Some Physical Constants

Quantity	Symbol	Value[a]
Atomic mass unit	u	$1.660\ 538\ 73\ (13) \times 10^{-27}$ kg 931.494 013 (37) MeV/c^2
Avogadro's number	N_A	$6.022\ 141\ 99\ (47) \times 10^{23}$ particles/mol
Bohr magneton	$\mu_B = \dfrac{e\hbar}{2m_e}$	$9.274\ 008\ 99\ (37) \times 10^{-24}$ J/T
Bohr radius	$a_0 = \dfrac{\hbar^2}{m_e e^2 k_e}$	$5.291\ 772\ 083\ (19) \times 10^{-11}$ m
Boltzmann's constant	$k_B = \dfrac{R}{N_A}$	$1.380\ 650\ 3\ (24) \times 10^{-23}$ J/K
Compton wavelength	$\lambda_C = \dfrac{h}{m_e c}$	$2.426\ 310\ 215\ (18) \times 10^{-12}$ m
Coulomb constant	$k_e = \dfrac{1}{4\pi\epsilon_0}$	$8.987\ 551\ 788 \times 10^9$ N·m^2/C^2 (exact)
Deuteron mass	m_d	$3.343\ 583\ 09\ (26) \times 10^{-27}$ kg 2.013 553 212 71 (35) u
Electron mass	m_e	$9.109\ 381\ 88\ (72) \times 10^{-31}$ kg $5.485\ 799\ 110\ (12) \times 10^{-4}$ u 0.510 998 902 (21) MeV/c^2
Electron volt	eV	$1.602\ 176\ 462\ (63) \times 10^{-19}$ J
Elementary charge	e	$1.602\ 176\ 462\ (63) \times 10^{-19}$ C
Gas constant	R	8.314 472 (15) J/mol·K
Gravitational constant	G	$6.673\ (10) \times 10^{-11}$ N·m^2/kg^2
Josephson frequency–voltage ratio	$\dfrac{2e}{h}$	$4.835\ 978\ 98\ (19) \times 10^{14}$ Hz/V
Magnetic flux quantum	$\Phi_0 = \dfrac{h}{2e}$	$2.067\ 833\ 636\ (81) \times 10^{-15}$ T·m^2
Neutron mass	m_n	$1.674\ 927\ 16\ (13) \times 10^{-27}$ kg 1.008 664 915 78 (55) u 939.565 330 (38) MeV/c^2
Nuclear magneton	$\mu_n = \dfrac{e\hbar}{2m_p}$	$5.050\ 783\ 17\ (20) \times 10^{-27}$ J/T
Permeability of free space	μ_0	$4\pi \times 10^{-7}$ T·m/A (exact)
Permittivity of free space	$\epsilon_0 = \dfrac{1}{\mu_0 c^2}$	$8.854\ 187\ 817 \times 10^{-12}$ C^2/N·m^2 (exact)
Planck's constant	h	$6.626\ 068\ 76\ (52) \times 10^{-34}$ J·s
	$\hbar = \dfrac{h}{2\pi}$	$1.054\ 571\ 596\ (82) \times 10^{-34}$ J·s
Proton mass	m_p	$1.672\ 621\ 58\ (13) \times 10^{-27}$ kg 1.007 276 466 88 (13) u 938.271 998 (38) MeV/c^2
Rydberg constant	R_H	$1.097\ 373\ 156\ 854\ 9\ (83) \times 10^7$ m^{-1}
Speed of light in vacuum	c	$2.997\ 924\ 58 \times 10^8$ m/s (exact)

Note: These constants are the values recommended in 1998 by CODATA, based on a least-squares adjustment of data from different measurements. For a more complete list, see P. J. Mohr and B. N. Taylor, "CODATA recommended values of the fundamental physical constants: 1998." *Rev. Mod. Phys.* 72:351, 2000.

[a]The numbers in parentheses for the values represent the uncertainties of the last two digits.

Solar System Data

Body	Mass (kg)	Mean Radius (m)	Period (s)	Distance from the Sun (m)
Mercury	3.18×10^{23}	2.43×10^{6}	7.60×10^{6}	5.79×10^{10}
Venus	4.88×10^{24}	6.06×10^{6}	1.94×10^{7}	1.08×10^{11}
Earth	5.98×10^{24}	6.37×10^{6}	3.156×10^{7}	1.496×10^{11}
Mars	6.42×10^{23}	3.37×10^{6}	5.94×10^{7}	2.28×10^{11}
Jupiter	1.90×10^{27}	6.99×10^{7}	3.74×10^{8}	7.78×10^{11}
Saturn	5.68×10^{26}	5.85×10^{7}	9.35×10^{8}	1.43×10^{12}
Uranus	8.68×10^{25}	2.33×10^{7}	2.64×10^{9}	2.87×10^{12}
Neptune	1.03×10^{26}	2.21×10^{7}	5.22×10^{9}	4.50×10^{12}
Pluto	$\approx 1.4 \times 10^{22}$	$\approx 1.5 \times 10^{6}$	7.82×10^{9}	5.91×10^{12}
Moon	7.36×10^{22}	1.74×10^{6}	—	—
Sun	1.991×10^{30}	6.96×10^{8}	—	—

Physical Data Often Used

Average Earth–Moon distance	3.84×10^{8} m
Average Earth–Sun distance	1.496×10^{11} m
Average radius of the Earth	6.37×10^{6} m
Density of air (20°C and 1 atm)	1.20 kg/m^3
Density of water (20°C and 1 atm)	1.00×10^{3} kg/m^3
Free-fall acceleration	9.80 m/s^2
Mass of the Earth	5.98×10^{24} kg
Mass of the Moon	7.36×10^{22} kg
Mass of the Sun	1.99×10^{30} kg
Standard atmospheric pressure	1.013×10^{5} Pa

Note: These values are the ones used in the text.

Some Prefixes for Powers of Ten

Power	Prefix	Abbreviation	Power	Prefix	Abbreviation
10^{-24}	yocto	y	10^{1}	deka	da
10^{-21}	zepto	z	10^{2}	hecto	h
10^{-18}	atto	a	10^{3}	kilo	k
10^{-15}	femto	f	10^{6}	mega	M
10^{-12}	pico	p	10^{9}	giga	G
10^{-9}	nano	n	10^{12}	tera	T
10^{-6}	micro	μ	10^{15}	peta	P
10^{-3}	milli	m	10^{18}	exa	E
10^{-2}	centi	c	10^{21}	zetta	Z
10^{-1}	deci	d	10^{24}	yotta	Y

PRINCIPLES OF PHYSICS

A CALCULUS-BASED TEXT FOURTH EDITION

Raymond A. Serway

Emeritus, James Madison University

John W. Jewett, Jr.

California State Polytechnic University—Pomona

Australia · Brazil · Japan · Korea · Mexico · Singapore · Spain · United Kingdom · United States

Principles of Physics: A Calculus-Based Text, Fourth Edition,
Raymond A. Serway, John W. Jewett, Jr.

Physics Acquisitions Editor: Chris Hall
Publisher: David Harris
Editor-in-Chief: Michelle Julet
Senior Developmental Editor: Susan Dust Pashos
Assistant Editor: Sarah Lowe
Editorial Assistant: Seth Dobrin
Technology Project Manager: Sam Subity
Marketing Managers: Erik Evans, Julie Conover
Marketing Assistant: Leyla Jowza
Advertising Project Manager: Stacey Purviance
Senior Project Manager, Editorial Production: Teri Hyde
Print/Media Buyer: Barbara Britton
Permissions Editor: Joohee Lee
Production Service: Progressive Publishing Alternatives
Text Designer: John Walker
Art Director: Rob Hugel
Photo Researcher: Dena Digilio Betz
Copy Editor: Kathleen Lafferty
Illustrator: Progressive Information Technologies
Cover Designer: John Walker
Cover Image: Transrapid International, Berlin, Germany
Compositor: Progressive Information Technologies

Library of Congress Control Number: 2004113842

ISBN-13: 978-0-534-49143-7
ISBN-10: 0-534-49143-X

Brooks/Cole
10 Davis Drive
Belmont, CA 94002-3098
USA

Printed in China by China Translation & Printing Services Limited
6 7 8 9 10 11 10

Welcome to PhysicsNow, your fully integrated system for physics tutorials and self-assessment on the web. To get started, just follow these simple instructions.

Your first visit to PhysicsNow

1. Go to **http://www.pop4e.com** and click the **Login** button.
2. The first time you visit, you will be asked to select your school. Choose your state from the drop-down menu, then type in your school's name in the box provided and click **Search**. A list of schools with names similar to what you entered will show on the right. Find your school and click on it.
3. On the next screen, enter the access code from the card that came with your textbook in the "Content or Course Access Code" box*. Enter your email address in the next box and click **Submit.**

   * PhysicsNow access codes may be purchased separately. Should you need to purchase an access code, go back to the sign-in screen and click the **Buy Now** button.

PhysicsNow - Microsoft Internet Explorer

Physics Now™

First Time Users: REGISTER. Register here with the access code from your book to setup your account. View Quick Start Instructions

Returning Users: LOGIN. Forgot your username or password?

Need an Access code? BUY DEMO

Additional Resources for instructors and students. Select your book below.

Serway *Physics for Scientists and Engineers*

Serway *College Physics*

Serway *Principles of Physics*

Students: Need help? Technical questions? Call us at (800) 354-9706 or email academic.cengage.com/support

Instructors: Get an account, set up training, or ask a question. Call us at (888) 281-2990 or email SupportTeam@ilrn-support.com

4. On the next screen, choose a password and click **Submit.**
5. Lastly, fill out the registration form and click **Register and Enter iLrn.** This information will only be used to contact you if there is a problem with your account.
6. You should now see the **PhysicsNow** homepage. Select a chapter and begin!

Note: Your account information will be sent to the email address that you entered in Step 3, so be sure to enter a valid email address. You will use your email address as your username the next time you login.

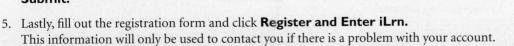

Second and later visits

1. Go to **http://www.pop4e.com** and click the **Login** button.
2. Enter your user name (the email address you entered when you registered) and your password and then click **Login.**

SYSTEM REQUIREMENTS:
(Please see the System Requirements link at* www.ilrn.com *for complete list.)
PC: Windows 98 or higher, Internet Explorer 5.5 or higher
Mac: OS X or higher, Mozilla browser 1.2.1 or higher

TECHNICAL SUPPORT:
For online help, click on **Technical Support** in the upper right corner of the screen, or contact us at:

1-800-354-9706 Monday–Friday • 8:30 A.M. to 6:00 P.M. EST
academic.cengage.com/support

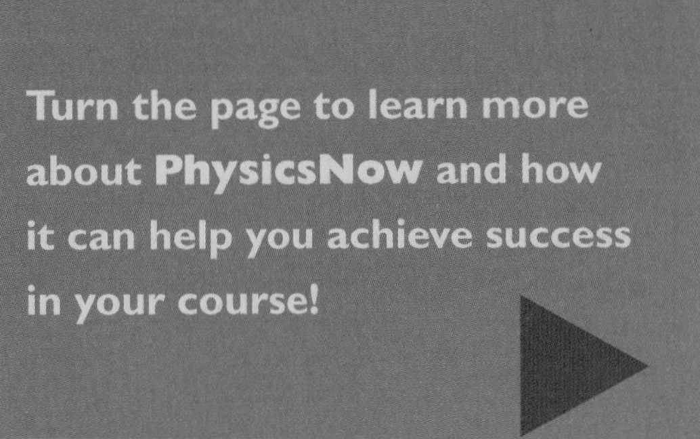

What do you need to learn now?

Take charge of your learning with **PhysicsNow**™, a powerful student-learning tool for physics! This interactive resource helps you gauge your unique study needs, then gives you a *Personalized Learning Plan* that will help you focus in on the concepts and problems that will most enhance your understanding. With **PhysicsNow**, you have the resources you need to take charge of your learning!

The access code card included with this new copy of ***Principles of Physics*** is your ticket to all of the resources in **PhysicsNow.** (See the previous page for login instructions.)

Interact at every turn with the POWER and SIMPLICITY of PhysicsNow!

PhysicsNow combines Serway and Jewett's best-selling *Principles of Physics* with carefully crafted media resources that will help you learn. This dynamic resource and the Fourth Edition of the text were developed in concert, to enhance each other and provide you with a seamless, integrated learning system.

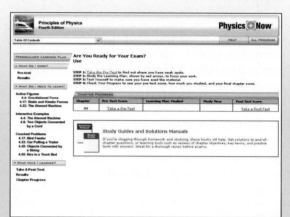

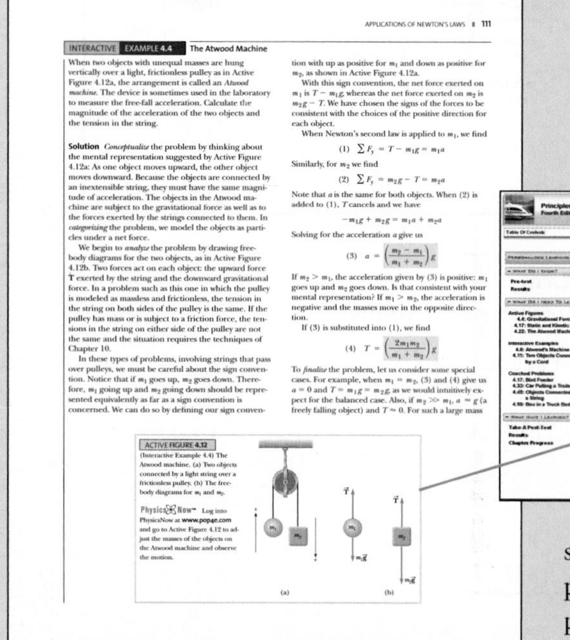

APPLICATIONS OF NEWTON'S LAWS ■ 111

INTERACTIVE EXAMPLE 4.4 The Atwood Machine

When two objects with unequal masses are hung vertically over a light, frictionless pulley as in Active Figure 4.12a, the arrangement is called an *Atwood machine*. The device is sometimes used in the laboratory to measure the free-fall acceleration. Calculate the magnitude of the acceleration of the two objects and the tension in the string.

Solution *Conceptualize* the problem by thinking about the mental representation suggested by Active Figure 4.12a: As one object moves upward, the other object moves downward. Because the objects are connected by an inextensible string, they must have the same magnitude of acceleration. The objects in the Atwood machine are subject to the gravitational force as well as to the forces exerted by the strings connected to them. In *categorizing* the problem, we model the objects as particles under a net force.

We begin to *analyze* the problem by drawing free-body diagrams for the two objects, as in Active Figure 4.12b. Two forces act on each object: the upward force $\vec{T}$ exerted by the string and the downward gravitational force. In a problem such as this one in which the pulley is modeled as massless and frictionless, the tension in the string on both sides of the pulley is the same. If the pulley has mass or is subject to a friction force, the tensions in the string on either side of the pulley are not the same and the situation requires the techniques of Chapter 10.

In these types of problems, involving strings that pass over pulleys, we must be careful about the sign convention. Notice that if m_1 goes up, m_2 goes down. Therefore, m_1 going up and m_2 going down should be represented equivalently as far as a sign convention is concerned. We can do so by defining our sign convention with up as positive for m_1 and down as positive for m_2, as shown in Active Figure 4.12a.

With this sign convention, the net force exerted on m_1 is $T - m_1g$, whereas the net force exerted on m_2 is $m_2g - T$. We have chosen the signs of the forces to be consistent with the choices of the positive direction for each object.

When Newton's second law is applied to m_1, we find

$$(1)\quad \sum F_y = T - m_1g = m_1a$$

Similarly, for m_2 we find

$$(2)\quad \sum F_y = m_2g - T = m_2a$$

Note that a is the same for both objects. When (2) is added to (1), T cancels and we have

$$-m_1g + m_2g = m_1a + m_2a$$

Solving for the acceleration a give us

$$(3)\quad a = \left(\frac{m_2 - m_1}{m_1 + m_2}\right)g$$

If $m_2 > m_1$, the acceleration given by (3) is positive: m_1 goes up and m_2 goes down. Is that consistent with your mental representation? If $m_1 > m_2$, the acceleration is negative and the masses move in the opposite direction.

If (3) is substituted into (1), we find

$$(4)\quad T = \left(\frac{2m_1m_2}{m_1 + m_2}\right)g$$

To *finalize* the problem, let us consider some special cases. For example, when $m_1 = m_2$, (3) and (4) give us $a = 0$ and $T = m_1g = m_2g$, as we would intuitively expect for the balanced case. Also, if $m_2 \gg m_1$, $a \approx g$ (a freely falling object) and $T \approx 0$. For such a large mass

ACTIVE FIGURE 4.12

(Interactive Example 4.4) The Atwood machine. (a) Two objects connected by a light string over a frictionless pulley. (b) The free-body diagrams for m_1 and m_2.

PhysicsNow™ Log into PhysicsNow at www.pop4e.com and go to Active Figure 4.12 to adjust the masses of the objects on the Atwood machine and observe the motion.

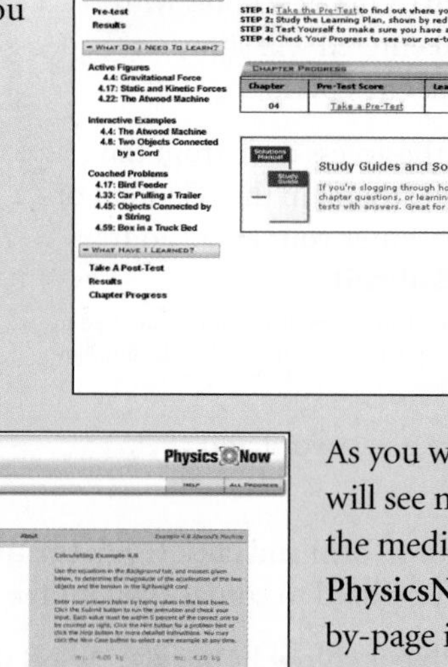

As you work through the text, you will see notes that direct you to the media-enhanced activities in **PhysicsNow.** This precise page-by-page integration means you'll spend less time flipping through pages or navigating websites looking for useful exercises. These multimedia exercises will make all the difference when you're studying and taking exams . . . after all, it's far easier to understand physics if it's seen in action, and **PhysicsNow** enables you to become a part of the action!

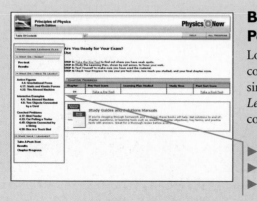

Begin at http://www.pop4e.com and build your own Personalized Learning Plan now!

Log into PhysicsNow at http://www.pop4e.com by using the free access code packaged with the text. You'll immediately notice the system's simple, browser-based format. You can build a complete *Personalized Learning Plan* for yourself by taking advantage of all three powerful components found on PhysicsNow:

- **What I Know**
- **What I Need to Learn**
- **What I've Learned**

The best way to maximize the system and optimize your time is to start by taking the *Pre-Test* ▶▶▶

What I Know

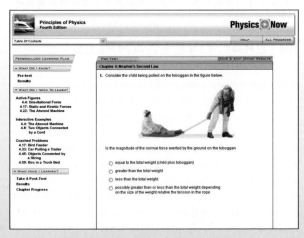

▲ You take a *Pre-Test* to measure your level of comprehension after reading a chapter. Each *Pre-Test* includes approximately 15 questions. The *Pre-Test* is your first step in creating your custom-tailored *Personalized Learning Plan.*

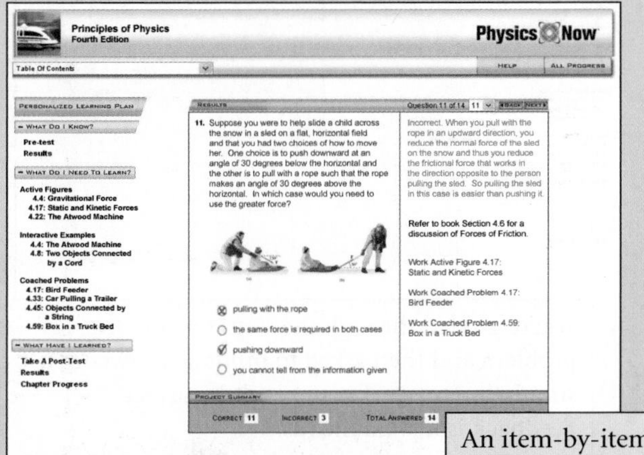

An item-by-item analysis gives you feedback on each of your answers.

▲ Once you've completed the "What I Know" *Pre-Test,* you are presented with a detailed *Personalized Learning Plan,* with text references that outline the elements you need to review in order to master the chapter's most essential concepts. This roadmap to concept mastery guides you to exercises designed to improve skills and to increase your understanding of the basic concepts.

At each stage, the *Personalized Learning Plan* refers to ***Principles of Physics*** to reinforce the connection between text and technology as a powerful learning tool.

What I Need to Learn

Once you've completed the *Pre-Test,* you're ready to work through tutorials and exercises that will help you master the concepts that are essential to your success in the course.

ACTIVE FIGURES

A remarkable bank of more than 200 animated figures helps you visualize physics in action. Taken straight from illustrations in the text, these *Active Figures* help you master key concepts from the book. By interacting with the animations and accompanying quiz questions, you come to an even greater understanding of the concepts you need to learn from each chapter. ▼

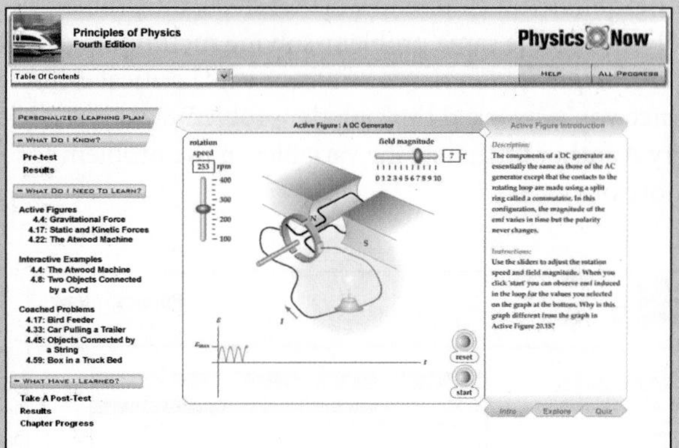

Each figure is titled so you can easily identify the concept you are seeing. The final tab features a *Quiz*. The *Explore* tab guides you through the animation so you understand what you should be seeing and learning.

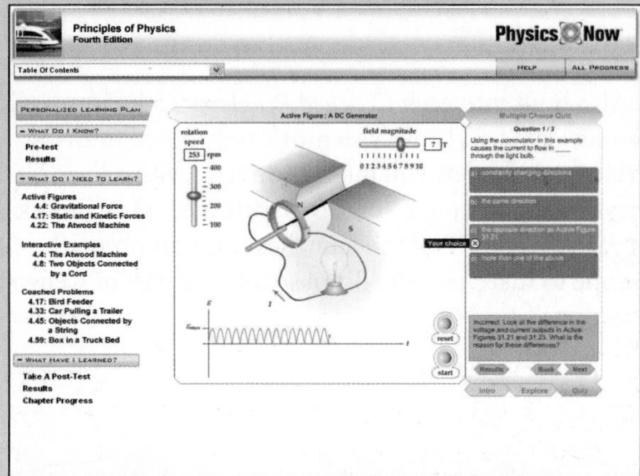

▲ The brief *Quiz* ensures that you mastered the concept played out in the animation—and gives you feedback on each response.

Continued on the next page

What I Need to Learn *Continued*

COACHED PROBLEMS

Engaging *Coached Problems* reinforce the lessons in the text by taking a step-by-step approach to problem-solving methodology. Each *Coached Problem* gives you the option of breaking down a problem from the text into steps with feedback to 'coach' you toward the solution. There are approximately five *Coached Problems* per chapter.

You can choose to work through the *Coached Problems* by inputting an answer directly or working in steps with the program. If you choose to work in steps, the problem is solved with the same problem-solving methodology used in ***Principles of Physics*** to reinforce these critical skills. Once you've worked through the problem, you can click **Try Another** to change the variables in the problem for more practice.

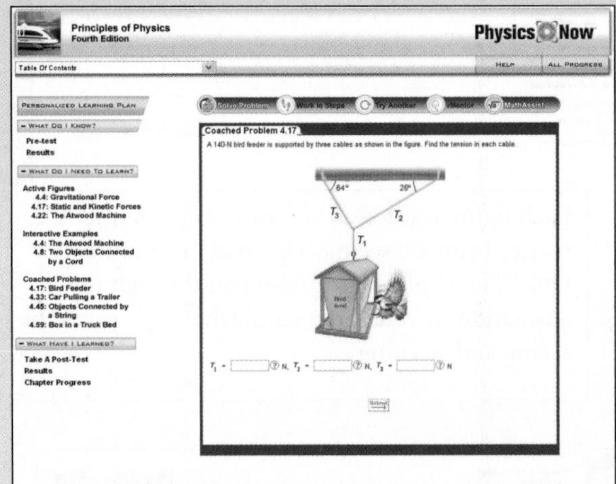

▲ Also built into each *Coached Problem* is a link to Brooks/Cole's exclusive **vMentor**™ web-based tutoring service site that lets you interact directly with a live physics tutor. If you're stuck on math, a *MathAssist* link on each *Coached Problem* launches tutorials on math specific to that problem. Brooks/Cole is a part of Cengage Learning.

INTERACTIVE EXAMPLES

You'll strengthen your problem-solving and visualization skills with *Interactive Examples.* Extending selected examples from the text, *Interactive Examples* utilize the proven and trusted problem-solving methodology presented in ***Principles of Physics.*** These animated learning modules give you all the tools you need to solve a problem type—you're then asked to apply what you have learned to different scenarios. You will find approximately two *Interactive Examples* for each chapter of the text.▼

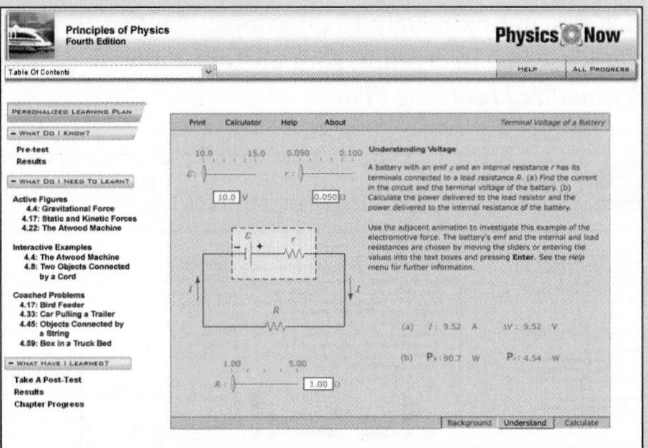

You're guided through the steps to solve the problem and then asked to input an answer in a simulation to see if your result is correct. Feedback is instantaneous.

What I've Learned

► After working through the problems highlighted in your *Personalized Learning Plan,* you move on to a *Post-Test,* about 15 questions per chapter.

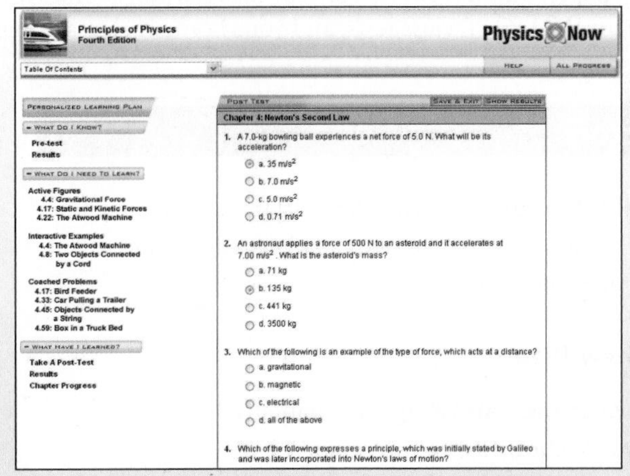

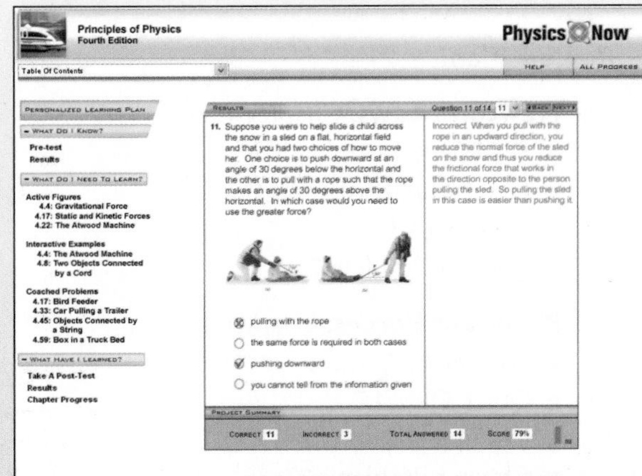

◄ Once you've completed the *Post-Test*, you receive your percentage score and specific feedback on each answer. The *Post-Tests* give you a new set of questions with each attempt, so you can take them over and over as you continue to build your knowledge and skills and master concepts.

Also available to help you succeed in your course

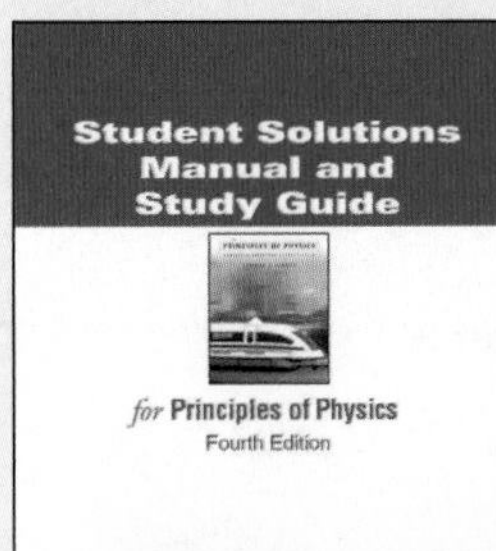

Student Solutions Manual and Study Guide

Volume I (Ch. 1–15) ISBN: 0-534-49145-6
Volume II (Ch. 16–31) ISBN: 0-534-49147-2

These manuals contain detailed solutions to approximately 20-percent of the end-of-chapter problems. These problems are indicated in the textbook with boxed problem numbers. Each manual also features a skills section, important notes from key sections of the text, and a list of important equations and concepts.

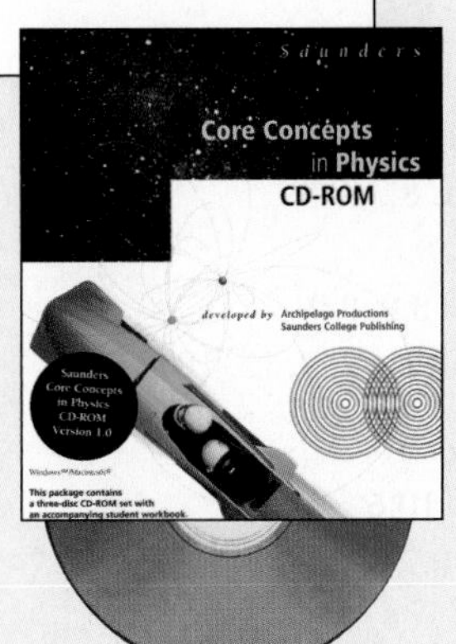

Core Concepts in Physics CD-ROM, Version 2.0

ISBN: 0-03-033731-3

Explore the core of physics with this powerful CD-ROM/workbook program! Content screens provide in-depth coverage of abstract and often difficult principles, building connections between physical concepts and mathematics. The presentation contains more than 350 movies—both animated and live video—including laboratory demonstrations, real-world examples, graphic models, and step-by-step explanations of essential mathematics. An accompanying workbook contains practical physics problems directly related to the presentation, along with worked solutions. Package includes three discs and a workbook.

Welcome to your MCAT Test Preparation Guide

The MCAT Test Preparation Guide makes your copy of *Principles of Physics,* **Fourth Edition,** the most comprehensive MCAT study tool and classroom resource in introductory physics. The grid, which begins below and continues on the next two pages, outlines twelve concept-based **study courses** for the physics part of your MCAT exam. Use it to prepare for the MCAT, class tests, and your homework assignments.

Vectors

Skill Objectives: To calculate distance, calculate angles between vectors, calculate magnitudes, and to understand vectors.

Review Plan:

Distance and Angles: Chapter 1
- Section 1.6
- Active Figure 1.4
- Chapter Problem 33

Using Vectors: Chapter 1
- Sections 1.7–1.9
- Quick Quizzes 1.4–1.8
- Examples 1.6–1.8
- Active Figures 1.9, 1.16
- Chapter Problems 37, 38, 45, 47, 51, 53

Motion

Skill Objectives: To understand motion in two dimensions, to calculate speed and velocity, to calculate centripetal acceleration, and acceleration in free fall problems.

Review Plan:

Motion in 1 Dimension: Chapter 2
- Sections 2.1, 2.2, 2.4, 2.6, 2.7
- Quick Quizzes 2.3–2.6
- Examples 2.1, 2.2, 2.4–2.10
- Active Figure 2.12
- Chapter Problems 3, 5,13, 19, 21, 29, 31, 33

Motion in 2 Dimensions: Chapter 3
- Sections 3.1–3.3
- Quick Quizzes 3.2, 3.3
- Examples 3.1–3.4
- Active Figures 3.4, 3.5, 3.8
- Chapter Problems 1, 7, 15

Centripetal Acceleration: Chapter 3
- Sections 3.4, 3.5
- Quick Quizzes 3.4, 3.5
- Example 3.5
- Active Figure 3.12
- Chapter Problems 23, 31

Force

Skill Objectives: To know and understand Newton's Laws, to calculate resultant forces and weight.

Review Plan:

Newton's Laws: Chapter 4
- Sections 4.1–4.6
- Quick Quizzes 4.1–4.6
- Example 4.1
- Chapter Problem 7

Resultant Forces: Chapter 4
- Section 4.7
- Quick Quiz 4.7
- Example 4.6
- Chapter Problems 27, 35

Gravity: Chapter 11
- Section 11.1
- Quick Quiz 11.1
- Chapter Problem 3

Equilibrium

Skill Objectives: To calculate momentum and impulse, center of gravity, and torque.

Review Plan:

Momentum: Chapter 8
- Section 8.1
- Quick Quiz 8.2
- Examples 8.2, 8.3

Impulse: Chapter 8
- Sections 8.2, 8.3
- Quick Quizzes 8.3, 8.4
- Examples 8.4, 8.6
- Active Figures 8.8, 8.9
- Chapter Problems 7, 9, 15, 19, 21

Torque: Chapter 10
- Sections 10.5, 10.6
- Quick Quiz 10.7
- Example 10.8
- Chapter Problems 21, 27

Work

Skill Objectives: To calculate friction, work, kinetic energy, power, and potential energy.

Review Plan:

Friction: Chapter 5
- Section 5.1
- Quick Quizzes 5.1, 5.2

Work: Chapter 6
- Section 6.2
- Chapter Problems 1, 3

Kinetic Energy: Chapter 6
- Section 6.5
- Example 6.4

Power: Chapter 6
- Section 6.8
- Chapter Problem 35

Potential Energy: Chapter 7
- Sections 7.1, 7.2
- Quick Quizzes 7.1, 7.2
- Chapter Problem 5

Waves

Skill Objectives: To understand interference of waves, to calculate basic properties of waves, properties of springs, and properties of pendulums.

Review Plan:

Wave Properties: Chapters 12, 13
- Sections 12.1, 12.2, 13.1-13.3
- Quick Quiz 13.1
- Examples 12.1, 13.2
- Active Figures 12.1, 12.2, 12.4, 12.6, 12.10

Chapter 13
- Problem 9

Pendulum: Chapter 12
- Sections 12.4, 12.5
- Quick Quizzes 12.3, 12.4
- Examples 12.5, 12.6
- Active Figure 12.11
- Chapter Problem 23

Interference: Chapter 14
- Sections 14.1–14.3
- Quick Quiz 14.1
- Active Figures 14.1–14.3

Matter

Skill Objectives: To calculate density, pressure, specific gravity, and flow rates.

Review Plan:

Density: Chapters 1, 15
- Sections 1.1, 15.2

Pressure: Chapter 15
- Sections 15.1–15.4
- Quick Quizzes 15.1–15.4
- Examples 15.1, 15.3
- Chapter Problems 3, 7, 19, 23, 27

Flow rates: Chapter 15
- Section 15.6
- Quick Quiz 15.5

Sound

Skill Objectives: To understand interference of waves, calculate properties of waves, the speed of sound, Doppler shifts, and intensity.

Review Plan:

Sound Properties: Chapters 13, 14
- Sections 13.3, 13.4, 13.7, 13.8, 14.4
- Quick Quizzes 13.2, 13.3, 13.6
- Example 14.3
- Active Figures 13.6–13.8, 13.21, 13.22

Chapter 13
- Problems 3, 17, 23, 29, 35, 37

Chapter 14
- Problem 23

Interference/Beats: Chapter 14
- Sections 14.1, 14.2, 14.6
- Quick Quiz 14.6
- Active Figures 14.1–14.3, 14.12
- Chapter Problems 5, 39, 41

Light

Skill Objectives: To understand mirrors and lenses, to calculate the angles of reflection, to use the index of refraction, and to find focal lengths.

Review Plan:

Reflection: Chapter 25
- Sections 25.1–25.3
- Example 25.1
- Active Figure 25.5

Refraction: Chapter 25
- Sections 25.4, 25.5
- Quick Quizzes 25.2–25.5
- Example 25.2
- Chapter Problems 7, 13

Mirrors and Lenses: Chapter 26
- Sections 26.1–26.4
- Quick Quizzes 26.1–26.6
- Examples 26.1–26.7
- Active Figures 26.2, 26.24
- Chapter Problems 23, 27, 31, 35

Electrostatics

Skill Objectives: To understand and calculate the electric field, the electrostatic force, and the electric potential.

Review Plan:

Coulomb's Law: Chapter 19
- Section 19.2–19.4
- Quick Quiz 19.1–19.3
- Examples 19.1, 19.2
- Active Figure 19.7
- Chapter Problems 3, 5

Electric Field: Chapter 19
- Sections 19.5, 19.6
- Quick Quizzes 19.4, 19.5
- Active Figures 19.10, 19.19, 19.21

Potential: Chapter 20
- Sections 20.1–20.3
- Examples 20.1, 20.2
- Active Figure 20.6
- Chapter Problems 1, 5, 11, 13

Circuits

Skill Objectives: To understand and calculate current, resistance, voltage, and power, and to use circuit analysis.

Review Plan:

Ohm's Law: Chapter 21
- Sections 21.1, 21.2
- Quick Quizzes 21.1, 21.2
- Examples 21.1, 21.2
- Chapter Problem 7

Power and energy: Chapter 21
- Section 21.5
- Quick Quiz 21.4
- Example 21.5
- Active Figure 21.10
- Chapter Problems 17, 19, 23

Circuits: Chapter 21
- Section 21.6–21.8
- Quick Quizzes 21.5–21.8
- Example 21.7–21.9
- Active Figures 21.13, 21.14, 21.16
- Chapter Problems 25, 29, 35

Atoms

Skill Objectives: To understand decay processes and nuclear reactions and to calculate half-life.

Review Plan:

Atoms: Chapter 30
- Sections 30.1
- Quick Quizzes 30.1, 30.2
- Active Figure 30.1

Decays: Chapter 30
- Sections 30.3, 30.4
- Quick Quizzes 30.3–30.6
- Examples 30.3–30.6
- Active Figures 30.11–30.14, 30.16, 30.17
- Chapter Problems 13, 19, 23

Nuclear reactions: Chapter 30
- Sections 30.5
- Active Figure 30.21
- Chapter Problems 27, 29

DEDICATION

IN MEMORY OF

Emily and Fargo Serway

Two hard working and dedicated parents, for their unforgettable love, vision, and wisdom.

John W. Jewett

Marvin V. Schober

These fathers and fathers-in-law provided models for hard work, inspiration for creativity, and motivation for excellence.

They are sincerely missed.

BRIEF CONTENTS

VOLUME 1

VOLUME 2

CONTENTS

VOLUME 1

ABOUT THE AUTHORS

RAYMOND A. SERWAY received his doctorate at Illinois Institute of Technology and is Professor Emeritus at James Madison University. In 1990, he received the Madison Scholar Award at James Madison University, where he taught for 17 years. Dr. Serway began his teaching career at Clarkson University, where he conducted research and taught from 1967 to 1980. He was the recipient of the Distinguished Teaching Award at Clarkson University in 1977 and of the Alumni Achievement Award from Utica College in 1985. As Guest Scientist at the IBM Research Laboratory in Zurich, Switzerland, he worked with K. Alex Müller, 1987 Nobel Prize recipient. Dr. Serway also was a visiting scientist at Argonne National Laboratory, where he collaborated with his mentor and friend, Sam Marshall. In addition to earlier editions of this textbook, Dr. Serway is the co-author of *Physics for Scientists and Engineers,* Sixth Edition; *College Physics,* Seventh Edition; and *Modern Physics,* Third Edition. He also is the author of the high-school textbook *Physics,* published by Holt, Rinehart, & Winston. In addition, Dr. Serway has published more than 40 research papers in the field of condensed matter physics and has given more than 70 presentations at professional meetings. Dr. Serway and his wife Elizabeth enjoy traveling, golfing, and spending quality time with their four children and seven grandchildren.

JOHN W. JEWETT, JR. earned his doctorate at Ohio State University, specializing in optical and magnetic properties of condensed matter. Dr. Jewett began his academic career at Richard Stockton College of New Jersey, where he taught from 1974 to 1984. He is currently Professor of Physics at California State Polytechnic University, Pomona. Throughout his teaching career, Dr. Jewett has been active in promoting science education. In addition to receiving four National Science Foundation grants, he helped found and direct the Southern California Area Modern Physics Institute (SCAMPI). He also directed Science IMPACT (Institute for Modern Pedagogy and Creative Teaching), which works with teachers and schools to develop effective science curricula. Dr. Jewett's honors include the Stockton Merit Award at Richard Stockton College in 1980, the Outstanding Professor Award at California State Polytechnic University for 1991–1992, and the Excellence in Undergraduate Physics Teaching Award from the American Association of Physics Teachers (AAPT) in 1998. He has given over 80 presentations at professional meetings, including presentations at international conferences in China and Japan. In addition to his work on this textbook, he is co-author of *Physics for Scientists and Engineers,* Sixth Edition with Dr. Serway and author of *The World of Physics . . . Mysteries, Magic, and Myth.* Dr. Jewett enjoys playing keyboard with his all-physicist band, traveling, and collecting antiques that can be used as demonstration apparatus in physics lectures. Most importantly, he relishes spending time with his wife Lisa and their children and grandchildren.

PREFACE

Principles of Physics is designed for a one-year introductory calculus-based physics course for engineering and science students and for premed students taking a rigorous physics course. This fourth edition contains many new pedagogical features—most notably, an integrated Web-based learning system and a structured problem-solving strategy that uses a modeling approach. Based on comments from users of the third edition and reviewers' suggestions, a major effort was made to improve organization, clarity of presentation, precision of language, and accuracy throughout.

This project was conceived because of well-known problems in teaching the introductory calculus-based physics course. The course content (and hence the size of textbooks) continues to grow, while the number of contact hours with students has either dropped or remained unchanged. Furthermore, traditional one-year courses cover little if any physics beyond the 19th century.

In preparing this textbook, we were motivated by the spreading interest in reforming the teaching and learning of physics through physics education research. One effort in this direction was the Introductory University Physics Project (IUPP), sponsored by the American Association of Physics Teachers and the American Institute of Physics. The primary goals and guidelines of this project are to

- Reduce course content following the "less may be more" theme;
- Incorporate contemporary physics naturally into the course;
- Organize the course in the context of one or more "story lines";
- Treat all students equitably.

Recognizing a need for a textbook that could meet these guidelines several years ago, we studied the various proposed IUPP models and the many reports from IUPP committees. Eventually, one of us (RAS) became actively involved in the review and planning of one specific model, initially developed at the U.S. Air Force Academy, entitled "A Particles Approach to Introductory Physics." Part of the summer of 1990 was spent at the Academy working with Colonel James Head and Lt. Col. Rolf Enger, the primary authors of the Particles model, and other members of that department. This most useful collaboration was the starting point of this project.

The other author (JWJ) became involved with the IUPP model called "Physics in Context," developed by John Rigden (American Institute of Physics), David Griffiths (Oregon State University), and Lawrence Coleman (University of Arkansas at Little Rock). This involvement led to the contextual overlay that is used in this book and described in detail later in the Preface.

The combined IUPP approach in this book has the following features:

- It is an evolutionary approach (rather than a revolutionary approach), which should meet the current demands of the physics community.
- It deletes many topics in classical physics (such as alternating current circuits and optical instruments) and places less emphasis on rigid object motion, optics, and thermodynamics.
- Some topics in contemporary physics, such as special relativity, energy quantization, and the Bohr model of the hydrogen atom, are introduced early in the textbook.
- A deliberate attempt is made to show the unity of physics.
- As a motivational tool, the textbook connects physics principles to interesting social issues, natural phenomena, and technological advances.

OBJECTIVES

This introductory physics textbook has two main objectives: to provide the student with a clear and logical presentation of the basic concepts and principles of physics, and to strengthen an understanding of the concepts and principles through a broad range of interesting applications to the real world. To meet these objectives, we have emphasized sound physical arguments and problem-solving methodology. At the same time, we have attempted to motivate the student through practical examples that demonstrate the role of physics in other disciplines, including engineering, chemistry, and medicine.

CHANGES IN THE FOURTH EDITION

A number of changes and improvements have been made in the fourth edition of this text. Many of these are in response to recent findings in physics education research and to comments and suggestions provided by the reviewers of the manuscript and instructors using the first three editions. The following represent the major changes in the fourth edition:

New Context The context overlay approach is described below under "Text Features." The fourth edition introduces a new Context for Chapters 2–7, "Alternative-Fuel Vehicles." This context addresses the current social issue of the depletion of our supply of petroleum and the efforts being made to develop new fuels and new types of automobiles to respond to this situation.

Active Figures Many diagrams from the text have been animated to form **Active Figures,** part of the new ***PhysicsNow*™** integrated Web-based learning system. There are over 150 Active Figures available at **www.pop4e.com.** By visualizing phenomena and processes that cannot be fully represented on a static page, students greatly increase their conceptual understanding. An addition to the figure caption, marked with the Physics Now™ icon, describes briefly the nature and contents of the animation. In addition to viewing animations of the figures, students can change variables to see the effects, conduct suggested explorations of the principles involved in the figure, and take and receive feedback on quizzes related to the figure.

Interactive Examples Sixty-seven of the worked examples have been identified as interactive. As part of the ***PhysicsNow*™** Web-based learning system, students can engage in an extension of the problem solved in the example. This often includes elements of both visualization and calculation, and may also involve prediction and intuition-building. Interactive Examples are available at **www.pop4e.com.**

Quick Quizzes Quick Quizzes have been cast in an objective format, including multiple choice, true-false, and ranking. Quick Quizzes provide students with opportunities to test their understanding of the physical concepts presented. The questions require students to make decisions on the basis of sound reasoning, and some of them have been written to help students overcome common misconceptions. Answers to all Quick Quiz questions are found at the end of each chapter. Additional Quick Quizzes that can be used in classroom teaching are available on the instructor's companion Web site. Many instructors choose to use such questions in a "peer instruction" teaching style, but they can be used in standard quiz format as well. To support the use of classroom response systems, we have coded the Quick Quiz questions so that they may be used within the response system of your choice.

General Problem-Solving Strategy A general strategy to be followed by the student is outlined at the end of Chapter 1 and provides students with a structured process for solving problems. In the remaining chapters, the steps of the Strategy appear explicitly in one example per chapter so that students are encouraged throughout the course to follow the procedure.

Line-by-Line Revision The text has been carefully edited to improve clarity of presentation and precision of language. We hope that the result is a book both accurate and enjoyable to read.

Problems In an effort to improve variety, clarity and quality, the end-of-chapter problems were substantially revised. Approximately 15% of the problems (about 300) are new to this edition. The new problems especially are chosen to include interesting applications, notably biological applications. As in previous editions, many problems require students to make order-of-magnitude calculations. More problems now explicitly ask students to design devices and to change among different representations of a situation. All problems have been carefully edited and reworded where necessary. Solutions to approximately 20% of the end-of-chapter problems are included in the *Student Solutions Manual and Study Guide*. Boxed numbers identify these problems. A smaller subset of problems will be available with coached solutions as part of the *PhysicsNow*™ Web-based learning system and will be accessible to students and instructors using *Principles of Physics.* These coached problems are identified with the PhysicsNow™ icon.

Biomedical Applications For biology and premed students, icons point the way to various practical and interesting applications of physical principles to biology and medicine. Where possible, an effort was made to include more problems that would be relevant to these disciplines.

TEXT FEATURES

Most instructors would agree that the textbook selected for a course should be the student's primary guide for understanding and learning the subject matter. Furthermore, the textbook should be easily accessible as well as styled and written to facilitate instruction and learning. With these points in mind, we have included many pedagogical features that are intended to enhance the textbook's usefulness to both students and instructors. These features are as follows:

Style To facilitate rapid comprehension, we have attempted to write the book in a clear, logical, and engaging style. The somewhat informal and relaxed writing style is intended to increase reading enjoyment. New terms are carefully defined, and we have tried to avoid the use of jargon.

Organization We have incorporated a "context overlay" scheme into the textbook, in response to the "Physics in Context" approach in the IUPP. This feature adds interesting applications of the material to real issues. We have developed this feature to be flexible, so that the instructor who does not wish to follow the contextual approach can simply ignore the additional contextual features without sacrificing complete coverage of the existing material. We believe, though, that the benefits students will gain from this approach will be many.

The context overlay organization divides the text into nine sections, or "Contexts," after Chapter 1, as follows:

Context Number	Context	Physics Topics	Chapters
1	Alternative-Fuel Vehicles	Classical mechanics	2–7
2	Mission to Mars	Classical mechanics	8–11
3	Earthquakes	Vibrations and waves	12–14
4	Search for the *Titanic*	Fluids	15
5	Global Warming	Thermodynamics	16–18
6	Lightning	Electricity	19–21
7	Magnetic Levitation Vehicles	Magnetism	22–23
8	Lasers	Optics	24–27
9	The Cosmic Connection	Modern physics	28–31

Each Context begins with an introduction, leading to a "central question" that motivates study within the Context. The final section of each chapter is a "Context Connection," which discusses how the material in the chapter relates to the Context and to the central question. The final chapter in each Context is followed by a "Context Conclusion." Each Conclusion uses the principles learned in the Context to respond fully to the central question. Each chapter, as well as the Context Conclusions, includes problems related to the context material.

Pitfall Prevention These features are placed in the margins of the text and address common student misconceptions and situations in which students often follow unproductive paths. Over 140 Pitfall Preventions are provided to help students avoid common mistakes and misunderstandings.

Modeling A modeling approach, based on four types of models commonly used by physicists, is introduced to help students understand they are solving problems that approximate reality. They must then learn how to test the validity of the model. This approach also helps students see the unity in physics, as a large fraction of problems can be solved with a small number of models. The modeling approach is introduced in Chapter 1.

Alternative Representations We emphasize alternative representations of information, including mental, pictorial, graphical, tabular, and mathematical representations. Many problems are easier to solve if the information is presented in alternative ways, to reach the many different methods students use to learn.

Problem-Solving Strategies We have included specific strategies for solving the types of problems featured both in the examples and in the end-of-chapter problems. These specific strategies are structured according to the steps in the General Problem-Solving Strategy introduced in Chapter 1. This feature helps students identify necessary steps in solving problems and eliminate any uncertainty they might have.

Worked Examples A large number of worked examples of varying difficulty are presented to promote students' understanding of concepts. In many cases, the examples serve as models for solving the end-of-chapter problems. Because of the increased emphasis on understanding physical concepts, many examples are conceptual in nature. The examples are set off in boxes, and the answers to examples with numerical solutions are highlighted with a tan screen.

Thinking Physics We have included many Thinking Physics examples throughout each chapter. These questions relate the physics concepts to common experiences or extend the concepts beyond what is discussed in the textual material. Immediately following each of these questions is a "Reasoning" section that responds to the question. Ideally, the student will use these features to better understand physical concepts before being presented with quantitative examples and working homework problems.

Previews Most chapters begin with a brief preview that includes a discussion of the particular chapter's objectives and content.

Important Statements and Equations Most important statements and definitions are set in boldface type or are highlighted with a blue outline for added emphasis and ease of review. Similarly, important equations are highlighted with a tan background screen to facilitate location.

Marginal Notes Comments and notes appearing in the margin can be used to locate important statements, equations, and concepts in the text.

Illustrations and Tables The readability and effectiveness of the text material and worked examples are enhanced by the large number of figures, diagrams, photographs, and tables. Full color adds clarity to the artwork and makes illustrations as realistic as possible. For example, vectors are color coded, and curves in graphs are drawn in color. The color photographs have been carefully selected, and their accompanying captions have been written to serve as an added instructional tool.

Mathematical Level We have introduced calculus gradually, keeping in mind that students often take introductory courses in calculus and physics concurrently. Most steps are shown when basic equations are developed, and reference is often made to mathematical appendices at the end of the textbook. Vector products are discussed in detail later in the text, where they are needed in physical applications. The dot product is introduced in Chapter 6, which addresses work and energy; the cross product is introduced in Chapter 10, which deals with rotational dynamics.

Significant Figures Significant figures in both worked examples and end-of-chapter problems have been handled with care. Most numerical examples and problems are worked out to either two or three significant figures, depending on the accuracy of the data provided.

Questions Questions requiring verbal responses are provided at the end of each chapter. Over 540 questions are included in the text. Some questions provide the student with a means of self-testing the concepts presented in the chapter. Others could serve as a basis for initiating classroom discussions. Answers to selected questions are included in the *Student Solutions Manual and Study Guide.*

Problems The end-of-chapter problems are more numerous in this edition and more varied (in all, over 1980 problems are given throughout the text). For the convenience of both the student and the instructor, about two thirds of the problems are keyed to specific sections of the chapter, including Context Connection sections. The remaining problems, labeled "Additional Problems," are not keyed to specific sections. The icon identifies problems dealing with applications to the life sciences and medicine. One or more problems in each chapter ask students to make an order-of-magnitude calculation based on their own estimated data. Other types of problems are described in more detail below. Answers to odd-numbered problems are provided at the end of the book.

Usually, the problems within a given section are presented so that the straightforward problems (those with black problem numbers) appear first. For ease of identification, the numbers of intermediate-level problems are printed in blue, and those of challenging problems are printed in magenta.

Solutions to approximately 20% of the problems in each chapter are in the *Student Solutions Manual and Study Guide.* Among these, selected problems are identified with PhysicsNow™ icons and have coached solutions available at **www.pop4e.com.**

Review Problems Many chapters include review problems requiring the student to relate concepts covered in the chapter to those discussed in previous chapters. These problems can be used by students in preparing for tests and by instructors in routine or special assignments and for classroom discussions.

Paired Problems As an aid for students learning to solve problems symbolically, paired numerical and symbolic problems are included in Chapters 1 through 4 and 16 through 21. Paired problems are identified by a common background screen.

Computer- and Calculator-Based Problems Many chapters include one or more problems whose solution requires the use of a computer or graphing calculator. Modeling of physical phenomena enables students to obtain graphical representations of variables and to perform numerical analyses.

Units The international system of units (SI) is used throughout the text. The U.S. customary system of units is used only to a limited extent in the chapters on mechanics and thermodynamics.

Summaries Each chapter contains a summary that reviews the important concepts and equations discussed in that chapter.

Appendices and Endpapers Several appendices are provided at the end of the textbook. Most of the appendix material represents a review of mathematical concepts and techniques used in the text, including scientific notation, algebra, geometry, trigonometry, differential calculus, and integral calculus. Reference to these appendices is made throughout the text. Most mathematical review sections in the appendices include worked examples and exercises with answers. In addition to the mathematical reviews, the appendices contain tables of physical data, conversion factors, atomic masses, and the SI units of physical quantities, as well as a periodic table of the elements and a list of Nobel Prize recipients. Other useful information, including fundamental constants and physical data, planetary data, a list of standard prefixes, mathematical symbols, the Greek alphabet, and standard abbreviations of units of measure, appears on the endpapers.

ANCILLARIES

The ancillary package has been updated substantially and streamlined in response to suggestions from users of the third edition. The most essential parts of the student package are the two-volume *Student Solutions Manual and Study Guide* with a tight focus on problem-solving and the Web-based ***PhysicsNow*™** learning system. Instructors will find increased support for their teaching efforts with new electronic materials.

Student Ancillaries

Student Solutions Manual and Study Guide by John R. Gordon, Ralph McGrew, and Raymond A. Serway. This two-volume manual features detailed solutions to approximately 20% of the end-of-chapter problems from the textbook. Boxed numbers identify those problems in the textbook whose complete solutions are found in the manual. The manual also features a summary of important chapter notes, a list of important equations and concepts, a short list of important study skills and strategies as well as answers to selected end-of-chapter conceptual questions.

PhysicsNow™ Students log into ***PhysicsNow*™** at **www.pop4e.com** by using the free access code packaged with this text.* The ***PhysicsNow*™** system is made up of three interrelated parts:

- How much do you know?
- What do you need to learn?
- What have you learned?

Students maximize their success by starting with the Pre-Test for the relevant chapter. Each Pre-Test is a mix of conceptual and numerical questions. After completing the Pre-Test, each student is presented with a detailed Learning Plan. The Learning Plan outlines elements to review in the text and Web-based media (Active Figures, Interactive Examples, and Coached Problems) in order to master the chapter's most essential concepts. After working through these materials, students move on to a multiple-choice Post-Test presenting them with questions similar to those that might appear on an exam. Results can be e-mailed to instructors.

WebTutor™ on WebCT and Blackboard WebTutor™ offers students real-time access to a full array of study tools, including a glossary of terms and a selection of animations.

The Brooks/Cole Physics Resource Center You'll find additional online quizzes, Web links, and animations at **academic.cengage.com/physics.**

Instructor's Ancillaries

The following ancillaries are available to qualified adopters. Please contact your local Cengage Learning sales representative for details.

Instructor's Solutions Manual by Ralph McGrew. This single manual contains worked solutions to all the problems in the textbook (Volumes 1 and 2) and answers to the end-of-chapter questions. The solutions to problems new to the fourth edition are marked for easy identification by the instructor.

Test Bank by Edward Adelson. Contains approximately 2,000 multiple-choice questions. It is provided in print form for the instructor who does not have access to a computer. The questions in the *Test Bank* are also available in electronic format with complete answers and solutions in iLrn Computerized Testing. The number of conceptual questions has been increased for the 4th edition.

Multimedia Manager This easy-to-use multimedia lecture tool allows you to quickly assemble art and database files with notes to create fluid lectures. The CD-ROM set (Volume 1, Chapters 1–15; Volume 2, Chapters 16–31) includes a database of animations, video clips, and digital art from the text as well as PowerPoint lectures and electronic files of the *Instructor's Solutions Manual* and *Test Bank.*

PhysicsNow™ ***PhysicsNow*™ Course Management Tools** This extension to the student tutorial environment of ***PhysicsNow*™** allows instructors to deliver online assignments in an environment that is familiar to students. This powerful system is your gateway to managing on-line homework, testing, and course administration all in one shell with the proven content to make your course a success. ***PhysicsNow*™** is a fully integrated testing, tutorial, and course management software accessible by instructors and students anytime, anywhere. To see a demonstration of this powerful system, contact your Thomson representative or go to **www.pop4e.com**.

PhysicsNow™ ***PhysicsNow*™ Homework Management** ***PhysicsNow*™** gives you a rich array of problem types and grading options. Its library of assignable questions includes all of the end-of-chapter problems from the text so that you can select the problems you want to

*Free access codes are only available with new copies of *Principles of Physics,* 4th edition.

include in your online homework assignments. These well-crafted problems are algorithmically generated so that you can assign the same problem with different variables for each student. A flexible grading tolerance feature allows you to specify a percentage range of correct answers so that your students are not penalized for rounding errors. You can give students the option to work an assignment multiple times and record the highest score or limit the times they are able to attempt it. In addition, you can create your own problems to complement the problems from the text. Results flow automatically to an exportable grade book so that instructors are better able to assess student understanding of the material, even prior to class or to an actual test.

iLrn Computerized Testing Extend the student experience with **PhysicsNow™** into a testing or quizzing environment. The test item file from the text is included to give you a bank of well-crafted questions that you can deliver online or print out. As with the homework problems, you can use the program's friendly interface to craft your own questions to complement the Serway/Jewett questions. You have complete control over grading, deadlines, and availability and can create multiple tests based on the same material.

WebTutor™ on WebCT and Blackboard With **WebTutor™**'s text-specific, pre-formatted content and total flexibility, instructors can easily create and manage their own personal Web site. **WebTutor™**'s course management tool gives instructors the ability to provide virtual office hours, post syllabi, set up threaded discussions, track student progress with the quizzing material, and much more. **WebTutor™** also provides robust communication tools, such as a course calendar, asynchronous discussion, real-time chat, a whiteboard, and an integrated e-mail system.

Additional Options for Online Homework

WebAssign: A Web-Based Homework System WebAssign is the most utilized homework system in physics. Designed by physicists for physicists, this system is a trusted companion to your teaching. An enhanced version of WebAssign is available for *Principles of Physics.* This enhanced version includes animations with conceptual questions and tutorial problems with feedback and hints to guide student content mastery. Take a look at this new innovation from the most trusted name in physics homework at **www.webassign.net.**

LON-CAPA: A Computer-Assisted Personalized Approach LON-CAPA is a Web-based course management system. For more information, visit the LON-CAPA Web site at **www.lon-capa.org.**

University of Texas Homework Service With this service, instructors can browse problem banks, select those problems they wish to assign to their students, and then let the Homework Service take over the delivery and grading. Details about and a demonstration of this service are available at **http://hw.ph.utexas.edu/hw.html.**

TEACHING OPTIONS

Although some topics found in traditional textbooks have been omitted from this textbook, instructors may find that the current text still contains more material than can be covered in a two-semester sequence. For this reason, we would like to offer the following suggestions. If you wish to place more emphasis on contemporary topics in physics, you should consider omitting parts or all of Chapters 15, 16, 17, 18, 24, 25, and 26. On the other hand, if you wish to follow a more traditional approach that places more emphasis on classical physics, you could omit Chapters 9, 11, 28, 29, 30, and 31. Either approach can be used without any loss in continuity. Other teaching options would fall somewhere between these two extremes by choosing to omit some or all of the following sections, which can be considered optional:

3.6 Relative Velocity
7.7 Energy Diagrams and Stability of Equilibrium
9.9 General Relativity
10.11 Rolling Motion of Rigid Objects
12.6 Damped Oscillations
12.7 Forced Oscillations
14.7 Nonsinusoidal Wave Patterns

15.8 Other Applications of Fluid Dynamics
16.6 Distribution of Molecular Speeds
17.7 Molar Specific Heats of Ideal Gases
17.8 Adiabatic Processes for an Ideal Gas
17.9 Molar Specific Heats and the Equipartition of Energy
20.10 Capacitors with Dielectrics
22.11 Magnetism in Matter
27.9 Diffraction of X-Rays by Crystals
28.13 Tunneling Through a Potential Energy Barrier

ACKNOWLEDGMENTS

The fourth edition of this textbook was prepared with the guidance and assistance of many professors who reviewed part or all of the manuscript, the pre-revision text, or both. We wish to acknowledge the following scholars and express our sincere appreciation for their suggestions, criticisms, and encouragement:

Anthony Aguirre, *University of California at Santa Cruz*

Royal Albridge, *Vanderbilt University*

Billy E. Bonner, *Rice University*

Richard Cardenas, *St. Mary's University*

Christopher R. Church, *Miami University (Ohio)*

Athula Herat, *Northern Kentucky University*

Huan Z. Huang, *University of California at Los Angeles*

George Igo, *University of California at Los Angeles*

Edwin Lo

Michael J. Longo, *University of Michigan*

Rafael Lopez-Mobilia, *University of Texas at San Antonio*

Ian S. McLean, *University of California at Los Angeles*

Richard Rolleigh, *Hendrix College*

Gregory Severn, *University of San Diego*

Satinder S. Sidhu, *Washington College*

Fiona Waterhouse, *University of California at Berkeley*

Principles of Physics, fourth edition was carefully checked for accuracy by James E. Rutledge (University of California at Irvine), Harry W. K. Tom (University of California at Riverside), Gregory Severn (University of San Diego), Bruce Mason (University of Oklahoma at Norman), and Ralf Rapp (Texas A&M University). We thank them for their dedication and vigilance.

We thank the following people for their suggestions and assistance during the preparation of earlier editions of this textbook:

Edward Adelson, *Ohio State University;* Yildirim M. Aktas, *University of North Carolina—Charlotte;* Alfonso M. Albano, *Bryn Mawr College;* Subash Antani, *Edgewood College;* Michael Bass, *University of Central Florida;* Harry Bingham, *University of California, Berkeley;* Anthony Buffa, *California Polytechnic State University, San Luis Obispo;* James Carolan, *University of British Columbia;* Kapila Clara Castoldi, *Oakland University;* Ralph V. Chamberlin, *Arizona State University;* Gary G. DeLeo, *Lehigh University;* Michael Dennin, *University of California, Irvine;* Alan J. DeWeerd, *Creighton University;* Madi Dogariu, *University of Central Florida;* Gordon Emslie, *University of Alabama at Huntsville;* Donald Erbsloe, *United States Air Force Academy;* William Fairbank, *Colorado State University;* Marco Fatuzzo, *University of Arizona;* Philip Fraundorf, *University of Missouri—St. Louis;* Patrick Gleeson, *Delaware State University;* Christopher M. Gould, *University of Southern California;* James D. Gruber, *Harrisburg Area Community College;* John B. Gruber, *San Jose State University;* Todd Hann, *United States Military Academy;* Gail Hanson, *Indiana University;* Gerald Hart, *Moorhead State University;* Dieter H. Hartmann, *Clemson University;* Richard W. Henry, *Bucknell University;* Laurent Hodges, *Iowa State University;* Michael J. Hones, *Villanova University;* Joey Huston, *Michigan State University;* Herb Jaeger, *Miami University;* David Judd, *Broward Community College;* Thomas H. Keil, *Worcester Polytechnic Institute;* V. Gordon Lind, *Utah State University;* Roger M. Mabe, *United States Naval Academy;*

David Markowitz, *University of Connecticut;* Thomas P. Marvin, *Southern Oregon University;* Martin S. Mason, *College of the Desert;* Wesley N. Mathews, Jr., *Georgetown University;* John W. McClory, *United States Military Academy;* L. C. McIntyre, Jr., *University of Arizona;* Alan S. Meltzer, *Rensselaer Polytechnic Institute;* Ken Mendelson, *Marquette University;* Roy Middleton, *University of Pennsylvania;* Allen Miller, *Syracuse University;* Clement J. Moses, *Utica College of Syracuse University;* John W. Norbury, *University of Wisconsin—Milwaukee;* Anthony Novaco, *Lafayette College;* Romulo Ochoa, *The College of New Jersey;* Melvyn Oremland, *Pace University;* Desmond Penny, *Southern Utah University;* Steven J. Pollock, *University of Colorado—Boulder;* Prabha Ramakrishnan, *North Carolina State University;* Rex D. Ramsier, *The University of Akron;* Rogers Redding, *University of North Texas;* Charles R. Rhyner, *University of Wisconsin—Green Bay;* Perry Rice, *Miami University;* Dennis Rioux, *University of Wisconsin—Oshkosh;* Janet E. Seger, *Creighton University;* Gregory D. Severn, *University of San Diego;* Antony Simpson, *Dalhousie University;* Harold Slusher, *University of Texas at El Paso;* J. Clinton Sprott, *University of Wisconsin at Madison;* Shirvel Stanislaus, *Valparaiso University;* Randall Tagg, *University of Colorado at Denver;* Cecil Thompson, *University of Texas at Arlington;* Chris Vuille, *Embry–Riddle Aeronautical University;* Robert Watkins, *University of Virginia;* James Whitmore, *Pennsylvania State University*

We are indebted to the developers of the IUPP models, "A Particles Approach to Introductory Physics" and "Physics in Context," upon which much of the pedagogical approach in this textbook is based.

Ralph McGrew coordinated the end-of-chapter problems. Problems new to this edition were written by Edward Adelson, Michael Browne, Andrew Duffy, Robert Forsythe, Perry Ganas, John Jewett, Randall Jones, Boris Korsunsky, Edwin Lo, Ralph McGrew, Clement Moses, Raymond Serway, and Jerzy Wrobel. Daniel Fernandez, David Tamres, and Kevin Kilty made corrections in problems from the previous edition.

We are grateful to John R. Gordon and Ralph McGrew for writing the *Student Solutions Manual and Study Guide,* to Ralph McGrew for preparing an excellent *Instructor's Solutions Manual,* and to Edward Adelson of Ohio State University for preparing the *Test Bank.* We thank M & N Toscano for the attractive layout of these volumes. During the development of this text, the authors benefited from many useful discussions with colleagues and other physics instructors, including Robert Bauman, William Beston, Don Chodrow, Jerry Faughn, John R. Gordon, Kevin Giovanetti, Dick Jacobs, Harvey Leff, Clem Moses, Dorn Peterson, Joseph Rudmin, and Gerald Taylor.

Special thanks and recognition go to the professional staff at Brooks/Cole, a part of Cengage Learning—in particular, Susan Pashos, Jay Campbell, Sarah Lowe, Seth Dobrin, Teri Hyde, Michelle Julet, David Harris, and Chris Hall—for their fine work during the development and production of this textbook. We are most appreciative of Sam Subity's masterful management of the ***PhysicsNow***™ media program. Julie Conover is our enthusiastic Marketing Manager, and Stacey Purviance coordinates our marketing communications. We recognize the skilled production service provided by Donna King and the staff at Progressive Publishing Alternatives and the dedicated photo research efforts of Dena Betz.

Finally, we are deeply indebted to our wives and children for their love, support, and long-term sacrifices.

RAYMOND A. SERWAY
St. Petersburg, Florida

JOHN W. JEWETT, JR.
Pomona, California

TO THE STUDENT

It is appropriate to offer some words of advice that should benefit you, the student. Before doing so, we assume you have read the Preface, which describes the various features of the text that will help you through the course.

HOW TO STUDY

Very often instructors are asked, "How should I study physics and prepare for examinations?" There is no simple answer to this question, but we would like to offer some suggestions based on our own experiences in learning and teaching over the years.

First and foremost, maintain a positive attitude toward the subject matter, keeping in mind that physics is the most fundamental of all natural sciences. Other science courses that follow will use the same physical principles, so it is important that you understand and are able to apply the various concepts and theories discussed in the text.

The Contexts in the text will help you understand how the physical principles relate to real issues, phenomena, and applications. Be sure to read the Context Introductions, Context Connection sections in each chapter, and Context Conclusions. These will be most helpful in motivating your study of physics.

CONCEPTS AND PRINCIPLES

It is essential that you understand the basic concepts and principles before attempting to solve assigned problems. You can best accomplish this goal by carefully reading the textbook before you attend your lecture on the covered material. When reading the text, you should jot down those points that are not clear to you. We've purposely left wide margins in the text to give you space for doing this. Also be sure to make a diligent attempt at answering the questions in the Quick Quizzes as you come to them in your reading. We have worked hard to prepare questions that help you judge for yourself how well you understand the material. Pay careful attention to the many Pitfall Preventions throughout the text. These will help you avoid misconceptions, mistakes, and misunderstandings as well as maximize the efficiency of your time by minimizing adventures along fruitless paths. During class, take careful notes and ask questions about those ideas that are unclear to you. Keep in mind that few people are able to absorb the full meaning of scientific material after only one reading.

After class, several readings of the text and your notes may be necessary. Be sure to take advantage of the features available in the ***PhysicsNow***™ learning system, such as the Active Figures, Interactive Examples, and Coached Problems. Your lectures and laboratory work supplement your reading of the textbook and should clarify some of the more difficult material. You should minimize your memorization of material. Successful memorization of passages from the text, equations, and derivations does not necessarily indicate that you understand the material.

Your understanding of the material will be enhanced through a combination of efficient study habits, discussions with other students and with instructors, and your ability to solve the problems presented in the textbook. Ask questions whenever you feel clarification of a concept is necessary.

STUDY SCHEDULE

It is important for you to set up a regular study schedule, preferably a daily one. Make sure you read the syllabus for the course and adhere to the schedule set by your instructor. The lectures will be much more meaningful if you read the corresponding textual material before attending them. As a general rule, you should devote about two hours of study time for every hour you are in class. If you are having trouble with the course, seek the advice of the instructor or other students who have taken the course. You may find it necessary to seek further instruction from experienced students. Very often, instructors offer review sessions in addition to regular class periods. It is important that you avoid the practice of delaying study until a day or two before an exam. More often than not, this approach has disastrous results. Rather than undertake an all-night study session, briefly review the basic concepts and equations and get a good night's rest. If you feel you need additional help in understanding the concepts, in preparing for exams, or in problem-solving, we suggest that you acquire a copy of the *Student Solutions Manual and Study Guide* that accompanies this textbook; this manual should be available at your college bookstore.

USE THE FEATURES

You should make full use of the various features of the text discussed in the preface. For example, marginal notes are useful for locating and describing important equations and concepts, and **boldfaced** type indicates important statements and definitions. Many useful tables are contained in the Appendices, but most tables are incorporated in the text where they are most often referenced. Appendix B is a convenient review of mathematical techniques.

Answers to odd-numbered problems are given at the end of the textbook, answers to Quick Quizzes are located at the end of each chapter, and answers to selected end-of-chapter questions are provided in the *Student Solutions Manual and Study Guide.* Problem-Solving Strategies are included in selected chapters throughout the text and give you additional information about how you should solve problems. The Table of Contents provides an overview of the entire text, while the Index enables you to locate specific material quickly. Footnotes sometimes are used to supplement the text or to cite other references on the subject discussed.

After reading a chapter, you should be able to define any new quantities introduced in that chapter and to discuss the principles and assumptions used to arrive at certain key relations. The chapter summaries and the review sections of the *Student Solutions Manual and Study Guide* should help you in this regard. In some cases, it may be necessary for you to refer to the index of the text to locate certain topics. You should be able to correctly associate with each physical quantity the symbol used to represent that quantity and the unit in which the quantity is specified. Furthermore, you should be able to express each important relation in a concise and accurate prose statement.

PROBLEM-SOLVING

R. P. Feynman, Nobel laureate in physics, once said, "You do not know anything until you have practiced." In keeping with this statement, we strongly advise that you develop the skills necessary to solve a wide range of problems. Your ability to solve problems will be one of the main tests of your knowledge of physics; therefore, you should try to solve as many problems as possible. It is essential that you understand basic concepts and principles before attempting to solve problems. It is good practice to try to find alternative solutions to the same problem. For example, you can solve problems in mechanics using Newton's laws, but very often an alternative method that draws on energy considerations is more direct. You should not deceive yourself into thinking you understand a problem merely because you have seen it solved in class. You must be able to solve the problem and similar problems on your own.

The approach to solving problems should be carefully planned. A systematic plan is especially important when a problem involves several concepts. First, read the problem several times until you are confident you understand what is being asked. Look for any key words that will help you interpret the problem and perhaps allow you to make certain assumptions. Your ability to interpret a question properly is an integral part of problem-solving. Second, you should acquire the habit of writing down the information given in a problem and those quantities that need to be found; for example, you might construct a table listing both the quantities given and the quantities to be found. This procedure is sometimes used in the worked examples of the textbook. After you have decided on the method you feel is appropriate for a given problem, proceed with your solution. Finally, check your results to see if they are reasonable and consistent with your initial understanding of the problem. General problem-solving strategies of this type are included in the text and are set off in their own boxes. We have also developed a General Problem-Solving Strategy, making use of models, to help guide you through complex problems. This strategy is located at the end of Chapter 1. If you follow the steps of this procedure, you will find it easier to come up with a solution and also gain more from your efforts.

Often, students fail to recognize the limitations of certain equations or physical laws in a particular situation. It is very important that you understand and remember the assumptions underlying a particular theory or formalism. For example, certain equations in kinematics apply only to a particle moving with constant acceleration. These equations are not valid for describing motion whose acceleration is not constant, such as the motion of an object connected to a spring or the motion of an object through a fluid.

EXPERIMENTS

Physics is a science based on experimental observations. In view of this fact, we recommend that you try to supplement the text by performing various types of "hands-on" experiments, either at home or in the laboratory. For example, the common Slinky™ toy is excellent for studying traveling waves; a ball swinging on the end of a long string can be used to investigate pendulum motion; various masses attached to the end of a vertical spring or rubber band can be used to determine their elastic nature; an old pair of Polaroid sunglasses and some discarded lenses and a magnifying glass are the components of various experiments in optics; and the approximate measure of the free-fall acceleration can be determined simply by measuring with a stopwatch the time it takes for a ball to drop from a known height. The list of such experiments is endless. When physical models are not available, be imaginative and try to develop models of your own.

NEW MEDIA

We strongly encourage you to use the ***PhysicsNow***™ Web-based learning system that accompanies this textbook. It is far easier to understand physics if you see it in action, and these new materials will enable you to become a part of that action. ***PhysicsNow***™ media described in the Preface are accessed at the URL **www.pop4e.com,** and feature a three-step learning process consisting of a Pre-Test, a personalized learning plan, and a Post-Test.

In addition to the Coached Problems identified with icons, ***PhysicsNow***™ includes the following Active Figures and Interactive Examples:

Chapter 1

Active Figures 1.4, 1.9, and 1.16

Interactive Example 1.8

Chapter 2

Active Figures 2.1, 2.2, 2.8, 2.11, and 2.12

Interactive Examples 2.8 and 2.10

Chapter 3

Active Figures 3.4, 3.5, 3.8, and 3.12

Interactive Examples 3.2 and 3.6

Chapter 4

Active Figures 4.12 and 4.13

Interactive Examples 4.4 and 4.5

Chapter 5

Active Figures 5.1, 5.9, 5.15, and 5.18

Interactive Examples 5.7 and 5.8

Chapter 6

Active Figure 6.8

Interactive Examples 6.6 and 6.7

Chapter 7

Active Figures 7.3, 7.6, and 7.15

Interactive Examples 7.1 and 7.2

Chapter 8

Active Figures 8.8, 8.9, 8.11, 8.13, and 8.14

Interactive Examples 8.2 and 8.8

Chapter 9

Active Figures 9.3, 9.5, and 9.8

Interactive Example 9.5

Chapter 10

Active Figures 10.4, 10.11, 10.12, 10.21, and 10.28

Interactive Examples 10.5, 10.8, and 10.9

Chapter 11

Active Figures 11.1, 11.5, 11.7, 11.19, and 11.20

Interactive Examples 11.1 and 11.3

Chapter 12

Active Figures 12.1, 12.2, 12.4, 12.6, 12.9, 12.10, 12.11, and 12.14

Interactive Example 12.1

Chapter 13

Active Figures 13.6, 13.7, 13.8, 13.14, 13.15, 13.21, 13.22, and 13.24

Interactive Examples 13.5 and 13.7

Chapter 14

Active Figures 14.1, 14.2, 14.3, 14.8, 14.9, 14.12, 14.15, and 14.16

Interactive Examples 14.1 and 14.3

Chapter 15

Active Figures 15.9 and 15.10

Interactive Examples 15.4 and 15.7

Chapter 16

Active Figures 16.9, 16.14, 16.16, 16.17, and 16.18

Interactive Example 16.4

Chapter 17

Active Figures 17.5, 17.6, 17.8, and 17.13

Interactive Example 17.9

Chapter 18

Active Figures 18.1, 18.5, 18.6, and 18.7

Interactive Example 18.3

Chapter 19

Active Figures 19.7, 19.10, 19.19, 19.21, 19.26, and 19.31

Interactive Examples 19.1, 19.7, and 19.10

Chapter 20

Active Figures 20.6, 20.20, 20.23, and 20.24

Interactive Examples 20.2, 20.3, 20.8, and 20.9

Chapter 21

Active Figures 21.4, 21.10, 21.13, 21.14, 21.16, 21.25, and 21.27

Interactive Examples 21.2, 21.6, 21.8, and 21.9

Chapter 22

Active Figures 22.1, 22.7, 22.8, 22.11, 22.12, 22.20, 22.27, and 22.28

Interactive Examples 22.3 and 22.6

Chapter 23

Active Figures 23.2, 23.3, 23.11, 23.14, 23.23, 23.24, 23.26, and 23.27

Interactive Examples 23.3, 23.4 and 23.8

Chapter 24

Active Figures 24.3, 24.8, 24.14, and 24.16

Interactive Examples 24.1 and 24.4

Chapter 25

Active Figures 25.2, 25.5, 25.8, 25.9, 25.16, 25.22, 25.28, and 25.30

Interactive Examples 25.1 and 25.3

Chapter 26

Active Figures 26.2, 26.12, 26.17, and 26.24

Interactive Examples 26.2, 26.3, 26.7 and 26.8

Chapter 27

Active Figures 27.2, 27.14, 27.21, and 27.22

Interactive Examples 27.1, 27.3, 27.5, and 27.7

Chapter 28

Active Figures 28.2, 28.7, 28.8, 28.9, 28.16, 28.17, 28.19, 28.23 and 28.24

Interactive Examples 28.3, 28.4, 28.9, and 28.12

Chapter 29

Active Figure 29.6

Interactive Example 29.6

Chapter 30

Active Figures 30.1, 30.11, 30.12, 30.13, 30.14, 30.16, 30.17, and 30.21

Interactive Examples 30.3 and 30.6

Chapter 31

Active Figure 31.11

Interactive Example 31.2

It is our sincere hope that you too will find physics an exciting and enjoyable experience and that you will profit from this experience, regardless of your chosen profession. Welcome to the exciting world of physics!

> *The scientist does not study nature because it is useful; he studies it because he delights in it, and he delights in it because it is beautiful. If nature were not beautiful, it would not be worth knowing, and if nature were not worth knowing, life would not be worth living.*
>
> **Henri Poincaré**

List of Life Science Applications and Problems

An Invitation to Physics

(© Stockbyte)

Technicians use electronic devices to test motherboards for computer systems. The principles of physics are involved in the design, manufacturing, and testing of these motherboards. ■

Physics, the most fundamental physical science, is concerned with the basic principles of the universe. It is the foundation on which engineering, technology, and the other sciences—astronomy, biology, chemistry, and geology—are based. The beauty of physics lies in the simplicity of its fundamental theories and in the manner in which just a small number of basic concepts, equations, and assumptions can alter and expand our view of the world around us.

Classical physics, developed prior to 1900, includes the theories, concepts, laws, and experiments in classical mechanics, thermodynamics, electromagnetism, and optics. For example, Galileo Galilei (1564–1642) made significant contributions to classical mechanics through his work on the laws of motion with constant acceleration. In the same era, Johannes Kepler (1571–1630) used astronomical observations to develop empirical laws for the motions of planetary bodies.

The most important contributions to classical mechanics, however, were provided by Isaac Newton (1642–1727), who developed classical mechanics as a systematic theory and was one of the originators of calculus as a mathematical tool. Although major developments in classical physics continued in the 18th century, thermodynamics and electromagnetism were not developed until the latter part of the 19th century, principally because the apparatus for controlled experiments was either too crude or unavailable until then. Although many electric and magnetic phenomena had been studied earlier, the work of James Clerk Maxwell (1831–1879) provided a unified theory of electromagnetism. In this text, we shall treat the various disciplines of classical physics in separate sections; we will see, however, that the disciplines of mechanics and electromagnetism are basic to all the branches of physics.

A major revolution in physics, usually referred to as *modern physics,* began near the end of the 19th century. Modern physics developed mainly because many physical phenomena could not be explained by classical physics. The two most important developments in this modern era were the theories of relativity and quantum mechanics. Albert Einstein's theory of relativity completely revolutionized the traditional concepts of space, time, and energy. This theory correctly describes the motion of objects moving at speeds comparable to the speed of light. The theory of relativity also shows that the speed of light is the upper limit of the speed of an object and that mass and energy are related. Quantum mechanics was formulated by a number of distinguished scientists to provide descriptions of physical phenomena at the atomic level.

Scientists continually work at improving our understanding of fundamental laws, and new discoveries are made every day. In many research areas, a great deal of overlap exists among physics, chemistry, and biology. Evidence for this overlap is seen in the names of some subspecialties in science: biophysics, biochemistry, chemical physics, biotechnology, and so on. Numerous technological advances in recent times are the result of the efforts of many scientists, engineers, and technicians. Some of the most notable developments in the latter half of the 20th century were (1) space missions to the Moon and other planets, (2) microcircuitry and high-speed computers, (3) sophisticated imaging techniques used in scientific research and medicine, and (4) several remarkable accomplishments in genetic engineering. The impact of such developments and discoveries on society has indeed been great, and future discoveries and developments will very likely be exciting, challenging, and of great benefit to humanity.

To investigate the impact of physics on developments in our society, we will use a *contextual* approach to the study of the content in this textbook. The book is divided into nine *Contexts,* which relate the physics to social issues, natural phenomena, or technological applications, as outlined here:

Chapters	Context
2–7	Alternative-Fuel Vehicles
8–11	Mission to Mars
12–14	Earthquakes
15	Search for the *Titanic*
16–18	Global Warming
19–21	Lightning
22–23	Magnetic Levitation Vehicles
24–27	Lasers
28–31	The Cosmic Connection

The Contexts provide a story line for each section of the text, which will help provide relevance and motivation for studying the material.

Each Context begins with a discussion of the topic, culminating in a *central question,* which forms the focus for the study of the physics in the Context. The final section of each chapter is a Context Connection, in which the material in the chapter is explored with the central question in mind. At the end of each Context, a

A technician works on the H1 detector in the Hadron Electron Accelerator Ring at the Deutsche Elektronen Synchrotron near Hamburg, Germany. Technicians educated in the physical sciences contribute their skills in many areas of modern technology. ■

Context Conclusion brings together all the principles necessary to respond as fully as possible to the central question.

In Chapter 1, we investigate some of the mathematical fundamentals and problem-solving strategies that we will use in our study of physics. The first Context, *Alternative-Fuel Vehicles,* is introduced just before Chapter 2; in this Context, the principles of classical mechanics are applied to the problem of designing, developing, producing, and marketing a vehicle that will help to reduce dependence on foreign oil and emit fewer harmful by-products into the atmosphere than current gasoline engines.

Introduction and Vectors

These controls in the cockpit of a commercial aircraft assist the pilot in maintaining control over the *velocity* of the aircraft—how fast it is traveling and in what direction it is traveling—allowing it to land safely. Quantities that are defined by both a magnitude and a direction, such as velocity, are called *vectors*.

(Mark Wagner/Stone/Getty Images)

CHAPTER OUTLINE

The goal of physics is to provide a quantitative understanding of certain basic phenomena that occur in our Universe. Physics is a science based on experimental observations and mathematical analyses. The main objectives behind such experiments and analyses are to develop theories that explain the phenomenon being studied and to relate those theories to other established theories. Fortunately, it is possible to explain the behavior of various physical systems using relatively few fundamental laws. Analytical procedures require the expression of those laws in the language of mathematics, the tool that provides a bridge between theory and experiment. In this chapter, we shall discuss a few mathematical concepts and techniques that will be used throughout the text. In addition, we will outline an effective problem-solving strategy that should be adopted and used in your problem-solving activities throughout the text.

PhysicsNow™ This icon throughout the text indicates an opportunity for you to test yourself on key concepts and explore animations and interactions on the PhysicsNow Web site at **http://www.pop4e.com.**

1.1 STANDARDS OF LENGTH, MASS, AND TIME

If we measure a certain quantity and wish to describe it to someone, a unit for the quantity must be specified and defined. For example, it would be meaningless for a visitor from another planet to talk to us about a length of 8 "glitches" if we did not know the meaning of the unit glitch. On the other hand, if someone familiar with our system of measurement reports that a wall is 2.0 meters high and our unit of length is defined to be 1.0 meter, we then know that the height of the wall is twice our fundamental unit of length. An international committee has agreed on a system of definitions and standards to describe fundamental physical quantities. It is called the **SI system** (Système International) of units. Its units of length, mass, and time are the meter, kilogram, and second, respectively.

Length

In A.D. 1120, King Henry I of England decreed that the standard of length in his country would be the yard and that the yard would be precisely equal to the distance from the tip of his nose to the end of his outstretched arm. Similarly, the original standard for the foot adopted by the French was the length of the royal foot of King Louis XIV. This standard prevailed until 1799, when the legal standard of length in France became the **meter,** defined as one ten-millionth of the distance from the equator to the North Pole.

Many other systems have been developed in addition to those just discussed, but the advantages of the French system have caused it to prevail in most countries and in scientific circles everywhere. Until 1960, the length of the meter was defined as the distance between two lines on a specific bar of platinum–iridium alloy stored under controlled conditions. This standard was abandoned for several reasons, a principal one being that the limited accuracy with which the separation between the lines can be determined does not meet the current requirements of science and technology. The definition of the meter was modified to be equal to 1 650 763.73 wavelengths of orange–red light emitted from a krypton-86 lamp. In October 1983, the meter was redefined to be **the distance traveled by light in a vacuum during a time interval of 1/299 792 458 second.** This value arises from the establishment of the speed of light in a vacuum as exactly 299 792 458 meters per second. We will use the standard scientific notation for numbers with more than three digits in which groups of three digits are separated by spaces rather than commas. Therefore, 1 650 763.73 and 299 792 458 in this paragraph are the same as the more popular American cultural notations of 1,650,763.73 and 299,792,458. Similarly, $\pi = 3.14159265$ is written as 3.141 592 65.

■ Definition of the meter

Mass

Mass represents a measure of the resistance of an object to changes in its motion. The SI unit of mass, the **kilogram,** is defined as **the mass of a specific platinum–iridium alloy cylinder kept at the International Bureau of Weights and Measures at Sèvres, France.** At this point, we should add a word of caution. Many beginning students of physics tend to confuse the physical quantities called *weight* and *mass.* For the present we shall not discuss the distinction between them; they will be clearly defined in later chapters. For now you should note that they are distinctly different quantities.

■ Definition of the kilogram

Time

Before 1960, the standard of time was defined in terms of the average length of a solar day in the year 1900. (A solar day is the time interval between successive appearances of the Sun at the highest point it reaches in the sky each day.) The basic

(Courtesy of National Institute of Standards and Technology, U.S. Department of Commerce)

FIGURE 1.1 The nation's primary time standard is a cesium fountain atomic clock developed at the National Institute of Standards and Technology laboratories in Boulder, Colorado. The clock will neither gain nor lose a second in 20 million years.

unit of time, the **second,** was defined to be $(1/60)(1/60)(1/24) = 1/86\,400$ of the average solar day. In 1967, the second was redefined to take advantage of the great precision obtainable with a device known as an atomic clock (Fig. 1.1), which uses the characteristic frequency of the cesium-133 atom as the "reference clock." The second is now defined as **9 192 631 770 times the period of oscillation of radiation from the cesium atom.** It is possible today to purchase clocks and watches that receive radio signals from an atomic clock in Colorado, which the clock or watch uses to continuously reset itself to the correct time.

Definition of the second

Approximate Values for Length, Mass, and Time

Approximate values of various lengths, masses, and time intervals are presented in Tables 1.1, 1.2, and 1.3, respectively. Note the wide range of values for these quantities.[1] You should study the tables and begin to generate an intuition for what is meant by a mass of 100 kilograms, for example, or by a time interval of 3.2×10^7 seconds.

Systems of units commonly used in science, commerce, manufacturing, and everyday life are (1) the *SI system,* in which the units of length, mass, and time are the meter (m), kilogram (kg), and second (s), respectively; and (2) the *U.S. customary system,* in which the units of length, mass, and time are the foot (ft), slug, and second, respectively. Throughout most of this text we shall use SI units because they are almost universally accepted in science and industry. We will make limited use of U.S. customary units in the study of classical mechanics.

Some of the most frequently used prefixes for the powers of ten and their abbreviations are listed in Table 1.4. For example, 10^{-3} m is equivalent to 1 millimeter (mm), and 10^3 m is 1 kilometer (km). Likewise, 1 kg is 10^3 grams (g), and 1 megavolt (MV) is 10^6 volts (V).

The variables length, time, and mass are examples of *fundamental quantities.* A much larger list of variables contains *derived quantities,* or quantities that can be expressed as a mathematical combination of fundamental quantities. Common examples are *area,* which is a product of two lengths, and *speed,* which is a ratio of a length to a time interval.

PITFALL PREVENTION 1.1

Reasonable values The generation of intuition about typical values of quantities suggested here is critical. An important step in solving problems is to think about your result at the end of a problem and determine if it seems reasonable. If you are calculating the mass of a housefly and arrive at a value of 100 kg, this value is *unreasonable;* there is an error somewhere. If you are calculating the length of a spacecraft on a launch pad and end up with a value of 10 cm, this value is *unreasonable;* and you should look for a mistake.

[1] If you are unfamiliar with the use of powers of ten (scientific notation), you should review Appendix B.1.

TABLE 1.1 Approximate Values of Some Measured Lengths

	Length (m)
Distance from the Earth to the most remote quasar known	1.4×10^{26}
Distance from the Earth to the most remote normal galaxies known	4×10^{25}
Distance from the Earth to the nearest large galaxy (M 31, the Andromeda galaxy)	2×10^{22}
Distance from the Sun to the nearest star (Proxima Centauri)	4×10^{16}
One lightyear	9.46×10^{15}
Mean orbit radius of the Earth	1.5×10^{11}
Mean distance from the Earth to the Moon	3.8×10^{8}
Distance from the equator to the North Pole	1×10^{7}
Mean radius of the Earth	6.4×10^{6}
Typical altitude of an orbiting Earth satellite	2×10^{5}
Length of a football field	9.1×10^{1}
Length of this textbook	2.8×10^{-1}
Length of a housefly	5×10^{-3}
Size of smallest visible dust particles	1×10^{-4}
Size of cells of most living organisms	1×10^{-5}
Diameter of a hydrogen atom	1×10^{-10}
Diameter of a uranium nucleus	1.4×10^{-14}
Diameter of a proton	1×10^{-15}

TABLE 1.2 Masses of Various Objects (Approximate Values)

	Mass (kg)
Visible Universe	10^{52}
Milky Way galaxy	10^{42}
Sun	2×10^{30}
Earth	6×10^{24}
Moon	7×10^{22}
Shark	3×10^{2}
Human	7×10^{1}
Frog	1×10^{-1}
Mosquito	1×10^{-5}
Bacterium	1×10^{-15}
Hydrogen atom	1.67×10^{-27}
Electron	9.11×10^{-31}

Another example of a derived quantity is **density.** The density ρ (Greek letter rho; a table of the letters in the Greek alphabet is provided at the back of the book) of any substance is defined as its *mass per unit volume:*

$$\rho \equiv \frac{m}{V} \qquad [1.1]$$

■ Definition of density

which is a ratio of mass to a product of three lengths. For example, aluminum has a density of 2.70×10^3 kg/m^3, and lead has a density of 11.3×10^3 kg/m^3. An extreme difference in density can be imagined by thinking about holding a 10-centimeter (cm) cube of Styrofoam in one hand and a 10-cm cube of lead in the other.

TABLE 1.3 Approximate Values of Some Time Intervals

	Time Interval (s)
Age of the Universe	5×10^{17}
Age of the Earth	1.3×10^{17}
Time interval since the fall of the Roman empire	5×10^{12}
Average age of a college student	6.3×10^{8}
One year	3.2×10^{7}
One day (time interval for one revolution of the Earth about its axis)	8.6×10^{4}
One class period	3.0×10^{3}
Time interval between normal heartbeats	8×10^{-1}
Period of audible sound waves	1×10^{-3}
Period of typical radio waves	1×10^{-6}
Period of vibration of an atom in a solid	1×10^{-13}
Period of visible light waves	2×10^{-15}
Duration of a nuclear collision	1×10^{-22}
Time interval for light to cross a proton	3.3×10^{-24}

TABLE 1.4 Some Prefixes for Powers of Ten

Power	Prefix	Abbreviation
10^{-24}	yocto	y
10^{-21}	zepto	z
10^{-18}	atto	a
10^{-15}	femto	f
10^{-12}	pico	p
10^{-9}	nano	n
10^{-6}	micro	μ
10^{-3}	milli	m
10^{-2}	centi	c
10^{-1}	deci	d
10^{3}	kilo	k
10^{6}	mega	M
10^{9}	giga	G
10^{12}	tera	T
10^{15}	peta	P
10^{18}	exa	E
10^{21}	zetta	Z
10^{24}	yotta	Y

1.2 DIMENSIONAL ANALYSIS

The word *dimension* has a special meaning in physics. It denotes the physical nature of a quantity. Whether a distance is measured in units of feet or meters or miles, it is a distance. We say its dimension is *length.*

The symbols used in this book to specify the dimensions[2] of length, mass, and time are L, M, and T, respectively. We shall often use square brackets [ ] to denote the dimensions of a physical quantity. For example, in this notation the dimensions of velocity v are written $[v] = \mathrm{L/T}$, and the dimensions of area A are $[A] = \mathrm{L}^2$. The dimensions of area, volume, velocity, and acceleration are listed in Table 1.5, along with their units in the two common systems. The dimensions of other quantities, such as force and energy, will be described as they are introduced in the text.

In many situations, you may be faced with having to derive or check a specific equation. Although you may have forgotten the details of the derivation, a useful and powerful procedure called *dimensional analysis* can be used as a consistency check, to assist in the derivation, or to check your final expression. Dimensional analysis makes use of the fact that **dimensions can be treated as algebraic quantities.** For example, quantities can be added or subtracted only if they have the same dimensions. Furthermore, the terms on both sides of an equation must have the same dimensions. By following these simple rules, you can use dimensional analysis to help determine whether an expression has the correct form because the relationship can be correct only if the dimensions on the two sides of the equation are the same.

To illustrate this procedure, suppose you wish to derive an expression for the position x of a car at a time t if the car starts from rest at $t = 0$ and moves with constant acceleration a. In Chapter 2, we shall find that the correct expression for this special case is $x = \frac{1}{2}at^2$. Let us check the validity of this expression from a dimensional analysis approach.

The quantity x on the left side has the dimension of length. For the equation to be dimensionally correct, the quantity on the right side must also have the dimension of length. We can perform a dimensional check by substituting the basic dimensions for acceleration, $\mathrm{L/T}^2$ (Table 1.5), and time, T, into the equation $x = \frac{1}{2}at^2$. That is, the dimensional form of the equation $x = \frac{1}{2}at^2$ can be written as

$$[x] = \frac{\mathrm{L}}{\cancel{\mathrm{T}^2}} \cdot \cancel{\mathrm{T}^2} = \mathrm{L}$$

The dimensions of time cancel as shown, leaving the dimension of length, which is the correct dimension for the position x. Notice that the number $\frac{1}{2}$ in the equation has no units, so it does not enter into the dimensional analysis.

QUICK QUIZ 1.1 True or false: Dimensional analysis can give you the numerical value of constants of proportionality that may appear in an algebraic expression.

TABLE 1.5 Units of Area, Volume, Velocity, and Acceleration

System	Area (L^2)	Volume (L^3)	Velocity (L/T)	Acceleration ($\mathrm{L/T}^2$)
SI	m^2	m^3	m/s	$\mathrm{m/s}^2$
U.S. customary	ft^2	ft^3	ft/s	$\mathrm{ft/s}^2$

[2] The *dimensions* of a variable will be symbolized by a capitalized, nonitalic letter, such as, in the case of length, L. The *symbol* for the variable itself will be italicized, such as L for the length of an object or t for time.

EXAMPLE 1.1 **Analysis of an Equation**

Show that the expression $v_f = v_i + at$ is dimensionally correct, where v_f and v_i represent velocities at two instants of time, a is acceleration, and t is an instant of time.

Solution The dimensions of the velocities are

$$[v_f] = [v_i] = \frac{\text{L}}{\text{T}}$$

and the dimensions of acceleration are L/T^2. Therefore, the dimensions of at are

$$[at] = \frac{\text{L}}{\text{T}^2}\cdot\text{T} = \frac{\text{L}}{\text{T}}$$

and the expression is dimensionally correct. On the other hand, if the expression were given as $v_f = v_i + at^2$, it would be dimensionally *incorrect.* Try it and see!

1.3 CONVERSION OF UNITS

Sometimes it is necessary to convert units from one system to another or to convert within a system, for example, from kilometers to meters. Equalities between SI and U.S. customary units of length are as follows:

$$1 \text{ mile (mi)} = 1\,609 \text{ m} = 1.609 \text{ km} \qquad 1 \text{ ft} = 0.304\,8 \text{ m} = 30.48 \text{ cm}$$

$$1 \text{ m} = 39.37 \text{ in.} = 3.281 \text{ ft} \qquad 1 \text{ inch (in.)} = 0.025\,4 \text{ m} = 2.54 \text{ cm}$$

A more complete list of equalities can be found in Appendix A.

Units can be treated as algebraic quantities that can cancel each other. To perform a conversion, a quantity can be multiplied by a **conversion factor,** which is a fraction equal to 1, with numerator and denominator having different units, to provide the desired units in the final result. For example, suppose we wish to convert 15.0 in. to centimeters. Because 1 in. = 2.54 cm, we multiply by a conversion factor that is the appropriate ratio of these equal quantities and find that

$$15.0 \text{ in.} = (15.0 \text{ in.})\left(\frac{2.54 \text{ cm}}{1 \text{ in.}}\right) = 38.1 \text{ cm}$$

where the ratio in parentheses is equal to 1. Notice that we put the unit of an inch in the denominator and that it cancels with the unit in the original quantity. The remaining unit is the centimeter, which is our desired result.

PITFALL PREVENTION 1.2

ALWAYS INCLUDE UNITS When performing calculations, make it a habit to include the units with every quantity and carry the units through the entire calculation. Avoid the temptation to drop the units during the calculation steps and then apply the expected unit to the number that results for an answer. By including the units in every step, you can detect errors if the units for the answer are incorrect.

QUICK QUIZ 1.2 The distance between two cities is 100 mi. The number of kilometers in the distance between the two cities is (**a**) smaller than 100, (**b**) larger than 100, (**c**) equal to 100.

EXAMPLE 1.2 **Is He Speeding?**

On an interstate highway in a rural region of Wyoming, a car is traveling at a speed of 38.0 m/s.

A Is this car exceeding the speed limit of 75.0 mi/h?

Solution We first convert meters to miles:

$$(38.0 \text{ m/s})\left(\frac{1 \text{ mi}}{1\,609 \text{ m}}\right) = 2.36 \times 10^{-2} \text{ mi/s}$$

Now we convert seconds to hours:

$$(2.36 \times 10^{-2} \text{ mi/s})\left(\frac{60 \text{ s}}{1 \text{ min}}\right)\left(\frac{60 \text{ min}}{1 \text{ h}}\right) = 85.0 \text{ mi/h}$$

Therefore, the car is exceeding the speed limit and should slow down.

B What is the speed of the car in kilometers per hour?

Solution We convert our answer in part A to the appropriate units:

$$(85.0 \text{ mi/h})\left(\frac{1.609 \text{ km}}{1 \text{ mi}}\right) = 137 \text{ km/h}$$

Figure 1.2 shows the speedometer of an automobile, with speeds in both miles per hour and kilometers per hour. Can you check the conversion we just performed using this photograph?

(Phil Boorman/Getty Images)

FIGURE 1.2 (Example 1.2) The speedometer of this vehicle shows speeds in both miles per hour and kilometers per hour.

1.4 ORDER-OF-MAGNITUDE CALCULATIONS

It is often useful to compute an approximate answer to a given physical problem even when little information is available. This answer can then be used to determine whether a more precise calculation is necessary. Such an approximation is usually based on certain assumptions, which must be modified if greater precision is needed. Therefore, we will sometimes refer to an *order of magnitude* of a certain quantity as the power of ten of the number that describes that quantity. Usually, when an order-of-magnitude calculation is made, the results are reliable to within about a factor of 10. If a quantity increases in value by three orders of magnitude, its value increases by a factor of $10^3 = 1\,000$. We use the symbol ~ for "is on the order of." Therefore,

$$0.008\,6 \sim 10^{-2} \qquad 0.0021 \sim 10^{-3} \qquad 700 \sim 10^3$$

EXAMPLE 1.3 The Number of Atoms in a Solid

Estimate the number of atoms in 1 cm^3 of a solid.

Solution From Table 1.1 we note that the diameter d of an atom is about 10^{-10} m. Let us assume that the atoms in the solid are spheres of this diameter. Then the volume of each sphere is about 10^{-30} m^3 (more precisely, volume $= 4\pi r^3/3 = \pi d^3/6$, where $r = d/2$). Therefore, because 1 $cm^3 = 10^{-6}$ m^3, the number of atoms in the solid is on the order of $10^{-6}/10^{-30} = 10^{24}$ atoms.

A more precise calculation would require additional knowledge that we could find in tables. Our estimate, however, agrees with the more precise calculation to within a factor of 10.

EXAMPLE 1.4 How Much Gas Do We Use?

Estimate the number of gallons of gasoline used by all cars in the United States each year.

Solution Because there are about 280 million people in the United States, an estimate of the number of cars in the country is 7×10^7 (assuming one car and four people per family). We can also estimate that the average distance traveled per year is 1×10^4 miles. If we assume gasoline consumption of 0.05 gal/mi (equivalent to 20 miles per gallon), each car uses about 5×10^2 gal/year. Multiplying this number by the total number of cars in the United States gives an estimated total consumption of about 10^{11} gal, which corresponds to a yearly consumer expenditure on the order of 10^2 billion dollars. This estimate is probably low because we haven't accounted for commercial consumption.

1.5 SIGNIFICANT FIGURES

When certain quantities are measured, the measured values are known only to within the limits of the experimental uncertainty. The value of the uncertainty can depend on various factors, such as the quality of the apparatus, the skill of the experimenter, and the number of measurements performed. The number of **significant figures** in a measurement can be used to express something about the uncertainty.

As an example of significant figures, consider the population of New York State, as reported in a published road atlas: 18 976 457. Notice that this number reports the population *to the level of one individual.* We would describe this number as having eight significant figures. Can the population really be this accurate? First of all, is the census process accurate enough to measure the population to one individual? By the time this number was actually published, had the number of births and immigrations into the state balanced the number of deaths and emigrations out of the state, so that the change in the population is exactly zero?

The claim that the population is measured and known to the level of one individual is unjustified. We would describe it by saying that *there are too many significant figures in the measurement.* To account for the inherent uncertainty in the census-taking process and the inevitable changes in population by the time the number is read in the road atlas, it might be better to report the population as something like 19.0 million. This number has three significant figures rather than the eight significant figures in the published population.

Let us look at a more scientific example. Suppose we are asked in a laboratory experiment to measure the area of a rectangular plate using a meter stick as a measuring instrument. Let us assume that the accuracy to which we can measure a particular dimension of the plate is ± 0.1 cm. If the length of the plate is measured to be 16.3 cm, we can claim only that its length lies somewhere between 16.2 cm and 16.4 cm. In this case, we say that the measured value has three significant figures. Likewise, if its width is measured to be 4.5 cm, the actual value lies between 4.4 cm and 4.6 cm. This measured value has only two significant figures. Note that the significant figures include the first estimated digit. Therefore, we could write the measured values as 16.3 ± 0.1 cm and 4.5 ± 0.1 cm.

Suppose we would now like to find the area of the plate by multiplying the two measured values. If we were to claim that the area is $(16.3 \text{ cm})(4.5 \text{ cm}) = 73.35 \text{ cm}^2$, our answer would be unjustifiable because it contains four significant figures, which is greater than the number of significant figures in either of the measured lengths. The following is a good rule of thumb to use in determining the number of significant figures that can be claimed:

> When multiplying several quantities, the number of significant figures in the final answer is the same as the number of significant figures in the quantity having the lowest number of significant figures. The same rule applies to division.

Applying this rule to the previous multiplication example, we see that the answer for the area can have only two significant figures because the length of 4.5 cm has only two significant figures. Therefore, all we can claim is that the area is 73 cm^2, realizing that the value can range between $(16.2 \text{ cm})(4.4 \text{ cm}) = 71 \text{ cm}^2$ and $(16.4 \text{ cm})(4.6 \text{ cm}) = 75 \text{ cm}^2$.

Zeros may or may not be significant figures. Those used to position the decimal point in such numbers as 0.03 and 0.007 5 are not significant. Therefore, there are one and two significant figures, respectively, in these two values. When the positioning of zeros comes after other digits, however, there is the possibility of misinterpretation. For example, suppose the mass of an object is given as 1 500 g. This value is ambiguous because we do not know whether the two zeros are being used to locate the decimal point or whether they represent significant figures in the measurement. To remove this ambiguity, it is common to use scientific notation to indicate the number of significant figures. In this case, we would express the mass as 1.5×10^3 g if the measured value has two significant figures, 1.50×10^3 g if it has three significant figures, and 1.500×10^3 g if it has four significant figures. Likewise, 0.000 150 should be expressed in scientific notation as 1.5×10^{-4} if it has two significant figures or as 1.50×10^{-4} if it has three significant figures. The three zeros between the decimal point and the digit 1 in the number 0.000 150 are not counted as significant figures because they are present only to locate the decimal point. In general, a **significant figure** in a measurement is a reliably known digit (other than a zero used to locate the decimal point) or the first estimated digit.

For addition and subtraction, the number of decimal places must be considered when you are determining how many significant figures to report.

> When numbers are added or subtracted, the number of decimal places in the result should equal the smallest number of decimal places of any term in the sum.

For example, if we wish to compute 123 + 5.35, the answer is 128 and not 128.35. If we compute the sum 1.000 1 + 0.000 3 = 1.000 4, the result has the correct number of decimal places; consequently, it has five significant figures even though one of the terms in the sum, 0.000 3, has only one significant figure. Likewise, if we perform the subtraction 1.002 − 0.998 = 0.004, the result has only one significant figure even though one term has four significant figures and the other has three. In this book, **most of the numerical examples and end-of-chapter problems will yield answers having three significant figures.**

If the number of significant figures in the result of an addition or subtraction must be reduced, a general rule for rounding numbers states that the last digit retained is to be increased by 1 if the last digit dropped is greater than 5. If the last digit dropped is less than 5, the last digit retained remains as it is. If the last digit dropped is equal to 5, the last digit retained should be rounded to the nearest even number. (This rule helps avoid accumulation of errors in long arithmetic processes.)

EXAMPLE 1.5 The Area of a Dish 

A biologist is filling a rectangular dish with growth culture and wishes to know the area of the dish. The length of the dish is measured to be 12.71 cm (four significant figures), and the width is measured to be 7.46 cm (three significant figures). Find the area of the dish.

Solution If you multiply 12.71 cm by 7.46 cm on your calculator, you will obtain an answer of 94.816 6 cm^2. How many of these numbers should you claim? Our rule of thumb for multiplication tells us that you can claim only the number of significant figures in the quantity with the smallest number of significant figures. In this example, that number is three (in the width 7.46 cm), so we should express our final answer as 94.8 cm^2.

1.6 COORDINATE SYSTEMS

Many aspects of physics deal in some way or another with locations in space. For example, the mathematical description of the motion of an object requires a method for specifying the object's position. Therefore, we first discuss how to describe the position of a point in space by means of coordinates in a graphical representation. A point on a line can be located with one coordinate, a point in a plane is located with two coordinates, and three coordinates are required to locate a point in space.

A coordinate system used to specify locations in space consists of

- A fixed reference point O, called the origin
- A set of specified axes or directions with an appropriate scale and labels on the axes
- Instructions that tell us how to label a point in space relative to the origin and axes

One convenient coordinate system that we will use frequently is the *Cartesian coordinate system,* sometimes called the *rectangular coordinate system.* Such a system in two dimensions is illustrated in Figure 1.3. An arbitrary point in this system is labeled with the coordinates (x, y). Positive x is taken to the right of the origin, and positive y is upward from the origin. Negative x is to the left of the origin, and negative y is downward from the origin. For example, the point P, which has coordinates (5, 3), may be reached by going first 5 m to the right of the origin and then 3 m above the origin (*or* by going 3 m above the origin and then 5 m to the right). Similarly, the point Q has coordinates $(-3, 4)$, which correspond to going 3 m to the left of the origin and 4 m above the origin.

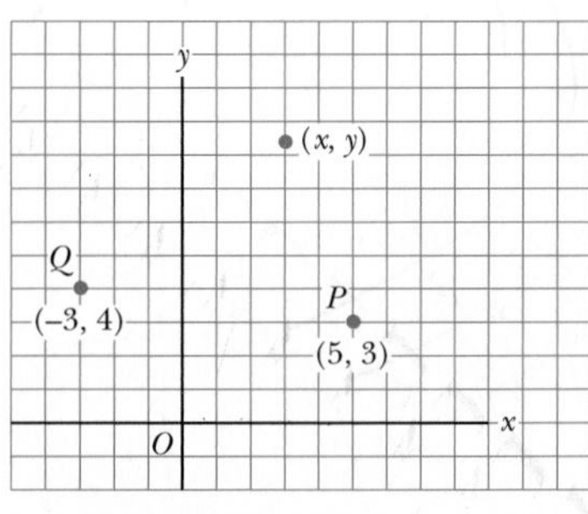

FIGURE 1.3 Designation of points in a Cartesian coordinate system. Each square in the xy plane is 1 m on a side. Every point is labeled with coordinates (x, y).

Sometimes it is more convenient to represent a point in a plane by its *plane polar coordinates* (r, θ), as in Active Figure 1.4a. In this coordinate system, r is the length of the line from the origin to the point, and θ is the angle between that line and a fixed axis, usually the positive x axis, with θ measured counterclockwise. From the right triangle in Active Figure 1.4b, we find that $\sin\theta = y/r$ and $\cos\theta = x/r$. (A review of trigonometric functions is given in Appendix B.4.) Therefore, starting with plane polar coordinates, one can obtain the Cartesian coordinates through the equations

$$x = r\cos\theta \qquad [1.2]$$

$$y = r\sin\theta \qquad [1.3]$$

Furthermore, it follows that

$$\tan\theta = \frac{y}{x} \qquad [1.4]$$

and

$$r = \sqrt{x^2 + y^2} \qquad [1.5]$$

You should note that these expressions relating the coordinates (x, y) to the coordinates (r, θ) apply only when θ is defined as in Active Figure 1.4a, where positive θ is an angle measured *counterclockwise* from the positive x axis. Other choices are

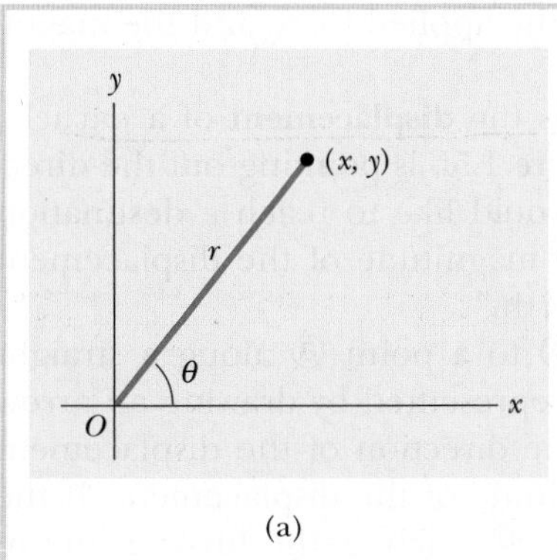

(a)

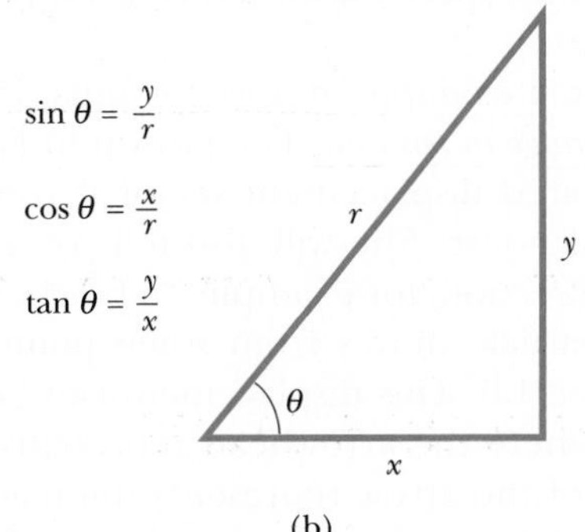

(b)

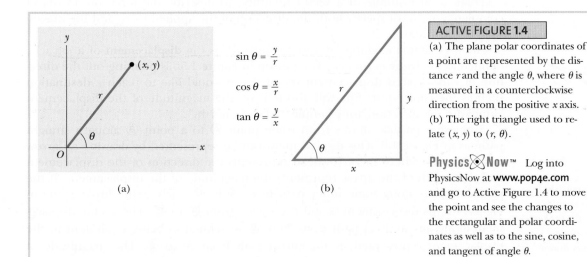

ACTIVE FIGURE 1.4
(a) The plane polar coordinates of a point are represented by the distance r and the angle θ, where θ is measured in a counterclockwise direction from the positive x axis. (b) The right triangle used to relate (x, y) to (r, θ).

PhysicsNow™ Log into PhysicsNow at **www.pop4e.com** and go to Active Figure 1.4 to move the point and see the changes to the rectangular and polar coordinates as well as to the sine, cosine, and tangent of angle θ.

(a)

(b)

FIGURE 1.5 (a) The number of grapes in this bunch is one example of a scalar quantity. Can you think of other examples? (b) This helpful person pointing in the correct direction tells us to travel five blocks north to reach the courthouse. A vector is a physical quantity that is specified by both magnitude and direction.

made in navigation and astronomy. If the reference axis for the polar angle θ is chosen to be other than the positive x axis or if the sense of increasing θ is chosen differently, the corresponding expressions relating the two sets of coordinates will change.

1.7 VECTORS AND SCALARS

Each of the physical quantities that we shall encounter in this text can be placed in one of two categories, either a scalar or a vector. A **scalar** is a quantity that is completely specified by a positive or negative number with appropriate units. On the other hand, a **vector** is a physical quantity that must be specified by both magnitude and direction.

The number of grapes in a bunch (Fig. 1.5a) is an example of a scalar quantity. If you are told that there are 38 grapes in the bunch, this statement completely specifies the information; no specification of direction is required. Other examples of scalars are temperature, volume, mass, and time intervals. The rules of ordinary arithmetic are used to manipulate scalar quantities; they can be freely added and subtracted (assuming that they have the same units!), multiplied and divided.

Force is an example of a vector quantity. To describe the force on an object completely, we must specify both the direction of the applied force and the magnitude of the force.

Another simple example of a vector quantity is the **displacement** of a particle, defined as its *change in position.* The person in Figure 1.5b is pointing out the direction of your desired displacement vector if you would like to reach a destination such as the courthouse. She will also tell you the magnitude of the displacement along with the direction, for example, "5 blocks north."

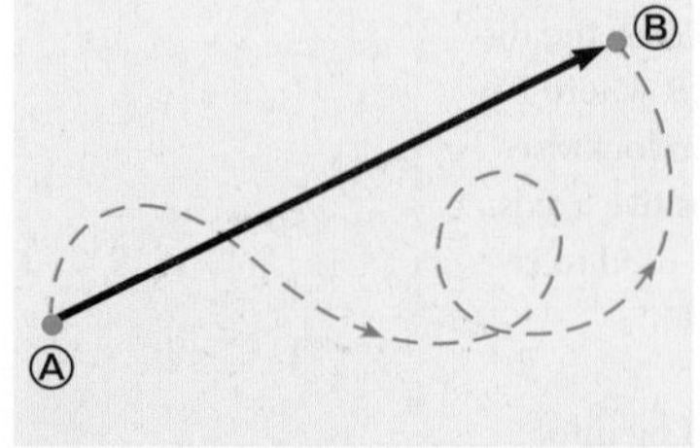

FIGURE 1.6 After a particle moves from Ⓐ to Ⓑ along an arbitrary path represented by the broken line, its displacement is a vector quantity shown by the arrow drawn from Ⓐ to Ⓑ.

Suppose a particle moves from some point Ⓐ to a point Ⓑ along a straight path, as in Figure 1.6. This displacement can be represented by drawing an arrow from Ⓐ to Ⓑ, where the arrowhead represents the direction of the displacement and the length of the arrow represents the magnitude of the displacement. If the particle travels along some other path from Ⓐ to Ⓑ, such as the broken line in Figure 1.6, its displacement is still the vector from Ⓐ to Ⓑ. The vector displacement along any indirect path from Ⓐ to Ⓑ is defined as being equivalent to the displacement represented by the direct path from Ⓐ to Ⓑ. The magnitude of the displacement is the shortest distance between the end points. Therefore, **the**

displacement of a particle is completely known if its initial and final coordinates are known. The path need not be specified. In other words, the **displacement is independent of the path** if the end points of the path are fixed.

Note that the **distance** traveled by a particle is distinctly different from its displacement. The distance traveled (a scalar quantity) is the length of the path, which in general can be much greater than the magnitude of the displacement. In Figure 1.6, the length of the curved red path is much larger than the magnitude of the black displacement vector.

If the particle moves along the x axis from position x_i to position x_f, as in Figure 1.7, its displacement is given by $x_f - x_i$. (The indices i and f refer to the initial and final values.) We use the Greek letter delta (Δ) to denote the *change* in a quantity. Therefore, we define the change in the position of the particle (the displacement) as

$$\Delta x \equiv x_f - x_i \qquad [1.6]$$

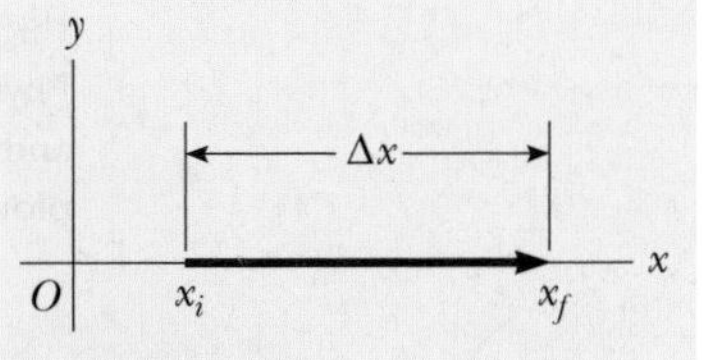

FIGURE 1.7 A particle moving along the x axis from x_i to x_f undergoes a displacement $\Delta x = x_f - x_i$.

From this definition we see that Δx is positive if x_f is greater than x_i and negative if x_f is less than x_i. For example, if a particle changes its position from $x_i = -5$ m to $x_f = 3$ m, its displacement is $\Delta x = +8$ m.

Many physical quantities in addition to displacement are vectors. They include velocity, acceleration, force, and momentum, all of which will be defined in later chapters. In this text, we will use boldface letters with an arrow over the letter, such as $\vec{\mathbf{A}}$, to represent vectors. Another common notation for vectors with which you should be familiar is a simple boldface character: **A**.

The magnitude of the vector $\vec{\mathbf{A}}$ is written with an italic letter A or, alternatively, $|\vec{\mathbf{A}}|$. The magnitude of a vector is always positive and carries the units of the quantity that the vector represents, such as meters for displacement or meters per second for velocity. Vectors combine according to special rules, which will be discussed in Sections 1.8 and 1.9.

QUICK QUIZ 1.3 Which of the following are scalar quantities and which are vector quantities? **(a)** your age **(b)** acceleration **(c)** velocity **(d)** speed **(e)** mass

■ Thinking Physics 1.1

Consider your commute to work or school in the morning. Which is larger, the distance you travel or the magnitude of the displacement vector?

Reasoning Unless you have a very unusual commute, the distance traveled *must* be larger than the magnitude of the displacement vector. The distance includes all the twists and turns you make in following the roads from home to work or school. On the other hand, the magnitude of the displacement vector is the length of a straight line from your home to work or school. This length is often described informally as "the distance as the crow flies." The only way that the distance could be the same as the magnitude of the displacement vector is if your commute is a perfect straight line, which is highly unlikely! The distance could *never* be less than the magnitude of the displacement vector because the shortest distance between two points is a straight line. ■

1.8 SOME PROPERTIES OF VECTORS

Equality of Two Vectors

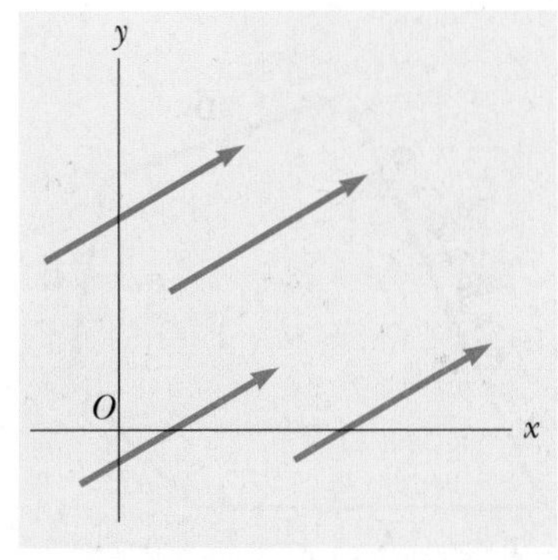

FIGURE 1.8 These four representations of vectors are equal because all four vectors have the same magnitude and point in the same direction.

Two vectors $\vec{\mathbf{A}}$ and $\vec{\mathbf{B}}$ are defined to be equal if they have the same units, the same magnitude, and the same direction. That is, $\vec{\mathbf{A}} = \vec{\mathbf{B}}$ only if $A = B$ *and* $\vec{\mathbf{A}}$ and $\vec{\mathbf{B}}$ point in the same direction. For example, all the vectors in Figure 1.8 are equal even

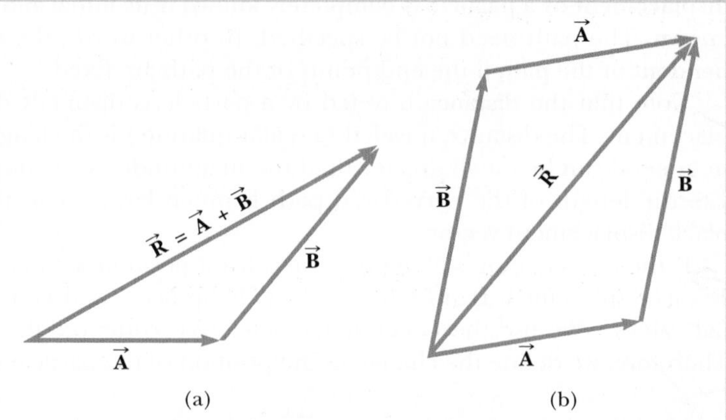

ACTIVE FIGURE 1.9
(a) When vector $\vec{\mathbf{B}}$ is added to vector $\vec{\mathbf{A}}$, the resultant $\vec{\mathbf{R}}$ is the vector that runs from the tail of $\vec{\mathbf{A}}$ to the tip of $\vec{\mathbf{B}}$. (b) This construction shows that $\vec{\mathbf{A}} + \vec{\mathbf{B}} = \vec{\mathbf{B}} + \vec{\mathbf{A}}$; vector addition is commutative.

PhysicsNow™ Log into PhysicsNow at **www.pop4e.com** and go to Active Figure 1.9 to explore the addition of two vectors.

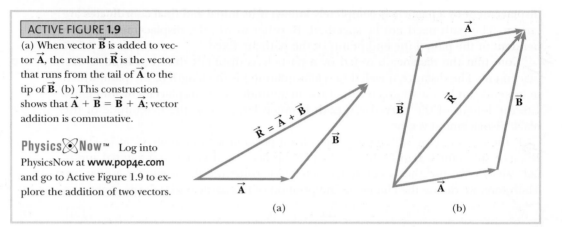

though they have different starting points. This property allows us to translate a vector parallel to itself in a diagram without affecting the vector.

Addition

When two or more vectors are added together, they must *all* have the same units. For example, it would be meaningless to add a velocity vector to a displacement vector because they are different physical quantities. Scalars obey the same rule. For example, it would be meaningless to add time intervals and temperatures.

The rules for vector sums are conveniently described using geometry. To add vector $\vec{\mathbf{B}}$ to vector $\vec{\mathbf{A}}$, first draw a diagram of vector $\vec{\mathbf{A}}$ on graph paper, with its magnitude represented by a convenient scale, and then draw vector $\vec{\mathbf{B}}$ to the same scale with its tail starting from the tip of $\vec{\mathbf{A}}$, as in Active Figure 1.9a. The *resultant vector* $\vec{\mathbf{R}} = \vec{\mathbf{A}} + \vec{\mathbf{B}}$ is the vector drawn from the tail of $\vec{\mathbf{A}}$ to the tip of $\vec{\mathbf{B}}$. If these vectors are displacements, $\vec{\mathbf{R}}$ is the single displacement that has the same effect as the displacements $\vec{\mathbf{A}}$ and $\vec{\mathbf{B}}$ performed one after the other. This process is known as the *triangle method of addition* because the three vectors can be geometrically modeled as the sides of a triangle.

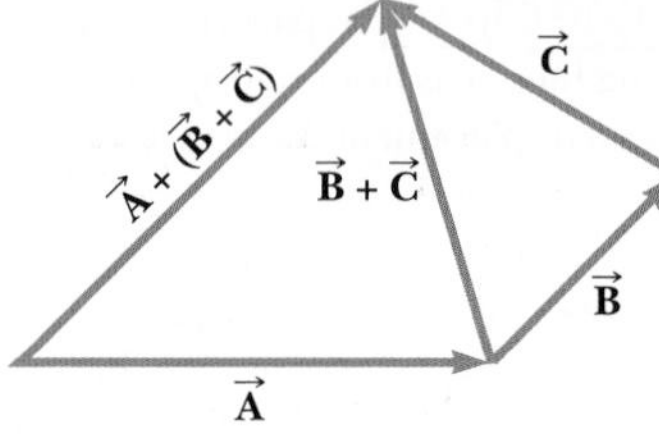

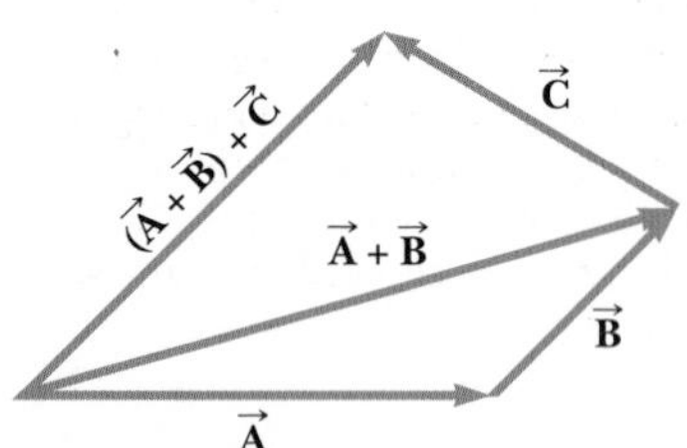

FIGURE 1.10 Geometric constructions for verifying the associative law of addition.

When vectors are added, the sum is independent of the order of the addition. This independence can be seen for two vectors from the geometric construction in Active Figure 1.9b and is known as the **commutative law of addition:**

$$\vec{\mathbf{A}} + \vec{\mathbf{B}} = \vec{\mathbf{B}} + \vec{\mathbf{A}} \qquad [1.7]$$

If three or more vectors are added, their sum is independent of the way in which they are grouped. A geometric demonstration of this property for three vectors is given in Figure 1.10. It is called the **associative law of addition:**

$$\vec{\mathbf{A}} + (\vec{\mathbf{B}} + \vec{\mathbf{C}}) = (\vec{\mathbf{A}} + \vec{\mathbf{B}}) + \vec{\mathbf{C}} \qquad [1.8]$$

Geometric constructions can also be used to add more than three vectors, as shown in Figure 1.11 for the case of four vectors. The resultant vector $\vec{\mathbf{R}} = \vec{\mathbf{A}} + \vec{\mathbf{B}} + \vec{\mathbf{C}} + \vec{\mathbf{D}}$ is the *vector that closes the polygon formed by the vectors being added.* In other words, $\vec{\mathbf{R}}$ is the *vector drawn from the tail of the first vector to the tip of the last vector.* Again, the order of the summation is unimportant.

We conclude that a vector is a quantity that has both magnitude and direction and also obeys the laws of vector addition described in Figures 1.9 to 1.11.

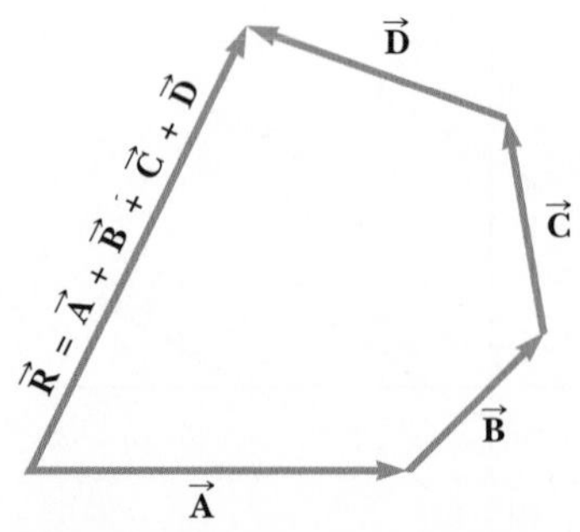

FIGURE 1.11 Geometric construction for summing four vectors. The resultant vector $\vec{\mathbf{R}}$ closes the polygon and points from the tail of the first vector to the tip of the final vector.

Negative of a Vector

The negative of the vector $\vec{\mathbf{A}}$ is defined as the vector that, when added to $\vec{\mathbf{A}}$, gives zero for the vector sum. That is, $\vec{\mathbf{A}} + (-\vec{\mathbf{A}}) = 0$. The vectors $\vec{\mathbf{A}}$ and $-\vec{\mathbf{A}}$ have the same magnitude but opposite directions.

Subtraction of Vectors

The operation of vector subtraction makes use of the definition of the negative of a vector. We define the operation $\vec{\mathbf{A}} - \vec{\mathbf{B}}$ as vector $-\vec{\mathbf{B}}$ added to vector $\vec{\mathbf{A}}$:

$$\vec{\mathbf{A}} - \vec{\mathbf{B}} = \vec{\mathbf{A}} + (-\vec{\mathbf{B}}) \qquad [1.9]$$

A diagram for subtracting two vectors is shown in Figure 1.12.

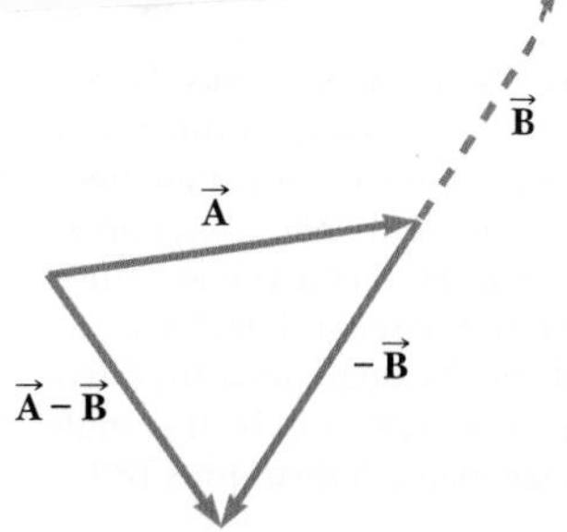

FIGURE 1.12 This construction shows how to subtract vector $\vec{\mathbf{B}}$ from vector $\vec{\mathbf{A}}$: Add the vector $-\vec{\mathbf{B}}$ to vector $\vec{\mathbf{A}}$. The vector $-\vec{\mathbf{B}}$ is equal in magnitude and opposite to the vector $\vec{\mathbf{B}}$.

Multiplication of a Vector by a Scalar

If a vector $\vec{\mathbf{A}}$ is multiplied by a positive scalar quantity s, the product $s\vec{\mathbf{A}}$ is a vector that has the same direction as $\vec{\mathbf{A}}$ and magnitude sA. If s is a negative scalar quantity, the vector $s\vec{\mathbf{A}}$ is directed opposite to $\vec{\mathbf{A}}$. For example, the vector $5\vec{\mathbf{A}}$ is five times greater in magnitude than $\vec{\mathbf{A}}$ and has the same direction as $\vec{\mathbf{A}}$. On the other hand, the vector $-\frac{1}{3}\vec{\mathbf{A}}$ has one third the magnitude of $\vec{\mathbf{A}}$ and points in the direction opposite $\vec{\mathbf{A}}$ (because of the negative sign).

Multiplication of Two Vectors

Two vectors $\vec{\mathbf{A}}$ and $\vec{\mathbf{B}}$ can be multiplied in two different ways to produce either a scalar or a vector quantity. The **scalar product** (or dot product) $\vec{\mathbf{A}} \cdot \vec{\mathbf{B}}$ is a scalar quantity equal to $AB \cos \theta$, where θ is the angle between $\vec{\mathbf{A}}$ and $\vec{\mathbf{B}}$. The **vector product** (or cross product) $\vec{\mathbf{A}} \times \vec{\mathbf{B}}$ is a vector quantity whose magnitude is equal to $AB \sin \theta$. We shall discuss these products more fully in Chapters 6 and 10, where they are first used.

PITFALL PREVENTION 1.3

VECTOR ADDITION VERSUS SCALAR ADDITION Keep in mind that $\vec{\mathbf{A}} + \vec{\mathbf{B}} = \vec{\mathbf{C}}$ is very different from $A + B = C$. The first is a vector sum, which must be handled carefully, such as with the graphical method described in Active Figure 1.9. The second is a simple algebraic addition of numbers that is handled with the normal rules of arithmetic.

QUICK QUIZ 1.4 The magnitudes of two vectors $\vec{\mathbf{A}}$ and $\vec{\mathbf{B}}$ are $A = 12$ units and $B = 8$ units. Which of the following pairs of numbers represents the *largest* and *smallest* possible values for the magnitude of the resultant vector $\vec{\mathbf{R}} = \vec{\mathbf{A}} + \vec{\mathbf{B}}$? (a) 14.4 units, 4 units (b) 12 units, 8 units (c) 20 units, 4 units (d) none of these answers

QUICK QUIZ 1.5 If vector $\vec{\mathbf{B}}$ is added to vector $\vec{\mathbf{A}}$, under what condition does the resultant vector $\vec{\mathbf{A}} + \vec{\mathbf{B}}$ have magnitude $A + B$? (a) $\vec{\mathbf{A}}$ and $\vec{\mathbf{B}}$ are parallel and in the same direction. (b) $\vec{\mathbf{A}}$ and $\vec{\mathbf{B}}$ are parallel and in opposite directions. (c) $\vec{\mathbf{A}}$ and $\vec{\mathbf{B}}$ are perpendicular.

1.9 COMPONENTS OF A VECTOR AND UNIT VECTORS

The geometric method of adding vectors is not the recommended procedure for situations in which great precision is required or in three-dimensional problems because we are forced to represent them on two-dimensional paper. In this section, we describe a method of adding vectors that makes use of the *projections* of a vector along the axes of a rectangular coordinate system.

Consider a vector $\vec{\mathbf{A}}$ lying in the xy plane and making an arbitrary angle θ with the positive x axis, as in Figure 1.13a. The vector $\vec{\mathbf{A}}$ can be represented by its rectangular **components,** A_x and A_y. The component A_x represents the projection of $\vec{\mathbf{A}}$ along the x axis, and A_y represents the projection of $\vec{\mathbf{A}}$ along the y axis. The components of a vector, which are scalar quantities, can be positive or negative. For example, in Figure 1.13a, A_x and A_y are both positive. The absolute values of the components are the magnitudes of the associated **component vectors** $\vec{\mathbf{A}}_x$ and $\vec{\mathbf{A}}_y$.

Figure 1.13b shows the component vectors again, but with the y component vector shifted so that it is added vectorially to the x component vector. This diagram

PITFALL PREVENTION 1.4

Tangents on Calculators Generally, the inverse tangent function on calculators provides an angle between −90° and +90°. As a consequence, if the vector you are studying lies in the second or third quadrant, the angle measured from the positive x axis will be the angle your calculator returns plus 180°.

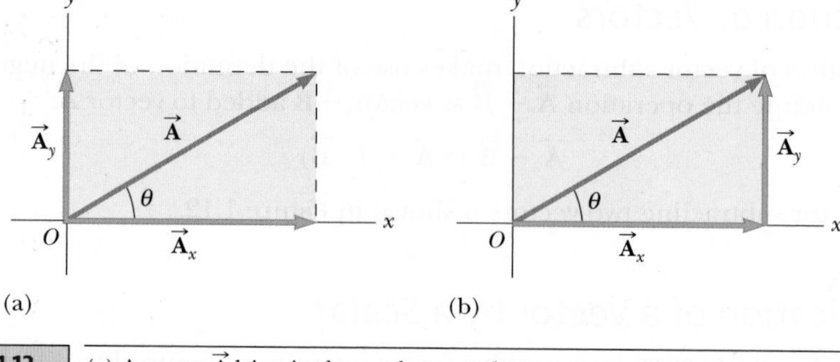

FIGURE 1.13 (a) A vector $\vec{\mathbf{A}}$ lying in the xy plane can be represented by its component vectors $\vec{\mathbf{A}}_x$ and $\vec{\mathbf{A}}_y$. (b) The y component vector $\vec{\mathbf{A}}_y$ can be moved to the right so that it adds to $\vec{\mathbf{A}}_x$. The vector sum of the component vectors is $\vec{\mathbf{A}}$. These three vectors form a right triangle.

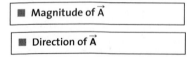

	y	
A_x negative A_y positive	A_x positive A_y positive	
A_x negative A_y negative	A_x positive A_y negative	x

FIGURE 1.14 The signs of the components of a vector $\vec{\mathbf{A}}$ depend on the quadrant in which the vector is located.

shows us two important features. First, a vector is equal to the sum of its component vectors. Therefore, the combination of the component vectors is a valid substitute for the actual vector. The second feature is that the vector and its component vectors form a right triangle. Therefore, we can let the triangle be a model for the vector and can use right triangle trigonometry to analyze the vector. The legs of the triangle are of lengths proportional to the components (depending on what scale factor you have chosen), and the hypotenuse is of a length proportional to the magnitude of the vector.

From Figure 1.13b and the definition of the sine and cosine of an angle, we see that $\cos\theta = A_x/A$ and $\sin\theta = A_y/A$. Hence, the components of $\vec{\mathbf{A}}$ are given by

$$A_x = A\cos\theta \qquad \text{and} \qquad A_y = A\sin\theta \tag{1.10}$$

When using these component equations, θ must be measured counterclockwise from the positive x axis. From our triangle, it follows that the magnitude of $\vec{\mathbf{A}}$ and its direction are related to its components through the Pythagorean theorem and the definition of the tangent function:

■ Magnitude of $\vec{A}$

$$A = \sqrt{A_x^{\,2} + A_y^{\,2}} \tag{1.11}$$

■ Direction of $\vec{A}$

$$\tan\theta = \frac{A_y}{A_x} \tag{1.12}$$

To solve for θ, we can write $\theta = \tan^{-1}(A_y/A_x)$, which is read "$\theta$ equals the angle whose tangent is the ratio A_y/A_x." *Note that the signs of the components A_x and A_y depend on the angle θ.* For example, if $\theta = 120°$, A_x is negative and A_y is positive. On the other hand, if $\theta = 225°$, both A_x and A_y are negative. Figure 1.14 summarizes the signs of the components when $\vec{\mathbf{A}}$ lies in the various quadrants.

If you choose reference axes or an angle other than those shown in Figure 1.13, the components of the vector must be modified accordingly. In many applications, it is more convenient to express the components of a vector in a coordinate system having axes that are not horizontal and vertical but are still perpendicular to each other. Suppose a vector $\vec{\mathbf{B}}$ makes an angle θ' with the x' axis defined in Figure 1.15. The components of $\vec{\mathbf{B}}$ along these axes are given by $B_{x'} = B\cos\theta'$ and $B_{y'} = B\sin\theta'$, as in Equation 1.10. The magnitude and direction of $\vec{\mathbf{B}}$ are obtained from expressions equivalent to Equations 1.11 and 1.12. Therefore, we can express the components of a vector in *any* coordinate system that is convenient for a particular situation.

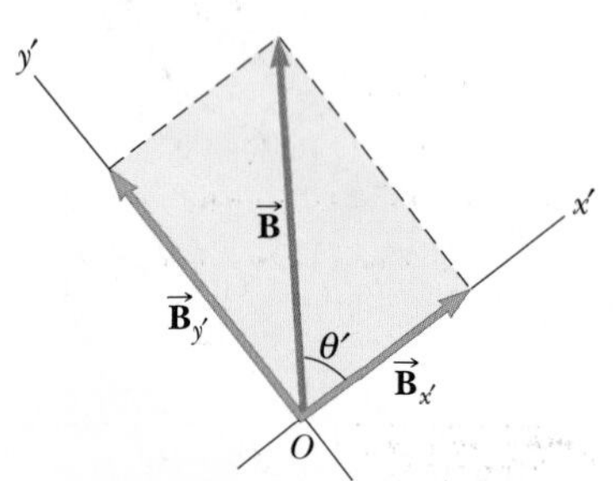

FIGURE 1.15 The component vectors of vector $\vec{\mathbf{B}}$ in a coordinate system that is tilted.

QUICK QUIZ 1.6 Choose the correct response to make the sentence true: A component of a vector is (**a**) always, (**b**) never, or (**c**) sometimes larger than the magnitude of the vector.

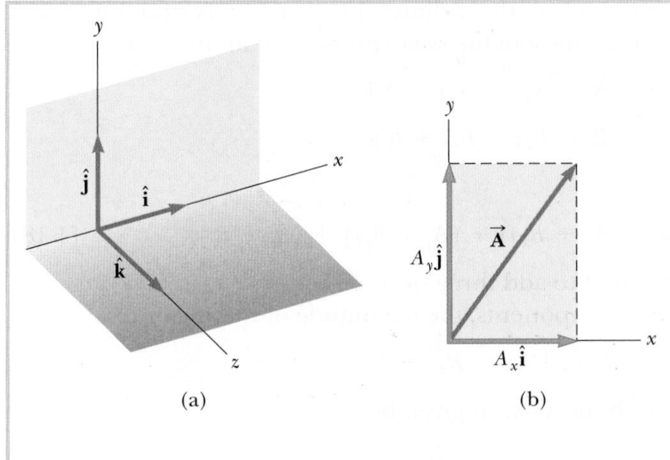

ACTIVE FIGURE 1.16

(a) The unit vectors $\hat{\mathbf{i}}$, $\hat{\mathbf{j}}$, and $\hat{\mathbf{k}}$ are directed along the *x*, *y*, and *z* axes, respectively. (b) A vector $\vec{\mathbf{A}}$ lying in the *xy* plane has component vectors $A_x\hat{\mathbf{i}}$ and $A_y\hat{\mathbf{j}}$, where A_x and A_y are the components of $\vec{\mathbf{A}}$.

Physics Now™ Log into PhysicsNow at **www.pop4e.com** and go to Active Figure 1.16 to rotate the coordinate axes in three-dimensional space and view a representation of vector $\vec{\mathbf{A}}$ in three dimensions.

Vector quantities are often expressed in terms of unit vectors. **A unit vector is a dimensionless vector with a magnitude of 1 and is used to specify a given direction.** Unit vectors have no other physical significance. They are used simply as a bookkeeping convenience when describing a direction in space. We will use the symbols $\hat{\mathbf{i}}$, $\hat{\mathbf{j}}$, and $\hat{\mathbf{k}}$ to represent unit vectors pointing in the *x*, *y*, and *z* directions, respectively. The "hat" over the letters is a common notation for a unit vector; for example, $\hat{\mathbf{i}}$ is called "i-hat." The unit vectors $\hat{\mathbf{i}}$, $\hat{\mathbf{j}}$, and $\hat{\mathbf{k}}$ form a set of mutually perpendicular vectors as shown in Active Figure 1.16a, where the magnitude of each unit vector equals 1; that is, $|\hat{\mathbf{i}}| = |\hat{\mathbf{j}}| = |\hat{\mathbf{k}}| = 1$.

Consider a vector $\vec{\mathbf{A}}$ lying in the *xy* plane, as in Active Figure 1.16b. The product of the component A_x and the unit vector $\hat{\mathbf{i}}$ is the component vector $\vec{\mathbf{A}}_x = A_x\hat{\mathbf{i}}$ parallel to the *x* axis with magnitude A_x. Likewise, $A_y\hat{\mathbf{j}}$ is a component vector of magnitude A_y parallel to the *y* axis. When using the unit-vector form of a vector, we are simply multiplying a vector (the unit vector) by a scalar (the component). Therefore, the unit-vector notation for the vector $\vec{\mathbf{A}}$ is written

$$\vec{\mathbf{A}} = A_x\hat{\mathbf{i}} + A_y\hat{\mathbf{j}} \qquad [1.13]$$

Now suppose we wish to add vector $\vec{\mathbf{B}}$ to vector $\vec{\mathbf{A}}$, where $\vec{\mathbf{B}}$ has components B_x and B_y. The procedure for performing this sum is simply to add the *x* and *y* components separately. The resultant vector $\vec{\mathbf{R}} = \vec{\mathbf{A}} + \vec{\mathbf{B}}$ is therefore

$$\vec{\mathbf{R}} = (A_x + B_x)\hat{\mathbf{i}} + (A_y + B_y)\hat{\mathbf{j}} \qquad [1.14]$$

From this equation, the components of the resultant vector are given by

$$R_x = A_x + B_x \qquad R_y = A_y + B_y \qquad [1.15]$$

The magnitude of $\vec{\mathbf{R}}$ and the angle it makes with the *x* axis can then be obtained from its components using the relationships

$$R = \sqrt{R_x^2 + R_y^2} = \sqrt{(A_x + B_x)^2 + (A_y + B_y)^2} \qquad [1.16]$$

$$\tan\theta = \frac{R_y}{R_x} = \frac{A_y + B_y}{A_x + B_x} \qquad [1.17]$$

The procedure just described for adding two vectors $\vec{\mathbf{A}}$ and $\vec{\mathbf{B}}$ using the component method can be checked using a diagram like Figure 1.17.

PITFALL PREVENTION 1.5

x **Components** Equation 1.10 for the *x* and *y* components of a vector associates the cosine of the angle with the *x* component and the sine of the angle with the *y* component. This association occurs *solely* because we chose to measure the angle with respect to the *x* axis, so don't memorize these equations. Invariably, you will face a problem in the future in which the angle is measured with respect to the *y* axis, and the equations will be incorrect. It is much better to always think about which side of the triangle containing the components is adjacent to the angle and which side is opposite, and then assign the sine and cosine accordingly.

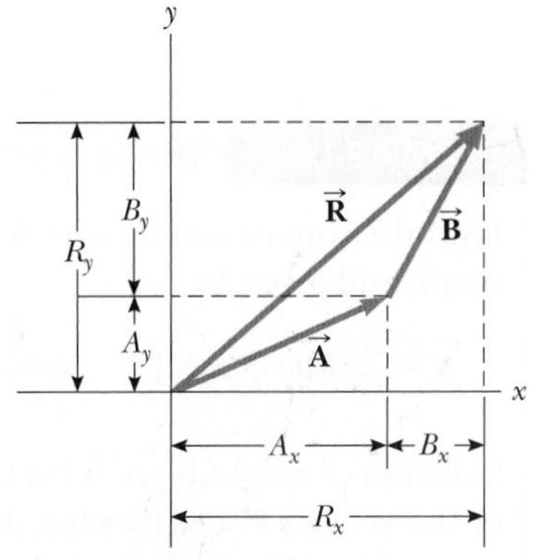

FIGURE 1.17 A geometric construction showing the relation between the components of the resultant $\vec{\mathbf{R}}$ of two vectors and the individual components.

The extension of these methods to three-dimensional vectors is straightforward. If $\vec{\mathbf{A}}$ and $\vec{\mathbf{B}}$ both have x, y, and z components, we express them in the form

$$\vec{\mathbf{A}} = A_x\hat{\mathbf{i}} + A_y\hat{\mathbf{j}} + A_z\hat{\mathbf{k}}$$

$$\vec{\mathbf{B}} = B_x\hat{\mathbf{i}} + B_y\hat{\mathbf{j}} + B_z\hat{\mathbf{k}}$$

The sum of $\vec{\mathbf{A}}$ and $\vec{\mathbf{B}}$ is

$$\vec{\mathbf{R}} = \vec{\mathbf{A}} + \vec{\mathbf{B}} = (A_x + B_x)\hat{\mathbf{i}} + (A_y + B_y)\hat{\mathbf{j}} + (A_z + B_z)\hat{\mathbf{k}} \qquad [1.18]$$

The same procedure can be used to add three or more vectors.

If a vector $\vec{\mathbf{R}}$ has x, y, and z components, the magnitude of the vector is

$$R = \sqrt{R_x^2 + R_y^2 + R_z^2}$$

The angle θ_x that $\vec{\mathbf{R}}$ makes with the x axis is given by

$$\cos\theta_x = \frac{R_x}{R}$$

with similar expressions for the angles with respect to the y and z axes.

QUICK QUIZ 1.7 If at least one component of a vector is a positive number, the vector cannot (**a**) have any component that is negative, (**b**) be zero, (**c**) have three dimensions.

QUICK QUIZ 1.8 If $\vec{\mathbf{A}} + \vec{\mathbf{B}} = 0$, the corresponding components of the two vectors $\vec{\mathbf{A}}$ and $\vec{\mathbf{B}}$ must be (**a**) equal, (**b**) positive, (**c**) negative, (**d**) of opposite sign.

■ Thinking Physics 1.2

You may have asked someone directions to a destination in a city and been told something like, "Walk 3 blocks east and then 5 blocks south." If so, are you experienced with vector components?

Reasoning Yes, you are! Although you may not have thought of vector component language when you heard these directions, that is exactly what the directions represent. The perpendicular streets of the city reflect an xy coordinate system; we can assign the x axis to the east–west streets, and the y axis to the north–south streets. Thus, the comment of the person giving you directions can be translated as, "Undergo a displacement vector that has an x component of $+3$ blocks and a y component of -5 blocks." You would arrive at the same destination by undergoing the y component first, followed by the x component, demonstrating the commutative law of addition. ■

EXAMPLE 1.6 The Sum of Two Vectors

Find the sum of two vectors $\vec{\mathbf{A}}$ and $\vec{\mathbf{B}}$ lying in the xy plane and given by

$$\vec{\mathbf{A}} = 2.00\hat{\mathbf{i}} + 3.00\hat{\mathbf{j}} \quad \text{and} \quad \vec{\mathbf{B}} = 5.00\hat{\mathbf{i}} - 4.00\hat{\mathbf{j}}$$

Solution It might be helpful for you to draw a diagram of the vectors to clarify what they look like on the xy plane. Using the rule given by Equation 1.14, we solve this problem mathematically as follows. Note that $A_x = 2.00$, $A_y = 3.00$, $B_x = 5.00$, and $B_y = -4.00$. Therefore, the resultant vector $\vec{\mathbf{R}}$ is

$$\vec{\mathbf{R}} = \vec{\mathbf{A}} + \vec{\mathbf{B}} = (2.00 + 5.00)\hat{\mathbf{i}} + (3.00 - 4.00)\hat{\mathbf{j}}$$

$$= 7.00\,\hat{\mathbf{i}} - 1.00\hat{\mathbf{j}}$$

or

$$R_x = 7.00, \qquad R_y = -1.00$$

The magnitude of $\vec{\mathbf{R}}$ is

$$R = \sqrt{R_x^2 + R_y^2}$$

$$= \sqrt{(7.00)^2 + (-1.00)^2} = \sqrt{50.0} = 7.07$$

EXAMPLE 1.7 The Resultant Displacement

A particle undergoes three consecutive displacements: $\Delta\vec{\mathbf{r}}_1 = (1.50\hat{\mathbf{i}} + 3.00\hat{\mathbf{j}} - 1.20\hat{\mathbf{k}})$ cm, $\Delta\vec{\mathbf{r}}_2 = (2.30\hat{\mathbf{i}} - 1.40\hat{\mathbf{j}} - 3.60\hat{\mathbf{k}})$ cm, and $\Delta\vec{\mathbf{r}}_3 = (-1.30\hat{\mathbf{i}} + 1.50\hat{\mathbf{j}})$ cm. Find the components of the resultant displacement and its magnitude.

Solution We use Equation 1.18 for three vectors:

$$\vec{\mathbf{R}} = \Delta\vec{\mathbf{r}}_1 + \Delta\vec{\mathbf{r}}_2 + \Delta\vec{\mathbf{r}}_3 = (1.50 + 2.30 - 1.30)\hat{\mathbf{i}}\,\text{cm} + (3.00 - 1.40 + 1.50)\hat{\mathbf{j}}\,\text{cm} + (-1.20 - 3.60 + 0)\hat{\mathbf{k}}\,\text{cm}$$

$$= (2.50\hat{\mathbf{i}} + 3.10\hat{\mathbf{j}} - 4.80\hat{\mathbf{k}})\text{ cm}$$

That is, the resultant displacement has components $R_x = 2.50$ cm, $R_y = 3.10$ cm, and $R_z = -4.80$ cm. Its magnitude is

$$R = \sqrt{R_x^2 + R_y^2 + R_z^2}$$

$$= \sqrt{(2.50\text{ cm})^2 + (3.10\text{ cm})^2 + (-4.80\text{ cm})^2}$$

$$= 6.24\text{ cm}$$

INTERACTIVE EXAMPLE 1.8 Taking a Hike

A hiker begins a two-day trip by first walking 25.0 km due southeast from her car. She stops and sets up her tent for the night. On the second day she walks 40.0 km in a direction 60.0° north of east, at which point she discovers a forest ranger's tower.

A Determine the components of the hiker's displacements on the first and second days.

Solution If we denote the displacement vectors on the first and second days by $\vec{\mathbf{A}}$ and $\vec{\mathbf{B}}$, respectively, and use the car as the origin of coordinates, we obtain the vectors shown in the diagram in Figure 1.18. Notice that the resultant vector $\vec{\mathbf{R}}$ can be drawn in the diagram to provide an approximation of the final result of the two hikes.

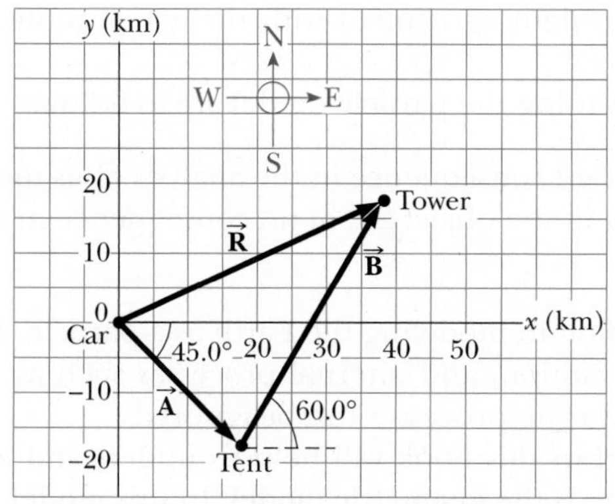

FIGURE 1.18 (Interactive Example 1.8) The total displacement of the hiker is the vector $\vec{\mathbf{R}} = \vec{\mathbf{A}} + \vec{\mathbf{B}}$.

Displacement $\vec{\mathbf{A}}$ has a magnitude of 25.0 km and is 45.0° southeast. Its components are

$$A_x = A\cos(-45.0°) = (25.0\text{ km})(0.707) = 17.7\text{ km}$$

$$A_y = A\sin(-45.0°) = (25.0\text{ km})(-0.707) = -17.7\text{ km}$$

The positive value of A_x indicates that the x coordinate increased in this displacement. The negative value of A_y indicates that the y coordinate decreased in this displacement. Notice in the diagram of Figure 1.18 that vector $\vec{\mathbf{A}}$ lies in the fourth quadrant, consistent with the signs of the components we calculated.

The second displacement $\vec{\mathbf{B}}$ has a magnitude of 40.0 km and is 60.0° north of east. Its components are

$$B_x = B\cos 60.0° = (40.0\text{ km})(0.500) = 20.0\text{ km}$$

$$B_y = B\sin 60.0° = (40.0\text{ km})(0.866) = 34.6\text{ km}$$

B Determine the components of the hiker's total displacement for the trip.

Solution The resultant displacement vector for the trip, $\vec{\mathbf{R}} = \vec{\mathbf{A}} + \vec{\mathbf{B}}$, has components given by

$$R_x = A_x + B_x = 17.7\text{ km} + 20.0\text{ km} = 37.7\text{ km}$$

$$R_y = A_y + B_y = -17.7\text{ km} + 34.6\text{ km} = 16.9\text{ km}$$

In unit-vector form, we can write the total displacement as

$$\vec{\mathbf{R}} = (37.7\hat{\mathbf{i}} + 16.9\hat{\mathbf{j}})\text{ km}$$

PhysicsNow™ Investigate this vector addition situation by logging into PhysicsNow at www.pop4e.com and going to Interactive Example 1.8.

1.10 MODELING, ALTERNATIVE REPRESENTATIONS, AND PROBLEM-SOLVING STRATEGY

Most courses in general physics require the student to learn the skills of problem solving, and examinations usually include problems that test such skills. This section describes some useful ideas that will enable you to enhance your understanding of physical concepts, increase your accuracy in solving problems, eliminate initial panic or lack of direction in approaching a problem, and organize your work.

One of the primary problem-solving methods in physics is to form an appropriate **model** of the problem. **A model is a simplified substitute for the real problem that allows us to solve the problem in a relatively simple way.** As long as the predictions of the model agree to our satisfaction with the actual behavior of the real system, the model is valid. If the predictions do not agree, the model must be refined or replaced with another model. The power of modeling is in its ability to reduce a wide variety of very complex problems to a limited number of classes of problems that can be approached in similar ways.

In science, a model is very different from, for example, an architect's scale model of a proposed building, which appears as a smaller version of what it represents. A scientific model is a theoretical construct and may have no visual similarity to the physical problem. A simple application of modeling is presented in Example 1.9, and we shall encounter many more examples of models as the text progresses.

Models are needed because the actual operation of the Universe is extremely complicated. Suppose, for example, we are asked to solve a problem about the Earth's motion around the Sun. The Earth is very complicated, with many processes occurring simultaneously. These processes include weather, seismic activity, and ocean movements as well as the multitude of processes involving human activity. Trying to maintain knowledge and understanding of all these processes is an impossible task.

The modeling approach recognizes that none of these processes affects the motion of the Earth around the Sun to a measurable degree. Therefore, these details are all ignored. In addition, as we shall find in Chapter 11, the size of the Earth does not affect the gravitational force between the Earth and the Sun; only the masses of the Earth and Sun and the distance between them determine this force. In a simplified model, the Earth is imagined to be a particle, an object with mass but zero size. This replacement of an extended object by a particle is called the **particle model,** which is used extensively in physics. By analyzing the motion of a particle with the mass of the Earth in orbit around the Sun, we find that the predictions of the particle's motion are in excellent agreement with the actual motion of the Earth.

The two primary conditions for using the particle model are as follows:

- The size of the actual object is of no consequence in the analysis of its motion.
- Any internal processes occurring in the object are of no consequence in the analysis of its motion.

Both of these conditions are in action in modeling the Earth as a particle. Its radius is not a factor in determining its motion, and internal processes such as thunderstorms, earthquakes, and manufacturing processes can be ignored.

Four categories of models used in this book will help us understand and solve physics problems. The first category is the **geometric model.** In this model, we form a geometric construction that represents the real situation. We then set aside the real problem and perform an analysis of the geometric construction. Consider a popular problem in elementary trigonometry, as in the following example.

EXAMPLE 1.9 Finding the Height of a Tree

You wish to find the height of a tree but cannot measure it directly. You stand 50.0 m from the tree and determine that a line of sight from the ground to the top of the tree makes an angle of 25.0° with the ground. How tall is the tree?

Solution Figure 1.19 shows the tree and a right triangle corresponding to the information in the problem superimposed over it. (We assume that the tree is exactly perpendicular to a perfectly flat ground.) In the triangle, we know the length of the horizontal leg and the angle between the hypotenuse and the horizontal leg. We can find the height of the tree by calculating the length of the vertical leg. We do so with the tangent function:

$$\tan\theta = \frac{\text{opposite side}}{\text{adjacent side}} = \frac{h}{50.0\text{ m}}$$

$$h = (50.0\text{ m})\tan\theta = (50.0\text{ m})\tan 25.0° = \boxed{23.3\text{ m}}$$

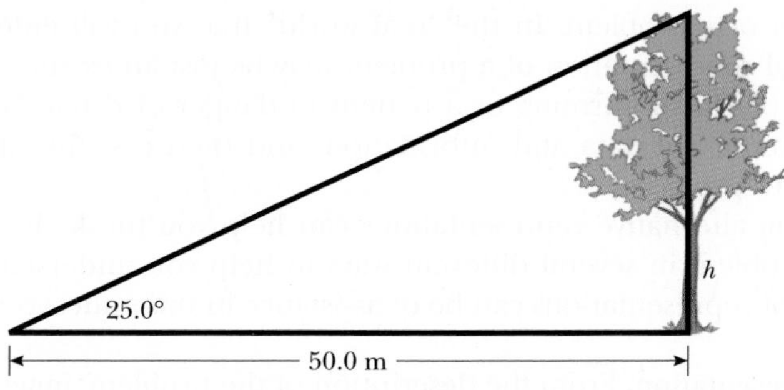

FIGURE 1.19 (Example 1.9) The height of a tree can be found by measuring the distance from the tree and the angle of sight to the top above the ground. This problem is a simple example of geometrically *modeling* the actual problem.

You may have solved a problem very similar to Example 1.9 but never thought about the notion of modeling. From the modeling approach, however, once we draw the triangle in Figure 1.19, the triangle is a geometric model of the real problem; it is a *substitute.* Until we reach the end of the problem, we do not imagine the problem to be about a tree but to be about a triangle. We use trigonometry to find the vertical leg of the triangle, leading to a value of 23.3 m. Because this leg *represents* the height of the tree, we can now return to the original problem and claim that the height of the tree is 23.3 m.

Other examples of geometric models include modeling the Earth as a perfect sphere, a pizza as a perfect disk, a meter stick as a long rod with no thickness, and an electric wire as a long, straight, cylinder.

The particle model is an example of the second category of models, which we will call **simplification models.** In a simplification model, details that are not significant in determining the outcome of the problem are ignored. When we study rotation in Chapter 10, objects will be modeled as *rigid objects.* All the molecules in a rigid object maintain their exact positions with respect to one another. We adopt this simplification model because a spinning rock is much easier to analyze than a spinning block of gelatin, which is *not* a rigid object. Other simplification models will assume that quantities such as friction forces are negligible, remain constant, or are proportional to some power of the object's speed.

The third category is that of **analysis models,** which are general types of problems that we have solved before. An important technique in problem solving is to cast a new problem into a form similar to one we have already solved and which can be used as a model. As we shall see, there are about two dozen analysis models that can be used to solve most of the problems you will encounter. We will see our first analysis models in Chapter 2, where we will discuss them in more detail.

The fourth category of models is **structural models.** These models are generally used to understand the behavior of a system that is far different in scale from our macroscopic world—either much smaller or much larger—so that we cannot interact with it directly. As an example, the notion of a hydrogen atom as an electron in a circular orbit around a proton is a structural model of the atom. We will discuss this model and structural models in general in Chapter 11.

Intimately related to the notion of modeling is that of forming **alternative representations** of the problem. **A representation is a method of viewing or presenting the information related to the problem.** Scientists must be able to communicate complex ideas to individuals without scientific backgrounds. The best representation to use in conveying the information successfully will vary from one individual to the next. Some will be convinced by a well-drawn graph, and others will require a picture. Physicists are often persuaded to agree with a point of view by examining an equation, but nonphysicists may not be convinced by this mathematical representation of the information.

A word problem, such as those at the ends of the chapters in this book, is one representation of a problem. In the "real world" that you will enter after graduation, the initial representation of a problem may be just an existing situation, such as the effects of global warming or a patient in danger of dying. You may have to identify the important data and information, and then cast the situation into an equivalent word problem!

Considering alternative representations can help you think about the information in the problem in several different ways to help you understand and solve it. Several types of representations can be of assistance in this endeavor:

- **Mental representation.** From the description of the problem, imagine a scene that describes what is happening in the word problem, then let time progress so that you understand the situation and can predict what changes will occur in the situation. This step is critical in approaching *every* problem.
- **Pictorial representation.** Drawing a picture of the situation described in the word problem can be of great assistance in understanding the problem. In Example 1.9, the pictorial representation in Figure 1.19 allows us to identify the triangle as a geometric model of the problem. In architecture, a blueprint is a pictorial representation of a proposed building.

  Generally, a pictorial representation describes *what you would see* if you were observing the situation in the problem. For example, Figure 1.20 shows a pictorial representation of a baseball player hitting a short pop foul. Any coordinate axes included in your pictorial representation will be in two dimensions: x and y axes.

FIGURE 1.20 A pictorial representation of a pop foul being hit by a baseball player.

- **Simplified pictorial representation.** It is often useful to redraw the pictorial representation without complicating details by applying a simplification model. This process is similar to the discussion of the particle model described earlier. In a pictorial representation of the Earth in orbit around the Sun, you might draw the Earth and the Sun as spheres, with possibly some attempt to draw continents to identify which sphere is the Earth. In the simplified pictorial representation, the Earth and the Sun would be drawn simply as dots, representing particles. Figure 1.21 shows a simplified pictorial representation corresponding to the pictorial representation of the baseball trajectory in Figure 1.20. The notations v_x and v_y refer to the components of the velocity vector for the baseball. We shall use such simplified pictorial representations throughout the book.

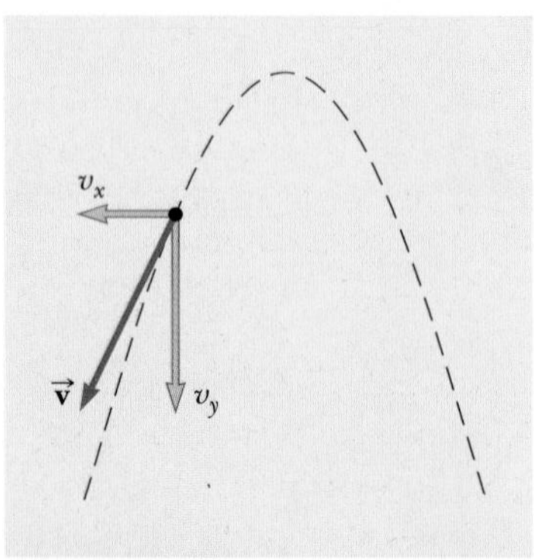

FIGURE 1.21 A simplified pictorial representation for the situation shown in Figure 1.20.

- **Graphical representation.** In some problems, drawing a graph that describes the situation can be very helpful. In mechanics, for example, position–time graphs can be of great assistance. Similarly, in thermodynamics, pressure–volume graphs are essential to understanding. Figure 1.22 shows a graphical representation of the position as a function of time of a block on the end of a vertical spring as it oscillates up and down. Such a graph is helpful for understanding simple harmonic motion, which we study in Chapter 12.

  A graphical representation is different from a pictorial representation, which is also a two-dimensional display of information but whose axes, if any, represent *length* coordinates. In a graphical representation, the axes may represent any two related variables. For example, a graphical representation may have axes for temperature and time. Therefore, in comparison to a pictorial representation, a graphical representation is generally *not* something you would see when observing the situation in the problem with your eyes.

- **Tabular representation.** It is sometimes helpful to organize the information in tabular form to help make it clearer. For example, some students find that making tables of known quantities and unknown quantities is helpful. The periodic table is an extremely useful tabular representation of information in chemistry and physics.
- **Mathematical representation.** The ultimate goal in solving a problem is often the mathematical representation. You want to move from the information contained in the word problem, through various representations of the problem that allow you to understand what is happening, to one or more equations that represent the situation in the problem and that can be solved mathematically for the desired result.

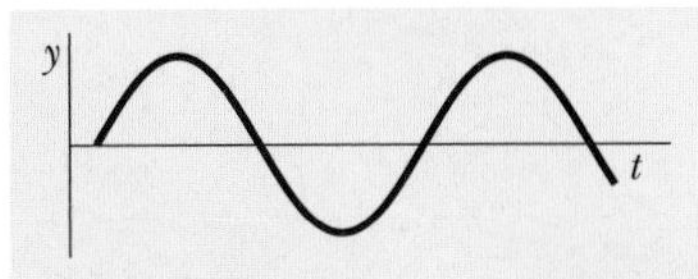

FIGURE 1.22 A graphical representation of the position as a function of time of a block hanging from a spring and oscillating.

GENERAL PROBLEM-SOLVING STRATEGY

An important way to become a skilled problem solver is to adopt a problem-solving strategy. This General Problem-Solving Strategy provides useful steps for solving numerical problems.

Conceptualize

- Read the problem carefully at least twice. Be sure you understand the nature of the problem before proceeding further. Imagine a movie, running in your mind, of what happens in the problem. This step allows you to set up the mental representation of the problem.
- Draw a suitable diagram with appropriate labels and coordinate axes, if needed. This process provides the pictorial representation. If appropriate, generate a graphical representation. If you find it helpful, generate a tabular representation.
- Now focus on the expected result of solving the problem. Exactly what is the question asking? Will the final result be numerical or algebraic? Do you know what units to expect?
- Don't forget to incorporate information from your own experiences and common sense. What should a reasonable answer look like? For example, you wouldn't expect to calculate the speed of an automobile to be 5×10^6 m/s.

Categorize

- Once you have a good idea of what the problem is about, you need to *simplify* the problem by drawing a simplified pictorial representation. Use a simplification model to remove additional unnecessary details if the conditions for the model are satisfied. If it helps you solve the problem, identify a useful geometric model from the diagrams.
- Once the problem is simplified, it is important to *categorize* the problem. Is it a simple *plug-in problem,* such that numbers can be simply substituted into a definition? If so, the problem is likely to be finished when this substitution is done. If not, you face an *analysis problem,* and the situation must be analyzed more deeply to reach a solution.
- Once you have eliminated the unnecessary details and have simplified the problem to its fundamental level, identify an analysis model for the problem. (We will see how to identify analysis models as we introduce them throughout the book.)

Analyze

- Now you must analyze the problem and strive for a mathematical representation of the problem. From the analysis model, identify the basic physical principle or principles that are involved, listing the knowns and unknowns. Select relevant equations that apply to the model.
- Use algebra (and calculus, if necessary) to solve symbolically for the unknown variable in terms of what is given. Substitute in the appropriate numbers, calculate the result, and round it to the proper number of significant figures.

Finalize

- This final step is the most important part. Examine your numerical answer. Does it have the correct units? Is it of reasonable value? Does it meet your expectations from your conceptualization of the problem? What about the algebraic form of the result, before you substituted numerical values? Does it make sense? Examine the variables in the problem to see whether the answer would change in a physically meaningful way if the variables were drastically increased, decreased, or even became zero. Looking at limiting cases to see whether they yield expected values is a very useful way to make sure that you are obtaining reasonable results.

Although this problem-solving strategy may look complicated, it may not be necessary to perform all the steps for a given problem. Examples in this text focus on how to apply these steps explicitly to help you become an effective problem solver. Many chapters include a section labeled "Problem-Solving Strategy" that should help you through the rough spots. These sections are organized according to the General Problem-Solving Strategy and tailor this strategy to the specific types of problems addressed in individual chapters. Once you have developed an organized system for examining problems and extracting relevant information, you will become a more confident problem solver in physics as well as in other areas.

SUMMARY

PhysicsNow™ Take a practice test by logging into PhysicsNow at www.pop4e.com and clicking on the Pre-Test link for this chapter.

Mechanical quantities can be expressed in terms of three fundamental quantities—**length, mass,** and **time**—which in the SI system have the units **meters** (m), **kilograms** (kg), and **seconds** (s), respectively. It is often useful to use the method of **dimensional analysis** to check equations and to assist in deriving expressions.

The **density** of a substance is defined as its mass per unit volume:

$$\rho \equiv \frac{m}{V} \qquad [1.1]$$

Vectors are quantities that have both magnitude and direction and obey the vector law of addition. **Scalars** are quantities that add algebraically.

Two vectors $\vec{\mathbf{A}}$ and $\vec{\mathbf{B}}$ can be added using the triangle method. In this method (see Fig. 1.9), the vector $\vec{\mathbf{R}} = \vec{\mathbf{A}} + \vec{\mathbf{B}}$ runs from the tail of $\vec{\mathbf{A}}$ to the tip of $\vec{\mathbf{B}}$.

The x component A_x of the vector $\vec{\mathbf{A}}$ is equal to its projection along the x axis of a coordinate system, where $A_x = A\cos\theta$ and where θ is the angle $\vec{\mathbf{A}}$ makes with the x axis. Likewise, the y component A_y of $\vec{\mathbf{A}}$ is its projection along the y axis, where $A_y = A\sin\theta$.

If a vector $\vec{\mathbf{A}}$ has an x component equal to A_x and a y component equal to A_y, the vector can be expressed in unit-vector form as $\vec{\mathbf{A}} = (A_x\hat{\mathbf{i}} + A_y\hat{\mathbf{j}})$. In this notation, $\hat{\mathbf{i}}$ is a unit vector in the positive x direction and $\hat{\mathbf{j}}$ is a unit vector in the positive y direction. Because $\hat{\mathbf{i}}$ and $\hat{\mathbf{j}}$ are unit vectors, $|\hat{\mathbf{i}}| = |\hat{\mathbf{j}}| = 1$. In three dimensions, a vector can be expressed as $\vec{\mathbf{A}} = (A_x\hat{\mathbf{i}} + A_y\hat{\mathbf{j}} + A_z\hat{\mathbf{k}})$, where $\hat{\mathbf{k}}$ is a unit vector in the z direction.

The resultant of two or more vectors can be found by resolving all vectors into their x, y, and z components and adding their components:

$$\vec{\mathbf{R}} = \vec{\mathbf{A}} + \vec{\mathbf{B}} = (A_x + B_x)\hat{\mathbf{i}} + (A_y + B_y)\hat{\mathbf{j}} + (A_z + B_z)\hat{\mathbf{k}} \qquad [1.18]$$

Problem-solving skills and physical understanding can be improved by **modeling** the problem and by constructing **alternative representations** of the problem. Models helpful in solving problems include **geometric, simplification,** and **analysis models.** Scientists use **structural models** to understand systems larger or smaller in scale than those with which we normally have direct experience. Helpful representations include the **mental, pictorial, simplified pictorial, graphical, tabular,** and **mathematical representations.**

QUESTIONS

☐ = answer available in the *Student Solutions Manual and Study Guide*

1. What types of natural phenomena could serve as time standards?

2. ☐ Suppose the three fundamental standards of the metric system were length, *density,* and time rather than length, *mass,* and time. The standard of density in this system is to be defined as that of water. What considerations about water would you need to address to make sure that the standard of density is as accurate as possible?

3. ☐ Express the following quantities using the prefixes given in Table 1.4: (a) 3×10^{-4} m, (b) 5×10^{-5} s, (c) 72×10^{2} g.

4. Suppose two quantities A and B have different dimensions. Determine which of the following arithmetic operations *could* be physically meaningful: (a) $A + B$, (b) A/B, (c) $B - A$, (d) AB.

5. If an equation is dimensionally correct, does that mean that the equation must be true? If an equation is not dimensionally correct, does that mean that the equation cannot be true?

6. Find the order of magnitude of your age in seconds.

7. What level of precision is implied in an order-of-magnitude calculation?

8. In reply to a student's question, a guard in a natural history museum says of the fossils near his station, "When I started work here twenty-four years ago, they were eighty million years old, so you can add it up." What should the student conclude about the age of the fossils?

9. Can the magnitude of a particle's displacement be greater than the distance traveled? Explain.

10. Which of the following are vectors and which are not: force, temperature, the volume of water in a can, the

ratings of a TV show, the height of a building, the velocity of a sports car, the age of the Universe?

11. A vector $\vec{A}$ lies in the xy plane. For what orientations of $\vec{A}$ will both of its components be negative? For what orientations will its components have opposite signs?

12. A book is moved once around the perimeter of a tabletop with the dimensions 1.0 m × 2.0 m. If the book ends up at its initial position, what is its displacement? What is the distance traveled?

13. While traveling along a straight interstate highway you notice that the mile marker reads 260. You travel until you reach the 150-mile marker and then retrace your path to the 175-mile marker. What is the magnitude of your resultant displacement from mile marker 260?

14. If the component of vector $\vec{A}$ along the direction of vector $\vec{B}$ is zero, what can you conclude about the two vectors?

15. Can the magnitude of a vector have a negative value? Explain.

16. Under what circumstances would a nonzero vector lying in the xy plane have components that are equal in magnitude?

17. Is it possible to add a vector quantity to a scalar quantity? Explain.

18. In what circumstance is the x component of a vector given by the magnitude of the vector multiplied by the sine of its direction angle?

19. Identify the type of model (geometrical, simplification, or structural) represented by each of the following. (a) In its orbit around the Sun, the Earth is treated as a particle. (b) The distance the Earth travels around the Sun is calculated as 2π multiplied by the Earth–Sun distance. (c) The atomic structure of a solid material is imagined to consist of small objects (atoms) connected to neighboring identical objects by springs. (d) For an object you drop, air resistance is ignored. (e) The volume of water in a bottle is estimated by calculating the volume of a cylinder. (f) A bat hits a baseball. In studying the motion of the baseball, any distortion of the ball while it is in contact with the bat is not considered. (g) In the early 20th century, the atom was proposed to consist of electrons in orbit around a very small but massive nucleus.

PROBLEMS

1, 2, 3 = straightforward, intermediate, challenging
☐ = full solution available in the *Student Solutions Manual and Study Guide*
PhysicsNow™ = coached problem with hints available at www.pop4e.com
= computer useful in solving problem
= paired numerical and symbolic problems
= biomedical application

Note: Consult the endpapers, appendices, and tables in the text whenever necessary in solving problems. For this chapter, Appendix B.3 and Table 15.1 may be particularly useful. Answers to odd-numbered problems appear in the back of the book.

Section 1.1 ■ Standards of Length, Mass, and Time

1. Use information on the endpapers of this book to calculate the average density of the Earth. Where does the value fit among those listed in Table 15.1? Look up the density of a typical surface rock like granite in another source and compare the density of the Earth to it.

2. A major motor company displays a die-cast model of its first automobile, made from 9.35 kg of iron. To celebrate its hundredth year in business, a worker will recast the model in gold from the original dies. What mass of gold is needed to make the new model? The density of iron is 7.86×10^3 kg/m^3, and that of gold is 19.3×10^3 kg/m^3.

3. What mass of a material with density ρ is required to make a hollow spherical shell having inner radius r_1 and outer radius r_2?

4. Two spheres are cut from a certain uniform rock. One has radius 4.50 cm. The mass of the other is five times greater. Find its radius.

Section 1.2 ■ Dimensional Analysis

5. The position of a particle moving under uniform acceleration is some function of time and the acceleration. Suppose we write this position as $x = ka^m t^n$, where k is a dimensionless constant. Show by dimensional analysis that this expression is satisfied if $m = 1$ and $n = 2$. Can this analysis give the value of k?

6. Figure P1.6 shows a *frustrum of a cone.* Of the following mensuration (geometrical) expressions, which describes (a) the total circumference of the flat circular faces, (b) the volume, and (c) the area of the curved surface? (i) $\pi(r_1 + r_2)[h^2 + (r_1 - r_2)^2]^{1/2}$ (ii) $2\pi(r_1 + r_2)$ (iii) $\pi h(r_1{}^2 + r_1 r_2 + r_2{}^2)$

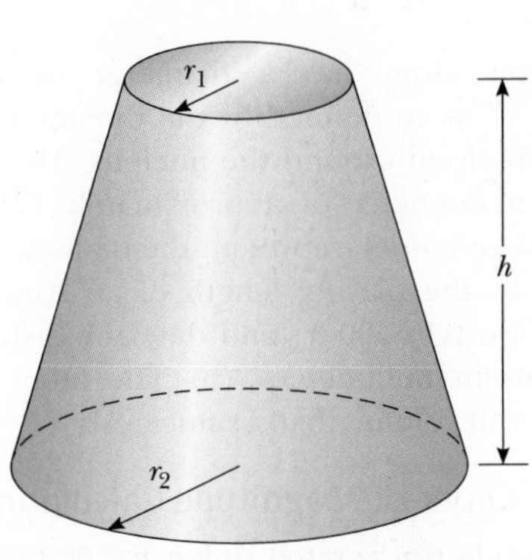

FIGURE P1.6

7. Which of the following equations are dimensionally correct? (a) $v_f = v_i + ax$ (b) $y = (2\text{ m})\cos(kx)$, where $k = 2\text{ m}^{-1}$.

Section 1.3 ▪ Conversion of Units

8. Suppose your hair grows at the rate $\frac{1}{32}$ in. per day. Find the rate at which it grows in nanometers (nm) per second. Because the distance between atoms in a molecule is on the order of 0.1 nm, your answer suggests how rapidly layers of atoms are assembled in this protein synthesis.

9. Assume it takes 7.00 minutes to fill a 30.0-gal gasoline tank. (a) Calculate the rate at which the tank is filled in gallons per second. (b) Calculate the flow rate of the gasoline in cubic meters per second. (c) Determine the time interval, in hours, required to fill a 1.00-m^3 volume at the same rate. (1 U.S. gal = 231 in.^3)

10. A *section* of land has an area of 1 square mile and contains 640 acres. Determine the number of square meters in 1 acre.

11. An ore loader moves 1 200 tons/h from a mine to the surface. Convert this rate to pounds per second, using 1 ton = 2 000 lb.

12. At the time of this book's printing, the U.S. national debt is about \$7 trillion. (a) If payments were made at the rate of \$1 000 per second, how many years would it take to pay off the debt assuming that no interest were charged? (b) A one-dollar bill is about 15.5 cm long. If seven trillion one-dollar bills were laid end to end around the Earth's equator, how many times would they encircle the planet? Take the radius of the Earth at the equator to be 6 378 km. (*Note:* Before doing any of these calculations, try to guess at the answers. You may be very surprised.)

13. **PhysicsNow™** One gallon of paint with a volume of $3.78 \times 10^{-3}\text{ m}^3$ covers an area of 25.0 m^2. What is the thickness of the paint on the wall?

14. The mass of the Sun is 1.99×10^{30} kg, and the mass of an atom of hydrogen, of which the Sun is mostly composed, is 1.67×10^{-27} kg. How many atoms are in the Sun?

15. **PhysicsNow™** One cubic meter (1.00 m^3) of aluminum has a mass of 2.70×10^3 kg, and 1.00 m^3 of iron has a mass of 7.86×10^3 kg. Find the radius of a solid aluminum sphere that will balance a solid iron sphere of radius 2.00 cm on an equal-arm balance.

16. Let ρ_{Al} represent the density of aluminum and ρ_{Fe} that of iron. Find the radius of a solid aluminum sphere that balances a solid iron sphere of radius r_{Fe} on an equal-arm balance.

17. A hydrogen atom has a diameter of approximately 1.06×10^{-10} m, as defined by the diameter of the spherical electron cloud around the nucleus. The hydrogen nucleus has a diameter of approximately 2.40×10^{-15} m. (a) For a scale model, represent the diameter of the hydrogen atom by the playing length of an American football field (100 yards = 300 ft) and determine the diameter of the nucleus in millimeters. (b) The atom is how many times larger in volume than its nucleus?

Section 1.4 ▪ Order-of-Magnitude Calculations

18. An automobile tire is rated to last for 50 000 miles. To an order of magnitude, through how many revolutions will it turn? In your solution, state the quantities you measure or estimate and the values you take for them.

19. **PhysicsNow™** Estimate the number of Ping-Pong balls that would fit into a typical-size room (without being crushed). In your solution, state the quantities you measure or estimate and the values you take for them.

20. Compute the order of magnitude of the mass of a bathtub half full of water. Compute the order of magnitude of the mass of a bathtub half full of pennies. In your solution, list the quantities you take as data and the value you measure or estimate for each.

21. To an order of magnitude, how many piano tuners are in New York City? The physicist Enrico Fermi was famous for asking questions like this one on oral Ph.D. qualifying examinations. His own facility in making order-of-magnitude calculations is exemplified in Problem 30.58.

22. Soft drinks are commonly sold in aluminum containers. To an order of magnitude, how many such containers are thrown away or recycled each year by U.S. consumers? How many tons of aluminum does this number represent? In your solution, state the quantities you measure or estimate and the values you take for them.

Section 1.5 ▪ Significant Figures

23. How many significant figures are in the following numbers: (a) 78.9 ± 0.2, (b) 3.788×10^9, (c) 2.46×10^{-6}, (d) 0.005 3?

24. Carry out the following arithmetic operations: (a) the sum of the measured values 756, 37.2, 0.83, and 2.5; (b) the product $0.003\,2 \times 356.3$; (c) the product $5.620 \times \pi$.

25. The *tropical year,* the time interval from vernal equinox to vernal equinox, is the basis for our calendar. It contains 365.242 199 days. Find the number of seconds in a tropical year.

> *Note:* Appendix B.8 on propagation of uncertainty may be useful in solving the next two problems.

26. The radius of a sphere is measured to be (6.50 ± 0.20) cm, and its mass is measured to be (1.85 ± 0.02) kg. The sphere is solid. Determine its density in kilograms per cubic meter and the uncertainty in the density.

27. A sidewalk is to be constructed around a swimming pool that measures (10.0 ± 0.1) m by (17.0 ± 0.1) m. If the sidewalk is to measure (1.00 ± 0.01) m wide by (9.0 ± 0.1) cm thick, what volume of concrete is needed and what is the approximate uncertainty of this volume?

> *Note:* The next four problems call upon mathematical skills that will be useful throughout the course.

28. **Review problem.** Prove that one solution of the equation

$$2.00x^4 - 3.00x^3 + 5.00x = 70.0$$

is $x = -2.22$.

29. **Review problem.** Find every angle θ between 0 and 360° for which $\sin\theta$ is equal to -3.00 multiplied by $\cos\theta$.

30. **Review problem.** A highway curve forms a section of a circle. A car goes around the curve. Its dashboard compass

shows that the car is initially heading due east. After it travels 840 m, it is heading 35.0° south of east. Find the radius of curvature of its path.

31. Review problem. From the set of equations

$$p = 3q$$
$$pr = qs$$
$$\tfrac{1}{2}pr^2 + \tfrac{1}{2}qs^2 = \tfrac{1}{2}qt^2$$

involving the unknowns p, q, r, s, and t, find the value of t/r.

Section 1.6 ▪ Coordinate Systems

32. The polar coordinates of a point are $r = 5.50$ m and $\theta = 240°$. What are the Cartesian coordinates of this point?

33. A fly lands on one wall of a room. The lower left corner of the wall is selected as the origin of a two-dimensional Cartesian coordinate system. If the fly is located at the point having coordinates (2.00, 1.00) m, (a) how far is it from the corner of the room? (b) What is its location in polar coordinates?

34. Two points in the xy plane have Cartesian coordinates (2.00, − 4.00) m and (− 3.00, 3.00) m. Determine (a) the distance between these points and (b) their polar coordinates.

35. Let the polar coordinates of the point (x, y) be (r, θ). Determine the polar coordinates for the points (a) $(-x, y)$, (b) $(-2x, -2y)$, and (c) $(3x, -3y)$.

Section 1.7 ▪ Vectors and Scalars

Section 1.8 ▪ Some Properties of Vectors

36. A plane flies from base camp to Lake A, 280 km away in the direction 20.0° north of east. After dropping off supplies, it flies to Lake B, which is 190 km at 30.0° west of north from Lake A. Graphically determine the distance and direction from Lake B to the base camp.

37. **Physics Now™** A skater glides along a circular path of radius 5.00 m. Assuming he coasts around one half of the circle, find (a) the magnitude of the displacement vector and (b) how far the person skated. (c) What is the magnitude of the displacement if he skates all the way around the circle?

38. Each of the displacement vectors $\vec{\mathbf{A}}$ and $\vec{\mathbf{B}}$ shown in Figure P1.38 has a magnitude of 3.00 m. Find graphically (a) $\vec{\mathbf{A}} + \vec{\mathbf{B}}$, (b) $\vec{\mathbf{A}} - \vec{\mathbf{B}}$, (c) $\vec{\mathbf{B}} - \vec{\mathbf{A}}$, and (d) $\vec{\mathbf{A}} - 2\vec{\mathbf{B}}$. Report all angles counterclockwise from the positive x axis.

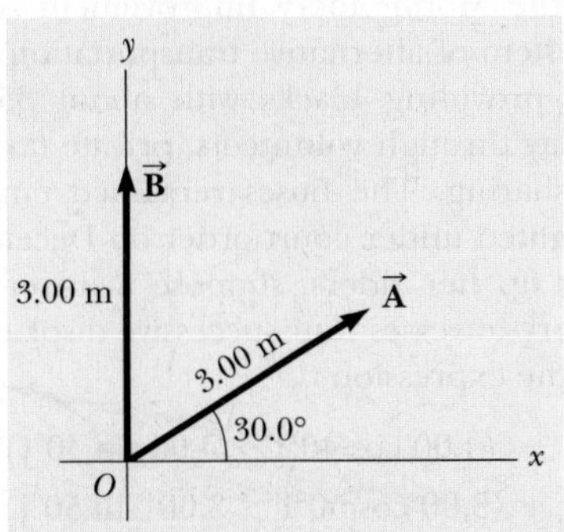

FIGURE **P1.38** Problems 1.38 and 1.48

39. A roller coaster car moves 200 ft horizontally and then rises 135 ft at an angle of 30.0° above the horizontal. It then travels 135 ft at an angle of 40.0° downward. What is its displacement from its starting point? Use graphical techniques.

Section 1.9 ▪ Components of a Vector and Unit Vectors

40. Find the horizontal and vertical components of the 100-m displacement of a superhero who flies from the top of a tall building following the path shown in Figure P1.40.

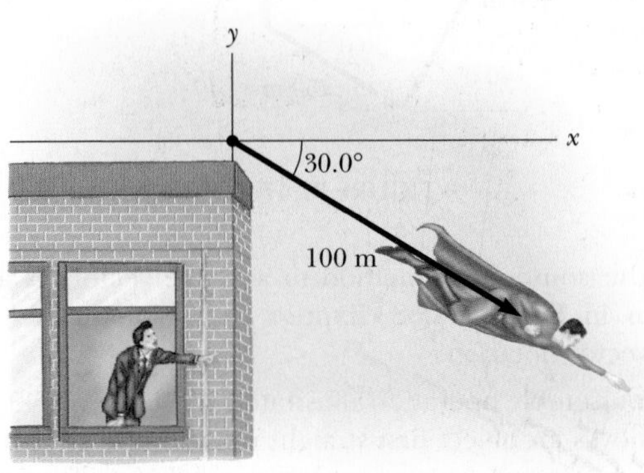

FIGURE **P1.40**

41. A vector has an x component of − 25.0 units and a y component of 40.0 units. Find the magnitude and direction of this vector.

42. For the vectors $\vec{\mathbf{A}} = 2.00\hat{\mathbf{i}} + 6.00\hat{\mathbf{j}}$ and $\vec{\mathbf{B}} = 3.00\hat{\mathbf{i}} - 2.00\hat{\mathbf{j}}$, (a) draw the vector sum $\vec{\mathbf{C}} = \vec{\mathbf{A}} + \vec{\mathbf{B}}$ and the vector difference $\vec{\mathbf{D}} = \vec{\mathbf{A}} - \vec{\mathbf{B}}$. (b) Calculate $\vec{\mathbf{C}}$ and $\vec{\mathbf{D}}$, first in terms of unit vectors and then in terms of polar coordinates, with angles measured with respect to the positive x axis.

43. A man pushing a mop across a floor causes it to undergo two displacements. The first has a magnitude of 150 cm and makes an angle of 120° with the positive x axis. The resultant displacement has a magnitude of 140 cm and is directed at an angle of 35.0° to the positive x axis. Find the magnitude and direction of the second displacement.

44. Vector $\vec{\mathbf{A}}$ has x and y components of − 8.70 cm and 15.0 cm, respectively; vector $\vec{\mathbf{B}}$ has x and y components of 13.2 cm and − 6.60 cm, respectively. If $\vec{\mathbf{A}} - \vec{\mathbf{B}} + 3\vec{\mathbf{C}} = 0$, what are the components of $\vec{\mathbf{C}}$?

45. Consider the two vectors $\vec{\mathbf{A}} = 3\hat{\mathbf{i}} - 2\hat{\mathbf{j}}$ and $\vec{\mathbf{B}} = -\hat{\mathbf{i}} - 4\hat{\mathbf{j}}$. Calculate (a) $\vec{\mathbf{A}} + \vec{\mathbf{B}}$, (b) $\vec{\mathbf{A}} - \vec{\mathbf{B}}$, (c) $|\vec{\mathbf{A}} + \vec{\mathbf{B}}|$, (d) $|\vec{\mathbf{A}} - \vec{\mathbf{B}}|$, and (e) the directions of $\vec{\mathbf{A}} + \vec{\mathbf{B}}$ and $\vec{\mathbf{A}} - \vec{\mathbf{B}}$.

46. Consider the three displacement vectors $\vec{\mathbf{A}} = (3\hat{\mathbf{i}} + 3\hat{\mathbf{j}})$ m, $\vec{\mathbf{B}} = (\hat{\mathbf{i}} - 4\hat{\mathbf{j}})$ m, and $\vec{\mathbf{C}} = (-2\hat{\mathbf{i}} + 5\hat{\mathbf{j}})$ m. Use the component method to determine (a) the magnitude and direction of the vector $\vec{\mathbf{D}} = \vec{\mathbf{A}} + \vec{\mathbf{B}} + \vec{\mathbf{C}}$ and (b) the magnitude and direction of $\vec{\mathbf{E}} = -\vec{\mathbf{A}} - \vec{\mathbf{B}} + \vec{\mathbf{C}}$.

47. A person going for a walk follows the path shown in Figure P1.47. The total trip consists of four straight-line paths. At the end of the walk, what is the person's resultant displacement measured from the starting point?

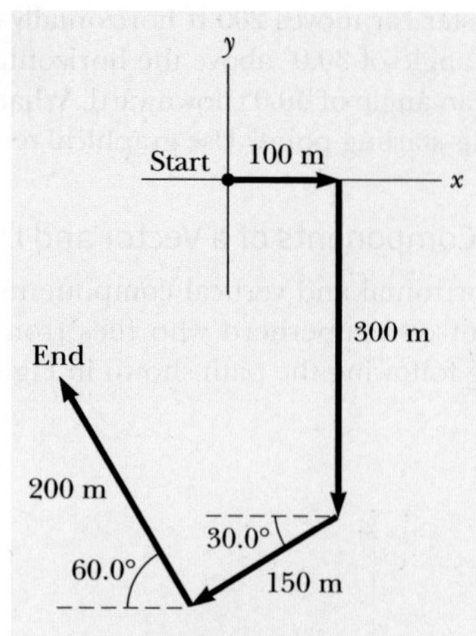

FIGURE P1.47

48. Use the component method to add the vectors $\vec{\mathbf{A}}$ and $\vec{\mathbf{B}}$ shown in Figure P1.38. Express the resultant $\vec{\mathbf{A}} + \vec{\mathbf{B}}$ in unit-vector notation.

49. In an assembly operation illustrated in Figure P1.49, a robot moves an object first straight upward and then also to the east, around an arc forming one quarter of a circle of radius 4.80 cm that lies in an east–west vertical plane. The robot then moves the object upward and to the north, through a quarter of a circle of radius 3.70 cm that lies in a north–south vertical plane. Find (a) the magnitude of the total displacement of the object and (b) the angle the total displacement makes with the vertical.

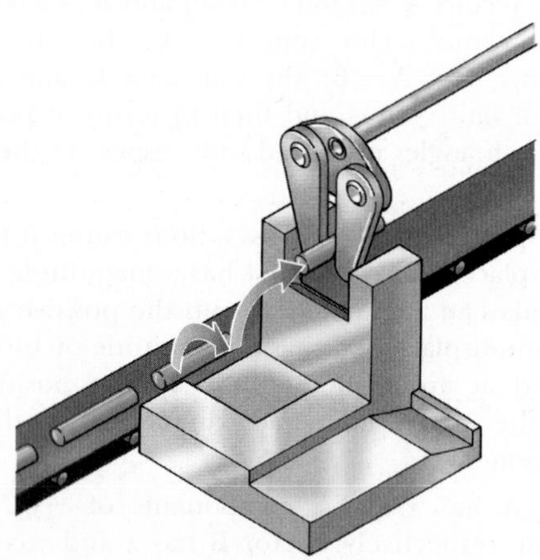

FIGURE P1.49

50. Vector $\vec{\mathbf{B}}$ has x, y, and z components of 4.00, 6.00, and 3.00 units, respectively. Calculate the magnitude of $\vec{\mathbf{B}}$ and the angles that $\vec{\mathbf{B}}$ makes with the coordinate axes.

51. The vector $\vec{\mathbf{A}}$ has x, y, and z components of 8.00, 12.0, and -4.00 units, respectively. (a) Write a vector expression for $\vec{\mathbf{A}}$ in unit-vector notation. (b) Obtain a unit-vector expression for a vector $\vec{\mathbf{B}}$ one fourth the length of $\vec{\mathbf{A}}$ pointing in the same direction as $\vec{\mathbf{A}}$. (c) Obtain a unit-vector expression for a vector $\vec{\mathbf{C}}$ three times the length of $\vec{\mathbf{A}}$ pointing in the direction opposite the direction of $\vec{\mathbf{A}}$.

52. (a) Vector $\vec{\mathbf{E}}$ has magnitude 17.0 cm and is directed 27.0° counterclockwise from the $+x$ axis. Express it in unit-vector notation. (b) Vector $\vec{\mathbf{F}}$ has magnitude 17.0 cm and is directed 27.0° counterclockwise from the $+y$ axis. Express it in unit-vector notation. (c) Vector $\vec{\mathbf{G}}$ has magnitude 17.0 cm and is directed 27.0° clockwise from the $-y$ axis. Express it in unit-vector notation.

53. **PhysicsNow™** Three displacement vectors of a croquet ball are shown in Figure P1.53, where $|\vec{\mathbf{A}}| = 20.0$ units, $|\vec{\mathbf{B}}| = 40.0$ units, and $|\vec{\mathbf{C}}| = 30.0$ units. Find (a) the resultant in unit-vector notation and (b) the magnitude and direction of the resultant displacement.

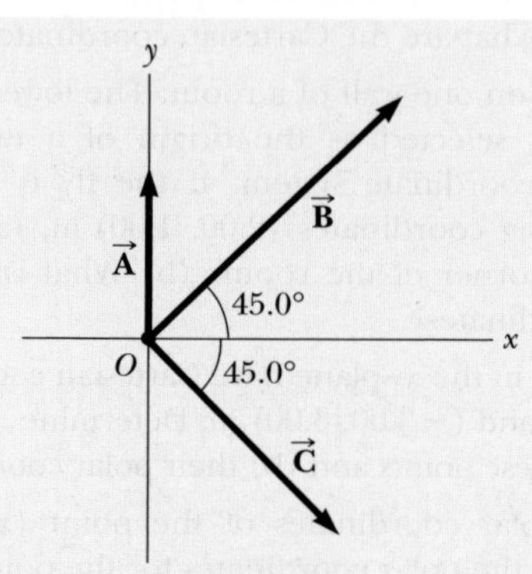

FIGURE P1.53

54. Taking $\vec{\mathbf{A}} = (6.00\hat{\mathbf{i}} - 8.00\hat{\mathbf{j}})$ units, $\vec{\mathbf{B}} = (-8.00\hat{\mathbf{i}} + 3.00\hat{\mathbf{j}})$ units, and $\vec{\mathbf{C}} = (26.0\hat{\mathbf{i}} + 19.0\hat{\mathbf{j}})$ units, determine a and b such that $a\vec{\mathbf{A}} + b\vec{\mathbf{B}} + \vec{\mathbf{C}} = 0$.

Section 1.10 ▪ Modeling, Alternative Representations, and Problem-Solving Strategy

55. A surveyor measures the distance across a straight river by the following method. Starting directly across from a tree on the opposite bank, she walks 100 m along the riverbank to establish a baseline. Then she sights across to the tree. The angle from her baseline to the tree is 35.0°. How wide is the river?

56. On December 1, 1955, Rosa Parks stayed seated in her bus seat when a white man demanded it. Police in Montgomery, Alabama, arrested her. On December 5, blacks began refusing to use all city buses. Under the leadership of the Montgomery Improvement Association, an efficient system of alternative transportation sprang up immediately, providing blacks with about 35 000 essential trips per day through volunteers, private taxis, carpooling, and ride sharing. The buses remained empty until they were integrated under court order on December 21, 1956. In picking up her riders, suppose a driver in downtown Montgomery traverses four successive displacements represented by the expression

$$(-6.30\hat{\mathbf{i}})\text{b} - (4.00\cos 40°\hat{\mathbf{i}} + 4.00\sin 40°\hat{\mathbf{j}})\text{b} + (3.00\cos 50°\hat{\mathbf{i}} - 3.00\sin 50°\hat{\mathbf{j}})\text{b} - (5.00\hat{\mathbf{j}})\text{b}$$

Here b represents one city block, a convenient unit of distance of uniform size; $\hat{\mathbf{i}}$ = east and $\hat{\mathbf{j}}$ = north. (a) Draw a map of the successive displacements. (b) What total distance did she travel? (c) Compute the magnitude and direction of her total displacement. The logical structure of this problem and of several problems in later chapters was suggested by Alan Van Heuvelen and David Maloney, *American Journal of Physics* 67(3) (March 1999) 252–256.

57. A crystalline solid consists of atoms stacked up in a repeating lattice structure. Consider a crystal as shown in Figure P1.57a. The atoms reside at the corners of cubes of side $L = 0.200$ nm. One piece of evidence for the regular arrangement of atoms comes from the flat surfaces along which a crystal separates, or cleaves, when it is broken. Suppose this crystal cleaves along a face diagonal, as shown in Figure P1.57b. Calculate the spacing d between two adjacent atomic planes that separate when the crystal cleaves.

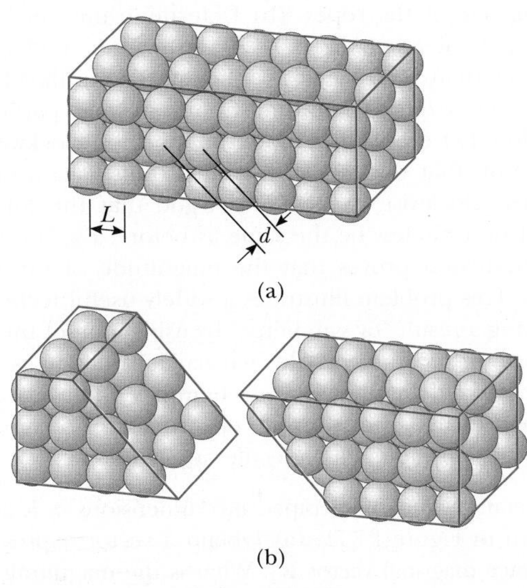

FIGURE **P1.57**

Additional Problems

58. In a situation where data are known to three significant figures, we write 6.379 m = 6.38 m and 6.374 m = 6.37 m. When a number ends in 5, we arbitrarily choose to write 6.375 m = 6.38 m. We could equally well write 6.375 m = 6.37 m, "rounding down" instead of "rounding up," because we would change the number 6.375 by equal increments in both cases. Now consider an order-of-magnitude estimate. Here factors of change, rather than increments, are important. We write 500 m $\sim 10^3$ m because 500 differs from 100 by a factor of 5 whereas it differs from 1 000 by only a factor of 2. We write 437 m $\sim 10^3$ m and 305 m $\sim 10^2$ m. What distance differs from 100 m and from 1 000 m by equal factors, so that we could equally well choose to represent its order of magnitude either as $\sim 10^2$ m or as $\sim 10^3$ m?

59. The basic function of the carburetor of an automobile is to "atomize" the gasoline and mix it with air to promote rapid combustion. As an example, assume that 30.0 cm^3 of gasoline is atomized into N spherical droplets, each with a radius of 2.00×10^{-5} m. What is the total surface area of these N spherical droplets?

60. The consumption of natural gas by a company satisfies the empirical equation $V = 1.50t + 0.008\,00t^2$, where V is the volume in millions of cubic feet and t the time in months. Express this equation in units of cubic feet and seconds. Assign proper units to the coefficients. Assume that a month is 30.0 days.

61. There are nearly $\pi \times 10^7$ s in one year. Find the percentage error in this approximation, where "percentage error" is defined as

$$\text{Percentage error} = \frac{|\text{assumed value} - \text{true value}|}{\text{true value}} \times 100\%$$

62. In physics, it is important to use mathematical approximations. Demonstrate that for small angles (<20°)

$$\tan \alpha \approx \sin \alpha \approx \alpha = \pi\alpha'/180°$$

where α is in radians and α' is in degrees. Use a calculator to find the largest angle for which $\tan \alpha$ may be approximated by α with an error less than 10.0%.

63. A child loves to watch as you fill a transparent plastic bottle with shampoo. Every horizontal cross-section is a circle, but the diameters of the circles have different values, so the bottle is much wider in some places than others. You pour in bright green shampoo with constant volume flow rate 16.5 cm^3/s. At what rate is its level in the bottle rising (a) at a point where the diameter of the bottle is 6.30 cm and (b) at a point where the diameter is 1.35 cm?

64. One cubic centimeter of water has a mass of 1.00×10^{-3} kg. (a) Determine the mass of 1.00 m^3 of water. (b) Biological substances are 98% water. Assume that they have the same density as water to estimate the masses of a cell that has a diameter of 1.00 μm, a human kidney, and a fly. Model the kidney as a sphere with a radius of 4.00 cm and the fly as a cylinder 4.00 mm long and 2.00 mm in diameter.

65. The distance from the Sun to the nearest star is 4×10^{16} m. The Milky Way galaxy is roughly a disk of diameter $\sim 10^{21}$ m and thickness $\sim 10^{19}$ m. Find the order of magnitude of the number of stars in the Milky Way. Assume that the distance between the Sun and our nearest neighbor is typical.

66. Two vectors $\vec{\mathbf{A}}$ and $\vec{\mathbf{B}}$ have precisely equal magnitudes. For the magnitude of $\vec{\mathbf{A}} + \vec{\mathbf{B}}$ to be larger than the magnitude of $\vec{\mathbf{A}} - \vec{\mathbf{B}}$ by the factor n, what must be the angle between them?

67. The helicopter view in Figure P1.67 shows two people pulling on a stubborn mule. (a) Find the single force that is equivalent to the two forces shown. The forces are measured in units of newtons (symbolized N). (b) Find the force that a third person would have to exert on the mule to make the resultant force equal to zero.

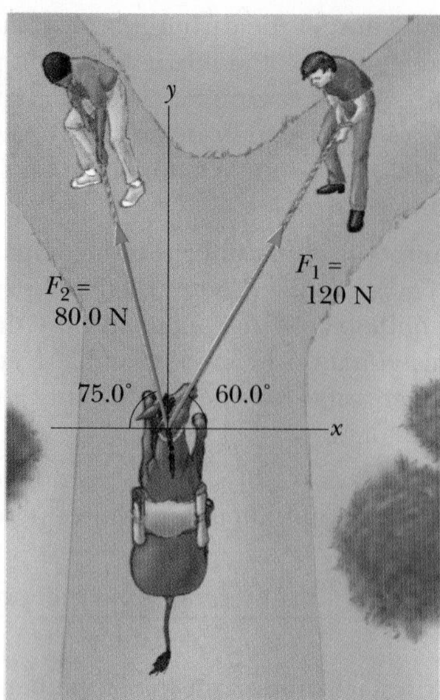

FIGURE **P1.67**

68. An air-traffic controller observes two aircraft on his radar screen. The first is at altitude 800 m, horizontal distance 19.2 km, and 25.0° south of west. The second aircraft is at altitude 1 100 m, horizontal distance 17.6 km, and 20.0° south of west. What is the distance between the two aircraft? (Place the x axis west, the y axis south, and the z axis vertical.)

69. Long John Silver, a pirate, has buried his treasure on an island with five trees, located at the following points: (30.0 m, − 20.0 m), (60.0 m, 80.0 m), (− 10.0 m, − 10.0 m), (40.0 m, − 30.0 m), and (− 70.0 m, 60.0 m), all measured relative to some origin, as shown in Figure P1.69. His ship's log instructs you to start at tree A and move toward tree B, but to cover only one half of the distance between A and B. Then move toward tree C, covering one third of the distance between your current location and C. Next move toward D, covering one fourth of the distance between where you are and D. Finally, move toward E, covering one fifth of the distance between you and E, stop, and dig. (a) Assume that you have correctly determined the order in which the pirate labeled the trees as A, B, C, D, and E, as shown in the figure. What are the coordinates of the point where his treasure is buried? (b) What if you do not really know the way the pirate labeled the trees? Rearrange the order of the trees [for instance, B(30 m, − 20 m), A(60 m, 80 m), E(− 10 m, − 10 m), C(40 m, − 30 m), and D(− 70 m, 60 m)] and repeat the calculation to show that the answer does not depend on the order in which the trees are labeled.

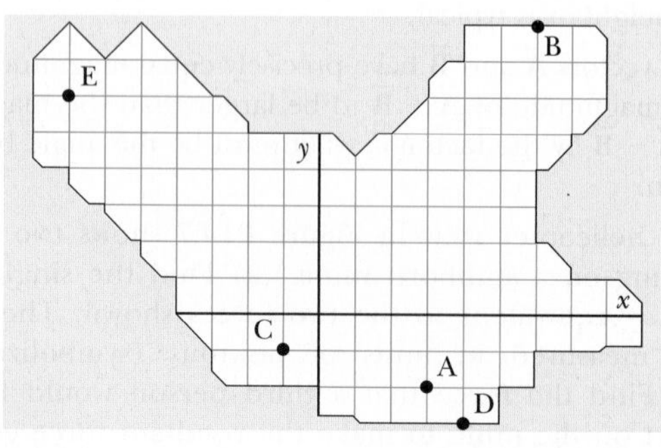

FIGURE **P1.69**

70. Consider a game in which N children position themselves at equal distances around the circumference of a circle. At the center of the circle is a rubber tire. Each child holds a rope attached to the tire and, at a signal, pulls on his or her rope. All children exert forces of the same magnitude F. In the case $N = 2$, it is easy to see that the net force on the tire will be zero because the two oppositely directed force vectors add to zero. Similarly, if $N = 4$, 6, or any even integer, the resultant force on the tire must be zero because the forces exerted by each pair of oppositely positioned children will cancel. When an odd number of children are around the circle, it is not as obvious whether the total force on the central tire will be zero. (a) Calculate the net force on the tire in the case $N = 3$ by adding the components of the three force vectors. Choose the x axis to lie along one of the ropes. (b) Determine the net force for the general case where N is any integer, odd or even, greater than 1. Proceed as follows: Assume that the total force is not zero. Then it must point in some particular direction. Let every child move one position clockwise. Give a reason that the total force must then have a direction turned clockwise by $360°/N$. Argue that the total force must nevertheless be the same as before. Explain that the contradiction proves that the magnitude of the force is zero. This problem illustrates a widely useful technique of proving a result "by symmetry," by using a bit of the mathematics of *group theory*. The particular situation is actually encountered in physics and chemistry when an array of electric charges (ions) exerts electric forces on an atom at a central position in a molecule or in a crystal.

71. A rectangular parallelepiped has dimensions a, b, and c, as shown in Figure P1.71. (a) Obtain a vector expression for the face diagonal vector $\vec{\mathbf{R}}_1$. What is the magnitude of this vector? (b) Obtain a vector expression for the body diagonal vector $\vec{\mathbf{R}}_2$. Note that $\vec{\mathbf{R}}_1$, $c\hat{\mathbf{k}}$, and $\vec{\mathbf{R}}_2$ make a right triangle and prove that the magnitude of $\vec{\mathbf{R}}_2$ is $\sqrt{a^2 + b^2 + c^2}$.

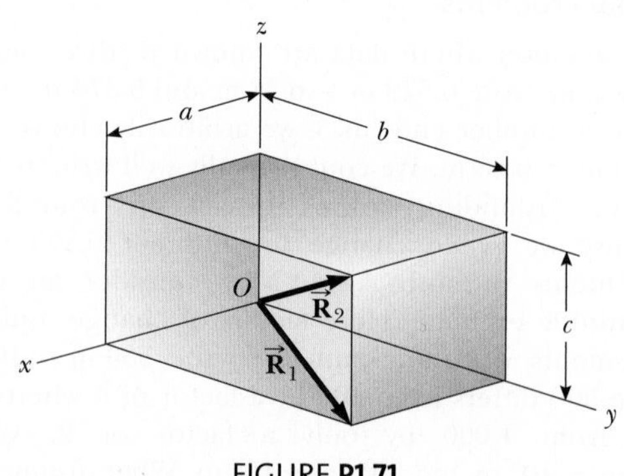

FIGURE **P1.71**

72. Vectors $\vec{\mathbf{A}}$ and $\vec{\mathbf{B}}$ have equal magnitudes of 5.00. The sum of $\vec{\mathbf{A}}$ and $\vec{\mathbf{B}}$ is the vector $6.00\hat{\mathbf{j}}$. Determine the angle between $\vec{\mathbf{A}}$ and $\vec{\mathbf{B}}$.

ANSWERS TO QUICK QUIZZES

1.1 False. Dimensional analysis gives the units of the proportionality constant but provides no information about its numerical value. Determining its numerical value requires either experimental data or mathematical reasoning. For example, in the generation of the equation $x = \frac{1}{2}at^2$, because the factor $\frac{1}{2}$ is dimensionless, there is no way of determining it using dimensional analysis.

1.2 (b). Because kilometers are shorter than miles, it takes a larger number of kilometers than miles to represent a given distance.

1.3 Scalars: (a), (d), (e). None of these quantities has a direction. Vectors: (b), (c). For these quantities, the direction is important to completely specify the quantity.

1.4 (c). The resultant has its maximum magnitude $A + B = 12 + 8 = 20$ units when vector $\vec{\mathbf{A}}$ is oriented in the same direction as vector $\vec{\mathbf{B}}$. The resultant vector has its minimum magnitude $A - B = 12 - 8 = 4$ units when vector $\vec{\mathbf{A}}$ is oriented in the direction opposite vector $\vec{\mathbf{B}}$.

1.5 (a). The resultant has magnitude $A + B$ when $\vec{\mathbf{A}}$ is oriented in the same direction as $\vec{\mathbf{B}}$.

1.6 (b). From the Pythagorean theorem, the magnitude of a vector is always larger than the absolute value of each component, unless there is only one nonzero component, in which case the magnitude of the vector is equal to the absolute value of that component.

1.7 (b). From the Pythagorean theorem, we see that the magnitude of a vector is nonzero if at least one component is nonzero.

1.8 (d). Each set of components, for example, the two x components A_x and B_x, must add to zero, so the components must be of opposite sign.

CONTEXT 1

Alternative-Fuel Vehicles

The idea of self-propelled vehicles has been part of the human imagination for centuries. Leonardo da Vinci drew plans for a vehicle powered by a wound spring in 1478. This vehicle was never built although models have been constructed from his plans and appear in museums. Isaac Newton developed a vehicle in 1680 that operated by ejecting steam out the back, similar to a rocket engine. This invention did not develop into a useful device. Despite these and other attempts, self-propelled vehicles did not succeed; that is, they did not begin to replace the horse as a primary means of transportation until the 19th century.

The history of *successful* self-propelled vehicles begins in 1769 with the invention of a military tractor by Nicolas Joseph Cugnot in France. This vehicle, as well as Cugnot's follow-up vehicles, was powered by a steam engine. During the remainder of the 18th century and for most of the 19th century, additional steam-driven vehicles were developed in France, Great Britain, and the United States.

After the invention of the electric battery by Italian Alessandro Volta at the beginning of the 19th century and its further development over three decades came the invention of early electric vehicles in the 1830s. The development in 1859 of the storage battery, which could be recharged, provided significant impetus to the development of electric vehicles. By the early 20th century, electric cars with a range of about 20 miles and a top speed of 15 miles per hour had been developed.

An internal combustion engine was designed but never built by Dutch physicist Christiaan Huygens in 1680. The invention of modern gasoline-powered internal combustion vehicles is generally credited to Gottlieb Daimler in 1885 and Karl Benz in 1886. Several earlier vehicles, dating back to 1807, however, used internal combustion engines operating on various fuels, including coal gas and primitive gasoline.

At the beginning of the 20th century, steam-powered, gasoline-powered, and electric cars shared the roadways in the United States. Electric cars did not possess the vibration, smell, and noise of gasoline-powered cars and did not suffer from the long start-up time intervals, up to 45 minutes, of steam-powered cars on cold mornings. Electric cars were especially preferred by women, who did not enjoy the difficult task of cranking a gasoline-powered car to start the engine. The limited range of electric cars was not a significant problem because the only roads that existed were in highly populated areas and cars were primarily used for short trips in town.

The end of electric cars in the early 20th century began with the following developments:

(Courtesy of The Exhibition Alliance, Hamilton, N.Y.)

FIGURE 1 A model of a spring-drive car designed by Leonardo da Vinci.

(© Bettmann/CORBIS)

FIGURE 2 This magazine advertisement for an electric car is typical of this popular type of car in the early 20th century.

- 1901: A major discovery of crude oil in Texas reduced prices of gasoline to widely affordable levels.
- 1912: The electric starter for gasoline engines was invented, removing the physical task of cranking the engine.
- During the 1910s: Henry Ford successfully introduced mass production of internal combustion vehicles, resulting in a drop in the price of these vehicles to significantly less than that of an electric car.
- By the early 1920s: Roadways in the United States were of much better quality than previously and connected cities, requiring vehicles with a longer range than that of electric cars.

Because of these factors, the roadways were ruled by gasoline-powered cars almost exclusively by the 1920s. Gasoline, however, is a finite and short-lived commodity. We are approaching the end of our ability to use gasoline in transportation; some experts predict that diminishing supplies of crude oil will push the cost of gasoline to prohibitively high levels within two more decades. Furthermore, gasoline and diesel fuel result in serious tailpipe emissions that are harmful to the environment. As we look for a replacement for gasoline, we also want to pursue fuels that will be kinder to the atmosphere. Such fuels will help reduce the effects of global warming, which we will study in Context 5.

What do the steam engine, the electric motor, and the internal combustion engine have in common? That is, what do they each extract from a source, be it a type of fuel or an electric battery? The answer to this question is *energy*. Regardless of the type of automobile, some source of energy must be

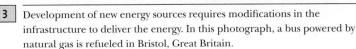

(© Martin Bond/Photo Researchers, Inc.)

FIGURE 3 Development of new energy sources requires modifications in the infrastructure to deliver the energy. In this photograph, a bus powered by natural gas is refueled in Bristol, Great Britain.

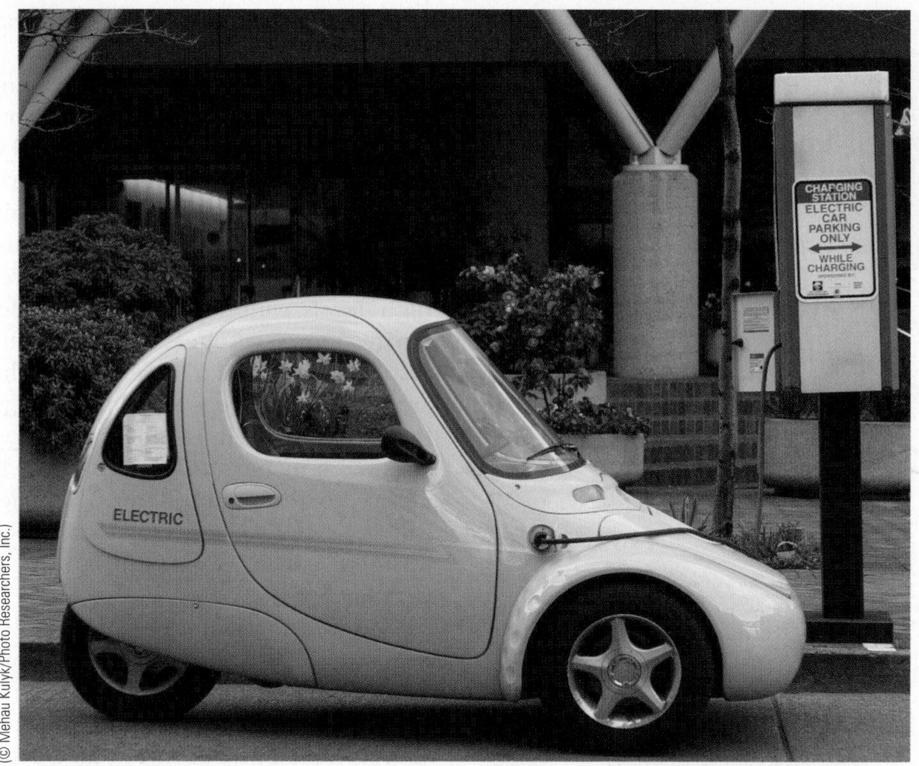

(© Mehau Kulyk/Photo Researchers, Inc.)

FIGURE 4 Modern electric cars can take advantage of an infrastructure set up in some localities to provide charging stations in parking lots.

provided. Energy is one of the physical concepts that we will investigate in this Context. A fuel such as gasoline contains energy due to its chemical composition and its ability to undergo a combustion process. The battery in an electric car also contains energy, again related to chemical composition, but in this case it is associated with an ability to produce an electric current.

One difficult social aspect of developing a new energy source for automobiles is that there must be a synchronized development of the new automobile along with the infrastructure for delivering the new source of energy. This aspect requires close cooperation between automotive corporations and energy manufacturers and suppliers. For example, electric cars cannot be used to travel long distances unless an infrastructure of charging stations develops in parallel with the development of electric cars.

As we draw near to the time when we run out of gasoline, our central question in this first Context is an important one for our future development:

What source besides gasoline can be used to provide energy for an automobile while reducing environmentally damaging emissions?

Motion in One Dimension

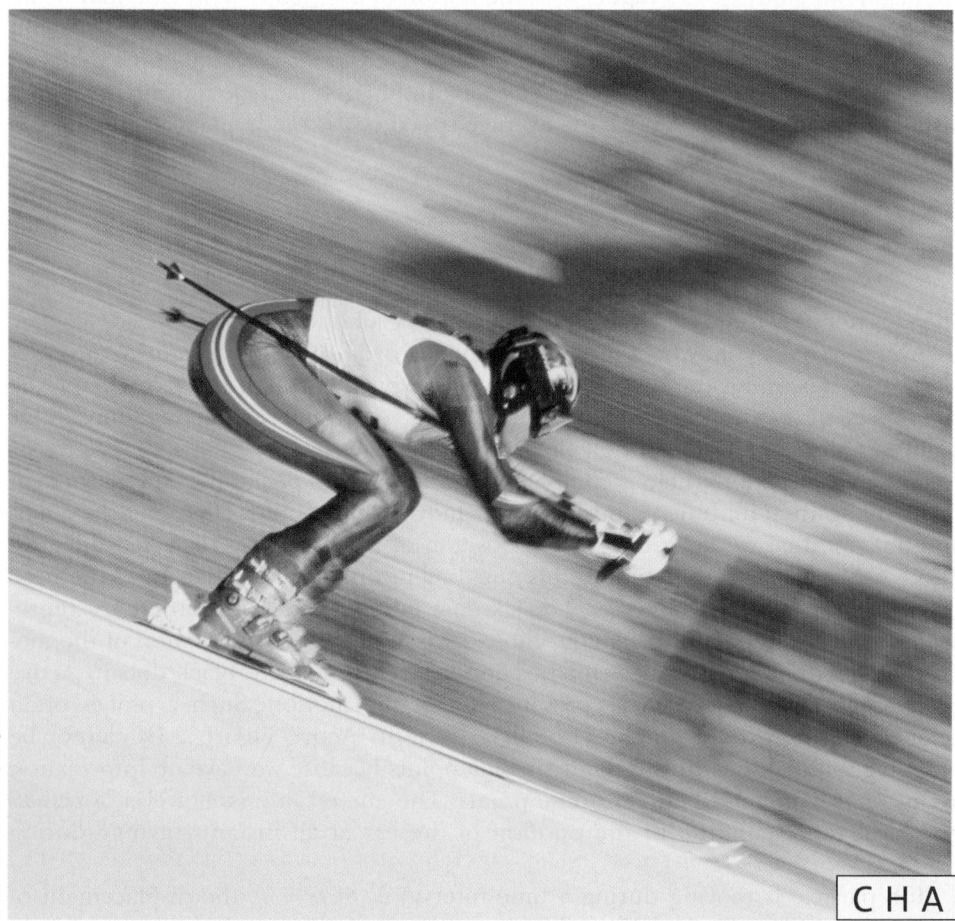

(Jean Y. Ruszniewski/Getty Images)

One of the physical quantities we will study in this chapter is the velocity of an object moving in a straight line. Downhill skiers can reach velocities with a magnitude greater than 100 km/h.

CHAPTER OUTLINE

To begin our study of motion, it is important to be able to *describe* motion using the concepts of space and time without regard to the causes of the motion. This portion of mechanics is called *kinematics.* In this chapter, we shall consider motion along a straight line, that is, one-dimensional motion. Chapter 3 extends our discussion to two-dimensional motion.

From everyday experience we recognize that motion represents continuous change in the position of an object. For example, if you are driving from your home to a destination, your position on the Earth's surface is changing.

The movement of an object through space (translation) may be accompanied by the rotation or vibration of the object. Such motions can be quite complex. It is often possible to simplify matters, however, by temporarily ignoring rotation and internal motions of the moving object. The result is the simplification model that we call the particle model, discussed in Chapter 1. In many situations, an object can be treated as a particle if the only motion being considered is translation through space. We will use the particle model extensively throughout this book.

2.1 AVERAGE VELOCITY

We begin our study of kinematics with the notion of average velocity. You may be familiar with a similar notion, average speed, from experiences with driving. If you drive your car 100 miles according to your odometer and it takes 2.0 hours to do so, your average speed is (100 mi)/(2.0 h) = 50 mi/h. For a particle moving through a distance d in a time interval Δt, the **average speed** v_{avg} is mathematically defined as

■ Definition of average speed

$$v_{\text{avg}} \equiv \frac{d}{\Delta t} \qquad [2.1]$$

Speed is not a vector, so there is no direction associated with average speed.

Average velocity may be a little less familiar to you due to its vector nature. Let us start by imagining the motion of a particle, which, through the particle model, can represent the motion of many types of objects. We shall restrict our study at this point to one-dimensional motion along the x axis.

The motion of a particle is completely specified if the position of the particle in space is known at all times. Consider a car moving back and forth along the x axis and imagine that we take data on the position of the car every 10 s. Active Figure 2.1a is a *pictorial* representation of this one-dimensional motion that shows the positions of the car at 10-s intervals. The six data points we have recorded are represented by the letters Ⓐ through Ⓕ. Table 2.1 is a *tabular* representation of the motion. It lists the data as entries for position at each time. The black dots in Active Figure 2.1b show a *graphical* representation of the motion. Such a plot is often called a **position–time graph.** The curved line in Active Figure 2.1b cannot be unambiguously drawn through our six data points because we have no information about what happened between these points. The curved line is, however, a *possible* graphical representation of the position of the car at all instants of time during the 50 s.

If a particle is moving during a time interval $\Delta t = t_f - t_i$, the displacement of the particle is described as $\Delta\vec{\mathbf{x}} = \vec{\mathbf{x}}_f - \vec{\mathbf{x}}_i = (x_f - x_i)\hat{\mathbf{i}}$. (Recall that displacement is defined as the change in the position of the particle, which is equal to its final position value minus its initial position value.) Because we are considering only one-dimensional motion in this chapter, we shall drop the vector notation at this point and pick it up again in Chapter 3. The direction of a vector in this chapter will be indicated by means of a positive or negative sign.

The **average velocity** $v_{x,\text{avg}}$ of the particle is defined as the ratio of its displacement Δx to the time interval Δt during which the displacement takes place:

PITFALL PREVENTION 2.1

AVERAGE SPEED AND AVERAGE VELOCITY The magnitude of the average velocity is *not* the average speed. Consider a particle moving from the origin to $x = 10$ m and then back to the origin in a time interval of 4.0 s. The magnitude of the average velocity is zero because the particle ends the time interval at the same position at which it started; the displacement is zero. The average speed, however, is the total distance divided by the time interval: 20 m/4.0 s = 5.0 m/s.

■ Definition of average velocity

$$v_{x,\text{avg}} \equiv \frac{\Delta x}{\Delta t} = \frac{x_f - x_i}{t_f - t_i} \qquad [2.2]$$

where the subscript x indicates motion along the x axis. From this definition we see that average velocity has the dimensions of length divided by time: meters per second in SI units and feet per second in U.S. customary units. The average velocity is *independent* of the path taken between the initial and final points. This independence is a major difference from the average speed discussed at the beginning of this section. The average velocity is independent of path because it is proportional

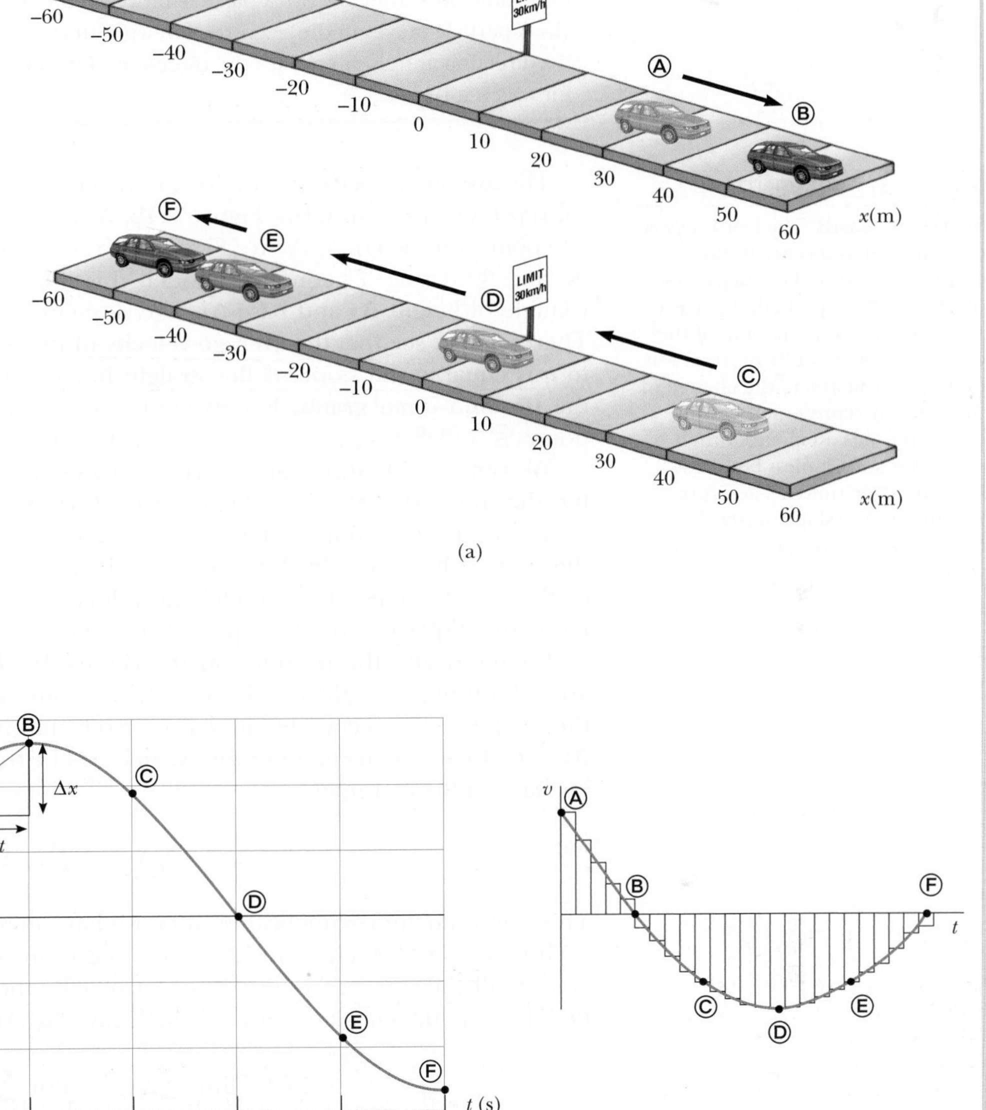

ACTIVE FIGURE 2.1

(a) A pictorial representation of the motion of a car. The positions of the car at six instants of time are shown and labeled. (b) A graphical representation, known as a position–time graph, of the car's motion in part (a). The average velocity $v_{x,\text{avg}}$ in the interval $t = 0$ to $t = 10$ s is obtained from the slope of the straight line connecting points Ⓐ and Ⓑ. (c) A velocity–time graph of the motion of the car in part (a).

Physics Now™ Log into PhysicsNow at **www.pop4e.com** and go to Active Figure 2.1. You can move each of the six points Ⓐ through Ⓕ and observe the car's motion in both a pictorial and a graphical representation as the car follows a smooth path through the six points.

to the displacement Δx, which depends only on the initial and final coordinates of the particle. Average speed (a scalar) is found by dividing the *distance* traveled by the time interval, whereas average velocity (a vector) is the *displacement* divided by the time interval. Therefore, average velocity gives us no details of the motion; rather, it only gives us the result of the motion. Finally, note that the average velocity in one dimension can be positive or negative, depending on the sign of the displacement. (The time interval Δt is always positive.) If the x coordinate of the particle increases during the time interval (i.e., if $x_f > x_i$), Δx is positive and $v_{x,\text{avg}}$ is positive, which corresponds to an average velocity in the positive x direction. On the other hand, if the coordinate decreases over time ($x_f < x_i$), Δx is negative; hence, $v_{x,\text{avg}}$ is negative, which corresponds to an average velocity in the negative x direction.

TABLE 2.1 Positions of the Car at Various Times

Position	t(s)	x (m)
Ⓐ	0	30
Ⓑ	10	52
Ⓒ	20	38
Ⓓ	30	0
Ⓔ	40	−37
Ⓕ	50	−53

QUICK QUIZ 2.1 Under which of the following conditions is the magnitude of the average velocity of a particle moving in one dimension smaller than the average speed over some time interval? (**a**) A particle moves in the $+x$ direction without reversing. (**b**) A particle moves in the $-x$ direction without reversing. (**c**) A particle moves in the $+x$ direction and then reverses the direction of its velocity. (**d**) There are no conditions for which it is true.

PITFALL PREVENTION 2.2

SLOPES OF GRAPHS The word *slope* is often used in reference to the graphs of physical data. Regardless of what data are plotted, the word *slope* will represent the ratio of the change in the quantity represented on the vertical axis to the change in the quantity represented on the horizontal axis. Remember that *a slope has units* (unless both axes have the same units). Therefore, the units of the slope in Active Figure 2.1b are m/s, the units of velocity.

The average velocity can also be interpreted geometrically, as seen in the graphical representation in Active Figure 2.1b. A straight line can be drawn between any two points on the curve. Active Figure 2.1b shows such a line drawn between points Ⓐ and Ⓑ. Using a geometric model, this line forms the hypotenuse of a right triangle of height Δx and base Δt. The slope of the hypotenuse is the ratio $\Delta x/\Delta t$. Therefore, we see that **the average velocity of the particle during the time interval t_i to t_f is equal to the slope of the straight line joining the initial and final points on the position–time graph.** For example, the average velocity of the car between points Ⓐ and Ⓑ is $v_{x,\text{avg}} = (52\text{ m} - 30\text{ m})/(10\text{ s} - 0) = 2.2\text{ m/s}$.

We can also identify a geometric interpretation for the total displacement during the time interval. Active Figure 2.1c shows the velocity–time graphical representation of the motion in Active Figures 2.1a and 2.1b. The total time interval for the motion has been divided into small increments of duration Δt_n. During each of these increments, if we model the velocity as constant during the short increment, the displacement of the particle is given by $\Delta x_n = v_n\,\Delta t_n$.

Geometrically, the product on the right side of this expression represents the area of a thin rectangle associated with each time increment in Active Figure 2.1c; the height of the rectangle (measured from the time axis) is v_n, and the width is Δt_n. The total displacement of the particle will be the sum of the displacements during each of the increments:

$$\Delta x \approx \sum_n \Delta x_n = \sum_n v_n\,\Delta t_n$$

This sum is an approximation because we have modeled the velocity as constant in each increment, which is not the case. The term on the right represents the total area of all the thin rectangles. Now let us take the limit of this expression as the time increments shrink to zero, in which case the approximation becomes exact:

$$\Delta x = \lim_{\Delta t_n \to 0} \sum_n \Delta x_n = \lim_{\Delta t_n \to 0} \sum_n v_n\,\Delta t_n$$

In this limit, the sum of the areas of all the very thin rectangles becomes equal to the total area under the curve. Therefore, **the displacement of a particle during the time interval t_i to t_f is equal to the area under the curve between the initial and final points on the velocity–time graph.** We will make use of this geometric interpretation in Section 2.6.

EXAMPLE 2.1 Calculate the Average Velocity

A particle moving along the x axis is located at $x_i = 12$ m at $t_i = 1$ s and at $x_f = 4$ m at $t_f = 3$ s. Find its displacement and average velocity during this time interval.

Solution First, establish the mental representation. Imagine the particle moving along the axis. Based on the information in the problem, which way is it moving? You may find it useful to draw a pictorial representation, but for this simple example, we will go straight to the mathematical representation. The displacement is

$$\Delta x = x_f - x_i = 4\text{ m} - 12\text{ m} = -8\text{ m}$$

The average velocity is, according to Equation 2.2,

$$v_{x,\text{avg}} = \frac{\Delta x}{\Delta t} = \frac{4\text{ m} - 12\text{ m}}{3\text{ s} - 1\text{ s}} = -4\text{ m/s}$$

Because the displacement is negative for this time interval, we conclude that the particle has moved to the left, toward decreasing values of x. Is this conclusion consistent with your mental representation? Keep in mind that it may not have *always* been moving to the left. We only have information about its location at two points in time. After $t_i = 1$ s, it could have moved to the right, turned around, and ended up farther to the left than its original position by the time $t_f = 3$ s. To be completely confident that we know the motion of the particle, we would need to have information about its location at *every* instant of time.

EXAMPLE 2.2 **Motion of a Jogger**

A jogger runs in a straight line, with a magnitude of average velocity of 5.00 m/s for 4.00 min and then with a magnitude of average velocity of 4.00 m/s for 3.00 min.

[A] What is the magnitude of the final displacement from her initial position?

Solution That this problem involves a jogger is not important; we model the jogger as a particle. We have data for two separate portions of the motion, so we use these data to find the displacement for each portion, using Equation 2.2:

$$v_{x,\text{avg}} = \frac{\Delta x}{\Delta t} \quad \rightarrow \quad \Delta x = v_{x,\text{avg}}\,\Delta t$$

$$\Delta x_{\text{portion 1}} = (5.00\text{ m/s})(4.00\text{ min})\left(\frac{60\text{ s}}{1\text{ min}}\right) = 1.20 \times 10^3\text{ m}$$

$$\Delta x_{\text{portion 2}} = (4.00\text{ m/s})(3.00\text{ min})\left(\frac{60\text{ s}}{1\text{ min}}\right) = 7.20 \times 10^2\text{ m}$$

We add these two displacements to find the total displacement of 1.92×10^3 m.

[B] What is the magnitude of her average velocity during this entire time interval of 7.00 min?

Solution We now have the data we need to find the average velocity for the entire time interval using Equation 2.2:

$$v_{x,\text{avg}} = \frac{\Delta x}{\Delta t} = \frac{1.92 \times 10^3\text{ m}}{7.00\text{ min}}\left(\frac{1\text{ min}}{60\text{ s}}\right) = 4.57\text{ m/s}$$

Notice that the average velocity is *not* calculated as the simple arithmetic mean of the two velocities given in the problem.

2.2 INSTANTANEOUS VELOCITY

Suppose you drive your car through a displacement of magnitude 40 miles and it takes exactly 1 hour to do so, from 1:00:00 P.M. to 2:00:00 P.M. Then the magnitude of your average velocity is 40 mi/h for the 1-h interval. How fast, though, were you going at the particular *instant* of time 1:20:00 P.M.? It is likely that your velocity varied during the trip, owing to hills, traffic lights, slow drivers ahead of you, and the like, so that there was not a single velocity maintained during the entire hour of travel. The velocity of a particle at any instant of time is called the **instantaneous velocity.**

Consider again the motion of the car shown in Active Figure 2.1a. Active Figure 2.2a is the graphical representation again, with two blue lines representing average velocities over very different time intervals. One blue line represents the average velocity we calculated earlier over the interval from Ⓐ to Ⓑ. The second blue line represents the average velocity over the much longer interval Ⓐ to Ⓕ. How well does either of these represent the instantaneous velocity at point Ⓐ? In Active Figure 2.1a, the car begins to move to the right, which we identify as a positive velocity. The average velocity from Ⓐ to Ⓕ is *negative* (because the slope of the line from Ⓐ to Ⓕ is negative), so this velocity clearly is not an accurate representation of the instantaneous velocity at Ⓐ. The average velocity from interval Ⓐ to Ⓑ is *positive,* so this velocity at least has the right sign.

In Active Figure 2.2b, we show the result of drawing the lines representing the average velocity of the car as point Ⓑ is brought closer and closer to point Ⓐ. As

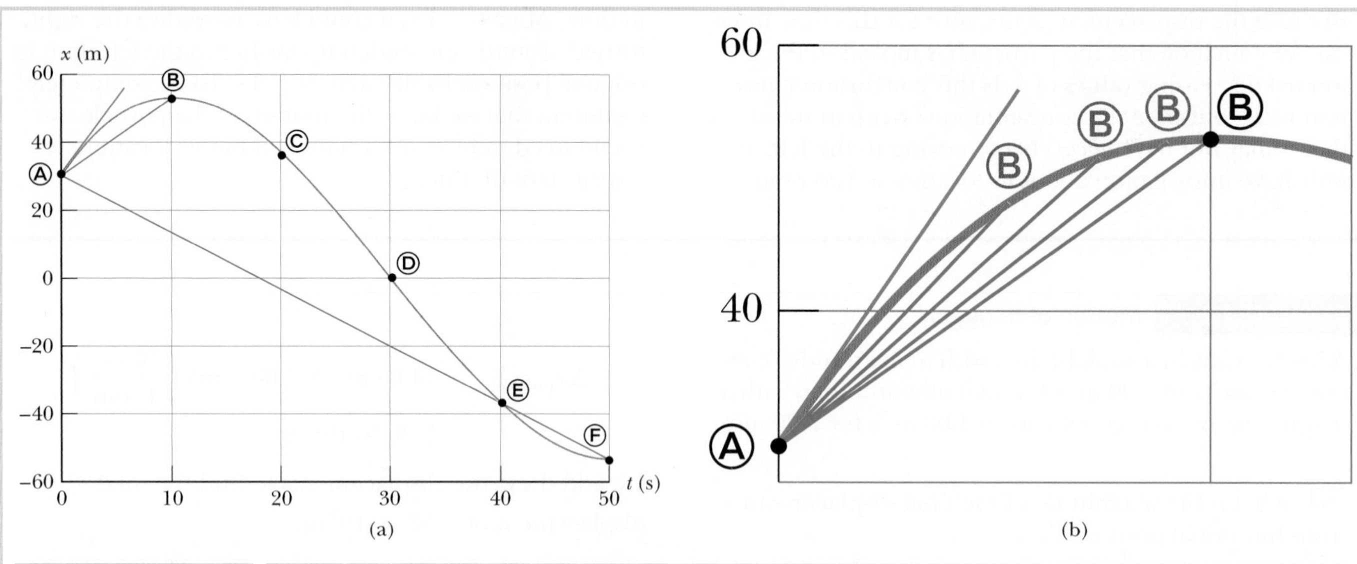

ACTIVE FIGURE 2.2 (a) Position–time graph for the motion of the car in Active Figure 2.1. (b) An enlargement of the upper left-hand corner of the graph in part (a) shows how the blue line between positions Ⓐ and Ⓑ approaches the green tangent line as point Ⓑ is moved closer to point Ⓐ.

PhysicsNow™ Log into PhysicsNow at **www.pop4e.com** and go to Active Figure 2.2. You can move point Ⓑ as suggested in part (b) and observe the blue line approaching the green tangent line.

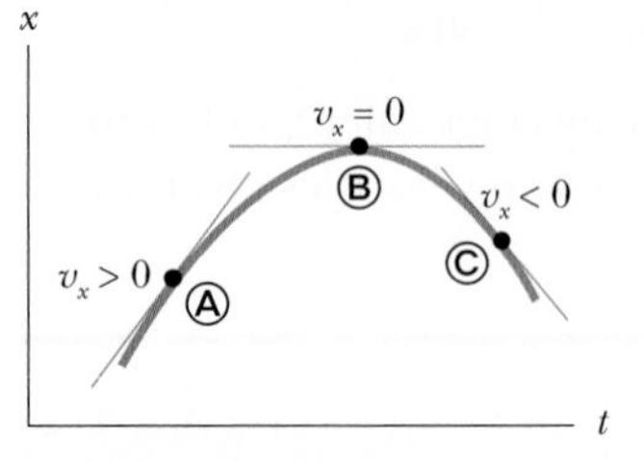

FIGURE 2.3 In the position–time graph shown, the velocity is positive at Ⓐ, where the slope of the tangent line is positive; the velocity is zero at Ⓑ, where the slope of the tangent line is zero; and the velocity is negative at Ⓒ, where the slope of the tangent line is negative.

■ Definition of instantaneous velocity

that occurs, the slope of the blue line approaches that of the green line, which is the line drawn tangent to the curve at point Ⓐ. As Ⓑ approaches Ⓐ, the time interval that includes point Ⓐ becomes infinitesimally small. Therefore, the average velocity over this interval as the interval shrinks to zero can be interpreted as the instantaneous velocity at point Ⓐ. Furthermore, the slope of the line tangent to the curve at Ⓐ is the instantaneous velocity at the time t_A. In other words, **the instantaneous velocity v_x equals the limiting value of the ratio $\Delta x/\Delta t$ as Δt approaches zero:**[1]

$$v_x \equiv \lim_{\Delta t \to 0} \frac{\Delta x}{\Delta t}$$

In calculus notation, this limit is called the *derivative* of x with respect to t, written dx/dt:

$$v_x \equiv \lim_{\Delta t \to 0} \frac{\Delta x}{\Delta t} = \frac{dx}{dt} \qquad [2.3]$$

The instantaneous velocity can be positive, negative, or zero. When the slope of the position–time graph is positive, such as at point Ⓐ in Figure 2.3, v_x is positive. At point Ⓒ, v_x is negative because the slope is negative. Finally, the instantaneous velocity is zero at the peak Ⓑ (the turning point), where the slope is zero. From here on, we shall usually use the word *velocity* to designate instantaneous velocity.

The **instantaneous speed** of a particle is defined as the magnitude of the instantaneous velocity vector. Hence, by definition, *speed* can never be negative.

[1] Note that the displacement Δx also approaches zero as Δt approaches zero. As Δx and Δt become smaller and smaller, however, the ratio $\Delta x/\Delta t$ approaches a value equal to the *true* slope of the line tangent to the x versus t curve.

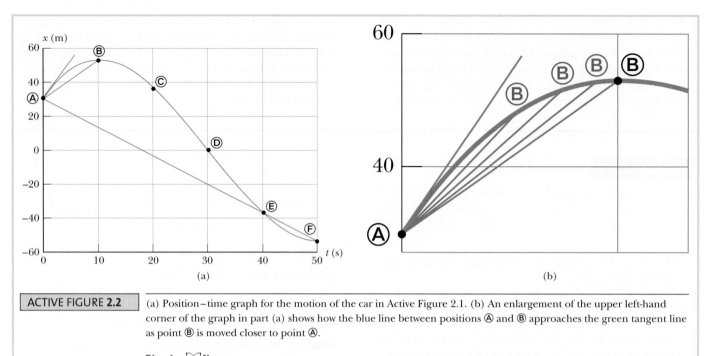

QUICK QUIZ 2.2 Are members of the highway patrol more interested in (**a**) your average speed or (**b**) your instantaneous speed as you drive?

If you are familiar with calculus, you should recognize that specific rules exist for taking the derivatives of functions. These rules, which are listed in Appendix B.6, enable us to evaluate derivatives quickly.

Suppose x is proportional to some power of t, such as

$$x = At^n$$

where A and n are constants. (This equation is a very common functional form.) The derivative of x with respect to t is

$$\frac{dx}{dt} = nAt^{n-1}$$

For example, if $x = 5t^3$, we see that $dx/dt = 3(5)t^{3-1} = 15t^2$.

Thinking Physics 2.1

Consider the following motions of an object in one dimension. (**a**) A ball is thrown directly upward, rises to its highest point, and falls back into the thrower's hand. (**b**) A race car starts from rest and speeds up to 100 m/s along a straight line. (**c**) A spacecraft on the way to another star drifts through empty space at constant velocity. Are there any instants of time in the motion of these objects at which the instantaneous velocity at the instant and the average velocity over the entire interval are the same? If so, identify the point(s).

Reasoning (**a**) The average velocity over the entire interval for the thrown ball is zero; the ball returns to the starting point at the end of the time interval. There is one point—at the top of the motion—at which the instantaneous velocity is zero. (**b**) The average velocity for the motion of the race car cannot be evaluated unambiguously with the information given, but its magnitude must be some value between 0 and 100 m/s. Because the magnitude of the instantaneous velocity of the car will have every value between 0 and 100 m/s at some time during the interval, there must be some instant at which the instantaneous velocity is equal to the average velocity over the entire interval. (**c**) Because the instantaneous velocity of the spacecraft is constant, its instantaneous velocity at *any* time and its average velocity over *any* time interval are the same. ■

EXAMPLE 2.3 The Limiting Process

The position of a particle moving along the x axis varies in time according to the expression[2] $x = 3t^2$, where x is in meters and t is in seconds. Find the velocity in terms of t at any time.

Solution The position–time graphical representation for this motion is shown in Figure 2.4. We can compute the velocity at any time t by using the definition of the instantaneous velocity. If the initial coordinate of the particle at time t is $x_i = 3t^2$, the coordinate at a later time $t + \Delta t$ is

$$x_f = 3(t + \Delta t)^2 = 3[t^2 + 2t\,\Delta t + (\Delta t)^2]$$
$$= 3t^2 + 6t\,\Delta t + 3(\Delta t)^2$$

Therefore, the displacement in the time interval Δt is

$$\Delta x = x_f - x_i = (3t^2 + 6t\,\Delta t + 3(\Delta t)^2) - (3t^2)$$
$$= 6t\,\Delta t + 3(\Delta t)^2$$

[2] Simply to make it easier to read, we write the equation as $x = 3t^2$ rather than as $x = (3.00\text{ m/s}^2)t^{2.00}$. When an equation summarizes measurements, consider its coefficients to have as many significant digits as other data quoted in a problem. Also consider its coefficients to have the units required for dimensional consistency. When we start our clocks at $t = 0$, we usually do not mean to limit precision to a single digit. Consider any zero value in this book to have as many significant figures as you need.

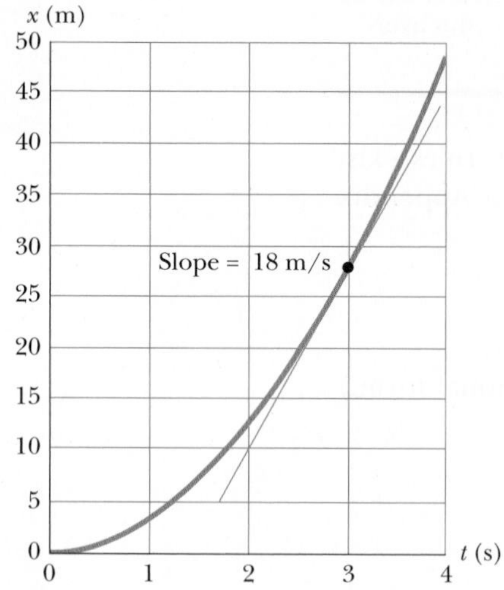

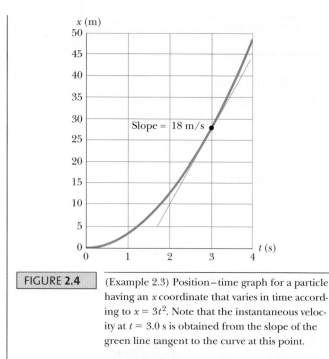

FIGURE 2.4 (Example 2.3) Position–time graph for a particle having an x coordinate that varies in time according to $x = 3t^2$. Note that the instantaneous velocity at $t = 3.0$ s is obtained from the slope of the green line tangent to the curve at this point.

The average velocity in this time interval is

$$v_{x,\,\text{avg}} = \frac{\Delta x}{\Delta t} = \frac{6t\,\Delta t + 3(\Delta t)^2}{\Delta t} = 6t + 3\,\Delta t$$

To find the instantaneous velocity, we take the limit of this expression as Δt approaches zero. In doing so, we see that the term $3\,\Delta t$ goes to zero; therefore,

$$v_x = \lim_{\Delta t \to 0} \frac{\Delta x}{\Delta t} = 6t$$

Notice that this expression gives us the velocity at *any* general time t. It tells us that v_x is increasing linearly in time. It is then a straightforward matter to find the velocity at some specific time from the expression $v_x = 6t$ by substituting the value of the time. For example, at $t = 3.0$ s, the velocity is $v_x = 6(3) = 18$ m/s. Again, this answer can be checked from the slope at $t = 3.0$ s (the green line in Fig. 2.4).

We can also find v_x by taking the first derivative of x with respect to time, as in Equation 2.3. In this example, $x = 3t^2$, and we see that $v_x = dx/dt = 6t$, in agreement with our result from taking the limit explicitly.

EXAMPLE 2.4 Average and Instantaneous Velocity

A particle moves along the x axis. Its x coordinate varies with time according to the expression $x = -4t + 2t^2$, where x is in meters and t is in seconds. The position–time graph for this motion is shown in Figure 2.5.

A Determine the displacement of the particle in the time intervals $t = 0$ to $t = 1$ s and $t = 1$ s to $t = 3$ s.

Solution This problem provides a graphical representation of the motion in Figure 2.5. In your mental representation, note that the particle moves in the negative x direction for the first second of motion, stops instantaneously at $t = 1$ s, and then heads back in the positive x direction for $t > 1$ s. Remember that it is a one-dimensional problem, so the curve in Figure 2.5 does not represent the path the particle follows through space; be sure not to confuse a graphical representation with a pictorial representation of the motion in space (see Active Fig. 2.1 for a comparison). In your mental representation, you should imagine the particle moving to the left and then to the right, with all the motion taking place along a single line.

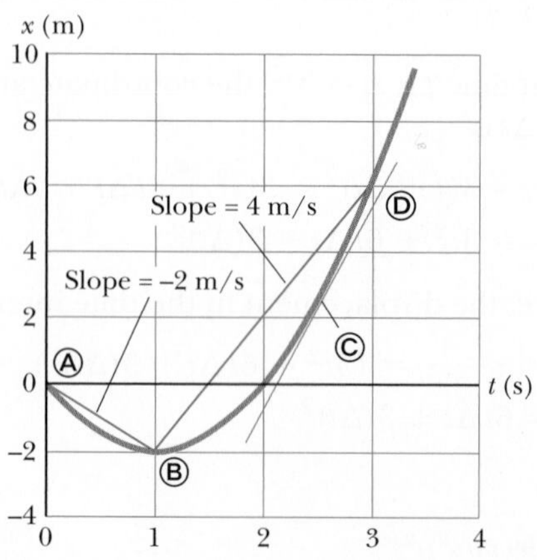

FIGURE 2.5 (Example 2.4) Position–time graph for a particle having an x coordinate that varies in time according to $x = -4t + 2t^2$.

In the first time interval (Ⓐ to Ⓑ), we set $t_i = 0$ and $t_f = 1$ s. Because $x = -4t + 2t^2$, the displacement during the first time interval is

$$\begin{aligned}\Delta x_{\text{AB}} &= x_f - x_i = -4(1) + 2(1)^2 - [-4(0) + 2(0)^2] \\ &= -2\text{ m}\end{aligned}$$

Likewise, in the second time interval (Ⓑ to Ⓓ), we can set $t_i = 1$ s and $t_f = 3$ s. Therefore, the displacement in this interval is

$$\begin{aligned}\Delta x_{\text{BD}} &= x_f - x_i = -4(3) + 2(3)^2 - [-4(1) + 2(1)^2] \\ &= 8\text{ m}\end{aligned}$$

These displacements can also be read directly from the position–time graph (see Fig. 2.5).

B Calculate the average velocity in the time intervals $t = 0$ to $t = 1$ s and $t = 1$ s to $t = 3$ s.

Solution In the first time interval, $\Delta t = t_f - t_i = 1$ s. Therefore, using Equation 2.2 and the result from part A gives

$$v_{x,\text{avg}} = \frac{\Delta x_{AB}}{\Delta t} = \frac{-2\text{ m}}{1\text{ s}} = -2\text{ m/s}$$

Likewise, in the second time interval, $\Delta t = 2$ s; therefore,

$$v_{x,\text{avg}} = \frac{\Delta x_{BD}}{\Delta t} = \frac{8\text{ m}}{2\text{ s}} = 4\text{ m/s}$$

These values agree with the slopes of the lines joining these points in Figure 2.5.

C Find the instantaneous velocity of the particle at $t = 2.5$ s (point ©).

Solution We can find the instantaneous velocity at any time t by taking the first derivative of x with respect to t:

$$v_x = \frac{dx}{dt} = \frac{d}{dt}(-4t + 2t^2) = -4 + 4t$$

Therefore, at $t = 2.5$ s, we find that

$$v_x = -4 + 4(2.5) = 6\text{ m/s}$$

We can also obtain this result by measuring the slope of the position–time graph at $t = 2.5$ s. Do you see any symmetry in the motion? For example, are there points at which the speed is the same? Is the velocity the same at these points?

2.3 ANALYSIS MODELS—THE PARTICLE UNDER CONSTANT VELOCITY

As mentioned in Section 1.10, the third category of models used in this book is that of **analysis models.** Such models help us analyze the situation in a physics problem and guide us toward the solution. **An analysis model is a problem we have solved before. It is a description of either (1) the behavior of some physical entity or (2) the interaction between that entity and the environment.** When you encounter a new problem, you should identify the fundamental details of the problem and attempt to recognize which, if any, of the types of problems you have already solved might be used as a model for the new problem. For example, suppose an automobile is moving along a straight freeway at a constant speed. Is it important that it is an automobile? Is it important that it is a freeway? If the answers to both questions are "no," we model the situation as a *particle under constant velocity,* which we will discuss in this section.

This method is somewhat similar to the common practice in the legal profession of finding "legal precedents." If a previously resolved case can be found that is very similar legally to the present one, it is offered as a model and an argument is made in court to link them logically. The finding in the previous case can then be used to sway the finding in the present case. We will do something similar in physics. For a given problem, we search for a "physics precedent," a model with which we are already familiar and that can be applied to the present problem.

We shall generate analysis models based on four fundamental simplification models. The first simplification model is the particle model discussed in Chapter 1. We will look at a particle under various behaviors and environmental interactions. Further analysis models are introduced in later chapters based on simplification models of a *system,* a *rigid object,* and a *wave.* Once we have introduced these analysis models, we shall see that they appear over and over again later in the book in different situations.

Let us use Equation 2.2 to build our first analysis model for solving problems. We imagine a particle moving with a constant velocity. The particle under constant velocity model can be applied in *any* situation in which an entity that can be modeled as a particle is moving with constant velocity. This situation occurs frequently, so it is an important model.

If the velocity of a particle is constant, its instantaneous velocity at any instant during a time interval is the same as the average velocity over the interval, $v_x = v_{x,\text{avg}}$. Therefore, we start with Equation 2.2 to generate an equation to be

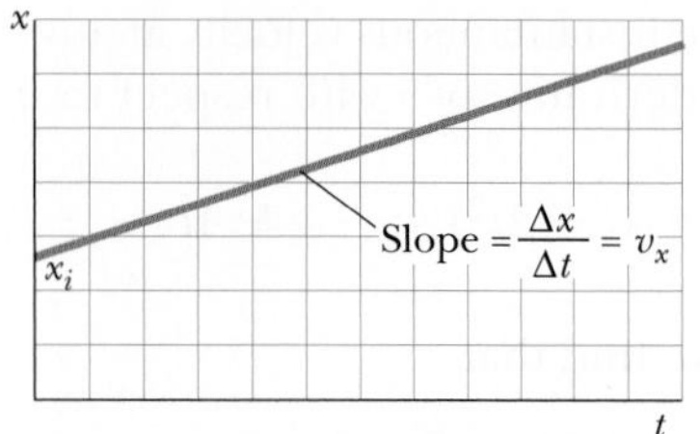

FIGURE 2.6 Position–time graph for a particle under constant velocity. The value of the constant velocity is the slope of the line.

used in the mathematical representation of this situation:

$$v_x = v_{x,\text{avg}} = \frac{\Delta x}{\Delta t} \qquad [2.4]$$

Remembering that $\Delta x = x_f - x_i$, we see that $v_x = (x_f - x_i)/\Delta t$, or

$$x_f = x_i + v_x \Delta t$$

This equation tells us that the position of the particle is given by the sum of its original position x_i plus the displacement $v_x \Delta t$ that occurs during the time interval Δt. In practice, we usually choose the time at the beginning of the interval to be $t_i = 0$ and the time at the end of the interval to be $t_f = t$, so our equation becomes

■ Position of a particle under constant velocity

$$x_f = x_i + v_x t \quad \text{(for constant } v_x\text{)} \qquad [2.5]$$

Equations 2.4 and 2.5 are the primary equations used in the model of a particle under constant velocity. They can be applied to particles or objects that can be modeled as particles.

Figure 2.6 is a graphical representation of the particle under constant velocity. On the position–time graph, the slope of the line representing the motion is constant and equal to the velocity. It is consistent with the mathematical representation, Equation 2.5, which is the equation of a straight line. The slope of the straight line is v_x and the y intercept is x_i in both representations.

EXAMPLE 2.5 Modeling a Runner as a Particle

A scientist is studying the biomechanics of the human body. She determines the velocity of an experimental subject while he runs at a constant rate. The scientist starts the stopwatch at the moment the runner passes a given point and stops it at the moment the runner passes another point 20 m away. The time interval indicated on the stopwatch is 4.4 s.

A What is the runner's velocity?

Solution We model the runner as a particle, as we did in Example 2.2, because the size of the runner and the movement of arms and legs are unnecessary details. This choice, in combination with the velocity being constant, allows us to use Equation 2.4 to find the velocity:

$$v_x = \frac{\Delta x}{\Delta t} = \frac{x_f - x_i}{\Delta t} = \frac{20\text{ m} - 0}{4.4\text{ s}} = 4.5\text{ m/s}$$

B What is the position of the runner after 10 s has passed?

Solution In this part of the problem, we use Equation 2.5 to find the position of the particle at the time $t = 10$ s. Using the velocity found in part A,

$$x_f = x_i + v_x t = 0 + (4.5\text{ m/s})(10\text{ s}) = 45\text{ m}$$

The mathematical manipulations for the particle under constant velocity stem from Equation 2.4 and its descendent, Equation 2.5. These equations can be used to solve for any variable in the equations that happens to be unknown if the other variables are known. For example, in part B of Example 2.5, we find the position when the velocity and the time are known. Similarly, if we know the velocity and the final position, we could use Equation 2.5 to find the time at which the runner is at this position. We shall present more examples of a particle under constant velocity in Chapter 3.

A particle under constant velocity moves with a constant speed along a straight line. Now consider a particle moving with a constant speed along a curved path. It can be represented with the *particle under constant speed* model. The primary equation for this model is Equation 2.1, with the average speed v_{avg} replaced by the constant speed v. As an example, imagine a particle moving at a constant speed in a circular path. If the speed is 5.00 m/s and the radius of the path is 10.0 m, we can calculate the time interval required to complete one trip around the circle:

$$v = \frac{d}{\Delta t} \quad \rightarrow \quad \Delta t = \frac{d}{v} = \frac{2\pi r}{v} = \frac{2\pi(10.0 \text{ m})}{5.00 \text{ m/s}} = 12.6 \text{ s}$$

2.4 ACCELERATION

When the velocity of a particle changes with time, the particle is said to be *accelerating.* For example, the speed of a car increases when you "step on the gas," the car slows down when you apply the brakes, and it changes direction when you turn the wheel; these changes are all accelerations. We will need a precise definition of acceleration for our studies of motion.

Suppose a particle moving along the x axis has a velocity v_{xi} at time t_i and a velocity v_{xf} at time t_f. The **average acceleration** $a_{x,\text{avg}}$ of the particle in the time interval $\Delta t = t_f - t_i$ is defined as the ratio $\Delta v_x/\Delta t$, where $\Delta v_x = v_{xf} - v_{xi}$ is the *change* in velocity of the particle in this time interval:

$$a_{x,\text{avg}} \equiv \frac{v_{xf} - v_{xi}}{t_f - t_i} = \frac{\Delta v_x}{\Delta t} \qquad [2.6]$$

■ Definition of average acceleration

Therefore, **acceleration is a measure of how rapidly the velocity is changing.** Acceleration is a vector quantity having dimensions of length divided by (time)^2, or L/T^2. Some of the common units of acceleration are meters per second per second (m/s^2) and feet per second per second (ft/s^2). For example, an acceleration of 2 m/s^2 means that the velocity changes by 2 m/s during each second of time that passes.

In some situations, the value of the average acceleration may be different for different time intervals. It is therefore useful to define the **instantaneous acceleration** as the limit of the average acceleration as Δt approaches zero, analogous to the definition of instantaneous velocity discussed in Section 2.2:

$$a_x \equiv \lim_{\Delta t \to 0} \frac{\Delta v_x}{\Delta t} = \frac{dv_x}{dt} \qquad [2.7]$$

■ Definition of instantaneous acceleration

That is, the instantaneous acceleration equals the derivative of the velocity with respect to time, which by definition is the slope of the velocity–time graph. Note that if a_x is positive, the acceleration is in the positive x direction, whereas negative a_x implies acceleration in the negative x direction. A negative acceleration does not necessarily mean that the particle is *moving* in the negative x direction, a point we shall address in more detail shortly. From now on, we use the term *acceleration* to mean instantaneous acceleration.

Because $v_x = dx/dt$, the acceleration can also be written

$$a_x = \frac{dv_x}{dt} = \frac{d}{dt}\left(\frac{dx}{dt}\right) = \frac{d^2x}{dt^2} \qquad [2.8]$$

This equation shows that the acceleration equals the *second derivative* of the position with respect to time.

Figure 2.7 shows how the acceleration–time curve in a graphical representation can be derived from the velocity–time curve. In these diagrams, the acceleration of a particle at any time is simply the slope of the velocity–time graph at that time. Positive values of the acceleration correspond to those points (between t_A and t_B)

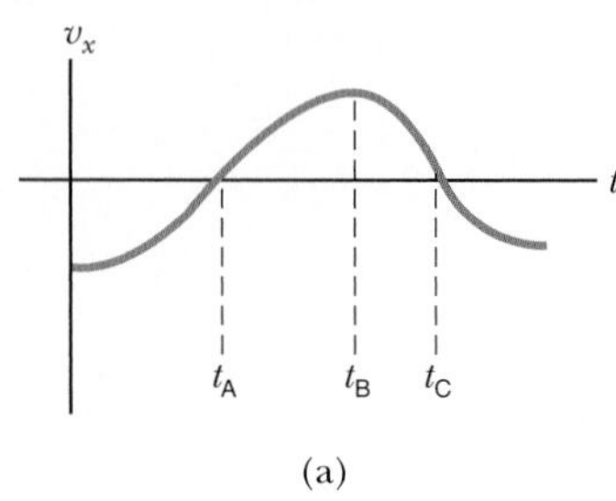

(a)

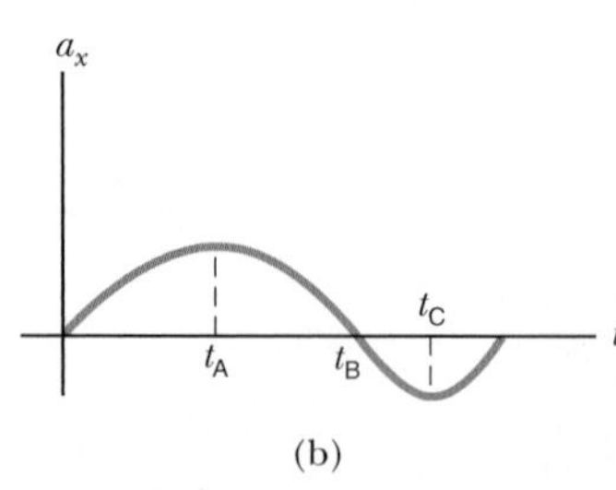

(b)

FIGURE 2.7 The instantaneous acceleration can be obtained from the velocity–time graph (a). At each instant the acceleration in the a_x versus t graph (b) equals the slope of the line tangent to the v_x versus t curve.

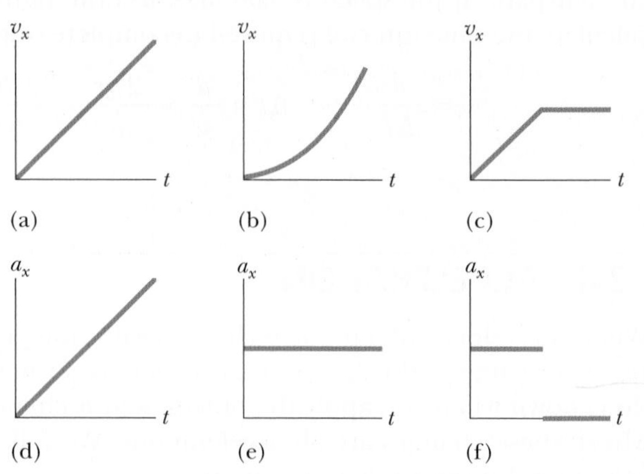

ACTIVE FIGURE 2.8 (Quick Quiz 2.3) Parts (a), (b), and (c) are velocity–time graphs of objects in one-dimensional motion. The possible acceleration–time graphs of each object are shown in scrambled order in parts (d), (e), and (f).

Physics Now™ Log into PhysicsNow at **www.pop4e.com** and go to Active Figure 2.8 to practice matching appropriate velocity versus time graphs and acceleration versus time graphs.

where the velocity in the positive x direction is increasing in magnitude (the particle is speeding up) or those points (between $t = 0$ and t_A) where the velocity in the negative x direction is decreasing in magnitude (the particle is slowing down). The acceleration reaches a maximum at time t_A, when the slope of the velocity–time graph is a maximum. The acceleration then goes to zero at time t_B, when the velocity is a maximum (i.e., when the velocity is momentarily not changing and the slope of the v versus t graph is zero). Finally, the acceleration is negative when the velocity in the positive x direction is decreasing in magnitude (between t_B and t_C) or when the velocity in the negative direction is increasing in magnitude (after t_C).

PITFALL PREVENTION 2.3

NEGATIVE ACCELERATION Keep in mind that *negative acceleration does not necessarily mean that an object is slowing down.* If the acceleration is negative and the velocity is negative, the object is speeding up!

QUICK QUIZ 2.3 Using Active Figure 2.8, match each of the velocity–time graphical representations on the top with the acceleration–time graphical representation on the bottom that best describes the motion.

PITFALL PREVENTION 2.4

DECELERATION The word *deceleration* has a common popular connotation as *slowing down.* When combined with the misconception in Pitfall Prevention 2.3 that negative acceleration means slowing down, the situation can be further confused by the use of the word *deceleration.* We will not use this word in this text.

As an example of the computation of acceleration, consider the pictorial representation of a car's motion in Figure 2.9. In this case, the velocity of the car has changed from an initial value of 30 m/s to a final value of 15 m/s in a time interval of 2.0 s. The average acceleration during this time interval is

$$a_{x,\text{avg}} = \frac{15 \text{ m/s} - 30 \text{ m/s}}{2.0 \text{ s}} = -7.5 \text{ m/s}^2$$

The negative sign in this example indicates that the acceleration vector is in the negative x direction (to the left in Figure 2.9). For the case of motion in a straight line, the direction of the velocity of an object and the direction of its acceleration are related as follows. **When the object's velocity and acceleration are in the same direction, the object is speeding up in that direction. On the other hand, when the object's velocity and acceleration are in opposite directions, the speed of the object decreases in time.**

To help with this discussion of the signs of velocity and acceleration, let us take a peek ahead to Chapter 4, where we shall relate the acceleration of an object to the *force* on the object. We will save the details until that later discussion, but for now, let us borrow the notion that **force is proportional to acceleration:**

$$\vec{\mathbf{F}} \propto \vec{\mathbf{a}}$$

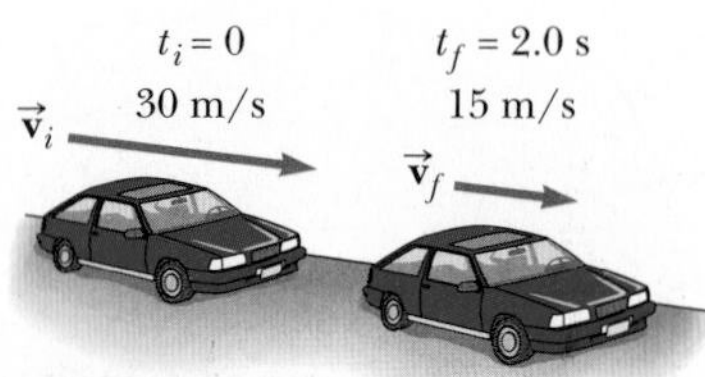

FIGURE 2.9 The velocity of the car decreases from 30 m/s to 15 m/s in a time interval of 2.0 s.

This proportionality indicates that acceleration is caused by force. What's more, as indicated by the vector notation in the proportionality, force and acceleration are in the same direction. Therefore, let us think about the signs of velocity and acceleration by forming a mental representation in which a force is applied to the object to cause the acceleration. Again consider the case in which the velocity and acceleration are in the same direction. This situation is equivalent to an object moving in a given direction and experiencing a force that pulls on it in the same direction. It is clear in this case that the object speeds up! If the velocity and acceleration are in opposite directions, the object moves one way and a force pulls in the opposite direction. In this case, the object slows down! It is very useful to equate the direction of the acceleration in these situations to the direction of a force because it is easier from our everyday experience to think about what effect a force will have on an object than to think only in terms of the direction of the acceleration.

QUICK QUIZ 2.4 If a car is traveling eastward and slowing down, what is the direction of the force on the car that causes it to slow down? **(a)** eastward **(b)** westward **(c)** neither of these directions

EXAMPLE 2.6 Average and Instantaneous Acceleration

The velocity of a particle moving along the x axis varies in time according to the expression $v_x = 40 - 5t^2$, where t is in seconds.

A Find the average acceleration in the time interval $t = 0$ to $t = 2.0$ s.

Solution Build your mental representation from the mathematical expression given for the velocity. For example, which way is the particle moving at $t = 0$? How does the velocity change in the first few seconds? Does it move faster or slower? The velocity–time graphical representation for this function is given in Figure 2.10. The velocities at $t_i = t_A = 0$ and $t_f = t_B = 2.0$ s are found by substituting these values of t into the expression given for the velocity:

$$v_{xA} = 40 - 5t_A^2 = 40 - 5(0)^2 = 40 \text{ m/s}$$

$$v_{xB} = 40 - 5t_B^2 = 40 - 5(2.0)^2 = 20 \text{ m/s}$$

Therefore, the average acceleration in the specified time interval $\Delta t = t_B - t_A$ is

$$a_{x,\text{avg}} = \frac{20 \text{ m/s} - 40 \text{ m/s}}{2.0 \text{ s}} = -10 \text{ m/s}^2$$

The negative sign is consistent with the negative slope of the line joining the initial and final points on the velocity–time graph.

B Determine the acceleration at $t = 2.0$ s.

Solution Because this question refers to a specific instant of time, it is asking for an instantaneous acceleration. The velocity at time t is $v_{xi} = 40 - 5t^2$, and the velocity at time $t + \Delta t$ is

$$v_{xf} = 40 - 5(t + \Delta t)^2 = 40 - 5t^2 - 10t\,\Delta t - 5(\Delta t)^2$$

Therefore, the change in velocity over the time interval Δt is

$$\Delta v_x = v_{xf} - v_{xi} = -10t\,\Delta t - 5(\Delta t)^2$$

Dividing this expression by Δt and taking the limit of the result as Δt approaches zero gives the acceleration at *any* time t:

$$a_x = \lim_{\Delta t \to 0} \frac{\Delta v_x}{\Delta t} = \lim_{\Delta t \to 0} (-10t - 5\Delta t) = -10t$$

Therefore, at $t = 2.0$ s we find that

$$a_x = (-10)(2.0) \text{ m/s}^2 = -20 \text{ m/s}^2$$

This result can also be obtained by measuring the slope of the velocity–time graph at $t = 2.0$ s (see Fig. 2.10) or by taking the derivative of the velocity expression.

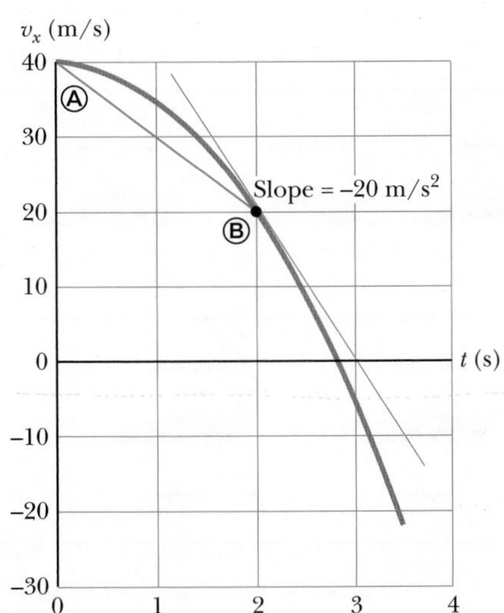

FIGURE 2.10 (Example 2.6) The velocity–time graph for a particle moving along the x axis according to the relation $v_x = 40 - 5t^2$. The acceleration at $t = 2.0$ s is obtained from the slope of the green tangent line at that time.

2.5 MOTION DIAGRAMS

The concepts of velocity and acceleration are often confused with each other, but in fact they are quite different quantities. It is instructive to make use of the specialized pictorial representation called a **motion diagram** to describe the velocity and acceleration vectors while an object is in motion.

A *stroboscopic photograph* of a moving object shows several images of the object taken as the strobe light flashes at a constant rate. Active Figure 2.11 represents three sets of strobe photographs of cars moving along a straight roadway in a single direction, from left to right. The time intervals between flashes of the stroboscope are equal in each part of the diagram. To distinguish between the two vector quantities, we use red for velocity vectors and violet for acceleration vectors in Active Figure 2.11. The vectors are sketched at several instants during the motion of the object. Let us describe the motion of the car in each diagram.

In Active Figure 2.11a, the images of the car are equally spaced, and the car moves through the same displacement in each time interval. Therefore, the car moves with *constant positive velocity* and has *zero acceleration*. We could model the car as a particle and describe it as a particle under constant velocity.

In Active Figure 2.11b, the images of the car become farther apart as time progresses. In this case, the velocity vector increases in time because the car's displacement between adjacent positions increases as time progresses. Therefore, the car is moving with a *positive velocity* and a *positive acceleration*. The velocity and acceleration are in the same direction. In terms of our earlier force discussion, imagine a force pulling on the car in the same direction it is moving: it speeds up.

In Active Figure 2.11c, we interpret the car as slowing down as it moves to the right because its displacement between adjacent positions decreases as time progresses. In this case, the car moves initially to the right with a *positive velocity* and a *negative acceleration*. The velocity vector decreases in time and eventually reaches zero. (This type of motion is exhibited by a car that skids to a stop after its brakes are applied.) From this diagram we see that the acceleration and velocity vectors are *not* in the same direction. The velocity and acceleration are in opposite directions. In terms of our earlier force discussion, imagine a force pulling on the car opposite to the direction it is moving: it slows down.

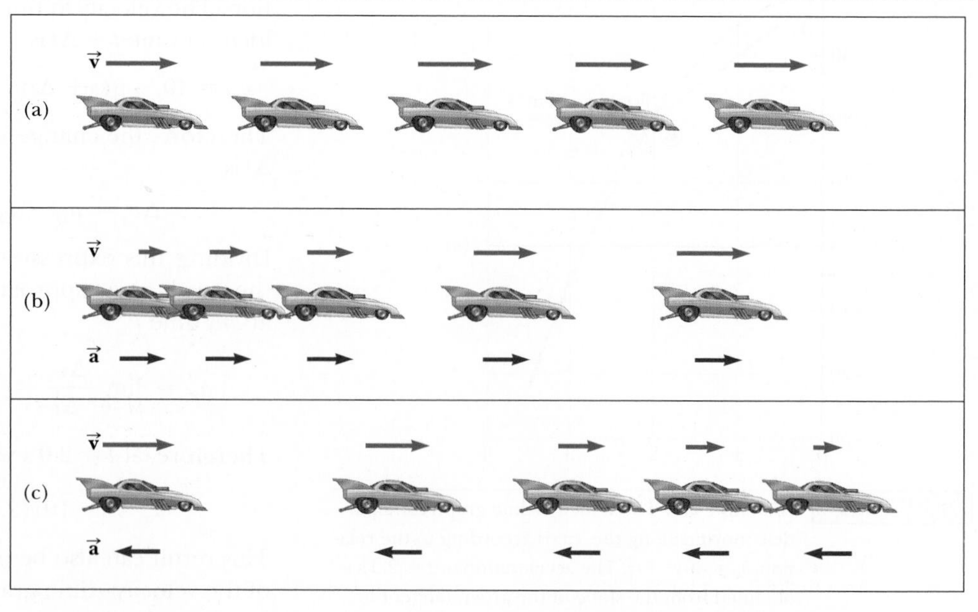

ACTIVE FIGURE **2.11**
(a) Motion diagram for a car moving at constant velocity.
(b) Motion diagram for a car whose constant acceleration is in the direction of its velocity. The velocity vector at each instant is indicated by a red arrow, and the constant acceleration vector is indicated by a violet arrow.
(c) Motion diagram for a car whose constant acceleration is in the direction *opposite* the velocity at each instant.

PhysicsNow™ Log into PhysicsNow at **www.pop4e.com** and go to Active Figure 2.11 to select the constant acceleration and initial velocity of the car and observe pictorial and graphical representations of its motion.

The violet acceleration vectors in Active Figures 2.11b and 2.11c are all the same length. Therefore, these diagrams represent a motion with constant acceleration. This important type of motion is discussed in the next section.

QUICK QUIZ 2.5 Which of the following is true? (**a**) If a car is traveling eastward, its acceleration is eastward. (**b**) If a car is slowing down, its acceleration must be negative. (**c**) A particle with constant acceleration can never stop and stay stopped.

2.6 THE PARTICLE UNDER CONSTANT ACCELERATION

If the acceleration of a particle varies in time, the motion may be complex and difficult to analyze. A very common and simple type of one-dimensional motion occurs when the acceleration is constant, such as for the motion of the cars in Active Figures 2.11b and 2.11c. In this case, the average acceleration over any time interval equals the instantaneous acceleration at any instant of time within the interval. Consequently, the velocity increases or decreases at the same rate throughout the motion. The *particle under constant acceleration* model is a common analysis model that we can apply to appropriate problems. It is often used to model situations such as falling objects and braking cars.

If we replace $a_{x,\text{avg}}$ with the constant a_x in Equation 2.6, we find that

$$a_x = \frac{v_{xf} - v_{xi}}{t_f - t_i}$$

For convenience, let $t_i = 0$ and t_f be any arbitrary time t. With this notation, we can solve for v_{xf}:

$$v_{xf} = v_{xi} + a_x t \qquad \text{(for constant } a_x\text{)} \qquad [2.9]$$

■ **Velocity as a function of time for a particle under constant acceleration**

This expression enables us to predict the velocity at *any* time t if the initial velocity and constant acceleration are known. It is the first of four equations that can be used to solve problems using the particle under constant acceleration model. A graphical representation of position versus time for this motion is shown in Active Figure 2.12a. The velocity–time graph shown in Active Figure 2.12b is a straight line, the slope of which is the constant acceleration a_x. The straight line on this

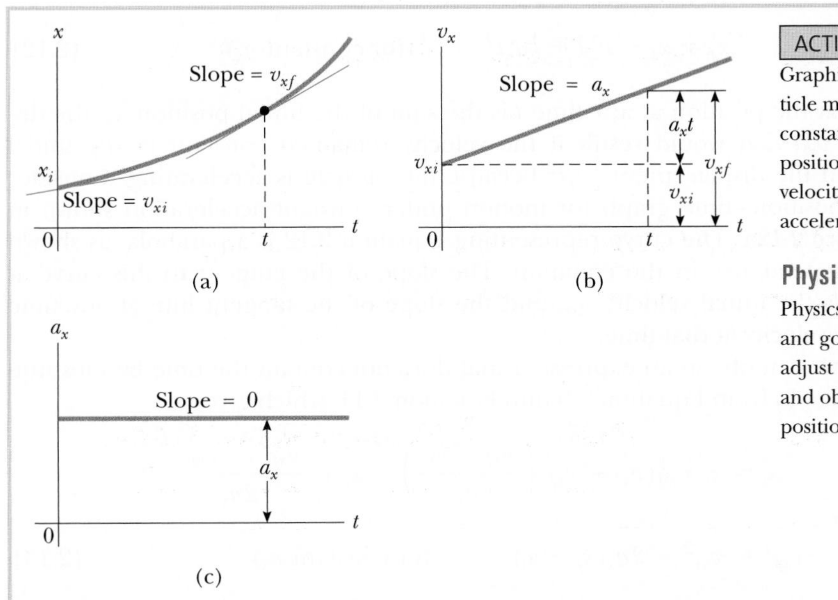

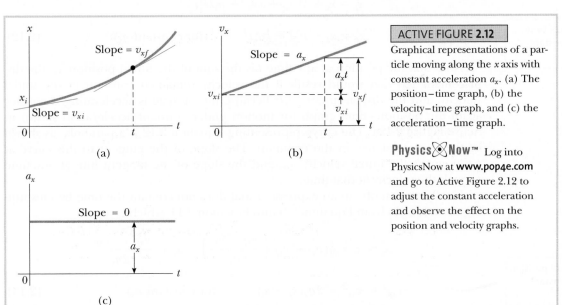

ACTIVE FIGURE 2.12
Graphical representations of a particle moving along the x axis with constant acceleration a_x. (a) The position–time graph, (b) the velocity–time graph, and (c) the acceleration–time graph.

PhysicsNow™ Log into PhysicsNow at **www.pop4e.com** and go to Active Figure 2.12 to adjust the constant acceleration and observe the effect on the position and velocity graphs.

graph is consistent with $a_x = dv_x/dt$ being a constant. From this graph and from Equation 2.9, we see that the velocity at any time t is the sum of the initial velocity v_{xi} and the change in velocity $a_x t$ due to the acceleration. The graph of acceleration versus time (Active Fig. 2.12c) is a straight line with a slope of zero because the acceleration is constant. If the acceleration were negative, the slope of Active Figure 2.12b would be negative and the horizontal line in Active Figure 2.12c would be below the time axis.

We can generate another equation for the particle under constant acceleration model by recalling a result from Section 2.1 that the displacement of a particle is the area under the curve on a velocity–time graph. Because the velocity varies linearly with time (see Active Fig. 2.12b), the area under the curve is the sum of a rectangular area (under the horizontal dashed line in Active Fig. 2.12b) and a triangular area (from the horizontal dashed line upward to the curve). Therefore,

$$\Delta x = v_{xi}\,\Delta t + \tfrac{1}{2}(v_{xf} - v_{xi})\Delta t$$

which can be simplified as follows:

$$\Delta x = (v_{xi} + \tfrac{1}{2}v_{xf} - \tfrac{1}{2}v_{xi})\Delta t = \tfrac{1}{2}(v_{xi} + v_{xf})\Delta t$$

In general, from Equation 2.2, the displacement for a time interval is

$$\Delta x = v_{x,\text{avg}}\,\Delta t$$

Comparing these last two equations, we find that the average velocity in any time interval is the arithmetic mean of the initial velocity v_{xi} and the final velocity v_{xf}:

■ Average velocity for a particle under constant acceleration

$$v_{x,\text{avg}} = \tfrac{1}{2}(v_{xi} + v_{xf}) \qquad \text{(for constant } a_x\text{)} \qquad [2.10]$$

Remember that this expression is valid only when the acceleration is constant, that is, when the velocity varies linearly with time.

We now use Equations 2.2 and 2.10 to obtain the position as a function of time. Again we choose $t_i = 0$, at which time the initial position is x_i, which gives

$$\Delta x = v_{x,\text{avg}}\,\Delta t = \tfrac{1}{2}(v_{xi} + v_{xf})\,t$$

■ Position as a function of velocity and time for a particle under constant acceleration

$$x_f = x_i + \tfrac{1}{2}(v_{xi} + v_{xf})\,t \qquad \text{(for constant } a_x\text{)} \qquad [2.11]$$

We can obtain another useful expression for the position by substituting Equation 2.9 for v_{xf} in Equation 2.11:

$$x_f = x_i + \tfrac{1}{2}[v_{xi} + (v_{xi} + a_x t)]\,t$$

■ Position as a function of time for a particle under constant acceleration

$$x_f = x_i + v_{xi}t + \tfrac{1}{2}a_x t^2 \qquad \text{(for constant } a_x\text{)} \qquad [2.12]$$

Note that the position at any time t is the sum of the initial position x_i, the displacement $v_{xi}t$ that would result if the velocity remained constant at the initial velocity, and the displacement $\frac{1}{2}a_x t^2$ because the particle is accelerating. Consider again the position–time graph for motion under constant acceleration shown in Active Figure 2.12a. The curve representing Equation 2.12 is a parabola, as shown by the t^2 dependence in the equation. The slope of the tangent to this curve at $t = 0$ equals the initial velocity v_{xi}, and the slope of the tangent line at any time t equals the velocity at that time.

Finally, we can obtain an expression that does not contain the time by substituting the value of t from Equation 2.9 into Equation 2.11, which gives

$$x_f = x_i + \tfrac{1}{2}(v_{xi} + v_{xf})\left(\frac{v_{xf} - v_{xi}}{a_x}\right) = x_i + \frac{v_{xf}^{\,2} - v_{xi}^{\,2}}{2a_x}$$

■ Velocity as a function of position for a particle under constant acceleration

$$v_{xf}^{\,2} = v_{xi}^{\,2} + 2a_x(x_f - x_i) \qquad \text{(for constant } a_x\text{)} \qquad [2.13]$$

TABLE 2.2 Kinematic Equations for Motion in a Straight Line Under Constant Acceleration

Equation	Information Given by Equation
$v_{xf} = v_{xi} + a_x t$	Velocity as a function of time
$x_f = x_i + \frac{1}{2}(v_{xf} + v_{xi})t$	Position as a function of velocity and time
$x_f = x_i + v_{xi}t + \frac{1}{2}a_x t^2$	Position as a function of time
$v_{xf}{}^2 = v_{xi}{}^2 + 2a_x(x_f - x_i)$	Velocity as a function of position

Note: Motion is along the x axis. At $t = 0$, the position of the particle is x_i and its velocity is v_{xi}.

This expression is *not* an independent equation because it arises from combining Equations 2.9 and 2.11. It is useful, however, for those problems in which a value for the time is not involved.

If motion occurs in which the constant value of the acceleration is *zero,* Equations 2.9 and 2.12 become

$$\left.\begin{aligned} v_{xf} &= v_{xi} \\ x_f &= x_i + v_{xi}t \end{aligned}\right\} \quad \text{when } a_x = 0$$

That is, when the acceleration is zero, the velocity remains constant and the position changes linearly with time. In this case, the particle under constant *acceleration* becomes the particle under constant *velocity* (Equation 2.5).

Equations 2.9, 2.11, 2.12, and 2.13 are four **kinematic equations** that may be used to solve any problem in one-dimensional motion of a particle (or an object that can be modeled as a particle) under constant acceleration. Keep in mind that these relationships were derived from the definitions of velocity and acceleration together with some simple algebraic manipulations and the requirement that the acceleration be constant. It is often convenient to choose the initial position of the particle as the origin of the motion so that $x_i = 0$ at $t = 0$. We will see cases, however, in which we must choose the value of x_i to be something other than zero.

The four kinematic equations for the particle under constant acceleration are listed in Table 2.2 for convenience. The choice of which kinematic equation or equations you should use in a given situation depends on what is known beforehand. Sometimes it is necessary to use two of these equations to solve for two unknowns, such as the position and velocity at some instant. You should recognize that the quantities that vary during the motion are velocity v_{xf}, position x_f, and time t. The other quantities—x_i, v_{xi}, and a_x—are *parameters* of the motion and remain constant.

PROBLEM-SOLVING STRATEGY Particle Under Constant Acceleration

The following procedure is recommended for solving problems that involve an object undergoing a constant acceleration. As mentioned in Chapter 1, individual strategies such as this one will follow the outline of the General Problem-Solving Strategy from Chapter 1, with specific hints regarding the application of the general strategy to the material in the individual chapters.

1. Conceptualize Think about what is going on physically in the problem. Establish the mental representation.

2. Categorize Simplify the problem as much as possible. Confirm that the problem involves either a particle or an object that can be modeled as a particle and that it is moving with a constant acceleration. Construct an appropriate pictorial representation, such as a motion diagram, or a graphical representation. Make sure all the units in the problem are consistent. That is, if positions are measured in meters, be sure that velocities have units of m/s and accelerations have units of m/s^2. Choose a coordinate system to be used throughout the problem.

3. Analyze Set up the mathematical representation. Choose an instant to call the "initial" time $t = 0$ and another to call the "final" time t. Let your choice be guided by what you know about the particle and what you want to know about it. The initial instant need not be when the particle starts to move, and the final instant will only rarely be when the particle stops moving. Identify all the quantities given in the problem and a separate list of those to be determined. A tabular representation of these quantities may be helpful to you. Select from the list of kinematic equations the one or ones that will enable you to determine the unknowns. Solve the equations.

4. Finalize Once you have determined your result, check to see if your answers are consistent with the mental and pictorial representations and that your results are realistic.

EXAMPLE 2.7 Accelerating an Electron

An electron in the cathode-ray tube of a television set enters a region in which it accelerates uniformly in a straight line from a speed of 3.00×10^4 m/s to a speed of 5.00×10^6 m/s in a distance of 2.00 cm. For what time interval is the electron accelerating?

Solution For this example, we shall identify the individual steps in the General Problem-Solving Strategy in Chapter 1. In subsequent examples, you should identify the portions of the solution that correspond to each step. For step 1 (*Conceptualize*), think about the electron moving through space. Note that it is moving faster at the end of the interval than before, so imagine it speeding up as it covers the 2.00-cm displacement. In step 2 (*Categorize*), ignore that it is an electron and that it is in a television. The electron is easily modeled as a particle, and the phrase "accelerates uniformly" tells us that it is a particle under constant acceleration. All the parts of Active Figure 2.12 represent the motion of the particle as a function of time, although you may want to graph velocity versus position because no time is given in the problem. Note that all units are metric, although we must convert 2.00 cm to meters to put all units in SI. We make the simple choice of the x axis lying along the straight line mentioned in the text of the problem.

We are now ready to move on to step 3 (*Analyze*), in which we develop the mathematical representation of the problem. Notice that no acceleration is given in the problem and that the time interval is requested, which provides a hint that we should use an equation that does not involve acceleration. We can find the time at which the particle is at the end of the 2.00-cm distance from Equation 2.11:

$$x_f = x_i + \tfrac{1}{2}(v_{xi} + v_{xf})t \quad \rightarrow \quad t = \frac{2(x_f - x_i)}{v_{xi} + v_{xf}}$$

$$t = \frac{2(0.020\,0 \text{ m})}{3.00 \times 10^4 \text{ m/s} + 5.00 \times 10^6 \text{ m/s}}$$

$$= 7.95 \times 10^{-9} \text{ s}$$

Finally, we check if the answer is reasonable (step 4, *Finalize*). The average speed is on the order of 10^6 m/s. Let us estimate the time interval required to move 1 cm at this speed:

$$\Delta t = \frac{\Delta x}{v} \approx \frac{0.01 \text{ m}}{10^6 \text{ m/s}} \approx 10^{-8} \text{ s} = 10 \times 10^{-9} \text{ s}$$

This result is the same order of magnitude as our answer, providing confidence that our answer is reasonable.

INTERACTIVE EXAMPLE 2.8 Watch Out for the Speed Limit!

A car traveling at a constant velocity of magnitude 45.0 m/s passes a trooper hidden behind a billboard. One second after the speeding car passes the billboard, the trooper sets out from the billboard to catch it, accelerating at a constant rate of 3.00 m/s^2. How long does it take her to overtake the speeding car?

Solution We will point out again in this example steps in the General Problem-Solving Strategy. A pictorial representation of the situation is shown in Figure 2.13. Establish the mental representation (*Conceptualize*) of this situation for yourself; in the following solution, we will go straight to the mathematical representation. As you become more proficient at solving physics problems, a quick thought about the mental representation may be enough to allow you to skip pictorial representations and go right to the mathematics. Let us model the speeding car as a particle under constant velocity and the trooper's motorcycle as a particle under constant acceleration (*Categorize*). We shall ignore that they are vehicles and instead will imagine the speeder and the trooper as point particles undergoing the motion described in the problem.

Note that all units are in the same system. To solve this problem algebraically, we will write an expression for the position of each vehicle as a function of time. It is convenient to choose the origin at the position of the billboard and take $t_B = 0$ as the time the trooper

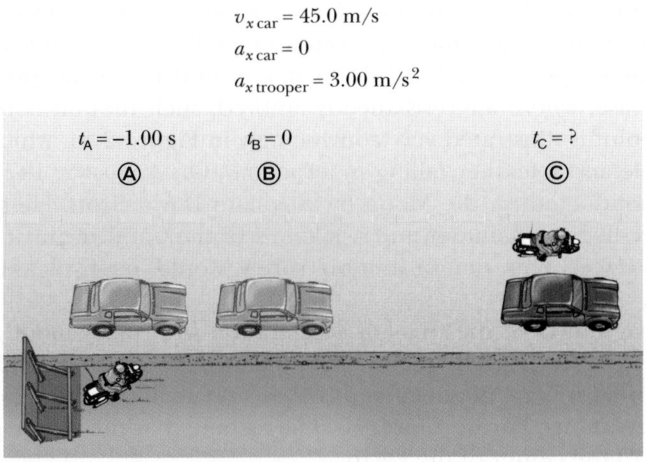

$v_{x\,\text{car}} = 45.0$ m/s

$a_{x\,\text{car}} = 0$

$a_{x\,\text{trooper}} = 3.00$ m/s^2

$t_A = -1.00$ s　　$t_B = 0$　　$t_C = ?$

Ⓐ　Ⓑ　Ⓒ

FIGURE 2.13 (Interactive Example 2.8) A speeding car passes a hidden trooper. The trooper catches up to the car at point Ⓒ.

begins moving. At that instant, the speeding car has already traveled a distance of 45.0 m because it has traveled at a constant speed of $v_x = 45.0$ m/s for 1.00 s; it is at point Ⓑ in Figure 2.13. Therefore, the initial position of the speeding car is $x_i = x_B = 45.0$ m. We do *not* choose $t = 0$ as the time at which the car passes the trooper (point Ⓐ in Fig. 2.13), because then the acceleration of the trooper is not constant during the problem. Her acceleration is $a_x = 0$ for the first second and then 3.00 m/s^2 after that. Therefore, we could not model the trooper as a particle under constant acceleration with this choice.

Now we set up the mathematical representation (*Analyze*). Because the car moves with constant velocity, its acceleration is zero, and applying Equation 2.5 gives us

$$x_f = x_B + v_x t \quad \rightarrow \quad x_{\text{car}} = 45.0\text{ m} + (45.0\text{ m/s})t$$

Note that at $t = 0$, this expression gives the car's correct initial position, $x_{\text{car}} = 45.0$ m.

For the trooper, who starts from the origin at $t = 0$, we have $x_i = 0$, $v_{xi} = 0$, and $a_x = 3.00$ m/s^2. Hence, from Equation 2.12 for a particle under constant acceleration, the position of the trooper as a function of time is

$$x_f = x_i + v_{xi}t + \tfrac{1}{2}a_x t^2 \quad \rightarrow \quad x_{\text{trooper}} = \tfrac{1}{2}a_x t^2 = \tfrac{1}{2}(3.00\text{ m/s}^2)t^2$$

The trooper overtakes the car at the instant that $x_{\text{trooper}} = x_{\text{car}}$, which is at position Ⓒ in Figure 2.13:

$$\tfrac{1}{2}(3.00\text{ m/s}^2)t^2 = 45.0\text{ m} + (45.0\text{ m/s})t$$

This result gives the quadratic equation (dropping the units)

$$1.50t^2 - 45.0t - 45.0 = 0$$

whose positive solution is $t = 31.0$ s.

From your everyday experience, is this value reasonable (*Finalize*)? (For help in solving quadratic equations, see Appendix B.2.).

Physics Now™ You can study the motion of the car and the trooper for various velocities of the car by logging into PhysicsNow at www.pop4e.com and going to Interactive Example 2.8.

2.7 FREELY FALLING OBJECTS

It is well known that all objects, when dropped, fall toward the Earth with nearly constant acceleration. Legend has it that Galileo Galilei first discovered this fact by observing that two different weights dropped simultaneously from the Leaning Tower of Pisa hit the ground at approximately the same time. (Air resistance plays a role in the falling of an object, but for now we shall model falling objects as if they are falling through a vacuum; this is a simplification model.) Although there is some doubt that this particular experiment was actually carried out, it is well established that Galileo did perform many systematic experiments on objects moving on inclined planes. Through careful measurements of distances and time intervals, he was able to show that the displacement from an origin of an object starting from rest is proportional to the square of the time interval during which the object is in motion. This observation is consistent with one of the kinematic equations we derived for a particle under constant acceleration (Eq. 2.12, with $v_{xi} = 0$). Galileo's achievements in mechanics paved the way for Newton in his development of the laws of motion.

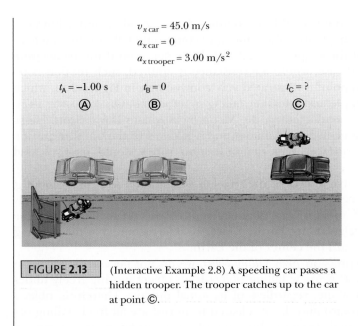

If a coin and a crumpled-up piece of paper are dropped simultaneously from the same height, there will be a small time difference between their arrivals at the

(North Wind Picture Archive)

GALILEO GALILEI (1564–1642)

Italian physicist and astronomer Galileo formulated the laws that govern the motion of objects in free-fall. He also investigated the motion of an object on an inclined plane, established the concept of relative motion, invented the thermometer, and discovered that the motion of a swinging pendulum could be used to measure time intervals. After designing and constructing his own telescope, he discovered four of Jupiter's moons, found that the Moon's surface is rough, discovered sunspots and the phases of Venus, and showed that the Milky Way consists of an enormous number of stars. Galileo publicly defended Nicolaus Copernicus's assertion that the Sun is at the center of the Universe (the heliocentric system). He published *Dialogue Concerning Two New World Systems* to support the Copernican model, a view that the Catholic Church declared to be heretical. After being taken to Rome in 1633 on a charge of heresy, he was sentenced to life imprisonment and later was confined to his villa at Arcetri, near Florence, where he died.

floor. If this same experiment could be conducted in a good vacuum, however, where air friction is truly negligible, the paper and coin would fall with the same acceleration, regardless of the shape or weight of the paper, even if the paper were still flat. In the idealized case, where air resistance is ignored, such motion is referred to as *free-fall.* This point is illustrated very convincingly in Figure 2.14, which is a photograph of an apple and a feather falling in a vacuum. On August 2, 1971, such an experiment was conducted on the Moon by astronaut David Scott. He simultaneously released a geologist's hammer and a falcon's feather, and in unison they fell to the lunar surface. This demonstration surely would have pleased Galileo!

We shall denote the magnitude of the free-fall acceleration with the symbol g, representing a vector acceleration $\vec{\mathbf{g}}$. At the surface of the Earth, g is approximately 9.80 m/s^2, or 980 cm/s^2, or 32 ft/s^2. Unless stated otherwise, we shall use the value 9.80 m/s^2 when doing calculations. Furthermore, we shall assume that the vector $\vec{\mathbf{g}}$ is directed downward toward the center of the Earth.

When we use the expression *freely falling object,* we do not necessarily mean an object dropped from rest. A freely falling object is an object moving freely under the influence of gravity alone, regardless of its initial motion. Therefore, objects thrown upward or downward and those released from rest are all freely falling objects once they are released! Because the value of g is constant as long as we are close to the surface of the Earth, we can model a freely falling object as a particle under constant acceleration.

In previous examples in this chapter, the particles were undergoing constant acceleration, as stated in the problem. Therefore, it may have been difficult to understand the need for modeling. We can now begin to see the need for modeling; we are *modeling* a real falling object with an analysis model. Notice that we are (1) ignoring air resistance and (2) assuming that the free-fall acceleration is constant. Therefore, the model of a particle under constant acceleration is a *replacement* for the real problem, which could be more complicated. If air resistance and any variation in g are small, however, the model should make predictions that agree closely with the real situation.

The equations developed in Section 2.6 for objects moving with constant acceleration can be applied to the falling object. The only necessary modification that we need to make in these equations for freely falling objects is to note that the

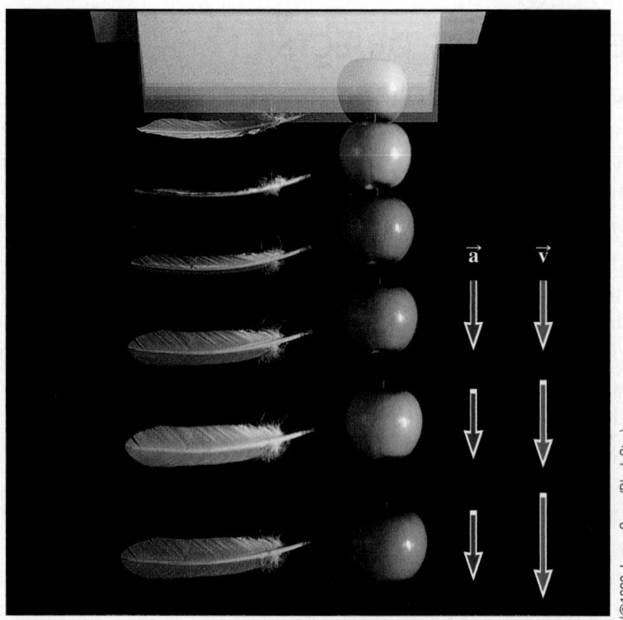

(©1993 James Sugar/Black Star)

FIGURE 2.14 An apple and a feather, released from rest in a vacuum chamber, fall at the same rate, regardless of their masses. Ignoring air resistance, all objects fall to the Earth with the same acceleration of magnitude 9.80 m/s^2, as indicated by the violet arrows in this multiflash photograph. The velocity of the two objects increases linearly with time, as indicated by the series of red arrows.

motion is in the vertical direction, so we will use y instead of x, and that the acceleration is downward and of magnitude 9.80 m/s^2. Therefore, for a freely falling object we commonly take $a_y = -g = -9.80$ m/s^2, where the negative sign indicates that the acceleration of the object is downward. The choice of negative for the downward direction is arbitrary, but common.

PITFALL PREVENTION 2.5

Acceleration at the Top of the Motion Imagine throwing a baseball straight up into the air. It is a common misconception that the acceleration of a projectile at the top of its trajectory is zero. This misconception generally arises owing to confusion between velocity and acceleration. Although the velocity at the top of the motion of an object thrown upward momentarily goes to zero, *the acceleration is still that due to gravity* at this point. Remember that acceleration is proportional to force and that the gravitational force still acts at the moment that the object has stopped. If the velocity and acceleration were both zero, the projectile would stay at the top!

QUICK QUIZ 2.6 A ball is thrown upward. While the ball is in free-fall, does its acceleration (**a**) increase, (**b**) decrease, (**c**) increase and then decrease, (**d**) decrease and then increase, or (**e**) remain constant?

■ Thinking Physics 2.2

A sky diver steps out of a stationary helicopter. A few seconds later, another sky diver steps out, so that both sky divers fall along the same vertical line. Ignore air resistance, so that both sky divers fall with the same acceleration, and model the sky divers as particles under constant acceleration. Does the vertical separation distance between them stay the same? Does the difference in their speeds stay the same?

Reasoning At any given instant of time, the speeds of the sky divers are definitely different, because one had a head start over the other. In any time interval, however, each sky diver increases his or her speed by the same amount, because they have the same acceleration. Therefore, the difference in speeds remains the same. The first sky diver will always be moving with a higher speed than the second. In a given time interval, then, the first sky diver will have a larger displacement than the second. Therefore, the separation distance between them increases. ■

PITFALL PREVENTION 2.6

The Sign of g Keep in mind that g is a *positive number*. It is tempting to substitute -9.80 m/s^2 for g, but resist the temptation. That the gravitational acceleration is downward is indicated explicitly by stating the acceleration as $a_y = -g$.

EXAMPLE 2.9 Try to Catch the Dollar

Emily challenges David to catch a dollar bill as follows. She holds the bill vertically, as in Figure 2.15, with the center of the bill between David's index finger and thumb. David must catch the bill after Emily releases it without moving his hand downward. The reaction time of most people is at best about 0.2 s. Who would you bet on?

Solution Place your bets on Emily. There is a time delay between the instant Emily releases the bill and the time David reacts and closes his fingers. We model the bill as a particle. When released, the bill will probably flutter downward to the floor due to the effects of the air, but for the very early part of its motion, we will assume that it can be modeled as a particle falling through a vacuum. Because the bill is in free-fall and undergoes a downward acceleration of magnitude 9.80 m/s^2, in 0.2 s it falls a distance of $y = \frac{1}{2}gt^2 \approx 0.2$ m $= 20$ cm. This distance is about twice the distance between the center of the bill and its top edge (≈ 8 cm). Therefore, David will be unsuccessful.

You might want to try this "trick" on one of your friends.

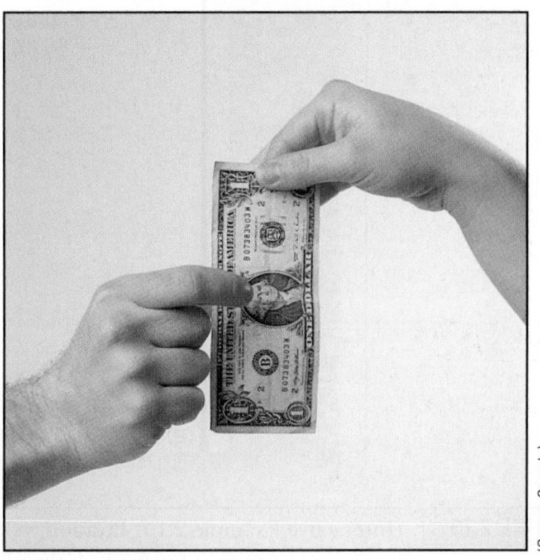

(George Semple)

FIGURE 2.15 (Example 2.9)

INTERACTIVE EXAMPLE 2.10 Not a Bad Throw for a Rookie!

A stone is thrown at point Ⓐ from the top of a building with an initial velocity of 20.0 m/s straight upward. The building is 50.0 m high, and the stone just misses the edge of the roof on its way down, as in the pictorial representation of Figure 2.16.

A Determine the time at which the stone reaches its maximum height.

Solution Think about the mental representation: the stone rises upward, slowing down. It stops momentarily (point Ⓑ) and then begins to fall downward again. During the entire motion, it is accelerating downward because the gravitational force is always pulling downward on it. Ignoring air resistance, we model the stone as a particle under constant acceleration.

To begin the mathematical representation, we consider the portion of the motion from Ⓐ to Ⓑ and find the time at which the stone reaches the maximum height, point Ⓑ. We use the vertical modification of Equation 2.9, noting that $v_{yf} = 0$ at the maximum height:

$$v_{yf} = v_{yi} + a_y t \quad \rightarrow \quad 0 = 20.0 \text{ m/s} + (-9.80 \text{ m/s}^2)t_B$$

$$t_B = \frac{20.0 \text{ m/s}}{9.80 \text{ m/s}^2} = \boxed{2.04 \text{ s}}$$

B Determine the maximum height of the stone above the roof top.

Solution The value of time from part A can be substituted into Equation 2.12 to give the maximum height measured from the position of the thrower:

$$\begin{aligned} y_{\max} = y_B &= y_A + v_{yA}t_B + \tfrac{1}{2}a_y t_B^2 \\ &= 0 + (20.0 \text{ m/s})(2.04 \text{ s}) + \tfrac{1}{2}(-9.80 \text{ m/s}^2)(2.04 \text{ s})^2 \\ &= \boxed{20.4 \text{ m}} \end{aligned}$$

C Determine the time at which the stone returns to the level of the thrower.

Solution Now we identify the initial point of the motion as Ⓐ and the final point as Ⓒ. When the stone is back at the height of the thrower, the y coordinate is zero. From Equation 2.12, letting $y_f = y_C = 0$, we obtain the expression

$$y_C = y_A + v_{yA}t_C + \tfrac{1}{2}a_y t_C^2 \quad \rightarrow \quad 0 = 20.0t_C - 4.90t_C^2$$

This result is a quadratic equation and has two solutions for t_C. The equation can be factored to give

$$t_C(20.0 - 4.90t_C) = 0$$

One solution is $t_C = 0$, corresponding to the time the stone starts its motion. The other solution—the one we are after—is $\boxed{t_C = 4.08 \text{ s}}$. Note that this value is twice the value for t_B. The fall from Ⓑ to Ⓒ is the reverse of the rise from Ⓐ to Ⓑ, and the stone requires exactly the same time interval to undergo each part of the motion.

D Determine the velocity of the stone at this instant.

Solution The value for t_C found in part C can be inserted into Equation 2.9 to give

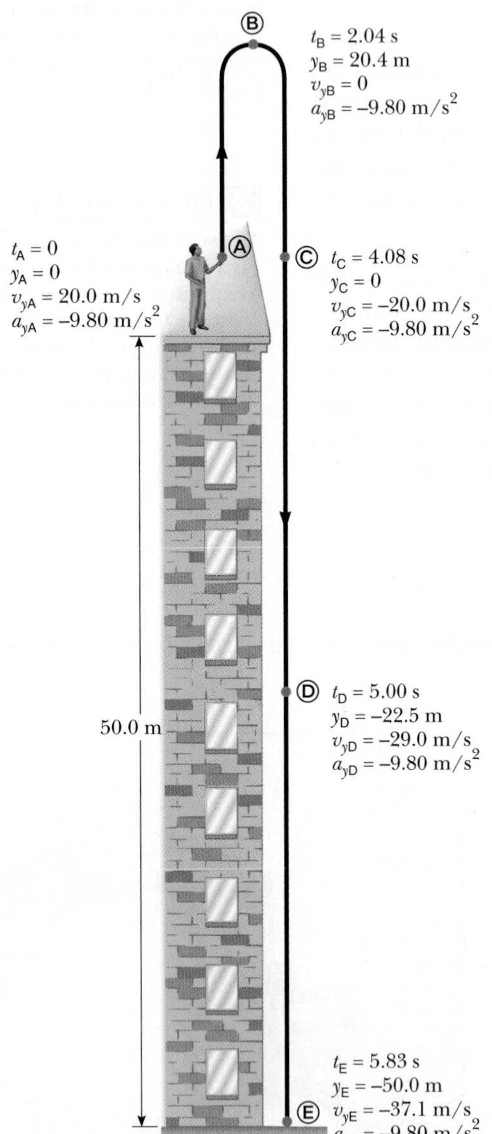

FIGURE 2.16 (Interactive Example 2.10) Position, velocity, and acceleration at various instants of time for a freely falling particle initially thrown upward with a velocity $v_y = 20.0$ m/s.

$$v_{yC} = v_{yA} + a_y t_C$$
$$= 20.0 \text{ m/s} + (-9.80 \text{ m/s}^2)(4.08 \text{ s})$$
$$= -20.0 \text{ m/s}$$

Note that the velocity of the stone when it arrives back at its original height is equal in magnitude to its initial velocity but opposite in direction. This equal magnitude, along with the equal time intervals noted at the end of part C, indicates that the motion to this point is symmetric.

E Determine the velocity and position of the stone at $t = 5.00$ s.

Solution For this part of the problem, we analyze the portion of the motion from Ⓐ to Ⓓ. From Equation 2.9, the velocity at Ⓓ after 5.00 s is

$$v_{yD} = v_{yA} + a_y t_D$$
$$= 20.0 \text{ m/s} + (-9.80 \text{ m/s}^2)(5.00 \text{ s})$$
$$= -29.0 \text{ m/s}$$

We can use Equation 2.12 to find the position of the stone at $t = 5.00$ s:

$$y_D = y_A + v_{yA} t_D + \tfrac{1}{2} a_y t_D^2$$
$$= 0 + (20.0 \text{ m/s})(5.00 \text{ s}) + \tfrac{1}{2}(-9.80 \text{ m/s}^2)(5.00 \text{ s})^2$$
$$= -22.5 \text{ m}$$

F Determine the position of the stone at $t = 6.00$ s. How does the model fail this last part of the problem?

Solution We use Equation 2.12 again to find the position of the stone at $t = 6.00$ s:

$$y = 0 + (20.0 \text{ m/s})(6.00 \text{ s}) + \tfrac{1}{2}(-9.80 \text{ m/s}^2)(6.00 \text{ s})^2$$
$$= -56.4 \text{ m}$$

The failure of the model is that the building is only 50.0 m high, so the stone cannot be at a position 6.4 m below ground. Our model does not include that the ground exists at $y = -50.0$ m, so the mathematical representation gives us an answer that is not consistent with our expectations in this case.

PhysicsNow™ You can study the motion of the thrown ball by logging into PhysicsNow at www.pop4e.com and going to Interactive Example 2.10.

2.8 ACCELERATION REQUIRED BY CONSUMERS

CONTEXT CONNECTION

We now have our first opportunity to address a Context in a closing section, as we will do in each remaining chapter. Our present Context is *Alternative-Fuel Vehicles*, and our central question is, *What source besides gasoline can be used to provide energy for an automobile while reducing environmentally damaging emissions?*

Consumers have been driving gasoline-powered vehicles for decades and have become used to a certain amount of acceleration, such as that required to enter a freeway on-ramp. This experience raises the question as to what kind of acceleration today's consumer would expect for an alternative-fuel vehicle that might replace a gasoline-powered vehicle. In turn, developers of alternative-fuel vehicles should strive for such an acceleration so as to satisfy consumer expectations and hope to generate a demand for the new vehicle.

If we consider published time intervals for accelerations from 0 to 60 mi/h for a number of automobile models, we find the data shown in the middle column of Table 2.3. The average acceleration of each vehicle is calculated from these data using Equation 2.6. It is clear from the upper part of this table (*Performance vehicles*) that acceleration upward of 10 mi/h · s is very expensive. The highest accelerations are 16.7 mi/h · s and cost either $480,000 for the Ferrari F50 or a bargain at $292,000 for the Lamborghini Diablo GT. For the less affluent driver, the accelerations in the middle part of the table (*Traditional vehicles*) have an average value of 7.8 mi/h · s. This number is typical of consumer-oriented gasoline-powered vehicles and provides an approximate standard for the acceleration desired in an alternative-fuel vehicle.

In the lower part of Table 2.3, we see data for three alternative vehicles. The General Motors EV1 is an electric car that was discontinued in 2001, even though it was a technological success. Notice that its acceleration is similar to those in the

TABLE 2.3 Accelerations of Various Vehicles, 0–60 mi/h

Automobile	Model Year	Time Interval, 0–60 mi/h (s)	Average Acceleration (mi/h · s)	Price
Performance vehicles				
Aston Martin DB7 Vantage	2001	5.0	12.0	$170,000
BMW Z8	2001	4.6	13.0	$134,000
Chevrolet Corvette	2000	4.6	13.0	$46,000
Dodge Viper GTS-R	1998	4.2	14.3	$92,000
Ferrari F50	1997	3.6	16.7	$480,000
Ferrari 360 Spider F1	2000	4.6	13.0	$171,000
Lamborghini Diablo GT	2000	3.6	16.7	$292,000
Porsche 911 GT2	2002	4.0	15.0	$182,000
Traditional vehicles				
Acura Integra GS	2000	7.9	7.6	$22,000
BMW Mini Cooper S	2003	6.9	8.7	$17,500
Cadillac Escalade (SUV)	2002	8.6	7.0	$51,000
Dodge Stratus	2002	7.5	8.0	$22,000
Lexus ES300	1997	8.6	7.0	$29,000
Mitsubishi Eclipse GT	2000	7.0	8.6	$23,000
Nissan Maxima	2000	6.7	9.0	$25,000
Pontiac Grand Prix	2003	8.5	7.1	$25,000
Toyota Sienna (SUV)	2004	8.3	7.2	$23,000
Volkswagen Beetle	1999	7.6	7.9	$19,000
Alternative vehicles				
GM EV1	1998	7.6	7.9	(lease only) $399/month
Toyota Prius	2004	12.7	4.7	$21,000
Honda Insight	2001	11.6	5.2	$21,000

Note: Data given in this table as well as in similar tables in Chapters 3 through 6 were gathered from a number of websites. Other data, such as the accelerations in this table, were calculated from the raw data.

middle part of Table 2.3. This acceleration is sufficiently large that it satisfies consumer demand for a car with "get-up-and-go."

The Toyota Prius and Honda Insight are *hybrid vehicles,* which we will discuss further in the Context Conclusion. These vehicles combine a gasoline engine and an electric motor. The accelerations for these vehicles are the lowest in the table. The disadvantage of the low acceleration is offset by other factors. These vehicles obtain relatively high gas mileage, have very low emissions, and do not require recharging as does a pure electric vehicle.

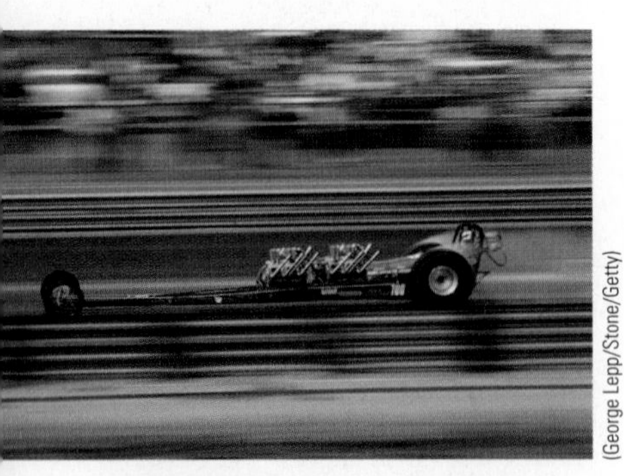

(George Lepp/Stone/Getty)

FIGURE 2.17 In drag racing, acceleration is a highly desired quantity. In a distance of 1/4 mile, speeds of over 320 mi/h are reached, with the entire distance being covered in under 5 s.

In comparison to the vehicles in the upper part of the table, consider the acceleration of an even higher-level "performance vehicle," a typical drag racer, as shown in Figure 2.17. Typical data show that such a vehicle covers a distance of 0.25 mi in 5.0 s, starting from rest. We can find the acceleration from Equation 2.12:

$$x_f = x_i + v_i t + \tfrac{1}{2}a_x t^2 = 0 + 0(t) + \tfrac{1}{2}(a_x)(t)^2 \quad \rightarrow \quad a_x = \frac{2x_f}{t^2}$$

$$= \frac{2(0.25 \text{ mi})}{(5.0 \text{ s})^2} = 0.020 \text{ mi/s}^2\left(\frac{3\,600 \text{ s}}{1 \text{ h}}\right) = 72 \text{ mi/h} \cdot \text{s}$$

This value is much larger than any accelerations in the table, as would be expected. We can show that the acceleration due to gravity has the following value in units of mi/h · s:

$$g = 9.80 \text{ m/s}^2 = 21.9 \text{ mi/h} \cdot \text{s}$$

Therefore, the drag racer is moving horizontally with 3.3 times as much acceleration as it would move vertically if you pushed it off a cliff! (Of course, the horizontal acceleration can only be maintained for a very short time interval.)

As we investigate two-dimensional motion in the next chapter, we shall consider a different type of acceleration for vehicles, that associated with the vehicle turning in a sharp circle at high speed. ■

SUMMARY

PhysicsNow™ Take a practice test by logging into PhysicsNow at www.pop4e.com and clicking on the Pre-Test link for this chapter.

The **average speed** of a particle during some time interval is equal to the ratio of the distance d traveled by the particle and the time interval Δt:

$$v_{\text{avg}} \equiv \frac{d}{\Delta t} \qquad [2.1]$$

The **average velocity** of a particle moving in one dimension during some time interval is equal to the ratio of the displacement Δx and the time interval Δt:

$$v_{x,\text{avg}} \equiv \frac{\Delta x}{\Delta t} \qquad [2.2]$$

The **instantaneous velocity** of a particle is defined as the limit of the ratio $\Delta x/\Delta t$ as Δt approaches zero:

$$v_x \equiv \lim_{\Delta t \to 0} \frac{\Delta x}{\Delta t} = \frac{dx}{dt} \qquad [2.3]$$

The **instantaneous speed** of a particle is defined as the magnitude of the instantaneous velocity vector.

If the velocity v_x is constant, the preceding equations can be modified and used to solve problems describing the motion of a *particle under constant velocity:*

$$v_x = \frac{\Delta x}{\Delta t} \qquad [2.4]$$

$$x_f = x_i + v_x t \qquad [2.5]$$

The **average acceleration** of a particle moving in one dimension during some time interval is defined as the ratio of the change in its velocity Δv_x and the time interval Δt:

$$a_{x,\text{avg}} \equiv \frac{\Delta v_x}{\Delta t} \qquad [2.6]$$

The **instantaneous acceleration** is equal to the limit of the ratio $\Delta v_x/\Delta t$ as $\Delta t \to 0$. By definition, this limit equals the derivative of v_x with respect to t, or the time rate of change of the velocity:

$$a_x \equiv \lim_{\Delta t \to 0} \frac{\Delta v_x}{\Delta t} = \frac{dv_x}{dt} \qquad [2.7]$$

The slope of the tangent to the x versus t curve at any instant gives the instantaneous velocity of the particle.

The slope of the tangent to the v versus t curve gives the instantaneous acceleration of the particle.

The **kinematic equations** for a *particle under constant acceleration* a_x (constant in magnitude and direction) are

$$v_{xf} = v_{xi} + a_x t \qquad [2.9]$$

$$x_f = x_i + \tfrac{1}{2}(v_{xi} + v_{xf})t \qquad [2.11]$$

$$x_f = x_i + v_{xi}t + \tfrac{1}{2}a_x t^2 \qquad [2.12]$$

$$v_{xf}^2 = v_{xi}^2 + 2a_x(x_f - x_i) \qquad [2.13]$$

An object falling freely experiences an acceleration directed toward the center of the Earth. If air friction is ignored and if the altitude of the motion is small compared with the Earth's radius, one can assume that the magnitude of the free-fall acceleration g is constant over the range of motion, where g is equal to 9.80 m/s^2, or 32 ft/s^2. Assuming y to be positive upward, the acceleration is given by $-g$, and the equations of kinematics for an object in free-fall are the same as those already given, with the substitutions $x \to y$ and $a_y \to -g$.

QUESTIONS

☐ = answer available in the *Student Solutions Manual and Study Guide*

1. The speed of sound in air is 331 m/s. During the next thunderstorm, try to estimate your distance from a lightning bolt by measuring the time lag between the flash and the thunderclap. You can ignore the time interval it takes for the light flash to reach you. Why?

2. The average velocity of a particle moving in one dimension has a positive value. Is it possible for the instantaneous velocity to have been negative at any time in the interval? Suppose the particle started at the origin $x = 0$. If its average velocity is positive, could the particle ever have been in the $-x$ region of the axis?

3. If the average velocity of an object is zero in some time interval, what can you say about the displacement of the object for that interval?

4. Can the instantaneous velocity of an object at an instant of time ever be greater in magnitude than the average velocity over a time interval containing the instant? Can it ever be less?

5. If an object's average velocity is nonzero over some time interval, does that mean that its instantaneous velocity is never zero during the interval? Explain your answer.

6. An object's average velocity is zero over some time interval. Show that its instantaneous velocity must be zero at some time during the interval. It may be useful in your proof to sketch a graph of x versus t and to note that $v_x(t)$ is a continuous function.

7. If the velocity of a particle is nonzero, can its acceleration be zero? Explain.

8. If the velocity of a particle is zero, can its acceleration be nonzero? Explain.

9. Two cars are moving in the same direction in parallel lanes along a highway. At some instant, the velocity of car A exceeds the velocity of car B. Does that mean that the acceleration of A is greater than that of B? Explain.

10. Is it possible for the velocity and the acceleration of an object to have opposite signs? If not, state a proof. If so, give an example of such a situation and sketch a velocity–time graph to prove your point.

11. Consider the following combinations of signs and values for velocity and acceleration of a particle with respect to a one-dimensional x axis:

Velocity	Acceleration
a. Positive	Positive
b. Positive	Negative
c. Positive	Zero
d. Negative	Positive
e. Negative	Negative
f. Negative	Zero
g. Zero	Positive
h. Zero	Negative

Describe what a particle is doing in each case and give a real-life example for an automobile on an east–west one-dimensional axis, with east considered the positive direction.

12. Can the kinematic equations (Eqs. 2.9 through 2.13) be used in a situation where the acceleration varies in time? Can they be used when the acceleration is zero?

13. A child throws a marble into the air with an initial speed v_i. Another child drops a ball at the same instant. Compare the accelerations of the two objects while they are in flight.

14. An object falls freely from height h. It is released at time zero and strikes the ground at time t. (a) When the object is at height $0.5h$, is the time earlier than $0.5t$, equal to $0.5t$, or later than $0.5t$? (b) When the time is $0.5t$, is the height of the object greater than $0.5h$, equal to $0.5h$, or less than $0.5h$? Give reasons for your answers.

15. A student at the top of a building of height h throws one ball upward with a speed of v_i and then throws a second ball downward with the same initial speed. How do the final velocities of the balls compare when they reach the ground?

16. You drop a ball from a window on an upper floor of a building. It strikes the ground with speed v. You now repeat the drop, but you have a friend down on the street who throws another ball upward at speed v. Your friend throws the ball upward at precisely the same time that you drop yours from the window. At some location, the balls pass each other. Is this location *at* the halfway point between window and ground, *above* this point, or *below* this point?

PROBLEMS

1, 2, 3 = straightforward, intermediate, challenging
☐ = full solution available in the *Student Solutions Manual and Study Guide*
Physics Now™ = coached problem with hints available at www.pop4e.com
= computer useful in solving problem
= paired numerical and symbolic problems
= biomedical application

Section 2.1 ■ Average Velocity

1. The position of a pinewood derby car was observed at various times; the results are summarized in the following table. Find the average velocity of the car for (a) the first second, (b) the last 3 s, and (c) the entire period of observation.

t (s)	0	1.0	2.0	3.0	4.0	5.0
x (m)	0	2.3	9.2	20.7	36.8	57.5

2. A particle moves according to the equation $x = 10t^2$, where x is in meters and t is in seconds. (a) Find the average velocity for the time interval from 2.00 s to 3.00 s. (b) Find the average velocity for the time interval from 2.00 s to 2.10 s.

3. The position versus time for a certain particle moving along the x axis is shown in Figure P2.3. Find the average velocity in the time intervals (a) 0 to 2 s, (b) 0 to 4 s, (c) 2 s to 4 s, (d) 4 s to 7 s, and (e) 0 to 8 s.

4. A person walks first at a constant speed of 5.00 m/s along a straight line from point A to point B and then back along the line from B to A at a constant speed of 3.00 m/s. (a) What is her average speed over the entire trip? (b) What is her average velocity over the entire trip?

Section 2.2 ■ Instantaneous Velocity

5. **Physics Now™** A position–time graph for a particle moving along the x axis is shown in Figure P2.5. (a) Find

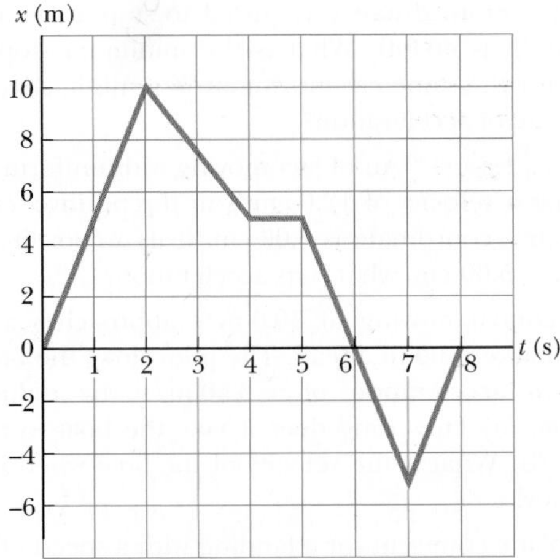

FIGURE P2.3 Problems 2.3 and 2.8.

the average velocity in the time interval $t = 1.50$ s to $t = 4.00$ s. (b) Determine the instantaneous velocity at $t = 2.00$ s by measuring the slope of the tangent line shown in the graph. (c) At what value of t is the velocity zero?

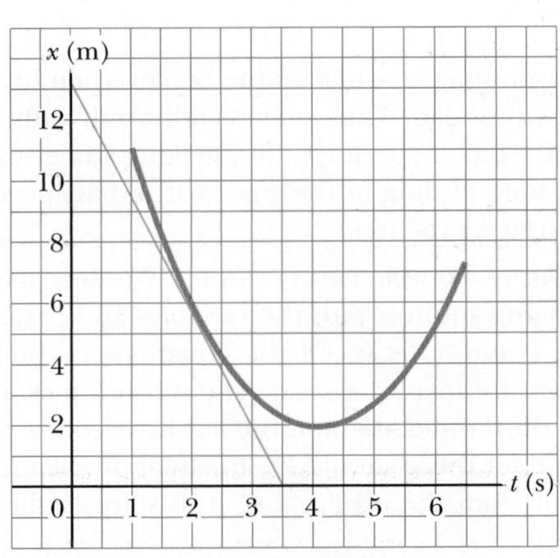

FIGURE P2.5

6. The position of a particle moving along the x axis varies in time according to the expression $x = 3t^2$, where x is in meters and t is in seconds. Evaluate its position (a) at $t = 3.00$ s and (b) at 3.00 s $+ \Delta t$. (c) Evaluate the limit of $\Delta x/\Delta t$ as Δt approaches zero to find the velocity at $t = 3.00$ s.

7. (a) Use the data in Problem 2.1 to construct a smooth graph of position versus time. (b) By constructing tangents to the $x(t)$ curve, find the instantaneous velocity of the car at several instants. (c) Plot the instantaneous velocity versus time and, from this information, determine the average acceleration of the car. (d) What was the initial velocity of the car?

8. Find the instantaneous velocity of the particle described in Figure P2.3 at the following times: (a) $t = 1.0$ s, (b) $t = 3.0$ s, (c) $t = 4.5$ s, (d) $t = 7.5$ s.

Section 2.3 ■ Analysis Models—The Particle Under Constant Velocity

9. A hare and a tortoise compete in a race over a course 1.00 km long. The tortoise crawls straight and steadily at its maximum speed of 0.200 m/s toward the finish line. The hare runs at its maximum speed of 8.00 m/s toward the goal for 0.800 km and then stops to taunt the tortoise. How close to the goal can the hare let the tortoise approach before resuming the race, which the tortoise wins in a photo finish? Assume that, when moving, both animals move steadily at their respective maximum speeds.

Section 2.4 ■ Acceleration

10. A 50.0-g superball traveling at 25.0 m/s bounces off a brick wall and rebounds at 22.0 m/s. A high-speed camera records this event. If the ball is in contact with the wall for 3.50 ms, what is the magnitude of the average acceleration of the ball during this time interval? (*Note:* 1 ms $= 10^{-3}$ s.)

11. A particle starts from rest and accelerates as shown in Figure P2.11. Determine (a) the particle's speed at $t = 10.0$ s and at $t = 20.0$ s, and (b) the distance traveled in the first 20.0 s.

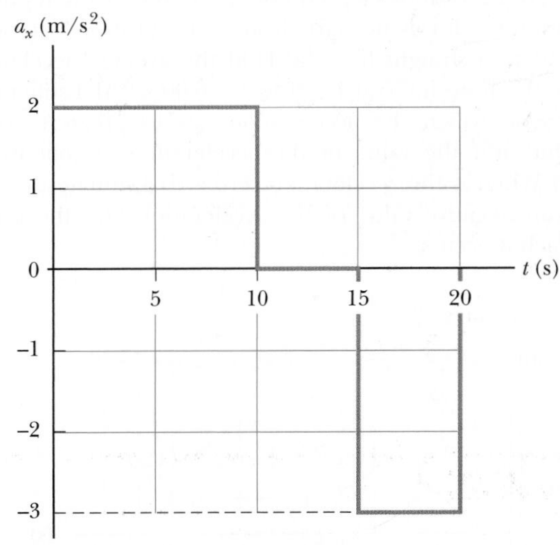

FIGURE P2.11

12. An object moves along the x axis according to the equation $x(t) = (3.00t^2 - 2.00t + 3.00)$m, where t is in seconds. Determine (a) the average speed between $t = 2.00$ s and $t = 3.00$ s, (b) the instantaneous speed at $t = 2.00$ s and at $t = 3.00$ s, (c) the average acceleration between $t = 2.00$ s and $t = 3.00$ s, and (d) the instantaneous acceleration at $t = 2.00$ s and $t = 3.00$ s.

13. **PhysicsNow™** A particle moves along the x axis according to the equation $x = 2.00 + 3.00t - 1.00t^2$, where x is in meters and t is in seconds. At $t = 3.00$ s, find (a) the position of the particle, (b) its velocity, and (c) its acceleration.

14. A student drives a moped along a straight road as described by the velocity versus time graph in Figure P2.14. Sketch this graph in the middle of a sheet of graph paper. (a) Directly above your graph, sketch a graph of the position versus time, aligning the time coordinates of the two graphs. (b) Sketch a graph of the acceleration versus time directly below the v_x-t graph, again aligning the time coordinates. On each graph, show the numerical values of x and a_x for all points of inflection. (c) What is the acceleration at $t = 6$ s? (d) Find the position (relative to the starting point) at $t = 6$ s. (e) What is the moped's final position at $t = 9$ s?

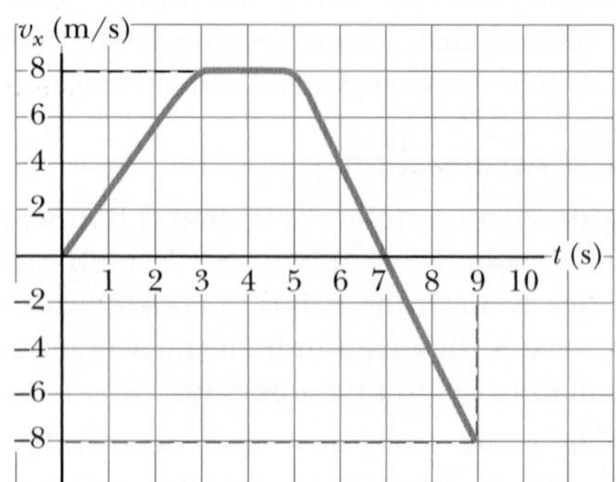

FIGURE P2.14

15. Figure P2.15 shows a graph of v_x versus t for the motion of a motorcyclist as he starts from rest and moves along the road in a straight line. (a) Find the average acceleration for the time interval $t = 0$ to $t = 6.00$ s. (b) Estimate the time at which the acceleration has its greatest positive value and the value of the acceleration at that instant. (c) When is the acceleration zero? (d) Estimate the maximum negative value of the acceleration and the time at which it occurs.

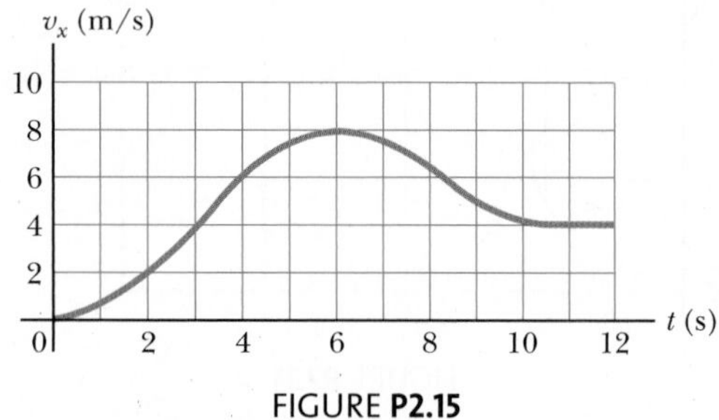

FIGURE P2.15

Section 2.5 ■ Motion Diagrams

16. Draw motion diagrams for (a) an object moving to the right at constant speed, (b) an object moving to the right and speeding up at a constant rate, (c) an object moving to the right and slowing down at a constant rate, (d) an object moving to the left and speeding up at a constant rate, and (e) an object moving to the left and slowing down at a constant rate. (f) How would your drawings change if the changes in speed were not uniform, that is, if the speed were not changing at a constant rate?

Section 2.6 ■ The Particle Under Constant Acceleration

17. A truck covers 40.0 m in 8.50 s while smoothly slowing down to a final speed of 2.80 m/s. (a) Find its original speed. (b) Find its acceleration.

18. The minimum distance required to stop a car moving at 35.0 mi/h is 40.0 ft. What is the minimum stopping distance for the same car moving at 70.0 mi/h, assuming the same rate of acceleration?

19. **PhysicsNow™** An object moving with uniform acceleration has a velocity of 12.0 cm/s in the positive x direction when its x coordinate is 3.00 cm. If its x coordinate 2.00 s later is -5.00 cm, what is its acceleration?

20. A speedboat moving at 30.0 m/s approaches a no-wake buoy marker 100 m ahead. The pilot slows the boat with a constant acceleration of -3.50 m/s^2 by reducing the throttle. (a) How long does it take the boat to reach the buoy? (b) What is the velocity of the boat when it reaches the buoy?

21. A jet plane comes in for a landing with a speed of 100 m/s and can accelerate at a maximum rate of -5.00 m/s^2 as it comes to rest. (a) From the instant the plane touches the runway, what is the minimum time interval needed before it can come to rest? (b) Can this plane land at a small tropical island airport where the runway is 0.800 km long?

22. A particle moves along the x axis. Its position is given by the equation $x = 2 + 3t - 4t^2$ with x in meters and t in seconds. Determine (a) its position when it changes direction and (b) its velocity when it returns to the position it had at $t = 0$.

23. The driver of a car slams on the brakes when he sees a tree blocking the road. The car slows uniformly with an acceleration of -5.60 m/s^2 for 4.20 s, making straight skid marks 62.4 m long ending at the tree. With what speed does the car then strike the tree?

24. *Help! One of our equations is missing!* We describe constant-acceleration motion with the variables and parameters v_{xi}, v_{xf}, a_x, t, and $x_f - x_i$. Of the equations in Table 2.2, the first does not involve $x_f - x_i$. The second does not contain a_x, the third omits v_{xf}, and the last leaves out t. So, to complete the set there should be an equation *not* involving v_{xi}. Derive it from the others. Use it to solve Problem 2.23 in one step.

25. A truck on a straight road starts from rest, accelerating at 2.00 m/s^2 until it reaches a speed of 20.0 m/s. Then the truck travels for 20.0 s at constant speed until the brakes are applied, stopping the truck in a uniform manner in an additional 5.00 s. (a) How long is the truck in motion? (b) What is the average velocity of the truck for the motion described?

26. An electron in a cathode-ray tube accelerates uniformly from 2.00×10^4 m/s to 6.00×10^6 m/s over 1.50 cm. (a) In what time interval does the electron travel this 1.50 cm? (b) What is its acceleration?

27. Speedy Sue, driving at 30.0 m/s, enters a one-lane tunnel. She then observes a slow-moving van 155 m ahead traveling at 5.00 m/s. Sue applies her brakes but can accelerate only at -2.00 m/s^2 because the road is wet. Will there be a collision? If yes, determine how far into the tunnel and at what time the collision occurs. If no, determine the distance of closest approach between Sue's car and the van.

Section 2.7 ■ Freely Falling Objects

Note: In all problems in this section, ignore the effects of air resistance.

28. In a classic clip on *America's Funniest Home Videos,* a sleeping cat rolls gently off the top of a warm TV set. Ignoring air resistance, calculate the position and velocity of the cat after (a) 0.100 s, (b) 0.200 s, and (c) 0.300 s.

29. A baseball is hit so that it travels straight upward after being struck by the bat. A fan observes that it takes 3.00 s for the ball to reach its maximum height. Find (a) its initial velocity and (b) the height it reaches.

30. *Every morning at seven o'clock*
There's twenty terriers drilling on the rock.
The boss comes around and he says, "Keep still
And bear down heavy on the cast-iron drill
And drill, ye terriers, drill." And drill, ye terriers, drill.
It's work all day for sugar in your tea
Down beyond the railway. And drill, ye terriers, drill.

The foreman's name was John McAnn.
By God, he was a blamed mean man.
One day a premature blast went off
And a mile in the air went big Jim Goff. And drill . . .
Then when next payday came around
Jim Goff a dollar short was found.
When he asked what for, came this reply:
"You were docked for the time you were up in the sky."
And drill . . .

—American folksong

What was Goff's hourly wage? State the assumptions you make in computing it.

31. **Physics Now™** A student throws a set of keys vertically upward to her sorority sister, who is in a window 4.00 m above. The keys are caught 1.50 s later by the sister's outstretched hand. (a) With what initial velocity were the keys thrown? (b) What was the velocity of the keys just before they were caught?

32. A ball is thrown directly downward, with an initial speed of 8.00 m/s, from a height of 30.0 m. After what time interval does the ball strike the ground?

33. **Physics Now™** A daring ranch hand sitting on a tree limb wishes to drop vertically onto a horse galloping under the tree. The constant speed of the horse is 10.0 m/s, and the distance from the limb to the level of the saddle is 3.00 m. (a) What must be the horizontal distance between the saddle and limb when the ranch hand makes his move? (b) How long is he in the air?

34. It is possible to shoot an arrow at a speed as high as 100 m/s. (a) If friction can be ignored, how high would an arrow launched at this speed rise if shot straight up? (b) How long would the arrow be in the air?

Section 2.8 ■ Context Connection—Acceleration Required by Consumers

35. (a) Show that the largest and smallest average accelerations in Table 2.3 are correctly computed from the measured time intervals required for the cars to speed up from 0 to 60 mi/h. (b) Convert both of these accelerations to the standard SI unit. (c) Modeling each acceleration as constant, find the distance traveled by both cars as they speed up. (d) If an automobile were able to maintain an acceleration of magnitude $a = g = 9.80 \text{ m/s}^2$ on a horizontal roadway, what time interval would be required to accelerate from zero to 60.0 mi/h?

36. A certain automobile manufacturer claims that its deluxe sports car will accelerate from rest to a speed of 42.0 m/s in 8.00 s. (a) Determine the average acceleration of the car. (b) Assume that the car moves with constant acceleration. Find the distance the car travels in the first 8.00 s. (c) What is the speed of the car 10.0 s after it begins its motion if it can continue to move with the same acceleration?

37. A steam catapult launches a jet aircraft from the aircraft carrier *John C. Stennis,* giving it a speed of 175 mi/h in 2.50 s. (a) Find the average acceleration of the plane. (b) Modeling the acceleration as constant, find the distance the plane moves in this time interval.

38. *Vroom—vroom!* As soon as a traffic light turns green, a car speeds up from rest to 50.0 mi/h with constant acceleration 9.00 mi/h · s. In the adjoining bike lane, a cyclist speeds up from rest to 20.0 mi/h with constant acceleration 13.0 mi/h · s. Each vehicle maintains constant velocity after reaching its cruising speed. (a) For what time interval is the bicycle ahead of the car? (b) By what maximum distance does the bicycle lead the car?

Additional Problems

Note: The human body can undergo brief accelerations up to 15 times the free-fall acceleration without injury or with only strained ligaments. Acceleration of long duration can do damage by preventing circulation of blood. Acceleration of larger magnitude can cause severe internal injuries, such as by tearing the aorta away from the heart. Problems 2.35, 2.37, and 2.39 through 2.41 deal with variously large accelerations of the human body that you can compare with the $15g$ datum.

39. For many years Colonel John P. Stapp, USAF, held the world's land speed record. He participated in studying whether a jet pilot could survive emergency ejection. On March 19, 1954, he rode a rocket-propelled sled that moved down a track at 632 mi/h. He and the sled were safely brought to rest in 1.40 s (Fig. P2.39). Determine (a) the negative acceleration he experienced and (b) the distance he traveled during this negative acceleration, assumed to be constant.

40. A woman is reported to have fallen 144 ft from the 17th floor of a building, landing on a metal ventilator box that she crushed to a depth of 18.0 in. She suffered only minor injuries. Ignoring air resistance, calculate (a) the speed of the woman just before she collided with the ventilator and (b) her average acceleration while in contact with the box. (c) Modeling her acceleration as constant, calculate the time interval it took to crush the box.

41. Jules Verne in 1865 suggested sending people to the Moon by firing a space capsule from a 220-m-long cannon with a final velocity of 10.97 km/s. What would have been

(Courtesy of the U.S. Air Force)

(Photri, Inc.)

FIGURE **P2.39** (*Left*) Col. John Stapp on the rocket sled. (*Right*) Col. Stapp's face is contorted by the stress of rapid negative acceleration.

the unrealistically large acceleration experienced by the space travelers during launch? Compare your answer with the free-fall acceleration 9.80 m/s^2.

42. **Review problem.** The biggest stuffed animal in the world is a snake 420 m long constructed by Norwegian children. Suppose the snake is laid out in a park as shown in Figure P2.42, forming two straight sides of a 105° angle, with one side 240 m long. Olaf and Inge run a race they invent. Inge runs directly from the tail of the snake to its head and Olaf starts from the same place at the same time but runs along the snake. If both children run steadily at 12.0 km/h, Inge reaches the head of the snake how much earlier than Olaf?

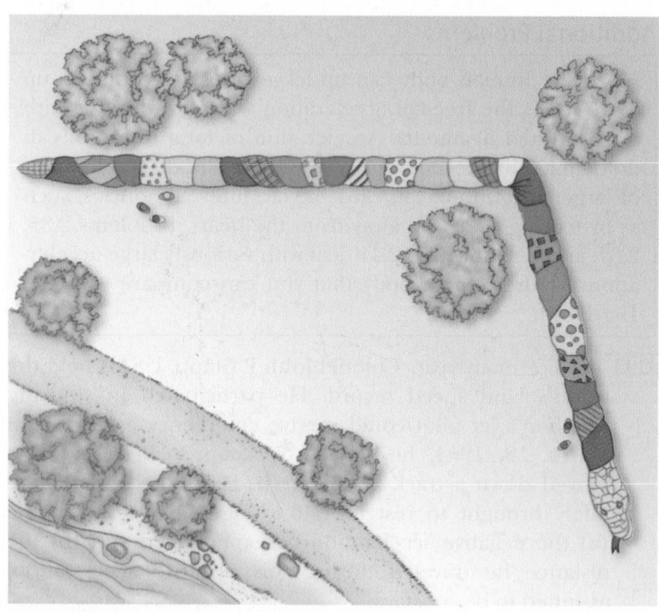

FIGURE **P2.42**

43. A ball starts from rest and accelerates at 0.500 m/s^2 while moving down an inclined plane 9.00 m long. When it reaches the bottom, the ball rolls up another plane, where it comes to rest after moving 15.0 m on that plane. (a) What is the speed of the ball at the bottom of the first plane? (b) During what time interval does the ball roll down the first plane? (c) What is the acceleration along the second plane? (d) What is the ball's speed 8.00 m along the second plane?

44. A glider on an air track carries a flag of length ℓ through a stationary photogate, which measures the time interval Δt_d during which the flag blocks a beam of infrared light passing across the photogate. The ratio $v_d = \ell/\Delta t_d$ is the average velocity of the glider over this part of its motion. Assume that the glider moves with constant acceleration. (a) Argue for or against the idea that v_d is equal to the instantaneous velocity of the glider when it is halfway through the photogate in space. (b) Argue for or against the idea that v_d is equal to the instantaneous velocity of the glider when it is halfway through the photogate in time.

45. Liz rushes down onto a subway platform to find her train already departing. She stops and watches the cars go by. Each car is 8.60 m long. The first moves past her in 1.50 s and the second in 1.10 s. Find the constant acceleration of the train.

46. The Acela is the Porsche of American trains. Shown in Figure P2.46a, the electric train whose name is pronounced ah-SELL-ah is in service on the Washington–New York–Boston run. With two power cars and six coaches, it can carry 304 passengers at 170 mi/h. The carriages tilt as much as 6° from the vertical to prevent passengers from feeling pushed to the side as they go around curves. Its braking mechanism uses electric generators to recover its energy of motion. A velocity–time graph for the Acela is shown in Figure P2.46b. (a) Describe the motion of the train in each successive time interval. (b) Find the peak positive acceleration of the train in the motion graphed. (c) Find the train's displacement in miles between $t = 0$ and $t = 200$ s.

47. A test rocket is fired vertically upward from a well. A catapult gives it initial speed 80.0 m/s at ground level. Its engines then fire and it accelerates upward at 4.00 m/s^2 until it reaches an altitude of 1 000 m. At that point its engines fail and the rocket goes into free-fall, with an acceleration of $-$ 9.80 m/s^2. (a) How long is the rocket in motion above the ground? (b) What is its maximum altitude? (c) What is its velocity just before it collides with the Earth? (You will need to consider the motion while the engine is operating separately from the free-fall motion.)

48. A motorist drives along a straight road at a constant speed of 15.0 m/s. Just as she passes a parked motorcycle police officer, the officer starts to accelerate at 2.00 m/s^2 to overtake her. Assuming that the officer maintains this acceleration, (a) determine the time interval required for the

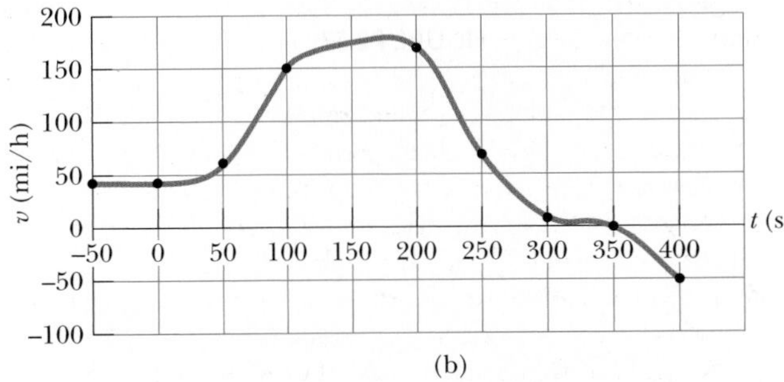

FIGURE P2.46 (a) The Acela, 1 171 000 lb of cold steel thundering along at 150 mi/h. (b) Velocity versus time graph for the Acela.

police officer to reach the motorist. Find (b) the speed and (c) the total displacement of the officer as he overtakes the motorist.

49. Setting a world record in a 100-m race, Maggie and Judy cross the finish line in a dead heat, both taking 10.2 s. Accelerating uniformly, Maggie took 2.00 s and Judy 3.00 s to attain maximum speed, which they maintained for the rest of the race. (a) What was the acceleration of each sprinter? (b) What were their respective maximum speeds? (c) Which sprinter was ahead at the 6.00-s mark and by how much?

50. A commuter train travels between two downtown stations. Because the stations are only 1.00 km apart, the train never reaches its maximum possible cruising speed. During rush hour the engineer minimizes the time interval Δt between two stations by accelerating at a rate $a_1 = 0.100\ \text{m/s}^2$ for a time interval Δt_1 and then immediately braking with acceleration $a_2 = -0.500\ \text{m/s}^2$ for a time interval Δt_2. Find the minimum time interval of travel Δt and the time interval Δt_1.

51. An inquisitive physics student and mountain climber climbs a 50.0-m cliff that overhangs a calm pool of water. He throws two stones vertically downward, 1.00 s apart, and observes that they cause a single splash. The first stone has an initial speed of 2.00 m/s. (a) How long after release of the first stone do the two stones hit the water? (b) What initial velocity must the second stone have if the two stones are to hit simultaneously? (c) What is the speed of each stone at the instant the two hit the water?

52. A hard rubber ball, released at chest height, falls to the pavement and bounces back to nearly the same height. When it is in contact with the pavement, the lower side of the ball is temporarily flattened. Suppose the maximum depth of the dent is on the order of 1 cm. Compute an order-of-magnitude estimate for the maximum acceleration of the ball while it is in contact with the pavement. State your assumptions, the quantities you estimate, and the values you estimate for them.

53. To protect his food from hungry bears, a Boy Scout raises his food pack with a rope that is thrown over a tree limb at height h above his hands. He walks away from the vertical rope with constant velocity v_{boy}, holding the free end of the rope in his hands (Fig. P2.53). (a) Show that the speed v of the food pack is given by $x(x^2 + h^2)^{-1/2}\, v_{\text{boy}}$ where x is the distance he has walked away from the vertical rope. (b) Show that the acceleration a of the food pack is $h^2(x^2 + h^2)^{-3/2} v_{\text{boy}}^2$. (c) What values do the acceleration and velocity v have shortly after the boy leaves the point under the pack ($x = 0$)? (d) What values do the pack's velocity and acceleration approach as the distance x continues to increase?

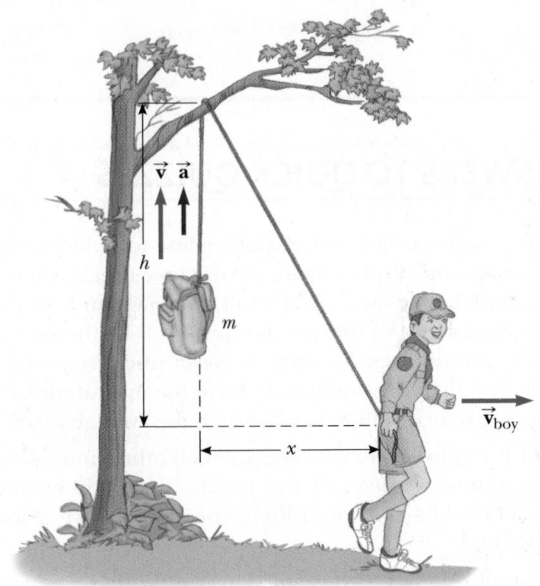

FIGURE P2.53 Problems 2.53 and 2.54.

54. In Problem 2.53, let the height h equal 6.00 m and the speed v_{boy} equal 2.00 m/s. Assume that the food pack starts from rest. (a) Tabulate and graph the speed–time graph. (b) Tabulate and graph the acceleration–time graph. Let the range of time be from 0 s to 5.00 s and the time intervals be 0.500 s.

55. A rock is dropped from rest into a well. (a) The sound of the splash is heard 2.40 s after the rock is released from rest. How far below the top of the well is the surface of the water? The speed of sound in air (at the ambient temperature) is 336 m/s. (b) If the travel time for the sound is ignored, what percentage error is introduced when the depth of the well is calculated?

56. Astronauts on a distant planet toss a rock into the air. With the aid of a camera that takes pictures at a steady rate, they record the height of the rock as a function of time as given in the Table P2.56. (a) Find the average velocity of the rock in the time interval between each measurement and the next. (b) Using these average velocities to approximate instantaneous velocities at the midpoints of the time intervals, make a graph of velocity as a function of time. Does the rock move with constant acceleration? If so, plot a straight line of best fit on the graph and calculate its slope to find the acceleration.

TABLE P2.56 Height of a Rock Versus Time

Time (s)	Height (m)	Time (s)	Height (m)
0.00	5.00	2.75	7.62
0.25	5.75	3.00	7.25
0.50	6.40	3.25	6.77
0.75	6.94	3.50	6.20
1.00	7.38	3.75	5.52
1.25	7.72	4.00	4.73
1.50	7.96	4.25	3.85
1.75	8.10	4.50	2.86
2.00	8.13	4.75	1.77
2.25	8.07	5.00	0.58
2.50	7.90		

57. Two objects, A and B, are connected by a rigid rod that has a length L. The objects slide along perpendicular guide rails, as shown in Figure P2.57. If A slides to the left with a constant speed v, find the velocity of B when $\alpha = 60.0°$.

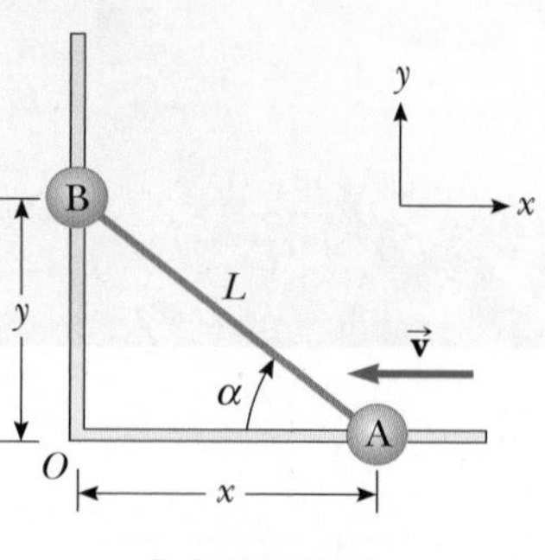

FIGURE **P2.57**

ANSWERS TO QUICK QUIZZES

2.1 (c). If the particle moves along a line without changing direction, the displacement and distance over any time interval will be the same. As a result, the magnitude of the average velocity and the average speed will be the same. If the particle reverses direction, however, the displacement will be less than the distance. In turn, the magnitude of the average velocity will be smaller than the average speed.

2.2 (b). Regardless of your speeds at all other times, if your instantaneous speed at the instant that it is measured is higher than the speed limit, you may receive a speeding ticket.

2.3 Graph (a) has a constant slope, indicating a constant acceleration; this situation is represented by graph (e). Graph (b) represents a speed that is increasing constantly but not at a uniform rate. Therefore, the acceleration must be increasing, and the graph that best indicates this situation is (d). Graph (c) depicts a velocity that first increases at a constant rate, indicating constant acceleration. Then the velocity stops increasing and becomes constant, indicating zero acceleration. The best match to this situation is graph (f).

2.4 (b). If the car is slowing down, a force must be acting in the direction opposite to its velocity.

2.5 (c). If a particle with constant acceleration stops and its acceleration remains constant, it must begin to move again in the opposite direction. If it did not, the acceleration would change from its original constant value to zero. Choice (a) is not correct because the direction of acceleration is independent of the direction of the velocity. Choice (b) is not correct either. For example, a car moving in the negative x direction and slowing down has a positive acceleration.

2.6 (e). For the entire time interval the ball is in free-fall, the acceleration is that due to gravity.

Motion in Two Dimensions

(© Arndt/Premium Stock/PictureQuest)

Lava spews from a volcanic eruption. Notice the parabolic paths of embers projected into the air. We will find in this chapter that all projectiles follow a parabolic path in the absence of air resistance.

CHAPTER OUTLINE

In this chapter, we shall study the kinematics of an object that can be modeled as a particle moving in a plane. This motion is two dimensional. Some common examples of motion in a plane are the motions of satellites in orbit around the Earth, projectiles such as a thrown baseball, and the motion of electrons in uniform electric fields. We shall also study a particle in uniform circular motion and discuss various aspects of particles moving in curved paths.

3.1 THE POSITION, VELOCITY, AND ACCELERATION VECTORS

In Chapter 2, we found that the motion of a particle moving along a straight line is completely specified if its position is known as a function of time. Now let us extend this idea to motion in the *xy* plane. We will find equations for position and velocity that are the same as those in Chapter 2 except for their vector nature.

We begin by describing the position of a particle with a **position vector** $\vec{\mathbf{r}}$, drawn from the origin of a coordinate system to the location of the particle in the *xy* plane, as in Figure 3.1. At time t_i, the particle is at the point Ⓐ, and at some later time t_f, the particle is at Ⓑ, where the subscripts i and f refer to initial and final values. As the particle moves from Ⓐ to Ⓑ in the time interval $\Delta t = t_f - t_i$, the position vector changes from $\vec{\mathbf{r}}_i$ to $\vec{\mathbf{r}}_f$. As we learned in Chapter 2, the displacement of a particle is the difference between its final position and its initial position:

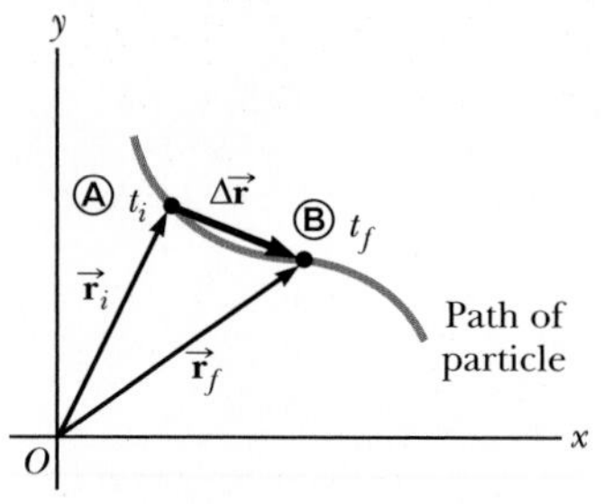

FIGURE 3.1 A particle moving in the *xy* plane is located with the position vector $\vec{\mathbf{r}}$ drawn from the origin to the particle. The displacement of the particle as it moves from Ⓐ to Ⓑ in the time interval $\Delta t = t_f - t_i$ is equal to the vector $\Delta\vec{\mathbf{r}} \equiv \vec{\mathbf{r}}_f - \vec{\mathbf{r}}_i$.

$$\Delta\vec{\mathbf{r}} \equiv \vec{\mathbf{r}}_f - \vec{\mathbf{r}}_i \qquad [3.1]$$

The direction of $\Delta\vec{\mathbf{r}}$ is indicated in Figure 3.1.

The **average velocity** $\vec{\mathbf{v}}_{avg}$ of the particle during the time interval Δt is defined as the ratio of the displacement to the time interval:

■ Definition of average velocity

$$\vec{\mathbf{v}}_{avg} \equiv \frac{\Delta\vec{\mathbf{r}}}{\Delta t} \qquad [3.2]$$

Because displacement is a vector quantity and the time interval is a scalar quantity, we conclude that the average velocity is a *vector* quantity directed along $\Delta\vec{\mathbf{r}}$. The average velocity between points Ⓐ and Ⓑ is *independent of the path* between the two points. That is because the average velocity is proportional to the displacement, which in turn depends only on the initial and final position vectors and not on the path taken between those two points. As with one-dimensional motion, if a particle starts its motion at some point and returns to this point via any path, its average velocity is zero for this trip because its displacement is zero.

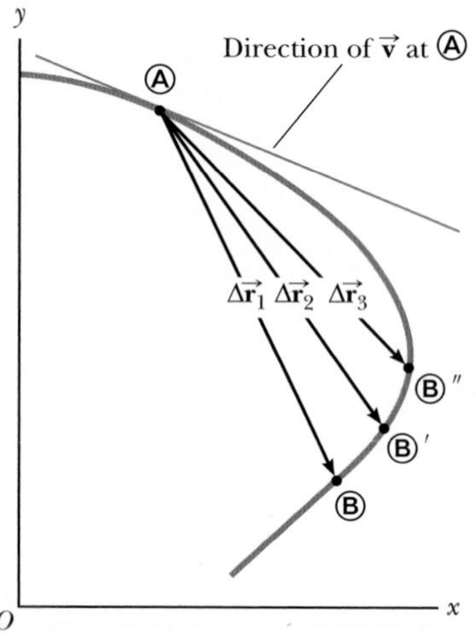

FIGURE 3.2 As a particle moves between two points, its average velocity is in the direction of the displacement vector $\Delta\vec{\mathbf{r}}$. As the end point of the path is moved from Ⓑ to Ⓑ′ to Ⓑ″, the respective displacements and corresponding time intervals become smaller and smaller. In the limit that the end point approaches Ⓐ, Δt approaches zero and the direction of $\Delta\vec{\mathbf{r}}$ approaches that of the line tangent to the curve at Ⓐ. By definition, the instantaneous velocity at Ⓐ is in the direction of this tangent line.

Consider again the motion of a particle between two points in the *xy* plane, as shown in Figure 3.2. As the time intervals over which we observe the motion become smaller and smaller, the direction of the displacement approaches that of the line tangent to the path at the point Ⓐ.

The **instantaneous velocity** $\vec{\mathbf{v}}$ is defined as the limit of the average velocity $\Delta\vec{\mathbf{r}}/\Delta t$ as Δt approaches zero:

$$\vec{\mathbf{v}} \equiv \lim_{\Delta t \to 0} \frac{\Delta\vec{\mathbf{r}}}{\Delta t} = \frac{d\vec{\mathbf{r}}}{dt} \qquad [3.3]$$

That is, the instantaneous velocity equals the derivative of the position vector with respect to time. The direction of the instantaneous velocity vector at any point in a particle's path is along a line that is tangent to the path at that point and in the direction of motion. The magnitude of the instantaneous velocity is called the *speed.*

As a particle moves from point Ⓐ to point Ⓑ along some path as in Figure 3.3, its instantaneous velocity changes from $\vec{\mathbf{v}}_i$ at time t_i to $\vec{\mathbf{v}}_f$ at time t_f. The **average acceleration** $\vec{\mathbf{a}}_{avg}$ of a particle over a time interval is defined as the ratio of the change in the instantaneous velocity $\Delta\vec{\mathbf{v}}$ to the time interval Δt:

■ Definition of average acceleration

$$\vec{\mathbf{a}}_{avg} \equiv \frac{\vec{\mathbf{v}}_f - \vec{\mathbf{v}}_i}{t_f - t_i} = \frac{\Delta\vec{\mathbf{v}}}{\Delta t} \qquad [3.4]$$

Because the average acceleration is the ratio of a vector quantity $\Delta\vec{\mathbf{v}}$ and a scalar quantity Δt, we conclude that $\vec{\mathbf{a}}_{avg}$ is a vector quantity directed along $\Delta\vec{\mathbf{v}}$. As

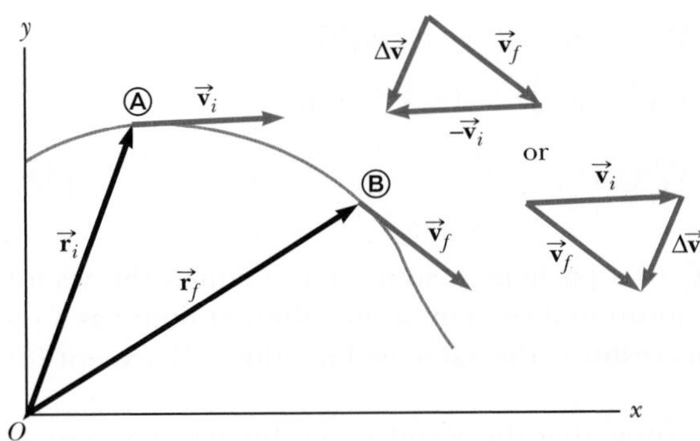

FIGURE 3.3 A particle moves from position Ⓐ to position Ⓑ. Its velocity vector changes from $\vec{\mathbf{v}}_i$ at time t_i to $\vec{\mathbf{v}}_f$ at time t_f. The vector addition diagrams at the upper right show two ways of determining the vector $\Delta\vec{\mathbf{v}}$ from the initial and final velocities.

PITFALL PREVENTION 3.1

VECTOR ADDITION The vector addition that was discussed in Chapter 1 involves displacement vectors. Because we are familiar with movements through space in our everyday experience, the addition of displacement vectors can be understood easily. The notion of vector addition can be applied to *any* type of vector quantity. Figure 3.3, for example, shows the addition of velocity vectors using the tip-to-tail approach.

indicated in Figure 3.3, the direction of $\Delta\vec{\mathbf{v}}$ is found by adding the vector $-\vec{\mathbf{v}}_i$ (the negative of $\vec{\mathbf{v}}_i$) to the vector $\vec{\mathbf{v}}_f$ because by definition $\Delta\vec{\mathbf{v}} = \vec{\mathbf{v}}_f - \vec{\mathbf{v}}_i$.

The **instantaneous acceleration** $\vec{\mathbf{a}}$ is defined as the limiting value of the ratio $\Delta\vec{\mathbf{v}}/\Delta t$ as Δt approaches zero:

$$\vec{\mathbf{a}} \equiv \lim_{\Delta t \to 0} \frac{\Delta\vec{\mathbf{v}}}{\Delta t} = \frac{d\vec{\mathbf{v}}}{dt} \quad [3.5]$$

▪ **Definition of instantaneous acceleration**

That is, the instantaneous acceleration equals the derivative of the velocity vector with respect to time.

It is important to recognize that various changes can occur that represent a particle undergoing an acceleration. First, the magnitude of the velocity vector (the speed) may change with time as in straight-line (one-dimensional) motion. Second, the direction of the velocity vector may change with time as its magnitude remains constant. Finally, both the magnitude and the direction of the velocity vector may change.

QUICK QUIZ 3.1 Consider the following controls in an automobile: gas pedal, brake, steering wheel. The controls in this list that cause an acceleration of the car are (**a**) all three controls, (**b**) the gas pedal and the brake, (**c**) only the brake, or (**d**) only the gas pedal.

3.2 TWO-DIMENSIONAL MOTION WITH CONSTANT ACCELERATION

Let us consider two-dimensional motion during which the magnitude and direction of the acceleration remain unchanged. In this situation, we shall investigate motion as a two-dimensional version of the analysis in Section 2.6.

The motion of a particle can be determined if its position vector $\vec{\mathbf{r}}$ is known at all times. The position vector for a particle moving in the xy plane can be written

$$\vec{\mathbf{r}} = x\hat{\mathbf{i}} + y\hat{\mathbf{j}} \quad [3.6]$$

where x, y, and $\vec{\mathbf{r}}$ change with time as the particle moves. If the position vector is known, the velocity of the particle can be obtained from Equations 3.3 and 3.6:

$$\vec{\mathbf{v}} = \frac{d\vec{\mathbf{r}}}{dt} = \frac{dx}{dt}\hat{\mathbf{i}} + \frac{dy}{dt}\hat{\mathbf{j}} = v_x\hat{\mathbf{i}} + v_y\hat{\mathbf{j}} \quad [3.7]$$

Because we are assuming that $\vec{\mathbf{a}}$ is constant in this discussion, its components a_x and a_y are also constants. Therefore, we can apply the equations of kinematics to the x and y components of the velocity vector separately. Substituting $v_x = v_{xf} = v_{xi} + a_x t$ and $v_y = v_{yf} = v_{yi} + a_y t$ into Equation 3.7 gives

$$\vec{\mathbf{v}}_f = (v_{xi} + a_x t)\hat{\mathbf{i}} + (v_{yi} + a_y t)\hat{\mathbf{j}}$$
$$= (v_{xi}\hat{\mathbf{i}} + v_{yi}\hat{\mathbf{j}}) + (a_x\hat{\mathbf{i}} + a_y\hat{\mathbf{j}})t$$

■ Velocity vector as a function of time for a particle under constant acceleration

$$\vec{\mathbf{v}}_f = \vec{\mathbf{v}}_i + \vec{\mathbf{a}}t \qquad [3.8]$$

This result states that the velocity $\vec{\mathbf{v}}_f$ of a particle at some time t equals the vector sum of its initial velocity $\vec{\mathbf{v}}_i$ and the additional velocity $\vec{\mathbf{a}}t$ acquired at time t as a result of its constant acceleration. This result is the same as Equation 2.9, except for its vector nature.

Similarly, from Equation 2.12 we know that the x and y coordinates of a particle moving with constant acceleration are

$$x_f = x_i + v_{xi}t + \tfrac{1}{2}a_x t^2 \qquad \text{and} \qquad y_f = y_i + v_{yi}t + \tfrac{1}{2}a_y t^2$$

Substituting these expressions into Equation 3.6 gives

$$\vec{\mathbf{r}}_f = (x_i + v_{xi}t + \tfrac{1}{2}a_x t^2)\hat{\mathbf{i}} + (y_i + v_{yi}t + \tfrac{1}{2}a_y t^2)\hat{\mathbf{j}}$$
$$= (x_i\hat{\mathbf{i}} + y_i\hat{\mathbf{j}}) + (v_{xi}\hat{\mathbf{i}} + v_{yi}\hat{\mathbf{j}})t + \tfrac{1}{2}(a_x\hat{\mathbf{i}} + a_y\hat{\mathbf{j}})t^2$$

■ Position vector as a function of time for a particle under constant acceleration

$$\vec{\mathbf{r}}_f = \vec{\mathbf{r}}_i + \vec{\mathbf{v}}_i t + \tfrac{1}{2}\vec{\mathbf{a}}t^2 \qquad [3.9]$$

This equation implies that the final position vector $\vec{\mathbf{r}}_f$ is the vector sum of the initial position vector $\vec{\mathbf{r}}_i$ plus a displacement $\vec{\mathbf{v}}_i t$, arising from the initial velocity of the particle, and a displacement $\frac{1}{2}\vec{\mathbf{a}}t^2$, resulting from the uniform acceleration of the particle. It is the same as Equation 2.12 except for its vector nature.

Pictorial representations of Equations 3.8 and 3.9 are shown in Active Figures 3.4a and 3.4b. Note from Active Figure 3.4b that $\vec{\mathbf{r}}_f$ is generally not along the direction of $\vec{\mathbf{v}}_i$ or $\vec{\mathbf{a}}$ because the relationship between these quantities is a vector expression. For the same reason, from Active Figure 3.4a we see that $\vec{\mathbf{v}}_f$ is generally not along the direction of $\vec{\mathbf{v}}_i$ or $\vec{\mathbf{a}}$. Finally, if we compare the two figures, we see that $\vec{\mathbf{v}}_f$ and $\vec{\mathbf{r}}_f$ are not in the same direction.

Because Equations 3.8 and 3.9 are *vector* expressions, we may also write their x and y component equations:

$$\vec{\mathbf{v}}_f = \vec{\mathbf{v}}_i + \vec{\mathbf{a}}t \quad \rightarrow \quad \begin{cases} v_{xf} = v_{xi} + a_x t \\ v_{yf} = v_{yi} + a_y t \end{cases}$$

$$\vec{\mathbf{r}}_f = \vec{\mathbf{r}}_i + \vec{\mathbf{v}}_i t + \tfrac{1}{2}\vec{\mathbf{a}}t^2 \quad \rightarrow \quad \begin{cases} x_f = x_i + v_{xi}t + \frac{1}{2}a_x t^2 \\ y_f = y_i + v_{yi}t + \frac{1}{2}a_y t^2 \end{cases}$$

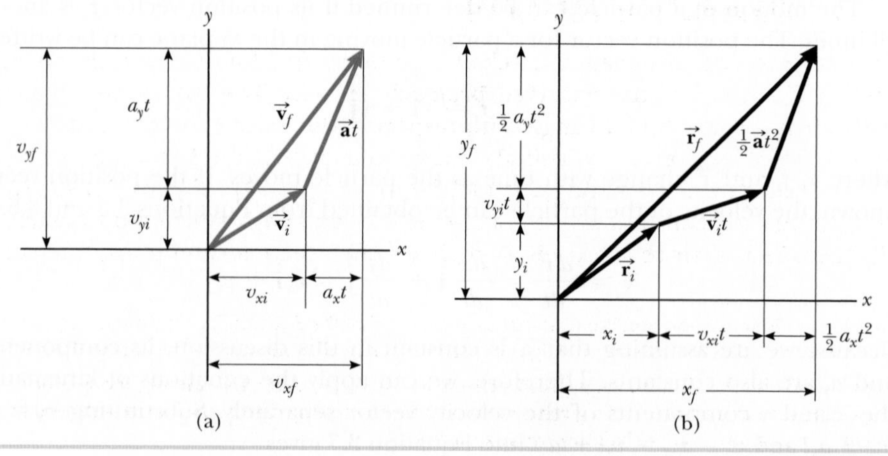

ACTIVE FIGURE 3.4
Vector representations and components of (a) the velocity and (b) the position of a particle under constant acceleration $\vec{\mathbf{a}}$.

Physics Now™ Log into PhysicsNow at **www.pop4e.com** and go to Active Figure 3.4 to investigate the effect of different initial positions and velocities on the final position and velocity (for constant acceleration).

These components are illustrated in Active Figure 3.4. In other words, **two-dimensional motion having constant acceleration is equivalent to two *independent* motions in the x and y directions having constant accelerations a_x and a_y.** Motion in the x direction does not affect motion in the y direction and vice versa. Therefore, there is no new model for a particle under two-dimensional constant acceleration; the appropriate model is just the one-dimensional particle under constant acceleration applied twice, in the x and y directions separately!

EXAMPLE 3.1 Motion in a Plane

A particle moves through the origin of an xy coordinate system at $t = 0$ with initial velocity $\vec{\mathbf{v}}_i = (20\hat{\mathbf{i}} - 15\hat{\mathbf{j}})$ m/s. The particle moves in the xy plane with an acceleration $\vec{\mathbf{a}} = 4.0\hat{\mathbf{i}}$ m/s^2.

A Determine the components of velocity as a function of time and the total velocity vector at any time.

Solution *Conceptualize* by establishing the mental representation and thinking about what the particle is doing. From the given information we see that the particle starts off moving to the right and downward and accelerates only toward the right. What will the particle do under these conditions? It may help if you draw a pictorial representation. To *categorize* consider that because the acceleration is only in the x direction, the moving particle can be modeled as one under constant acceleration in the x direction and one under constant velocity in the y direction.

To *analyze* the situation, we identify $v_{xi} = 20$ m/s and $a_x = 4.0$ m/s^2. The equations of kinematics give us, for the x direction,

$$v_{xf} = v_{xi} + a_x t = (20 + 4.0t)$$

Also, with $v_{yi} = -15$ m/s and $a_y = 0$,

$$v_{yf} = v_{yi} + a_y t = -15 \text{ m/s}$$

Therefore, using these results and noting that the velocity vector $\vec{\mathbf{v}}_f$ has two components, we find

$$\vec{\mathbf{v}}_f = v_{xf}\hat{\mathbf{i}} + v_{yf}\hat{\mathbf{j}} = [(20 + 4.0t)\hat{\mathbf{i}} - 15\hat{\mathbf{j}}]$$

Note that only the x component varies in time, reflecting that acceleration occurs only in the x direction.

B Calculate the velocity and speed of the particle at $t = 5.0$ s.

Solution At $t = 5.0$ s, the velocity expression from part A gives

$$\vec{\mathbf{v}}_f = \{[20 + 4(5.0)]\hat{\mathbf{i}} - 15\hat{\mathbf{j}}\} \text{ m/s} = (40\hat{\mathbf{i}} - 15\hat{\mathbf{j}}) \text{ m/s}$$

That is, at $t = 5.0$ s, $v_{xf} = 40$ m/s and $v_{yf} = -15$ m/s. To determine the angle θ that $\vec{\mathbf{v}}_f$ makes with the x axis, use $\tan\theta = v_{yf}/v_{xf}$, or

$$\theta = \tan^{-1}\left(\frac{v_{yf}}{v_{xf}}\right) = \tan^{-1}\left(\frac{-15 \text{ m/s}}{40 \text{ m/s}}\right) = -21°$$

The speed is the magnitude of $\vec{\mathbf{v}}_f$:

$$v_f = |\vec{\mathbf{v}}_f| = \sqrt{v_{xf}^2 + v_{yf}^2} = \sqrt{(40)^2 + (-15)^2} \text{ m/s}$$

$$= 43 \text{ m/s}$$

Now we *finalize.* In examining our result, we find that $v_f > v_i$. Does that make sense to you? Is it consistent with your mental representation?

3.3 PROJECTILE MOTION

Anyone who has observed a baseball in motion (or, for that matter, any object thrown into the air) has observed projectile motion. The ball moves in a curved path when thrown at some angle with respect to the Earth's surface. This very common form of motion is surprisingly simple to analyze if the following two assumptions are made when building a model for these types of problems: (1) the free-fall acceleration g is constant over the range of motion and is directed downward,[1] and (2) the effect of air resistance is negligible.[2] With these assumptions, the path of a

[1]In effect, this approximation is equivalent to assuming that the Earth is flat within the range of motion considered and that the maximum height of the object is small compared to the radius of the Earth.

[2]This approximation is often *not* justified, especially at high velocities. In addition, the spin of a projectile, such as a baseball, can give rise to some very interesting effects associated with aerodynamic forces (for example, a curve ball thrown by a pitcher).

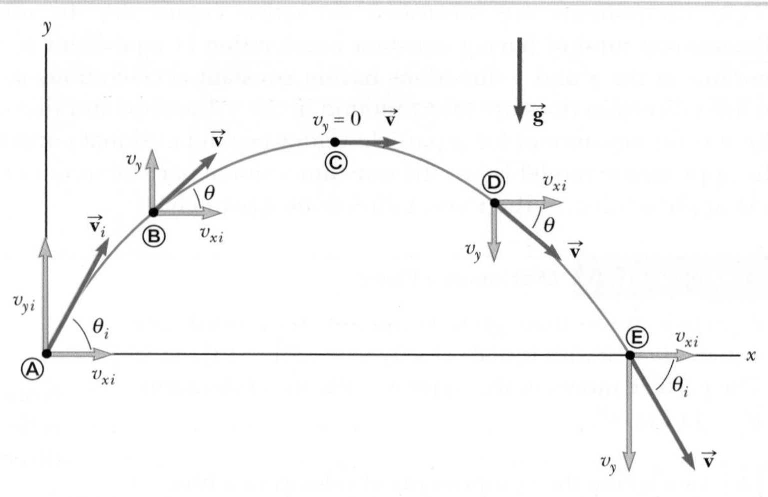

ACTIVE FIGURE 3.5
The parabolic path of a projectile that leaves the origin (point Ⓐ) with a velocity $\vec{v}_i$. The velocity vector $\vec{v}$ changes with time in both magnitude and direction. The change in the velocity vector is the result of acceleration in the negative *y* direction. The *x* component of velocity remains constant in time because no acceleration occurs in the horizontal direction. The *y* component of velocity is zero at the peak of the path (point Ⓒ).

PhysicsNow™ Log into PhysicsNow at **www.pop4e.com** and go to Active Figure 3.5 to change the launch angle and initial speed. You can also observe the changing components of velocity along the trajectory of the projectile.

projectile, called its *trajectory,* is *always* a parabola. **We shall use a simplification model based on these assumptions throughout this chapter.**

If we choose our reference frame such that the *y* direction is vertical and positive upward, $a_y = -g$ (as in one-dimensional free-fall) and $a_x = 0$ (because the only possible horizontal acceleration is due to air resistance, and it is ignored). Furthermore, let us assume that at $t = 0$, the projectile leaves the origin (point Ⓐ, $x_i = y_i = 0$) with speed v_i, as in Active Figure 3.5. If the vector $\vec{v}_i$ makes an angle θ_i with the horizontal, we can identify a right triangle in the diagram as a geometric model, and from the definitions of the cosine and sine functions we have

$$\cos\theta_i = \frac{v_{xi}}{v_i} \qquad \text{and} \qquad \sin\theta_i = \frac{v_{yi}}{v_i}$$

Therefore, the initial *x* and *y* components of velocity are

$$v_{xi} = v_i\cos\theta_i \qquad \text{and} \qquad v_{yi} = v_i\sin\theta_i$$

Substituting these expressions into Equations 3.8 and 3.9 with $a_x = 0$ and $a_y = -g$ gives the velocity components and position coordinates for the projectile at any time t:

$$v_{xf} = v_{xi} = v_i\cos\theta_i = \text{constant} \qquad [3.10]$$

$$v_{yf} = v_{yi} - gt = v_i\sin\theta_i - gt \qquad [3.11]$$

$$x_f = x_i + v_{xi}t = (v_i\cos\theta_i)\,t \qquad [3.12]$$

$$y_f = y_i + v_{yi}t - \tfrac{1}{2}gt^2 = (v_i\sin\theta_i)\,t - \tfrac{1}{2}gt^2 \qquad [3.13]$$

A welder cuts holes through a heavy metal construction beam with a hot torch. The sparks generated in the process follow parabolic paths. ■

From Equation 3.10 we see that v_{xf} remains constant in time and is equal to v_{xi}; there is no horizontal component of acceleration. Therefore, we model the horizontal motion as that of a particle under constant velocity. For the *y* motion, note that the equations for v_{yf} and y_f are similar to Equations 2.9 and 2.12 for freely falling objects. Therefore, we can apply the model of a particle under constant acceleration to the *y* component. In fact, *all* the equations of kinematics developed in Chapter 2 are applicable to projectile motion.

If we solve for t in Equation 3.12 and substitute this expression for t into Equation 3.13, we find that

$$y_f = (\tan \theta_i) x_f - \left(\frac{g}{2v_i^2 \cos^2 \theta_i}\right) x_f^2 \qquad [3.14]$$

which is valid for angles in the range $0 < \theta_i < \pi/2$. This expression is of the form $y = ax - bx^2$, which is the equation of a parabola that passes through the origin. Thus, we have proven that the trajectory of a projectile can be geometrically modeled as a parabola. The trajectory is *completely* specified if v_i and θ_i are known.

The vector expression for the position of the projectile as a function of time follows directly from Equation 3.9, with $\vec{\mathbf{a}} = \vec{\mathbf{g}}$:

$$\vec{\mathbf{r}}_f = \vec{\mathbf{r}}_i + \vec{\mathbf{v}}_i t + \tfrac{1}{2}\vec{\mathbf{g}} t^2$$

This equation gives the same information as the combination of Equations 3.12 and 3.13 and is plotted in Figure 3.6. Note that this expression for $\vec{\mathbf{r}}_f$ is consistent with Equation 3.13 because the expression for $\vec{\mathbf{r}}_f$ is a vector equation and $\vec{\mathbf{a}} = \vec{\mathbf{g}} = -g\hat{\mathbf{j}}$ when the upward direction is taken to be positive.

The position of a particle can be considered the sum of its original position $\vec{\mathbf{r}}_i$, the term $\vec{\mathbf{v}}_i t$, which would be the displacement if no acceleration were present, and the term $\frac{1}{2}\vec{\mathbf{g}} t^2$, which arises from the acceleration caused by gravity. In other words, if no gravitational acceleration occurred, the particle would continue to move along a straight path in the direction of $\vec{\mathbf{v}}_i$.

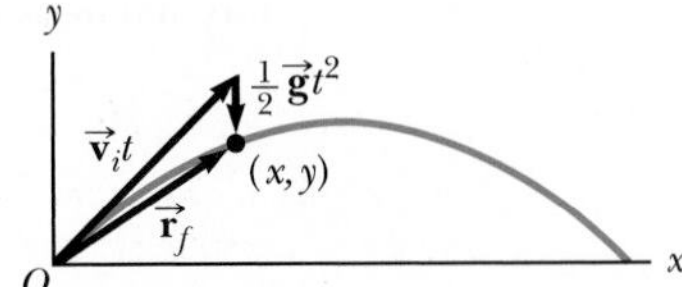

FIGURE 3.6 The position vector $\vec{\mathbf{r}}_f$ of a projectile whose initial velocity at the origin is $\vec{\mathbf{v}}_i$. The vector $\vec{\mathbf{v}}_i t$ would be the position vector of the projectile if gravity were absent and the vector $\frac{1}{2}\vec{\mathbf{g}}t^2$ is the particle's vertical displacement due to its downward gravitational acceleration.

QUICK QUIZ 3.2 As a projectile thrown upward moves in its parabolic path (such as in Figure 3.6), at what point along its path are the velocity and acceleration vectors for the projectile perpendicular to each other? **(a)** nowhere **(b)** the highest point **(c)** the launch point At what point are the velocity and acceleration vectors for the projectile parallel to each other? **(d)** nowhere **(e)** the highest point **(f)** the launch point

Horizontal Range and Maximum Height of a Projectile

Let us assume that a projectile is launched over flat ground from the origin at $t = 0$ with a positive v_y component, as in Figure 3.7. There are two special points that are interesting to analyze: the peak point Ⓐ, which has Cartesian coordinates $(R/2, h)$, and the landing point Ⓑ, having coordinates $(R, 0)$. The distance R is called the *horizontal range* of the projectile, and h is its *maximum height*. Because of the symmetry of the trajectory, the projectile is at the maximum height h when its x position is half the range R. Let us find h and R in terms of v_i, θ_i, and g.

We can determine h by noting that at the peak $v_{yA} = 0$. Therefore, Equation 3.11 can be used to determine the time t_A at which the projectile reaches the peak:

$$t_A = \frac{v_i \sin \theta_i}{g}$$

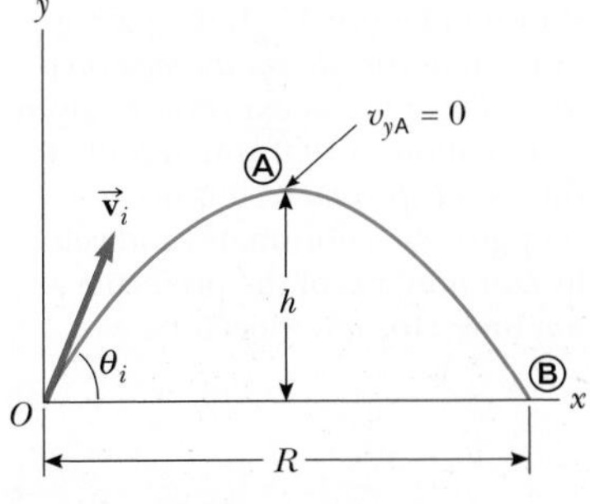

FIGURE 3.7 A projectile launched from the origin at $t = 0$ with an initial velocity $\vec{\mathbf{v}}_i$. The maximum height of the projectile is h, and its horizontal range is R. At Ⓐ, the peak of the trajectory, the projectile has coordinates $(R/2, h)$.

Substituting this expression for t_A into Equation 3.13 and replacing y_f with h gives h in terms of v_i and θ_i:

$$h = (v_i \sin \theta_i)\frac{v_i \sin \theta_i}{g} - \tfrac{1}{2}g\left(\frac{v_i \sin \theta_i}{g}\right)^2$$

$$h = \frac{v_i^2 \sin^2 \theta_i}{2g} \qquad [3.15]$$

Notice from the mathematical representation how you could increase the maximum height h: You could launch the projectile with a larger initial velocity, at a higher angle, or at a location with lower free-fall acceleration, such as on the Moon. Is that consistent with your mental representation of this situation?

The range R is the horizontal distance traveled in twice the time interval required to reach the peak. Equivalently, we are seeking the position of the projectile

ACTIVE FIGURE 3.8

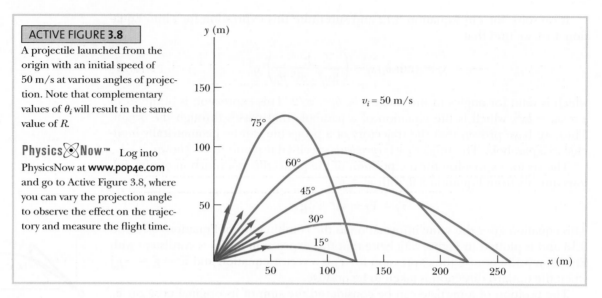

A projectile launched from the origin with an initial speed of 50 m/s at various angles of projection. Note that complementary values of θ_i will result in the same value of R.

Physics Now™ Log into PhysicsNow at **www.pop4e.com** and go to Active Figure 3.8, where you can vary the projection angle to observe the effect on the trajectory and measure the flight time.

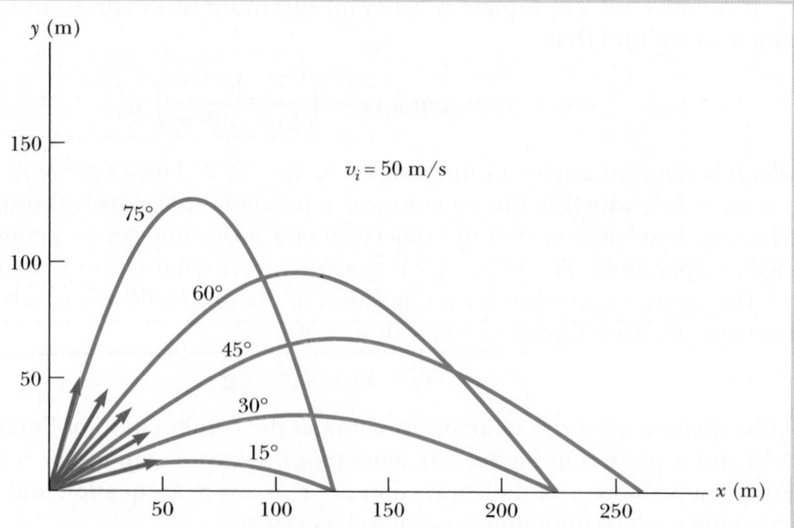

at a time $2t_A$. Using Equation 3.12 and noting that $x_f = R$ at $t = 2t_A$, we find that

$$R = (v_i \cos \theta_i)2t_A = (v_i \cos \theta_i)\frac{2v_i \sin \theta_i}{g} = \frac{2v_i^2 \sin \theta_i \cos \theta_i}{g}$$

Because $\sin 2\theta = 2 \sin \theta \cos \theta$, R can be written in the more compact form

$$R = \frac{v_i^2 \sin 2\theta_i}{g} \qquad [3.16]$$

PITFALL PREVENTION 3.2

THE HEIGHT AND RANGE EQUATIONS Keep in mind that Equations 3.15 and 3.16 are useful for calculating h and R only for a symmetric path, as shown in Figure 3.7. If the path is not symmetric, *do not use these equations*. The general expressions given by Equations 3.10 through 3.13 are the *more important* results because they give the coordinates and velocity components of the projectile at *any* time t for *any* trajectory.

Notice from the mathematical expression how you could increase the range R: You could launch the projectile with a larger initial velocity or at a location with lower free-fall acceleration, such as on the Moon. Is that consistent with your mental representation of this situation?

The range also depends on the angle of the initial velocity vector. The maximum possible value of R from Equation 3.16 is given by $R_{\max} = v_i^2/g$. This result follows from the maximum value of $\sin 2\theta_i$ being unity, which occurs when $2\theta_i = 90°$. Therefore, R is a maximum when $\theta_i = 45°$.

Active Figure 3.8 illustrates various trajectories for a projectile of a given initial speed. As you can see, the range is a maximum for $\theta_i = 45°$. In addition, for any θ_i other than 45°, a point with coordinates $(R, 0)$ can be reached by using either one of two complementary values of θ_i, such as 75° and 15°. Of course, the maximum height and the time of flight will be different for these two values of θ_i.

QUICK QUIZ 3.3 Rank the launch angles for the five paths in Active Figure 3.8 with respect to time of flight, from the shortest time of flight to the longest.

PROBLEM-SOLVING STRATEGY Projectile Motion

We suggest that you use the following approach when solving projectile motion problems:

1. Conceptualize Think about what is going on physically in the problem. Establish the mental representation by imagining the projectile moving along its trajectory.

2. Categorize Confirm that the problem involves a particle in free-fall and that air resistance is neglected. Select a coordinate system with x in the horizontal direction and y in the vertical direction.

3. Analyze If the initial velocity vector is given, resolve it into x and y components. Treat the horizontal motion and the vertical motion independently. Analyze the horizontal motion of the projectile as a particle under constant velocity. Analyze the vertical motion of the projectile as a particle under constant acceleration.

4. Finalize Once you have determined your result, check to see if your answers are consistent with the mental and pictorial representations and that your results are realistic.

■ Thinking Physics 3.1

A home run is hit in a baseball game. The ball is hit from home plate into the stands along a parabolic path. What is the acceleration of the ball (**a**) while it is rising, (**b**) at the highest point of the trajectory, and (**c**) while it is descending after reaching the highest point? Ignore air resistance.

Reasoning The answers to all three parts are the same: the acceleration is that due to gravity, $a_y = -9.80 \text{ m/s}^2$, because the gravitational force is pulling downward on the ball during the entire motion. During the rising part of the trajectory, the downward acceleration results in the decreasing positive values of the vertical component of the ball's velocity. During the falling part of the trajectory, the downward acceleration results in the increasing negative values of the vertical component of the velocity. ■

INTERACTIVE EXAMPLE 3.2 That's Quite an Arm

A stone is thrown from the top of a building at an angle of 30.0° to the horizontal and with an initial speed of 20.0 m/s, as in Figure 3.9.

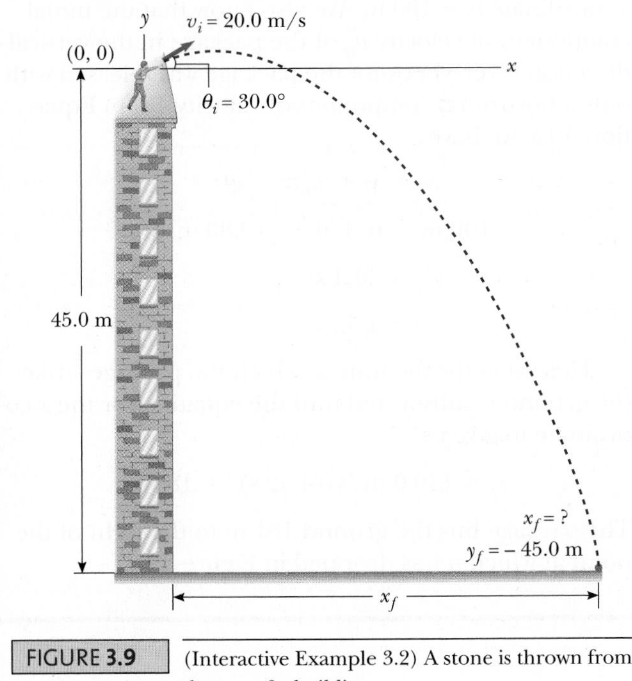

FIGURE 3.9 (Interactive Example 3.2) A stone is thrown from the top of a building.

A If the height of the building is 45.0 m, how long is the stone "in flight"?

Solution Looking at the pictorial representation in Figure 3.9, it is clear that this trajectory is not symmetric. Therefore, we cannot use Equations 3.15 and 3.16. We use the more general approach described by the Problem-Solving Strategy and represented by Equations 3.10 to 3.13.

The initial x and y components of the velocity are

$$v_{xi} = v_i \cos \theta_i = (20.0 \text{ m/s})(\cos 30.0°) = 17.3 \text{ m/s}$$

$$v_{yi} = v_i \sin \theta_i = (20.0 \text{ m/s})(\sin 30.0°) = 10.0 \text{ m/s}$$

To find t, we use the vertical motion, in which we model the stone as a particle under constant acceleration. We use Equation 3.13 with $y_f = -45.0$ m and $v_{yi} = 10.0$ m/s (we have chosen the top of the building as the origin, as in Figure 3.9):

$$y_f = y_i + v_{yi}t - \tfrac{1}{2}gt^2$$

$$-45.0 \text{ m} = 0 + (10.0 \text{ m/s})t - \tfrac{1}{2}(9.80 \text{ m/s}^2)t^2$$

Solving the quadratic equation for t gives, for the positive root, $t = 4.22$ s. Does the negative root have any physical meaning? (Can you think of another way of finding t from the information given?)

B What is the speed of the stone just before it strikes the ground?

Solution The y component of the velocity just before the stone strikes the ground can be obtained using Equation 3.11, with $t = 4.22$ s:

$$v_{yf} = v_{yi} - gt$$

$$= 10.0 \text{ m/s} - (9.80 \text{ m/s}^2)(4.22 \text{ s}) = -31.4 \text{ m/s}$$

In the horizontal direction, the appropriate model is the particle under constant velocity. Because $v_{xf} = v_{xi} = 17.3$ m/s, the speed as the stone strikes the ground is

$$v_f = \sqrt{v_{xf}^2 + v_{yf}^2}$$

$$= \sqrt{(17.3)^2 + (-31.4)^2} \text{ m/s} = 35.9 \text{ m/s}$$

PhysicsNow™ Investigate this projectile situation by logging into PhysicsNow at **www.pop4e.com** and going to Interactive Example 3.2.

EXAMPLE 3.3 **The Stranded Explorers**

An Alaskan rescue plane drops a package of emergency rations to a stranded party of explorers, as shown in the pictorial representation in Figure 3.10. If the plane is traveling horizontally at 40.0 m/s at a height of 100 m above the ground, where does the package strike the ground relative to the point at which it is released?

Solution We ignore air resistance, so we model this problem as a particle in two-dimensional free-fall, which, as we have seen, is modeled by a combination of a particle under constant velocity in the x direction and a particle under constant acceleration in the y direction. The coordinate system for this problem is selected as shown in Figure 3.10, with the positive x direction to the right and the positive y direction upward.

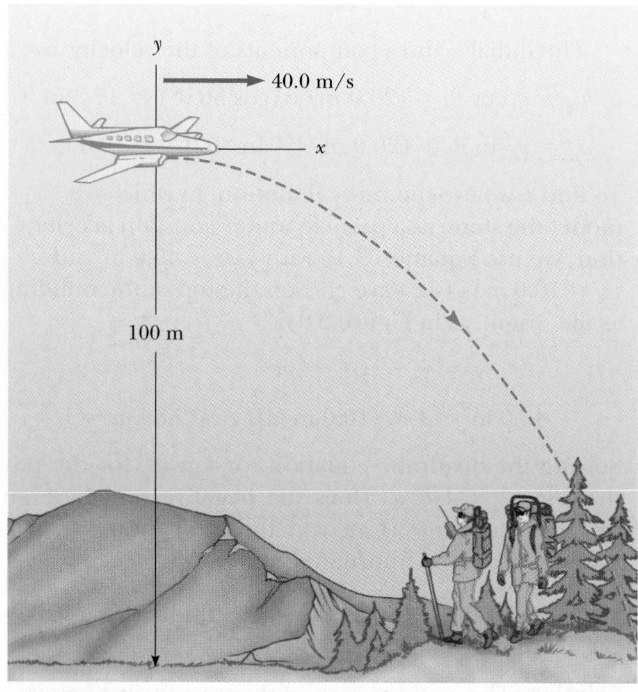

FIGURE 3.10 (Example 3.3) A package of emergency supplies is dropped from a plane to stranded explorers.

Consider first the horizontal motion of the package. From Equation 3.12, the position is given by $x_f = x_i + v_{xi}t$. The initial x component of the package velocity is the same as that of the plane when the package is released, 40.0 m/s. We define the initial position $x_i = 0$ right under the plane at the instant the package is released. Therefore,

$$x_f = x_i + v_{xi}t = 0 + (40.0 \text{ m/s})t$$

If we know t, the time at which the package strikes the ground, we can determine x_f, the final position and therefore the distance traveled by the package in the horizontal direction. At present, however, we have no information about t. To find t, we turn to the equations for the vertical motion of the package, modeling the package as a particle under constant acceleration. We know that at the instant the package hits the ground, its y coordinate is -100 m. We also know that the initial component of velocity v_{yi} of the package in the vertical direction is zero because the package was released with only a horizontal component of velocity. From Equation 3.13, we have

$$y_f = y_i + v_{yi}t - \tfrac{1}{2}gt^2$$
$$-100 \text{ m} = 0 + 0 - \tfrac{1}{2}(9.80 \text{ m/s}^2)t^2$$
$$t^2 = 20.4 \text{ s}^2$$
$$t = 4.52 \text{ s}$$

This value for the time at which the package strikes the ground is substituted into the equation for the x coordinate to give us

$$x_f = (40.0 \text{ m/s})(4.52 \text{ s}) = \boxed{181 \text{ m}}$$

The package hits the ground 181 m to the right of the point at which it was dropped in Figure 3.10.

EXAMPLE 3.4 **Javelin Throwing at the Olympics**

An athlete throws a javelin a distance of 80.0 m at the Olympics held at the equator, where $g = 9.78 \text{ m/s}^2$. Four years later the Olympics are held at the North Pole, where $g = 9.83 \text{ m/s}^2$. Assuming that the thrower provides the javelin with exactly the same initial velocity as she did at the equator, how far does the javelin travel at the North Pole?

Solution In the absence of any information about how the javelin is affected by moving through the air, we adopt the free-fall model for the javelin. Track and field events are normally held on flat fields. Therefore, we surmise that the javelin returns to the same vertical position from which it was thrown and therefore that the trajectory is symmetric. These assumptions allow us to use Equations 3.15 and 3.16 to analyze the motion. The difference in range is due to the difference in the free-fall acceleration at the two locations.

To solve this problem, we will set up a ratio based on the range of the projectile being mathematically

(Tony Duffy/Getty Images)

A javelin can be thrown over a very long distance by a world class athlete. ■

related to the acceleration due to gravity. This technique of solving by ratios is very powerful and should be studied and understood so that it can be applied in future problem solving. We use Equation 3.16 to express the range of the particle at each of the two locations:

$$R_{\text{North Pole}} = \frac{v_i^2 \sin 2\theta_i}{g_{\text{North Pole}}}$$

$$R_{\text{equator}} = \frac{v_i^2 \sin 2\theta_i}{g_{\text{equator}}}$$

We divide the first equation by the second to establish a relationship between the ratio of the ranges and the ratio of the free-fall accelerations. Because the problem states that the same initial velocity is provided to the javelin at both locations, v_i and θ_i are the same in the numerator and denominator of the ratio, which gives us

$$\frac{R_{\text{North Pole}}}{R_{\text{equator}}} = \frac{\left(\dfrac{v_i^2 \sin 2\theta_i}{g_{\text{North Pole}}}\right)}{\left(\dfrac{v_i^2 \sin 2\theta_i}{g_{\text{equator}}}\right)} = \frac{g_{\text{equator}}}{g_{\text{North Pole}}}$$

We can now solve this equation for the range at the North Pole and substitute the numerical values:

$$R_{\text{North Pole}} = \frac{g_{\text{equator}}}{g_{\text{North Pole}}} R_{\text{equator}} = \frac{9.78 \text{ m/s}^2}{9.83 \text{ m/s}^2}(80.0 \text{ m})$$

$$= 79.6 \text{ m}$$

Notice one of the advantages of this powerful technique of setting up ratios; we do not need to know the magnitude (v_i) nor the direction (θ_i) of the initial velocity. As long as they are the same at both locations, they cancel in the ratio.

3.4 THE PARTICLE IN UNIFORM CIRCULAR MOTION

Figure 3.11a shows a car moving in a circular path with *constant speed v.* Such motion is called **uniform circular motion** and serves as the basis for a new group of problems we can solve.

It is often surprising to students to find that **even though an object moves at a constant speed in a circular path, it still has an acceleration.** To see why, consider the defining equation for average acceleration, $\vec{\mathbf{a}}_{\text{avg}} = \Delta\vec{\mathbf{v}}/\Delta t$ (Eq. 3.4). The acceleration depends on *the change in the velocity vector.* Because velocity is a vector quantity, an acceleration can be produced in two ways, as mentioned in Section 3.1: by a change in the *magnitude* of the velocity or by a change in the *direction* of the velocity. The latter situation is occurring for an object moving with constant speed in a circular path. The velocity vector is always tangent to the path of the object and perpendicular to the radius of the circular path. We now show that the acceleration vector in uniform circular motion is always perpendicular to the path and always points toward the center of the circle. An acceleration of this nature is called a **centripetal acceleration** (*centripetal* means *center seeking*), and its magnitude is

$$a_c = \frac{v^2}{r} \quad [3.17]$$

■ Magnitude of centripetal acceleration

where r is the radius of the circle. The subscript on the acceleration symbol reminds us that the acceleration is centripetal.

PITFALL PREVENTION 3.3

Acceleration of a Particle in Uniform Circular Motion Many students have trouble with the notion of a particle moving in a circular path at constant speed and yet having an acceleration because the everyday interpretation of acceleration means *speeding up* or *slowing down.* Remember, though, that acceleration is defined as a change in the *velocity,* not a change in the *speed.* In circular motion, the velocity vector is changing in direction, so there is indeed an acceleration.

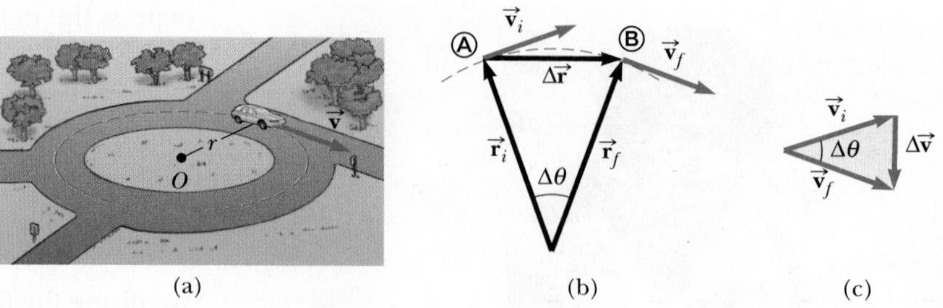

FIGURE 3.11 (a) A car moving along a circular path at constant speed is in uniform circular motion. (b) As the particle moves from Ⓐ to Ⓑ, its velocity vector changes from $\vec{v}_i$ to $\vec{v}_f$. (c) The construction for determining the direction of the change in velocity $\Delta\vec{v}$, which is toward the center of the circle for small $\Delta\theta$.

Let us first argue conceptually that the acceleration must be perpendicular to the path followed by the particle. If not, there would be a component of the acceleration parallel to the path and therefore parallel to the velocity vector. Such an acceleration component would lead to a change in the speed of the object, which we model as a particle, along the path. This change, however, is inconsistent with our setup of the problem in which the particle moves with constant speed along the path. Therefore, for *uniform* circular motion, the acceleration vector can only have a component perpendicular to the path, which is toward the center of the circle.

To derive Equation 3.17, consider the pictorial representation of the position and velocity vectors in Figure 3.11b. In addition, the figure shows the vector representing the change in position, $\Delta\vec{r}$. The particle follows a circular path, part of which is shown by the dashed curve. The particle is at Ⓐ at time t_i, and its velocity at that time is $\vec{v}_i$; it is at Ⓑ at some later time t_f, and its velocity at that time is $\vec{v}_f$. Let us also assume that $\vec{v}_i$ and $\vec{v}_f$ differ only in direction; their magnitudes are the same (i.e., $v_i = v_f = v$, because it is *uniform* circular motion). To calculate the acceleration of the particle, let us begin with the defining equation for average acceleration (Eq. 3.4):

$$\vec{a}_{avg} = \frac{\vec{v}_f - \vec{v}_i}{t_f - t_i} = \frac{\Delta\vec{v}}{\Delta t}$$

In Figure 3.11c, the velocity vectors in Figure 3.11b have been redrawn tail to tail. The vector $\Delta\vec{v}$ connects the tips of the vectors, representing the vector addition, $\vec{v}_f = \vec{v}_i + \Delta\vec{v}$. In Figures 3.11b and 3.11c, we can identify triangles that can serve as geometric models to help us analyze the motion. The angle $\Delta\theta$ between the two position vectors in Figure 3.11b is the same as the angle between the velocity vectors in Figure 3.11c because the velocity vector $\vec{v}$ is always perpendicular to the position vector $\vec{r}$. Therefore, the two triangles are *similar.* (Two triangles are similar if the angle between any two sides is the same for both triangles and if the ratio of the lengths of these sides is the same.) This similarity enables us to write a relationship between the lengths of the sides for the two triangles:

$$\frac{|\Delta\vec{v}|}{v} = \frac{|\Delta\vec{r}|}{r}$$

where $v = v_i = v_f$ and $r = r_i = r_f$. This equation can be solved for $|\Delta\vec{v}|$ and the expression so obtained can be substituted into $\vec{a}_{avg} = \Delta\vec{v}/\Delta t$ (Eq. 3.4) to give the magnitude of the average acceleration over the time interval for the particle to move from Ⓐ to Ⓑ:

$$|\vec{a}_{avg}| = \frac{v}{r}\frac{|\Delta\vec{r}|}{\Delta t}$$

Now imagine that we bring points Ⓐ and Ⓑ in Figure 3.11b very close together. As Ⓐ and Ⓑ approach each other, Δt approaches zero and the ratio $|\Delta\vec{r}|/\Delta t$

approaches the speed v. In addition, the average acceleration becomes the instantaneous acceleration at point Ⓐ. Hence, in the limit $\Delta t \rightarrow 0$, the magnitude of the acceleration is

$$a_c = \frac{v^2}{r}$$

Therefore, in uniform circular motion, the acceleration is directed inward toward the center of the circle and has magnitude v^2/r.

In many situations, it is convenient to describe the motion of a particle moving with constant speed in a circle of radius r in terms of the **period** T, which is defined as the time interval required for one complete revolution. In the time interval T, the particle moves a distance of $2\pi r$, which is equal to the circumference of the particle's circular path. Therefore, because its speed is equal to the circumference of the circular path divided by the period, or $v = 2\pi r/T$, it follows that

$$T = \frac{2\pi r}{v} \qquad [3.18]$$

■ Period of a particle in uniform circular motion

The *particle in uniform circular motion* is a very common physical situation and is useful as an analysis model for problem solving.

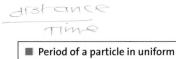

PITFALL PREVENTION 3.4

CENTRIPETAL ACCELERATION IS NOT CONSTANT We derived the magnitude of the centripetal acceleration vector and found it to be constant for uniform circular motion, but *the centripetal acceleration vector is not constant.* It always points toward the center of the circle, but it continuously changes direction as the particle moves around the circular path.

QUICK QUIZ 3.4 Which of the following correctly describes the centripetal acceleration vector for a particle moving in a circular path? **(a)** constant and always perpendicular to the velocity vector for the particle **(b)** constant and always parallel to the velocity vector for the particle **(c)** of constant magnitude and always perpendicular to the velocity vector for the particle **(d)** of constant magnitude and always parallel to the velocity vector for the particle

■ Thinking Physics 3.2

An airplane travels from Los Angeles to Sydney, Australia. After cruising altitude is reached, the instruments on the plane indicate that the ground speed holds rock-steady at 700 km/h and that the heading of the airplane does not change. Is the velocity of the airplane constant during the flight?

Reasoning The velocity is not constant because of the curvature of the Earth. Even though the speed does not change and the heading is always toward Sydney (is that actually true?), the airplane travels around a significant portion of the Earth's circumference. Therefore, the direction of the velocity vector does indeed change. We could extend this situation by imagining that the airplane passes over Sydney and continues (assuming it has enough fuel!) around the Earth until it arrives at Los Angeles again. It is impossible for an airplane to have a constant velocity (relative to the Universe, not to the Earth's surface) and return to its starting point. ■

EXAMPLE 3.5 The Centripetal Acceleration of the Earth

What is the centripetal acceleration of the Earth as it moves in its orbit around the Sun?

Solution We shall model the Earth as a particle and approximate the Earth's orbit as circular (it's actually slightly elliptical, as we discuss in Chapter 11). Although we don't know the orbital speed of the Earth, with the help of Equation 3.18 we can recast Equation 3.17 in terms of the period of the Earth's orbit, which we know is one year:

$$a_c = \frac{v^2}{r} = \frac{\left(\frac{2\pi r}{T}\right)^2}{r} = \frac{4\pi^2 r}{T^2}$$

$$= \frac{4\pi^2(1.5 \times 10^{11}\text{ m})}{(1\text{ yr})^2}\left(\frac{1\text{ yr}}{3.16 \times 10^7\text{ s}}\right)^2$$

$$= 5.9 \times 10^{-3}\text{ m/s}^2$$

Note that this small acceleration can also be expressed as $6.0 \times 10^{-4}\,g$.

3.5 TANGENTIAL AND RADIAL ACCELERATION

Let us consider the motion of a particle along a curved path where the velocity changes both in direction and in magnitude, as described in Active Figure 3.12. In this situation, the velocity vector is always tangent to the path; the acceleration vector $\vec{\mathbf{a}}$, however, is at some angle to the path. At each of three points Ⓐ, Ⓑ, and Ⓒ in Active Figure 3.12, we draw dashed circles that form geometric models of circular paths for the actual path at each point. The radius of the model circle is equal to the radius of curvature of the path at each point.

As the particle moves along the curved path in Active Figure 3.12, the direction of the total acceleration vector $\vec{\mathbf{a}}$ changes from point to point. This vector can be resolved into two components based on an origin at the center of the model circle: a radial component a_r along the radius of the model circle and a tangential component a_t perpendicular to this radius. The *total* acceleration vector $\vec{\mathbf{a}}$ can be written as the vector sum of the component vectors:

$$\vec{\mathbf{a}} = \vec{\mathbf{a}}_r + \vec{\mathbf{a}}_t \qquad [3.19]$$

The tangential acceleration arises from the change in the speed of the particle and is given by

■ Tangential acceleration

$$a_t = \frac{d|\vec{\mathbf{v}}|}{dt} \qquad [3.20]$$

The radial acceleration is a result of the change in direction of the velocity vector and is given by

■ Radial acceleration

$$a_r = -a_c = -\frac{v^2}{r}$$

where r is the radius of curvature of the path at the point in question, which is the radius of the model circle. We recognize the radial component of the acceleration as the centripetal acceleration discussed in Section 3.4. The negative sign indicates that the direction of the centripetal acceleration is toward the center of the model circle, opposite the direction of the radial unit vector $\hat{\mathbf{r}}$, which always points away from the center of the circle.

Because $\vec{\mathbf{a}}_r$ and $\vec{\mathbf{a}}_t$ are perpendicular component vectors of $\vec{\mathbf{a}}$, it follows that $a = \sqrt{a_r^2 + a_t^2}$. At a given speed, a_r is large when the radius of curvature is small (as at points Ⓐ and Ⓑ in Active Fig. 3.12) and small when r is large (such as at

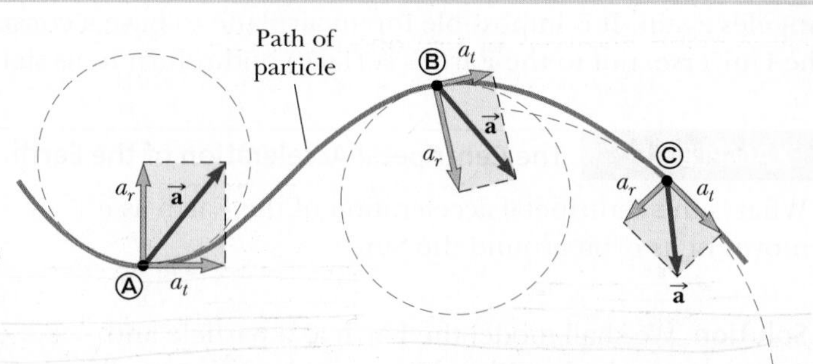

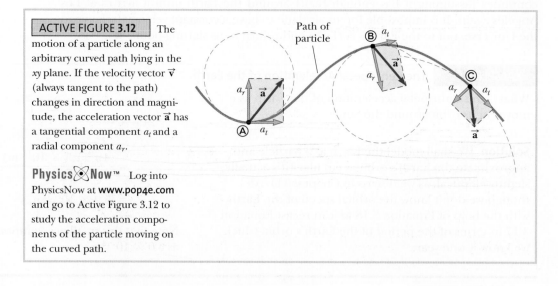

ACTIVE FIGURE 3.12 The motion of a particle along an arbitrary curved path lying in the *xy* plane. If the velocity vector $\vec{\mathbf{v}}$ (always tangent to the path) changes in direction and magnitude, the acceleration vector $\vec{\mathbf{a}}$ has a tangential component a_t and a radial component a_r.

Physics Now™ Log into PhysicsNow at **www.pop4e.com** and go to Active Figure 3.12 to study the acceleration components of the particle moving on the curved path.

point ©). The direction of $\vec{\mathbf{a}}_t$ is either in the same direction as $\vec{\mathbf{v}}$ (if v is increasing) or opposite $\vec{\mathbf{v}}$ (if v is decreasing).

In the case of uniform circular motion, where v is constant, $a_t = 0$ and the acceleration is always radial, as described in Section 3.4. In other words, uniform circular motion is a special case of motion along a curved path. Furthermore, if the direction of $\vec{\mathbf{v}}$ doesn't change, no radial acceleration occurs and the motion is one dimensional ($a_r = 0$, but a_t may not be zero).

QUICK QUIZ 3.5 A particle moves along a path and its speed increases with time. **(i)** In which of the following cases are its acceleration and velocity vectors parallel? **(a)** The path is circular. **(b)** The path is straight. **(c)** The path is a parabola. **(d)** Never. **(ii)** From the same choices, in which case are its acceleration and velocity vectors perpendicular everywhere along the path?

3.6 RELATIVE VELOCITY

In Section 1.6, we discussed the need for a fixed reference point as the origin of a coordinate system used to locate the position of a point. We have made observations of position, velocity, and acceleration of a particle with respect to this reference point. Now imagine that we have two observers making measurements of a particle located in space and that one of them moves with respect to the other at constant velocity. Each observer can define a coordinate system with an origin fixed with respect to him or her. The origins of the two coordinate systems are in motion with respect to each other. In this section, we explore how we relate the measurements of one observer to that of the other.

As an example, consider two cars, a red one and a blue one, moving on a highway in the same direction, both with speeds of 60 mi/h, as in Figure 3.13. We identify the red car as a particle to be observed, and an observer on the side of the road measures a speed for this car of 60 mi/h. Now consider an observer riding in the blue car. This observer looks out the window and sees that the red car is always in the same position with respect to the blue car. Therefore, this observer measures a speed for the red car of *zero*. This simple example demonstrates that speed measurements differ in different frames of reference. Both observers look at the same particle (the red car) and arrive at different values for its speed. Both are correct; the difference in their measurements is a result of the relative velocity of their frames of reference.

Let us now generate a mathematical representation that will allow us to calculate one observer's measurements from the other's. Consider a particle located at point P in an xy plane, as shown in Figure 3.14. Imagine that the motion of this particle is being observed by two observers. Observer O is in reference frame S. Observer O' is in reference frame S′, which moves with velocity $\vec{\mathbf{v}}_{O'O}$ with respect to S,

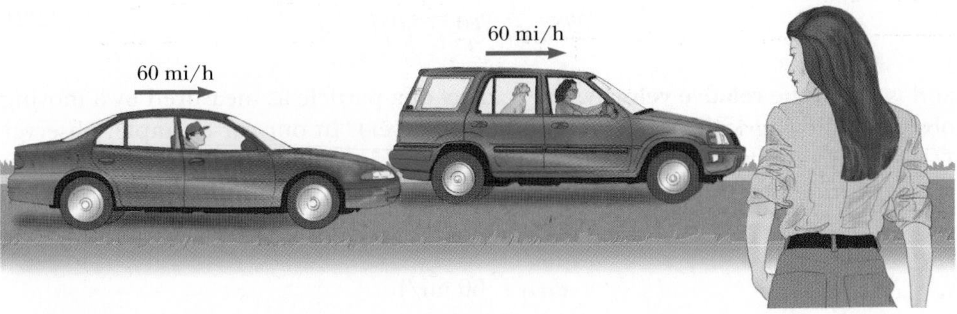

FIGURE 3.13 Two observers measure the speed of the red car. Observer O is standing on the ground beside the highway. Observer O' is in the blue car.

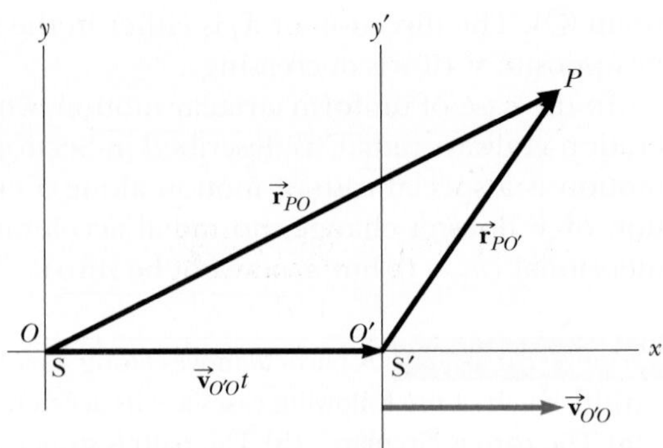

FIGURE 3.14 Position vectors for an event occurring at point P for two observers. Observer O' is moving to the right at speed $v_{O'O}$ with respect to observer O.

where the first subscript describes what is being observed and the second describes who is doing the observing. Therefore, $\vec{\mathbf{v}}_{O'O}$ is the velocity of observer O' as measured by observer O. At $t = 0$, the origins of the reference frames coincide. Therefore, when modeling the origin of S′ as a particle under constant velocity, the origins of the two reference frames are separated by a displacement $\vec{\mathbf{v}}_{O'O}t$ at time t. This displacement is shown in Figure 3.14. Also shown in the figure are the position vectors $\vec{\mathbf{r}}_{PO}$ and $\vec{\mathbf{r}}_{PO'}$ for point P from each of the two origins. They are the position vectors that the two observers would use to describe the location of point P, using the same subscript notation. From the diagram, we see that these three vectors form a vector addition triangle:

$$\vec{\mathbf{r}}_{PO} = \vec{\mathbf{r}}_{PO'} + \vec{\mathbf{v}}_{O'O}t$$

Notice the order of subscripts in this expression. The subscripts on the left side are the same as the first and last subscripts on the right. The second and third subscripts on the right are both O'. These subscripts are helpful in analyzing these types of situations. On the left, we are looking at the position vector that points directly to P from O, as described by the subscripts. On the right, the same point P is located by first going to P from O' and then describing where O' is relative to O, again as suggested by the subscripts.

Let us now differentiate this expression with respect to time to find an expression for the velocity of a particle located at point P:

$$\frac{d}{dt}(\vec{\mathbf{r}}_{PO}) = \frac{d}{dt}(\vec{\mathbf{r}}_{PO'} + \vec{\mathbf{v}}_{O'O}t) \quad \rightarrow \quad \vec{\mathbf{v}}_{PO} = \vec{\mathbf{v}}_{PO'} + \vec{\mathbf{v}}_{O'O} \qquad [3.21]$$

This expression relates the velocity of the particle as measured by O to that measured by O' and the relative velocity of the two reference frames.

In the one-dimensional case, this equation reduces to

$$v_{PO} = v_{PO'} + v_{O'O}$$

Often, this equation is expressed in terms of the observer O' as

$$v_{PO'} = v_{PO} - v_{O'O} \qquad [3.22]$$

and is called the **relative velocity,** the velocity of a particle as measured by a moving observer (moving with respect to another observer). In our car example, observer O is standing on the side of the road. Observer O' is in the blue car. Both observers are measuring the speed of the red car, which is located at point P. Therefore,

$$v_{PO} = 60 \text{ mi/h}$$

$$v_{O'O} = 60 \text{ mi/h}$$

and, using Equation 3.22,

$$v_{PO'} = v_{PO} - v_{O'O} = 60 \text{ mi/h} - 60 \text{ mi/h} = 0$$

The result of our calculation agrees with our previous intuitive discussion. This equation will be used in Chapter 9, when we discuss special relativity. We shall find that this simple expression is valid for low-speed particles but is no longer valid when the particle or observers are moving at speeds close to the speed of light.

INTERACTIVE EXAMPLE 3.6 A Boat Crossing a River

A boat heading due north crosses a wide river with a speed of 10.0 km/h relative to the water. The river has a current such that the water moves with uniform speed of 5.00 km/h due east relative to the ground.

A What is the velocity of the boat relative to a stationary observer on the side of the river?

Solution It is often useful to use subscripts other than *O* and *P* that make it easy to identify the observers and the object being observed. Observer *O* is standing on the side of the river. Because he is at rest with respect to the Earth, we will use the subscript E for this observer. Let us identify an imaginary observer O' at rest in the water, floating with the current. Because he is at rest with respect to the water, we will use the subscript w for this observer. Both observers are looking at the boat, denoted by the subscript b. We can identify the velocity of the boat relative to the water as $\vec{\mathbf{v}}_{bw} = 10.0\hat{\mathbf{j}}$ km/h. The velocity of the water relative to the Earth is that of the current in the river, $\vec{\mathbf{v}}_{wE} = 5.00\hat{\mathbf{i}}$ km/h.

We are looking for the velocity of the boat relative to the Earth, so, from Equation 3.21,

$$\vec{\mathbf{v}}_{bE} = \vec{\mathbf{v}}_{bw} + \vec{\mathbf{v}}_{wE} = (10.0\hat{\mathbf{j}} + 5.00\hat{\mathbf{i}})\ \text{km/h}$$

This vector addition is shown in Figure 3.15a. The speed of the boat relative to the observer on shore is found from the Pythagorean theorem:

$$v_{bE} = \sqrt{v_{bw}^2 + v_{wE}^2} = \sqrt{(10.0)^2 + (5.00)^2}\ \text{km/h}$$
$$= 11.2\ \text{km/h}$$

The direction of the velocity vector can be found with the inverse tangent function:

$$\theta = \tan^{-1}\left(\frac{v_{wE}}{v_{bw}}\right) = \tan^{-1}\left(\frac{5.00}{10.0}\right) = 26.6°$$

B At what angle should the boat be headed if it is to travel directly north across the river, and what is the speed of the boat relative to the Earth?

Solution We now want $\vec{\mathbf{v}}_{bE}$ to be pointed due north, as shown in Figure 3.15b. From the vector triangle,

$$\theta = \sin^{-1}\left(\frac{v_{wE}}{v_{bw}}\right) = \sin^{-1}\left(\frac{5.00\ \text{km/h}}{10.0\ \text{km/h}}\right) = 30.0°$$

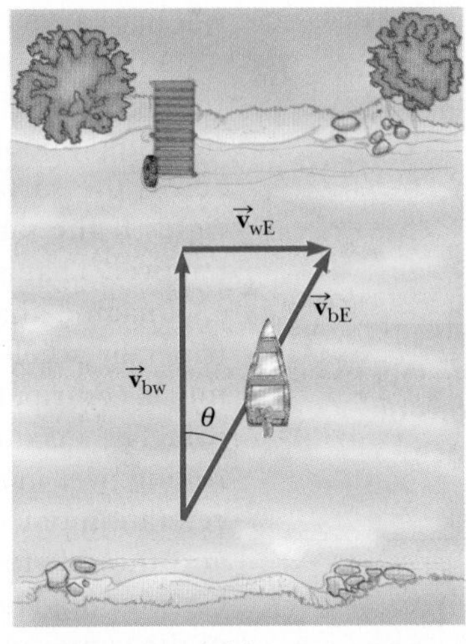

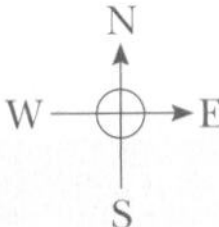

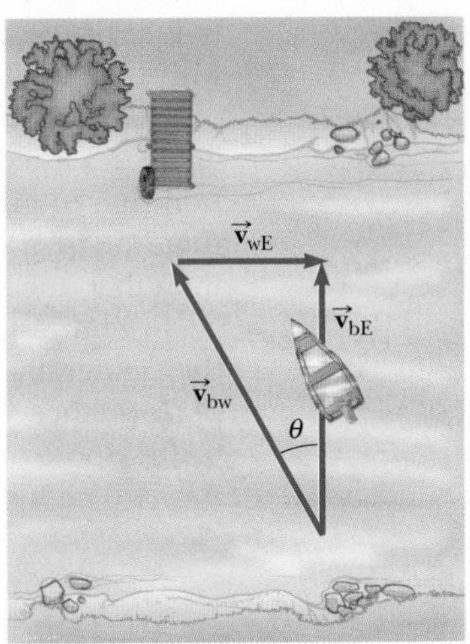

FIGURE 3.15 (Interactive Example 3.6) (a) A boat aims directly across a river and ends up downstream. (b) To move directly across the river, the boat must aim upstream.

The speed of the boat relative to the Earth is

$$v_{bE} = \sqrt{v_{bw}^2 - v_{wE}^2} = \sqrt{(10.0)^2 - (5.00)^2}\ \text{km/h}$$
$$= 8.66\ \text{km/h}$$

PhysicsNow™ Investigate the crossing of the river for various boat speeds and current speeds by logging into PhysicsNow at www.pop4e.com and going to Interactive Example 3.6.

3.7 LATERAL ACCELERATION OF AUTOMOBILES

CONTEXT CONNECTION

An automobile does not travel in a straight line. It follows a two-dimensional path on a flat Earth surface and a three-dimensional path if there are hills and valleys. Let us restrict our thinking at this point to an automobile traveling in two dimensions on a flat roadway. During a turn, the automobile can be modeled as following an arc of a circular path at each point in its motion. Consequently, the automobile will have a centripetal acceleration.

A desired characteristic of automobiles is that they can negotiate a curve without rolling over. This characteristic depends on the centripetal acceleration. Imagine standing a book upright on a strip of sandpaper. If the sandpaper is moved slowly across the surface of a table with a very small acceleration, the book will stay upright. If the sandpaper is moved with a large acceleration, however, the book will fall over. That is what we would like to avoid in a car.

Imagine that instead of accelerating a book in one dimension we are centripetally accelerating a car in a circular path. The effect is the same. If there is too much centripetal acceleration, the car will "fall over" and will go into a sideways roll. The maximum possible centripetal acceleration that a car can exhibit without rolling over in a turn is called *lateral acceleration.* Two contributions to the lateral acceleration of a car are the height of the center of mass of the car above the ground and the side-to-side distance between the wheels. (We will study center of mass in Chapter 8.) The book in our demonstration has a relatively large ratio of the height of the center of mass to the width of the book upon which it is sitting, so it falls over relatively easily at low accelerations. An automobile has a much lower ratio of the height of the center of mass to the distance between the wheels. Therefore, it can withstand higher accelerations.

Consider the documented lateral acceleration of the performance vehicles from Table 2.3 listed in Table 3.1. These values are given as multiples of g, the acceleration due to gravity. Notice that all the vehicles have a lateral acceleration close to that due to gravity and that the lateral acceleration of the Ferrari F50 is 20% larger than that due to gravity. The Ferrari is a very stable vehicle!

In contrast, the lateral acceleration of nonperformance cars is lower because they generally are not designed to travel around turns at such a high speed as the performance cars. For example, the Honda Insight has a lateral acceleration of $0.80g$. Sport utility vehicles have lateral accelerations as low as $0.62g$. As a result, they are highly prone to rollovers in emergency maneuvers. ■

TABLE 3.1

Lateral Accelerations of Various Performance Vehicles

Automobile	Lateral Acceleration
Aston Martin DB7 Vantage	$0.90g$
BMW Z8	$0.92g$
Chevrolet Corvette	$1.00g$
Dodge Viper GTS-R	$0.98g$
Ferrari F50	$1.20g$
Ferrari 360 Spider F1	$0.94g$
Lamborghini Diablo GT	$0.99g$
Porsche 911 GT2	$0.96g$

SUMMARY

Physics Now™ Take a practice test by logging into PhysicsNow at www.pop4e.com and clicking on the Pre-Test link for this chapter.

If a particle moves with *constant* acceleration $\vec{\mathbf{a}}$ and has velocity $\vec{\mathbf{v}}_i$ and position $\vec{\mathbf{r}}_i$ at $t = 0$, its velocity and position vectors at some later time t are

$$\vec{\mathbf{v}}_f = \vec{\mathbf{v}}_i + \vec{\mathbf{a}}t \qquad [3.8]$$

$$\vec{\mathbf{r}}_f = \vec{\mathbf{r}}_i + \vec{\mathbf{v}}_i t + \tfrac{1}{2}\vec{\mathbf{a}}t^2 \qquad [3.9]$$

For two-dimensional motion in the xy plane under constant acceleration, these vector expressions are equivalent to two component expressions, one for the motion along x and one for the motion along y.

Projectile motion is a special case of two-dimensional motion under constant acceleration, where $a_x = 0$ and $a_y = -g$. In this case, the horizontal components of Equations 3.8 and 3.9 reduce to those of a particle under constant velocity:

$$v_{xf} = v_{xi} = \text{constant} \qquad [3.10]$$

$$x_f = x_i + v_{xi}t \qquad [3.12]$$

The vertical components of Equations 3.8 and 3.9 are those of a particle under constant acceleration:

$$v_{yf} = v_{yi} - gt \qquad [3.11]$$

$$y_f = y_i + v_{yi}t - \tfrac{1}{2}gt^2 \qquad [3.13]$$

where $v_{xi} = v_i \cos \theta_i$, $v_{yi} = v_i \sin \theta_i$, v_i is the initial speed of the projectile, and θ_i is the angle $\vec{\mathbf{v}}_i$ makes with the positive x axis.

A particle moving in a circle of radius r with constant speed v undergoes a **centripetal acceleration** because the direction of $\vec{\mathbf{v}}$ changes in time. The magnitude of this acceleration is

$$a_c = \frac{v^2}{r} \qquad [3.17]$$

and its direction is always toward the center of the circle.

If a particle moves along a curved path in such a way that the magnitude and direction of $\vec{\mathbf{v}}$ change in time, the particle has an acceleration vector that can be described by two components: (1) a radial component $a_r = -a_c$ arising from the change in direction of $\vec{\mathbf{v}}$ and (2) a tangential component a_t arising from the change in magnitude of $\vec{\mathbf{v}}$.

If an observer O' is moving with velocity $\vec{\mathbf{v}}_{O'O}$ with respect to observer O, their measurements of the velocity of a particle located at point P are related according to

$$\vec{\mathbf{v}}_{PO} = \vec{\mathbf{v}}_{PO'} + \vec{\mathbf{v}}_{O'O} \qquad [3.21]$$

The velocity $\vec{\mathbf{v}}_{PO'}$ is called the **relative velocity,** the velocity of a particle as measured by a moving observer (moving at constant velocity with respect to another observer).

QUESTIONS

☐ = answer available in the *Student Solutions Manual and Study Guide*

1. [boxed] If you know the position vectors of a particle at two points along its path and also know the time interval it took to move from one point to the other, can you determine the particle's instantaneous velocity? Its average velocity? Explain.
2. Construct motion diagrams showing the velocity and acceleration of a projectile at several points along its path, assuming that (a) the projectile is launched horizontally and (b) the projectile is launched at an angle θ with the horizontal.
3. A baseball is thrown such that its initial x and y components of velocity are known. Ignoring air resistance, describe how you would calculate, at the instant the ball reaches the top of its trajectory, (a) its coordinates, (b) its velocity, and (c) its acceleration. How would these results change if air resistance were taken into account?
4. A ball is projected horizontally from the top of a building. One second later another ball is projected horizontally from the same point with the same velocity. At what point in the motion will the balls be closest to each other? Will the first ball always be traveling faster than the second ball? What will be the time interval between the moments when the two balls hit the ground? Can the horizontal projection velocity of the second ball be changed so that the balls arrive at the ground at the same time?
5. [boxed] A spacecraft drifts through space at a constant velocity. Suddenly a gas leak in the side of the spacecraft gives it a constant acceleration in a direction perpendicular to the initial velocity. The orientation of the spacecraft does not change, so the acceleration remains perpendicular to the original direction of the velocity. What is the shape of the path followed by the spacecraft in this situation?
6. State which of the following quantities, if any, remain constant as a projectile moves through its parabolic trajectory: (a) speed, (b) acceleration, (c) horizontal component of velocity, (d) vertical component of velocity.
7. [boxed] A projectile is launched at some angle to the horizontal with some initial speed v_i, and air resistance is negligible. Is the projectile a freely falling body? What is its acceleration in the vertical direction? What is its acceleration in the horizontal direction?
8. The maximum range of a projectile occurs when it is launched at an angle of 45.0° with the horizontal, if air resistance is ignored. If air resistance is not ignored, will the optimum angle be greater or less than 45.0°? Explain.
9. A projectile is launched on the Earth with some initial velocity. Another projectile is launched on the Moon with the same initial velocity. If air resistance can be ignored, which projectile has the greater range? Which reaches the greater altitude? (Note that the free-fall acceleration on the Moon is about 1.6 m/s^2.)
10. Correct the following statement: "The racing car rounds the turn at a constant velocity of 90 miles per hour."
11. [boxed] Explain whether or not the following particles have an acceleration: (a) a particle moving in a straight line with constant speed, (b) a particle moving around a curve with constant speed.
12. An object moves in a circular path with constant speed v. (a) Is the velocity of the object constant? (b) Is its acceleration constant? Explain.
13. Describe how a driver can steer a car traveling at constant speed so that (a) the acceleration is zero or (b) the magnitude of the acceleration remains constant.
14. An ice skater is executing a figure eight, consisting of two equal, tangent circular paths. Throughout the first loop she increases her speed uniformly, and during the second loop she moves at a constant speed. Draw a motion diagram showing her velocity and acceleration vectors at several points along the path of motion.
15. A sailor drops a wrench from the top of a sailboat's mast while the boat is moving rapidly and steadily in a straight line. Where will the wrench hit the deck? (Galileo posed this question.)
16. A ball is thrown upward in the air by a passenger on a train that is moving with constant velocity. (a) Describe the path of the ball as seen by the passenger. Describe the path as seen by an observer standing by the tracks outside the train. (b) How would these observations change if the train were accelerating along the track?

PROBLEMS

1, 2, 3 = straightforward, intermediate, challenging
☐ = full solution available in the *Student Solutions Manual and Study Guide*
PhysicsNow™ = coached problem with hints available at www.pop4e.com
= computer useful in solving problem
= paired numerical and symbolic problems
= biomedical application

Section 3.1 ■ The Position, Velocity, and Acceleration Vectors

1. PhysicsNow™ A motorist drives south at 20.0 m/s for 3.00 min, then turns west and travels at 25.0 m/s for 2.00 min, and finally travels northwest at 30.0 m/s for 1.00 min. For this 6.00-min trip, find (a) the total vector displacement, (b) the average speed, and (c) the average velocity. Let the positive x axis point east.

2. Suppose the position vector for a particle is given as a function of time by $\vec{\mathbf{r}}(t) = x(t)\hat{\mathbf{i}} + y(t)\hat{\mathbf{j}}$, with $x(t) = at + b$ and $y(t) = ct^2 + d$, where $a = 1.00$ m/s, $b = 1.00$ m, $c = 0.125$ m/s^2, and $d = 1.00$ m. (a) Calculate the average velocity during the time interval from $t = 2.00$ s to $t = 4.00$ s. (b) Determine the velocity and the speed at $t = 2.00$ s.

Section 3.2 ■ Two-Dimensional Motion with Constant Acceleration

3. A fish swimming in a horizontal plane has velocity $\vec{\mathbf{v}}_i = (4.00\hat{\mathbf{i}} + 1.00\hat{\mathbf{j}})$ m/s at a point in the ocean where the position relative to a certain rock is $\vec{\mathbf{r}}_i = (10.0\hat{\mathbf{i}} - 4.00\hat{\mathbf{j}})$ m. After the fish swims with constant acceleration for 20.0 s, its velocity is $\vec{\mathbf{v}}_f = (20.0\hat{\mathbf{i}} - 5.00\hat{\mathbf{j}})$ m/s. (a) What are the components of the acceleration? (b) What is the direction of the acceleration with respect to unit vector $\hat{\mathbf{i}}$? (c) If the fish maintains constant acceleration, where is it at $t = 25.0$ s and in what direction is it moving?

4. At $t = 0$, a particle moving in the xy plane with constant acceleration has a velocity of $\vec{\mathbf{v}}_i = (3.00\hat{\mathbf{i}} - 2.00\hat{\mathbf{j}})$ m/s and is at the origin. At $t = 3.00$ s, the particle's velocity is $\vec{\mathbf{v}}_f = (9.00\hat{\mathbf{i}} + 7.00\hat{\mathbf{j}})$ m/s. Find (a) the acceleration of the particle and (b) its coordinates at any time t.

5. A particle initially located at the origin has an acceleration of $\vec{\mathbf{a}} = 3.00\hat{\mathbf{j}}$ m/s^2 and an initial velocity of $\vec{\mathbf{v}}_i = 5.00\hat{\mathbf{i}}$ m/s. Find (a) the vector position and velocity at any time t and (b) the coordinates and speed of the particle at $t = 2.00$ s.

6. It is not possible to see very small objects, such as viruses, using an ordinary light microscope. An electron microscope, however, can view such objects using an electron beam instead of a light beam. Electron microscopy has proved invaluable for investigations of viruses, cell membranes and subcellular structures, bacterial surfaces, visual receptors, chloroplasts, and the contractile properties of muscles. The "lenses" of an electron microscope consist of electric and magnetic fields that control the electron beam. As an example of the manipulation of an electron beam, consider an electron traveling away from the origin along the x axis in the xy plane with initial velocity $\vec{\mathbf{v}}_i = v_i\hat{\mathbf{i}}$. As it passes through the region $x = 0$ to $x = d$, the electron experiences acceleration $\vec{\mathbf{a}} = a_x\hat{\mathbf{i}} + a_y\hat{\mathbf{j}}$, where a_x and a_y are constants. Taking $v_i = 1.80 \times 10^7$ m/s, $a_x = 8.00 \times 10^{14}$ m/s^2, and $a_y = 1.60 \times 10^{15}$ m/s^2, determine at $x = d = 0.010\,0$ m (a) the position of the electron, (b) the velocity of the electron, (c) the speed of the electron, and (d) the direction of travel of the electron (i.e., the angle between its velocity and the x axis).

Section 3.3 ■ Projectile Motion

Note: Ignore air resistance in all problems and take $g = 9.80$ m/s^2 at the Earth's surface.

7. PhysicsNow™ In a local bar, a customer slides an empty beer mug down the counter for a refill. The bartender is just deciding to go home and rethink his life. He does not see the mug, which slides off the counter and strikes the floor 1.40 m from the base of the counter. If the height of the counter is 0.860 m, (a) with what velocity did the mug leave the counter and (b) what was the direction of the mug's velocity just before it hit the floor?

8. In a local bar, a customer slides an empty beer mug down the counter for a refill. The bartender is momentarily distracted and does not see the mug, which slides off the counter and strikes the floor at distance d from the base of the counter. The height of the counter is h. (a) With what velocity did the mug leave the counter? (b) What was the direction of the mug's velocity just before it hit the floor?

9. Mayan kings and many school sports teams are named for the puma, cougar, or mountain lion *Felis concolor*, the best jumper among animals. It can jump to a height of 12.0 ft when leaving the ground at an angle of 45.0°. With what speed, in SI units, does it leave the ground to make this leap?

10. An astronaut on a strange planet finds that she can jump a maximum horizontal distance of 15.0 m if her initial speed is 3.00 m/s. What is the free-fall acceleration on the planet?

11. A cannon with a muzzle speed of 1 000 m/s is used to start an avalanche on a mountain slope. The target is 2 000 m from the cannon horizontally and 800 m above the cannon. At what angle, above the horizontal, should the cannon be fired?

12. A ball is tossed from an upper-story window of a building. The ball is given an initial velocity of 8.00 m/s at an angle of 20.0° below the horizontal. It strikes the ground 3.00 s later. (a) How far horizontally from the base of the building does the ball strike the ground? (b) Find the height from which the ball was thrown. (c) How long does it take the ball to reach a point 10.0 m below the level of launching?

13. The speed of a projectile when it reaches its maximum height is one half its speed when it is at half its maximum

height. What is the initial projection angle of the projectile?

14. The small archerfish (length 20 to 25 cm) lives in brackish waters of Southeast Asia from India to the Philippines. This aptly named creature captures its prey by shooting a stream of water drops at an insect, either flying or at rest. The bug falls into the water and the fish gobbles it up. The archerfish has high accuracy at distances of 1.2 m to 1.5 m, and it sometimes makes hits at distances up to 3.5 m. A groove in the roof of its mouth, along with a curled tongue, forms a tube that enables the fish to impart high velocity to the water in its mouth when it suddenly closes its gill flaps. Suppose the archerfish shoots at a target that is 2.00 m away, measured along a line at an angle of 30.0° above the horizontal. With what velocity must the water stream be launched if it is not to drop more than 3.00 cm vertically on its path to the target?

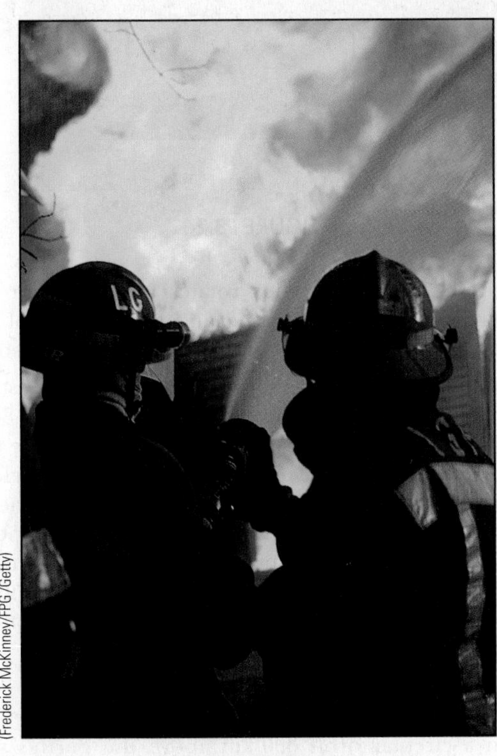

(Frederick McKinney/FPG/Getty)

FIGURE P3.16

15. **PhysicsNow™** A placekicker must kick a football from a point 36.0 m (about 40 yards) from the goal, and half the crowd hopes the ball will clear the crossbar, which is 3.05 m high. When kicked, the ball leaves the ground with a speed of 20.0 m/s at an angle of 53.0° to the horizontal. (a) By how much does the ball clear or fall short of clearing the crossbar? (b) Does the ball approach the crossbar while still rising or while falling?

16. A firefighter a distance d from a burning building directs a stream of water from a fire hose at angle θ_i above the horizontal as shown in Figure P3.16. If the initial speed of the stream is v_i, at what height h does the water strike the building?

17. A playground is on the flat roof of a city school, 6.00 m above the street below. The vertical wall of the building is 7.00 m high, forming a 1 m-high railing around the playground. A ball has fallen to the street below, and a passerby returns it by launching it at an angle of 53.0° above the horizontal at a point 24.0 m from the base of the building wall. The ball takes 2.20 s to reach a point vertically above the wall. (a) Find the speed at which the ball was launched. (b) Find the vertical distance by which the ball clears the wall. (c) Find the distance from the wall to the point on the roof where the ball lands.

18. The motion of a human body through space can be precisely modeled as the motion of a particle at the body's center of mass, as we will study in Chapter 8. The components of the displacement of an athlete's center of mass from the beginning to the end of a certain jump are described by the two equations

$$x_f = 0 + (11.2\ \text{m/s})(\cos 18.5°)t$$

$$0.360\ \text{m} = 0.840\ \text{m} + (11.2\ \text{m/s})(\sin 18.5°)t - \tfrac{1}{2}(9.80\ \text{m/s}^2)t^2$$

where t is the time at which the athlete lands after taking off at time $t = 0$. Identify (a) his position and (b) his vector velocity at the takeoff point. (c) The world long jump record is 8.95 m. How far did the athlete in this problem jump? (d) Make a sketch of the motion of his center of mass.

19. A soccer player kicks a rock horizontally off a 40.0-m-high cliff into a pool of water. If the player hears the sound of the splash 3.00 s later, what was the initial speed given to the rock? Assume that the speed of sound in air is 343 m/s.

20. A basketball star covers 2.80 m horizontally in a jump to dunk the ball (Fig. P3.20). His motion through space can be modeled precisely as that of a particle at his *center of mass*, which we will define in Chapter 8. His center of mass is at elevation 1.02 m when he leaves the floor. It reaches a maximum height of 1.85 m above the floor and is at elevation 0.900 m when he touches down again. Determine (a) his time of flight (his "hang time"), (b) his horizontal and (c) vertical velocity components at the instant of takeoff, and (d) his takeoff angle. (e) For comparison, determine the hang time of a whitetail deer making a jump with center of mass elevations $y_i = 1.20$ m, $y_{max} = 2.50$ m, and $y_f = 0.700$ m.

(Top, Jed Jacobsohn/Getty Images; bottom, Bill Lee/Dembinsky Photo Associates)

FIGURE **P3.20**

21. A fireworks rocket explodes at height h, the peak of its vertical trajectory. It throws out burning fragments in all directions, but all at the same speed v. Pellets of solidified metal fall to the ground without air resistance. Find the smallest angle that the final velocity of an impacting fragment makes with the horizontal.

Section 3.4 ■ The Particle in Uniform Circular Motion

22. From information on the endsheets of this book, compute the radial acceleration of a point on the surface of the Earth at the equator owing to the rotation of the Earth about its axis.

23. **Physics Now™** The athlete shown in Figure P3.23 rotates a 1.00-kg discus along a circular path of radius 1.06 m. The maximum speed of the discus is 20.0 m/s. Determine the magnitude of the maximum radial acceleration of the discus.

(Sam Sargent/Liaison International)

FIGURE **P3.23**

24. Casting of molten metal is important in many industrial processes. *Centrifugal casting* is used for manufacturing pipes, bearings, and many other structures. A variety of sophisticated techniques have been invented, but the basic idea is as illustrated in Figure P3.24. A cylindrical enclosure is rotated rapidly and steadily about a horizontal axis. Molten metal is poured into the rotating cylinder and then cooled, forming the finished product. Turning the cylinder at a high rotation rate forces the solidifying metal strongly to the outside. Any bubbles are displaced toward the axis, so unwanted voids will not be present in the casting. Sometimes it is desirable to form a composite casting, such as for a bearing. Here a strong steel outer surface is poured and then inside it a lining of special low-friction metal. In some applications, a very strong metal is given a coating of corrosion-resistant metal. Centrifugal casting results in strong bonding between the layers.

Suppose a copper sleeve of inner radius 2.10 cm and outer radius 2.20 cm is to be cast. To eliminate bubbles and give high structural integrity, the centripetal acceleration of each bit of metal should be at least $100g$. What rate of rotation is required? State the answer in revolutions per minute.

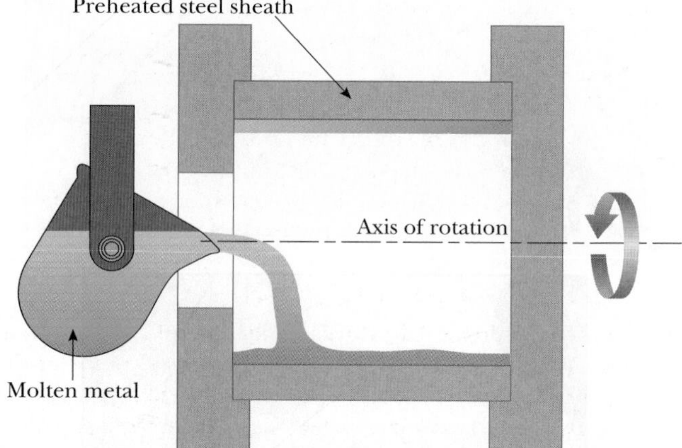

FIGURE **P3.24**

25. A tire 0.500 m in radius rotates at a constant rate of 200 rev/min. Find the speed and acceleration of a small stone lodged in the tread of the tire (on its outer edge).

26. As their booster rockets separate, Space Shuttle astronauts typically feel accelerations up to $3g$, where $g = 9.80\ \text{m/s}^2$. In their training, astronauts ride in a device where they experience such an acceleration as a centripetal acceleration. Specifically, the astronaut is fastened securely at the end of a mechanical arm, which then turns at constant speed in a horizontal circle. Determine the rotation rate, in revolutions per second, required to give an astronaut a centripetal acceleration of $3.00g$ while in circular motion with radius 9.45 m.

27. The astronaut orbiting the Earth in Figure P3.27 is preparing to dock with a Westar VI satellite. The satellite is in a circular orbit 600 km above the Earth's surface, where the free-fall acceleration is $8.21\ \text{m/s}^2$. Take the radius of the Earth as 6 400 km. Determine the speed of

the satellite and the time interval required to complete one orbit around the Earth, which is the period of the satellite.

(Courtesy of NASA)

FIGURE **P3.27**

Section 3.5 ■ Tangential and Radial Acceleration

28. A point on a rotating turntable 20.0 cm from the center accelerates from rest to a final speed of 0.700 m/s in 1.75 s. At $t = 1.25$ s, find the magnitude and direction of (a) the radial acceleration, (b) the tangential acceleration, and (c) the total acceleration of the point.

29. A train slows down as it rounds a sharp horizontal turn, slowing from 90.0 km/h to 50.0 km/h in the 15.0 s that it takes to round the bend. The radius of the curve is 150 m. Compute the acceleration at the moment the train speed reaches 50.0 km/h. Assume that it continues to slow down at this time at the same rate.

30. A ball swings in a vertical circle at the end of a rope 1.50 m long. When the ball is 36.9° past the lowest point on its way up, its total acceleration is $(-22.5\hat{\mathbf{i}} + 20.2\hat{\mathbf{j}})$ m/s^2. At that instant, (a) sketch a vector diagram showing the components of its acceleration, (b) determine the magnitude of its radial acceleration, and (c) determine the speed and velocity of the ball.

31. Figure P3.31 represents the total acceleration of a particle moving clockwise in a circle of radius 2.50 m at a certain instant of time. At this instant, find (a) the radial acceleration, (b) the speed of the particle, and (c) its tangential acceleration.

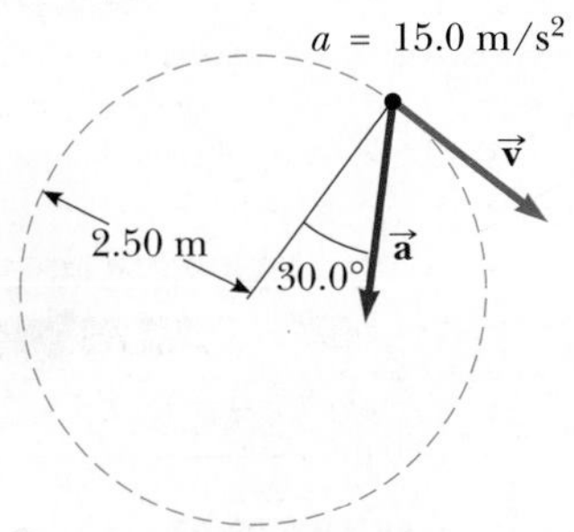

FIGURE **P3.31**

Section 3.6 ■ Relative Velocity

32. How long does it take an automobile traveling in the left lane at 60.0 km/h to pull alongside a car traveling in the right lane at 40.0 km/h if the cars' front bumpers are initially 100 m apart?

33. A river has a steady speed of 0.500 m/s. A student swims upstream a distance of 1.00 km and swims back to the starting point. If the student can swim at a speed of 1.20 m/s in still water, how long does the trip take? Compare this answer with the time interval the trip would take if the water were still.

34. A car travels due east with a speed of 50.0 km/h. Raindrops are falling at constant speed vertically with respect to the Earth. The traces of the rain on the side windows of the car make an angle of 60.0° with the vertical. Find the velocity of the rain with respect to (a) the car and (b) the Earth.

35. The pilot of an airplane notes that the compass indicates a heading due west. The airplane's speed relative to the air is 150 km/h. The air is moving in a wind at 30.0 km/h toward the north. Find the velocity of the airplane relative to the ground.

36. Two swimmers, Alan and Beth, start together at the same point on the bank of a wide stream that flows with a speed v. Both move at the same speed c ($c > v$) relative to the water. Alan swims downstream a distance L and then upstream the same distance. Beth swims so that her motion relative to the Earth is perpendicular to the banks of the stream. She swims the distance L and then back the same distance, so that both swimmers return to the starting point. Which swimmer returns first? (*Note:* First, guess the answer.)

37. A science student is riding on a flatcar of a train traveling along a straight horizontal track at a constant speed of 10.0 m/s. The student throws a ball into the air along a path that he judges to make an initial angle of 60.0° with the horizontal and to be in line with the track. The student's professor, who is standing on the ground nearby, observes the ball to rise vertically. How high does she see the ball rise?

38. A Coast Guard cutter detects an unidentified ship at a distance of 20.0 km in the direction 15.0° east of north. The ship is traveling at 26.0 km/h on a course at 40.0° east of north. The Coast Guard wishes to send a speedboat to intercept the vessel and investigate it. If the speedboat travels 50.0 km/h, in what direction should it head? Express the direction as a compass bearing with respect to due north.

Section 3.7 ■ Context Connection—Lateral Acceleration of Automobiles

39. The cornering performance of an automobile is evaluated on a skid pad, where the maximum speed a car can maintain around a circular path on a dry, flat surface is measured. Then the magnitude of the centripetal acceleration, also called the lateral acceleration, is calculated as a multiple of the free-fall acceleration g. Along with the height and width of the car, factors affecting its performance are the tire characteristics and the suspension system. A Dodge Viper GTS-R can negotiate a skid pad of radius 156 m at 139 km/h. Calculate its maximum lateral acceleration from these data to verify the corresponding entry in Table 3.1.

40. A certain light truck can go around an unbanked curve having a radius of 150 m with a maximum speed of 32.0 m/s. With what maximum speed can it go around a curve having a radius of 75.0 m?

Additional Problems

41. *The "Vomit Comet."* In zero-gravity astronaut training and equipment testing, NASA flies a KC135A aircraft along a parabolic flight path. As shown in Figure P3.41, the aircraft climbs from 24 000 ft to 31 000 ft, where it enters the zero-g parabola with a velocity of 143 m/s at 45.0° nose high and exits with velocity 143 m/s at 45.0° nose low. During this portion of the flight, the aircraft and objects inside its padded cabin are in free-fall; they have gone ballistic. The aircraft then pulls out of the dive with an upward acceleration of $0.800g$, moving in a vertical circle with radius 4.13 km. (During this portion of the flight, occupants of the plane perceive an acceleration of $1.800g$.) What are the aircraft (a) speed and (b) altitude at the top of the maneuver? (c) What is the time interval spent in zero gravity? (d) What is the speed of the aircraft at the bottom of the flight path?

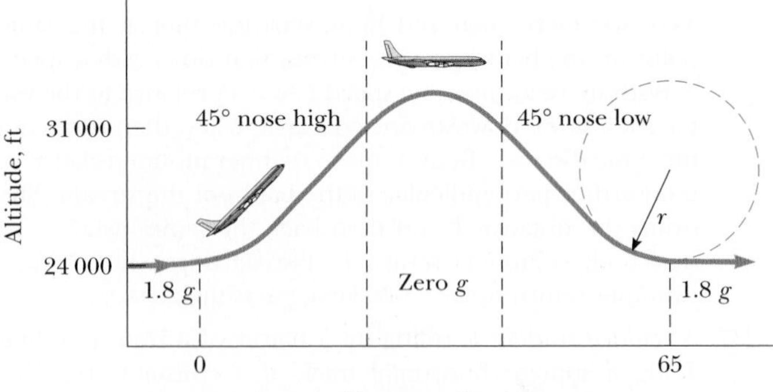

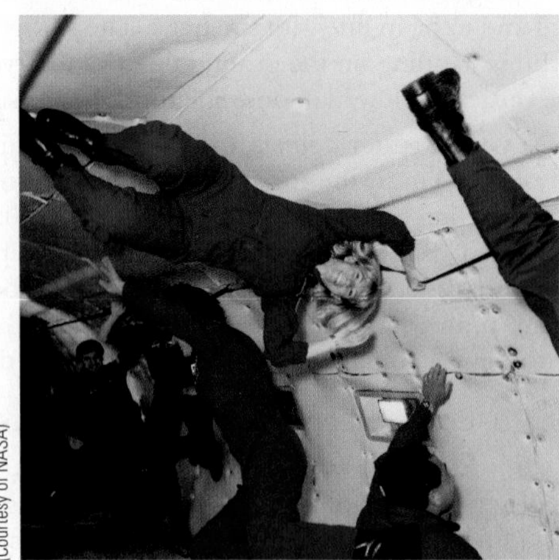

FIGURE P3.41

42. A landscape architect is planning an artificial waterfall in a city park. Water flowing at 1.70 m/s will leave the end of a horizontal channel at the top of a vertical wall 2.35 m high and from there fall into a pool. (a) Will there be a wide enough space for a walkway on which people can go behind the waterfall? (b) To sell her plan to the city council, the architect wants to build a model to standard scale, one-twelfth actual size. How fast should the water in the channel flow in the model?

43. A ball on the end of a string is whirled around in a horizontal circle of radius 0.300 m. The plane of the circle is 1.20 m above the ground. The string breaks and the ball lands 2.00 m (horizontally) away from the point on the ground directly beneath the ball's location when the string breaks. Find the radial acceleration of the ball during its circular motion.

44. A projectile is fired up an incline (incline angle ϕ) with an initial speed v_i at an angle θ_i with respect to the horizontal ($\theta_i > \phi$), as shown in Figure P3.44. (a) Show that the projectile travels a distance d up the incline, where

$$d = \frac{2v_i^2 \cos\theta_i \sin(\theta_i - \phi)}{g\cos^2\phi}$$

(b) For what value of θ_i is d a maximum, and what is that maximum value?

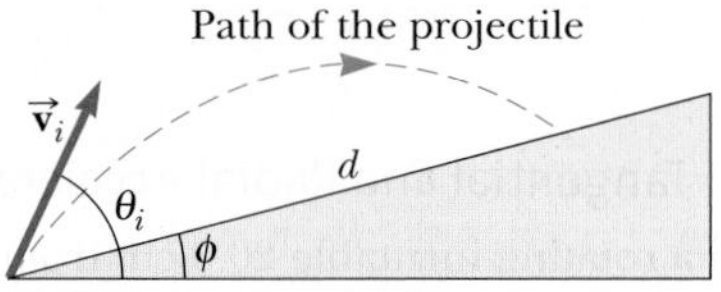

FIGURE P3.44

45. Barry Bonds hits a home run so that the baseball just clears the top row of bleachers, 21.0 m high, located 130 m from home plate. The ball is hit at an angle of 35.0° to the horizontal, and air resistance is negligible. Find (a) the initial speed of the ball, (b) the time interval that elapses before the ball reaches the top row, and (c) the velocity components and the speed of the ball when it passes over the top row. Assume that the ball is hit at a height of 1.00 m above the ground.

46. An astronaut on the surface of the Moon fires a cannon to launch an experiment package, which leaves the barrel moving horizontally. (a) What must be the muzzle speed of the probe so that it travels completely around the Moon and returns to its original location? (b) How long does this trip around the Moon take? Assume that the free-fall acceleration on the Moon is one sixth that on the Earth.

47. A basketball player who is 2.00 m tall is standing on the floor 10.0 m from the basket, as shown in Figure P3.47. If he shoots the ball at a 40.0° angle with the horizontal, at

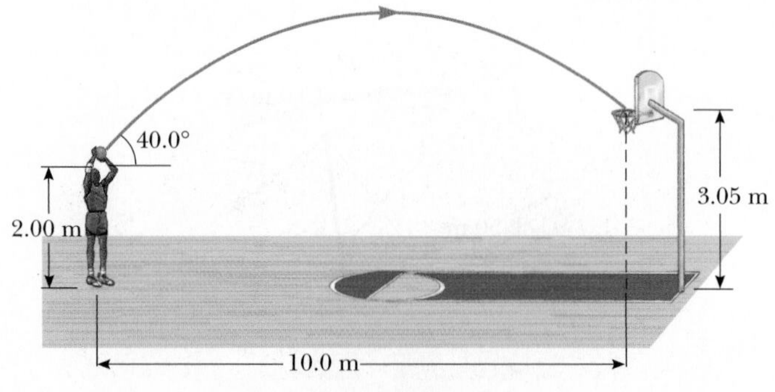

FIGURE P3.47

what initial speed must he throw so that it goes through the hoop without striking the backboard? The basket height is 3.05 m.

48. When baseball players throw the ball in from the outfield, they usually allow it to take one bounce before it reaches the infield, on the theory the ball arrives sooner that way. Suppose the angle at which a bounced ball leaves the ground is the same as the angle at which the outfielder threw it, as shown in Figure P3.48, but that the ball's speed after the bounce is one half what it was before the bounce. (a) Assume that the ball is always thrown with the same initial speed. At what angle θ should the fielder throw the ball to make it go the same distance D with one bounce (blue path) as a ball thrown upward at 45.0° with no bounce (green path)? (b) Determine the ratio of the time intervals required for the one-bounce and no-bounce throws.

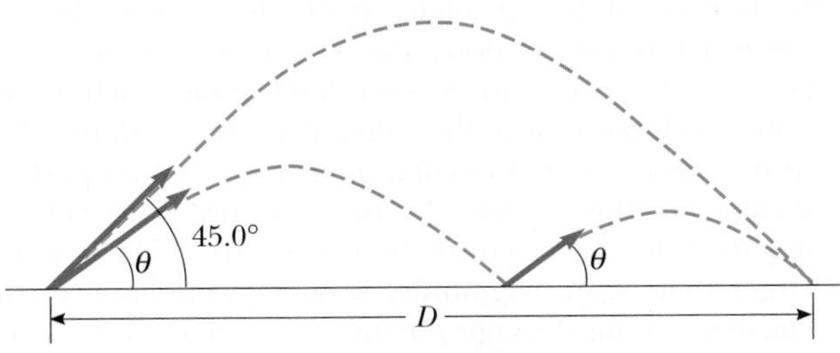

FIGURE **P3.48**

49. Your grandfather is copilot of a bomber, flying horizontally over level terrain, with a speed of 275 m/s relative to the ground at an altitude of 3 000 m. (a) The bombardier releases one bomb. How far will the bomb travel horizontally between its release and its impact on the ground? Ignore the effects of air resistance. (b) Firing from the people on the ground suddenly incapacitates the bombardier before he can call, "Bombs away!" Consequently, the pilot maintains the plane's original course, altitude, and speed through a storm of flak. Where will the plane be when the bomb hits the ground? (c) The plane has a telescopic bomb sight set so that the bomb hits the target seen in the sight at the moment of release. At what angle from the vertical was the bomb sight set?

50. A person standing at the top of a hemispherical rock of radius R kicks a ball (initially at rest on the top of the rock) to give it horizontal velocity $\vec{\mathbf{v}}_i$ as shown in Figure P3.50.

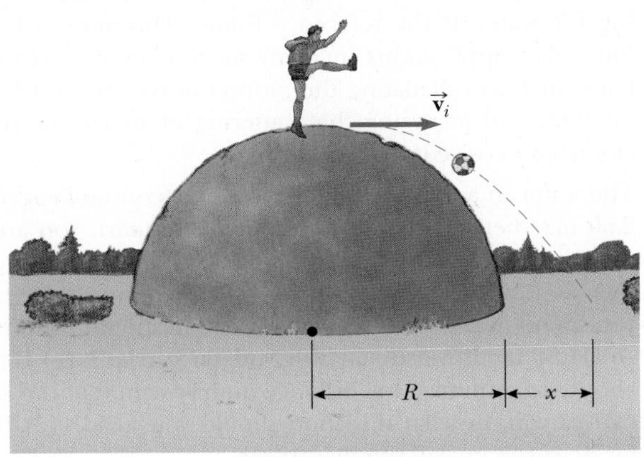

FIGURE **P3.50**

(a) What must be its minimum initial speed if the ball is never to hit the rock after it is kicked? (b) With this initial speed, how far from the base of the rock does the ball hit the ground?

51. **Physics Now™** A car is parked on a steep incline overlooking the ocean, where the incline makes an angle of 37.0° below the horizontal. The negligent driver leaves the car in neutral, and the parking brakes are defective. The car rolls from rest down the incline with a constant acceleration of 4.00 m/s^2, traveling 50.0 m to the edge of a vertical cliff. The cliff is 30.0 m above the ocean. Find (a) the speed of the car when it reaches the edge of the cliff and the time interval it takes to get there, (b) the velocity of the car when it lands in the ocean, (c) the total time interval during which the car is in motion, and (d) the position of the car when it lands in the ocean, relative to the base of the cliff.

52. A truck loaded with cannonball watermelons stops suddenly to avoid running over the edge of a washed-out bridge (Fig. P3.52). The quick stop causes a number of melons to fly off the truck. One melon rolls over the edge with an initial speed v_i = 10.0 m/s in the horizontal direction. A cross-section of the bank has the shape of the bottom half of a parabola with its vertex at the edge of the road and with the equation $y^2 = 16x$, where x and y are measured in meters. What are the x and y coordinates of the melon when it splatters on the bank?

FIGURE **P3.52**

53. The determined coyote is out once more in pursuit of the elusive roadrunner. The coyote wears a pair of Acme jet-powered roller skates, which provide a constant horizontal acceleration of 15.0 m/s^2 (Fig. P3.53). The coyote starts at rest 70.0 m from the brink of a cliff at the instant the roadrunner zips past him in the direction of the cliff. (a) The roadrunner moves with constant speed. Determine the minimum speed he must have so as to reach the cliff before the coyote. At the edge of the cliff, the roadrunner escapes by making a sudden turn, while the coyote continues straight ahead. The coyote's skates remain horizontal and continue to operate while he is in flight so that his acceleration while in the air is $(15.0\hat{\mathbf{i}} - 9.80\hat{\mathbf{j}})$ m/s^2. (b) The cliff is 100 m above the flat floor of a wide canyon. Determine

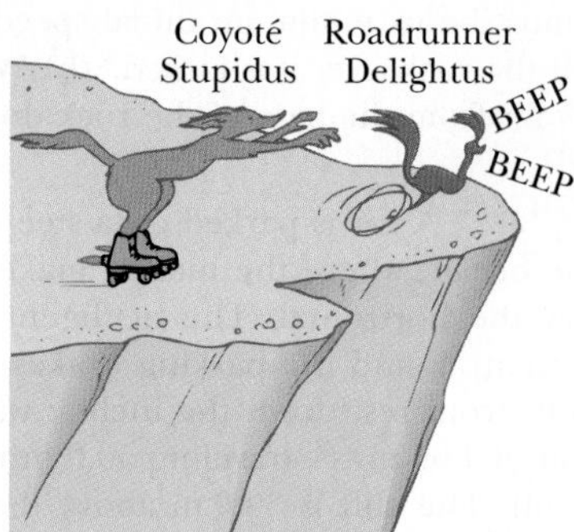

FIGURE P3.53

where the coyote lands in the canyon. (c) Determine the components of the coyote's impact velocity.

54. A ball is thrown with an initial speed v_i at an angle θ_i with the horizontal. The horizontal range of the ball is R, and the ball reaches a maximum height $R/6$. In terms of R and g, find (a) the time interval during which the ball is in motion, (b) the ball's speed at the peak of its path, (c) the initial vertical component of its velocity, (d) its initial speed, and (e) the angle θ_i. (f) Suppose the ball is thrown at the same initial speed found in (d) but at the angle appropriate for reaching the greatest height that it can. Find this height. (g) Suppose the ball is thrown at the same initial speed but at the angle for greatest possible range. Find this maximum horizontal range.

55. A catapult launches a rocket at an angle of 53.0° above the horizontal with an initial speed of 100 m/s. The rocket engine immediately starts a burn, and for 3.00 s the rocket moves along its initial line of motion with an acceleration of 30.0 m/s^2. Then its engine fails, and the rocket proceeds to move in free-fall. Find (a) the maximum altitude reached by the rocket, (b) its total time of flight, and (c) its horizontal range.

56. Do not hurt yourself; do not strike your hand against anything. Within these limitations, describe what you do to give your hand a large acceleration. Compute an order-of-magnitude estimate of this acceleration, stating the quantities you measure or estimate and their values.

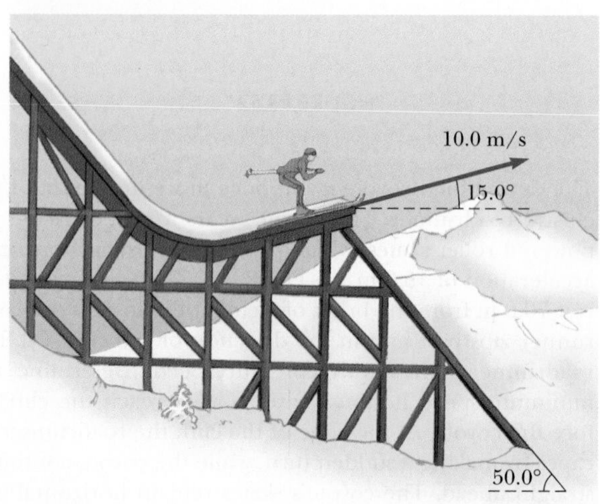

FIGURE P3.57

57. A skier leaves the ramp of a ski jump with a velocity of 10.0 m/s, 15.0° above the horizontal, as shown in Figure P3.57. The slope is inclined at 50.0°, and air resistance is negligible. Find (a) the distance from the ramp to where the jumper lands and (b) the velocity components just before the landing. (How do you think the results might be affected if air resistance were included? Note that jumpers lean forward in the shape of an airfoil, with their hands at their sides, to increase their distance. Why does this method work?)

58. In a television picture tube (a cathode-ray tube), electrons are emitted with velocity $\vec{\mathbf{v}}_i$ from a source at the origin of coordinates. The initial velocities of different electrons make different angles θ with the x axis. As they move a distance D along the x axis, the electrons are acted on by a constant electric field, giving each a constant acceleration $\vec{\mathbf{a}}$ in the x direction. At $x = D$, the electrons pass through a circular aperture, oriented perpendicular to the x axis. At the aperture, the velocity imparted to the electrons by the electric field is much larger than $\vec{\mathbf{v}}_i$ in magnitude. Show that velocities of the electrons going through the aperture radiate from a certain point on the x axis, which is not the origin. Determine the location of this point. This point is called a *virtual source,* and it is important in determining where the electron beam hits the screen of the tube.

59. An angler sets out upstream from Metaline Falls on the Pend Oreille River in northwestern Washington State. His small boat, powered by an outboard motor, travels at a constant speed v in still water. The water flows at a lower constant speed v_w. He has traveled upstream for 2.00 km when his ice chest falls out of the boat. He notices that the chest is missing only after he has gone upstream for another 15.0 minutes. At that point, he turns around and heads back downstream, all the time traveling at the same speed relative to the water. He catches up with the floating ice chest just as it is about to go over the falls at his starting point. How fast is the river flowing? Solve this problem in two ways. (a) First, use the Earth as a reference frame. With respect to the Earth, the boat travels upstream at speed $v - v_w$ and downstream at $v + v_w$. (b) A second much simpler and more elegant solution is obtained by using the water as the reference frame. This approach has important applications in many more complicated problems, such as calculating the motion of rockets and Earth satellites and analyzing the scattering of subatomic particles from massive targets.

60. The water in a river flows uniformly at a constant speed of 2.50 m/s between parallel banks 80.0 m apart. You are to deliver a package directly across the river, but you can swim only at 1.50 m/s. (a) If you choose to minimize the time you spend in the water, in what direction should you head? (b) How far downstream will you be carried? (c) If you choose to minimize the distance downstream that the river carries you, in what direction should you head? (d) How far downstream will you be carried?

61. An enemy ship is on the east side of a mountain island, as shown in Figure P3.61. The enemy ship has maneuvered to within 2 500 m of the 1 800-m-high mountain peak and can shoot projectiles with an initial speed of 250 m/s. If the western shoreline is horizontally 300 m from the peak, what are the distances from the western shore at which a ship can be safe from the bombardment of the enemy ship?

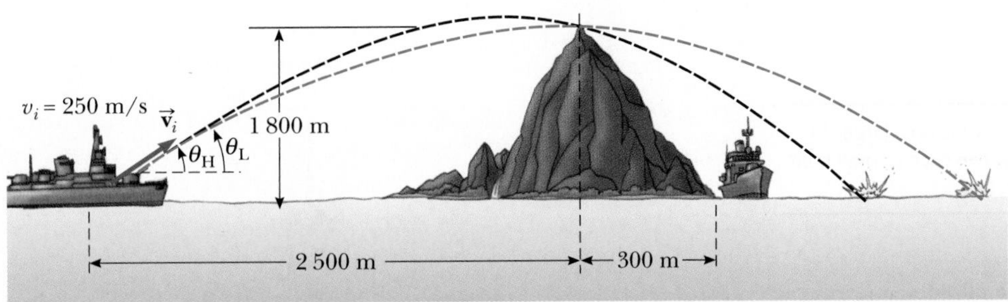

FIGURE P3.61 View looking south.

ANSWERS TO QUICK QUIZZES

3.1 (a) Because acceleration occurs whenever the velocity changes in any way—with an increase or decrease in speed, a change in direction, or both—all three controls are accelerators. The gas pedal causes the car to speed up; the brake pedal causes the car to slow down. The steering wheel changes the direction of the velocity vector.

3.2 (b), (d). At only one point—the peak of the trajectory—are the velocity and acceleration vectors perpendicular to each other. The velocity vector is horizontal at that point and the acceleration vector is downward. The acceleration vector is always directed downward. The velocity vector is never vertical if the object follows a path such as that in Figure 3.6.

3.3 15°, 30°, 45°, 60°, 75°. The greater the maximum height, the longer it takes the projectile to reach that altitude and then fall back down from it. So, as the launch angle increases, the time of flight increases.

3.4 (c). We cannot choose (a) or (b) because the centripetal acceleration vector is not constant; it continuously changes in direction. Of the remaining choices, only (c) gives the correct perpendicular relationship between $\vec{a}_c$ and $\vec{v}$.

3.5 (i), (b). The velocity vector is tangent to the path. If the acceleration vector is to be parallel to the velocity vector, it must also be tangent to the path. To be tangent requires that the acceleration vector have no component perpendicular to the path. If the path were to change direction, the acceleration vector would have a radial component, perpendicular to the path. Therefore, the path must remain straight. (ii), (d). If the acceleration vector is to be perpendicular to the velocity vector, it must have no component tangent to the path. On the other hand, if the speed is changing, there *must* be a component of the acceleration tangent to the path. Therefore, the velocity and acceleration vectors are never perpendicular in this situation. They can only be perpendicular if there is no change in the speed.

CHAPTER 4

The Laws of Motion

A small tugboat exerts a force on a large ship, causing it to move. How can such a small boat move such a large object?

(Steve Raymer/CORBIS)

CHAPTER OUTLINE

In the preceding two chapters on kinematics, we described the motion of particles based on the definitions of position, velocity, and acceleration. Aside from our discussion of gravity for objects in free-fall, we did not address what causes an object to move as it does. We would like to be able to answer general questions related to the causes of motion, such as "What mechanism causes changes in motion?" and "Why do some objects accelerate at higher rates than others?" In this first chapter on *dynamics,* we shall discuss the causes of the change in motion of particles using the concepts of force and mass. We will discuss the three fundamental laws of motion, which are based on experimental observations and were formulated about three centuries ago by Sir Isaac Newton.

4.1 THE CONCEPT OF FORCE

As a result of everyday experiences, everyone has a basic understanding of the concept of force. When you push or pull an object, you exert a force on it. You exert a force when you throw or kick a ball. In these examples, the word *force* is associated with the result of muscular activity and with some change in the state of motion of an object. Forces do not always cause an object to move, however. For example, as you sit reading this book, the gravitational force acts on your body and yet you remain stationary. You can push on a heavy block of stone and yet fail to move it.

This chapter is concerned with the relation between the force on an object and the change in motion of that object. If you pull on a spring, as in Figure 4.1a, the spring stretches. If the spring is calibrated, the distance it stretches can be used to measure the strength of the force. If a child pulls on a wagon, as in Figure 4.1b, the wagon moves. When a football is kicked, as in Figure 4.1c, it is both deformed and set in motion. These examples all show the results of a class of forces called *contact forces.* That is, these forces represent the result of physical contact between two objects.

There exist other forces that do not involve physical contact between two objects. These forces, known as *field forces,* can act through empty space. The gravitational force between two objects that causes the free-fall acceleration described in Chapters 2 and 3 is an example of this type of force and is illustrated in Figure 4.1d. This gravitational force keeps objects bound to the Earth and gives rise to what we commonly call the *weight* of an object. The planets of our solar system are bound to the Sun under the action of gravitational forces. Another common example of a field force is the electric force that one electric charge exerts on another electric

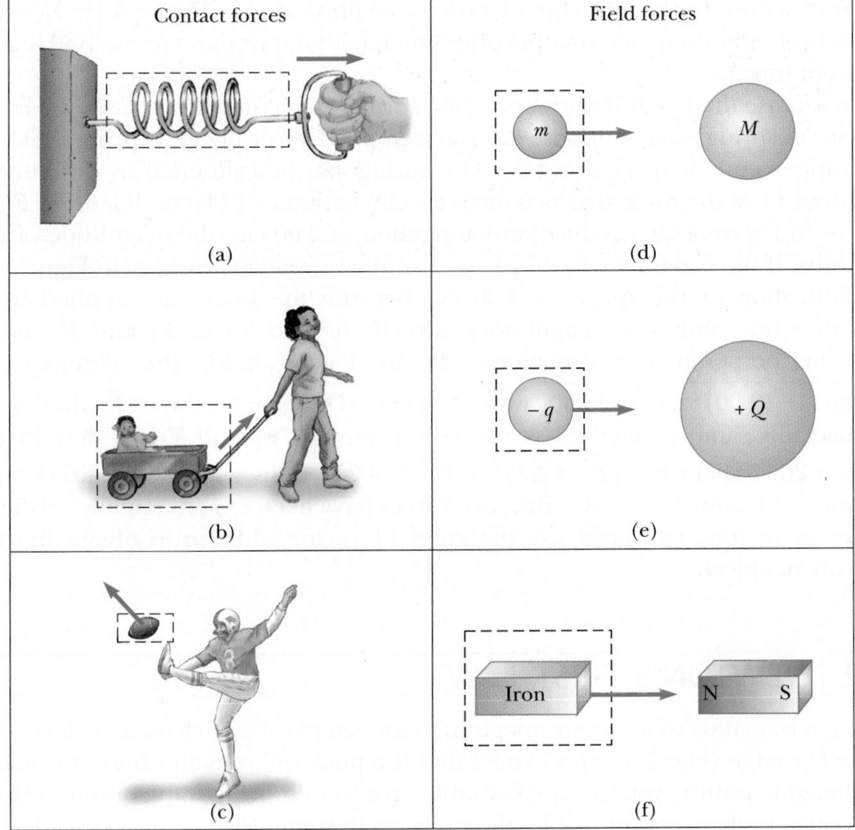

FIGURE 4.1 Some examples of forces applied to various objects. In each case, a force is exerted on the particle or object within the boxed area. The environment external to the boxed area provides this force.

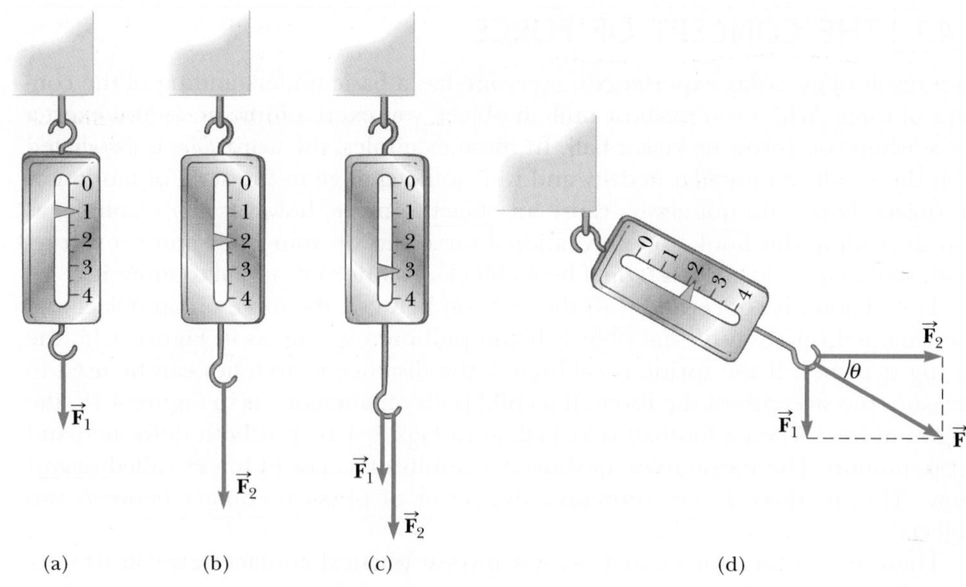

FIGURE 4.2 The vector nature of a force is tested with a spring scale. (a) A downward vertical force $\vec{F}_1$ elongates the spring 1.00 cm. (b) A downward vertical force $\vec{F}_2$ elongates the spring 2.00 cm. (c) When $\vec{F}_1$ and $\vec{F}_2$ are applied simultaneously, the spring elongates by 3.00 cm. (d) When $\vec{F}_1$ is downward and $\vec{F}_2$ is horizontal, the combination of the two forces elongates the spring $\sqrt{(1.00\text{ cm})^2 + (2.00\text{ cm})^2} = \sqrt{5.00}$ cm.

charge, as in Figure 4.1e. These charges might be an electron and proton forming a hydrogen atom. A third example of a field force is the force that a bar magnet exerts on a piece of iron, as shown in Figure 4.1f.

The distinction between contact forces and field forces is not as sharp as you may have been led to believe by the preceding discussion. At the atomic level, all the forces classified as contact forces turn out to be caused by electric (field) forces similar in nature to the attractive electric force illustrated in Figure 4.1e. Nevertheless, in understanding macroscopic phenomena, it is convenient to use both classifications of forces.

We can use the linear deformation of a spring to measure force, as in the case of a common spring scale. Suppose a force is applied vertically to a spring that has a fixed upper end, as in Figure 4.2a. The spring can be calibrated by defining the unit force $\vec{F}_1$ as the force that produces an elongation of 1.00 cm. If a force $\vec{F}_2$, applied as in Figure 4.2b, produces an elongation of 2.00 cm, the magnitude of $\vec{F}_2$ is 2.00 units. If the two forces $\vec{F}_1$ and $\vec{F}_2$ are applied simultaneously, as in Figure 4.2c, the elongation of the spring is 3.00 cm because the forces are applied in the same direction and their magnitudes add. If the two forces $\vec{F}_1$ and $\vec{F}_2$ are applied in perpendicular directions, as in Figure 4.2d, the elongation is $\sqrt{(1.00)^2 + (2.00)^2}$ cm $= \sqrt{5.00}$ cm $= 2.24$ cm. The single force $\vec{F}$ that would produce this same elongation is the vector sum of $\vec{F}_1$ and $\vec{F}_2$, as described in Figure 4.2d. That is, $|\vec{F}| = \sqrt{F_1^{\,2} + F_2^{\,2}} = 2.24$ units, and its direction is $\theta = \tan^{-1}(-0.500) = -26.6°$. **Because forces have been experimentally verified to behave as vectors, you must use the rules of vector addition to obtain the total force on an object.**

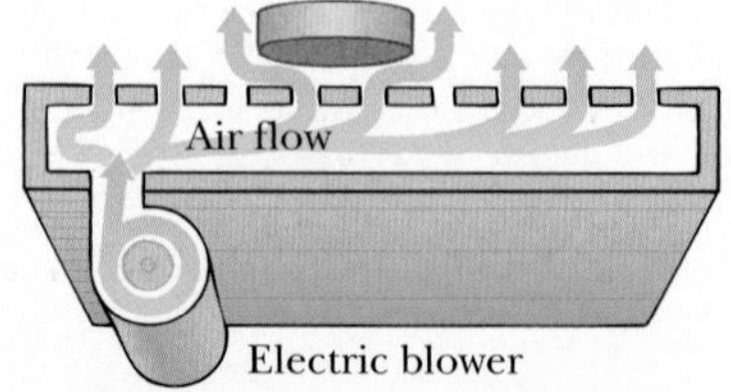

FIGURE 4.3 On an air hockey table, air blown through holes in the surface allows the puck to move almost without friction. If the table is not accelerating, a puck placed on the table will remain at rest with respect to the table if there are no horizontal forces acting on it.

4.2 NEWTON'S FIRST LAW

We begin our study of forces by imagining that you place a puck on a perfectly level air hockey table (Fig. 4.3). You expect that the puck will remain where it is placed. Now imagine putting your air hockey table on a train moving with constant velocity. If the puck is placed on the table, the puck again remains where it is placed. If the train were to accelerate, however, the puck would start moving along the table, just as a set of papers on your dashboard falls onto the front seat of your car when you step on the gas.

As we saw in Section 3.6, a moving object can be observed from any number of reference frames. **Newton's first law of motion,** sometimes called the *law of inertia,* defines a special set of reference frames called *inertial frames.* This law can be stated as follows:

■ Newton's first law

> If an object does not interact with other objects, it is possible to identify a reference frame in which the object has zero acceleration.

■ Inertial frame of reference

Such a reference frame is called an **inertial frame of reference.** When the puck is on the air hockey table located on the ground, you are observing it from an inertial reference frame; there are no horizontal interactions of the puck with any other objects, and you observe it to have zero acceleration in the horizontal direction. When you are on the train moving at constant velocity, you are also observing the puck from an inertial reference frame. **Any reference frame that moves with constant velocity relative to an inertial frame is itself an inertial frame.** When the train accelerates, however, you are observing the puck from a **noninertial reference frame** because you and the train are accelerating relative to the inertial reference frame of the surface of the Earth. Although the puck appears to be accelerating according to your observations, we can identify a reference frame in which the puck has zero acceleration. For example, an observer standing outside the train on the ground sees the puck moving with the same velocity as the train had before it started to accelerate (because there is almost no friction to "tie" the puck and the train together). Therefore, Newton's first law is still satisfied even though your observations say otherwise.

(Giraudon/Art Resource)

Isaac Newton (1642–1727)
Newton, an English physicist and mathematician, was one of the most brilliant scientists in history. Before the age of 30, he formulated the basic concepts and laws of mechanics, discovered the law of universal gravitation, and invented the mathematical methods of calculus. As a consequence of his theories, Newton was able to explain the motions of the planets, the ebb and flow of the tides, and many special features of the motions of the Moon and the Earth. His contributions to physical theories dominated scientific thought for two centuries and remain important today.

A reference frame that moves with constant velocity relative to the distant stars is the best approximation of an inertial frame, and for our purposes we can consider the Earth as being such a frame. The Earth is not really an inertial frame because of its orbital motion around the Sun and its rotational motion about its own axis, both of which are related to centripetal accelerations. These accelerations, however, are small compared with g and can often be neglected. (This is a simplification model.) For this reason, we assume that the Earth is an inertial frame, as is any other frame attached to it.

Let us assume that we are observing an object from an inertial reference frame. Before about 1600, scientists believed that the natural state of matter was the state of rest. Observations showed that moving objects eventually stopped moving. Galileo was the first to take a different approach to motion and the natural state of matter. He devised thought experiments and concluded that it is not the nature of an object to stop once set in motion; rather, it is its nature to *resist changes in its motion.* In his words, "Any velocity once imparted to a moving body will be rigidly maintained as long as the external causes of retardation are removed."

Given our assumption of observations made from inertial reference frames, we can pose a more practical statement of Newton's first law of motion:

■ Another statement of Newton's first law

> In the absence of external forces, when viewed from an inertial reference frame, an object at rest remains at rest and an object in motion continues in motion with a constant velocity (that is, with a constant speed in a straight line).

In simpler terms, we can say that **when no force acts on an object, the acceleration of the object is zero.** If nothing acts to change the object's motion, its velocity does not change. From the first law, we conclude that any *isolated object* (one that does not interact with its environment) is either at rest or moving with constant

PITFALL PREVENTION 4.1

NEWTON'S FIRST LAW Newton's first law does *not* say what happens for an object with *zero net force,* that is, multiple forces that cancel; it says what happens *in the absence of a force.* This subtle but important difference allows us to define force as that which causes a change in the motion. The description of an object under the effect of forces that balance is covered by Newton's second law.

velocity. The tendency of an object to resist any attempt to change its velocity is called **inertia.**

Consider a spacecraft traveling in space, far removed from any planets or other matter. The spacecraft requires some propulsion system to change its velocity. If the propulsion system is turned off when the spacecraft reaches a velocity $\vec{\mathbf{v}}$, however, the spacecraft "coasts" in space with that velocity and the astronauts enjoy a "free ride" (i.e., no propulsion system is required to keep them moving at the velocity $\vec{\mathbf{v}}$).

Finally, recall our discussion in Chapter 2 about the proportionality between force and acceleration:

$$\vec{\mathbf{F}} \propto \vec{\mathbf{a}}$$

Newton's first law tells us that the velocity of an object remains constant if no force acts on an object; the object maintains its state of motion. The preceding proportionality tells us that if a force *does* act, a change does occur in the motion, measured by the acceleration. This notion will form the basis of Newton's second law, and we shall provide more details on this concept shortly.

QUICK QUIZ 4.1 Which of the following statements is most correct? **(a)** It is possible for an object to have motion in the absence of forces on the object. **(b)** It is possible to have forces on an object in the absence of motion of the object. **(c)** Neither (a) nor (b) is correct. **(d)** Both (a) and (b) are correct.

4.3 MASS

Imagine playing catch with either a basketball or a bowling ball. Which ball is more likely to keep moving when you try to catch it? Which ball has the greater tendency to remain motionless when you try to throw it? The bowling ball is more resistant to changes in its velocity than the basketball. How can we quantify this concept?

▮ Definition of mass

Mass is that property of an object that specifies how much resistance an object exhibits to changes in its velocity, and as we learned in Section 1.1, the SI unit of mass is the kilogram. The greater the mass of an object, the less that object accelerates under the action of a given applied force.

To describe mass quantitatively, we begin by experimentally comparing the accelerations a given force produces on different objects. Suppose a force acting on an object of mass m_1 produces an acceleration $\vec{\mathbf{a}}_1$ and the *same force* acting on an object of mass m_2 produces an acceleration $\vec{\mathbf{a}}_2$. The ratio of the two masses is defined as the *inverse* ratio of the magnitudes of the accelerations produced by the force:

$$\frac{m_1}{m_2} \equiv \frac{a_2}{a_1} \qquad [4.1]$$

For example, if a given force acting on a 3-kg object produces an acceleration of 4 m/s^2, the same force applied to a 6-kg object produces an acceleration of 2 m/s^2. If one object has a known mass, the mass of the other object can be obtained from acceleration measurements.

Mass is an inherent property of an object and is independent of the object's surroundings and of the method used to measure it. Also, **mass is a scalar quantity** and therefore obeys the rules of ordinary arithmetic. That is, several masses can be combined in simple numerical fashion. For example, if you combine a 3-kg mass with a 5-kg mass, their total mass is 8 kg. We can verify this result experimentally by comparing the acceleration that a known force gives to several objects separately with the acceleration that the same force gives to the same objects combined as one unit.

Mass should not be confused with weight. **Mass and weight are two different quantities.** As we shall see later in this chapter, the weight of an object is equal to the magnitude of the gravitational force exerted on the object and varies with location. For example, a person who weighs 180 lb on the Earth weighs only about 30 lb on the Moon. On the other hand, the mass of an object is the same everywhere. An object having a mass of 2 kg on Earth also has a mass of 2 kg on the Moon.

■ Mass and weight are different quantities

4.4 NEWTON'S SECOND LAW—THE PARTICLE UNDER A NET FORCE

Newton's first law explains what happens to an object when no force acts on it: It either remains at rest or moves in a straight line with constant speed. This law allows us to define an inertial frame of reference. It also allows us to identify force as that which changes motion. Newton's second law answers the question of what happens to an object that has a nonzero net force acting on it, based on our discussion of mass in the preceding section.

Imagine you are pushing a block of ice across a frictionless horizontal surface. When you exert some horizontal force $\vec{\mathbf{F}}$, the block moves with some acceleration $\vec{\mathbf{a}}$. Experiments show that if you apply a force twice as large to the same object, the acceleration doubles. If you increase the applied force to $3\vec{\mathbf{F}}$, the original acceleration is tripled, and so on. From such observations, we conclude that **the acceleration of an object is directly proportional to the net force acting on it.** We alluded to this proportionality in our discussion of acceleration in Chapter 2. We are now ready to extend that discussion.

These observations and those in Section 4.3 relating mass and acceleration are summarized in **Newton's second law:**

> The acceleration of an object is directly proportional to the net force acting on it and inversely proportional to its mass.

■ Newton's second law

We write this law as

$$\vec{\mathbf{a}} \propto \frac{\sum \vec{\mathbf{F}}}{m}$$

where $\Sigma\vec{\mathbf{F}}$ is the **net force,** which is the vector sum of *all* forces acting on the object of mass m. If the object consists of a system of individual elements, the net force is the vector sum of all forces *external* to the system. Any *internal* forces—that is, forces between elements of the system—are not included because they do not affect the motion of the entire system. The net force is sometimes called the *resultant force,* the *sum of the forces,* the *total force,* or the *unbalanced force.*

Newton's second law in mathematical form is a statement of this relationship that makes the preceding proportionality an equality:[1]

$$\sum \vec{\mathbf{F}} = m\vec{\mathbf{a}} \qquad [4.2]$$

■ Mathematical representation of Newton's second law

Note that Equation 4.2 is a *vector* expression and hence is equivalent to the following three component equations:

$$\sum F_x = ma_x \qquad \sum F_y = ma_y \qquad \sum F_z = ma_z \qquad [4.3]$$

■ Newton's second law in component form

Newton's second law introduces us to a new analysis model, the particle under a net force. If a particle, or an object that can be modeled as a particle, is under the

PITFALL PREVENTION 4.2

FORCE IS THE CAUSE OF CHANGES IN MOTION Be sure that you are clear on the role of force. Many times, students make the mistake of thinking that force is the cause of motion. We can, though, have motion in the absence of forces, as described in Newton's first law. Be sure to understand that force is the cause of *changes* in motion.

PITFALL PREVENTION 4.3

$m\vec{\mathbf{a}}$ IS NOT A FORCE Equation 4.2 does *not* say that the product $m\vec{\mathbf{a}}$ is a force. All forces on an object are added vectorially to generate the net force on the left side of the equation. This net force is then equated to the product of the mass of the object and the acceleration that results from the net force. Do *not* include an "$m\vec{\mathbf{a}}$ force" in your analysis.

[1] Equation 4.2 is valid only when the speed of the object is much less than the speed of light. We will treat the relativistic situation in Chapter 9.

influence of a net force, Equation 4.2, the mathematical statement of Newton's second law, can be used to describe its motion. The acceleration is constant if the net force is constant. Therefore, the particle under a constant net force will have its motion described as a particle under constant acceleration. Of course, not all forces are constant, and when they are not, the particle cannot be modeled as one under constant acceleration. We shall investigate situations in this chapter and the next involving both constant and varying forces.

QUICK QUIZ 4.2 An object experiences no acceleration. Which of the following *cannot* be true for the object? **(a)** A single force acts on the object. **(b)** No forces act on the object. **(c)** Forces act on the object, but the forces cancel.

QUICK QUIZ 4.3 You push an object, initially at rest, across a frictionless floor with a constant force for a time interval Δt, resulting in a final speed of v for the object. You repeat the experiment, but with a force that is twice as large. What time interval is now required to reach the same final speed v? (a) $4\,\Delta t$ (b) $2\,\Delta t$ (c) Δt (d) $\Delta t/2$ (e) $\Delta t/4$

Unit of Force

The SI unit of force is the **newton,** which is defined as the force that, when acting on a 1-kg mass, produces an acceleration of 1 m/s^2.

From this definition and Newton's second law, we see that the newton can be expressed in terms of the fundamental units of mass, length, and time:

■ Definition of the newton

$$1\ \text{N} \equiv 1\ \text{kg}\cdot\text{m/s}^2 \qquad [4.4]$$

The units of mass, acceleration, and force are summarized in Table 4.1. Most of the calculations we shall make in our study of mechanics will be in SI units. Equalities between units in the SI and U.S. customary systems are given in Appendix A.

■ Thinking Physics 4.1

In a train, the cars are connected by *couplers.* The couplers between the cars exert forces on the cars as the train is pulled by the locomotive in the front. Imagine that the train is speeding up in the forward direction. As you imagine moving from the locomotive to the last car, does the force exerted by the couplers *increase, decrease,* or *stay the same*? What if the engineer applies the brakes? How does the force vary from locomotive to last car in this case? (Assume that the only brakes applied are those on the engine.)

Reasoning The force *decreases* from the front of the train to the back. The coupler between the locomotive and the first car must apply enough force to accelerate all the remaining cars. As we move back along the train, each coupler is accelerating less mass behind it. The last coupler only has to accelerate the last car, so it exerts the smallest force. If the brakes are applied, the force decreases from front to back of the train also. The first coupler, at the back of the locomotive, must apply a large force to slow down all the remaining cars. The final coupler must only apply a force large enough to slow down the mass of the last car. ■

TABLE 4.1 Units of Mass, Acceleration, and Force

System of Units	Mass (M)	Acceleration (L/T^2)	Force (ML/T^2)
SI	kg	m/s^2	N = kg · m/s^2
U.S. customary	slug	ft/s^2	lb = slug · ft/s^2

EXAMPLE 4.1 An Accelerating Hockey Puck

A 0.30-kg hockey puck slides on the horizontal frictionless surface of an ice rink. It is struck simultaneously by two different hockey sticks. The two constant forces that act on the puck as a result of the hockey sticks are parallel to the ice surface and are shown in the pictorial representation in Figure 4.4. The force $\vec{\mathbf{F}}_1$ has a magnitude of 5.0 N, and $\vec{\mathbf{F}}_2$ has a magnitude of 8.0 N. Determine the acceleration of the puck while it is in contact with the two sticks.

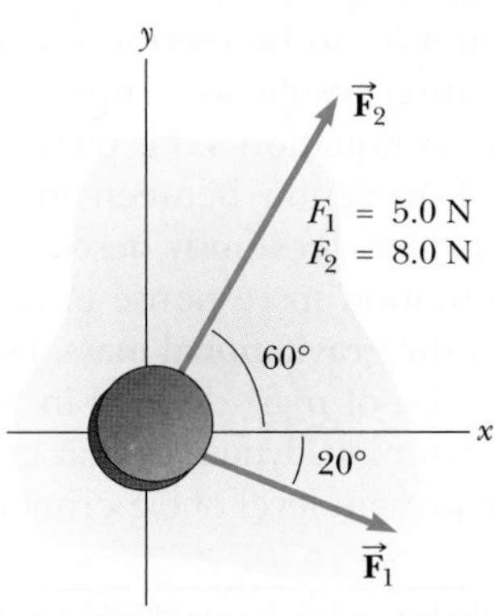

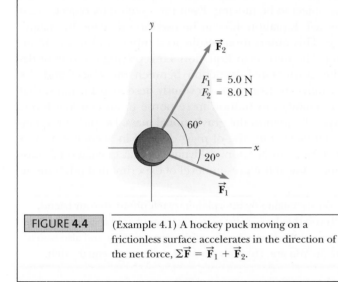

FIGURE 4.4 (Example 4.1) A hockey puck moving on a frictionless surface accelerates in the direction of the net force, $\Sigma\vec{\mathbf{F}} = \vec{\mathbf{F}}_1 + \vec{\mathbf{F}}_2$.

Solution The puck is modeled as a particle under a net force. We first find the components of the net force. The component of the net force in the x direction is

$$\sum F_x = F_{1x} + F_{2x} = F_1 \cos 20° + F_2 \cos 60°$$
$$= (5.0\text{ N})(0.940) + (8.0\text{ N})(0.500) = 8.7\text{ N}$$

The component of the net force in the y direction is

$$\sum F_y = F_{1y} + F_{2y} = -F_1 \sin 20° + F_2 \sin 60°$$
$$= -(5.0\text{ N})(0.342) + (8.0\text{ N})(0.866) = 5.2\text{ N}$$

Now we use Newton's second law in component form to find the x and y components of acceleration:

$$a_x = \frac{\sum F_x}{m} = \frac{8.7\text{ N}}{0.30\text{ kg}} = 29\text{ m/s}^2$$

$$a_y = \frac{\sum F_y}{m} = \frac{5.2\text{ N}}{0.30\text{ kg}} = 17\text{ m/s}^2$$

The acceleration has a magnitude of

$$a = \sqrt{(29\text{ m/s}^2)^2 + (17\text{ m/s}^2)^2} = 34\text{ m/s}^2$$

and its direction is

$$\theta = \tan^{-1}\frac{a_y}{a_x} = \tan^{-1}\frac{17\text{ m/s}^2}{29\text{ m/s}^2} = 30°$$

relative to the positive x axis.

4.5 THE GRAVITATIONAL FORCE AND WEIGHT

We are well aware that all objects are attracted to the Earth. The force exerted by the Earth on an object is the **gravitational force** $\vec{\mathbf{F}}_g$. This force is directed toward the center of the Earth.[2] The magnitude of the gravitational force is called the **weight** F_g of the object.

We have seen in Chapters 2 and 3 that a freely falling object experiences an acceleration $\vec{\mathbf{g}}$ directed toward the center of the Earth. A freely falling object has only one force on it, the gravitational force, so the net force on the object in this situation is equal to the gravitational force:

$$\sum \vec{\mathbf{F}} = \vec{\mathbf{F}}_g$$

Because the acceleration of a freely falling object is equal to the free-fall acceleration $\vec{\mathbf{g}}$, it follows that

$$\sum \vec{\mathbf{F}} = m\vec{\mathbf{a}} \rightarrow \vec{\mathbf{F}}_g = m\vec{\mathbf{g}}$$

or, in magnitude,

$$F_g = mg \qquad [4.5]$$

■ Relation between mass and weight of an object

[2] This statement represents a simplification model in that it ignores that the mass distribution of the Earth is not perfectly spherical.

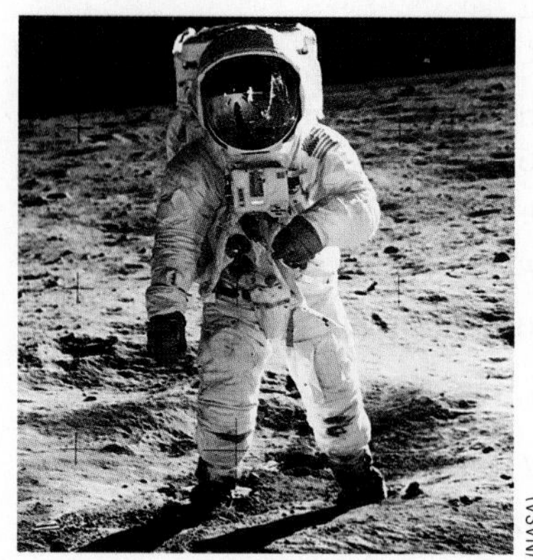
(NASA)

Astronaut Edwin E. Aldrin Jr., walking on the Moon after the *Apollo 11* lunar landing. Aldrin's weight on the Moon is less than it is on the Earth, but his mass is the same in both places. ■

Because it depends on g, weight varies with location, as we mentioned in Section 4.3. Objects weigh less at higher altitudes than at sea level because g decreases with increasing distance from the center of the Earth. Hence, weight, unlike mass, is not an inherent property of an object. For example, if an object has a mass of 70 kg, its weight in a location where $g = 9.80\ \text{m/s}^2$ is $mg = 686$ N. At the top of a mountain where $g = 9.76\ \text{m/s}^2$, the object's weight would be 683 N. Therefore, if you want to lose weight without going on a diet, climb a mountain or weigh yourself at 30 000 ft during an airplane flight.

Because $F_g = mg$, we can compare the masses of two objects by measuring their weights with a spring scale. At a given location (so that g is fixed) the ratio of the weights of two objects equals the ratio of their masses.

Equation 4.5 quantifies the gravitational force on the object, but notice that this equation does not require the object to be moving. Even for a stationary object, or an object on which several forces act, Equation 4.5 can be used to calculate the magnitude of the gravitational force. This observation results in a subtle shift in the interpretation of m in the equation. The mass m in Equation 4.5 is playing the role of determining the strength of the gravitational attraction between the object and the Earth. This role is completely different from that previously described for mass, that of measuring the resistance to changes in motion in response to an external force. Therefore, we call m in this type of equation the **gravitational mass.** Despite this quantity being different from inertial mass (the type of mass defined in Section 4.3), it is one of the experimental conclusions in Newtonian dynamics that gravitational mass and inertial mass have the same value at the present level of experimental refinement.

PITFALL PREVENTION 4.4

DIFFERENTIATE BETWEEN g AND g Be sure not to confuse the italicized letter g that we use for the magnitude of the free-fall acceleration with the abbreviation g that is used for grams.

QUICK QUIZ 4.4 Suppose you are talking by interplanetary telephone to your friend, who lives on the Moon. He tells you that he has just won a newton of gold in a contest. Excitedly, you tell him that you entered the Earth version of the same contest and also won a newton of gold! Who is richer? (**a**) You are. (**b**) Your friend is. (**c**) You are equally rich.

4.6 NEWTON'S THIRD LAW

Newton's third law conveys the notion that forces are always interactions between two objects: **If two objects interact, the force $\vec{\mathbf{F}}_{12}$ exerted by object 1 on object 2 is equal in magnitude but opposite in direction to the force $\vec{\mathbf{F}}_{21}$ exerted by object 2 on object 1:**

■ Newton's third law

$$\vec{\mathbf{F}}_{12} = -\vec{\mathbf{F}}_{21} \qquad [4.6]$$

When it is important to designate forces as interactions between two objects, we will use this subscript notation, where $\vec{\mathbf{F}}_{ab}$ means "the force exerted *by* a *on* b." The third law, illustrated in Figure 4.5a, is equivalent to stating that **forces always occur in pairs** or that **a single isolated force cannot exist.** The force that object 1 exerts on object 2 may be called the *action force,* and the force of object 2 on object 1 may be called the *reaction force.* In reality, either force can be labeled the action or reaction force. **The action force is equal in magnitude to the reaction force and opposite in direction. In all cases, the action and reaction forces act on different objects and must be of the same type.** For example, the force acting on a freely falling projectile is the gravitational force exerted by the Earth on the projectile $\vec{\mathbf{F}}_g = \vec{\mathbf{F}}_{Ep}$ (E = Earth, p = projectile), and the magnitude of this force is mg. The reaction to this force is the gravitational force exerted by the projectile on the Earth $\vec{\mathbf{F}}_{pE} = -\vec{\mathbf{F}}_{Ep}$. The reaction force $\vec{\mathbf{F}}_{pE}$ must accelerate the Earth toward the projectile just as the action force $\vec{\mathbf{F}}_{Ep}$ accelerates the projectile toward the Earth. Because the Earth has such a large mass, however, its acceleration as a result of this reaction force is negligibly small.

PITFALL PREVENTION 4.5

NEWTON'S THIRD LAW Newton's third law is such an important and often misunderstood notion that it is repeated here in a Pitfall Prevention. In Newton's third law, action and reaction forces act on *different* objects. Two forces acting on the same object, even if they are equal in magnitude and opposite in direction, *cannot* be an action–reaction pair.

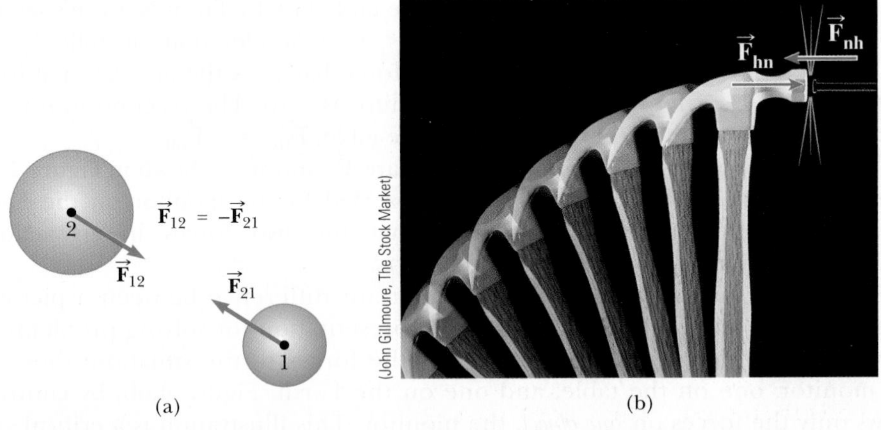

FIGURE 4.5 Newton's third law. (a) The force $\vec{F}_{12}$ exerted by object 1 on object 2 is equal in magnitude and opposite in direction to the force $\vec{F}_{21}$ exerted by object 2 on object 1. (b) The force $\vec{F}_{hn}$ exerted by the hammer on the nail is equal in magnitude and opposite in direction to the force $\vec{F}_{nh}$ exerted by the nail on the hammer.

Another example of Newton's third law in action is shown in Figure 4.5b. The force $\vec{F}_{hn}$ exerted by the hammer on the nail (the action) is equal in magnitude and opposite the force $\vec{F}_{nh}$ exerted by the nail on the hammer (the reaction). This latter force stops the forward motion of the hammer when it strikes the nail.

The Earth exerts a gravitational force $\vec{F}_g$ on any object. If the object is a computer monitor at rest on a table, as in the pictorial representation in Figure 4.6a, the reaction force to $\vec{F}_g = \vec{F}_{Em}$ is the force exerted by the monitor on the Earth $\vec{F}_{mE} = -\vec{F}_{Em}$. The monitor does not accelerate because it is held up by the table. The table exerts on the monitor an upward force $\vec{n} = \vec{F}_{tm}$, called the **normal force.**[3] This force prevents the monitor from falling through the table; it can have

■ Normal force

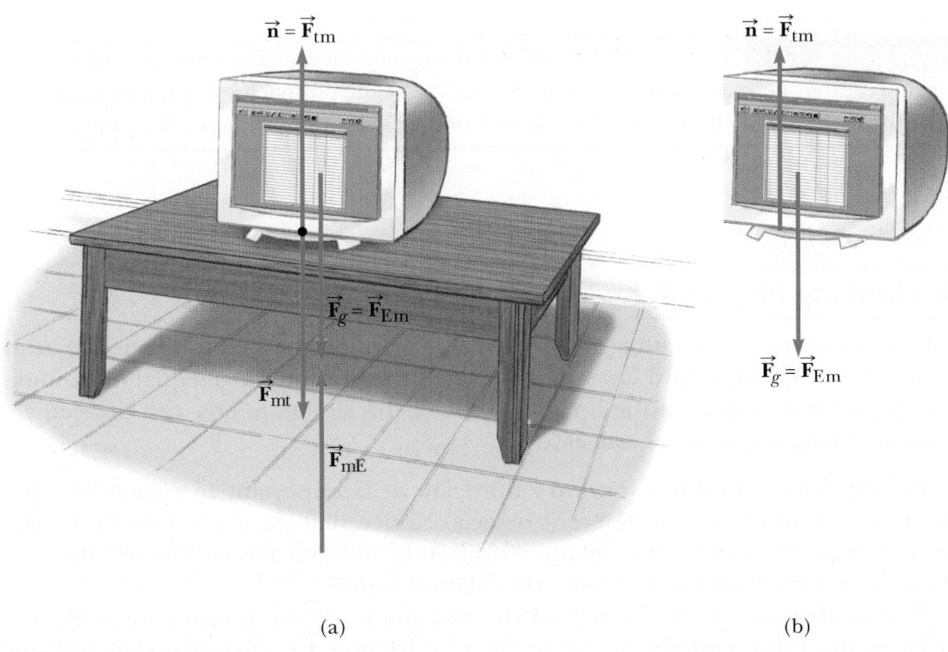

FIGURE 4.6 (a) When a computer monitor is sitting on a table, several forces are acting. (b) The free-body diagram for the monitor. The forces acting on the monitor are the normal force $\vec{n} = \vec{F}_{tm}$ and the gravitational force $\vec{F}_g = \vec{F}_{Em}$.

[3] The word *normal* is used because the direction of $\vec{n}$ is always *perpendicular* to the surface.

any value needed, up to the point at which the table breaks. From Newton's second law we see that, because the monitor has zero acceleration, it follows that $\Sigma\vec{\mathbf{F}} = \vec{\mathbf{n}} - m\vec{\mathbf{g}} = 0$, or $n = mg$. The normal force balances the gravitational force on the monitor, so the net force on the monitor is zero. The reaction to **n** is the force exerted by the monitor downward on the table, $\vec{\mathbf{F}}_{\text{mt}} = -\vec{\mathbf{F}}_{\text{tm}}$.

Note that the forces acting on the monitor are $\vec{\mathbf{F}}_g$ and $\vec{\mathbf{n}}$, as shown in Figure 4.6b. The two reaction forces $\vec{\mathbf{F}}_{\text{mE}}$ and $\vec{\mathbf{F}}_{\text{mt}}$ are exerted by the monitor on the Earth and the table, respectively. Remember that the two forces in an action–reaction pair always act on two different objects.

Figure 4.6 illustrates an extremely important difference between a pictorial representation and a simplified pictorial representation for solving problems involving forces. Figure 4.6a shows many of the forces in the situation: those on the monitor, one on the table, and one on the Earth. Figure 4.6b, by contrast, shows only the forces on *one object,* the monitor. This illustration is a critical simplified pictorial representation called a **free-body diagram.** When analyzing a particle under a net force, we are interested in the net force on one object, an object of mass m, which we will model as a particle. Therefore, a free-body diagram helps us isolate only those forces on the object and eliminate the other forces from our analysis. The free-body diagram can be simplified further, if you wish, by representing the object, such as the monitor in this case, as a particle by simply drawing a dot.

PITFALL PREVENTION 4.6

n DOES NOT ALWAYS EQUAL mg In the situation shown in Figure 4.6, we find that $n = mg$. There are many situations in which the normal force has the same magnitude as the gravitational force, but *do not* adopt this equality as a general rule (a common student pitfall). If the problem involves an object on an incline, if there are applied forces with vertical components, or if there is a vertical acceleration of the system, $n \neq mg$. *Always* apply Newton's second law to find the relationship between n and mg.

PITFALL PREVENTION 4.7

FREE-BODY DIAGRAMS The *most important* step in solving a problem using Newton's laws is to draw a proper simplified pictorial representation, the free-body diagram. Be sure to draw only those forces that act on the object you are isolating. Be sure to draw *all* forces acting on the object, including any field forces, such as the gravitational force. Do not include velocity, position, or acceleration vectors. Do not include a vector for the net force or for $m\vec{\mathbf{a}}$.

QUICK QUIZ 4.5 If a fly collides with the windshield of a fast-moving bus, which experiences an impact force with a larger magnitude? **(a)** The fly does. **(b)** The bus does. **(c)** The same force is experienced by both. Which experiences the greater acceleration? **(d)** The fly does. **(e)** The bus does. **(f)** The same acceleration is experienced by both.

QUICK QUIZ 4.6 Which of the following is the reaction force to the gravitational force acting on your body as you sit in your desk chair? **(a)** the normal force from the chair **(b)** the force you apply downward on the seat of the chair **(c)** neither of these forces

■ Thinking Physics 4.2

A horse pulls on a sled with a horizontal force, causing the sled to accelerate as in Figure 4.7a. Newton's third law says that the sled exerts a force of equal magnitude and opposite direction on the horse. In view of this situation, how can the sled accelerate? Don't these forces cancel?

Reasoning When applying Newton's third law, it is important to remember that the forces involved act on different objects. Notice that the force exerted by the horse acts *on the sled,* whereas the force exerted by the sled acts *on the horse.* Because these forces act on different objects, they cannot cancel.

The horizontal forces exerted on the *sled* alone are the forward force $\vec{\mathbf{F}}_{\text{hs}}$ exerted by the horse and the backward force of friction $\vec{\mathbf{f}}_{\text{sled}}$ between sled and surface (Fig. 4.7b). When $\vec{\mathbf{F}}_{\text{hs}}$ exceeds $\vec{\mathbf{f}}_{\text{sled}}$, the sled accelerates to the right.

The horizontal forces exerted on the *horse* alone are the forward friction force $\vec{\mathbf{f}}_{\text{horse}}$ from the ground and the backward force $\vec{\mathbf{F}}_{\text{sh}}$ exerted by the sled (Fig. 4.7c). The resultant of these two forces causes the horse to accelerate. When $\vec{\mathbf{f}}_{\text{horse}}$ exceeds $\vec{\mathbf{F}}_{\text{sh}}$, the horse accelerates to the right. ■

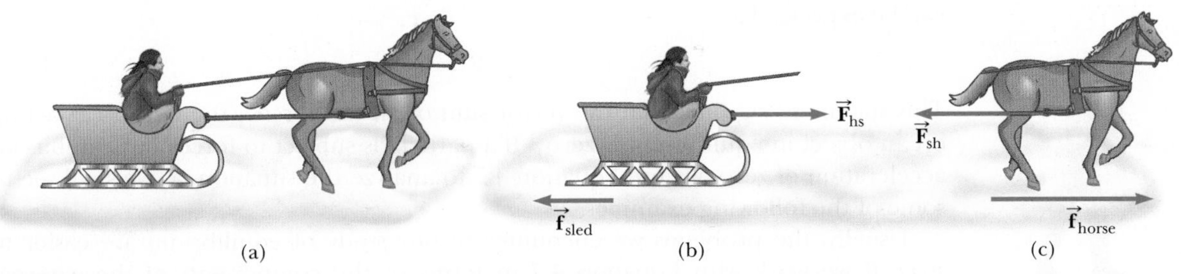

FIGURE 4.7 (Thinking Physics 4.2) (a) A horse pulls a sled through the snow. (b) The forces on the sled. (c) The forces on the horse.

4.7 APPLICATIONS OF NEWTON'S LAWS

In this section, we present some simple applications of Newton's laws to objects that are either in equilibrium ($\vec{a} = 0$) or are accelerating under the action of constant external forces. We shall assume that the objects behave as particles so that we need not worry about rotational motion or other complications. In this section, we also apply some additional simplification models. We ignore the effects of friction for those problems involving motion, which is equivalent to stating that the surfaces are *frictionless.* We usually ignore the masses of any ropes or strings involved. In this approximation, the magnitude of the force exerted at any point along a string is the same. In problem statements, the terms *light* and *of negligible mass* are used to indicate that a mass is to be ignored when you work the problem. These two terms are synonymous in this context.

When an object such as a block is being pulled by a rope or string attached to it, the rope exerts a force $\vec{T}$ on the object. Its direction is along the rope, away from the object. The magnitude T of this force is called the **tension** in the rope.

▮ Tension

Consider a crate being pulled to the right on a frictionless, horizontal surface, as in Figure 4.8a. Suppose you are asked to find the acceleration of the crate and the force the floor exerts on it. Note that the horizontal force $\vec{T}$ being applied to the crate acts through the rope.

Because we are interested only in the motion of the crate, we must be able to *identify any and all external forces acting on it.* These forces are illustrated in the free-body diagram in Figure 4.8b. In addition to the force $\vec{T}$, the free-body diagram for the crate includes the gravitational force $\vec{F}_g$ and the normal force **n** exerted by the floor on the crate. The *reactions* to the forces we have listed—namely, the force exerted by the crate on the rope, the force exerted by the crate on the Earth, and the force exerted by the crate on the floor—are not included in the free-body diagram because they act on *other* objects and not on the crate.

Now let us apply Newton's second law to the crate. First, we must choose an appropriate coordinate system. In this case, it is convenient to use the coordinate system shown in Figure 4.8b, with the x axis horizontal and the y axis vertical. We can apply Newton's second law in the x direction, y direction, or both, depending on what we are asked to find in the problem. In addition, we may be able to use the equations of motion for the particle under constant acceleration that we discussed in Chapter 2. You should use these equations only when the acceleration is constant, however, which is the case if the net force is constant. For example, if the force $\vec{T}$ in Figure 4.8 is constant, the acceleration in the x direction is also constant because $\vec{a} = \vec{T}/m$.

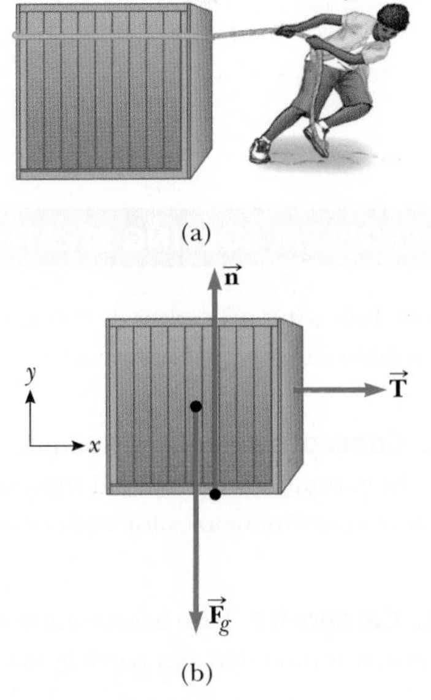

FIGURE 4.8 (a) A crate being pulled to the right on a frictionless surface. (b) The free-body diagram that represents the external forces on the crate.

The Particle in Equilibrium

Objects that are either at rest or moving with constant velocity are said to be in **equilibrium.** From Newton's second law with $\vec{a} = 0$, this condition of equilibrium can be expressed as

$$\sum \vec{\mathbf{F}} = 0 \qquad [4.7]$$

This statement signifies that the vector sum of all the forces (the net force) acting on an object in equilibrium is zero.[4] If a particle is subject to forces but exhibits an acceleration of zero, we use Equation 4.7 to analyze the situation, as we shall see in some of the following examples.

Usually, the problems we encounter in our study of equilibrium are easier to solve if we work with Equation 4.7 in terms of the components of the external forces acting on an object. In other words, in a two-dimensional problem, the sum of all the external forces in the x and y directions must separately equal zero; that is,

$$\sum F_x = 0 \qquad \sum F_y = 0 \qquad [4.8]$$

The extension of Equations 4.8 to a three-dimensional situation can be made by adding a third component equation, $\Sigma F_z = 0$.

In a given situation, we may have balanced forces on an object in one direction but unbalanced forces in the other. Therefore, for a given problem, we may need to model the object as a particle in equilibrium for one component and a particle under a net force for the other.

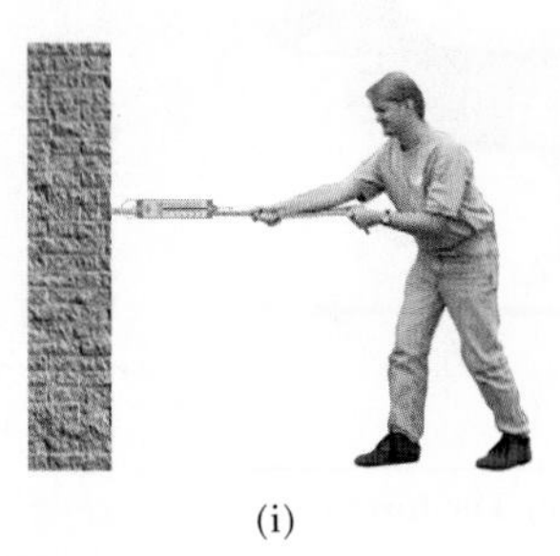

(i)

(ii)

FIGURE 4.9 (Quick Quiz 4.7) (i) An individual pulls with a force of magnitude F on a spring scale attached to a wall. (ii) Two individuals pull with forces of magnitude F in opposite directions on a spring scale attached between two ropes.

QUICK QUIZ 4.7 Consider the two situations shown in Figure 4.9, in which no acceleration occurs. In both cases, all individuals pull with a force of magnitude F on a rope attached to a spring scale. Is the reading on the spring scale in part (i) of the figure **(a)** greater than, **(b)** less than, or **(c)** equal to the reading in part (ii)?

The Particle Under a Net Force

In a situation in which a nonzero net force is acting on an object, the object is accelerating. We use Newton's second law to determine the features of the motion:

$$\sum \vec{\mathbf{F}} = m\vec{\mathbf{a}}$$

In practice, this equation is broken into components so that two (or three) equations can be handled independently. The representative suggestions and problems that follow should help you solve problems of this kind.

PROBLEM-SOLVING STRATEGY Applying Newton's Laws

The following procedure is recommended when dealing with problems involving Newton's law.

1. Conceptualize Draw a simple, neat diagram of the system to help establish the mental representation. Establish convenient coordinate axes for each object in the system.

2. Categorize If an acceleration component for an object is zero, it is modeled as a particle in equilibrium in this direction and $\Sigma F = 0$. If not, the object is modeled as a particle under a net force in this direction and $\Sigma F = ma$.

3. Analyze Isolate the object whose motion is being analyzed. Draw a free-body diagram for this object. For systems containing more than one object, draw *separate* free-body diagrams for each object. *Do not* include in the free-body diagram forces exerted by the object on its surroundings.

Find the components of the forces along the coordinate axes. Apply Newton's second law, $\Sigma\vec{\mathbf{F}} = m\vec{\mathbf{a}}$, in component

[4] This statement is only one condition of equilibrium for an object. An object that can be modeled as a particle moving through space is said to be in translational motion. If the object is spinning, it is said to be in rotational motion. A second condition of equilibrium is a statement of rotational equilibrium. This condition will be discussed in Chapter 10 when we discuss spinning objects. Equation 4.7 is sufficient for analyzing objects in translational motion, which are those of interest to us at this point.

form. Check your dimensions to make sure all terms have units of force.

Solve the component equations for the unknowns. Remember that to obtain a complete solution, you must have as many independent equations as you have unknowns.

4. Finalize Make sure your results are consistent with the free-body diagram. Also check the predictions of your solutions for extreme values of the variables. By doing so, you can often detect errors in your results.

We now embark on a series of examples that demonstrate how to solve problems involving a particle in equilibrium or a particle under a net force. You should read and study these examples very carefully.

EXAMPLE 4.2 A Traffic Light at Rest

A traffic light weighing 122 N hangs from a cable tied to two other cables fastened to a support, as in Figure 4.10a. The upper cables make angles of 37.0° and 53.0° with the horizontal. These upper cables are not as strong as the vertical cable and will break if the tension in them exceeds 100 N. Does the traffic light remain in this situation, or will one of the cables break?

Solution Let us assume that the cables do not break, so no acceleration of any sort occurs in any direction. Therefore, we use the model of a particle in equilibrium for both x and y components for any part of the system. We shall construct two free-body diagrams. The first is for the traffic light, shown in Figure 4.10b; the second is for the knot that holds the three cables together, as in Figure 4.10c. The knot is a convenient point to choose because all the forces in which we are interested act through this point. Because the acceleration of the system is zero, we can use the equilibrium conditions that the net force on the light is zero and that the net force on the knot is zero.

Considering Figure 4.10b, we apply the equilibrium condition in the y direction, $\Sigma F_y = 0 \rightarrow T_3 - F_g = 0$, which leads to $T_3 = F_g = 122$ N. Thus, the force $\vec{\mathbf{T}}_3$ exerted by the vertical cable balances the weight of the light.

Considering the knot next, we choose the coordinate axes as shown in Figure 4.10c and resolve the forces into their x and y components, as shown in the following tabular representation:

Force	x component	y component
$\vec{\mathbf{T}}_1$	$-T_1 \cos 37.0°$	$T_1 \sin 37.0°$
$\vec{\mathbf{T}}_2$	$T_2 \cos 53.0°$	$T_2 \sin 53.0°$
$\vec{\mathbf{T}}_3$	0	-122 N

Equations 4.8 give us

$$(1)\ \sum F_x = T_2 \cos 53.0° - T_1 \cos 37.0° = 0$$

$$(2)\ \sum F_y = T_1 \sin 37.0° + T_2 \sin 53.0° - 122\ \text{N} = 0$$

We solve (1) for T_2 in terms of T_1 to give

$$T_2 = T_1 \left(\frac{\cos 37.0°}{\cos 53.0°} \right) = 1.33 T_1$$

This value for T_2 is substituted into (2) to give

$$T_1 \sin 37.0° + (1.33 T_1)(\sin 53.0°) - 122\ \text{N} = 0$$

$$T_1 = 73.4\ \text{N}$$

We then calculate T_2:

$$T_2 = 1.33 T_1 = 97.4\ \text{N}$$

Both of these values are less than 100 N (just barely for T_2!), so the cables do not break.

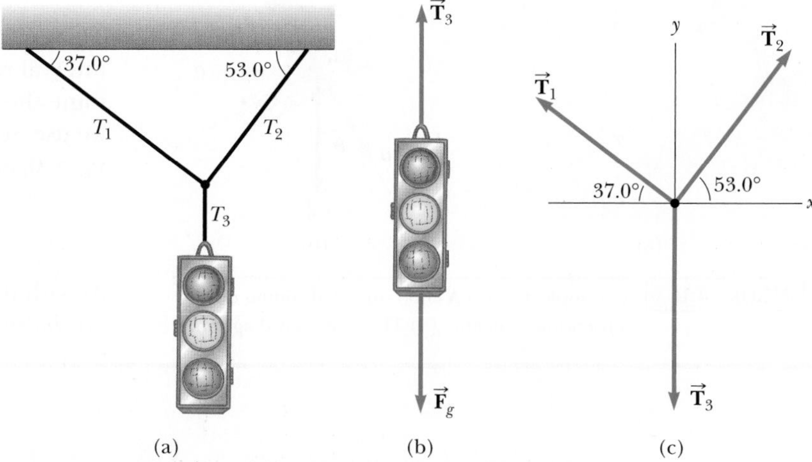

FIGURE 4.10 (Example 4.2) (a) A traffic light suspended by cables. (b) The free-body diagram for the traffic light. (c) The free-body diagram for the knot in the cable.

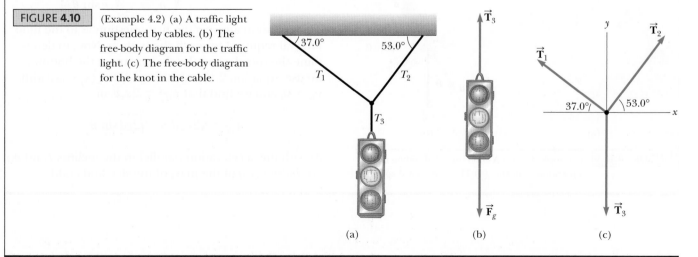

EXAMPLE 4.3 A Sled on Frictionless Snow

A child on a sled is released on a frictionless hill of angle θ, as in Figure 4.11a.

A Determine the acceleration of the sled after it is released.

Solution We identify the combination of the sled and the child as our object of interest. We model the object as a particle of mass m. Newton's second law can be used to determine the acceleration of the particle. First, we construct the free-body diagram for the particle as in Figure 4.11b. The only forces on the particle are the normal force $\vec{n}$ acting perpendicularly to the incline and the gravitational force $m\vec{g}$ acting vertically downward. For problems of this type involving inclines, it is convenient to choose the coordinate axes with x *along* the incline and y *perpendicular* to it. Then, we replace $m\vec{g}$ by a combination of a component vector of magnitude $mg \sin \theta$ along the *positive* x axis (down the incline) and one of magnitude $mg \cos \theta$ in the *negative* y direction.

Applying Newton's second law in component form to the particle and noting that $a_y = 0$ gives

$$(1)\quad \sum F_x = mg \sin \theta = ma_x$$

$$(2)\quad \sum F_y = n - mg \cos \theta = 0$$

From (1) we see that the acceleration along the incline is provided by the component of the gravitational force parallel to the incline, which gives us

$$(3)\quad a_x = g \sin \theta$$

Note that the acceleration given by (3) is *independent* of the mass of the particle; it depends only on the angle of inclination and on g. From (2) we conclude that the component of the gravitational force perpendicular to the incline is *balanced* by the normal force; that is, $n = mg \cos \theta$. (Notice, as pointed out in Pitfall Prevention 4.6, that n does not equal mg in this case.)

Special Cases When $\theta = 90°$, (3) gives us $a_x = g$ and (2) gives us $n = 0$. This case corresponds to the particle in free-fall. (For our choice of coordinate system, positive x is in the downward direction when $\theta = 90°$; hence, the acceleration is $+g$ rather than $-g$.) When $\theta = 0°$, $a_x = 0$ and $n = mg$ (its maximum value), which corresponds to the situation in which the particle is on a level surface and not accelerating.

This technique of looking at special cases of limiting situations is often useful in checking an answer. In this situation, if the angle θ goes to 90°, we know intuitively that the object should be falling parallel to the surface of the incline. That (3) mathematically reduces to $a_x = g$ when $\theta = 90°$ gives us confidence in our answer. It doesn't prove that the answer is correct, but if the acceleration does not reduce to g, it would tell us that the answer is incorrect.

B Suppose the sled is released from rest at the top of the hill and the distance from the front of the sled to the bottom of the hill is d. How long does it take the front of the sled to reach the bottom, and what is its speed just as it arrives at that point?

Solution In part A, we found $a_x = g \sin \theta$, which is constant. Hence, we can model the system as a particle under constant acceleration for the motion parallel to the incline. We use Equation 2.12, $x_f = x_i + v_{xi}t + \frac{1}{2}a_x t^2$, to describe the position of the sled's front edge. We define the initial position as $x_i = 0$ and the final position as $x_f = d$. Because the sled starts sliding from rest, $v_{xi} = 0$. With these values, Equation 2.12 becomes simply $d = \frac{1}{2}a_x t^2$, or

$$t = \sqrt{\frac{2d}{a_x}} = \sqrt{\frac{2d}{g \sin \theta}}$$

This equation answers the first question as to the time interval required to reach the bottom. Now, to determine the speed when the sled arrives at the bottom, we use Equation 2.13, $v_{xf}^2 = v_{xi}^2 + 2a_x(x_f - x_i)$ with $v_{xi} = 0$, and we find that $v_{xf}^2 = 2a_x d$, or

$$v_{xf} = \sqrt{2a_x d} = \sqrt{2gd \sin \theta}$$

As with the acceleration parallel to the incline, t and v_{xf} are *independent* of the mass of the sled and child.

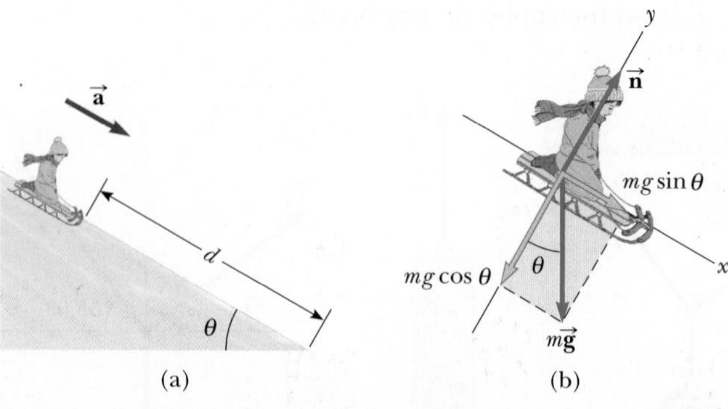

FIGURE 4.11 (Example 4.3) (a) A child on a sled sliding down a frictionless incline. (b) The free-body diagram.

INTERACTIVE EXAMPLE 4.4 The Atwood Machine

When two objects with unequal masses are hung vertically over a light, frictionless pulley as in Active Figure 4.12a, the arrangement is called an *Atwood machine.* The device is sometimes used in the laboratory to measure the free-fall acceleration. Calculate the magnitude of the acceleration of the two objects and the tension in the string.

Solution *Conceptualize* the problem by thinking about the mental representation suggested by Active Figure 4.12a: As one object moves upward, the other object moves downward. Because the objects are connected by an inextensible string, they must have the same magnitude of acceleration. The objects in the Atwood machine are subject to the gravitational force as well as to the forces exerted by the strings connected to them. In *categorizing* the problem, we model the objects as particles under a net force.

We begin to *analyze* the problem by drawing free-body diagrams for the two objects, as in Active Figure 4.12b. Two forces act on each object: the upward force $\vec{\mathbf{T}}$ exerted by the string and the downward gravitational force. In a problem such as this one in which the pulley is modeled as massless and frictionless, the tension in the string on both sides of the pulley is the same. If the pulley has mass or is subject to a friction force, the tensions in the string on either side of the pulley are not the same and the situation requires the techniques of Chapter 10.

In these types of problems, involving strings that pass over pulleys, we must be careful about the sign convention. Notice that if m_1 goes up, m_2 goes down. Therefore, m_1 going up and m_2 going down should be represented equivalently as far as a sign convention is concerned. We can do so by defining our sign convention with up as positive for m_1 and down as positive for m_2, as shown in Active Figure 4.12a.

With this sign convention, the net force exerted on m_1 is $T - m_1g$, whereas the net force exerted on m_2 is $m_2g - T$. We have chosen the signs of the forces to be consistent with the choices of the positive direction for each object.

When Newton's second law is applied to m_1, we find

$$(1) \quad \sum F_y = T - m_1 g = m_1 a$$

Similarly, for m_2 we find

$$(2) \quad \sum F_y = m_2 g - T = m_2 a$$

Note that a is the same for both objects. When (2) is added to (1), T cancels and we have

$$-m_1 g + m_2 g = m_1 a + m_2 a$$

Solving for the acceleration a give us

$$(3) \quad a = \left(\frac{m_2 - m_1}{m_1 + m_2}\right) g$$

If $m_2 > m_1$, the acceleration given by (3) is positive: m_1 goes up and m_2 goes down. Is that consistent with your mental representation? If $m_1 > m_2$, the acceleration is negative and the masses move in the opposite direction.

If (3) is substituted into (1), we find

$$(4) \quad T = \left(\frac{2m_1 m_2}{m_1 + m_2}\right) g$$

To *finalize* the problem, let us consider some special cases. For example, when $m_1 = m_2$, (3) and (4) give us $a = 0$ and $T = m_1 g = m_2 g$, as we would intuitively expect for the balanced case. Also, if $m_2 \gg m_1$, $a \approx g$ (a freely falling object) and $T \approx 0$. For such a large mass

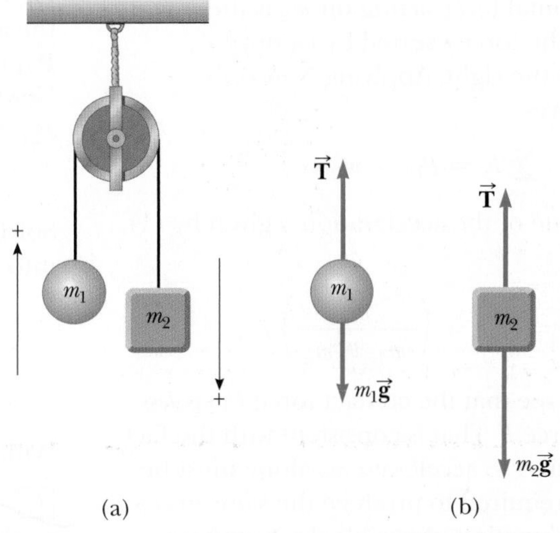

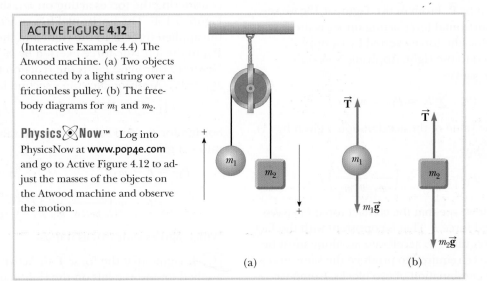

ACTIVE FIGURE 4.12
(Interactive Example 4.4) The Atwood machine. (a) Two objects connected by a light string over a frictionless pulley. (b) The free-body diagrams for m_1 and m_2.

PhysicsNow™ Log into PhysicsNow at **www.pop4e.com** and go to Active Figure 4.12 to adjust the masses of the objects on the Atwood machine and observe the motion.

m_2, we would expect m_1 to have little effect so that m_2 is simply falling. Our results are consistent with our intuitive predictions in both of these limiting situations.

PhysicsNow™ Investigate the motion of the Atwood machine for different masses by logging into PhysicsNow at www.pop4e.com and going to Interactive Example 4.4.

INTERACTIVE EXAMPLE 4.5 One Block Pushes Another

Two blocks of masses m_1 and m_2, with $m_1 > m_2$, are placed in contact with each other on a frictionless, horizontal surface, as in Active Figure 4.13a. A constant horizontal force $\vec{\mathbf{F}}$ is applied to m_1 as shown.

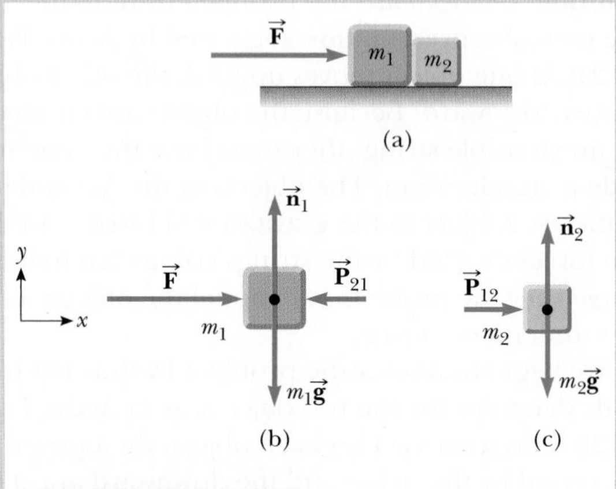

ACTIVE FIGURE 4.13

(Interactive Example 4.5) A force is applied to a block of mass m_1, which pushes on a second block of mass m_2. (b) The free-body diagram for m_1. (c) The free-body diagram for m_2.

PhysicsNow™ Log into PhysicsNow at www.pop4e.com and go to Active Figure 4.13 to study the forces involved in this two-block system.

A Find the magnitude of the acceleration of the system of two blocks.

Solution Both blocks must experience the *same* acceleration because they are in contact with each other and remain in contact with each other. We model the combination of both blocks as a particle under a net force. Because $\vec{\mathbf{F}}$ is the only horizontal force exerted on the particle, we have

$$\sum F_x \text{ (system)} = F = (m_1 + m_2)a$$

$$(1) \quad a = \frac{F}{m_1 + m_2}$$

B Determine the magnitude of the contact force between the two blocks.

Solution The contact force is internal to the combination of two blocks. Therefore, we cannot find this force by modeling the combination as a single particle. We now need to treat each of the two blocks individually as a particle under a net force. We first construct a free-body diagram for each block, as shown in Active Figures 4.13b and 4.13c, where the contact force is denoted by $\vec{\mathbf{P}}$. From Active Figure 4.13c we see that the only horizontal force acting on m_2 is the contact force $\vec{\mathbf{P}}_{12}$ (the force exerted by m_1 on m_2), which is directed to the right. Applying Newton's second law to m_2 gives

$$(2) \quad \sum F_x = P_{12} = m_2 a$$

Substituting the value of the acceleration a given by (1) into (2) gives

$$(3) \quad P_{12} = m_2 a = \left(\frac{m_2}{m_1 + m_2}\right) F$$

From this result we see that the contact force P_{12} is *less* than the applied force F. That is consistent with the fact that the force required to accelerate m_2 alone must be less than the force required to produce the same acceleration for the combination of two blocks. Compare this result with the forces in the couplers in the train of Thinking Physics 4.1.

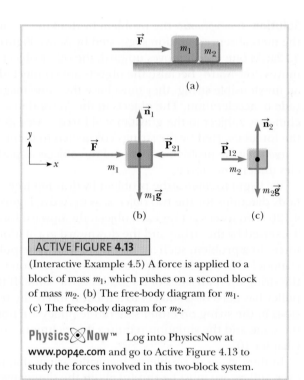

It is instructive to check this expression for P_{12} by considering the forces acting on m_1, shown in Active Figure 4.13b. The horizontal forces acting on m_1 are the applied force $\vec{\mathbf{F}}$ to the right and the contact force $\vec{\mathbf{P}}_{21}$ to the left (the force exerted by m_2 on m_1). From Newton's third law, $\vec{\mathbf{P}}_{21}$ is the reaction to $\vec{\mathbf{P}}_{12}$, so $P_{21} = P_{12}$. Applying Newton's second law to m_1 gives

$$(4) \quad \sum F_x = F - P_{21} = F - P_{12} = m_1 a$$

Solving for P_{12} and substituting the value of a from (1) into (4) gives

$$P_{12} = F - m_1 a$$

$$= F - m_1\left(\frac{F}{m_1 + m_2}\right) = \left(\frac{m_2}{m_1 + m_2}\right) F$$

Which agrees with (3), as it must.

C Imagine that the force $\vec{\mathbf{F}}$ in Active Figure 4.13 is applied toward the left on the right-hand block of mass

m_2. Is the magnitude of the force $\vec{\mathbf{P}}_{12}$ the same as it was when the force was applied toward the right on m_1?

Solution With the force applied toward the left on m_2, the contact force must accelerate m_1. In the original situation, the contact force accelerates m_2. Because $m_1 > m_2$, more force is required, so the magnitude of $\vec{\mathbf{P}}_{12}$ is greater.

PhysicsNow™ Investigate the motion of the blocks for different mass combinations and applied forces by logging into PhysicsNow at **www.pop4e.com** and going to Interactive Example 4.5.

EXAMPLE 4.6 Weighing a Fish in an Elevator

A person weighs a fish on a spring scale attached to the ceiling of an elevator, as shown in Figure 4.14. Show that if the elevator accelerates, the spring scale reads an apparent weight different from the fish's true weight.

Solution An observer on the accelerating elevator is not in an inertial frame. We need to analyze this situation in an inertial frame, so let us imagine observing it from the stationary ground. We model the fish as a particle under a net force. The external forces acting on the fish are the downward gravitational force $\vec{\mathbf{F}}_g$ and the upward force $\vec{\mathbf{T}}$ exerted on it by the hook hanging from the bottom of the scale. (It might be more fruitful in your mental representation to imagine that the hook is a string connecting the fish to the spring in the scale.) Because the tension is the same everywhere in the hook supporting the fish, the hook pulls downward with a force of magnitude T on the spring scale. Therefore, the tension T in the hook is also the reading of the spring scale.

If the elevator is either at rest or moves at constant velocity, the fish is not accelerating and is a particle in equilibrium, which gives us $\Sigma F_y = T - mg = 0$

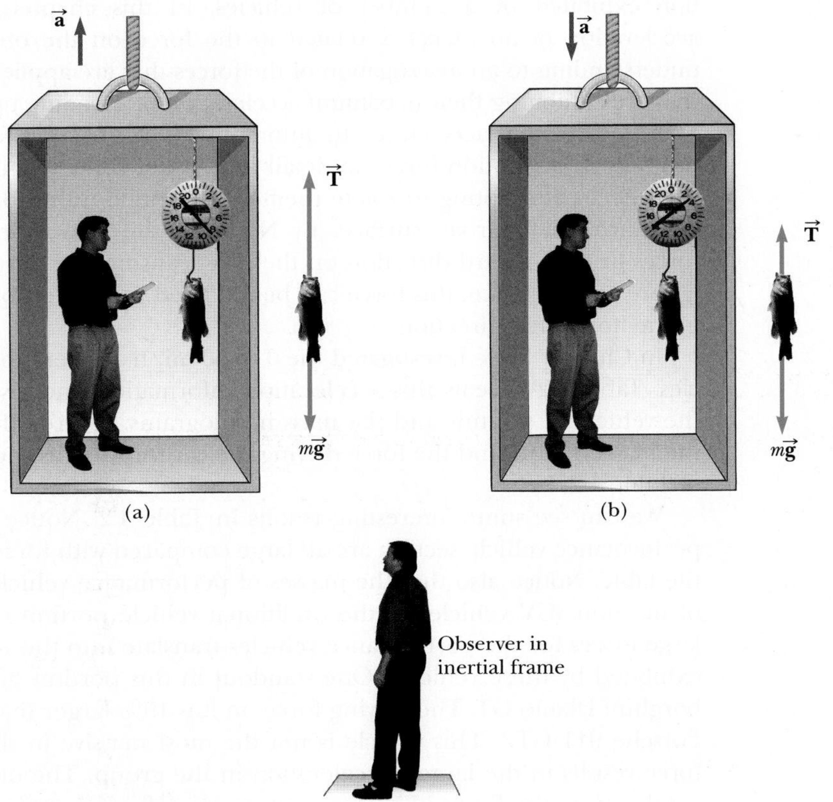

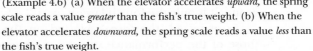

FIGURE 4.14 (Example 4.6) (a) When the elevator accelerates *upward,* the spring scale reads a value *greater* than the fish's true weight. (b) When the elevator accelerates *downward,* the spring scale reads a value *less* than the fish's true weight.

→ $T = mg$. If the elevator accelerates either up or down, however, the tension is no longer equal to the weight of the fish because $T - mg$ does not equal zero.

If the elevator accelerates with an acceleration $\vec{\mathbf{a}}$ relative to an observer in an inertial frame outside the elevator, Newton's second law applied to the fish in the vertical direction gives us

$$\sum F_y = T - mg = ma_y$$

which leads to

$$(1) \quad T = mg + ma_y$$

We conclude from (1) that the scale reading T is greater than the weight mg if $\vec{\mathbf{a}}$ is upward as in Figure 4.14a. Furthermore, we see that T is less than mg if $\vec{\mathbf{a}}$ is downward as in Figure 4.14b. For example, if the weight of the fish is 40.0 N and $\vec{\mathbf{a}}$ is upward with $a_y = 2.00\text{ m/s}^2$, the scale reading is

$$T = mg + ma_y = mg\left(1 + \frac{a_y}{g}\right)$$

$$= (40.0\text{ N})\left(1 + \frac{2.00\text{ m/s}^2}{9.80\text{ m/s}^2}\right) = 48.2\text{ N}$$

If $a_y = -2.00\text{ m/s}^2$ so that $\vec{\mathbf{a}}$ is downward,

$$T = mg\left(1 + \frac{a_y}{g}\right)$$

$$= (40.0\text{ N})\left(1 + \frac{-2.00\text{ m/s}^2}{9.80\text{ m/s}^2}\right) = 31.8\text{ N}$$

Hence, if you buy a fish in an elevator, make sure the fish is weighed while the elevator is at rest or is accelerating downward!

Special Case If the cable breaks, the elevator falls freely so that $a_y = -g$, and from (1) we see that the tension T is zero; that is, the fish appears to be weightless.

4.8 FORCES ON AUTOMOBILES

CONTEXT CONNECTION

In the Context Connections of Chapters 2 and 3, we focused on two types of acceleration exhibited by a number of vehicles. In this chapter, we learned how the acceleration of an object is related to the force on the object. Let us apply this understanding to an investigation of the forces that are applied to automobiles when they are exhibiting their maximum acceleration in speeding up from rest to 60 mi/h.

The force that accelerates an automobile is the friction force from the ground. (We will study friction forces in detail in Chapter 5.) The engine applies a force to the wheels, attempting to rotate them so that the bottoms of the tires apply forces backward on the road surface. By Newton's third law, the road surface applies forces in the forward direction on the tires, causing the car to move forward. If we ignore air resistance, this force can be modeled as the net force on the automobile in the horizontal direction.

In Chapter 2, we investigated the 0 to 60 mi/h acceleration of a number of vehicles. Table 4.2 repeats this acceleration information and also shows the weight of the vehicle in pounds and the mass in kilograms. With both the acceleration and the mass, we can find the force driving the car forward, as shown in the last column of Table 4.2.

We can see some interesting results in Table 4.2. Notice that the forces in the performance vehicle section are all large compared with forces in the other parts of the table. Notice also that the masses of performance vehicles are similar to those of the non-SUV vehicles in the traditional vehicle portion of the table. Thus, the large forces for the performance vehicles translate into the very large accelerations exhibited by these vehicles. One standout in this portion of the table is the Lamborghini Diablo GT. The driving force on it is 15% larger than the next largest, the Porsche 911 GT2. This vehicle is not the most massive in the group, so the large force results in the largest acceleration in the group. The other car with the same acceleration, the Ferrari F50, has a mass only 81% of that of the Lamborghini. Consequently, although the force on the Ferrari is higher than the average in the group, it is only the fourth largest.

As expected, the forces exerted on the traditional vehicles are smaller than those of the performance vehicles, corresponding to the smaller accelerations of

TABLE 4.2 Driving Forces on Various Vehicles

Automobile	Model Year	Acceleration (mi/h · s)	Weight (lb)	Mass (kg)	Force (N)
Performance vehicles:					
Aston Martin DB7 Vantage	2001	12.0	3 285	1 493	8.01×10^3
BMW Z8	2001	13.0	3 215	1 461	8.52×10^3
Chevrolet Corvette	2000	13.0	3 115	1 416	8.25×10^3
Dodge Viper GTS-R	1998	14.3	2 865	1 302	8.32×10^3
Ferrari F50	1997	16.7	2 655	1 207	8.99×10^3
Ferrari 360 Spider F1	2000	13.0	3 400	1 545	9.01×10^3
Lamborghini Diablo GT	2000	16.7	3 285	1 493	11.12×10^3
Porsche 911 GT2	2002	15.0	3 175	1 443	9.68×10^3
Traditional vehicles:					
Acura Integra GS	2000	7.6	2 725	1 239	4.20×10^3
BMW Mini Cooper S	2003	8.7	2 678	1 217	4.73×10^3
Cadillac Escalade (SUV)	2002	7.0	5 542	2 519	7.86×10^3
Dodge Stratus	2002	8.0	3 192	1 451	5.19×10^3
Lexus ES300	1997	7.0	3 296	1 498	4.67×10^3
Mitsubishi Eclipse GT	2000	8.6	3 186	1 448	5.55×10^3
Nissan Maxima	2000	9.0	3 221	1 464	5.86×10^3
Pontiac Grand Prix	2003	7.1	3 384	1 538	4.85×10^3
Toyota Sienna (SUV)	2004	7.2	3 912	1 778	5.74×10^3
Volkswagen Beetle	1999	7.9	2 771	1 260	4.44×10^3
Alternative vehicles:					
GM EV1	1998	7.9	2 970	1 350	4.76×10^3
Toyota Prius	2004	4.7	2 765	1 257	2.65×10^3
Honda Insight	2001	5.2	1 967	894	2.07×10^3

this group. Notice, however, that the forces for the two SUVs are large. Because these two vehicles have accelerations that are somewhat similar to those of the other vehicles in this portion of the table, we can identify these large forces as being required to accelerate the larger mass of the SUVs.

Also as expected, the forces driving the two hybrid vehicles, the Toyota Prius and the Honda Insight, are the lowest in the table. This finding is consistent with the accelerations of these vehicles being much lower than those elsewhere in the table. ■

SUMMARY

PhysicsNow™ Take a practice test by logging into PhysicsNow at www.pop4e.com and clicking on the Pre-Test link for this chapter.

Newton's first law states that if an object does not interact with other objects, it is possible to identify a reference frame in which the object has zero acceleration. Thus, if we observe an object from such a frame and no force is exerted on the object, an object at rest remains at rest and an object in uniform motion in a straight line maintains that motion.

Newton's first law defines an **inertial frame of reference,** which is a frame in which Newton's first law is valid.

Newton's second law states that the acceleration of an object is directly proportional to the net force acting on the object and inversely proportional to the object's mass. Therefore, the net force on an object equals the product of the mass of the object and its acceleration, or

$$\sum \vec{\mathbf{F}} = m\vec{\mathbf{a}} \qquad [4.2]$$

The **weight** of an object is equal to the product of its mass (a scalar quantity) and the magnitude of the free-fall acceleration, or

$$F_g = mg \qquad [4.5]$$

If the acceleration of an object is zero, the object is modeled as a particle in equilibrium, with the appropriate equations being

$$\sum F_x = 0 \qquad \sum F_y = 0 \qquad [4.8]$$

Newton's third law states that if two objects interact, the force exerted by object 1 on object 2 is equal in magnitude but opposite in direction to the force exerted by object 2 on object 1. Therefore, an isolated force cannot exist in nature.

QUESTIONS

☐ = answer available in the *Student Solutions Manual and Study Guide*

1. A ball is held in a person's hand. (a) Identify all the external forces acting on the ball and the reaction to each. (b) If the ball is dropped, what force is exerted on it while it is falling? Identify the reaction force in this case.

2. What is wrong with the statement, "Because the car is at rest, there are no forces acting on it"? How would you correct this sentence?

3. In the motion picture *It Happened One Night* (Columbia Pictures, 1934), Clark Gable is standing inside a stationary bus in front of Claudette Colbert, who is seated. The bus suddenly starts moving forward and Clark falls into Claudette's lap. Why did that happen?

4. As you sit in a chair, the chair pushes up on you with a normal force. The force is equal to your weight and in the opposite direction. Is this force the Newton's third law reaction to your weight?

5. A passenger sitting in the rear of a bus claims that she was injured as the driver slammed on the brakes, causing a suitcase to come flying toward her from the front of the bus. If you were the judge in this case, what disposition would you make? Why?

6. A space explorer is moving through space in a space ship far from any planet or star. She notices a large rock, taken as a specimen from an alien planet, floating around the cabin of the ship. Should she push it gently or kick it toward the storage compartment? Why?

7. A rubber ball is dropped onto the floor. What force causes the ball to bounce?

8. While a football is in flight, what forces act on it? What are the action–reaction pairs while the football is being kicked and while it is in flight?

9. If gold were sold by weight, would you rather buy it in Denver or in Death Valley? If it were sold by mass, at which of the two locations would you prefer to buy it? Why?

10. If you hold a horizontal metal bar several centimeters above the ground and move it through grass, each leaf of grass bends out of the way. If you increase the speed of the bar, each leaf of grass will bend more quickly. How then does a rotary power lawn mower manage to cut grass? How can it exert enough force on a leaf of grass to shear it off?

11. A weightlifter stands on a bathroom scale. He pumps a barbell up and down. What happens to the reading on the bathroom scale as he does so? What if he is strong enough to actually *throw* the barbell upward? How does the reading on the scale vary now?

12. The mayor of a city decides to fire some city employees because they will not remove the obvious sags from the cables that support the city traffic lights. If you were a lawyer, what defense would you give on behalf of the employees? Who do you think would win the case in court?

13. Suppose a truck loaded with sand accelerates along a highway. If the driving force on the truck remains constant, what happens to the truck's acceleration if its trailer leaks sand at a constant rate through a hole in its bottom?

14. As a rocket is fired from a launching pad, its speed and acceleration increase with time as its engines continue to operate. Explain why that occurs even though the thrust of the engines remains constant.

15. Twenty people participate in a tug-of-war. The two teams of ten people are so evenly matched that neither team wins. After the game they notice that one person's car is mired in mud. They attach the tug-of-war rope to the bumper of the car, and all the people pull on the rope. The heavy car has just moved a couple of decimeters when the rope breaks. Why did the rope break in this situation when it did not break when the same twenty people pulled on it in a tug-of-war?

16. "When the locomotive in Figure Q4.16 broke through the wall of the train station, the force exerted by the locomotive on the wall was greater than the force the wall could exert on the locomotive." Is this statement true or in need of correction? Explain your answer.

(Roger Viollet, Mill Valley, CA, University Science Books, 1982)

FIGURE Q4.16

17. An athlete grips a light rope that passes over a low-friction pulley attached to the ceiling of a gym. A sack of sand precisely equal in weight to the athlete is tied to the rope's other end. Both the sand and the athlete are initially at rest. The athlete climbs the rope, sometimes speeding up and slowing down as he does so. What happens to the sack of sand? Explain.

18. If action and reaction forces are always equal in magnitude and opposite in direction to each other, doesn't the net vector force on any object necessarily add up to zero? Explain your answer.

19. Can an object exert a force on itself? Argue for your answer.

PROBLEMS

1, 2, 3 = straightforward, intermediate, challenging
□ = full solution available in the *Student Solutions Manual and Study Guide*
Physics Now™ = coached problem with hints available at www.pop4e.com
= computer useful in solving problem
= paired numerical and symbolic problems
= biomedical application

Section 4.3 ▪ Mass

1. A force $\vec{\mathbf{F}}$ applied to an object of mass m_1 produces an acceleration of 3.00 m/s^2. The same force applied to a second object of mass m_2 produces an acceleration of 1.00 m/s^2. (a) What is the value of the ratio m_1/m_2? (b) If m_1 and m_2 are combined, find their acceleration under the action of the force $\vec{\mathbf{F}}$.

2. (a) A car with a mass of 850 kg is moving to the right with a constant speed of 1.44 m/s. What is the total force on the car? (b) What is the total force on the car if it is moving to the left?

Section 4.4 ▪ Newton's Second Law—The Particle Under a Net Force

3. A 3.00-kg object undergoes an acceleration given by $\vec{\mathbf{a}} = (2.00\hat{\mathbf{i}} + 5.00\hat{\mathbf{j}})$ m/s^2. Find the resultant force acting on it and the magnitude of the resultant force.

4. Two forces, $\vec{\mathbf{F}}_1 = (-6\hat{\mathbf{i}} - 4\hat{\mathbf{j}})$ N and $\vec{\mathbf{F}}_2 = (-3\hat{\mathbf{i}} + 7\hat{\mathbf{j}})$ N, act on a particle of mass 2.00 kg that is initially at rest at coordinates (− 2.00 m, + 4.00 m). (a) What are the components of the particle's velocity at t = 10.0 s? (b) In what direction is the particle moving at t = 10.0 s? (c) What displacement does the particle undergo during the first 10.0 s? (d) What are the coordinates of the particle at t = 10.0 s?

5. Physics Now™ To model a spacecraft, a toy rocket engine is securely fastened to a large puck that can glide with negligible friction over a horizontal surface, taken as the xy plane. The 4.00-kg puck has a velocity of $3.00\hat{\mathbf{i}}$ m/s at one instant. Eight seconds later, its velocity is to be $(8.00\hat{\mathbf{i}} + 10.0\hat{\mathbf{j}})$ m/s. Assuming that the rocket engine exerts a constant horizontal force, find (a) the components of the force and (b) its magnitude.

6. A 3.00-kg object is moving in a plane, with its x and y coordinates given by $x = 5t^2 - 1$ and $y = 3t^3 + 2$, where x and y are in meters and t is in seconds. Find the magnitude of the net force acting on this object at t = 2.00 s.

7. Two forces $\vec{\mathbf{F}}_1$ and $\vec{\mathbf{F}}_2$ act on a 5.00-kg object. If F_1 = 20.0 N and F_2 = 15.0 N, find the accelerations in (a) and (b) of Figure P4.7.

8. Three forces, given by $\vec{\mathbf{F}}_1 = (-2.00\hat{\mathbf{i}} + 2.00\hat{\mathbf{j}})$ N, $\vec{\mathbf{F}}_2 = (5.00\hat{\mathbf{i}} - 3.00\hat{\mathbf{j}})$ N, and $\vec{\mathbf{F}}_3 = (-45.0\hat{\mathbf{i}})$ N, act on an object to give it an acceleration of magnitude 3.75 m/s^2.

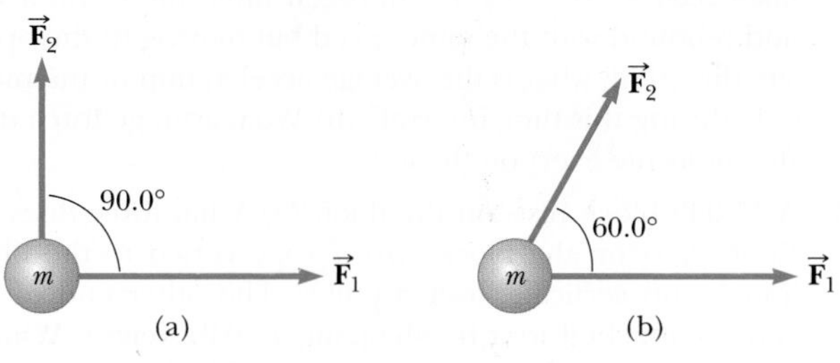

FIGURE P4.7

(a) What is the direction of the acceleration? (b) What is the mass of the object? (c) If the object is initially at rest, what is its speed after 10.0 s? (d) What are the velocity components of the object after 10.0 s?

Section 4.5 ▪ The Gravitational Force and Weight

9. A woman weighs 120 lb. Determine (a) her weight in newtons and (b) her mass in kilograms.

10. If a man weighs 900 N on the Earth, what would he weigh on Jupiter, where the free-fall acceleration is 25.9 m/s^2?

11. The distinction between mass and weight was discovered after Jean Richer transported pendulum clocks from Paris, France, to Cayenne, French Guiana in 1671. He found that they quite systematically ran slower in Cayenne than in Paris. The effect was reversed when the clocks returned to Paris. How much weight would you personally lose in traveling from Paris, where g = 9.809 5 m/s^2, to Cayenne, where g = 9.780 8 m/s^2? (We will consider how the free-fall acceleration influences the period of a pendulum in Section 12.4.)

12. The gravitational force on a baseball is $-F_g\hat{\mathbf{j}}$. A pitcher throws the baseball with velocity $v\hat{\mathbf{i}}$ by uniformly accelerating it straight forward horizontally for a time interval $\Delta t = t - 0 = t$. If the ball starts from rest, (a) through what distance does it accelerate before its release? (b) What force does the pitcher exert on the ball?

13. An electron of mass 9.11×10^{-31} kg has an initial speed of 3.00×10^5 m/s. It travels in a straight line, and its speed increases to 7.00×10^5 m/s in a distance of 5.00 cm. Assuming that its acceleration is constant, (a) determine the net force exerted on the electron and (b) compare this force with the weight of the electron.

14. Besides its weight, a 2.80-kg object is subjected to one other constant force. The object starts from rest and in 1.20 s experiences a displacement of $(4.20\hat{\mathbf{i}} - 3.30\hat{\mathbf{j}})$ m, where the direction of $\hat{\mathbf{j}}$ is the upward vertical direction. Determine the other force.

Section 4.6 ▪ Newton's Third Law

15. You stand on the seat of a chair and then hop off. (a) During the time you are in flight down to the floor, the Earth is lurching up toward you with an acceleration of what order of magnitude? In your solution, explain your logic. Model the Earth as a perfectly solid object. (b) The Earth moves up through a distance of what order of magnitude?

16. The average speed of a nitrogen molecule in air is about 6.70×10^2 m/s, and its mass is 4.68×10^{-26} kg. (a) If it takes 3.00×10^{-13} s for a nitrogen molecule to hit a wall and rebound with the same speed but moving in the opposite direction, what is the average acceleration of the molecule during this time interval? (b) What average force does the molecule exert on the wall?

17. A 15.0-lb block rests on the floor. (a) What force does the floor exert on the block? (b) A rope is tied to the block and is run vertically over a pulley. The other end of the rope is attached to a free-hanging 10.0-lb weight. What is the force exerted by the floor on the 15.0-lb block? (c) If we replace the 10.0-lb weight in part (b) with a 20.0-lb weight, what is the force exerted by the floor on the 15.0-lb block?

Section 4.7 ■ Applications of Newton's Laws

18. A bag of cement of weight 325 N hangs in equilibrium from three wires as suggested in Figure P4.18. Two of the wires make angles $\theta_1 = 60.0°$ and $\theta_2 = 25.0°$ with the horizontal. Find the tensions T_1, T_2, and T_3 in the wires.

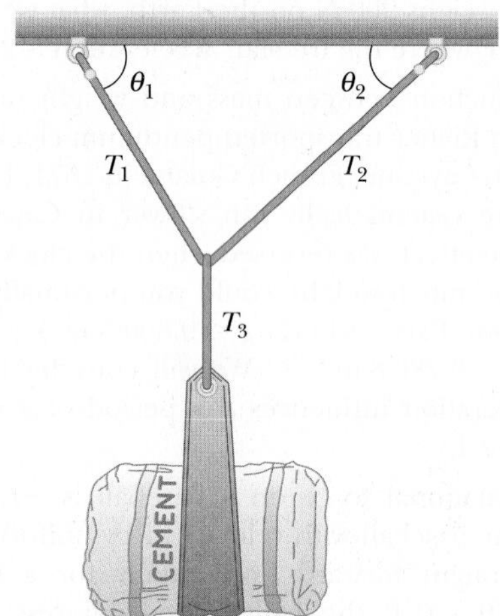

FIGURE P4.18 Problems 4.18 and 4.19.

19. A bag of cement of weight F_g hangs in equilibrium from three wires as shown in Figure P4.18. Two of the wires make angles θ_1 and θ_2 with the horizontal. Show that the tension in the left-hand wire is

$$T_1 = \frac{F_g \cos \theta_2}{\sin(\theta_1 + \theta_2)}$$

20. Figure P4.20 shows a worker poling a boat—a very efficient mode of transportation—across a shallow lake. He pushes parallel to the length of the light pole, exerting on the bottom of the lake a force of 240 N. The pole lies in the vertical plane containing the keel of the boat. At one moment the pole makes an angle of 35.0° with the vertical and the water exerts a horizontal drag force of 47.5 N on the boat, opposite to its forward motion at 0.857 m/s. The mass of the boat including its cargo and the worker is 370 kg. (a) The water exerts a buoyant force vertically upward on the boat. Find the magnitude of this force. (b) Model the forces as constant over a short interval of time to find the velocity of the boat 0.450 s after the moment described.

FIGURE P4.20

21. You are a judge in a children's kite-flying contest, and two children will win prizes for the kites that pull most strongly and least strongly on their strings. To measure string tensions, you borrow a weight hanger, some slotted weights, and a protractor from your physics teacher, and you use the following protocol, illustrated in Figure P4.21: Wait for a child to get her kite well controlled, hook the hanger onto the kite string about 30 cm from her hand, pile on weight until that section of string is horizontal, record the mass required, and record the angle between the horizontal and the string running up to the kite. (a) Explain how this method works. As you construct your explanation, imagine that the children's parents ask you about your method, that they might make false assumptions about your ability without concrete evidence, and that your

FIGURE P4.21

explanation is an opportunity to give them confidence in your evaluation technique. (b) Find the string tension assuming that the mass is 132 g and the angle of the kite string is 46.3°.

22. The systems shown in Figure P4.22 are in equilibrium. If the spring scales are calibrated in newtons, what do they read? (Ignore the masses of the pulleys and strings, and assume that the incline is frictionless.)

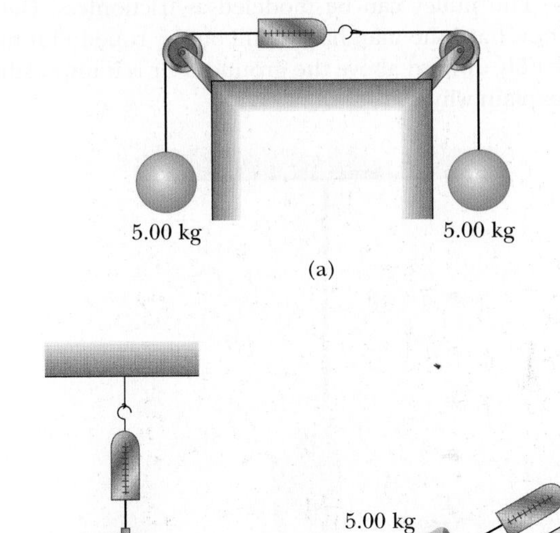

FIGURE **P4.22**

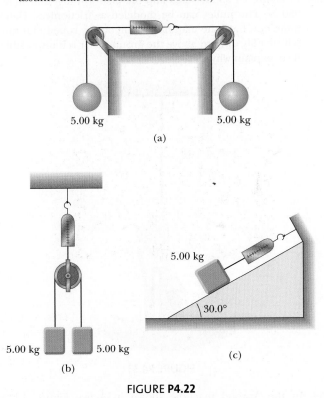

23. A simple accelerometer is constructed inside a car by suspending an object of mass m from a string of length L that is tied to the car's ceiling. As the car accelerates the string–object system makes a constant angle of θ with the vertical. (a) Assuming that the string mass is negligible compared with m, derive an expression for the car's acceleration in terms of θ and show that it is independent of the mass m and the length L. (b) Determine the acceleration of the car when $\theta = 23.0°$.

24. Figure P4.24 shows loads hanging from the ceiling of an elevator that is moving at constant velocity. Find the tension in each of the three strands of cord supporting each load.

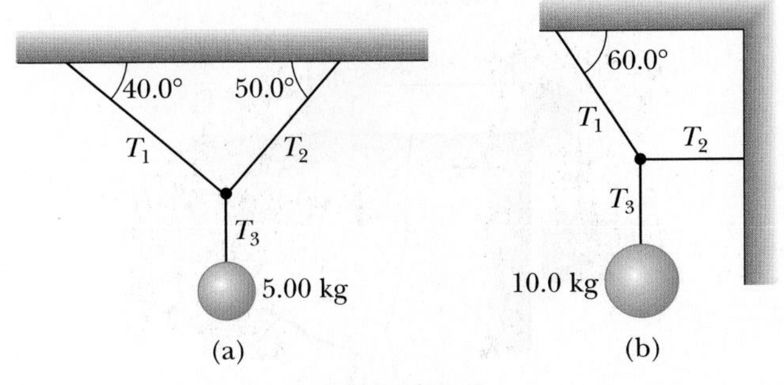

FIGURE **P4.24**

25. Two people pull as hard as they can on horizontal ropes attached to a boat that has a mass of 200 kg. If they pull in the same direction, the boat has an acceleration of 1.52 m/s^2 to the right. If they pull in opposite directions, the boat has an acceleration of 0.518 m/s^2 to the left. What is the magnitude of the force each person exerts on the boat? Disregard any other horizontal forces on the boat.

26. Draw a free-body diagram of a block that slides down a frictionless plane having an inclination of $\theta = 15.0°$ (Fig. P4.26). Assuming that the block starts from rest at the top and that the length of the incline is 2.00 m, find (a) the acceleration of the block and (b) its speed when it reaches the bottom of the incline.

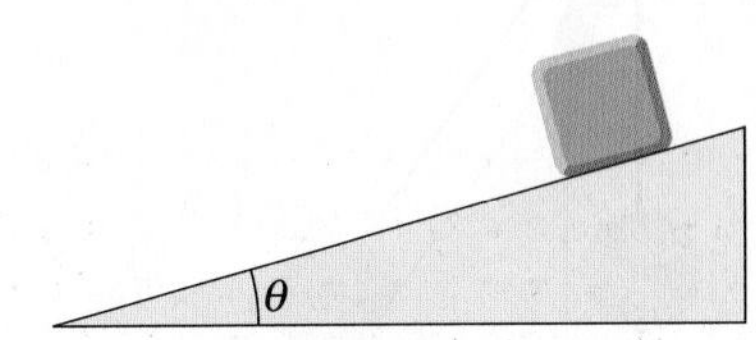

FIGURE **P4.26** Problems 4.26, 4.29, and 4.46.

27. **Physics Now™** A 1.00-kg object is observed to accelerate at 10.0 m/s^2 in a direction 30.0° north of east (Fig. P4.27). The force $\vec{\mathbf{F}}_2$ acting on the object has magnitude 5.00 N and is directed north. Determine the magnitude and direction of the force $\vec{\mathbf{F}}_1$ acting on the object.

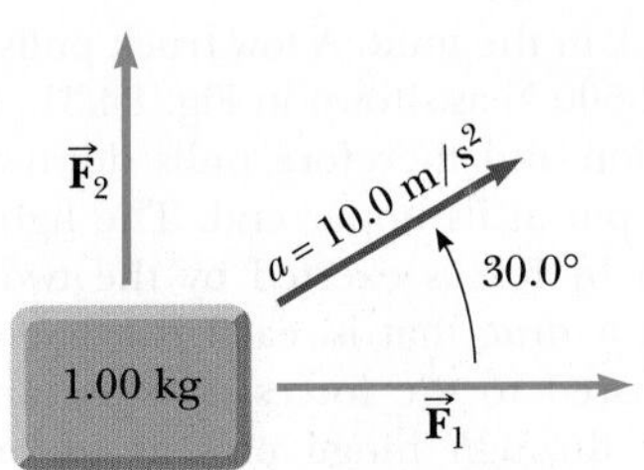

FIGURE **P4.27**

28. A 5.00-kg object placed on a frictionless, horizontal table is connected to a cable that passes over a pulley and then is fastened to a hanging 9.00-kg object as shown in Figure P4.28. Draw free-body diagrams of both objects. Find the acceleration of the two objects and the tension in the string.

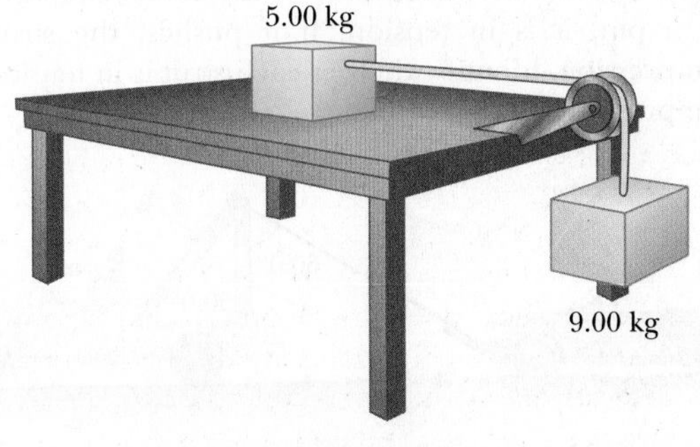

FIGURE **P4.28**

29. **Physics Now™** A block is given an initial velocity of 5.00 m/s up a frictionless 20.0° incline (Fig. P4.26). How far up the incline does the block slide before coming to rest?

30. Two objects are connected by a light string that passes over a frictionless pulley as shown in Figure P4.30. Draw free-body diagrams of both objects. The incline is frictionless, and $m_1 = 2.00$ kg, $m_2 = 6.00$ kg, and $\theta = 55.0°$. Find (a) the accelerations of the objects, (b) the tension in the string, and (c) the speed of each of the objects 2.00 s after they are released simultaneously from rest.

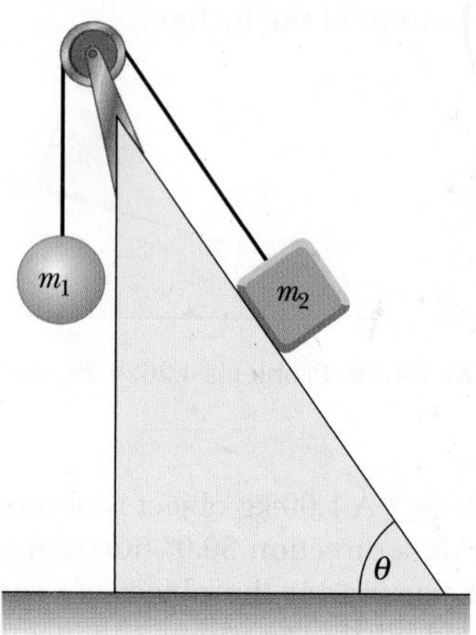

FIGURE **P4.30**

31. A car is stuck in the mud. A tow truck pulls on the car with a force of 2 500 N as shown in Fig. P4.31. The tow cable is under tension and therefore pulls downward and to the left on the pin at its upper end. The light pin is held in equilibrium by forces exerted by the two bars A and B. Each bar is a *strut;* that is, each is a bar whose weight is small compared to the forces it exerts and which exerts forces only through hinge pins at its ends. Each strut exerts a force directed parallel to its length. Determine the force of tension or compression in each strut. Proceed as follows. Make a guess as to which way (pushing or pulling) each force acts on the top pin. Draw a free-body diagram of the pin. Use the condition for equilibrium of the pin to translate the free-body diagram into equations. From the equations calculate the forces exerted by struts A and B. If you obtain a positive answer, you correctly guessed the direction of the force. A negative answer means that the direction should be reversed, but the absolute value correctly gives the magnitude of the force. If a strut pulls on a pin, it is in tension. If it pushes, the strut is in compression. Identify whether each strut is in tension or in compression.

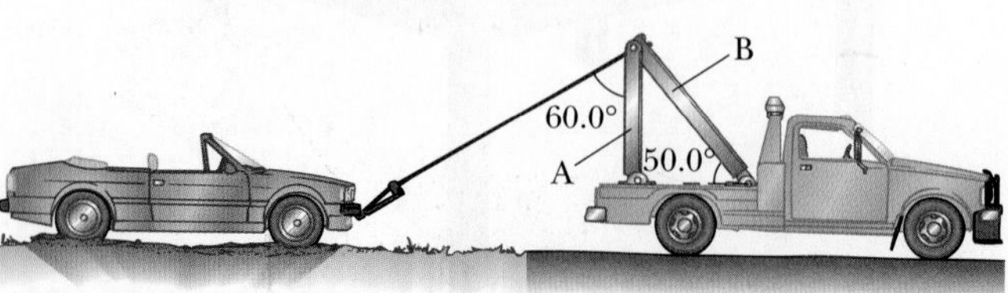

FIGURE **P4.31**

32. Two objects with masses of 3.00 kg and 5.00 kg are connected by a light string that passes over a light frictionless pulley to form an Atwood machine as shown in Active Figure 4.12a. Determine (a) the tension in the string, (b) the acceleration of each object, and (c) the distance each object will move in the first second of motion if they start from rest.

33. In Figure P4.33, the man and the platform together weigh 950 N. The pulley can be modeled as frictionless. Determine how hard the man has to pull on the rope to lift himself steadily upward above the ground. (Or is it impossible? If so, explain why.)

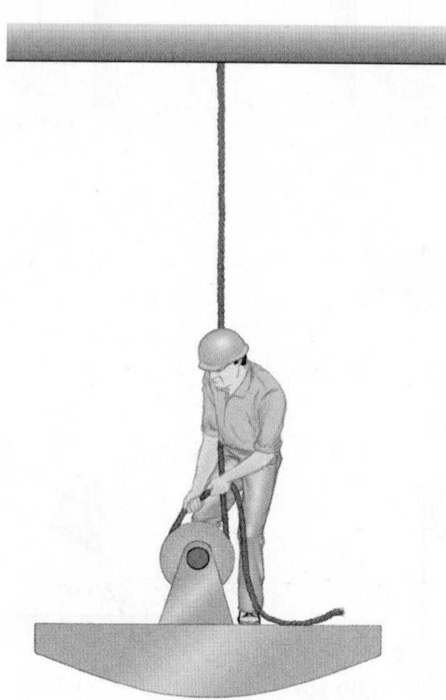
FIGURE **P4.33**

34. In the Atwood machine shown in Active Figure 4.12a, $m_1 = 2.00$ kg and $m_2 = 7.00$ kg. The masses of the pulley and string are negligible by comparison. The pulley turns without friction, and the string does not stretch. The lighter object is released with a sharp push that sets it into motion at $v_i = 2.40$ m/s downward. (a) How far will m_1 descend below its initial level? (b) Find the velocity of m_1 after 1.80 s.

35. **Physics Now™** In the system shown in Figure P4.35, a horizontal force $\vec{\mathbf{F}}_x$ acts on the 8.00-kg object. The horizontal surface is frictionless. (a) For what values of F_x does the 2.00-kg object accelerate upward? (b) For what values of F_x is the tension in the cord zero? (c) Plot the acceleration of the 8.00-kg object versus F_x. Include values of F_x from − 100 N to + 100 N.

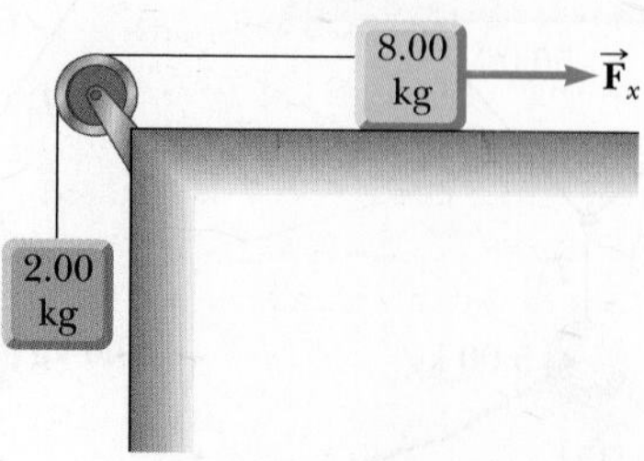

FIGURE **P4.35**

36. A frictionless plane is 10.0 m long and inclined at 35.0°. A sled starts at the bottom with an initial speed of 5.00 m/s up the incline. When it reaches the point at which it momentarily stops, a second sled is released from the top of this incline with an initial speed v_i. Both sleds reach the bottom of the incline at the same moment. (a) Determine the distance that the first sled traveled up the incline. (b) Determine the initial speed of the second sled.

37. A 72.0-kg man stands on a spring scale in an elevator. Starting from rest, the elevator ascends, attaining its maximum speed of 1.20 m/s in 0.800 s. It travels with this constant speed for the next 5.00 s. The elevator then undergoes a uniform acceleration in the negative y direction for 1.50 s and comes to rest. What does the spring scale register (a) before the elevator starts to move, (b) during the first 0.800 s, (c) while the elevator is traveling at constant speed, and (d) during the time it is slowing down?

38. An object of mass m_1 on a frictionless horizontal table is connected to an object of mass m_2 through a very light pulley P_1 and a light fixed pulley P_2 as shown in Figure P4.38. (a) If a_1 and a_2 are the accelerations of m_1 and m_2, respectively, what is the relation between these accelerations? Express (b) the tensions in the strings and (c) the accelerations a_1 and a_2 in terms of g and the masses m_1 and m_2.

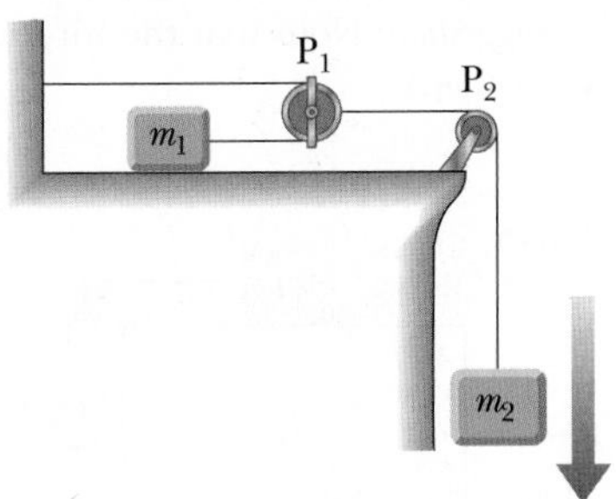

FIGURE **P4.38**

Section 4.8 ■ Context Connection—Forces on Automobiles

39. A young woman buys an inexpensive used car for stock car racing. It can attain highway speed with an acceleration of 8.40 mi/h·s. By making changes to its engine, she can increase the net horizontal force on the car by 24.0%. With much less expense, she can remove material from the body of the car to decrease its mass by 24.0%. (a) Which of these two changes, if either, will result in the greater increase in the car's acceleration? (b) If she makes both changes, what acceleration can she attain?

40. A 1 000-kg car is pulling a 300-kg trailer. Together the car and trailer move forward with an acceleration of 2.15 m/s^2. Ignore any force of air drag on the car and all frictional forces on the trailer. Determine (a) the net force on the car, (b) the net force on the trailer, (c) the force exerted by the trailer on the car, and (d) the resultant force exerted by the car on the road.

Additional Problems

41. An inventive child named Pat wants to reach an apple in a tree without climbing the tree. While sitting in a chair connected to a rope that passes over a frictionless pulley (Fig. P4.41), Pat pulls on the loose end of the rope with such a force that the spring scale reads 250 N. Pat's true weight is 320 N, and the chair weighs 160 N. (a) Draw free-body diagrams for Pat and the chair considered as separate systems and another diagram for Pat and the chair considered as one system. (b) Show that the acceleration of the system is *upward* and find its magnitude. (c) Find the force Pat exerts on the chair.

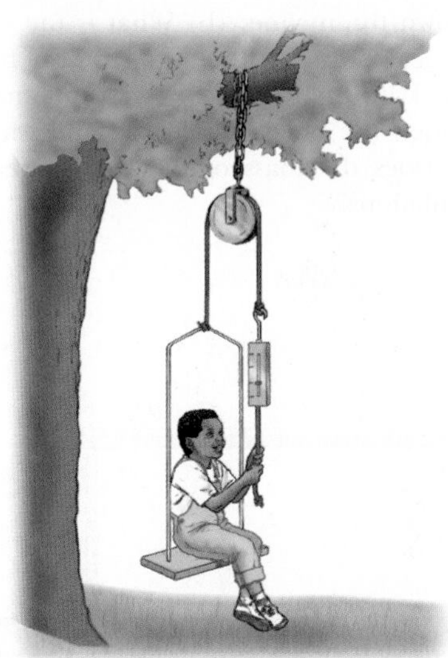

FIGURE **P4.41** Problems 4.41 and 4.42.

42. In the situation described in Problem 4.41 and Figure P4.41, the masses of the rope, spring balance, and pulley are negligible. Pat's feet are not touching the ground. (a) Assume that Pat is momentarily at rest when he stops pulling down on the rope and passes the end of the rope to another child, of weight 440 N, who is standing on the ground next to him. The rope does not break. Describe the ensuing motion. (b) Instead, assume that Pat is momentarily at rest when he ties the rope to a strong hook projecting from the tree trunk. Explain why this action can make the rope break.

43. Three blocks are in contact with one another on a frictionless, horizontal surface as shown in Figure P4.43. A horizontal force $\vec{\mathbf{F}}$ is applied to m_1. Taking $m_1 = 2.00$ kg, $m_2 = 3.00$ kg, $m_3 = 4.00$ kg, and $F = 18.0$ N, draw a separate free-body diagram for each block and find (a) the acceleration of the blocks, (b) the *resultant* force on each block, and (c) the magnitudes of the contact forces between the blocks. (d) You are working on a construction project. A coworker is nailing up plasterboard on one side of a light partition, and you are on the opposite side, providing "backing" by leaning against the wall with your back pushing on it. Every hammer blow makes your back sting.

FIGURE **P4.43**

The supervisor helps you put a heavy block of wood between the wall and your back. Using the situation analyzed in parts (a), (b), and (c) as a model, explain how this change works to make your job more comfortable.

44. **Review problem.** A block of mass $m = 2.00$ kg is released from rest at $h = 0.500$ m above the surface of a table, at the top of a $\theta = 30.0°$ incline as shown in Figure P4.44. The frictionless incline is fixed on a table of height $H = 2.00$ m. (a) Determine the acceleration of the block as it slides down the incline. (b) What is the velocity of the block as it leaves the incline? (c) How far from the table will the block hit the floor? (d) What time interval elapses between when the block is released and when it hits the floor? (e) Does the mass of the block affect any of the above calculations?

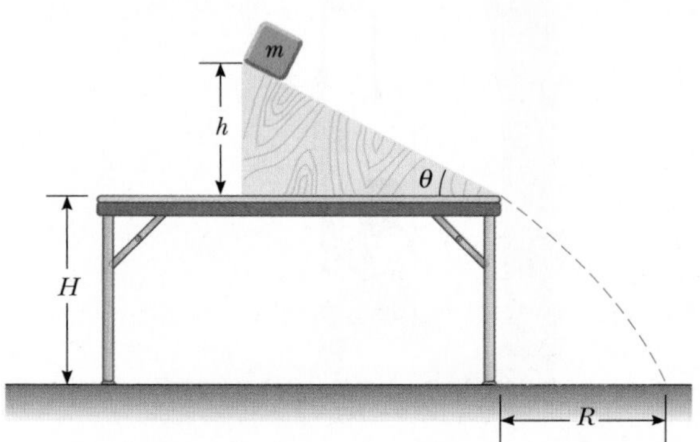

FIGURE **P4.44** Problems 4.44 and 4.55.

45. Physics Now™ An object of mass M is held in place by an applied force $\vec{\mathbf{F}}$ and a pulley system as shown in Figure P4.45. The pulleys are massless and frictionless. Find (a) the tension in each section of rope, T_1, T_2, T_3, T_4, and T_5 and (b) the magnitude of $\vec{\mathbf{F}}$. *Suggestion:* Draw a free-body diagram for each pulley.

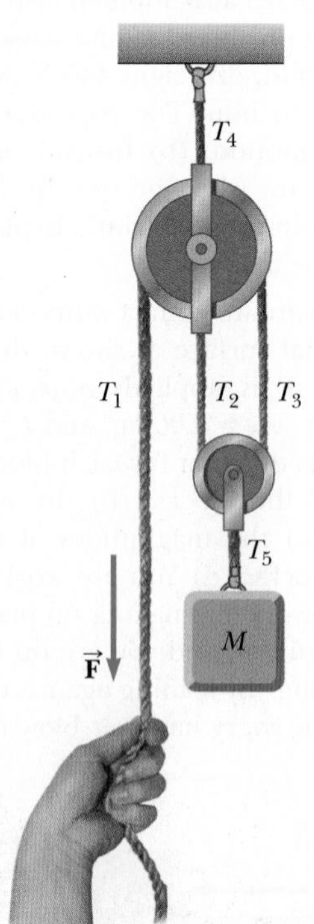

FIGURE **P4.45**

46. A student is asked to measure the acceleration of a cart on a "frictionless" inclined plane as shown in Figure P4.26 and analyzed in Example 4.3, using an air track, a stopwatch, and a meter stick. The height of the incline is measured to be 1.774 cm, and the total length of the incline is measured to be $d = 127.1$ cm. Hence, the angle of inclination θ is determined from the relation $\sin \theta = 1.774/127.1$. The cart is released from rest at the top of the incline, and its position x along the incline is measured as a function of time, where $x = 0$ refers to the initial position of the cart. For x values of 10.0 cm, 20.0 cm, 35.0 cm, 50.0 cm, 75.0 cm, and 100 cm, the measured times at which these positions are reached (averaged over five runs) are 1.02 s, 1.53 s, 2.01 s, 2.64 s, 3.30 s, and 3.75 s, respectively. Construct a graph of x versus t^2 and perform a linear least-squares fit to the data. Determine the acceleration of the cart from the slope of this graph and compare it with the value you would get using $a = g \sin \theta$, where $g = 9.80 \text{ m/s}^2$.

47. What horizontal force must be applied to the cart shown in Figure P4.47 so that the blocks remain stationary relative to the cart? Assume that all surfaces, wheels, and pulley are frictionless. (*Suggestion:* Note that the force exerted by the string accelerates m_1.)

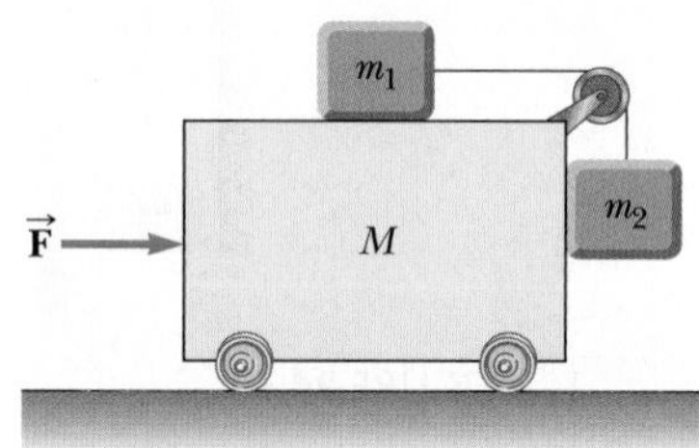

FIGURE **P4.47** Problems 4.47 and 4.48.

48. Initially, the system of objects shown in Figure P4.47 is held motionless. The pulley and all surfaces and wheels are frictionless. Let the force $\vec{\mathbf{F}}$ be zero and assume that m_2 can move only vertically. At the instant after the system of objects is released, find (a) the tension T in the string, (b) the acceleration of m_2, (c) the acceleration of M, and (d) the acceleration of m_1. (*Note:* The pulley accelerates along with the cart.)

49. A 1.00-kg glider on a horizontal air track is pulled by a string at an angle θ. The taut string runs over a pulley and is attached to a hanging object of mass 0.500 kg as shown in Figure P4.49. (a) Show that the speed v_x of the glider and the speed v_y of the hanging object are related by $v_x = uv_y$, where $u = z(z^2 - h_0^2)^{-1/2}$. (b) The glider is released from rest. Show that at that instant the acceleration a_x of the glider and the acceleration a_y of the hanging object are related by $a_x = ua_y$. (c) Find the tension in the string at the instant the glider is released for $h_0 = 80.0$ cm and $\theta = 30.0°$.

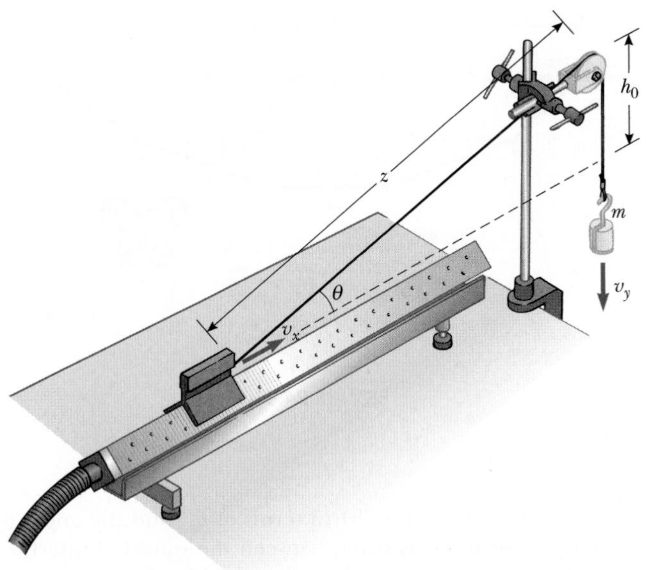

FIGURE **P4.49**

50. Cam mechanisms are used in many machines. For example, cams open and close the valves in your car engine to admit gasoline vapor to each cylinder and to allow the escape of exhaust. The principle is illustrated in Figure P4.50, showing a follower rod (also called a pushrod) of mass m resting on a wedge of mass M. The sliding wedge duplicates the function of a rotating eccentric disk on a car's camshaft. Assume that there is no friction between the wedge and the base, between the pushrod and the wedge, or between the rod and the guide through which it slides. When the wedge is pushed to the left by the force $\vec{\mathbf{F}}$, the rod moves upward and does something such as opening a valve. By varying the shape of the wedge, the motion of the follower rod could be made quite complex, but assume that the wedge makes a constant angle of $\theta = 15.0°$. Suppose you want the wedge and the rod to start from rest and move with constant acceleration, with the rod moving upward 1.00 mm in 8.00 ms. Take $m = 0.250$ kg and $M = 0.500$ kg. What force F must be applied to the wedge?

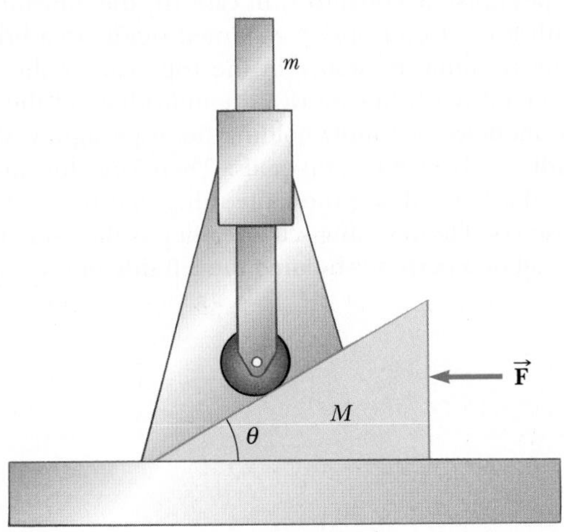

FIGURE **P4.50**

51. If you jump from a desktop and land stiff-legged on a concrete floor, you run a significant risk that you will break a leg. To see how that happens, consider the average force stopping your body when you drop from rest from a height of 1.00 m and stop in a much shorter distance d. Your leg is likely to break at the point where the cross-sectional area of the bone (the tibia) is smallest. This point is just above the ankle, where the cross-sectional area of one bone is about 1.60 cm^2. A bone will fracture when the compressive stress on it exceeds about 1.60×10^8 N/m^2. If you land on both legs, the maximum force that your ankles can safely exert on the rest of your body is then about

$$2(1.60 \times 10^8\ \text{N/m}^2)(1.60 \times 10^{-4}\ \text{m}^2) = 5.12 \times 10^4\ \text{N}.$$

Calculate the minimum stopping distance d that will not result in a broken leg if your mass is 60.0 kg. Don't try it! Bend your knees!

52. Any device that allows you to increase the force you exert is a kind of *machine.* Some machines, such as the prybar or the inclined plane, are very simple. Some machines do not even look like machines. For example, your car is stuck in the mud and you can't pull hard enough to get it out. You do, however, have a long cable that you connect taut between your front bumper and the trunk of a stout tree. You now pull sideways on the cable at its midpoint, exerting a force f. Each half of the cable is displaced through a small angle θ from the straight line between the ends of the cable. (a) Deduce an expression for the force acting on the car. (b) Evaluate the cable tension for the case where $\theta = 7.00°$ and $f = 100$ N.

53. A van accelerates down a hill (Fig. P4.53), going from rest to 30.0 m/s in 6.00 s. During the acceleration, a toy ($m = 0.100$ kg) hangs by a string from the van's ceiling. The acceleration is such that the string remains perpendicular to the ceiling. Determine (a) the angle θ and (b) the tension in the string.

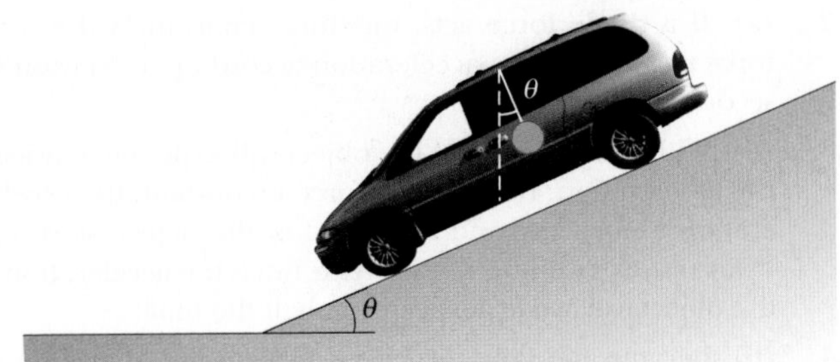

FIGURE **P4.53**

54. Two blocks of mass 3.50 kg and 8.00 kg are connected by a massless string that passes over a frictionless pulley (Fig. P4.54). The inclines are frictionless. Find (a) the

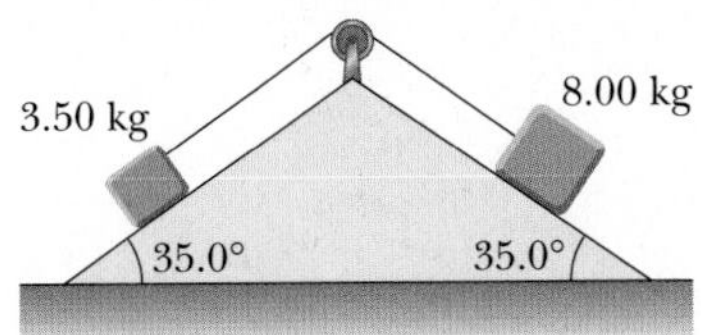

FIGURE **P4.54** Problems 4.54 and 5.41.

magnitude of the acceleration of each block and (b) the tension in the string.

55. In Figure P4.44, the incline has mass M and is fastened to the stationary horizontal tabletop. The block of mass m is placed near the bottom of the incline and is released with a quick push that sets it sliding upward. It stops near the top of the incline, as shown in the figure, and then slides down again, always without friction. Find the force that the tabletop exerts on the incline throughout this motion.

56. An 8.40-kg object slides down a fixed, frictionless inclined plane. Use a computer to determine and tabulate the normal force exerted on the object and its acceleration for a series of incline angles (measured from the horizontal) ranging from 0° to 90° in 5° increments. Plot a graph of the normal force and the acceleration as functions of the incline angle. In the limiting cases of 0° and 90°, are your results consistent with the known behavior?

57. A mobile is formed by supporting four metal butterflies of equal mass m from a string of length L. The points of support are evenly spaced a distance ℓ apart as shown in Figure P4.57. The string forms an angle θ_1 with the ceiling at each end point. The center section of string is horizontal. (a) Find the tension in each section of string in terms of θ_1, m, and g. (b) In terms of θ_1, find the angle θ_2 that the sections of string between the outside butterflies and the inside butterflies form with the horizontal. (c) Show that the distance D between the end points of the string is

$$D = \frac{L}{5}(2\cos\theta_1 + 2\cos[\tan^{-1}(\tfrac{1}{2}\tan\theta_1)] + 1)$$

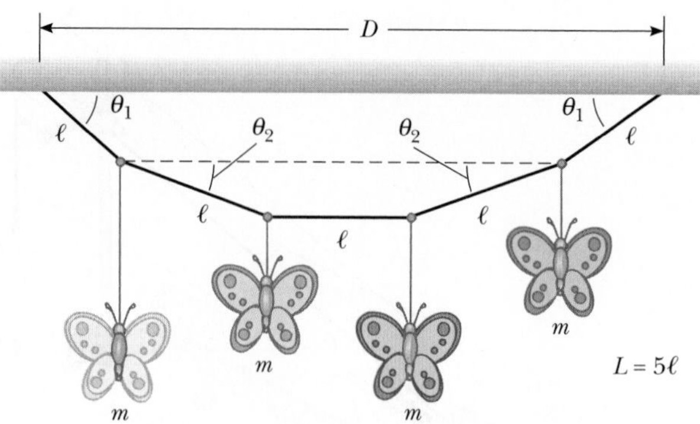

FIGURE **P4.57**

ANSWERS TO QUICK QUIZZES

4.1 (d). Choice (a) is true. Newton's first law tells us that motion requires no force: An object in motion continues to move at constant velocity in the absence of external forces. Choice (b) is also true: A stationary object can have several forces acting on it, but if the vector sum of all these external forces is zero, there is no net force and the object remains stationary.

4.2 (a). If a single force acts, this force constitutes the net force and there is an acceleration according to Newton's second law.

4.3 (d). With twice the force, the object will experience twice the acceleration. Because the force is constant, the acceleration is constant, and the speed of the object, starting from rest, is given by $v = at$. With twice the acceleration, the object will arrive at speed v at half the time.

4.4 (b). Because the value of g is smaller on the Moon than on the Earth, more mass of gold would be required to represent 1 N of weight on the Moon. Therefore, your friend on the Moon is richer, by about a factor of 6!

4.5 (c), (d). In accordance with Newton's third law, the fly and the bus experience forces that are equal in magnitude but opposite in direction. Because the fly has such a small mass, Newton's second law tells us that it undergoes a very large acceleration. The huge mass of the bus means that it more effectively resists any change in its motion and exhibits a small acceleration.

4.6 (c). The reaction force to the gravitational force on you is an upward gravitational force on the Earth caused by you.

4.7 (c). The scale is in equilibrium in both situations, so it experiences a net force of zero. Because each individual pulls with a force F and there is no acceleration, each individual is in equilibrium. Therefore, the tension in the ropes must be equal to F. In case (i), the individual pulls with force F on a spring mounted rigidly to a brick wall. The resulting tension F in the rope causes the scale to read a force F. In case (ii), the individual on the left can be modeled as simply holding the rope tightly while the individual on the right pulls. Therefore, the individual on the left is doing the same thing that the wall does in case (i). The resulting scale reading is the same whether a wall or a person is holding the left side of the scale.

CHAPTER 5

More Applications of Newton's Laws

(© Paul Hardy/CORBIS)

The London Eye, a ride on the River Thames in downtown London. Riders travel in a large vertical circle for a breathtaking view of the city. In this chapter, we will study the forces involved in circular motion.

CHAPTER OUTLINE

In Chapter 4, we introduced Newton's laws of motion and applied them to situations in which we ignored friction. In this chapter, we shall expand our investigation to objects moving in the presence of friction, which will allow us to model situations more realistically. Such objects include those sliding on rough surfaces and those moving through viscous media such as liquids and air. We also apply Newton's laws to the dynamics of circular motion so that we can understand more about objects moving in circular paths under the influence of various types of forces.

5.1 FORCES OF FRICTION

When an object moves either on a surface or through a viscous medium such as air or water, there is resistance to the motion because the object interacts with its surroundings. We call such resistance a **force of friction.** Forces of friction are very important in our everyday lives. They allow us to walk or run and are necessary for the motion of wheeled vehicles.

Imagine you are working in your garden and have filled a trash can with yard clippings. You then try to drag the trash can across the surface of your concrete patio as in Active Figure 5.1a. The patio surface is *real,* not an idealized, frictionless surface in a simplification model. If we apply an external horizontal force $\vec{\mathbf{F}}$ to the trash can, acting to the right, the trash can remains stationary if $\vec{\mathbf{F}}$ is small. The force that counteracts $\vec{\mathbf{F}}$ and keeps the trash can from moving acts to the left and is called the **force of static friction** $\vec{\mathbf{f}}_s$. As long as the trash can is not moving, it is modeled as a particle in equilibrium and $f_s = F$. Therefore, if $\vec{\mathbf{F}}$ is increased in magnitude, the magnitude of $\vec{\mathbf{f}}_s$ also increases. Likewise, if $\vec{\mathbf{F}}$ decreases, $\vec{\mathbf{f}}_s$ also decreases. Experiments show that the friction force arises from the nature of the two surfaces; because of their roughness, contact is made only at a few points, as shown in the magnified surface view in Active Figure 5.1a.

If we increase the magnitude of $\vec{\mathbf{F}}$, as in Active Figure 5.1b, the trash can eventually slips. When the trash can is on the verge of slipping, f_s is a maximum as shown in Active Figure 5.1c. If F exceeds $f_{s,\text{max}}$, the trash can moves and accelerates to the right. While the trash can is in motion, the friction force is less than $f_{s,\text{max}}$ (Active Fig. 5.1c). We call the friction force for an object in motion the **force of**

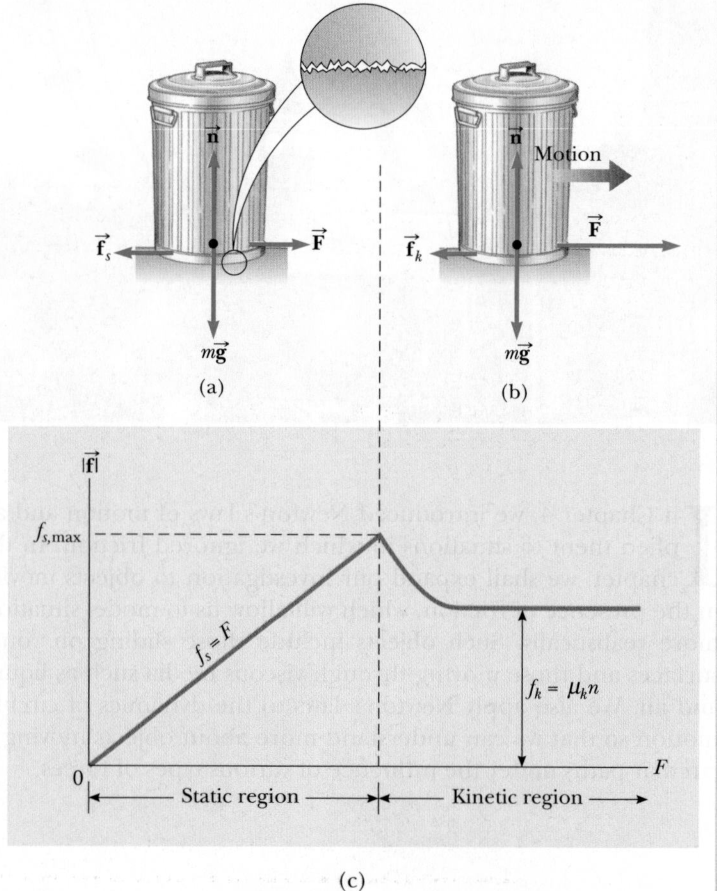

ACTIVE FIGURE 5.1

(a) The force of static friction $\vec{\mathbf{f}}_s$ between a trash can and a concrete patio is opposite the applied force $\vec{\mathbf{F}}$. The magnitude of the force of static friction equals that of the applied force. (b) When the magnitude of the applied force exceeds the magnitude of the force of kinetic friction $\vec{\mathbf{f}}_k$, the trash can accelerates to the right. (c) A graph of the magnitude of the friction force versus that of the applied force. In our model, the force of kinetic friction is independent of the applied force and the relative speed of the surfaces. Note that $f_{s,\text{max}} > f_k$.

Physics Now™ You can vary the load in the trash can and practice sliding it on surfaces of varying roughness by logging into PhysicsNow at **www.pop4e.com** and going to Active Figure 5.1. Note the effect on the trash can's motion and the corresponding behavior of the graph in (c).

kinetic friction $\vec{\mathbf{f}}_k$. The net force $F - f_k$ in the x direction produces an acceleration to the right, according to Newton's second law. If we reduce the magnitude of $\vec{\mathbf{F}}$ so that $F = f_k$, the acceleration is zero and the trash can moves to the right with constant speed. If the applied force is removed, the friction force acting to the left provides an acceleration of the trash can in the $-x$ direction and eventually brings it to rest.

Experimentally, one finds that, to a good approximation, both $f_{s,\max}$ and f_k for an object on a surface are proportional to the normal force exerted by the surface on the object; thus, we adopt a simplification model in which this approximation is assumed to be exact. The assumptions in this simplification model can be summarized as follows:

- The magnitude of the force of static friction between any two surfaces in contact can have the values

$$f_s \leq \mu_s n \qquad [5.1]$$

■ Force of static friction

where the dimensionless constant μ_s is called the **coefficient of static friction** and n is the magnitude of the normal force. The equality in Equation 5.1 holds when the surfaces are on the verge of slipping, that is, when $f_s = f_{s,\max} \equiv \mu_s n$. This situation is called *impending motion.* The inequality holds when the component of the applied force parallel to the surfaces is less than this value.

- The magnitude of the force of kinetic friction acting between two surfaces is

$$f_k = \mu_k n \qquad [5.2]$$

■ Force of kinetic friction

where μ_k is the **coefficient of kinetic friction.** In our simplification model, this coefficient is independent of the relative speed of the surfaces.

- The values of μ_k and μ_s depend on the nature of the surfaces, but μ_k is generally less than μ_s. Table 5.1 lists some measured values.
- The direction of the friction force on an object is opposite to the actual motion (kinetic friction) or the impending motion (static friction) of the object relative to the surface with which it is in contact.

The approximate nature of Equations 5.1 and 5.2 is easily demonstrated by trying to arrange for an object to slide down an incline at constant speed. Especially at low speeds, the motion is likely to be characterized by alternate stick and slip episodes. The simplification model described in the bulleted list above has been developed so that we can solve problems involving friction in a relatively straightforward way.

PITFALL PREVENTION 5.1

The equal sign is used in limited situations In Equation 5.1, the equal sign is used *only* when the surfaces are just about to break free and begin sliding. Do not fall into the common trap of using $f_s = \mu_s n$ in *any* static situation.

PITFALL PREVENTION 5.2

The direction of the friction force Sometimes, an incorrect statement about the friction force between an object and a surface is made—"The friction force on an object is opposite to its motion or impending motion"—rather than the correct phrasing, "The friction force on an object is opposite to its motion or impending motion *relative to the surface.*" Think carefully about Quick Quiz 5.2.

TABLE 5.1 Coefficients of Friction

	μ_s	μ_k
Steel on steel	0.74	0.57
Aluminum on steel	0.61	0.47
Copper on steel	0.53	0.36
Rubber on concrete	1.0	0.8
Wood on wood	0.25–0.5	0.2
Glass on glass	0.94	0.4
Waxed wood on wet snow	0.14	0.1
Waxed wood on dry snow	—	0.04
Metal on metal (lubricated)	0.15	0.06
Ice on ice	0.1	0.03
Teflon on Teflon	0.04	0.04
Synovial joints in humans	0.01	0.003

Note: All values are approximate.

Now that we have identified the characteristics of the friction force, we can include the friction force in the net force on an object in the model of a particle under a net force.

QUICK QUIZ 5.1 You press your physics textbook flat against a vertical wall with your hand, which applies a normal force perpendicular to the book. What is the direction of the friction force on the book due to the wall? **(a)** downward **(b)** upward **(c)** out from the wall **(d)** into the wall

QUICK QUIZ 5.2 A crate is located in the center of a flatbed truck. The truck accelerates to the east and the crate moves with it, not sliding at all. What is the direction of the friction force exerted by the truck on the crate? **(a)** It is to the west. **(b)** It is to the east. **(c)** No friction force exists because the crate is not sliding.

QUICK QUIZ 5.3 You are playing with your daughter in the snow. She sits on a sled and asks you to slide her across a flat, horizontal field. You have a choice of **(a)** pushing her from behind, by applying a force downward on her shoulders at 30° below the horizontal (Fig. 5.2a) or **(b)** attaching a rope to the front of the sled and pulling with a force at 30° above the horizontal (Fig 5.2b). Which would require less force for a given acceleration of the daughter?

Thinking Physics 5.1

In the motion picture *The Abyss* (Twentieth Century Fox, 1989), an underwater oil exploration rig is located at the ocean bottom in very deep water. It is connected to a ship on the ocean surface by a cable called an "umbilical cord" as suggested in Figure 5.3a. On the ship, the umbilical cord is attached to a gantry. During a hurricane, the gantry structure breaks loose from the ship, falls into the water, and sinks to the bottom, passing over the edge of an extremely deep abyss. As a result, the rig is dragged by the umbilical cord along the ocean bottom as described in Figure 5.3b. As the rig approaches the edge of the abyss, however, it is not pulled over the edge but rather, stops just short of the edge as shown in Figure 5.3c. Is this scenario purely a cinematic edge-of-the-seat situation, or do the principles of physics suggest why the moving rig does not topple over the edge?

Reasoning Physics can explain this phenomenon. While the rig is being pulled across the ocean floor (Fig. 5.3b), it is pulled by the section of the umbilical cord that is almost horizontal and therefore almost parallel to the ocean floor. Therefore, the rig is subject to two horizontal forces: the tension in the umbilical cord

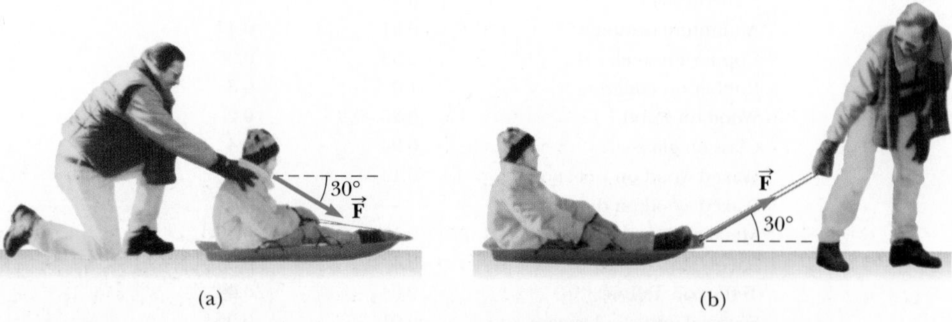

FIGURE 5.2 (Quick Quiz 5.3) A father tries to slide his daughter on a sled over snow by (a) pushing downward on her shoulders or (b) pulling upward on a rope attached to the sled. Which is easier?

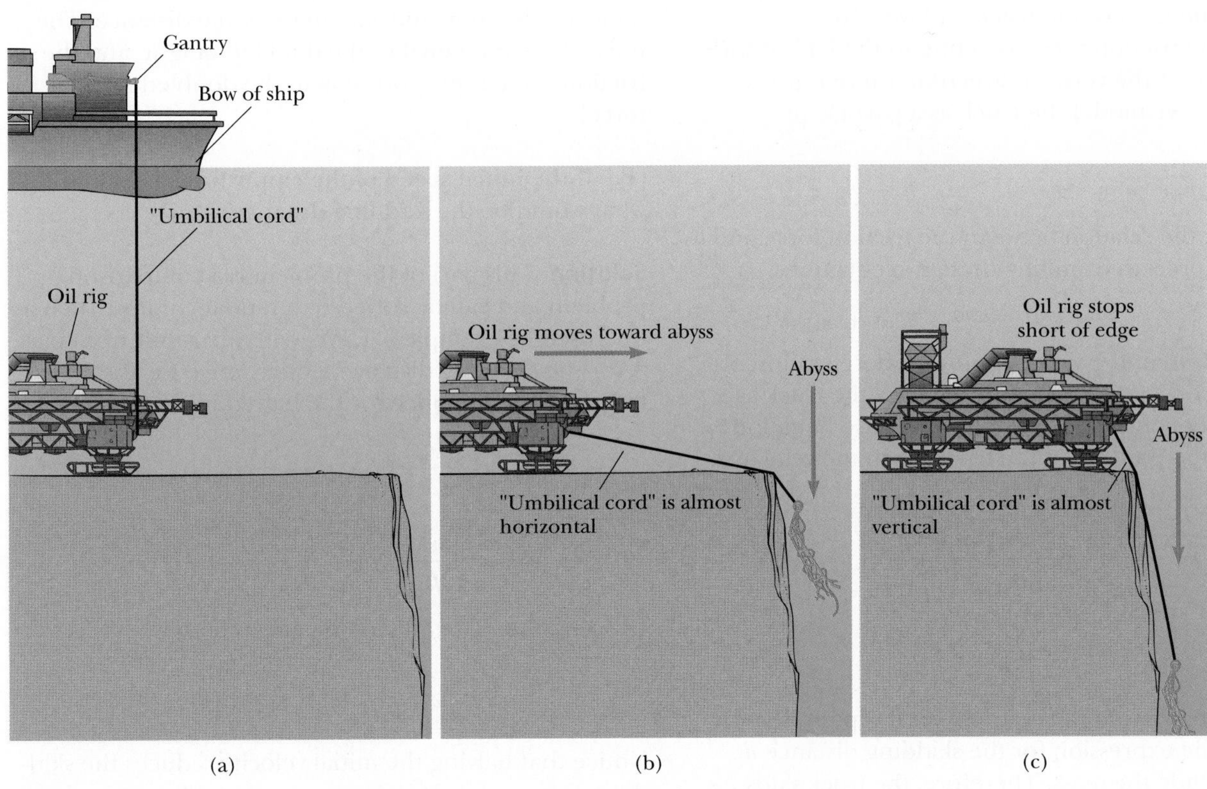

FIGURE 5.3 (Thinking Physics 5.1) An oil rig at the bottom of the ocean is dragged by a cable.

pulling it forward and friction with the ocean floor pulling back. Let us assume that these forces are equal in magnitude so that the rig moves with constant speed. As the rig nears the edge of the abyss, the angle the umbilical cord makes with the horizontal increases. As a result, the component of the force from the cord parallel to the ocean floor decreases and the downward vertical component increases. As a result of the increased vertical force, the rig is pulled downward more strongly to the ocean floor, increasing the normal force on it and, in turn, increasing the friction force between the rig and the ocean floor. Therefore, with less force pulling it forward (from the umbilical cord) and more force opposing the motion (as a result of friction), the rig slows down. By the time the rig reaches the edge of the abyss, the force from the umbilical cord is almost *straight down* (Fig. 5.3c), resulting in little forward force. Furthermore, this large downward force pulls the rig into the ocean floor, resulting in a very large friction force that stops the rig. ■

EXAMPLE 5.1 The Skidding Truck

The driver of an empty speeding truck slams on the brakes and skids to a stop through a distance d.

A If the truck carries a heavy load such that the moving mass is doubled, what would be its skidding distance if it starts from the same initial speed?

Solution Figure 5.4 shows a free-body diagram for the skidding truck. The only force in the horizontal direction is the friction force, which is assumed to be independent of speed in our simplification model for friction. Therefore, from Newton's second law,

$$\sum F_x = -f_k = ma$$

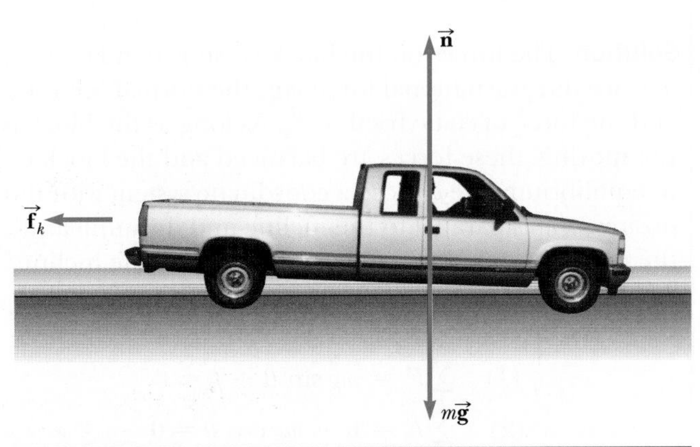

FIGURE 5.4 (Example 5.1) A truck skids to a stop.

where m is the mass of the truck and we have expressed the friction force as acting to the left, in the $-x$ direction. In the vertical direction, there is no acceleration, so we model the truck as a particle in equilibrium:

$$\sum F_y = n - mg = 0 \quad \rightarrow \quad n = mg$$

Finally, from the relation between the friction force and the normal force, we combine these two equations:

$$f_k = \mu_k n \quad \rightarrow \quad -\mu_k(mg) = ma \quad \rightarrow \quad a = -\mu_k g$$

Because both μ_k and g are constant, the acceleration of the truck is constant. We therefore model the truck as a particle under constant acceleration. We use Equation 2.13 to find the position of the truck when the velocity is zero:

$$v_{xf}{}^2 = v_{xi}{}^2 + 2a_x(x_f - x_i)$$

$$0 = v_{xi}{}^2 - 2(\mu_k g)(x_f - 0)$$

$$x_f = d = \frac{v_{xi}{}^2}{2\mu_k g}$$

We can argue from the mathematical representation as follows. The expression for the skidding distance d does not include the mass. Therefore, the truck skids the same distance regardless of the mass of the load. Conceptually, we can argue that the truck with twice the mass requires twice the friction force to exhibit the same acceleration and stop in the same distance. The normal force is equal to the doubled weight, and the friction force is proportional to the doubled normal force!

B If the initial speed of the empty truck is halved, what would be the skidding distance?

Solution This part of the problem is a comparison problem and can be solved by a ratio technique such as that used in Example 3.4. We write the result from part A for the skidding distance d twice, once for the original situation and once for the halved initial velocity:

$$d_1 = \frac{v_{1xi}^2}{2\mu_k g}$$

$$d_2 = \frac{v_{2xi}^2}{2\mu_k g} = \frac{\left(\frac{1}{2}v_{1xi}\right)^2}{2\mu_k g} = \frac{1}{4}\frac{v_{1xi}^2}{2\mu_k g}$$

Dividing the first equation by the second, we have

$$\frac{d_1}{d_2} = 4 \quad \rightarrow \quad d_2 = \tfrac{1}{4}d_1$$

Notice that halving the initial velocity reduces the skidding distance by 75%! This important safety consideration is associated with the possibility of an accident when driving at high speed.

EXAMPLE 5.2 Experimental Determination of μ_s and μ_k

The following is a simple method of measuring coefficients of friction. Suppose a block is placed on a rough surface inclined relative to the horizontal, as shown in Figure 5.5. The incline angle θ is increased until the block starts to move.

A How is the coefficient of static friction related to the critical angle θ_c at which the block begins to move?

Solution The forces on the block, as shown in Figure 5.5, are the gravitational force $m\vec{g}$, the normal force $\vec{n}$, and the force of static friction $\vec{f}_s$. As long as the block is not moving, these forces are balanced and the block is in equilibrium. We choose a coordinate system with the positive x axis parallel to the incline and downhill and the positive y axis upward perpendicular to the incline. Applying Newton's second law in component form to the block gives

$$(1) \quad \sum F_x = mg\sin\theta - f_s = 0$$

$$(2) \quad \sum F_y = n - mg\cos\theta = 0$$

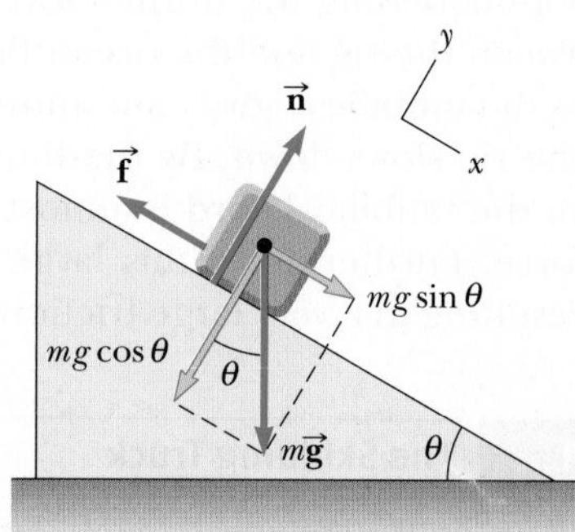

FIGURE 5.5 (Example 5.2) A block on an adjustable incline is used to determine the coefficients of friction.

These equations are valid for any angle of inclination θ. At the critical angle θ_c at which the block is on the verge of slipping, the friction force has its maximum magnitude $\mu_s n$, so we rewrite (1) and (2) for this condition as

$$(3) \quad mg\sin\theta_c = \mu_s n$$

$$(4) \quad mg\cos\theta_c = n$$

Dividing (3) by (4), we have

$$\tan \theta_c = \mu_s$$

Therefore, the coefficient of static friction is equal to the tangent of the angle of the incline at which the block begins to slide.

B How could we find the coefficient of kinetic friction?

Solution Once the block begins to move, the magnitude of the friction force is the kinetic value $\mu_k n$, which is smaller than that of the force of static friction. As a result, if the angle is maintained at the critical angle, the block accelerates down the incline. To restore the equilibrium situation in Equation (1), with f_s replaced by f_k, the angle must be reduced to a value θ'_c such that the block slides down the incline at *constant speed.* In this situation, Equations (3) and (4), with θ_c replaced by θ'_c and μ_s by μ_k, give us

$$\tan \theta'_c = \mu_k$$

EXAMPLE 5.3 Connected Objects

A ball and a cube are connected by a light string that passes over a frictionless light pulley, as in Figure 5.6a. The coefficient of kinetic friction between the cube and the surface is 0.30. Find the acceleration of the two objects and the tension in the string.

Solution To *conceptualize* the problem, imagine the ball moving downward and the cube sliding to the right, both accelerating from rest. We recognize that there are two objects that are accelerating, so we *categorize* this problem as one involving particles under a net force, where one of the forces to be included is the friction force. To begin to *analyze* the problem, we set up a simplified pictorial representation by drawing the free-body diagrams for the two objects as in Figures 5.6b and 5.6c. For the ball, no forces are exerted in the horizontal direction, and we apply Newton's second law in the vertical direction. For the cube, the acceleration is horizontal, so we know the cube is in equilibrium in the vertical direction. We use the fact that the magnitude of the force of kinetic friction acting on the cube is proportional to the normal force according to $f_k = \mu_k n$. Because the pulley is light (massless) and frictionless, the tension in the string is the same on both sides of the pulley. Because the tension acts on both objects, it is the common quantity that applies to both objects and allows us to combine separate equations for the two objects into one equation.

Let us address the cube of mass m_1 first. Newton's second law applied to the cube in component form, with the positive x direction to the right, gives

$$\sum F_x = m_1 a \quad \rightarrow \quad T - f_k = m_1 a$$

$$\sum F_y = 0 \quad \rightarrow \quad n - m_1 g = 0$$

where T is the tension in the string. Because $f_k = \mu_k n$ and $n = m_1 g$ from the equilibrium equation for the y direction, we have $f_k = \mu_k m_1 g$. Therefore, from the equation for the x direction,

$$(1) \quad T = \mu_k m_1 g + m_1 a$$

Now we apply Newton's second law to the ball moving in the vertical direction. Because the ball moves downward when the cube moves to the right, we choose the positive direction downward for the ball:

$$(2) \quad \sum F_y = m_2 a \quad \rightarrow \quad m_2 g - T = m_2 a$$

Substituting the expression for T from (1) into (2) gives us

$$m_2 g - (\mu_k m_1 g + m_1 a) = m_2 a$$

$$a = \frac{m_2 - \mu_k m_1}{m_1 + m_2} g$$

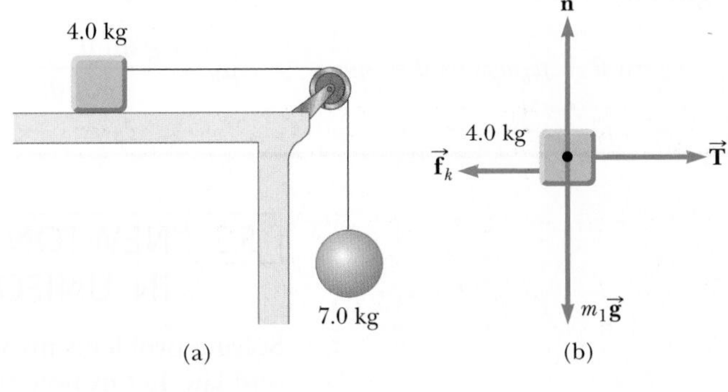

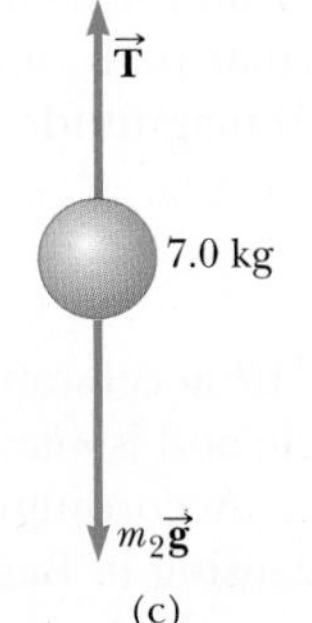

FIGURE 5.6 (Example 5.3) (a) Two objects connected by a light string that passes over a frictionless pulley. (b) Free-body diagram for the sliding cube. (c) Free-body diagram for the hanging ball.

Now, substituting the known values,

$$a = \frac{7.0\text{ kg} - 0.30(4.0\text{ kg})}{7.0\text{ kg} + 4.0\text{ kg}}(9.80\text{ m/s}^2) = \boxed{5.2\text{ m/s}^2}$$

which is the *magnitude* of the acceleration of each of the two objects. For the ball, the acceleration *vector* is downward and the vector is toward the right for the cube. When the magnitude of the acceleration is substituted into (1), we find the tension:

$$T = 0.30(4.0\text{ kg})(9.80\text{ m/s}^2) + (4.0\text{ kg})(5.2\text{ m/s}^2) = \boxed{33\text{ N}}$$

To *finalize* the problem, note that the acceleration is smaller than that due to gravity. That does not tell us that the answer is correct, but if the acceleration were larger than g, it would tell is that we have made an error. Note also that the tension in the string is smaller than $m_2g = (7.0\text{ kg})(9.80\text{ m/s}^2) = 69\text{ N}$, which is consistent with m_2 accelerating downward.

EXAMPLE 5.4 **The Sliding Crate**

A warehouse worker places a crate on a sloped surface that is inclined at 30.0° with respect to the horizontal (Fig. 5.7a). If the crate slides down the incline with an acceleration of magnitude $g/3$, determine the coefficient of kinetic friction between the crate and the surface.

Solution Figure 5.7b shows the forces acting on the crate. The x axis is chosen parallel to the incline and the y axis perpendicular. From Newton's second law,

$$(1)\quad \sum F_x = ma \quad\rightarrow\quad mg\sin\theta - f_k = ma$$

$$(2)\quad \sum F_y = 0 \quad\rightarrow\quad n - mg\cos\theta = 0$$

The kinetic friction force is $f_k = \mu_k n$ and, from (2), we find that $n = mg\cos\theta$. Therefore, the friction force can be expressed as $f_k = \mu_k mg\cos\theta$. Substituting into (1) gives us

$$mg\sin\theta - \mu_k mg\cos\theta = ma \quad\rightarrow\quad \mu_k = \frac{g\sin\theta - a}{g\cos\theta}$$

Substituting the known values, we have

$$\mu_k = \frac{g\sin 30.0° - \frac{1}{3}g}{g\cos 30.0°} = \frac{(0.500 - 0.333)}{0.867} = \boxed{0.192}$$

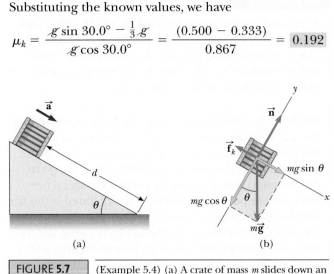

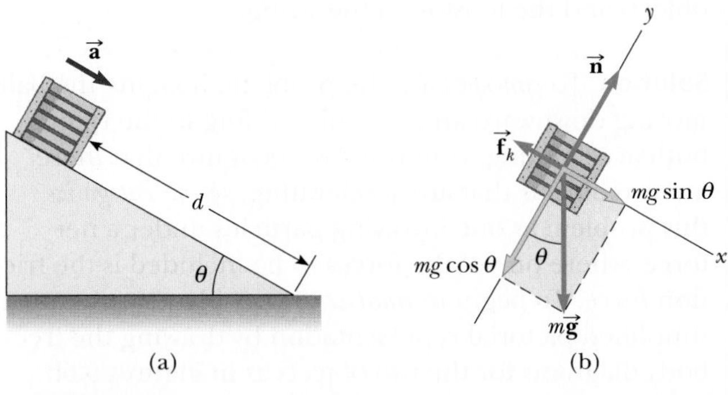

FIGURE 5.7 (Example 5.4) (a) A crate of mass m slides down an incline. (b) Free-body diagram for the sliding crate.

5.2 NEWTON'S SECOND LAW APPLIED TO A PARTICLE IN UNIFORM CIRCULAR MOTION

Solving problems involving friction is just one of many applications of Newton's second law. Let us now consider another common situation, associated with a particle in uniform circular motion. In Chapter 3, we found that a particle moving in a circular path of radius r with uniform speed v experiences a centripetal acceleration of magnitude

■ Centripetal acceleration

$$a_c = \frac{v^2}{r}$$

The acceleration vector with this magnitude is directed toward the center of the circle and is *always* perpendicular to $\vec{\mathbf{v}}$.

According to Newton's second law, if an acceleration occurs, a net force must be causing it. Because the acceleration is toward the center of the circle, the net force must be toward the center of the circle. Therefore, when a particle travels in a circular path, a force must be acting *inward* on the particle that causes the circular motion. We investigate the forces causing this type of acceleration in this section.

Consider an object of mass m tied to a string of length r and being whirled in a horizontal circular path on a frictionless table top as in the overhead view in

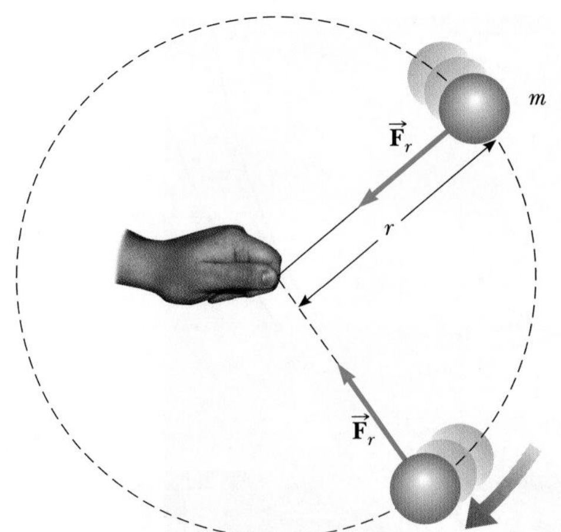

FIGURE 5.8 Overhead view of a ball moving in a circular path in a horizontal plane. A force $\vec{\mathbf{F}}_r$ directed toward the center of the circle keeps the ball moving in its circular path.

Figure 5.8. Let us assume that the object moves with constant speed. The natural tendency of the object is to move in a straight-line path, according to Newton's first law; the string, however, prevents this motion along a straight line by exerting a radial force $\vec{\mathbf{F}}_r$ on the object to make it follow a circular path. This force, whose magnitude is the tension in the string, is directed along the length of the string toward the center of the circle as shown in Figure 5.8.

In this discussion, the tension in the string causes the circular motion. Other forces also cause objects to move in circular paths. For example, friction forces cause automobiles to travel around curved roadways and the gravitational force causes a planet to orbit the Sun.

Regardless of the nature of the force acting on the particle in circular motion, we can apply Newton's second law to the particle along the radial direction:

$$\sum F = ma_c = m\frac{v^2}{r} \qquad [5.3]$$

In general, an object can move in a circular path under the influence of various types of forces, or a *combination* of forces, as we shall see in some of the examples that follow.

If the force acting on an object vanishes, the object no longer moves in its circular path; instead, it moves along a straight-line path tangent to the circle. This idea is illustrated in Active Figure 5.9 for the case of the ball whirling in a circle at the

PITFALL PREVENTION 5.3

Centripetal force The force causing centripetal acceleration is called *centripetal force* in some textbooks. Giving the force causing circular motion a name leads many students to consider it as a new *kind* of force rather than a new *role* for force. A common mistake is to draw the forces in a free-body diagram and then add another vector for the centripetal force. Yet it is not a separate force; it is one of our familiar forces *acting in the role of causing a circular motion.* For the motion of the Earth around the Sun, for example, the "centripetal force" is *gravity.* For a rock whirled on the end of a string, the "centripetal force" is the *tension* in the string. After this discussion, we shall no longer use the phrase *centripetal force.*

ACTIVE FIGURE 5.9
An overhead view of a ball moving in a circular path in a horizontal plane. When the string breaks, the ball moves in the direction tangent to the circular path.

PhysicsNow™ Log into PhysicsNow at **www.pop4e.com** and go to Active Figure 5.9 to "break" the string yourself and observe the effect on the ball's motion.

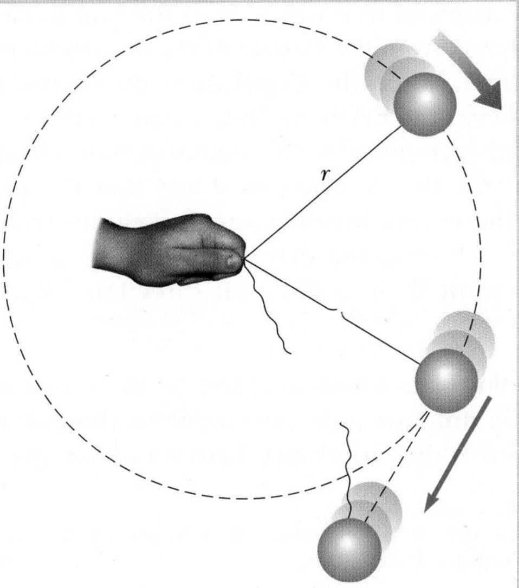

PITFALL PREVENTION 5.4

Direction of travel when the string is cut Study Active Figure 5.9 carefully. Many students have a misconception that the ball moves *radially* away from the center of the circle when the string is cut. The velocity of the ball is *tangent* to the circle. By Newton's first law, the ball simply continues to move in the direction that it is moving just as the force from the string disappears.

The cars of a corkscrew roller coaster must travel in tight loops. The normal force from the track contributes to the centripetal acceleration. The gravitational force, because it remains constant in direction, is sometimes in the same direction as the normal force, but is sometimes in the opposite direction. ■

end of a string. If the string breaks at some instant, the ball moves along the straight-line path tangent to the circle at the point on the circle at which the ball is located at that instant.

PITFALL PREVENTION 5.5

Centrifugal force The commonly heard phrase "centrifugal force" is described as a force pulling *outward* on an object moving in a circular path. If you are experiencing a "centrifugal force" on a rotating carnival ride, what is the other object with which you are interacting? You cannot identify another object because it is a fictitious force that occurs as a result of your being in a noninertial reference frame.

QUICK QUIZ 5.4 You are riding on a Ferris wheel (Fig. 5.10) that is rotating with constant speed. The car in which you are riding always maintains its correct upward orientation; it does not invert. **(i)** What is the direction of the normal force on you from the seat when you are at the top of the wheel? **(a)** upward **(b)** downward **(c)** impossible to determine. **(ii)** What is the direction of the net force on you when you are at the top of the wheel? **(a)** upward **(b)** downward **(c)** impossible to determine

FIGURE 5.10 (Quick Quiz 5.4) A Ferris wheel located on Navy Pier in Chicago, Illinois.

■ Thinking Physics 5.2

The Copernican theory of the solar system is a structural model in which the planets are assumed to travel around the Sun in circular orbits. Historically, this theory was a break from the Ptolemaic theory, a structural model in which the Earth was at the center. When the Copernican theory was proposed, a natural question arose: What keeps the Earth and other planets moving in their paths around the Sun? An interesting response to this question comes from Richard Feynman[1]: "In those days, one of the theories proposed was that the planets went around because behind them there were invisible angels, beating their wings and driving the planets forward. . . . It turns out that in order to keep the planets going around, the invisible angels must fly in a different direction." What did Feynman mean by this statement?

Reasoning The question asked by those at the time of Copernicus indicates that they did not have a proper understanding of inertia as described by Newton's first law. At that time in history, before Galileo and Newton, the interpretation was that

[1] R. P. Feynman, R. B. Leighton, and M. Sands, *The Feynman Lectures on Physics,* Vol. 1, (Reading, MA: Addison-Wesley, 1963), p. 7-2.

motion was caused by force. This interpretation is different from our current understanding that *changes in motion* are caused by force. Therefore, it was natural for Copernicus's contemporaries to ask what force propelled a planet in its orbit. According to our current understanding, it is equally natural for us to realize that no force tangent to the orbit is necessary, that the motion simply continues owing to inertia.

Therefore, in Feynman's imagery, the angels do not have to push the planet *from behind.* The angels must push *inward,* to provide the centripetal acceleration associated with the orbital motion of the planet. Of course, the angels are not real from a scientific point of view, but are a metaphor for the *gravitational force.* ■

EXAMPLE 5.5 How Fast Can It Spin?

An object of mass 0.500 kg is attached to the end of a cord whose length is 1.50 m. The object is whirled in a horizontal circle as in Figure 5.8. If the cord can withstand a maximum tension of 50.0 N, what is the maximum speed the object can have before the cord breaks?

Solution Because the magnitude of the force that provides the centripetal acceleration of the object in this case is the tension T exerted by the cord on the object, Newton's second law gives us for the inward radial direction

$$\sum F_r = ma_c \quad \rightarrow \quad T = m\frac{v^2}{r}$$

Solving for the speed v, we have

$$v = \sqrt{\frac{Tr}{m}}$$

The maximum speed that the object can have corresponds to the maximum value of the tension. Hence, we find

$$v_{\text{max}} = \sqrt{\frac{T_{\text{max}}r}{m}} = \sqrt{\frac{(50.0 \text{ N})(1.50 \text{ m})}{0.500 \text{ kg}}} = 12.2 \text{ m/s}$$

EXAMPLE 5.6 The Conical Pendulum

A small object of mass m is suspended from a string of length L. The object revolves in a horizontal circle of radius r with constant speed v, as in Figure 5.11a. (Because the string sweeps out the surface of a cone, the system is known as a *conical pendulum.*)

A Find the speed of the object.

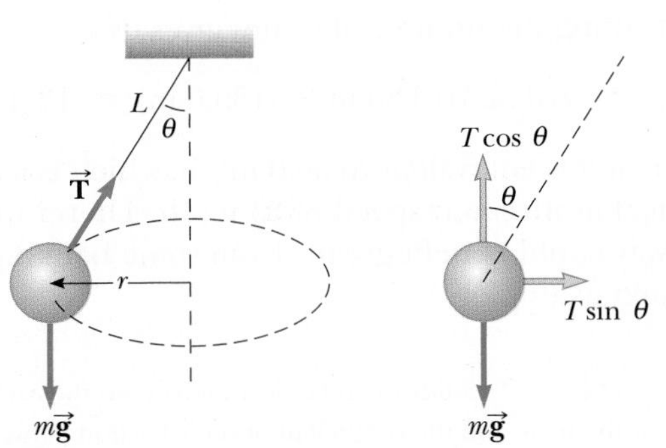

FIGURE 5.11 (Example 5.6) The conical pendulum and its free-body diagram.

Solution The free-body diagram for the object of mass m is shown in Figure 5.11b, where the force $\vec{\mathbf{T}}$ exerted by the string has been resolved into a vertical component $T\cos\theta$ and a horizontal component $T\sin\theta$ acting toward the center of rotation. Because the object does not accelerate in the vertical direction, we model it as a particle in equilibrium in the vertical direction:

$$\sum F_y = 0 \quad \rightarrow \quad T\cos\theta - mg = 0$$

$$(1) \quad T\cos\theta = mg$$

In the horizontal direction, we have a centripetal acceleration so we model the object as a particle under a net force. Because the force that provides the centripetal acceleration in this example is the component $T\sin\theta$, from Newton's second law we have

$$(2) \quad \sum F_r = ma_c \quad \rightarrow \quad T\sin\theta = m\frac{v^2}{r}$$

By dividing (2) by (1), we eliminate T and find that

$$\tan\theta = \frac{v^2}{rg} \quad \rightarrow \quad v = \sqrt{rg\tan\theta}$$

From a triangle we can construct in the pictorial representation in Figure 5.11a, we note that $r = L\sin\theta$; therefore,

$$v = \sqrt{Lg\sin\theta\tan\theta}$$

B Find the period of revolution, defined as the time interval required to complete one revolution.

Solution The object is traveling at constant speed around its circular path. Because the object travels a distance of $2\pi r$ (the circumference of the circular path) in a time interval Δt equal to the period of revolution, we find

$$(3)\quad \Delta t = \frac{2\pi r}{v} = \frac{2\pi r}{\sqrt{rg\tan\theta}} = 2\pi\sqrt{\frac{L\cos\theta}{g}}$$

The intermediate algebraic steps used in obtaining (3) are left to the reader. Note that the period is independent of the mass m!

INTERACTIVE EXAMPLE 5.7 What Is the Maximum Speed of the Car?

A 1 500-kg car moving on a flat, horizontal road negotiates a curve whose radius is 35.0 m (Fig. 5.12a). If the coefficient of static friction between the tires and the dry pavement is 0.523, find the maximum speed the car can have to make the turn successfully.

Solution In the rolling motion of each tire, the bit of rubber meeting the road is instantaneously at rest relative to the road. It is prevented from skidding radially outward by a static friction force that acts radially inward, enabling the car to move in its circular path. The car is an extended object with four friction forces acting on it, one on each wheel, but we shall model it as a particle with only one net friction force. Figure 5.12b shows a free-body diagram for the car. From Newton's second law in the horizontal direction, we have

$$(1)\quad \sum F_x = ma \quad\rightarrow\quad f_s = m\frac{v^2}{r}$$

The maximum speed that the car can have around the curve corresponds to the speed at which it is on the verge of skidding toward the side of the road. At this point, the friction force has its maximum value

$$f_{s,\max} = \mu_s n$$

In the vertical direction, no acceleration occurs, so

$$\sum F_y = 0 \quad\rightarrow\quad n - mg = 0$$

Therefore, the magnitude of the normal force equals the weight in this case, and we find

$$f_{s,\max} = \mu_s mg$$

Substituting this expression into (1), we find the maximum speed:

$$\mu_s mg = m\frac{v_{\max}^2}{r} \quad\rightarrow\quad v_{\max} = \sqrt{\mu_s gr}$$

Substituting the numerical values gives us

$$v_{\max} = \sqrt{(0.523)(9.80\text{ m/s}^2)(35.0\text{ m})} = 13.4\text{ m/s}$$

This result is equivalent to 30.0 mi/h, which is less than a typical nonfreeway speed of 35 mi/h. Therefore, this roadway could benefit greatly from some banking, as in the next example!

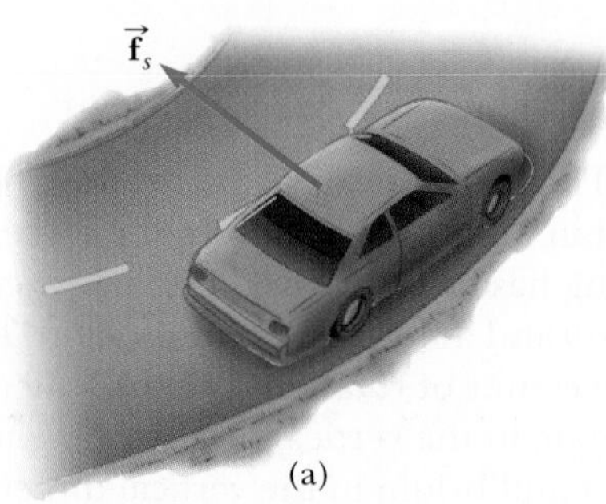

(a)

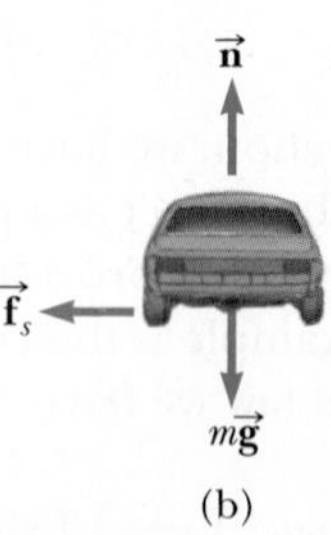

(b)

FIGURE 5.12 (Interactive Example 5.7) (a) The force of static friction directed toward the center of the curve keeps the car moving in a circular path. (b) Free-body diagram for the car.

PhysicsNow™ Study the relationship between the car's speed, radius of the turn, and the coefficient of static friction between road and tires by logging into PhysicsNow at www.pop4e.com and going to Interactive Example 5.7.

INTERACTIVE EXAMPLE 5.8 The Banked Roadway

A civil engineer wishes to redesign the curved roadway in Interactive Example 5.7 in such a way that a car will not have to rely on friction to round the curve without skidding. In other words, a car moving at the designated speed can negotiate the curve even when the road is covered with ice. Such a curve is usually *banked,* meaning that the roadway is tilted toward the inside of the curve. Suppose the designated speed for the curve is to be 13.4 m/s (30.0 mi/h) and the radius of the curve is 35.0 m. At what angle should the curve be banked?

Solution On a level (unbanked) road, the force that causes the centripetal acceleration is the force of static friction between car and road, as we saw in the previous example. If the road is banked at an angle θ, however, as in Figure 5.13, the normal force $\vec{\mathbf{n}}$ has a horizontal component $n_x = n \sin \theta$ pointing toward the center of the curve. Because the curve is to be designed so that the force of static friction is zero, only the component $n \sin \theta$ causes the centripetal acceleration. Hence, Newton's second law for the radial direction gives

$$(1) \quad \sum F_r = n \sin\theta = \frac{mv^2}{r}$$

The car is in equilibrium in the vertical direction. Therefore, from $\Sigma F_y = 0$ we have

$$(2) \quad n \cos \theta = mg$$

Dividing (1) by (2) gives

$$(3) \quad \tan \theta = \frac{v^2}{rg}$$

$$\theta = \tan^{-1}\left(\frac{(13.4 \text{ m/s})^2}{(35.0 \text{ m})(9.80 \text{ m/s}^2)}\right) = 27.6°$$

If a car rounds the curve at a speed less than 13.4 m/s, friction is needed to keep it from sliding down the bank (to the left in Fig. 5.13). A driver who attempts to negotiate the curve at a speed greater than 13.4 m/s has to depend on friction to keep from sliding up the bank (to the right in Fig. 5.13). The banking angle is independent of the mass of the vehicle negotiating the curve.

PhysicsNow™ Adjust the turn radius and the speed to see the effect on the banking angle by logging into PhysicsNow at www.pop4e.com and going to Interactive Example 5.8.

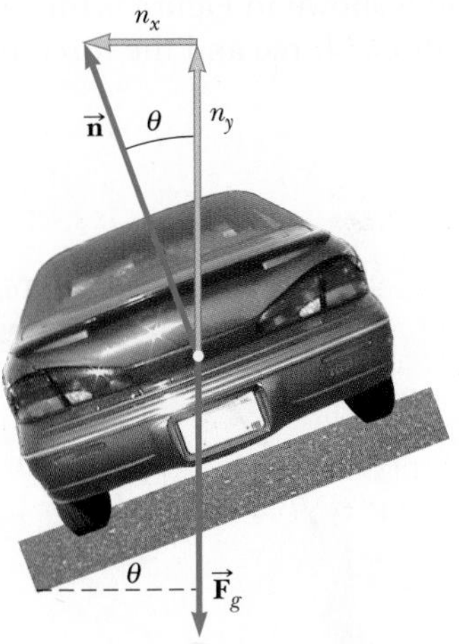

FIGURE 5.13 (Interactive Example 5.8) A car rounding a curve on a road banked at an angle θ to the horizontal. In the absence of friction the force that causes the centripetal acceleration and keeps the car moving in its circular path is the horizontal component of the normal force.

EXAMPLE 5.9 Let's Go Loop-the-Loop

A pilot of mass m in a jet aircraft executes a "loop-the-loop" maneuver as illustrated in Figure 5.14a. The aircraft moves in a vertical circle of radius 2.70 km at a *constant speed* of 225 m/s.

A Determine the force exerted by the seat on the pilot at the bottom of the loop. Express the answer in terms of the weight mg of the pilot.

Solution This example is the first numerical one we have seen in which the force causing the centripetal acceleration is a *combination* of forces rather than a single force. We shall model the pilot as a particle under a net force and analyze the situation at the bottom and top of the circular path.

The free-body diagram for the pilot at the bottom of the loop is shown in Figure 5.14b. The forces acting on the pilot are the downward gravitational force $m\vec{\mathbf{g}}$ and the upward normal force $\vec{\mathbf{n}}_{\text{bot}}$ exerted by the seat on the pilot. Because the net upward force at the bottom that provides the centripetal acceleration has a magnitude $n_{\text{bot}} - mg$, Newton's second law for the radial (upward) direction gives

$$\sum F_y = ma \quad \rightarrow \quad n_{\text{bot}} - mg = m\frac{v^2}{r}$$

$$n_{\text{bot}} = mg + m\frac{v^2}{r} = mg\left[1 + \frac{v^2}{rg}\right]$$

Substituting the values given for the speed and radius gives

$$n_{\text{bot}} = mg\left[1 + \frac{(225\text{ m/s})^2}{(2.70 \times 10^3\text{ m})(9.80\text{ m/s}^2)}\right]$$

$$= 2.91\,mg$$

Therefore, the force exerted by the seat on the pilot at the bottom of the loop is *greater* than the pilot's weight by a factor of 2.91.

B Determine the force exerted by the seat on the pilot at the top of the loop. Express the answer in terms of the weight mg of the pilot.

Solution The free-body diagram for the pilot at the top of the loop is shown in Figure 5.14c. At this point, both the gravitational force and the force $\vec{\mathbf{n}}_{\text{top}}$ exerted by the seat on the pilot act *downward,* so the net force downward that provides the centripetal acceleration has a magnitude $n_{\text{top}} + mg$. Applying Newton's second law gives

$$\sum F_y = ma \quad \rightarrow \quad n_{\text{top}} + mg = m\frac{v^2}{r}$$

$$n_{\text{top}} = m\frac{v^2}{r} - mg = mg\left[\frac{v^2}{rg} - 1\right]$$

$$n_{\text{top}} = mg\left[\frac{(225\text{ m/s})^2}{(2.70 \times 10^3\text{ m})(9.80\text{ m/s}^2)} - 1\right]$$

$$= 0.911\,mg$$

In this case, the force exerted by the seat on the pilot is *less* than the weight by a factor of 0.911. Therefore, the pilot feels lighter at the top of the loop.

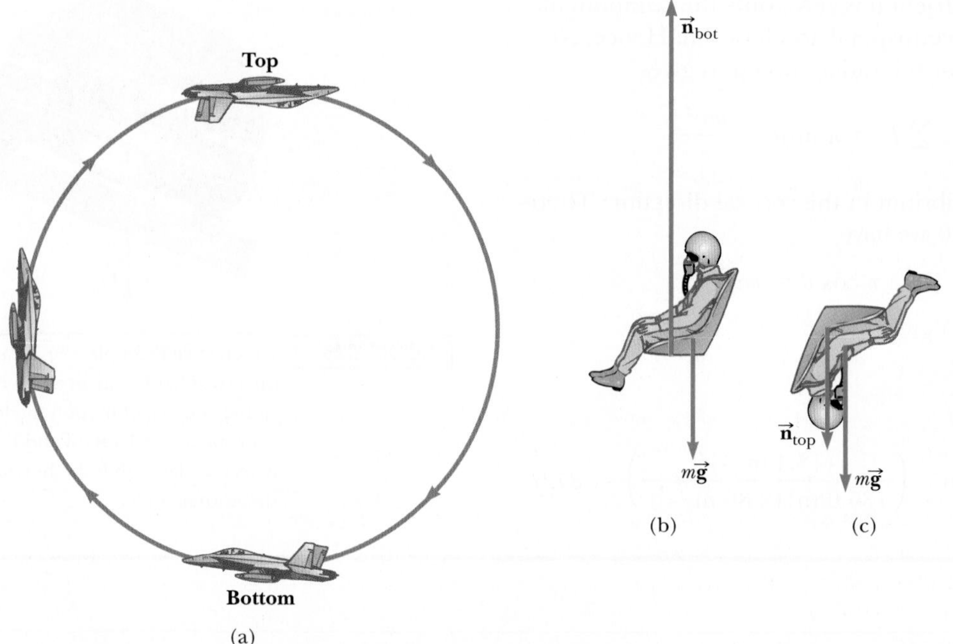

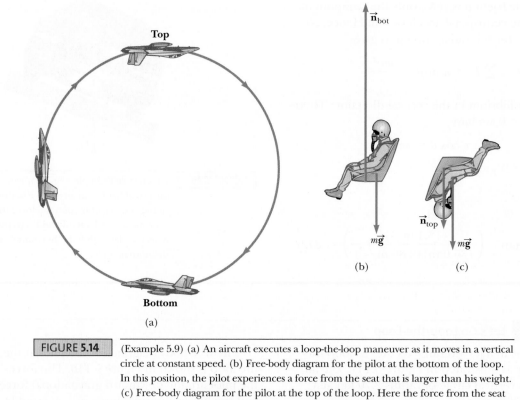

FIGURE 5.14 (Example 5.9) (a) An aircraft executes a loop-the-loop maneuver as it moves in a vertical circle at constant speed. (b) Free-body diagram for the pilot at the bottom of the loop. In this position, the pilot experiences a force from the seat that is larger than his weight. (c) Free-body diagram for the pilot at the top of the loop. Here the force from the seat could be smaller than his weight or larger, depending on the speed of the aircraft.

5.3 NONUNIFORM CIRCULAR MOTION

In Chapter 3, we found that if a particle moves with varying speed in a circular path, there is, in addition to the radial component of acceleration, a tangential component of magnitude dv/dt. Therefore, the net force acting on the particle must also have a radial and a tangential component as shown in Active Figure 5.15.

That is, because the total acceleration is $\vec{\mathbf{a}} = \vec{\mathbf{a}}_r + \vec{\mathbf{a}}_t$, the total force exerted on the particle is $\sum \vec{\mathbf{F}} = \sum \vec{\mathbf{F}}_r + \sum \vec{\mathbf{F}}_t$. The component vector $\sum \vec{\mathbf{F}}_r$ is directed toward the center of the circle and is responsible for the centripetal acceleration. The component vector $\sum \vec{\mathbf{F}}_t$ tangent to the circle is responsible for the tangential acceleration, which causes the speed of the particle to change with time.

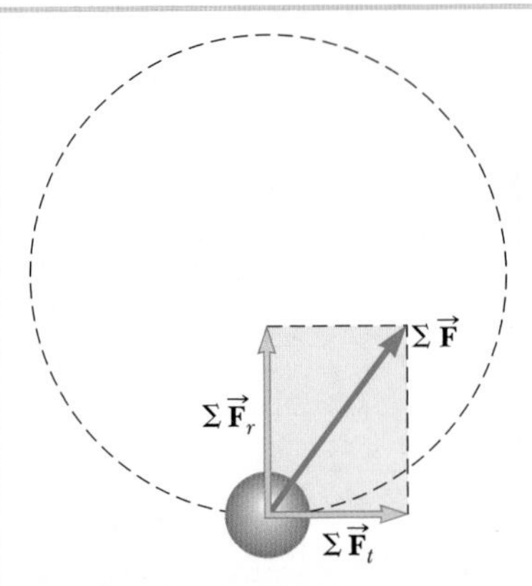

ACTIVE FIGURE 5.15

When the net force acting on a particle moving in a circular path has a tangential component vector $\sum \vec{\mathbf{F}}_t$, its speed changes. The total force on the particle also has a component vector $\sum \vec{\mathbf{F}}_r$ directed toward the center of the circular path. Therefore, the total force is $\sum \vec{\mathbf{F}} = \sum \vec{\mathbf{F}}_r + \sum \vec{\mathbf{F}}_t$.

Physics Now™ Log into PhysicsNow at www.pop4e.com and go to Active Figure 5.15 to adjust the initial position of the particle. Compare the component forces acting on the particle to those for a child swinging on a swing set.

QUICK QUIZ 5.5 Which of the following is *impossible* for a car moving in a circular path? Assume that the car is never at rest. **(a)** The car has tangential acceleration but no centripetal acceleration. **(b)** The car has centripetal acceleration but no tangential acceleration. **(c)** The car has both centripetal acceleration and tangential acceleration.

QUICK QUIZ 5.6 A bead slides freely along a *horizontal,* curved wire at constant speed, as shown in Figure 5.16. **(a)** Draw the vectors representing the force exerted by the wire on the bead at points Ⓐ, Ⓑ, and Ⓒ. **(b)** Suppose the bead in Figure 5.16 speeds up with constant tangential acceleration as it moves toward the right. Draw the vectors representing the force on the bead at points Ⓐ, Ⓑ, and Ⓒ.

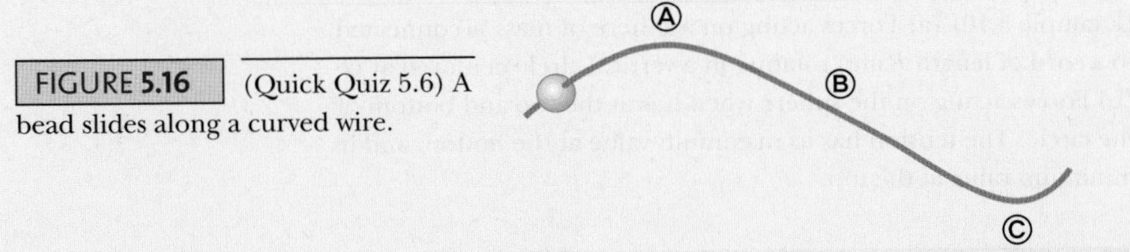

FIGURE 5.16 (Quick Quiz 5.6) A bead slides along a curved wire.

EXAMPLE 5.10 Follow the Rotating Ball

A small sphere of mass m is attached to the end of a cord of length R, which rotates under the influence of the gravitational force and the force exerted by the cord in a *vertical* circle about a fixed point O, as in Figure 5.17a. Let us determine the tension in the cord at any instant when the speed of the sphere is v and the cord makes an angle θ with the vertical.

Solution First, note that the speed is *not* uniform because a tangential component of acceleration arises from the gravitational force on the sphere. Although this example is similar to Example 5.9, it is not identical. From the free-body diagram in Figure 5.17a, we see that the only forces acting on the sphere are the gravitational force $m\vec{\mathbf{g}}$ and the force $\vec{\mathbf{T}}$ exerted by the cord.

We resolve $m\vec{\mathbf{g}}$ into a tangential component $mg \sin\theta$ and a radial component $mg\cos\theta$. Applying Newton's second law for the tangential direction gives

$$\sum F_t = ma_t \quad \rightarrow \quad mg\sin\theta = ma_t$$

$$a_t = g\sin\theta$$

This component causes v to change in time because $a_t = dv/dt$.

Applying Newton's second law to the forces in the radial direction (for which the outward direction is positive), we find

$$\sum F_r = ma_r \quad \rightarrow \quad mg\cos\theta - T = -m\frac{v^2}{r}$$

$$T = m\left(\frac{v^2}{R} + g\cos\theta\right)$$

At the bottom of the path, where $\cos\theta = \cos 0 = 1$, we see that

$$T_{\text{bot}} = m\left(\frac{v_{\text{bot}}^2}{R} + g\right)$$

which is the *maximum* value of T; as the sphere passes through the bottom point, the string is under the most tension. This property is of interest to trapeze artists because their support wires must withstand this largest tension at the bottom of the swing as well as to Tarzan when he chooses a nice, strong vine on which to swing to withstand this force.

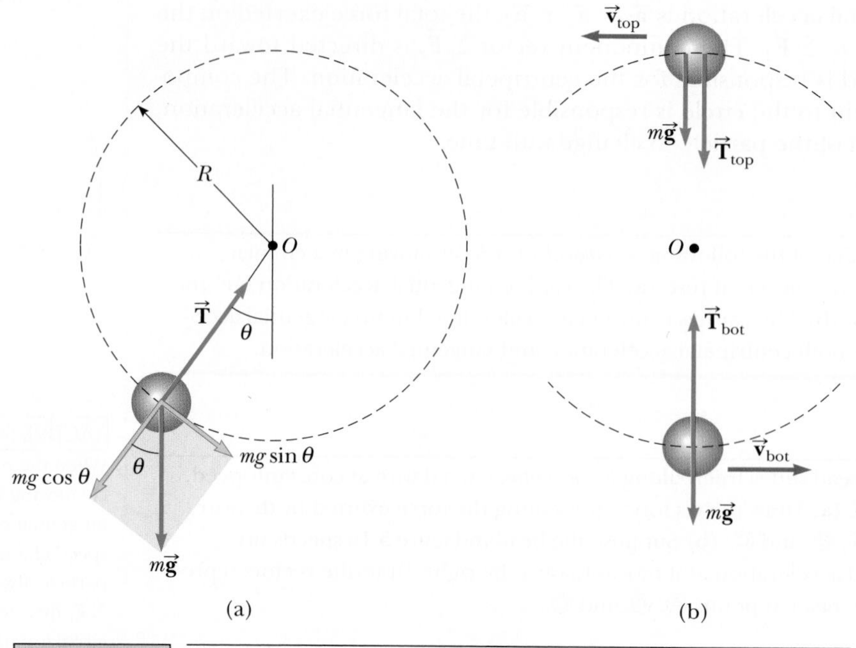

FIGURE 5.17 (Example 5.10) (a) Forces acting on a sphere of mass m connected to a cord of length R and rotating in a vertical circle centered at O. (b) Forces acting on the sphere when it is at the top and bottom of the circle. The tension has its maximum value at the bottom and its minimum value at the top.

5.4 MOTION IN THE PRESENCE OF VELOCITY-DEPENDENT RESISTIVE FORCES

Earlier, we described the friction force between a moving object and the surface along which it moves. So far, we have ignored any interaction between the object and the *medium* through which it moves. Let us now consider the effect of a medium such as a liquid or gas. The medium exerts a **resistive force $\vec{\mathbf{R}}$** on the object moving through it. You feel this force if you ride in a car at high speed with your hand out the window; the force you feel pushing your hand backward is the resistive force of the air rushing past the car. The magnitude of this force depends on the relative speed between the object and the medium, and the direction of $\vec{\mathbf{R}}$ on the object is always opposite the direction of the object's motion relative to the medium. Some examples are the air resistance associated with moving vehicles (sometimes called air drag), the force of the wind on the sails of a sailboat, and the viscous forces that act on objects sinking through a liquid.

Generally, the magnitude of the resistive force increases with increasing speed. The resistive force can have a complicated speed dependence. In the following discussions, we consider two simplification models that allow us to analyze these situations. The first model assumes that the resistive force is proportional to the velocity, which is approximately the case for objects that fall through a liquid with low speed and for very small objects, such as dust particles, that move through air. The second model treats situations for which we assume that the magnitude of the resistive force is proportional to the square of the speed of the object. Large objects, such as a sky diver moving through air in free-fall, experience such a force.

Model 1: Resistive Force Proportional to Object Velocity

At low speeds, the resistive force acting on an object that is moving through a viscous medium is effectively modeled as being proportional to the object's velocity.

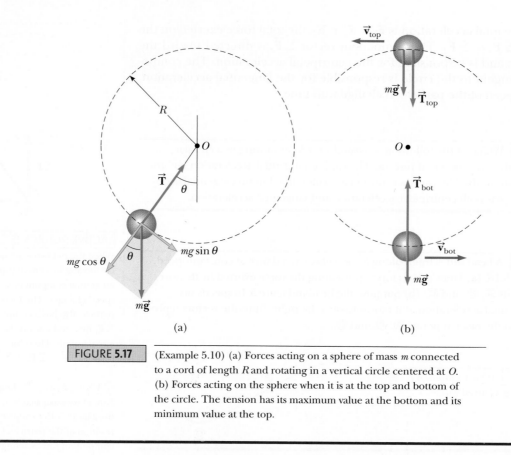

The mathematical representation of the resistive force can be expressed as

$$\vec{\mathbf{R}} = -b\vec{\mathbf{v}} \tag{5.4}$$

where $\vec{\mathbf{v}}$ is the velocity of the object relative to the medium and b is a constant that depends on the properties of the medium and on the shape and dimensions of the object. The negative sign represents that the resistive force is opposite the velocity of the object relative to the medium.

Consider a sphere of mass m released from rest in a liquid, as in Active Figure 5.18a. We assume that the only forces acting on the sphere are the resistive force $\vec{\mathbf{R}}$ and the weight $m\vec{\mathbf{g}}$, and we describe its motion using Newton's second law.[2] Considering the vertical motion and choosing the downward direction to be positive, we have

$$\sum F_y = ma_y \quad \rightarrow \quad mg - bv = m\frac{dv}{dt}$$

Dividing this equation by the mass m gives

$$\frac{dv}{dt} = g - \frac{b}{m}v \tag{5.5}$$

Equation 5.5 is called a *differential equation;* it includes both the speed v and the derivative of the speed. The methods of solving such an equation may not be familiar to you as yet. Note, however, that if we define $t = 0$ when $v = 0$, the resistive force is zero at this time and the acceleration dv/dt is simply g. As t increases, the speed increases, the resistive force increases, and the acceleration decreases. Thus, this problem is one in which neither the velocity nor the acceleration of the particle is constant.

The acceleration becomes zero when the increasing resistive force eventually balances the weight. At this point, the object reaches its **terminal speed** v_T and from then on it continues to move with zero acceleration. After this point, the motion is that of a particle under constant velocity. The terminal speed can be obtained from Equation 5.5 by setting $a = dv/dt = 0$, which gives

$$mg - bv_T = 0 \quad \rightarrow \quad v_T = \frac{mg}{b}$$

The expression for v that satisfies Equation 5.5 with $v = 0$ at $t = 0$ is

$$v = \frac{mg}{b}(1 - e^{-bt/m}) = v_T(1 - e^{-t/\tau}) \tag{5.6}$$

where $v_T = mg/b$, $\tau = m/b$, and $e = 2.718\ 28$ is the base of the natural logarithm. This expression for v can be verified by substituting it back into Equation 5.5. (Try it!) This function is plotted in Active Figure 5.18b.

The mathematical representation of the motion (Eq. 5.6) indicates that the terminal speed is never reached because the exponential function is never exactly equal to zero. For all practical purposes, however, when the exponential function is very small at large values of t, the speed of the particle can be approximated as being constant and equal to the terminal speed.

We cannot compare different objects by means of the time interval required to reach terminal speed because, as we have just discussed, this time interval is infinite for all objects! We need some means to compare these exponential behaviors for different objects. We do so with a parameter called the **time constant.** The time constant $\tau = m/b$ that appears in Equation 5.6 is the time interval required for the factor in parentheses in Equation 5.6 to become equal to $1 - e^{-1} = 0.632$. Therefore, the time constant represents the time interval required for the object to reach 63.2% of its terminal speed (Active Fig. 5.18b).

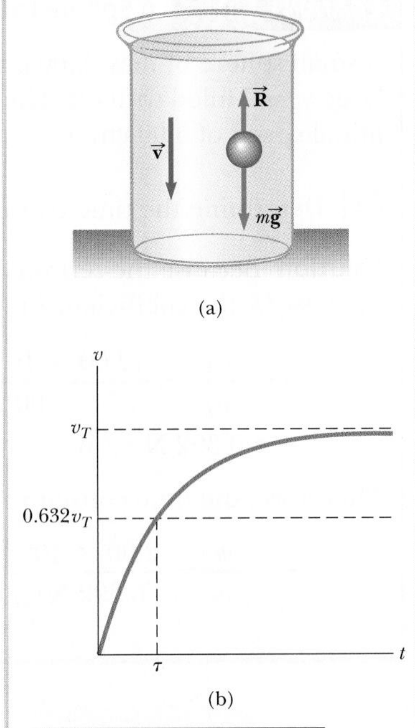

ACTIVE FIGURE 5.18

(a) A small sphere falling through a viscous fluid. (b) The speed–time graph for an object falling through a viscous medium. The object approaches a terminal speed v_T, and the time constant τ is the time interval required to reach $0.632v_T$.

PhysicsNow™ Log into PhysicsNow at www.pop4e.com and go to Active Figure 5.18 to vary the size and mass of the sphere and the viscosity (resistance to flow) of the surrounding medium. Observe the effects on the sphere's motion and its speed–time graph.

[2] A *buoyant* force also acts on any object surrounded by a fluid. This force is constant and equal to the weight of the displaced fluid, as will be discussed in Chapter 15. The effect of this force can be modeled by changing the apparent weight of the sphere by a constant factor, so we can ignore it here.

EXAMPLE 5.11 A Sphere Falling in Oil

A small sphere of mass 2.00 g is released from rest in a large vessel filled with oil. The sphere approaches a terminal speed of 5.00 cm/s.

A Determine the time constant τ.

Solution Because the terminal speed is given by $v_T = mg/b$, the coefficient b is

$$b = \frac{mg}{v_T} = \frac{(2.00 \times 10^{-3}\text{ kg})(9.80\text{ m/s}^2)}{5.00 \times 10^{-2}\text{ m/s}}$$
$$= 0.392\text{ N}\cdot\text{s/m}$$

Therefore, the time constant τ is

$$\tau = \frac{m}{b} = \frac{2.00 \times 10^{-3}\text{ kg}}{0.392\text{ N}\cdot\text{s/m}} = 5.1 \times 10^{-3}\text{ s}$$

B Determine the time interval required for the sphere to reach 90.0% of its terminal speed.

Solution The speed of the sphere as a function of time is given by Equation 5.6. To find the time t at which the sphere is traveling at a speed of $0.900v_T$, we set $v = 0.900v_T$, substitute into Equation 5.6, and solve for t:

$$0.900v_T = v_T(1 - e^{-t/\tau})$$
$$1 - e^{-t/\tau} = 0.900$$
$$e^{-t/\tau} = 0.100$$
$$-\frac{t}{\tau} = \ln 0.100 = -2.30$$
$$t = 2.30\tau = 2.30(5.10 \times 10^{-3}\text{ s})$$
$$= 11.7 \times 10^{-3}\text{ s} = 11.7\text{ ms}$$

Model 2: Resistive Force Proportional to Object Speed Squared

For large objects moving at high speeds through air, such as airplanes, sky divers, and baseballs, the magnitude of the resistive force is modeled as being proportional to the square of the speed:

$$R = \tfrac{1}{2}D\rho A v^2 \qquad [5.7]$$

where ρ is the density of air, A is the cross-sectional area of the moving object measured in a plane perpendicular to its velocity, and D is a dimensionless empirical quantity called the *drag coefficient.* The drag coefficient has a value of about 0.5 for spherical objects moving through air but can be as high as 2 for irregularly shaped objects.

Consider an airplane in flight that experiences such a resistive force. Equation 5.7 shows that the force is proportional to the density of air and hence decreases with decreasing air density. Because air density decreases with increasing altitude, the resistive force on a jet airplane flying at a given speed will decrease with increasing altitude. Therefore, airplanes tend to fly at very high altitudes to take advantage of this reduced resistive force, which allows them to fly faster for a given engine thrust. Of course, this higher speed *increases* the resistive force, in proportion to the square of the speed, so a balance is struck between fuel economy and higher speed.

Now let us analyze the motion of a falling object subject to an upward air resistive force whose magnitude is given by Equation 5.7. Suppose an object of mass m is released from rest, as in Figure 5.19, from the position $y = 0$. The object experiences two external forces: the downward gravitational force $m\vec{\mathbf{g}}$ and the upward resistive force $\vec{\mathbf{R}}$. Hence, using Newton's second law,

$$\sum F = ma \quad \rightarrow \quad mg - \tfrac{1}{2}D\rho A v^2 = ma \qquad [5.8]$$

Solving for a, we find that the object has a downward acceleration of magnitude

$$a = g - \left(\frac{D\rho A}{2m}\right)v^2 \qquad [5.9]$$

Because $a = dv/dt$, Equation 5.9 is another differential equation that provides us with the speed as a function of time.

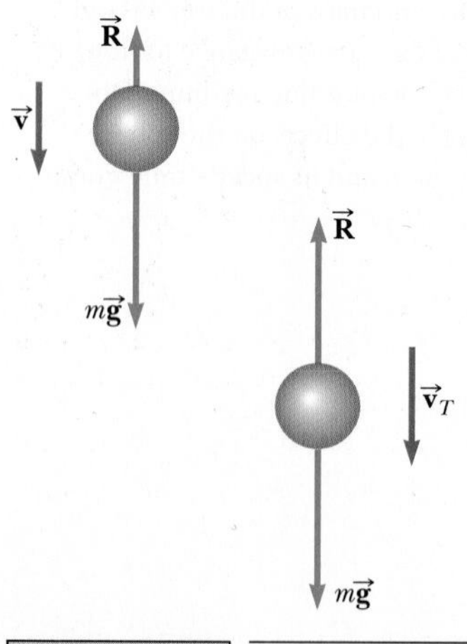

FIGURE 5.19 An object falling through air experiences a resistive drag force $\vec{\mathbf{R}}$ and a gravitational force $\vec{\mathbf{F}}_g = m\vec{\mathbf{g}}$. The object reaches terminal speed (on the right) when the net force acting on it is zero, that is, when $\vec{\mathbf{R}} = -\vec{\mathbf{F}}_g$, or $R = mg$. Before that occurs, the acceleration varies with speed according to Equation 5.9.

TABLE 5.2 Terminal Speeds for Various Objects Falling Through Air

Object	Mass (kg)	Cross-sectional Area (m^2)	v_T (m/s)[a]
Sky diver	75	0.70	60
Baseball (radius 3.7 cm)	0.145	4.2×10^{-3}	33
Golf ball (radius 2.1 cm)	0.046	1.4×10^{-3}	32
Hailstone (radius 0.50 cm)	4.8×10^{-4}	7.9×10^{-5}	14
Raindrop (radius 0.20 cm)	3.4×10^{-5}	1.3×10^{-5}	9.0

[a]The drag coefficient D is assumed to be 0.5 in each case.

Again, we can calculate the terminal speed v_T because when the gravitational force is balanced by the resistive force, the net force is zero and therefore the acceleration is zero. Setting $a = 0$ in Equation 5.9 gives

$$g - \left(\frac{D\rho A}{2m}\right) v_T^2 = 0$$

$$v_T = \sqrt{\frac{2mg}{D\rho A}} \qquad [5.10]$$

Table 5.2 lists the terminal speeds for several objects falling through air, all computed on the assumption that the drag coefficient is 0.5.

QUICK QUIZ 5.7 Consider a sky surfer falling through air, as in Figure 5.20, before reaching her terminal speed. As the speed of the sky surfer increases, the magnitude of her acceleration **(a)** remains constant, **(b)** decreases until it reaches a constant nonzero value, or **(c)** decreases until it reaches zero.

(Jump Run Productions/Image Bank)

FIGURE 5.20 (Quick Quiz 5.7) A sky surfer takes advantage of the upward force of the air on her board.

5.5 THE FUNDAMENTAL FORCES OF NATURE

We have described a variety of forces experienced in our everyday activities, such as the gravitational force acting on all objects at or near the Earth's surface and the force of friction as one surface slides over another. Newton's second law tells us how to relate the forces to the object's or particle's acceleration.

In addition to these familiar macroscopic forces in nature, forces also act in the atomic and subatomic world. For example, atomic forces within the atom are responsible for holding its constituents together and nuclear forces act on different parts of the nucleus to keep its parts from separating.

Until recently, physicists believed that there were four fundamental forces in nature: the gravitational force, the electromagnetic force, the strong force, and the weak force. We shall discuss these forces individually and then consider the current view of fundamental forces.

The Gravitational Force

The **gravitational force** is the mutual force of attraction between any two objects in the Universe. It is interesting and rather curious that although the gravitational force can be very strong between macroscopic objects, it is inherently the weakest of all the fundamental forces. For example, the gravitational force between the electron and proton in the hydrogen atom has a magnitude on the order of 10^{-47} N, whereas the electromagnetic force between these same two particles is on the order of 10^{-7} N.

■ Newton's law of universal gravitation

In addition to his contributions to the understanding of motion, Newton studied gravity extensively. **Newton's law of universal gravitation** states that every particle in the Universe attracts every other particle with a force that is directly proportional to the product of the masses of the particles and inversely proportional to the square of the distance between them. If the particles have masses m_1 and m_2 and are separated by a distance r, as in Figure 5.21, the magnitude of the gravitational force is

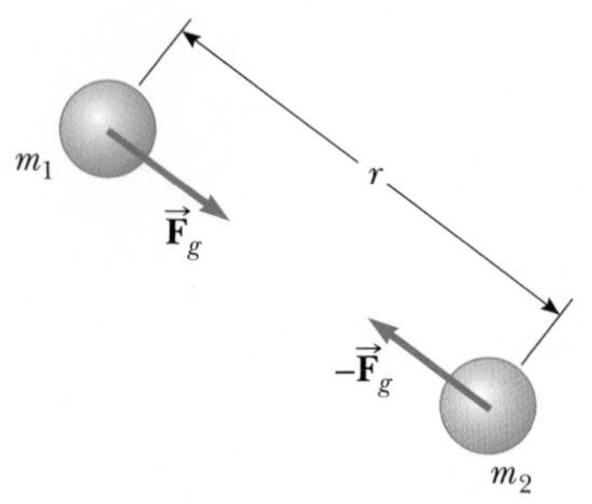

FIGURE 5.21 Two particles with masses m_1 and m_2 attract each other with a force of magnitude Gm_1m_2/r^2.

$$F_g = G\frac{m_1 m_2}{r^2} \qquad [5.11]$$

where $G = 6.67 \times 10^{-11}\ \text{N}\cdot\text{m}^2/\text{kg}^2$ is the **universal gravitational constant.** More detail on the gravitational force will be provided in Chapter 11.

The Electromagnetic Force

The **electromagnetic force** is the force that binds atoms and molecules in compounds to form ordinary matter. It is much stronger than the gravitational force. The force that causes a rubbed comb to attract bits of paper and the force that a magnet exerts on an iron nail are electromagnetic forces. Essentially all forces at work in our macroscopic world, apart from the gravitational force, are manifestations of the electromagnetic force. For example, friction forces, contact forces, tension forces, and forces in elongated springs are consequences of electromagnetic forces between charged particles in proximity.

The electromagnetic force involves two types of particles: those with positive charge and those with negative charge. (More information on these two types of charge is provided in Chapter 19.) Unlike the gravitational force, which is always an attractive interaction, the electromagnetic force can be either attractive or repulsive, depending on the charges on the particles.

Coulomb's law

Coulomb's law expresses the magnitude of the *electrostatic force*[3] F_e between two charged particles separated by a distance r:

$$F_e = k_e\frac{q_1 q_2}{r^2} \qquad [5.12]$$

where q_1 and q_2 are the charges on the two particles, measured in units called *coulombs* (C), and k_e $(= 8.99 \times 10^9\ \text{N}\cdot\text{m}^2/\text{C}^2)$ is the **Coulomb constant.** Note that the electrostatic force has the same mathematical form as Newton's law of universal gravitation (see Eq. 5.11), with charge playing the mathematical role of mass and the Coulomb constant being used in place of the universal gravitational constant. The electrostatic force is attractive if the two charges have opposite signs and is repulsive if the two charges have the same sign, as indicated in Figure 5.22.

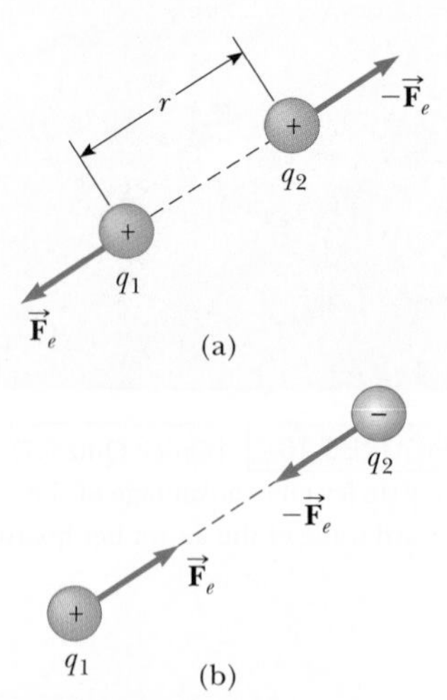

FIGURE 5.22 Two point charges separated by a distance r exert an electrostatic force on each other given by Coulomb's law. (a) When the charges are of the same sign, the charges repel each other. (b) When the charges are of opposite sign, the charges attract each other.

The smallest amount of isolated charge found in nature (so far) is the charge on an electron or proton. This fundamental unit of charge is given the symbol e and has the magnitude $e = 1.60 \times 10^{-19}$ C. An electron has charge $-e$, whereas a proton has charge $+e$. Theories developed in the latter half of the 20th century propose that protons and neutrons are made up of smaller particles called **quarks,** which have charges of either $\frac{2}{3}e$ or $-\frac{1}{3}e$ (discussed further in Chapter 31). Although experimental evidence has been found for such particles inside nuclear matter, free quarks have never been detected.

The Strong Force

An atom, as we currently model it, consists of an extremely dense positively charged nucleus surrounded by a cloud of negatively charged electrons, with the electrons attracted to the nucleus by the electric force. All nuclei except those of hydrogen are combinations of positively charged protons and neutral neutrons (collectively

[3]The electrostatic force is the electromagnetic force between two electric charges that are at rest. If the charges are moving, magnetic forces are also present; these forces will be studied in Chapter 22.

called nucleons), yet why does the repulsive electrostatic force between the protons not cause nuclei to break apart? Clearly, there must be an attractive force that counteracts the strong electrostatic repulsive force and is responsible for the stability of nuclei. This force that binds the nucleons to form a nucleus is called the **nuclear force.** It is one manifestation of the **strong force,** which is the force between particles formed from quarks, which we will discuss in Chapter 31. Unlike the gravitational and electromagnetic forces, which depend on distance in an inverse-square fashion, the nuclear force is extremely short range; its strength decreases very rapidly outside the nucleus and is negligible for separations greater than approximately 10^{-14} m.

The Weak Force

The **weak force** is a short-range force that tends to produce instability in certain nuclei. It was first observed in naturally occurring radioactive substances and was later found to play a key role in most radioactive decay reactions. The weak force is about 10^{36} times stronger than the gravitational force and about 10^3 times weaker than the electromagnetic force.

The Current View of Fundamental Forces

For years, physicists have searched for a simplification scheme that would reduce the number of fundamental forces needed to describe physical phenomena. In 1967, physicists predicted that the electromagnetic force and the weak force, originally thought to be independent of each other and both fundamental, are in fact manifestations of one force, now called the **electroweak** force. This prediction was confirmed experimentally in 1984. We shall discuss it more fully in Chapter 31.

We also now know that protons and neutrons are not fundamental particles; current models of protons and neutrons theorize that they are composed of simpler particles called quarks, as mentioned previously. The quark model has led to a modification of our understanding of the nuclear force. Scientists now define the strong force as the force that binds the quarks to one another in a nucleon (proton or neutron). This force is also referred to as a **color force,** in reference to a property of quarks called "color," which we shall investigate in Chapter 31. The previously defined nuclear force, the force that acts between nucleons, is now interpreted as a secondary effect of the strong force between the quarks.

Scientists believe that the fundamental forces of nature are closely related to the origin of the Universe. The Big Bang theory states that the Universe began with a cataclysmic explosion about 14 billion years ago. According to this theory, the first moments after the Big Bang saw such extremes of energy that all the fundamental forces were unified into one force. Physicists are continuing their search for connections among the known fundamental forces, connections that could eventually prove that the forces are all merely different forms of a single superforce. This fascinating search continues to be at the forefront of physics.

5.6 DRAG COEFFICIENTS OF AUTOMOBILES

CONTEXT CONNECTION

In the Context Connection of Chapter 4, we ignored air resistance and assumed that the driving force on the tires was the only force on the vehicle in the horizontal direction. Given our understanding of velocity-dependent forces from Section 5.4, we should understand now that air resistance could be a significant factor in the design of an automobile.

Table 5.3 shows the drag coefficients for the vehicles that we have investigated in previous chapters. Notice that the coefficients for the performance and traditional vehicles vary from 0.30 to 0.43, with the average coefficient in the two portions of the table almost the same. A look at the lower part of the table shows that this parameter

TABLE 5.3 Drag Coefficients of Various Vehicles

Automobile	Model Year	Drag Coefficient
Performance vehicles		
Aston Martin DB7 Vantage	2001	0.31
BMW Z8	2001	0.43
Chevrolet Corvette	2000	0.29
Dodge Viper GTS-R	1998	0.40
Ferrari F50	1997	0.37
Ferrari 360 Spider F1	2000	0.33
Lamborghini Diablo GT	2000	0.31
Porsche 911 GT2	2002	0.34
Traditional vehicles		
Acura Integra GS	2000	0.34
BMW Mini Cooper S	2003	0.35
Cadillac Escalade (SUV)	2002	0.42
Dodge Stratus	2002	0.34
Lexus ES300	1997	0.32
Mitsubishi Eclipse GT	2000	0.30
Nissan Maxima	2000	0.31
Pontiac Grand Prix	2003	0.31
Toyota Sienna (SUV)	2004	0.31
Volkswagen Beetle	1999	0.36
Alternative vehicles		
GM EV1	1998	0.19
Toyota Prius	2004	0.26
Honda Insight	2001	0.25

is where the alternative vehicles shine. All three vehicles have drag coefficients lower than all others in the table, and the GM EV1 has a remarkable coefficient of just 0.19.

Designers of alternative-fuel vehicles try to squeeze every last mile of travel out of the energy that is stored in the vehicle in the form of fuel or an electric battery. A significant method of doing so is to reduce the force of air resistance so that the net force driving the car forward is as large as possible.

A number of techniques can be used to reduce the drag coefficient. Two factors that help are a small frontal area and smooth curves from the front of the vehicle to the back. For example, the Chevrolet Corvette shown in Figure 5.23a exhibits a streamlined shape that contributes to its low drag coefficient. As a comparison, consider a large, boxy vehicle, such as the Hummer H2 in Figure 5.23b. The drag coefficient for this vehicle is 0.57. Another factor includes elimination or minimization of

(a)

(b)

(a, Courtesy of GM; b, Courtesy of GM–Hummer)

FIGURE 5.23 (a) The Chevrolet Corvette has a streamlined shape that contributes to its low drag coefficient of 0.29. (b) The Hummer H2 is not streamlined like the Corvette and consequently has a much higher drag coefficient of 0.57.

as many irregularities in the surfaces as possible, including door handles that project from the body, windshield wipers, wheel wells, and rough surfaces on headlamps and grills. An important consideration is the underside of the carriage. As air rushes beneath the car, there are many irregular surfaces associated with brakes, drive trains, suspension components, and so on. The drag coefficient can be made lower by assuring that the overall surface of the car's undercarriage is as smooth as possible. ■

SUMMARY

Physics Now™ Take a practice test by logging into PhysicsNow at www.pop4e.com and clicking on the Pre-Test link for this chapter.

Forces of friction are complicated, but we design a simplification model for friction that allows us to analyze motion that includes the effects of friction. The **maximum force of static friction** $f_{s,\max}$ between two surfaces is proportional to the normal force between the surfaces. This maximum force occurs when the surfaces are on the verge of slipping. In general, $f_s \leq \mu_s n$, where μ_s is the **coefficient of static friction** and n is the magnitude of the normal force. When an object slides over a rough surface, the **force of kinetic friction** $\vec{\mathbf{f}}_k$ is opposite the direction of the velocity of the object relative to the surface and its magnitude is proportional to the magnitude of the normal force on the object. The magnitude is given by $f_k = \mu_k n$, where μ_k is the **coefficient of kinetic friction.** Usually, $\mu_k < \mu_s$.

Newton's second law, applied to a particle moving in uniform circular motion, states that the net force in the inward radial direction must equal the product of the mass and the centripetal acceleration:

$$\sum F = ma_c = m\frac{v^2}{r} \qquad [5.3]$$

An object moving through a liquid or gas experiences a **resistive force** that is velocity dependent. This resistive force, which is opposite the velocity of the object relative to the medium, generally increases with speed. The force depends on the object's shape and on the properties of the medium through which the object is moving. In the limiting case for a falling object, when the resistive force balances the weight ($a = 0$), the object reaches its **terminal speed.**

The fundamental forces existing in nature can be expressed as the following four: the gravitational force, the electromagnetic force, the strong force, and the weak force.

QUESTIONS

☐ = answer available in the *Student Solutions Manual and Study Guide.*

1. Draw a free-body diagram for each of the following objects: (a) a projectile in motion in the presence of air resistance, (b) a rocket leaving the launch pad with its engines operating, (c) an athlete running along a horizontal track.
2. What force causes (a) an automobile, (b) a propeller-driven airplane, and (c) a rowboat to move?
3. Identify the action-reaction pairs in the following situations: a man takes a step, a snowball hits a girl in the back, a baseball player catches a ball, a gust of wind strikes a window.
4. In a contest of National Football League behemoths, teams from the Rams and the 49ers engage in a tug-of-war, pulling in opposite directions on a strong rope. The Rams exert a force of 9 200 N and they are winning, making the center of the light rope move steadily toward themselves. Is it possible to know the tension in the rope from the information stated? Is it larger or smaller than 9 200 N? How hard are the 49ers pulling on the rope? Would it change your answer if the 49ers were winning or if the contest were even? The stronger team wins by exerting a larger force, on what? Explain your answers.
5. Suppose you are driving a classic car. Why should you avoid slamming on your brakes when you want to stop in the shortest possible distance? (Many cars have antilock brakes that avoid this problem.)
6. A book is given a brief push to make it slide up a rough incline. It comes to a stop and slides back down to the starting point. Does it take the same time interval to go up as to come down? What if the incline is frictionless?
7. Describe a few examples in which the force of friction exerted on an object is in the direction of motion of the object.
8. An object executes circular motion with constant speed whenever a net force of constant magnitude acts perpendicular to the velocity. What happens to the speed if the force is not perpendicular to the velocity?
9. What causes a rotary lawn sprinkler to turn?
10. It has been suggested that rotating cylinders about 10 miles in length and 5 miles in diameter be placed in space and used as colonies. The purpose of the rotation is to simulate gravity for the inhabitants. Explain this concept for producing an effective imitation of gravity.
11. A pail of water can be whirled in a vertical path such that none is spilled. Why does the water stay in the pail, even when the pail is upside down above your head?
12. Why does a pilot tend to black out when pulling out of a steep dive?
13. If someone told you that astronauts are weightless in orbit because they are beyond the pull of gravity, would you accept the statement? Explain.

14. A falling sky diver reaches terminal speed with her parachute closed. After the parachute is opened, what parameters change to decrease this terminal speed?

15. On long journeys, jet aircraft usually fly at high altitudes of about 30 000 ft. What is the main advantage from an economic viewpoint of flying at these altitudes?

16. Consider a small raindrop and a large raindrop falling through the atmosphere. Compare their terminal speeds. What are their accelerations when they reach terminal speed?

17. "If the current position and velocity of every particle in the Universe were known, together with the laws describing the forces that particles exert on one another, then the whole future of the Universe could be calculated. The future is determinate and preordained. Free will is an illusion." Do you agree with this thesis? Argue for or against it.

PROBLEMS

1, 2, 3 = straightforward, intermediate, challenging
☐ = full solution available in the *Student Solutions Manual and Study Guide*
Physics Now™ = coached problem with hints available at www.pop4e.com
= computer useful in solving problem
= paired numerical and symbolic problems
= biomedical application

Section 5.1 ■ Forces of Friction

1. A 25.0-kg block is initially at rest on a horizontal surface. A horizontal force of 75.0 N is required to set the block in motion. After it is in motion, a horizontal force of 60.0 N is required to keep the block moving with constant speed. Find the coefficients of static and kinetic friction from this information.

2. A car is traveling at 50.0 mi/h on a horizontal highway. (a) If the coefficient of static friction between road and tires on a rainy day is 0.100, what is the minimum distance in which the car will stop? (b) What is the stopping distance when the surface is dry and $\mu_s = 0.600$?

3. Before 1960, it was believed that the maximum attainable coefficient of static friction for an automobile tire was less than 1. Then around 1962, three companies independently developed racing tires with coefficients of 1.6. Since then, tires have improved, as illustrated in this problem. According to the 1990 *Guinness Book of Records,* the shortest time interval in which a piston-engine car initially at rest has covered a distance of one-quarter mile is 4.96 s. This record was set by Shirley Muldowney in September 1989. (a) Assume that, as shown in Figure P5.3, the rear wheels lifted the front wheels off the pavement. What minimum value of μ_s is necessary to achieve the record time? (b) Suppose Muldowney were able to double her engine power, keeping other things equal. How would this change affect the elapsed time?

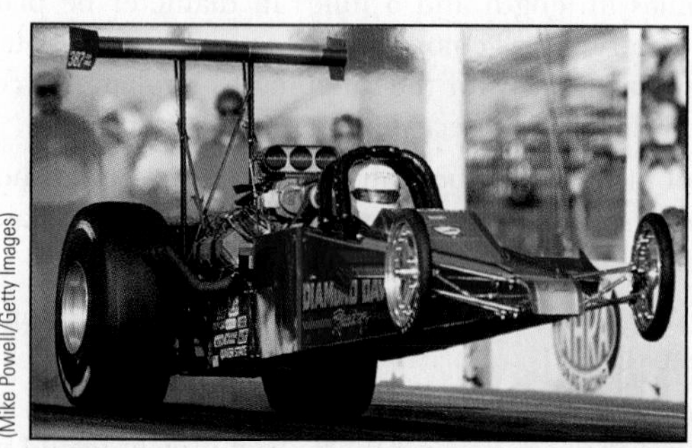
(Mike Powell/Getty Images)

FIGURE **P5.3**

4. The person in Figure P5.4 weighs 170 lb. As seen from the front, each light crutch makes an angle of 22.0° with the vertical. Half of the person's weight is supported by the crutches. The other half is supported by the vertical forces of the ground on the person's feet. Assuming that the person is moving with constant velocity and the force exerted by the ground on the crutches acts along the crutches, determine (a) the smallest possible coefficient of friction between crutches and ground and (b) the magnitude of the compression force in each crutch.

FIGURE **P5.4**

5. To meet a U.S. Postal Service requirement, footwear must have a coefficient of static friction of 0.5 or more on a specified tile surface. A typical athletic shoe has a coefficient of 0.800. In an emergency, what is the minimum time interval in which a person starting from rest can move 3.00 m on a tile surface if she is wearing (a) footwear meeting the Postal Service minimum and (b) a typical athletic shoe?

6. Consider a large truck carrying a heavy load, such as steel beams. A significant hazard for the driver is that the load may slide forward, crushing the cab, if the truck stops suddenly in an accident or even in braking. Assume, for example, that a 10 000-kg load sits on the flatbed of a 20 000-kg truck moving at 12.0 m/s. Assume that the load is not tied down to the truck, but has a coefficient of friction of 0.500 with the flatbed of the truck. (a) Calculate the minimum stopping distance for which the load will not slide forward relative to the truck. (b) Is any piece of data unnecessary for the solution?

7. To determine the coefficients of friction between rubber and various surfaces, a student uses a rubber eraser and an

incline. In one experiment, the eraser begins to slip down the incline when the angle of inclination is 36.0° and then moves down the incline with constant speed when the angle is reduced to 30.0°. From these data, determine the coefficients of static and kinetic friction for this experiment.

8. A woman at an airport is towing her 20.0-kg suitcase at constant speed by pulling on a strap at an angle θ above the horizontal (Fig. P5.8). She pulls on the strap with a 35.0-N force, and the friction force on the suitcase is 20.0 N. Draw a free-body diagram of the suitcase. (a) What angle does the strap make with the horizontal? (b) What normal force does the ground exert on the suitcase?

FIGURE **P5.8**

9. **Physics Now™** A 3.00-kg block starts from rest at the top of a 30.0° incline and slides a distance of 2.00 m down the incline in 1.50 s. Find (a) the magnitude of the acceleration of the block, (b) the coefficient of kinetic friction between block and plane, (c) the friction force acting on the block, and (d) the speed of the block after it has slid 2.00 m.

10. A 9.00-kg hanging block is connected by a string over a pulley to a 5.00-kg block that is sliding on a flat table (Fig. P5.10). The string is light and does not stretch; the pulley is light and turns without friction. The coefficient of kinetic friction between the sliding block and the table is 0.200. Find the tension in the string.

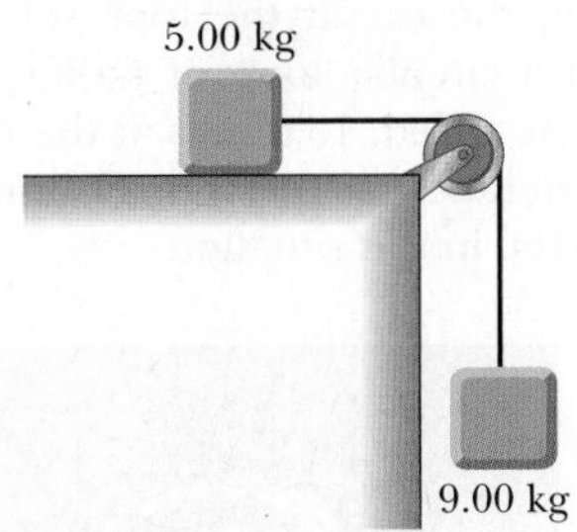

FIGURE **P5.10**

11. Two blocks connected by a rope of negligible mass are being dragged by a horizontal force $\vec{\mathbf{F}}$ (Fig. P5.11). Suppose $F = 68.0$ N, $m_1 = 12.0$ kg, $m_2 = 18.0$ kg, and the coefficient of kinetic friction between each block and the surface is 0.100. (a) Draw a free-body diagram for each block. (b) Determine the tension T and the magnitude of the acceleration of the system.

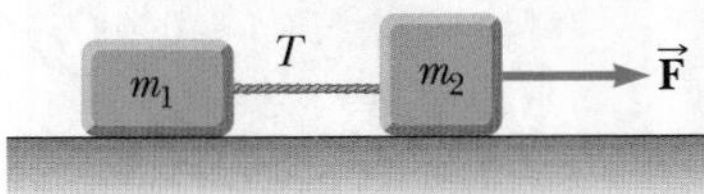

FIGURE **P5.11**

12. Three objects are connected on the table as shown in Figure P5.12. The table is rough and has a coefficient of kinetic friction of 0.350. The objects have masses 4.00 kg, 1.00 kg, and 2.00 kg, as shown, and the pulleys are frictionless. Draw a free-body diagram for each object. (a) Determine the acceleration of each object and their directions. (b) Determine the tensions in the two cords.

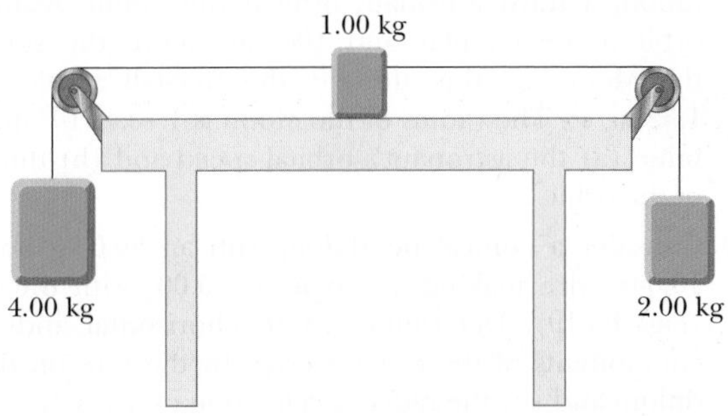

FIGURE **P5.12**

13. A block of mass 3.00 kg is pushed up against a wall by a force $\vec{\mathbf{P}}$ that makes a 50.0° angle with the horizontal as shown in Figure P5.13. The coefficient of static friction between the block and the wall is 0.250. Determine the possible values for the magnitude of $\vec{\mathbf{P}}$ that allow the block to remain stationary.

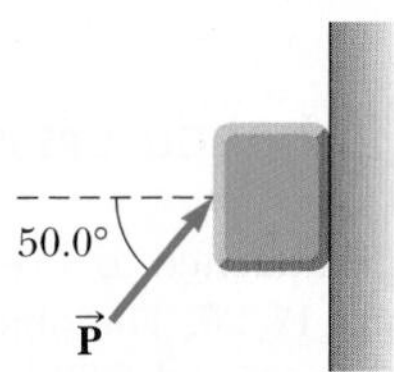

FIGURE **P5.13**

14. **Review problem.** One side of the roof of a building slopes up at 37.0°. A student throws a Frisbee onto the roof. It strikes with a speed of 15.0 m/s and does not bounce, but instead slides straight up the incline. The coefficient of kinetic friction between the plastic Frisbee and the roof is 0.400. The Frisbee slides 10.0 m up the roof to its peak, where it goes into free-fall, following a parabolic trajectory with negligible air resistance. Determine the maximum height the Frisbee reaches above the point where it struck the roof.

Section 5.2 ▪ Newton's Second Law Applied to a Particle in Uniform Circular Motion

15. A light string can support a stationary hanging load of 25.0 kg before breaking. A 3.00-kg object attached to the string rotates on a horizontal, frictionless table in a circle of radius 0.800 m, and the other end of the string is held fixed. What range of speeds can the object have before the string breaks?

16. In the Bohr model of the hydrogen atom, the speed of the electron is approximately 2.20×10^6 m/s. Find (a) the force acting on the electron as it revolves in a circular orbit of radius 0.530×10^{-10} m and (b) the centripetal acceleration of the electron.

17. A crate of eggs is located in the middle of the flatbed of a pickup truck as the truck negotiates an unbanked curve in the road. The curve may be regarded as an arc of a circle of radius 35.0 m. If the coefficient of static friction between crate and truck is 0.600, how fast can the truck be moving without the crate sliding?

18. Whenever two *Apollo* astronauts were on the surface of the Moon, a third astronaut orbited the Moon. Assume the orbit to be circular and 100 km above the surface of the Moon. At this altitude, the free-fall acceleration is 1.52 m/s^2. The radius of the Moon is 1.70×10^6 m. Determine (a) the astronaut's orbital speed and (b) the period of the orbit.

19. Consider a conical pendulum with an 80.0-kg bob on a 10.0-m wire making an angle $\theta = 5.00°$ with the vertical (Fig. P5.19). Determine (a) the horizontal and vertical components of the force exerted by the wire on the pendulum and (b) the radial acceleration of the bob.

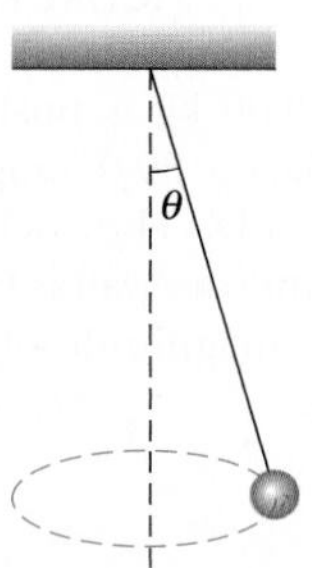

FIGURE **P5.19**

20. A 4.00-kg object is attached to a vertical rod by two strings as shown in Figure P5.20. The object rotates in a horizontal circle at constant speed 6.00 m/s. Find the tension in (a) the upper string and (b) the lower string.

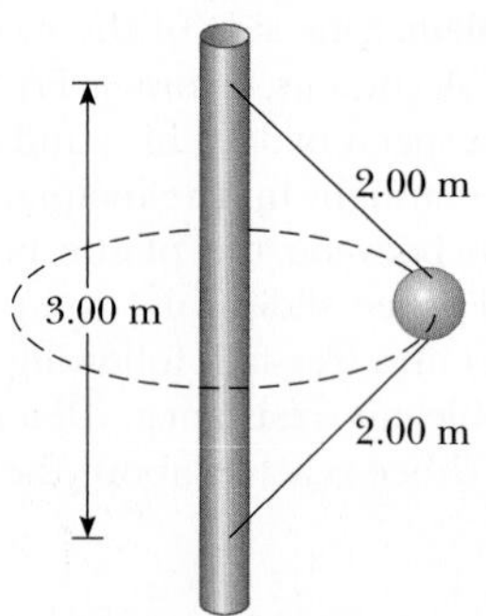

FIGURE **P5.20**

Section 5.3 ▪ Nonuniform Circular Motion

21. **Physics Now™** Tarzan (m = 85.0 kg) tries to cross a river by swinging from a vine. The vine is 10.0 m long, and his speed at the bottom of the swing (as he just clears the water) will be 8.00 m/s. Tarzan doesn't know that the vine has a breaking strength of 1 000 N. Does he make it safely across the river?

22. We will study the most important work of Nobel laureate Arthur Compton in Chapter 28. Disturbed by speeding cars outside the physics building at Washington University in St. Louis, he designed a speed bump and had it installed. Suppose a car of mass m passes over a bump in a road that follows the arc of a circle of radius R as shown in Figure P5.22. (a) What force does the road exert on the car as the car passes the highest point of the bump if the car travels at a speed v? (b) What is the maximum speed the car can have as it passes this highest point without losing contact with the road?

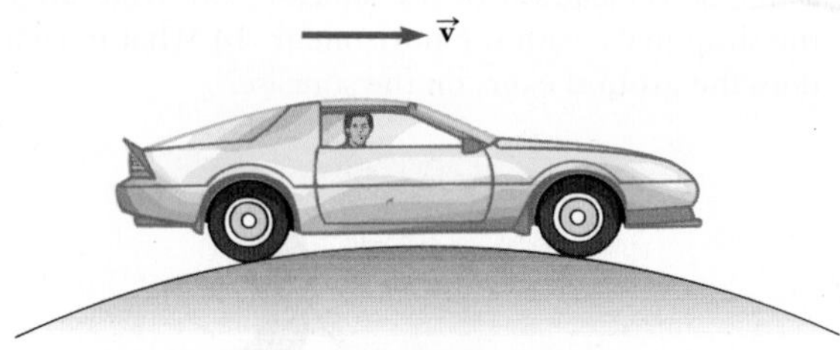

FIGURE **P5.22**

23. **Physics Now™** A pail of water is rotated in a vertical circle of radius 1.00 m. What is the minimum speed of the pail, upside down at the top of the circle, if no water is to spill out?

24. A roller coaster at Six Flags Great America amusement park in Gurnee, Illinois, incorporates some clever design technology and some basic physics. Each vertical loop, instead of being circular, is shaped like a teardrop (Fig. P5.24). The cars ride on the inside of the loop at the top, and the speeds are high enough to ensure that the cars remain on the track. The biggest loop is 40.0 m high, with a maximum speed of 31.0 m/s (nearly 70 mi/h) at the bottom. Suppose the speed at the top is 13.0 m/s and the corresponding centripetal acceleration is $2g$. (a) What is the radius of the arc of the teardrop at the top? (b) If the total mass of a car plus the riders is M, what force does the rail exert on the car at the top? (c) Suppose the roller coaster had a circular loop of radius 20.0 m. If the cars have the same speed, 13.0 m/s at the top, what is the centripetal acceleration at the top? Comment on the normal force at the top in this situation.

FIGURE **P5.24**

Section 5.4 ■ Motion in the Presence of Velocity-Dependent Resistive Forces

25. A small piece of Styrofoam packing material is dropped from a height of 2.00 m above the ground. Until it reaches terminal speed, the magnitude of its acceleration is given by $a = g - bv$. After falling 0.500 m, the Styrofoam effectively reaches terminal speed and then takes 5.00 s more to reach the ground. (a) What is the value of the constant b? (b) What is the acceleration at $t = 0$? (c) What is the acceleration when the speed is 0.150 m/s?

26. (a) Calculate the terminal speed of a wooden sphere (density 0.830 g/cm^3) falling through air if its radius is 8.00 cm and its drag coefficient is 0.500. (b) From what height would a freely falling object reach this speed in the absence of air resistance?

27. A small, spherical bead of mass 3.00 g is released from rest at $t = 0$ in a bottle of liquid shampoo. The terminal speed is observed to be $v_T = 2.00$ cm/s. Find (a) the value of the constant b in Equation 5.4, (b) the time τ at which it reaches $0.632v_T$, and (c) the value of the resistive force when the bead reaches terminal speed.

28. A 9.00-kg object starting from rest falls through a viscous medium and experiences a resistive force $\vec{\mathbf{R}} = -b\vec{\mathbf{v}}$, where $\vec{\mathbf{v}}$ is the velocity of the object. The object reaches one-half its terminal speed in 5.54 s. (a) Determine the terminal speed. (b) At what time is the speed of the object three-fourths the terminal speed? (c) How far has the object traveled in the first 5.54 s of motion?

29. **Physics Now™** A motorboat cuts its engine when its speed is 10.0 m/s and coasts to rest. The equation describing the motion of the motorboat during this period is $v = v_i e^{-ct}$, where v is the speed at time t, v_i is the initial speed, and c is a constant. At $t = 20.0$ s, the speed is 5.00 m/s. (a) Find the constant c. (b) What is the speed at $t = 40.0$ s? (c) Differentiate the expression for $v(t)$ and thus show that the acceleration of the boat is proportional to the speed at any time.

30. Consider an object on which the net force is a resistive force proportional to the square of its speed. For example, assume that the resistive force acting on a speed skater is $f = -kmv^2$, where k is a constant and m is the skater's mass. The skater crosses the finish line of a straight-line race with speed v_0 and then slows down by coasting on his skates. Show that the skater's speed at any time t after crossing the finish line is $v(t) = v_0/(1 + ktv_0)$.

Section 5.5 ■ The Fundamental Forces of Nature

31. Two identical isolated particles, each of mass 2.00 kg, are separated by a distance of 30.0 cm. What is the magnitude of the gravitational force exerted by one particle on the other?

32. Find the order of magnitude of the gravitational force that you exert on another person 2 m away. In your solution, state the quantities you measure or estimate and their values.

33. When a falling meteor is at a distance above the Earth's surface of 3.00 times the Earth's radius, what is its free-fall acceleration caused by the gravitational force exerted on it?

34. In a thundercloud, there may be electric charges of +40.0 C near the top of the cloud and −40.0 C near the bottom of the cloud. These charges are separated by 2.00 km. What is the electric force on the top charge?

Section 5.6 ■ Context Connection—Drag Coefficients of Automobiles

35. The mass of a sports car is 1 200 kg. The shape of the body is such that the aerodynamic drag coefficient is 0.250 and the frontal area is 2.20 m^2. Ignoring all other sources of friction, calculate the initial acceleration of the car assuming that it has been traveling at 100 km/h and is now shifted into neutral and allowed to coast.

36. Consider a 1 300-kg car presenting front-end area 2.60 m^2 and having drag coefficient 0.340. It can achieve instantaneous acceleration 3.00 m/s^2 when its speed is 10.0 m/s. Ignore any force of rolling resistance. Assume that the only horizontal forces on the car are static friction forward exerted by the road on the drive wheels and resistance exerted by the surrounding air, with density 1.20 kg/m^3. (a) Find the friction force exerted by the road. (b) Suppose the car body could be redesigned to have a drag coefficient of 0.200. If nothing else changes, what will be the car's acceleration? (c) Assume that the force exerted by the road remains constant. Then what maximum speed could the car attain with $D = 0.340$? (d) With $D = 0.200$?

Additional Problems

37. Consider the three connected objects shown in Figure P5.37. Assume first that the inclined plane is frictionless and that the system is in equilibrium. In terms of m, g, and θ, find (a) the mass M and (b) the tensions T_1 and T_2. Now assume that the value of M is double the value found in part (a). Find (c) the acceleration of each object and (d) the tensions T_1 and T_2. Next, assume that the coefficient of static friction between m and $2m$ and the inclined plane is μ_s and that the system is in equilibrium. Find (e) the maximum value of M and (f) the minimum value of M. (g) Compare the values of T_2 when M has its minimum and maximum values.

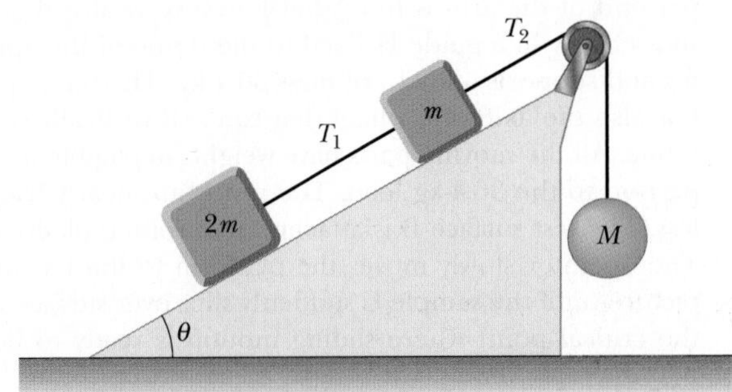

FIGURE **P5.37**

38. A 2.00-kg aluminum block and a 6.00-kg copper block are connected by a light string over a frictionless pulley. They sit on a steel surface, as shown in Figure P5.38, where $\theta = 30.0°$. When they are released from rest, will they start

to move? If so, determine (a) their acceleration and (b) the tension in the string. If not, determine the sum of the magnitudes of the forces of friction acting on the blocks.

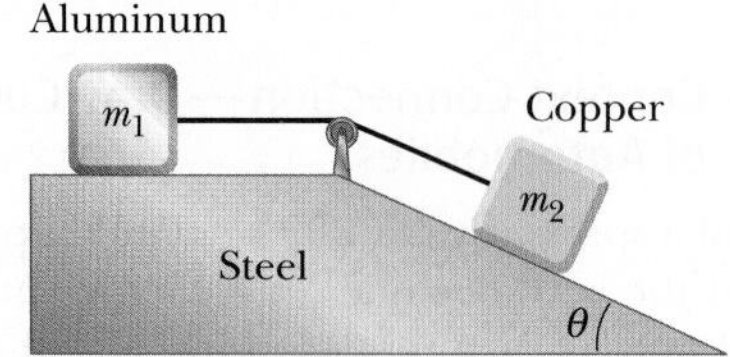

FIGURE P5.38

39. A crate of weight F_g is pushed by a force $\vec{\mathbf{P}}$ on a horizontal floor. (a) Assuming that the coefficient of static friction is μ_s and that $\vec{\mathbf{P}}$ is directed at angle θ below the horizontal, show that the minimum value of P that will move the crate is given by

$$P = \frac{\mu_s F_g \sec \theta}{1 - \mu_s \tan \theta}$$

(b) Find the minimum value of P that can produce motion when $\mu_s = 0.400$, $F_g = 100$ N, and $\theta = 0°$, 15.0°, 30.0°, 45.0°, and 60.0°.

40. A 1.30-kg toaster is not plugged in. The coefficient of static friction between the toaster and a horizontal countertop is 0.350. To make the toaster start moving you carelessly pull on its electric cord. (a) For the cord tension to be as small as possible, you should pull at what angle above the horizontal? (b) With this angle, how large must the tension be?

41. The system shown in Figure P4.54 (Chapter 4) has an acceleration of magnitude 1.50 m/s^2. Assume that the coefficient of kinetic friction between block and incline is the same for both inclines. Find (a) the coefficient of kinetic friction and (b) the tension in the string.

42. Materials such as automobile tire rubber and shoe soles are tested for coefficients of static friction with an apparatus called a James tester. The pair of surfaces for which μ_s is to be measured are labeled B and C in Figure P5.42. Sample C is attached to a foot D at the lower end of a pivoting arm E that makes angle θ with the vertical. The upper end of the arm is hinged at F to a vertical rod G that slides freely in a guide H fixed to the frame of the apparatus and supports a load I of mass 36.4 kg. The hinge pin at F is also the axle of a wheel that can roll vertically on the frame. All the moving parts have weights negligible in comparison to the 36.4-kg load. The pivots are nearly frictionless. The test surface B is attached to a rolling platform A. The operator slowly moves the platform to the left in the picture until the sample C suddenly slips over surface B. At the critical point where sliding motion is ready to begin, the operator notes the angle θ_s of the pivoting arm. (a) Make a free-body diagram of the pin at F. It is in equilibrium under three forces: the weight of the load I, a horizontal normal force exerted by the frame, and a force of compression directed upward along the arm E. (b) Draw a free-body diagram of the foot D and sample C, considered as one system. (c) Determine the normal force that the test surface B exerts on the sample for any angle θ. (d) Show that $\mu_s = \tan \theta_s$. (e) The protractor on the tester can record angles as large as 50.2°. What is the greatest coefficient of friction it can measure?

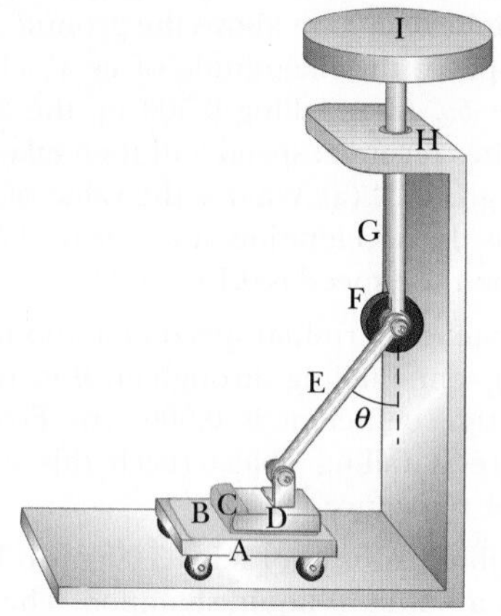

FIGURE P5.42

43. A block of mass $m = 2.00$ kg rests on the left edge of a block of mass $M = 8.00$ kg. The coefficient of kinetic friction between the two blocks is 0.300, and the surface on which the 8.00-kg block rests is frictionless. A constant horizontal force of magnitude $F = 10.0$ N is applied to the 2.00-kg block, setting it in motion as shown in Figure P5.43a. If the distance L that the leading edge of the smaller block travels on the larger block is 3.00 m, (a) in what time interval will the smaller block make it to the right side of the 8.00-kg block as shown in Figure P5.43b? (*Note:* Both blocks are set into motion when $\vec{\mathbf{F}}$ is applied.) (b) How far does the 8.00-kg block move in the process?

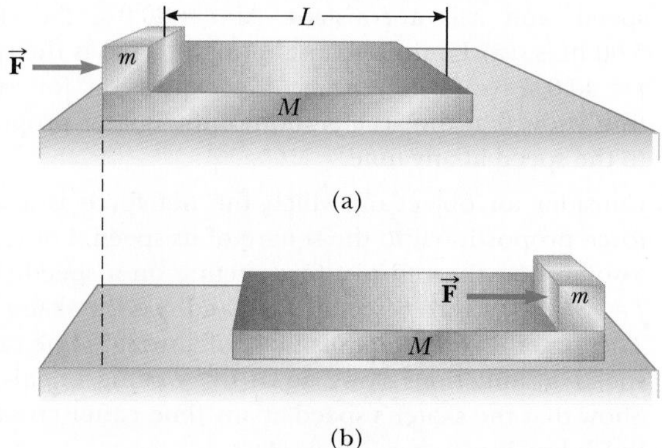

FIGURE P5.43

44. A 5.00-kg block is placed on top of a 10.0-kg block (Fig. P5.44). A horizontal force of 45.0 N is applied to the 10-kg block, and the 5-kg block is tied to the wall. The coefficient of kinetic friction between all moving surfaces is 0.200. (a) Draw a free-body diagram for each block and identify the action-reaction forces between the blocks. (b) Determine the tension in the string and the magnitude of the acceleration of the 10-kg block.

45. A car rounds a banked curve as in Figure 5.13. The radius of curvature of the road is R, the banking angle is θ, and the coefficient of static friction is μ_s. (a) Determine the range of speeds the car can have without slipping up or down the road. (b) Find the minimum value for μ_s such

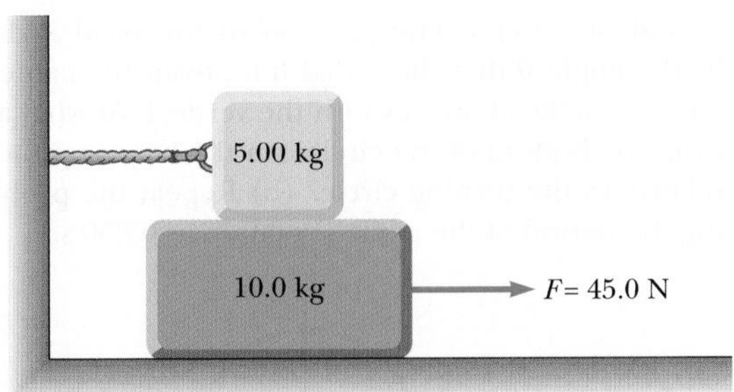

FIGURE **P5.44**

that the minimum speed is zero. (c) What is the range of speeds possible if $R = 100$ m, $\theta = 10.0°$, and $\mu_s = 0.100$ (slippery conditions)?

46. The following equations describe the motion of a system of two objects.

$$n - (6.50 \text{ kg})(9.80 \text{ m/s}^2) \cos 13.0° = 0$$

$$f_k = 0.360n$$

$$T + (6.50 \text{ kg})(9.80 \text{ m/s}^2) \sin 13.0° - f_k = (6.50 \text{ kg})a$$

$$-T + (3.80 \text{ kg})(9.80 \text{ m/s}^2) = (3.80 \text{ kg})a$$

(a) Solve the equations for a and T. (b) Describe a situation to which these equations apply. Draw free-body diagrams for both objects.

47. In a home laundry dryer, a cylindrical tub containing wet clothes is rotated steadily about a horizontal axis as shown in Figure P5.47. The clothes are made to tumble so that they will dry uniformly. The rate of rotation of the smooth-walled tub is chosen so that a small piece of cloth will lose contact with the tub when the cloth is at an angle of 68.0° above the horizontal. If the radius of the tub is 0.330 m, what rate of revolution is needed?

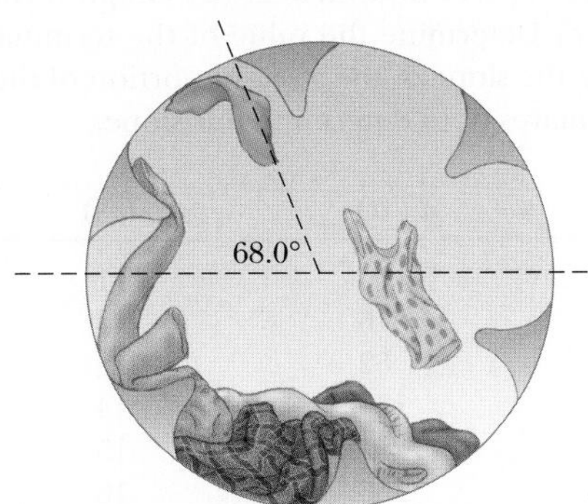

FIGURE **P5.47**

48. A student builds and calibrates an accelerometer and uses it to determine the speed of her car around a certain unbanked highway curve. The accelerometer is a plumb bob with a protractor that she attaches to the roof of her car. A friend riding in the car with the student observes that the plumb bob hangs at an angle of 15.0° from the vertical when the car has a speed of 23.0 m/s. (a) What is the centripetal acceleration of the car rounding the curve? (b) What is the radius of the curve? (c) What is speed of the car if the plumb bob deflection is 9.00° while rounding the same curve?

49. **Physics Now™** Because the Earth rotates about its axis, a point on the equator experiences a centripetal acceleration of 0.033 7 m/s^2, whereas a point at the poles experiences no centripetal acceleration. (a) Show that at the equator the gravitational force on an object must exceed the normal force required to support the object. That is, show that the object's true weight exceeds its apparent weight. (b) What is the apparent weight at the equator and at the poles of a person having a mass of 75.0 kg? (Assume that the Earth is a uniform sphere and take $g = 9.800$ m/s^2.)

50. An air puck of mass m_1 is tied to a string and allowed to revolve in a circle of radius R on a frictionless horizontal table. The other end of the string passes through a hole in the center of the table, and a counterweight of mass m_2 is tied to it (Fig. P5.50). The suspended object remains in equilibrium while the puck on the tabletop revolves. What are (a) the tension in the string, (b) the radial force acting on the puck, and (c) the speed of the puck?

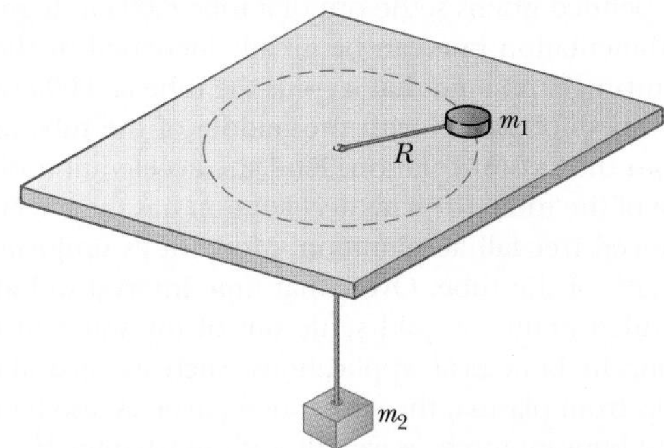

FIGURE **P5.50**

51. A Ferris wheel rotates four times each minute. It carries each car around a circle of diameter of 18.0 m. (a) What is the centripetal acceleration of a rider? (b) What force does the seat exert on a 40.0-kg rider at the lowest point of the ride? (c) At the highest point of the ride? (d) What force (magnitude and direction) does the seat exert on a rider when the rider is halfway between top and bottom?

52. An amusement park ride consists of a rotating circular platform 8.00 m in diameter from which 10.0-kg seats are suspended at the end of 2.50-m massless chains (Fig. P5.52). When the system rotates, the chains make an

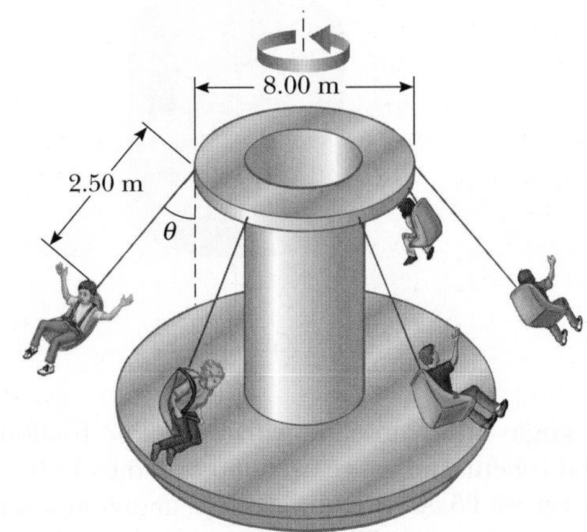

FIGURE **P5.52**

angle $\theta = 28.0°$ with the vertical. (a) What is the speed of each seat? (b) Draw a free-body diagram of a 40.0-kg child riding in a seat and find the tension in the chain.

53. A space station, in the form of a wheel 120 m in diameter, rotates to provide an "artificial gravity" of 3.00 m/s^2 for persons who walk around on the inner wall of the outer rim. Find the rate of rotation of the wheel (in revolutions per minute) that will produce this effect.

54. *Sedimentation and centrifugation.* According to Stokes's law, water exerts on a slowly moving immersed spherical object a resistive force described by

$$\vec{\mathbf{R}} = -(0.018\,8\ \text{N}\cdot\text{s/m}^2)r\vec{\mathbf{v}}$$

where r is the radius of the sphere and $\vec{\mathbf{v}}$ is its velocity. (a) Consider a spherical grain of gold dust with density 19.3×10^3 kg/m^3 and radius 0.500 μm. Ignore the buoyant force on the grain. Find the terminal speed at which the grain falls in water. (b) Over what time interval will all such suspended grains settle out of a tube 8.00 cm high? (c) The sedimentation rate can be greatly increased by the use of a centrifuge. Assume that it spins the tube at 3 000 rev/min in a horizontal plane, with the middle of the tube at 9.00 cm from the axis of rotation. Find the acceleration of the middle of the tube. (d) This acceleration has the effect of an enhanced free-fall acceleration. Model it as uniform over the length of the tube. Over what time interval will all the suspended grains of gold settle out of the water in this situation? In biological applications, such as separating blood cells from plasma, the suspended particles also feel a significant buoyant force, as we will study in Chapter 15.

55. An amusement park ride consists of a large vertical cylinder that spins about its axis sufficiently fast that any person inside is held up against the wall when the floor drops away (Fig. P5.55). The coefficient of static friction between person and wall is μ_s, and the radius of the cylinder is R. (a) Show that the maximum period of revolution necessary to keep the person from falling is $T = (4\pi^2 R\mu_s/g)^{1/2}$. (b) Obtain a numerical value for T assuming that $R = 4.00$ m and $\mu_s = 0.400$. How many revolutions per minute does the cylinder make?

FIGURE **P5.55**

56. A single bead can slide with negligible friction on a wire that is bent into a circular loop of radius 15.0 cm as shown in Figure P5.56. (a) The circle is always in a vertical plane and rotates steadily about its vertical diameter with a period of 0.450 s. The position of the bead is described by the angle θ that the radial line, from the center of the loop to the bead, makes with the vertical. At what angle up from the bottom of the circle can the bead stay motionless relative to the turning circle? (b) Repeat the problem taking the period of the circle's rotation as 0.850 s.

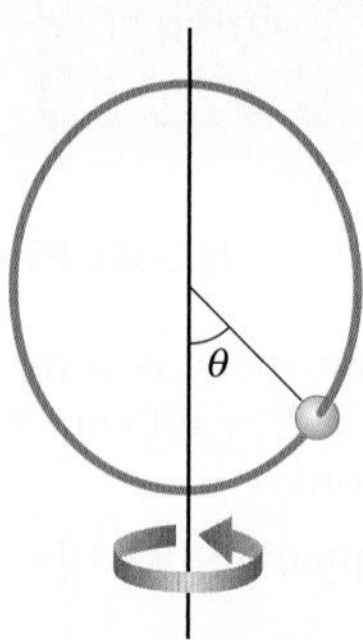

FIGURE **P5.56**

57. The expression $F = arv + br^2v^2$ gives the magnitude of the resistive force (in newtons) exerted on a sphere of radius r (in meters) by a stream of air moving at speed v (in meters per second), where a and b are constants with appropriate SI units. Their numerical values are $a = 3.10 \times 10^{-4}$ and $b = 0.870$. Using this expression, find the terminal speed for water droplets falling under their own weight in air, taking the following values for the drop radii: (a) 10.0 μm, (b) 100 μm, (c) 1.00 mm. Note that for (a) and (c) you can obtain accurate answers without solving a quadratic equation by considering which of the two contributions to the air resistance is dominant and ignoring the lesser contribution.

58. Members of a skydiving club were given the following data to use in planning their jumps. In the table, d is the distance fallen from rest by a sky diver in a "free-fall stable spread position" versus the time of fall t. (a) Convert the distances in feet into meters. (b) Graph d (in meters) versus t. (c) Determine the value of the terminal speed v_T by finding the slope of the straight portion of the curve. Use a least-squares fit to determine this slope.

t (s)	d (ft)	t (s)	d (ft)
0	0	11	1 309
1	16	12	1 483
2	62	13	1 657
3	138	14	1 831
4	242	15	2 005
5	366	16	2 179
6	504	17	2 353
7	652	18	2 527
8	808	19	2 701
9	971	20	2 875
10	1 138		

59. A model airplane of mass 0.750 kg flies in a horizontal circle at the end of a 60.0-m control wire with a speed of 35.0 m/s. Compute the tension in the wire assuming that it makes a constant angle of 20.0° with the horizontal. The forces exerted on the airplane are the pull of the control wire, the gravitational force, and aerodynamic lift, which acts at 20.0° inward from the vertical as shown in Figure P5.59.

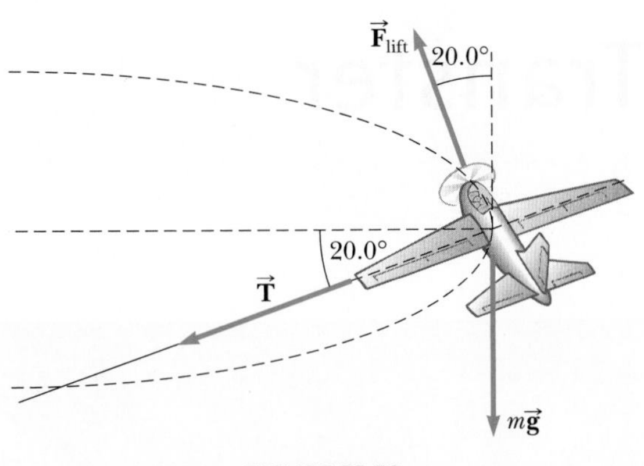

FIGURE **P5.59**

60. If a single constant force acts on an object that moves on a straight line, the object's velocity is a linear function of time. The equation $v = v_i + at$ gives its velocity v as a function of time, where a is its constant acceleration. What if velocity is instead a linear function of position? Assume that as a particular object moves through a resistive medium, its speed decreases as described by the equation $v = v_i - kx$, where k is a constant coefficient and x is the position of the object. Find the law describing the total force acting on this object.

ANSWERS TO QUICK QUIZZES

5.1 (b). The friction force acts opposite to the weight of the book to keep the book in equilibrium. Because the weight is downward, the friction force must be upward.

5.2 (b). The crate accelerates to the east. Because the only horizontal force acting on it is the force of static friction between its bottom surface and the truck bed, that force must also be directed to the east.

5.3 (b). When pulling with the rope, there is a component of your applied force that is upward, which reduces the normal force between the sled and the snow. In turn, the friction force between the sled and the snow is reduced, making the sled easier to move. If you push from behind, with a force with a downward component, the normal force is larger, the friction force is larger, and the sled is harder to move.

5.4 (i), (a). The normal force is always perpendicular to the surface that applies the force. Because your car maintains its orientation at all points on the ride, the normal force is always upward. (ii), (b). Your centripetal acceleration is downward toward the center of the circle, so the net force on you must be downward.

5.5 (a). If the car is moving in a circular path, it must have centripetal acceleration given by Equation 3.17.

5.6 (a) Because the speed is constant, the only direction the force can have is that of the centripetal acceleration. The force is larger at Ⓒ than at Ⓐ because the radius at Ⓒ is smaller. There is no force at Ⓑ because the wire is straight. (b) In addition to the forces in the centripetal direction in (a), there are now tangential forces to provide the tangential acceleration. The tangential force is the same at all three points because the tangential acceleration is constant.

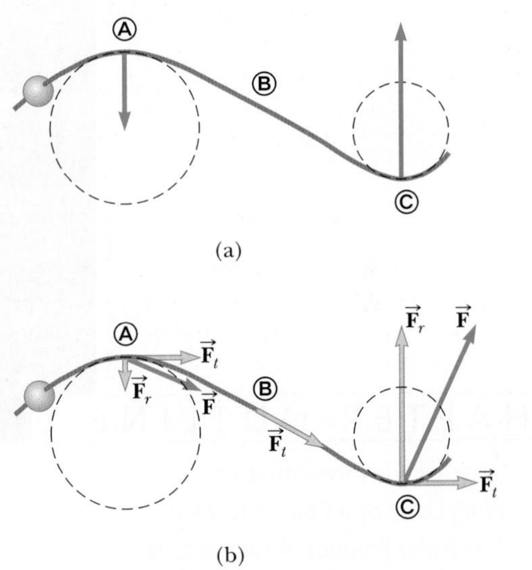

FIGURE **QQA.5.6**

5.7 (c). When the downward gravitational force $m\vec{g}$ and the upward force of air resistance $\vec{R}$ have the same magnitude, she reaches terminal speed and her acceleration is zero.

Energy and Energy Transfer

On a wind farm, a technician inspects one of the windmills. Moving air does work on the blades of the windmills, causing the blades and the rotor of an electrical generator to rotate. Energy is transferred out of the system of the windmill by means of electricity.

(Billy Hustace/Getty Images)

CHAPTER OUTLINE

In the preceding chapters, we analyzed the motion of an object using quantities such as *position, velocity, acceleration,* and *force,* with which you are familiar from everyday life. We developed a number of models using these notions that allow us to solve a variety of problems. Some problems that, in theory, could be solved with these models are very difficult to solve in practice, but they can be made much simpler with a different approach. In this and the following two chapters, we shall investigate this new approach, which will introduce us to new analysis models for problem solving. This approach includes definitions of quantities that may not be familiar to you. You may be familiar with some quantities, but they may have more specific meanings in physics than in everyday life. We begin this discussion by exploring *energy.*

Energy is present in the Universe in various forms. ***Every* physical process in the Universe involves energy and energy transfers**

or transformations. Therefore, energy is an extremely important concept to understand. Unfortunately, despite its importance, it cannot be easily defined. The variables in previous chapters were relatively concrete; we have everyday experience with velocities and forces, for example. Although the *notion* of energy is more abstract, we do have *experiences* with energy, such as running out of gasoline or losing our electrical service if we forget to pay the bill.

The concept of energy can be applied to the dynamics of a mechanical system without resorting to Newton's laws. This "energy approach" to describing motion is especially useful when the force acting on a particle is not constant; in such a case, the acceleration is not constant and we cannot apply the particle under constant acceleration model we developed in Chapter 2. Particles in nature are often subject to forces that vary with the particles' positions. These forces include gravitational forces and the force exerted on an object attached to a spring. We will develop a global approach to problems involving energy and energy transfer. This approach extends well beyond physics and can be applied to biological organisms, technological systems, and engineering situations.

6.1 SYSTEMS AND ENVIRONMENTS

All our analysis models in the earlier chapters were based on the motion of a particle or an object modeled as a particle. We begin our study of our new approach by identifying a **system.** A system is a simplification model in that we focus our attention on a small region of the Universe—the system—and ignore details of the rest of the Universe outside the system. A critical skill in applying the energy approach to problems in the next three chapters is *correctly identifying the system.* A system may

■ A system

- be a single object or particle
- be a collection of objects or particles
- be a region of space (e.g., the interior of an automobile engine combustion cylinder)
- vary in size and shape (e.g., a rubber ball that deforms upon striking a wall)

A **system boundary,** which is an imaginary surface (often but not necessarily coinciding with a physical surface), divides the Universe between the system and the **environment** of the system.

PITFALL PREVENTION 6.1

IDENTIFY THE SYSTEM One of the most important steps to take in solving a problem using the energy approach is to identify the system of interest correctly. Be sure this step is the *first* step you take in solving a problem.

As an example, imagine a force applied to an object in empty space. We can define the object as the system as in the first item in the bulleted list above. The force applied to it is an influence on the system from the environment and acts across the system boundary. We will see how to analyze this situation using a system approach in a subsequent section of this chapter.

Another example occurs in Example 5.3. Here the system can be defined as the combination of the ball, the cube, and the string, consistent with the second item of the bulleted list. The influences from the environment include the gravitational forces on the ball and the cube, the normal and friction forces on the cube, and the force of the pulley on the string. The forces exerted by the string on the ball and the cube are internal to the system and therefore are not included as influences from the environment.

6.2 WORK DONE BY A CONSTANT FORCE

Let us begin our analysis of systems by introducing a term whose meaning in physics is distinctly different from its everyday meaning. This new term is **work.** Imagine that you are trying to push a heavy sofa across your living room floor. If you push on the sofa *and* it moves through a displacement, you have done work on the sofa.

Consider a particle, which we identify as the system, that undergoes a displacement $\Delta\vec{\mathbf{r}}$ along a straight line while acted on by a constant force $\vec{\mathbf{F}}$ that makes

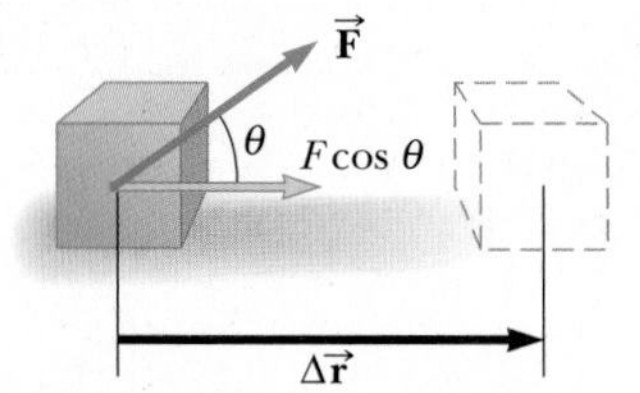

FIGURE 6.1 If an object undergoes a displacement $\Delta\vec{r}$, the work done by the constant force $\vec{F}$ on the object is $(F\cos\theta)\Delta r$.

an angle θ with $\Delta\vec{r}$, as in Figure 6.1. The force has accomplished something—it has moved the particle—so we say that work was done by the force on the particle.

Notice that we know only the force and the displacement given in the description of the situation. We have no information about how long it took for this displacement to occur, nor any information about velocities or accelerations. The absence of this information provides a hint of the power of the energy approach as well as a hint of how different it will be from our approach in previous chapters. We do not need this information to find the work done. Let us now formally define the work done on a system if the force is constant:

■ Work done by a constant force

The **work** W done on a system by an external agent exerting a constant force on the system is the product of the magnitude F of the force, the magnitude Δr of the displacement of the point of application of the force, and $\cos\theta$, where θ is the angle between the force and displacement vectors:

$$W \equiv F\Delta r\cos\theta \qquad [6.1]$$

Work is a scalar quantity; no direction is associated with it. Its units are those of force multiplied by length; therefore, the SI unit of work is the **newton · meter** (N · m). The newton · meter, when it refers to work or energy, is called the **joule** (J).

From the definition in Equation 6.1, we see that a force does no work on a system if the point of application of the force does not move. In the mathematical representation, if $\Delta r = 0$, Equation 6.1 gives $W = 0$. In the mental representation, imagine pushing on the sofa mentioned earlier. If it doesn't move, no work has been done on the sofa. Of course, the work is also zero if the applied force is zero. If you don't push on the sofa, no work is done on it!

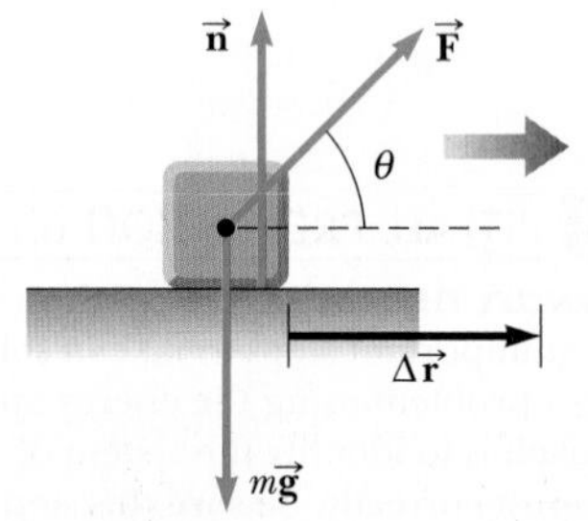

FIGURE 6.2 When an object is displaced horizontally on a flat table, the normal force $\vec{n}$ and the gravitational force $m\vec{g}$ do no work.

Also note from Equation 6.1 that the work done by a force is zero when the force is perpendicular to the displacement. That is, if $\theta = 90°$, then $\cos 90° = 0$ and $W = 0$. For example, consider the free-body diagram for a block moving across a frictionless surface in Figure 6.2. The work done by the normal force and the gravitational force on the block during its horizontal displacement are both zero for the same reason: they are both perpendicular to the displacement.

For now, we restrict our attention to systems consisting of a single particle or a small number of particles. In the case of a force applied to a particle, the displacement of the point of application of the force is necessarily the same as the displacement of the particle. In Chapter 17, we will consider work done in compressing a gas, which is modeled as a system consisting of a large number of particles. In this process, the displacement of the point of application of the force is very different from the displacement of the system.

In general, a particle may be moving under the influence of several forces. In that case, because work is a scalar quantity, the total work done as the particle undergoes some displacement is the algebraic sum of the work done by each of the forces.

The sign of the work depends on the direction of $\vec{F}$ relative to $\Delta\vec{r}$. The work done by the applied force is positive when the vector component of magnitude $F\cos\theta$ is in the *same direction* as the displacement. For example, when an object is lifted, the work done by the lifting force on the object is positive because the lifting force is upward, that is, in the same direction as the displacement. When the vector component of magnitude $F\cos\theta$ is in the direction *opposite* the displacement, W is *negative*. In the case of the object being lifted, for instance, the work done by the gravitational force on the object is negative.

If a constant applied force $\vec{F}$ acts parallel to the direction of the displacement, $\theta = 0$ and $\cos 0 = 1$. In this case, Equation 6.1 gives

$$W = F\Delta r \qquad [6.2]$$

Both Equations 6.1 and 6.2 are special cases of a more generalized definition of work. Both equations assume a constant force, and Equation 6.2 assumes that the force is parallel to the displacement. In the next two sections, we shall consider the situation in which a force is not parallel to the displacement and the more general case of a varying force.

PITFALL PREVENTION 6.2

WORK IS DONE BY . . . ON . . . Not only must you identify the system, you must also identify the interaction of the system with the environment. When discussing work, always use the phrase, "the work done by . . . on . . ." After "by" insert the part of the environment that is interacting directly with the system. After "on" insert the system. For example, "the work done by the hammer on the nail" identifies the nail as the system and the force from the hammer represents the interaction with the environment. This wording is similar to our use in Chapter 4 of "the force exerted by . . . on . . . ."

QUICK QUIZ 6.1 Figure 6.3 shows four situations in which a force is applied to an object. In all four cases, the force has the same magnitude and the displacement of the object is to the right and of the same magnitude. Rank the situations in order of the work done by the force on the object, from most positive to most negative.

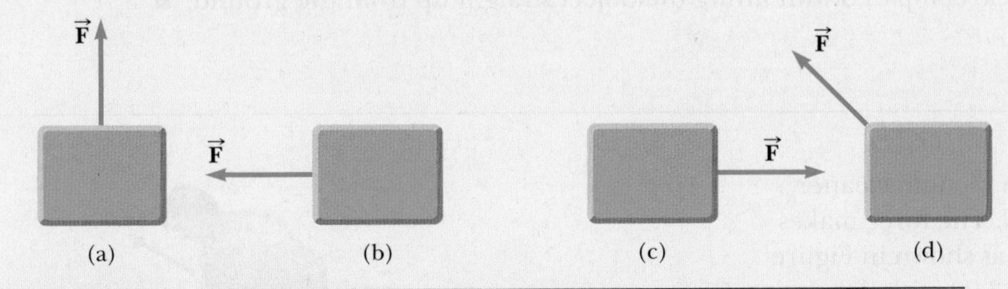

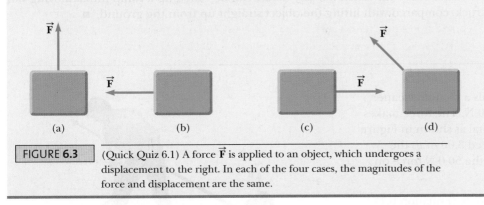

FIGURE 6.3 (Quick Quiz 6.1) A force $\vec{F}$ is applied to an object, which undergoes a displacement to the right. In each of the four cases, the magnitudes of the force and displacement are the same.

PITFALL PREVENTION 6.3

CAUSE OF THE DISPLACEMENT We can calculate the work done by a force on an object, but that force is *not* necessarily the cause of the object's displacement. For example, if you lift an object, work is done by the gravitational force, although gravity is not the cause of the object moving upward!

■ Thinking Physics 6.1

A person slowly lifts a heavy box of mass m a vertical height h and then walks horizontally at constant velocity a distance d while holding the box as in Figure 6.4. Determine the work done **(a)** by the person and **(b)** by the gravitational force on the box in this process.

Reasoning **(a)** Assume that the person lifts the box with a force of magnitude equal to the weight of the box mg. In this case, the work done by the person on the box during the vertical displacement is $W = F\,\Delta r = (mg)(h) = mgh$, which is positive because the lifting force is in the same direction as the displacement. For the horizontal displacement, we assume that the acceleration of the box is approximately zero. As a result, the work done by the person on the box during the horizontal displacement of the box is zero because the horizontal force is approximately zero, and the force supporting the box's weight in this process is perpendicular to the displacement. Therefore, the net work done by the person on the box during the complete process is mgh.

(b) The work done by the gravitational force on the box during the vertical displacement of the box is $-mgh$, which is negative because this force is opposite the displacement. The work done by the gravitational force is zero during the horizontal displacement because this force is perpendicular to the displacement. Hence, the net work done by the gravitational force for the complete process is $-mgh$. The net work done by all forces on the box is zero, because $+mgh + (-mgh) = 0$. ■

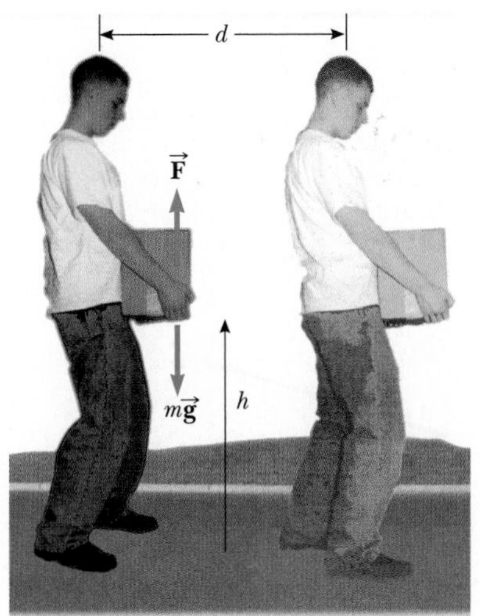

FIGURE 6.4 (Thinking Physics 6.1) A person lifts a heavy box of mass m a vertical distance h and then walks horizontally at constant velocity a distance d.

■ Thinking Physics 6.2

Roads going up mountains are formed into *switchbacks,* with the road weaving back and forth along the face of the slope so that any portion of the roadway has only a gentle rise. Do switchbacks require that an automobile climbing the mountain do any less work than if it were driving on a roadway that runs straight up the slope? Why are the switchbacks used?

Reasoning If we ignore the effects of rolling friction on the tires of the car, the same amount of work would be done in driving up the switchbacks and driving straight up the mountain because the weight of the car is moved upward against the gravitational force by the same vertical distance in each case. So why do we use the switchbacks? The answer lies in the force required, not the work. The force needed from the engine to follow a gentle rise is much less than that required to drive straight up the hill. Roadways running straight uphill would require redesigning engines so as to enable them to apply much larger forces. This situation is similar to the ease with which a heavy object can be rolled up a ramp into a moving van truck, compared with lifting the object straight up from the ground. ■

EXAMPLE 6.1 Mr. Clean

A man cleaning his apartment pulls a vacuum cleaner with a force of magnitude $F = 50.0$ N. The force makes an angle of 30.0° with the horizontal as shown in Figure 6.5. The vacuum cleaner is displaced 3.00 m to the right. Calculate the work done by the 50.0-N force on the vacuum cleaner.

Solution Using the definition of work (Equation 6.1), we have

$$W = (F\cos\theta)\,\Delta r = (50.0\text{ N})(\cos 30.0°)(3.00\text{ m})$$

$$= 130\text{ N}\cdot\text{m} = \boxed{130\text{ J}}$$

Note that the normal force $\vec{\mathbf{n}}$, the gravitational force $m\vec{\mathbf{g}}$, and the upward component of the applied force do *no* work because they are perpendicular to the displacement.

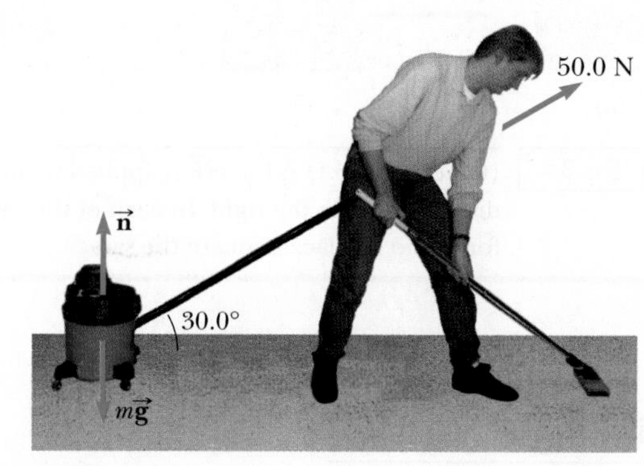

FIGURE 6.5 (Example 6.1) A vacuum cleaner being pulled at an angle of 30.0° with the horizontal.

6.3 THE SCALAR PRODUCT OF TWO VECTORS

Based on Equation 6.1, it is convenient to express the definition of work in terms of a **scalar product** of the two vectors $\vec{\mathbf{F}}$ and $\Delta\vec{\mathbf{r}}$. The scalar product was introduced briefly in Section 1.8. We formally provide its definition here:

The scalar product of any two vectors $\vec{\mathbf{A}}$ and $\vec{\mathbf{B}}$ is a scalar quantity equal to the product of the magnitudes of the two vectors and the cosine of the angle θ between them:

$$\vec{\mathbf{A}}\cdot\vec{\mathbf{B}} \equiv AB\cos\theta \qquad [6.3]$$

where θ is the angle between $\vec{\mathbf{A}}$ and $\vec{\mathbf{B}}$ as in Figure 6.6.

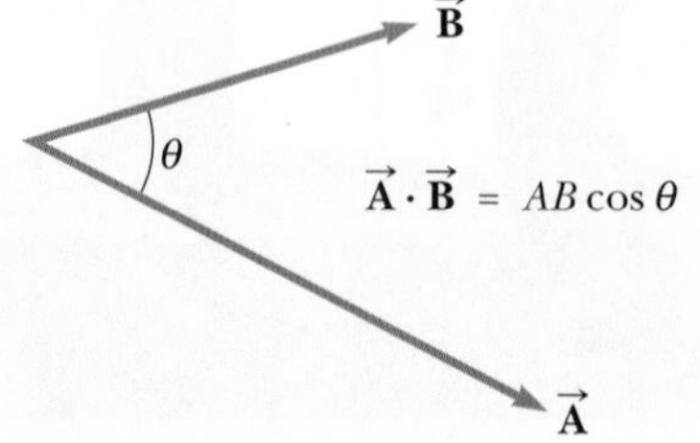

FIGURE 6.6 The scalar product $\vec{\mathbf{A}}\cdot\vec{\mathbf{B}}$ equals the magnitude of $\vec{\mathbf{A}}$ multiplied by the magnitude of $\vec{\mathbf{B}}$ and the cosine of the angle between $\vec{\mathbf{A}}$ and $\vec{\mathbf{B}}$.

Note that $\vec{\mathbf{A}}$ and $\vec{\mathbf{B}}$ need not have the same units. The units of the scalar product are simply the product of the units of the two vectors. Because of the dot symbol, the scalar product is often called the *dot product.*

Notice that the right-hand side of Equation 6.3 has the same mathematical structure as the right-hand side of Equation 6.1. Consequently, we can write the definition of work as the scalar product $\vec{\mathbf{F}} \cdot \Delta\vec{\mathbf{r}}$. Therefore, we can express Equation 6.1 as

$$W = \vec{\mathbf{F}} \cdot \Delta\vec{\mathbf{r}} = F\Delta r\cos\theta \qquad [6.4]$$

■ Work expressed as a scalar product

Before continuing with our discussion of work, let us investigate some properties of the scalar product because we will need to use it later in the book as well. From Equation 6.3 we see that the scalar product is *commutative.* That is,

$$\vec{\mathbf{A}} \cdot \vec{\mathbf{B}} = \vec{\mathbf{B}} \cdot \vec{\mathbf{A}} \qquad [6.5]$$

In addition, the scalar product obeys the *distributive law of multiplication,* so that

$$\vec{\mathbf{A}} \cdot (\vec{\mathbf{B}} + \vec{\mathbf{C}}) = \vec{\mathbf{A}} \cdot \vec{\mathbf{B}} + \vec{\mathbf{A}} \cdot \vec{\mathbf{C}} \qquad [6.6]$$

The scalar product is simple to evaluate from Equation 6.3 when $\vec{\mathbf{A}}$ is either perpendicular or parallel to $\vec{\mathbf{B}}$. If $\vec{\mathbf{A}}$ is perpendicular to $\vec{\mathbf{B}}$ ($\theta = 90°$), then $\vec{\mathbf{A}} \cdot \vec{\mathbf{B}} = 0$. (The equality $\vec{\mathbf{A}} \cdot \vec{\mathbf{B}} = 0$ also holds in the more trivial case when either $\vec{\mathbf{A}}$ or $\vec{\mathbf{B}}$ is zero.) If $\vec{\mathbf{A}}$ and $\vec{\mathbf{B}}$ point in the same direction ($\theta = 0$), then $\vec{\mathbf{A}} \cdot \vec{\mathbf{B}} = AB$. If $\vec{\mathbf{A}}$ and $\vec{\mathbf{B}}$ point in opposite directions ($\theta = 180°$), then $\vec{\mathbf{A}} \cdot \vec{\mathbf{B}} = -AB$. The scalar product is negative when $90° < \theta < 180°$.

The unit vectors $\hat{\mathbf{i}}$, $\hat{\mathbf{j}}$, and $\hat{\mathbf{k}}$, which were defined in Chapter 1, lie in the positive x, y, and z directions, respectively, of a right-handed coordinate system. Therefore, it follows from the definition of $\vec{\mathbf{A}} \cdot \vec{\mathbf{B}}$ that the scalar products of these unit vectors are given by

$$\hat{\mathbf{i}} \cdot \hat{\mathbf{i}} = \hat{\mathbf{j}} \cdot \hat{\mathbf{j}} = \hat{\mathbf{k}} \cdot \hat{\mathbf{k}} = 1 \qquad [6.7]$$

$$\hat{\mathbf{i}} \cdot \hat{\mathbf{j}} = \hat{\mathbf{i}} \cdot \hat{\mathbf{k}} = \hat{\mathbf{j}} \cdot \hat{\mathbf{k}} = 0 \qquad [6.8]$$

■ Scalar products of unit vectors

Two vectors $\vec{\mathbf{A}}$ and $\vec{\mathbf{B}}$ can be expressed in component form as

$$\vec{\mathbf{A}} = A_x\hat{\mathbf{i}} + A_y\hat{\mathbf{j}} + A_z\hat{\mathbf{k}} \qquad \vec{\mathbf{B}} = B_x\hat{\mathbf{i}} + B_y\hat{\mathbf{j}} + B_z\hat{\mathbf{k}}$$

Therefore, using these expressions, Equations 6.7 and 6.8 reduce the scalar product of $\vec{\mathbf{A}}$ and $\vec{\mathbf{B}}$ to

$$\vec{\mathbf{A}} \cdot \vec{\mathbf{B}} = A_xB_x + A_yB_y + A_zB_z \qquad [6.9]$$

where we have used the distributive law (Eq. 6.6) to simplify the result. This equation and Equation 6.3 are alternative but equivalent expressions for the scalar product. Equation 6.3 is useful if you know the magnitudes and directions of the vectors, and Equation 6.9 is useful if you know the components of the vectors. In the special case where $\vec{\mathbf{A}} = \vec{\mathbf{B}}$, we see that

$$\vec{\mathbf{A}} \cdot \vec{\mathbf{A}} = A_x^2 + A_y^2 + A_z^2 = A^2$$

QUICK QUIZ 6.2 Which of the following statements is true about the relationship between the scalar product of two vectors and the product of the magnitudes of the vectors? **(a)** $\vec{\mathbf{A}} \cdot \vec{\mathbf{B}}$ is larger than AB. **(b)** $\vec{\mathbf{A}} \cdot \vec{\mathbf{B}}$ is smaller than AB. **(c)** $\vec{\mathbf{A}} \cdot \vec{\mathbf{B}}$ could be larger or smaller than AB, depending on the angle between the vectors. **(d)** $\vec{\mathbf{A}} \cdot \vec{\mathbf{B}}$ could be equal to AB.

PITFALL PREVENTION 6.4

Work is a scalar Although Equation 6.4 defines the work in terms of two vectors, *work is a scalar;* there is no direction associated with it. *All* types of energy and energy transfer are scalars. This property is a major advantage of the energy approach because we don't need vector calculations!

EXAMPLE 6.2 **The Scalar Product**

The vectors $\vec{\mathbf{A}}$ and $\vec{\mathbf{B}}$ are given by $\vec{\mathbf{A}} = 2\hat{\mathbf{i}} + 3\hat{\mathbf{j}}$ and $\vec{\mathbf{B}} = -\hat{\mathbf{i}} + 2\hat{\mathbf{j}}$.

A Determine the scalar product $\vec{\mathbf{A}} \cdot \vec{\mathbf{B}}$.

Solution We can evaluate the scalar product directly using the unit vector notation:

$$\begin{aligned}\vec{\mathbf{A}} \cdot \vec{\mathbf{B}} &= (2\hat{\mathbf{i}} + 3\hat{\mathbf{j}}) \cdot (-\hat{\mathbf{i}} + 2\hat{\mathbf{j}}) \\ &= -2\hat{\mathbf{i}} \cdot \hat{\mathbf{i}} + 2\hat{\mathbf{i}} \cdot 2\hat{\mathbf{j}} - 3\hat{\mathbf{j}} \cdot \hat{\mathbf{i}} + 3\hat{\mathbf{j}} \cdot 2\hat{\mathbf{j}} \\ &= -2 + 6 = 4\end{aligned}$$

where we have used that $\hat{\mathbf{i}} \cdot \hat{\mathbf{i}} = \hat{\mathbf{j}} \cdot \hat{\mathbf{j}} = 1$ and $\hat{\mathbf{i}} \cdot \hat{\mathbf{j}} = \hat{\mathbf{j}} \cdot \hat{\mathbf{i}} = 0$. The same result is obtained using Equation 6.9 directly, where $A_x = 2$, $A_y = 3$, $B_x = -1$, and $B_y = 2$. Note that the result has no units because no units were specified on the original vectors $\vec{\mathbf{A}}$ and $\vec{\mathbf{B}}$.

B Find the angle θ between $\vec{\mathbf{A}}$ and $\vec{\mathbf{B}}$.

Solution The magnitudes of $\vec{\mathbf{A}}$ and $\vec{\mathbf{B}}$ are given by

$$A = \sqrt{A_x^2 + A_y^2} = \sqrt{(2)^2 + (3)^2} = \sqrt{13}$$

$$B = \sqrt{B_x^2 + B_y^2} = \sqrt{(-1)^2 + (2)^2} = \sqrt{5}$$

Using Equation 6.3 and the result from part A gives

$$\cos\theta = \frac{\vec{\mathbf{A}} \cdot \vec{\mathbf{B}}}{AB} = \frac{4}{\sqrt{13}\,\sqrt{5}} = \frac{4}{\sqrt{65}} = 0.496$$

$$\theta = \cos^{-1}(0.496) = 60.3^\circ$$

6.4 WORK DONE BY A VARYING FORCE

Consider a particle being displaced along the x axis under the action of a force with an x component F_x that varies with position, as in the graphical representation in Figure 6.7. The particle is displaced in the direction of increasing x from $x = x_i$ to $x = x_f$. In such a situation, we cannot use Equation 6.1 to calculate the work done by the force because this relationship applies only when $\vec{\mathbf{F}}$ is constant in magnitude and direction. As seen in Figure 6.7, we do not have a *single* value of the force to substitute into Equation 6.1. If, however, we imagine that the point of application of the force undergoes a *small* displacement in the x direction so that $\Delta r = \Delta x$, as shown in Figure 6.7a, the x component F_x of the force is approximately constant over this interval. We can then approximate the work done by the force on the particle for this small displacement as

$$W_1 \approx F_x\,\Delta x \qquad [6.10]$$

This quantity is just the area of the shaded geometric model rectangle in Figure 6.7a. If we imagine that the curve described by F_x versus x is divided into a large number of such intervals, the total work done for the displacement from x_i to x_f is approximately equal to the sum of a large number of such terms:

$$W \approx \sum_{x_i}^{x_f} F_x\,\Delta x$$

If the displacements Δx are allowed to approach zero, the number of terms in the sum increases without limit, but the value of the sum approaches a definite value equal to the area under the curve bounded by F_x and the x axis in Figure 6.7b. As you probably have learned in calculus, this limit of the sum is called an *integral* and is represented by

$$\lim_{\Delta x \to 0} \sum_{x_i}^{x_f} F_x\,\Delta x = \int_{x_i}^{x_f} F_x\,dx$$

The limits on the integral $x = x_i$ to $x = x_f$ define what is called a **definite integral.** (An *indefinite integral* is the limit of a sum over an unspecified interval. Appendix B.7 gives a brief description of integration.) This definite integral is numerically

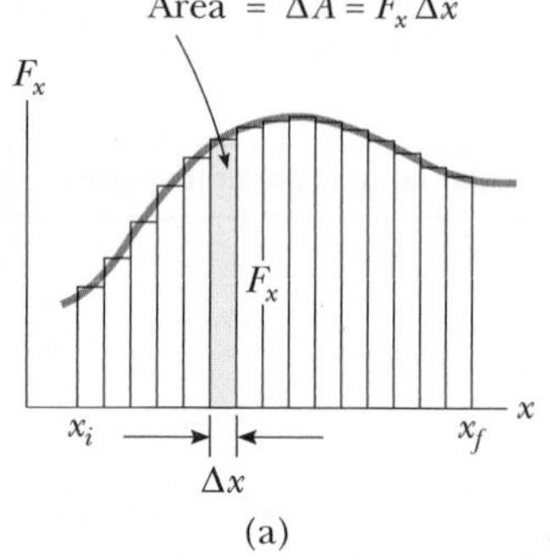

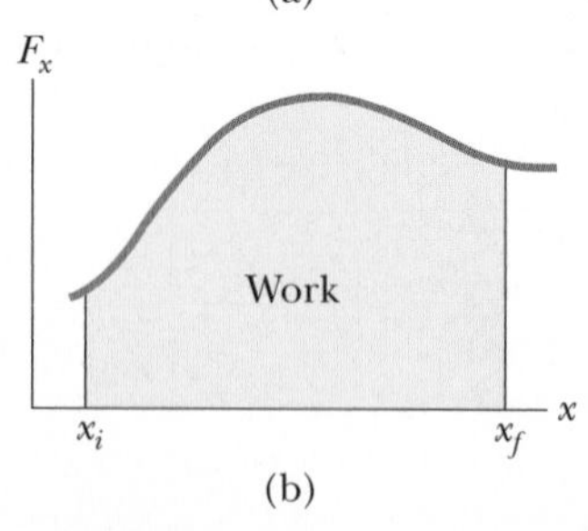

FIGURE 6.7 (a) The work done by a force of magnitude F_x for the small displacement Δx is $F_x\,\Delta x$, which equals the area of the shaded rectangle. The total work done for the displacement from x_i to x_f is approximately equal to the sum of the areas of all the rectangles. (b) The work done by the variable force F_x as the particle moves from x_i to x_f is *exactly* equal to the area under this curve.

equal to the area under the curve of F_x versus x between x_i and x_f. Therefore, we can express the work done by F_x for the displacement from x_i to x_f as

$$W = \int_{x_i}^{x_f} F_x \, dx \qquad [6.11]$$

This equation reduces to Equation 6.1 when $F_x = F\cos\theta$ is constant and $x_f - x_i = \Delta x$.

If more than one force acts on a system *and the system can be modeled as a particle,* the total work done on the system is just the work done by the net force. If we express the x component of the net force as $\sum F_x$, the total work, or *net work,* done on the particle as it moves from x_i to x_f is

$$\sum W = W_{\text{net}} = \int_{x_i}^{x_f} \left(\sum F_x\right) dx$$

For the general case of a particle moving along an arbitrary path while acted on by a net force $\sum \vec{\mathbf{F}}$, we use the scalar product:

$$\sum W = W_{\text{net}} = \int \left(\sum \vec{\mathbf{F}}\right) \cdot d\vec{\mathbf{r}} \qquad [6.12]$$

■ Work done by a variable net force

where the integral is calculated over the path that the particle takes through space.

If the system cannot be modeled as a particle (for example, if the system consists of multiple particles that can move with respect to each other), we cannot use Equation 6.12 because different forces on the system may move through different displacements. In that case, we must evaluate the work done by each force separately and then add the works algebraically.

Work Done by a Spring

A common physical system for which the force varies with position is shown in Active Figure 6.8. A block on a horizontal, frictionless surface is connected to a spring. If the block is located at a position x relative to its equilibrium position $x = 0$, the stretched or compressed spring exerts a force on the block given by

$$F_s = -kx \qquad [6.13]$$

■ Hooke's law

where k is a positive constant called the *force constant* (or *spring constant* or *stiffness constant*) of the spring. This force law for springs is known as **Hooke's law.** For many springs, Hooke's law can describe the behavior very accurately provided that the displacement from equilibrium is not too large. The value of k is a measure of the stiffness of the spring. Stiff springs have larger k values, and weak springs have smaller k values. We shall employ a simplification model in which all springs obey Hooke's law unless specified otherwise.

The negative sign in Equation 6.13 signifies that the force exerted by the spring on the block is always directed *opposite* the displacement from the equilibrium position $x = 0$. For example, when $x > 0$, such that the block is pulled to the right and the spring is stretched as in Active Figure 6.8a, the spring force is to the left, or negative. When $x < 0$, and the spring is compressed as in Active Figure 6.8c, the spring force is to the right, or positive. Of course, when $x = 0$, as in Active Figure 6.8b, the spring is unstretched and $F_s = 0$. Because the spring force always acts toward the equilibrium position, it is sometimes called a *restoring force.*

If the block is displaced to a position $-x_{\text{max}}$ and then released, it moves from $-x_{\text{max}}$ through zero to $+x_{\text{max}}$ (assuming a frictionless surface) and then turns around and returns to $-x_{\text{max}}$. The details of this oscillating motion will be

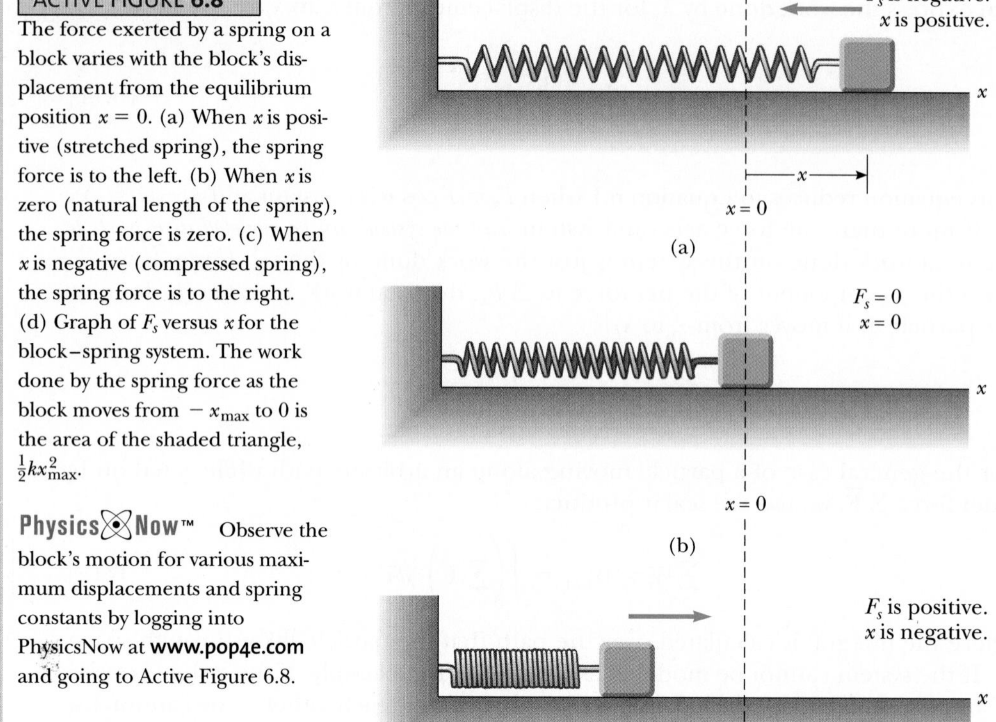

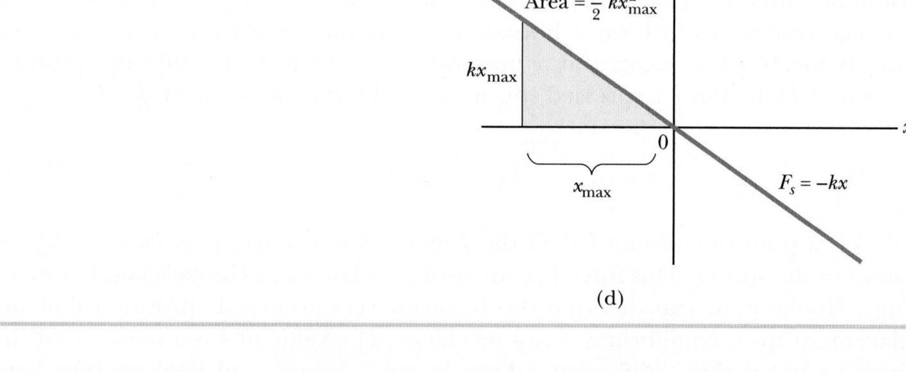

ACTIVE FIGURE 6.8
The force exerted by a spring on a block varies with the block's displacement from the equilibrium position $x = 0$. (a) When x is positive (stretched spring), the spring force is to the left. (b) When x is zero (natural length of the spring), the spring force is zero. (c) When x is negative (compressed spring), the spring force is to the right. (d) Graph of F_s versus x for the block–spring system. The work done by the spring force as the block moves from $-x_{max}$ to 0 is the area of the shaded triangle, $\frac{1}{2}kx_{max}^2$.

PhysicsNow™ Observe the block's motion for various maximum displacements and spring constants by logging into PhysicsNow at **www.pop4e.com** and going to Active Figure 6.8.

discussed in Chapter 12. For our purposes here, let us calculate the work done by the spring force on the block as the block moves from $x_i = -x_{max}$ to $x_f = 0$. Applying the particle model to the block and using Equation 6.11, we have

$$W_s = \int_{x_i}^{x_f} F_s\, dx = \int_{-x_{max}}^{0} (-kx)\, dx = \tfrac{1}{2}kx_{max}^2 \qquad [6.14]$$

The work done by the spring force on the block is positive because the spring force is in the same direction as the displacement (both are to the right).

If we consider the work done by the spring force on the block as the block continues to move from $x_i = 0$ to $x_f = x_{max}$, we find that $W_s = -\frac{1}{2}kx_{max}^2$. This work is negative because for this part of the motion the displacement is to the right and the spring force is to the left. Therefore, the *net* work done by the spring force on the block as it moves from $x_i = -x_{max}$ to $x_f = x_{max}$ is *zero.*

If we plot F_s versus x, as in Active Figure 6.8d, we arrive at the same results. The work calculated in Equation 6.14 is equal to the area of the shaded triangle in Active Figure 6.8d, with base x_{max} and height kx_{max}. This area is $\frac{1}{2}kx_{max}^2$.

If the block undergoes an *arbitrary* displacement from $x = x_i$ to $x = x_f$, the work done by the spring force is

$$W_s = \int_{x_i}^{x_f} (-kx)\, dx = \tfrac{1}{2}kx_i^2 - \tfrac{1}{2}kx_f^2 \qquad [6.15]$$

■ Work done by a spring

From this equation we see that the work done by the spring force on the block is zero for any motion that ends where it began ($x_i = x_f$). We shall make use of this important result in Chapter 7, where we describe the motion of this system in more detail. Equation 6.15 also shows that the work done by the spring force is zero when the block moves between any two symmetric locations, $x_i = -x_f$. Consider the curve representing the spring force in Active Figure 6.8d; if the block moves from $x = -x_{max}$ to $x = +x_{max}$, the total work is zero because we are adding a positive area (for $-x_{max} < x < 0$) to a negative area (for $0 < x < +x_{max}$) of equal magnitude.

Equations 6.14 and 6.15 describe the work done by the spring force on the block. Now consider the work done by an *external agent* on the block as the agent applies a force to the spring and stretches it *very slowly* from $x_i = -x_{max}$ to $x_f = 0$ as in Figure 6.9. This work can be easily calculated by noting that the applied force $\vec{\mathbf{F}}_{app}$ is of equal magnitude and opposite direction to the spring force $\vec{\mathbf{F}}_s$ at any value of the position (because the block is not accelerating), so that $F_{app} = -(-kx) = +kx$. The work done by this applied force (the external agent) on the block is therefore

$$W_{F_{app}} = \int_{-x_{max}}^{0} F_{app}\, dx = \int_{-x_{max}}^{0} kx\, dx = -\tfrac{1}{2}kx_{max}^2$$

Note that this work is equal to the negative of the work done by the spring force on the block for this displacement (Eq. 6.14). The work is negative because the external agent must push to the left on the spring in Figure 6.9 to prevent it from expanding, and this direction is opposite the direction of the displacement as the block moves from $-x_{max}$ to 0.

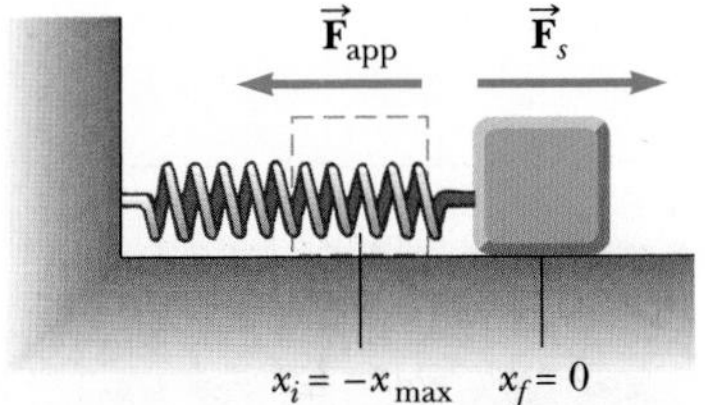

FIGURE 6.9 A block moves from $x_i = -x_{max}$ to $x_f = 0$ on a frictionless surface as a force $\vec{\mathbf{F}}_{app}$ is applied to the block. If the process is carried out very slowly, the applied force is equal in magnitude and opposite in direction to the spring force at all times.

QUICK QUIZ 6.3 A dart is loaded into a spring-loaded toy dart gun by pushing the spring in by a distance d. For the next loading, the spring is compressed a distance $2d$. How much work is required to load the second dart compared to that required to load the first? **(a)** four times as much **(b)** two times as much **(c)** the same **(d)** half as much **(e)** one-fourth as much

EXAMPLE 6.3 Work Required to Stretch a Spring

One end of a horizontal spring ($k = 80$ N/m) is held fixed while an external force is applied to the free end, stretching it slowly from $x_A = 0$ to $x_B = 4.0$ cm.

A Find the work done by the external force on the spring.

Solution Because we have not been told otherwise, we assume that the spring obeys Hooke's law. We place the zero reference of the coordinate axis at the free end of the unstretched spring. The applied force is $F_{app} = kx = (80 \text{ N/m})(x)$. The work done by F_{app} is the area of the triangle from 0 to 4.0 cm in Figure 6.10:

$$W = \tfrac{1}{2}kx_B^2 = \tfrac{1}{2}(80 \text{ N/m})(0.040 \text{ m})^2 = \boxed{0.064 \text{ J}}$$

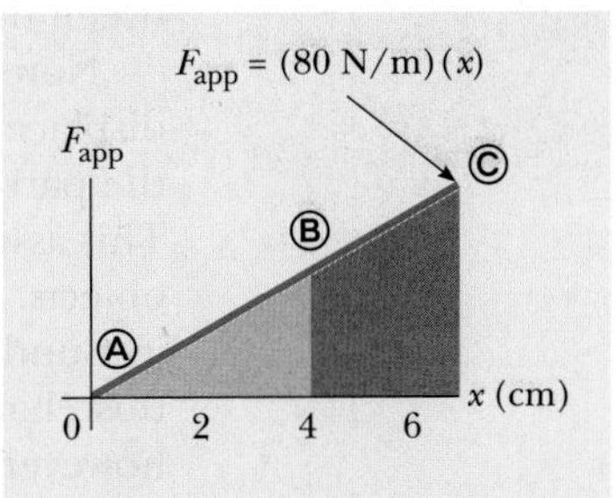

FIGURE 6.10 (Example 6.3) A graph of the applied force required to stretch a spring that obeys Hooke's law versus the elongation of the spring.

Note that the work is positive because the applied force and the displacement are in the same direction.

B Find the additional work done in stretching the spring from $x_B = 4.0$ cm to $x_C = 7.0$ cm.

Solution The work done in stretching the spring the additional amount is the darker shaded area between these limits in Figure 6.10. Geometrically, it is the difference in area between the large and small triangles:

$$W = \frac{1}{2}kx_C^2 - \frac{1}{2}kx_B^2$$
$$= \frac{1}{2}(80 \text{ N/m}) [(0.070 \text{ m})^2 - (0.040 \text{ m})^2] = 0.13 \text{ J}$$

Using calculus, we find the same result:

$$W = \int_{x_B}^{x_C} F_{app}\, dx = \int_{0.040 \text{ m}}^{0.070 \text{ m}} (80 \text{ N/m})\, x\, dx$$
$$= \frac{1}{2}(80 \text{ N/m})\, (x^2)\Big|_{0.040 \text{ m}}^{0.070 \text{ m}}$$
$$W = \frac{1}{2}(80 \text{ N/m}) [(0.070 \text{ m})^2 - (0.040 \text{ m})^2] = 0.13 \text{ J}$$

6.5 KINETIC ENERGY AND THE WORK–KINETIC ENERGY THEOREM

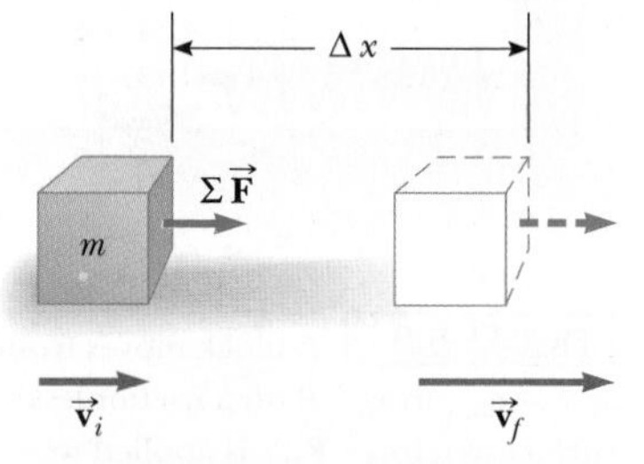

FIGURE 6.11 An object modeled as a particle undergoes a displacement of magnitude Δx and a change in speed under the action of a constant net force $\sum \vec{F}$.

Now that we have explored various means of evaluating the work done by a force on a system, let us explore the significance and benefits of the energy approach. As we shall see in this section, if the work done by the net force on a particle can be calculated for a given displacement, the change in the particle's speed is easy to evaluate. Let's see how it is done.

Figure 6.11 shows an object modeled as a particle of mass m moving to the right along the x axis under the action of a net force $\sum \vec{F}$, also to the right. If the point of application of the force moves through a displacement $\Delta x = x_f - x_i$, the work done by the force $\sum \vec{F}$ on the particle is

$$W_{net} = \int_{x_i}^{x_f} \sum F\, dx \qquad [6.16]$$

Using Newton's second law, we can substitute for the magnitude of the net force $\sum F = ma$ and then perform the following chain-rule manipulations on the integrand:

$$W_{net} = \int_{x_i}^{x_f} ma\, dx = \int_{x_i}^{x_f} m\frac{dv}{dt}\, dx = \int_{x_i}^{x_f} m\frac{dv}{dx}\frac{dx}{dt}\, dx = \int_{v_i}^{v_f} mv\, dv$$

$$W_{net} = \frac{1}{2}mv_f^2 - \frac{1}{2}mv_i^2 \qquad [6.17]$$

This equation was generated for the specific situation of one-dimensional motion, but it can also be used for two- or three-dimensional motion. It tells us that the work done by the net force on a particle of mass m is equal to the difference between the initial and final values of a quantity $\frac{1}{2}mv^2$.

Note that in deriving Equation 6.17, the dx we used to calculate the work was the displacement of the particle. In other words, we assumed that the displacement of the particle is the same as the displacement of the point of application of the force. This assumption is necessarily true for particles, but it may not be true for extended objects. It will only be true if the object is perfectly rigid, so that all parts of the object undergo the same displacement. Most of the situations that we will consider in this chapter and the next will satisfy this requirement. One important exception, however—objects subject to kinetic friction—will be explored in Section 6.7.

The quantity $\frac{1}{2}mv^2$ in Equation 6.17 is so important that we give it a special name. The **kinetic energy** K of an object of mass m moving with a speed v is defined as

▪ Kinetic energy of an object

$$K \equiv \frac{1}{2}mv^2 \qquad [6.18]$$

Kinetic energy is a scalar quantity and has the same units as work. For example, an object of mass 2.0 kg moving with a speed of 4.0 m/s has a kinetic energy of 16 J.

It is often convenient to write Equation 6.17 in the form

$$W_{\text{net}} = K_f - K_i = \Delta K \qquad [6.19]$$

■ The work–kinetic energy theorem

Equation 6.19 is an important result known as the **work–kinetic energy theorem:**

> When work is done on a system and the only change in the system is in its speed, the work done by the net force equals the change in kinetic energy of the system.

The work–kinetic energy theorem indicates that the speed of a particle increases if the net work done on it is positive because the final kinetic energy will be greater than the initial kinetic energy. The speed decreases if the net work is negative because the final kinetic energy will be less than the initial kinetic energy.

The work–kinetic energy theorem will clarify some results we saw earlier in this chapter that may have seemed odd. In Thinking Physics 6.1, a person lifts a block and moves it horizontally. At the end of the Reasoning, we mentioned that the net work done by all forces on the block is zero. That may seem strange, but it is correct. If we choose the block as the system, the net force on the system is zero because the upward lifting force is modeled as being equal in magnitude to the gravitational force. Therefore, the net force is zero and zero net work is done, which is consistent because the kinetic energy of the block does not change. It may seem incorrect that no work was done because something changed—the block was lifted—but that is correct *because we chose the block as the system.* If we had chosen the block and the Earth as the system, we would have a different result because the work done on this system is not zero. We will explore this idea in the next chapter.

In Section 6.4, we also saw a result of zero work done, when a block on a spring moved from $x_i = -x_{\max}$ to $x_f = x_{\max}$. The work is zero here for a different reason from that for lifting the block. It is the result of the combination of positive work and an equal amount of negative work done by the *same* force. It is also different from the lifting example in that the speed of the block on the spring is continually changing. The work–kinetic energy theorem refers only to the initial and final points for the speeds; it does not depend on details of the path followed between these points. We shall use this concept often in the remainder of this chapter and in the next chapter.

PITFALL PREVENTION 6.5

CONDITIONS FOR THE WORK–KINETIC ENERGY THEOREM Always remember the special conditions for the work–kinetic energy theorem. We will see many situations in which other changes occur in the system besides its speed, and there are other interactions with the environment besides work. The work–kinetic energy theorem is important, but it is limited in its application and is not a general principle. We shall present a general principle involving energy in Section 6.6.

QUICK QUIZ 6.4 A dart is loaded into a spring-loaded toy dart gun by pushing the spring in by a distance d. For the next loading, the spring is compressed a distance $2d$. How much faster does the second dart leave the gun compared with the first? **(a)** four times as fast **(b)** two times as fast **(c)** the same **(d)** half as fast **(e)** one-fourth as fast

EXAMPLE 6.4 A Block Pulled on a Frictionless Surface

A 6.00-kg block initially at rest is pulled to the right along a horizontal frictionless surface by a constant, horizontal force $\vec{\mathbf{F}}$ of magnitude 12.0 N as in Figure 6.12. Find the speed of the block after it has moved 3.00 m.

Solution The block is the system, and three external forces interact with it. Neither the gravitational force nor the normal force does work on the block because these forces are vertical and the displacement of the block is horizontal. There is no friction, so the only

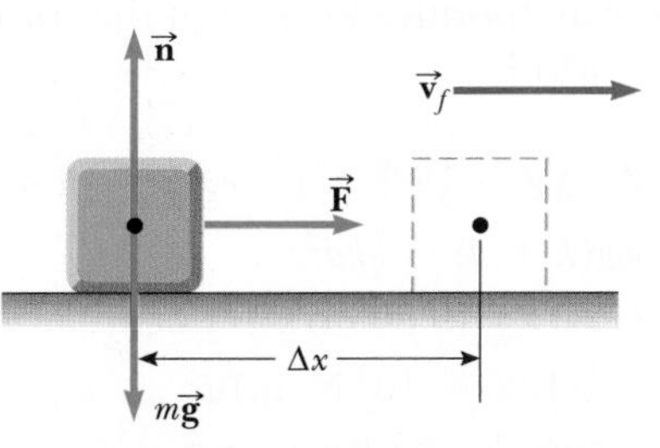

FIGURE 6.12 (Example 6.4) A block on a frictionless surface is pulled to the right by a constant horizontal force.

external force that we must consider in the calculation is the 12.0-N force.

The work done by the 12.0-N force is

$$W = F\Delta x = (12.0 \text{ N})(3.00 \text{ m}) = 36.0 \text{ N}\cdot\text{m} = 36.0 \text{ J}$$

Using the work–kinetic energy theorem and noting that the initial kinetic energy is zero, we find

$$W = K_f - K_i = \tfrac{1}{2}mv_f^2 - 0$$

$$v_f = \sqrt{\frac{2W}{m}} = \sqrt{\frac{2(36.0 \text{ J})}{6.00 \text{ kg}}} = \boxed{3.46 \text{ m/s}}$$

Notice that an energy calculation such as this one gives only the speed of the particle, not the velocity. In many cases, that is all you need. If you want to find the direction of the velocity vector, you may need to analyze the pictorial representation or perform other calculations. In this example, it is clear that $\vec{\mathbf{v}}_f$ is directed to the right.

EXAMPLE 6.5 Dropping a Block onto a Spring

A massless spring that has a force constant of 1.00×10^3 N/m is placed on a table in a vertical position as in Figure 6.13. A block of mass 1.60 kg is held 1.00 m above the free end of the spring. The block is dropped from rest so that it falls vertically onto the spring. By what maximum distance does the spring compress?

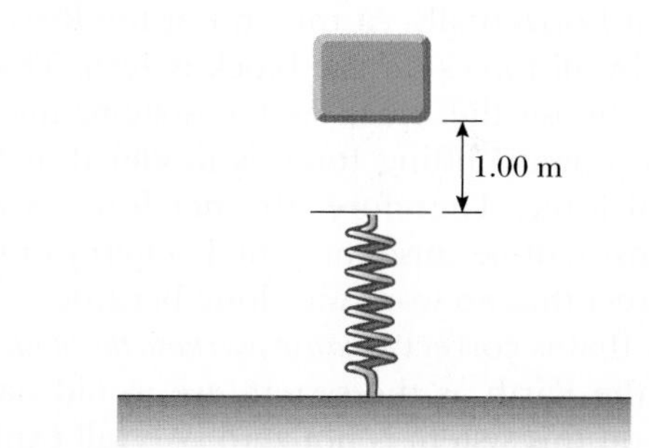

FIGURE 6.13 (Example 6.5) A block is dropped onto a vertical spring, causing the spring to compress.

Solution Conceptualize the problem by imagining the block dropping on the spring and compressing the spring by some distance. The block is at rest momentarily before the compressed spring begins to move the block upward again. We want to focus on that instant of time at which the block is at rest. We identify the block as the system. We identify the initial condition as the release of the block from the height $y_i = h = 1.00$ m above the free end of the spring. The final condition occurs when the block is momentarily at rest with the spring compressed its maximum distance. For this condition, the block is located at $y_f = -d$, where d is the maximum distance by which the spring is compressed. Because both the gravitational force and the spring force are doing work on the block, we categorize the problem as one that can be addressed with the work–kinetic energy theorem. To analyze the problem, we determine that the net work done on the block during its displacement between the initial and final positions by gravity (positive work) and the spring force (negative work) is

$$\begin{aligned} W_{\text{net}} &= \vec{\mathbf{F}}_g \cdot \Delta\vec{\mathbf{r}} - \tfrac{1}{2}kd^2 = (-mg)\hat{\mathbf{j}} \cdot (-d - h)\hat{\mathbf{j}} - \tfrac{1}{2}kd^2 \\ &= mg(h + d) - \tfrac{1}{2}kd^2 \\ &= (1.60 \text{ kg})(9.80 \text{ m/s}^2)(1.00 \text{ m} + d) \\ &\quad - \tfrac{1}{2}(1.00 \times 10^3 \text{ N/m})d^2 \\ &= -500d^2 + 15.7d + 15.7 \end{aligned}$$

The change in kinetic energy of the block is zero because it is at rest at both the initial and final conditions. Therefore, from the work–kinetic energy theorem, the work done by the net force must be equal to zero:

$$-500d^2 + 15.7d + 15.7 = 0$$

This quadratic equation can be solved, and the solutions are $d = 0.19$ m and $d = -0.16$ m. Because we have chosen the value of d as a positive number by claiming that $y = -d$ is *below* the initial position of the end of the spring, we must choose the positive root, $\boxed{d = 0.19 \text{ m}}$.

To finalize the problem, let us be sure that we can interpret the negative root. The negative root gives the position for the final condition as $y = -d = -(-0.16 \text{ m}) = +0.16$ m, which is the position above the *initial* position $y = 0$ at which the block again comes to rest in its oscillation, assuming that the block remains attached to the spring. These two positions are symmetric around $y = -0.016$ m, which is where the block would rest in equilibrium on the spring, according to Hooke's law.

6.6 THE NONISOLATED SYSTEM

We have seen a number of examples in which an object, modeled as a particle, is acted on by various forces, with the result that there is a change in its kinetic energy. This very simple situation is the first example of the **nonisolated system,** which is an important new analysis model for us. Physical problems for which this model is appropriate involve systems that interact with or are influenced by their environment, causing some kind of change in the system.

The work–kinetic energy theorem is our first introduction to the nonisolated system. The interaction is the work done by the external force and the quantity related to the system that changes is its kinetic energy. Because the energy of the system changes, we conceptualize work as a means of **energy transfer; work has the effect of transferring energy between the system and the environment.** If positive work is done on the system, energy is transferred to the system, whereas negative work indicates that energy is transferred from the system to the environment.

So far, we have discussed kinetic energy as the only type of energy in a system. We now argue the existence of a second type of energy. Consider a situation in which an object slides along a surface with friction. Clearly, work is done by the friction force because there is a force and a displacement of the object on which the force acts. Keep in mind, however, that our equations for work involve the displacement of the *point of application of the force.* If an object is perfectly rigid, the displacement of the point of application of the force is the same as the displacement of the object. For a nonrigid object, however, these displacements are not the same. Imagine, for example, a block of gelatin sitting on a plate. Suppose the block is pushed with a horizontal force applied to a vertical side so that the block deforms but does not slide on the plate. There has been a displacement of the object because most of the particles in the object, except for those along the stationary bottom edge, have moved horizontally through various displacements. The displacement of the point of application of the friction force between the block and the plate is zero, however, because the bottom of the block has not moved.

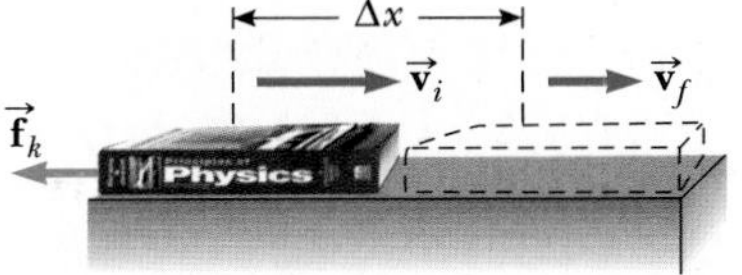

FIGURE 6.14 A book sliding to the right on a horizontal surface slows down in the presence of a force of kinetic friction acting to the left. The initial velocity of the book is $\vec{\mathbf{v}}_i$, and its final velocity is $\vec{\mathbf{v}}_f$. The normal force and gravitational force are not included in the diagram because they are perpendicular to the direction of motion and therefore do not influence the speed of the book.

On a microscopic scale, real objects are deformable; it is the deformation and interaction of the surfaces in contact that cause the friction force. In general, the displacement of the point of application of the friction force (assuming that we could calculate it!) is not the same as the displacement of the object.[1]

Let us imagine the book in Figure 6.14 sliding to the right on the surface of a heavy table and slowing down as a result of the friction force. Suppose the *surface* is the system. The sliding book exerts a friction force to the right on the surface. As a result, many atoms on the surface move slightly to the right under the influence of this force. Consequently, the points of application of the friction force move to the right and the friction force does positive work on the surface. The surface, however, is not moving after the book has stopped. Positive work has been done on the surface, yet the kinetic energy of the surface does not increase. Is this situation a violation of the work–kinetic energy theorem?

It is not so much a violation as a misapplication because this situation does not fit the description of the conditions given for the work–kinetic energy theorem. The theorem requires that the only change in the system is in its speed, which is not the case here. Work is done on the system of the surface by the book, but the result of that work is *not* an increase in kinetic energy. From your everyday experience with sliding over surfaces with friction, you can probably guess that the surface will be *warmer* after the book slides over it (rub your hands together briskly to experience that!). Therefore, the work done has gone into warming the surface rather than causing it to increase in speed. We use the phrase **internal energy** E_{int} for the

[1]For more details on energy transfer situations involving forces of kinetic friction, see B. A. Sherwood and W. H. Bernard, *American Journal of Physics* 52:1001, 1984; and R. P. Bauman, *The Physics Teacher* 30:264, 1992.

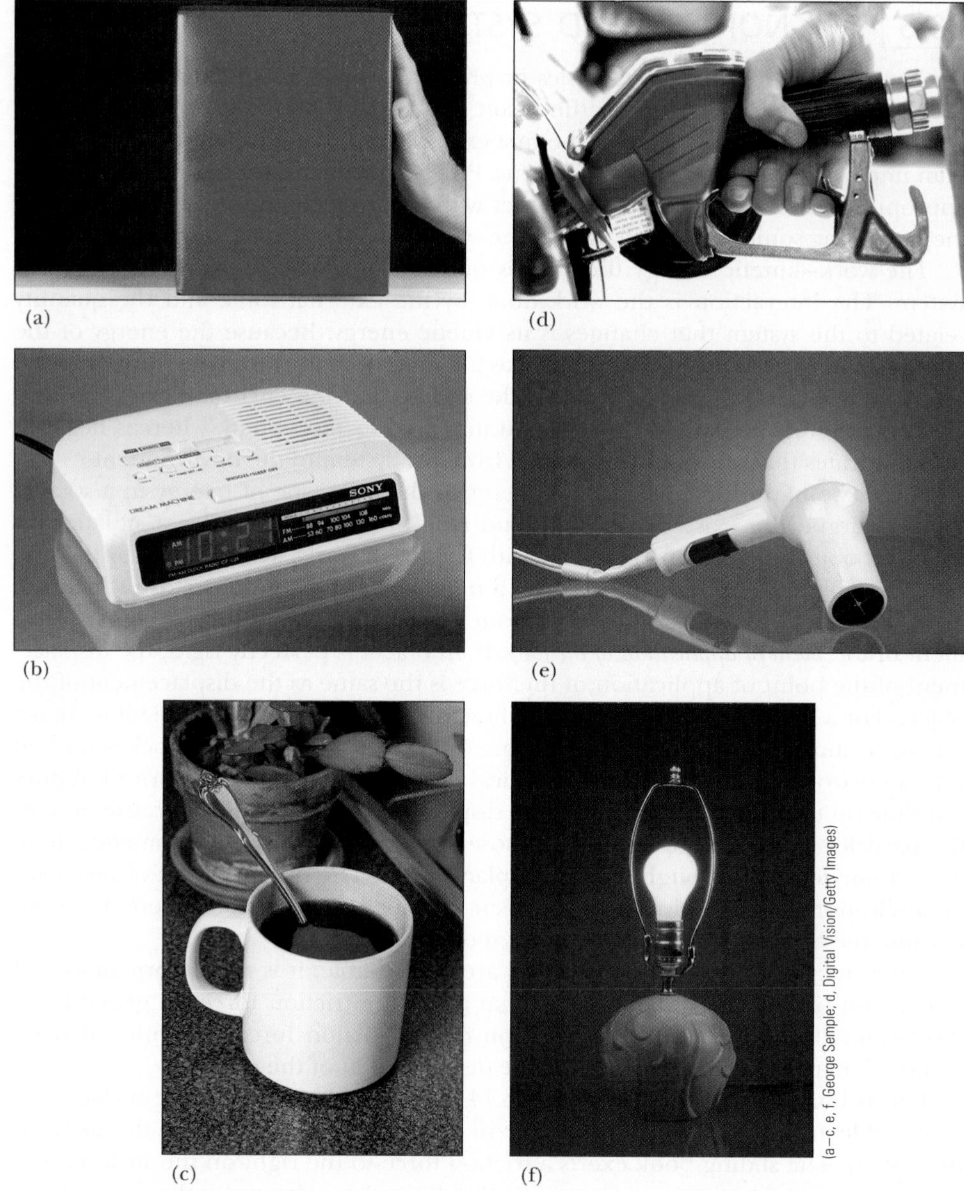

(a) (d) (b) (e) (c) (f)

FIGURE 6.15 Energy transfer mechanisms. (a) Energy is transferred to the block by *work,* (b) energy leaves the radio by *mechanical waves,* (c) energy transfers up the handle of the spoon by *heat,* (d) energy enters the automobile gas tank by *matter transfer,* (e) energy enters the hair dryer by *electrical transmission,* and (f) energy leaves the light bulb by *electromagnetic radiation.*

energy associated with an object's temperature. (We will see a more general definition for internal energy in Chapter 17.) In this case, the work done by the book on the surface does indeed represent energy transferred into the system, but it appears in the system as internal energy rather than kinetic energy.

We have now seen two methods of storing energy in a system: kinetic energy, related to motion of the system, and internal energy, related to its temperature. We have seen only one way to transfer energy into the system so far: work. Next, we introduce a few other ways to transfer energy into or out of a system, which will be studied in detail in other sections of the book. We will focus on the following six methods (Fig. 6.15) for transferring energy between the environment and the system.

■ Methods of energy transfer

Work (this chapter) is a method of transferring energy to a system by the application of a force to the system and a displacement of the point of application of the force, as we have seen in the previous sections (Fig. 6.15a).

Mechanical waves (Chapter 13) are a means of transferring energy by allowing a disturbance to propagate through air or another medium. This method is the one by which energy leaves a radio (Fig. 6.15b) through the loudspeaker—sound—and by which energy enters your ears to stimulate the hearing process. Mechanical waves also include seismic waves and ocean waves.

Heat (Chapter 17) is a method of transferring energy by means of microscopic collisions; for example, the end of a metal spoon in a cup of coffee becomes hot because fast-moving electrons and atoms in the bowl of the spoon bump into slower ones in the nearby part of the handle (Fig. 6.15c). These particles move faster because of the collisions and bump into the next group of slow particles. Therefore, the internal energy of the handle end of the spoon rises from energy transfer as a result of this bumping process. This process, also called *thermal conduction,* is caused by a temperature difference between two regions in space.[2]

In **matter transfer** (Chapter 17), matter physically crosses the boundary of the system, carrying energy with it. Examples include filling the system of your automobile tank with gasoline (Fig. 6.15d) and carrying energy to the rooms of your home by means of circulating warm air from the furnace. Matter transfer occurs in several situations and is introduced in Chapter 17 by means of one example, *convection.*

Electrical transmission (Chapter 21) involves energy transfer by means of electric currents. That is how energy transfers into your stereo system or any other electrical device such as a hair dryer (Fig. 6.15e).

Electromagnetic radiation (Chapter 24) refers to electromagnetic waves such as light, microwaves, and radio waves (Fig. 6.15f). Examples of this method of transfer include energy going into your baked potato in your microwave oven and light energy traveling from the Sun to the Earth through space.[3]

PITFALL PREVENTION 6.6

Heat is not a form of energy The word *heat* is one of the most misused words in our popular language. In this text, heat is a method of *transferring* energy across a system boundary, *not* a form of stored energy. Therefore, phrases such as "heat content," "the heat of the summer," and "the heat escaped" all represent uses of this word that are inconsistent with our physics definition. See Chapter 17.

The central feature of the energy approach is the notion that **we can neither create nor destroy energy; energy is *conserved.*** Therefore, **if the amount of energy in a system changes, it can *only* be because energy has crossed the boundary by a transfer mechanism such as those listed above.** This general statement of the principle of **conservation of energy** can be described mathematically as follows:

■ Conservation of energy: the continuity equation for energy

$$\Delta E_{\text{system}} = \sum T \qquad [6.20]$$

where E_{system} is the total energy of the system, including all methods of energy storage (kinetic, internal, and another to be discussed in Chapter 7) and T (for *transfer*) is the amount of energy transferred across the system boundary by a transfer mechanism. Two of our transfer mechanisms have well-established symbolic notations. For work, $T_{\text{work}} = W$, as we have seen in this chapter, and for heat, $T_{\text{heat}} = Q$, which we will see in detail in Chapter 17. The other four members of our list do not have established symbols, so we will call them T_{MW} (mechanical waves), T_{MT} (matter transfer), T_{ET} (electrical transmission), and T_{ER} (electromagnetic radiation).

In this chapter, we have seen how to calculate work. The other types of transfers will be discussed in subsequent chapters. Equation 6.20 is called the **continuity**

[2]Many textbooks use the term *heat* to include *conduction, convection,* and *radiation.* Conduction is the only one of these three processes driven by a temperature difference alone, so we will restrict heat to this process in this book. Convection and radiation are included in other types of energy transfer in our list of six.

[3]Electromagnetic radiation and work done by field forces are the only energy transfer mechanisms that do not require molecules of the environment to be available at the system boundary. Therefore, systems surrounded by a vacuum (such as planets) can only exchange energy with the environment by means of these two possibilities.

equation for energy. A continuity equation arises in any situation in which the change in a quantity in a system occurs solely because of transfers across the boundary (because the quantity is conserved), several examples of which occur in various areas of physics, as we shall see.

The full expansion of Equation 6.20, with kinetic and internal energy as the storage mechanisms, is

$$\Delta K + \Delta E_{\text{int}} = W + T_{\text{MW}} + Q + T_{\text{MT}} + T_{\text{ET}} + T_{\text{ER}}$$

This equation is the primary mathematical representation of the energy analysis of the nonisolated system. In most cases, it reduces to a much simpler equation because some of the terms are zero. If, for a given system, all terms on the right side of the continuity equation for energy are zero, the system is an *isolated system,* which we study in the next chapter.

The concept described by Equation 6.20 is no more complicated in theory than is that of balancing your checking account statement. If your account is the system, the change in the account balance for a given month is the sum of all the transfers: deposits, withdrawals, fees, interest, and checks written. It may be useful for you to think of energy as the *currency of nature*!

Suppose a force is applied to a nonisolated system and the point of application of the force moves through a displacement. Further, suppose the only effect on the system is to increase its speed. Then the only transfer mechanism is work (so that $\sum T$ in Equation 6.20 reduces to just W) and the only kind of energy in the system that changes is the kinetic energy (so that ΔE_{system} reduces to just ΔK). Equation 6.20 then becomes

$$\Delta K = W$$

which is the work–kinetic energy theorem, Equation 6.19. This theorem is a special case of the more general continuity equation for energy. In future chapters, we shall see several more examples of other special cases of the continuity equation for energy.

Bioluminescence

Equation 6.20 is not restricted to phenomena commonly described as belonging to the area of physics. For example, Figure 6.16 shows a glow worm whose last three segments of the abdomen glow with *bioluminescence.* In this process, chemical energy in the worm is transformed such that energy leaves the worm by electromagnetic radiation in the form of visible light. For this process, Equation

(Sinclair Stammers/Science Photo Library/Photo Researchers, Inc.)

FIGURE 6.16 The glow worm *Lampyris noctiluca* is found in Great Britain and parts of continental Europe. It exhibits the phenomenon of *bioluminescence.* The light leaving the last three segments of its abdomen represents a transfer of energy out of the system of the worm.

6.20 can be written

$$\Delta E_{\text{chem}} = T_{\text{ER}}$$

Chemical energy is a form of potential energy, which we will study in Chapter 7. Chemical energy is stored in any organism by means of food ingested by the organism. Therefore, the source of the light leaving the worm in Figure 6.16 is food ingested earlier by the worm.

QUICK QUIZ 6.5 By what transfer mechanisms does energy enter and leave **(a)** your television set, **(b)** Your gasoline-powered lawn mower, and **(c)** your hand-cranked pencil sharpener?

QUICK QUIZ 6.6 Consider a block sliding over a horizontal surface with friction. Ignore any sound the sliding might make. If we consider the system to be the *block,* this system is **(a)** isolated or **(b)** nonisolated. If we consider the system to be the *surface,* this system is **(c)** isolated or **(d)** nonisolated. If we consider the system to be the *block and the surface,* this system is **(e)** isolated or **(f)** nonisolated.

■ Thinking Physics 6.3

A toaster is turned on. Discuss the forms of energy and energy transfer occurring in the coils of the toaster.

Reasoning We identify the coils as the system. The energy that changes in the system is *internal energy* because the temperature of the coils rises. The energy transfer mechanism for energy coming into the coils is *electrical transmission* through the wire plugged into the wall. Energy is transferring out of the coils by *electromagnetic radiation* because the coils are hot and glowing. Some transfer of energy also occurs by *heat* from the hot surfaces of the coils into the air. We could express this process in terms of the continuity equation for energy as

$$\Delta E_{\text{int}} = Q + T_{\text{ET}} + T_{\text{ER}}$$

After a short warm-up period, the temperature of the coils reaches a constant value and the internal energy will no longer change. In this situation, the energy input and output are balanced:

$$0 = Q + T_{\text{ET}} + T_{\text{ER}} \quad \rightarrow \quad -T_{\text{ET}} = Q + T_{\text{ER}}$$

Note that Q and T_{ER} are both negative because they represent energy leaving the system; T_{ET} is positive because energy continues to enter the system by electrical transmission. ■

6.7 SITUATIONS INVOLVING KINETIC FRICTION

In the preceding section, we discussed the nature of the friction force and the situation with deformable objects. Let us see how to handle problems with friction forces such as that on our block in Figure 6.11 sliding on the surface.

Consider a situation in which forces, including friction, are applied to the block as it follows an arbitrary path in space and let us follow a similar procedure to that in generating Equation 6.17. We start by writing Equation 6.12 for all forces other than friction:

$$\sum W_{\text{other forces}} = \int \left(\sum \vec{\mathbf{F}}_{\text{other forces}}\right) \cdot d\vec{\mathbf{r}} \qquad [6.21]$$

The $d\vec{\mathbf{r}}$ in this equation is the displacement of the object because for forces other than friction, under the assumption that these forces do not deform the object, this displacement is the same as that of the point of application of the forces. To each side of Equation 6.21 let us add the integral of the scalar product of the force of kinetic friction and $d\vec{\mathbf{r}}$:

$$\sum W_{\text{other forces}} + \int \vec{\mathbf{f}}_k \cdot d\vec{\mathbf{r}} = \int \left(\sum \vec{\mathbf{F}}_{\text{other forces}} \right) \cdot d\vec{\mathbf{r}} + \int \vec{\mathbf{f}}_k \cdot d\vec{\mathbf{r}}$$
$$= \int \left(\sum \vec{\mathbf{F}}_{\text{other forces}} + \vec{\mathbf{f}}_k \right) \cdot d\vec{\mathbf{r}}$$

The integrand on the right side of this equation is the net force $\sum \vec{\mathbf{F}}$, so,

$$\sum W_{\text{other forces}} + \int \vec{\mathbf{f}}_k \cdot d\vec{\mathbf{r}} = \int \sum \vec{\mathbf{F}} \cdot d\vec{\mathbf{r}}$$

Incorporating Newton's second law $\sum \vec{\mathbf{F}} = m\vec{\mathbf{a}}$, gives us

$$\sum W_{\text{other forces}} + \int \vec{\mathbf{f}}_k \cdot d\vec{\mathbf{r}} = \int m\vec{\mathbf{a}} \cdot d\vec{\mathbf{r}} = \int m \frac{d\vec{\mathbf{v}}}{dt} \cdot d\vec{\mathbf{r}} = \int_{t_i}^{t_f} m \frac{d\vec{\mathbf{v}}}{dt} \cdot \vec{\mathbf{v}}\, dt \quad [6.22]$$

where we have used Equation 3.5 to rewrite $d\vec{\mathbf{r}}$ as $\vec{\mathbf{v}}\,dt$. The scalar product obeys the product rule for differentiation (See Eq. B.30 in Appendix B.6), so the derivative of the scalar product of $\vec{\mathbf{v}}$ with itself can be written

$$\frac{d}{dt}(\vec{\mathbf{v}} \cdot \vec{\mathbf{v}}) = \frac{d\vec{\mathbf{v}}}{dt} \cdot \vec{\mathbf{v}} + \vec{\mathbf{v}} \cdot \frac{d\vec{\mathbf{v}}}{dt} = 2 \frac{d\vec{\mathbf{v}}}{dt} \cdot \vec{\mathbf{v}}$$

where we have used the commutative property of the scalar product to justify the final expression in this equation. Consequently,

$$\frac{d\vec{\mathbf{v}}}{dt} \cdot \vec{\mathbf{v}} = \frac{1}{2} \frac{d}{dt}(\vec{\mathbf{v}} \cdot \vec{\mathbf{v}}) = \frac{1}{2} \frac{dv^2}{dt}$$

Substituting this result into Equation 6.22, we find that

$$\sum W_{\text{other forces}} + \int \vec{\mathbf{f}}_k \cdot d\vec{\mathbf{r}} = \int_{t_i}^{t_f} m \left(\frac{1}{2} \frac{dv^2}{dt} \right) dt = \frac{1}{2} m \int_{v_i}^{v_f} d(v^2)$$
$$= \frac{1}{2} m v_f^2 - \frac{1}{2} m v_i^2 = \Delta K$$

Looking at the left side of this equation, we realize that in the inertial frame of the surface, $\vec{\mathbf{f}}_k$ and $d\vec{\mathbf{r}}$ will be in opposite directions for every increment $d\vec{\mathbf{r}}$ of the path followed by the object. Therefore, $\vec{\mathbf{f}}_k \cdot d\vec{\mathbf{r}} = -f_k\, dr$. The previous expression now becomes

$$\sum W_{\text{other forces}} - \int f_k\, dr = \Delta K$$

If the kinetic friction force is constant, f_k can be brought out of the integral. The remaining integral $\int dr$ is simply the sum of increments of length along the path, which is the total path length d. Therefore,

■ The change in kinetic energy of an object due to friction and other forces

$$\sum W_{\text{other forces}} - f_k d = \Delta K \quad [6.23]$$

This equation can be considered to be a modification of the work–kinetic energy theorem to be used when a constant friction force acts on an object. The change in kinetic energy is equal to the work done by all forces other than friction minus a term $f_k d$ associated with the friction force.

Now consider the larger system consisting of the block *and* the surface as the block slows down under the influence of a friction force alone. No work is done across the boundary of this system; the system does not interact with the environment, so there is no work done by other forces beside friction. In this case, Equation 6.23 becomes $-f_k d = \Delta K$. For this situation, Equation 6.20 becomes

$$\Delta K + \Delta E_{\text{int}} = 0$$

The change in kinetic energy of this system is the same as the change in kinetic energy of the system of the block because the block is the only part of the block–surface system that is moving. Therefore,

$$-f_k d + \Delta E_{\text{int}} = 0$$

$$\Delta E_{\text{int}} = f_k d \qquad [6.24]$$

■ **The increase in internal energy of a system due to friction**

Therefore, the increase in internal energy of the system is equal to the product of the friction force and the path length through which the block moves. In summary, **a friction force transforms kinetic energy in a system to internal energy, and for a system in which the friction force alone acts, the increase in internal energy of the system is equal to its decrease in kinetic energy.**

QUICK QUIZ 6.7 You are traveling along a freeway at 65 mi/h. You suddenly skid to a stop because of congestion in traffic. Where is the energy that your car once had as kinetic energy before you stopped? **(a)** It is all in internal energy in the road. **(b)** It is all in internal energy in the tires. **(c)** Some of it has transformed to internal energy and some of it transferred away by mechanical waves. **(d)** It all transferred away from your car by various mechanisms.

INTERACTIVE EXAMPLE 6.6 A Block Pulled on a Rough Surface

A block of mass 6.00 kg initially at rest is pulled to the right by a constant horizontal force with magnitude $F = 12.0$ N (Fig. 6.17a). The coefficient of kinetic friction between the block and the surface is 0.150.

A Find the speed of the block after it has moved 3.00 m. (This question is Example 6.4 modified so that the surface is no longer frictionless.)

Solution We define the system as the block. Because the block moves in a straight line without reversing direction, the displacement Δx of the block and the distance d through which it moves are equal. We apply Equation 6.23:

$$\Delta K = -f_k d + \sum W_{\text{other forces}} = -\mu_k n d + F d$$

The block is modeled as a particle in equilibrium in the vertical direction so that $n = mg$. Therefore,

$$\Delta K = -\mu_k mgd + Fd$$

Evaluating ΔK, we have

$$K = -(0.150)(6.00\text{ kg})(9.80\text{ m/s}^2)(3.00\text{ m}) + (12.0\text{ N})(3.00\text{ m}) = 9.54\text{ J}$$

Now, we find v_f

$$\Delta K = \tfrac{1}{2}mv_f^2 - \tfrac{1}{2}mv_i^2$$

$$v_f = \sqrt{\left(\frac{2}{m}\right)\left(\Delta K + \tfrac{1}{2}mv_i^2\right)}$$

Substituting the numerical values, we find

$$v_f = \sqrt{\left(\frac{2}{6.00\text{ kg}}\right)(9.54\text{ J} + 0)} = 1.78\text{ m/s}$$

Notice that this value is less than that calculated in Example 6.4 because of the effect of the friction force.

B Suppose the force $\vec{\mathbf{F}}$ is applied at an angle θ as shown in Figure 6.17b. At what angle should the force be applied to achieve the largest possible speed after the block has moved 3.00 m to the right?

Solution At first, we might guess that $\theta = 0$ is the optimal angle to transfer the maximum energy to the

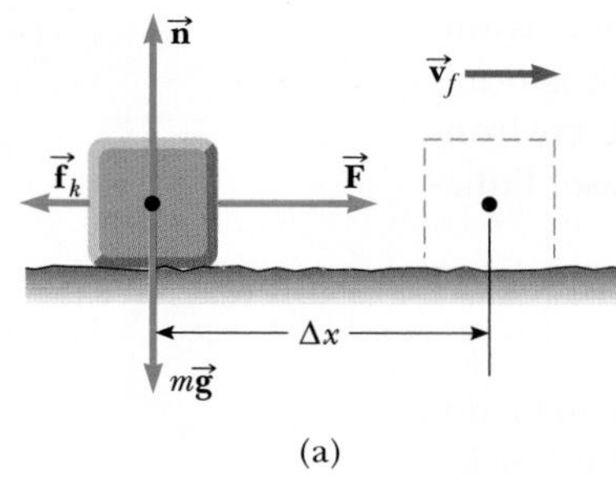

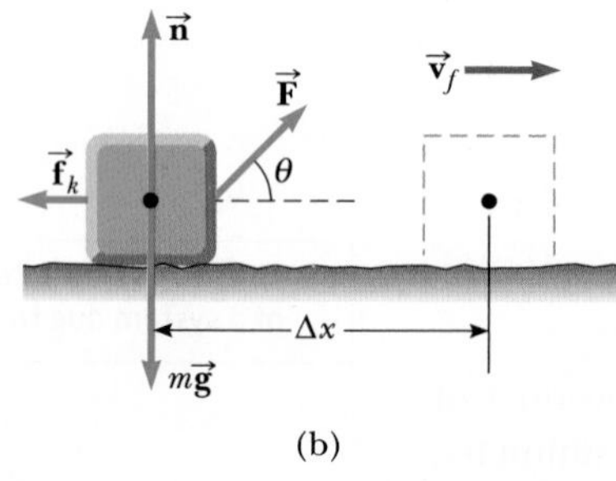

FIGURE 6.17 (Interactive Example 6.6) (a) A block is pulled to the right by a constant horizontal force on a surface with friction. (b) The applied force is at an angle θ to the horizontal.

block. That would indeed be the case when pulling the block on a frictionless surface. With friction, however, pulling the block at some angle $\theta \neq 0$ reduces the normal force on the block, which in turn reduces the friction force. As a result, more energy can be transferred by work by pulling at some nonzero angle. For a nonzero angle θ, the work done by the applied force is

$$W = F\Delta x \cos\theta = Fd\cos\theta$$

The block is in equilibrium in the vertical direction, so

$$\sum F_y = n + F\sin\theta - mg = 0$$

and

$$n = mg - F\sin\theta$$

Because $K_i = 0$, Equation 6.23 can be written as

$$\begin{aligned} K_f &= -f_k d + \sum W_{\text{other forces}} \\ &= -\mu_k n d + Fd\cos\theta \\ &= -\mu_k(mg - F\sin\theta)d + Fd\cos\theta \end{aligned}$$

Maximizing the speed is equivalent to maximizing the final kinetic energy. Consequently, we differentiate K_f with respect to θ and set the result equal to zero:

$$\frac{d(K_f)}{d\theta} = -\mu_k(0 - F\cos\theta)d - Fd\sin\theta = 0$$

$$\mu_k\cos\theta - \sin\theta = 0$$

$$\tan\theta = \mu_k$$

For $\mu_k = 0.150$, we have

$$\theta = \tan^{-1}(\mu_k) = \tan^{-1}(0.150) = 8.53^\circ$$

If we test this result by examining the second derivative of K_f, we find indeed that this angle gives a maximum value.

Physics Now™ Try out the effects of pulling the block at various angles by logging into PhysicsNow at www.pop4e.com and going to Interactive Example 6.6.

INTERACTIVE EXAMPLE 6.7 A Block–Spring System

A block of mass 1.6 kg is attached to a horizontal spring that has a force constant of 1.0×10^3 N/m as shown in Active Figure 6.8. The spring is compressed 2.0 cm and is then released from rest.

A Calculate the speed of the block as it passes through the equilibrium position $x = 0$ if the surface is frictionless.

Solution In this situation, the block starts with $v_i = 0$ at $x_i = -2.0$ cm and we want to find v_f at $x_f = 0$. We use Equation 6.14 to find the work done by the spring with $x_{\max} = x_i = -2.0 \text{ cm} = -2.0 \times 10^{-2}$ m:

$$\begin{aligned} W_s &= \tfrac{1}{2}kx_{\max}^2 = \tfrac{1}{2}(1.0 \times 10^3 \text{ N/m})(-2.0 \times 10^{-2}\text{ m})^2 \\ &= 0.20\text{ J} \end{aligned}$$

Using the work–kinetic energy theorem with $v_i = 0$, we obtain the change in kinetic energy of the block as a result of the work done on it by the spring:

$$W_s = \tfrac{1}{2}mv_f^2 - \tfrac{1}{2}mv_i^2$$

$$\begin{aligned} v_f &= \sqrt{v_i^2 + \frac{2}{m}W_s} \\ &= \sqrt{0 + \frac{2}{1.6\text{ kg}}(0.20\text{ J})} \\ &= 0.50\text{ m/s} \end{aligned}$$

B Calculate the speed of the block as it passes through the equilibrium position if a constant friction force of 4.0 N retards the block's motion from the moment it is released.

Solution Certainly, the answer has to be less than what we found in part A because the friction force retards the motion. We use Equation 6.23:

$$\sum W_{\text{other forces}} - f_k d = \Delta K$$

$$W_s - f_k d = \tfrac{1}{2}mv_f^2 - \tfrac{1}{2}mv_i^2$$

$$v_f = \sqrt{v_i^2 + \frac{2}{m}(W_s - f_k d)}$$

Substituting the numerical values, we find

$$v_f = \sqrt{0 + \frac{2}{1.6\text{ kg}}[0.20\text{ J} - (4.0\text{ N})(2.0 \times 10^{-2}\text{ m})]}$$

$$= 0.39\text{ m/s}$$

As expected, this value is somewhat less than the 0.50 m/s we found in part A.

PhysicsNow™ Investigate the role of the spring constant, amount of spring compression, and surface friction by logging into PhysicsNow at www.pop4e.com and going to Interactive Example 6.7.

6.8 POWER

We discussed transfers of energy across the boundary of a system by a number of methods. From a practical viewpoint, it is interesting to know not only the amount of energy transferred to a system but also the *rate* at which the energy is transferred. The time rate of energy transfer is called **power.**

We shall focus on work as our particular energy transfer method in this discussion, but keep in mind that the notion of power is valid for *any* means of energy transfer. If an external force is applied to an object (for which we will adopt the particle model) and if the work done by this force is W in the time interval Δt, the **average power** during this interval is defined as

$$\mathcal{P}_{\text{avg}} \equiv \frac{W}{\Delta t} \qquad [6.25]$$

The **instantaneous power** $\mathcal{P}$ at a particular point in time is the limiting value of the average power as Δt approaches zero:

$$\mathcal{P} \equiv \lim_{\Delta t \to 0} \frac{W}{\Delta t} = \frac{dW}{dt} \qquad [6.26]$$

where we represent the infinitesimal value of the work done by dW. We know from Equation 6.4 that we can write the infinitesimal amount of work done over a displacement $d\vec{\mathbf{r}}$ as $dW = \vec{\mathbf{F}} \cdot d\vec{\mathbf{r}}$. Therefore, the instantaneous power can be written

$$\mathcal{P} = \frac{dW}{dt} = \vec{\mathbf{F}} \cdot \frac{d\vec{\mathbf{r}}}{dt} = \vec{\mathbf{F}} \cdot \vec{\mathbf{v}} \qquad [6.27]$$

where we have used $\vec{\mathbf{v}} = d\vec{\mathbf{r}}/dt$.

In general, power is defined for any type of energy transfer. The most general expression for power is therefore

$$\mathcal{P} = \frac{dE}{dt} \qquad [6.28]$$

■ General expression for power

where dE/dt is the rate at which energy is crossing the boundary of the system by transfer mechanisms.

The SI unit of power is joules per second (J/s), also called a **watt** (W) (after James Watt):

$$1\text{ W} = 1\text{ J/s} = 1\text{ kg}\cdot\text{m}^2/\text{s}^3$$

The unit of power in the U.S. customary system is the **horsepower** (hp):

$$1\text{ hp} \equiv 550\text{ ft}\cdot\text{lb/s} \equiv 746\text{ W}$$

PITFALL PREVENTION 6.7

BE CAREFUL WITH POWER Do not confuse the symbol W for the watt with the italic symbol W for work. Also, remember that the watt already represents a rate of energy transfer, so we do not want to say something like "watts per second" for power. The watt is *the same as* a joule per second.

A new unit of energy can now be defined in terms of the unit of power. One **kilowatt-hour** (kWh) is the energy transferred in a time interval of 1 h at the constant rate of 1 kW. The numerical value of 1 kWh of energy is

$$1 \text{ kWh} = (10^3 \text{ W})(3\,600 \text{ s}) = 3.60 \times 10^6 \text{ J}$$

It is important to realize that a kilowatt-hour is a unit of energy, not power. When you pay your electric bill, you are buying energy, and the amount of energy transferred by electrical transmission into a home during the period represented by the electric bill is usually expressed in kilowatt-hours. For example, your bill may state that you used 900 kWh of energy during a month and that you are being charged a rate of 10¢ per kWh. Your obligation is then $90 for this amount of energy. As another example, suppose an electric bulb is rated at 100 W. In 1.00 h of operation, it will have energy transferred to it by electrical transmission in the amount of $(0.100 \text{ kW})(1.00 \text{ h}) = 0.100 \text{ kWh} = 3.60 \times 10^5 \text{ J}$.

EXAMPLE 6.8 Power Delivered by an Elevator Motor

A 1 000-kg elevator carries a maximum load of 800 kg. A constant friction force of 4 000 N retards its motion upward as in Figure 6.18.

A What is the minimum power delivered by the motor to lift the elevator at a constant speed of 3.00 m/s?

Solution We use two analysis models for the elevator. First, we model it as a particle in equilibrium because it moves at constant speed. The motor must supply the force $\vec{\mathbf{T}}$ that results in the tension in the cable that pulls the elevator upward. From Newton's second law and from $a = 0$ because v is constant, we have

$$T - f - Mg = 0$$

where M is the *total* mass (elevator plus load), equal to 1 800 kg. Therefore,

$$\begin{aligned} T &= f + Mg \\ &= 4.00 \times 10^3 \text{ N} + (1.80 \times 10^3 \text{ kg})(9.80 \text{ m/s}^2) \\ &= 2.16 \times 10^4 \text{ N} \end{aligned}$$

We now model the elevator as a nonisolated system. Work is being done on it by the tension force (as well as other forces). We can use Equation 6.27 to evaluate the power delivered by the motor, which is the rate at which work is done on the elevator by the tension force. Because $\vec{\mathbf{T}}$ is in the same direction as $\vec{\mathbf{v}}$, we have

$$\begin{aligned} \mathcal{P} &= \vec{\mathbf{T}} \cdot \vec{\mathbf{v}} = Tv \\ &= (2.16 \times 10^4 \text{ N})(3.00 \text{ m/s}) = 6.48 \times 10^4 \text{ W} \\ &= 64.8 \text{ kW} \end{aligned}$$

Because $\vec{\mathbf{T}}$ is the force the motor applies to the cable, the preceding result represents the rate at which energy is being transferred out of the motor by doing work on the cable.

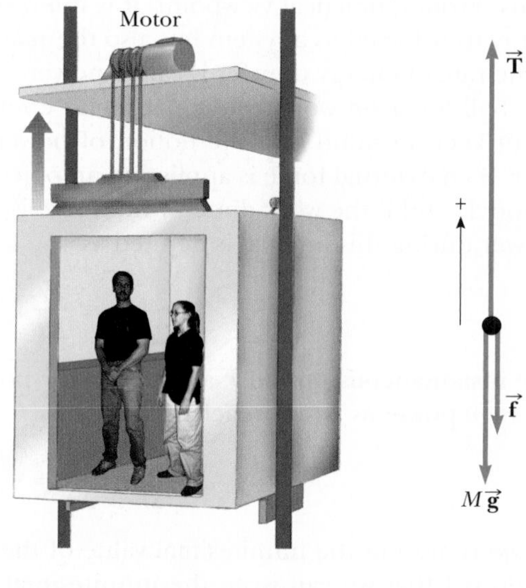

FIGURE 6.18 (Example 6.8) (a) A motor lifts an elevator car. (b) Free-body diagram for the elevator. The motor exerts an upward force $\vec{\mathbf{T}}$ on the supporting cables. The magnitude of this force is T, the tension in the cables, which is applied in the upward direction on the elevator. The downward forces on the elevator are the friction force $\vec{\mathbf{f}}$ and the gravitational force $\vec{\mathbf{F}}_g = M\vec{\mathbf{g}}$.

B What power must the motor deliver at any instant if it is designed to provide an upward acceleration of 1.00 m/s²?

Solution In this case, we expect the tension to be larger than in part A because the cable must now cause an upward acceleration of the elevator. Modeling the elevator as a particle under a net force, we apply Newton's second law, which gives

$$T - f - Mg = Ma$$

$$T = M(a + g) + f$$

$$= (1.80 \times 10^3 \text{ kg})(1.00 \text{ m/s}^2 + 9.80 \text{ m/s}^2) + 4.00 \times 10^3 \text{ N}$$

$$= 2.34 \times 10^4 \text{ N}$$

Therefore, using Equation 6.27, we have for the required power

$$\mathcal{P} = Tv = (2.34 \times 10^4 v)$$

where v is the instantaneous speed of the elevator in meters per second. Hence, the power required increases with increasing speed.

6.9 HORSEPOWER RATINGS OF AUTOMOBILES

CONTEXT CONNECTION

As discussed in Section 4.8, an automobile moves because of Newton's third law. The engine attempts to rotate the wheels in such a direction as to push the Earth toward the back of the car because of the friction force between the wheels and the roadway. By Newton's third law, the Earth pushes in the opposite direction on the wheels, which is toward the front of the car. Because the Earth is much more massive than the car, the Earth remains stationary while the car moves forward.

This principle is the same one humans use for walking. By pushing your leg backward while your foot is on the ground, you apply a friction force backward on the surface of the Earth. By Newton's third law, the surface applies a forward friction force on you, which causes your body to move forward.

The strength of the friction force $\vec{\mathbf{f}}$ exerted on a car by the roadway is related to the rate at which energy is transferred to the wheels to set them into rotation, which is the power of the engine:

$$\mathcal{P}_{\text{avg}} = \frac{\Delta E}{\Delta t} = \frac{f\Delta x}{\Delta t} = fv \quad \rightarrow \quad \mathcal{P} \leftrightarrow f$$

where the symbol $\leftrightarrow$ implies a relationship between the variables that is not necessarily an exact proportionality. In turn, the magnitude of the driving force is related to the acceleration of the car owing to Newton's second law:

$$f = ma \quad \rightarrow \quad f \propto a$$

Consequently, there should be a close relationship between the power rating of a vehicle and the possible acceleration of the vehicle:

$$\mathcal{P} \leftrightarrow a$$

Let us see if this relationship exists for actual data. For automobiles, a common unit for power is the *horsepower* (hp), defined in Section 6.8. Table 6.1 shows the gasoline-powered automobiles we have studied in the preceding chapters. The fourth column provides the published horsepower rating of each vehicle. The final column shows the ratio of the horsepower rating to the acceleration. Consider first the *Performance vehicles* section of the table. The ratio of power to acceleration is similar for all these vehicles, demonstrating the relationship between power and acceleration that we proposed.

In the second part of the table, under *Traditional vehicles,* there is a wider range of ratios of power to acceleration. This range is correlated to the range of vehicle masses in this listing. Notice that the BMW Mini Cooper S, Acura Integra GS, and Volkswagen Beetle have relatively low ratios and are cars with relatively small masses. It takes less power to accelerate this much mass to 60 mi/h than for a heavier car. Conversely, the two SUVs in this listing, the Cadillac Escalade and the Toyota Sienna, have the highest ratios of power to acceleration in this part of the table, 49 hp/mi/h · s and 32 hp/mi/h · s, respectively.

TABLE 6.1 Horsepower Ratings and Accelerations of Various Vehicles

Automobile	Time Interval, 0 to 60 mi/h (s)	Acceleration (mi/h · s)	Horsepower Rating (hp)	Ratio of Horsepower Rating to Acceleration (hp/mi/h · s)
Performance vehicles				
Aston Martin DB7 Vantage	5.0	12.0	414	35
BMW Z8	4.6	13.0	394	30
Chevrolet Corvette	4.6	13.0	385	30
Dodge Viper GTS-R	4.2	14.3	460	32
Ferrari F50	3.6	16.7	513	31
Ferrari 360 Spider F1	4.6	13.0	395	30
Lamborghini Diablo GT	3.6	16.7	567	34
Porsche 911 GT2	4.0	15.0	456	30
Traditional vehicles				
Acura Integra GS	7.9	7.6	140	18
BMW Mini Cooper S	6.9	8.7	163	19
Cadillac Escalade (SUV)	8.6	7.0	345	49
Dodge Stratus	7.5	8.0	200	25
Lexus ES300	8.6	7.0	200	29
Mitsubishi Eclipse GT	7.0	8.6	205	24
Nissan Maxima	6.7	9.0	222	25
Pontiac Grand Prix	8.5	7.1	200	28
Toyota Sienna (SUV)	8.3	7.2	230	32
Volkswagen Beetle	7.6	7.9	150	19

SUMMARY

Physics Now™ Take a practice test by logging intoPhysics-Now at www.pop4e.com and clicking on the Pre-Test link for this chapter.

A **system** can be a single particle, a collection of particles, or a region of space. A **system boundary** separates the system from the **environment.** Many physics problems can be solved by considering the interaction of a system with its environment.

The **work** done by a *constant* force $\vec{\mathbf{F}}$ on a particle is defined as the product of the magnitude F of the force, the magnitude Δr of the displacement of the point of application of the force, and $\cos\theta$, where θ is the angle between the force vector and the displacement vector $\Delta\vec{\mathbf{r}}$:

$$W \equiv F\,\Delta r\,\cos\theta \qquad [6.1]$$

The **scalar** or **dot product** of any two vectors $\vec{\mathbf{A}}$ and $\vec{\mathbf{B}}$ is defined by the relationship

$$\vec{\mathbf{A}}\cdot\vec{\mathbf{B}} \equiv AB\cos\theta \qquad [6.3]$$

where the result is a scalar quantity and θ is the angle between the directions of the two vectors. The scalar product obeys the commutative and distributive laws.

The scalar product allows us to write the work done by a constant force $\vec{\mathbf{F}}$ on a particle as

$$W = \vec{\mathbf{F}}\cdot\Delta\vec{\mathbf{r}} \qquad [6.4]$$

The work done by a *varying* force acting on a particle moving along the x axis from x_i to x_f is

$$W = \int_{x_i}^{x_f} F_x\,dx \qquad [6.11]$$

where F_x is the component of force in the x direction. If several forces act on the particle, the net work done by all forces is the sum of the individual amounts of work done by each force.

The **kinetic energy** of a particle of mass m moving with a speed v is

$$K \equiv \tfrac{1}{2}mv^2 \qquad [6.18]$$

The **work–kinetic energy theorem** states that when work is done on a system and the only change in the system is in its speed, the net work done on the system by external forces equals the change in kinetic energy of the system:

$$W_{\text{net}} = K_f - K_i = \Delta K \qquad [6.19]$$

For a nonisolated system, we can equate the change in the total energy stored in the system to the sum of all the transfers of energy across the system boundary:

$$\Delta E_{\text{system}} = \sum T \qquad [6.20]$$

which is the **continuity equation for energy.** Methods of energy transfer (T) include **work** ($T = W$), **mechanical waves** (T_{MW}), **heat** ($T = Q$), **matter transfer** (T_{MT}), **electrical transmission** (T_{ET}), and **electromagnetic radiation** (T_{ER}). Storage mechanisms (E_{system}) seen in this chapter include **kinetic energy** K and **internal energy** E_{int}. The continuity equation arises because **energy is conserved;** we can neither create nor destroy energy. The work–kinetic energy theorem is a special case of the continuity equation for energy in situations in which work is the only transfer mechanism and kinetic energy is the only type of energy storage in the system.

In the case of an object sliding through a distance d over a surface with friction, the change in kinetic energy of the system is found from

$$\sum W_{\text{other forces}} - f_k d = \Delta K \qquad [6.23]$$

where f_k is the force of kinetic friction and $\Sigma\, W_{\text{other forces}}$ is the work done by all forces other than friction.

Average power is the time rate of energy transfer. If we use work as the energy transfer mechanism,

$$\mathcal{P}_{\text{avg}} \equiv \frac{W}{\Delta t} \qquad [6.25]$$

If an agent applies a force $\vec{\mathbf{F}}$ to an object moving with a velocity $\vec{\mathbf{v}}$, the **instantaneous power** delivered by that agent is

$$\mathcal{P} \equiv \frac{dW}{dt} = \vec{\mathbf{F}} \cdot \vec{\mathbf{v}} \qquad [6.27]$$

Because power is defined for any type of energy transfer, the general expression for power is

$$\mathcal{P} = \frac{dE}{dt} \qquad [6.28]$$

QUESTIONS

☐ = answer available in the *Student Solutions Manual and Study Guide*

1. When a particle rotates in a circle, a force acts on it directed toward the center of rotation. Why is it that this force does no work on the particle?
2. When a punter kicks a football, is he doing any work on the ball while his toe is in contact with it? Is he doing any work on the ball after it loses contact with his toe? Are any forces doing work on the ball while it is in flight?
3. Cite two examples in which a force is exerted on an object without doing any work on the object.
4. Discuss the work done by a pitcher throwing a baseball. What is the approximate distance through which the force acts as the ball is thrown?
5. [boxed] As a simple pendulum swings back and forth, the forces acting on the suspended object are the gravitational force, the tension in the supporting cord, and air resistance. (a) Which of these forces, if any, does no work on the pendulum? (b) Which of these forces does negative work at all times during the pendulum's motion? (c) Describe the work done by the gravitational force while the pendulum is swinging.
6. If the scalar product of two vectors is positive, does that imply that the vectors must have positive rectangular components?
7. For what values of θ is the scalar product (a) positive and (b) negative?
8. A certain uniform spring has spring constant k. Now the spring is cut in half. What is the relationship between k and the spring constant k' of each resulting smaller spring? Explain your reasoning.
9. [boxed] Can kinetic energy be negative? Explain.
10. Two sharpshooters fire 0.30-caliber rifles using identical shells. A force exerted by expanding gases in the barrels accelerates the bullets. The barrel of rifle A is 2.00 cm longer than the barrel of rifle B. Which rifle will have the higher muzzle speed?
11. [boxed] One bullet has twice the mass of a second bullet. If both are fired so that they have the same speed, which has more kinetic energy? What is the ratio of the kinetic energies of the two bullets?
12. You are reshelving books in a library. You lift a book from the floor to the top shelf. The kinetic energy of the book on the floor was zero and the kinetic energy of the book sitting on the top shelf is zero, so no change occurs in the kinetic energy. Yet you did some work in lifting the book. Is the work–kinetic energy theorem violated?
13. [boxed] (a) If the speed of a particle is doubled, what happens to its kinetic energy? (b) What can be said about the speed of a particle if the net work done on it is zero?
14. A car salesperson claims that a souped-up 300-hp engine is a necessary option in a compact car in place of the conventional 130-hp engine. Suppose you intend to drive the car within speed limits ($\leq$ 65 mi/h) on flat terrain. How would you counter this sales pitch?
15. Can the average power over a time interval ever be equal to the instantaneous power at an instant within the interval? Explain.
16. Words given quantitative definitions in physics are sometimes used in popular literature in interesting ways. For example, a rock falling from the top of a cliff is said to be "gathering force as it falls to the beach below." What does the phrase "gathering force" mean, and can you repair this phrase?
17. In most circumstances, the normal force acting on an object and the force of static friction do zero work on the object. The reason that the work is zero is different for the two cases, however. Explain why each does zero work.
18. "A level air track can do no work." Argue for or against this statement.

PROBLEMS

1, 2, 3 = straightforward, intermediate, challenging
□ = full solution available in the *Student Solutions Manual and Study Guide*
PhysicsNow™ = coached problem with hints available at www.pop4e.com
= computer useful in solving problem
= paired numerical and symbolic problems
= biomedical application

Section 6.2 ■ Work Done by a Constant Force

1. A block of mass 2.50 kg is pushed 2.20 m along a frictionless horizontal table by a constant 16.0-N force directed 25.0° below the horizontal. Determine the work done on the block by (a) the applied force, (b) the normal force exerted by the table, and (c) the gravitational force. (d) Determine the total work done on the block.

2. A shopper in a supermarket pushes a cart with a force of 35.0 N directed at an angle of 25.0° downward from the horizontal. Find the work done by the shopper on the cart as he moves down an aisle 50.0 m long.

3. PhysicsNow™ Batman, whose mass is 80.0 kg, is dangling on the free end of a 12.0-m rope, the other end of which is fixed to a tree limb above. He is able to get the rope in motion as only Batman knows how, eventually getting it to swing enough that he can reach a ledge when the rope makes a 60.0° angle with the vertical. How much work was done by the gravitational force on Batman in this maneuver?

4. A raindrop of mass 3.35×10^{-5} kg falls vertically at constant speed under the influence of gravity and air resistance. Model the drop as a particle. As it falls 100 m, what is the work done on the raindrop (a) by the gravitational force and (b) by air resistance?

Section 6.3 ■ The Scalar Product of Two Vectors

In Problems 6.5 through 6.9, calculate numerical answers to three significant figures as usual.

5. Find the scalar product of the vectors in Figure P6.5.

6. For any two vectors $\vec{\mathbf{A}}$ and $\vec{\mathbf{B}}$, show that $\vec{\mathbf{A}} \cdot \vec{\mathbf{B}} = A_xB_x + A_yB_y + A_zB_z$. (*Suggestion:* Write $\vec{\mathbf{A}}$ and $\vec{\mathbf{B}}$ in unit vector form and use Equations 6.7 and 6.8.)

7. PhysicsNow™ A force $\vec{\mathbf{F}} = (6\hat{\mathbf{i}} - 2\hat{\mathbf{j}})$ N acts on a particle that undergoes a displacement $\Delta\vec{\mathbf{r}} = (3\hat{\mathbf{i}} + \hat{\mathbf{j}})$ m. Find (a) the work done by the force on the particle and (b) the angle between $\vec{\mathbf{F}}$ and $\Delta\vec{\mathbf{r}}$.

8. For $\vec{\mathbf{A}} = 3\hat{\mathbf{i}} + \hat{\mathbf{j}} - \hat{\mathbf{k}}$, $\vec{\mathbf{B}} = -\hat{\mathbf{i}} + 2\hat{\mathbf{j}} + 5\hat{\mathbf{k}}$, and $\vec{\mathbf{C}} = 2\hat{\mathbf{j}} - 3\hat{\mathbf{k}}$, find $\vec{\mathbf{C}} \cdot (\vec{\mathbf{A}} - \vec{\mathbf{B}})$.

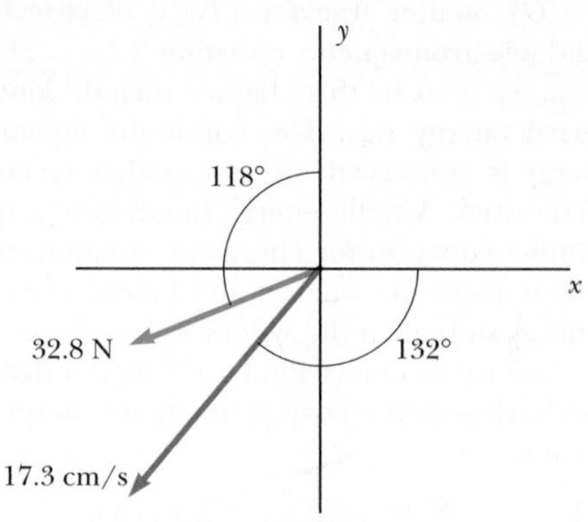

FIGURE P6.5

9. Using the definition of the scalar product, find the angles between (a) $\vec{\mathbf{A}} = 3\hat{\mathbf{i}} - 2\hat{\mathbf{j}}$ and $\vec{\mathbf{B}} = 4\hat{\mathbf{i}} - 4\hat{\mathbf{j}}$ (b) $\vec{\mathbf{A}} = -2\hat{\mathbf{i}} + 4\hat{\mathbf{j}}$ and $\vec{\mathbf{B}} = 3\hat{\mathbf{i}} - 4\hat{\mathbf{j}} + 2\hat{\mathbf{k}}$, and (c) $\vec{\mathbf{A}} = \hat{\mathbf{i}} - 2\hat{\mathbf{j}} + 2\hat{\mathbf{k}}$ and $\vec{\mathbf{B}} = 3\hat{\mathbf{j}} + 4\hat{\mathbf{k}}$.

Section 6.4 ■ Work Done by a Varying Force

10. The force acting on a particle is $F_x = (8x - 16)$ N, where x is in meters. (a) Make a plot of this force versus x from $x = 0$ to $x = 3.00$ m. (b) From your graph, find the net work done by this force on the particle as it moves from $x = 0$ to $x = 3.00$ m.

11. PhysicsNow™ A particle is subject to a force F_x that varies with position as shown in Figure P6.11. Find the work done by the force on the particle as it moves (a) from $x = 0$ to $x = 5.00$ m, (b) from $x = 5.00$ m to $x = 10.0$ m, and (c) from $x = 10.0$ m to $x = 15.0$ m. (d) What is the total work done by the force over the distance $x = 0$ to $x = 15.0$ m?

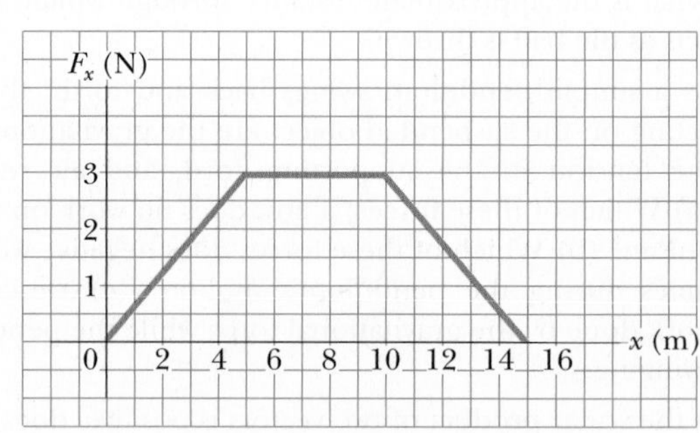

FIGURE P6.11 Problems 11 and 24.

12. A 6 000-kg freight car rolls along rails with negligible friction. The car is brought to rest by a combination of two coiled springs as illustrated in Figure P6.12. Both springs obey Hooke's law with $k_1 = 1\,600$ N/m and $k_2 = 3\,400$ N/m. After the first spring compresses a distance of 30.0 cm, the second spring acts with the first to increase the force as additional compression occurs as shown in the graph. The car comes to rest 50.0 cm after first contacting the two-spring system. Find the car's initial speed.

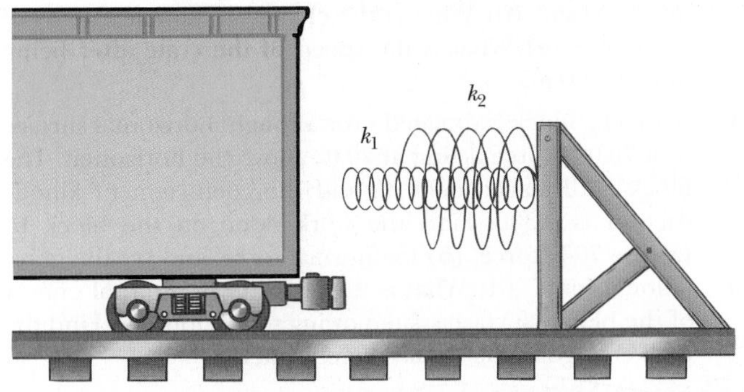

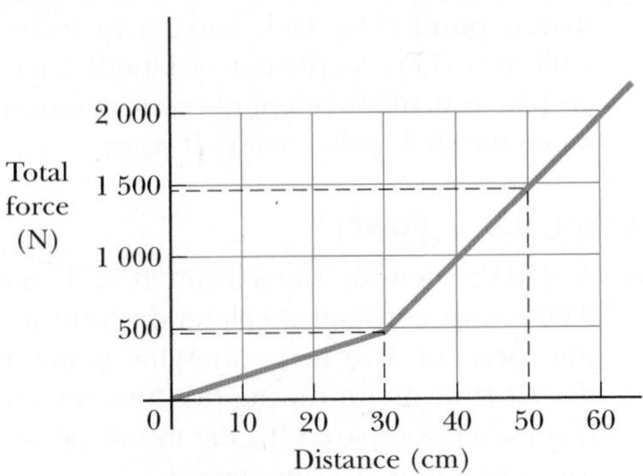

FIGURE P6.12

13. When a 4.00-kg object is hung vertically on a certain light spring that obeys Hooke's law, the spring stretches 2.50 cm. If the 4.00-kg object is removed, (a) how far will the spring stretch if a 1.50-kg block is hung on it and (b) how much work must an external agent do to stretch the same spring 4.00 cm from its unstretched position?

14. A force $\vec{\mathbf{F}} = (4x\hat{\mathbf{i}} + 3y\hat{\mathbf{j}})$ N acts on an object as the object moves in the x direction from the origin to $x = 5.00$ m. Find the work $W = \int \vec{\mathbf{F}} \cdot d\vec{\mathbf{r}}$ done on the object by the force.

15. An archer pulls her bowstring back 0.400 m by exerting a force that increases uniformly from zero to 230 N. (a) What is the equivalent spring constant of the bow? (b) How much work does the archer do in drawing the bow?

16. A 100-g bullet is fired from a rifle having a barrel 0.600 m long. Choose the origin to be at the location where the bullet begins to move. Then the force (in newtons) exerted by the expanding gas on the bullet is $15\,000 + 10\,000x - 25\,000x^2$, where x is in meters. (a) Determine the work done by the gas on the bullet as the bullet travels the length of the barrel. (b) If the barrel is 1.00 m long, how much work is done and how does this value compare to the work calculated in (a)?

17. It takes 4.00 J of work to stretch a Hooke's-law spring 10.0 cm from its unstressed length. Determine the extra work required to stretch it an additional 10.0 cm.

18. A cafeteria tray dispenser supports a stack of trays on a shelf that hangs from four identical spiral springs under tension, one near each corner of the shelf. Each tray is rectangular, 45.3 cm by 35.6 cm, is 0.450 cm thick, and has mass 580 g. Demonstrate that the top tray in the stack can always be at the same height above the floor, however many trays are in the dispenser. Find the spring constant each spring should have for the dispenser to function in this convenient way. Is any piece of data unnecessary for this determination?

19. A small particle of mass m moves at constant speed as it is pulled to the top of a frictionless half-cylinder (of radius R) by a cord that passes over the top of the cylinder as illustrated in Figure P6.19. (a) Show that $F = mg\cos\theta$. (*Note:* If the particle moves at constant speed, the component of its acceleration tangent to the cylinder must be zero at all times.) (b) By directly integrating $W = \int \vec{\mathbf{F}} \cdot d\vec{\mathbf{r}}$, find the work done by the force in moving the particle at constant speed from the bottom to the top of the half-cylinder.

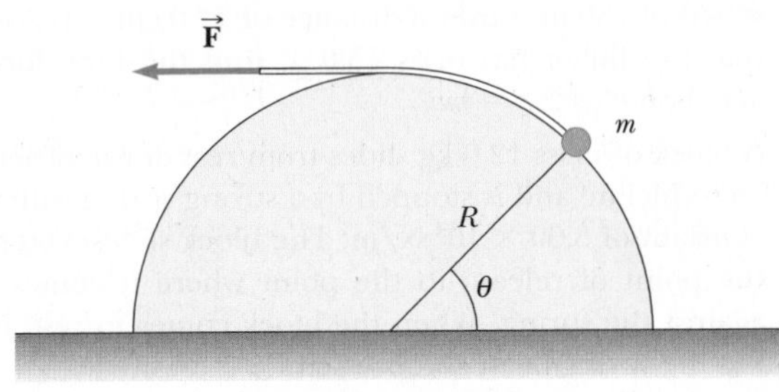

FIGURE P6.19

20. A light spring with spring constant k_1 is hung from an elevated support. From its lower end a second light spring that has spring constant k_2 is hung. An object of mass m is hung at rest from the lower end of the second spring. (a) Find the total extension distance of the pair of springs. (b) Find the effective spring constant of the pair of springs as a system. We describe these springs as *in series.*

Section 6.5 ▪ Kinetic Energy and the Work–Kinetic Energy Theorem

Section 6.6 ▪ The Nonisolated System

21. A 0.600-kg particle has a speed of 2.00 m/s at point Ⓐ and kinetic energy of 7.50 J at point Ⓑ. (a) What is its kinetic energy at Ⓐ? (b) What is its speed at Ⓑ? (c) What is the total work done on the particle as it moves from Ⓐ to Ⓑ?

22. A 0.300-kg ball has a speed of 15.0 m/s. (a) What is its kinetic energy? (b) If its speed were doubled, what would be its kinetic energy?

23. A 3.00-kg object has an initial velocity $(6.00\hat{\mathbf{i}} - 2.00\hat{\mathbf{j}})$ m/s. (a) What is its kinetic energy at this time? (b) Find the total work done on the object as its velocity changes to $(8.00\hat{\mathbf{i}} + 4.00\hat{\mathbf{j}})$ m/s. (*Note:* From the definition of the scalar product, $v^2 = \vec{\mathbf{v}} \cdot \vec{\mathbf{v}}$.)

24. A 4.00-kg particle is subject to a total force that varies with position as shown in Figure P6.11. The particle starts from rest at $x = 0$. What is its speed at (a) $x = 5.00$ m, (b) $x = 10.0$ m, and (c) $x = 15.0$ m?

25. A 2 100-kg pile driver is used to drive a steel I-beam into the ground. The pile driver falls 5.00 m before coming into contact with the top of the beam, and it drives the beam 12.0 cm farther into the ground before coming to rest. Using energy considerations, calculate the average

force the beam exerts on the pile driver while the pile driver is brought to rest.

26. You can think of the work–kinetic energy theorem as a second theory of motion, parallel to Newton's laws in describing how outside influences affect the motion of an object. In this problem, do parts (a) and (b) separately from parts (c) and (d) to compare the predictions of the two theories. In a rifle barrel, a 15.0-g bullet is accelerated from rest to a speed of 780 m/s. (a) Find the work that is done on the bullet. (b) Assuming that the rifle barrel is 72.0 cm long, find the magnitude of the average total force that acted on it as $F = W/(\Delta r \cos \theta)$. (c) Find the constant acceleration of a bullet that starts from rest and gains a speed of 780 m/s over a distance of 72.0 cm. (d) Assuming that the bullet has mass 15.0 g, find the total force that acted on it as $\Sigma F = ma$.

27. A block of mass 12.0 kg slides from rest down a frictionless 35.0° incline and is stopped by a strong spring with a force constant of 3.00×10^4 N/m. The block slides 3.00 m from the point of release to the point where it comes to rest against the spring. When the block comes to rest, how far has the spring been compressed?

28. In the neck of the picture tube of a certain black-and-white television set, an electron gun contains two charged metallic plates 2.80 cm apart. An electric force accelerates each electron in the beam from rest to 9.60% of the speed of light over this distance. (a) Determine the kinetic energy of the electron as it leaves the electron gun. Electrons carry this energy to a phosphorescent material on the inner surface of the television screen, making it glow. For an electron passing between the plates in the electron gun, determine (b) the magnitude of the constant electric force acting on the electron, (c) the acceleration, and (d) the time of flight.

Section 6.7 ▪ Situations Involving Kinetic Friction

29. A 40.0-kg box initially at rest is pushed 5.00 m along a rough, horizontal floor with a constant applied horizontal force of 130 N. The coefficient of friction between box and floor is 0.300. Find (a) the work done by the applied force, (b) the increase in internal energy in the box–floor system as a result of friction, (c) the work done by the normal force, (d) the work done by the gravitational force, (e) the change in kinetic energy of the box, and (f) the final speed of the box.

30. A 2.00-kg block is attached to a spring of force constant 500 N/m as shown in Active Figure 6.8. The block is pulled 5.00 cm to the right of equilibrium and released from rest. Find the speed the block has as it passes through equilibrium if (a) the horizontal surface is frictionless and (b) the coefficient of friction between block and surface is 0.350.

31. A crate of mass 10.0 kg is pulled up a rough incline with an initial speed of 1.50 m/s. The pulling force is 100 N parallel to the incline, which makes an angle of 20.0° with the horizontal. The coefficient of kinetic friction is 0.400, and the crate is pulled 5.00 m. (a) How much work is done by the gravitational force on the crate? (b) Determine the increase in internal energy of the crate–incline system owing to friction. (c) How much work is done by the 100-N force on the crate? (d) What is the change in kinetic energy of the crate? (e) What is the speed of the crate after being pulled 5.00 m?

32. A 15.0-kg block is dragged over a rough, horizontal surface by a 70.0-N force acting at 20.0° above the horizontal. The block is displaced 5.00 m, and the coefficient of kinetic friction is 0.300. Find the work done on the block by (a) the 70-N force, (b) the normal force, and (c) the gravitational force. (d) What is the increase in internal energy of the block–surface system owing to friction? (e) Find the total change in the block's kinetic energy.

33. **PhysicsNow™** A sled of mass m is given a kick on a frozen pond. The kick imparts to it an initial speed of 2.00 m/s. The coefficient of kinetic friction between sled and ice is 0.100. Use energy considerations to find the distance the sled moves before it stops.

Section 6.8 ▪ Power

34. A 650-kg elevator starts from rest. It moves upward for 3.00 s with constant acceleration until it reaches its cruising speed of 1.75 m/s. (a) What is the average power of the elevator motor during this time interval? (b) How does this power compare with the motor power when the elevator moves at its cruising speed?

35. **PhysicsNow™** A 700-N Marine in basic training climbs a 10.0-m vertical rope at a constant speed in 8.00 s. What is his power output?

36. A skier of mass 70.0 kg is pulled up a slope by a motor-driven cable. (a) How much work is required to pull the skier a distance of 60.0 m up a 30.0° slope (assumed frictionless) at a constant speed of 2.00 m/s? (b) A motor of what power is required to perform this task?

37. An energy-efficient lightbulb, taking in 28.0 W of power, can produce the same level of brightness as a conventional lightbulb operating at power 100 W. The lifetime of the energy-efficient bulb is 10 000 h and its purchase price is \$17.0, whereas the conventional bulb has lifetime 750 h and costs \$0.420 per bulb. Determine the total savings obtained by using one energy-efficient bulb over its lifetime as opposed to using conventional bulbs over the same time interval. Assume an energy cost of \$0.080 0 per kilowatt-hour.

38. Energy is conventionally measured in Calories as well as in joules. One Calorie in nutrition is one kilocalorie, defined as 1 kcal = 4 186 J. Metabolizing 1 g of fat can release 9.00 kcal. A student decides to try to lose weight by exercising. She plans to run up and down the stairs in a football stadium as fast as she can and as many times as necessary. Is this plan in itself a practical way to lose weight? To evaluate the program, suppose she runs up a flight of 80 steps, each 0.150 m high, in 65.0 s. For simplicity, ignore the energy she uses in coming down (which is small). Assume that a typical efficiency for human muscles is 20.0%. Therefore when your body converts 100 J from metabolizing fat, 20 J goes into doing mechanical work (here, climbing stairs) and the remainder goes into extra internal energy. Assume that the student's mass is 50.0 kg. (a) How many times must she run the flight of stairs to lose 1 lb of fat? (b) What is her average power output, in watts and in horsepower, as she is running up the stairs?

39\. For saving energy, bicycling and walking are far more efficient means of transportation than is travel by automobile. For example, when riding at 10.0 mi/h, a cyclist uses food energy at a rate of about 400 kcal/h above what he would use if merely sitting still. (In exercise physiology, power is often measured in kcal/h rather than in watts. Here 1 kcal = 1 nutritionist's Calorie = 4 186 J.) Walking at 3.00 mi/h requires about 220 kcal/h. It is interesting to compare these values with the energy consumption required for travel by car. Gasoline yields about 1.30×10^8 J/gal. Find the fuel economy in equivalent miles per gallon for a person (a) walking and (b) bicycling.

Section 6.9 ■ Context Connection—Horsepower Ratings of Automobiles

40. Make an order-of-magnitude estimate of the output power a car engine contributes to speeding the car up to highway speed. For concreteness, consider your own car, if you use one, and make the calculation as precise as you wish. In your solution, state the physical quantities you take as data and the values you measure or estimate for them. The mass of the vehicle is given in the owner's manual. If you do not wish to estimate for a car, consider a bus or truck that you specify.

41. A certain automobile engine delivers 2.24×10^4 W (30.0 hp) to its wheels when moving at a constant speed of 27.0 m/s (≈ 60 mi/h). What is the resistive force acting on the automobile at that speed?

Additional Problems

42\. A baseball outfielder throws a 0.150-kg baseball at a speed of 40.0 m/s and an initial angle of 30.0°. What is the kinetic energy of the baseball at the highest point of its trajectory?

43\. While running, a person transforms about 0.600 J of chemical energy to mechanical energy per step per kilogram of body mass. If a 60.0-kg runner transforms energy at a rate of 70.0 W during a race, how fast is the person running? Assume that a running step is 1.50 m long.

44\. In bicycling for aerobic exercise, a woman wants her heart rate to be between 136 and 166 beats per minute. Assume that her heart rate is directly proportional to her mechanical power output within the range relevant here. Ignore all forces on the woman-plus-bicycle system except for static friction forward on the drive wheel of the bicycle and an air resistance force proportional to the square of her speed. When her speed is 22.0 km/h, her heart rate is 90.0 beats per minute. In what range should her speed be so that her heart rate will be in the range she wants?

45\. A 4.00-kg particle moves along the x axis. Its position varies with time according to $x = t + 2.0t^3$, where x is in meters and t is in seconds. Find (a) the kinetic energy at any time t, (b) the acceleration of the particle and the force acting on it at time t, (c) the power being delivered to the particle at time t, and (d) the work done on the particle in the interval $t = 0$ to $t = 2.00$ s.

46\. A bead at the bottom of a bowl is one example of an object in a stable equilibrium position. When a physical system is displaced by an amount x from stable equilibrium, a restoring force acts on it, tending to return the system to its equilibrium configuration. The magnitude of the restoring force can be a complicated function of x. For example, when an ion in a crystal is displaced from its lattice site, the restoring force may not be a simple function of x. In such cases, we can generally imagine the function $F(x)$ to be expressed as a power series in x as $F(x) = -(k_1x + k_2x^2 + k_3x^3 + \cdots)$. The first term here is just Hooke's law, which describes the force exerted by a simple spring for small displacements. For small excursions from equilibrium we generally ignore the higher-order terms, but in some cases it may be desirable to keep the second term as well. If we model the restoring force as $F = -(k_1x + k_2x^2)$, how much work is done in displacing the system from $x = 0$ to $x = x_{max}$ by an applied force equal in magnitude to the restoring force?

47\. A traveler at an airport takes an escalator up one floor as shown in Figure P6.47. The moving staircase would itself carry him upward with vertical velocity component v between entry and exit points separated by height h. While the escalator is moving, however, the hurried traveler climbs the steps of the escalator at a rate of n steps/s. Assume that the height of each step is h_s. (a) Determine the amount of chemical energy converted into mechanical energy by the traveler's leg muscles during his escalator ride given that his mass is m. (b) Determine the work the escalator motor does on this person.

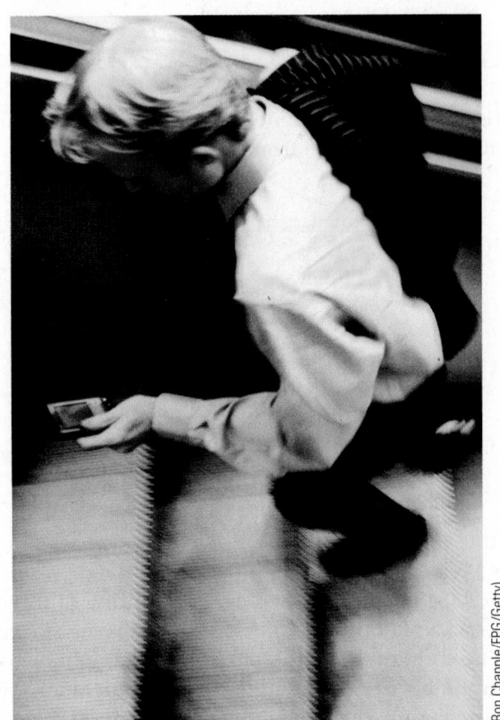

(Ron Chapple/FPG/Getty)

FIGURE **P6.47**

48. A 5.00-kg steel ball is dropped onto a copper plate from a height of 10.0 m. If the ball leaves a dent 3.20 mm deep, what is the average force exerted by the plate on the ball during the impact?

49. In a control system, an accelerometer consists of a 4.70-g object sliding on a horizontal rail. A low-mass spring attaches the object to a flange at one end of the rail. Grease on the rail makes static friction negligible, but rapidly damps out vibrations of the sliding object. When subject to a steady acceleration of $0.800g$, the object is to assume a location 0.500 cm away from its equilibrium position. Find the force constant required for the spring.

50. A light spring with force constant 3.85 N/m is compressed by 8.00 cm as it is held between a 0.250-kg block on the left and a 0.500-kg block on the right, both resting on a horizontal surface. The spring exerts a force on each block, tending to push the blocks apart. The blocks are simultaneously released from rest. Find the acceleration with which each block starts to move if the coefficient of kinetic friction between each block and the surface is (a) 0, (b) 0.100, and (c) 0.462.

51. A single constant force $\vec{\mathbf{F}}$ acts on a particle of mass m. The particle starts at rest at $t = 0$. (a) Show that the instantaneous power delivered by the force at any time t is $(F^2/m)t$. (b) If $F = 20.0$ N and $m = 5.00$ kg, what is the power delivered at $t = 3.00$ s?

52. A particle is attached between two identical springs on a horizontal frictionless table. Both springs have spring constant k and are initially unstressed. (a) The particle is pulled a distance x along a direction perpendicular to the initial configuration of the springs as shown in Figure P6.52. Show that the force exerted by the springs on the particle is

$$\vec{\mathbf{F}} = -2kx\left(1 - \frac{L}{\sqrt{x^2 + L^2}}\right)\hat{\mathbf{i}}$$

(b) Determine the amount of work done by this force in moving the particle from $x = A$ to $x = 0$.

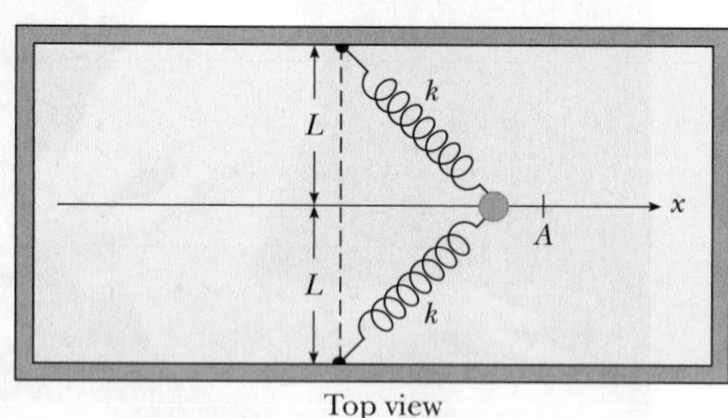

FIGURE **P6.52**

53. A 200-g block is pressed against a spring of force constant 1.40 kN/m until the block compresses the spring 10.0 cm. The spring rests at the bottom of a ramp inclined at 60.0° to the horizontal. Using energy considerations, determine how far up the incline the block moves before it stops (a) if there is no friction between the block and the ramp and (b) if the coefficient of kinetic friction is 0.400.

54. **Review problem.** Two constant forces act on a 5.00-kg object moving in the xy plane as shown in Figure P6.54. Force $\vec{\mathbf{F}}_1$ is 25.0 N at 35.0°, whereas $\vec{\mathbf{F}}_2$ is 42.0 N at 150°. At time $t = 0$, the object is at the origin and has velocity $(4.00\hat{\mathbf{i}} + 2.50\hat{\mathbf{j}})$ m/s. (a) Express the two forces in unit-vector notation. Use unit-vector notation for your other answers. (b) Find the total force on the object. (c) Find the object's acceleration. Now, considering the instant $t = 3.00$ s, (d) find the object's velocity, (e) its location, (f) its kinetic energy from $\frac{1}{2}mv_f^2$, and (g) its kinetic energy from $\frac{1}{2}mv_i^2 + \sum \vec{\mathbf{F}} \cdot \Delta\vec{\mathbf{r}}$.

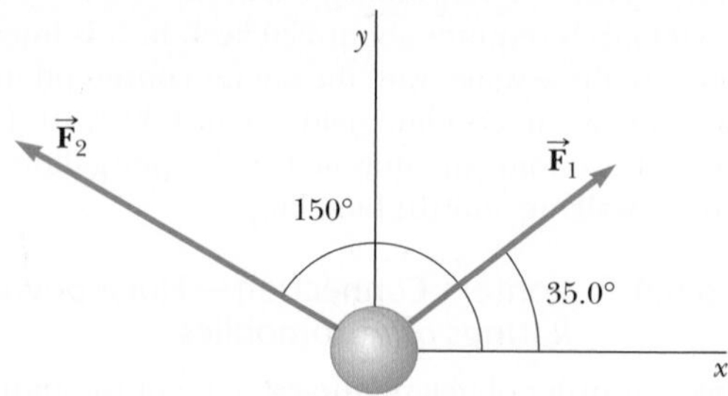

FIGURE **P6.54**

55. The ball launcher in a classic pinball machine has a spring that has a force constant of 1.20 N/cm (Fig. P6.55). The surface on which the ball moves is inclined 10.0° with respect to the horizontal. The spring is initially compressed 5.00 cm. Find the launching speed of a 100-g ball when the plunger is released. Friction and the mass of the plunger are negligible.

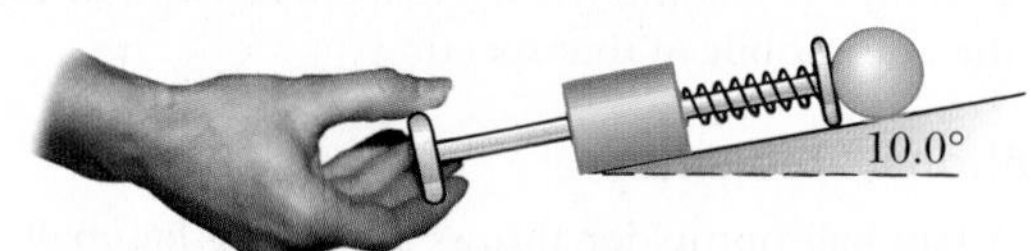

FIGURE **P6.55**

56. When objects with different weights are hung on a spring, the spring stretches to different lengths as shown in the following table. (a) Make a graph of the applied force versus the extension of the spring. By least-squares fitting, determine the straight line that best fits the data. (You may not want to use all the data points.) (b) From the slope of the best-fit line, find the spring constant k. (c) If the spring is extended to 105 mm, what force does it exert on the object it suspends?

F (N)	*L* (mm)	*F* (N)	*L* (mm)
2.0	15	14	112
4.0	32	16	126
6.0	49	18	149
8.0	64	20	175
10	79	22	190
12	98		

57. In diatomic molecules, the constituent atoms exert attractive forces on each other at large distances and repulsive forces at short distances. For many molecules, the Lennard–Jones law is a good approximation to the magnitude of these forces:

$$F = F_0\left[2\left(\frac{\sigma}{r}\right)^{13} - \left(\frac{\sigma}{r}\right)^{7}\right]$$

where r is the center-to-center distance between the atoms in the molecule, σ is a length parameter, and F_0 is the force when $r = \sigma$. For an oxygen molecule, $F_0 = 9.60 \times 10^{-11}$ N and $\sigma = 3.50 \times 10^{-10}$ m. Determine the work done by this force as the atoms are pulled apart from $r = 4.00 \times 10^{-10}$ m to $r = 9.00 \times 10^{-10}$ m.

58. A 0.400-kg particle slides around a horizontal track. The track has a smooth vertical outer wall forming a circle with a radius of 1.50 m. The particle is given an initial speed of 8.00 m/s. After one revolution, its speed has dropped to 6.00 m/s because of friction with the rough floor of the track. (a) Find the energy transformed from mechanical to internal in the system owing to friction in one revolution. (b) Calculate the coefficient of kinetic friction. (c) What is the total number of revolutions the particle makes before stopping?

59. A particle moves along the x axis from $x = 12.8$ m to $x = 23.7$ m under the influence of a force

$$F = \frac{375}{x^3 + 3.75x}$$

where F is in newtons and x is in meters. Using numerical integration, determine the total work done by this force on the particle during this displacement. Your result should be accurate to within 2%.

60. As it plows a parking lot, a snowplow pushes an ever-growing pile of snow in front of it. Suppose a car moving through the air is similarly modeled as a cylinder pushing a growing plug of air in front of it. The originally stationary air is set into motion at the constant speed v of the cylinder as shown in Figure P6.60. In a time interval Δt, a new disk of air of mass Δm must be moved a distance $v\,\Delta t$ and hence must be given a kinetic energy $\frac{1}{2}(\Delta m)v^2$. Using this model, show that the automobile's power loss owing to air resistance is $\frac{1}{2}\rho A v^3$ and that the resistive force acting on the car is $\frac{1}{2}\rho A v^2$, where ρ is the density of air. Compare this result with the empirical expression $\frac{1}{2}D\rho A v^2$ for the resistive force.

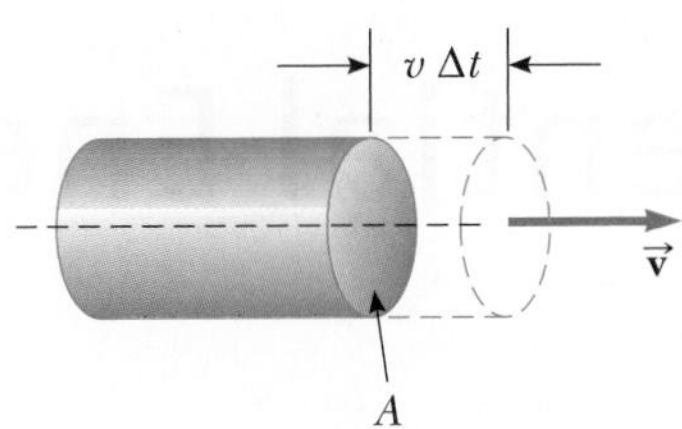

FIGURE **P6.60**

61. A windmill, such as that in the opening photograph of this chapter, turns in response to a force of high-speed air resistance, $R = \frac{1}{2}D\rho A v^2$. The power available is $\mathcal{P} = Rv = \frac{1}{2}D\rho\pi r^2 v^3$, where v is the wind speed and we have assumed a circular face for the windmill of radius r. Take the drag coefficient as $D = 1.00$ and the density of air from the front endpaper. For a home windmill with $r = 1.50$ m, calculate the power available if (a) $v = 8.00$ m/s and (b) $v = 24.0$ m/s. The power delivered to the generator is limited by the efficiency of the system, about 25%. For comparison, a typical home needs about 3 kW of electric power.

62. Consider the block–spring–surface system in part (b) of Interactive Example 6.7. (a) At what position x of the block is its speed a maximum? (b) Explore the effect of an increased friction force of 10.0 N. At what position of the block does its maximum speed occur in this situation?

ANSWERS TO QUICK QUIZZES

6.1 c, a, d, b. The work in (c) is positive and of the largest possible value because the angle between the force and the displacement is zero. The work done in (a) is zero because the force is perpendicular to the displacement. In (d) and (b), negative work is done by the applied force because in neither case is there a component of the force in the direction of the displacement. Situation (b) is the most negative value because the angle between the force and the displacement is 180°.

6.2 (d). Because of the range of values of the cosine function, $\vec{\mathbf{A}} \cdot \vec{\mathbf{B}}$ has values that range from AB to $-AB$.

6.3 (a). Because the work done in compressing a spring is proportional to the square of the compression distance x, doubling the value of x causes the work to increase fourfold.

6.4 (b). Because the work is proportional to the square of the compression distance x and the kinetic energy is proportional to the square of the speed v, doubling the compression distance doubles the speed.

6.5 (a) For the television set, energy enters by electrical transmission (through the power cord) and electromagnetic radiation (the television signal). Energy leaves by heat (from hot surfaces into the air), mechanical waves (sound from the speaker), and electromagnetic radiation (from the screen). (b) For the gasoline-powered lawn mower, energy enters by matter transfer (gasoline). Energy leaves by work (on the blades of grass), mechanical waves (sound), and heat (from hot surfaces into the air). (c) For the hand-cranked pencil sharpener, energy enters by work (from your hand turning the crank). Energy leaves by work (done on the pencil), mechanical waves (sound), and heat resulting from the temperature increase from friction.

6.6 (b), (d), (e). For the block, the friction force from the surface represents an interaction with the environment. For the surface, the friction force from the block represents an interaction with the environment. For the block and the surface, the friction force is internal to the system, so there are no interactions with the environment.

6.7 (c). The brakes and the roadway are warmer, so their internal energy has increased. In addition, the sound of the skid represents transfer of energy away by mechanical waves.

CHAPTER 7

Potential Energy

A strobe photograph of a pole vaulter. In the system of the pole vaulter and the Earth, several types of energy transformations occur during this process.

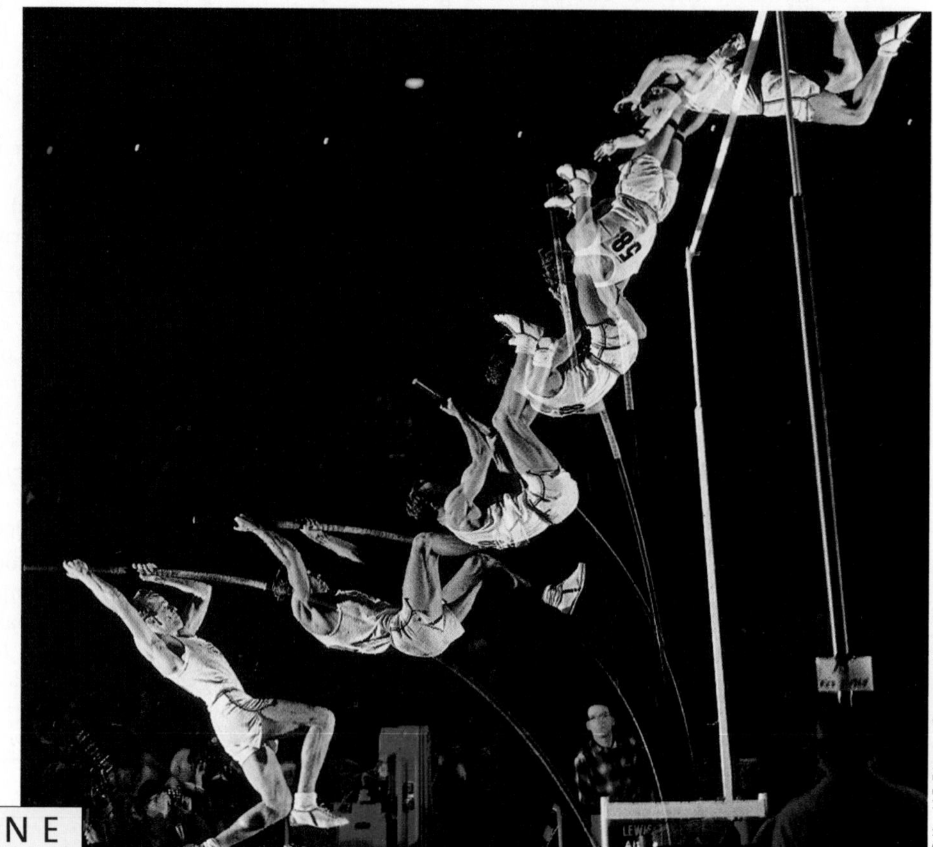

(© Harold E. Edgerton/Courtesy of Palm Press, Inc.)

CHAPTER OUTLINE

In Chapter 6, we introduced the concepts of kinetic energy, which is associated with the motion of an object or a particle, and internal energy, which is associated with the temperature of a system. In this chapter, we introduce another form of energy for a system, called *potential energy*, which is associated with the configuration of a system of two or more interacting objects or particles. This new type of energy will provide us with a powerful and universal fundamental principle for an isolated system.

7.1 POTENTIAL ENERGY OF A SYSTEM

In Chapter 6, we defined a system in general but focused our attention on single particles under the influence of an external force. In this chapter, we consider systems of two or more objects or particles interacting via a force that is *internal* to the system. The kinetic energy of such a system is the algebraic sum of the kinetic energies of all members of the system. In some systems, however, one object may be so massive that it can be modeled as stationary and its kinetic energy can be ignored. For example, if we consider a ball–Earth system as the ball falls to the ground, the kinetic energy of the system can be considered as only the kinetic energy of the ball. The Earth moves toward the ball so slowly in this process that we can ignore its kinetic energy. (We will justify this claim in Chapter 8.) On the other hand, the kinetic energy of a system of two electrons must include the kinetic energies of both particles.

Let us imagine a system consisting of a book and the Earth, interacting via the gravitational force. We do positive work on the system by lifting the book slowly through a height $\Delta y = y_b - y_a$ as in Figure 7.1. According to the continuity equation for energy (Eq. 6.20) introduced in Chapter 6, this work on the system must appear as an increase in energy of the system. The book is at rest before we perform the work and is at rest after we perform the work; therefore, the kinetic energy of the system does not change. There is no reason to suspect that the temperature of the book or of the Earth should change, so the internal energy of the system experiences no change.

Because the energy change of the system is not in the form of kinetic energy or internal energy, it must appear as some other form of energy storage. After lifting the book, suppose we release it and let it fall to the ground. Notice that the book (and therefore the system) now has kinetic energy, and its origin is in the work that was done in lifting the book. While the book is at the highest point, the energy of the system has the *potential* to become kinetic energy, but does not do so until the book is allowed to fall. Therefore, we call the energy storage mechanism before we release the book **potential energy.** We will find that a potential energy can be associated with a number of types of forces. In this particular case, we are discussing **gravitational potential energy.**

Let us now derive an expression for the gravitational potential energy associated with an object at a given location above the Earth's surface. To do so, consider an external agent lifting an object of mass m from an initial height y_a above the ground to a final height y_b as in Figure 7.1. We assume that the lifting is done slowly, with no acceleration, so that the lifting force can be modeled as equal to the weight of the object; the object is in equilibrium and moving at constant velocity. The work done by the external agent on the system (the object and the Earth) as the object undergoes this upward displacement is given by the product of the upward force $\vec{\mathbf{F}} = -m\vec{\mathbf{g}}$ and the displacement $\Delta\vec{\mathbf{r}} = \Delta y\hat{\mathbf{j}}$:

$$W = (-m\vec{\mathbf{g}})\cdot\Delta\vec{\mathbf{r}} = [-m(-g\hat{\mathbf{j}})]\cdot[(y_b - y_a)\hat{\mathbf{j}}] = mgy_b - mgy_a \qquad [7.1]$$

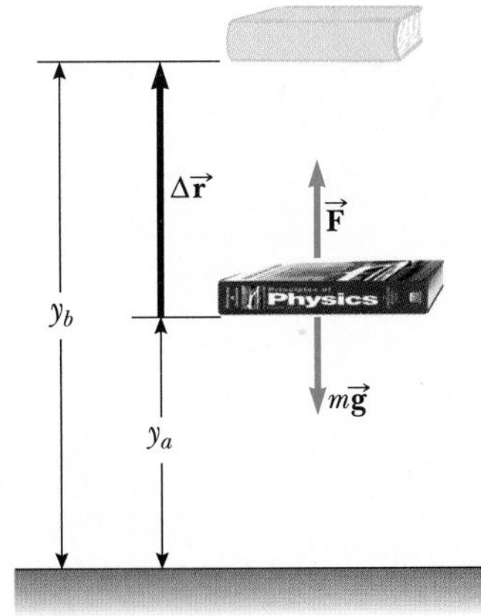

FIGURE 7.1 The work done by an external agent on the system of the book and the Earth as the book is lifted from y_a to y_b is equal to $mgy_b - mgy_a$.

PITFALL PREVENTION 7.1

POTENTIAL ENERGY BELONGS TO A SYSTEM Keep in mind that potential energy is always associated with a *system* of two or more interacting objects. In the gravitational case, in which a small object moves near the surface of the Earth, we may sometimes refer to the potential energy "associated with the object" rather than the more proper "associated with the system" because the Earth does not move significantly. We will not, however, refer to the potential energy "of the object" because this wording clearly ignores the role of the Earth in the potential energy.

We have discussed in Chapter 6 that work is a means of transferring energy into a system. Consequently, the expression on the right in Equation 7.1 must represent a change in the energy of the system, equal to the amount of work done on the system. Notice how similar Equation 7.1 is to Equation 6.17 in the preceding chapter. In each equation, the work done on a system equals a difference between the final and initial values of a quantity. Of course, both equations are nothing more than special cases of the continuity equation for energy, Equation 6.20. In Equation 6.17, the work represents a transfer of energy into the system and the increase in energy of the system is kinetic in form. In Equation 7.1, the work represents a transfer of energy into the system and the system energy appears in a different form, which we call gravitational potential energy.

Therefore, we can represent the quantity mgy to be the gravitational potential energy U_g of the object–Earth system:

$$U_g \equiv mgy \qquad [7.2]$$

■ Gravitational potential energy

The units of gravitational potential energy are joules, the same as those of work and kinetic energy. Potential energy, like work and kinetic energy, is a scalar quantity. Note that Equation 7.2 is valid only for objects near the surface of the Earth, where g is approximately constant.

Using our definition of gravitational potential energy, we can now rewrite Equation 7.1 as

$$W = \Delta U_g$$

which mathematically describes that the work done on the system by the external agent in this situation appears as a change in the gravitational potential energy of the system.

The gravitational potential energy depends only on the vertical height of the object above the Earth's surface. Therefore, the same amount of work is done on an object–Earth system whether the object is lifted vertically from the Earth or whether it starts at the same point and is pushed up a frictionless incline, ending up at the same height. This concept can be shown in a mathematical representation by reperforming the work calculation in Equation 7.1 with a displacement having both vertical and horizontal components:

$$W = (-m\vec{\mathbf{g}}) \cdot \Delta\vec{\mathbf{r}} = [-m(-g\hat{\mathbf{j}})] \cdot [(x_b - x_a)\hat{\mathbf{i}} + (y_b - y_a)\hat{\mathbf{j}}] = mgy_b - mgy_a$$

Note that no term involving x appears in the final result because $\hat{\mathbf{j}} \cdot \hat{\mathbf{i}} = 0$.

In solving problems, it is necessary to choose a reference configuration for which to set the gravitational potential energy equal to some reference value, which is normally zero. The choice of this configuration is completely arbitrary because the important quantity is the *difference* in potential energy, and this difference is independent of the choice of reference configuration.

It is often convenient to choose an object located at the surface of the Earth as the reference configuration for zero gravitational potential energy, but this choice is not essential. Often, the statement of the problem suggests a convenient configuration to use.

QUICK QUIZ 7.1 Choose the correct answer. The gravitational potential energy of a system (**a**) is always positive, (**b**) is always negative, or (**c**) can be negative or positive.

QUICK QUIZ 7.2 An object falls off a table to the floor. We wish to analyze the situation in terms of kinetic and potential energy. In discussing the potential energy of the system, we identify the system as (**a**) both the object and the Earth, (**b**) only the object, or (**c**) only the Earth.

7.2 THE ISOLATED SYSTEM

The introduction of potential energy allows us to generate a powerful and universally applicable principle for solving problems that are difficult to solve with Newton's laws. Let us develop this new principle by thinking about the book–Earth system in Figure 7.1 again. After we have lifted the book, there is gravitational potential energy stored in the system, which we can calculate from the work done by the external agent on the system using $W = \Delta U_g$.

Let us now shift our focus to the book alone as the system and let the book fall (Fig. 7.2). As the book falls from y_b to y_a, the work done by the gravitational force on the book is

$$W_{\text{on book}} = (m\vec{\mathbf{g}}) \cdot \Delta\vec{\mathbf{r}} = (-mg\hat{\mathbf{j}}) \cdot (y_a - y_b)\hat{\mathbf{j}} = mgy_b - mgy_a \qquad [7.3]$$

From the work–kinetic energy theorem of Chapter 6, the work done on the book is also

$$W_{\text{on book}} = \Delta K_{\text{book}}$$

Therefore, equating these two expressions for the work done on the book gives

$$\Delta K_{\text{book}} = mgy_b - mgy_a \qquad [7.4]$$

Now, let us relate each side of this equation to the *system* of the book and the Earth. For the right-hand side,

$$mgy_b - mgy_a = U_{gi} - U_{gf} = -\Delta U_g$$

where U_g is the gravitational potential energy of the system. Because the book is the only part of the system that is moving, the left-hand side of Equation 7.4 becomes

$$\Delta K_{\text{book}} = \Delta K$$

where K is the kinetic energy of the system. Therefore, by replacing each side of Equation 7.4 with its system equivalent, the equation becomes

$$\Delta K = -\Delta U_g \qquad [7.5]$$

This equation can be manipulated to provide a very important result for a new analysis model. First, we bring the change in potential energy to the left side of the equation:

$$\Delta K + \Delta U_g = 0 \qquad [7.6]$$

Notice that this equation is in the form of the continuity equation for energy, Equation 6.20. On the left, we have a sum of changes of the energy stored in the system. The right-hand side in the continuity equation is the sum of the transfers across the boundary of the system. This sum is equal to zero in this case because our book–Earth system is isolated from the environment.

Let us now write the changes in energy in Equation 7.6 explicitly:

$$(K_f - K_i) + (U_{gf} - U_{gi}) = 0 \quad \rightarrow \quad K_f + U_{gf} = K_i + U_{gi} \qquad [7.7]$$

In general, we define the sum of kinetic and potential energies of a system as the **total mechanical energy of the system.** Therefore, Equation 7.7 is a statement of **conservation of mechanical energy** for an **isolated system.** An isolated system is one for which no energy transfers occur across the boundary. Therefore, the energy in the system is conserved and the sum of the kinetic and potential energies remains constant. **Equation 7.7 is only true when no friction acts between members of the system.** In Section 7.3, we shall see how this equation must be modified to include the effects of friction.

For the falling book situation that we are describing in this discussion, Equation 7.7 can be written as

$$\tfrac{1}{2}mv_f^{\,2} + mgy_f = \tfrac{1}{2}mv_i^{\,2} + mgy_i$$

As the book falls to the Earth, the book–Earth system loses potential energy and gains kinetic energy such that the total of the two types of energy always remains constant. The transformation of one type of energy to another is the result of the process of work done by the gravitational force on the book. Note that this

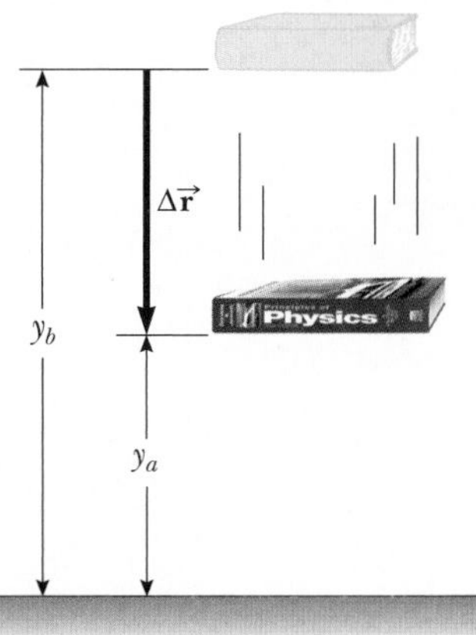

FIGURE 7.2 The work done by the gravitational force on the book as the book falls from y_b to y_a is equal to $mgy_b - mgy_a$.

PITFALL PREVENTION 7.2

ISOLATED SYSTEMS The isolated system model goes far beyond Equation 7.7. This equation is only the *mechanical energy* version of this model. We will see shortly how to include internal energy. In later chapters, we will see other isolated systems and generate new versions (and associated equations) related to such quantities as momentum, angular momentum, and electric charge.

■ Conservation of mechanical energy for an isolated system

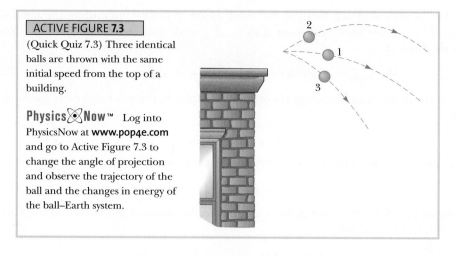

ACTIVE FIGURE 7.3
(Quick Quiz 7.3) Three identical balls are thrown with the same initial speed from the top of a building.

Physics Now™ Log into PhysicsNow at **www.pop4e.com** and go to Active Figure 7.3 to change the angle of projection and observe the trajectory of the ball and the changes in energy of the ball–Earth system.

work is *internal* to the system; it is not work done *on* the system from the environment.

We will see other types of potential energy besides gravitational, so we can write the general form of the definition for mechanical energy as

$$E_{\text{mech}} \equiv K + U \quad [7.8]$$

where U without a subscript refers to the total potential energy of the system, including all types. In addition, K in general refers to the sum of the kinetic energies of all particles in the system.

QUICK QUIZ 7.3 Three identical balls are thrown from the top of a building, all with the same initial speed. The first is thrown horizontally, the second at some angle above the horizontal, and the third at some angle below the horizontal, as shown in Active Figure 7.3. Neglecting air resistance, rank the speeds of the balls at the instant each hits the ground.

■ Thinking Physics 7.1

You have graduated from college and are designing roller coasters for a living. You design a roller coaster in which a car is pulled to the top of a hill of height h and then, starting from a momentary rest, rolls freely down the hill and upward toward the peak of the next hill, which is at height $1.1h$. Will you have a long career in this business?

Reasoning Your career will probably not be long because this roller coaster will not work! At the top of the first hill, the roller coaster car has no kinetic energy and the gravitational potential energy of the car–Earth system is that associated with a height for the car of h. If the car were to reach the top of the next hill, the system would have higher potential energy, that associated with height $1.1h$. This situation would violate the principle of conservation of mechanical energy. If this coaster were actually built, the car would move upward on the second hill to a height h (ignoring the effects of friction), stop short of the peak, and then start rolling backward, becoming trapped between the two hills. ■

INTERACTIVE EXAMPLE 7.1 Ball in Free-Fall

A ball of mass m is dropped from rest at a height h above the ground as in Figure 7.4. Ignore air resistance.

A Determine the speed of the ball when it is at a height y above the ground.

Solution The ball and the Earth do not experience any forces from the environment because we ignore air resistance. The ball–Earth system is isolated and we use the principle of conservation of mechanical energy. Note that the system has potential energy and no kinetic energy at the beginning of our time interval of interest. As the ball falls, the total mechanical energy of the system (the sum of kinetic and potential energies) remains constant and equal to its initial potential energy. The potential energy of the system decreases, and the kinetic energy of the system (which is due only to the ball) increases.

Before the ball is released from rest at a height h above the ground, the kinetic energy of the system is $K_i = 0$ and the potential energy is $U_i = mgh$, where the y coordinate is measured from ground level. When the ball is at an arbitrary position y above the ground, its kinetic energy is $K_f = \frac{1}{2}mv_f^2$ and the potential energy of the system is $U_f = mgy$. Applying Equation 7.7, we have

$$K_f + U_{gf} = K_i + U_{gi}$$

$$\tfrac{1}{2}mv_f^2 + mgy = 0 + mgh$$

$$v_f^2 = 2g(h - y)$$

$$v_f = \sqrt{2g(h - y)}$$

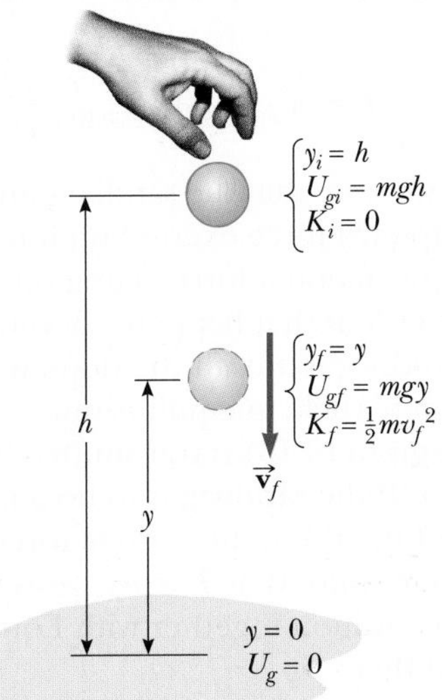

FIGURE 7.4 (Interactive Example 7.1) A ball is dropped from rest at a height h above the ground. Initially, the total energy of the ball–Earth system is gravitational potential energy, equal to mgh when $h = 0$ is at the ground. When the ball is at elevation y, the total system energy is the sum of kinetic and potential energies.

B Determine the speed of the ball at y if it is given an initial speed v_i at the initial altitude h.

Solution In this case, the initial energy includes kinetic energy of the ball equal to $\frac{1}{2}mv_i^2$ and Equation 7.7 gives

$$\tfrac{1}{2}mv_f^2 + mgy = \tfrac{1}{2}mv_i^2 + mgh$$

$$v_f^2 = v_i^2 + 2g(h - y)$$

$$v_f = \sqrt{v_i^2 + 2g(h - y)}$$

Note that this result is consistent with Equation 2.13 (Chapter 2), $v_{yf}^2 = v_{yi}^2 - 2g(y_f - y_i)$, for a particle under constant acceleration, where $y_i = h$. Furthermore, this result is valid even if the initial velocity is at an angle to the horizontal (the projectile situation), as discussed in Quick Quiz 7.3.

PhysicsNow™ Compare the effect of upward, downward, and zero initial velocities by logging into PhysicsNow at **www.pop4e.com** and going to Interactive Example 7.1.

INTERACTIVE EXAMPLE 7.2 A Grand Entrance

You are designing apparatus to support an actor of mass 65 kg who is to "fly" down to the stage during the performance of a play. You attach the actor's harness to a 130-kg sandbag by means of a lightweight steel cable running smoothly over two frictionless pulleys as in Figure 7.5a. You need 3.0 m of cable between the harness and the nearest pulley so that the pulley can be hidden behind a curtain. For the apparatus to work successfully, the sandbag must never lift above the floor as the actor swings from above the stage to the floor. Let us identify the angle θ as the angle that the actor's cable makes with the vertical when he begins his motion from rest. What is the maximum value θ can have such that the sandbag does not lift off the floor during the actor's swing?

Solution We must draw on several concepts to solve this problem. To conceptualize the problem, imagine what happens as the actor approaches the bottom of the swing. At the bottom, the rope is vertical and must

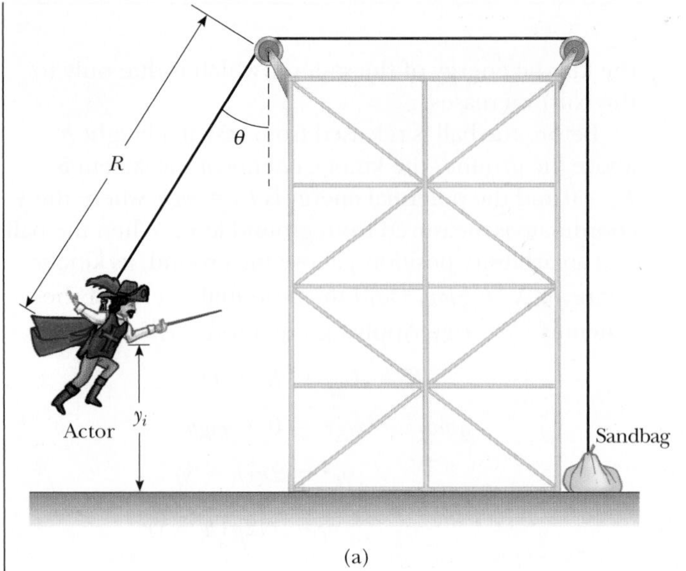

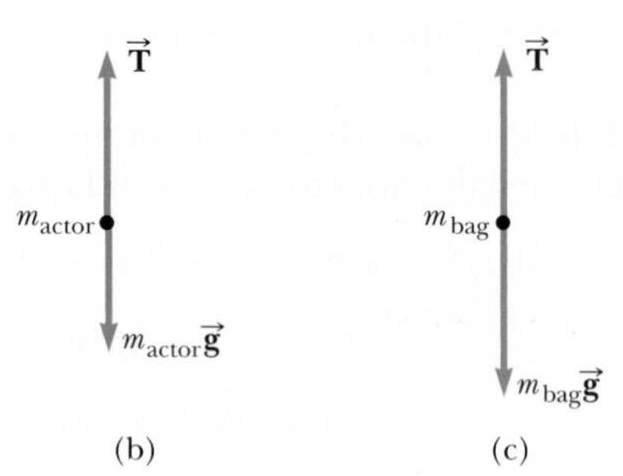

FIGURE 7.5 (Interactive Example 7.2) (a) An actor uses some clever staging to make his entrance. (b) Free-body diagram for the actor at the bottom of the circular path. (c) Free-body diagram for the sandbag when it is just lifted from the floor.

support his weight as well as provide centripetal acceleration of his body in the upward direction. At this point, the tension in the rope is the highest and the sandbag is most likely to lift off the floor. Looking first at the swinging of the actor from the initial point to the lowest point, we categorize this problem as an energy problem involving an isolated system, the actor and the Earth. To analyze this part of the problem we use the principle of conservation of mechanical energy for the system to find the actor's speed as he arrives at the floor as a function of the initial angle θ and the radius R of the circular path through which he swings.

Applying Equation 7.7 to the actor–Earth system gives

$$K_f + U_{gf} = K_i + U_{gi}$$

$$(1) \quad \tfrac{1}{2}m_{\text{actor}}v_f^2 + 0 = 0 + m_{\text{actor}}gy_i$$

where y_i is the initial height of the actor above the floor and v_f is the speed of the actor at the instant before he lands. (Note that $K_i = 0$ because he starts from rest and that $U_f = 0$ because we define the configuration of the actor at the floor as having a gravitational potential energy of zero.) From the geometry in Figure 7.5a, we see that $y_i = R - R\cos\theta = R(1 - \cos\theta)$. Using this relationship in Equation (1), we obtain

$$(2) \quad v_f^2 = 2gR(1 - \cos\theta)$$

Next, we focus on the instant the actor is at the lowest point. Because the tension in the cable is transferred to the sandbag by means of force, we categorize the actor at this instant as a particle under a net force and a particle in uniform circular motion. To analyze, we apply Newton's second law to the actor at the bottom of his path, using the free-body diagram in Figure 7.5b as a guide:

$$\sum F_y = T - m_{\text{actor}}g = m_{\text{actor}}\frac{v_f^2}{R}$$

$$(3) \quad T = m_{\text{actor}}g + m_{\text{actor}}\frac{v_f^2}{R}$$

Finally, we note that the sandbag lifts off the floor when the upward force exerted on it by the cable exceeds the gravitational force acting on it; the normal force is zero when that happens. Because we do not want the sandbag to lift off the floor, we categorize the sandbag as a particle in equilibrium. A force T of the magnitude given by (3) is transmitted by the cord to the sandbag. If the sandbag is to be on the verge of being lifted off the floor, the normal force on it becomes zero and we require that $T = m_{\text{bag}}g$ as in Figure 7.5c. Using this condition together with Equations (2) and (3), we find that

$$m_{\text{bag}}g = m_{\text{actor}}g + m_{\text{actor}}\frac{2gR(1 - \cos\theta)}{R}$$

Solving for $\cos\theta$ and substituting in the given parameters, we obtain

$$\cos\theta = \frac{3m_{\text{actor}} - m_{\text{bag}}}{2m_{\text{actor}}} = \frac{3(65\text{ kg}) - 130\text{ kg}}{2(65\text{ kg})} = \frac{1}{2}$$

$$\theta = 60°$$

To finalize the problem, note that we had to combine techniques from different areas of our study: energy and Newton's second law. Furthermore, the length R of the cable from the actor's harness to the leftmost pulley did not appear in the final algebraic equation. Therefore, the final answer is independent of R.

PhysicsNow™ Let the actor fly or crash without injury by logging into PhysicsNow at **www.pop4e.com** and going to Interactive Example 7.2.

7.3 CONSERVATIVE AND NONCONSERVATIVE FORCES

In the preceding section, we showed that the mechanical energy of a system is conserved in a process in which the force between members of the system is the gravitational force. The gravitational force is one example of a category of forces for which the mechanical energy of a system is conserved. These forces are called **conservative forces.** The other possibility for energy storage in a system besides kinetic and potential is internal energy. Therefore, a conservative force, for our purposes in mechanics, is a **force between members of a system that causes no transformation of mechanical energy to internal energy within the system.** If no energy is transformed to internal energy, the mechanical energy of the system is conserved, as described by Equation 7.7.

If a force is conservative, the work done by such a force has a special property as the members of the system move in response either to the force itself or to an external force: **The work done by a conservative force is independent of the path followed by the members of the system and depends only on the initial and final configurations of the system.**

■ A conservative force

From this property, it follows that **the work done by a conservative force when a member of the system is moved through a closed path is equal to zero.**

These statements can be mathematically demonstrated and serve as general mathematical definitions of conservative forces. Both statements can be seen for the gravitational force from Equation 7.3. The work done is expressed only in terms of the initial and final heights, with no indication of what path is followed. If the path is closed, the initial and final heights are the same in Equation 7.3 and the work is equal to zero.

Another example of a conservative force is the force of a spring on an object attached to the spring, where the spring force is given by Hooke's law, $F_s = -kx$. As we learned in Chapter 6 (Eq. 6.15), the work done by the spring force is

$$W_s = \tfrac{1}{2}kx_i^{\,2} - \tfrac{1}{2}kx_f^{\,2}$$

where the initial and final positions of the object are measured from its equilibrium position $x = 0$, at which the spring is unstretched. Again we see that W_s depends only on the initial and final coordinates of the object and is zero for any closed path. Hence, the spring force is conservative.

In Section 7.1, we discussed the notion of an external agent lifting a book and storing energy as potential energy in the book–Earth system. In Section 6.4, we discussed an external agent pulling a block attached to a spring from $x = 0$ to $x = x_{\text{max}}$ and calculated the work done on the system as $\frac{1}{2}kx_{\text{max}}^2$. This situation is another one, like that of the book in Section 7.1, in which work is done on a system but there is no change in kinetic energy of the system. Therefore, the energy must be stored in the block–spring system as potential energy. The **elastic potential energy** associated with the spring force is defined by

$$U_s \equiv \tfrac{1}{2}kx^2 \qquad [7.9]$$

■ Elastic potential energy

The elastic potential energy can be considered as the energy stored in the deformed spring (one that is either compressed or stretched from its equilibrium position). The elastic potential energy stored in the spring is zero whenever the spring is undeformed ($x = 0$). Because elastic potential energy is proportional to x^2, we see that U_s is always positive in a deformed spring.

Consider Active Figure 7.6a, which shows an undeformed spring on a frictionless, horizontal surface. When the block is pushed against the spring (Active Fig. 7.6b), compressing the spring a distance x, the elastic potential energy stored in the spring is $\frac{1}{2}kx^2$. When the block is released, the spring returns to its original

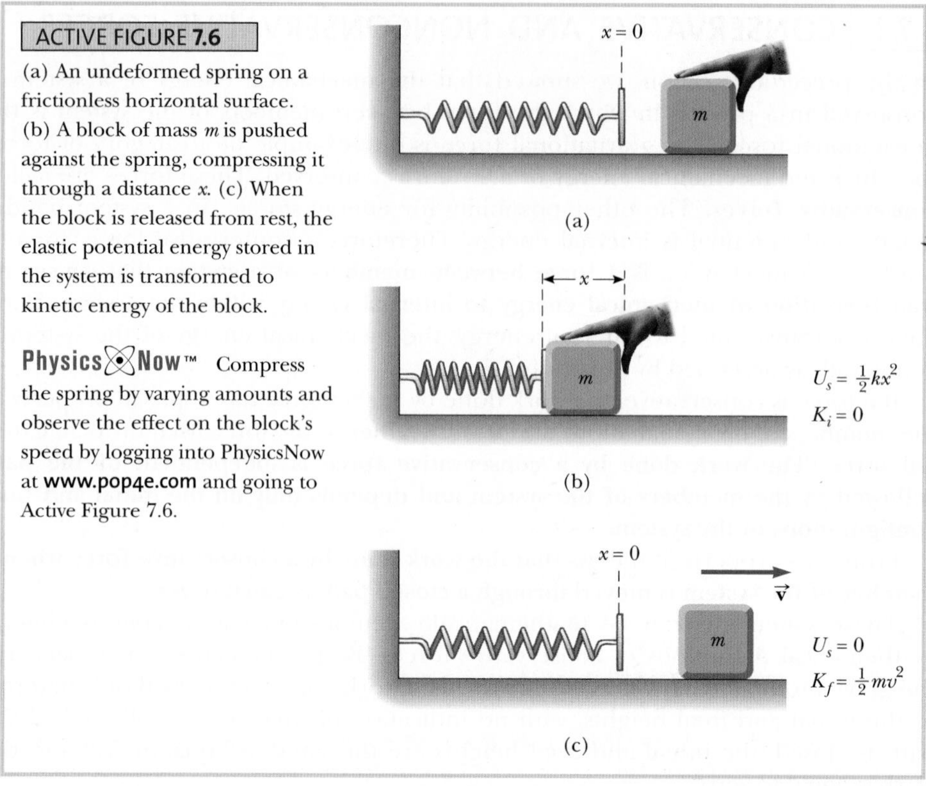

ACTIVE FIGURE 7.6 (a) An undeformed spring on a frictionless horizontal surface. (b) A block of mass m is pushed against the spring, compressing it through a distance x. (c) When the block is released from rest, the elastic potential energy stored in the system is transformed to kinetic energy of the block.

Physics Now™ Compress the spring by varying amounts and observe the effect on the block's speed by logging into PhysicsNow at **www.pop4e.com** and going to Active Figure 7.6.

length, applying a force to the block. This force does work on the block, resulting in kinetic energy of the block (Active Fig. 7.6c).

■ A nonconservative force

In comparison to a conservative force, a **nonconservative force** in mechanics is **a force between members of a system that is not conservative;** that is, it causes transformation of mechanical energy to internal energy within the system. A common nonconservative force in mechanics is the friction force. If we consider a system consisting of a block and a surface and imagine an initially sliding block coming to rest because of friction, we see the result of a nonconservative force. Initially, the system has kinetic energy (of the block). Afterward, nothing is moving so the final kinetic energy is zero. The friction force between the block and the surface transforms the mechanical energy into internal energy; the block and surface are both slightly warmer than before.

Let us return to the notion of the work done over a path. The two statements claimed for conservative forces are not true for nonconservative forces. For nonconservative forces, the work done depends on the path taken between the initial and final configurations, and the work done over a closed path is not zero. As an example, consider Figure 7.7. Suppose you displace a book between two points on a table. If the book is displaced in a straight line along the blue path between points Ⓐ and Ⓑ in Figure 7.7, you do a certain amount of work against the kinetic friction force to keep the book moving at a constant speed. Now, imagine that you push the book along the brown semicircular path in Figure 7.7. You perform more work against friction along this longer path than along the straight path. The work done depends on the path, so the friction force cannot be conservative.

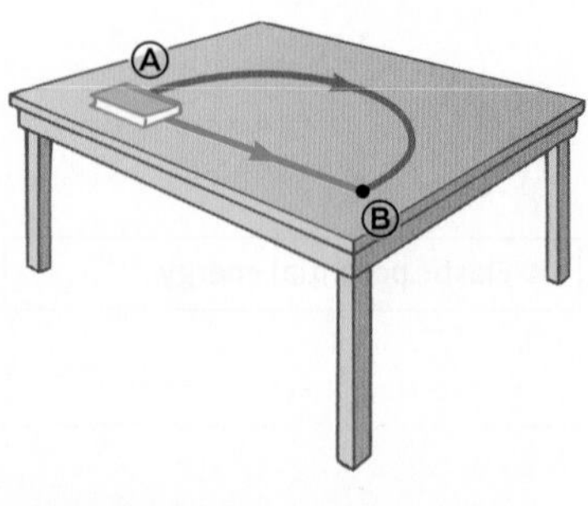

FIGURE 7.7 The work done against the force of friction depends on the path taken as the book is moved from Ⓐ to Ⓑ; hence, friction is a nonconservative force. The work required is greater along the brown path than the blue path.

In Chapter 6, we discussed that we cannot calculate the work done by friction on an object because the displacement of the point of application of the friction force is not the same as the displacement of the object. In the case of an object subject to a force of kinetic friction, the particle model is not valid. Another example of a situation in which we cannot use the particle model is seen for deformable objects. For example, suppose a rubber ball is flattened against a brick wall. The ball deforms during the pushing process. If the ball is suddenly released, it jumps away from the wall.

The force causing the ball to accelerate is the normal force of the wall on the ball. The point of application of this force, however, does not move through space; it stays fixed at the point of contact between the ball and the wall. Therefore, no work is done on the ball. Yet the ball has kinetic energy afterward. The work–kinetic energy theorem does not describe this situation correctly. It is more valuable to apply the isolated system model to this situation. With the ball as the system, there is no transfer of energy across the boundary as the ball springs off the wall. Rather, there is a *transformation* of energy, from elastic potential energy (stored in the ball when it was flattened) to kinetic energy. In the same way, if a skateboarder pushes off a wall to start rolling, no work is done by the force from the wall; the kinetic energy is transformed within the system from potential energy stored in the body from previous meals.

We also discussed in Chapter 6 the difficulties associated with the nonconservative force of friction in energy calculations. Recall that the result of a friction force is to transform kinetic energy in a system to internal energy and that the increase in internal energy is equal to the decrease in kinetic energy. If a potential energy is associated with the system, the decrease in *mechanical* energy, equals the increase in internal energy in the isolated system. Therefore, for a constant friction force,

$$-f_k d = \Delta K + \Delta U = \Delta E_{\text{mech}} = -\Delta E_{\text{int}} \qquad [7.10]$$

We can recast this expression by putting the changes in all forms of energy storage on one side of the equation:

$$\Delta K + \Delta U + \Delta E_{\text{int}} = \Delta E_{\text{system}} = 0 \qquad [7.11]$$

This gives us the most general expression of the continuity equation for energy for an isolated system. Note that ΔK may represent more than one term if two or more parts of the system are moving. Also, ΔU may represent more than one term if different types of potential energy (e.g., gravitational and elastic) are associated with the system. Equation 7.11 is equivalent to

$$K + U + E_{\text{int}} = \text{constant} \qquad [7.12]$$

■ Conservation of energy for an isolated system

which tells us that **the total energy (kinetic, potential, and internal) of an isolated system is conserved, regardless of whether the forces acting within the system are conservative or nonconservative. No violation of this critical conservation principle has ever been observed.** If we consider the Universe as an isolated system, this statement claims that there is a fixed amount of energy in our Universe and that all processes within the Universe represent transformations of energy from one type to another.

PITFALL PREVENTION 7.3

Complicated systems For simplicity, our discussion leading to Equation 7.10 assumes that only one object in the system is sliding on a surface. If there are two or more objects sliding on surfaces in the system, a term $-f_k d$ must be included for each object, with d representing the distance the object slides relative to the surface with which it is in contact.

QUICK QUIZ 7.4 A ball is connected to a light spring suspended vertically. When displaced downward from its equilibrium position and released, the ball oscillates up and down. **(i)** In the system of *the ball, the spring, and the Earth,* what forms of energy are there during the motion? **(ii)** In the system of *the ball and the spring,* what forms of energy are there during the motion? **(a)** kinetic and elastic potential **(b)** kinetic and gravitational potential **(c)** kinetic, elastic potential, and gravitational potential **(d)** elastic potential and gravitational potential

PROBLEM-SOLVING STRATEGY Isolated Systems

Many problems in physics can be solved using the principle of conservation of energy for an isolated system. The following procedure should be used when you apply this principle:

1. **Conceptualize** Define your system, which may consist of more than one object and may or may not include springs or other possibilities for storage of potential energy. Choose configurations to represent the initial and final conditions of the system.
2. **Categorize** Determine if any energy transfers occur across the boundary of your system. If so, use the nonisolated system model, $\Delta E_{\text{system}} = \Sigma T$. If not, use the isolated system model, $\Delta E_{\text{system}} = 0$.

Determine whether any nonconservative forces are present. Remember that if friction or air resistance is present, *mechanical* energy is not conserved but the *total* energy of an isolated system is.

3. **Analyze** For each object that changes elevation, select a reference position for the object that will define the zero configuration of gravitational potential energy for the system. For a spring, the zero configuration for elastic potential energy is when the spring is neither compressed nor extended from its equilibrium position. If there is more than one conservative force, write an expression for the potential energy associated with each force.

   If mechanical energy is conserved, write the total initial mechanical energy E_i of the system for some configuration as the sum of the kinetic and potential energy associated with the configuration. Then write a similar expression for the total mechanical energy E_f of the system for the final configuration that is of interest. Because mechanical energy is conserved, equate the two total energies and solve for the quantity that is unknown.

   If nonconservative forces are present (and therefore mechanical energy is not conserved), first write expressions for the total initial and total final mechanical energies. In this case, the difference between the total final mechanical energy and the total initial mechanical energy equals the energy transformed to or from internal energy by the nonconservative forces.

4. **Finalize** Make sure your results are consistent with your mental representation. Also make sure that the values of your results are reasonable and consistent with connections to everyday experience.

EXAMPLE 7.3 Crate Sliding Down a Ramp

A 3.00-kg crate slides down a ramp at a loading dock. The ramp is 1.00 m in length and is inclined at an angle of 30.0° as shown in Figure 7.8. The crate starts from rest at the top and experiences a constant friction force of magnitude 5.00 N. Use energy methods to determine the speed of the crate when it reaches the bottom of the ramp.

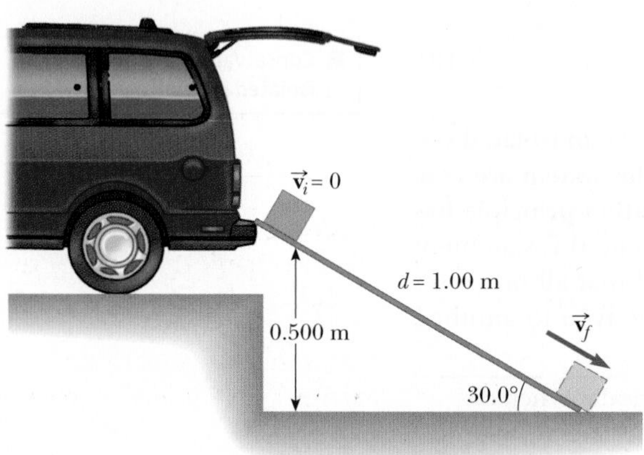

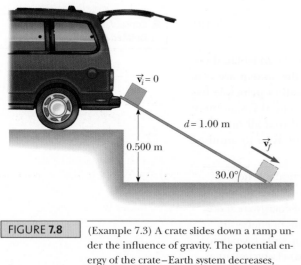

FIGURE 7.8 (Example 7.3) A crate slides down a ramp under the influence of gravity. The potential energy of the crate–Earth system decreases, whereas the kinetic energy of the crate increases.

Solution We define the system as the crate, the Earth, and the ramp. This system is isolated. If we had chosen the crate and the Earth as the system, we would need to use the nonisolated system model because the friction force between the crate and the ramp is an external influence. There would be work done across the boundary as well as flow of energy by heat between the crate and the ramp. This problem would be difficult to solve. In general, if a friction force acts, it is easiest to define the system so that the friction force is an internal force.

Because $v_i = 0$ for the crate, the initial kinetic energy of the system is zero. If the y coordinate is measured from the bottom of the ramp, $y_i = (1.00 \text{ m}) \sin 30° = 0.500$ m for the crate. The total mechanical energy of the crate–Earth–ramp system when the crate is at the top is therefore the gravitational potential energy:

$$E_i = U_i = mgy_i$$

When the crate reaches the bottom, the gravitational potential energy of the system is *zero* because the elevation of the crate is $y_f = 0$. The total mechanical energy when the crate is at the bottom is therefore kinetic energy:

$$E_f = K_f = \tfrac{1}{2}mv_f^2$$

We cannot say that $E_f = E_i$ in this case, however, because a nonconservative force—the force of friction—reduces the mechanical energy of the system. In this case, the change in mechanical energy for the system is $\Delta E_{\text{mech}} = -f_k d$, where $d = 1.00$ m. Because $\Delta E_{\text{mech}} = \Delta K + \Delta U = \frac{1}{2}mv_f^2 - mgy_i$ in this situation, Equation 7.10 gives

$$-f_k d = \tfrac{1}{2}mv_f^2 - mgy_i$$

$$v_f = \sqrt{2gy_i - 2\frac{f_k d}{m}}$$

$$= \sqrt{2(9.80 \text{ m/s}^2)(0.500 \text{ m}) - 2\frac{(5.00 \text{ N})(1.00 \text{ m})}{(3.00 \text{ kg})}}$$

$$= 2.54 \text{ m/s}$$

EXAMPLE 7.4 **Motion on a Curved Track**

A child of mass m takes a ride on an irregularly curved slide of height $h = 2.00$ m as in Figure 7.9. The child starts from rest at the top.

A Determine the speed of the child at the bottom, assuming that no friction is present.

Solution We will define the system as the child and the Earth and will model the child as a particle. The normal force $\vec{\mathbf{n}}$ does no work on the system because this force is always perpendicular to each element of the displacement. Furthermore, because no friction is present, no work is done by friction across the boundary of the system. Therefore, we use the isolated system model with no friction forces, for which mechanical energy is conserved; that is, $K + U =$ constant.

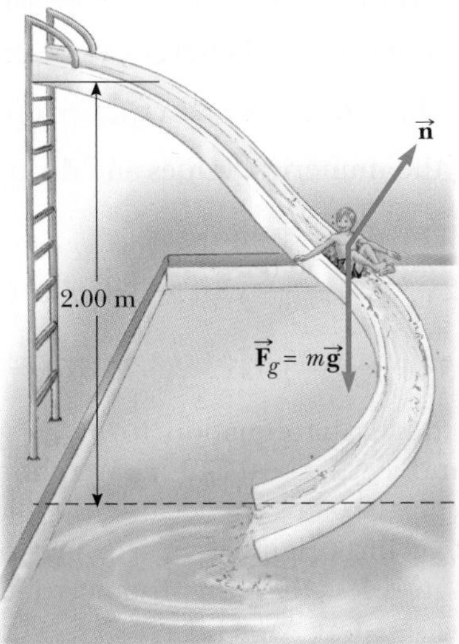

FIGURE 7.9 (Example 7.4) If the slide is frictionless, the speed of the child at the bottom depends only on the height of the slide.

If we measure the y coordinate for the child from the bottom of the slide, $y_i = h$, $y_f = 0$, and we have for the system

$$K_f + U_f = K_i + U_i$$

$$\tfrac{1}{2}mv_f^2 + 0 = 0 + mgh$$

$$v_f = \sqrt{2gh}$$

The result for the speed is the same as if the child simply fell vertically through a distance h! In this example, $h = 2.00$ m, giving

$$v_f = \sqrt{2gh} = \sqrt{2(9.80\text{ m/s}^2)(2.00\text{ m})} = \boxed{6.26\text{ m/s}}$$

B If a friction force acts on the 20.0-kg child and he arrives at the bottom of the slide with a speed $v_f = 3.00$ m/s, by how much does the mechanical energy of the system decrease as a result of this force?

Solution We define the system as the child, the Earth, and the slide. In this case, a nonconservative force acts within the system and mechanical energy is *not* conserved. We can find the change in mechanical energy as a result of friction, given that the final speed at the bottom is known:

$$\Delta E_{\text{mech}} = K_f + U_f - K_i - U_i = \tfrac{1}{2}mv_f^2 + 0 - 0 - mgh$$

$$\Delta E_{\text{mech}} = \tfrac{1}{2}(20.0\text{ kg})(3.00\text{ m/s})^2 - (20.0\text{ kg})(9.80\text{ m/s}^2)(2.00\text{ m})$$

$$= \boxed{-302\text{ J}}$$

The change in mechanical energy ΔE_{mech} is negative because friction reduces the mechanical energy of the system. The change in internal energy in the system is $+302$ J.

EXAMPLE 7.5 **Block–Spring Collision**

A block of mass 0.800 kg is given an initial velocity $v_A = 1.20$ m/s to the right and collides with a light spring of force constant $k = 50.0$ N/m as in Figure 7.10.

A If the surface is frictionless, calculate the maximum compression of the spring after the collision.

Solution We define the system as the block and the spring. No transfers of energy occur across the boundary of this system, so we use the isolated system model. Before the collision, when the block is at Ⓐ, for example, the system has kinetic energy due to the moving block and the spring is uncompressed, so the potential energy stored in the system is zero. Therefore, the total energy of the system before the collision is $\frac{1}{2}mv_A^2$. After the collision, and when the spring is fully compressed at point Ⓒ, the block is momentarily at rest and has zero kinetic energy, whereas the potential energy stored in the spring has its maximum value $\frac{1}{2}kx_{\text{max}}^2$. The total mechanical energy of the system is conserved because no nonconservative forces act within the system.

Because the mechanical energy of the system is conserved,

$$\tfrac{1}{2}mv_A^2 + 0 = 0 + \tfrac{1}{2}kx_{\text{max}}^2$$

$$x_{\text{max}} = \sqrt{\frac{m}{k}}\,v_A = \sqrt{\frac{0.800\text{ kg}}{50.0\text{ N/m}}}(1.20\text{ m/s}) = \boxed{0.152\text{ m}}$$

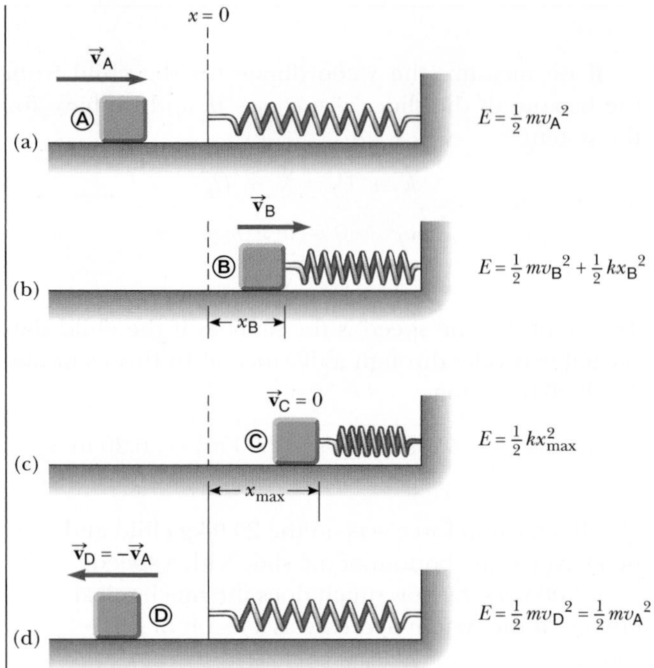

FIGURE 7.10 (Example 7.5) A block sliding on a smooth, horizontal surface collides with a light spring. (a) Initially, the mechanical energy of the system is all kinetic energy. (b) The mechanical energy is the sum of the kinetic energy of the block and the elastic potential energy in the spring. (c) The mechanical energy is entirely potential energy. (d) The mechanical energy is transformed back to the kinetic energy of the block. The total energy of the block–spring system remains constant throughout the motion.

B If a constant force of kinetic friction acts between the block and the surface with $\mu_k = 0.500$ and if the speed of the block just as it collides with the spring is $v_A = 1.20$ m/s, what is the maximum compression in the spring?

Solution We define the system as the block, the spring, and the surface. In this case, mechanical energy of the system is *not* conserved because a friction force acts between members of the system. The magnitude of the friction force is

$$f_k = \mu_k n = \mu_k mg = 0.500(0.800 \text{ kg})(9.80 \text{ m/s}^2)$$
$$= 3.92 \text{ N}$$

where we have used $n = mg$ from Newton's second law in the vertical direction. Therefore, the decrease in mechanical energy due to friction as the block is displaced through a straight line from $x_i = 0$ to the point $x_f = x_{max}$ at which the block stops is

$$\Delta E_{mech} = -f_k x_{max} = -3.92x_{max}$$

The change in mechanical energy can be expressed as

$$\Delta E_{mech} = E_f - E_i = (0 + \tfrac{1}{2}kx_{max}^2) - (\tfrac{1}{2}mv_A^2 + 0)$$

Substituting the numerical values and dropping the units, we have

$$-3.92x_{max} = \frac{50.0}{2}x_{max}^2 - \tfrac{1}{2}(0.800)(1.20)^2$$

$$25.0x_{max}^2 + 3.92x_{max} - 0.576 = 0$$

Solving the quadratic equation for x_{max} gives $x_{max} = 0.092\,4$ m and $x_{max} = -0.249$ m. We choose the positive root $x_{max} = 0.092\,4$ m because the block must be to the right of the origin when it comes to rest. Note that 0.092 4 m is less than the distance obtained in the frictionless case (part A). This result is what we expect because friction retards the motion of the system.

7.4 CONSERVATIVE FORCES AND POTENTIAL ENERGY

Let us return to the falling book discussed in Section 7.2. We found that the work done within the book–Earth system by the gravitational force on the book can be expressed as the negative of the difference between two quantities that we called the initial and final potential energies of the system:

$$W_{\text{on book}} = mgy_b - mgy_a = -\Delta U \qquad [7.13]$$

This expression is the hallmark of a conservative force: we can identify a potential energy function such that the work done by the force on a member of the system in which the force acts depends only on the difference in the function's initial and final values. Such a function does not exist for a nonconservative force because the work done depends on the particular path followed between the initial and final points.

For a conservative force, this notion allows us to generate a mathematical relationship between a force and its potential energy function. From the definition of work, we can write Equation 7.13 for a general force in the x direction as

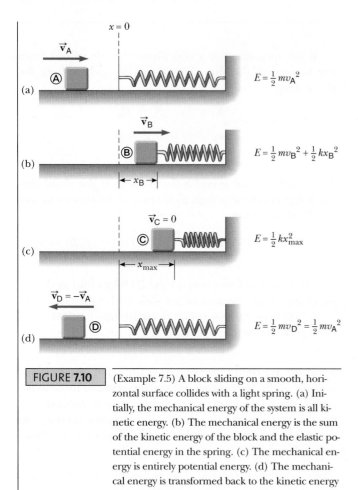

$$W = \int_{x_i}^{x_f} F_x\,dx = -\Delta U = -(U_f - U_i) = -U_f + U_i \qquad [7.14]$$

Therefore, the potential energy function can be written as

$$U_f = -\int_{x_i}^{x_f} F_x \, dx + U_i \qquad [7.15]$$

■ **Finding the potential energy of a system associated with a force between members of the system**

This expression allows us to calculate the potential energy function associated with a conservative force if we know the force function. The value of U_i is often taken to be zero at some arbitrary reference point. It really doesn't matter what value we assign to U_i because any value simply shifts U_f by a constant, and it is the *change* in potential energy that is physically meaningful.

As an example, let us calculate the potential energy function for the spring force. We model the spring as obeying Hooke's law, so the force the spring exerts is $F_s = -kx$. The potential energy stored in a block-spring system is

$$U_f = -\int_{x_i}^{x_f} (-kx)\, dx + U_i = \tfrac{1}{2}kx_f^2 - \tfrac{1}{2}kx_i^2 + U_i$$

As mentioned earlier, we can choose the configuration representing the zero of potential energy arbitrarily. Let us choose $U_i = 0$ when the block is at the position $x_i = 0$. Then,

$$U_f = \tfrac{1}{2}kx_f^2 - \tfrac{1}{2}kx_i^2 + U_i = \tfrac{1}{2}kx_f^2 - 0 + 0 \quad \rightarrow \quad U_f = U_s = \tfrac{1}{2}kx^2$$

which is the potential energy function we have already recognized (see Eq. 7.9) for a spring that obeys Hooke's law.

In the preceding discussion, we have seen how to find a potential energy function if we know the force function. Let us now turn this process around. Suppose we know the potential energy function. Can we find the force function? We start from the basic definition of work done by a conservative force for an infinitesimal displacement $d\vec{\mathbf{r}} = dx\hat{\mathbf{i}}$ in the x direction:

$$dW = \vec{\mathbf{F}} \cdot d\vec{\mathbf{r}} = \vec{\mathbf{F}} \cdot dx\hat{\mathbf{i}} = F_x \, dx = -\, dU$$

This equation can be rewritten as

$$F_x = -\frac{dU}{dx} \qquad [7.16]$$

■ **Finding the force between members of the system from the potential energy of the system**

In general, **the conservative force acting between parts of a system equals the negative derivative of the potential energy associated with that system.**[1]

In the case of an object located a distance y above some reference point, the gravitational potential energy function is given by $U_g = mgy$, and it follows from Equation 7.16 that (considering the y direction rather than x)

$$F_y = -\frac{dU_g}{dy} = -\frac{d}{dy}(mgy) = -mg$$

which is the correct expression for the vertical component of the gravitational force.

---

[1]In three dimensions, the appropriate expression is

$$\vec{\mathbf{F}} = -\hat{\mathbf{i}}\frac{\partial U}{\partial x} - \hat{\mathbf{j}}\frac{\partial U}{\partial y} - \hat{\mathbf{k}}\frac{\partial U}{\partial z}$$

where $\partial U/\partial x$ and so on are *partial derivatives*. In the language of vector calculus, $\vec{\mathbf{F}}$ equals the negative of the *gradient* of the scalar potential energy function $U(x, y, z)$.

7.5 THE NONISOLATED SYSTEM IN STEADY STATE

We have seen two approaches related to systems so far. In a nonisolated system, the energy stored in the system changes due to transfers across the boundaries of the system. Therefore, nonzero terms occur on both sides of the continuity equation for energy, $\Delta E_{\text{system}} = \Sigma T$. For an isolated system, no energy transfer takes place across the boundary, so the right-hand side of the continuity equation is zero; that is, $\Delta E_{\text{system}} = 0$.

Another possibility exists that we have not yet addressed. It is possible for no change to occur in the energy of the system even though nonzero terms are present on the right-hand side of the continuity equation, $0 = \Sigma T$. This situation can only occur if the rate at which energy is entering the system is equal to the rate at which it is leaving. In this case, the system is in steady state under the effects of two or more competing transfers, which we describe as a **nonisolated system in steady state.** The system is nonisolated because it is interacting with the environment, but it is in steady state because the system energy remains constant.

We could identify a number of examples of this type of situation. First, consider your home as a nonisolated system. Ideally, you would like to keep the temperature of your home constant for the comfort of the occupants. Therefore, your goal is to keep the internal energy in the home fixed.

The energy transfer mechanisms for the home are numerous, as we can see in Figure 7.11. Solar electromagnetic radiation is absorbed by the roof and walls of the home and enters the home through the windows. Energy enters by electrical transmission to operate electrical devices. Leaks in the walls, windows, and doors allow warm or cold air to enter and leave, carrying energy across the boundary of the system by matter transfer. Matter transfer also occurs if any devices in the home operate from natural gas because energy is carried in with the gas. Energy transfer by heat occurs through the walls, windows, floor, and roof as a result of temperature differences between the inside and outside of the home. Therefore, we have a variety of transfers, but the energy in the home remains constant in the idealized case. In reality, the home is a system in *quasi-steady state* because some small temperature variations actually occur over a 24-h period, but we can imagine an idealized situation that conforms to the nonisolated system in steady-state model.

As a second example, consider the Earth and its atmosphere as a system. Because this system is located in the vacuum of space, the only possible types of

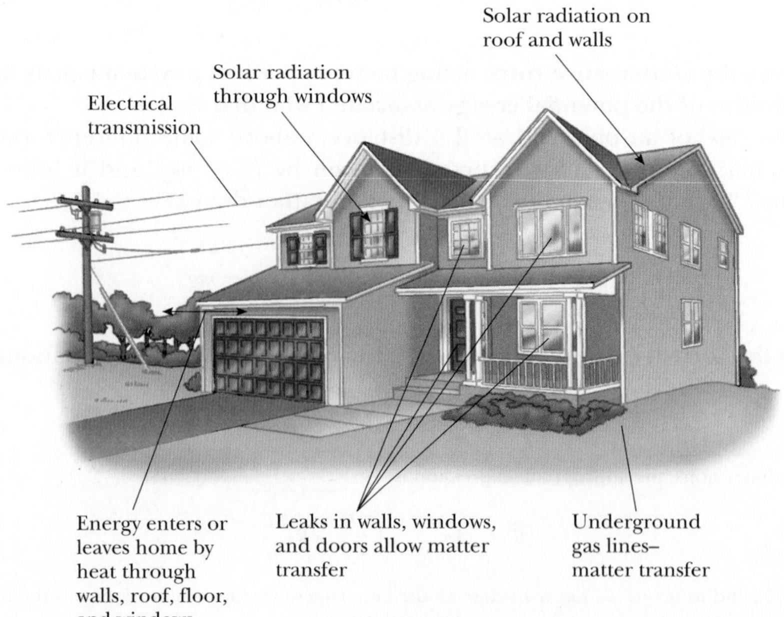

FIGURE 7.11 Energy enters and leaves a home by several mechanisms. The home can be modeled as a nonisolated system in steady state.

energy transfers are those that involve no contact between the system and external molecules in the environment. As mentioned in the footnote on page 172, only two types of transfer do not depend on contact with molecules: work done by field forces and electromagnetic radiation. The Earth–atmosphere system exchanges energy with the rest of the Universe only by means of electromagnetic radiation (ignoring work done by field forces and ignoring some small matter transfer as a result of cosmic ray particles and meteoroids entering the system and spacecraft leaving the system!). The primary input radiation is that from the Sun, and the output radiation is primarily infrared radiation emitted from the atmosphere and the ground. Ideally, these transfers are balanced so that the Earth maintains a constant temperature. In reality, however, the transfers are not *exactly* balanced, so the Earth is in quasi-steady state; measurements of the temperature show that it does appear to be changing. The change in temperature is very gradual and currently appears to be in the positive direction. This change is the essence of the social issue of global warming. (See Context 5, beginning on page 497.)

The human body as a nonisolated system

If we consider a time interval of several days, the human body can be modeled as another nonisolated system in steady state. If the body is at rest at the beginning and end of the time interval, there is no change in kinetic energy. Assuming that no major weight gain or loss occurs during this time interval, the amount of potential energy stored in the body as food in the stomach and fat remains constant on the average. If no fevers are experienced during this time interval, the internal energy of the body remains constant. Therefore, the change in the energy of the system is zero. Energy transfer methods during this time interval include work (you apply forces on objects which move), heat (your body is warmer than the surrounding air), matter transfer (breathing, eating), mechanical waves (you speak and hear), and electromagnetic radiation (you see, as well as absorb and emit radiation from your skin).

7.6 POTENTIAL ENERGY FOR GRAVITATIONAL AND ELECTRIC FORCES

Earlier in this chapter we introduced the concept of gravitational potential energy, that is, the energy associated with a system of objects interacting via the gravitational force. We emphasized that the gravitational potential energy function, Equation 7.2, is valid only when the object of mass m is near the Earth's surface. We would like to find a more general expression for the gravitational potential energy that is valid for all separation distances. Because the free-fall acceleration g varies as $1/r^2$, it follows that the general dependence of the potential energy function of the system on separation distance is more complicated than our simple expression, Equation 7.2.

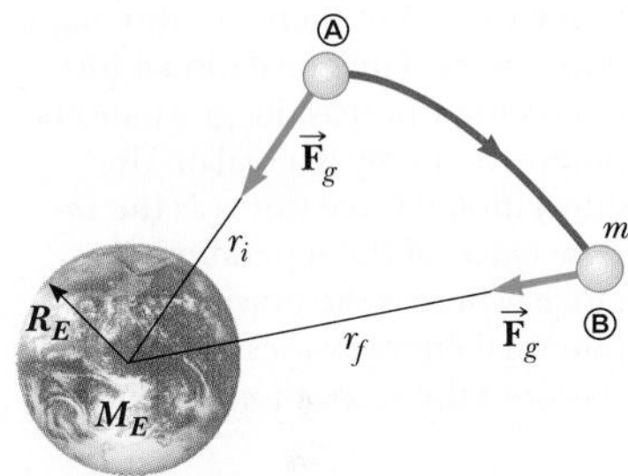

FIGURE 7.12 As a particle of mass m moves from Ⓐ to Ⓑ above the Earth's surface, the potential energy of the particle–Earth system, given by Equation 7.19, changes because of the change in the particle–Earth separation distance r from r_i to r_f.

Consider a particle of mass m moving between two points Ⓐ and Ⓑ above the Earth's surface as in Figure 7.12. The gravitational force on the particle due to the Earth, first introduced in Section 5.6, can be written in vector form as

$$\vec{\mathbf{F}}_g = -\frac{GM_E m}{r^2}\hat{\mathbf{r}} \qquad [7.17]$$

where $\hat{\mathbf{r}}$ is a unit vector directed from the Earth toward the particle and the negative sign indicates that the force is downward toward the Earth. This expression shows that the gravitational force depends on the radial coordinate r. Furthermore, the gravitational force is conservative. Equation 7.15 gives

$$U_f = -\int_{r_i}^{r_f} F(r)\,dr + U_i = GM_E m\int_{r_i}^{r_f}\frac{dr}{r^2} + U_i = GM_E m\left(-\frac{1}{r}\right)\Bigg|_{r_i}^{r_f} + U_i$$

or

$$U_f = -GM_E m\left(\frac{1}{r_f} - \frac{1}{r_i}\right) + U_i \qquad [7.18]$$

PITFALL PREVENTION 7.4

What is r? In Section 5.5, we discussed the gravitational force between two *particles*. In Equation 7.17, we present the gravitational force between a particle and an extended object, the Earth. We could also express the gravitational force between two extended objects, such as the Earth and the Sun. In these kinds of situations, remember that r is measured *between the centers* of the objects. Be sure *not* to measure r from the surface of the Earth.

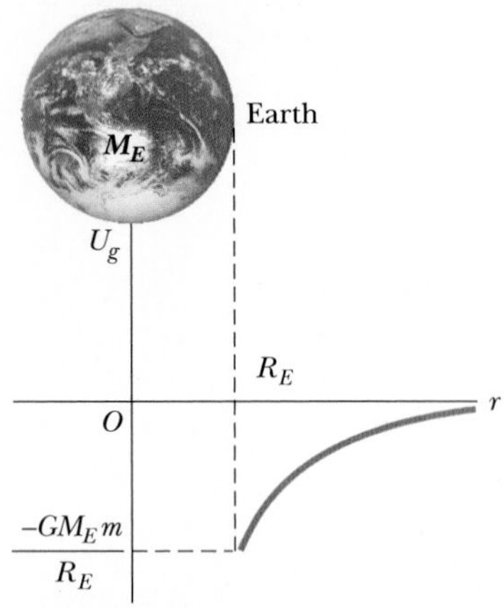

FIGURE 7.13 Graph of the gravitational potential energy U_g versus r for a particle above the Earth's surface. The potential energy of the system goes to zero as r approaches infinity.

PITFALL PREVENTION 7.5

GRAVITATIONAL POTENTIAL ENERGY Be careful! Equation 7.20 looks similar to Equation 5.14 for the gravitational force, but there are two major differences. The gravitational force is a vector, whereas the gravitational potential energy is a scalar. The gravitational force varies as the *inverse square* of the separation distance, whereas the gravitational potential energy varies as the simple *inverse* of the separation distance.

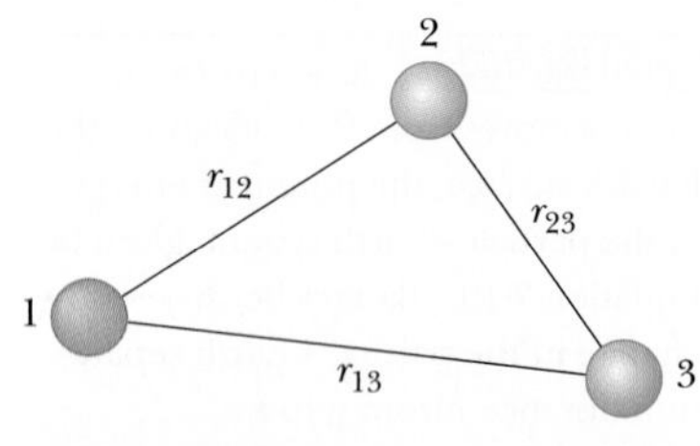

FIGURE 7.14 Three interacting particles.

As always, the choice of a reference point for the potential energy is completely arbitrary. It is customary to locate the reference point where the force is zero. Letting $U_i \rightarrow 0$ as $r_i \rightarrow \infty$, we obtain the important result

$$U_g = -\frac{GM_E m}{r} \quad [7.19]$$

for separation distances $r > R_E$, the radius of the Earth. Because of our choice of the reference point for zero potential energy, the function U_g is always negative (Fig. 7.13).

Although Equation 7.19 was derived for the particle–Earth system, it can be applied to *any* two particles. For *any pair* of particles of masses m_1 and m_2 separated by a distance r, the gravitational force of attraction is given by Equation 5.11 and the gravitational potential energy of the system of two particles is

$$U_g = -\frac{Gm_1 m_2}{r} \quad [7.20]$$

This expression also applies to larger objects *if their mass distributions are spherically symmetric,* as first shown by Newton. In this case, r is measured between the centers of the spherical objects.

Equation 7.20 shows that the gravitational potential energy for any pair of particles varies as $1/r$ (whereas the force between them varies as $1/r^2$). Furthermore, the potential energy is *negative* because the force is attractive and we have chosen the potential energy to be zero when the particle separation is infinity. Because the force between the particles is attractive, we know that an external agent must do positive work to increase the separation between the two particles. The work done by the external agent produces an increase in the potential energy as the two particles are separated. That is, U_g becomes less negative as r increases.

We can extend this concept to three or more particles. In this case, the total potential energy of the system is the sum over all *pairs* of particles. Each pair contributes a term of the form given by Equation 7.20. For example, if the system contains three particles, as in Figure 7.14, we find that

$$U_{\text{total}} = U_{12} + U_{13} + U_{23} = -G\left(\frac{m_1 m_2}{r_{12}} + \frac{m_1 m_3}{r_{13}} + \frac{m_2 m_3}{r_{23}}\right) \quad [7.21]$$

The absolute value of U_{total} represents the work needed to separate all three particles by an infinite distance.

■ Thinking Physics 7.2

Why is the Sun hot?

Reasoning The Sun was formed when a cloud of gas and dust coalesced, because of gravitational attraction, into a massive astronomical object. Let us define this cloud as our system and model the gas and dust as particles. Initially, the particles of the system were widely scattered, representing a large amount of gravitational potential energy. As the particles moved together to form the Sun, the gravitational potential energy of the system decreased. According to the isolated system model, this potential energy was transformed to kinetic energy as the particles fell toward the center. As the speeds of the particles increased, many collisions occurred between particles, randomizing their motion and transforming the kinetic energy to internal energy, which represented an increase in temperature. As the particles came together, the temperature rose to a point at which nuclear reactions occurred. These reactions release huge amounts of energy that maintain the high temperature of the Sun. This process has occurred for every star in the Universe. ■

EXAMPLE 7.6 **The Change in Potential Energy**

A particle of mass m is displaced through a small vertical distance Δy near the Earth's surface. Show that the general expression for the change in gravitational potential energy reduces to the familiar relationship $\Delta U_g = mg\,\Delta y$.

Solution We can express Equation 7.18 in the form

$$\Delta U_g = -GM_E m\left(\frac{1}{r_f} - \frac{1}{r_i}\right) = GM_E m\left(\frac{r_f - r_i}{r_i r_f}\right)$$

If both the initial and final positions of the particle are close to the Earth's surface, $r_f - r_i = \Delta y$ and $r_i r_f \approx R_E^2$. (Recall that r is measured from the center of the Earth.) The change in potential energy therefore becomes

$$\Delta U_g \approx \frac{GM_E m}{R_E^2}\,\Delta y = F_g \Delta y = mg\,\Delta y$$

where we have used Equation 7.17 to express $GM_E m/R_E^2$ as the magnitude of the gravitational force F_g on an object of mass m at the Earth's surface and then Equation 4.5 to express F_g as mg.

In Chapter 5, we discussed the electrostatic force between two point particles, which is given by Coulomb's law,

$$F_e = k_e \frac{q_1 q_2}{r^2} \qquad [7.22]$$

Because this expression looks so similar to Newton's law of universal gravitation, we would expect that the generation of a potential energy function for this force would proceed in a similar way. That is indeed the case, and this procedure results in the **electric potential energy** function,

$$U_e = k_e \frac{q_1 q_2}{r} \qquad [7.23]$$

As with the gravitational potential energy, the electric potential energy is defined as zero when the charges are infinitely far apart. Comparing this expression with that for the gravitational potential energy, we see the obvious differences in the constants and the use of charges instead of masses, but there is one more difference. The gravitational expression has a negative sign, but the electrical expression doesn't. For systems of objects that experience an attractive force, the potential energy decreases as the objects are brought closer together. Because we have defined zero potential energy at infinite separation, all real separations are finite and the energy must decrease from a value of zero. Therefore, all potential energies for systems of objects that attract must be negative. In the gravitational case, attraction is the only possibility. The constant, the masses, and the separation distance are all positive, so the negative sign must be included explicitly, as it is in Equation 7.20.

The electric force can be either attractive or repulsive. Attraction occurs between charges of opposite sign. Therefore, for the two charges in Equation 7.23, one is positive and one is negative if the force is attractive. The product of the charges provides the negative sign for the potential energy mathematically, and we do not need an explicit negative sign in the potential energy expression. In the case of charges with the same sign, either a product of two negative charges or two positive charges will be positive, leading to a positive potential energy. This conclusion is reasonable because to cause repelling particles to move together from infinite separation requires work to be done on the system, so the potential energy increases.

7.7 ENERGY DIAGRAMS AND STABILITY OF EQUILIBRIUM

The motion of a system can often be understood qualitatively by analyzing a graphical representation of the system's potential energy curve. An **energy diagram** shows the potential energy of the system as a function of the position of one of the members of the system (or as a function of the separation distance between two members of the system). Consider the potential energy function for the block–spring system, given by $U_s = \frac{1}{2}kx^2$. This function is plotted versus x in Active Figure 7.15a.

The spring force is related to U_s through Equation 7.16:

$$F_s = -\frac{dU_s}{dx} = -kx$$

That is, the force is equal to the negative of the *slope* of the U_s versus x curve. When the block is placed at rest at the equilibrium position ($x = 0$), where $F_s = 0$, it will remain there unless some external force acts on it. If the spring in Active Figure 7.15b is stretched to the right from equilibrium, x is positive and the slope dU_s/dx is positive; therefore, F_s is negative and the block accelerates back toward $x = 0$. If the spring is compressed, x is negative and the slope is negative; therefore, F_s is positive and again the block accelerates toward $x = 0$.

From this analysis, we conclude that the $x = 0$ position is one of **stable equilibrium.** That is, any movement away from this position results in a force directed back toward $x = 0$. (We described this type of force in Chapter 6 as a *restoring force.*) In general, **positions of stable equilibrium correspond to those values of x for which $U(x)$ has a relative minimum value on an energy diagram.**

From Active Figure 7.15 we see that if the block is given an initial displacement x_{max} and is released from rest, the total initial energy of the system is the potential energy stored in the spring, given by $\frac{1}{2}kx_{max}^2$. As motion commences, the system acquires kinetic energy and loses an equal amount of potential energy. From an energy viewpoint, the energy of the system cannot exceed $\frac{1}{2}kx_{max}^2$; therefore, the block must stop at the points $x = \pm x_{max}$ and, because of the spring force, accelerate toward $x = 0$. The block oscillates between the two points $x = \pm x_{max}$, called the *turning points.* The block cannot be farther from equilibrium than $\pm x_{max}$ because the potential energy of the system beyond these points would be larger than the total energy, an

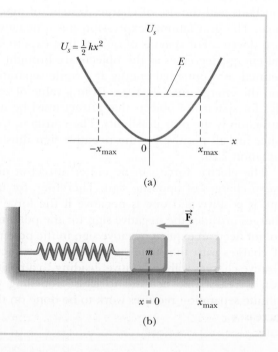

ACTIVE FIGURE 7.15
(a) Potential energy as a function of x for the block–spring system shown in part (b). The block oscillates between the turning points, which have the coordinates $x = \pm x_{max}$. The restoring force exerted by the spring always acts toward $x = 0$, the position of stable equilibrium.

PhysicsNow™ Log into PhysicsNow at **www.pop4e.com** and go to Active Figure 7.15 to observe the block oscillate between its turning points and trace the corresponding points on the potential energy curve for varying values of k.

impossible situation in classical physics. Because there is no transformation of mechanical energy to internal energy (no friction), the block oscillates between $-x_{max}$ and $+x_{max}$ forever. (We shall discuss these oscillations further in Chapter 12.)

Now consider an example in which the curve of U versus x is as shown in Figure 7.16. In this case, $F_x = 0$ at $x = 0$, and so the particle is in equilibrium at this point. This point, however, is a position of **unstable equilibrium** for the following reason. Suppose the particle is displaced to the *right* of the origin. Because the slope is negative for $x > 0$, $F_x = -dU/dx$ is positive and the particle accelerates away from $x = 0$. Now suppose the particle is displaced to the *left* of the origin. In this case, the force is negative because the slope is positive for $x < 0$, and the particle again accelerates away from the equilibrium position. The $x = 0$ position in this situation is called a position of unstable equilibrium because, for any displacement from this point, the force pushes the particle farther away from equilibrium. In fact, the force pushes the particle toward a position representing lower potential energy of the system. A ball placed on the top of an inverted spherical bowl is in a position of unstable equilibrium. If the ball is displaced slightly from the top and released, it will roll off the bowl. In general, **positions of unstable equilibrium correspond to those values of x for which $U(x)$ has a relative maximum value on an energy diagram.**[2]

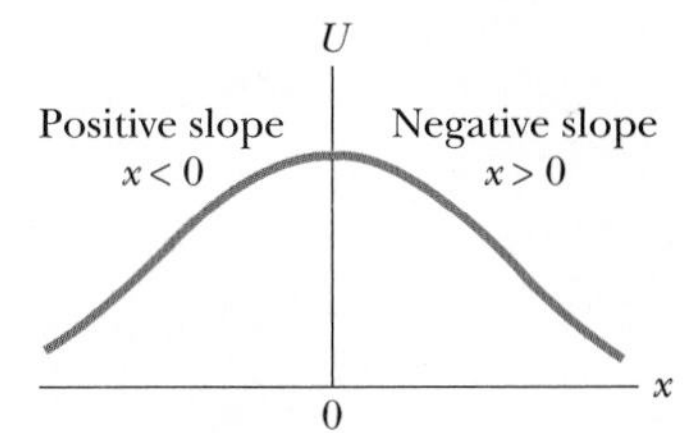

FIGURE 7.16 A plot of U versus x for a particle that has a position of unstable equilibrium, located at $x = 0$. For any finite displacement of the particle, the force on the particle is directed away from $x = 0$.

Finally, a situation may arise in which U is constant over some region, and hence $F = 0$. A point in this region is called a position of **neutral equilibrium.** Small displacements from this position produce neither restoring nor disrupting forces. A ball lying on a flat horizontal surface is an example of an object in neutral equilibrium.

7.8 POTENTIAL ENERGY IN FUELS

CONTEXT CONNECTION

Fuel represents a storage mechanism for potential energy to be used to make a vehicle move. The standard fuel for automobiles for several decades has been *gasoline.* Gasoline is refined from crude oil that is present in the Earth. This oil represents the decay products of plant life that existed on the Earth, primarily from 100 to 600 million years ago. The source of energy in crude oil is hydrocarbons produced from molecules in the ancient plants.

The primary chemical reactions occurring in an internal combustion engine involve the oxidation of carbon and hydrogen:

$$C + O_2 \rightarrow CO_2$$

$$4H + O_2 \rightarrow 2H_2O$$

Both reactions release energy that is used to operate the automobile.

Notice the final products in these reactions. One is water, which is not harmful to the environment. Carbon dioxide, however, contributes to the greenhouse effect, which leads to global warming, which we will study in Context 5. The incomplete combustion of carbon and oxygen can form CO, carbon monoxide, which is a poisonous gas. Because air contains other elements besides oxygen, other harmful emission products, such as oxides of nitrogen, exist.

The amount of potential energy stored in a fuel and available from the fuel is typically called the *heat of combustion,* even though this term is a misuse of the word *heat.* For automotive gasoline, this value is about 44 MJ/kg. Because the efficiency of the engine is not 100%, only part of this energy eventually finds its way into kinetic energy of the car. We will study efficiencies of engines in Context 5.

Another common fuel is *diesel fuel.* The heat of combustion for diesel fuel is 42.5 MJ/kg, slightly lower than that for gasoline. Diesel engines, however, operate at a higher efficiency than gasoline engines, so they can extract a larger percentage of the available energy.

[2] You can mathematically test whether an extreme of U represents stable or unstable equilibrium by examining the sign of d^2U/dx^2.

A number of additional fuels have been developed to operate internal combustion engines with minimal modifications. They are described briefly below.

Ethanol

Ethanol is the most widely used alternative fuel and is used primarily on commercial fleet vehicles. It is an alcohol made from such crops as corn, wheat, and barley. Because these crops can be grown, ethanol is renewable. The use of ethanol reduces carbon monoxide and carbon dioxide emissions compared with the use of normal gasoline.

Ethanol is mixed with gasoline to form the following mixtures:

E10: 10% ethanol, 90% gasoline

E85: 85% ethanol, 15% gasoline

The energy content of E85 is about 70% of that for gasoline, so the miles per gallon ratio will be lower than that for a vehicle powered by straight gasoline. On the other hand, the renewable nature of ethanol counteracts this disadvantage significantly.

Biodiesel

Biodiesel fuel is formed by a chemical reaction between alcohol and oils from field crops as well as vegetable oil, fat, and grease from commercial sources. Pacific Biodiesel in Hawaii makes biodiesel from used restaurant cooking oil, providing a usable fuel as well as diverting this used oil from landfills.

Biodiesel is available in the following forms:

B20: 20% biodiesel, 80% gasoline

B100: 100% biodiesel

B100 is nontoxic and biodegradable. The use of biodiesel reduces environmentally harmful tailpipe emissions significantly. Furthermore, tests have shown that the emission of cancer-causing particulate matter is reduced by 94% with the use of pure biodiesel.

The energy content of B100 is about 90% of that for conventional diesel. As with ethanol, the renewable nature of biodiesel counteracts this disadvantage significantly.

Natural Gas

Natural gas is a fossil fuel, originating from gas wells or as a by-product of the refining process for crude oil. It is primarily methane (CH_4), with smaller amounts of nitrogen, ethane, propane, and other gases. It burns cleanly and generates much lower amounts of harmful tailpipe emissions than gasoline. Natural gas vehicles are used in many fleets of buses, delivery trucks, and refuse haulers.

Although ethanol and biodiesel mixtures can be used in conventional engines with minimal modifications, a natural gas engine is much more heavily modified. In addition, the gas must be carried on board the vehicle in one of two ways that require higher-level technology than a simple fuel tank. One possibility is to liquefy the gas, requiring a well-insulated storage container to keep the gas at $-190°C$. The other possibility is to compress the gas to about 200 times atmospheric pressure and carry it in the vehicle in a high-pressure storage tank.

The energy content of natural gas is 48 MJ/kg, a bit higher than that for gasoline. Note that natural gas, like gasoline, is *not* a renewable source.

Propane

Propane is available commercially as liquefied petroleum gas, which is actually a mixture of propane, propylene, butane, and butylenes. It is a by-product of natural gas processing and refining of crude oil. Propane is the most widely accessible alternative fuel, with fueling facilities in all states of the United States.

Tailpipe emissions for propane-fueled vehicles are significantly lower than those for gasoline-powered vehicles. Tests show that carbon monoxide is reduced by 30% to 90%.

As with natural gas, high-pressure tanks are necessary to carry the fuel. In addition propane is a nonrenewable resource. The energy content of propane is 46 MJ/kg, slightly higher than that of gasoline.

Electric Vehicles

In the Context introduction before Chapter 2, we discussed the electric cars that were on the roadways in the early part of the twentieth century. As mentioned, these electric cars virtually disappeared around the 1920s due to several factors. One was that oil was plentiful during the twentieth century and there was little incentive to operate vehicles on anything other than gasoline or diesel.

In the early 1970s, difficulties arose with regard to the availability of oil from the Middle East, leading to shortages at gas stations. At this time, interest arose anew in electric-powered vehicles. An early attempt to market a new electric vehicle was the Electrovette, an electric version of the Chevrolet Chevette.

Although the oil crisis eased somewhat, political instabilities in the Middle East created uncertainty in the availability of oil and interest in electric cars continued, albeit on a small scale. In the late 1980s, General Motors developed a prototype called the Impact, an electric car that could accelerate from 0 to 60 in 8 s and had a drag coefficient of 0.19, much lower than that of traditional cars. The Impact was the hit of the 1990 Los Angeles Auto Show. In the 1990s, the Impact became commercially available as the EV1.

Although the EV1 was a very successful electric car in terms of quality and performance, it was difficult to convince consumers that oil was in short supply and not many consumers chose to drive the car. A few other manufacturers also developed electric cars, and consumer response was similar. Two major disadvantages of electric cars were the limited range, 70 to 100 mi, on a single charging of the batteries and the several hours of time required to recharge the batteries. These difficulties, as well as a federal court ruling that relaxed emissions standards, led General Motors to cancel the EV1 program in 2001. An additional contribution to the demise of contemporary electric cars is the development of hybrid electric vehicles, which will be discussed in the Context Conclusion.

SUMMARY

PhysicsNow™ Take a practice test by logging into PhysicsNow at **www.pop4e.com** and clicking on the Pre-Test link for this chapter.

If a particle of mass m is elevated a distance y from a reference point $y = 0$ near the Earth's surface, the **gravitational potential energy** of the particle–Earth system can be defined as

$$U_g \equiv mgy \qquad [7.2]$$

The **total mechanical energy of a system** is defined as the sum of the kinetic energy and potential energy:

$$E_{\text{mech}} \equiv K + U \qquad [7.8]$$

If no energy transfers occur across the boundary of the system, the system is modeled as an **isolated system.** In this model, the principle of **conservation of mechanical energy** states that the total mechanical energy of the system is constant if all of the forces in the system are conservative. For example, if a system involves gravitational forces,

$$K_f + U_{gf} = K_i + U_{gi} \qquad [7.7]$$

A force is **conservative** if the work it does on a particle is independent of the path the particle takes between two given points. A conservative force in mechanics does not cause a transformation of mechanical energy to internal energy. A force that does not meet these criteria is said to be **nonconservative.**

The **elastic potential energy** stored in a spring of force constant k is

$$U_s \equiv \tfrac{1}{2}kx^2 \qquad [7.9]$$

If some of the forces acting within a system are not conservative, the mechanical energy of the system does not remain constant. In the case of a common nonconservative force, a constant force of friction, the change in mechanical energy of the system when an object in the system moves is equal to the product of the kinetic friction force and the distance through which the object moves:

$$-f_k d = \Delta K + \Delta U \qquad [7.10]$$

This decrease in mechanical energy in the system is equal to the increase in internal energy:

$$\Delta E_{\text{int}} = f_k d \qquad [7.10]$$

A potential energy function U can be associated only with a conservative force. If a conservative force $\vec{\mathbf{F}}$ acts within a system on a particle that moves along the x axis from x_i to x_f, the potential energy function can be written

$$U_f = -\int_{x_i}^{x_f} F_x\, dx + U_i \qquad [7.15]$$

If we know the potential energy function, the component of a conservative force is given by the negative of the derivative of the potential energy function:

$$F_x = -\frac{dU}{dx} \qquad [7.16]$$

In some situations, a system may have energy crossing the boundary with no change in the energy stored in the system. In such a case, the energy input in any time interval equals the energy output, and we describe this system as a **nonisolated system in steady state.**

The **gravitational potential energy** associated with a system of two particles or uniform spherical distributions of mass separated by a distance r is

$$U_g = -\frac{Gm_1m_2}{r} \qquad [7.20]$$

where U_g is taken to approach zero as $r \rightarrow \infty$.

The **electric potential energy** associated with two charged particles separated by a distance r is

$$U_e = k_e\frac{q_1q_2}{r} \qquad [7.23]$$

where U_e is taken to approach zero as $r \rightarrow \infty$.

In an energy diagram, a point of **stable equilibrium** is one at which the potential energy is a minimum. A point of **unstable equilibrium** is one at which the potential energy is a maximum. **Neutral equilibrium** exists if the potential energy function is constant.

QUESTIONS

☐ = answer available in the *Student Solutions Manual and Study Guide.*

1. If the height of a playground slide is kept constant, will the length of the slide or the presence of bumps make any difference in the final speed of children playing on it? Assume that the slide is slick enough to be considered frictionless. Repeat this question assuming that friction is present.
2. Explain why the total energy of a system can be either positive or negative, whereas the kinetic energy is always positive.
3. One person drops a ball from the top of a building, while another person at the bottom observes its motion. Will these two people agree on the value of the gravitational potential energy of the ball–Earth system? On the change in potential energy? On the kinetic energy?
4. Discuss the changes in mechanical energy of an object–Earth system in (a) lifting the object, (b) holding the object at a fixed position, and (c) lowering the object slowly. Include the muscles in your discussion.
5. In Chapter 6, the work–kinetic energy theorem, $W = \Delta K$, was introduced. This equation states that work done on a system appears as a change in kinetic energy. It is a special-case equation, valid if there are no changes in any other type of energy, such as potential or internal. Give some examples in which work is done on a system but the change in energy of the system is not that of kinetic energy.
6. If three conservative forces and one nonconservative force act within a system, how many potential energy terms appear in the equation that describes the system?
7. If only one external force acts on a particle, (a) does it necessarily change the particle's kinetic energy? (b) Does it change the particle's velocity?
8. A driver brings an automobile to a stop. If the brakes lock so that the car skids, where is the original kinetic energy of the car and in what form is it after the car stops? Answer the same question for the case in which the brakes do not lock but the wheels continue to turn.
9. You ride a bicycle. In what sense is your bicycle solar-powered?
10. In an earthquake, a large amount of energy is "released" and spreads outward, potentially causing severe damage. In what form does this energy exist before the earthquake, and by what energy transfer mechanism does it travel?
11. A bowling ball is suspended from the ceiling of a lecture hall by a strong cord. The ball is drawn away from its equilibrium position and released from rest at the tip of the demonstrator's nose as shown in Figure Q7.11. Assuming

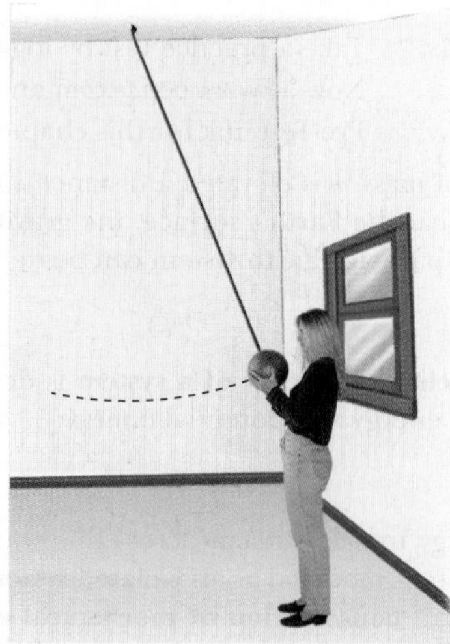

FIGURE Q7.11

that the demonstrator remains stationary, explain why the ball does not strike her on its return swing. Would this demonstrator be safe if the ball were given a push from its starting position at her nose?

12. A ball is thrown straight up into the air. At what position is its kinetic energy a maximum? At what position is the gravitational potential energy of the ball–Earth system a maximum?

13. A pile driver is a device used to drive objects into the Earth by repeatedly dropping a heavy weight on them. By how much does the energy of the pile driver–Earth system increase when the weight it drops is doubled? Assume that the weight is dropped from the same height each time.

14. Our body muscles exert forces when we lift, push, run, jump, and so forth. Are these forces conservative?

15. A block is connected to a spring that is suspended from the ceiling. Assuming that the block is set in motion and that air resistance can be ignored, describe the energy transformations that occur within the system consisting of the block, Earth, and spring.

16. Discuss the energy transformations that occur during the operation of an automobile.

17. What would the curve of U versus x look like if a particle were in a region of neutral equilibrium?

18. A ball rolls on a horizontal surface. Is the ball in stable, unstable, or neutral equilibrium?

PROBLEMS

1, 2, 3 = straightforward, intermediate, challenging

☐ = full solution available in the *Student Solutions Manual and Study Guide*

PhysicsNow™ = coached problem with hints available at www.pop4e.com

 = computer useful in solving problem

 = paired numerical and symbolic problems

 = biomedical application

Section 7.1 ▪ Potential Energy of a System

1. A 1 000-kg roller coaster train is initially at the top of a rise, at point Ⓐ. It then moves 135 ft, at an angle of 40.0° below the horizontal, to a lower point Ⓑ. (a) Choose the train at point Ⓑ to be the zero configuration for gravitational potential energy. Find the potential energy of the roller coaster–Earth system at points Ⓐ and Ⓑ, and the change in potential energy as the coaster moves. (b) Repeat part (a), setting the zero configuration when the train is at point Ⓐ.

2. A 400-N child is in a swing that is attached to ropes 2.00 m long. Find the gravitational potential energy of the child–Earth system relative to the child's lowest position when (a) the ropes are horizontal, (b) the ropes make a 30.0° angle with the vertical, and (c) the child is at the bottom of the circular arc.

3. A person with a remote mountain cabin plans to install her own hydroelectric plant. A nearby stream is 3.00 m wide and 0.500 m deep. Water flows at 1.20 m/s over the brink of a waterfall 5.00 m high. The manufacturer promises only 25.0% efficiency in converting the potential energy of the water–Earth system into electric energy. Find the power she can generate. (Large-scale hydroelectric plants, with a much larger drop, can be more efficient.)

Section 7.2 ▪ The Isolated System

4. At 11:00 A.M. on September 7, 2001, more than one million British school children jumped up and down for one minute. The curriculum focus of the "Giant Jump" was on earthquakes, but it was integrated with many other topics, such as exercise, geography, cooperation, testing hypotheses, and setting world records. Children built their own seismographs that registered local effects. (a) Find the mechanical energy released in the experiment. Assume that 1 050 000 children of average mass 36.0 kg jump 12 times each, raising their centers of mass by 25.0 cm each time and briefly resting between one jump and the next. The free-fall acceleration in Britain is 9.81 m/s^2. (b) Most of the energy is converted very rapidly into internal energy within the bodies of the children and the floors of the school buildings. Of the energy that propagates into the ground, most produces high frequency "microtremor" vibrations that are rapidly damped and cannot travel far. Assume that 0.01% of the energy is carried away by a long-range seismic wave. The magnitude of an earthquake on the Richter scale is given by

$$M = \frac{\log E - 4.8}{1.5}$$

where E is the seismic wave energy in joules. According to this model, what is the magnitude of the demonstration quake? It did not register above background noise overseas or on the seismograph of the Wolverton Seismic Vault, Hampshire.

5. A bead slides without friction around a loop-the-loop (Fig. P7.5). The bead is released from a height $h = 3.50R$. (a) What is its speed at point Ⓐ? (b) How large is the normal force on it if its mass is 5.00 g?

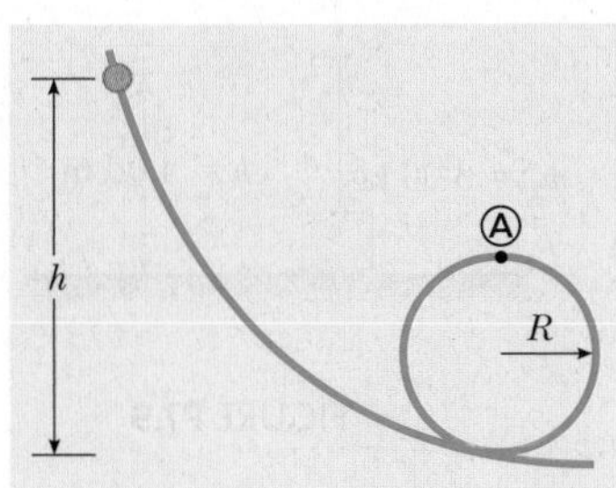

FIGURE P7.5

6. **Review problem.** A particle of mass 0.500 kg is shot from P as shown in Figure P7.6. The particle has an initial velocity $\vec{\mathbf{v}}_i$ with a horizontal component of 30.0 m/s. The particle rises to a maximum height of 20.0 m above P. Using the law of conservation of energy, determine (a) the vertical component of $\vec{\mathbf{v}}_i$, (b) the work done by the gravitational force on the particle during its motion from P to B, and (c) the horizontal and the vertical components of the velocity vector when the particle reaches B.

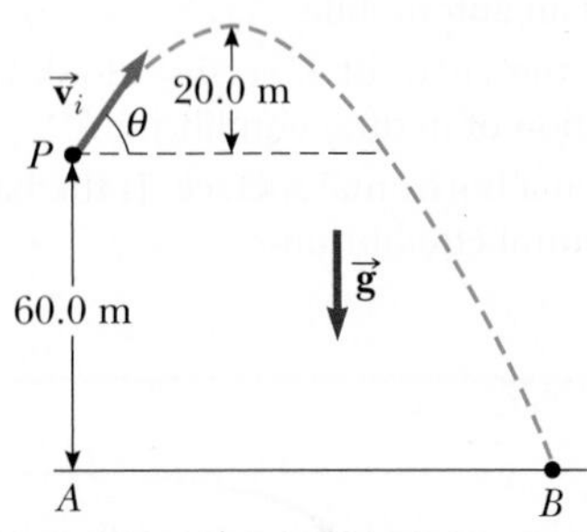

FIGURE P7.6

7. Dave Johnson, the bronze medallist at the 1992 Olympic decathlon in Barcelona, leaves the ground at the high jump with vertical velocity component 6.00 m/s. How far does his center of mass move up as he makes the jump?

8. A simple pendulum, which you will consider in detail in Chapter 12, consists of an object suspended by a string. The object is assumed to be a particle. The string, with its top end fixed, has negligible mass and does not stretch. In the absence of air friction, the system oscillates by swinging back and forth in a vertical plane. The string is 2.00 m long and makes an initial angle of 30.0° with the vertical. Calculate the speed of the particle (a) at the lowest point in its trajectory and (b) when the angle is 15.0°.

9. Two objects are connected by a light string passing over a light frictionless pulley as shown in Figure P7.9. The 5.00-kg object is released from rest. Using the principle of conservation of energy, (a) determine the speed of the 3.00-kg object just as the 5.00-kg object hits the ground, and (b) find the maximum height to which the 3.00-kg object rises.

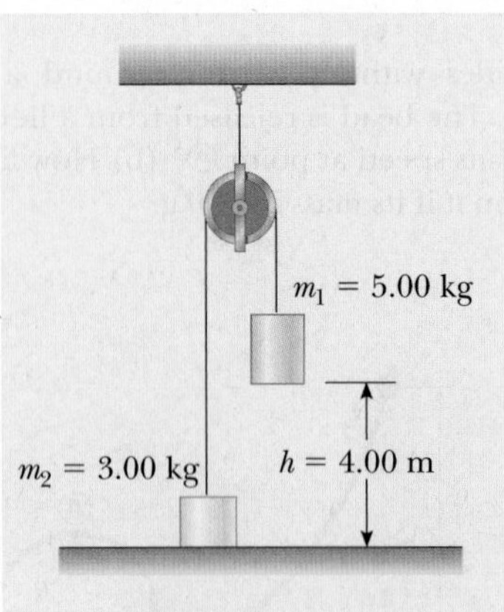

FIGURE P7.9

10. A particle of mass m = 5.00 kg is released from point Ⓐ and slides on the frictionless track shown in Figure P7.10. Determine (a) the particle's speed at points Ⓑ and Ⓒ and (b) the net work done by the gravitational force as the particle moves from Ⓐ to Ⓒ.

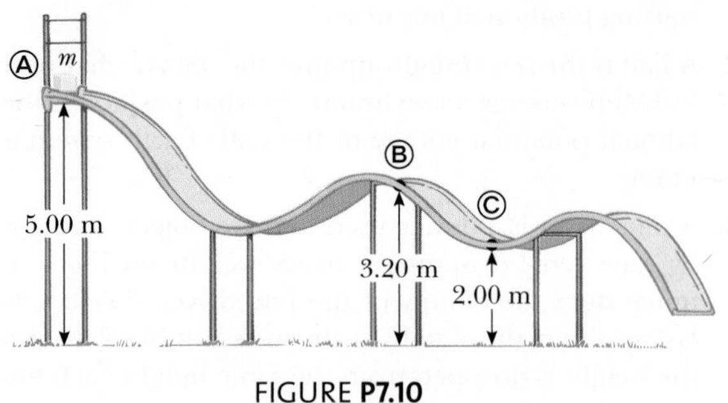

FIGURE P7.10

11. A circus trapeze consists of a bar suspended by two parallel ropes, each of length ℓ, allowing performers to swing in a vertical circular arc (Fig. P7.11). Suppose a performer with mass m holds the bar and steps off an elevated platform, starting from rest with the ropes at an angle θ_i with respect to the vertical. Suppose the size of the performer's body is small compared to the length ℓ, she does not pump the trapeze to swing higher, and air resistance is negligible. (a) Show that when the ropes make an angle θ with the vertical, the performer must exert a force

$$mg(3 \cos \theta - 2 \cos \theta_i)$$

so as to hang on. (b) Determine the angle θ_i for which the force needed to hang on at the bottom of the swing is twice the performer's weight.

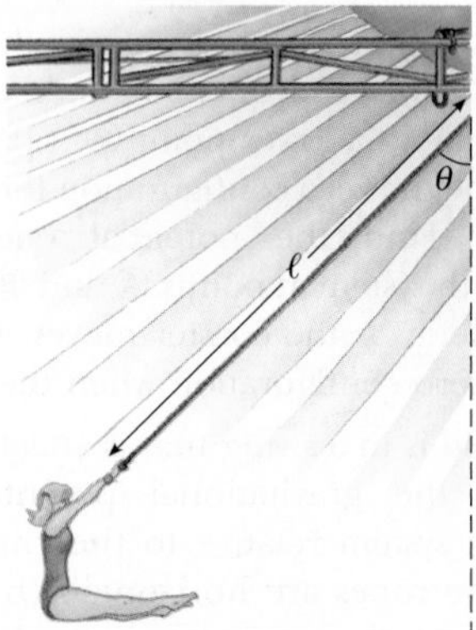

FIGURE P7.11

12. A light rigid rod is 77.0 cm long. Its top end is pivoted on a low-friction horizontal axle. The rod hangs straight down at rest, with a small massive ball attached to its bottom end. You strike the ball, suddenly giving it a horizontal velocity so that it swings around in a full circle. What minimum speed at the bottom is required to make the ball go over the top of the circle?

13. Columnist Dave Barry poked fun at the name "The Grand Cities" adopted by Grand Forks, North Dakota, and East Grand Forks, Minnesota. Residents of the prairie towns then named a sewage pumping station for him. At the Dave Barry Lift Station No. 16, untreated sewage is raised vertically by 5.49 m, in the amount 1 890 000 L each day. The waste has density 1 050 kg/m^3. It enters and leaves the

pump at atmospheric pressure, through pipes of equal diameter. (a) Find the output power of the lift station. (b) Assume that an electric motor continuously operating with average power 5.90 kW runs the pump. Find its efficiency. Barry attended the outdoor January dedication of the lift station and a festive potluck supper to which the residents of the different Grand Forks sewer districts brought casseroles, Jell-O salads, and "bars" (desserts).

Section 7.3 ■ Conservative and Nonconservative Forces

14. (a) Suppose a constant force acts on an object. The force does not vary with time nor with the position or the velocity of the object. Start with the general definition for work done by a force

$$W = \int_i^f \vec{\mathbf{F}} \cdot d\vec{\mathbf{r}}$$

and show that the force is conservative. (b) As a special case, suppose the force $\vec{\mathbf{F}} = (3\hat{\mathbf{i}} + 4\hat{\mathbf{j}})$N acts on a particle that moves from O to C in Figure P7.14. Calculate the work the force $\vec{\mathbf{F}}$ does on the particle as it moves along each one of the three paths OAC, OBC, and OC. (Your three answers should be identical.)

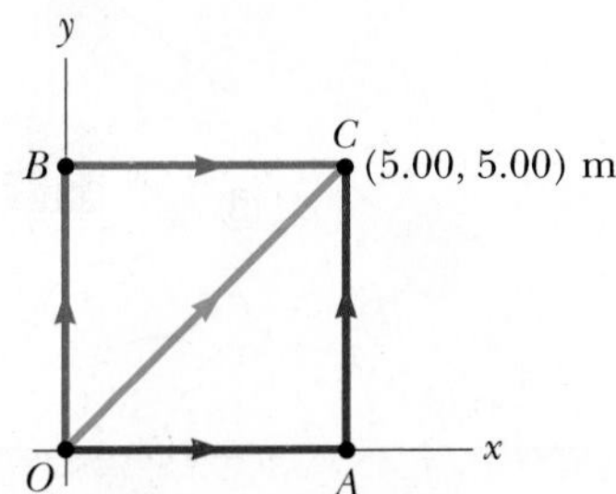

FIGURE P7.14 Problems 7.14 and 7.15.

15. A force acting on a particle moving in the xy plane is given by $\vec{\mathbf{F}} = (2y\hat{\mathbf{i}} + x^2\hat{\mathbf{j}})$N, where x and y are in meters. The particle moves from the origin to a final position having coordinates $x = 5.00$ m and $y = 5.00$ m as shown in Figure P7.14. Calculate the work done by $\vec{\mathbf{F}}$ on the particle as it moves along (a) OAC, (b) OBC, and (c) OC. (d) Is $\vec{\mathbf{F}}$ conservative or nonconservative? Explain.

16. An object of mass m starts from rest and slides a distance d down a frictionless incline of angle θ. While sliding, it contacts an unstressed spring of negligible mass as shown in Figure P7.16. The object slides an additional distance x as it is brought momentarily to rest by compression of the spring (of force constant k). Find the initial separation d between the object and the spring.

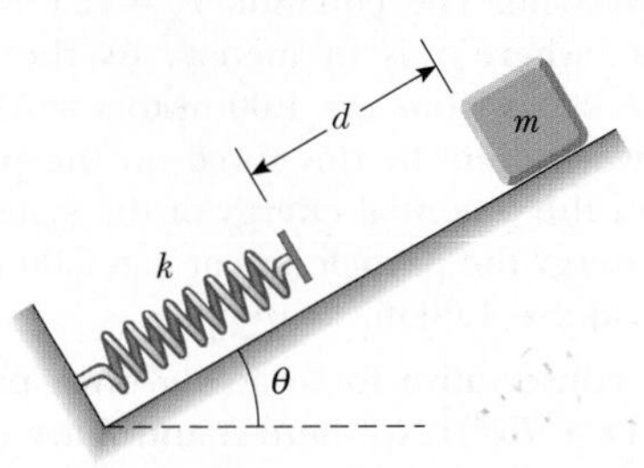

FIGURE P7.16

17. A block of mass 0.250 kg is placed on top of a light vertical spring of force constant 5 000 N/m and pushed downward so that the spring is compressed by 0.100 m. After the block is released from rest it travels upward and then leaves the spring. To what maximum height above the point of release does it rise?

18. A daredevil plans to bungee jump from a balloon 65.0 m above a carnival midway (Fig. P7.18). He will use a uniform elastic cord, tied to a harness around his body, to stop his fall at a point 10.0 m above the ground. Model his body as a particle. Assume that the cord has negligible mass and is described by Hooke's force law. In a preliminary test, hanging at rest from a 5.00-m length of the cord, he finds that his body weight stretches it by 1.50 m. He will drop from rest at the point where the top end of a longer section of the cord is attached to the stationary balloon. (a) What length of cord should he use? (b) What maximum acceleration will he experience?

FIGURE P7.18 Problems 7.18 and 7.64.

19. At time t_i, the kinetic energy of a particle is 30.0 J and the potential energy of the system to which it belongs is 10.0 J. At some later time t_f, the kinetic energy of the particle is 18.0 J. (a) If only conservative forces act on the particle, what are the potential energy and the total energy of the system at time t_f? (b) If the potential energy of the system at time t_f is 5.00 J, are there any nonconservative forces acting on the particle? Explain.

20. Heedless of danger, a child leaps onto a pile of old mattresses to use them as a trampoline. His motion between two particular points is described by the energy conservation equation

$$\tfrac{1}{2}(46.0\text{ kg})(2.40\text{ m/s})^2 + (46.0\text{ kg})(9.80\text{ m/s}^2)(2.80\text{ m} + x) = \tfrac{1}{2}(1.94 \times 10^4\text{ N/m})x^2$$

(a) Solve the equation for x. (b) Compose the statement of a problem, including data, for which this equation gives the solution. Identify the physical meaning of the value of x.

21. In her hand, a softball pitcher swings a ball of mass 0.250 kg around a vertical circular path of radius 60.0 cm before releasing it from her hand. The pitcher maintains a component of force on the ball of constant magnitude 30.0 N in the direction of motion around the complete path. The ball's speed at the top of the circle is 15.0 m/s. If she releases the ball at the bottom of the circle, what is its speed upon release?

22. In a needle biopsy, a narrow strip of tissue is extracted from a patient using a hollow needle. Rather than being

pushed by hand, to ensure a clean cut the needle can be fired into the patient's body by a spring. Assume that the needle has mass 5.60 g, the light spring has force constant 375 N/m, and the spring is originally compressed 8.10 cm to project the needle horizontally without friction. After the needle leaves the spring, the tip of the needle moves through 2.40 cm of skin and soft tissue, which exerts on it a resistive force of 7.60 N. Next, the needle cuts 3.50 cm into an organ, which exerts on it a backward force of 9.20 N. Find (a) the maximum speed of the needle and (b) the speed at which a flange on the back end of the needle runs into a stop that is set to limit the penetration to 5.90 cm.

23. **Physics Now™** The coefficient of friction between the 3.00-kg block and the surface in Figure P7.23 is 0.400. The system starts from rest. What is the speed of the 5.00-kg ball when it has fallen 1.50 m?

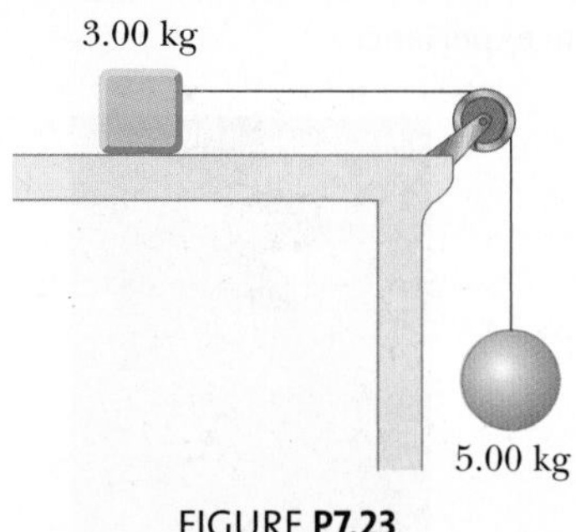

FIGURE **P7.23**

24. A boy in a wheelchair (total mass 47.0 kg) wins a race with a skateboarder. He has speed 1.40 m/s at the crest of a slope 2.60 m high and 12.4 m long. At the bottom of the slope his speed is 6.20 m/s. Assuming that air resistance and rolling resistance can be modeled as a constant friction force of 41.0 N, find the work he did in pushing forward on his wheels during the downhill ride.

25. A 5.00-kg block is set into motion up an inclined plane with an initial speed of 8.00 m/s (Fig. P7.25). The block comes to rest after traveling 3.00 m along the plane, which is inclined at an angle of 30.0° to the horizontal. For this motion, determine (a) the change in the block's kinetic energy, (b) the change in the potential energy of the block–Earth system, and (c) the friction force exerted on the block (assumed to be constant). (d) What is the coefficient of kinetic friction?

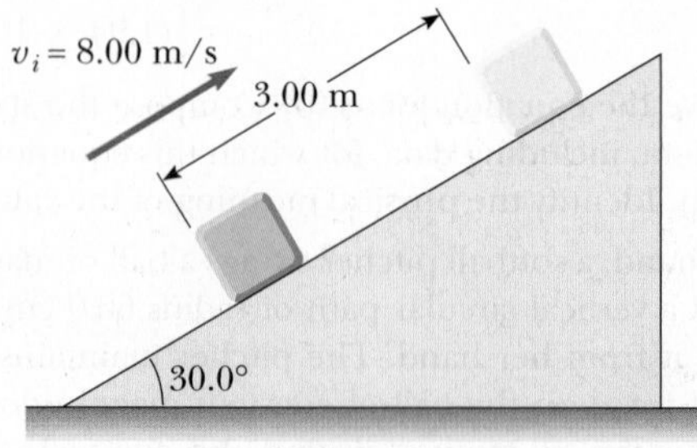

FIGURE **P7.25**

26. An 80.0-kg sky diver jumps out of a balloon at an altitude of 1 000 m and opens the parachute at an altitude of 200 m. (a) Assuming that the total retarding force on the diver is constant at 50.0 N with the parachute closed and constant at 3 600 N with the parachute open, find the speed of the sky diver when he lands on the ground. (b) Do you think the sky diver will be injured? Explain. (c) At what height should the parachute be opened so that the final speed of the sky diver when he hits the ground is 5.00 m/s? (d) How realistic is the assumption that the total retarding force is constant? Explain.

27. A toy cannon uses a spring to project a 5.30-g soft rubber ball. The spring is originally compressed by 5.00 cm and has a force constant of 8.00 N/m. When it is fired, the ball moves 15.0 cm through the horizontal barrel of the cannon and the barrel exerts a constant friction force of 0.032 0 N on the ball. (a) With what speed does the projectile leave the barrel of the cannon? (b) At what point does the ball have maximum speed? (c) What is this maximum speed?

28. A 50.0-kg block and 100-kg block are connected by a string as shown in Figure P7.28. The pulley is frictionless and of negligible mass. The coefficient of kinetic friction between the 50-kg block and incline is 0.250. Determine the change in the kinetic energy of the 50-kg block as it moves from Ⓐ to Ⓑ, a distance of 20.0 m.

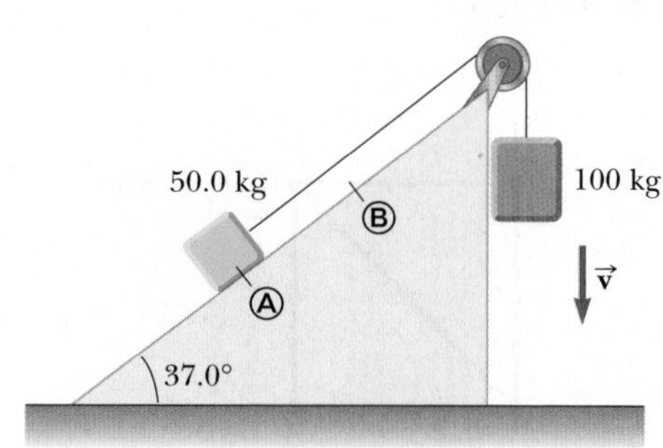

FIGURE **P7.28**

29. A 1.50-kg object is held 1.20 m above a relaxed massless vertical spring with a force constant of 320 N/m. The object is dropped onto the spring. (a) How far does it compress the spring? (b) How far does it compress the spring if the same experiment is performed on the Moon, where $g = 1.63$ m/s^2? (c) Repeat part (a), but now assume that a constant air-resistance force of 0.700 N acts on the object during its motion.

30. A 75.0-kg sky surfer is falling straight down with terminal speed 60.0 m/s. Determine the rate at which the sky surfer–Earth system is losing mechanical energy.

Section 7.4 ■ Conservative Forces and Potential Energy

31. **Physics Now™** A single conservative force acts on a 5.00-kg particle. The equation $F_x = (2x + 4)$ N describes the force, where x is in meters. As the particle moves along the x axis from $x = 1.00$ m to $x = 5.00$ m, calculate (a) the work done by this force on the particle, (b) the change in the potential energy of the system, and (c) the kinetic energy the particle has at $x = 5.00$ m if its speed is 3.00 m/s at $x = 1.00$ m.

32. A single conservative force acting on a particle varies as $\vec{\mathbf{F}} = (-Ax + Bx^2)\hat{\mathbf{i}}$ N, where A and B are constants and x is in meters. (a) Calculate the potential energy function

$U(x)$ associated with this force, taking $U = 0$ at $x = 0$. (b) Find the change in potential energy of the system and the change in kinetic energy of the particle as it moves from $x = 2.00$ m to $x = 3.00$ m.

33. **Physics Now™** The potential energy of a system of two particles separated by a distance r is given by $U(r) = A/r$, where A is a constant. Find the radial force $\vec{\mathbf{F}}$ that each particle exerts on the other.

34. A potential energy function for a two-dimensional force is of the form $U = 3x^3y - 7x$. Find the force that acts at the point (x, y).

Section 7.6 ▪ Potential Energy for Gravitational and Electric Forces

35. A satellite of the Earth has a mass of 100 kg and is at an altitude of 2.00×10^6 m. (a) What is the potential energy of the satellite–Earth system? (b) What is the magnitude of the gravitational force exerted by the Earth on the satellite? (c) What force does the satellite exert on the Earth?

36. How much energy is required to move a 1 000-kg object from the Earth's surface to an altitude twice the Earth's radius?

37. At the Earth's surface, a projectile is launched straight up at a speed of 10.0 km/s. To what height will it rise? Ignore air resistance.

38. A system consists of three particles, each of mass 5.00 g, located at the corners of an equilateral triangle with sides of 30.0 cm. (a) Calculate the potential energy describing the gravitational interactions internal to the system. (b) If the particles are released simultaneously, where will they collide?

Section 7.7 ▪ Energy Diagrams and Stability of Equilibrium

39. For the potential energy curve shown in Figure P7.39, (a) determine whether the force F_x is positive, negative, or zero at the five points indicated. (b) Indicate points of stable, unstable, and neutral equilibrium. (c) Sketch the curve for F_x versus x from $x = 0$ to $x = 9.5$ m.

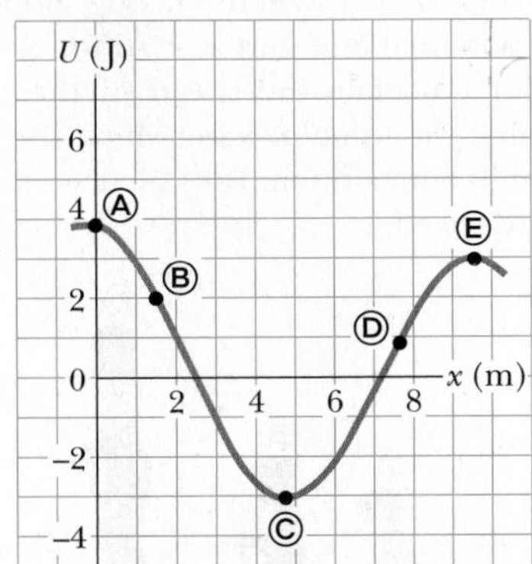

FIGURE P7.39

40. A particle moves along a line where the potential energy of its system depends on its position r as graphed in Figure P7.40. In the limit as r increases without bound, $U(r)$ approaches +1 J. (a) Identify each equilibrium position for this particle. Indicate whether each is a point of stable, unstable, or neutral equilibrium. (b) The particle will be bound if the total energy of the system is in what range? Now suppose the system has energy −3 J. Determine (c) the range of positions where the particle can be found, (d) its maximum kinetic energy, (e) the location where it has maximum kinetic energy, and (f) the *binding energy* of the system, that is, the additional energy that it would have to be given for the particle to move out to $r \to \infty$.

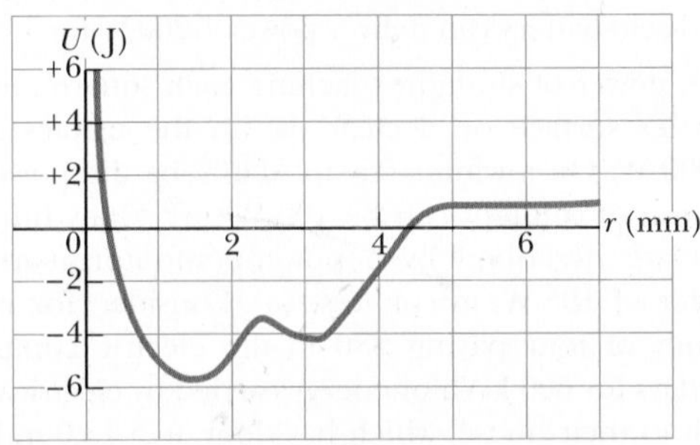

FIGURE P7.40

41. A particle of mass 1.18 kg is attached between two identical springs on a horizontal, frictionless tabletop. The springs have force constant k and each is initially unstressed. (a) The particle is pulled a distance x along a direction perpendicular to the initial configuration of the springs as shown in Figure P7.41. Show that the potential energy of the system is

$$U(x) = kx^2 + 2kL(L - \sqrt{x^2 + L^2})$$

(*Suggestion:* See Problem 6.52 in Chapter 6.) (b) Make a plot of $U(x)$ versus x and identify all equilibrium points. Assume that $L = 1.20$ m and $k = 40.0$ N/m. (c) If the particle is pulled 0.500 m to the right and then released, what is its speed when it reaches the equilibrium point $x = 0$?

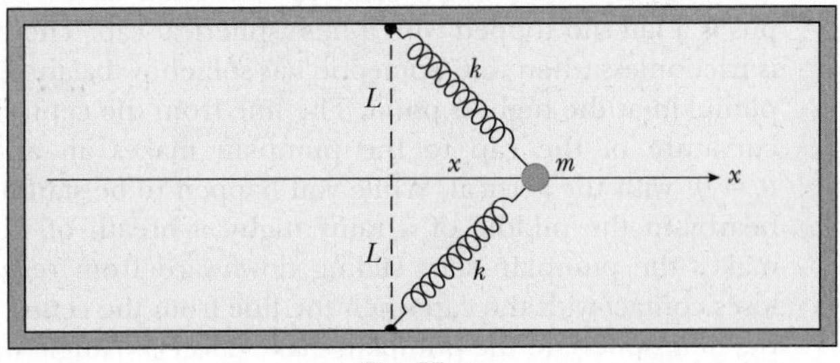

FIGURE P7.41

Section 7.8 ▪ Context Connection—Potential Energy in Fuels

42. **Review problem.** The mass of a car is 1 500 kg. The shape of the body is such that its aerodynamic drag coefficient is $D = 0.330$ and the frontal area is 2.50 m^2. Assuming that the drag force is proportional to v^2 and ignoring other sources of friction, calculate the power required to maintain a speed of 100 km/h as the car climbs a long hill sloping at 3.20°.

43. In considering the energy supply for an automobile, the energy per unit mass of the energy source is an important parameter. As the chapter text points out, the "heat of combustion" or stored energy per mass is quite similar for gasoline, ethanol, diesel fuel, cooking oil, methane, and propane. For a broader perspective, compare the energy per mass in joules per kilogram for gasoline, lead-acid batteries, hydrogen, and hay. Rank the four in order of increasing energy density and state the factor of increase between each one and the next. Hydrogen has "heat of combustion" 142 MJ/kg. For wood, hay, and dry vegetable matter in general, this parameter is 17 MJ/kg. A fully charged 16.0-kg lead-acid battery can deliver power 1 200 W for 1.0 hr.

44. The power of sunlight reaching each square meter of the Earth's surface on a clear day in the tropics is close to 1 000 W. On a winter day in Manitoba the power concentration of sunlight can be 100 W/m^2. Many human activities are described by a power-per-footprint-area on the order of 10^2 W/m^2 or less. (a) Consider, for example, a family of four paying $80 to the electric company every 30 days for 600 kWh of energy carried by electrical transmission to their house, which has floor area 13.0 m by 9.50 m. Compute the power-per-area measure of this energy use. (b) Consider a car 2.10 m wide and 4.90 m long traveling at 55.0 mi/h using gasoline having "heat of combustion" 44.0 MJ/kg with fuel economy 25.0 mi/gal. One gallon of gasoline has a mass of 2.54 kg. Find the power-per-area measure of the car's energy use. It can be similar to that of a steel mill where rocks are melted in blast furnaces. (c) Explain why direct use of solar energy is not practical for a conventional automobile.

Additional Problems

45. Make an order-of-magnitude estimate of your power output as you climb stairs. In your solution, state the physical quantities you take as data and the values you measure or estimate for them. Do you consider your peak power or your sustainable power?

46. Assume that you attend a state university that was founded as an agricultural college. Close to the center of the campus is a tall silo topped with a hemispherical cap. The cap is frictionless when wet. Someone has somehow balanced a pumpkin at the highest point. The line from the center of curvature of the cap to the pumpkin makes an angle $\theta_i = 0°$ with the vertical. While you happen to be standing nearby in the middle of a rainy night, a breath of wind makes the pumpkin start sliding downward from rest. It loses contact with the cap when the line from the center of the hemisphere to the pumpkin makes a certain angle with the vertical. What is this angle?

47. **Review problem.** The system shown in Figure P7.47 consists of a light inextensible cord, light frictionless pulleys, and blocks of equal mass. It is initially held at rest so that the blocks are at the same height above the ground. The blocks are then released. Find the speed of block A at the moment when the vertical separation of the blocks is h.

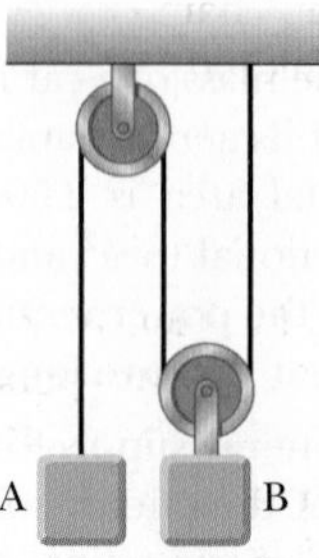

FIGURE **P7.47**

48. A 200-g particle is released from rest at point Ⓐ along the horizontal diameter on the inside of a frictionless, hemispherical bowl of radius $R = 30.0$ cm (Fig. P7.48). Calculate (a) the gravitational potential energy of the particle–Earth system when the particle is at point Ⓐ relative to point Ⓑ, (b) the kinetic energy of the particle at point Ⓑ, (c) its speed at point Ⓑ, and (d) its kinetic energy and the potential energy when the particle is at point Ⓒ.

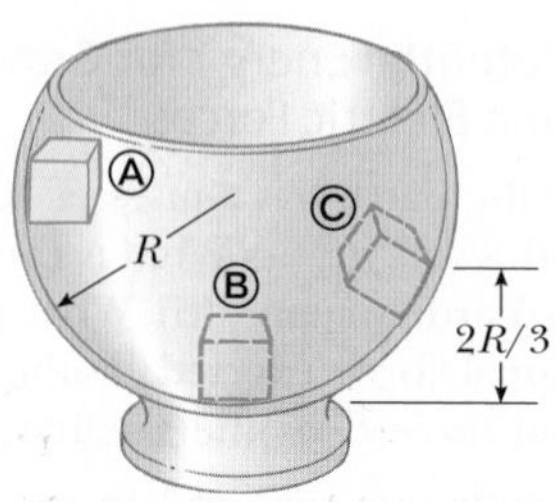

FIGURE **P7.48** Problems 7.48 and 7.49.

49. **Physics Now™** The particle described in Problem 7.48 (Fig. P7.48) is released from rest at Ⓐ, and the surface of the bowl is rough. The speed of the particle at Ⓑ is 1.50 m/s. (a) What is its kinetic energy at Ⓑ? (b) How much mechanical energy is transformed into internal energy as the particle moves from Ⓐ to Ⓑ? (c) Is it possible to determine the coefficient of friction from these results in any simple manner? Explain.

50. A child's pogo stick (Fig. P7.50) stores energy in a spring with a force constant of 2.50×10^4 N/m. At position Ⓐ ($x_A = -0.100$ m), the spring compression is a maximum and the child is momentarily at rest. At position Ⓑ ($x_B = 0$), the spring is relaxed and the child is moving upward. At position Ⓒ, the child is again momentarily at rest at the top of the jump. The combined mass of child and pogo stick is 25.0 kg. (a) Calculate the total energy of the child–stick–Earth system, taking both gravitational and elastic potential energies as zero for $x = 0$. (b) Determine x_C. (c) Calculate the speed of the child at $x = 0$. (d) Determine the value of x for which the kinetic energy of the system is a maximum. (e) Calculate the child's maximum upward speed.

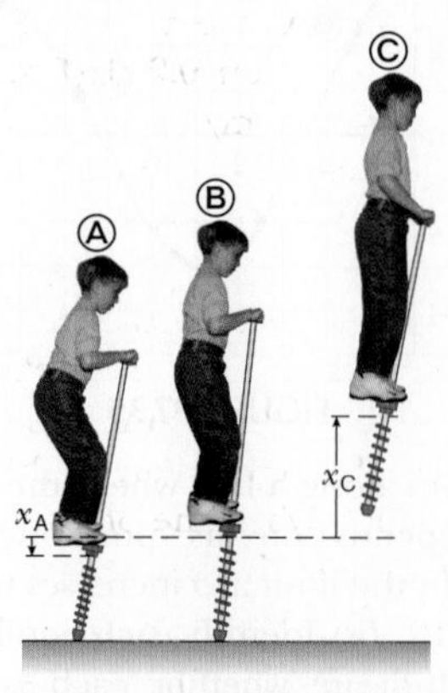

FIGURE **P7.50**

51. A 10.0-kg block is released from point Ⓐ in Figure P7.51. The track is frictionless except for the portion between points Ⓑ and Ⓒ, which has a length of 6.00 m. The block travels down the track, hits a spring of force constant 2 250 N/m, and compresses the spring 0.300 m from its equilibrium position before coming to rest momentarily. Determine the coefficient of kinetic friction between the block and the rough surface between Ⓑ and Ⓒ.

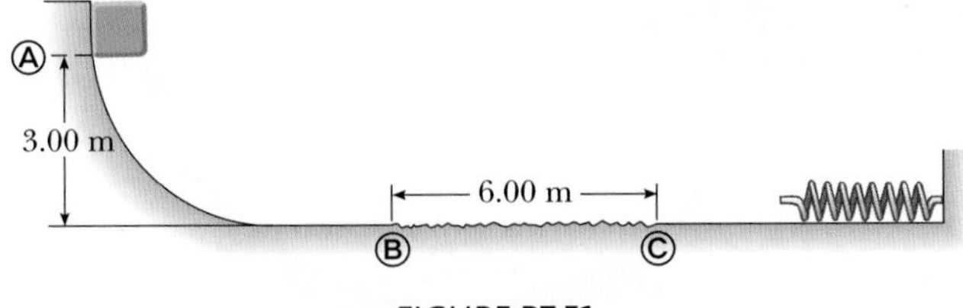

FIGURE **P7.51**

52. The potential energy function for a system is given by $U(x) = -x^3 + 2x^2 + 3x$. (a) Determine the force F_x as a function of x. (b) For what values of x is the force equal to zero? (c) Plot $U(x)$ versus x and F_x versus x, and indicate points of stable and unstable equilibrium.

53. A 20.0-kg block is connected to a 30.0-kg block by a string that passes over a light frictionless pulley. The 30.0-kg block is connected to a spring that has negligible mass and a force constant of 250 N/m as shown in Figure P7.53. The spring is unstretched when the system is as shown in the figure, and the incline is frictionless. The 20.0-kg block is pulled 20.0 cm down the incline (so that the 30.0-kg block is 40.0 cm above the floor) and released from rest. Find the speed of each block when the 30.0-kg block is 20.0 cm above the floor (that is, when the spring is unstretched).

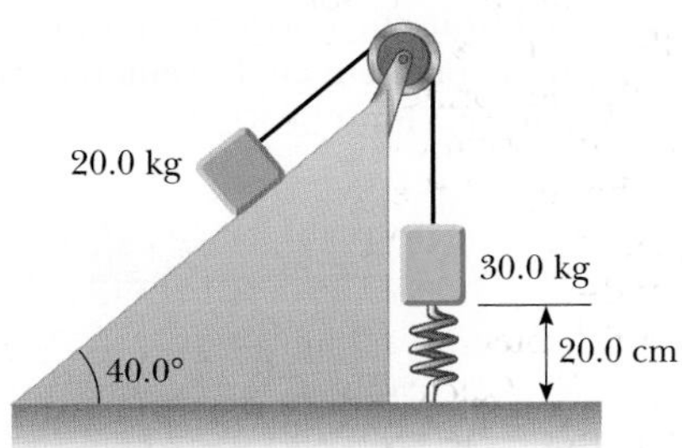

FIGURE **P7.53**

54. A 1.00-kg object slides to the right on a surface having a coefficient of kinetic friction 0.250 (Fig. P7.54). The object has a speed of $v_i = 3.00$ m/s when it makes contact with a light spring that has a force constant of 50.0 N/m. The object comes to rest after the spring has been compressed a distance d. The object is then forced toward the left by the spring and continues to move in that direction beyond the spring's unstretched position. The object finally comes to rest a distance D to the left of the unstretched spring. Find (a) the distance of compression d, (b) the speed v at the unstretched position when the object is moving to the left, and (c) the distance D where the object comes to rest.

55. **Physics Now™** A block of mass 0.500 kg is pushed against a horizontal spring of negligible mass until the spring is compressed a distance x (Fig. P7.55). The force constant of

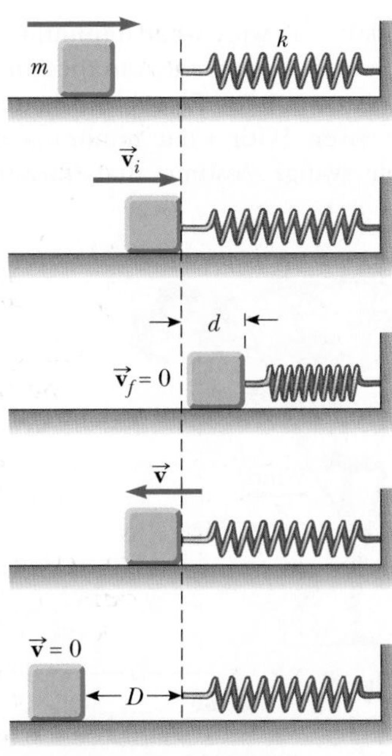

FIGURE **P7.54**

the spring is 450 N/m. When it is released, the block travels along a frictionless, horizontal surface to point B, the bottom of a vertical circular track of radius $R = 1.00$ m, and continues to move up the track. The speed of the block at the bottom of the track is $v_B = 12.0$ m/s, and the block experiences an average friction force of 7.00 N while sliding up the track. (a) What is x? (b) What speed do you predict for the block at the top of the track? (c) Does the block actually reach the top of the track, or does it fall off before reaching the top?

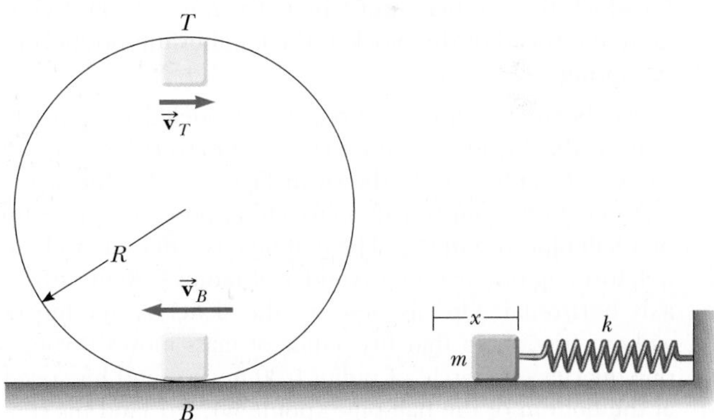

FIGURE **P7.55**

56. A uniform chain of length 8.00 m initially lies stretched out on a horizontal table. (a) Assuming that the coefficient of static friction between chain and table is 0.600, show that the chain will begin to slide off the table if at least 3.00 m of it hangs over the edge of the table. (b) Determine the speed of the chain as it all leaves the table, given that the coefficient of kinetic friction between the chain and the table is 0.400.

57. Jane, whose mass is 50.0 kg, needs to swing across a river (having width D) filled with person-eating crocodiles to save Tarzan from danger. She must swing into a wind exerting constant horizontal force $\vec{\mathbf{F}}$, on a vine having length L and initially making an angle θ with the vertical (Fig. P7.57). Taking $D = 50.0$ m, $F = 110$ N, $L = 40.0$ m,

and $\theta = 50.0°$, (a) with what minimum speed must Jane begin her swing to just make it to the other side? (b) Once the rescue is complete, Tarzan and Jane must swing back across the river. With what minimum speed must they begin their swing? Assume that Tarzan has a mass of 80.0 kg.

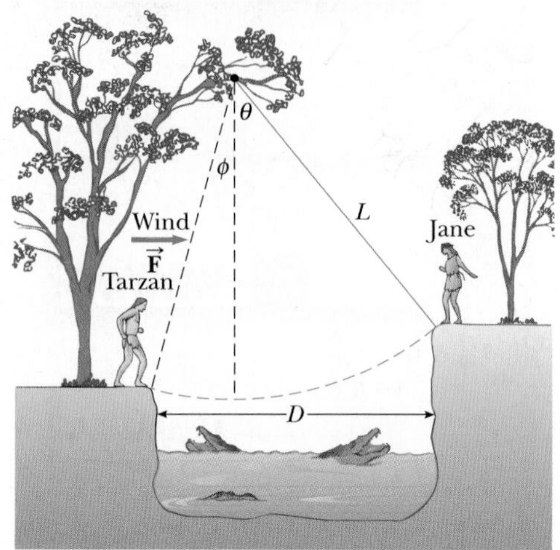

FIGURE **P7.57**

58. A 5.00-kg block free to move on a horizontal, frictionless surface is attached to one end of a light horizontal spring. The other end of the spring is held fixed. The spring is compressed 0.100 m from equilibrium and released. The speed of the block is 1.20 m/s when it passes the equilibrium position of the spring. The same experiment is now repeated with the frictionless surface replaced by a surface for which the coefficient of kinetic friction is 0.300. Determine the speed of the block at the equilibrium position of the spring.

59. A skateboarder with his board can be modeled as a particle of mass 76.0 kg, located at his center of mass (which we will study in Chapter 8). As shown in Figure P7.59, the skateboarder starts from rest in a crouching position at one lip of a half-pipe (point Ⓐ). The half-pipe is a dry water channel, forming one half of a cylinder of radius 6.80 m with its axis horizontal. On his descent, the skateboarder moves without friction so that his center of mass moves through one quarter of a circle of radius 6.30 m. (a) Find his speed at the bottom of the half-pipe (point Ⓑ). (b) Find his centripetal acceleration. (c) Find the normal force n_B acting on the skateboarder at point Ⓑ. Immediately after passing point Ⓑ, he stands up and raises his arms, lifting his center of mass from 0.500 m to 0.950 m above the concrete (point Ⓒ). To account for the conversion of chemical into mechanical energy, model his legs as doing work by pushing him vertically up with a constant force equal to the normal force n_B over a distance of 0.450 m. (You will be able to solve this problem with a more accurate model in Chapter 10.) (d) What is the work done on the skateboarder's body in this process? Next, the skateboarder glides upward with his center of mass moving in a quarter circle of radius 5.85 m. His body is horizontal when he passes point Ⓓ, the far lip of the half-pipe. (e) Find his speed at this location. At last he goes ballistic, twisting around while his center of mass moves vertically. (f) How high above point Ⓓ does he rise? (g) Over what time interval is he airborne before he touches down, 2.34 m below the level of point Ⓓ? (*Caution:* Do not try this yourself without the required skill and protective equipment or in a drainage channel to which you do not have legal access.)

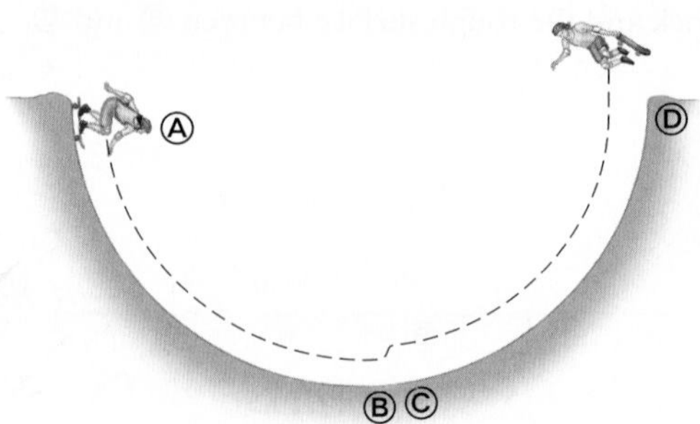

FIGURE **P7.59**

60. A block of mass M rests on a table. It is fastened to the lower end of a light vertical spring. The upper end of the spring is fastened to a block of mass m. The upper block is pushed down by an additional force $3mg$ so that the spring compression is $4mg/k$. In this configuration, the upper block is released from rest. The spring lifts the lower block off the table. In terms of m, what is the greatest possible value for M?

61. A pendulum, comprising a light string of length L and a small sphere, swings in a vertical plane. The string hits a peg located a distance d below the point of suspension (Fig. P7.61). (a) Show that if the sphere is released from a height below that of the peg, it will return to this height after the string strikes the peg. (b) Show that if the pendulum is released from the horizontal position ($\theta = 90°$) and is to swing in a complete circle centered on the peg, the minimum value of d must be $3L/5$.

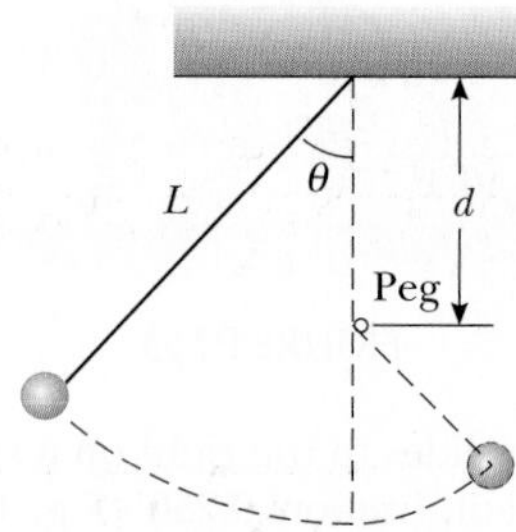

FIGURE **P7.61**

62. A roller coaster car is released from rest at the top of the first rise and then moves freely with negligible friction. The roller coaster shown in Figure P7.62 has a circular loop of radius R in a vertical plane. (a) First, suppose the car barely makes it around the loop; at the top of the loop the riders are upside down and feel weightless. Find the required height of the release point above the bottom of the loop, in terms of R. (b) Now assume that the release point is at or above the minimum required height. Show that the normal force on the car at the bottom of the loop exceeds the normal force at the top of the loop by six times the

weight of the car. The normal force on each rider follows the same rule. Such a large normal force is dangerous and very uncomfortable for the riders. Roller coasters are therefore not built with circular loops in vertical planes. Figure P5.24 and the photograph on page 134 show two actual designs.

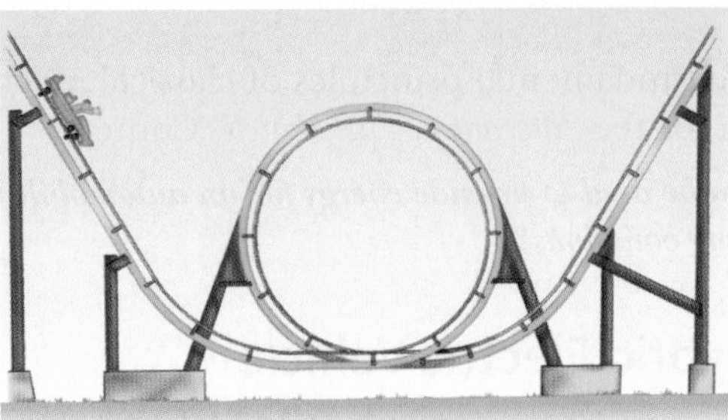

FIGURE P7.62

63. **Review problem.** In 1887 in Bridgeport, Connecticut, C. J. Belknap built the water slide shown in Figure P7.63. A rider on a small sled, of total mass 80.0 kg, pushed off to start at the top of the slide (point Ⓐ) with a speed of 2.50 m/s. The chute was 9.76 m high at the top, 54.3 m long, and 0.51 m wide. Along its length, 725 wheels made friction negligible. Upon leaving the chute horizontally at its bottom end (point Ⓒ), the rider skimmed across the water of Long Island Sound for as much as 50 m, "skipping along like a flat pebble," before at last coming to rest and swimming ashore, pulling his sled after him. According to *Scientific American,* "The facial expression of novices taking their first adventurous slide is quite remarkable, and the sensations felt are correspondingly novel and peculiar." (a) Find the speed of the sled and rider at point Ⓒ. (b) Model the force of water friction as a constant retarding force acting on a particle. Find the work done by water friction in stopping the sled and rider. (c) Find the magnitude of the force the water exerts on the sled. (d) Find the magnitude of the force the chute exerts on the sled at point Ⓑ. (e) At point Ⓒ the chute is horizontal but curving in the vertical plane. Assume that its radius of curvature is 20.0 m. Find the force the chute exerts on the sled at point Ⓒ.

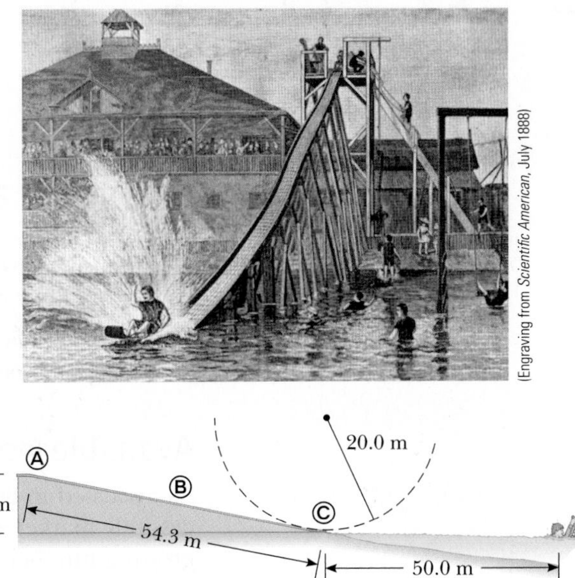

FIGURE P7.63

64. Starting from rest, a 64.0-kg person bungee jumps from a tethered balloon 65.0 m above the ground (Fig. P7.18). The bungee cord has negligible mass and unstretched length 25.8 m. One end is tied to the basket of the balloon and the other end to a harness around the person's body. The cord is described by Hooke's law with a spring constant of 81.0 N/m. The balloon does not move. (a) Model the person's body as a particle. Express the gravitational potential energy of the person–Earth system as a function of the person's variable height y above the ground. (b) Express the elastic potential energy of the cord as a function of y. (c) Express the total potential energy of the person–cord–Earth system as a function of y. (d) Plot a graph of the gravitational, elastic, and total potential energies as functions of y. (e) Assume that air resistance is negligible. Determine the minimum height of the person above the ground during his plunge. (f) Does the potential energy graph show any equilibrium position or positions? If so, at what elevations? Are they stable or unstable? (g) Determine the jumper's maximum speed.

ANSWERS TO QUICK QUIZZES

7.1 (c). The sign of the gravitational potential energy depends on your choice of zero configuration. If the two objects in the system are closer together than in the zero configuration, the potential energy is negative. If they are farther apart, the potential energy is positive.

7.2 (a). We must include the Earth if we are going to work with gravitational potential energy.

7.3 $v_1 = v_2 = v_3$. The first and third balls speed up after they are thrown, whereas the second ball initially slows down but then speeds up after reaching its peak. The paths of all three balls are parabolas, and the balls take different time intervals to reach the ground because they have different initial velocities. All three balls, however, have the same speed at the moment they hit the ground because all start with the same kinetic energy and because the ball–Earth system undergoes the same change in gravitational potential energy in all three cases.

7.4 (i), (c). This system exhibits changes in kinetic energy as well as in both types of potential energy. (ii), (a). Because the Earth is not included in the system, there is no gravitational potential energy associated with the system.

Present and Future Possibilities

Now that we have explored some fundamental principles of classical mechanics, let us return to our central question for the *Alternative-Fuel Vehicles* Context:

What source besides gasoline can be used to provide energy for an automobile while reducing environmentally damaging emissions?

Available Now—The Hybrid Electric Vehicle

As discussed in Section 7.8, electric vehicles such as the GM EV1 have not been successfully marketed and are falling by the wayside. Currently taking their place are a growing number of **hybrid electric vehicles.** In these automobiles, a gasoline engine and an electric motor are combined to increase the fuel economy of the vehicle and reduce its emissions. Currently available models include the Toyota Prius and Honda Insight, which are originally designed hybrid vehicles, as well as other existing models that have been modified with a hybrid drive system, such as the Honda Civic.

Two major categories of hybrid vehicles are the **parallel hybrid** and the **series hybrid.** In a parallel hybrid, both the engine and the motor are connected to the transmission, so either one can provide propulsion energy for the car. In a series hybrid, the gasoline engine does not provide propulsion energy to the transmission directly. The engine turns a generator, which in turn either charges the batteries or powers the electric motor. Only the electric motor is connected directly to the transmission to propel the car.

(Courtesy of Honda Motor Co., Inc.)

FIGURE 1 The Honda Insight.

The Honda Insight (Fig. 1) is a parallel hybrid. Both the engine and the motor provide power to the transmission, and the engine is running at all times while the car is moving. The goal of the development of this hybrid is maximum mileage, which is achieved through a number of design features. Because the engine is small, the Insight has lower emissions than a traditional gasoline-powered vehicle. Because the engine is running at all vehicle speeds, however, its emissions are not as low as those of the Toyota Prius.

Figure 2 shows the engine compartment of the Toyota Prius. In this parallel hybrid, power to the wheels can come from either the gasoline engine or the electric motor. The vehicle has some aspects of a series hybrid, however, in that the electric motor alone accelerates the vehicle from rest until it is moving at a speed of about 15 mph (24 kph). During this acceleration period, the engine is not running, so gasoline is not used and there is no emission. As a result, the average tailpipe emissions are lower than those of the Insight, although the gasoline mileage is not quite as high.

When a hybrid vehicle brakes, the motor acts as a generator and returns some of the kinetic energy of the vehicle back to the battery as electric potential energy. In a normal vehicle, this kinetic energy is not recoverable because it is transformed to internal energy in the brakes and roadway.

Gas mileage for hybrid vehicles is in the range of 45 to 60 mi/gal and emissions are far below those of a standard gasoline engine. A hybrid vehicle does not need to be charged like a purely electric vehicle. The battery that drives the electric motor is charged while the gasoline engine is running. Consequently, even though the hybrid vehicle has an electric motor like a pure electric vehicle, it can simply be filled at a gas station like a normal vehicle.

Hybrid electric vehicles are not strictly alternative-fuel vehicles because they use the same fuel as normal vehicles, gasoline. They do, however, represent an important step toward more efficient cars with lower emissions, and the increased mileage helps conserve crude oil.

(Photo by Brent Romans/www.Edmunds.com)

FIGURE 2 The engine compartment of the Toyota Prius.

In the Future — The Fuel Cell Vehicle

In an internal combustion engine, the chemical potential energy in the fuel is transformed to internal energy during an explosion initiated by a spark plug. The resulting expanding gases do work on pistons, directing energy to the wheels of the vehicle. In current development is the **fuel cell,** in which the conversion of the energy in the fuel to internal energy is not required. The fuel (hydrogen) is oxidized, and energy leaves the fuel cell by electrical transmission. The energy is used by an electric motor to drive the vehicle.

The advantages of this type of vehicle are many. There is no internal combustion engine to generate harmful emissions, so the vehicle is emission-free. Other than the energy used to power the vehicle, the only by-products are internal energy and water. The fuel is hydrogen, which is the most abundant element in the universe. The efficiency of a fuel cell is much higher than that of an internal combustion engine, so more of the potential energy in the fuel can be extracted.

That is all good news. The bad news is that fuel cell vehicles are still only in the early prototype stage (Fig. 3). It will be many years before fuel cell vehicles are available to consumers. During these years, fuel cells must be perfected to operate in weather extremes, manufacturing infrastructure must be set up to supply the hydrogen, and a fueling infrastructure must be established to allow transfer of hydrogen into individual vehicles.

(© Adam Hart-Davis/Photo Researchers, Inc.)

FIGURE 3 A hydrogen fuel cell in Europe's first hydrogen-powered taxi. Fuel cells convert the energy from a chemical reaction into electricity.

Problems

1. When a conventional car brakes to a stop, all (100%) its kinetic energy is converted into internal energy. None of this energy is available to get the car moving again. Consider a hybrid electric car of mass 1 300 kg moving at 22.0 m/s. (a) Calculate its kinetic energy. (b) The car uses its regenerative braking system to come to a stop at a red light. Assume that the motor-generator converts 70.0% of the car's kinetic energy into energy delivered to the battery by electrical transmission. The other 30.0% becomes internal energy. Compute the amount of energy charging up the battery. (c) Assume that the battery can give back 85.0% of the energy chemically stored in it. Compute the amount of this energy. The other 15.0% becomes internal energy. (d) When the light turns green, the car's motor-generator runs as a motor to convert 68.0% of the energy from the battery into kinetic energy of the car. Compute the amount of this energy and (e) the speed at which the car will be set moving with no other energy input. (f) Compute the overall efficiency of the braking-and-starting process. (g) Compute the net amount of internal energy produced.

2. In both a conventional car and a hybrid electric car, the gasoline engine is the original source of all the energy the car uses to push through the air and against rolling resistance of the road. In city traffic, a conventional gasoline engine must run at a wide variety of rotation rates and fuel inputs. That is, it must run at a wide variety of tachometer and throttle settings. It is almost never running at its maximum-efficiency point. In a hybrid electric car, on the other hand, the gasoline engine can run at maximum efficiency whenever it is on. A simple model can reveal the distinction numerically. Assume that the two cars both do 66.0 MJ of "useful" work in making the same trip to the drugstore. Let the conventional car run at 7.00% efficiency as it puts out useful energy 33.0 MJ and let it run at 30.0% efficiency as it puts out 33.0 MJ. Let the hybrid car run at 30.0% efficiency all the time. Compute (a) the required energy input for each car and (b) the overall efficiency of each.

C O N T E X T 2

Mission to Mars

In this Context, we shall investigate the physics necessary to send a spacecraft from Earth to Mars. If the two planets were sitting still in space, millions of kilometers apart, it would be a difficult enough proposition, but keep in mind that we are launching the spacecraft from a moving object, the Earth, and are aiming at a moving target, Mars. Furthermore, the spacecraft's motion is influenced by gravitational forces from the Earth, the Sun, and Mars as well as from any other massive objects in the vicinity. Despite these apparent difficulties, we can use the principles of physics to plan a successful mission.

Travel in space began in the early 1960s, with the launch of human-occupied spacecraft in both the United States and the Soviet Union. The first human to ride into space was Yury Gagarin, who made a one-orbit trip in 1961 in the Soviet spacecraft *Vostok.* Competition between the two countries resulted in a "space race," which led to the successful landing of American astronauts on the Moon in 1969.

(Japanese Aerospace Exploration Agency (JAXA))

FIGURE 1 The *Nozomi* is the first Mars orbiter to be launched by Japan. This photo shows its launch on July 4, 1998 from Kagoshima Space Center. Unfortunately, the *Nozomi* mission was unsuccessful because of technical difficulties, and the spacecraft did not achieve orbit around Mars.

In the 1970s, the Viking Project landed spacecraft on Mars to analyze the soil for signs of life. These tests were inconclusive.

U.S. efforts in the 1980s focused on the development and implementation of the space shuttle system, a reusable space transportation system. The shuttle has been used extensively in moving supplies and personnel to the International Space Station, which was begun in 1998 and continues to develop. It has also been an important means of performing scientific experiments in space and delivering satellites into orbit.

The United States returned to Mars in the 1990s with the Mars Global Surveyor, designed to perform careful mapping of the Martian surface, and Mars Pathfinder, which landed on Mars and deployed a roving robot to analyze rocks and soil. Not all trips have been successful. In 1999, Mars Polar Lander was launched to land near the polar ice cap and search for water. As it entered the Martian atmosphere, it sent its last data and was never heard from again. Mars Climate Orbiter was also lost in 1999 due to communication errors between the builder of the spacecraft and the mission control team.

In late 2003 and early 2004, arrivals of spacecraft at Mars were expected by three space agencies, the National Aerodynamics and Space Administration (NASA) in the United States, the European Space Agency (ESA) in Europe, and the Japanese Aerospace Exploration Agency (JAXA) in Japan. The extreme difficulties associated with

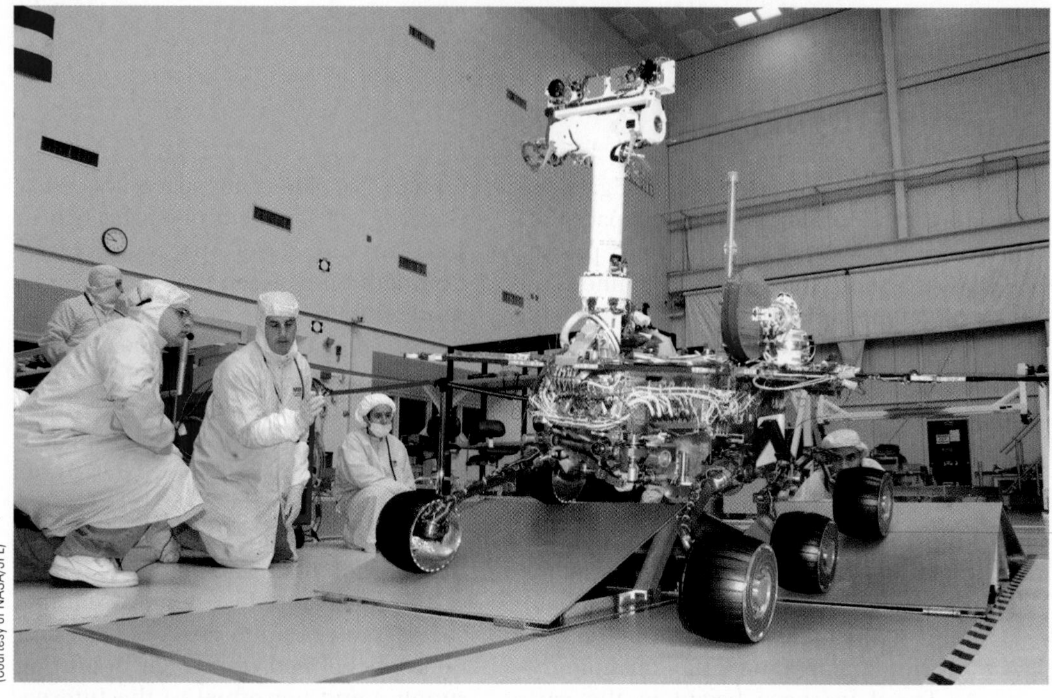

(Courtesy of NASA/JPL)

FIGURE 2 The Mars rover *Spirit* is tested in a clean room at the Jet Propulsion Laboratory in Pasadena, California.

such an endeavor can be appreciated by examining the results of these simultaneous missions. The Japanese mission ended in failure when a stuck valve and electrical circuit problems affected a critical midcourse correction, resulting in the inability of the spacecraft, named *Nozomi,* to achieve an orbit around Mars. It passed about 1 000 km above the Martian surface on December 14, 2003, and then left the planet to continue its orbit around the Sun.

(Courtesy of NASA/JPL/Cornell)

FIGURE 3 An image from a camera on the Mars rover *Opportunity* shows a rock called the "Berry Bowl." The "berries" are sphere-like grains containing hematite, which scientists used to confirm the earlier presence of water on the surface. The circular area on the rock is the result of using the rover's rock abrasion tool to remove a layer of dust. In this way, a clean surface of the rock was available for spectral analysis by the rover's spectrometers.

The European effort resulted in a successful injection of their *Mars Express* spacecraft into an orbit around Mars. A lander, named *Beagle 2,* descended to the surface. Unfortunately, no signals from the lander have been detected and it is presumed lost. The *Mars Express* orbiter continues to send data and is equipped to perform scientific analyses from orbit.

The NASA effort was the most successful of the three missions, with the *Spirit* rover landing successfully on the surface of Mars on January 4, 2004. Its twin, *Opportunity,* also landed successfully, on January 24, 2004, on the opposite side of the planet from *Spirit.* Amazingly, *Opportunity* landed inside a crater, providing scientists with a

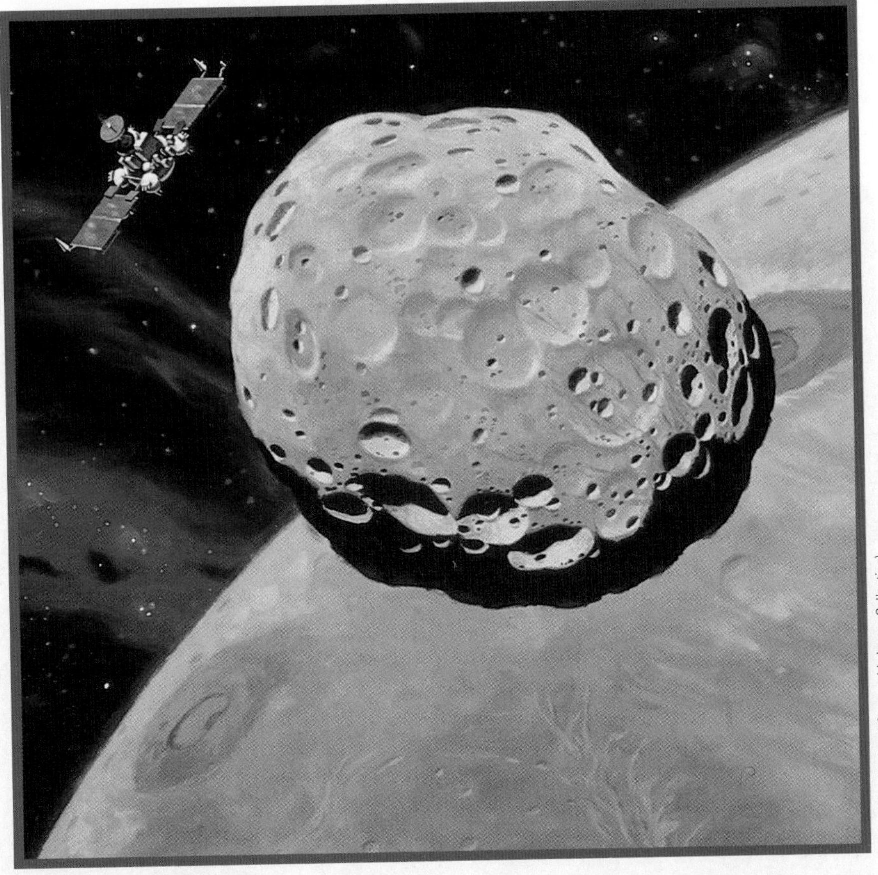

(Pierre Mion/National Geographic Image Collection)

FIGURE 4 In this Context, we shall investigate the details of the challenging task of sending a spacecraft from the Earth to Mars.

wonderful opportunity to study the geology of an impact crater. Aside from a computer glitch that was successfully repaired, both rovers performed excellently and sent back very high-quality photographs of the Martian surface as well as large amounts of data including verification of water that once existed on the surface.

Many individuals dream of one day establishing colonies on Mars. This dream is far in the future; we are still learning much about Mars today and have yet taken only a handful of trips to the planet. Travel to Mars is still not an everyday occurrence, although we learn more from each mission. In this Context, we address the central question,

How can we undertake a successful transfer of a spacecraft from Earth to Mars?

CHAPTER 8

Momentum and Collisions

A golf ball is struck by a club and begins to leave the tee. Note the deformation of the ball as a result of the large force exerted on it by the club.

CHAPTER OUTLINE

Consider what happens when a golf ball is struck by a club as in the opening photograph for this chapter. The ball changes its motion from being at rest to having a very large velocity as a result of the collision; consequently, it is able to travel a large distance through the air. Because the ball experiences this change in velocity over a very short time interval, the average force on it during the collision is very large. By Newton's third law, the club experiences a reaction force equal in magnitude and opposite to the force on the ball. This reaction force produces a change in the velocity of the club. Because the club is much more massive than the ball, however, the change in the club's velocity is much less than the change in the ball's velocity.

One main objective of this chapter is to enable you to understand and analyze such events. As a first step, we shall introduce the concept of *momentum,* a term used to describe objects in motion. The concept of momentum leads us to a new conservation law and momentum approaches for treating isolated and nonisolated systems. This conservation law is especially useful for treating problems that involve collisions between objects.

8.1 LINEAR MOMENTUM AND ITS CONSERVATION

In the preceding two chapters, we studied situations that are difficult to analyze with Newton's laws. We were able to solve problems involving these situations by applying a conservation principle, conservation of energy. Consider another situation. A 60-kg archer stands on frictionless ice and fires a 0.50-kg arrow horizontally at 50 m/s. From Newton's third law, we know that the force that the bow exerts on the arrow will be matched by a force in the opposite direction on the bow (and the archer). This force will cause the archer to begin to slide backward on the ice. But with what speed? We cannot answer this question using *either* Newton's second law or an energy approach because there is not enough information.

Despite our inability to solve the archer problem using our techniques learned so far, this problem is very simple to solve if we introduce a new quantity that describes motion. To motivate this new quantity, let us apply the General Problem-Solving Strategy from Chapter 1 and *conceptualize* an isolated system of two particles (Fig. 8.1) with masses m_1 and m_2 and moving with velocities $\vec{\mathbf{v}}_1$ and $\vec{\mathbf{v}}_2$ at an instant of time. Because the system is isolated, the only force on one particle is that from the other particle, and we can *categorize* this situation as one in which Newton's laws can be applied. If a force from particle 1 (for example, a gravitational force) acts on particle 2, there must be a second force—equal in magnitude but opposite in direction—that particle 2 exerts on particle 1. That is, the forces form a Newton's third law action–reaction pair so that $\vec{\mathbf{F}}_{12} = -\vec{\mathbf{F}}_{21}$. We can express this condition as a statement about the *system* of two particles as follows:

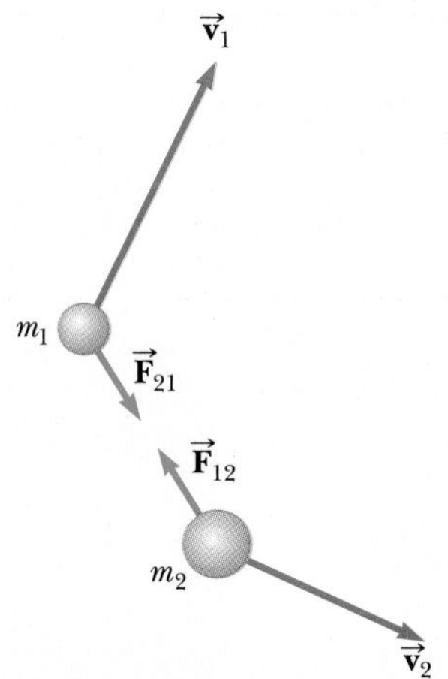

FIGURE 8.1 Two particles interact with each other. According to Newton's third law, we must have $\vec{\mathbf{F}}_{12} = -\vec{\mathbf{F}}_{21}$.

$$\vec{\mathbf{F}}_{21} + \vec{\mathbf{F}}_{12} = 0$$

Let us further *analyze* this situation by incorporating Newton's second law. Over some time interval, the interacting particles in the system will accelerate. Therefore, replacing each force with $m\vec{\mathbf{a}}$ gives

$$m_1\vec{\mathbf{a}}_1 + m_2\vec{\mathbf{a}}_2 = 0$$

Now we replace the acceleration with its definition from Equation 3.5:

$$m_1\frac{d\vec{\mathbf{v}}_1}{dt} + m_2\frac{d\vec{\mathbf{v}}_2}{dt} = 0$$

If the masses m_1 and m_2 are constant, we can bring them into the derivatives, which gives

$$\frac{d(m_1\vec{\mathbf{v}}_1)}{dt} + \frac{d(m_2\vec{\mathbf{v}}_2)}{dt} = 0$$

$$\frac{d}{dt}(m_1\vec{\mathbf{v}}_1 + m_2\vec{\mathbf{v}}_2) = 0 \qquad [8.1]$$

To *finalize* this discussion, note that the derivative of the sum $m_1\vec{\mathbf{v}}_1 + m_2\vec{\mathbf{v}}_2$ with respect to time is zero. Consequently, this sum must be constant. We learn from this discussion that the quantity $m\vec{\mathbf{v}}$ for a particle is important in that the sum of the values of this quantity for the particles in an isolated system is conserved. We call this quantity *linear momentum:*

The **linear momentum** $\vec{\mathbf{p}}$ of a particle or an object that can be modeled as a particle of mass m moving with a velocity $\vec{\mathbf{v}}$ is defined to be the product of the mass and velocity:[1]

■ Definition of linear momentum of a particle

$$\vec{\mathbf{p}} \equiv m\vec{\mathbf{v}} \tag{8.2}$$

Because momentum equals the product of a scalar m and a vector $\vec{\mathbf{v}}$, it is a vector quantity. Its direction is the same as that for $\vec{\mathbf{v}}$, and it has dimensions ML/T. In the SI system, momentum has the units kg · m/s.

If an object is moving in an arbitrary direction in three-dimensional space, $\vec{\mathbf{p}}$ has three components and Equation 8.2 is equivalent to the component equations

$$p_x = mv_x \qquad p_y = mv_y \qquad p_z = mv_z \tag{8.3}$$

As you can see from its definition, the concept of momentum provides a quantitative distinction between objects of different masses moving at the same velocity. For example, the momentum of a truck moving at 2 m/s is much greater in magnitude than that of a Ping-Pong ball moving at the same speed. Newton called the product $m\vec{\mathbf{v}}$ the *quantity of motion,* perhaps a more graphic description than *momentum,* which comes from the Latin word for movement.

QUICK QUIZ 8.1 Two objects have equal kinetic energies. How do the magnitudes of their momenta compare? **(a)** $p_1 < p_2$ **(b)** $p_1 = p_2$ **(c)** $p_1 > p_2$ **(d)** not enough information to determine the answer

QUICK QUIZ 8.2 Your physical education teacher throws a baseball to you at a certain speed and you catch it. The teacher is next going to throw you a medicine ball whose mass is ten times the mass of the baseball. You are given the following choices. You can have the medicine ball thrown with **(a)** the same speed as the baseball, **(b)** the same momentum, or **(c)** the same kinetic energy. Rank these choices from easiest to hardest to catch.

Let us use the particle model for an object in motion. By using Newton's second law of motion, we can relate the linear momentum of a particle to the net force acting on the particle. In Chapter 4, we learned that Newton's second law can be written as $\sum \vec{\mathbf{F}} = m\vec{\mathbf{a}}$. This form applies only when the mass of the particle remains constant, however. In situations where the mass is changing with time, one must use an alternative statement of Newton's second law: **The time rate of change of momentum of a particle is equal to the net force acting on the particle,** or

■ Newton's second law for a particle

$$\sum \vec{\mathbf{F}} = \frac{d\vec{\mathbf{p}}}{dt} \tag{8.4}$$

If the mass of the particle is constant, the preceding equation reduces to our previous expression for Newton's second law:

$$\sum \vec{\mathbf{F}} = \frac{d\vec{\mathbf{p}}}{dt} = \frac{d(m\vec{\mathbf{v}})}{dt} = m\frac{d\vec{\mathbf{v}}}{dt} = m\vec{\mathbf{a}}$$

It is difficult to imagine a particle whose mass is changing, but if we consider objects, a number of examples emerge. These examples include a rocket that is ejecting its

[1]This expression is nonrelativistic and is valid only when $v \ll c$, where c is the speed of light. In the next chapter, we discuss momentum for high-speed particles.

fuel as it operates, a snowball rolling down a hill and picking up additional snow, and a watertight pickup truck whose bed is collecting water as it moves in the rain.

From Equation 8.4 we see that if the net force on an object is zero, the time derivative of the momentum is zero and therefore the momentum of the object must be constant. This conclusion should sound familiar because it is the case of a particle in equilibrium, expressed in terms of momentum. Of course, if the particle is *isolated* (that is, if it does not interact with its environment), no forces act on it and $\vec{\mathbf{p}}$ remains unchanged, which is Newton's first law.

Momentum and Isolated Systems

Using the definition of momentum, Equation 8.1 can be written as

$$\frac{d}{dt}(\vec{\mathbf{p}}_1 + \vec{\mathbf{p}}_2) = 0$$

Because the time derivative of the total system momentum $\vec{\mathbf{p}}_{\text{tot}} = \vec{\mathbf{p}}_1 + \vec{\mathbf{p}}_2$ is *zero*, we conclude that the *total* momentum $\vec{\mathbf{p}}_{\text{tot}}$ must remain constant:

$$\vec{\mathbf{p}}_{\text{tot}} = \text{constant} \qquad [8.5]$$

■ **Conservation of momentum for an isolated system**

or, equivalently,

$$\vec{\mathbf{p}}_{1i} + \vec{\mathbf{p}}_{2i} = \vec{\mathbf{p}}_{1f} + \vec{\mathbf{p}}_{2f} \qquad [8.6]$$

where $\vec{\mathbf{p}}_{1i}$ and $\vec{\mathbf{p}}_{2i}$ are initial values and $\vec{\mathbf{p}}_{1f}$ and $\vec{\mathbf{p}}_{2f}$ are final values of the momentum during a period over which the particles interact. Equation 8.6 in component form states that the momentum components of the isolated system in the *x*, *y*, and *z* directions are all *independently constant;* that is,

$$\sum_{\text{system}} p_{ix} = \sum_{\text{system}} p_{fx} \qquad \sum_{\text{system}} p_{iy} = \sum_{\text{system}} p_{fy} \qquad \sum_{\text{system}} p_{iz} = \sum_{\text{system}} p_{fz} \qquad [8.7]$$

This result, known as the law of **conservation of linear momentum,** is the mathematical representation of the momentum version of the isolated system model. It is considered one of the most important laws of mechanics. We have generated this law for a system of two interacting particles, but it can be shown to be true for a system of any number of particles. We can state it as follows: **The total momentum of an isolated system remains constant.**

Notice that we have made no statement concerning the nature of the forces acting between members of the system. The only requirement is that the forces must be *internal* to the system. Therefore, momentum is conserved for an isolated system regardless of the nature of the internal forces, *even if the forces are nonconservative.*

PITFALL PREVENTION 8.1

MOMENTUM OF A SYSTEM IS CONSERVED Remember that the momentum of an isolated *system* is conserved. The momentum of one particle within an isolated system is not necessarily conserved because other particles in the system may be interacting with it. Always apply conservation of momentum to an isolated *system*.

EXAMPLE 8.1 Can We Really Ignore the Kinetic Energy of the Earth?

In Section 7.1, we claimed that we can ignore the kinetic energy of the Earth when considering the energy of a system consisting of the Earth and a dropped ball. Verify this claim.

Solution We will verify this claim by setting up a ratio of the kinetic energy of the Earth to that of the ball:

$$(1) \qquad \frac{K_E}{K_b} = \frac{\frac{1}{2}m_E v_E^{\,2}}{\frac{1}{2}m_b v_b^{\,2}} = \left(\frac{m_E}{m_b}\right)\left(\frac{v_E}{v_b}\right)^2$$

where v_E and v_b are the speeds of the Earth and the ball, respectively, after the ball has fallen through some distance. Now we find a relationship between these two speeds by considering conservation of momentum in the vertical direction for the system of the ball and the Earth. The initial momentum of the system is zero, so the final momentum must also be zero:

$$p_i = p_f \quad \to \quad 0 = m_b v_b + m_E v_E$$

$$\to \quad \frac{v_E}{v_b} = -\frac{m_b}{m_E}$$

Substituting for v_E/v_b in (1), we have

$$\frac{K_E}{K_b} = \left(\frac{m_E}{m_b}\right)\left(-\frac{m_b}{m_E}\right)^2 = \frac{m_b}{m_E}$$

Substituting order-of-magnitude numbers for the masses, this ratio becomes

$$\frac{K_E}{K_b} = \frac{m_b}{m_E} \sim \frac{1 \text{ kg}}{10^{25} \text{ kg}} \sim 10^{-25}$$

The kinetic energy of the Earth is a very small fraction of the kinetic energy of the ball, so we are justified in ignoring it in the kinetic energy of the system.

INTERACTIVE EXAMPLE 8.2 The Archer

Let us consider the situation proposed at the beginning of this section. A 60-kg archer stands at rest on frictionless ice and fires a 0.50-kg arrow horizontally at 50 m/s (Fig. 8.2). With what velocity does the archer move across the ice after firing the arrow?

FIGURE 8.2 (Interactive Example 8.2) An archer fires an arrow horizontally. Because he is standing on frictionless ice, he will begin to slide across the ice.

Solution We *cannot* solve this problem using Newton's second law, $\sum \vec{\mathbf{F}} = m\vec{\mathbf{a}}$, because we have no information about the force on the arrow or its acceleration. We *cannot* solve this problem using an energy approach because we do not know how much work is done in pulling the bow back or how much potential energy is stored in the bow. We *can,* however, solve this problem very easily with conservation of momentum because momentum does not depend on any of these quantities that we do not know.

Let us take the system to consist of the archer (including the bow) and the arrow. The system is not isolated because the gravitational force and the normal force act on the system. These forces, however, are vertical and perpendicular to the motion of the system. Therefore, there are no external forces in the horizontal direction, and we can consider the system to be isolated in terms of momentum components in this direction.

The total horizontal momentum of the system before the arrow is fired is zero ($m_1\vec{\mathbf{v}}_{1i} + m_2\vec{\mathbf{v}}_{2i} = 0$), where the archer is particle 1 and the arrow is particle 2. Therefore, the total horizontal momentum of the system after the arrow is fired must be zero; that is,

$$m_1\vec{\mathbf{v}}_{1f} + m_2\vec{\mathbf{v}}_{2f} = 0$$

We choose the direction of firing of the arrow as the positive *x* direction. With $m_1 = 60$ kg, $m_2 = 0.50$ kg, and $\vec{\mathbf{v}}_{2f} = 50\hat{\mathbf{i}}$ m/s, solving for $\vec{\mathbf{v}}_{1f}$ we find the recoil velocity of the archer to be

$$\vec{\mathbf{v}}_{1f} = -\frac{m_2}{m_1}\vec{\mathbf{v}}_{2f} = -\left(\frac{0.50 \text{ kg}}{60 \text{ kg}}\right)(50\hat{\mathbf{i}} \text{ m/s})$$

$$= -0.42\hat{\mathbf{i}} \text{ m/s}$$

The negative sign for $\vec{\mathbf{v}}_{1f}$ indicates that the archer is moving to the left after the arrow is fired, in the direction opposite the direction of the arrow's motion, in accordance with Newton's third law. Because the archer is much more massive than the arrow, his acceleration and consequent velocity are much smaller than the arrow's acceleration and velocity.

PhysicsNow™ Log into PhysicsNow at **www.pop4e.com** and go to Interactive Example 8.2 to change the arrow's speed and the masses of the archer and the arrow.

EXAMPLE 8.3 Decay of the Kaon at Rest

One type of nuclear particle, called the *neutral kaon* (K^0), decays into a pair of other particles called *pions* (π^+ and π^-), which are oppositely charged but equal in mass, as in Figure 8.3. Assuming that the kaon is initially at rest, prove that the two pions must have momenta that are equal in magnitude and opposite in direction.

Solution The isolated system is the kaon before the decay and the two pions afterward. The decay of the kaon, represented in Figure 8.3, can be written

$$K^0 \rightarrow \pi^+ + \pi^-$$

If we let $\vec{\mathbf{p}}^+$ be the momentum of the positive pion and $\vec{\mathbf{p}}^-$ be the momentum of the negative pion after

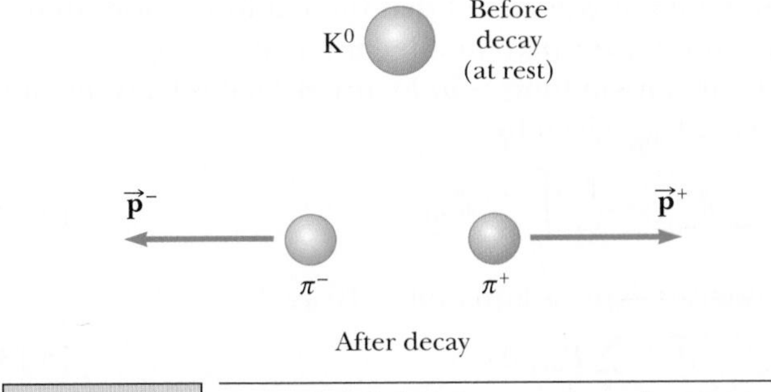

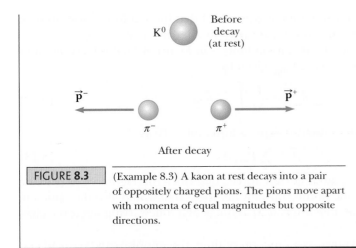

FIGURE 8.3 (Example 8.3) A kaon at rest decays into a pair of oppositely charged pions. The pions move apart with momenta of equal magnitudes but opposite directions.

the decay, the final momentum $\vec{\mathbf{p}}_f$ of the isolated system of two pions can be written

$$\vec{\mathbf{p}}_f = \vec{\mathbf{p}}^+ + \vec{\mathbf{p}}^-$$

Because the kaon is at rest before the decay, we know that the initial system momentum $\vec{\mathbf{p}}_i = 0$. Furthermore, because the momentum of the isolated system is conserved, $\vec{\mathbf{p}}_i = \vec{\mathbf{p}}_f = 0$, so that $\vec{\mathbf{p}}^+ + \vec{\mathbf{p}}^- = 0$ or

$$\vec{\mathbf{p}}^+ = -\vec{\mathbf{p}}^-$$

Therefore, we see that the two momentum vectors of the pions are equal in magnitude and opposite in direction.

8.2 IMPULSE AND MOMENTUM

As described by Equation 8.4, the momentum of a particle changes if a net force acts on the particle. Let us assume that a net force $\sum \vec{\mathbf{F}}$ acts on a particle and that this force may vary with time. According to Equation 8.4,

$$d\vec{\mathbf{p}} = \sum \vec{\mathbf{F}}\, dt \qquad [8.8]$$

We can integrate this expression to find the change in the momentum of a particle during the time interval $\Delta t = t_f - t_i$. Integrating Equation 8.8 gives

$$\Delta\vec{\mathbf{p}} = \vec{\mathbf{p}}_f - \vec{\mathbf{p}}_i = \int_{t_i}^{t_f} \sum \vec{\mathbf{F}}\, dt \qquad [8.9]$$

The integral of a force over the time interval during which it acts is called the **impulse** of the force. The impulse of the net force $\sum \vec{\mathbf{F}}$ is a vector defined by

$$\vec{\mathbf{I}} \equiv \int_{t_i}^{t_f} \sum \vec{\mathbf{F}}\, dt \qquad [8.10]$$

■ Impulse of a net force

The direction of the impulse vector is the same as the direction of the change in momentum. Impulse has the dimensions of momentum, ML/T.

Based on this definition, Equation 8.9 tells us that the total impulse of the net force $\sum \vec{\mathbf{F}}$ on a particle equals the change in the momentum of the particle: $\vec{\mathbf{I}} = \Delta\vec{\mathbf{p}}$. This statement, known as the **impulse–momentum theorem,** is equivalent to Newton's second law. It also applies to a system of particles for which the net external force on the system causes a change in the total momentum of the system:

$$\vec{\mathbf{I}} \equiv \int_{t_i}^{t_f} \sum \vec{\mathbf{F}}_{\text{ext}}\, dt = \Delta\vec{\mathbf{p}}_{\text{tot}} \qquad [8.11]$$

■ Impulse–momentum theorem

Impulse is an interaction between the system and its environment. As a result of this interaction, the momentum of the system changes. This idea is an analog to the continuity equation for energy, which relates an interaction with the environment to the change in the energy of the system. Therefore, when we say that an impulse is given to a system, we imply that momentum is transferred from an external agent to that system. In many situations, the system can be modeled as a particle, so Equation 8.10 can be used rather than the more general Equation 8.11.

From the definition, we see that impulse is a vector quantity having a magnitude equal to the area under the curve of the magnitude of the net force versus time, as

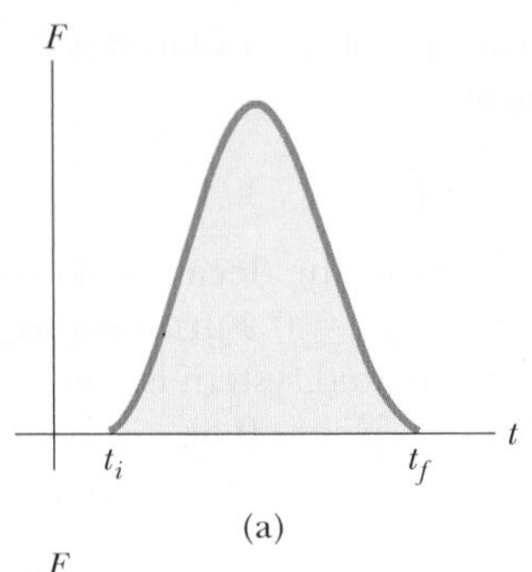

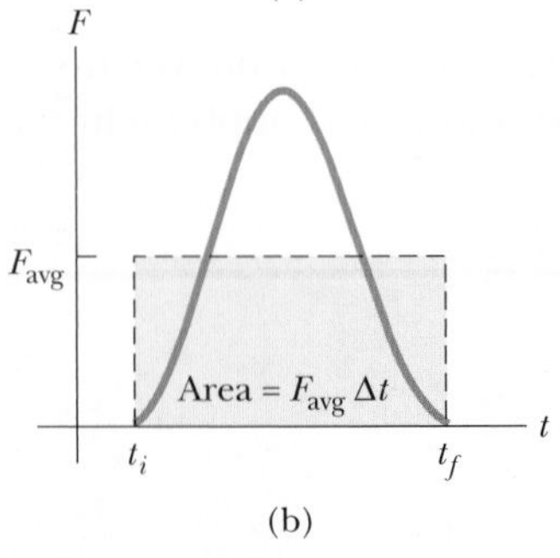

FIGURE 8.4 (a) A net force acting on a particle may vary in time. The impulse is the area under the curve of the magnitude of the net force versus time. (b) The average force (*horizontal dashed line*) gives the same impulse to the particle in the time interval Δt as the time-varying force described in part (a). The area of the rectangle is the same as the area under the curve.

illustrated in Figure 8.4. In this figure, it is assumed that the net force varies in time in the general manner shown and is nonzero in the time interval $\Delta t = t_f - t_i$.

Because the force can generally vary in time as in Figure 8.4a, it is convenient to define a time-averaged net force $\sum \vec{\mathbf{F}}_{avg}$ given by

$$\sum \vec{\mathbf{F}}_{avg} \equiv \frac{1}{\Delta t}\int_{t_i}^{t_f} \sum \vec{\mathbf{F}}\, dt \qquad [8.12]$$

where $\Delta t = t_f - t_i$. Therefore, we can express Equation 8.10 as

$$\vec{\mathbf{I}} = \sum \vec{\mathbf{F}}_{avg}\,\Delta t \qquad [8.13]$$

The magnitude of this average net force, described in Figure 8.4b, can be thought of as the magnitude of the constant net force that would give the same impulse to the particle in the time interval Δt as the actual time-varying net force gives over this same interval.

In principle, if $\sum \vec{\mathbf{F}}$ is known as a function of time, the impulse can be calculated from Equation 8.10. The calculation becomes especially simple if the net force acting on the particle is constant. In this case, $\sum \vec{\mathbf{F}}_{avg}$ over a time interval is the same as the constant $\sum \vec{\mathbf{F}}$ at any instant within the interval, and Equation 8.13 becomes

$$\vec{\mathbf{I}} = \Delta \vec{\mathbf{p}} = \sum \vec{\mathbf{F}}\,\Delta t \qquad [8.14]$$

In many physical situations, we shall use what is called the **impulse approximation: We assume that one of the forces exerted on a particle acts for a short time but is much greater than any other force present.** This simplification model allows us to ignore the effects of other forces because these effects are small for the short time interval during which the large force acts. This approximation is especially useful in treating collisions in which the duration of the collision is very short. When this approximation is made, we refer to the force that is greater as an *impulsive force.* For example, when a baseball is struck with a bat, the duration of the collision is about 0.01 s and the average force the bat exerts on the ball during this time interval is typically several thousand newtons. This average force is much greater than the gravitational force, so we ignore any change in velocity related to the gravitational force during the collision. It is important to remember that $\vec{\mathbf{p}}_i$ and $\vec{\mathbf{p}}_f$ represent the momenta *immediately* before and after the collision, respectively. Therefore in the impulse approximation, very little motion of the particle takes place during the collision.

Advantages of air bags in reducing injury

The concept of impulse helps us understand the value of air bags in stopping a passenger in an automobile accident (Fig. 8.5). The passenger experiences the same change in momentum and therefore the same impulse in a collision whether the car has air bags or not. The air bag allows the passenger to experience that change in momentum over a longer time interval, however, reducing the peak force on the passenger and increasing the chances of escaping without injury. Without the air bag, the passenger's head could move forward and be brought to rest in a short time interval by the steering wheel or the dashboard. In this case, the passenger undergoes the same change in momentum, but the short time interval results in a very large force that could cause severe head injury. Such injuries often result in spinal cord nerve damage where the nerves enter the base of the brain.

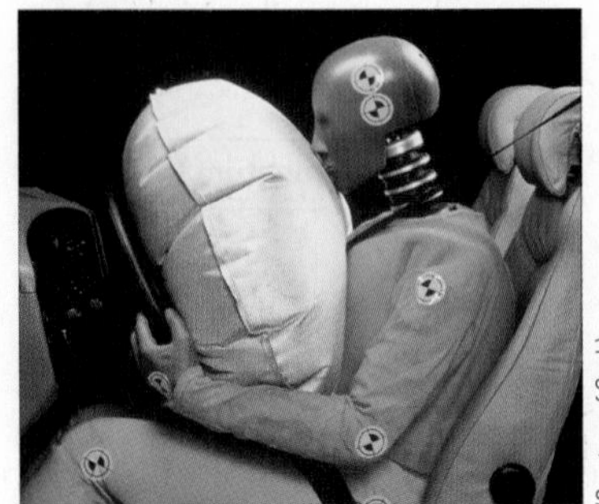
(Courtesy of Saab)

FIGURE 8.5 A test dummy is brought to rest by an air bag in an automobile.

QUICK QUIZ 8.3 Two objects are at rest on a frictionless surface. Object 1 has a greater mass than object 2. (i) When a constant force is applied to object 1, it accelerates through a distance *d*. The force is removed from object 1 and is applied to object 2. At the moment when object 2 has accelerated through the same distance *d*, which statements are true? **(a)** $p_1 < p_2$ **(b)** $p_1 = p_2$ **(c)** $p_1 > p_2$ **(d)** $K_1 < K_2$ **(e)** $K_1 = K_2$ **(f)** $K_1 > K_2$ **(ii)** When a constant force is applied to object 1, it accelerates for a time interval Δt. The force is removed from object 1 and is applied to object 2. After object 2 has accelerated for the same time interval Δt, which statements are true? **(a)** $p_1 < p_2$ **(b)** $p_1 = p_2$ **(c)** $p_1 > p_2$ **(d)** $K_1 < K_2$ **(e)** $K_1 = K_2$ **(f)** $K_1 > K_2$

EXAMPLE 8.4 How Good Are the Bumpers?

In a crash test, an automobile of mass 1 500 kg collides with a wall as in Figure 8.6. The initial and final velocities of the automobile are $\vec{\mathbf{v}}_i = -15.0\hat{\mathbf{i}}$ m/s and $\vec{\mathbf{v}}_f = 2.60\hat{\mathbf{i}}$ m/s. If the collision lasts for 0.150 s, find the impulse due to the collision and the average force exerted on the automobile.

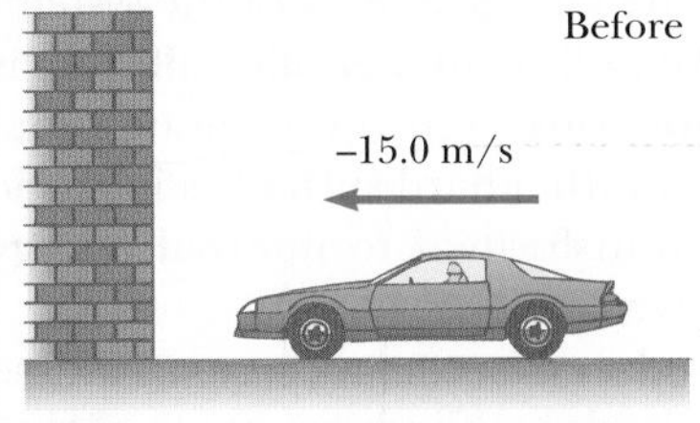

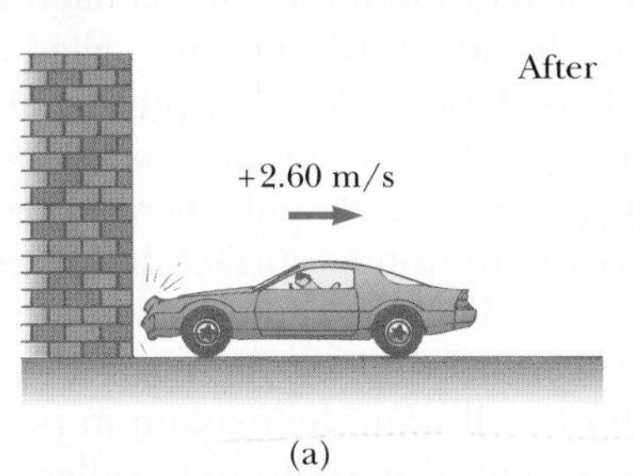

(a)

(b)

(Tim Wright/CORBIS)

FIGURE 8.6 (Example 8.4) (a) The car's momentum changes as a result of its collision with the wall. (b) In a crash test, the large force exerted by the wall on the car produces extensive damage to the car's front end.

Solution We identify the automobile as the system. The initial and final momenta of the automobile are

$$\vec{\mathbf{p}}_i = m\vec{\mathbf{v}}_i = (1\,500 \text{ kg})(-15.0\hat{\mathbf{i}} \text{ m/s})$$
$$= -2.25 \times 10^4\hat{\mathbf{i}} \text{ kg}\cdot\text{m/s}$$
$$\vec{\mathbf{p}}_f = m\vec{\mathbf{v}}_f = (1\,500 \text{ kg})(2.60\hat{\mathbf{i}} \text{ m/s})$$
$$= 0.390 \times 10^4\hat{\mathbf{i}} \text{ kg}\cdot\text{m/s}$$

Hence, the impulse is

$$\vec{\mathbf{I}} = \Delta\vec{\mathbf{p}} = \vec{\mathbf{p}}_f - \vec{\mathbf{p}}_i$$
$$= 0.390 \times 10^4\hat{\mathbf{i}} \text{ kg}\cdot\text{m/s} - (-2.25 \times 10^4\hat{\mathbf{i}} \text{ kg}\cdot\text{m/s})$$
$$\vec{\mathbf{I}} = 2.64 \times 10^4\hat{\mathbf{i}} \text{ kg}\cdot\text{m/s}$$

The average force exerted on the automobile is

$$\vec{\mathbf{F}}_{\text{avg}} = \frac{\Delta\vec{\mathbf{p}}}{\Delta t} = \frac{2.64 \times 10^4\hat{\mathbf{i}} \text{ kg}\cdot\text{m/s}}{0.150 \text{ s}} = 1.76 \times 10^5\hat{\mathbf{i}} \text{ N}$$

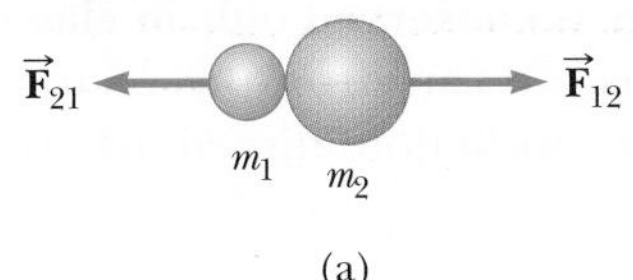

(a)

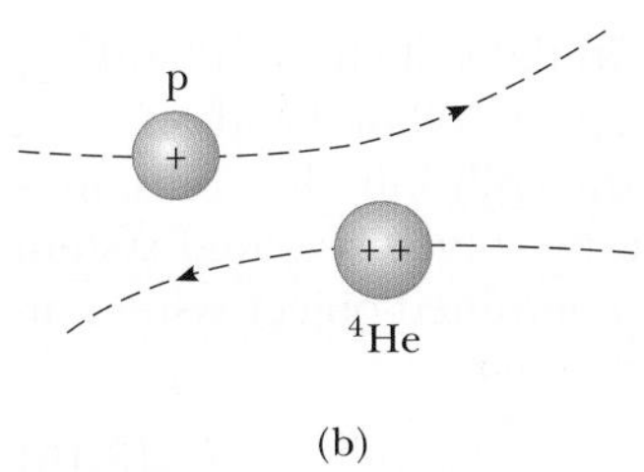

(b)

FIGURE 8.7 (a) A collision between two objects as the result of direct contact. (b) A "collision" between two charged particles that do not make contact.

8.3 COLLISIONS

In this section, we use the law of conservation of momentum to describe what happens when two objects collide. The forces due to the collision are assumed to be much larger than any external forces present, so we use the simplification model we call the impulse approximation. The general goal in collision problems is to relate the final conditions of the system to the initial conditions.

A collision may be the result of physical contact between two objects, as described in Figure 8.7a. This observation is common when two macroscopic objects collide, such as two billiard balls or a baseball and a bat.

The notion of what we mean by *collision* must be generalized because "contact" on a microscopic scale is ill defined. To understand the distinction between macroscopic and microscopic collisions, consider the collision of a proton with an alpha particle (the nucleus of the helium atom), illustrated in Figure 8.7b. Because the two particles are positively charged, they repel each other. A collision has occurred, but the colliding particles were never in "contact."

When two particles of masses m_1 and m_2 collide, the collision forces may vary in time in a complicated way, as seen in Figure 8.4. As a result, an analysis of the situation with Newton's second law could be very complicated. We find, however, that the momentum concept is similar to the energy concept in Chapters 6 and 7 in that it provides us with a much easier method to solve problems involving isolated systems.

According to Equation 8.5, the momentum of an isolated system is conserved during some interaction event, such as a collision. The kinetic energy of the system, however, is generally *not* conserved in a collision. We define an **inelastic collision** as one in which the kinetic energy of the system is not conserved (even though momentum is conserved). The collision of a rubber ball with a hard surface is inelastic because some of the kinetic energy of the ball is transformed to internal energy when the ball is deformed while in contact with the surface.

Glaucoma testing

A practical example of an inelastic collision is used to detect glaucoma, a disease in which the pressure inside the eye builds up and leads to blindness by damaging the cells of the retina. In this application, medical professionals use a device called a *tonometer* to measure the pressure inside the eye. This device releases a puff of air against the outer surface of the eye and measures the speed of the air after reflection from the eye. At normal pressure, the eye is slightly spongy and the pulse is reflected at low speed. As the pressure inside the eye increases, the outer surface becomes more rigid and the speed of the reflected pulse increases. Therefore, the speed of the reflected puff of air is used to measure the internal pressure of the eye.

When two objects collide and stick together after a collision, the maximum possible fraction of the initial kinetic energy is transformed; this collision is called a **perfectly inelastic collision.** For example, if two vehicles collide and become entangled, they move with some common velocity after the perfectly inelastic collision. If a meteorite collides with the Earth, it becomes buried in the ground and the collision is perfectly inelastic.

PITFALL PREVENTION 8.2

PERFECTLY INELASTIC COLLISIONS Keep in mind the distinction between inelastic and perfectly inelastic collisions. If the colliding particles stick together, the collision is perfectly inelastic. If they bounce off each other (and kinetic energy is not conserved), the collision is inelastic. Generally, inelastic collisions are hard to analyze unless additional information is provided. This difficulty appears in the mathematical representation as having more unknowns than equations.

An **elastic collision** is defined as one in which the kinetic energy of the system is conserved (as well as momentum). Real collisions in the macroscopic world, such as those between billiard balls, are only approximately elastic because some transformation of kinetic energy takes place and some energy leaves the system by mechanical waves, sound. Imagine a billiard game with truly elastic collisions. The opening break would be completely silent! Truly elastic collisions do occur between atomic and subatomic particles. Elastic and perfectly inelastic collisions are *limiting* cases; a large number of collisions fall in the range between them.

In the remainder of this section, we treat collisions in one dimension and consider the two extreme cases: perfectly inelastic collisions and elastic collisions. The important distinction between these two types of collisions is that **the momentum of the system is conserved in all cases, but the kinetic energy is conserved only in elastic collisions.** When analyzing one-dimensional collisions, we can drop the vector notation and use positive and negative signs for velocities to denote directions, as we did in Chapter 2.

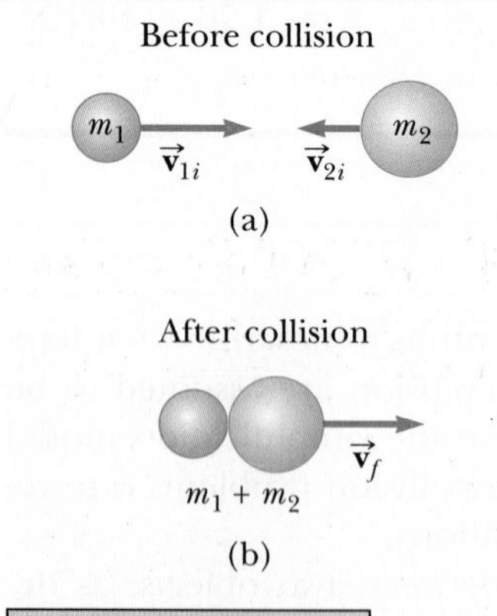

ACTIVE FIGURE 8.8
A perfectly inelastic head-on collision between two particles: (a) before the collision and (b) after the collision.

Physics Now™ Log into PhysicsNow at **www.pop4e.com** and go to Active Figure 8.8 to adjust the masses and velocities of the colliding objects and see the effect on the final velocity.

One-Dimensional Perfectly Inelastic Collisions

Consider two objects of masses m_1 and m_2 moving with initial velocities v_{1i} and v_{2i} along a straight line as in Active Figure 8.8. If the two objects collide head-on, stick together, and move with some common velocity v_f after the collision, the collision is perfectly inelastic. Because the total momentum of the two-object isolated system before the collision equals the total momentum of the combined-object system after the collision, we have

$$m_1 v_{1i} + m_2 v_{2i} = (m_1 + m_2) v_f \qquad [8.15]$$

$$v_f = \frac{m_1 v_{1i} + m_2 v_{2i}}{m_1 + m_2} \qquad [8.16]$$

Therefore, if we know the initial velocities of the two objects, we can use this single equation to determine the final common velocity.

One-Dimensional Elastic Collisions

Now consider two objects that undergo an elastic head-on collision (Active Fig. 8.9) in one dimension. In this collision, both momentum and kinetic energy are conserved; therefore, we can write[2]

$$m_1 v_{1i} + m_2 v_{2i} = m_1 v_{1f} + m_2 v_{2f} \qquad [8.17]$$

$$\tfrac{1}{2} m_1 v_{1i}^2 + \tfrac{1}{2} m_2 v_{2i}^2 = \tfrac{1}{2} m_1 v_{1f}^2 + \tfrac{1}{2} m_2 v_{2f}^2 \qquad [8.18]$$

In a typical problem involving elastic collisions, two unknown quantities occur (such as v_{1f} and v_{2f}), and Equations 8.17 and 8.18 can be solved simultaneously to find them. An alternative approach, employing a little mathematical manipulation of Equation 8.18, often simplifies this process. Let us cancel the factor of $\frac{1}{2}$ in Equation 8.18 and rewrite the equation as

$$m_1 (v_{1i}^2 - v_{1f}^2) = m_2 (v_{2f}^2 - v_{2i}^2)$$

Here we have moved the terms containing m_1 to one side of the equation and those containing m_2 to the other. Next, let us factor both sides:

$$m_1 (v_{1i} - v_{1f})(v_{1i} + v_{1f}) = m_2 (v_{2f} - v_{2i})(v_{2f} + v_{2i}) \qquad [8.19]$$

We now separate the terms containing m_1 and m_2 in the equation for conservation of momentum (Eq. 8.17) to obtain

$$m_1 (v_{1i} - v_{1f}) = m_2 (v_{2f} - v_{2i}) \qquad [8.20]$$

To obtain our final result, we divide Equation 8.19 by Equation 8.20 and obtain

$$v_{1i} + v_{1f} = v_{2f} + v_{2i}$$

or, gathering initial and final values on opposite sides of the equation,

$$v_{1i} - v_{2i} = -(v_{1f} - v_{2f}) \qquad [8.21]$$

This equation, in combination with the condition for conservation of momentum, Equation 8.17, can be used to solve problems dealing with one-dimensional elastic collisions between two objects. According to Equation 8.21, the relative speed[3] $v_{1i} - v_{2i}$ of the two objects before the collision equals the negative of their relative speed after the collision, $-(v_{1f} - v_{2f})$.

Suppose the masses and the initial velocities of both objects are known. Equations 8.17 and 8.21 can be solved for the final velocities in terms of the initial values because we have two equations and two unknowns:

$$v_{1f} = \left(\frac{m_1 - m_2}{m_1 + m_2}\right) v_{1i} + \left(\frac{2m_2}{m_1 + m_2}\right) v_{2i} \qquad [8.22]$$

$$v_{2f} = \left(\frac{2m_1}{m_1 + m_2}\right) v_{1i} + \left(\frac{m_2 - m_1}{m_1 + m_2}\right) v_{2i} \qquad [8.23]$$

It is important to remember that the appropriate signs for the velocities v_{1i} and v_{2i} must be included in Equations 8.22 and 8.23. For example, if m_2 is moving to the left initially, as in Active Figure 8.9a, v_{2i} is negative.

Let us consider some special cases. If $m_1 = m_2$, Equations 8.22 and 8.23 show us that $v_{1f} = v_{2i}$ and $v_{2f} = v_{1i}$. That is, the objects exchange speeds if they have equal

[2]Notice that the kinetic energy of the system is the sum of the kinetic energies of the two particles. In our energy conservation examples in Chapter 7 involving a falling object and the Earth, we ignored the kinetic energy of the Earth because it is so small. Therefore, the kinetic energy of the *system* is just the kinetic energy of the falling *object*. That is a special case in which the mass of one of the objects (the Earth) is so immense that ignoring its kinetic energy introduces no measurable error. For problems such as those described here, however, and for the particle decay problems we will see in Chapters 30 and 31, we need to include the kinetic energies of *all* particles in the system.

[3]See Section 3.6 for a review of relative speed.

PITFALL PREVENTION 8.3

Momentum and Kinetic Energy in Collisions Linear momentum of an isolated system is conserved in *all* collisions. Kinetic energy of an isolated system is conserved *only* in elastic collisions. These statements are true because kinetic energy can be transformed into several types of energy or can be transferred out of the system (so that the system may *not* be isolated in terms of energy during the collision), but there is only one type of linear momentum.

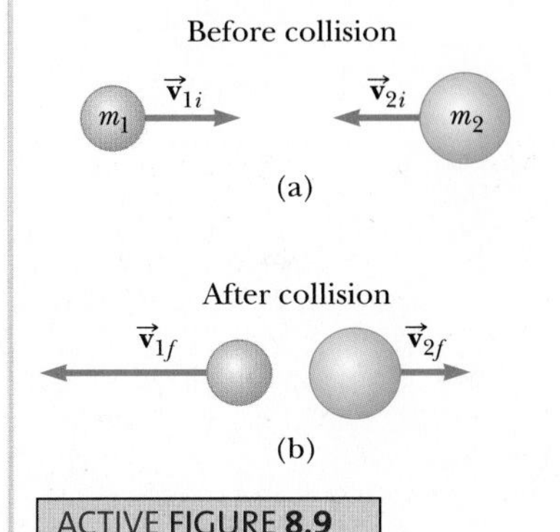

ACTIVE FIGURE 8.9
An elastic head-on collision between two particles: (a) before the collision and (b) after the collision.

PhysicsNow™ Log into PhysicsNow at **www.pop4e.com** and go to Active Figure 8.9 to adjust the masses and velocities of the colliding objects and see the effect on the final velocities.

PITFALL PREVENTION 8.4

Not a General Equation We have spent some effort on deriving Equation 8.21, but remember that it can be used only in a very *specific* situation: a one-dimensional, elastic collision between two objects. The *general* concept is conservation of momentum (and conservation of kinetic energy if the collision is elastic) for an isolated system.

masses. That is what one observes in head-on billiard ball collisions, assuming there is no spin on the ball: The initially moving ball stops and the initially stationary ball moves away with approximately the same speed.

If m_2 is initially at rest, $v_{2i} = 0$ and Equations 8.22 and 8.23 become

■ Elastic collision in one dimension: particle 2 initially at rest

$$v_{1f} = \left(\frac{m_1 - m_2}{m_1 + m_2}\right) v_{1i} \qquad [8.24]$$

$$v_{2f} = \left(\frac{2m_1}{m_1 + m_2}\right) v_{1i} \qquad [8.25]$$

If m_1 is very large compared with m_2, we see from Equations 8.24 and 8.25 that $v_{1f} \approx v_{1i}$ and $v_{2f} \approx 2v_{1i}$. That is, when a very heavy object collides head-on with a very light one initially at rest, the heavy object continues its motion unaltered after the collision but the light object rebounds with a speed equal to about twice the initial speed of the heavy object. An example of such a collision is that of a moving heavy atom, such as uranium, with a light atom, such as hydrogen.

If m_2 is much larger than m_1 and if m_2 is initially at rest, we find from Equations 8.24 and 8.25 that $v_{1f} \approx -v_{1i}$ and $v_{2f} \approx 0$. That is, when a very light object collides head-on with a very heavy object initially at rest, the velocity of the light object is reversed and the heavy object remains approximately at rest. For example, imagine what happens when a marble hits a stationary bowling ball.

QUICK QUIZ 8.4 A Ping-Pong ball is thrown at a stationary bowling ball hanging from a wire. The Ping-Pong ball makes a one-dimensional elastic collision and bounces back along the same line. After the collision, the Ping-Pong ball has, compared with the bowling ball, **(a)** a larger magnitude of momentum and more kinetic energy, **(b)** a smaller magnitude of momentum and more kinetic energy, **(c)** a larger magnitude of momentum and less kinetic energy, **(d)** a smaller magnitude of momentum and less kinetic energy, or **(e)** the same magnitude of momentum and the same kinetic energy

PROBLEM-SOLVING STRATEGY One-Dimensional Collisions

We suggest that you use the following approach when solving collision problems in one dimension:

1. **Conceptualize** Establish the mental representation by imagining the collision occurring in your mind. Draw simple diagrams of the particles before and after the collision with appropriate velocity vectors. You may have to guess for now at the directions of final velocity vectors.
2. **Categorize** Is the system of particles truly isolated? If so, categorize the collision as elastic, inelastic, or perfectly inelastic.
3. **Analyze** Set up the appropriate mathematical representation for the problem. If the collision is perfectly inelastic, use Equation 8.15. If the collision is elastic, use Equations 8.17 and 8.21. If the collision is inelastic, use Equation 8.17. To find the final velocities in this case, you will need some additional piece of information.
4. **Finalize** Once you have determined your result, check to see that your answers are consistent with the mental and pictorial representations and that your results are realistic.

EXAMPLE 8.5 Kinetic Energy in a Perfectly Inelastic Collision

We claimed that the maximum amount of kinetic energy was transformed to other forms in a perfectly inelastic collision. Prove this statement mathematically for a one-dimensional two-particle collision.

Solution We will assume that the maximum kinetic energy is transformed and prove that the collision must be perfectly inelastic. We set up the fraction f of the final kinetic energy after the collision to the initial kinetic energy:

$$f = \frac{K_f}{K_i} = \frac{\frac{1}{2}m_1 v_{1f}^2 + \frac{1}{2}m_2 v_{2f}^2}{\frac{1}{2}m_1 v_{1i}^2 + \frac{1}{2}m_2 v_{2i}^2} = \frac{m_1 v_{1f}^2 + m_2 v_{2f}^2}{m_1 v_{1i}^2 + m_2 v_{2i}^2}$$

The *maximum* amount of energy transformed to other forms corresponds to the *minimum* value of f. For fixed initial conditions, we imagine that the final velocities v_{1f} and v_{2f} are variables. We minimize the fraction f by taking the derivative of f with respect to v_{1f} and setting the result equal to zero:

$$\frac{df}{dv_{1f}} = \frac{d}{dv_{1f}}\left(\frac{m_1 v_{1f}^{\,2} + m_2 v_{2f}^{\,2}}{m_1 v_{1i}^{\,2} + m_2 v_{2i}^{\,2}}\right)$$

$$= \frac{2m_1 v_{1f} + 2m_2 v_{2f}\dfrac{dv_{2f}}{dv_{1f}}}{m_1 v_{1i}^{\,2} + m_2 v_{2i}^{\,2}} = 0$$

$$(1) \quad \rightarrow \quad m_1 v_{1f} + m_2 v_{2f}\frac{dv_{2f}}{dv_{1f}} = 0$$

From the conservation of momentum condition, we can evaluate the derivative in (1). We differentiate Equation 8.17 with respect to v_{1f}:

$$\frac{d}{dv_{1f}}(m_1 v_{1i} + m_2 v_{2i}) = \frac{d}{dv_{1f}}(m_1 v_{1f} + m_2 v_{2f})$$

$$\rightarrow \quad 0 = m_1 + m_2 \frac{dv_{2f}}{dv_{1f}} \quad \rightarrow \quad \frac{dv_{2f}}{dv_{1f}} = -\frac{m_1}{m_2}$$

Substituting this expression for the derivative into (1), we find

$$m_1 v_{1f} - m_2 v_{2f}\frac{m_1}{m_2} = 0 \quad \rightarrow \quad v_{1f} = v_{2f}$$

If the particles come out of the collision with the same velocities, they are joined together and it is a perfectly inelastic collision, which is what we set out to prove.

EXAMPLE 8.6 Carry Collision Insurance

An 1 800-kg car stopped at a traffic light is struck from the rear by a 900-kg car and the two become entangled. If the smaller car was moving at 20.0 m/s before the collision, what is the speed of the entangled cars after the collision?

Solution The total momentum of the system (the two cars) before the collision equals the total momentum of the system after the collision because the system is isolated (in the impulse approximation). Notice that we ignore friction with the road in the impulse approximation. Therefore, the result we obtain for the final speed will only be approximately true just after the collision. For longer time intervals after the collision, we would use Newton's second law to describe the slowing down of the system as a result of friction. Because the cars "become entangled," it is a perfectly inelastic collision.

The magnitude of the total momentum of the system before the collision is equal to that of only the smaller car because the larger car is initially at rest:

$$p_i = m_1 v_i = (900 \text{ kg})(20.0 \text{ m/s}) = 1.80 \times 10^4 \text{ kg}\cdot\text{m/s}$$

After the collision, the mass that moves is the sum of the masses of the cars. The magnitude of the momentum of the combination is

$$p_f = (m_1 + m_2)v_f = (2\,700 \text{ kg})v_f$$

Equating the initial momentum to the final momentum and solving for v_f, the speed of the entangled cars, we have

$$v_f = \frac{p_f}{m_1 + m_2} = \frac{p_i}{m_1 + m_2} = \frac{1.80 \times 10^4 \text{ kg}\cdot\text{m/s}}{2\,700 \text{ kg}}$$

$$= 6.67 \text{ m/s}$$

EXAMPLE 8.7 Slowing Down Neutrons by Collisions

In a nuclear reactor, neutrons are produced when $^{235}_{92}\text{U}$ atoms split in a process called *fission*. These neutrons are moving at about 10^7 m/s and must be slowed down to about 10^3 m/s before they take part in another fission event. They are slowed down by being passed through a solid or liquid material called a *moderator*. The slowing-down process involves elastic collisions. Let us show that a neutron can lose most of its kinetic energy if it collides elastically with a moderator containing light nuclei, such as deuterium (in "heavy water," D_2O).

Solution We identify the system as the neutron and a moderator nucleus. Because the momentum and kinetic energy of this system are conserved in an elastic collision, Equations 8.24 and 8.25 can be applied to a one-dimensional collision of these two particles.

Let us assume that the moderator nucleus of mass m_m is at rest initially and that the neutron of mass m_n and initial speed v_{ni} collides head-on with it. The initial kinetic energy of the neutron is

$$K_{ni} = \tfrac{1}{2} m_n v_{ni}^{\,2}$$

After the collision, the neutron has kinetic energy $\frac{1}{2}m_n v_{nf}^2$, where v_{nf} is given by Equation 8.24:

$$K_{nf} = \tfrac{1}{2}m_n v_{nf}^2 = \tfrac{1}{2}m_n\left(\frac{m_n - m_m}{m_n + m_m}\right)^2 v_{ni}^2$$

Therefore, the fraction of the total kinetic energy possessed by the neutron after the collision is

$$(1)\quad f_n = \frac{K_{nf}}{K_{ni}} = \frac{\frac{1}{2}m_n\left(\frac{m_n - m_m}{m_n + m_m}\right)^2 v_{ni}^2}{\frac{1}{2}m_n v_{ni}^2} = \left(\frac{m_n - m_m}{m_n + m_m}\right)^2$$

From this result, we see that the final kinetic energy of the neutron is small when m_m is close to m_n and is zero when $m_m = m_n$.

We can calculate the kinetic energy of the moderator nucleus after the collision using Equation 8.25:

$$K_{mf} = \tfrac{1}{2}m_m v_{mf}^2 = \frac{2m_n^2 m_m}{(m_n + m_m)^2} v_{ni}^2$$

Hence, the fraction of the total kinetic energy transferred to the moderator nucleus is

$$(2)\quad f_{\text{trans}} = \frac{K_{mf}}{K_{ni}} = \frac{\frac{2m_n^2 m_m}{(m_n + m_m)^2} v_{ni}^2}{\frac{1}{2}m_n v_{ni}^2} = \frac{4m_n m_m}{(m_n + m_m)^2}$$

If $m_m \approx m_n$, we see that $f_{\text{trans}} \approx 1 = 100\%$. Because the system's kinetic energy is conserved, (2) can also be obtained from (1) with the condition that $f_n + f_m = 1$, so that $f_m = 1 - f_n$.

For collisions of the neutrons with deuterium nuclei in D_2O ($m_m = 2m_n$), $f_n = 1/9$ and $f_{\text{trans}} = 8/9$. That is, 89% of the neutron's kinetic energy is transferred to the deuterium nucleus. In practice, the moderator efficiency is reduced because head-on collisions are very unlikely to occur.

INTERACTIVE EXAMPLE 8.8 Two Blocks and a Spring

A block of mass $m_1 = 1.60$ kg, initially moving to the right with a speed of 4.00 m/s on a frictionless horizontal track, collides with a massless spring attached to a second block of mass $m_2 = 2.10$ kg, moving to the left with a speed of 2.50 m/s as in Figure. 8.10a. The spring has a spring constant of 600 N/m.

A At the instant when m_1 is moving to the right with a speed of 3.00 m/s as in Figure 8.10b, determine the speed of m_2.

Solution Figure 8.10 helps conceptualize the problem. Because the blocks move along a frictionless straight track, we categorize this problem as one involving a one-dimensional collision between objects forming an isolated system. We identify the system as the two blocks and the spring and identify the collision as elastic because the force from the spring is conservative. Because the total momentum of the isolated system is conserved, we analyze the problem by recognizing that

$$m_1 v_{1i} + m_2 v_{2i} = m_1 v_{1f} + m_2 v_{2f}$$

$$(1.60\text{ kg})(4.00\text{ m/s}) + (2.10\text{ kg})(-2.50\text{ m/s}) = (1.60\text{ kg})(3.00\text{ m/s}) + (2.10\text{ kg})v_{2f}$$

$$v_{2f} = -1.74\text{ m/s}$$

Note that the initial velocity of m_2 is -2.50 m/s because its direction is to the left. The negative value for v_{2f} means that m_2 is still moving to the left at the instant we are considering.

B Determine the distance the spring is compressed at that instant.

Solution Because the system is isolated, we can also analyze this problem from the energy version of the

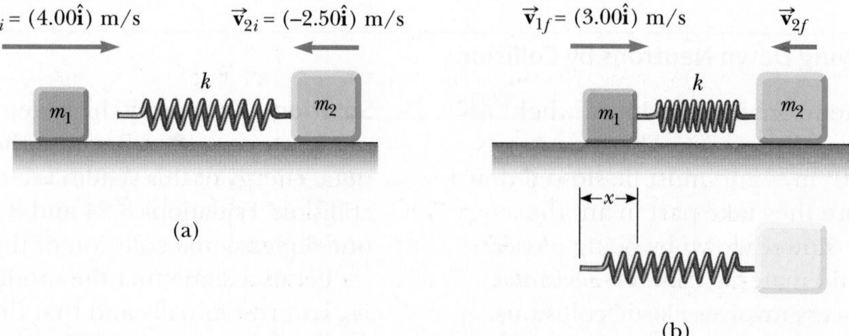

FIGURE 8.10 (Interactive Example 8.8) A moving block collides with another moving block with a spring attached: (a) before the collision and (b) at one instant during the collision.

isolated system model to determine the compression x in the spring shown in Figure 8.10b. No nonconservative forces are acting within the system, so the mechanical energy of the system is conserved:

$$E_i = E_f$$

$$\tfrac{1}{2}m_1v_{1i}^2 + \tfrac{1}{2}m_2v_{2i}^2 = \tfrac{1}{2}m_1v_{1f}^2 + \tfrac{1}{2}m_2v_{2f}^2 + \tfrac{1}{2}kx^2$$

Substituting the given values and the result to part A into this expression gives

$$x = 0.173 \text{ m}$$

C Determine the maximum distance by which the spring is compressed during the collision.

Solution The maximum compression of the spring occurs when the two blocks are not moving relative to each other. For their relative velocity to be zero they must be moving with the same velocity in our reference frame as we watch the collision. Therefore, we can model the collision *up to this point* as a perfectly inelastic collision:

$$m_1v_{1i} + m_2v_{2i} = (m_1 + m_2)v_f$$

$$(1.60 \text{ kg})(4.00 \text{ m/s}) + (2.10 \text{ kg})(-2.50 \text{ m/s}) = (1.60 \text{ kg} + 2.10 \text{ kg})v_f$$

$$v_f = 0.311 \text{ m/s}$$

As in part B, mechanical energy is conserved, so we set up a conservation of mechanical energy expression:

$$E_i = E_f$$

$$\tfrac{1}{2}m_1v_{1i}^2 + \tfrac{1}{2}m_2v_{2i}^2 = \tfrac{1}{2}(m_1 + m_2)v_f^2 + \tfrac{1}{2}kx^2$$

Substituting the values into this expression gives

$$x = 0.253 \text{ m}$$

To finalize the problem, note that the value for x in part C is larger than that in part B. We can argue that this result is consistent with our mental representation. In part C, the blocks are moving at the same speed so that their relative speed is zero. In parts A and B, note that the blocks are moving with speeds $v_{1f} = 3.00$ m/s and $v_{2f} = -1.74$ m/s at the instant of interest. Therefore, the blocks are moving toward each other with a relative speed of 4.74 m/s. As a result, the spring will continue to compress, and the ultimate maximum value of x will be larger than that value found in part B.

PhysicsNow™ Investigate this situation with a variety of masses and initial speeds of the blocks by logging into PhysicsNow at **www.pop4e.com** and going to Interactive Example 8.8.

8.4 TWO-DIMENSIONAL COLLISIONS

In Section 8.1, we showed that the total momentum of a system is conserved when the system is isolated (i.e., when no external forces act on the system). For a general collision of two objects in three-dimensional space, the principle of conservation of momentum implies that the total momentum in each direction is conserved. An important subset of collisions takes place in a plane. The game of billiards is a familiar example involving multiple collisions of objects moving on a two-dimensional surface. Let us restrict our attention to a single two-dimensional collision between two objects that takes place in a plane. For such collisions, we obtain two component equations for the conservation of momentum:

$$m_1v_{1ix} + m_2v_{2ix} = m_1v_{1fx} + m_2v_{2fx}$$

$$m_1v_{1iy} + m_2v_{2iy} = m_1v_{1fy} + m_2v_{2fy}$$

where we use three subscripts in this general equation to represent, respectively, (1) the identification of the object, (2) initial and final values, and (3) the velocity component in the x or y direction.

Consider a two-dimensional problem in which an object of mass m_1 collides with an object of mass m_2 that is initially at rest as in Active Figure 8.11. After the collision, m_1 moves at an angle θ with respect to the horizontal and m_2 moves at an angle ϕ with respect to the horizontal. This collision is called a *glancing* collision. Applying the law of conservation of momentum in component form and noting that the initial y component of momentum is zero, we have

$$x \text{ component:} \quad m_1v_{1i} + 0 = m_1v_{1f}\cos\theta + m_2v_{2f}\cos\phi \qquad [8.26]$$

$$y \text{ component:} \quad 0 + 0 = m_1v_{1f}\sin\theta - m_2v_{2f}\sin\phi \qquad [8.27]$$

ACTIVE FIGURE **8.11**

A glancing collision between two particles.

Physics Now™ Log into PhysicsNow at **www.pop4e.com** and go to Active Figure 8.11 to adjust the speed and position of the blue particle and the masses of both particles to see the effects.

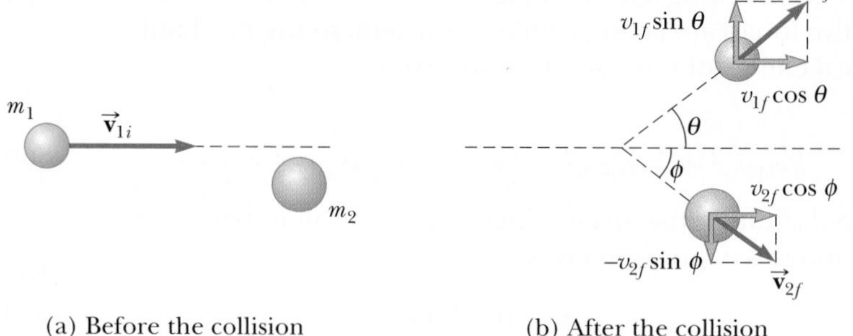

If the collision is elastic, we can write a third equation for conservation of kinetic energy in the form

$$\tfrac{1}{2}m_1{v_{1i}}^2 = \tfrac{1}{2}m_1{v_{1f}}^2 + \tfrac{1}{2}m_2{v_{2f}}^2 \qquad [8.28]$$

If we know the initial velocity v_{1i} and the masses, we are left with four unknowns (v_{1f}, v_{2f}, θ, and ϕ). Because we have only three equations, one of the four remaining quantities must be given to determine the motion after the collision from conservation principles alone.

If the collision is inelastic, kinetic energy is *not* conserved and Equation 8.28 does *not* apply.

PROBLEM-SOLVING STRATEGY Two-Dimensional Collisions

The following procedure is recommended when dealing with problems involving collisions between two objects.

1. **Conceptualize** Imagine the collisions occurring in your mind and predict the approximate directions in which the particles will move after the collision. Set up a coordinate system and define your velocities with respect to that system. It is convenient to have the *x* axis coincide with one of the initial velocities. Sketch the coordinate system, draw and label all velocity vectors, and include all the given information.
2. **Categorize** Is the system of particles truly isolated? If so, categorize the collision as elastic, inelastic, or perfectly inelastic.
3. **Analyze** Write expressions for the *x* and *y* components of the momentum of each object before and after the collision. Remember to include the appropriate signs for the components of the velocity vectors. It is essential that you pay careful attention to signs.

   Write expressions for the *total* momentum in the *x* direction *before* and *after* the collision and equate the two. Repeat this procedure for the total momentum in the *y* direction.

   Proceed to solve the momentum equations for the unknown quantities. If the collision is inelastic, kinetic energy is *not* conserved and additional information is probably required. If the collision is perfectly inelastic, the final velocities of the two objects are equal.

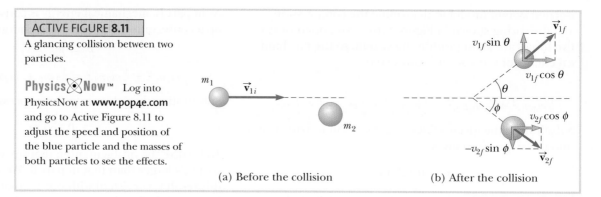

   If the collision is elastic, kinetic energy is conserved and you can equate the total kinetic energy before the collision to the total kinetic energy after the collision. This step provides an additional relationship between the velocity magnitudes.
4. **Finalize** Once you have determined your result, check to see that your answers are consistent with the mental and pictorial representations and that your results are realistic.

EXAMPLE 8.9 Proton–Proton Collision

A proton collides elastically with another proton that is initially at rest. The incoming proton has an initial speed of 3.5×10^5 m/s and makes a glancing collision with the second proton as in Active Figure 8.11. (At close separations, the protons exert a repulsive electrostatic force on each other.) After the collision, one proton moves off at an angle of 37° to the original direction of motion and the second deflects at an angle of ϕ to the same axis. Find the final speeds of the two protons and the angle ϕ.

Solution The isolated system is the pair of protons. Both momentum and kinetic energy of the system are conserved in this glancing elastic collision. Because

$m_1 = m_2$, $\theta = 37°$, and we are given that $v_{1i} = 3.5 \times 10^5$ m/s, Equations 8.26, 8.27, and 8.28 become

$$(1) \qquad v_{1f} \cos 37° + v_{2f} \cos \phi = 3.5 \times 10^5 \text{ m/s}$$

$$(2) \qquad v_{1f} \sin 37° - v_{2f} \sin \phi = 0$$

$$(3) \quad {v_{1f}}^2 + {v_{2f}}^2 = (3.5 \times 10^5 \text{ m/s})^2 = 1.2 \times 10^{11} \text{ m}^2/\text{s}^2$$

We rewrite (1) and (2) as follows:

$$v_{2f} \cos \phi = 3.5 \times 10^5 \text{ m/s} - v_{1f} \cos 37°$$

$$v_{2f} \sin \phi = v_{1f} \sin 37°$$

Now we square these two equations and add them:

$$\begin{aligned} {v_{2f}}^2 \cos^2 \phi + {v_{2f}}^2 \sin^2 \phi = 1.2 \times 10^{11} \text{ m}^2/\text{s}^2 \\ - (7.0 \times 10^5 \text{ m/s}) v_{1f} \cos 37° \\ + {v_{1f}}^2 \cos^2 37° + {v_{1f}}^2 \sin^2 37° \end{aligned}$$

$$\rightarrow \quad {v_{2f}}^2 = 1.2 \times 10^{11} - (5.6 \times 10^5) v_{1f} + {v_{1f}}^2$$

Substituting this expression into (3) gives

$${v_{1f}}^2 + [1.2 \times 10^{11} - (5.6 \times 10^5) v_{1f} + {v_{1f}}^2] = 1.2 \times 10^{11}$$

$$\rightarrow \quad 2{v_{1f}}^2 - (5.6 \times 10^5) v_{1f} = (2v_{1f} - 5.6 \times 10^5) v_{1f} = 0$$

One possibility for the solution of this equation is $v_{1f} = 0$, which corresponds to a head-on collision; the first proton stops and the second continues with the same speed in the same direction. This result is not what we want. The other possibility is

$$2v_{1f} - 5.6 \times 10^5 = 0 \quad \rightarrow \quad v_{1f} = 2.8 \times 10^5 \text{ m/s}$$

From (3),

$$v_{2f} = \sqrt{1.2 \times 10^{11} - {v_{1f}}^2} = \sqrt{1.2 \times 10^{11} - (2.8 \times 10^5)^2}$$

$$= 2.1 \times 10^5 \text{ m/s}$$

and from (2),

$$\phi = \sin^{-1}\left(\frac{v_{1f} \sin 37°}{v_{2f}}\right) = \sin^{-1}\left(\frac{(2.8 \times 10^5) \sin 37°}{2.1 \times 10^5}\right)$$

$$= 53°$$

It is interesting that $\theta + \phi = 90°$. This result is *not* accidental. Whenever two objects of equal mass collide elastically in a glancing collision and one of them is initially at rest, their final velocities are at right angles to each other.

EXAMPLE 8.10 **Collision at an Intersection**

A 1 500-kg car traveling east with a speed of 25.0 m/s collides at an intersection with a 2 500-kg van traveling north at a speed of 20.0 m/s as shown in Figure 8.12. Find the direction and magnitude of the velocity of the wreckage after the collision, assuming that the vehicles undergo a perfectly inelastic collision (i.e., they stick together).

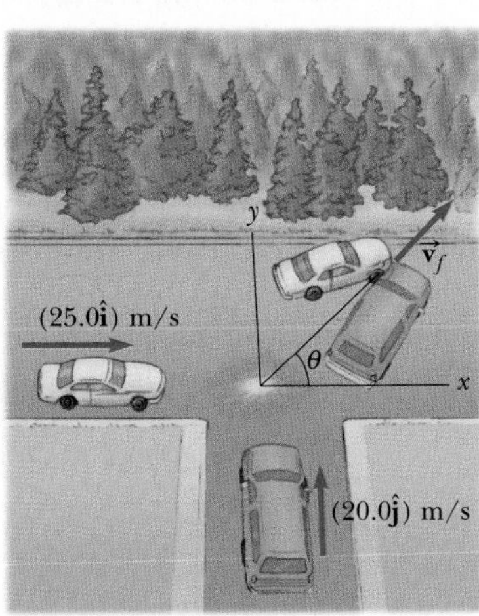

FIGURE **8.12** (Example 8.10) An eastbound car colliding with a northbound van.

Solution Let us choose east to be along the positive *x* direction and north to be along the positive *y* direction as in Figure 8.12. Before the collision, the only object having momentum in the *x* direction is the car. Therefore, the magnitude of the total initial momentum of the system (car plus van) in the *x* direction is

$$\sum p_{xi} = (1\,500 \text{ kg})(25.0 \text{ m/s}) = 3.75 \times 10^4 \text{ kg}\cdot\text{m/s}$$

The wreckage moves at an angle θ and speed v_f after the collision. The magnitude of the total momentum in the *x* direction after the collision is

$$\sum p_{xf} = (4\,000 \text{ kg}) v_f \cos \theta$$

Because the total momentum in the *x* direction is conserved, we can equate these two equations to obtain

$$(1) \quad 3.75 \times 10^4 \text{ kg}\cdot\text{m/s} = (4\,000 \text{ kg})\, v_f \cos \theta$$

Similarly, the total initial momentum of the system in the *y* direction is that of the van, whose magnitude is equal to (2 500 kg)(20.0 m/s). Applying conservation of momentum to the *y* direction, we have

$$\sum p_{yi} = \sum p_{yf}$$

$$(2\,500 \text{ kg})(20.0 \text{ m/s}) = (4\,000 \text{ kg}) v_f \sin \theta$$

$$(2) \qquad 5.00 \times 10^4 \text{ kg}\cdot\text{m/s} = (4\,000 \text{ kg}) v_f \sin \theta$$

If we divide (2) by (1), we find that

$$\tan\theta = \frac{5.00 \times 10^4}{3.75 \times 10^4} = 1.33$$

$$\theta = 53.1^\circ$$

When this angle is substituted into (2), the value of v_f is

$$v_f = \frac{5.00 \times 10^4\ \text{kg}\cdot\text{m/s}}{(4\,000\ \text{kg})\sin 53.1^\circ} = 15.6\ \text{m/s}$$

8.5 THE CENTER OF MASS

In this section, we describe the overall motion of a system of particles in terms of a very special point called the **center of mass** of the system. This notion gives us confidence in the particle model because we will see that the center of mass accelerates as if all the system's mass were concentrated at that point and all external forces act there.

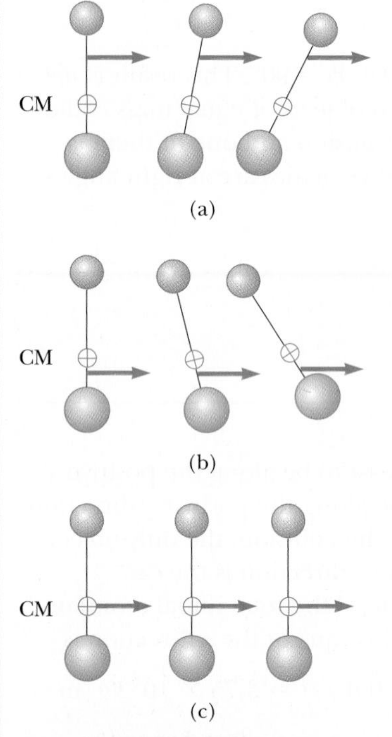

ACTIVE FIGURE 8.13
Two particles of unequal mass are connected by a light, rigid rod. (a) The system rotates clockwise when a force is applied between the less massive particle and the center of mass. (b) The system rotates counterclockwise when a force is applied between the more massive particle and the center of mass. (c) The system moves in the direction of the force without rotating when a force is applied at the center of mass.

Physics Now™ Log into PhysicsNow at **www.pop4e.com** and go to Active Figure 8.13 to choose the point at which to apply the force.

Consider a system consisting of a pair of particles connected by a light, rigid rod (Active Fig. 8.13). The center of mass as indicated in the figure is located on the rod and is closer to the larger mass in the figure; we will see why soon. If a single force is applied at some point on the rod that is above the center of mass, the system rotates clockwise (Active Fig. 8.13a) as it translates through space. If the force is applied at a point on the rod below the center of mass, the system rotates counterclockwise (Active Fig. 8.13b). If the force is applied exactly at the center of mass, the system moves in the direction of $\vec{\mathbf{F}}$ without rotating (Active Fig. 8.13c) as if the system is behaving as a particle. Therefore, in theory, the center of mass can be located with this experiment.

If we were to analyze the motion in Active Figure 8.13c, we would find that the system moves as if all its mass were concentrated at the center of mass. Furthermore, if the external net force on the system is $\Sigma\vec{\mathbf{F}}$ and the total mass of the system is M, the center of mass moves with an acceleration given by $\vec{\mathbf{a}} = \Sigma\vec{\mathbf{F}}/M$. That is, the system moves as if the resultant external force were applied to a single particle of mass M located at the center of mass, which justifies our particle model for extended objects. We have ignored all rotational effects for extended objects so far, implicitly assuming that forces were provided at just the right position so as to cause no rotation. We will study rotational motion in Chapter 10, where we will apply forces that do not pass through the center of mass.

The position of the center of mass of a system can be described as being the *average position* of the system's mass. For example, the center of mass of the pair of particles described in Active Figure 8.14 is located on the x axis, somewhere between the particles. The x coordinate of the center of mass in this case is

$$x_{CM} = \frac{m_1 x_1 + m_2 x_2}{m_1 + m_2} \qquad [8.29]$$

For example, if $x_1 = 0$, $x_2 = d$, and $m_2 = 2m_1$, we find that $x_{CM} = \frac{2}{3}d$. That is, the center of mass lies closer to the more massive particle. If the two masses are equal, the center of mass lies midway between the particles.

We can extend the concept of center of mass to a system of many particles in three dimensions. The x coordinate of the center of mass of n particles is defined to be

$$x_{CM} \equiv \frac{m_1 x_1 + m_2 x_2 + m_3 x_3 + \cdots + m_n x_n}{m_1 + m_2 + m_3 + \cdots + m_n} = \frac{\sum_i m_i x_i}{\sum_i m_i} = \frac{\sum_i m_i x_i}{M} \qquad [8.30]$$

where x_i is the x coordinate of the ith particle and M is the *total mass* of the system. The y and z coordinates of the center of mass are similarly defined by the equations

$$y_{CM} \equiv \frac{\sum_i m_i y_i}{M} \quad \text{and} \quad z_{CM} \equiv \frac{\sum_i m_i z_i}{M} \qquad [8.31]$$

The center of mass can also be located by its position vector, $\vec{\mathbf{r}}_{CM}$. The rectangular coordinates of this vector are x_{CM}, y_{CM}, and z_{CM}, defined in Equations 8.30 and 8.31. Therefore,

$$\vec{\mathbf{r}}_{CM} = x_{CM}\hat{\mathbf{i}} + y_{CM}\hat{\mathbf{j}} + z_{CM}\hat{\mathbf{k}} = \frac{\sum_i m_i x_i \hat{\mathbf{i}} + \sum_i m_i y_i \hat{\mathbf{j}} + \sum_i m_i z_i \hat{\mathbf{k}}}{M}$$

$$\vec{\mathbf{r}}_{CM} = \frac{\sum_i m_i \vec{\mathbf{r}}_i}{M} \qquad [8.32]$$

where $\vec{\mathbf{r}}_i$ is the position vector of the ith particle, defined by

$$\vec{\mathbf{r}}_i \equiv x_i\hat{\mathbf{i}} + y_i\hat{\mathbf{j}} + z_i\hat{\mathbf{k}}$$

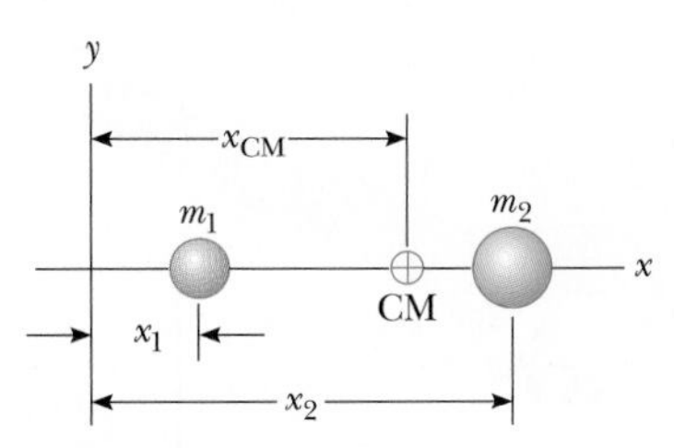

ACTIVE FIGURE 8.14
The center of mass of two particles having unequal mass is located on the x axis at x_{CM}, a point between the particles, closer to the one having the larger mass.

PhysicsNow™ Log into PhysicsNow at **www.pop4e.com** and go to Active Figure 8.14 to adjust the masses of the particles and see the effect on the location of the center of mass.

Although locating the center of mass for an extended object is somewhat more cumbersome than locating the center of mass of a system of particles, this location is based on the same fundamental ideas. We can model the extended object as a system containing a large number of elements (Fig. 8.15). Each element is modeled as a particle of mass Δm_i, with coordinates x_i, y_i, z_i. The particle separation is very small, so this model is a good representation of the continuous mass distribution of the object. The x coordinate of the center of mass of the particles representing the object, and therefore of the approximate center of mass of the object, is

$$x_{CM} \approx \frac{\sum_i x_i \Delta m_i}{M}$$

with similar expressions for y_{CM} and z_{CM}. If we let the number of elements approach infinity (and, as a consequence, the size and mass of each element approach zero), the model becomes indistinguishable from the continuous mass distribution and x_{CM} is given precisely. In this limit, we replace the sum by an integral and Δm_i by the differential element dm:

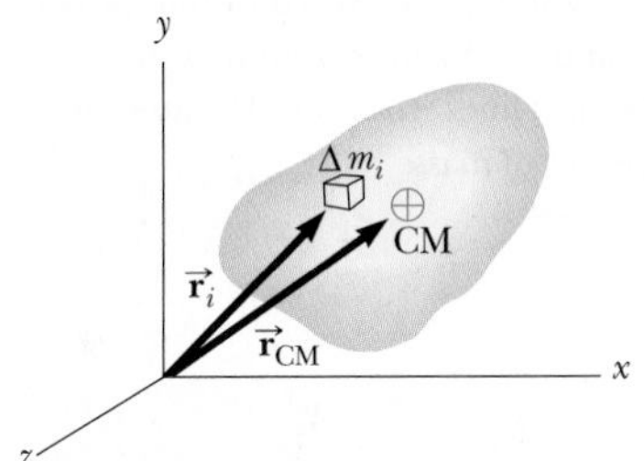

FIGURE 8.15 An extended object can be modeled as a distribution of small elements of mass Δm_i. The center of mass of the object is located at the vector position $\vec{\mathbf{r}}_{CM}$, which has coordinates x_{CM}, y_{CM}, and z_{CM}.

$$x_{CM} = \lim_{\Delta m_i \to 0} \frac{\sum_i x_i \Delta m_i}{M} = \frac{1}{M}\int x\, dm \qquad [8.33]$$

where the integration is over the length of the object in the x direction. Likewise, for y_{CM} and z_{CM} we obtain

$$y_{CM} = \frac{1}{M}\int y\, dm \quad \text{and} \quad z_{CM} = \frac{1}{M}\int z\, dm \qquad [8.34]$$

We can express the vector position of the center of mass of an extended object as

$$\vec{\mathbf{r}}_{CM} = \frac{1}{M}\int \vec{\mathbf{r}}\, dm \qquad [8.35]$$

▪ Center of mass of a continuous mass distribution

which is equivalent to the three expressions in Equations 8.33 and 8.34.

The center of mass of a homogeneous, symmetric object must lie on an axis of symmetry. For example, the center of mass of a homogeneous rod must lie midway between the ends of the rod. The center of mass of a homogeneous sphere or a homogeneous cube must lie at the geometric center of the object.

The center of mass of a system is often confused with the **center of gravity** of a system. Each portion of a system is acted on by the gravitational force. The net effect of all these forces is equivalent to the effect of a single force $M\vec{\mathbf{g}}$ acting at a

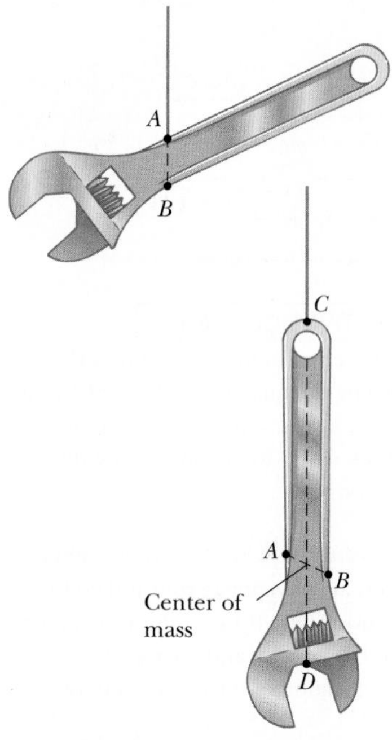

FIGURE 8.16 An experimental technique for determining the center of mass of a wrench. The wrench is hung freely from two different pivots, A and C. The intersection of the two vertical lines AB and CD locates the center of mass.

special point called the center of gravity. The center of gravity is the average position of the gravitational forces on all parts of the object. If $\vec{\mathbf{g}}$ is uniform over the system, the center of gravity coincides with the center of mass. If the gravitational field over the system is not uniform, the center of gravity and the center of mass are different. In most cases, for objects or systems of reasonable size, the two points can be considered to be coincident.

One can experimentally determine the center of gravity of an irregularly shaped object, such as a wrench, by suspending the wrench from two different points (Fig. 8.16). An object of this size has virtually no variation in the gravitational field over its dimensions, so this method also locates the center of mass. The wrench is first hung from point A, and a vertical line AB is drawn (which can be established with a plumb bob) when the wrench is in equilibrium. The wrench is then hung from point C, and a second vertical line CD is drawn. The center of mass coincides with the intersection of these two lines. In fact, if the wrench is hung freely from any point, the vertical line through that point will pass through the center of mass.

QUICK QUIZ 8.5 A baseball bat is made of wood of uniform density. The bat is cut at the location of its center of mass as shown in Figure 8.17. Which piece has the smaller mass? **(a)** the piece on the right **(b)** the piece on the left **(c)** both pieces have the same mass **(d)** impossible to determine

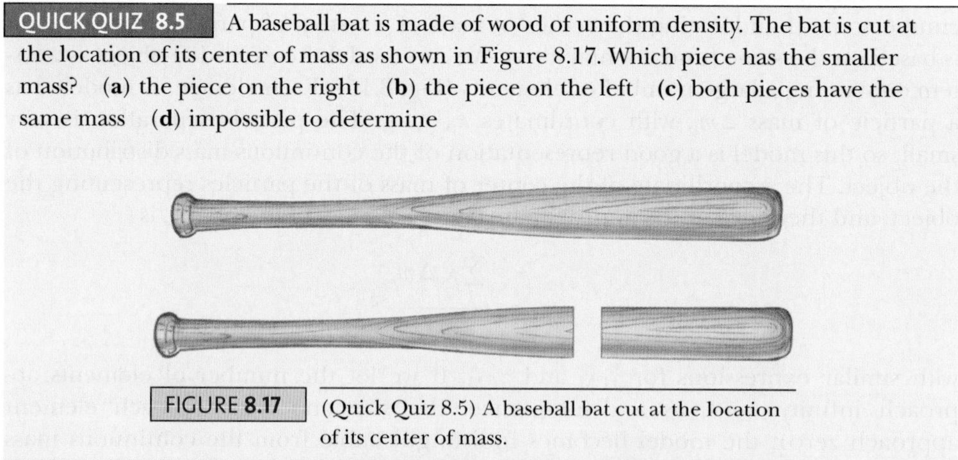

FIGURE 8.17 (Quick Quiz 8.5) A baseball bat cut at the location of its center of mass.

EXAMPLE 8.11 The Center of Mass of Three Particles

A system consists of three particles located at the corners of a right triangle as in Figure 8.18. Find the center of mass of the system.

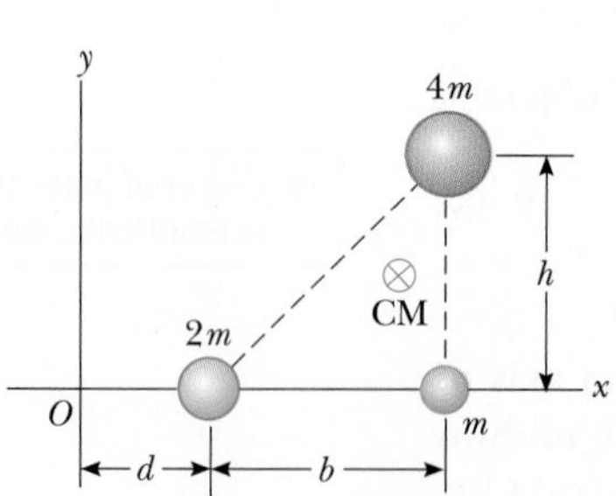

FIGURE 8.18 (Example 8.11) Locating the center of mass for a system of three particles.

Solution Using the basic defining equations for the coordinates of the center of mass and noting that $z_{\text{CM}} = 0$, we have

$$x_{\text{CM}} = \frac{\sum_i m_i x_i}{M} = \frac{2md + m(d+b) + 4m(d+b)}{7m}$$

$$= d + \tfrac{5}{7}b$$

$$y_{\text{CM}} = \frac{\sum_i m_i y_i}{M} = \frac{2m(0) + m(0) + 4mh}{7m} = \tfrac{4}{7}h$$

Therefore, we can express the position vector to the center of mass measured from the origin as

$$\vec{\mathbf{r}}_{\text{CM}} = x_{\text{CM}}\hat{\mathbf{i}} + y_{\text{CM}}\hat{\mathbf{j}} + z_{\text{CM}}\hat{\mathbf{k}} = (d + \tfrac{5}{7}b)\hat{\mathbf{i}} + \tfrac{4}{7}h\hat{\mathbf{j}}$$

EXAMPLE 8.12 **The Center of Mass of a Right Triangle**

You have been asked to hang a metal sign from a single vertical wire. The sign is of the triangular shape shown in Figure 8.19a. The bottom of the sign is to be parallel to the ground. At what distance from the left end of the sign should you attach the wire?

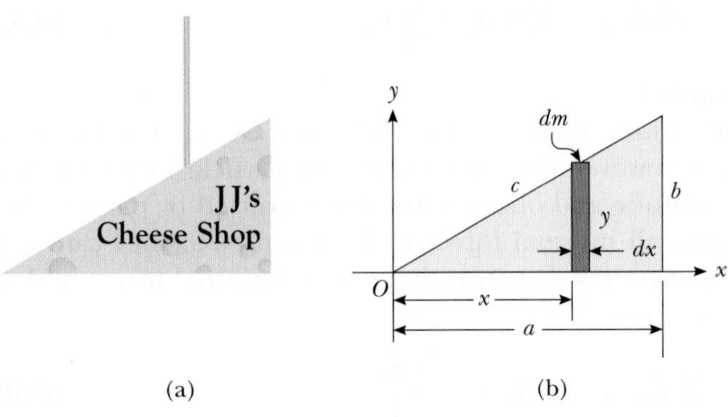

FIGURE 8.19 (Example 8.12) (a) A triangular sign to be hung from a single wire. (b) Geometric construction for locating the center of mass.

Solution We will need to attach the wire at a point directly above the center of gravity of the sign, which is the same as its center of mass because it is in a uniform gravitational field. We model the sign as a perfect triangle. We assume that the sign has a uniform density and total mass M. Because the sign is a continuous distribution of mass, we will need to use the integral expression in Equation 8.33 to find the x coordinate of the center of mass.

We divide the triangle into narrow strips of width dx and height y as shown in Figure 8.19b, where y is the height to the hypotenuse of the triangle above the x axis for a given value of x. The mass of each strip is the product of the volume of the strip and the density ρ of the material from which the sign is made: $dm = \rho yt\, dx$, where t is the thickness of the metal sign. The density of the material is the total mass of the sign divided by its total volume (area of the triangle times thickness), so

$$dm = \rho yt\, dx = \left(\frac{M}{\frac{1}{2}abt}\right) yt\, dx = \frac{2My}{ab}\, dx$$

Using Equation 8.33 to find the x coordinate of the center of mass gives

$$x_{\text{CM}} = \frac{1}{M}\int x\, dm = \frac{1}{M}\int_0^a x\,\frac{2My}{ab}\, dx = \frac{2}{ab}\int_0^a xy\, dx$$

To proceed further and evaluate the integral, we must express y in terms of x. The line representing the hypotenuse of the triangle in Figure 8.19b has a slope of b/a and passes through the origin, so the equation of this line is $y = (b/a)x$. With this substitution for y in the integral, we have

$$x_{\text{CM}} = \frac{2}{ab}\int_0^a x\left(\frac{b}{a}x\right) dx = \frac{2}{a^2}\int_0^a x^2\, dx = \frac{2}{a^2}\left[\frac{x^3}{3}\right]_0^a$$

$$= \frac{2}{3}a$$

Therefore, the wire must be attached to the sign at a distance two thirds of the length of the bottom edge from the left end. We could also find the y coordinate of the center of mass of the sign, but that is not needed to determine where the wire should be attached.

8.6 MOTION OF A SYSTEM OF PARTICLES

We can begin to understand the physical significance and utility of the center of mass concept by taking the time derivative of the position vector $\vec{\mathbf{r}}_{\text{CM}}$ of the center of mass, given by Equation 8.32. Assuming that M remains constant—that is, no particles enter or leave the system—we find the following expression for the **velocity of the center of mass:**

$$\vec{\mathbf{v}}_{\text{CM}} = \frac{d\vec{\mathbf{r}}_{\text{CM}}}{dt} = \frac{1}{M}\sum_i m_i \frac{d\vec{\mathbf{r}}_i}{dt} = \frac{1}{M}\sum_i m_i \vec{\mathbf{v}}_i \qquad [8.36]$$

■ Velocity of the center of mass for a system of particles

where $\vec{\mathbf{v}}_i$ is the velocity of the ith particle. Rearranging Equation 8.36 gives

$$M\vec{\mathbf{v}}_{\text{CM}} = \sum_i m_i\vec{\mathbf{v}}_i = \sum_i \vec{\mathbf{p}}_i = \vec{\mathbf{p}}_{\text{tot}} \qquad [8.37]$$

This result tells us that **the total momentum of the system equals its total mass multiplied by the velocity of its center of mass.** In other words, the total momentum of the system is equal to the momentum of a single particle of mass M moving with a velocity $\vec{\mathbf{v}}_{\text{CM}}$; this is the particle model.

If we now differentiate Equation 8.36 with respect to time, we find the **acceleration of the center of mass:**

■ Acceleration of the center of mass for a system of particles

$$\vec{\mathbf{a}}_{CM} = \frac{d\vec{\mathbf{v}}_{CM}}{dt} = \frac{1}{M}\sum_i m_i \frac{d\vec{\mathbf{v}}_i}{dt} = \frac{1}{M}\sum_i m_i \vec{\mathbf{a}}_i \quad [8.38]$$

Rearranging this expression and using Newton's second law, we have

$$M\vec{\mathbf{a}}_{CM} = \sum_i m_i \vec{\mathbf{a}}_i = \sum_i \vec{\mathbf{F}}_i \quad [8.39]$$

where $\vec{\mathbf{F}}_i$ is the force on particle i.

The forces on any particle in the system may include both external and internal forces. By Newton's third law, however, the force exerted by particle 1 on particle 2, for example, is equal in magnitude and opposite the force exerted by particle 2 on particle 1. When we sum over all internal forces in Equation 8.39, they cancel in pairs. Therefore, the net force on the system is due *only* to external forces and we can write Equation 8.39 in the form

■ Newton's second law for a system of particles

$$\sum \vec{\mathbf{F}}_{ext} = M\vec{\mathbf{a}}_{CM} = \frac{d\vec{\mathbf{p}}_{tot}}{dt} \quad [8.40]$$

That is, the external net force on the system of particles equals the total mass of the system multiplied by the acceleration of the center of mass, or the time rate of change of the momentum of the system. If we compare this statement to Newton's second law for a single particle, we see that the center of mass moves like an imaginary particle of mass M under the influence of the external net force on the system. In the absence of external forces, the center of mass moves with uniform velocity as in the case of the translating and rotating wrench in Figure 8.20. If the net force acts along a line through the center of mass of an extended object such as the wrench, the object is accelerated without rotation. If the net force does not act through the center of mass, the object will undergo rotation in addition to translation. The linear acceleration of the center of mass is the same in either case, as given by Equation 8.40.

Finally, we see that if the external net force is zero, from Equation 8.40 it follows that

$$\frac{d\vec{\mathbf{p}}_{tot}}{dt} = M\vec{\mathbf{a}}_{CM} = 0$$

so that

$$\vec{\mathbf{p}}_{tot} = M\vec{\mathbf{v}}_{CM} = \text{constant} \qquad \left(\text{when } \sum \vec{\mathbf{F}}_{ext} = 0\right) \quad [8.41]$$

That is, the total linear momentum of a system of particles is constant if no external forces act on the system. It follows that, for an *isolated* system of particles, the total momentum is conserved. The law of conservation of momentum that was derived in Section 8.1 for a two-particle system is thus generalized to a many-particle system.

(Richard Megna, Fundamental Photographs)

FIGURE 8.20 Strobe photograph showing an overhead view of a wrench moving on a horizontal surface. The center of mass of the wrench (marked with a white dot) moves in a straight line as the wrench rotates about this point. The wrench moves from left to right in the photograph and is slowing down due to friction between the wrench and the supporting surface. (*Note* The decreasing distance between the white dots.)

FIGURE 8.21 (Thinking Physics 8.1) A boy takes a step in a canoe. What happens to the canoe?

■ Thinking Physics 8.1

A boy stands at one end of a canoe that is stationary relative to the shore (Fig. 8.21). He then walks to the opposite end of the canoe, away from the shore. Does the canoe move?

Reasoning Yes, the canoe moves toward the shore. Ignoring friction between the canoe and water, no horizontal force acts on the system consisting of the boy and canoe. The center of mass of the system therefore remains fixed relative to the shore (or any stationary point). As the boy moves away from the shore, the canoe must move toward the shore such that the center of mass of the system remains fixed in position. ■

QUICK QUIZ 8.6 The vacationers on a cruise ship are eager to arrive at their next destination. They decide to try to speed up the cruise ship by gathering at the bow (the front) and running all at once toward the stern (the back) of the ship. **(i)** While they are running toward the stern, what is the speed of the ship? **(a)** higher than it was before **(b)** unchanged **(c)** lower than it was before **(d)** impossible to determine **(ii)** The vacationers stop running when they reach the stern of the ship. After they have all stopped running, what is the speed of the ship? **(a)** higher than it was before they started running **(b)** unchanged from what it was before they started running **(c)** lower than it was before they started running **(d)** impossible to determine

EXAMPLE 8.13 An Exploding Projectile

A projectile is fired into the air and suddenly explodes into several fragments (Fig. 8.22). What can be said about the motion of the center of mass of the system made up of all the fragments after the explosion?

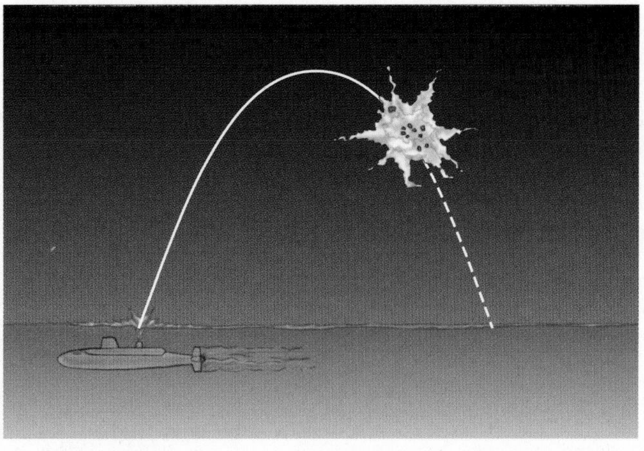

FIGURE 8.22 (Example 8.13) When a projectile explodes into several fragments, where does the center of mass of the fragments land?

Solution Neglecting air resistance, the only external force on the projectile is the gravitational force. Therefore, if the projectile did not explode, it would continue to move along the parabolic path indicated by the dashed line in Figure 8.22. Because the forces caused by the explosion are internal, they do not affect the motion of the center of mass of the system (the fragments). Therefore, after the explosion, the center of mass follows the same parabolic path the projectile would have followed if there had been no explosion.

8.7 ROCKET PROPULSION

CONTEXT CONNECTION

On our trip to Mars, we will need to control our spacecraft by firing the rocket engines. When ordinary vehicles, such as the automobiles in Context 1, are propelled, the driving force for the motion is the friction force exerted by the road on the car. A rocket moving in space, however, has no road to "push" against. The source of the propulsion of a rocket must therefore be different. **The operation of a rocket depends on the law of conservation of momentum as applied to a system, where the system is the rocket plus its ejected fuel.**

The propulsion of a rocket can be understood by first considering the archer on ice in Interactive Example 8.2. As an arrow is fired from the bow, the arrow receives momentum $m\vec{\mathbf{v}}$ in one direction and the archer receives a momentum of equal magnitude in the opposite direction. As additional arrows are fired, the archer moves faster, so a large velocity of the archer can be established by firing many arrows.

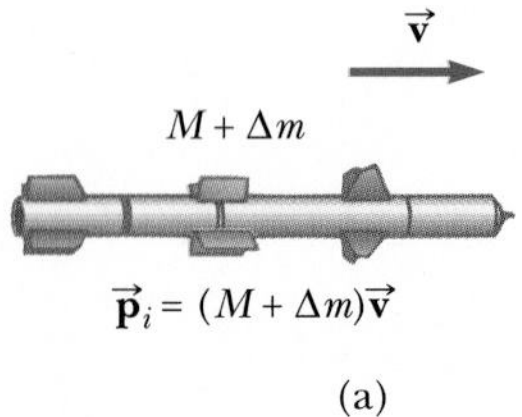

(a)

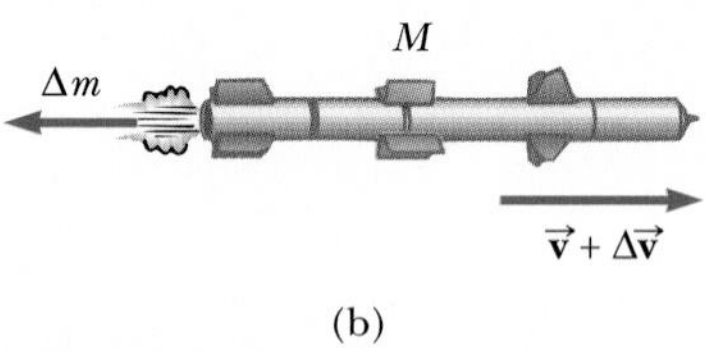

(b)

FIGURE 8.23 Rocket propulsion. (a) The initial mass of the rocket and fuel is $M + \Delta m$ at a time t, and its speed is v. (b) At a time $t + \Delta t$, the rocket's mass has been reduced to M, and an amount of fuel Δm has been ejected. The rocket's speed increases by an amount Δv.

In a similar manner, as a rocket moves in free space (a vacuum), its momentum changes when some of its mass is released in the form of ejected gases. Because the ejected gases acquire some momentum, the rocket receives a compensating momentum in the opposite direction. The rocket therefore is accelerated as a result of the "push," or thrust, from the exhaust gases. Note that the rocket represents the *inverse* of an inelastic collision; that is, momentum is conserved, but the kinetic energy of the system is *increased* (at the expense of energy stored in the fuel of the rocket).

Suppose at some time t the magnitude of the momentum of the rocket plus the fuel is $(M + \Delta m)v$ (Fig. 8.23a). During a short time interval Δt, the rocket ejects fuel of mass Δm and the rocket's speed therefore increases to $v + \Delta v$ (Fig. 8.22b). If the fuel is ejected with velocity $\vec{\mathbf{v}}_e$ *relative to the rocket,* the speed of the fuel relative to a stationary frame of reference is $v - v_e$ according to our discussion of relative velocity in Section 3.6. Therefore, if we equate the total initial momentum of the system with the total final momentum, we have

$$(M + \Delta m)v = M(v + \Delta v) + \Delta m(v - v_e)$$

Simplifying this expression gives

$$M\,\Delta v = \Delta m(v_e)$$

If we now take the limit as Δt goes to zero, $\Delta v \rightarrow dv$ and $\Delta m \rightarrow dm$. Furthermore, the increase dm in the exhaust mass corresponds to an equal decrease in the rocket mass, so $dm = -dM$. Note that the negative sign is introduced into the equation because dM represents a decrease in mass. Using this fact, we have

$$M\,dv = -v_e\,dM \qquad [8.42]$$

Integrating this equation and taking the initial mass of the rocket plus fuel to be M_i and the final mass of the rocket plus its remaining fuel to be M_f, we have

$$\int_{v_i}^{v_f} dv = -v_e \int_{M_i}^{M_f} \frac{dM}{M}$$

■ Velocity change in rocket propulsion

$$v_f - v_i = v_e \ln\left(\frac{M_i}{M_f}\right) \qquad [8.43]$$

which is the basic expression for rocket propulsion. It tells us that the increase in speed is proportional to the exhaust speed v_e. The exhaust speed should therefore be very high.

The **thrust** on the rocket is the force exerted on the rocket by the ejected exhaust gases. We can obtain an expression for the instantaneous thrust from Equation 8.42:

■ Rocket thrust

$$\text{Instantaneous thrust} = Ma = M\frac{dv}{dt} = \left|v_e \frac{dM}{dt}\right| \qquad [8.44]$$

Here we see that the thrust increases as the exhaust speed increases and as the rate of change of mass (burn rate) increases.

We can now determine the amount of fuel needed to set us on our journey to Mars. The fuel requirements are well within the capabilities of current technology, as evidenced by the several missions to Mars that have already been accomplished. What if we wanted to visit another *star,* however, rather than another *planet?* This question raises many new technological challenges, including the requirement to consider the effects of relativity, which we investigate in the next chapter.

■ Thinking Physics 8.2

When Robert Goddard proposed the possibility of rocket-propelled vehicles, the *New York Times* agreed that such vehicles would be useful and successful within the Earth's atmosphere ("Topics of the Times," *New York Times,* January 13, 1920, p. 12). The *Times,* however, balked at the idea of using such a rocket in the vacuum of space, noting that "its flight would be neither accelerated nor maintained by the explosion of the charges it then might have left. To claim that it would be is to deny a fundamental law of dynamics, and only Dr. Einstein and his chosen dozen, so few and fit, are licensed to do that. . . . That Professor Goddard, with his 'chair' in Clark College and the countenancing of the Smithsonian Institution, does not know the relation of action to reaction, and of the need to have something better than a vacuum against which to react—to say that would be absurd. Of course, he only seems to lack the knowledge ladled out daily in high schools." What did the writer of this passage overlook?

Reasoning The writer of this passage was making a common mistake in believing that a rocket works by expelling gases that push on something, propelling the rocket forward. With this belief, it is impossible to see how a rocket fired in empty space would work.

Gases do not need to push on anything; it is the act itself of expelling the gases that pushes the rocket forward. This point can be argued from Newton's third law: The rocket pushes the gases backward, resulting in the gases pushing the rocket forward. It can also be argued from conservation of momentum: As the gases gain momentum in one direction, the rocket must gain momentum in the opposite direction to conserve the original momentum of the rocket–gas system.

The *New York Times* did publish a retraction 49 years later ("A Correction," *New York Times,* July 17, 1969, p. 43) while the *Apollo 11* astronauts were on their way to the Moon. It appeared on a page with two other articles entitled "Fundamentals of Space Travel" and "Spacecraft, Like Squid, Maneuver by 'Squirts'" and contained the following passages: "an editorial feature of the *New York Times* dismissed the notion that a rocket could function in a vacuum and commented on the ideas of Robert H. Goddard. . . . Further investigation and experimentation have confirmed the findings of Isaac Newton in the 17th century, and it is now definitely established that a rocket can function in a vacuum as well as in an atmosphere. The *Times* regrets the error." ■

EXAMPLE 8.14 A Rocket in Space

A rocket in free space has a speed of 3.0×10^3 m/s relative to the Earth. Its engines are turned on, and fuel is ejected in a direction opposite the rocket's motion at a speed of 5.0×10^3 m/s relative to the rocket.

A What is the speed of the rocket relative to the Earth once its mass is reduced to one half its mass before ignition?

Solution Applying Equation 8.43, we have

$$v_f = v_i + v_e \ln\left(\frac{M_i}{M_f}\right)$$

$$= 3.0 \times 10^3 \text{ m/s} + (5.0 \times 10^3 \text{ m/s}) \ln\left(\frac{M_i}{0.5M_i}\right)$$

$$= 6.5 \times 10^3 \text{ m/s}$$

B What is the thrust on the rocket if it burns fuel at the rate of 50 kg/s?

Solution Using Equation 8.44, we have

$$\text{Thrust} = \left| v_e \frac{dM}{dt} \right| = (5.0 \times 10^3 \text{ m/s})(50 \text{ kg/s})$$

$$= 2.5 \times 10^5 \text{ N}$$

SUMMARY

Physics Now™ Take a practice test by logging into PhysicsNow at **www.pop4e.com** and clicking on the Pre-Test link for this chapter.

The linear momentum of any object of mass m moving with a velocity $\vec{\mathbf{v}}$ is

$$\vec{\mathbf{p}} \equiv m\vec{\mathbf{v}} \qquad [8.2]$$

Conservation of linear momentum applied to two interacting objects states that if the two objects form an isolated system, the total momentum of the system at all times equals its initial total momentum:

$$\vec{\mathbf{p}}_{1i} + \vec{\mathbf{p}}_{2i} = \vec{\mathbf{p}}_{1f} + \vec{\mathbf{p}}_{2f} \qquad [8.6]$$

The **impulse** of a net force $\Sigma\vec{\mathbf{F}}$ is defined as the integral of the force over the time interval during which it acts. The total impulse on any system is equal to the change in the momentum of the system and is given by

$$\vec{\mathbf{I}} = \int_{t_i}^{t_f} \sum \vec{\mathbf{F}}_{\text{ext}}\, dt = \Delta\vec{\mathbf{p}}_{\text{tot}} \qquad [8.11]$$

This is known as the **impulse–momentum theorem.**

When two objects collide, the total momentum of the isolated system before the collision always equals the total momentum after the collision, regardless of the nature of the collision. An **inelastic collision** is one in which kinetic energy is not conserved. A **perfectly inelastic collision** is one in which the colliding objects stick together after the collision. An **elastic collision** is one in which both momentum and kinetic energy are conserved.

In a two- or three-dimensional collision, the components of momentum in each of the directions are conserved independently.

The vector position of the center of mass of a system of particles is defined as

$$\vec{\mathbf{r}}_{\text{CM}} = \frac{\sum_i m_i \vec{\mathbf{r}}_i}{M} \qquad [8.32]$$

where M is the total mass of the system and $\vec{\mathbf{r}}_i$ is the position vector of the ith particle.

The **velocity of the center of mass for a system of particles** is

$$\vec{\mathbf{v}}_{\text{CM}} = \frac{1}{M}\sum_i m_i \vec{\mathbf{v}}_i \qquad [8.36]$$

The total momentum of a system of particles equals the total mass multiplied by the velocity of the center of mass; that is, $\vec{\mathbf{p}}_{\text{tot}} = M\vec{\mathbf{v}}_{\text{CM}}$.

Newton's second law applied to a system of particles is

$$\sum \vec{\mathbf{F}}_{\text{ext}} = M\vec{\mathbf{a}}_{\text{CM}} = \frac{d\vec{\mathbf{p}}_{\text{tot}}}{dt} \qquad [8.40]$$

where $\vec{\mathbf{a}}_{\text{CM}}$ is the acceleration of the center of mass and the sum is over all external forces. The center of mass therefore moves like an imaginary particle of mass M under the influence of the resultant external force on the system.

QUESTIONS

☐ = answer available in the *Student Solutions Manual and Study Guide*

1. Does a large force always produce a larger impulse on an object than a smaller force does? Explain.
2. If the speed of a particle is doubled, by what factor is its momentum changed? By what factor is its kinetic energy changed?
3. If two particles have equal kinetic energies, are their momenta necessarily equal? Explain.
4. While in motion, a pitched baseball carries kinetic energy and momentum. (a) Can we say that it carries a force that it can exert on any object it strikes? (b) Can the baseball deliver more kinetic energy to the object it strikes than the ball carries initially? (c) Can the baseball deliver to the object it strikes more momentum than the ball carries initially? Explain your answers.
5. You are watching a movie about a superhero and notice that the superhero hovers in the air and throws a piano at some villains while remaining stationary in the air. What is wrong with this scenario?
6. If two objects collide and one is initially at rest, is it possible for both to be at rest after the collision? Is it possible for one to be at rest after the collision? Explain.
7. Explain how linear momentum is conserved when a ball bounces from a floor.

8. A bomb, initially at rest, explodes into several pieces. (a) Is linear momentum of the system conserved? (b) Is kinetic energy of the system conserved? Explain.

9. You are standing perfectly still and then you take a step forward. Before the step your momentum was zero, but afterward you have some momentum. Is the principle of conservation of momentum violated in this case?

10. Consider a perfectly inelastic collision between a car and a large truck. Which vehicle experiences a larger change in kinetic energy as a result of the collision?

11. A sharpshooter fires a rifle while standing with the butt of the gun against her shoulder. If the forward momentum of a bullet is the same as the backward momentum of the gun, why isn't it as dangerous to be hit by the gun as by the bullet?

12. A pole-vaulter falls from a height of 6.0 m onto a foam rubber pad. Can you calculate his speed just before he reaches the pad? Can you calculate the force exerted on him by the pad? Explain.

13. Firefighters must apply large forces to hold a fire hose steady (Fig. Q8.13). What factors related to the projection of the water determine the magnitude of the force needed to keep the end of the fire hose stationary?

(© Bill Stormont/The Stock Market)

FIGURE Q8.13 Firefighters attack a burning house with a hose line.

14. A large bedsheet is held vertically by two students. A third student, who happens to be the star pitcher on the baseball team, throws a raw egg at the sheet. Explain why the egg does not break when it hits the sheet, regardless of its initial speed. (If you try this demonstration, make sure the pitcher hits the sheet near its center, and do not allow the egg to fall on the floor after being caught.)

15. NASA often uses a planet's gravity to "slingshot" a probe on its way to a more distant planet. The interaction of the planet and the spacecraft is a collision in which the objects do not touch. How can the probe have its speed increased in this manner?

16. Can the center of mass of an object be located at a position at which there is no mass? If so, give examples.

17. A juggler juggles three balls in a continuous cycle. Any one ball is in contact with his hands for one fifth of the time. Describe the motion of the center of mass of the three balls. What average force does the juggler exert on one ball while he is touching it?

18. Explain how you could use a balloon to demonstrate the mechanism responsible for rocket propulsion.

19. Does the center of mass of a rocket in free space accelerate? Explain. Can the speed of a rocket exceed the exhaust speed of the fuel? Explain.

20. On the subject of the following positions, state your own view and argue to support it. (a) The best theory of motion is that force causes acceleration. (b) The true measure of a force's effectiveness is the work it does, and the best theory of motion is that work on an object changes its energy. (c) The true measure of a force's effect is impulse, and the best theory of motion is that impulse imparted to an object changes its momentum.

PROBLEMS

1, 2, 3 = straightforward, intermediate, challenging
☐ = full solution available in the *Student Solutions Manual and Study Guide*
PhysicsNow™ = coached problem with hints available at www.pop4e.com
= computer useful in solving problem
= paired numerical and symbolic problems
= biomedical application

Section 8.1 ■ Linear Momentum and Its Conservation

1. A 3.00-kg particle has a velocity of $(3.00\hat{\mathbf{i}} - 4.00\hat{\mathbf{j}})$ m/s. (a) Find its x and y components of momentum. (b) Find the magnitude and direction of its momentum.

2. How fast can you set the Earth moving? In particular, when you jump straight up as high as you can, what is the order of magnitude of the maximum recoil speed that you give to the Earth? Model the Earth as a perfectly solid object. In your solution, state the physical quantities you take as data and the values you measure or estimate for them.

3. In research in cardiology and exercise physiology, it is often important to know the mass of blood pumped by a person's heart in one stroke. This information can be obtained by means of a *ballistocardiograph.* The instrument works as follows. The subject lies on a horizontal pallet floating on a film of air. Friction on the pallet is negligible. Initially, the momentum of the system is zero. When the heart beats, it expels a mass m of blood into the aorta with speed v, and the body and platform move in the opposite direction with speed V. The blood velocity can be determined independently (e.g., by observing the Doppler shift of ultrasound). Assume that it is 50.0 cm/s in one typical trial. The mass of the subject plus the pallet is 54.0 kg. The pallet moves 6.00×10^{-5} m in 0.160 s after one heartbeat. Calculate the mass of blood that leaves the heart. Assume

that the mass of blood is negligible compared with the total mass of the person. (This simplified example illustrates the principle of ballistocardiography, but in practice a more sophisticated model of heart function is used.)

4. (a) A particle of mass m moves with momentum p. Show that the kinetic energy of the particle is given by $K = p^2/2m$. (b) Express the magnitude of the particle's momentum in terms of its kinetic energy and mass.

5. Two blocks with masses M and $3M$ are placed on a horizontal, frictionless surface. A light spring is attached to one of them, and the blocks are pushed together with the spring between them (Fig. P8.5). A cord initially holding the blocks together is burned; after this, the block of mass $3M$ moves to the right with a speed of 2.00 m/s. (a) What is the speed of the block of mass M? (b) Find the original elastic potential energy in the spring, taking $M = 0.350$ kg.

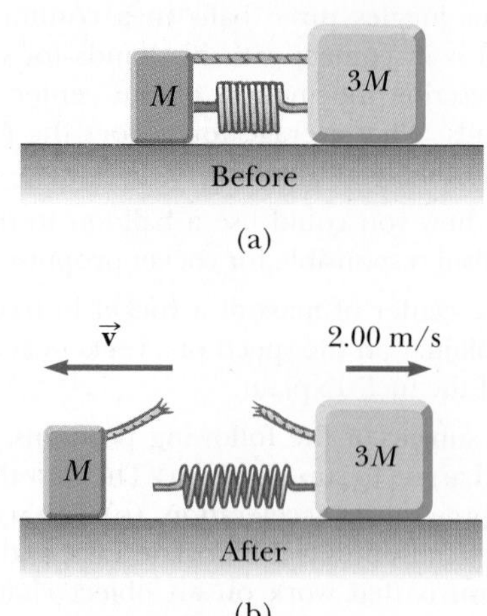

FIGURE P8.5

Section 8.2 ▪ Impulse and Momentum

6. A friend claims that as long as he has his seat belt on, he can hold on to a 12.0-kg child in a 60.0 mi/h head-on collision with a brick wall in which the car passenger compartment comes to a stop in 0.050 0 s. Show that the violent force during the collision will tear the child from his arms. (A child should always be in a toddler seat secured with a seat belt in the back seat of a car.)

7. An estimated force–time curve for a baseball struck by a bat is shown in Figure P8.7. From this curve, determine (a) the impulse delivered to the ball, (b) the average force exerted on the ball, and (c) the peak force exerted on the ball.

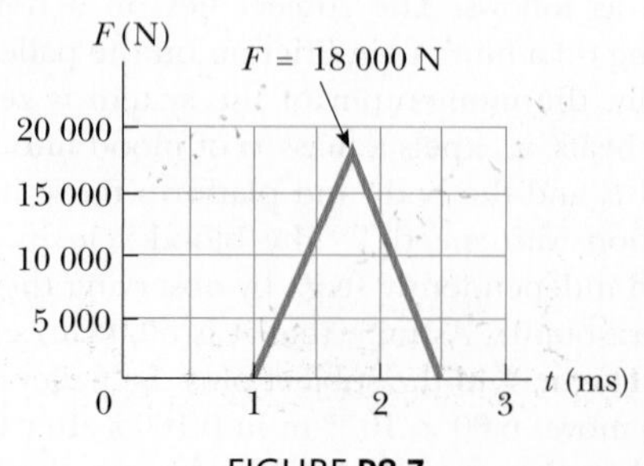

FIGURE P8.7

8. A tennis player receives a shot with the ball (0.060 0 kg) traveling horizontally at 50.0 m/s and returns the shot with the ball traveling horizontally at 40.0 m/s in the opposite direction. (a) What is the impulse delivered to the ball by the racquet? (b) What work does the racquet do on the ball?

9. **PhysicsNow™** A 3.00-kg steel ball strikes a wall with a speed of 10.0 m/s at an angle of 60.0° with the surface. It bounces off with the same speed and angle (Fig. P8.9). If the ball is in contact with the wall for 0.200 s, what is the average force exerted on the ball by the wall?

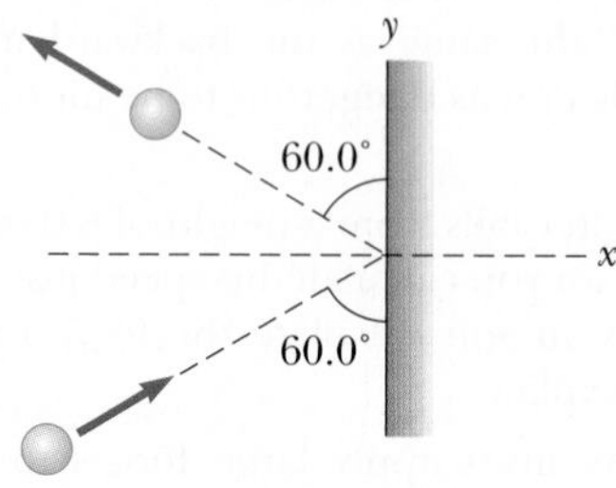

FIGURE P8.9

10. In a slow-pitch softball game, a 0.200-kg softball crosses the plate at 15.0 m/s at an angle of 45.0° below the horizontal. The batter hits the ball toward center field, giving it a velocity of 40.0 m/s at 30.0° above the horizontal. (a) Determine the impulse delivered to the ball. (b) If the force on the ball increases linearly for 4.00 ms, holds constant for 20.0 ms, and then decreases linearly to zero in another 4.00 ms, what is the maximum force on the ball?

11. A garden hose is held as shown in Figure P8.11. The hose is originally full of motionless water. What additional force is necessary to hold the nozzle stationary after the water flow is turned on if the discharge rate is 0.600 kg/s with a speed of 25.0 m/s?

FIGURE P8.11

12. A glider of mass m is free to slide along a horizontal air track. It is pushed against a launcher at one end of the track. Model the launcher as a light spring of force constant k, compressed by a distance x. The glider is released from rest. (a) Show that the glider attains a speed $v = x(k/m)^{1/2}$. (b) Does a glider of large or of small mass attain a greater speed? (c) Show that the impulse imparted to the glider is given by the expression $x(km)^{1/2}$. (d) Is a

greater impulse imparted to a large or a small mass? (e) Is more work done on a large or a small mass?

Section 8.3 ■ Collisions

13. A railroad car of mass 2.50×10^4 kg is moving with a speed of 4.00 m/s. It collides and couples with three other coupled railroad cars, each of the same mass as the single car and moving in the same direction with an initial speed of 2.00 m/s. (a) What is the speed of the four cars after the collision? (b) How much mechanical energy is lost in the collision?

14. Four railroad cars, each of mass 2.50×10^4 kg, are coupled together and coasting along horizontal tracks at speed v_i toward the south. A very strong but foolish movie actor, riding on the second car, uncouples the front car and gives it a big push, increasing its speed to 4.00 m/s southward. The remaining three cars continue moving south, now at 2.00 m/s. (a) Find the initial speed of the four cars. (b) How much work did the actor do? (c) State the relationship between the process described here and the process in Problem 8.13.

15. A 45.0-kg girl is standing on a plank that has a mass of 150 kg. The plank, originally at rest, is free to slide on a frozen lake, which is a flat, frictionless supporting surface. The girl begins to walk along the plank at a constant speed of 1.50 m/s relative to the plank. (a) What is her speed relative to the ice surface? (b) What is the speed of the plank relative to the ice surface?

16. Two blocks are free to slide along the frictionless wooden track *ABC* shown in Figure P8.16. A block of mass $m_1 = 5.00$ kg is released from *A*. Protruding from its front end is the north pole of a strong magnet, repelling the north pole of an identical magnet embedded in the back end of the block of mass $m_2 = 10.0$ kg, initially at rest. The two blocks never touch. Calculate the maximum height to which m_1 rises after the elastic collision.

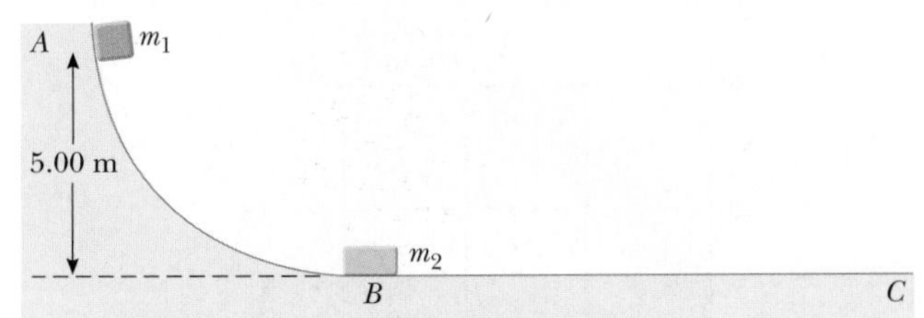

FIGURE **P8.16**

17. Most of us know intuitively that in a head-on collision between a large dump truck and a subcompact car, you are better off being in the truck than in the car. Why? Many people imagine that the collision force exerted on the car is much greater than that experienced by the truck. To substantiate this view, they point out that the car is crushed, whereas the truck is only dented. This idea of unequal forces, of course, is false. Newton's third law tells us that both objects experience forces of the same magnitude. The truck suffers less damage because it is made of stronger metal. What about the two drivers? Do they experience the same forces? To answer this question, suppose each vehicle is initially moving at 8.00 m/s and they undergo a perfectly inelastic head-on collision. Each driver has mass 80.0 kg. Including the drivers, the total vehicle masses are 800 kg for the car and 4 000 kg for the truck. If the collision time is 0.120 s, what force does the seat belt exert on each driver?

18. As shown in Figure P8.18, a bullet of mass m and speed v passes completely through a pendulum bob of mass M. The bullet emerges with a speed of $v/2$. The pendulum bob is suspended by a stiff rod of length ℓ and negligible mass. What is the minimum value of v such that the pendulum bob will barely swing through a complete vertical circle?

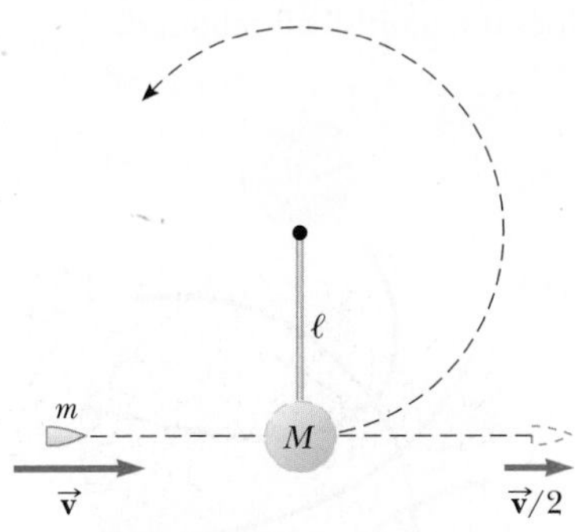

FIGURE **P8.18**

19. **PhysicsNow™** A neutron in a nuclear reactor makes an elastic head-on collision with the nucleus of a carbon atom initially at rest. (a) What fraction of the neutron's kinetic energy is transferred to the carbon nucleus? (b) Assume that the initial kinetic energy of the neutron is 1.60×10^{-13} J. Find its final kinetic energy and the kinetic energy of the carbon nucleus after the collision. (The mass of the carbon nucleus is nearly 12.0 times the mass of the neutron.)

20. A 7.00-g bullet, when fired from a gun into a 1.00-kg block of wood held in a vise, penetrates the block to a depth of 8.00 cm. This block of wood is next placed on a frictionless horizontal surface, and a second 7.00-g bullet is fired from the gun into the block. To what depth will the bullet penetrate the block in this case?

21. **PhysicsNow™** A 12.0-g wad of sticky clay is hurled horizontally at a 100-g wooden block initially at rest on a horizontal surface. The clay sticks to the block. After impact, the block slides 7.50 m before coming to rest. If the coefficient of friction between the block and the surface is 0.650, what was the speed of the clay immediately before impact?

22. (a) Three carts of masses 4.00 kg, 10.0 kg, and 3.00 kg move on a frictionless, horizontal track with speeds of 5.00 m/s, 3.00 m/s, and 4.00 m/s, respectively, as shown in Figure P8.22. Velcro couplers make the carts stick together after colliding. Find the final velocity of the train of three

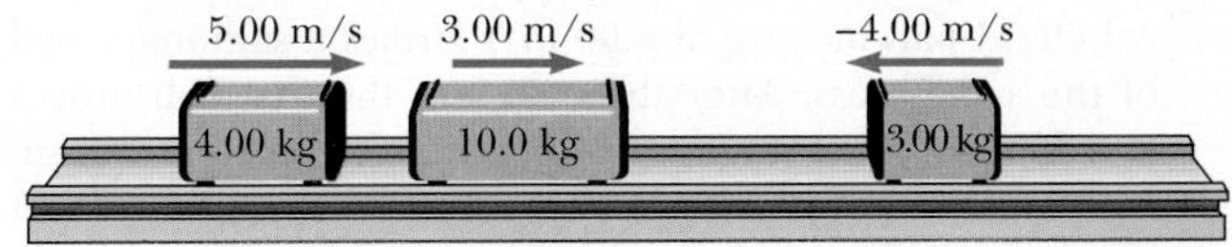

FIGURE **P8.22**

carts. (b) Does your answer require that all the carts collide and stick together at the same time? What if they collide in a different order?

23. A tennis ball of mass 57.0 g is held just above a basketball of mass 590 g. With their centers vertically aligned, both are released from rest at the same time, to fall through a distance of 1.20 m, as shown in Figure P8.23. (a) Find the magnitude of the downward velocity with which the basketball reaches the ground. Assume that an elastic collision with the ground instantaneously reverses the velocity of the basketball while the tennis ball is still moving down. Next, the two balls meet in an elastic collision. (b) To what height does the tennis ball rebound?

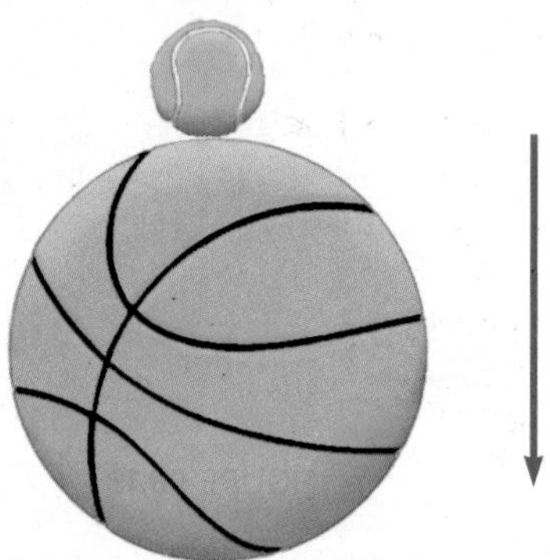

FIGURE P8.23

Section 8.4 ■ Two-Dimensional Collisions

24. A 90.0-kg fullback running east with a speed of 5.00 m/s is tackled by a 95.0-kg opponent running north with a speed of 3.00 m/s. Noting that the collision is perfectly inelastic, (a) calculate the speed and direction of the players just after the tackle and (b) determine the mechanical energy lost as a result of the collision. Account for the missing energy.

25. Two shuffleboard disks of equal mass, one orange and the other yellow, are involved in an elastic, glancing collision. The yellow disk is initially at rest and is struck by the orange disk moving with a speed v_i. After the collision, the orange disk moves along a direction that makes an angle θ with its initial direction of motion. The velocities of the two disks are perpendicular after the collision. Determine the final speed of each disk.

26. Two automobiles of equal mass approach an intersection. One vehicle is traveling with velocity 13.0 m/s toward the east, and the other is traveling north with speed v_{2i}. Neither driver sees the other. The vehicles collide in the intersection and stick together, leaving parallel skid marks at an angle of 55.0° north of east. The speed limit for both roads is 35 mi/h, and the driver of the northward-moving vehicle claims that he was within the speed limit when the collision occurred. Is he telling the truth?

27. A billiard ball moving at 5.00 m/s strikes a stationary ball of the same mass. After the collision, the first ball moves at 4.33 m/s, at an angle of 30.0° with respect to the original line of motion. Assuming an elastic collision (and ignoring friction and rotational motion), find the struck ball's velocity.

28. A proton, moving with a velocity of $v_i\hat{\mathbf{i}}$, collides elastically with another proton that is initially at rest. Assuming that the two protons have equal speeds after the collision, find (a) the speed of each proton after the collision in terms of v_i and (b) the direction of the velocity vectors after the collision.

29. An object of mass 3.00 kg, with an initial velocity of $5.00\hat{\mathbf{i}}$ m/s, collides with and sticks to an object of mass 2.00 kg, with an initial velocity of $-3.00\hat{\mathbf{j}}$ m/s. Find the final velocity of the composite object.

30. A 0.300-kg puck, initially at rest on a horizontal, frictionless surface, is struck by a 0.200-kg puck moving initially along the x axis with a speed of 2.00 m/s. After the collision, the 0.200-kg puck has a speed of 1.00 m/s at an angle of $\theta = 53.0°$ to the positive x axis (see Active Fig. 8.11). (a) Determine the velocity of the 0.300-kg puck after the collision. (b) Find the fraction of kinetic energy lost in the collision.

31. **PhysicsNow™** An unstable atomic nucleus of mass 17.0×10^{-27} kg initially at rest disintegrates into three particles. One of the particles, of mass 5.00×10^{-27} kg, moves along the y axis with a velocity of 6.00×10^{6} m/s. Another particle, of mass 8.40×10^{-27} kg, moves along the x axis with a speed of 4.00×10^{6} m/s. Find (a) the velocity of the third particle and (b) the total kinetic energy increase in the process.

Section 8.5 ■ The Center of Mass

32. Four objects are situated along the y axis as follows: a 2.00-kg object is at +3.00 m, a 3.00-kg object is at +2.50 m, a 2.50-kg object is at the origin, and a 4.00-kg object is at −0.500 m. Where is the center of mass of these objects?

33. A uniform piece of sheet steel is shaped as shown in Figure P8.33. Compute the x and y coordinates of the center of mass of the piece.

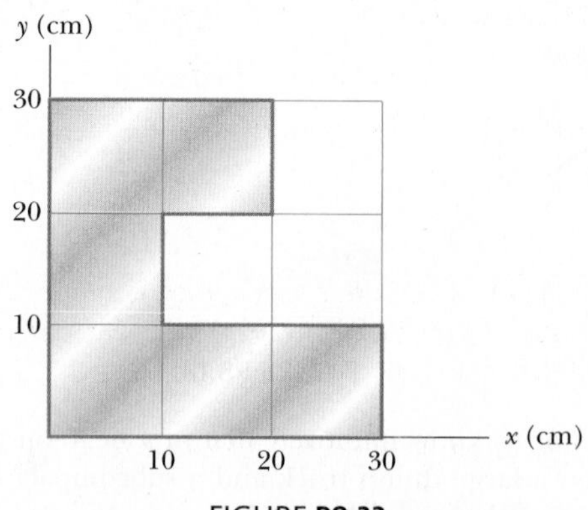

FIGURE P8.33

34. A water molecule consists of an oxygen atom with two hydrogen atoms bound to it (Fig. P8.34). The angle between the two bonds is 106°. If the bonds are 0.100 nm long, where is the center of mass of the molecule?

35. (a) Consider an extended object whose different portions have different elevations. Assume that the free fall acceleration is uniform over the object. Prove that the gravitational potential energy of the object–Earth system is given by $U_g = Mgy_{CM}$, where M is the total mass of the object and

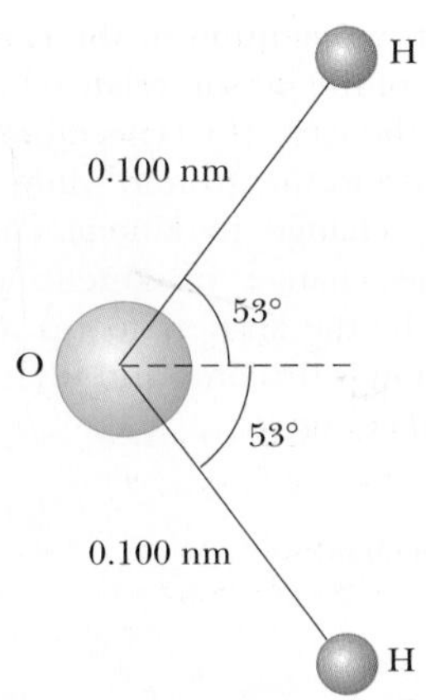

FIGURE **P8.34**

y_{CM} is the elevation of its center of mass above the chosen reference level. (b) Calculate the gravitational potential energy associated with a ramp constructed on level ground with stone with density 3 800 kg/m^3 and everywhere 3.60 m wide (Fig. P8.35). In a side view, the ramp appears as a right triangle with height 15.7 m at the top end and base 64.8 m.

FIGURE **P8.35**

36. A rod of length 30.0 cm has linear density (mass-per-length) given by

$$\lambda = 50.0 \text{ g/m} + 20.0x \text{ g/m}^2$$

where x is the distance from one end, measured in meters. (a) What is the mass of the rod? (b) How far from the $x = 0$ end is its center of mass?

Section 8.6 ■ Motion of a System of Particles

37. A 2.00-kg particle has a velocity $(2.00\hat{\mathbf{i}} - 3.00\hat{\mathbf{j}})$ m/s, and a 3.00-kg particle has a velocity $(1.00\hat{\mathbf{i}} + 6.00\hat{\mathbf{j}})$ m/s. Find (a) the velocity of the center of mass and (b) the total momentum of the system.

38. Consider a system of two particles in the xy plane: $m_1 = 2.00$ kg is at the location $\vec{\mathbf{r}}_1 = (1.00\hat{\mathbf{i}} + 2.00\hat{\mathbf{j}})$ m and has a velocity of $(3.00\hat{\mathbf{i}} + 0.500\hat{\mathbf{j}})$ m/s; $m_2 = 3.00$ kg is at $\vec{\mathbf{r}}_2 = (-4.00\hat{\mathbf{i}} - 3.00\hat{\mathbf{j}})$ m and has velocity $(3.00\hat{\mathbf{i}} - 2.00\hat{\mathbf{j}})$ m/s. (a) Plot these particles on a grid or graph paper. Draw their position vectors and show their velocities. (b) Find the position of the center of mass of the system and mark it on the grid. (c) Determine the velocity of the center of mass and also show it on the diagram. (d) What is the total linear momentum of the system?

39. Romeo (77.0 kg) entertains Juliet (55.0 kg) by playing his guitar from the rear of their boat at rest in still water, 2.70 m away from Juliet who is in the front of the boat. After the serenade, Juliet carefully moves to the rear of the boat (away from shore) to plant a kiss on Romeo's cheek. How far does the 80.0-kg boat move toward the shore it is facing?

40. A ball of mass 0.200 kg has a velocity of $1.50\hat{\mathbf{i}}$ m/s; a ball of mass 0.300 kg has a velocity of $-0.400\hat{\mathbf{i}}$ m/s. They meet in a head-on elastic collision. (a) Find their velocities after the collision. (b) Find the velocity of their center of mass before and after the collision.

Section 8.7 ■ Context Connection—Rocket Propulsion

41. **PhysicsNow**™ The first stage of a *Saturn V* space vehicle consumed fuel and oxidizer at the rate of 1.50×10^4 kg/s, with an exhaust speed of 2.60×10^3 m/s. (a) Calculate the thrust produced by these engines. (b) Find the acceleration of the vehicle just as it lifted off the launch pad on the Earth, taking the vehicle's initial mass as 3.00×10^6 kg. You must include the gravitational force to solve part (b).

42. Model rocket engines are sized by thrust, thrust duration, and total impulse, among other characteristics. A size C5 model rocket engine has an average thrust of 5.26 N, a fuel mass of 12.7 g, and an initial mass of 25.5 g. The duration of its burn is 1.90 s. (a) What is the average exhaust speed of the engine? (b) If this engine is placed in a rocket body of mass 53.5 g, what is the final velocity of the rocket if it is fired in outer space? Assume that the fuel burns at a constant rate.

43. A rocket for use in deep space is to be capable of boosting a total load (payload plus rocket frame and engine) of 3.00 metric tons to a speed of 10 000 m/s. (a) It has an engine and fuel designed to produce an exhaust speed of 2 000 m/s. How much fuel plus oxidizer is required? (b) If a different fuel and engine design could give an exhaust speed of 5 000 m/s, what amount of fuel and oxidizer would be required for the same task?

44. *Rocket science.* A rocket has total mass $M_i = 360$ kg, including 330 kg of fuel and oxidizer. In interstellar space, it starts from rest at the position $x = 0$, turns on its engine at time $t = 0$, and puts out exhaust with relative speed $v_e = 1\,500$ m/s at the constant rate $k = 2.50$ kg/s. The fuel will last for an actual burn time of 330 kg/(2.5 kg/s) = 132 s, but define a "projected depletion time" as $T_p = M_i/k = 360$ kg/(2.5 kg/s) = 144 s (which would be the burn time if the rocket could use its payload and fuel tanks, and even the walls of the combustion chamber, as fuel.) (a) Show that during the burn the velocity of the rocket is given as a function of time by

$$v(t) = -v_e \ln\left(1 - \frac{t}{T_p}\right)$$

(b) Make a graph of the velocity of the rocket as a function of time for times running from 0 to 132 s. (c) Show that the acceleration of the rocket is

$$a(t) = \frac{v_e}{T_p - t}$$

(d) Graph the acceleration as a function of time. (e) Show that the position of the rocket is

$$x(t) = v_e(T_p - t)\ln\left(1 - \frac{t}{T_p}\right) + v_e t$$

(f) Graph the position during the burn as a function of time.

45. An orbiting spacecraft is described not as a "zero-g" but rather as a "microgravity" environment for its occupants and for onboard experiments. Astronauts experience slight lurches due to the motions of equipment and other astronauts and as a result of venting of materials from the craft. Assume that a 3 500-kg spacecraft undergoes an acceleration of $2.50\ \mu g = 2.45 \times 10^{-5}\ \text{m/s}^2$ due to a leak from one of its hydraulic control systems. The fluid is known to escape with a speed of 70.0 m/s into the vacuum of space. How much fluid will be lost in 1.00 h if the leak is not stopped?

Additional Problems

46. Two gliders are set in motion on an air track. A spring of force constant k is attached to the near side of one glider. The first glider of mass m_1 has velocity $\vec{\mathbf{v}}_1$, and the second glider of mass m_2 moves more slowly, with velocity $\vec{\mathbf{v}}_2$, as shown in Figure P8.46. When m_1 collides with the spring attached to m_2 and compresses the spring to its maximum compression x_{max}, the velocity of the gliders is $\vec{\mathbf{v}}$. In terms of $\vec{\mathbf{v}}_1$, $\vec{\mathbf{v}}_2$, m_1, m_2, and k, find (a) the velocity $\vec{\mathbf{v}}$ at maximum compression, (b) the maximum compression x_{max}, and (c) the velocity of each glider after m_1 has lost contact with the spring.

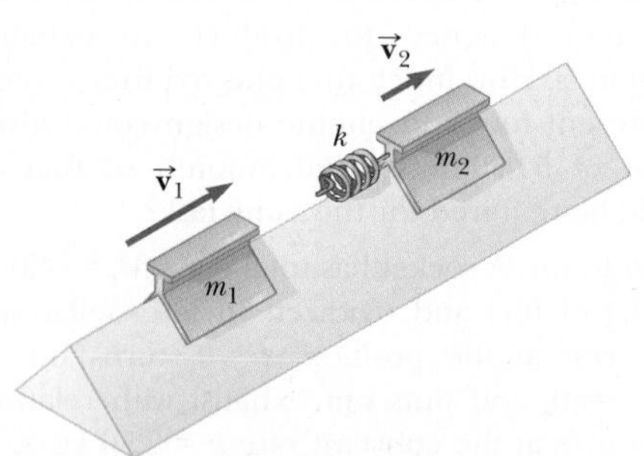

FIGURE **P8.46**

47. Review problem. A 60.0-kg person running at an initial speed of 4.00 m/s jumps onto a 120-kg cart initially at rest (Fig. P8.47). The person slides on the cart's top surface and finally comes to rest relative to the cart. The coefficient of kinetic friction between the person and the cart is 0.400. Friction between the cart and ground can be ignored. (a) Find the final velocity of the person and cart relative to the ground. (b) Find the friction force acting on the person while he is sliding across the top surface of the cart. (c) How long does the friction force act on the person? (d) Find the change in momentum of the person and the change in momentum of the cart. (e) Determine the displacement of the person relative to the ground while he is sliding on the cart. (f) Determine the displacement of the cart relative to the ground while the person is sliding. (g) Find the change in kinetic energy of the person. (h) Find the change in kinetic energy of the cart. (i) Explain why the answers to (g) and (h) differ. (What kind of collision is this one, and what accounts for the loss of mechanical energy?)

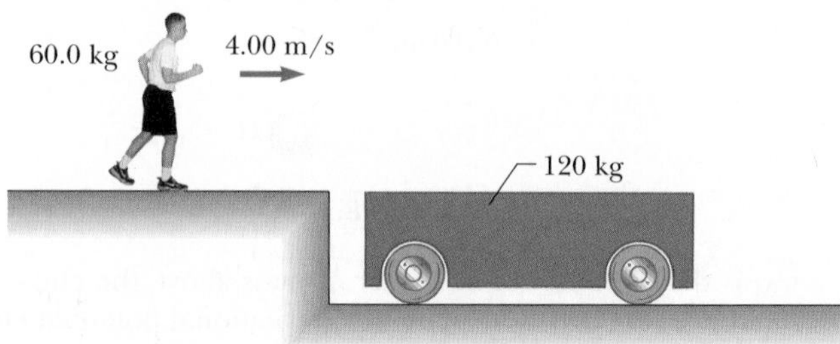

FIGURE **P8.47**

48. A bullet of mass m is fired into a block of mass M initially at rest at the edge of a frictionless table of height h (Fig. P8.48). The bullet remains in the block, and after impact the block lands a distance d from the bottom of the table. Determine the initial speed of the bullet.

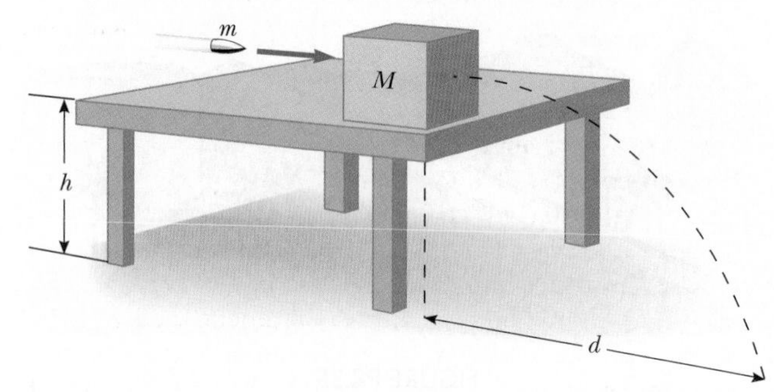

FIGURE **P8.48**

49. When it is threatened, a squid can escape by expelling a jet of water, sometimes colored with camouflaging ink. Consider a squid originally at rest in ocean water of constant density 1 030 kg/m³. Its original mass is 90.0 kg, of which a significant fraction is water inside its mantle. It expels this water through its siphon, a circular opening of diameter 3.00 cm, at a speed of 16.0 m/s. (a) As the squid is just starting to move, the surrounding water exerts no drag force on it. Find the squid's initial acceleration. (b) To estimate the maximum speed of the escaping squid, model the drag force of the surrounding water as described by Equation 5.7. Assume that the squid has a drag coefficient of 0.300 and a cross-sectional area of 800 cm². Find the speed at which the drag force counterbalances the thrust of its jet.

50. Pursued by ferocious wolves, you are in a sleigh with no horses, gliding without friction across an ice-covered lake. You take an action described by these equations:

$$(270\ \text{kg})(7.50\ \text{m/s})\hat{\mathbf{i}} = (15.0\ \text{kg})(-v_{1f}\hat{\mathbf{i}}) + (255\ \text{kg})(v_{2f}\hat{\mathbf{i}})$$
$$v_{1f} + v_{2f} = 8.00\ \text{m/s}$$

(a) Complete the statement of the problem, giving the data and identifying the unknowns. (b) Find the values of v_{1f} and v_{2f}. (c) Find the work you do.

51. A small block of mass $m_1 = 0.500$ kg is released from rest at the top of a curve-shaped, frictionless wedge of mass $m_2 = 3.00$ kg, which sits on a frictionless, horizontal surface as shown in Figure P8.51a. When the block leaves the wedge, its velocity is measured to be 4.00 m/s to the right as shown in Figure P8.51b. (a) What is the velocity of the wedge after the block reaches the horizontal surface? (b) What is the height h of the wedge?

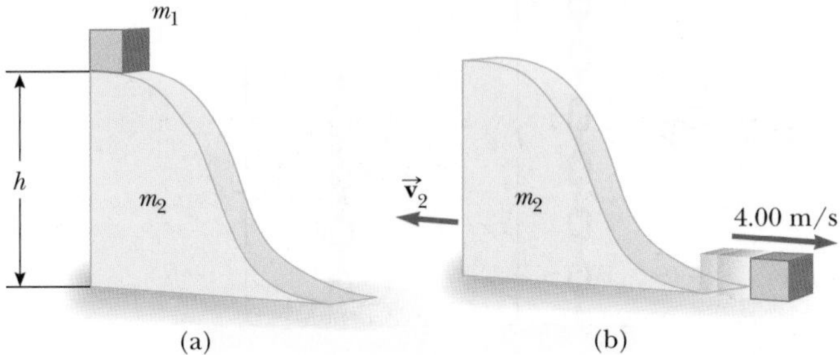

FIGURE **P8.51**

52. A jet aircraft is traveling at 500 mi/h (223 m/s) in horizontal flight. The engine takes in air at a rate of 80.0 kg/s and burns fuel at a rate of 3.00 kg/s. The exhaust gases are ejected at 600 m/s relative to the aircraft. Find the thrust of the jet engine and the delivered power.

53. **Review problem.** A light spring of force constant 3.85 N/m is compressed by 8.00 cm and held between a 0.250-kg block on the left and a 0.500-kg block on the right. Both blocks are at rest on a horizontal surface. The blocks are released simultaneously so that the spring tends to push them apart. Find the maximum velocity each block attains if the coefficient of kinetic friction between each block and the surface is (a) 0, (b) 0.100, and (c) 0.462. Assume that the coefficient of static friction is larger than that for kinetic friction.

54. **Review problem.** There are (one can say) three coequal theories of motion: Newton's second law, stating that the total force on an object causes its acceleration; the work–kinetic energy theorem, stating that the total work on an object causes its change in kinetic energy; and the impulse–momentum theorem, stating that the total impulse on an object causes its change in momentum. In this problem, you compare predictions of the three theories in one particular case. A 3.00-kg object has velocity $7.00\hat{\mathbf{j}}$ m/s. Then, a total force $12.0\hat{\mathbf{i}}$ N acts on the object for 5.00 s. (a) Calculate the object's final velocity, using the impulse–momentum theorem. (b) Calculate its acceleration from $\vec{\mathbf{a}} = (\vec{\mathbf{v}}_f - \vec{\mathbf{v}}_i)/\Delta t$. (c) Calculate its acceleration from $\vec{\mathbf{a}} = \Sigma \vec{\mathbf{F}}/m$. (d) Find the object's vector displacement from $\Delta\vec{\mathbf{r}} = \vec{\mathbf{v}}_i t + \frac{1}{2}\vec{\mathbf{a}}t^2$. (e) Find the work done on the object from $W = \vec{\mathbf{F}}\cdot\Delta\vec{\mathbf{r}}$. (f) Find the final kinetic energy from $\frac{1}{2}mv_f^2 = \frac{1}{2}m\vec{\mathbf{v}}_f\cdot\vec{\mathbf{v}}_f$. (g) Find the final kinetic energy from $\frac{1}{2}mv_i^2 + W$.

55. Two particles with masses m and $3m$ are moving toward each other along the x axis with the same initial speeds v_i. The particle with mass m is traveling to the left, and particle $3m$ is traveling to the right. They undergo a head-on elastic collision and each rebounds along the same line as it approached. Find the final speeds of the particles.

56. Two particles with masses m and $3m$ are moving toward each other along the x axis with the same initial speeds v_i. Particle m is traveling to the left, and particle $3m$ is traveling to the right. They undergo an elastic glancing collision such that particle m is moving downward after the collision at a right angle to its initial direction. (a) Find the final speeds of the two particles. (b) What is the angle θ at which the particle $3m$ is scattered?

57. George of the Jungle, with mass m, swings on a light vine hanging from a stationary tree branch. A second vine of equal length hangs from the same point, and a gorilla of larger mass M swings in the opposite direction on it. Both vines are horizontal when the primates start from rest at the same moment. George and the gorilla meet at the lowest point of their swings. Each is afraid that one vine will break, so they grab each other and hang on. They swing upward together, reaching a point where the vines make an angle of 35.0° with the vertical. (a) Find the value of the ratio m/M. (b) Try this experiment at home. Tie a small magnet and a steel screw to opposite ends of a string. Hold the center of the string fixed to represent the tree branch and reproduce a model of the motions of George and the gorilla. What changes in your analysis will make it apply to this situation? Assume next that the magnet is strong so that it noticeably attracts the screw over a distance of a few centimeters. Then the screw will be moving faster just before it sticks to the magnet. Does this change make a difference?

58. A cannon is rigidly attached to a carriage, which can move along horizontal rails but is connected to a post by a large spring, initially unstretched and with force constant $k = 2.00 \times 10^4$ N/m, as shown in Figure P8.58. The cannon fires a 200-kg projectile at a velocity of 125 m/s directed 45.0° above the horizontal. (a) Assuming that the mass of the cannon and its carriage is 5 000 kg, find the recoil speed of the cannon. (b) Determine the maximum extension of the spring. (c) Find the maximum force the spring exerts on the carriage. (d) Consider the system consisting of the cannon, carriage, and projectile. Is the momentum of this system conserved during the firing? Why or why not?

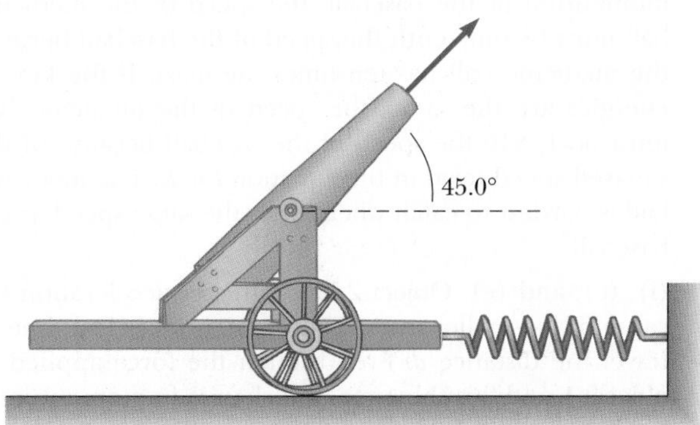

FIGURE **P8.58**

59. Sand from a stationary hopper falls onto a moving conveyor belt at the rate of 5.00 kg/s as shown in Figure P8.59. The conveyor belt is supported by frictionless rollers and moves at a constant speed of 0.750 m/s under the action of a constant horizontal external force $\vec{\mathbf{F}}_{\text{ext}}$ supplied by the

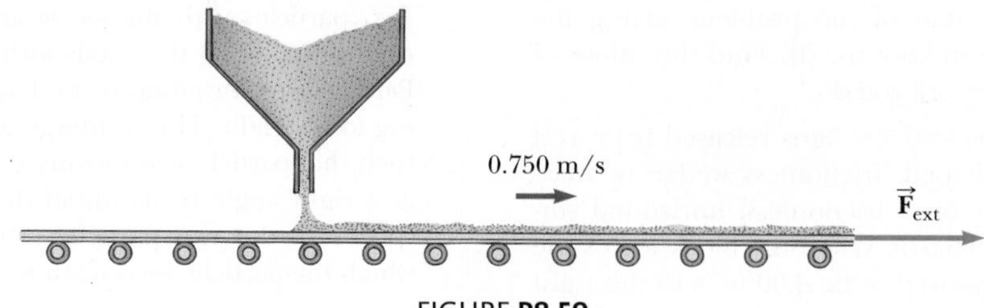

FIGURE **P8.59**

motor that drives the belt. Find (a) the sand's rate of change of momentum in the horizontal direction, (b) the force of friction exerted by the belt on the sand, (c) the external force $\vec{\mathbf{F}}_{ext}$, (d) the work done by $\vec{\mathbf{F}}_{ext}$ in 1 s, and (e) the kinetic energy acquired by the falling sand each second due to the change in its horizontal motion. (f) Why are the answers to (d) and (e) different?

60. A chain of length L and total mass M is released from rest with its lower end just touching the top of a table as shown in Figure P8.60a. Find the force exerted by the table on the chain after the chain has fallen through a distance x as shown in Figure P8.60b. (Assume that each link comes to rest the instant it reaches the table.)

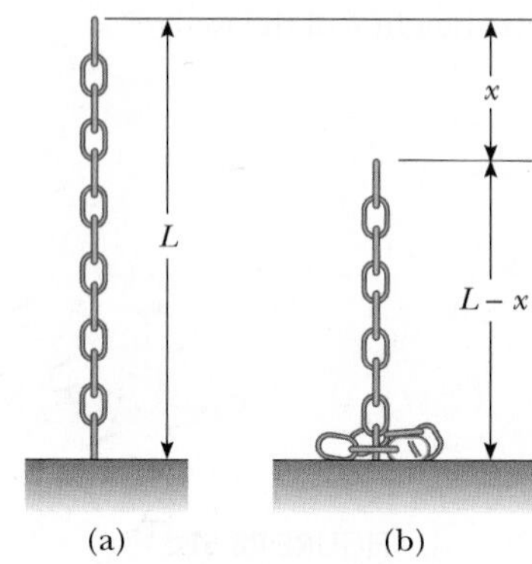

FIGURE **P8.60**

ANSWERS TO QUICK QUIZZES

8.1 (d). Two identical objects ($m_1 = m_2$) traveling at the same speed ($v_1 = v_2$) have the same kinetic energies and the same magnitudes of momentum. It also is possible, however, for particular combinations of masses and velocities to satisfy $K_1 = K_2$ but not $p_1 = p_2$. For example, a 1-kg object moving at 2 m/s has the same kinetic energy as a 4-kg object moving at 1 m/s, but the two clearly do not have the same momenta. Because we have no information about masses and speeds, we cannot choose among (a), (b), or (c).

8.2 (b), (c), (a). The slower the ball, the easier it is to catch. If the momentum of the medicine ball is the same as the momentum of the baseball, the speed of the medicine ball must be one-tenth the speed of the baseball because the medicine ball has ten times the mass. If the kinetic energies are the same, the speed of the medicine ball must be $1/\sqrt{10}$ the speed of the baseball because of the squared speed term in the equation for K. The medicine ball is hardest to catch when it has the same speed as the baseball.

8.3 (i), (c), and (e). Object 2 has a greater acceleration because of its smaller mass. Therefore, it takes less time to travel the distance d. Even though the force applied to objects 1 and 2 is the same, the change in momentum is less for object 2 because Δt is smaller. The work $W = Fd$ done on both objects is the same, because both F and d are the same in the two cases. Therefore, $K_1 = K_2$. (ii), (b) and (d). The same impulse is applied to both objects, so they experience the same change in momentum. Object 2 has a larger acceleration because of its smaller mass. Therefore, the distance that object 2 covers in the time interval Δt is larger than that for object 1. As a result, more work is done on object 2 and $K_2 > K_1$.

8.4 (b). Because momentum of the two-ball system is conserved, $\vec{\mathbf{p}}_{Pi} + 0 = \vec{\mathbf{p}}_{Pf} + \vec{\mathbf{p}}_B$. Because the Ping-Pong ball bounces back from the much more massive bowling ball with approximately the same speed, $\vec{\mathbf{p}}_{Pf} = -\vec{\mathbf{p}}_{Pi}$. As a consequence, $\vec{\mathbf{p}}_B = 2\vec{\mathbf{p}}_{Pi}$. Kinetic energy can be expressed as $K = p^2/2m$. Because of the much larger mass of the bowling ball, its kinetic energy is much smaller than that of the Ping-Pong ball.

8.5 (b). The piece with the handle will have less mass than the piece made up of the end of the bat. To see why, take the origin of coordinates as the center of mass before the bat was cut. Replace each cut piece by a small sphere located at the center of mass for each piece. The sphere representing the handle piece is farther from the origin, but the product of less mass and greater distance balances the product of greater mass and less distance for the end piece, as shown.

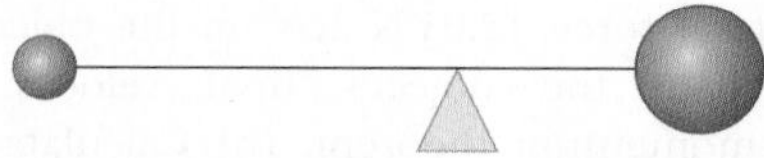

8.6 (i), (a). The vessel–passengers system is isolated. If the passengers all start running one way, the speed of the vessel increases (a *small* amount!) the other way, so the speed of the center of mass of the system remains constant. (ii), (b). Once they stop running, the momentum of the system is the same as it was before they started running; you cannot change the momentum of an isolated system by means of internal forces. In case you are thinking that the passengers could run from bow to stern over and over to take advantage of the speed increase *while* they are running, remember that they will slow the ship down every time they return to the bow!

Relativity

(Emily Serway)

Standing on the shoulders of a giant. David Serway, son of one of the authors, watches over his children, Nathan and Kaitlyn, as they frolic in the arms of Albert Einstein at the Einstein memorial in Washington, D.C. It is well known that Einstein, the principal architect of relativity, was very fond of children.

CHAPTER OUTLINE

Our everyday experiences and observations are associated with objects that move at speeds much less than that of light in a vacuum, $c = 3.00 \times 10^8$ m/s. Models based on Newtonian mechanics and early concepts of space and time were formulated to describe the motion of such objects. This formalism is very successful in describing a wide range of phenomena that occur at low speeds, as we have seen in previous chapters. It fails, however, when applied to objects whose speeds approach that of light. Experimentally, the predictions of Newtonian theory can be tested by accelerating electrons or other particles to very high speeds. For example, it is possible to accelerate an electron to a speed of $0.99c$. According to the Newtonian definition of kinetic energy, if the energy transferred to such an electron were increased by a factor of 4, the electron speed should double to $1.98c$. Relativistic calculations, however, show that the speed of the electron—as well as the speeds of all other objects in the Universe—remains less than the speed of light. Because it places no upper limit on speed, Newtonian mechanics is contrary to modern theoretical predictions and experimental results, and the Newtonian models that we have developed are limited to objects moving much slower than the speed of light. Because Newtonian mechanics does not correctly predict the results of experiments carried out on objects moving at high speeds, we need a new formalism that is valid for these objects.

In 1905, at the age of only 26, Albert Einstein published his *special theory of relativity*, which is the subject of most of this chapter. Regarding the theory, Einstein wrote:

> The relativity theory arose from necessity, from serious and deep contradictions in the old theory from which there seemed no escape. The strength of the new theory lies in the consistency and simplicity with which it solves all these difficulties, using only a few very convincing assumptions.[1]

Although Einstein made many important contributions to science, special relativity alone represents one of the greatest intellectual achievements of the 20th century. With special relativity, experimental observations can be correctly predicted for objects over the range of all possible speeds, from rest to speeds approaching the speed of light. This chapter gives an introduction to special relativity, with emphasis on some of its consequences.

9.1 THE PRINCIPLE OF NEWTONIAN RELATIVITY

We will begin by considering the notion of relativity at low speeds. This discussion was actually begun in Section 3.6 when we discussed relative velocity. At that time, we discussed the importance of the observer. In a similar way here, we will generate equations that allow us to express one observer's measurements in terms of the other's. This process will lead to some rather unexpected and startling results about our understanding of space and time.

As we have mentioned previously, it is necessary to establish a frame of reference when describing a physical event. You should recall from Chapter 4 that an inertial frame is one in which an object is measured to have no acceleration if no forces act on it. Furthermore, any frame moving with constant velocity with respect to an inertial frame must also be an inertial frame. The laws predicting the results of an experiment performed in a vehicle moving with uniform velocity will be identical for the driver of the vehicle and a hitchhiker on the side of the road. The formal statement of this result is called the **principle of Newtonian relativity:**

■ Principle of Newtonian relativity

The laws of mechanics must be the same in all inertial frames of reference.

The following observation illustrates the equivalence of the laws of mechanics in different inertial frames. Consider a pickup truck moving with a constant velocity as in Figure 9.1a. If a passenger in the truck throws a ball straight up in the air, the passenger observes that the ball moves in a vertical path (ignoring air resistance). The motion of the ball appears to be precisely the same as if the ball were thrown by a person at rest on the Earth and observed by that person. The kinematic equations of Chapter 3 describe the results correctly whether the truck is at rest or in uniform motion. Now consider the ball thrown in the truck as viewed by an observer at rest on the Earth. This observer sees the path of the ball as a parabola as in Figure 9.1b. Furthermore, according to this observer, the ball has a horizontal component of velocity equal to the speed of the truck. Although the two observers measure different velocities and see different paths of the ball, they see the same forces on the ball and agree on the validity of Newton's laws as well as classical principles such as conservation of energy and conservation of momentum. Their measurements differ, but the measurements they make satisfy the same laws. All differences between the two views stem from the relative motion of one frame with respect to the other.

[1]A. Einstein and L. Infeld, *The Evolution of Physics* (New York, Simon and Schuster, 1966), p. 192.

FIGURE 9.1 (a) The observer in the truck sees the ball move in a vertical path when thrown upward. (b) The Earth observer sees the path of the ball to be a parabola.

Suppose some physical phenomenon, which we call an **event,** occurs. The event's location in space and time of occurrence can be specified by an observer with the coordinates (x, y, z, t). We would like to be able to transform these coordinates from one inertial frame to another moving with uniform relative velocity, which will allow us to express one observer's measurements in terms of the other's.

Consider two inertial frames S and S′ (Fig. 9.2). The frame S′ moves with a constant velocity $\vec{\mathbf{v}}$ along the common x and x' axes, where $\vec{\mathbf{v}}$ is measured relative to S. We assume that the origins of S and S′ coincide at $t = 0$. Therefore, at time t, the origin of frame S′ is to the right of the origin of S by a distance vt. An event occurs at point P. An observer in S describes the event with space–time coordinates (x, y, z, t), and an observer in S′ describes the same event with coordinates (x', y', z', t'). As we can see from Figure 9.2, a simple geometric argument shows that the space coordinates are related by the equations

$$x' = x - vt \qquad y' = y \qquad z' = z \tag{9.1}$$

Time is assumed to be the same in both inertial frames. That is, within the framework of classical mechanics, all clocks run at the same rate, regardless of their velocity, so that the time at which an event occurs for an observer in S is the same as the time for the same event in S′:

$$t' = t \tag{9.2}$$

Equations 9.1 and 9.2 constitute what is known as the **Galilean transformation of coordinates.**

Now suppose a particle moves through a displacement dx in a time interval dt as measured by an observer in S. It follows from the first of Equations 9.1 that the corresponding displacement dx' measured by an observer in S′ is $dx' = dx - v dt$. Because $dt = dt'$ (Eq. 9.2), we find that

$$\frac{dx'}{dt'} = \frac{dx}{dt} - v$$

or

$$u_x' = u_x - v \tag{9.3}$$

where u_x and u_x' are the instantaneous x components of velocity of the particle[2] relative to S and S′, respectively. This result, which is called the **Galilean velocity**

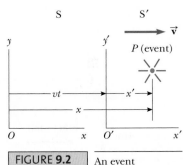

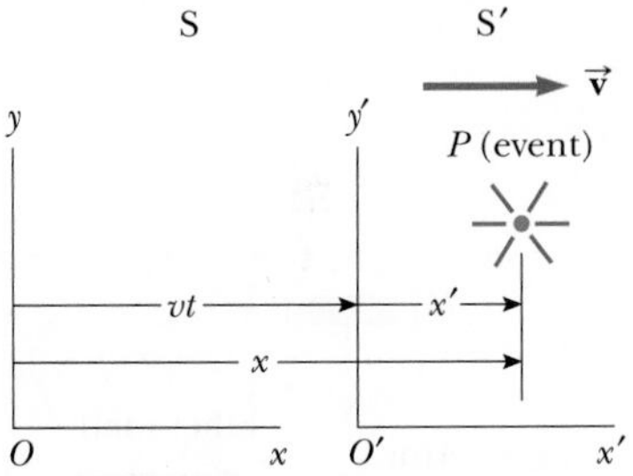

FIGURE 9.2 An event occurs at a point P. The event is seen by two observers O and O' in inertial frames S and S′, where S′ moves with a velocity $\vec{\mathbf{v}}$ relative to S.

PITFALL PREVENTION 9.1

THE RELATIONSHIP BETWEEN THE S AND S′ FRAMES Keep in mind the relationship between the S and S′ frames. Otherwise, many of the mathematical representations in this chapter could be misinterpreted. We choose the time $t = 0$ to be the instant at which the origins of the two coordinate systems coincide. The x and x' axes coincide except that their origins are different at all times other than $t = 0$. The y and y' axes (and the z and z' axes) are parallel, but they do not coincide for $t \neq 0$ because of the displacement of the origin of S′ with respect to that of S. If the S′ frame is moving in the positive x direction relative to S, v is positive; otherwise, it is negative.

[2]We have used v for the speed of the S′ frame relative to the S frame. To avoid confusion, we will use u for the speed of an object or particle.

transformation, is used in everyday observations and is consistent with our intuitive notion of time and space. It is the same equation we generated in Section 3.6 (Eq. 3.22) when we first discussed relative velocity in one dimension. We find, however, that it leads to serious contradictions when applied to objects moving at high speeds.

9.2 THE MICHELSON–MORLEY EXPERIMENT

Many experiments similar to throwing the ball in the pickup truck, described in the preceding section, show us that the laws of classical mechanics are the same in all inertial frames of reference. When similar inquiries are made into the laws of other branches of physics, however, the results are contradictory. In particular, the laws of electricity and magnetism are found to depend on the frame of reference used. It might be argued that these laws are wrong, but that is difficult to accept because the laws are in total agreement with known experimental results. The Michelson–Morley experiment was one of many attempts to investigate this dilemma.

The experiment stemmed from a misconception early physicists had concerning the manner in which light propagates. The properties of mechanical waves, such as water and sound waves, were well known, and all these waves require a *medium* to support the propagation of the disturbance, as we shall discuss in Chapter 13. For sound from your stereo system, the medium is the air, and for ocean waves, the medium is the water surface. In the 19th century, physicists subscribed to a model for light in which electromagnetic waves also require a medium through which to propagate. They proposed that such a medium exists, filling all space, and they named it the **luminiferous ether.** The ether would define an **absolute frame of reference** in which the speed of light is c.

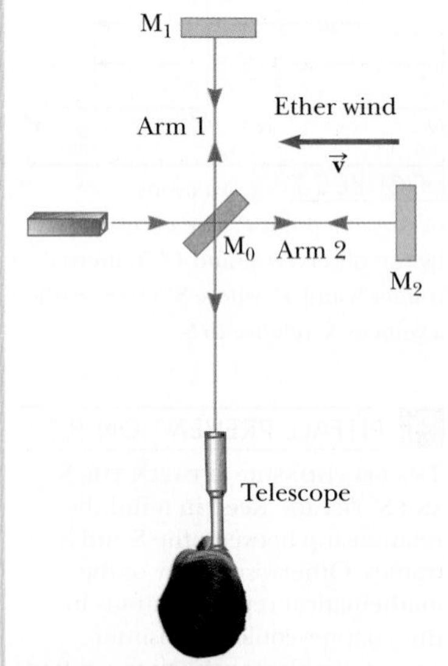

ACTIVE FIGURE 9.3
In the Michelson interferometer, the ether theory claims that the time of travel for a light beam traveling from the beam splitter to mirror M_1 and back will be different from that for a light beam traveling from the beam splitter to mirror M_2 and back. The interferometer is sufficiently sensitive to detect this time difference.

PhysicsNow™ Log into PhysicsNow at **www.pop4e.com** and go to Active Figure 9.3 to adjust the speed of the ether wind and see the effect on the light beams if there were an ether.

The most famous experiment designed to show the presence of the ether was performed in 1887 by A. A. Michelson (1852–1931) and E. W. Morley (1838–1923). The objective was to determine the speed of the Earth through space with respect to the ether, and the experimental tool used was a device called the *interferometer*, shown schematically in Active Figure 9.3.

Light from the source at the left encounters a beam splitter M_0, which is a partially silvered mirror. Part of the light passes through toward mirror M_2, and the other part is reflected upward toward mirror M_1. Both mirrors are the same distance from the beam splitter. After reflecting from these mirrors, the light returns to the beam splitter, and part of each light beam propagates toward the observer at the bottom.

Suppose one arm of the interferometer (Arm 2, in Active Fig. 9.3) is aligned along the direction of the velocity $\vec{v}$ of the Earth through space and therefore through the ether. The "ether wind" blowing in the direction opposite the Earth's motion should cause the speed of light, as measured in the Earth's frame of reference, to be $c - v$ as the light approaches mirror M_2 in Active Figure 9.3 and $c + v$ after reflection.

The other arm (Arm 1) is perpendicular to the ether wind. For light to travel in this direction, the vector $\vec{c}$ must be aimed "upstream" so that the vector addition of $\vec{c}$ and $\vec{v}$ gives the speed of the light perpendicular to the ether wind as $\sqrt{c^2 - v^2}$. This situation is similar to Example 3.6, in which a boat crosses a river with a current. The boat is a model for the light beam in the Michelson–Morley experiment, and the river current is a model for the ether wind.

Because they travel in perpendicular directions with different speeds, light beams leaving the beam splitter simultaneously will arrive back at the beam splitter at different times. The interferometer is designed to detect this time difference. Measurements failed, however, to show any time difference! The Michelson–Morley experiment was repeated by other researchers under varying conditions and at different locations, but the results were always the same: *No time difference of the magnitude required was ever observed.*[3]

The negative result of the Michelson–Morley experiment not only contradicted the ether hypothesis, but it also meant that it was impossible to measure the absolute speed of the Earth with respect to the ether frame. From a theoretical viewpoint, it was impossible to find the absolute frame. As we shall see in the next section, however, Einstein offered a postulate that places a different interpretation on the negative result. In later years, when more was known about the nature of light, the idea of an ether that permeates all space was abandoned. **Light is now understood to be an electromagnetic wave that requires no medium for its propagation.** As a result, an ether through which light travels is an unnecessary construct.

Modern versions of the Michelson–Morley experiment have placed an upper limit of about 5 cm/s = 0.05 m/s on ether wind velocity. We can show that the speed of the Earth in its orbit around the Sun is 2.97×10^4 m/s, six orders of magnitude larger than the upper limit of ether wind velocity! These results have shown quite conclusively that the motion of the Earth has no effect on the measured speed of light.

(AIP Emilio Segrè Visual Archives, Michelson Collection)

ALBERT A. MICHELSON (1852–1931)

Michelson was born in Prussia in a town that later became part of Poland. He moved to the United States as a small child and spent much of his adult life making accurate measurements of the speed of light. In 1907, he was the first American to be awarded the Nobel Prize in Physics, which he received for his work in optics. His most famous experiment, conducted with Edward Morley in 1887, indicated that it was impossible to measure the absolute velocity of the Earth with respect to the ether.

9.3 EINSTEIN'S PRINCIPLE OF RELATIVITY

In the preceding section, we noted the failure of experiments to measure the relative speed between the ether and the Earth. Einstein proposed a theory that boldly removed these difficulties and at the same time completely altered our notion of space and time.[4] He based his relativity theory on two postulates:

1. **The principle of relativity:** All the laws of physics are the same in all inertial reference frames.
2. **The constancy of the speed of light:** The speed of light in vacuum has the same value in all inertial frames, regardless of the velocity of the observer or the velocity of the source emitting the light.

These postulates form the basis of **special relativity,** which is the relativity theory applied to observers moving with constant velocity. The first postulate asserts that *all* the laws of physics—those dealing with mechanics, electricity and magnetism, optics, thermodynamics, and so on—are the same in all reference frames moving with constant velocity relative to each other. This postulate is a sweeping generalization of the principle of Newtonian relativity that only refers to the laws of mechanics. From an experimental point of view, Einstein's principle of relativity means that any kind of experiment performed in a laboratory at rest must agree with the same laws of physics as when performed in a laboratory moving at constant velocity relative to the first one. Hence, no preferred inertial reference frame exists and it is impossible to detect absolute motion.

Note that postulate 2, the principle of the constancy of the speed of light, is required by postulate 1: If the speed of light were not the same in all inertial frames, it would be possible to experimentally distinguish between inertial frames and a

[3]From an Earth observer's point of view, changes in the Earth's speed and direction of motion in the course of a year are viewed as ether wind shifts. Even if the speed of the Earth with respect to the ether were zero at some time, six months later the Earth is moving in the opposite direction, the speed of the Earth with respect to the ether would be nonzero, and a clear time difference should be detected. None has ever been observed, however.

[4]A. Einstein, "On the Electrodynamics of Moving Bodies," *Ann. Physik* 17:891, 1905. For an English translation of this article and other publications by Einstein, see the book by H. Lorentz, A. Einstein, H. Minkowski, and H. Weyl, *The Principle of Relativity* (New York: Dover, 1958).

(AIP Niels Bohr Library)

ALBERT EINSTEIN (1879–1955)

Einstein, one of the greatest physicists of all times, was born in Ulm, Germany. He left Germany in 1932 for the United States and became a U.S. citizen in 1940. In 1905, at the age of 26, he published four scientific papers that revolutionized physics. Two of these papers were concerned with what is now considered his most important contribution of all, the special theory of relativity.

In 1916, Einstein published his work on the general theory of relativity. The most dramatic prediction of this theory is the degree to which light is deflected by a gravitational field. Measurements made by astronomers on bright stars in the vicinity of the eclipsed sun in 1919 confirmed Einstein's prediction, and Einstein suddenly became a world celebrity.

Einstein was deeply disturbed by the development of quantum mechanics in the 1920s despite his own role as a scientific revolutionary. In particular, he could never accept the probabilistic view of events in nature that is a central feature of quantum theory. The last few decades of his life were devoted to an unsuccessful search for a unified theory that would combine gravitation and electromagnetism into one picture.

preferred, absolute frame in which the speed of light is c, in contradiction to postulate 1. Postulate 2 also eliminates the problem of measuring the speed of the ether by denying the existence of the ether and boldly asserting that light always moves with speed c relative to all inertial observers.

9.4 CONSEQUENCES OF SPECIAL RELATIVITY

If we accept the postulates of special relativity, we must conclude that relative motion is unimportant when measuring the speed of light, which is the lesson of the Michelson–Morley experiment. At the same time, we must alter our commonsense notion of space and time and be prepared for some very unexpected consequences, as we shall see now.

Simultaneity and the Relativity of Time

A basic premise of Newtonian mechanics is that a universal time scale exists that is the same for all observers. In fact, Newton wrote, "Absolute, true, and mathematical time, of itself, and from its own nature, flows equably without relation to anything external." Thus, Newton and his followers simply took simultaneity for granted. In his development of special relativity, Einstein abandoned the notion that two events that appear simultaneous to one observer appear simultaneous to all observers. According to Einstein, **a time measurement depends on the reference frame in which the measurement is made.**

Einstein devised the following thought experiment to illustrate this point. A boxcar moves with uniform velocity and two lightning bolts strike its ends, as in Figure 9.4a, leaving marks on the boxcar and on the ground. The marks on the boxcar are labeled A' and B', and those on the ground are labeled A and B. An observer at O' moving with the boxcar is midway between A' and B', and a ground observer at O is midway between A and B. The events recorded by the observers are the arrivals of light signals from the lightning bolts.

The two light signals reach observer O at the same time as indicated in Figure 9.4b. As a result, O concludes that the events at A and B occurred simultaneously. Now consider the same events as viewed by the observer on the boxcar at O'. From our frame of reference, at rest with respect to the tracks in Figure 9.4, we see the lightning strikes occur as A' passes A, O' passes O, and B' passes B. By the time the light has reached observer O, observer O' has moved as indicated in Figure 9.4b. Therefore, the light signal from B' has already swept past O' because it had less distance to travel, but the light from A' has not yet reached O'. According to Einstein, **observer O' must find that light travels at the same speed as that measured by observer O.** Observer O' therefore concludes that the lightning struck the front of the boxcar before it struck the back. This thought experiment clearly demonstrates that the two events, which appear to be simultaneous to observer O, do not appear to be simultaneous to observer O'. In general, two events separated in space and observed to be simultaneous in one reference frame are not observed to be simultaneous in a second frame moving relative to the first. That is, simultaneity is not an absolute concept but one that depends on the state of motion of the observer.

Einstein's thought experiment demonstrates that two observers can disagree on the simultaneity of two events. **This disagreement, however, depends on the transit time of light to the observers and therefore does *not* demonstrate the deeper meaning of relativity.** In relativistic analyses of high-speed situations, relativity shows that **simultaneity is relative even when the transit time is subtracted out.** In fact, all the relativistic effects that we will discuss from here on will assume that we are ignoring differences caused by the transit time of light to the observers.

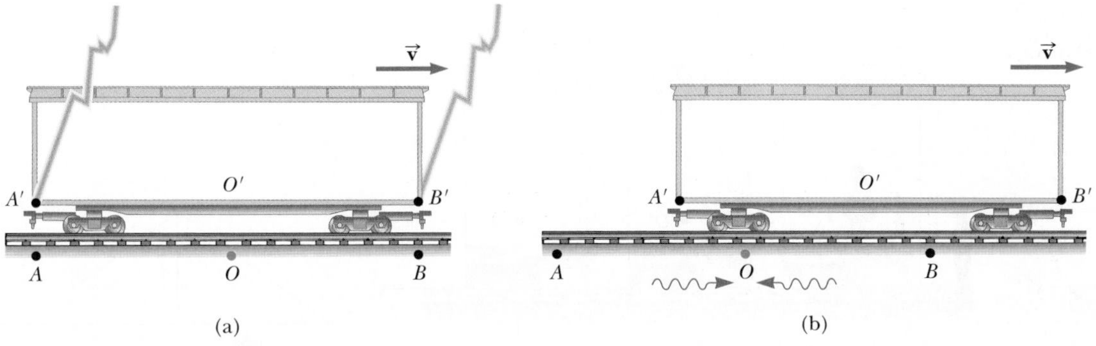

FIGURE 9.4 (a) Two lightning bolts strike the ends of a moving boxcar. (b) The events appear to be simultaneous to the observer at O, who is standing on the ground midway between A and B. The events do not appear to be simultaneous to the observer O' riding on the boxcar, who claims that the front of the car is struck before the rear. Note that the leftward-traveling light signal from B' has already passed observer O', but the rightward-traveling light signal from A' has not yet reached O'.

Time Dilation

According to the preceding paragraph, observers in different inertial frames measure different time intervals between a pair of events, independent of the transit time of the light. This situation can be illustrated by considering a vehicle moving to the right with a speed v as in the pictorial representation in Active Figure 9.5a. A mirror is fixed to the ceiling of the vehicle, and observer O', at rest in a frame attached to the vehicle, holds a flashlight a distance d below the mirror. At some instant, the flashlight is turned on momentarily and emits a pulse of light (event 1) directed toward the mirror. At some later time after reflecting from the mirror, the pulse arrives back at the flashlight (event 2). Observer O' carries a clock that she uses to measure the time interval Δt_p between these two events. (The subscript p stands for "proper," as will be discussed shortly.) Because the light pulse has a constant speed c, the time interval required for the pulse to travel from O' to the mirror and back to O' (a distance of $2d$) can be found by modeling the light pulse as a particle under constant speed as discussed in Chapter 2:

$$\Delta t_p = \frac{2d}{c} \quad [9.4]$$

This time interval Δt_p is measured by O', for whom the two events occur at the same spatial position.

Now consider the same pair of events as viewed by observer O at rest with respect to a second frame attached to the ground as in Active Figure 9.5b. According to this observer, the mirror and flashlight are moving to the right with a speed v. The geometry appears to be entirely different as viewed by this observer. By the time the light from the flashlight reaches the mirror, the mirror has moved horizontally a distance $v\,\Delta t/2$, where Δt is the time interval required for the light to travel from the flashlight to the mirror and back to the flashlight as measured by observer O. In other words, the second observer concludes that because of the motion of the vehicle, if the light is to hit the mirror, it must leave the flashlight at an angle with respect to the vertical direction. Comparing Active Figures 9.5a and 9.5b, we see that the light must travel farther when observed in the second frame than in the first frame.

According to the second postulate of special relativity, both observers must measure c for the speed of light. Because the light travels farther in the second frame but at the same speed, it follows that the time interval Δt measured by the observer in the second frame is longer than the time interval Δt_p measured by the observer in the first frame. To obtain a relationship between these two time intervals, it is

PITFALL PREVENTION 9.2

WHO'S RIGHT? At this point, you might wonder which observer in Figure 9.4 is correct concerning the two events. *Both are correct* because the principle of relativity states that *no inertial frame of reference is preferred.* Although the two observers reach different conclusions, both are correct in their own reference frame because the concept of simultaneity is not absolute. In fact, the central point of relativity is that any uniformly moving frame of reference can be used to describe events and do physics.

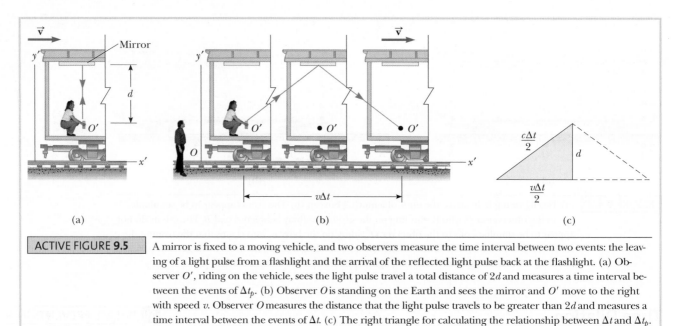

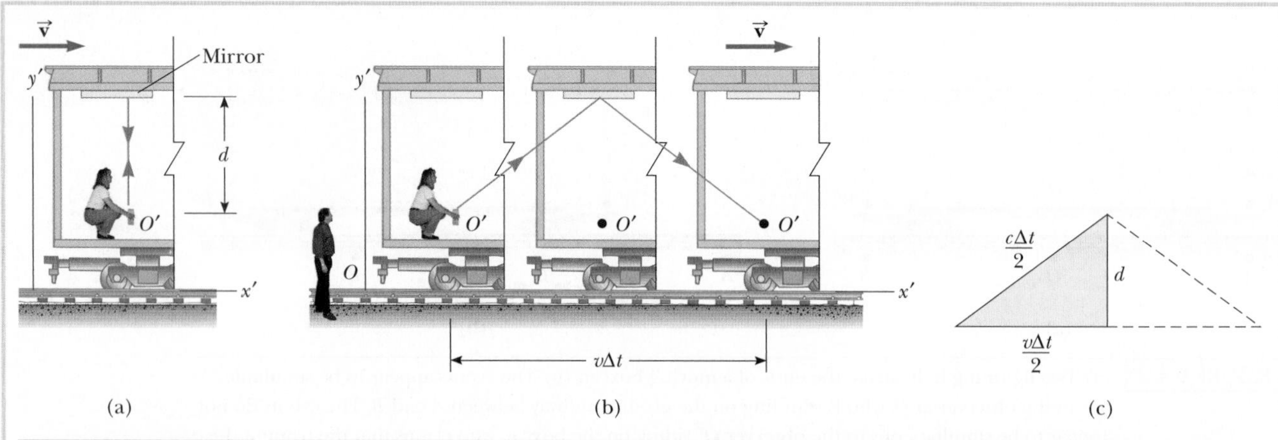

ACTIVE FIGURE 9.5 A mirror is fixed to a moving vehicle, and two observers measure the time interval between two events: the leaving of a light pulse from a flashlight and the arrival of the reflected light pulse back at the flashlight. (a) Observer O', riding on the vehicle, sees the light pulse travel a total distance of $2d$ and measures a time interval between the events of Δt_p. (b) Observer O is standing on the Earth and sees the mirror and O' move to the right with speed v. Observer O measures the distance that the light pulse travels to be greater than $2d$ and measures a time interval between the events of Δt. (c) The right triangle for calculating the relationship between Δt and Δt_p.

PhysicsNow™ By logging into PhysicsNow at **www.pop4e.com** and going to Active Figure 9.5 you can observe the bouncing of the light pulse for various speeds of the train.

convenient to use the right triangle geometric model shown in Active Figure 9.5c. The Pythagorean theorem applied to the triangle gives

$$\left(\frac{c\,\Delta t}{2}\right)^2 = \left(\frac{v\,\Delta t}{2}\right)^2 + d^2$$

Solving for Δt gives

$$\Delta t = \frac{2d}{\sqrt{c^2 - v^2}} = \frac{2d}{c\sqrt{1 - \dfrac{v^2}{c^2}}} \qquad [9.5]$$

Because $\Delta t_p = 2d/c$, we can express Equation 9.5 as

$$\Delta t = \frac{\Delta t_p}{\sqrt{1 - \dfrac{v^2}{c^2}}} = \gamma\,\Delta t_p \qquad [9.6]$$

where $\gamma = (1 - v^2/c^2)^{-1/2}$. This result says that **the time interval Δt measured by O is longer than the time interval Δt_p measured by O'** because γ is always greater than one. That is, $\Delta t > \Delta t_p$. This effect is known as **time dilation.**

We can see that time dilation is not observed in our everyday lives by considering the factor γ. This factor deviates significantly from a value of 1 only for very high speeds, as shown in Table 9.1. For example, for a speed of $0.1c$, the value of γ is 1.005. Therefore, a time dilation of only 0.5% occurs at one-tenth the speed of light. Speeds we encounter on an everyday basis are far slower than that, so we do not see time dilation in normal situations.

The time interval Δt_p in Equation 9.6 is called the **proper time interval.** In general, the proper time interval is defined as **the time interval between two events as measured by an observer for whom the events occur at the same point in space.** In our case, observer O' measures the proper time interval. For us to be able to use Equation 9.6, the events must occur at the same spatial position in *some* inertial

TABLE 9.1

Approximate Values for γ at Various Speeds

v/c	γ
0.001 0	1.000 000 5
0.010	1.000 05
0.10	1.005
0.20	1.021
0.30	1.048
0.40	1.091
0.50	1.155
0.60	1.250
0.70	1.400
0.80	1.667
0.90	2.294
0.92	2.552
0.94	2.931
0.96	3.571
0.98	5.025
0.99	7.089
0.995	10.01
0.999	22.37

frame. Therefore, for instance, this equation cannot be used to relate the measurements made by the two observers in the lightning example described at the beginning of this section because the lightning strikes occur at different positions for both observers.

If a clock is moving with respect to you, the time interval between ticks of the moving clock is measured to be longer than the time interval between ticks of an identical clock in your reference frame. Therefore, it is often said that a moving clock is measured to run more slowly than a clock in your reference frame by a factor γ. That is true for mechanical clocks as well as for the light clock just described. We can generalize this result by stating that all physical processes, including chemical and biological ones, slow down relative to a stationary clock when those processes occur in a frame moving with respect to the clock. For example, the heartbeat of an astronaut moving through space would keep time with a clock inside the spaceship. Both the astronaut's clock and heartbeat would be measured to be slowed down according to an observer on the Earth comparing time intervals with his own clock at rest with respect to him (although the astronaut would have no sensation of life slowing down in the spaceship).

Time dilation is a verifiable phenomenon; let us look at one situation in which the effects of time dilation can be observed and that served as an important historical confirmation of the predictions of relativity. Muons are unstable elementary particles that have a charge equal to that of the electron and a mass 207 times that of the electron. Muons can be produced as a result of collisions of cosmic radiation with atoms high in the atmosphere. Slow-moving muons in the laboratory have a lifetime measured to be the proper time interval $\Delta t_p = 2.2\ \mu\text{s}$. If we assume that the speed of atmospheric muons is close to the speed of light, we find that these particles can travel a distance during their lifetime of approximately $(3.0 \times 10^8\ \text{m/s})(2.2 \times 10^{-6}\ \text{s}) \approx 6.6 \times 10^2\ \text{m}$ before they decay (Fig. 9.6a). Hence, they are unlikely to reach the surface of the Earth from high in the atmosphere where they are produced; nonetheless, experiments show that a large number of muons *do* reach the surface. The phenomenon of time dilation explains this effect. As measured by an observer on the Earth, the muons have a dilated lifetime equal to $\gamma\Delta t_p$. For example, for $v = 0.99c$, $\gamma \approx 7.1$ and $\gamma\,\Delta t_p \approx 16\ \mu\text{s}$. Hence, the average distance traveled by the muons in this time interval as measured by an observer on the Earth is approximately $(3.0 \times 10^8\ \text{m/s})(16 \times 10^{-6}\ \text{s}) \approx 4.8 \times 10^3\ \text{m}$, as shown in Figure 9.6b.

The results of an experiment reported by J. C. Hafele and R. E. Keating provided direct evidence of time dilation.[5] The experiment involved the use of very stable atomic clocks. Time intervals measured with four such clocks in jet flight were compared with time intervals measured by reference clocks located at the U.S. Naval Observatory. Their results were in good agreement with the predictions of special relativity and can be explained in terms of the relative motion between the Earth's rotation and the jet aircraft. In their paper, Hafele and Keating report the following: "Relative to the atomic time scale of the U.S. Naval Observatory, the flying clocks lost 59 ± 10 ns during the eastward trip and gained 273 ± 7 ns during the westward trip."

Muon is created

$\approx 6.6 \times 10^2$ m

Muon decays

(a)

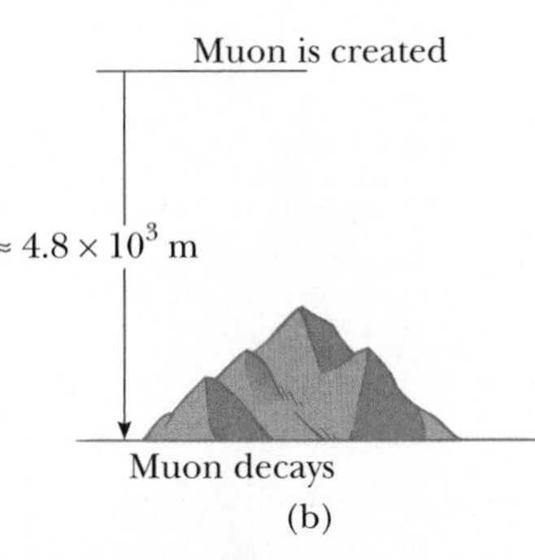

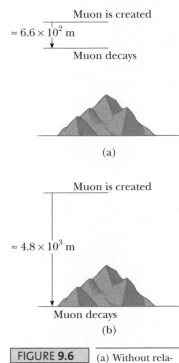

FIGURE 9.6 (a) Without relativistic considerations, muons created in the atmosphere and traveling downward with a speed of $0.99c$ would travel only about 6.6×10^2 m before decaying with an average lifetime of 2.2 μs. Therefore, very few muons would reach the surface of the Earth. (b) With relativistic considerations, the muon's lifetime is dilated according to an observer on Earth. As a result, according to this observer, the muon can travel about 4.8×10^3 m before decaying, which results in many of them arriving at the surface.

QUICK QUIZ 9.1 Suppose the observer O' on the train in Active Figure 9.5 aims her flashlight at the far wall of the boxcar and turns it on and off, sending a pulse of light toward the far wall. Both O' and O measure the time interval between when the pulse leaves the flashlight and when it hits the far wall. Which observer measures the proper time interval between these two events? **(a)** O' **(b)** O **(c)** both observers **(d)** neither observer

[5]J. C. Hafele and R. E. Keating, "Around the World Atomic Clocks: Relativistic Time Gains Observed," *Science,* July 14, 1972, p. 168.

QUICK QUIZ 9.2 A crew watches a movie that is 2 h long in a spacecraft that is moving at high speed through space. Will an Earth-bound observer, who is watching the movie through a powerful telescope, measure the duration of the movie to be (**a**) longer than, (**b**) shorter than, or (**c**) equal to 2 h?

The Twin Paradox

An intriguing consequence of time dilation is the so-called twin paradox (Fig. 9.7). Consider an experiment involving a set of twins named Speedo and Goslo. At age 20, Speedo, the more adventuresome of the two, sets out on an epic journey to Planet X, located 20 lightyears (ly) from the Earth. (Note that 1 ly is the distance light travels through free space in 1 year. It is equal to 9.46×10^{15} m.) Furthermore, his spaceship is capable of reaching a speed of $0.95c$ relative to the inertial frame of his twin brother back home. After reaching Planet X, Speedo becomes homesick and immediately returns to the Earth at the same speed $0.95c$. Upon his return, Speedo is shocked to discover that Goslo has aged 42 yr and is now 62 yr old. Speedo, on the other hand, has aged only 13 yr.

At this point, it is fair to raise the following question: Which twin is the traveler and which is really younger as a result of this experiment? From Goslo's frame of reference, he was at rest while his brother traveled at a high speed away from him and then came back. According to Speedo, however, he himself remained stationary while Goslo and the Earth raced away from him and then headed back. There is an apparent contradiction due to the apparent symmetry of the observations. Which twin has developed signs of excess aging?

 Varying rates of aging in relativity

The situation in this problem is actually not symmetrical. To resolve this apparent paradox, recall that the special theory of relativity describes observations made in inertial frames of reference moving relative to each other. Speedo, the space traveler, must experience a series of accelerations during his journey because he must fire his rocket engines to slow down and start moving back toward the Earth. As a result, his speed is not always uniform, and consequently he is not always in an inertial frame. Therefore, there is no paradox because only Goslo, who is always in a single inertial frame, can make correct predictions based on special relativity. During each passing year noted by Goslo, slightly less than 4 months elapses for Speedo.

Only Goslo, who is in a single inertial frame, can apply the simple time dilation formula to Speedo's trip. Therefore, Goslo finds that instead of aging 42 yr, Speedo ages only $(1 - v^2/c^2)^{1/2}(42 \text{ yr}) = 13$ yr. According to both twins, Speedo spends 6.5 yr traveling to Planet X and 6.5 yr returning, for a total travel time of 13 yr.

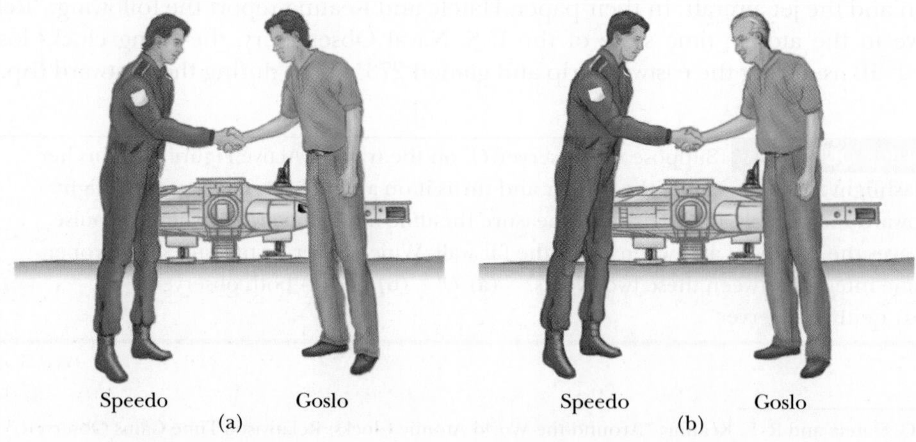

FIGURE 9.7 (a) As Speedo leaves his twin brother, Goslo, on the Earth, both are the same age. (b) When Speedo returns from his journey to Planet X, he is younger than Goslo.

QUICK QUIZ 9.3 Suppose astronauts are paid according to the amount of time they spend traveling in space. After a long voyage traveling at a speed approaching c, would a crew rather be paid according to (**a**) an Earth-based clock, (**b**) their spacecraft's clock, or (**c**) either clock?

■ Thinking Physics 9.1

Suppose a student explains time dilation with the following argument: If I start running away from a clock at 12:00 at a speed very close to the speed of light, I would not see the time change, because the light from the clock representing 12:01 would never reach me. What is the flaw in this argument?

Reasoning The implication in this argument is that the velocity of light relative to the runner is approximately *zero* because "the light . . . would never reach me." In this Galilean point of view, the relative velocity is a simple subtraction of running velocity from the light velocity. From the point of view of special relativity, one of the fundamental postulates is that the speed of light is the same for all observers, *including one running away from the light source at the speed of light.* Therefore, the light from 12:01 will move toward the runner at the speed of light, as measured by all observers, including the runner. ■

EXAMPLE 9.1 **What Is the Heart Rate of the Astronaut?**

An astronaut at rest on the Earth has a heartbeat rate of 70 beats/min.

A When the astronaut is traveling on a spacecraft at $0.95c$, what will this rate be as measured by another observer on the spacecraft?

Solution The proper time interval between beats of the heart is the period $T_p = 60 \text{ s}/70 \text{ beats} = 0.86 \text{ s/beat}$. The observer in the spacecraft will measure this time interval because two successive beats of the heart take place at the same position according to this observer.

B When the astronaut is traveling on a spacecraft at $0.95c$, what will this rate be as measured by an observer at rest on the Earth?

Solution For the time interval measured by the observer on the Earth, Equation 9.6 gives

$$T = \gamma T_p = \frac{1}{\sqrt{1 - \dfrac{(0.95c)^2}{c^2}}} T_p$$

$$= (3.2)(0.86 \text{ s}) = 2.7 \text{ s}$$

Therefore, the heart rate as measured by this observer is

$$\text{Heart rate} = \frac{1}{2.7 \text{ s/beat}} = 0.36 \text{ beat/s}\left(\frac{60 \text{ s}}{1 \text{ min}}\right)$$

$$= 22 \text{ beats/min}$$

That is, measurements by this observer show that the heart rate is measured to slow down compared with that measured by an observer on the spacecraft.

Length Contraction

The measured distance between two points also depends on the frame of reference. The **proper length** of an object is defined as **the distance in space between the end points of the object measured by someone who is at rest relative to the object.** An observer in a reference frame that is moving with respect to the object will measure a length along the direction of the velocity that is always less than the proper length. This effect is known as **length contraction.** Although we have introduced this effect through the mental representation of an object, the object is not

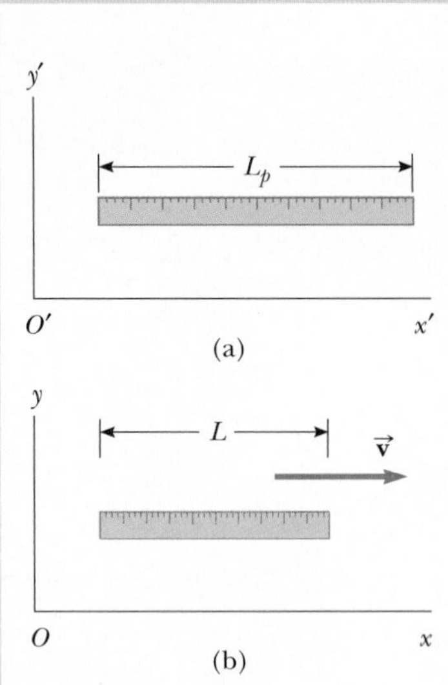

ACTIVE FIGURE 9.8
(a) According to an observer in a frame attached to the meter stick (that is, both the stick and the frame have the same velocity), the stick is measured to have its proper length L_p. (b) According to an observer in a frame in which the meter stick has a velocity $\vec{v}$ relative to the frame, the stick is measured to be *shorter* than the proper length L_p by a factor $(1 - v^2/c^2)^{1/2}$.

Physics Now™ By logging into PhysicsNow at **www.pop4e.com** and going to Active Figure 9.8 you can view the meter stick from the points of view of two observers to compare the measured length of the stick.

necessary. **The distance between *any* two points in space is measured by an observer to be contracted along the direction of the velocity of the observer relative to the points.**

Consider a spacecraft traveling with a speed v from one star to another. We will consider the time interval between two events: (1) the leaving of the spacecraft from the first star and (2) the arrival of the spacecraft at the second star. There are two observers: one on the Earth and the other in the spacecraft. The observer at rest on the Earth (and also at rest with respect to the two stars) measures the distance between the stars to be L_p, the proper length. According to this observer, the time interval required for the spacecraft to complete the voyage is $\Delta t = L_p/v$. What does an observer in the moving spacecraft measure for the distance between the stars? This observer measures the proper time interval because the passage of each of the two stars by his spacecraft occurs at the same position in his reference frame, at his spacecraft. Therefore, because of time dilation, the time interval required to travel between the stars as measured by the space traveler will be smaller than that for the Earth-bound observer. Using the time dilation expression, the proper time interval between events is $\Delta t_p = \Delta t/\gamma$. The space traveler claims to be at rest and sees the destination star moving toward the spacecraft with speed v. Because the space traveler reaches the star in the time interval $\Delta t_p < \Delta t$, he concludes that the distance L between the stars is shorter than L_p. This distance measured by the space traveler is

$$L = v\,\Delta t_p = v\frac{\Delta t}{\gamma}$$

Because $L_p = v\,\Delta t$, we see that

$$L = \frac{L_p}{\gamma} = L_p\left(1 - \frac{v^2}{c^2}\right)^{1/2} \qquad [9.7]$$

Because $(1 - v^2/c^2)^{1/2}$ is less than 1, the space traveler measures a length that is shorter than the proper length. Therefore, an observer in motion with respect to two points in space measures the length L between the points (along the direction of motion) to be shorter than the length L_p measured by an observer at rest with respect to the points (the proper length).

Note that **length contraction takes place only along the direction of motion.** For example, suppose a meter stick moves past an Earth observer with speed v as in Active Figure 9.8. The length of the meter stick as measured by an observer in a frame attached to the stick is the proper length L_p as in Active Figure 9.8a. The length L of the stick measured by the Earth observer is shorter than L_p by the factor $(1 - v^2/c^2)^{1/2}$, but the width is the same. Furthermore, length contraction is a symmetric effect. If the stick is at rest on the Earth, an observer in the moving frame would also measure its length to be shorter by the same factor $(1 - v^2/c^2)^{1/2}$.

It is important to emphasize that the proper length and proper time interval are defined differently. The proper length is measured by an observer at rest with respect to the end points of the length. The proper time interval between two events is measured by someone for whom the events occur at the same position. Often, the proper time interval and the proper length are not measured by the same observer. As an example, let us return to the decaying muons moving at speeds close to the speed of light. An observer in the muon's reference frame would measure the proper lifetime, and an Earth-based observer would measure the proper length (the distance from creation to decay in Fig. 9.6). In the muon's reference frame, no time dilation occurs, but the distance of travel is observed to be shorter when measured in this frame. Likewise, in the Earth observer's reference frame, a time dilation does occur, but the distance of travel is measured to be the proper length. Therefore, when calculations on the muon are performed in both frames, the outcome of the experiment in one frame is the same as the outcome in the other frame: More muons reach the surface than would be predicted without relativistic calculations.

QUICK QUIZ 9.4 You are packing for a trip to another star. During the journey, you will be traveling at $0.99c$. You are trying to decide whether you should buy smaller sizes of your clothing because you will be thinner on your trip as a result of length contraction. You are also considering saving money by reserving a smaller cabin to sleep in because you will be shorter when you lie down. Should you **(a)** buy smaller sizes of clothing, **(b)** reserve a smaller cabin, **(c)** do neither, or **(d)** do both?

EXAMPLE 9.2 A Voyage to Sirius

An astronaut takes a trip to Sirius, located 8.00 ly from the Earth. The astronaut measures the time interval for the one-way journey to be 6.00 yr. If the spacecraft moves at a constant speed of $0.800c$, how can the 8.00-ly distance be reconciled with the 6.00-yr duration measured by the astronaut?

Solution The 8.00 ly represents the proper length (the distance from the Earth to Sirius) measured by an observer for whom both the Earth and Sirius are at rest. The astronaut measures Sirius to be approaching her at $0.800c$ but also measures the distance contracted to

$$L = \frac{L_p}{\gamma} = \frac{8.00 \text{ ly}}{\gamma} = (8.00 \text{ ly})\sqrt{1 - \frac{v^2}{c^2}}$$

$$= (8.00 \text{ ly})\sqrt{1 - \frac{(0.800c)^2}{c^2}} = 4.80 \text{ ly}$$

So the travel time measured on her clock is

$$\Delta t = \frac{L}{v} = \frac{4.80 \text{ ly}}{0.800c} = \frac{4.80 \text{ ly}}{0.800(1.00 \text{ ly/yr})} = 6.00 \text{ yr}$$

Notice that we have used the speed of light as $c = 1.00$ ly/yr to determine this last result.

EXAMPLE 9.3 The Triangular Spacecraft

A spacecraft in the form of a triangle flies by an observer on the Earth with a speed of $0.95c$ along the x direction. According to an observer on the spacecraft (Fig. 9.9a), the distances L_p and y are measured to be 52 m and 25 m, respectively. What are the dimensions of the spacecraft as measured by the Earth observer when the spacecraft is in motion along the direction shown in Figure 9.9b?

Solution In Figure 9.9a, we show the shape of the spacecraft as measured[6] by the observer on the spacecraft. The proper length along the direction of motion

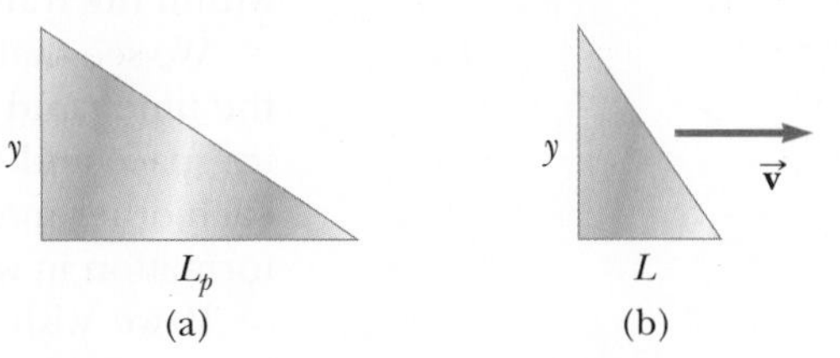

FIGURE 9.9 (Example 9.3) (a) When the spacecraft is at rest, its shape is measured as shown. (b) The spacecraft is measured to have this shape when it moves to the right with a speed v. Note that only its x dimension is contracted in this case.

[6]Notice that we are careful here to say the shape "as measured" by an observer rather than "as seen" by an observer. What an observer *sees* when looking at an object is the set of light rays entering the eye at a given instant. These rays left different parts of the object at different times because different parts of the object are at different distances from the eye. Figure 9.9b is a representation of the light rays leaving different parts of the object *simultaneously*. Viewing an object moving at high speed introduces additional changes to the object besides length contraction, including apparent rotations of the object.

is $L_p = 52$ m. The Earth observer watching the moving spacecraft measures the horizontal length of the spacecraft to be contracted to

$$L = L_p\sqrt{1 - \frac{v^2}{c^2}} = (52 \text{ m})\sqrt{1 - \frac{(0.95c)^2}{c^2}} = 16 \text{ m}$$

The 25-m vertical height is unchanged because it is perpendicular to the direction of relative motion between observer and spacecraft. Figure 9.9b represents the spacecraft's shape as measured by the Earth observer.

9.5 THE LORENTZ TRANSFORMATION EQUATIONS

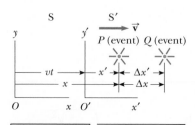

FIGURE 9.10 Events occur at points P and Q and are observed by an observer at rest in the S frame and another in the S′ frame, which is moving to the right with a speed v.

Suppose an event that occurs at some point P is reported by two observers: one at rest in a frame S and another in a frame S′ that is moving to the right with speed v as in Figure 9.10. The observer in S reports the event with space–time coordinates (x, y, z, t), and the observer in S′ reports the same event using the coordinates (x', y', z', t'). If two events occur at P and Q in Figure 9.10, Equation 9.1 predicts that $\Delta x = \Delta x'$; that is, the distance between the two points in space at which the events occur does not depend on the motion of the observer. Because this notion is contradictory to that of length contraction, the Galilean transformation is not valid when v approaches the speed of light. In this section, we state the correct transformation equations that apply for all speeds in the range $0 \le v < c$.

The equations that relate these measurements and enable us to transform coordinates from S to S′ are the **Lorentz transformation equations:**

■ Lorentz transformation for S→S′

$$\begin{aligned} x' &= \gamma(x - vt) \\ y' &= y \\ z' &= z \\ t' &= \gamma\left(t - \frac{v}{c^2}x\right) \end{aligned} \qquad [9.8]$$

These transformation equations were developed by Hendrik A. Lorentz (1853–1928) in 1890 in connection with electromagnetism. Einstein, however, recognized their physical significance and took the bold step of interpreting them within the framework of special relativity.

We see that the value for t' assigned to an event by observer O' depends both on the time t and on the coordinate x as measured by observer O. Therefore, in relativity, space and time are not separate concepts but rather are closely interwoven with each other in what we call **space–time.** This case is unlike that of the Galilean transformation in which $t = t'$.

If we wish to transform coordinates in the S′ frame to coordinates in the S frame, we simply replace v by $-v$ and interchange the primed and unprimed coordinates in Equation 9.8:

■ Inverse Lorentz transformation for S′→S

$$\begin{aligned} x &= \gamma(x' + vt') \\ y &= y' \\ z &= z' \\ t &= \gamma\left(t' + \frac{v}{c^2}x'\right) \end{aligned} \qquad [9.9]$$

When $v \ll c$, the Lorentz transformation reduces to the Galilean transformation. To check, note that if $v \ll c$, $v^2/c^2 \ll 1$, so γ approaches 1 and Equation 9.8 reduces in this limit to Equations 9.1 and 9.2:

$$x' = x - vt \qquad y' = y \qquad z' = z \qquad t' = t$$

Lorentz Velocity Transformation

Let us now derive the **Lorentz velocity transformation,** which is the relativistic counterpart of the Galilean velocity transformation, Equation 9.3. Once again S′ is a frame of reference that moves at a speed v relative to another frame S along the common x and x' axes. Suppose an object is measured in S′ to have an instantaneous velocity component u'_x given by

$$u'_x = \frac{dx'}{dt'} \qquad [9.10]$$

Using Equations 9.8, we have

$$dx' = \gamma(dx - v\,dt) \qquad \text{and} \qquad dt' = \gamma\left(dt - \frac{v}{c^2}\,dx\right)$$

Substituting these values into Equation 9.10 gives

$$u'_x = \frac{dx'}{dt'} = \frac{dx - v\,dt}{dt - \dfrac{v}{c^2}\,dx} = \frac{\dfrac{dx}{dt} - v}{1 - \dfrac{v}{c^2}\,\dfrac{dx}{dt}}$$

Note, though, that dx/dt is the velocity component u_x of the object measured in S, so this expression becomes

$$u'_x = \frac{u_x - v}{1 - \dfrac{u_x v}{c^2}} \qquad [9.11]$$

■ Lorentz velocity transformation for S → S′

Similarly, if the object has velocity components along y and z, the components in S′ are

$$u'_y = \frac{u_y}{\gamma\left(1 - \dfrac{u_x v}{c^2}\right)} \qquad \text{and} \qquad u'_z = \frac{u_z}{\gamma\left(1 - \dfrac{u_x v}{c^2}\right)} \qquad [9.12]$$

When u_x or v is much smaller than c (the nonrelativistic case), the denominator of Equation 9.11 approaches unity and so $u'_x \approx u_x - v$. This result corresponds to the Galilean velocity transformations. In the other extreme, when $u_x = c$, Equation 9.11 becomes

$$u'_x = \frac{c - v}{1 - \dfrac{cv}{c^2}} = \frac{c\left(1 - \dfrac{v}{c}\right)}{1 - \dfrac{v}{c}} = c$$

From this result, we see that an object whose speed approaches c relative to an observer in S also has a speed approaching c relative to an observer in S′, independent of the relative motion of S and S′. Note that this conclusion is consistent with Einstein's second postulate, namely, that the speed of light must be c relative to all inertial frames of reference.

To obtain u_x in terms of u'_x, we replace v by $-v$ in Equation 9.11 and interchange the roles of primed and unprimed variables:

$$u_x = \frac{u'_x + v}{1 + \dfrac{u'_x v}{c^2}} \qquad [9.13]$$

■ Inverse Lorentz velocity transformation for S′ → S

PITFALL PREVENTION 9.3

What can the observers agree on? We have seen several measurements on which the two observers O and O' do not agree. These measurements include (1) the time interval between events that take place in the same position in one of the frames, (2) the distance between two points that remain fixed in one of their frames, (3) the velocity components of a moving particle, and (4) whether two events occurring at different locations in both frames are simultaneous. It is worth noting here what the two observers *can* agree on: (1) the relative speed v with which they move with respect to each other, (2) the speed c of any ray of light, and (3) the simultaneity of two events taking place at the same position and time in some frame.

QUICK QUIZ 9.5 You are driving on a freeway at a relativistic speed. Straight ahead of you, a technician standing on the ground turns on a searchlight and a beam of light moves exactly vertically upward, as seen by the technician. As you observe the beam of light, you measure the magnitude of the vertical component of its velocity as **(a)** equal to c, **(b)** greater than c, or **(c)** less than c. If the technician aims the searchlight directly at you instead of upward, you measure the magnitude of the horizontal component of its velocity as **(d)** equal to c, **(e)** greater than c, or **(f)** less than c.

EXAMPLE 9.4 Relative Velocity of Spacecraft

Two spacecraft A and B are moving directly toward each other as in Figure 9.11. An observer on the Earth measures the speed of A to be $0.750c$ and the speed of B to be $0.850c$. Find the velocity of B with respect to A.

Solution Conceptualize the problem by studying Figure 9.11. Note that the spacecraft are approaching each other, so the speed of one as measured by an observer in the other will be larger than the speed of either as measured by an observer on the Earth.

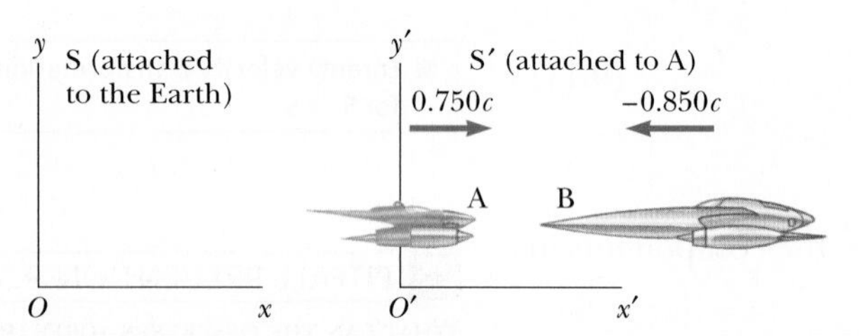

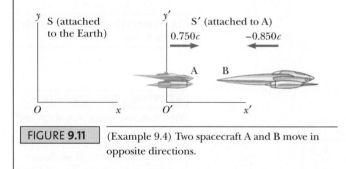

FIGURE 9.11 (Example 9.4) Two spacecraft A and B move in opposite directions.

Because the speeds of the spacecraft are large fractions of the speed of light, we categorize the problem as a relativistic one. We analyze the problem by taking the S′ frame as being attached to A so that $v = 0.750c$ relative to the Earth observer (in the S frame). Spacecraft B can be considered as an object moving with a velocity component $u_x = -0.850c$ relative to the Earth observer. Hence, the velocity of B with respect to A is the velocity of B as measured by the observer O' on A, which can be obtained by using Equation 9.11:

$$u'_x = \frac{u_x - v}{1 - \dfrac{u_x v}{c^2}} = \frac{-0.850c - 0.750c}{1 - \dfrac{(-0.850c)(0.750c)}{c^2}}$$

$$= -0.980c$$

To finalize the problem, note that the negative sign in the result indicates that spacecraft B is moving in the negative x direction as observed by A. Also note that the relative speed is larger than each of the individual speeds but is smaller than the speed of light, as it must be.

INTERACTIVE EXAMPLE 9.5 Relativistic Leaders of the Pack

Two motorcycle pack leaders named David and Emily are racing at relativistic speeds along perpendicular paths as in Figure 9.12. How fast does Emily recede over David's right shoulder as seen by David?

Solution To determine Emily's speed of recession as seen by David, we take S′ to move in the x direction along with David. Figure 9.12 represents the situation as seen by a police officer at rest in frame S who observes the following:

David: $v = 0.75c$

Emily: $u_x = 0$ $\quad u_y = -0.90c$

We calculate u'_x and u'_y for Emily using Equations 9.11 and 9.12:

$$u'_x = \frac{u_x - v}{1 - \dfrac{u_x v}{c^2}} = \frac{0 - 0.75c}{1 - \dfrac{(0)(0.75c)}{c^2}} = -0.75c$$

$$u'_y = \frac{u_y}{\gamma\left(1 - \dfrac{u_x v}{c^2}\right)} = \frac{\sqrt{1 - \dfrac{(0.75c)^2}{c^2}}\,(-0.90c)}{\left(1 - \dfrac{(0)(0.75c)}{c^2}\right)}$$

$$= -0.60c$$

Therefore, the speed of Emily as observed by David is

$$u' = \sqrt{(u'_x)^2 + (u'_y)^2} = \sqrt{(-0.75c)^2 + (-0.60c)^2}$$

$$= 0.96c$$

PhysicsNow™ Investigate this situation with various speeds of David and Emily by logging into PhysicsNow at **www.pop4e.com** and going to Interactive Example 9.5.

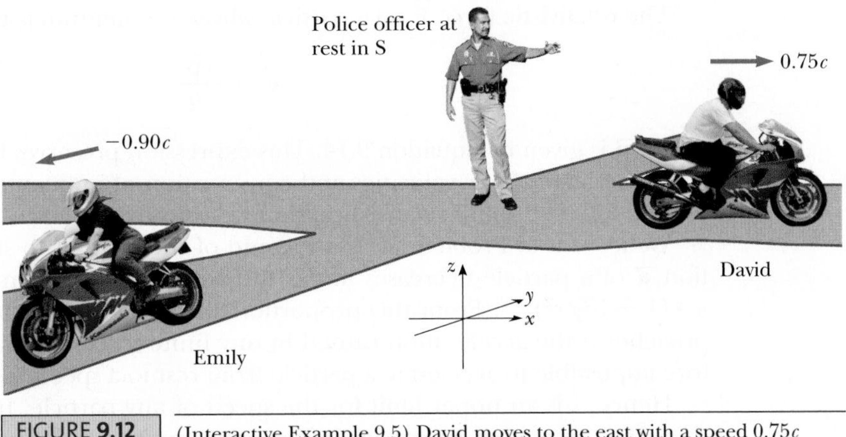

FIGURE 9.12 (Interactive Example 9.5) David moves to the east with a speed $0.75c$ relative to the police officer, and Emily travels south at a speed $0.90c$ relative to the officer.

9.6 RELATIVISTIC MOMENTUM AND THE RELATIVISTIC FORM OF NEWTON'S LAWS

We have seen that to describe the motion of particles within the framework of special relativity properly, the Galilean transformation must be replaced by the Lorentz transformation. Because the laws of physics must remain unchanged under the Lorentz transformation, we must generalize Newton's laws and the definitions of momentum and energy to conform to the Lorentz transformation and the principle of relativity. These generalized definitions should reduce to the classical (nonrelativistic) definitions for $v \ll c$ or $u \ll c$. (As we have done previously, we will use v for the speed of one reference frame relative to another and u for the speed of a particle.)

First, recall that the total momentum of an isolated system of particles is conserved. Suppose a collision between two particles is described in a reference frame S in which the momentum of the system is measured to be conserved. If the velocities in a second reference frame S′ are calculated using the Lorentz transformation and the Newtonian definition of momentum, $\vec{\mathbf{p}} = m\vec{\mathbf{u}}$, is used, it is found that the momentum of the system is *not* measured to be conserved in the second reference frame. This finding violates one of Einstein's postulates: The laws of physics are the same in all inertial frames. Therefore, assuming the Lorentz transformation to be correct, we must modify the definition of momentum.

The relativistic equation for the momentum of a particle of mass m that maintains the principle of conservation of momentum is

Definition of relativistic momentum

$$\vec{\mathbf{p}} \equiv \frac{m\vec{\mathbf{u}}}{\sqrt{1 - \dfrac{u^2}{c^2}}} \qquad [9.14]$$

where $\vec{\mathbf{u}}$ is the velocity of the particle. When u is much less than c, the denominator of Equation 9.14 approaches unity, so $\vec{\mathbf{p}}$ approaches $m\vec{\mathbf{u}}$. Therefore, the relativistic equation for $\vec{\mathbf{p}}$ reduces to the classical expression when u is small compared with c. Equation 9.14 is often written in simpler form as

$$\vec{\mathbf{p}} = \gamma m\vec{\mathbf{u}} \qquad [9.15]$$

using our previously defined expression[7] for γ.

PITFALL PREVENTION 9.4

WATCH OUT FOR "RELATIVISTIC MASS" Some older treatments of relativity maintained the conservation of momentum principle at high speeds by using a model in which the mass of a particle increases with speed. You might still encounter this notion of "relativistic mass" in your outside reading, especially in older books. Be aware that this notion is no longer widely accepted; today, mass is considered as *invariant,* independent of speed. The mass of an object in all frames is considered to be the mass as measured by an observer at rest with respect to the object.

[7]We defined γ previously in terms of the speed v of one frame relative to another frame. The same symbol is also used for $(1 - u^2/c^2)^{-1/2}$, where u is the speed of a particle.

The relativistic force $\vec{\mathbf{F}}$ on a particle whose momentum is $\vec{\mathbf{p}}$ is defined as

$$\vec{\mathbf{F}} \equiv \frac{d\vec{\mathbf{p}}}{dt} \qquad [9.16]$$

where $\vec{\mathbf{p}}$ is given by Equation 9.14. This expression preserves both classical mechanics in the limit of low velocities and conservation of momentum for an isolated system ($\sum \vec{\mathbf{F}}_{ext} = 0$) both relativistically and classically.

We leave it to Problem 9.57 at the end of the chapter to show that the acceleration $\vec{\mathbf{a}}$ of a particle decreases under the action of a constant force, in which case $a \propto (1 - u^2/c^2)^{3/2}$. From this proportionality, note that as the particle's speed approaches c, the acceleration caused by any finite force approaches zero. It is therefore impossible to accelerate a particle from rest to a speed $u \geq c$.

Hence, c is an upper limit for the speed of any particle. In fact, it is possible to show that **no matter, energy, or information can travel through space faster than *c*.** Note that the relative speeds of the two spacecraft in Example 9.4 and the two motorcyclists in Interactive Example 9.5 were both less than c. If we had attempted to solve these examples with Galilean transformations, we would have obtained relative speeds larger than c in both cases.

EXAMPLE 9.6 Momentum of an Electron

An electron, which has a mass of 9.11×10^{-31} kg, moves with a speed of $0.750c$. Find its relativistic momentum and compare it with the momentum calculated from the classical expression.

Solution Using Equation 9.14 with $u = 0.750c$, we have

$$p = \frac{m_e u}{\sqrt{1 - \frac{u^2}{c^2}}}$$

$$p = \frac{(9.11 \times 10^{-31}\text{ kg})(0.750 \times 3.00 \times 10^8\text{ m/s})}{\sqrt{1 - \frac{(0.750c)^2}{c^2}}}$$

$$= 3.10 \times 10^{-22}\text{ kg}\cdot\text{m/s}$$

The classical expression, if used (inappropriately) for this high-speed particle, gives

$$p = m_e u = 2.05 \times 10^{-22}\text{ kg}\cdot\text{m/s}$$

Hence, the correct relativistic result is more than 50% larger than the classical result!

9.7 RELATIVISTIC ENERGY

We have seen that the definition of momentum requires generalization to make it compatible with the principle of relativity. We find that the definition of kinetic energy must also be modified.

To derive the relativistic form of the work–kinetic energy theorem, let us start with the definition of the work done by a force of magnitude F on a particle initially at rest. Recall from Chapter 6 that the work–kinetic energy theorem states that the work done by a net force acting on a particle equals the change in kinetic energy of the particle. Because the initial kinetic energy is zero, we conclude that the work W done in accelerating a particle from rest is equivalent to the relativistic kinetic energy K of the particle:

$$W = \Delta K = K - 0 = K = \int_{x_1}^{x_2} F\,dx = \int_{x_1}^{x_2} \frac{dp}{dt}\,dx \qquad [9.17]$$

where we are considering the special case of force and displacement vectors along the x axis for simplicity. To perform this integration and find the relativistic kinetic energy as a function of u, we first evaluate dp/dt, using Equation 9.14:

$$\frac{dp}{dt} = \frac{d}{dt}\frac{mu}{\sqrt{1-\dfrac{u^2}{c^2}}} = \frac{m(du/dt)}{\left(1-\dfrac{u^2}{c^2}\right)^{3/2}}$$

Substituting this expression for dp/dt and $dx = u\,dt$ into Equation 9.17 gives

$$K = \int_0^t \frac{m(du/dt)\,u\,dt}{\left(1-\dfrac{u^2}{c^2}\right)^{3/2}} = m\int_0^u \frac{u}{\left(1-\dfrac{u^2}{c^2}\right)^{3/2}}\,du$$

Evaluating the integral, we find that

$$K = \frac{mc^2}{\sqrt{1-\dfrac{u^2}{c^2}}} - mc^2 = \gamma mc^2 - mc^2 = (\gamma - 1)\,mc^2 \qquad [9.18]$$

■ **Relativistic kinetic energy**

At low speeds, where $u/c \ll 1$, Equation 9.18 should reduce to the classical expression $K = \frac{1}{2}mu^2$. We can show this reduction by using the binomial expansion $(1 - x^2)^{-1/2} \approx 1 + \frac{1}{2}x^2 + \cdots$ for $x \ll 1$, where the higher-order powers of x are ignored in the expansion because they are so small. In our case, $x = u/c$, so

$$\gamma = \frac{1}{\sqrt{1-\dfrac{u^2}{c^2}}} = \left(1-\frac{u^2}{c^2}\right)^{-1/2} \approx 1 + \frac{1}{2}\frac{u^2}{c^2} + \cdots$$

Substituting into Equation 9.18 gives

$$K \approx \left(1 + \frac{1}{2}\frac{u^2}{c^2} + \cdots\right)mc^2 - mc^2 = \tfrac{1}{2}mu^2$$

which agrees with the classical result. Figure 9.13 shows a comparison of the speed–kinetic energy relationships for a particle using the nonrelativistic expression for K (the blue curve) and the relativistic expression for K (the brown curve). The curves are in good agreement at low speeds, but deviate at higher speeds. The nonrelativistic expression indicates a violation of special relativity because it suggests that sufficient energy can be added to the particle to accelerate it to a speed larger than c. In the relativistic case, the particle speed never exceeds c, regardless

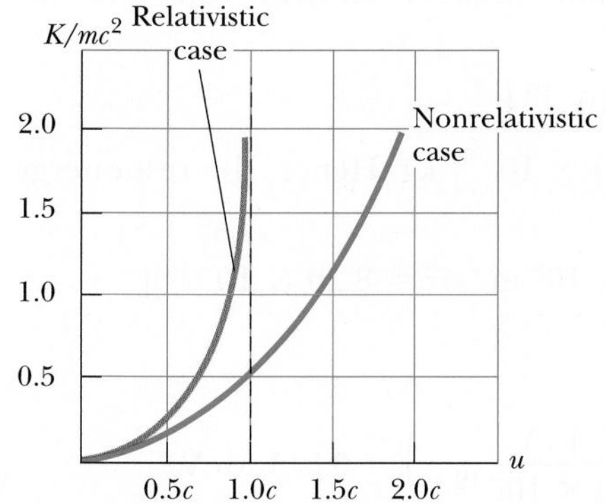

FIGURE 9.13 A graph comparing relativistic and nonrelativistic kinetic energy of a particle. The energies are plotted as a function of speed u. In the relativistic case, u is always less than c.

of the kinetic energy, which is consistent with experimental results. When an object's speed is less than one-tenth the speed of light, the classical kinetic energy equation differs by less than 1% from the relativistic equation (which is experimentally verified at all speeds). Therefore, for practical calculations it is valid to use the classical equation when the object's speed is less than $0.1c$.

The constant term mc^2 in Equation 9.18, which is independent of the speed, is called the **rest energy** E_R of the particle:

■ Rest energy

$$E_R = mc^2 \qquad [9.19]$$

The term γmc^2 in Equation 9.18 depends on the particle speed and is the sum of the kinetic and rest energies. We define γmc^2 to be the **total energy** E; that is, total energy = kinetic energy + rest energy:

$$E = \gamma mc^2 = K + mc^2 = K + E_R \qquad [9.20]$$

or, when γ is replaced by its equivalent,

■ Total energy of a relativistic particle

$$E = \frac{mc^2}{\sqrt{1 - \dfrac{u^2}{c^2}}} \qquad [9.21]$$

The relation $E_R = mc^2$ shows that **mass is a manifestation of energy.** It also shows that a small mass corresponds to an enormous amount of energy. This concept is fundamental to much of the field of nuclear physics.

In many situations, the momentum or energy of a particle is measured rather than its speed. It is therefore useful to have an expression relating the total energy E to the relativistic momentum p, which is accomplished by using the expressions $E = \gamma mc^2$ and $p = \gamma mu$. By squaring these equations and subtracting, we can eliminate u (see Problem 9.37). The result, after some algebra, is

■ Energy–momentum relationship for a relativistic particle

$$E^2 = p^2c^2 + (mc^2)^2 \qquad [9.22]$$

When the particle is at rest, $p = 0$, and so $E = E_R = mc^2$. That is, the total energy equals the rest energy.

For the case of particles that have zero mass, such as photons (massless, chargeless particles of light to be discussed further in Chapter 28), we set $m = 0$ in Equation 9.22 and see that

$$E = pc \qquad [9.23]$$

This equation is an exact expression relating energy and momentum for photons, which always travel at the speed of light.

When dealing with subatomic particles, it is convenient to express their energy in a unit called an *electron volt* (eV). The equality between electron volts and our standard energy unit is

$$1 \text{ eV} = 1.60 \times 10^{-19}\text{ J}$$

For example, the mass of an electron is 9.11×10^{-31} kg. Hence, the rest energy of the electron is

$$E_R = m_ec^2 = (9.11 \times 10^{-31}\text{ kg})(3.00 \times 10^8\text{ m/s})^2 = 8.20 \times 10^{-14}\text{ J}$$

Converting to eV, we have

$$E_R = m_ec^2 = (8.20 \times 10^{-14}\text{ J})\left(\frac{1\text{ eV}}{1.60 \times 10^{-19}\text{ J}}\right) = 0.511\text{ MeV}$$

QUICK QUIZ 9.6 The following *pairs* of energies represent the rest energy and total energy of three different particles: particle 1: E, $2E$; particle 2: E, $3E$; particle 3: $2E$, $4E$. Rank the particles, from greatest to least, according to their (**a**) mass, (**b**) kinetic energy, and (**c**) speed.

EXAMPLE 9.7 The Energy of a Speedy Proton

Let us consider the relativistic motion of a proton.

A Find the proton's rest energy in electron volts.

Solution To find the rest energy, we use Equation 9.19,

$$E_R = m_pc^2 = (1.67 \times 10^{-27}\text{ kg})(3.00 \times 10^8\text{ m/s})^2$$

$$= (1.50 \times 10^{-10}\text{ J})\left(\frac{1.00\text{ eV}}{1.60 \times 10^{-19}\text{ J}}\right)$$

$$= 938\text{ MeV}$$

B The total energy of a proton is three times its rest energy. With what speed is the proton moving?

Solution Because the total energy E is three times the rest energy, $E = \gamma mc^2$ (Eq. 9.20) gives

$$E = 3m_pc^2 = \frac{m_pc^2}{\sqrt{1 - \dfrac{u^2}{c^2}}}$$

$$3 = \frac{1}{\sqrt{1 - \dfrac{u^2}{c^2}}}$$

Solving for u gives

$$\left(1 - \frac{u^2}{c^2}\right) = \frac{1}{9} \quad \text{or} \quad \frac{u^2}{c^2} = \frac{8}{9}$$

$$u = \frac{\sqrt{8}}{3}c = 2.83 \times 10^8\text{ m/s}$$

C Determine the kinetic energy of the proton in part B in electron volts.

Solution We use Equation 9.20:

$$K = E - m_pc^2 = 3m_pc^2 - m_pc^2 = 2m_pc^2$$

Because $m_pc^2 = 938$ MeV, $K =$ 1.88 GeV.

D What is the magnitude of the proton's momentum in part B?

Solution We can use Equation 9.22 to calculate the momentum with $E = 3m_pc^2$:

$$E^2 = p^2c^2 + (m_pc^2)^2 = (3m_pc^2)^2$$

$$p^2c^2 = 9(m_pc^2)^2 - (m_pc^2)^2 = 8(m_pc^2)^2$$

$$p = \sqrt{8}\,\frac{m_pc^2}{c} = \sqrt{8}\,\frac{(938\text{ MeV})}{c}$$

$$= 2.65 \times 10^3\,\frac{\text{MeV}}{c}$$

The unit of momentum is written MeV/c, which is a momentum unit often used in particle studies.

9.8 MASS AND ENERGY

Equation 9.20, $E = \gamma mc^2$, which represents the total energy of a particle, suggests that even when a particle is at rest ($\gamma = 1$) it still possesses enormous energy through its mass. The clearest experimental proof of the equivalence of mass and energy occurs in nuclear and elementary particle interactions in which the conversion of mass into kinetic energy takes place. Hence, we cannot use the principle of conservation of energy in relativistic situations exactly as it is outlined in Chapter 7. We must include rest energy as another form of energy storage.

This concept is important in atomic and nuclear processes, in which the change in mass during the process is on the order of the initial mass. For example, in a conventional nuclear reactor, the uranium nucleus undergoes *fission,* a reaction that results in several lighter fragments having considerable kinetic energy. In the case of a ^{235}U atom, which is used as fuel in nuclear power plants, the fragments are two lighter nuclei and a few neutrons. The total mass of the fragments is less than that of the ^{235}U by an amount Δm. The corresponding energy Δmc^2 associated with this mass difference is exactly equal to the total kinetic energy of the fragments. The kinetic energy is transferred by collisions with water molecules as the fragments move through water, raising the internal energy of the water. This internal energy is used to produce steam for the generation of electric power.

Next, consider a basic *fusion* reaction in which two deuterium atoms combine to form one helium atom. The decrease in mass that results from the creation of one helium atom from two deuterium atoms is $\Delta m = 4.25 \times 10^{-29}$ kg. Hence, the corresponding energy that results from one fusion reaction is calculated to be $\Delta mc^2 = 3.83 \times 10^{-12}$ J $= 23.9$ MeV. To appreciate the magnitude of this result, consider that if 1 g of deuterium is converted to helium, the energy released is on the order of 10^{12} J! At the year 2006 cost of electric energy, this energy would be worth about $20 000. We will see more details of these nuclear processes in Chapter 30.

EXAMPLE 9.8 Mass Change in a Radioactive Decay

The ^{216}Po nucleus is unstable and exhibits radioactivity, which we will study in detail in Chapter 30. It decays to ^{212}Pb by emitting an alpha particle, which is a helium nucleus, ^{4}He.

A Find the mass change in this decay.

Solution Using values in Table A.3, we see that the initial and final masses are

$$m_i = m(^{216}\text{Po}) = 216.001\,905 \text{ u}$$

$$m_f = m(^{212}\text{Pb}) + m(^{4}\text{He})$$
$$= 211.991\,888 \text{ u} + 4.002\,603 \text{ u}$$
$$= 215.994\,491 \text{ u}$$

Therefore, the mass change is

$$\Delta m = 216.001\,905 \text{ u} - 215.994\,491 \text{ u} = 0.007\,414 \text{ u}$$
$$= 1.23 \times 10^{-29} \text{ kg}$$

B Find the energy that this mass change represents.

Solution The energy associated with this mass change is

$$E = \Delta mc^2 = (1.23 \times 10^{-29} \text{ kg})(3.00 \times 10^8 \text{ m/s})^2$$
$$= 1.11 \times 10^{-12} \text{ J} = 6.92 \text{ MeV}$$

This energy appears as the kinetic energies of the alpha particle and the ^{212}Pb nucleus after the decay.

9.9 GENERAL RELATIVITY

Up to this point, we have sidestepped a curious puzzle. Mass has two seemingly different properties: It determines a force of mutual gravitational attraction between two objects (Newton's law of universal gravitation), and it also represents the resistance of a single object to being accelerated (Newton's second law), regardless of the type of force producing the acceleration. How can one quantity have two such different properties? An answer to this question, which puzzled Newton and many other physicists over the years, was provided when Einstein published his theory of gravitation, known as *general relativity*, in 1916. Because it is a mathematically complex theory, we offer merely a hint of its elegance and insight.

In Einstein's view, the dual behavior of mass was evidence for a very intimate and basic connection between the two behaviors. He pointed out that no mechanical experiment (e.g., dropping an object) could distinguish between the two situations illustrated in Figures 9.14a and 9.14b. In Figure 9.14a, a person is standing in an elevator on the surface of a planet and feels pressed into the floor due to the gravitational force. If he releases his briefcase, he observes it moving toward the floor with acceleration $\vec{\mathbf{g}} = -g\hat{\mathbf{j}}$. In Figure 9.14b, the person is in an elevator in empty space accelerating upward with $\vec{\mathbf{a}}_{\text{el}} = +g\hat{\mathbf{j}}$. The person feels pressed into the floor with the same force as in Figure 9.14a. If he releases his briefcase, he observes it moving toward the floor with acceleration g, just as in the previous situation. In each case, an object released by the observer undergoes a downward

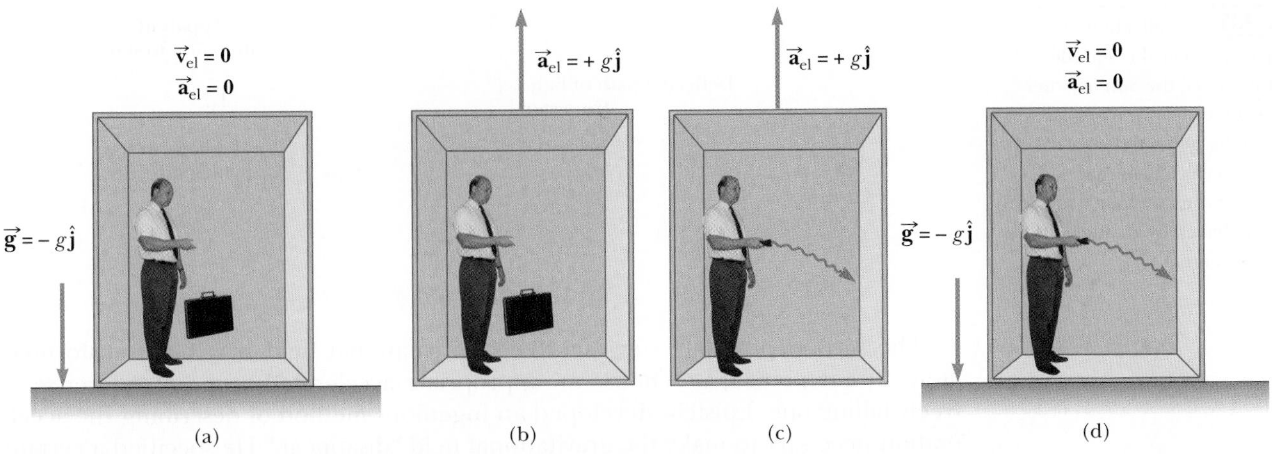

FIGURE 9.14 (a) The observer is at rest in an elevator in a uniform gravitational field $\vec{g} = -g\hat{j}$, directed downward. The observer drops his briefcase, which moves downward with acceleration g. (b) The observer is in a region where gravity is negligible, but the elevator moves upward with an acceleration $\vec{a}_{el} = +g\hat{j}$. The observer releases his briefcase, which moves downward (according to the observer) with acceleration g relative to the floor of the elevator. According to Einstein, the frames of reference in parts (a) and (b) are equivalent in every way. No local experiment can distinguish any difference between the two frames. (c) In the accelerating frame, a ray of light would appear to bend downward due to the acceleration. (d) If parts (a) and (b) are truly equivalent, as Einstein proposed, part (c) suggests that a ray of light would bend downward in a gravitational field.

acceleration of magnitude g relative to the floor. In Figure 9.14a, the person is at rest in an inertial frame in a gravitational field due to the planet. (A gravitational field exists around any object with mass, such as a planet. We will define the gravitational field formally in Chapter 11.) In Figure 9.14b, the person is in a noninertial frame accelerating in gravity-free space. Einstein's claim is that these two situations are completely equivalent.

Einstein carried this idea further and proposed that *no* experiment, mechanical or otherwise, could distinguish between the two cases. This extension to include all phenomena (not just mechanical ones) has interesting consequences. For example, suppose a light pulse is sent horizontally across the elevator as in Figure 9.14c, in which the elevator is accelerating upward in empty space. From the point of view of an observer in an inertial frame outside the elevator, the light travels in a straight line while the floor of the elevator accelerates upward. According to the observer on the elevator, however, the trajectory of the light pulse bends downward as the floor of the elevator (and the observer) accelerates upward. Therefore, based on the equality of parts (a) and (b) of the figure for all phenomena, Einstein proposed that **a beam of light should also be bent downward by a gravitational field,** as in Figure 9.14d.

The two postulates of Einstein's **general theory of relativity** are as follows:

■ Postulates of general relativity

- All the laws of nature have the same form for observers in any frame of reference, whether accelerated or not.
- In the vicinity of any given point, a gravitational field is equivalent to an accelerated frame of reference in the absence of gravitational effects. (This postulate is known as the **principle of equivalence.**)

One interesting effect predicted by general relativity is that the passage of time is altered by gravity. A clock in the presence of gravity runs more slowly than one for which gravity is negligible. Consequently, the frequencies of radiation emitted by atoms in the presence of a strong gravitational field are shifted to lower values compared with the same emissions in a weak field. This gravitational shift has been detected in light emitted by atoms in massive stars. It has also been verified on the Earth by comparing the frequencies of gamma rays (a high-energy form of electromagnetic radiation) emitted from nuclei separated vertically by about 20 m.

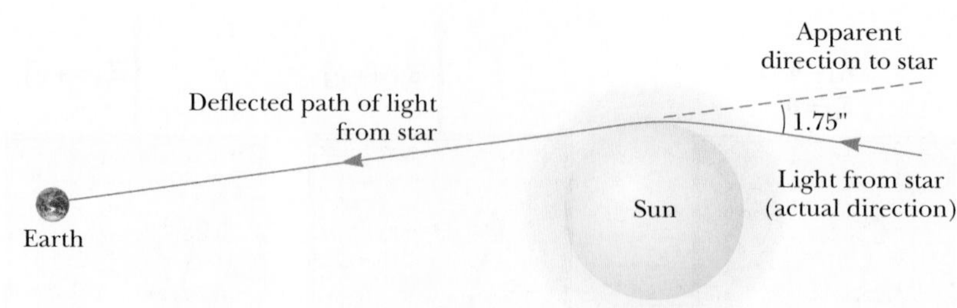

FIGURE 9.15 Deflection of starlight passing near the Sun. Because of this effect, the Sun or other remote objects can act as a *gravitational lens.* In his general theory of relativity, Einstein calculated that starlight just grazing the Sun's surface should be deflected by an angle of 1.75 seconds of arc.

The second postulate suggests that a gravitational field may be "transformed away" at any point if we choose an appropriate accelerated frame of reference, a freely falling one. Einstein developed an ingenious method of describing the acceleration necessary to make the gravitational field "disappear." He specified a certain quantity, the *curvature of space–time,* that describes the gravitational effect of a mass. In fact, the curvature of space–time completely replaces Newton's gravitational theory. According to Einstein, there is no such thing as a gravitational force. Rather, the presence of a mass causes a curvature of space–time in the vicinity of the mass, and this curvature dictates the space-time path that all freely moving objects must follow.

One important test of general relativity is the prediction that a light ray passing near the Sun should be deflected by some angle. This prediction was confirmed by astronomers as the bending of starlight during a total solar eclipse shortly following World War I (Fig. 9.15).

As an example of the effects of curved space–time, imagine two travelers moving on parallel paths a few meters apart on the surface of the Earth and maintaining an exact northward heading along two longitude lines. As they observe each other near the equator, they will claim that their paths are exactly parallel. As they approach the North Pole, however, they will notice that they are moving closer together and that they will actually meet at the North Pole. Thus, they will claim that they moved along parallel paths, but moved toward each other, *as if there were an attractive force between them.* They will make this conclusion based on their everyday experience of moving on flat surfaces. From our mental representation, however, we realize that they are walking on a curved surface, and the geometry of the curved surface, rather than an attractive force, causes them to converge. In a similar way, general relativity replaces the notion of forces with the movement of objects through curved space–time.

If a concentration of mass in space becomes very great, as is believed to occur when a large star exhausts its nuclear fuel and collapses to a very small volume, a **black hole** may form. Here the curvature of space–time is so extreme that, within a certain distance from the center of the black hole, all matter and light become trapped. We will say more about black holes in Chapter 11.

■ Thinking Physics 9.2

Atomic clocks are extremely accurate; in fact, an error of 1 s in 3 million years is typical. This error can be described as about 1 part in 10^{14}. On the other hand, the atomic clock in Boulder, Colorado, near Denver, is often 15 ns faster than the one in Washington, D.C., after only one day. This error is one of about 1 part in 6×10^{12}, which is about 17 times larger than the previously expressed error. If atomic clocks are so accurate, why does a clock in Boulder not remain in synchronization with one in Washington, D.C.? (*Hint:* Denver is known as the Mile High City.)

Reasoning According to the general theory of relativity, the passage of time depends on gravity. Time is measured to run more slowly in strong gravitational fields. Washington, D.C., is at an elevation very close to sea level, but Boulder is about a mile higher in altitude. This difference results in a weaker gravitational field at Boulder than at Washington, D.C. As a result, time is measured to run more rapidly in Boulder than in Washington, D.C. ■

9.10 FROM MARS TO THE STARS

CONTEXT CONNECTION

In this chapter, we have discussed the strange effects of traveling at high speeds. Do we need to consider these effects in our planned mission to Mars?

To answer this question, let us consider a typical spacecraft speed necessary to travel from the Earth to Mars. This speed is on the order of 10^4 m/s. Let us evaluate γ for this speed:

$$\gamma = \frac{1}{\sqrt{1 - \dfrac{u^2}{c^2}}} = \frac{1}{\sqrt{1 - \dfrac{(10^4 \text{ m/s})^2}{(3.00 \times 10^8 \text{ m/s})^2}}} = 1.000\,000\,000\,6$$

where we have completely ignored the rules of significant figures so that we could find the first nonzero digit to the right of the decimal place!

It is clear from this result that relativistic considerations are not important for our trip to Mars. Yet what about deeper travels into space? Suppose we wish to travel to another star. This distance is several orders of magnitude larger. The nearest star is about 4.2 ly from the Earth. In comparison, Mars is 4.0×10^{-5} ly at its farthest from the Earth. Therefore, we are talking about a distance to the nearest star that is five orders of magnitude larger than the distance to Mars. Very long travel times will be needed to reach even the nearest star. At the escape speed from the Sun, for example, assuming that this speed is maintained during the entire trip, the travel time is 30 000 years to the nearest star. This time period is clearly prohibitive, especially if we would like the people who leave the Earth to be the same people who arrive at the star!

We can use the principles of relativity to reduce this travel time significantly by traveling at very high speeds. Suppose our spacecraft travels at a constant speed of $0.99c$. The travel time as measured by an observer on the Earth then is

$$\Delta t = \frac{L_p}{u} = \frac{4.2 \text{ ly}}{0.99(1.0 \text{ ly/yr})} = 4.2 \text{ yr}$$

where the distance between the Earth and the destination star is the proper length L_p.

Because the spacecraft occupants see both the Earth and the destination star moving, the distance between them is measured to be shorter than that measured by observers on the Earth. We can use length contraction to calculate the distance from the Earth to the star as measured by the spacecraft occupants:

$$L = \frac{L_p}{\gamma} = L_p\sqrt{1 - \frac{u^2}{c^2}} = (4.2 \text{ ly})\sqrt{1 - \frac{(0.99c)^2}{c^2}} = 0.59 \text{ ly}$$

The time interval required to reach the star is now

$$\Delta t = \frac{L}{u} = \frac{0.59 \text{ ly}}{0.99(1.0 \text{ ly/yr})} = 0.60 \text{ yr}$$

which is clearly a reduction in travel time from the low-speed trip!

There are three major problems with this scenario, however. The first is the technological challenge of designing and building a spacecraft and rocket engine assembly that can attain a speed of $0.99c$. Second is the design of a safety system that will provide early warnings about running into asteroids, meteoroids, or other bits of matter while traveling at almost light speed through space. Even a small piece of rock could be disastrous if struck at $0.99c$. The third problem is related to the twin paradox discussed earlier in this chapter. During the trip to the star, 4.2 yr will pass on the Earth. If the travelers return to the Earth, another 4.2 yr will pass. Therefore, the travelers will have aged by only 2(0.6 yr) = 1.2 yr, but 8.4 yr will have passed on the Earth. For stars farther away than the nearest star, these effects could result in the personnel assisting with the liftoff from the Earth no longer being alive when the travelers return. In conclusion, we see that travel to the stars will be an enormous challenge!

SUMMARY

PhysicsNow™ Take a practice test by logging into PhysicsNow at www.pop4e.com and clicking on the Pre-Test link for this chapter.

The two basic postulates of **special relativity** are:

- All the laws of physics are the same in all inertial reference frames.
- The speed of light in vacuum has the same value, in all inertial frames, regardless of the velocity of the observer or the velocity of the source emitting the light.

Three consequences of special relativity are:

- Events that are simultaneous for one observer may not be simultaneous for another observer who is in motion relative to the first.
- Clocks in motion relative to an observer are measured to be slowed down by a factor γ. This phenomenon is known as **time dilation.**
- Lengths of objects in motion are measured to be shorter in the direction of motion. This phenomenon is known as **length contraction.**

To satisfy the postulates of special relativity, the Galilean transformations must be replaced by the **Lorentz transformation equations:**

$$\begin{aligned} x' &= \gamma(x - vt) \\ y' &= y \\ z' &= z \\ t' &= \gamma\left(t - \frac{v}{c^2}x\right) \end{aligned} \qquad [9.8]$$

where $\gamma = (1 - v^2/c^2)^{-1/2}$.

The relativistic form of the **Lorentz velocity transformation** is

$$u_x' = \frac{u_x - v}{1 - \dfrac{u_x v}{c^2}} \qquad [9.11]$$

where u_x is the speed of an object as measured in the S frame and u_x' is its speed measured in the S′ frame.

The relativistic expression for the momentum of a particle moving with a velocity $\vec{\mathbf{u}}$ is

$$\vec{\mathbf{p}} \equiv \frac{m\vec{\mathbf{u}}}{\sqrt{1 - \dfrac{u^2}{c^2}}} = \gamma m\vec{\mathbf{u}} \qquad [9.14, 9.15]$$

The relativistic expression for the kinetic energy of a particle is

$$K = \gamma mc^2 - mc^2 = (\gamma - 1)mc^2 \qquad [9.18]$$

where $E_R = mc^2$ is the **rest energy** of the particle.

The **total energy** E of a particle is given by the expression

$$E = \frac{mc^2}{\sqrt{1 - \dfrac{u^2}{c^2}}} \qquad [9.21]$$

The total energy of a particle is the sum of its rest energy and its kinetic energy: $E = E_R + K$.

The relativistic momentum of a particle is related to its total energy through the equation

$$E^2 = p^2c^2 + (mc^2)^2 \qquad [9.22]$$

The **general theory of relativity** claims that no experiment can distinguish between a gravitational field and an accelerating reference frame. It correctly predicts that the path of light is affected by a gravitational field.

QUESTIONS

☐ = answer available in the *Student Solutions Manual and Study Guide*

1. On what two speed measurements do two observers in relative motion always agree?
2. A spacecraft with the shape of a sphere moves past an observer on Earth with a speed 0.5c. What shape does the observer measure for the spacecraft as it moves past?
3. The speed of light in water is 230 Mm/s. Suppose an electron is moving through water at 250 Mm/s. Does that violate the principle of relativity?
4. Two identical clocks are synchronized. One is then put in orbit directed eastward around the Earth, and the other remains on the Earth. According to an observer on the Earth, which clock runs more slowly? When the moving clock returns to the Earth, are the two still synchronized?

5. Explain why it is necessary, when defining the length of a rod, to specify that the positions of the ends of the rod are to be measured simultaneously.

6. A train is approaching you at very high speed as you stand next to the tracks. Just as an observer on the train passes you, you both begin to play the same Beethoven symphony on portable compact disc players. (a) According to you, whose CD player finishes the symphony first? (b) According to the observer on the train, whose CD player finishes the symphony first? (c) Whose CD player really finishes the symphony first?

7. List some ways our day-to-day lives would change if the speed of light were only 50 m/s.

8. A particle is moving at a speed less than $c/2$. If the speed of the particle is doubled, what happens to its momentum?

9. Give a physical argument that shows that it is impossible to accelerate an object of mass m to the speed of light, even with a continuous force acting on it.

10. The upper limit of the speed of an electron is the speed of light c. Does that mean that the momentum of the electron has an upper limit?

11. Because mass is a measure of energy, can we conclude that the mass of a compressed spring is greater than the mass of the same spring when it is not compressed?

12. It is said that Einstein, in his teenage years, asked the question, "What would I see in a mirror if I carried it in my hands and ran at the speed of light?" How would you answer this question?

13. Some distant astronomical objects, called quasars, are receding from us at half the speed of light (or greater). What is the speed of the light we receive from these quasars?

14. Photons of light have zero mass. How is it possible that they have momentum?

15. "Newtonian mechanics correctly describes objects moving at ordinary speeds, and relativistic mechanics correctly describes objects moving very fast." "Relativistic mechanics must make a smooth transition as it reduces to Newtonian mechanics in a case where the speed of an object becomes small compared with the speed of light." Argue for or against each of these two statements.

16. Two cards have straight edges. Suppose the top edge of one card crosses the bottom edge of another card at a small angle as shown in Figure Q9.16a. A person slides the cards together at a moderately high speed. In what direction does the intersection point of the edges move? Show that it can move at a speed greater than the speed of light.

A small flashlight is suspended in a horizontal plane and set into rapid rotation. Show that the spot of light it produces on a distant screen can move across the screen at a speed greater than the speed of light. (If you use a laser pointer, as shown in Fig. Q9.16b, make sure that the direct laser light cannot enter a person's eyes.) Argue that these experiments do not invalidate the principle that no material, no energy, and no information can move faster than light moves in a vacuum.

17. With regard to reference frames, how does general relativity differ from special relativity?

18. Two identical clocks are in the same house, one upstairs in a bedroom and the other downstairs in the kitchen. Which clock runs more slowly? Explain.

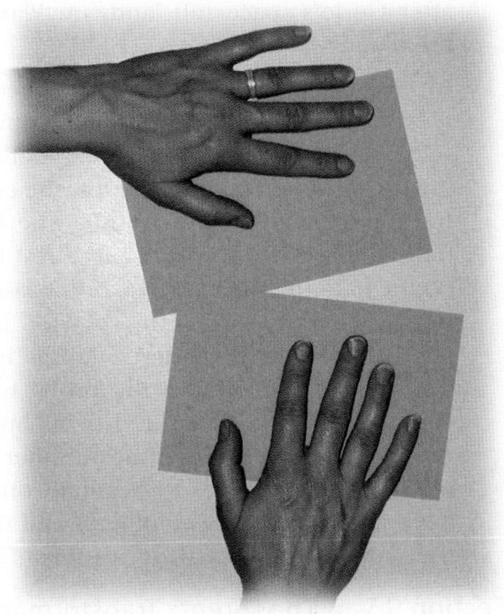

(a)

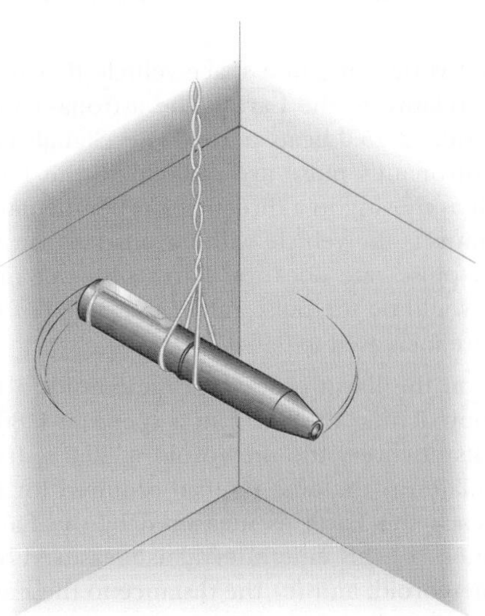

(b)

FIGURE **Q9.16**

PROBLEMS

1, 2, 3 = straightforward, intermediate, challenging
☐ = full solution available in the *Student Solutions Manual and Study Guide*
PhysicsNow™ = coached problem with hints available at www.pop4e.com
= computer useful in solving problem
= paired numerical and symbolic problems
= biomedical application

Section 9.1 ▪ The Principle of Newtonian Relativity

1. In a laboratory frame of reference, an observer notes that Newton's second law is valid. Show that it is also valid for an observer moving at a constant speed, small compared with the speed of light, relative to the laboratory frame.

2. Show that Newton's second law is *not* valid in a reference frame moving past the laboratory frame of Problem 9.1 with a constant acceleration.

3. A 2 000-kg car moving at 20.0 m/s collides and locks together with a 1 500-kg car at rest at a stop sign. Show that momentum is conserved in a reference frame moving at 10.0 m/s in the direction of the moving car.

Section 9.2 ▪ The Michelson–Morley Experiment
Section 9.3 ▪ Einstein's Principle of Relativity
Section 9.4 ▪ Consequences of Special Relativity

Problem 3.36 in Chapter 3 can be assigned with this section.

4. How fast must a meter stick be moving if its length is measured to shrink to 0.500 m?

5. At what speed does a clock move if it is measured to run at a rate that is one-half the rate of a clock at rest with respect to an observer?

6. An astronaut is traveling in a space vehicle that has a speed of $0.500c$ relative to the Earth. The astronaut measures her pulse rate at 75.0 beats per minute. Signals generated by the astronaut's pulse are radioed to the Earth when the vehicle is moving in a direction perpendicular to the line that connects the vehicle with an observer on the Earth. (a) What pulse rate does the Earth observer measure? (b) What would be the pulse rate if the speed of the space vehicle were increased to $0.990c$?

7. An astronomer on the Earth observes a meteoroid in the southern sky approaching the Earth at a speed of $0.800c$. At the time of its discovery the meteoroid is 20.0 ly from the Earth. Calculate (a) the time interval required for the meteoroid to reach the Earth as measured by the Earth-bound astronomer, (b) this time interval as measured by a tourist on the meteoroid, and (c) the distance to the Earth as measured by the tourist.

8. A muon formed high in the Earth's atmosphere travels at speed $v = 0.990c$ for a distance of 4.60 km before it decays into an electron, a neutrino, and an antineutrino ($\mu^- \rightarrow e^- + \nu + \bar{\nu}$). (a) How long does the muon live, as measured in its reference frame? (b) How far does the Earth travel, as measured in the frame of the muon?

9. An atomic clock moves at 1 000 km/h for 1.00 h as measured by an identical clock on the Earth. How many nanoseconds slow will the moving clock be compared with the Earth clock at the end of the 1.00-h interval?

10. For what value of v does $\gamma = 1.010\,0$? Observe that for speeds lower than this value, time dilation and length contraction are effects amounting to less than 1%.

11. PhysicsNow™ A spacecraft with a proper length of 300 m takes 0.750 μs to pass an Earth observer. Determine the speed of the spacecraft as measured by the Earth observer.

12. (a) An object of proper length L_p takes a time interval Δt to pass an Earth observer. Determine the speed of the object as measured by the Earth observer. (b) A column of tanks, 300 m long, takes 75.0 s to pass a child waiting at a street corner on her way to school. Determine the speed of the armored vehicles. (c) Show that the answer to part (a) includes the answer to Problem 9.11 as a special case and includes the answer to part (b) as another special case.

13. A friend passes by you in a spacecraft traveling at a high speed. He tells you that his craft is 20.0 m long and that the identically constructed craft you are sitting in is 19.0 m long. According to your observations, (a) how long is your spacecraft, (b) how long is your friend's craft, and (c) what is the speed of your friend's craft?

14. The identical twins Speedo and Goslo join a migration from the Earth to Planet X. It is 20.0 ly away in a reference frame in which both planets are at rest. The twins, of the same age, depart at the same time on different spacecraft. Speedo's craft travels steadily at $0.950c$, and Goslo's travels at $0.750c$. Calculate the age difference between the twins after Goslo's spacecraft lands on Planet X. Which twin is the older?

15. An interstellar space probe is launched from the Earth. After a brief period of acceleration it moves with a constant velocity, with a magnitude of 70.0% of the speed of light. Its nuclear-powered batteries supply the energy to keep its data transmitter active continuously. The batteries have a lifetime of 15.0 yr as measured in a rest frame. (a) How long do the batteries on the space probe last as measured by Mission Control on the Earth? (b) How far is the probe from the Earth when its batteries fail as measured by Mission Control? (c) How far is the probe from the Earth when its batteries fail as measured by its built-in trip odometer? (d) For what total time interval after launch are data received from the probe by Mission Control? Note that radio waves travel at the speed of light and fill the space between the probe and the Earth at the time of battery failure.

Section 9.5 ▪ The Lorentz Transformation Equations

16. Suzanne observes two light pulses to be emitted from the same location, but separated in time by 3.00 μs. Mark sees

the emission of the same two pulses separated in time by 9.00 μs. (a) How fast is Mark moving relative to Suzanne? (b) According to Mark, what is the separation in space of the two pulses?

17. A moving rod is measured to have a length of 2.00 m and to be oriented at an angle of 30.0° with respect to the direction of motion as shown in Figure P9.17. The rod has a speed of 0.995*c*. (a) What is the proper length of the rod? (b) What is the orientation angle in the proper frame?

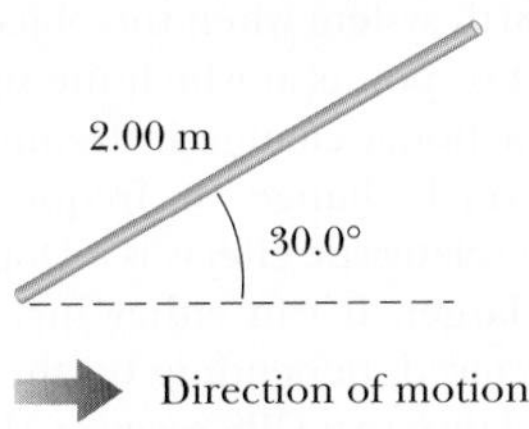

FIGURE **P9.17**

18. An observer in reference frame S measures two events as simultaneous. Event A occurs at the point (50.0 m, 0, 0) at the instant 9:00:00 Universal time on January 15, 2005. Event B occurs at the point (150 m, 0, 0) at the same moment. A second observer, moving past with a velocity of $0.800c\hat{\mathbf{i}}$, also observes the two events. In her reference frame S′, which event occurred first and what time interval elapsed between the events?

19. A red light flashes at position $x_R = 3.00$ m and time $t_R = 1.00 \times 10^{-9}$ s, and a blue light flashes at $x_B = 5.00$ m and $t_B = 9.00 \times 10^{-9}$ s, all measured in the S reference frame. Reference frame S′ has its origin at the same point as S at $t = t' = 0$; frame S′ moves uniformly to the right. Both flashes are observed to occur at the same place in S′. (a) Find the relative speed between S and S′. (b) Find the location of the two flashes in frame S′. (c) At what time does the red flash occur in the S′ frame?

20. A Klingon spacecraft moves away from the Earth at a speed of 0.800*c* (Fig. P9.20). The starship *Enterprise* pursues at a speed of 0.900*c* relative to the Earth. Observers on the Earth measure the *Enterprise* overtaking the Klingon craft at a relative speed of 0.100*c*. With what speed is the *Enterprise* overtaking the Klingon craft as measured by the crew of the *Enterprise*?

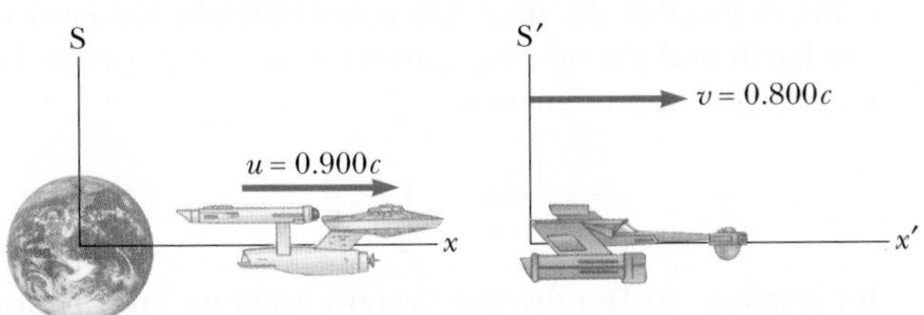

FIGURE **P9.20**

21. **Physics Now™** Two jets of material from the center of a radio galaxy are ejected in opposite directions. Both jets move at 0.750*c* relative to the galaxy. Determine the speed of one jet relative to the other.

22. A spacecraft is launched from the surface of the Earth with a velocity of 0.600*c* at an angle of 50.0° above the horizontal positive *x* axis. Another spacecraft is moving past with a velocity of 0.700*c* in the negative *x* direction. Determine the magnitude and direction of the velocity of the first spacecraft as measured by the pilot of the second spacecraft.

Section 9.6 ■ Relativistic Momentum and the Relativistic Form of Newton's Laws

23. Calculate the momentum of an electron moving with a speed of (a) 0.010 0*c*, (b) 0.500*c*, and (c) 0.900*c*.

24. The nonrelativistic expression for the momentum of a particle, $p = mu$, agrees with experimental results if $u \ll c$. For what speed does the use of this equation give an error in the momentum of (a) 1.00% and (b) 10.0%?

25. A golf ball travels with a speed of 90.0 m/s. By what fraction does its relativistic momentum magnitude p differ from its classical value mu? That is, find the ratio $(p - mu)/mu$.

26. The speed limit on a certain roadway is 90.0 km/h. Suppose speeding fines are made proportional to the amount by which a vehicle's momentum exceeds the momentum it would have when traveling at the speed limit. The fine for driving at 190 km/h (that is, 100 km/h more than the speed limit) is $80.0. What, then, will be the fine for traveling at (a) 1 090 km/h and (b) 1 000 000 090 km/h?

27. **Physics Now™** An unstable particle at rest breaks into two fragments of unequal mass. The mass of one fragment is 2.50×10^{-28} kg and that of the other is 1.67×10^{-27} kg. If the lighter fragment has a speed of 0.893*c* after the breakup, what is the speed of the heavier fragment?

28. Show that the speed of an object having momentum of magnitude p and mass m is

$$u = \frac{c}{\sqrt{1 + (mc/p)^2}}$$

Section 9.7 ■ Relativistic Energy

29. Determine the energy required to accelerate an electron from (a) 0.500*c* to 0.900*c* and (b) 0.900*c* to 0.990*c*.

30. Show that, for any object moving at less than one-tenth the speed of light, the relativistic kinetic energy agrees with the result of the classical equation $K = \frac{1}{2}mu^2$ to within less than 1%. Therefore, for most purposes the classical equation is good enough to describe these objects, whose motion we call *nonrelativistic.*

31. An electron has a kinetic energy five times greater than its rest energy. Find (a) its total energy and (b) its speed.

32. Find the kinetic energy of a 78.0-kg spacecraft launched out of the solar system with speed 106 km/s by using (a) the classical equation $K = \frac{1}{2}mu^2$ and (b) the relativistic equation.

33. **Physics Now™** A proton moves at 0.950*c*. Calculate its (a) rest energy, (b) total energy, and (c) kinetic energy.

34. A cube of steel has a volume of 1.00 cm³ and a mass of 8.00 g when at rest on the Earth. If this cube is now given a speed $u = 0.900c$, what is its density as measured by a stationary observer? Note that relativistic density is defined as E_R/c^2V.

35. The rest energy of an electron is 0.511 MeV. The rest energy of a proton is 938 MeV. Assume that both particles have kinetic energies of 2.00 MeV. Find the speed of (a) the electron and (b) the proton. (c) By how much does the speed of the electron exceed that of the proton? (d) Repeat the calculations assuming that both particles have kinetic energies of 2 000 MeV.

36. An unstable particle with a mass of 3.34×10^{-27} kg is initially at rest. The particle decays into two fragments that fly off along the x axis with velocity components $0.987c$ and $-0.868c$. Find the masses of the fragments. (*Suggestion:* Conserve both energy and momentum.)

37. Show that the energy–momentum relationship $E^2 = p^2c^2 + (mc^2)^2$ follows from the expressions $E = \gamma mc^2$ and $p = \gamma mu$.

38. An object having mass 900 kg and traveling at speed $0.850c$ collides with a stationary object having mass 1 400 kg. The two objects stick together. Find (a) the speed and (b) the mass of the composite object.

39. A pion at rest ($m_\pi = 273m_e$) decays to a muon ($m_\mu = 207m_e$) and an antineutrino ($m_{\bar{\nu}} \approx 0$). The reaction is written $\pi^- \rightarrow \mu^- + \bar{\nu}$. Find the kinetic energy of the muon and the energy of the antineutrino in electron volts. (*Suggestion:* Conserve both energy and momentum.)

Section 9.8 ▪ Mass and Energy

40. When 1.00 g of hydrogen combines with 8.00 g of oxygen, 9.00 g of water is formed. During this chemical reaction, 2.86×10^5 J of energy is released. How much mass do the constituents of this reaction lose? Is the loss of mass likely to be detectable?

41. The power output of the Sun is 3.85×10^{26} W. How much mass is converted to energy in the Sun each second?

42. In a nuclear power plant, the fuel rods last 3 yr before they are replaced. If a plant with rated thermal power 1.00 GW operates at 80.0% capacity for the 3.00 yr, what is the loss of mass of the fuel?

43. A gamma ray (a high-energy photon) can produce an electron (e^-) and a positron (e^+) when it enters the electric field of a heavy nucleus: $\gamma \rightarrow e^+ + e^-$. What minimum gamma-ray energy is required to accomplish this task? (*Note:* The masses of the electron and the positron are equal.)

44. (a) In a crash test, two 1 500-kg cars, both moving at 20.0 m/s, collide head-on and stick together. Consider the whole quantity of wreckage before it loses any energy by such processes as thermal radiation. Is its mass greater or less than the original total mass of the two cars? By how much? (b) Repeat the problem for a relativistic crash test in which two 1 500-kg space vehicles, both moving at 200 Mm/s, meet head-on in a completely inelastic collision.

Section 9.9 ▪ General Relativity

45. An Earth satellite used in the Global Positioning System (GPS) moves in a circular orbit with radius 2.66×10^7 m and period 11 h 58 min. (a) Determine its speed. (b) The satellite contains an oscillator producing the principal nonmilitary GPS signal. Its frequency is 1 575.42 MHz in the reference frame of the satellite. When it is received on the Earth's surface, what is the fractional change in this frequency due to time dilation, as described by special relativity? (c) The gravitational "blueshift" of the frequency according to general relativity is a separate effect. It is called a blueshift to indicate a change to a higher frequency. The magnitude of that fractional change is given by

$$\frac{\Delta f}{f} = \frac{\Delta U_g}{mc^2}$$

where ΔU_g is the change in gravitational potential energy of an object–Earth system when the object of mass m is moved between the two points at which the signal is observed. Calculate this fractional change in frequency. (d) What is the overall fractional change in frequency? Superposed on both of these relativistic effects is a Doppler shift that is generally much larger. It can either increase or decrease the frequency received, depending on the motion of a particular satellite relative to a GPS receiver (Fig. P9.45).

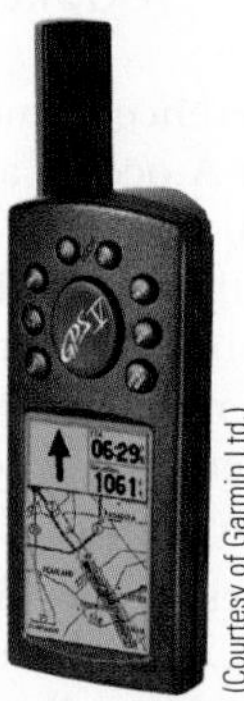
(Courtesy of Garmin Ltd.)

FIGURE P9.45 This Global Positioning System (GPS) receiver incorporates relativistically corrected time calculations in its analysis of signals it receives from orbiting satellites, allowing the unit to determine its position on the Earth's surface to within a few meters. If these corrections were not made, the location error would be about 1 km.

Section 9.10 ▪ Context Connection—From Mars to the Stars

46. In 1963, Mercury astronaut Gordon Cooper orbited the Earth 22 times. The press stated that for each orbit he aged 2 millionths of a second less than he would have if he had remained on the Earth. (a) Assuming that he was 160 km above the Earth moving at 7.82 km/s in a circular orbit, determine the time difference between someone on the Earth and the orbiting astronaut for the 22 orbits. You may use the approximation

$$\frac{1}{\sqrt{1-x}} \approx 1 + \frac{x}{2}$$

for small x. (b) Did the press report accurate information? Explain.

47. An astronaut wishes to visit the Andromeda galaxy, making a one-way trip that will take 30.0 yr in the spacecraft's frame of reference. Assume that the galaxy is 2.00 million ly away and that the astronaut's speed is constant. (a) How fast must he travel relative to the Earth? (b) What will be the kinetic energy of his 1 000-metric-ton spacecraft? (c) What is the cost of this energy if it is purchased at a typical consumer price for energy from the electric company of \$0.130/kWh?

Additional Problems

48. An electron has a speed of $0.750c$. (a) Find the speed of a proton that has the same kinetic energy as the electron. (b) Find the speed of a proton that has the same momentum as the electron.

49. **PhysicsNow™** The cosmic rays of highest energy are protons that have kinetic energy on the order of 10^{13} MeV. (a) How long would it take a proton of this energy to travel across the Milky Way galaxy, having a diameter on the order of 10^5 ly, as measured in the proton's frame? (b) From the point of view of the proton, how many kilometers across is the galaxy?

50. Ted and Mary are playing a game of catch in frame S′, which is moving at $0.600c$ with respect to frame S, while Jim, at rest in frame S, watches the action (Fig. P9.50). Ted throws the ball to Mary at $0.800c$ (according to Ted), and their separation (measured in S′) is 1.80×10^{12} m. (a) According to Mary, how fast is the ball moving? (b) According to Mary, how long does it take the ball to reach her? (c) According to Jim, how far apart are Ted and Mary, and how fast is the ball moving? (d) According to Jim, how long does it take the ball to reach Mary?

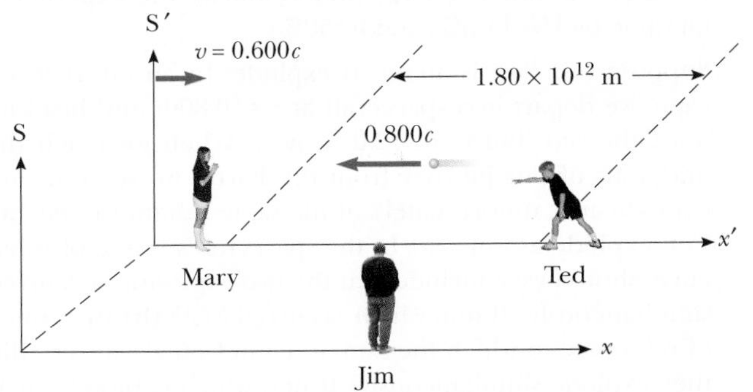

FIGURE **P9.50**

51. The net nuclear fusion reaction inside the Sun can be written as $4\,^1\text{H} \rightarrow {}^4\text{He} + E$. The rest energy of each hydrogen atom is 938.78 MeV and the rest energy of the helium-4 atom is 3 728.4 MeV. Calculate the percentage of the starting mass that is transformed to other forms of energy.

52. An object disintegrates into two fragments. One of the fragments has mass $1.00\ \text{MeV}/c^2$ and momentum $1.75\ \text{MeV}/c$ in the positive x direction. The other fragment has mass $1.50\ \text{MeV}/c^2$ and momentum $2.00\ \text{MeV}/c$ in the positive y direction. Find (a) the mass and (b) the speed of the original object.

53. An alien spaceship traveling at $0.600c$ toward the Earth launches a landing craft with an advance guard of purchasing agents and physics teachers. The lander travels in the same direction with a speed of $0.800c$ relative to the mother ship. As observed on the Earth, the spaceship is 0.200 ly from the Earth when the lander is launched. (a) What speed do the Earth observers measure for the approaching lander? (b) What is the distance to the Earth at the time of lander launch as observed by the aliens? (c) How long does it take the lander to reach the Earth as observed by the aliens on the mother ship? (d) If the lander has a mass of 4.00×10^5 kg, what is its kinetic energy as observed in the Earth reference frame?

54. A physics professor on the Earth gives an exam to her students, who are in a spacecraft traveling at speed v relative to the Earth. The moment the craft passes the professor, she signals the start of the exam. She wishes her students to have a time interval T_0 (spacecraft time) to complete the exam. Show that she should wait a time interval (Earth time) of

$$T = T_0\sqrt{\frac{1 - v/c}{1 + v/c}}$$

before sending a light signal telling them to stop. (*Suggestion:* Remember that it takes some time for the second light signal to travel from the professor to the students.)

55. A supertrain (proper length 100 m) travels at a speed of $0.950c$ as it passes through a tunnel (proper length 50.0 m). As seen by a trackside observer, is the train ever completely within the tunnel? If so, with how much space to spare?

56. Energy reaches the upper atmosphere of the Earth from the Sun at the rate of 1.79×10^{17} W. If all this energy were absorbed by the Earth and not re-emitted, how much would the mass of the Earth increase in 1.00 yr?

57. A particle with electric charge q moves along a straight line in a uniform electric field $\vec{\mathbf{E}}$ with a speed of u. The electric force exerted on the charge is $q\vec{\mathbf{E}}$. The motion and the electric field are both in the x direction. (a) Show that the acceleration of the particle in the x direction is given by

$$a = \frac{du}{dt} = \frac{qE}{m}\left(1 - \frac{u^2}{c^2}\right)^{3/2}$$

(b) Discuss the significance of the dependence of the acceleration on the speed. (c) If the particle starts from rest at $x = 0$ at $t = 0$, how could you proceed to find the speed of the particle and its position at time t?

58. Imagine that the entire Sun collapses to a sphere of radius R_g such that the work required to remove a small mass m from the surface would be equal to its rest energy mc^2. This radius is called the *gravitational radius* for the Sun. Find R_g. (It is believed that the ultimate fate of very massive stars is to collapse beyond their gravitational radii into black holes.)

59. The creation and study of new elementary particles is an important part of contemporary physics. Especially interesting is the discovery of a very massive particle. To create a particle of mass M requires an energy Mc^2. With enough energy, an exotic particle can be created by allowing a fast-moving particle of ordinary matter, such as a proton, to collide with a similar target particle. Let us consider a perfectly inelastic collision between two protons in which an incident proton with mass m_p, kinetic energy K, and momentum magnitude p joins with an originally stationary target proton to form a single product particle of mass M. You might think that the creation of a new product particle, nine times more massive than in a previous experiment, would require just nine times more energy for the incident proton. Unfortunately, not all the kinetic energy of the incoming proton is available to create the product particle because conservation of momentum requires that after the collision the system as a whole still must have some kinetic energy. Only a fraction of the energy of the incident particle is thus available to create a new particle. Determine how the energy available for particle creation depends on the energy of the moving proton. In particular, show that the energy available to create a product particle is given by

$$Mc^2 = 2m_pc^2\sqrt{1 + \frac{K}{2m_pc^2}}$$

From this result, when the kinetic energy K of the incident proton is large compared with its rest energy m_pc^2, we see that M approaches $(2m_pK)^{1/2}/c$. Thus, if the energy of the incoming proton is increased by a factor of nine, the mass you can create increases only by a factor of three. This disappointing result is the main reason that most modern accelerators, such as those at CERN (in Europe), at Fermilab (near Chicago), at SLAC (at Stanford), and at DESY (in Germany), use *colliding beams*. Here the total momentum of a pair of interacting particles can be zero. The center of mass can be at rest after the collision, so in principle all the initial kinetic energy can be used for particle creation, according to

$$Mc^2 = 2mc^2 + K = 2mc^2\left(1 + \frac{K}{2mc^2}\right)$$

where K is the total kinetic energy of two identical colliding particles. Here, if $K >> mc^2$, we have M directly proportional to K, as we would desire. These machines are difficult to build and to operate, but they open new vistas in physics.

60. An observer in a coasting spacecraft moves toward a mirror at speed v relative to the reference frame labeled by S in Figure P9.60. The mirror is stationary with respect to S. A light pulse emitted by the spacecraft travels toward the mirror and is reflected back to the craft. The front of the craft is a distance d from the mirror (as measured by observers in S) at the moment the light pulse leaves the craft. What is the total travel time of the pulse as measured by observers in (a) the S frame and (b) the front of the spacecraft?

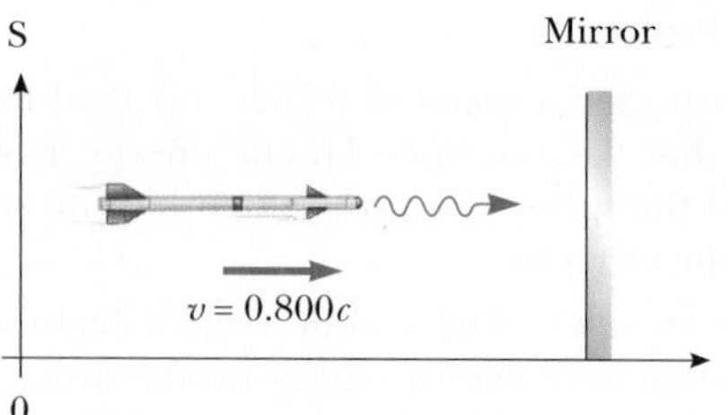

FIGURE **P9.60**

61. A rod of length L_0 moving with a speed v along the horizontal direction makes an angle θ_0 with respect to the x' axis. (a) Show that the length of the rod as measured by a stationary observer is $L = L_0[1 - (v^2/c^2)\cos^2\theta_0]^{1/2}$. (b) Show that the angle that the rod makes with the x axis is given by $\tan\theta = \gamma\tan\theta_0$. These results show that the rod is both contracted and rotated. (Take the lower end of the rod to be at the origin of the primed coordinate system.)

62. Prepare a graph of the relativistic kinetic energy and the classical kinetic energy, both as a function of speed, for an object with a mass of your choice. At what speed does the classical kinetic energy underestimate the experimental value by 1%, by 5%, and by 50%?

63. Suppose our Sun is about to explode. In an effort to escape, we depart in a spacecraft at $v = 0.800c$ and head toward the star Tau Ceti, 12.0 ly away. When we reach the midpoint of our journey from the Earth, we see our Sun explode and, unfortunately, at the same instant we see Tau Ceti explode as well. (a) In the spacecraft's frame of reference, should we conclude that the two explosions occurred simultaneously? If not, which occurred first? (b) In a frame of reference in which the Sun and Tau Ceti are at rest, did they explode simultaneously? If not, which exploded first?

ANSWERS TO QUICK QUIZZES

9.1 (d). The two events (the pulse leaving the flashlight and the pulse hitting the far wall) take place at different locations for both observers, so neither measures the proper time interval.

9.2 (a). The two events are the beginning and the end of the movie, both of which take place at rest with respect to the spacecraft crew. Therefore, the crew measures the proper time interval of 2 h. Any observer in motion with respect to the spacecraft, which includes the observer on Earth, will measure a longer time interval because of time dilation.

9.3 (a). If their on-duty time is based on clocks that remain on the Earth, they will have larger paychecks. A shorter time interval will have passed for the astronauts in their frame of reference than for their employer back on the Earth.

9.4 (c). Both your body and your sleeping cabin are at rest in your reference frame; therefore, they will have their proper length according to you. There will be no change in measured lengths of objects, including yourself, within your spacecraft.

9.5 (c), (d). Because of your motion toward the source of the light, the light beam has a horizontal component of velocity as measured by you. The magnitude of the vector sum of the horizontal and vertical component vectors must be equal to c, so the magnitude of the vertical component must be smaller than c. When the searchlight is aimed directly toward you, there is only a horizontal component of the velocity of the light and you must measure a speed of c.

9.6 (a) $m_3 > m_2 = m_1$; the rest energy of particle 3 is $2E$, whereas it is E for particles 1 and 2. (b) $K_3 = K_2 > K_1$; the kinetic energy is the difference between the total energy and the rest energy. The kinetic energy is $4E - 2E = 2E$ for particle 3, $3E - E = 2E$ for particle 2, and $2E - E = E$ for particle 1. (c) $u_2 > u_3 = u_1$; from Equation 9.21, $E = \gamma E_R$. Solving for the square of the particle speed u, we find that $u^2 = c^2[1 - (E_R/E)^2]$. Therefore, the particle with the smallest ratio of rest energy to total energy will have the largest speed. Particles 1 and 3 have the same ratio as each other, and the ratio of particle 2 is smaller.

Rotational Motion

The Malaysian pastime of *gasing* involves the spinning of tops that can have masses up to 5 kg. Professional spinners can spin their tops so that they might rotate for 1 to 2 h before stopping. We will study the rotational motion of objects such as these tops in this chapter.

(Courtesy of Tourism Malaysia)

CHAPTER OUTLINE

When an extended object, such as a wheel, rotates about its axis, the motion cannot be analyzed by treating the object as a particle because at any given time different parts of the object are moving with different speeds and in different directions. We can, however, analyze the motion by considering an extended object to be composed of a collection of moving particles.

In dealing with a rotating object, analysis is greatly simplified by assuming that the object is rigid. A **rigid object** is one that is nondeformable; that is, it is an object in which the relative locations of all particles of which the object is composed remain constant. All real objects are deformable to some extent; our rigid-object model, however, is useful in many situations in which deformation is negligible.

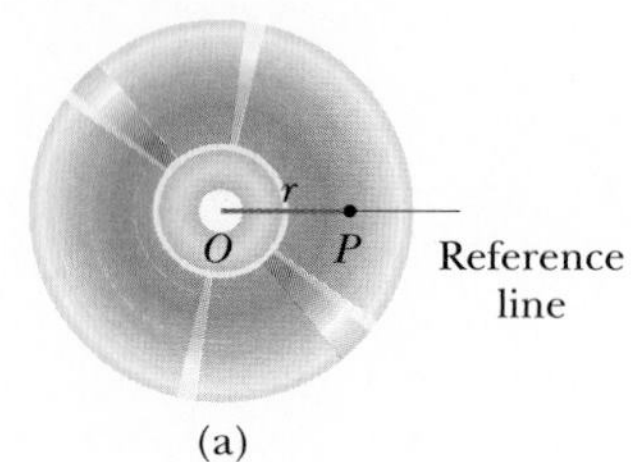

(a)

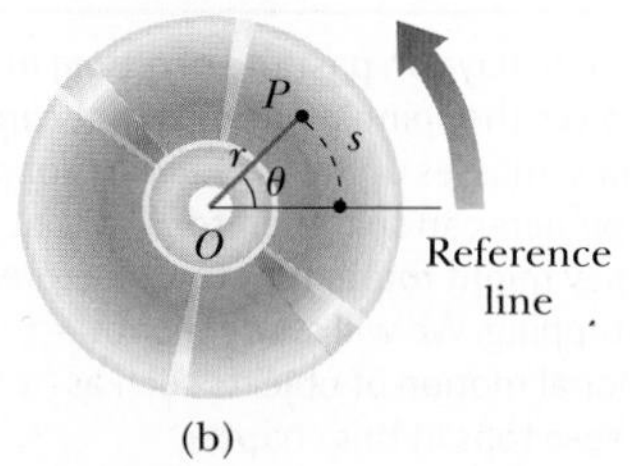

(b)

FIGURE 10.1 A compact disc rotating about a fixed axis through O perpendicular to the plane of the figure. (a) To define angular position for the disc, a fixed reference line is chosen. A particle at P is located at a radial distance r from the rotation axis at O. (b) As the disc rotates, point P moves through an arc length s on a circular path of radius r. The angular position of P is θ.

10.1 ANGULAR POSITION, SPEED, AND ACCELERATION

We began our study of translational motion in Chapter 2 by defining the terms *position, velocity,* and *acceleration.* For example, we locate a particle in one-dimensional space with the position variable x. In this chapter, we will insert the word *translational* before our previously studied kinematic variables to distinguish them from the analogous *rotational* variables that we will develop.

Let us think about a rotating object. How would we describe its position in its rotational motion? We do so by describing its *orientation* relative to some fixed reference direction. For example, imagine two soldiers performing a military about-face maneuver. They both begin by facing due north. One soldier, who has been practicing diligently, ends up after the maneuver with his body facing due south. The second, who has not been practicing, ends up facing southeast. We could describe their respective rotational positions after the maneuver by reporting the angle through which each turned from the original direction. The first soldier turned through 180°, but the second turned through only 135°. Thus, we can use an angle measured from a reference direction as a measure of **rotational position,** or **angular position,** which is our starting point for our description of rotational motion.

Figure 10.1 illustrates an overhead view of a rotating compact disc. The disc is rotating about a fixed axis through O. The axis is perpendicular to the plane of the figure. Let us investigate the motion of only one of the millions of "particles" making up the disc. A particle at P is at a fixed distance r from the origin and rotates about it in a circle of radius r. (In fact, *every* particle on the disc undergoes circular motion about O.) It is convenient to represent the position of P with its polar coordinates (r, θ), where r is the distance from the origin to P and θ is measured *counterclockwise* from some reference line shown in Figure 10.1. In this representation, the only coordinate for the particle that changes in time is the angle θ; r remains constant. As the particle moves along the circle from the reference line ($\theta = 0$) to an angular position θ, it moves through an arc of length s as in Figure 10.1b. The arc length s is related to the angle θ through the relationship

PITFALL PREVENTION 10.1

Remember the Radian Keep in mind that Equation 10.1b defines an angle expressed in *radians.* Don't fall into the trap of using this equation for angles measured in degrees. Also, be sure to set your calculator in radian mode when doing problems in rotation.

$$s = r\theta \quad [10.1a]$$

$$\theta = \frac{s}{r} \quad [10.1b]$$

It is important to note the units of θ in Equation 10.1b. Because θ is the ratio of an arc length and the radius of the circle, it is a pure number. Nonetheless, we commonly give θ the artificial unit **radian** (rad), where

▪ The radian

one radian is the angle subtended by an arc length equal to the radius of the arc.

Because the circumference of a circle is $2\pi r$, it follows from Equation 10.1b that 360° corresponds to an angle of $(2\pi r/r)$ rad $= 2\pi$ rad. (Also note that 2π rad corresponds to one complete revolution.) Hence, 1 rad $= 360°/2\pi \approx 57.3°$. To convert an angle in degrees to an angle in radians, we use π rad $= 180°$, so

$$\theta\ (\text{rad}) = \frac{\pi}{180°}\theta\ (\text{deg})$$

For example, 60° equals $\pi/3$ rad, and 45° equals $\pi/4$ rad.

Because the disc in Figure 10.1 is a rigid object, as the particle moves along the circle from the reference line every other particle on the object rotates through the same angle θ. Therefore, **we can associate the angle θ with the entire rigid object as well as with an individual particle,** which allows us to define the angular position of a rigid object in its rotational motion. We choose a radial line on the object, such as a line connecting O and a chosen particle on the object. The angular position of the rigid object is the angle θ between this radial line on the object and the fixed reference line in space, which is often chosen as the x axis. This process is similar to the way we identify the position of an object in translational motion as the distance x between the object and the reference position, which is the origin, $x = 0$.

As a particle on a rigid object travels from position Ⓐ to position Ⓑ in a time interval Δt as in Figure 10.2, the radial line of length r sweeps out an angle $\Delta\theta = \theta_f - \theta_i$. This quantity $\Delta\theta$ is defined as the **angular displacement** of the rigid object:

$$\Delta\theta \equiv \theta_f - \theta_i$$

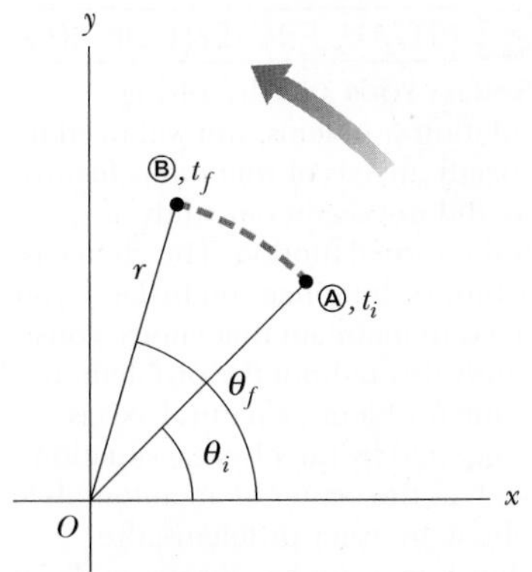

FIGURE 10.2 A particle on a rotating rigid object moves from Ⓐ to Ⓑ along the arc of a circle. In the time interval $\Delta t = t_f - t_i$, the radial line of length r sweeps out an angle $\Delta\theta = \theta_f - \theta_i$.

The rate at which this angular displacement occurs can vary. If the rigid object spins rapidly, this displacement can occur in a short time interval. If it rotates slowly, this displacement occurs in a longer time interval. These different rotation rates can be quantified by introducing *angular speed.* We define the **average angular speed** ω_{avg} as the ratio of the angular displacement of a rigid object to the time interval Δt during which the displacement occurs:

■ Average angular speed

$$\omega_{avg} \equiv \frac{\theta_f - \theta_i}{t_f - t_i} = \frac{\Delta\theta}{\Delta t} \qquad [10.2]$$

In analogy to linear speed, the **instantaneous angular speed** ω is defined as the limit of the ratio $\Delta\theta/\Delta t$ as Δt approaches zero:

■ Instantaneous angular speed

$$\omega \equiv \lim_{\Delta t \to 0} \frac{\Delta\theta}{\Delta t} = \frac{d\theta}{dt} \qquad [10.3]$$

Angular speed has units of rad/s (or s^{-1} because radians are not dimensional). Let us adopt the convention that the fixed axis of rotation for an object is the z axis, which is directed out of the page in Figures 10.1 and 10.2. We shall take ω to be positive when θ is increasing (counterclockwise motion in Figs. 10.1 and 10.2) and negative when θ is decreasing (clockwise motion).

If the instantaneous angular speed of a particle changes from ω_i to ω_f in the time interval Δt, the particle has an angular acceleration. The **average angular acceleration** α_{avg} of a particle moving in a circular path is defined as the ratio of the change in the angular speed to the time interval Δt:

■ Average angular acceleration

$$\alpha_{avg} \equiv \frac{\omega_f - \omega_i}{t_f - t_i} = \frac{\Delta\omega}{\Delta t} \qquad [10.4]$$

In analogy to linear acceleration, the **instantaneous angular acceleration** is defined as the limit of the ratio $\Delta\omega/\Delta t$ as Δt approaches zero:

■ Instantaneous angular acceleration

$$\alpha \equiv \lim_{\Delta t \to 0} \frac{\Delta\omega}{\Delta t} = \frac{d\omega}{dt} \qquad [10.5]$$

Angular acceleration has units of rad/s^2 or s^{-2}.

As pointed out in the introduction, we will focus much of our attention in this chapter on rigid objects. Approximating a real object as a rigid object is a simplification model, similar to the particle model, which we call the **rigid object model.** If we were to imagine a rotating block of gelatin, which is *not* a rigid object, the motion is very complicated because of the combination of rotation of the particles

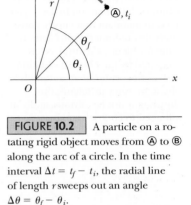

PITFALL PREVENTION 10.2

SPECIFY YOUR AXIS In solving rotation problems, you will need to specify an axis of rotation, a feature we did not see in our study of translational motion. The choice is arbitrary, but once you make it, you need to maintain that choice consistently throughout the problem. In some problems, a natural axis is suggested by the physical situation, such as the center of an automobile wheel. In other problems, the choice may not be obvious, and you will need to choose an axis.

and the movement of particles within the deformable block relative to one another. Another example of a nonrigid object is our own Sun; the region of the Sun near the solar equator is moving with a higher angular speed than the region near the poles. We shall not analyze such messy problems, however. Instead, our analyses will address only rigid objects. As we investigate rotational motion, we shall develop a number of analysis models for rigid objects that have analogs in our analysis models for particles.

With our simplification model of a rigid object, we can make a statement about the various particles in the rigid object: When a rigid object is rotating about a *fixed* axis, **every particle on the object rotates about that axis through the same angle in a given time interval and has the same angular speed and the same angular acceleration.** That is, the quantities θ, ω, and α characterize the rotational motion of the entire rigid object as well as individual particles in the object. Using these quantities, we can greatly simplify the analysis of rigid-object rotation.

The angular position θ, angular speed ω, and angular acceleration α of a rigid object are analogous to translational position x, translational speed v, and translational acceleration a, respectively, for the corresponding one-dimensional motion of a particle discussed in Chapter 2. The variables θ, ω, and α differ dimensionally from the variables x, v, and a only by a length factor, as we shall see shortly.

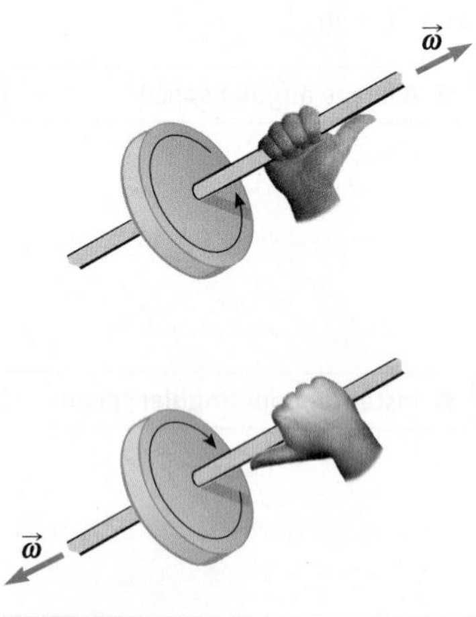

FIGURE 10.3 The orange disk rotates in the directions indicated. The right-hand rule determines the direction of the angular velocity vector.

We have not associated any direction with the angular speed and angular acceleration.[1] Strictly speaking, these variables are the magnitudes of the angular velocity and angular acceleration vectors $\vec{\omega}$ and $\vec{\alpha}$. Because we are considering rotation about a fixed axis, we can indicate the directions of these vectors by assigning a positive or negative sign to ω and α, as discussed for ω after Equation 10.3. For rotation about a fixed axis, the only direction in space that uniquely specifies the rotational motion is the direction along the axis, but we still must specify one of the two directions along this axis as positive.

The direction of $\vec{\omega}$ is along the axis of rotation, which is the z axis in Figure 10.1. By convention, we take the direction of $\vec{\omega}$ to be *out of* the plane of the diagram when the rotation is counterclockwise and *into* the plane of the diagram when the rotation is clockwise. To further illustrate this convention, it is convenient to use the **right-hand rule** illustrated by Figure 10.3. The four fingers of the right hand are wrapped in the direction of the rotation. The extended right thumb points in the direction of $\vec{\omega}$.

The direction of $\vec{\alpha}$ follows from its vector definition as $d\vec{\omega}/dt$. For rotation about a fixed axis, the direction of $\vec{\alpha}$ is the same as $\vec{\omega}$ if the angular speed (the magnitude of $\vec{\omega}$) is increasing in time and is antiparallel to $\vec{\omega}$ if the angular speed is decreasing in time.

The full vector treatment of rotational motion is beyond the scope of this book and not necessary for our level of understanding, so we will not use vector notation for most of this chapter.

QUICK QUIZ 10.1 A rigid object is rotating in a counterclockwise sense around a fixed axis. Each of the following pairs of quantities represents an initial angular position and a final angular position of the rigid object. **(i)** Which of the sets can *only* occur if the rigid object rotates through more than 180°? **(a)** 3 rad, 6 rad **(b)** -1 rad, 1 rad **(c)** 1 rad, 5 rad **(ii)** If each of the displacements occurs during the same time interval, which choice represents the lowest average angular speed?

[1]Although we do not verify it here, the instantaneous angular velocity and instantaneous angular acceleration are vector quantities, but the corresponding average values are not because angular displacements do not add as vector quantities for finite rotations.

10.2 ROTATIONAL KINEMATICS: THE RIGID OBJECT UNDER CONSTANT ANGULAR ACCELERATION

In our study of one-dimensional motion, we found that the simplest accelerated motion to analyze is motion under constant translational acceleration (Chapter 2). Likewise, for rotational motion about a fixed axis, the simplest accelerated motion to analyze is motion of a rigid object under constant angular acceleration. We will identify this situation as an analysis model that can be used to solve a wide variety of rotational problems.

If we write Equation 10.5 in the form $d\omega = \alpha\, dt$ and let $\omega = \omega_i$ at $t_i = 0$, we can integrate this expression directly to find the final angular speed ω_f of the rigid object as a function of time:

$$\omega_f = \omega_i + \alpha t \qquad \text{(for constant } \alpha) \qquad [10.6]$$

Likewise, if we rewrite Equation 10.3 and substitute Equation 10.6, we can integrate once more (with $\theta = \theta_i$ at $t_i = 0$) to find the angular position of the rigid object as a function of time:

$$\theta_f = \theta_i + \omega_i t + \tfrac{1}{2}\alpha t^2 \qquad \text{(for constant } \alpha) \qquad [10.7]$$

If we eliminate t from Equations 10.6 and 10.7, we obtain

$$\omega_f^{\,2} = \omega_i^{\,2} + 2\alpha(\theta_f - \theta_i) \qquad \text{(for constant } \alpha) \qquad [10.8]$$

If we eliminate α, we find

$$\theta_f = \theta_i + \tfrac{1}{2}(\omega_i + \omega_f)\,t \qquad \text{(for constant } \alpha) \qquad [10.9]$$

Notice that these kinematic expressions for rotational motion of a rigid object under constant angular acceleration are of the *same mathematical form* as those for translational motion of a particle under constant acceleration, with the substitutions $x \rightarrow \theta$, $v \rightarrow \omega$, and $a \rightarrow \alpha$. The similarities between rotational and translational kinematic equations are shown in Table 10.1.

QUICK QUIZ 10.2 Consider again the pairs of angular positions for the rigid object in Quick Quiz 10.1. If the object starts from rest at the initial angular position, moves counterclockwise with constant angular acceleration, and arrives at the final angular position with the same angular speed in all three cases, for which choice is the angular acceleration the highest?

TABLE 10.1 A Comparison of Equations for Rotational and Translational Motion: Kinematic Equations

Rotational Motion About a Fixed Axis with α = Constant Variables: θ_f and ω_f	Translational Motion with a = Constant Variables: x_f and v_f
$\omega_f = \omega_i + \alpha t$	$v_f = v_i + at$
$\theta_f = \theta_i + \omega_i t + \frac{1}{2}\alpha t^2$	$x_f = x_i + v_i t + \frac{1}{2}at^2$
$\theta_f = \theta_i + \frac{1}{2}(\omega_i + \omega_f)t$	$x_f = x_i + \frac{1}{2}(v_i + v_f)t$
$\omega_f^{\,2} = \omega_i^{\,2} + 2\alpha(\theta_f - \theta_i)$	$v_f^{\,2} = v_i^{\,2} + 2a(x_f - x_i)$

PITFALL PREVENTION 10.3

Just like Translation? Table 10.1 suggests that rotational kinematics is just like translational kinematics. That is almost true, but keep in mind two differences that you must address. (1) In rotational kinematics, as suggested in Pitfall Prevention 10.2, you need to specify a rotation axis. (2) In rotational motion, the object keeps returning to its original orientation; therefore, you may be asked for the number of revolutions made by a rigid object, a concept that has no meaning in translational motion.

EXAMPLE 10.1 Rotating Wheel

A wheel rotates with a constant angular acceleration of 3.50 rad/s^2.

A If the angular speed of the wheel is 2.00 rad/s at $t = 0$, through what angle does the wheel rotate between $t = 0$ and $t = 2.00$ s?

Solution We assume that the wheel is perfectly rigid, so we can use the rigid object model. Because the angular acceleration in the problem is given as constant, we model the wheel as a rigid object under constant angular acceleration and use the rotational kinematic equations. We use Equation 10.7, setting $\theta_i = 0$, and obtain

$$\theta_f = \theta_i + \omega_i t + \tfrac{1}{2}\alpha t^2$$
$$= 0 + (2.00 \text{ rad/s})(2.00 \text{ s}) + \tfrac{1}{2}(3.50 \text{ rad/s}^2)(2.00 \text{ s})^2$$
$$= \boxed{11.0 \text{ rad}}$$

which is equivalent to 11.0 rad/$(2\pi$ rad/rev$)$ = 1.75 rev.

B What is the angular speed of the wheel at $t = 2.00$ s?

Solution We use Equation 10.6:

$$\omega_f = \omega_i + \alpha t = 2.00 \text{ rad/s} + (3.50 \text{ rad/s}^2)(2.00 \text{ s})$$
$$= \boxed{9.00 \text{ rad/s}}$$

We could also obtain this result using Equation 10.8 and the results of part A. Try it!

10.3 RELATIONS BETWEEN ROTATIONAL AND TRANSLATIONAL QUANTITIES

In this section, we shall derive some useful relations between the angular speed and angular acceleration of a single particle on a rotating rigid object and its translational speed and translational acceleration. Keep in mind that when a rigid object rotates about a fixed axis, *every* particle of the object moves in a circle whose center is on the axis of rotation.

Consider a particle on a rotating rigid object, moving in a circle of radius r about the z axis, as in Active Figure 10.4. Because the particle moves along a circular path, its translational velocity vector $\vec{\mathbf{v}}$ is always tangent to the path; hence, we often call this quantity **tangential velocity.** The magnitude of the tangential velocity of the particle is, by definition, the **tangential speed,** given by $v = ds/dt$, where s is the distance traveled by the particle along the circular path. Recalling from Equation 10.1a that $s = r\theta$ and noting that r is a constant, we have

$$v = \frac{ds}{dt} = r\frac{d\theta}{dt}$$

$$\boxed{v = r\omega} \quad [10.10]$$

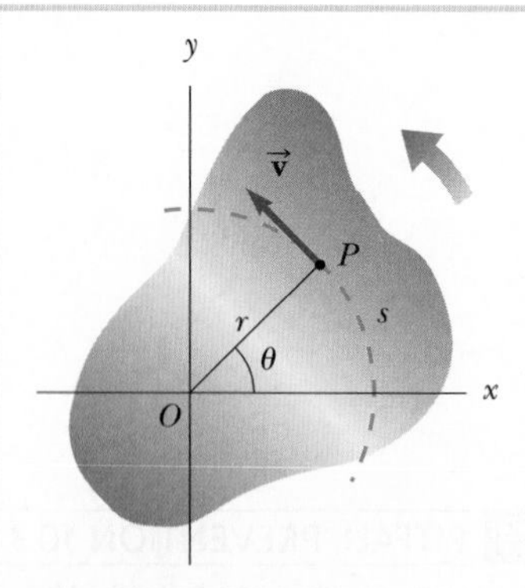

ACTIVE FIGURE 10.4
As a rigid object rotates about the fixed axis through O, the point P has a tangential velocity $\vec{\mathbf{v}}$ that is always tangent to the circular path of radius r.

Physics Now™ Log into PhysicsNow at **www.pop4e.com** and go to Active Figure 10.4 to move point P and see the change in the tangential velocity.

That is, the tangential speed of the particle equals the distance of the particle from the axis of rotation multiplied by the particle's angular speed.

We can relate the angular acceleration of the particle to its tangential acceleration a_t—which is the component of its acceleration tangent to the path of motion—by taking the time derivative of v:

$$a_t = \frac{dv}{dt} = r\frac{d\omega}{dt}$$

$$\boxed{a_t = r\alpha} \quad [10.11]$$

That is, the tangential component of the translational acceleration of a particle undergoing circular motion equals the distance of the particle from the axis of rotation multiplied by the angular acceleration.

In Chapter 3, we found that a particle rotating in a circular path undergoes a centripetal, or radial, acceleration of magnitude v^2/r directed toward the center of rotation (Fig. 10.5). Because $v = r\omega$, we can express the centripetal acceleration of the particle in terms of the angular speed as

$$a_c = \frac{v^2}{r} = r\omega^2 \qquad [10.12]$$

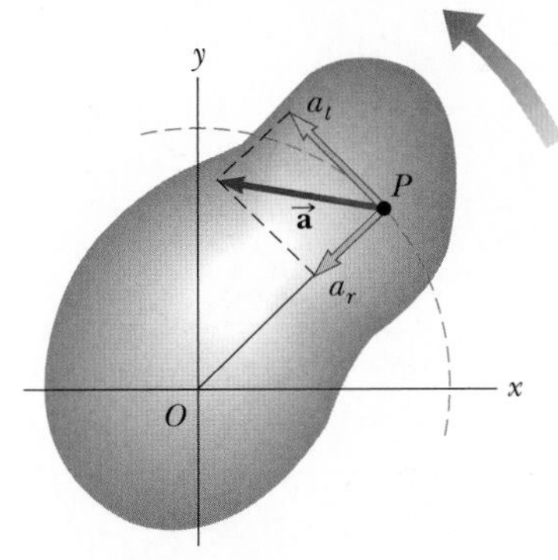

FIGURE 10.5 As a rigid object rotates about a fixed axis through O, a particle at point P experiences a tangential component a_t and a radial component a_r of translational acceleration. The total translational acceleration of this particle is $\vec{\mathbf{a}} = \vec{\mathbf{a}}_t + \vec{\mathbf{a}}_r$, where $\vec{\mathbf{a}}_r = -a_c\hat{\mathbf{r}}$.

The *total translational acceleration* of the particle is $\vec{\mathbf{a}} = \vec{\mathbf{a}}_t + \vec{\mathbf{a}}_r$. The magnitude of the total translational acceleration of the particle is therefore

$$a = \sqrt{a_t^2 + a_r^2} = \sqrt{r^2\alpha^2 + r^2\omega^4} = r\sqrt{\alpha^2 + \omega^4} \qquad [10.13]$$

QUICK QUIZ 10.3 Benjamin and Torrey are riding on a merry-go-round. Benjamin rides on a horse at the outer rim of the circular platform, twice as far from the center of the circular platform as Torrey, who rides on an inner horse. **(i)** When the merry-go-round is rotating at a constant angular speed, what is Benjamin's angular speed? **(a)** twice Torrey's **(b)** the same as Torrey's **(c)** half of Torrey's **(d)** impossible to determine. **(ii)** When the merry-go-round is rotating at a constant angular speed, what is Benjamin's tangential speed from the same list of choices?

■ Thinking Physics 10.1

A phonograph record (LP, for *long-playing*) rotates at a constant *angular* speed. A compact disc (CD) rotates so that the surface sweeps past the laser at a constant *tangential* speed. Consider two circular grooves of information on an LP, one near the outer edge and one near the inner edge. Suppose the outer groove "contains" 1.8 s of music. Does the inner groove also contain 1.8 s of music? And for the CD, do the inner and outer "grooves" contain the same time interval of music?

Reasoning On the LP the inner and outer grooves must both rotate once in the same time interval. Therefore, each groove, regardless of where it is on the record, contains the same time interval of information. Of course, on the inner grooves, this same information must be compressed into a smaller circumference. On a CD, the constant tangential speed requires that no such compression occur; the digital pits representing the information are spaced uniformly everywhere on the surface. Therefore, there is more information in an outer "groove," because of its larger circumference and, as a result, a longer time interval of music than in the inner "groove." ■

■ Thinking Physics 10.2

The launch area for the European Space Agency is not in Europe, but rather in South America. Why?

Reasoning Placing a satellite in Earth orbit requires providing a large tangential speed to the satellite, which is the task of the rocket propulsion system. Anything that reduces the requirements on the propulsion system is a welcome contribution. The surface of the Earth is already traveling toward the east at a high speed due to the rotation of the Earth. Therefore, if rockets are launched toward the east, the rotation of the Earth provides some initial tangential speed, reducing somewhat the requirements on the propulsion system. If rockets were launched from Europe, which is at a relatively large latitude, the contribution of the Earth's rotation is relatively small because the distance between Europe and the rotation axis of the Earth is relatively small. The ideal place for launching is at the equator, which is as far as one can be from the rotation axis of the Earth and still be on the surface of the Earth. This location results in the largest possible tangential speed due to the Earth's rotation. The European Space Agency exploits this advantage by launching from French Guiana, which is only a few degrees north of the equator.

A second advantage of this location is that launching toward the east takes the spacecraft over water. In the event of an accident or a failure, the wreckage will fall into the ocean rather than into populated areas as it would if launched to the east from Europe. Similarly, the United States launches spacecraft from Florida rather than California, despite the more favorable weather conditions in California. ■

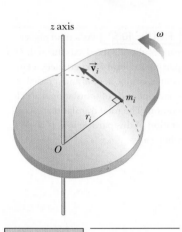

FIGURE 10.6 A rigid object rotating about the z axis with angular speed ω. The kinetic energy of the particle of mass m_i is $\frac{1}{2}m_iv_i^2$. The kinetic energy of the rigid object is called its rotational kinetic energy.

10.4 ROTATIONAL KINETIC ENERGY

Imagine that you begin a workout session on a stationary exercise bicycle. You apply a force with your feet on the pedals, moving them through a displacement; as a result, you have done work. The result of this work is the spinning of the wheel. This rotational motion represents kinetic energy because an object with mass is in motion. In this section, we will investigate this kinetic energy for rotating objects. In a later section, we will consider the work done in rotational motion and develop a rotational version of the work–kinetic energy theorem.

Let us consider a rigid object as a collection of particles and assume that it rotates about a fixed z axis with an angular speed ω (Fig. 10.6). Each particle of the object is in motion and therefore has some kinetic energy, determined by its mass and tangential speed. If the mass of the ith particle is m_i and its tangential speed is v_i, the kinetic energy of this particle is

$$K_i = \tfrac{1}{2}m_iv_i^2$$

We can express the *total* kinetic energy K_R of the rotating rigid object as the sum of the kinetic energies of the individual particles. Therefore, incorporating Equation 10.10,

$$K_R = \sum_i K_i = \sum_i \tfrac{1}{2}m_iv_i^2 = \tfrac{1}{2}\sum_i m_ir_i^2\omega^2$$
$$= \tfrac{1}{2}\left(\sum_i m_ir_i^2\right)\omega^2$$

where we have factored ω^2 from the sum because it is common to every particle in the object. The quantity in parentheses is called the **moment of inertia** I of the rigid object:

■ Moment of inertia for a system of particles

$$I = \sum_i m_ir_i^2 \qquad [10.14]$$

Therefore, we can express the kinetic energy of the rotating rigid object around the z axis as

PITFALL PREVENTION 10.4

No single moment of inertia We have pointed out that moment of inertia is analogous to mass, but there is one major difference. Mass is an inherent property of an object and has a single value. The moment of inertia of an object depends on your choice of rotation axis; therefore, an object has no single value of the moment of inertia. An object does have a minimum value of the moment of inertia, which is that calculated around an axis passing through the center of mass of the object.

$$K_R = \frac{1}{2}I\omega^2 \qquad [10.15]$$

■ **Kinetic energy of a rotating rigid object**

From the definition of moment of inertia, we see that it has dimensions of ML^2 ($kg \cdot m^2$ in SI units). The moment of inertia is a measure of an object's *resistance to change in its angular speed.* Therefore, it plays a role in rotational motion identical to the role mass plays in translational motion. Notice that moment of inertia depends not only on the mass of the rigid object but also on *how the mass is distributed around the rotation axis.*

Although we shall commonly refer to the quantity $\frac{1}{2}I\omega^2$ as the **rotational kinetic energy,** it is not a new form of energy. It is ordinary kinetic energy because it was derived from a sum over individual kinetic energies of the particles contained in the rigid object. It is a new role for kinetic energy for us, however, because we have only considered kinetic energy associated with translation through space so far. **On the storage side of the continuity equation for energy (see Eq. 6.20), we should now consider that the kinetic energy term should be the sum of the changes in both translational and rotational kinetic energy.** Therefore, in energy versions of system models, we should keep in mind the possibility of rotational kinetic energy.

Equation 10.14 gives the moment of inertia of a collection of particles. For an extended, continuous object, we can calculate the moment of inertia by dividing the object into many small elements with mass Δm_i. Then, the moment of inertia is approximately $I \approx \sum r_i^2 \, \Delta m_i$, where r_i is the perpendicular distance of the element of mass Δm_i from the rotation axis. Now we take the limit as $\Delta m_i \rightarrow 0$, in which case the sum becomes an integral:

$$I = \lim_{\Delta m_i \rightarrow 0} \sum_i r_i^2 \, \Delta m_i = \int r^2 \, dm \qquad [10.16]$$

It is usually easier to calculate moments of inertia in terms of the volume of the elements rather than their mass, and we can easily make this change by using Equation 1.1, $\rho = m/V$, where ρ is the density of the object and V is its volume. We can express the mass of an element by writing Equation 1.1 in differential form, $dm = \rho \, dV$. Using this form, Equation 10.16 becomes

$$I = \int \rho r^2 \, dV \qquad [10.17]$$

■ **Moment of inertia for an extended, continuous object**

If the object is homogenous, the density ρ is constant and the integral can be evaluated for a given geometry. If ρ is not uniform over the volume of the object, its variation with position must be known in order to perform the integration.

For symmetric objects, the moment of inertia can be expressed in terms of the total mass of the object and one or more dimensions of the object. Table 10.2 shows the moments of inertia of various common symmetric objects.

QUICK QUIZ 10.4 A section of hollow pipe and a solid cylinder have the same radius, mass, and length. They both rotate about their long central axes with the same angular speed. Which object has the higher rotational kinetic energy? **(a)** the pipe **(b)** the solid cylinder **(c)** they have the same rotational kinetic energy **(d)** impossible to determine

TABLE 10.2 Moments of Inertia of Homogeneous Rigid Objects With Different Geometries

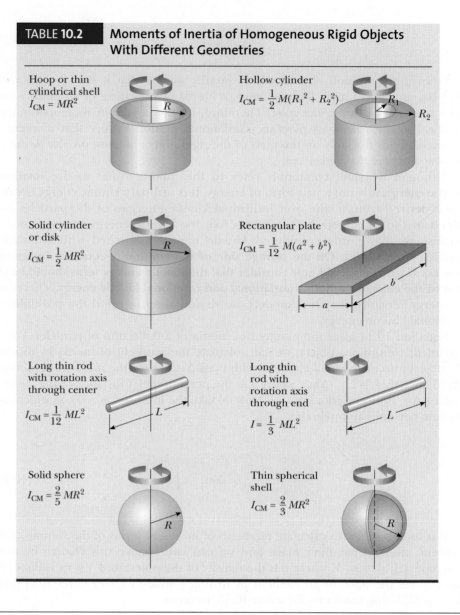

Hoop or thin cylindrical shell
$I_{CM} = MR^2$

Hollow cylinder
$I_{CM} = \frac{1}{2}M(R_1^2 + R_2^2)$

Solid cylinder or disk
$I_{CM} = \frac{1}{2}MR^2$

Rectangular plate
$I_{CM} = \frac{1}{12}M(a^2 + b^2)$

Long thin rod with rotation axis through center
$I_{CM} = \frac{1}{12}ML^2$

Long thin rod with rotation axis through end
$I = \frac{1}{3}ML^2$

Solid sphere
$I_{CM} = \frac{2}{5}MR^2$

Thin spherical shell
$I_{CM} = \frac{2}{3}MR^2$

EXAMPLE 10.2 The Oxygen Molecule

Consider the diatomic oxygen molecule O_2, which is rotating in the *xy* plane about the *z* axis passing through its center, perpendicular to its length. The mass of each oxygen atom is 2.66×10^{-26} kg, and at room temperature, the average separation between the two oxygen atoms is $d = 1.21 \times 10^{-10}$ m.

A Calculate the moment of inertia of the molecule about the *z* axis.

Solution We model the molecule as a rigid object, consisting of two particles (the two oxygen atoms), in rotation. Because the distance of each particle from the *z* axis is $d/2$, the moment of inertia about the *z* axis is

$$I = \sum_i m_i r_i^2 = m\left(\frac{d}{2}\right)^2 + m\left(\frac{d}{2}\right)^2 = \frac{md^2}{2}$$

$$= \frac{(2.66 \times 10^{-26}\ \text{kg})(1.21 \times 10^{-10}\ \text{m})^2}{2}$$

$$= 1.95 \times 10^{-46}\ \text{kg} \cdot \text{m}^2$$

B A typical angular speed of a molecule is 4.60×10^{12} rad/s. If the oxygen molecule is rotating with this angular speed about the *z* axis, what is its rotational kinetic energy?

Solution We use Equation 10.15:

$$K_R = \tfrac{1}{2}I\omega^2$$

$$= \tfrac{1}{2}(1.95 \times 10^{-46}\ \text{kg} \cdot \text{m}^2)(4.60 \times 10^{12}\ \text{rad/s})^2$$

$$= 2.06 \times 10^{-21}\ \text{J}$$

EXAMPLE 10.3 **Four Rotating Objects**

Four small spheres are fastened to the corners of a frame of negligible mass lying in the *xy* plane (Fig. 10.7).

A If the rotation of the system occurs about the *y* axis, as in Figure 10.7a, with an angular speed ω, find the moment of inertia I_y about the *y* axis and the rotational kinetic energy about this axis.

Solution Because the spheres are small, we will model them as particles. First, note that the two spheres of mass m that lie on the *y* axis do not contribute to I_y. Because they are modeled as particles, $r_i = 0$ for these spheres about this axis. Applying Equation 10.14, we have for the two spheres on the *x* axis

$$I_y = \sum_i m_i r_i^2 = Ma^2 + Ma^2 = 2Ma^2$$

Therefore, the rotational kinetic energy about the *y* axis is

$$K_R = \frac{1}{2} I_y \omega^2 = \frac{1}{2}(2Ma^2)\omega^2 = Ma^2\omega^2$$

That the spheres of mass m do not enter into this result makes sense because they have no motion about the chosen axis of rotation; hence, they have no kinetic energy.

B Suppose the system rotates in the *xy* plane about an axis (the *z* axis) through O (Fig. 10.7b). Calculate the moment of inertia about the *z* axis and the rotational energy about this axis.

Solution Because r_i in Equation 10.14 is the perpendicular distance to the axis of rotation, we have

$$I_z = \sum_i m_i r_i^2 = Ma^2 + Ma^2 + mb^2 + mb^2 = 2Ma^2 + 2mb^2$$

$$K_R = \frac{1}{2} I_z \omega^2 = \frac{1}{2}(2Ma^2 + 2mb^2)\omega^2 = (Ma^2 + mb^2)\omega^2$$

Comparing the results for parts A and B, we see explicitly that the moment of inertia and therefore the rotational energy associated with a given angular speed depend on the axis of rotation. In part B, we expect the result to include all masses and distances because all four spheres are in motion for rotation in the *xy* plane. Furthermore, that the rotational energy in part A is smaller than in part B indicates that there is less resistance to changes in rotational motion about the *y* axis than about the *z* axis.

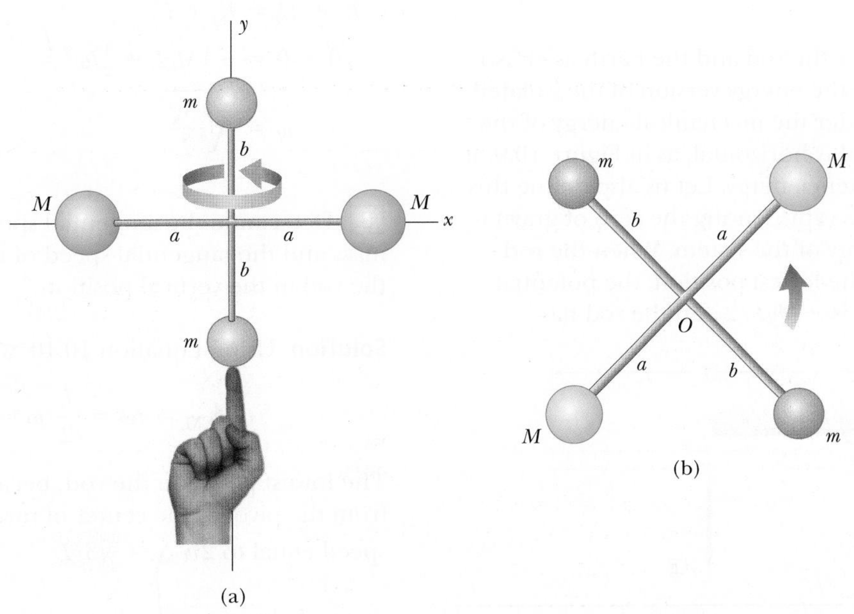

FIGURE 10.7 (Example 10.3) Four spheres form an unusual baton. (a) The baton is rotated about the *y* axis. (b) The baton is rotated about the *z* axis.

EXAMPLE 10.4 Moment of Inertia of a Uniform Solid Cylinder

A uniform solid cylinder has a radius R, mass M, and length L. Calculate its moment of inertia about its central axis (the z axis shown in Fig. 10.8).

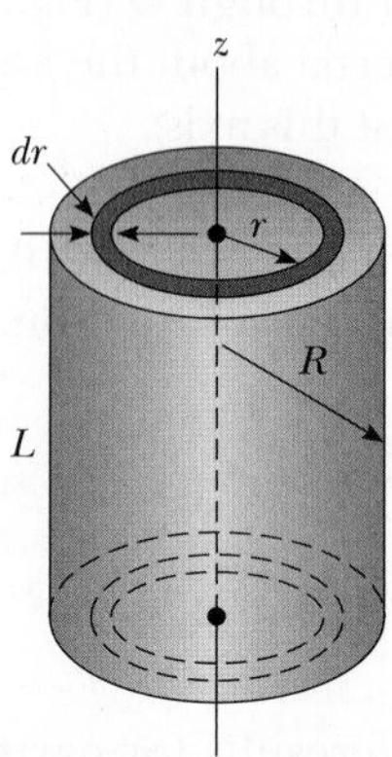

FIGURE 10.8 (Example 10.4) The geometry for calculating the moment of inertia about the central axis of a uniform solid cylinder.

Solution The integral in Equation 10.17 can be evaluated relatively simply by dividing the cylinder into many cylindrical shells of radius r, thickness dr, and length L as shown in Figure 10.8. The volume dV of a shell is its cross-sectional area multiplied by its length: $dV = (dA)L = (2\pi r\, dr)L$. Equation 10.17 gives the moment of inertia:

$$I = \int \rho r^2\, dV = \int_0^R \rho r^2 (2\pi r L)\, dr$$

$$= 2\pi\rho L \int_0^R r^3\, dr = \tfrac{1}{2}\pi\rho L R^4$$

The volume of the entire cylinder is $\pi R^2 L$, so the density is $\rho = M/V = M/\pi R^2 L$. Substituting this value of ρ in the above result gives

$$I = \tfrac{1}{2}\pi\left(\frac{M}{\pi R^2 L}\right) L R^4 = \boxed{\tfrac{1}{2}MR^2}$$

Note that this result, which appears in Table 10.2, does not depend on L. Therefore, it applies equally well to a long cylinder and a flat disk.

INTERACTIVE EXAMPLE 10.5 Rotating Rod

A uniform rod of length L and mass M is free to rotate on a frictionless pin through one end (Fig. 10.9). The rod is released from rest in the horizontal position.

A What is the angular speed of the rod at its lowest position?

Solution We consider the rod and the Earth as an isolated system and use the energy version of the isolated system model. Consider the mechanical energy of the system. When the rod is horizontal, as in Figure 10.9, it has no rotational kinetic energy. Let us also define this position of the rod as representing the zero of gravitational potential energy of the system. When the rod's center of mass is at the lowest position, the potential energy of the system is $-MgL/2$ and the rod has rotational kinetic energy $\frac{1}{2}I\omega^2$, where I is the moment of inertia about the pivot.

Because $I = \frac{1}{3}ML^2$ (see Table 10.2) for a geometric model of a long, thin rod and because mechanical energy of the isolated system is conserved, we have

$$K_i + U_i = K_f + U_f$$

$$0 + 0 = -\tfrac{1}{2}MgL + \tfrac{1}{2}I\omega^2 = -\tfrac{1}{2}MgL + \tfrac{1}{2}(\tfrac{1}{3}ML^2)\omega^2$$

$$\omega = \boxed{\sqrt{\frac{3g}{L}}}$$

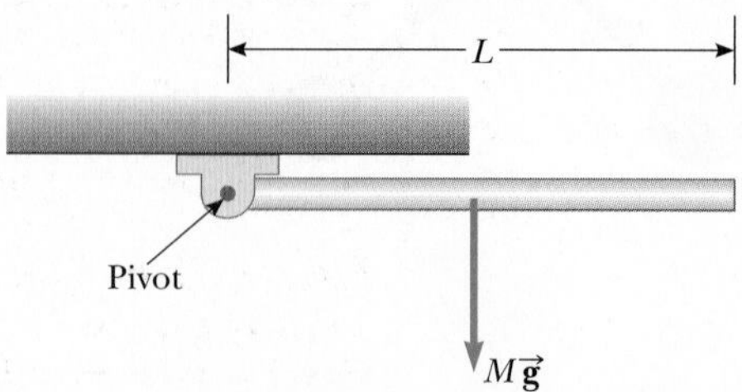

FIGURE 10.9 (Interactive Example 10.5) A uniform rod rotates freely under the influence of gravity around a pivot at the left end.

B Determine the tangential speed of the center of mass and the tangential speed of the lowest point on the rod in the vertical position.

Solution Using Equation 10.10, we have

$$v_{CM} = r\omega = \frac{L}{2}\,\omega = \boxed{\tfrac{1}{2}\sqrt{3gL}}$$

The lowest point on the rod, because it is twice as far from the pivot as the center of mass, has a tangential speed equal to $2v_{CM} = \boxed{\sqrt{3gL}}$.

PhysicsNow™ By logging into PhysicsNow at **www.pop4e.com** and going to Interactive Example 10.5 you can alter the mass and length of the rod and see the effect on the velocity at the lowest point.

10.5 TORQUE AND THE VECTOR PRODUCT

Recall our stationary exercise bicycle from the preceding section. We caused the rotational motion of the wheel by applying forces to the pedals. When a net force is exerted on a rigid object pivoted about some axis and the line of action[2] of the force does not pass through the pivot, the object tends to rotate about that axis. For example, when you push on a door, the door rotates about an axis through the hinges. The tendency of a force to rotate an object about some axis is measured by a vector quantity called **torque.** Torque is the cause of changes in rotational motion and is analogous to force, which causes changes in translational motion. Consider the wrench pivoted about the axis through O in Figure 10.10. The applied force $\vec{\mathbf{F}}$ generally can act at an angle ϕ with respect to the position vector $\vec{\mathbf{r}}$ locating the point of application of the force. We define the torque τ resulting from the force $\vec{\mathbf{F}}$ with the expression[3]

$$\tau \equiv rF\sin\phi \tag{10.18}$$

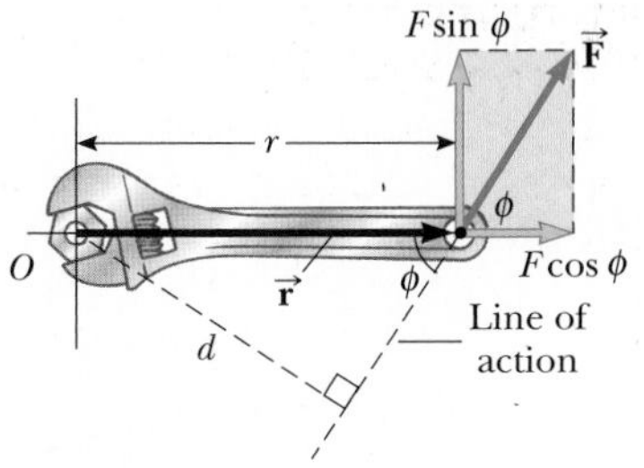

FIGURE 10.10 A force $\vec{\mathbf{F}}$ is applied to a wrench in an effort to loosen a bolt. The force has a greater rotating tendency about O as F increases and as the moment arm d increases. The component $F \sin \phi$ tends to rotate the system about O.

It is very important to recognize that **torque is defined only when a reference axis is specified,** from which the distance r is determined. We can interpret Equation 10.18 in two different ways. Looking at the force components in Figure 10.10, we see that the component $F \cos \phi$ parallel to $\vec{\mathbf{r}}$ will not cause a rotation of the wrench around the pivot point because its line of action passes right through the pivot point. Similarly, you cannot open a door by pushing on the hinges! Therefore, only the perpendicular component $F \sin \phi$ causes a rotation of the wrench about the pivot. In this case, we can write Equation 10.18 as

$$\tau = r(F\sin\phi)$$

so that the torque is the product of the distance to the point of application of the force and the perpendicular component of the force. In some problems, this method is the easiest way to interpret the calculation of the torque.

The second way to interpret Equation 10.18 is to associate the sine function with the distance r so that we can write

$$\tau = F(r\sin\phi) = Fd$$

The quantity $d = r \sin \phi$, called the **moment arm** (or *lever arm*) of the force $\vec{\mathbf{F}}$, represents the perpendicular distance from the rotation axis to the line of action of $\vec{\mathbf{F}}$. In some problems, this approach to the calculation of the torque is easier than that of resolving the force into components.

If two or more forces are acting on a rigid object, as in Active Figure 10.11, each has a tendency to produce a rotation about the pivot at O. For example, if the object is initially at rest, $\vec{\mathbf{F}}_2$ tends to rotate the object clockwise and $\vec{\mathbf{F}}_1$ tends to rotate the object counterclockwise. We shall use the convention that the sign of the torque resulting from a force is positive if its turning tendency is counterclockwise around the rotation axis and negative if its turning tendency is clockwise. For example, in Active Figure 10.11, the torque resulting from $\vec{\mathbf{F}}_1$, which has a moment arm of d_1, is *positive* and equal to $+F_1d_1$; the torque from $\vec{\mathbf{F}}_2$ is *negative* and equal to $-F_2d_2$. Hence, the *net* torque acting on the rigid object about an axis through O is

$$\tau_{\text{net}} = \tau_1 + \tau_2 = F_1d_1 - F_2d_2$$

PITFALL PREVENTION 10.5

TORQUE DEPENDS ON YOUR CHOICE Like moment of inertia, torque has no unique value. Its value depends on your choice of rotation axis.

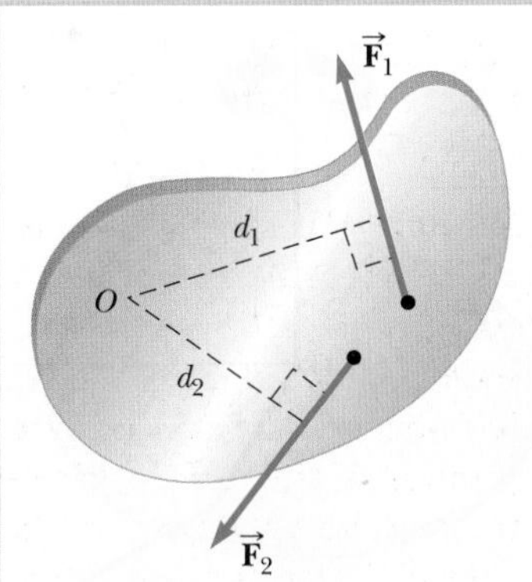

ACTIVE FIGURE 10.11 The force $\vec{\mathbf{F}}_1$ tends to rotate the object counterclockwise about an axis through O, and $\vec{\mathbf{F}}_2$ tends to rotate the object clockwise.

PhysicsNow™ Change the magnitudes, directions, and points of application of forces $\vec{\mathbf{F}}_1$ and $\vec{\mathbf{F}}_2$ to see how the object accelerates under the action of the two forces by logging into PhysicsNow at www.pop4e.com and going to Active Figure 10.11.

[2] The line of action of a force is an imaginary line colinear with the force vector and extending to infinity in both directions.

[3] In general, torque is a vector. For rotation about a fixed axis, however, we will use italic, nonbold notation and specify the direction with a positive or a negative sign as we did for angular speed and acceleration in Section 10.1. We will treat the vector nature of torque briefly in a short while.

From the definition of torque, we see that the rotating tendency increases as F increases and as d increases. For example, we cause more rotation of a door if (a) we push harder or (b) we push at the doorknob rather than at a point close to the hinges. **Torque should not be confused with force.** Torque *depends* on force, but it also depends on *where the force is applied.* Torque has units of force times length, or newton · meters (N · m) in SI units.[4]

So far, we have not discussed the vector nature of torque aside from assigning a positive or negative value to τ. Consider a force $\vec{\mathbf{F}}$ acting on a particle of a rigid object located at the vector position $\vec{\mathbf{r}}$ (Active Fig. 10.12). The *magnitude* of the torque due to this force relative to an axis through the origin is $|rF\sin\phi|$, where ϕ is the angle between $\vec{\mathbf{r}}$ and $\vec{\mathbf{F}}$. The axis about which $\vec{\mathbf{F}}$ would tend to produce rotation of the object is perpendicular to the plane formed by $\vec{\mathbf{r}}$ and $\vec{\mathbf{F}}$. If the force lies in the *xy* plane, as in Active Figure 10.12, the torque is represented by a vector parallel to the *z* axis. The force in Active Figure 10.12 creates a torque that tends to rotate the object counterclockwise when we are looking down the *z* axis. We define the direction of torque such that the vector $\vec{\boldsymbol{\tau}}$ is in the positive *z* direction (i.e., coming toward your eyes). If we reverse the direction of $\vec{\mathbf{F}}$ in Active Figure 10.12, $\vec{\boldsymbol{\tau}}$ is in the negative *z* direction. With this choice, the torque vector can be defined to be equal to the **vector product,** or **cross product,** of $\vec{\mathbf{r}}$ and $\vec{\mathbf{F}}$:

■ Definition of torque using the cross product

$$\vec{\boldsymbol{\tau}} \equiv \vec{\mathbf{r}} \times \vec{\mathbf{F}} \qquad [10.19]$$

We now give a formal definition of the vector product, first introduced in Section 1.8. Given any two vectors $\vec{\mathbf{A}}$ and $\vec{\mathbf{B}}$, the vector product $\vec{\mathbf{A}} \times \vec{\mathbf{B}}$ is defined as a third vector $\vec{\mathbf{C}}$, the *magnitude* of which is $AB\sin\theta$, where θ is the angle between $\vec{\mathbf{A}}$ and $\vec{\mathbf{B}}$:

$$\vec{\mathbf{C}} = \vec{\mathbf{A}} \times \vec{\mathbf{B}} \qquad [10.20]$$

$$C = |\vec{\mathbf{C}}| \equiv AB\sin\theta \qquad [10.21]$$

Note that the quantity $AB\sin\theta$ is equal to the area of the parallelogram formed by $\vec{\mathbf{A}}$ and $\vec{\mathbf{B}}$, as shown in Figure 10.13. The *direction* of $\vec{\mathbf{A}} \times \vec{\mathbf{B}}$ is perpendicular to the plane formed by $\vec{\mathbf{A}}$ and $\vec{\mathbf{B}}$ and is determined by the right-hand rule illustrated in Figure 10.13. The four fingers of the right hand are pointed along $\vec{\mathbf{A}}$ and then "wrapped" into $\vec{\mathbf{B}}$ through the angle θ. The direction of the upright thumb is the direction of $\vec{\mathbf{A}} \times \vec{\mathbf{B}}$. Because of the notation, $\vec{\mathbf{A}} \times \vec{\mathbf{B}}$ is often read "$\vec{\mathbf{A}}$ cross $\vec{\mathbf{B}}$," hence the term *cross product.*

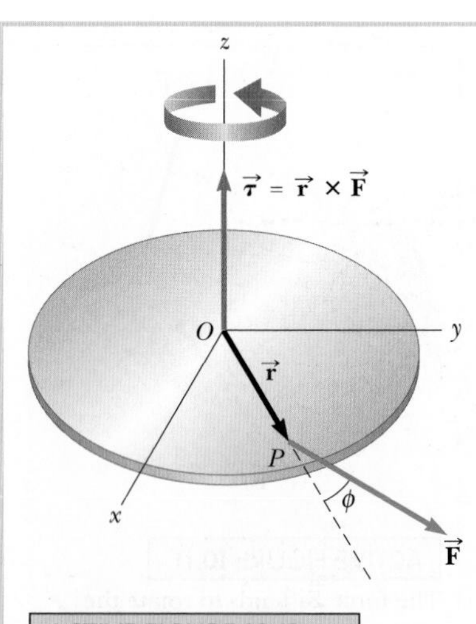

ACTIVE FIGURE 10.12
The torque vector $\vec{\boldsymbol{\tau}}$ lies in a direction perpendicular to the plane formed by the position vector $\vec{\mathbf{r}}$ and the applied force vector $\vec{\mathbf{F}}$.

Physics Now™ By logging into PhysicsNow at **www.pop4e.com** and going to Active Figure 10.12 you can move point P and change the force vector $\vec{\mathbf{F}}$ to see the effect on the torque vector.

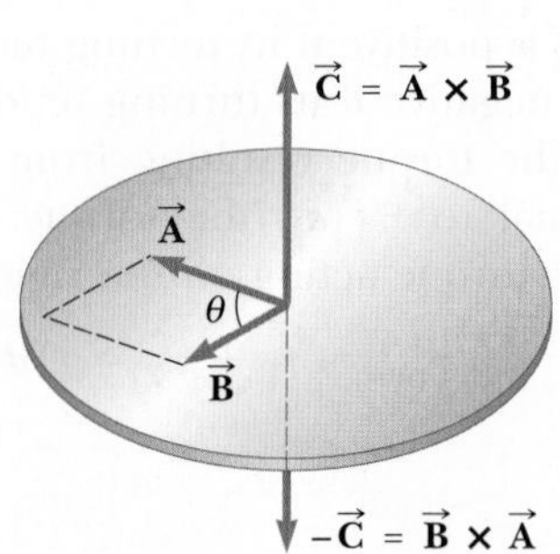

FIGURE 10.13 The vector product $\vec{\mathbf{A}} \times \vec{\mathbf{B}}$ is a third vector $\vec{\mathbf{C}}$ having a magnitude $AB\sin\theta$ equal to the area of the parallelogram shown. The vector $\vec{\mathbf{C}}$ is perpendicular to the plane formed by $\vec{\mathbf{A}}$ and $\vec{\mathbf{B}}$, and its direction is determined by the right-hand rule.

[4]In Chapter 6, we saw the product of newtons and meters when we defined work and called this product a *joule.* We do not use this term here because the joule is only to be used when discussing energy. For torque, the unit is simply the newton · meter, or N · m.

Some properties of the vector product follow from its definition:

- Unlike the case of the scalar product, the vector product is not commutative; in fact,

$$\vec{\mathbf{A}} \times \vec{\mathbf{B}} = -\vec{\mathbf{B}} \times \vec{\mathbf{A}} \qquad [10.22]$$

  Therefore, if you change the order of the vector product, you must change the sign. One can easily verify this relation with the right-hand rule (see Fig. 10.13).
- If $\vec{\mathbf{A}}$ is parallel to $\vec{\mathbf{B}}$ ($\theta = 0°$ or $180°$), then $\vec{\mathbf{A}} \times \vec{\mathbf{B}} = 0$; therefore, it follows that $\vec{\mathbf{A}} \times \vec{\mathbf{A}} = 0$.
- If $\vec{\mathbf{A}}$ is perpendicular to $\vec{\mathbf{B}}$, then $|\vec{\mathbf{A}} \times \vec{\mathbf{B}}| = AB$. It is left to Problem 10.25 to show, from Equations 10.20 and 10.21 and the definition of unit vectors, that the vector products of the unit vectors $\hat{\mathbf{i}}$, $\hat{\mathbf{j}}$, and $\hat{\mathbf{k}}$ obey the following expressions:

$$\begin{aligned} \hat{\mathbf{i}} \times \hat{\mathbf{i}} &= \hat{\mathbf{j}} \times \hat{\mathbf{j}} = \hat{\mathbf{k}} \times \hat{\mathbf{k}} = 0 \\ \hat{\mathbf{i}} \times \hat{\mathbf{j}} &= -\hat{\mathbf{j}} \times \hat{\mathbf{i}} = \hat{\mathbf{k}} \\ \hat{\mathbf{j}} \times \hat{\mathbf{k}} &= -\hat{\mathbf{k}} \times \hat{\mathbf{j}} = \hat{\mathbf{i}} \\ \hat{\mathbf{k}} \times \hat{\mathbf{i}} &= -\hat{\mathbf{i}} \times \hat{\mathbf{k}} = \hat{\mathbf{j}} \end{aligned} \qquad [10.23]$$

Signs are interchangeable. For example, $\hat{\mathbf{i}} \times (-\hat{\mathbf{j}}) = -\hat{\mathbf{i}} \times \hat{\mathbf{j}} = -\hat{\mathbf{k}}$.

QUICK QUIZ 10.5 If you are trying to loosen a stubborn screw from a piece of wood with a screwdriver and fail, should you find a screwdriver for which the handle is **(a)** longer or **(b)** fatter? If you are trying to loosen a stubborn bolt from a piece of metal with a wrench and fail, should you find a wrench for which the handle is **(c)** longer or **(d)** fatter?

EXAMPLE 10.6 The Net Torque on a Cylinder

A one-piece cylinder is shaped as in Figure 10.14, with a core section protruding from the larger drum. The cylinder is free to rotate around the central axis shown in the drawing. A rope wrapped around the drum, of radius R_1, exerts a force $\vec{\mathbf{T}}_1$ to the right on the cylinder. A rope wrapped around the core, of radius R_2, exerts a force $\vec{\mathbf{T}}_2$ downward on the cylinder.

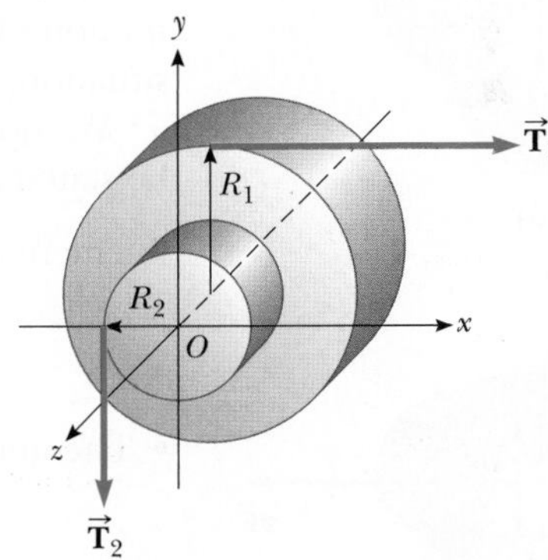

FIGURE 10.14 (Example 10.6) A solid cylinder pivoted about the z axis through O. The moment arm of $\vec{\mathbf{T}}_1$ is R_1, and the moment arm of $\vec{\mathbf{T}}_2$ is R_2.

A What is the net torque acting on the cylinder about the rotation axis (which is the z axis in Fig. 10.14)?

Solution The torque due to $\vec{\mathbf{T}}_1$ is $-R_1T_1$. It is negative because it tends to produce a clockwise rotation from the point of view in Figure 10.14. The torque due to $\vec{\mathbf{T}}_2$ is $+R_2T_2$ and is positive because it tends to produce a counterclockwise rotation. Therefore, the net torque about the rotation axis is

$$\tau_{\text{net}} = \tau_1 + \tau_2 = R_2T_2 - R_1T_1$$

B Suppose $T_1 = 5.0$ N, $R_1 = 1.0$ m, $T_2 = 6.0$ N, and $R_2 = 0.50$ m. What is the net torque about the rotation axis and which way does the cylinder rotate if it starts from rest?

Solution We substitute numerical values in the result from part A:

$$\tau_{\text{net}} = (6.0\ \text{N})(0.50\ \text{m}) - (5.0\ \text{N})(1.0\ \text{m}) = -2.0\ \text{N}\cdot\text{m}$$

Because the net torque is negative, the cylinder rotates clockwise from rest.

EXAMPLE 10.7 The Vector Product

Two vectors lying in the *xy* plane are given by the equations $\vec{\mathbf{A}} = 2\hat{\mathbf{i}} + 3\hat{\mathbf{j}}$ and $\vec{\mathbf{B}} = -\hat{\mathbf{i}} + 2\hat{\mathbf{j}}$. Find $\vec{\mathbf{A}} \times \vec{\mathbf{B}}$ and verify explicitly that $\vec{\mathbf{A}} \times \vec{\mathbf{B}} = -\vec{\mathbf{B}} \times \vec{\mathbf{A}}$.

Solution Using Equation 10.23 for the vector product of unit vectors gives

$$\vec{\mathbf{A}} \times \vec{\mathbf{B}} = (2\hat{\mathbf{i}} + 3\hat{\mathbf{j}}) \times (-\hat{\mathbf{i}} + 2\hat{\mathbf{j}})$$

$$= 2\hat{\mathbf{i}} \times 2\hat{\mathbf{j}} + 3\hat{\mathbf{j}} \times (-\hat{\mathbf{i}}) = 4\hat{\mathbf{k}} + 3\hat{\mathbf{k}} = 7\hat{\mathbf{k}}$$

(We have omitted the terms containing $\hat{\mathbf{i}} \times \hat{\mathbf{i}}$ and $\hat{\mathbf{j}} \times \hat{\mathbf{j}}$ because, as Equation 10.23 shows, they are equal to zero.)

We can show that $\vec{\mathbf{A}} \times \vec{\mathbf{B}} = -\vec{\mathbf{B}} \times \vec{\mathbf{A}}$, because

$$\vec{\mathbf{B}} \times \vec{\mathbf{A}} = (-\hat{\mathbf{i}} + 2\hat{\mathbf{j}}) \times (2\hat{\mathbf{i}} + 3\hat{\mathbf{j}})$$

$$= -\hat{\mathbf{i}} \times 3\hat{\mathbf{j}} + 2\hat{\mathbf{j}} \times 2\hat{\mathbf{i}} = -3\hat{\mathbf{k}} - 4\hat{\mathbf{k}} = -7\hat{\mathbf{k}}$$

Therefore, $\vec{\mathbf{A}} \times \vec{\mathbf{B}} = -\vec{\mathbf{B}} \times \vec{\mathbf{A}}$.

10.6 THE RIGID OBJECT IN EQUILIBRIUM

We have defined a rigid object and have discussed torque as the cause of changes in rotational motion of a rigid object. We can now establish models for a rigid object subject to torques that are analogous to those for a particle subject to forces. We begin by imagining a rigid object with balanced torques, which will give us an analysis model that we call the rigid object in equilibrium.

Consider two forces of equal magnitude and opposite direction applied to an object as shown in Figure 10.15a. The force directed to the right tends to rotate the object clockwise about an axis perpendicular to the diagram through *O*, whereas the force directed to the left tends to rotate it counterclockwise about that axis. Because the forces are of equal magnitude and act at the same perpendicular distance from *O*, their torques are equal in magnitude. Therefore, the net torque on the rigid object is zero. The situation shown in Figure 10.15b is another case in which the net torque about *O* is zero (although the net *force* on the object is not zero), and we can devise many more cases.

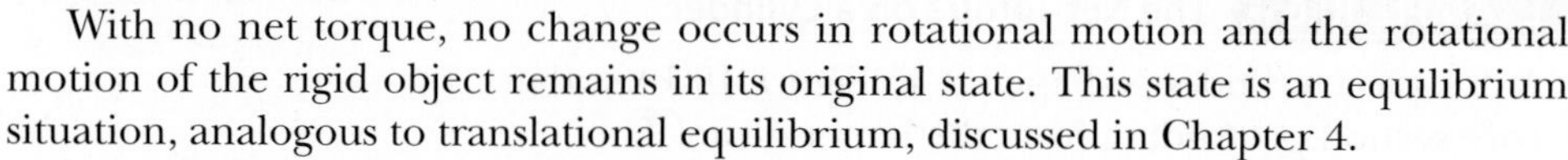

With no net torque, no change occurs in rotational motion and the rotational motion of the rigid object remains in its original state. This state is an equilibrium situation, analogous to translational equilibrium, discussed in Chapter 4.

We now have **two conditions for complete equilibrium of an object,** which can be stated as follows:

- The net external force must equal zero:

$$\sum \vec{\mathbf{F}} = 0 \qquad [10.24]$$

- The net external torque must be zero about *any* axis:

$$\sum \vec{\boldsymbol{\tau}} = 0 \qquad [10.25]$$

The first condition is a statement of translational equilibrium. The second condition is a statement of rotational equilibrium. In the special case of **static equilibrium,** the object is at rest, so it has no translational or angular speed (i.e., $v_{\text{CM}} = 0$ and $\omega = 0$).

The two vector expressions given by Equations 10.24 and 10.25 are equivalent, in general, to six scalar equations: three from the first condition of equilibrium and three from the second (corresponding to *x*, *y*, and *z* components). Hence, in a complex system involving several forces acting in various directions, you would be faced with solving a set of equations with many unknowns. Here, we restrict our discussion to situations in which all the forces on an object lie in the *xy* plane. (Forces whose vector representations are in the same plane are said to be *coplanar.*) With this restriction, we need to deal with only three scalar equations. Two of them

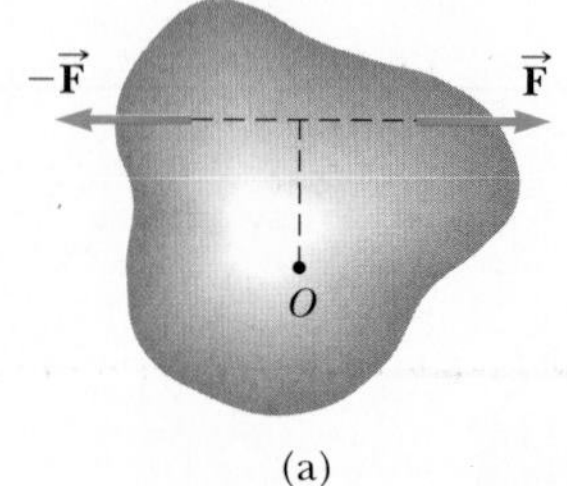

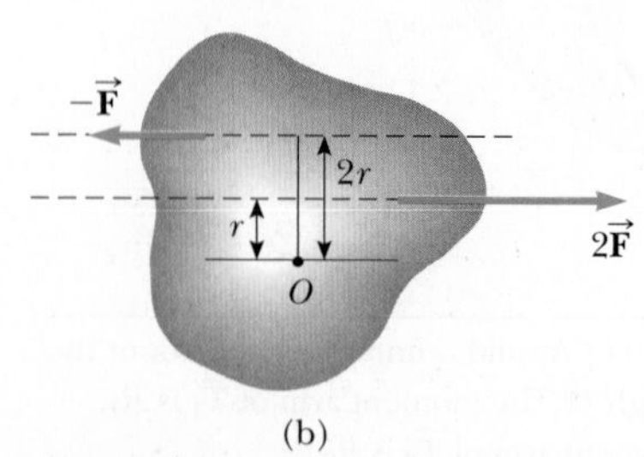

FIGURE 10.15 (a) The two forces acting on the object are equal in magnitude and opposite in direction. Because they also act along the same line of action, the net torque is zero and the object is in equilibrium. (b) Another situation in which two forces act on an object to produce zero net torque about *O* (but *not* zero net force).

come from balancing the forces on the object in the x and y directions. The third comes from the torque equation, namely, that the net torque on the object about an axis through *any* point in the xy plane must be zero. Hence, the two conditions of equilibrium provide the equations

$$\sum F_x = 0 \qquad \sum F_y = 0 \qquad \sum \tau_z = 0 \qquad [10.26]$$

where the axis of the torque equation is arbitrary.

In working static equilibrium problems, it is important to recognize all external forces acting on the object. Failure to do so will result in an incorrect analysis. The following procedure is recommended when analyzing an object in equilibrium under the action of several external forces:

PROBLEM-SOLVING STRATEGY Rigid Object in Equilibrium

1. **Conceptualize** Think about the object that is in equilibrium and identify the forces on it. Imagine what effect each force would have on the rotation of the object if it were the only force acting.
2. **Categorize** Confirm that the object under consideration is indeed a rigid object in equilibrium.
3. **Analyze** Draw a free-body diagram and label all external forces acting on the object. Try to guess the correct direction for each force.

   Resolve all forces into rectangular components, choosing a convenient coordinate system. Then apply the first condition for equilibrium, Equation 10.24. Remember to keep track of the signs of the various force components.

   Choose a convenient axis for calculating the net torque on the rigid object. Remember that the choice of the axis for the torque equation is arbitrary; therefore, choose an axis that will simplify your calculation as much as possible. Usually, the most convenient axis for calculating torques is one through a point at which several forces act, so their torques around this axis are zero. If you don't know a force or don't need to know a force, it is often beneficial to choose an axis through the point at which this force acts. Apply the second condition for equilibrium, Equation 10.25.

   Solve the simultaneous equations for the unknowns in terms of the known quantities.
4. **Finalize** Make sure your results are consistent with the free-body diagram. If you selected a direction that leads to a negative sign in your solution for a force, do not be alarmed; it merely means that the direction of the force is the opposite of what you guessed. Add up the vertical and horizontal forces on the object and confirm that each set of components adds to zero. Add up the torques on the object and confirm that the sum equals zero.

INTERACTIVE EXAMPLE 10.8 Standing on a Horizontal Beam

A uniform horizontal beam of length 8.00 m and weight 200 N is attached to a wall by a pin connection. Its far end is supported by a cable that makes an angle of 53.0° with the horizontal (Fig. 10.16a). If a 600-N man stands 2.00 m from the wall, find the tension in the cable and the force exerted by the wall on the beam at the pivot.

Solution The beam–man system is at rest and remains at rest, so it is clearly in static equilibrium. First, we must identify all the external forces acting on the system, which we do in the free-body diagram in Figure 10.16b. These forces are the gravitational forces on the beam and the man, the force $\vec{\mathbf{T}}$ exerted by the cable, and the force $\vec{\mathbf{R}}$ exerted by the wall at the pivot (the direction of this force is unknown). (The force between the man and the beam is internal to the system, so it is not included in the free-body diagram.) Notice that we have imagined the gravitational force on the beam as acting at its center of gravity. Because the beam is uniform, the center of gravity is at the geometric center. If we resolve $\vec{\mathbf{T}}$ and $\vec{\mathbf{R}}$ into horizontal and vertical components (Fig. 10.16c) and apply the first condition for equilibrium for the beam, we have

$$(1) \quad \sum F_x = R\cos\theta - T\cos 53.0° = 0$$

$$(2) \quad \sum F_y = R\sin\theta + T\sin 53.0° - 600\text{ N} - 200\text{ N} = 0$$

Because we have three unknowns—R, T, and θ—we cannot obtain a solution from these two expressions alone.

To generate a third expression, let us invoke the condition for rotational equilibrium because the beam can be modeled as a rigid object in equilibrium. A convenient axis to choose for our torque equation is the one that passes through the pivot at the wall. The feature that makes this point so convenient is that the

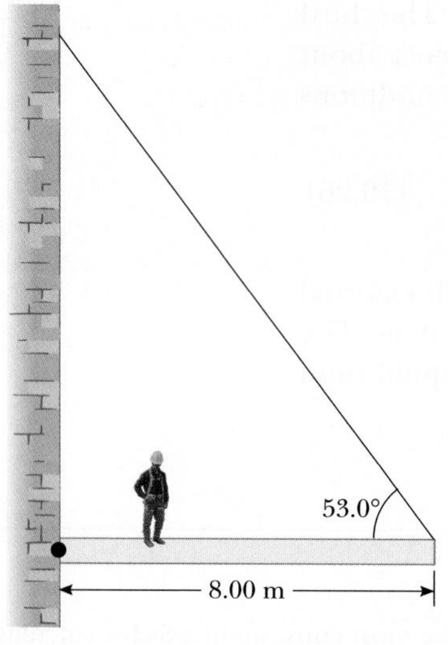

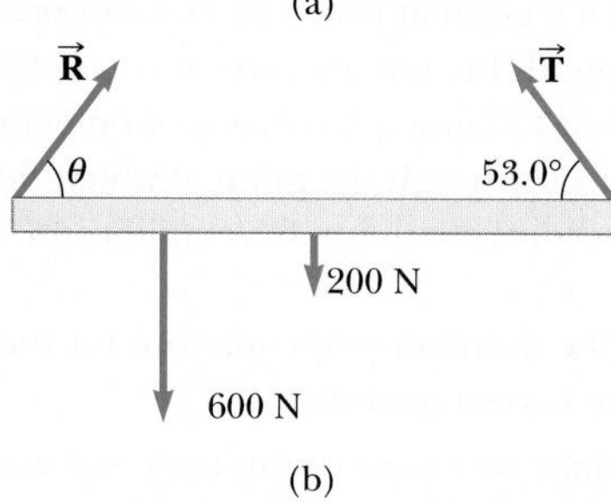

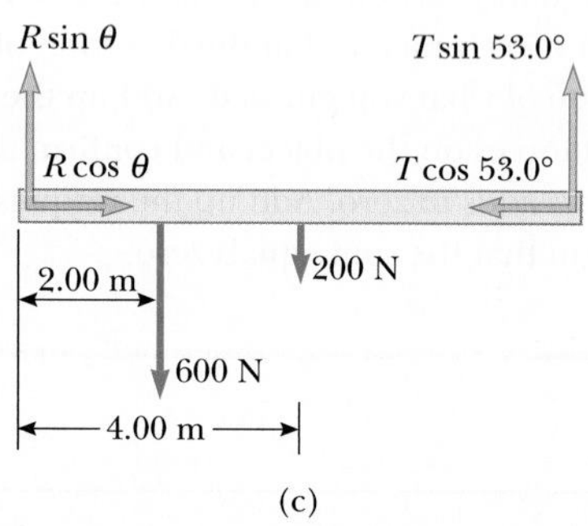

FIGURE 10.16 (Interactive Example 10.8) (a) A uniform beam supported by a cable. A man walks out on the beam. (b) The free-body diagram for the beam–man system. (c) The free-body diagram with forces resolved into horizontal and vertical components.

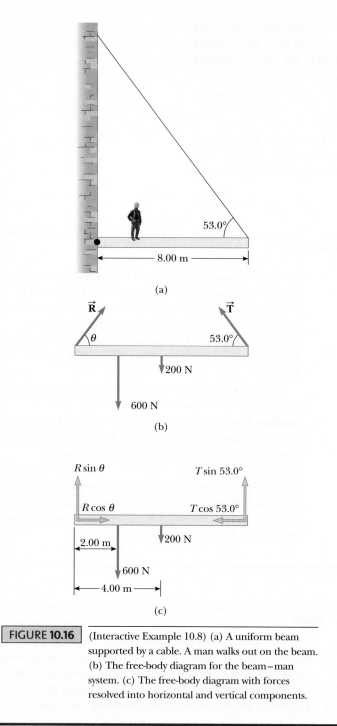

force $\vec{\mathbf{R}}$ and the horizontal component of $\vec{\mathbf{T}}$ both have a lever arm of zero, and hence zero torque, about this pivot. Recalling our convention for the sign of the torque about an axis and noting that the lever arms of the 600-N, 200-N, and $T \sin 53°$ forces are 2.00 m, 4.00 m, and 8.00 m, respectively, we have

$$\sum \tau = (T \sin 53.0°)(8.00 \text{ m}) - (600 \text{ N})(2.00 \text{ m}) - (200 \text{ N})(4.00 \text{ m}) = 0$$

$$T = \boxed{313 \text{ N}}$$

The torque equation gives us one of the unknowns directly, thanks to our judicious choice of the axis! This value is substituted into (1) and (2) to give

$$R \cos \theta = 188 \text{ N}$$

$$R \sin \theta = 550 \text{ N}$$

We divide these two equations to find

$$\tan \theta = \frac{550 \text{ N}}{188 \text{ N}} = 2.93$$

$$\theta = 71.1°$$

Finally,

$$R = \frac{188 \text{ N}}{\cos \theta} = \frac{188 \text{ N}}{\cos 71.1°} = \boxed{581 \text{ N}}$$

If we had selected some other axis for the torque equation, the results would have been the same although the details of the solution would be somewhat different. For example, if we had chosen to have the axis pass through the center of gravity of the beam, the torque equation would involve both T and R. This equation, coupled with (1) and (2), however could still be solved for the unknowns T, R, and θ, yielding the same results. Try it!

Physics Now™ Adjust the position of the person and observe the effect on the forces by logging into PhysicsNow at **www.pop4e.com** and going to Interactive Example 10.8.

INTERACTIVE EXAMPLE 10.9 The Leaning Ladder

A uniform ladder of length ℓ and mass m rests against a smooth, vertical wall (Fig. 10.17a). If the coefficient of static friction between ladder and ground is $\mu_s = 0.40$, find the minimum angle $\theta_{\min}$ such that the ladder does not slip.

Solution The ladder is at rest and remains at rest, so we model it as a rigid object in equilibrium. The free-body diagram showing all the external forces acting on the ladder is illustrated in Figure 10.17b. The force exerted by the ground on the ladder is the vector sum of a normal force $\vec{\mathbf{n}}$ and the force of static friction $\vec{\mathbf{f}}_s$. The reaction force $\vec{\mathbf{P}}$ exerted by the wall on the ladder is horizontal because the wall is smooth, meaning that it is frictionless. Therefore, $\vec{\mathbf{P}}$ is simply the normal force on the ladder from the wall.

From the first condition of equilibrium applied to the ladder, we have

$$\sum F_x = f_s - P = 0$$
$$\sum F_y = n - mg = 0$$

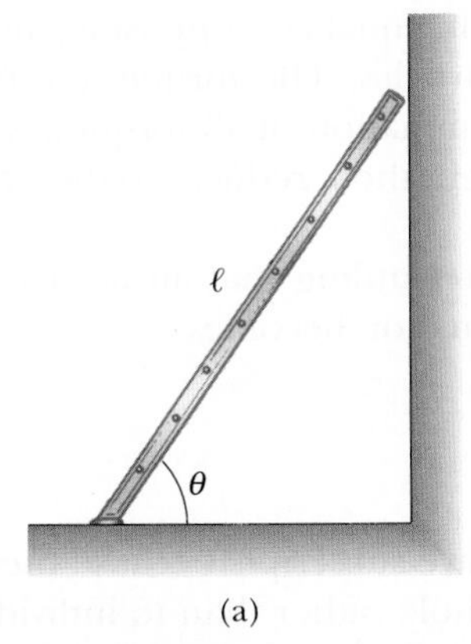

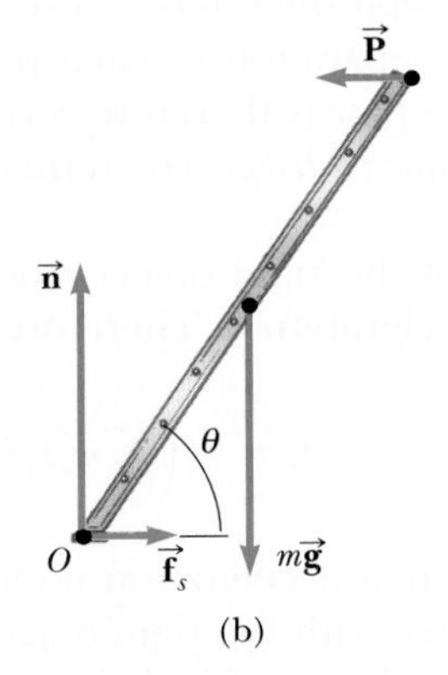

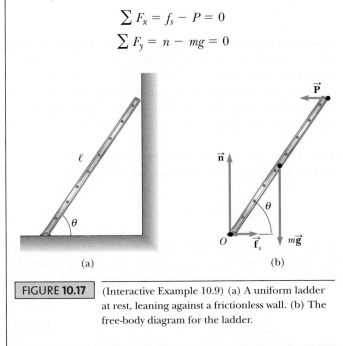

FIGURE 10.17 (Interactive Example 10.9) (a) A uniform ladder at rest, leaning against a frictionless wall. (b) The free-body diagram for the ladder.

We see from the second equation that $n = mg$. Furthermore, when the ladder is on the verge of slipping, the force of static friction must be a maximum, given by $f_{s,\,\max} = \mu_s n$.

To find θ, we use the second condition of equilibrium. When the torques are taken about the origin O at the bottom of the ladder, we have

$$\sum \tau_O = P\ell \sin\theta - mg\frac{\ell}{2}\cos\theta = 0$$

This expression gives

$$\tan\theta_{\min} = \frac{mg}{2P} = \frac{n}{2f_{s,\,\max}} = \frac{n}{2(\mu_s n)} = \frac{1}{2(0.40)} = 1.25$$

$$\theta_{\min} = 51°$$

It is interesting that the result does not depend on ℓ or m. The answer depends only on μ_s.

PhysicsNow™ Adjust the angle of the ladder and watch what happens when it is released by logging into PhysicsNow at www.pop4e.com and going to Interactive Example 10.9.

10.7 THE RIGID OBJECT UNDER A NET TORQUE

In the preceding section, we investigated the equilibrium situation in which the net torque on a rigid object is zero. What if the net torque on a rigid object is not zero? In analogy with Newton's second law for translational motion, we should expect the angular speed of the rigid object to change. The net torque will cause angular acceleration of the rigid object. We describe this situation with a new analysis model, the rigid object under a net torque, and investigate this model in this section.

Let us imagine a rotating rigid object again as a collection of particles. The rigid object will be subject to a number of forces applied at various locations on the rigid object at which individual particles will be located. Therefore, we can imagine that the forces on the rigid object are exerted on individual particles of the rigid object. We will calculate the net torque on the object due to the torques resulting from these forces around the rotation axis of the rotating object. Any applied force can be represented by its radial component and its tangential component. The radial component of an applied force provides no torque because its line of action goes through the rotation axis. Therefore, only the tangential component of an applied force contributes to the torque.

On any given particle, described by index variable i, within the rigid object, we can use Newton's second law to describe the tangential acceleration of the particle:

$$F_{ti} = m_i a_{ti}$$

where the t subscript refers to tangential components. Let us multiply both sides of this expression by r_i, the distance of the particle from the rotation axis:

$$r_i F_{ti} = r_i m_i a_{ti}$$

Using Equation 10.11 and recognizing the definition of torque ($\tau = rF\sin\phi = rF_t$ in this case), we can rewrite this expression as

$$\tau_i = m_i r_i^2 \alpha_i$$

Now, let us add up the torques on all particles of the rigid object:

$$\sum_i \tau_i = \sum_i m_i r_i^2 \alpha_i$$

The left side is the net torque on all particles of the rigid object. The net torque associated with *internal* forces is zero, however. To understand why, recall that Newton's third law tells us that the internal forces occur in equal and opposite pairs that lie along the line of separation of each pair of particles. The torque due to each action–reaction force pair is therefore zero. On summation of all torques, we see that the *net internal torque vanishes.* The term on the left, then, reduces to the net *external* torque.

On the right, we adopt the rigid object model by demanding that all particles have the same angular acceleration α. Therefore, this equation becomes

$$\sum \tau = \left(\sum_i m_i r_i^2\right) \alpha$$

where the torque and angular acceleration no longer have subscripts because they refer to quantities associated with the rigid object as a whole rather than to individual particles. We recognize the quantity in parentheses as the moment of inertia I. Therefore,

Rotational analog to Newton's second law

$$\sum \tau = I\alpha \qquad [10.27]$$

That is, **the net torque acting on the rigid object is proportional to its angular acceleration,** and the proportionality constant is the moment of inertia. It is important to note that $\sum \tau = I\alpha$ is the rotational analog of Newton's second law of motion, $\sum F = ma$.

QUICK QUIZ 10.6 You turn off your electric drill and find that the time interval for the rotating bit to come to rest due to frictional torque in the drill is Δt. You replace the drill bit with a larger one that results in a doubling of the moment of inertia of the drill's entire rotating mechanism. When this larger bit is rotated at the same angular speed as the first and turned off, the frictional torque remains the same as that for the previous situation. What is the time interval for this second bit to come to rest? **(a)** $4\Delta t$ **(b)** $2\Delta t$ **(c)** Δt **(d)** $0.5\Delta t$ **(e)** $0.25\Delta t$ **(f)** impossible to determine

EXAMPLE 10.10 An Atwood Machine with a Massive Pulley

In Example 4.4, we analyzed an Atwood machine in which two objects with unequal masses hang from a string that passes over a light, frictionless pulley. Suppose the pulley, which is modeled as a disk, has mass M and radius R, and suppose the pulley surface is not frictionless so that the string does not slide on the pulley (Fig. 10.18a). We will assume that the frictional torque acting at the bearing of the pulley is negligible. Calculate the magnitude of the acceleration of the two objects.

Solution Conceptualize the problem by imagining the motion of the two objects in Figure 10.18, as we did in Example 4.4. The difference here is that the pulley is not considered to be massless. Because this problem involves a massive object in rotation as well as two other objects in translational motion, we categorize it as one involving the rigid object under a net torque model (for the pulley) and the particle under a net force model (for the hanging objects).

To analyze the problem, we first set up a coordinate system. Because counterclockwise angular acceleration of the pulley is defined as positive, we define the positive directions for m_1 and m_2 as shown in Figure 10.18 so that all accelerations, translational and rotational, are positive if m_1 accelerates downward. If the pulley has mass and friction, the tensions T_1 and T_2 in the string on either side of the pulley are not equal in magnitude as they are in Example 4.4. Indeed, it is the

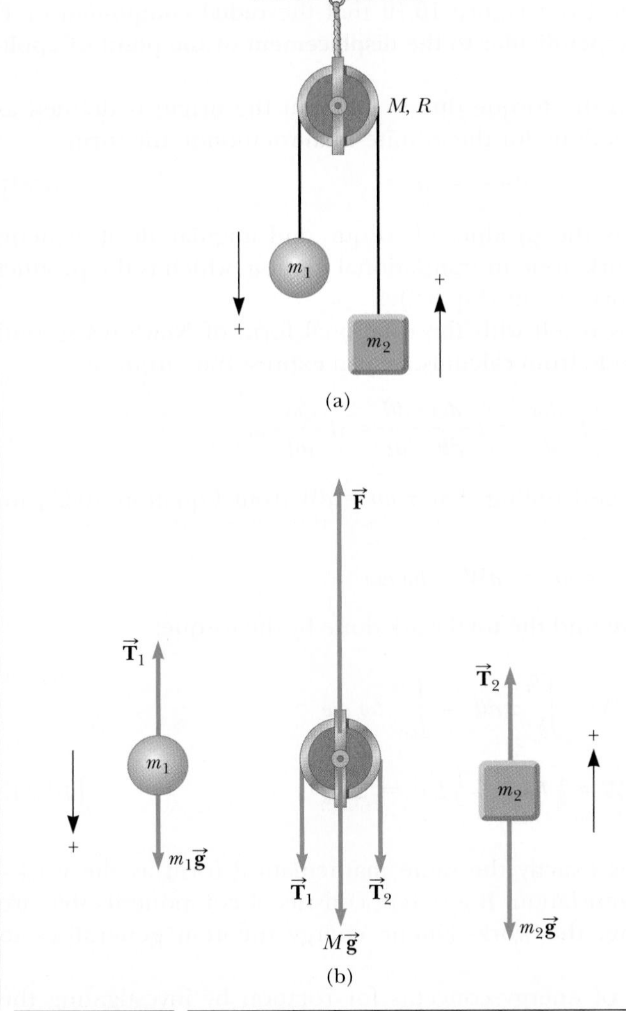

FIGURE 10.18 (Example 10.10) (a) An Atwood machine with a massive pulley. The pulley is modeled as a disk. (b) Free-body diagrams for the two hanging objects and the pulley.

difference in torque due to these different tensions that provides the net torque to cause the angular acceleration of the pulley (Fig. 10.18b). Consequently, some of what we do here will look similar to Example 4.4, with the exception of incorporating T_1 and T_2 in our mathematical representation instead of just a single T. The forces $M\vec{g}$ and $\vec{F}$ (the force supporting the pulley) both act through the pulley axle, so these forces do not contribute to the torque on the pulley.

With the help of the free-body diagrams in Figure 10.18b, we apply Newton's second law to m_1 so that

$$(1)\quad \sum F_y = m_1 g - T_1 = m_1 a$$

and for m_2,

$$(2)\quad \sum F_y = T_2 - m_2 g = m_2 a$$

We cannot solve these two equations for a, as is done in Example 4.4, because we have three unknowns: a, T_1, and T_2. We can find a third equation by applying Equation 10.27 to the pulley (Fig. 10.18b):

$$\sum \tau = T_1 R - T_2 R = I\alpha = (\tfrac{1}{2}MR^2)\left(\frac{a}{R}\right) = \tfrac{1}{2}MRa$$

$$(3)\quad T_1 - T_2 = \tfrac{1}{2}Ma$$

We substitute into (3) expressions for T_1 and T_2 from (1) and (2):

$$(m_1 g - m_1 a) - (m_2 a + m_2 g) = \tfrac{1}{2}Ma$$

$$a = \left(\frac{m_1 - m_2}{m_1 + m_2 + \frac{1}{2}M}\right) g$$

To finalize this problem, notice that this result differs from the result for Example 4.4 only in the extra term $\frac{1}{2}M$ in the denominator. If the pulley mass $M \rightarrow 0$, this expression reduces to that in Example 4.4.

Work and Energy in Rotational Motion

In translational motion, we found energy concepts, and in particular the reduction of the continuity equation for energy called the work–kinetic energy theorem, to be extremely useful in describing the motion of a system. Energy concepts can be equally useful in simplifying the analysis of rotational motion, as we saw in the isolated system analysis in Interactive Example 10.5. From the continuity equation for energy, we expect that for rotation of an object about a fixed axis, the work done by external forces on the object will equal the change in the rotational kinetic energy as long as energy is not stored by any other means. To show that this case is in fact true, we begin by finding an expression for the work done by a torque.

Consider a rigid object pivoted at the point O in Figure 10.19. Suppose a single external force $\vec{F}$ is applied at the point P and $d\vec{s}$ is the displacement of the point of application of the force. The small amount of work dW done on the object by $\vec{F}$ as the point of application rotates through an infinitesimal distance $ds = r\,d\theta$ in a time interval dt is

$$dW = \vec{F} \cdot d\vec{s} = (F \sin\phi) r\, d\theta$$

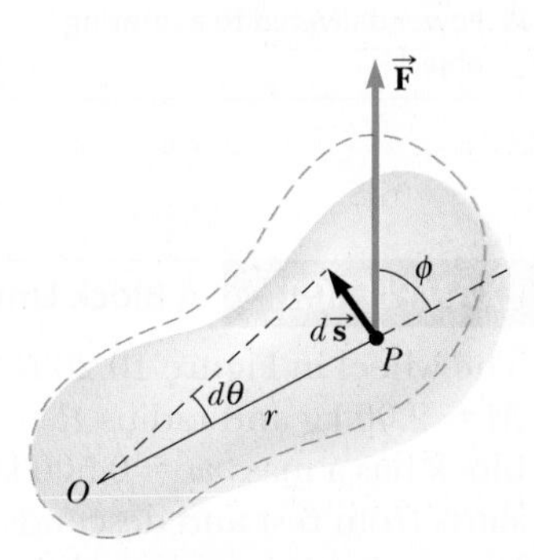

FIGURE 10.19 A rigid object rotates about an axis through O under the action of an external force $\vec{F}$ applied at P.

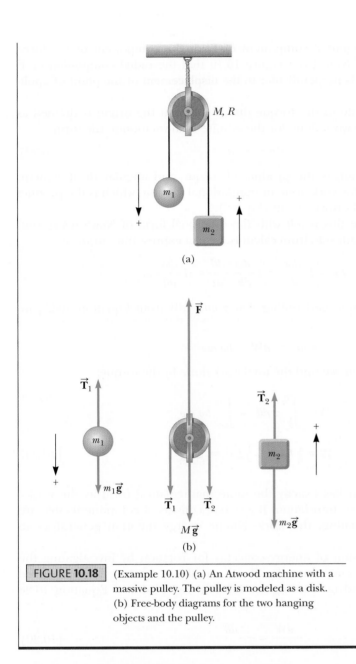

where $F \sin \phi$ is the tangential component of $\vec{\mathbf{F}}$, or the component of the force along the displacement. Note from Figure 10.19 that the **radial component of $\vec{\mathbf{F}}$ does no work because it is perpendicular to the displacement of the point of application of the force.**

Because the magnitude of the torque due to $\vec{\mathbf{F}}$ about the origin is defined as $rF \sin \phi$, we can write the work done for the infinitesimal rotation in the form

$$dW = \tau \, d\theta \qquad [10.28]$$

Notice that this expression is the product of torque and angular displacement, making it analogous to the work done in translational motion, which is the product of force and translational displacement (Eq. 6.2).

Now, we will combine this result with the rotational form of Newton's second law, $\tau = I\alpha$. Using the chain rule from calculus, we can express the torque as

$$\tau = I\alpha = I\frac{d\omega}{dt} = I\frac{d\omega}{d\theta}\frac{d\theta}{dt} = I\frac{d\omega}{d\theta}\,\omega$$

Rearranging this expression and noting that $\tau \, d\theta = dW$ from Equation 10.28, we have

$$\tau \, d\theta = dW = I\omega \, d\omega$$

Integrating this expression, we find the total work done by the torque:

$$W = \int_{\theta_i}^{\theta_f} \tau \, d\theta = \int_{\omega_i}^{\omega_f} I\omega \, d\omega$$

■ Work–kinetic energy theorem for pure rotation

$$W = \tfrac{1}{2} I\omega_f^{\,2} - \tfrac{1}{2} I\omega_i^{\,2} = \Delta K_R \qquad [10.29]$$

Notice that this equation has exactly the same mathematical form as the work–kinetic energy theorem for translation. If a system consists of components that are both translating and rotating, the work–kinetic energy theorem generalizes to $W = \Delta K + \Delta K_R$.

We finish this discussion of energy concepts for rotation by investigating the *rate* at which work is being done by $\vec{\mathbf{F}}$ on an object rotating about a fixed axis. This rate is obtained by dividing the left and right sides of Equation 10.28 by dt:

$$\frac{dW}{dt} = \tau \frac{d\theta}{dt} \qquad [10.30]$$

The quantity dW/dt is, by definition, the instantaneous power $\mathcal{P}$ delivered by the force. Furthermore, because $d\theta/dt = \omega$, Equation 10.30 reduces to

■ Power delivered to a rotating object

$$\mathcal{P} = \tau\omega \qquad [10.31]$$

This expression is analogous to $\mathcal{P} = Fv$ in the case of translational motion.

EXAMPLE 10.11 A Block Unwinding from a Wheel

The wheel in Figure 10.20 is a solid disk of mass $M = 2.00$ kg and radius $R = 30.0$ cm. The suspended block has a mass $m = 0.500$ kg. If the suspended block starts from rest and descends to a position 1.00 m lower, what is its speed when it is at this position?

Solution The work done on the system of the block and the wheel is due to the gravitational force $m\vec{\mathbf{g}}$ acting on the hanging block. From the continuity equation for energy, the work must be equal to the change in kinetic energy of the system:

$$W = \Delta K + \Delta K_R$$

where K is the translational kinetic energy of the block and K_R is the rotational kinetic energy of the wheel.

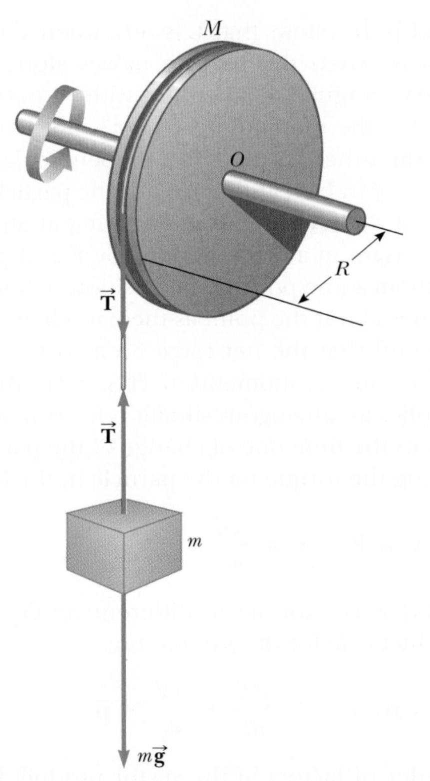

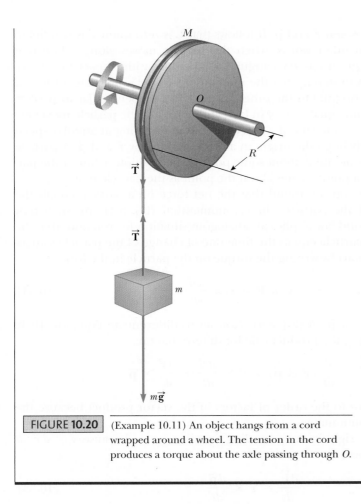

FIGURE 10.20 (Example 10.11) An object hangs from a cord wrapped around a wheel. The tension in the cord produces a torque about the axle passing through O.

The system begins from rest, so we can write this expression as

$$W = \vec{F} \cdot \Delta\vec{r} = (-mg\hat{j}) \cdot (-\Delta y\hat{j}) = mg(\Delta y) = \tfrac{1}{2}mv^2 + \tfrac{1}{2}I\omega^2$$

where v is the speed of the block at its final position. It is also the speed of the string at this instant as well as the speed of a point on the rim of the wheel at this instant. Therefore, $\omega = v/R$. In addition, because the wheel is a solid disk, its moment of inertia is $I = \frac{1}{2}MR^2$. Consequently,

$$mg(\Delta y) = \tfrac{1}{2}mv^2 + \tfrac{1}{2}(\tfrac{1}{2}MR^2)\left(\frac{v}{R}\right)^2 = \tfrac{1}{2}mv^2 + \tfrac{1}{4}Mv^2$$

Solving for v, we find that

$$v = \sqrt{\frac{mg(\Delta y)}{\frac{1}{2}m + \frac{1}{4}M}}$$

$$= \sqrt{\frac{(0.500\text{ kg})(9.80\text{ m/s}^2)(1.00\text{ m})}{\frac{1}{2}(0.500\text{ kg}) + \frac{1}{4}(2.00\text{ kg})}} = 2.56\text{ m/s}$$

10.8 ANGULAR MOMENTUM

Imagine an object rotating in space with no motion of its center of mass. Each particle in the object is moving in a circular path, so momentum is associated with the motion of each particle. Although the object has no linear momentum (its center of mass is not moving through space), a "quantity of motion" is associated with its rotation. We will investigate the **angular momentum** that the object has in this section.

Let us first consider a particle of mass m, situated at the vector position $\vec{r}$ and moving with a momentum $\vec{p}$, as shown in Active Figure 10.21. For now, we don't consider it as a particle on a rigid object; it is any particle moving with momentum $\vec{p}$. We will apply the result to a rotating rigid object shortly. The **instantaneous angular momentum** $\vec{L}$ of the particle relative to the origin O is defined by the vector product of its instantaneous position vector $\vec{r}$ and the instantaneous linear momentum $\vec{p}$:

$$\vec{L} \equiv \vec{r} \times \vec{p} \qquad [10.32]$$

The SI units of angular momentum are $\text{kg} \cdot \text{m}^2/\text{s}$. Note that both the magnitude and the direction of $\vec{L}$ depend on the choice of origin. The direction of $\vec{L}$ is perpendicular to the plane formed by $\vec{r}$ and $\vec{p}$, and the sense of $\vec{L}$ is governed by the right-hand rule. For example, in Active Figure 10.21, $\vec{r}$ and $\vec{p}$ are assumed to be in the xy plane and $\vec{L}$ points in the z direction. Because $\vec{p} = m\vec{v}$, the magnitude of $\vec{L}$ is

$$L = mvr\sin\phi \qquad [10.33]$$

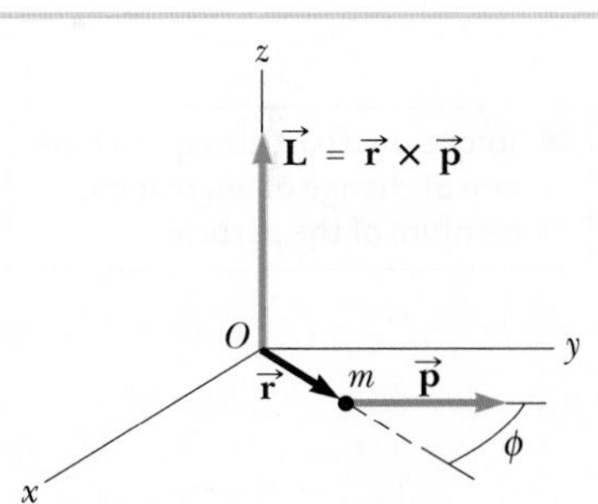

ACTIVE FIGURE 10.21 The angular momentum $\vec{L}$ of a particle of mass m and linear momentum $\vec{p}$ located at the position $\vec{r}$ is given by $\vec{L} = \vec{r} \times \vec{p}$. The value of $\vec{L}$ depends on the origin about which it is measured and is a vector perpendicular to both $\vec{r}$ and $\vec{p}$.

PhysicsNow™ By logging into PhysicsNow at **www.pop4e.com** and going to Active Figure 10.21 you can change the position vector $\vec{r}$ and the momentum vector $\vec{p}$ to see the effect on the angular momentum vector.

PITFALL PREVENTION 10.6

IS ROTATION NECESSARY FOR ANGULAR MOMENTUM? We can define angular momentum even if the particle is not moving in a circular path. Even a particle moving in a straight line has angular momentum about any axis displaced from the path of the particle.

where ϕ is the angle between $\vec{\mathbf{r}}$ and $\vec{\mathbf{p}}$. It follows that $\vec{\mathbf{L}}$ is zero when $\vec{\mathbf{r}}$ is parallel to $\vec{\mathbf{p}}$ ($\phi = 0°$ or $180°$). In other words, when the particle moves along a line that passes through the origin, it has zero angular momentum with respect to the origin, which is equivalent to stating that the momentum vector is not tangent to *any* circle drawn about the origin. On the other hand, if $\vec{\mathbf{r}}$ is perpendicular to $\vec{\mathbf{p}}$ ($\phi = 90°$), L is a maximum and equal to mvr. In fact, at that instant the particle moves exactly as though it were on the rim of a wheel of radius r rotating at angular speed $\omega = v/r$ about an axis through the origin in a plane defined by $\vec{\mathbf{r}}$ and $\vec{\mathbf{p}}$. A particle has nonzero angular momentum about some point if the position vector of the particle measured from that point rotates about the point as the particle moves.

For translational motion, we found that the net force on a particle equals the time rate of change of the particle's linear momentum (Eq. 8.4). We shall now show that Newton's second law implies an analogous situation for rotation: that the net torque acting on a particle equals the time rate of change of the particle's angular momentum. Let us start by writing the torque on the particle in the form

$$\vec{\boldsymbol{\tau}} = \vec{\mathbf{r}} \times \vec{\mathbf{F}} = \vec{\mathbf{r}} \times \frac{d\vec{\mathbf{p}}}{dt} \qquad [10.34]$$

where we have used $\vec{\mathbf{F}} = d\vec{\mathbf{p}}/dt$ (Eq. 8.4). Now let us differentiate Equation 10.32 with respect to time, using the product rule for differentiation:

$$\frac{d\vec{\mathbf{L}}}{dt} = \frac{d}{dt}(\vec{\mathbf{r}} \times \vec{\mathbf{p}}) = \vec{\mathbf{r}} \times \frac{d\vec{\mathbf{p}}}{dt} + \frac{d\vec{\mathbf{r}}}{dt} \times \vec{\mathbf{p}}$$

It is important to adhere to the order of factors in the vector product because the vector product is not commutative.

The last term on the right in the preceding equation is zero because $\vec{\mathbf{v}} = d\vec{\mathbf{r}}/dt$ is parallel to $\vec{\mathbf{p}}$. Therefore,

$$\frac{d\vec{\mathbf{L}}}{dt} = \vec{\mathbf{r}} \times \frac{d\vec{\mathbf{p}}}{dt} \qquad [10.35]$$

Comparing Equations 10.34 and 10.35, we see that

■ Torque on a particle equals time rate of change of angular momentum of the particle

$$\vec{\boldsymbol{\tau}} = \frac{d\vec{\mathbf{L}}}{dt} \qquad [10.36]$$

This result is the rotational analog of Newton's second law, $\vec{\mathbf{F}} = d\vec{\mathbf{p}}/dt$. Equation 10.36 says that the torque acting on a particle is equal to the time rate of change of the particle's angular momentum. Note that Equation 10.36 is valid only if the axes used to define $\vec{\boldsymbol{\tau}}$ and $\vec{\mathbf{L}}$ are the *same.* Equation 10.36 is also valid when several forces are acting on the particle, in which case $\vec{\boldsymbol{\tau}}$ is the *net* torque on the particle. Of course, the same origin must be used in calculating all torques as well as the angular momentum.

Now, let us apply these ideas to a system of particles. The total angular momentum $\vec{\mathbf{L}}$ of the system of particles about some point is defined as the vector sum of the angular momenta of the individual particles:

$$\vec{\mathbf{L}} = \vec{\mathbf{L}}_1 + \vec{\mathbf{L}}_2 + \cdots + \vec{\mathbf{L}}_n = \sum_i \vec{\mathbf{L}}_i$$

where the vector sum is over all the n particles in the system.

Because the individual angular momenta of the particles may change in time, the total angular momentum may also vary in time. In fact, from Equations 10.34 and 10.35 we find that the time rate of change of the total angular momentum of the system equals the vector sum of *all* torques, including those associated with internal forces between particles and those associated with external forces.

As we found in our discussion of the rigid object under a net torque, however, the sum of the internal torques is zero. Therefore, we conclude that the total angular momentum can vary with time *only* if there is a net *external* torque on the system, so that we have

$$\sum \vec{\boldsymbol{\tau}}_{\text{ext}} = \sum_i \frac{d\vec{\mathbf{L}}_i}{dt} = \frac{d}{dt}\sum_i \vec{\mathbf{L}}_i = \frac{d\vec{\mathbf{L}}_{\text{tot}}}{dt} \qquad [10.37]$$

■ **Net external torque on a system equals time rate of change of angular momentum of the system**

That is, the time rate of change of the total angular momentum of the system about some origin in an inertial frame equals the net external torque acting on the system about that origin. Note that Equation 10.37 is the rotational analog of $\sum \vec{\mathbf{F}}_{\text{ext}} = d\vec{\mathbf{p}}_{\text{tot}}/dt$ (Eq. 8.40) for a system of particles.

This result is valid for a system of particles that change their positions with respect to one another, that is, a nonrigid object. In this discussion of angular momentum of a system of particles, notice that we never imposed the rigid-object condition.

Equation 10.37 is the primary equation in the angular momentum version of the nonisolated system model. The system's angular momentum changes in response to an interaction with the environment, described by means of the net torque on the system.

One final result can be obtained for angular momentum, which will serve as an analog to the definition of linear momentum. Let us imagine a rigid object rotating about an axis. Each particle of mass m_i in the rigid object moves in a circular path of radius r_i, with a tangential speed v_i. Therefore, the total angular momentum of the rigid object is

$$L = \sum_i m_i v_i r_i$$

Let us now replace the tangential speed with the product of the radial distance and the angular speed (Eq. 10.10):

$$L = \sum_i m_i v_i r_i = \sum_i m_i (r_i\omega) r_i = \left(\sum_i m_i {r_i}^2\right)\omega$$

We recognize the combination in the parentheses as the moment of inertia, so we can write the angular momentum of the rigid object as

$$L = I\omega$$

■ **Angular momentum of an object with moment of inertia *I***

which is the rotational analog to $p = mv$. Table 10.3 is a continuation of Table 10.1, with additional translational and rotational analogs that we have developed in the past few sections.

TABLE 10.3 A Comparison of Equations for Rotational and Translational Motion: Dynamic Equations

	Rotational Motion About a Fixed Axis	Translational Motion
Kinetic energy	$K_R = \frac{1}{2}I\omega^2$	$K = \frac{1}{2}mv^2$
Equilibrium	$\sum \vec{\boldsymbol{\tau}} = 0$	$\sum \vec{\mathbf{F}} = 0$
Newton's second law	$\sum \tau = I\alpha$	$\sum \vec{\mathbf{F}} = m\vec{\mathbf{a}}$
Newton's second law	$\vec{\boldsymbol{\tau}} = \frac{d\vec{\mathbf{L}}}{dt}$	$\vec{\mathbf{F}} = \frac{d\vec{\mathbf{p}}}{dt}$
Momentum	$L = I\omega$	$\vec{\mathbf{p}} = m\vec{\mathbf{v}}$
Conservation principle	$\vec{\mathbf{L}}_i = \vec{\mathbf{L}}_f$	$\vec{\mathbf{p}}_i = \vec{\mathbf{p}}_f$
Power	$\mathcal{P} = \tau\omega$	$\mathcal{P} = Fv$

Note: Equations in translation motion expressed in terms of vectors have rotational analogs in terms of vectors. Because the full vector treatment of rotation is beyond the scope of this book, however, some rotational equations are given in nonvector form.

QUICK QUIZ 10.7 A solid sphere and a hollow sphere have the same mass and radius. They are rotating with the same angular speed. Which is the one with the higher angular momentum? **(a)** the solid sphere **(b)** the hollow sphere **(c)** both have the same angular momentum **(d)** impossible to determine

EXAMPLE 10.12 The Atwood Machine Once Again

Consider again the Atwood machine with the massive pulley in Example 10.10. Determine the acceleration of the two objects using an angular momentum approach.

Solution This example is of a nonrigid object experiencing a net torque, so we use the nonisolated system model. We will evaluate the angular momentum of the system at any time and then differentiate the angular momentum, setting it equal to the net external torque. We will solve the resulting expression for the acceleration of the objects.

Let us first calculate the angular momentum of the system, which consists of the two objects plus the pulley. At the instant m_1 and m_2 have a speed v, the angular momentum of m_1 around the axle of the pulley is m_1vR and that of m_2 is m_2vR. At the same instant, the angular momentum of the pulley around its center is $L = I\omega = Iv/R$. Therefore, the total angular momentum of the system is

$$(1) \quad L = m_1vR + m_2vR + I\frac{v}{R}$$

$$= m_1vR + m_2vR + (\tfrac{1}{2}MR^2)\frac{v}{R}$$

$$= (m_1 + m_2 + \tfrac{1}{2}M)vR$$

Now let us evaluate the total external torque on the system about the axle. The weight of the pulley and the force of the axle upward on the pulley have zero moment arm around the center of the pulley, so they do not contribute to the torque. The external forces on the system that produce torques about the axle are $m_1\vec{\mathbf{g}}$, with a torque of m_1gR, and $m_2\vec{\mathbf{g}}$, with a torque of $-m_2gR$. Combining the net external torque with (1) and Equation 10.37 gives us

$$\tau_{ext} = \frac{dL}{dt}$$

$$m_1gR - m_2gR = \frac{d}{dt}[(m_1 + m_2 + \tfrac{1}{2}M)vR]$$

$$(2) \quad m_1gR - m_2gR = (m_1 + m_2 + \tfrac{1}{2}M)R\frac{dv}{dt}$$

Because $dv/dt = a$, we can solve Equation (2) for a to find

$$a = \left(\frac{m_1 - m_2}{m_1 + m_2 + \frac{1}{2}M}\right)g$$

which is the same result as that obtained in Example 10.10. You may wonder why we did not include the tension forces that the cord exerts on the objects in evaluating the net torque about the axle. The reason is that these forces are *internal* to the system under consideration. Only the *external* torques contribute to the change in angular momentum.

10.9 CONSERVATION OF ANGULAR MOMENTUM

In Chapter 8, we found that the total linear momentum of a system of particles is conserved when the net external force acting on the system is zero. In rotational motion, we have an analogous conservation law that states that **the total angular momentum of a system is conserved if the net external torque acting on the system is zero.**

Because the net external torque acting on the system equals the time rate of change of the system's angular momentum, we see from Equation 10.37 that if

$$\sum \vec{\boldsymbol{\tau}}_{ext} = \frac{d\vec{\mathbf{L}}_{tot}}{dt} = 0 \qquad [10.38]$$

then

■ Conservation of angular momentum for an isolated system

$$\vec{\mathbf{L}}_{tot} = \text{constant} \rightarrow \vec{\mathbf{L}}_{tot,\,i} = \vec{\mathbf{L}}_{tot,\,f} \qquad [10.39]$$

Equation 10.39 represents a third conservation law to add to our list of fundamental conservation principles. We can now state that **the total energy, linear**

momentum, and angular momentum of an isolated system are all conserved. We have focused our attention in this chapter on rigid objects; the conservation of angular momentum principle, however, is a general result of the isolated system model. Therefore, **the angular momentum of an isolated system is conserved whether the system is a rigid object or not.**

At any instant of time, the angular momentum of a system of particles about a fixed axis has a magnitude given by $L = I\omega$, where I is the moment of inertia of the system about the axis. In this case, if the net external torque on the system is zero, we can express the conservation of angular momentum principle as $I\omega$ = constant. Imagine a situation in which a rotating system undergoes a change in moment of inertia. Because of the principle of conservation of angular momentum, there must be a corresponding change in the angular speed.

Many examples can be used to demonstrate this effect; some of them should be familiar to you. You may have observed a figure skater spinning (Fig. 10.22). The angular speed of the skater is large when his hands and feet are close to the trunk of his body. Ignoring friction between skater and ice, we see that there are no external torques on the skater. The moment of inertia of his body increases as his hands and feet are moved away from his body at the finish of the spin. According to the principle of conservation of angular momentum, his angular speed must decrease.

An interesting astrophysical example of conservation of angular momentum occurs when, at the end of its lifetime, a massive star uses up all its fuel and collapses under the influence of gravitational forces, causing a gigantic outburst of energy called a supernova explosion. The best-studied example of a remnant of a supernova explosion is the Crab Nebula, a chaotic, expanding mass of gas (Fig. 10.23). In a supernova, part of the star's mass is released into space, where it eventually condenses into new stars and planets. Most of what is left behind typically collapses into a **neutron star,** an extremely dense sphere of matter with a diameter of about 10 km in comparison with the 10^6-km diameter of the original star and containing a large fraction of the star's original mass. As the moment of inertia of the system decreases during the collapse, the star's rotational speed increases, similar to the change in speed of the skater in Figure 10.22. More than 700 rapidly rotating neutron stars have been identified since the first discovery of such astronomical bodies in 1967, with periods of rotation ranging from a millisecond to several seconds. The neutron star—an object with a mass greater than the Sun, rotating about its axis many times each second—is a most dramatic system!

(© Stuart Franklin/Getty Images)

FIGURE 10.22 Angular momentum is conserved as Russian figure skater Evgeni Plushenko performs during the 2004 World Figure Skating Championships. When his arms and legs are close to his body, his moment of inertia is small and his angular speed is large. To slow down for the finish of his spin, he moves his arms and legs outward, increasing his moment of inertia.

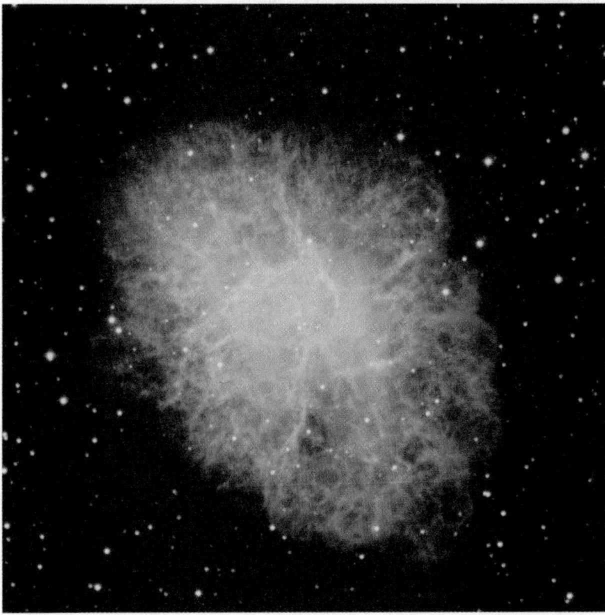

(David Malin, Anglo-Australian Observatory)

FIGURE 10.23 The Crab Nebula, in the constellation Taurus. This nebula is the remnant of a supernova explosion, which was seen on Earth in the year A.D. 1054. It is located some 6 300 lightyears away and is approximately 6 lightyears in diameter, still expanding outward.

QUICK QUIZ 10.8 (i) A competitive diver leaves the diving board and falls toward the water with her body straight and rotating slowly. She pulls her arms and legs into a tight tuck position. What happens to her angular speed? **(a)** It increases. **(b)** It decreases. **(c)** It stays the same. **(d)** It is impossible to determine. **(ii)** From the same list of choices, what happens to the rotational kinetic energy of her body?

EXAMPLE 10.13 A Revolving Puck on a Horizontal, Frictionless Surface

A puck of mass m on a horizontal, frictionless table is connected to a string that passes through a small hole in the table. The puck is set into circular motion of radius R, at which time its speed is v_i (Fig. 10.24).

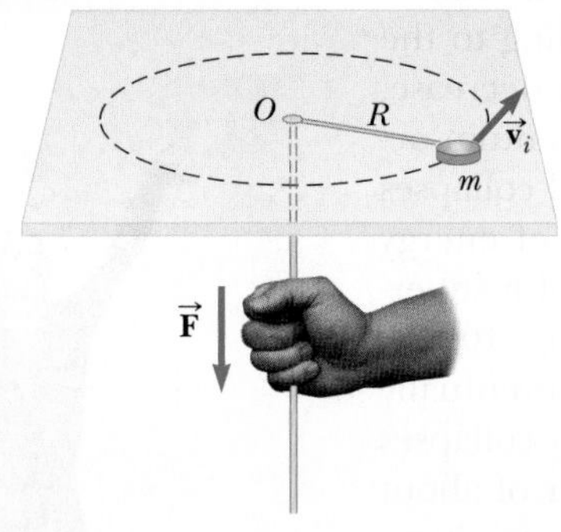

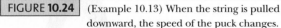

FIGURE 10.24 (Example 10.13) When the string is pulled downward, the speed of the puck changes.

A If the string is pulled from the bottom so that the radius of the circular path is decreased to r, what is the final speed v_f of the puck?

Solution We identify the system as the puck. We will calculate torque about the center of rotation O. Note that the gravitational force acting on the puck is balanced by the upward normal force, so these forces cancel, resulting in zero net torque from these forces. The force $\vec{\mathbf{F}}$ of the string on the puck acts toward the center of rotation, and the vector position $\vec{\mathbf{r}}$ is directed away from O. Therefore, we see that $\vec{\boldsymbol{\tau}} = \vec{\mathbf{r}} \times \vec{\mathbf{F}} = 0$, so no torque is applied on the puck due to this force. Three forces are acting on the puck, but zero net torque occurs. Therefore, $\vec{\mathbf{L}}$ is a constant of the motion. The puck can be modeled as a particle moving in a circular path, so $L = mv_iR = mv_fr$, or

$$v_f = \frac{v_iR}{r}$$

From this result we see that as r decreases, the speed v increases.

B Is the kinetic energy of the puck conserved in this process?

Solution We set up the ratio of the final kinetic energy to the initial kinetic energy:

$$\frac{K_f}{K_i} = \frac{\frac{1}{2}mv_f^2}{\frac{1}{2}mv_i^2} = \frac{1}{v_i^2}\left(\frac{v_iR}{r}\right)^2 = \frac{R^2}{r^2}$$

Because this ratio is not equal to 1, kinetic energy is not conserved. Furthermore, because $R > r$, the kinetic energy of the puck has increased. This increase corresponds to energy entering the system of the puck by means of the work done by the person pulling the string.

EXAMPLE 10.14 Rotation Period of a Neutron Star

A star undergoes a supernova explosion. The material left behind forms a sphere of radius 8.0×10^6 m just after the explosion with a rotation period of 15 h. This remaining material collapses into a neutron star of radius 8.0 km. What is the rotation period T of the neutron star?

Solution We model the star as an isolated system. As the moment of inertia of the stellar core decreases during its collapse, the angular speed increases. From conservation of angular momentum,

$$I_i\omega_i = I_f\omega_f$$

We have no information about the variation in the density of material with radius in either the initial star or the neutron star, so we choose a simplification model in which the density is uniform. Our result will most likely not be entirely accurate, but we can consider it an estimate. Using the moment of inertia of a sphere of uniform density (see Table 10.2),

$$(\tfrac{2}{5}MR_i^{\,2})\omega_i = (\tfrac{2}{5}MR_f^{\,2})\omega_f \quad\rightarrow\quad \omega_f = \left(\frac{R_i}{R_f}\right)^2 \omega_i$$

Now, because $\omega = 2\pi/T$, we have

$$\frac{2\pi}{T_f} = \left(\frac{R_i}{R_f}\right)^2 \frac{2\pi}{T_i} \quad\rightarrow\quad T_f = \left(\frac{R_f}{R_i}\right)^2 T_i$$

Substituting numerical values, we have

$$T_f = \left(\frac{8.0 \times 10^3 \text{ m}}{8.0 \times 10^6 \text{ m}}\right)^2 (15 \text{ h}) = 1.5 \times 10^{-5} \text{ h} = 0.054 \text{ s}$$

10.10 PRECESSIONAL MOTION OF GYROSCOPES

Angular momentum is the basis of the operation of a **gyroscope,** which is a spinning object used to control or maintain the orientation in space of the object or a system containing the object. As an example, consider a quarterback passing a football. If he imparts no spin to the ball, there is no angular momentum to be conserved and forces from the air might cause the ball to tumble as it moves through its trajectory. If a spin is imparted to the ball along the long axis of the football, however, the angular momentum vector stays fixed in direction and the football maintains its orientation throughout the trajectory, resulting in much less air resistance and a longer pass. In this application, the football is acting as a gyroscope to maintain its own orientation in space.

An unusual and fascinating type of motion you probably have observed is that of a top spinning rapidly about its axis of symmetry, as shown in Figure 10.25a. The top is acting as a gyroscope and one might expect the orientation to remain fixed in space. If the top is leaning over, however, it is observed that the symmetry axis rotates about the z axis, sweeping out a cone (see Fig. 10.25b). This phenomenon is called **precessional motion.** The angular speed of the symmetry axis about the vertical is usually slow relative to the angular speed of the top about the symmetry axis.

It is quite natural to wonder why the top does not maintain its direction of spin. Because the center of mass of the top is not directly above the pivot point O, a net torque is acting on the top about O, a torque resulting from the gravitational force $M\vec{\mathbf{g}}$. The top would certainly fall over if it were not spinning. Because it is spinning, however, it has an angular momentum $\vec{\mathbf{L}}$ directed along its symmetry axis. As we shall show, the motion of this symmetry axis about the z axis (the precessional motion) occurs because the torque produces a change in the *direction* of the symmetry axis. This motion is an excellent example of the importance of the directional nature of angular momentum.

The two forces acting on the top are the downward gravitational force $M\vec{\mathbf{g}}$ and the normal force $\vec{\mathbf{n}}$ acting upward at the pivot point O. The normal force produces no torque about the pivot because its moment arm through that point is zero. The gravitational force, however, produces a torque $\vec{\boldsymbol{\tau}} = \vec{\mathbf{r}} \times M\vec{\mathbf{g}}$ about O, where the direction of $\vec{\boldsymbol{\tau}}$ is perpendicular to the plane formed by $\vec{\mathbf{r}}$ and $M\vec{\mathbf{g}}$. By necessity, the vector $\vec{\boldsymbol{\tau}}$ lies in a horizontal plane perpendicular to the angular momentum vector. The net torque and angular momentum of the top are related through Equation 10.37:

$$\sum \vec{\boldsymbol{\tau}} = \frac{d\vec{\mathbf{L}}}{dt}$$

From this expression we see that the nonzero torque produces a change in angular momentum $d\vec{\mathbf{L}}$, a change that is in the same direction as $\sum \vec{\boldsymbol{\tau}}$. Therefore, like the torque vector, $d\vec{\mathbf{L}}$ must also be perpendicular to $\vec{\mathbf{L}}$. Figure 10.25b illustrates the resulting precessional motion of the symmetry axis of the top. In a time interval Δt, the change in angular momentum is $\Delta\vec{\mathbf{L}} = \vec{\mathbf{L}}_f - \vec{\mathbf{L}}_i = \sum \vec{\boldsymbol{\tau}}\Delta t$. Because $\Delta\vec{\mathbf{L}}$ is

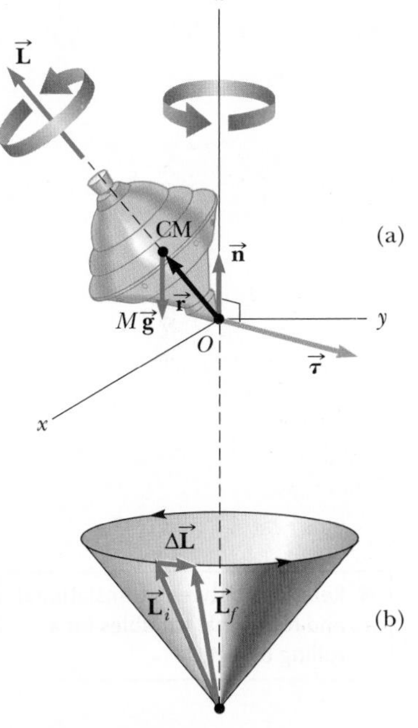

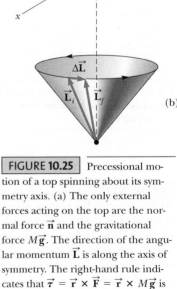

FIGURE 10.25 Precessional motion of a top spinning about its symmetry axis. (a) The only external forces acting on the top are the normal force $\vec{\mathbf{n}}$ and the gravitational force $M\vec{\mathbf{g}}$. The direction of the angular momentum $\vec{\mathbf{L}}$ is along the axis of symmetry. The right-hand rule indicates that $\vec{\boldsymbol{\tau}} = \vec{\mathbf{r}} \times \vec{\mathbf{F}} = \vec{\mathbf{r}} \times M\vec{\mathbf{g}}$ is in the xy plane. (b). The direction of $\Delta\vec{\mathbf{L}}$ is parallel to that of $\vec{\boldsymbol{\tau}}$ in part (a). That $\vec{\mathbf{L}}_f = \vec{\mathbf{L}}_i + \Delta\vec{\mathbf{L}}$ indicates that the top precesses about the z axis.

perpendicular to $\vec{\mathbf{L}}$, the magnitude of $\vec{\mathbf{L}}$ does not change ($|\vec{\mathbf{L}}_i| = |\vec{\mathbf{L}}_f|$). Rather, what is changing is the *direction* of $\vec{\mathbf{L}}$. Because the change in angular momentum $\Delta\vec{\mathbf{L}}$ is in the direction of $\sum \vec{\boldsymbol{\tau}}$, which lies in the *xy* plane, the top undergoes precessional motion.

With careful manufacturing tolerances, precession due to gravitational torque can be made very small and gyroscopes can be used for guidance systems in vehicles, whereby a change in the direction of the velocity of a vehicle is detected as a change between the direction of the angular momentum of the gyroscope and a reference direction attached to the vehicle. With proper electronic feedback, the deviation from the desired direction of motion can be removed, bringing the angular momentum back in line with the reference direction. Precession rates for highly specialized military gyroscopes are as low as 0.02° per day.

10.11 ROLLING MOTION OF RIGID OBJECTS

In this section, we shall investigate the special case of rotational motion in which a round object rolls on a surface. Many everyday examples exist for such motion, including automobile tires rolling on roads and bowling balls rolling toward the pins.

Suppose a cylinder is rolling on a straight path as in Figure 10.26. The center of mass moves in a straight line, but a point on the rim moves in a more complex path called a *cycloid*. Let us further assume that the cylinder of radius R is uniform and rolls on a surface with friction. We make a rather odd, but valid, simplification model here for rolling objects. The surfaces must exert friction forces on each other; otherwise, the cylinder would simply slide rather than roll. If the friction force on the cylinder is large enough, the cylinder rolls without slipping. In this situation, the friction force is static rather than kinetic because the contact point of the cylinder with the surface is at rest relative to the surface at any instant. The static friction force acts through no displacement, so it does no work on the cylinder and causes no decrease in mechanical energy of the cylinder. In real rolling objects, deformations of the surfaces result in some rolling resistance. If both surfaces are hard, however, they will deform very little, and rolling resistance can be negligibly small. Therefore, we can model the rolling motion as maintaining constant mechanical energy. The wheel was a great invention!

As the cylinder rotates through an angle θ, its center of mass moves a distance of $s = r\theta$. Therefore, the speed and acceleration of the center of mass for **pure rolling motion** are

■ Relations between translational and rotational variables for a rolling object

$$v_{\text{CM}} = \frac{ds}{dt} = R\frac{d\theta}{dt} = R\omega \qquad [10.40]$$

$$a_{\text{CM}} = \frac{dv_{\text{CM}}}{dt} = R\frac{d\omega}{dt} = R\alpha \qquad [10.41]$$

(Courtesy of Henry Leap and Jim Lehman)

FIGURE 10.26 Light sources at the center and rim of a rolling cylinder illustrate the different paths these points take. The center moves in a straight line *(green line)*, whereas a point on the rim moves in the path of a cycloid *(red curve)*.

The translational velocities of various points on the rolling cylinder are illustrated in Figure 10.27. Note that the translational velocity of any point is in a direction perpendicular to the line from that point to the contact point. At any instant, the point P is at rest relative to the surface because sliding does not occur.

We can express the **total kinetic energy** of a rolling object of mass M and moment of inertia I as the combination of the rotational kinetic energy around the center of mass plus the translational kinetic energy of the center of mass:

■ Total kinetic energy of a rolling object

$$K = \tfrac{1}{2}I_{CM}\omega^2 + \tfrac{1}{2}Mv_{CM}^2 \qquad [10.42]$$

A useful theorem called the **parallel axis theorem** enables us to express this energy in terms of the moment of inertia I_p through any axis parallel to the axis through the center of mass of an object. This theorem states that

$$I_p = I_{CM} + MD^2 \qquad [10.43]$$

where D is the distance from the center-of-mass axis to the parallel axis and M is the total mass of the object. Let us use this theorem to express the moment of inertia around an axis passing through the contact point P between the rolling object and the surface. The distance from this point to the center of mass of the symmetric object is its radius, so

$$I_p = I_{CM} + MR^2$$

If we write the translational speed of the center of mass of the object in Equation 10.42 in terms of the angular speed, we have

$$K = \tfrac{1}{2}I_{CM}\omega^2 + \tfrac{1}{2}MR^2\omega^2 = \tfrac{1}{2}(I_{CM} + MR^2)\,\omega^2 = \tfrac{1}{2}I_p\omega^2 \qquad [10.44]$$

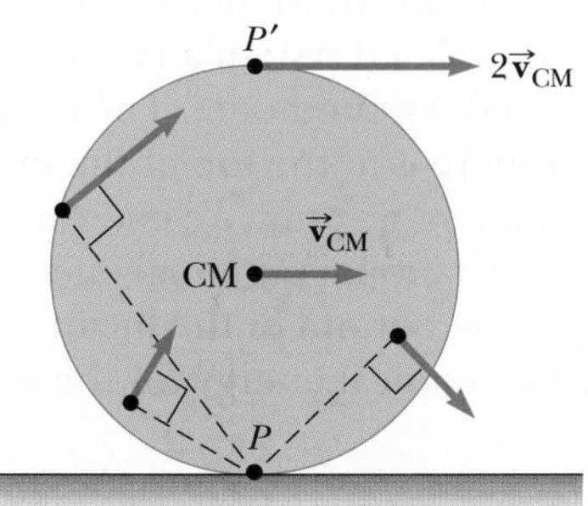

FIGURE 10.27 All points on a rolling object move in a direction perpendicular to a line through the instantaneous point of contact P. The center of the object moves with a velocity $\vec{\mathbf{v}}_{CM}$, whereas the point P' moves with a velocity $2\vec{\mathbf{v}}_{CM}$.

Therefore, the kinetic energy of the rolling object can be considered as equivalent to a purely rotational kinetic energy of the object rotating around its contact point.

We can use the energy version of the isolated system model to treat a class of problems concerning the rolling motion of a rigid object down a rough incline. In these types of problems, gravitational potential energy of the object–Earth system decreases as the rotational and translational kinetic energies of the object increase. For example, consider a sphere rolling without slipping after being released from rest at the top of an incline. Note that accelerated rolling motion is possible only if a friction force is present between the sphere and the incline to produce a net torque about the center of mass. Despite the presence of friction, no loss of mechanical energy occurs because the contact point is at rest relative to the surface at any instant. (On the other hand, if the sphere were to slip, mechanical energy of the sphere–incline–Earth system would be lost due to the nonconservative force of kinetic friction.)

Using $v_{CM} = R\omega$ for pure rolling motion, we can express Equation 10.42 as

$$K = \tfrac{1}{2}I_{CM}\left(\frac{v_{CM}}{R}\right)^2 + \tfrac{1}{2}Mv_{CM}^2$$

$$K = \tfrac{1}{2}\left(\frac{I_{CM}}{R^2} + M\right)v_{CM}^2 \qquad [10.45]$$

For the system of the sphere and the Earth, we define the zero configuration of gravitational potential energy to be when the sphere is at the bottom of the incline. Therefore, conservation of mechanical energy gives us

$$K_f + U_f = K_i + U_i$$

$$\tfrac{1}{2}\left(\frac{I_{CM}}{R^2} + M\right)v_{CM}^2 + 0 = 0 + Mgh$$

$$v_{CM} = \left(\frac{2gh}{1 + I_{CM}/MR^2}\right)^{1/2} \qquad [10.46]$$

QUICK QUIZ 10.9 Two items A and B are placed at the top of an incline and released from rest. For *each* of the three pairs of items in (i), (ii), and (iii), which item arrives at the bottom of the incline first? **(i)** a ball A rolling without slipping and a box B sliding on a frictionless portion of the incline **(ii)** a sphere A that has twice the mass and twice the radius of a sphere B, where both roll without slipping **(iii)** a sphere A that has the same mass and radius as a sphere B, but sphere A is solid while sphere B is hollow and both roll without slipping. Choose from the following list for each of the three pairs of items. **(a)** item A **(b)** item B **(c)** items A and B arrive at the same time **(d)** impossible to determine

EXAMPLE 10.15 Sphere Rolling Down an Incline

A If the object in Active Figure 10.28 is a solid sphere, calculate the speed of its center of mass at the bottom.

Solution We shall consider the sphere and the Earth as an isolated system and use the energy version of the isolated system model. The energy of the system when the sphere is at the top of the incline is gravitational potential energy only. We choose the zero configuration of gravitational potential energy to be when the sphere is at the bottom of the incline. Therefore, conservation of mechanical energy for the system gives us

$$K_f + U_f = K_i + U_i$$

$$(\tfrac{1}{2}Mv_{\text{CM},f}^2 + \tfrac{1}{2}I_{\text{CM}}\omega_f^{\,2}) + 0 = 0 + Mgh$$

Using Equation 10.40 to relate the translational and angular speeds, and substituting the moment of inertia for a sphere, we have

$$\tfrac{1}{2}Mv_{\text{CM},f}^2 + \tfrac{1}{2}(\tfrac{2}{5}MR^2)\frac{v_{\text{CM},f}^2}{R^2} = Mgh$$

$$\tfrac{1}{2}Mv_{\text{CM},f}^2 + \tfrac{1}{5}Mv_{\text{CM},f}^2 = \tfrac{7}{10}Mv_{\text{CM},f}^2 = Mgh$$

$$v_{\text{CM},f} = \sqrt{\tfrac{10}{7}gh}$$

B Determine the magnitude of the translational acceleration of the center of mass.

Solution To find the acceleration, let us recognize that the constant gravitational force should cause a constant acceleration of the center of mass of the sphere. From Equation 2.13,

$$v_{\text{CM},f}^2 = v_{\text{CM},i}^2 + 2a_{\text{CM}}(x_{\text{CM},f} - x_{\text{CM},i})$$

we can solve for the acceleration

$$a_{\text{CM}} = \frac{v_{\text{CM},f}^2 - v_{\text{CM},i}^2}{2(x_{\text{CM},f} - x_{\text{CM},i})} = \frac{\tfrac{10}{7}gh - 0}{2\left(\dfrac{h}{\sin\theta}\right)} = \tfrac{5}{7}g\sin\theta$$

These results are quite interesting in that both the speed and the acceleration of the center of mass are independent of the mass and radius of the sphere. That is, *all* homogeneous solid spheres experience the same speed and acceleration on a given incline!

If we repeated the calculations for a hollow sphere, a solid cylinder, or a hoop, we would obtain similar results with different numerical factors appearing in the expressions for $v_{\text{CM},f}$ and a_{CM}. These factors depend only on the moment of inertia about the center of mass for the specific object. In all cases, the acceleration of the center of mass is less than $g\sin\theta$, the value it would have if the plane were frictionless and no rolling occurred.

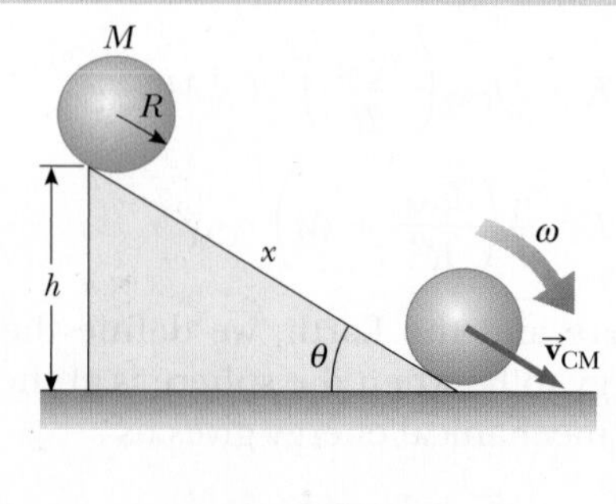

ACTIVE FIGURE 10.28 (Example 10.15) A round object rolling down an incline. Mechanical energy of the object–surface–Earth system is conserved if no slipping occurs and there is no rolling resistance.

PhysicsNow™ Roll several objects down the hill and see the effect on the final speed by logging into PhysicsNow at **www.pop4e.com** and going to Active Figure 10.28.

10.12 TURNING THE SPACECRAFT

CONTEXT CONNECTION

In the Context Connection of Chapter 8, we discussed how to make a spacecraft move in empty space by firing its rocket engines. Let us know consider how to make the spacecraft turn in empty space.

One way to change the orientation of a spacecraft is to have small rocket engines that fire perpendicularly out the side of the spacecraft, providing a torque around its center of mass. This torque causes an angular acceleration around the center of mass of the spacecraft and therefore an angular speed. This rotation can be stopped to give the spacecraft the desired final orientation by firing the sideward-mounted rocket engines in the opposite direction. This option is desirable, and many spacecraft have such sideward-mounted rocket engines. An undesirable feature of this technique is that it consumes nonrenewable fuel on the spacecraft, both to initiate and to stop the rotation.

Let us consider another possibility related to angular momentum. Suppose the spacecraft carries a gyroscope that is not rotating, as in Figure 10.29a. In this case, the angular momentum of the spacecraft about its center of mass is zero. Suppose the gyroscope is set into rotation. Now, it would appear that the spacecraft system has a nonzero angular momentum because of the rotation of the gyroscope. Yet there is no external torque on the system, so the angular momentum of the isolated system must remain zero according to the principle of conservation of angular momentum. This principle can be satisfied by realizing that the spacecraft will turn in the direction opposite to that of the gyroscope so that the angular momentum vectors of the gyroscope and the spacecraft cancel, resulting in no angular momentum of the system. The result of rotating the gyroscope, as in Figure 10.29b, is that the spacecraft turns! By including three gyroscopes with mutually perpendicular axles, any desired rotation in space can be achieved. Once the desired orientation is achieved, the rotation of the gyroscope is halted.

This effect occurred in an undesirable situation with the *Voyager 2* spacecraft during its flight. The spacecraft carried a tape recorder whose reels rotated at high speeds. Each time the tape recorder was turned on, the reels acted as gyroscopes and the spacecraft started an undesirable rotation in the opposite direction. This rotation had to be counteracted by Mission Control by using the sideward-firing jets to stop the rotation!

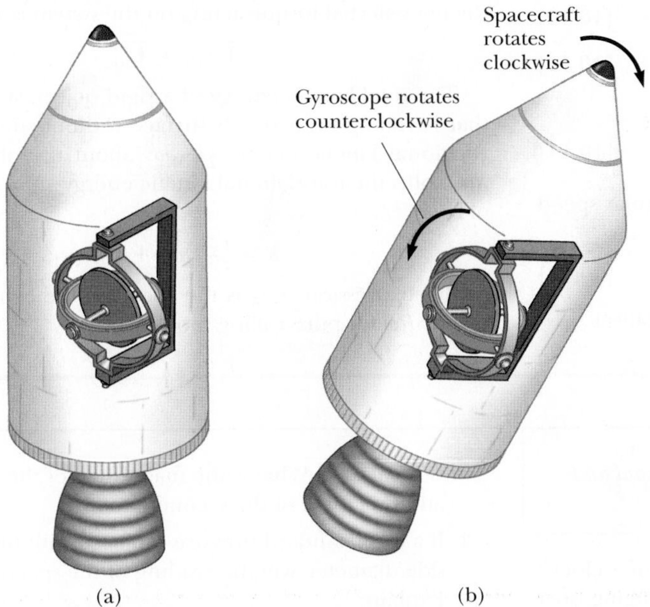

FIGURE 10.29 (a) A spacecraft carries a gyroscope that is not spinning. (b) When the gyroscope is set into rotation, the spacecraft turns the other way so that the angular momentum of the system is conserved.

SUMMARY

PhysicsNow™ Take a practice test by logging into PhysicsNow at www.pop4e.com and clicking on the Pre-Test link for this chapter.

The **instantaneous angular speed** of a particle rotating in a circle or of a rigid object rotating about a fixed axis is

$$\omega \equiv \frac{d\theta}{dt} \qquad [10.3]$$

where ω is in rad/s or s^{-1}.

The **instantaneous angular acceleration** of a particle rotating in a circle or of a rigid object rotating about a fixed axis is

$$\alpha \equiv \frac{d\omega}{dt} \qquad [10.5]$$

and has units of rad/s^2 or s^{-2}.

When a rigid object rotates about a fixed axis, every part of the object has the same angular speed and the same angular acceleration. Different parts of the object, in general, have different translational speeds and different translational accelerations, however.

If a particle (or object) undergoes rotational motion about a fixed axis under constant angular acceleration α, one can apply equations of kinematics by analogy with kinematic equations for translational motion with constant translational acceleration:

$$\omega_f = \omega_i + \alpha t \qquad [10.6]$$

$$\theta_f = \theta_i + \omega_i t + \tfrac{1}{2}\alpha t^2 \qquad [10.7]$$

$$\omega_f^2 = \omega_i^2 + 2\alpha(\theta_f - \theta_i) \qquad [10.8]$$

$$\theta_f = \theta_i + \tfrac{1}{2}(\omega_i + \omega_f)t \qquad [10.9]$$

When a particle rotates about a fixed axis, the angular position, the angular speed, and the angular acceleration are related to the tangential position, the tangential speed, and the tangential acceleration through the relationships

$$s = r\theta \qquad [10.1a]$$

$$v = r\omega \qquad [10.10]$$

$$a_t = r\alpha \qquad [10.11]$$

The **moment of inertia** of a system of particles is

$$I = \sum_i m_i r_i^2 \qquad [10.14]$$

If a rigid object rotates about a fixed axis with angular speed ω, its **rotational kinetic energy** can be written

$$K_R = \tfrac{1}{2}I\omega^2 \qquad [10.15]$$

where I is the moment of inertia about the axis of rotation.

The moment of inertia of a continuous object of density ρ is

$$I = \int \rho r^2 dV \qquad [10.17]$$

The **torque** $\vec{\tau}$ due to a force $\vec{\mathbf{F}}$ about an origin in an inertial frame is defined to be

$$\vec{\tau} \equiv \vec{\mathbf{r}} \times \vec{\mathbf{F}} \qquad [10.19]$$

where $\vec{\mathbf{r}}$ is the position vector of the point of application of the force.

Given two vectors $\vec{\mathbf{A}}$ and $\vec{\mathbf{B}}$, their **vector product** or **cross product** $\vec{\mathbf{A}} \times \vec{\mathbf{B}}$ is a vector $\vec{\mathbf{C}}$ having the magnitude

$$C \equiv AB\sin\theta \qquad [10.21]$$

where θ is the angle between $\vec{\mathbf{A}}$ and $\vec{\mathbf{B}}$. The direction of $\vec{\mathbf{C}}$ is perpendicular to the plane formed by $\vec{\mathbf{A}}$ and $\vec{\mathbf{B}}$, and is determined by the right-hand rule.

The net torque acting on an object is proportional to the angular acceleration of the object, and the proportionality constant is the moment of inertia I:

$$\sum \tau = I\alpha \qquad [10.27]$$

The **angular momentum** $\vec{\mathbf{L}}$ of a particle with linear momentum $\vec{\mathbf{p}} = m\vec{\mathbf{v}}$ is

$$\vec{\mathbf{L}} \equiv \vec{\mathbf{r}} \times \vec{\mathbf{p}} \qquad [10.32]$$

where $\vec{\mathbf{r}}$ is the vector position of the particle relative to the origin. If ϕ is the angle between $\vec{\mathbf{r}}$ and $\vec{\mathbf{p}}$, the magnitude of $\vec{\mathbf{L}}$ is

$$L = mvr\sin\phi \qquad [10.33]$$

The net external torque acting on a system is equal to the time rate of change of its angular momentum:

$$\sum \vec{\tau}_{ext} = \frac{d\vec{\mathbf{L}}_{tot}}{dt} \qquad [10.37]$$

The law of conservation of angular momentum states that the total angular momentum of a system remains constant if the net external torque acting on the system is zero:

$$\vec{\mathbf{L}}_{tot,i} = \vec{\mathbf{L}}_{tot,f} \qquad [10.39]$$

The **total kinetic energy** of a rigid object, such as a cylinder, that is rolling on a rough surface without slipping equals the rotational kinetic energy $\frac{1}{2}I_{CM}\omega^2$ about the object's center of mass plus the translational kinetic energy $\frac{1}{2}Mv_{CM}^2$ of the center of mass:

$$K = \tfrac{1}{2}I_{CM}\omega^2 + \tfrac{1}{2}Mv_{CM}^2 \qquad [10.42]$$

In this expression, v_{CM} is the speed of the center of mass and $v_{CM} = R\omega$ for pure rolling motion.

QUESTIONS

☐ = answer available in the *Student Solutions Manual and Study Guide*

1. What is the angular speed of the second hand of a clock? What is the direction of $\vec{\omega}$ as you view a clock hanging on a vertical wall? What is the magnitude of the angular acceleration vector $\vec{\alpha}$ of the second hand?

2. If a car's standard tires are replaced with tires of larger outside diameter, will the reading of the speedometer change? Explain.

3. Suppose just two external forces act on a stationary rigid object and the two forces are equal in magnitude and opposite in direction. Under what condition does the object start to rotate?

4. Suppose you remove two eggs from the refrigerator, one hard-boiled and the other uncooked. You wish to determine which is the hard-boiled egg without breaking the eggs. You can do so by spinning the two eggs on the floor and comparing the rotational motions. Which egg spins faster? Which rotates more uniformly? Explain.

5. If you see an object rotating, is there necessarily a net torque acting on it?

6. Which of the entries in Table 10.2 applies to finding the moment of inertia of a long, straight sewer pipe rotating about its axis of symmetry? of an embroidery hoop rotating about an axis through its center and perpendicular to its plane? of a uniform door turning on its hinges? of a coin turning about an axis through its center and perpendicular to its faces?

7. In a tape recorder, the tape is pulled past the read-and-write heads at a constant speed by the drive mechanism. Consider the reel from which the tape is pulled. As the tape is pulled from it, the radius of the roll of remaining tape decreases. How does the torque on the reel change with time? How does the angular speed of the reel change in time? If the drive mechanism is switched on so that the tape is suddenly jerked with a large force, is the tape more likely to break when it is being pulled from a nearly full reel or from a nearly empty reel?

8. Vector $\vec{\mathbf{A}}$ is in the negative y direction and vector $\vec{\mathbf{B}}$ is in the negative x direction. What are the directions of (a) $\vec{\mathbf{A}} \times \vec{\mathbf{B}}$ and (b) $\vec{\mathbf{B}} \times \vec{\mathbf{A}}$?

9. For a helicopter to be stable as it flies, it must have at least two propellers. Why?

10. Often when a high diver wants to turn a flip in midair, she draws her legs up against her chest. Why does this movement make her rotate faster? What should she do when she wants to come out of her flip?

11. Why does a long pole help a tightrope walker stay balanced?

12. In some motorcycle races, the riders drive over small hills and the motorcycle becomes airborne for a short time. If the motorcycle racer keeps the throttle open while leaving the hill and going into the air, the motorcycle tends to nose upward. Why?

13. If global warming continues over the next one hundred years, it is likely that some polar ice will melt and the water will be distributed closer to the Equator. How would that change the moment of inertia of the Earth? Would the length of the day (one revolution) increase or decrease?

14. Two uniform solid spheres, a large, massive sphere and a small sphere with low mass, are rolled down a hill. Which one reaches the bottom of the hill first? Next, we roll a large, low-density sphere and a small high-density sphere of equal mass. Which one wins in this case?

15. In a soapbox derby race, the cars have no engines; they simply coast down a hill to race with one another. Suppose you are designing a car for a coasting race. Do you want to use large wheels or small wheels? Do you want to use solid disk-like wheels or hoop-like wheels? Should the wheels be heavy or light?

16. Stand with your back against a wall. Why can't you put your heels firmly against the wall and then bend forward without falling?

17. A ladder stands on the ground, leaning against a wall. Would you feel safer climbing up the ladder if you were told that the ground is frictionless but the wall is rough, or that the wall is frictionless but the ground is rough? Justify your answer.

18. (a) Give an example in which the net force acting on an object is zero and yet the net torque is nonzero. (b) Give an example in which the net torque acting on an object is zero and yet the net force is nonzero.

PROBLEMS

1, 2, 3 = straightforward, intermediate, challenging
□ = full solution available in the *Student Solutions Manual and Study Guide*
Physics Now™ = coached problem with hints available at www.pop4e.com
= computer useful in solving problem
= paired numerical and symbolic problems
= biomedical application

Section 10.1 ▪ Angular Position, Speed, and Acceleration

1. During a certain period of time, the angular position of a swinging door is described by $\theta = 5.00 + 10.0t + 2.00t^2$, where θ is in radians and t is in seconds. Determine the angular position, angular speed, and angular acceleration of the door (a) at $t = 0$ and (b) at $t = 3.00$ s.

Section 10.2 ▪ Rotational Kinematics: The Rigid Object Under Constant Angular Acceleration

2. A dentist's drill starts from rest. After 3.20 s of constant angular acceleration, it turns at a rate of 2.51×10^4 rev/min. (a) Find the drill's angular acceleration. (b) Determine the angle (in radians) through which the drill rotates during this period.

3. Physics Now™ An electric motor rotating a grinding wheel at 100 rev/min is switched off. The wheel then moves with a constant negative angular acceleration of magnitude 2.00 rad/s^2. (a) During what time interval does the wheel come to rest? (b) Through how many radians does it turn while it is slowing down?

4. A centrifuge in a medical laboratory rotates at an angular speed of 3 600 rev/min. When switched off, it rotates through 50.0 revolutions before coming to rest. Find the constant angular acceleration of the centrifuge.

5. The tub of a washer goes into its spin cycle, starting from rest and gaining angular speed steadily for 8.00 s, at which time it is turning at 5.00 rev/s. At this point, the person doing the laundry opens the lid and a safety switch turns off the washer. The tub smoothly slows to rest in 12.0 s. Through how many revolutions does the tub turn while it is in motion?

6. A rotating wheel requires 3.00 s to rotate through 37.0 rev. Its angular speed at the end of the 3.00-s interval is 98.0 rad/s. What is the constant angular acceleration of the wheel?

7. (a) Find the angular speed of the Earth's rotation on its axis. As the Earth turns toward the east, we see the sky turning toward the west at this same rate.
(b) *The rainy Pleiads wester*
And seek beyond the sea
The head that I shall dream of
That shall not dream of me.
—A. E. Housman (© Robert E. Symons)

Cambridge, England, is at longitude 0° and Saskatoon, Saskatchewan, is at longitude 107° west. How much time elapses after the Pleiades set in Cambridge until these stars fall below the western horizon in Saskatoon?

Section 10.3 ■ Relations Between Rotational and Translational Quantities

8. Make an order-of-magnitude estimate of the number of revolutions through which a typical automobile tire turns in 1 yr. State the quantities you measure or estimate and their values.

9. **PhysicsNow™** A disk 8.00 cm in radius rotates at a constant rate of 1 200 rev/min about its central axis. Determine (a) its angular speed, (b) the tangential speed at a point 3.00 cm from its center, (c) the radial acceleration of a point on the rim, and (d) the total distance a point on the rim moves in 2.00 s.

10. A wheel 2.00 m in diameter lies in a vertical plane and rotates with a constant angular acceleration of 4.00 rad/s^2. The wheel starts at rest at $t = 0$, and the radius vector of a certain point P on the rim makes an angle of 57.3° with the horizontal at this time. At $t = 2.00$ s, find (a) the angular speed of the wheel, (b) the tangential speed and the total acceleration of the point P, and (c) the angular position of the point P.

11. Figure P10.11 shows the drivetrain of a bicycle that has wheels 67.3 cm in diameter and pedal cranks 17.5 cm long. The cyclist pedals at a steady cadence of 76.0 rev/min. The chain engages with a front sprocket 15.2 cm in diameter and a rear sprocket 7.00 cm in diameter. (a) Calculate the speed of a link of the chain relative to the bicycle frame. (b) Calculate the angular speed of the bicycle wheels. (c) Calculate the speed of the bicycle relative to the road. (d) What pieces of data, if any, are not necessary for the calculations?

12. A digital audio compact disc carries data, each bit of which occupies 0.6 μm along a continuous spiral track from the inner circumference of the disc to the outside edge. A CD player turns the disc to carry the track counterclockwise above a lens at a constant speed of 1.30 m/s. Find the required angular speed (a) at the beginning of the recording, where the spiral has a radius of 2.30 cm, and (b) at the end of the recording, where the spiral has a radius of 5.80 cm. (c) A full-length recording lasts for 74 min 33 s. Find the average angular acceleration of the disc. (d) Assuming that the acceleration is constant, find the total angular displacement of the disc as it plays. (e) Find the total length of the track.

FIGURE **P10.11**

13. A car traveling on a flat (unbanked) circular track accelerates uniformly from rest with a tangential acceleration of 1.70 m/s^2. The car makes it one fourth of the way around the circle before it skids off the track. Determine the coefficient of static friction between the car and track from these data.

Section 10.4 ■ Rotational Kinetic Energy

14. Rigid rods of negligible mass lying along the y axis connect three particles (Fig. P10.14). The system rotates about the x axis with an angular speed of 2.00 rad/s. Find (a) the moment of inertia about the x axis and the total rotational kinetic energy evaluated from $\frac{1}{2}I\omega^2$ and (b) the tangential speed of each particle and the total kinetic energy evaluated from $\Sigma \frac{1}{2}m_i v_i^2$.

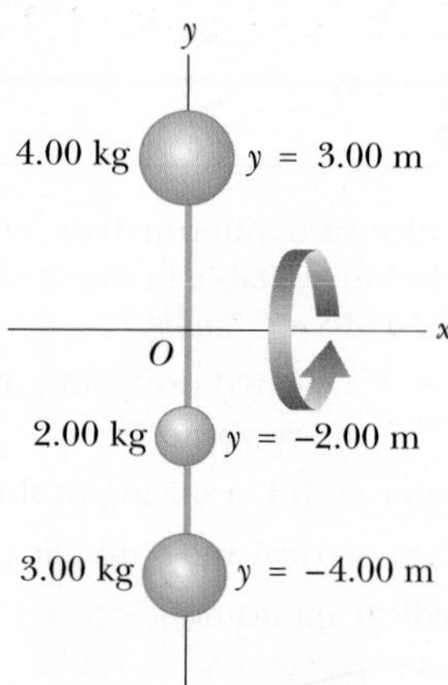

FIGURE **P10.14**

15. This problem describes one experimental method for determining the moment of inertia of an irregularly shaped object such as the payload for a satellite. Figure P10.15 shows a counterweight of mass m suspended by a cord wound around a spool of radius r, forming part of a

turntable supporting the object. The turntable can rotate without friction. When the counterweight is released from rest, it descends through a distance h, acquiring a speed v. Show that the moment of inertia I of the rotating apparatus (including the turntable) is $mr^2(2gh/v^2 - 1)$.

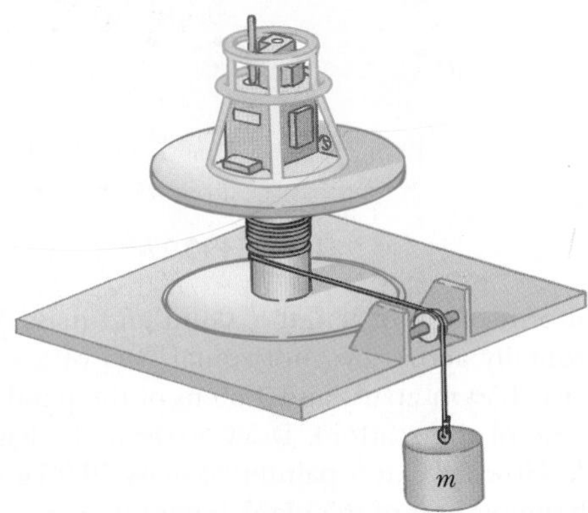

FIGURE **P10.15**

16. Big Ben, the Parliament tower clock in London, has an hour hand 2.70 m long with a mass of 60.0 kg and a minute hand 4.50 m long with a mass of 100 kg (Fig. P10.16). Calculate the total rotational kinetic energy of the two hands about the axis of rotation. (You may model the hands as uniform long, thin rods.)

FIGURE **P10.16** Problems 10.16, 10.42, and 10.64.

17. Consider two objects with $m_1 > m_2$ connected by a light string that passes over a pulley having a moment of inertia of I about its axis of rotation as shown in Figure P10.17. The string does not slip on the pulley or stretch. The pulley turns without friction. The two objects are released from rest separated by a vertical distance $2h$. (a) Use the principle of conservation of energy to find the translational speeds of the objects as they pass each other. (b) Find the angular speed of the pulley at this time.

18. As a gasoline engine operates, a flywheel turning with the crankshaft stores energy after each fuel explosion, providing the energy required to compress the next charge of fuel and air. For the engine of a certain lawn tractor, suppose a flywheel must be no more than 18.0 cm in diameter. Its thickness, measured along its axis of rotation, must be no larger than 8.00 cm. The flywheel must release energy 60.0 J when its angular speed drops from 800 rev/min to 600 rev/min. Design a sturdy, steel flywheel to meet these requirements with the smallest mass that you can reasonably attain. Assume that the material has the density listed for iron in Table 15.1. Specify the shape and mass of the flywheel.

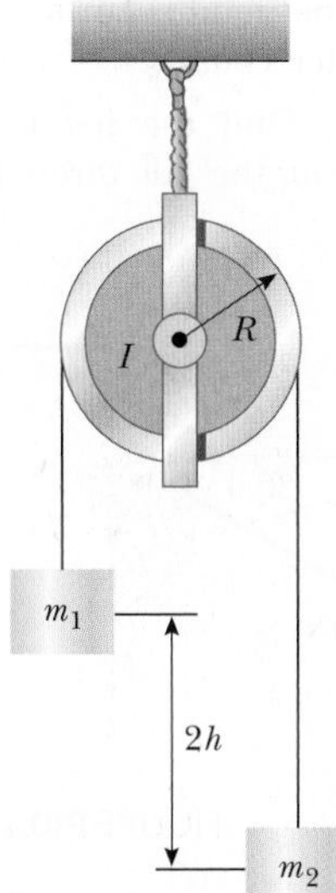

FIGURE **P10.17**

19. A *war-wolf* or *trebuchet* is a device used during the Middle Ages to throw rocks at castles and sometimes now used to fling pianos as a sport. A simple trebuchet is shown in Figure P10.19. Model it as a stiff rod of negligible mass, 3.00 m long, joining particles of mass 60.0 kg and 0.120 kg at its ends. It can turn on a frictionless horizontal axle perpendicular to the rod and 14.0 cm from the large-mass particle. The rod is released from rest in a horizontal orientation. Find the maximum speed that the small-mass object attains.

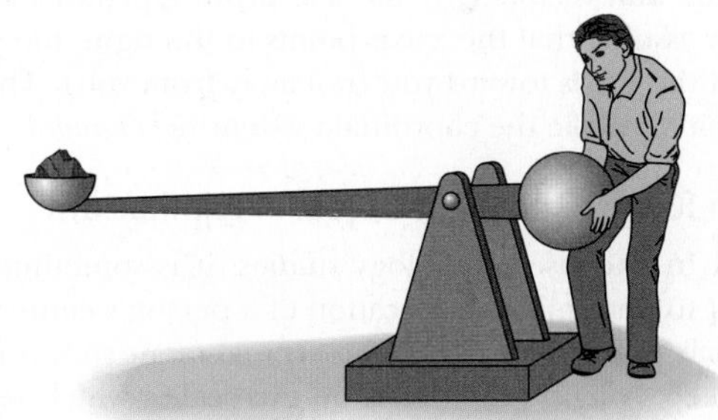

FIGURE **P10.19**

Section 10.5 ■ Torque and the Vector Product

20. The fishing pole in Figure P10.20 makes an angle of 20.0° with the horizontal. What is the torque exerted by the fish

FIGURE **P10.20**

about an axis perpendicular to the page and passing through the angler's hand?

21. **Physics Now™** Find the net torque on the wheel in Figure P10.21 about the axle through O, taking $a = 10.0$ cm and $b = 25.0$ cm.

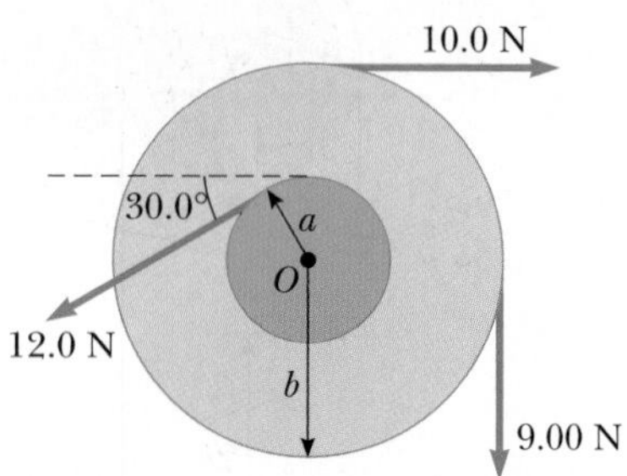

FIGURE **P10.21**

22. Given $\vec{\mathbf{M}} = 6\hat{\mathbf{i}} + 2\hat{\mathbf{j}} - \hat{\mathbf{k}}$ and $\vec{\mathbf{N}} = 2\hat{\mathbf{i}} - \hat{\mathbf{j}} - 3\hat{\mathbf{k}}$, calculate the vector product $\vec{\mathbf{M}} \times \vec{\mathbf{N}}$.

23. A force of $\vec{\mathbf{F}} = (2.00\hat{\mathbf{i}} + 3.00\hat{\mathbf{j}})$ N is applied to an object that is pivoted about a fixed axle aligned along the z coordinate axis. The force is applied at the point $\vec{\mathbf{r}} = (4.00\hat{\mathbf{i}} + 5.00\hat{\mathbf{j}})$ m. Find (a) the magnitude of the net torque about the z axis and (b) the direction of the torque vector $\vec{\boldsymbol{\tau}}$.

24. Two vectors are given by $\vec{\mathbf{A}} = -3\hat{\mathbf{i}} + 7\hat{\mathbf{j}} - 4\hat{\mathbf{k}}$ and $\vec{\mathbf{B}} = 6\hat{\mathbf{i}} - 10\hat{\mathbf{j}} + 9\hat{\mathbf{k}}$. Evaluate the following quantities: (a) $\cos^{-1}[\vec{\mathbf{A}} \cdot \vec{\mathbf{B}}/AB]$ and (b) $\sin^{-1}[|\vec{\mathbf{A}} \times \vec{\mathbf{B}}|/AB]$. (c) Which give(s) the angle between the vectors?

25. Use the definition of the vector product and the definitions of the unit vectors $\hat{\mathbf{i}}$, $\hat{\mathbf{j}}$, and $\hat{\mathbf{k}}$ to prove Equations 10.23. You may assume that the x axis points to the right, the y axis up, and the z axis toward you (not away from you). This choice is said to make the coordinate system *right-handed.*

Section 10.6 ■ The Rigid Object in Equilibrium

26. In exercise physiology studies, it is sometimes important to determine the location of a person's center of mass, which can be done with the arrangement shown in Figure P10.26. A light plank rests on two scales, which read $F_{g1} = 380$ N and $F_{g2} = 320$ N. A distance of 2.00 m separates the scales. How far from the woman's feet is her center of mass?

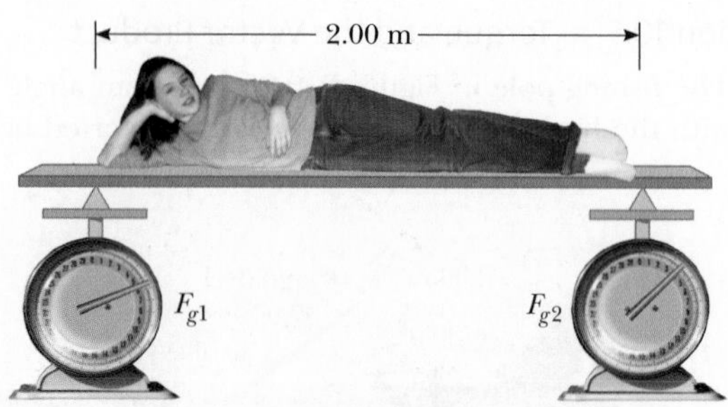

FIGURE **P10.26**

27. A uniform beam of mass m_b and length ℓ supports blocks with masses m_1 and m_2 at two positions as shown in Figure P10.27. The beam rests on two knife edges. For what value of x will the beam be balanced at P such that the normal force at O is zero?

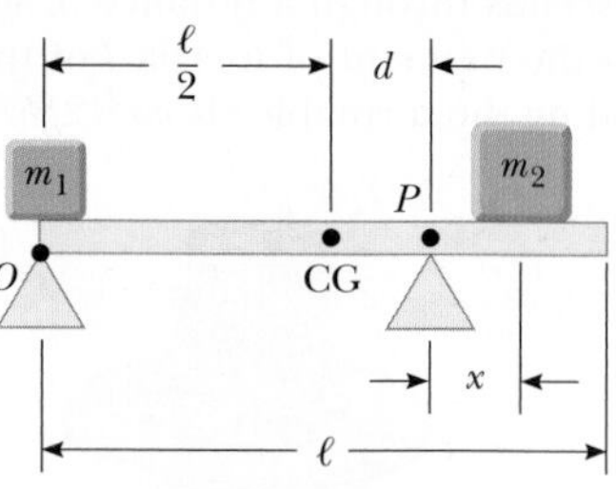

FIGURE **P10.27**

28. A uniform plank of length 6.00 m and mass 30.0 kg rests horizontally across two horizontal bars of a scaffold. The bars are 4.50 m apart, and 1.50 m of the plank hangs over one side of the scaffold. Draw a free-body diagram of the plank. How far can a painter of mass 70.0 kg walk on the overhanging part of the plank before it tips?

29. Figure P10.29 shows a claw hammer as it is being used to pull a nail out of a horizontal board. A force of 150 N is exerted horizontally as shown. Find (a) the force exerted by the hammer claws on the nail and (b) the force exerted by the surface on the point of contact with the hammer head. Assume that the force the hammer exerts on the nail is parallel to the nail.

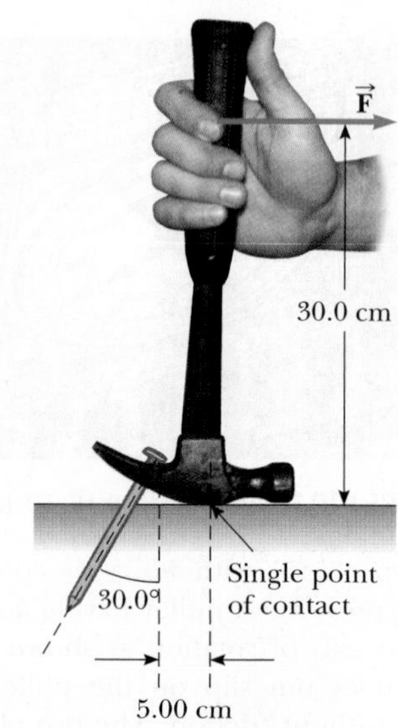

FIGURE **P10.29**

30. A uniform ladder of length L and mass m_1 rests against a frictionless wall. The ladder makes an angle θ with the horizontal. (a) Find the horizontal and vertical forces the ground exerts on the base of the ladder when a firefighter of mass m_2 is a distance x from the bottom. (b) If the ladder is just on the verge of slipping when the firefighter is a distance d from the bottom, what is the coefficient of static friction between ladder and ground?

31. **Physics Now™** A uniform sign of weight F_g and width $2L$ hangs from a light, horizontal beam hinged at the wall and supported by a cable (Fig. P10.31). Determine (a) the tension in the cable and (b) the components of the reaction

force exerted by the wall on the beam, in terms of F_g, d, L, and θ.

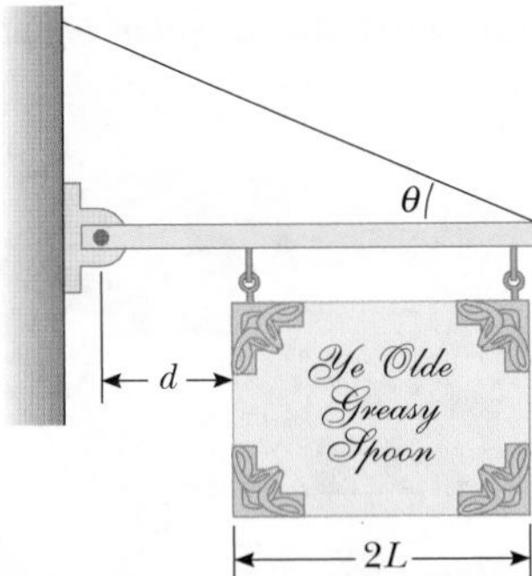

FIGURE **P10.31**

32. A crane of mass 3 000 kg supports a load of 10 000 kg as shown in Figure P10.32. The crane is pivoted with a frictionless pin at A and rests against a smooth support at B. Find the reaction forces at A and B.

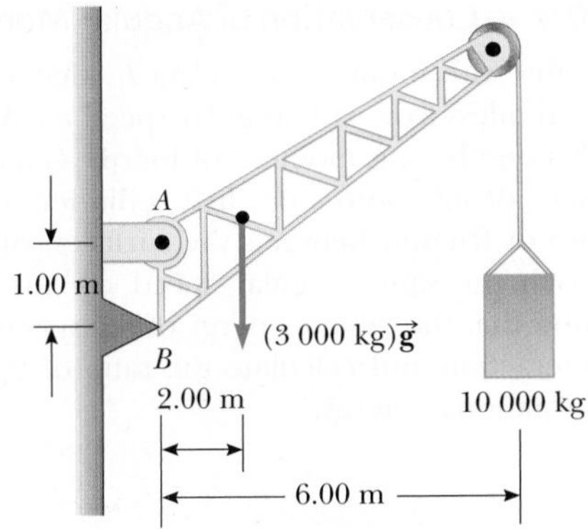

FIGURE **P10.32**

Section 10.7 ▪ The Rigid Object Under a Net Torque

33. The combination of an applied force and a friction force produces a constant total torque of 36.0 N·m on a wheel rotating about a fixed axis. The applied force acts for 6.00 s. During this time the angular speed of the wheel increases from 0 to 10.0 rad/s. The applied force is then removed, and the wheel comes to rest in 60.0 s. Find (a) the moment of inertia of the wheel, (b) the magnitude of the frictional torque, and (c) the total number of revolutions of the wheel.

34. A potter's wheel—a thick stone disk of radius 0.500 m and mass 100 kg—is freely rotating at 50.0 rev/min. The potter can stop the wheel in 6.00 s by pressing a wet rag against the rim and exerting a radially inward force of 70.0 N. Find the effective coefficient of kinetic friction between wheel and rag.

35. An electric motor turns a flywheel through a drive belt that joins a pulley on the motor and a pulley that is rigidly attached to the flywheel, as shown in Figure P10.35. The flywheel is a solid disk with a mass of 80.0 kg and a diameter of 1.25 m. It turns on a frictionless axle. Its pulley has much smaller mass and a radius of 0.230 m. The tension in the upper (taut) segment of the belt is 135 N, and the flywheel has a clockwise angular acceleration of 1.67 rad/s^2. Find the tension in the lower (slack) segment of the belt.

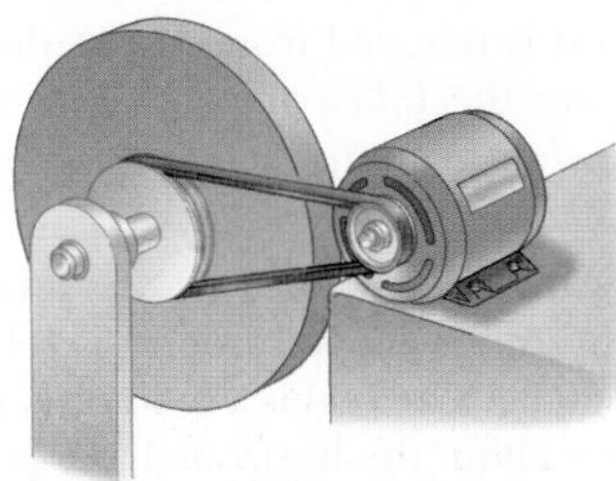

FIGURE **P10.35**

36. In Figure P10.36, the sliding block has a mass of 0.850 kg, the counterweight has a mass of 0.420 kg, and the pulley is a hollow cylinder with a mass of 0.350 kg, an inner radius of 0.020 0 m, and an outer radius of 0.030 0 m. The coefficient of kinetic friction between the block and the horizontal surface is 0.250. The pulley turns without friction on its axle. The light cord does not stretch and does not slip on the pulley. The block has a velocity of 0.820 m/s toward the pulley when it passes through a photogate. (a) Use energy methods to predict its speed after it has moved to a second photogate, 0.700 m away. (b) Find the angular speed of the pulley at the same moment.

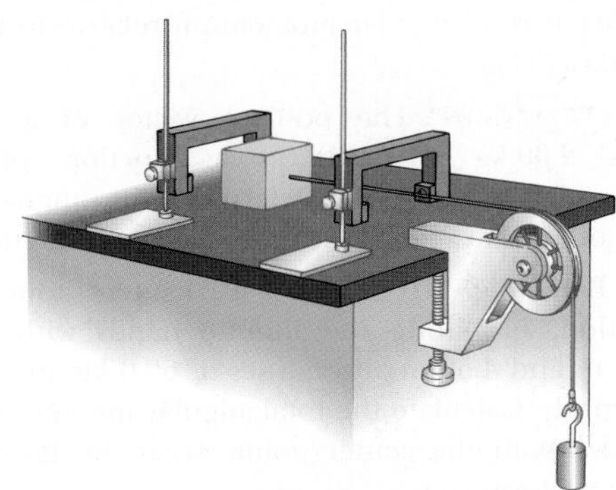

FIGURE **P10.36**

37. Two blocks, as shown in Figure P10.37, are connected by a string of negligible mass passing over a pulley of radius 0.250 m and moment of inertia I. The block on the frictionless incline is moving up with a constant acceleration of 2.00 m/s^2. (a) Determine T_1 and T_2, the tensions in the

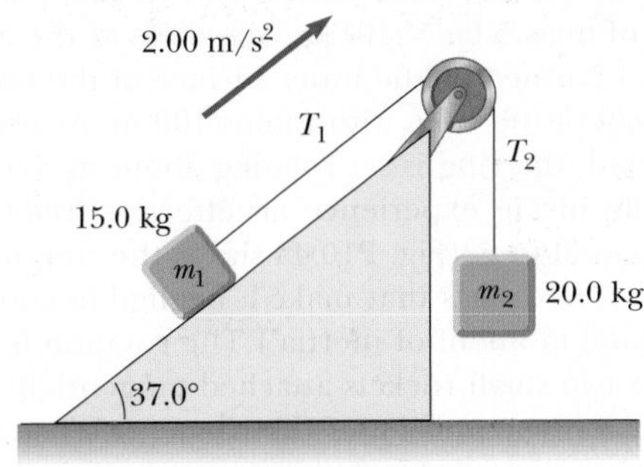

FIGURE **P10.37**

two parts of the string. (b) Find the moment of inertia of the pulley.

38. A uniform rod of length L and mass M is free to rotate about a frictionless pivot at one end as shown in Figure 10.9. The rod is released from rest in the horizontal position. What are the *initial* angular acceleration of the rod and the *initial* translational acceleration of the right end of the rod?

39. An object with a weight of 50.0 N is attached to the free end of a light string wrapped around a reel of radius 0.250 m and mass 3.00 kg. The reel is a solid disk, free to rotate in a vertical plane about the horizontal axis passing through its center. The suspended object is released 6.00 m above the floor. (a) Determine the tension in the string, the acceleration of the object, and the speed with which the object hits the floor. (b) Verify your last answer by using the principle of conservation of energy to find the speed with which the object hits the floor.

Section 10.8 ■ Angular Momentum

40. Heading straight toward the summit of Pikes Peak, an airplane of mass 12 000 kg flies over the plains of Kansas at nearly constant altitude 4.30 km with constant velocity 175 m/s west. (a) What is the airplane's vector angular momentum relative to a wheat farmer on the ground directly below the airplane? (b) Does this value change as the airplane continues its motion along a straight line? (c) What is its angular momentum relative to the summit of Pikes Peak?

41. **PhysicsNow™** The position vector of a particle of mass 2.00 kg is given as a function of time by $\vec{\mathbf{r}} = (6.00\,\hat{\mathbf{i}} + 5.00t\,\hat{\mathbf{j}})$ m. Determine the angular momentum of the particle about the origin as a function of time.

42. Big Ben (Fig. P10.16), the Parliament tower clock in London, has hour and minute hands with lengths of 2.70 m and 4.50 m and masses of 60.0 kg and 100 kg, respectively. Calculate the total angular momentum of these hands about the center point. Treat the hands as long, thin, uniform rods.

43. A particle of mass 0.400 kg is attached to the 100-cm mark of a meter stick of mass 0.100 kg. The meter stick rotates on a horizontal, frictionless table with an angular speed of 4.00 rad/s. Calculate the angular momentum of the system when the stick is pivoted about an axis (a) perpendicular to the table through the 50.0-cm mark and (b) perpendicular to the table through the 0-cm mark.

44. A space station is constructed in the shape of a hollow ring of mass 5.00×10^4 kg. Members of the crew walk on a deck formed by the inner surface of the outer cylindrical wall of the ring, with radius 100 m. At rest when constructed, the ring is set rotating about its axis so that the people inside experience an effective free-fall acceleration equal to g. (Fig. P10.44 shows the ring together with some other parts that make a negligible contribution to the total moment of inertia.) The rotation is achieved by firing two small rockets attached tangentially to opposite points on the outside of the ring. (a) What angular momentum does the space station acquire? (b) How long must the rockets be fired if each exerts a thrust of 125 N? (c) Prove that the total torque on the ring, multiplied by the time interval found in part (b), is equal to the change in angular momentum, found in part (a). This equality represents the *angular impulse–angular momentum theorem.*

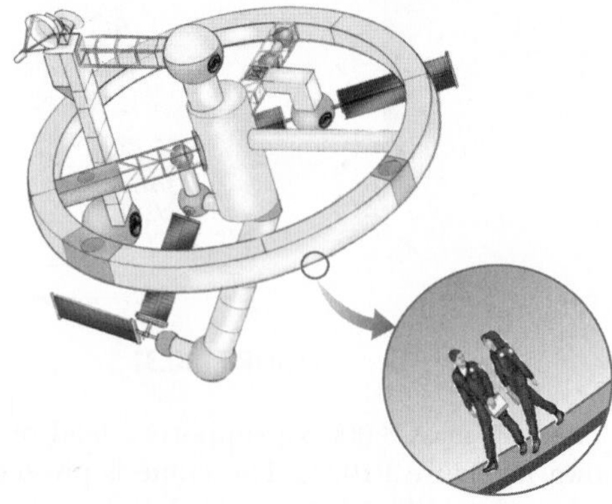

FIGURE P10.44 Problems 10.44 and 10.50.

Section 10.9 ■ Conservation of Angular Momentum

45. A cylinder with moment of inertia I_1 rotates about a vertical, frictionless axle with angular speed ω_i. A second cylinder, this one having moment of inertia I_2 and initially not rotating, drops onto the first cylinder (Fig. P10.45). Because of friction between the surfaces, the two eventually reach the same angular speed ω_f. (a) Calculate ω_f. (b) Show that the kinetic energy of the system decreases in this interaction and calculate the ratio of the final to the initial rotational energy.

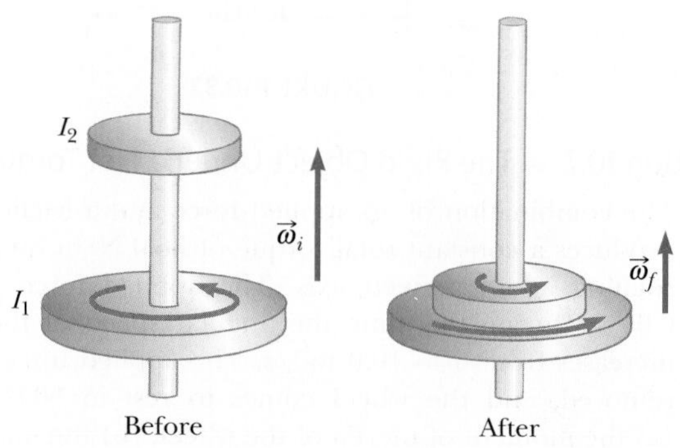

FIGURE P10.45

46. A playground merry-go-round of radius $R = 2.00$ m has a moment of inertia $I = 250$ kg·m^2 and is rotating at 10.0 rev/min about a frictionless vertical axle. Facing the axle, a 25.0-kg child hops onto the merry-go-round and manages to sit down on the edge. What is the new angular speed of the merry-go-round?

47. A 60.0-kg woman stands at the rim of a horizontal turntable having a moment of inertia of 500 kg·m^2 and a radius of 2.00 m. The turntable is initially at rest and is free to rotate about a frictionless, vertical axle through its center. The woman then starts walking around the rim clockwise (as viewed from above the system) at a constant speed of 1.50 m/s relative to the Earth. (a) In what direction and

with what angular speed does the turntable rotate? (b) How much work does the woman do to set herself and the turntable into motion?

48. A student sits on a freely rotating stool holding two weights, each of mass 3.00 kg (Fig. P10.48). When his arms are extended horizontally, the weights are 1.00 m from the axis of rotation and he rotates with an angular speed of 0.750 rad/s. The moment of inertia of the student plus stool is 3.00 kg · m^2 and is assumed to be constant. The student pulls the weights inward horizontally to a position 0.300 m from the rotation axis. (a) Find the new angular speed of the student. (b) Find the kinetic energy of the rotating system before and after he pulls the weights inward.

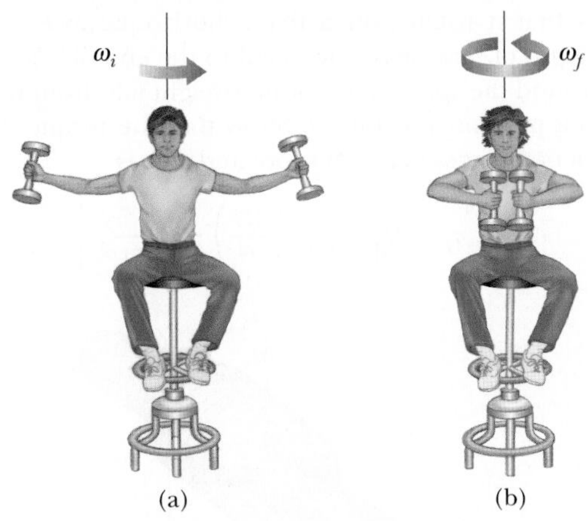

FIGURE **P10.48**

49. A puck of mass 80.0 g and radius 4.00 cm slides along an air table at a speed of 1.50 m/s as shown in Figure P10.49a. It makes a glancing collision with a second puck of radius 6.00 cm and mass 120 g (initially at rest) such that their rims just touch. Because their rims are coated with instant-acting glue, the pucks stick together and spin after the collision (Fig. P10.49b). (a) What is the angular momentum of the system relative to the center of mass? (b) What is the angular speed about the center of mass?

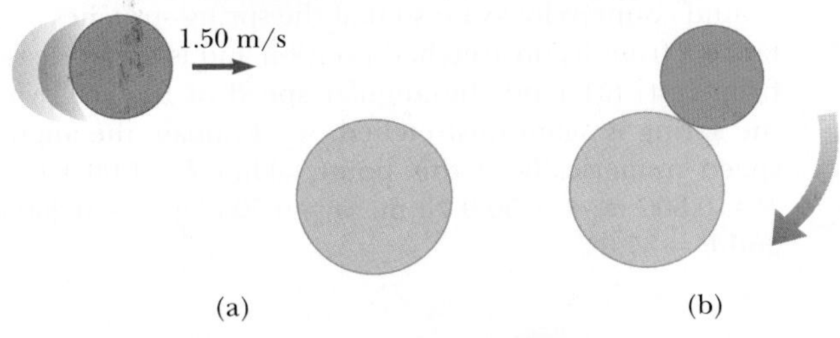

FIGURE **P10.49**

50. A space station shaped like a giant wheel has a radius of 100 m and a moment of inertia of 5.00 × 10^8 kg·m^2. A crew of 150 is living on the rim, and the station's rotation causes the crew to experience an apparent free-fall acceleration of g (Fig. P10.44). When 100 people move to the center of the station for a union meeting, the angular speed changes. Assume that the average mass for each inhabitant is 65.0 kg. What apparent free-fall acceleration is experienced by the managers remaining at the rim?

51. The puck in Figure 10.24 has a mass of 0.120 kg. The distance of the puck from the center of rotation is originally 40.0 cm, and the puck is sliding with a speed of 80.0 cm/s. The string is pulled downward 15.0 cm through the hole in the frictionless table. Determine the work done on the puck. (*Suggestion:* Consider the change of kinetic energy.)

Section 10.10 ▪ Precessional Motion of Gyroscopes

52. The angular momentum vector of a precessing gyroscope sweeps out a cone as shown in Figure 10.25b. Its angular speed, called its precessional frequency, is given by $\omega_p = \tau/L$, where τ is the magnitude of the torque on the gyroscope and L is the magnitude of its angular momentum. In the motion called *precession of the equinoxes,* represented in Figure P10.52, the Earth's axis of rotation precesses about the perpendicular to its orbital plane with a period of 2.58 × 10^4 yr. Model the Earth as a uniform sphere and calculate the torque on the Earth that is causing this precession.

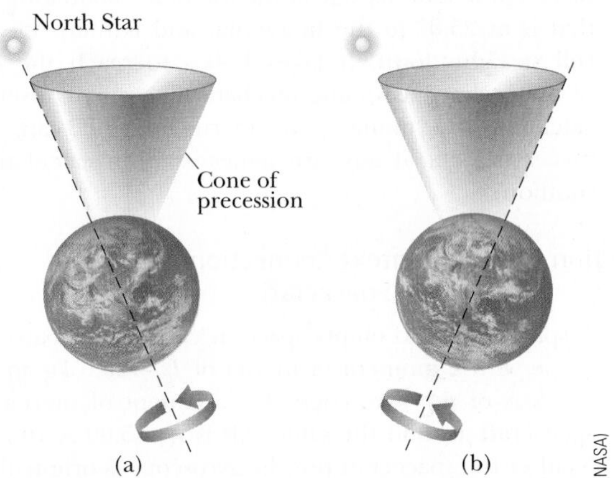

FIGURE **P10.52** (a) At present, the spin axis of the Earth points toward the North Star. (b) Torque on the spinning Earth will cause it to precess, so the spin axis will no longer be pointing in this direction in the future.

Section 10.11 ▪ Rolling Motion of Rigid Objects

53. A cylinder of mass 10.0 kg rolls without slipping on a horizontal surface. At a certain instant its center of mass has a speed of 10.0 m/s. Determine (a) the translational kinetic energy of its center of mass, (b) the rotational kinetic energy about its center of mass, and (c) its total energy.

54. A uniform solid disk and a uniform hoop are placed side by side at the top of an incline of height h. If they are released from rest and roll without slipping, which object reaches the bottom first? Verify your answer by calculating their speeds when they reach the bottom in terms of h.

55. A tennis ball is a hollow sphere with a thin wall. It is set rolling without slipping at 4.03 m/s on a horizontal section of a track as shown in Figure P10.55. It rolls around the inside of a vertical circular loop 90.0 cm in diameter and

finally leaves the track at a point 20.0 cm below the horizontal section. (a) Find the speed of the ball at the top of the loop. Demonstrate that it will not fall from the track. (b) Find its speed as it leaves the track. (c) Suppose static friction between ball and track were negligible so that the ball slid instead of rolling. Would its speed then be higher, lower, or the same at the top of the loop? Explain.

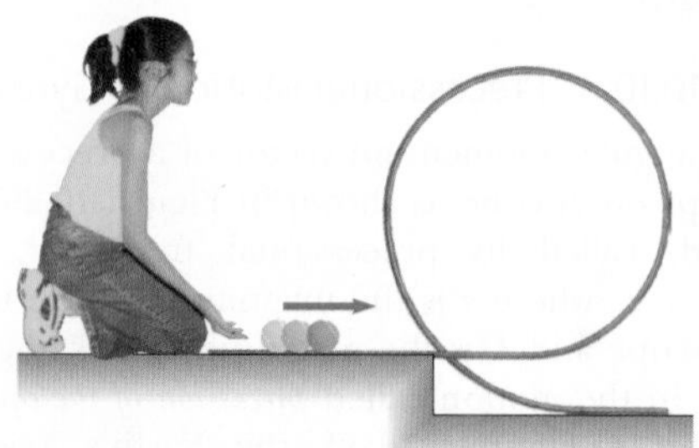

FIGURE **P10.55**

56. A metal can containing condensed mushroom soup has mass 215 g, height 10.8 cm, and diameter 6.38 cm. It is placed at rest on its side at the top of a 3.00-m-long incline that is at 25.0° to the horizontal and is then released to roll straight down. It takes 1.50 s to reach the bottom of the incline. Assuming mechanical energy conservation, calculate the moment of inertia of the can. Which pieces of data, if any, are unnecessary in calculating the solution?

Section 10.12 ■ Context Connection—Turning the Spacecraft

57. A spacecraft is in empty space. It carries on board a gyroscope with a moment of inertia of $I_g = 20.0\ \text{kg}\cdot\text{m}^2$ about the axis of the gyroscope. The moment of inertia of the spacecraft around the same axis is $I_s = 5.00 \times 10^5\ \text{kg}\cdot\text{m}^2$. Neither the spacecraft nor the gyroscope is originally rotating. The gyroscope can be powered up in a negligible period of time to an angular speed of $100\ \text{s}^{-1}$. If the orientation of the spacecraft is to be changed by 30.0°, for how long should the gyroscope be operated?

Additional Problems

58. **Review problem.** A mixing beater consists of three thin rods, each 10.0 cm long. The rods diverge from a central hub, separated from each other by 120°, and all turn in the same plane. A ball is attached to the end of each rod. Each ball has cross-sectional area 4.00 cm^2 and is so shaped that it has a drag coefficient of 0.600. Calculate the power input required to spin the beater at 1 000 rev/min (a) in air and (b) in water.

59. A long uniform rod of length L and mass M is pivoted about a horizontal, frictionless pin through one end. The rod is released from rest in a vertical position as shown in Figure P10.59. At the instant the rod is horizontal, find (a) its angular speed, (b) the magnitude of its angular acceleration, (c) the x and y components of the acceleration of its center of mass, and (d) the components of the reaction force at the pivot.

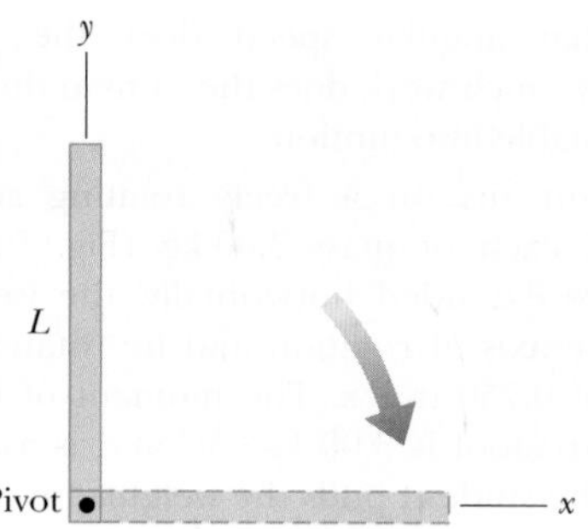

FIGURE **P10.59**

60. A uniform, hollow, cylindrical spool has inside radius $R/2$, outside radius R, and mass M (Fig. P10.60). It is mounted so that it rotates on a fixed, horizontal axle. A counterweight of mass m is connected to the end of a string wound around the spool. The counterweight falls from rest at $t = 0$ to a position y at time t. Show that the torque due to the friction forces between spool and axle is

$$\tau_f = R\left[m\left(g - \frac{2y}{t^2}\right) - M\frac{5y}{4t^2}\right]$$

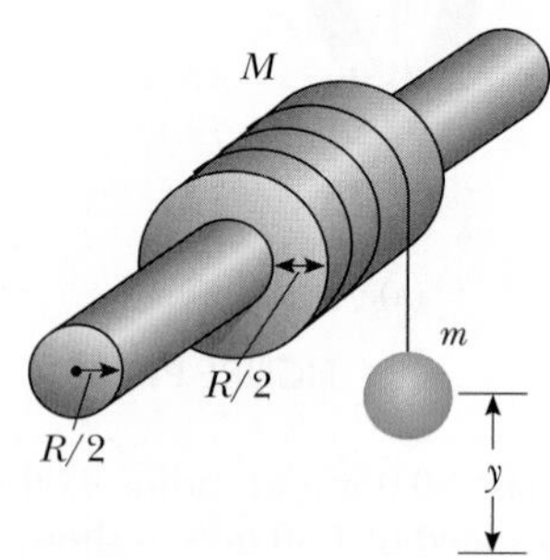

FIGURE **P10.60**

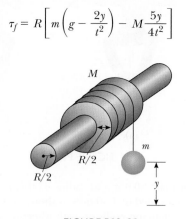

61. The reel shown in Figure P10.61 has radius R and moment of inertia I. One end of the block of mass m is connected to a spring of force constant k, and the other end is fastened to a cord wrapped around the reel. The reel axle and the incline are frictionless. The reel is wound counterclockwise so that the spring stretches a distance d from its unstretched position and is then released from rest. (a) Find the angular speed of the reel when the spring is again unstretched. (b) Evaluate the angular speed numerically at this point taking $I = 1.00\ \text{kg}\cdot\text{m}^2$, $R = 0.300$ m, $k = 50.0$ N/m, $m = 0.500$ kg, $d = 0.200$ m, and $\theta = 37.0°$.

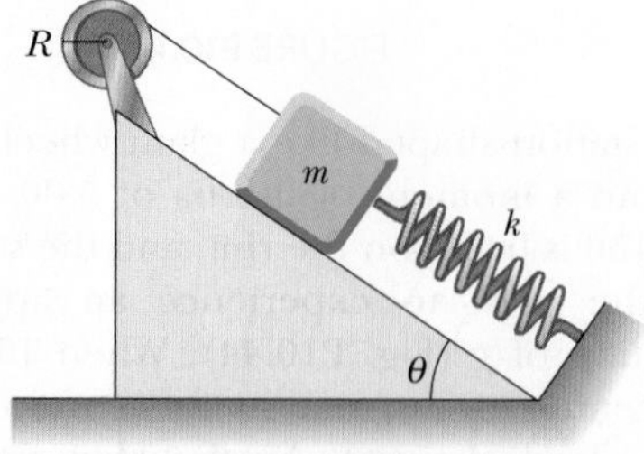

FIGURE **P10.61**

62. A block of mass $m_1 = 2.00$ kg and a block of mass $m_2 = 6.00$ kg are connected by a massless string over a pulley in the shape of a solid disk having radius $R = 0.250$ m and mass $M = 10.0$ kg. These blocks are allowed to move on a fixed block-wedge of angle $\theta = 30.0°$ as shown in Figure P10.62. The coefficient of kinetic friction is 0.360 for both blocks. Draw free-body diagrams of both blocks and of the pulley. Determine (a) the acceleration of the two blocks and (b) the tensions in the string on both sides of the pulley.

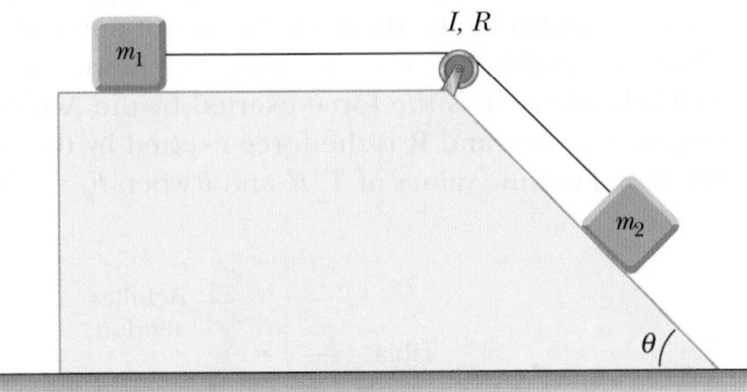

FIGURE **P10.62**

63. A common demonstration, illustrated in Figure P10.63, consists of a ball resting at one end of a uniform board of length ℓ, hinged at the other end, and elevated at an angle θ. A light cup is attached to the board at r_c so that it will catch the ball when the support stick is suddenly removed. (a) Show that the ball will lag behind the falling board when θ is less than 35.3°. (b) Assume that the board is 1.00 m long and is supported at this limiting angle. Show that the cup must be 18.4 cm from the moving end.

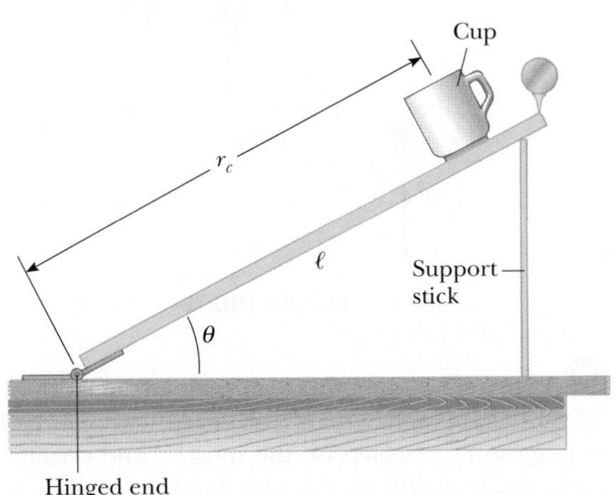

FIGURE **P10.63**

64. The hour hand and the minute hand of Big Ben, the Parliament tower clock in London, are 2.70 m and 4.50 m long and have masses of 60.0 kg and 100 kg, respectively (see Fig. P10.16). (a) Determine the total torque due to the weight of these hands about the axis of rotation when the time reads (i) 3:00, (ii) 5:15, (iii) 6:00, (iv) 8:20, and (v) 9:45. (You may model the hands as long, thin, uniform rods.) (b) Determine all times when the total torque about the axis of rotation is zero. Determine the times to the nearest second, solving a transcendental equation numerically.

65. A string is wound around a uniform disk of radius R and mass M. The disk is released from rest with the string vertical and its top end tied to a fixed bar (Fig. P10.65). Show that (a) the tension in the string is one-third the weight of the disk, (b) the magnitude of the acceleration of the center of mass is $2g/3$, and (c) the speed of the center of mass is $(4gh/3)^{1/2}$ after the disk has descended through distance h. Verify your answer to (c) using the energy approach.

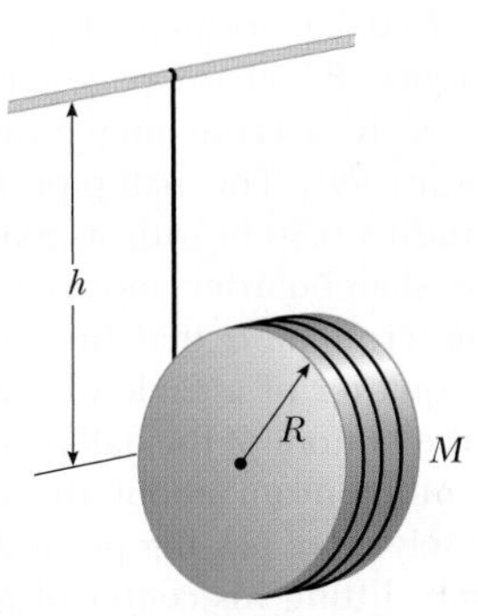

FIGURE **P10.65**

66. A new General Electric stove has a mass of 68.0 kg and the dimensions shown in Figure P10.66. The stove comes with a warning that it can tip forward if a person stands or sits on the oven door when it is open. What can you conclude about the weight of such a person? Could it be a child? List the assumptions you make in solving this problem. The stove is supplied with a wall bracket to prevent the accident.

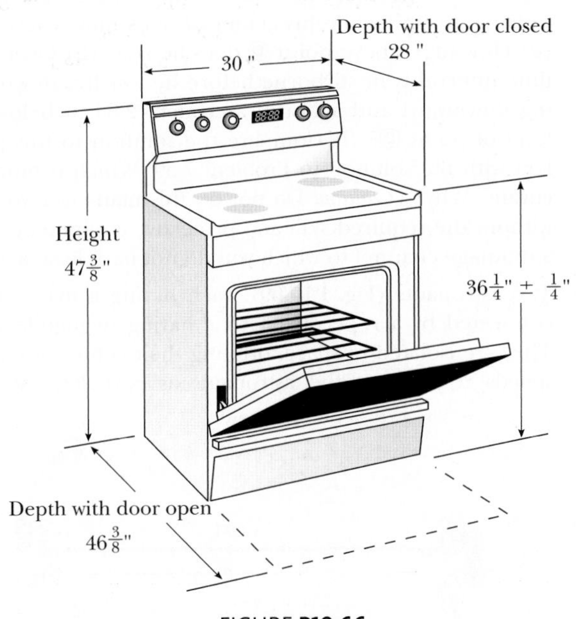

FIGURE **P10.66**

67. (a) Without the wheels, a bicycle frame has a mass of 8.44 kg. Each of the wheels can be roughly modeled as a

uniform solid disk with a mass of 0.820 kg and a radius of 0.343 m. Find the kinetic energy of the whole bicycle when it is moving forward at 3.35 m/s. (b) Before the invention of a wheel turning on an axle, ancient people moved heavy loads by placing rollers under them. (Modern people use rollers, too. Any hardware store will sell you a roller bearing for a lazy Susan.) A stone block of mass 844 kg moves forward at 0.335 m/s, supported by two uniform cylindrical tree trunks each of mass 82.0 kg and radius 0.343 m. No slipping occurs between the block and the rollers or between the rollers and the ground. Find the total kinetic energy of the moving objects.

68. A skateboarder with his board can be modeled as a particle of mass 76.0 kg, located at his center of mass. As shown in Figure P7.59 on page 218, the skateboarder starts from rest in a crouching position at one lip of a half-pipe (point Ⓐ). The half-pipe forms one half of a cylinder of radius 6.80 m with its axis horizontal. On his descent, the skateboarder moves without friction and maintains his crouch so that his center of mass moves through one quarter of a circle of radius 6.30 m. (a) Find his speed at the bottom of the half-pipe (point Ⓑ). (b) Find his angular momentum about the center of curvature. (c) Immediately after passing point Ⓑ, he stands up and raises his arms, lifting his center of gravity from 0.500 m to 0.950 m above the concrete (point Ⓒ). Explain why his angular momentum is constant in this maneuver, whereas his linear momentum and his mechanical energy are not constant. (d) Find his speed immediately after he stands up, when his center of mass is moving in a quarter circle of radius 5.85 m. (e) What work did the skateboarder's legs do on his body as he stood up? Next, the skateboarder glides upward with his center of mass moving in a quarter circle of radius 5.85 m. His body is horizontal when he passes point Ⓓ, the far lip of the half-pipe. (f) Find his speed at this location. At last he goes ballistic, twisting around while his center of mass moves vertically. (g) How high above point Ⓓ does he rise? (h) Over what time interval is he airborne before he touches down, facing downward and again in a crouch, 2.34 m below the level of point Ⓓ? (i) Compare the solution to this problem with the solution to Problem 7.59. Which is more accurate? Why? (*Caution*: Do not try this maneuver yourself without the required skill and protective equipment, or in a drainage channel to which you do not have legal access.)

69. Two astronauts (Fig. P10.69), each having a mass M, are connected by a rope of length d having negligible mass. They are isolated in space, orbiting their center of mass at speeds v. Treating the astronauts as particles, calculate (a) the magnitude of the angular momentum of the system and (b) the rotational energy of the system. By pulling on the rope, one of the astronauts shortens the distance between them to $d/2$. (c) What is the new angular momentum of the system? (d) What are the astronauts' new speeds? (e) What is the new rotational energy of the system? (f) How much work does the astronaut do in shortening the rope?

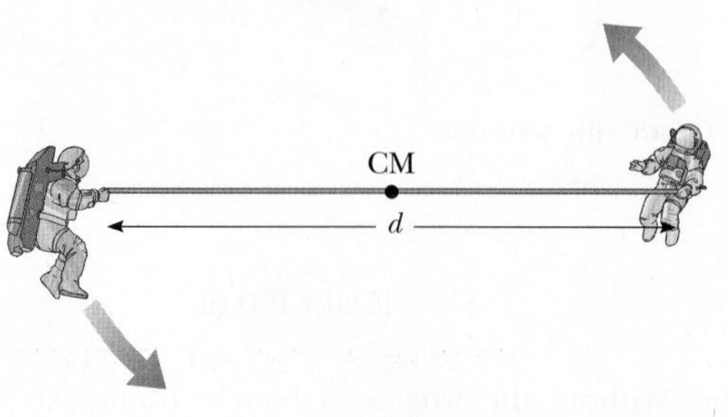

FIGURE P10.69

70. When a person stands on tiptoe (a strenuous position), the position of the foot is as shown in Figure P10.70a. The total gravitational force on the body $\vec{\mathbf{F}}_g$ is supported by the force $\vec{\mathbf{n}}$ exerted by the floor on the toes of one foot. A mechanical model for the situation is shown in Figure P10.70b, where $\vec{\mathbf{T}}$ is the force exerted by the Achilles tendon on the foot and $\vec{\mathbf{R}}$ is the force exerted by the tibia on the foot. Find the values of T, R, and θ when $F_g = 700$ N.

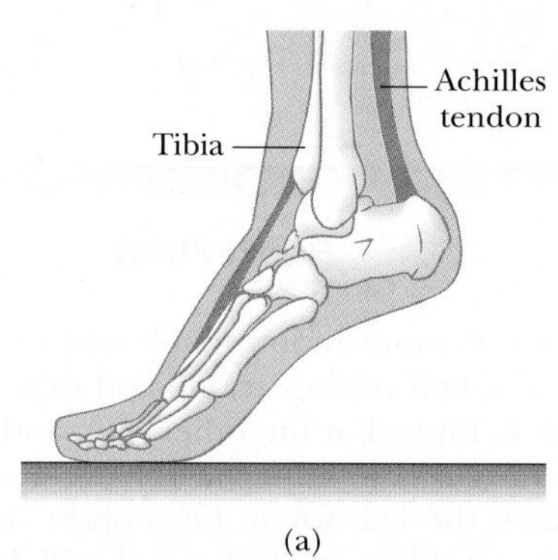

(a)

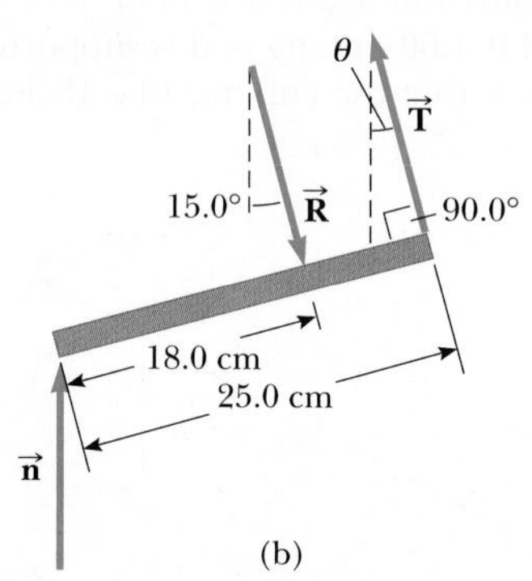

(b)

FIGURE P10.70

71. A person bending forward to lift a load "with his back" (Fig. P10.71a) rather than "with his knees" can be injured by large forces exerted on the muscles and vertebrae. The spine pivots mainly at the fifth lumbar vertebra, with the principal supporting force provided by the erector spinalis muscle in the back. To see the magnitude of the forces involved and to understand why back problems are common among humans, consider the model shown in Fig. P10.71b for a person bending forward to lift a 200-N object. The spine and upper body are represented as a uniform horizontal rod of weight 350 N, pivoted at the base of the spine. The erector spinalis muscle, attached at a point two thirds of the way up the spine, maintains the position of the back. The angle between the spine and this muscle is 12.0°. Find the tension

in the back muscle and the compressional force in the spine.

FIGURE **P10.71**

72. A wad of sticky clay with mass m and velocity $\vec{v}_i$ is fired at a solid cylinder of mass M and radius R (Fig. P10.72). The cylinder is initially at rest and is mounted on a fixed horizontal axle that runs through its center of mass. The line of motion of the projectile is perpendicular to the axle and at a distance $d < R$ from the center. (a) Find the angular speed of the system just after the clay strikes and sticks to the surface of the cylinder. (b) Is mechanical energy of the clay–cylinder system conserved in this process? Explain your answer.

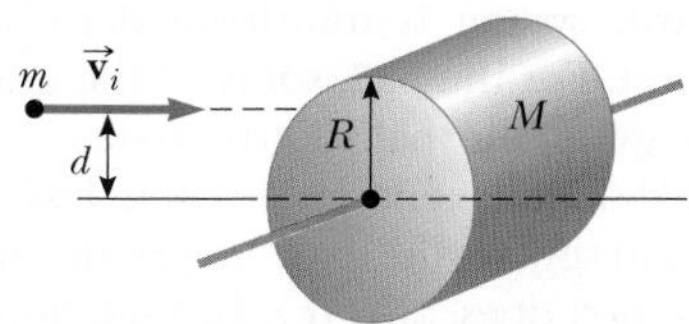

FIGURE **P10.72**

73. A force acts on a rectangular cabinet weighing 400 N as shown in Figure P10.73. (a) Assuming that the cabinet slides with constant speed when $F = 200$ N and $h = 0.400$ m, find the coefficient of kinetic friction and the position of the resultant normal force. (b) Taking $F = 300$ N, find the value of h for which the cabinet just begins to tip.

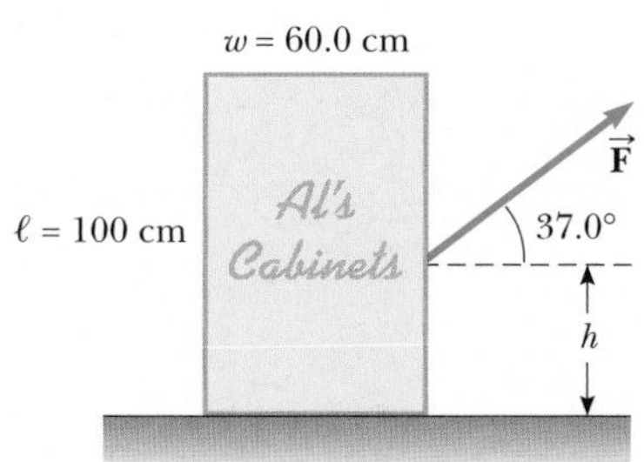

FIGURE **P10.73**

74. The following equations are obtained from a free-body diagram of a rectangular farm gate, supported by two hinges on the left-hand side. A bucket of grain is hanging from the latch.

$$-A + C = 0$$

$$+B - 392\text{ N} - 50.0\text{ N} = 0$$

$$A(0) + B(0) + C(1.80\text{ m}) - 392\text{ N}(1.50\text{ m}) - 50.0\text{ N}(3.00\text{ m}) = 0$$

(a) Draw the free-body diagram and complete the statement of the problem, specifying the unknowns. (b) Determine the values of the unknowns and state the physical meaning of each.

75. A stepladder of negligible weight is constructed as shown in Figure P10.75. A painter of mass 70.0 kg stands on the ladder 3.00 m from the bottom. Assuming that the floor is frictionless, find (a) the tension in the horizontal bar connecting the two halves of the ladder, (b) the normal forces at A and B, and (c) the components of the reaction force at the single hinge C that the left half of the ladder exerts on the right half. (*Suggestion:* Treat the ladder as a single object, but also treat each half of the ladder separately.)

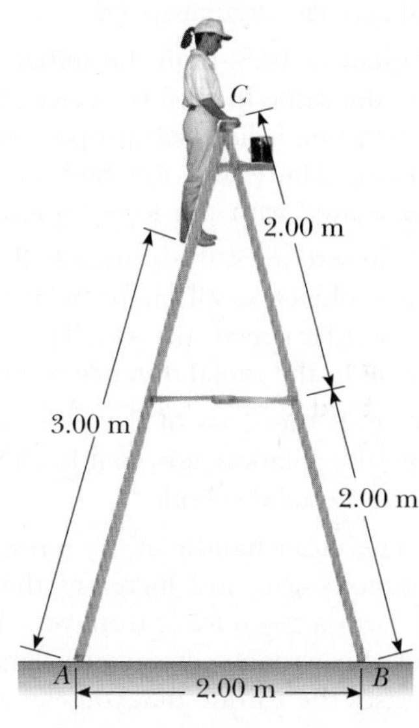

FIGURE **P10.75**

76. A solid sphere of mass m and radius r rolls without slipping along the track shown in Figure P10.76. It starts from rest with the lowest point of the sphere at height h above the

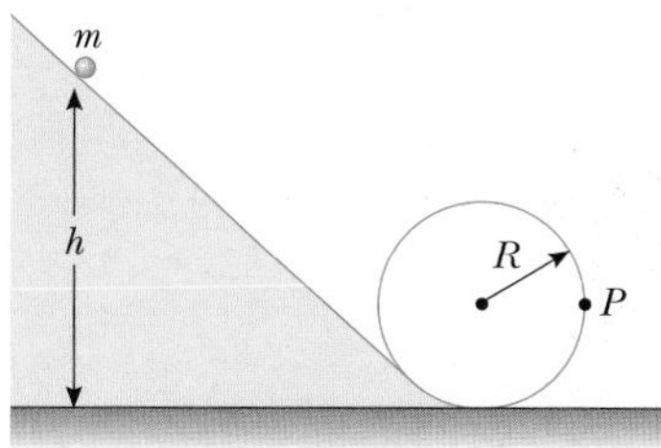

FIGURE **P10.76**

bottom of the loop of radius R, much larger than r. (a) What is the minimum value of h (in terms of R) such that the sphere completes the loop? (b) What are the force components on the sphere at the point P if $h = 3R$?

77. Figure P10.77 shows a vertical force applied tangentially to a uniform cylinder of weight F_g. The coefficient of static friction between the cylinder and all surfaces is 0.500. In terms of F_g, find the maximum force P that can be applied that does not cause the cylinder to rotate. (*Suggestion:* When the cylinder is on the verge of slipping, both friction forces are at their maximum values. Why?)

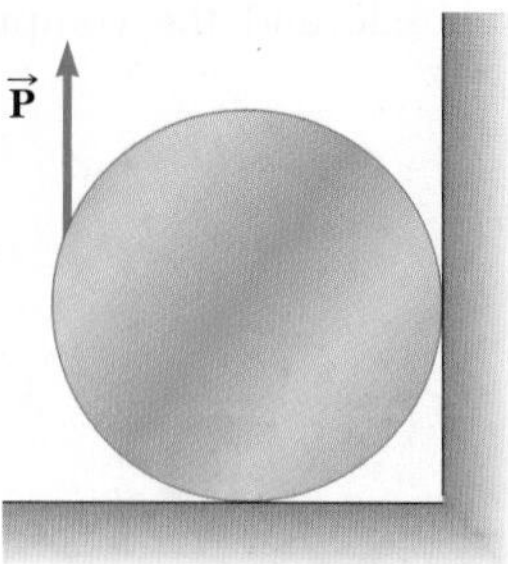

FIGURE **P10.77**

ANSWERS TO QUICK QUIZZES

10.1 (i), (c). For a rotation of more than 180°, the angular displacement must be larger than $\pi = 3.14$ rad. The angular displacements in the three choices are (a) 6 rad − 3 rad = 3 rad, (b) 1 rad − (−1) rad = 2 rad, and (c) 5 rad − 1 rad = 4 rad. (ii), (b). Because all angular displacements occur in the same time interval, the displacement with the lowest value will be associated with the lowest average angular speed.

10.2 (b). In Equation 10.8, both the initial and final angular speeds are the same in all three cases. As a result, the angular acceleration is inversely proportional to the angular displacement. Therefore, the highest angular acceleration is associated with the lowest angular displacement.

10.3 (i), (b). The system of the platform, Benjamin, and Torrey is a rigid object, so all points on the rigid object have the same angular speed. (ii), (a). The tangential speed is proportional to the radial distance from the rotation axis.

10.4 (a). Almost all the mass of the pipe is at the same distance from the rotation axis, so it has a larger moment of inertia than the solid cylinder.

10.5 (b), (c). The fatter handle of the screwdriver gives you a larger moment arm and increases the torque that you can apply with a given force from your hand. The longer handle of the wrench gives you a larger moment arm and increases the torque that you can apply with a given force from your hand.

10.6 (b). With twice the moment of inertia and the same frictional torque, there is half the angular acceleration. With half the angular acceleration, it will require twice as long to change the speed to zero.

10.7 (b). The hollow sphere has a larger moment of inertia than the solid sphere because much of its mass is far from the rotation axis. Because $L = I\omega$ and ω is the same for both objects, the hollow sphere has a larger angular momentum.

10.8 (i), (a). The diver is an isolated system, so the product $I\omega$ remains constant. As the moment of inertia of the diver decreases, the angular speed increases by the same factor. For example, if I goes down by a factor of 2, ω goes up by a factor of 2. (ii), (a). The rotational kinetic energy varies as the square of ω. If I is halved, ω^2 increases by a factor of 4 and the energy increases by a factor of 2.

10.9 (i), (b). All the gravitational potential energy of the box–Earth system is transformed to kinetic energy of translation. For the ball, some of the gravitational potential energy of the ball–Earth system is transformed to rotational kinetic energy, leaving less for translational kinetic energy, so the ball moves downhill more slowly than the box does. (ii), (c). In Equation 10.46, I_{CM} for a sphere is $\frac{2}{5}MR^2$. Therefore, MR^2 will cancel and the remaining expression on the right-hand side of the equation is independent of mass and radius. (iii), (a). The moment of inertia of the hollow sphere B is larger than that of sphere A. As a result, Equation 10.46 tells us that sphere B will have a smaller speed of the center of mass, so sphere A should arrive first.

Gravity, Planetary Orbits, and the Hydrogen Atom

The Rosette Nebula is a region of gas and dust surrounding an open cluster of stars. Bundles of matter in the Universe such as this one interact with other bundles of matter by means of the gravitational force. The red color is due to hydrogen atoms, excited by light from the stars in the cluster, making transitions from the $n = 3$ quantum state to the $n = 2$ state. In this chapter, we will study both the gravitational force and the origin of the red color in the hydrogen atoms.

CHAPTER OUTLINE

At the beginning of our discussion of mechanics in Chapter 1, we introduced the notion of modeling and defined four categories of models: geometric, simplification, analysis, and structural. We can apply our analysis models to two very common *structural models*. In this chapter we shall discuss a structural model for a large system—the Solar System—and a structural model for a small system—the hydrogen atom.

We return to Newton's law of universal gravitation—one of the fundamental force laws in nature—and show how it, together with our analysis models, enables us to understand the motions of planets, moons, and artificial Earth satellites.

We conclude this chapter with a discussion of Niels Bohr's model of the hydrogen atom, which represents an interesting mixture of classical and nonclassical physics. Despite the hybrid nature of the model, some of its predictions agree with experimental measurements made on hydrogen atoms. This discussion will be our first major venture into the area of *quantum physics,* which we will continue in Chapter 28.

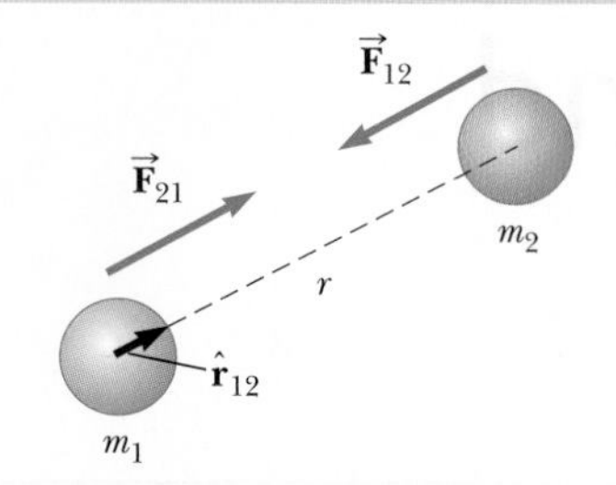

ACTIVE FIGURE 11.1
The gravitational force between two particles is attractive. The unit vector $\hat{\mathbf{r}}_{12}$ is directed from particle 1 toward particle 2. Note that $\vec{\mathbf{F}}_{21} = -\vec{\mathbf{F}}_{12}$.

Physics Now™ By logging into PhysicsNow at **www.pop4e.com** and going to Active Figure 11.1 you can change the separation distance between the particles to see the effect on the gravitational force.

11.1 NEWTON'S LAW OF UNIVERSAL GRAVITATION REVISITED

Prior to 1686, many data had been collected on the motions of the Moon and the planets, but a clear understanding of the forces involved with the motions was not yet attainable. In that year, Isaac Newton provided the key that unlocked the secrets of the heavens. He knew, from the first law of motion, that a net force had to be acting on the Moon. If not, the Moon would move in a straight-line path rather than in its almost circular orbit. Newton reasoned that this force between the Moon and the Earth was an attractive force. He also concluded that there could be nothing special about the Earth–Moon system or the Sun and its planets that would cause gravitational forces to act on them alone.

As you should recall from Chapter 5, every particle in the Universe attracts every other particle with a force that is directly proportional to the product of their masses and inversely proportional to the square of the distance between them. If two particles have masses m_1 and m_2 and are separated by a distance r, the magnitude of the gravitational force between them is

$$F_g = G\,\frac{m_1 m_2}{r^2} \qquad [11.1]$$

where G is the **universal gravitational constant** whose value in SI units is

$$G = 6.673 \times 10^{-11}\ \mathrm{N \cdot m^2/kg^2} \qquad [11.2]$$

PITFALL PREVENTION 11.1

BE CLEAR ON g AND G Be sure you understand the difference between g and G. The symbol g represents the magnitude of the free-fall acceleration near a planet. At the surface of the Earth, g has the value $9.80\ \mathrm{m/s^2}$. On the other hand, G is a universal constant that has the same value everywhere in the Universe.

The force law given by Equation 11.1 is often referred to as an **inverse-square law** because the magnitude of the force varies as the inverse square of the separation of the particles. We can express this attractive force in vector form by defining a unit vector $\hat{\mathbf{r}}_{12}$ directed from m_1 toward m_2 as shown in Active Figure 11.1. The force exerted by m_1 on m_2 is

$$\vec{\mathbf{F}}_{12} = -G\,\frac{m_1 m_2}{r^2}\,\hat{\mathbf{r}}_{12} \qquad [11.3]$$

where the negative sign indicates that particle 1 is attracted toward particle 2. Likewise, by Newton's third law, the force exerted by m_2 on m_1, designated $\vec{\mathbf{F}}_{21}$, is equal in magnitude to $\vec{\mathbf{F}}_{12}$ and in the opposite direction. That is, these forces form an action–reaction pair, and $\vec{\mathbf{F}}_{21} = -\vec{\mathbf{F}}_{12}$.

As Newton demonstrated, **the gravitational force exerted by a finite-sized, spherically symmetric mass distribution on a particle outside the distribution is the same as if the entire mass of the distribution were concentrated at its center.** For example, the force on a particle of mass m at the Earth's surface has the magnitude

$$F_g = G\,\frac{M_E m}{R_E^2}$$

where M_E is the Earth's mass and R_E is the Earth's radius. This force is directed toward the center of the Earth.

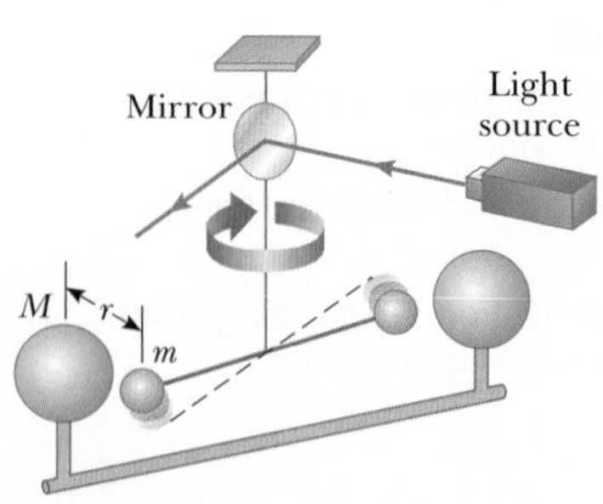

FIGURE 11.2 Schematic diagram of the Cavendish apparatus for measuring G. As the small spheres of mass m are attracted to the large spheres of mass M, the rod rotates through a small angle. A light beam reflected from a mirror on the rotating apparatus measures the angle of rotation. The dashed line represents the original position of the rod. (In reality, the length of wire above the mirror is much larger than that below it.)

Measurement of the Gravitational Constant

The universal gravitational constant G was first measured in an important experiment by Sir Henry Cavendish in 1798. The apparatus he used consists of two small spheres, each of mass m, fixed to the ends of a light horizontal rod suspended by a thin wire as in Figure 11.2. Two large spheres, each of mass M, are then placed near the smaller spheres. The attractive force between the smaller and larger spheres causes the rod to rotate and twist the wire. If the system is oriented as shown in Figure 11.2, the rod rotates clockwise when viewed from the top. The angle through which it rotates is measured by the deflection of a light beam that is reflected from a mirror attached to the wire. The experiment is carefully repeated with different masses at various separations. In addition to providing a value for G, the results confirm that the force is attractive, proportional to the product mM, and inversely proportional to the square of the distance r.

It is interesting that G is the least well known of the fundamental constants, with a percentage uncertainty thousands of times larger than those for other constants such as the speed of light c and the fundamental electric charge e. Several measurements of G made in the 1990s varied significantly from the previous value and from one another! The search for a more precise value of G continues to be an area of active research.

QUICK QUIZ 11.1 A planet has two moons of equal mass. Moon 1 is in a circular orbit of radius r. Moon 2 is in a circular orbit of radius $2r$. What is the magnitude of the gravitational force exerted by the planet on Moon 2? (**a**) four times as large as that on Moon 1 (**b**) twice as large as that on Moon 1 (**c**) equal to that on Moon 1 (**d**) half as large as that on Moon 1 (**e**) one-fourth as large as that on Moon 1

■ Thinking Physics 11.1

The novel *Icebound,* by Dean Koontz (Bantam Books, 2000), is a story of a group of scientists trapped on a floating iceberg near the North Pole. One of the devices the scientists have with them is a transmitter with which they can fix their position with "the aid of a geosynchronous polar satellite." Can a satellite in a *polar* orbit be *geosynchronous*?

Reasoning A geosynchronous satellite is one that stays over one location on the Earth's surface at all times. Therefore, an antenna on the surface that receives signals from the satellite, such as a television dish, can stay pointed in a fixed direction toward the sky. The satellite must be in an orbit with the correct radius such that its orbital period is the same as that of the Earth's rotation. This orbit results in the satellite appearing to have no east–west motion relative to the observer at the chosen location. Another requirement is that a geosynchronous satellite *must be in orbit over the equator.* Otherwise it would appear to undergo a north–south oscillation during one orbit. Therefore, it would be impossible to have a geosynchronous satellite in a *polar* orbit. Even if such a satellite were at the proper distance from the Earth, it would be moving rapidly in the north–south direction, resulting in the necessity of accurate tracking equipment. What's more, it would be below the horizon for long periods of time, making it useless for determining one's position. ■

The Gravitational Field

When Newton first published his theory of gravitation, his contemporaries found it difficult to accept the concept of a force that one object could exert on another without anything happening in the space between them. They asked how it was possible for two objects with mass to interact even though they were not in contact with each other. Although Newton himself could not answer this question, his theory was considered a success because it satisfactorily explained the motions of the planets.

An alternative mental representation of the gravitational force is to think of the gravitational interaction as a two-step process involving a *field,* as discussed in Section 4.1. First, one object (a *source mass*) creates a **gravitational field** $\vec{\mathbf{g}}$ throughout the space around it. Then, a second object (a *test mass*) of mass m residing in this field experiences a force $\vec{\mathbf{F}}_g = m\vec{\mathbf{g}}$. In other words, we model the *field* as exerting a force on the test mass rather than the source mass exerting the force directly. The gravitational field is defined by

$$\vec{\mathbf{g}} \equiv \frac{\vec{\mathbf{F}}_g}{m} \qquad [11.4]$$

■ Gravitational field

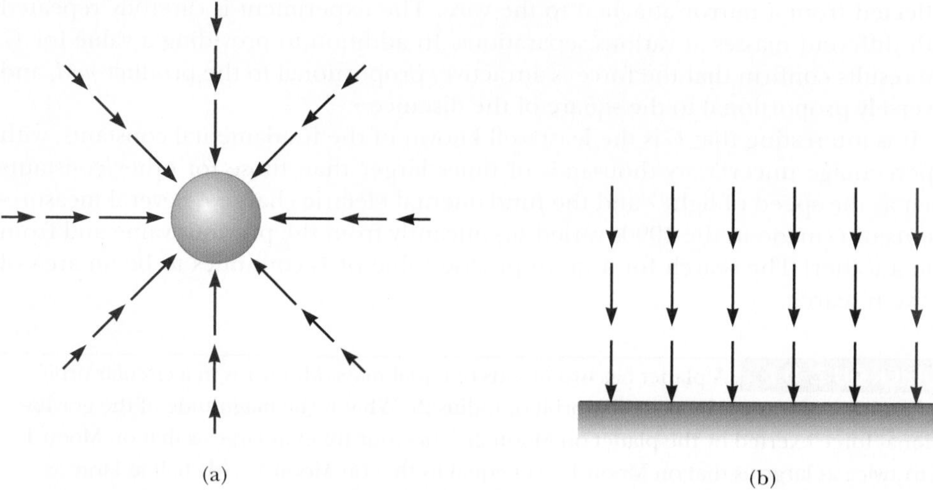

FIGURE 11.3 (a) The gravitational field vectors in the vicinity of a uniform spherical mass vary in both direction and magnitude. (b) The gravitational field vectors in a small region near the Earth's surface are uniform; that is, they all have the same direction and magnitude.

That is, the gravitational field at a point in space equals the gravitational force that a test mass m experiences at that point divided by the mass. Consequently, if $\vec{\mathbf{g}}$ is known at some point in space, a particle of mass m experiences a gravitational force $\vec{\mathbf{F}}_g = m\vec{\mathbf{g}}$ when placed at that point. We will also see the model of a particle in a field for electricity and magnetism in later chapters, where it plays a much larger role than it does for gravity.

As an example, consider an object of mass m near the Earth's surface. The gravitational force on the object is directed toward the center of the Earth and has a magnitude mg. Therefore, we see that the gravitational field experienced by the object at some point has a magnitude equal to the free-fall acceleration at that point. Because the gravitational force on the object has a magnitude $GM_E m/r^2$ (where M_E is the mass of the Earth), the field $\vec{\mathbf{g}}$ at a distance r from the center of the Earth is given by

$$\vec{\mathbf{g}} = \frac{\vec{\mathbf{F}}_g}{m} = -\frac{GM_E}{r^2}\hat{\mathbf{r}} \qquad [11.5]$$

where $\hat{\mathbf{r}}$ is a unit vector pointing radially outward from the Earth and the negative sign indicates that the field vector points toward the center of the Earth as shown in Figure 11.3a. Note that the field vectors at different points surrounding the spherical mass vary in both direction and magnitude. In a small region near the Earth's surface, $\vec{\mathbf{g}}$ is approximately constant and the downward field is uniform as indicated in Figure 11.3b. Equation 11.5 is valid at all points outside the Earth's surface, assuming that the Earth is spherical and that rotation can be neglected. At the Earth's surface, where $r = R_E$, $\vec{\mathbf{g}}$ has a magnitude of 9.80 m/s^2.

INTERACTIVE EXAMPLE 11.1 An Earth Satellite

A satellite of mass m moves in a circular orbit about the Earth with a constant speed v and at a height of $h = 1\,000$ km above the Earth's surface as in Figure 11.4. (For clarity, this figure is not drawn to scale.) Find the orbital speed of the satellite.

Solution The only external force on the satellite is the gravitational force exerted by the Earth. This force is directed toward the center of the satellite's circular path. We apply Newton's second law to the satellite modeled as a particle in uniform circular motion. Because the magnitude of the gravitational force between the Earth and the satellite is $GM_E m/r^2$ we find that

$$F_g = G\frac{M_E m}{r^2} = m\frac{v^2}{r}$$

$$v = \sqrt{\frac{GM_E}{r}}$$

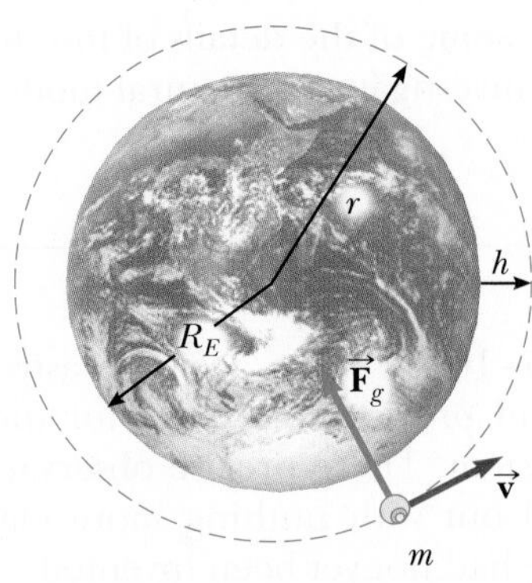

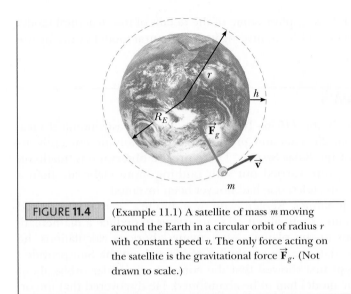

FIGURE 11.4 (Example 11.1) A satellite of mass m moving around the Earth in a circular orbit of radius r with constant speed v. The only force acting on the satellite is the gravitational force $\vec{F}_g$. (Not drawn to scale.)

In this expression, the distance r is the Earth's radius plus the height of the satellite; that is, $r = R_E + h = 6.37 \times 10^6 + 1.00 \times 10^6 = 7.37 \times 10^6$ m, so that

$$v = \sqrt{\frac{(6.67 \times 10^{-11}\ \mathrm{N \cdot m^2/kg^2})(5.98 \times 10^{24}\ \mathrm{kg})}{7.37 \times 10^6\ \mathrm{m}}}$$

$$= 7.36 \times 10^3\ \mathrm{m/s} \approx 16\,400\ \mathrm{mi/h}$$

Note that v is independent of the mass of the satellite!

PhysicsNow™ You can adjust the altitude of the satellite and observe the orbit by logging into PhysicsNow at www.pop4e.com and going to Interactive Example 11.1.

11.2 STRUCTURAL MODELS

In Chapter 1, we mentioned that we would discuss four categories of models. The fourth category is **structural models.** In these models, we propose theoretical structures in an attempt to understand the behavior of a system with which we cannot interact directly because it is far different in scale—either much smaller or much larger—from our macroscopic world.

One of the earliest structural models to be explored was that of the place of the Earth in the Universe. The movements of the planets, stars, and other celestial bodies have been observed by people for thousands of years. Early in history, scientists regarded the Earth as the center of the Universe because it appeared that objects in the sky moved around the Earth. This organization of the Earth and other objects is a structural model for the Universe called the *geocentric model.* It was elaborated and formalized by the Greek astronomer Claudius Ptolemy in the second century A.D. and was accepted for the next 1400 years. In 1543, Polish astronomer Nicolaus Copernicus (1473–1543) offered a different structural model in which the Earth is part of a local Solar System, suggesting that the Earth and the other planets revolve in perfectly circular orbits about the Sun (the *heliocentric model*).

In general, a structural model contains the following features:

■ Features of structural models

1. A description of the physical components of the system; in the heliocentric model, the components are the planets and the Sun.
2. A description of where the components are located relative to one another and how they interact; in the heliocentric model, the planets are in orbit around the Sun and they interact via the gravitational force.
3. A description of the time evolution of the system; the heliocentric model assumes a steady-state Solar System, with planets revolving in orbits around the Sun with fixed periods.
4. A description of the agreement between predictions of the model and actual observations and, possibly, predictions of new effects that have not yet been observed; the heliocentric model predicts Earth-based observations of Mars that are in agreement with historical and present measurements. The geocentric model was also able to find agreement between predictions and observations, but only at the expense of a very complicated structural model in which the planets moved in circles built on other circles. The heliocentric model, along with Newton's law of universal gravitation, predicted that a spacecraft could be sent from the Earth to Mars long before it was actually first done in the 1970s.

In Sections 11.3 and 11.4, we explore some of the details of the structural model of the Solar System. In Section 11.5, we investigate a structural model of the hydrogen atom.

(Art Resource)

JOHANNES KEPLER (1571–1630)

Kepler, a German astronomer, is best known for developing the laws of planetary motion based on the careful observations of Tycho Brahe.

11.3 KEPLER'S LAWS

Danish astronomer Tycho Brahe (1546–1601) made accurate astronomical measurements over a period of 20 years and provided the basis for the currently accepted structural model of the Solar System. These precise observations, made on the planets and 777 stars, were carried out with nothing more elaborate than a large sextant and compass; the telescope had not yet been invented.

German astronomer Johannes Kepler, who was Brahe's assistant, acquired Brahe's astronomical data and spent about 16 years trying to deduce a mathematical model for the motions of the planets. After many laborious calculations, he found that Brahe's precise data on the revolution of Mars about the Sun provided the answer. Kepler's analysis first showed that the concept of circular orbits about the Sun in the heliocentric model had to be abandoned. He discovered that the orbit of Mars could be accurately described by a curve called an *ellipse.* He then generalized this analysis to include the motions of all planets. The complete analysis is summarized in three statements, known as **Kepler's laws of planetary motion,** each of which is discussed in the following sections.

Newton demonstrated that these laws are consequences of the gravitational force that exists between any two masses. Newton's law of universal gravitation, together with his laws of motion, provides the basis for a full mathematical representation of the motion of planets and satellites.

Kepler's First Law

We are familiar with circular orbits of objects around gravitational force centers from Interactive Example 11.1. Kepler's first law indicates that the circular orbit is a very special case and that elliptical orbits are the general situation:[1]

Each planet in the Solar System moves in an elliptical orbit with the Sun at one focus.

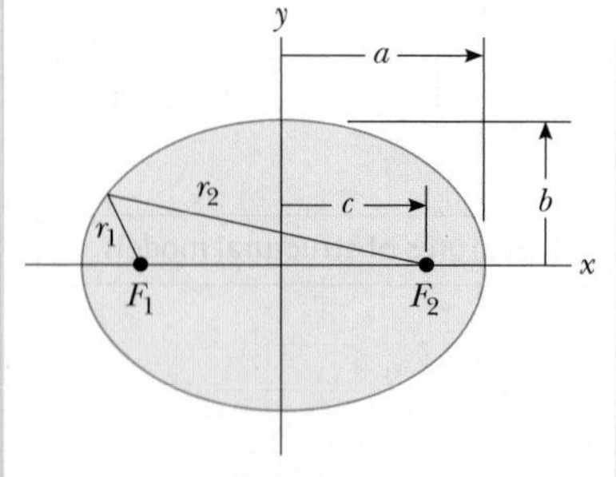

ACTIVE FIGURE 11.5
Plot of an ellipse. The semimajor axis has length a, and the semiminor axis has length b. A focus is located at a distance c from the center on each side of the center.

Physics Now™ Log into PhysicsNow at www.pop4e.com and go to Active Figure 11.5 to move the focal points or enter values for a, b, c, and e and see the resulting elliptical shape.

Active Figure 11.5 shows the geometry of an ellipse, which serves as our geometric model for the elliptical orbit of a planet.[2] An ellipse is mathematically defined by choosing two points, F_1 and F_2, each of which is a called a **focus,** and then drawing a curve through points for which the sum of the distances r_1 and r_2 from F_1 and F_2 is a constant. The longest distance through the center between points on the ellipse (and passing through both foci) is called the **major axis,** and this distance is $2a$. In Active Figure 11.5, the major axis is drawn along the x direction. The distance a is called the **semimajor axis.** Similarly, the shortest distance through the center between points on the ellipse is called the **minor axis** of length $2b$, where the distance b is the **semiminor axis.** Either focus of the ellipse is located at a distance c from the center of the ellipse, where $a^2 = b^2 + c^2$. In the elliptical orbit of a planet around the Sun, the Sun is at one focus of the ellipse. Nothing is at the other focus.

[1] We choose a simplification model in which a body of mass m is in orbit around a body of mass M, with $M >> m$. In this way, we model the body of mass M to be stationary. In reality, that is not true; both M and m move around the center of mass of the system of two objects. That is how we indirectly detect planets around other stars; we see the "wobbling" motion of the star as the planet and the star rotate about the center of mass.

[2] Actual orbits show perturbations due to moons in orbit around the planet and passages of the planet near other planets. We will ignore these perturbations and adopt a simplification model in which the planet follows a perfectly elliptical orbit.

The **eccentricity** of an ellipse is defined as $e \equiv c/a$ and describes the general shape of the ellipse. For a circle, $c = 0$ and the eccentricity is therefore zero. The smaller b is than a, the shorter the ellipse is along the y direction compared with its extent in the x direction in Active Figure 11.5. As b decreases, c increases and the eccentricity e increases. Therefore, higher values of eccentricity correspond to longer and thinner ellipses. The range of values of the eccentricity for an ellipse is $0 < e < 1$. Eccentricities higher than 1 correspond to hyperbolas.

Eccentricities for planetary orbits vary widely in the Solar System. The eccentricity of the Earth's orbit is 0.017, which makes it nearly circular. On the other hand, the eccentricity of Pluto's orbit is 0.25, the highest of all the nine planets. Figure 11.6a shows an ellipse with the eccentricity of that of Pluto's orbit. Notice that even this highest eccentricity orbit is difficult to distinguish from a circle, which is why Kepler's first law is an admirable accomplishment.

The eccentricity of the orbit of Comet Halley is 0.97, describing an orbit whose major axis is much longer than its minor axis as shown in Figure 11.6b. As a result, Comet Halley spends much of its 76-year period far from the Sun and invisible from the Earth. It is only visible to the naked eye during a small part of its orbit when it is near the Sun.

Let us imagine now a planet in an elliptical orbit such as that shown in Active Figure 11.5 with the Sun at focus F_2. When the planet is at the far left in the diagram, the distance between the planet and the Sun is $a + c$. This point is called the *aphelion*, where the planet is the farthest away from the Sun that it can be in the orbit (for an object in orbit around the Earth, this point is called the *apogee*). Conversely, when the planet is at the right end of the ellipse, the point is called the *perihelion* (for an Earth orbit, the *perigee*), and the distance between the planet and the Sun is $a - c$.

Kepler's first law is a direct result of the inverse-square nature of the gravitational force. We have discussed circular and elliptical orbits, which are the allowed shapes of orbits for objects that are *bound* to the gravitational force center. These objects include planets, asteroids, and comets that move repeatedly around the Sun, as well as moons orbiting a planet. *Unbound* objects might also occur, such as a meteoroid from deep space that might pass by the Sun once and then never return. The gravitational force between the Sun and these objects also varies as the inverse square of the separation distance, and the allowed paths for these objects are parabolas and hyperbolas.

PITFALL PREVENTION 11.2

WHERE IS THE SUN? The Sun is located at one focus of the elliptical orbit of a planet. It is *not* located at the center of the ellipse.

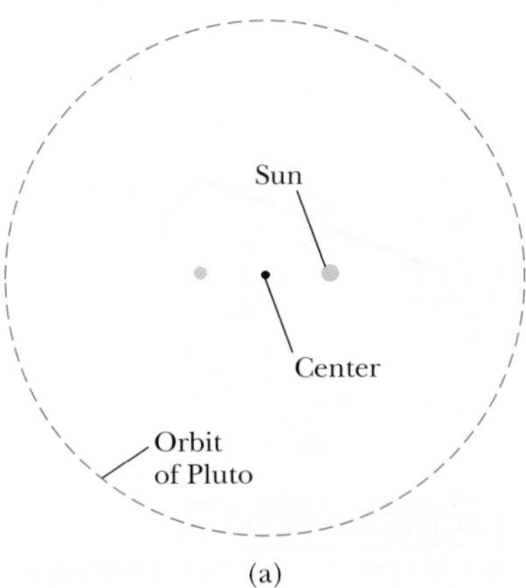

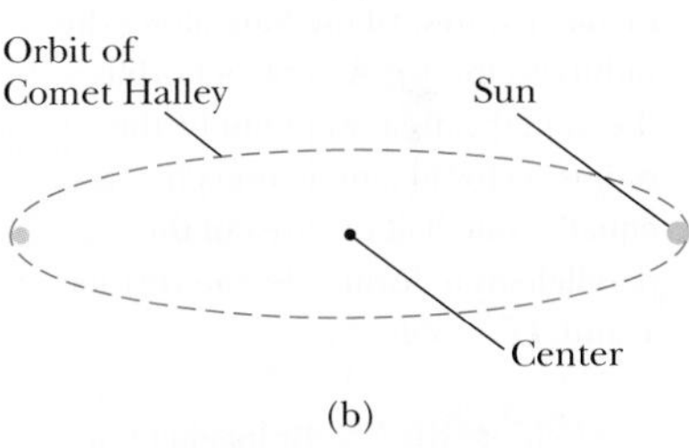

FIGURE 11.6 (a) The shape of the orbit of Pluto, which has the highest eccentricity ($e = 0.25$) among the planets in the Solar System. The Sun is located at the large yellow dot, which is a focus of the ellipse. Nothing physical is located at the center of the orbit (the small dot) or the other focus (the blue dot). (b) The shape of the orbit of Comet Halley.

Kepler's Second Law

Let us now look at the second of Kepler's laws:

▮ Kepler's second law

> The radius vector drawn from the Sun to any planet sweeps out equal areas in equal time intervals.

This law can be shown to be a consequence of angular momentum conservation as follows. Consider a planet of mass M_p moving about the Sun in an elliptical orbit (Active Fig. 11.7a). Let us consider the planet as a system. We shall assume that the Sun is much more massive than the planet, so the Sun does not move. The gravitational force acting on the planet is a central force, that is, a force that is always directed along the radius vector. Therefore, the force on the planet is directed toward the Sun. The torque on the planet due to this central force is zero because $\vec{\mathbf{F}}$ is parallel to $\vec{\mathbf{r}}$. That is,

$$\vec{\boldsymbol{\tau}} \equiv \vec{\mathbf{r}} \times \vec{\mathbf{F}} = \vec{\mathbf{r}} \times F(r)\hat{\mathbf{r}} = 0$$

Recall that the external net torque on a system equals the time rate of change of angular momentum of the system; that is, $\vec{\boldsymbol{\tau}} = d\vec{\mathbf{L}}/dt$. Therefore, because $\vec{\boldsymbol{\tau}} = 0$ for the planet, the angular momentum $\vec{\mathbf{L}}$ of the planet is a constant of the motion:

$$\vec{\mathbf{L}} = \vec{\mathbf{r}} \times \vec{\mathbf{p}} = M_p\vec{\mathbf{r}} \times \vec{\mathbf{v}} = \text{constant}$$

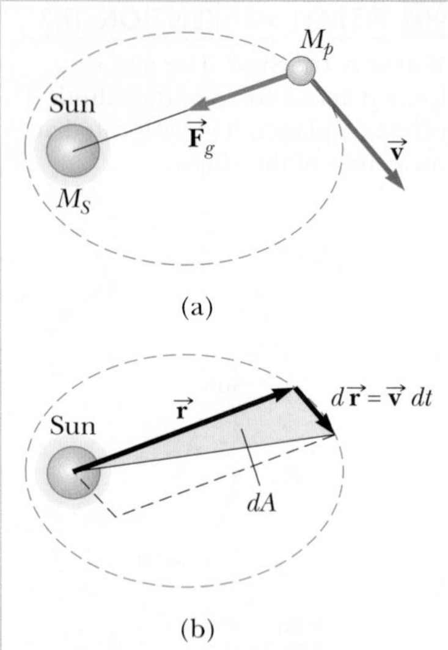

ACTIVE FIGURE 11.7
(a) The gravitational force acting on a planet acts toward the Sun, along the radius vector. (b) As a planet orbits the Sun, the area swept out by the radius vector in a time interval dt is equal to one-half the area of the parallelogram formed by the vectors $\vec{r}$ and $d\vec{r} = \vec{v}dt$.

PhysicsNow™ By logging into PhysicsNow at **www.pop4e.com** and going to Active Figure 11.7 you can assign a value of the eccentricity and see the resulting motion of the planet around the Sun.

We can relate this result to the following geometric consideration. In a time interval dt, the radius vector $\vec{r}$ in Active Figure 11.7b sweeps out the area dA, which equals one-half the area $|\vec{r} \times d\vec{r}|$ of the parallelogram formed by the vectors $\vec{r}$ and $d\vec{r}$. Because the displacement of the planet in the time interval dt is given by $d\vec{r} = \vec{v}dt$, we have

$$dA = \tfrac{1}{2}|\vec{r} \times d\vec{r}| = \tfrac{1}{2}|\vec{r} \times \vec{v}dt| = \frac{L}{2M_p}\,dt$$

$$\frac{dA}{dt} = \frac{L}{2M_p} = \text{constant} \qquad [11.6]$$

where L and M_p are both constants. Therefore, we conclude that the radius vector from the Sun to any planet sweeps out equal areas in equal times.

It is important to recognize that this result is a consequence of the gravitational force being a central force, which in turn implies that angular momentum of the planet is constant. Therefore, the law applies to *any* situation that involves a central force, whether inverse-square or not.

■ Thinking Physics 11.2

The Earth is closer to the Sun when it is winter in the Northern Hemisphere than when it is summer. July and January both have 31 days. In which month, if either, does the Earth move through a longer distance in its orbit?

Reasoning The Earth is in a slightly elliptical orbit around the Sun. Because of angular momentum conservation, the Earth moves more rapidly when it is close to the Sun and more slowly when it is farther away. Therefore, because it is closer to the Sun in January, it is moving faster and will cover more distance in its orbit than it will in July. ■

Kepler's Third Law

Kepler's third law reads as follows:

> The square of the orbital period of any planet is proportional to the cube of the semimajor axis of the elliptical orbit.

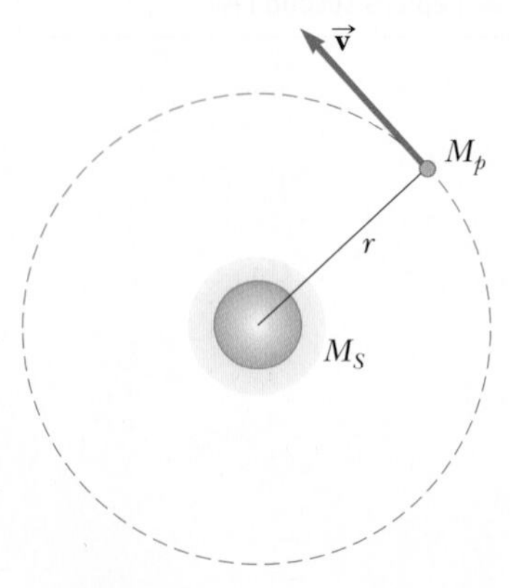

FIGURE 11.8 A planet of mass M_p moving in a circular orbit about the Sun. Kepler's third law relates the period of the orbit to the radius.

This law can be shown easily for circular orbits. Consider a planet of mass M_p that is assumed to be moving about the Sun (mass M_S) in a circular orbit as in Figure 11.8. Because the gravitational force provides the centripetal acceleration of the planet as it moves in a circle, we use Newton's second law for a particle in uniform circular motion:

$$\frac{GM_SM_p}{r^2} = \frac{M_pv^2}{r}$$

The orbital speed of the planet is $2\pi r/T$, where T is the period; therefore, the preceding expression becomes

$$\frac{GM_S}{r^2} = \frac{(2\pi r/T)^2}{r}$$

$$T^2 = \left(\frac{4\pi^2}{GM_S}\right)r^3 = K_Sr^3$$

TABLE 11.1 Useful Planetary Data

Body	Mass (kg)	Mean Radius (m)	Period (s)	Average Distance from Sun (m)	$\frac{T^2}{a^3}$ (s^2/m^3)
Mercury	3.18×10^{23}	2.43×10^6	7.60×10^6	5.79×10^{10}	2.97×10^{-19}
Venus	4.88×10^{24}	6.06×10^6	1.94×10^7	1.08×10^{11}	2.99×10^{-19}
Earth	5.98×10^{24}	6.37×10^6	3.156×10^7	1.496×10^{11}	2.97×10^{-19}
Mars	6.42×10^{23}	3.37×10^6	5.94×10^7	2.28×10^{11}	2.98×10^{-19}
Jupiter	1.90×10^{27}	6.99×10^7	3.74×10^8	7.78×10^{11}	2.97×10^{-19}
Saturn	5.68×10^{26}	5.85×10^7	9.35×10^8	1.43×10^{12}	2.99×10^{-19}
Uranus	8.68×10^{25}	2.33×10^7	2.64×10^9	2.87×10^{12}	2.95×10^{-19}
Neptune	1.03×10^{26}	2.21×10^7	5.22×10^9	4.50×10^{12}	2.99×10^{-19}
Pluto	$\approx 1.4 \times 10^{22}$	$\approx 1.5 \times 10^6$	7.82×10^9	5.91×10^{12}	2.96×10^{-19}
Moon	7.36×10^{22}	1.74×10^6	—	—	—
Sun	1.991×10^{30}	6.96×10^8	—	—	—

Note: For a more complete set of data, see, for example, the *Handbook of Chemistry and Physics* (Boca Raton, FL: CRC Press, published annually).

where K_S is a constant given by

$$K_S = \frac{4\pi^2}{GM_S} = 2.97 \times 10^{-19} \text{ s}^2/\text{m}^3$$

For elliptical orbits, Kepler's third law is expressed by starting with $T^2 = K_S r^3$ and replacing r with the length a of the semimajor axis (see Fig. 11.5):

$$T^2 = \left(\frac{4\pi^2}{GM_S}\right) a^3 = K_S a^3 \qquad [11.7]$$

■ Kepler's third law

Equation 11.7 is Kepler's third law. Because the semimajor axis of a circular orbit is its radius, Equation 11.7 is valid for both circular and elliptical orbits. Note that the constant of proportionality K_S is independent of the mass of the planet. Equation 11.7 is therefore valid for *any* planet. If we were to consider the orbit of a satellite about the Earth, such as the Moon, the constant would have a different value, with the Sun's mass replaced by the Earth's mass; that is, $K_E = 4\pi^2/GM_E$.

Table 11.1 is a collection of useful planetary data. The last column verifies that the ratio T^2/a^3 is constant. The small variations in the values in this column are because of uncertainties in the data measured for the periods and semimajor axes of the planets.

QUICK QUIZ 11.2 A comet is in a highly elliptical orbit around the Sun. The period of the comet's orbit is 90 days. Which of the following statements is true about the possibility of a collision between this comet and the Earth? **(a)** Collision is not possible. **(b)** Collision is possible. **(c)** Not enough information is available to determine whether a collision is possible.

11.4 ENERGY CONSIDERATIONS IN PLANETARY AND SATELLITE MOTION

So far we have approached orbital mechanics from the point of view of forces and angular momentum. Let us now investigate the motion of planets in orbit from the *energy* point of view.

Consider an object of mass m moving with a speed v in the vicinity of a massive object of mass $M \gg m$. This two-object system might be a planet moving around the Sun, a satellite orbiting the Earth, or a comet making a one-time flyby past the Sun. We will treat the two objects of mass m and M as an isolated system. If we assume that M is at rest in an inertial reference frame (because $M \gg m$), the total mechanical energy E of the two-object system is the sum of the kinetic energy of the object of mass m and the gravitational potential energy of the system:

$$E = K + U_g$$

Recall from Chapter 7 that the gravitational potential energy U_g associated with *any pair* of particles of masses m_1 and m_2 separated by a distance r is given by

$$U_g = -\frac{Gm_1m_2}{r}$$

where we have defined $U_g \rightarrow 0$ as $r \rightarrow \infty$; therefore, in our case, the mechanical energy of the system of m and M is

$$E = \tfrac{1}{2}mv^2 - \frac{GMm}{r} \qquad [11.8]$$

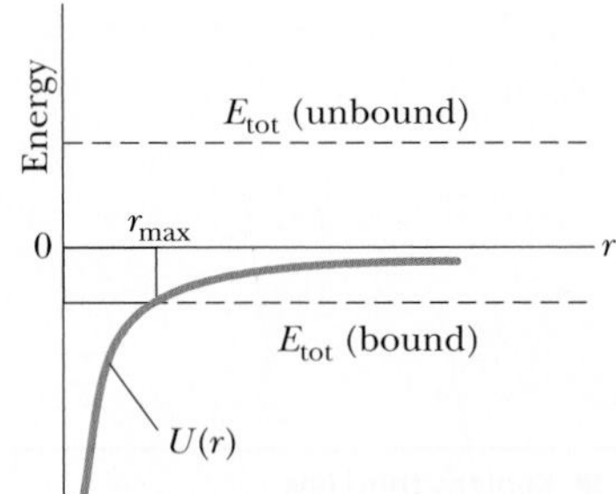

FIGURE 11.9 The lower total energy line represents a bound system. The separation distance r between the two gravitationally bound objects never exceeds r_{max}. The upper total energy line represents an unbound system of two objects interacting gravitationally. The separation distance r between the two objects can have any value.

Equation 11.8 shows that E may be positive, negative, or zero, depending on the value of v at a particular separation distance r. If we consider the energy diagram method of Section 7.7, we can show the potential and total energies of the system as a function of r as in Figure 11.9. A planet moving around the Sun and a satellite in orbit around the Earth are *bound* systems, such as those we discussed in Section 11.3; the Earth will always stay near the Sun and the satellite near the Earth. In Figure 11.9, these systems are represented by a total energy that is negative. The point at which the total energy line intersects the potential energy curve is a turning point, the maximum separation distance r_{max} between the two bound objects.

A one-time meteoroid flyby represents an unbound system. The meteoroid interacts with the Sun but is not bound to it. Therefore, the meteoroid can in theory move infinitely far away from the Sun as represented in Figure 11.9 by a total energy line in the positive region of the graph. This line never intersects the potential energy curve, so all values of r are possible.

For a bound system, such as the Earth and Sun, E is necessarily less than zero because we have chosen the convention that $U_g \rightarrow 0$ as $r \rightarrow \infty$. We can easily establish that $E < 0$ for the system consisting of an object of mass m moving in a circular orbit about an object of mass $M \gg m$. Applying Newton's second law to the object of mass m in uniform circular motion gives

$$\sum F = ma \quad \rightarrow \quad \frac{GMm}{r^2} = \frac{mv^2}{r}$$

Many artificial satellites have been placed in orbit about the Earth. This diagram shows a plot of all known unclassified satellites and satellite debris larger in size than a baseball as of 1995. Note the large number of geosynchronous satellites (see Thinking Physics 11.1) that form a visible circle above the Earth's equator.

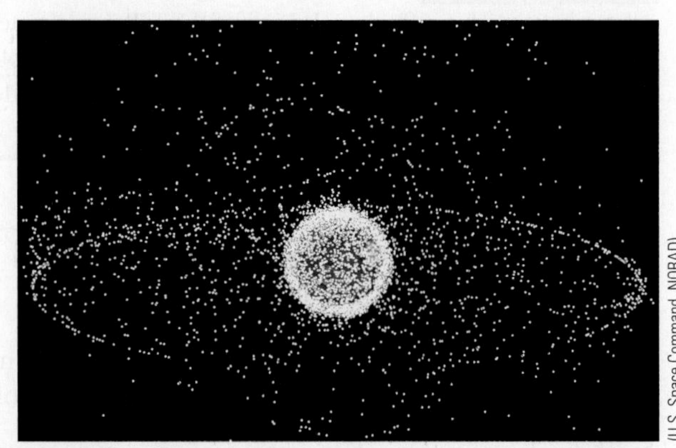

(U.S. Space Command, NORAD)

Multiplying both sides by r and dividing by 2 gives

$$\tfrac{1}{2}mv^2 = \frac{GMm}{2r} \qquad [11.9]$$

Substituting this result into Equation 11.8, we obtain

$$E = \frac{GMm}{2r} - \frac{GMm}{r}$$

$$E = -\frac{GMm}{2r} \qquad [11.10]$$

This result clearly shows that **the total mechanical energy must be negative in the case of circular orbits.** Furthermore, Equation 11.9 shows that **the kinetic energy of an object in a circular orbit is equal to one-half the magnitude of the potential energy of the system** (when the potential energy is chosen to be zero at infinite separation).

The total mechanical energy is also negative in the case of elliptical orbits. The expression for E for elliptical orbits is the same as Equation 11.10, with r replaced by the semimajor axis a:

$$E = -\frac{GMm}{2a} \qquad [11.11]$$

▮ Total energy of a planet–star system

The total energy, the total angular momentum, and the total linear momentum of a planet–star system are constants of the motion, according to the isolated system model.

QUICK QUIZ 11.3 A comet moves in an elliptical orbit around the Sun. Which point in its orbit represents the highest value of (**a**) the speed of the comet, (**b**) the potential energy of the comet–Sun system, (**c**) the kinetic energy of the comet, and (**d**) the total energy of the comet–Sun system?

EXAMPLE 11.2 A Satellite in an Elliptical Orbit

A satellite moves in an elliptical orbit about the Earth as in Figure 11.10. The minimum and maximum distances from the surface of the Earth are 400 km and 3 000 km, respectively. Find the speeds of the satellite at apogee and perigee.

Solution Figure 11.10 helps conceptualize the motion of the satellite. Gravity is a central force, so there is zero torque exerted on the satellite. There is also zero torque on the Earth for the same reason. Consequently, we categorize the problem as one involving an isolated system for which the angular momentum is conserved. Because the mass of the satellite is negligible compared with the Earth's mass, we take the center of mass of the Earth to be at rest. Therefore, we only need to consider the angular momentum of the satellite. To analyze the problem, we assign subscripts a and p for the apogee and perigee positions, and apply the principle of conservation of angular momentum for the satellite at these two positions. The result is $L_p = L_a$, or

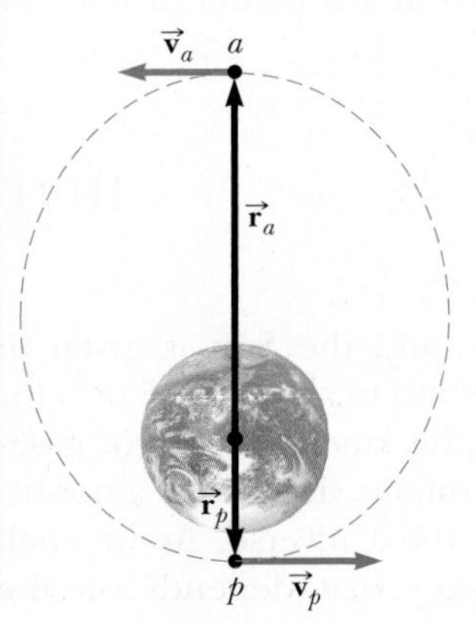

FIGURE 11.10 (Example 11.2) A satellite in an elliptical orbit about the Earth.

$$mv_pr_p = mv_ar_a$$

$$v_pr_p = v_ar_a$$

Using the Earth's radius of 6.37×10^6 m and the given data, we find that $r_a = 9.37 \times 10^6$ m and $r_p = 6.77 \times 10^6$ m. Therefore,

$$(1) \quad \frac{v_p}{v_a} = \frac{r_a}{r_p} = \frac{9.37 \times 10^6 \text{ m}}{6.77 \times 10^6 \text{ m}} = 1.38$$

Because the satellite and the Earth form an isolated system, we can apply conservation of energy for the system and obtain $E_p = E_a$, or

$$U_p + K_p = U_a + K_a$$

$$-G\frac{M_E m}{r_p} + \frac{1}{2}mv_p^2 = -G\frac{M_E m}{r_a} + \frac{1}{2}mv_a^2$$

$$(2) \quad 2GM_E\left(\frac{1}{r_a} - \frac{1}{r_p}\right) = (v_a^2 - v_p^2)$$

Because we know the numerical values of G, M_E, r_p, and r_a, we can use Equations (1) and (2) to determine the two unknowns v_p and v_a. Solving the equations simultaneously, we obtain

$$v_p = \boxed{8.27 \text{ km/s}} \qquad v_a = \boxed{5.98 \text{ km/s}}$$

To finalize the problem, note that $v_p > v_a$, as we would expect.

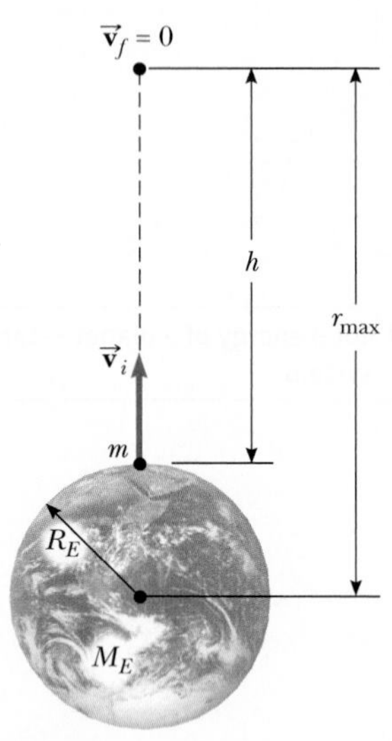

FIGURE 11.11 An object of mass m projected upward from the Earth's surface with an initial speed v_i reaches a maximum altitude $h = r_{max} - R_E$.

Escape Speed

Suppose an object of mass m is projected vertically from the Earth's surface with an initial speed v_i as in Figure 11.11. We can use energy considerations to find the minimum value of the initial speed such that the object will continue to move away from the Earth forever. Equation 11.8 gives the total energy of the object–Earth system at any point when the speed of the object and its distance from the center of the Earth are known. At the surface of the Earth, $r_i = R_E$. When the object reaches its maximum altitude, $v_f = 0$ and $r_f = r_{max}$. Because the total energy of the system is conserved, substitution of these conditions into Equation 11.8 gives

$$\frac{1}{2}mv_i^2 - \frac{GM_E m}{R_E} = -\frac{GM_E m}{r_{max}}$$

Solving for v_i^2 gives

$$v_i^2 = 2GM_E\left(\frac{1}{R_E} - \frac{1}{r_{max}}\right) \qquad [11.12]$$

If the initial speed is known, this expression can therefore be used to calculate the maximum altitude h because we know that $h = r_{max} - R_E$.

We are now in a position to calculate the minimum speed the object must have at the Earth's surface to continue to move away forever. This **escape speed** v_{esc} results in the speed asymptotically approaching *zero*. Letting $r_{max} \rightarrow \infty$ in Equation 11.12 and setting $v_i = v_{esc}$, we have

$$v_{esc} = \sqrt{\frac{2GM_E}{R_E}} \qquad [11.13]$$

Note that this expression for v_{esc} is independent of the mass of the object projected from the Earth. For example, a spacecraft has the same escape speed as a molecule. Furthermore, the result is independent of the *direction* of the velocity.

Note also that Equations 11.12 and 11.13 can be applied to objects projected from *any* planet. That is, in general, the escape speed from any planet of mass M and radius R is

$$v_{esc} = \sqrt{\frac{2GM}{R}} \qquad [11.14]$$

A list of escape speeds for the planets, the Moon, and the Sun is given in Table 11.2. Note that the values vary from 1.1 km/s for Pluto to about 618 km/s for the Sun. These results, together with some ideas from the kinetic theory of gases (Chapter 16), explain why our atmosphere does not contain significant amounts of hydrogen, which is the most abundant element in the Universe. As we shall see later, gas molecules have an average kinetic energy that depends on the

PITFALL PREVENTION 11.3

YOU CAN'T REALLY ESCAPE Although Equation 11.13 provides the escape speed, remember that this conventional name is misleading. It is impossible to escape *completely* from the Earth's gravitational influence because the gravitational force is of infinite range. No matter how far away you are, you will always feel some gravitational force due to the Earth. In practice, however, this force will be much smaller than forces due to other astronomical objects closer to you, so the gravitational force from the Earth can be ignored.

TABLE 11.2 Escape Speeds from the Surfaces of the Planets, the Moon, and the Sun

Planet	v_{esc} (km/s)
Mercury	4.3
Venus	10.3
Earth	11.2
Mars	5.0
Jupiter	60
Saturn	36
Uranus	22
Neptune	24
Pluto	1.1
Moon	2.3
Sun	618

temperature of the gas. Lighter molecules in an atmosphere have translational speeds that are closer to the escape speed than more massive molecules, so they have a higher probability of escaping from the planet and the lighter molecules diffuse into space. This mechanism explains why the Earth does not retain hydrogen molecules and helium atoms in its atmosphere but does retain much heavier molecules, such as oxygen and nitrogen. On the other hand, Jupiter has a very large escape speed (60 km/s), which enables it to retain hydrogen, the primary constituent of its atmosphere.

Black Holes

In Chapter 10, we briefly described a rare event called a supernova, the catastrophic explosion of a very massive star. The material that remains in the central core of such an object continues to collapse, and the core's ultimate fate depends on its mass. If the core has a mass less than 1.4 times the mass of our Sun, it gradually cools down and ends its life as a white dwarf star. If, however, the core's mass is greater than that, it may collapse further due to gravitational forces. What remains is a neutron star, discussed in Chapter 10, in which the mass of a star is compressed to a radius of about 10 km. (On the Earth, a teaspoon of this material would weigh about 5 billion tons!)

An even more unusual star death may occur when the core has a mass greater than about three solar masses. The collapse may continue until the star becomes a very small object in space, commonly referred to as a **black hole.** In effect, black holes are the remains of stars that have collapsed under their own gravitational force. If an object such as a spacecraft comes close to a black hole, it experiences an extremely strong gravitational force and is trapped forever.

The escape speed from any spherical body depends on the mass and radius of the body. The escape speed for a black hole is very high because of the concentration of the star's mass into a sphere of very small radius. If the escape speed exceeds the speed of light c, radiation from the body (e.g. visible light) cannot escape and the body appears to be black, hence the origin of the term *black hole.* The critical radius R_S at which the escape speed is c is called the **Schwarzschild radius** (Fig. 11.12). The imaginary surface of a sphere of this radius surrounding the black hole is called the **event horizon,** which is the limit of how close you can approach the black hole and hope to be able to escape.

Although light from a black hole cannot escape, light from events taking place near the black hole should be visible. For example, it is possible for a binary star system to consist of one normal star and one black hole. Material surrounding the ordinary star can be pulled into the black hole, forming an **accretion disk** around the

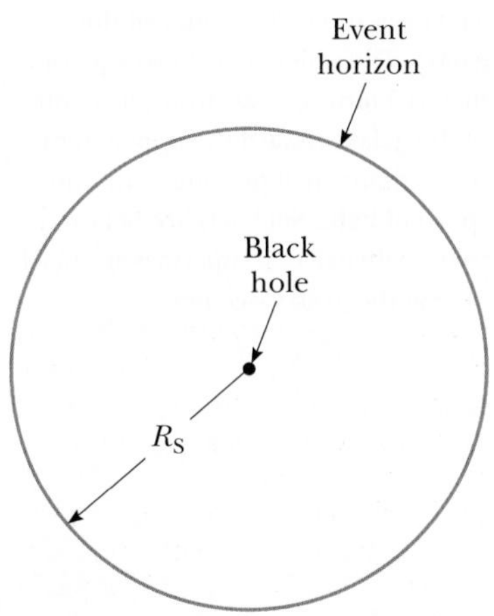

FIGURE 11.12 A black hole. The distance R_S equals the Schwarzschild radius. Any event occurring within the boundary of radius R_S, called the event horizon, is invisible to an outside observer.

FIGURE 11.13 A binary star system consisting of an ordinary star on the left and a black hole on the right. Matter pulled from the ordinary star forms an accretion disk around the black hole, in which matter is raised to very high temperatures, resulting in the emission of x-rays.

black hole as suggested in Figure 11.13. Friction among particles in the accretion disk results in transformation of mechanical energy into internal energy. As a result, the orbital height of the material above the event horizon decreases and the temperature rises. This high-temperature material emits a large amount of radiation, extending well into the x-ray region of the electromagnetic spectrum. These x-rays are characteristic of a black hole. Several possible candidates for black holes have been identified by observation of these x-rays.

Evidence also supports the existence of supermassive black holes at the centers of galaxies, with masses very much larger than the Sun. (The evidence is strong for a supermassive black hole of mass 2 to 3 million solar masses at the center of our galaxy.) Theoretical models for these bizarre objects predict that jets of material should be evident along the rotation axis of the black hole. Figure 11.14 shows a Hubble Space Telescope photograph of the galaxy M87. The jet of material coming from this galaxy is believed to be evidence for a supermassive black hole at the center of the galaxy.

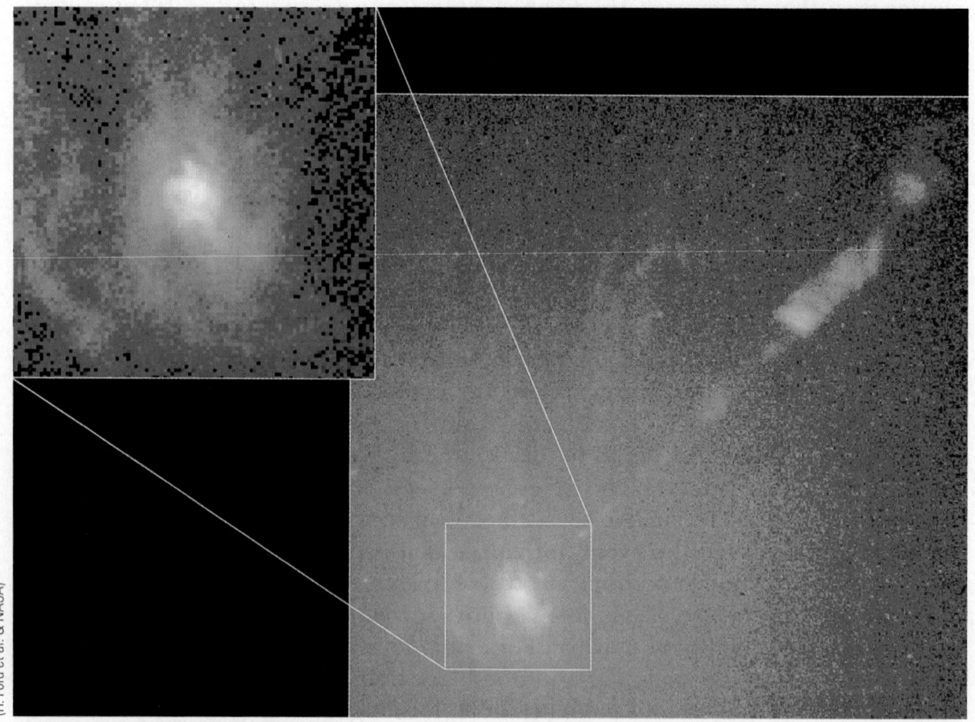

FIGURE 11.14 Hubble Space Telescope images of the galaxy M87. The inset shows the center of the galaxy. The wider view shows a jet of material moving away from the center of the galaxy toward the upper right of the figure at about one-tenth the speed of light. Such jets are believed to be evidence of a supermassive black hole at the galaxy's center.

(H. Ford et al. & NASA)

Black holes are of considerable interest to those searching for **gravity waves,** which are ripples in space–time caused by changes in a gravitational system. These ripples can be caused by a star collapsing into a black hole, a binary star consisting of a black hole and a visible companion, and supermassive black holes at a galaxy center. A gravity wave detector, the Laser Interferometer Gravitational Wave Observatory (LIGO), is currently being built and tested in the United States, and hopes are high for detecting gravitational waves with this instrument.

11.5 ATOMIC SPECTRA AND THE BOHR THEORY OF HYDROGEN

In the preceding sections, we described a structural model for a large-scale system, the Solar System. Let us now do the same for a very small-scale system, the hydrogen atom. We shall find that a Solar System model of the atom, with a few extra features, provides explanations for some of the experimental observations made on the hydrogen atom.

As you may have already learned in a chemistry course, the hydrogen atom is the simplest known atomic system and an especially important one to understand. Much of what is learned about the hydrogen atom (which consists of one proton and one electron) can be extended to single-electron ions such as He^+ and Li^{2+}. Furthermore, a thorough understanding of the physics underlying the hydrogen atom can then be used to describe more complex atoms and the periodic table of the elements.

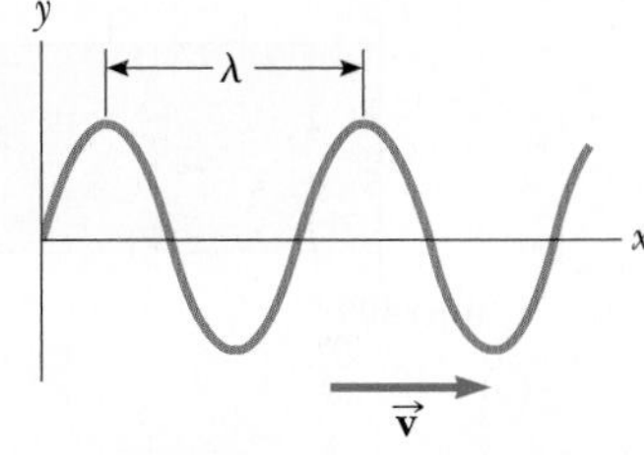

FIGURE 11.15 A sinusoidal wave traveling to the right with wave speed v. Any point on the wave moves a distance of one wavelength λ in a time interval equal to the period T of the wave.

Atomic systems can be investigated by observing *electromagnetic waves* emitted from the atom. Our eyes are sensitive to visible light, one type of electromagnetic wave. The wave will be one of our four simplification models around which we will identify analysis models, as we have done for a particle, a system, and a rigid object. A common form of periodic wave is the sinusoidal wave, whose shape is depicted in Figure 11.15. If this graph represents an electromagnetic wave, the vertical axis represents the magnitude of the electric field. (We will study electric fields in Chapter 19.) The horizontal axis is position in the direction of travel of the wave. The distance between two consecutive crests of the wave is called the **wavelength** λ. As the wave travels to the right with a speed v, any point on the wave travels a distance of one wavelength in a time interval of one period T (the time interval for one cycle), so the wave speed is given by $v = \lambda/T$. The inverse of the period, $1/T$, is called the **frequency** f of the wave; it represents the number of cycles per second. Therefore, the speed of the wave is often written as $v = \lambda f$. In this section, because we shall deal with electromagnetic waves—which travel at the speed of light c—the appropriate relation is

$$c = \lambda f \qquad [11.15]$$

▪ Relation between wavelength, frequency, and wave speed

Suppose an evacuated glass tube is filled with hydrogen (or some other gas). If a voltage applied between metal electrodes in the tube is great enough to produce an electric current in the gas, the tube emits light with colors that are characteristic of the gas. (That is how a neon sign works.) When the emitted light is analyzed with a device called a spectroscope, in which the light passes through a narrow slit, a series of discrete **spectral lines** is observed, each line corresponding to a different wavelength, or color, of light. Such a series of spectral lines is commonly referred to as an **emission spectrum.** The wavelengths contained in a given spectrum are characteristic of the element emitting the light. Figure 11.16 is a semigraphical representation of the spectra of various elements. It is semigraphical because the horizontal axis is linear in wavelength, but the vertical axis has no significance. Because no two elements emit the same line spectrum, this phenomenon represents a marvelous and reliable technique for identifying elements in a substance.

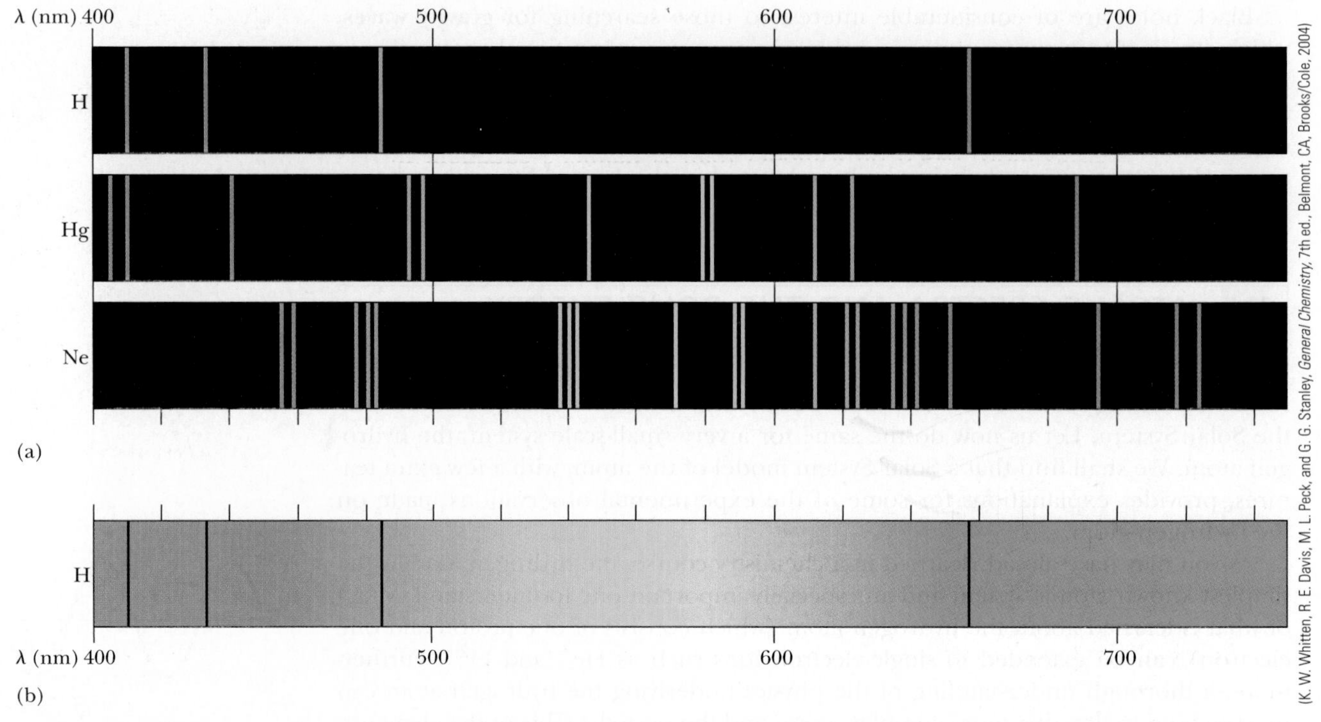

FIGURE 11.16 Visible spectra. (a) Line spectra produced by emission in the visible range for the elements hydrogen, mercury, and neon. (b) The absorption spectrum for hydrogen. The dark absorption lines occur at the same wavelengths as the emission lines for hydrogen shown in (a).

In addition to emitting light at specific wavelengths, an element can also absorb light at specific wavelengths. The spectral lines corresponding to this process form what is known as an **absorption spectrum.** An absorption spectrum can be obtained by passing a continuous radiation spectrum (one containing all wavelengths) through a vapor of the element being analyzed. The absorption spectrum consists of a series of dark lines superimposed on the otherwise continuous spectrum (Fig. 11.16b).

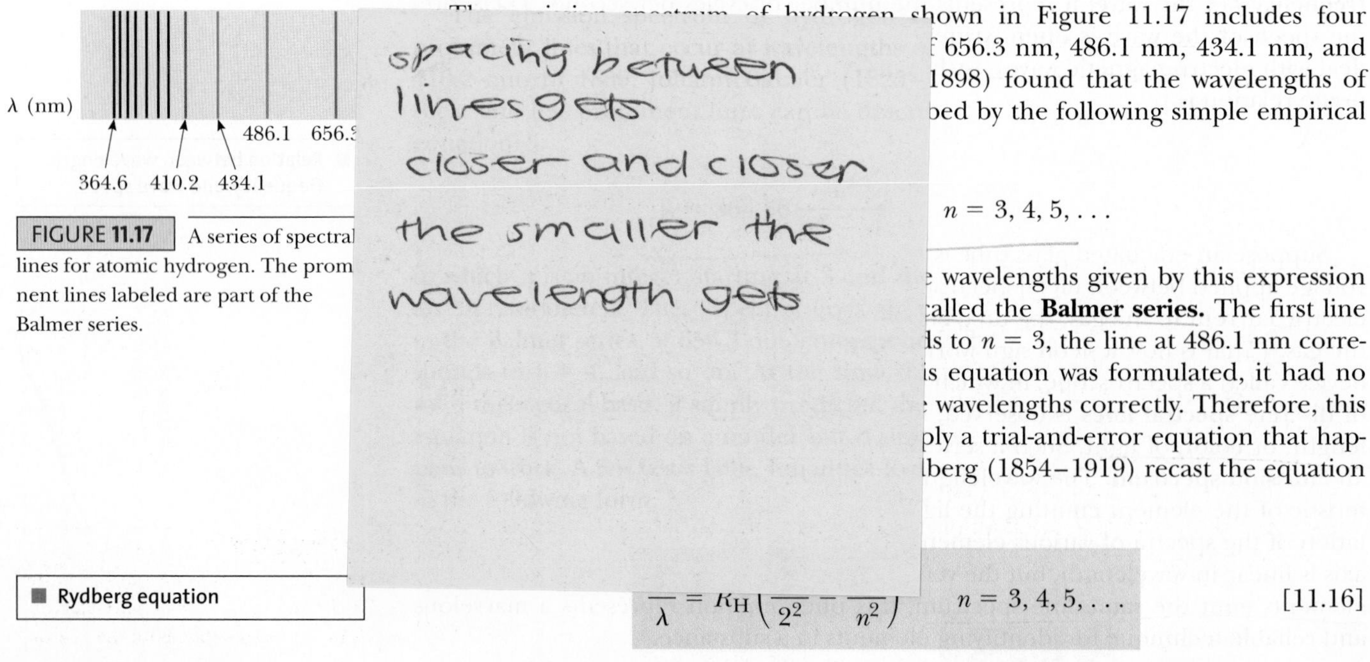

FIGURE 11.17 A series of spectral lines for atomic hydrogen. The prom nent lines labeled are part of the Balmer series.

shown in Figure 11.17 includes four f 656.3 nm, 486.1 nm, 434.1 nm, and 1898) found that the wavelengths of bed by the following simple empirical

$n = 3, 4, 5, \ldots$

wavelengths given by this expression called the **Balmer series.** The first line ds to $n = 3$, the line at 486.1 nm corre- is equation was formulated, it had no wavelengths correctly. Therefore, this ply a trial-and-error equation that hap- lberg (1854–1919) recast the equation

■ **Rydberg equation**

$$\frac{1}{\lambda} = R_H\left(\frac{1}{2^2} - \frac{1}{n^2}\right) \qquad n = 3, 4, 5, \ldots \qquad [11.16]$$

where n may have integral values of 3, 4, 5, . . . and R_{H} is a constant, now called the **Rydberg constant,** which has the value

$$R_{\mathrm{H}} = 1.097\,373\,2 \times 10^{7}\ \mathrm{m}^{-1}$$

Equation 11.16 is no more based on a model than is Balmer's equation. In this form, however, we can compare it with the predictions of a structural model of the hydrogen atom that is described below.

At the beginning of the 20th century, scientists were perplexed by the failure of classical physics to explain the characteristics of atomic spectra. Why did atoms of a given element emit only certain wavelengths of radiation so that the emission spectrum displayed discrete lines? Furthermore, why did the atoms absorb many of the same wavelengths that they emitted? In 1913, Niels Bohr provided an explanation of atomic spectra that includes some features of the currently accepted theory. Using the simplest atom, hydrogen, Bohr described a structural model for the atom. His model of the hydrogen atom contains some classical features that can be related to our analysis models as well as some revolutionary postulates that could not be justified within the framework of classical physics. The basic assumptions of the Bohr model as it applies to the hydrogen atom are as follows:

(Photo courtesy of AIP Niels Bohr Library, Margarethe Bohr Collection)

NIELS BOHR (1885–1962)
Bohr, a Danish physicist, was an active participant in the early development of quantum mechanics and provided much of its philosophical framework. During the 1920s and 1930s, Bohr headed the Institute for Advanced Studies in Copenhagen. The institute was a magnet for many of the world's best physicists and provided a forum for the exchange of ideas. Bohr was awarded the 1922 Nobel Prize in Physics for his investigation of the structure of atoms and of the radiation emanating from them.

1. The electron moves in a circular orbit about the proton under the influence of the electric force of attraction as in Figure 11.18. This notion is purely classical and is very similar to our previous discussion of planets in orbit around the Sun in our structural model of the Solar System.
2. Only certain electron orbits are stable, and they are the only orbits in which we find the electron. In these orbits, the hydrogen atom does not emit energy in the form of radiation. Hence, the total energy of the atom remains constant, and classical mechanics can be used to describe the electron's motion. This restriction to certain orbits is a new idea that is not consistent with classical physics. As we shall see in Chapter 24, an accelerating electron should emit energy by electromagnetic radiation. Therefore, according to the continuity equation for energy, the emission of radiation from the atom should result in a decrease in the energy of the atom. Bohr's postulate boldly claims that this radiation simply does not happen.
3. Radiation is emitted by the hydrogen atom when the atom makes a transition from a more energetic initial state to a lower state. The transition cannot be visualized or treated classically. In particular, the frequency f of the radiation emitted in the transition is related to the change in the atom's energy. The frequency of the emitted radiation is found from

$$E_i - E_f = hf \qquad [11.17]$$

where E_i is the energy of the initial state, E_f is the energy of the final state, h is **Planck's constant** ($h = 6.63 \times 10^{-34}\ \mathrm{J\cdot s}$; we will see Planck's constant extensively in our studies of modern physics), and $E_i > E_f$. The notion of energy being emitted only when a transition occurs is nonclassical. Given this notion, however, Equation 11.17 is simply the continuity equation for energy, $\Delta E = \Sigma T \rightarrow E_f - E_i = -hf$. On the left is the change in energy of the system—the atom—and on the right is the energy transferred out of the system by electromagnetic radiation.
4. The size of the allowed electron orbits is determined by a condition imposed on the electron's orbital angular momentum. The allowed orbits are those for which the electron's orbital angular momentum about the nucleus is an integral multiple of $\hbar \equiv h/2\pi$:

$$m_e vr = n\hbar \qquad n = 1, 2, 3, \ldots \qquad [11.18]$$

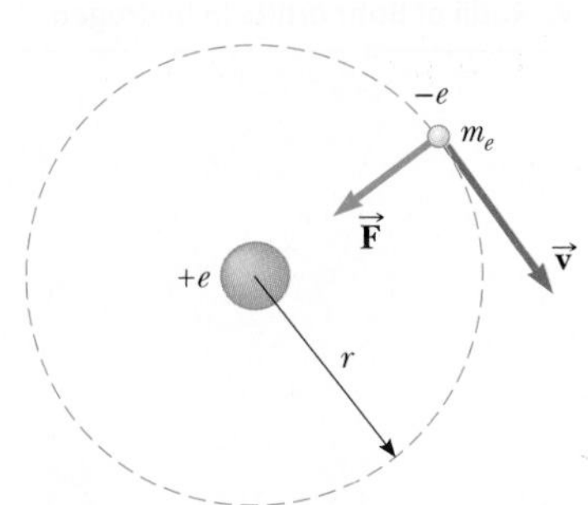

FIGURE 11.18 A pictorial representation of Bohr's model of the hydrogen atom, in which the electron is in a circular orbit about the proton.

This new idea cannot be related to any of the models we have developed so far. It can be related, however, to a model that will be developed in later chapters, and we shall return to this idea at that time to see how it is predicted by the model. This concept is our first introduction to a notion from **quantum mechanics,** which describes the behavior of microscopic particles.

Using these four assumptions, Bohr built a structural model that explains the emission wavelengths of the hydrogen atom. The electric potential energy of the system shown in Figure 11.18 is given by Equation 7.23, $U_e = -k_e e^2/r$, where k_e is the electric constant, e is the charge on the electron, and r is the electron–proton separation. Therefore, the total energy of the atom, which contains both kinetic and potential energy terms, is

$$E = K + U_e = \tfrac{1}{2}m_e v^2 - k_e\frac{e^2}{r} \qquad [11.19]$$

According to assumption 2, the energy of the system remains constant; the system is isolated because the structural model does not allow for electromagnetic radiation for a given orbit.

Applying Newton's second law to this system, we see that the magnitude of the attractive electric force on the electron, $k_e e^2/r^2$ (Eq. 5.15), is equal to the product of its mass and its centripetal acceleration ($a_c = v^2/r$):

$$\frac{k_e e^2}{r^2} = \frac{m_e v^2}{r}$$

From this expression, the kinetic energy of the electron is found to be

$$K = \tfrac{1}{2}m_e v^2 = \frac{k_e e^2}{2r} \qquad [11.20]$$

Substituting this value of K into Equation 11.19 gives the following expression for the total energy E of the hydrogen atom:

■ Total energy of the hydrogen atom

$$E = -\frac{k_e e^2}{2r} \qquad [11.21]$$

Note that the total energy is negative,[3] indicating a bound electron–proton system. Therefore, energy in the amount of $k_e e^2/2r$ must be added to the atom just to separate the electron and proton by an infinite distance and make the total energy zero.[4] An expression for r, the radius of the allowed orbits, can be obtained by eliminating v by substitution between Equations 11.18 and 11.20:

■ Radii of Bohr orbits in hydrogen

$$r_n = \frac{n^2\hbar^2}{m_e k_e e^2} \qquad n = 1, 2, 3 \ldots \qquad [11.22]$$

This result shows that the radii have discrete values, or are *quantized.* The integer n is called a **quantum number** and specifies the particular allowed **quantum state** of the atomic system.

---

[3] Compare this expression with Equation 11.11 for a gravitational system.

[4] This process is called *ionizing* the atom. In theory, ionization requires separating the electron and proton by an infinite distance. In reality, however, the electron and proton are in an environment with huge numbers of other particles. Therefore, ionization means separating the electron and proton by a distance large enough so that the interaction of these particles with other entities in their environment is larger than the remaining interaction between them.

The orbit for which $n = 1$ has the smallest radius; it is called the **Bohr radius** a_0 and has the value

$$a_0 = \frac{\hbar^2}{m_e k_e e^2} = 0.052\,9 \text{ nm} \qquad [11.23]$$

■ The Bohr radius

The first three Bohr orbits are shown to scale in Active Figure 11.19.

The quantization of the orbit radii immediately leads to quantization of the energy of the atom, which can be seen by substituting $r_n = n^2 a_0$ into Equation 11.21. The allowed energies of the atom are

$$E_n = -\frac{k_e e^2}{2a_0}\left(\frac{1}{n^2}\right) \qquad n = 1, 2, 3, \ldots \qquad [11.24]$$

Insertion of numerical values into Equation 11.24 gives

$$E_n = -\frac{13.606 \text{ eV}}{n^2} \qquad n = 1, 2, 3, \ldots \qquad [11.25]$$

■ Energies of quantum states of the hydrogen atom

(Recall from Section 9.7 that $1 \text{ eV} = 1.60 \times 10^{-19}$ J.) The lowest quantum state, corresponding to $n = 1$, is called the **ground state** and has an energy of $E_1 = -13.606$ eV. The next state, the **first excited state,** has $n = 2$ and an energy of $E_2 = E_1/2^2 = -3.401$ eV. Active Figure 11.20 is an **energy level diagram** showing the energies of these discrete energy states and the corresponding quantum numbers. This diagram is another semigraphical representation. The vertical axis is linear in energy, but the horizontal axis has no significance. The horizontal lines correspond to the allowed energies. The atomic system cannot have any energies other than those represented by the lines. The vertical lines with arrowheads represent transitions between states, during which energy is emitted.

The upper limit of the quantized levels, corresponding to $n \rightarrow \infty$(or $r \rightarrow \infty$) and $E \rightarrow 0$, represents the state for which the electron is removed from the atom.[5] Above this energy is a continuum of available states for the ionized atom. The minimum energy required to ionize the atom is called the **ionization energy.** As can be seen from Active Figure 11.20, the ionization energy for hydrogen, based on Bohr's calculation, is 13.6 eV. This finding constituted a major achievement for the Bohr theory because the ionization energy for hydrogen had already been measured to be 13.6 eV.

Active Figure 11.20 also shows various transitions of the atom from one state to a lower state, as referred to in Bohr's assumption 3. As the energy of the atom decreases in a transition, the difference in energy between the states is carried away by electromagnetic radiation. Those transitions ending on $n = 2$ are shown in color, corresponding to the color of the light they represent. The transitions ending on $n = 2$ form the Balmer series of spectral lines, the wavelengths of which are correctly predicted by the Rydberg equation (see Eq. 11.16). Active Figure 11.20 also shows other spectral series (the Lyman series and the Paschen series) that were found after Balmer's discovery.

Equation 11.24, together with Bohr's third postulate, can be used to calculate the frequency of the radiation that is emitted when the atom makes a transition[6] from a high-energy state to a low-energy state:

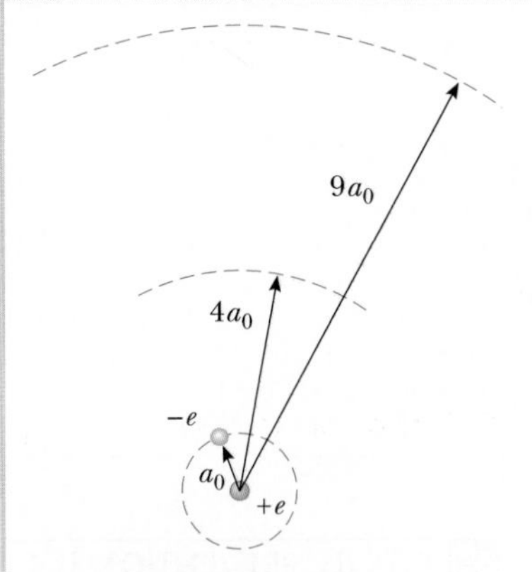

ACTIVE FIGURE 11.19
The first three circular orbits predicted by the Bohr model for hydrogen.

PhysicsNow™ Log into PhysicsNow at www.pop4e.com and go to Active Figure 11.19 to choose the initial and final states of the hydrogen atom and observe the transitions in this figure and in Active Figure 11.20.

[5] The phrase "the electron is removed from the atom" is very commonly used, but, of course, we realize that we mean that the electron and proton are separated *from each other.*

[6] The phrase "the electron makes a transition" is also commonly used, but we will use "the atom makes a transition" to emphasize that the energy belongs to the system of the atom, not just to the electron. This wording is similar to our discussion in Chapter 7 of gravitational potential energy belonging to the system of an object and the Earth, not to the object alone.

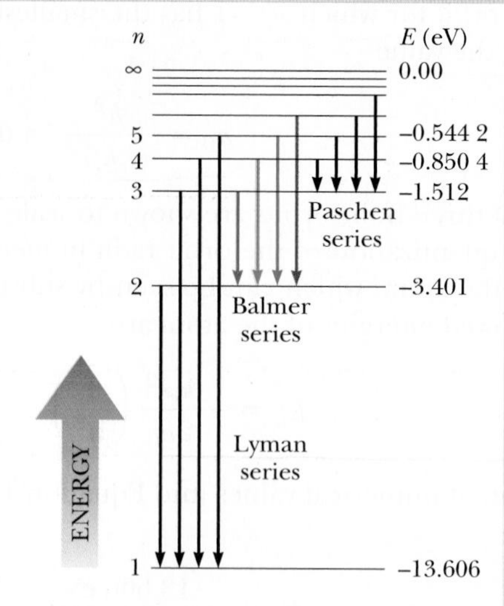

ACTIVE FIGURE 11.20 An energy level diagram for hydrogen. The discrete allowed energies are plotted on the vertical axis. Nothing is plotted on the horizontal axis, but the horizontal extent of the diagram is made large enough to show allowed transitions. Quantum numbers are given on the left and energies (in electron volts) on the right. Vertical arrows represent the four lowest-energy transitions in each of the spectral series shown. The colored arrows for the Balmer series indicate that this series results in visible light.

PhysicsNow™ Log into PhysicsNow at **www.pop4e.com** and go to Active Figure 11.20 to choose the initial and final states of the hydrogen atom and observe the transitions in this figure and in Active Figure 11.19.

■ Frequency of radiation emitted from hydrogen

$$f = \frac{E_i - E_f}{h} = \frac{k_e e^2}{2a_0 h}\left(\frac{1}{n_f^2} - \frac{1}{n_i^2}\right) \qquad [11.26]$$

Because the quantity expressed in the Rydberg equation is wavelength, it is convenient to convert frequency to wavelength, using $c = f\lambda$, to obtain

■ Emission wavelengths of hydrogen

$$\frac{1}{\lambda} = \frac{f}{c} = \frac{k_e e^2}{2a_0 hc}\left(\frac{1}{n_f^2} - \frac{1}{n_i^2}\right) \qquad [11.27]$$

Notice that the *theoretical* expression, Equation 11.27, is identical to the *empirical* Rydberg equation (Equation 11.16), provided that the combination of constants $k_e e^2/2a_0 hc$ is equal to the experimentally determined Rydberg constant and that $n_f = 2$. After Bohr demonstrated the agreement of the constants in these two equations to a precision of about 1%, it was soon recognized as the crowning achievement of his structural model of the atom.

One question remains: What is the significance of $n_f = 2$? Its importance is simply because those transitions ending on $n_f = 2$ result in radiation that happens to lie in the visible; therefore, they were easily observed! As seen in Active Figure 11.20, other series of lines end on other final states. These lines lie in regions of the spectrum not visible to the eye, the infrared and ultraviolet. The generalized Rydberg equation for any initial and final states is

$$\frac{1}{\lambda} = R_{\mathrm{H}}\left(\frac{1}{n_f^2} - \frac{1}{n_i^2}\right) \qquad [11.28]$$

In this equation, different series correspond to different values of n_f and different lines within a series correspond to varying values of n_i.

Bohr immediately extended his structural model for hydrogen to other elements in which all but one electron had been removed. Ionized elements such as He^+, Li^{2+}, and Be^{3+} were suspected to exist in hot stellar atmospheres, where frequent atomic collisions occur with enough energy to completely remove one or

PITFALL PREVENTION 11.4

THE BOHR MODEL IS GREAT, BUT . . . The Bohr model correctly predicts the ionization energy for hydrogen, but it cannot account for the spectra of more complex atoms and is unable to predict many subtle spectral details of hydrogen and other simple atoms. Despite its cultural popularity, the notion of electrons in well-defined orbits about the nucleus is *not* consistent with current models of the atom.

more atomic electrons. Bohr showed that many mysterious lines observed in the Sun and several stars could not be due to hydrogen, but were correctly predicted by his theory if attributed to singly ionized helium.

QUICK QUIZ 11.4 A hydrogen atom makes a transition from the $n = 3$ level to the $n = 2$ level. It then makes a transition from the $n = 2$ level to the $n = 1$ level. Which transition results in emission of the longest-wavelength photon? **(a)** the first transition **(b)** the second transition **(c)** neither, because the wavelengths are the same for both transitions

INTERACTIVE EXAMPLE 11.3 An Electronic Transition in Hydrogen

A hydrogen atom makes a transition from the $n = 2$ state to the ground state (corresponding to $n = 1$). Find the wavelength and frequency of the emitted radiation.

Solution We can use Equation 11.28 directly to obtain λ, with $n_i = 2$ and $n_f = 1$:

$$\frac{1}{\lambda} = R_{\mathrm{H}}\left(\frac{1}{n_f^2} - \frac{1}{n_i^2}\right) = R_{\mathrm{H}}\left(\frac{1}{1^2} - \frac{1}{2^2}\right) = \frac{3R_{\mathrm{H}}}{4}$$

$$\lambda = \frac{4}{3R_{\mathrm{H}}} = \frac{4}{3(1.097 \times 10^7\ \mathrm{m}^{-1})}$$

$$= 1.215 \times 10^{-7}\ \mathrm{m} = 121.5\ \mathrm{nm\ (ultraviolet)}$$

Because $c = f\lambda$, the frequency of the radiation is

$$f = \frac{c}{\lambda} = \frac{3.00 \times 10^8\ \mathrm{m/s}}{1.215 \times 10^{-7}\ \mathrm{m}} = 2.47 \times 10^{15}\ \mathrm{s}^{-1}$$

PhysicsNow™ Investigate transitions of the atom between states by logging into PhysicsNow at **www.pop4e.com** and going to Interactive Example 11.3.

11.6 CHANGING FROM A CIRCULAR TO AN ELLIPTICAL ORBIT

CONTEXT CONNECTION

In Interactive Example 11.1, we discussed a spacecraft in a circular orbit around the Earth. From our studies of Kepler's laws in this chapter, we are also aware that an elliptical orbit is possible for our spacecraft. Let us investigate how the motion of our spacecraft can be changed from a circular to an elliptical orbit, which will set us up for the conclusion to our *Mission to Mars* Context.

Let us identify the system as the spacecraft and the Earth, *but not the portion of the fuel in the spacecraft that we use to change the orbit.* In a given orbit, the mechanical energy of the spacecraft–Earth system is given by Equation 11.10,

$$E = -\frac{GMm}{2r}$$

This energy includes the kinetic energy of the spacecraft and the potential energy associated with the gravitational force between the spacecraft and the Earth. If the rocket engines are fired, the exhausted fuel can be seen as doing work on the spacecraft–Earth system because the thrust force moves through a displacement. As a result, the mechanical energy of the spacecraft–Earth system increases.

The spacecraft has a new, higher energy but is constrained to be in an orbit that includes the original starting point. It cannot be in a higher-energy circular orbit having a larger radius because this orbit would not contain the starting point. The only possibility is that the orbit is elliptical. Figure 11.21 shows the change from the original circular orbit to the new elliptical orbit for our spacecraft.

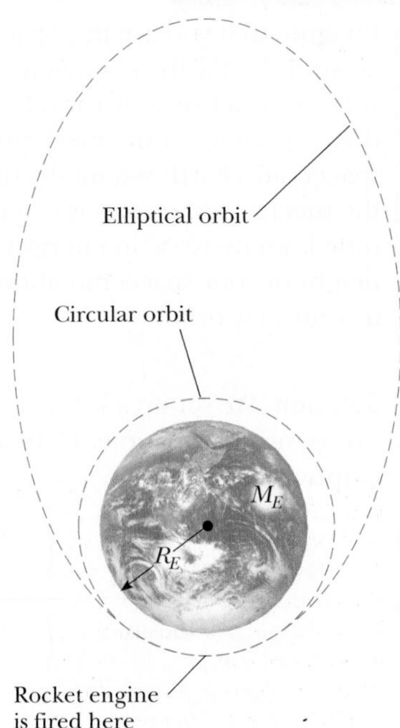

FIGURE 11.21 A spacecraft, originally in a circular orbit about the Earth, fires its engines and enters an elliptical orbit about the Earth.

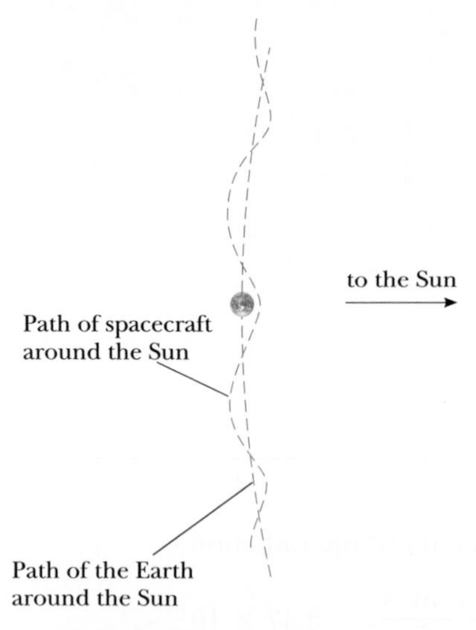

FIGURE 11.22 A spacecraft in orbit about the Earth can be modeled as one in a circular orbit about the Sun, with its orbit about the Earth appearing as small perturbations from the circular orbit.

Equation 11.11 gives the energy of the spacecraft–Earth system for an elliptical orbit. Therefore, if we know the new energy of the orbit, we can find the semimajor axis of the elliptical orbit. Conversely, if we know the semimajor axis of an elliptical orbit we would like to achieve, we can calculate how much additional energy is required from the rocket engines. This information can then be converted to a required burn time for the rockets.

Larger amounts of energy increase supplied by the rocket engines will move the spacecraft into elliptical orbits with larger semimajor axes. What happens if the burn time of the engines is so long that the total mechanical energy of the spacecraft–Earth system becomes positive? A positive energy refers to an *unbound* system. Therefore, in this case, the spacecraft will *escape* from the Earth, going into a hyperbolic path that would not bring it back to the Earth.

This process is the essence of what must be done to transfer to Mars. Our rocket engines must be fired to leave the circular parking orbit and escape the Earth. At this point, our thinking must shift to a spacecraft–Sun system rather than a spacecraft–Earth system. From this point of view, the spacecraft in orbit around the Earth can also be considered to be in a circular orbit around the Sun, moving along with the Earth, as shown in Figure 11.22. The orbit is not a perfect circle because there are perturbations corresponding to its extra motion around the Earth, but these perturbations are small compared with the radius of the orbit around the Sun. When our engines are fired to escape from the Earth, our orbit around the Sun changes from a circular orbit (ignoring the perturbations) to an elliptical one with the Sun at one focus. We shall choose the semimajor axis of our elliptical orbit so that it intersects the orbit of Mars! In the Context 2 Conclusion, we shall look at more details of this process.

EXAMPLE 11.4 How High Do We Go?

Imagine that you are in a spacecraft in circular orbit around the Earth, at a height of 300 km from the surface. You fire your rocket engines, and as a result the magnitude of the mechanical energy of the spacecraft–Earth system decreases by 10.0%. (Because the mechanical energy is negative, a decrease in magnitude is an increase in energy.) What is the greatest height of your spacecraft above the surface of the Earth in your new orbit?

Solution We set up a ratio of the energies of the two orbits, using Equations 11.10 and 11.11 for circular and elliptical orbits:

$$\frac{E_{\text{elliptical}}}{E_{\text{circular}}} = \frac{\left(-\dfrac{GMm}{2a}\right)}{\left(-\dfrac{GMm}{2r}\right)} = \frac{r}{a}$$

The ratio on the left is 0.900 because of the 10.0% decrease in magnitude of the mechanical energy. Therefore,

$$0.900 = \frac{r}{a} \quad \rightarrow \quad a = \frac{r}{0.900} = 1.11r$$

From this equation, we can find the semimajor axis for the elliptical orbit:

$$a = 1.11r = 1.11(6.37 \times 10^3 \text{ km} + 3.00 \times 10^2 \text{ km})$$
$$= 7.40 \times 10^3 \text{ km}$$

The maximum distance from the center of the Earth will occur when the spacecraft is at apogee and is given by

$$r_{\text{max}} = 2a - r = 2(7.40 \times 10^3 \text{ km}) - (6.67 \times 10^3 \text{ km})$$
$$= 8.14 \times 10^3 \text{ km}$$

Now, if we subtract the radius of the Earth, we have the maximum height above the surface:

$$h_{\text{max}} = r_{\text{max}} - R_E = 8.14 \times 10^3 \text{ km} - 6.37 \times 10^3 \text{ km}$$
$$= 1.77 \times 10^3 \text{ km}$$

SUMMARY

Physics Now™ Take a practice test by logging into PhysicsNow at www.pop4e.com and clicking on the Pre-Test link for this chapter.

Newton's law of universal gravitation states that the gravitational force of attraction between any two particles of masses m_1 and m_2 separated by a distance r has the magnitude

$$F_g = G\frac{m_1 m_2}{r^2} \quad [11.1]$$

where G is the **universal gravitational constant** whose value is $6.673 \times 10^{-11}\ \text{N}\cdot\text{m}^2/\text{kg}^2$.

Rather than considering the gravitational force as a direct interaction between two objects, we can imagine that one object sets up a **gravitational field** in space:

$$\vec{\mathbf{g}} \equiv \frac{\vec{\mathbf{F}}_g}{m} \quad [11.4]$$

A second object in this field experiences a force $\vec{\mathbf{F}}_g = m\vec{\mathbf{g}}$ when placed in this field.

Kepler's laws of planetary motion state the following:

1. Each planet in the Solar System moves in an elliptical orbit with the Sun at one focus.
2. The radius vector drawn from the Sun to any planet sweeps out equal areas in equal time intervals.
3. The square of the orbital period of any planet is proportional to the cube of the semimajor axis of the elliptical orbit.

Kepler's first law is a consequence of the inverse-square nature of the law of universal gravitation. The **semimajor axis** of an ellipse is a, where $2a$ is the longest dimension of the ellipse. The **semiminor axis** of the ellipse is b, where $2b$ is the shortest dimension of the ellipse. The **eccentricity** of the ellipse is $e = c/a$, where c is the distance between the center and a focus and $a^2 = b^2 + c^2$.

Kepler's second law is a consequence of the gravitational force being a central force. For a central force, the angular momentum of the planet is conserved.

Kepler's third law is a consequence of the inverse-square nature of the universal law of gravitation. Newton's second law, together with the force law given by Equation 11.1, verifies that the period T and semimajor axis a of the orbit of a planet about the Sun are related by

$$T^2 = \left(\frac{4\pi^2}{GM_S}\right)a^3 \quad [11.7]$$

where M_S is the mass of the Sun.

If an isolated system consists of a particle of mass m moving with a speed v in the vicinity of a massive body of mass M, the *total energy* of the system is constant and is

$$E = \tfrac{1}{2}mv^2 - \frac{GMm}{r} \quad [11.8]$$

If m moves in an elliptical orbit of major axis $2a$ about M, where $M >> m$, the total energy of the system is

$$E = -\frac{GMm}{2a} \quad [11.11]$$

The total energy is negative for any bound system, that is, one in which the orbit is closed, such as a circular or an elliptical orbit.

The Bohr model of the atom successfully describes the spectra of atomic hydrogen and hydrogen-like ions. One basic assumption of this structural model is that the electron can exist only in discrete orbits such that the angular momentum $m_e vr$ is an integral multiple of $\hbar \equiv h/2\pi$. Assuming circular orbits and a simple electrical attraction between the electron and proton, the energies of the quantum states for hydrogen are calculated to be

$$E_n = -\frac{k_e e^2}{2a_0}\left(\frac{1}{n^2}\right) \qquad n = 1, 2, 3, \ldots \quad [11.24]$$

where k_e is the Coulomb constant, e is the fundamental electric charge, n is a positive integer called a **quantum number,** and $a_0 = 0.052\,9$ nm is the **Bohr radius.**

If the hydrogen atom makes a transition from a state whose quantum number is n_i to one whose quantum number is n_f, where $n_f < n_i$, the frequency of the radiation emitted by the atom is

$$f = \frac{k_e e^2}{2a_0 h}\left(\frac{1}{n_f^2} - \frac{1}{n_i^2}\right) \quad [11.26]$$

Using $E_i - E_f = hf = hc/\lambda$, one can calculate the wavelengths of the radiation for various transitions. The calculated wavelengths are in excellent agreement with those in observed atomic spectra.

QUESTIONS

☐ = answer available in the *Student Solutions Manual and Study Guide*

1. If the gravitational force on an object is directly proportional to its mass, why don't objects with large masses fall with greater acceleration than small ones?

2. The gravitational force exerted by the Sun on you is downward into the Earth at night and upward into the sky during the day. If you had a sensitive enough bathroom scale, would you expect to weigh more at night than during the day? Note also that you are farther away from the Sun at night than during the day. Would you expect to weigh less?

3. The gravitational force that the Sun exerts on the Moon is about twice as great as the gravitational force that the Earth exerts on the Moon. Why doesn't the Sun pull the Moon away from the Earth during a total eclipse of the Sun?

4. A satellite in orbit is not truly traveling through a vacuum. It is moving through very thin air. Does the resulting air friction cause the satellite to slow down?

5. Explain why it takes more fuel for a spacecraft to travel from the Earth to the Moon than for the return trip. Estimate the difference.

6. Explain why no work is done on a planet as it moves in a circular orbit around the Sun, even though a gravitational force is acting on the planet. What is the net work done on a planet during each revolution as it moves around the Sun in an elliptical orbit?

7. Why don't we put a geosynchronous weather satellite in orbit around the 45th parallel? Wouldn't that be more useful in the United States than one in orbit around the equator?

8. If a hole could be dug to the center of the Earth, would the force on an object of mass m still obey Equation 11.1 there? What do you think the force on m would be at the center of the Earth?

9. At what position in its elliptical orbit is the speed of a planet a maximum? At what position is the speed a minimum?

10. Each *Voyager* spacecraft was accelerated toward escape speed from the Sun by Jupiter's gravitational force exerted on the spacecraft. How is that possible?

11. In his 1798 experiment, Cavendish was said to have "weighed the Earth." Explain this statement.

12. The *Apollo 13* spacecraft developed trouble in the oxygen system about halfway to the Moon. Why did the mission continue on around the Moon and then return home, rather than immediately turn back to the Earth?

13. Suppose the system of a hydrogen atom obeyed classical mechanics rather than quantum mechanics. Why should such a hypothetical atom emit a continuous spectrum rather than the observed line spectrum?

14. Can the electron in the ground state of hydrogen absorb a photon of energy (a) less than 13.6 eV and (b) greater than 13.6 eV?

15. Explain why, in the Bohr model, the total energy of the atom is negative.

16. Let $-E$ represent the energy of a hydrogen atom. What is the kinetic energy of the electron? What is the potential energy of the atom?

PROBLEMS

1, 2, 3 = straightforward, intermediate, challenging
☐ = full solution available in the *Student Solutions Manual and Study Guide*
Physics Now™ = coached problem with hints available at www.pop4e.com
= computer useful in solving problem
= paired numerical and symbolic problems
= biomedical application

Section 11.1 ▪ Newton's Law of Universal Gravitation Revisited

Problems 5.31 through 5.33 in Chapter 5 can be assigned with this section.

1. Two ocean liners, each with a mass of 40 000 metric tons, are moving on parallel courses, 100 m apart. What is the magnitude of the acceleration of one of the liners toward the other due to their mutual gravitational attraction? Treat the ships as particles.

2. A 200-kg object and a 500-kg object are separated by 0.400 m. (a) Find the net gravitational force exerted by these objects on a 50.0-kg object placed midway between them. (b) At what position (other than an infinitely remote one) can the 50.0-kg object be placed so as to experience a net force of zero?

3. Physics Now™ In introductory physics laboratories, a typical Cavendish balance for measuring the gravitational constant G uses lead spheres with masses of 1.50 kg and 15.0 g whose centers are separated by about 4.50 cm. Calculate the gravitational force between these spheres, treating each as a particle located at the center of the sphere.

4. Three uniform spheres of mass 2.00 kg, 4.00 kg, and 6.00 kg are placed at the corners of a right triangle as shown in Figure P11.4. Calculate the resultant gravitational force on the 4.00-kg object, assuming that the spheres are isolated from the rest of the Universe.

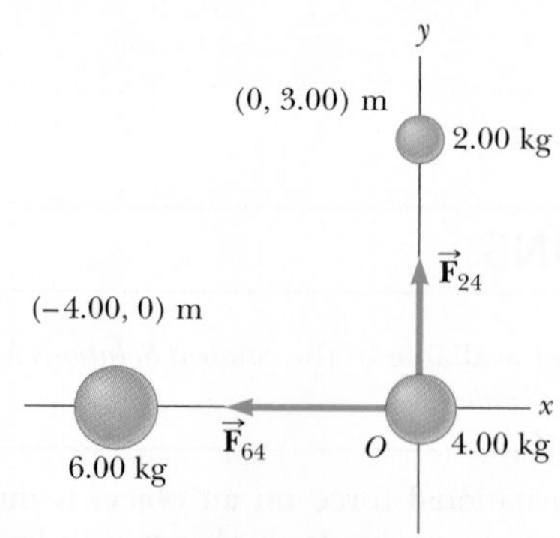

FIGURE P11.4

5. During a solar eclipse, the Moon, the Earth, and the Sun all lie on the same line, with the Moon between the Earth and the Sun. (a) What force is exerted by the Sun on the Moon? (b) What force is exerted by the Earth on the Moon? (c) What force is exerted by the Sun on the Earth?

6. A student proposes to measure the gravitational constant G by suspending two spherical objects from the ceiling of a tall cathedral and measuring the deflection of the cables from the vertical. Draw a free-body diagram of one of the objects. If two 100.0-kg objects are suspended at the lower ends of cables 45.00 m long and the cables are attached to the ceiling 1.000 m apart, what is the separation of the objects?

7. The free-fall acceleration on the surface of the Moon is about one-sixth that on the surface of the Earth. Assuming that the radius of the Moon is about $0.250R_E$, find the ratio of their average densities, ρ_{Moon}/ρ_{Earth}.

8. On the way to the Moon, the *Apollo* astronauts passed a point after which the Moon's gravitational pull became stronger than the Earth's. (a) Determine the distance of this point from the center of the Earth. (b) What is the acceleration due to the Earth's gravitation at this point?

9. **Review problem.** Miranda, a satellite of Uranus, is shown in Figure P11.9a. It can be modeled as a sphere of radius 242 km and mass 6.68×10^{19} kg. (a) Find the free-fall acceleration on its surface. (b) A cliff on Miranda is 5 000 m high. It appears on the limb at the 11 o'clock position in Figure P11.9a and is magnified in Figure P11.9b. If a devotee of extreme sports runs horizontally off the top of the cliff at 8.50 m/s, for what time interval will he be in flight? (Or will he be in orbit?) (c) How far from the base of the vertical cliff will he strike the icy surface of Miranda? (d) What will be his vector impact velocity?

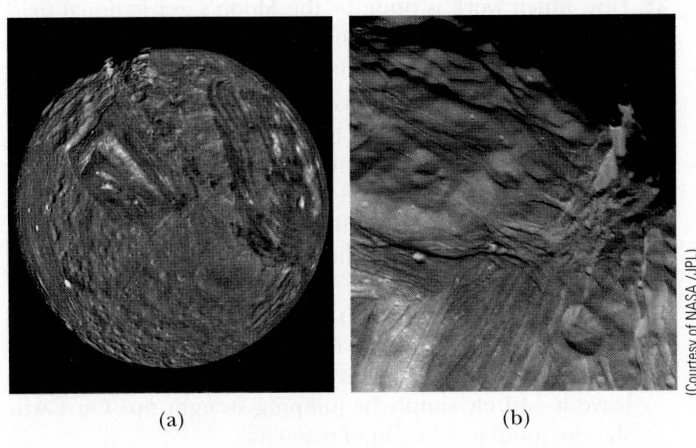

FIGURE P11.9 (a) Miranda, a moon of Uranus. (b) A magnified image of a 5 000-m cliff on Miranda

10. A spacecraft in the shape of a long cylinder has a length of 100 m, and its mass with occupants is 1 000 kg. It has strayed too close to a black hole having a mass 100 times that of the Sun (Fig. P11.10). The nose of the spacecraft points toward the black hole, and the distance between the nose and the center of the black hole is 10.0 km. (a) Determine the total force on the spacecraft. (b) What is the difference in the gravitational fields acting on the occupants in the nose of the ship and on those in the rear of the ship, farthest from the black hole? This difference in accelerations grows rapidly as the ship approaches the black hole. It puts the body of the ship under extreme tension and eventually tears it apart.

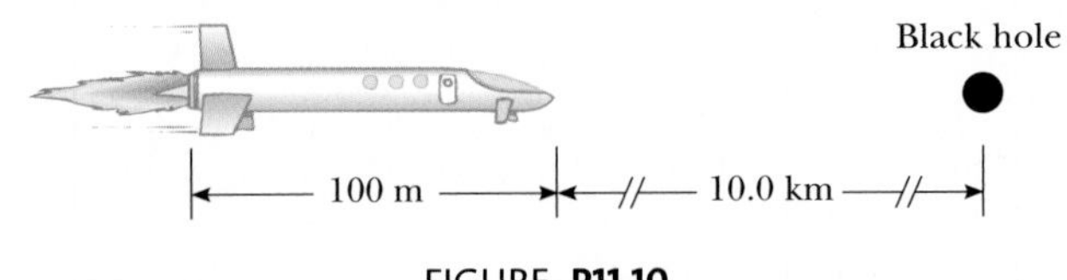

FIGURE P11.10

11. Compute the magnitude and direction of the gravitational field at a point P on the perpendicular bisector of the line joining two objects of equal mass separated by a distance $2a$, as shown in Figure P11.11.

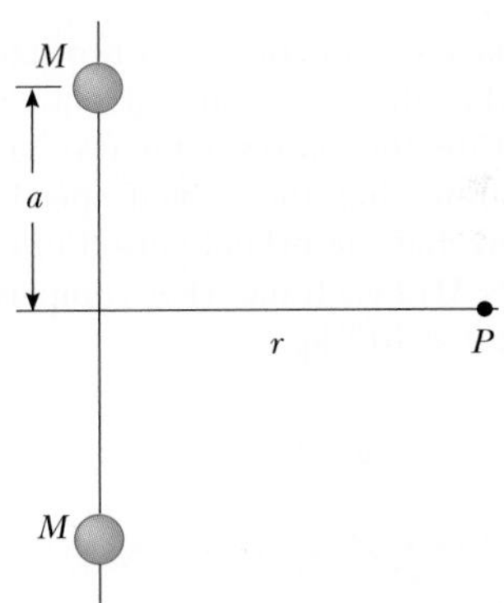

FIGURE P11.11

12. A satellite of mass 300 kg is in a circular orbit around the Earth at an altitude equal to the Earth's mean radius. Find (a) the satellite's orbital speed, (b) the period of its revolution, and (c) the gravitational force acting on it.

Section 11.3 ▪ Kepler's Laws

13. A communication satellite in geosynchronous orbit remains above a single point on the Earth's equator as the planet rotates on its axis. (a) Calculate the radius of its orbit. (b) The satellite relays a radio signal from a transmitter near the North Pole to a receiver, also near the North Pole. Traveling at the speed of light, how long is the radio wave in transit?

14. The *Explorer VIII* satellite, placed into orbit November 3, 1960, to investigate the ionosphere, had the following orbit parameters: perigee, 459 km; apogee, 2 289 km (both distances above the Earth's surface); period, 112.7 min. Find the ratio v_p/v_a of the speed at perigee to that at apogee.

15. **PhysicsNow™** Io, a satellite of Jupiter, has an orbital period of 1.77 days and an orbital radius of 4.22×10^5 km. From these data, determine the mass of Jupiter.

16. Comet Halley approaches the Sun to within 0.570 AU (Fig. P11.16), and its orbital period is 75.6 yr. (AU is the symbol for astronomical unit, where 1 AU = 1.50×10^{11} m is the mean Earth–Sun distance.) How far from the Sun will this comet travel before it starts its return journey?

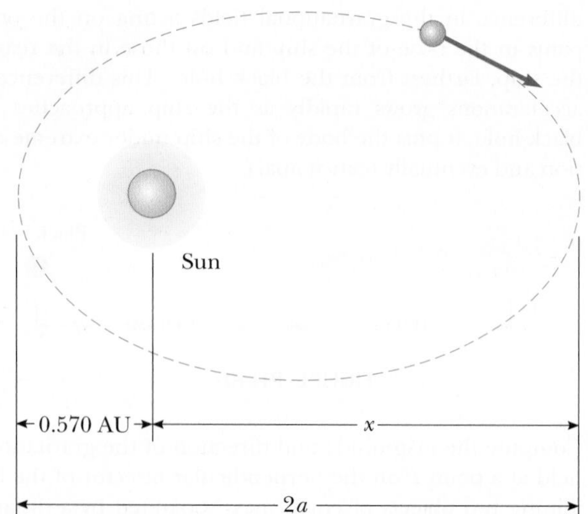

FIGURE P11.16 The elliptical orbit of Comet Halley (not to scale).

17. Plaskett's binary system consists of two stars that revolve in a circular orbit about a center of mass midway between them. Therefore, the masses of the two stars are equal (Fig. P11.17). Assume that the orbital speed of each star is 220 km/s and that the orbital period of each is 14.4 days. Find the mass M of each star. (For comparison, the mass of our Sun is 1.99×10^{30} kg.)

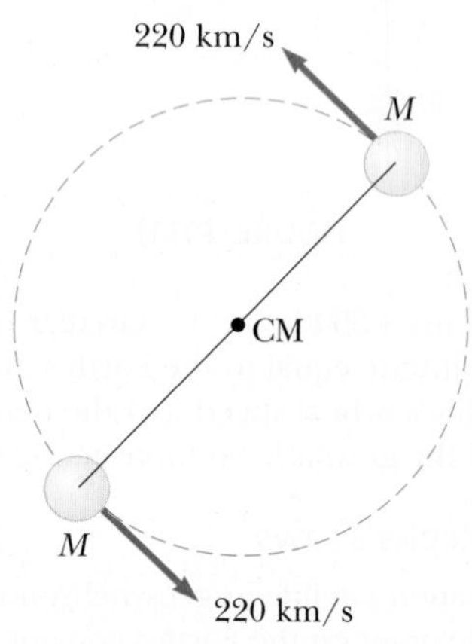

FIGURE P11.17

18. Two planets X and Y travel counterclockwise in circular orbits about a star as shown in Figure P11.18. The radii of their orbits are in the ratio 3:1. At some time, they are aligned as shown in Figure P11.18a, making a straight line with the star. During the next five years, the angular displacement of planet X is 90.0°, as shown in Figure P11.18b. Where is planet Y at this time?

19. Suppose the Sun's gravity were switched off. The planets would leave their nearly circular orbits and fly away in straight lines, as described by Newton's first law. Would Mercury ever be farther from the Sun than Pluto? If so, find how long it would take for Mercury to achieve this passage. If not, give a convincing argument that Pluto is always farther from the Sun than is Mercury.

20. As thermonuclear fusion proceeds in its core, the Sun loses mass at a rate of 3.64×10^9 kg/s. During the 5 000-yr period of recorded history, by how much has the length of the year changed due to the loss of mass from the Sun? (*Suggestions:* Assume that the Earth's orbit is circular. No external torque acts on the Earth–Sun system, so angular momentum is conserved. If x is small compared to 1, then $(1 + x)^n$ is nearly equal to $1 + nx$.)

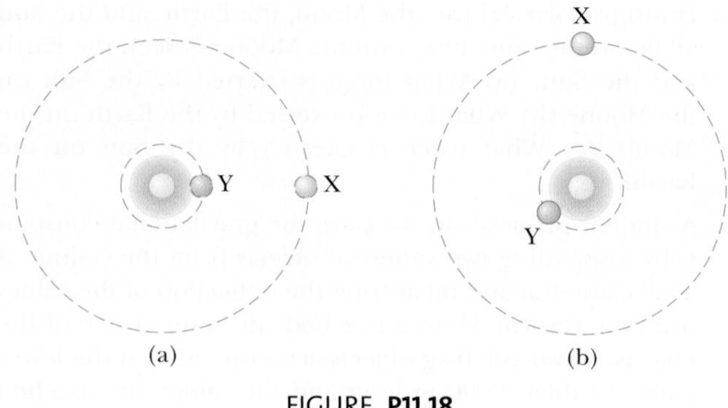

FIGURE P11.18

Section 11.4 ■ Energy Considerations in Planetary and Satellite Motion

Problems 7.35 through 7.38 in Chapter 7 can be assigned with this section.

21. After our Sun exhausts its nuclear fuel, its ultimate fate may be to collapse to a *white dwarf* state, in which it has approximately the same mass as it has now but a radius equal to the radius of the Earth. Calculate (a) the average density of the white dwarf, (b) the free-fall acceleration, and (c) the gravitational potential energy associated with a 1.00-kg object at its surface.

22. How much work is done by the Moon's gravitational field as a 1 000-kg meteor comes in from outer space and impacts on the Moon's surface?

23. An asteroid is on a collision course with Earth. An astronaut lands on the rock to bury explosive charges that will blow the asteroid apart. Most of the small fragments will miss the Earth, and those that fall into the atmosphere will produce only a beautiful meteor shower. The astronaut finds that the density of the spherical asteroid is equal to the average density of the Earth. To ensure its pulverization, she incorporates into the explosives the rocket fuel and oxidizer intended for her return journey. What maximum radius can the asteroid have for her to be able to leave it entirely simply by jumping straight up? On Earth she can jump to a height of 0.500 m.

24. (a) Determine the amount of work that must be done on a 100-kg payload to elevate it to a height of 1 000 km above the Earth's surface. (b) Determine the amount of additional work that is required to put the payload into circular orbit at this elevation.

25. **PhysicsNow™** A space probe is fired up from the Earth's surface with an initial speed of 2.00×10^4 m/s. What will its speed be when it is very far from the Earth? Ignore friction and the rotation of the Earth.

B.C. by John Hart

By permission of John Hart FLP, and Creators Syndicate, Inc.

FIGURE **P11.27**

26. (a) What is the minimum speed, relative to the Sun, necessary for a spacecraft to escape the solar system if it starts at the Earth's orbit? (b) *Voyager 1* achieved a maximum speed of 125 000 km/h on its way to photograph Jupiter. Beyond what distance from the Sun is this speed sufficient to escape the solar system?

27. A "treetop satellite" (Fig. P11.27) moves in a circular orbit just above the surface of a planet, assumed to offer no air resistance. Show that its orbital speed v and the escape speed from the planet are related by the expression $v_{\text{esc}} = \sqrt{2}v$.

28. An object is released from rest at an altitude h above the surface of the Earth. (a) Show that its speed at a distance r from the Earth's center, where $R_E \le r \le R_E + h$, is given by

$$v = \sqrt{2GM_E\left(\frac{1}{r} - \frac{1}{R_E + h}\right)}$$

(b) Assume that the release altitude is 500 km. Perform the integral

$$\Delta t = \int_i^f dt = \int_i^f -\frac{dr}{v}$$

to find the time of fall during which the object moves from the release point to the Earth's surface. The negative sign appears because the object is moving opposite to the radial direction, so its speed is $v = -dr/dt$. Perform the integral numerically.

29. A 500-kg satellite is in a circular orbit at an altitude of 500 km above the Earth's surface. Because of air friction, the satellite eventually falls to the Earth's surface, where it hits the ground with a speed of 2.00 km/s. How much energy was transformed into internal energy by means of air friction?

30. A satellite of mass m, originally on the surface of the Earth, is placed into Earth orbit at an altitude h. (a) With a circular orbit, how long does the satellite take to complete one orbit? (b) What is the satellite's speed? (c) What is the minimum energy input necessary to place this satellite in orbit? Ignore air resistance but include the effect of the planet's daily rotation. At what location on the Earth's surface and in what direction should the satellite be launched to minimize the required energy investment? Represent the mass and radius of the Earth as M_E and R_E.

31. An object is fired vertically upward from the surface of the Earth (of radius R_E) with an initial speed v_i that is comparable to but less than the escape speed v_{esc}. (a) Show that the object attains a maximum height h given by

$$h = \frac{R_E v_i^2}{v_{\text{esc}}^2 - v_i^2}$$

(b) A space vehicle is launched vertically upward from the Earth's surface with an initial speed of 8.76 km/s, which is less than the escape speed of 11.2 km/s. What maximum height does it attain? (c) A meteorite falls toward the Earth. It is essentially at rest with respect to the Earth when it is at a height of 2.51×10^7 m. With what speed does the meteorite strike the Earth? (d) Assume that a baseball is tossed up with an initial speed that is very small compared with the escape speed. Show that the equation from part (a) is consistent with Equation 3.15.

32. Derive an expression for the work required to move an Earth satellite of mass m from a circular orbit of radius $2R_E$ to one of radius $3R_E$.

33. A comet of mass 1.20×10^{10} kg moves in an elliptical orbit around the Sun. Its distance from the Sun ranges between 0.500 AU and 50.0 AU. (a) What is the eccentricity of its orbit? (b) What is its period? (c) At aphelion, what is the potential energy of the comet–Sun system? (*Note:* 1 AU = one astronomical unit = the average distance from the Sun to the Earth = 1.496×10^{11} m.)

Section 11.5 ■ Atomic Spectra and the Bohr Theory of Hydrogen

34. Within the Rosette Nebula shown in the photograph opening this chapter, a hydrogen atom emits light as it undergoes a transition from the $n = 3$ state to the $n = 2$ state. Calculate (a) the energy, (b) the wavelength, and (c) the frequency of the radiation.

35. (a) What value of n_i is associated with the 94.96-nm spectral line in the Lyman series of hydrogen? (b) Could this wavelength be associated with the Paschen series or the Balmer series?

36. For a hydrogen atom in its ground state, use the Bohr model to compute (a) the orbital speed of the electron, (b) the kinetic energy of the electron, and (c) the electric potential energy of the atom.

37. Four possible transitions for a hydrogen atom are as follows:

(i) $n_i = 2;\ n_f = 5$ (ii) $n_i = 5;\ n_f = 3$

(iii) $n_i = 7;\ n_f = 4$ (iv) $n_i = 4;\ n_f = 7$

(a) In which transition is light of the shortest wavelength emitted? (b) In which transition does the atom gain the most energy? (c) In which transition(s) does the atom lose energy?

38. How much energy is required to ionize hydrogen (a) when it is in the ground state and (b) when it is in the state for which $n = 3$?

39. **PhysicsNow™** A hydrogen atom is in its first excited state ($n = 2$). Using the Bohr theory of the atom, calculate (a) the radius of the orbit, (b) the linear momentum of the electron, (c) the angular momentum of the electron, (d) the kinetic energy, (e) the potential energy, and (f) the total energy.

40. Show that the speed of the electron in the nth Bohr orbit in hydrogen is given by

$$v_n = \frac{k_e e^2}{n\hbar}$$

41. Two hydrogen atoms collide head-on and end up with zero kinetic energy. Each atom then emits light with a wavelength of 121.6 nm ($n = 2$ to $n = 1$ transition). At what speed were the atoms moving before the collision?

Section 11.6 ■ Context Connection—Changing From a Circular to an Elliptical Orbit

42. A spacecraft of mass 1.00×10^4 kg is in a circular orbit at an altitude of 500 km above the Earth's surface. Mission Control wants to fire the engines so as to put the spacecraft in an elliptical orbit around the Earth with an apogee of 2.00×10^4 km. How much energy must be used from the fuel to achieve this orbit? (Assume that all the fuel energy goes into increasing the orbital energy. This model will give a lower limit to the required energy because some of the energy from the fuel will appear as internal energy in the hot exhaust gases and engine parts.)

43. A spacecraft is approaching Mars after a long trip from the Earth. Its velocity is such that it is traveling along a parabolic trajectory under the influence of the gravitational force from Mars. The distance of closest approach will be 300 km above the Martian surface. At this point of closest approach, the engines will be fired to slow down the spacecraft and place it in a circular orbit 300 km above the surface. (a) By what percentage must the speed of the spacecraft be reduced to achieve the desired orbit? (b) How would the answer to part (a) change if the distance of closest approach and the desired circular orbit altitude were 600 km instead of 300 km? (*Note:* The energy of the spacecraft–Mars system for a parabolic orbit is $E = 0$.)

Additional Problems

44. The Solar and Heliospheric Observatory (SOHO) spacecraft has a special orbit, chosen so that its view of the Sun is never eclipsed and it is always close enough to the Earth to transmit data easily. It moves in a near-circle around the Sun that is smaller than the Earth's circular orbit. Its period, however, is not less than 1 yr, but just equal to 1 yr. It is always located between the Earth and the Sun along the line joining them. Both objects exert gravitational forces on the observatory. Show that the spacecraft's distance from the Earth must be between 1.47×10^9 m and 1.48×10^9 m. In 1772, Joseph Louis Lagrange determined theoretically the special location allowing this orbit. The SOHO spacecraft took this position on February 14, 1996. (*Suggestions:* Use data that are precise to four digits. The mass of the Earth is 5.983×10^{24} kg.)

45. Let Δg_M represent the difference in the gravitational fields produced by the Moon at the points on the Earth's surface nearest to and farthest from the Moon. Find the fraction $\Delta g_M/g$, where g is the Earth's gravitational field. (This difference is responsible for the occurrence of the *lunar tides* on the Earth.)

46. **Review problem.** Two identical hard spheres, each of mass m and radius r, are released from rest in otherwise empty space with their centers separated by the distance R. They are allowed to collide under the influence of their gravitational attraction. (a) Show that the magnitude of the impulse received by each sphere before they make contact is given by $[Gm^3(1/2r - 1/R)]^{1/2}$. (b) Find the magnitude of the impulse each receives if they collide elastically.

47. (a) Show that the rate of change of the free-fall acceleration with distance above the Earth's surface is

$$\frac{dg}{dr} = -\frac{2GM_E}{R_E^3}$$

This rate of change over distance is called a *gradient.* (b) Assuming that h is small in comparison to the radius of the Earth, show that the difference in free-fall acceleration between two points separated by vertical distance h is

$$|\Delta g| = \frac{2GM_E h}{R_E^3}$$

(c) Evaluate this difference for $h = 6.00$ m, a typical height for a two-story building.

48. A ring of matter is a familiar structure in planetary and stellar astronomy. Examples include Saturn's rings and a ring nebula. Consider a uniform ring of mass 2.36×10^{20} kg and radius 1.00×10^8 m. An object of mass 1 000 kg is placed at a point A on the axis of the ring, 2.00×10^8 m from the center of the ring (Fig. P11.48). When the object is released, the attraction of the ring makes the object move along the axis toward the center of the ring (point B). (a) Calculate the gravitational potential energy of the object–ring system when the object is at A. (b) Calculate the gravitational potential energy of the system when the object is at B. (c) Calculate the speed of the object as it passes through B.

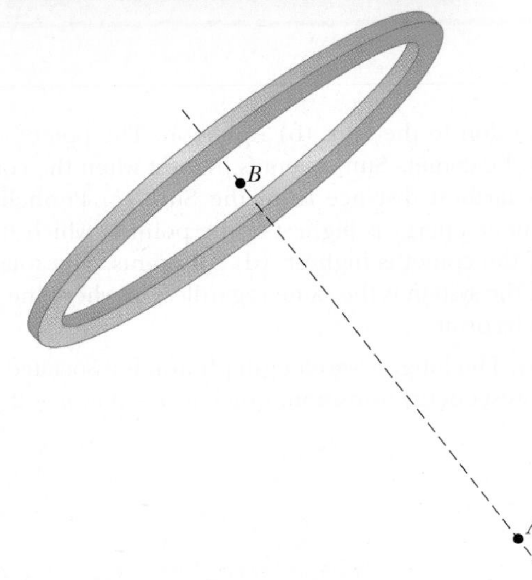

FIGURE **P11.48**

49. *Voyager 1* and *Voyager 2* surveyed the surface of Jupiter's moon Io and photographed active volcanoes spewing liquid sulfur to heights of 70 km above the surface of this moon. Find the speed with which the liquid sulfur left the volcano. Io's mass is 8.9×10^{22} kg, and its radius is 1 820 km.

50. As an astronaut, you observe a small planet to be spherical. After landing on the planet, you set off, walking always straight ahead, and find yourself returning to your spacecraft from the opposite side after completing a lap of 25.0 km. You hold a hammer and a falcon feather at a height of 1.40 m, release them, and observe that they fall together to the surface in 29.2 s. Determine the mass of the planet.

51. Many people assume that air resistance acting on a moving object will always make the object slow down. It can actually be responsible for making the object speed up. Consider a 100-kg Earth satellite in a circular orbit at an altitude of 200 km. A small force of air resistance makes the satellite drop into a circular orbit with an altitude of 100 km. (a) Calculate its initial speed. (b) Calculate its final speed in this process. (c) Calculate the initial energy of the satellite–Earth system. (d) Calculate the final energy of the system. (e) Show that the system has lost mechanical energy and find the amount of the loss due to friction. (f) What force makes the satellite's speed increase? You will find a free-body diagram useful in explaining your answer.

52. The maximum distance from the Earth to the Sun (at our aphelion) is 1.521×10^{11} m, and the distance of closest approach (at perihelion) is 1.471×10^{11} m. The Earth's orbital speed at perihelion is 3.027×10^4 m/s. Determine (a) the Earth's orbital speed at aphelion, (b) the kinetic and potential energies of the Earth–Sun system at perihelion, and (c) the kinetic and potential energies at aphelion. Is the total energy constant? (Ignore the effect of the Moon and other planets.)

53. **PhysicsNow™** Two hypothetical planets of masses m_1 and m_2 and radii r_1 and r_2, respectively, are nearly at rest when they are an infinite distance apart. Because of their gravitational attraction, they head toward each other on a collision course. (a) When their center-to-center separation is d, find expressions for the speed of each planet and for their relative speed. (b) Find the kinetic energy of each planet just before they collide, taking $m_1 = 2.00 \times 10^{24}$ kg, $m_2 = 8.00 \times 10^{24}$ kg, $r_1 = 3.00 \times 10^6$ m, and $r_2 = 5.00 \times 10^6$ m. (*Note:* Both energy and momentum of the system are conserved.)

54. Assume that you are agile enough to run across a horizontal surface at 8.50 m/s, independently of the value of the gravitational field. What would be (a) the radius and (b) the mass of an airless spherical asteroid of uniform density 1.10×10^3 kg/m^3 on which you could launch yourself into orbit by running? (c) What would be your period? (d) Take your mass as 90.0 kg. If the asteroid were originally stationary, your running would set it into rotation with what period?

55. Studies of the relationship of the Sun to its galaxy—the Milky Way—have revealed that the Sun is located near the outer edge of the galactic disc, about 30 000 ly from the center. The Sun has an orbital speed of approximately 250 km/s around the galactic center. (a) What is the period of the Sun's galactic motion? (b) What is the order of magnitude of the mass of the Milky Way galaxy? Suppose the galaxy is made mostly of stars of which the Sun is typical. What is the order of magnitude of the number of stars in the Milky Way?

56. The oldest artificial satellite in orbit is *Vanguard I*, launched March 3, 1958. Its mass is 1.60 kg. In its initial orbit, its minimum distance from the center of the Earth was 7.02 Mm and its speed at this perigee point was 8.23 km/s. (a) Find the total energy of the satellite–Earth system. (b) Find the magnitude of the angular momentum of the satellite. (c) Find its speed at apogee and its maximum (apogee) distance from the center of the Earth. (d) Find the semimajor axis of its orbit. (e) Determine its period.

57. Astronomers detect a distant meteoroid moving along a straight line that, if extended, would pass at a distance $3R_E$ from the center of the Earth, where R_E is the radius of the Earth. What minimum speed must the meteoroid have if the Earth's gravitation is not to deflect the meteoroid to make it strike the Earth?

58. A spherical planet has uniform density ρ. Show that the minimum period for a satellite in orbit around it is

$$T_{\text{min}} = \sqrt{\frac{3\pi}{G\rho}}$$

independent of the radius of the planet.

59. Two stars of masses M and m, separated by a distance d, revolve in circular orbits about their center of mass (Fig. P11.59). Show that each star has a period given by

$$T^2 = \frac{4\pi^2 d^3}{G(M + m)}$$

Proceed as follows: Apply Newton's second law to each star. Note that the center-of-mass condition requires that $Mr_2 = mr_1$, where $r_1 + r_2 = d$.

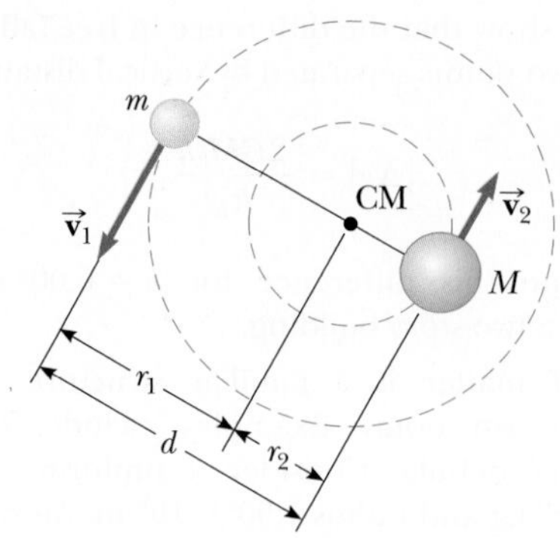

FIGURE **P11.59**

60. (a) A 5.00-kg object is released 1.20×10^7 m from the center of the Earth. It moves with what acceleration relative to the Earth? (b) A 2.00×10^{24} kg object is released 1.20×10^7 m from the center of the Earth. It moves with what acceleration relative to the Earth? Assume that the objects behave as pairs of particles, isolated from the rest of the Universe.

61. The positron is the antiparticle to the electron. It has the same mass and a positive electric charge of the same magnitude as the electron charge. Positronium is a hydrogen-like atom consisting of a positron and an electron revolving around each other. Using the Bohr model, find the allowed distances between the two particles and the allowed energies of the system.

ANSWERS TO QUICK QUIZZES

11.1 (e). The gravitational force follows an inverse-square behavior, so doubling the distance causes the force to be one-fourth as large.

11.2 (a). From Kepler's third law and the given period, the major axis of the asteroid can be calculated. It is found to be 1.2×10^{11} m. Because this axis is smaller than the Earth–Sun distance, the asteroid cannot possibly collide with the Earth.

11.3 (a) Perihelion. Because of conservation of angular momentum, the speed of the comet is highest at its closest position to the Sun. (b) Aphelion. The potential energy of the comet–Sun system is highest when the comet is at its farthest distance from the Sun. (c) Perihelion. The kinetic energy is highest at the point at which the speed of the comet is highest. (d) All points. The total energy of the system is the same regardless of where the comet is in its orbit.

11.4 (a). The longest-wavelength photon is associated with the lowest-energy transition, which is $n = 3$ to $n = 2$.

A Successful Mission Plan

Now that we have explored the physics of classical mechanics, let us return to our central question for the *Mission to Mars* Context:

How can we undertake a successful transfer of a spacecraft from the Earth to Mars?

We make use of the physical principles that we now understand and apply them to our journey from the Earth to Mars.

Let us start with a more modest proposition. Suppose a spacecraft is in a circular orbit around the Earth and you are a passenger on the spacecraft. If you toss a wrench in the direction of travel, tangent to the circular path, what orbital path will the wrench follow?

Let us adopt a simplification model in which the spacecraft is much more massive than the wrench. Conservation of momentum for the isolated system of the wrench and the spacecraft tells us that the spacecraft must slow down slightly once the wrench is thrown. Because of the mass difference between the wrench and spacecraft, however, we can ignore the small change in the spacecraft's speed. The wrench now enters a new orbit, from its perigee position, and the wrench–Earth system has more energy than it had when the wrench was in the circular orbit. Because the orbital energy is related to the major axis, the wrench is injected into an elliptical orbit as discussed in the Context Connection of Chapter 11 and as shown in Figure 1. Therefore, the path of the wrench is changed from a circular orbit to an elliptical orbit by providing the wrench–Earth system with extra energy. The energy is provided by the force you apply to the wrench tangent to the circular orbit because you have done work on the system. The elliptical orbit will take the wrench farther from the Earth than the circular orbit. If there were another spacecraft in a higher circular orbit than your spacecraft, you could throw the wrench so that it transfers from one spacecraft to another as shown in Figure 2. For that to occur, the elliptical orbit of the wrench must intersect with the higher spacecraft orbit. Furthermore, the wrench and the second spacecraft must arrive at the same point at the same time.

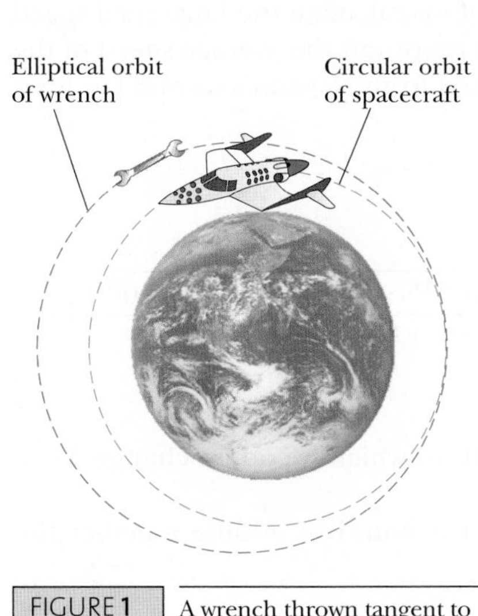

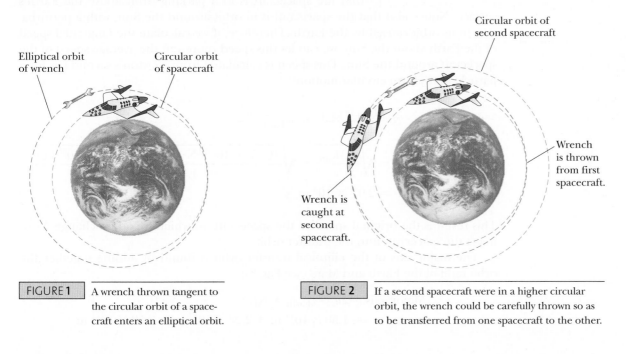

FIGURE 1 A wrench thrown tangent to the circular orbit of a spacecraft enters an elliptical orbit.

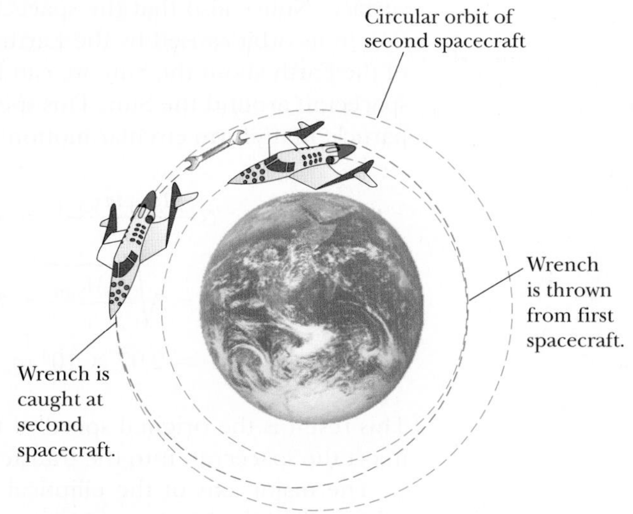

FIGURE 2 If a second spacecraft were in a higher circular orbit, the wrench could be carefully thrown so as to be transferred from one spacecraft to the other.

This scenario is the essence of our planned mission from the Earth to Mars. Rather than transferring a wrench between two spacecraft in orbit around the Earth, we will transfer a spacecraft between two planets in orbit around the Sun. Kinetic energy is added to the wrench–Earth system by throwing the wrench. Kinetic energy is added to the spacecraft–Sun system by firing the engines.

What if you were to throw the wrench harder and harder in the previous example? The wrench would be placed in a larger and larger elliptical orbit around the Earth. As you increased the launch velocity, you could inject the wrench into a *hyperbolic* escape orbit, relative to the Earth, and into an *elliptical* orbit around the *Sun*. This approach is the one we will take for the trip from the Earth to Mars; we will break free from a circular *parking* orbit around the Earth and move into an elliptical *transfer* orbit around the Sun. The spacecraft will then continue on its journey to Mars, where it will enter a new parking orbit.

Now let us focus our attention on the transfer orbit part of the journey. One simple transfer orbit is called a **Hohmann transfer,** the type of transfer imparted to the wrench shown in Figure 2. The Hohmann transfer involves the least energy expenditure and thus requires the smallest amount of fuel. As might be expected for a lowest-energy transfer, the transfer time for a Hohmann transfer is longer than for other types of orbits. We shall investigate the Hohmann transfer because of its simplicity and its general usefulness in planetary transfers.

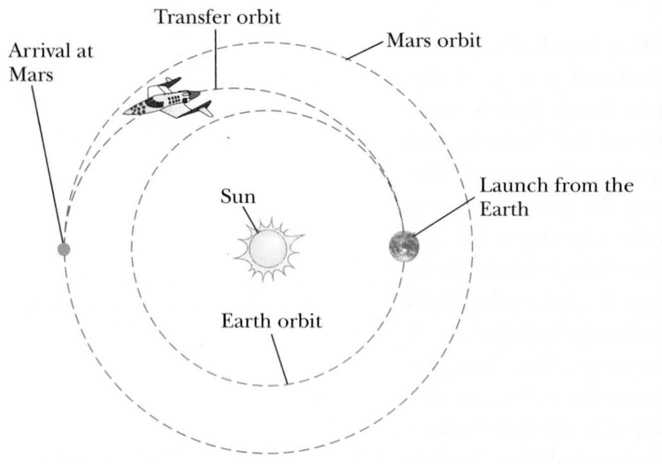

FIGURE 3 The Hohmann transfer orbit from the Earth to Mars. It is similar to transferring the wrench from one spacecraft to another in Figure 2, but here we are transferring a spacecraft from one planet to another.

The rocket engine on the spacecraft is fired from the parking orbit such that the spacecraft enters an elliptical orbit around the Sun at its perihelion and encounters the planet at the spacecraft's aphelion. Therefore, the spacecraft makes exactly one half of a revolution about its elliptical path during the transfer as shown in Figure 3.

This process is energy efficient because fuel is expended only at the beginning and the end. The movement between parking orbits around the Earth and Mars is free; the spacecraft simply follows Kepler's laws while in an elliptical orbit around the Sun.

Let us perform a simple numerical calculation to see how to apply the mechanical laws to this process. We assume that the spacecraft is in a parking orbit above the Earth's surface. Notice also that the spacecraft is in orbit around the Sun, with a perturbation in its orbit caused by the Earth. Therefore, if we calculate the tangential speed of the Earth about the Sun, we can let this speed represent the average speed of the spacecraft around the Sun. This speed is calculated from Newton's second law for a particle in uniform circular motion:

$$F = ma \quad \rightarrow \quad G\frac{M_{\text{Sun}}\, m_{\text{Earth}}}{r^2} = m_{\text{Earth}}\frac{v^2}{r}$$

$$\rightarrow \quad v = \sqrt{\frac{GM_{\text{Sun}}}{r}} = \sqrt{\frac{(6.67 \times 10^{-11}\,\text{N}\cdot\text{m}^2/\text{kg}^2)(1.99 \times 10^{30}\,\text{kg})}{1.50 \times 10^{11}\,\text{m}}}$$

$$= 2.97 \times 10^4\ \text{m/s}$$

This result is the original speed of the spacecraft, to which we add a change Δv to inject the spacecraft into the transfer orbit.

The major axis of the elliptical transfer orbit is found by adding together the orbit radii of the Earth and Mars (see Fig. 3):

$$\text{Major axis} = 2a = r_{\text{Earth}} + r_{\text{Mars}}$$

$$= 1.50 \times 10^{11}\ \text{m} + 2.28 \times 10^{11}\ \text{m} = 3.78 \times 10^{11}\ \text{m}$$

Therefore

$$a = 1.89 \times 10^{11}\ \text{m}$$

From this value, Kepler's third law is used to find the travel time, which is one half of the period of the orbit:

$$\begin{aligned}\Delta t_{\text{travel}} &= \tfrac{1}{2}T = \tfrac{1}{2}\sqrt{\frac{4\pi^2}{GM_{\text{Sun}}}a^3}\\ &= \tfrac{1}{2}\sqrt{\frac{4\pi^2}{(6.67 \times 10^{-11}\ \text{N}\cdot\text{m}^2/\text{kg}^2)(1.99 \times 10^{30}\ \text{kg})}(1.89 \times 10^{11}\ \text{m})^3}\\ &= 2.24 \times 10^7\ \text{s} = 0.711\ \text{yr} = 260\ \text{d}\end{aligned}$$

Therefore, the journey to Mars will require 260 Earth days. We can also determine where in their orbits Mars and the Earth must be so that the planet will be there when the spacecraft arrives. Mars has an orbital period of 687 Earth days. During the transfer time, the angular position *change* of Mars is

$$\Delta\theta_{\text{Mars}} = \frac{260\ \text{d}}{687\ \text{d}}(2\pi) = 2.38\ \text{rad} = 136°$$

Therefore, for the spacecraft and Mars to arrive at the same point at the same time, the spacecraft must be launched when Mars is 180° − 136° = 44° ahead of the Earth in its orbit. This geometry is shown in Figure 4.

With relatively simple mathematics, that is as far as we can go in describing the details of a trip to Mars. We have found the desired path, the time for the trip, and the position of Mars at launch time. Another important issue for the spacecraft captain would be that of the amount of fuel required for the trip. This question is related to the speed changes necessary to put us into a transfer orbit. These types of calculations involve energy considerations and are explored in Problem 3.

Although many considerations for a successful mission to Mars have not been addressed, we have successfully designed a transfer orbit from the Earth to Mars that is consistent with the laws of mechanics. We consequently declare success for our endeavor and bring this *Mission to Mars* Context to a close.

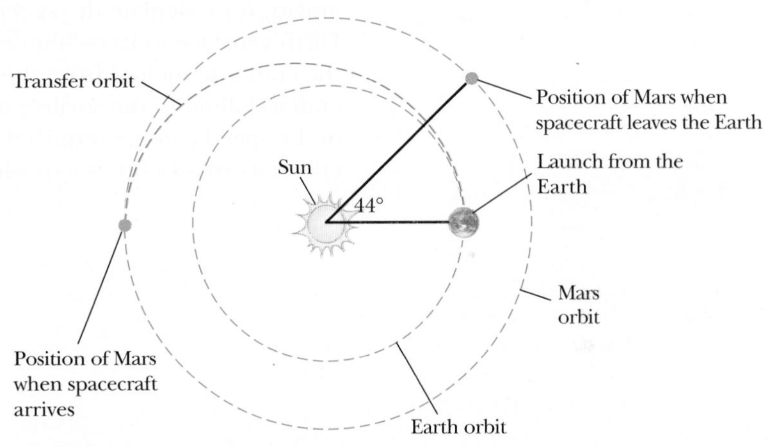

FIGURE 4 The spacecraft must be launched when Mars is 44° ahead of the Earth in its orbit.

Questions

1. Some science fiction stories describe a twin planet to the Earth. It is exactly 180° ahead of us in the same orbit as the Earth, so we will never see it because it is on the other side of the Sun. Assuming you are in a spacecraft in orbit around the Earth, describe conceptually how you could visit this planet by altering your orbit.
2. You are in an orbiting spacecraft. Another spacecraft is in precisely the same orbit but is 1 km ahead of you, moving in the same direction around the circle. Through an oversight, your food supplies have been exhausted, but there is more than enough food in the other spacecraft. The commander of the other spacecraft is going to throw, from her spacecraft to yours, a picnic basket full of sandwiches. Give a qualitative description of how she should throw it.

Problems

1. Consider a Hohmann transfer from the Earth to Venus. (a) How long will this transfer take? (b) Should Venus be ahead of or behind the Earth in its orbit

when the spacecraft leaves the Earth on its way to the rendezvous? How many degrees is Venus ahead or behind the Earth?

2. You are on a space station in a circular orbit 500 km above the surface of the Earth. Your passenger and guest is a large, strong, intelligent extraterrestrial. You cannot answer her penetrating questions about bigotry and war, so you try to teach her to play golf. Walking on the space station surface with magnetic shoes, you demonstrate a drive. The alien tees up a golf ball and hits it with incredible power, sending it off with speed Δv, relative to the space station, in a direction parallel to the instantaneous velocity vector of the space station. You notice that after you then complete precisely 2.00 orbits of the Earth, the golf ball also returns to the same location, so you reach up and catch the ball as it is passing the space station. With what speed Δv was the golf ball hit?

3. Investigate what the engine has to do to make a spacecraft follow the Hohmann transfer orbit from the Earth to Mars described in the text. Short-duration burns of our rocket engine are required to change the speed of our spacecraft whenever we alter our orbit. There are no brakes in space, so fuel is required both to increase and to decrease the speed of the spacecraft. First, ignore the gravitational attraction between the spacecraft and the planets. (a) Calculate the speed change required for switching the craft from a circular orbit around the Sun at the Earth's distance to the transfer orbit to Mars. (b) Calculate the speed change required for switching from the transfer orbit to a circular orbit around the Sun at the distance of Mars. Now consider the effects of the two planets' gravity. (c) Calculate the speed change required to carry the craft from the Earth's surface to its own independent orbit around the Sun. You may suppose the craft is launched from the Earth's equator toward the east. (d) Model the craft as falling to the surface of Mars from solar orbit. Calculate the magnitude of the speed change required to make a soft landing on Mars at the end of the fall. Mars rotates on its axis with a period of 24.6 h.

CONTEXT 3

Earthquakes

Earthquakes result in massive movement of the ground, as evidenced by the accompanying photograph of railroad tracks in Mexico, damaged severely by an earthquake in 1985. Anyone who has experienced a serious earthquake can attest to the violent shaking it produces. In this Context, we shall focus on earthquakes as an application of our study of the physics of vibrations and waves.

The cause of an earthquake is a release of energy within the Earth at a point called the *focus,* or *hypocenter,* of the earthquake. The point on the Earth's surface radially above the focus is called the *epicenter.* As the energy from the focus reaches the surface, it spreads out along the surface of the Earth. We might expect that the risk of damage in an earthquake decreases as one moves farther from the epicenter, and over long distances that assumption is correct. For example, structures in Kansas are not affected by earthquakes in California. In regions close to the earthquake, however, the notion of decrease in risk with distance is not consistent. Consider, for example, the following quotations describing damage in two different earthquakes.

After the Northridge, California, earthquake, January 17, 1994:[1]

> Although the city [Santa Monica] sits 25 kilometers from the shock's epicenter, it suffered more [damage] than did other areas less than one-third that distance from the jolt.

After the Michoacán earthquake, September 19, 1985:[2]

> An earthquake rattled the coast of Mexico in the state of Michoacán, about 400 kilometers west of Mexico City. Near the coast, the shaking of the ground was mild and caused little damage. As the seismic waves raced

(Guillermo Aldana Espinosa/National Geographic Image Collection)

FIGURE 1 In this Context, we shall study the physics of vibrations and waves by investigating earthquakes. These deformed railroad tracks suggest the large amounts of energy released in an earthquake.

(Joseph Sohm/ChromoSohm Inc./CORBIS)

FIGURE 2 The Northridge earthquake in California in 1994 caused billions of dollars in damage.

[1] *Science News,* Volume 145, 1994, p. 287.

[2] *American Scientist,* November–December 1992, p. 566.

FIGURE 3 Severe damage occurred in localized regions of Mexico City in 1985, even though the epicenter of the Michoacán earthquake was hundreds of kilometers away.

inland, the ground shook even less, and by the time the waves were 100 kilometers from Mexico City, the shaking had nearly subsided. Nevertheless, the seismic waves induced severe shaking in the city, and some areas continued to shake for several minutes after the seismic waves had passed. Some 300 buildings collapsed and more than 20,000 people died.

It is clear from these quotations that the notion of a simple decrease in risk with distance is misleading. We will use these quotations as motivation in our study of the physics of vibrations and waves so that we can better analyze the risk of damage to structures in an earthquake. Our study here will also be important when we investigate electromagnetic waves in Chapters 24 through 27. In this Context, we shall address the central question:

How can we choose locations and build structures to minimize the risk of damage in an earthquake?

CHAPTER 12

Oscillatory Motion

(Courtesy of Motioneering Inc.)

(© Simon Kwong/Reuters/CORBIS)

To reduce swaying in tall buildings because of wind, tuned dampers are placed near the top of the building. These mechanisms include an object of large mass that oscillates under computer control at the same frequency as the building, reducing the swaying. The large sphere in the photograph on the left is part of the tuned damper system of the building in the photograph on the right, called Taipei 101, in Taiwan. The building, also called the Taipei Financial Center, was completed in 2004, at which time it held the record for the world's tallest building.

CHAPTER OUTLINE

You are most likely familiar with several examples of *periodic* motion, such as the oscillations of an object on a spring, the motion of a pendulum, and the vibrations of a stringed musical instrument. Numerous other systems exhibit periodic behavior. For example, the molecules in a solid oscillate about their equilibrium positions; electromagnetic waves, such as light waves, radar, and radio waves, are characterized by oscillating electric and magnetic field vectors; in alternating-current circuits, such as in your household electrical service, voltage and current vary periodically with time. In this chapter, we will investigate mechanical systems that exhibit periodic motion.

We have experienced a number of situations in which the net force on a particle is constant. In these situations, the acceleration of the particle is also constant and we can describe the motion of the particle using the kinematic equations of Chapter 2. If a force acting on a particle varies in time, the acceleration of the particle also changes with time and so the kinematic equations cannot be used.

A special kind of periodic motion occurs when the force that acts on a particle is always directed toward an equilibrium position and is proportional to the position of the particle relative to the

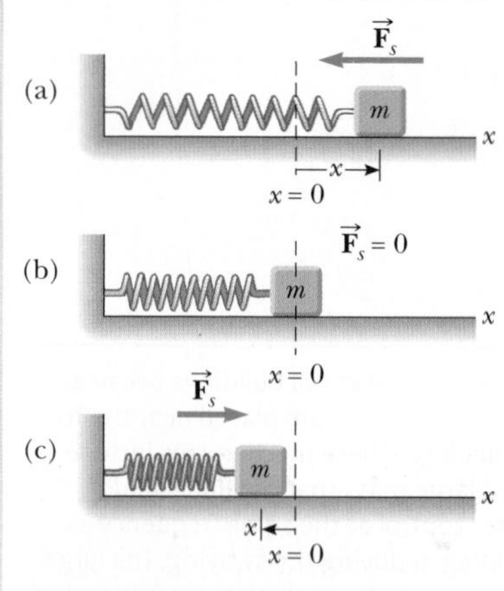

ACTIVE FIGURE 12.1

A block attached to a spring on a frictionless track moves in simple harmonic motion. (a) When the block is displaced to the right of equilibrium, the position is positive and the force and acceleration are negative. (b) At the equilibrium position $x = 0$, the force and acceleration of the block are zero but the speed is a maximum. (c) When the position is negative, the force and acceleration of the block are positive.

PhysicsNow™ By logging into PhysicsNow at www.pop4e.com and going to Active Figure 12.1 you can choose the spring constant and the initial position and velocities of the block to see the resulting simple harmonic motion.

PITFALL PREVENTION 12.1

THE ORIENTATION OF THE SPRING Active Figure 12.1 shows the pictorial representation we will use to study the behavior of systems with springs: a *horizontal* spring with an attached block sliding on a frictionless surface. Another possible pictorial representation is a block hanging from a *vertical* spring. All the results that we discuss for the horizontal spring will be the same for the vertical spring, except for one difference. When the block is placed on the vertical spring, its weight will cause the spring to extend. If the position of the block at which it hangs at rest on the spring is defined as $x = 0$, the results of this chapter will apply to this system also.

equilibrium position. We shall study this special type of varying force in this chapter. When this type of force acts on a particle, the particle exhibits *simple harmonic motion,* which will serve as an analysis model for a large class of oscillation problems.

12.1 MOTION OF A PARTICLE ATTACHED TO A SPRING

At this point in your study of physics, you have probably started to develop a set of mental models associated with the analysis models we have developed. By mental model, we mean a typical physical situation that comes to your mind each time you identify the analysis model to be used in a problem. For example, a rock falling in the absence of air resistance is a possible mental model for the analysis model of a particle under constant acceleration. Collisions between two billiard balls represent a mental model to help understand the momentum version of the isolated system model. A bicycle wheel might offer a mental model for the rigid object under a net torque model.

In the case of oscillatory motion, a useful mental model is an object of mass m attached to a horizontal spring as in Active Figure 12.1. If the spring is unstretched, the object is at rest on a frictionless surface at its **equilibrium position,** which is defined as $x = 0$ (Active Fig. 12.1b). If the object is pulled to the side to position x and released, it will oscillate back and forth as we discussed in Section 6.4. We will use the particle model to ignore the object's size and analyze a particle on a spring as our system.

Recall from Chapter 6 that when the particle attached to an idealized massless spring is located at a position x, the spring exerts a force F_s on it given by **Hooke's law,**

$$F_s = -kx \qquad [12.1]$$

where k is the **force constant** or the **spring constant** of the spring. We call the force in Equation 12.1 a **linear restoring force** because it is proportional to the position of the particle relative to the equilibrium position and is always directed toward the equilibrium position. That is, when the particle is displaced to the right in Active Figure 12.1a, x is positive and the spring force is negative, to the left. When the particle is displaced to the left of $x = 0$ (Active Fig. 12.1c), x is negative and the spring force is positive, to the right. When a particle is under the effect of a linear restoring force, the motion it follows is a special type of oscillatory motion called **simple harmonic motion.** You can test whether or not a particle will undergo simple harmonic motion by seeing if the force on the particle is linear in x. A system undergoing simple harmonic motion is called a **simple harmonic oscillator,** and we describe the motion of the particle with an analysis model called the particle in simple harmonic motion.

Let us imagine a particle subject to a linear restoring force such as that given by Equation 12.1. Applying Newton's second law in the x direction to the particle gives us

$$\sum F = F_s = ma \quad \rightarrow \quad -kx = ma$$

$$a = -\frac{k}{m}x \qquad [12.2]$$

That is, **the acceleration of a particle in simple harmonic motion is proportional to the position of the particle relative to the equilibrium position and is in the opposite direction.** If the particle is released from rest at position $x = A$, its *initial* acceleration is $-kA/m$. When the particle passes through the equilibrium position $x = 0$, its acceleration is zero. At this instant, its speed is a maximum because the acceleration changes sign. The particle then continues to travel to the left of the equilibrium position with a positive acceleration and finally reaches $x = -A$, at which time its acceleration is $+kA/m$ and its speed is again zero. The particle completes a full cycle of its motion by returning to the original position, again passing through $x = 0$ with maximum speed. Thus, we see that the particle oscillates between the turning points $x = \pm A$. In the absence of friction, this motion will continue forever because the force exerted by the spring is conservative. Real systems are generally subject to friction and so cannot oscillate forever. We explore the details of the situation with friction in Section 12.6.

QUICK QUIZ 12.1 A block on the end of a spring is pulled to position $x = A$ and released. In one full cycle of its motion, through what total distance does it travel? **(a)** $A/2$ **(b)** A **(c)** $2A$ **(d)** $4A$

12.2 MATHEMATICAL REPRESENTATION OF SIMPLE HARMONIC MOTION

Let us now develop a mathematical representation of the motion we described in the preceding section. Recall that, by definition, $a = dv/dt = d^2x/dt^2$, so we can express Equation 12.2 as

$$\frac{d^2x}{dt^2} = -\frac{k}{m}x \qquad [12.3]$$

We denote the ratio k/m with the symbol ω^2 (we choose ω^2 rather than ω to make the solution simpler in form),

$$\omega^2 = \frac{k}{m} \qquad [12.4]$$

and Equation 12.3 can be written in the form

$$\frac{d^2x}{dt^2} = -\omega^2 x \qquad [12.5]$$

What we now require is a mathematical solution to Equation 12.5, that is, a function $x(t)$ that satisfies this second-order differential equation. This function will be a mathematical representation of the particle's position as a function of time. We seek a function $x(t)$ such that the second derivative of the function is the same as the original function with a negative sign and multiplied by ω^2. The trigonometric functions sine and cosine exhibit this behavior, so we can build a solution around one or both of them. The following cosine function is a solution to the differential equation:

$$x(t) = A\cos(\omega t + \phi) \qquad [12.6]$$

■ Position of a particle in simple harmonic motion

where A, ω, and ϕ are constants.[1] To see explicitly that this expression is a solution to Equation 12.5, note that

$$\frac{dx}{dt} = A\frac{d}{dt}\cos(\omega t + \phi) = -\omega A\sin(\omega t + \phi) \qquad [12.7]$$

$$\frac{d^2x}{dt^2} = -\omega A\frac{d}{dt}\sin(\omega t + \phi) = -\omega^2 A\cos(\omega t + \phi) \qquad [12.8]$$

PITFALL PREVENTION 12.2

A NONCONSTANT ACCELERATION Notice that the acceleration of the particle in simple harmonic motion is not constant; Equation 12.2 shows that it varies with position x. Therefore, as pointed out in the introduction to the chapter, we *cannot* apply the kinematic equations of Chapter 2 in this situation. We now explore the correct approach in Section 12.2.

[1] In earlier chapters, we saw many examples in which we evaluated a trigonometric function of an angle. The argument of a trigonometric function, such as sine or cosine, *must* be a pure number. The radian is a pure number because it is a ratio of lengths. Degrees are pure simply because the degree is a completely artificial "unit"; it is not related to measurements of lengths. The notion of requiring a pure number for a trigonometric function is important in Equation 12.6, where the angle is expressed in terms of other measurements. Therefore, ω *must* be expressed in radians per second (and not, for example, in revolutions per second) if t is expressed in seconds. Furthermore, the argument of other types of functions must also be pure numbers, including logarithms and exponential functions.

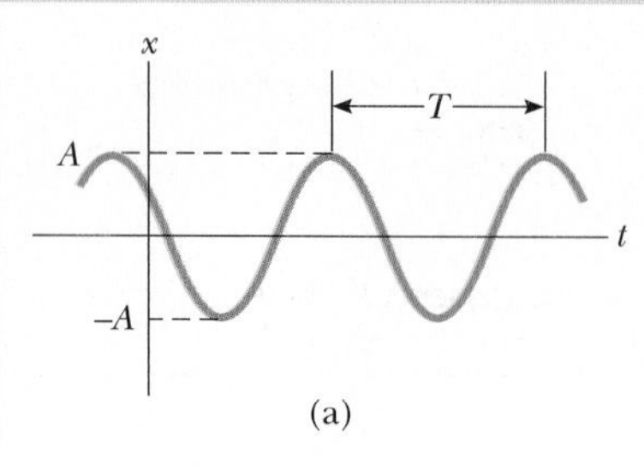

(a)

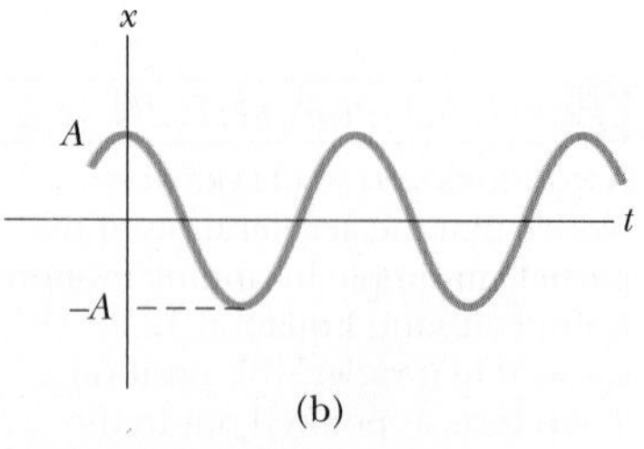

(b)

ACTIVE FIGURE 12.2

(a) A graphical representation (position versus time) for the system in Active Figure 12.1, a particle in simple harmonic motion. The amplitude of the motion is A and the period is T. (b) The x-t curve in the special case in which $x = A$ and $v = 0$ at $t = 0$.

PhysicsNow™ By logging into PhysicsNow at **www.pop4e.com** and going to Active Figure 12.2 you can adjust the graphical representation and see the resulting simple harmonic motion of the block in Active Figure 12.1.

Comparing Equations 12.6 and 12.8, we see that $d^2x/dt^2 = -\omega^2 x$ and that Equation 12.5 is satisfied.

The parameters A, ω, and ϕ are constants of the motion. To give physical significance to these constants, it is convenient to form a graphical representation of the motion by plotting x as a function of t as in Active Figure 12.2a. First, we note that A, called the **amplitude** of the motion, is simply the **maximum value of the position of the particle in either the positive or negative x direction.** The constant ω is called the **angular frequency** and has units of radians per second (rad/s). From Equation 12.4, the angular frequency is

$$\omega = \sqrt{\frac{k}{m}} \qquad [12.9]$$

The constant angle ϕ is called the **phase constant** (or phase angle) and, along with the amplitude A, is determined uniquely by the position and velocity of the particle at $t = 0$. If the particle is at its maximum position $x = A$ at $t = 0$, the phase constant is $\phi = 0$ and the graphical representation of the motion is shown in Active Figure 12.2b. The quantity $(\omega t + \phi)$ is called the **phase** of the motion. Note that the function $x(t)$ is periodic and that its value is the same each time ωt increases by 2π rad.

Equations 12.1, 12.5, and 12.6 form the basis for the analysis model of the particle in simple harmonic motion. We can be assured that a particle is undergoing simple harmonic motion if (1) we analyze the situation and find that the force on the particle is of the mathematical form of Equation 12.1, (2) we analyze the situation and find that it is described by a differential equation of the form of Equation 12.5, or (3) we analyze the situation and find that the position of the particle is described by Equation 12.6.

Let us investigate further the mathematical description of the motion. The **period** T of the motion is the time interval required for the particle to go through one full cycle of its motion (see Active Fig. 12.2a). That is, the values of x and v for the particle at time t equal the values of x and v at time $t + T$. We can relate the period to the angular frequency by noting that the phase increases by 2π rad in a time interval of T:

$$[\omega(t + T) + \phi] - (\omega t + \phi) = 2\pi$$

Simplifying this expression, we see that $\omega T = 2\pi$, or

■ Relation of period to angular frequency

$$T = \frac{2\pi}{\omega} \qquad [12.10]$$

The inverse of the period is called the **frequency** f of the motion. Whereas the period is the time interval per oscillation, the frequency represents **the number of oscillations the particle makes per unit time interval:**

$$f = \frac{1}{T} = \frac{\omega}{2\pi} \qquad [12.11]$$

The units of f are cycles per second, or **hertz** (Hz). Rearranging Equation 12.11 gives

$$\omega = 2\pi f = \frac{2\pi}{T} \qquad [12.12]$$

We can use Equations 12.9, 12.10, and 12.11 to express the period and frequency of the motion for the particle–spring system in terms of the characteristics m and k of the system as

PITFALL PREVENTION 12.3

TWO KINDS OF FREQUENCY We identify two kinds of frequency for a simple harmonic oscillator: f, called simply the *frequency*, is measured in hertz, and ω, the *angular frequency*, is measured in radians per second. Be sure you are clear about which frequency is being discussed or requested in a given problem. Equations 12.11 and 12.12 show the relationship between the two frequencies.

$$T = \frac{2\pi}{\omega} = 2\pi\sqrt{\frac{m}{k}} \qquad [12.13]$$

■ Period in terms of system parameters

$$f = \frac{1}{T} = \frac{1}{2\pi}\sqrt{\frac{k}{m}} \qquad [12.14]$$

■ Frequency in terms of system parameters

That is, the period and frequency depend *only* on the mass of the particle and the force constant of the spring and *not* on the parameters of the motion, such as A or ϕ. As we might expect, the frequency is larger for a stiffer spring (larger value of k) and decreases with increasing mass of the particle.

We can obtain the velocity and acceleration[2] of a particle undergoing simple harmonic motion from Equations 12.7 and 12.8:

$$v = \frac{dx}{dt} = -\omega A \sin(\omega t + \phi) \qquad [12.15]$$

■ Velocity of a particle in simple harmonic motion

$$a = \frac{d^2x}{dt^2} = -\omega^2 A \cos(\omega t + \phi) \qquad [12.16]$$

■ Acceleration of a particle in simple harmonic motion

From Equation 12.15 we see that because the sine and cosine functions oscillate between ± 1, the extreme values of v are $\pm \omega A$. Likewise, Equation 12.16 tells us that the extreme values of the acceleration are $\pm \omega^2 A$. Therefore, the *maximum* values of the magnitudes of the speed and acceleration are

$$v_{\max} = \omega A = \sqrt{\frac{k}{m}}\,A \qquad [12.17]$$

■ Maximum values of speed and acceleration of a particle in simple harmonic motion

$$a_{\max} = \omega^2 A = \frac{k}{m}\,A \qquad [12.18]$$

Figure 12.3a plots position versus time for an arbitrary value of the phase constant. The associated velocity–time and acceleration–time curves are illustrated in Figures 12.3b and 12.3c. They show that the phase of the velocity differs from the

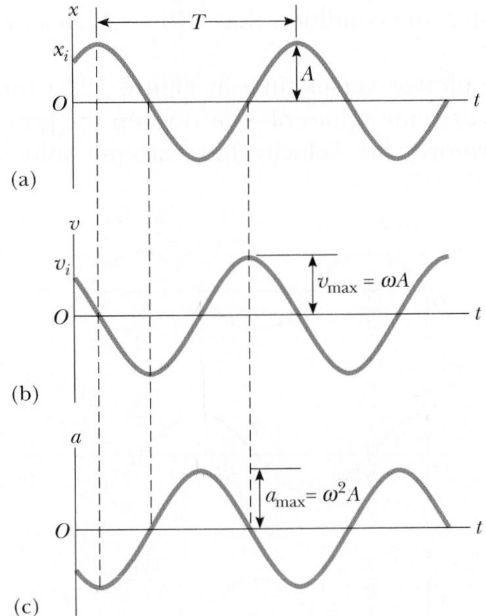

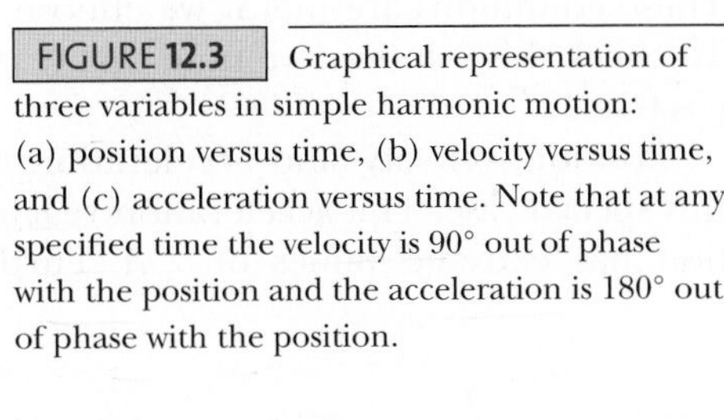
FIGURE 12.3 Graphical representation of three variables in simple harmonic motion: (a) position versus time, (b) velocity versus time, and (c) acceleration versus time. Note that at any specified time the velocity is 90° out of phase with the position and the acceleration is 180° out of phase with the position.

[2]Because the motion of a simple harmonic oscillator takes place in one dimension, we will denote velocity as v and acceleration as a, with the direction indicated by a positive or negative sign, as in Chapter 2.

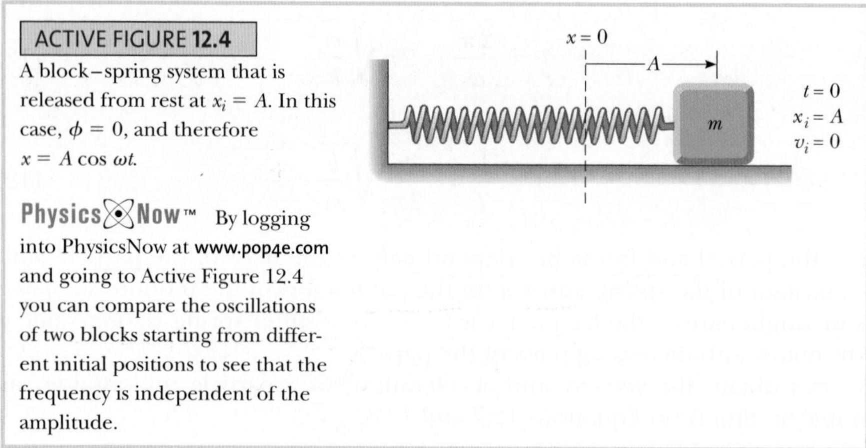

ACTIVE FIGURE 12.4 A block–spring system that is released from rest at $x_i = A$. In this case, $\phi = 0$, and therefore $x = A \cos \omega t$.

Physics Now™ By logging into PhysicsNow at www.pop4e.com and going to Active Figure 12.4 you can compare the oscillations of two blocks starting from different initial positions to see that the frequency is independent of the amplitude.

phase of the position by $\pi/2$ rad, or 90°. That is, when x is a maximum or a minimum, the velocity is zero. Likewise, when x is zero, the speed is a maximum. Furthermore, note that the phase of the acceleration differs from the phase of the position by π rad, or 180°. For example, when x is a maximum, a has a maximum magnitude in the opposite direction.

Equation 12.6 describes simple harmonic motion of a particle in general. Let us now see how to evaluate the constants of the motion. The angular frequency ω is evaluated using Equation 12.9. The constants A and ϕ are evaluated from the initial conditions, that is, the state of the oscillator at $t = 0$.

Suppose we initiate the motion by pulling the particle from equilibrium by a distance A and releasing it from rest at $t = 0$ as in Active Figure 12.4. We must then require that our solutions for $x(t)$ and $v(t)$ (Eqs. 12.6 and 12.15) obey the initial conditions that $x(0) = A$ and $v(0) = 0$:

$$x(0) = A \cos \phi = A$$

$$v(0) = -\omega A \sin \phi = 0$$

These conditions are met if we choose $\phi = 0$, giving $x = A \cos \omega t$ as our solution. To check this solution, we note that it satisfies the condition that $x(0) = A$ because $\cos 0 = 1$.

Position, velocity, and acceleration are plotted versus time in Figure 12.5a for this special case. The acceleration reaches extreme values of $\mp \omega^2 A$ when the position has extreme values of $\pm A$. Furthermore, the velocity has extreme values

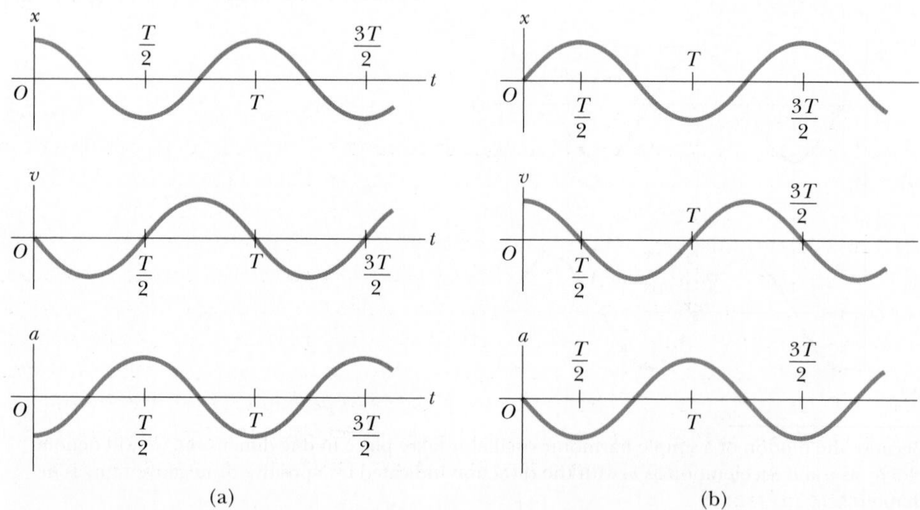

FIGURE 12.5 (a) Position, velocity, and acceleration versus time for a block undergoing simple harmonic motion under the initial conditions that at $t = 0$, $x(0) = A$ and $v(0) = 0$. (b) Position, velocity, and acceleration versus time for a block undergoing simple harmonic motion under the initial conditions that at $t = 0$, $x(0) = 0$ and $v(0) = v_i$.

of $\pm\,\omega A$, which both occur at $x = 0$. Hence, the quantitative solution agrees with our qualitative description of this system.

Let us consider another possibility. Suppose the system is oscillating and we define $t = 0$ as the instant that the particle passes through the unstretched position of the spring while moving to the right (Active Fig. 12.6) with speed v_i. We must then require that our solutions for $x(t)$ and $v(t)$ obey the initial conditions that $x(0) = 0$ and $v(0) = v_i$:

$$x(0) = A\cos\phi = 0$$

$$v(0) = -\omega A\sin\phi = v_i$$

The first of these conditions tells us that $\phi = -\pi/2$. With this value for ϕ, the second condition tells us that $A = v_i/\omega$. Hence, the solution is given by

$$x = \frac{v_i}{\omega}\cos\left(\omega t - \frac{\pi}{2}\right)$$

Figure 12.5b shows the graphs of position, velocity, and acceleration versus time for this choice of $t = 0$. Note that these curves are the same as those in Figure 12.5a, but shifted to the right by one fourth of a cycle. This shift is described mathematically by the phase constant $\phi = -\pi/2$, which is one fourth of a full cycle of 2π.

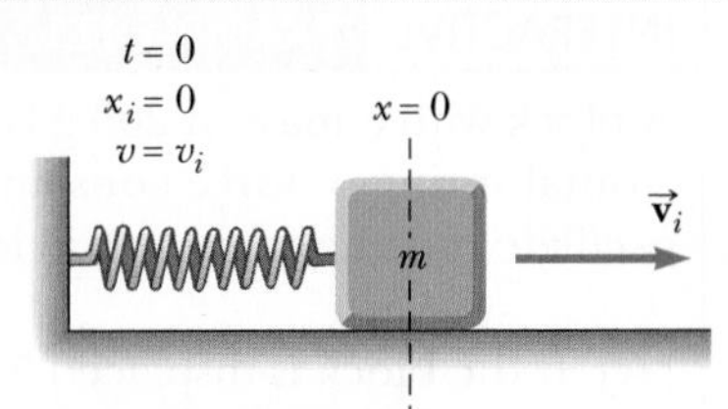

ACTIVE FIGURE 12.6
The block–spring system is undergoing oscillation, and $t = 0$ is defined at an instant when the block passes through the equilibrium position $x = 0$ and is moving to the right with speed v_i.

PhysicsNow™ By logging into PhysicsNow at **www.pop4e.com** and going to Active Figure 12.6 you can compare the oscillations of two blocks with different velocities at $t = 0$ to see that the frequency is independent of the amplitude.

QUICK QUIZ 12.2 Consider a graphical representation (Fig. 12.7) of simple harmonic motion as described mathematically in Equation 12.6. **(i)** When the object is at point Ⓐ on the graph, what are, respectively, its position and velocity? **(a)** both positive **(b)** both negative **(c)** positive and zero **(d)** negative and zero **(e)** positive and negative **(f)** negative and positive **(ii)** From the same list of choices, what are the respective signs of the velocity and acceleration when the object is at position Ⓐ on the graph?

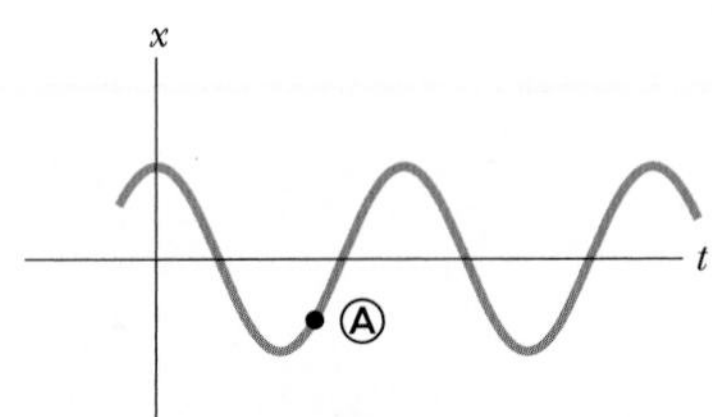

FIGURE 12.7 (Quick Quiz 12.2) An x-t graph for an object undergoing simple harmonic motion. At a particular time, the object's position is indicated by Ⓐ in the diagram.

■ Thinking Physics 12.1

We know that the period of oscillation of an object attached to a spring is proportional to the square root of the mass of the object (Eq. 12.13). Therefore, if we perform an experiment in which we place objects with a range of masses on the end of a spring and measure the period of oscillation of each object–spring system, a graph of the square of the period versus the mass will result in a straight line as suggested in Figure 12.8. We find, however, that the line does not go through the origin. Why not?

Reasoning The line does not go through the origin because the spring itself has mass. Therefore, the resistance to changes in motion of the system is a combination of the mass of the object on the end of the spring and the mass of the oscillating spring coils. The entire mass of the spring is not oscillating in the same way, however. The coil of the spring attached to the object is oscillating over the same amplitude as the object, but the coil at the fixed end of the spring is not oscillating at all. For a cylindrical spring, energy arguments can be used to show that the effective additional mass representing the oscillations of the spring is one third of the mass of the spring. The square of the period is proportional to the total oscillating mass, but the graph in Figure 12.8 shows the square of the period versus only the mass of the object on the spring. A graph of period squared versus total mass (mass of the object on the spring plus the effective oscillating mass of the spring) would pass through the origin. ■

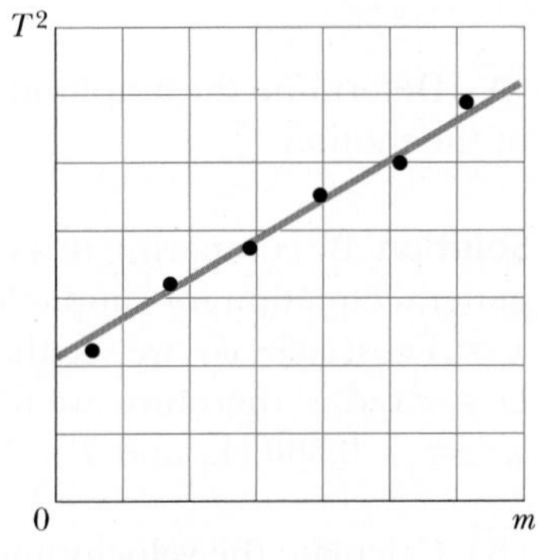

FIGURE 12.8 (Thinking Physics 12.1) A graph of experimental data: the square of the period versus mass of a block in a block–spring system.

INTERACTIVE EXAMPLE 12.1 **A Block–Spring System**

A block with a mass of 200 g is connected to a light horizontal spring of force constant 5.00 N/m and is free to oscillate on a horizontal, frictionless surface.

A If the block is displaced 5.00 cm from equilibrium and released from rest as in Active Figure 12.4, find the period of its motion.

Solution The situation (we assume an ideal spring) tells us to use the simple harmonic motion model. Using Equation 12.13,

$$T = 2\pi\sqrt{\frac{m}{k}} = 2\pi\sqrt{\frac{200 \times 10^{-3}\text{ kg}}{5.00\text{ N/m}}} = 1.26\text{ s}$$

B Determine the maximum speed and maximum acceleration of the block.

Solution Using Equations 12.17 and 12.18, with $A = 5.00 \times 10^{-2}$ m, we have

$$v_{\text{max}} = \omega A = \frac{2\pi}{T}A = \left(\frac{2\pi}{1.26\text{ s}}\right)(5.00 \times 10^{-2}\text{ m})$$
$$= 0.250\text{ m/s}$$

$$a_{\text{max}} = \omega^2 A = \left(\frac{2\pi}{T}\right)^2 A = \left(\frac{2\pi}{1.26\text{ s}}\right)^2 (5.00 \times 10^{-2}\text{ m})$$
$$= 1.25\text{ m/s}^2$$

C Express the position, velocity, and acceleration of this object as functions of time, assuming that $\phi = 0$.

Solution From Equations 12.6, 12.15, and 12.16,

$$x = A\cos\omega t = (0.050\,0\text{ m})\cos 5.00t$$

$$v = -\omega A\sin\omega t = -(0.250\text{ m/s})\sin 5.00t$$

$$a = -\omega^2 A\cos\omega t = -(1.25\text{ m/s}^2)\cos 5.00t$$

Physics Now™ You can adjust the mass of the object, the force constant of the spring, and the starting position by logging into PhysicsNow at **www.pop4e.com** and going to Interactive Example 12.1.

EXAMPLE 12.2 **An Oscillating Particle**

A particle oscillates with simple harmonic motion along the *x* axis. Its position varies with time according to the equation

$$x = (4.00\text{ m})\cos\left(\pi t + \frac{\pi}{4}\right)$$

where *t* is in seconds.

A Determine the amplitude, frequency, and period of the motion.

Solution By comparing this equation with the general equation for simple harmonic motion, $x = A\cos(\omega t + \phi)$, we see that $A = 4.00$ m and $\omega = \pi$ rad/s; therefore, we find that $f = \omega/2\pi = \pi/2\pi = 0.500$ Hz and $T = 1/f = 2.00$ s.

B Calculate the velocity and acceleration of the particle at any time *t*.

Solution Using Equations 12.15 and 12.16,

$$v = \frac{dx}{dt} = -(4.00\pi\text{ m/s})\sin\left(\pi t + \frac{\pi}{4}\right)$$

$$a = \frac{dv}{dt} = -(4.00\pi^2\text{ m/s}^2)\cos\left(\pi t + \frac{\pi}{4}\right)$$

C What are the position and the velocity of the particle at time $t = 0$?

Solution The position function is given in the text of the problem. Evaluating this expression at $t = 0$ gives us

$$x = (4.00\text{ m})\cos\left(\frac{\pi}{4}\right) = 2.83\text{ m}$$

From part B, we evaluate the velocity function at $t = 0$:

$$v = -(4.00\pi\text{ m/s})\sin\left(\frac{\pi}{4}\right)$$
$$= -8.89\text{ m/s}$$

EXAMPLE 12.3 Initial Conditions

Suppose the initial position x_i and initial velocity v_i of a harmonic oscillator of known angular frequency are given; that is, $x(0) = x_i$ and $v(0) = v_i$. Find general expressions for the amplitude and the phase constant in terms of these initial parameters.

Solution With these initial conditions, Equations 12.6 and 12.15 give us at $t = 0$

$$x_i = A\cos\phi \quad \text{and} \quad v_i = -\omega A\sin\phi$$

Dividing these two equations eliminates A, giving $v_i/x_i = -\omega\tan\phi$, or

$$\tan\phi = -\frac{v_i}{\omega x_i}$$

Furthermore, if we take the sum $x_i^2 + (v_i/\omega)^2 = A^2\cos^2\phi + A^2\sin^2\phi = A^2$ (where we have used Eqs. 12.6 and 12.15) and solve for A, we find that

$$A = \sqrt{x_i^2 + \left(\frac{v_i}{\omega}\right)^2}$$

12.3 ENERGY CONSIDERATIONS IN SIMPLE HARMONIC MOTION

If an object attached to a spring slides on a frictionless surface, we can consider the combination of the spring and the attached object to be an isolated system. As a result, we can apply the energy version of the isolated system model to the system. Let us examine the mechanical energy of the system described in Active Figure 12.1. Because the surface is frictionless, the total mechanical energy of the system is constant. We model the object as a particle. The kinetic energy, in the simplification model in which the spring is massless, is associated only with the motion of the particle of mass m. We use Equation 12.15 to express the kinetic energy as

$$K = \tfrac{1}{2}mv^2 = \tfrac{1}{2}m\omega^2A^2\sin^2(\omega t + \phi) \qquad [12.19]$$

■ Kinetic energy of a simple harmonic oscillator

Elastic potential energy in this system is associated with the spring and for any position x of the particle is, as we found in Chapter 7, $U = \frac{1}{2}kx^2$. Using Equation 12.6, we have

$$U = \tfrac{1}{2}kx^2 = \tfrac{1}{2}kA^2\cos^2(\omega t + \phi) \qquad [12.20]$$

■ Potential energy of a simple harmonic oscillator

We see that K and U are always positive quantities and that each varies with time. We can express the **total energy** of the simple harmonic oscillator as

$$E = K + U = \tfrac{1}{2}m\omega^2A^2\sin^2(\omega t + \phi) + \tfrac{1}{2}kA^2\cos^2(\omega t + \phi)$$

Because $\omega^2 = k/m$, we can write this expression as

$$E = \tfrac{1}{2}kA^2[\sin^2(\omega t + \phi) + \cos^2(\omega t + \phi)]$$

Because $\sin^2\theta + \cos^2\theta = 1$ for any angle θ, this equation reduces to

$$E = \tfrac{1}{2}kA^2 \qquad [12.21]$$

■ Total energy of a simple harmonic oscillator

That is, the total energy of an isolated simple harmonic oscillator is a constant of the motion and proportional to the square of the amplitude. In fact, the total energy is just equal to the maximum potential energy stored in the spring when $x = \pm A$. At these points, $v = 0$ and there is no kinetic energy. At the equilibrium position, $x = 0$ and $U = 0$, so the total energy is all in the form of kinetic energy of the particle,

$$K_{\text{max}} = \tfrac{1}{2}mv_{\text{max}}^2 = \tfrac{1}{2}kA^2$$

These results are appropriate for the simplification model in which we consider the spring to be massless. In the real situation in which the spring has mass, additional kinetic energy is associated with the motion of the spring. In numerical problems in this book, we will consider only massless springs unless otherwise noted.

ACTIVE FIGURE 12.9
(a) Kinetic energy and potential energy versus time for a simple harmonic oscillator with $\phi = 0$. (b) Kinetic energy and potential energy versus position for a simple harmonic oscillator. In either plot, note that $K + U =$ constant.

Physics Now™ By logging into PhysicsNow at www.pop4e.com and going to Active Figure 12.9 you can compare the physical oscillation of a block with energy graphs in this figure as well as with energy bar graphs.

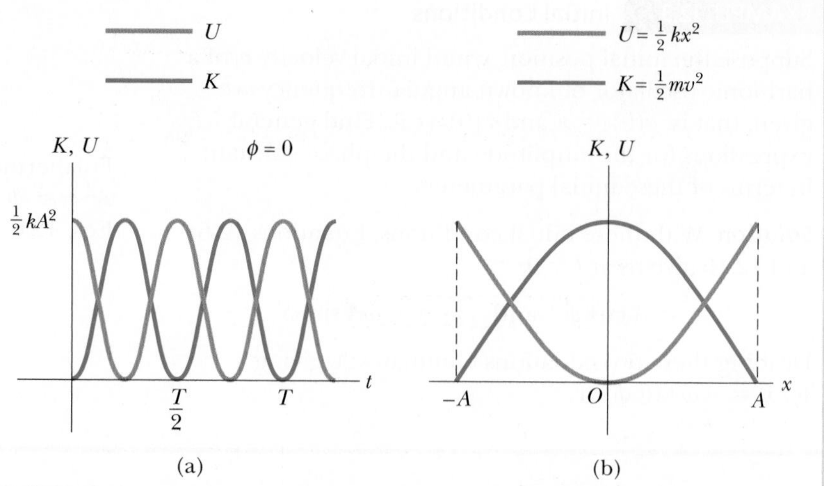

Graphical representations of the kinetic and potential energies versus time for the system of a particle on a massless spring are shown in Active Figure 12.9a, where $\phi = 0$. In this situation, the sum of the kinetic and potential energies at all times is a constant equal to $\frac{1}{2}kA^2$, the total energy of the system. The variations of K and U with position are plotted in Active Figure 12.9b. Energy in the system is continuously being transformed between potential energy (in the spring) and kinetic energy (of the object attached to the spring). Active Figure 12.10 illustrates the position, velocity, acceleration, kinetic energy, and potential energy of the particle–spring system for one full period of the motion. Most of the ideas discussed so far for simple harmonic motion are incorporated in this important figure. We suggest that you study it carefully.

Finally, we can use conservation of mechanical energy for an isolated system to obtain the velocity for an arbitrary position x of the particle, expressing the total energy as

■ Velocity as a function of position for a particle in simple harmonic motion

$$E = K + U = \tfrac{1}{2}mv^2 + \tfrac{1}{2}kx^2 = \tfrac{1}{2}kA^2$$

$$v = \pm\sqrt{\frac{k}{m}(A^2 - x^2)} = \pm\,\omega\sqrt{A^2 - x^2} \qquad [12.22]$$

This expression confirms that the speed is a maximum at $x = 0$ and is zero at the turning points $x = \pm A$.

■ Thinking Physics 12.2

An object oscillating on the end of a horizontal spring slides back and forth over a frictionless surface. During one oscillation, you set an identical object at the maximum displacement point, with instant-acting glue on its surface. Just as the oscillating object reaches its largest displacement and is momentarily at rest, it adheres to the new object by means of the glue and the two objects continue the oscillation together. Does the period of the oscillation change? Does the amplitude of oscillation change? Does the energy of the oscillation change?

Reasoning The period of oscillation changes because the period depends on the mass that is oscillating (Eq. 12.13). The amplitude does not change. Because the new object was added under the special condition that the original object was at rest, the combined objects are at rest at this point also, defining the amplitude as the same as in the original oscillation. The energy does not change either. At the maximum displacement point, the energy is all potential energy stored in the spring, which depends only on the force constant and the amplitude, not on

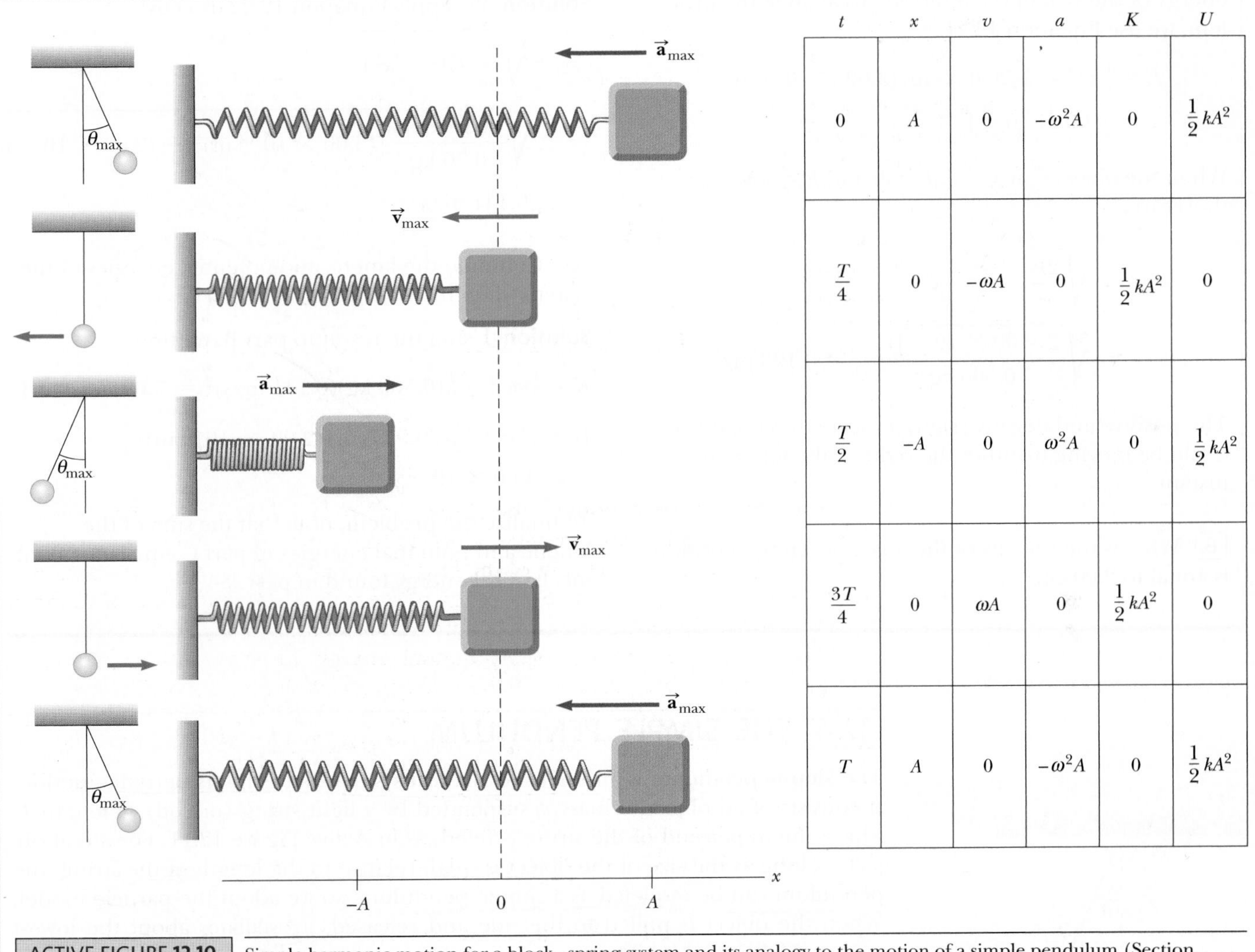

t	x	v	a	K	U
0	A	0	$-\omega^2 A$	0	$\frac{1}{2}kA^2$
$\frac{T}{4}$	0	$-\omega A$	0	$\frac{1}{2}kA^2$	0
$\frac{T}{2}$	$-A$	0	$\omega^2 A$	0	$\frac{1}{2}kA^2$
$\frac{3T}{4}$	0	ωA	0	$\frac{1}{2}kA^2$	0
T	A	0	$-\omega^2 A$	0	$\frac{1}{2}kA^2$

ACTIVE FIGURE 12.10 Simple harmonic motion for a block–spring system and its analogy to the motion of a simple pendulum (Section 12.4). The parameters in the table at the right refer to the block–spring system, assuming that at $t = 0$, $x = A$ so that $x = A \cos \omega t$.

Physics Now™ By logging into PhysicsNow at **www.pop4e.com** and going to Active Figure 12.10 you can set the initial position of the block and see the block–spring system and the analogous pendulum in motion.

the mass of the object. The object of increased mass will pass through the equilibrium point with lower speed than in the original oscillation but with the same kinetic energy. Another approach is to think about how energy could be transferred into the oscillating system. No work was done on the system (nor did any other form of energy transfer occur), so the energy in the system cannot change. ■

EXAMPLE 12.4 Oscillations on a Horizontal Surface

A 0.500-kg object connected to a massless spring of force constant 20.0 N/m oscillates on a horizontal, frictionless track.

A Calculate the total energy of the system and the maximum velocity of the object if the amplitude of the motion is 3.00 cm.

Solution Conceptualize the problem by studying the block–spring system in Active Figure 12.10. Because the object slides on a frictionless surface, we can categorize the problem as one involving an isolated system of the object and the spring. Because only conservative forces are acting within the system, the mechanical energy of the system is conserved. To analyze the problem, we use Equation 12.21:

$$E = \tfrac{1}{2}kA^2 = \tfrac{1}{2}(20.0\ \text{N/m})(3.00 \times 10^{-2}\ \text{m})^2$$
$$= 9.00 \times 10^{-3}\ \text{J}$$

When the object is at $x = 0$, $U = 0$ and $E = \frac{1}{2}mv_{\text{max}}^2$; therefore,

$$v_{\text{max}} = \sqrt{\frac{2E}{m}}$$
$$= \sqrt{\frac{2(9.00 \times 10^{-3}\ \text{J})}{0.500\ \text{kg}}} = \pm 0.190\ \text{m/s}$$

The positive and negative signs indicate that the object could be moving to either the right or the left at this instant.

B What is the velocity of the object when the position is equal to 2.00 cm?

Solution We apply Equation 12.22 directly:

$$v = \pm\sqrt{\frac{k}{m}(A^2 - x^2)}$$
$$= \pm\sqrt{\frac{20.0\ \text{N/m}}{0.50\ \text{kg}}[(3.00 \times 10^{-2}\ \text{m})^2 - (2.00 \times 10^{-2}\ \text{m})^2]}$$
$$= \pm 0.141\ \text{m/s}$$

C Compute the kinetic and potential energies of the system when the position equals 2.00 cm.

Solution Using the result to part B, we find

$$K = \tfrac{1}{2}mv^2 = \tfrac{1}{2}(0.500\ \text{kg})(0.141\ \text{m/s})^2 = 5.00 \times 10^{-3}\ \text{J}$$

$$U = \tfrac{1}{2}kx^2 = \tfrac{1}{2}(20.0\ \text{N/m})(2.00 \times 10^{-2}\ \text{m})^2$$
$$= 4.00 \times 10^{-3}\ \text{J}$$

To finalize the problem, note that the sum of the kinetic and potential energies in part C equals the total mechanical energy found in part A.

12.4 THE SIMPLE PENDULUM

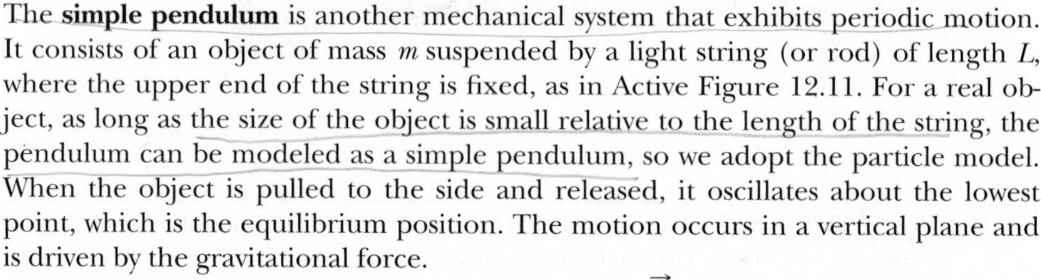

The **simple pendulum** is another mechanical system that exhibits periodic motion. It consists of an object of mass m suspended by a light string (or rod) of length L, where the upper end of the string is fixed, as in Active Figure 12.11. For a real object, as long as the size of the object is small relative to the length of the string, the pendulum can be modeled as a simple pendulum, so we adopt the particle model. When the object is pulled to the side and released, it oscillates about the lowest point, which is the equilibrium position. The motion occurs in a vertical plane and is driven by the gravitational force.

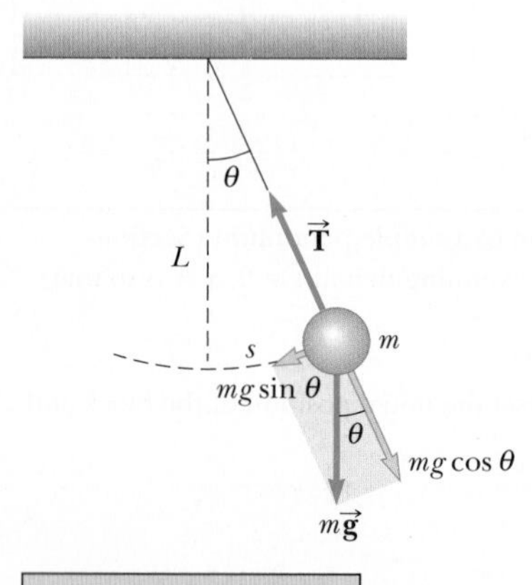

ACTIVE FIGURE **12.11**

When θ is small, the oscillation of the simple pendulum can be modeled as simple harmonic motion about the equilibrium position ($\theta = 0$). The restoring force is $mg \sin \theta$, the component of the gravitational force tangent to the arc.

Physics Now™ By logging into PhysicsNow at **www.pop4e.com** and going to Active Figure 12.11 you can adjust the mass of the bob, the length of the string, and the initial angle and see the resulting oscillation of the pendulum.

The forces acting on the object are the force $\vec{T}$ acting along the string and the gravitational force $m\vec{g}$. The tangential component of the gravitational force $mg \sin \theta$ always acts toward $\theta = 0$, opposite the displacement. The gravitational force is therefore a restoring force, and we can use Newton's second law to write the equation of motion in the tangential direction as

$$F_t = ma_t \quad \rightarrow \quad -mg \sin \theta = m\frac{d^2s}{dt^2}$$

where s is the position measured along the circular arc in Active Figure 12.11 and the negative sign indicates that F_t acts toward the equilibrium position. Because $s = L\theta$ (Eq. 10.1a) and L is constant, this equation reduces to

$$\frac{d^2\theta}{dt^2} = -\frac{g}{L}\sin \theta$$

Compare this equation to Equation 12.5, which is of a similar, but not identical, mathematical form. The right side is proportional to $\sin \theta$ rather than to θ; hence, we conclude that the motion is *not* simple harmonic motion because the equation describing the motion is not of the form of Equation 12.5. If we assume that θ is *small* (less than about 10° or 0.2 rad), however, we can use a simplification model called the **small angle approximation,** in which $\sin \theta \approx \theta$, where θ is measured in radians. Table 12.1 shows angles, in degrees and radians, and the sines of these

TABLE 12.1 Angles and Sines of Angles

Angle in Degrees	Angle in Radians	Sine of Angle	Percent Difference
0°	0.000 0	0.000 0	0.0%
1°	0.017 5	0.017 5	0.0%
2°	0.034 9	0.034 9	0.0%
3°	0.052 4	0.052 3	0.0%
5°	0.087 3	0.087 2	0.1%
10°	0.174 5	0.173 6	0.5%
15°	0.261 8	0.258 8	1.2%
20°	0.349 1	0.342 0	2.1%
30°	0.523 6	0.500 0	4.7%

angles. As long as θ is less than about 10°, the angle in radians and its sine are the same, at least to within an accuracy of less than 1.0%.

Therefore, for small angles, the equation of motion becomes

$$\frac{d^2\theta}{dt^2} = -\frac{g}{L}\theta \qquad [12.23]$$

PITFALL PREVENTION 12.4

Not true simple harmonic motion Remember that the pendulum *does not* exhibit true simple harmonic motion for *any* angle. If the angle is less than about 10°, the motion can be *modeled* as simple harmonic.

Now we have an expression with exactly the same mathematical form as Equation 12.5, with $\omega^2 = g/L$, and so we conclude that the motion is approximately simple harmonic motion for small amplitudes. Modeling the solution after Equation 12.6, θ can therefore be written as $\theta = \theta_{max}\cos(\omega t + \phi)$, where θ_{max} is the **maximum angular position** and the angular frequency ω is

$$\omega = \sqrt{\frac{g}{L}} \qquad [12.24]$$

▪ Angular frequency for a simple pendulum

The period of the motion is

$$T = \frac{2\pi}{\omega} = 2\pi\sqrt{\frac{L}{g}} \qquad [12.25]$$

▪ Period for a simple pendulum

In other words, **the period and frequency of a simple pendulum oscillating at small angles depend only on the length of the string and the free-fall acceleration.** Because the period is *independent* of the mass, we conclude that *all* simple pendulums of equal length at the same location oscillate with equal periods. Experiments show that this conclusion is correct. The analogy between the motion of a simple pendulum and the particle–spring system is illustrated in Active Figure 12.10.

QUICK QUIZ 12.3 A grandfather clock depends on the period of a pendulum to keep correct time. **(i)** Suppose a grandfather clock is calibrated correctly and then a mischievous child slides the bob of the pendulum downward on the oscillating rod. Does the grandfather clock run **(a)** slow, **(b)** fast, or **(c)** correctly? **(ii)** Suppose the grandfather clock is calibrated correctly at sea level and is then taken to the top of a very tall mountain. Does the grandfather clock run **(a)** slow, **(b)** fast, or **(c)** correctly?

▪ Thinking Physics 12.3

You set up two oscillating systems: a simple pendulum and a block hanging from a vertical spring. You carefully adjust the length of the pendulum so that both oscillators have the same period. You now take the two oscillators to the Moon. Will they still have the same period as each other? What happens if you observe the two oscillators in an orbiting spacecraft? (Assume that the spring is one with open space between the coils when it is unstretched, so the spring can be both stretched and compressed.)

Reasoning The block hanging from the spring will have the same period on the Moon that it had on the Earth because the period depends on the mass of the block and the force constant of the spring, neither of which have changed. The pendulum's period on the Moon will be different from its period on the Earth because the period of the pendulum depends on the value of g. Because g is smaller on the Moon than on the Earth, the pendulum will oscillate with a longer period.

In the orbiting spacecraft, the block–spring system will oscillate with the same period as on the Earth when it is set into motion because the period does not depend on gravity. The pendulum will not oscillate at all; if you pull it to the side from a direction you define as "vertical" and release it, it stays there. Because the spacecraft is in free-fall while in orbit around the Earth, the effective gravity is zero and there is no restoring force on the pendulum. ■

EXAMPLE 12.5 **A Measure of Height**

A man enters a tall tower, needing to know its height. He notes that a long pendulum extends from the ceiling almost to the floor and that its period is 12.0 s. How tall is the tower?

Solution We adopt a simplification model in which the height of the tower is equal to the length of the pendulum. If we use $T = 2\pi\sqrt{L/g}$ and solve for L, we have

$$L = \frac{gT^2}{4\pi^2} = \frac{(9.80\ \text{m/s}^2)(12.0\ \text{s})^2}{4\pi^2} = 35.7\ \text{m}$$

12.5 THE PHYSICAL PENDULUM

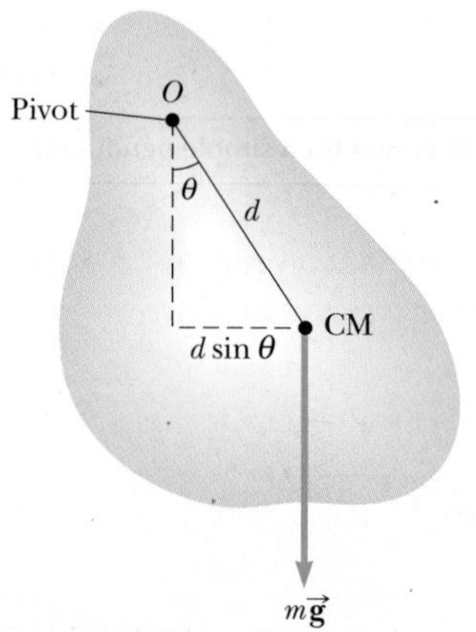

FIGURE 12.12 The physical pendulum consists of a rigid object pivoted at the point O, which is not at the center of mass.

If a hanging object that cannot be modeled as a particle oscillates about a fixed axis that does not pass through its center of mass, it must be treated as a **physical, or compound, pendulum.** For the simple pendulum we generated Equation 12.23 from the particle under a net force model. For the physical pendulum we will need to use the rigid object under a net torque model from Chapter 10. Consider a rigid object pivoted at a point O that is a distance d from the center of mass (Fig. 12.12). The torque about O is provided by the gravitational force, and its magnitude is $mgd \sin \theta$. Using Newton's second law for rotation, $\sum \tau = I\alpha$ (Eq. 10.27), where I is the moment of inertia of the object about the axis through O, we have

$$-mgd \sin \theta = I\frac{d^2\theta}{dt^2}$$

The negative sign on the left indicates that the torque about O tends to decrease θ. That is, the gravitational force produces a restoring torque.

If we again assume that θ is small, the small angle approximation $\sin \theta \approx \theta$ is valid and the equation of motion reduces to

$$\frac{d^2\theta}{dt^2} = -\left(\frac{mgd}{I}\right)\theta \qquad [12.26]$$

Note that this equation has the same mathematical form as Equation 12.5, with $\omega^2 = mgd/I$, and so the motion of the object is approximately simple harmonic motion for small amplitudes. That is, the solution of Equation 12.26 is $\theta = \theta_{max} \cos(\omega t + \phi)$, where θ_{max} is the maximum angular position and

■ Angular frequency for a physical pendulum

$$\omega = \sqrt{\frac{mgd}{I}}$$

The period is

$$T = \frac{2\pi}{\omega} = 2\pi\sqrt{\frac{I}{mgd}} \qquad [12.27]$$

■ **Period for a physical pendulum**

One can use this result to measure the moment of inertia of a rigid object. If the location of the center of mass and hence the distance d are known, the moment of inertia can be obtained through a measurement of the period. Finally, note that Equation 12.27 becomes the equation for the period of a simple pendulum (Eq. 12.25) when $I = md^2$—that is, when all the mass is concentrated at a point—and the physical pendulum reduces to the simple pendulum.

QUICK QUIZ 12.4 Two students, Alex and Brian, are in a museum watching the swinging of a pendulum with a large bob. Alex says, "I'm going to sneak past the fence and stick some chewing gum on the top of the pendulum bob to change its period of oscillation." Brian says, "That won't change the period. The period of a pendulum is independent of mass." Which student is correct? **(a)** Alex **(b)** Brian

EXAMPLE 12.6 A Swinging Sign

A circular sign of mass M and radius R is hung on a nail from a small loop located at one edge (Fig. 12.13). After it is placed on the nail, the sign oscillates in a vertical plane. Find the period of oscillation if the amplitude of the motion is small.

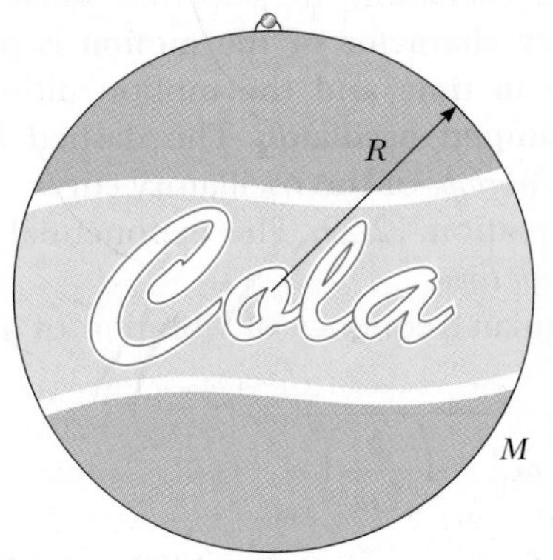

FIGURE 12.13 (Example 12.6) A circular sign oscillating about a pivot as a physical pendulum.

Solution The moment of inertia of a disk about an axis through the center is $\frac{1}{2}MR^2$ (see Table 10.2). The pivot point for the sign is through a point on the rim, so we use the parallel axis theorem (see Eq. 10.43) to find the moment of inertia about the pivot:

$$I_p = \tfrac{1}{2}MR^2 + MR^2 = \tfrac{3}{2}MR^2$$

The distance d from the pivot to the center of mass is the radius R. Substituting these quantities into Equation 12.27 gives

$$T = 2\pi\sqrt{\frac{\frac{3}{2}MR^2}{MgR}} = 2\pi\sqrt{\frac{3R}{2g}}$$

12.6 DAMPED OSCILLATIONS

The oscillatory motions we have considered so far have occurred under the simplification model of an ideal frictionless system, that is, one that oscillates indefinitely under the action of only a linear restoring force. In many realistic systems, resistive forces, such as friction, are present and retard the motion of the system. Consequently, the mechanical energy of the system diminishes in time, and the motion is described as a **damped oscillation.**

Consider an object moving through a medium such as a liquid or a gas. One common type of resistive force on the object, which we discussed in Chapter 5, is proportional to the velocity of the object and acts in the direction opposite that of the object's velocity relative to the medium. This type of force is often observed

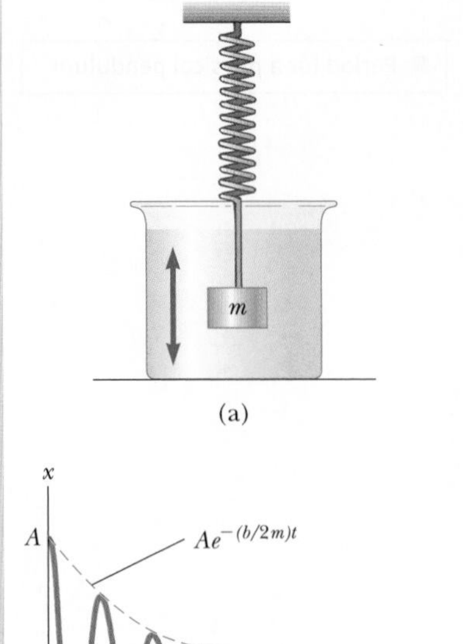

ACTIVE FIGURE 12.14

(a) One example of a damped oscillator is an object attached to a spring and submerged in a viscous liquid. (b) Graph of the position versus time for a damped oscillator with small damping. Note the decrease in amplitude with time.

Physics Now™ By logging into PhysicsNow at www.pop4e.com and going to Active Figure 12.14 you can adjust the spring constant, the mass of the object, and the damping constant and see the resulting damped oscillation of the object.

when an object is oscillating slowly in air, for instance. Because the resistive force can be expressed as $\vec{\mathbf{R}} = -b\vec{\mathbf{v}}$, where b is a constant related to the strength of the resistive force, and the restoring force exerted on the system is $-kx$, Newton's second law gives us

$$\sum F_x = -kx - bv = ma_x$$

$$-kx - b\frac{dx}{dt} = m\frac{d^2x}{dt^2} \qquad [12.28]$$

The solution of this differential equation requires mathematics that may not yet be familiar to you, so it will simply be stated without proof. When the parameters of the system are such that $b < \sqrt{4mk}$ so that the resistive force is small, the solution to Equation 12.28 is

$$x = (Ae^{-(b/2m)t})\cos(\omega t + \phi) \qquad [12.29]$$

where the angular frequency of the motion is

$$\omega = \sqrt{\frac{k}{m} - \left(\frac{b}{2m}\right)^2} \qquad [12.30]$$

This result can be verified by substituting Equation 12.29 into Equation 12.28. Notice that Equation 12.29 is similar to Equation 12.6, with the new feature that the amplitude (in the parentheses before the cosine function) depends on time. In Active Figure 12.14a, we see one example of a damped system. The object suspended from the spring experiences both a force from the spring and a resistive force from the surrounding liquid. Active Figure 12.14b shows the position as a function of time for such a damped oscillator. We see that **when the resistive force is relatively small, the oscillatory character of the motion is preserved but the amplitude of vibration decreases in time** and the motion ultimately ceases. This system is known as an **underdamped oscillator.** The dashed blue lines in Active Figure 12.14b, which form the *envelope* of the oscillatory curve, represent the exponential factor that appears in Equation 12.29. The exponential factor shows that the *amplitude decays exponentially with time.*

It is convenient to express the angular frequency of vibration of a damped system (Eq. 12.30) in the form

$$\omega = \sqrt{\omega_0^2 - \left(\frac{b}{2m}\right)^2}$$

where $\omega_0 = \sqrt{k/m}$ represents the angular frequency of oscillation in the absence of a resistive force (the undamped oscillator). In other words, when $b = 0$, the resistive force is zero and the system oscillates with angular frequency ω_0, called the **natural frequency.**[3] As the magnitude of the resistive force increases, the oscillations dampen more rapidly. When b reaches a critical value b_c, so that $b_c/2m = \omega_0$, the system does not oscillate and is said to be **critically damped.** In this case, it returns to equilibrium in an exponential manner with time, as in Figure 12.15.

If the medium is highly viscous and the parameters meet the condition that $b/2m > \omega_0$, the system is **overdamped.** Again, the displaced system does not oscillate but simply returns to its equilibrium position. As the damping increases, the time interval required for the particle to approach the equilibrium position also increases, as indicated in Figure 12.15. In any case, when a resistive force is present, the mechanical energy of the oscillator eventually falls to zero. The mechanical energy is transformed into internal energy in the oscillating system and the resistive medium.

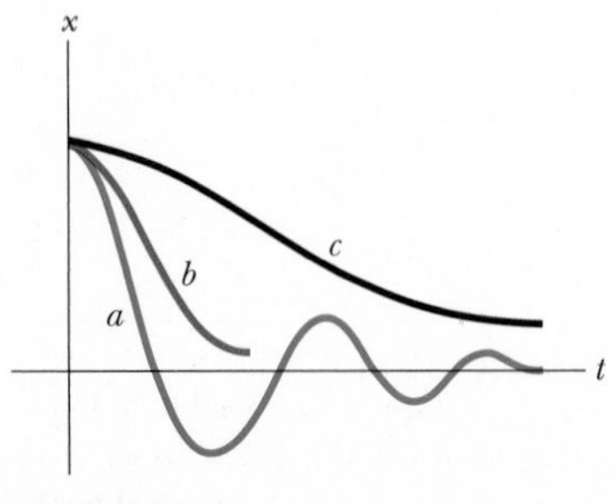

FIGURE 12.15 Plots of position versus time for an underdamped oscillator (a), a critically damped oscillator (b), and an overdamped oscillator (c).

[3]In practice, both ω_0 and $f_0 = \omega_0/2\pi$ are described as the natural frequency. The context of the discussion will help you determine which frequency is being discussed.

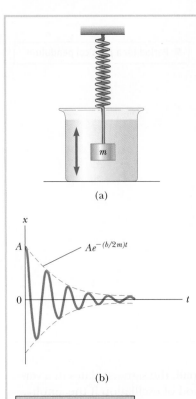

12.7 FORCED OSCILLATIONS

We have seen that the mechanical energy of a damped oscillator decreases in time as a result of the resistive force. It is possible to compensate for this energy decrease by applying an external force that does positive work on the system. Such an oscillator then undergoes **forced oscillations.** At any instant, energy can be transferred into the system by an applied force that acts in the direction of motion of the oscillator. For example, a child on a swing can be kept in motion by appropriately timed "pushes." The amplitude of motion remains constant if the energy input per cycle of motion exactly equals the decrease in mechanical energy in each cycle that results from resistive forces.

A common example of a forced oscillator is a damped oscillator driven by an external force that varies periodically, such as $F(t) = F_0 \sin \omega t$, where ω is the angular frequency of the driving force and F_0 is a constant. In general, the frequency ω of the driving force is different from the natural frequency ω_0 of the oscillator. Newton's second law in this situation gives

$$\sum F_x = ma_x \quad \rightarrow \quad F_0 \sin \omega t - b\frac{dx}{dt} - kx = m\frac{d^2x}{dt^2} \qquad [12.31]$$

Again, the solution of this equation is rather lengthy and will not be presented. After the driving force on an initially stationary object begins to act, the amplitude of the oscillation will increase. After a sufficiently long time interval, when the energy input per cycle from the driving force equals the amount of mechanical energy transformed to internal energy for each cycle, a steady-state condition is reached in which the oscillations proceed with constant amplitude. In this case, Equation 12.31 has the solution

$$x = A\cos(\omega t + \phi) \qquad [12.32]$$

where

$$A = \frac{F_0/m}{\sqrt{\left(\omega^2 - \omega_0^{\,2}\right)^2 + \left(\frac{b\omega}{m}\right)^2}} \qquad [12.33]$$

and where $\omega_0 = \sqrt{k/m}$ is the natural frequency of the undamped oscillator ($b = 0$).

Equation 12.33 shows that the amplitude of the forced oscillator is constant for a given driving force because it is being driven in steady state by an external force. For small damping the amplitude becomes large when the frequency of the driving force is near the natural frequency of oscillation, or when $\omega \approx \omega_0$ as can be seen in Equation 12.33. The dramatic increase in amplitude near the natural frequency is called **resonance,** and the natural frequency ω_0 is called the **resonance frequency** of the system.

Figure 12.16 is a graph of amplitude as a function of frequency for the forced oscillator, with varying resistive forces. Note that the amplitude increases with decreasing damping ($b \rightarrow 0$) and that the resonance curve flattens as the damping increases. In the absence of a damping force ($b = 0$), we see from Equation 12.33 that the steady-state amplitude approaches infinity as $\omega \rightarrow \omega_0$. In other words, if there are no resistive forces in the system and we continue to drive an oscillator with a sinusoidal force at the resonance frequency, the amplitude of motion will build up without limit. This situation does not occur in practice because some damping is always present in real oscillators.

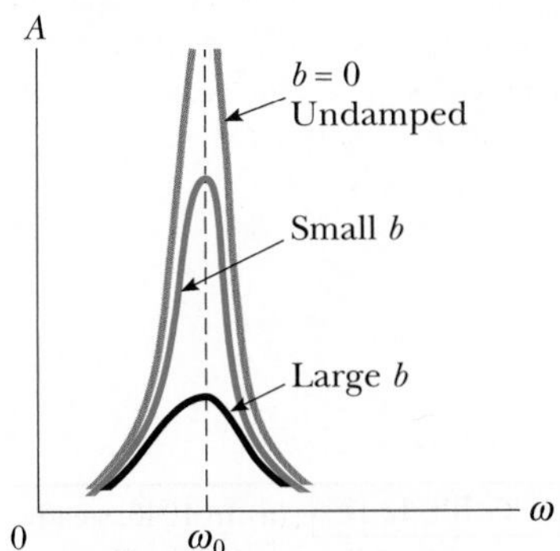

FIGURE 12.16 Graph of amplitude versus frequency for a damped oscillator when a periodic driving force is present. When the frequency of the driving force equals the natural frequency ω_0, resonance occurs. Note that the shape of the resonance curve depends on the size of the damping coefficient b.

Resonance appears in many areas of physics. For example, certain electric circuits have resonance frequencies. This fact is exploited in radio tuners, which allow you to select the station you wish to hear. Vibrating strings and columns of air also have resonance frequencies, which allow them to be used for musical instruments, which we shall discuss in Chapter 14.

12.8 RESONANCE IN STRUCTURES

CONTEXT CONNECTION

In the preceding section, we investigated the phenomenon of resonance in which an oscillating system exhibits its maximum response to a periodic driving force when the frequency of the driving force matches the oscillator's natural frequency. We now apply this understanding to the interaction between the shaking of the ground during an earthquake and structures attached to the ground. The structure is the oscillator. It has a set of natural frequencies, determined by its stiffness, its mass, and the details of its construction. The periodic driving force is supplied by the shaking of the ground.

A disastrous result can occur if a natural frequency of the building matches a frequency contained in the ground shaking. In this case, the resonance vibrations of the building can build to a very large amplitude, large enough to damage or destroy the building. This result can be avoided in two ways. The first involves designing the structure so that natural frequencies of the building lie outside the range of earthquake frequencies. (A typical range of earthquake frequencies is 0–15 Hz.) Such a building can be designed by varying its size or mass structure. The second method involves incorporating sufficient damping in the building. This method may not change the resonance frequency significantly, but it will lower the response to the natural frequency as in Figure 12.16. It will also flatten the resonance curve, so the building will respond to a wide range of frequencies but with relatively small amplitude at any given frequency.

We now describe two examples involving resonance excitations in bridge structures. Soldiers are commanded to break step when marching across a bridge. This command takes into account resonance; if the marching frequency of the soldiers matches that of the bridge, the bridge could be set into resonance oscillation. If the amplitude becomes large enough, the bridge could actually collapse. Just such a situation occurred on April 14, 1831, when the Broughton suspension bridge in England collapsed while troops marched over it. Investigations after the accident showed that the bridge was near failure, and the resonance vibration induced by the marching soldiers caused it to fail sooner than it otherwise might have.

The second example of such a structural resonance occurred in 1940, when the Tacoma Narrows Bridge in Washington State was destroyed by resonant vibrations (Fig. 12.17). The winds were not particularly strong on that occasion, but the bridge still collapsed because vortices (turbulences) generated by the wind blowing through the bridge occurred at a frequency that matched a natural frequency of the bridge. The flapping of this wind across the roadway (similar to the flapping of

(a)

(b)

FIGURE 12.17 (a) In 1940, steady winds set up vibrations in the Tacoma Narrows Bridge, causing it to oscillate at a frequency near one of the natural frequencies of the bridge structure. (b) Once established, this resonance condition led to the bridge's collapse.

a flag in a strong breeze) provided the periodic driving force that brought the bridge down into the river.

Resonance gives us our first clue to responding to the central question for this Context. Suppose a building is far from the epicenter of an earthquake so that the ground shaking is small. If the shaking frequency matches a natural frequency of the building, a very effective energy coupling occurs between the ground and the building. Therefore, even for relatively small shaking, the ground, by resonance, can feed energy into the building efficiently enough to cause the failure of the structure. The structure must be carefully designed so as to reduce the resonance response.

SUMMARY

PhysicsNow™ Take a practice test by logging into PhysicsNow at **www.pop4e.com** and clicking on the Pre-Test link for this chapter.

The particle in simple harmonic motion model is used for a particle experiencing a linear restoring force, expressed by **Hooke's law,**

$$F_s = -kx \qquad [12.1]$$

where k is the **force constant** of the spring. The motion caused by such a force is called **simple harmonic motion,** and the system is called a **simple harmonic oscillator.** The position of a particle in simple harmonic motion varies periodically in time according to the relation

$$x(t) = A\cos(\omega t + \phi) \qquad [12.6]$$

where A is the **amplitude** of the motion, ω is the **angular frequency,** and ϕ is the **phase constant.** The values of A and ϕ depend on the initial position and velocity of the particle.

The time for one complete oscillation is called the **period** T of the motion. The inverse of the period is the **frequency** f of the motion, which equals the number of oscillations per second:

$$f = \frac{1}{T} = \frac{\omega}{2\pi} \qquad [12.11]$$

A particle–spring system oscillating without friction exhibits simple harmonic motion with the period

$$T = \frac{2\pi}{\omega} = 2\pi\sqrt{\frac{m}{k}} \qquad [12.13]$$

where m is the mass of the particle attached to the spring of force constant k.

The velocity and acceleration of a particle in simple harmonic oscillation are

$$v = \frac{dx}{dt} = -\omega A\sin(\omega t + \phi) \qquad [12.15]$$

$$a = \frac{d^2x}{dt^2} = -\omega^2 A\cos(\omega t + \phi) \qquad [12.16]$$

Therefore, the maximum speed of the particle is ωA and its maximum acceleration is of magnitude $\omega^2 A$. The speed is zero when the particle is at its turning points, $x = \pm A$, and the speed is a maximum at the equilibrium position, $x = 0$. The magnitude of the acceleration is a maximum at the turning points and is zero at the equilibrium position.

The kinetic energy and potential energy of a simple harmonic oscillator vary with time and are given by

$$K = \tfrac{1}{2}mv^2 = \tfrac{1}{2}m\omega^2A^2\sin^2(\omega t + \phi) \qquad [12.19]$$

$$U = \tfrac{1}{2}kx^2 = \tfrac{1}{2}kA^2\cos^2(\omega t + \phi) \qquad [12.20]$$

The **total energy** of a simple harmonic oscillator is a constant of the motion and is

$$E = \tfrac{1}{2}kA^2 \qquad [12.21]$$

The potential energy of a simple harmonic oscillator is a maximum when the particle is at its turning points (maximum displacement from equilibrium) and is zero at the equilibrium position. The kinetic energy is zero at the turning points and is a maximum at the equilibrium position.

A **simple pendulum** of length L exhibits simple harmonic motion for small angular displacements from the vertical, with a period of

$$T = 2\pi\sqrt{\frac{L}{g}} \qquad [12.25]$$

The period of a simple pendulum is independent of the mass of the suspended object.

A **physical pendulum** exhibits simple harmonic motion for small angular displacements from equilibrium about a pivot that does not go through the center of mass. The period of this motion is

$$T = 2\pi\sqrt{\frac{I}{mgd}} \qquad [12.27]$$

where I is the moment of inertia about an axis through the pivot and d is the distance from the pivot to the center of mass.

Damped oscillations occur in a system in which a resistive force opposes the motion of the oscillating object. If such a system is set in motion and then left to itself, its mechanical energy decreases in time because of the presence of the nonconservative resistive force. It is possible to compensate for this transformation of energy by driving the system with an external periodic force. The oscillator in this case is undergoing **forced oscillations.** When the frequency of the driving force matches the natural frequency of the *undamped* oscillator, energy is efficiently transferred to the oscillator and its steady-state amplitude is a maximum. This situation is called **resonance.**

QUESTIONS

☐ = answer available in the *Student Solutions Manual and Study Guide*

1. Is a bouncing ball an example of simple harmonic motion? Is the daily movement of a student from home to school and back simple harmonic motion? Why or why not?
2. Does the displacement of an oscillating particle between $t = 0$ and a later time t necessarily equal the position of the particle at time t? Explain.
3. If the position of a particle varies as $x = -A \cos \omega t$, what is the phase constant in Equation 12.6? At what position is the particle at $t = 0$?
4. Can the amplitude A and phase constant ϕ be determined for an oscillator if only the position is specified at $t = 0$? Explain.
5. Determine whether or not the following quantities can be in the same direction for a simple harmonic oscillator: (a) position and velocity, (b) velocity and acceleration, (c) position and acceleration.
6. A block is hung on a spring and the frequency f of the oscillation of the system is measured. The block, a second identical block, and the spring are carried into space in a spacecraft. The two blocks are attached to opposite ends of the spring, and the system is taken out into space on a space walk. The spring is extended, and the system is released to oscillate while floating in space. What is the frequency of oscillation for this system in terms of f?
7. A block–spring system undergoes simple harmonic motion with amplitude A. Does the total energy change if the mass is doubled but the amplitude is not changed? Do the kinetic and potential energies depend on the mass? Explain.
8. The equations listed in Table 2.2 give position as a function of time, velocity as a function of time, and velocity as function of position for an object moving in a straight line with constant acceleration. The quantity v_{xi} appears in every equation. Do any of these equations apply to an object moving in a straight line with simple harmonic motion? Using a similar format, make a table of equations describing simple harmonic motion. Include equations giving acceleration as a function of time and acceleration as a function of position. State the equations in such a form that they apply equally to a block–spring system, to a pendulum, and to other vibrating systems. What quantity appears in every equation?
9. What happens to the period of a simple pendulum if the pendulum's length is doubled? What happens to the period if the mass of the suspended bob is doubled?
10. If a grandfather clock were running slow, how could we adjust the pendulum's length to correct the time?
11. Will damped oscillations occur for any values of b and k? Explain.
12. You stand on the end of a diving board and bounce to set it into oscillation. You find a maximum response, in terms of the amplitude of oscillation of the end of the board, when you bounce at frequency f. You now move to the middle of the board and repeat the experiment. Is the resonance frequency for forced oscillations at this point higher, lower, or the same as f? Why?
13. Is it possible to have damped oscillations when a system is at resonance? Explain.
14. You are looking at a small tree. You do not notice any breeze, and most of the leaves on the tree are motionless. One leaf, however, is fluttering back and forth wildly. After you wait a while, that leaf stops moving and you notice a different leaf moving much more than all the others. Explain what could cause the large motion of one particular leaf.

PROBLEMS

1, 2, 3 = straightforward, intermediate, challenging
☐ = full solution available in the *Student Solutions Manual and Study Guide*
Physics Now™ = coached problem with hints available at www.pop4e.com
= computer useful in solving problem
= paired numerical and symbolic problems
= biomedical application

Note: Ignore the mass of every spring except in problems 12.54 and 12.56.

Section 12.1 ■ Motion of a Particle Attached to a Spring

Problems 6.13, 6.15, 6.17, 6.18, 6.30, and 6.56 in Chapter 6 can also be assigned with this section.

1. A ball dropped from a height of 4.00 m makes an elastic collision with the ground. Assuming that no mechanical energy is lost due to air resistance, (a) show that the ensuing motion is periodic and (b) determine the period of the motion. (c) Is the motion simple harmonic? Explain.

Section 12.2 ■ Mathematical Representation of Simple Harmonic Motion

2. In an engine, a piston oscillates with simple harmonic motion so that its position varies according to the expression

$$x = (5.00 \text{ cm}) \cos(2t + \pi/6)$$

where x is in centimeters and t is in seconds. At $t = 0$, find (a) the position of the piston, (b) its velocity, and (c) its acceleration. (d) Find the period and amplitude of the motion.

3. The position of a particle is given by the expression $x = (4.00 \text{ m}) \cos(3.00\pi t + \pi)$, where x is in meters and t is in

seconds. Determine (a) the frequency and period of the motion, (b) the amplitude of the motion, (c) the phase constant, and (d) the position of the particle at $t = 0.250$ s.

4. A particle moves in simple harmonic motion with a frequency of 3.00 Hz and an amplitude of 5.00 cm. (a) Through what total distance does the particle move during one cycle of its motion? (b) What is its maximum speed? Where does this maximum speed occur? (c) Find the maximum acceleration of the particle. Where in the motion does the maximum acceleration occur?

5. **Physics Now™** A particle moving along the x axis in simple harmonic motion starts from its equilibrium position, the origin, at $t = 0$ and moves to the right. The amplitude of its motion is 2.00 cm and the frequency is 1.50 Hz. (a) Show that the position of the particle is given by

$$x = (2.00 \text{ cm}) \sin(3.00\pi t)$$

Determine (b) the maximum speed and the earliest time ($t > 0$) at which the particle has this speed, (c) the maximum acceleration and the earliest time ($t > 0$) at which the particle has this acceleration, and (d) the total distance traveled between $t = 0$ and $t = 1.00$ s.

6. **Review problem.** A particle moves along the x axis. It is initially at the position 0.270 m, moving with velocity 0.140 m/s and acceleration -0.320 m/s^2. First, assume that it moves with constant acceleration for 4.50 s. Find (a) its position and (b) its velocity at the end of this time interval. Next, assume that it moves with simple harmonic motion for 4.50 s and that $x = 0$ is its equilibrium position. Find (c) its position and (d) its velocity at the end of this time interval.

7. The initial position, velocity, and acceleration of an object moving in simple harmonic motion are x_i, v_i, and a_i; the angular frequency of oscillation is ω. (a) Show that the position and velocity of the object for all time can be written as

$$x(t) = x_i \cos \omega t + \left(\frac{v_i}{\omega}\right) \sin \omega t$$

$$v(t) = -x_i \omega \sin \omega t + v_i \cos \omega t$$

(b) Using A to represent the amplitude of the motion, show that

$$v^2 - ax = v_i^2 - a_i x_i = \omega^2 A^2$$

8. A simple harmonic oscillator takes 12.0 s to undergo five complete vibrations. Find (a) the period of its motion, (b) the frequency in hertz, and (c) the angular frequency in radians per second.

9. A 7.00 kg object is hung from the bottom end of a vertical spring fastened to an overhead beam. The object is set into vertical oscillations having a period of 2.60 s. Find the force constant of the spring.

10. A vibration sensor, used in testing a washing machine, consists of a cube of aluminum 1.50 cm on edge mounted on one end of a strip of spring steel (like a hacksaw blade) that lies in a vertical plane. The strip's mass is small compared with that of the cube, but the strip's length is large compared with the size of the cube. The other end of the strip is clamped to the frame of the washing machine that is not operating. A horizontal force of 1.43 N applied to the cube is required to hold it 2.75 cm away from its equilibrium position. If it is released, what is its frequency of vibration?

11. A 0.500-kg object attached to a spring with a force constant of 8.00 N/m vibrates in simple harmonic motion with an amplitude of 10.0 cm. Calculate (a) the maximum value of its speed and acceleration, (b) the speed and acceleration when the object is 6.00 cm from the equilibrium position, and (c) the time interval required for the object to move from $x = 0$ to $x = 8.00$ cm.

12. A 1.00-kg glider attached to a spring with a force constant of 25.0 N/m oscillates on a horizontal, frictionless air track. At $t = 0$, the glider is released from rest at $x = -3.00$ cm (that is, the spring is compressed by 3.00 cm). Find (a) the period of its motion, (b) the maximum values of its speed and acceleration, and (c) the position, velocity, and acceleration as functions of time.

Section 12.3 ▪ Energy Considerations in Simple Harmonic Motion

13. **Physics Now™** An automobile having a mass of 1 000 kg is driven into a brick wall in a safety test. The bumper behaves like a spring of force constant 5.00×10^6 N/m and compresses 3.16 cm as the car is brought to rest. What was the speed of the car before impact, assuming that no mechanical energy is lost during impact with the wall?

14. A 200-g block is attached to a horizontal spring and executes simple harmonic motion with a period of 0.250 s. The total energy of the system is 2.00 J. Find (a) the force constant of the spring and (b) the amplitude of the motion.

15. A block of unknown mass is attached to a spring with a spring constant of 6.50 N/m and undergoes simple harmonic motion with an amplitude of 10.0 cm. When the block is halfway between its equilibrium position and the end point, its speed is measured to be 30.0 cm/s. Calculate (a) the mass of the block, (b) the period of the motion, and (c) the maximum acceleration of the block.

16. A block–spring system oscillates with an amplitude of 3.50 cm. The spring constant is 250 N/m and the mass of the block is 0.500 kg. Determine (a) the mechanical energy of the system, (b) the maximum speed of the block, and (c) the maximum acceleration.

17. A 50.0-g object connected to a spring with a force constant of 35.0 N/m oscillates on a horizontal, frictionless surface with an amplitude of 4.00 cm. Find (a) the total energy of the system and (b) the speed of the object when the position is 1.00 cm. Find (c) the kinetic energy and (d) the potential energy when the position is 3.00 cm.

18. A 2.00-kg object is attached to a spring and placed on a horizontal, smooth surface. A horizontal force of 20.0 N is required to hold the object at rest when it is pulled 0.200 m from its equilibrium position (the origin of the x axis). The object is now released from rest with an initial position of $x_i = 0.200$ m, and it subsequently undergoes simple harmonic oscillations. Find (a) the force constant of the spring, (b) the frequency of the oscillations, and (c) the maximum speed of the object. Where does this

maximum speed occur? (d) Find the maximum acceleration of the object. Where does it occur? (e) Find the total energy of the oscillating system. Find (f) the speed and (g) the acceleration of the object when its position is equal to one third of the maximum value.

19. The amplitude of a system moving in simple harmonic motion is doubled. Determine the change in (a) the total energy, (b) the maximum speed, (c) the maximum acceleration, and (d) the period.

20. A 65.0-kg bungee jumper steps off a bridge with a light bungee cord tied to her and to the bridge (Figure P12.20). The unstretched length of the cord is 11.0 m. She reaches the bottom of her motion 36.0 m below the bridge before bouncing back. Her motion can be separated into an 11.0-m free-fall and a 25.0-m section of simple harmonic oscillation. (a) For what time interval is she in free-fall? (b) Use the principle of conservation of energy to find the spring constant of the bungee cord. (c) What is the location of the equilibrium point where the spring force balances the gravitational force acting on the jumper? Note that this point is taken as the origin in our mathematical description of simple harmonic oscillation. (d) What is the angular frequency of the oscillation? (e) What time interval is required for the cord to stretch by 25.0 m? (f) What is the total time interval for the entire 36.0-m drop?

(Telegraph Colour Library/FPG International)

FIGURE P12.20 Problems 12.20 and 12.44.

21. A particle executes simple harmonic motion with an amplitude of 3.00 cm. At what position does its speed equal half of its maximum speed?

Section 12.4 ▪ The Simple Pendulum

Section 12.5 ▪ The Physical Pendulum

Problem 1.62 in Chapter 1 can also be assigned with this section.

22. A "seconds pendulum" is one that moves through its equilibrium position once each second. (The period of the pendulum is precisely 2 s.) The length of a seconds pendulum is 0.992 7 m at Tokyo, Japan, and 0.994 2 m at Cambridge, England. What is the ratio of the free-fall accelerations at these two locations?

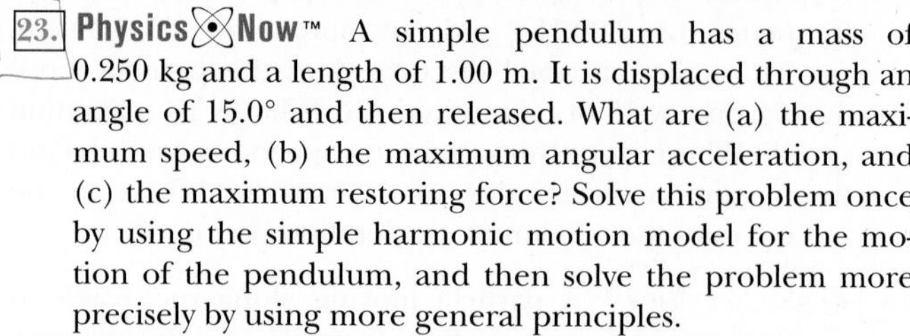

23. **PhysicsNow™** A simple pendulum has a mass of 0.250 kg and a length of 1.00 m. It is displaced through an angle of 15.0° and then released. What are (a) the maximum speed, (b) the maximum angular acceleration, and (c) the maximum restoring force? Solve this problem once by using the simple harmonic motion model for the motion of the pendulum, and then solve the problem more precisely by using more general principles.

24. The angular position of a pendulum is represented by the equation $\theta = (0.032\ 0 \text{ rad}) \cos \omega t$, where θ is in radians and $\omega = 4.43$ rad/s. Determine the period and length of the pendulum.

25. A particle of mass m slides without friction inside a hemispherical bowl of radius R. Show that if it starts from rest with a small displacement from equilibrium, the particle moves in simple harmonic motion with an angular frequency equal to that of a simple pendulum of length R (that is, $\omega = \sqrt{g/R}$).

26. A small object is attached to the end of a string to form a simple pendulum. The period of its harmonic motion is measured for small angular displacements and three lengths, each time by clocking the motion with a stopwatch for 50 oscillations. For lengths of 1.000 m, 0.750 m, and 0.500 m, total time intervals of 99.8 s, 86.6 s, and 71.1 s are measured for 50 oscillations. (a) Determine the period of motion for each length. (b) Determine the mean value of g obtained from these three independent measurements, and compare it with the accepted value. (c) Plot T^2 versus L, and obtain a value for g from the slope of your best-fit straight-line graph. Compare this value with that obtained in part (b).

27. A physical pendulum in the form of a planar object moves in simple harmonic motion with a frequency of 0.450 Hz. The pendulum has a mass of 2.20 kg, and the pivot is located 0.350 m from the center of mass. Determine the moment of inertia of the pendulum about the pivot point.

28. A very light rigid rod with a length of 0.500 m extends straight out from one end of a meter stick. The stick is suspended from a pivot at the far end of the rod and is set into oscillation. (a) Determine the period of oscillation. (*Suggestion:* Use the parallel-axis theorem from Section 10.11.) (b) By what percentage does the period differ from the period of a simple pendulum 1.00 m long?

29. Consider the physical pendulum of Figure 12.12. (a) Representing its moment of inertia about an axis passing through its center of mass and parallel to the axis passing through its pivot point as I_{CM}, show that its period is

$$T = 2\pi\sqrt{\frac{I_{CM} + md^2}{mgd}}$$

where d is the distance between the pivot point and center of mass. (b) Show that the period has a minimum value when d satisfies $md^2 = I_{CM}$.

Section 12.6 ▪ Damped Oscillations

30. Show that the time rate of change of mechanical energy for a damped, undriven oscillator is given by $dE/dt = -bv^2$ and hence is always negative. (*Suggestion:* Differentiate the expression for the mechanical energy of an oscillator, $E = \frac{1}{2}mv^2 + \frac{1}{2}kx^2$, and use Equation 12.28.)

31. A pendulum with a length of 1.00 m is released from an initial angle of 15.0°. After 1 000 s, its amplitude has been reduced by friction to 5.50°. What is the value of $b/2m$?

32. Show that Equation 12.29 is a solution of Equation 12.28 provided that $b^2 < 4mk$.

Section 12.7 ▪ Forced Oscillations

33. A 2.00-kg object attached to a spring moves without friction and is driven by an external force $F = (3.00 \text{ N}) \sin(2\pi t)$. Assuming that the force constant of the spring is 20.0 N/m, determine (a) the period and (b) the amplitude of the motion.

34. The front of her sleeper wet from teething, a baby rejoices in the day by crowing and bouncing up and down in her crib. Her mass is 12.5 kg, and the crib mattress can be modeled as a light spring with force constant 4.30 kN/m. (a) The baby soon learns to bounce with maximum amplitude and minimum effort by bending her knees at what frequency? (b) She learns to use the mattress as a trampoline—losing contact with it for part of each cycle—when her amplitude exceeds what value?

35. Considering an undamped, forced oscillator ($b = 0$), show that Equation 12.32 is a solution of Equation 12.31, with an amplitude given by Equation 12.33.

36. Damping is negligible for a 0.150-kg object hanging from a light 6.30-N/m spring. A sinusoidal force with an amplitude of 1.70 N drives the system. At what frequency will the force make the object vibrate with an amplitude of 0.440 m?

37. You are a research biologist. You take your emergency pager along to a fine restaurant even though its batteries are getting low. You switch the small pager to vibrate instead of beep, and you put it into a side pocket of your suit coat. The arm of your chair presses the light cloth against your body at one spot. Fabric with a length of 8.21 cm hangs freely below that spot, with the pager at the bottom. A coworker urgently needs instructions and calls you from your laboratory. The pager's motion makes the hanging part of your coat swing back and forth with remarkably large amplitude. The waiter, maître d', wine steward, and nearby diners notice immediately and fall silent. Your daughter pipes up and says, "Daddy, look! Your cockroaches must have gotten out again!" Find the frequency at which your pager vibrates.

Section 12.8 ▪ Context Connection—Resonance in Structures

38. Four people, each with a mass of 72.4 kg, are in a car with a mass of 1 130 kg. An earthquake strikes. The vertical oscillations of the ground surface make the car bounce up and down on its suspension springs, but the driver manages to pull off the road and stop. When the frequency of the shaking is 1.80 Hz, the car exhibits a maximum amplitude of vibration. The earthquake ends and the four people leave the car as fast as they can. By what distance does the car's undamaged suspension lift the car's body as the people get out?

39. People who ride motorcycles and bicycles learn to look out for bumps in the road and especially for *washboarding*, a condition in which many equally spaced ridges are worn into the road. What is so bad about washboarding? A motorcycle has several springs and shock absorbers in its suspension, but you can model it as a single spring supporting a block. You can estimate the force constant by thinking about how far the spring compresses when a large biker sits down on the seat. A motorcyclist traveling at highway speed must be particularly careful of washboard bumps that are a certain distance apart. What is the order of magnitude of their separation distance? State the quantities you take as data and the values you measure or estimate for them.

Additional Problems

40. An object of mass $m_1 = 9.00$ kg is in equilibrium while connected to a light spring of constant $k = 100$ N/m that is fastened to a wall as shown in Figure P12.40a. A second object, $m_2 = 7.00$ kg, is slowly pushed up against m_1, compressing the spring by the amount $A = 0.200$ m (see Fig. P12.40b). The system is then released and both objects start moving to the right on the frictionless surface. (a) When m_1 reaches the equilibrium point, m_2 loses contact with m_1 (see Fig. P12.40c) and moves to the right with speed v. Determine the value of v. (b) How far apart are the objects when the spring is fully stretched for the first time (D in Fig. P12.40d)? (*Suggestion:* First determine the period of oscillation and the amplitude of the m_1–spring system after m_2 loses contact with m_1.)

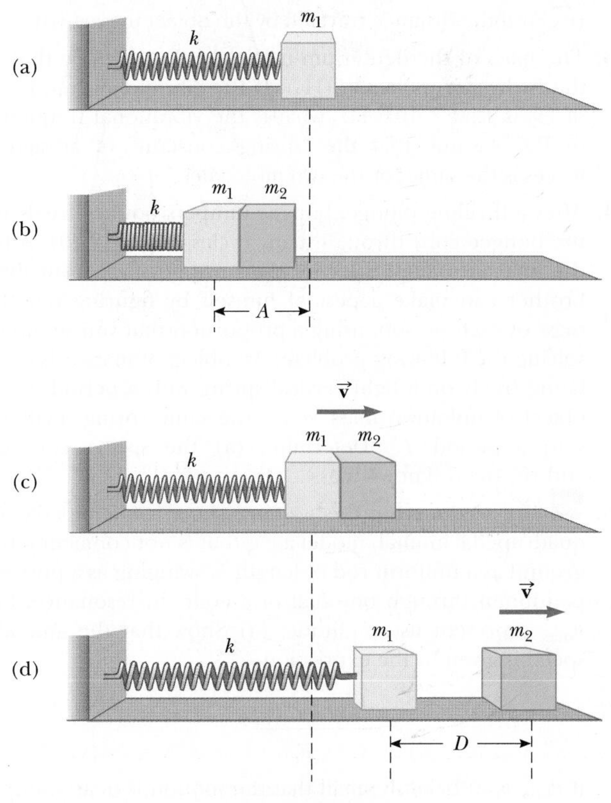

FIGURE P12.40

41. **Physics**⊗**Now**™ A large block P executes horizontal simple harmonic motion as it slides across a frictionless surface with a frequency $f = 1.50$ Hz. Block B rests on it, as shown in Figure P12.41, and the coefficient of static friction between the two is $\mu_s = 0.600$. What maximum amplitude of oscillation can the system have if block B is not to slip?

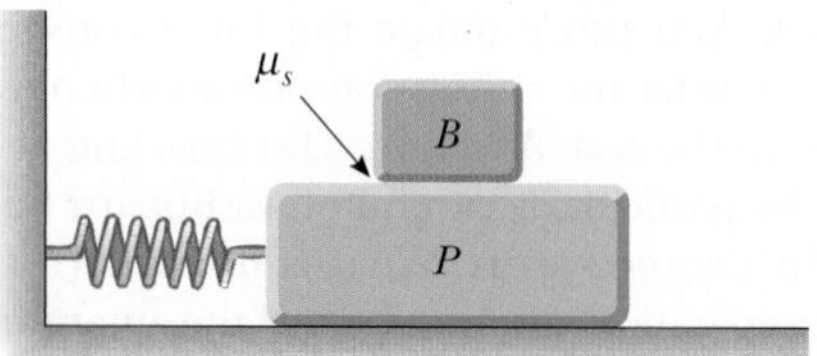

FIGURE **P12.41**

42. (a) A hanging spring stretches by 35.0 cm when an object of mass 450 g is hung on it at rest. In this situation, we define its position as $x = 0$. The object is pulled down an additional 18.0 cm and released from rest to oscillate without friction. What is its position x at a time 84.4 s later? (b) A hanging spring stretches by 35.5 cm when an object of mass 440 g is hung on it at rest. We define this new position as $x = 0$. This object is also pulled down an additional 18.0 cm and released from rest to oscillate without friction. Find its position 84.4 s later. (c) Why are the answers to parts (a) and (b) different by such a large percentage when the data are so similar? Does this circumstance reveal a fundamental difficulty in calculating the future? (d) Find the distance traveled by the vibrating object in part (a). (e) Find the distance traveled by the object in part (b).

43. The mass of the deuterium molecule (D_2) is twice that of the hydrogen molecule (H_2). If the vibrational frequency of H_2 is 1.30×10^{14} Hz, what is the vibrational frequency of D_2? Assume that the "spring constant" of attracting forces is the same for the two molecules.

44. After a thrilling plunge, bungee jumpers bounce freely on the bungee cord through many cycles (Fig. P12.20). After the first few cycles, the cord does not go slack. Your little brother can make a pest of himself by figuring out the mass of each person, using a proportion that you set up by solving the following problem. An object of mass m is oscillating freely on a light vertical spring with a period T. An object of unknown mass m' on the same spring oscillates with a period T'. Determine (a) the spring constant and (b) the unknown mass.

45. To account for the walking speed of a bipedal or quadrupedal animal, model a leg that is not contacting the ground as a uniform rod of length ℓ, swinging as a physical pendulum through one half of a cycle, in resonance. Let θ_{max} represent its amplitude. (a) Show that the animal's speed is given by the expression

$$\frac{\sqrt{6g\ell}\,\sin\theta_{max}}{\pi}$$

if θ_{max} is sufficiently small that the motion is nearly simple harmonic. An empirical relationship that is based on the same model and applies over a wider range of angles is

$$\frac{\sqrt{6g\ell\cos(\theta_{max}/2)}\,\sin\theta_{max}}{\pi}$$

(b) Evaluate the walking speed of a human with leg length 0.850 m and leg-swing amplitude 28.0°. (c) What leg length would give twice the speed for the same angular amplitude?

46. **Review problem.** The problem extends the reasoning of Problem 8.46 in Chapter 8. Two gliders are set in motion on an air track. Glider 1 has mass $m_1 = 0.240$ kg and velocity $0.740\hat{\mathbf{i}}$ m/s. It will have a rear-end collision with glider 2, of mass $m_2 = 0.360$ kg, which has original velocity $0.120\hat{\mathbf{i}}$ m/s. A light spring of force constant 45.0 N/m is attached to the back end of glider 2 as shown in Figure P8.46. When glider 1 touches the spring, superglue instantly and permanently makes it stick to its end of the spring. (a) Find the common velocity the two gliders have when the spring compression is a maximum. (b) Find the maximum spring compression distance. (c) Argue that the motion after the gliders become attached consists of the center of mass of the two-glider system moving with the constant velocity found in part (a) while both gliders oscillate in simple harmonic motion relative to the center of mass. (d) Find the energy of the center-of-mass motion. (e) Find the energy of the oscillation.

47. A pendulum of length L and mass M has a spring of force constant k connected to it at a distance h below its point of suspension (Fig. P12.47). Find the frequency of vibration of the system for small values of the amplitude (small θ). Assume that the vertical suspension of length L is rigid, but ignore its mass.

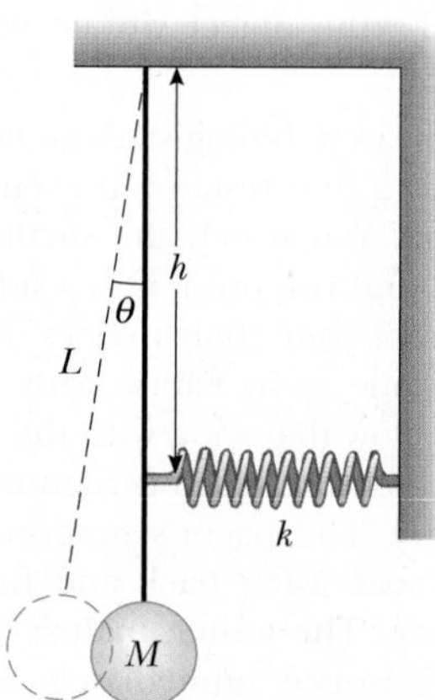

FIGURE **P12.47**

48. A particle with a mass of 0.500 kg is attached to a spring with a force constant of 50.0 N/m. At time $t = 0$ the particle has its maximum speed of 20.0 m/s and is moving to the left. (a) Determine the particle's equation of motion, specifying its position as a function of time. (b) Where in the motion is the potential energy three times the kinetic energy? (c) Find the length of a simple pendulum with the same period. (d) Find the minimum time interval required for the particle to move from $x = 0$ to $x = 1.00$ m.

49. A horizontal plank of mass m and length L is pivoted at one end. The plank's other end is supported by a spring of force constant k (Fig. P12.49). The moment of inertia of the plank about the pivot is $\frac{1}{3}mL^2$. The plank is displaced by a small angle θ from its horizontal equilibrium position and released. (a) Show that it moves with simple harmonic motion with an angular frequency $\omega = \sqrt{3k/m}$. (b) Evaluate the frequency, assuming that the mass is 5.00 kg and that the spring has a force constant of 100 N/m.

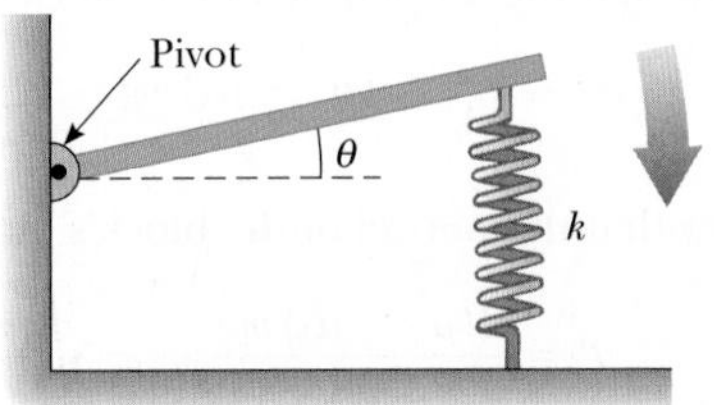

FIGURE **P12.49**

50. **Review problem.** A particle of mass 4.00 kg is attached to a spring with a force constant of 100 N/m. It is oscillating on a horizontal, frictionless surface with an amplitude of 2.00 m. A 6.00-kg object is dropped vertically on top of the 4.00-kg object as it passes through its equilibrium point. The two objects stick together. (a) By how much does the amplitude of the vibrating system change as a result of the collision? (b) By how much does the period change? (c) By how much does the mechanical energy of the system change? (d) Account for the change in energy.

51. A simple pendulum with a length of 2.23 m and a mass of 6.74 kg is given an initial speed of 2.06 m/s at its equilibrium position. Assume that it undergoes simple harmonic motion and determine its (a) period, (b) total energy, and (c) maximum angular displacement.

52. **Review problem.** One end of a light spring with force constant 100 N/m is attached to a vertical wall. A light string is tied to the other end of the horizontal spring. The string changes from horizontal to vertical as it passes over a solid pulley of diameter 4.00 cm. The pulley is free to turn on a fixed, smooth axle. The vertical section of the string supports a 200-g object. The string does not slip at its contact with the pulley. Find the frequency at which the object oscillates if the mass of the pulley is (a) negligible, (b) 250 g, and (c) 750 g.

53. **Physics Now™** A ball of mass m is connected to two rubber bands of length L, each under tension T, as shown in Figure P12.53. The ball is displaced by a small distance y perpendicular to the length of the rubber bands. Assuming that the tension does not change, show that (a) the restoring force is $-(2T/L)y$ and (b) the system exhibits simple harmonic motion with an angular frequency $\omega = \sqrt{2T/mL}$.

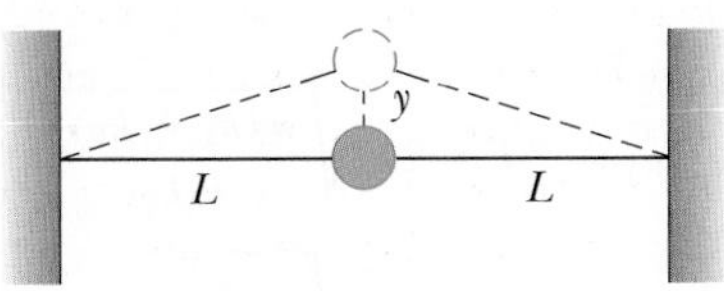

FIGURE **P12.53**

54. A block of mass M is connected to a spring of mass m and oscillates in simple harmonic motion on a horizontal, frictionless track (Fig. P12.54). The force constant of the spring is k and the equilibrium length is ℓ. Assume that all portions of the spring oscillate in phase and that the velocity of a segment dx is proportional to the distance x from the fixed end; that is, $v_x = (x/\ell)v$. Also, note that the mass of a segment of the spring is $dm = (m/\ell)dx$, Find (a) the kinetic energy of the system when the block has a speed v and (b) the period of oscillation.

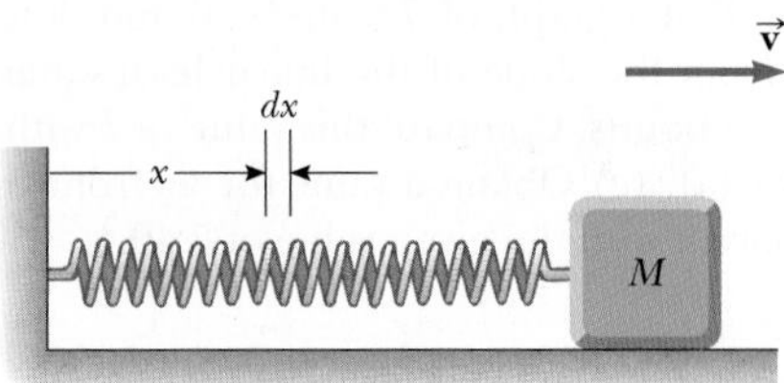

FIGURE **P12.54**

55. A smaller disk of radius r and mass m is attached rigidly to the face of a second larger disk of radius R and mass M as shown in Figure P12.55. The center of the small disk is located at the edge of the large disk. The large disk is mounted at its center on a frictionless axle. The assembly is rotated through a small angle θ from its equilibrium position and released. (a) Show that the speed of the center of the small disk as it passes through the equilibrium position is

$$v = 2\left[\frac{Rg(1 - \cos\theta)}{(M/m) + (r/R)^2 + 2}\right]^{1/2}$$

(b) Show that the period of the motion is

$$T = 2\pi\left[\frac{(M + 2m)R^2 + mr^2}{2mgR}\right]^{1/2}$$

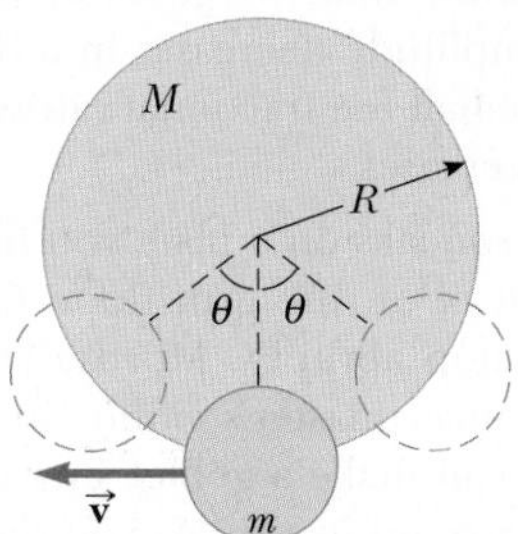

FIGURE **P12.55**

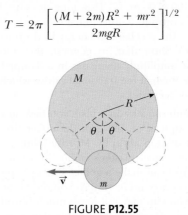

56. When a block of mass M, connected to the end of a spring of mass $m_s = 7.40$ g and force constant k, is set into simple harmonic motion, the period of its motion is

$$T = 2\pi\sqrt{\frac{M + (m_s/3)}{k}}$$

A two-part experiment is conducted with the use of blocks of various masses suspended vertically from the spring as shown in Figure P12.56. (a) Static extensions of 17.0, 29.3, 35.3, 41.3, 47.1, and 49.3 cm are measured for M values of

20.0, 40.0, 50.0, 60.0, 70.0, and 80.0 g, respectively. Construct a graph of Mg versus x and perform a linear least-squares fit to the data. From the slope of your graph, determine a value for k for this spring. (b) The system is now set into simple harmonic motion, and periods are measured with a stopwatch. With $M = 80.0$ g, the total time interval for ten oscillations is measured to be 13.41 s. The experiment is repeated with M values of 70.0, 60.0, 50.0, 40.0, and 20.0 g, with corresponding time intervals for ten oscillations of 12.52, 11.67, 10.67, 9.62, and 7.03 s. Compute the experimental value for T from each of these measurements. Plot a graph of T^2 versus M and determine a value for k from the slope of the linear least-squares fit through the data points. Compare this value of k with that obtained in part (a). (c) Obtain a value for m_s from your graph and compare it with the given value of 7.40 g.

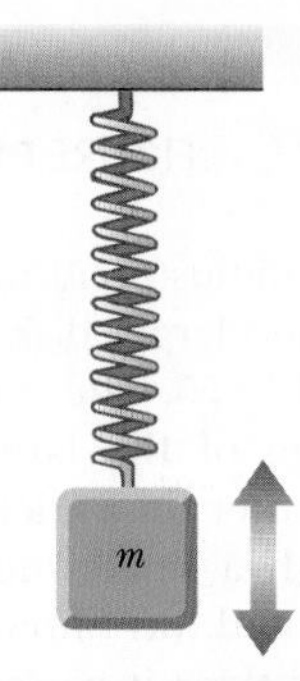

FIGURE **P12.56**

57. An object is hung from a spring, set into vertical vibration, and immersed in a beaker of oil. Its motion is graphed in Active Figure 12.14b. The object has mass 375 g, the spring has force constant 100 N/m, and the damping coefficient is $b = 0.100$ N·s/m. (a) How long does it take for the amplitude to drop to half its initial value? (b) How long does it take for the mechanical energy to drop to half its initial value? (c) Show that, in general, the fractional rate at which the amplitude decreases in a damped harmonic oscillator is one-half the fractional rate at which the mechanical energy decreases.

58. Your thumb squeaks on a plate you have just washed. Your sneakers squeak on the gym floor. Car tires squeal when you start or stop abruptly. Mortise joints groan in an old barn. The concertmaster's violin sings out over a full orchestra. You can make a goblet sing by wiping your moistened finger around its rim. As you slide it across the table, a Styrofoam cup may not make much sound, but it makes the surface of some water inside it dance in a complicated resonance vibration. When chalk squeaks on a blackboard, you can see that it makes a row of regularly spaced dashes. As these examples suggest, vibration commonly results when friction acts on a moving elastic object. The oscillation is not simple harmonic motion, but is called *stick-and-slip*. This problem models stick-and-slip motion.

A block of mass m is attached to a fixed support by a horizontal spring with force constant k and negligible mass (Fig. P12.58). Hooke's law describes the spring both in extension and in compression. The block sits on a long horizontal board with which it has coefficient of static friction μ_s and a smaller coefficient of kinetic friction μ_k. The board moves to the right at constant speed v. Assume that the block spends most of its time sticking to the board and moving to the right, so the speed v is small in comparison to the average speed the block has as it slips back toward the left. (a) Show that the maximum extension of the spring from its unstressed position is very nearly given by $\mu_s mg/k$. (b) Show that the block oscillates around an equilibrium position at which the spring is stretched by $\mu_k mg/k$. (c) Graph the block's position versus time. (d) Show that the amplitude of the block's motion is

$$A = \frac{(\mu_s - \mu_k)\,mg}{k}$$

(e) Show that the period of the block's motion is

$$T = \frac{2(\mu_s - \mu_k)\,mg}{vk} + \pi\sqrt{\frac{m}{k}}$$

(f) Evaluate the frequency of the motion assuming that $\mu_s = 0.400$, $\mu_k = 0.250$, $m = 0.300$ kg, $k = 12.0$ N/m, and $v = 2.40$ cm/s. (g) What happens to the frequency if the mass increases? (h) What happens if the spring constant increases? (i) What happens if the speed of the board increases? (j) What happens if the coefficient of static friction increases relative to the coefficient of kinetic friction? Note that it is the excess of static over kinetic friction that is important for the vibration. "The squeaky wheel gets the grease" because even a viscous fluid cannot exert a force of static friction.

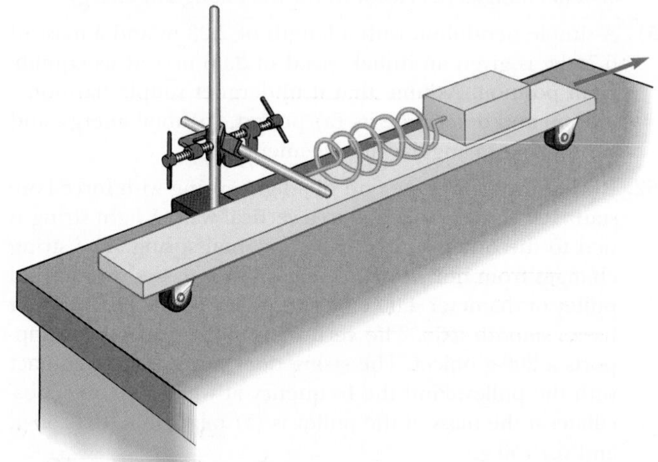

FIGURE **P12.58**

59. A block of mass m is connected to two springs of force constants k_1 and k_2 in two ways as shown in Figure P12.59. In both cases, the block moves on a frictionless table after it is displaced from equilibrium and released. Show that in the two cases the block exhibits simple harmonic motion with periods

$$\text{(a)}\quad T = 2\pi\sqrt{\frac{m(k_1 + k_2)}{k_1 k_2}}$$

$$\text{(b)}\quad T = 2\pi\sqrt{\frac{m}{k_1 + k_2}}$$

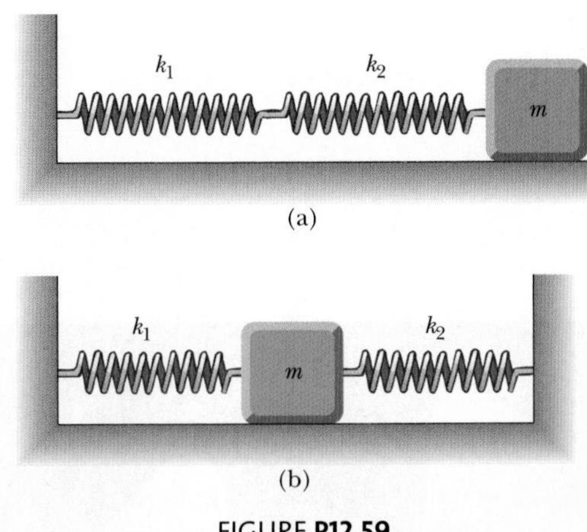

FIGURE **P12.59**

60. **Review problem.** Imagine that a hole is drilled through the center of the Earth to the other side. An object of mass m at a distance r from the center of the Earth is pulled toward the center of the Earth only by the mass within the sphere of radius r (the reddish region in Fig. P12.60). (a) Write Newton's law of gravitation for an object at the distance r from the center of the Earth and show that the force on it is of Hooke's law form, $F = -kr$, where the effective force constant is $k = \frac{4}{3}\pi\rho Gm$. Here ρ is the density of the Earth, assumed uniform, and G is the gravitational constant. (b) Show that a sack of mail dropped into the hole will execute simple harmonic motion if it moves without friction. How long does it take to arrive at the other side of the Earth? (c) At the same time as the sack of mail is dropped in the hole, a golf ball is struck so that it becomes a treetop satellite. (See Problem 11.27 in Chapter 11.) Which object, sack of mail or golf ball, arrives first at the end of the hole halfway around the Earth?

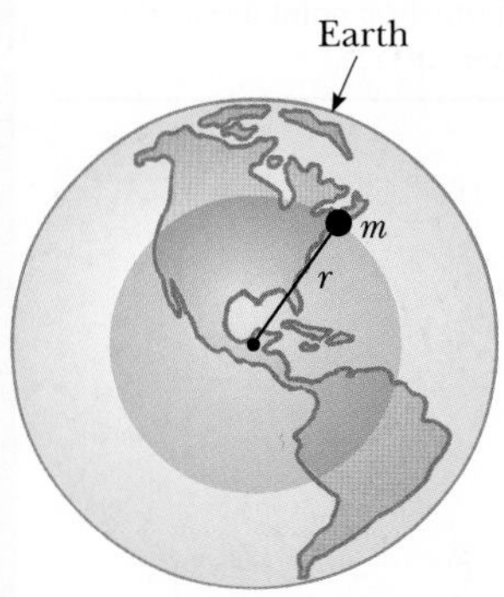

FIGURE **P12.60**

ANSWERS TO QUICK QUIZZES

12.1 (d). From its maximum positive position to the equilibrium position, the block travels a distance A. It then goes an equal distance past the equilibrium position to its maximum negative position. It then repeats these two motions in the reverse direction to return to its original position and complete one cycle.

12.2 (i), (f). The object is in the region $x < 0$, so the position is negative. Because the object is moving back toward the origin in this region, the velocity is positive. (ii), (a). The velocity is positive, as in (i). Because the spring is pulling the object toward equilibrium from the negative x region, the acceleration is also positive.

12.3 (i), (a). With a longer length, the period of the pendulum will increase. Therefore, it will take slightly longer to execute each swing, so each second according to the clock will take longer than an actual second and the clock will run *slow.* (ii), (a). At the top of the mountain, the value of g is less than that at sea level. As a result, the period of the pendulum will increase slightly and the clock will run slow.

12.4 (a). Although changing the mass of a simple pendulum does not change the frequency, because the bob is large means that we must model the pendulum as a physical pendulum rather than a simple pendulum. When the gum is placed on top of the bob, the moment of inertia and the center of mass of the physical pendulum are altered slightly. According to Equation 12.27, these changes alter the period of the pendulum.

CHAPTER 13

Mechanical Waves

Drops of water fall from a leaf into a pond. The disturbance caused by the falling water moves away from the drop point as circular ripples on the water surface.

(Don Bonsey/Stone Getty)

CHAPTER OUTLINE

Most of us experienced waves as children when we dropped pebbles into a pond. The disturbance created by a pebble manifests itself as ripples that move outward from the point at which the pebble lands in the water, like the ripples from the falling water drops in the opening photograph. If you were to carefully examine the motion of a leaf floating near the point where the pebble enters the water, you would see that the leaf moves up and down and back and forth about its original position but does not undergo any net displacement away from or toward the source of the disturbance. The *disturbance* in the water moves over a long distance, but a given small *element of the water* oscillates only over a very small distance. This behavior is the essence of wave motion.

The world is full of other kinds of waves, including sound waves, waves on strings, seismic waves, radio waves, and x-rays. Most waves can be placed in one of two categories. **Mechanical**

waves are waves that disturb and propagate through a medium; the ripple in the water because of the pebble and a sound wave, for which air is the medium, are examples of mechanical waves. **Electromagnetic waves** are a special class of waves that do not require a medium to propagate, as discussed with regard to the absence of the ether in Section 9.2; light waves and radio waves are two familiar examples. In this chapter, we shall confine our attention to the study of mechanical waves, deferring our study of electromagnetic waves to Chapter 24.

13.1 PROPAGATION OF A DISTURBANCE

In the introduction, we alluded to the essence of wave motion: the transfer of a *disturbance* through space without the accompanying transfer of *matter*. The propagation of the disturbance also represents a transfer of energy; thus, we can view waves as a means of energy transfer. In the list of energy transfer mechanisms in Section 6.6, we see two entries that depend on waves: mechanical waves and electromagnetic radiation. These entries are to be contrasted with another entry—matter transfer—in which the energy transfer is accompanied by a movement of matter through space.

All waves carry energy, but the amount of energy transmitted through a medium and the mechanism responsible for the energy transport differ from case to case. For instance, the power of ocean waves during a storm is much greater than that of sound waves generated by a musical instrument.

All mechanical waves require (1) some source of disturbance, (2) a medium that can be disturbed, and (3) some physical mechanism through which elements of the medium can influence one another. This final requirement assures that a disturbance to one element will cause a disturbance to the next so that the disturbance will indeed propagate through the medium.

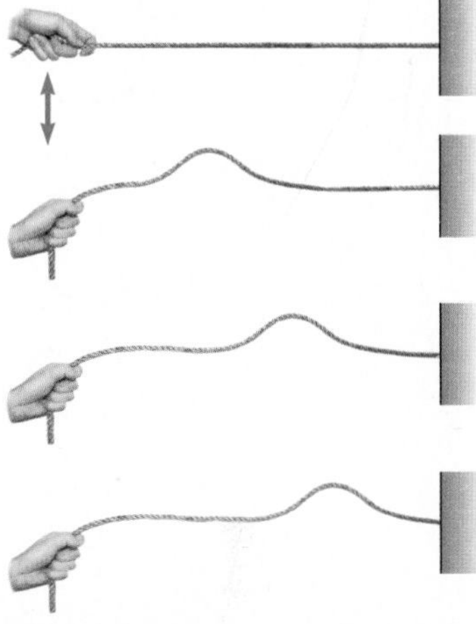

FIGURE 13.1 A pulse traveling down a stretched rope. The shape of the pulse is approximately unchanged as it travels along the rope.

One way to demonstrate wave motion is to flip the free end of a long rope that is under tension and has its opposite end fixed as in Figure 13.1. In this manner, a single **pulse** is formed and travels (to the right in Fig. 13.1) with a definite speed. The rope is the medium through which the pulse travels. Figure 13.1 represents consecutive "snapshots" of the traveling pulse. The shape of the pulse changes very little as it travels along the rope.

As the pulse travels, each rope element that is disturbed moves in a direction perpendicular to the direction of propagation. Figure 13.2 illustrates this point for a particular element, labeled P. Note that there is no motion of any part of the rope that is in the direction of the wave. A disturbance such as this one in which the elements of the disturbed medium move perpendicularly to the direction of propagation is called a **transverse wave.**

In another class of waves, called **longitudinal waves,** the elements of the medium undergo displacements *parallel* to the direction of propagation. Sound waves in air, for instance, are longitudinal. Their disturbance corresponds to a series of high- and low-pressure regions that may travel through air or through any material medium with a certain speed. A longitudinal pulse can be easily produced in a stretched spring as in Figure 13.3. A group of coils at the free end is pushed forward and pulled back. This action produces a pulse in the form of a compressed region of coils that travels along the spring.

So far, we have provided pictorial representations of a traveling pulse and hope you have begun to develop a mental representation of such a pulse. Let us now

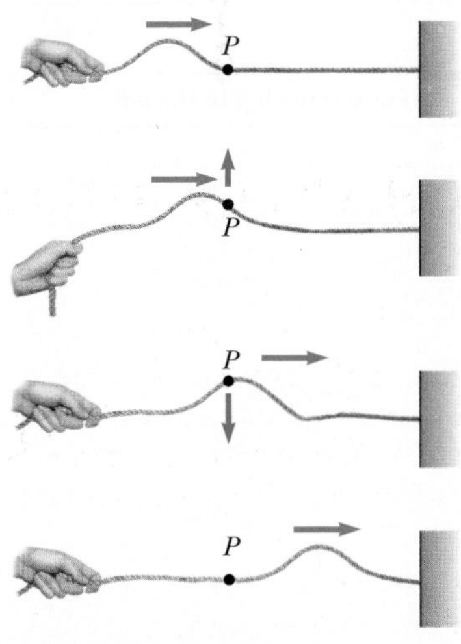

FIGURE 13.2 A pulse traveling on a stretched rope is a transverse disturbance. That is, any element of the rope, such as that at P, moves *(blue arrows)* in a direction perpendicular to the propagation of the pulse *(red arrows)*.

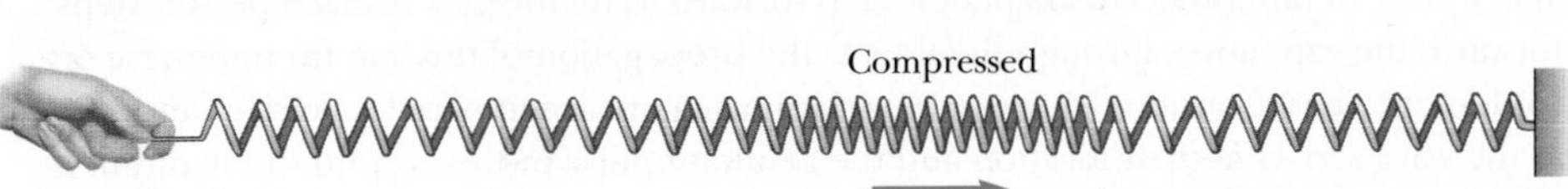

FIGURE 13.3 A longitudinal pulse along a stretched spring. The displacement of the coils is in the direction of the wave motion. The compressed region moves to the right along the spring.

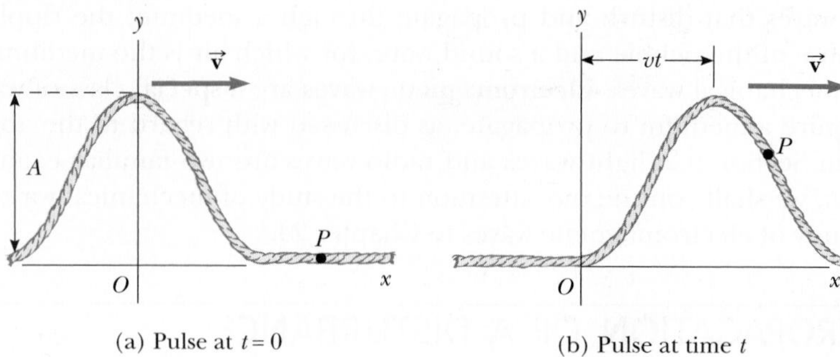

FIGURE 13.4 A one-dimensional pulse traveling to the right with a speed v. (a) At $t = 0$, the shape of the pulse is given by $y = f(x)$. (b) At some later time t, the shape remains unchanged and the vertical position of any element of the medium is given by $y = f(x - vt)$.

develop a mathematical representation for the propagation of this pulse. Consider a pulse traveling to the right with constant speed v on a long, stretched string as in Figure 13.4. The pulse moves along the x axis (the axis of the string), and the transverse (up-and-down) displacement of the elements of the string is described by means of the position y.

Figure 13.4a represents the shape and position of the pulse at time $t = 0$. At this time, the shape of the pulse, whatever it may be, can be represented by some mathematical function that we will write as $y(x, 0) = f(x)$. This function describes the vertical position y of the element of the string located at each value of x at time $t = 0$. Because the speed of the pulse is v, the pulse has traveled to the right a distance vt at time t (Fig. 13.4b). We adopt a simplification model in which the shape of the pulse does not change with time.[1] Therefore, at time t, the shape of the pulse is the same as it was at time $t = 0$, as in Figure 13.4a. Consequently, an element of the string at x at this time has the same y position as an element located at $x - vt$ had at time $t = 0$:

$$y(x, t) = y(x - vt, 0)$$

In general, then, we can represent the position y for all values of x and t, measured in a stationary frame with the origin at O, as

■ Pulse traveling to the right

$$y(x, t) = f(x - vt) \qquad [13.1a]$$

If the pulse travels to the left, the position of an element of the string is described by

■ Pulse traveling to the left

$$y(x, t) = f(x + vt) \qquad [13.1b]$$

The function y, sometimes called the **wave function,** depends on the two variables x and t. For this reason, it is often written $y(x, t)$, which is read "y as a function of x and t."

It is important to understand the meaning of y. Consider a point P on the string, identified by a particular value of its x coordinate as in Figure 13.4. As the pulse passes through P, the y coordinate of this point increases, reaches a maximum, and then decreases to zero. **The wave function $y(x, t)$ represents the y position of any element of string located at position x at any time t.** Furthermore, if t is fixed (e.g., in the case of taking a snapshot of the pulse), the wave function y as a function of x, sometimes called the **waveform,** defines a curve representing the actual geometric shape of the pulse at that time.

QUICK QUIZ 13.1 In a long line of people waiting to buy tickets, the first person leaves and a pulse of motion occurs as people step forward to fill the gap. As each person steps forward, the gap moves through the line. Is the propagation of this gap **(a)** transverse or **(b)** longitudinal? Consider "the wave" at a baseball game when people stand up and shout as the wave arrives at their location and the resultant pulse moves around the stadium. Is this wave **(c)** transverse or **(d)** longitudinal?

[1] In reality, the pulse changes its shape and gradually spreads out during the motion. This effect, called *dispersion,* is common to many mechanical waves, but we adopt a simplification model that ignores this effect.

EXAMPLE 13.1 A Pulse Moving to the Right

A pulse moving to the right along the x axis is represented by the wave function

$$y(x, t) = \frac{2.0}{(x - 3.0t)^2 + 1}$$

where x and y are measured in centimeters and t is in seconds. Let us plot the waveform at $t = 0$, $t = 1.0$ s, and $t = 2.0$ s.

Solution First, note that this function is of the form $y = f(x - vt)$. By inspection, we see that the speed of the wave is $v = 3.0$ cm/s. The location of the peak of the pulse occurs at the value of x for which the denominator is a minimum, that is, where $(x - 3.0t) = 0$. Therefore, the peaks occur at $x = 0.0$ cm at $t = 0$, at $x = 3.0$ cm at $t = 1.0$ s, and $x = 6.0$ cm at $t = 2.0$ s.

At times $t = 0$, $t = 1.0$ s, and $t = 2.0$ s, the wave function expressions are

$$y(x, 0) = \frac{2.0}{x^2 + 1} \quad \text{at } t = 0$$

$$y(x, 1.0) = \frac{2.0}{(x - 3.0)^2 + 1} \quad \text{at } t = 1.0 \text{ s}$$

$$y(x, 2.0) = \frac{2.0}{(x - 6.0)^2 + 1} \quad \text{at } t = 2.0 \text{ s}$$

We can now use these expressions to plot the wave function versus x at these times. For example, let us evaluate $y(x, 0)$ at $x = 0.50$ cm:

$$y(0.50, 0) = \frac{2.0}{(0.50)^2 + 1} = 1.6 \text{ cm}$$

Likewise, $y(1.0, 0) = 1.0$ cm, $y(2.0, 0) = 0.40$ cm, and so on. A continuation of this procedure for other values of x yields the waveform shown in Figure 13.5a. In a similar manner, one obtains the graphs of $y(x, 1.0)$ and $y(x, 2.0)$, shown in Figures 13.5b and 13.5c, respectively. These snapshots show that the pulse moves to the right without changing its shape and has a constant speed of 3.0 cm/s.

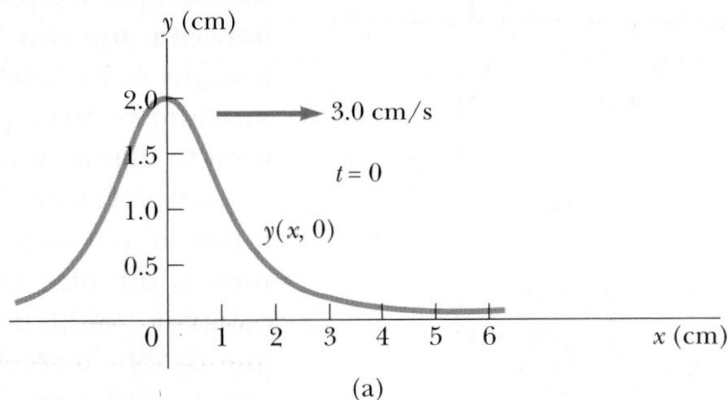

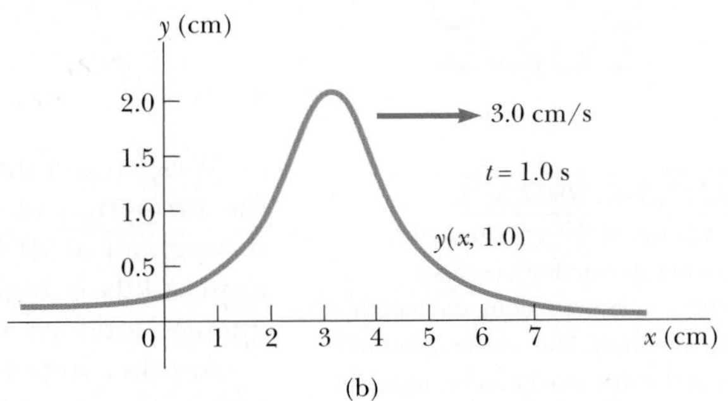

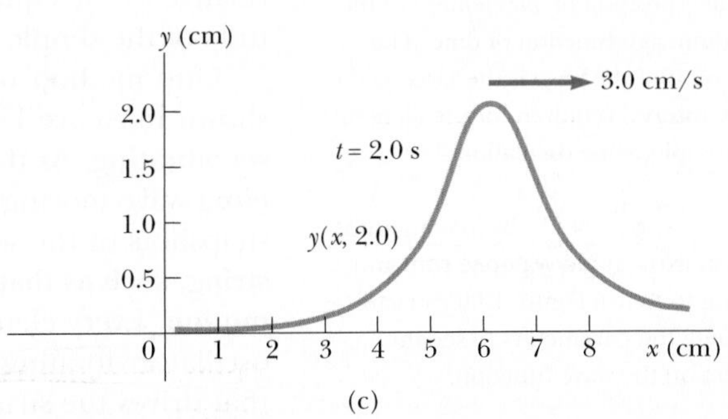

FIGURE 13.5 (Example 13.1) Graphs of the function $y(x, t) = 2.0/[(x - 3.0t)^2 + 1]$ at (a) $t = 0$, (b) $t = 1.0$ s, and (c) $t = 2.0$ s.

13.2 THE WAVE MODEL

We have discussed creating a disturbance moving through a medium such as a stretched string by a simple up-and-down displacement of the end of the string. This action results in a pulse moving along the medium. A continuous wave is created by shaking the end of the string in simple harmonic motion, which we studied in Chapter 12. If we do that, the string will take on the shape shown by the curve in the graph in Active Figure 13.6a, with this shape remaining the same but moving toward the right. This shape is what we call a **sinusoidal** wave because the waveform in Active Figure 13.6a is that of a sine wave. The point with the largest positive displacement of the string is called the **crest** of the wave. The lowest point is called the **trough.** The crest and trough move along with the wave, and a particular point on

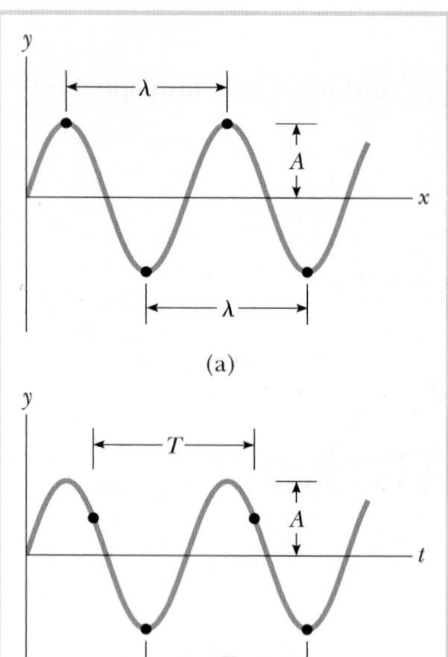

ACTIVE FIGURE 13.6
(a) A graph of the y position of elements of a medium versus x position, measured along the length of the medium. The wavelength λ of a wave is the distance between adjacent crests or adjacent troughs. (b) A graph of the y position of *one* element of the medium as a function of time. The period T of the wave is the same as the time interval required for the element to complete one oscillation.

Physics Now™ By logging into PhysicsNow at **www.pop4e.com** and going to Active Figure 13.6 you can change the parameters to see the effect on the wave function.

the string alternates between locations on a crest and a trough. In idealized wave motion in an idealized medium, each element of the medium undergoes simple harmonic motion around its equilibrium position.

Three physical characteristics are important in describing a sinusoidal wave: **wavelength, frequency,** and **wave speed. One wavelength is the minimum distance between any two identical points on a wave** such as adjacent crests or adjacent troughs as in Active Figure 13.6a, which is a graph of y position of elements of the medium versus x position for a sinusoidal wave at a specific time. The symbol λ is used to denote wavelength.

Active Figure 13.6b shows y position versus time for a single element of the medium as a sinusoidal wave is passing through its position x. The **period** T of the wave is the time interval required for an element of the medium to undergo one complete oscillation. The **frequency** f of sinusoidal waves is the same as the frequency of simple harmonic motion of an element of the medium. The period is equal to the inverse of the frequency:

$$T = \frac{1}{f} \qquad [13.2]$$

Waves travel through the medium with a specific **wave speed,** which depends on the properties of the medium being disturbed. For instance, sound waves travel through air at 20°C with a speed of about 343 m/s, whereas the speed of sound in most solids is higher than 343 m/s. We will learn more about wavelength, frequency, and wave speed in the next section.

Another important parameter for the wave in Active Figure 13.6 is the **amplitude** of the wave. Amplitude is the maximum position of an element of the medium relative to the equilibrium position. It is denoted by A and is the same as the amplitude of the simple harmonic motion of the elements of the medium.

One method of producing a traveling sinusoidal wave on a very long string is shown in Active Figure 13.7. One end of the string is connected to a blade that is set vibrating. As the blade oscillates vertically with simple harmonic motion, a traveling wave moving to the right is set up on the string. Active Figure 13.7 represents snapshots of the wave at intervals of one quarter of a period. **Each element of the string, such as that at P, oscillates vertically in the y direction with simple harmonic motion.** Every element of the string can therefore be treated as a simple harmonic oscillator vibrating with a frequency equal to the frequency of vibration of the blade that drives the string. Although each element oscillates in the y direction, the wave (or disturbance) travels in the x direction with a speed v. Of course, this situation is the definition of a transverse wave. In this case, the energy carried by the traveling wave is supplied by the vibrating blade.

ACTIVE FIGURE 13.7
One method for producing a sinusoidal wave on a continuous string. The left end of the string is connected to a blade that is set into vibration. Every element of the string, such as the one at point P, oscillates with simple harmonic motion in the vertical direction.

Physics Now™ Log into PhysicsNow at **www.pop4e.com** and go to Active Figure 13.7 to adjust the frequency of the blade.

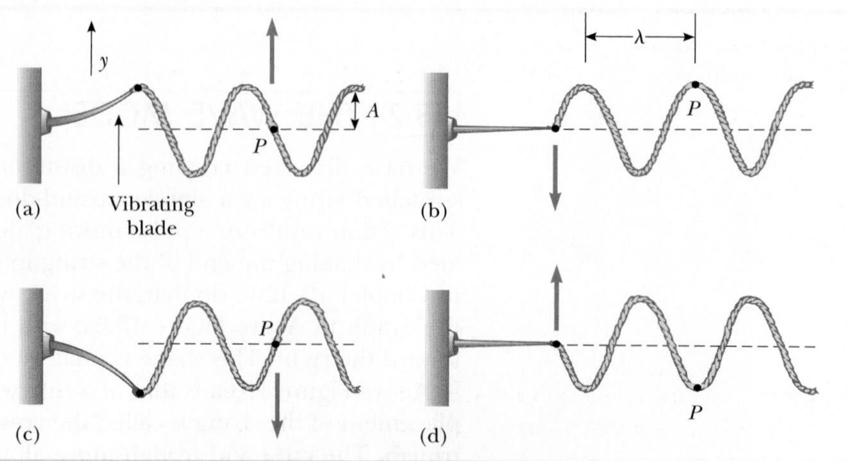

In the early chapters of this book, we developed several analysis models based on the particle model. With our introduction to waves, we can develop a new simplification model, the **wave model,** which will allow us to explore more analysis models for solving problems. An ideal particle has zero size. We can build physical objects with nonzero size as combinations of particles. Thus, the particle can be considered a basic building block. An ideal wave has a single frequency and is infinitely long; that is, the wave exists throughout the Universe. (It is beyond the mathematical scope of this text at this point to prove this fact, but a wave of finite length must necessarily have a mixture of frequencies.) We will find that we can combine ideal waves, just as we combined particles, and thus the ideal wave can be considered as a basic building block. We will explore this concept in Section 14.7.

The wave model starts with an ideal wave having a single frequency, wavelength, wave speed, and amplitude. From this beginning, we describe waves in a variety of situations that serve as analysis models to help us solve problems. In the next section, we further develop our mathematical representation for the wave model.

PITFALL PREVENTION 13.1

WHAT'S THE DIFFERENCE BETWEEN ACTIVE FIGURES 13.6a AND 13.6b? The curves in both parts of the figure are the same, but (a) is a graph of vertical position versus horizontal position and (b) is vertical position versus time. Part (a) can also be interpreted as a pictorial representation of the wave *for a series of elements of the medium,* which is what you would see at an instant of time. Part (b) is a graphical representation of the position of *one element of the medium* as a function of time.

13.3 THE TRAVELING WAVE

Let us investigate further the mathematics of a sinusoidal wave (Active Fig. 13.8). The brown curve represents a snapshot of a sinusoidal wave at $t = 0$, and the blue curve represents a snapshot of the wave at some later time t. In what follows, we will develop the principal features and mathematical representations of the model of a **traveling wave.** This analysis model is used in situations in which a wave moves through space without interacting with any other waves or particles.

At $t = 0$, the brown curve in Active Figure 13.8 can be described mathematically as

$$y = A \sin\left(\frac{2\pi}{\lambda}x\right) \quad [13.3]$$

where the amplitude A, as usual, represents the maximum value of the position of an element relative to the equilibrium position and λ is the wavelength as defined in Active Figure 13.6a. Therefore, we see that the value of y is the same when x is increased by an integral multiple of λ. If the wave moves to the right with a speed of v, the wave function at some later time t is

$$y = A \sin\left[\frac{2\pi}{\lambda}(x - vt)\right] \quad [13.4]$$

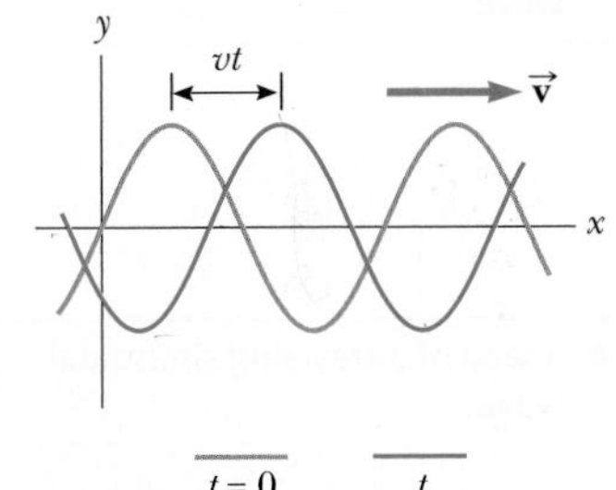

ACTIVE FIGURE 13.8

A one-dimensional sinusoidal wave traveling to the right with a speed v. The brown curve represents a snapshot of the wave at $t = 0$, and the blue curve represents a snapshot at some later time t.

PhysicsNow™ By logging into PhysicsNow at **www.pop4e.com** and going to Active Figure 13.8 you can watch the wave move and take snapshots of it at various times.

That is, the sinusoidal wave has moved to the right a distance of vt at time t as in Active Figure 13.8. Note that the wave function has the form $f(x - vt)$ and represents a wave traveling to the right. If the wave were traveling to the left, the quantity $x - vt$ would be replaced by $x + vt$, just as in the case of the traveling pulse described by Equations 13.1a and 13.1b.

Because the period T is the time interval required for the wave to travel a distance of one wavelength, the speed, wavelength, and period are related by

$$v = \frac{\lambda}{T} \quad [13.5]$$

Substituting Equation 13.5 into Equation 13.4, we find that

$$y = A \sin\left[2\pi\left(\frac{x}{\lambda} - \frac{t}{T}\right)\right] \quad [13.6]$$

This form of the wave function shows the periodic nature of y in both space and time. That is, at any given time t (a snapshot of the wave), y has the same value at the positions x, $x + \lambda$, $x + 2\lambda$, and so on. Furthermore, at any given position x (at which a single element of the medium is undergoing simple harmonic motion), the values of y at times t, $t + T$, $t + 2T$, and so on are the same.

We can express the sinusoidal wave function in a compact form by defining two other quantities: **angular wave number** k (often called simply the **wave number**) and **angular frequency** ω:

■ Angular wave number

$$k \equiv \frac{2\pi}{\lambda} \tag{13.7}$$

■ Angular frequency

$$\omega \equiv \frac{2\pi}{T} = 2\pi f \tag{13.8}$$

Note that in Equation 13.8 we use the definition of frequency, $f = 1/T$. Using these definitions, Equation 13.6 can be written in the more compact form

■ Wave function for a sinusoidal wave

$$y = A\sin(kx - \omega t) \tag{13.9}$$

We shall use this form most frequently.

Using Equations 13.7 and 13.8, we can express the wave speed v (Eq. 13.5) in the alternative forms

■ Speed of a traveling sinusoidal wave

$$v = \frac{\omega}{k} \tag{13.10}$$

$$v = \lambda f \tag{13.11}$$

The wave function given by Equation 13.9 assumes that the position y is zero at $x = 0$ and $t = 0$, but that need not be the case. If the transverse position of an element is not zero at $x = 0$ and $t = 0$, we generally express the wave function in the form

$$y = A\sin(kx - \omega t + \phi) \tag{13.12}$$

where ϕ is called the **phase constant** and can be determined from the initial conditions.

QUICK QUIZ 13.2 A sinusoidal wave of frequency f is traveling along a stretched string. The string is brought to rest and a second traveling wave of frequency $2f$ is established on the same string. **(i)** What is the wave speed of the second wave? **(a)** twice that of the first wave **(b)** half that of the first wave **(c)** the same as that of the first wave **(d)** impossible to determine **(ii)** What is the wavelength of the second wave? **(a)** twice that of the first wave **(b)** half that of the first wave **(c)** the same as that of the first wave **(d)** impossible to determine **(iii)** What is the amplitude of the second wave? **(a)** twice that of the first wave **(b)** half that of the first wave **(c)** the same as that of the first wave **(d)** impossible to determine

EXAMPLE 13.2 A Traveling Sinusoidal Wave

A sinusoidal wave traveling in the positive x direction has an amplitude of 15.0 cm, a wavelength of 40.0 cm, and a frequency of 8.00 Hz. The vertical position of an element of the medium at $t = 0$ and $x = 0$ is also 15.0 cm as shown in Figure 13.9.

A Find the angular wave number, period, angular frequency, and speed of the wave.

Solution This problem is a simple one in which we apply the traveling wave model. Using Equations 13.7,

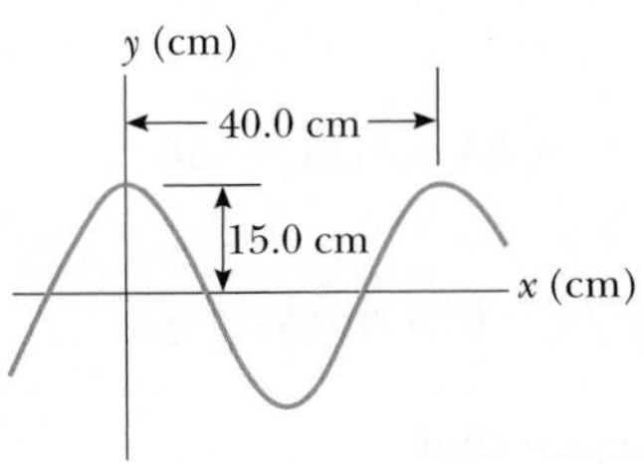

FIGURE 13.9 (Example 13.2) A sinusoidal wave of wavelength $\lambda = 40.0$ cm and amplitude $A = 15.0$ cm. The wave function can be written in the form $y = A\cos(kx - \omega t)$.

13.8, and 13.11, we find the following:

$$k = \frac{2\pi}{\lambda} = \frac{2\pi \text{ rad}}{40.0 \text{ cm}} = 0.157 \text{ rad/cm}$$

$$T = \frac{1}{f} = \frac{1}{8.00 \text{ s}^{-1}} = 0.125 \text{ s}$$

$$\omega = 2\pi f = 2\pi(8.00 \text{ s}^{-1}) = 50.3 \text{ rad/s}$$

$$v = f\lambda = (8.00 \text{ s}^{-1})(40.0 \text{ cm}) = 320 \text{ cm/s}$$

B Determine the phase constant ϕ and write a general expression for the wave function.

Solution Because $A = 15.0$ cm and because it is given that $y = 15.0$ cm at $x = 0$ and $t = 0$, substitution into Equation 13.12 gives

$$15.0 = 15.0 \sin\phi \quad \text{or} \quad \sin\phi = 1$$

We see that $\phi = \pi/2$ rad (or 90°). Hence, the wave function is of the form

$$y = A\sin\left(kx - \omega t + \frac{\pi}{2}\right) = A\cos(kx - \omega t)$$

As we can see by inspection, the wave function must have this form because the cosine argument is displaced by 90° from the sine function. Substituting the values for A, k, and ω into this expression gives

$$y = (15.0 \text{ cm}) \cos(0.157x - 50.3t)$$

The Linear Wave Equation

If the waveform at $t = 0$ is as described in Active Figure 13.7b, the wave function can be written

$$y = A\sin(kx - \omega t)$$

We can use this expression to describe the motion of any element of the string. The element at point P (or any other point on the string) moves only vertically, so its x coordinate *remains constant.* The **transverse velocity** v_y of the element and its **transverse acceleration** a_y are therefore

$$v_y = \left.\frac{dy}{dt}\right|_{x=\text{constant}} = \frac{\partial y}{\partial t} = -\omega A\cos(kx - \omega t) \qquad [13.13]$$

$$a_y = \left.\frac{dv_y}{dt}\right|_{x=\text{constant}} = \frac{\partial v_y}{\partial t} = \frac{\partial^2 y}{\partial t^2} = -\omega^2 A\sin(kx - \omega t) \qquad [13.14]$$

The maximum values of these quantities are simply the absolute values of the coefficients of the cosine and sine functions:

$$v_{y,\,\text{max}} = \omega A \qquad [13.15]$$

$$a_{y,\,\text{max}} = \omega^2 A \qquad [13.16]$$

You should recognize from Equations 13.13 and 13.14 that the transverse velocity and transverse acceleration of any element of the string do not reach their maximum values simultaneously. In fact, the transverse velocity reaches its maximum value (ωA) when position $y = 0$, whereas the transverse acceleration reaches its maximum magnitude ($\omega^2 A$) when $y = \pm A$. These relationships are due to the sine and cosine functions differing by a phase constant of $\pi/2$. Finally, note that Equations 13.15 and 13.16 are identical to the corresponding equations for simple harmonic motion (Eq. 12.17 and Eq. 12.18).

Let us take derivatives of our wave function with respect to position at a fixed time, similar to the process by which we took derivatives with respect to time in

PITFALL PREVENTION 13.2

Two kinds of speed/velocity Be sure to differentiate between v, the speed of the wave as it propagates through the medium, and v_y, the transverse velocity of an element of the string. The speed v is constant for a uniform medium, whereas v_y varies sinusoidally.

Equations 13.13 and 13.14:

$$\left.\frac{dy}{dx}\right|_{t=\text{constant}} = \frac{\partial y}{\partial x} = -kA\cos(kx - \omega t) \qquad [13.17]$$

$$\left.\frac{d^2y}{dx^2}\right|_{t=\text{constant}} = \frac{\partial^2 y}{\partial x^2} = -k^2A\sin(kx - \omega t) \qquad [13.18]$$

Comparing Equations 13.14 and 13.18, we see that

$$A\sin(kx - \omega t) = -\frac{1}{k^2}\frac{\partial^2 y}{\partial x^2} = -\frac{1}{\omega^2}\frac{\partial^2 y}{\partial t^2} \quad \rightarrow \quad \frac{\partial^2 y}{\partial x^2} = \frac{k^2}{\omega^2}\frac{\partial^2 y}{\partial t^2}$$

Using Equation 13.10, we can rewrite this expression as

■ Linear wave equation

$$\frac{\partial^2 y}{\partial x^2} = \frac{1}{v^2}\frac{\partial^2 y}{\partial t^2} \qquad [13.19]$$

which is known as the **linear wave equation.** If we analyze a situation and find this kind of relationship between derivatives of a function describing the situation, wave motion is occurring. Equation 13.19 is a differential equation representation of the traveling wave model. The solutions to the equation describe **linear mechanical waves.** We have developed the linear wave equation from a sinusoidal mechanical wave traveling through a medium, but it is much more general. The linear wave equation successfully describes waves on strings, sound waves, and also electromagnetic waves.[2] What's more, although the sinusoidal wave that we have studied is a solution to Equation 13.19, the general solution to the equation is *any* function of the form $y(x, t) = f(x \pm vt)$ as discussed in Section 13.1.

Nonlinear waves are more difficult to analyze, but they are an important area of current research, especially in optics. An example of a nonlinear mechanical wave is one for which the amplitude is not small compared with the wavelength.

EXAMPLE 13.3 A Solution to the Linear Wave Equation

Verify that the wave function presented in Example 13.1 is a solution to the linear wave equation.

Solution The wave function is

$$y(x, t) = \frac{2.0}{(x - 3.0t)^2 + 1}$$

By taking partial derivatives of this function with respect to x and to t, we find that

$$\frac{\partial^2 y}{\partial x^2} = \frac{12(x - 3.0t)^2 - 4.0}{[(x - 3.0t)^2 + 1]^3}$$

$$\frac{\partial^2 y}{\partial t^2} = \frac{108(x - 3.0t)^2 - 36}{[(x - 3.0t)^2 + 1]^3} = 9.0\,\frac{[12(x - 3.0t)^2 - 4.0]}{[(x - 3.0t)^2 + 1]^3}$$

Comparing these two expressions, we see that

$$\frac{\partial^2 y}{\partial x^2} = \frac{1}{9.0}\frac{\partial^2 y}{\partial t^2}$$

Comparing this result with Equation 13.19, we see that the wave function is a solution to the linear wave equation if the speed at which the pulse moves is 3.0 cm/s. We have already determined in Example 13.1 that this speed is indeed the speed of the pulse, so we have proven what we set out to do.

13.4 THE SPEED OF TRANSVERSE WAVES ON STRINGS

An aspect of the behavior of linear mechanical waves is that **the wave speed depends only on the properties of the medium through which the wave travels.** Waves for which the amplitude A is small relative to the wavelength λ are well represented

[2]In the case of electromagnetic waves, y is interpreted to represent an electric field, which we will study in Chapter 24.

as linear waves. In this section, we determine the speed of a transverse wave traveling on a stretched string.

Let us use a mechanical analysis to derive the expression for the speed of a pulse traveling on a stretched string under tension T. Consider a pulse moving to the right with a uniform speed v, measured relative to a stationary (with respect to the Earth) inertial reference frame. Recall from Chapter 9 that Newton's laws are valid in any inertial reference frame. Therefore, let us view this pulse from a different inertial reference frame, one that moves along with the pulse at the same speed so that the pulse appears to be at rest in the frame as in Figure 13.10a. In this reference frame, the pulse remains fixed and each element of the string moves to the left through the pulse shape.

A short element of the string, of length Δs, forms an arc with a radius of curvature R as shown in Figure 13.10a and magnified in Figure 13.10b. We use a simplification model in which this arc is an arc of a perfect circle. In our moving frame of reference, the element of the string moves to the left with speed v through the arc. As it travels through the arc, we can model the element as a particle in uniform circular motion. This element has a centripetal acceleration of v^2/R, which is supplied by components of the force $\vec{\mathbf{T}}$ in the string at each end of the element. The force $\vec{\mathbf{T}}$ acts on each side of the element, tangent to the arc, as in Figure 13.10b. The horizontal components of $\vec{\mathbf{T}}$ cancel, and each vertical component $T \sin \theta$ acts radially inward toward the center of the arc. Hence, the magnitude of the total radial force on the element is $2T \sin \theta$. Because the element is small, θ is small and we can use the small-angle approximation $\sin \theta \approx \theta$. Therefore, the magnitude of the total radial force can be expressed as

$$F_r = 2T \sin \theta \approx 2T\theta$$

The element has mass $m = \mu \Delta s$, where μ is the mass per unit length of the string. Because the element forms part of a circle and subtends an angle of 2θ at the center, $\Delta s = R(2\theta)$, and hence

$$m = \mu \, \Delta s = 2\mu R\theta$$

The radial component of Newton's second law applied to the element gives

$$F_r = \frac{mv^2}{R} \quad \rightarrow \quad 2T\theta = \frac{2\mu R\theta v^2}{R} \quad \rightarrow \quad T = \mu v^2$$

where F_r is the force that supplies the centripetal acceleration of the element. Solving for v gives

$$v = \sqrt{\frac{T}{\mu}} \qquad [13.20]$$

■ Speed of a wave on a stretched string

Notice that this derivation is based on the linear wave assumption that the pulse height is small relative to the length of the pulse. Using this assumption, we were able to use the approximation $\sin \theta \approx \theta$. Furthermore, the model assumes that the tension T is not affected by the presence of the pulse, so T is the same at all points on the string. Finally, this proof does *not* assume any particular shape for the pulse. We therefore conclude that a pulse of *any shape* will travel on the string with speed $v = \sqrt{T/\mu}$, without changing its shape.

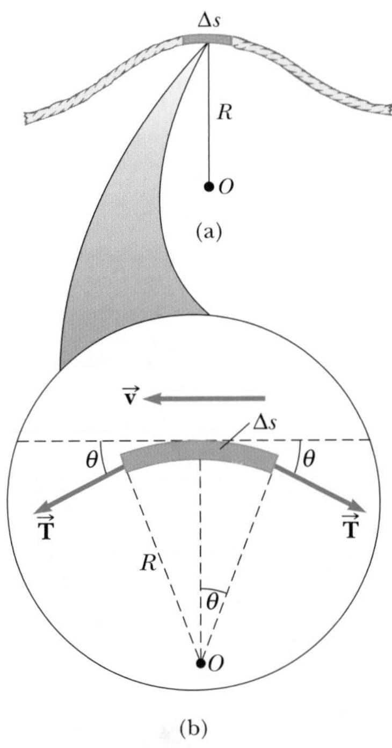

FIGURE 13.10 (a) To obtain the speed v of a wave on a stretched string, it is convenient to describe the motion of a small element of the string in a moving frame of reference. (b) The horizontal components of the force $\vec{\mathbf{T}}$ on a small element of length Δs cancel. The radial components add, so there is a net force in the radial direction.

PITFALL PREVENTION 13.3

Multiple T**'s** Be careful not to confuse the T for the magnitude of the tension in this discussion with the T we are using in this chapter for the period of a wave. The context of the equation should help you to identify which one it is. The alphabet simply doesn't have enough letters to allow us to assign a unique letter to each variable!

QUICK QUIZ 13.3 Suppose you create a pulse by moving the free end of a taut string up and down once with your hand. The string is attached at its other end to a distant wall. The pulse reaches the wall in a time interval Δt. Which of the following actions, taken by itself, decreases the time interval required for the pulse to reach the wall? More than one choice may be correct. **(a)** Moving your hand more quickly, but still only up and down once by the same amount. **(b)** Moving your hand more slowly, but still only up and down once by the same amount. **(c)** Moving your hand a greater distance up and down in the same amount of time. **(d)** Moving your hand a smaller distance up and down in the same amount of time. **(e)** Using a heavier string of the same length and under the same tension. **(f)** Using a lighter string of the same length and under the same tension. **(g)** Using a string of the same linear mass density but under decreased tension. **(h)** Using a string of the same linear mass density but under increased tension.

■ Thinking Physics 13.1

A secret agent is trapped in a building on top of an elevator car at a lower floor. He attempts to signal a fellow agent on the roof by tapping a message in Morse code on the elevator cable so that transverse pulses move upward on the cable. As the pulses move up the cable toward the accomplice, does the speed with which they move stay the same, increase, or decrease? If the pulses are sent 1 s apart, are they received 1 s apart by the agent on the roof?

Reasoning The elevator cable can be modeled as a vertical string. The speed of waves on the cable is a function of the tension in the cable. As the waves move higher on the cable, they encounter increased tension because each higher point on the cable must support the weight of all the cable below it (and the elevator). Therefore, the speed of the pulses increases as they move higher on the cable. The frequency of the pulses will not be affected because each pulse takes the same time interval to reach the top. They will still arrive at the top of the cable at intervals of 1 s. ■

EXAMPLE 13.4 The Speed of a Pulse on a Cord

A uniform cord has a mass of 0.300 kg and a total length of 6.00 m. Tension is maintained in the cord by suspending an object of mass 2.00 kg from one end (Fig. 13.11). Find the speed of a pulse on this cord.

Solution Because the suspended object can be modeled

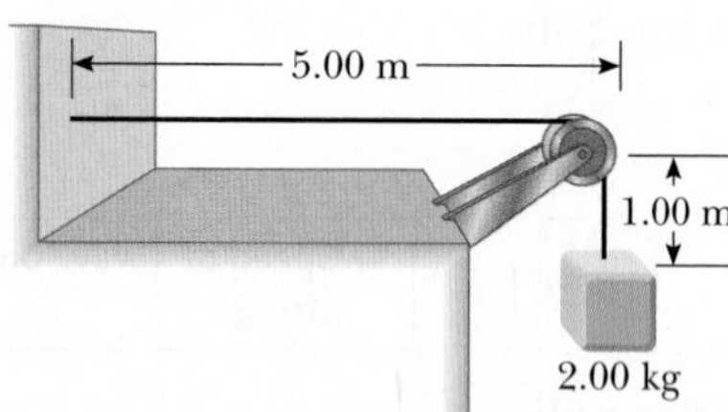

FIGURE 13.11 (Example 13.4) The tension T in the cord is maintained by the suspended object. The wave speed is given by the expression $v = \sqrt{T/\mu}$.

as a particle in equilibrium, the tension T in the cord is equal to the weight of the suspended 2.00-kg object:

$$T = mg = (2.00 \text{ kg})(9.80 \text{ m/s}^2) = 19.6 \text{ N}$$

(This calculation of the tension neglects the small mass of the cord. Strictly speaking, the horizontal portion of the cord can never be exactly straight—it will sag slightly—and therefore the tension is not uniform.)

The mass per unit length μ is

$$\mu = \frac{m}{\ell} = \frac{0.300 \text{ kg}}{6.00 \text{ m}} = 0.050\,0 \text{ kg/m}$$

Therefore, the wave speed is

$$v = \sqrt{\frac{T}{\mu}} = \sqrt{\frac{19.6 \text{ N}}{0.050\,0 \text{ kg/m}}} = 19.8 \text{ m/s}$$

INTERACTIVE EXAMPLE 13.5 Rescuing the Hiker

An 80.0-kg hiker is trapped on a mountain ledge following a storm. A helicopter rescues the hiker by hovering above him and lowering a cable to him. The mass of the cable is 8.00 kg and its length is 15.0 m. A chair of mass 70.0 kg is attached to the end of the cable. The hiker attaches himself to the chair and the helicopter then accelerates upward. Terrified by hanging from the cable in midair, the hiker tries to signal the pilot by sending transverse pulses up the cable. A pulse takes 0.250 s to travel the length of the cable. What is the acceleration of the helicopter?

Solution To conceptualize this problem, imagine the effect of the helicopter's acceleration on the cable. The higher the upward acceleration, the larger the tension in the cable. In turn, the larger the tension, the higher the speed of pulses on the cable. Therefore, we categorize this problem as a combination of one involving Newton's laws and one involving the speed of pulses on a string. To analyze the problem, we use the time interval for the pulse to travel from the hiker to the helicopter to find the speed of the pulses on the cable:

$$v = \frac{\Delta x}{\Delta t} = \frac{15.0 \text{ m}}{0.250 \text{ s}} = 60.0 \text{ m/s}$$

The speed of pulses on the cable is given by Equation 13.20, which allows us to find the tension in the cable:

$$v = \sqrt{\frac{T}{\mu}} \quad \rightarrow \quad T = \mu v^2 = \left(\frac{8.00 \text{ kg}}{15.0 \text{ m}}\right)(60.0 \text{ m/s})^2$$

$$= 1.92 \times 10^3 \text{ N}$$

Newton's second law relates the tension in the cable to the acceleration of the hiker and the chair, which is the same as the acceleration of the helicopter (we ignore the mass of the cable relative to that of the hiker and the chair):

$$\sum F = ma \quad \rightarrow \quad T - mg = ma$$

$$a = \frac{T}{m} - g = \frac{1.92 \times 10^3 \text{ N}}{150.0 \text{ kg}} - 9.80 \text{ m/s}^2$$

$$= 3.00 \text{ m/s}^2$$

To finalize this problem, note that a real cable has stiffness in addition to tension. Stiffness tends to return a cable or a wire to its original straight-line shape even when it is not under tension. For example, a piano wire, which has stiffness, will straighten if released from a curved shape, whereas normal package wrapping string will not.

Stiffness represents a restoring force in addition to tension, which tends to increase the speed of waves on the cable over that due to tension alone. Consequently, for a real cable, the speed of 60.0 m/s that we determined is most likely associated with a tension lower than 1.92×10^3 N and a correspondingly smaller acceleration of the helicopter.

PhysicsNow™ Investigate the rescue situation by logging into PhysicsNow at **www.pop4e.com** and going to Interactive Example 13.5.

13.5 REFLECTION AND TRANSMISSION OF WAVES

So far, we have only considered a wave traveling through a medium with no changes in the medium and no interactions with anything other than the elements of the medium. This model is the traveling wave model. This situation is similar to a particle traveling through empty space and obeying Newton's first law. Although these situations demonstrate important physics, things become more interesting when particles and waves interact with something. Let us see what happens when a wave encounters a boundary between two media.

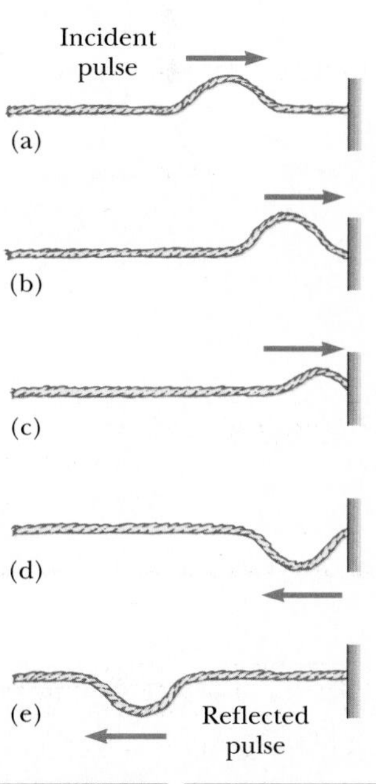

FIGURE 13.12 The reflection of a traveling pulse at the fixed end of a stretched string. The reflected pulse is inverted, but its shape remains the same.

For simplicity, consider a single pulse once again. When a traveling pulse reaches a boundary, part or all of the pulse is *reflected*. Any part not reflected is said to be *transmitted* through the boundary. Suppose a pulse travels on a string that is fixed at one end (Fig. 13.12). When the pulse reaches the fixed boundary, it is reflected. In the simplification model in which the support attaching the string to the wall is rigid, none of the pulse is transmitted through the fixed end.

Note that the reflected pulse (Figs. 13.12d and 13.12e) has exactly the same amplitude as the incoming pulse but is inverted. The inversion can be explained as follows. The pulse is created initially with an upward and then downward force on the free end of the string. As the pulse arrives at the fixed end of the string, the string first produces an upward force on the support. By Newton's third law, the support exerts a reaction force in the opposite direction on the string. Therefore, the positive shape of the pulse results in a downward and then upward force on the string as the entirety of the pulse encounters the rigid end. This situation is equivalent to a person replacing the fixed support and applying a downward and then an upward force to the string. Therefore, reflection at a rigid end causes the pulse to invert on reflection.

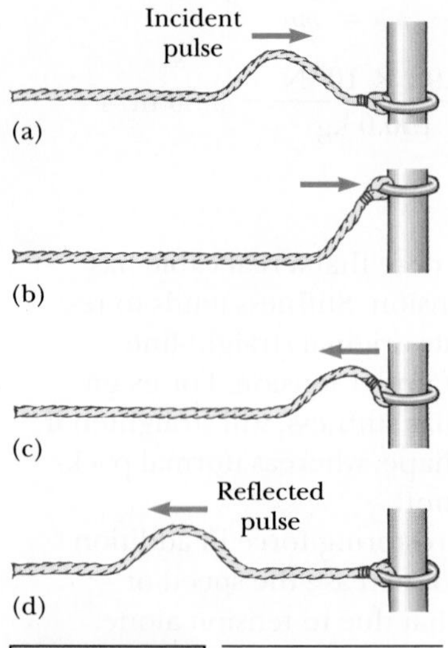

FIGURE 13.13 The reflection of a traveling pulse at the free end of a stretched string. In this case, the reflected pulse is not inverted.

Now consider a second idealized situation in which reflection is total and transmission is zero. In this simplification model, the pulse arrives at the end of a string that is perfectly free to move vertically, as in Figure 13.13. The tension at the free end is maintained by tying the string to a ring of negligible mass that is free to slide vertically on a frictionless post. Again, the pulse is reflected, but this time it is not inverted. As the pulse reaches the post, it exerts a force on the free end, causing the ring to accelerate upward. In the process, the ring reaches the top of its motion and is then returned to its original position by the downward component of the tension force. Therefore, the ring experiences the same motion as if it were raised and lowered by hand. This motion produces a reflected pulse that is not inverted and whose amplitude is the same as that of the incoming pulse.

Finally, in some situations the boundary is intermediate between these two extreme cases; that is, it is neither completely rigid nor completely free. In that case, part of the wave is transmitted and part is reflected. For instance, suppose a string is attached to a denser string as in Active Figure 13.14. When a pulse traveling on the first string reaches the boundary between the two strings, part of the pulse is reflected and inverted and part is transmitted to the denser string. Both the reflected and transmitted pulses have smaller amplitude than the incident pulse. The inversion in the reflected pulse is similar to the behavior of a pulse meeting a rigid boundary. As the pulse travels from the initial string to the denser string, the junction acts more like a rigid end than a free end. Therefore, the reflected pulse is inverted.

When a pulse traveling on a dense string strikes the boundary of a less dense string, as in Active Figure 13.15, again part is reflected and part transmitted. This time, however, the reflected pulse is not inverted. As the pulse travels from the dense string to the less dense one, the junction acts more like a free end than a rigid end.

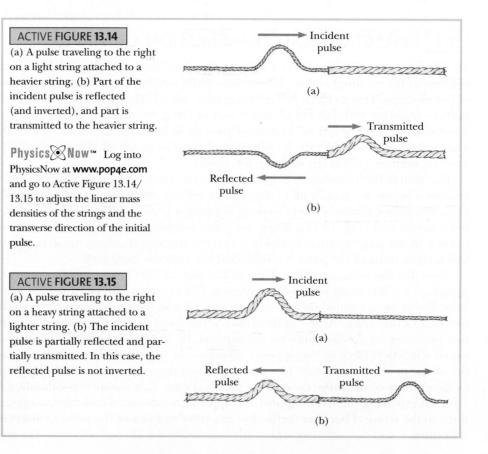

ACTIVE FIGURE 13.14 (a) A pulse traveling to the right on a light string attached to a heavier string. (b) Part of the incident pulse is reflected (and inverted), and part is transmitted to the heavier string.

Physics Now™ Log into PhysicsNow at **www.pop4e.com** and go to Active Figure 13.14/13.15 to adjust the linear mass densities of the strings and the transverse direction of the initial pulse.

Incident pulse
(a)
Transmitted pulse
Reflected pulse
(b)

ACTIVE FIGURE 13.15 (a) A pulse traveling to the right on a heavy string attached to a lighter string. (b) The incident pulse is partially reflected and partially transmitted. In this case, the reflected pulse is not inverted.

Incident pulse
(a)
Reflected pulse
Transmitted pulse
(b)

The limiting value between the two cases in Figures 13.14 and 13.15 would be that in which both strings have the same linear mass density. In this case, no boundary exists between the two media. Both strings are identical. As a result, no reflection occurs and transmission is total.

In the preceding section, we found that the speed of a wave on a string increases as the mass per unit length of the string decreases. In other words, a pulse travels more slowly on a dense string than on a less dense one if both are under the same tension. This comparison is illustrated by the lengths of the red velocity vectors in Figures 13.14 and 13.15.

This discussion has focused on pulses arriving at a boundary. If a sinusoidal wave on a string arrives at a rigid end, the inversion of the waveform is equivalent to shifting the entire wave by half a wavelength. This equivalence can be seen by looking back at Active Figure 13.7. The wave in Active Figure 13.7d is the inversion of the wave in Active Figure 13.7b. Notice, however, that Active Figure 13.7b would look like Active Figure 13.7d if the wave were shifted to the right or left by half a wavelength. Therefore, because a full wavelength can be associated with an angle of 360°, we describe the inversion of a wave at a rigid end as a **180° phase shift.** We will see this effect again in Chapter 27 when we discuss reflection of light waves from materials.

13.6 RATE OF ENERGY TRANSFER BY SINUSOIDAL WAVES ON STRINGS

As waves propagate through a medium, they transport energy. This fact is easily demonstrated by hanging an object on a stretched string and sending a pulse down the string as in Figure 13.16. When the pulse meets the suspended object, the object is momentarily displaced as in Figure 13.16b. In the process, energy is transferred to the object because work must be done in moving it upward. This section examines the rate at which energy is transferred along a string. We shall assume a one-dimensional sinusoidal wave in the calculation of the energy transferred.

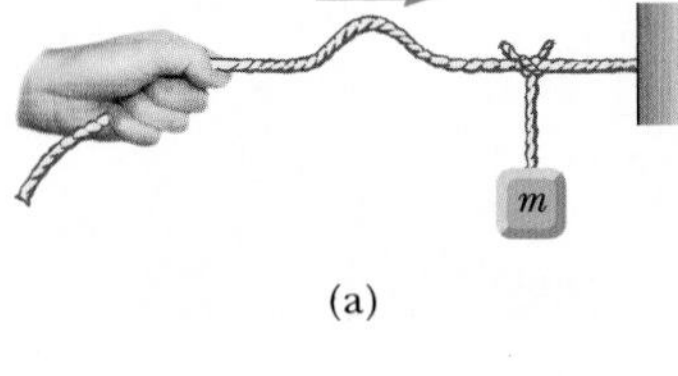

(a)

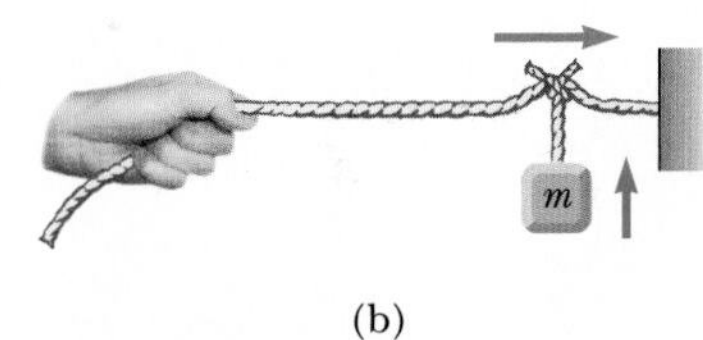

(b)

FIGURE 13.16 (a) A pulse traveling to the right on a stretched string on which an object has been suspended. (b) Energy is transmitted to the suspended object when the pulse arrives.

Consider a sinusoidal wave traveling on a string (Fig. 13.17). The source of the energy is some external agent at the left end of the string, which does work in producing the oscillations. We can consider the string to be a nonisolated system. As the external agent performs work on the end of the string, moving it up and down, energy enters the system of the string and propagates along its length. Let us focus our attention on an element of the string of length Δx and mass Δm. Each such element moves vertically with simple harmonic motion. Therefore, we can model each element of the string as a simple harmonic oscillator, with the oscillation in the y direction. All elements have the same angular frequency ω and the same amplitude A. The kinetic energy K associated with a particle in simple harmonic motion is $K = \frac{1}{2}mv^2$, where v varies sinusoidally during the oscillation. If we apply this equation to an element of length Δx, we see that the kinetic energy ΔK of this element is

$$\Delta K = \tfrac{1}{2}(\Delta m)\, v_y^{\,2}$$

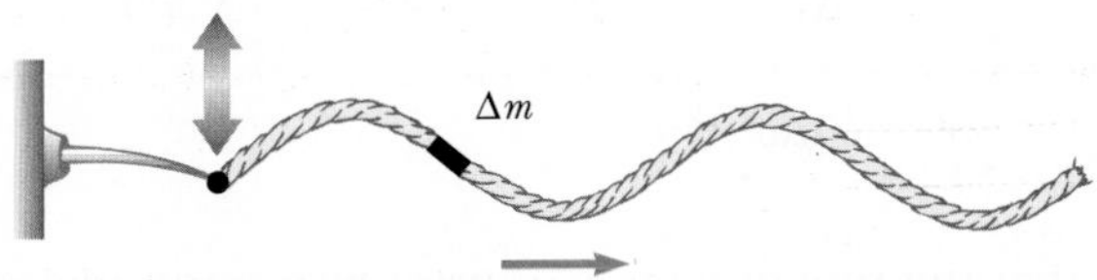

FIGURE 13.17 A sinusoidal wave traveling along the x axis on a stretched string. Every element, such as the one labeled with its mass Δm, moves vertically, and each element has the same total energy. The average power transmitted by the wave equals the energy contained in one wavelength divided by the period of the wave.

If μ is the mass per unit length of the string, the element of length Δx has a mass Δm that is equal to $\mu\,\Delta x$. Hence, we can express the kinetic energy of an element of the string as

$$\Delta K = \tfrac{1}{2}(\mu\,\Delta x)\,v_y^2 \qquad [13.21]$$

As the length of the element of the string shrinks to zero, this expression becomes a differential relationship:

$$dK = \tfrac{1}{2}(\mu\,dx)\,v_y^2$$

We substitute for the general velocity of an element of the string using Equation 13.13:

$$\begin{aligned} dK &= \tfrac{1}{2}\mu[\omega A\cos(kx - \omega t)]^2\,dx \\ &= \tfrac{1}{2}\mu\omega^2A^2\cos^2(kx - \omega t)\,dx \end{aligned}$$

If we take a snapshot of the wave at time $t = 0$, the kinetic energy of a given element is

$$dK = \tfrac{1}{2}\mu\omega^2A^2\cos^2 kx\,dx$$

Let us integrate this expression over all the string elements in a wavelength of the wave, which will give us the kinetic energy in one wavelength:

$$\begin{aligned} K_\lambda &= \int_0^\lambda \tfrac{1}{2}\mu\omega^2A^2\cos^2 kx\,dx = \tfrac{1}{2}\mu\omega^2A^2\int_0^\lambda \cos^2 kx\,dx \\ &= \tfrac{1}{2}\mu\omega^2A^2\left[\tfrac{1}{2}x + \frac{1}{4k}\sin 2kx\right]_0^\lambda = \tfrac{1}{2}\mu\omega^2A^2\left[\tfrac{1}{2}\lambda\right] = \tfrac{1}{4}\mu\omega^2A^2\lambda \end{aligned}$$

In addition to this kinetic energy, there is potential energy associated with each element of the string due to its displacement from the equilibrium position. A similar analysis as that above for the total potential energy in a wavelength gives the same result:

$$U_\lambda = \tfrac{1}{4}\mu\omega^2A^2\lambda$$

The total energy in one wavelength of the wave is the sum of the kinetic and potential energies:

$$E_\lambda = K_\lambda + U_\lambda = \tfrac{1}{2}\mu\omega^2A^2\lambda \qquad [13.22]$$

As the wave moves along the string, this amount of energy passes by a given point on the string during one period of the oscillation. Therefore, the **power,** or rate of energy transfer, associated with the wave is

$$\mathcal{P} = \frac{E_\lambda}{\Delta t} = \frac{\tfrac{1}{2}\mu\omega^2A^2\lambda}{T} = \tfrac{1}{2}\mu\omega^2A^2\left(\frac{\lambda}{T}\right)$$

■ Rate of energy transfer for a wave

$$\mathcal{P} = \tfrac{1}{2}\mu\omega^2A^2v \qquad [13.23]$$

This result shows that the rate of energy transfer by a sinusoidal wave on a string is proportional to (a) the square of the angular frequency, (b) the square of the amplitude, and (c) the wave speed. In fact, *all* sinusoidal waves have the following general property: **The rate of energy transfer in any sinusoidal wave is proportional to the square of the angular frequency and to the square of the amplitude.**

QUICK QUIZ 13.4 Which of the following, taken by itself, would be most effective in increasing the rate at which energy is transferred by a wave traveling along a string? **(a)** reducing the linear mass density of the string by one half **(b)** doubling the wavelength of the wave **(c)** doubling the tension in the string **(d)** doubling the amplitude of the wave

EXAMPLE 13.6 **Power Supplied to a Vibrating String**

A string having a linear mass density of $\mu = 5.00 \times 10^{-2}$ kg/m is under a tension of 80.0 N. How much power must be supplied to the string to generate sinusoidal waves at a frequency of 60.0 Hz and an amplitude of 6.00 cm?

Solution The wave speed on the string is

$$v = \sqrt{\frac{T}{\mu}} = \left(\frac{80.0\text{ N}}{5.00 \times 10^{-2}\text{ kg/m}}\right)^{1/2} = 40.0\text{ m/s}$$

Because $f = 60.0$ Hz, the angular frequency ω of the sinusoidal waves on the string has the value

$$\omega = 2\pi f = 2\pi(60.0\text{ Hz}) = 377\text{ s}^{-1}$$

Using these values in Equation 13.23 for the power, with $A = 6.00 \times 10^{-2}$ m, gives

$$\begin{aligned}\mathcal{P} &= \tfrac{1}{2}\mu\omega^2A^2v \\ &= \tfrac{1}{2}(5.00 \times 10^{-2}\text{ kg/m})(377\text{ s}^{-1})^2 \\ &\quad\times (6.00 \times 10^{-2}\text{ m})^2(40.0\text{ m/s}) = 512\text{ W}\end{aligned}$$

13.7 SOUND WAVES

Let us turn our attention from transverse waves to longitudinal waves. As stated in Section 13.2, for longitudinal waves the elements of the medium undergo displacements parallel to the direction of wave motion. Sound waves in air are the most important examples of longitudinal waves. Sound waves can travel through any material medium, however, and their speed depends on the properties of that medium. Table 13.1 provides examples of the speed of sound in different media.

The displacements accompanying a sound wave in air are longitudinal displacements of small elements of air from their equilibrium positions. Such displacements result if the source of the waves, such as the diaphragm of a loudspeaker, oscillates in air. If the oscillation of the diaphragm is described by simple harmonic motion, a sinusoidal sound wave propagates away from the loudspeaker. For instance, a one-dimensional sound wave can be produced in a long, narrow tube containing a gas by means of a vibrating piston at one end, as in Figure 13.18.

It is difficult to draw a pictorial representation of longitudinal waves because the displacements of the elements of the medium are in the same direction as that of the propagation of the wave. Figure 13.18 is one way to represent these types of waves. The darker color in the figure represents a region where the gas is compressed; consequently, the density and pressure are *above* their equilibrium values. Such a compressed region of gas, called a **compression,** is formed when the piston is being pushed into the tube. The compression moves along the tube, continuously compressing the layers in front of it. When the piston is withdrawn from the tube, the gas in front of it expands, and consequently the pressure and density in this region fall below their equilibrium values. These low-pressure regions, called **rarefactions,** are represented by the lighter areas in Figure 13.18. The rarefactions also propagate along the tube, following the compressions. Both regions move with a speed equal to the speed of sound in that medium.

As the piston oscillates back and forth in a sinusoidal fashion, regions of compression and rarefaction are continuously set up. The distance between two successive compressions (or two successive rarefactions) equals the wavelength λ. As these regions travel along the tube, any small element of the medium moves with simple harmonic motion parallel to the direction of the wave (in other words,

TABLE 13.1

Speed of Sound in Various Media

Medium	v (m/s)
Gases	
Hydrogen (0°C)	1 286
Helium (0°C)	972
Air (20°C)	343
Air (0°C)	331
Oxygen (0°C)	317
Liquids at 25°C	
Glycerol	1 904
Sea water	1 533
Water	1 493
Mercury	1 450
Kerosene	1 324
Methyl alcohol	1 143
Carbon tetrachloride	926
Solids[a]	
Pyrex glass	5 640
Iron	5 950
Aluminum	6 420
Brass	4 700
Copper	5 010
Gold	3 240
Lucite	2 680
Lead	1 960
Rubber	1 600

[a]Values given are for propagation of longitudinal waves in bulk media. Speeds for longitudinal waves in thin rods are smaller, and speeds of transverse waves in bulk are smaller yet.

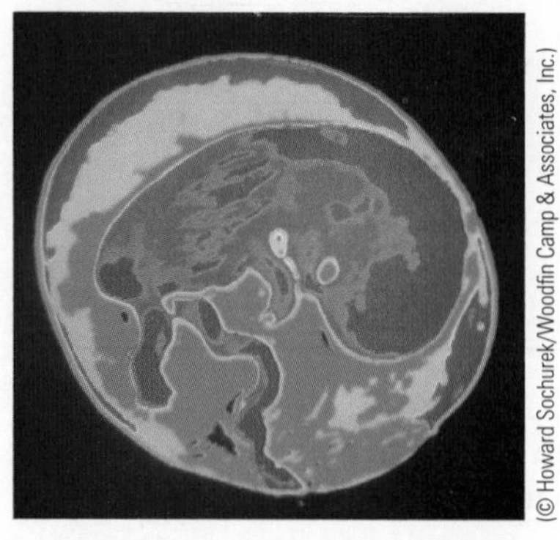

An ultrasound image showing a young human fetus and umbilical cord. Ultrasound refers to sound waves that are higher in frequency than those audible to humans. The sound waves transmit through the body of the mother and reflect from the skin of the fetus. The reflected sound waves are organized by the electronics of the ultrasonic imaging system into a visual image of the fetus. ■

longitudinally). If $s(x, t)$ is the position of a small element measured relative to its equilibrium position, we can express this position function as

$$s(x, t) = s_{max} \sin(kx - \omega t) \qquad [13.24]$$

where s_{max} is the **maximum position relative to equilibrium,** often called the **displacement amplitude.** Equation 13.24 represents the **displacement wave,** where k is the wave number and ω is the angular frequency of the piston. The variation ΔP in the pressure[3] of the gas measured from its equilibrium value is also sinusoidal; it is given by

$$\Delta P = \Delta P_{max} \cos(kx - \omega t) \qquad [13.25]$$

The pressure amplitude ΔP_{max} is the maximum change in pressure from the equilibrium value, and Equation 13.25 represents the **pressure wave.** The pressure amplitude is proportional to the displacement amplitude s_{max}:

$$\Delta P_{max} = \rho v \omega s_{max} \qquad [13.26]$$

where ρ is the density of the medium, v is the wave speed, and ωs_{max} is the maximum longitudinal speed of an element of the medium. It is these pressure variations in a sound wave that result in an oscillating force on the eardrum, leading to the sensation of hearing.

Therefore, we see that a sound wave may be considered as either a displacement wave or a pressure wave. A comparison of Equations 13.24 and 13.25 shows that **the pressure wave is 90° out of phase with the displacement wave.** Graphs of these functions are shown in Figure 13.19. Note that the change in pressure from equilibrium is a maximum when the displacement is zero, whereas the displacement is a maximum when the pressure change is zero.

Note that Figure 13.19 presents two graphical representations of the longitudinal wave: one for position of the elements of the medium and the other for pressure variation. They are *not* pictorial representations for longitudinal waves, however. For transverse waves, the element displacement is perpendicular to the direction of propagation and the pictorial and graphical representations look the same because the perpendicularity of the oscillations and propagation is matched by the perpendicularity of x and y axes. For longitudinal waves, the oscillations and propagation exhibit no perpendicularity, so those pictorial representations look like Figure 13.18.

The speed of a sound wave in air depends only on the temperature of the air. For a small range of temperatures around room temperature, the speed of sound is described by

$$v = 331 \text{ m/s} + (0.6 \text{ m/s}\cdot{}^\circ\text{C})\, T_C \qquad [13.27]$$

where T_C is the temperature in degrees Celsius and the speed of sound at 0°C is 331 m/s.

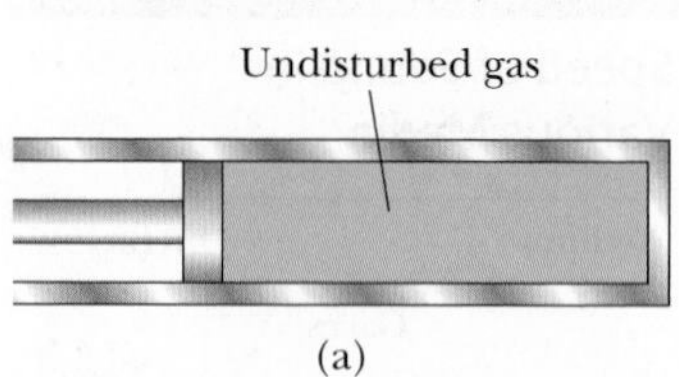

(a)

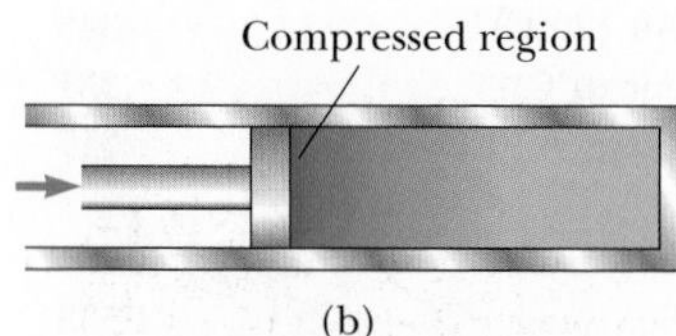

(b)

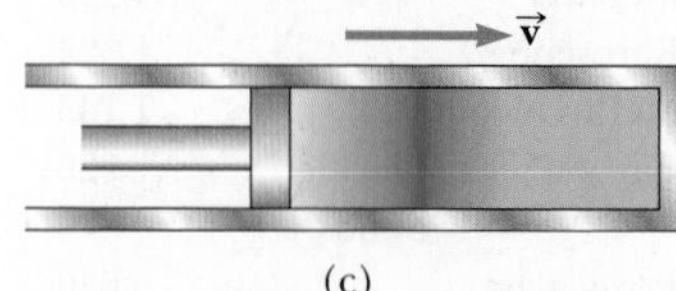

(c)

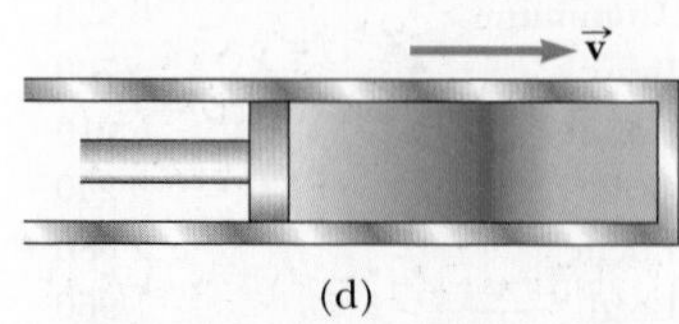

(d)

FIGURE 13.18 A longitudinal wave propagating along a tube filled with a compressible gas. The source of the wave is a vibrating piston at the left. The high- and low-pressure regions are dark and light, respectively.

■ Thinking Physics 13.2

Why does thunder produce an extended "rolling" sound when its source, a lightning strike, occurs in a fraction of a second? How does lightning produce thunder in the first place?

Reasoning Let us assume that we are at ground level and ignore ground reflections. When cloud-to-ground lightning strikes, a channel of ionized air carries a very large electric current from the cloud to the ground. (We will study electric

[3] We will formally introduce pressure in Chapter 15. In the case of longitudinal waves in a gas, each compressed area is a region of higher-than-average pressure and density, and each stretched region is a region of lower-than-average pressure and density.

current in Chapter 21.) The result is a very rapid temperature increase of this channel of air as it carries the current. The temperature increase causes a sudden expansion of the air. This expansion is so sudden and so intense that a tremendous disturbance is produced in the air: thunder. The thunder rolls because the lightning channel is a long, extended source; the entire length of the channel produces the sound at essentially the same instant of time. Sound produced at the end of the channel nearest you reaches you first, but sounds from progressively farther portions of the channel reach you shortly thereafter. If the lightning channel were a perfectly straight line, the resulting sound might be a steady roar, but the zigzagged shape of the path results in the rolling variation in loudness. ■

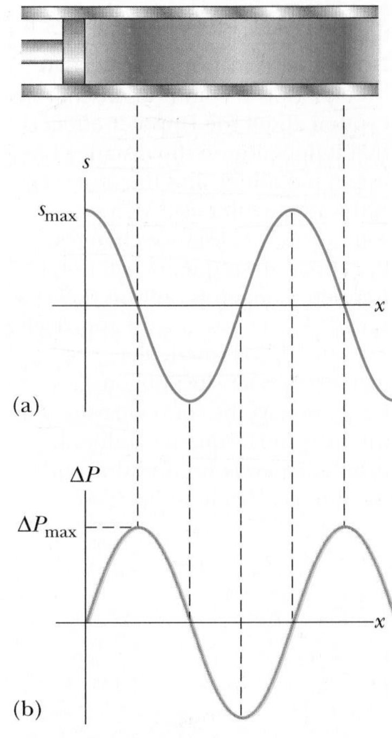

FIGURE 13.19 (a) Displacement versus position and (b) pressure versus position for a sinusoidal longitudinal wave. The displacement wave is 90° out of phase with the pressure wave.

13.8 THE DOPPLER EFFECT

When someone honks the horn of a vehicle as it travels along a highway, the frequency of the sound you hear is higher as the vehicle approaches you than it is as the vehicle moves away from you. This change is one example of the **Doppler effect,** named after Christian Johann Doppler (1803–1853), an Austrian physicist.

The Doppler effect for sound is experienced whenever there is relative motion between the source of sound and the observer. Motion of the source or observer toward the other results in the observer's hearing a frequency that is higher than the true frequency of the source. Motion of the source or observer away from the other results in the observer hearing a frequency that is lower than the true frequency of the source.

Although we shall restrict our attention to the Doppler effect for sound waves, it is associated with waves of all types. The Doppler effect for electromagnetic waves is used in police radar systems to measure the speeds of motor vehicles. Likewise, astronomers use the effect to determine the relative motions of stars, galaxies, and other celestial objects. In 1842, Doppler first reported the frequency shift in connection with light emitted by two stars revolving about each other in double-star systems. In the early 20th century, the Doppler effect for light from galaxies was used to argue for the expansion of the Universe, which led to the Big Bang theory, discussed in Chapter 31.

To see what causes this apparent frequency change, imagine you are in a boat lying at anchor on a gentle sea where the waves have a period of $T = 2.0$ s. Thus, every 2.0 s a crest hits your boat. Figure 13.20a shows this situation with the water waves moving toward the left. If you start a stopwatch at $t = 0$ just as one crest hits, the stopwatch reads 2.0 s when the next crest hits, 4.0 s when the third crest hits, and so on. From these observations you conclude that the wave frequency is $f = 1/T = 0.50$ Hz. Now suppose you start your motor and head directly into the oncoming waves as shown in Figure 13.20b. Again you set your stopwatch to $t = 0$ as a crest hits the bow of your boat. This time, however, because you are moving toward the next wave crest as it moves toward you, it hits you less than 2.0 s after the first hit. In other words, the period you observe is shorter than the 2.0-s period you observed when you were stationary. Because $f = 1/T$, you observe a higher wave frequency than when you were at rest.

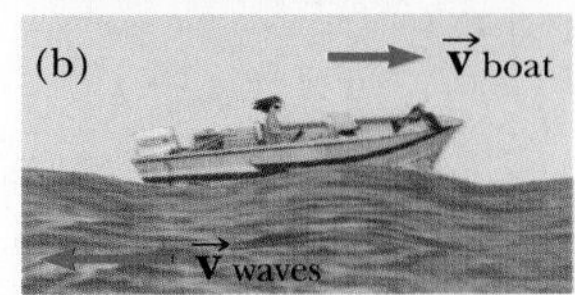

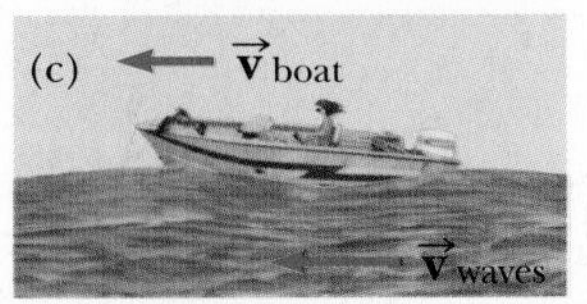

FIGURE 13.20 (a) Waves moving toward a stationary boat. The waves travel to the left and their source is far to the right of the boat, out of the frame of the drawing. (b) The boat moving toward the wave source. (c) The boat moving away from the wave source.

If you turn around and move in the same direction as the waves (Fig. 13.20c), you observe the opposite effect. You set your watch to $t = 0$ as a crest hits the stern of the boat. Because you are now moving away from the next crest, more than 2.0 s has elapsed on your watch by the time that crest catches you. Therefore, you observe a lower frequency than when you were at rest.

These effects occur because the relative speed between your boat and the crest of a wave depends on the direction of travel and on the speed of your boat. When you are moving toward the right in Figure 13.20b, this relative speed is higher than that of the wave speed, which leads to the observation of an increased frequency. When you turn around and move to the left, the relative speed is lower, as is the observed frequency of the water waves.

PITFALL PREVENTION 13.4

DOPPLER EFFECT DOES NOT DEPEND ON DISTANCE A common misconception about the Doppler effect is that it depends on the distance between the source and the observer. Although the intensity of a sound will vary as the distance changes, the apparent frequency will not; the frequency depends only on the speed. As you listen to an approaching source, you will detect increasing intensity but constant frequency. As the source passes, you will hear the frequency suddenly drop to a new constant value and the intensity begin to decrease.

Let us now examine an analogous situation with sound waves in which we replace the water waves with sound waves, the water surface becomes the air, and the person on the boat becomes an observer listening to the sound. In this case, an observer O is moving with a speed of v_O and a sound source S is stationary. For simplicity, we assume that the air is also stationary and that the observer moves directly toward the source.

The red lines in Active Figure 13.21 represent circles connecting the crests of sound waves moving away from the source. Therefore, the radial distance between adjacent red lines is one wavelength. We shall take the frequency of the source to be f, the wavelength to be λ, and the speed of sound to be v. A stationary observer would detect a frequency f, where $f = v/\lambda$ (i.e., when the source and observer are both at rest, the observed frequency must equal the true frequency of the source). If the observer moves toward the source with the speed v_O, however, the relative speed of sound experienced by the observer is higher than the speed of sound in air. Using our relative speed discussion of Section 3.6, if the sound is coming toward the observer at v and the observer is moving toward the sound at v_O, the relative speed of sound as measured by the observer is

$$v_{\text{rel}} = v + v_O$$

The frequency of sound heard by the observer is based on this apparent speed of sound:

$$f' = \frac{v_{\text{rel}}}{\lambda} = \frac{v + v_O}{\lambda} = f\left(\frac{v + v_O}{v}\right) \quad \text{(observer moving toward source)} \quad [13.28]$$

Now consider the situation in which the source moves with a speed of v_S relative to the medium and the observer is at rest. Active Figure 13.22a shows this situation. Because the source is moving, the crest of each new wave is emitted from the source to the right of the position of the emission of the previous crest a distance $v_S T$, where T is the period of the wave being generated by the source. Therefore, the center of each colored circle (indicated by the identically colored dot) in Active Figure 13.22a is shifted to the right by this distance relative to the circle representing the previous crest. If the source moves directly toward observer A in Active Figure 13.22a, the crests detected by the observer along a line between the source and observer are closer to one another than they would be if the source were at rest. As a result, the wavelength λ' measured by observer A is shorter than the true wavelength λ of the source. The wavelength is *shortened* by the distance $v_S T$, and the observed wavelength has the value $\lambda' = \lambda - v_S/f$. Because $\lambda = v/f$, the frequency heard by observer A is

$$f' = \frac{v}{\lambda'} = f\left(\frac{v}{v - v_S}\right) \quad \text{(source moving toward observer)} \quad [13.29]$$

ACTIVE FIGURE 13.21
An observer O (the cyclist) moving with a speed v_O toward a stationary point source S, the horn of a parked car. The observer hears a frequency f' that is greater than the source frequency.

Physics Now™ Log into PhysicsNow at **www.pop4e.com** and go to Active Figure 13.21 to adjust the speed of the observer.

That is, the frequency is *increased* when the source moves toward the observer. In a similar manner, if the source moves away from observer B at rest, the sign of v_S is reversed in Equation 13.29 and the frequency is lower.

In Equation 13.29, notice that the denominator approaches zero when the speed of the source approaches the speed of sound, resulting in the frequency f' approaching infinity. Such a situation results in waves that cannot escape from the source in the direction of motion of the source. This concentration of energy in front of the source results in a *shock wave.* Such a disturbance is noted when a jet aircraft flying at a speed equal to or greater than the speed of sound produces a *sonic boom.*

(Courtesy of U.S. Navy. Photo by Ensign John Gay)

A shock wave due to a jet traveling at the speed of sound is made visible as a fog of water vapor. The large pressure variation in the shock wave causes the water in the air to condense into water droplets. ■

Finally, if both the source and the observer are in motion, the following general equation for the observed frequency is found:

$$f' = f\left(\frac{v + v_O}{v - v_S}\right) \quad [13.30]$$

In this expression, the signs for the values substituted for v_O and v_S depend on the direction of the velocity. A positive value is used for motion of the observer or the source *toward* the other, and a negative sign is used for motion of one *away from* the other.

When working with any Doppler effect problem, remember the following rule concerning signs: The word *toward* is associated with an *increase* in the observed frequency, and the words *away from* are associated with a *decrease* in the observed frequency.

Doppler measurements of blood flow

The Doppler effect is used in medicine to measure the speed of blood flow. In ultrasound Doppler procedures, an ultrasonic sound wave is sent into the skin from a transducer. The sound waves reflect from moving blood cells, undergoing a frequency shift based on the speed of the cells. The instrumentation detects the reflected sound waves and converts the frequency information to a speed of flow of the blood. It is also possible to use the Doppler shift of light to measure the speed of blood flow. That can be done for blood vessels just under the skin by shining light from a low power laser onto the skin surface and monitoring the reflected light. The procedure can be performed on internal blood vessels by means of optical fibers (see Section 25.8) entering the body through natural openings or small incisions.

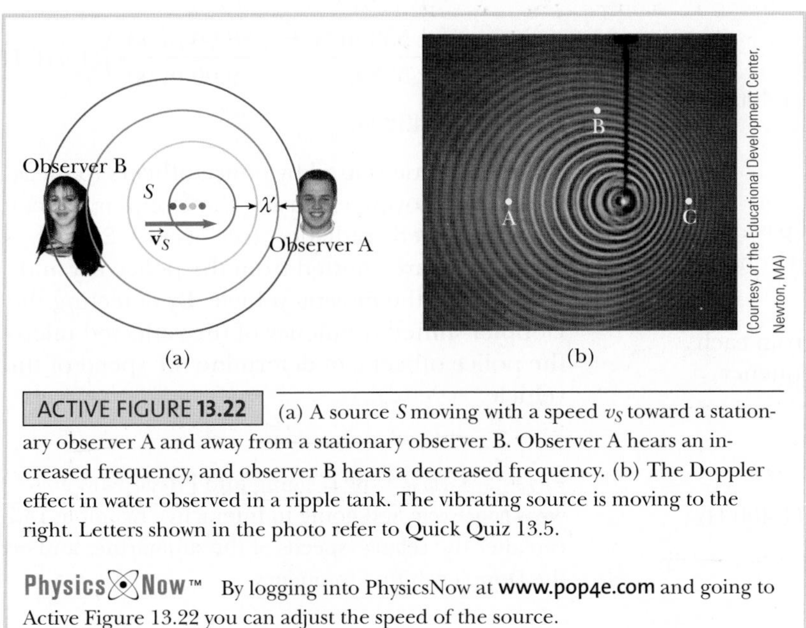

ACTIVE FIGURE 13.22 (a) A source S moving with a speed v_S toward a stationary observer A and away from a stationary observer B. Observer A hears an increased frequency, and observer B hears a decreased frequency. (b) The Doppler effect in water observed in a ripple tank. The vibrating source is moving to the right. Letters shown in the photo refer to Quick Quiz 13.5.

PhysicsNow™ By logging into PhysicsNow at **www.pop4e.com** and going to Active Figure 13.22 you can adjust the speed of the source.

QUICK QUIZ 13.5 Consider detectors of water waves at three locations A, B, and C in Active Figure 13.22b. Which of the following statements is true? **(a)** The wave speed is highest at location A. **(b)** The wave speed is highest at location C. **(c)** The detected wavelength is largest at location B. **(d)** The detected wavelength is largest at location C. **(e)** The detected frequency is highest at location C. **(f)** The detected frequency is highest at location A.

QUICK QUIZ 13.6 You stand on a platform at a train station and listen to a train approaching the station at a constant velocity. While the train approaches, but before it arrives, you hear **(a)** the intensity and the frequency of the sound both increasing, **(b)** the intensity and the frequency of the sound both decreasing, **(c)** the intensity increasing and the frequency decreasing, **(d)** the intensity decreasing and the frequency increasing, **(e)** the intensity increasing and the frequency remaining the same, or **(f)** the intensity decreasing and the frequency remaining the same.

INTERACTIVE EXAMPLE 13.7 Doppler Submarines

Submarines A and B are traveling toward each other under water. Sub A travels through the water at a speed of 8.00 m/s, emitting a sonar wave at a frequency of 1 400 Hz. Sub B travels through the water at a speed of 9.00 m/s. The speed of sound in the water is 1 533 m/s.

A What frequency is detected by an observer riding on sub B as the subs approach each other?

Solution We use Equation 13.30 to find the Doppler-shifted frequency. As the two submarines approach each other, the observer in sub B hears the frequency

$$f' = \left(\frac{v + v_O}{v - v_S}\right) f$$

$$= \left(\frac{1\,533 \text{ m/s} + (+9.00 \text{ m/s})}{1\,533 \text{ m/s} - (+8.00 \text{ m/s})}\right)(1\,400 \text{ Hz})$$

$$= 1\,416 \text{ Hz}$$

B The subs barely miss each other and pass. What frequency is detected by an observer riding on sub B as the subs recede from each other?

Solution As the two submarines recede from each other, the observer in sub B hears the frequency

$$f' = \left(\frac{v + v_O}{v - v_S}\right) f$$

$$= \left(\frac{1\,533 \text{ m/s} + (-9.00 \text{ m/s})}{1\,533 \text{ m/s} - (-8.00 \text{ m/s})}\right)(1\,400 \text{ Hz})$$

$$= 1\,385 \text{ Hz}$$

C While the subs are approaching each other, some of the sound from sub A will reflect from sub B and return to sub A. If this sound were to be detected by an observer on sub A, what is its frequency?

Solution The sound of apparent frequency 1 416 Hz found in part A will be reflected from a moving source (sub B) and then detected by a moving observer (sub A). Therefore, the frequency detected by sub A is

$$f'' = \left(\frac{v + v_O}{v - v_S}\right) f'$$

$$= \left(\frac{1\,533 \text{ m/s} + (+8.00 \text{ m/s})}{1\,533 \text{ m/s} - (+9.00 \text{ m/s})}\right)(1\,416 \text{ Hz})$$

$$= 1\,432 \text{ Hz}$$

This technique is used by police officers to measure the speed of a moving car, using the Doppler effect for electromagnetic radiation (see Section 24.3). Microwaves are emitted from the police car and reflected by the moving vehicle. By detecting the Doppler-shifted frequency of the reflected microwaves, the police officer can determine the speed of the vehicle.

PhysicsNow™ By Logging into PhysicsNow at www.pop4e.com and going to Interactive Example 13.7 you can alter the relative speeds of the submarines and observe the Doppler-shifted frequency.

13.9 SEISMIC WAVES

CONTEXT CONNECTION

When an earthquake occurs, a sudden release of energy takes place at a location called the **focus** or **hypocenter** of the earthquake. The **epicenter** is the point on the Earth's surface radially above the hypocenter. The released energy will propagate away from the focus of the earthquake by means of **seismic waves.** Seismic waves are like the sound waves that we have studied in the later sections of this chapter in that they are mechanical disturbances moving through a medium.

In discussing mechanical waves in this chapter, we identified two types: transverse and longitudinal. In the case of mechanical waves moving through air, we have only a longitudinal possibility. For mechanical waves moving through a solid, however, both possibilities are available because of the strong interatomic forces between elements of the solid. Therefore, in the case of seismic waves, energy propagates away from the focus both by longitudinal and transverse waves.

In the language used in earthquake studies, these two types of waves are named according to the order of their arrival at a seismograph. The longitudinal wave travels at a higher speed than the transverse wave. As a result, the longitudinal wave arrives at a seismograph first and is thus called the ***P* wave,** where *P* stands for *primary.* The slower moving transverse wave arrives next, so it is called the ***S* wave,** or *secondary* wave.

Let us see why longitudinal waves travel faster than transverse waves. The speed of *all* mechanical waves follows an expression of the general form

$$v = \sqrt{\frac{\text{elastic property}}{\text{inertial property}}} \qquad [13.31]$$

For a wave traveling on a string, we have seen the speed given by Equation 13.20:

$$v = \sqrt{\frac{T}{\mu}}$$

where the elastic property is the tension in the string. It is the tension in the string that returns a displaced element of the string to equilibrium. The appropriate inertial property is the linear mass density of the string.

For a transverse wave moving in a bulk solid, the elastic property is the *shear modulus S* of the material.[4] The shear modulus is a parameter that measures the deformation of a solid to a shear force, a force in the sideways direction. For example, lay your textbook down on a table and place your hand flat on the cover. Now, move your hand in a direction away from the book spine. The book will deform so that its cross-section changes from a rectangle to a parallelogram. The amount by which the book deforms under a given force from your hand is related to the shear modulus of the book. The speed of a transverse wave (an S wave) in a bulk solid is

$$v_S = \sqrt{\frac{S}{\rho}} \qquad [13.32]$$

where ρ is the density and S is the shear modulus of the material.

For a longitudinal wave moving in a gas or liquid, the elastic property in Equation 13.31 is the *bulk modulus B* of the material. The bulk modulus is a parameter that measures the change in volume of a sample of material due to a force compressing it that is uniform over a surface area. The speed of sound in a gas is given by

$$v = \sqrt{\frac{B}{\rho}} \qquad [13.33]$$

where B is the bulk modulus of the gas and ρ is the gas density.

[4]For details on various elastic moduli for materials see R. A. Serway and J. W. Jewett Jr., *Physics for Scientists and Engineers,* 6th ed. (Brooks-Cole, Belmont, CA: 2004), Section 12.4.

Now we consider longitudinal waves moving through a bulk solid. As a wave passes through a sample of the material, the material is compressed, so the wave speed should depend on the bulk modulus. As the material is compressed along the direction of travel of the wave, however, it is also distorted in the perpendicular direction. (Imagine a partially inflated balloon that is pressed downward against a table. It spreads out in the direction parallel to the table.) The result is a shear distortion of the sample of material. Therefore, the wave speed should depend on both the bulk modulus and the shear modulus! Careful analysis shows that this wave speed is

$$v_P = \sqrt{\frac{B + \frac{4}{3}S}{\rho}} \qquad [13.34]$$

Notice that this equation for the speed of a *P* wave gives a value that is larger than that for the *S* wave in Equation 13.32.

The wave speed for a seismic wave depends on the medium through which it travels. Typical values are 8 km/s for a *P* wave and 5 km/s for an *S* wave. Figure 13.23 shows a typical seismograph trace of a distant earthquake, with the *S* wave clearly arriving after the *P* wave.

The *P* and *S* waves move through the body of the Earth and can be detected by seismographs at various locations around the planet. Once these waves reach the surface, the energy can propagate by additional types of waves along the surface. In a *Rayleigh wave,* the motion of the elements of the medium at the surface is a combination of longitudinal and transverse displacements so that the net motion of a point on the surface is circular or elliptical. This motion is similar to the path followed by elements of water on the ocean surface as a wave passes by, as in Active Figure 13.24. The *Love wave* is a transverse surface wave in which the transverse oscillations are parallel to the surface. Therefore, no vertical displacement of the surface occurs in a Love wave.

It is possible to use the *P* and *S* waves traveling through the body of the Earth to gain information about the structure of the Earth's interior. Measurements of a given earthquake by seismographs at various locations on the surface indicate that the Earth has an interior region that allows the passage of *P* waves but not *S* waves. This fact can be understood if this particular region is modeled as having liquid

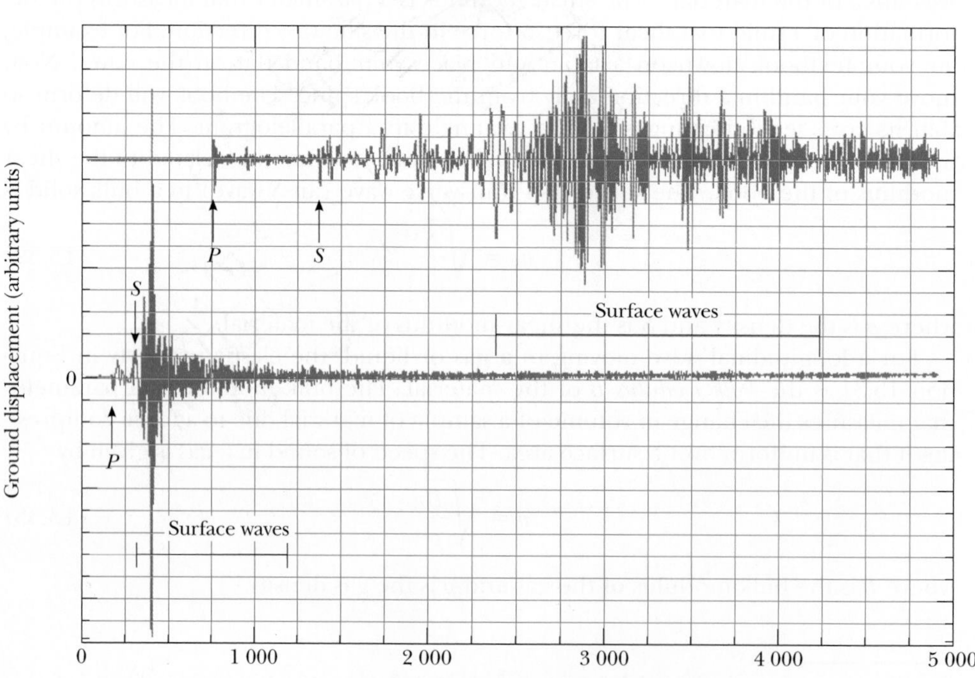

FIGURE 13.23 A seismograph trace, showing the arrival of *P* and *S* waves from the Northridge, California, earthquake at San Pablo, Spain (*top trace*) and Albuquerque, New Mexico (*bottom trace*). The *P* wave arrives first because it travels the fastest, followed by the slower moving *S* wave. The farther the seismograph station is from the epicenter, the longer the time interval between the arrivals of the *P* and *S* waves.

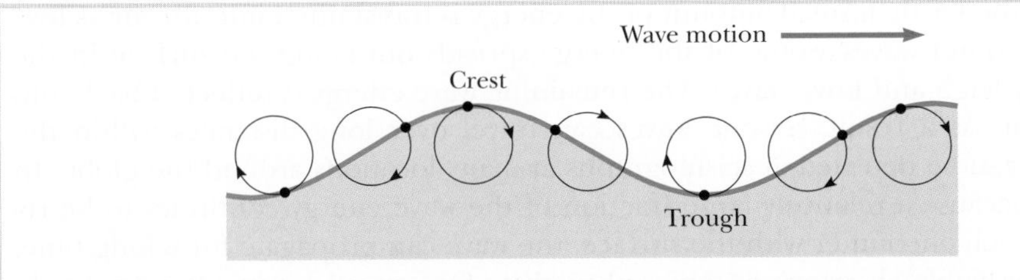

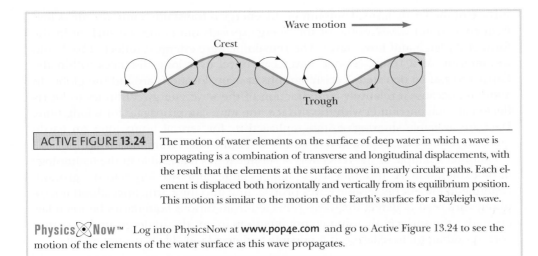

ACTIVE FIGURE 13.24 The motion of water elements on the surface of deep water in which a wave is propagating is a combination of transverse and longitudinal displacements, with the result that the elements at the surface move in nearly circular paths. Each element is displaced both horizontally and vertically from its equilibrium position. This motion is similar to the motion of the Earth's surface for a Rayleigh wave.

PhysicsNow™ Log into PhysicsNow at **www.pop4e.com** and go to Active Figure 13.24 to see the motion of the elements of the water surface as this wave propagates.

characteristics. Similar to a gas, a liquid cannot sustain a transverse force. Therefore, the transverse *S* waves cannot pass through this region. This information leads us to a structural model in which the Earth has a **liquid core** between radii of approximately 1.2×10^3 km and 3.5×10^3 km.

Other measurements of seismic waves allow additional interpretations of layers within the interior of the Earth, including a **solid core** at the center, a rocky region called the **mantle,** and a relatively thin outer layer called the **crust.** Figure 13.25 shows this structure. Using x-rays or ultrasound in medicine to provide information about the interior of the human body is somewhat similar to using seismic waves to provide information about the interior of the Earth.

As *P* and *S* waves propagate in the interior of the Earth, they will encounter variations in the medium. At each boundary at which the properties of the medium change, reflection and transmission occur. When the seismic wave arrives at the

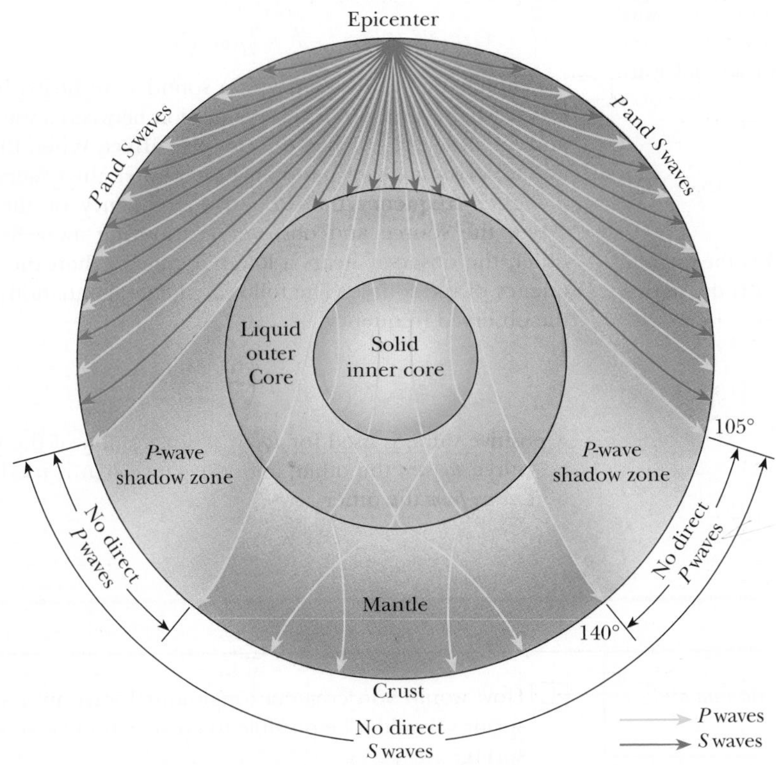

FIGURE 13.25 Cross-section of the Earth showing paths of waves produced by an earthquake. Only *P* waves (yellow) can propagate in the liquid core. The *S* waves (blue) do not enter the liquid core. When the *P* waves transmit from one region to another, such as from the mantle to the liquid core, they experience *refraction,* a change in the direction of propagation. We will study refraction for light in Chapter 25. Because of the refraction for seismic waves, there is a "shadow" zone between 105° and 140° from the epicenter in which no waves following a direct path (i.e., a path with no reflections) arrive.

surface of the Earth, a small amount of the energy is transmitted into the air as low-frequency sound waves. Some of the energy spreads out along the surface in the form of Rayleigh and Love waves. The remaining wave energy is reflected back into the interior. As a result, seismic waves can travel over long distances within the Earth and can be detected at seismographs at many locations around the globe. In addition, because a relatively large fraction of the wave energy continues to be reflected at each encounter with the surface, the wave can propagate for a long time. Data are available showing seismograph activity for several hours after an earthquake, a result of the repeated reflections of seismic waves from the surface.

Another example of the reflection of seismic waves is available in the technology of oil exploration. A "thumper truck" applies large impulsive forces to the ground, resulting in low-energy seismic waves propagating into the Earth. Specialized microphones are used to detect the waves reflected from various boundaries between layers under the surface. By using computers to map out the underground structure corresponding to these layers, it is possible to detect layers likely to contain oil.

SUMMARY

PhysicsNow™ Take a practice test by logging into PhysicsNow at www.pop4e.com and clicking on the Pre-Test link for this chapter.

A **transverse wave** is a wave in which the elements of the medium move in a direction perpendicular to the direction of the wave velocity. An example is a wave moving along a stretched string.

Longitudinal waves are waves in which the elements of the medium move back and forth parallel to the direction of the wave velocity. Sound waves in air are longitudinal.

Any one-dimensional wave traveling with a speed of v in the positive x direction can be represented by a **wave function** of the form $y = f(x - vt)$. Likewise, the wave function for a wave traveling in the negative x direction has the form $y = f(x + vt)$.

The wave function for a one-dimensional sinusoidal wave traveling to the right can be expressed as

$$y = A \sin\left[\frac{2\pi}{\lambda}(x - vt)\right] = A \sin(kx - \omega t) \quad [13.4, 13.9]$$

where A is the **amplitude,** λ is the **wavelength,** k is the **angular wave number,** and ω is the **angular frequency.** If T is the **period** and f is the **frequency,** v, k, and ω can be written as

$$v = \frac{\lambda}{T} = \lambda f \quad [13.5, 13.11]$$

$$k \equiv \frac{2\pi}{\lambda} \quad [13.7]$$

$$\omega \equiv \frac{2\pi}{T} = 2\pi f \quad [13.8]$$

The speed of a transverse wave traveling on a stretched string of mass per unit length μ and tension T is

$$v = \sqrt{\frac{T}{\mu}} \quad [13.20]$$

When a pulse traveling on a string meets a fixed end, the pulse is reflected and inverted. If the pulse reaches a free end, it is reflected but not inverted.

The **power** transmitted by a sinusoidal wave on a stretched string is

$$\mathcal{P} = \tfrac{1}{2}\mu\omega^2 A^2 v \quad [13.23]$$

The change in frequency of a sound wave heard by an observer whenever there is relative motion between a wave source and the observer is called the **Doppler effect.** When the source and observer are moving toward each other, the observer hears a higher frequency than the true frequency of the source. When the source and observer are moving away from each other, the observer hears a lower frequency than the true frequency of the source. The following general equation provides the observed frequency:

$$f' = f\left(\frac{v + v_O}{v - v_S}\right) \quad [13.30]$$

A positive value is used for v_O or v_S for motion of the observer or source *toward* the other, and a negative sign is used for motion *away from* the other.

QUESTIONS

☐ = answer available in the *Student Solutions Manual and Study Guide*

1. How would you create a longitudinal wave in a stretched spring? Would it be possible to create a transverse wave in a spring?

2. By what factor would you have to multiply the tension in a stretched string so as to double the wave speed?

3. When a pulse travels on a taut string, does it always invert upon reflection? Explain.

4. Consider a wave traveling on a taut rope. What is the difference, if any, between the speed of the wave and the speed of a small element of the rope?

5. What happens to the wavelength of a wave on a string when the frequency is doubled? Assume that the tension in the string remains constant.

6. What happens to the speed of a wave on a taut string when the frequency is doubled? Assume that the tension in the string remains constant.

7. If you stretch a rubber hose and pluck it, you can observe a pulse traveling up and down the hose. What happens to the speed of the pulse if you stretch the hose more tightly? What happens to the speed if you fill the hose with water?

8. If one end of a heavy rope is attached to one end of a light rope, the speed of a wave will change as the wave goes from the heavy rope to the light one. Will it increase or decrease? What happens to the frequency? What happens to the wavelength?

9. A vibrating source generates a sinusoidal wave on a string under constant tension. If the power delivered to the string is doubled, by what factor does the amplitude change? Does the wave speed change under these circumstances?

10. Why are sound waves characterized as longitudinal?

11. If an alarm clock is placed in a good vacuum and then activated, no sound is heard. Explain.

12. If the wavelength of sound is reduced by a factor of 2, what happens to its frequency? What happens to its speed?

13. By listening to a band or orchestra, how can you determine that the speed of sound is the same for all frequencies?

14. *The Tunguska event.* On June 30, 1908, a meteor burned up and exploded in the atmosphere above the Tunguska River valley in Siberia. It knocked down trees over thousands of square kilometers and started a forest fire, but apparently caused no human casualties. A witness sitting on his doorstep outside the zone of falling trees recalled events in the following sequence. He saw a moving light in the sky, brighter than the sun and descending at a low angle to the horizon. He felt his face become warm. He felt the ground shake. An invisible agent picked him up and immediately dropped him about a meter farther away from where the light had been. He heard a very loud protracted rumbling. Suggest an explanation for these observations and for the order in which they happened.

15. How can an object move with respect to an observer so that the sound from it is not shifted in frequency?

16. Suppose the wind blows. Does that cause a Doppler effect for sound propagating through the air? Is it like a moving source or a moving observer?

17. In an earthquake, both S (transverse) and P (longitudinal) waves propagate from the focus of the earthquake. The focus is in the ground below the epicenter on the surface. Assume that the waves move in straight lines through uniform material. The S waves travel through the Earth more slowly than the P waves (at about 5 km/s versus 8 km/s). By detecting the time of arrival of the waves, how can one determine the distance to the focus of the quake? How many detection stations are necessary to locate the focus unambiguously?

PROBLEMS

1, 2, 3 = straightforward, intermediate, challenging
☐ = full solution available in the *Student Solutions Manual and Study Guide*
Physics Now™ = coached problem with hints available at www.pop4e.com
= computer useful in solving problem
= paired numerical and symbolic problems
= biomedical application

Section 13.1 ■ Propagation of a Disturbance

1. At $t = 0$, a transverse pulse in a wire is described by the function

$$y = \frac{6}{x^2 + 3}$$

where x and y are in meters. Write the function $y(x, t)$ that describes this pulse if it is traveling in the positive x direction with a speed of 4.50 m/s.

2. Ocean waves with a crest-to-crest distance of 10.0 m can be described by the wave function

$$y(x, t) = (0.800 \text{ m}) \sin[0.628(x - vt)]$$

where $v = 1.20$ m/s. (a) Sketch $y(x, t)$ at $t = 0$. (b) Sketch $y(x, t)$ at $t = 2.00$ s. Note that the entire wave form has shifted 2.40 m in the positive x direction in this time interval.

Section 13.2 ■ The Wave Model

Section 13.3 ■ The Traveling Wave

3. A sinusoidal wave is traveling along a rope. The oscillator that generates the wave completes 40.0 vibrations in 30.0 s. Also, a given maximum travels 425 cm along the rope in 10.0 s. What is the wavelength?

4. For a certain transverse wave, the distance between two successive crests is 1.20 m and eight crests pass a given point along the direction of travel every 12.0 s. Calculate the wave speed.

5. The wave function for a traveling wave on a taut string is (in SI units)

$$y(x, t) = (0.350 \text{ m}) \sin(10\pi t - 3\pi x + \pi/4)$$

(a) What are the speed and direction of travel of the wave? (b) What is the vertical position of an element of the string at $t = 0$, $x = 0.100$ m? (c) What are the wavelength and frequency of the wave? (d) What is the maximum transverse speed of an element of the string?

6. A wave is described by $y = (2.00 \text{ cm}) \sin(kx - \omega t)$, where $k = 2.11$ rad/m, $\omega = 3.62$ rad/s, x is in meters, and t is in seconds. Determine the amplitude, wavelength, frequency, and speed of the wave.

7. The string shown in Active Figure 13.8 is driven at a frequency of 5.00 Hz. The amplitude of the motion is 12.0 cm and the wave speed is 20.0 m/s. Furthermore, the wave is such that $y = 0$ at $x = 0$ and $t = 0$. Determine (a) the angular frequency and (b) wave number for this wave. (c) Write an expression for the wave function. Calculate (d) the maximum transverse speed and (e) the maximum transverse acceleration of an element of the string.

8. Consider the sinusoidal wave of Example 13.2, with the wave function

$$y = (15.0 \text{ cm}) \cos(0.157x - 50.3t)$$

At a certain instant, let point A be at the origin and point B be the first point along the x axis where the wave is 60.0° out of phase with point A. What is the coordinate of point B?

9. **Physics Now™** (a) Write the expression for y as a function of x and t for a sinusoidal wave traveling along a rope in the *negative* x direction with the following characteristics: $A = 8.00$ cm, $\lambda = 80.0$ cm, $f = 3.00$ Hz, and $y(0, t) = 0$ at $t = 0$. (b) Write the expression for y as a function of x and t for the wave in part (a) assuming that $y(x, 0) = 0$ at the point $x = 10.0$ cm.

10. A transverse wave on a string is described by the wave function

$$y = (0.120 \text{ m}) \sin\left(\frac{\pi}{8}x + 4\pi t\right)$$

(a) Determine the transverse speed and acceleration of an element of the string at $t = 0.200$ s for the point on the string located at $x = 1.60$ m. (b) What are the wavelength, period, and speed of propagation of this wave?

11. A transverse sinusoidal wave on a string has a period $T = 25.0$ ms and travels in the negative x direction with a speed of 30.0 m/s. At $t = 0$, an element of the string at $x = 0$ has a transverse position of 2.00 cm and is traveling downward with a speed of 2.00 m/s. (a) What is the amplitude of the wave? (b) What is the initial phase angle? (c) What is the maximum transverse speed of an element of the string? (d) Write the wave function for the wave.

12. Show that the wave function $y = e^{b(x-vt)}$ is a solution of the linear wave equation (Eq. 13.19), where b is a constant.

Section 13.4 ▪ The Speed of Transverse Waves on Strings

13. A telephone cord is 4.00 m long and has a mass of 0.200 kg. A transverse pulse is produced by plucking one end of the taut cord. The pulse makes four trips down and back along the cord in 0.800 s. What is the tension in the cord?

14. An astronaut on the Moon wishes to measure the local value of the free-fall acceleration by timing pulses traveling down a wire that has an object of large mass suspended from it. Assume that a wire has a mass of 4.00 g and a length of 1.60 m and that a 3.00-kg object is suspended from it. A pulse requires 36.1 ms to traverse the length of the wire. Calculate g_{Moon} from these data. (You may ignore the mass of the wire when calculating the tension in it.)

15. Transverse waves travel with a speed of 20.0 m/s in a string under a tension of 6.00 N. What tension is required for a wave speed of 30.0 m/s in the same string?

16. **Review problem.** A light string with a mass per unit length of 8.00 g/m has its ends tied to two walls separated by a distance equal to three-fourths the length of the string (Fig. P13.16). An object of mass m is suspended from the center of the string, putting a tension in the string. (a) Find an expression for the transverse wave speed in the string as a function of the mass of the hanging object. (b) What should be the mass of the object suspended from the string so as to produce a wave speed of 60.0 m/s?

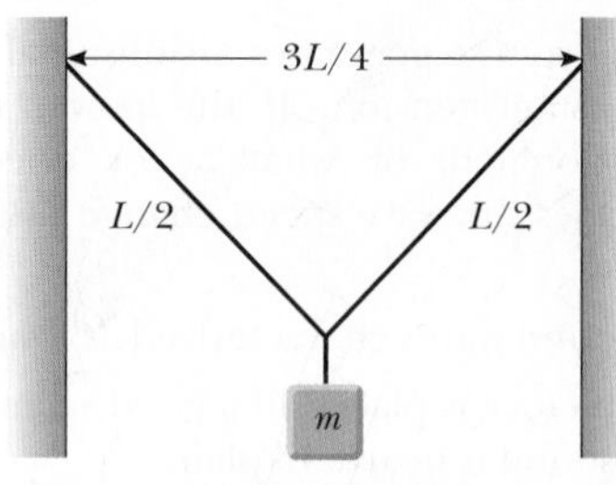

FIGURE **P13.16**

17. **Physics Now™** A 30.0-m steel wire and a 20.0-m copper wire, both with 1.00-mm diameters, are connected end to end and stretched to a tension of 150 N. How long does it take a transvers e wave to travel the entire length of the two wires?

Section 13.5 ▪ Reflection and Transmission of Waves

18. A series of pulses, each of amplitude 0.150 m, are sent down a string that is attached to a post at one end. The pulses are reflected at the post and travel back along the string without loss of amplitude. When two waves are present on the same string, the net displacement of a particular element of the string is the sum of the displacements of the individual waves at that point. What is the net displacement of an element at a point on the string where two pulses are crossing (a) if the string is rigidly attached to the post and (b) if the end at which reflection occurs is free to slide up and down?

Section 13.6 ▪ Rate of Energy Transfer by Sinusoidal Waves on Strings

19. A taut rope has a mass of 0.180 kg and a length of 3.60 m. What power must be supplied to the rope so as to generate sinusoidal waves having an amplitude of 0.100 m and a wavelength of 0.500 m and traveling with a speed of 30.0 m/s?

20. It is found that a 6.00-m segment of a long string contains four complete waves and has a mass of 180 g. The string is vibrating sinusoidally with a frequency of 50.0 Hz and a peak-to-valley displacement of 15.0 cm. (The peak-to-valley distance is the vertical distance from the farthest positive position to the farthest negative position.) (a) Write the function that describes this wave traveling in the positive x direction. (b) Determine the power being supplied to the string.

21. Physics Now™ Sinusoidal waves 5.00 cm in amplitude are to be transmitted along a string that has a linear mass density of 4.00×10^{-2} kg/m. If the source can deliver a maximum power of 300 W and the string is under a tension of 100 N, what is the highest frequency at which the source can operate?

22. A horizontal string can transmit a maximum power $\mathcal{P}_0$ (without breaking) if a wave with amplitude A and angular frequency ω is traveling along it. To increase this maximum power, a student folds the string and uses this "double string" as a medium. Determine the maximum power that can be transmitted along the "double string," assuming that the tension is constant.

Section 13.7 ▪ Sound Waves

Note: Use the following values as needed unless otherwise specified. The equilibrium density of air at 20°C is $\rho = 1.20$ kg/m^3. The speed of sound in air is $v = 343$ m/s at 20°C. Pressure variations ΔP are measured relative to atmospheric pressure, 1.013×10^5 N/m^2.

Problem 2.55 in Chapter 2 can also be assigned with this section.

23. Suppose you hear a clap of thunder 16.2 s after seeing the associated lightning stroke. The speed of sound waves in air is 343 m/s and the speed of light in air is 3.00×10^8 m/s. How far are you from the lightning stroke?

24. A dolphin in sea water at a temperature of 25°C emits sound directed toward the bottom of the ocean 150 m below. How much time passes before it hears an echo?

25. Many artists sing very high notes in *ad lib* ornaments and cadenzas. The highest note written for a singer in a published score was F-sharp above high C, 1.480 kHz, for Zerbinetta in the original version of Richard Strauss's opera *Ariadne auf Naxos.* (a) Find the wavelength of this sound in air. (b) In response to complaints, Strauss later transposed the note down to F above high C, 1.397 kHz. By what increment did the wavelength change?

26. A bat (Fig. P13.26) can detect very small objects, such as an insect whose length is approximately equal to one wavelength of the sound the bat makes. If a bat emits chirps at a frequency of 60.0 kHz and the speed of sound in air is 340 m/s, what is the smallest insect the bat can detect?

27. An ultrasonic tape measure uses frequencies above 20 MHz to determine dimensions of structures such as buildings. It does so by emitting a pulse of ultrasound into air and then measuring the time interval for an echo to return from a reflecting surface whose distance away is to be measured. The distance is displayed as a digital readout. For a tape measure that emits a pulse of ultrasound with a frequency of 22.0 MHz, (a) what is the distance to an object from which the echo pulse returns after 24.0 ms when the air temperature is 26°C? (b) What should be the duration of the emitted pulse if it is to include ten cycles of the ultrasonic wave? (c) What is the spatial length of such a pulse?

(Joe McDonald/Visuals Unlimited)

FIGURE P13.26 Problems 13.26 and 13.59.

28. Ultrasound is used in medicine both for diagnostic imaging and for therapy. For diagnosis, short pulses of ultrasound are passed through the patient's body. An echo reflected from a structure of interest is recorded, and from the time interval for the return of the echo the distance to the structure can be determined. A single transducer emits and detects the ultrasound. An image of the structure is obtained by reducing the data with a computer. With sound of low intensity, this technique is noninvasive and harmless. It is used to examine fetuses, tumors, aneurysms, gallstones, hearts, and many other structures. To reveal detail, the wavelength of the reflected ultrasound must be small compared with the size of the object reflecting the wave. (a) What is the wavelength of ultrasound with a frequency of 2.40 MHz, used in echo cardiography to map the beating heart? (b) In the whole set of imaging techniques, frequencies in the range 1.00 to 20.0 MHz are used. What is the range of wavelengths corresponding to this range of frequencies? The speed of ultrasound in human tissue is about 1 500 m/s (nearly the same as the speed of sound in water).

29. Physics Now™ An experimenter wishes to generate in air a sound wave that has a displacement amplitude of 5.50×10^{-6} m. The pressure amplitude is to be limited to 0.840 N/m^2. What is the minimum wavelength the sound wave can have?

30. A sinusoidal sound wave is described by the displacement wave function

$$s(x, t) = (2.00\ \mu\text{m})\cos[(15.7\ \text{m}^{-1})x - (858\ \text{s}^{-1})t]$$

(a) Find the amplitude, wavelength, and speed of this wave. (b) Determine the instantaneous displacement from equilibrium of the elements of the medium at the position $x = 0.050\,0$ m at $t = 3.00$ ms. (c) Determine the maximum speed of the element's oscillatory motion.

31. Write an expression that describes the pressure variation as a function of position and time for a sinusoidal sound wave in air, taking $\lambda = 0.100$ m and $\Delta P_{max} = 0.200$ N/m^2.

32. Calculate the pressure amplitude of a 2.00-kHz sound wave in air, assuming that the displacement amplitude is equal to 2.00×10^{-8} m.

Section 13.8 ■ The Doppler Effect

33. A driver travels northbound on a highway at a speed of 25.0 m/s. A police car, traveling southbound at a speed of 40.0 m/s, approaches with its siren producing sound at a frequency of 2 500 Hz. (a) What frequency does the driver observe as the police car approaches? (b) What frequency does the driver detect after the police car passes him? (c) Repeat parts (a) and (b) for the case when the police car is traveling northbound.

34. Expectant parents are thrilled to hear their unborn baby's heartbeat, revealed by an ultrasonic motion detector. Suppose the fetus's ventricular wall moves in simple harmonic motion with an amplitude of 1.80 mm and a frequency of 115 per minute. (a) Find the maximum linear speed of the heart wall. Suppose the motion detector in contact with the mother's abdomen produces sound at 2 000 000.0 Hz that travels through tissue at 1.50 km/s. (b) Find the maximum frequency at which sound arrives at the wall of the baby's heart. (c) Find the maximum frequency at which reflected sound is received by the motion detector. By electronically "listening" for echoes at a frequency different from the broadcast frequency, the motion detector can produce beeps of audible sound in synchronization with the fetal heartbeat.

35. **Physics Now™** Standing at a crosswalk, you hear a frequency of 560 Hz from the siren of an approaching ambulance. After the ambulance passes, the observed frequency of the siren is 480 Hz. Determine the ambulance's speed from these observations.

36. A block with a speaker bolted to it is connected to a spring having spring constant $k = 20.0$ N/m as shown in Figure P13.36. The total mass of the block and speaker is 5.00 kg, and the amplitude of this unit's motion is 0.500 m. The speaker emits sound waves of frequency 440 Hz. Determine the highest and lowest frequencies heard by the person to the right of the speaker. Assume that the speed of sound is 343 m/s.

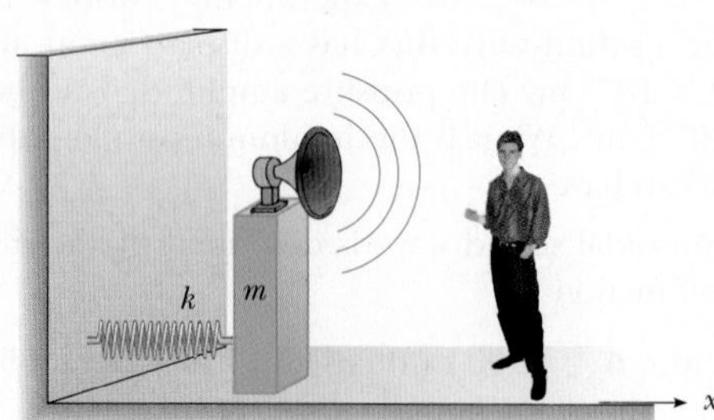

FIGURE P13.36

37. A tuning fork vibrating at 512 Hz falls from rest and accelerates at 9.80 m/s^2. How far below the point of release is the tuning fork when waves of frequency 485 Hz reach the release point? Take the speed of sound in air to be 340 m/s.

38. At the Winter Olympics, an athlete rides her luge down the track while a bell just above the wall of the chute rings continuously. When her sled passes the bell, she hears the frequency of the bell fall by the musical interval called a minor third. That is, the frequency she hears drops to five sixths of its original value. (a) Find the speed of sound in air at the ambient temperature −10.0°C. (b) Find the speed of the athlete.

39. A siren mounted on the roof of a firehouse emits sound at a frequency of 900 Hz. A steady wind is blowing with a speed of 15.0 m/s. Taking the speed of sound in calm air to be 343 m/s, find the wavelength of the sound (a) upwind of the siren and (b) downwind of the siren. Firefighters are approaching the siren from various directions at 15.0 m/s. What frequency does a firefighter hear (c) if she is approaching from an upwind position so that she is moving in the direction in which the wind is blowing and (d) if she is approaching from a downwind position and moving against the wind?

Section 13.9 ■ Context Connection—Seismic Waves

40. Two points A and B on the surface of the Earth are at the same longitude and 60.0° apart in latitude. Suppose an earthquake at point A creates a P wave that reaches point B by traveling straight through the body of the Earth at a constant speed of 7.80 km/s. The earthquake also radiates a Rayleigh wave, which travels along the surface of the Earth at 4.50 km/s. (a) Which of these two seismic waves arrives at B first? (b) What is the time difference between the arrivals of the two waves at B? Take the radius of the Earth to be 6 370 km.

41. A seismographic station receives S and P waves from an earthquake, 17.3 s apart. Assume that the waves have traveled over the same path at speeds of 4.50 km/s and 7.80 km/s. Find the distance from the seismograph to the hypocenter of the quake.

Additional Problems

42. "The wave" is a particular type of pulse that can propagate through a large crowd gathered at a sports arena (Fig. P13.42). The elements of the medium are the spectators, with zero position corresponding to their being seated and maximum position corresponding to their standing and raising their arms. When a large fraction of the spectators participate in the wave motion, a somewhat stable pulse shape can develop. The wave speed depends on people's reaction time, which is typically on the order of 0.1 s. Estimate the order of magnitude, in minutes, of the time interval required for such a pulse to make one circuit around a large sports stadium. State the quantities you measure or estimate and their values.

43. **Review problem.** A block of mass M, supported by a string, rests on a frictionless incline making an angle θ with the horizontal (Fig. P13.43). The length of the string is L and its mass is $m \ll M$. Derive an expression for the time interval required for a transverse wave to travel from one end of the string to the other.

FIGURE P13.42

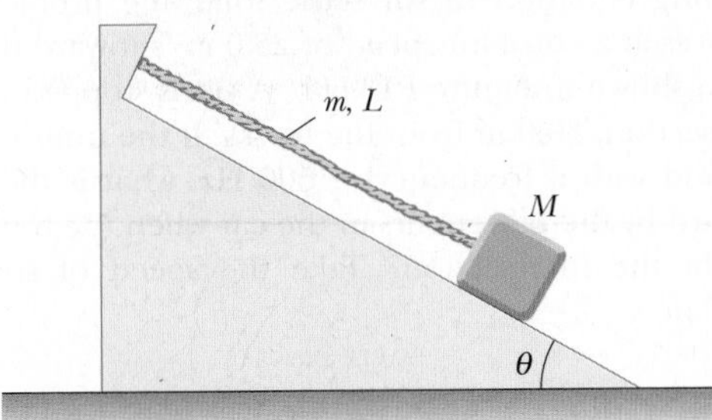

FIGURE P13.43

44. **Review problem.** A block of mass M hangs from a rubber cord. The block is supported so that the cord is not stretched. The unstretched length of the cord is L_0 and its mass is m, much less than M. The "spring constant" for the cord is k. The block is released and stops at the lowest point. (a) Determine the tension in the cord when the block is at this lowest point. (b) What is the length of the cord in this "stretched" position? (c) Find the speed of a transverse wave in the cord, assuming that the block is held in this lowest position.

45. **Review problem.** A block of mass 0.450 kg is attached to one end of a cord of mass 0.003 20 kg; the other end of the cord is attached to a fixed point. The block rotates with constant angular speed in a circle on a horizontal, frictionless table. Through what angle does the block rotate in the time interval required for a transverse wave to travel along the string from the center of the circle to the block?

46. (a) Show that the speed of longitudinal waves along a spring of force constant k is $v = \sqrt{kL/\mu}$, where L is the unstretched length of the spring and μ is the mass per unit length. (b) A spring with a mass of 0.400 kg has an unstretched length of 2.00 m and a force constant of 100 N/m. Using the result you obtained in part (a), determine the speed of longitudinal waves along this spring.

47. A rope of total mass m and length L is suspended vertically. Show that a transverse pulse travels the length of the rope in a time interval $\Delta t = 2\sqrt{L/g}$. (*Suggestion:* First find an expression for the wave speed at any point a distance x from the lower end by considering the tension in the rope as resulting from the weight of the segment below that point.)

48. Assume that an object of mass M is suspended from the bottom of the rope in Problem 13.47. (a) Show that the time interval for a transverse pulse to travel the length of the rope is

$$\Delta t = 2\sqrt{\frac{L}{mg}}\left(\sqrt{M+m} - \sqrt{M}\right)$$

(b) Show that this expression reduces to the result of Problem 13.47 when $M = 0$. (c) Show that for $m \ll M$, the expression in part (a) reduces to

$$\Delta t = \sqrt{\frac{mL}{Mg}}$$

49. A pulse traveling along a string of linear mass density μ is described by the wave function

$$y = [A_0 e^{-bx}] \sin(kx - \omega t)$$

where the factor in brackets is said to be the amplitude. (a) What is the power $\mathcal{P}(x)$ carried by this wave at a point x? (b) What is the power carried by this wave at the origin? (c) Compute the ratio $\mathcal{P}(x)/\mathcal{P}(0)$.

50. An earthquake on the ocean floor in the Gulf of Alaska produces a *tsunami* (sometimes incorrectly called a "tidal wave") that reaches Hilo, Hawaii, 4 450 km away, in a time interval of 9 h 30 min. Tsunamis have enormous wavelengths (100 to 200 km), and the propagation speed for these waves is $v \approx \sqrt{g\bar{d}}$, where $\bar{d}$ is the average depth of the water. From the information given, find the average wave speed and the average ocean depth between Alaska and Hawaii. (This method was used in 1856 to estimate the average depth of the Pacific Ocean long before soundings were made to give a direct determination.)

51. A string on a musical instrument is held under tension T and extends from the point $x = 0$ to the point $x = L$. The string is overwound with wire in such a way that its mass per unit length $\mu(x)$ increases uniformly from μ_0 at $x = 0$ to μ_L at $x = L$. (a) Find an expression for $\mu(x)$ as a function of x over the range $0 \le x \le L$. (b) Show that the time interval required for a transverse pulse to travel the length of the string is given by

$$\Delta t = \frac{2L(\mu_L + \mu_0 + \sqrt{\mu_L \mu_0})}{3\sqrt{T}(\sqrt{\mu_L} + \sqrt{\mu_0})}$$

52. A flowerpot is knocked off a balcony 20.0 m above the sidewalk and falls toward an unsuspecting 1.75-m-tall man who is standing below. How close to the sidewalk can the

flowerpot fall before it is too late for a warning shouted from the balcony to reach the man in time? Assume that the man below requires 0.300 s to respond to the warning.

53. A sound wave in a cylinder is described by Equations 13.24 through 13.26. Show that $\Delta P = \pm \rho v \omega \sqrt{s_{\max}^2 - s^2}$.

54. On a Saturday morning, pickup trucks and sport utility vehicles carrying garbage to the town landfill form a nearly steady procession on a country road, all traveling at 19.7 m/s. From one direction, two trucks arrive at the dump every 3 min. A bicyclist is also traveling toward the landfill, at 4.47 m/s. (a) With what frequency do the trucks pass him? (b) A hill does not slow down the trucks, but makes the out-of-shape cyclist's speed drop to 1.56 m/s. How often do noisy, smelly, inefficient, garbage-dripping, roadhogging trucks whiz past him now?

55. The ocean floor is underlain by a layer of basalt that constitutes the crust, or uppermost layer, of the Earth in that region. Below this crust is found denser periodotite rock that forms the Earth's mantle. The boundary between these two layers is called the Mohorovicic discontinuity ("Moho" for short). If an explosive charge is set off at the surface of the basalt, it generates a seismic wave that is reflected back out at the Moho. If the speed of this wave in basalt is 6.50 km/s and the two-way travel time is 1.85 s, what is the thickness of this oceanic crust?

56. A train whistle (f = 400 Hz) sounds higher or lower in frequency depending on whether it approaches or recedes. (a) Prove that the difference in frequency between the approaching and receding train whistle is

$$\Delta f = \frac{2u/v}{1 - u^2/v^2} f$$

where u is the speed of the train and v is the speed of sound. (b) Calculate this difference for a train moving at a speed of 130 km/h. Take the speed of sound in air to be 340 m/s.

57. To permit measurement of her speed, a sky diver carries a buzzer emitting a steady tone at 1 800 Hz. A friend on the ground at the landing site directly below listens to the amplified sound he receives. Assume that the air is calm and that the sound speed is 343 m/s, independent of altitude. While the sky diver is falling at terminal speed, her friend on the ground receives waves of frequency 2 150 Hz. (a) What is the sky diver's speed of descent? (b) Suppose the sky diver can hear the sound of the buzzer reflected from the ground. What frequency does she receive?

58. A police car is traveling east at 40.0 m/s along a straight road, overtaking a car ahead of it moving east at 30.0 m/s. The police car has a malfunctioning siren that is stuck at 1 000 Hz. (a) Sketch the appearance of the wave fronts of the sound produced by the siren. Show the wave fronts both to the east and to the west of the police car. (b) What would be the wavelength in air of the siren sound if the police car were at rest? (c) What is the wavelength in front of the car? (d) What is it behind the police car? (e) What is the frequency heard by the driver being chased?

59. A bat, moving at 5.00 m/s, is chasing a flying insect (Fig. P13.26). If the bat emits a 40.0-kHz chirp and receives back an echo at 40.4 kHz, at what speed is the insect moving toward or away from the bat? Take the speed of sound in air to be v = 340 m/s.

60. The Doppler Equation 13.30 is valid when the motion between the observer and the source occurs on a straight line so that the source and observer are moving either directly toward or directly away from each other. If this restriction is relaxed, one must use the more general Doppler equation

$$f' = \left(\frac{v + v_O \cos \theta_O}{v - v_S \cos \theta_S}\right) f$$

where θ_O and θ_S are defined in Figure P13.60a. (a) Show that if the observer and source are moving away from each other, the preceding equation reduces to Equation 13.30 with negative values for both v_O and v_S. (b) Use the preceding equation to solve the following problem. A train moves at a constant speed of 25.0 m/s toward the intersection shown in Figure P13.60b. A car is stopped near the intersection, 30.0 m from the tracks. If the train's horn emits sound with a frequency of 500 Hz, what is the frequency heard by the passengers in the car when the train is 40.0 m from the intersection? Take the speed of sound to be 343 m/s.

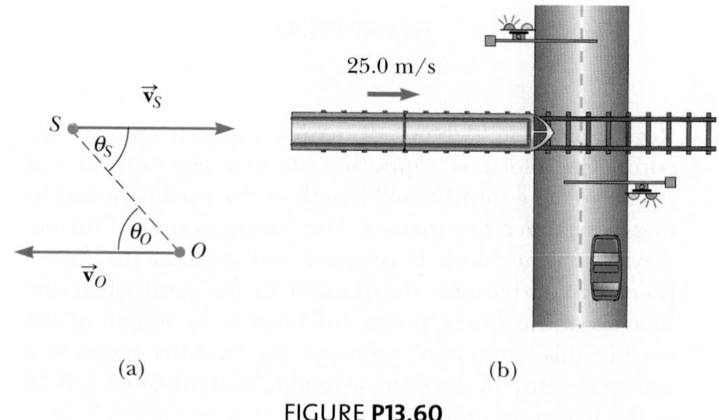

FIGURE **P13.60**

ANSWERS TO QUICK QUIZZES

13.1 (b), (c). The movement of the people in the line is parallel to the direction of propagation of the gap. The fans participating in the "wave" stand up vertically as the wave sweeps past them horizontally.

13.2 (i), (c). The wave speed is determined by the medium, so it is unaffected by changing the frequency. (ii), (b). Because the wave speed remains the same, the result of doubling the frequency is that the wavelength is half as large. (iii), (d). The amplitude of a wave is unrelated to the frequency, so we cannot determine the new amplitude without further information.

13.3 Only choices (f) and (h) are correct. Choices (a) and (b) affect the transverse speed of an element of the string but not the wave speed along the string. Choices (c) and (d) change the amplitude. Choices (e) and (g) increase the time interval by decreasing the wave speed.

13.4 (d). Doubling the amplitude of the wave causes the power to be larger by a factor of 4. In (a), halving the linear mass density of the string causes the power to change by a factor of 0.71; the rate decreases. In (b), doubling the wavelength of the wave halves the frequency and causes the power to change by a factor of 0.25; the rate decreases. In (c), doubling the tension in the string changes the wave speed and causes the power to change by a factor of 1.4, which is not as large as in part (d).

13.5 (e). The wave speed cannot be changed by moving the source, so (a) and (b) are incorrect. The detected wavelength is largest at A, so (c) and (d) are incorrect. Choice (f) is incorrect because the detected frequency is lowest at location A. Choice (e) is correct because at location C the wavelength is the smallest, so the frequency must be the largest.

13.6 (e). The intensity of the sound increases because the train is moving closer to you. Because the train moves at a constant velocity, the Doppler-shifted frequency remains fixed.

CHAPTER 14

Superposition and Standing Waves

The rich sound of a piano is due to standing waves on strings under tension. Many such strings can be seen in this photograph. Waves also travel on the soundboard, which is visible below the strings.

(Kathy Ferguson Johnson/PhotoEdit/PictureQuest)

CHAPTER OUTLINE

In Chapter 13, we introduced the wave model. We have seen that waves are very different from particles. An ideal particle is of zero size, but an ideal wave is of infinite length. Another important difference between waves and particles is that we can explore the possibility of two or more waves combining at one point in the same medium. We can combine particles to form extended objects, but the particles must be at different locations. In contrast, two waves can both be present at a given location, and the ramifications of this possibility are explored in this chapter.

One ramification of the combination of waves is that only certain allowed frequencies can exist on systems with boundary conditions; that is, the frequencies are *quantized.* In Chapter 11, we learned about quantized energies of the hydrogen atom. Quantization is at the heart of quantum mechanics, a subject that is introduced formally in Chapter 28. We shall see that waves under boundary conditions explain many of the quantum phenomena. For our present purposes in this chapter, quantization enables us to understand the behavior of the wide array of musical instruments that are based on strings and air columns.

14.1 THE PRINCIPLE OF SUPERPOSITION

Many interesting wave phenomena in nature cannot be described by a single wave. Instead, one must analyze complex waveforms in terms of a combination of traveling waves. To analyze such wave combinations, we make use of the **principle of superposition:**

> If two or more traveling waves are moving through a medium and combine at a given point, the resultant position of the element of the medium at that point is the sum of the positions due to the individual waves.

■ Principle of superposition

This rather striking property is exhibited by many waves in nature, including waves on strings, sound waves, and surface water waves. It is also exhibited by electromagnetic waves, for which the electric fields of the combined waves are added. Waves that obey this principle are called *linear waves.* In general, linear waves have an amplitude that is small relative to their wavelength. Waves that violate the superposition principle are called *nonlinear waves* and, as mentioned in Chapter 13, are often characterized by large amplitudes. In this book, we shall deal only with linear waves.

A simple pictorial representation of the superposition principle is obtained by considering two pulses traveling in opposite directions on a stretched string as in Active Figure 14.1. The wave function for the pulse moving to the right is y_1, and the wave function for the pulse moving to the left is y_2. The pulses have the same speed but different shapes. Each pulse is assumed to be symmetric (although that is not a necessary condition), and in both cases displacements of the elements of the string in the vertical direction are taken to be positive. When the waves overlap, the resulting waveform is given by $y_1 + y_2$. After the time interval during which the pulses combine, they separate and continue moving in their original directions (Active Fig. 14.1d). Note that the final waveforms remain unchanged as if the two

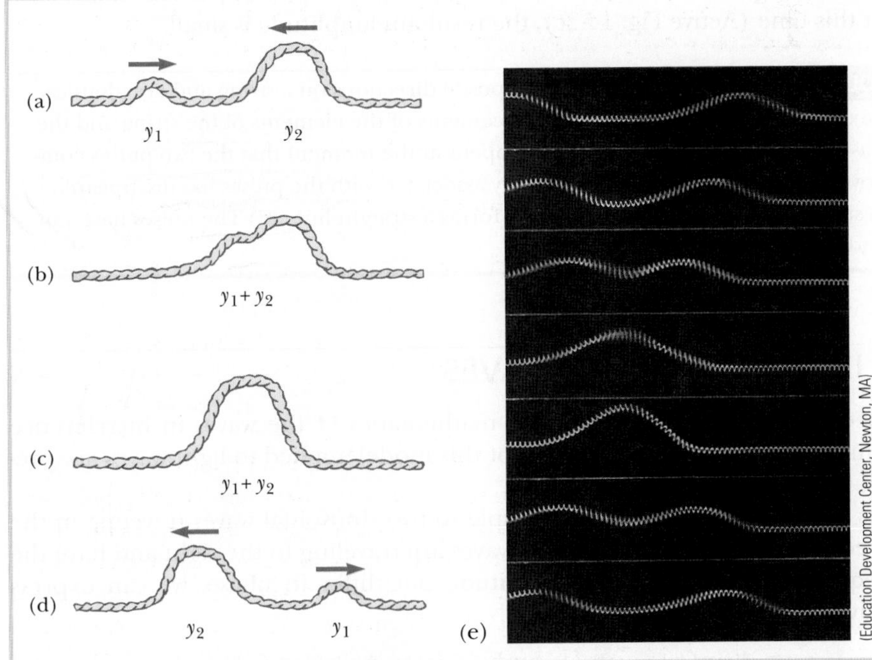

ACTIVE FIGURE 14.1 (*Left*) Two pulses traveling on a stretched string in opposite directions pass through each other. When the pulses overlap, as in (b) and (c), the net displacement of each element of the string equals the sum of the displacements produced by each pulse. Because each pulse produces positive displacements of the string, we refer to their superposition as *constructive interference.* (*Right*) Photograph of the superposition of two equal and symmetric pulses traveling in opposite directions on a stretched spring.

PhysicsNow™ By logging into PhysicsNow at www.pop4e.com and going to Active Figure 14.1 you can choose the amplitude and orientation of each of the pulses and observe the interference as they pass each other.

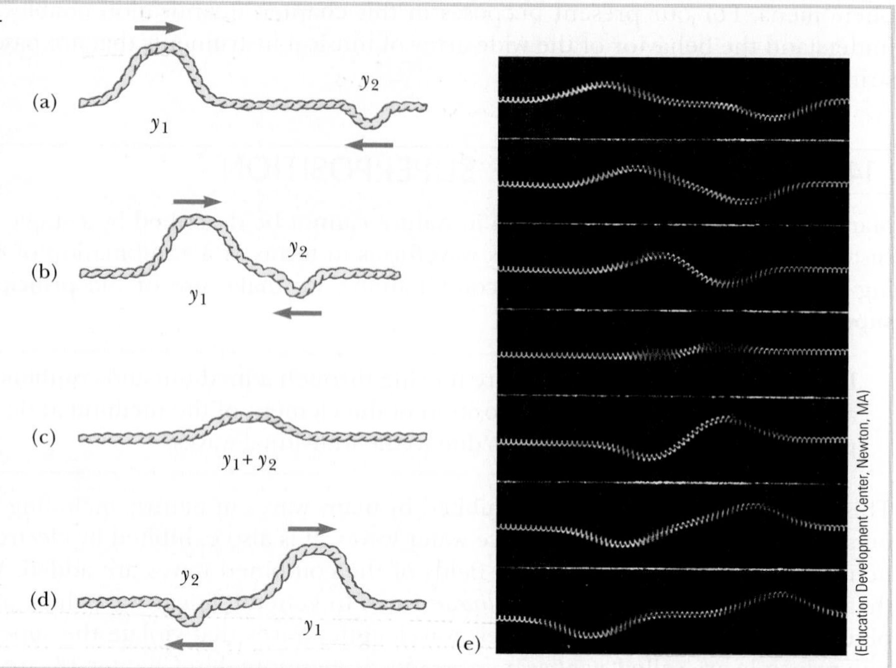

ACTIVE FIGURE 14.2 (*Left*) Two pulses traveling in opposite directions with displacements that are inverted relative to each other. When the two overlap as in (c), their displacements subtract from each other. (*Right*) Photograph of the superposition of two symmetric pulses traveling in opposite directions, where one is inverted relative to the other.

Physics Now™ By logging into PhysicsNow at **www.pop4e.com** and going to Active Figure 14.2 you can choose the amplitude and orientation of each of the pulses and observe the interference as they pass each other.

pulses had never met! The combination of separate waves in the same region of space to produce a resultant wave is called **interference.** Notice that the interference exists only while the waves are in the same region of space, and there is no permanent effect on the pulses after they separate.

For the two pulses shown in Active Figure 14.1, the vertical displacements are in the same direction and so the resultant waveform (when the pulses overlap) exhibits an amplitude greater than those of the individual pulses. Now consider two identical pulses, again traveling in opposite directions on a stretched string, but this time one pulse is inverted relative to the other as in Active Figure 14.2. In this case, when the pulses begin to overlap, the resultant waveform is the sum of the two separate waveforms again, but one of the displacements is negative. Again, the two pulses pass through each other. When they exactly overlap, they partially *cancel* each other. At this time (Active Fig. 14.2c), the resultant amplitude is small.

PITFALL PREVENTION 14.1

DO WAVES REALLY *INTERFERE*? In popular usage, the term *interfere* implies that an agent affects a situation in some way so as to preclude something from happening. For example, in American football, *pass interference* means that a defending player has affected the receiver so that he is unable to catch the ball. This usage is very different from that in physics, in which waves pass through one another and interfere but do not affect one another in any way. In physics, we will consider interference to be similar to the notion of *combination*, as described in this chapter, as opposed to the popular usage.

QUICK QUIZ 14.1 Two pulses move in opposite directions on a string and are identical in shape except that one has positive displacements of the elements of the string and the other has negative displacements. What happens at the moment that the two pulses completely overlap on the string? **(a)** The energy associated with the pulses has disappeared. **(b)** The string is not moving. **(c)** The string forms a straight line. **(d)** The pulses have vanished and will not reappear.

14.2 INTERFERENCE OF WAVES

In this section, we shall investigate the mathematics of the waves in interference analysis model. Additional applications of this model applied to light waves are presented in Chapter 27.

Let us apply the superposition principle to two sinusoidal waves traveling in the same direction in a medium. If the two waves are traveling to the right and have the same frequency, wavelength, and amplitude but differ in phase, we can express their individual wave functions as

$$y_1 = A \sin(kx - \omega t) \qquad \text{and} \qquad y_2 = A \sin(kx - \omega t + \phi)$$

where ϕ is the phase difference between the two waves. Let us imagine that these two waves coincide in the medium. For example, these expressions might represent two waves traveling along the same string. In this situation, the resultant wave function y is, according to the principle of superposition,

$$y = y_1 + y_2 = A[\sin(kx - \omega t) + \sin(kx - \omega t + \phi)]$$

To simplify this expression, it is convenient to use the trigonometric identity

$$\sin a + \sin b = 2\cos\left(\frac{a-b}{2}\right)\sin\left(\frac{a+b}{2}\right)$$

If we let $a = kx - \omega t$ and $b = kx - \omega t + \phi$, the resultant wave function y reduces to

$$y = \left(2A\cos\frac{\phi}{2}\right)\sin\left(kx - \omega t + \frac{\phi}{2}\right) \quad [14.1]$$

This mathematical representation of the resultant wave has several important features. The resultant wave function y is also a sinusoidal wave and has the *same* frequency and wavelength as the individual waves. The amplitude of the resultant wave is $2A\cos(\phi/2)$ and the phase angle is $\phi/2$. If the phase angle ϕ equals 0, $\cos(\phi/2) = \cos 0 = 1$ and the amplitude of the resultant wave is $2A$. In other words, the amplitude of the resultant wave is twice the amplitude of either individual wave. In this case, the waves are said to be everywhere *in phase* ($\phi = 0$) and to **interfere constructively.** That is, the crests of the individual waves occur at the same positions, as is shown by the blue line in Active Figure 14.3a. In general, constructive interference occurs when $\cos(\phi/2) = \pm 1$ or when $\phi = 0, 2\pi, 4\pi, \ldots$.

■ Constructive interference

On the other hand, if ϕ is equal to π radians or to any *odd* multiple of π, $\cos(\phi/2) = \cos(\pi/2) = 0$ and the resultant wave has *zero* amplitude everywhere. In this case, the two waves **interfere destructively.** That is, the crest of one wave coincides with the trough of the second (Active Fig. 14.3b) and their displacements cancel at every point. Finally, when the phase constant has a value between 0 and π, as in Active Figure 14.3c, the resultant wave has an amplitude whose value is somewhere between 0 and $2A$.

■ Destructive interference

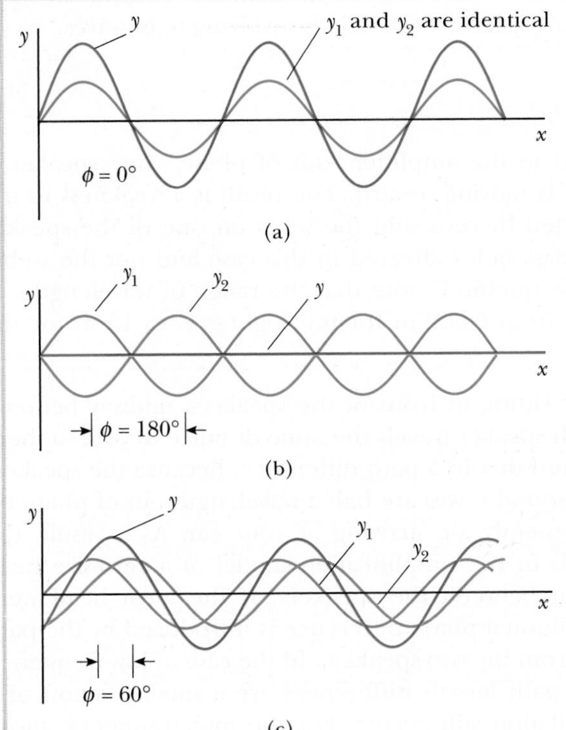

ACTIVE FIGURE 14.3
The superposition of two identical waves y_1 and y_2. (a) When the two waves are in phase, the result is constructive interference. (b) When the two waves are π rad out of phase, the result is destructive interference. (c) When the phase angle has a value other than 0 or π rad, the resultant wave y falls somewhere between the extremes shown in (a) and (b).

PhysicsNow™ Log into PhysicsNow at **www.pop4e.com** and go to Active Figure 14.3 to change the phase relationship between the waves and observe the wave representing the superposition.

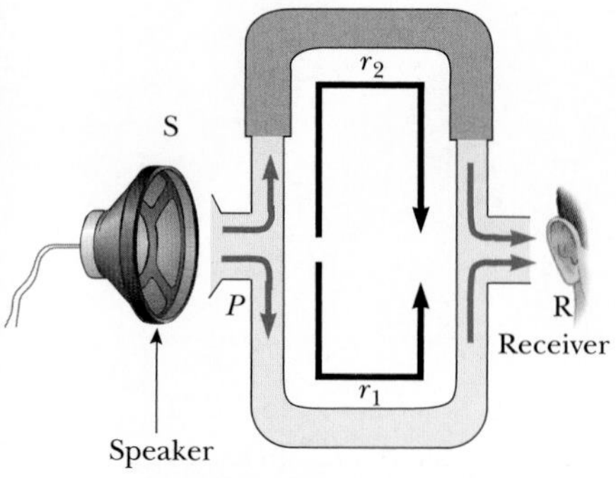

FIGURE 14.4 An acoustical system for demonstrating interference of sound waves. Sound waves from the speaker propagate into the tube and the energy splits into two parts at point P. The waves from the two paths, which combine at the opposite side, are detected at the receiver R. The upper path length r_2 can be varied by sliding the upper section.

Although we have used waves having the same amplitude in the preceding discussion, waves of differing amplitudes will interfere in a similar way. If the waves are in phase, the combined amplitude is the sum of the individual amplitudes. If they are 180° out of phase, the combined amplitude is the difference between the individual amplitudes.

Figure 14.4 shows a simple device for demonstrating interference of sound waves. Sound from speaker S is sent into a tube at P, where there is a T-shaped junction. Half the sound energy travels in one direction and half in the opposite direction. Therefore, the sound waves that reach receiver R at the other side can travel along either of two paths. The total distance from speaker to receiver is called the **path length** r. The length of the lower path is fixed at r_1. The upper path length r_2 can be varied by sliding the U-shaped tube (similar to that on a slide trombone). When the difference in the path lengths $\Delta r = |r_2 - r_1|$ is either zero or some integral multiple of the wavelength λ, the two waves reaching the receiver are in phase and interfere constructively as in Active Figure 14.3a. In this case, a maximum in the sound intensity is detected at the receiver. If path length r_2 is adjusted so that Δr is $\lambda/2$, $3\lambda/2$, . . . , $n\lambda/2$ (for n odd), the two waves are exactly 180° out of phase at the receiver and hence cancel each other. In this case of completely destructive interference, no sound is detected at the receiver. This simple experiment is a striking illustration of interference. In addition, it demonstrates that a phase difference may arise between two waves generated by the same source when they travel along paths of unequal lengths.

It is often useful to express a path difference in terms of the phase angle ϕ between the two waves. Because a path difference of one wavelength corresponds to a phase angle of 2π rad, we obtain the ratio $\phi/2\pi = \Delta r/\lambda$ or

■ Relationship between path difference and phase angle

$$\Delta r = \frac{\phi}{2\pi}\lambda \qquad [14.2]$$

Therefore, for example, a phase difference of 180° or π rad corresponds to a shift of $\lambda/2$. Conversely, a one-quarter-wavelength shift corresponds to a 90° phase difference.

Nature provides many examples of interference phenomena. Later in the text we shall describe several interesting interference effects involving light waves.

■ Thinking Physics 14.1

If stereo speakers are connected to the amplifier "out of phase," one speaker is moving outward when the other is moving inward. The result is a weakness in the bass notes, which can be corrected by reversing the wires on one of the speaker connections. Why are only the bass notes affected in this case and not the treble notes? For help in answering this question, note that the range of wavelengths of sound from a standard piano is from 0.082 m for the highest C to 13 m for the lowest A.

Reasoning Imagine that you are sitting in front of the speakers, midway between them. Then, the sound from each speaker travels the same distance to you, so there is no phase difference in the sound due to a path difference. Because the speakers are connected out of phase, the sound waves are half a wavelength out of phase on leaving the speaker and, consequently, on arriving at your ear. As a result, the sound for all frequencies cancels in the simplification model of a zero-size head located exactly on the midpoint between the speakers. If the ideal head were moved off the centerline, an additional phase difference is introduced by the path length difference for the sound from the two speakers. In the case of low-frequency, long-wavelength bass notes, the path length differences are a small fraction of a wavelength, so significant cancellation still occurs. For the high-frequency, short-wavelength treble notes, a small movement of the ideal head results in a much larger fraction of a wavelength in path length difference or even multiple wavelengths. Therefore, the treble notes could be in phase with this head movement. If we now add that the head is not of zero size and that it has two ears, we can see that complete cancellation is not possible and, with even small movements of the head, one or both ears will be at or near maxima for the treble notes. The size of the head is much smaller than bass wavelengths, however, so the bass notes are significantly weakened over much of the region in front of the speakers. ■

INTERACTIVE EXAMPLE 14.1 Two Speakers Driven by the Same Source

Two speakers placed 3.00 m apart are driven in phase by the same oscillator (Fig. 14.5). A listener is originally at point *O*, which is located 8.00 m from the center of the line connecting the two speakers. The listener then moves to point *P*, which is a perpendicular distance 0.350 m from *O*, at which the first cancellation of waves occurs, resulting in a minimum in sound intensity. What is the frequency of the oscillator?

Solution The first cancellation occurs when the two waves reaching the listener at *P* are 180° out of phase

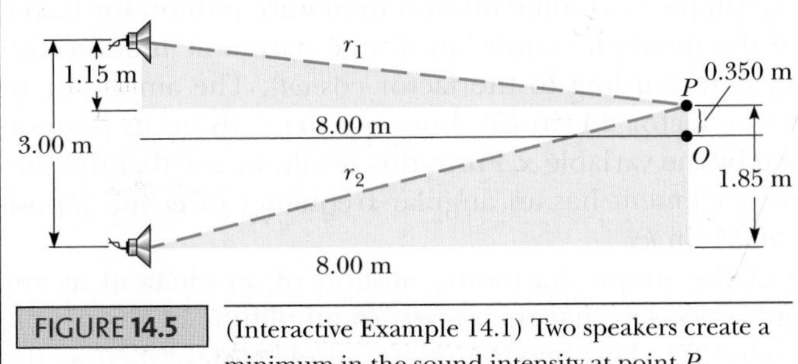

FIGURE 14.5 (Interactive Example 14.1) Two speakers create a minimum in the sound intensity at point *P*.

or, in other words, when their path difference equals $\lambda/2$. To calculate the path difference, we must first find the path lengths r_1 and r_2. Consider the two geometric model triangles shaded in Figure 14.5. Making use of these triangles, we find the path lengths to be

$$r_1 = \sqrt{(8.00 \text{ m})^2 + (1.15 \text{ m})^2} = 8.08 \text{ m}$$

$$r_2 = \sqrt{(8.00 \text{ m})^2 + (1.85 \text{ m})^2} = 8.21 \text{ m}$$

Hence, the path difference is $r_2 - r_1 = 0.13$ m. Because we require that this path difference be equal to $\lambda/2$ for the first minimum, we find that $\lambda = 0.26$ m.

To obtain the oscillator frequency, we use $v = \lambda f$, where v is the speed of sound in air, 343 m/s:

$$f = \frac{v}{\lambda} = \frac{343 \text{ m/s}}{0.26 \text{ m}} = 1.3 \text{ kHz}$$

PhysicsNow™ By logging into PhysicsNow at www.pop4e.com and going to Interactive Example 14.1 you can vary the point *P* at which the first minimum occurs to determine the frequency of the sound waves.

14.3 STANDING WAVES

The sound waves from the speakers in Interactive Example 14.1 leave the speakers in the forward direction, and we considered interference at a point in front of the speakers. Suppose we turn the speakers so that they face each other as in Figure 14.6 and then have them emit sound of the same frequency and amplitude. In this situation, two identical waves travel in opposite directions in the same medium. These waves combine in accordance with the superposition principle.

We can analyze such a situation by considering wave functions for two transverse sinusoidal waves having the same amplitude, frequency, and wavelength but traveling in opposite directions in the same medium:

$$y_1 = A \sin(kx - \omega t) \quad \text{and} \quad y_2 = A \sin(kx + \omega t)$$

where y_1 represents a wave traveling in the $+x$ direction and y_2 represents a wave traveling in the $-x$ direction. According to the principle of superposition, adding these two functions gives the resultant wave function y:

$$y = y_1 + y_2 = A \sin(kx - \omega t) + A \sin(kx + \omega t)$$

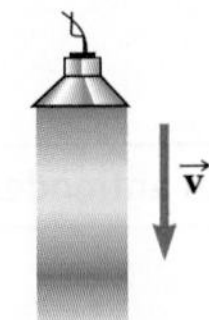

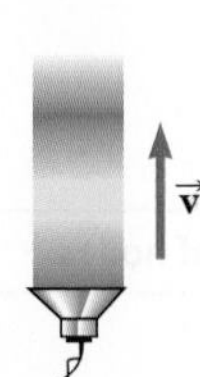

FIGURE 14.6 Two speakers emit sound waves toward each other. Between the speakers, identical waves traveling in opposite directions combine to form standing waves.

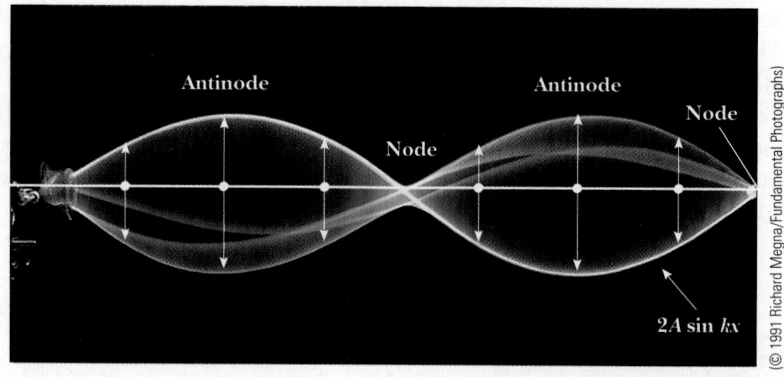

FIGURE 14.7 Multiflash photograph of a standing wave on a string. The vertical displacement from equilibrium of an individual element of the string is proportional to cos ωt. That is, each element vibrates at an angular frequency ω. The amplitude of the vertical oscillation of any element on the string depends on the horizontal position of the element. Each element vibrates within the confines of the envelope function $2A \sin kx$.

PITFALL PREVENTION 14.2

THREE TYPES OF AMPLITUDE We need to distinguish carefully here between the **amplitude of the individual waves,** which is A, and the **amplitude of the simple harmonic motion of the elements of the medium,** which is $2A \sin kx$. A given element in a standing wave vibrates within the constraints of the *envelope* function $2A \sin kx$, where x is that element's position in the medium. That vibration is in contrast to traveling sinusoidal waves, in which all elements oscillate with the same amplitude and the same frequency and the amplitude A of the wave is the same as the amplitude A of the simple harmonic motion of the elements. Furthermore, we can identify the **amplitude of the standing wave** as $2A$.

Using the trigonometric identity $\sin(a \pm b) = \sin a \cos b \pm \cos a \sin b$, this expression reduces to

$$y = (2A \sin kx) \cos \omega t \qquad [14.3]$$

Notice that this function does not look mathematically like a traveling wave because there is no function of $kx - \omega t$. Equation 14.3 represents the wave function of a **standing wave** such as that shown in Figure 14.7. A standing wave is an oscillation pattern that results from two waves traveling in opposite directions. Mathematically, this equation looks more like simple harmonic motion than wave motion for traveling waves. Every element of the medium vibrates in simple harmonic motion with the same angular frequency ω (according to the factor $\cos \omega t$). The amplitude of motion of a given element (the factor $2A \sin kx$), however, depends on its position along the medium, described by the variable x. From this result, we see that the simple harmonic motion of every element has an angular frequency of ω and a position-dependent amplitude of $2A \sin kx$.

Because the amplitude of the simple harmonic motion of an element at any value of x is equal to $2A \sin kx$, we see that the *maximum* amplitude of the simple harmonic motion has the value $2A$. This maximum amplitude is described as the amplitude of the standing wave. It occurs when the coordinate x for an element satisfies the condition $\sin kx = 1$, or when

$$kx = \frac{\pi}{2}, \frac{3\pi}{2}, \frac{5\pi}{2}, \ldots$$

Because $k = 2\pi/\lambda$, the positions of maximum amplitude, called **antinodes,** are

■ Positions of antinodes

$$x = \frac{\lambda}{4}, \frac{3\lambda}{4}, \frac{5\lambda}{4}, \ldots = \frac{n\lambda}{4} \qquad [14.4]$$

where $n = 1, 3, 5, \ldots$. Note that **adjacent antinodes are separated by a distance $\lambda/2$.**

Similarly, the simple harmonic motion has a *minimum* amplitude of zero when x satisfies the condition $\sin kx = 0$, or when $kx = \pi, 2\pi, 3\pi, \ldots$, giving

■ Positions of nodes

$$x = \frac{\lambda}{2}, \lambda, \frac{3\lambda}{2}, \ldots = \frac{n\lambda}{2} \qquad [14.5]$$

where $n = 1, 2, 3, \ldots$. **These points of zero amplitude, called nodes, are also spaced by $\lambda/2$.** The distance between a node and an adjacent antinode is $\lambda/4$. The standing wave patterns produced at various times by two waves traveling in opposite directions are represented graphically in Active Figure 14.8. The upper part of each figure represents the individual traveling waves and the lower part represents the standing wave patterns. The nodes of the standing wave are labeled N and the

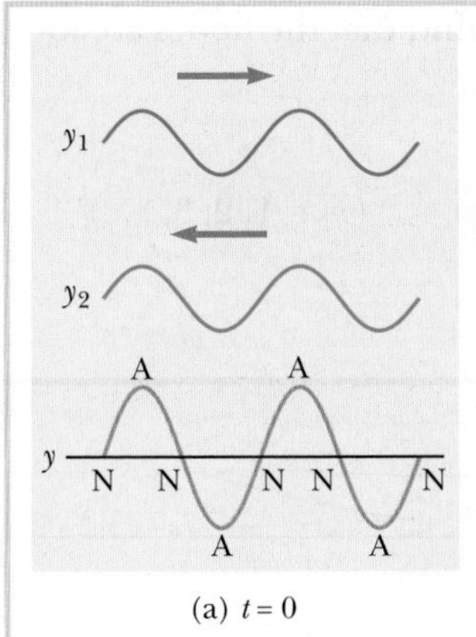

(a) $t = 0$

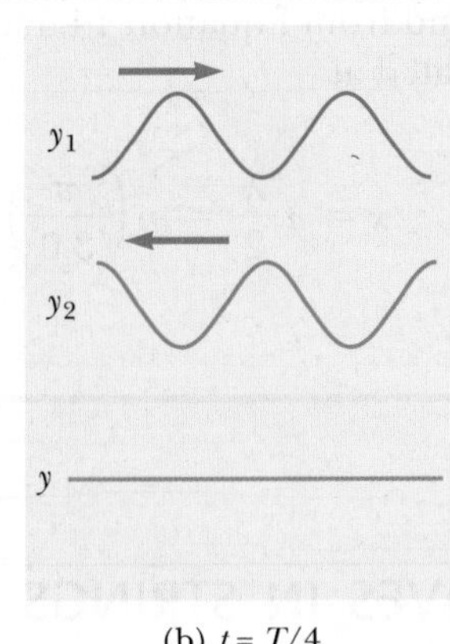

(b) $t = T/4$

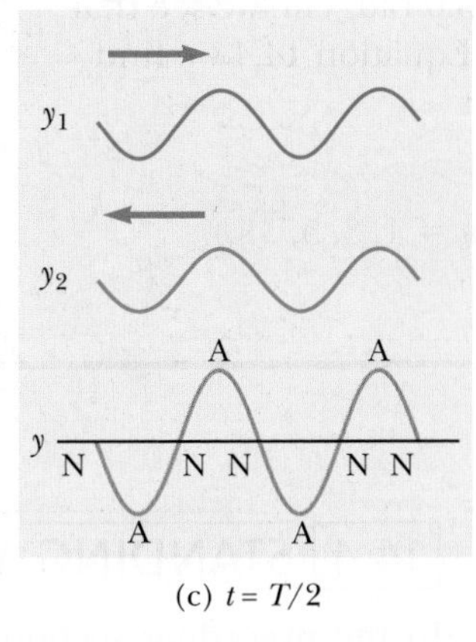

(c) $t = T/2$

ACTIVE FIGURE 14.8
Standing wave patterns at various times produced by two waves of equal amplitude traveling in opposite directions. For the resultant wave y, the nodes (N) are points of zero displacement and the antinodes (A) are points of maximum displacement.

Physics Now™ Log into PhysicsNow at **www.pop4e.com** and go to Active Figure 14.8 to choose the wavelength of the waves and see the standing wave that results.

antinodes are labeled A. At $t = 0$ (Active Fig. 14.8a), the two waves are in phase, giving a wave pattern with amplitude $2A$. One quarter of a period later, at $t = T/4$ (Active Fig. 14.8b), the individual waves have moved one quarter of a wavelength (one to the right and the other to the left). At this time, the waves are 180° out of phase. The individual displacements of the elements of the medium from their equilibrium positions are of equal magnitude and opposite direction for all values of x; hence, the resultant wave has zero displacement everywhere. At $t = T/2$ (Active Fig. 14.8c), the individual waves are again in phase, producing a wave pattern that is inverted relative to the $t = 0$ pattern.

QUICK QUIZ 14.2 Consider Active Figure 14.8 as representing a standing wave on a string. Define the velocity of elements of the string as positive if they are moving upward in the figure. **(i)** At the moment the string has the shape shown at the bottom of Active Figure 14.8a, the instantaneous velocity of elements along the string **(a)** is zero for all elements, **(b)** is positive for all elements, **(c)** is negative for all elements, or **(d)** varies with the position of the element. **(ii)** From the same set of choices, choose the best answer at the moment the string has the shape shown at the bottom of Active Figure 14.8b.

EXAMPLE 14.2 Formation of a Standing Wave

Two transverse waves traveling in opposite directions produce a standing wave. The individual wave functions are

$$y_1 = (4.0 \text{ cm}) \sin(3.0x - 2.0t)$$

$$y_2 = (4.0 \text{ cm}) \sin(3.0x + 2.0t)$$

where x and y are in centimeters.

A Find the maximum transverse position of an element of the medium at $x = 2.3$ cm.

Solution When the two waves are summed, the result is a standing wave whose mathematical representation is given by Equation 14.3, with $A = 4.0$ cm and $k = 3.0$ rad/cm:

$$y = (2A \sin kx) \cos \omega t = [(8.0 \text{ cm}) \sin 3.0x] \cos 2.0t$$

Therefore, the maximum transverse position of an element at the position $x = 2.3$ cm is

$$y_{\max} = [(8.0 \text{ cm}) \sin 3.0x]_{x=2.3 \text{ cm}}$$
$$= (8.0 \text{ cm}) \sin(6.9 \text{ rad}) = 4.6 \text{ cm}$$

B Find the positions of the nodes and antinodes.

Solution Because $k = 2\pi/\lambda = 3.0$ rad/cm, we see that $\lambda = 2\pi/3$ cm. Therefore, from Equation 14.4 we find that the antinodes are located at

$$x = n\left(\frac{\pi}{6.0}\right) \text{cm} \qquad (n = 1, 3, 5, \ldots)$$

and from Equation 14.5 we find that the nodes are located at

$$x = n\frac{\lambda}{2} = n\left(\frac{\pi}{3.0}\right) \text{cm} \qquad (n = 1, 2, 3, \ldots)$$

14.4 STANDING WAVES IN STRINGS

In the preceding section, we discussed standing waves formed by identical waves moving in opposite directions in the same medium. One way to establish a standing wave on a string is to combine incoming and reflected waves from a rigid end. If a string is stretched between *two* rigid supports (Active Fig. 14.9a) and waves are established on the string, standing waves will be set up in the string by the continuous superposition of the waves incident on and reflected from the ends. This physical system is a model for the source of sound in any stringed instrument, such as the guitar, the violin, and the piano. The string has a number of natural patterns of vibration, called **normal modes,** and each mode has a characteristic frequency.

This discussion is our first introduction to an important analysis model, the **wave under boundary conditions.** When boundary conditions are applied to a wave, we find very interesting behavior that has no analog in the physics of particles. The most prominent aspect of this behavior is **quantization.** We shall find that only certain waves—those that satisfy the boundary conditions—are allowed. The notion of quantization was introduced in Chapter 11 when we discussed the Bohr model of the atom. In that model, angular momentum was quantized. As we shall see in Chapter 29, this quantization is just an application of the wave under boundary conditions model.

In the standing wave pattern on a stretched string, the ends of the string must be nodes because these points are fixed, establishing the boundary condition on the waves. The rest of the pattern can be built from this boundary condition along with the requirement that nodes and antinodes are equally spaced and separated by one fourth of a wavelength. The simplest pattern that satisfies these conditions

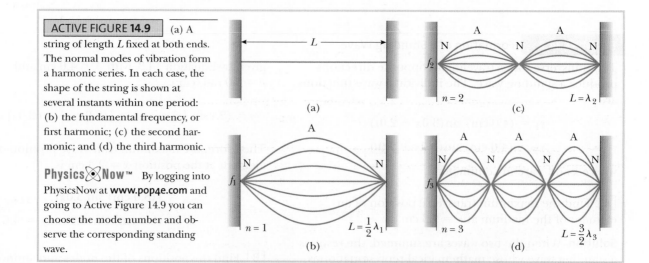

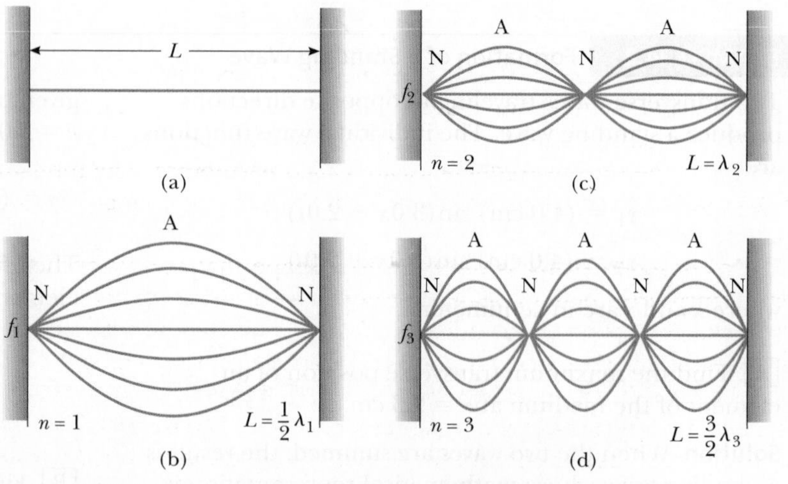

ACTIVE FIGURE 14.9 (a) A string of length L fixed at both ends. The normal modes of vibration form a harmonic series. In each case, the shape of the string is shown at several instants within one period: (b) the fundamental frequency, or first harmonic; (c) the second harmonic; and (d) the third harmonic.

Physics Now™ By logging into PhysicsNow at **www.pop4e.com** and going to Active Figure 14.9 you can choose the mode number and observe the corresponding standing wave.

has the required nodes at the ends of the string and an antinode at the center point (Active Fig. 14.9b). For this normal mode, the length of the string equals $\lambda/2$ (the distance between adjacent nodes):

$$L = \frac{\lambda_1}{2} \quad \text{or} \quad \lambda_1 = 2L$$

The next normal mode, of wavelength λ_2 (Active Fig. 14.9c), occurs when the length of the string equals one wavelength, that is, when $\lambda_2 = L$. In this mode, the two halves of the string are moving in opposite directions at a given instant, and we sometimes say that two *loops* occur. The third normal mode (Active Fig. 14.9d) corresponds to the case when the length equals $3\lambda/2$; therefore, $\lambda_3 = 2L/3$. In general, the wavelengths of the various normal modes can be conveniently expressed as

$$\lambda_n = \frac{2L}{n} \qquad (n = 1, 2, 3, \ldots) \qquad [14.6]$$

▪ Wavelengths of normal modes

where the index n refers to the nth mode of vibration. The natural frequencies associated with these modes are obtained from the relationship $f = v/\lambda$, where the wave speed v is determined by the tension T and linear mass density μ of the string and therefore is the same for all frequencies. Using Equation 14.6, we find that the frequencies of the normal modes are

$$f_n = \frac{v}{\lambda_n} = \frac{n}{2L}v \qquad (n = 1, 2, 3, \ldots) \qquad [14.7]$$

▪ Frequencies of normal modes as functions of wave speed and length of string

Because $v = \sqrt{T/\mu}$ (Equation 13.20), we can express the natural frequencies of a stretched string as

$$f_n = \frac{n}{2L}\sqrt{\frac{T}{\mu}} \qquad (n = 1, 2, 3, \ldots) \qquad [14.8]$$

▪ Frequencies of normal modes as functions of string tension and linear mass density

Equation 14.8 demonstrates the quantization that we mentioned as a feature of the wave under boundary conditions model. The frequencies are quantized because only certain frequencies of waves satisfy the boundary conditions and can exist on the string. The lowest frequency, corresponding to $n = 1$, is called the **fundamental frequency** f_1 and is

$$f_1 = \frac{1}{2L}\sqrt{\frac{T}{\mu}} \qquad [14.9]$$

▪ Fundamental frequency of a stretched string

Equation 14.8 shows that the frequencies of the higher modes are integral multiples of the fundamental frequency, that is, $2f_1$, $3f_1$, $4f_1$, and so on. These higher natural frequencies, together with the fundamental frequency, form a **harmonic series** and the various frequencies are called **harmonics.** The fundamental f_1 is the first harmonic; the frequency $f_2 = 2f_1$ is the second harmonic; and the frequency f_n is the nth harmonic.

If a stretched string is distorted to a shape that corresponds to any one of its harmonics, after being released it will vibrate at the frequency of that harmonic. If the string is plucked, bowed, or struck, however, as occurs when playing a stringed instrument, the resulting vibration will include frequencies of many modes, including the fundamental. In effect, the string "selects" a mixture of normal-mode frequencies when disturbed by a finger or a bow. The frequency of the combination is that of the fundamental because that is the rate at which the waveform repeats; the frequency associated with the string by a listener is that of the fundamental.

The frequency of a given string on a stringed instrument can be changed either by varying the string's tension T or by changing the length L of the vibrating portion of the string. For example, the tension in the strings of guitars and violins is adjusted by a screw mechanism or by tuning pegs on the neck of the instrument. As the tension increases, the frequencies of the normal modes increase according to Equation 14.8. Once the instrument is "tuned," the player varies the frequency by moving his or her fingers along the neck, thereby changing the length of the vibrating portion of the string. As this length is reduced, the frequency increases because the normal-mode frequencies are inversely proportional to the length of the vibrating portion of the string.

Imagine that we have several strings of the same length under the same tension but varying linear mass density μ. The strings will have different wave speeds and therefore different fundamental frequencies. The linear mass density can be changed either by varying the diameter of the string or by wrapping extra mass around the string. Both of these possibilities can be seen on the guitar, on which the higher-frequency strings vary in diameter and the lower-frequency strings have additional wire wrapped around them.

QUICK QUIZ 14.3 Which of the following is true when a standing wave is set up on a string fixed at both ends? **(a)** The number of nodes is equal to the number of antinodes. **(b)** The wavelength is equal to the length of the string divided by an integer. **(c)** The frequency is equal to the number of nodes times the fundamental frequency. **(d)** The center of the string is either a node or an antinode.

INTERACTIVE EXAMPLE 14.3 Give Me a C Note

A middle C string on a piano has a fundamental frequency of 262 Hz, and the A note has a fundamental frequency of 440 Hz.

A Calculate the frequencies of the next two harmonics of the C string.

Solution Because the higher frequencies are integer multiples of the fundamental frequency,

$$f_2 = 2f_1 = \boxed{524 \text{ Hz}}$$

$$f_3 = 3f_1 = \boxed{786 \text{ Hz}}$$

B If the strings for the A and C notes are assumed to have the same mass per unit length and the same length, determine the ratio of tensions in the two strings.

Solution Using Equation 14.9 for the two strings vibrating at their fundamental frequencies gives

$$\left.\begin{aligned} f_{1\text{A}} &= \frac{1}{2L}\sqrt{\frac{T_\text{A}}{\mu}} \\ f_{1\text{C}} &= \frac{1}{2L}\sqrt{\frac{T_\text{C}}{\mu}} \end{aligned}\right\} \rightarrow \frac{f_{1\text{A}}}{f_{1\text{C}}} = \sqrt{\frac{T_\text{A}}{T_\text{C}}}$$

$$\frac{T_\text{A}}{T_\text{C}} = \left(\frac{f_{1\text{A}}}{f_{1\text{C}}}\right)^2 = \left(\frac{440 \text{ Hz}}{262 \text{ Hz}}\right)^2 = \boxed{2.82}$$

C In a real piano, the assumption we made in part B is only partially true. The string densities are equal, but the A string is 64% as long as the C string. What is the ratio of their tensions?

Solution We start from the same point as in part B, but the string lengths do not cancel in the ratio:

$$\left.\begin{aligned} f_{1\text{A}} &= \frac{1}{2L_\text{A}}\sqrt{\frac{T_\text{A}}{\mu}} \\ f_{1\text{C}} &= \frac{1}{2L_\text{C}}\sqrt{\frac{T_\text{C}}{\mu}} \end{aligned}\right\} \rightarrow \frac{f_{1\text{A}}}{f_{1\text{C}}} = \frac{L_\text{C}}{L_\text{A}}\sqrt{\frac{T_\text{A}}{T_\text{C}}}$$

$$\frac{T_\text{A}}{T_\text{C}} = \left(\frac{L_\text{A}}{L_\text{C}}\right)^2\left(\frac{f_{1\text{A}}}{f_{1\text{C}}}\right)^2 = (0.64)^2\left(\frac{440 \text{ Hz}}{262 \text{ Hz}}\right)^2$$

$$= \boxed{1.16}$$

PhysicsNow™ Investigate this situation for different combinations of notes by logging into PhysicsNow at **www.pop4e.com** and going to Interactive Example 14.3.

14.5 STANDING WAVES IN AIR COLUMNS

We have discussed musical instruments that use strings, which include guitars, violins, and pianos. What about instruments classified as brasses or woodwinds? These instruments produce music using a column of air. Standing longitudinal waves can be set up in an air column, such as an organ pipe or a clarinet, as the result of interference between longitudinal sound waves traveling in opposite directions. Whether a node or an antinode occurs at the end of an air column depends on whether that end is open or closed. **The closed end of an air column is a displacement node,** just as the fixed end of a vibrating string is a displacement node. Furthermore, because the pressure wave is 90° out of phase with the displacement wave (Section 13.7), **the closed end of an air column corresponds to a pressure antinode** (i.e., a point of maximum pressure variation). On the other hand, **the open end of an air column is approximately a displacement antinode and a pressure node.**

You may wonder how a sound wave can reflect from an open end because there may not appear to be a change in the medium at this point. It is indeed true that the medium through which the sound wave moves is air both inside and outside the pipe. Sound is a pressure wave, however, and a compression region of the sound wave is constrained by the sides of the pipe as long as the region is inside the pipe. As the compression region exits at the open end of the pipe, the constraint of the pipe is removed and the compressed air is free to expand into the atmosphere. Therefore, there is a change in the *character* of the medium between the inside of the pipe and the outside even though there is no change in the *material* of the medium. This change in character is sufficient to allow some reflection.

Strictly speaking, the open end of an air column is not exactly an antinode. A compression in the sound wave does not reach full expansion until it passes somewhat beyond the open end. Therefore, to calculate frequencies of the normal modes accurately, an **end correction** must be added to the length of the air column at each open end. For a thin-walled tube of circular cross-section, this end correction is about $0.6R$, where R is the tube's radius. Hence, the effective acoustical length of the tube is somewhat greater than the physical length L.

We can determine the modes of vibration of an air column by applying the appropriate boundary condition at the end of the column, along with the requirement that nodes and antinodes be separated by one fourth of a wavelength. We shall find that the frequency for sound waves in air columns is quantized, similar to the results found for waves on strings under boundary conditions.

The first three modes of vibration of an air column that is open at both ends are shown in Figure 14.10a. Note that the ends are displacement antinodes (approximately). In the fundamental mode, the wavelength is twice the length of the air column; hence, the frequency of the fundamental f_1 is $v/2L$. Similarly, the frequencies of the higher harmonics are $2f_1, 3f_1, \ldots$. Therefore, in an air column that is open at both ends, the natural frequencies of vibration form a harmonic series; that is, the higher harmonics are integral multiples of the fundamental frequency. Because all harmonics are present, we can express the natural frequencies of vibration as

$$f_n = n\frac{v}{2L} \qquad (n = 1, 2, 3, \ldots) \qquad [14.10]$$

▮ Natural frequencies of an air column open at both ends

where v is the speed of sound in air.

If an air column is closed at one end and open at the other, the closed end is a displacement node and the open end is a displacement antinode (Fig. 14.10b). In this case, the wavelength for the fundamental mode is four times the length of the column. Hence, the fundamental frequency f_1 is equal to $v/4L$, and the frequencies

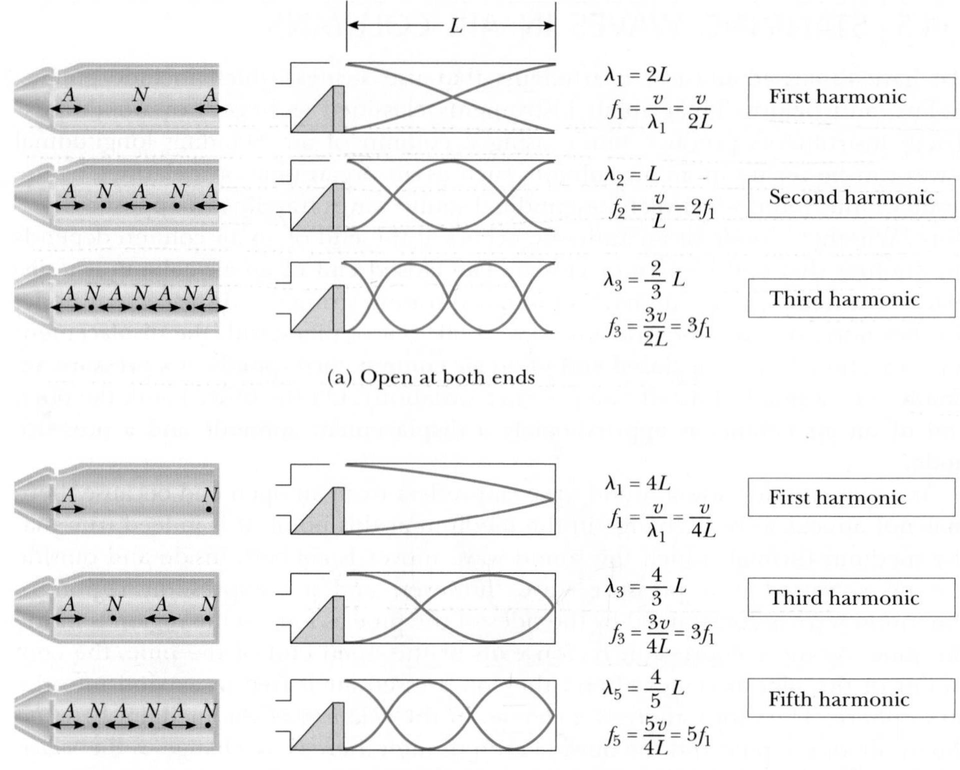

FIGURE 14.10 Motion of elements of air in standing longitudinal waves in an air column, along with graphical representations of the displacements of the elements. (a) In an air column open at both ends, the harmonic series created consists of all integer multiples of the fundamental frequency: $f_1, 2f_1, 3f_1, \ldots$. (b) In an air column closed at one end and open at the other, the harmonic series consists of only odd-integer multiples of the fundamental frequency: $f_1, 3f_1, 5f_1, \ldots$.

of the higher harmonics are equal to $3f_1, 5f_1, \ldots$. That is, **in an air column that is closed at one end, only odd harmonics are present,** and the frequencies are

■ **Natural frequencies of an air column closed at one end and open at the other**

$$f_n = n\frac{v}{4L} \qquad (n = 1, 3, 5, \ldots) \qquad [14.11]$$

Standing waves in air columns are the primary sources of the sounds produced by wind instruments. In a woodwind instrument, a key is pressed, which opens a hole in the side of the column. This hole defines the end of the vibrating column of air (because the hole acts as an open end at which pressure can be released), so that the column is effectively shortened and the fundamental frequency rises. In a brass instrument, the length of the air column is changed by an adjustable section, as in a trombone, or by adding segments of tubing, as is done in a trumpet when a valve is pressed.

PITFALL PREVENTION 14.3

SOUND WAVES IN AIR ARE NOT TRANSVERSE Note that the standing longitudinal waves are drawn as transverse waves in Figure 14.10. It is difficult to draw longitudinal displacements because they are in the same direction as the propagation. Therefore, it is best to interpret the curves in Figure 14.10 as a graphical representation of the waves (our diagrams of string waves are pictorial representations), with the vertical axis representing horizontal position of the elements of the medium.

QUICK QUIZ 14.4 Standing waves in a pipe open at both ends are excited at a fundamental frequency f_{open}. When one end is closed and the pipe is again excited, the fundamental frequency is f_{closed}. Which of the following expressions describes how these two frequencies that are heard compare? **(a)** $f_{\text{closed}} = f_{\text{open}}$ **(b)** $f_{\text{closed}} = \frac{1}{2}f_{\text{open}}$ **(c)** $f_{\text{closed}} = 2f_{\text{open}}$ **(d)** $f_{\text{closed}} = \frac{3}{2}f_{\text{open}}$

QUICK QUIZ 14.5 Balboa Park in San Diego has an outdoor organ. When the air temperature increases, what happens to the fundamental frequency of one of the organ pipes? **(a)** It stays the same. **(b)** It goes down. **(c)** It goes up. **(d)** It is impossible to determine.

■ Thinking Physics 14.2

A bugle has no valves, keys, slides, or finger holes. How can it play a song?

Reasoning Songs for the bugle are limited to harmonics of the fundamental frequency because the bugle has no control over frequencies by means of valves, keys, slides, or finger holes. The player obtains different notes by changing the tension in the lips as the bugle is played to excite different harmonics. The normal playing range of a bugle is among the third, fourth, fifth, and sixth harmonics of the fundamental. As examples, "Reveille" is played with just the three notes D (294 Hz), G (392 Hz), and B (490 Hz), and "Taps" is played with these same three notes and the D one octave above the lower D (588 Hz). Note that the frequencies of these four notes are, respectively, three, four, five, and six times the fundamental of 98 Hz. ■

■ Thinking Physics 14.3

If an orchestra doesn't warm up before a performance, the strings go flat and the wind instruments go sharp during the performance. Why?

Reasoning Without warming up, all the instruments will be at room temperature at the beginning of the concert. As the wind instruments are played, they fill with warm air from the player's exhalation. The increase in temperature of the air in the instrument causes an increase in the speed of sound, which raises the fundamental frequencies of the air columns. As a result, the wind instruments go sharp. The strings on the stringed instruments also increase in temperature due to the friction of rubbing with the bow. This increase in temperature results in thermal expansion, which causes a decrease in the tension in the strings. (We will study thermal expansion in Chapter 16.) With a decrease in tension, the wave speed on the strings drops and the fundamental frequencies decrease. Therefore, the stringed instruments go flat. ■

EXAMPLE 14.4 Harmonics in a Pipe

A pipe has a length of 1.23 m.

A Determine the frequencies of the first three harmonics if the pipe is open at each end. Take $v = 343$ m/s as the speed of sound in air.

Solution The first harmonic of a pipe open at both ends is

$$f_1 = \frac{v}{2L} = \frac{343 \text{ m/s}}{2(1.23 \text{ m})} = \boxed{139 \text{ Hz}}$$

Because all harmonics are possible for a pipe open at both ends, the second and third harmonics are $f_2 = 2f_1 = \boxed{278 \text{ Hz}}$ and $f_3 = 3f_1 = \boxed{417 \text{ Hz}}$.

B What are the three frequencies requested in part A if the pipe is closed at one end?

Solution The fundamental frequency of a pipe closed at one end is

$$f_1 = \frac{v}{4L} = \frac{343 \text{ m/s}}{4(1.23 \text{ m})} = \boxed{69.7 \text{ Hz}}$$

In this case, only odd harmonics are present, so the next two harmonics have frequencies $f_3 = 3f_1 = \boxed{209 \text{ Hz}}$ and $f_5 = 5f_1 = \boxed{349 \text{ Hz}}$.

C For the pipe open at both ends, how many harmonics are present in the normal human hearing range (20–20 000 Hz)?

Solution Because all harmonics are present, $f_n = nf_1$. For $f_n = 20\,000$ Hz, we have $n = 20\,000/139 = 144$, so $\boxed{144}$ harmonics are present in the audible range. Actually, only the first few harmonics have sufficient amplitude to be heard.

EXAMPLE 14.5 Measuring the Frequency of a Tuning Fork

A simple apparatus for demonstrating standing waves in a tube is described in Figure 14.11a. A long, vertical tube open at both ends is partially submerged in a beaker of water, and a vibrating tuning fork of unknown frequency is placed near the top. The length L of the air column is adjusted by moving the tube vertically. The sound waves generated by the fork excite a resonance response in the air column when its length equals that associated with one of the harmonic frequencies of the tube.

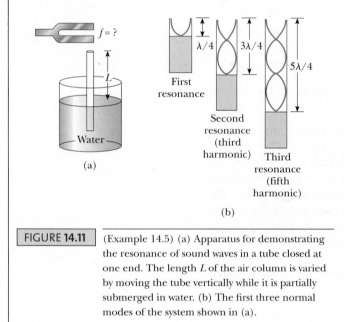

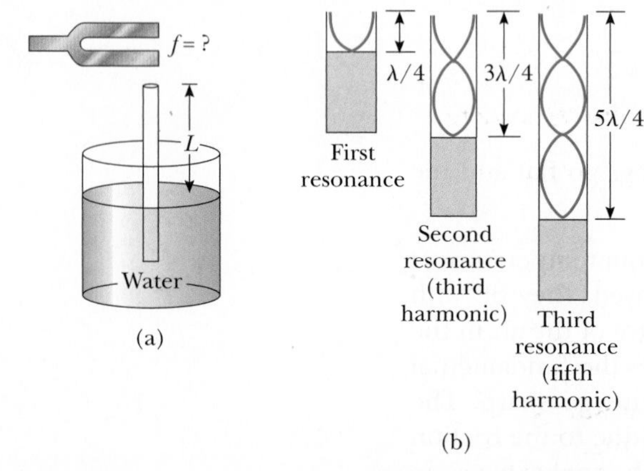

FIGURE 14.11 (Example 14.5) (a) Apparatus for demonstrating the resonance of sound waves in a tube closed at one end. The length L of the air column is varied by moving the tube vertically while it is partially submerged in water. (b) The first three normal modes of the system shown in (a).

For a certain tube, the smallest value of L for which a peak occurs in the sound intensity is 9.00 cm. From this measurement, determine the frequency of the tuning fork and the value of L for the next two resonant modes.

Solution To conceptualize the problem, we realize that although the tube is open at the bottom end to allow the water in, the water surface acts like a rigid barrier at that end. This setup can therefore be categorized as an air column closed at one end, and the fundamental has frequency $v/4L$, where L is the length of the tube from the open end to the water surface (Fig. 14.11b). To analyze the problem, we take $v = 343$ m/s for the speed of sound in air and $L = 0.090\,0$ m. Then, from Equation 14.11, we have

$$f_1 = \frac{v}{4L} = \frac{343 \text{ m/s}}{4(0.090\,0 \text{ m})} = 953 \text{ Hz}$$

From the information about the fundamental mode, we see that the wavelength is $\lambda = 4L = 0.360$ m. Because the frequency of the source is constant, the next two natural modes (Fig. 14.11b) correspond to lengths of $3\lambda/4 = 0.270$ m and $5\lambda/4 = 0.450$ m.

To finalize the problem, note in Figure 14.11 that the wavelength of the sound remains fixed because it is determined by the tuning fork. A resonance response occurs whenever the level of the water coincides with a node of the standing wave. In this condition, the driving frequency of the tuning fork matches the natural frequency of the air column, and the amplitude of the sound increases.

14.6 BEATS: INTERFERENCE IN TIME

The interference phenomena that we have discussed so far involve the superposition of two or more waves with the same frequency. Because the resultant displacement of an element in the medium in this case depends on the position of the element, we can refer to the phenomenon as **spatial interference.** Standing waves in strings and air columns are common examples of spatial interference.

We now consider another type of interference effect, one that results from the superposition of two waves with slightly *different* frequencies. In this case, when the two waves of amplitudes A_1 and A_2 are observed at a given point, they are alternately in and out of phase. We refer to this phenomenon as **interference in time** or **temporal interference.** When the waves are in phase, the combined amplitude is $A_1 + A_2$. When they are out of phase, the combined amplitude is $|A_1 - A_2|$. The combination therefore varies between small and large amplitudes, resulting in what we call **beats.**

Although beats occur for all types of waves, they are particularly noticeable for sound waves. For example, if two tuning forks of slightly different frequencies are struck, you hear a sound of pulsating intensity.

The number of beats you hear per second, the **beat frequency,** equals the difference in frequency between the two sources. The maximum beat frequency that the human ear can detect is about 20 beats/s. When the beat frequency exceeds this value, it blends with the sounds producing the beats.

One can use beats to tune a stringed instrument, such as a piano, by beating a note against a reference tone of known frequency. The frequency of the string can then be adjusted to equal the frequency of the reference by changing the string's tension until the beats disappear; the two frequencies are then the same.

Let us look at the mathematical representation of beats. Consider two waves with equal amplitudes traveling through a medium with slightly different frequencies f_1 and f_2. We can represent the position of an element of the medium associated with each wave at a fixed point, which we choose as $x = 0$, as

$$y_1 = A\cos 2\pi f_1 t \qquad \text{and} \qquad y_2 = A\cos 2\pi f_2 t$$

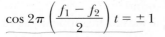

Using the superposition principle, we find that the resultant position is given by

$$y = y_1 + y_2 = A(\cos 2\pi f_1 t + \cos 2\pi f_2 t)$$

It is convenient to write this expression in a form that uses the trigonometric identity

$$\cos a + \cos b = 2\cos\left(\frac{a-b}{2}\right)\cos\left(\frac{a+b}{2}\right)$$

Letting $a = 2\pi f_1 t$ and $b = 2\pi f_2 t$, we find that

$$y = \left[2A\cos 2\pi\left(\frac{f_1 - f_2}{2}\right)t\right]\cos 2\pi\left(\frac{f_1 + f_2}{2}\right)t \qquad [14.12]$$

Graphs demonstrating the individual waves as well as the resultant wave are shown in Active Figure 14.12. From the factors in Equation 14.12, we see that the resultant wave has an effective frequency equal to the average frequency $(f_1 + f_2)/2$ and an amplitude of

$$A_{x=0} = 2A\cos 2\pi\left(\frac{f_1 - f_2}{2}\right)t \qquad [14.13]$$

That is, the *amplitude varies in time* with a frequency of $(f_1 - f_2)/2$. When f_1 is close to f_2, this amplitude variation is slow compared with the frequency of the individual waves, as illustrated by the envelope (broken line) of the resultant wave in Active Figure 14.12b.

Note that a maximum in amplitude will be detected whenever

$$\cos 2\pi\left(\frac{f_1 - f_2}{2}\right)t = \pm 1$$

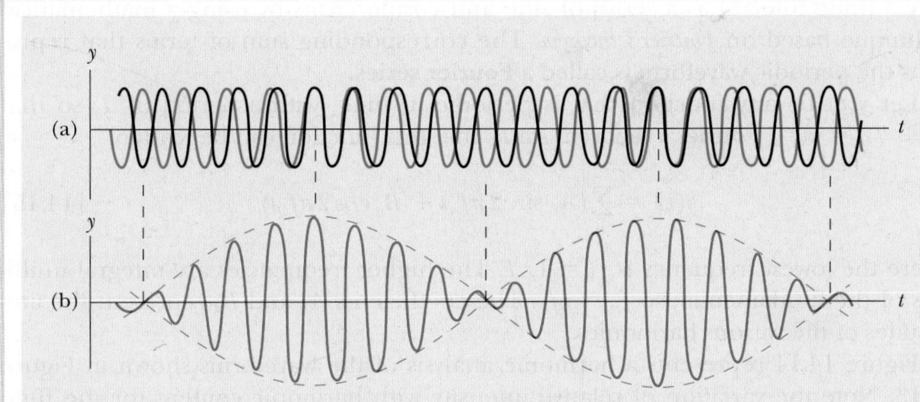

ACTIVE FIGURE 14.12 Beats are formed by the combination of two waves of slightly different frequencies. (a) The blue and black curves represent the individual waves. (b) The combined wave has an amplitude (broken line) that oscillates in time.

PhysicsNow™ By logging into PhysicsNow at **www.pop4e.com** and going to Active Figure 14.12 you can choose the two frequencies and observe the corresponding beats.

That is, the amplitude maximizes twice in each cycle of the function on the left in the preceding expression. Therefore, the number of beats per second, or the beat frequency f_b, is twice the frequency of this function:

■ Beat frequency

$$f_b = |f_1 - f_2| \qquad [14.14]$$

For instance, if two tuning forks vibrate individually at frequencies of 438 Hz and 442 Hz, respectively, the resultant sound wave of the combination has a frequency of $(f_1 + f_2)/2 = 440$ Hz (the musical note A) and a beat frequency of $|f_1 - f_2| = 4$ Hz. That is, the listener hears the 440-Hz sound wave go through an intensity maximum four times every second.

QUICK QUIZ 14.6 You are tuning a guitar by comparing the sound of the string with that of a standard tuning fork. You notice a beat frequency of 5 Hz when both sounds are present. You tighten the guitar string and the beat frequency rises to 8 Hz. To tune the string exactly to the tuning fork, what should you do? **(a)** Continue to tighten the string. **(b)** Loosen the string. **(c)** It is impossible to determine.

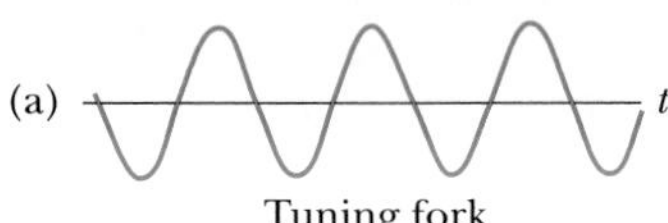

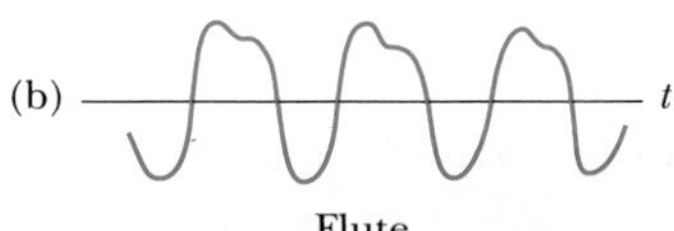

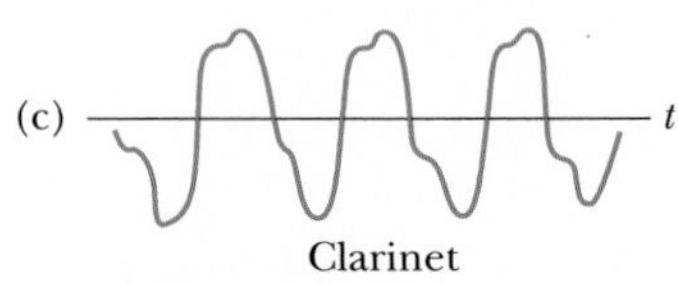

FIGURE 14.13 Waveforms of sound produced by (a) a tuning fork, (b) a flute, and (c) a clarinet, each at approximately the same frequency.

14.7 NONSINUSOIDAL WAVE PATTERNS

The sound wave patterns produced by most instruments are not sinusoidal. Some characteristic waveforms produced by a tuning fork, a flute, and a clarinet are shown in Figure 14.13. Although each instrument has its own characteristic pattern, Figure 14.13 shows that all three waveforms are periodic. A struck tuning fork produces primarily one harmonic (the fundamental), whereas the flute and clarinet produce many frequencies, which include the fundamental and various harmonics. The nonsinusoidal waveforms produced by a violin or clarinet, and the corresponding richness of musical tones, are the result of the superposition of various harmonics.

This phenomenon is in contrast to a percussive musical instrument, such as the drum, in which the combination of frequencies does not form a harmonic series. When frequencies that are integer multiples of a fundamental frequency are combined, the result is a *musical* sound. A listener can assign a pitch to the sound based on the fundamental frequency. Pitch is a psychological reaction to a sound that allows the listener to place the sound on a scale of low to high (bass to treble). Combinations of frequencies that are not integer multiples of a fundamental result in a *noise* rather than a musical sound. It is much harder for a listener to assign a pitch to a noise than to a musical sound.

PITFALL PREVENTION 14.4

PITCH VERSUS FREQUENCY A very common mistake made in speech when talking about sound is to use the term *pitch* when one means *frequency*. Frequency is the physical measurement of the number of oscillations per second, as we have defined. Pitch is a psychological reaction of humans to sound that enables a human to place the sound on a scale from high to low or from treble to bass. Therefore, frequency is the stimulus and pitch is the response. Although pitch is related mostly (but not completely) to frequency, they are not the same. A phrase such as "the pitch of the sound" is incorrect because pitch is not a physical property of the sound.

Analysis of nonsinusoidal waveforms appears at first sight to be a formidable task. If the waveform is periodic, however, it can be represented with arbitrary precision by the combination of a sufficiently large number of sinusoidal waves that form a harmonic series. In fact, one can represent any periodic function or any function over a finite interval as a series of sine and cosine terms by using a mathematical technique based on *Fourier's theorem.* The corresponding sum of terms that represents the periodic waveform is called a **Fourier series.**

Let $y(t)$ be any function that is periodic in time, with a period of T, so that $y(t + T) = y(t)$. **Fourier's theorem** states that this function can be written

■ Fourier's theorem

$$y(t) = \sum_n (A_n \sin 2\pi f_n t + B_n \cos 2\pi f_n t) \qquad [14.15]$$

where the lowest frequency is $f_1 = 1/T$. The higher frequencies are integral multiples of the fundamental, so $f_n = nf_1$. The coefficients A_n and B_n represent the amplitudes of the various harmonics.

Figure 14.14 represents a harmonic analysis of the waveforms shown in Figure 14.13. Note the variation of relative intensity with harmonic content for the flute and the clarinet. In general, any musical sound contains components that are members of a harmonic series with varying relative intensities.

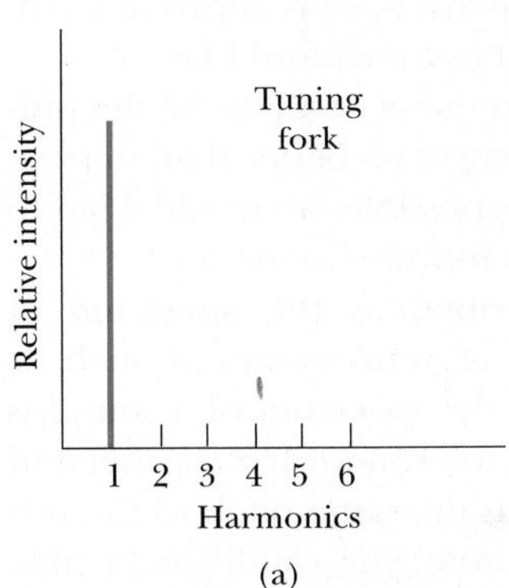

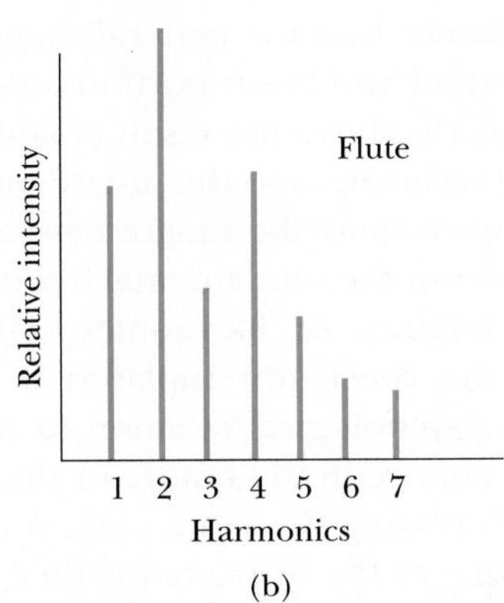

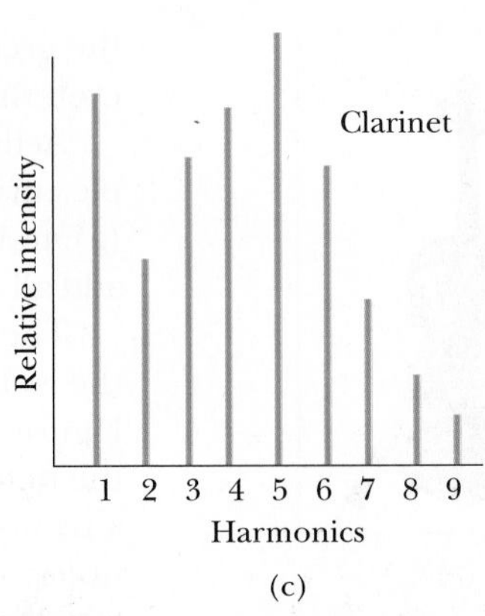

FIGURE 14.14 Harmonics of the waveforms shown in Figure 14.13. Note the variations in intensity of the various harmonics.

We have discussed the *analysis* of a wave pattern using Fourier's theorem. The analysis involves determining the coefficients of the trigonometric functions in Equation 14.15 from a knowledge of the wave pattern. We can also perform the reverse process, *Fourier synthesis.* In this process, the various harmonics are added together to form a resultant wave pattern. As an example of Fourier synthesis, consider the building of a square wave as shown in Active Figure 14.15. The symmetry of the square wave results in only odd multiples of the fundamental combining in the synthesis. In Active Figure 14.15a, the brown curve shows the combination of f and $3f$. In Active Figure 14.15b, we have added $5f$ to the combination and obtained

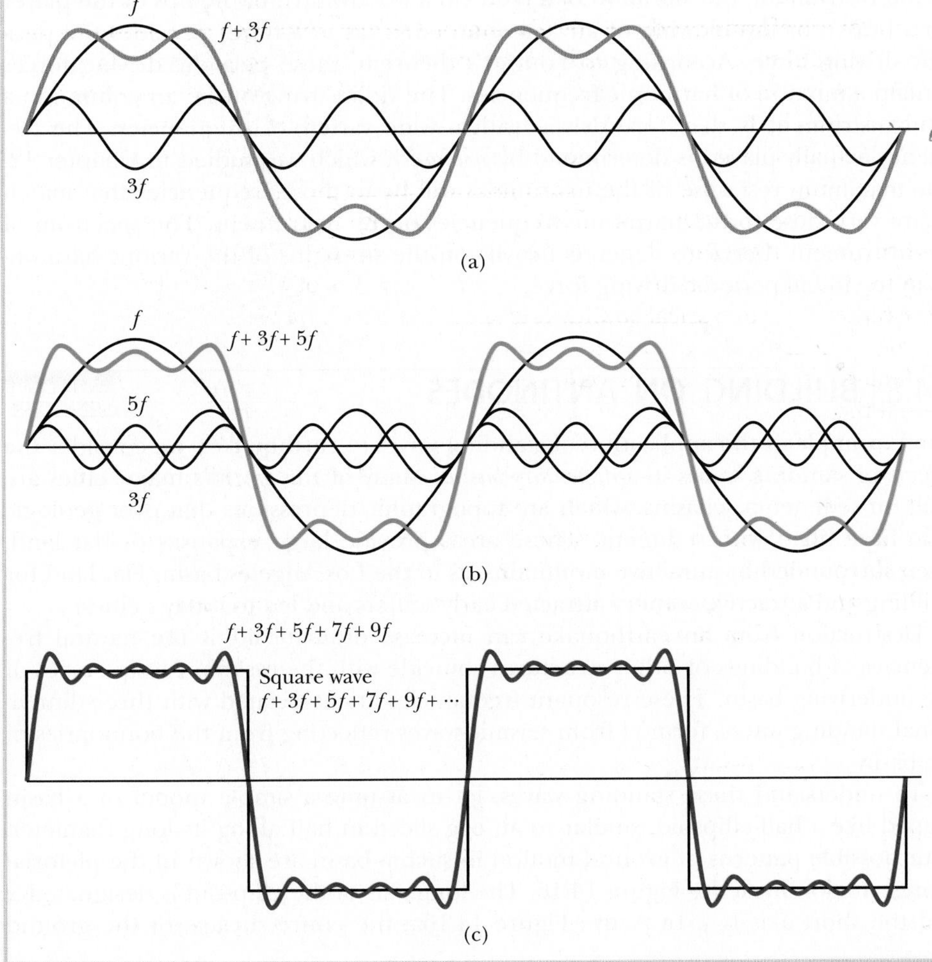

ACTIVE FIGURE 14.15 Fourier synthesis of a square wave represented by the sum of odd multiples of the first harmonic, which has frequency f. (a) Waves of frequency f and $3f$ are added. (b) One more odd frequency of $5f$ is added. (c) The synthesis curve approaches the square wave when odd frequencies up to $9f$ are combined.

PhysicsNow™ By logging into PhysicsNow at **www.pop4e.com** and going to Active Figure 14.15 you can add in harmonics with frequencies higher than $9f$ to try to synthesize a square wave.

(a)

(b)

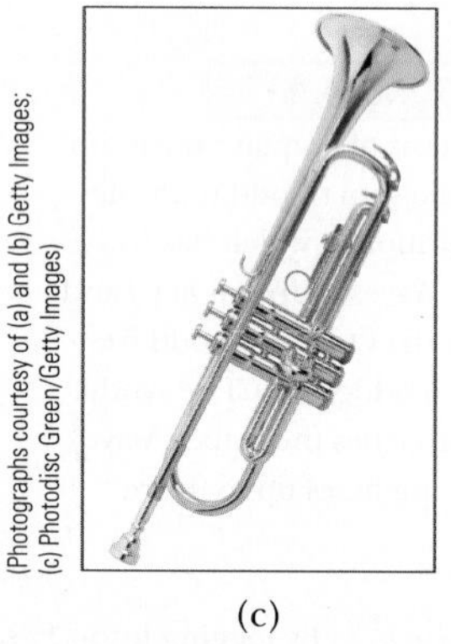

(c)

(Photographs courtesy of (a) and (b) Getty Images; (c) Photodisc Green/Getty Images)

Each musical instrument has its own characteristic sound and mixture of harmonics. Instruments shown are (a) the violin, (b) the saxophone, and (c) the trumpet. ■

the green curve. Notice how the general shape of the square wave is approximated, even though the upper and lower portions are not as flat as they should be.

Active Figure 14.15c shows the result of adding odd frequencies up to $9f$, the purple curve. This approximation to the square wave (black curve) is better than in parts (a) and (b). To approximate the square wave as closely as possible, we would need to add all odd multiples of the fundamental frequency up to infinite frequency.

The physical mixture of harmonics can be described as the **spectrum** of the sound, with the spectrum displayed in a graphical representation such as Figure 14.14. The psychological reaction to changes in the spectrum of a sound is the detection of a change in the **timbre** or the **quality** of the sound. If a clarinet and a trumpet are both playing the same note, you will assign the same pitch to the two notes. Yet if only one of the instruments then plays the note, you will likely be able to tell which instrument is playing. The sounds you hear from the two instruments differ in timbre because of a different physical mixture of harmonics. For example, the timbre due to the sound of a trumpet is different from that of a clarinet. You have probably developed words to describe timbres of various instruments, such as "brassy," "mellow," and "tinny."

Fourier's theorem allows us to understand the excitation process of musical instruments. In a stringed instrument that is plucked, such as a guitar, the string is pulled aside and released. After release, the string oscillates almost freely; a small damping causes the amplitude to decay to zero eventually. The mixture of harmonic frequencies depends on the length of the string, its linear mass density, and the plucking point.

On the other hand, a bowed stringed instrument, such as a violin, or a wind instrument is a forced oscillator. In the case of the violin, the alternate sticking and slipping of the bow on the string provides the periodic driving force. In the case of a wind instrument, the vibration of a reed (in a woodwind), of the lips of the player (in a brass), or the blowing of air across an edge (as in a flute) provides the periodic driving force. According to Fourier's theorem, these periodic driving forces contain a mixture of harmonic frequencies. The violin string or the air column in a wind instrument is therefore driven with a wide variety of frequencies. The frequency actually played is determined by *resonance,* which we studied in Chapter 12. The maximum response of the instrument will be to those frequencies that match or are very close to the harmonic frequencies of the instrument. The spectrum of the instrument therefore depends heavily on the strengths of the various harmonics in the initial periodic driving force.

14.8 BUILDING ON ANTINODES

CONTEXT CONNECTION

As an example of the application of standing waves to earthquakes, we consider the effects of standing waves in *sedimentary basins.* Many of the world's major cities are built on sedimentary basins, which are topographic depressions that over geologic time have filled with sediment. These areas provide large expanses of flat land, often surrounded by attractive mountains, as in the Los Angeles basin. Flat land for building and attractive scenery attracted early settlers and led to today's cities.

Destruction from an earthquake can increase dramatically if the natural frequencies of buildings or other structures coincide with the resonant frequencies of the underlying basin. These resonant frequencies are associated with three-dimensional standing waves, formed from seismic waves reflecting from the boundaries of the basin.

To understand these standing waves, let us assume a simple model of a basin shaped like a half-ellipsoid, similar to an egg sliced in half along its long diameter. Four possible patterns of ground motion in such a basin are shown in the pictorial representation in Active Figure 14.16. The long axis of the ellipsoid is designated x and the short axis is y. In Active Figure 14.16a, the entire surface of the ground

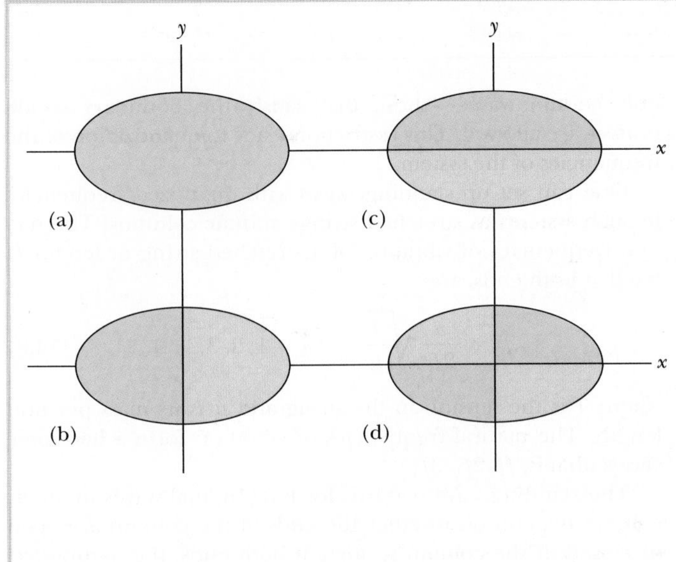

ACTIVE FIGURE 14.16 Overhead views of standing waves in a basin shaped like a half-ellipsoid. In each case, if the blue element is above the plane of the page at an instant of time, the yellow element is below the plane of the page. (a) In the fundamental mode of oscillation, the only nodal line is the rim of the basin and the entire surface moves up and down together. (b) The surface oscillates in two halves, with a nodal line along the y axis. (c) The surface oscillates in two halves, with a nodal line along the x axis. (d) The surface oscillates in four quarters, with nodal lines along both the x and y axes.

Physics Now™ By logging into PhysicsNow at **www.pop4e.com** and going to Active Figure 14.16 you can observe a three-dimensional view of the oscillations for the four modes shown in the figure.

moves up and down (that is, in and out of the page) except at a nodal curve running around the edge of the basin.

In Figures 14.16b and 14.16c, half the ground surface lies above and half lies below the equilibrium position, and each half oscillates up and down on either side of a nodal line. The nodal line is along the y axis in Active Figure 14.16b and along the x axis in Active Figure 14.16c. In Active Figure 14.16d, nodal lines occur along both the x and y axes and the surface oscillates in four segments, with two above the equilibrium position at any time and the other two below.

The standing wave patterns in a basin arise from seismic waves traveling horizontally between the boundaries of the basin. For structures built on sedimentary basins, the degree of seismic risk will depend on the standing wave modes excited by the interference of seismic waves trapped in the basin. It is clear that structures built on regions of maximum ground motion (i.e., the antinodes) will suffer maximum shaking, whereas structures residing near nodes will experience relatively mild ground motion. These considerations appear to have played an important role in the selective destruction that occurred in Mexico City in the Michoacán earthquake in 1985 and in the 1989 Loma Prieta earthquake, which caused the collapse of a section of the Nimitz Freeway in Oakland, California.

A similar effect occurs in bounded bodies of water, such as harbors and bays. A standing wave pattern established in such a body of water is called a **seiche.** This wave pattern can result in variations in the water level that exhibit a period of several minutes, superposed on the longer-period tidal variations. Seiches can be caused by earthquakes, tsunamis, winds, or weather disturbances. You can create a seiche in your bathtub by sliding back and forth at just the right frequency such that the water sloshes back and forth at such a large amplitude that much of it spills out onto the floor.

During the Northridge earthquake of 1994, swimming pools throughout southern California overflowed as a result of seiches set up by the shaking of the ground. In a more dramatic example, seismic waves from the 1964 Alaska earthquake caused severe seiches in the bays and bayous of Louisiana, some causing the water level to shift by 2 m.

We have now considered the role of standing waves in the damage caused by an earthquake. In the Context Conclusion, we will gather together the principles of vibrations and waves that we have learned to respond more fully to the central question of this Context. ■

SUMMARY

PhysicsNow™ Take a practice test by logging into PhysicsNow at www.pop4e.com and clicking on the Pre-Test link for this chapter.

The **principle of superposition** states that if two or more traveling waves are moving through a medium and combine at a given point, the resultant position of the element of the medium at that point is the sum of the positions due to the individual waves.

When two waves with equal amplitudes and frequencies superpose, the resultant wave has an amplitude that depends on the phase angle ϕ between the two waves. **Constructive interference** occurs when the two waves are *in phase* everywhere, corresponding to $\phi = 0, 2\pi, 4\pi, \ldots$. **Destructive interference** occurs when the two waves are 180° out of phase everywhere, corresponding to $\phi = \pi, 3\pi, 5\pi, \ldots$.

Standing waves are formed from the superposition of two sinusoidal waves that have the same frequency, amplitude, and wavelength but are traveling in *opposite* directions. The resultant standing wave is described by the wave function

$$y = (2A \sin kx) \cos \omega t \qquad [14.3]$$

The maximum amplitude points (called **antinodes**) are separated by a distance $\lambda/2$. Halfway between antinodes are points of zero amplitude (called **nodes**).

The **wave under boundary conditions** model tells us that when boundary conditions are applied to a wave, we find that only certain waves—those that satisfy the boundary conditions—are allowed. This restriction leads to **quantization** of the frequencies of the system.

One can set up standing waves with quantized frequencies in such systems as stretched strings and air columns. The natural frequencies of vibration of a stretched string of length L, fixed at both ends, are

$$f_n = \frac{n}{2L}\sqrt{\frac{T}{\mu}} \qquad (n = 1, 2, 3, \ldots) \qquad [14.8]$$

where T is the tension in the string and μ is its mass per unit length. The natural frequencies of vibration form a **harmonic series,** that is, $f_1, 2f_1, 3f_1, \ldots$.

The standing wave patterns for longitudinal waves in an air column depend on whether the ends of the column are open or closed. If the column is open at both ends, the natural frequencies of vibration form a harmonic series. If one end is closed, only odd harmonics of the fundamental are present.

The phenomenon of **beats** occurs as a result of the superposition of two traveling waves of slightly different frequencies. For sound waves at a given point, one hears an alternation in sound intensity with time.

Any periodic waveform can be represented by the combination of sinusoidal waves that form a harmonic series. The process is based on **Fourier's theorem.**

QUESTIONS

☐ = answer available in the *Student Solutions Manual and Study Guide*

1. Does the phenomenon of wave interference apply only to sinusoidal waves?
2. Can two pulses traveling in opposite directions on the same string reflect from each other? Explain.
3. When two waves interfere, can the amplitude of the resultant wave be greater than either of the two original waves? If so, under what conditions can that happen?
4. For certain positions of the movable section shown in Figure 14.4, no sound is detected at the receiver, a situation corresponding to destructive interference. This situation suggests that energy is somehow lost. What happens to the energy transmitted by the speaker?
5. When two waves interfere constructively or destructively, is there any gain or loss in energy? Explain.
6. What limits the amplitude of motion of a real vibrating system that is driven at one of its resonant frequencies?
7. Explain why your voice seems to sound better than usual when you sing in the shower.
8. What is the purpose of the slide on a trombone or of the valves on a trumpet?
9. Why does a vibrating guitar string sound louder when placed on the instrument than it would if allowed to vibrate in air while off the instrument?
10. Explain why all harmonics are present in an organ pipe open at both ends, but only the odd harmonics are present in a pipe closed at one end.
11. An archer shoots an arrow from a bow. Does the string of the bow exhibit standing waves after the arrow leaves? If so, and if the bow is perfectly symmetric so that the arrow leaves from the center of the string, what harmonics are excited?
12. Explain how a musical instrument such as a piano may be tuned by using the phenomenon of beats.
13. An airplane mechanic notices that the sound from a twin-engine aircraft rapidly varies in loudness when both engines are running. What could be causing this variation from loud to soft?
14. Despite a reasonably steady hand, a person often spills his coffee when carrying it to his seat. Discuss resonance as a possible cause of this difficulty and devise a means for solving the problem.
15. You have a standard tuning fork whose frequency is 262 Hz and a second tuning fork with an unknown frequency. When you tap both of them on the heel of one of your sneakers, you hear beats with a frequency of 4 per second.

Thoughtfully chewing your gum, you wonder whether the unknown frequency is 258 Hz or 266 Hz. How can you decide?

16. When the base of a vibrating tuning fork is placed against a chalkboard, the sound that it emits becomes louder because the vibrations of the tuning fork are transmitted to the chalkboard. Because it has a larger area than the tuning fork, the vibrating chalkboard sets more air into vibration. Therefore, the chalkboard is a better radiator of sound than the tuning fork. How does that affect the time interval during which the fork vibrates? Does that agree with the principle of conservation of energy?

PROBLEMS

1, 2, 3 = straightforward, intermediate, challenging
☐ = full solution available in the *Student Solutions Manual and Study Guide*
PhysicsNow = coached problem with hints available at www.pop4e.com
= computer useful in solving problem
= paired numerical and symbolic problems
= biomedical application

Section 14.1 ▪ The Principle of Superposition

1. Two waves in one string are described by the wave functions

$$y_1 = 3.0 \cos(4.0x - 1.6t) \quad \text{and} \quad y_2 = 4.0 \sin(5.0x - 2.0t)$$

where y and x are in centimeters and t is in seconds. Find the superposition of the waves $y_1 + y_2$ at the points (a) $x = 1.00$, $t = 1.00$; (b) $x = 1.00$, $t = 0.500$; and (c) $x = 0.500$, $t = 0$. (Remember that the arguments of the trigonometric functions are in radians.)

2. Two pulses A and B are moving in opposite directions along a taut string with a speed of 2.00 cm/s. The amplitude of A is twice the amplitude of B. The pulses are shown in Figure P14.2 at $t = 0$. Sketch the shape of the string at $t = 1$, 1.5, 2, 2.5, and 3 s.

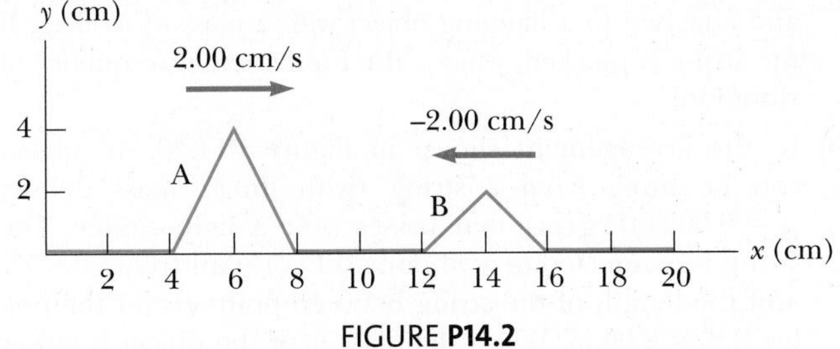

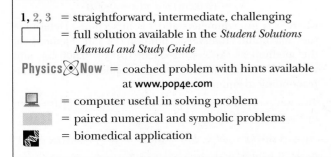

FIGURE P14.2

3. Two pulses traveling on the same string are described by

$$y_1 = \frac{5}{(3x - 4t)^2 + 2} \quad \text{and} \quad y_2 = \frac{-5}{(3x + 4t - 6)^2 + 2}$$

(a) In which direction does each pulse travel? (b) At what time do the two cancel everywhere? (c) At what point do the two pulses always cancel?

Section 14.2 ▪ Interference of Waves

4. Two waves are traveling in the same direction along a stretched string. The waves are 90.0° out of phase. Each wave has an amplitude of 4.00 cm. Find the amplitude of the resultant wave.

5. PhysicsNow™ Two traveling sinusoidal waves are described by the wave functions

$$y_1 = (5.00 \text{ m}) \sin[\pi(4.00x - 1\,200t)]$$

and

$$y_2 = (5.00 \text{ m}) \sin[\pi(4.00x - 1\,200t - 0.250)]$$

where x, y_1, and y_2 are in meters and t is in seconds. (a) What is the amplitude of the resultant wave? (b) What is the frequency of the resultant wave?

6. Two identical sinusoidal waves with wavelengths of 3.00 m travel in the same direction at a speed of 2.00 m/s. The second wave originates from the same point as the first, but at a later time. The amplitude of the resultant wave is the same as that of each of the two initial waves. Determine the minimum possible time interval between the starting moments of the two waves.

7. A tuning fork generates sound waves with a frequency of 246 Hz. The waves travel in opposite directions along a hallway, are reflected by end walls, and return. The hallway is 47.0 m long and the tuning fork is located 14.0 m from one end. What is the phase difference between the reflected waves when they meet at the tuning fork? The speed of sound in air is 343 m/s.

8. Two loudspeakers are placed on a wall 2.00 m apart. A listener stands 3.00 m from the wall directly in front of one of the speakers. A single oscillator is driving the speakers at a frequency of 300 Hz. (a) What is the phase difference between the two waves when they reach the observer? (b) What is the frequency closest to 300 Hz to which the oscillator may be adjusted such that the observer hears minimal sound?

9. Two sinusoidal waves in a string are defined by the functions

$$y_1 = (2.00 \text{ cm}) \sin(20.0x - 32.0t)$$

and

$$y_2 = (2.00 \text{ cm}) \sin(25.0x - 40.0t)$$

where y and x are in centimeters and t is in seconds. (a) What is the phase difference between these two waves at the point $x = 5.00$ cm at $t = 2.00$ s? (b) What is the positive x value closest to the origin for which the two phases differ by $\pm\pi$ at $t = 2.00$ s? (This location is where the two waves add to zero.)

10. Two speakers are driven by the same oscillator of frequency f. They are located a distance d from each other on a vertical pole. A man walks straight toward the lower speaker in a direction perpendicular to the pole as shown in Figure P14.10. (a) How many times will he hear a minimum in sound intensity? (b) How far is he from the pole at these moments? Let v represent the speed of sound and assume that the ground does not reflect sound.

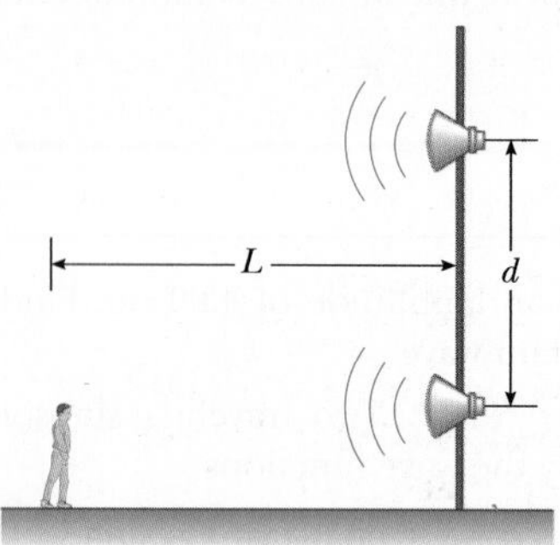

FIGURE **P14.10**

11. Two identical speakers 10.0 m apart are driven by the same oscillator with a frequency of $f = 21.5$ Hz (Fig. P14.11). (a) Explain why a receiver at point A records a minimum in sound intensity from the two speakers. (b) If the receiver is moved in the plane of the speakers, what path should it take so that the intensity remains at a minimum? That is, determine the relationship between x and y (the coordinates of the receiver) that causes the receiver to record a minimum in sound intensity. Take the speed of sound to be 344 m/s.

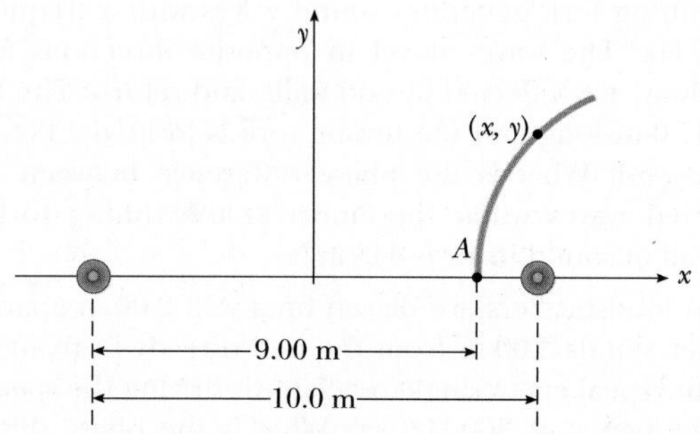

FIGURE **P14.11**

Section 14.3 ▪ Standing Waves

12. Two sinusoidal waves traveling in opposite directions interfere to produce a standing wave with the wave function

$$y = (1.50 \text{ m}) \sin(0.400x) \cos(200t)$$

where x is in meters and t is in seconds. Determine the wavelength, frequency, and speed of the interfering waves.

13. PhysicsNow™ Two speakers are driven in phase by a common oscillator at 800 Hz and face each other at a distance of 1.25 m. Locate the points along a line joining the two speakers where relative minima of sound pressure amplitude would be expected. (Use $v = 343$ m/s.)

14. Verify by direct substitution that the wave function for a standing wave given in Equation 14.3,

$$y = (2A \sin kx) \cos \omega t$$

is a solution of the general linear wave equation, Equation 13.19:

$$\frac{\partial^2 y}{\partial x^2} = \frac{1}{v^2} \frac{\partial^2 y}{\partial t^2}$$

15. Two sinusoidal waves combining in a medium are described by the wave functions

$$y_1 = (3.0 \text{ cm}) \sin \pi(x + 0.60t)$$

and

$$y_2 = (3.0 \text{ cm}) \sin \pi(x - 0.60t)$$

where x is in centimeters and t is in seconds. Determine the maximum transverse position of an element of the medium at (a) $x = 0.250$ cm, (b) $x = 0.500$ cm, and (c) $x = 1.50$ cm. (d) Find the three smallest values of x corresponding to antinodes.

16. Two waves simultaneously present in a long string are given by the wave functions

$$y_1 = A \sin(kx - \omega t + \phi) \quad \text{and} \quad y_2 = A \sin(kx + \omega t)$$

In the case $\phi = 0$, the chapter text shows that they add to a standing wave. Demonstrate (a) that the addition of the arbitrary phase constant ϕ changes only the position of the nodes and, in particular, (b) that the distance between nodes is still one half the wavelength.

Section 14.4 ▪ Standing Waves in Strings

17. Find the fundamental frequency and the next three frequencies that could cause standing wave patterns on a string that is 30.0 m long, has a mass per length of 9.00×10^{-3} kg/m, and is stretched to a tension of 20.0 N.

18. A standing wave is established in a 120-cm-long string fixed at both ends. The string vibrates in four segments when driven at 120 Hz. (a) Determine the wavelength. (b) What is the fundamental frequency of the string?

19. A string with a mass of 8.00 g and a length of 5.00 m has one end attached to a wall; the other end is draped over a pulley and attached to a hanging object with a mass of 4.00 kg. If the string is plucked, what is the fundamental frequency of vibration?

20. In the arrangement shown in Figure P14.20, an object can be hung from a string (with linear mass density $\mu = 0.002\,00$ kg/m) that passes over a light pulley. The string is connected to a vibrator (of constant frequency f), and the length of the string between point P and the pulley is $L = 2.00$ m. When the mass m of the object is either 16.0 kg or 25.0 kg, standing waves are observed, but no standing waves are observed with any mass between these

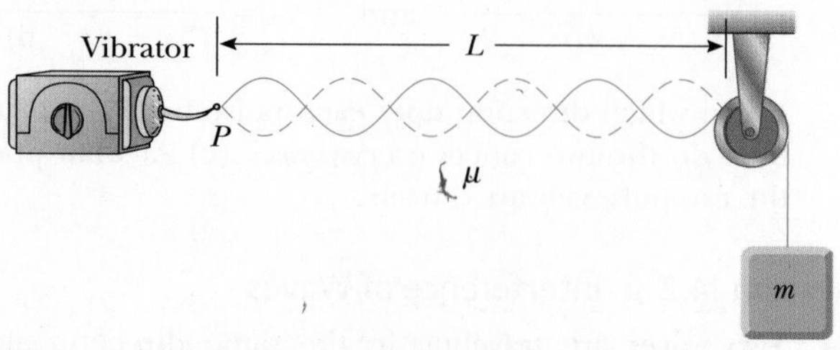

FIGURE **P14.20**

values. (a) What is the frequency of the vibrator? (*Note:* The greater the tension in the string, the smaller the number of nodes in the standing wave.) (b) What is the largest object mass for which standing waves could be observed?

21. A string of length L, mass per unit length μ, and tension T is vibrating at its fundamental frequency. What effect will the following have on the fundamental frequency? (a) The length of the string is doubled, with all other factors held constant. (b) The mass per unit length is doubled, with all other factors held constant. (c) The tension is doubled, with all other factors held constant.

22. The top string of a guitar has a fundamental frequency of 330 Hz when it is allowed to vibrate as a whole, along all its 64.0-cm length from the neck to the bridge. A fret is provided for limiting vibration to just the lower two thirds of the string. (a) If the string is pressed down at this fret and plucked, what is the new fundamental frequency? (b) The guitarist can play a "natural harmonic" by gently touching the string at the location of this fret and plucking the string at about one sixth of the way along its length from the bridge. What frequency will be heard then?

23. The A string on a cello vibrates in its first normal mode with a frequency of 220 Hz. The vibrating segment is 70.0 cm long and has a mass of 1.20 g. (a) Find the tension in the string. (b) Determine the frequency of vibration when the string vibrates in three segments.

24. A violin string has a length of 0.350 m and is tuned to concert G, with $f_G = 392$ Hz. Where must the violinist place her finger to play concert A, with $f_A = 440$ Hz? If this position is to remain correct to one-half the width of a finger (that is, to within 0.600 cm), what is the maximum allowable percentage change in the string tension?

25. **Review problem.** A sphere of mass M is supported by a string that passes over a light horizontal rod of length L (Fig. P14.25). Given that the angle is θ and that f represents the fundamental frequency of standing waves in the portion of the string above the rod, determine the mass of this portion of the string.

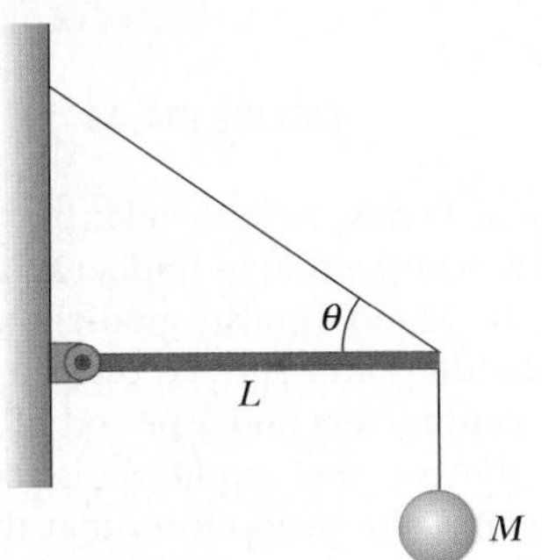

FIGURE **P14.25**

26. A standing wave pattern is observed in a thin wire with a length of 3.00 m. The equation of the wave is

$$y = (0.002 \text{ m}) \sin(\pi x) \cos(100\pi t)$$

where x is in meters and t is in seconds. (a) How many loops does this pattern exhibit? (b) What is the fundamental frequency of vibration of the wire? (c) If the original frequency is held constant and the tension in the wire is increased by a factor of nine, how many loops are present in the new pattern?

Section 14.5 ▪ Standing Waves in Air Columns

Note: Unless otherwise specified, assume that the speed of sound in air is 343 m/s at 20°C and is described by

$$v = 331 \text{ m/s} + (0.6 \text{ m/s}\cdot{}^\circ\text{C})T_C$$

at any Celsius temperature T_C.

27. Calculate the length of a pipe that has a fundamental frequency of 240 Hz if the pipe is (a) closed at one end and (b) open at both ends.

28. The overall length of a piccolo is 32.0 cm. The resonating air column vibrates as in a pipe open at both ends. (a) Find the frequency of the lowest note that a piccolo can play, assuming that the speed of sound in air is 340 m/s. (b) Opening holes in the side effectively shortens the length of the resonant column. Assuming that the highest note a piccolo can sound is 4 000 Hz, find the distance between adjacent antinodes for this mode of vibration.

29. The windpipe of one typical whooping crane is 5.00 feet long. What is the fundamental resonant frequency of the bird's trachea, modeled as a narrow pipe closed at one end? Assume a temperature of 37°C.

30. The fundamental frequency of an open organ pipe corresponds to middle C (261.6 Hz on the chromatic musical scale). The third resonance of a closed organ pipe has the same frequency. What is the length of each of the two pipes?

31. **PhysicsNow™** A shower stall has dimensions 86.0 cm × 86.0 cm × 210 cm. If you were singing in this shower, which frequencies would sound the richest (because of resonance)? Assume that the shower stall acts as a pipe closed at both ends, with nodes at opposite sides, and that the voices of various singers range from 130 Hz to 2 000 Hz. Let the speed of sound in the hot air be 355 m/s.

32. Do not stick anything into your ear! Estimate the length of your ear canal, from its opening at the external ear to the eardrum. If you regard the canal as a narrow tube that is open at one end and closed at the other, at approximately what fundamental frequency would you expect your hearing to be most sensitive? Explain why you can hear especially soft sounds just around this frequency.

33. **PhysicsNow™** Two adjacent natural frequencies of an organ pipe are determined to be 550 Hz and 650 Hz. Calculate the fundamental frequency and length of this pipe. (Use $v = 340$ m/s.)

34. As shown in Figure P14.34, water is pumped into a tall vertical cylinder at a volume flow rate R. The radius of the cylinder is r, and at the open top of the cylinder a tuning fork is vibrating with a frequency f. As the water rises, how much time elapses between successive resonances?

35. A glass tube (open at both ends) of length L is positioned near an audio speaker of frequency $f = 680$ Hz. For what values of L will the tube resonate with the speaker?

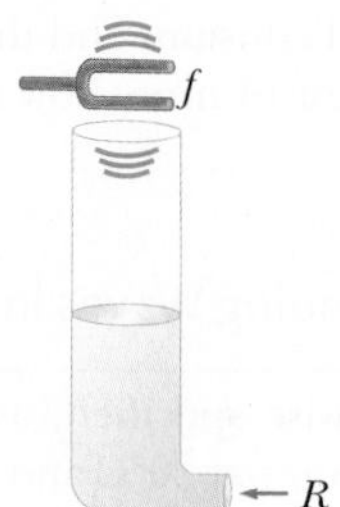

FIGURE P14.34

36. A tuning fork with a frequency of 512 Hz is placed near the top of the tube shown in Figure 14.11a. The water level is lowered so that the length L slowly increases from an initial value of 20.0 cm. Determine the next two values of L that correspond to resonant modes.

37. An air column in a glass tube is open at one end and closed at the other by a movable piston. The air in the tube is warmed above room temperature and a 384-Hz tuning fork is held at the open end. Resonance is heard when the piston is 22.8 cm from the open end and again when it is 68.3 cm from the open end. (a) What speed of sound is implied by these data? (b) How far from the open end will the piston be when the next resonance is heard?

38. With a particular fingering, a flute plays a note with frequency 880 Hz at 20.0°C. The flute is open at both ends. (a) Find the air column length. (b) Find the frequency it produces at the beginning of the halftime performance at a late-season American football game, when the ambient temperature is −5.00°C and the player has not had a chance to warm up the flute.

Section 14.6 ▪ Beats: Interference in Time

39. **PhysicsNow™** In certain ranges of a piano keyboard, more than one string is tuned to the same note to provide extra loudness. For example, the note at 110 Hz has two strings at this frequency. If one string slips from its normal tension of 600 N to 540 N, what beat frequency is heard when the hammer strikes the two strings simultaneously?

40. While attempting to tune the note C at 523 Hz, a piano tuner hears 2.00 beats/s between a reference oscillator and the string. (a) What are the possible frequencies of the string? (b) When she tightens the string slightly, she hears 3.00 beats/s. What is the frequency of the string now? (c) By what percentage should the piano tuner now change the tension in the string to bring it into tune?

41. A student holds a tuning fork oscillating at 256 Hz. He walks toward a wall at a constant speed of 1.33 m/s. (a) What beat frequency does he observe between the tuning fork and its echo? (b) How fast must he walk away from the wall to observe a beat frequency of 5.00 Hz?

Section 14.7 Nonsinusoidal Wave Patterns

42. Suppose a flutist plays a 523-Hz C note with first harmonic displacement amplitude $A_1 = 100$ nm. From Figure 14.14b read, by proportion, the displacement amplitudes of harmonics 2 through 7. Take these values as the coefficients A_2 through A_7 in the Fourier analysis of the sound and assume that $B_1 = B_2 = \cdots = B_7 = 0$. Construct a graph of the waveform of the sound. Your waveform will not look exactly like the flute waveform in Figure 14.13b because you simplify by ignoring cosine terms; nevertheless, it produces the same sensation to human hearing.

43. An A-major chord consists of the notes called A, C$^{\#}$, and E. It can be played on a piano by simultaneously striking strings with fundamental frequencies of 440.00 Hz, 554.37 Hz, and 659.26 Hz. The rich consonance of the chord is associated with near equality of the frequencies of some of the higher harmonics of the three tones. Consider the first five harmonics of each string and determine which harmonics show near equality.

Section 14.8 ▪ Context Connection—Building on Antinodes

44. An earthquake can produce a seiche in a lake in which the water sloshes back and forth from end to end with remarkably large amplitude and long period. Consider a seiche produced in a rectangular farm pond as shown in the cross-sectional view of Figure P14.44. (The figure is not drawn to scale.) Suppose the pond is 9.15 m long and of uniform width and depth. You measure that a pulse produced at one end reaches the other end in 2.50 s. (a) What is the wave speed? (b) To produce the seiche, several people stand on the bank at one end and paddle together with snow shovels, moving them in simple harmonic motion. What should be the frequency of this motion?

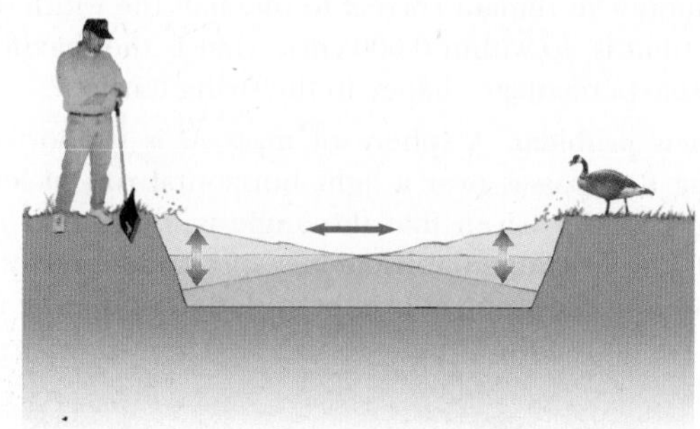

FIGURE P14.44

45. The Bay of Fundy, Nova Scotia, has the highest tides in the world. Assume that in midocean and at the mouth of the bay, the Moon's gravity gradient and the Earth's rotation make the water surface oscillate with an amplitude of a few centimeters and a period of 12 h 24 min. At the head of the bay, the amplitude is several meters. Argue for or against the proposition that the tide is magnified by standing wave resonance. Assume that the bay has a length of 210 km and a uniform depth of 36.1 m. The speed of long-wavelength water waves is given by $\sqrt{gd}$, where d is the water's depth.

Additional Problems

46. Figure P14.46a is a photograph of a vibrating wine glass. A special technique makes black and white stripes appear where the glass is moving, with closer spacing where the amplitude is larger. Six nodes and six antinodes alternate

around the rim of the glass in the vibration photographed, but consider instead the case of a standing wave vibration with four nodes and four antinodes equally spaced around the 20.0-cm circumference of the rim of a goblet. If transverse waves move around the glass at 900 m/s, an opera singer would have to produce a high harmonic with what frequency to shatter the glass with a resonant vibration as shown in Figure P14.46b?

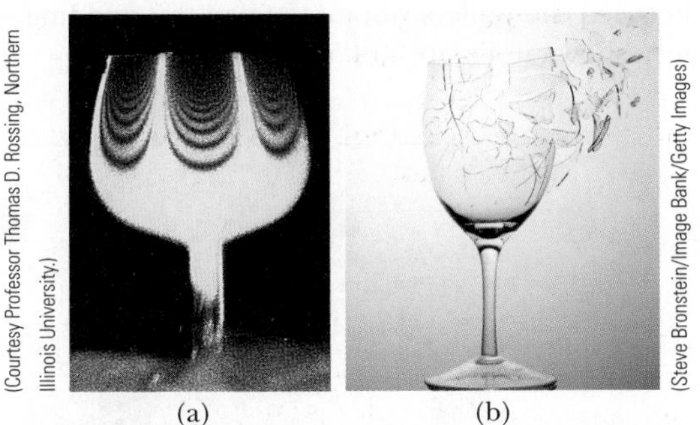

FIGURE **P14.46** (a) Nodes (in white) and antinodes (where the stripes converge to black) alternate around the rim of a vibrating wine glass. (b) A glass shatters when vibrating with large amplitude.

47. On a marimba (Fig. P14.47), the wooden bar that sounds a tone when struck vibrates in a transverse standing wave having three antinodes and two nodes. The lowest-frequency note is 87.0 Hz, produced by a bar 40.0 cm long. (a) Find the speed of transverse waves on the bar. (b) A resonant pipe suspended vertically below the center of the bar enhances the loudness of the emitted sound. If the pipe is open at the top end only and the speed of sound in air is 340 m/s, what is the length of the pipe required to resonate with the bar in part (a)?

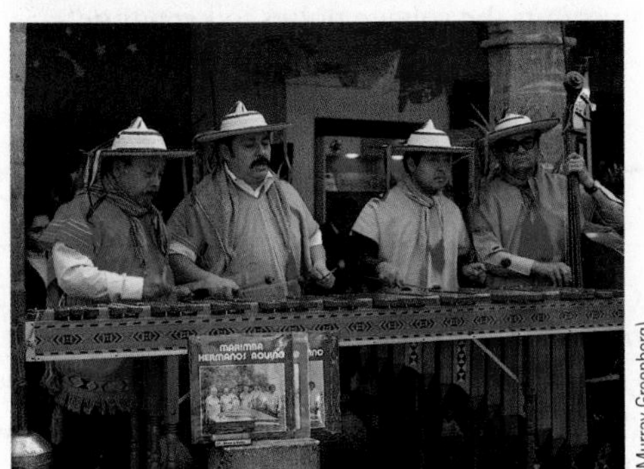

FIGURE **P14.47** Marimba players in Mexico City.

48. A nylon string has mass 5.50 g and length 86.0 cm. One end is tied to the floor and the other end to a small magnet, with a mass negligible compared with the string. A magnetic field (which we will study in Chapter 22) exerts an upward force of 1.30 N on the magnet, wherever the magnet is located. At equilibrium, the string is vertical and motionless, with the magnet at the top. When it is carrying a small-amplitude wave, you may assume the string is always under uniform tension 1.30 N. (a) Find the speed of transverse waves on the string. (b) The string's vibration possibilities are a set of standing wave states, each with a node at the fixed bottom end and an antinode at the free top end. Find the node–antinode distances for each one of the three simplest states. (c) Find the frequency of each of these states.

49. Two train whistles have identical frequencies of 180 Hz. When one train is at rest in the station and the other is moving nearby, a commuter standing on the station platform hears beats with a frequency of 2.00 beats/s when the whistles sound at the same time. What are the two possible speeds and directions that the moving train can have?

50. A loudspeaker at the front of a room and an identical loudspeaker at the rear of the room are being driven by the same oscillator at 456 Hz. A student walks at a uniform rate of 1.50 m/s along the length of the room. She hears a single tone repeatedly becoming louder and softer. (a) Model these variations as beats between the Doppler-shifted sounds the student receives. Calculate the number of beats the student hears each second. (b) Model the two speakers as producing a standing wave in the room and the student as walking between antinodes. Calculate the number of intensity maxima the student hears each second.

51. A student uses an audio oscillator of adjustable frequency to measure the depth of a water well. The student hears two successive resonances at 51.5 Hz and 60.0 Hz. How deep is the well?

52. A string fixed at both ends and having a mass of 4.80 g, a length of 2.00 m, and a tension of 48.0 N vibrates in its second ($n = 2$) normal mode. What is the wavelength in air of the sound emitted by this vibrating string?

53. Two wires are welded together end to end. The wires are made of the same material, but the diameter of one is twice that of the other. They are subjected to a tension of 4.60 N. The thin wire has a length of 40.0 cm and a linear mass density of 2.00 g/m. The combination is fixed at both ends and vibrated in such a way that two antinodes are present, with the node between them being right at the weld. (a) What is the frequency of vibration? (b) How long is the thick wire?

54. A string is 0.400 m long and has a mass per unit length of 9.00×10^{-3} kg/m. What must be the tension in the string if its second harmonic has the same frequency as the second resonance mode of a 1.75-m-long pipe open at one end?

55. A standing wave is set up in a string of variable length and tension by a vibrator of variable frequency. Both ends of the string are fixed. When the vibrator has a frequency f, in a string of length L and under tension T, n antinodes are set up in the string. (a) If the length of the string is doubled, by what factor should the frequency be changed so that the same number of antinodes is produced? (b) If the frequency and length are held constant, what tension will produce $n + 1$ antinodes? (c) If the frequency is tripled and the length of the string is halved, by what factor should the tension be changed so that twice as many antinodes are produced?

56. **Review problem.** For the arrangement shown in Figure P14.56, $\theta = 30.0°$, the inclined plane and the small pulley are frictionless, the string supports the object of mass M at the bottom of the plane, and the string has mass m that is small compared with M. The system is in equilibrium and the vertical part of the string has a length h. Standing

waves are set up in the vertical section of the string. (a) Find the tension in the string. (b) Model the shape of the string as one leg and the hypotenuse of a right triangle. Find the whole length of the string. (c) Find the mass per unit length of the string. (d) Find the speed of waves on the string. (e) Find the lowest frequency for a standing wave. (f) Find the period of the standing wave having three nodes. (g) Find the wavelength of the standing wave having three nodes. (h) Find the frequency of the beats resulting from the interference of the sound wave of lowest frequency generated by the string with another sound wave having a frequency that is 2.00% greater.

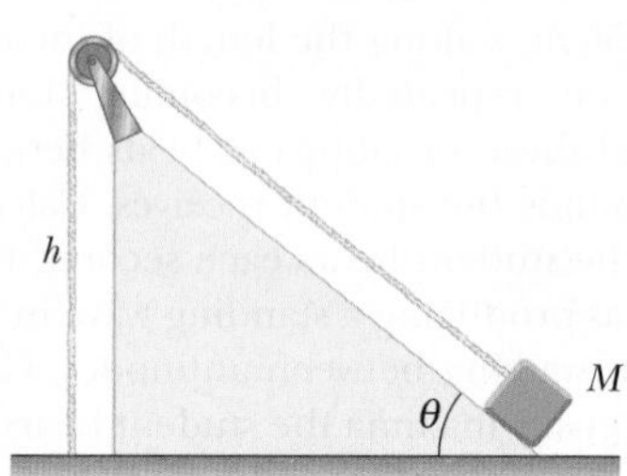

FIGURE **P14.56**

57. Two waves are described by the wave functions

$$y_1(x, t) = 5.0 \sin(2.0x - 10t)$$

$$y_2(x, t) = 10 \cos(2.0x - 10t)$$

where y_1, y_2, and x are in meters and t is in seconds. Show that the wave resulting from their superposition is sinusoidal. Determine the amplitude and phase of this sinusoidal wave.

58. A 0.010 0-kg wire, 2.00 m long, is fixed at both ends and vibrates in its simplest mode under a tension of 200 N. When a vibrating tuning fork is placed near the wire, a beat frequency of 5.00 Hz is heard. (a) What could be the frequency of the tuning fork? (b) What should the tension in the wire be if the beats are to disappear?

59. **Review problem.** A 12.0-kg object hangs in equilibrium from a string with a total length of $L = 5.00$ m and a linear mass density of $\mu = 0.001\ 00$ kg/m. The string is wrapped around two light, frictionless pulleys that are separated by a distance of $d = 2.00$ m (Fig. P14.59a). (a) Determine the tension in the string. (b) At what frequency must the string between the pulleys vibrate to form the standing wave pattern shown in Figure P14.59b?

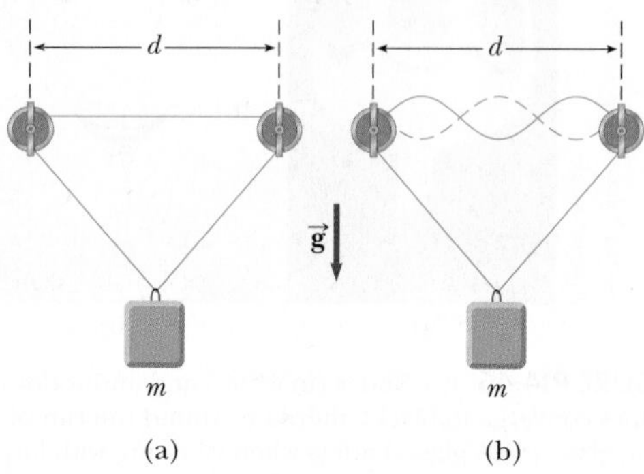

FIGURE **P14.59**

60. A quartz watch contains a crystal oscillator in the form of a block of quartz that vibrates by contracting and expanding. Two opposite faces of the block, 7.05 mm apart, are antinodes, moving alternately toward each other and away from each other. The plane halfway between these two faces is a node of the vibration. The speed of sound in quartz is 3.70 km/s. Find the frequency of the vibration. An oscillating electric voltage accompanies the mechanical oscillation; the quartz is described as *piezoelectric.* An electric circuit feeds in energy to maintain the oscillation and also counts the voltage pulses to keep time.

ANSWERS TO QUICK QUIZZES

14.1 (c). The pulses completely cancel each other in terms of displacement of elements of the string from equilibrium, but the string is still moving. A short time later, the string will be displaced again and the pulses will have passed each other.

14.2 (i), (a). The pattern shown at the bottom of Active Figure 14.8a corresponds to the extreme position of the string. All elements of the string have momentarily come to rest. Notice that the time derivative of Equation 14.3 gives the transverse velocity of each element of the string: $v(t) = dy/dt = -(2\omega A \sin kx) \sin \omega t$. Whenever $\omega t = n\pi$, the velocity of every element of the string is equal to zero. (ii), (d). Near a nodal point, elements on one side of the point are moving upward at this instant and elements on the other side are moving downward.

14.3 (d). Choice (a) is incorrect because the number of nodes is one greater than the number of antinodes. Choice (b) is only true for half of the modes; it is not true for any odd-numbered mode. Choice (c) would be correct if we replace the word *nodes* with *antinodes.*

14.4 (b). With both ends open, the pipe has a fundamental frequency given by Equation 14.10: $f_{\text{open}} = v/2L$. With one end closed, the pipe has a fundamental frequency given by Equation 14.11:

$$f_{\text{closed}} = \frac{v}{4L} = \frac{1}{2}\left(\frac{v}{2L}\right) = \frac{1}{2} f_{\text{open}}$$

14.5 (c). The increase in temperature causes the speed of sound to go up. According to Equation 14.10, this effect will result in an increase in the fundamental frequency of a given organ pipe.

14.6 (b). Tightening the string has caused the frequencies to be farther apart based on the increase in the beat frequency, so you want to loosen the string.

Minimizing the Risk

We have explored the physics of vibrations and waves. Let us now return to our central question for this *Earthquakes* Context:

> ***How can we choose locations and build structures to minimize the risk of damage in an earthquake?***

To answer this question, we shall use the physical principles that we now understand more clearly and apply them to our choices of locations and structural design.

In our discussion of simple harmonic oscillation, we learned about resonance. Designers of structures in earthquake-prone areas need to pay careful attention to resonance vibrations from shaking of the ground. The design features to be considered include ensuring that the resonance frequencies of the building do not match typical earthquake frequencies. In addition, the structural details should include sufficient damping to ensure that the amplitude of resonance vibration does not destroy the structure.

In Chapter 13, we discussed the role of the medium in the propagation of a wave. For seismic waves moving across the surface of the Earth, the soil on the surface is the medium. Because soil varies from one location to another, the speed of seismic waves will vary at different locations. A particularly dangerous situation exists for structures built on loose soil or mudfill. In these types of media, the interparticle forces are much weaker than in a more solid foundation such as granite bedrock. As a result, the wave speed is less in loose soil than in bedrock.

Consider Equation 13.23, which provides an expression for the rate of energy transfer by waves. This equation was derived for waves on strings, but the proportionality to the square of the amplitude and the speed is general. Because of conservation of energy, the rate of energy transfer for a wave must remain constant regardless of the medium. Thus, according to Equation 13.23, **if the wave speed decreases, as it does for seismic waves moving from rock into loose soil, the amplitude must increase.** As a result, the shaking of structures built on loose soil is of larger magnitude than for those built on solid bedrock.

This factor contributed to the collapse of the Nimitz Freeway during the Loma Prieta earthquake, near San Francisco, in 1989. Figure 1 shows the results of the earthquake on the freeway. The portion of the freeway that collapsed was built on mudfill, but the surviving portion was built on bedrock. The amplitude of oscillation in the portion built on mudfill was more than five times as large as the amplitude of other portions.

(Paul X. Scott/SYGMA)

FIGURE 1 Portions of the double-decked Nimitz Freeway in Oakland, California, collapsed during the Loma Prieta earthquake of 1989.

Another danger for structures on loose soil is the possibility of

(Photo by M. Hamada; used by permission from the Multidisciplinary Center for Earthquake Engineering Research, Univ. of Buffalo)

FIGURE 2 The collapse of this crane during the Kobe, Japan, earthquake of 1995 was caused by liquefaction of the underlying soil.

liquefaction of the soil. When soil is shaken, the elements of soil can move with respect to one another and the soil tends to act like a liquid rather than a solid. It is possible for the structure to sink into the soil during an earthquake. If the liquefaction is not uniform over the foundation of the structure, the structure can tip over, as seen in Figure 2. As a result, even if the earthquake is not sufficient to damage the structure, it will be unusable in its tipped-over orientation.

As discussed in Chapter 14, building structures where standing seismic waves can be established is dangerous. Such construction was a factor in the Michoacán earthquake of 1985. The shape of the bedrock under Mexico City resulted in standing waves, with severe damage to buildings located at antinodes.

In summary, to minimize risk of damage in an earthquake, architects and engineers must design structures to prevent destructive resonances, avoid building on loose soil, and pay attention to the underground rock formations so as to be aware of possible standing wave patterns. Other precautions can also be taken. For example, buildings can be constructed with **seismic isolation** from the ground. This method involves mounting the structure on **isolation dampers,** heavy-duty bearings that dampen the oscillations of the building, resulting in reduced amplitude of vibration.

We have not addressed many other considerations for earthquake safety in structures, but we have been able to apply many of our concepts from oscillations and waves so as to understand some aspects of logical choices in locating and designing structures.

Problems

1. For seismic waves spreading out from a point (the epicenter) on the surface of the Earth, the intensity of the waves decreases with distance according to an inverse proportionality to distance. That is, the wave intensity is proportional to $1/r$, where r is the distance from the epicenter to the observation point. This rule applies if the medium is uniform. The intensity of the wave is proportional to the rate of energy transfer for the wave. Furthermore, we have shown that the energy of vibration of an oscillator is proportional to the square of the amplitude of the vibration. Assume that a particular earthquake causes ground shaking with an amplitude of 5.0 cm at a distance of 10 km from the epicenter. If the medium is uniform, what is the amplitude of the ground shaking at a point 20 km from the epicenter?
2. As mentioned in the text, the amplitude of oscillation during the Loma Prieta earthquake of 1989 was five times greater in areas of mudfill than in areas of bedrock. From this information, find the factor by which the seismic wave speed changed as the waves moved from the bedrock to the mudfill. Ignore any reflection of wave energy and any change in density between the two media.

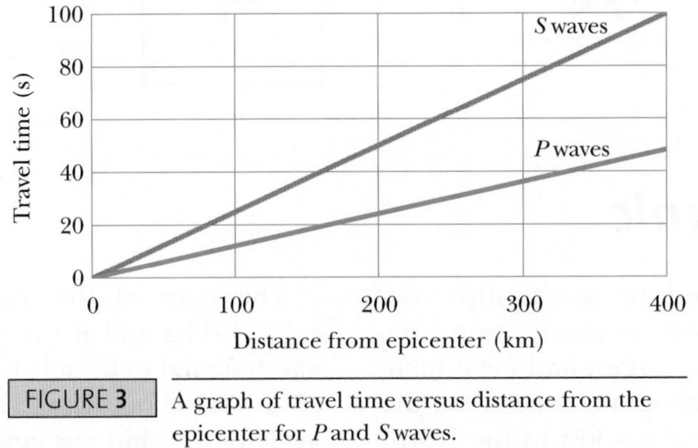

FIGURE 3 A graph of travel time versus distance from the epicenter for *P* and *S* waves.

3. Figure 3 is a graphical representation of the travel time for *P* and *S* waves from the epicenter of an earthquake to a seismograph as a function of the distance of travel. The following table shows the measured times of day for arrival of *P* waves from a particular earthquake at three seismograph locations. In the last column, fill in the times of day for the arrival of the *S* waves at the three seismograph locations.

Seismograph Station	Distance from Epicenter (km)	*P* Wave Arrival Time	*S* Wave Arrival Time
#1	200	15:46:06	
#2	160	15:46:01	
#3	105	15:45:54	

CONTEXT 4

Search for the *Titanic*

The *Titanic* and its sister ships, the *Olympic* and the *Britannic,* were designed to be the largest and most luxurious liners sailing the ocean at the time. The *Titanic* was lost in the North Atlantic on its maiden voyage from Southampton, England, to New York, on April 15, 1912. It was claimed by many (although not by the White Star Line, which operated the *Titanic*) that the ship was unsinkable. This claim was proven to be untrue when the *Titanic* struck an iceberg at 11:40 P.M. on April 14 and sank less than 3 h later. This event was one of the worst maritime disasters of all time, with more than 1 500 lives lost because of a severe shortage of lifeboats. Amazingly, British ships of the time were not required to carry enough lifeboats for all passengers on board. It is fortunate that the *Titanic* was not completely full on its maiden voyage, as there would have been only enough lifeboats for a third of its passengers. As it was, there were enough lifeboats for a little more than half of the 2 200 passengers on board, but only 705 were saved due to the partial filling of the boats, which occurred for a number of reasons.

The mass of the *Titanic* was over 4.2×10^7 kg and it was 269 m long. It was designed to be able to travel at 24 to 25 knots (about 12–13 m/s), and the safety of the ship was ensured by lateral bulkheads at several places across the ship with electrically operated watertight doors. It is ironic to note that the design of the bulkheads actually worked against the safety of the ship, and the closing of the watertight doors, according to some experts after the disaster, caused the ship to sink more rapidly than if they had been left open.

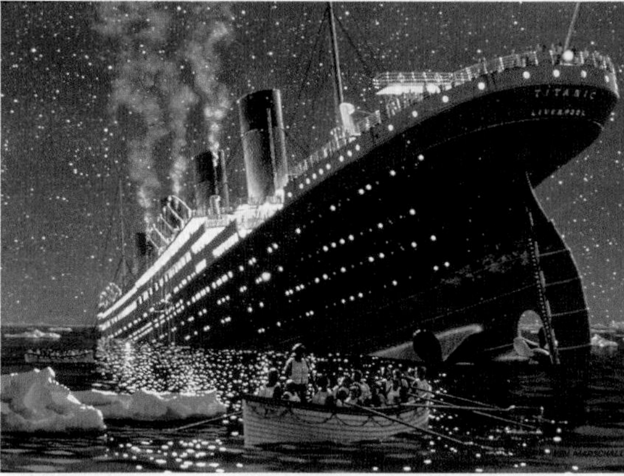

FIGURE 2 During the sinking process in the early morning of April 15, 1912, the bow of the *Titanic* is under water and the stern is lifted out of the water. The huge forces necessary to hold the stern aloft caused the *Titanic* to split in the middle before sinking.

The accidental sinking of the *Titanic* resulted from a remarkable confluence of bad luck, complacency, and poor policy. Several events occurred during the last day of its voyage that would not have led to the foundering of the ship if they had happened in just a slightly different way.

The *Titanic* has captured the interest of the public for many years. Many theatrical movies related to the ship have

(Illustrations by Ken Marschall © 1992 from *Titanic: An Illustrated History,* a Hyperion/Madison Press Book)

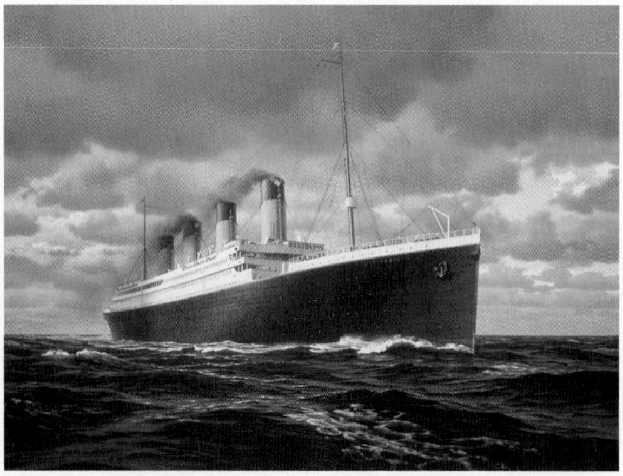

FIGURE 1 The *Titanic* on its way from Southampton to New York.

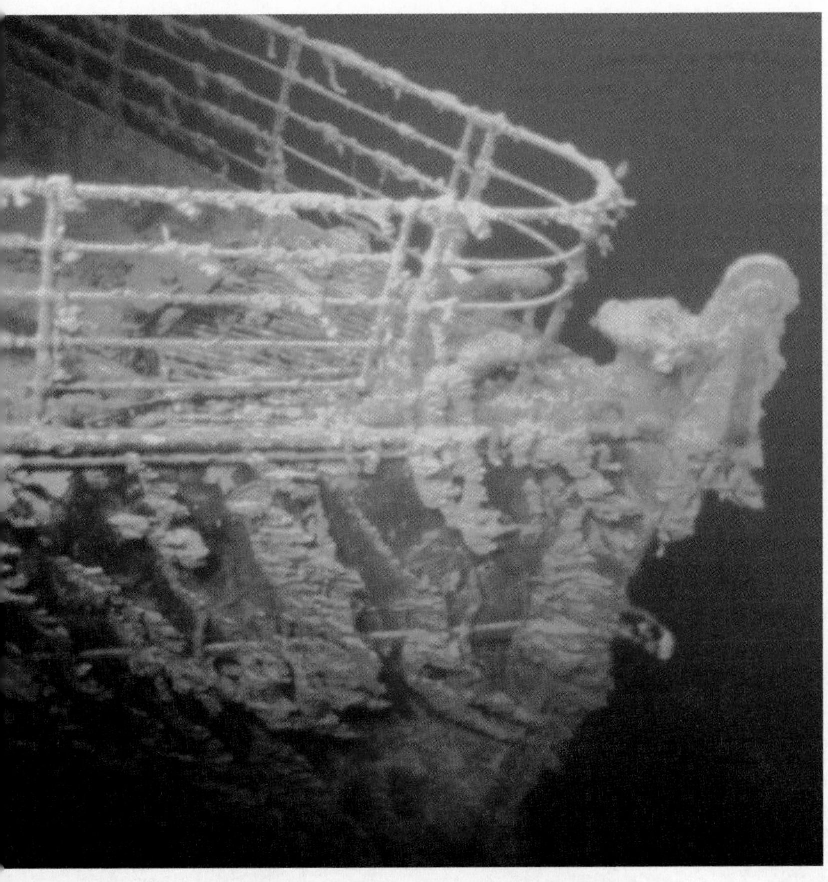

FIGURE 3 The bow of the *Titanic* as it rests on the ocean floor 80 years after it sank in the North Atlantic.

(Courtesy of RMS Titanic, Inc.)

FIGURE 4 In 1998, a 20-ton section of the *Titanic*'s hull was raised to the surface and is now part of a touring exhibit of *Titanic* artifacts. Called the "Big Piece," the hull section contains four portholes, three of which still contain the original glass.

been produced, including *Titanic* (1953), *A Night to Remember* (1958), *Raise the Titanic* (1980), *Titanic* (1997), and *Ghosts of the Abyss* (2003). A musical play, *Titanic* (1997), has been produced, and the *Titanic* plays a role in the Broadway and movie musical *The Unsinkable Molly Brown* (1964). A large number of books have been written on the disaster, many of which were reissued when the *Titanic* became wildly popular in response to the 1997 film.

Finding the *Titanic* on the ocean floor was a dream of many individuals ever since the ship was lost. The wreck of the *Titanic* was discovered in 1985 by a research team from Woods Hole Oceanographic Institute, led by Dr. Robert Ballard. This discovery led to interesting ethical questions because subsequent expeditions to the site salvaged items from the wreck, making them available for exhibition and sale to the public. We will use the *Titanic*, the search for its wreckage, and the underwater visits to its gravesite in this short Context on fluids as we address the central question:

How can we safely visit the wreck of the *Titanic*?

CHAPTER 15

Fluid Mechanics

Icebergs float in the cold waters of the North Atlantic. Although the visible portion of an iceberg may tower over a passing ship, only about 11% of the iceberg is above water.

(Norman Tobley/Taxi/Getty Images)

CHAPTER OUTLINE

Matter is normally classified as being in one of three states: solid, liquid, or gas. Everyday experience tells us that a solid has a definite volume and shape. A brick maintains its familiar shape and size over a long time. We also know that a liquid has a definite volume but no definite shape. For example, a cup of liquid water has a fixed volume but assumes the shape of its container. Finally, an unconfined gas has neither definite volume nor definite shape. For example, if there is a leak in the natural gas supply in your home, the escaping gas continues to expand into the surrounding atmosphere. These definitions help us picture the states of matter, but they are somewhat artificial. For example, asphalt, glass, and plastics are normally considered solids, but over a long time interval they tend to flow like liquids. Likewise, most substances can be a solid, liquid, or gas (or combinations of these states), depending on the temperature and pressure. In general, the time interval required for a particular substance to change its shape in response to an external force determines whether we treat the substance as a solid, liquid, or gas.

A **fluid** is a collection of molecules that are randomly arranged and held together by weak cohesive forces between molecules and forces exerted by the walls of a container. Both liquids and gases are fluids. In our treatment of the mechanics of fluids, we shall see that no new physical principles are needed to explain such effects as the buoyant force on a submerged object and vascular flutter in an artery. In this chapter, we shall apply a number of familiar analysis models to the physics of fluids.

15.1 PRESSURE

Our first task in understanding the physics of fluids is to define a new quantity to describe fluids. Imagine applying a force to the surface of an object, with the force having components both parallel to and perpendicular to the surface. If the object is a solid at rest on a table, the force component perpendicular to the surface may cause the object to flatten, depending on how hard the object is. Assuming that the object does not slide on the table, the component of the force parallel to the surface of the object will cause the object to distort. As an example, suppose you place your physics book flat on a table and apply a force with your hand parallel to the front cover and perpendicular to the spine. The book will distort, with the bottom pages staying fixed at their original location and the top pages shifting horizontally by some distance. The cross-section of the book changes from a rectangle to a parallelogram. This kind of force parallel to the surface is called a *shearing force.*

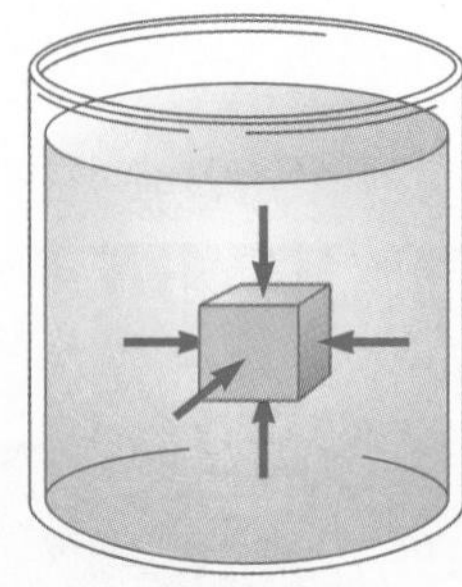

FIGURE 15.1 The force exerted by the fluid on a submerged object at any point is perpendicular to the surface of the object. The force exerted by the fluid on the walls of the container is perpendicular to the walls at all points.

We shall adopt a simplification model in which the fluids we study will be nonviscous; that is, no friction exists between adjacent layers of the fluid. Nonviscous fluids and static fluids do not sustain shearing forces. If you imagine placing your hand on a water surface and pushing parallel to the surface, your hand simply slides over the water; you cannot distort the water as you did the book. This phenomenon occurs because the interatomic forces in a fluid are not strong enough to lock atoms in place with respect to one another. The fluid cannot be modeled as a rigid object as in Chapter 10. If we try to apply a shearing force, the molecules of the fluid simply slide past one another.

Therefore, the only type of force that can exist in a fluid is one that is perpendicular to a surface. For example, the forces exerted by the fluid on the object in Figure 15.1 are everywhere perpendicular to the surfaces of the object.

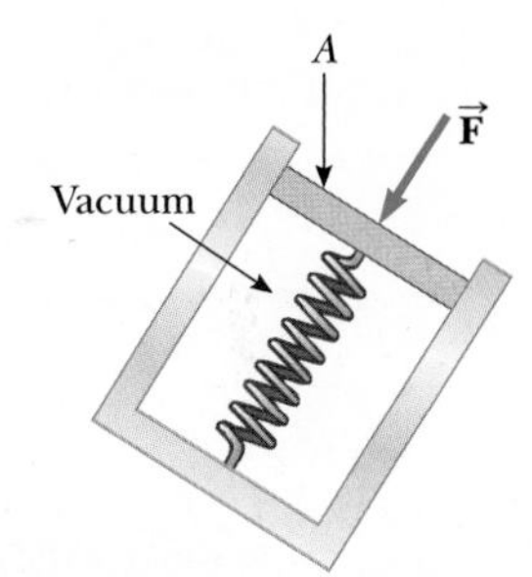

FIGURE 15.2 A simple device for measuring pressure in a fluid.

The force that a fluid exerts on a surface originates in the collisions of molecules of the fluid with the surface. Each collision results in the reversal of the component of the velocity vector of the molecule perpendicular to the surface. By the impulse–momentum theorem and Newton's third law, each collision results in a force on the surface. A huge number of these impulsive forces occur every second, resulting in a constant macroscopic force on the surface. This force is spread out over the area of the surface and is related to a new quantity called **pressure.**

PITFALL PREVENTION 15.1

FORCE AND PRESSURE Equation 15.1 makes a clear distinction between force and pressure. Another important distinction is that *force is a vector* and *pressure is a scalar.* No direction is associated with pressure, but the direction of the force associated with the pressure is perpendicular to the surface of interest.

The pressure at a specific point in a fluid can be measured with the device pictured in Figure 15.2. The device consists of an evacuated cylinder enclosing a light piston connected to a spring. As the device is submerged in a fluid, the fluid presses in on the top of the piston and compresses the spring until the inward force of the fluid is balanced by the outward force of the spring. The force exerted on the piston by the fluid can be measured if the spring is calibrated in advance.

If F is the magnitude of the force exerted by the fluid on the piston and A is the surface area of the piston, **the pressure P of the fluid at the level to which the device has been submerged is defined as the ratio of force to area:**

■ Definition of pressure

$$P \equiv \frac{F}{A} \qquad [15.1]$$

Although we have defined pressure in terms of our device in Figure 15.2, the definition is general. Because pressure is force per unit area, it has units of newtons per square meter in the SI system. Another name for the SI unit of pressure is the **pascal** (Pa):

■ The pascal

$$1 \text{ Pa} \equiv 1 \text{ N/m}^2 \quad [15.2]$$

Hypodermic needles

Notice that pressure and force are different quantities. We can have a very large pressure from a relatively small force by making the area over which the force is applied small. Such is the case with hypodermic needles. The area of the tip of the needle is very small, so a small force pushing on the needle is sufficient to cause a pressure large enough to puncture the skin. We can also create a small pressure from a large force by enlarging the area over which the force acts. Such is the principle behind the design of snowshoes. If a person were to walk on deep snow with regular shoes, it is possible for his or her feet to break through the snow and sink. Snowshoes, however, allow the force on the snow due to the weight of the person to spread out over a larger area, reducing the pressure enough so that the snow surface is not broken (Fig. 15.3).

(Royalty-free/CORBIS)

FIGURE 15.3 Snowshoes keep you from sinking into soft snow because they spread the downward force you exert on the snow over a large area, reducing the pressure on the snow surface.

The atmosphere exerts a pressure on the surface of the Earth and all objects at the surface. This pressure is responsible for the action of suction cups, drinking straws, vacuum cleaners, and many other devices. In our calculations and end-of-chapter problems, we usually take atmospheric pressure to be

$$P_0 = 1.00 \text{ atm} \approx 1.013 \times 10^5 \text{ Pa} \quad [15.3]$$

QUICK QUIZ 15.1 Suppose you are standing directly behind someone who steps back and accidentally stomps on your foot with the heel of one shoe. Would you be better off if that person were **(a)** a large male professional basketball player wearing sneakers or **(b)** a petite woman wearing spike-heeled shoes?

■ Thinking Physics 15.1

Suction cups can be used to hold objects onto surfaces. Why don't astronauts use suction cups to hold onto the outside surface of an orbiting spacecraft?

Reasoning A suction cup works because air is pushed out from under the cup when it is pressed against a surface. When the cup is released, it tends to spring back a bit, causing the trapped air under the cup to expand. This expansion causes a reduced pressure inside the cup. Thus, the difference between the atmospheric pressure on the outside of the cup and the reduced pressure inside provides a net force pushing the cup against the surface. For astronauts in orbit around the Earth, almost no air exists outside the surface of the spacecraft. Therefore, if a suction cup were to be pressed against the outside surface of the spacecraft, the pressure differential needed to press the cup to the surface is not present. ■

15.2 VARIATION OF PRESSURE WITH DEPTH

The study of fluid mechanics involves the density of a substance, defined in Equation 1.1 as the mass per unit volume for the substance. Table 15.1 lists the densities of various substances. These values vary slightly with temperature because the volume of a substance is temperature-dependent (as we shall see in Chapter 16). Note that under standard conditions (0°C and atmospheric pressure) the densities of gases are on the order of 1/1 000 the densities of solids and liquids. This difference implies that the average molecular spacing in a gas under these conditions is about ten times greater in each dimension than in a solid or liquid.

As divers know well, the pressure in the sea or a lake increases as they dive to greater depths. Likewise, atmospheric pressure decreases with increasing altitude.

TABLE 15.1 Densities of Some Common Substances at Standard Temperature (0°C) and Pressure (Atmospheric)

Substance	ρ (kg/m^3)	Substance	ρ (kg/m^3)
Air	1.29	Ice	0.917×10^3
Aluminum	2.70×10^3	Iron	7.86×10^3
Benzene	0.879×10^3	Lead	11.3×10^3
Copper	8.92×10^3	Mercury	13.6×10^3
Ethyl alcohol	0.806×10^3	Oak	0.710×10^3
Fresh water	1.00×10^3	Oxygen gas	1.43
Glycerin	1.26×10^3	Pine	0.373×10^3
Gold	19.3×10^3	Platinum	21.4×10^3
Helium gas	1.79×10^{-1}	Sea water	1.03×10^3
Hydrogen gas	8.99×10^{-2}	Silver	10.5×10^3

For this reason, aircraft flying at high altitudes must have pressurized cabins to provide sufficient oxygen for the passengers.

We now show mathematically how the pressure in a liquid increases with depth. Consider a liquid of density ρ at rest as in Figure 15.4. Let us select a sample of the liquid contained within an imaginary cylinder of cross-sectional area A extending from depth d to depth $d + h$. This sample of liquid is in equilibrium and at rest. Therefore, according to Newton's second law, the net force on the sample must be equal to zero. We will investigate the forces on the sample related to the pressure on it.

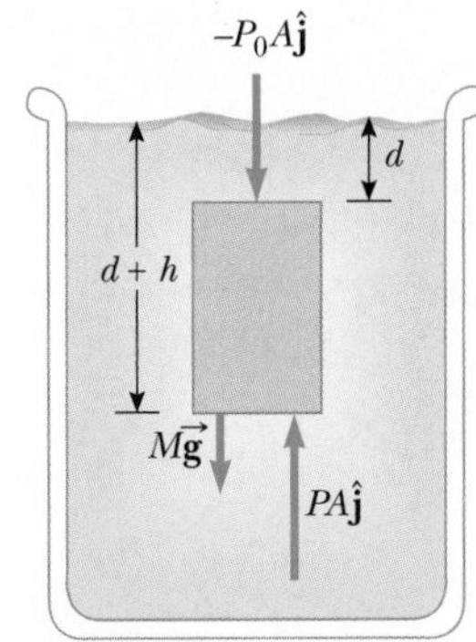

FIGURE 15.4 The net force on the sample of liquid within the darker region must be zero because the sample is in equilibrium.

The liquid external to our sample exerts forces at all points on the sample's surface, perpendicular to it. On the sides of the sample of liquid in Figure 15.4, forces due to the pressure act horizontally and cancel in pairs on opposite sides of the sample for a net horizontal force of zero. The pressure exerted by the liquid on the sample's bottom face is P and the pressure on the top face is P_0. Therefore, from Equation 15.1, the magnitude of the upward force exerted by the liquid on the bottom of the sample is PA, and the magnitude of the downward force exerted by the liquid on the top is P_0A. In addition, a gravitational force is exerted on the sample. Because the sample is in equilibrium, the net force in the vertical direction must be zero:

$$\sum F_y = 0 \quad \rightarrow \quad PA - P_0A - Mg = 0$$

Because the mass of liquid in the sample is $M = \rho V = \rho Ah$, the gravitational force on the liquid in the sample is $Mg = \rho gAh$. Therefore,

$$PA = P_0A + \rho gAh$$

or

$$P = P_0 + \rho gh \qquad [15.4]$$

▪ Variation of pressure with depth in a liquid

If the top surface of our sample is at $d = 0$ so that it is open to the atmosphere, P_0 is atmospheric pressure. Equation 15.4 indicates that the pressure in a liquid depends only on the depth h within the liquid. The pressure is therefore the same at all points having the same depth, independent of the shape of the container.

In view of Equation 15.4, any increase in pressure at the surface must be transmitted to every point in the liquid. This behavior was first recognized by French scientist Blaise Pascal (1623–1662) and is called **Pascal's law: A change in the pressure applied to an enclosed fluid is transmitted undiminished to every point of the fluid and to the walls of the container.** You use Pascal's law when you squeeze the sides of your toothpaste tube. The increase in pressure on the sides of the tube increases the pressure everywhere, which pushes a stream of toothpaste out of the opening.

▪ Pascal's law

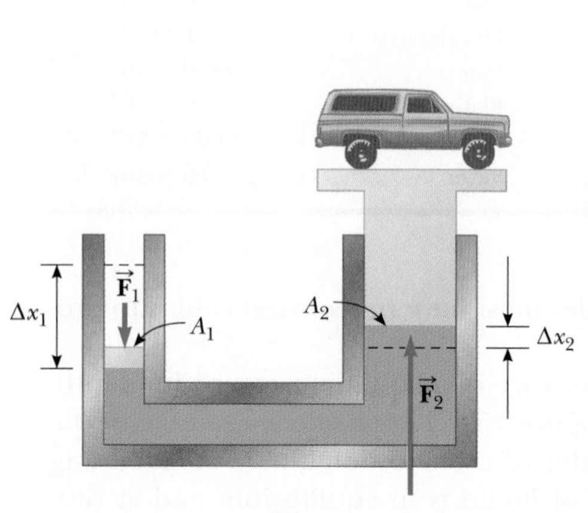

(a)

(b)

(David Frazier)

FIGURE 15.5 (a) Diagram of a hydraulic press. Because the increase in pressure is the same at the left and right sides, a small force $\vec{F}_1$ at the left produces a much larger force $\vec{F}_2$ at the right. (b) A vehicle under repair is supported by a hydraulic lift in a garage.

An important application of Pascal's law is the hydraulic press illustrated by Figure 15.5. A force $\vec{F}_1$ is applied to a small piston of area A_1. The pressure is transmitted through a liquid to a larger piston of area A_2, and force $\vec{F}_2$ is exerted by the liquid on this piston. Because the pressure is the same at both pistons, we see that $P = F_1/A_1 = F_2/A_2$. The force magnitude F_2 is therefore larger than F_1 by the multiplying factor A_2/A_1. Hydraulic brakes, car lifts, hydraulic jacks, and forklifts all make use of this principle.

QUICK QUIZ 15.2 The pressure at the bottom of a filled glass of water ($\rho = 1\,000$ kg/m^3) is P. The water is poured out and the glass is filled with ethyl alcohol ($\rho = 806$ kg/m^3). What is the pressure at the bottom of the glass? **(a)** smaller than P **(b)** equal to P **(c)** larger than P **(d)** indeterminate

■ Thinking Physics 15.2

 Measuring blood pressure

Blood pressure is normally measured with the cuff of the sphygmomanometer around the arm. Suppose the blood pressure were measured with the cuff around the calf of the leg of a standing person. Would the reading of the blood pressure be the same here as it is for the arm?

Reasoning The blood pressure measured at the calf would be larger than that measured at the arm. If we imagine the vascular system of the body to be a vessel containing a liquid (blood), the pressure in the liquid will increase with depth. The blood at the calf is deeper in the liquid than that at the arm and is at a higher pressure.

Blood pressures are normally taken at the arm because it is at approximately the same height as the heart. If blood pressures at the calf were used as a standard, adjustments would need to be made for the height of the person and the blood pressure would be different if the person were lying down. ■

EXAMPLE 15.1 The Car Lift

In a car lift used in a service station, compressed air exerts a force on a small piston of circular cross-section having a radius of 5.00 cm. This pressure is transmitted by an incompressible liquid to a second piston of radius 15.0 cm.

A What force must the compressed air exert to lift a car weighing 13 300 N?

Solution Because the pressure exerted by the compressed air is transmitted undiminished throughout the liquid, we have

$$F_1 = \left(\frac{A_1}{A_2}\right) F_2 = \frac{\pi(5.00 \times 10^{-2}\text{ m})^2}{\pi(15.0 \times 10^{-2}\text{ m})^2}(1.33 \times 10^4\text{ N})$$
$$= 1.48 \times 10^3\text{ N}$$

B What air pressure will produce this force?

Solution The air pressure that will produce this force is

$$P = \frac{F_1}{A_1} = \frac{1.48 \times 10^3\text{ N}}{\pi(5.00 \times 10^{-2}\text{ m})^2} = 1.88 \times 10^5\text{ Pa}$$

This pressure is approximately twice atmospheric pressure.

C Consider the lift as a nonisolated system and show that the input energy transfer is equal in magnitude to the output energy transfer.

Solution The energy input and output are by means of work done by the forces as the pistons move. To determine the work done, we must find the magnitude of the displacement through which each force acts. Because the liquid is modeled to be incompressible, the volume of the cylinder through which the input piston moves must equal that through which the output piston moves. The lengths of these cylinders are the magnitudes Δx_1 and Δx_2 of the displacements of the forces (see Fig. 15.5a). Setting the volumes equal, we have

$$V_1 = V_2 \quad \rightarrow \quad A_1\Delta x_1 = A_2\Delta x_2$$
$$\frac{A_1}{A_2} = \frac{\Delta x_2}{\Delta x_1}$$

Evaluating the ratio of the input work to the output work, we find that

$$\frac{W_1}{W_2} = \frac{F_1\,\Delta x_1}{F_2\,\Delta x_2} = \left(\frac{F_1}{F_2}\right)\left(\frac{\Delta x_1}{\Delta x_2}\right) = \left(\frac{A_1}{A_2}\right)\left(\frac{A_2}{A_1}\right) = 1$$

which verifies that the work input and output are the same, as they must be to conserve energy.

EXAMPLE 15.2 The Force on a Dam

Water is filled to a height H behind a dam of width w (Fig. 15.6). Determine the resultant force on the dam.

Solution We cannot calculate the force on the dam by simply multiplying the area by the pressure because the pressure varies with depth. The problem can be solved by finding the force dF on a narrow horizontal strip at depth h and then integrating the expression over the height of the dam to find the total force.

The pressure at the depth h beneath the surface at the red strip in Figure 15.6 is

$$P = \rho g h = \rho g(H - y)$$

(We have not included atmospheric pressure in our calculation because it acts on both sides of the dam, resulting in a net contribution of zero to the total force.) From Equation 15.1, we find the force on the red strip of area dA:

$$F = PA \quad \rightarrow \quad dF = P\,dA$$

Because $dA = w\,dy$, we have

$$dF = P\,dA = \rho g(H - y)w\,dy$$

Therefore, the total force on the dam is

$$F = \int_0^H \rho g(H - y)w\,dy = \tfrac{1}{2}\rho g w H^2$$

Note that because the pressure increases with depth, the dam is designed such that its thickness increases with depth as in Figure 15.6.

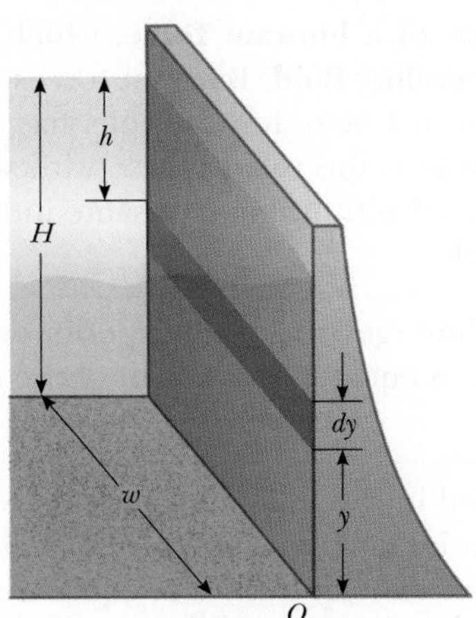

FIGURE 15.6 (Example 15.2) The total force on a dam is obtained from the expression $F = \int P\,dA$, where dA is the area of the red strip.

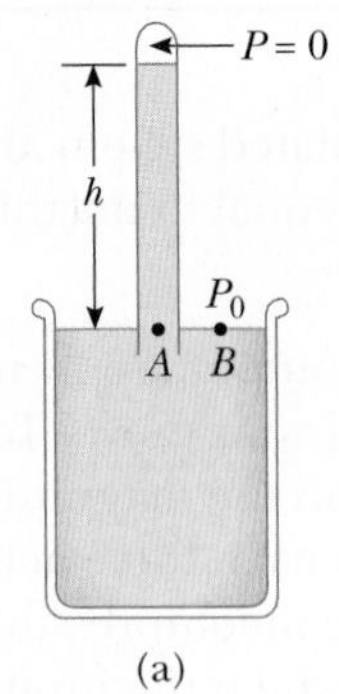

(a)

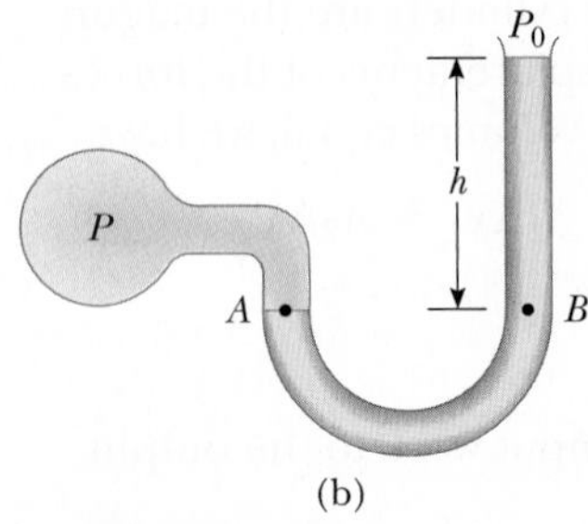

(b)

FIGURE 15.7 Two devices for measuring pressure: (a) a mercury barometer and (b) an open-tube manometer.

15.3 PRESSURE MEASUREMENTS

During the weather report on a television news program, the *barometric pressure* is often provided. Barometric pressure is the current pressure of the atmosphere, which varies over a small range from the standard value provided in Equation 15.3. How is this pressure measured?

One instrument used to measure atmospheric pressure is the common barometer, invented by Evangelista Torricelli (1608–1647). A long tube closed at one end is filled with mercury and then inverted into a dish of mercury (Fig. 15.7a). The closed end of the tube is nearly a vacuum, so the pressure at the top of the mercury column can be taken as zero. In Figure 15.7a, the pressure at point A due to the column of mercury must equal the pressure at point B due to the atmosphere. If that were not the case, a net force would move mercury from one point to the other until equilibrium was established. It therefore follows that $P_0 = \rho_{\text{Hg}} g h$, where ρ_{Hg} is the density of the mercury and h is the height of the mercury column. As atmospheric pressure varies, the height of the mercury column varies, so the height can be calibrated to measure atmospheric pressure. Let us determine the height of a mercury column for one atmosphere of pressure, $P_0 = 1\text{ atm} = 1.013 \times 10^5\text{ Pa}$:

$$P_0 = \rho_{\text{Hg}} g h \quad \rightarrow \quad h = \frac{P_0}{\rho_{\text{Hg}} g} = \frac{1.013 \times 10^5\text{ Pa}}{(13.6 \times 10^3\text{ kg/m}^3)(9.80\text{ m/s}^2)} = 0.760\text{ m}$$

Based on a calculation such as this one, one atmosphere of pressure is defined as the pressure equivalent of a column of mercury that is exactly 0.760 0 m in height at 0°C.

The open-tube manometer illustrated in Figure 15.7b is a device for measuring the pressure of a gas contained in a vessel. One end of a U-shaped tube containing a liquid is open to the atmosphere, and the other end is connected to a system of unknown pressure P. The pressures at points A and B must be the same (otherwise, the curved portion of the liquid would experience a net force and would accelerate), and the pressure at A is the unknown pressure of the gas. Therefore, equating the unknown pressure P to the pressure at point B, we see that $P = P_0 + \rho g h$. The difference in pressure $P - P_0$ is equal to $\rho g h$. Pressure P is called the **absolute pressure,** and the difference $P - P_0$ is called the **gauge pressure.** For example, the pressure you measure in your bicycle tire is gauge pressure.

(© Hulton Deutsch Collection/CORBIS)

ARCHIMEDES (ca. 287–212 B.C.)

Archimedes, a Greek mathematician, physicist, and engineer, was perhaps the greatest scientist of antiquity. He was the first to compute accurately the ratio of a circle's circumference to its diameter, and he showed how to calculate the volume and surface area of spheres, cylinders, and other geometric shapes. He is well known for discovering the nature of the buoyant force.

15.4 BUOYANT FORCES AND ARCHIMEDES'S PRINCIPLE

In this section, we investigate the origin of a **buoyant force,** which is **an upward force exerted on an object by the surrounding fluid.** Buoyant forces are evident in many situations. Anyone who has ridden in a boat, for example, has experienced a buoyant force. Another common example is the relative ease with which you can lift someone in a swimming pool compared with lifting that same individual on dry land. According to **Archimedes's principle:**

> Any object completely or partially submerged in a fluid experiences an upward buoyant force whose magnitude is equal to the weight of the fluid displaced by the object.

Archimedes's principle can be verified in the following manner. Suppose we focus our attention on a small parcel of a larger fluid such as the indicated cube of fluid in the container of Figure 15.8. This cube of fluid is in equilibrium under the action of the forces exerted on it by the fluid surrounding it. One of these forces in the vertical direction is the gravitational force. Because the cube is in equilibrium, the net force on it in the vertical direction must be zero. What cancels the downward gravitational force so that the cube remains in equilibrium? Apparently, the rest of the fluid inside the container is applying an upward force, the buoyant force. Therefore, the magnitude B of the buoyant force must be exactly equal to the weight of the fluid inside the cube:

$$\sum F_y = 0 \quad \rightarrow \quad B - F_g = 0 \quad \rightarrow \quad B = Mg$$

where M is the mass of the fluid in the cube.

Now imagine that the cube of fluid is replaced by a cube of steel of the same dimensions. What is the buoyant force on the steel? The fluid surrounding a cube behaves in the same way whether it is exerting pressure on a cube of fluid or a cube of steel; therefore, **the buoyant force acting on the steel is the same as the buoyant force acting on a cube of fluid of the same dimensions.** This result applies for a submerged object of any shape, size, or density.

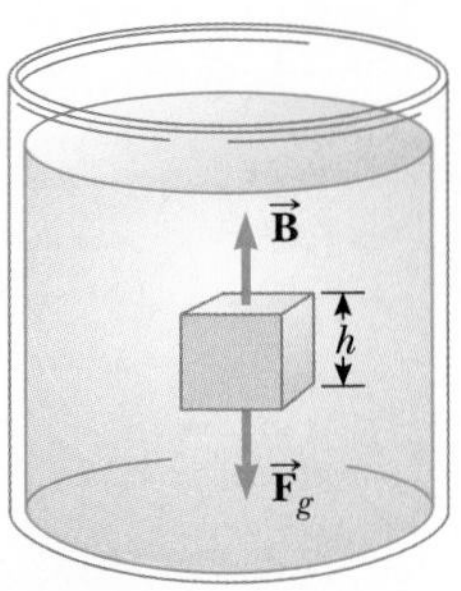

FIGURE 15.8 The external forces on the cube of liquid are the gravitational force $\vec{\mathbf{F}}_g$ and the buoyant force $\vec{\mathbf{B}}$. Under equilibrium conditions, $B = F_g$.

Let us now show more explicitly that the magnitude of the buoyant force is equal to the weight of the displaced fluid. Although that is true for both liquids and gases, we will perform the derivation for a liquid. On the sides of the cube of liquid in Figure 15.8, forces due to the pressure act horizontally and cancel in pairs on opposite sides of the cube for a net horizontal force of zero. In a liquid, the pressure at the bottom of the cube is greater than the pressure at the top by an amount $\rho_f gh$, where ρ_f is the density of the liquid and h is the height of the cube. Therefore, the upward force F_{bot} on the bottom is greater than the downward force F_{top} on the top of the cube. The net vertical force *exerted by the liquid* (we are ignoring the gravitational force for now) is

$$\sum F_{\text{liquid}} = B = F_{\text{bot}} - F_{\text{top}}$$

Expressing the forces in terms of pressure gives us

$$B = P_{\text{bot}}A - P_{\text{top}}A = \Delta PA = \rho_f ghA$$

$$B = \rho_f gV \qquad [15.5]$$

■ Archimedes's principle

where $V = hA$ is the volume of the cube. Because the mass of the liquid in the cube is $M = \rho_f V$, we see that

$$B = Mg$$

which is the weight of the displaced liquid.

Before proceeding with a few examples, it is instructive to compare two common cases: the buoyant force acting on a totally submerged object and that acting on a floating object.

Case I: A Totally Submerged Object

When an object is totally submerged in a liquid of density ρ_f, the magnitude of the upward buoyant force is $B = \rho_f gV$, where V is the volume of the liquid displaced by the object. Because the object is totally submerged, the volume V_O of the object and the volume V of liquid displaced by the object are the same, $V = V_O$. If the object has a density ρ_O, its weight is $Mg = \rho_O V_O g$. Therefore, the net force on it is $\sum F = B - Mg = (\rho_f - \rho_O)V_O g$. We see that if the density of the object is less than the density of the liquid as in Active Figure 15.9a, the net force is positive and the unsupported object accelerates upward. If the density of the object is greater than the density of the liquid as in Active Figure 15.9b, the net force is negative and the unsupported object sinks.

The same behavior is exhibited by an object immersed in a gas, such as the air in the atmosphere.[1] If the object is less dense than air, like a helium-filled balloon, the object floats upward. If it is denser, like a rock, it falls downward.

PITFALL PREVENTION 15.2

BUOYANT FORCE IS EXERTED BY THE FLUID Notice the important point in this discussion of the buoyant force: **it is exerted by the fluid.** It is not determined by properties of the object except for the amount of fluid displaced by the object. Therefore, if several objects of different densities but the same volume are immersed in a fluid, they will all experience the same buoyant force. Whether they sink or float will be determined by the relationship between the buoyant force and the weight.

[1]The general behavior is the same, but the buoyant force varies with height in the atmosphere due to the variation in density of the air.

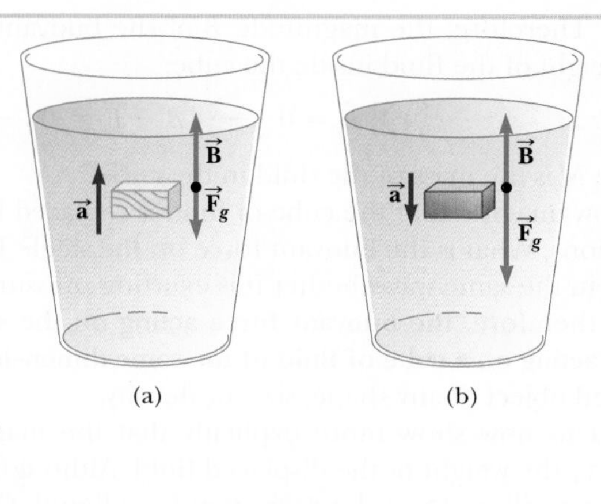

ACTIVE FIGURE 15.9
(a) A totally submerged object that is less dense than the fluid in which it is submerged experiences a net upward force. (b) A totally submerged object that is denser than the fluid sinks.

Physics Now™ Log into PhysicsNow at **www.pop4e.com** and go to Active Figure 15.9 to move the object to new positions as well as change the density of the object to see the results.

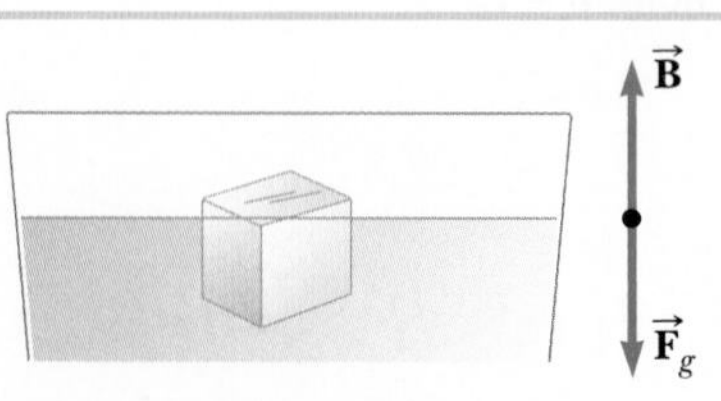

ACTIVE FIGURE 15.10 An object floating on the surface of a liquid experiences two forces, the gravitational force $\vec{\mathbf{F}}_g$ and the buoyant force $\vec{\mathbf{B}}$. Because the object floats in equilibrium, $B = F_g$.

Physics Now™ By logging into PhysicsNow at **www.pop4e.com** and going to Active Figure 15.10 you can change the densities of the object and the liquid.

Case II: A Floating Object

Now consider an object in static equilibrium floating on the surface of a liquid, that is, an object that is only partially submerged such as the ice cube floating in water in Active Figure 15.10. Because it is only partially submerged, the volume V of liquid displaced by the object is only a fraction of the total volume V_O of the object. The volume of the liquid displaced by the object corresponds to that volume of the object beneath the liquid surface. Because the object is in equilibrium, the upward buoyant force is balanced by the downward gravitational force exerted on the object. The buoyant force has a magnitude $B = \rho_f gV$. Because the weight of the object is $Mg = \rho_O V_O g$ and because Newton's second law tells us that $\Sigma F = 0$ in the vertical direction, $Mg = B$. We see that $\rho_f gV = \rho_O V_O g$, or

$$\frac{\rho_O}{\rho_f} = \frac{V}{V_O} \qquad [15.6]$$

Therefore, the fraction of the volume of the object under the liquid surface is equal to the ratio of the object density to the liquid density.

Let us consider examples of both cases. Under normal conditions, the average density of a fish is slightly greater than the density of water. That being the case, a fish would sink if it did not have some mechanism to counteract the net downward force. The fish does so by internally regulating the size of its swim bladder, a gas-filled cavity within the fish's body. Increasing its size increases the amount of water displaced, which increases the buoyant force. In this manner, fish are able to swim to various depths. Because the fish is totally submerged in the water, this example illustrates Case I.

As an example of Case II, imagine a large cargo ship. When the ship is at rest, the upward buoyant force from the water balances the weight so that the ship is in equilibrium. Only part of the volume of the ship is under water. If the ship is loaded with heavy cargo, it sinks deeper into the water. The increased weight of the ship due to the cargo is balanced by the extra buoyant force related to the extra volume of the ship that is now beneath the water surface.

(Richard Megna/Fundamental Photographs)

These hot-air balloons float on air because they are filled with air at high temperature. The buoyant force on a balloon due to the surrounding air is equal to the weight of the balloon, resulting in a net force of zero. ■

QUICK QUIZ 15.3 An apple is held completely submerged just below the surface of a container of water. The apple is then moved to a deeper point in the water. Compared with the force needed to hold the apple just below the surface, what is the force needed to hold it at a deeper point? **(a)** larger **(b)** the same **(c)** smaller **(d)** impossible to determine

QUICK QUIZ 15.4 You are shipwrecked and floating in the middle of the ocean on a raft. Your cargo on the raft includes a treasure chest full of gold that you found before your ship sank and the raft is just barely afloat. To keep you floating as high as possible in the water, should you **(a)** leave the treasure chest on top of the raft, **(b)** secure the treasure chest to the underside of the raft, or **(c)** hang the treasure chest in the water with a rope attached to the raft? (Assume that throwing the treasure chest overboard is not an option you wish to consider!)

Thinking Physics 15.3

A florist delivery person is delivering a flower basket to a home. The basket includes an attached helium-filled balloon, which suddenly comes loose from the basket and begins to accelerate upward toward the sky. Startled by the release of the balloon, the delivery person drops the flower basket. As the basket falls, the basket–Earth system experiences an increase in kinetic energy and a decrease in gravitational potential energy, consistent with conservation of mechanical energy. The balloon–Earth system, however, experiences an increase in *both* gravitational potential energy and kinetic energy. Is that consistent with the principle of conservation of mechanical energy? If not, from where is the extra energy coming?

Reasoning In the case of the system of the flower basket and the Earth, a good approximation to the motion of the basket can be made by ignoring the effects of the air. Therefore, the basket–Earth system can be analyzed with the isolated system model and mechanical energy is conserved. For the balloon–Earth system, we cannot ignore the effects of the air because it is the buoyant force of the air that causes the balloon to rise. Therefore, the balloon–Earth system is analyzed with the nonisolated system model. The buoyant force of the air does work across the boundary of the system, and that work results in an increase in both the kinetic and gravitational potential energies of the system. ■

EXAMPLE 15.3 Eureka!

Archimedes supposedly was asked to determine whether a crown made for the king consisted of pure gold. Legend has it that Archimedes solved this problem by weighing the crown first in air and then in water as shown in Figure 15.11. Suppose the scale reads 7.84 N in air and 6.84 N in water. What should Archimedes have told the king?

Solution Our strategy will be based on determining the density of the crown and comparing it with the density of gold. Figure 15.11 helps us conceptualize the problem. Because of our understanding of the buoyant force, we realize that the scale reading will be smaller in Figure 15.11b than in Figure 15.11a. The scale reading is a measure of one of the forces on the crown and we recognize that the crown is stationary. Therefore, we can categorize this problem as one in which we model the crown as a particle in equilibrium. To analyze the problem, note that when the crown is suspended in air, the scale reads the true weight $T_1 = F_g$ (neglecting the buoyancy of air). When it is immersed in water, the buoyant force $\vec{\mathbf{B}}$ reduces the scale reading to an *apparent*

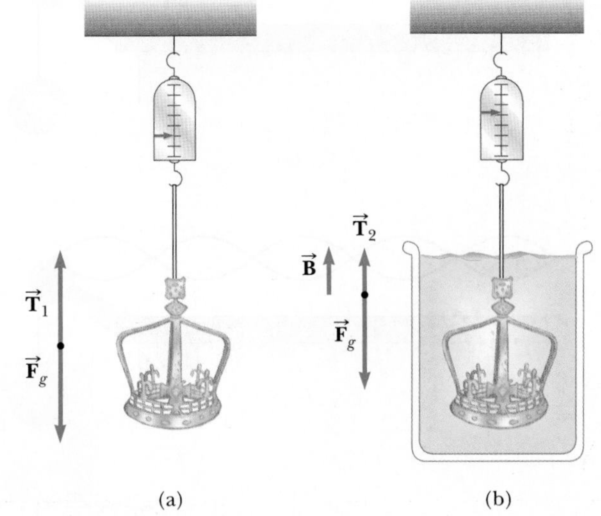

FIGURE 15.11 (Example 15.3) (a) When the crown is suspended in air, the scale reads its true weight because $T_1 = F_g$ (the buoyancy due to air is negligible). (b) When the crown is immersed in water, the buoyant force $\vec{\mathbf{B}}$ reduces the scale reading to $T_2 = F_g - B$.

weight of $T_2 = F_g - B$. Because the crown is in equilibrium, the net force on it is zero. When the crown is in water, then,

$$\sum F = B + T_2 - F_g = 0$$

so that

$$B = F_g - T_2 = 7.84 \text{ N} - 6.84 \text{ N} = 1.00 \text{ N}$$

Because this buoyant force is equal in magnitude to the weight of the displaced water, we have $\rho_w g V_w = 1.00$ N, where V_w is the volume of the displaced water and ρ_w is its density. Also, the volume of the crown V_c is equal to the volume of the displaced water because the crown is completely submerged. Therefore,

$$V_c = V_w = \frac{1.00 \text{ N}}{\rho_w g} = \frac{1.00 \text{ N}}{(1\,000 \text{ kg/m}^3)(9.80 \text{ m/s}^2)} = 1.02 \times 10^{-4} \text{ m}^3$$

Finally, the density of the crown is

$$\rho_c = \frac{m_c}{V_c} = \frac{m_c g}{V_c g} = \frac{7.84 \text{ N}}{(1.02 \times 10^{-4} \text{ m}^3)(9.80 \text{ m/s}^2)} = 7.84 \times 10^3 \text{ kg/m}^3$$

To finalize the problem, from Table 15.1 we see that the density of gold is 19.3×10^3 kg/m^3. Therefore, Archimedes should have told the king that he had been cheated. Either the crown was hollow or it was not made of pure gold.

INTERACTIVE EXAMPLE 15.4 Changing String Vibration with Water

One end of a horizontal string is attached to a vibrating blade and the other end passes over a pulley as in Figure 15.12a. A sphere of mass 2.00 kg hangs on the end of the string. The string is vibrating in its second harmonic. A container of water is raised under the sphere so that the sphere is completely submerged. After that is done, the string vibrates in its fifth harmonic as shown in Figure 15.12b. What is the radius of the sphere?

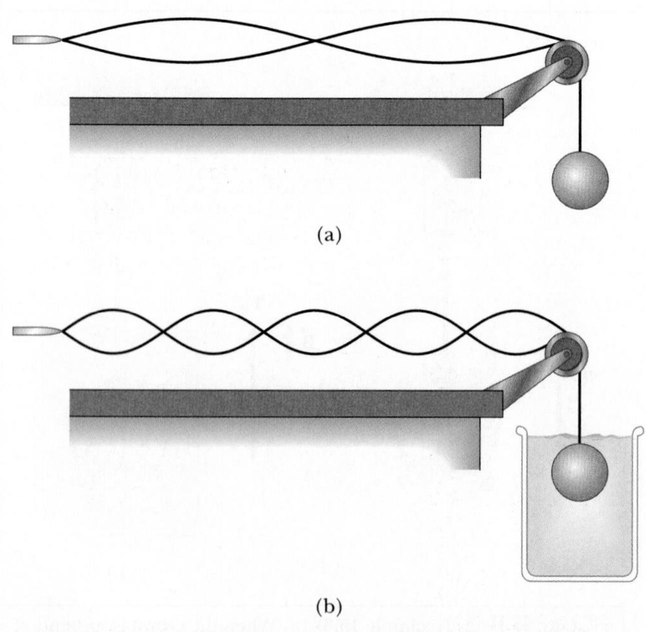

FIGURE 15.12 (Interactive Example 15.4) (a) When the sphere hangs in air, the string vibrates in its second harmonic. (b) When the sphere is immersed in water, the string vibrates in its fifth harmonic.

Solution In Figure 15.12a, Newton's second law applied to the sphere tells us that the initial tension T_i in the string is equal to the weight of the sphere:

$$T_i - mg = 0 \quad \rightarrow \quad T_i = mg$$

$$T_i = (2.00 \text{ kg})(9.80 \text{ m/s}^2) = 19.6 \text{ N}$$

where the subscript i is used to indicate initial variables before we immerse the sphere in water. Once the sphere is immersed in water, the tension in the string will decrease to T_f. Applying Newton's second law to the sphere again in this situation, we have

$$(1) \quad T_f + B - mg = 0 \quad \rightarrow \quad B = mg - T_f$$

The desired quantity, the radius of the sphere, will appear in the expression for the buoyant force B. Before we can proceed in this direction, however, we need to evaluate T_f. We do so from the standing wave information. We write the equation for the frequency of a standing wave on a string (Equation 14.8) twice: once before we immerse the sphere and once after, and divide the equations:

$$(2) \quad \left.\begin{aligned} f &= \frac{n_i}{2L}\sqrt{\frac{T_i}{\mu}} \\ f &= \frac{n_f}{2L}\sqrt{\frac{T_f}{\mu}} \end{aligned}\right\} \quad \rightarrow \quad 1 = \frac{n_i}{n_f}\sqrt{\frac{T_i}{T_f}}$$

where the frequency f is the same in both cases because it is determined by the vibrating blade. In addition, the linear mass density μ and the length L of the vibrating portion of the string are the same in both cases. Solving (2) for T_f gives

$$T_f = \left(\frac{n_i}{n_f}\right)^2 T_i = \left(\frac{2}{5}\right)^2 (19.6\text{ N}) = 3.14\text{ N}$$

Substituting this value into equation (1), we can evaluate the buoyant force on the sphere:

$$B = mg - T_f = 19.6\text{ N} - 3.14\text{ N} = 16.5\text{ N}$$

Finally, expressing the buoyant force in terms of the radius of the sphere, we solve for the radius,

$$B = \rho_{\text{water}} g V_{\text{sphere}} = \rho_{\text{water}} g \left(\tfrac{4}{3}\pi r^3\right)$$

$$r = \left(\frac{3B}{4\pi\rho_{\text{water}} g}\right)^{1/3}$$

$$= \left(\frac{3(16.5\text{ N})}{4\pi(1\,000\text{ kg/m}^3)(9.80\text{ m/s}^2)}\right)^{1/3}$$

$$= 7.38 \times 10^{-2}\text{ m} = \boxed{7.38\text{ cm}}$$

PhysicsNow™ You can adjust the mass of the sphere by logging into PhysicsNow at **www.pop4e.com** and going to Interactive Example 15.4.

15.5 FLUID DYNAMICS

Thus far, our study of fluids has been restricted to fluids at rest, or **fluid statics.** We now turn our attention to **fluid dynamics,** the study of fluids in motion. Instead of trying to study the motion of each particle of the fluid as a function of time, we describe the properties of the fluid as a whole.

Flow Characteristics

When fluid is in motion, its flow is of one of two main types. The flow is said to be **steady,** or **laminar,** if each particle of the fluid follows a smooth path so that the paths of different particles never cross each other as in Figure 15.13. Therefore, in steady flow, the velocity of the fluid at any point remains constant in time.

Above a certain critical speed, fluid flow becomes **turbulent.** Turbulent flow is an irregular flow characterized by small, whirlpool-like regions as in Figure 15.14. As an example, the flow of water in a river becomes turbulent in regions where rocks and other obstructions are encountered, often forming "white-water" rapids.

(Andy Sacks/Tony Stone Images/Getty Images)

FIGURE 15.13 An illustration of steady flow around an automobile in a test wind tunnel. The streamlines in the airflow are made visible by smoke particles.

FIGURE 15.14 Hot gases from a cigarette made visible by smoke particles. The smoke first moves in laminar flow at the bottom and then in turbulent flow above.

The term **viscosity** is commonly used in fluid flow to characterize the degree of internal friction in the fluid. This internal friction, or viscous force, is associated with the resistance of two adjacent layers of the fluid against moving relative to each other. Because viscosity represents a nonconservative force, part of a fluid's kinetic energy is converted to internal energy when layers of fluid slide past one another. This conversion is similar to the mechanism by which an object sliding on a rough horizontal surface experiences a transformation of kinetic energy to internal energy.

Because the motion of a real fluid is very complex and not yet fully understood, we adopt a simplification model. As we shall see, many features of real fluids in motion can be understood by considering the behavior of an ideal fluid. In our simplification model, we make the following four assumptions:

1. *Nonviscous fluid.* In a nonviscous fluid, internal friction is ignored. An object moving through the fluid experiences no viscous force.
2. *Incompressible fluid.* The density of the fluid is assumed to remain constant regardless of the pressure in the fluid.
3. *Steady flow.* In steady flow, we assume that the velocity of the fluid at each point remains constant in time.
4. *Irrotational flow.* Fluid flow is irrotational if the fluid has no angular momentum about any point. If a small paddle wheel placed anywhere in the fluid does not rotate about the wheel's center of mass, the flow is irrotational. (If the wheel were to rotate, as it would if turbulence were present, the flow would be rotational.)

The first two assumptions in our simplification model are properties of our ideal fluid. The last two are descriptions of the way that the fluid flows.

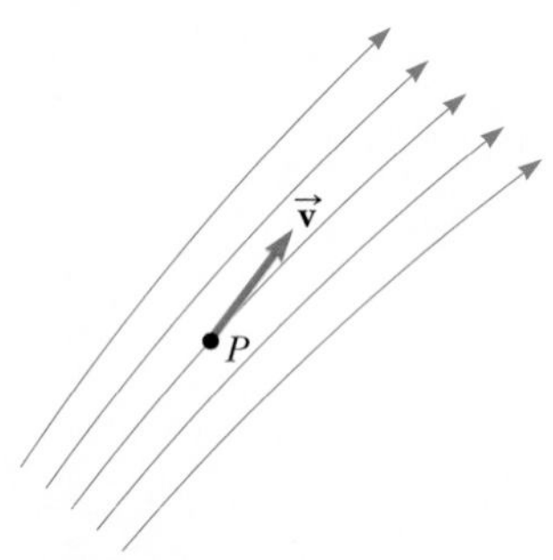

FIGURE 15.15 This diagram represents a set of streamlines (*blue lines*). A particle at *P* follows one of these streamlines, and its velocity is tangent to the streamline at each point along its path.

15.6 STREAMLINES AND THE CONTINUITY EQUATION FOR FLUIDS

If you are watering your garden and your garden hose is too short, you might do one of two things to help you reach the garden with the water (before you look for a longer hose!). You might attach a nozzle to the end of the hose, or, in the absence of a nozzle, you might place your thumb over the end of the hose, allowing the water to come out of a narrower opening. Why does either of these techniques cause the water to come out faster so that it can be projected over a longer range? We shall see the answer to this question in this section.

The path taken by a particle of the fluid under steady flow is called a **streamline.** The velocity of the particle is always tangent to the streamline as shown in Figure 15.15. No two streamlines can cross each other; if they did, a particle could move either way at the crossover point and then the flow would not be steady.

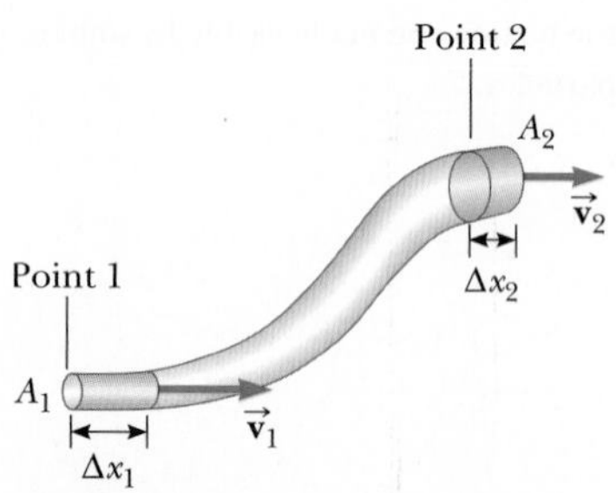

FIGURE 15.16 A fluid moving with steady flow through a pipe of varying cross-sectional area. The volume of fluid flowing through A_1 in a time interval Δt must equal the volume flowing through A_2 in the same time interval.

Consider an ideal fluid flowing through a pipe of nonuniform size as in Figure 15.16. The particles in the fluid move along the streamlines in steady flow. Let us analyze this situation using the nonisolated system in steady-state model. We have seen this model used for energy in Chapter 7, but we mentioned at that time that the model can be used for any conserved quantity. The volume of an incompressible fluid is a conserved quantity. Assuming no leaks in our pipe, we can neither create nor destroy fluid, just as we could not create nor destroy energy in Chapters 6 and 7.

We choose as our system the region of space in the pipe from point 1 to point 2 in Figure 15.16. Let us assume that this region is filled with fluid at all times. As the fluid flows in the pipe, fluid enters the system at point 1 and leaves the system at point 2. Imagine that the fluid moves through a displacement Δx_1 at point 1 and moves through a displacement Δx_2 at point 2 as it leaves the system. The volume of fluid entering the system at point 1 is $A_1\,\Delta x_1$ and the volume leaving at point 2 is $A_2\,\Delta x_2$. Because the volume of an incompressible fluid is a conserved quantity, these two volumes must be equal for the system to be in steady state. If that were not true, the volume of fluid in the system would be changing. Therefore,

$$A_1\,\Delta x_1 = A_2\,\Delta x_2$$

Let us divide this equation by the time interval during which the fluid moves:

$$\frac{A_1\,\Delta x_1}{\Delta t} = \frac{A_2\,\Delta x_2}{\Delta t}$$

In the limit as the time interval shrinks to zero, the ratio of the displacement of the fluid to the time interval is the instantaneous speed of the fluid, so we can write this expression as

$$A_1 v_1 = A_2 v_2 \qquad [15.7]$$

■ Continuity equation for fluids

The product Av, which has the dimensions of volume per time, is called the **volume flow rate.** Equation 15.7, called the **continuity equation for fluids,** says that **the product of the area and the fluid speed at all points along the pipe is a constant.** Therefore, the speed is high where the tube is constricted and low where the tube is wide. Hence, a nozzle or your thumb over the garden hose allows you to project the water farther. By reducing the area through which the water flows, you increase its speed. Therefore, you project the water from the hose with a high initial velocity, resulting in a large value of the range, as discussed for projectiles in Chapter 3.

QUICK QUIZ 15.5 You tape two different sodas straws together end to end to make a longer straw with no leaks. The two straws have radii of 3 mm and 5 mm. You drink a soda through your combination straw. In which straw is the speed of the liquid higher? **(a)** It is higher in whichever one is nearest your mouth. **(b)** It is higher in the one of radius 3 mm. **(c)** It is higher in the one of radius 5 mm. **(d)** Neither, because the speed is the same in both straws.

EXAMPLE 15.5 Watering a Garden

A water hose 2.50 cm in diameter is used by a gardener to fill a 30.0-L bucket. The gardener notes that it takes 1.00 min to fill the bucket. A nozzle with an opening of cross-sectional area 0.500 cm^2 is then attached to the hose. The nozzle is held so that water is projected horizontally from a point 1.00 m above the ground. Over what horizontal distance can the water be projected?

Solution We identify point 1 within the hose and point 2 at the exit of the nozzle. We first find the speed of the water in the hose from the bucket-filling information. The cross-sectional area of the hose is

$$A_1 = \pi r^2 = \pi \frac{d^2}{4} = \pi\left[\frac{(2.50\text{ cm})^2}{4}\right] = 4.91\text{ cm}^2$$

According to the data given, the volume flow rate is equal to 30.0 L/min:

$$A_1 v_1 = 30.0\text{ L/min} = \frac{30.0 \times 10^3\text{ cm}^3}{60.0\text{ s}} = 500\text{ cm}^3/\text{s}$$

$$v_1 = \frac{500\text{ cm}^3/\text{s}}{4.91\text{ cm}^2} = 102\text{ cm/s} = 1.02\text{ m/s}$$

Now we use the continuity equation for fluids to find the speed $v_2 = v_{xi}$ with which the water exits the nozzle. The subscript i anticipates that this speed will be the *initial* velocity component of the water projected from the hose, and the subscript x indicates that the initial velocity vector of the projected water is in the horizontal direction. So,

$$A_1 v_1 = A_2 v_2 = A_2 v_{xi}$$

$$v_{xi} = \frac{A_1}{A_2} v_1 = \frac{4.91\text{ cm}^2}{0.500\text{ cm}^2}(1.02\text{ m/s}) = 10.0\text{ m/s}$$

We now shift our thinking away from fluids and to projectile motion because the water is in free-fall once it exits the nozzle. An element of the water is modeled as a particle under constant acceleration as it falls through a vertical distance of 1.00 m starting from rest at $t = 0$. We find the time at which the water strikes the ground. From Equation 3.13,

$$y_f = y_i + v_{yi}t - \tfrac{1}{2}gt^2$$

$$-1.00\text{ m} = 0 + 0 - \tfrac{1}{2}(9.80\text{ m/s}^2)t^2$$

$$t = \sqrt{\frac{2(1.00\text{ m})}{9.80\text{ m/s}^2}} = 0.452\text{ s}$$

In the horizontal direction, the element of water is modeled as a particle under constant velocity. We apply Equation 3.12 to find the horizontal position as the water strikes the ground:

$$x_f = x_i + v_{xi}t = 0 + (10.0 \text{ m/s})(0.452 \text{ s}) = 4.52 \text{ m}$$

(© Bettmann/CORBIS)

DANIEL BERNOULLI (1700–1782)

Bernoulli, a Swiss physicist and mathematician, made important discoveries in fluid dynamics. His most famous work, *Hydrodynamica,* published in 1738, is both a theoretical and a practical study of equilibrium, pressure, and speed in fluids. In this publication, Bernoulli also attempted the first explanation of the behavior of gases with changing pressure and temperature; this effort was the beginning of the kinetic theory of gases, which we will study in Chapter 16.

15.7 BERNOULLI'S EQUATION

You have probably had the experience of driving on a highway and having a large truck pass by you at high speed. In that situation, you may have had the frightening feeling that your car was being pulled in toward the truck as it passed. We will see the origin for this effect in this section.

As a fluid moves through a region where its speed, elevation above the Earth's surface, or both change, the pressure in the fluid varies with these changes. The relationship between fluid speed, pressure, and elevation was first derived in 1738 by Swiss physicist Daniel Bernoulli. Consider the flow of a segment of an ideal fluid through a nonuniform pipe in a time interval Δt as illustrated in Figure 15.17. At the beginning of the time interval, the segment of fluid consists of the blue shaded portion (portion 1) at the left and the unshaded portion. During the time interval, the left end of the segment moves to the right through a displacement Δx_1, which is the length of the blue shaded portion at the left. Meanwhile, the right end of the segment moves to the right through a displacement Δx_2, which is the length of the blue shaded portion (portion 2) at the upper right of Figure 15.17. Therefore, at the end of the time interval, the segment of fluid consists of the unshaded portion and the blue shaded portion at the upper right.

Now consider forces exerted on this segment by fluid to the left and the right of the segment. The force exerted by the fluid on the left end has a magnitude P_1A_1. The work done by this force on the segment in a time interval Δt is $W_1 = F_1\,\Delta x_1 = P_1A_1\,\Delta x_1 = P_1V$, where V is the volume of portion 1. In a similar manner, the work done by the fluid to the right of the segment in the same time interval Δt is $W_2 = -P_2A_2\,\Delta x_2 = -P_2V$. (The volume of portion 1 equals the volume of portion 2.) This work is negative because the force on the segment of fluid is to the left and the displacement of the point of application of the force is to the right. Therefore, the net work done on the segment by these forces in the time interval Δt is

$$W = (P_1 - P_2)V \qquad [15.8]$$

Part of this work goes into changing the kinetic energy of the segment of fluid and part goes into changing the gravitational potential energy of the segment–Earth system. Because we are assuming streamline flow, the kinetic energy of the unshaded portion of the segment in Figure 15.17 is unchanged during the time interval. The only change is that before the time interval we have portion 1 traveling at v_1, whereas after the time interval we have portion 2 traveling at v_2. Therefore, the change in the kinetic energy of the segment of fluid is

$$\Delta K = \tfrac{1}{2}mv_2^2 - \tfrac{1}{2}mv_1^2 \qquad [15.9]$$

where m is the mass of either portion 1 or portion 2. Because the volumes of both portions are the same, they also have the same mass.

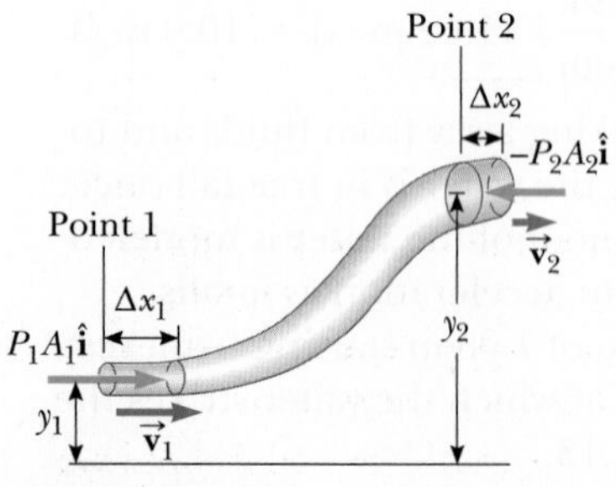

FIGURE 15.17 A fluid in laminar flow through a constricted pipe. The volume of the shaded portion on the left is equal to the volume of the shaded portion on the right.

Considering the gravitational potential energy of the segment–Earth system, once again there is no change during the time interval for the unshaded portion of the fluid. The net change is that the mass of the fluid in portion 1 has effectively been moved to the location of portion 2. Consequently, the change in gravitational potential energy of the system is

$$\Delta U = mgy_2 - mgy_1 \qquad [15.10]$$

The total work done on the segment–Earth system by the fluid outside the segment is equal to the change in mechanical energy of the system: $W = \Delta K + \Delta U$. Substituting for each of these terms gives

$$(P_1 - P_2)V = \tfrac{1}{2}mv_2^2 - \tfrac{1}{2}mv_1^2 + mgy_2 - mgy_1 \qquad [15.11]$$

If we divide each term by the portion volume V and recall that $\rho = m/V$, this expression reduces to

$$P_1 - P_2 = \tfrac{1}{2}\rho v_2^2 - \tfrac{1}{2}\rho v_1^2 + \rho g y_2 - \rho g y_1$$

Rearranging terms, we obtain

$$P_1 + \tfrac{1}{2}\rho v_1^2 + \rho g y_1 = P_2 + \tfrac{1}{2}\rho v_2^2 + \rho g y_2 \qquad [15.12]$$

■ Bernoulli's equation

which is **Bernoulli's equation** applied to an ideal fluid. It is often expressed as

$$P + \tfrac{1}{2}\rho v^2 + \rho g y = \text{constant} \qquad [15.13]$$

Bernoulli's equation says that the sum of the pressure P, the kinetic energy per unit volume $\frac{1}{2}\rho v^2$, and gravitational potential energy per unit volume $\rho g y$ has the same value at all points along a streamline.

When the fluid is at rest, $v_1 = v_2 = 0$ and Equation 15.12 becomes

$$P_1 - P_2 = \rho g(y_2 - y_1) = \rho g h$$

which agrees with Equation 15.4.

Although Equation 15.13 was derived for an incompressible fluid, the general behavior of pressure with speed is true even for gases: as the speed increases, the pressure decreases. This *Bernoulli effect* explains the experience with the truck on the highway at the opening of this section. As air passes between your car and the truck, it must pass through a relatively narrow channel. According to the continuity equation, the speed of the air is higher. According to the Bernoulli effect, this higher-speed air exerts less pressure on your car than the slower-moving air on the other side of your car. Thus, there is a net force pushing you toward the truck.

QUICK QUIZ 15.6 You observe two helium balloons floating next to each other at the ends of strings secured to a table. The facing surfaces of the balloons are separated by 1 to 2 cm. You blow through the opening between the balloons. What happens to the balloons? **(a)** They move toward each other. **(b)** They move away from each other. **(c)** They are unaffected.

EXAMPLE 15.6 Sinking the Cruise Ship

A scuba diver is hunting for fish with a spear gun. He accidentally fires the gun so that a spear punctures the side of a cruise ship. The hole is located at a depth of 10.0 m below the water surface. With what speed does the water enter the cruise ship through the hole?

Solution We identify point 1 as the water surface outside the ship, which we will assign as $y = 0$. At this point, the water is static, so $v_1 = 0$. We identify point 2 as a point just inside the hole in the interior of the ship because that is the point at which we wish to evaluate the speed of the water. This point is at a depth $y = -h = -10.0$ m below the water surface. We use Bernoulli's equation to compare these two points. At both points, the water is open to atmospheric pressure, so $P_1 = P_2 = P_0$.

Based on this argument, Bernoulli's equation becomes

$$P_0 + \tfrac{1}{2}\rho(0)^2 + \rho g(0) = P_0 + \tfrac{1}{2}\rho v_2^2 + \rho g(-h) \rightarrow$$

$$v_2 = \sqrt{2gh} = \sqrt{2(9.80\text{ m/s}^2)(10.0\text{ m})} = 14\text{ m/s}$$

INTERACTIVE EXAMPLE 15.7 Torricelli's Law

An enclosed tank containing a liquid of density ρ has a hole in its side at a distance y_1 from the tank's bottom (Fig. 15.18). The hole is open to the atmosphere, and its diameter is much smaller than the diameter of the tank. The air above the liquid is maintained at a pressure P.

A Determine the speed of the liquid as it leaves the hole when the liquid's level is a distance h above the hole.

Solution Because $A_2 >> A_1$, the liquid is approximately at rest at the top of the tank, where the pressure is P. Applying Bernoulli's equation to points 1 and 2 and noting that at the hole P_1 is equal to atmospheric pressure P_0, we find that

$$P_0 + \tfrac{1}{2}\rho v_1^2 + \rho g y_1 = P + \rho g y_2$$

In this case, $y_2 - y_1 = h$; therefore, this expression reduces to

$$v_1 = \sqrt{\frac{2(P - P_0)}{\rho} + 2gh}$$

When P is much greater than P_0 and $P/\rho >> 2gh$ (so that the term $2gh$ can be neglected), the exit speed of the water is mainly a function of P. If the tank is open to the atmosphere, $P = P_0$ and $v_1 = \sqrt{2gh}$. In other words, for an open tank the speed of liquid coming out through a hole a distance h below the surface is equal to that acquired by an object falling freely through a vertical distance h. This phenomenon is known as **Torricelli's law.**

B Suppose the position of the hole in Figure 15.18 could be adjusted vertically. If the tank is open to the atmosphere and sitting on a table, what position of the hole would cause the water to land on the table at the farthest distance from the tank?

Solution Because the tank is open to the atmosphere, the pressure at both points 1 and 2 is atmospheric pressure. Therefore, Bernoulli's equation becomes

$$P_0 + \tfrac{1}{2}\rho v_1^2 + \rho g y_1 = P_0 + \rho g y_2 \quad \rightarrow \quad v_1 = \sqrt{2g(y_2 - y_1)}$$

We model a parcel of water exiting the hole as a projectile. We find the time at which the parcel strikes the table from a hole at an arbitrary position:

$$y_f = y_i + v_{yi}t - \tfrac{1}{2}gt^2$$

$$0 = y_1 + 0 - \tfrac{1}{2}gt^2$$

$$t = \sqrt{\frac{2y_1}{g}}$$

Therefore, the horizontal position of the parcel at the time it strikes the table is

$$x_f = x_i + v_{xi}t = 0 + \sqrt{2g(y_2 - y_1)}\sqrt{\frac{2y_1}{g}} = 2\sqrt{(y_2 y_1 - y_1^2)}$$

Now we maximize the horizontal position by taking the derivative of x_f with respect to y_1 (because y_1, the height of the hole, is the variable that can be adjusted) and setting it equal to zero:

$$\frac{dx_f}{dy_1} = \tfrac{1}{2}(2)[(y_2 y_1 - y_1^2)]^{-1/2}(y_2 - 2y_1) = 0$$

This expression is satisfied if

$$y_1 = \tfrac{1}{2}y_2$$

Therefore, the hole should be halfway between the bottom of the tank and the upper surface of the water to maximize the horizontal distance. Below this location, the water is projected at a higher speed but falls for a short time interval, reducing the horizontal range. Above this point, the water spends more time in the air but is projected with a smaller horizontal speed.

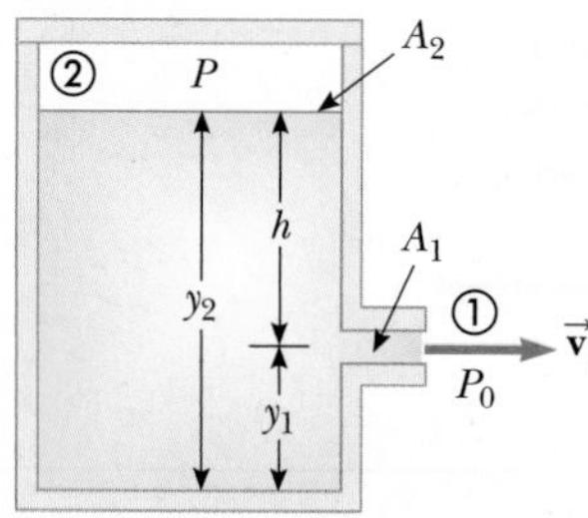

FIGURE 15.18 (Interactive Example 15.7) A liquid leaves a hole in a tank at speed v_1.

PhysicsNow™ By logging into PhysicsNow at www.pop4e.com and going to Interactive Example 15.7 you can move the hole vertically to see where the water lands.

15.8 OTHER APPLICATIONS OF FLUID DYNAMICS

Consider the streamlines that flow around an airplane wing as shown in Figure 15.19. Let us assume that the airstream approaches the wing horizontally from the right. The tilt of the wing causes the airstream to be deflected downward. Because the airstream is deflected by the wing, the wing must exert a force on the airstream. According to Newton's third law, the airstream must exert an equal and opposite force $\vec{\mathbf{F}}$ on the wing. This force has a vertical component called the **lift** (or aerodynamic lift) and a horizontal component called **drag.** The lift depends on several factors, such as the speed of the airplane, the area of the wing, its curvature, and the angle between the wing and the horizontal. As this angle increases, turbulent flow can set in above the wing to reduce the lift.

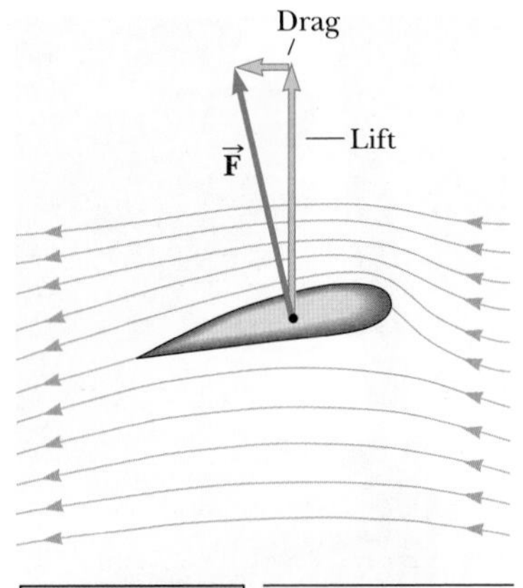

FIGURE 15.19 Streamline flow around a moving airplane wing. The air approaching from the right is deflected downward by the wing.

In general, an object experiences lift by any effect that causes the fluid to change its direction as it flows past the object. Some factors that influence lift are the shape of the object, its orientation with respect to the fluid flow, spinning motion (for example, a curve ball thrown in a baseball game due to the spinning of the baseball), and the texture of the object's surface.

A number of devices operate in a manner similar to the *atomizer* in Figure 15.20. A stream of air passing over an open tube reduces the pressure above the tube. This reduction in pressure causes the liquid to rise into the air stream. The liquid is then dispersed into a fine spray of droplets. This type of system is used in perfume bottles and paint sprayers.

Vascular flutter

Bernoulli's principle explains one symptom of advanced arteriosclerosis called *vascular flutter.* The artery is constricted as a result of an accumulation of plaque on its inner walls (Fig. 15.21). Plaque is a combination of fat, cell debris, connective tissue, and sometimes calcium that forms a flat patch inside a blood vessel. The blood speed through the constriction is higher than elsewhere according to the continuity equation for fluids. According to Bernoulli's principle, the pressure in the constriction is lower than elsewhere. If the blood speed is sufficiently high in the constricted region, the artery may collapse under the larger external pressure, causing a momentary interruption in blood flow. At this point, the speed of the blood goes to zero, its pressure rises again, and the vessel reopens. As the blood rushes through the constricted artery, the internal pressure drops and again the artery closes. Such variations in blood flow can be heard with a stethoscope.

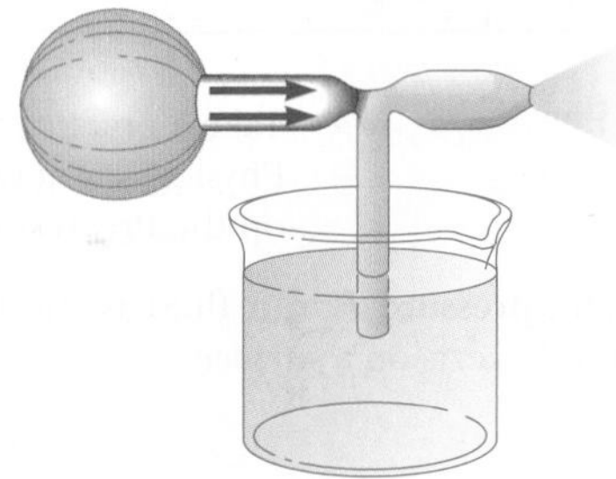

FIGURE 15.20 A stream of air passing over a tube dipped into a liquid will cause the liquid to rise in the tube.

15.9 A NEAR MISS EVEN BEFORE LEAVING SOUTHAMPTON

CONTEXT CONNECTION

As the *Titanic* began its maiden voyage, it experienced a potentially disastrous incident before it left Southampton harbor. (If this disaster had occurred, however, it is highly likely that it would have been much less disastrous in terms of lives lost than the incident that actually occurred later in the voyage.) The *Titanic* passed closely by the *New York,* which was tied securely next to the *Oceanic* at the dock, with the keels of the two ships parallel. As the *Titanic* passed by, the *New York* was forced toward it, the *New York*'s mooring ropes snapped, and its stern swung out toward the *Titanic.* It was only quick thinking by the harbor pilot on the *Titanic,* who reversed the engines, causing the *Titanic* to slow and allow the *New York* to pass by safely, that saved the two ships from a collision. As it was, a collision was averted by only a few feet, and the *Titanic* was delayed by over an hour in her departure. Figure 15.22 is a photograph taken from the *Titanic,* showing how close the ships came to colliding.

It is ironic that the captain of the *Titanic,* E. J. Smith, who watched the near miss from the bridge, was captain on one of the *Titanic*'s sister ships, the *Olympic,* when a similar incident occurred seven months before the *New York* incident. In this case, the cruiser *Hawke* was pulled toward the *Olympic* and a collision was not averted. The *Hawke*'s bow was seriously damaged in the collision, and the hull of the *Olympic* was punctured above and below the waterline. Both ships were able to return to port but needed extensive repairs.

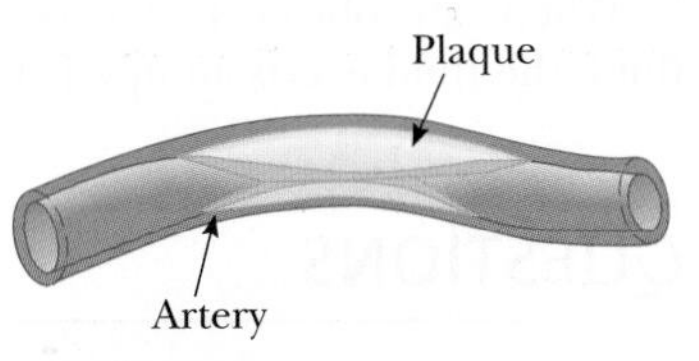

FIGURE 15.21 Blood must travel faster than normal through a constricted region of an artery.

(Courtesy of The Father Browne S.J. Collection)

FIGURE 15.22 While leaving Southampton harbor, the *Titanic (left)* experienced a near miss with the *New York (right)* due to Bernoulli's principle. If this accident had actually occurred, it may have changed the timing enough that the *Titanic,* once under way, might not have been sunk by an iceberg.

Why did these events occur? The answer lies in Bernoulli's principle. As ships move through the water, they push water out of the way and the water moves around the sides of the ship. Imagine now that a ship such as the *Titanic* passes near another ship, such as the *New York,* with their keels parallel. The water moving around the side of the *Titanic* toward the *New York* is forced into a narrow channel between the ships. Because water is incompressible, its volume remains constant when it is squeezed into a narrow channel. The initial tendency is for the compressed water to rise into the air between the ships because the air above the water offers little resistance to being compressed. As soon as the water level between the ships rises, however, the water will begin to flow in a direction parallel to the keels toward the lower-level water near the bow and stern of the ships. Therefore, the *water between the ships is moving at a higher speed than the water on the opposite sides of the ships.* According to Bernoulli's principle, this rapidly moving water exerts less pressure on the sides of the ships than the slower moving water on the outer sides. The result is a *net force pushing the two ships toward each other.*

Therefore, captains of boats and ships are advised not to pass too close by other boats moving in a parallel direction. If that does occur, the boats could be pushed into each other. This effect occurs for air, explaining the effect of the passing truck in Section 15.7.

In this Context Connection section, we investigated an application of Bernoulli's principle. In the Context Conclusion, we shall explore the difficulties in visiting the *Titanic* because of its great depth under the ocean surface. ■

SUMMARY

PhysicsNow™ Take a practice test by logging into PhysicsNow at **www.pop4e.com** and clicking on the Pre-Test link for this chapter.

The **pressure** P in a fluid is the force per unit area that the fluid exerts on a surface:

$$P \equiv \frac{F}{A} \qquad [15.1]$$

In the SI system, pressure has units of newtons per square meter, and $1\ \text{N/m}^2 = 1$ pascal (Pa).

The pressure in a liquid varies with depth h according to the expression

$$P = P_0 + \rho g h \qquad [15.4]$$

where P_0 is the pressure at the surface of the liquid and ρ is the density of the liquid, assumed uniform.

Pascal's law states that when a change in pressure is applied to a fluid, the change in pressure is transmitted undiminished to every point in the fluid and to every point on the walls of the container.

When an object is partially or fully submerged in a fluid, the fluid exerts an upward force on the object called the **buoyant force.** According to **Archimedes's principle,** the buoyant force is equal to the weight of the fluid displaced by the object.

Various aspects of fluid dynamics can be understood by adopting a simplification model in which the fluid is nonviscous and incompressible and the fluid motion is a steady flow with no turbulence.

Using this model, two important results regarding fluid flow through a pipe of nonuniform size can be obtained:

1. The flow rate through the pipe is a constant, which is equivalent to stating that the product of the cross-sectional area A and the speed v at any point is a constant. This behavior is described by the **continuity equation for fluids:**

$$A_1 v_1 = A_2 v_2 = \text{constant} \qquad [15.7]$$

2. The sum of the pressure, kinetic energy per unit volume, and gravitational potential energy per unit volume has the same value at all points along a streamline. This behavior is described by **Bernoulli's equation:**

$$P_1 + \tfrac{1}{2}\rho v_1^2 + \rho g y_1 = P_2 + \tfrac{1}{2}\rho v_2^2 + \rho g y_2 \qquad [15.12]$$

QUESTIONS

☐ = answer available in the *Student Solutions Manual and Study Guide*

1. Two drinking glasses having equal weights but different shapes and different cross-sectional areas are filled to the same level with water. According to the expression

$P = P_0 + \rho g h$, the pressure is the same at the bottom of both glasses. In view of this fact, why does one weigh more than the other?

2. Figure Q15.2 shows aerial views from directly above two dams. Both dams are equally wide (the vertical dimension in the diagram) and equally high (into the page in the diagram). The dam on the left holds back a very large lake, whereas the dam on the right holds back a narrow river. Which dam has to be built stronger?

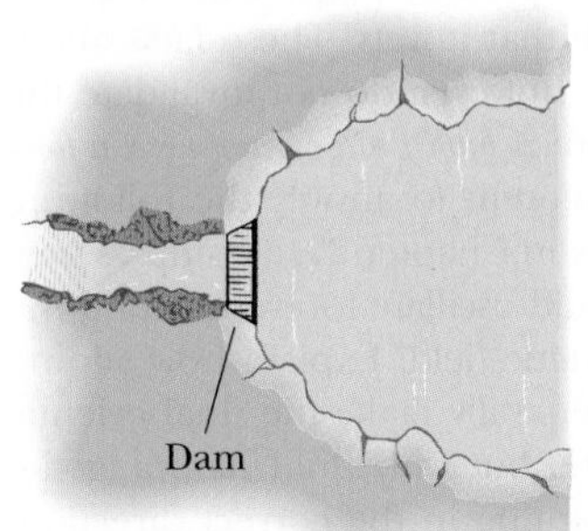

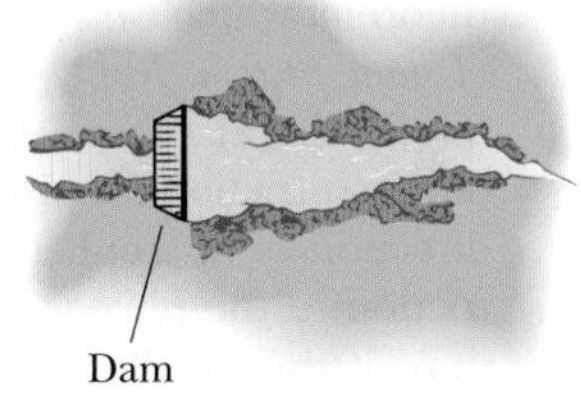

FIGURE Q15.2

3. Some physics students attach a long tube to the opening of a hot water bottle made of strong rubber. Leaving the hot water bottle on the ground, they hoist the other end of the tube to the roof of a multistory campus building. Students at the top of the building pour water into the tube. The students on the ground watch the bottle fill with water. On the roof, the students are surprised to see that the tube never seems to fill up: they can continue to pour more and more water down the tube. On the ground, the hot water bottle swells up like a balloon and bursts, drenching the students. Explain these observations.

4. Suppose a damaged ship can just barely keep afloat in the ocean. It is towed toward shore and into a river, heading toward a dry dock for repair. As it is pulled up the river, it sinks. Why?

5. A fish rests on the bottom of a bucket of water while the bucket is being weighed on a scale. When the fish begins to swim around, does the scale reading change?

6. Lead has a greater density than iron, and both are denser than water. Is the buoyant force on a lead object greater than, less than, or equal to the buoyant force on an iron object of the same volume?

7. Is the buoyant force a conservative force? Is a potential energy associated with it? Explain your answers.

8. If the air stream from a hair dryer is directed over a Ping-Pong ball, the ball can be levitated. Explain.

9. The water supply for a city is often provided from reservoirs built on high ground. Water flows from the reservoir, through pipes, and into your home when you turn the tap on your faucet. Why is the water flow more rapid out of a faucet on the first floor of a building than in an apartment on a higher floor?

10. When ski jumpers are airborne (Fig. Q15.10), why do they bend their bodies forward and keep their hands at their sides?

(© TempSport/CORBIS)

FIGURE Q15.10

11. Explain why a sealed bottle partially filled with a liquid can float in a basin of the same liquid.

12. When is the buoyant force on a swimmer greater, after exhaling or after inhaling?

13. A barge is carrying a load of gravel along a river. As it approaches a low bridge the captain realizes that the top of the pile of gravel is not going to make it under the bridge. The captain orders the crew to shovel gravel quickly from the pile into the water. Is that a good decision?

14. A person in a boat floating in a small pond throws an anchor overboard. Does the level of the pond rise, fall, or remain the same?

15. An empty metal soap dish barely floats in water. A bar of Ivory soap floats in water. When the soap is stuck in the soap dish, the combination sinks. Explain why.

16. A piece of unpainted porous wood barely floats in a container partly filled with water. If the container is sealed and pressurized above atmospheric pressure, does the wood rise, fall, or remain at the same level?

17. Because atmospheric pressure is about 10^5 N/m^2 and the area of a person's chest is about 0.13 m^2, the force of the atmosphere on one's chest is around 13 000 N. In view of this enormous force, why don't our bodies collapse?

18. A small piece of steel is tied to a block of wood. When the wood is placed in a tub of water with the steel on top, half of the block is submerged. If the block is inverted so that the steel is under water, does the amount of the block submerged increase, decrease, or remain the same? What happens to the water level in the tub when the block is inverted?

19. An unopened can of diet cola floats when placed in a tank of water, whereas a can of regular cola of the same brand sinks in the tank. What do you suppose could explain this behavior?

20. Prairie dogs (Fig. Q15.20) ventilate their burrows by building a mound around one entrance, which is open to a stream of air when wind blows from any direction. A second entrance at ground level is open to almost stagnant air. How does this construction create an airflow through the burrow?

(Pamela Zilly/The Image Bank/Getty)

FIGURE Q15.20

21. You are a passenger on a spacecraft. For your survival and comfort, the interior contains air just like that at the surface of the Earth. The craft is coasting through a very empty region of space. That is, a nearly perfect vacuum exists just outside the wall. Suddenly, a meteoroid pokes a hole, about the size of a large coin, right through the wall next to your seat. What will happen? Is there anything you can or should do about it?

22. Consider a stationary fluid in contact with a solid surface. If the force exerted by the fluid is entirely characterized by a pressure, the force must be perpendicular to the surface. That is, a bulk fluid exerts a normal force but cannot exert a force of static friction. In contrast, a *moving* fluid can have the effect of exerting kinetic friction if it possesses viscosity; for example, think of the force that molasses exerts on a stirring spoon. In Chapter 5, we modeled this drag force as possibly proportional to the first or to the second power of the speed of an object moving through the fluid. Now we are saying that the drag force must go to zero when the speed approaches zero. A *thin film,* as opposed to a bulk fluid, can exert a force parallel to a solid surface. For example, a droplet or a thin film of water can temporarily support its weight by adhering to a vertical surface that it wets. These facts about moving fluids and thin films, however, do not invalidate the theorem, which we state again: *A bulk fluid cannot exert a force of static friction.* Apply this theorem to answer the following questions. (a) As we will study in Chapter 22, a compass needle on a frictionless pivot would oscillate forever around the direction of an applied magnetic field. Explain how adding friction to the pivot would generally make the final orientation of the needle inaccurate, but with fluid damping the needle will approach rest pointing in the correct direction. (b) A carpenter's level can consist of a bubble of air in water within a tube forming an arc of a circle. Explain how, after a quick calibration, it can be a very accurate level. (c) Assume that just after you step out of a shower the whole bathtub and also the bar of soap are thoroughly wet, covered with more than a thin film of water. If you drop the soap into the tub, will it come to rest? If so, where? Explain.

PROBLEMS

1, 2, 3 = straightforward, intermediate, challenging
□ = full solution available in the *Student Solutions Manual and Study Guide*
PhysicsNow™ = coached problem with hints available at www.pop4e.com
= computer useful in solving problem
= paired numerical and symbolic problems
= biomedical application

Section 15.1 ■ Pressure

1. Calculate the mass of a solid iron sphere that has a diameter of 3.00 cm.

2. The four tires of an automobile are inflated to a gauge pressure of 200 kPa. Each tire has an area of 0.024 0 m^2 in contact with the ground. Determine the weight of the automobile.

3. A 50.0-kg woman balances on one heel of a pair of high-heeled shoes. If the heel is circular and has a radius of 0.500 cm, what pressure does she exert on the floor?

4. What is the total mass of the Earth's atmosphere? (The radius of the Earth is 6.37×10^6 m, and atmospheric pressure at the surface is 1.013×10^5 N/m^2.)

Section 15.2 ■ Variation of Pressure with Depth

5. The spring of the pressure gauge shown in Figure 15.2 has a force constant of 1 000 N/m and the piston has a diameter of 2.00 cm. As the gauge is lowered into water, what change in depth causes the piston to move in by 0.500 cm?

6. (a) Calculate the absolute pressure at an ocean depth of 1 000 m. Assume that the density of sea water is 1 024 kg/m^3 and that the air above exerts a pressure of 101.3 kPa. (b) At this depth, what force must the frame around a circular submarine porthole having a diameter of 30.0 cm exert to counterbalance the force exerted by the water?

7. PhysicsNow™ What must be the contact area between a suction cup (completely exhausted) and a ceiling if the cup is to support the weight of an 80.0-kg student?

8. (a) A very powerful vacuum cleaner has a hose 2.86 cm in diameter (Fig. P15.8a). With no nozzle on the hose, what is the weight of the heaviest brick that the cleaner can lift? (b) A very powerful octopus uses one sucker of diameter 2.86 cm on each of the two shells of a clam in an attempt to pull the shells apart (Fig. P15.8b). Find the greatest force the octopus can exert in salt water 32.3 m deep. (*Caution:* Experimental verification can be interesting, but do not drop a brick on your foot. Do not overheat the motor of a vacuum cleaner. Do not get an octopus mad at you.)

(a)

(b)

FIGURE **P15.8**

9. For the cellar of a new house, a hole is dug in the ground, with vertical sides going down 2.40 m. A concrete foundation wall is built all the way across the 9.60-m width of the excavation. This foundation wall is 0.183 m away from the front of the cellar hole. During a rainstorm, drainage from the street fills up the space in front of the concrete wall, but not the cellar behind the wall. The water does not soak into the clay soil. Find the force the water causes on the foundation wall. For comparison, the weight of the water is given by $2.40 \text{ m} \times 9.60 \text{ m} \times 0.183 \text{ m} \times 1\,000 \text{ kg/m}^3 \times 9.80 \text{ m/s}^2 = 41.3 \text{ kN}$.

10. A swimming pool has dimensions $30.0 \text{ m} \times 10.0 \text{ m}$ and a flat bottom. When the pool is filled to a depth of 2.00 m with fresh water, what is the force caused by the water on the bottom? On each end? On each side?

11. **Review problem.** Piston ① in Figure P15.11 has a diameter of 0.250 in. Piston ② has a diameter of 1.50 in. Determine the magnitude F of the force necessary to support the 500 lb load in the absence of friction.

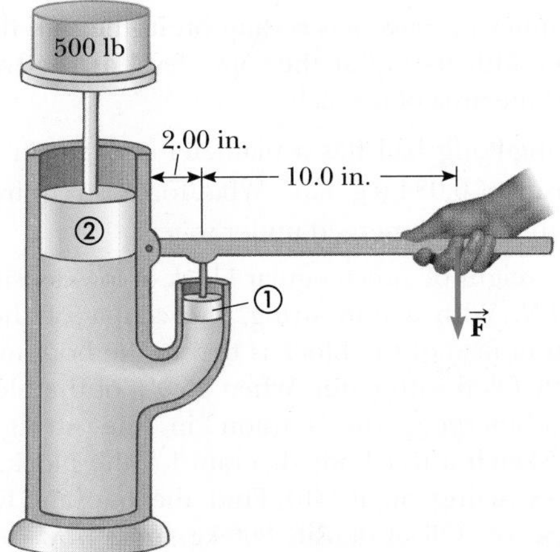

FIGURE **P15.11**

Section 15.3 ■ Pressure Measurements

12. Figure P15.12 shows Superman attempting to drink water through a very long straw. With his great strength he achieves maximum possible suction. The walls of the tubular straw do not collapse. (a) Find the maximum height through which he can lift the water. (b) Still thirsty, the Man of Steel repeats his attempt on the Moon, which has no atmosphere. Find the difference between the water levels inside and outside the straw.

FIGURE **P15.12**

13. **Physics Now™** Blaise Pascal duplicated Torricelli's barometer using a red Bordeaux wine, of density 984 kg/m^3, as the working liquid (Fig. P15.13). What was the height h of the wine column for normal atmospheric pressure? Would you expect the vacuum above the column to be as good as for mercury?

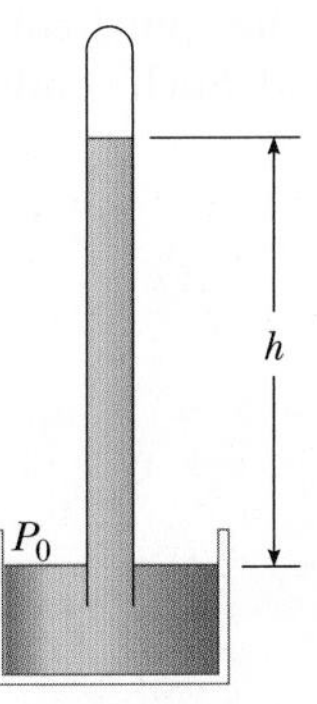

FIGURE **P15.13**

14. Mercury is poured into a U-tube as shown in Figure P15.14a. The left arm of the tube has cross-sectional area A_1 of 10.0 cm^2 and the right arm has a cross-sectional area A_2 of 5.00 cm^2. One hundred grams of water are then poured into the right arm as shown in Figure P15.14b. (a) Determine the length of the water column in the right arm of the U-tube. (b) Given that the density of mercury is 13.6 g/cm^3, what distance h does the mercury rise in the left arm?

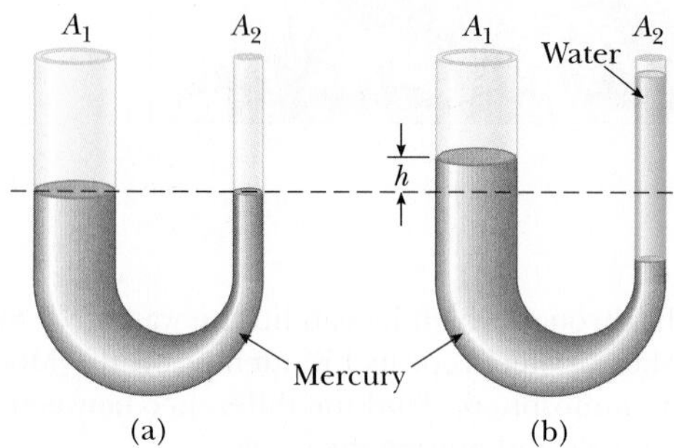

FIGURE **P15.14**

15. Normal atmospheric pressure is 1.013×10^5 Pa. The approach of a storm causes the height of a mercury barometer to drop by 20.0 mm from the normal height. What is the atmospheric pressure? (The density of mercury is 13.59 g/cm^3.)

16. The human brain and spinal cord are immersed in the cerebrospinal fluid. The fluid is normally continuous between the cranial and spinal cavities. It normally exerts a pressure of 100 to 200 mm of H_2O above the prevailing atmospheric pressure. In medical work, pressures are often measured in millimeters of H_2O because body fluids, including the cerebrospinal fluid, typically have the same density as water. The pressure of the cerebrospinal fluid can be measured by means of a *spinal tap* as illustrated in Figure P15.16. A hollow tube is inserted into the spinal column, and the height to which the fluid rises is observed. If the fluid rises to a height of 160 mm, we write its gauge pressure as 160 mm H_2O. (a) Express this pressure in pascals, in atmospheres, and in millimeters of mercury. (b) Sometimes it is necessary to determine whether an accident victim has suffered a crushed vertebra that is blocking flow of the cerebrospinal fluid in the spinal column. In other cases, a physician may suspect that a tumor or other growth is blocking the spinal column and inhibiting flow of cerebrospinal fluid. Such conditions can be investigated by means of *Queckenstedt's test*. In this procedure, the veins in the patient's neck are compressed so as to make the blood pressure rise in the brain. The increase in pressure in the blood vessels is transmitted to the cerebrospinal fluid. What should be the normal effect on the height of the fluid in the spinal tap? (c) Suppose compressing the veins had no effect on the fluid level. What might account for that?

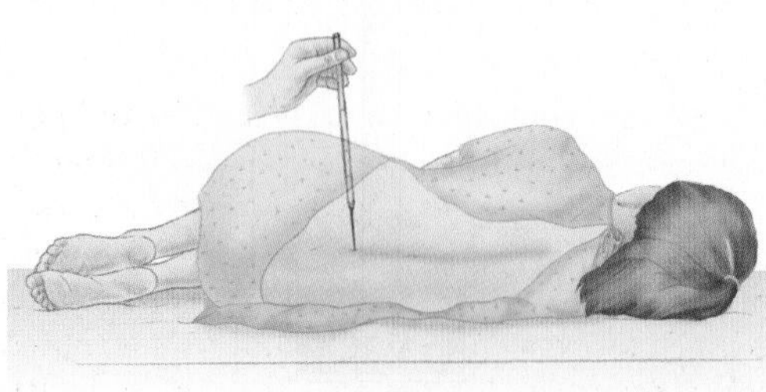

FIGURE **P15.16**

17. Mercury is a poison. The liquid and mercury vapor can enter the body through the skin and mucous membranes. Do not carry out the procedure described here without appropriate safety precautions. Assume that a barometer is constructed as follows. A rigid, thin-walled plastic tube, closed at one end, has mass 480 g, inner diameter 2.10 cm, and height 160 cm. Mercury is poured into it to fill the tube. (a) Find the mass of the metal. With the open end covered, the tube is inverted and held just above the flat bottom of an originally empty pan (mass 320 g) on a table. The tube is hung from a cord attached between its closed end and the ceiling. Next, the bottom end of the tube is uncovered. It is found that more than half, but not all, of the mercury runs out of the tube into the pan. The flow stops when the open end of the tube is below the level of the mercury in the pan and the level of mercury in the tube is 76.0 cm higher than the level in the pan. We say that the barometer reading is 76.0 cm. (b) Find the tension in the cord. (c) Find the normal force exerted by the table on the pan. (d) If another barometer were placed into the space within the tube above the mercury, what would it read? (This question was discussed by town philosophical societies in colonial America.) (e) A slow leak at the closed end of the tube allows air to enter over a period of days. What happens to the tension in the cord? What happens to the normal force exerted by the table?

Section 15.4 ■ Buoyant Forces and Archimedes's Principle

18. A Styrofoam slab has thickness h and density ρ_s. When a swimmer of mass m is resting on it, the slab floats in fresh water with its top at the same level as the water surface. Find the area of the slab.

19. A Ping-Pong ball has a diameter of 3.80 cm and average density of 0.084 0 g/cm^3. What force is required to hold it completely submerged under water?

20. The weight of a rectangular block of low-density material is 15.0 N. With a thin string, the center of the horizontal bottom face of the block is tied to the bottom of a beaker partly filled with water. When 25.0% of the block's volume is submerged, the tension in the string is 10.0 N. (a) Sketch a free-body diagram for the block, showing all forces acting on it. (b) Find the buoyant force on the block. (c) Oil of density 800 kg/m^3 is now steadily added to the beaker, forming a layer above the water and surrounding the block. The oil exerts forces on each of the four side walls of the block that the oil touches. What are the directions of these forces? (d) What happens to the string tension as the oil is added? Explain how the oil has this effect on the string tension. (e) The string breaks when its tension reaches 60.0 N. At this moment, 25.0% of the block's volume is still below the waterline. What addi-

tional fraction of the block's volume is below the top surface of the oil? (f) After the string breaks, the block comes to a new equilibrium position in the beaker. It is now in contact only with the oil. What fraction of the block's volume is submerged?

21. A 10.0-kg block of metal measuring 12.0 cm × 10.0 cm × 10.0 cm is suspended from a scale and immersed in water as shown in Figure P15.21. The 12.0-cm dimension is vertical and the top of the block is 5.00 cm below the surface of the water. (a) What are the forces acting on the top and on the bottom of the block? (Use $P_0 = 1.013\ 0 \times 10^5\ \text{N/m}^2$.) (b) What is the reading of the spring scale? (c) Show that the buoyant force equals the difference between the forces at the top and the bottom of the block.

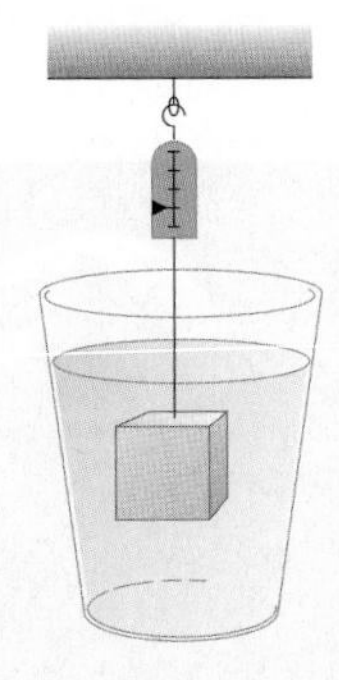

FIGURE **P15.21**

22. To an order of magnitude, how many helium-filled toy balloons would be required to lift you? Because helium is an irreplaceable resource, develop a theoretical answer rather than an experimental answer. In your solution, state what physical quantities you take as data and the values you measure or estimate for them.

23. **Physics Now™** A cube of wood having an edge dimension of 20.0 cm and a density of 650 kg/m^3 floats on water. (a) What is the distance from the horizontal top surface of the cube to the water level? (b) What mass of lead should be placed on top of the cube so that its top will be just level with the water?

24. Determination of the density of a fluid has many important applications. A car battery contains sulfuric acid, for which density is a measure of concentration. For the battery to function properly, the density must be within a range specified by the manufacturer. Similarly, the effectiveness of antifreeze in your car's engine coolant depends on the density of the mixture (usually ethylene glycol and water). When you donate blood to a blood bank, its screening includes determination of the density of your blood because higher density correlates with higher hemoglobin content. A *hydrometer* is an instrument used to determine liquid density. A simple one is sketched in Figure P15.24. The bulb of a syringe is squeezed and released to let the atmosphere lift a sample of the liquid of interest into a tube containing a calibrated rod of known density. The rod, of length L and average density ρ_0, floats partially immersed in the liquid of density ρ. A length h of the rod protrudes above the surface of the liquid. Show that the density of the liquid is given by

$$\rho = \frac{\rho_0 L}{L - h}$$

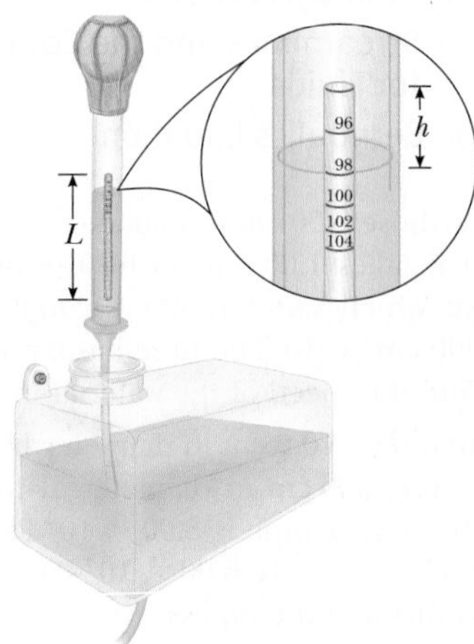

FIGURE **P15.24**

25. How many cubic meters of helium are required to lift a balloon with a 400-kg payload to a height of 8 000 m? (Take $\rho_{\text{He}} = 0.180\ \text{kg/m}^3$.) Assume that the balloon maintains a constant volume and that the density of air decreases with the altitude z according to the expression $\rho_{\text{air}} = \rho_0 e^{-z/8\ 000}$, where z is in meters and the density of air at sea level is $\rho_0 = 1.25\ \text{kg/m}^3$.

26. A bathysphere used for deep-sea exploration has a radius of 1.50 m and a mass of 1.20×10^4 kg. To dive, this submarine takes on mass in the form of sea water. Determine the amount of mass the submarine must take on if it is to descend at a constant speed of 1.20 m/s, when the resistive force on it is 1 100 N in the upward direction. The density of sea water is $1.03 \times 10^3\ \text{kg/m}^3$.

27. A plastic sphere floats in water with 50.0% of its volume submerged. This same sphere floats in glycerin with 40.0% of its volume submerged. Determine the densities of the glycerin and the sphere.

28. **Review problem.** A long, cylindrical rod of radius r is weighted on one end so that it floats upright in a fluid having a density ρ. It is pushed down a distance x from its equilibrium position and released. Show that the rod will execute simple harmonic motion if the resistive effects of the fluid are negligible and determine the period of the oscillations.

29. Decades ago, it was thought that huge herbivorous dinosaurs such as *Apatosaurus* and *Brachiosaurus* habitually walked on the bottom of lakes, extending their long necks up to the surface to breathe. *Brachiosaurus* had its nostrils on the top of its head. In 1977, Knut Schmidt-Nielsen pointed out that breathing would be too much work for such a creature. For a simple model, consider a sample consisting of 10.0 L of air at absolute pressure 2.00 atm, with density 2.40 kg/m^3, located at the surface of a freshwater lake. Find the work required to transport it to a depth of 10.3 m, with its temperature, volume, and pressure remaining constant. This energy investment is greater than the energy that can be obtained by metabolism of food with the oxygen in that quantity of air.

Section 15.5 ■ Fluid Dynamics
Section 15.6 ■ Streamlines and the Continuity Equation for Fluids
Section 15.7 ■ Bernoulli's Equation

30. (a) A water hose 2.00 cm in diameter is used to fill a 20.0-L bucket. If it takes 1.00 min to fill the bucket, what is the speed v at which water moves through the hose? (*Note:* 1 L = 1 000 cm^3.) (b) The hose has a nozzle 1.00 cm in diameter. Find the speed of the water at the nozzle.

31. A horizontal pipe 10.0 cm in diameter has a smooth reduction to a pipe 5.00 cm in diameter. If the pressure of the water in the larger pipe is 8.00×10^4 Pa and the pressure in the smaller pipe is 6.00×10^4 Pa, at what rate does water flow through the pipes?

32. Water flows through a fire hose of diameter 6.35 cm at a rate of 0.012 0 m^3/s. The fire hose ends in a nozzle of inner diameter 2.20 cm. What is the speed with which the water exits the nozzle?

33. **PhysicsNow™** A large storage tank with an open top is filled to a height h_0. The tank is punctured at a height h above the bottom of the tank (Fig. P15.33). Find an expression for how far from the tank the exiting stream lands.

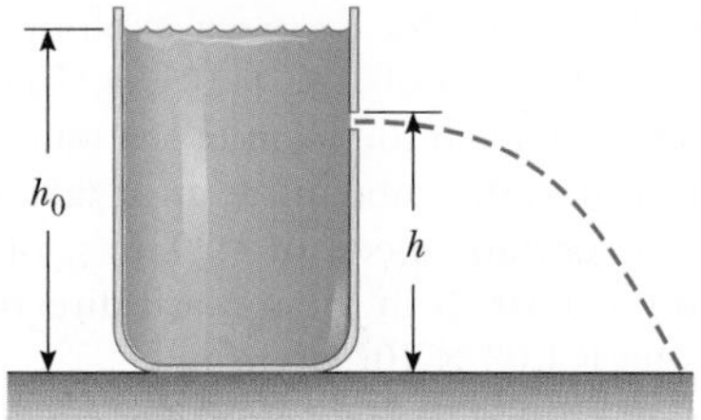

FIGURE **P15.33** Problems 15.33 and 15.34.

34. A large storage tank, open at the top and filled with water, develops a small hole in its side (Fig. P15.33) at a point 16.0 m below the water level. If the rate of flow from the leak is 2.50×10^{-3} m^3/min, determine (a) the speed at which the water leaves the hole and (b) the diameter of the hole.

35. A village maintains a large tank with an open top, containing water for emergencies. The water can drain from the tank through a hose of diameter 6.60 cm. The hose ends with a nozzle of diameter 2.20 cm. A rubber stopper is inserted into the nozzle. The water level in the tank is kept 7.50 m above the nozzle. (a) Calculate the friction force exerted by the nozzle on the stopper. (b) The stopper is removed. What mass of water flows from the nozzle in 2.00 h? (c) Calculate the gauge pressure of the flowing water in the hose just behind the nozzle.

36. Water falls over a dam of height h with a mass flow rate of R, in kilograms per second. (a) Show that the power available from the water is

$$\mathcal{P} = Rgh$$

where g is the free-fall acceleration. (b) Each hydroelectric unit at the Grand Coulee Dam takes in water at a rate of 8.50×10^5 kg/s from a height of 87.0 m. The power developed by the falling water is converted to electric power with an efficiency of 85.0%. How much electric power does each hydroelectric unit produce?

37. Figure P15.37 shows a stream of water in steady flow from a kitchen faucet. At the faucet the diameter of the stream is 0.960 cm. The stream fills a 125-cm^3 container in 16.3 s. Find the diameter of the stream 13.0 cm below the opening of the faucet.

FIGURE **P15.37**

38. A legendary Dutch boy saved Holland by plugging a hole in a dike with his finger, 1.20 cm in diameter. If the hole was 2.00 m below the surface of the North Sea (density 1 030 kg/m^3), (a) what was the force on his finger? (b) If he pulled his finger out of the hole, during what time interval would the released water fill 1 acre of land to a depth of 1 ft? Assume that the hole remained constant in size. (A typical U.S. family of four uses 1 acre-foot of water, 1 234 m^3, in 1 year.)

39. Water is pumped up from the Colorado River to supply Grand Canyon Village, located on the rim of the canyon. The river is at an elevation of 564 m and the village is at an elevation of 2 096 m. Imagine that water is pumped through a single, long pipe 15.0 cm in diameter, driven by a single pump at the bottom end. (a) What is the minimum pressure at which the water must be pumped if it is to arrive at the village? (b) If 4 500 m^3 of water are pumped per day, what is the speed of the water in the pipe? (c) What additional pressure is necessary to deliver this flow? (*Note:* You may assume that the free-fall acceleration and the density of air are constant over this range of elevations. The pressures you calculate are too high for an ordinary pipe. In fact, the water is lifted in stages by several pumps through shorter pipes.)

40. Old Faithful Geyser in Yellowstone Park (Fig. P15.40) erupts at approximately 1-h intervals and the height of the

water column reaches 40.0 m. (a) Model the rising stream as a series of separate drops. Analyze the free-fall motion of one of the drops to determine the speed at which the water leaves the ground. (b) Model the rising stream as an ideal fluid in streamline flow. Use Bernoulli's equation to determine the speed of the water as it leaves ground level. (c) What is the pressure (above atmospheric) in the heated underground chamber if its depth is 175 m? You may assume that the chamber is large compared with the geyser's vent.

(Stan Osolinski/Dembinsky Photo Associates)

FIGURE **P15.40**

41. An airplane is cruising at altitude 10 km. The pressure outside the craft is 0.287 atm; within the passenger compartment the pressure is 1.00 atm and the temperature is 20°C. A small leak occurs in one of the window seals in the passenger compartment. Model the air as an ideal fluid to find the speed of the stream of air flowing through the leak.

Section 15.8 ■ Other Applications of Fluid Dynamics

42. An airplane has a mass of 1.60×10^4 kg and each wing has an area of 40.0 m^2. During level flight, the pressure on the lower wing surface is 7.00×10^4 Pa. Determine the pressure on the upper wing surface.

43. A siphon is used to drain water from a tank as illustrated in Figure P15.43. The siphon has a uniform diameter. Assume steady flow without friction. (a) Assuming that the distance $h = 1.00$ m, find the speed of outflow at the end of the siphon. (b) What is the limitation on the height of the top of the siphon above the water surface? (For the flow of the liquid to be continuous, the pressure must not drop below the vapor pressure of the liquid.)

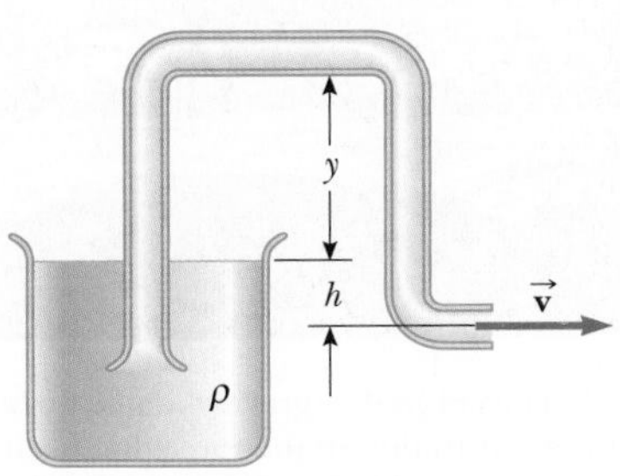

FIGURE **P15.43**

44. The Bernoulli effect can have important consequences for the design of buildings. For example, wind can blow around a skyscraper at remarkably high speed, creating low pressure. The higher atmospheric pressure in the still air inside the buildings can cause windows to pop out. As originally constructed, the John Hancock Building in Boston popped window panes, which fell many stories to the sidewalk below. (a) Suppose a horizontal wind blows in streamline flow with a speed of 11.2 m/s outside a large pane of plate glass with dimensions 4.00 m × 1.50 m. Assume that the density of the air is 1.30 kg/m^3. The air inside the building is at atmospheric pressure. What is the total force exerted by air on the window pane? (b) If a second skyscraper is built nearby, the air speed can be especially high where wind passes through the narrow separation between the buildings. Solve part (a) again, this time taking the wind speed as 22.4 m/s, twice as high.

45. A hypodermic syringe contains a medicine with the density of water (Fig. P15.45). The barrel of the syringe has a cross-sectional area $A = 2.50 \times 10^{-5}$ m^2 and the needle has a cross-sectional area $a = 1.00 \times 10^{-8}$ m^2. In the absence of a force on the plunger, the pressure everywhere is 1 atm. A force $\vec{F}$ of magnitude 2.00 N acts on the plunger, making medicine squirt horizontally from the needle. Determine the speed of the medicine as it leaves the needle's tip.

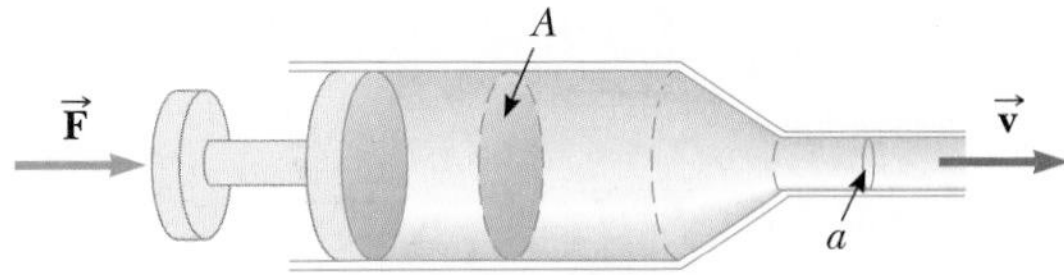

FIGURE **P15.45**

Section 15.9 ■ Context Connection—A Near Miss Even Before Leaving Southampton

46. According to the caption of the chapter-opening photograph, about 11% of an iceberg is above water. (a) Confirm this value mathematically. (b) Suppose an iceberg were floating in fresh water rather than in sea water. Would a larger or smaller percentage be above the waterline? Calculate this percentage.

47. The *Titanic* is docked in Southampton harbor just before boarding. You, as a ticket agent for White Star Lines, notice where the water level is on a scale of numbers marked on the side of the vessel. Because your ticket-collecting job is so boring that you need something to occupy your mind, you make a note of the water level and check it again after everyone boards. Over an interval of 2 h, 2 205 passengers, of average mass 75.0 kg, board the *Titanic.* You notice that the ship has sunk 1.00 cm deeper in the water with the passengers on board. What is the horizontal area enclosed by the waterline of the *Titanic*?

48. **Review problem.** Assume that the *Titanic* is drifting in Southampton harbor before its fateful journey and the captain wishes to stop the drift by dropping an anchor. The iron anchor has a mass of 2 000 kg. It is attached to a

massless rope. The rope is wrapped around a reel in the form of a solid disk of radius 0.250 m and mass 300 kg that rotates on a frictionless axle. (a) Find the angular displacement of the reel when the anchor moves down 15.0 m. (b) Find the acceleration of the anchor as it falls through the air, which offers negligible resistance. (c) While the anchor continues to drop through the water, the water exerts a drag force of 2 500 N on it. With what acceleration does the anchor move through the water? (d) While the anchor drops through the water, what torque is exerted on the reel?

Additional Problems

49. The true weight of an object can be measured in a vacuum, where buoyant forces are absent. An object of volume V is weighed in air on an equal-arm balance with the use of counterweights of density ρ. Let the density of air be ρ_{air} and the balance reading be F'_g. Show that the true weight F_g is

$$F_g = F'_g + \left(V - \frac{F'_g}{\rho g}\right)\rho_{air} g$$

50. Water is forced out of a fire extinguisher by air pressure as shown in Figure P15.50. How much gauge air pressure in the tank (above atmospheric) is required for the water jet to have a speed of 30.0 m/s when the water level is 0.500 m below the nozzle?

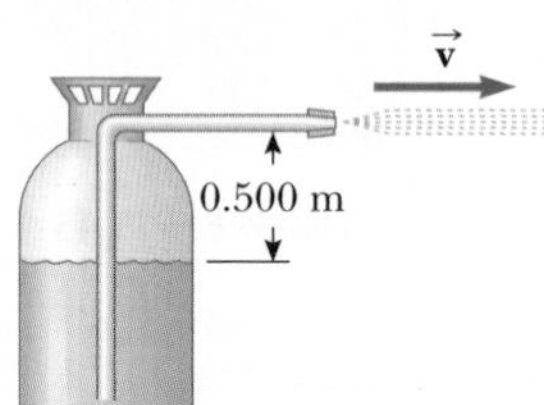

FIGURE P15.50

51. A light spring of constant k = 90.0 N/m is attached vertically to a table (Fig. P15.51a). A 2.00-g balloon is filled with helium (density = 0.180 kg/m^3) to a volume of 5.00 m^3 and is then connected to the spring, causing it to stretch as shown in Figure P15.51b. Determine the extension distance L when the balloon is in equilibrium.

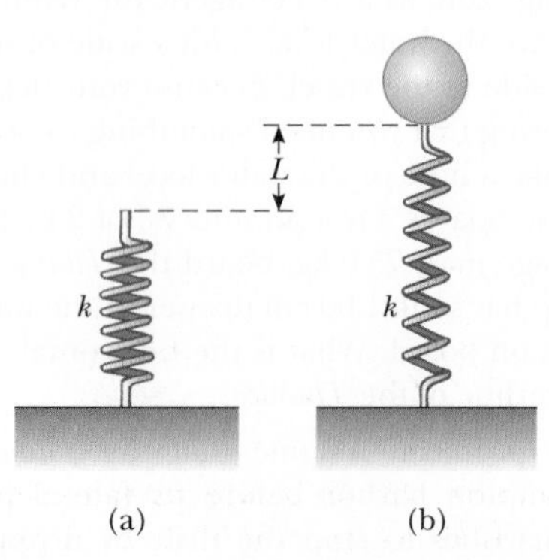

FIGURE P15.51

52. Evangelista Torricelli was the first person to realize that we live at the bottom of an ocean of air. He correctly surmised that the pressure of our atmosphere is attributable to the weight of the air. The density of air at 0°C at the Earth's surface is 1.29 kg/m^3. The density decreases with increasing altitude (as the atmosphere thins). On the other hand, if we assume that the density is constant at 1.29 kg/m^3 up to some altitude h and is zero above that altitude, h would represent the depth of the ocean of air. Use this model to determine the value of h that gives a pressure of 1.00 atm at the surface of the Earth. Would the peak of Mount Everest rise above the surface of such an atmosphere?

53. **Physics Now™ Review problem.** With reference to Figure 15.6, show that the total torque exerted by the water behind the dam about a horizontal axis through O is $\frac{1}{6}\rho g w H^3$. Show that the effective line of action of the total force exerted by the water is at a distance $\frac{1}{3}H$ above O.

54. In about 1657, Otto von Guericke, inventor of the air pump, evacuated a sphere made of two brass hemispheres. Two teams of eight horses each could pull the hemispheres apart only on some trials, and then "with greatest difficulty," with the resulting sound likened to a cannon firing (Fig. P15.54). (a) Show that the force F required to pull the evacuated hemispheres apart is $\pi R^2(P_0 - P)$, where R is the radius of the hemispheres and P is the pressure inside the hemispheres, which is much less than P_0. (b) Determine the force, taking $P = 0.100P_0$ and $R = 0.300$ m.

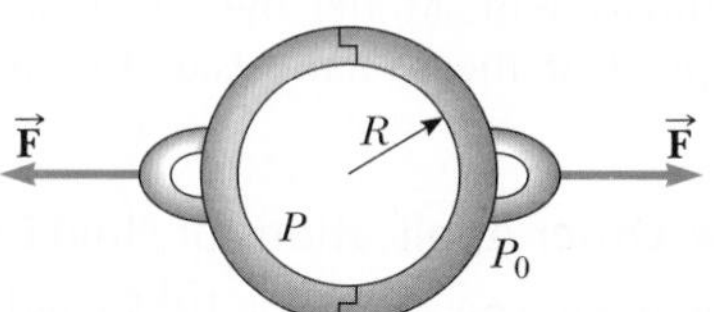

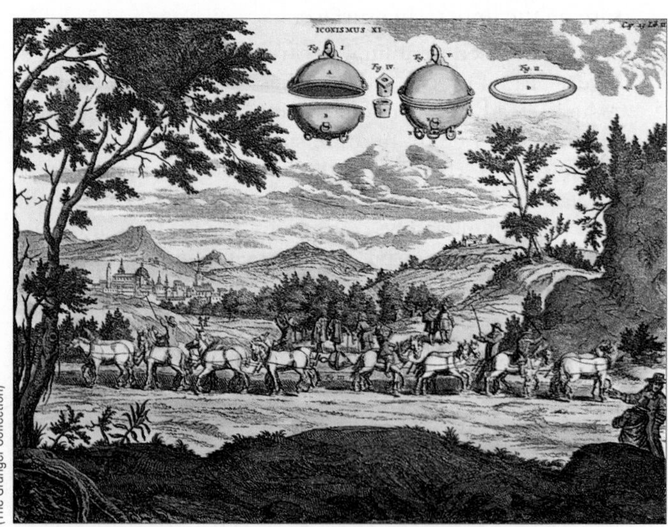

FIGURE P15.54 The colored engraving, dated 1672, illustrates Otto von Guericke's demonstration of the force due to air pressure as it might have been performed before Emperor Ferdinand III in about 1657.

55. A beaker of mass m_b containing oil of mass m_0 and density ρ_0 rests on a scale. A block of iron of mass m_{Fe} is suspended from a spring scale and completely submerged in the oil as shown in Figure P15.55. Determine the equilibrium readings of both scales.

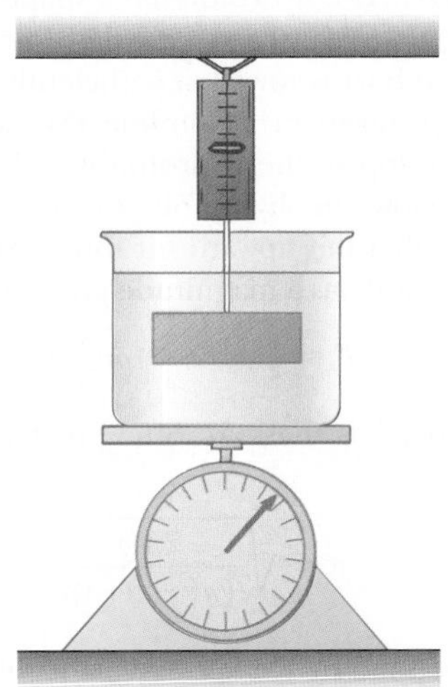

FIGURE **P15.55**

56. **Review problem.** A copper cylinder hangs at the bottom of a steel wire of negligible mass. The top end of the wire is fixed. When the wire is struck, it emits sound with a fundamental frequency of 300 Hz. The copper cylinder is then submerged in water so that half its volume is below the waterline. Determine the new fundamental frequency.

57. **Review problem.** This problem extends the reasoning of Problem 5.54 in Chapter 5 on sedimentation and centrifugation. According to Stokes's law, water exerts on a slowly moving immersed spherical object a resistive force described by

$$\vec{\mathbf{R}} = (I) - 0.018\,8\ \mathrm{N \cdot s/m^2}\ r\vec{\mathbf{v}}$$

where r is the radius of the sphere and $\vec{\mathbf{v}}$ is its velocity. (a) Spherical cells of average density $1.02 \times 10^3\ \mathrm{kg/m^3}$ and radius 8.00 μm are suspended in water. Find the terminal speed with which the cells drift down. (b) Over what time interval will all the cells settle out of a tube 8.00 cm high? (c) The sedimentation rate can be greatly increased by the use of a centrifuge. Assume that it spins the tube at 3 000 rev/min in a horizontal plane, with the middle of the tube at 9.00 cm from the axis of rotation. Find the acceleration of the middle of the tube. (d) This acceleration has the effect of an enhanced free-fall acceleration. Model it as uniform over the length of the tube. Over what time interval will all the suspended cells settle out the water in this situation?

58. Show that the variation of atmospheric pressure with altitude is given by $P = P_0 e^{-\alpha y}$, where $\alpha = \rho_0 g / P_0$, P_0 is atmospheric pressure at some reference level $y = 0$, and ρ_0 is the atmospheric density at this level. Assume that the decrease in atmospheric pressure over an infinitesimal change in altitude (so that the density is approximately uniform) is given by $dP = -\rho g\ dy$ and that the density of air is proportional to the pressure.

59. An incompressible, nonviscous fluid is initially at rest in the vertical portion of the pipe shown in Figure P15.59a, where $L = 2.00$ m. When the valve is opened, the fluid flows into the horizontal section of the pipe. What is the speed of the fluid when it is all in the horizontal section as shown in Figure P15.59b? Assume that the cross-sectional area of the entire pipe is constant.

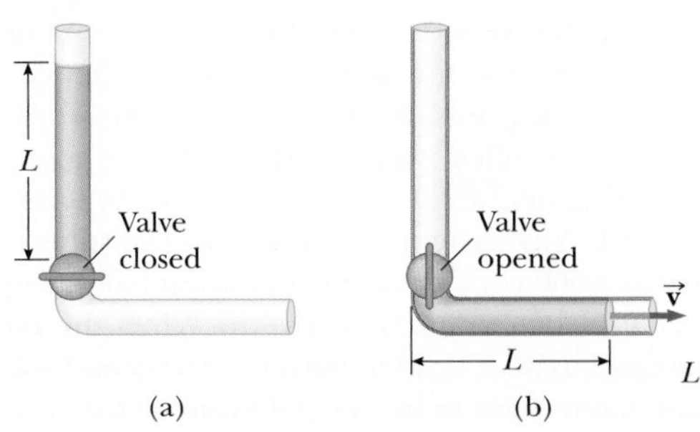

FIGURE **P15.59**

60. A cube of ice whose edges measure 20.0 mm is floating in a glass of ice-cold water with one of its faces parallel to the water's surface. (a) How far below the water surface is the bottom face of the block? (b) Ice-cold ethyl alcohol is gently poured onto the water surface to form a layer 5.00 mm thick above the water. The alcohol does not mix with the water. When the ice cube again attains hydrostatic equilibrium, what will be the distance from the top of the water to the bottom face of the block? (c) Additional cold ethyl alcohol is poured onto the water's surface until the top surface of the alcohol coincides with the top surface of the ice cube (in hydrostatic equilibrium). How thick is the required layer of ethyl alcohol?

61. A U-tube open at both ends is partially filled with water (Fig. P15.61a). Oil having a density 750 kg/m^3 is then poured into the right arm and forms a column $L = 5.00$ cm high (Fig. P15.61b). (a) Determine the difference h in the heights of the two liquid surfaces. (b) The right arm is then shielded from any air motion while air is blown across the top of the left arm until the surfaces of the two liquids are at the same height (Fig. P15.61c). Determine the speed of the air being blown across the left arm. Take the density of air as 1.29 kg/m^3.

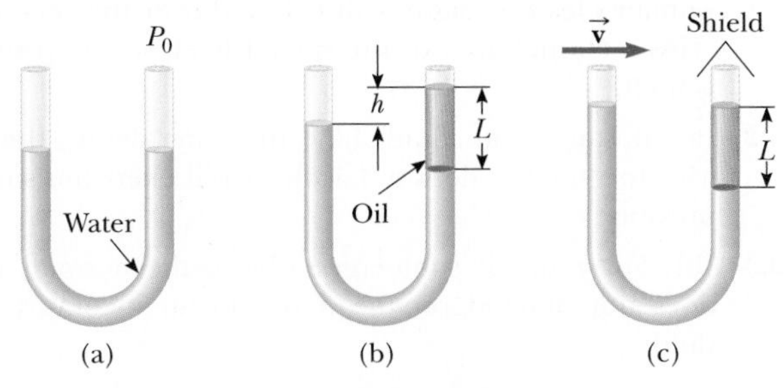

FIGURE **P15.61**

62. The water supply of a building is fed through a main pipe 6.00 cm in diameter. A 2.00-cm-diameter faucet tap, located 2.00 m above the main pipe, is observed to fill a

25.0 L container in 30.0 s. (a) What is the speed at which the water leaves the faucet? (b) What is the gauge pressure in the 6-cm main pipe? (Assume that the faucet is the only "leak" in the building.)

63. The *spirit-in-glass thermometer,* invented in Florence, Italy, around 1654, consists of a tube of liquid (the spirit) containing a number of submerged glass spheres with slightly different masses (Fig. P15.63). At sufficiently low temperatures, all the spheres float, but as the temperature rises, the spheres sink one after another. The device is a crude but interesting tool for measuring temperature. Suppose the tube is filled with ethyl alcohol, whose density is 0.789 45 g/cm^3 at 20.0°C and decreases to 0.780 97 g/cm^3 at 30.0°C. (a) Assuming that one of the spheres has a radius of 1.000 cm and is in equilibrium halfway up the tube at 20.0°C, determine its mass. (b) When the temperature increases to 30.0°C, what mass must a second sphere of the same radius have to be in equilibrium at the halfway point?

FIGURE **P15.63**

(c) At 30.0°C, the first sphere has fallen to the bottom of the tube. What upward force does the bottom of the tube exert on this sphere?

64. The hull of an experimental boat is to be lifted above the water by a hydrofoil mounted below its keel as shown in Figure P15.64. The hydrofoil has a shape like that of an airplane wing. Its area projected onto a horizontal surface is A. When the boat is towed at sufficiently high speed, water of density ρ moves in streamline flow so that its average speed at the top of the hydrofoil is n times larger than its speed v_b below the hydrofoil. (a) Ignoring the buoyant force, show that the upward lift force exerted by the water on the hydrofoil has a magnitude given by

$$F \approx \frac{1}{2}(n^2 - 1)\rho v_b^{\,2} A$$

(b) The boat has mass M. Show that the liftoff speed is given by

$$v \approx \sqrt{\frac{2Mg}{(n^2 - 1)A\rho}}$$

(c) Assume that an 800-kg boat is to lift off at 9.50 m/s. Evaluate the area A required for the hydrofoil if its design yields n = 1.05.

FIGURE **P15.64**

ANSWERS TO QUICK QUIZZES

15.1 (a). Because the basketball player's weight is distributed over the larger surface area of the shoe, the pressure (F/A) that he applies is relatively small. The woman's lesser weight is distributed over the very small cross-sectional area of the spiked heel, so the pressure is high.

15.2 (a). Because both fluids have the same depth, the one with the smaller density (alcohol) will exert the smaller pressure.

15.3 (b). For a totally submerged object, the buoyant force does not depend on the depth in an incompressible fluid.

15.4 (b) or (c). In all three cases, the weight of the treasure chest causes a downward force on the raft that makes it sink into the water. In (b) and (c), however, the treasure chest also displaces water, which provides a buoyant force in the upward direction, reducing the effect of the weight of the chest on the raft.

15.5 (b). The liquid moves at the highest speed in the straw with the smallest cross sectional area.

15.6 (a). The high-speed air between the balloons results in low pressure in this region. The higher pressure on the outer surfaces of the balloons pushes them toward each other.

Finding and Visiting the *Titanic*

We have now investigated the physics of fluids and can respond to our central question for the *Search for the Titanic* Context:

How can we safely visit the wreck of the Titanic?

Many individuals believed that the *Titanic* was unsinkable. One factor in this belief was the series of watertight bulkheads that divided the hull of the ship into several watertight compartments. Even if the hull were breached so that a compartment became flooded, the incoming water could be isolated to that compartment by closing watertight doors in the bulkhead.

According to the design of the ship, the *Titanic* could be kept afloat if its four forwardmost compartments were flooded. Unfortunately, the collision with the iceberg caused a breach in the first five compartments. As these forward compartments filled, the extra weight of the water in the bow of the ship resulted in the bow sinking into the water and the stern lifting out of the water (Fig. 1).

Despite the shipbuilders' pride in their watertight compartments, they were not watertight at the top. The bulkheads only went up to a certain height in the ship and then ended. Therefore, as the *Titanic* tilted forward, water from one compartment simply spilled over the top of the bulkhead into the next compartment and the compartments filled one by one.

(Illustrations by Ken Marschall © 1992 from *Titanic: An Illustrated History*, a Hyperion/Madison Press Book)

FIGURE 1 The *Titanic* struck the iceberg near the bow, so the forward compartments filled with water and sank, lifting the stern of the ship above the water.

Some experts after the disaster claimed that opening the watertight doors in the bulkheads would have kept the *Titanic* afloat longer, with an increased possibility of another ship arriving in time to save those who were not able to leave in the lifeboats. According to this hypothesis, if the water entering the forward compartments had been allowed to distribute evenly along the ship by passing through the doors in the bulkheads, the ship would not have tilted so that water could spill over the tops of the bulkheads. The sinking of a ship is a complicated event, however, and this hypothesis is not universally accepted.

In the region of the sinking of the *Titanic,* the depth of the ocean is about 4 km. When the wreckage was located and visited, the depth was measured to be 3 784 m.

Let us use Equation 15.4 to calculate the pressure at this depth of sea water:

$$\begin{aligned} P &= P_0 + \rho g h \\ &= 1.013 \times 10^5 \text{ Pa} + (1.03 \times 10^3 \text{ kg/m}^3)(9.80 \text{ m/s}^2)(3\,784 \text{ m}) \\ &= 3.83 \times 10^7 \text{ Pa} = 378 \text{ atm} \end{aligned}$$

Therefore, the pressure is 378 times that at the surface! A human being could not survive at this pressure.

Plans for finding and possibly salvaging the *Titanic* began immediately after it sank. Families of some of the wealthy victims contacted salvage companies with requests for a salvage operation. One plan suggested filling the *Titanic* with Ping-Pong balls so that its overall density would be less than that of water and the ship would float to the surface! This plan, of course, ignores the obvious problems of the Ping-Pong balls' failure to withstand the tremendous pressure at that depth.

An early expedition to find the *Titanic* occurred in 1980 and met with failure. After unsuccessful searches were carried out by a number of teams, Dr. Robert Ballard of Woods Hole Oceanographic Institute discovered the wreck in 1985, in cooperation with a team from IFREMER, the French National Institute of Oceanography. The search began by towing a sonar device, which emitted sound waves through the water and analyzed the reflection of the waves from solid objects such as the hull of the *Titanic*. The search pattern was a tedious back-and-forth sweeping of the area near the reported location of the sinking of the ship, looking for sonar reflections and checking possible sites with a magnetometer for the presence of an iron hull.

After failing to find the *Titanic* with the sonar system, Ballard switched to a visual search using an underwater video system called Argo. After three more grueling weeks with no reward, the searchers saw one of the *Titanic*'s boilers in the early morning of September 1, 1985. After this first evidence of the wreck was found, the remainder was located quickly.

The visual evidence indicated clearly that the *Titanic* had split in two as had been reported by some of the survivors in 1912. As it tilted steeply in the water due to the sinking of the bow, the midsection was subjected to forces that it was not designed to sustain. After the break occurred, but while the two sections were still connected, the stern section settled back into the water, with the bow section hanging from it underwater. As more water entered the bow section, it pulled the stern section into a vertical orientation and then broke free, beginning its trip to the bottom. The stern bobbed for a while as it filled with water and then sank into the ocean.

The two sections of the *Titanic* lie about 600 m apart on the ocean floor. The bow section (Fig. 2) is fairly intact, but the stern section (Fig. 3) is tremendously damaged. As the bow section sank, it was already filled with water. As the pressure of the water outside the bow section increased during the plummet to the bottom, the pressure inside the section increased. On the other hand, the stern section spent its time in the air before sinking. Therefore, as it sank, a significant volume of air was still

FIGURE 2 The bow section of the *Titanic* rests on the ocean floor relatively intact.

trapped inside the stern section. As the water pressure increased while the stern section sank, the air pressure inside could not increase along with the external water pressure because many air pockets existed in the relatively sealed sections of the structure. Therefore, some areas of the hull of the stern section experienced very large pressure on the outside surface, with relatively low pressure on the inside surface. This extreme imbalance in pressures possibly caused an implosion of the stern section at some depth, causing severe destruction of the structure. Further damage was caused by the sudden impact of hitting the bottom. With little structural integrity left, the decks pancaked downward as the stern hit the ocean floor.

(Illustrations by Ken Marschall © 1992 from Titanic: *An Illustrated History*, a Hyperion/Madison Press Book)

FIGURE 3 The stern section of the *Titanic* is heavily damaged and lies in pieces on the ocean floor.

The *Titanic* has been visited by a number of teams for purposes of research, salvage, and even filmmaking, by James Cameron, the director of the 1997 version of the film *Titanic* and the 2003 IMAX film *Ghosts of the Abyss*. What is necessary to travel to such depths? Our calculation of the pressure at the location of the *Titanic* indicates that special submarines must be used that can withstand such high pressure while maintaining normal atmospheric pressure inside for the human occupants. That was first done by Ballard in the summer of 1986 using a deep-sea submersible called *Alvin* with a remote-controlled robot named *Jason Junior*. Figure 4 shows the structure of *Alvin*.

Alvin has a titanium alloy hull that can withstand the pressure at the depth of the *Titanic*. The submersible has room for three occupants, although they are quite cramped. A number of air tanks on the craft can be flooded with water. Blocks of iron can also be jettisoned. When the tanks are filled with air and the iron blocks are attached, *Alvin* floats on water. When the air tanks are flooded with water, the submersible sinks. In visiting the *Titanic*, the first step is to fill the air tanks with water and then wait 2.5 h to sink to the bottom. After visiting the wreckage, the iron blocks are jettisoned. After doing so, the buoyant force on *Alvin* is larger than its weight and it starts upward on another long journey to the surface. The descent and ascent of *Alvin* are effective examples of applications of the physics described in Case I in Section 15.4.

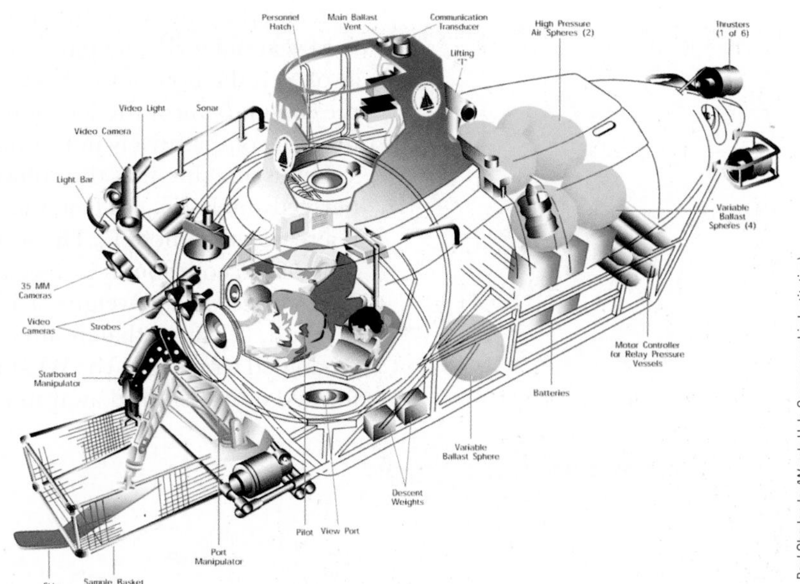

(E. Paul Oberlander/Woods Hole Oceanographic Institution)

FIGURE 4 The submersible *Alvin*, which can carry three scientists to the great depths at which the *Titanic* currently lies.

Once *Alvin* reaches the *Titanic*, the only means of viewing it are by visual inspection through the portholes or by video, using *Jason Junior.* It is impossible to don scuba gear and exit the submersible because of the tremendous pressure. There is much more to the story of the *Titanic,* but we need to return to our investigations into physics.

Problems

1. When the *Titanic* is in its normal sailing position, the torque due to the gravitational force about a horizontal axis through the midpoint of the ship is zero. Imagine now that the bow of the ship is under water and the stern is in the air above the water during the sinking process as shown in Figure 1. The waterline is at the midpoint of the ship and the keel makes a 45.0° angle with the horizontal. The net torque is zero here also because the *Titanic* as a whole is in equilibrium. If we take as our system the stern half of the ship, the torque counteracting the weight of the stern section must be applied by the framework of the ship at the midpoint. Calculate the torque about the midpoint of the *Titanic* required to hold the stern section in the air. Model the ship as a uniform rod of length 269 m and mass 4.2×10^7 kg. This torque caused the *Titanic* to split near the middle of the ship during the sinking process.
2. The *Titanic* had two almost identical sister ships, the *Britannic* and the *Olympic.* The *Britannic* sank in 1916 off the coast of Athens, Greece, possibly due to a mine planted during World War I. It now sits below 119 m of sea water. (a) What is the pressure at the location of the *Britannic*? (b) Look up a practical limit for scuba diving and determine whether a person can visit the *Britannic* by scuba diving.
3. The submersible *Alvin* requires 2.5 h to sink to the location of the *Titanic.* (a) What is the average speed during descent? (b) Assume that the speed of *Alvin* remains constant during the entire descent. Is the average density of *Alvin* greater than, less than, or equal to the density of sea water? (*Note:* Include in your analysis the resistive force on *Alvin* as it moves through the water.)
4. The *Titanic* is only one of many maritime disasters. In 1956, despite the use of radar, which was invented after the time of the *Titanic,* a collision occurred between the Italian luxury liner *Andrea Doria* and the Swedish liner *Stockholm.* Deaths were relatively few because of the long time interval between the collision and the sinking of the *Andrea Doria,* but a remarkable event occurred. A 14-year-old girl, asleep in her bed on the *Andrea Doria* before the collision, awoke on the bow of the *Stockholm.* The bow of the latter ship pierced the hull of the *Andrea Doria* at the location of the girl's berth and miraculously scooped her up with comparatively minor injuries.

   Let us imagine that the collision between the *Andrea Doria* and the *Stockholm* is perfectly inelastic. (In fact, the *Stockholm* drew away after the collision, but ours is a reasonable model.) The weight of the *Andrea Doria* is 29 100 tons. At the time of the collision, it is traveling at full speed of 23 knots at 15° south of west. The *Stockholm* has a weight of 12 165 tons and is traveling at 18 knots at 30° east of south. (a) Immediately after the collision, what is the velocity, in knots, of the entangled wreckage? (b) What fraction of the initial kinetic energy was transformed or transferred away in the collision?

CONTEXT 5

Global Warming

Numerous news stories have detailed the increase in temperature of the Earth and its subsequent results, including melting of ice from the polar ice caps and changes in climate and the corresponding effects on vegetation. Data taken over the past few decades are interpreted by some scientists as showing a measurable global temperature increase. Life on this planet depends on a delicate balance that keeps the global temperature in a narrow range necessary for our survival. How is this temperature determined? What factors need to be in balance to keep the temperature constant? If we can devise an adequate structural model to calculate the correct surface temperature of the Earth, we can use the model to predict changes in the temperature as we vary the parameters.

You most likely have an intuitive sense for the temperature of an object, and as long as the object is small (and the object is not undergoing combustion or some other rapid process) no significant temperature variation occurs between different points on the object. What about a huge object like the Earth, though? It is clear that no single temperature describes the entire planet; we know that it is summer in Australia when it is winter in Canada. The polar ice caps clearly have different temperatures from the tropical regions. Variations also occur in temperature within a single large body of water such as an ocean. Temperature varies greatly with altitude in a relatively local region, such as in and near Palm Springs, California, as shown in Figure 1. Thus, when we speak of the temperature of the Earth, we will refer to an *average* surface temperature, taking into account all the variations across the surface. It is this average temperature that we would like to calculate by building a structural model of the atmosphere and comparing its prediction with the measured surface temperature.

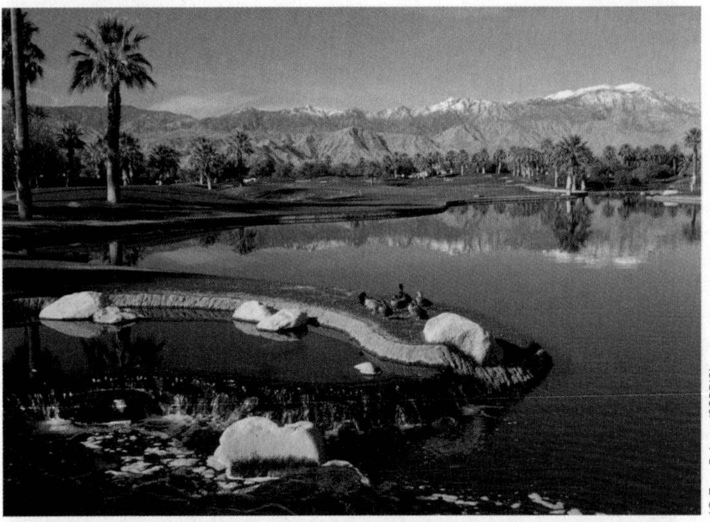

(© Tony Roberts/CORBIS)

FIGURE 1 Temperature variations with altitude can exist in a local region on the Earth. Here in Palm Springs, California, palm trees grow in the city while snow is present at the top of the local mountains.

A primary factor in determining the surface temperature of the Earth is the

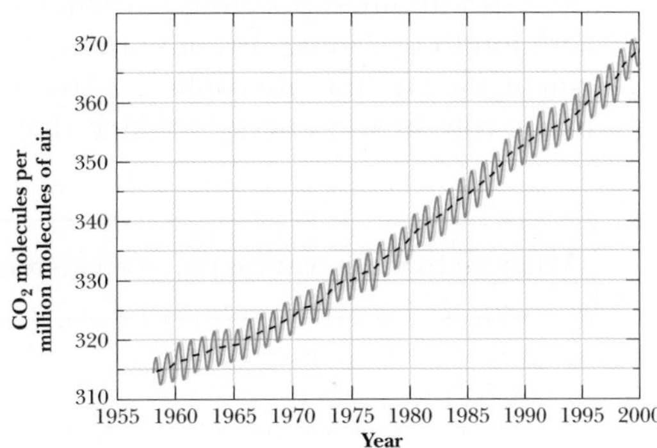

FIGURE 2 The concentration of atmospheric carbon dioxide in parts per million (ppm) of dry air as a function of time during the latter part of the 20th century. These data were recorded at the Mauna Loa Observatory in Hawaii. The yearly variations (*red curve*) coincide with growing seasons because vegetation absorbs carbon dioxide from the air. The steady increase in the average concentration (*black curve*) is of concern to scientists.

FIGURE 3 In this Context, we explore the energy balance of the Earth, which is a large, nonisolated system that interacts with its environment solely by means of electromagnetic radiation.

existence of our atmosphere. The atmosphere is a relatively thin (compared with the radius of the Earth) layer of gas above the surface that provides us with life-supporting oxygen. In addition to providing this important element for life, the atmosphere plays a major role in the energy balance that determines the average temperature. As we proceed with this Context, we shall focus on the physics of gases and apply the principles we learn to the atmosphere.

One important component of the global warming problem is the concentration of carbon dioxide in the atmosphere. Carbon dioxide plays an important role in absorbing energy and raising the temperature of the atmosphere. As seen in Figure 2, the amount of carbon dioxide in the atmosphere has been steadily increasing since the middle of the 20th century. This graph shows hard data that indicate that the atmosphere is undergoing a distinct change, although not all scientists agree on the interpretation of what that change means in terms of global temperatures.

In addition to its scientific aspects, global warming is a social issue with many facets. These aspects encompass international politics and economics, because global warming is a worldwide problem. Changing our policies requires real costs to solve the problem. Global warming also has technological aspects, and new methods of manufacturing, transportation, and energy supply must be designed to slow down or reverse the increase in temperature. We shall restrict our attention to the physical aspects of global warming as we address this central question:

What factors determine the average temperature at the Earth's surface?

CHAPTER 16

Temperature and the Kinetic Theory of Gases

Why would someone designing a pipeline include these strange loops? Pipelines carrying liquids often contain loops such as these to allow for expansion and contraction as the temperature changes. We will study thermal expansion in this chapter.

CHAPTER OUTLINE

Our study thus far has focused mainly on Newtonian mechanics, which explains a wide range of phenomena such as the motion of baseballs, rockets, and planets. We have applied these principles to oscillating systems, the propagation of mechanical waves through a medium, and the properties of fluids at rest and in motion. In Chapter 6, we introduced the notions of temperature and internal energy. We now extend the study of such notions as we focus on **thermodynamics,** which is concerned with the concepts of energy transfers between a system and its environment and the resulting variations in temperature or changes of state. As we shall see, thermodynamics explains the bulk properties of matter and the correlation between these properties and the mechanics of atoms and molecules.

Have you ever wondered how a refrigerator cools or what types of transformations occur in an automobile engine or why a bicycle pump becomes warm as you inflate a tire? The laws of thermodynamics enable us to answer such questions. In general, thermodynamics deals with the physical and chemical transformations of matter in all its states: solid, liquid, gas, and plasma.

This chapter concludes with a study of ideal gases, which we shall approach on two levels. The first examines ideal gases on the macroscopic scale. Here we shall be concerned with the relationships among such quantities as pressure, volume, and temperature of the gas. On the second level, we shall examine gases on a microscopic (molecular) scale, using a structural model that treats the gas as a collection of particles. The latter approach will help us understand how behavior on the atomic level affects such macroscopic properties as pressure and temperature.

16.1 TEMPERATURE AND THE ZEROTH LAW OF THERMODYNAMICS

We often associate the concept of temperature with how hot or cold an object feels to the touch. Our sense of touch provides us with a qualitative indication of temperature. Our senses are unreliable and often misleading, however. For example, if you stand with one bare foot on a tile floor and the other on an adjacent carpeted floor, the tile floor feels colder to your foot than the carpet even though the two are at the same temperature. That is because the properties of the tile are such that the transfer of energy (by heat) to the tile floor from your foot is more rapid than to the carpet. Your skin is sensitive to the *rate* of energy transfer—power—not the temperature of the object. Of course, the larger the difference in temperature between that of the object and that of your hand, the faster the energy transfer, so temperature and your sense of touch are related in *some* way. What we need is a reliable and reproducible method for establishing the relative "hotness" or "coldness" of objects that is related solely to the temperature of the object. Scientists have developed a variety of thermometers for making such quantitative measurements.

Sense of warm and cold

We are all familiar with experiences in which two objects at different initial temperatures eventually reach some intermediate temperature when placed in contact with each other. For example, if you combine hot water and cold water in a bathtub from separate faucets, the combined water reaches an equilibrium temperature between the temperatures of the hot water and the cold water. Likewise, if an ice cube is placed in a cup of hot coffee, the ice eventually melts and the temperature of the coffee decreases.

We shall use these familiar examples to develop the scientific notion of temperature. Imagine two objects placed in an insulated container so that they form an isolated system. If the objects are at different temperatures, energy can be exchanged between them by, for example, heat or electromagnetic radiation. Objects that can exchange energy with each other in this way are said to be in **thermal contact.** Eventually, the temperatures of the two objects will become the same, one becoming warmer and the other cooler, as in our preceding examples. **Thermal equilibrium** is the situation in which two objects in thermal contact cease to have any exchange of energy by heat or electromagnetic radiation.

Using these ideas, we can develop a formal definition of temperature. Consider two objects A and B that are not in thermal contact and a third object C that will be our **thermometer,** a device calibrated to measure the temperature of an object. We wish to determine whether A and B would be in thermal equilibrium if they were placed in thermal contact. The thermometer is first placed in thermal contact with A and its reading is recorded, as shown in Figure 16.1a. The thermometer is then placed in thermal contact with B and its reading is recorded (Fig. 16.1b). If the two readings are the same, A and B are in thermal equilibrium with each other. If they are placed in thermal contact with each other, as in Figure 16.1c, there is no net transfer of energy between them.

We can summarize these results in a statement known as the **zeroth law of thermodynamics:**

Zeroth law of thermodynamics

If objects A and B are separately in thermal equilibrium with a third object C, then A and B are in thermal equilibrium with each other.

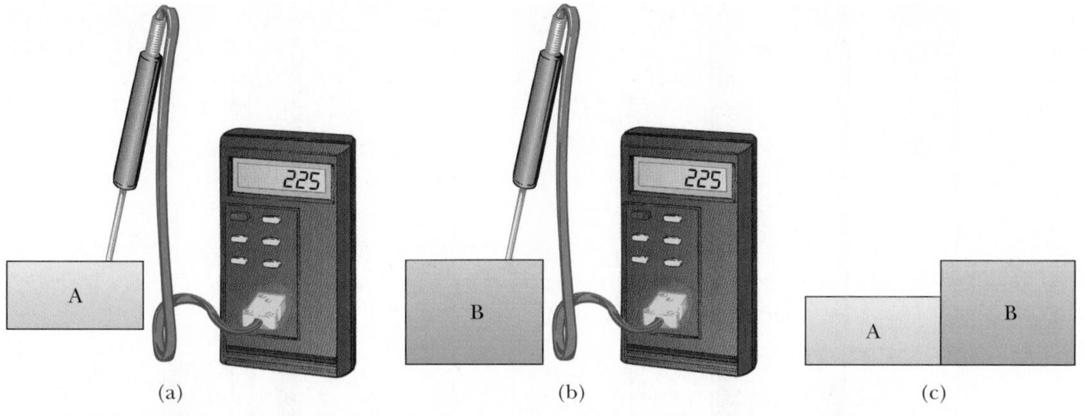

FIGURE 16.1 The zeroth law of thermodynamics. (a) and (b) If the temperatures of A and B are measured to be the same by placing them in thermal contact with a thermometer (object C), no energy will be exchanged between them when they are placed in thermal contact with each other (c).

This statement, elementary as it may seem, is very important because it can be used to define the notion of temperature and is easily proved experimentally. We can think of temperature as the property that determines whether an object is in thermal equilibrium with other objects. **Two objects in thermal equilibrium with each other are at the same temperature.**

16.2 THERMOMETERS AND TEMPERATURE SCALES

In our discussion of the zeroth law, we mentioned a thermometer. Thermometers are devices used to measure the temperature of an object or a system with which the thermometer is in thermal equilibrium. All thermometers make use of some physical property that exhibits a change with temperature that can be calibrated to make the temperature measurable. Some of the physical properties used are (1) the volume of a liquid, (2) the length of a solid, (3) the pressure of a gas held at constant volume, (4) the volume of a gas held at constant pressure, (5) the electric resistance of a conductor, and (6) the color of a hot object.

A common thermometer in everyday use consists of a liquid—usually mercury or alcohol—that expands into a glass capillary tube when its temperature rises (Fig. 16.2). In this case, the physical property that changes is the volume of a liquid. Because the cross-sectional area of the capillary tube is uniform, the change in volume of the liquid varies linearly with its length along the tube. We can then define a temperature to be related to the length of the liquid column.

The thermometer can be calibrated by placing it in thermal contact with some environments that remain at constant temperature and marking the end of the liquid column on the thermometer. One such environment is a mixture of water and ice in thermal equilibrium with each other at atmospheric pressure. Once we have marked the ends of the liquid column for our chosen environments on our thermometer, we need to define a scale of numbers associated with various temperatures. One such scale is the **Celsius temperature scale.** On the Celsius scale, the temperature of the ice–water mixture is defined as zero degrees Celsius, written 0°C; this temperature is called the **ice point** or **freezing point** of water. Another commonly used environment is a mixture of water and steam in thermal equilibrium with each other at atmospheric pressure. On the Celsius scale, this temperature is defined as 100°C, the **steam point** or **boiling point** of water. Once the ends of the liquid column in the thermometer have been marked at these two points, the distance between the marks is divided into 100 equal segments, each denoting a change in temperature of one degree Celsius.

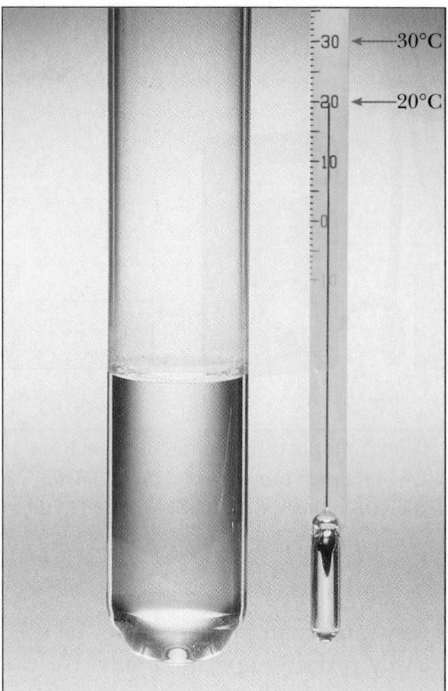

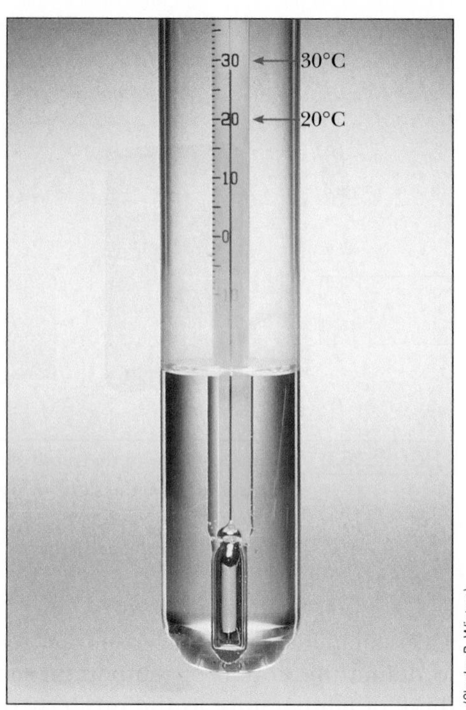

(Charles D. Winters)

FIGURE 16.2 As a result of thermal expansion, the level of the mercury in the thermometer rises as the mercury is heated by water in the test tube.

Thermometers calibrated in this way present problems when extremely accurate readings are needed. For instance, an alcohol thermometer calibrated at the ice and steam points of water might agree with a mercury thermometer only at the calibration points. Because mercury and alcohol have different thermal expansion properties (the expansion may not be perfectly linear with temperature), when one indicates a given temperature, the other may indicate a slightly different value. The discrepancies between different types of thermometers are especially large when the temperatures to be measured are far from the calibration points.

The Constant-Volume Gas Thermometer and the Kelvin Scale

Although practical devices such as the mercury thermometer can measure temperature, they do not define it in a fundamental way. Only one thermometer, the **gas thermometer,** offers a way to define temperature and relate it to internal energy directly. In a gas thermometer, the temperature readings are nearly independent of the substance used in the thermometer. One type of gas thermometer is the constant-volume example shown in Figure 16.3. The behavior observed in this device is the pressure variation with temperature of a fixed volume of gas.

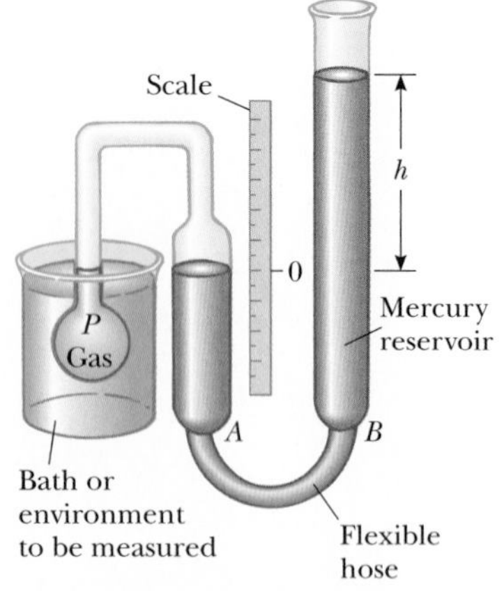

FIGURE 16.3 A constant-volume gas thermometer measures the pressure of the gas contained in the flask immersed in the bath. The volume of gas in the flask is kept constant by raising or lowering reservoir B such that the mercury level in column A remains constant.

When the constant-volume gas thermometer was developed, it was calibrated using the ice and steam points of water as follows. (A different calibration procedure, to be discussed shortly, is now used.) The gas flask is inserted into an ice bath, and mercury reservoir B is raised or lowered until the volume of the confined gas is at some value, indicated by the zero point on the scale. The height h, the difference between the levels in the reservoir and column A, indicates the pressure in the flask at 0°C, according to Equation 15.4. The flask is inserted into water at the steam point, and reservoir B is readjusted until the height in column A is again brought to zero on the scale, ensuring that the gas volume is the same as it had been in the ice bath (hence the designation "constant-volume"). A measure of the new value for h gives a value for the pressure at 100°C. These pressure and temperature values are then plotted on a graph, as in Figure 16.4. Based on experimental observations that the pressure of a gas varies linearly with its temperature, which is discussed in more detail in Section 16.4, we draw a straight line through our two points. The line connecting the two points serves as a calibration curve for measuring unknown temperatures. If we want to measure the temperature of a substance, we place the gas flask in thermal contact with the substance and adjust the column of mercury until the level in column A again returns to zero. The height of the mercury column tells us the pressure of the gas, and we can then find the temperature of the substance from the calibration curve.

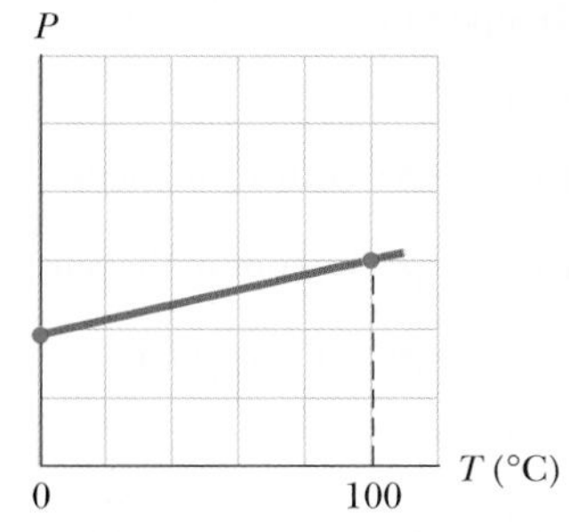

FIGURE 16.4 A typical graph of pressure versus temperature taken with a constant-volume gas thermometer. The two dots represent known reference temperatures (the ice and steam points of water).

Now suppose temperatures are measured with various gas thermometers containing different gases. Experiments show that the thermometer readings are nearly independent of the type of gas used as long as the gas pressure is low and the temperature is well above the point at which the gas liquefies.

We can also perform the temperature measurements with the gas in the flask at different starting pressures at 0°C. As long as the pressure is low, we will generate straight-line calibration curves for each different starting pressure, as shown for three experimental trials (solid lines) in Figure 16.5.

If the curves in Figure 16.5 are extended back toward negative temperatures, we find a startling result. **In every case, regardless of the type of gas or the value of the low starting pressure, the pressure extrapolates to zero when the temperature is −273.15°C.** This result suggests that this particular temperature is universal in its importance because it does not depend on the substance used in the thermometer. In addition, because the lowest possible pressure is $P = 0$, which would be a perfect vacuum, this temperature must represent a lower bound for physical processes. Therefore, we define this temperature as **absolute zero.** Some interesting effects occur at temperatures near absolute zero, such as the phenomenon of *superconductivity,* which we shall study in Chapter 21.

This significant temperature is used as the basis for the **Kelvin temperature scale,** which sets −273.15°C as its zero point (0 K). The size of a degree on the Kelvin scale is chosen to be identical to the size of a degree on the Celsius scale. Therefore, the following relationship enables conversion between these temperatures:

PITFALL PREVENTION 16.1

A MATTER OF DEGREE Note that notations for temperatures in the Kelvin scale do not use the degree sign. The unit for a Kelvin temperature is simply "kelvins" and not "degrees Kelvin."

$$T_C = T - 273.15 \qquad [16.1]$$

where T_C is the Celsius temperature and T is the Kelvin temperature (sometimes called the **absolute temperature**). The primary difference between these two temperature scales is a shift in the zero of the scale. The zero of the Celsius scale is arbitrary; it depends on a property associated with only one substance, water. The zero on the Kelvin scale is not arbitrary because it is characteristic of a behavior associated with all substances. Consequently, when an equation contains T as a variable, the absolute temperature must be used. Similarly, a ratio of temperatures is only meaningful if the temperatures are expressed on the Kelvin scale.

Equation 16.1 shows that the Celsius temperature T_C is shifted from the absolute temperature T by 273.15°. Because the size of a degree is the same on the

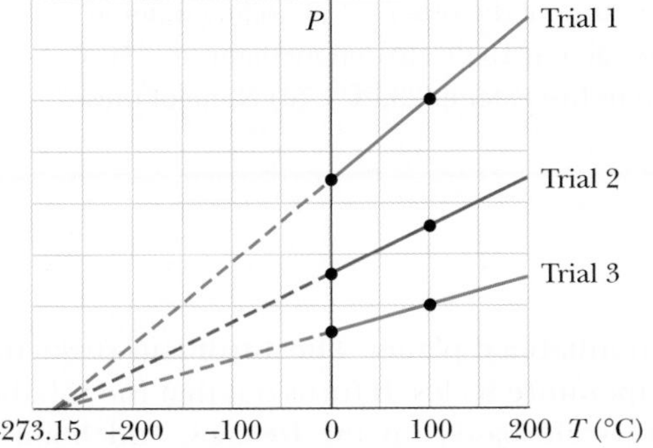

FIGURE 16.5 Pressure versus temperature for experimental trials in which gases have different pressures in a constant-volume gas thermometer. Note that the pressure extrapolates to zero at the temperature of −273.15°C for all three trials.

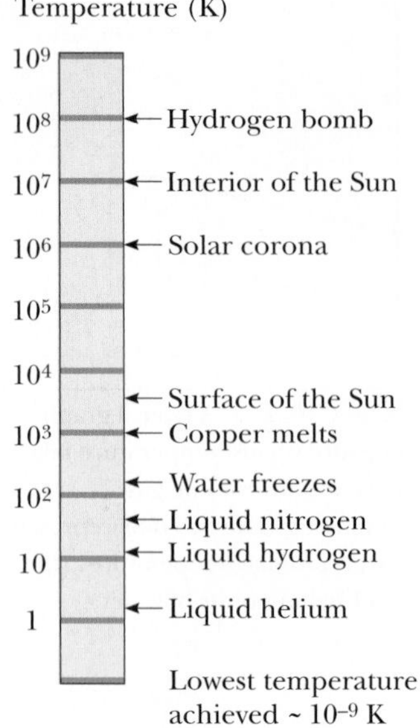

FIGURE 16.6 Absolute temperatures at which various selected physical processes occur. Note that the scale is logarithmic.

two scales, a temperature difference of 5°C is equal to a temperature difference of 5 K. The two scales differ only in the choice of the zero point. Therefore, the ice point (273.15 K) corresponds to 0.00°C, and the steam point (373.15 K) is equivalent to 100.00°C.

Early gas thermometers made use of ice and steam points according to the procedure just described. These points are experimentally difficult to duplicate, however. For this reason, a new procedure based on two new points was adopted in 1954 by the International Committee on Weights and Measures. The first point is absolute zero. The second point is the **triple point of water,** which corresponds to the **single temperature and pressure at which water, water vapor, and ice can coexist in equilibrium.** This point is a convenient and reproducible reference temperature for the Kelvin scale. It occurs at a temperature of 0.01°C and a very low pressure of 4.58 mm of mercury. The temperature at the triple point of water on the Kelvin scale has a value of 273.16 K. Therefore, the SI unit of temperature, the **kelvin,** is defined as **1/273.16 of the temperature of the triple point of water.**

Figure 16.6 shows the Kelvin temperatures for various physical processes and conditions. As the figure reveals, absolute zero has never been achieved, although laboratory experiments have created conditions that are very close to absolute zero.

What would happen to a gas if its temperature could reach 0 K? As Figure 16.5 indicates (if we ignore the liquefaction and solidification of the substance), the pressure it would exert on the container's walls would be zero. In Section 16.5, we shall show that the pressure of a gas is proportional to the kinetic energy of the molecules of that gas. Therefore, according to classical physics, the kinetic energy of the gas would go to zero and there would be no motion at all of the individual components of the gas; hence, the molecules would settle out on the bottom of the container. Quantum theory, to be discussed in Chapter 28, modifies this statement to indicate that there would be some residual energy, called the *zero-point energy,* at this low temperature.

The Fahrenheit Scale

The most common temperature scale in everyday use in the United States is the **Fahrenheit scale.** This scale sets the temperature of the ice point at 32°F and the temperature of the steam point at 212°F. The relationship between the Celsius and Fahrenheit temperature scales is

$$T_F = \frac{9}{5} T_C + 32°\text{F} \qquad [16.2]$$

Equation 16.2 can easily be used to find a relationship between changes in temperature on the Celsius and Fahrenheit scales. It is left as a problem for you to show that if the Celsius temperature changes by ΔT_C, the Fahrenheit temperature changes by an amount ΔT_F given by

$$\Delta T_F = \frac{9}{5}\, \Delta T_C \qquad [16.3]$$

QUICK QUIZ 16.1 Consider the following pairs of materials. Which pair represents two materials in which one has twice the temperature of the other? **(a)** boiling water at 100°C, a glass of water at 50°C **(b)** boiling water at 100°C, frozen methane at −50°C **(c)** an ice cube at −20°C, flames from a circus fire-eater at 233°C **(d)** None of these pairs

■ Thinking Physics 16.1

A group of future astronauts lands on an inhabited planet. The astronauts strike up a conversation with the aliens about temperature scales. It turns out that the inhabitants of this planet have a temperature scale based on the freezing and boiling points of water, which are separated by 100 of the inhabitants' degrees. Would these two temperatures on this planet be the same as those on the Earth? Would the size of the aliens' degrees be the same as ours? Suppose the aliens have also devised a scale similar to the Kelvin scale. Would their absolute zero be the same as ours?

Reasoning The values of 0°C and 100°C for the freezing and boiling points of water are defined at atmospheric pressure. On another planet, it is unlikely that atmospheric pressure would be exactly the same as that on the Earth. Therefore, water would freeze and boil at different temperatures on the alien planet. The aliens may call these temperatures 0° and 100°, but they would not be the same temperatures as our 0°C and 100°C. If the aliens did assign values of 0° and 100° for these temperatures, their degrees would not be the same size as our Celsius degrees (unless their atmospheric pressure were the same as ours). For an alien version of the Kelvin scale, the absolute zero would be the same as ours because it is based on a natural, universal definition rather than being associated with a particular substance or a given atmospheric pressure. ■

EXAMPLE 16.1 **Converting Temperatures**

On a day when the temperature reaches 50°F, what is the temperature in degrees Celsius and in kelvins?

Solution Solving Equation 16.2 for the Celsius temperature and substituting $T_F = 50°\text{F}$, we have

$$T_C = \frac{5}{9}(T_F - 32°\text{F}) = \frac{5}{9}(50°\text{F} - 32°\text{F}) = \boxed{10°\text{C}}$$

From Equation 16.1, we find that

$$T = T_C + 273.15 = 10°\text{C} + 273.15 = \boxed{283\text{ K}}$$

16.3 THERMAL EXPANSION OF SOLIDS AND LIQUIDS

Our discussion of the liquid thermometer makes use of one of the best known changes that occur in most substances, that as the temperature of a substance increases, its volume increases. This phenomenon, known as **thermal expansion,** plays an important role in numerous applications. For example, thermal expansion joints (Fig. 16.7) must be included in buildings, concrete highways, railroad tracks, and bridges to compensate for changes in dimensions with temperature variations.

The overall thermal expansion of an object is a consequence of the change in the average separation between its constituent atoms or molecules. To understand this concept, consider how the atoms in a solid substance behave. These atoms are located at fixed equilibrium positions; if an atom is pulled away from its position, a restoring force pulls it back. We can build a structural model in which we imagine

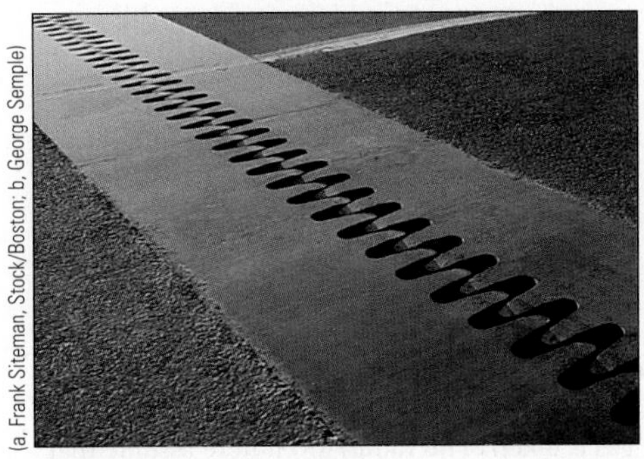

(a)

(b)

(a, Frank Siteman, Stock/Boston; b, George Semple)

FIGURE 16.7 (a) Thermal expansion joints are used to separate sections of roadways on bridges. Without these joints, the surfaces would buckle due to thermal expansion on very hot days or crack due to contraction on very cold days. (b) The long, vertical joint in a wall is filled with a soft material that allows the wall to expand and contract as the temperature of the bricks changes.

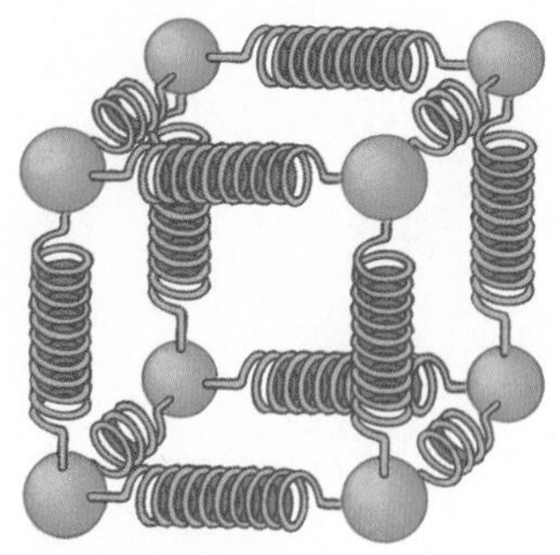

FIGURE 16.8 A structural model of the atomic configuration in a solid. The atoms (spheres) are imagined to be connected to one another by springs that reflect the elastic nature of the interatomic forces.

(Wide World Photos)

Thermal expansion. The extremely high temperature of a July day in Asbury Park, New Jersey, caused these railroad tracks to buckle. ■

that the atoms are particles at their equilibrium positions connected by springs to their neighboring atoms (Fig. 16.8). If an atom is pulled away from its equilibrium position, the distortion of the springs provides a restoring force. If the atom is released, it oscillates, and we can apply the simple harmonic motion model to it. A number of macroscopic properties of the substance can be understood with this type of structural model on the atomic level.

In Chapter 6, we introduced the notion of internal energy and pointed out that it is related to the temperature of a system. For a solid, the internal energy is associated with the kinetic and potential energy of the vibrations of the atoms around their equilibrium positions. At ordinary temperatures, the atoms vibrate with an amplitude of about 10^{-11} m, and the average spacing between the atoms is about 10^{-10} m. As the temperature of the solid increases, the average separation between atoms increases. The increase in average separation with increasing temperature (and subsequent thermal expansion) is the result of a breakdown in the model of simple harmonic motion. Active Figure 7.15 in Chapter 7 shows the potential energy curve for an ideal simple harmonic oscillator. The potential energy curve for atoms in a solid is similar but not exactly the same as that one; it is slightly asymmetric around the equilibrium position. It is this asymmetry that leads to thermal expansion.

If the thermal expansion of an object is sufficiently small compared with the object's initial dimensions, the change in any dimension is, to a good approximation, dependent on the first power of the temperature change. For most situations, we can adopt a simplification model in which this dependence is true. Suppose an object has an initial length L_i along some direction at some temperature. The length increases by ΔL for a change in temperature ΔT. Experiments show that when ΔT is small enough, ΔL is proportional to ΔT and to L_i:

$$\Delta L = \alpha L_i \Delta T \qquad [16.4]$$

or

$$L_f - L_i = \alpha L_i (T_f - T_i) \qquad [16.5]$$

where L_f is the final length, T_f is the final temperature, and the proportionality constant α is called the **average coefficient of linear expansion** for a given material and has units of inverse degrees Celsius, or $(°C)^{-1}$.

Table 16.1 lists the average coefficient of linear expansion for various materials. Note that for these materials α is positive, indicating an increase in length with

TABLE 16.1 Average Expansion Coefficients for Some Materials Near Room Temperature

Material	Average Coefficient of Linear Expansion (α) $(°C^{-1})$	Material	Average Coefficient of Volume Expansion (β) $(°C^{-1})$
Aluminum	24×10^{-6}	Acetone	1.5×10^{-4}
Brass and bronze	19×10^{-6}	Alcohol, ethyl	1.12×10^{-4}
Copper	17×10^{-6}	Benzene	1.24×10^{-4}
Glass (ordinary)	9×10^{-6}	Gasoline	9.6×10^{-4}
Glass (Pyrex)	3.2×10^{-6}	Glycerin	4.85×10^{-4}
Lead	29×10^{-6}	Mercury	1.82×10^{-4}
Steel	11×10^{-6}	Turpentine	9.0×10^{-4}
Invar (Ni–Fe alloy)	0.9×10^{-6}	Air[a] at 0°C	3.67×10^{-3}
Concrete	12×10^{-6}	Helium[a]	3.665×10^{-3}

[a]Gases do not have a specific value for the volume expansion coefficient because the amount of expansion depends on the type of process through which the gas is taken. The values given here assume that the gas undergoes an expansion at constant pressure.

increasing temperature, but that is not always the case. For example, some substances, such as calcite ($CaCO_3$), expand along one dimension (positive α) and contract along another (negative α) with increasing temperature.

It may be helpful to think of thermal expansion as a magnification or a photographic enlargement. For example, as a metal washer is heated (Active Figure 16.9), all dimensions, including the radius of the hole, increase according to Equation 16.4. Because the linear dimensions of an object change with temperature, it follows that volume and surface area also change with temperature. Consider a cube having an initial edge length L_i and therefore an initial volume $V_i = L_i^3$. As the temperature is increased, the length of each side increases to

$$L_f = L_i + \alpha L_i \Delta T$$

The new volume, $V_f = L_f^3$, is

$$L_f^3 = (L_i + \alpha L_i \Delta T)^3 = L_i^3 + 3\alpha L_i^3 \Delta T + 3\alpha^2 L_i^3 (\Delta T)^2 + \alpha^3 L_i^3 (\Delta T)^3$$

The last two terms in this expression contain the quantity $\alpha \Delta T$ raised to the second and third powers. Because $\alpha \Delta T$ is a pure number much less than 1, raising it to a power makes it even smaller. Therefore, we can ignore these terms to obtain a simpler expression:

$$V_f = L_f^3 = L_i^3 + 3\alpha L_i^3 \Delta T = V_i + 3\alpha V_i \Delta T$$

or

$$\Delta V = V_f - V_i = \beta V_i \Delta T \qquad [16.6]$$

where $\beta = 3\alpha$. The quantity β is called the **average coefficient of volume expansion.** We considered a cubic shape in deriving this equation, but Equation 16.6 describes a sample of any shape as long as the average coefficient of linear expansion is the same in all directions.

By a similar procedure, we can show that the *increase in area* of an object accompanying an increase in temperature is

$$\Delta A = \gamma A_i \Delta T \qquad [16.7]$$

where γ, the **average coefficient of area expansion,** is given by $\gamma = 2\alpha$.

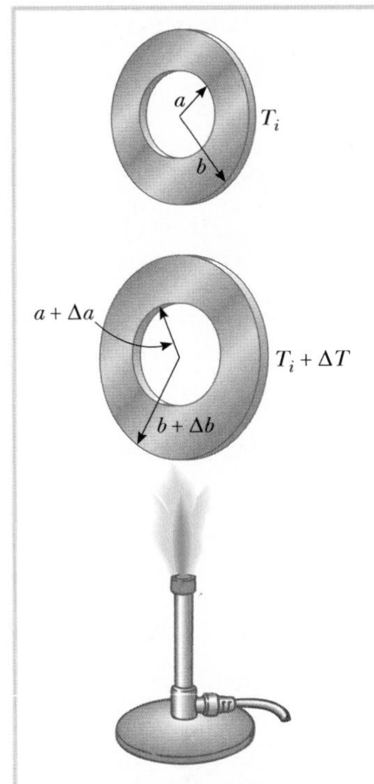

ACTIVE FIGURE **16.9**

Thermal expansion of a homogeneous metal washer. Note that as the washer is heated, all dimensions increase. (The expansion is exaggerated in this figure.)

Physics Now™ Log into PhysicsNow at **www.pop4e.com** and go to Active Figure 16.9 to compare expansions for various temperatures of the burner and materials from which the washer is made.

QUICK QUIZ **16.2** Two spheres are made of the same metal and have the same radius, but one is hollow and the other is solid. The spheres are taken through the same temperature increase. Which sphere expands more? **(a)** The solid sphere does. **(b)** The hollow sphere does. **(c)** They expand by the same amount. **(d)** There is not enough information to say.

∎ Thinking Physics 16.2

As a homeowner is painting a ceiling, a drop of paint falls from the brush onto an operating incandescent lightbulb. The bulb breaks. Why?

Reasoning The glass envelope of an incandescent lightbulb receives energy on the inside surface by electromagnetic radiation from the very hot filament. In addition, because the bulb contains gas, the glass envelope receives energy by matter transfer related to the movement of the hot gas near the filament to the colder glass. Therefore, the glass can become very hot. If a drop of relatively cold paint falls onto the glass, that portion of the glass envelope suddenly becomes colder than the other portions, and the contraction of this region can cause thermal stresses that might break the glass. ∎

As Table 16.1 indicates, each substance has its own characteristic coefficients of expansion. For example, when the temperatures of a brass rod and a steel rod of

PITFALL PREVENTION 16.2

DO HOLES BECOME LARGER OR SMALLER? When an object's temperature is raised, every linear dimension increases in size. Included are any holes in the material, which expand in the same way as if the hole were filled with the material, as shown in Active Figure 16.9. Keep in mind the notion of thermal expansion as being similar to a photographic enlargement.

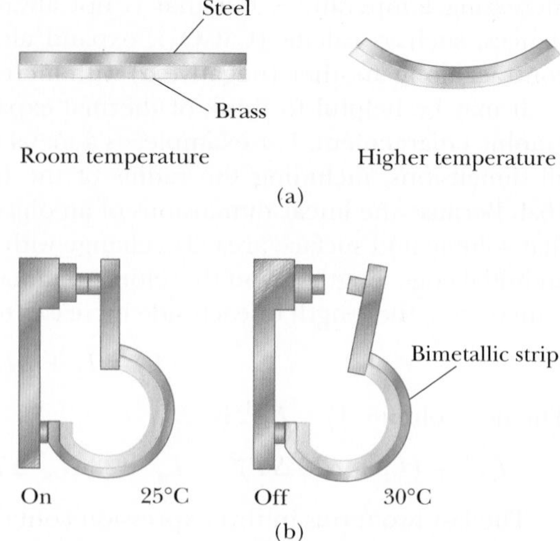

FIGURE 16.10 (a) A bimetallic strip bends as the temperature changes because the two metals have different expansion coefficients. (b) A bimetallic strip used in a thermostat to make or break electrical contact.

equal length are raised by the same amount from some common initial value, the brass rod expands more than the steel rod because brass has a larger coefficient of expansion than steel. A simple device called a bimetallic strip that demonstrates this principle is found in practical devices such as thermostats in home furnace systems. The strip is made by securely bonding two different metals together along their surfaces. As the temperature of the strip increases, the two metals expand by different amounts and the strip bends as in Figure 16.10.

EXAMPLE 16.2 The Thermal Electrical Short

An electronic device has been poorly designed so that two bolts attached to different parts of the device almost touch in the interior of the device, as in Figure 16.11. The steel and brass bolts are at different electric potentials and if they touch, a short circuit will develop, damaging the device. (We will study electric potential in Chapter 20.) If the initial gap between the ends of the bolts is 5.0 μm at 27°C, at what temperature will the bolts touch? Assume that the frame supporting the bolts is not affected by the temperature change.

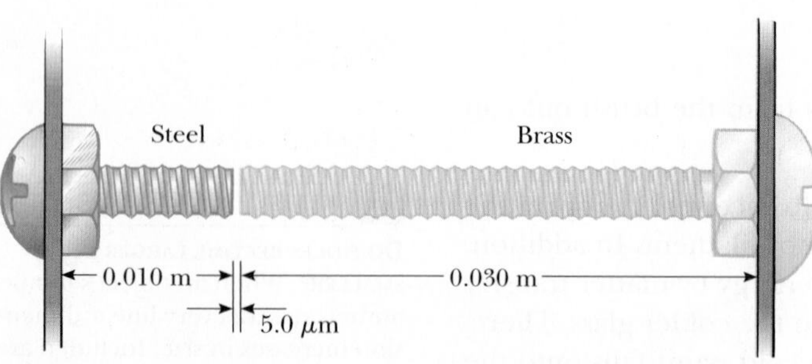

FIGURE 16.11 (Example 16.2) Two bolts attached to different parts of an electrical device are almost touching when the temperature is 27°C. As the temperature increases, the ends of the bolts move toward each other.

Solution We can conceptualize the situation by imagining that the ends of both bolts expand into the gap between them as the temperature rises. We categorize this problem as a thermal expansion one in which the *sum* of the changes in length of the two bolts must equal the length of the initial gap between the ends. To analyze the problem, we write this condition mathematically:

$$\Delta L_{\text{Br}} + \Delta L_{\text{St}} = \alpha_{\text{Br}} L_{i,\text{Br}} \Delta T + \alpha_{\text{St}} L_{i,\text{St}} \Delta T = 5.0 \times 10^{-6} \text{ m}$$

Solving for ΔT, we find,

$$\Delta T = \frac{5.0 \times 10^{-6} \text{ m}}{\alpha_{\text{Br}} L_{i,\text{Br}} + \alpha_{\text{St}} L_{i,\text{St}}}$$

$$= \frac{5.0 \times 10^{-6} \text{ m}}{(19 \times 10^{-6} \text{ °C}^{-1})(0.030 \text{ m}) + (11 \times 10^{-6} \text{ °C}^{-1})(0.010 \text{ m})}$$

$$= 7.4\text{°C}$$

Therefore, the temperature at which the bolts touch is 27°C + 7.4°C = 34°C. To finalize this problem, note that this temperature is possible if the air conditioning in the building housing the device fails for a long period on a very hot summer day.

The Unusual Behavior of Water

Liquids generally increase in volume with increasing temperature and have volume expansion coefficients on the order ten times greater than those of solids. Water is an exception to this rule over a small temperature range, as we can see from its density versus temperature curve in Figure 16.12. As the temperature increases from 0°C to 4°C, water contracts and thus its density increases. Above 4°C, water exhibits the expected expansion with increasing temperature. Therefore, the density of water reaches a maximum value of 1 000 kg/m^3 at 4°C.

We can use this unusual thermal expansion behavior of water to explain why a pond freezes at the surface. When the atmospheric temperature drops from 7°C to 6°C, for example, the water at the surface of the pond also cools and consequently decreases in volume. Hence, the surface water is denser than the water below it, which has not cooled and has not decreased in volume. As a result, the surface water sinks and warmer water from below moves to the surface to be cooled in a process called upwelling. When the atmospheric temperature is between 4°C and 0°C, however, the surface water expands as it cools, becoming less dense than the water below it. The sinking process stops, and eventually the surface water freezes. As the water freezes, the ice remains on the surface because ice is less dense than water. The ice continues to build up on the surface, while water near the bottom of the pool remains at 4°C. If that did not happen, fish and other forms of marine life would not survive through the winter.

Survival of fish in winter

A vivid example of the dangers of the absence of the upwelling and mixing processes is the sudden and deadly release of carbon dioxide gas from Lake Monoun in August 1984 and Lake Nyos in August 1986 (Fig. 16.13). Both lakes are located in the rain forest country of Cameroon in Africa. More than 1 700 natives of Cameroon died in these events.

In a lake located in a temperate zone such as the United States, significant temperature variations occur during the day and during the entire year. For example, imagine the Sun going down in the evening. As the temperature of the surface water drops because of the absence of sunlight, the sinking process tends to mix the upper and lower layers of water.

This mixing process does not normally occur in Lake Monoun and Lake Nyos because of two characteristics that contributed significantly to the disasters. First, the lakes are very deep, so mixing the various layers of water over such a large vertical distance is difficult. This factor also results in such very large pressure at the bottom of the lake that a large amount of carbon dioxide from local rocks and deep springs dissolves into the water. Second, both lakes are located in an equatorial rain forest region where the temperature variation is much smaller than in temperate zones, which results in little driving force to mix the layers of water in the lakes.

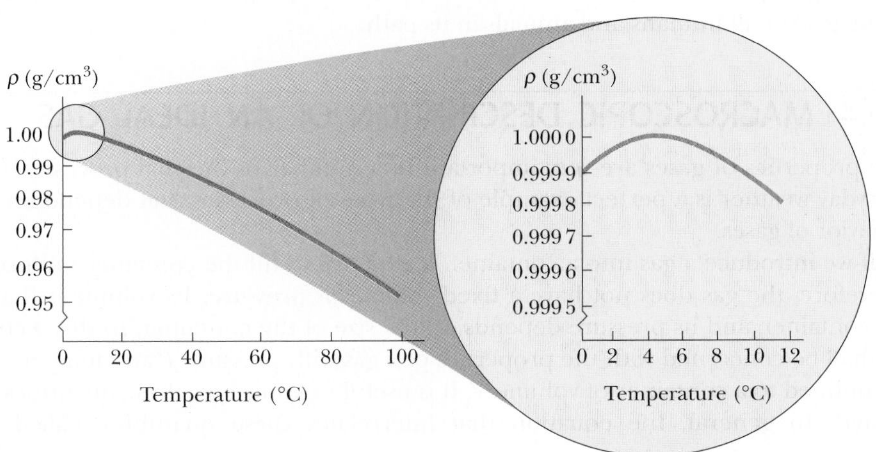

FIGURE 16.12 The variation of density with temperature for water at atmospheric pressure. The inset at the right shows that the maximum density of water occurs at 4°C.

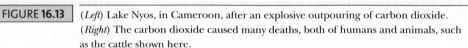

(© Peter Turnley/Corbis)

FIGURE 16.13 (*Left*) Lake Nyos, in Cameroon, after an explosive outpouring of carbon dioxide. (*Right*) The carbon dioxide caused many deaths, both of humans and animals, such as the cattle shown here.

Water near the bottom of the lake stays there for a long time and collects a large amount of dissolved carbon dioxide. In the absence of a mixing process, this carbon dioxide cannot be brought to the surface and released safely. It simply continues to increase in concentration.

Suffocation by explosive release of carbon dioxide

The situation described is explosive. If the carbon dioxide–laden water is brought to the surface where the pressure is much lower, the gas expands and comes out of the solution rapidly. Once the carbon dioxide comes out of the solution, bubbles of carbon dioxide rise through the water and cause more mixing of layers.

Suppose the temperature of the surface water were to decrease; this water would become denser and sink, possibly triggering the release of carbon dioxide and the beginning of the explosive situation just described. The monsoon season in Cameroon occurs in August. Monsoon clouds block the sunlight, resulting in lower surface water temperatures, which may be the reason the disasters occurred in August. Climate data for Cameroon show lower than normal temperatures and higher than normal rainfall in the mid-1980s. The resulting decrease in surface temperature could explain why these events occurred in 1984 and 1986. The exact reasons for the sudden release of carbon dioxide are unknown and remain an area of active research.

Finally, once the carbon dioxide was released from the lakes, it stayed near the ground because carbon dioxide is denser than air. Therefore, a layer of carbon dioxide gas spread out over the land around the lake, representing a deadly suffocating gas for all humans and animals in its path.

16.4 MACROSCOPIC DESCRIPTION OF AN IDEAL GAS

The properties of gases are very important in a number of thermal processes. Our everyday weather is a perfect example of the types of processes that depend on the behavior of gases.

If we introduce a gas into a container, it expands to fill the container uniformly. Therefore, the gas does not have a fixed volume or pressure. Its volume is that of the container, and its pressure depends on the size of the container. In this section, we shall be concerned with the properties of a gas with pressure P and temperature T, confined to a container of volume V. It is useful to know how these quantities are related. In general, the equation that interrelates these quantities, called the

equation of state, can be complicated. If the gas is maintained at a very low pressure (or low density), however, the equation of state is found experimentally to be relatively simple. Such a low-density gas is commonly referred to as an **ideal gas.** Most gases at room temperature and atmospheric pressure behave approximately as ideal gases. We shall adopt a simplification model, called the **ideal gas model,** for these types of studies. An ideal gas is **a collection of atoms or molecules that move randomly, exert no long-range forces on one another, and are so small that they occupy a negligible fraction of the volume of their container.**

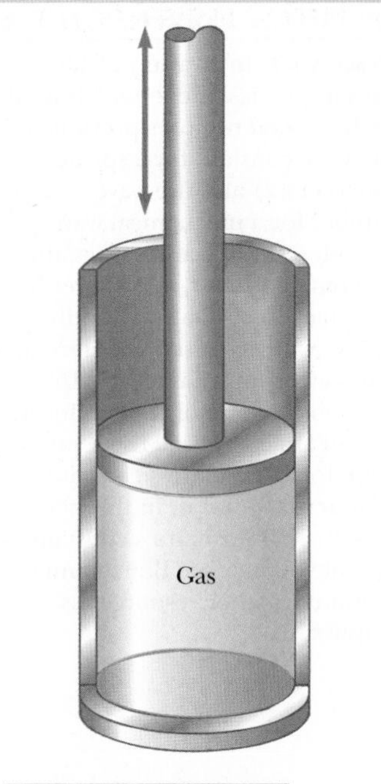

ACTIVE FIGURE 16.14 An ideal gas confined to a cylinder whose volume can be varied with a movable piston.

Physics Now™ Log into PhysicsNow at **www.pop4e.com** and go to Active Figure 16.14. You can choose to keep either the temperature or the pressure constant and verify Boyle's law and Charles's law.

It is convenient to express the amount of gas in a given volume in terms of the number of moles. One **mole** of any substance is that mass of the substance that contains **Avogadro's number,** $N_A = 6.022 \times 10^{23}$, of molecules. The number of moles n of a substance in a sample is related to its mass m through the expression

$$n = \frac{m}{M} \tag{16.8}$$

where M is the **molar mass** of the substance, usually expressed in grams per mole. For example, the molar mass of molecular oxygen O_2 is 32.0 g/mol. The mass of one mole of oxygen is therefore 32.0 g. We can calculate the mass m_0 of one molecule by dividing the molar mass by the number of molecules, which is Avogadro's number. Therefore, for oxygen,

$$m_0 = \frac{M}{N_A} = \frac{32.0 \times 10^{-3} \text{ kg/mol}}{6.02 \times 10^{23} \text{ molecule/mol}} = 5.32 \times 10^{-26} \text{ kg/molecule}$$

Now suppose an ideal gas is confined to a cylindrical container whose volume can be varied by means of a movable piston, as in Active Figure 16.14. We shall assume that the cylinder does not leak, so the number of moles of gas remains constant. For such a system, experiments provide the following information:

- When the gas is kept at a constant temperature, its pressure is inversely proportional to the volume. (This principle is known historically as Boyle's law.)
- When the pressure of the gas is kept constant, the volume is directly proportional to the temperature. (This principle is known historically as Charles's law.)
- When the volume of the gas is kept constant, the pressure is directly proportional to the temperature. (This principle is known historically as Gay-Lussac's law.)

These observations can be summarized by the following equation of state, known as the **ideal gas law:**

▮ Ideal gas law

$$PV = nRT \tag{16.9}$$

In this expression, R is a constant for a specific gas that can be determined from experiments and T is the absolute temperature in kelvins. Experiments on several gases show that as the pressure approaches zero, the quantity PV/nT approaches the same value of R for all gases. For this reason, R is called the **universal gas constant.** In SI units, where pressure is expressed in pascals and volume in cubic meters, R has the value

▮ The universal gas constant

$$R = 8.314 \text{ J/mol}\cdot\text{K} \tag{16.10}$$

If the pressure is expressed in atmospheres and the volume in liters ($1 \text{ L} = 10^3 \text{ cm}^3 = 10^{-3} \text{ m}^3$), R has the value

$$R = 0.082\,1 \text{ L}\cdot\text{atm/mol}\cdot\text{K}$$

Using this value of R and Equation 16.9, one finds that the volume occupied by 1 mol of any gas at atmospheric pressure and 0°C (273 K) is 22.4 L.

PITFALL PREVENTION 16.3

SO MANY *k*'S In a variety of situations in physics, the letter k is used. We have seen two uses previously, the force constant for a spring (Chapter 12) and the wave number for a mechanical wave (Chapter 13). We also saw k_e, the Coulomb constant, in Chapter 5. Boltzmann's constant is another k, and we will see k used for thermal conductivity in Chapter 17. To make some sense of this confusing state of affairs, we will use a subscript for Boltzmann's constant to help us recognize it. In this book, we will see Boltzmann's constant as k_B, but you may see Boltzmann's constant in other resources as simply k.

The ideal gas law is often expressed in terms of the total number of molecules N. Because the total number of molecules equals the product of the number of moles and Avogadro's number N_A, we can write Equation 16.9 as

$$PV = nRT = \frac{N}{N_A} RT$$

$$PV = Nk_BT \qquad [16.11]$$

where k_B is called **Boltzmann's constant** and has the value

$$k_B = \frac{R}{N_A} = 1.38 \times 10^{-23}\ \text{J/K} \qquad [16.12]$$

QUICK QUIZ 16.3 A common material for cushioning objects in packages is made by trapping bubbles of air between sheets of plastic. This material is more effective at keeping the contents of the package from moving around inside the package on **(a)** a hot day, **(b)** a cold day, or **(c)** either hot or cold days.

QUICK QUIZ 16.4 On a winter day, you turn on your furnace and the temperature of the air inside your home increases. Assuming that your home has the normal amount of leakage between inside air and outside air, the number of moles of air in your room at the higher temperature is **(a)** larger than before, **(b)** smaller than before, or **(c)** the same as before.

EXAMPLE 16.3 Squeezing a Tank of Gas

Pure helium gas is admitted into a tank containing a movable piston. The initial volume, pressure, and temperature of the gas are $15.0 \times 10^{-3}\ \text{m}^3$, 200 kPa, and 300 K, respectively. If the volume is decreased to $12.0 \times 10^{-3}\ \text{m}^3$ and the pressure increased to 350 kPa, find the final temperature of the gas.

Solution Let us model the helium as an ideal gas. If no gas escapes from the tank, the number of moles of gas remains constant; therefore, using $PV = nRT$ at the initial and final conditions gives

$$nR = \frac{P_iV_i}{T_i} = \frac{P_fV_f}{T_f}$$

where i and f refer to the initial and final values, respectively, for the variables. Solving for T_f, we find

$$T_f = \frac{P_fV_f}{P_iV_i}\,T_i = \frac{(350\ \text{kPa})(12.0 \times 10^{-3}\ \text{m}^3)}{(200\ \text{kPa})(15.0 \times 10^{-3}\ \text{m}^3)}(300\ \text{K})$$

$$= 420\ \text{K}$$

INTERACTIVE EXAMPLE 16.4 Heating a Spray Can

A spray can containing a propellant gas at twice atmospheric pressure (202 kPa) and having a volume of $125\ \text{cm}^3$ is at 22°C. It is then tossed into an open fire.

A When the temperature of the gas in the can reaches 195°C, what is the pressure inside the can? Assume that any change in the volume of the can is negligible.

Solution We employ the same approach we used in Example 16.3, starting with the expression

$$(1) \quad \frac{P_iV_i}{T_i} = \frac{P_fV_f}{T_f}$$

Because the initial and final volumes of the gas are assumed to be equal, this expression reduces to

$$\frac{P_i}{T_i} = \frac{P_f}{T_f}$$

Solving for P_f gives

$$(2) \quad P_f = \frac{T_f}{T_i}\,P_i = \left(\frac{468\ \text{K}}{295\ \text{K}}\right)(202\ \text{kPa}) = 320\ \text{kPa}$$

where we have substituted the temperatures on the Kelvin scale to evaluate their ratio. Obviously, the higher the temperature, the higher the pressure exerted by the trapped gas. Of course, if the pressure increases high enough, the can will explode. Because of this possibility, you should never dispose of spray cans in a fire.

B Suppose we include a volume change due to thermal expansion of the steel can as the temperature increases. Does that alter our answer for the final pressure significantly?

Solution Because the thermal expansion coefficient of steel is very small, we do not expect much of an effect on our final answer. Let find the change in the volume of the can, using Equation 16.6 and the value for α for steel from Table 16.1:

$$\Delta V = \beta V_i \, \Delta T = 3\alpha V_i \Delta T$$
$$= 3(11 \times 10^{-6}\ ^\circ\text{C}^{-1})(125\ \text{cm}^3)(173^\circ\text{C})$$
$$= 0.71\ \text{cm}^3$$

So the final volume of the can is 125.71 cm^3. Starting from (1) again, the equation for the final pressure becomes

$$P_f = \left(\frac{T_f}{T_i}\right)\left(\frac{V_i}{V_f}\right) P_i$$

This equation differs from (2) only in the factor V_i/V_f. Let us evaluate this factor:

$$\frac{V_i}{V_f} = \frac{125\ \text{cm}^3}{125.71\ \text{cm}^3} = 0.994 = 99.4\%$$

Therefore, the final pressure will differ by only 0.6% from the value we calculated without considering the thermal expansion of the can. Taking 99.4% of the previous final pressure, the final pressure including thermal expansion is 318 kPa.

PhysicsNow™ Put the can in the fire by logging into PhysicsNow at **www.pop4e.com** and going to Interactive Example 16.4.

16.5 THE KINETIC THEORY OF GASES

In the preceding section, we discussed the macroscopic properties of an ideal gas using such quantities as pressure, volume, number of moles, and temperature. From a macroscopic point of view, the mathematical representation of the ideal gas model is the ideal gas law. In this section, we consider the microscopic point of view of the ideal gas model. We shall show that the macroscopic properties can be understood on the basis of what is happening on the atomic scale.

Using the ideal gas model, we shall build a structural model of a gas enclosed in a container. The mathematical structure and the predictions made by this model constitute what is known as the **kinetic theory of gases.** With this theory, we shall interpret the pressure and temperature of an ideal gas in terms of microscopic variables. In our structural model, we make the following assumptions:

1. **The number of molecules in the gas is large, and the average separation between them is large compared with their dimensions.** Thus, the molecules occupy a negligible volume in the container. This assumption is consistent with the ideal gas model, in which we imagine the molecules to be point-like.
2. **The molecules obey Newton's laws of motion, but as a whole their motion is isotropic.** By "isotropic" we mean that any molecule can move in any direction with any speed.
3. **The molecules interact only by short-range forces during elastic collisions.** This assumption is consistent with the ideal gas model, in which the molecules exert no long-range forces on one another.
4. **The molecules make elastic collisions with the walls.**
5. **The gas under consideration is a pure substance; that is, all molecules are identical.**

Although we often picture an ideal gas as consisting of single atoms, molecular gases exhibit equally good approximations to ideal gas behavior at low pressures. Effects associated with molecular structure have no influence on the motions considered here. Therefore, we can apply the results of the following development to molecular gases as well as to monatomic gases.

(Courtesy of AIP Niels Bohr Library, Lande Collection)

LUDWIG BOLTZMANN (1844–1906)
Austrian theoretical physicist Boltzmann made many important contributions to the development of the kinetic theory of gases, electromagnetism, and thermodynamics. His pioneering work in the field of kinetic theory led to the branch of physics known as statistical mechanics.

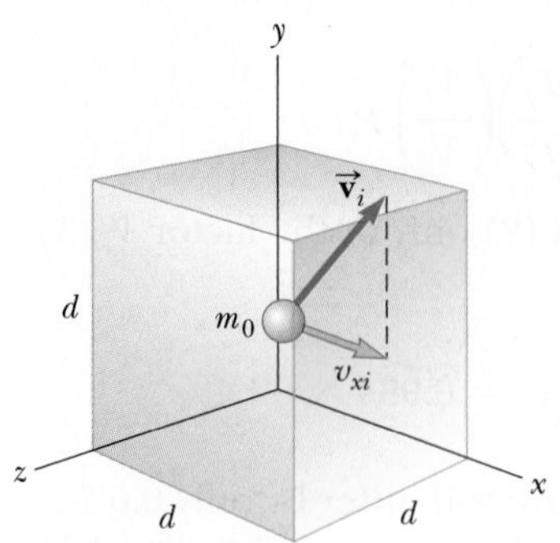

FIGURE 16.15 A cubical box with sides of length d containing an ideal gas. The molecule shown moves with velocity $\vec{\mathbf{v}}_i$.

Molecular Interpretation of the Pressure of an Ideal Gas

For our first application of kinetic theory, let us derive an expression for the pressure of N molecules of an ideal gas in a container of volume V in terms of microscopic quantities. The container is a cube with edges of length d (Fig. 16.15). We shall focus our attention on one of these molecules of mass m_0 and assumed to be moving so that its component of velocity in the x direction is v_{xi} as in Active Figure 16.16. (The subscript i here refers to the ith molecule, not to an initial value. We will combine the effects of all of the molecules shortly.) As the molecule collides elastically with any wall (assumption 4), its velocity component perpendicular to the wall is reversed because the mass of the wall is far greater than the mass of the molecule. Because the momentum component p_{xi} of the molecule is $m_0 v_{xi}$ before the collision and $-m_0 v_{xi}$ after the collision, the change in momentum of the molecule in the x-direction is

$$\Delta p_{xi} = -m_0 v_{xi} - (m_0 v_{xi}) = -2m_0 v_{xi}$$

Applying the impulse–momentum theorem (Eq. 8.11) to the molecule gives

$$\overline{F}_{i,\text{on molecule}}\,\Delta t_{\text{collision}} = \Delta p_{xi} = -2m_0 v_{xi}$$

where $\overline{F}_{i,\text{on molecule}}$ is the average force component,[1] perpendicular to the wall, for the force that the wall exerts on the molecule during the collision and $\Delta t_{\text{collision}}$ is the duration of the collision. For the molecule to make another collision with the same wall after this first collision, it must travel a distance of $2d$ in the x direction (across the container and back). The time interval between two collisions with the same wall is therefore

$$\Delta t = \frac{2d}{v_{xi}}$$

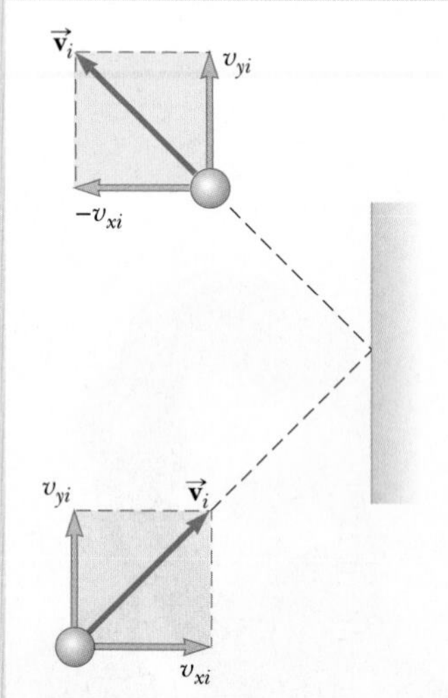

ACTIVE FIGURE 16.16 A molecule makes an elastic collision with the wall of the container. Its x component of momentum is reversed whereas its y component remains unchanged. In this construction, we assume that the molecule moves in the xy plane.

PhysicsNow™ Log into PhysicsNow at **www.pop4e.com** and go to Active Figure 16.16 to observe molecules within a container making collisions with the walls of the container and with each other.

The force that causes the change in momentum of the molecule in the collision with the wall occurs only during the collision. We can, however, average the force over the time interval for the molecule to move across the cube and back. Sometime during this time interval the collision occurs, so the change in momentum for this time interval is the same as that for the short duration of the collision. Therefore, we can rewrite the impulse–momentum theorem as

$$\overline{F}_i\,\Delta t = -2m_0 v_{xi}$$

where $\overline{F}_i$ is interpreted as the average force component on the molecule over the time for the molecule to move across the cube and back. Because exactly one collision occurs for each such time interval, it is also the long-term average force component on the molecule, over long time intervals containing any number of multiples of Δt.

The substitution of Δt into the impulse–momentum equation enables us to express the long-term average force component of the wall on the molecule:

$$\overline{F}_i = \frac{-2m_0 v_{xi}}{\Delta t} = \frac{-2m_0 v_{xi}^2}{2d} = \frac{-m_0 v_{xi}^2}{d}$$

Now, by Newton's third law, the force component of the molecule on the wall is equal in magnitude and opposite in direction:

$$\overline{F}_{i,\text{on wall}} = -\overline{F}_i = -\left(\frac{-m_0 v_{xi}^2}{d}\right) = \frac{m_0 v_{xi}^2}{d}$$

[1] For this discussion, we will use a bar over a variable to represent the average value of the variable, such as $\overline{F}$ for the average force, rather than the subscript "avg" that we have used before. This notation saves confusion because we will already have a number of subscripts on variables.

The magnitude of the total average force $\overline{F}$ exerted on the wall by the gas is found by adding the average force components exerted by the individual molecules. We add terms such as those shown in the preceding equations for all molecules:

$$\overline{F} = \sum_{i=1}^{N} \frac{m_0 v_{xi}^{\,2}}{d} = \frac{m_0}{d} \sum_{i=1}^{N} v_{xi}^{\,2}$$

where we have factored out the length of the box and the mass m_0 because assumption 5 tells us that all the molecules are the same. We now impose assumption 1, that the number of molecules is large. For a small number of molecules, the actual force on the wall would vary with time. It would be nonzero during the short interval of a collision of a molecule with the wall and zero when no molecule happens to be hitting the wall. For a very large number of molecules, however, such as Avogadro's number, these variations in force are smoothed out, so the average force is the same over *any* time interval. Therefore, the *constant* force F on the wall due to the molecular collisions is the same as the average force $\overline{F}$ and is of magnitude

$$F = \frac{m_0}{d} \sum_{i=1}^{N} v_{xi}^{\,2}$$

To proceed further, let us consider how we express the average value of the square of the x component of the velocity for the N molecules. The traditional average of a value is the sum of the values over the number of values:

$$\overline{v_x^{\,2}} = \frac{\sum_{i=1}^{N} v_{xi}^{\,2}}{N}$$

The numerator of this expression is contained in the right-hand side of the previous equation. Therefore, by combining the two expressions the total force on the wall can be written

$$F = \frac{m_0}{d} N \overline{v_x^{\,2}}$$

Now let us focus again on one molecule with velocity components v_{xi}, v_{yi}, and v_{zi}. The Pythagorean theorem relates the square of the speed of the molecule to the squares of the velocity components:

$$v_i^{\,2} = v_{xi}^{\,2} + v_{yi}^{\,2} + v_{zi}^{\,2}$$

If we take an average of both sides of this equation (sum over all particles and divide by N), the average value of v^2 for all the molecules in the container is related to the average values of $v_x^{\,2}$, $v_y^{\,2}$, and $v_z^{\,2}$ according to the expression

$$\overline{v^2} = \overline{v_x^{\,2}} + \overline{v_y^{\,2}} + \overline{v_z^{\,2}}$$

Now we assume that the motion is completely isotropic (assumption 2), which implies that no direction is preferred. On the average, the x, y, and z directions are equivalent, so

$$\overline{v_x^{\,2}} = \overline{v_y^{\,2}} = \overline{v_z^{\,2}}$$

which allows us to write

$$\overline{v^2} = 3\overline{v_x^{\,2}}$$

Therefore, the total force on the wall is

$$F = \frac{m_0}{d} N \left(\tfrac{1}{3}\overline{v^2}\right) = \frac{N}{3}\left(\frac{m_0 \overline{v^2}}{d}\right)$$

From this expression, we can find the pressure exerted on the wall by dividing this force by the area of the wall:

$$P = \frac{F}{A} = \frac{F}{d^2} = \frac{1}{3}\frac{N}{d^3}(m_0\overline{v^2}) = \frac{1}{3}\left(\frac{N}{V}\right)(m_0\overline{v^2})$$

▪ Pressure of an ideal gas

$$P = \frac{2}{3}\left(\frac{N}{V}\right)\left(\tfrac{1}{2}m_0\overline{v^2}\right) \qquad [16.13]$$

This result shows that **the pressure is proportional to the number of molecules per unit volume and to the average translational kinetic energy of the molecules,** $\frac{1}{2}m_0\overline{v^2}$. With this structural model of an ideal gas, we have arrived at an important result that relates the macroscopic quantity of pressure to a microscopic quantity, the average value of the molecular translational kinetic energy. Thus, we have a key link between the atomic world and the large-scale world.

Equation 16.13 verifies some features of pressure that are probably familiar to you. One way to increase the pressure inside a container is to increase the number of molecules per unit volume in the container (N/V). You do so when you add air to a tire. The pressure in the tire can also be increased by increasing the average translational kinetic energy of the molecules in the tire. As we shall see shortly, that can be accomplished by increasing the temperature of the gas inside the tire. Hence, the pressure inside a tire increases as the tire warms up during long trips. The continuous flexing of the tires as they move along the road surface results in work done as parts of the tire distort and in an increase in internal energy of the rubber. The increased temperature of the rubber results in transfer of energy into the air by heat, increasing the average translational kinetic energy of the molecules, which in turn produces an increase in pressure.

Molecular Interpretation of the Temperature of an Ideal Gas

We have related the pressure to the average kinetic energy of molecules; let us now relate temperature to a microscopic description of the gas. We can obtain some insight into the meaning of temperature by first writing Equation 16.13 in the form

$$PV = \tfrac{2}{3}N\left(\tfrac{1}{2}m_0\overline{v^2}\right)$$

Let us now compare this equation with the equation of state for an ideal gas:

$$PV = Nk_BT$$

The left-hand sides of these two equations are identical. Equating the right-hand sides of these expressions, we find that

▪ Temperature is proportional to average kinetic energy

$$T = \frac{2}{3k_B}\left(\tfrac{1}{2}m_0\overline{v^2}\right) \qquad [16.14]$$

which tells us that **the temperature of a gas is a direct measure of average translational molecular kinetic energy.** Therefore, as the temperature of a gas increases, the molecules move with higher average kinetic energy.

By rearranging Equation 16.14, we can relate the average translational molecular kinetic energy to the temperature:

▪ Average kinetic energy per molecule

$$\tfrac{1}{2}m_0\overline{v^2} = \tfrac{3}{2}k_BT \qquad [16.15]$$

That is, the average translational kinetic energy per molecule is $\frac{3}{2}k_BT$. Because $\overline{v_x^2} = \frac{1}{3}\overline{v^2}$, it follows that

$$\tfrac{1}{2}m_0\overline{v_x^2} = \tfrac{1}{2}k_BT \qquad [16.16]$$

In a similar manner, for the y and z motions we find that

$$\tfrac{1}{2} m_0 \overline{v_y^{\,2}} = \tfrac{1}{2} k_B T \qquad \text{and} \qquad \tfrac{1}{2} m_0 \overline{v_z^{\,2}} = \tfrac{1}{2} k_B T$$

Therefore, each translational degree of freedom contributes an equal amount of energy to the gas, namely $\frac{1}{2} k_B T$ per molecule. (In general, the phrase *degrees of freedom* refers to the number of independent means by which a molecule can possess energy.) A generalization of this result, known as the **theorem of equipartition of energy,** states that **the energy of a system in thermal equilibrium is equally divided among all degrees of freedom.** Furthermore, each degree of freedom contributes the same amount of average energy to the total, $\frac{1}{2} k_B T$ per molecule.

The total translational kinetic energy of N molecules of gas is simply N times the average translational kinetic energy per molecule, which is given by Equation 16.15:

$$E_{\text{total}} = N(\tfrac{1}{2} m_0 \overline{v^2}) = \tfrac{3}{2} N k_B T = \tfrac{3}{2} nRT \qquad [16.17]$$

▮ Total kinetic energy of N molecules

where we have used $k_B = R/N_A$ for Boltzmann's constant and $n = N/N_A$ for the number of moles of gas. From this result we see that **the total translational kinetic energy of a system of molecules is proportional to the absolute temperature of the system.**

For a monatomic gas, translational kinetic energy is the only type of energy the particles of the gas can have. Therefore, Equation 16.17 gives the **internal energy for a monatomic gas:**

$$E_{\text{int}} = \tfrac{3}{2} nRT \qquad \text{(monatomic gas)} \qquad [16.18]$$

This equation mathematically justifies our claim that internal energy is related to the temperature of a system, which we introduced in Chapter 6. For diatomic and polyatomic molecules, additional possibilities for energy storage are available in the vibration and rotation of the molecule, but a proportionality between E_{int} and T remains.

▮ Root-mean-square speed

The square root of $\overline{v^2}$ is called the **root-mean-square** (rms) **speed** of the molecules. From Equation 16.15 we find for the rms speed that

$$v_{\text{rms}} = \sqrt{\overline{v^2}} = \sqrt{\frac{3k_B T}{m_0}} = \sqrt{\frac{3RT}{M}} \qquad [16.19]$$

where M is the molar mass in kilograms per mole. This expression shows that, at a given temperature, lighter molecules move faster, on the average, than heavier molecules. For example, hydrogen, with a molar mass of 2.0×10^{-3} kg/mol, moves four times as fast as oxygen, whose molar mass is 32×10^{-3} kg/mol. If we calculate the rms speed for hydrogen at room temperature ($\approx$ 300 K), we find that

$$v_{\text{rms}} = \sqrt{\frac{3RT}{M}} = \sqrt{\frac{3(8.31 \text{ J/mol}\cdot\text{K})(300 \text{ K})}{2.0 \times 10^{-3} \text{ kg/mol}}} = 1.9 \times 10^3 \text{ m/s}$$

This value is about 17% of the escape speed for the Earth, which we calculated in Chapter 11. Because this value is an average speed, a large number of molecules having speeds much higher than the average can escape from the Earth's atmosphere. Thus, the Earth's atmosphere does not at present contain hydrogen because it has all bled off into space.

Table 16.2 lists the rms speeds for various molecules at 20°C.

TABLE 16.2

Some rms Speeds

Gas	Molar mass (g/mol)	v_{rms} at 20°C (m/s)
H_2	2.02	1 902
He	4.00	1 352
H_2O	18.0	637
Ne	20.2	602
N_2 or CO	28.0	511
NO	30.0	494
O_2	32.0	478
CO_2	44.0	408
SO_2	64.1	338

QUICK QUIZ 16.5 Two containers of ideal gas are at the same temperature. Both containers hold the same type of gas, but container B has twice the volume of container A. **(i)** The average translational kinetic energy per molecule in container B is **(a)** twice that for container A, **(b)** the same as that for container A, **(c)** half that for container A, or **(d)** impossible to determine. **(ii)** The internal energy of the gas in container B is related in what way to that in A, from the same list of choices? **(iii)** The rms speed of the gas molecules in container B is related in what way to that in A, from the same list of choices?

EXAMPLE 16.5 A Tank of Helium

A tank of volume 0.300 m^3 contains 2.00 mol of helium gas at 20.0°C.

A Assuming that the helium behaves like an ideal gas, find the total internal energy of the gas.

Solution Helium is a monatomic gas, so Equation 16.18 can be used for the internal energy. With $n = 2.00$ mol and $T = 293$ K, we have

$$E_{\text{int}} = \tfrac{3}{2}nRT = \tfrac{3}{2}(2.00 \text{ mol})(8.31 \text{ J/mol}\cdot\text{K})(293 \text{ K})$$
$$= 7.30 \times 10^3 \text{ J}$$

B What is the average kinetic energy per molecule?

Solution From Equation 16.15, we see that the average kinetic energy per molecule is

$$\tfrac{1}{2}m_0\overline{v^2} = \tfrac{3}{2}k_BT = \tfrac{3}{2}(1.38 \times 10^{-23} \text{ J/K})(293 \text{ K})$$
$$= 6.07 \times 10^{-21} \text{ J}$$

16.6 DISTRIBUTION OF MOLECULAR SPEEDS

In the preceding section, we derived an expression for the average speed of a gas molecule but made no mention of the actual distribution of molecular speeds among all possible values. In 1860, James Clerk Maxwell (1831–1879) derived an expression that describes this distribution of molecular speeds. His work and developments by other scientists shortly thereafter were highly controversial because experiments at that time could not directly detect molecules. About 60 years later, however, experiments confirmed Maxwell's predictions.

Consider a container of gas whose molecules have some distribution of speeds. Suppose we want to determine how many gas molecules have a speed in the range from, for example, 400 to 410 m/s. Intuitively, we expect that the speed distribution depends on temperature. Furthermore, we expect that the distribution peaks in the vicinity of v_{rms}. That is, few molecules are expected to have speeds much less than or much greater than v_{rms} because these extreme speeds will result only from an unlikely chain of collisions.

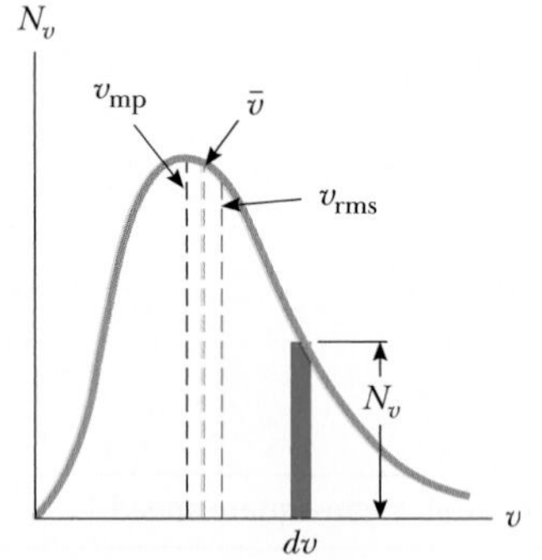

ACTIVE FIGURE 16.17 The speed distribution of gas molecules at some temperature. The number of molecules having speeds in the range v to $v + dv$ is equal to the area of the shaded rectangle, $N_v\,dv$. The function N_v approaches zero as v approaches infinity.

Physics Now™ Log into PhysicsNow at **www.pop4e.com** and go to Active Figure 16.17. You can move the blue rectangle and measure the number of molecules having speeds within a small range.

The observed speed distribution of gas molecules in thermal equilibrium is shown in Active Figure 16.17. The quantity N_v, called the **Maxwell–Boltzmann distribution function,** is defined as follows. If N is the total number of molecules, the number of molecules with speeds between v and $v + dv$ is $dN = N_v\,dv$. This number is also equal to the area of the shaded rectangle in Active Figure 16.17. Furthermore, the fraction of molecules with speeds between v and $v + dv$ is $N_v dv/N$. This fraction is also equal to the *probability* that a molecule has a speed in the range v to $v + dv$.

The fundamental expression that describes the distribution of speeds of N gas molecules is

■ Maxwell–Boltzmann distribution function

$$N_v = 4\pi N\left(\frac{m_0}{2\pi k_BT}\right)^{3/2} v^2 e^{-m_0v^2/2k_BT} \qquad [16.20]$$

where m_0 is the mass of a gas molecule, k_B is Boltzmann's constant, and T is the absolute temperature.[2]

As indicated in Active Figure 16.17, the average speed $\overline{v}$ is somewhat lower than the rms speed. The most probable speed v_{mp} is the speed at which the distribution

[2] For the derivation of this expression, see a text on thermodynamics such as that by R. P. Bauman, *Modern Thermodynamics and Statistical Mechanics* (New York: Macmillan, 1992).

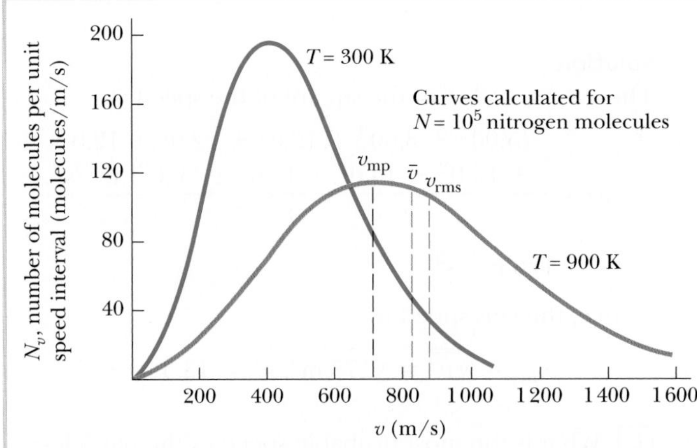

ACTIVE FIGURE 16.18 The speed distribution function for 10^5 nitrogen molecules at 300 K and 900 K. The total area under either curve is equal to the total number of molecules, which in this case equals 10^5. Note that $v_{rms} > \bar{v} > v_{mp}$.

PhysicsNow™ Log into PhysicsNow at **www.pop4e.com** and go to Active Figure 16.18 to set a desired temperature and see the effect on the distribution curve.

curve reaches a peak. Using Equation 16.20, we find that

$$v_{rms} = \sqrt{\overline{v^2}} = \sqrt{3k_BT/m_0} = 1.73\sqrt{k_BT/m_0} \qquad [16.21]$$

$$\bar{v} = \sqrt{8k_BT/\pi m_0} = 1.60\sqrt{k_BT/m_0} \qquad [16.22]$$

$$v_{mp} = \sqrt{2k_BT/m_0} = 1.41\sqrt{k_BT/m_0} \qquad [16.23]$$

From these equations we see that $v_{rms} > \bar{v} > v_{mp}$.

Active Figure 16.18 represents speed distribution curves for nitrogen molecules. The curves were obtained using Equation 16.20 to evaluate the distribution function at various speeds and at two temperatures. Note that the peak in the curve shifts to the right as T increases, indicating that the average speed increases with increasing temperature, as expected. In addition, the overall width of the curve increases with temperature. The shape of the curves is asymmetrical because the lowest speed possible is zero, whereas the upper classical limit of the speed is infinity.

The speed distribution curve for molecules in a liquid is similar to those shown in Active Figure 16.18. The phenomenon of evaporation of a liquid can be understood from this distribution in speeds because some molecules in the liquid are more energetic than others. Some of the faster-moving molecules in the liquid penetrate the surface and leave the liquid even at temperatures well below the boiling point. The molecules that escape the liquid by evaporation are those that have sufficient energy to overcome the attractive forces of the molecules in the liquid phase. Consequently, the molecules left behind in the liquid phase have a lower average kinetic energy, causing the temperature of the liquid to decrease. Hence, evaporation is a cooling process. For example, an alcohol-soaked cloth is often placed on a feverish head to cool and comfort the patient.

Cooling a patient with an alcohol-soaked cloth

QUICK QUIZ 16.6 Consider the qualitative shapes of the two curves in Active Figure 16.18, without regard for the numerical values or labels in the graph. Suppose you have two containers of gas *at the same temperature.* Container A has 10^5 nitrogen molecules and container B has 10^5 hydrogen molecules. What is the correct qualitative matching between the containers and the two curves in Active Figure 16.18? **(a)** Container A corresponds to the blue curve and container B to the brown curve. **(b)** Container B corresponds to the blue curve and container A to the brown curve. **(c)** Both containers will correspond to the same curve.

EXAMPLE 16.6 A System of Nine Particles

Nine particles have speeds of 5.00, 8.00, 12.0, 12.0, 12.0, 14.0, 14.0, 17.0, and 20.0 m/s.

A Find the average speed.

Solution We cannot use Equations 16.21, 16.22, and 16.23 to calculate average speeds because these equations are only valid for a large number of gas particles. On the other hand, because we have so few particles, we can calculate the averages directly.

The average speed is the sum of the speeds divided by the total number of particles:

$$\overline{v} = \frac{(5.00 + 8.00 + 12.0 + 12.0 + 12.0 + 14.0 + 14.0 + 17.0 + 20.0)\ \text{m/s}}{9}$$

$$= \boxed{12.7\ \text{m/s}}$$

B What is the rms speed?

Solution

The average value of the square of the speed is

$$\overline{v^2} = \frac{(5.00^2 + 8.00^2 + 12.0^2 + 12.0^2 + 12.0^2 + 14.0^2 + 14.0^2 + 17.0^2 + 20.0^2)\ \text{m}^2/\text{s}^2}{9}$$

$$= 178\ \text{m}^2/\text{s}^2$$

Hence, the rms speed is

$$v_{\text{rms}} = \sqrt{\overline{v^2}} = \sqrt{178\ \text{m}^2/\text{s}^2} = \boxed{13.3\ \text{m/s}}$$

C What is the most probable speed of the particles?

Solution

Three of the particles have a speed of 12.0 m/s, two have a speed of 14.0 m/s, and the remaining particles have different speeds. Hence, we see that the most probable speed v_{mp} is $\boxed{12.0\ \text{m/s}}$.

16.7 THE ATMOSPHERIC LAPSE RATE

CONTEXT CONNECTION

We have discussed the temperature of a gas with the assumption that all parts of the gas are at the same temperature. For small volumes of gas, this assumption is relatively good. What about a *huge* volume of gas, however, such as the atmosphere? It is clear that the assumption of a uniform temperature throughout the gas is not valid in this case. When it is a hot summer day in Los Angeles, it is a cold winter day in Melbourne; different parts of the atmosphere are clearly at different temperatures.

We can address this question, as we discussed in the opening section of this Context, by considering the global average of the air temperature at the surface of the Earth. Yet variations also occur in temperature at different *heights* in the atmosphere. It is this variation of temperature with height that we explore here.

Figure 16.19 shows graphical representations of average air temperature in January at various heights in four states.[3] These data are taken at locations on the surface of the Earth, but at varying elevations, such as at sea level and on mountains. For all four states we see a clear indication that the temperature decreases as we move to higher elevations. Of course, one look at snow-capped mountains tells us that is the case. We also see that the data tend to lie along straight lines, although some of the data are scattered, suggesting that the temperature decreases approximately linearly with height above the surface.

We can argue conceptually why the temperature decreases with height. Imagine a parcel of air moving upward along the slope of a mountain. As this parcel rises into higher elevations, the pressure on it from the surrounding air decreases. The pressure difference between the interior and the exterior of the parcel causes the parcel to expand. In doing so, the parcel is pushing the surrounding air outward,

[3] Data are taken from National Oceanic and Atmospheric Administration, *Climates of the States* (Port Washington, NY: U.S. Department of Commerce, Water Information Center Inc., 1974.)

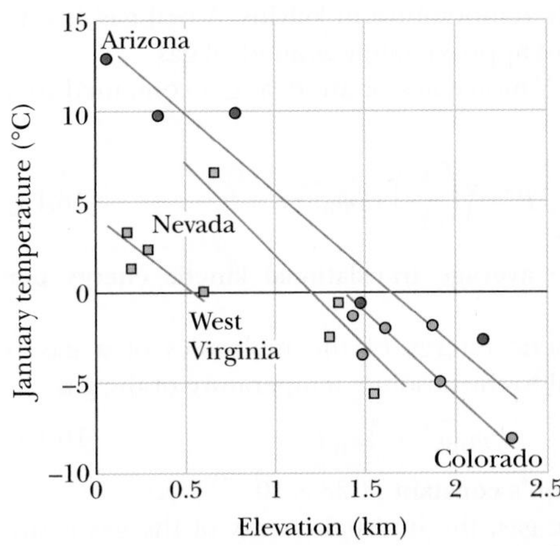

FIGURE 16.19 Variation of average temperature with elevation for four states. Note that the slopes of all four lines are approximately the same.

doing work on it. Because the system (the parcel of air) is doing work on the environment, the energy in the parcel decreases. The decreased energy is manifested as a decrease in temperature.

If this process is reversed so that the parcel moves toward lower elevations, work is done on the parcel, which increases its energy so that it becomes warmer. This situation occurs during Santa Ana wind conditions in the Los Angeles basin in which air is pushed from the mountains down into the low elevations of the basin, resulting in hot, dry winds. Similar conditions go by other names in other regions, such as the *chinook* from the Rocky Mountains and the *foehn* from the Swiss Alps.

Notice in Figure 16.19 that the slopes of all four lines are relatively close together. This closeness suggests that the decrease in temperature with height—called the **atmospheric lapse rate**—is similar at various locations across the surface of the Earth, so we might define an average lapse rate for the entire surface.

That is indeed the case, and we find that the average global lapse rate is about $-6.5°C/km$. If you determine the slopes of the lines in Figure 16.19, you will find that they are close to this value.

The linear decrease with temperature only occurs in the lower part of the atmosphere called the **troposphere,** the part of the atmosphere in which weather occurs and airplanes fly. Above the troposphere is the **stratosphere,** with an imaginary boundary called the **tropopause** separating the two layers. In the stratosphere, temperature tends to be relatively constant with height.

The decrease in temperature with height in the troposphere is one component to a structural model of the atmosphere that will allow us to predict the surface temperature of the Earth. If we can find the temperature of the stratosphere and the height of the tropopause, we can extrapolate to the surface, using the lapse rate to find the temperature at the surface. The lapse rate and the height of the tropopause can be measured. To find the temperature of the stratosphere, we need to know more about energy exchanges in the Earth's atmosphere, which we will investigate in the next chapter.

SUMMARY

PhysicsNow™ Take a practice test by logging into PhysicsNow at www.pop-4e.com and clicking on the Pre-Test link for this chapter.

The **zeroth law of thermodynamics** states that if two objects, A and B, are separately in thermal equilibrium with a third object, A and B are in thermal equilibrium with each other.

The relationship between T_C, the **Celsius temperature,** and T, the **Kelvin (absolute) temperature,** is

$$T_C = T - 273.15 \qquad [16.1]$$

The relationship between the **Fahrenheit** and Celsius temperatures is

$$T_F = \frac{9}{5}T_C + 32°F \qquad [16.2]$$

When the temperature of a substance is raised, it generally expands. If an object has an initial length of L_i at some temperature and undergoes a change in temperature ΔT, its length changes by the amount ΔL, which is proportional to the object's initial length and the temperature change:

$$\Delta L = \alpha L_i \Delta T \qquad [16.4]$$

The parameter α is called the **average coefficient of linear expansion.**

The change in volume of most substances is proportional to the initial volume V_i and the temperature change ΔT:

$$\Delta V = \beta V_i \Delta T \qquad [16.6]$$

where β is the **average coefficient of volume expansion** and is equal to 3α.

The change in area of a substance is given by

$$\Delta A = \gamma A_i \Delta T \qquad [16.7]$$

where γ is the **average coefficient of area expansion** and is equal to 2α.

The **ideal gas model** refers to a collection of gas molecules that move randomly and are of negligible size. An ideal gas obeys the equation

$$PV = nRT \qquad [16.9]$$

where P is the pressure of the gas, V is its volume, n is the number of moles of gas, R is the universal gas constant (8.314 J/mol·K), and T is the absolute temperature in kelvins. A real gas at very low pressures behaves approximately as an ideal gas.

The pressure of N molecules of an ideal gas contained in a volume V is given by

$$P = \frac{2}{3}\left(\frac{N}{V}\right)\left(\frac{1}{2}m_0\overline{v^2}\right) \qquad [16.13]$$

where $\frac{1}{2}m_0\overline{v^2}$ is the **average translational kinetic energy per molecule.**

The average kinetic energy of the molecules of a gas is directly proportional to the absolute temperature of the gas:

$$\frac{1}{2}m_0\overline{v^2} = \frac{3}{2}k_B T \qquad [16.15]$$

where k_B is **Boltzmann's constant** (1.38×10^{-23} J/K).

For a monatomic gas, the internal energy of the gas is the total translational kinetic energy

$$E_{int} = \frac{3}{2}nRT \quad \text{(monatomic gas)} \qquad [16.18]$$

The **root-mean-square** (rms) **speed** of the molecules of a gas is

$$v_{rms} = \sqrt{\overline{v^2}} = \sqrt{\frac{3k_BT}{m_0}} = \sqrt{\frac{3RT}{M}} \qquad [16.19]$$

The **Maxwell–Boltzmann distribution function** describes the distribution of speeds of N gas molecules:

$$N_v = 4\pi N\left(\frac{m_0}{2\pi k_B T}\right)^{3/2} v^2 e^{-m_0 v^2/2k_B T} \qquad [16.20]$$

where m_0 is the mass of a gas molecule, k_B is Boltzmann's constant, and T is the absolute temperature.

QUESTIONS

☐ = answer available in the *Student Solutions Manual and Study Guide*

1. A piece of copper is dropped into a beaker of water. If the water's temperature rises, what happens to the temperature of the copper? Under what conditions are the water and copper in thermal equilibrium?

2. In describing his upcoming trip to the Moon, and as portrayed in the movie *Apollo 13* (Universal, 1995), astronaut Jim Lovell said, "I'll be walking in a place where there's a 400-degree difference between sunlight and shadow." What is it that is hot in sunlight and cold in shadow? Suppose an astronaut standing on the Moon holds a thermometer in his gloved hand. Is the thermometer reading the temperature of the vacuum at the Moon's surface? Does it read any temperature? If so, what object or substance has that temperature?

3. Why should the amalgam used in dental fillings have the same average coefficient of expansion as a tooth? What would occur if they were mismatched?

4. Markings to indicate length are placed on a steel tape in a room that has a temperature of 22°C. Are measurements made with the tape on a day when the temperature is 27°C too long, too short, or accurate? Defend your answer.

5. Use a periodic table of the elements, as in Appendix C, to determine the number of grams in one mole of the following gases: (a) hydrogen, (b) helium, (c) carbon monoxide.

6. What does the ideal gas law predict about the volume of a sample of gas at absolute zero? Why is this prediction incorrect?

7. An inflated rubber balloon filled with air is immersed in a flask of liquid nitrogen that is at 77 K. Describe what happens to the balloon, assuming that it remains flexible while being cooled.

8. Two identical cylinders at the same temperature each contain the same kind of gas and the same number of moles of gas. If the volume of cylinder A is three times greater than the volume of cylinder B, what can you say about the relative pressures in the cylinders?

9. After food is cooked in a pressure cooker, why is it very important to cool off the container with cold water before attempting to remove the lid?

10. The shore of the ocean is very rocky at a particular place. The rocks form a cave sloping upward from an underwater opening, as shown in Figure Q16.10a. Inside the cave is a pocket of trapped air. As the level of the ocean rises and falls with the tides, will the level of water in the cave rise and fall? If so, will it have the same amplitude as that of the

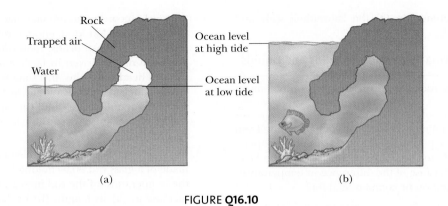

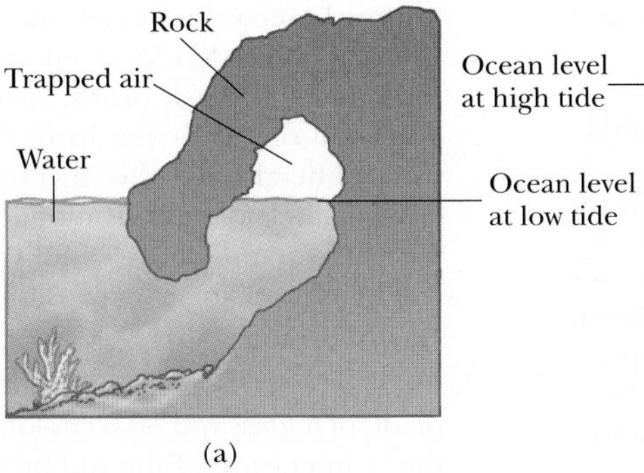

Rock
Trapped air
Water
Ocean level at high tide
Ocean level at low tide

(a)

(b)

FIGURE **Q16.10**

ocean? (b) Now suppose that the cave is deeper in the water so that it is completely submerged and filled with water at high tide, as shown in Figure Q16.10b. At low tide, will the level of the water in the cave be the same as that of the ocean?

11. When the metal ring and metal sphere in Figure Q16.11 are both at room temperature, the sphere can just be passed through the ring. After the sphere is heated, it cannot be passed through the ring. Explain. What if the ring is heated and the sphere is left at room temperature? Does the sphere pass through the ring?

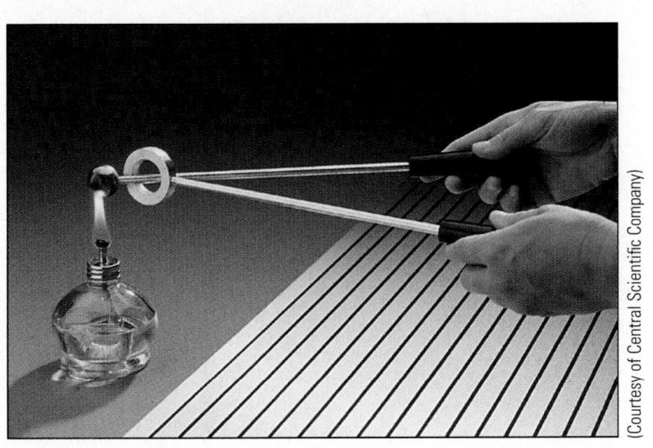

FIGURE **Q16.11**

12. Metal lids on glass jars can often be loosened by running hot water over them. How is that possible?

13. When alcohol is rubbed on your body, it lowers your skin temperature. Explain this effect.

14. Dalton's law of partial pressures states that the total pressure of a mixture of gases is equal to the sum of the partial pressures of gases making up the mixture. Give a convincing argument of this law based on the kinetic theory of gases.

15. If a helium-filled balloon initially at room temperature is placed in a freezer, will its volume increase, decrease, or remain the same?

16. Which is denser, dry air or air saturated with water vapor? Explain.

17. What happens to a helium-filled balloon released into the air? Will it expand or contract? Will it stop rising at some height?

18. An ideal gas is contained in a vessel at 300 K. If the temperature is increased to 900 K, determine the factor by which each of the following changes: (a) the average kinetic energy of the molecules, (b) the rms molecular speed, (c) the average momentum change of one molecule in a collision with a wall, (d) the rate of collisions of molecules with walls, (e) the pressure of the gas.

PROBLEMS

1, 2, 3 = straightforward, intermediate, challenging
□ = full solution available in the *Student Solutions Manual and Study Guide*
PhysicsNow™ = coached problem with hints available at www.pop4e.com
= computer useful in solving problem
= paired numerical and symbolic problems
= biomedical application

Section 16.2 ▪ Thermometers and Temperature Scales

1. **PhysicsNow™** A constant-volume gas thermometer is calibrated in dry ice (that is, evaporating carbon dioxide in the solid state, with a temperature of −80.0°C) and in boiling ethyl alcohol (78.0°C). The two pressures are 0.900 atm and 1.635 atm, respectively. (a) What Celsius value of absolute zero does the calibration yield? What is the pressure at (b) the freezing point of water and (c) the boiling point of water?

2. Convert the following to equivalent temperatures on the Celsius and Kelvin scales: (a) the normal human body temperature, 98.6°F; (b) the air temperature on a cold day, −5.00°F.

3. Liquid nitrogen has a boiling point of −195.81°C at atmospheric pressure. Express this temperature in (a) degrees Fahrenheit and (b) kelvins.

4. The temperature difference between the inside and the outside of an automobile engine is 450°C. Express this temperature difference on (a) the Fahrenheit scale and (b) the Kelvin scale.

Section 16.3 ■ Thermal Expansion of Solids and Liquids

Note: Table 16.1 is available for use in solving problems in this section.

5. The Trans-Alaska pipeline is 1 300 km long, reaching from Prudhoe Bay to the port of Valdez. It experiences temperatures from $-73°C$ to $+35°C$. How much does the steel pipeline expand because of the difference in temperature? How can this expansion be compensated for?

6. A pair of eyeglass frames is made of epoxy plastic. At room temperature (20.0°C), the frames have circular lens holes 2.20 cm in radius. To what temperature must the frames be heated if lenses 2.21 cm in radius are to be inserted in them? The average coefficient of linear expansion for epoxy is $1.30 \times 10^{-4}\ (°C)^{-1}$.

7. Each year thousands of children are badly burned by hot tap water. Figure P16.7 shows a cross-sectional view of an antiscalding faucet attachment designed to prevent such accidents. Within the device, a spring made of material with a high coefficient of thermal expansion controls a movable plunger. When the water temperature rises above a preset safe value, the expansion of the spring causes the plunger to shut off the water flow. Assuming that the initial length L of the unstressed spring is 2.40 cm and its coefficient of linear expansion is $22.0 \times 10^{-6}\ (°C)^{-1}$, determine the increase in length of the spring when the water temperature rises by 30.0°C. (You will find the increase in length to be small. Therefore, to provide a greater variation in valve opening for the temperature change anticipated, actual devices have a more complicated mechanical design.)

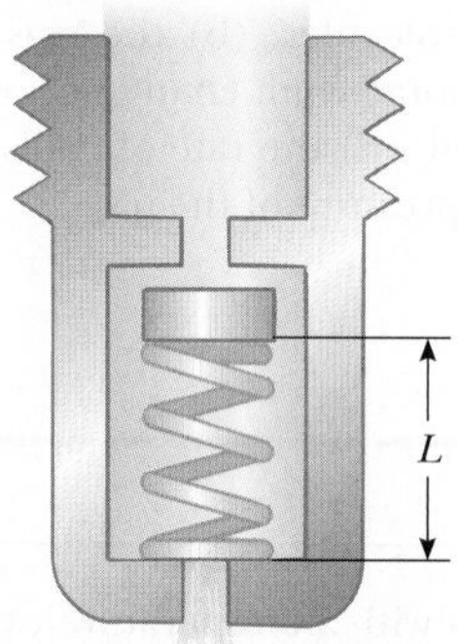

FIGURE **P16.7**

8. Vinyl siding for houses is manufactured with horizontal slots 2.80 cm long for nailing. Assume that the manufacturers chose this length to allow for the expansion that would occur in the following situation. The wall of a house is 15.0 m long. Its vinyl siding is nailed through insulating panels to studs that stay at nearly constant temperature throughout the year. The siding is installed in midwinter, when its temperature is $-5°C$. Adjacent pieces of vinyl are butted snug, end to end against each other. The nails holding them to the wall are placed at the centers of the slots in the vinyl. According to the manufacturer's directions, most of the nails are left loose enough to allow the vinyl slots to slide under the nail heads. Near one end of the wall, however, one nail is driven in farther than the others so that the nail head holds the vinyl stationary relative to the insulating panel at that location. On a summer day the sun raises the temperature of all the vinyl to 38.0°C. Find the coefficient of linear expansion for vinyl implied by these assumptions.

9. Physics Now™ The active element of a certain laser is made of a glass rod 30.0 cm long by 1.50 cm in diameter. If the temperature of the rod increases by 65.0°C, what is the increase in (a) its length, (b) its diameter, and (c) its volume? Assume that the average coefficient of linear expansion of the glass is $9.00 \times 10^{-6}\ (°C)^{-1}$.

10. Review problem. Inside the wall of a house, an L-shaped section of hot water pipe consists of a straight horizontal piece 28.0 cm long, an elbow, and a straight vertical piece 134 cm long (Fig. P16.10). A stud and a second-story floorboard hold stationary the ends of this section of copper pipe. Find the magnitude and direction of the displacement of the pipe elbow when the water flow is turned on, raising the temperature of the pipe from 18.0°C to 46.5°C.

FIGURE **P16.10**

11. A square hole 8.00 cm along each side is cut in a sheet of copper. (a) Calculate the change in the area of this hole resulting when the temperature of the sheet is increased by 50.0 K. (b) Does this change represent an increase or a decrease in the area enclosed by the hole?

12. The average coefficient of volume expansion for carbon tetrachloride is $5.81 \times 10^{-4}\ (°C)^{-1}$. If a 50.0-gal steel container is filled completely with carbon tetrachloride when the temperature is 10.0°C, how much will spill over when the temperature rises to 30.0°C?

13. A hollow aluminum cylinder 20.0 cm deep has an internal capacity of 2.000 L at 20.0°C. It is completely filled with turpentine and then slowly warmed to 80.0°C. (a) How much turpentine overflows? (b) If the cylinder is then cooled back to 20.0°C, how far below the cylinder's rim does the turpentine's surface recede?

14. At 20.0°C, an aluminum ring has an inner diameter of 5.000 0 cm and a brass rod has a diameter of 5.050 0 cm. (a) If only the ring is heated, what temperature must it reach so that it will just slip over the rod? (b) If both are heated together, what temperature must they both reach so that the ring just slips over the rod? Would this latter process work?

Section 16.4 ▪ Macroscopic Description of an Ideal Gas

15. On your wedding day your lover gives you a gold ring of mass 3.80 g. Fifty years later its mass is 3.35 g. On the average, how many atoms were abraded from the ring during each second of your marriage? The molar mass of gold is 197 g/mol.

16. Use the definition of Avogadro's number to find the mass of a helium atom.

17. An automobile tire is inflated with air originally at 10.0°C and normal atmospheric pressure. During the process, the air is compressed to 28.0% of its original volume and the temperature is increased to 40.0°C. (a) What is the tire pressure? (b) After the car is driven at high speed, the tire air temperature rises to 85.0°C and the interior volume of the tire increases by 2.00%. What is the new tire pressure (absolute) in pascals?

18. A rigid tank having a volume of 0.100 m^3 contains helium gas at 150 atm. How many balloons can be inflated by opening the valve at the top of the tank? Each filled balloon is a sphere 0.300 m in diameter at an absolute pressure of 1.20 atm.

19. An auditorium has dimensions 10.0 m × 20.0 m × 30.0 m. How many molecules of air fill the auditorium at 20.0°C and a pressure of 101 kPa?

20. Your father and your little brother are confronted with the same puzzle. Your father's garden sprayer and your brother's water cannon both have tanks with a capacity of 5.00 L (Fig. P16.20). Your father introduces a negligible amount of concentrated insecticide into his tank. Your father and your brother both pour in 4.00 L of water and seal up their tanks so that the tanks also contain air at atmospheric pressure. Next, each uses a hand-operated piston pump to inject more air until the absolute pressure in the tank reaches 2.40 atm and it becomes too difficult to move the pump handle. Now each uses his device to spray out water—not air—until the stream becomes feeble, as it does when the pressure in the tank reaches 1.20 atm. Then each device must be pumped up again, sprayed again, and so on. To spray out all the water, each finds that he must pump up the tank three times. Here is the puzzle: most of the water sprays out as a result of the second pumping. The first and the third pumping-up processes seem just as difficult, but they result in a disappointingly small amount of water coming out. Account for this phenomenon.

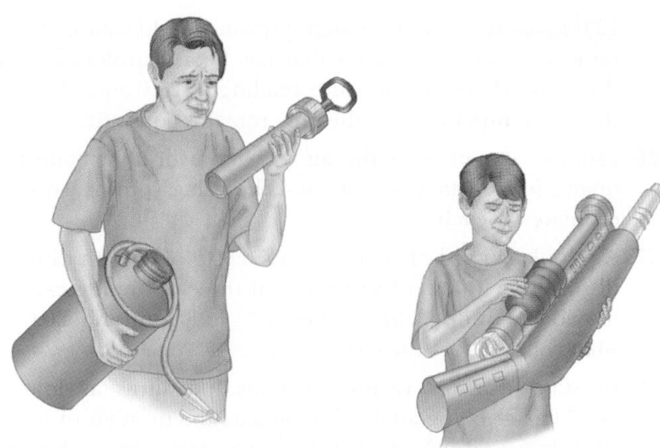

FIGURE P16.20

21. **Physics Now™** The mass of a hot-air balloon and its cargo (not including the air inside) is 200 kg. The air outside is at 10.0°C and 101 kPa. The volume of the balloon is 400 m^3. To what temperature must the air in the balloon be heated before the balloon will lift off? (Air density at 10.0°C is 1.25 kg/m^3.)

22. A cube 10.0 cm on each edge contains air (with equivalent molar mass 28.9 g/mol) at atmospheric pressure and temperature 300 K. Find (a) the mass of the gas, (b) its weight, and (c) the force it exerts on each face of the cube. (d) Comment on the physical reason why such a small sample can exert such a great force.

23. How deep will a shearwater dive? To measure how far below the ocean surface the bird goes to catch a fish, Will Mackin used a method originated by Lord Kelvin for soundings by the British Navy. It was adapted for zoological use by Alan Burger and Rory Wilson. With powdered sugar Mackin dusted the interiors of very thin plastic tubes and then sealed one end of each tube with a cigarette lighter. Charging around on a rocky beach at night with a miner's headlamp, he would grab an Audubon's shearwater in its nest and attach a tube to its back. He caught the same bird the next night and removed the tube. After hundreds of captures the birds thoroughly disliked him but were not permanently frightened away from the rookery. Assume that in one trial, with a tube 6.50 cm long, he found that water had entered the tube to wash away the sugar over a distance of 2.70 cm from the open end. (a) Find the greatest depth to which the shearwater dove, assuming that the air in the tube stayed at constant temperature. (b) Must the tube be attached to the bird in any particular orientation for this method to work? (Audubon's shearwater can dive to more than twice the depth you calculate, and larger species can dive nearly ten times deeper.)

24. At 25.0 m below the surface of the sea (density = 1 025 kg/m^3), where the temperature is 5.00°C, a diver exhales an air bubble having a volume of 1.00 cm^3. If the surface temperature of the sea is 20.0°C, what is the volume of the bubble just before it breaks the surface?

25. The pressure gauge on a tank registers the gauge pressure, which is the difference between the interior and exterior pressure. When the tank is full of oxygen (O_2), it contains

12.0 kg of the gas at a gauge pressure of 40.0 atm. Determine the mass of oxygen that has been withdrawn from the tank when the pressure reading is 25.0 atm. Assume that the temperature of the tank remains constant.

26. Estimate the mass of the air in your bedroom. State the quantities you take as data and the value you measure or estimate for each.

27. A popular brand of cola contains 6.50 g of carbon dioxide dissolved in 1.00 L of soft drink. If the evaporating carbon dioxide is trapped in a cylinder at 1.00 atm and 20.0°C, what volume does the gas occupy?

28. In state-of-the-art vacuum systems, pressures as low as 10^{-9} Pa are being attained. Calculate the number of molecules in a 1.00-m^3 vessel at this pressure assuming that the temperature is 27.0°C.

29. A room of volume V contains air having equivalent molar mass M (in g/mol). If the temperature of the room is raised from T_1 to T_2, what mass of air will leave the room? Assume that the air pressure in the room is maintained at P_0.

Section 16.5 ▪ The Kinetic Theory of Gases

30. **Review problem.** In a time interval $\Delta t = t - 0 = t$, N hailstones strike a glass window of area A at an angle θ to the window surface. Each hailstone has a mass m and a speed v. If the collisions are elastic, what are the average force and pressure on the window?

31. In a period of 1.00 s, 5.00×10^{23} nitrogen molecules strike a wall with an area of 8.00 cm^2. If the molecules move with a speed of 300 m/s and strike the wall head-on in elastic collisions, what is the pressure exerted on the wall? (The mass of one N_2 molecule is 4.68×10^{-26} kg.)

32. A 5.00-L vessel contains nitrogen gas at 27.0°C and 3.00 atm. Find (a) the total translational kinetic energy of the gas molecules and (b) the average kinetic energy per molecule.

33. (a) How many atoms of helium gas fill a balloon of diameter 30.0 cm at 20.0°C and 1.00 atm? (b) What is the average kinetic energy of the helium atoms? (c) What is the root-mean-square speed of the helium atoms?

34. *Brownian motion.* Molecular motion is invisible in itself. When a small particle is suspended in a fluid, bombardment by molecules makes the particle jitter about at random. Robert Brown discovered this motion in 1827 while studying plant fertilization. Albert Einstein analyzed it in 1905 and Jean Perrin used it for an early measurement of Avogadro's number. The visible particle's average kinetic energy can be taken as $\frac{3}{2}k_BT$, the same as that of a molecule in an ideal gas. Consider a spherical particle of density 1 000 kg/m^3 in water at 20°C. (a) For a particle of diameter 3.00 μm, evaluate the rms speed. (b) The particle's actual motion is a random walk, but imagine that it moves with constant velocity equal in magnitude to its rms speed. In what time interval would it move by a distance equal to its own diameter? (c) Repeat parts (a) and (b) for a particle of mass 70.0 kg, modeling your own body. (d) Find the diameter of a particle whose rms speed is equal to its own diameter divided by 1 s. (*Note:* You can solve all parts of this problem most efficiently by first finding a symbolic relationship between the particle size and its rms speed.)

35. **Physics Now™** A cylinder contains a mixture of helium and argon gas in equilibrium at 150°C. (a) What is the average kinetic energy for each type of gas molecule? (b) What is the root-mean-square speed of each type of molecule?

Section 16.6 ▪ Distribution of Molecular Speeds

36. Fifteen identical particles have various speeds. One has a speed of 2.00 m/s, two have speeds of 3.00 m/s, three have speeds of 5.00 m/s, four have speeds of 7.00 m/s, three have speeds of 9.00 m/s, and two have speeds of 12.0 m/s. Find (a) the average speed, (b) the rms speed, and (c) the most probable speed of these particles.

37. From the Maxwell–Boltzmann speed distribution, show that the most probable speed of a gas molecule is given by Equation 16.23. Note that the most probable speed corresponds to the point at which the slope of the speed distribution curve dN_v/dv is zero.

38. Helium gas is in thermal equilibrium with liquid helium at 4.20 K. Even though it is on the point of condensation, model the gas as ideal and determine the most probable speed of a helium atom (mass = 6.64×10^{-27} kg) in it.

39. **Review problem.** At what temperature would the average speed of helium atoms equal (a) the escape speed from the Earth, 1.12×10^4 m/s; and (b) the escape speed from the Moon, 2.37×10^3 m/s? (See Chapter 11 for a discussion of escape speed; also note that the mass of a helium atom is 6.64×10^{-27} kg.)

Section 16.7 ▪ Context Connection—The Atmospheric Lapse Rate

40. The summit of Mount Whitney, in California, is 3 660 m higher than a point in the foothills. Assume that the atmospheric lapse rate in the Mount Whitney area is the same as the global average of −6.5°C/km. What is the temperature of the summit of Mount Whitney when eager hikers depart from the foothill location at a temperature of 30°C?

41. The theoretical lapse rate for dry air (no water vapor) in an atmosphere is given by

$$\frac{dT}{dy} = -\frac{\gamma - 1}{\gamma}\frac{gM}{R}$$

where g is the acceleration due to gravity, M is the molar mass of the uniform ideal gas in the atmosphere, R is the gas constant, and γ is the ratio of molar specific heats, which we will study in Chapter 17. (a) Calculate the theoretical lapse rate on the Earth given that $\gamma = 1.40$ and the effective molar mass of air is 28.9 g/mol. (b) Why is this value different from the value of −6.5°C/km given in the text? (c) The atmosphere of Mars is mostly dry carbon dioxide, with a molar mass of 44.0 g/mol and a ratio of molar specific heats of $\gamma = 1.30$. The mass of Mars is 6.42×10^{23} kg and the radius is 3.37×10^6 m. What is the lapse rate for the Martian troposphere? (d) A typical surface atmospheric temperature on Mars is −40.0°C. Using the lapse rate calculated in part (c), find the height in the Martial troposphere at which the temperature is −60.0°C.

(e) Data from the *Mariner* flights in 1969 indicated a lapse rate in the Martian troposphere of about $-1.5°C/km$. The *Viking* missions in 1976 gave measured lapse rates of about $-2°C/km$. These values deviate from the ideal value calculated in part (c) because of *dust* in the Martian atmosphere. Why would dust affect the lapse rate? Which mission occurred in dustier conditions, *Mariner* or *Viking*?

Additional Problems

42. A student measures the length of a brass rod with a steel tape at 20.0°C. The reading is 95.00 cm. What will the tape indicate for the length of the rod when the rod and the tape are at (a) $-15.0°C$ and (b) 55.0°C?

43. The density of gasoline is 730 kg/m^3 at 0°C. Its average coefficient of volume expansion is 9.60×10^{-4} (°C)$^{-1}$. If 1.00 gal of gasoline occupies 0.003 80 m^3, how many extra kilograms of gasoline would you get if you bought 10.0 gal of gasoline at 0°C rather than at 20.0°C from a pump that is not temperature compensated?

44. A liquid with a coefficient of volume expansion β just fills a spherical shell of volume V_i at a temperature of T_i (Fig. P16.44). The shell is made of a material that has an average coefficient of linear expansion α. The liquid is free to expand into an open capillary of area A projecting from the top of the sphere. (a) The temperature increases by ΔT. Show that the liquid rises in the capillary by the amount Δh given by the equation $\Delta h = (V_i/A)(\beta - 3\alpha)\,\Delta T$. (b) For a typical system, such as a mercury thermometer, why is it a good approximation to ignore the expansion of the shell?

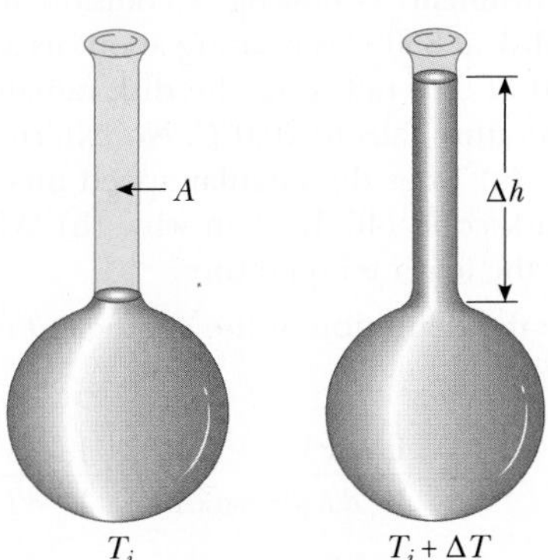

FIGURE **P16.44**

45. A liquid has a density ρ. (a) Show that the fractional change in density for a change in temperature ΔT is $\Delta\rho/\rho = -\beta\,\Delta T$. What does the negative sign signify? (b) Fresh water has a maximum density of 1.000 0 g/cm^3 at 4.0°C. At 10.0°C, its density is 0.999 7 g/cm^3. What is β for water over this temperature interval?

46. Long-term space missions require reclamation of the oxygen in the carbon dioxide exhaled by the crew. In one method of reclamation, 1.00 mol of carbon dioxide produces 1.00 mol of oxygen and 1.00 mol of methane as a byproduct. The methane is stored in a tank under pressure and is available to control the attitude of the spacecraft by controlled venting. A single astronaut exhales 1.09 kg of carbon dioxide each day. If the methane generated in the respiration recycling of three astronauts during one week of flight is stored in an originally empty 150-L tank at $-45.0°C$, what is the final pressure in the tank?

47. **Physics Now™** A vertical cylinder of cross-sectional area A is fitted with a tight-fitting, frictionless piston of mass m (Fig. P16.47). (a) If n moles of an ideal gas are in the cylinder at a temperature of T, what is the height h at which the piston is in equilibrium under its own weight? (b) What is the value for h if $n = 0.200$ mol, $T = 400$ K, $A = 0.008\ 00$ m^2, and $m = 20.0$ kg?

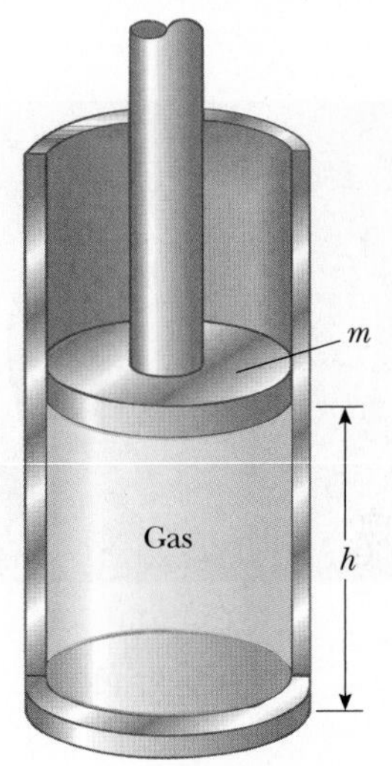

FIGURE **P16.47**

48. A bimetallic strip is made of two ribbons of dissimilar metals bonded together. (a) First assume that the strip is originally straight. As the metals are heated, the metal with the greater average coefficient of expansion expands more than the other, forcing the strip into an arc, with the outer radius having a greater circumference (Fig. P16.48a). Derive an expression for the angle of bending θ as a function of the initial length of the strips, their average coefficients of linear expansion, the change in temperature, and the separation of the centers of the strips ($\Delta r = r_2 - r_1$). (b) Show that the angle of bending decreases to zero when ΔT decreases to zero and also when the two average coefficients of expansion become equal. (c) What happens if the strip is cooled? (d) Figure P16.48b shows a compact spiral bimetallic strip in a home thermostat. The equation from part (a) applies to it as well if θ is interpreted as the angle of additional bending caused by a change in temperature. The inner end of the spiral strip is fixed and the outer end is free to move. Assume that the metals are bronze and invar, the thickness of the strip is $2\Delta r = 0.500$ mm, and the overall length of the spiral strip is 20.0 cm. Find the angle through which the free end of the strip turns when the temperature changes by 1°C. The free end of the strip supports a capsule partly filled with mercury, visible above the strip in Figure P16.48b. When the capsule tilts, the mercury shifts from one end to the other, to make or break an electrical contact switching the furnace on or off.

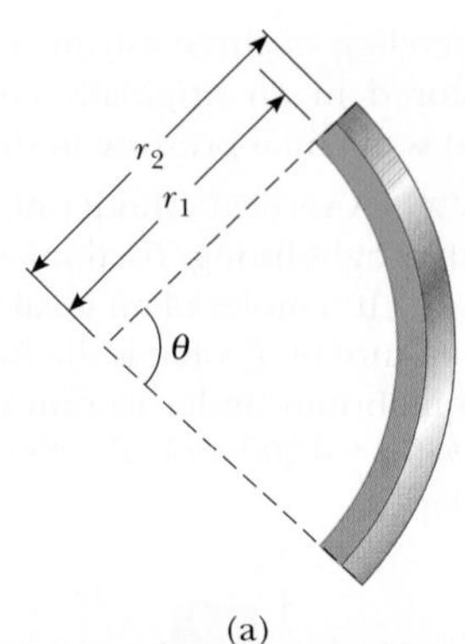

(a)

(b)

FIGURE P16.48

49. The rectangular plate shown in Figure P16.49 has an area A_i equal to ℓw. If the temperature increases by ΔT, each dimension increases according to the equation $\Delta L = \alpha L_i \Delta T$, where α is the average coefficient of linear expansion. Show that the increase in area is $\Delta A = 2\alpha A_i \Delta T$. What approximation does this expression assume?

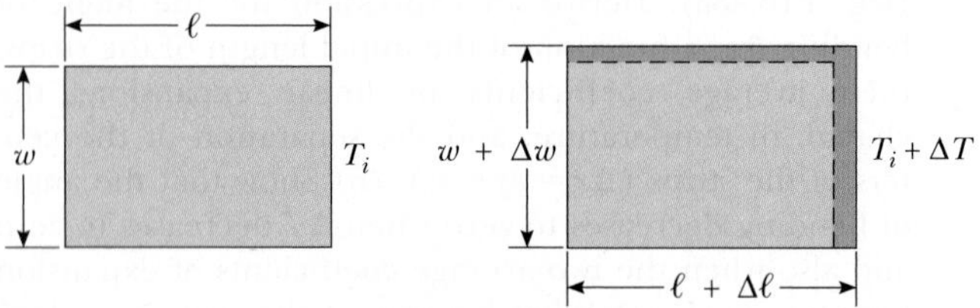

FIGURE P16.49

50. **Review problem.** A clock with a brass pendulum has a period of 1.000 s at 20.0°C. If the temperature increases to 30.0°C, (a) by how much does the period change and (b) how much time does the clock gain or lose in one week?

51. **Review problem.** Consider an object with any one of the shapes displayed in Table 10.2. What is the percentage increase in the moment of inertia of the object when it is heated from 0°C to 100°C if it is composed of (a) copper or (b) aluminum? Assume that the average linear expansion coefficients shown in Table 16.1 do not vary between 0°C and 100°C.

52. (a) Derive an expression for the buoyant force on a spherical balloon, submerged in water, as a function of the depth below the surface, the volume of the balloon at the surface, the pressure at the surface, and the density of the water. (Assume that the water temperature does not change with depth.) (b) Does the buoyant force increase or decrease as the balloon is submerged? (c) At what depth is the buoyant force one-half the surface value?

53. Two concrete spans of a 250-m-long bridge are placed end to end so that no room is allowed for expansion (Fig. P16.53a). If a temperature increase of 20.0°C occurs, what is the height y to which the spans rise when they buckle (Fig. P16.53b)?

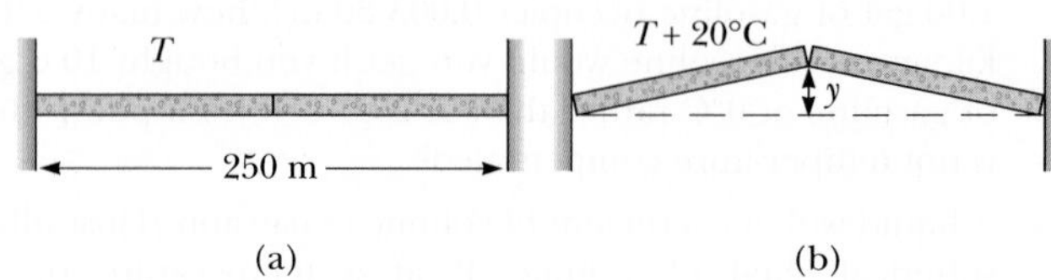

FIGURE P16.53 Problems 16.53 and 16.54.

54. Two concrete spans of a bridge of length L are placed end to end so that no room is allowed for expansion (Fig. P16.53a). If a temperature increase of ΔT occurs, what is the height y to which the spans rise when they buckle (Fig. P16.53b)?

55. **Review problem.** Following a collision in outer space, a copper disk at 850°C is rotating about its axis with an angular speed of 25.0 rad/s. As the disk radiates infrared light, its temperature falls to 20.0°C. No external torque acts on the disk. (a) Does the angular speed increase or decrease as the disk cools off? Explain why. (b) What is its angular speed at the lower temperature?

56. (a) Take the definition of the coefficient of volume expansion to be

$$\beta = \frac{1}{V}\frac{dV}{dT}\bigg|_{P=\text{constant}} = \frac{1}{V}\frac{\partial V}{\partial T}$$

Use the equation of state for an ideal gas to show that the coefficient of volume expansion for an ideal gas at constant pressure is given by $\beta = 1/T$, where T is the absolute temperature. (b) What value does this expression predict for β at 0°C? Compare this result with the experimental values for helium and air in Table 16.1. Note that these are much larger than the coefficients of volume expansion for most liquids and solids.

57. (a) Show that the density of an ideal gas occupying a volume V is given by $\rho = PM/RT$, where M is the molar mass. (b) Determine the density of oxygen gas at atmospheric pressure and 20.0°C.

58. A cylinder that has a 40.0-cm radius and is 50.0 cm deep is filled with air at 20.0°C and 1.00 atm (Fig. P16.58a). A 20.0-kg piston is now lowered into the cylinder, compressing the air trapped inside (Fig. P16.58b). Finally, a 75.0-kg

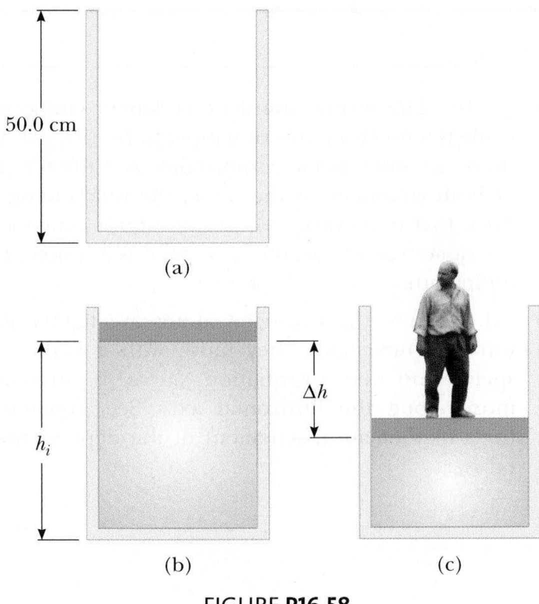

FIGURE P16.58

man stands on the piston, further compressing the air, which remains at 20°C (Fig. P16.58c). (a) How far down (Δh) does the piston move when the man steps onto it? (b) To what temperature should the gas be heated to raise the piston and man back to h_i?

59. In a chemical processing plant, a reaction chamber of fixed volume V_0 is connected to a reservoir chamber of fixed volume $4V_0$ by a passage containing a thermally insulating porous plug. The plug permits the chambers to be at different temperatures. The plug allows gas to pass from either chamber to the other, ensuring that the pressure is the same in both. At one point in the processing, both chambers contain gas at a pressure of 1.00 atm and a temperature of 27.0°C. Intake and exhaust valves to the pair of chambers are closed. The reservoir is maintained at 27.0°C while the reaction chamber is heated to 400°C. What is the pressure in both chambers after that is done?

60. A 1.00-km steel railroad rail is fastened securely at both ends when the temperature is 20.0°C. As the temperature increases, the rail begins to buckle. Assuming that its shape is an arc of a vertical circle, find the height h of the center of the rail when the temperature is 25.0°C. You will need to solve a transcendental equation.

61. **Review problem.** A perfectly plane house roof makes an angle θ with the horizontal. When its temperature changes, between T_c before dawn each day to T_h in the middle of each afternoon, the roof expands and contracts uniformly with a coefficient of thermal expansion α_1. Resting on the roof is a flat, rectangular, metal plate with expansion coefficient α_2, greater than α_1. The length of the plate is L, measured up the slope of the roof. The component of the plate's weight perpendicular to the roof is supported by a normal force uniformly distributed over the area of the plate. The coefficient of kinetic friction between the plate and the roof is μ_k. The plate is always at the same temperature as the roof, so we assume that its temperature is continuously changing. Because of the difference in expansion coefficients, each bit of the plate is moving relative to the roof below it, except for points along a certain horizontal line running across the plate, called the stationary line. If the temperature is rising, parts of the plate below the stationary line are moving down relative to the roof, and feel a force of kinetic friction acting up the roof. Elements of area above the stationary line are sliding up the roof, and on them kinetic friction acts downward parallel to the roof. The stationary line occupies no area, so we assume that no force of static friction acts on the plate while the temperature is changing. The plate as a whole is very nearly in equilibrium, so the net frictional force on it must be equal to the component of its weight acting down the incline. (a) Prove that the stationary line is at a distance of

$$\frac{L}{2}\left(1-\frac{\tan\theta}{\mu_k}\right)$$

below the top edge of the plate. (b) Analyze the forces that act on the plate when the temperature is falling and prove that the stationary line is at that same distance above the bottom edge of the plate. (c) Show that the plate steps down the roof like an inchworm, moving each day by the distance

$$L(\alpha_2-\alpha_1)(T_h-T_c)\frac{\tan\theta}{\mu_k}$$

(d) Evaluate the distance an aluminum plate moves each day if its length is 1.20 m, if the temperature cycles between 4.00°C and 36.0°C, and if the roof has slope 18.5°, coefficient of linear expansion 1.50×10^{-5} $(°C)^{-1}$, and coefficient of friction 0.420 with the plate. (e) What if the expansion coefficient of the plate is less than that of the roof? Will the plate creep up the roof?

62. **Review problem.** Oxygen at pressures much greater than 1 atm is toxic to lung cells. Assume that a deep-sea diver breathes a mixture of oxygen (O_2) and helium (He). By weight, what ratio of helium to oxygen must be used if the diver is at an ocean depth of 50.0 m?

63. For a Maxwellian gas, use a computer or programmable calculator to find the numerical value of the ratio $N_v(v)/N_v(v_{mp})$ for the following values of v: $v = (v_{mp}/50)$, $(v_{mp}/10)$, $(v_{mp}/2)$, v_{mp}, $2v_{mp}$, $10v_{mp}$, $50v_{mp}$. Give your results to three significant figures.

64. A vessel contains 1.00×10^4 oxygen molecules at 500 K. (a) Make an accurate graph of the Maxwell–Boltzmann speed distribution function versus speed with points at speed intervals of 100 m/s. (b) Determine the most probable speed from this graph. (c) Calculate the average and rms speeds for the molecules and label these points on your graph. (d) From the graph, estimate the fraction of molecules with speeds in the range 300 m/s to 600 m/s.

ANSWERS TO QUICK QUIZZES

16.1 (c). A ratio of temperatures must be calculated using temperatures on the Kelvin scale. When the given temperatures are converted to kelvins, only those in part (c) are in the correct ratio.

16.2 (c). A cavity in a material expands in the same way as if it were filled with material.

16.3 (a). On a cold day, the trapped air in the bubbles is reduced in volume, according to the ideal gas law. Therefore, the smaller bubbles can allow the package contents to shift more.

16.4 (b). Because of the increased temperature, the air expands. Consequently, some of the air leaks to the outside, leaving less air in the house.

16.5 (i), (b). The average translational kinetic energy per molecule is a function only of temperature. (ii), (a). Because there are twice as many molecules and the temperature of both containers is the same, the total energy in B is twice that in A. (iii), (b). Because both containers hold the same type of gas, the rms speed is a function only of temperature.

16.6 (a). Because the hydrogen atoms are lighter than the nitrogen molecules, they move with a higher average speed and the distribution curve is stretched out more along the horizontal axis. See Equation 16.20 for a mathematical statement of the dependence of N_v on m_0.

CHAPTER 17

Energy in Thermal Processes: The First Law of Thermodynamics

(Jon Eisberg/Taxi/Getty)

In this photograph of Moraine Lake in Banff National Park, Alberta, we see evidence of water in all three phases. In the lake is liquid water, and solid water in the form of snow appears on the mountains. The clouds in the sky consist of liquid water droplets that have condensed from the gaseous water vapor in the air. Changes of a substance from one phase to another are a result of energy transfer.

CHAPTER OUTLINE

In Chapters 6 and 7, we introduced the relationship between energy in mechanics and energy in thermodynamics. We discussed the transformation of mechanical energy to internal energy in cases in which a nonconservative force such as friction is acting. In Chapter 16, we discussed additional concepts of the relationship between internal energy and temperature. In this chapter, we extend these discussions into a complete treatment of energy in thermal processes.

Until around 1850, the fields of thermodynamics and mechanics were considered to be two distinct branches of science, and the law of conservation of energy seemed to describe only certain kinds of mechanical systems. Mid-19th-century experiments performed by English physicist James Joule (1818–1889) and others showed that energy may enter or leave a system by heat and by work. Today, as we discussed in Chapter 6, internal energy is treated as a form of energy that can be transformed into mechanical energy and vice versa. Once the concept of energy was broadened to include internal energy, the law of conservation of energy emerged as a universal law of nature.

(By kind permission of the President and Council of the Royal Society)

JAMES PRESCOTT JOULE (1818–1889)

Joule, an English physicist, received some formal education in mathematics, philosophy, and chemistry but was in large part self-educated. His research led to the establishment of the principle of conservation of energy. His study of the quantitative relationship among electrical, mechanical, and chemical effects of heat culminated in his announcement in 1843 of the amount of work required to produce a unit of internal energy, called the mechanical equivalent of heat.

This chapter focuses on developing the concept of heat, extending our concept of work to thermal processes, introducing the first law of thermodynamics, and investigating some important applications.

17.1 HEAT AND INTERNAL ENERGY

A major distinction must be made between internal energy and heat because these terms tend to be used interchangeably in everyday communication. You should read the following descriptions carefully and try to use these terms correctly because they are not interchangeable. They have very different meanings.

We introduced internal energy in Chapter 6, and we formally define it here:

Internal energy E_{int} is the energy associated with the microscopic components of a system—atoms and molecules—when viewed from a reference frame at rest with respect to the system. It includes kinetic and potential energy associated with the random translational, rotational, and vibrational motion of the atoms or molecules that make up the system as well as intermolecular potential energy.

In Chapter 16, we showed that the internal energy of a monatomic ideal gas is associated with the translational motion of its atoms. In this special case, the internal energy is simply the total translational kinetic energy of the atoms; the higher the temperature of the gas, the greater the kinetic energy of the atoms and the greater the internal energy of the gas. For more complicated diatomic and polyatomic gases, internal energy includes other forms of molecular energy, such as rotational kinetic energy and the kinetic and potential energy associated with molecular vibrations.

Heat was introduced in Chapter 6 as one possible method of energy transfer, and we provide a formal definition here:

Heat is a mechanism by which energy is transferred between a system and its environment because of a temperature difference between them. It is also the amount of energy Q transferred by this mechanism.

PITFALL PREVENTION 17.1

HEAT, TEMPERATURE, AND INTERNAL ENERGY ARE DIFFERENT As you read the newspaper or listen to the radio, be alert for incorrectly used phrases including the word *heat* and think about the proper word to be used in place of it. "As the truck braked to a stop, a large amount of heat was generated by friction" and "The heat of a hot summer day . . ." are two examples.

Figure 17.1 shows a pan of water in contact with a gas flame. Energy enters the water by heat from the hot gases in the flame, and the internal energy of the water increases as a result. It is *incorrect* to say that the water has more heat as time goes by.

(Charles D. Winters)

FIGURE 17.1 A pan of boiling water is warmed by a gas flame. Energy enters the water through the bottom of the pan by heat.

As further clarification of the use of the word *heat,* consider the distinction between work and energy. The work done on (or by) a system is a measure of the amount of energy transferred between the system and its surroundings, whereas the mechanical energy of the system (kinetic or potential) is a consequence of its motion and coordinates. Thus, when a person does work on a system, energy is transferred from the person to the system. It makes no sense to talk about the work *in* a system; one refers only to the work done *on* or *by* a system when some process has occurred in which energy has been transferred to or from the system. Likewise, it makes no sense to use the term *heat* unless energy has been transferred as a result of a temperature difference.

Units of Heat

Early in the development of thermodynamics, before scientists recognized the connection between thermodynamics and mechanics, heat was defined in terms of the temperature changes it produced in an object, and a separate unit of energy, the calorie, was used for heat. The **calorie** (cal) was defined as **the heat necessary to**

raise the temperature of 1 g of water[1] **from 14.5°C to 15.5°C.** (The "Calorie," with a capital C, used in describing the energy content of foods, is actually a kilocalorie.) Likewise, the unit of heat in the U.S. customary system, the **British thermal unit** (Btu), was defined as **the heat required to raise the temperature of 1 lb of water from 63°F to 64°F.**

In 1948, scientists agreed that because heat (like work) is a measure of the transfer of energy, its SI unit should be the joule. The calorie is now defined to be exactly 4.186 J:

$$1 \text{ cal} \equiv 4.186 \text{ J} \qquad [17.1]$$

■ Mechanical equivalent of heat

Note that this definition makes no reference to the heating of water. The calorie is a general energy unit. We could have used it in Chapter 6 for the kinetic energy of an object, for example. It is introduced here for historical reasons, but we shall make little use of it as an energy unit. The definition in Equation 17.1 is known as the **mechanical equivalent of heat.**

EXAMPLE 17.1 Losing Weight the Hard Way

A student eats a dinner containing 2 000 Calories of energy. He wishes to do an equivalent amount of work in the gymnasium by lifting a 50.0-kg object. How many times must he raise the object to expend this much energy? Assume that he raises it a distance of 2.00 m each time.

Solution The student desires to transfer 2 000 Calories of energy from his body by doing work on the object–Earth system. Because 1 Calorie = 1.00×10^3 cal, the total work required is 2.00×10^6 cal. Converting to joules, we have for the total work required

$$W = (2.00 \times 10^6 \text{ cal})(4.186 \text{ J/cal}) = 8.37 \times 10^6 \text{ J}$$

The work done in lifting the object of mass m a distance h is equal to mgh, and the work done in lifting it n times is $nmgh$. We equate this expression to the total work required:

$$W = nmgh = 8.37 \times 10^6 \text{ J}$$

Solving for n,

$$n = \frac{8.37 \times 10^6 \text{ J}}{(50.0 \text{ kg})(9.80 \text{ m/s}^2)(2.00 \text{ m})} = 8.54 \times 10^3 \text{ times}$$

If the student is in good shape and lifts the weight once every 5 s, it will take him about 12 h to perform this feat. Clearly, it is much easier to lose weight by dieting than by lifting weights!

In reality, the human body is not 100% efficient. Thus, not all the energy transformed within the body from the dinner transfers out of the body by work done on the barbell. Some of this energy is used to pump blood and perform other functions within the body. Thus, the 2 000 Calories can be worked off in less time than 12 hours when these other energy requirements are included.

17.2 SPECIFIC HEAT

The definition of the calorie indicates the amount of energy necessary to raise the temperature of 1 g of a specific substance—water—by 1°C, which is 4.186 J. To raise the temperature of 1 kg of water by 1°C, we need to transfer 4 186 J of energy. The quantity of energy required to raise the temperature of 1 kg of an arbitrary substance by 1°C varies with the substance. For example, the energy required to raise the temperature of 1 kg of copper by 1°C is 387 J, which is significantly less than that required for water. Every substance requires a unique amount of energy per unit mass to change the temperature of that substance by 1°C.

[1]Originally, the calorie was defined as the heat necessary to raise the temperature of 1 g of water by 1°C at any initial temperature. Careful measurements, however, showed that the energy required depends somewhat on temperature; hence, a more precise definition evolved.

PITFALL PREVENTION 17.2

AN UNFORTUNATE CHOICE OF TERMINOLOGY The name *specific heat* is an unfortunate holdover from the days when thermodynamics and mechanics developed separately. A better name would be *specific energy transfer*, but the existing term is too entrenched to be replaced.

Suppose a quantity of energy Q is transferred to a mass m of a substance, thereby changing its temperature by ΔT. The **specific heat** c of the substance is defined as

$$c \equiv \frac{Q}{m\,\Delta T} \qquad [17.2]$$

The units of specific heat are joules per kilogram-degree Celsius, or J/kg · °C. Table 17.1 lists specific heats for several substances. From the definition of the calorie, the specific heat of water is 4 186 J/kg · °C.

From this definition, we can express the energy Q transferred between a system of mass m and its surroundings in terms of the resulting temperature change ΔT as

$$Q = mc\,\Delta T \qquad [17.3]$$

For example, the energy required to raise the temperature of 0.500 kg of water by 3.00°C is $Q = (0.500 \text{ kg})(4\,186 \text{ J/kg}\cdot{}^\circ\text{C})(3.00^\circ\text{C}) = 6.28 \times 10^3$ J. Note that when the temperature increases, ΔT and Q are taken to be *positive,* corresponding to energy flowing *into* the system. When the temperature decreases, ΔT and Q are *negative* and energy flows *out* of the system. These sign conventions are consistent with those in our discussion of the continuity equation for energy, Equation 6.20.

Table 17.1 shows that water has a high specific heat relative to most other common substances (the specific heats of hydrogen and helium are higher). The high specific heat of water is responsible for the moderate temperatures found in regions near large bodies of water. As the temperature of a body of water decreases during the winter, the water transfers energy to the air, which carries the energy landward when prevailing winds are toward the land. For example, the prevailing winds off the western coast of the United States are toward the land, and the energy liberated by the Pacific Ocean as it cools keeps coastal areas much warmer than they would be otherwise. Thus, the western coastal states generally have warmer

TABLE 17.1 Specific Heats of Some Substances at 25°C and Atmospheric Pressure

Substance	Specific Heat c J/kg · °C	Specific Heat c cal/g · °C
Elemental Solids		
Aluminum	900	0.215
Beryllium	1 830	0.436
Cadmium	230	0.055
Copper	387	0.092 4
Germanium	322	0.077
Gold	129	0.030 8
Iron	448	0.107
Lead	128	0.030 5
Silicon	703	0.168
Silver	234	0.056
Other Solids		
Brass	380	0.092
Glass	837	0.200
Ice (−5°C)	2 090	0.50
Marble	860	0.21
Wood	1 700	0.41
Liquids		
Alcohol (ethyl)	2 400	0.58
Mercury	140	0.033
Water (15°C)	4 186	1.00
Gas		
Steam (100°C)	2 010	0.48

winter weather than the eastern coastal states, where the winds do not transfer energy toward land.

That the specific heat of water is higher than that of sand accounts for the pattern of air flow at a beach. During the day, the Sun adds roughly equal amounts of energy to beach and water, but the lower specific heat of sand causes the beach to reach a higher temperature than the water. As a result, the air above the land reaches a higher temperature than the air above the water. The denser cold air pushes the less dense hot air upward (due to Archimedes's principle), which results in a breeze from ocean to land during the day. Because the hot air gradually cools as it rises, it subsequently sinks, setting up the circulating pattern shown in Figure 17.2. During the night, the sand cools more quickly than the water, and the circulating pattern reverses because the hotter air is now over the water. These offshore and onshore breezes are well known to sailors.

FIGURE 17.2 Circulation of air at the beach. On a hot day, the air above the warm sand warms faster than the air above the cooler water. The warmer air floats upward due to Archimedes's principle, resulting in the movement of cooler air toward the beach.

QUICK QUIZ 17.1 Imagine you have 1.00 kg each of iron, glass, and water and that all three samples are at 10.0°C. **(a)** Rank them from lowest to highest temperature after 100 J of energy is added to each. **(b)** Rank them from lowest to greatest amount of energy transfer by heat if each increases in temperature by 20.0°C.

Calorimetry

One technique for measuring the specific heat of a solid or liquid is to raise the temperature of the substance to some value, place it into a vessel containing water of known mass and temperature, and measure the temperature of the combination after equilibrium is reached. Let us define the system as the substance and the water. If the vessel is assumed to be a good insulator so that energy does not leave the system by heat (nor by any other means), we can use the isolated system model. A vessel having this property is called a **calorimeter,** and the analysis performed by using such a vessel is called **calorimetry.**

The principle of conservation of energy for this isolated system requires that the energy leaving by heat from the warmer substance (of unknown specific heat) equals the energy entering the water.[2] Thus, we can write

$$Q_{\text{cold}} = -Q_{\text{hot}} \qquad [17.4]$$

To see how to set up a calorimetry problem, suppose m_x is the mass of a substance whose specific heat we wish to determine, c_x its specific heat, and T_x its initial temperature. Let m_w, c_w, and T_w represent corresponding values for the water. If T is the final equilibrium temperature after the substance and the water are combined, from Equation 17.3 we find that the energy gained by the water is $m_w c_w(T - T_w)$ and that the energy lost by the substance of unknown specific heat is $m_x c_x(T - T_x)$. Substituting these values into Equation 17.4, we have

$$m_w c_w(T - T_w) = -m_x c_x(T - T_x)$$

Solving for c_x gives

$$c_x = \frac{m_w c_w(T - T_w)}{m_x(T_x - T)}$$

By substituting the known values in the right-hand side, we can calculate the specific heat of the substance.

PITFALL PREVENTION 17.3

Remember the Negative Sign It is *critical* to include the negative sign in Equation 17.4. The negative sign in the equation is necessary for consistency with our sign convention for energy transfer. The energy transfer Q_{hot} is negative because energy is leaving the hot substance. The negative sign in the equation ensures that the right-hand side is a positive number, consistent with the left-hand side, which is positive because energy is entering the cold substance.

PITFALL PREVENTION 17.4

Celsius versus Kelvin In equations in which T appears (e.g., the ideal gas law), the Kelvin temperature *must* be used. In equations involving ΔT, such as calorimetry equations, it is *possible* to use Celsius temperatures because a change in temperature is the same on both scales. It is *safest,* however, to use Kelvin temperatures *consistently* in all equations involving T or ΔT.

[2]For precise measurements, the container holding the water should be included in the calculations because it also changes temperature. Doing so would require a knowledge of its mass and composition. If, however, the mass of the water is large compared with that of the container, we can adopt a simplification model in which we ignore the energy gained by the container.

■ Thinking Physics 17.1

The equation $Q = mc\,\Delta T$ indicates the relationship between energy Q transferred to an object of mass m and specific heat c by means of heat and the resulting temperature change ΔT. In reality, the energy transfer on the left-hand side of the equation could be made by any method, not just heat. Give a few examples in which this equation could be used to calculate a temperature change of an object due to an energy transfer process other than heat.

Reasoning The following are a few of several possible examples.

> During the first few seconds after turning on a toaster, the temperature of the electrical coils rises. The transfer mechanism here is *electrical transmission* of energy through the power cord.
>
> The temperature of a potato in a microwave oven increases due to the absorption of microwaves. In this case, the energy transfer mechanism is by *electromagnetic radiation,* the microwaves.
>
> A carpenter attempts to use a dull drill bit to bore a hole in a piece of wood. The bit fails to make much headway but becomes very warm. The increase in temperature in this case is due to *work* done on the bit by the wood.

In each of these cases, as well as many other possibilities, the Q on the left of the equation of interest is not a measure of heat but, rather, is replaced with the energy transferred or transformed by other means. Even though heat is not involved, the equation can still be used to calculate the temperature change. ■

EXAMPLE 17.2 Cooling a Hot Ingot

The temperature of a 0.050 0-kg ingot of metal is raised to 200.0°C and the ingot is then dropped into a light, insulated beaker containing 0.400 kg of water initially at 20.0°C. If the final equilibrium temperature of the mixed system is 22.4°C, find the specific heat of the metal.

Solution Using Equations 17.3 and 17.4, we can write

$$Q_{\text{cold}} = -Q_{\text{hot}}$$

$$m_w c_w(T - T_w) = -m_x c_x(T - T_x)$$

Substituting numerical values, we have

$$(0.400\text{ kg})(4\,186\text{ J/kg}\cdot{}^\circ\text{C})(22.4^\circ\text{C} - 20.0^\circ\text{C}) = -(0.050\,0\text{ kg})(c_x)(22.4^\circ\text{C} - 200.0^\circ\text{C})$$

from which we find that

$$c_x = 453\text{ J/kg}\cdot{}^\circ\text{C}$$

The ingot is most likely iron, as can be seen by comparing this result with the data in Table 17.1.

17.3 LATENT HEAT AND PHASE CHANGES

A substance often undergoes a change in temperature when energy is transferred between the substance and its environment. In some situations, however, the transfer of energy does not result in a change in temperature. That can occur when the physical characteristics of the substance change from one form to another, commonly referred to as a **phase change.** Some common phase changes are solid to liquid (melting), liquid to gas (boiling), and a change in crystalline structure of a solid. All such phase changes involve a change in internal energy but no change in temperature. The energy that enters the substance during melting and boiling appears as increased intermolecular potential energy as bonds are broken rather than as an increase in random motion of the molecules.

In Chapter 11, we discussed two types of bound systems. An object is gravitationally bound to the Earth, and a certain amount of energy must be added to the object–Earth system to separate the Earth and the object by an infinite distance. Knowing this energy allowed us to calculate the escape speed (Section 11.4). The input of energy breaks the bond between the object and the Earth. In Section 11.5, we discussed the bound system of an electron and a proton in the hydrogen atom. The minimum energy required to separate the electron and the proton of a hydrogen atom was called the ionization energy. An input of energy of at least the ionization energy breaks the bond between the electron and the proton. The energy input during a phase change is similar to these examples. During phase changes, the energy added to the system of all the molecules of a substance modifies or breaks the bonds between the molecules.

The energy transfer required to change the phase of a given mass m of a pure substance is

$$Q = \pm mL \quad [17.5]$$

■ Latent heat

where L is called the **latent heat**[3] of the substance and depends on the nature of the phase change as well as on the substance. The proper sign in Equation 17.5 is chosen according to the direction of the energy flow. When an ice cube melts into water, we express Equation 17.5 as $Q = mL$, but in the case of liquid water freezing into ice, because we are removing energy from the water, we express Equation 17.5 with the negative sign: $Q = -mL$.

Heat of fusion L_f is the term used when the phase change occurs during melting or freezing, and **heat of vaporization** L_v is the term used when the phase change occurs during boiling or condensing. For example, the latent heat of fusion for water at atmospheric pressure is 3.33×10^5 J/kg, and the latent heat of vaporization of water is 2.26×10^6 J/kg. The latent heats of different substances vary considerably, as is seen in Table 17.2.

The latent heat of fusion is the energy required to modify all the intermolecular bonds in 1 kg of a substance so as to convert the solid phase to the liquid phase. The latent heat of vaporization is the energy that must be added to 1 kg of the liquid phase of a substance to break all the liquid bonds so as to form a gas.

As you can see from Table 17.2, the latent heat of vaporization for a given substance is usually larger than the latent heat of fusion. In the change from solid to liquid phase, solid bonds between molecules are transformed into somewhat weaker liquid bonds. In the change from liquid to gas phase, however, liquid bonds

PITFALL PREVENTION 17.5

SIGNS ARE CRITICAL Sign errors occur very often when students perform calorimetry equations, so we will make this point once again. For phase changes, use the correct explicit sign in Equation 17.5, depending on whether you are adding or removing energy from the substance. Equation 17.3 has no explicit sign to consider, but be sure that your ΔT is *always* the final temperature minus the initial temperature. In addition, make sure that you *always* include the negative sign on the right-hand side of Equation 17.4.

TABLE 17.2 Latent Heats of Fusion and Vaporization

Substance	Melting Point (°C)	Latent Heat of Fusion (J/kg)	Boiling Point (°C)	Latent Heat of Vaporization (J/kg)
Helium	−269.65	5.23×10^3	−268.93	2.09×10^4
Nitrogen	−209.97	2.55×10^4	−195.81	2.01×10^5
Oxygen	−218.79	1.38×10^4	−182.97	2.13×10^5
Ethyl alcohol	−114	1.04×10^5	78	8.54×10^5
Water	0.00	3.33×10^5	100.00	2.26×10^6
Sulfur	119	3.81×10^4	444.60	3.26×10^5
Lead	327.3	2.45×10^4	1 750	8.70×10^5
Aluminum	660	3.97×10^5	2 450	1.14×10^7
Silver	960.80	8.82×10^4	2 193	2.33×10^6
Gold	1 063.00	6.44×10^4	2 660	1.58×10^6
Copper	1 083	1.34×10^5	1 187	5.06×10^6

[3]The word *latent* is from the Latin *latere,* meaning hidden or concealed. Notice that this phrase uses the term *heat* incorrectly, but we will still use it because it is well ingrained in the terminology of physics.

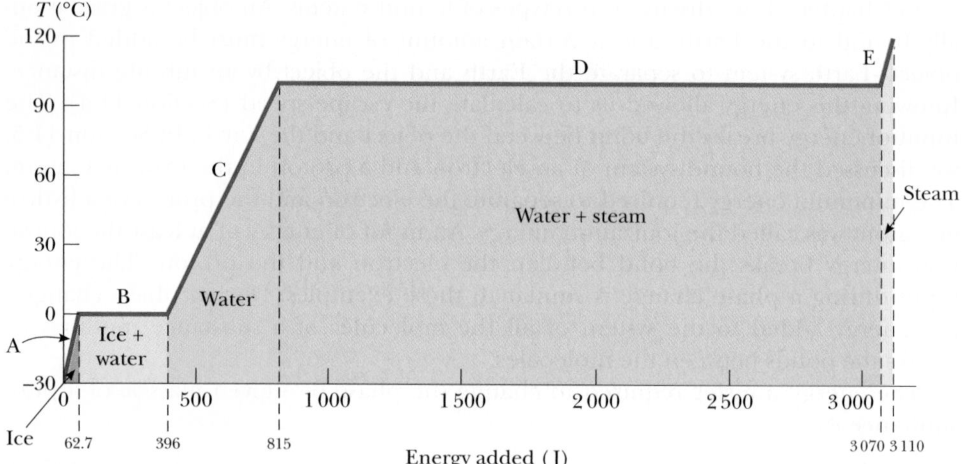

FIGURE 17.3 A plot of temperature versus energy added when 1 g of ice initially at −30.0°C is converted to steam at 120.0°C.

are broken, creating a situation in which the molecules of the gas have essentially no bonding to one another. More energy is therefore required to vaporize a given mass of substance than to melt it.

With our knowledge of latent heat, we can understand the full behavior of a substance as energy is added to it. Consider, for example, the addition of energy to a system of a 1-g block of ice at − 30.0°C in a container held at constant pressure. Suppose this energy results in the ice turning to steam (water vapor) at 120.0°C. Figure 17.3 indicates the experimental measurement of temperature as energy is added to the system. Let us examine each portion of the curve separately.

Part A During this portion of the curve, the temperature of the ice changes from − 30.0°C to 0.0°C. Because the specific heat of ice is 2 090 J/kg · °C, we can calculate the amount of energy added from Equation 17.3:

$$Q = mc_{\text{ice}}\,\Delta T = (1.00 \times 10^{-3}\text{ kg})(2\,090\text{ J/kg}\cdot{}^\circ\text{C})(30.0^\circ\text{C}) = 62.7\text{ J}$$

The transferred energy appears in the system as internal energy associated with random motion of the molecules of the ice, as represented by the increasing temperature of the ice.

Part B When the ice reaches 0.0°C, the ice–water mixture remains at this temperature—even though energy is being added—until all the ice melts to become water at 0.0°C. According to Equation 17.5, the energy required to melt 1.00 g of ice at 0.0°C is

$$Q = mL_f = (1.00 \times 10^{-3}\text{ kg})(3.33 \times 10^5\text{ J/kg}) = 333\text{ J}$$

During this process, the transferred energy appears in the system as internal energy associated with the increase in intermolecular potential energy as the bonds between water molecules in the ice break.

Part C Between 0.0°C and 100.0°C, no phase change occurs. The energy added to the water is used to increase its temperature, as it was in part A. The amount of energy necessary to increase the temperature from 0.0°C to 100.0°C is

$$Q = mc_{\text{water}}\,\Delta T = (1.00 \times 10^{-3}\text{ kg})(4.19 \times 10^3\text{ J/kg}\cdot{}^\circ\text{C})(100.0^\circ\text{C}) = 419\text{ J}$$

Part D At 100.0°C, another phase change occurs as the water changes to steam at 100.0°C. Similar to the behavior in part B, the water–steam mixture remains at a constant temperature, this time at 100.0°C—and the energy goes into breaking bonds so that the gas molecules move far apart—until all the liquid has been converted to steam. The energy required to convert 1.00 g of water to steam at 100.0°C is

$$Q = mL_v = (1.00 \times 10^{-3}\text{ kg})(2.26 \times 10^6\text{ J/kg}) = 2.26 \times 10^3\text{ J}$$

Part E On this portion of the curve, as in parts A and C, no phase change occurs, so all energy added is used to increase the temperature of the steam. The energy that must be added to raise the temperature of the steam to 120.0°C is

$$Q = mc_{\text{steam}}\,\Delta T = (1.00 \times 10^{-3}\text{ kg})(2.01 \times 10^{3}\text{ J/kg}\cdot{}^{\circ}\text{C})(20.0^{\circ}\text{C}) = 40.2\text{ J}$$

The total amount of energy that must be added to change 1 g of ice at −30.0°C to steam at 120.0°C is the sum of the results from all five parts of the curve, 3.11×10^{3} J. Conversely, to cool 1 g of steam at 120.0°C to ice at −30.0°C, we must remove 3.11×10^{3} J of energy.

QUICK QUIZ 17.2 Suppose the same process of adding energy to the ice cube is performed but we graph the internal energy of the system as a function of energy input. What would this graph look like?

QUICK QUIZ 17.3 Calculate the slopes for the A, C, and E portions of Figure 17.3. Rank the slopes from least steep to steepest, and explain what this ordering means.

EXAMPLE 17.3 **Boiling Liquid Helium**

Liquid helium has a very low boiling point, 4.2 K, and a very low heat of vaporization, 2.09×10^{4} J/kg (see Table 17.2). Energy is transferred at a constant rate of 10.0 W to a container of liquid helium from an immersed electric heater. At this rate, how long does it take to vaporize 1.00 kg of liquid helium?

Solution We divide Equation 17.5 by our desired time interval and recognize the ratio of the energy transfer to the time interval as the power:

$$Q = mL_v \quad \rightarrow \quad \mathcal{P} = \frac{Q}{\Delta t} = \frac{mL_v}{\Delta t}$$

Solving for the time interval, we have

$$\Delta t = \frac{mL_v}{\mathcal{P}}$$

Now, substituting the numerical values gives

$$\Delta t = \frac{(1.00\text{ kg})(2.09 \times 10^{4}\text{ J/kg})}{10.0\text{ W}} = 2.09 \times 10^{3}\text{ s}$$

$$\approx 35\text{ min}$$

17.4 WORK IN THERMODYNAMIC PROCESSES

In the macroscopic approach to thermodynamics, we describe the *state* of a system with such quantities as pressure, volume, temperature, and internal energy. As a result, these quantities belong to a category called **state variables.** For any given condition of the system, we can identify values of the state variables. It is important to note, however, that a macroscopic state of a system can be specified only if the system is in internal thermal equilibrium. In the case of a gas in a container, internal thermal equilibrium requires that every part of the gas be at the same pressure and temperature. If the temperature varies from one part of the gas to another, for example, we cannot specify a single temperature for the entire gas to be used in the ideal gas law.

▪ State variables

A second category of variables in situations involving energy is **transfer variables.** These variables only have a value if a process occurs in which energy is transferred across the boundary of the system. Because a transfer of energy across the boundary represents a change in the system, transfer variables are not associated with a given state of the system, but with a *change* in the state of the system. In the previous sections, we discussed heat as a transfer variable. For a given set of conditions of a system, the heat has no defined value. We can assign a value to the heat only if energy crosses the boundary by heat, resulting in a change in the system.

▪ Transfer variables

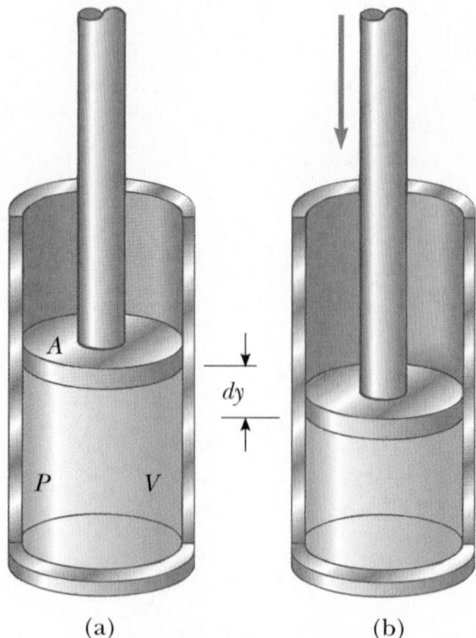

FIGURE 17.4 Work is done on a gas contained in a cylinder at pressure P as the piston is pushed downward so that the gas is compressed.

State variables are characteristic of a system in internal thermal equilibrium. Transfer variables are characteristic of a process in which energy is transferred between a system and its environment.

We have seen this notion before, but we have not used the language of state variables and transfer variables. In the continuity equation for energy, $\Delta E_{\text{system}} = \Sigma T$, we can identify the terms on the right-hand side as transfer variables: work, heat, mechanical waves, matter transfer, electromagnetic radiation, and electrical transmission. The left side of the continuity equation represents *changes* in state variables: kinetic energy, potential energy, and internal energy. For a gas, we have additional state variables, such as pressure, volume, and temperature.

In this section, we study an important transfer variable for thermodynamic systems, work. Work performed on particles was studied extensively in Chapter 6, and here we investigate the work done on a deformable system, a gas. Consider a gas contained in a cylinder fitted with a frictionless, movable piston of face area A (Fig. 17.4) and in thermal equilibrium. The gas occupies a volume V and exerts a uniform pressure P on the cylinder walls and the piston. Now let us adopt a simplification model in which the gas is compressed in a **quasi-static process,** that is, slowly enough to allow the system to remain in thermal equilibrium at all times. As the piston is pushed inward by an external force $\vec{\mathbf{F}}_{\text{ext}}$ through a displacement $d\vec{\mathbf{r}} = dy\hat{\mathbf{j}}$ (Fig. 17.4b), the work done on the gas is, according to our definition of work in Chapter 6,

$$dW = \vec{\mathbf{F}}_{\text{ext}} \cdot d\vec{\mathbf{r}} = \vec{\mathbf{F}}_{\text{ext}} \cdot dy\hat{\mathbf{j}}$$

Because the piston is in equilibrium at all times during the process, the external force has the same magnitude as the force exerted by the gas but is in the opposite direction:

$$\vec{\mathbf{F}}_{\text{ext}} = -\vec{\mathbf{F}}_{\text{gas}} = -PA\hat{\mathbf{j}}$$

where we have set the magnitude of the force exerted by the gas equal to PA. The work done by the external force can now be expressed as

$$dW = -PA\hat{\mathbf{j}} \cdot dy\hat{\mathbf{j}} = -PA\,dy$$

Because $A\,dy$ is the change in volume of the gas dV, we can express the work done *on* the gas as

$$dW = -P\,dV \qquad [17.6]$$

If the gas is compressed, dV is negative and the work done on the gas is positive. If the gas expands, dV is positive and the work done on the gas is negative. If the volume remains constant, the work done on the gas is zero. The total **work** done on the gas as its volume changes from V_i to V_f is given by the integral of Equation 17.6:

$$W = -\int_{V_i}^{V_f} P\,dV \qquad [17.7]$$

To evaluate this integral, one must know how the pressure varies with volume during the expansion process.

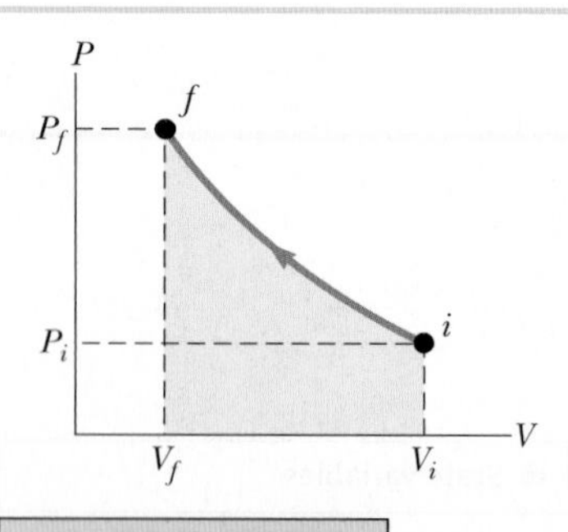

ACTIVE FIGURE 17.5
A gas is compressed quasi-statically from state i to state f. The work done on the gas equals the negative of the area under the PV curve. Note that the area is negative because $V_f < V_i$, so the work done on the gas is positive.

PhysicsNow™ Log into PhysicsNow at www.pop4e.com and go to Active Figure 17.5. You can compress the piston in Figure 17.4 and see the result on the PV diagram of this figure.

In general, the pressure is not constant during a process that takes a gas from its initial state to its final state but depends on the volume and temperature. If the pressure and volume are known at each step of the process, the state of the gas at each step can be plotted on a specialized graphical representation—a ***PV* diagram,** as in Active Figure 17.5—that is very important in thermodynamics. This type of diagram allows us to visualize a process through which a gas is progressing. The curve on such a graphical representation is called the *path* taken between the initial and final states.

Considering the integral in Equation 17.7 and recognizing the significance of the integral as an area under a curve, we can identify an important use for PV diagrams:

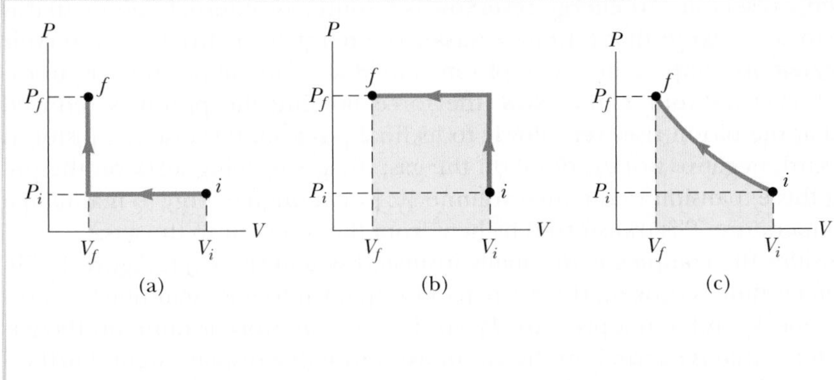

ACTIVE FIGURE 17.6 The work done on a gas as it is taken from an initial state to a final state depends on the path between these states.

Physics Now™ Log into PhysicsNow at **www.pop4e.com** and go to Active Figure 17.6. You can choose one of the three paths and see the movement of the piston in Figure 17.4 and of a point on the *PV* diagram of this figure.

The work done on a gas in a quasi-static process that takes the gas from an initial state to a final state is the negative of the area under the curve on a *PV* diagram, evaluated between the initial and final states.

For our process of compressing a gas in the cylinder, as Active Figure 17.5 suggests, the work done depends on the particular path taken between the initial and final states. To illustrate this important point, consider several different paths connecting i and f (Active Fig. 17.6). In the process depicted in Active Figure 17.6a, the volume of the gas is first reduced from V_i to V_f at constant pressure P_i and the pressure of the gas then increases from P_i to P_f by heating at constant volume V_f. The work done on the gas along this path is $-P_i(V_f - V_i)$. In Active Figure 17.6b, the pressure of the gas is increased from P_i to P_f at constant volume V_i and then the volume of the gas is reduced from V_i to V_f at constant pressure P_f. The work done on the gas along this path is $-P_f(V_f - V_i)$, which is greater in magnitude than that for the process described in Active Figure 17.6a because the piston is displaced through the same distance by a larger force than for the situation in Active Figure 17.6a. Finally, for the process described in Active Figure 17.6c, where both P and V change continuously, the work done on the gas has some value intermediate between the values obtained in the first two processes.

In a similar manner, the energy transferred by heat into or out of the gas is also found to depend on the process, which can be demonstrated by the situations depicted in Figure 17.7. In each case, the gas has the same initial volume, temperature, and pressure and is assumed to be ideal. In Figure 17.7a, the gas is thermally insulated from its surroundings except at the bottom where it is in thermal contact

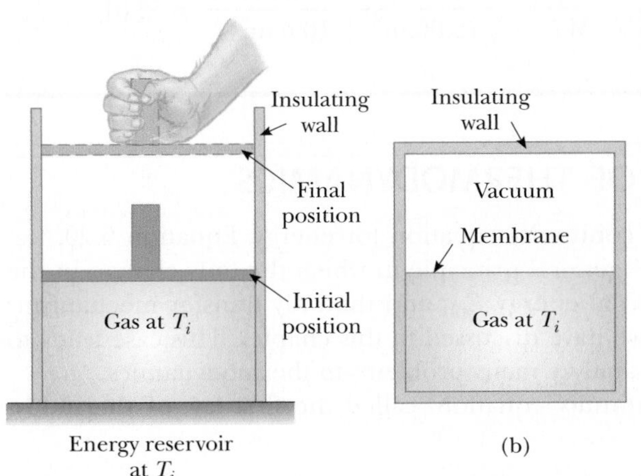

FIGURE 17.7 (a) A gas at temperature T_i expands slowly while absorbing energy from a reservoir so as to maintain a constant temperature. (b) A gas expands rapidly into an evacuated region after a membrane is broken.

with an energy reservoir. An **energy reservoir** is a source of internal energy that is considered to be so large that a finite transfer of energy to or from the reservoir does not change its temperature. The piston is held at its initial position by an external agent, such as your hand. Now, the force holding the piston is reduced slightly so that the piston rises very slowly to its final position. Because the piston is moving upward, negative work is done on the gas; the gas is doing work on the piston. During this expansion to the final volume V_f, just enough energy to maintain a constant temperature T_i is transferred by heat from the reservoir to the gas.

Now consider the completely thermally insulated system shown in Figure 17.7b. When the membrane is broken, the gas expands rapidly into the vacuum until it occupies a volume V_f and is at a pressure P_f. In this case, no work is done on the gas because no force that is exerted on the gas moves through a displacement. Furthermore, no energy is transferred by heat through the insulating wall. As we shall show in Section 17.6, the initial and final temperatures of the gas are the same in this type of expansion.

The initial and final states of the ideal gas in Figure 17.7a are identical to the initial and final states in Figure 17.7b, but the paths are different. In the first case, the gas does work on the piston and energy is transferred slowly to the gas. In the second case, no energy is transferred and the work done is zero. We therefore conclude that **energy transfer by heat, like work done, depends on the process followed between the initial and final states of the system.**

EXAMPLE 17.4 Comparing Processes

An ideal gas is taken through two processes in which $P_f = 1.00 \times 10^5$ Pa, $V_f = 2.00$ m^3, $P_i = 0.200 \times 10^5$ Pa, and $V_i = 10.0$ m^3. For process 1 shown in Active Figure 17.5, the temperature remains constant. For process 2 shown in Active Figure 17.6a, the pressure remains constant and then the volume remains constant. What is the ratio of the work W_1 done on the gas in the first process to the work W_2 done in the second process?

Solution For process 1, the pressure as a function of volume can be expressed using the ideal gas law:

$$P = \frac{nRT}{V}$$

For process 2, no work is done during the portion at constant volume because the piston does not move through a displacement. During the first part of the process, the pressure is constant at $P = P_i$. Let us use these results in the ratio of the work done:

$$\frac{W_1}{W_2} = \frac{-\int_{\text{process 1}} P\,dV}{-\int_{\text{process 2}} P\,dV} = \frac{\int_{V_i}^{V_f} \frac{nRT}{V}\,dV}{\int_{V_i}^{V_f} P_i\,dV} = \frac{nRT\int_{V_i}^{V_f} \frac{dV}{V}}{P_i\int_{V_i}^{V_f} dV}$$

$$= \frac{nRT\ln\left(\frac{V_f}{V_i}\right)}{P_i(V_f - V_i)} = \frac{P_iV_i\ln\left(\frac{V_f}{V_i}\right)}{P_i(V_f - V_i)} = \frac{V_i\ln\left(\frac{V_f}{V_i}\right)}{V_f - V_i}$$

Substituting the numerical values for the initial and final volumes gives us

$$\frac{W_1}{W_2} = \frac{(10.0\text{ m}^3)\ln\left(\frac{2.00\text{ m}^3}{10.0\text{ m}^3}\right)}{(2.00\text{ m}^3 - 10.0\text{ m}^3)} = 2.01$$

17.5 THE FIRST LAW OF THERMODYNAMICS

In Chapter 6, we discussed the continuity equation for energy, Equation 6.20. Let us consider a special case of this general principle in which the only change in the energy of a system is in its internal energy E_{int} and the only transfer mechanisms are heat Q and work W, which we have discussed in this chapter. This case leads to an equation that can be used to analyze many problems in thermodynamics.

This special case of the continuity equation, called the **first law of thermodynamics,** can be written as

■ First law of thermodynamics

$$\Delta E_{\text{int}} = Q + W \qquad [17.8]$$

This equation indicates that the change in the internal energy of a system is equal to the sum of the energy transferred across the system boundary by heat and the energy transferred by work.

Active Figure 17.8 shows the energy transfers and change in internal energy for a gas in a cylinder consistent with the first law. Equation 17.8 can be used in a variety of problems in which the only energy considerations are internal energy, heat, and work. We shall consider several examples shortly. Some problems may not fit the conditions of the first law. For example, the internal energy of the coils in your toaster does not increase due to heat or work, but rather due to electrical transmission. Keep in mind that the first law is a special case of the continuity equation for energy and the latter is the more general equation that covers the widest range of possible situations.

When a system undergoes an infinitesimal change in state, such that a small amount of energy dQ is transferred by heat and a small amount of work dW is done on the system, the internal energy also changes by a small amount dE_{int}. Thus, for infinitesimal processes we can express the first law as[4]

$$dE_{\text{int}} = dQ + dW \qquad [17.9]$$

No practical distinction exists between the results of heat and work on a microscopic scale. Each can produce a change in the internal energy of a system. Although the macroscopic quantities Q and W are *not* properties of a system, they are related to changes of the internal energy of a stationary system through the first law of thermodynamics. Once a process or path is defined, Q and W can be either calculated or measured, and the change in internal energy can be found from the first law.

PITFALL PREVENTION 17.6

Dual sign conventions Some physics and engineering textbooks present the first law as $\Delta E_{\text{int}} = Q - W$, with a minus sign between the heat and work. The reason is that work is defined in these treatments as the work done *by* the gas rather than *on* the gas, as in our treatment. The equivalent equation to Equation 17.7 in these treatments defines work as $W = \int_{V_i}^{V_f} P\,dV$. Thus, if positive work is done by the gas, energy is leaving the system, leading to the negative sign in the first law.

In your studies in other chemistry or engineering courses, or in your reading of other physics textbooks, be sure to note which sign convention is being used for the first law.

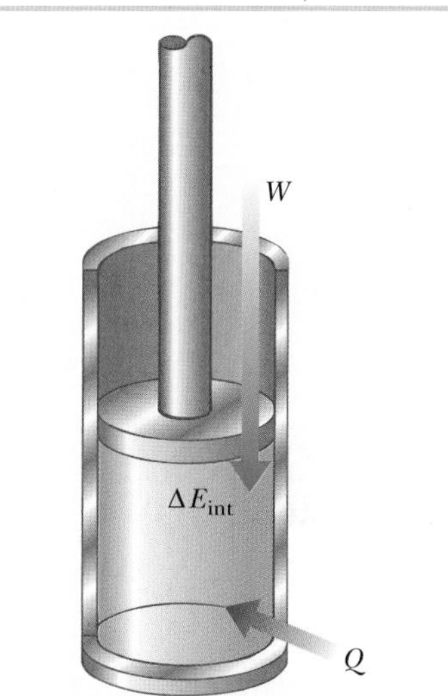

ACTIVE FIGURE 17.8
The first law of thermodynamics equates the change in internal energy in a system to the net energy transfer to the system by heat Q and work W.

Physics Now™ Log into PhysicsNow at www.pop4e.com and go to Active Figure 17.8. You can choose one of the four processes for the gas discussed in Section 17.6 and see the movement of the piston and of a point on a *PV* diagram.

QUICK QUIZ 17.4 Fill in the spaces in the following table with −, +, or 0 for each of the three terms in the first law of thermodynamics. For each situation, the system to be considered is identified.

Situation	System	Q	W	ΔE_{int}
(a) Rapidly pumping up a bicycle tire	Air in the pump			
(b) Pan of room-temperature water sitting on a hot stove	Water in the pan			
(c) Air quickly leaking out of a balloon	Air originally in the balloon			

Thinking Physics 17.2

In the late 1970s, casino gambling was approved in Atlantic City, New Jersey, which can become quite cold in the winter. Energy projections that were performed for the design of the casinos showed that the air-conditioning system would need to operate in the casino even in the middle of a very cold January. Why?

Reasoning If we consider the air in the casino to be the gas to which we apply the first law, imagine a simplification model in which there is no air conditioning and no ventilation so that this gas simply stays in the room. No work is being done on the gas, so we focus on the energy transferred by heat. A casino contains a large number of people, many of whom are active (throwing dice, cheering, etc.) and many of whom are in excited states (celebration, frustration, panic, etc.). As a result, these people have large rates of energy flow by heat from their bodies into the air. This energy results in an increase in internal energy of the air in the casino.

[4]It should be noted that dQ and dW are not true differential quantities because Q and W are not state variables, although dE_{int} is a true differential. For further details on this point, see R. P. Bauman, *Modern Thermodynamics and Statistical Mechanics* (New York, Macmillan, 1992).

With the large number of excited people in a casino (along with the large number of machines and incandescent lights), the temperature of the gas can rise quickly, and to a very high value. To keep the temperature at a comfortable level, energy must be transferred out of the air to compensate for the energy input. Calculations show that energy transfer by heat through the walls even on a 10°F January day is not sufficient to provide the required energy transfer, so the air-conditioning system must be in almost continuous use throughout the year. ■

17.6 SOME APPLICATIONS OF THE FIRST LAW OF THERMODYNAMICS

To apply the first law of thermodynamics to specific systems, it is useful to first define some common thermodynamic processes. We shall identify four special processes used as simplification models to approximate real processes. For each of the following processes, we build a mental representation by imagining that the process occurs for the gas in Active Figure 17.8.

During an **adiabatic process,** no energy enters or leaves the system by heat; that is, $Q = 0$. For the piston in Active Figure 17.8, imagine that all surfaces of the piston are perfect insulators so that energy transfer by heat does not exist. (Another way to achieve an adiabatic process is to perform the process very rapidly because energy transfer by heat tends to be relatively slow.) Applying the first law in this case, we see that

$$\Delta E_{\text{int}} = W \qquad [17.10]$$

From this result, we see that when a gas is compressed adiabatically, both W and ΔE_{int} are positive; work is done on the gas, representing a transfer of energy into the system, so the internal energy increases. Conversely, when the gas expands adiabatically, ΔE_{int} is negative.

Adiabatic processes are very important in engineering practice. Common applications include the expansion of hot gases in an internal combustion engine, the liquefaction of gases in a cooling system, and the compression stroke in a diesel engine. We study adiabatic processes in more detail in Section 17.8.

The **free expansion** depicted in Figure 17.7b is a unique adiabatic process in which no work is done on the gas. Because $Q = 0$ and $W = 0$, we see from the first law that $\Delta E_{\text{int}} = 0$ for this process. That is, **the initial and final internal energies of a gas are equal in a free expansion.** As we saw in Chapter 16, the internal energy of an ideal gas depends only on its temperature. Thus, we expect no change in temperature during an adiabatic free expansion, which is in accord with experiments performed at low pressures. Experiments with real gases at high pressures show a slight increase or decrease in temperature after the expansion because of interactions between molecules.

A process that occurs at constant pressure is called an **isobaric process.** In Active Figure 17.8, as long as the piston is perfectly free to move, the pressure of the gas inside the cylinder is due to atmospheric pressure and the weight of the piston. Hence, the piston can be modeled as a particle in equilibrium. When such a process occurs, the work done on the gas is simply the negative of the constant pressure multiplied by the change in volume, or $-P(V_f - V_i)$. On a PV diagram, an isobaric process appears as a horizontal line, such as the first portion of the process in Active Figure 17.6a or the second portion of the process in Active Figure 17.6b.

A process that takes place at constant volume is called an **isovolumetric process.** In Active Figure 17.8, an isovolumetric process is created by locking the piston in place so that it cannot move. In such a process, the work done is zero because the volume does not change. Hence, the first law applied to an isovolumetric process gives

$$\Delta E_{\text{int}} = Q \qquad [17.11]$$

This equation tells us that if energy is added by heat to a system kept at constant volume, all the energy goes into increasing the internal energy of the system and none enters or leaves the system by work. For example, when an aerosol can is thrown into a fire, energy enters the system (the gas in the can) by heat through the metal walls of the can. Consequently, the temperature and pressure of the gas rise until the can possibly explodes. On a *PV* diagram, an isovolumetric process appears as a vertical line, such as the second portion of the process in Active Figure 17.6a or the first portion of the process in Active Figure 17.6b.

A process that occurs at constant temperature is called an **isothermal process.** Because the internal energy of an ideal gas is a function of temperature only, in an isothermal process for an ideal gas, $\Delta E_{\text{int}} = 0$. Hence, the first law applied to an isothermal process gives

$$Q = -W$$

Whatever energy enters the gas by work leaves the gas by heat in an isothermal process so that the internal energy remains fixed. On a *PV* diagram, an isothermal process appears as a curved line such as that in Figure 17.9. The path of the isothermal process in Figure 19.7 follows along the blue curve, which is an **isotherm,** defined as the curve passing through all points on the *PV* diagram for which the gas has the same temperature. The work done on the ideal gas in an isothermal process was calculated in Example 17.4:

$$W = -nRT\ln\left(\frac{V_f}{V_i}\right) \qquad [17.12]$$

■ Work done on a gas in an isothermal process

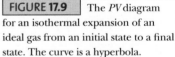

The isothermal process can be analyzed as a model of a nonisolated system in steady state. There is a transfer of energy across the boundary of the system, but no change occurs in the internal energy of the system. The adiabatic, isobaric, and isovolumetric processes are examples of the nonisolated system model.

Next consider the case in which a nonisolated system is taken through a **cyclic process,** that is, one that originates and ends at the same state. In this case, the change in the internal energy must be zero because internal energy is a state variable and the initial and final states are identical. The energy added by heat to the system must therefore equal the negative of the work done on the system during the cycle. That is, in a cyclic process,

$$\Delta E_{\text{int}} = 0 \qquad \text{and} \qquad Q = -W$$

The net work done per cycle equals the area enclosed by the path representing the process on a *PV* diagram. As we shall see in Chapter 18, cyclic processes are very important in describing the thermodynamics of **heat engines,** thermal devices in which a fraction of the energy added by heat to the system is extracted by mechanical work.

PITFALL PREVENTION 17.7

$Q \neq 0$ IN AN ISOTHERMAL PROCESS
Do not fall into the common trap of thinking that no energy is transferred by heat if the temperature does not change, as is the case in an isothermal process. Because the cause of temperature change can be either heat *or* work, the temperature can remain constant even if energy enters the gas by heat. That can only happen if the energy entering the system by heat leaves by work.

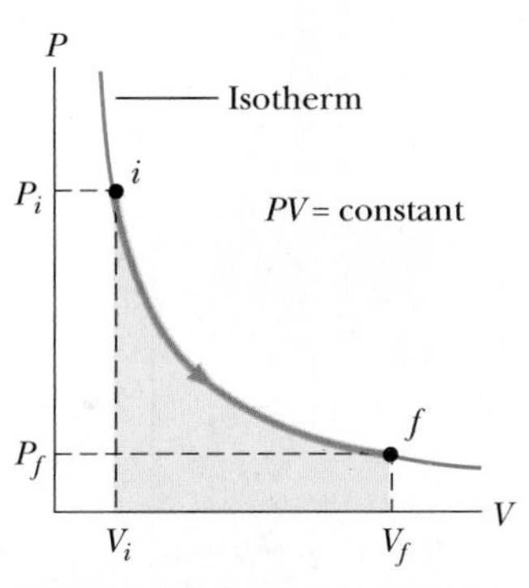

FIGURE 17.9 The *PV* diagram for an isothermal expansion of an ideal gas from an initial state to a final state. The curve is a hyperbola.

QUICK QUIZ 17.5 Characterize the paths in Figure 17.10 as isobaric, isovolumetric, isothermal, or adiabatic. Note that $Q = 0$ in path B.

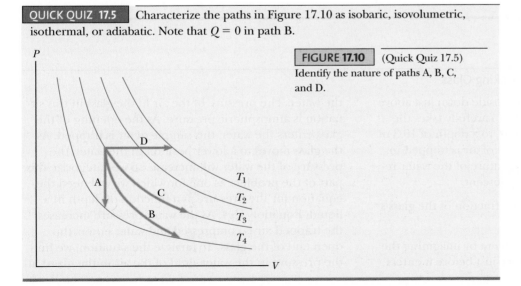

FIGURE 17.10 (Quick Quiz 17.5) Identify the nature of paths A, B, C, and D.

EXAMPLE 17.5 Cylinder in an Ice Water Bath

The cylinder in Figure 17.11a has thermally conducting walls and is immersed in an ice-water bath. The gas in the cylinder undergoes three processes: (1) the piston is rapidly pushed downward, compressing the gas in the cylinder; (2) the piston is held at the final position of the previous process while the gas returns to the temperature of the ice-water bath; and (3) the piston is very slowly raised back to the original position.

A For the system of the gas, what kind of special thermodynamic process does each of these represent?

Solution Because process 1 occurs rapidly, it is very close to an adiabatic compression. In process 2, the piston is held fixed, so this process is isovolumetric. In the very slow process 3, the gas and the ice-water bath can be approximated as remaining in thermal equilibrium at all times, so the process is very close to isothermal.

B Draw the complete cycle on a *PV* diagram.

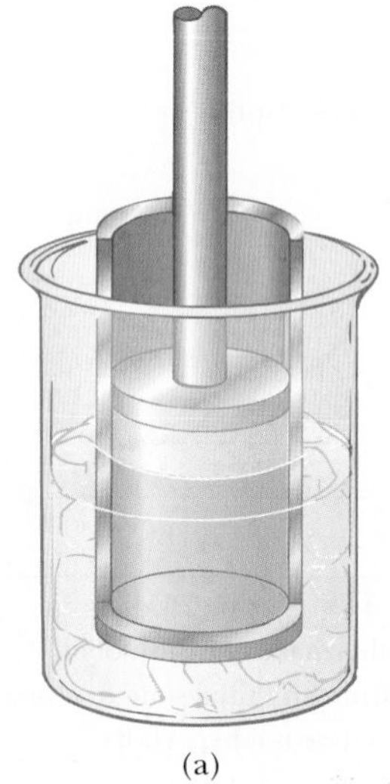

(a)

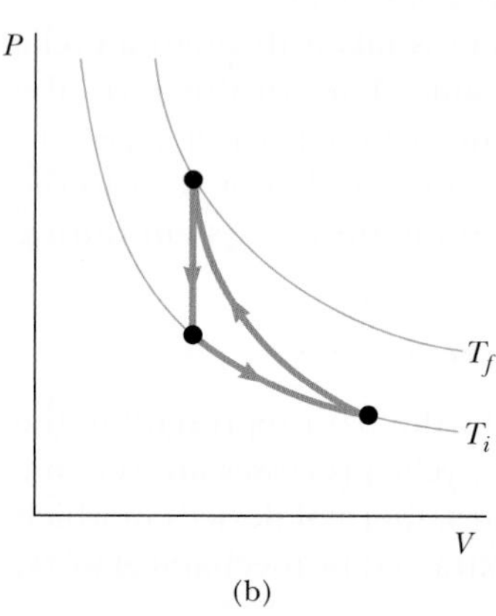

(b)

Solution The cycle is shown in Figure 17.11b.

C The work done on the gas during the cycle is 500 J. What mass of ice in the ice water bath melts during the cycle?

Solution For the entire cycle, the change in internal energy is zero. Thus, from the first law, the energy transfer by heat must equal the negative of the work done on the gas, $Q = -W = -500\text{ J}$. This equation indicates that energy leaves the system of the gas by heat during the cycle, entering the ice-water bath (so $Q_{\text{ice}} = +500\text{ J}$), where it melts some of the ice. The amount of ice that melts is found using Equation 17.5:

$$Q_{\text{ice}} = mL_f$$

$$m = \frac{Q_{\text{ice}}}{L_f} = \frac{500\text{ J}}{3.33 \times 10^5\text{ J/kg}} = 1.5 \times 10^{-3}\text{ kg}$$

$$= 1.50\text{ g}$$

FIGURE 17.11 (Example 17.5) (a) Cutaway view of a cylinder containing an ideal gas immersed in an ice water bath. (b) The *PV* diagram for the cycle described.

EXAMPLE 17.6 The Diving Drinking Glass

An empty drinking glass is held upside down just above the surface of water. A scuba diver carefully takes the glass, which remains upside down, to a depth of 10.3 m below the surface so that a sample of air is trapped in the glass. Assume that the temperature of the water remains fixed at 285 K during the descent.

A At the depth of 10.3 m, what fraction of the glass's volume is filled with air?

Solution Conceptualize the problem by imagining the glass held above the water surface just before it enters the water. The pressure of the air in the glass in this situation is atmospheric pressure. As the opening of the glass enters the water, this sample of air is trapped. As the glass moves to a lower position in the water, the pressure of the water will increase, so we categorize this part of the problem as one for which we will need the equation for the pressure as a function of depth in a liquid, Equation 15.4. As the water pressure increases, the trapped air is compressed and water enters the open end of the glass. To analyze the situation, we find the pressure in the water (and of the air in the glass) at the depth of 10.3 m:

$$P = P_{atm} + \rho g h = 1.013 \times 10^5 \text{ Pa} + (1\,000 \text{ kg/m}^3)(9.80 \text{ m/s}^2)(10.3 \text{ m}) = 2.02 \times 10^5 \text{ Pa}$$

Because the temperature remains fixed, we categorize the process occurring for the gas as isothermal. We calculate the ratio of volumes of the air in the glass for the final and initial conditions of this isothermal process from the ideal gas law:

$$P_i V_i = P_f V_f \rightarrow \frac{V_f}{V_i} = \frac{P_i}{P_f} = \frac{1.013 \times 10^5 \text{ Pa}}{2.02 \times 10^5 \text{ Pa}} = \boxed{0.500}$$

B There are 0.020 0 mol of air trapped in the glass. How much energy crosses the boundary of the system of the air trapped in the glass by heat during the process?

Solution Because the process is isothermal, the first law tells us that $\Delta E_{int} = 0$ and the energy flow by heat is equal to the negative of the work done on the gas; from Equation 17.12,

$$Q = -W = nRT \ln\left(\frac{V_f}{V_i}\right) = (0.020\,0 \text{ mol})(8.314 \text{ J/mol}\cdot\text{K})(285 \text{ K}) \ln(0.500) = \boxed{-32.8 \text{ J}}$$

To finalize the problem, note that because Q is negative, energy comes out of the air by heat. As the air is compressed, the tendency is for its temperature to increase as work is done on it by the surrounding water. Because its temperature is to remain fixed, however, energy must leave the air in the glass by heat so that the temperature does not rise.

17.7 MOLAR SPECIFIC HEATS OF IDEAL GASES

In Section 17.2, we considered the energy necessary to change the temperature of a mass m of a substance by ΔT. In this section, we focus our attention on ideal gases, and the amount of gas is measured by the number of moles n rather than the mass m. In doing so, some important new connections are found between thermodynamics and mechanics.

The energy transfer by heat required to raise the temperature of n moles of gas from T_i to T_f depends on the path taken between the initial and final states. To understand this concept, consider an ideal gas undergoing several processes such that the change in temperature is $\Delta T = T_f - T_i$ for all processes. The temperature change can be achieved by traveling along a variety of paths from one isotherm to another as in Figure 17.12. Because ΔT is the same for all paths, the change in internal energy ΔE_{int} is the same for all paths. From the first law, $Q = \Delta E_{int} - W$; we see, however, that the heat Q for each path is different because W (the negative of the area under the curves) is different for each path. Thus, the heat required to produce a given change in temperature does not have a unique value.

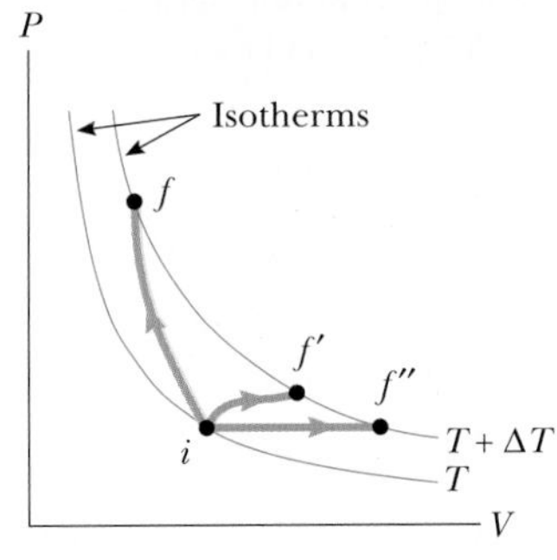

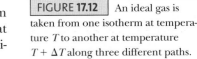

FIGURE 17.12 An ideal gas is taken from one isotherm at temperature T to another at temperature $T + \Delta T$ along three different paths.

This difficulty is addressed by defining specific heats for two processes from Section 17.6: isovolumetric and isobaric processes. Modifying Equation 17.3 so that the amount of gas is measured in moles, we define the **molar specific heats** associated with these processes with the following equations:

$$Q = nC_V \Delta T \quad \text{(constant volume)} \qquad [17.13]$$

$$Q = nC_P \Delta T \quad \text{(constant pressure)} \qquad [17.14]$$

where C_V is the **molar specific heat at constant volume** and C_P is the **molar specific heat at constant pressure.**

In Chapter 16, we found that the temperature of a monatomic gas is a measure of the average translational kinetic energy of the gas molecules. In view of this finding, let us first consider the simplest case of an ideal monatomic gas (i.e., a gas containing one atom per molecule), such as helium, neon, or argon. When energy is added to a monatomic gas in a container of fixed volume (e.g., by heating), all the added energy goes into increasing the translational kinetic energy of the atoms. There is no other way to store the energy in a monatomic gas. The constant-volume process from i to f is described in Active Figure 17.13, where ΔT is the temperature

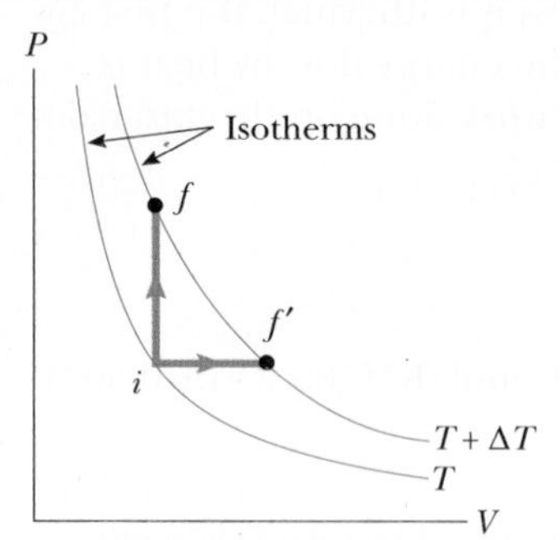

ACTIVE FIGURE 17.13
Energy is transferred by heat to an ideal gas in two ways. For the constant-volume path $i \rightarrow f$, all the energy goes into increasing the internal energy of the gas because no work is done. Along the constant-pressure path $i \rightarrow f'$, part of the energy transferred into the gas by heat is transferred out by work.

Physics Now™ Log into PhysicsNow at www.pop4e.com and go to Active Figure 17.13. You can choose initial and final temperatures for one mole of an ideal gas undergoing constant-volume and constant-pressure processes and measure Q, W, ΔE_{int}, C_V, and C_P.

difference between the two isotherms. From Equation 16.18, we see that the total internal energy E_{int} of N molecules (or n mol) of an ideal monatomic gas is

$$E_{\text{int}} = \tfrac{3}{2}nRT \qquad [17.15]$$

If energy is transferred by heat to the system at constant volume, the work done on the system is zero. That is, $W = -\int P\,dV = 0$ for a constant-volume process. Hence, from the first law of thermodynamics and Equation 17.15 we find that

$$Q = \Delta E_{\text{int}} = \tfrac{3}{2}nR\,\Delta T \qquad [17.16]$$

Substituting the value for Q given by Equation 17.13 into Equation 17.16, we have

$$nC_V\,\Delta T = \tfrac{3}{2}nR\,\Delta T$$

$$C_V = \tfrac{3}{2}R = 12.5\ \text{J/mol}\cdot\text{K} \qquad [17.17]$$

This expression predicts a value of $C_V = \frac{3}{2}R$ for *all* monatomic gases, regardless of the type of gas. This prediction is based on our structural model of kinetic theory, in which the atoms interact with one another only via short-range forces. The third column of Table 17.3 indicates that this prediction is in excellent agreement with measured values of molar specific heats for monatomic gases. It also indicates that this prediction is not in agreement with values of molar specific heats for diatomic and polyatomic gases. We address these types of gases shortly.

Because no work is done on an ideal gas undergoing an isovolumetric process, the energy transfer by heat is equal to the change in internal energy. Thus, the change in internal energy can be expressed as

$$\Delta E_{\text{int}} = nC_V\,\Delta T \qquad [17.18]$$

Because internal energy is a state function, the change in internal energy does not depend on the path followed between the initial and final states. Thus, **Equation 17.18 gives the change in internal energy of an ideal gas for *any* process in which the temperature change is ΔT, not just an isovolumetric process. Furthermore, it is true for monatomic, diatomic, and polyatomic gases.**

TABLE 17.3 Molar Specific Heats of Various Gases

	Molar Specific Heat[a] (J/mol · K)			
Gas	C_P	C_V	$C_P - C_V$	$\gamma = C_P/C_V$
Monatomic Gases				
He	20.8	12.5	8.33	1.67
Ar	20.8	12.5	8.33	1.67
Ne	20.8	12.7	8.12	1.64
Kr	20.8	12.3	8.49	1.69
Diatomic Gases				
H_2	28.8	20.4	8.33	1.41
N_2	29.1	20.8	8.33	1.40
O_2	29.4	21.1	8.33	1.40
CO	29.3	21.0	8.33	1.40
Cl_2	34.7	25.7	8.96	1.35
Polyatomic Gases				
CO_2	37.0	28.5	8.50	1.30
SO_2	40.4	31.4	9.00	1.29
H_2O	35.4	27.0	8.37	1.30
CH_4	35.5	27.1	8.41	1.31

[a]All values except that for water were obtained at 300 K.

In the case of infinitesimal changes, we can use Equation 17.18 to express the molar specific heat at constant volume as

$$C_V = \frac{1}{n}\frac{dE_{\text{int}}}{dT} \qquad [17.19]$$

Now suppose the gas is taken along the constant-pressure path $i \rightarrow f'$ in Active Figure 17.13. Along this path, the temperature again increases by ΔT. The energy transferred to the gas by heat in this process is $Q = nC_P\,\Delta T$. Because the volume changes in this process, the work done on the gas is $W = -P\,\Delta V$. Applying the first law to this process gives

$$\Delta E_{\text{int}} = Q + W = nC_P\,\Delta T - P\,\Delta V \qquad [17.20]$$

The change in internal energy for the process $i \rightarrow f'$ is equal to that for the process $i \rightarrow f$ because E_{int} depends only on temperature for an ideal gas and ΔT is the same for both processes. Because $PV = nRT$, for a constant-pressure process $P\Delta V = nR\,\Delta T$. Substituting this value for $P\,\Delta V$ into Equation 17.20 with $\Delta E_{\text{int}} = nC_V\,\Delta T$ (Eq. 17.18) gives

■ Relation between molar specific heats

$$nC_V\,\Delta T = nC_P\,\Delta T - nR\,\Delta T \quad \rightarrow \quad C_P - C_V = R \qquad [17.21]$$

This expression applies to *any* ideal gas. It shows that the molar specific heat of an ideal gas at constant pressure is greater than the molar specific heat at constant volume by an amount R, the universal gas constant. As shown by the fourth column in Table 17.3, this result is in good agreement with real gases regardless of the number of atoms in the molecule.

Because $C_V = \frac{3}{2}R$ for a monatomic ideal gas, Equation 17.21 predicts a value $C_P = \frac{5}{2}R = 20.8\text{ J/mol}\cdot\text{K}$ for the molar specific heat of a monatomic gas at constant pressure. The second column of Table 17.3 shows the validity of this prediction for monatomic gases.

The ratio of molar specific heats is a dimensionless quantity γ:

$$\gamma = \frac{C_P}{C_V} \qquad [17.22]$$

For a monatomic gas, this ratio has the value

$$\gamma = \frac{C_P}{C_V} = \frac{\frac{5}{2}R}{\frac{3}{2}R} = \frac{5}{3} = 1.67$$

The last column in Table 17.3 shows good agreement between this predicted value for γ and experimentally measured values for monatomic gases.

QUICK QUIZ 17.6 (i) How does the internal energy of an ideal gas change as it follows path $i \rightarrow f$ in Active Figure 17.13? **(a)** E_{int} increases. **(b)** E_{int} decreases. **(c)** E_{int} stays the same. **(d)** There is not enough information to determine how E_{int} changes. **(ii)** From the same list of choices, how does the internal energy of a gas change as it follows path $f \rightarrow f'$ along the isotherm labeled $T + \Delta T$ in Active Figure 17.13?

EXAMPLE 17.7 Heating a Cylinder of Helium

A cylinder contains 3.00 mol of helium gas at a temperature of 300 K.

A How much energy must be transferred to the gas by heat to increase its temperature to 500 K if it is heated at constant volume?

Solution For the constant-volume process,

$$Q = nC_V\,\Delta T = \tfrac{3}{2}nR\,\Delta T$$

$$Q = \tfrac{3}{2}(3.00\text{ mol})(8.31\text{ J/mol}\cdot\text{K})(500\text{ K} - 300\text{ K})$$

$$= 7.48 \times 10^3\text{ J}$$

B How much energy must be transferred to the gas by heat at constant pressure to raise the temperature to 500 K?

Solution For the constant-pressure process,

$$Q = nC_P\,\Delta T = \tfrac{5}{2}nR\,\Delta T$$

$$Q = \tfrac{5}{2}(3.00\text{ mol})(8.31\text{ J/mol}\cdot\text{K})(500\text{ K} - 300\text{ K})$$

$$= 12.5 \times 10^3\text{ J}$$

Notice that the energy transfer for part B is larger than that for part A because positive work must be done *by* the gas pushing the piston outward to maintain the constant pressure in an isobaric process, but no work is done in an isovolumetric process.

17.8 ADIABATIC PROCESSES FOR AN IDEAL GAS

In Section 17.6, we identified four special processes of interest for ideal gases. In three of them, a state variable is held constant: P = constant for an isobaric process, V = constant for an isovolumetric process, and T = constant for an isothermal process. What about our fourth special process, the adiabatic process? Is anything constant in this process? As you recall, an adiabatic process is one in which no energy is transferred by heat between a system and its surroundings. In reality, true adiabatic processes on the Earth cannot occur because there is no such thing as a perfect thermal insulator. Some processes, however, are nearly adiabatic. For example, if a gas is compressed (or expanded) very rapidly, very little energy flows out of (or into) the system by heat and so the process is nearly adiabatic.

Suppose an ideal gas undergoes a quasi-static adiabatic expansion. We find that all three variables in the ideal gas law—P, V, and T—change during an adiabatic process. At any time during the process, however, the ideal gas law $PV = nRT$ describes the correct relationship among these variables. Although none of the three variables alone is constant in this process, we find that a *combination* of some of these variables remains constant. This relationship is derived in the following discussion.

Imagine a gas expanding adiabatically in a thermally insulated cylinder so that $Q = 0$. Let us take the infinitesimal change in volume to be dV and the infinitesimal change in temperature to be dT. The work done on the gas is $-P\,dV$. The change in internal energy is given by the differential form of Equation 17.18, $dE_{\text{int}} = nC_V\,dT$. Hence, the first law of thermodynamics becomes

$$dE_{\text{int}} = dQ + dW \quad\rightarrow\quad nC_V\,dT = 0 - P\,dV \qquad [17.23]$$

Taking the differential of the equation of state for an ideal gas, $PV = nRT$, gives

$$P\,dV + V\,dP = nR\,dT$$

Eliminating $n\,dT$ from these last two equations, we find that

$$P\,dV + V\,dP = -\frac{R}{C_V}P\,dV$$

From Equation 17.21, we substitute $R = C_P - C_V$ and divide by PV to obtain

$$\frac{dV}{V} + \frac{dP}{P} = -\left(\frac{C_P - C_V}{C_V}\right)\frac{dV}{V} = (1-\gamma)\frac{dV}{V}$$

$$\frac{dP}{P} + \gamma\frac{dV}{V} = 0$$

Integrating this expression gives

$$\ln P + \gamma \ln V = \text{constant}$$

which we can write as

■ Relation between pressure and volume in an adiabatic process for an ideal gas

$$PV^{\gamma} = \text{constant} \qquad [17.24]$$

The PV diagram for an adiabatic expansion is shown in Figure 17.14. Because $\gamma > 1$, the PV curve is steeper than that for an isothermal expansion, in which

$PV =$ constant. Equation 17.23 shows that during an adiabatic expansion, ΔE_{int} is negative and so ΔT is also negative. Thus, the gas cools during an adiabatic expansion. Conversely, the temperature increases if the gas is compressed adiabatically. Equation 17.24 can be expressed in terms of initial and final states as

$$P_i V_i^{\gamma} = P_f V_f^{\gamma} \quad [17.25]$$

Using the ideal gas law, Equation 17.25 can also be expressed as

$$T_i V_i^{\gamma-1} = T_f V_f^{\gamma-1} \quad [17.26]$$

Given the relationship in Equation 17.24, it can be shown that the work done on a gas during an adiabatic process is

$$W = \frac{1}{\gamma - 1}(P_f V_f - P_i V_i) \quad [17.27]$$

Problem 17.72 invites you to derive this equation.

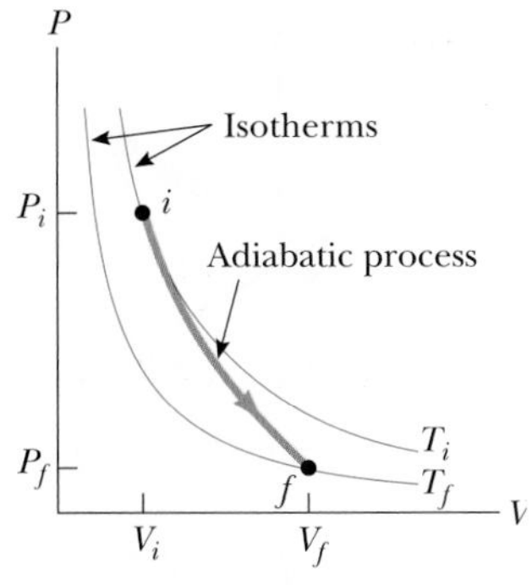

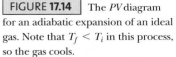

FIGURE 17.14 The PV diagram for an adiabatic expansion of an ideal gas. Note that $T_f < T_i$ in this process, so the gas cools.

EXAMPLE 17.8 **A Diesel Engine Cylinder**

The fuel–air mixture in the cylinder of a diesel engine at 20.0°C is compressed from an initial pressure of 1.00 atm and volume of 800 cm^3 to a volume of 60.0 cm^3. Assuming that the mixture behaves as an ideal gas with $\gamma = 1.40$ and that the compression is adiabatic, find the final pressure and temperature of the mixture.

Solution Using Equation 17.25, we find that

$$P_f = P_i\left(\frac{V_i}{V_f}\right)^{\gamma} = (1.00\text{ atm})\left(\frac{800\text{ cm}^3}{60.0\text{ cm}^3}\right)^{1.40}$$

$$= 37.6\text{ atm}$$

Because $PV = nRT$ for an ideal gas, we can compare initial and final conditions in another way as

$$\frac{P_i V_i}{T_i} = \frac{P_f V_f}{T_f}$$

By solving for the final temperature and substituting the given and calculated quantities,

$$T_f = \frac{P_f V_f}{P_i V_i} T_i = \frac{(37.6\text{ atm})(60.0\text{ cm}^3)}{(1.00\text{ atm})(800\text{ cm}^3)}(293\text{ K}) = 826\text{ K}$$

$$= 553°\text{C}$$

The high compression in a diesel engine raises the temperature of the fuel enough to cause its combustion without the need for spark plugs.

17.9 MOLAR SPECIFIC HEATS AND THE EQUIPARTITION OF ENERGY

We have found that predictions of molar specific heats based on kinetic theory agree quite well with the behavior of monatomic gases but not with the behavior of complex gases (Table 17.3). To explain the variations in C_V and C_P between monatomic gases and more complex gases, let us explore the origin of specific heat by extending our structural model of kinetic theory in Chapter 16. In Section 16.5, we discussed that the sole contribution to the internal energy of a monatomic gas is the translational kinetic energy of the molecules. We also discussed the theorem of equipartition of energy, which states that, at equilibrium, each degree of freedom contributes, on the average, $\frac{1}{2}k_BT$ of energy per molecule. The monatomic gas has three degrees of freedom, one associated with each of the independent directions of translational motion.

For more complex molecules, other types of motion exist in addition to translation. The internal energy of a diatomic or polyatomic gas includes contributions from the vibrational and rotational motion of the molecules in addition to translation. The rotational and vibrational motions of molecules with structure can be activated by collisions and therefore are "coupled" to the translational motion of the

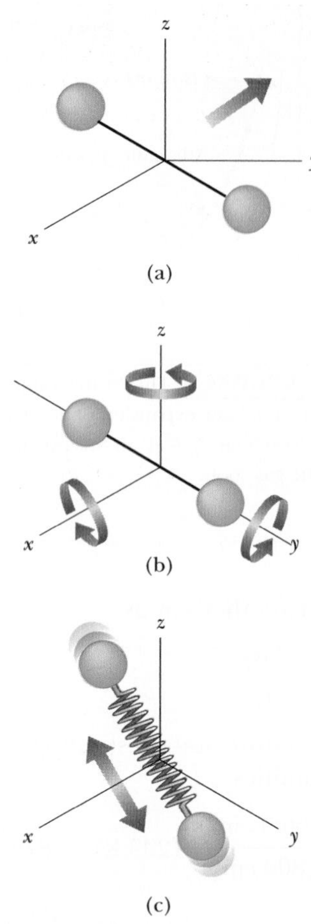

FIGURE 17.15 Possible motions of a diatomic molecule: (a) translational motion of the center of mass, (b) rotational motion about the various axes, and (c) vibrational motion along the molecular axis.

molecules. The branch of physics known as *statistical mechanics* suggests that the average energy for each of these additional degrees of freedom is the same as that for translation, which in turn suggests that the determination of a gas's internal energy is a simple matter of counting the degrees of freedom. We will find that this process works well, although the model must be modified with some notions from quantum physics for us to explain the experimental data completely.

Let us consider a diatomic gas, which we can model as consisting of dumbbell-shaped molecules (Fig. 17.15), and apply concepts that we studied in Chapter 10. In this model, the center of mass of the molecule can translate in the x, y, and z directions (Fig. 17.15a). In addition, the molecule can rotate about three mutually perpendicular axes (Fig. 17.15b). We can ignore the rotation about the y axis because the moment of inertia and the rotational energy $\frac{1}{2}I\omega^2$ about this axis are negligible compared with those associated with the x and z axes. Thus, there are five degrees of freedom: three associated with the translational motion and two associated with the rotational motion. Because each degree of freedom contributes, on average, $\frac{1}{2}k_BT$ of energy per molecule, the total internal energy for a diatomic gas consisting of N molecules and considering both translation and rotation is

$$E_{int} = 3N(\tfrac{1}{2}k_BT) + 2N(\tfrac{1}{2}k_BT) = \tfrac{5}{2}Nk_BT = \tfrac{5}{2}nRT$$

We can use this result and Equation 17.19 to predict the molar specific heat at constant volume:

$$C_V = \frac{1}{n}\frac{dE_{int}}{dT} = \frac{1}{n}\frac{d}{dT}(\tfrac{5}{2}nRT) = \tfrac{5}{2}R = 20.8\text{ J/mol}\cdot\text{K} \qquad [17.28]$$

From Equations 17.21 and 17.22, we find that the model predicts that

$$C_P = C_V + R = \tfrac{7}{2}R \qquad [17.29]$$

$$\gamma = \frac{C_P}{C_V} = \frac{\frac{7}{2}R}{\frac{5}{2}R} = \frac{7}{5} = 1.40 \qquad [17.30]$$

Let us now incorporate the vibration of the molecule in the model. We use the structural model for the diatomic molecule in which the two atoms are joined by an imaginary spring (Fig. 17.15c) and apply the concepts of Chapter 12. The vibrational motion has two types of energy associated with the vibrations along the length of the molecule—kinetic energy of the atoms and potential energy in the model spring—which adds two more degrees of freedom for a total of seven for translation, rotation, and vibration. Because each degree of freedom contributes $\frac{1}{2}k_BT$ of energy per molecule, the total internal energy for a diatomic gas consisting of N molecules and considering all types of motion is

$$E_{int} = 3N(\tfrac{1}{2}k_BT) + 2N(\tfrac{1}{2}k_BT) + 2N(\tfrac{1}{2}k_BT) = \tfrac{7}{2}Nk_BT = \tfrac{7}{2}nRT$$

Thus, the molar specific heat at constant volume is predicted to be

$$C_V = \frac{1}{n}\frac{dE_{int}}{dT} = \frac{1}{n}\frac{d}{dT}(\tfrac{7}{2}nRT) = \tfrac{7}{2}R = 29.1\text{ J/mol}\cdot\text{K} \qquad [17.31]$$

From Equations 17.21 and 17.22,

$$C_P = C_V + R = \tfrac{9}{2}R \qquad [17.32]$$

$$\gamma = \frac{C_P}{C_V} = \frac{\frac{9}{2}R}{\frac{7}{2}R} = \frac{9}{7} = 1.29 \qquad [17.33]$$

When we compare our predictions with the section of Table 17.3 corresponding to diatomic gases, we find a curious result. For the first four gases—hydrogen, nitrogen, oxygen, and carbon monoxide—the value of C_V is close to that predicted by Equation 17.28, which includes rotation but not vibration. The value for the fifth gas, chlorine, lies between the prediction including rotation and the prediction

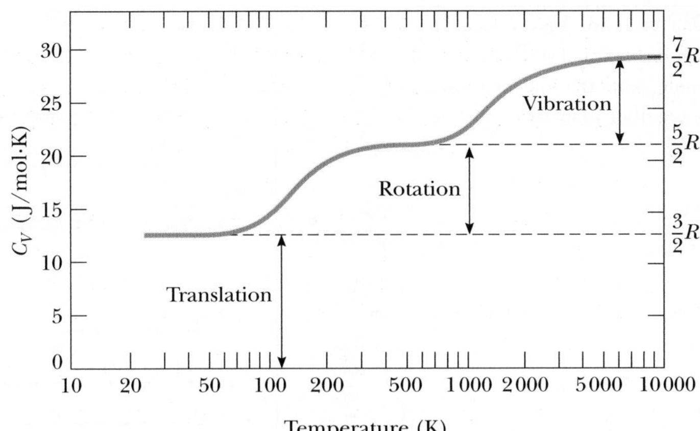

FIGURE 17.16 The molar specific heat of hydrogen as a function of temperature. The horizontal scale is logarithmic. Note that hydrogen liquefies at 20 K.

that includes rotation and vibration. None of the values agrees with Equation 17.31, which is based on the most complete model for motion of the diatomic molecule!

It might seem that our model is a failure for predicting molar specific heats for diatomic gases. We can claim success for our model, however, if measurements of molar specific heat are made over a wide temperature range rather than at the single temperature that gives us the values in Table 17.3. Figure 17.16 shows the molar specific heat of hydrogen as a function of temperature. The curve has three plateaus and they are at the values of the molar specific heat predicted by Equations 17.17, 17.28, and 17.31! For low temperatures, the diatomic hydrogen gas behaves like a monatomic gas. As the temperature rises to room temperature, its molar specific heat rises to a value for a diatomic gas, consistent with the inclusion of rotation but not vibration. For high temperatures, the molar specific heat is consistent with a model including all types of motion.

Before addressing the reason for this mysterious behavior, let us make a brief remark about polyatomic gases. For molecules with more than two atoms, the number of degrees of freedom is even larger and the vibrations are more complex than for diatomic molecules. These considerations result in an even higher predicted molar specific heat, which is in qualitative agreement with experiment. For the polyatomic gases shown in Table 17.3, we see that the molar specific heats are higher than those for diatomic gases. The more degrees of freedom available to a molecule, the more "ways" of storing energy are available, resulting in a higher molar specific heat.

A Hint of Energy Quantization

Our model for molar specific heats has been based so far on purely classical notions. It predicts a value of the specific heat for a diatomic gas that, according to Figure 17.16, only agrees with experimental measurements made at high temperatures. To explain why this value is only true at high temperatures and why the plateaus exist in Figure 17.16, we must go beyond classical physics and introduce some quantum physics into the model. In Section 11.5, we discussed energy quantization for the hydrogen atom. Only certain energies are allowed for the system, and an energy level diagram can be drawn to illustrate those allowed energies. For a molecule, quantum physics tells us that the rotational and vibrational energies are quantized. Figure 17.17 shows an energy level diagram for the rotational and vibrational quantum states of a diatomic molecule. Notice that vibrational states are separated by larger energy gaps than rotational states.

At low temperatures, the energy that a molecule gains in collisions with its neighbors is generally not large enough to raise it to the first excited state of either rotation or vibration. All molecules are in the ground state for rotation and vibration. Therefore, at low temperatures, the only contribution to the molecules'

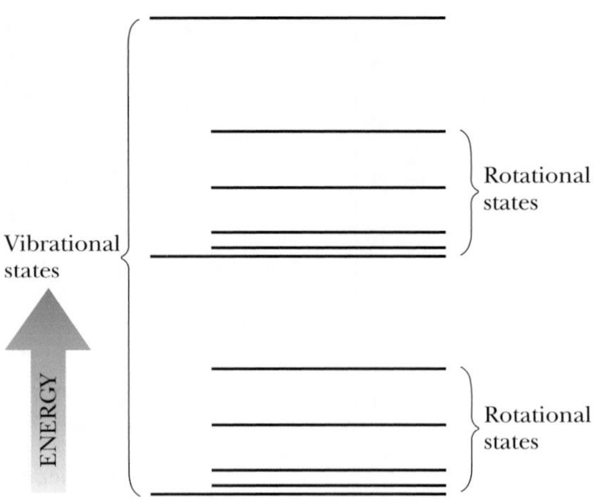

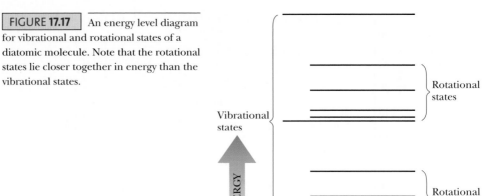

FIGURE 17.17 An energy level diagram for vibrational and rotational states of a diatomic molecule. Note that the rotational states lie closer together in energy than the vibrational states.

average energy is from translation, and the specific heat is that predicted by Equation 17.17.

As the temperature is raised, the average energy of the molecules increases. In some collisions, a molecule may have enough energy transferred to it from another molecule to excite the first rotational state. As the temperature is raised further, more molecules can be excited to this state. The result is that rotation begins to contribute to the internal energy and the molar specific heat rises. At about room temperature in Figure 17.16, the second plateau is reached and rotation contributes fully to the molar specific heat. The molar specific heat is now equal to the value predicted by Equation 17.28.

Vibration contributes nothing at room temperature because the vibrational states are farther apart in energy than the rotational states; the molecules are in the ground vibrational state. The temperature must be even higher to raise the molecules to the first excited vibrational state. That happens in Figure 17.16 between 1 000 K and 10 000 K. At 10 000 K on the right side of the figure, vibration is contributing fully to the internal energy and the molar specific heat has the value predicted by Equation 17.31.

The predictions of this structural model are supportive of the theorem of equipartition of energy. In addition, the inclusion in the model of energy quantization from quantum physics allows a full understanding of Figure 17.16. This excellent example shows the power of the modeling approach.

(Courtesy of Dr. Albert A. Bartlett, University of Colorado, Boulder)

The absence of snow on some parts of the roof show that energy is conducted from the inside of the residence to the exterior more rapidly on those parts of the roof. The dormer appears to have been added and insulated. The main roof does not appear to be well insulated. ■

17.10 ENERGY TRANSFER MECHANISMS IN THERMAL PROCESSES

In Chapter 6, we introduced the continuity equation for energy $\Delta E_{\text{system}} = \Sigma T$ as a principle allowing a global approach to energy considerations in physical processes. Earlier in this chapter, we discussed two of the terms on the right-hand side of the continuity equation: work and heat. In this section, we consider more details of heat and two other energy transfer methods that are often related to temperature changes: convection (a form of matter transfer) and electromagnetic radiation.

Conduction

The process of energy transfer by heat can also be called **conduction** or **thermal conduction.** In this process, the transfer mechanism can be viewed on an atomic scale as an exchange of kinetic energy between molecules in which the less energetic molecules gain energy by colliding with the more energetic molecules. For example, if you hold one end of a long metal bar and insert the other end into a flame, the temperature of the metal in your hand soon increases. The energy reaches your hand through conduction. How that happens can be understood by examining what is happening to the atoms in the metal. Initially, before the rod is inserted into the flame, the atoms are vibrating about their equilibrium positions. As the flame provides energy to the rod, those atoms near the flame begin to vibrate with larger and larger amplitudes. These atoms in turn collide with their neighbors and transfer some of their energy in the collisions. Slowly, metal atoms farther and farther from the flame increase their amplitude of vibration until eventually those in the metal near your hand are affected. This increased vibration represents an increase in temperature of the metal (and possibly a burned hand).

Although the transfer of energy through a material can be partially explained by atomic vibrations, the rate of conduction also depends on the properties of the substance. For example, it is possible to hold a piece of asbestos in a flame indefinitely, which implies that very little energy is being conducted through the asbestos. In general, metals are good thermal conductors because they contain large numbers of electrons that are relatively free to move through the metal and can transport energy from one region to another. Thus, in a good thermal conductor, such as copper, conduction takes place via the vibration of atoms and via the motion of free electrons. Materials such as asbestos, cork, paper, and fiberglass are poor thermal conductors. Gases also are poor thermal conductors because of the large distance between the molecules.

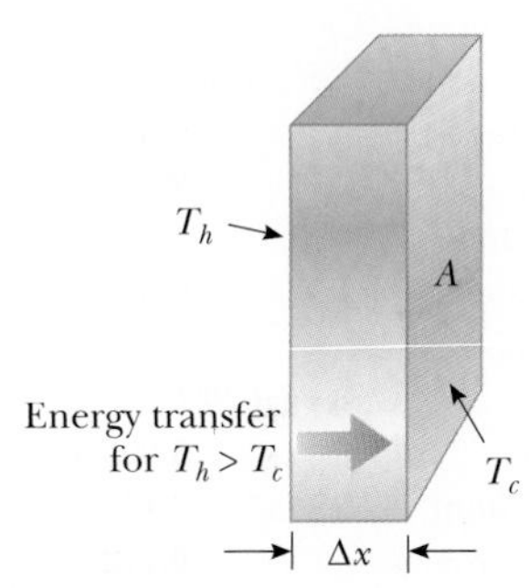

FIGURE 17.18 Energy transfer through a conducting slab with cross-sectional area A and thickness Δx. The opposite faces are at different temperatures T_c and T_h.

Conduction occurs only if the temperatures differ between two parts of the conducting medium. This temperature difference drives the flow of energy. Consider a slab of material of thickness Δx and cross-sectional area A with its opposite faces at different temperatures T_c and T_h, where $T_h > T_c$ (Fig. 17.18). The slab allows energy to transfer from the region of high temperature to that of low temperature by thermal conduction. The rate of energy transfer by heat, $\mathcal{P} = Q/\Delta t$, is proportional to the cross-sectional area of the slab and the temperature difference and inversely proportional to the thickness of the slab:

$$\mathcal{P} = \frac{Q}{\Delta t} \propto A\frac{\Delta T}{\Delta x}$$

Note that $\mathcal{P}$ has units of watts when Q is in joules and Δt is in seconds. That is not surprising because $\mathcal{P}$ is *power,* the rate of transfer of energy by heat. For a slab of infinitesimal thickness dx and temperature difference dT, we can write the **law of conduction** as

■ Law of conduction

$$\mathcal{P} = kA\left|\frac{dT}{dx}\right| \qquad [17.34]$$

where the proportionality constant k is called the **thermal conductivity** of the material and dT/dx is the **temperature gradient** (the variation of temperature with position). It is the higher thermal conductivity of tile relative to carpet that makes the tile floor feel colder than the carpeted floor in the discussion at the beginning of Chapter 16.

Suppose a substance is in the shape of a long uniform rod of length L as in Figure 17.19 and is insulated so that energy cannot escape by heat from its surface except at the ends, which are in thermal contact with reservoirs having temperatures T_c and T_h. When steady state is reached, the temperature at each point along the rod is constant in time. In this case, the temperature gradient is the same everywhere along the rod and is

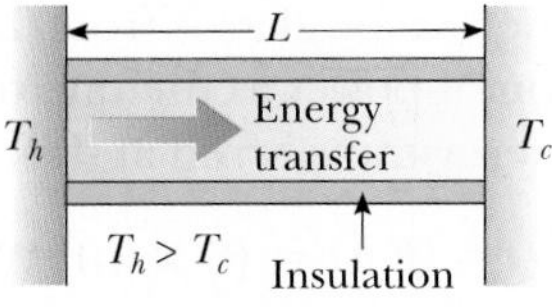

FIGURE 17.19 Conduction of energy through a uniform, insulated rod of length L. The opposite ends are in thermal contact with energy reservoirs at different temperatures.

$$\left|\frac{dT}{dx}\right| = \frac{T_h - T_c}{L}$$

TABLE 17.4

Thermal Conductivities

Substance	Thermal Conductivity (W/m·°C)
Metals (at 25°C)	
Aluminum	238
Copper	397
Gold	314
Iron	79.5
Lead	34.7
Silver	427
Nonmetals (approximate values)	
Asbestos	0.08
Concrete	0.8
Diamond	2 300
Glass	0.8
Ice	2
Rubber	0.2
Water	0.6
Wood	0.08
Gases (at 20°C)	
Air	0.023 4
Helium	0.138
Hydrogen	0.172
Nitrogen	0.023 4
Oxygen	0.023 8

Thus, the rate of energy transfer by heat is

$$\mathcal{P} = kA\frac{(T_h - T_c)}{L} \qquad [17.35]$$

Substances that are good thermal conductors have large thermal conductivity values, whereas good thermal insulators have low thermal conductivity values. Table 17.4 lists thermal conductivities for various substances.

QUICK QUIZ 17.7 You have two rods of the same length and diameter, but they are formed from different materials. The rods will be used to connect two regions of different temperature such that energy will transfer through the rods by heat. They can be connected in series, as in Figure 17.20a, or in parallel, as in Figure 17.20b. In which case is the rate of energy transfer by heat larger? **(a)** It is larger when the rods are in series. **(b)** It is larger when the rods are in parallel. **(c)** The rate is the same in both cases.

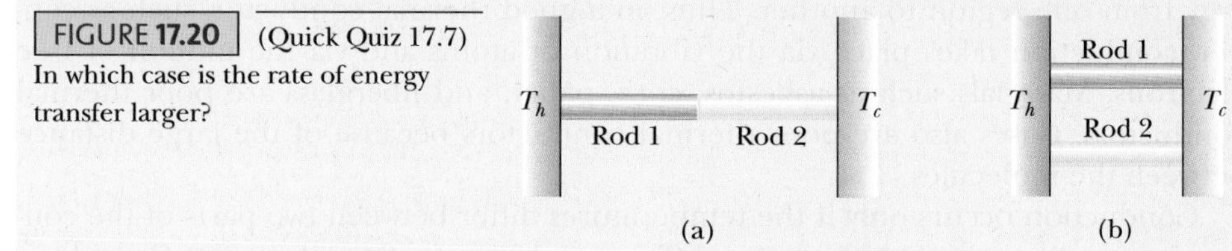

FIGURE 17.20 (Quick Quiz 17.7) In which case is the rate of energy transfer larger?

INTERACTIVE EXAMPLE 17.9 The Leaky Window

A window of area 2.0 m^2 is glazed with glass of thickness 4.0 mm. The window is in the wall of a house, and the outside temperature is 10°C. The temperature inside the house is 25°C.

A How much energy transfers through the window by heat in 1.0 h?

Solution We use Equation 17.35 to find the rate of energy transfer by heat:

$$\mathcal{P} = kA\frac{(T_h - T_c)}{L}$$

$$= (0.8 \text{ W/m}\cdot{}^\circ\text{C})(2.0 \text{ m}^2)\frac{(25^\circ\text{C} - 10^\circ\text{C})}{4.0 \times 10^{-3} \text{ m}}$$

$$= 6 \times 10^3 \text{ W}$$

where k for glass is from Table 17.4. From the definition of power as the rate of energy transfer, we find the energy transferred at this rate in 1.0 h:

$$Q = \mathcal{P}\,\Delta t = (6 \times 10^3 \text{ W})(3.6 \times 10^3 \text{ s}) = 2 \times 10^7 \text{ J}$$

B If electrical energy costs 12¢/kWh, how much does the transfer of energy in part A cost to replace with electrical heating?

Solution We cast the answer to part A in units of kilowatt-hours:

$$Q = \mathcal{P}\,\Delta t = (6 \times 10^3 \text{ W})(1.0 \text{ h}) = 6 \times 10^3 \text{ Wh} = 6 \text{ kWh}$$

Thus, the cost to replace the energy transferred through the window is (6 kWh)(12¢/kWh) ≈ 72¢.

If you imagine paying this much for each hour for each window in your home, your electric bill will be extremely high! For example, for ten such windows, your electric bill would be over $5 000 for one month. It seems like something is wrong here because electric bills are not that high. In reality, a thin layer of air adheres to each of the two surfaces of the window. This air provides additional insulation to that of the glass. As seen in Table 17.4, air is a much poorer thermal conductor than glass, so most of the insulation is performed by the air, not the glass, in a window!

PhysicsNow™ Determine the cost for replacing the energy conducting through windows by logging into PhysicsNow at **www.pop4e.com** and going to Interactive Example 17.9.

Convection

At one time or another you may have warmed your hands by holding them over an open flame. In this situation, the air directly above the flame is heated and expands. As a result, the density of the air decreases and the air rises. This warmed mass of air transfers energy by heat into your hands as it flows by. The transfer of energy from the flame to your hands is performed by matter transfer because the energy travels with the air. Energy transferred by the movement of a fluid is a process called **convection.** When the movement results from differences in density, as in the example of air around a fire, the process is called **natural convection.** When the fluid is forced to move by a fan or pump, as in some air and water heating systems, the process is called **forced convection.**

The circulating pattern of air flow at a beach (Fig 17.2) is an example of convection in nature. The mixing that occurs as water is cooled and eventually freezes at its surface (Section 16.3) is another example.

If it were not for convection currents, it would be very difficult to boil water. As water is heated in a teakettle, the lower layers are warmed first. These regions expand and rise to the top because their density is lower than that of the cooler water. At the same time, the denser cool water falls to the bottom of the kettle so that it can be heated.

The same process occurs near the surface of the Sun. Figure 17.21 shows a close-up view of the solar surface. The granulation that appears is because of *convection cells.* The bright center of a cell is the location at which hot gases rise to the surface, just like the hot water rises to the surface in a pan of boiling water. As the gases cool, they sink back downward along the edges of the cell, forming the darker outline of each cell. The sinking gases appear dark because they are cooler than the gases in the center of the cell. Although the sinking gases emit a tremendous amount of radiation, the filter used to take the photograph in Figure 17.21 makes these areas appear dark relative to the warmer center of the cell.

Convection occurs when a room is heated by a radiator. The radiator warms the air in the lower regions of the room by heat at the interface between the radiator surface and the air. The warm air expands and floats to the ceiling because of its lower density, setting up the continuous air current pattern shown in Figure 17.22.

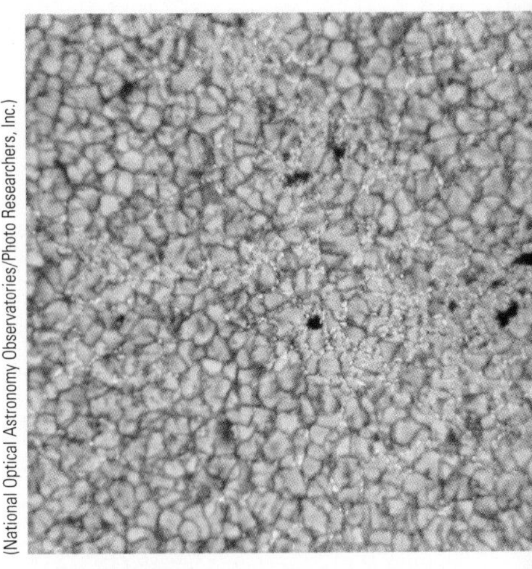

(National Optical Astronomy Observatories/Photo Researchers, Inc.)

FIGURE 17.21 The surface of the Sun shows *granulation,* due to the existence of separate convection cells, each carrying energy to the surface by convection.

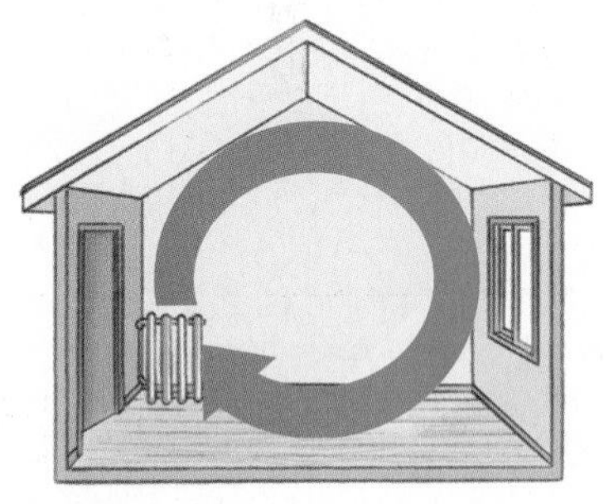

FIGURE 17.22 Convection currents are set up in a room heated by a radiator.

Radiation

Another method of transferring energy that can be related to a temperature change is **electromagnetic radiation.** All objects radiate energy continuously in the form of electromagnetic waves. As we shall find out in Chapter 24, electromagnetic radiation arises from accelerating electric charges. From our discussion of temperature, we know that temperature corresponds to random motion of molecules that are constantly changing direction and therefore are accelerating. Because the molecules contain electric charges, the charges also accelerate. Thus, any object emits electromagnetic radiation because of the thermal motion of its molecules. This radiation is called **thermal radiation.**

Through electromagnetic radiation, approximately 1 370 J of energy from the Sun strikes each square meter at the top of the Earth's atmosphere every second. Some of this energy is reflected back into space and some is absorbed by the atmosphere, but enough arrives at the surface of the Earth each day to supply all our energy needs on this planet hundreds of times over, if it could be captured and used efficiently. The growth in the number of solar houses in the United States is one example of an attempt to make use of this abundant energy.

The rate at which an object emits energy by thermal radiation from its surface is proportional to the fourth power of its absolute surface temperature. This principle, known as **Stefan's law,** is expressed in equation form as

■ Stefan's law

$$\mathcal{P} = \sigma A e T^4 \qquad [17.36]$$

where $\mathcal{P}$ is the power radiated by the object in watts, σ is the **Stefan–Boltzmann constant,** equal to $5.669\ 6 \times 10^{-8}\ \mathrm{W/m^2 \cdot K^4}$, A is the surface area of the object in square meters, e is a constant called the **emissivity,** and T is the surface temperature of the body in kelvins. The value of e can vary between zero and unity, depending on the properties of the surface. The emissivity of a surface is equal to its absorptivity, or the fraction of incoming radiation that the surface absorbs.

At the same time as it radiates, the object also absorbs electromagnetic radiation from the environment. If the latter process did not occur, an object would continuously radiate its energy and its temperature would eventually decrease spontaneously to absolute zero. If an object is at a temperature T and its environment is at a temperature T_0, the net rate of energy change for the object as a result of radiation is

$$\mathcal{P}_{\text{net}} = \sigma A e (T^4 - T_0{}^4) \qquad [17.37]$$

When an object is in equilibrium with its environment, it radiates and absorbs energy at the same rate and so its temperature remains constant, which is the nonisolated system in steady-state model. When an object is hotter than its surroundings, it radiates more energy than it absorbs and so it cools, which is the nonisolated system model.

■ Thinking Physics 17.3

If you sit in front of a fire with your eyes closed, you can feel significant warmth in your eyelids. If you now put on a pair of eyeglasses and repeat this activity, your eyelids will not feel nearly so warm. Why?

Reasoning Much of the warmth you feel is because of electromagnetic radiation from the fire. A large fraction of this radiation is in the infrared part of the electromagnetic spectrum. (We will study the electromagnetic spectrum in detail in Chapter 24.) Your eyelids are particularly sensitive to infrared radiation. On the other hand, glass is very opaque to infrared radiation. Therefore, when you put on the glasses, you block much of the radiation from reaching your eyelids and they feel cooler. ■

■ Thinking Physics 17.4

If you inspect a lightbulb that has been operating for a long time, a dark region appears on the inner surface of the bulb. This region is located on the highest parts of the bulb's glass envelope. What is the origin of this dark region, and why is it located at the high point?

Reasoning The dark region is tungsten that vaporized from the filament of the lightbulb and collected on the inner surface of the glass. Many lightbulbs contain a gas that allows convection to occur within the bulb. The gas near the filament is at a very high temperature, causing it to expand and float upward due to Archimedes's principle. As it floats upward, it carries the vaporized tungsten with it, and the tungsten collects on the surface at the top of the lightbulb. ■

17.11 ENERGY BALANCE FOR THE EARTH

Let us follow up on our discussion of energy transfer by radiation for the Earth from Section 17.10. We will then perform an initial calculation of the temperature of the Earth.

As mentioned previously, energy arrives at the Earth by electromagnetic radiation from the Sun.[5] This energy is absorbed by the surface of the Earth and is reradiated out into space according to Stefan's law, Equation 17.36. The only type of energy in the system that can change due to radiation is internal energy. Let us assume that any change in temperature of the Earth is so small over a time interval that we can approximate the change in internal energy as zero. This assumption leads to the following reduction of the continuity equation, Equation 6.20:

$$0 = T_{\text{ER}}$$

Two energy transfer mechanisms occur by electromagnetic radiation, so we can write this equation as

$$0 = T_{\text{ER}}\text{ (in)} + T_{\text{ER}}\text{ (out)} \quad\rightarrow\quad T_{\text{ER}}\text{ (in)} = -T_{\text{ER}}\text{ (out)} \qquad [17.38]$$

where "in" and "out" refer to energy transfers across the boundary of the system of the Earth. The energy coming into the system is from the Sun, and the energy going out of the system is by thermal radiation emitted from the Earth's surface. Figure 17.23 depicts these energy exchanges. The energy coming in from the Sun comes from only one direction, but the energy radiated out from the Earth's surface leaves in all directions. This distinction will be important in setting up our calculation of the equilibrium temperature.

As mentioned in Section 17.10, the rate of energy transfer per unit area from the Sun is approximately 1 370 W/m^2 at the top of the atmosphere. The rate of energy transfer per area is called **intensity,** and the intensity of radiation from the Sun at the top of the atmosphere is called the **solar constant** I_S. A large amount of this energy is in the form of visible radiation, to which the atmosphere is transparent. The radiation emitted from the Earth's surface, however, is not visible. For a radiating object at the temperature of the Earth's surface, the radiation peaks in the infrared, with greatest intensity at a wavelength of about 10 μm. In general, objects with typical household temperatures have wavelength distributions in the infrared, so we do not see them glowing visibly. Only much hotter items emit enough radiation to be seen visibly. An example is a household electric stove burner. When turned off, it emits a small amount of radiation, mostly in the infrared. When turned to its highest setting, its much higher temperature results in significant radiation, with much of it in the visible. As a result, it appears to glow with a reddish color and is described as *red-hot.*

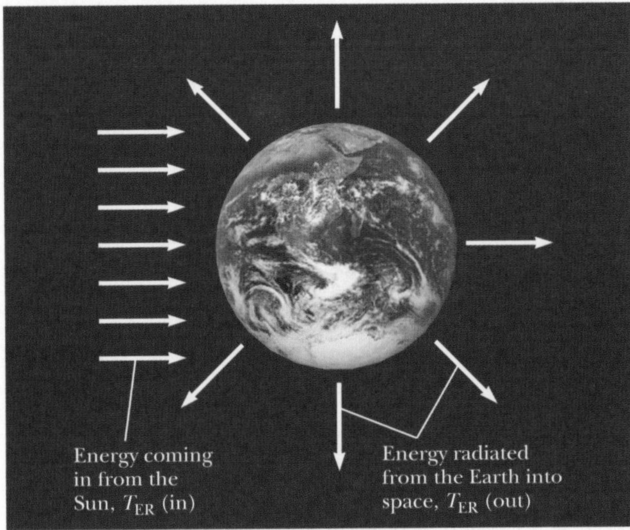

FIGURE 17.23 Energy exchanges by electromagnetic radiation for the Earth. The Sun is far to the left of the diagram and is not visible. When the rates of energy transfer due to these exchanges are equal, the temperature of the Earth remains constant.

[5]Some energy arrives at the surface of the Earth from the interior. The source of this energy is radioactive decay (Chapter 30) deep underground. We will ignore this energy because its contribution is much smaller than that due to electromagnetic radiation from the Sun.

Let us divide Equation 17.38 by the time interval Δt during which the energy transfer occurs, which gives us

$$\mathcal{P}_{\text{ER}}\,(\text{in}) = -\mathcal{P}_{\text{ER}}\,(\text{out}) \qquad [17.39]$$

We can express the rate of energy transfer into the top of the atmosphere of the Earth in terms of the solar constant I_S:

$$\mathcal{P}_{\text{ER}}\,(\text{in}) = I_S A_c$$

where A_c is the circular cross-sectional area of the Earth. Not all the radiation arriving at the top of the atmosphere reaches the ground. A fraction of it is reflected from clouds and the ground and escapes back into space. For the Earth, this fraction is about 30%, so only 70% of the incident radiation reaches the surface. Using this fact, we modify the input power, assuming that 70.0% reaches the surface:

$$\mathcal{P}_{\text{ER}}\,(\text{in}) = (0.700)\,I_S A_c$$

Stefan's law can be used to express the outgoing power, assuming that the Earth is a perfect emitter ($e = 1$):

$$\mathcal{P}_{\text{ER}}\,(\text{out}) = -\sigma A T^4$$

In this expression, A is the surface area of the Earth and the negative sign indicates that energy is leaving the Earth. Substituting the expressions for the input and output power into Equation 17.39, we have

$$(0.700)\,I_S A_c = -(-\sigma A T^4)$$

Solving for the temperature of the Earth's surface gives

$$T = \left(\frac{(0.700)\,I_S A_c}{\sigma A}\right)^{1/4}$$

Substituting the numbers, we find that

$$T = \left(\frac{(0.700)(1\,370\ \text{W/m}^2)(\pi R_E{}^2)}{(5.67 \times 10^{-8}\ \text{W/m}^2\cdot\text{K}^4)(4\pi R_E{}^2)}\right)^{1/4} = 255\ \text{K} \qquad [17.40]$$

Measurements show that the average global temperature at the surface of the Earth is 288 K, about 33 K warmer than the temperature from our calculation. This difference indicates that a major factor was left out of our analysis. The major factor is the thermodynamic effects of the atmosphere, which result in additional energy from the Sun being "trapped" in the system of the Earth and raising the temperature. This effect is not included in the simple energy balance calculation we performed. To evaluate it, we must incorporate into our model the principles of thermodynamics of gases for the air in the atmosphere. The details of this incorporation are explored in the Context Conclusion.

SUMMARY

PhysicsNow™ Take a practice test by logging into PhysicsNow at www.pop4e.com and clicking on the Pre-Test link for this chapter.

The **internal energy** E_{int} of a system is the total of the kinetic and potential energies of the system associated with its microscopic components. **Heat** is a process by which energy is transferred as a consequence of a temperature difference. It is also the amount of energy Q transferred by this process.

The energy required to change the temperature of a substance by ΔT is

$$Q = mc\,\Delta T \qquad [17.3]$$

where m is the mass of the substance and c is its **specific heat.**

The energy required to change the phase of a pure substance of mass m is

$$Q = \pm mL \qquad [17.5]$$

The parameter L is called the **latent heat** of the substance and depends on the nature of the phase change and the properties of the substance.

A **state variable** of a system is a quantity that is defined for a given condition of the system. State variables for a gas include pressure, volume, temperature, and internal energy.

A **quasi-static process** is one that proceeds slowly enough to allow the system to always be in a state of thermal equilibrium.

The **work** done on a gas as its volume changes from some initial value V_i to some final value V_f is

$$W = -\int_{V_i}^{V_f} P\, dV \qquad [17.7]$$

where P is the pressure, which may vary during the process.

The **first law of thermodynamics** is a special case of the continuity equation for energy, relating the internal energy of a system to energy transfer by heat and work:

$$\Delta E_{\text{int}} = Q + W \qquad [17.8]$$

where Q is the energy transferred across the boundary of the system by heat and W is the work done on the system. Although Q and W both depend on the path taken from the initial state to the final state, internal energy is a state variable, so the quantity ΔE_{int} is independent of the path taken between given initial and final states.

An **adiabatic process** is one in which no energy is transferred by heat between the system and its surroundings ($Q = 0$). In this case, the first law gives $\Delta E_{\text{int}} = W$.

An **isobaric process** is one that occurs at constant pressure. The work done on the gas in such a process is $-P(V_f - V_i)$.

An **isovolumetric process** is one that occurs at constant volume. No work is done in such a process.

An **isothermal process** is one that occurs at constant temperature. The work done on an ideal gas during an isothermal process is

$$W = -nRT \ln\left(\frac{V_f}{V_i}\right) \qquad [17.12]$$

In a **cyclic process** (one that originates and terminates at the same state), $\Delta E_{\text{int}} = 0$, and therefore $Q = -W$.

We define the **molar specific heats** of an ideal gas with the following equations:

$$Q = nC_V \Delta T \qquad \text{(constant volume)} \qquad [17.13]$$

$$Q = nC_P \Delta T \qquad \text{(constant pressure)} \qquad [17.14]$$

where C_V is the **molar specific heat at constant volume** and C_P is the **molar specific heat at constant pressure.**

The change in internal energy of an ideal gas for *any* process in which the temperature change is ΔT is

$$\Delta E_{\text{int}} = nC_V \Delta T \qquad [17.18]$$

The molar specific heat at constant volume is related to internal energy as follows:

$$C_V = \frac{1}{n}\frac{dE_{\text{int}}}{dT} \qquad [17.19]$$

The molar specific heat at constant volume and molar specific heat at constant pressure for all ideal gases are related as follows:

$$C_P - C_V = R \qquad [17.21]$$

For an ideal gas undergoing an adiabatic process, where

$$\gamma = \frac{C_P}{C_V} \qquad [17.22]$$

the pressure and volume are related as

$$PV^\gamma = \text{constant} \qquad [17.24]$$

The theorem of equipartition of energy can be used to predict the molar specific heat at constant volume for various types of gases. Monatomic gases can only store energy by means of translational motion of the molecules of the gas. Diatomic and polyatomic gases can also store energy by means of rotation and vibration of the molecules. For a given molecule, the rotational and vibrational energies are quantized, so their contribution does not enter into the internal energy until the temperature is raised to a sufficiently high value.

Thermal conduction is the transfer of energy by molecular collisions. It is driven by a temperature difference, and the rate of energy transfer is

$$\mathcal{P} = kA\left|\frac{dT}{dx}\right| \qquad [17.34]$$

where the constant k is called the **thermal conductivity** of the material and dT/dx is the **temperature gradient** (the variation of temperature with position).

Convection is energy transfer by means of a moving fluid.

All objects emit **electromagnetic radiation** continuously in the form of **thermal radiation,** which depends on temperature according to **Stefan's law:**

$$\mathcal{P} = \sigma A e T^4 \qquad [17.36]$$

QUESTIONS

☐ = answer available in the *Student Solutions Manual and Study Guide*

1. Clearly distinguish among temperature, heat, and internal energy.
2. Ethyl alcohol has about half the specific heat of water. Assuming that equal-mass samples of alcohol and water in separate beakers are supplied with the same amount of energy, compare the temperature increases of the two liquids.
3. A small metal crucible is taken from a 200°C oven and immersed in a tub full of water at room temperature (in a process often referred to as *quenching*). What is the approximate final equilibrium temperature?
4. What is a major problem that arises in measuring specific heats if a sample with a temperature above 100°C is placed in water?

5. What is wrong with the statement, "Given any two objects, the one with the higher temperature contains more heat"?

6. Why is a person able to remove a piece of dry aluminum foil from a hot oven with bare fingers, whereas a burn results if there is moisture on the foil?

7. The air temperature above coastal areas is profoundly influenced by the large specific heat of water. One reason is that the energy released when 1 m^3 of water cools by 1°C will raise the temperature of a much larger volume of air by 1°C. Find this volume of air. The specific heat of air is approximately 1 kJ/kg · °C. Take the density of air to be 1.3 kg/m^3.

8. When a sealed Thermos bottle full of hot coffee is shaken, what are the changes, if any, in (a) the temperature of the coffee and (b) the internal energy of the coffee?

9. Using the first law of thermodynamics, explain why the *total* energy of an isolated system is always constant.

10. The U.S. penny was formerly made mostly of copper and is now made of copper-coated zinc. Can a calorimetric experiment be devised to test for the metal content in a collection of pennies? If so, describe the procedure you would use.

11. A tile floor in a bathroom may feel uncomfortably cold to your bare feet, but a carpeted floor in an adjoining room at the same temperature will feel warm. Why?

12. Why can potatoes be baked more quickly when a metal skewer has been inserted through them?

13. Why do heavy draperies over the windows help keep a home cool in the summer as well as warm in the winter?

14. Pioneers stored fruits and vegetables in underground cellars. Discuss the advantages of this choice for a storage site.

15. The pioneers referred to in the last question found that a large tub of water placed in a storage cellar would prevent their food from freezing on really cold nights. Explain why.

16. Why is it more comfortable to hold a cup of hot tea by the handle rather than by wrapping your hands around the cup itself?

17. You need to pick up a very hot cooking pot in your kitchen. You have a pair of hot pads. To be able to pick up the pot most comfortably, should you soak them in cold water or keep them dry?

18. Suppose you pour hot coffee for your guests, and one of them wants to drink it with cream, several minutes later, and then as warm as possible. To have the warmest coffee, should the person add the cream just after the coffee is poured or just before drinking? Explain.

19. A warning sign often seen on highways just before a bridge is "Caution—Bridge surface freezes before road surface." Which of the three energy transfer processes discussed in Section 17.10 is most important in causing a bridge surface to freeze before a road surface on very cold days?

20. A professional physics teacher drops one marshmallow into a flask of liquid nitrogen, waits for the most energetic boiling to stop, fishes it out with tongs, shakes it off, pops it into his mouth, chews it up, and swallows it. Clouds of ice crystals issue from his mouth as he crunches noisily and comments on the sweet taste. How can he do that without injury? (*Caution:* Liquid nitrogen can be a dangerous substance and you should *not* try this experiment yourself. The teacher might be badly injured if he did not shake the marshmallow off, if he touched the tongs to a tooth, or if he did not start with a mouthful of saliva.)

21. In 1801, Humphry Davy rubbed together pieces of ice inside an icehouse. He took care that nothing in his environment was at a higher temperature than the rubbed pieces. He observed the production of drops of liquid water. Make a table listing this and other experiments or processes to illustrate each of the following situations. (a) A system can absorb energy by heat, increase in internal energy, and increase in temperature. (b) A system can absorb energy by heat and increase in internal energy, without an increase in temperature. (c) A system can absorb energy by heat without increasing in temperature or in internal energy. (d) A system can increase in internal energy and in temperature, without absorbing energy by heat. (e) A system can increase in internal energy without absorbing energy by heat or increasing in temperature. (f) If a system's temperature increases, is it necessarily true that its internal energy increases?

22. A liquid partially fills a container. Explain why the temperature of the liquid decreases if the container is then partially evacuated. (Using this technique, it is possible to freeze water at temperatures above 0°C.)

PROBLEMS

1, 2, 3 = straightforward, intermediate, challenging
☐ = full solution available in the *Student Solutions Manual and Study Guide*
Physics Now™ = coached problem with hints available at www.pop4e.com
= computer useful in solving problem
= paired numerical and symbolic problems
= biomedical application

Section 17.1 ■ Heat and Internal Energy

1. On his honeymoon, James Joule traveled from England to Switzerland. He attempted to verify his idea of the interconvertibility of mechanical energy and internal energy by measuring the increase in temperature of water that fell in a waterfall. For the waterfall near Chamonix in the French Alps, which has a 120-m drop, what maximum temperature rise could Joule expect? He did not succeed in measuring it, partly because evaporation cooled the falling water and also because his thermometer was not sufficiently sensitive.

2. Consider Joule's apparatus diagrammed in Figure P17.2. The mass of each of the two blocks is 1.50 kg, and the insulated tank is filled with 200 g of water. What is the increase in the temperature of the water after the blocks fall through a distance of 3.00 m?

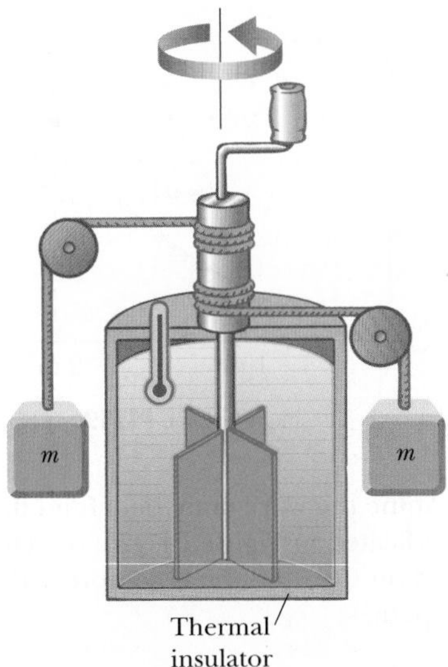

FIGURE **P17.2** The falling weights rotate the paddles, causing the temperature of the water to increase.

Section 17.2 ■ Specific Heat

3. A 50.0-g sample of copper is at 25.0°C. If 1 200 J of energy is added to it by heat, what is the final temperature of the copper?

4. Systematic use of solar energy can yield a large saving in the cost of winter space heating for a typical house in the north central United States. If the house has good insulation, you may model it as losing energy by heat steadily at the rate 6 000 W on a day in April when the average exterior temperature is 4°C and when the conventional heating system is not used at all. The passive solar energy collector can consist simply of very large windows in a room facing south. Sunlight shining in during the daytime is absorbed by the floor, interior walls, and objects in the room, raising their temperature to 38°C. As the sun goes down, insulating draperies or shutters are closed over the windows. During the period between 5:00 P.M. and 7:00 A.M., the temperature of the house will drop and a sufficiently large "thermal mass" is required to keep it from dropping too far. The thermal mass can be a large quantity of stone (with specific heat 850 J/kg · °C) in the floor and the interior walls exposed to sunlight. What mass of stone is required if the temperature is not to drop below 18°C overnight?

5. A 1.50-kg iron horseshoe initially at 600°C is dropped into a bucket containing 20.0 kg of water at 25.0°C. What is the final temperature? (Ignore the heat capacity of the container and assume that a negligible amount of water boils away.)

6. An aluminum cup of mass 200 g contains 800 g of water in thermal equilibrium at 80.0°C. The combination of cup and water is cooled uniformly so that the temperature decreases by 1.50°C per minute. At what rate is energy being removed by heat? Express your answer in watts.

7. An electric drill with a steel drill bit of mass 27.0 g and diameter 0.635 cm is used to drill into a cubical steel block of mass 240 g. Assume that steel has the same properties as iron. The cutting process can be modeled as happening at one point on the circumference of the bit. This point moves in a helix at constant speed 40.0 m/s and exerts a force of constant magnitude 3.20 N on the block. As shown in Figure P17.7, a groove in the bit carries the chips up to the top of the block, where they form a pile around the hole. The block is held in a clamp made of material of low thermal conductivity, and the drill bit is held in a chuck also made of this material. We consider turning the drill on for a time interval of 15.0 s. This time is sufficiently short that the steel objects lose only a negligible amount of energy by conduction, convection, and radiation into their environment, but 15.0 s is long enough for conduction within the steel to bring it all to a uniform temperature. The temperature is promptly measured with a thermometer probe, shown in the side of the block in the figure. (a) First suppose the drill bit is sharp and cuts three quarters of the way through the block during 15.0 s. Find the temperature change of the whole quantity of steel. (b) Now suppose the drill bit is dull and cuts only one eighth of the way through the block. Identify the temperature change of the whole quantity of steel in this case.

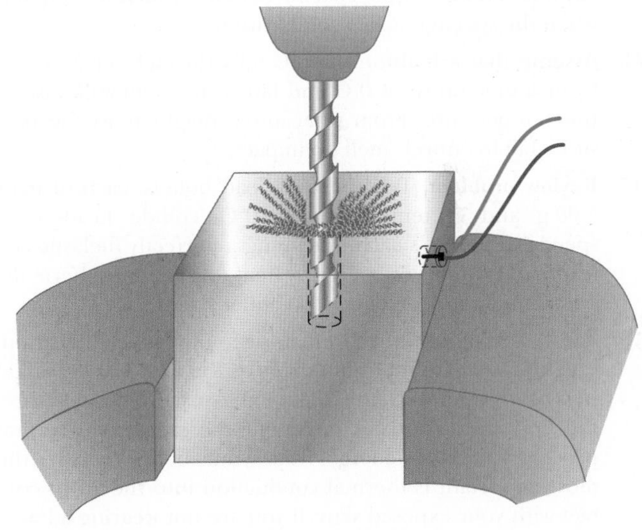

FIGURE **P17.7**

8. An aluminum calorimeter with a mass of 100 g contains 250 g of water. The calorimeter and water are in thermal equilibrium at 10.0°C. Two metallic blocks are placed into the water. One is a 50.0-g piece of copper at 80.0°C. The other block has a mass of 70.0 g and is originally at a temperature of 100°C. The entire system stabilizes at a final temperature of 20.0°C. (a) Determine the specific heat of the unknown sample. (b) Guess the material of the unknown, using the data in Table 17.1.

9. A combination of 0.250 kg of water at 20.0°C, 0.400 kg of aluminum at 26.0°C, and 0.100 kg of copper at 100°C is

mixed in an insulated container and allowed to come to thermal equilibrium. Ignore any energy transfer to or from the container and determine the final temperature of the mixture.

10. If water with a mass m_h at temperature T_h is poured into an aluminum cup of mass m_{Al} containing mass m_c of water at T_c, where $T_h > T_c$, what is the equilibrium temperature of the system?

Section 17.3 ▪ Latent Heat and Phase Changes

11. How much energy is required to change a 40.0-g ice cube from ice at −10.0°C to steam at 110°C?

12. A 50.0-g copper calorimeter contains 250 g of water at 20.0°C. How much steam must be condensed into the water if the final temperature of the system is to reach 50.0°C?

13. A 3.00-g lead bullet at 30.0°C is fired at a speed of 240 m/s into a large block of ice at 0°C, in which it becomes embedded. What quantity of ice melts?

14. A 1.00-kg block of copper at 20.0°C is dropped into a large vessel of liquid nitrogen at 77.3 K. How many kilograms of nitrogen boil away by the time the copper reaches 77.3 K? (The specific heat of copper is 0.092 0 cal/g · °C. The latent heat of vaporization of nitrogen is 48.0 cal/g.)

15. **PhysicsNow™** In an insulated vessel, 250 g of ice at 0°C is added to 600 g of water at 18.0°C. (a) What is the final temperature of the system? (b) How much ice remains when the system reaches equilibrium?

16. Assume that a hailstone at 0°C falls through air at a uniform temperature of 0°C and lands on a sidewalk also at this temperature. From what initial height must the hailstone fall to entirely melt on impact?

17. **Review problem.** Two speeding lead bullets, each of mass 5.00 g and at temperature 20.0°C, collide head-on at speeds of 500 m/s each. Assuming a perfectly inelastic collision and no loss of energy by heat to the atmosphere, describe the final state of the two-bullet system.

18. A resting adult of average size converts chemical energy in food into internal energy at the rate 120 W, called her *basal metabolic rate.* To stay at constant temperature, the body must put out energy at the same rate. Several processes exhaust energy from your body. Usually, the most important is thermal conduction into the air in contact with your exposed skin. If you are not wearing a hat, a convection current of warm air rises vertically from your head like a plume from a smokestack. Your body also loses energy by electromagnetic radiation, by your exhaling warm air, and by evaporation of perspiration. In this problem, consider still another pathway for energy loss: moisture in exhaled breath. Suppose you breathe out 22.0 breaths per minute, each with a volume of 0.600 L. Assume that you inhale dry air and exhale air at 37°C containing water vapor with a vapor pressure of 3.20 kPa. The vapor came from evaporation of liquid water in your body. Model the water vapor as an ideal gas. Assume that its latent heat of evaporation at 37°C is the same as its heat of vaporization at 100°C. Calculate the rate at which you lose energy by exhaling humid air.

Section 17.4 ▪ Work in Thermodynamic Processes

19. **PhysicsNow™** A sample of ideal gas is expanded to twice its original volume of 1.00 m^3 in a quasi-static process for which $P = \alpha V^2$, with $\alpha = 5.00$ atm/m^6, as shown in Figure P17.19. How much work is done on the expanding gas?

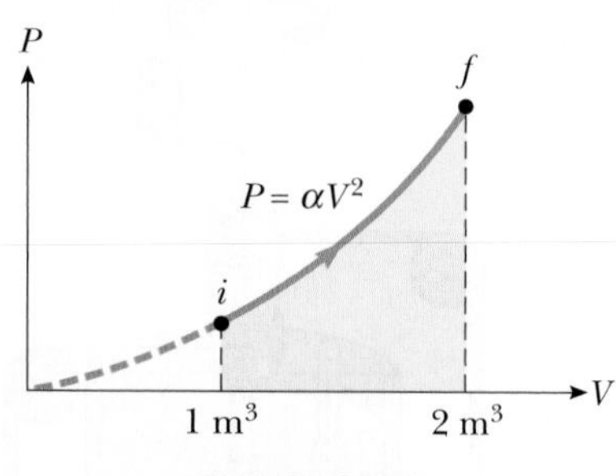

FIGURE **P17.19**

20. (a) Determine the work done on a fluid that expands from i to f as indicated in Figure P17.20. (b) How much work is performed on the fluid if it is compressed from f to i along the same path?

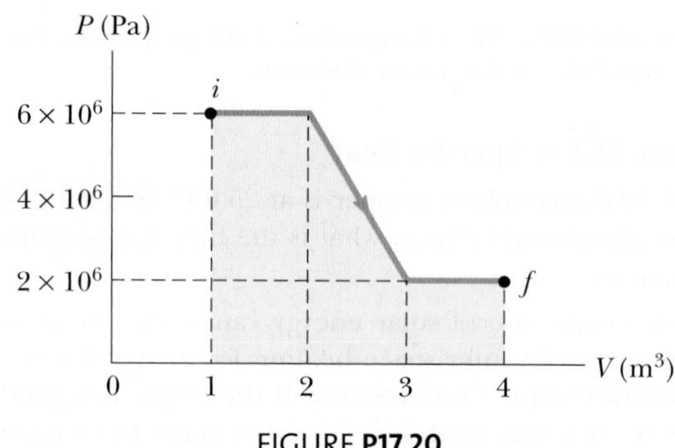

FIGURE **P17.20**

21. An ideal gas is enclosed in a cylinder with a movable piston on top of it. The piston has a mass of 8 000 g and an area of 5.00 cm^2 and is free to slide up and down, keeping the pressure of the gas constant. How much work is done on the gas as the temperature of 0.200 mol of the gas is raised from 20.0°C to 300°C?

22. An ideal gas is enclosed in a cylinder that has a movable piston on top. The piston has a mass m and an area A and is free to slide up and down, keeping the pressure of the gas constant. How much work is done on the gas as the temperature of n mol of the gas is raised from T_1 to T_2?

Section 17.5 ▪ The First Law of Thermodynamics

23. A thermodynamic system undergoes a process in which its internal energy decreases by 500 J. At the same time, 220 J of work is done on the system. Find the energy transferred to or from it by heat.

24. A gas is taken through the cyclic process described in Figure P17.24. (a) Find the net energy transferred to the system by heat during one complete cycle. (b) If the cycle is reversed—that is, the process follows the path *ACBA*—what is the net energy input per cycle by heat?

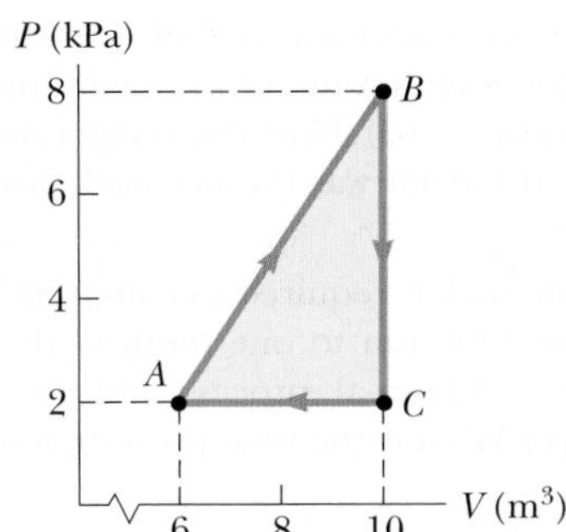

FIGURE **P17.24** Problems 17.24 and 17.25.

25. Consider the cyclic process depicted in Figure P17.24. If Q is negative for the process BC, and ΔE_{int} is negative for the process CA, what are the signs of Q, W, and ΔE_{int} that are associated with each process?

Section 17.6 ■ Some Applications of the First Law of Thermodynamics

26. One mole of an ideal gas does 3 000 J of work on its surroundings as it expands isothermally to a final pressure of 1.00 atm and volume of 25.0 L. Determine (a) the initial volume and (b) the temperature of the gas.

27. An ideal gas initially at 300 K undergoes an isobaric expansion at 2.50 kPa. If the volume increases from 1.00 m^3 to 3.00 m^3 and 12.5 kJ is transferred to the gas by heat, what are (a) the change in its internal energy and (b) its final temperature?

28. A 1.00-kg block of aluminum is heated at atmospheric pressure so that its temperature increases from 22.0°C to 40.0°C. Find (a) the work done on the aluminum, (b) the energy added to it by heat, and (c) the change in its internal energy.

29. How much work is done on the steam when 1.00 mol of water at 100°C boils and becomes 1.00 mol of steam at 100°C at 1.00 atm pressure? Assuming the steam to behave as an ideal gas, determine the change in internal energy of the material as it vaporizes.

30. An ideal gas initially at P_i, V_i, and T_i is taken through a cycle as shown in Figure P17.30. (a) Find the net work done on the gas per cycle. (b) What is the net energy added by heat to the system per cycle? (c) Obtain a numerical value for the net work done per cycle for 1.00 mol of gas initially at 0°C.

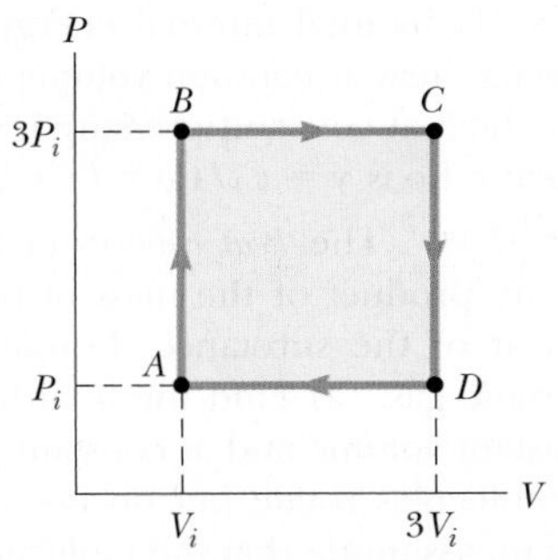

FIGURE **P17.30**

31. A 2.00-mol sample of helium gas initially at 300 K and 0.400 atm is compressed isothermally to 1.20 atm. Noting that the helium behaves as an ideal gas, find (a) the final volume of the gas, (b) the work done on the gas, and (c) the energy transferred by heat.

32. In Figure P17.32, the change in internal energy of a gas that is taken from A to C is $+800$ J. The work done on the gas along path ABC is -500 J. (a) How much energy must be added to the system by heat as it goes from A through B to C? (b) If the pressure at point A is five times that of point C, what is the work done on the system in going from C to D? (c) What is the energy exchanged with the surroundings by heat as the cycle goes from C to A along the green path? (d) If the change in internal energy in going from point D to point A is $+500$ J, how much energy must be added to the system by heat as it goes from point C to point D?

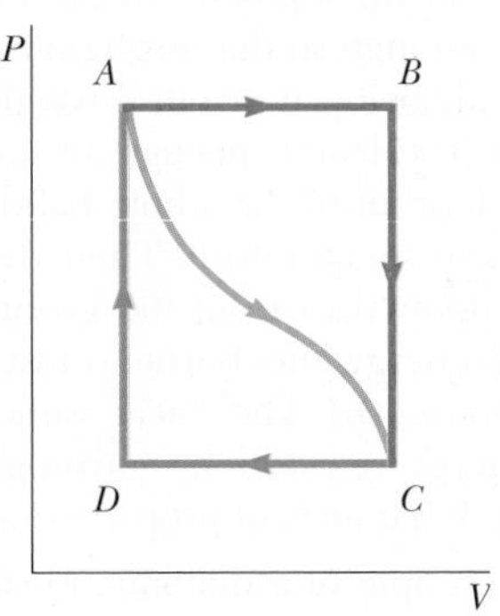

FIGURE **P17.32**

Section 17.7 ■ Molar Specific Heats of Ideal Gases

Note: You may use data in Table 17.3 about particular gases. Here we define a "monatomic ideal gas" to have molar specific heats $C_V = 3R/2$ and $C_P = 5R/2$, and a "diatomic ideal gas" to have $C_V = 5R/2$ and $C_P = 7R/2$.

33. **PhysicsNow™** A 1.00-mol sample of hydrogen gas is heated at constant pressure from 300 K to 420 K. Calculate (a) the energy transferred to the gas by heat, (b) the increase in its internal energy, and (c) the work done on the gas.

34. Calculate the change in internal energy of 3.00 mol of helium gas when its temperature is increased by 2.00 K.

35. In a constant-volume process, 209 J of energy is transferred by heat to 1.00 mol of an ideal monatomic gas initially at 300 K. Find (a) the increase in internal energy of the gas, (b) the work done on it, and (c) its final temperature.

36. A vertical cylinder with a heavy piston contains air at a temperature of 300 K. The initial pressure is 200 kPa and the initial volume is 0.350 m^3. Take the molar mass of air as 28.9 g/mol and assume that $C_V = 5R/2$. (a) Find the specific heat of air at constant volume in units of J/kg · °C. (b) Calculate the mass of the air in the cylinder. (c) Suppose the piston is held fixed. Find the energy input required to raise the temperature of the air to 700 K. (d) Assume again the conditions of the initial state and that the heavy piston is free to move. Find the energy input required to raise the temperature to 700 K.

37. A 1-L Thermos bottle is full of tea at 90°C. You pour out one cup and immediately screw the stopper back on. Make an order-of-magnitude estimate of the change in temperature of the tea remaining in the flask that results from the admission of air at room temperature. State the quantities you take as data and the values you measure or estimate for them.

38. **Review problem.** This problem is a continuation of Problem 16.21 in Chapter 16. A hot-air balloon consists of an envelope of constant volume 400 m^3. Not including the air inside, the balloon and cargo have mass 200 kg. The air outside and originally inside is a diatomic ideal gas at 10.0°C and 101 kPa, with density 1.25 kg/m^3. A propane burner at the center of the spherical envelope injects energy into the air inside. The air inside stays at constant pressure. Hot air, at just the temperature required to make the balloon lift off, starts to fill the envelope at its closed top, rapidly enough so that negligible energy flows by heat to the cool air below it or out through the wall of the balloon. Air at 10°C leaves through an opening at the bottom of the envelope until the whole balloon is filled with hot air at uniform temperature. Then the burner is shut off and the balloon rises from the ground. (a) Evaluate the quantity of energy the burner must transfer to the air in the balloon. (b) The "heat value" of propane—the internal energy released by burning each kilogram—is 50.3 MJ/kg. What mass of propane must be burned?

39. A 1.00-mol sample of a diatomic ideal gas has pressure P and volume V. When the gas is heated, its pressure triples and its volume doubles. This heating process includes two steps, the first at constant pressure and the second at constant volume. Determine the amount of energy transferred to the gas by heat.

Section 17.8 ■ Adiabatic Processes for an Ideal Gas

40. During the compression stroke of a certain gasoline engine, the pressure increases from 1.00 atm to 20.0 atm. If the process is adiabatic and the fuel–air mixture behaves as a diatomic ideal gas, (a) by what factor does the volume change and (b) by what factor does the temperature change? (c) Assuming that the compression starts with 0.016 0 mol of gas at 27.0°C, find the values of Q, W, and ΔE_{int} that characterize the process.

41. A 2.00-mol sample of a diatomic ideal gas expands slowly and adiabatically from a pressure of 5.00 atm and a volume of 12.0 L to a final volume of 30.0 L. (a) What is the final pressure of the gas? (b) What are the initial and final temperatures? (c) Find Q, W, and ΔE_{int}.

42. Air in a thundercloud expands as it rises. If its initial temperature is 300 K and no energy is lost by thermal conduction on expansion, what is its temperature when the initial volume has doubled?

43. A 4.00-L sample of a diatomic ideal gas with specific heat ratio 1.40, confined to a cylinder, is carried through a closed cycle. The gas is initially at 1.00 atm and at 300 K. First, its pressure is tripled under constant volume. Then, it expands adiabatically to its original pressure. Finally, the gas is compressed isobarically to its original volume. (a) Draw a PV diagram of this cycle. (b) Determine the volume of the gas at the end of the adiabatic expansion. (c) Find the temperature of the gas at the start of the adiabatic expansion. (d) Find the temperature at the end of the cycle. (e) What was the net work done on the gas for this cycle?

44. How much work is required to compress 5.00 mol of air at 20.0°C and 1.00 atm to one tenth of the original volume (a) by an isothermal process and (b) by an adiabatic process? (c) What is the final pressure in each of these two cases?

45. During the power stroke in a four-stroke automobile engine, the piston is forced down as the mixture of combustion products and air undergoes an adiabatic expansion (Fig. P17.45). Assume that (1) the engine is running at 2 500 cycles/min; (2) the gauge pressure right before the expansion is 20.0 atm; (3) the volumes of the mixture right before and after the expansion are 50.0 and 400 cm^3, respectively; (4) the time interval for the expansion is one-fourth that of the total cycle; and (5) the mixture behaves like an ideal gas with specific heat ratio 1.40. Find the average power generated during the expansion stroke.

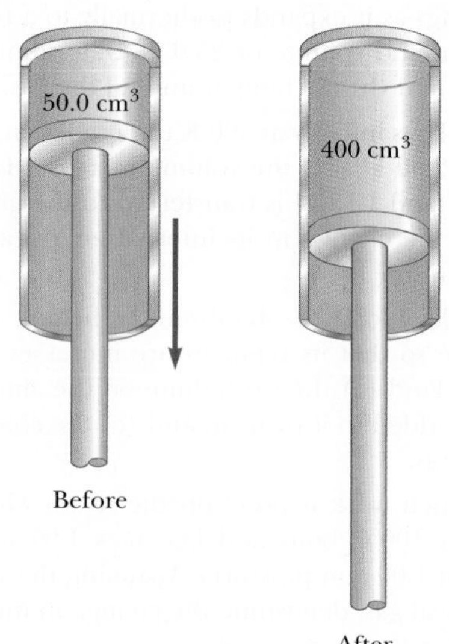

FIGURE P17.45

Section 17.9 ■ Molar Specific Heats and the Equipartition of Energy

46. A certain molecule has f degrees of freedom. Show that an ideal gas consisting of such molecules has the following properties: (1) its total internal energy is $fnRT/2$, (2) its molar specific heat at constant volume is $fR/2$, (3) its molar specific heat at constant pressure is $(f + 2)R/2$, (4) its specific heat ratio is $\gamma = C_P/C_V = (f + 2)/f$.

47. PhysicsNow™ The *heat capacity* of a sample of a substance is the product of the mass of the sample and the specific heat of the substance. Consider 2.00 mol of an ideal diatomic gas. (a) Find the total heat capacity of the gas at constant volume and at constant pressure, assuming that the molecules rotate but do not vibrate. (b) Repeat the problem, assuming that the molecules both rotate and vibrate.

Section 17.10 ■ Energy Transfer Mechanisms in Thermal Processes

48. A box with a total surface area of $1.20\ m^2$ and a wall thickness of 4.00 cm is made of an insulating material. A 10.0-W electric heater inside the box maintains the inside temperature at 15.0°C above the outside temperature. Find the thermal conductivity k of the insulating material.

49. A bar of gold is in thermal contact with a bar of silver of the same length and area (Fig. P17.49). One end of the compound bar is maintained at 80.0°C, and the opposite end is at 30.0°C. When the energy transfer reaches steady state, what is the temperature at the junction?

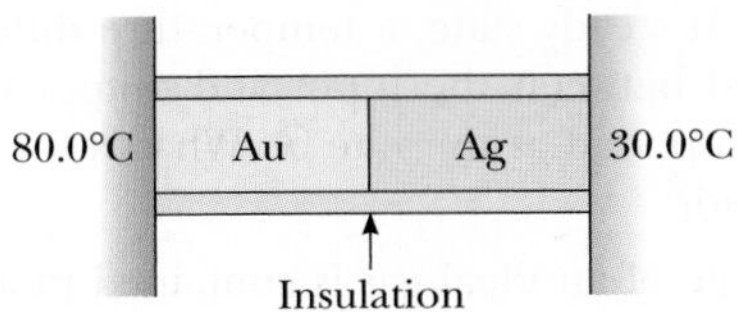

FIGURE **P17.49**

50. A power transistor is a solid-state electronic device. Assume that energy entering the device at the rate of 1.50 W by electrical transmission causes the internal energy of the device to increase. The surface area of the transistor is so small that it tends to overheat. To prevent overheating, the transistor is attached to a larger metal heat sink with fins. The temperature of the heat sink remains constant at 35.0°C under steady-state conditions. The transistor is electrically insulated from the heat sink by a rectangular sheet of mica measuring 8.25 mm by 6.25 mm, and 0.085 2 mm thick. Assume the thermal conductivity of mica to be 0.075 3 W/m · °C. What is the operating temperature of the transistor?

51. The surface of the Sun has a temperature of about 5 800 K. The radius of the Sun is 6.96×10^8 m. Calculate the total energy radiated by the Sun each second. Assume that the emissivity is 0.986.

52. The human body must maintain its core temperature inside a rather narrow range around 37°C. Metabolic processes, notably muscular exertion, convert chemical energy into internal energy deep in the interior. From the interior, energy must flow out to the skin or lungs to be expelled to the environment. During moderate exercise, an 80-kg man can metabolize food energy at the rate 300 kcal/h, do 60 kcal/h of mechanical work, and put out the remaining 240 kcal/h of energy by heat. Most of the energy is carried from the body interior out to the skin by forced convection (as a plumber would say), whereby blood is warmed in the interior and then cooled at the skin, which is a few degrees cooler than the body core. Without blood flow, living tissue is a good thermal insulator, with thermal conductivity about 0.210 W/m · °C. Show that blood flow is essential to cool the man's body by calculating the rate of energy conduction in kcal/h through the tissue layer under his skin. Assume that its area is $1.40\ m^2$, its thickness is 2.50 cm, and it is maintained at 37.0°C on one side and at 34.0°C on the other side.

53. A student is trying to decide what to wear. His bedroom is at 20°C. His skin temperature is 35°C. The area of his exposed skin is $1.50\ m^2$. People of all races have skin that is dark in the infrared, with emissivity about 0.900. Find the net energy loss from his body by radiation in 10.0 min.

Section 17.11 ■ Context Connection—Energy Balance for the Earth

54. At high noon, the Sun delivers 1 000 W to each square meter of a blacktop road. If the hot asphalt loses energy only by radiation, what is its equilibrium temperature?

55. The intensity of solar radiation reaching the top of the Earth's atmosphere is 1 370 W/m^2. The temperature of the Earth is affected by the so-called greenhouse effect of the atmosphere. That effect makes our planet's emissivity for visible light higher than its emissivity for infrared light. For comparison, consider a spherical object with no atmosphere, at the same distance from the Sun as the Earth. Assume that its emissivity is the same for all kinds of electromagnetic waves and that its temperature is uniform over its surface. Identify the projected area over which it absorbs sunlight and the surface area over which it radiates. Compute its equilibrium temperature. Chilly, isn't it? Your calculation applies to (a) the average temperature of the Moon, (b) astronauts in mortal danger aboard the crippled *Apollo 13* spacecraft, and (c) global catastrophe on the Earth if widespread fires caused a layer of soot to accumulate throughout the upper atmosphere so that most of the radiation from the Sun were absorbed there rather than at the surface below the atmosphere.

56. *A theoretical atmospheric lapse rate.* Section 16.7 described experimental data on the decrease in temperature with altitude in the Earth's atmosphere. Model the troposphere as an ideal gas, everywhere with equivalent molar mass M and ratio of specific heats γ. Absorption of sunlight at the Earth's surface warms the troposphere from below, so vertical convection currents are continually mixing the air. As a parcel of air rises, its pressure drops and it expands. The parcel does work on its surroundings, so its internal energy decreases and it drops in temperature. Assume that the vertical mixing is so rapid as to be adiabatic. (a) Show that the quantity $TP^{(1-\gamma)/\gamma}$ has a uniform value through the layers of the troposphere. (b) By differentiating with respect to altitude y, show that the lapse rate is given by

$$\frac{dT}{dy} = \frac{T}{P}\left(1 - \frac{1}{\gamma}\right)\frac{dP}{dy}$$

(c) A lower layer of air must support the weight of the layers above. From Equation 15.4, observe that mechanical equilibrium of the atmosphere requires that the pressure decrease with altitude according to $dP/dy = -\rho g$. The depth of the troposphere is small compared with the radius of the Earth, so you may assume that the free-fall acceleration is uniform. Proceed to prove that the lapse rate is

$$\frac{dT}{dy} = -\left(1 - \frac{1}{\gamma}\right)\frac{Mg}{R}$$

Problem 16.38 in Chapter 16 calls for evaluation of this theoretical lapse rate on the Earth and on Mars and for comparison with experimental results.

Additional Problems

57. For bacteriological testing of water supplies and in medical clinics, samples must routinely be incubated for 24 h at 37°C. A standard constant-temperature bath with electric heating and thermostatic control is not practical in third-world locations without continuously operating electric power lines. Peace Corps volunteer and MIT engineer Amy Smith invented a low-cost, low-maintenance incubator to fill the need. It consists of a foam-insulated box containing several packets of a waxy material that melts at 37.0°C, interspersed among tubes, dishes, or bottles containing the test samples and growth medium (bacteria food). Outside the box, the waxy material is first melted by a stove or solar energy collector. Then it is put into the box to keep the test samples warm as it solidifies. The heat of fusion of the phase-change material is 205 kJ/kg. Model the insulation as a panel with surface area 0.490 m^2, thickness 4.50 cm, and conductivity 0.012 0 W/m°C. Assume that the exterior temperature is 23.0°C for 12.0 h and 16.0°C for 12.0 h. (a) What mass of the waxy material is required to conduct the bacteriological test? (b) Explain why your calculation can be done without knowing the mass of the test samples or of the insulation.

58. A 75.0-kg cross-country skier moves across the snow (Fig. P17.58). The coefficient of friction between the skis and the snow is 0.200. Assume that all the snow beneath his skis is at 0°C and that all the internal energy generated by friction is added to the snow, which sticks to his skis until it melts. How far would he have to ski to melt 1.00 kg of snow?

FIGURE P17.58 A cross-country skier.

59. On a cold winter day, you buy roasted chestnuts from a street vendor. Into the pocket of your down parka you put the change he gives you: coins constituting 9.00 g of copper at − 12.0°C. Your pocket already contains 14.0 g of silver coins at 30.0°C. A short time later the temperature of the copper coins is 4.00°C and is increasing at a rate of 0.500°C/s. At this time, (a) what is the temperature of the silver coins and (b) at what rate is it changing?

60. An aluminum rod 0.500 m in length and with a cross-sectional area of 2.50 cm^2 is inserted into a thermally insulated vessel containing liquid helium at 4.20 K. The rod is initially at 300 K. (a) If one half of the rod is inserted into the helium, how many liters of helium boil off by the time the inserted half cools to 4.20 K? (Assume that the upper half does not yet cool.) (b) If the upper end of the rod is maintained at 300 K, what is the approximate boil-off rate of liquid helium after the lower half has reached 4.20 K? (Aluminum has thermal conductivity of 31.0 J/s · cm · K at 4.2 K; ignore its temperature variation. Aluminum has a specific heat of 0.210 cal/g · °C and density of 2.70 g/cm^3. The density of liquid helium is 0.125 g/cm^3.)

61. A *flow calorimeter* is an apparatus used to measure the specific heat of a liquid. The technique of flow calorimetry involves measuring the temperature difference between the input and output points of a flowing stream of the liquid while energy is added by heat at a known rate. A liquid of density ρ flows through the calorimeter with volume flow rate R. At steady state, a temperature difference ΔT is established between the input and output points when energy is supplied at the rate $\mathcal{P}$. What is the specific heat of the liquid?

62. One mole of an ideal gas is contained in a cylinder with a movable piston. The initial pressure, volume, and temperature are P_i, V_i, and T_i, respectively. Find the work done on the gas for the following processes and show each process on a PV diagram: (a) an isobaric compression in which the final volume is one-half the initial volume, (b) an isothermal compression in which the final pressure is four times the initial pressure, (c) an isovolumetric process in which the final pressure is three times the initial pressure.

63. **Review problem.** Continue the analysis of Problem 16.55 in Chapter 16. Following a collision between a large spacecraft and an asteroid, a copper disk of radius 28.0 m and thickness 1.20 m, at a temperature of 850°C, is floating in space, rotating about its axis with an angular speed of 25.0 rad/s. As the disk radiates infrared light, its temperature falls to 20.0°C. No external torque acts on the disk. (a) Find the change in kinetic energy of the disk. (b) Find the change in internal energy of the disk. (c) Find the amount of energy it radiates.

64. **Review problem.** A 670-kg meteorite happens to be composed of aluminum. When it is far from the Earth, its temperature is − 15°C and it moves with a speed of 14.0 km/s relative to the Earth. As it crashes into the Earth, assume that the resulting additional internal energy is shared equally between the meteor and the planet, and that all the material of the meteor rises momentarily to the same final temperature. Find this temperature. Assume that the specific heat of liquid and of gaseous aluminum is 1 170 J/kg · °C.

65. **PhysicsNow™** A solar cooker consists of a curved reflecting surface that concentrates sunlight onto the object to be warmed (Fig. P17.65). The solar power per unit area reaching the Earth's surface at the location is 600 W/m^2. The cooker faces the Sun and has a diameter of 0.600 m. Assume that 40.0% of the incident energy is transferred to 0.500 L of water in an open container, initially at 20.0°C. How long does it take to completely boil away the water? (Ignore the heat capacity of the container.)

66. An iron plate is held against an iron wheel so that a kinetic friction force of 50.0 N acts between the two pieces of metal. The relative speed at which the two surfaces slide over each other is 40.0 m/s. (a) Calculate the rate at which

FIGURE P17.65

mechanical energy is converted to internal energy. (b) The plate and the wheel each have a mass of 5.00 kg, and each receives 50.0% of the internal energy. If the system is run as described for 10.0 s and each object is then allowed to reach a uniform internal temperature, what is the resultant temperature increase?

67. (a) In air at 0°C, a 1.60-kg copper block at 0°C is set sliding at 2.50 m/s over a sheet of ice at 0°C. Friction brings the block to rest. Find the mass of the ice that melts. To describe the process of slowing down, identify the energy input Q, the work input W, the change in internal energy ΔE_{int}, and the change in mechanical energy ΔK for the block and also for the ice. (b) A 1.60-kg block of ice at 0°C is set sliding at 2.50 m/s over a sheet of copper at 0°C. Friction brings the block to rest. Find the mass of the ice that melts. Identify Q, W, ΔE_{int}, and ΔK for the block and for the metal sheet during the process. (c) A thin, 1.60-kg slab of copper at 20°C is set sliding at 2.50 m/s over an identical stationary slab at the same temperature. Friction quickly stops the motion. Assuming that no energy is lost to the environment by heat, find the change in temperature of both objects. Identify Q, W, ΔE_{int}, and ΔK for each object during the process.

68. A pond of water at 0°C is covered with a layer of ice 4.00 cm thick. If the air temperature stays constant at -10.0°C, how long does it take for the ice thickness to increase to 8.00 cm? (*Suggestion:* Use Equation 17.34 in the form

$$\frac{dQ}{dt} = kA\frac{\Delta T}{x}$$

and note that the incremental energy dQ extracted from the water through the thickness x of ice is the amount required to freeze a thickness dx of ice. That is, $dQ = L\rho A\,dx$, where ρ is the density of the ice, A is the area, and L is the latent heat of fusion.)

69. The average thermal conductivity of the walls (including the windows) and roof of the house depicted in Figure P17.69 is 0.480 W/m · °C, and their average thickness is 21.0 cm. The house is heated with natural gas having a heat of combustion (that is, the energy provided per cubic meter of gas burned) of 9 300 kcal/m^3. How many cubic meters of gas must be burned each day to maintain an inside temperature of 25.0°C if the outside temperature is 0.0°C? Disregard radiation and the energy lost by heat through the ground.

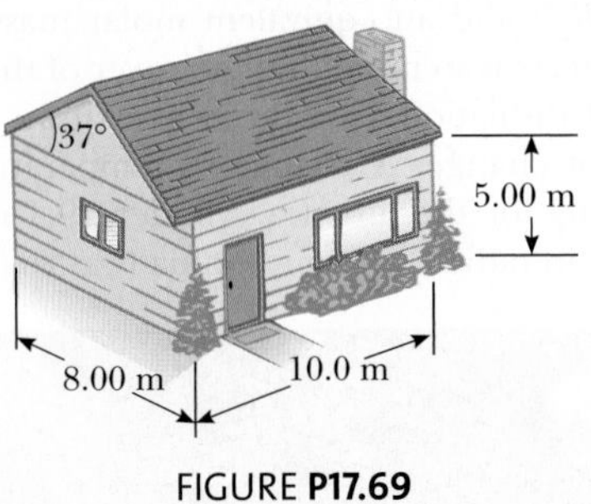

FIGURE P17.69

70. A student obtains the following data in a calorimetry experiment designed to measure the specific heat of aluminum:

Initial temperature of water and calorimeter	70°C
Mass of water	0.400 kg
Mass of calorimeter	0.040 kg
Specific heat of calorimeter	0.63 kJ/kg · °C
Initial temperature of aluminum	27°C
Mass of aluminum	0.200 kg
Final temperature of mixture	66.3°C

Use these data to determine the specific heat of aluminum. Your result should be within 15% of the value listed in Table 17.1.

71. The function $E_{\text{int}} = 3.50nRT$ describes the internal energy of a certain ideal gas. A 2.00-mol sample of the gas always starts at pressure 100 kPa and temperature 300 K. For each one of the following processes, determine the final pressure, volume, and temperature; the change in internal energy of the gas; the energy added to the gas by heat; and the work done on the gas. (a) The gas is heated at constant pressure to 400 K. (b) The gas is heated at constant volume to 400 K. (c) The gas is compressed at constant temperature to 120 kPa. (d) The gas is compressed adiabatically to 120 kPa.

72. A cylinder containing n mol of an ideal gas undergoes an adiabatic process. (a) Starting with the expression $W = -\int P\,dV$ and using the condition $PV^\gamma = $ constant, show that the work done on the gas is

$$W = \left(\frac{1}{\gamma - 1}\right)(P_fV_f - P_iV_i)$$

(b) Starting with the first law of thermodynamics in differential form, prove that the work done on the gas is also equal to $nC_V(T_f - T_i)$. Show that this result is consistent with the equation in part (a).

73. As a 1.00-mol sample of a monatomic ideal gas expands adiabatically, the work done on it is $-2\,500$ J. The initial temperature and pressure of the gas are 500 K and 3.60 atm, respectively. Calculate (a) the final temperature and (b) the final pressure. You may use the result of Problem 17.72.

74. *Smokin'!* A pitcher throws a 0.142-kg baseball at 47.2 m/s (Fig. P17.74). As it travels 19.4 m, the ball slows to 42.5 m/s because of air resistance. Find the change in temperature of the air through which it passes. To find the greatest possible temperature change, you may make the following assumptions. Air has a molar specific heat of $C_P = 7R/2$ and an equivalent molar mass of 28.9 g/mol. The process is so rapid that the cover of the baseball acts as thermal insulation, and the temperature of the ball itself does not change. A change in temperature happens initially only for the air in a cylinder 19.4 m in length and 3.70 cm in radius. This air is initially at 20.0°C.

FIGURE P17.74 John Lackey, the first rookie to win a World Series game 7 in 93 years, pitches for the Anaheim Angels during the final game of the 2002 World Series.

75. A sample of monatomic ideal gas occupies 5.00 L at atmospheric pressure and 300 K. Its state is represented by point A in Figure P17.75. It is heated at constant volume to 3.00 atm (point B). Then it is allowed to expand isothermally to 1.00 atm (point C) and at last compressed isobarically to its original state. (a) Find the number of moles in the sample. (b) Find the temperature at points B and C and the volume at point C. (c) Assuming that the specific heat does not depend on temperature, so that $E_{int} = 3nRT/2$, find the internal energy at points A, B, and C. (d) Tabulate P, V, T, and E_{int} at points A, B, and C. (e) Now consider the processes $A \to B$, $B \to C$, and $C \to A$. Describe just how to carry out each process experimentally. (f) Find Q, W, and ΔE_{int} for each of the processes. (g) For the whole cycle $A \to B \to C \to A$, find Q, W, and ΔE_{int}.

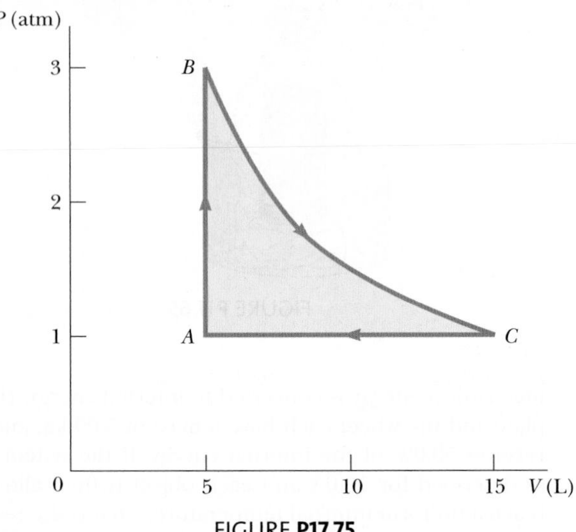

FIGURE P17.75

76. The rate at which a resting person converts food energy is called one's basal metabolic rate (BMR). Assume that the resulting internal energy leaves a person's body by radiation and convection of dry air. When you jog, most of the food energy you burn above your BMR becomes internal energy that would raise your body temperature if it were not eliminated. Assume that evaporation of perspiration is the mechanism for eliminating this energy. Suppose a person is jogging for "maximum fat burning," converting food energy at the rate 400 kcal/h above his BMR, and putting out energy by work at the rate 60.0 W. Assume that the heat of evaporation of water at body temperature is equal to its heat of vaporization at 100°C. (a) Determine the hourly rate at which water must evaporate from his skin. (b) When you metabolize fat, the hydrogen atoms in the fat molecule are transferred to oxygen to form water. Assume that metabolism of 1 g of fat generates 9.00 kcal of energy and produces 1 g of water. What fraction of the water the jogger needs is provided by fat metabolism?

ANSWERS TO QUICK QUIZZES

17.1 (a) Water, glass, iron. Because water has the highest specific heat (4 186 J/kg · °C), it undergoes the smallest change in temperature. Glass is next (837 J/kg · °C), and iron (448 J/kg · °C) is last. (b) Iron, glass, water. For a given temperature increase, the energy transfer by heat is proportional to the specific heat.

17.2 The figure on the next page shows a graphical representation of the internal energy of the ice in parts A to E as a function of energy added. Notice that this graph looks quite different from Figure 17.3 because it doesn't have the flat portions during the phase changes. Regardless of how the temperature is varying in Figure 17.3, the internal energy of the system simply increases linearly with energy input.

17.3 C, A, E. The slope is the ratio of the temperature change to the amount of energy input. Thus, the slope is proportional to the reciprocal of the specific heat. Liquid water, which has the highest specific heat, has the lowest slope.

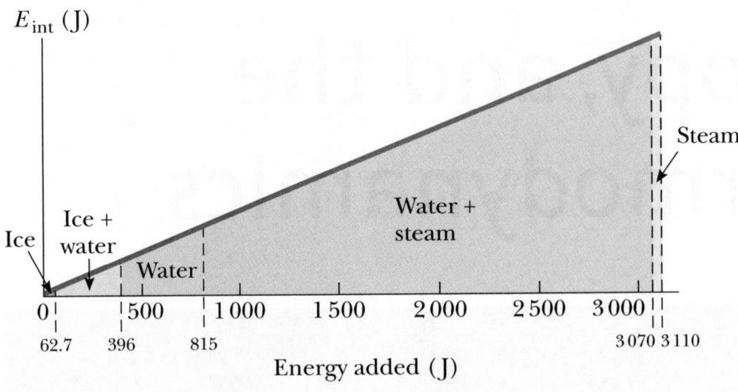

17.4

Situation	System	Q	W	ΔE_{int}
(a) Rapidly pumping up a bicycle tire	Air in the pump	0	+	+
(b) Pan of room-temperature water sitting on a hot stove	Water in the pan	+	0	+
(c) Air quickly leaking out of a balloon	Air originally in the balloon	0	−	−

(a) Because the pumping is rapid, no energy enters or leaves the system by heat. Because $W > 0$ when work is done *on* the system, it is positive here. Thus, $\Delta E_{int} = Q + W$ must be positive. The air in the pump is warmer. (b) There is no work done either on or by the system, but energy flows into the water by heat from the hot burner, making both Q and ΔE_{int} positive. (c) Again no energy flows into or out of the system by heat, but the air molecules escaping from the balloon do work on the surrounding air molecules as they push them out of the way. Thus, W is negative and ΔE_{int} is negative. The decrease in internal energy is evidenced by the escaping air becoming cooler.

17.5 Path A is isovolumetric, path B is adiabatic, path C is isothermal, and path D is isobaric.

17.6 (i), (a). According to Equation 17.15, E_{int} is a function of temperature only. Because the temperature increases, the internal energy increases. (ii), (c). Along an isotherm, T is constant by definition. Therefore, the internal energy of the gas does not change.

17.7 (b). In parallel, the rods present a larger area through which energy can transfer and a smaller length.

CHAPTER 18

Heat Engines, Entropy, and the Second Law of Thermodynamics

This computer artwork shows the interior of an automobile engine cylinder at the moment the spark plug (upper left) fires and ignites the air–fuel mixture. The expanding gases push downward on the piston (lower right), ultimately resulting in energy provided to the drive wheels of the automobile. An automobile engine is one example of a heat engine, which we study in this chapter.

(Roger Harris/Photo Researchers, Inc.)

CHAPTER OUTLINE

The first law of thermodynamics that we studied in Chapter 17 and the more general continuity equation for energy (Eq. 6.20) are statements of the principle of conservation of energy. This principle places no restrictions on the types of energy conversions that can occur. In reality, however, only certain types of energy conversions are observed to take place. Consider the following examples of processes that are consistent with the principle of conservation of energy in either direction but that proceed only in a particular direction in practice.

1. When two objects at different temperatures are placed in thermal contact with each other, energy transfer by heat always occurs from the warmer to the cooler object. We never see energy transfer from the cooler object to the warmer object.
2. A rubber ball dropped to the ground bounces several times and eventually comes to rest, the original gravitational potential energy having been transformed to internal energy in the ball and the ground. A ball lying on the ground, however,

never gathers internal energy up from the ground and begins bouncing on its own.

3. If oxygen and nitrogen are maintained in separate halves of a container by a membrane and the membrane is punctured, the oxygen and nitrogen molecules mix together. We never see a mixture of oxygen and nitrogen spontaneously separate into different sides of the container.

These situations all *illustrate irreversible processes;* that is, they occur naturally in only one direction. In this chapter, we investigate a new fundamental principle that allows us to understand why these processes occur in one direction only.[1] The second law of thermodynamics, which is the primary focus of this chapter, establishes which natural processes do and which do not occur.

LORD KELVIN (1824–1907)

Born William Thomson in Belfast, British physicist and mathematician Kelvin was the first to propose the use of an absolute scale of temperature. The Kelvin temperature scale is named in his honor. Kelvin's work in thermodynamics led to the idea that energy cannot pass spontaneously from a colder object to a hotter object.

18.1 HEAT ENGINES AND THE SECOND LAW OF THERMODYNAMICS

One device that is very useful in understanding the second law of thermodynamics is the heat engine. A **heat engine** is a device that takes in energy by heat[2] and, operating in a cycle, expels a fraction of that energy by work. In a typical process for producing electricity in a power plant, for instance, coal or some other fuel is burned and the resulting internal energy is used to convert water to steam. This steam is directed at the blades of a turbine, setting it into rotation. Finally, the mechanical energy associated with this rotation is used to drive an electric generator. In another heat engine, the internal combustion engine in your automobile, energy enters the engine by matter transfer as the fuel is injected into the cylinder and a fraction of this energy is converted to mechanical energy.

In general, a heat engine carries some working substance through cyclic processes[3] during which (1) the working substance absorbs energy by heat from an energy reservoir at a high temperature, (2) work is done by the engine, and (3) energy is expelled by heat to a reservoir at a lower temperature. This output energy is often called wasted energy, exhaust energy, or thermal pollution. As an example, consider the operation of a steam engine in which the working substance is water. The water in the engine is carried through a cycle in which it first evaporates into steam in a boiler and then expands against a piston. After the steam is condensed with cooling water, it is returned to the boiler, and the process is repeated.

It is useful to draw a heat engine schematically as in the pictorial representation in Active Figure 18.1. The engine absorbs a quantity of energy Q_h from the hot reservoir.[4] The engine does work W_{eng} (so that *negative* work $W = -W_{eng}$ is done *on* the engine), and then gives up energy Q_c to the cold reservoir. Because the working substance goes through a cycle, its initial and final internal energies are equal, so $\Delta E_{int} = 0$. The engine can be modeled as a nonisolated system in steady state. Hence, from the first law,

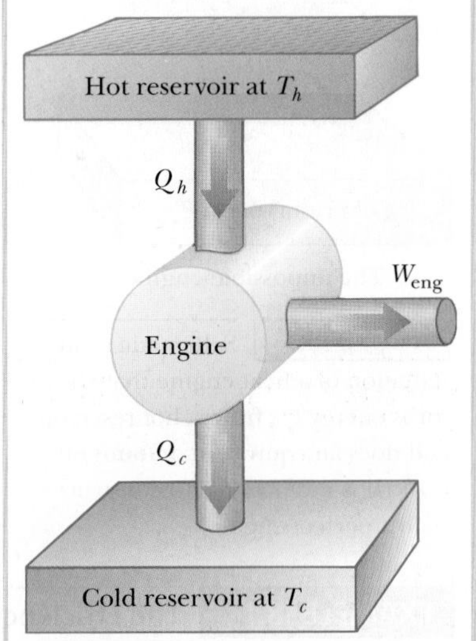

ACTIVE FIGURE 18.1
Schematic representation of a heat engine. The engine does work W_{eng}. The arrow at the top represents energy $Q_h > 0$ entering the engine. At the bottom, $Q_c < 0$ represents energy leaving the engine.

PhysicsNow™ Log into PhysicsNow at **www.pop4e.com** and go to Active Figure 18.1 to select the efficiency of the engine and observe the transfer of energy.

[1] As we shall see in this chapter, it is more proper to say that the set of events in the time-reversed sense is highly improbable. From this viewpoint, events in one direction are vastly more probable than those in the opposite direction.

[2] We will use heat as our model for energy transfer into a heat engine. Other methods of energy transfer are also possible in the model of a heat engine, however. For example, as we shall show in Section 18.9, the Earth's atmosphere can be modeled as a heat engine in which the input energy transfer is by means of electromagnetic radiation from the Sun. The output of the atmospheric heat engine causes the wind structure in the atmosphere.

[3] The automobile engine is not strictly a heat engine according to the cyclic process description because the substance (the air–fuel mixture) undergoes only one cycle and is then expelled through the exhaust system.

[4] We will adopt a simplification model in which we assume that the energy transfer from the reservoir is by heat, but will realize that other transfer mechanisms are possible. For example, as mentioned earlier, energy is brought into the cylinder of an automobile engine by matter transfer.

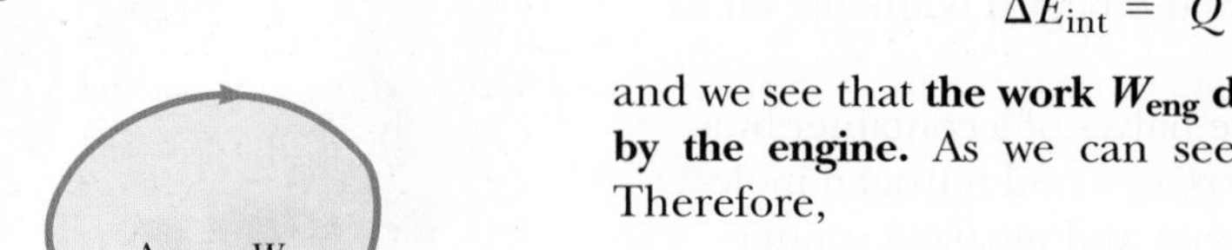

$$\Delta E_{\text{int}} = Q + W \quad \rightarrow \quad Q_{\text{net}} = -W = W_{\text{eng}}$$

and we see that **the work W_{eng} done by a heat engine equals the net energy absorbed by the engine.** As we can see from Active Figure 18.1, $Q_{\text{net}} = |Q_h| - |Q_c|$. Therefore,

$$W_{\text{eng}} = |Q_h| - |Q_c| \qquad [18.1]$$

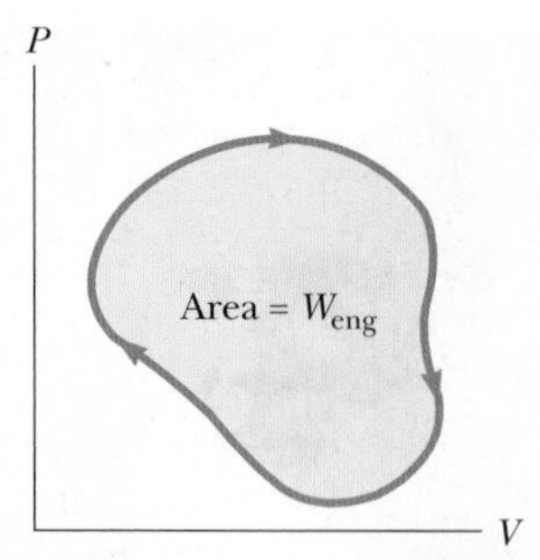

FIGURE 18.2 The *PV* diagram for an arbitrary cyclic process. The net work done by the engine equals the area enclosed by the curve.

If the working substance is a gas, **the net work done by the engine for a cyclic process is the area enclosed by the curve representing the process on a *PV* diagram.** This area is shown for an arbitrary cyclic process in Figure 18.2.

The **thermal efficiency** e of a heat engine is defined as the ratio of the work done by the engine to the energy absorbed at the higher temperature during one cycle:

$$e = \frac{W_{\text{eng}}}{|Q_h|} = \frac{|Q_h| - |Q_c|}{|Q_h|} = 1 - \frac{|Q_c|}{|Q_h|} \qquad [18.2]$$

We can think of the efficiency as the ratio of what you gain (energy transfer by work) to what you give (energy transfer from the high temperature reservoir). Equation 18.2 shows that a heat engine has 100% efficiency ($e = 1$) only if $Q_c = 0$ (i.e., if no energy is expelled to the cold reservoir). In other words, a heat engine with perfect efficiency would have to expel all the input energy by mechanical work.

The Kelvin–Planck statement of the **second law of thermodynamics** can be stated as follows:

> It is impossible to construct a heat engine that, operating in a cycle, produces no effect other than the absorption of energy by heat from a reservoir and the performance of an equal amount of work.

The essence of this form of the second law is that it is theoretically impossible to construct an engine such as that in Figure 18.3 that works with 100% efficiency. All engines must exhaust some energy Q_c to the environment.

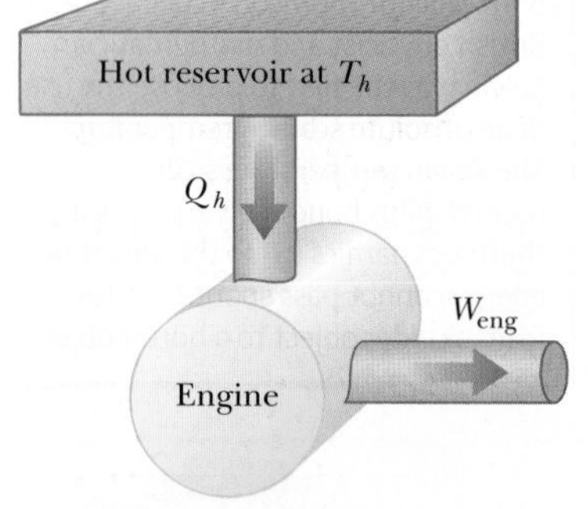

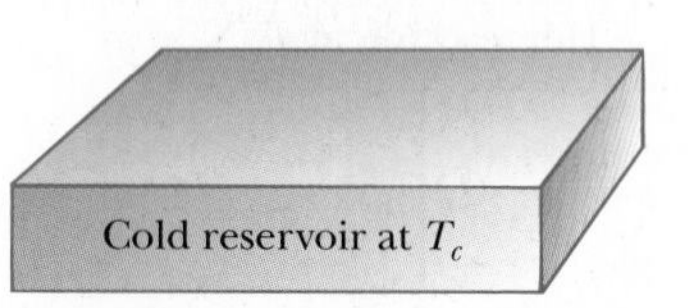

The impossible engine

FIGURE 18.3 Schematic representation of a heat engine that absorbs energy Q_h from a hot reservoir and does an equivalent amount of work. It is not possible to construct such a perfect engine.

QUICK QUIZ 18.1 The energy input to an engine is 3.00 times greater than the work it performs. **(i)** What is its thermal efficiency? **(a)** 3.00 **(b)** 1.00 **(c)** 0.333 **(d)** impossible to determine **(ii)** For this engine, what fraction of the energy input is expelled to the cold reservoir? **(a)** 0.333 **(b)** 0.667 **(c)** 1.00 **(d)** impossible to determine

EXAMPLE 18.1 The Efficiency of an Engine

An engine transfers 2.00×10^3 J of energy from a hot reservoir during a cycle and transfers 1.50×10^3 J to a cold reservoir.

A Find the efficiency of the engine.

Solution The efficiency of the engine is calculated using Equation 18.2:

$$e = 1 - \frac{|Q_c|}{|Q_h|} = 1 - \frac{1.50 \times 10^3\ \text{J}}{2.00 \times 10^3\ \text{J}} = 0.250, \text{ or } 25.0\%$$

B How much work does this engine do in one cycle?

Solution The work done is the difference between the input and output energies:

$$W_{\text{eng}} = |Q_h| - |Q_c| = 2.00 \times 10^3\ \text{J} - 1.50 \times 10^3\ \text{J} = 500\ \text{J}$$

C One cycle of this engine occurs in a time interval of 0.010 0 s. What is the power output of this engine?

Solution Power is defined as the rate of energy transfer per unit time interval:

$$\mathcal{P} = \frac{W}{\Delta t} = \frac{500\ \text{J}}{0.010\ 0\ \text{s}} = 5.00 \times 10^4\ \text{W}$$

18.2 REVERSIBLE AND IRREVERSIBLE PROCESSES

In the next section, we shall discuss a theoretical heat engine that is the most efficient engine possible. To understand its nature, we must first examine the meaning of reversible and irreversible processes. A **reversible** process is one for which the system can be returned to its initial conditions along the same path and for which every point along the path is an equilibrium state. A process that does not satisfy these requirements is **irreversible.**

Most natural processes are known to be irreversible; the reversible process is an idealization. The three processes described in the introduction to this chapter are irreversible, and we see them proceed in only one direction. The free expansion of a gas discussed in Section 17.6 is irreversible. When the membrane is removed, the gas rushes into the empty half of the vessel and the environment is not changed. No matter how long we watched, we would never see the gas in the full volume spontaneously rush back into only half the volume. The only way we could cause that to happen would be to interact with the gas, perhaps by pushing it inward with a piston, but that method would result in a change in the environment.

If a real process occurs very slowly so that the system is always very nearly in equilibrium, the process can be modeled as reversible. For example, imagine compressing a gas very slowly by dropping some grains of sand onto a frictionless piston as in Figure 18.4. The pressure, volume, and temperature of the gas are well defined during this isothermal compression. Each added grain of sand represents a small change to a new equilibrium state. The process can be reversed by the slow removal of grains of sand from the piston.

PITFALL PREVENTION 18.1

REAL PROCESSES ARE IRREVERSIBLE
The reversible process is an idealization. All real processes on Earth are irreversible.

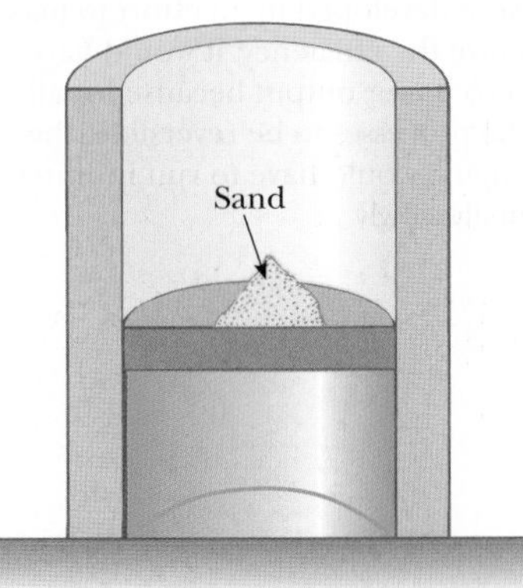

FIGURE 18.4 A gas in thermal contact with an energy reservoir is compressed slowly as individual grains of sand are dropped onto the piston. The compression is isothermal and reversible.

18.3 THE CARNOT ENGINE

In 1824, a French engineer named Sadi Carnot described a theoretical engine, now called a **Carnot engine,** that is of great importance from both practical and theoretical viewpoints. He showed that a heat engine operating in an ideal, reversible cycle—called a **Carnot cycle**—between two energy reservoirs is the most efficient engine possible. Such an ideal engine establishes an upper limit on the efficiencies of all real engines. That is, **the net work done by a working substance taken through the Carnot cycle is the greatest amount of work possible for a given amount of energy supplied to the substance at the upper temperature.**

To describe the Carnot cycle, we shall assume that the working substance in the engine is an ideal gas contained in a cylinder with a movable piston at one end. The cylinder walls and the piston are thermally nonconducting. Four stages of the Carnot cycle are shown in Active Figure 18.5; Active Figure 18.6 is the *PV* diagram for the cycle, which consists of two adiabatic and two isothermal processes, all reversible.

In the Carnot cycle, the following processes take place:

- The process $A \rightarrow B$ is an isothermal expansion at temperature T_h, in which the gas is placed in thermal contact with an energy reservoir at temperature T_h (Active Fig. 18.5a). During the process, the gas absorbs energy Q_h by heat from the reservoir and does work W_{AB} in raising the piston.
- In the process $B \rightarrow C$, the base of the cylinder is replaced by a thermally nonconducting wall and the gas expands adiabatically; that is, no energy enters or leaves the system by heat (Active Fig. 18.5b). During the process, the temperature falls from T_h to T_c and the gas does work W_{BC} in raising the piston.
- In the process $C \rightarrow D$, the gas is placed in thermal contact with an energy reservoir at temperature T_c (Active Fig. 18.5c) and is compressed isothermally at

(J.L. Charmet/SPL/Photo Researchers, Inc.)

SADI CARNOT (1796–1832)
French physicist Carnot was the first to show the quantitative relationship between work and heat. In 1824, he published his only work—*Reflections on the Motive Power of Heat,* which reviewed the industrial, political, and economic importance of the steam engine. In it, he defined work as "weight lifted through a height."

PITFALL PREVENTION 18.2

Don't Shop for a Carnot Engine The Carnot engine is an idealization, so do not expect a Carnot engine to be developed for commercial use. If a Carnot engine were developed in an effort to maximize the efficiency, it would have zero power output because for all the processes to be reversible, the engine would have to run infinitesimally slowly.

temperature T_c. During this time, the gas expels energy Q_c to the reservoir and the work done on the gas is W_{CD}.

- In the final process $D \rightarrow A$, the base of the cylinder is again replaced by a thermally nonconducting wall (Active Fig. 18.5d) and the gas is compressed adiabatically. The temperature of the gas increases to T_h, and the work done on the gas is W_{DA}.

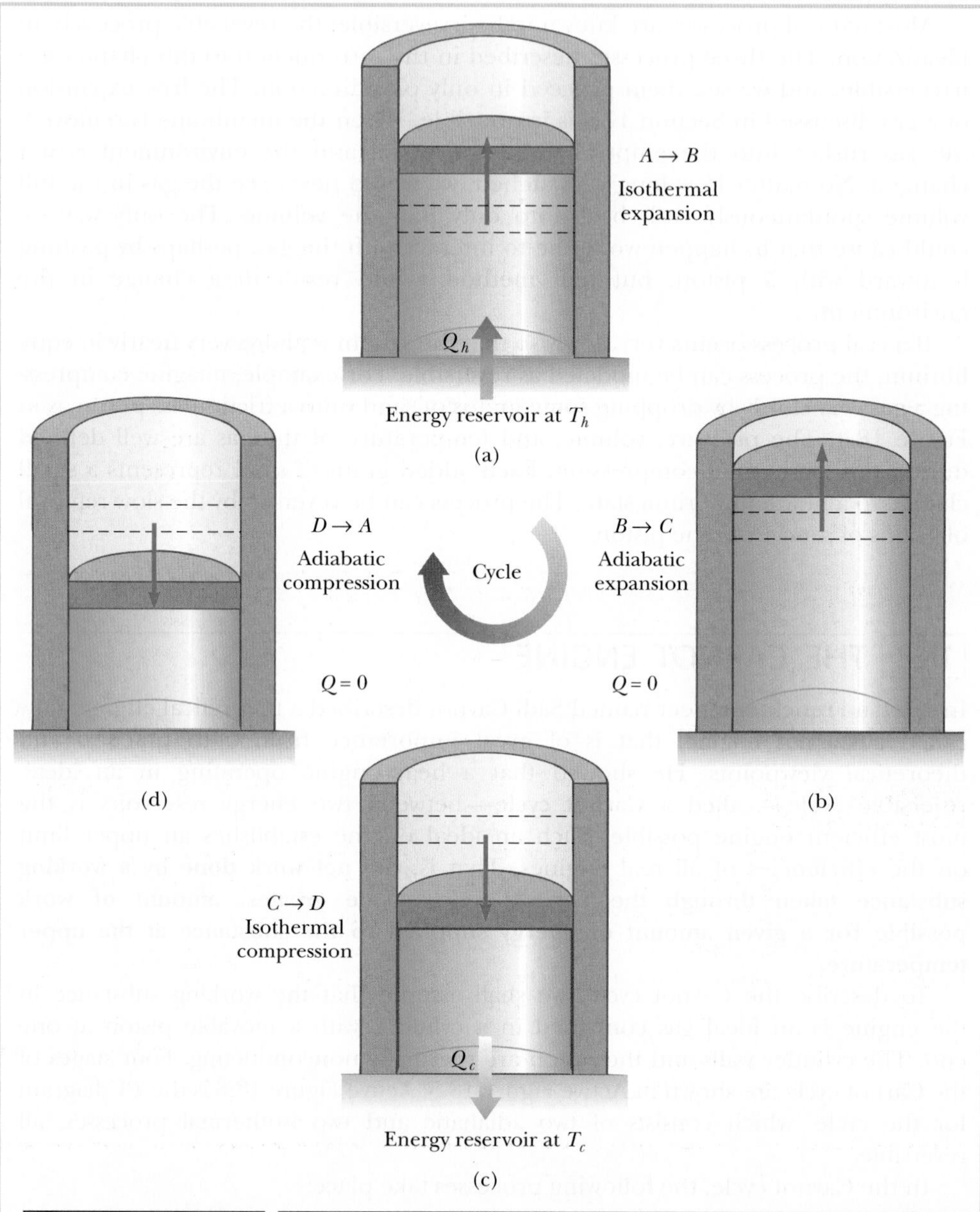

ACTIVE FIGURE 18.5 The Carnot cycle. (a) In process $A \rightarrow B$, the gas expands isothermally while in contact with an energy reservoir at T_h. (b) In process $B \rightarrow C$, the gas expands adiabatically ($Q = 0$). (c) In process $C \rightarrow D$, the gas is compressed isothermally while in contact with an energy reservoir at $T_c < T_h$. (d) In process $D \rightarrow A$, the gas is compressed adiabatically. The arrows on the piston indicate the direction of its motion in each process.

PhysicsNow™ Log into PhysicsNow at **www.pop4e.com** and go to Active Figure 18.5. You can observe the motion of the piston in the Carnot cycle while you also observe the cycle on the *PV* diagram of Active Figure 18.6.

Carnot showed that for this cycle,

$$\frac{|Q_c|}{|Q_h|} = \frac{T_c}{T_h} \qquad [18.3]$$

Therefore, using Equation 18.2, the thermal efficiency of a Carnot engine is

$$e_{\text{Carnot}} = 1 - \frac{T_c}{T_h} \qquad [18.4]$$

From this result, we see that all Carnot engines operating between the same two temperatures have the same efficiency.

Equation 18.4 can be applied to any working substance operating in a Carnot cycle between two energy reservoirs. According to this result, the efficiency is zero if $T_c = T_h$, as one would expect. The efficiency increases as T_c is lowered and as T_h is increased. The efficiency can be unity (100%), however, only if $T_c = 0$ K. It is impossible to reach absolute zero,[5] so such reservoirs are not available. Therefore, the maximum efficiency is always less than unity. In most practical cases, the cold reservoir is near room temperature, about 300 K. Therefore, one usually strives to increase the efficiency by raising the temperature of the hot reservoir. **All real engines are less efficient than the Carnot engine because they all operate irreversibly so as to complete a cycle in a brief time interval.**[6] In addition to this theoretical limitation, real engines are subject to practical difficulties, including friction, that reduce the efficiency further.

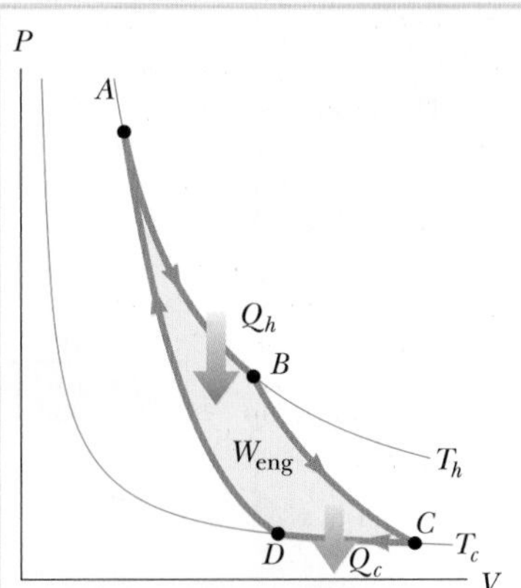

ACTIVE FIGURE 18.6
The *PV* diagram for the Carnot cycle. The net work done W_{eng} equals the net energy transferred into the Carnot engine in one cycle, $|Q_h| - |Q_c|$. Note that $\Delta E_{int} = 0$ for the cycle.

PhysicsNow™ Log into PhysicsNow at **www.pop4e.com** and go to Active Figure 18.6. You can observe the Carnot cycle on the *PV* diagram while you also observe the motion of the piston in Active Figure 18.5.

QUICK QUIZ 18.2 Three engines operate between reservoirs separated in temperature by 300 K. The reservoir temperatures are as follows: engine A: T_h = 1 000 K, T_c = 700 K; engine B: T_h = 800 K, T_c = 500 K; engine C: T_h = 600 K, T_c = 300 K. Rank the engines in order of theoretically possible efficiency, from highest to lowest.

EXAMPLE 18.2 The Steam Engine

A steam engine has a boiler that operates at 500 K. The energy resulting from the burning of the fuel changes water to steam, and this steam then drives the piston. The exhaust temperature is that of the outside air, approximately 300 K. What is the maximum thermal efficiency of this steam engine?

Solution From the expression for the efficiency of a Carnot engine, we find the maximum thermal efficiency for any engine operating between these temperatures:

$$e_{\text{Carnot}} = 1 - \frac{T_c}{T_h} = 1 - \frac{300\text{ K}}{500\text{ K}} = 0.400, \text{ or } 40.0\%$$

Note that this result is the highest theoretical efficiency of the engine. In practice, the efficiency is considerably lower.

[5] The inability to reach absolute zero is known as the *third law of thermodynamics.* It would require an infinite amount of energy to lower the temperature of a substance to absolute zero.

[6] For the processes in the Carnot cycle to be reversible, they must be carried out infinitesimally slowly. Therefore, although the Carnot engine is the most efficient engine possible, it has zero power output because it takes an infinite time interval to complete one cycle! For a real engine, the short time interval for each cycle results in the working substance reaching a high temperature lower than that of the hot reservoir and a low temperature higher than that of the cold reservoir. An engine undergoing a Carnot cycle between this narrower temperature range was analyzed by F. L. Curzon and B. Ahlborn (*Am. J. Phys.*, 43(1):22, 1975), who found that the efficiency at maximum power output depends only on the reservoir temperatures T_c and T_h, and is given by $e_{\text{C-A}} = 1 - (T_c/T_h)^{1/2}$. The Curzon–Ahlborn efficiency $e_{\text{C-A}}$ provides a closer approximation to the efficiencies of real engines than does the Carnot efficiency.

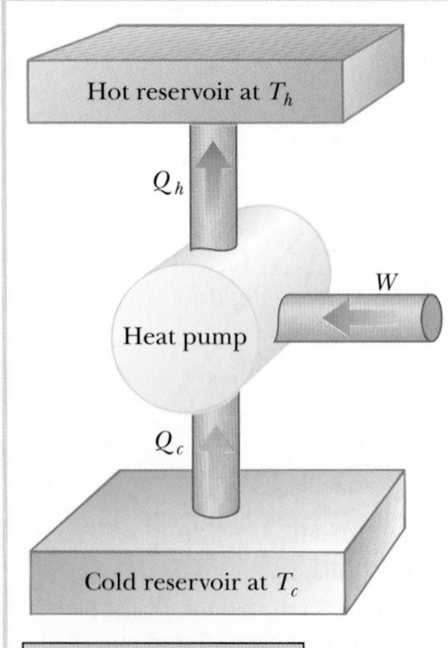

ACTIVE FIGURE 18.7
Schematic representation of a heat pump, which absorbs energy Q_c from a cold reservoir and expels energy Q_h to a hot reservoir. The work done on the heat pump is W.

Physics Now™ Log into PhysicsNow at **www.pop4e.com** and go to Active Figure 18.7 to select the COP of the heat pump and observe the transfer of energy.

18.4 HEAT PUMPS AND REFRIGERATORS

In a heat engine, the direction of energy transfer is from the hot reservoir to the cold reservoir, which is the natural direction. The role of the heat engine is to process the energy from the hot reservoir so as to do useful work. What if we wanted to transfer energy by heat from the cold reservoir to the hot reservoir? Because this direction is not the natural one, we must transfer some energy into a device to cause it to occur. Devices that perform this task are called **heat pumps** or **refrigerators.**

Active Figure 18.7 is a schematic representation of a heat pump. The cold reservoir temperature is T_c, the hot reservoir temperature is T_h, and the energy absorbed by the heat pump is Q_c. Energy is transferred into the system, which we model as work[7] W, and the energy transferred out of the pump is Q_h.

Heat pumps have long been popular for cooling in homes, where they are called *air conditioners,* and are now becoming increasingly popular for heating purposes as well. In the heating mode, a circulating coolant fluid absorbs energy from the outside air (the cold reservoir) and releases energy to the interior of the structure (the hot reservoir). The fluid is usually in the form of a low-pressure vapor when in the coils of the exterior part of the unit, where it absorbs energy from either the air or the ground by heat. This gas is then compressed into a hot, high-pressure vapor and enters the interior part of the unit, where it condenses to a liquid and releases its stored energy. An air conditioner is simply a heat pump installed backward, with "exterior" and "interior" interchanged. The inside of the home is the cold reservoir and the outside air is the hot reservoir.

The effectiveness of a heat pump is described in terms of a number called the **coefficient of performance** COP. In the heating mode, the COP is defined as the ratio of the energy transferred by heat into the hot reservoir to the work required to transfer that energy:

$$\text{COP (heat pump)} \equiv \frac{\text{energy transferred to hot reservoir}}{\text{work done on pump}} = \frac{|Q_h|}{W} \qquad [18.5]$$

As a practical example, if the outside temperature is $-4°C$ (25°F) or higher, the COP for a typical heat pump is about 4. That is, the energy transferred into the house is about four times greater than the work done by the compressor in the heat pump. As the outside temperature decreases, however, it becomes more difficult for the heat pump to extract sufficient energy from the air, and the COP therefore drops.

A Carnot cycle heat engine operating in reverse constitutes an ideal heat pump, the heat pump with the highest possible COP for the temperatures between which it operates. The maximum coefficient of performance is

$$\text{COP}_{\text{Carnot}}(\text{heat pump}) = \frac{T_h}{T_h - T_c}$$

Although heat pumps are relatively new products in heating, the refrigerator has been a standard appliance in homes for decades. The refrigerator cools its interior by pumping energy from the food storage compartments into the warmer air

[7] The traditional notation is to model the input energy as transferred by work, although most heat pumps operate from electricity, so the more appropriate transfer mechanism *into the device as the system* is *electrical transmission.* If we identify the refrigerant fluid in a heat pump as the system, the energy transfers into the fluid by work done by a piston attached to a compressor operated electrically. In keeping with tradition, we will schematicize the heat pump with input by work regardless of the choice of system.

outside. During its operation, a refrigerator removes energy Q_c from the interior of the refrigerator, and in the process its motor does work W. The COP of a refrigerator or of a heat pump used in its cooling cycle is

$$\text{COP (refrigerator)} = \frac{|Q_c|}{W} \qquad [18.6]$$

An efficient refrigerator is one that removes the greatest amount of energy from the cold reservoir with the least amount of work. Therefore, a good refrigerator should have a high coefficient of performance, typically 5 or 6.

The highest possible COP is again that of a refrigerator whose working substance is carried through the Carnot heat engine cycle in reverse:

$$\text{COP}_{\text{Carnot}}\text{ (refrigerator)} = \frac{T_c}{T_h - T_c}$$

As the difference between the temperatures of the two reservoirs approaches zero, the theoretical coefficient of performance of a Carnot heat pump approaches infinity. In practice, the low temperature of the cooling coils and the high temperature at the compressor limit the COP to values below 10.

QUICK QUIZ 18.3 The energy entering an electric heater by electrical transmission can be converted to internal energy with an efficiency of 100%. By what factor does the cost of heating your home change when you replace your electric heating system with an electric heat pump that has a COP of 4.00? Assume that the motor running the heat pump is 100% efficient. **(a)** 4.00 **(b)** 2.00 **(c)** 0.500 **(d)** 0.250

■ Thinking Physics 18.1

It is a sweltering summer day and your air-conditioning system is not operating. In your kitchen, you have a working refrigerator and an ice chest full of ice. Which should you open and leave open to cool the room more effectively?

Reasoning The high-temperature reservoir for your kitchen refrigerator is the air in the kitchen. If the refrigerator door were left open, energy would be drawn from the air in the kitchen, passed through the refrigeration system and transferred right back into the air. The result would be that the kitchen would become *warmer* because of the addition of the energy coming in by electricity to run the refrigeration system. If the ice chest were opened, energy in the air would enter the ice, raising its temperature and causing it to melt. The transfer of energy from the air would cause its temperature to drop. Therefore, it would be more effective to open the ice chest. ■

18.5 AN ALTERNATIVE STATEMENT OF THE SECOND LAW

Suppose you wish to cool off a hot piece of pizza by placing it on a block of ice. You will certainly be successful because in every similar situation, energy transfer has always taken place from a hot object to a cooler one. Yet nothing in the first law of thermodynamics says that this energy transfer could not proceed in the opposite direction. (Imagine your astonishment if someday you place a piece of hot pizza on ice and the pizza becomes warmer!) It is the second law that determines the directions of such natural phenomena.

An analogy can be made with the impossible sequence of events seen in a movie film running backward, such as a person rising out of a swimming pool and landing back on the diving board, an apple rising from the ground and latching onto the branch of a tree, or a pot of hot water becoming colder as it rests over an open flame. Such events occurring backward in time are impossible because they violate the second law of thermodynamics. **Real processes proceed in a preferred direction.**

■ Second law of thermodynamics; Clausius statement

The second law can be stated in several different ways, but all the statements can be shown to be equivalent. Which form you use depends on the application you have in mind. For example, if you were concerned about the energy transfer between pizza and ice, you might choose to concentrate on the **Clausius statement of the second law: Energy does not flow spontaneously by heat from a cold object to a hot object.** Figure 18.8 shows a heat pump that violates this statement of the second law. Energy is transferring from the cold reservoir to the hot reservoir without an input of work. At first glance, this statement of the second law seems to be radically different from that in Section 18.1, but the two are, in fact, equivalent in all respects. Although we shall not prove it here, it can be shown that if either statement of the second law is false, so is the other.

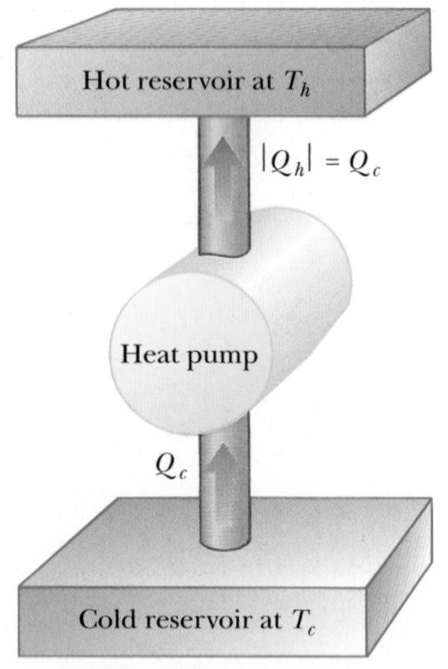

FIGURE 18.8 Schematic diagram of an impossible heat pump or refrigerator, that is, one that takes in energy from a cold reservoir and expels an equivalent amount of energy to a hot reservoir without the input of energy by work.

18.6 ENTROPY

The zeroth law of thermodynamics involves the concept of temperature and the first law involves the concept of internal energy. Temperature and internal energy are both state variables; that is, they can be used to describe the thermodynamic state of a system. Another state variable, this one related to the second law of thermodynamics, is **entropy** S. In this section, we define entropy on a macroscopic scale as German physicist Rudolf Clausius (1822–1888) first expressed it in 1865.

Equation 18.3, which describes the Carnot engine, can be rewritten as

$$\frac{|Q_c|}{T_c} = \frac{|Q_h|}{T_h}$$

Thus, the ratio of the energy transfer by heat in a Carnot cycle to the (constant) temperature at which the transfer takes place has the same magnitude for both isothermal processes. To generalize the current discussion beyond heat engines, let us drop the absolute value notation and revive our original sign convention, in which Q_c represents energy leaving the system of the gas and is therefore a negative number. Therefore, we need an explicit negative sign to maintain the equality:

$$-\frac{Q_c}{T_c} = \frac{Q_h}{T_h}$$

We can write this equation as

$$\frac{Q_h}{T_h} + \frac{Q_c}{T_c} = 0 \quad \rightarrow \quad \sum \frac{Q}{T} = 0 \qquad [18.7]$$

PITFALL PREVENTION 18.3

ENTROPY IS ABSTRACT Entropy is one of the most abstract notions in physics, so follow the discussion in this and the subsequent sections very carefully. Be sure that you do not confuse energy and entropy; even though the names sound similar, they are very different.

We have not specified a particular Carnot cycle in generating this equation, so it must be true for all Carnot cycles. Furthermore, by approximating a general reversible cycle with a series of Carnot cycles, we can show that this equation is true for *any* reversible cycle, which suggests that the ratio Q/T may have some special significance. Indeed it does, as we note in the following discussion.

Consider a system undergoing any infinitesimal process between two equilibrium states. If dQ_r is the energy transferred by heat as the system follows a reversible path between the states, the **change in entropy,** regardless of the actual path followed, is equal to this energy transferred by heat along the reversible path divided by the absolute temperature of the system:

■ Change in entropy for an infinitesimal process

$$dS = \frac{dQ_r}{T} \qquad [18.8]$$

The subscript r on the term dQ_r is a reminder that the heat is to be determined along a reversible path, even though the system may actually follow some irreversible path. Therefore, we *must* model a nonreversible process by a reversible process between the same initial and final states to calculate the entropy change. In this case, the model might not be close to the actual process at all, but that is not a concern because entropy is a state variable and the entropy change depends only on the initial and final states. The only requirements are that the model process must be reversible and must connect the given initial and final states.

When energy is absorbed by the system, dQ_r is positive and hence the entropy increases. When energy is expelled by the system, dQ_r is negative and the entropy decreases. Note that Equation 18.8 defines not entropy but rather the *change* in entropy. Hence, the meaningful quantity in a description of a process is the *change* in entropy.

With Equation 18.8, we have a mathematical representation of the change in entropy, but we have developed no mental representation of what entropy means. In this and the next few sections, we explore various aspects of entropy that will allow us to gain a conceptual understanding of it.

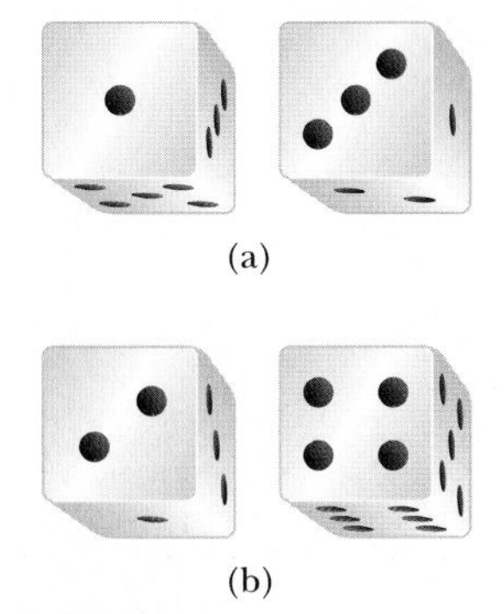

FIGURE 18.9 Two different microstates for a throw of two dice. These correspond to two macrostates, having values of (a) 4 and (b) 6.

Entropy originally found its place in thermodynamics, but its importance grew tremendously as the field of physics called *statistical mechanics* developed because this method of analysis provided an alternative way of interpreting entropy. In statistical mechanics, a substance's behavior is described in terms of the statistical behavior of its large number of atoms and molecules. Kinetic theory, which we studied in Chapter 16, is an excellent example of the statistical mechanics approach. A main outcome of this treatment is the principle that **isolated systems tend toward disorder, and entropy is a measure of that disorder.**

To understand this notion, we introduce the distinction between **microstates** and **macrostates** for a system. We can do so by looking at an example far removed from thermodynamics, the throwing of dice at a craps table in a casino. For two dice, a *microstate* is the particular combination of numbers on the upturned faces of the dice; for example, 1–3 and 2–4 are two different microstates (Fig. 18.9). The *macrostate* is the sum of the numbers. Therefore, the macrostates for the two example microstates in Figure 18.9 are 4 and 6. Now, here is the central notion that we will need to understand entropy: **The number of microstates associated with a given macrostate is not the same for all macrostates, and the most probable macrostate is that with the largest number of possible microstates.** A macrostate of 7 on our pair of dice has six possible microstates: 1–6, 2–5, 3–4, 4–3, 5–2, and 6–1 (Fig. 18.10a). For a macrostate of 2, there is only one possible microstate: 1–1 (Fig. 18.10b). Therefore, a macrostate of 7 has six times as many microstates as a macrostate of 2 and is therefore six times as probable. In fact, the macrostate of 7 is the most probable macrostate for two dice. The game of craps is built around these probabilities of various macrostates.

Consider the low probability macrostate 2. The *only* way of achieving it is to have a 1 on each die. We say that this macrostate has a high degree of *order;* we *must* have a 1 on each die for this macrostate to exist. Considering the possible microstates for a macrostate of 7, however, we see six possibilities. This macrostate is more *disordered* because several microstates are possible that will result in the same macrostate. Thus, we conclude that **high-probability macrostates are disordered macrostates and low-probability macrostates are ordered macrostates.**

FIGURE 18.10 Possible two-dice microstates for a macrostate of (a) 7 and (b) 2. The macrostate of 7 is more probable because there are more ways of achieving it; more microstates are associated with a 7 than with a 2.

As a more physical example, consider the molecules in the air in your room. Let us compare two possible macrostates. Macrostate 1 is the condition in which the oxygen and nitrogen molecules are mixed evenly throughout the room. Macrostate 2 is that in which the oxygen molecules are in the front half of the room and the nitrogen molecules are in the back half. From our everyday experience, it is *extremely unlikely* for macrostate 2 to exist. On the other hand, macrostate 1 is what we would normally expect to see. Let us relate this experience to the microstates, which correspond to the possible positions of molecules of each type. For macrostate 2 to exist, every molecule of oxygen would have to be in one half of the room and every molecule of nitrogen in the other half, which is a highly ordered and unlikely situation. The probability of this occurrence is infinitesimal. For macrostate 1 to exist, both types of molecules are simply distributed evenly around the room, which is a much lower level of order and a highly probable situation. Therefore, the mixed state is much more likely than the separated state, and that is what we normally see.

Let us look now at the notion that isolated systems tend toward disorder. The cause of this tendency toward disorder is easily seen. Let us assume that all microstates for the system are equally probable. When the possible macrostates associated with the microstates are examined, however, far more of them are disordered macrostates with many microstates than ordered macrostates with few microstates. Because each of the microstates is equally probable, it is highly probable that the actual macrostate will be one of the highly disordered macrostates simply because there are more microstates.

In physical systems, we are not talking about microstates of two entities like our pair of dice; we are talking about a number on the order of Avogadro's number of molecules. If you imagine throwing Avogadro's number of dice, the game of craps would be meaningless. You could make an almost perfect prediction of the result when the numbers on the faces are all added up (if the numbers on the face of the dice were added once per second, more than 19 thousand trillion years would be needed to tabulate the results for only one throw!) because you are dealing with the statistics of a huge number of dice. We face these kinds of statistics with Avogadro's number of molecules. The macrostate can be well predicted. Even if a system starts off in a very low probability state (e.g., the nitrogen and oxygen molecules separated in a room by a membrane that is then punctured), it quickly develops into a high-probability state (the molecules rapidly mix evenly throughout the room).

We can now present this as a general principle for physical processes: **All physical processes tend toward more probable macrostates for the system and its surroundings. The more probable macrostate is always one of higher disorder.**

QUICK QUIZ 18.4 **(i)** Suppose you select four cards at random from a standard deck of playing cards and end up with a macrostate of four deuces. How many microstates are associated with this macrostate? **(a)** 1 **(b)** 2 **(c)** 3 **(d)** 4 **(e)** 5 **(f)** 6 **(ii)** Suppose you pick up two cards and end up with a macrostate of two aces. From the same choices, how many microstates are associated with this macrostate? **(iii)** Which macrostate is more probable? **(a)** Four deuces are more probable. **(b)** Two aces are more probable. **(c)** They are of equal probability.

Now, what does all this talk about dice and states have to do with entropy? To answer this question, we can show that entropy is a measure of the disorder of a state. Then, we shall use these ideas to generate a new statement of the second law of thermodynamics.

As we have seen, entropy can be defined using the macroscopic concepts of heat and temperature. Entropy can also be treated from a microscopic viewpoint through statistical analysis of molecular motions. We can make a connection between entropy and the number of microstates associated with a given macrostate with the following expression:[8]

■ Entropy (microscopic definition)

$$S \equiv k_B \ln W \qquad [18.9]$$

where W is the number of microstates associated with the macrostate whose entropy is S.

[8] For a derivation of this expression, see Chapter 22 of R. A. Serway and J. W. Jewett Jr., *Physics for Scientists and Engineers*, 6th ed. (Belmont, CA: 2004), Brooks-Cole.

Because the more probable macrostates are the ones with larger numbers of microstates and the larger numbers of microstates are associated with more disorder, Equation 18.9 tells us that **entropy is a measure of microscopic disorder.**

INTERACTIVE **EXAMPLE 18.3** **Let's Play Marbles!**

Suppose you have a bag of 100 marbles. Fifty of the marbles are red and 50 are green. You are allowed to draw four marbles from the bag according to the following rules. Draw one marble, record its color, and return it to the bag. Shake the bag and then draw another marble. Continue this process until you have drawn and returned four marbles. What are the possible macrostates for this set of events? What is the most likely macrostate? What is the least likely macrostate?

Solution Because each marble is returned to the bag before the next one is drawn and because the bag is shaken, the probability of drawing a red marble is always the same as the probability of drawing a green one. All the possible microstates and macrostates are shown in Table 18.1. As this table indicates, there is only one way to draw a macrostate of four red marbles, and so there is only one microstate. There are, however, four possible microstates that correspond to the macrostate of one green marble and three red marbles, six microstates that correspond to two green marbles and two red marbles, four microstates that correspond to three green marbles and one red marble, and one microstate that corresponds to four green marbles. The most likely, and most disordered, macrostate—two red marbles and two green marbles—corresponds to the largest number of microstates. The least likely, most ordered macrostates—four red marbles or four green marbles—correspond to the smallest number of microstates.

TABLE **18.1** **Possible Results of Drawing Four Marbles from a Bag**

Macrostate	Possible Microstates	Total Number of Microstates
All R	RRRR	1
1G, 3R	RRRG, RRGR, RGRR, GRRR	4
2G, 2R	RRGG, RGRG, GRRG, RGGR, GRGR, GGRR	6
3G, 1R	GGGR, GGRG, GRGG, RGGG	4
All G	GGGG	1

PhysicsNow™ Explore the generation of microstates and macrostates by logging into PhysicsNow at **www.pop4e.com** and going to Interactive Example 18.3.

18.7 ENTROPY AND THE SECOND LAW OF THERMODYNAMICS

Because entropy is a measure of disorder and physical systems tend toward disordered macrostates, we can state that **the entropy of the Universe increases in all natural processes.** This statement is yet another way of stating the second law of thermodynamics.

■ Second law of thermodynamics; entropy statement

To calculate the change in entropy for a finite process, we must recognize that T is generally not constant. If dQ_r is the energy transferred reversibly by heat when the system is at temperature T, the change in entropy in an arbitrary reversible process between an initial state and a final state is

$$\Delta S = \int_i^f dS = \int_i^f \frac{dQ_r}{T} \qquad \text{(reversible path)} \qquad [18.10]$$

■ Change in entropy for a finite process

The change in entropy of a system depends only on the properties of the initial and final equilibrium states because entropy is a state variable, like internal energy, which is consistent with the relationship of entropy to disorder. For a given macrostate of a system, a given amount of disorder exists, measured by W (Eq. 18.9), the number of microstates corresponding to the macrostate. This number does not depend on the path followed as a system goes from one state to another.

In the case of a reversible, adiabatic process, no energy is transferred by heat between the system and its surroundings, and therefore $\Delta S = 0$. Because no change in entropy occurs, such a process is often referred to as an **isentropic process.**

Consider the changes in entropy that occur in a Carnot heat engine operating between the temperatures T_c and T_h. Equation 18.7 tells us that for a Carnot cycle,

$$\Delta S = 0$$

Now consider a system taken through an arbitrary reversible cycle. Because entropy is a state variable and hence depends only on the properties of a given equilibrium state, we conclude that $\Delta S = 0$ for *any* reversible cycle. In general, we can write this condition in the mathematical form

$$\oint \frac{dQ_r}{T} = 0 \qquad [18.11]$$

where the symbol $\oint$ indicates that the integration is over a *closed* path.

QUICK QUIZ 18.5 Which of the following is true for the entropy change of a system that undergoes a reversible, adiabatic process? **(a)** $\Delta S < 0$ **(b)** $\Delta S = 0$ **(c)** $\Delta S > 0$

QUICK QUIZ 18.6 An ideal gas is taken from an initial temperature T_i to a higher final temperature T_f along two different reversible paths starting from the same point on a *PV* diagram. Path A is at constant pressure, and path B is at constant volume. What is the relation between the entropy changes of the gas for these paths? **(a)** $\Delta S_A > \Delta S_B$ **(b)** $\Delta S_A = \Delta S_B$ **(c)** $\Delta S_A < \Delta S_B$

■ Thinking Physics 18.2

A box contains five gas molecules, spread throughout the box. At some time, all five are in one half of the box, which is a highly ordered situation. Does this situation violate the second law of thermodynamics? Is the second law valid for this system?

Reasoning Strictly speaking, this situation does violate the second law of thermodynamics. In response to the second question, however, the second law is not valid for small numbers of particles. The second law is based on collections of huge numbers of particles for which disordered states have astronomically higher probabilities than ordered states. Because the macroscopic world is built from these huge numbers of particles, the second law is valid as real processes proceed from order to disorder. In the five-molecule system, the general idea of the second law is valid in that there are more disordered states than ordered ones, but the relatively high probability of the ordered states results in their existence from time to time. ■

EXAMPLE 18.4 Change in Entropy—Melting Process

A solid substance that has a latent heat of fusion L_f melts at a temperature T_m. Calculate the change in entropy when a mass m of this substance is melted.

Solution Let us assume that the melting process occurs so slowly that it can be considered a reversible process; we can reverse the process by extracting energy very slowly to freeze the liquid into the solid form. In this case, the temperature can be regarded as constant and equal to T_m. Making use of Equation 18.10 and that the latent heat of fusion $Q = mL_f$, (Eq. 17.5), we find that

$$\Delta S = \int \frac{dQ_r}{T} = \frac{1}{T_m}\int dQ = \frac{Q}{T_m} = \frac{mL_f}{T_m}$$

Note that we are able to remove T_m from the integral because the process is isothermal.

18.8 ENTROPY CHANGES IN IRREVERSIBLE PROCESSES

So far, we have calculated changes in entropy using information about a reversible path connecting the initial and final equilibrium states. We can calculate entropy changes for irreversible processes by devising a reversible process (or a series of reversible processes) between the same two equilibrium states and computing $\int dQ_r/T$ for the reversible process. In irreversible processes, it is critically important to distinguish between Q, the actual energy transfer in the process, and Q_r, the energy that would have been transferred by heat along a reversible path between the same states. Only the second value gives the correct entropy change. For example, as we shall see, if an ideal gas expands adiabatically into a vacuum, $Q = 0$, but $\Delta S \neq 0$ because $Q_r \neq 0$. The reversible path between the same two states is the reversible, isothermal expansion that gives $\Delta S > 0$.

As we shall see in the following examples, the change in entropy for the system plus its environment is always positive for an irreversible process. In general, the total entropy (and disorder) always increases in irreversible processes. From these considerations, the second law of thermodynamics can be stated as follows: **The total entropy of an isolated system that undergoes a change cannot decrease.** Furthermore, if the process is *irreversible,* the total entropy of an isolated system always *increases.* On the other hand, in a reversible process, the total entropy of an isolated system remains constant.

When dealing with interacting objects that are not isolated from the environment, we must consider the change of entropy for the system *and* its environment. When two objects interact in an irreversible process, the increase in entropy of one part of the Universe is greater than the decrease in entropy of the other part. Hence, we conclude that the change in entropy of the Universe must be greater than zero for an irreversible process and equal to zero for a reversible process. Ultimately, the entropy of the Universe should reach a maximum value. At this point, the Universe will be in a state of uniform temperature and density. All physical, chemical, and biological processes will cease because a state of perfect disorder implies that no energy is available for doing work. This gloomy state of affairs is sometimes referred to as the "heat death" of the Universe.

QUICK QUIZ 18.7 True or false: The entropy change in an adiabatic process must be zero because $Q = 0$.

■ Thinking Physics 18.3

According to the entropy statement of the second law, the entropy of the Universe increases in irreversible processes. This statement sounds very different from the Kelvin–Planck and Clausius forms of the second law. Can these two statements be made consistent with the entropy interpretation of the second law?

Reasoning These three forms are consistent. In the Kelvin–Planck statement, the energy in the reservoir is disordered internal energy, the random motion of molecules. Performing work results in ordered energy, such as pushing a piston through a displacement. In this case, the motion of all molecules of the piston is in the same direction. If a heat engine absorbed energy by heat and performed an equal amount of work, it would have converted disorder into order, in violation of the entropy statement. In the Clausius statement, we start with an ordered system: higher temperature in the hot object, lower in the cold object. This separation of temperatures is an example of order. Energy transferring spontaneously from the cold object to the hot object, so that the temperatures spread even farther apart, would be an increase in order, in violation of the entropy statement. ■

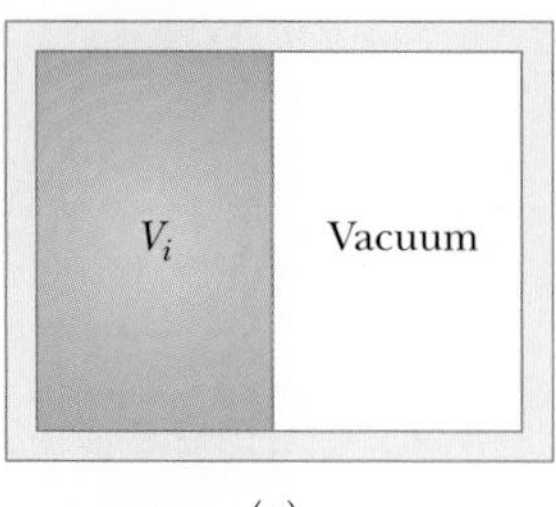

(a)

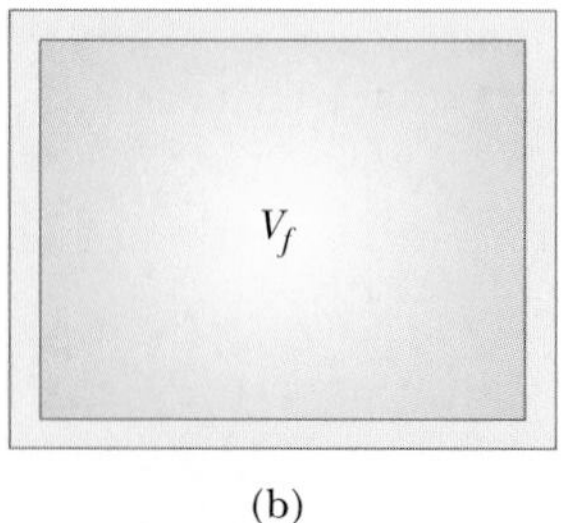

(b)

FIGURE 18.11 (a) A gas is initially restricted to an available volume V_i while the remainder of the volume is evacuated. (b) During a free expansion, the gas is allowed to expand into the evacuated volume.

Entropy Changes in a Free Expansion

An ideal gas in an insulated container initially occupies a volume of V_i (Fig. 18.11). A partition separating the gas from an evacuated region is broken so that the gas expands (irreversibly) to the volume V_f. Let us find the change in entropy of the gas and the Universe.

The process is neither reversible nor quasi-static. The work done on the gas is zero, and because the walls are insulating, no energy is transferred by heat during the expansion. That is, $W = 0$ and $Q = 0$. The first law tells us that the change in internal energy ΔE_{int} is zero; therefore, $E_{\text{int},i} = E_{\text{int},f}$. Because the gas is ideal, E_{int} depends on temperature only, so we conclude that $T_i = T_f$.

To apply Equation 18.10, we must find Q_r; that is, we must find an equivalent reversible path that shares the same initial and final states. A simple choice is an isothermal, reversible expansion in which the gas pushes slowly against a piston. Because T is constant in this process, Equation 18.10 gives

$$\Delta S = \int \frac{dQ_r}{T} = \frac{1}{T}\int_i^f dQ_r$$

Because we are considering an isothermal process, $\Delta E_{\text{int}} = 0$, so the first law of thermodynamics tells us that the energy input by heat is equal to the negative of the work done on the gas, $dQ_r = -dW = P\,dV$. Using this result, we find that

$$\Delta S = \frac{1}{T}\int dQ_r = \frac{1}{T}\int P\,dV = \frac{1}{T}\int \frac{nRT}{V}\,dV = nR\int_{V_i}^{V_f}\frac{dV}{V}$$

$$\Delta S = nR\ln\frac{V_f}{V_i} \qquad [18.12]$$

Because $V_f > V_i$, we conclude that ΔS is positive, and so both the entropy and the disorder of the gas (and the Universe) increase as a result of the irreversible, adiabatic expansion.

EXAMPLE 18.5 **Free Expansion of an Ideal Gas, Revisited**

Again consider the free expansion of an ideal gas. Let us verify that the macroscopic and microscopic approaches lead to the same conclusion. Suppose 1 mol of an ideal gas undergoes a free expansion to four times its initial volume. The initial and final temperatures are, as we have seen, the same.

A Using a macroscopic approach, calculate the entropy change of the gas.

Solution Let us conceptualize this situation from both a macroscopic and a microscopic point of view. Macroscopically, the gas has become more disordered by changing from a situation in which all molecules are in a small space to one in which the molecules are distributed within a space four times as large. Microscopically, there are more ways of placing the molecules in a large volume than in a small volume. From either approach, we see that the entropy of the gas must increase. The text of this first part of the problem tells us to categorize the problem as a macroscopic approach to entropy change. To analyze the problem, Equation 18.12 gives us

$$\Delta S = nR\ln\frac{V_f}{V_i} = (1)R\ln\left(\frac{4V_i}{V_i}\right) = R\ln 4$$

B The number of microstates associated with placing one molecule of volume V_m into a space of volume V is $w = V/V_m$. (The symbol w is used to represent the word "way"; one microstate is one "way" of achieving a macrostate.) The number of microstates associated with placing N molecules in this space is $W = w^N = (V/V_m)^N$. Find the number of ways the molecules of 1 mol of an ideal gas can be distributed in an initial volume V_i.

Solution The text of this part of the problem tells us to now categorize the problem as a microscopic approach to entropy change. To analyze the problem, as discussed in the text of the problem, the number of states available to a single molecule in the initial volume V_i is $w_i = (V_i/V_m)$. For 1 mol (N_A molecules), the number of available states is

$$W_i = w_i^{N_A} = \left(\frac{V_i}{V_m}\right)^{N_A}$$

C Using the considerations of part B, calculate the change in entropy for the free expansion to four times the initial volume and show that it agrees with part A.

Solution The number of states for all N_A molecules in the volume $V_f = 4V_i$ is

$$W_f = w_f^{N_A} = \left(\frac{V_f}{V_m}\right)^{N_A} = \left(\frac{4V_i}{V_m}\right)^{N_A}$$

From Equation 18.9, we obtain

$$\Delta S = k_B \ln W_f - k_B \ln W_i = k_B \ln\left(\frac{4V_i}{V_i}\right)^{N_A}$$

$$= k_B \ln(4)^{N_A} = N_A k_B \ln 4 = R \ln 4$$

To finalize the problem, note that this answer is the same as that in part A, which dealt with macroscopic parameters. This equality suggests that entropy can be approached both macroscopically and microscopically and the results should be the same.

18.9 THE ATMOSPHERE AS A HEAT ENGINE

CONTEXT CONNECTION

In Chapter 17, we predicted a global temperature based on the notion of energy balance between incoming visible radiation from the Sun and outgoing infrared radiation from the Earth. This model leads to a global temperature that is well below the measured temperature. This discrepancy results because atmospheric effects are not included in our model. In this section, we shall introduce some of these effects and show that the atmosphere can be modeled as a heat engine. In the Context Conclusion, we shall use concepts learned in the thermodynamics chapters to build a model that is more successful at predicting the correct temperature of the Earth.

What happens to the energy that enters the atmosphere by radiation from the Sun? Figure 18.12 helps answer this question by showing how the input energy undergoes various processes. If we identify the incoming energy as 100%, we find that 30% of it is reflected back into space, as we mentioned in Chapter 17. This 30% includes 6% back-scattered from air molecules, 20% reflected from clouds, and 4% reflected from the surface of the Earth. The remaining 70% is absorbed by either the air or the surface. Before reaching the surface, 20% of the original radiation is absorbed in the air; 4% by clouds; and 16% by water, dust particles, and ozone in

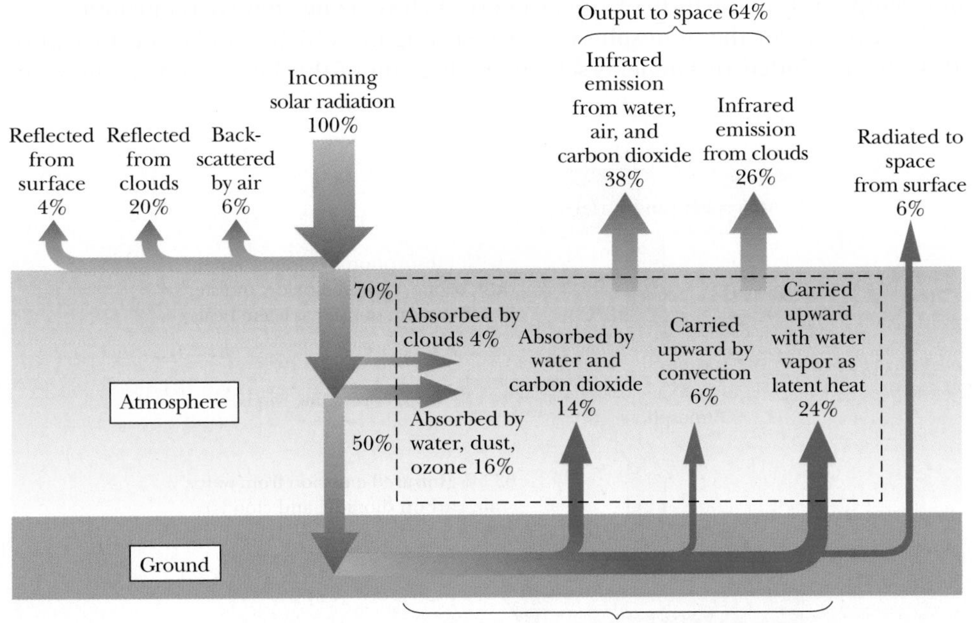

FIGURE 18.12 Energy input to the atmosphere from the Sun is divided into several components. Only 50% of the incident radiation is absorbed by the ground, and it is then reradiated into the atmosphere. The dashed line represents the atmospheric heat engine that is shown schematically in Figure 18.13.

the atmosphere. Of the original radiation striking the top of the atmosphere, the ground absorbs 50%.

The ground emits radiation upward and transfers energy to the atmosphere by several processes. Of the original 100% of the incoming energy, 6% simply passes back through the atmosphere into space (at the right in Fig. 18.12). In addition, 14% of the original incoming energy emitted as radiation from the ground is absorbed by water and carbon dioxide molecules. The air warmed by the surface rises upward by convection, carrying 6% of the original energy into the atmosphere. The hydrological cycle results in 24% of the original energy being carried upward as water vapor and released into the atmosphere when the water vapor condenses into liquid water.

These processes result in a total of 64% of the original energy being absorbed in the atmosphere, with another 6% from the surface passing back through into space. Because the atmosphere is in steady state, this 64% is also emitted from the atmosphere into space. The emission is divided into two types. The first is infrared radiation from molecules in the atmosphere, including water vapor, carbon dioxide, and the nitrogen and oxygen molecules of the air, which accounts for emission of 38% of the original energy. The remaining 26% is emitted as infrared radiation from clouds.

Figure 18.12 accounts for all the energy; the amount of energy input equals the amount of energy output, which is the premise used in the Context Connection of Chapter 17. A major difference from our discussion in that chapter, however, is the notion of absorption of the energy by the atmosphere. It is this absorption that creates thermodynamic processes in the atmosphere to raise the surface temperature above the value we determined in Chapter 17. We shall explore more about these processes and the temperature profile of the atmosphere in the Context Conclusion.

To close this chapter, let us discuss one more process that is not included in Figure 18.12. The various processes depicted in that figure result in a small amount of work done on the air, which appears as the kinetic energy of the prevailing winds in the atmosphere.

The amount of the original solar energy that is converted to kinetic energy of prevailing winds is about 0.5%. The process of generating the winds does not change the energy balance shown in Figure 18.12. The kinetic energy of the wind is converted to internal energy as masses of air move past one another. This internal energy produces an increased infrared emission of the atmosphere into space, so the 0.5% is only temporarily in the form of kinetic energy before being emitted as radiation.

We can model the atmosphere as a heat engine, which is indicated in Figure 18.12 by the dotted rectangle. A schematic diagram of this heat engine is shown in

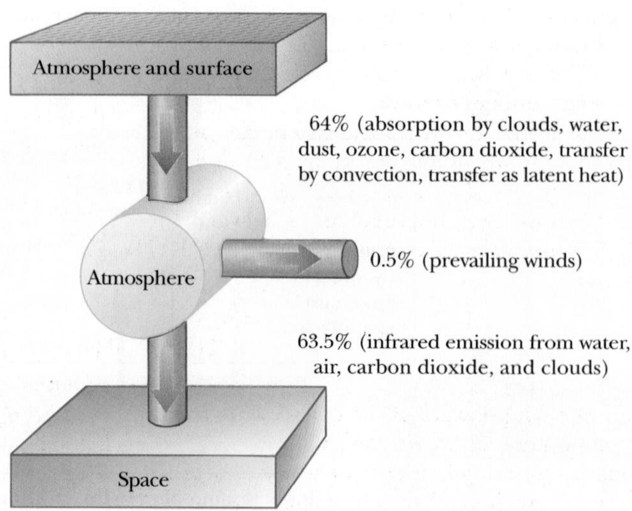

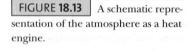

FIGURE 18.13 A schematic representation of the atmosphere as a heat engine.

Figure 18.13. The warm reservoir is the surface and the atmosphere, and the cold reservoir is empty space. We can calculate the efficiency of the atmospheric engine using Equation 18.2:

$$e = \frac{W_{\text{eng}}}{|Q_h|} = \frac{0.5\%}{64\%} = 0.008 = 0.8\%$$

which is a very low efficiency. Keep in mind, however, that a tremendous amount of energy enters the atmosphere from the Sun, so even a very small fraction of it can create a very complex and powerful wind system.

Notice that the output energy in Figure 18.13 is less than that in Figure 18.12 by 0.5%. As noted previously, the 0.5% transferred to the atmosphere by generating winds is eventually transformed to internal energy in the atmosphere by friction and then radiated into space as thermal radiation. We cannot separate the heat engine and the winds in the atmosphere in a diagram because the atmosphere *is* the heat engine and the winds are generated *in* the atmosphere!

We now have all the pieces we need to put together the puzzle of the temperature of the Earth. We shall discuss this subject in the Context Conclusion. ■

SUMMARY

Physics Now™ Take a practice test by logging into PhysicsNow at www.pop4e.com and clicking on the Pre-Test link for this chapter.

A **heat engine** is a device that takes in energy by heat and, operating in a cycle, expels a fraction of that energy by work. The net work done by a heat engine is

$$W_{\text{eng}} = |Q_h| - |Q_c| \qquad [18.1]$$

where Q_h is the energy absorbed from a hot reservoir and Q_c is the energy expelled to a cold reservoir.

The **thermal efficiency** e of a heat engine is defined as the ratio of the net work done to the energy absorbed per cycle from the higher temperature reservoir:

$$e = \frac{W_{\text{eng}}}{|Q_h|} = 1 - \frac{|Q_c|}{|Q_h|} \qquad [18.2]$$

The Kelvin–Planck statement of the **second law of thermodynamics** can be stated as follows: **It is impossible to construct a heat engine that, operating in a cycle, produces no effect other than the absorption of energy by heat from a reservoir and the performance of an equal amount of work.**

A **reversible** process is one for which the system can be returned to its initial conditions along the same path and for which every point along the path is an equilibrium state. A process that does not satisfy these requirements is **irreversible.**

The thermal efficiency of a heat engine operating in a **Carnot cycle** is given by

$$e_{\text{Carnot}} = 1 - \frac{T_c}{T_h} \qquad [18.4]$$

where T_c is the absolute temperature of the cold reservoir and T_h is the absolute temperature of the hot reservoir.

No real heat engine operating between the temperatures T_c and T_h can be more efficient than an engine operating reversibly in a Carnot cycle between the same two temperatures.

The **Clausius statement** of the second law states that **energy will not transfer spontaneously by heat from a cold object to a hot object.**

The second law of thermodynamics states that when real (irreversible) processes occur, the degree of disorder in the system plus the surroundings increases. The measure of disorder in a system is called **entropy** S.

The **change in entropy** dS of a system moving through an infinitesimal process between two equilibrium states is

$$dS = \frac{dQ_r}{T} \qquad [18.8]$$

where dQ_r is the energy transferred by heat in a reversible process between the same states.

From a microscopic viewpoint, the entropy S associated with a macrostate of a system is defined as

$$S \equiv k_B \ln W \qquad [18.9]$$

where k_B is Boltzmann's constant and W is the number of microstates corresponding to the macrostate whose entropy is S. Therefore, **entropy is a measure of microscopic disorder.** Because of the statistical tendency of systems to proceed toward states of greater probability and greater disorder, all natural processes are irreversible and result in an increase in entropy.

The change in entropy of a system moving between two general equilibrium states is

$$\Delta S = \int_i^f \frac{dQ_r}{T} \qquad [18.10]$$

The value of ΔS is the same for all paths connecting the initial and final states.

The change in entropy for any reversible, cyclic process is zero, and when such a process occurs, the entropy of the Universe remains constant. In an irreversible process, the total entropy of the Universe always increases.

QUESTIONS

☐ = answer available in the *Student Solutions Manual and Study Guide*

1. What are some factors that affect the efficiency of automobile engines?
2. Is it possible to construct a heat engine that creates no thermal pollution? What does your answer tell us about environmental considerations for an industrialized society?
3. A steam-driven turbine is one major component of an electric power plant. Why is it advantageous to have the temperature of the steam as high as possible?
4. Does the second law of thermodynamics contradict or correct the first law? Argue for your answer.
5. "The first law of thermodynamics says you can't really win, and the second law says you can't even break even." Explain how this statement applies to a particular device or process; alternatively, argue against the statement.
6. In solar ponds constructed in Israel, the Sun's energy is concentrated near the bottom of a salty pond. With the proper layering of salt in the water, convection is prevented and temperatures of 100°C may be reached. Can you estimate the maximum efficiency with which useful energy can be extracted from the pond?
7. Can a heat pump have a coefficient of performance less than unity? Explain.
8. Give various examples of irreversible processes that occur in nature. Give an example of a process in nature that is nearly reversible.
9. The device shown in Figure Q18.9, called a thermoelectric converter, uses a series of semiconductor cells to convert internal energy to electric potential energy, which we will study in Chapter 20. In the left photograph, both legs of the device are at the same temperature and no electric potential energy is produced. When one leg is at a higher temperature than the other, however, as shown in the right photograph, electric potential energy is produced as the device extracts energy from the hot reservoir and drives a small electric motor. (a) Why does the temperature differential produce electric potential energy in this demonstration? (b) In what sense does this intriguing experiment demonstrate the second law of thermodynamics?
10. A thermodynamic process occurs in which the entropy of a system changes by -8.0 J/K. According to the second law of thermodynamics, what can you conclude about the entropy change of the environment?
11. Discuss the change in entropy of a gas that expands (a) at constant temperature and (b) adiabatically.
12. How could you increase the entropy of 1 mol of a metal that is at room temperature? How could you decrease its entropy?
13. Suppose your roommate is "Mr. Clean" and tidies up your messy room after a big party. Because your roommate is creating more order, does this action represent a violation of the second law of thermodynamics?
14. "Energy is the mistress of the Universe and entropy is her shadow." Writing for an audience of general readers, argue for this statement with examples. Alternatively, argue for the view that entropy is like a decisive hands-on executive instantly determining what will happen, whereas energy is like a wretched back-office bookkeeper telling us how little we can afford.
15. If you shake a jar full of jelly beans of different sizes, the larger beans tend to appear near the top and the smaller ones tend to fall to the bottom. Why? Does this process violate the second law of thermodynamics?

(Courtesy of PASCO Scientific Company)

FIGURE Q18.9

PROBLEMS

1, 2, 3 = straightforward, intermediate, challenging
☐ = full solution available in the *Student Solutions Manual and Study Guide*
PhysicsNow™ = coached problem with hints available at www.pop4e.com
= computer useful in solving problem
= paired numerical and symbolic problems
= biomedical application

Section 18.1 ▪ Heat Engines and the Second Law of Thermodynamics

1. A heat engine takes in 360 J of energy from a hot reservoir and performs 25.0 J of work in each cycle. Find (a) the efficiency of the engine and (b) the energy expelled to the cold reservoir in each cycle.

2. A multicylinder gasoline engine in an airplane, operating at 2 500 rev/min, takes in energy 7.89×10^3 J and exhausts 4.58×10^3 J for each revolution of the crankshaft. (a) How many liters of fuel does it consume in 1.00 h of operation if the heat of combustion is 4.03×10^7 J/L? (b) What is the mechanical power output of the engine? Ignore friction and express the answer in horsepower. (c) What is the torque exerted by the crankshaft on the load? (d) What power must the exhaust and cooling system transfer out of the engine?

3. A particular heat engine has a useful power output of 5.00 kW and an efficiency of 25.0%. In each cycle, the engine expels 8 000 J of exhaust energy. Find (a) the energy taken in during each cycle and (b) the time interval for each cycle.

4. A gun is a heat engine. In particular, it is an internal combustion piston engine that does not operate in a cycle but, rather, comes apart during its adiabatic expansion process. A certain gun consists of 1.80 kg of iron. It fires one 2.40-g bullet at 320 m/s with an energy efficiency of 1.10%. Assume that the body of the gun absorbs all the energy exhaust—the other 98.9%—and increases uniformly in temperature for a short time interval before it loses any energy by heat into the environment. Find its temperature increase.

Section 18.2 ▪ Reversible and Irreversible Processes

Section 18.3 ▪ The Carnot Engine

5. One of the most efficient heat engines ever built is a steam turbine in the Ohio valley, operating between 430°C and 1 870°C on energy from West Virginia coal to produce electricity for the Midwest. (a) What is its maximum theoretical efficiency? (b) The actual efficiency of the engine is 42.0%. How much useful power does the engine deliver if it takes in 1.40×10^5 J of energy each second from its hot reservoir?

6. A Carnot engine has a power output of 150 kW. The engine operates between two reservoirs at 20.0°C and 500°C. (a) How much energy does it take in per hour? (b) How much energy is lost per hour in its exhaust?

7. PhysicsNow™ An ideal gas is taken through a Carnot cycle. The isothermal expansion occurs at 250°C, and the isothermal compression takes place at 50.0°C. The gas takes in 1 200 J of energy from the hot reservoir during the isothermal expansion. Find (a) the energy expelled to the cold reservoir in each cycle and (b) the net work done by the gas in each cycle.

8. The exhaust temperature of a Carnot heat engine is 300°C. What is the intake temperature if the efficiency of the engine is 30.0%?

9. A power plant operates at a 32.0% efficiency during the summer when the sea water used for cooling is at 20.0°C. The plant uses 350°C steam to drive turbines. If the plant's efficiency changes in the same proportion as the ideal efficiency, what would be the plant's efficiency in the winter, when the sea water is at 10.0°C?

10. An electric power plant that would make use of the temperature gradient in the ocean has been proposed. The system is to operate between 20.0°C (surface water temperature) and 5.00°C (water temperature at a depth of about 1 km). (a) What is the maximum efficiency of such a system? (b) If the useful power output of the plant is 75.0 MW, how much energy is taken in from the warm reservoir per hour? (c) In view of your answer to part (a), do you think such a system is worthwhile? Note that the "fuel" is free.

11. Here is a clever idea. Suppose you build a two-engine device such that the exhaust energy output from one heat engine is the input energy for a second heat engine. We say that the two engines are running *in series*. Let e_1 and e_2 represent the efficiencies of the two engines. (a) The overall efficiency of the two-engine device is defined as the total work output divided by the energy put into the first engine by heat. Show that the overall efficiency is given by

$$e = e_1 + e_2 - e_1 e_2$$

(b) Assume that the two engines are Carnot engines. Engine 1 operates between temperatures T_h and T_i. The gas in engine 2 varies in temperature between T_i and T_c. In terms of the temperatures, what is the efficiency of the combination engine? (c) What value of the intermediate temperature T_i will result in equal work being done by each of the two engines in series? (d) What value of T_i will result in each of the two engines in series having the same efficiency?

12. At point *A* in a Carnot cycle, 2.34 mol of a monatomic ideal gas has a pressure of 1 400 kPa, a volume of 10.0 L, and a temperature of 720 K. It expands isothermally to point *B* and then expands adiabatically to point *C*, where its volume is 24.0 L. An isothermal compression brings it to point *D*, where its volume is 15.0 L. An adiabatic process returns the gas to point *A*. (a) Determine all the unknown pressures, volumes, and temperatures as you fill in the following table:

	P	*V*	*T*
A	1 400 kPa	10.0 L	720 K
B			
C		24.0 L	
D		15.0 L	

(b) Find the energy added by heat, the work done by the engine, and the change in internal energy for each of the steps $A \rightarrow B$, $B \rightarrow C$, $C \rightarrow D$, and $D \rightarrow A$. (c) Calculate the efficiency W_{net}/Q_h. Show that it is equal to $1 - T_C/T_A$, the Carnot efficiency.

Section 18.4 ■ Heat Pumps and Refrigerators

13. A refrigerator has a coefficient of performance equal to 5.00. The refrigerator takes in 120 J of energy from a cold reservoir in each cycle. Find (a) the work required in each cycle and (b) the energy expelled to the hot reservoir.

14. A refrigerator has a coefficient of performance of 3.00. The ice tray compartment is at $-20.0°C$, and the room temperature is 22.0°C. The refrigerator can convert 30.0 g of water at 22.0°C to 30.0 g of ice at $-20.0°C$ each minute. What input power is required? Give your answer in watts.

15. In 1993, the U.S. government instituted a requirement that all room air conditioners sold in the United States must have an energy efficiency ratio (EER) of 10 or higher. The EER is defined as the ratio of the cooling capacity of the air conditioner, measured in British thermal units per hour, or Btu/h, to its electrical power requirement in watts. (a) Convert the EER of 10.0 to dimensionless form, using the conversion 1 Btu = 1 055 J. (b) What is the appropriate name for this dimensionless quantity? (c) In the 1970s, it was common to find room air conditioners with EERs of 5 or lower. Compare the operating costs for 10 000-Btu/h air conditioners with EERs of 5.00 and 10.0. Assume that each air conditioner operates for 1 500 h during the summer in a city where electricity costs 10.0¢ per kWh.

16. What is the coefficient of performance of a refrigerator that operates with Carnot efficiency between temperatures $-3.00°C$ and $+27.0°C$?

17. An ideal refrigerator or ideal heat pump is equivalent to a Carnot engine running in reverse. That is, energy Q_c is taken in from a cold reservoir and energy $|Q_h|$ is rejected to a hot reservoir. (a) Show that the work that must be supplied to run the refrigerator or heat pump is

$$W = \frac{T_h - T_c}{T_c} Q_c$$

(b) Show that the coefficient of performance of the ideal refrigerator is

$$\text{COP} = \frac{T_c}{T_h - T_c}$$

18. What is the maximum possible coefficient of performance of a heat pump that brings energy from outdoors at $-3.00°C$ into a 22.0°C house? Note that the work done to run the heat pump is also available to warm up the house.

19. A heat pump, shown in Figure P18.19, is essentially an air conditioner installed backward. It extracts energy from colder air outside and deposits it in a warmer room. Suppose the ratio of the actual energy entering the room to the work done by the device's motor is 10.0% of the theoretical maximum ratio. Determine the energy entering the room per joule of work done by the motor, given that the inside temperature is 20.0°C and the outside temperature is $-5.00°C$.

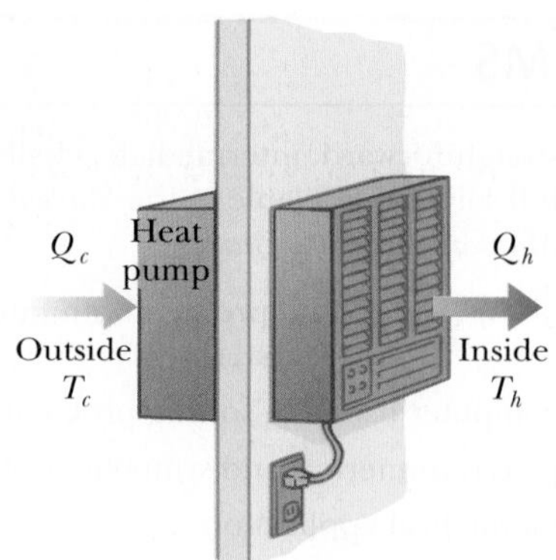

FIGURE P18.19

20. If a 35.0%-efficient Carnot heat engine (Active Fig. 18.1) is run in reverse so as to form a refrigerator (Active Fig. 18.7), what would be this refrigerator's coefficient of performance?

21. PhysicsNow™ How much work does an ideal Carnot refrigerator require to remove 1.00 J of energy from helium at 4.00 K and reject this energy to a room-temperature (293-K) environment?

22. Your father comes home to find that you have left a 400-W color television set turned on in an empty living room. The exterior temperature is 36°C. The room is cooled to 20°C by an air conditioner with coefficient of performance 4.50. The electric company charges you $0.150 per kilowatt-hour. Your father gets angry, and his metabolic rate increases from 120 W to 170 W. The excess internal energy he produces is carried away from his body into the room by convection and radiation. (a) Calculate the direct cost per minute of operating the television. (b) Calculate the per-minute cost of additional air conditioning attributable to the TV set before your father gets home. (c) Calculate the per-minute surcharge for still more air conditioning that your father blames on the TV set after he enters.

Section 18.6 ■ Entropy

Section 18.7 ■ Entropy and the Second Law of Thermodynamics

23. Calculate the change in entropy of 250 g of water warmed slowly from 20.0°C to 80.0°C. (*Suggestion:* Note that $dQ = mc\,dT$.)

24. An ice tray contains 500 g of liquid water at 0°C. Calculate the change in entropy of the water as it freezes slowly and completely at 0°C.

25. In making raspberry jelly, 900 g of raspberry juice is combined with 930 g of sugar. The mixture starts at room temperature, 23.0°C, and is slowly heated on a stove until it reaches 220°F. It is then poured into heated jars and allowed to cool. Assume that the juice has the same specific heat as water. The specific heat of sucrose is 0.299 cal/g·°C. Consider the heating process. (a) Which of the following terms describe(s) this process: adiabatic, isobaric, isothermal, isovolumetric, cyclic, reversible, isentropic? (b) How much energy does the mixture absorb? (c) What is the minimum change in entropy of the jelly while it is heated?

26. What change in entropy occurs when a 27.9-g ice cube at $-12°C$ is transformed into steam at 115°C?

27. If you toss two dice, what is the total number of ways in which you can obtain (a) a 12 and (b) a 7?

28. You toss a quarter, a dime, a nickel, and a penny into the air simultaneously and then record the results of your tosses in terms of the numbers of heads and tails that result. Prepare a table listing each macrostate and each of the microstates included in it. For example, the two microstates HHTH and HTHH, together with some others, are included in the macrostate of three heads and one tail. (a) On the basis of your table, what is the most probable result recorded for a toss? In terms of entropy, (b) what is the most ordered macrostate and (c) what is the most disordered?

29. A bag contains 50 red marbles and 50 green marbles. (a) You draw a marble at random from the bag, notice its color, return it to the bag, and repeat the process for a total of three draws. You record the result as the number of red marbles and the number of green marbles in the set of three. Construct a table listing each possible macrostate and the number of microstates within it. For example, RRG, RGR, and GRR are the three microstates constituting the macrostate 1G,2R. (b) Construct a table for the case in which you draw five marbles instead of three.

Section 18.8 ■ Entropy Changes in Irreversible Processes

30. The temperature at the surface of the Sun is approximately 5 700 K, and the temperature at the surface of the Earth is approximately 290 K. What entropy change occurs when 1 000 J of energy is transferred by radiation from the Sun to the Earth?

31. PhysicsNow™ A 1 500-kg car is moving at 20.0 m/s. The driver brakes to a stop. The brakes cool off to the temperature of the surrounding air, which is nearly constant at 20.0°C. What is the total entropy change?

32. A 1.00-kg iron horseshoe is taken from a forge at 900°C and dropped into 4.00 kg of water at 10.0°C. Assuming that no energy is lost by heat to the surroundings, determine the total entropy change of the horseshoe-plus-water system.

33. How fast are you personally making the entropy of the Universe increase right now? Compute an order-of-magnitude estimate, stating what quantities you take as data and the values you measure or estimate for them.

34. A 1.00-mol sample of H_2 gas is contained in the left-hand side of the container shown in Figure P18.34, which has equal volumes left and right. The right-hand side is evacuated. When the valve is opened, the gas streams into the right-hand side. What is the final entropy change of the gas? Does the temperature of the gas change? Assume that the container is so large that the hydrogen behaves as an ideal gas.

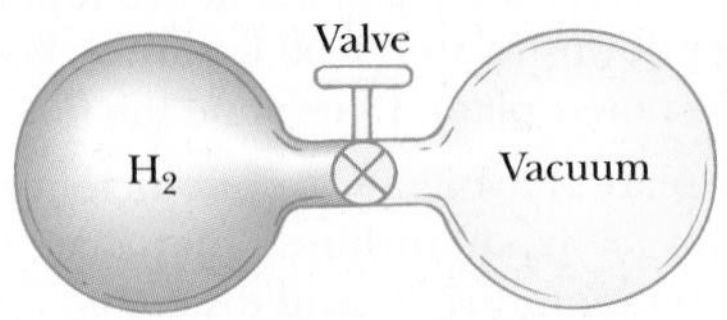

FIGURE P18.34

35. A 2.00-L container has a center partition that divides it into two equal parts, as shown in Figure P18.35. The left side contains H_2 gas, and the right side contains O_2 gas. Both gases are at room temperature and at atmospheric pressure. The partition is removed and the gases are allowed to mix. What is the entropy increase of the system?

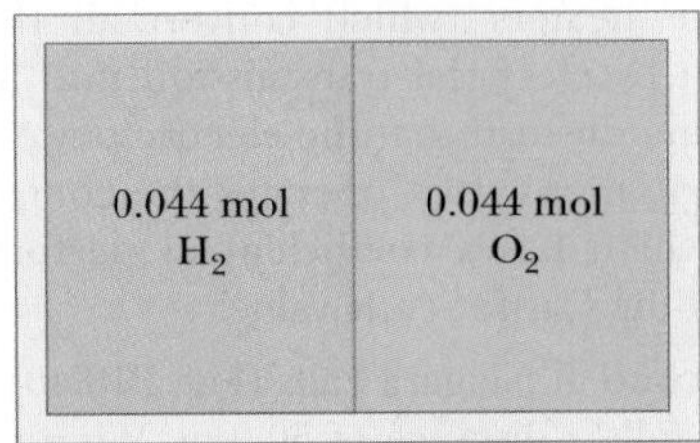

FIGURE P18.35

Section 18.9 ■ Context Connection—The Atmosphere as a Heat Engine

36. We found the efficiency of the atmospheric heat engine to be about 0.8%. Taking the intensity of incoming solar radiation to be 1 370 W/m^2 and assuming that 64% of this energy is absorbed in the atmosphere, find the "wind power," that is, the rate at which energy becomes available for driving the winds.

37. (a) Find the kinetic energy of the moving air in a hurricane, modeled as a disk 600 km in diameter and 11 km thick, with wind blowing at a uniform speed of 60 km/h. (b) Consider sunlight with an intensity of 1 000 W/m^2 falling perpendicularly on a circular area 600 km in diameter. During what time interval would the sunlight deliver the amount of energy computed in part (a)?

Additional Problems

38. A firebox is at 750 K, and the ambient temperature is 300 K. The efficiency of a Carnot engine doing 150 J of work as it transports energy between these constant-temperature baths is 60.0%. The Carnot engine must take in energy 150 J/0.600 = 250 J from the hot reservoir and must put out 100 J of energy by heat into the environment. To follow Carnot's reasoning, suppose some other heat engine S could have efficiency 70.0%. (a) Find the energy input and wasted energy output of engine S as it does 150 J of work. (b) Let engine S operate as in part (a) and run the Carnot engine in reverse. Find the total energy the firebox puts out as both engines operate together and the total energy transferred to the environment. Show that the Clausius statement of the second law of thermodynamics is violated. (c) Find the energy input and work output of engine S as it puts out exhaust energy of 100 J. (d) Let engine S operate as in part (c) and contribute 150 J of its work output to running the Carnot engine in reverse. Find the total energy the firebox puts out as both engines operate together, the total work output, and the total energy transferred to the environment. Show that the Kelvin–Planck statement of the second law is violated. Thus, our assumption about the efficiency of engine S must be false. (e) Let the engines operate

together through one cycle as in part (d). Find the change in entropy of the Universe. Show that the entropy statement of the second law is violated.

39. PhysicsNow™ A house loses energy through the exterior walls and roof at a rate of 5 000 J/s = 5.00 kW when the interior temperature is 22.0°C and the outside temperature is −5.00°C. Calculate the electric power required to maintain the interior temperature at 22.0°C for the following two cases. (a) The electric power is used in electric resistance heaters (which convert all the energy transferred in by electrical transmission into internal energy). (b) Assume instead that the electric power is used to drive an electric motor that operates the compressor of a heat pump, which has a coefficient of performance equal to 60.0% of the Carnot-cycle value.

40. Every second at Niagara Falls (Fig. P18.40), some 5 000 m^3 of water falls a distance of 50.0 m. What is the increase in entropy per second due to the falling water? Assume that the mass of the surroundings is so great that its temperature and that of the water stay nearly constant at 20.0°C. Suppose a negligible amount of water evaporates.

(CORBIS/Stock Market)

FIGURE **P18.40** Niagara Falls, a popular tourist attraction.

41. How much work is required, using an ideal Carnot refrigerator, to change 0.500 kg of tap water at 10.0°C into ice at −20.0°C? Assume that the freezer compartment is held at −20.0°C and that the refrigerator exhausts energy into a room at 20.0°C.

42. **Review problem.** This problem complements Problem 10.18 in Chapter 10. In the operation of a single-cylinder internal combustion piston engine, one charge of fuel explodes to drive the piston outward in the so-called power stroke. Part of the energy output is stored in a turning flywheel. This energy is then used to push the piston inward to compress the next charge of fuel and air. In this compression process, assume that an original volume of 0.120 L of a diatomic ideal gas at atmospheric pressure is compressed adiabatically to one-eighth of its original volume. (a) Find the work input required to compress the gas. (b) Assume that the flywheel is a solid disk of mass 5.10 kg and radius 8.50 cm, turning freely without friction between the power stroke and the compression stroke. How fast must the flywheel turn immediately after the power stroke? This situation represents the minimum angular speed at which the engine can operate because it is on the point of stalling. (c) When the engine's operation is well above the point of stalling, assume that the flywheel puts 5.00% of its maximum energy into compressing the next charge of fuel and air. Find its maximum angular speed in this case.

43. PhysicsNow™ In 1816, Robert Stirling, a Scottish clergyman, patented the *Stirling engine,* which has found a wide variety of applications ever since. Fuel is burned externally to warm one of the engine's two cylinders. A fixed quantity of inert gas moves cyclically between the cylinders, expanding in the hot one and contracting in the cold one. Figure P18.43 represents a model for its thermodynamic cycle. Consider n mol of an ideal monatomic gas being taken once through the cycle, consisting of two isothermal processes at temperatures $3T_i$ and T_i and two constant-volume processes. Determine, in terms of n, R, and T_i, (a) the net energy transferred by heat to the gas and (b) the efficiency of the engine. A Stirling engine is easier to manufacture than an internal combustion engine or a turbine. It can run on burning garbage. It can run on the energy of sunlight and produce no material exhaust.

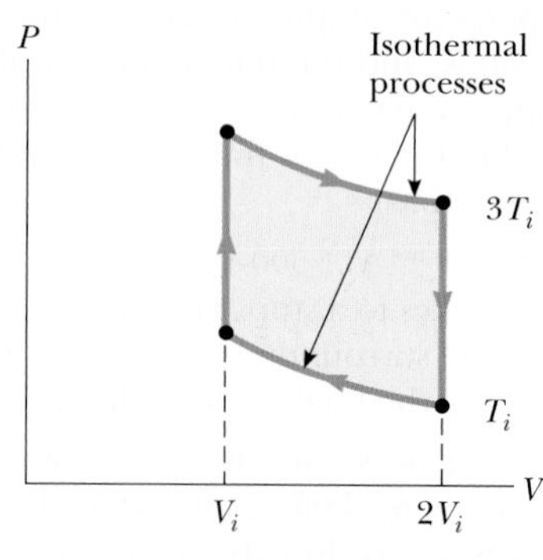

FIGURE **P18.43**

44. A heat engine operates between two reservoirs at $T_2 =$ 600 K and $T_1 =$ 350 K. It takes in 1 000 J of energy from the higher-temperature reservoir and performs 250 J of work. Find (a) the entropy change of the Universe ΔS_U for this process and (b) the work W that could have been done by an ideal Carnot engine operating between these two reservoirs. (c) Show that the difference between the amounts of work done in parts (a) and (b) is $T_1 \Delta S_U$.

45. A power plant, having a Carnot efficiency, produces 1 000 MW of electrical power from turbines that take in steam at 500 K and reject water at 300 K into a flowing river. The water downstream is 6.00 K warmer because of the output of the power plant. Determine the flow rate of the river.

46. A power plant, having a Carnot efficiency, produces electric power $\mathcal{P}$ from turbines that take in energy from steam at temperature T_h and discharge energy at temperature T_c through a heat exchanger into a flowing river.

The water downstream is warmer by ΔT because of the output of the power plant. Determine the flow rate of the river.

47. On the *PV* diagram for an ideal gas, one isothermal curve and one adiabatic curve pass through each point. Prove that the slope of the adiabat is steeper than the slope of the isotherm by the factor γ.

48. An athlete whose mass is 70.0 kg drinks 16 ounces (453.6 g) of refrigerated water. The water is at a temperature of 35.0°F. (a) Ignoring the temperature change of the human body that results from the water intake (so that the human body is regarded as a reservoir always at 98.6°F), find the entropy increase of the entire system. (b) Assume instead that the entire human body is cooled by the drink and that the average specific heat of a person is equal to the specific heat of liquid water. Ignoring any other energy transfers by heat and any metabolic energy release, find the athlete's temperature after she drinks the cold water, given an initial body temperature of 98.6°F. Under these assumptions, what is the entropy increase of the entire system? Compare this result with the one you obtained in part (a).

49. A 1.00-mol sample of an ideal monatomic gas is taken through the cycle shown in Figure P18.49. The process $A \rightarrow B$ is a reversible isothermal expansion. Calculate (a) the net work done by the gas, (b) the energy added to the gas by heat, (c) the energy exhausted from the gas by heat, and (d) the efficiency of the cycle.

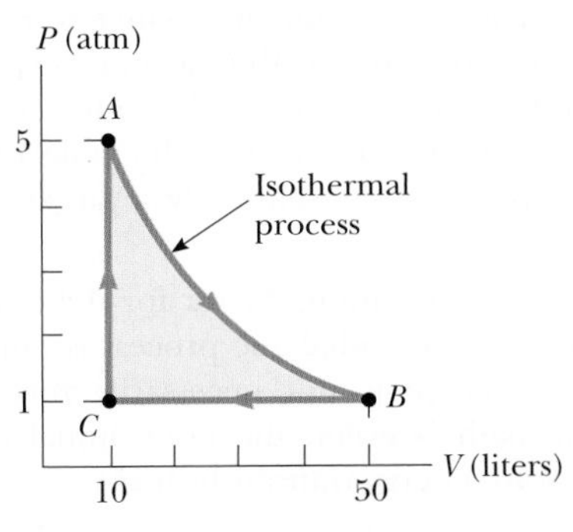

FIGURE **P18.49**

50. A biology laboratory is maintained at a constant temperature of 7.00°C by an air conditioner, which is vented to the air outside. On a typical hot summer day, the outside temperature is 27.0°C and the air-conditioning unit emits energy to the outside at a rate of 10.0 kW. Model the unit as having a coefficient of performance equal to 40.0% of the coefficient of performance of an ideal Carnot device. (a) At what rate does the air conditioner remove energy from the laboratory? (b) Calculate the power required for the work input. (c) Find the change in entropy produced by the air conditioner in 1.00 h. (d) The outside temperature increases to 32.0°C. Find the fractional change in the coefficient of performance of the air conditioner.

51. A 1.00-mol sample of a monatomic ideal gas is taken through the cycle shown in Figure P18.51. At point *A*, the pressure, volume, and temperature are P_i, V_i, and T_i, respectively. In terms of *R* and T_i, find (a) the total energy entering the system by heat per cycle, (b) the total energy leaving the system by heat per cycle, (c) the efficiency of an engine operating in this cycle, and (d) the efficiency of an engine operating in a Carnot cycle between the same temperature extremes.

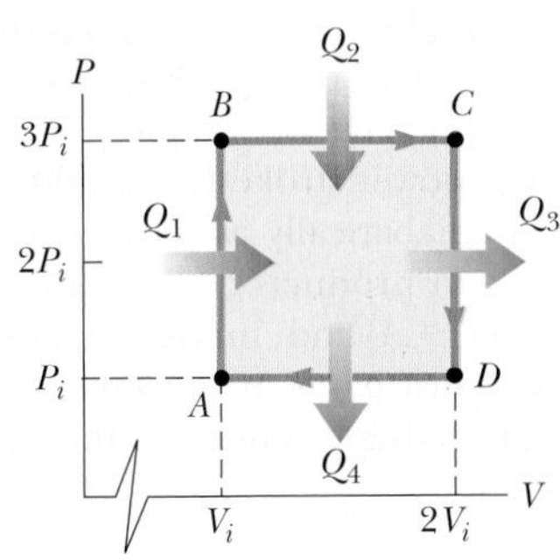

FIGURE **P18.51**

52. A sample consisting of *n* mol of an ideal gas undergoes a reversible isobaric expansion from volume V_i to volume $3V_i$. Find the change in entropy of the gas by calculating $\int_i^f dQ/T$, where $dQ = nC_P\,dT$.

53. A system consisting of *n* mol of an ideal gas undergoes two reversible processes. It starts with pressure P_i and volume V_i, expands isothermally, and then contracts adiabatically to reach a final state with pressure P_i and volume $3V_i$. (a) Find its change in entropy in the isothermal process. The entropy does not change in the adiabatic process. (b) Explain why the answer to part (a) must be the same as the answer to Problem 18.52.

54. Suppose you are working in a patent office and an inventor comes to you with the claim that her heat engine, which employs water as a working substance, has a thermodynamic efficiency of 0.61. She explains that it operates between energy reservoirs at 4°C and 0°C. It is a very complicated device, with many pistons, gears, and pulleys, and the cycle involves freezing and melting. Does her claim that $e = 0.61$ warrant serious consideration? Explain.

55. A 1.00-mol sample of an ideal gas ($\gamma = 1.40$) is carried through the Carnot cycle described in Active Figure 18.6. At point *A*, the pressure is 25.0 atm and the temperature is 600 K. At point *C*, the pressure is 1.00 atm and the temperature is 400 K. (a) Determine the pressures and volumes at points *A*, *B*, *C*, and *D*. (b) Calculate the net work done per cycle. (c) Determine the efficiency of an engine operating in this cycle.

56. An ideal (Carnot) freezer in a kitchen has a constant temperature of 260 K, whereas the air in the kitchen has a constant temperature of 300 K. Suppose the insulation for the freezer is not perfect but, rather, conducts energy into the freezer at a rate of 0.150 W. Determine the average power required for the freezer's motor to maintain the constant temperature in the freezer.

57. Calculate the increase in entropy of the Universe when you add 20.0 g of 5.00°C cream to 200 g of 60.0°C coffee. Assume that the specific heats of cream and coffee are both 4.20 J/g · °C.

58. The *Otto cycle* in Figure P18.58 models the operation of the internal combustion engine in an automobile. A mixture of gasoline vapor and air is drawn into a cylinder as the piston moves down during the intake stroke $O \rightarrow A$. The piston moves up toward the closed end of the cylinder to compress the mixture adiabatically in process $A \rightarrow B$. The ratio $r = V_1/V_2$ is the *compression ratio* of the engine. At B, the gasoline is ignited by the spark plug and the pressure rises rapidly as it burns in process $B \rightarrow C$. In the power stroke $C \rightarrow D$, the combustion products expand adiabatically as they drive the piston down. The combustion products cool further in an isovolumetric process $D \rightarrow A$ and in the exhaust stroke $A \rightarrow O$, when the exhaust gases are pushed out of the cylinder. Assume that a single value of the specific heat ratio γ characterizes both the fuel–air mixture and the exhaust gases after combustion. Prove that the efficiency of the engine is $1 - r^{1-\gamma}$.

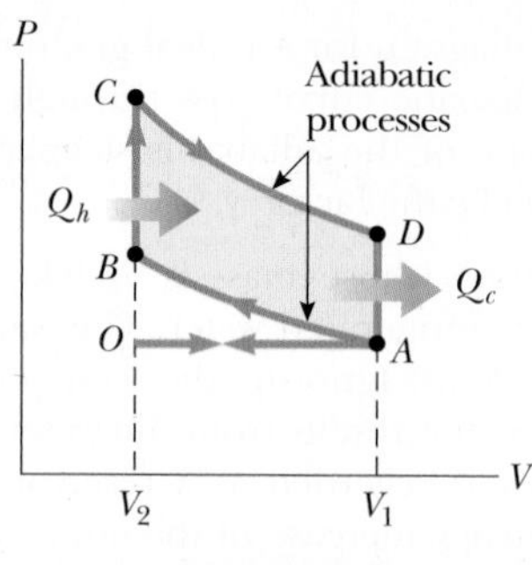

FIGURE **P18.58**

ANSWERS TO QUICK QUIZZES

18.1 (i), (c). The efficiency is the ratio of the work to the input energy, which is the inverse of the given value of 3.00. (ii), (b). The work represents one third of the input energy. The remainder, two thirds, must be expelled to the cold reservoir.

18.2 C, B, A. Although all three engines operate over a 300-K temperature difference, the efficiency depends on the ratio of temperatures, not on the difference.

18.3 (d). The COP of 4.00 for the heat pump means that the energy leaving the heat pump is four times as much as the energy entering by electrical transmission. With four times as much energy per unit of energy from electricity, you need only one-fourth as much electricity.

18.4 (i), (a). The only microstate is all four deuces. (ii), (f). The six microstates for two aces are club–diamond, club–heart, club–spade, diamond–heart, diamond–spade, and heart–spade. (iii), (b). The macrostate of two aces is more probable because this particular macrostate has six times as many microstates as the macrostate of four deuces. Therefore, in a hand of poker, two aces is less valuable than four deuces, even though a single ace is ranked higher than a single deuce.

18.5 (b). Because the process is reversible and adiabatic, $Q_r = 0$; therefore, $\Delta S = 0$.

18.6 (a). From the first law of thermodynamics, for these two reversible processes, $Q_r = \Delta E_{\text{int}} - W$. During the constant-volume process, $W = 0$; the work W is nonzero and negative during the constant-pressure expansion. Therefore, Q_r is larger for the constant-pressure process, leading to a larger value for the change in entropy. In terms of entropy as disorder, the gas must expand during the constant–pressure process. The increase in volume results in more ways of locating the molecules of the gas in a container and consequently a larger increase in entropy.

18.7 False. The determining factor for the entropy change is Q_r, not Q. If the adiabatic process is not reversible, the entropy change is not necessarily zero because a reversible path between the same initial and final states may involve energy transfer by heat.

Predicting the Earth's Surface Temperature

Now that we have investigated the principles of thermodynamics, we respond to our central question for the Context on global warming:

What factors determine the average temperature at the Earth's surface?

We discussed some of these factors—the energy input from the Sun and the energy output by thermal radiation from the surface of the Earth—in Chapter 17. In Chapter 18, we introduced the role of the atmosphere in absorbing radiation by means of various molecules. In the following discussion, we explore how the atmosphere modifies the temperature calculation performed in Chapter 17, which leads to a structural model that predicts a temperature in agreement with observations.

Modeling the Atmosphere

We first ask if the temperature of 255 K that we found in Chapter 17 is valid and, if so, what does it represent? The answer to the first question is *yes*. The energy balance concept is certainly valid, and the Earth, as a system, must emit energy at the same rate as it absorbs energy. The temperature of 255 K is representative of the radiation leaving the atmosphere. A space traveler outside our atmosphere who takes a reading of radiation from the Earth would determine that the temperature representing this radiation is indeed 255 K. This temperature is the one associated with radiation leaving the *top* of the atmosphere, however. It is not the temperature at the surface of the Earth.

As we have mentioned, the atmosphere is almost transparent to the visible radiation from the Sun but not to the infrared radiation emitted by the surface of the Earth. Let us build a model in which we assume that all radiation with wavelengths less than about 5 μm is allowed to pass through the atmosphere. Thus, almost *all* incoming radiation from the Sun (except for the 30% reflected) reaches the Earth's surface. In addition, let us assume that all radiation above about 5 μm (which is *infrared* radiation, including that emitted by the Earth surface) is absorbed by the atmosphere.

We can identify two layers to the atmosphere in our model (Fig. 1). The lower part of the atmosphere is the *troposphere.* In this layer, the density of the air is relatively high so that the probability of absorption of infrared radiation from the surface by molecules in the air is large. This absorption warms parcels of air near the surface, which then rise upward. As a parcel rises, it expands and its temperature drops. Therefore, the

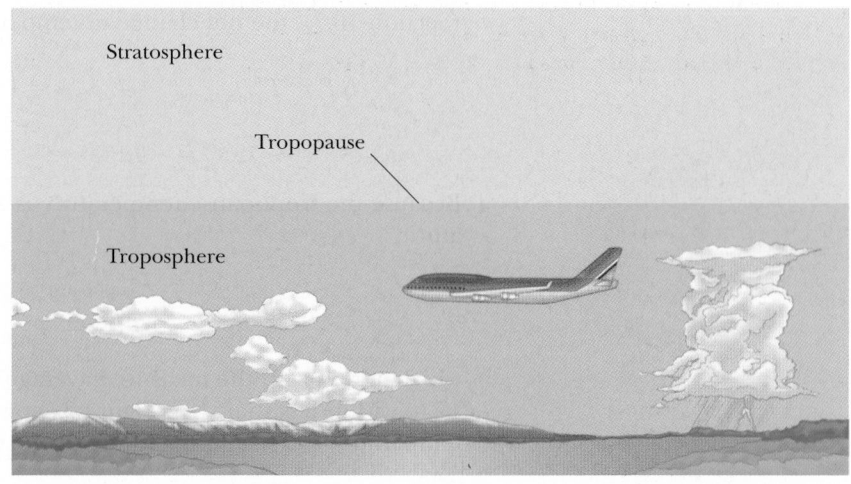

FIGURE 1 In our structural model of the atmosphere, we consider two layers. The upper layer is the stratosphere, in which the temperature is modeled as constant. The lower layer is the troposphere, in which the temperature increases linearly from the tropopause to the ground.

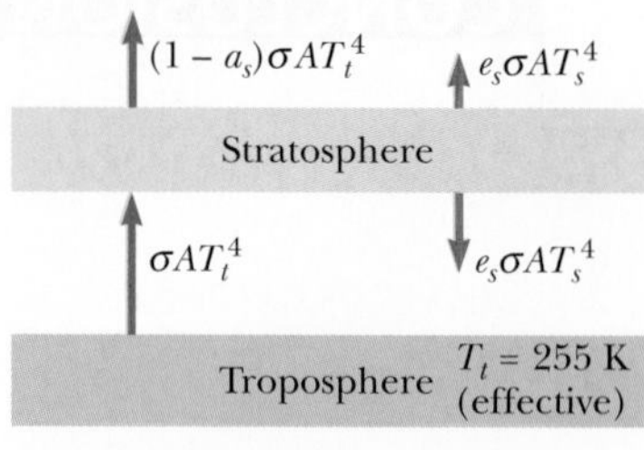

FIGURE 2 A portion of the stratosphere of area A is modeled as an object with a temperature, emitting thermal radiation from both upper and lower surfaces. The input of energy to the stratosphere is a portion of the radiation passing through it from the troposphere.

troposphere is the convective region in which the temperature decreases with height according to the lapse rate, as discussed in Section 16.7. It is also the region of the atmosphere in which our familiar weather occurs. Above the troposphere is the *stratosphere*. In this layer, the density of the air is relatively low so that the probability of absorption of infrared radiation is small. As a result, infrared radiation tends to pass through into space with little absorption. Without this absorption, the temperature in the stratosphere remains approximately constant with height. Between these two layers is the *tropopause,* which is about 11 km from the Earth's surface.[1] In reality, the tropopause is a thin region in which the primary energy transfer mechanism changes continuously from convection to radiation. In our model, we imagine the tropopause to be a sharp boundary.

The first task is to find the temperature, assumed constant, of the stratosphere. We appeal again to Stefan's law and consider the energy transfer into and out of the stratosphere, as indicated in Figure 2. Radiation from the troposphere (to which we assign an effective average temperature of $T_t = 255$ K so that it is the temperature associated with the radiation coming through the stratosphere to our imagined outer space observer) passes through the stratosphere, with a fraction a_s absorbed. The stratosphere, at temperature T_s, radiates both upward and downward, according to its emissivity e_s. Thus, because the stratosphere is in steady state, the power balance equation for the stratosphere is

$$\mathcal{P}_{\text{ER}}\text{ (in)} = -\mathcal{P}_{\text{ER}}\text{ (out)}$$

$$a_s\sigma A T_t^4 = 2e_s\sigma A T_s^4$$

where the factor of 2 arises from the output radiation of the stratosphere from both top and bottom surfaces. We can solve for the temperature of the stratosphere:

$$T_s = \left(\frac{a_s\sigma T_t^4}{2e_s\sigma}\right)^{1/4} = \left(\frac{a_s}{2e_s}\right)^{1/4} T_t = \left(\tfrac{1}{2}\right)^{1/4}(255\text{ K}) = 214\text{ K}$$

where we have used that the absorptivity and the emissivity of the stratosphere are the same number.

Now, we have all the pieces: the temperature of the stratosphere, the height of the tropopause, and the lapse rate. We simply need to extrapolate, using the lapse rate, from the temperature at the tropopause, which is the temperature of the stratosphere, to that at the Earth's surface.

If the tropopause is 11 km from the surface and the lapse rate is -6.5°C/km (Section 16.7), the net change in temperature from the surface to the tropopause is

$$\Delta T = T_{\text{tropopause}} - T_{\text{surface}} = \left(\frac{\Delta T}{\Delta y}\right)\Delta y = (-6.5\text{°C/km})(11\text{ km})$$
$$= -72\text{°C} = -72\text{ K}$$

Because the tropopause temperature is 214 K, we can now find the surface temperature:

$$\Delta T = T_{\text{tropopause}} - T_{\text{surface}}$$
$$-72\text{ K} = 214\text{ K} - T_{\text{surface}} \quad\rightarrow\quad T_{\text{surface}} = 286\text{ K}$$

which agrees with the measured average temperature of 288 K discussed in Chapter 17 to within less than 1%! Figure 3 shows a graphical representation (height versus temperature) of the temperature in the troposphere.

[1]The tropopause height of 11 km that we assume here is a simplification model in our structural model. In reality, the tropopause height varies with latitude and with season. At various latitudes and at different times of the year, the tropopause height can vary from less than 8 km to more than 17 km. The height of 11 km is a reasonable average for all latitudes over an entire year.

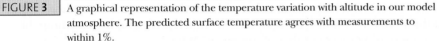

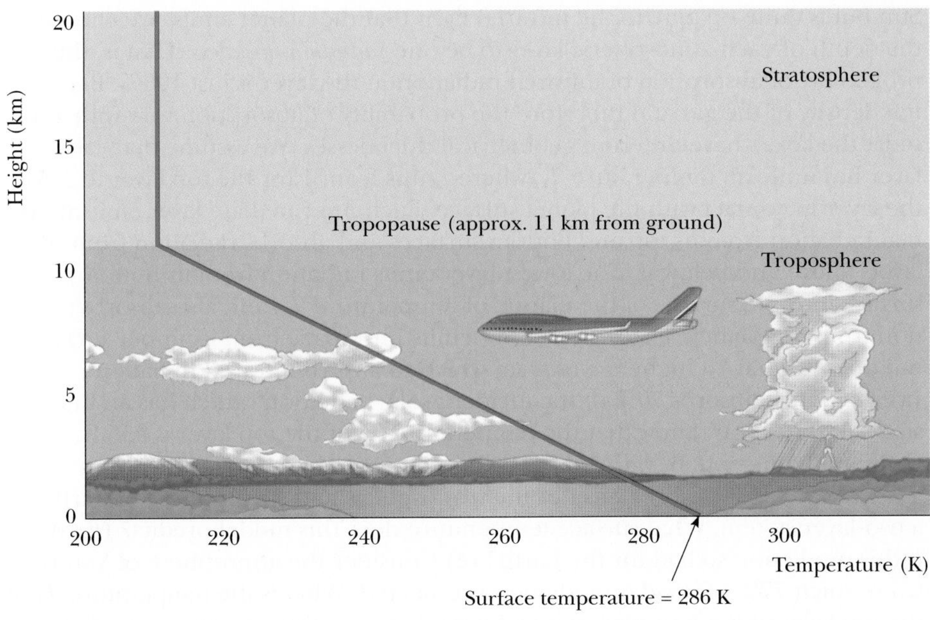

FIGURE 3 A graphical representation of the temperature variation with altitude in our model atmosphere. The predicted surface temperature agrees with measurements to within 1%.

The absorption of infrared radiation from the Earth's surface is dependent on molecules in the atmosphere. Our industrialized society is changing the atmospheric concentrations of molecules such as water, carbon dioxide, and methane. As a result, we are altering the energy balance and putting the Earth at risk of a change in temperature. Some data taken since the mid-19th century show a temperature increase of 0.5 to 1.0°C in the last 150 years. Although this increase may seem small, a slightly warmer Earth results in some melting of the polar ice caps and subsequent rises in the level of the oceans. Measured ocean levels show rises by as much as 50 cm in the 20th century. Further rises will cause severe problems for coastal populations. In addition, changes in temperature will have major effects on balanced ecosystems in various parts of the Earth.

The model we have described in this Context is successful in predicting the surface temperature. If we extend the model to predict changes in the surface temperature as we add more carbon dioxide to the atmosphere, we find that the predictions are not in agreement with more sophisticated models. The atmosphere is a very complicated entity, and the models used by atmospheric scientists are far more sophisticated than the one we have studied here. For our purposes, however, our successful prediction of the surface temperature is sufficient.

Problems

1. A simple model of absorption in the atmosphere shows that doubling the amount of carbon dioxide in the future will raise the altitude of the tropopause from 11 km to about 13 km. If the stratospheric temperature and the lapse rate remain the same, what is the surface temperature in this case? The result you obtain is much larger than the temperature predicted by sophisticated computer models. This disagreement displays a weakness in our simple model.
2. The stratosphere of Venus has a temperature of about 200 K. The lapse rate in the Venutian troposphere is about -8.8°C/km. The measured temperature on the surface of Venus is 732 K. What is the altitude of the Venutian tropopause?
3. Another atmospheric model is based on splitting the atmosphere into N layers of gas. We assume that the atmosphere is transparent to visible light from the

Sun but is quite opaque to the infrared light that the planet emits. We choose the depth of each atmospheric layer to be one *radiation thickness.* That is, the probability of absorption of infrared radiation in the layer is just 100%. Because the density of the gas and therefore the probability of absorption vary with altitude, the layers have different geometrical thicknesses. We assume that each layer has uniform temperature T_i, where i runs from 1 for the top layer to N for the layer in contact with the planet surface. Each intermediate layer emits thermal radiation from its top and bottom surfaces and absorbs radiation from the layers above and below it. The lowest layer emits radiation from its bottom surface into the surface of the planet, of temperature T_s, and also absorbs radiation from the planet. The highest layer emits into space from its upper surface but does not have a higher layer from which to absorb infrared radiation. (a) The Earth absorbs 70% of the incident solar radiation, which has an intensity of 1 370 W/m^2. Show that the temperature T_1 of the top layer is 255 K. (b) For an atmosphere with N layers, show that the surface temperature is $T_s = (N + 1)^{1/4} T_1$. (c) Consider the troposphere and stratosphere of Earth as a two-layer system. What surface temperature does this model predict? (d) Why is this prediction so bad for the Earth? (e) Consider the atmosphere of Venus, from which 77% of incident radiation is reflected. What is the temperature T_1 of the top layer of the Venutian atmosphere? (f) Given that the surface temperature of Venus is 732 K, how many layers are in the Venutian atmosphere? (g) Do you think that the multilayer model will be more successful in describing the atmosphere of Venus than that of the Earth? Why?

CONTEXT 6

Lightning

Lightning occurs all over the world, but more often in some places than others. Florida, for example, experiences lightning storms very often, but lightning is rare in Southern California. We begin this Context by looking at the details of a flash of lightning in a *qualitative* way. As we continue deeper into the Context, we will return to this description and attach more quantitative structure to it.

In general, we shall consider a flash of lightning to be an electric discharge occurring between a charged cloud and the ground or, in other words, an enormous spark. Lightning, however, can occur in *any* situation in which a large electric charge (which we discuss in Chapter 19) can result in electrical breakdown of the air, including snowstorms, sandstorms, and erupting volcanoes. If we consider lightning associated with clouds, we observe cloud-to-ground discharge, cloud-to-cloud discharge, intracloud discharge, and cloud-to-air discharge. In this Context, we shall consider only the most commonly described discharge, *cloud to ground.* Intracloud discharge actually occurs more often than cloud-to-ground discharge, but it is not the type of lightning that we regularly observe.

Because a flash of lightning occurs in a very short time, the structure of the process is hidden from normal human observation. A *flash* of lightning is comprised of a number of individual *strokes* of lightning, separated by tens of milliseconds. A typical number of strokes is 3 or 4, although as many as 26 strokes (for a total duration of 2 s) have been measured in a flash.

Although a stroke of lightning may appear as a sudden, single event, several steps are involved in the process. The process begins with an electrical breakdown in the air near the cloud that results in a column of negative charge, called a stepped leader, moving toward the ground at a typical speed of 10^5 m/s. The term *stepped leader* is used because the movement occurs in discrete steps of length about 50 m, with a delay of about 50 μs before the next step. A step occurs whenever the air becomes randomly ionized with sufficient free electrons in a short length of air to conduct electricity. The stepped leader is only faintly luminous and is not the bright flash we ordinarily think of as lightning. The radius of the channel of charge carried by the stepped leader is typically several meters.

As the tip of the stepped leader approaches the ground, it can initiate an

(Paul and LindaMarie Ambrose/Taxi/Getty Images)

FIGURE 1 Lightning electrically connects a cloud and the ground. In this Context, we shall learn about the details of such a lightning flash and find out how many lightning flashes occur on the Earth in a typical day.

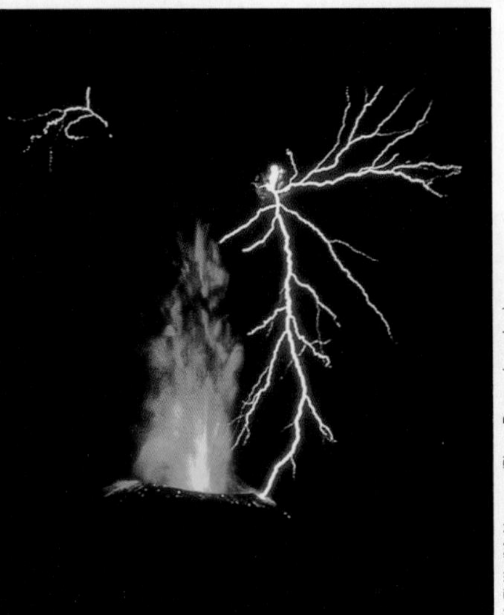
(M. Zhilin/M. Newman/Photo Researchers, Inc.)

FIGURE 2 During an eruption of the Sakurajima volcano in Japan, lightning is prevalent in the charged atmosphere above the volcano. Although lightning is possible in this as well as many other situations, in this Context we shall study the familiar lightning that occurs in a thunderstorm.

electrical breakdown in the air near the ground, often at the tip of a pointed object. Negative charges in the ground are repelled by the approaching tip of the column of negative charge in the stepped leader. As a result, the electrical breakdown in the air near the ground results in a column of positive charge beginning to move upward from the ground. (Electrons move downward in this column, which is equivalent to positive charges moving upward.) This process is the beginning of the *return stroke.* At 20 to 100 m above the ground, the return stroke meets the stepped leader, producing an effective short circuit between the cloud and the ground. Electrons pour downward into the ground at high speed, resulting in a very large electric current moving through a channel with a radius measured in centimeters. This high current rapidly raises the temperature of the air, ionizing atoms and providing the bright light flash we associate with lightning. Emission spectra of lightning show many spectral lines from oxygen and nitrogen, the major components of air.

After the return stroke, the conducting channel retains its conductivity for a short time (measured in tens of milliseconds). If more negative charge from the cloud is made available at the top of the conducting channel, this charge can move downward and result in a new stroke. In this case, because the conducting channel is "open," the leader does not move in a stepped fashion but, rather, moves downward smoothly and quickly. For this reason, it is called a *dart leader.* Once again, as the dart leader approaches the ground, a return stroke is initiated and a bright flash of light occurs.

Just after the current has passed through the conducting channel, the air is turned into a plasma at a typical temperature of 30 000 K. As a result, there is a sudden increase of pressure causing a rapid expansion of the plasma and generating a shock wave in the surrounding gas. This shock wave is the origin of the *thunder* associated with lightning.

(© Johnny Autery)

FIGURE 3 This photograph shows a lightning stroke as well as the individual components of the stroke. The bright channel represents a lightning stroke in progress, just after a stepped leader and a return stroke have connected and the channel becomes conducting. Several stepped leaders can be seen at the top of the photograph, branching off from the bright channel. They are less luminous than the bright channel because they have not yet connected with return strokes. A return stroke can be seen, just to the left of the bright channel, moving upward from the tree in search of a stepped leader. Another very faint return stroke can be seen leaving the top of the power pole at the left of the photograph.

Having taken this first qualitative step into the understanding of lightning, let us now seek more details. After investigating the physics of lightning, we shall respond to our central question:

How can we determine the number of lightning flashes on the Earth in a typical day?

CHAPTER 19

Electric Forces and Electric Fields

Mother and daughter are both enjoying the effects of electrically charging their bodies. Each individual hair on their heads becomes charged and exerts a repulsive force on the other hairs, resulting in the "stand-up" hairdos that you see here.

(Courtesy of Resonance Research Corporation)

CHAPTER OUTLINE

This chapter is the first of three on *electricity*. You are probably familiar with electrical effects, such as the static cling between articles of clothing removed from the dryer. You may also be familiar with the spark that jumps from your finger to a doorknob after you have walked across a carpet. Much of your daily experience involves working with devices that operate on energy transferred to the device by means of electrical transmission and provided by the electric power company. Even your own body is an electrochemical machine that uses electricity extensively. Nerves carry impulses as electrical signals, and electric forces are involved in the flow of materials across cell membranes.

This chapter begins with a review of some of the basic properties of the electrostatic force that we introduced in Chapter 5 as well as some properties of the electric field associated with stationary charged particles. Our study of electrostatics then continues with the concept of an electric field that is associated with a continuous charge distribution and the effect of this field on other charged particles. In these studies, we shall apply the models of a particle in a field and a particle under a net force that we have seen in earlier chapters.

19.1 HISTORICAL OVERVIEW

The laws of electricity and magnetism play a central role in the operation of devices such as radios, televisions, electric motors, computers, high-energy particle accelerators, and a host of electronic devices used in medicine. More fundamental, however, is that the interatomic and intermolecular forces responsible for the formation of solids and liquids are electric in origin. Furthermore, such forces as the pushes and pulls between objects in contact and the elastic force in a spring arise from electric forces at the atomic level.

Chinese documents suggest that magnetism was recognized as early as about 2000 B.C. The ancient Greeks observed electric and magnetic phenomena possibly as early as 700 B.C. They found that a piece of amber, when rubbed, attracted pieces of straw or feathers. The existence of magnetic forces was known from observations that pieces of a naturally occurring stone called *magnetite* (Fe_3O_4) were attracted to iron. (The word *electric* comes from the Greek word for amber, *elektron.* The word *magnetic* comes from *Magnesia,* a city on the coast of Turkey where magnetite was found.)

In 1600, Englishman William Gilbert discovered that electrification was not limited to amber but was a general phenomenon. Scientists went on to electrify a variety of objects, including people!

It was not until the early part of the 19th century that scientists established that electricity and magnetism are related phenomena. In 1820, Hans Oersted discovered that a compass needle, which is magnetic, is deflected when placed near an electric current. In 1831, Michael Faraday in England and, almost simultaneously, Joseph Henry in the United States showed that when a wire loop is moved near a magnet (or, equivalently, when a magnet is moved near a wire loop) an electric current is observed in the wire. In 1873, James Clerk Maxwell used these observations and other experimental facts as a basis for formulating the laws of electromagnetism as we know them today. Shortly thereafter (around 1888), Heinrich Hertz verified Maxwell's predictions by producing electromagnetic waves in the laboratory. This achievement was followed by such practical developments as radio and television.

Maxwell's contributions to the science of electromagnetism were especially significant because the laws he formulated are basic to *all* forms of electromagnetic phenomena. His work is comparable in importance to Newton's discovery of the laws of motion and the theory of gravitation.

(Charles D. Winters)

FIGURE 19.1 Rubbing a balloon against your hair on a dry day causes the balloon and your hair to become electrically charged.

19.2 PROPERTIES OF ELECTRIC CHARGES

A number of simple experiments demonstrate the existence of electrostatic forces. For example, after running a comb through your hair, you will find that the comb attracts bits of paper. The attractive electrostatic force is often strong enough to suspend the bits. The same effect occurs with other rubbed materials, such as glass or rubber.

Another simple experiment is to rub an inflated balloon with wool or across your hair (Fig. 19.1). On a dry day, the rubbed balloon will stick to the wall of a room, often for hours. When materials behave this way, they are said to have become electrically charged. You can give your body an electric charge by walking across a wool rug or by sliding across a car seat. You can then feel, and remove, the charge on your body by lightly touching another person or object. Under the right conditions, a visible spark is seen when you touch and a slight tingle is felt by both parties. (Such an experiment works best on a dry day

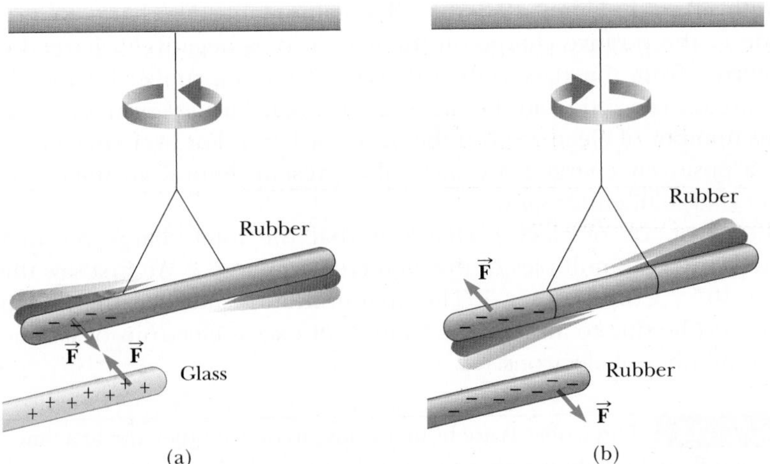

FIGURE 19.2 (a) A negatively charged rubber rod, suspended by an insulating thread, is attracted to a positively charged glass rod. (b) A negatively charged rubber rod is repelled by another negatively charged rubber rod.

because excessive moisture in the air can provide a pathway for charge to leak off a charged object.)

Experiments also demonstrate that there are two kinds of **electric charge,** given the names **positive** and **negative** by Benjamin Franklin (1706–1790). Figure 19.2 illustrates the interactions of the two kinds of charge. A hard rubber (or plastic) rod that has been rubbed with fur (or an acrylic material) is suspended by a piece of thread. When a glass rod that has been rubbed with silk is brought near the rubber rod, the rubber rod is attracted toward the glass rod (Fig. 19.2a). If two charged rubber rods (or two charged glass rods) are brought near each other, as in Figure 19.2b, the force between them is repulsive. This observation demonstrates that the rubber and glass have different kinds of charge. We use the convention suggested by Franklin; the electric charge on the glass rod is called positive and that on the rubber rod is called negative. On the basis of such observations, we conclude that **charges of the same sign repel each other and charges with opposite signs attract each other.**

We know that only two kinds of electric charge exist because any unknown charge that is found experimentally to be attracted to a positive charge is also repelled by a negative charge. No one has ever observed a charged object that is repelled by both a positive and a negative charge or that is attracted to both.

 Electrical attraction of contact lenses

Attractive electric forces are responsible for the behavior of a wide variety of commercial products. For example, the plastic in many contact lenses, *etafilcon,* is made up of molecules that electrically attract the protein molecules in human tears. These protein molecules are absorbed and held by the plastic so that the lens ends up being primarily composed of the wearer's tears. Therefore, the lens does not behave as a foreign object to the wearer's eye and can be worn comfortably. Many cosmetics also take advantage of electric forces by incorporating materials that are electrically attracted to skin or hair, causing the pigments or other chemicals to stay put once they are applied.

Another important characteristic of electric charge is that **the net charge in an isolated system is always conserved.** This represents the electric charge version of the isolated system model. We first introduced isolated system models in Chapter 7 when we discussed conservation of energy; we now see a principle of **conservation of electric charge** for an isolated system. When two initially neutral objects are charged by being rubbed together, charge is not created in the process. The objects become charged because *electrons are transferred* from one object to the other. One object gains some amount of negative charge from the electrons transferred to it while the other loses an equal amount of negative charge and hence is left with a positive charge. For the isolated system of the two objects, no transfer of charge occurs across the boundary of the system. The only change is that charge has been transferred between two members of the system. For example, when a glass rod is

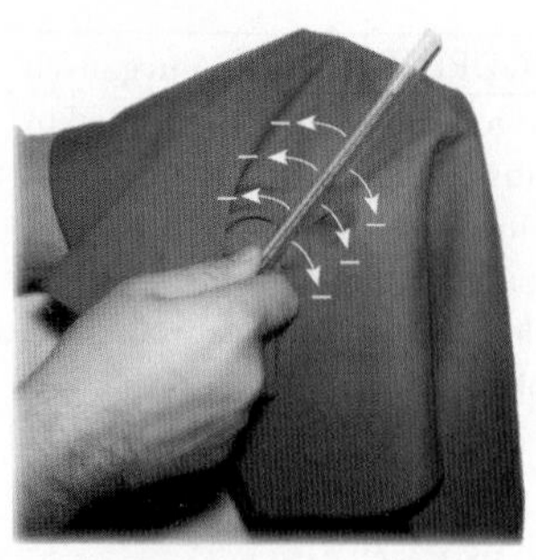

FIGURE 19.3 When a glass rod is rubbed with silk, electrons are transferred from the glass to the silk. Because of conservation of charge, each electron adds negative charge to the silk, and an equal positive charge is left behind on the rod. Also, because the charges are transferred in discrete bundles, the charges on the two objects are $\pm e$ or $\pm 2e$ or $\pm 3e$, and so on.

rubbed with silk, as in Figure 19.3, the silk obtains a negative charge that is equal in magnitude to the positive charge on the glass rod as negatively charged electrons are transferred from the glass to the silk. Likewise, when rubber is rubbed with fur, electrons are transferred from the fur to the rubber. An *uncharged object* contains an enormous number of electrons (on the order of 10^{23}). For every negative electron, however, a positively charged proton is also present; hence, an uncharged object has no net charge of either sign.

Another property of electric charge is that the total charge on an object is quantized as integral multiples of the elementary charge *e*. We first saw this charge $e = 1.60 \times 10^{-19}$ C in Chapter 5. The quantization results because the charge on an object must be due to an integral number of excess electrons or a deficiency of an integral number of electrons.

QUICK QUIZ 19.1 Three objects are brought close to one another, two at a time. When objects A and B are brought together, they repel. When objects B and C are brought together, they also repel. Which of the following statements are true? **(a)** Objects A and C possess charges of the same sign. **(b)** Objects A and C possess charges of opposite sign. **(c)** All three objects possess charges of the same sign. **(d)** One of the objects is neutral. **(e)** We need to perform additional experiments to determine the signs of the charges.

19.3 INSULATORS AND CONDUCTORS

We have discussed the transfer of charge from one object to another. It is also possible for electric charges to move from one location to another within an object; such motion of charge is called **electrical conduction.** It is convenient to classify substances in terms of the ability of charges to move within the substance:

Conductors are materials in which electric charges move relatively freely and **insulators** are materials in which electric charges do not move freely.

Materials such as glass, rubber, and Lucite are insulators. When such materials are charged by rubbing, only the rubbed area becomes charged; the charge does not tend to move to other regions of the material. In contrast, materials such as copper, aluminum, and silver are good conductors. When such materials are charged in some small region, the charge readily distributes itself over the entire surface of the material. If you hold a copper rod in your hand and rub it with wool or fur, it will not attract a small piece of paper, which might suggest that a metal cannot be charged. If you hold the copper rod by an insulating handle and then rub, however, the rod remains charged and attracts the piece of paper. In the first case, the electric charges produced by rubbing readily move from the copper through your body, which is a conductor, and finally to the Earth. In the second case, the insulating handle prevents the flow of charge to your hand.

Semiconductors are a third class of materials, and their electrical properties are somewhere between those of insulators and those of conductors. Charges can move somewhat freely in a semiconductor, but far fewer charges are moving through a semiconductor than in a conductor. Silicon and germanium are well-known examples of semiconductors that are widely used in the fabrication of a variety of electronic devices. The electrical properties of semiconductors can be changed over many orders of magnitude by adding controlled amounts of certain foreign atoms to the materials.

Charging by Induction

When a conductor is connected to the Earth by means of a conducting wire or pipe, it is said to be **grounded.** For present purposes, the Earth can be modeled as an infinite reservoir for electrons, which means that it can accept or supply an unlimited number of electrons. In this context, the Earth serves a purpose similar to our energy reservoirs introduced in Chapter 17. With that in mind, we can understand how to charge a conductor by a process known as **charging by induction.**

To understand how to charge a conductor by induction, consider a neutral (uncharged) metallic sphere insulated from the ground as shown in Figure 19.4a. There are an equal number of electrons and protons in the sphere if the charge on the sphere is exactly zero. When a negatively charged rubber rod is brought near the sphere, electrons in the region nearest the rod experience a repulsive force and migrate to the opposite side of the sphere. This migration leaves the side of the sphere near the rod with an effective positive charge because of the diminished number of electrons as in Figure 19.4b. (The left side of the sphere in Figure 19.4b is positively charged *as if* positive charges moved into this region, but in a metal it is only electrons that are free to move.) This migration occurs even if the rod never actually touches the sphere. If the same experiment is performed with a conducting wire connected from the sphere to the Earth (Fig. 19.4c), some of the electrons in the conductor are so strongly repelled by the presence of the negative charge in the rod that they move out of the sphere through the wire and into the Earth. The symbol ⏚ at the end of the wire in Figure 19.4c indicates that the wire is connected to **ground,** which means a reservoir such as the Earth. If the wire to ground is then removed (Fig. 19.4d), the conducting sphere contains an excess of *induced* positive charge because it has fewer electrons than it needs to cancel out the positive charge of the protons. When the rubber rod is removed from the vicinity of the sphere (Fig. 19.4e), this induced positive charge remains on the ungrounded sphere. Note that the rubber rod loses none of its negative charge during this process.

Charging an object by induction requires no contact with the object inducing the charge. This behavior is in contrast to charging an object by rubbing, which does require contact between the two objects.

A process similar to the first step in charging by induction in conductors takes place in insulators. In most neutral atoms and molecules, the average position of the positive charge coincides with the average position of the negative charge. In the presence of a charged object, however, these positions may shift slightly because of the attractive and repulsive forces from the charged object, resulting in more positive charge on one side of the molecule than on the other. This effect is known as **polarization.** The polarization of individual molecules produces a layer of charge on the surface of the insulator as shown in Figure 19.5a, in which a charged balloon on the left is placed against a wall on the right. In the figure, the negative charge layer in the wall is closer to the positively charged balloon than the positive

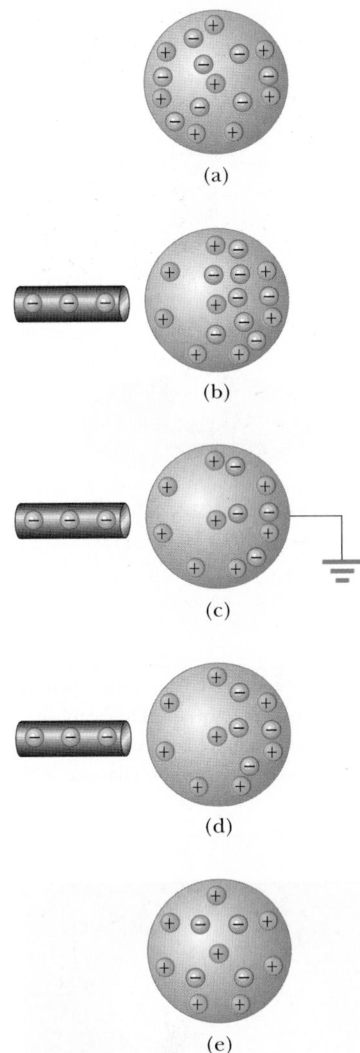

FIGURE 19.4 Charging a metallic object by *induction;* that is, the two objects never touch each other. (a) A neutral metallic sphere, with equal numbers of positive and negative charges. (b) The electrons on the neutral sphere are redistributed when a charged rubber rod is placed near the sphere. (c) When the sphere is grounded, some of its electrons leave through the ground wire. (d) When the ground connection is removed, the sphere has excess positive charge that is nonuniformly distributed. (e) When the rod is removed, the remaining electrons redistribute uniformly and there is a net uniform distribution of positive charge on the sphere.

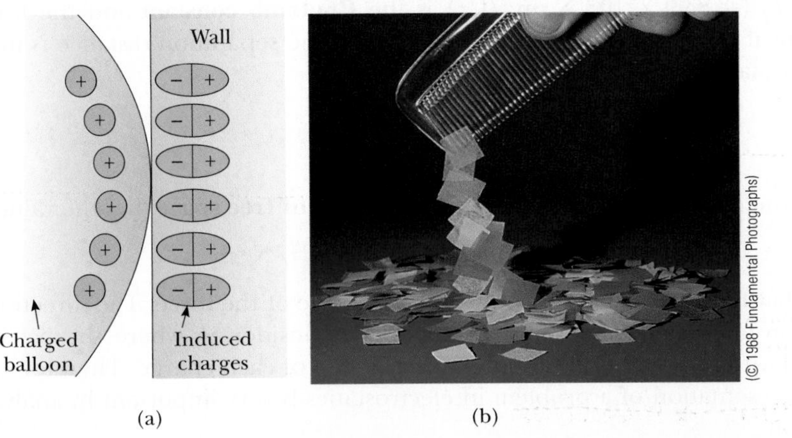

FIGURE 19.5 (a) The charged balloon on the left induces a charge distribution on the wall's surface due to realignment of charges in the molecules. (b) A charged comb attracts bits of paper because charges in the paper's molecules are realigned.

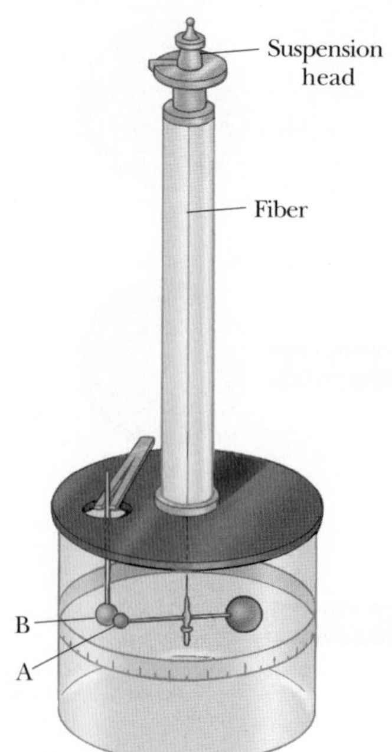

FIGURE 19.6 Coulomb's torsion balance, which was used to establish the inverse-square law for the electrostatic force between two charges.

(Photo courtesy of AIP Niels Bohr Library/E. Scott Barr Collection)

CHARLES COULOMB (1736–1806)

French physicist Coulomb's major contributions to science were in the areas of electrostatics and magnetism. During his lifetime, he also investigated the strengths of materials and determined the forces that affect objects on beams, thereby contributing to the field of structural mechanics. In the field of ergonomics, his research provided a fundamental understanding of the ways in which people and animals can best do work.

charges at the other ends of the molecules. Therefore, the attractive force between the positive and negative charges is larger than the repulsive force between the positive charges. The result is a net attractive force between the charged balloon and the neutral insulator. It is this polarization effect that explains why a comb that has been rubbed through hair attracts bits of neutral paper (Fig. 19.5b) or why a balloon that has been rubbed against your hair can stick to a neutral wall.

QUICK QUIZ 19.2 Three objects are brought close to one another, two at a time. When objects A and B are brought together, they attract. When objects B and C are brought together, they repel. From this experiment, what can we conclude? **(a)** Objects A and C possess charges of the same sign. **(b)** Objects A and C possess charges of opposite sign. **(c)** All three of the objects possess charges of the same sign. **(d)** One of the objects is neutral. **(e)** We need to perform additional experiments to determine information about the charges on the objects.

19.4 COULOMB'S LAW

Electric forces between charged objects were measured quantitatively by Charles Coulomb using the torsion balance, which he invented (Fig. 19.6). Coulomb confirmed that the electric force between two small charged spheres is proportional to the inverse square of their separation distance r, that is, $F_e \propto 1/r^2$. The operating principle of the torsion balance is the same as that of the apparatus used by Sir Henry Cavendish to measure the gravitational constant (Section 11.1), with the electrically neutral spheres replaced by charged ones. The electric force between charged spheres A and B in Figure 19.6 causes the spheres to either attract or repel each other, and the resulting motion causes the suspended fiber to twist. Because the restoring torque of the twisted fiber is proportional to the angle through which it rotates, a measurement of this angle provides a quantitative measure of the electric force of attraction or repulsion. Once the spheres are charged by rubbing, the electric force between them is very large compared with the gravitational attraction, and so the gravitational force can be ignored.

In Chapter 5, we introduced **Coulomb's law,** which describes the magnitude of the electrostatic force between two charged particles with charges q_1 and q_2 and separated by a distance r:

$$F_e = k_e \frac{|q_1||q_2|}{r^2} \qquad [19.1]$$

where k_e ($= 8.99 \times 10^9$ N·m^2/C^2) is the **Coulomb constant** and the force is in newtons if the charges are in coulombs and if the separation distance is in meters. The constant k_e is also written as

$$k_e = \frac{1}{4\pi\epsilon_0}$$

where the constant ϵ_0, known as the **permittivity of free space,** has the value

$$\epsilon_0 = 8.854\,2 \times 10^{-12} \text{ C}^2/\text{N}\cdot\text{m}^2$$

Note that Equation 19.1 gives only the magnitude of the force. The direction of the force on a given particle must be found by considering where the particles are located with respect to one another and the sign of each charge. Therefore, a pictorial representation of a problem in electrostatics is very important in analyzing the problem.

The charge of an electron is $q = -e = -1.60 \times 10^{-19}$ C, and the proton has a charge of $q = +e = 1.60 \times 10^{-19}$ C; therefore, 1 C of charge is equal to the

TABLE 19.1 Charge and Mass of the Electron, Proton, and Neutron

Particle	Charge (C)	Mass (kg)
Electron (e)	$-1.602\,176\,5 \times 10^{-19}$	$9.109\,38 \times 10^{-31}$
Proton (p)	$+1.602\,176\,5 \times 10^{-19}$	$1.672\,62 \times 10^{-27}$
Neutron (n)	0	$1.674\,93 \times 10^{-27}$

magnitude of the charge of $(1.60 \times 10^{-19})^{-1} = 6.25 \times 10^{18}$ electrons. (The elementary charge e was introduced in Section 5.5.) Note that 1 C is a substantial amount of charge. In typical electrostatic experiments, where a rubber or glass rod is charged by friction, a net charge on the order of 10^{-6} C ($= 1\ \mu$C) is obtained. In other words, only a very small fraction of the total available electrons (on the order of 10^{23} in a 1-cm^3 sample) are transferred between the rod and the rubbing material. The experimentally measured values of the charges and masses of the electron, proton, and neutron are given in Table 19.1.

When dealing with Coulomb's law, remember that force is a *vector* quantity and must be treated accordingly. Furthermore, **Coulomb's law applies exactly only to particles.**[1] The electrostatic force exerted by q_1 on q_2, written $\vec{\mathbf{F}}_{12}$, can be expressed in vector form as[2]

$$\vec{\mathbf{F}}_{12} = k_e \frac{q_1 q_2}{r^2} \hat{\mathbf{r}}_{12} \qquad [19.2]$$

where $\hat{\mathbf{r}}_{12}$ is a unit vector directed from q_1 toward q_2 as in Active Figure 19.7a. Equation 19.2 can be used to find the direction of the force in space, although a carefully drawn pictorial representation is needed to clearly identify the direction of $\hat{\mathbf{r}}_{12}$. From Newton's third law, we see that the electric force exerted by q_2 on q_1 is equal in magnitude to the force exerted by q_1 on q_2 and in the opposite direction; that is, $\vec{\mathbf{F}}_{21} = -\vec{\mathbf{F}}_{12}$. From Equation 19.2, we see that if q_1 and q_2 have the same sign, the product $q_1 q_2$ is positive and the force is repulsive as in Active Figure 19.7a. The force on q_2 is in the same direction as $\hat{\mathbf{r}}_{12}$ and is directed away from q_1. If q_1 and q_2 are of opposite sign as in Active Figure 19.7b, the product $q_1 q_2$ is negative and the force is attractive. In this case, the force on q_2 is in the direction opposite to $\hat{\mathbf{r}}_{12}$, directed toward q_1.

When more than two charged particles are present, the force between any pair is given by Equation 19.2. Therefore, **the resultant force on any one particle equals the *vector* sum of the individual forces due to all other particles.** This **principle of superposition** as applied to electrostatic forces is an experimentally observed fact and simply represents the traditional vector sum of forces introduced in Chapter 4. As an example, if four charged particles are present, the resultant force on particle 1 due to particles 2, 3, and 4 is given by the vector sum

$$\vec{\mathbf{F}}_1 = \vec{\mathbf{F}}_{21} + \vec{\mathbf{F}}_{31} + \vec{\mathbf{F}}_{41}$$

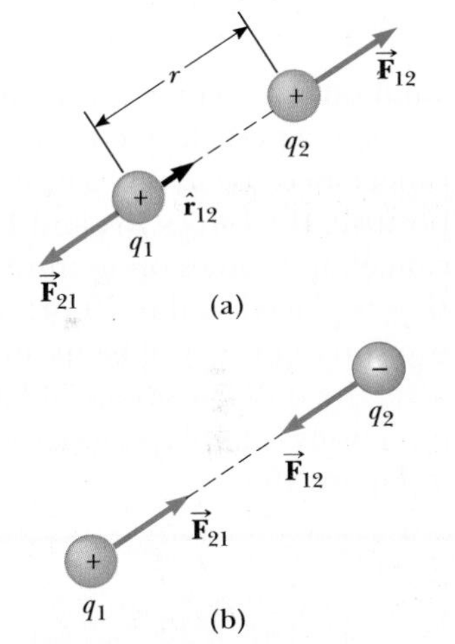

ACTIVE FIGURE 19.7
Two point charges separated by a distance r exert a force on each other given by Coulomb's law. Note that the force $\vec{\mathbf{F}}_{21}$ exerted by q_2 on q_1 is equal in magnitude and opposite in direction to the force $\vec{\mathbf{F}}_{12}$ exerted by q_1 on q_2. (a) When the charges are of the same sign, the force is repulsive. (b) When the charges are of opposite signs, the force is attractive.

PhysicsNow™ Log into PhysicsNow at **www.pop4e.com** and go to Active Figure 19.7 to move the charges to any position in two-dimensional space and observe the electric forces on them.

QUICK QUIZ 19.3 (i) Object A has a charge of $+2\ \mu$C, and object B has a charge of $+6\ \mu$C. Which of the following statements is true about the electric forces on the objects? **(a)** $F_{AB} = -3F_{BA}$ **(b)** $F_{AB} = -F_{BA}$ **(c)** $3F_{AB} = -F_{BA}$ **(d)** $F_{AB} = 3F_{BA}$ **(e)** $F_{AB} = F_{BA}$ **(f)** $3F_{AB} = F_{BA}$ **(ii)** Which of the following statements is true about the electric forces on the objects? **(a)** $\vec{\mathbf{F}}_{AB} = -3\vec{\mathbf{F}}_{BA}$ **(b)** $\vec{\mathbf{F}}_{AB} = -\vec{\mathbf{F}}_{BA}$ **(c)** $3\vec{\mathbf{F}}_{AB} = -\vec{\mathbf{F}}_{BA}$ **(d)** $\vec{\mathbf{F}}_{AB} = 3\vec{\mathbf{F}}_{BA}$ **(e)** $\vec{\mathbf{F}}_{AB} = \vec{\mathbf{F}}_{BA}$ **(f)** $3\vec{\mathbf{F}}_{AB} = \vec{\mathbf{F}}_{BA}$

[1] Coulomb's law can also be used for larger objects to which the particle model can be applied.

[2] Notice that we use "q_2" as shorthand notation for "the particle with charge q_2." This usage is common when discussing charged particles, similar to the use in mechanics of "m_2" for "the particle with mass m_2." The context of the sentence will tell you whether the symbol represents an amount of charge or a particle with that charge.

INTERACTIVE EXAMPLE 19.1 Where Is the Resultant Force Zero?

Three charged particles lie along the x axis as in Figure 19.8. The particle with charge $q_1 = +15.0\ \mu C$ is at $x = 2.00$ m, and the particle with charge $q_2 = +6.00\ \mu C$ is at the origin. Where on the x axis can a particle with negative charge q_3 be placed such that the resultant force on it is zero?

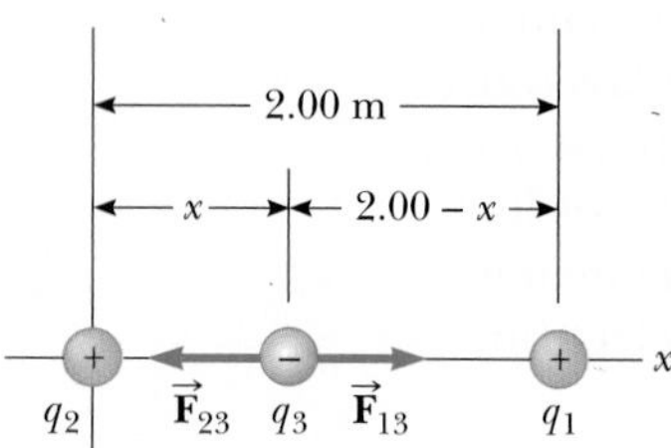

FIGURE 19.8 (Interactive Example 19.1) Three point charges are placed along the x axis. If the net force on q_3 is zero, the force $\vec{\mathbf{F}}_{13}$ exerted by q_1 on q_3 must be equal in magnitude and opposite in direction to the force $\vec{\mathbf{F}}_{23}$ exerted by q_2 on q_3.

Solution The requested resultant force of zero indicates that q_3 is a particle in equilibrium, so the two forces on q_3 cancel. Because q_3 is negative and both q_1 and q_2 are positive, the forces $\vec{\mathbf{F}}_{13}$ and $\vec{\mathbf{F}}_{23}$ are both attractive. To cancel, the forces on q_3 must be in opposite directions. If q_3 is placed to the left of q_2 or to the right of q_1, the two forces on q_3 will be in the same direction. Therefore, the only possibility of having forces in opposite directions is to place q_3 between q_1 and q_2, as indicated in Figure 19.8.

If we let x be the coordinate of q_3, the forces $\vec{\mathbf{F}}_{13}$ and $\vec{\mathbf{F}}_{23}$ can be written as

$$\vec{\mathbf{F}}_{13} = k_e \frac{q_1 q_3}{(2.00 - x)^2}\,\hat{\mathbf{r}}_{13} = -k_e \frac{q_1 q_3}{(2.00 - x)^2}\,\hat{\mathbf{i}}$$

$$\vec{\mathbf{F}}_{23} = k_e \frac{q_2 q_3}{x^2}\,\hat{\mathbf{r}}_{23} = k_e \frac{q_2 q_3}{x^2}\,\hat{\mathbf{i}}$$

where we have recognized that $\hat{\mathbf{r}}_{13} = -\hat{\mathbf{i}}$ for the force due to q_1 because q_1 is to the right of q_3. (Remember that q_3 is negative, so $\vec{\mathbf{F}}_{13}$ will be in the positive direction as shown in Figure 19.8 and $\vec{\mathbf{F}}_{23}$ will be in the negative direction.) We now add these two forces and set the resultant equal to zero:

$$\sum \vec{\mathbf{F}} = \vec{\mathbf{F}}_{13} + \vec{\mathbf{F}}_{23} = -k_e \frac{q_1 q_3}{(2.00 - x)^2}\,\hat{\mathbf{i}} + k_e \frac{q_2 q_3}{x^2}\,\hat{\mathbf{i}} = 0$$

Because k_e and q_3 are common to both terms, they cancel, and we can solve for x:

$$\frac{q_1}{(2.00 - x)^2} = \frac{q_2}{x^2}$$

$$(4.00 - 4.00x + x^2)(6.00 \times 10^{-6}\ \text{C}) = x^2(15.0 \times 10^{-6}\ \text{C})$$

which simplifies to

$$9.00x^2 + 24.0x - 24.0 = 0$$

Solving this quadratic equation for x, we find that $x = 0.775$ m. Why is the negative root not acceptable?

Physics Now™ By logging into PhysicsNow at www.pop4e.com and going to Interactive Example 19.1, you can predict where on the x axis the electric force is zero for random values of q_1 and q_2.

EXAMPLE 19.2 The Hydrogen Atom

The electron and proton of a hydrogen atom are separated (on average) by a distance of approximately 5.3×10^{-11} m. Find the magnitudes of the electrostatic force and the gravitational force that either particle exerts on the other.

Solution From Coulomb's law, we find that the magnitude of the attractive electrostatic force is

$$F_e = k_e \frac{e^2}{r^2} = (8.99 \times 10^9\ \text{N}\cdot\text{m}^2/\text{C}^2)\frac{(1.60 \times 10^{-19}\ \text{C})^2}{(5.3 \times 10^{-11}\ \text{m})^2}$$

$$= 8.2 \times 10^{-8}\ \text{N}$$

Using Newton's law of universal gravitation (Section 5.5) and Table 19.1 for the particle masses, we find that the magnitude of the gravitational force is

$$F_g = G\frac{m_e m_p}{r^2}$$

$$= (6.67 \times 10^{-11}\ \text{N}\cdot\text{m}^2/\text{kg}^2)\frac{(9.11 \times 10^{-31}\ \text{kg})(1.67 \times 10^{-27}\ \text{kg})}{(5.3 \times 10^{-11}\ \text{m})^2}$$

$$= 3.6 \times 10^{-47}\ \text{N}$$

The ratio $F_g/F_e \approx 4 \times 10^{-40}$. Therefore, the gravitational force between charged atomic particles is negligible compared with the electric force.

19.5 ELECTRIC FIELDS

The gravitational field $\vec{\mathbf{g}}$ at a point in space was defined in Section 11.1 to be equal to the gravitational force $\vec{\mathbf{F}}_g$ acting on a test particle of mass m_0 divided by the mass of the test particle: $\vec{\mathbf{g}} = \vec{\mathbf{F}}_g/m_0$. It represents the gravitational version of the model of a particle in a field. In a similar manner, an electric field at a point in space can be defined in terms of the electric force acting on a test particle with charge q_0 placed at that point. Because charge exists in two varieties, we must choose a convention for our test particle. We choose the convention that **a test particle always carries a positive electric charge.** With this convention, we can introduce the electric version of the particle in a field model. The **electric field $\vec{\mathbf{E}}$** at a point in space is defined as **the electric force $\vec{\mathbf{F}}_e$ acting on a test particle placed at that point divided by the charge q_0 of the test particle:**

$$\vec{\mathbf{E}} \equiv \frac{\vec{\mathbf{F}}_e}{q_0} \qquad [19.3]$$

■ Definition of electric field

Therefore, **an electric field exists at a point if a charged test particle placed at rest at that point experiences an electric force.** Because force is a vector, the electric field is also a vector. Note that $\vec{\mathbf{E}}$ is the field produced by some charged particle(s) separate from the test particle; it is *not* the field produced by the test particle. We call the particle(s) creating the electric field the **source particle(s).** The electric field set up by a source charge is analogous to the gravitational field set up by some massive object such as the Earth. This gravitational field exists whether a test particle of mass m_0 is present or not. Similarly, the electric field of the source particles is present whether or not we introduce a test particle into the field. The test particle is used only to measure the force and thus detect the existence of the field and evaluate its strength.

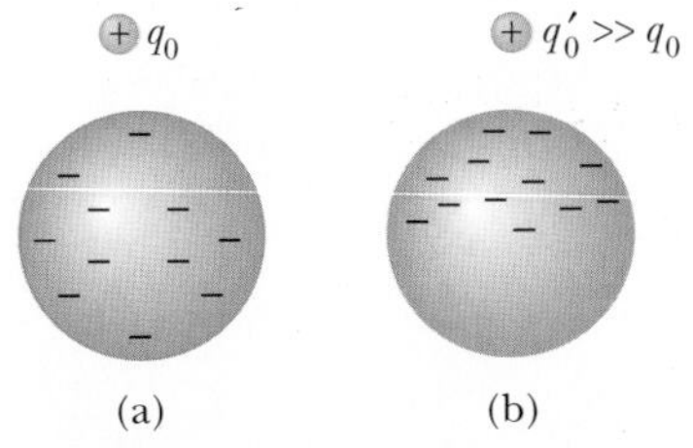

FIGURE 19.9 (a) For a small enough test charge q_0, the charge distribution on the sphere is undisturbed. (b) If the test charge q_0' were larger, the charge distribution on the sphere would be disturbed as a result of the proximity of q_0'.

When using Equation 19.3, we must assume that the test charge q_0 is small enough that it does not disturb the charge distribution responsible for the electric field. If a vanishingly small test charge q_0 is placed near a uniformly charged metallic sphere as in Figure 19.9a, the charge on the metallic sphere remains uniformly distributed. If the test charge is large enough ($q_0' >> q_0$) as in Figure 19.9b, the charge on the metallic sphere is redistributed and the ratio of the force to the test charge is different: ($F_e'/q_0' \neq F_e/q_0$). That is, because of this redistribution of charge on the metallic sphere, the electric field it sets up is different from the field it sets up in the presence of the much smaller q_0.

The vector $\vec{\mathbf{E}}$ has the SI units of newtons per coulomb (N/C), analogous to the units newtons per kilogram (N/kg) for the gravitational field. The direction of $\vec{\mathbf{E}}$ is the same as the direction of $\vec{\mathbf{F}}_e$ because we have used the convention of a positive charge on the test particle.

Once the electric field is known at some point, the force on *any* particle with charge q placed at that point can be calculated from a rearrangement of Equation 19.3:

$$\vec{\mathbf{F}}_e = q\vec{\mathbf{E}} \qquad [19.4]$$

PITFALL PREVENTION 19.1

Particles only Keep in mind that Equation 19.4 is only valid for a charged *particle*, an object of zero size. For a charged object of finite size in an electric field, the field may vary in magnitude and direction over the size of the object, so the corresponding force equation would be more complicated.

Once the electric force on a particle is evaluated, its motion can be determined from the particle under a net force model or the particle in equilibrium model (the electric force may have to be combined with other forces acting on the particle), and the techniques of earlier chapters can be used to find the motion of the particle.

Consider a point charge[3] q located a distance r from a test particle with charge q_0. According to Coulomb's law, the force exerted on the test particle by q is

[3]We have used the phrase "charged particle" so far. The phrase "point charge" is somewhat misleading because charge is a property of a particle, not a physical entity. It is similar to misleading phrasing in mechanics such as "a mass m is placed . . ." (which we have avoided) rather than "a particle with mass m is placed. . . ." This phrase is so ingrained in physics usage, however, that we will use it and hope that this footnote suffices to clarify its use.

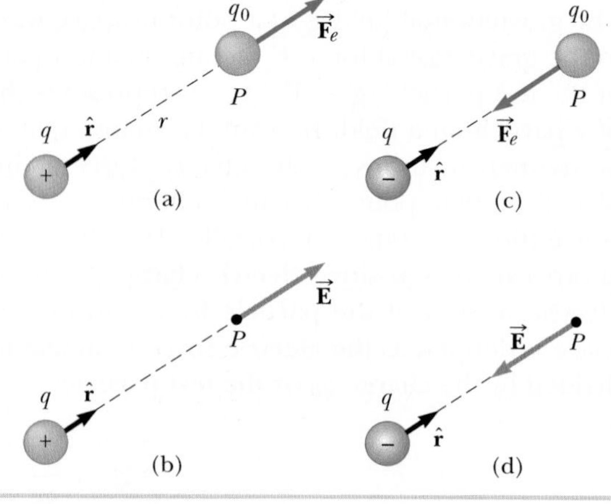

ACTIVE FIGURE 19.10

A test charge q_0 at point P is a distance r from a point charge q. (a) If q is positive, the force on the test charge is directed away from q. (b) For the positive source charge, the electric field at P points radially outward from q. (c) If q is negative, the force on the test charge is directed toward q. (d) For the negative source charge, the electric field at P points radially inward toward q.

Physics Now™ Log into PhysicsNow at **www.pop4e.com** and go to Active Figure 19.10 to move point P to any position in two-dimensional space and observe the electric field due to q.

$$\vec{\mathbf{F}}_e = k_e \frac{q q_0}{r^2} \hat{\mathbf{r}}$$

where $\hat{\mathbf{r}}$ is a unit vector directed from q toward q_0. This force in Active Figure 19.10a is directed away from the source charge q. Because the electric field at P, the position of the test charge, is defined by $\vec{\mathbf{E}} = \vec{\mathbf{F}}_e / q_0$, we find that at P, the electric field created by q is

■ Electric field due to a point charge

$$\vec{\mathbf{E}} = k_e \frac{q}{r^2} \hat{\mathbf{r}} \qquad [19.5]$$

If the source charge q is positive, Active Figure 19.10b shows the situation with the test charge removed; the source charge sets up an electric field at point P, directed away from q. If q is negative as in Active Figure 19.10c, the force on the test charge is toward the source charge, so the electric field at P is directed toward the source charge as in Active Figure 19.10d.

To calculate the electric field at a point P due to a group of point charges, we first calculate the electric field vectors at P individually using Equation 19.5 and then add them vectorially. In other words, **the total electric field at a point in space due to a group of charged particles equals the vector sum of the electric fields at that point due to all the particles.** This superposition principle applied to fields follows directly from the vector addition property of forces. Therefore, the electric field at point P of a group of source charges can be expressed as

■ Electric field due to a finite number of point charges

$$\vec{\mathbf{E}} = k_e \sum_i \frac{q_i}{r_i^2} \hat{\mathbf{r}}_i \qquad [19.6]$$

where r_i is the distance from the ith charge q_i to the point P (the location at which the field is to be evaluated) and $\hat{\mathbf{r}}_i$ is a unit vector directed from q_i toward P.

QUICK QUIZ 19.4 A test charge of $+3\ \mu$C is at a point P where an external electric field is directed to the right and has a magnitude of 4×10^6 N/C. If the test charge is replaced with another charge of $-3\ \mu$C, the external electric field at P **(a)** is unaffected, **(b)** reverses direction, or **(c)** changes in a way that cannot be determined.

EXAMPLE 19.3 Electric Field of a Dipole

An **electric dipole** consists of a point charge q and a point charge $-q$ separated by a distance of $2a$ as in Figure 19.11. As we shall see in later chapters, neutral atoms and molecules behave as dipoles when placed in an external electric field. Furthermore, many molecules, such as HCl, are permanent dipoles. (HCl can be effectively modeled as an H^+ ion combined with a Cl^- ion.) The effect of such dipoles on the behavior of materials subjected to electric fields is discussed in Chapter 20.

A Find the electric field $\vec{\mathbf{E}}$ due to the dipole along the *y* axis at the point *P*, which is a distance *y* from the origin.

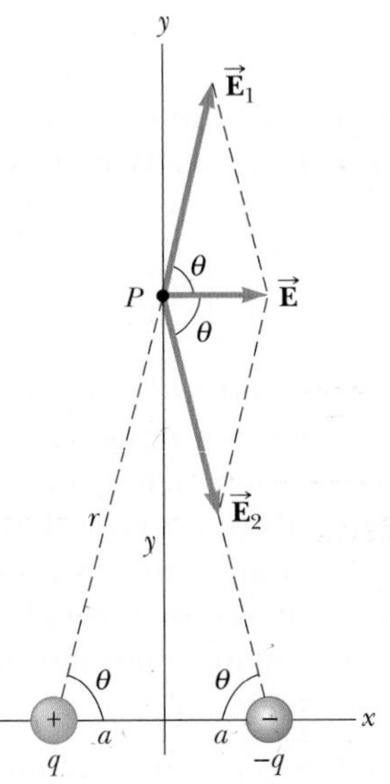

FIGURE 19.11 (Example 19.3) The total electric field $\vec{\mathbf{E}}$ at *P* due to two equal and opposite charges (an electric dipole) equals the vector sum $\vec{\mathbf{E}}_1 + \vec{\mathbf{E}}_2$. The field $\vec{\mathbf{E}}_1$ is due to the positive charge *q*, and $\vec{\mathbf{E}}_2$ is the field due to the negative charge $-q$.

Solution At *P*, the fields $\vec{\mathbf{E}}_1$ and $\vec{\mathbf{E}}_2$ due to the two particles are equal in magnitude because *P* is equidistant from the two charges. The total field at *P* is $\vec{\mathbf{E}} = \vec{\mathbf{E}}_1 + \vec{\mathbf{E}}_2$, where the magnitudes of the fields are

$$E_1 = E_2 = k_e \frac{q}{r^2} = k_e \frac{q}{y^2 + a^2}$$

The *y* components of $\vec{\mathbf{E}}_1$ and $\vec{\mathbf{E}}_2$ are equal in magnitude and opposite in sign, so they cancel. The *x* components are equal and add because they have the same sign. The total field $\vec{\mathbf{E}}$ is therefore parallel to the *x* axis and has a magnitude

$$E = 2k_e \frac{q}{y^2 + a^2} \cos\theta$$

From the geometry in Figure 19.11 we see that $\cos\theta = a/r = a/(y^2 + a^2)^{1/2}$. Therefore,

$$E = 2k_e \frac{q}{y^2 + a^2} \cos\theta = 2k_e \frac{q}{(y^2 + a^2)} \frac{a}{(y^2 + a^2)^{1/2}}$$

$$= k_e \frac{2qa}{(y^2 + a^2)^{3/2}}$$

B Find the electric field for points $y >> a$ far from the dipole.

Solution The preceding equation gives the value of the electric field on the *y* axis at all values of *y*. For points far from the dipole, for which $y >> a$, we can ignore a^2 in the denominator and write

$$E \approx k_e \frac{2qa}{y^3}$$

Therefore, we see that along the *y* axis the field of a dipole at a distant point varies as $1/r^3$, whereas the more slowly varying field of a point charge varies as $1/r^2$. (*Note:* In the geometry of this example, $r = y$.) At distant points, the fields of the two charges in the dipole almost cancel each other. The $1/r^3$ variation in *E* for the dipole is also obtained for a distant point along the *x* axis (Problem 19.16) and for a general distant point.

Electric Field Due to Continuous Charge Distributions

In most practical situations (e.g., an object charged by rubbing), the average separation between source charges is small compared with their distances from the point at which the field is to be evaluated. In such cases, the system of source charges can be modeled as *continuous.* That is, we imagine that the system of closely spaced charges is equivalent to a total charge that is continuously distributed through some volume or over some surface.

To evaluate the electric field of a continuous charge distribution, the following procedure is used. First, we divide the charge distribution into small elements, each of which contains a small amount of charge Δq as in Figure 19.12. Next, modeling the element as a point charge, we use Equation 19.5 to calculate the electric field $\Delta\vec{\mathbf{E}}$ at a point *P* due to one of these elements. Finally, we evaluate the total field at *P* due to the charge distribution by performing a vector sum of the contributions of all the charge elements (i.e., by applying the superposition principle).

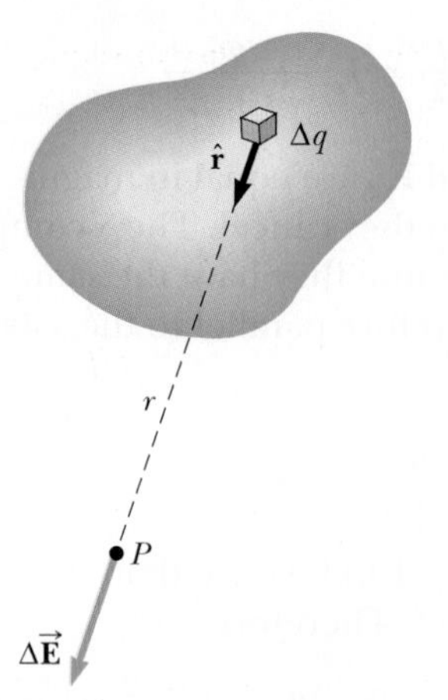

FIGURE 19.12 The electric field $\vec{\mathbf{E}}$ at P due to a continuous charge distribution is the vector sum of the fields $\Delta\vec{\mathbf{E}}$ due to all the elements Δq of the charge distribution.

The electric field at P due to one element of charge Δq_i is given by

$$\Delta\vec{\mathbf{E}}_i = k_e \frac{\Delta q_i}{r_i^{\,2}}\hat{\mathbf{r}}_i$$

where the index i refers to the ith element in the distribution, r_i is the distance from the element to point P, and $\hat{\mathbf{r}}_i$ is a unit vector directed from the element toward P. The total electric field $\vec{\mathbf{E}}$ at P due to all elements in the charge distribution is approximately

$$\vec{\mathbf{E}} \approx k_e \sum_i \frac{\Delta q_i}{r_i^{\,2}}\hat{\mathbf{r}}_i$$

Now, we apply the model in which the charge distribution is continuous, and we let the elements of charge become infinitesimally small. With this model, the total field at P in the limit $\Delta q_i \rightarrow 0$ becomes

$$\vec{\mathbf{E}} = \lim_{\Delta q_i \to 0} k_e \sum_i \frac{\Delta q_i}{r_i^{\,2}}\hat{\mathbf{r}}_i = k_e \int \frac{dq}{r^2}\hat{\mathbf{r}} \qquad [19.7]$$

where dq is an infinitesimal amount of charge and the integration is over all the charge creating the electric field. The integration is a *vector* operation and must be treated with caution. It can be evaluated in terms of individual components, or perhaps symmetry arguments can be used to reduce it to a scalar integral. We shall illustrate this type of calculation with several examples in which we assume that the charge is *uniformly* distributed on a line or a surface or throughout some volume. When performing such calculations, it is convenient to use the concept of a *charge density* along with the following notations:

- If a total charge Q is uniformly distributed throughout a volume V, the **volume charge density** ρ is defined by

■ Volume charge density

$$\rho \equiv \frac{Q}{V} \qquad [19.8]$$

where ρ has units of coulombs per cubic meter.

- If Q is uniformly distributed on a surface of area A, the **surface charge density** σ is defined by

■ Surface charge density

$$\sigma \equiv \frac{Q}{A} \qquad [19.9]$$

where σ has units of coulombs per square meter.

- If Q is uniformly distributed along a line of length ℓ, the **linear charge density** λ is defined by

■ Linear charge density

$$\lambda \equiv \frac{Q}{\ell} \qquad [19.10]$$

where λ has units of coulombs per meter.

PROBLEM-SOLVING STRATEGY Calculating the Electric Field

The following procedure is recommended for solving problems that involve the determination of an electric field due to individual charges or a charge distribution:

1. Conceptualize Think carefully about the individual charges or the charge distribution that you have in the problem. Imagine what type of electric field they would create and establish the mental representation. Appeal to any symmetry in the arrangement of charges to help you visualize the electric field.

2. Categorize Are you analyzing a group of individual charges or a continuous charge distribution? The answer to this question will tell you how to proceed in the *Analyze* step.

3. Analyze

(a) *If you are analyzing a group of individual charges,* use the superposition principle. When several point charges are present, the resultant field at a point in space is the *vector sum* of the individual fields due to the individual charges (Eq. 19.6). Example 19.3 demonstrated this procedure. Be very careful in the manipulation of vector quantities. It may be useful to review the material on vector addition in Chapter 1.

(b) *If you are analyzing a continuous charge distribution,* replace the vector sums for evaluating the total electric field from individual charges by vector integrals. The charge distribution is divided into infinitesimal pieces, and the vector sum is carried out by integrating over the entire charge distribution (Eq. 19.7). Examples 19.4 and 19.5 demonstrate such procedures.

Symmetry. Whenever dealing with either a distribution of point charges or a continuous charge distribution, take advantage of any symmetry in the system that you observed in the *Conceptualize* step to simplify your calculations. The cancellation of field components parallel to the y axis in Example 19.3 and perpendicular to the axis in Example 19.5 is an example of the application of symmetry.

4. Finalize Once you have determined your result, check to see if your field is consistent with the mental representation and that it reflects any symmetry that you noted previously. Imagine varying parameters such as the distance of the observation point from the charges or the radius of any circular or spherical objects to see if the mathematical result changes in a reasonable way.

EXAMPLE 19.4 The Electric Field Due to a Charged Rod

A rod of length ℓ has a uniform linear charge density λ and a total charge Q. Calculate the electric field at a point P along the axis of the rod, a distance a from one end (Fig. 19.13).

Solution Figure 19.13 helps us visualize the source of the electric field and conceptualize what the field might look like. We expect the field to be symmetric around the horizontal dimension of the rod and would expect the field to decrease for increasing values of a. We categorize this problem as one involving a continuous distribution of charge on the rod rather than a collection of individual charges. To analyze the problem, we choose an infinitesimal element of the charge distribution as indicated by the blue portion in Figure 19.13. Let us use dx to represent the length of one small segment of the rod and let dq be the charge on the segment. We express the charge dq of the element in terms of the other variables within the integral (in this example, there is one variable, x). The charge dq on the small segment is $dq = \lambda\, dx$.

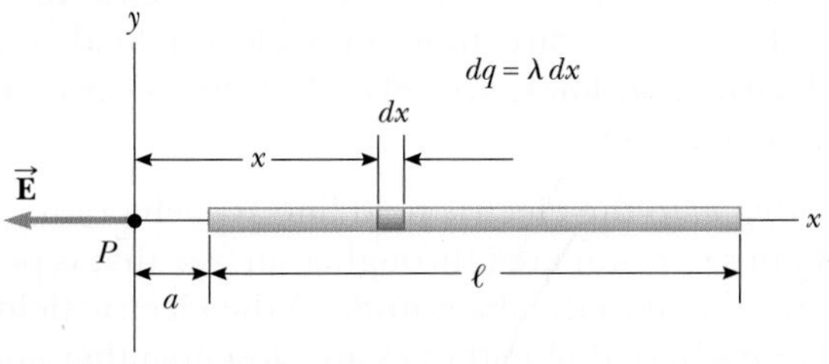

FIGURE 19.13 (Example 19.4) The electric field at P due to a uniformly charged rod lying along the x axis. The field at P due to the segment of charge dq is $k_e\, dq/x^2$. The total field at P is the vector sum over all segments of the rod.

The field $d\vec{\mathbf{E}}$ due to this segment at the point P is in the negative x direction, and its magnitude is

$$dE = k_e \frac{dq}{x^2} = k_e \frac{\lambda\, dx}{x^2}$$

Each element of the charge distribution produces a field at P in the negative x direction, so the vector sum of their contributions reduces to an algebraic sum. The total field at P due to all segments of the rod, which are at different distances from P, is given by Equation 19.7, which in this case becomes

$$E = \int_a^{\ell + a} k_e \lambda \frac{dx}{x^2}$$

where the limits on the integral extend from one end of the rod ($x = a$) to the other ($x = \ell + a$). Because k_e and λ are constants, they can be removed from the integral. Therefore, we find that

$$E = k_e \lambda \int_a^{\ell + a} \frac{dx}{x^2} = k_e \lambda \left[-\frac{1}{x} \right]_a^{\ell + a}$$

$$= k_e \lambda \left(\frac{1}{a} - \frac{1}{\ell + a} \right) = \frac{k_e Q}{a(\ell + a)}$$

where we have used that the linear charge density is $\lambda = Q/\ell$.

To finalize, note that E decreases as a increases, as we expected from our mental representation. If point P is very far from the rod ($a \gg \ell$), we can ignore the ℓ in the denominator, and $E \approx k_e Q/a^2$. This result is just the form you would expect for a point charge. Therefore, at large values of a, the charge distribution appears to be a point charge of magnitude Q as you should expect.

EXAMPLE 19.5 The Electric Field of a Uniform Ring of Charge

A ring of radius a has a uniform positive charge per unit length, with a total charge Q. Calculate the electric field at a point P on the axis of the ring at a distance x from the center of the ring (Fig. 19.14a).

Solution The magnitude of the electric field at P due to the segment of charge dq is

$$dE = k_e \frac{dq}{r^2}$$

This field has an x component $dE_x = dE \cos\theta$ along the axis of the ring and a component $dE_\perp$ perpendicular to the axis. The perpendicular component of any element is canceled by the perpendicular component of an element on the opposite side of the ring, as for the elements 1 and 2 in Figure 19.14b. Therefore, the perpendicular components of the field for the entire ring sum to zero and the resultant field at P must lie along the x axis. Because $r = (x^2 + a^2)^{1/2}$ and $\cos\theta = x/r$, we find that

$$dE_x = dE \cos\theta = \left(k_e \frac{dq}{r^2}\right)\left(\frac{x}{r}\right) = \frac{k_e x}{(x^2 + a^2)^{3/2}}\, dq$$

We integrate this expression to find the total field at P. In this case, all segments of the ring give the same contribution to the field at P because they are all equidistant from this point. Therefore,

$$E_x = \int \frac{k_e x}{(x^2 + a^2)^{3/2}}\, dq = \frac{k_e x}{(x^2 + a^2)^{3/2}} \int dq$$

$$= \frac{k_e x}{(x^2 + a^2)^{3/2}}\, Q$$

This result shows that the field is zero at the center point of the ring, $x = 0$. Does that surprise you?

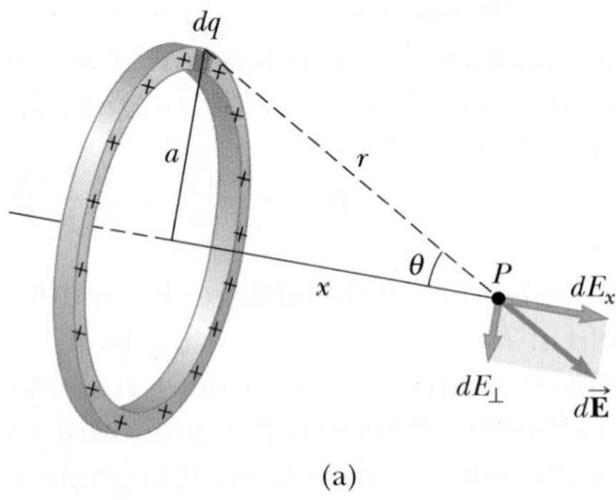

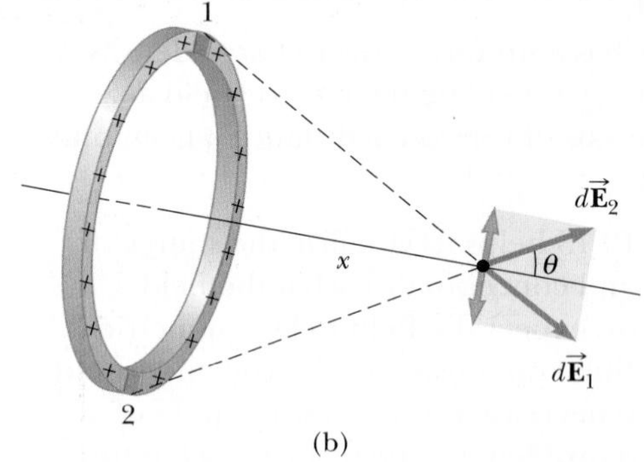

FIGURE 19.14 (Example 19.5) A uniformly charged ring of radius a. (a) The field at P on the x axis due to an element of charge dq. (b) The total electric field at P is along the x axis. The perpendicular component of the electric field at P due to segment 1 is canceled by the perpendicular component due to segment 2.

19.6 ELECTRIC FIELD LINES

A convenient specialized pictorial representation for visualizing electric field patterns is created by drawing lines showing the direction of the electric field vector at any point. These lines, called **electric field lines,** are related to the electric field in any region of space in the following manner:

- The electric field vector $\vec{\mathbf{E}}$ is *tangent* to the electric field line at each point.
- The number of electric field lines per unit area through a surface that is perpendicular to the lines is proportional to the magnitude of the electric field in that region. Therefore, E is large where the field lines are close together and small where they are far apart.

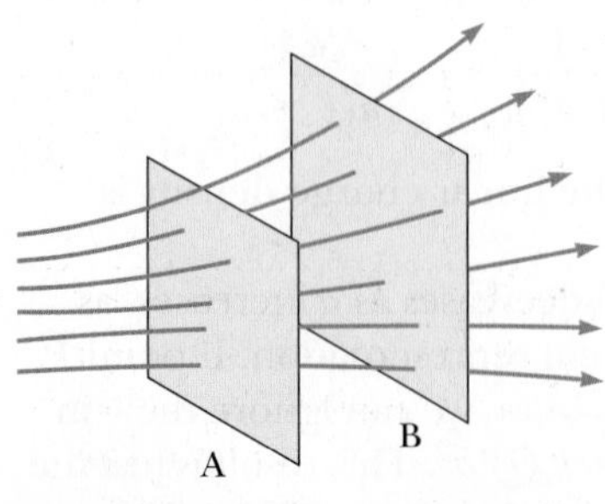

FIGURE 19.15 Electric field lines penetrating two surfaces. The magnitude of the field is greater on surface A than on surface B.

These properties are illustrated in Figure 19.15. The density of lines through surface A is greater than the density of lines through surface B. Therefore, the magnitude of the electric field on surface A is larger than on surface B. Furthermore, the field drawn in Figure 19.15 is nonuniform because the lines at different locations point in different directions.

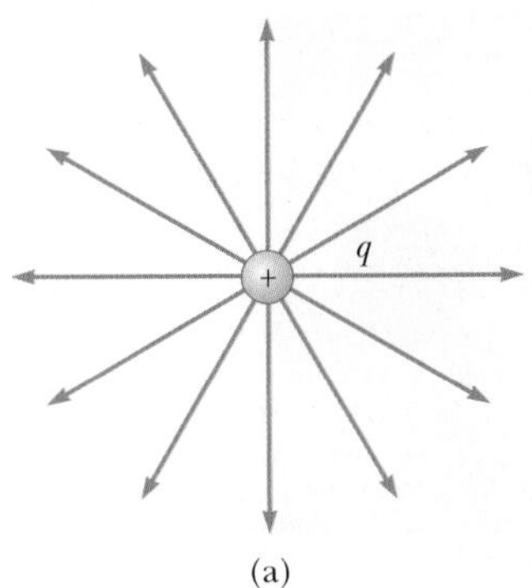

(a)

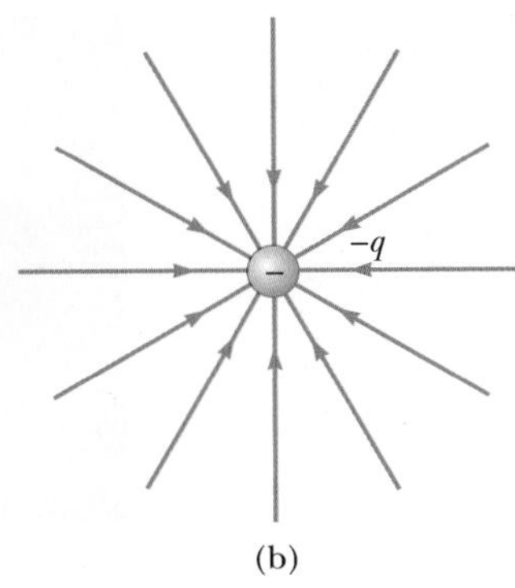

(b)

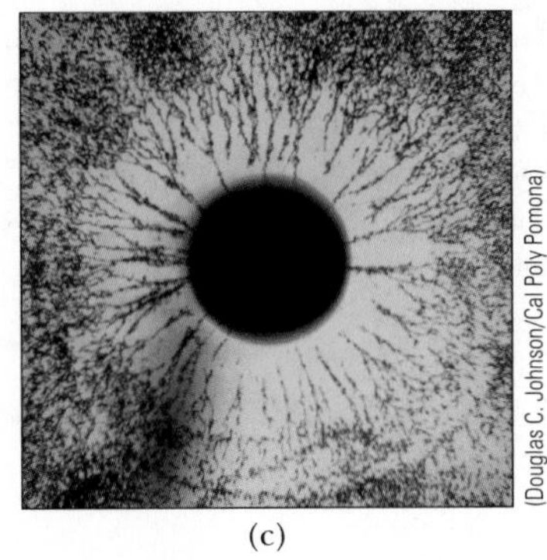

(c)

FIGURE 19.16 The electric field lines for a point charge. (a) For a positive point charge, the lines are directed radially outward. (b) For a negative point charge, the lines are directed radially inward. Note that the figures show only those field lines that lie in the plane containing the charge. (c) The dark areas are small particles suspended in oil, which align with the electric field produced by a small charged conductor at the center.

Some representative electric fiel[illegible] shown in Figure 19.16a. Note that i[illegible] the field lines that lie in the plane [illegible] radially outward in *all* directions from the charge, somewhat like the needles of a porcupine. Because a positively charged test particle placed in this field would be repelled by the charge q, the lines are directed radially away from q. Similarly, the electric field lines for a single negative point charge are directed toward the charge (Fig. 19.16b). In either case, the lines are radial and extend to infinity. Note that the lines are closer together as they come nearer to the charge, indicating that the magnitude of the field is increasing.

Is this visualization of the electric field in terms of field lines consistent with Equation 19.5? To answer this question, consider an imaginary spherical surface of radius r, concentric with the charge. From symmetry, we see that the magnitude of the electric field is the same everywhere on the surface of the sphere. The number of lines N emerging from the charge is equal to the number penetrating the spherical surface. Hence, the number of lines per unit area on the sphere is $N/4\pi r^2$ (where the surface area of the sphere is $4\pi r^2$). Because E is proportional to the number of lines per unit area, we see that E varies as $1/r^2$. This result is consistent with that obtained from Equation 19.5; that is, $E = k_e q/r^2$.

The rules for drawing electric field lines for any charge distribution are as follows:

- The lines for a group of point charges must begin on positive charges and end on negative ones. In the case of an excess of one type of charge, some lines will begin or end infinitely far away.
- The number of lines drawn beginning on a positive charge or ending on a negative one is proportional to the magnitude of the charge.
- Field lines cannot intersect.

Because charge is quantized, the number of lines leaving any positively charged object must be 0, ae, $2ae$, . . . , where a is an arbitrary (but fixed) proportionality constant chosen by the person drawing the lines. Once a is chosen, the number of lines is no longer arbitrary. For example, if object 1 has charge Q_1 and object 2 has charge Q_2, the ratio of the number of lines connected to object 2 to those connected to object 1 is $N_2/N_1 = Q_2/Q_1$.

The electric field lines for two point charges of equal magnitude but opposite signs (the electric dipole) are shown in Figure 19.17. In this case, the number of lines that begin at the positive charge must equal the number that terminate at the negative charge. At points very near the charges, the lines are nearly radial. The

PITFALL PREVENTION 19.2

Electric Field Lines Are Not Paths of Particles Electric field lines represent the field at various locations. Except in very special cases, they *do not* represent the path of a charged particle released in an electric field.

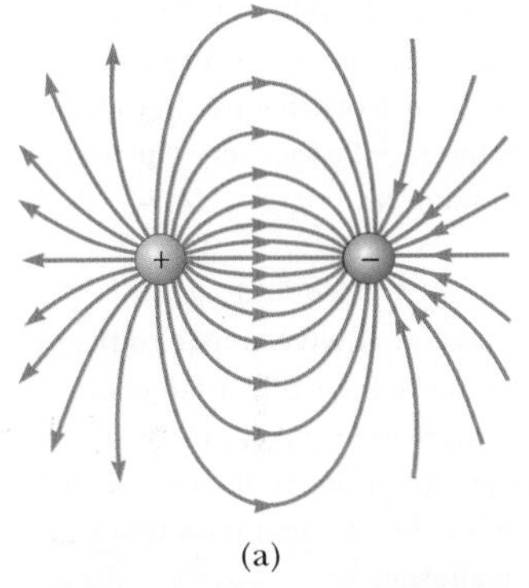

(a)

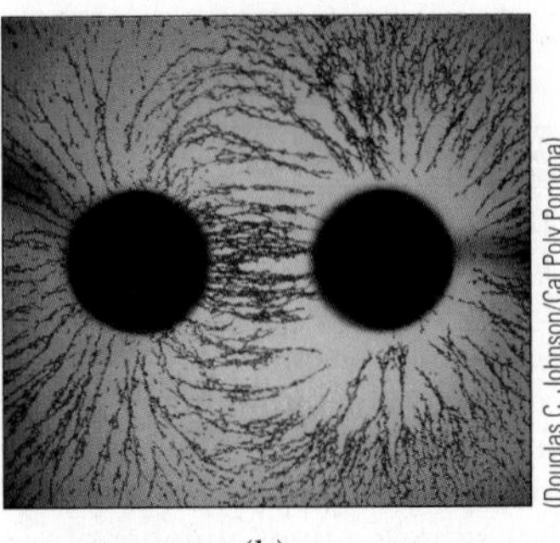

(b)

FIGURE 19.17 (a) The electric field lines for two charges of equal magnitude and opposite sign (an electric dipole). Note that the number of lines leaving the positive charge equals the number terminating at the negative charge. (b) Small particles suspended in oil align with the electric field.

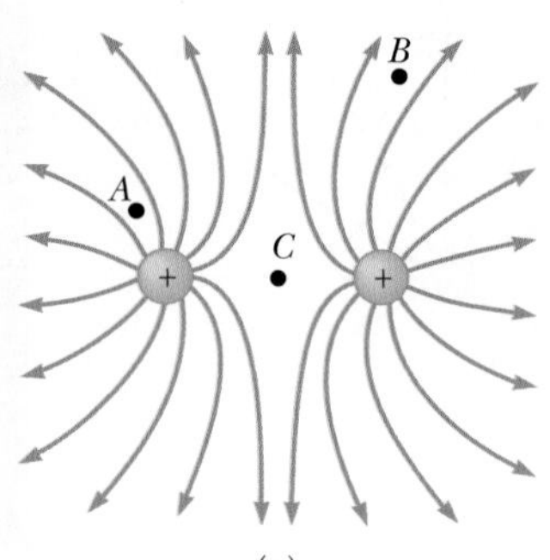

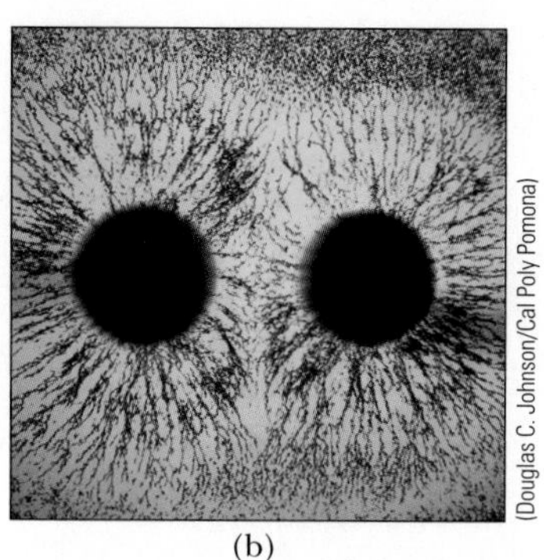

FIGURE 19.18 (a) The electric field lines for two positive point charges. (The locations A, B, and C are discussed in Quick Quiz 19.5.) (b) Small particles suspended in oil align with the electric field.

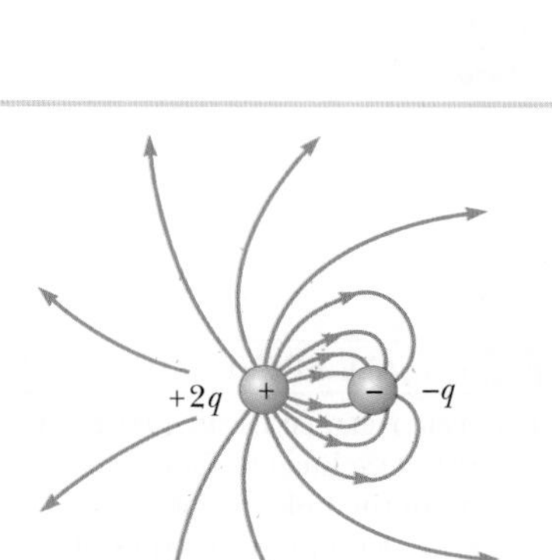

ACTIVE FIGURE 19.19 The electric field lines for a point charge $+2q$ and a second point charge $-q$. Note that two lines leave the charge $+2q$ for every one that terminates on $-q$.

Physics Now™ Log into PhysicsNow at www.pop4e.com and go to Active Figure 19.19 to choose the values and signs for the two charges and observe the electric field lines for the configuration that you have chosen.

high density of lines between the charges indicates a region of strong electric field. The attractive nature of the force between the particles is also suggested by Figure 19.17, with the lines from one particle ending on the other particle.

Figure 19.18 shows the electric field lines in the vicinity of two equal positive point charges. Again, close to either charge the lines are nearly radial. The same number of lines emerges from each particle because the charges are equal in magnitude. At great distances from the particles, the field is approximately equal to that of a single point charge of magnitude $2q$. The repulsive nature of the electric force between particles of like charge is suggested in the figure in that no lines connect the particles and that the lines bend away from the region between the charges.

Finally, we sketch the electric field lines associated with a positive point charge $+2q$ and a negative point charge $-q$ in Active Figure 19.19. In this case, the number of lines leaving $+2q$ is twice the number terminating on $-q$. Hence, only half the lines that leave the positive charge end at the negative charge. The remaining half terminate on hypothetical negative charges we assume to be located infinitely far away. At large distances from the particles (large compared with the particle separation), the electric field lines are equivalent to those of a single point charge $+q$.

QUICK QUIZ 19.5 Rank the magnitudes of the electric field at points A, B, and C in Figure 19.18a, largest magnitude first.

19.7 MOTION OF CHARGED PARTICLES IN A UNIFORM ELECTRIC FIELD

When a particle of charge q and mass m is placed in an electric field $\vec{\mathbf{E}}$, the electric force exerted on the charge is given by Equation 19.4, $\vec{\mathbf{F}}_e = q\vec{\mathbf{E}}$. If this force is the only force exerted on the particle, it is the net force. According to the particle under a net force model from Chapter 4, the net force causes the particle to accelerate. In this case, Newton's second law applied to the particle gives

$$\vec{\mathbf{F}}_e = q\vec{\mathbf{E}} = m\vec{\mathbf{a}}$$

The acceleration of the particle is therefore

$$\vec{\mathbf{a}} = \frac{q\vec{\mathbf{E}}}{m} \qquad [19.11]$$

If $\vec{\mathbf{E}}$ is uniform (i.e., constant in magnitude and direction), the acceleration is constant. If the particle has a positive charge, its acceleration is in the direction of the electric field. If the particle has a negative charge, its acceleration is in the direction opposite the electric field.

PITFALL PREVENTION 19.3

ELECTRIC FIELD LINES ARE NOT REAL Electric field lines are not material objects. They are used only as a pictorial representation to provide a qualitative description of the electric field. One problem with this representation is that one always draws a finite number of lines from each charge, which makes it appear as if the field were quantized and exists only in certain parts of space. The field, in fact, is continuous, existing at every point. Another problem with this representation is the danger of obtaining the wrong impression from a two-dimensional drawing of field lines used to describe a three-dimensional situation.

EXAMPLE 19.6 An Accelerating Positive Charge

A particle with positive charge q and mass m is released from rest in a uniform electric field $\vec{\mathbf{E}}$ directed along the x axis as in Figure 19.20. Describe its motion.

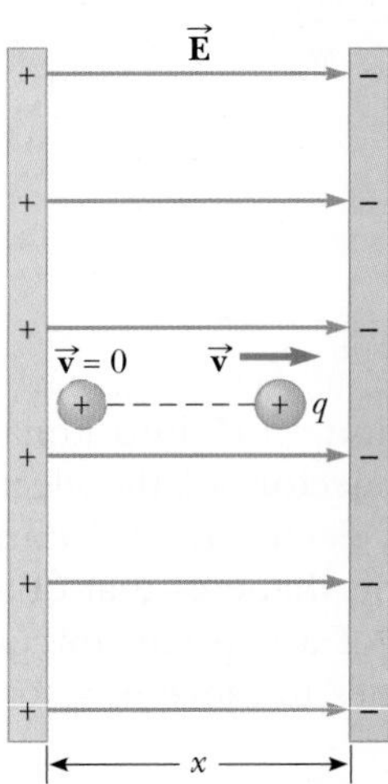

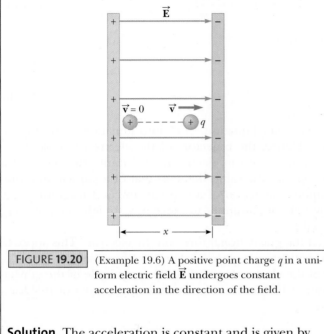

FIGURE 19.20 (Example 19.6) A positive point charge q in a uniform electric field $\vec{\mathbf{E}}$ undergoes constant acceleration in the direction of the field.

Solution The acceleration is constant and is given by $q\vec{\mathbf{E}}/m$ (Eq. 19.11). The motion is simple linear motion along the x axis. We can therefore apply the model of a particle under constant acceleration and use the equations of kinematics in one dimension (from Chapter 2):

$$x_f = x_i + v_i t + \tfrac{1}{2}at^2$$

$$v_f = v_i + at$$

$$v_f^2 = v_i^2 + 2a(x_f - x_i)$$

Choosing $x_i = 0$ and $v_i = 0$ gives

$$x_f = \tfrac{1}{2}at^2 = \frac{qE}{2m}t^2$$

$$v_f = at = \frac{qE}{m}t$$

$$v_f^2 = 2ax_f = \left(\frac{2qE}{m}\right)x_f$$

The kinetic energy of the particle after it has moved a distance $x = x_f - x_i$ is

$$K = \tfrac{1}{2}mv^2 = \tfrac{1}{2}m\left(\frac{2qE}{m}\right)x = qEx$$

This result can also be obtained by identifying the particle as a nonisolated system and applying the nonisolated system model. Energy is transferred from the environment (the electric field) by work, so the work–kinetic energy theorem gives the same result as the calculation above. Try it!

The electric field in the region between two oppositely charged flat metal plates is approximately uniform (Active Fig. 19.21). Suppose an electron of charge $-e$ is projected horizontally into this field with an initial velocity $v_i\hat{\mathbf{i}}$. Because the electric field $\vec{\mathbf{E}}$ in Active Figure 19.21 is in the positive y direction, the acceleration of the electron is in the negative y direction. That is,

$$\vec{\mathbf{a}} = -\frac{eE}{m_e}\hat{\mathbf{j}} \qquad [19.12]$$

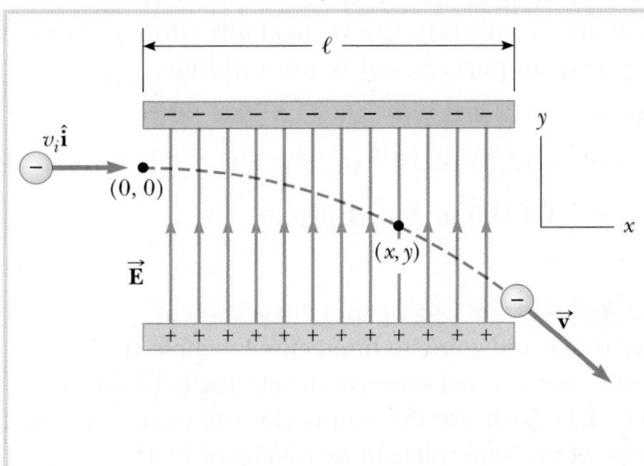

ACTIVE FIGURE 19.21
An electron is projected horizontally into a uniform electric field produced by two charged plates. The electron undergoes a downward acceleration (opposite $\vec{\mathbf{E}}$), and its motion is parabolic while it is between the plates.

PhysicsNow™ Log into PhysicsNow at **www.pop4e.com** and go to Active Figure 19.21 to choose the magnitude of the electric field and the mass and charge of the projected particle.

Because the acceleration is constant, we can apply the kinematic equations from Chapter 3 with $v_{xi} = v_i$ and $v_{yi} = 0$. At time t, the components of the velocity of the electron are

$$v_x = v_i = \text{constant} \quad [19.13]$$

$$v_y = a_y t = -\frac{eE}{m_e} t \quad [19.14]$$

Its position coordinates at time t are

$$x_f = v_i t \quad [19.15]$$

$$y_f = \tfrac{1}{2} a_y t^2 = -\tfrac{1}{2} \frac{eE}{m_e} t^2 \quad [19.16]$$

Substituting the value $t = x_f/v_i$ from Equation 19.15 into Equation 19.16, we see that y_f is proportional to x_f^2. Hence, the trajectory of the electron is a parabola. The trajectory of the electron in a uniform electric field $\vec{\mathbf{E}}$ under the action of a constant force of magnitude qE has the same shape as that of a particle in a uniform gravitational field $\vec{\mathbf{g}}$ under the action of a constant force of magnitude mg. After the electron leaves the field, it continues to move in a straight line, obeying Newton's first law, with a speed $v > v_i$.

Note that we have ignored the gravitational force on the electron. This approximation is valid when dealing with atomic particles. For an electric field of 10^4 N/C, the ratio of the magnitude of the electric force eE to the magnitude of the gravitational force mg is on the order of 10^{14} for an electron and on the order of 10^{11} for a proton.

INTERACTIVE EXAMPLE 19.7 An Accelerated Electron

An electron enters the region of a uniform electric field as in Active Figure 19.21, with $v_i = 3.00 \times 10^6$ m/s and $E = 200$ N/C. The horizontal length of the plates is $\ell = 0.100$ m.

A Find the acceleration of the electron while it is in the electric field.

Solution The charge on the electron is $-e$ and its mass is $m_e = 9.11 \times 10^{-31}$ kg. Therefore, Equation 19.12 gives

$$\vec{\mathbf{a}} = -\frac{eE}{m_e}\hat{\mathbf{j}} = -\frac{(1.60 \times 10^{-19}\ \text{C})(200\ \text{N/C})}{9.11 \times 10^{-31}\ \text{kg}}\hat{\mathbf{j}}$$

$$= -3.51 \times 10^{13}\hat{\mathbf{j}}\ \text{m/s}^2$$

B Find the time interval required for the electron to travel through the field.

Solution The horizontal distance through the field is $\ell = 0.100$ m. Modeling the electron as a particle under constant velocity in the horizontal direction, we find that the time interval spent in the electric field is

$$\Delta t = \frac{\ell}{v_i} = \frac{0.100\ \text{m}}{3.00 \times 10^6\ \text{m/s}} = 3.33 \times 10^{-8}\ \text{s}$$

C What is the vertical displacement Δy of the electron while it is in the field?

Solution Modeling the electron as a particle under constant acceleration in the vertical direction and using the results from parts A and B, we find that

$$\Delta y = y_f - y_i = \tfrac{1}{2} a_y t^2$$

$$= -\tfrac{1}{2}(3.51 \times 10^{13}\ \text{m/s}^2)(3.33 \times 10^{-8}\ \text{s})^2$$

$$= -0.0195\ \text{m} = -1.95\ \text{cm}$$

PhysicsNow™ By logging into PhysicsNow at www.pop4e.com and going to Interactive Example 19.7, you can predict, for random values of the electric field, the required initial velocity for the exiting electron to just miss the right edge of the lower plate in Active Figure 19.21.

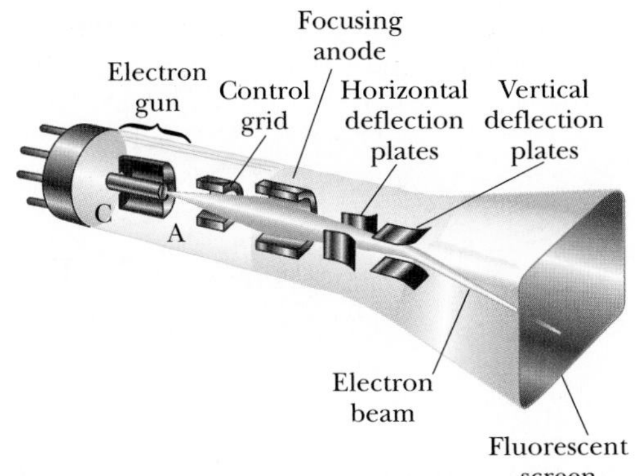

FIGURE 19.22 Schematic diagram of a cathode-ray tube. Electrons leaving the hot cathode C are accelerated to the anode A. In addition to accelerating electrons, the electron gun is also used to focus the beam of electrons, and the plates deflect the beam.

The Cathode-Ray Tube

The previous example describes a portion of a cathode-ray tube (CRT). This tube, illustrated in Figure 19.22, is commonly used to obtain a visual display of electronic information in oscilloscopes, radar systems, television receivers, and computer monitors. The CRT is a vacuum tube in which a beam of electrons is accelerated and deflected under the influence of electric or magnetic fields. The electron beam is produced by an *electron gun* located in the neck of the tube. The electrons travel through the *control grid,* which determines the number of electrons passing through (and therefore the brightness of the display). The focusing anode focuses the beam of electrons to a small spot on the display screen. The fluorescent screen is coated with a material that emits visible light when bombarded with electrons.

In an oscilloscope, the electrons are deflected in various directions by two sets of plates placed perpendicularly to each other in the neck of the tube. (A television CRT steers the beam with a magnetic field, which we will discuss in Chapter 22.) An external electric circuit is used to control the amount of charge present on the plates. Placing positive charge on one horizontal deflection plate and negative charge on the other creates an electric field between the plates and allows the beam to be steered from side to side. The vertical deflection plates act in the same way, except that changing the charge on them deflects the beam vertically.

19.8 ELECTRIC FLUX

Now that we have described the concept of electric field lines qualitatively, let us use a new concept, *electric flux,* to approach electric field lines on a quantitative basis. Electric flux is a quantity proportional to the number of electric field lines penetrating some surface. (We can define only a proportionality because the number of lines we choose to draw is arbitrary.)

First consider an electric field that is uniform in both magnitude and direction as in Figure 19.23. The field lines penetrate a plane rectangular surface of area A, which is perpendicular to the field. Recall that the number of lines per unit area is proportional to the magnitude of the electric field. The number of lines penetrating the surface of area A is therefore proportional to the product EA. The product of the electric field magnitude E and a surface area A perpendicular to the field is called the **electric flux** Φ_E:

$$\Phi_E = EA \qquad [19.17]$$

From the SI units of E and A, we see that electric flux has the units $\text{N} \cdot \text{m}^2/\text{C}$.

If the surface under consideration is not perpendicular to the field, the number of lines through it must be less than that given by Equation 19.17. This concept can

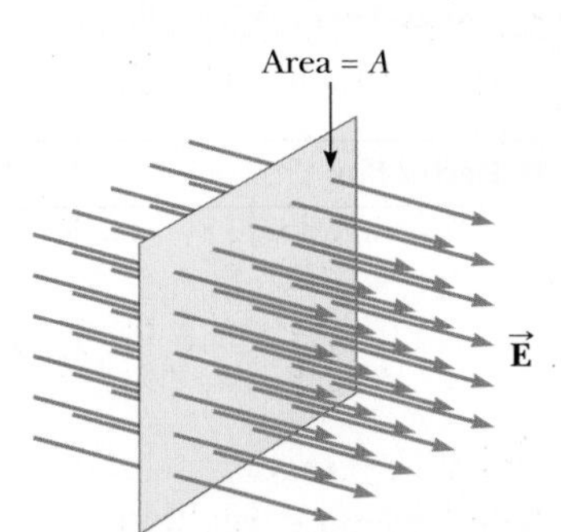

FIGURE 19.23 Field lines of a uniform electric field penetrating a plane of area A perpendicular to the field. The electric flux Φ_E through this area is equal to EA.

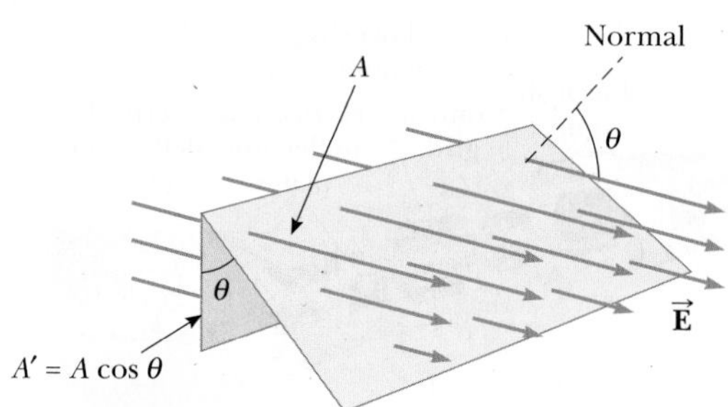

FIGURE 19.24 Field lines for a uniform electric field through an area A whose normal is at an angle θ to the field. Because the number of lines that go through the shaded area A' is the same as the number that go through A, we conclude that the total flux through A' is equal to the flux through A and is given by $\Phi_E = EA\cos\theta$.

be understood by considering Figure 19.24, where the normal to the surface of area A is at an angle of θ to the uniform electric field. Note that the number of lines that cross this area is equal to the number that cross the projected area A', which is perpendicular to the field. From Figure 19.24, we see that the two areas are related by $A' = A\cos\theta$. Because the flux through area A equals the flux through A', we conclude that the desired flux is

$$\Phi_E = EA\cos\theta \qquad [19.18]$$

...at the flux through a surface of fixed area has the ...he angle θ between the normal to the surface and the electric ... is zero. This situation occurs when the normal is parallel to the field and the surface is perpendicular to the field. The flux is zero when the surface is parallel to the field because the angle θ in Equation 19.18 is then 90°.

In more general situations, the electric field may vary in both magnitude and direction over the surface in question. Unless the field is uniform, our definition of flux given by Equation 19.18 therefore has meaning only over a small element of area. Consider a general surface divided up into a large number of small elements, each of area ΔA. The variation in the electric field over the element can be ignored if the element is small enough. It is convenient to define a vector $\Delta\vec{\mathbf{A}}_i$ whose magnitude represents the area of the ith element and whose direction is defined to be perpendicular to the surface as in Figure 19.25. The electric flux $\Delta\Phi_E$ through this small element is

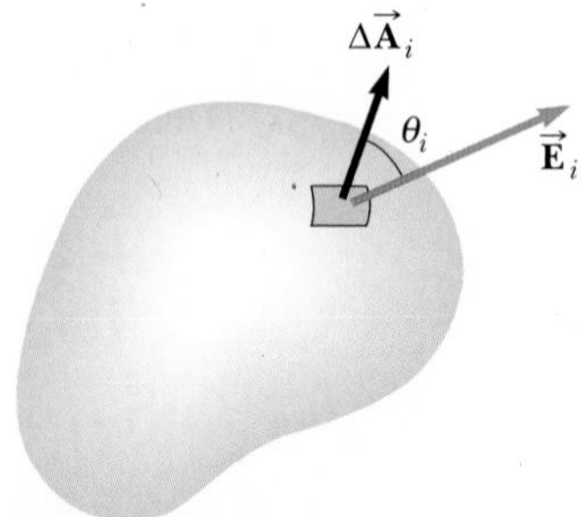

FIGURE 19.25 A small element of a surface of area ΔA_i. The electric field makes an angle θ_i with the normal to the surface (the direction of $\Delta\vec{\mathbf{A}}_i$), and the flux through the element is equal to $E_i\,\Delta A_i\cos\theta_i$.

$$\Delta\Phi_E = E_i\,\Delta A_i\cos\theta_i = \vec{\mathbf{E}}_i\cdot\Delta\vec{\mathbf{A}}_i$$

where we have used the definition of the scalar product of two vectors ($\vec{\mathbf{A}}\cdot\vec{\mathbf{B}} = AB\cos\theta$). By summing the contributions of all elements, we obtain the total flux through the surface. If we let the area of each element approach zero, the number of elements approaches infinity and the sum is replaced by an integral. The general definition of electric flux is therefore

■ Electric flux

$$\Phi_E \equiv \lim_{\Delta A_i\to 0}\sum \vec{\mathbf{E}}_i\cdot\Delta\vec{\mathbf{A}}_i = \int_{\text{surface}} \vec{\mathbf{E}}\cdot d\vec{\mathbf{A}} \qquad [19.19]$$

Equation 19.19 is a surface integral, which must be evaluated over the surface in question. In general, the value of Φ_E depends both on the field pattern and on the specified surface.

We shall often be interested in evaluating electric flux through a *closed surface*. A closed surface is defined as one that completely divides space into an inside region and an outside region so that movement cannot take place from one region to the other without penetrating the surface. This definition is similar to that of the system boundary in system models, in which the boundary divides space into a region inside the system and the outer region, the environment. The surface of a sphere is an example of a closed surface, whereas a drinking glass is an open surface.

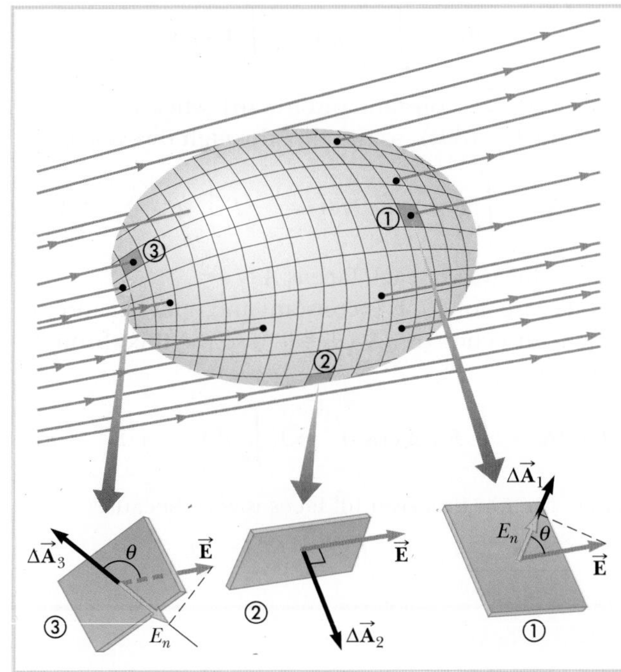

ACTIVE FIGURE 19.26

A closed surface in an electric field. The area vectors $\Delta\vec{\mathbf{A}}_i$ are, by convention, normal to the surface and point outward. The flux through an area element can be positive (element ①), zero (element ②), or negative (element ③).

PhysicsNow™ Log into PhysicsNow at **www.pop4e.com** and go to Active Figure 19.26 to select a segment on the surface and see the relationship between the electric field vector $\vec{\mathbf{E}}$ and the area vector $\Delta\vec{\mathbf{A}}_i$.

Consider the closed surface in Active Figure 19.26. Note that the vectors $\Delta\vec{\mathbf{A}}_i$ point in different directions for the various surface elements. At each point, these vectors are *perpendicular* to the surface and, by convention, always point *outward* from the inside region. At the element labeled ①, $\vec{\mathbf{E}}$ is outward and $\theta_i < 90°$; hence, the flux $\Delta\Phi_E = \vec{\mathbf{E}} \cdot \Delta\vec{\mathbf{A}}_i$ through this element is positive. For element ②, the field lines graze the surface (perpendicular to the vector $\Delta\vec{\mathbf{A}}_i$); therefore, $\theta_i = 90°$ and the flux is zero. For elements such as ③, where the field lines are crossing the surface from the outside to the inside, $180° > \theta_i > 90°$ and the flux is negative because $\cos\theta_i$ is negative. The net flux through the surface is proportional to the net number of lines penetrating the surface, where the net number means **the number leaving the volume surrounded by the surface minus the number entering the volume.** If more lines are leaving the surface than entering, the net flux is positive. If more lines enter than leave the surface, the net flux is negative. Using the symbol $\oint$ to represent an integral over a closed surface, we can write the net flux Φ_E through a closed surface as

$$\Phi_E = \oint \vec{\mathbf{E}} \cdot d\vec{\mathbf{A}} = \oint E_n \, dA \qquad [19.20]$$

where E_n represents the component of the electric field normal to the surface.

Evaluating the net flux through a closed surface can be very cumbersome. If the field is perpendicular or parallel to the surface at each point and constant in magnitude, however, the calculation is straightforward. The following example illustrates this point.

EXAMPLE 19.8 Flux Through a Cube

Consider a uniform electric field $\vec{\mathbf{E}}$ directed along the $+x$ axis. Find the net electric flux through the surface of a cube of edges ℓ oriented as shown in Figure 19.27.

Solution The net flux can be evaluated by summing up the fluxes through each face of the cube. First, note that the flux through four of the faces is zero because $\vec{\mathbf{E}}$ is perpendicular to $d\vec{\mathbf{A}}$ on these faces. In particular, the orientation of $d\vec{\mathbf{A}}$ is perpendicular to $\vec{\mathbf{E}}$ for the faces labeled ③ and ④ in Figure 19.27. Therefore, $\theta = 90°$, so $\vec{\mathbf{E}} \cdot d\vec{\mathbf{A}} = E\,dA\cos 90° = 0$. The flux through each face parallel to the xy plane is also zero for the same reason.

Now consider the faces labeled ① and ②. The net flux through these faces is

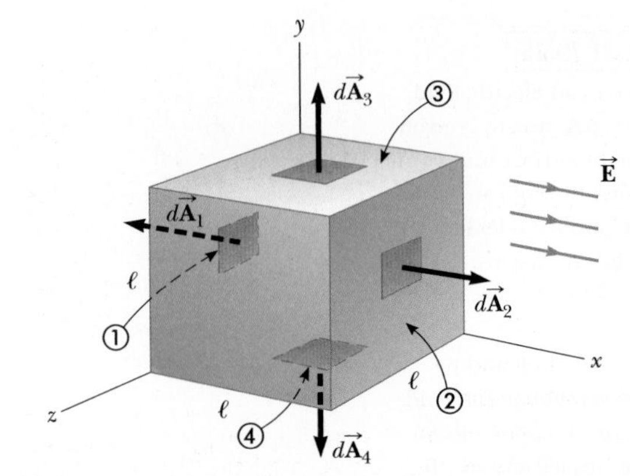

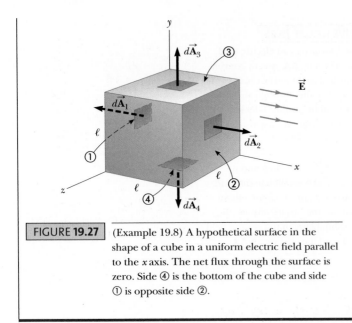

FIGURE 19.27 (Example 19.8) A hypothetical surface in the shape of a cube in a uniform electric field parallel to the *x* axis. The net flux through the surface is zero. Side ④ is the bottom of the cube and side ① is opposite side ②.

$$\Phi_E = \int_1 \vec{\mathbf{E}} \cdot d\vec{\mathbf{A}} + \int_2 \vec{\mathbf{E}} \cdot d\vec{\mathbf{A}}$$

For face ①, $\vec{\mathbf{E}}$ is constant and inward, whereas $d\vec{\mathbf{A}}$ is outward ($\theta = 180°$), so the flux through this face is

$$\int_1 \vec{\mathbf{E}} \cdot d\vec{\mathbf{A}} = \int_1 E\, dA \cos 180° = -E \int_1 dA = -EA = -E\ell^2$$

because the area of each face is $A = \ell^2$.

Likewise, for ②, $\vec{\mathbf{E}}$ is constant and outward and in the same direction as $d\vec{\mathbf{A}}$ ($\theta = 0°$), so the flux through this face is

$$\int_2 \vec{\mathbf{E}} \cdot d\vec{\mathbf{A}} = \int_2 E\, dA \cos 0° = E \int_2 dA = +EA = E\ell^2$$

Hence, the net flux over all faces is zero because

$$\Phi_E = -E\ell^2 + E\ell^2 = 0$$

19.9 GAUSS'S LAW

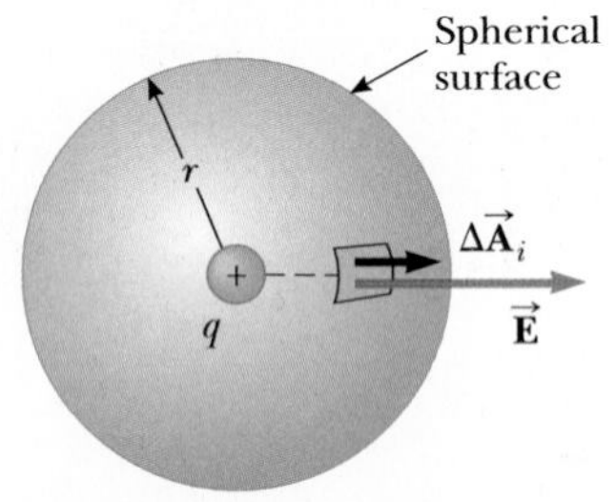

FIGURE 19.28 A spherical surface of radius *r* surrounding a point charge *q*. When the charge is at the center of the sphere, the electric field is normal to the surface and constant in magnitude everywhere on the surface.

In this section, we describe a general relation between the net electric flux through a closed surface and the charge *enclosed* by the surface. This relation, known as **Gauss's law,** is of fundamental importance in the study of electrostatic fields.

First, let us consider a positive point charge q located at the center of a spherical surface of radius r as in Figure 19.28. The field lines radiate outward and hence are perpendicular (or normal) to the surface at each point. That is, at each point on the surface, $\vec{\mathbf{E}}$ is parallel to the vector $\Delta\vec{\mathbf{A}}_i$ representing the local element of area ΔA_i. Therefore, at all points on the surface,

$$\vec{\mathbf{E}} \cdot \Delta\vec{\mathbf{A}}_i = E_n\, \Delta A_i = E\, \Delta A_i$$

and, from Equation 19.20, we find that the net flux through the surface is

$$\Phi_E = \oint E_n\, dA = \oint E\, dA = E \oint dA = EA$$

because E is constant over the surface. From Equation 19.5, we know that the magnitude of the electric field everywhere on the surface of the sphere is $E = k_e q/r^2$. Furthermore, for a spherical surface, $A = 4\pi r^2$ (the surface area of a sphere). Hence, the net flux through the surface is

$$\Phi_E = EA = \left(\frac{k_e q}{r^2}\right)(4\pi r^2) = 4\pi k_e q$$

Recalling that $k_e = 1/4\pi\epsilon_0$, we can write this expression in the form

$$\Phi_E = \frac{q}{\epsilon_0} \qquad [19.21]$$

This result, which is independent of *r*, says that the net flux through a spherical surface is proportional to the charge q at the center *inside* the surface. This result mathematically represents that (1) the net flux is proportional to the number of field lines, (2) the number of field lines is proportional to the charge inside the surface, and (3) every field line from the charge must pass through the surface. That the net flux is independent of the radius is a consequence of the inverse-square dependence of the electric field given by Equation 19.5. That is, E varies as $1/r^2$, but the area of the sphere varies as r^2. Their combined effect produces a flux that is independent of r.

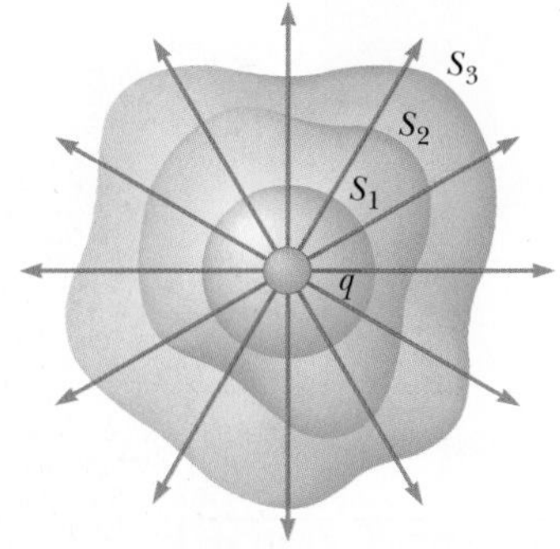

FIGURE 19.29 Closed surfaces of various shapes surrounding a charge q. The net electric flux through each surface is the same.

Now consider several closed surfaces surrounding a charge q as in Figure 19.29. Surface S_1 is spherical, whereas surfaces S_2 and S_3 are nonspherical. The flux that passes through surface S_1 has the value q/ϵ_0. As we discussed in Section 19.8, the flux is proportional to the number of electric field lines passing through that surface. The construction in Figure 19.29 shows that the number of electric field lines through the spherical surface S_1 is equal to the number of electric field lines through the nonspherical surfaces S_2 and S_3. It is therefore reasonable to conclude that the net flux through any closed surface is independent of the shape of that surface. (One can prove that conclusion using $E \propto 1/r^2$.) In fact, **the net flux through any closed surface surrounding the point charge q is given by q/ϵ_0.** Because we could choose a spherical surface surrounding the charge such that the charge is *not* at the center of the surface, **the flux through the surface is independent of the position of the charge within the surface.**

Now consider a point charge located *outside* a closed surface of arbitrary shape as in Figure 19.30. As you can see from this construction, electric field lines enter the surface and then leave it. Therefore, the number of electric field lines entering the surface equals the number leaving the surface. Consequently, we conclude that **the net electric flux through a closed surface that surrounds no net charge is zero.** If we apply this result to Example 19.8, we see that the net flux through the cube is zero because there was no charge inside the cube. If there were charge in the cube, the electric field could not be uniform throughout the cube as specified in the example.

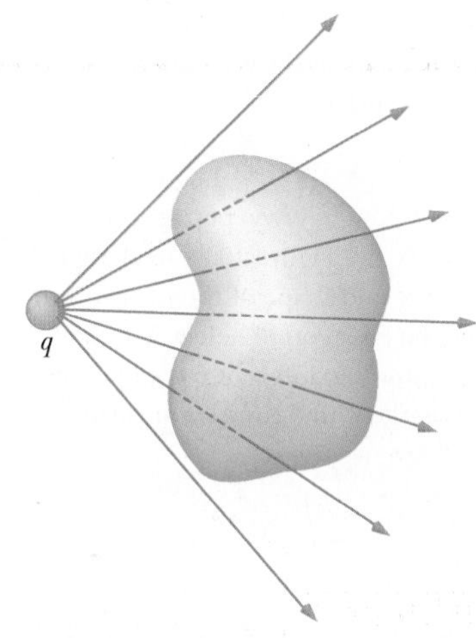

FIGURE 19.30 A point charge located *outside* a closed surface. The number of lines entering the surface equals the number leaving the surface.

Let us extend these arguments to the generalized case of many point charges. We shall again make use of the superposition principle. That is, we can express the net flux through any closed surface as

$$\oint \vec{\mathbf{E}} \cdot d\vec{\mathbf{A}} = \oint (\vec{\mathbf{E}}_1 + \vec{\mathbf{E}}_2 + \cdots) \cdot d\vec{\mathbf{A}}$$

where $\vec{\mathbf{E}}$ is the total electric field at any point on the surface and $\vec{\mathbf{E}}_1$, $\vec{\mathbf{E}}_2$, . . . are the fields produced by the individual charges at that point. Consider the system of charges shown in Active Figure 19.31. The surface S surrounds only one charge, q_1; hence, the net flux through S is q_1/ϵ_0. The flux through S due to the charges outside it is zero because each electric field line from these charges that enters S at one point leaves it at another. The surface S' surrounds charges q_2 and q_3; hence, the net flux through S' is $(q_2 + q_3)/\epsilon_0$. Finally, the net flux through surface S'' is zero because no charge exists inside this surface. That is, *all* electric field lines that enter S'' at one point leave S'' at another. Notice that charge q_4 does not contribute to the net flux through any of the surfaces because it is outside all the surfaces.

Gauss's law, which is a generalization of the foregoing discussion, states that the net flux through *any* closed surface is

$$\Phi_E = \oint \vec{\mathbf{E}} \cdot d\vec{\mathbf{A}} = \frac{q_{\text{in}}}{\epsilon_0} \qquad [19.22]$$

PITFALL PREVENTION 19.4

ZERO FLUX IS NOT ZERO FIELD
In this discussion, we see two possibilities in which there is zero flux through a closed surface: either no charged particles are enclosed by the surface, or charged particles are enclosed but the net charge is zero. For either possibility, *do not* fall into the trap of saying that because the flux is zero, the electric field is zero at the surfaces. Remember that Gauss's law states that the electric *flux* is proportional to the enclosed charge, not the electric *field*.

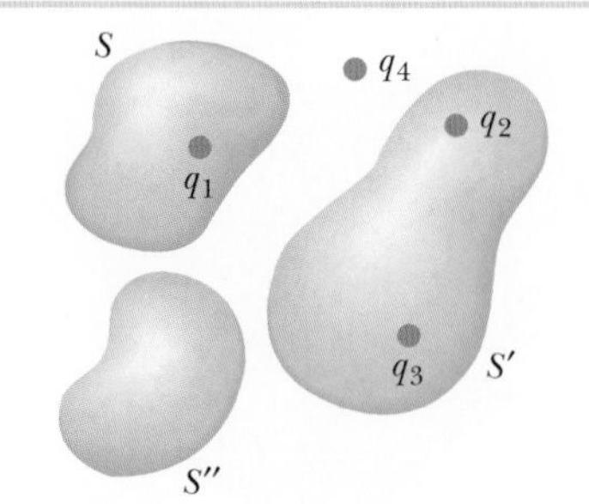

ACTIVE FIGURE 19.31

The net electric flux through any closed surface depends only on the charge *inside* that surface. The net flux through surface S is q_1/ϵ_0, the net flux through surface S' is $(q_2 + q_3)/\epsilon_0$, and the net flux through surface S'' is zero. Charge q_4 does not contribute to the flux through any surface because it is outside all surfaces.

PhysicsNow™ Log into PhysicsNow at **www.pop4e.com** and go to Active Figure 19.31 to change the size and shape of the surface and see the effect on the electric flux of surrounding different combinations of charge with a gaussian surface.

where q_{in} represents the *net charge inside the surface* and $\vec{\mathbf{E}}$ represents the electric field at any point on the surface. In words, Gauss's law states that **the net electric flux through any closed surface is equal to the net charge inside the surface divided by ϵ_0.** The closed surface used in Gauss's law is called a **gaussian surface.**

Gauss's law is valid for the electric field of any system of charges or continuous distribution of charge. In practice, however, **the technique is useful for calculating the electric field only in situations where the degree of symmetry is high.** As we shall see in the next section, **Gauss's law can be used to evaluate the electric field for charge distributions that have spherical, cylindrical, or plane symmetry.** We do so by choosing an appropriate gaussian surface that allows $\vec{\mathbf{E}}$ to be removed from the integral in Gauss's law and performing the integration. Note that a gaussian surface is a mathematical surface and need not coincide with any real physical surface.

QUICK QUIZ 19.6 For a gaussian surface through which the net flux is zero, the following four statements *could be true.* Which of the statements *must be true?* **(a)** No charges are inside the surface. **(b)** The net charge inside the surface is zero. **(c)** The electric field is zero everywhere on the surface. **(d)** The number of electric field lines entering the surface equals the number leaving the surface.

QUICK QUIZ 19.7 Consider the charge distribution shown in Active Figure 19.31. **(i)** What are the charges contributing to the total electric *flux* through surface S'? **(a)** q_1 only **(b)** q_4 only **(c)** q_2 and q_3 **(d)** all four charges **(e)** none of the charges **(ii)** What are the charges contributing to the total electric *field* at a chosen point on the surface S'? **(a)** q_1 only **(b)** q_4 only **(c)** q_2 and q_3 **(d)** all four charges **(e)** none of the charges

■ Thinking Physics 19.1

A spherical gaussian surface surrounds a point charge q. Describe what happens to the net flux through the surface if (a) the charge is tripled, (b) the volume of the sphere is doubled, (c) the surface is changed to a cube, and (d) the charge is moved to another location *inside* the surface.

Reasoning (a) If the charge is tripled, the flux through the surface is also tripled because the net flux is proportional to the charge inside the surface. (b) The net flux remains constant when the volume changes because the surface surrounds the same amount of charge, regardless of its volume. (c) The net flux does not change when the shape of the closed surface changes. (d) The net flux through the closed surface remains unchanged as the charge inside the surface is moved to another location as long as the new location remains inside the surface. ■

19.10 APPLICATION OF GAUSS'S LAW TO SYMMETRIC CHARGE DISTRIBUTIONS

As mentioned earlier, Gauss's law is useful in determining electric fields when the charge distribution has a high degree of symmetry. The following examples show ways of choosing the gaussian surface over which the surface integral given by Equation 19.22 can be simplified and the electric field determined. The surface should always be chosen to take advantage of the symmetry of the charge distribution so that we can remove E from the integral and solve for it. The crucial step in applying Gauss's law is to determine a useful gaussian surface. Such a surface should be a closed surface for which each portion of the surface satisfies one or more of the following conditions:

1. The value of the electric field can be argued by symmetry to be constant over the portion of the surface.
2. The dot product in Equation 19.22 can be expressed as a simple algebraic product $E\,dA$ because $\vec{\mathbf{E}}$ and $d\vec{\mathbf{A}}$ are parallel.
3. The dot product in Equation 19.22 is zero because $\vec{\mathbf{E}}$ and $d\vec{\mathbf{A}}$ are perpendicular.
4. The field can be argued to be zero everywhere on the portion of the surface.

Note that different portions of the gaussian surface can satisfy different conditions as long as every portion satisfies at least one condition. We will see all four of these conditions used in the examples through the remainder of this chapter.

EXAMPLE 19.9 The Electric Field Due to a Point Charge

Starting with Gauss's law, calculate the electric field due to an isolated point charge q.

Solution A single charge is the simplest possible charge distribution, and we will use this familiar example to show the technique of solving for the electric field with Gauss's law. We choose a spherical gaussian surface of radius r centered on the point charge as in Figure 19.32.

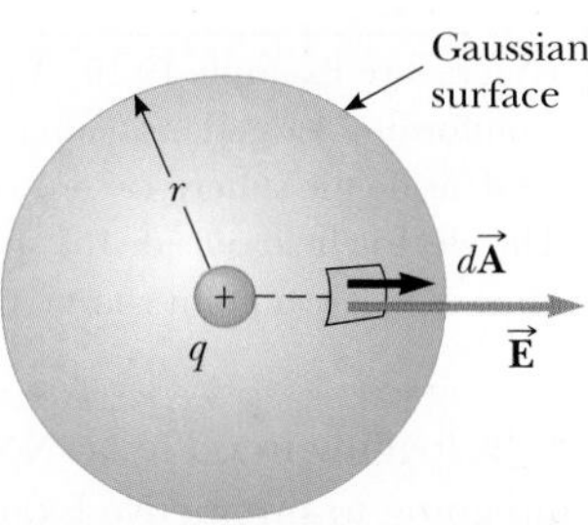

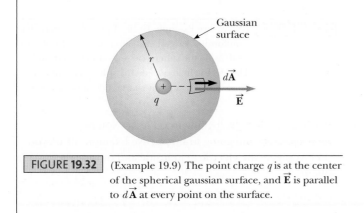

FIGURE 19.32 (Example 19.9) The point charge q is at the center of the spherical gaussian surface, and $\vec{\mathbf{E}}$ is parallel to $d\vec{\mathbf{A}}$ at every point on the surface.

The electric field of a positive point charge is radial outward by symmetry and is therefore normal to the surface at every point. As in condition 2, $\vec{\mathbf{E}}$ is therefore parallel to $d\vec{\mathbf{A}}$ at each point on the surface, so $\vec{\mathbf{E}} \cdot d\vec{\mathbf{A}} = E\,dA$ and Gauss's law gives

$$\Phi_E = \oint \vec{\mathbf{E}} \cdot d\vec{\mathbf{A}} = \oint E\,dA = \frac{q}{\epsilon_0}$$

By symmetry, E is constant everywhere on the surface, which satisfies condition 1, and so it can be removed from the integral. Therefore,

$$\oint E\;dA = E\oint dA = E(4\pi r^2) = \frac{q}{\epsilon_0}$$

where we have used that the surface area of a sphere is $4\pi r^2$. We now solve for the electric field:

$$E = \frac{q}{4\pi\epsilon_0 r^2} = k_e\frac{q}{r^2}$$

which is the familiar electric field of a point charge that we developed from Coulomb's law earlier in this chapter.

INTERACTIVE EXAMPLE 19.10 A Spherically Symmetric Charge Distribution

An insulating solid sphere of radius a has a uniform volume charge density ρ and carries a total positive charge Q (Fig. 19.33).

A Calculate the magnitude of the electric field at a point outside the sphere.

Solution Because the charge distribution is spherically symmetric, we again select a spherical gaussian surface of radius r, concentric with the sphere, as in Figure 19.33a. For this choice, conditions 1 and 2 are satisfied, as they were for the point charge in Example 19.9. Following the line of reasoning given in Example 19.9, we find that

$$E = k_e\frac{Q}{r^2} \qquad (\text{for } r > a)$$

Note that this result is identical to that obtained for a point charge. We therefore conclude that, for a uniformly charged sphere, the field in the region external to the sphere is *equivalent* to that of a point charge located at the center of the sphere.

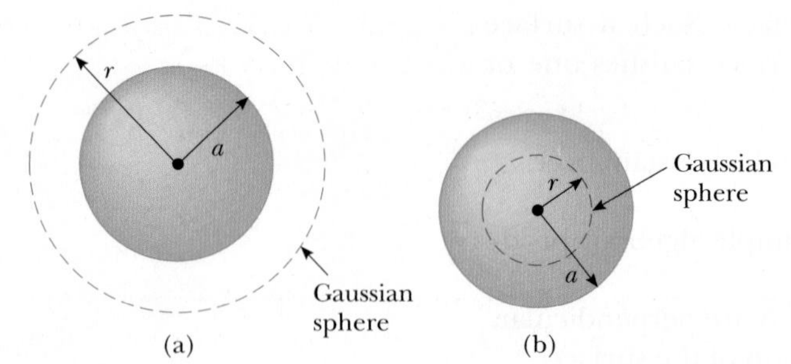

FIGURE 19.33 (Interactive Example 19.10) A uniformly charged insulating sphere of radius a and total charge Q. (a) For points outside the sphere, a large, spherical gaussian surface is drawn concentric with the sphere. In diagrams such as this one, the dotted line represents the intersection of the gaussian surface with the plane of the page. (b) For points inside the sphere, a spherical gaussian surface smaller than the sphere is drawn.

B Find the magnitude of the electric field at a point inside the sphere.

Solution In this case, we select a spherical gaussian surface having radius $r < a$, concentric with the insulating sphere (Fig. 19.33b). Let us denote the volume of this smaller sphere by V'. To apply Gauss's law in this situation, it is important to recognize that the charge q_{in} within the gaussian surface of volume V' is less than Q. To calculate q_{in}, we use that $q_{\text{in}} = \rho V'$:

$$q_{\text{in}} = \rho V' = \rho(\tfrac{4}{3}\pi r^3)$$

By symmetry, the magnitude of the electric field is constant everywhere on the spherical gaussian surface and the field is normal to the surface at each point, so both conditions 1 and 2 are satisfied. Gauss's law in the region $r < a$ therefore gives

$$\oint E\,dA = E\oint dA = E(4\pi r^2) = \frac{q_{\text{in}}}{\epsilon_0}$$

Solving for E gives

$$E = \frac{q_{\text{in}}}{4\pi\epsilon_0 r^2} = \frac{\rho(\frac{4}{3}\pi r^3)}{4\pi\epsilon_0 r^2} = \frac{\rho}{3\epsilon_0}r$$

Because $\rho = Q/\frac{4}{3}\pi a^3$ by definition and $k_e = 1/4\pi\epsilon_0$, this expression for E can be written as

$$E = \frac{Qr}{4\pi\epsilon_0 a^3} = \frac{k_e Q}{a^3}r \qquad \text{(for } r < a\text{)}$$

This result for E differs from that obtained in part A. It shows that $E \rightarrow 0$ as $r \rightarrow 0$. A plot of E versus r is shown in Figure 19.34. Note that the expressions for parts A and B match when $r = a$.

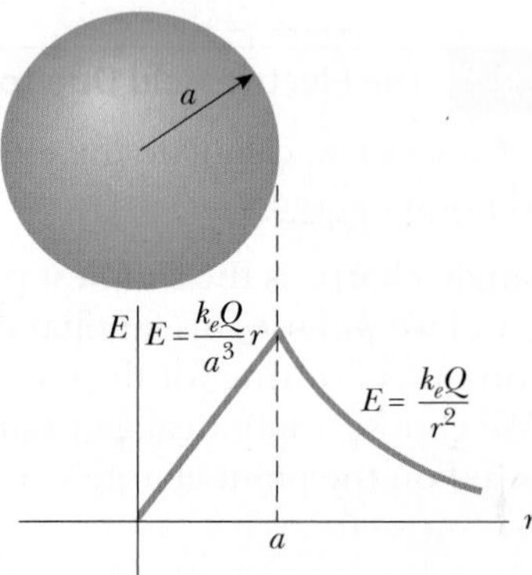

FIGURE 19.34 (Interactive Example 19.10) A plot of E versus r for a uniformly charged insulating sphere. The electric field inside the sphere ($r < a$) varies linearly with r. The electric field outside the sphere ($r > a$) is the same as that of a point charge Q located at $r = 0$.

PhysicsNow™ By logging into PhysicsNow at **www.pop4e.com** and going to Interactive Example 19.10, you can investigate the electric field inside and outside the sphere.

EXAMPLE 19.11 A Cylindrically Symmetric Charge Distribution

Find the electric field a distance r from a line of positive charge of infinite length and constant charge per unit length ℓ (Fig. 19.35a).

Solution The symmetry of the charge distribution requires that $\vec{\mathbf{E}}$ must be perpendicular to the line charge and directed outward as in Figure 19.35. To reflect the symmetry of the charge distribution, we select a cylindrical gaussian surface of radius r and length ℓ that is coaxial with the line charge. For the curved part of this surface, $\vec{\mathbf{E}}$ is constant in magnitude and perpendicular to the surface at each point (conditions 1 and 2). Furthermore, the flux through the ends of the gaussian cylinder is zero because $\vec{\mathbf{E}}$ is parallel to these surfaces (and therefore perpendicular to $d\vec{\mathbf{A}}$), which is the first application we have seen of condition 3.

The surface integral in Gauss's law is taken over the entire gaussian surface. Because of the zero value of $\vec{\mathbf{E}} \cdot d\vec{\mathbf{A}}$ for the ends of the cylinder, however, we can restrict our attention to only the curved surface of the cylinder.

The total charge inside our gaussian surface is $q_{\text{in}} = \lambda\ell$. Applying Gauss's law and applying conditions 1 and 2, we find, for the curved surface, that

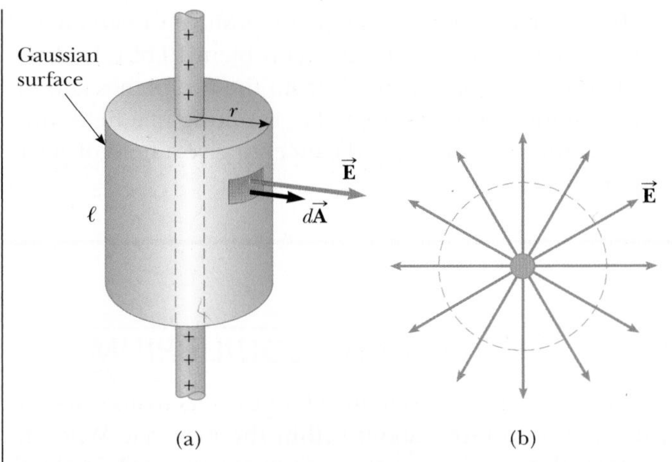

FIGURE 19.35 (Example 19.11) (a) An infinite line of charge surrounded by a cylindrical gaussian surface concentric with the line charge. (b) An end view shows that the electric field on the cylindrical surface is constant in magnitude and perpendicular to the surface.

$$\Phi_E = \oint \vec{\mathbf{E}} \cdot d\vec{\mathbf{A}} = E\oint dA = EA = \frac{q_{\text{in}}}{\epsilon_0} = \frac{\lambda \ell}{\epsilon_0}$$

The area of the curved surface is $A = 2\pi r\ell$. Therefore,

$$E(2\pi r\ell) = \frac{\lambda \ell}{\epsilon_0}$$

$$E = \frac{\lambda}{2\pi\epsilon_0 r} = 2k_e\frac{\lambda}{r} \qquad [19.23]$$

Therefore, we see that the electric field of a cylindrically symmetric charge distribution varies as $1/r$, whereas the field external to a spherically symmetric charge distribution varies as $1/r^2$. Equation 19.23 can also be obtained using Equation 19.7; the mathematical techniques necessary for this calculation, however, are more cumbersome.

If the line charge in this example were of finite length, the result for E is not that given by Equation 19.23. A finite line charge does not possess sufficient symmetry to use Gauss's law because the magnitude of the electric field is no longer constant over the surface of the gaussian cylinder; the field near the ends of the line would be different from that far from the ends. Therefore, condition 1 is not satisfied in this situation. Furthermore, $\vec{\mathbf{E}}$ is not perpendicular to the cylindrical surface at all points; the field vectors near the ends would have a component parallel to the line. Condition 2 is not satisfied. When the symmetry in the charge distribution is insufficient, as in this situation, it is necessary to calculate $\vec{\mathbf{E}}$ using Equation 19.7.

For points close to a finite line charge and far from the ends, Equation 19.23 gives a good approximation of the value of the field.

It is left as a problem (Problem 19.39) to show that the electric field inside a uniformly charged rod of finite thickness and infinite length is proportional to r.

EXAMPLE 19.12 A Nonconducting Plane Sheet of Charge

Find the electric field due to a nonconducting, infinite plane with uniform surface charge density σ.

Solution Symmetry tells us that $\vec{\mathbf{E}}$ must be perpendicular to the plane and that the field will have the same magnitude at points on opposite sides of the plane and equidistant from it. That the direction of $\vec{\mathbf{E}}$ is away from positive charges tells us that the direction of $\vec{\mathbf{E}}$ on one side of the plane must be opposite its direction on the other side as in Figure 19.36. A gaussian surface that reflects the symmetry is a small cylinder whose axis is perpendicular to the plane and whose ends each have an area A and are equidistant from the plane. Because $\vec{\mathbf{E}}$ is parallel to the curved surface and therefore perpendicular to $d\vec{\mathbf{A}}$ everywhere on the surface, condition 3 is satisfied and the curved surface makes no contribution to the surface integral. For the flat ends of the cylinder, conditions 1 and 2 are satisfied. The flux through each end of the cylinder is EA; hence, the total flux through the entire gaussian surface is just that through the ends, $\Phi_E = 2EA$.

Noting that the total charge inside the surface is $q_{\text{in}} = \sigma A$, we use Gauss's law to obtain

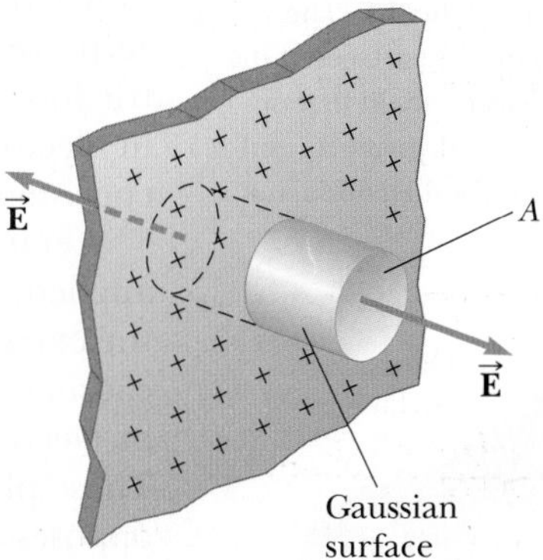

FIGURE 19.36 (Example 19.12) A cylindrical gaussian surface penetrating an infinite sheet of charge. The flux is EA through each end of the gaussian surface and zero through its curved surface.

$$\Phi_E = 2EA = \frac{q_{\text{in}}}{\epsilon_0} = \frac{\sigma A}{\epsilon_0}$$

$$E = \frac{\sigma}{2\epsilon_0} \qquad [19.24]$$

Because the distance of the flat end of the cylinder from the plane does not appear in Equation 19.24, we conclude that $E = \sigma/2\epsilon_0$ at *any* distance from the plane. That is, the field is uniform everywhere.

An important charge configuration related to this example is two parallel planes each with a surface charge density σ, with one plane positively charged and the other negatively charged (Problem 19.62). In this situation, the electric fields from the two planes add in the region between the planes, resulting in a field with a magnitude of σ/ϵ_0, and cancel to give a field of zero elsewhere.

19.11 CONDUCTORS IN ELECTROSTATIC EQUILIBRIUM

A good electrical conductor, such as copper, contains charges (electrons) that are not bound to any atom and are free to move about within the material. When no motion of charge occurs within the conductor, the conductor is in **electrostatic equilibrium.** In this situation, *every* charge in the conductor is a particle in equilibrium, experiencing zero net force. As we shall see, an isolated conductor (one that is insulated from ground) in electrostatic equilibrium has the following properties:

1. The electric field is zero everywhere inside the conductor.
2. If the isolated conductor carries a net charge, the net charge resides entirely on its surface.
3. The electric field just outside the charged conductor is perpendicular to the conductor surface and has a magnitude σ/ϵ_0, where σ is the surface charge density at that point.
4. On an irregularly shaped conductor, the surface charge density is highest at locations where the radius of curvature of the surface is smallest.

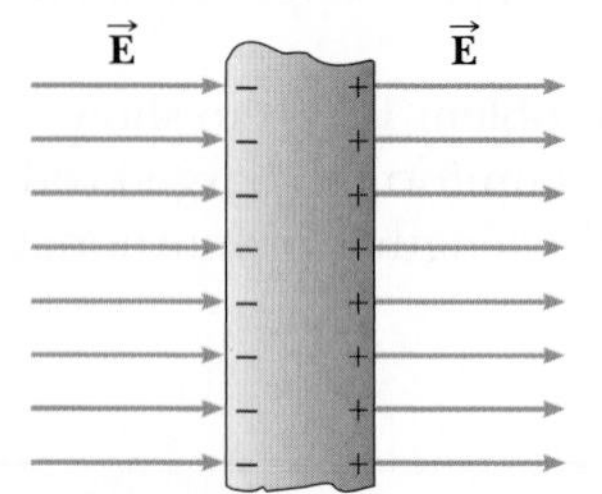

FIGURE 19.37 A conducting slab in an external electric field $\vec{E}$. The charges induced on the surfaces of the slab produce an electric field that opposes the external field, giving a resultant field of zero *inside* the conductor.

We will verify the first three properties in the following discussion. The fourth property is presented here so that we have a complete list of properties for conductors in electrostatic equilibrium. The verification of it, however, requires concepts from Chapter 20, so we will postpone its verification until then.

The first property can be understood by considering a conducting slab placed in an external field $\vec{E}$ (Fig. 19.37). The electric field inside the conductor *must* be zero under the assumption that we have electrostatic equilibrium. If the field were not zero, free charges in the conductor would accelerate under the action of the electric force. This motion of electrons, however, would mean that the conductor is not in electrostatic equilibrium. Therefore, the existence of electrostatic equilibrium is consistent only with a zero field in the conductor.

Let us investigate how this zero field is accomplished. Before the external field is applied, free electrons are uniformly distributed throughout the conductor. When the external field is applied, the free electrons accelerate to the left in Figure 19.37, causing a plane of negative charge to be present on the left surface. The movement of electrons to the left results in a plane of positive charge on the right surface. These planes of charge create an additional electric field inside the conductor that opposes the external field. As the electrons move, the surface charge density increases until the magnitude of the internal field equals that of the external field, giving a net field of zero inside the conductor.

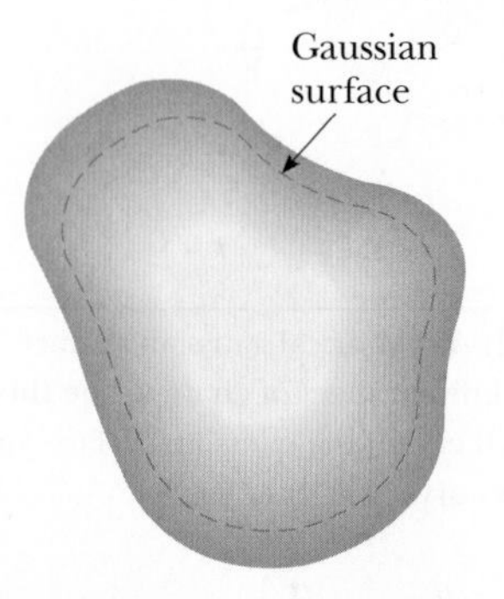

FIGURE 19.38 An isolated conductor of arbitrary shape. The broken line represents a gaussian surface just inside the physical surface of the conductor.

We can use Gauss's law to verify the second property of a conductor in electrostatic equilibrium. Figure 19.38 shows an arbitrarily shaped conductor. A gaussian surface is drawn just inside the conductor and can be as close to the surface as we wish. As we have just shown, the electric field everywhere inside a conductor in electrostatic equilibrium is zero. Therefore, the electric field must be zero at every point on the gaussian surface (condition 4 in Section 19.10). From this result and Gauss's law, we conclude that the net charge inside the gaussian surface is zero. Because there can be no net charge inside the gaussian surface (which is arbitrarily close to the conductor's surface), any net charge on the conductor must reside on its surface. Gauss's law does not tell us how this excess charge is distributed on the surface, only that it must reside on the surface.

Conceptually, we can understand the location of the charges on the surface by imagining placing many charges at the center of the conductor. The mutual repulsion of the charges causes them to move apart. They will move as far as they can, which is to various points on the surface.

To verify the third property, we can also use Gauss's law. We draw a gaussian surface in the shape of a small cylinder having its end faces parallel to the surface (Fig. 19.39). Part of the cylinder is just outside the conductor and part is inside. The field is normal to the surface because the conductor is in electrostatic equilibrium. If $\vec{\mathbf{E}}$ had a component parallel to the surface, an electric force would be exerted on the charges parallel to the surface, free charges would move along the surface, and so the conductor would not be in equilibrium. Therefore, we satisfy condition 3 in Section 19.10 for the curved part of the cylinder in that no flux exists through this part of the gaussian surface because $\vec{\mathbf{E}}$ is parallel to this part of the surface. No flux exists through the flat face of the cylinder inside the conductor because $\vec{\mathbf{E}} = 0$ (condition 4). Hence, the net flux through the gaussian surface is the flux through the flat face outside the conductor where the field is perpendicular to the surface. Using conditions 1 and 2 for this face, the flux is EA, where E is the electric field just outside the conductor and A is the area of the cylinder's face. Applying Gauss's law to this surface gives

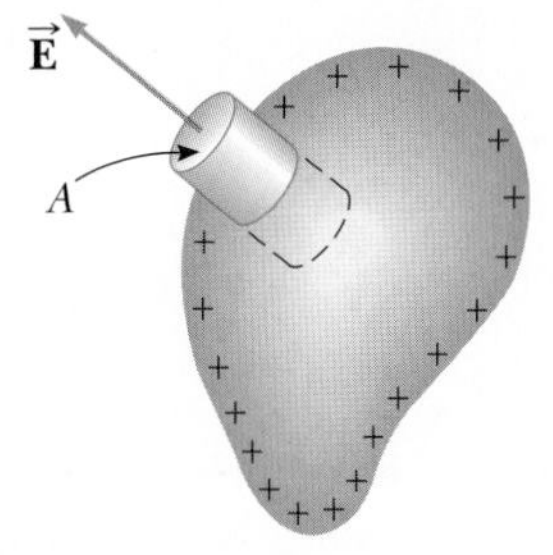

FIGURE 19.39 A gaussian surface in the shape of a small cylinder is used to calculate the electric field just outside a charged conductor. The flux through the gaussian surface is EA.

$$\Phi_E = \oint E\,dA = EA = \frac{q_{\text{in}}}{\epsilon_0} = \frac{\sigma A}{\epsilon_0}$$

where we have used that $q_{\text{in}} = \sigma A$. Solving for E gives

$$E = \frac{\sigma}{\epsilon_0} \quad [19.25]$$

■ Thinking Physics 19.2

Suppose a point charge $+Q$ is in empty space. We surround the charge with a spherical, uncharged conducting shell so that the charge is at the center of the shell. What effect does that have on the field lines from the charge?

Reasoning When the spherical shell is placed around the charge, the free charges in the shell adjust so as to satisfy the rules for a conductor in equilibrium and Gauss's law. A net charge of $-Q$ moves to the interior surface of the conductor, so the electric field within the conductor is zero (a spherical gaussian surface totally within the shell encloses no *net* charge). A net charge of $+Q$ resides on the outer surface, so a gaussian surface outside the sphere encloses a net charge of $+Q$, just as if the shell were not there. Therefore, the only change in the field lines from the initial situation is the absence of field lines over the thickness of the conducting shell. ■

19.12 THE ATMOSPHERIC ELECTRIC FIELD

CONTEXT CONNECTION

In this chapter, we discussed the electric field due to various charge distributions. On the surface of the Earth and in the atmosphere, a number of processes create charge distributions, resulting in an electric field in the atmosphere. These processes include cosmic rays entering the atmosphere, radioactive decay at the Earth's surface, and lightning, the focus of our study in this Context.

The result of these processes is an average negative charge distributed over the surface of the Earth of about 5×10^5 C, which is a tremendous amount of charge. (The Earth is neutral overall; the positive charges corresponding to this negative surface charge are spread through the atmosphere, as we shall discuss in

Chapter 20.) We can calculate the average surface charge density over the surface of the Earth:

$$\sigma_{\text{avg}} = \frac{Q}{A} = \frac{Q}{4\pi r^2} = \frac{5 \times 10^5\ \text{C}}{4\pi(6.37 \times 10^6\ \text{m})^2} \sim 10^{-9}\ \text{C/m}^2$$

Throughout this Context, we will be adopting a number of simplification models. Consequently, we will consider our calculations to be order-of-magnitude estimates of the actual values, as suggested by the ~ sign above.

The Earth is a good conductor. Therefore, we can use the third property of conductors in Section 19.11 to find the average magnitude of the electric field at the surface of the Earth:

$$E_{\text{avg}} = \frac{\sigma_{\text{avg}}}{\epsilon_0} = \frac{10^{-9}\ \text{C/m}^2}{8.85 \times 10^{-12}\ \text{C}^2/\text{N}\cdot\text{m}^2} \sim 10^2\ \text{N/C}$$

which is a typical value of the **fair-weather electric field** that exists in the absence of a thunderstorm. The direction of the field is downward because the charge on the Earth's surface is negative. During a thunderstorm, the electric field under the thundercloud is significantly higher than the fair-weather electric field, because of the charge distribution in the thundercloud.

Figure 19.40 shows a typical charge distribution in a thundercloud. The charge distribution can be modeled as a *tripole,* although the positive charge at the bottom of the cloud tends to be smaller than the other two charges. The mechanism of charging in thunderclouds is not well understood and continues to be an active area of research.

It is this high concentration of charge in the thundercloud that is responsible for the very strong electric fields that cause lightning discharge between the cloud and the ground. Typical electric fields during a thunderstorm are as high as 25 000 N/C. The distribution of negative charges in the center of the cloud in Figure 19.40 is the source of negative charge that moves downward in a lightning strike.

Atmospheric electric fields can be measured with an instrument called a **field mill.** Figure 19.41 shows the operation of the field mill. In Figure 19.41a, a metallic

FIGURE 19.40 A typical tripolar charge distribution in a thundercloud. The amounts of positive charge at the top and negative charge in the middle are approximately the same, but the amount of positive charge at the bottom is less. The dots indicate the average position of each charge distribution.

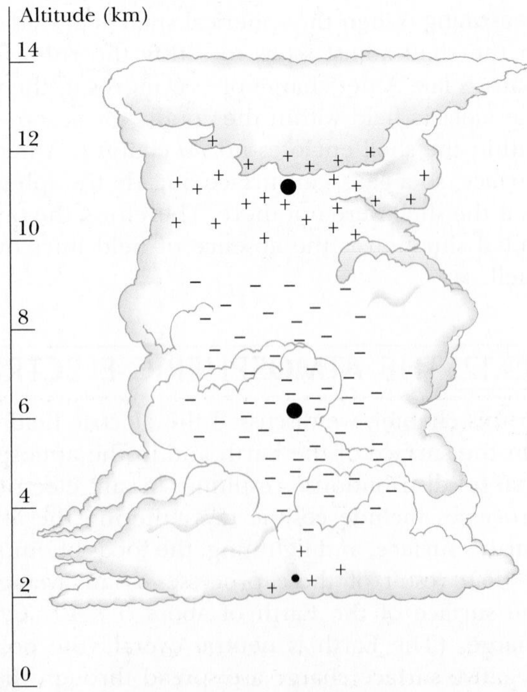

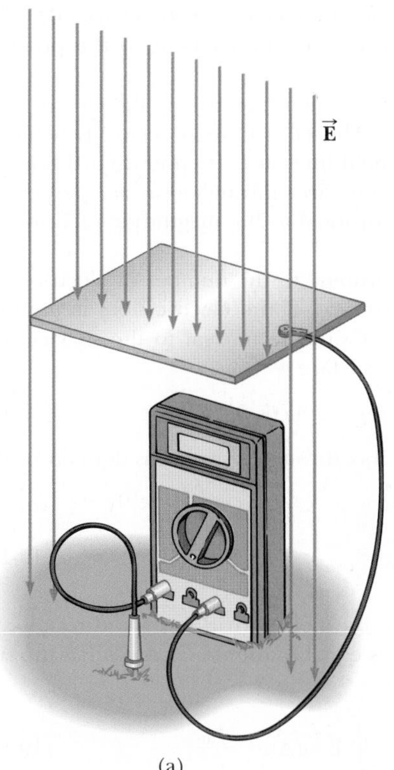

(a)

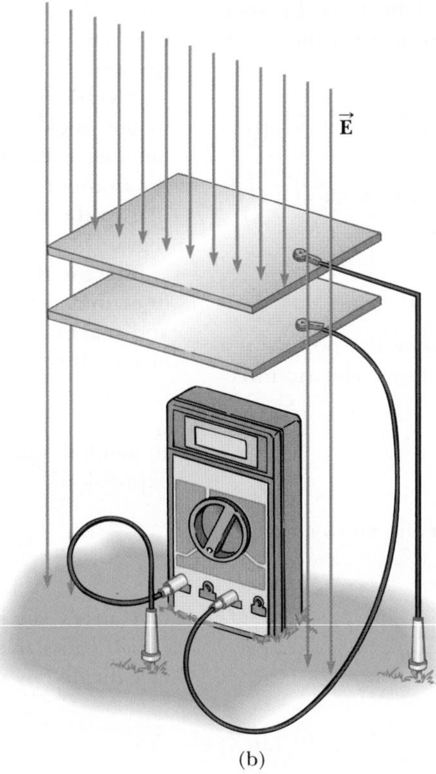

(b)

FIGURE 19.41 A field mill for measuring the atmospheric electric field. When the upper plate is moved over the lower plate, charges move through the meter.

plate is connected to the ground by a wire. A meter measures the flow of charge through the wire. Because the ground is negatively charged, electrons will flow from the ground into the metal plate. These electrons represent the ends of some of the electric field lines in the atmosphere.

Now, as shown in Figure 19.41b, this plate is covered with a second plate also attached to the ground. The electric field lines that previously ended on the lower plate now end on the upper plate. The charges on the lower plate are repelled by those in the upper plate and pass through the meter into the ground. The meter measures the amount of charge flowing through the wire. This charge is related to how much charge is on the lower plate, which, in turn, is related to the magnitude of the electric field. Therefore, the meter can be calibrated to measure the atmospheric electric field. In operation, the plates are similar to the blades of a fan. As one set of blades rotates over a second stationary set, charge pulses back and forth through the meter.

In this chapter, we have analyzed the atmosphere in terms of the electric field. In Chapter 20, we shall learn about electric potential and analyze the atmosphere again in terms of this new parameter. ■

SUMMARY

PhysicsNow™ Take a practice test by logging into PhysicsNow at www.pop4e.com and clicking on the Pre-Test link for this chapter.

Electric charges have the following important properties:

1. Two kinds of charges exist in nature, **positive** and **negative,** with the property that charges of opposite sign attract each other and charges of the same sign repel each other.
2. The force between charged particles varies as the inverse square of their separation distance.
3. Charge is conserved.
4. Charge is quantized.

Conductors are materials in which charges move relatively freely. **Insulators** are materials in which charges do not move freely.

Coulomb's law states that the electrostatic force between two stationary, charged particles separated by a distance r has the magnitude

$$F_e = k_e \frac{|q_1||q_2|}{r^2} \quad [19.1]$$

where the Coulomb constant $k_e = 8.99 \times 10^9\ \text{N} \cdot \text{m}^2/\text{C}^2$. The vector form of Coulomb's law is

$$\vec{\mathbf{F}}_{12} = k_e \frac{q_1 q_2}{r^2} \hat{\mathbf{r}}_{12} \quad [19.2]$$

An **electric field** exists at a point in space if a positive test charge q_0 placed at that point experiences an electric force. The electric field is defined as

$$\vec{\mathbf{E}} \equiv \frac{\vec{\mathbf{F}}_e}{q_0} \quad [19.3]$$

The force on a particle with charge q placed in an electric field $\vec{\mathbf{E}}$ is

$$\vec{\mathbf{F}}_e = q\vec{\mathbf{E}} \quad [19.4]$$

The electric field due to the point charge q at a distance r from the charge is

$$\vec{\mathbf{E}} = k_e \frac{q}{r^2} \hat{\mathbf{r}} \quad [19.5]$$

where $\hat{\mathbf{r}}$ is a unit vector directed from the charge toward the point in question. The electric field is directed radially outward from a positive charge and is directed toward a negative charge.

The electric field due to a group of charges can be obtained using the superposition principle. That is, the total electric field equals the vector sum of the electric fields of all the charges at some point:

$$\vec{\mathbf{E}} = k_e \sum_i \frac{q_i}{r_i^2} \hat{\mathbf{r}}_i \quad [19.6]$$

Similarly, the electric field of a continuous charge distribution at some point is

$$\vec{\mathbf{E}} = k_e \int \frac{dq}{r^2} \hat{\mathbf{r}} \quad [19.7]$$

where dq is the charge on one element of the charge distribution and r is the distance from the element to the point in question.

Electric field lines are useful for describing the electric field in any region of space. The electric field vector $\vec{\mathbf{E}}$ is always tangent to the electric field lines at every point. Furthermore, the number of lines per unit area through a surface perpendicular to the lines is proportional to the magnitude of $\vec{\mathbf{E}}$ in that region.

Electric flux is proportional to the number of electric field lines that penetrate a surface. If the electric field is uniform and makes an angle of θ with the normal to the surface, the electric flux through the surface is

$$\Phi_E = EA \cos \theta \quad [19.18]$$

In general, the electric flux through a surface is defined by the expression

$$\Phi_E \equiv \int_{\text{surface}} \vec{\mathbf{E}} \cdot d\vec{\mathbf{A}} \quad [19.19]$$

Gauss's law says that the net electric flux Φ_E through any closed gaussian surface is equal to the *net* charge *inside* the surface divided by ϵ_0:

$$\Phi_E = \oint \vec{\mathbf{E}} \cdot d\vec{\mathbf{A}} = \frac{q_{\text{in}}}{\epsilon_0} \quad [19.22]$$

Using Gauss's law, one can calculate the electric field due to various symmetric charge distributions.

A conductor in **electrostatic equilibrium** has the following properties:

1. The electric field is zero everywhere inside the conductor.
2. If the isolated conductor carries a net charge, the net charge resides entirely on its surface.
3. The electric field just outside the charged conductor is perpendicular to the conductor surface and has a magnitude σ/ϵ_0, where σ is the surface charge density at that point.
4. On an irregularly shaped conductor, the surface charge density is highest at locations where the radius of curvature of the surface is smallest.

QUESTIONS

☐ = answer available in the *Student Solutions Manual and Study Guide*

1. Explain what is meant by the term "a neutral atom." Explain what "a negatively charged atom" means.
2. Sparks are often seen or heard on a dry day when fabrics are removed from a clothes dryer in dim light. Explain.
3. Hospital personnel must wear special conducting shoes while working around oxygen in an operating room. Why? Contrast with what might happen if people wore rubber-soled shoes.
4. Explain the similarities and differences between Newton's law of universal gravitation and Coulomb's law.
5. A balloon is negatively charged by rubbing and then clings to a wall. Does that mean that the wall is positively charged? Why does the balloon eventually fall?
6. Is it possible for an electric field to exist in empty space? Explain. Consider point A in Figure 19.18a. Does charge exist at this point? Does a force exist at this point? Does a field exist at this point?
7. When is it valid to approximate a charge distribution by a point charge?
8. Figure 19.11 shows three electric field vectors at the same point. With a little extrapolation, Figure 19.16a would show many electric field lines at the same point. Is it really

true that "no two field lines can cross"? Are the diagrams drawn correctly? Explain your answers.

9. Would life be different if the electron were positively charged and the proton were negatively charged? Does the choice of signs have any bearing on physical and chemical interactions? Explain.

10. Consider two equal point charges separated by some distance d. At what point (other than ∞) would a third test charge experience no net force?

11. Consider two electric dipoles in empty space. Each dipole has zero net charge. Does an electric force exist between the dipoles? That is, can two objects with zero net charge exert electric forces on each other? If so, is the force one of attraction or of repulsion?

12. A particle with negative charge $-q$ is placed at the point P near the positively charged ring shown in Figure 19.14 (Example 19.5). Assuming that x is much less than a, describe the motion of the point charge after it is released from rest.

13. If more electric field lines leave a gaussian surface than enter it, what can you conclude about the net charge enclosed by that surface?

14. A uniform electric field exists in a region of space in which there are no charges. What can you conclude about the net electric flux through a gaussian surface placed in this region of space?

15. If the total charge inside a closed surface is known but the distribution of the charge is unspecified, can you use Gauss's law to find the electric field? Explain.

16. On the basis of the repulsive nature of the force between like charges and the freedom of motion of charge within a conductor, explain why excess charge on an isolated conductor must reside on its surface.

17. A person is placed in a large, hollow, metallic sphere that is insulated from ground. If a large charge is placed on the sphere, will the person be harmed upon touching the inside of the sphere? Explain what will happen if the person also has an initial charge whose sign is opposite that of the charge on the sphere.

18. Two solid spheres, both of radius R, carry identical total charges Q. One sphere is a good conductor, whereas the other is an insulator. If the charge on the insulating sphere is uniformly distributed throughout its interior volume, how do the electric fields outside these two spheres compare? Are the fields identical inside the two spheres?

19. A common demonstration involves charging a rubber balloon, which is an insulator, by rubbing it on your hair and then touching the balloon to a ceiling or wall, which is also an insulator. The electrical attraction between the charged balloon and the neutral wall results in the balloon sticking to the wall. Imagine now that we have two infinitely large, flat sheets of insulating material. One is charged and the other is neutral. If these sheets are brought into contact, will an attractive force exist between them, as there was for the balloon and the wall?

PROBLEMS

1, 2, 3 = straightforward, intermediate, challenging

☐ = full solution available in the *Student Solutions Manual and Study Guide*

Physics Now™ = coached problem with hints available at www.pop4e.com

= computer useful in solving problem

= paired numerical and symbolic problems

= biomedical application

Section 19.2 ■ Properties of Electric Charges

1. (a) Find to three significant digits the charge and the mass of an ionized hydrogen atom, represented as H^+. (*Suggestion:* Begin by looking up the mass of a neutral atom on the periodic table of the elements in Appendix C.) (b) Find the charge and the mass of Na^+, a singly ionized sodium atom. (c) Find the charge and the average mass of a chloride ion Cl^- that joins with the Na^+ to make one molecule of table salt. (d) Find the charge and the mass of $Ca^{++} = Ca^{2+}$, a doubly ionized calcium atom. (e) You can model the center of an ammonia molecule as an N^{3-} ion. Find its charge and mass. (f) The plasma in a hot star contains quadruply ionized nitrogen atoms, N^{4+}. Find their charge and mass. (g) Find the charge and the mass of the nucleus of a nitrogen atom. (h) Find the charge and the mass of the molecular ion H_2O^-.

2. (a) Calculate the number of electrons in a small, electrically neutral silver pin that has a mass of 10.0 g. Silver has 47 electrons per atom, and its molar mass is 107.87 g/mol. (b) Electrons are added to the pin until the net negative charge is 1.00 mC. How many electrons are added for every 10^9 electrons already present?

Section 19.4 ■ Coulomb's Law

3. Nobel laureate Richard Feynman once said that if two persons stood at arm's length from each other and each person had 1% more electrons than protons, the force of repulsion between them would be enough to lift a "weight" equal to that of the entire Earth. Carry out an order-of-magnitude calculation to substantiate this assertion.

4. Two protons in an atomic nucleus are typically separated by a distance of 2×10^{-15} m. The electric repulsion force between the protons is huge, but the attractive nuclear force is even stronger and keeps the nucleus from bursting apart. What is the magnitude of the electric force between two protons separated by 2.00×10^{-15} m?

5. Three point charges are located at the corners of an equilateral triangle as shown in Figure P19.5. Calculate the resultant electric force on the 7.00-μC charge.

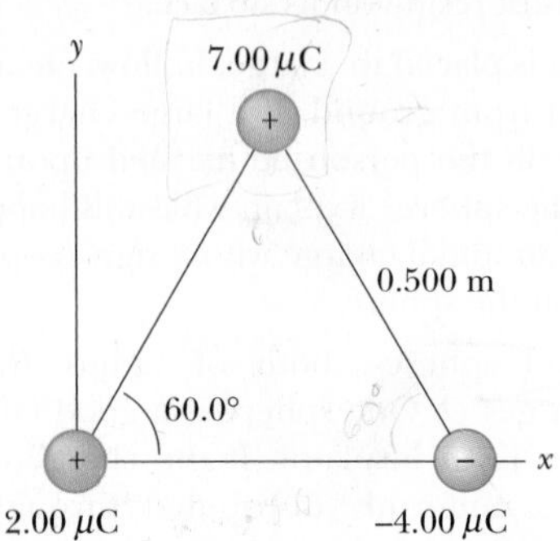

FIGURE P19.5

6. A charged particle A exerts a force of 2.62 μN to the right on charged particle B when the particles are 13.7 mm apart. Particle B moves straight away from A to make the distance between them 17.7 mm. What vector force does it then exert on A?

7. Two identical conducting small spheres are placed with their centers 0.300 m apart. One is given a charge of 12.0 nC and the other a charge of −18.0 nC. (a) Find the electric force exerted by one sphere on the other. (b) Next, the spheres are connected by a conducting wire. Find the electric force between the two after they have come to equilibrium.

8. Two small beads having positive charges $3q$ and q are fixed at the opposite ends of a horizontal, insulating rod, extending from the origin to the point $x = d$. As shown in Figure P19.8, a third small, charged bead is free to slide on the rod. At what position is the third bead in equilibrium? Can it be in stable equilibrium?

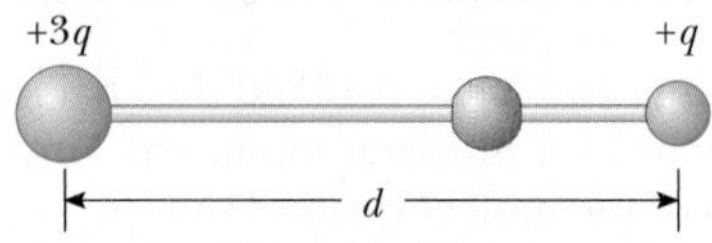

FIGURE P19.8

9. **Review problem.** In the Bohr theory of the hydrogen atom, an electron moves in a circular orbit about a proton, where the radius of the orbit is 0.529×10^{-10} m. (a) Find the magnitude of the electric force each exerts on the other. (b) If this force causes the centripetal acceleration of the electron, what is the speed of the electron?

Section 19.5 ■ Electric Fields

10. What are the magnitude and direction of the electric field that will balance the weight of (a) an electron and (b) a proton? (You may use the data in Table 19.1.)

11. In Figure P19.11, determine the point (other than infinity) at which the electric field is zero.

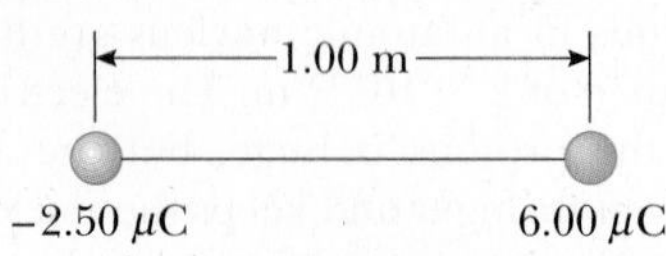

FIGURE P19.11

12. Two point charges are located on the x axis. The first is a charge $+Q$ at $x = -a$. The second is an unknown charge located at $x = +3a$. The net electric field these charges produce at the origin has a magnitude of $2k_eQ/a^2$. What are the two possible values of the unknown charge?

13. Three point charges are arranged as shown in Figure P19.13. (a) Find the vector electric field that the 6.00-nC and −3.00-nC charges together create at the origin. (b) Find the vector force on the 5.00-nC charge.

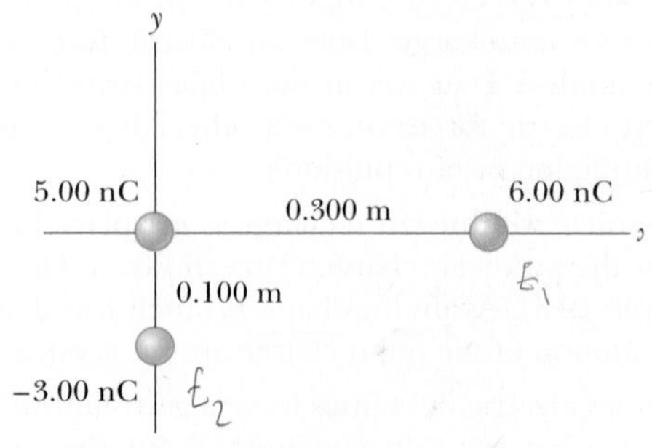

FIGURE P19.13

14. Two 2.00-μC point charges are located on the x axis. One is at $x = 1.00$ m, and the other is at $x = -1.00$ m. (a) Determine the electric field on the y axis at $y = 0.500$ m. (b) Calculate the electric force on a −3.00-μC charge placed on the y axis at $y = 0.500$ m.

15. Four point charges are at the corners of a square of side a as shown in Figure P19.15. (a) Determine the magnitude and direction of the electric field at the location of charge q. (b) What is the resultant force on q?

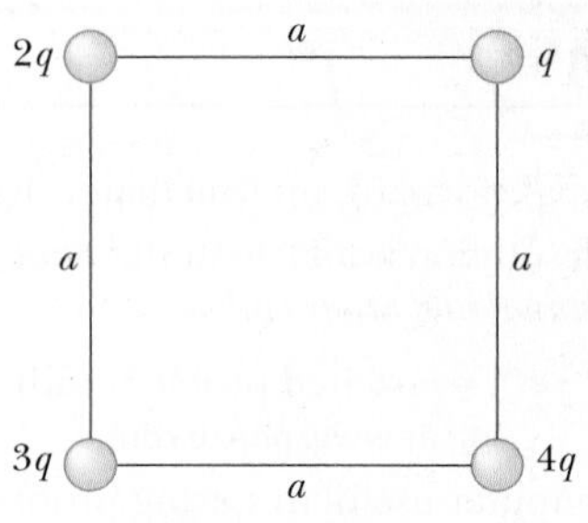

FIGURE P19.15

16. Consider the electric dipole shown in Figure P19.16. Show that the electric field at a *distant* point on the $+x$ axis is $E_x \approx 4k_eqa/x^3$.

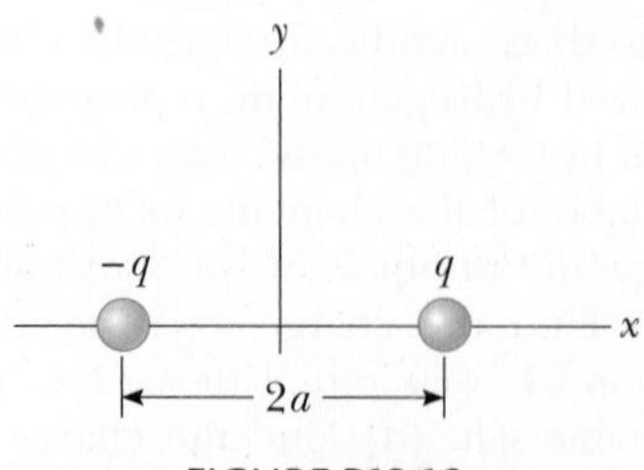

FIGURE P19.16

17. A rod 14.0 cm long is uniformly charged and has a total charge of −22.0 μC. Determine the magnitude and

direction of the electric field along the axis of the rod at a point 36.0 cm from its center.

18. A continuous line of charge lies along the x axis, extending from $x = +x_0$ to positive infinity. The line carries charge with a uniform linear charge density λ_0. What are the magnitude and direction of the electric field at the origin?

19. A uniformly charged ring of radius 10.0 cm has a total charge of 75.0 μC. Find the electric field on the axis of the ring at (a) 1.00 cm, (b) 5.00 cm, (c) 30.0 cm, and (d) 100 cm from the center of the ring.

20. Show that the maximum magnitude E_{max} of the electric field along the axis of a uniformly charged ring occurs at $x = a/\sqrt{2}$ (see Fig. 19.14) and has the value $Q/(6\sqrt{3}\pi\epsilon_0 a^2)$.

21. **PhysicsNow™** A uniformly charged insulating rod of length 14.0 cm is bent into the shape of a semicircle as shown in Figure P19.21. The rod has a total charge of -7.50 μC. Find the magnitude and direction of the electric field at O, the center of the semicircle.

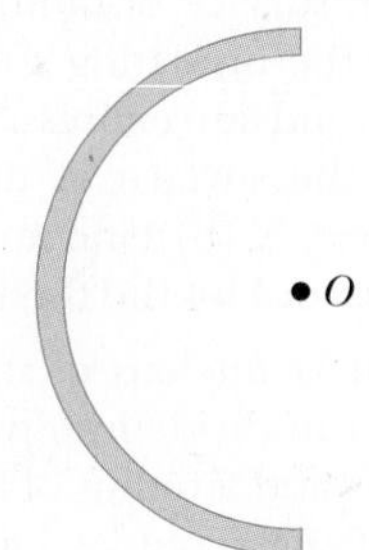

FIGURE P19.21 Problems 19.21 and 20.27.

22. A thin rod of length ℓ and uniform charge per unit length λ lies along the x axis as shown in Figure P19.22. (a) Show that the electric field at P, a distance y from the rod along its perpendicular bisector, has no x component and is given by $E = 2k_e\lambda \sin \theta_0/y$. (b) Using your result to part (a), show that the field of a rod of infinite length is $E = 2k_e\lambda/y$. (*Suggestion:* First, calculate the field at P due to an element of length dx, which has a charge $\lambda\, dx$. Then, change variables from x to θ, using the relationships $x = y \tan \theta$ and $dx = y \sec^2 \theta\, d\theta$, and integrate over θ.)

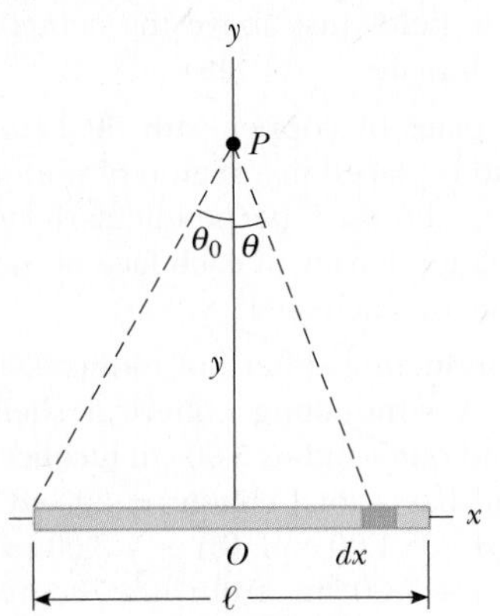

FIGURE P19.22

23. Three solid plastic cylinders all have radius 2.50 cm and length 6.00 cm. One (a) carries charge with uniform density 15.0 nC/m^2 everywhere on its surface. Another (b) carries charge with the same uniform density on its curved lateral surface only. The third (c) carries charge with uniform density 500 nC/m^3 throughout the plastic. Find the charge of each cylinder.

Section 19.6 ■ Electric Field Lines

24. Figure P19.24 shows the electric field lines for two point charges separated by a small distance. (a) Determine the ratio q_1/q_2. (b) What are the signs of q_1 and q_2?

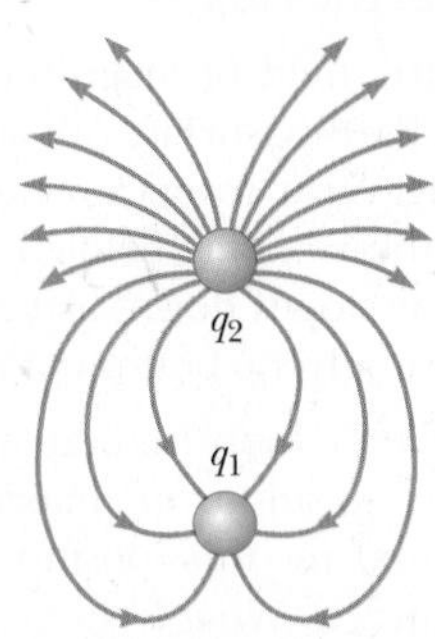

FIGURE P19.24

25. A negatively charged rod of finite length carries charge with a uniform charge per unit length. Sketch the electric field lines in a plane containing the rod.

26. Three equal positive charges q are at the corners of an equilateral triangle of side a as shown in Figure P19.26. (a) Assume that the three charges together create an electric field. Sketch the field lines in the plane of the charges. Find the location of a point (other than ∞) where the electric field is zero. (b) What are the magnitude and direction of the electric field at P due to the two charges at the base?

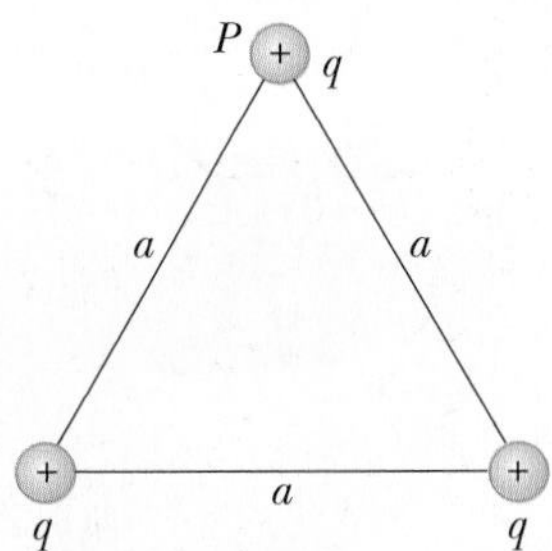

FIGURE P19.26 Problems 19.26 and 20.17.

Section 19.7 ■ Motion of Charged Particles in a Uniform Electric Field

27. A proton accelerates from rest in a uniform electric field of 640 N/C. At some later instant, its speed is 1.20×10^6 m/s (nonrelativistic, because v is much less than the speed of light). (a) Find the acceleration of the proton. (b) After what time interval does the proton reach this speed? (c) How far does the proton move in this time interval? (d) What is its kinetic energy at the end of this time interval?

28. The electrons in a particle beam each have a kinetic energy K. What are the magnitude and direction of the electric field that will stop these electrons in a distance d?

29. A proton moves at 4.50×10^5 m/s in the horizontal direction. It enters a uniform vertical electric field with a magnitude of 9.60×10^3 N/C. Ignoring any gravitational effects, find (a) the time interval required for the proton to travel 5.00 cm horizontally, (b) its vertical displacement during the time interval in which it travels 5.00 cm horizontally, and (c) the horizontal and vertical components of its velocity after it has traveled 5.00 cm horizontally.

Section 19.8 ▪ Electric Flux

30. A vertical electric field of magnitude 2.00×10^4 N/C exists above the Earth's surface on a day when a thunderstorm is brewing. A car covers a rectangle measuring 6.00 m by 3.00 m on the roadway below it, which is built on dry fill. The roadway slopes downward at 10.0°. Determine the electric flux through the bottom of the car.

31. A 40.0-cm-diameter loop is rotated in a uniform electric field until the position of maximum electric flux is found. The flux in this position is measured to be 5.20×10^5 N · m^2/C. What is the magnitude of the electric field?

Section 19.9 ▪ Gauss's Law

32. The electric field everywhere on the surface of a thin spherical shell of radius 0.750 m is measured to be 890 N/C and points radially toward the center of the sphere. (a) What is the net charge within the sphere's surface? (b) What can you conclude about the nature and distribution of the charge inside the spherical shell?

33. PhysicsNow™ A point charge Q is located just above the center of the flat face of a hemisphere of radius R as shown in Figure P19.33. What is the electric flux (a) through the curved surface and (b) through the flat face?

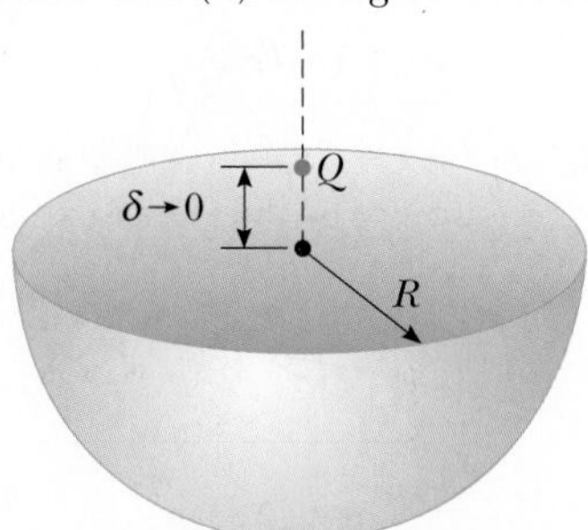

FIGURE **P19.33**

34. A charge of 170 μC is at the center of a cube of edge 80.0 cm. (a) Find the total flux through each face of the cube. (b) Find the flux through the whole surface of the cube. (c) Would your answers to parts (a) or (b) change if the charge were not at the center? Explain.

Section 19.10 ▪ Application of Gauss's Law to Symmetric Charge Distributions

35. A solid sphere of radius 40.0 cm has a total positive charge of 26.0 μC uniformly distributed throughout its volume. Calculate the magnitude of the electric field (a) 0 cm, (b) 10.0 cm, (c) 40.0 cm, and (d) 60.0 cm from the center of the sphere.

36. A 10.0-g piece of Styrofoam carries a net charge of -0.700 μC and floats above the center of a large horizontal sheet of plastic that has a uniform charge density on its surface. What is the charge per unit area on the plastic sheet?

37. A cylindrical shell of radius 7.00 cm and length 240 cm has its charge uniformly distributed on its curved surface. The magnitude of the electric field at a point 19.0 cm radially outward from its axis (measured from the midpoint of the shell) is 36.0 kN/C. Find (a) the net charge on the shell and (b) the electric field at a point 4.00 cm from the axis, measured radially outward from the midpoint of the shell.

38. Consider a thin spherical shell of radius 14.0 cm with a total charge of 32.0 μC distributed uniformly on its surface. Find the electric field (a) 10.0 cm and (b) 20.0 cm from the center of the charge distribution.

39. PhysicsNow™ Consider a long cylindrical charge distribution of radius R with a uniform charge density ρ. Find the electric field at distance r from the axis where $r < R$.

40. An insulating solid sphere of radius a has a uniform volume charge density and carries a total positive charge Q. A spherical gaussian surface of radius r, which shares a common center with the insulating sphere, is inflated starting from $r = 0$. (a) Find an expression for the electric flux passing through the surface of the gaussian sphere as a function of r for $r < a$. (b) Find an expression for the electric flux for $r > a$. (c) Plot the flux versus r.

41. In nuclear fission, a nucleus of uranium-238, which contains 92 protons, can divide into two smaller spheres, each having 46 protons and a radius of 5.90×10^{-15} m. What is the magnitude of the repulsive electric force pushing the two spheres apart?

Section 19.11 ▪ Conductors in Electrostatic Equilibrium

42. A long, straight metal rod has a radius of 5.00 cm and a charge per unit length of 30.0 nC/m. Find the electric field (a) 3.00 cm, (b) 10.0 cm, and (c) 100 cm from the axis of the rod, where distances are measured perpendicular to the rod.

43. A very large, thin, flat plate of aluminum of area A has a total charge Q uniformly distributed over its surfaces. Assuming that the same charge is spread uniformly over the *upper* surface of an otherwise identical glass plate, compare the electric fields just above the center of the upper surface of each plate.

44. A square plate of copper with 50.0-cm sides has no net charge and is placed in a region of uniform electric field of 80.0 kN/C directed perpendicularly to the plate. Find (a) the charge density of each face of the plate and (b) the total charge on each face.

45. A solid conducting sphere of radius 2.00 cm has a charge 8.00 μC. A conducting spherical shell of inner radius 4.00 cm and outer radius 5.00 cm is concentric with the solid sphere and has a total charge -4.00 μC. Find the electric field at (a) $r = 1.00$ cm, (b) $r = 3.00$ cm, (c) $r = 4.50$ cm, and (d) $r = 7.00$ cm from the center of this charge configuration.

46. The electric field on the surface of an irregularly shaped conductor varies from 56.0 kN/C to 28.0 kN/C. Calculate the local surface charge density at the point on the surface where the radius of curvature of the surface is (a) greatest and (b) smallest.

47. A long, straight wire is surrounded by a hollow metal cylinder whose axis coincides with that of the wire. The wire has a charge per unit length of λ, and the cylinder has a net charge per unit length of 2λ. From this information, use Gauss's law to find (a) the charge per unit length on the inner and outer surfaces of the cylinder and (b) the electric field outside the cylinder, a distance r from the axis.

48. Consider an electric field that is uniform in direction throughout a certain volume. Can it be uniform in magnitude? Must it be uniform in magnitude? Answer these questions (a) assuming that the volume is filled with an insulating material carrying charge described by a volume charge density and (b) assuming the volume is empty space. State reasoning to prove your answers.

49. PhysicsNow™ A thin, square conducting plate 50.0 cm on a side lies in the xy plane. A total charge of 4.00×10^{-8} C is placed on the plate. Find (a) the charge density on the plate, (b) the electric field just above the plate, and (c) the electric field just below the plate. You may assume that the charge density is uniform.

Section 19.12 ■ Context Connection—The Atmospheric Electric Field

50. In fair weather, the electric field in the air at a particular location just above the Earth's surface is 120 N/C directed downward. (a) What is the surface charge density on the ground surface? Is it positive or negative? (b) If the weather were fair everywhere and the surface charge density were uniform, what would be the charge of the whole surface of the Earth? How many excess electrons (or protons) would be on the entire surface of the Earth to produce an atmospheric field of 120 N/C down?

51. In the air over a particular region, at an altitude of 500 m above the ground, the electric field is 120 N/C directed downward. At 600 m above the ground, the electric field is 100 N/C downward. What is the average volume charge density in the layer of air between these two elevations? Is it positive or negative?

52. The electric field in the Earth's atmosphere suggests that the solid and liquid surface of the Earth has a charge of about -5×10^5 C. Imagine that the planet as a whole had a charge of -5.00×10^5 C and that the Moon, with 27.3% of the radius of the Earth, had a charge of -1.37×10^5 C. (a) Find the electric force that the Earth would then exert on the Moon. (b) Compare the answer to part (a) with the gravitational force that the Earth exerts on the Moon. As your calculation suggests, for the purpose of accounting for astronomical motions, we may treat the actual forces as purely gravitational. We may say that astronomical objects have negligible total charges.

Additional Problems

53. Two known charges, $-12.0\ \mu$C and $45.0\ \mu$C, and an unknown charge are located on the x axis. The charge $-12.0\ \mu$C is at the origin, and the charge $45.0\ \mu$C is at $x = 15.0$ cm. The unknown charge is to be placed so that each charge is in equilibrium under the action of the electric forces exerted by the other two charges. Is this situation possible? Is it possible in more than one way? Find the required location, magnitude, and sign of the unknown charge.

54. A small, 2.00-g plastic ball is suspended by a 20.0-cm-long string in a uniform electric field as shown in Figure P19.54. If the ball is in equilibrium when the string makes a 15.0° angle with the vertical, what is the net charge on the ball?

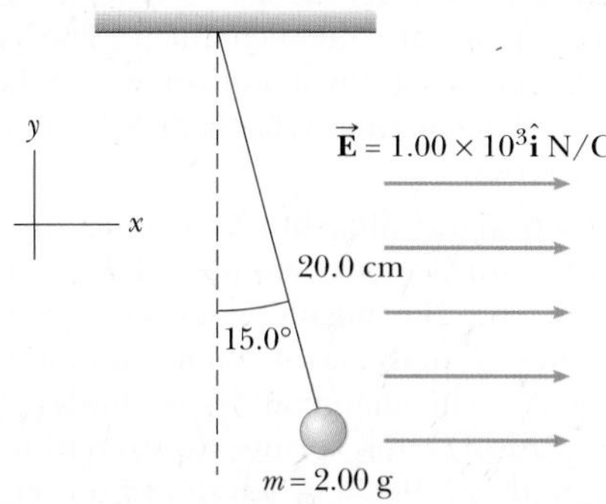

FIGURE P19.54

55. Four identical point charges ($q = +10.0\ \mu$C) are located on the corners of a rectangle as shown in Figure P19.55. The dimensions of the rectangle are $L = 60.0$ cm and $W = 15.0$ cm. Calculate the magnitude and direction of the resultant electric force exerted on the charge at the lower left corner by the other three charges.

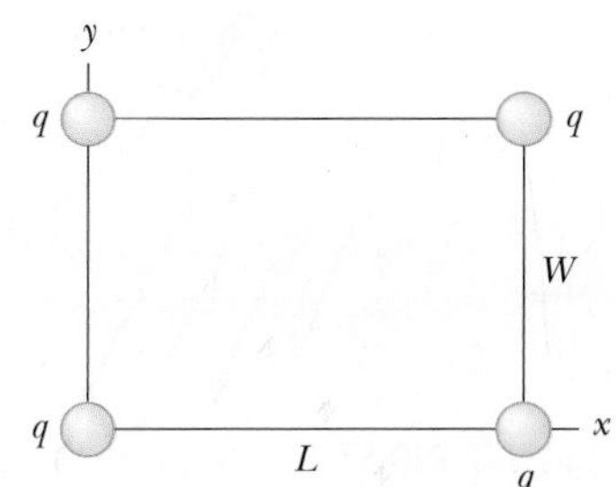

FIGURE P19.55 Problems 19.55 and 20.10.

56. Inez is putting up decorations for her sister's quinceañera (fifteenth birthday party). She ties three light silk ribbons together to the top of a gateway and hangs a rubber balloon from each ribbon (Fig. P19.56). To include the effects of

FIGURE P19.56

the gravitational and buoyant forces on it, each balloon can be modeled as a particle of mass 2.00 g, with its center 50.0 cm from the point of support. Inez wishes to show off the colors of the balloons. She rubs the whole surface of each balloon with her woolen scarf to make them hang separately with gaps between them. The centers of the hanging balloons form a horizontal equilateral triangle with sides 30.0 cm long. What is the common charge each balloon carries?

57. Two identical metallic blocks resting on a frictionless horizontal surface are connected by a light metallic spring having the spring constant 100 N/m and an unstretched length of 0.300 m as shown in Figure P19.57a. A total charge of Q is slowly placed on the system, causing the spring to stretch to an equilibrium length of 0.400 m as shown in Figure P19.57b. Determine the value of Q, assuming that all the charge resides on the blocks and modeling the blocks as point charges.

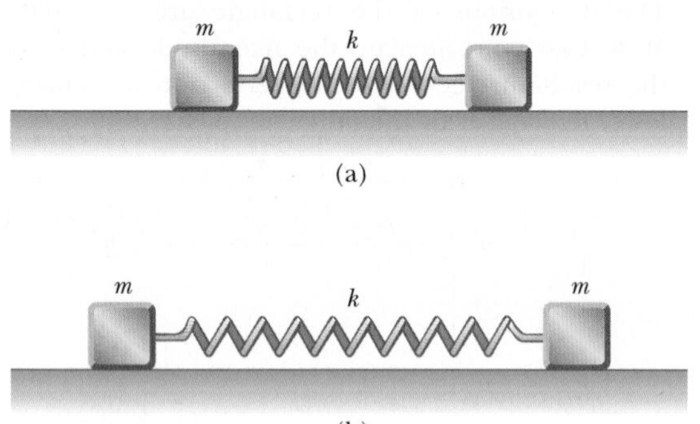

FIGURE **P19.57** Problems 19.57 and 19.58.

58. Two identical metallic blocks resting on a frictionless horizontal surface are connected by a light metallic spring having a spring constant k and an unstretched length L_i as shown in Figure P19.57a. A total charge Q is slowly placed on the system, causing the spring to stretch to an equilibrium length L as shown in Figure P19.57b. Determine the value of Q, assuming that all the charge resides on the blocks and modeling the blocks as point charges.

59. Two small spheres of mass m are suspended from strings of length ℓ that are connected at a common point. One sphere has charge Q, and the other has charge $2Q$. The strings make angles θ_1 and θ_2 with the vertical. (a) How are θ_1 and θ_2 related? (b) Assume that θ_1 and θ_2 are small. Show that the distance r between the spheres is given by

$$r \approx \left(\frac{4k_eQ^2\ell}{mg}\right)^{1/3}$$

60. Three charges of equal magnitude q are fixed in position at the vertices of an equilateral triangle (Fig. P19.60). A fourth charge Q is free to move along the positive x axis under the influence of the forces exerted by the three fixed charges. Find a value for s for which Q is in equilibrium. You will need to solve a transcendental equation.

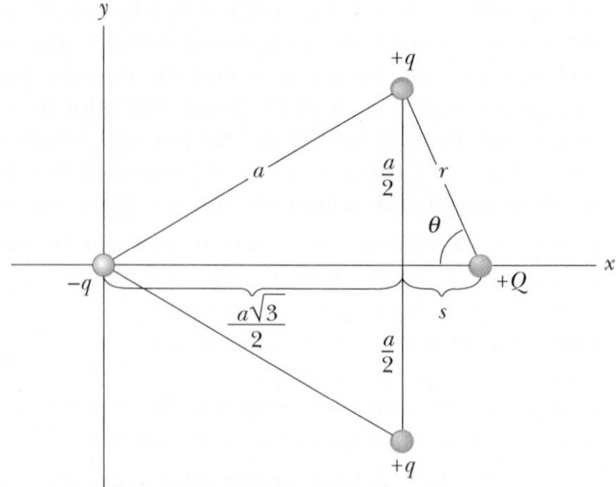

FIGURE **P19.60**

61. Consider the charge distribution shown in Figure P19.61. (a) Show that the magnitude of the electric field at the center of any face of the cube has a value of $2.18k_eq/s^2$. (b) What is the direction of the electric field at the center of the top face of the cube?

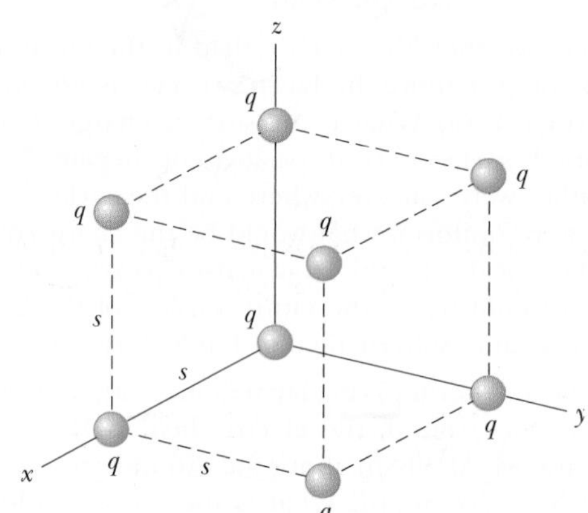

FIGURE **P19.61**

62. Two infinite, nonconducting sheets of charge are parallel to each other as shown in Figure P19.62. The sheet on the left has a uniform surface charge density σ, and the one on the right has a uniform charge density $-\sigma$. Calculate the electric field at points (a) to the left of, (b) in between, and (c) to the right of the two sheets.

FIGURE **P19.62**

63. Repeat the calculations for Problem 19.62 when both sheets have *positive* uniform surface charge densities of value σ.

64. A line of charge with uniform density 35.0 nC/m lies along the line $y = -15.0$ cm, between the points with coordinates $x = 0$ and $x = 40.0$ cm. Find the electric field it creates at the origin.

65. **PhysicsNow™** A solid, insulating sphere of radius a has a uniform charge density ρ and a total charge Q. Concentric with this sphere is an uncharged, conducting hollow sphere whose inner and outer radii are b and c as shown in Figure P19.65. (a) Find the magnitude of the electric field in the regions $r < a$, $a < r < b$, $b < r < c$, and $r > c$. (b) Determine the induced charge per unit area on the inner and outer surfaces of the hollow sphere.

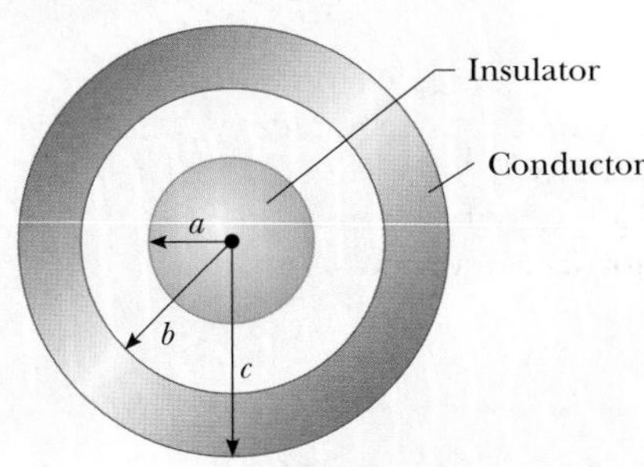

FIGURE **P19.65**

66. **Review problem.** A negatively charged particle $-q$ is placed at the center of a uniformly charged ring, where the ring has a total positive charge Q as shown in Example 19.5. The particle, confined to move along the x axis, is displaced a small distance x along the axis (where $x \ll a$) and released. Show that the particle oscillates in simple harmonic motion with a frequency given by

$$f = \frac{1}{2\pi}\left(\frac{k_e q Q}{m a^3}\right)^{1/2}$$

67. A sphere of radius $2a$ is made of a nonconducting material that has a uniform volume charge density ρ. (Assume that the material does not affect the electric field.) A spherical cavity of radius a is now removed from the sphere as shown in Figure P19.67. Show that the electric field within the cavity is uniform and is given by $E_x = 0$ and $E_y = \rho a/3\epsilon_0$. (*Suggestion:* The field within the cavity is the superposition of the field due to the original uncut sphere plus the field due to a sphere the size of the cavity with a uniform negative charge density $-\rho$.)

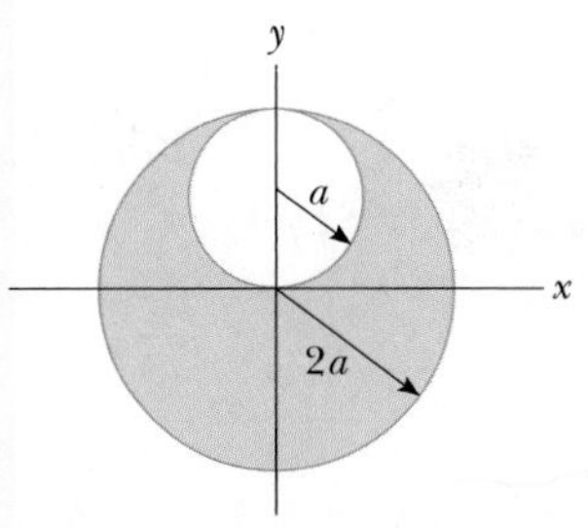

FIGURE **P19.67**

ANSWERS TO QUICK QUIZZES

19.1 (a), (c), and (e). The experiment shows that objects A and B have charges of the same sign, as do objects B and C. Therefore, all three objects have charges of the same sign. We cannot determine from this information, however, whether the charges are positive or negative.

19.2 (e). In the first experiment, objects A and B may have charges with opposite signs or one of the objects may be neutral. The second experiment shows that objects B and C have charges with opposite signs, so object B must be charged. We still do not know if object A is charged or neutral, however.

19.3 (i), (e). From Newton's third law, the electric force exerted by object B on object A is equal in magnitude to the force exerted by object A on object B. (ii), (b). From Newton's third law, the electric force exerted by object B on object A is equal in magnitude to the force exerted by object A on object B and in the opposite direction.

19.4 (a). There is no effect on the electric field if we assume that the source charge producing the field is not disturbed by our actions. Remember that the electric field is created by source charge(s) (unseen in this case), not the test charge(s).

19.5 *A*, *B*, and *C*. The field is greatest at point *A* because that is where the field lines are closest together. The absence of lines near point *C* indicates that the electric field there is zero.

19.6 (b) and (d). Statement (a) is not necessarily true because an equal number of positive and negative charges could be present inside the surface. Statement (c) is not necessarily true as can be seen from Figure 19.30 because a nonzero electric field exists everywhere on the surface, but the charge is not enclosed within the surface. Thus, the net flux is zero.

19.7 (i), (c). The charges q_1 and q_4 are outside the surface and contribute zero net flux through S'. (ii), (d). We don't need the surfaces to realize that any given point in space will experience an electric field due to all local source charges.

CHAPTER 20

Electric Potential and Capacitance

This device is a *variable capacitor,* used to tune radios to a selected station. When one set of metal plates is rotated so as to lie between a fixed set of plates, the *capacitance* of the device changes. Capacitance is a parameter that depends on *electric potential,* the primary topic of this chapter.

(George Semple)

CHAPTER OUTLINE

The concept of potential energy was introduced in Chapter 7 in connection with such conservative forces as gravity and the elastic force of a spring. By using the principle of conservation of mechanical energy in an isolated system, we are often able to avoid working directly with forces when solving mechanical problems. In this chapter, we shall use the energy concept in our study of electricity. Because the electrostatic force (given by Coulomb's law) is conservative, electrostatic phenomena can conveniently be described in terms of an *electric* potential energy function. This concept enables us to define a quantity called *electric potential,* which is a scalar quantity and which therefore leads to a simpler means of describing some electrostatic phenomena than the electric field method. As we shall see in subsequent chapters, the concept of electric potential is of great practical value in many applications.

This chapter also addresses the properties of capacitors, devices that store charge. The ability of a capacitor to store charge is measured by its *capacitance.* Capacitors are used in common applications such as frequency tuners in radio receivers, filters in power supplies, dampers to eliminate unwanted sparking in automobile ignition systems, and energy-storing devices in electronic flash units.

20.1 POTENTIAL DIFFERENCE AND ELECTRIC POTENTIAL

When a point charge q_0 is placed in an electric field $\vec{\mathbf{E}}$, the electric force on the particle is $q_0\vec{\mathbf{E}}$ (Eq. 19.4). This force is the vector sum of the individual forces exerted on q_0 by the various source charges producing the field $\vec{\mathbf{E}}$. It follows that the force $q_0\vec{\mathbf{E}}$ is conservative because the individual forces governed by Coulomb's law are conservative. (See Section 7.3 for a review of conservative forces.) Let us consider a system consisting of the point charge and all the source charges creating the electric field. Because the field represents the effect of the source charges, we can also consider the system to be the electric field and the charge q_0 that we place in the field, without referring specifically to the source charges. When the point charge moves in response to the electric force within the electric field, work is done on the particle by the field. For an infinitesimal displacement $d\vec{\mathbf{s}}$ of a point charge q_0, the work done by the electric field on the charge is $\vec{\mathbf{F}}_e \cdot d\vec{\mathbf{s}} = q_0\vec{\mathbf{E}} \cdot d\vec{\mathbf{s}}$.

The work done by the field on a point charge is similar to the work done by a gravitational field on a falling object. We found in Chapter 7 that the gravitational potential energy of an isolated object–field system changes by an amount equal to the negative of the work done within the system by the field on the object (Eq. 7.3). Similarly, the work done within the system by the electric field on a charged particle changes the potential energy of the isolated charge–field system by an amount $dU = -dW = -q_0\vec{\mathbf{E}} \cdot d\vec{\mathbf{s}}$. For a finite displacement of a test particle of charge q_0 between points A and B, the **change in potential energy** of the charge–field system is

$$\Delta U = U_B - U_A = -q_0 \int_A^B \vec{\mathbf{E}} \cdot d\vec{\mathbf{s}} \qquad [20.1]$$

■ Change in potential energy for a charge–field system

The integral in Equation 20.1 is performed over the path along which the particle moves from A to B. It is called either a **path integral** or a **line integral.** Because the force $q_0\vec{\mathbf{E}}$ is conservative, **this integral does not depend on the path taken between *A* and *B*.**

In Chapter 19, we recognized that the force between a test charge and a distribution of source charges depends on all the charges, whereas the electric field is defined as a quantity established only by the source charges. We do something similar in this discussion. The potential energy of the system of a test charge q_0 in an electric field $\vec{\mathbf{E}}$ depends on the test charge and all the source charges establishing the electric field. Let us remove the effect of the test charge by dividing the potential energy of the system by the test charge. The potential energy U of the system per unit charge q_0 is independent of the value of q_0 and has a unique value at every point in an electric field. The quantity U/q_0 is called the **electric potential** V (or simply the **potential**):

PITFALL PREVENTION 20.1

POTENTIAL AND POTENTIAL ENERGY *The potential is characteristic of the field only,* independent of a charged test particle that may be placed in the field. *Potential energy is characteristic of the charge–field system* due to an interaction between the field and a charged particle placed in the field.

$$V \equiv \frac{U}{q_0} \qquad [20.2]$$

■ Definition of electric potential

Because potential energy is a scalar, electric potential is also a scalar quantity. Note that potential is not a property of the charge–field system because we have divided the potential energy of the system by the charge. It is a property only of the field. Therefore, in the physical situation, we can imagine removing the test charge from the field. The potential still exists at the point the test charge occupied and is due to the source charges that establish the electric field.

The **potential difference** $\Delta V = V_B - V_A$ between the points A and B is defined as the change in potential energy of the charge–field system when the test particle is moved between the points divided by the charge q_0 on the test particle:

■ Potential difference between two points in an electric field

$$\Delta V = \frac{\Delta U}{q_0} = -\int_A^B \vec{\mathbf{E}} \cdot d\vec{\mathbf{s}} \qquad [20.3]$$

Potential difference should not be confused with potential energy difference. The potential difference between two points in an electric field is *proportional* to the potential energy difference of the charge–field system when the charge is at the two points, and we see from Equation 20.3 that the two are related by $\Delta U = q_0 \Delta V$.

Equation 20.3 defines potential difference only. The potential is often taken to be zero at some convenient point, sometimes called a *ground.* We usually set the potential due to one or more source charges at zero for a point at infinity (i.e., a point infinitely remote from the source charges producing the electric field). With this choice, we can say that **the electric potential at an arbitrary point due to source charges equals the work required by an external agent to bring a test particle from infinity to that point divided by the charge on the test particle.** Therefore, if we take $V_A = 0$ at infinity in Equation 20.3, the potential at any point P is

$$V_P = -\int_\infty^P \vec{\mathbf{E}} \cdot d\vec{\mathbf{s}} \qquad [20.4]$$

where $\vec{\mathbf{E}}$ is the electric field established by the source charges. In reality, V_P represents the potential difference between the point P and a point at infinity. (Note that Eq. 20.4 is a special case of Eq. 20.3.) When discussing potentials in an electric circuit, we shall set $V = 0$ at some selected point in the circuit.

Because potential is a measure of energy per unit charge, the SI units of potential are joules per coulomb, called the **volt** (V):

$$1 \text{ V} \equiv 1 \text{ J/C}$$

That is, if we release a particle with a charge of 1 C in an electric field and it moves from a point of high potential to a point of low potential through a potential difference of -1 V, it will have 1 J of work done on it by the field and therefore will attain a kinetic energy of 1 J. (From the continuity equation for energy, Eq. 6.20, for the system of the particle, $W = \Delta K$.) Alternatively, 1 J of work must be done by an external agent to take a particle with a charge of 1 C through a potential difference of $+1$ V at constant velocity. (From the continuity equation for energy, for the particle–field system, $W = \Delta U$.) Equation 20.3 shows that the potential difference also has the same units as the product of electric field and displacement. It therefore follows that the SI units of electric field, newtons per coulomb, can be expressed as volts per meter:

$$1 \text{ N/C} = 1 \text{ V/m}$$

which suggests that the electric field can be interpreted as the rate of change in space of the electric potential. A strong electric field corresponds to a potential that changes rapidly in space, whereas a weak field represents a slowly changing potential.

PITFALL PREVENTION 20.2

VOLTAGE In practice, a variety of phrases are used to describe the potential difference between two points, the most common being **voltage,** arising from the unit for potential. A voltage *applied* to a device, such as a television, or *across* a device has the same meaning as the potential difference across the device. For example, if we say that the voltage applied to a lightbulb is 120 V, we mean that the potential difference between the two electrical contacts on the lightbulb is 120 V.

As we learned in Section 9.7, a unit of energy commonly used in physics is the **electron volt** (eV):

■ The electron volt

$$1 \text{ eV} = (1e)(1 \text{ V}) = (1.60 \times 10^{-19} \text{ C})(1 \text{ J/C}) = 1.60 \times 10^{-19} \text{ J} \qquad [20.5]$$

One eV is the kinetic energy gained by a particle with charge e being accelerated by an electric field through a potential difference of magnitude 1 V. Equation 20.5 can be used to convert any energy in joules to electron volts. For instance, an electron in the beam of a typical TV picture tube may have a speed of 3.0×10^7 m/s. This speed corresponds to a kinetic energy of 4.1×10^{-16} J, which is equivalent to 2.6×10^3 eV. Such an electron has to be accelerated from rest through a potential difference of 2.6 kV to reach this speed.

20.2 POTENTIAL DIFFERENCES IN A UNIFORM ELECTRIC FIELD

In this section, we describe the potential difference between any two points in a *uniform* electric field. Consider a uniform electric field directed along the negative y axis as in Figure 20.1a. Let us calculate the potential difference between two points A and B, separated by a distance d, where d is measured parallel to the field lines. If we apply Equation 20.3 to this situation, we have

$$V_B - V_A = \Delta V = -\int_A^B \vec{\mathbf{E}} \cdot d\vec{\mathbf{s}} = -\int_A^B E \cos 0° \, ds = -\int_A^B E \, ds$$

Because the field is uniform, the magnitude E of the field is a constant and can be removed from the integral, giving

$$\Delta V = -E \int_A^B ds = -Ed \qquad [20.6]$$

■ Potential difference between two points in a uniform electric field

The negative sign results because point B is at a lower potential than point A; that is, $V_B < V_A$. In general, electric field lines always point in the direction of decreasing electric potential.

Now suppose a test particle with charge q_0 moves from A to B. The change in the electric potential energy of the charge–field system can be found from Equations 20.3 and 20.6:

$$\Delta U = q_0 \, \Delta V = -q_0 Ed \qquad [20.7]$$

From this result, we see that if q_0 is positive, ΔU is negative. Thus, **when a positive charge moves in the direction of the electric field, the electric potential energy of the charge–field system decreases.** This situation is analogous to the change in gravitational potential energy $-mgd$ of an object–field system when an object with mass m falls through a height d in a uniform gravitational field, as suggested in Figure 20.1b. If a particle with a positive charge q_0 is released from rest in the electric field, it experiences an electric force $q_0\vec{\mathbf{E}}$ in the direction of $\vec{\mathbf{E}}$ (downward in Fig. 20.1a). Therefore, it accelerates downward, gaining kinetic energy. **As the charged particle gains kinetic energy, the charge–field system loses an equal amount of potential energy.** This familiar result is similar to what we have seen for gravitational situations (Fig. 20.1b). The statement is simply the principle of conservation of mechanical energy in the isolated system model for electric fields.

If q_0 is negative, ΔU in Equation 20.7 is positive and the situation is reversed. If a negatively charged particle is released from rest in the field $\vec{\mathbf{E}}$, it accelerates in a direction opposite the electric field. **The charge–field system loses electric potential**

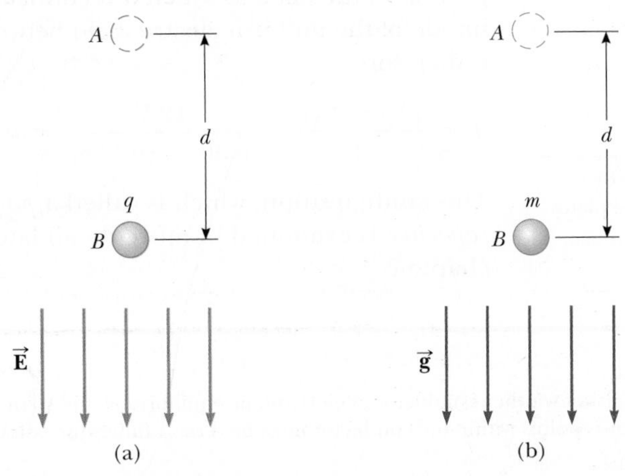

FIGURE 20.1 (a) When the electric field $\vec{\mathbf{E}}$ is directed downward, point B is at a lower electric potential than point A. When a positive test charge moves from A to B, the charge–field system loses electric potential energy. (b) A gravitational analogy: When an object with mass m moves downward in the direction of the gravitational field $\vec{\mathbf{g}}$, the object–field system loses gravitational potential energy.

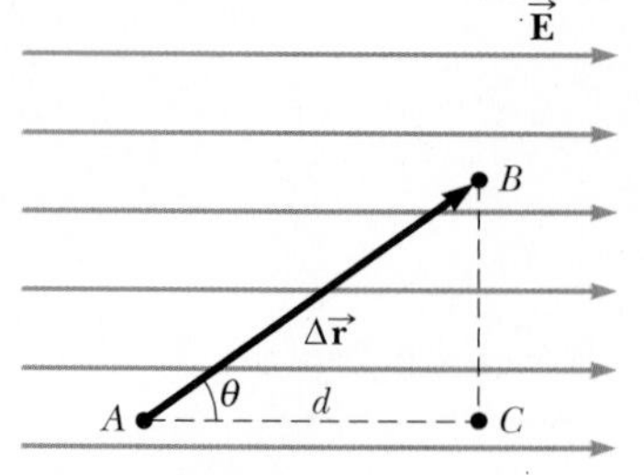

FIGURE 20.2 A particle is moved in a uniform electric field. Point B is at a lower potential than point A. Points B and C are at the *same* potential.

energy when a negative charge moves in the direction opposite to the electric field. We have no analog for this situation in the gravitational case because no negative mass has been observed.

Now consider the more general case of a charged particle moving between any two points in a uniform electric field as in Figure 20.2. If $\Delta\vec{\mathbf{r}}$ represents the displacement vector between points A and B, Equation 20.3 gives

$$\Delta V = -\int_A^B \vec{\mathbf{E}} \cdot d\vec{\mathbf{s}} = -\vec{\mathbf{E}} \cdot \int_A^B d\vec{\mathbf{s}} = -\vec{\mathbf{E}} \cdot \Delta\vec{\mathbf{r}} \qquad [20.8]$$

where again we are able to remove $\vec{\mathbf{E}}$ from the integral because the electric field is uniform. Furthermore, the change in electric potential energy of the charge–field system is

$$\Delta U = q_0\,\Delta V = -q_0\,\vec{\mathbf{E}} \cdot \Delta\vec{\mathbf{r}} \qquad [20.9]$$

Finally, our results show that all points in a plane *perpendicular* to a uniform electric field are at the same potential as can be seen in Figure 20.2, where the potential difference $V_B - V_A = -\vec{\mathbf{E}} \cdot \Delta\vec{\mathbf{r}} = -E\,\Delta r\cos\theta = -Ed = V_C - V_A$. Therefore, $V_B = V_C$. The name **equipotential surface** is given to any surface consisting of a continuous distribution of points having the same electric potential. Note that because $\Delta U = q_0\,\Delta V$, no work is required to move a test particle between any two points on an equipotential surface. The equipotential surfaces of a uniform electric field consist of a family of planes, all perpendicular to the field. Equipotential surfaces for fields with other symmetries will be described in later sections.

EXAMPLE 20.1 The Electric Field Between Two Parallel Plates of Opposite Charge

A 12-V battery is connected between two parallel plates as in Figure 20.3. The separation between the plates is 0.30 cm, and the electric field is assumed to be uniform. (This simplification model is reasonable if the plate separation is small relative to the plate size and if we do not consider points near the edges of the plates.) Find the magnitude of the electric field between the plates.

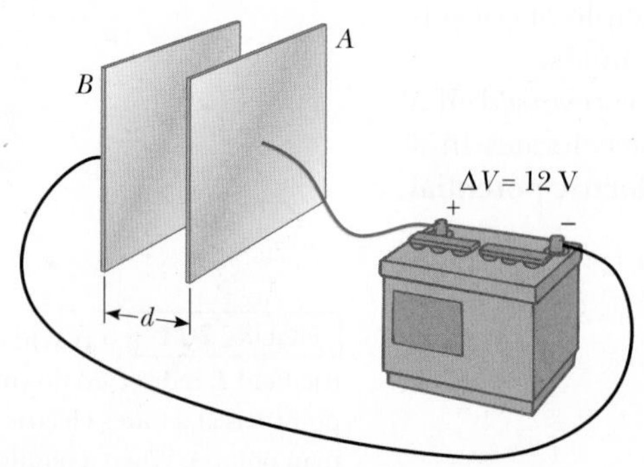

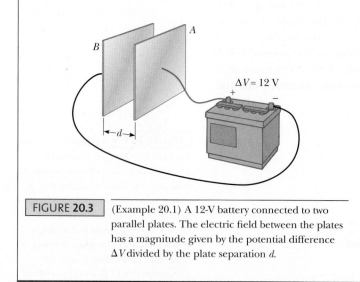

FIGURE 20.3 (Example 20.1) A 12-V battery connected to two parallel plates. The electric field between the plates has a magnitude given by the potential difference ΔV divided by the plate separation d.

Solution The electric field is directed from the positive plate A toward the negative plate B. The positive plate is at a higher potential than the negative plate. Note that the potential difference between plates must equal the potential difference between the battery terminals. This requirement can be understood by recognizing that all points on a conductor in equilibrium are at the same potential.[1] Hence, no potential difference occurs between a terminal of the battery and any portion of the plate to which it is connected. The magnitude of the uniform electric field between the plates is therefore

$$E = \frac{|V_B - V_A|}{d} = \frac{12\text{ V}}{0.30 \times 10^{-2}\text{ m}} = 4.0 \times 10^3\text{ V/m}$$

This configuration, which is called a *parallel-plate capacitor,* is examined in more detail later in this chapter.

[1]The electric field vanishes within a conductor in electrostatic equilibrium, and so the path integral $\int \vec{\mathbf{E}} \cdot d\vec{\mathbf{s}}$ between any two points within the conductor must be zero. A fuller discussion of this point is given in Section 20.6.

INTERACTIVE EXAMPLE 20.2 **Motion of a Proton in a Uniform Electric Field**

A proton is released from rest in a uniform electric field of magnitude 8.0×10^4 V/m directed along the positive x axis (Fig. 20.4). The proton undergoes a displacement of magnitude $d = 0.50$ m in the direction of $\vec{\mathbf{E}}$.

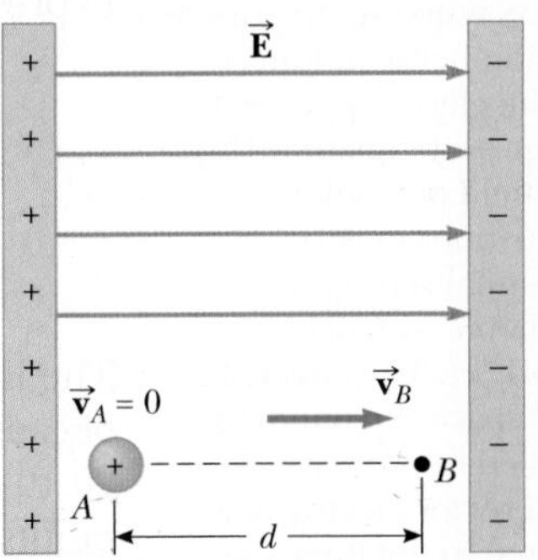

FIGURE 20.4 (Interactive Example 20.2) A proton accelerates from A to B in the direction of the electric field.

A Find the difference in the electric potential between the points A and B.

Solution The difference in electric potential does not depend on the presence of the proton. From Equation 20.6, we have

$$\Delta V = -Ed = -(8.0 \times 10^4 \text{ V/m})(0.50 \text{ m})$$
$$= -4.0 \times 10^4 \text{ V}$$

This negative result tells us that the electric potential decreases between points A and B.

B Find the change in potential energy of the charge–field system for this displacement.

Solution From Equation 20.3, we have

$$\Delta U = q\Delta V = e\Delta V$$
$$= (1.6 \times 10^{-19} \text{ C})(-4.0 \times 10^4 \text{ V}) = -6.4 \times 10^{-15} \text{ J}$$

The negative sign here means that the potential energy of the system decreases as the proton moves in the direction of the electric field. This decrease is consistent with conservation of energy in an isolated system: as the proton accelerates in the direction of the field, it gains kinetic energy and at the same time the system loses electric potential energy. The increase in kinetic energy of a charged particle in an electric field is exploited in many devices, including electron guns for TV picture tubes and particle accelerators for research in particle physics.

PhysicsNow™ Log into PhysicsNow at **www.pop4e.com** and go to Interactive Example 20.2 to predict and observe the speed of the proton as it arrives at the negative plate for random values of the electric field.

20.3 ELECTRIC POTENTIAL AND ELECTRIC POTENTIAL ENERGY DUE TO POINT CHARGES

In establishing the concept of electric potential, we imagined placing a test particle in an electric field set up by some undescribed source charges. As a simplification model, the field was assumed to be uniform in Section 20.2 so as to firmly plant the idea of electric potential in our minds. Let us now focus our attention on point charges, which we know set up electric fields that are not uniform.

Consider an isolated positive point charge q (Fig. 20.5). Recall that such a charge is a source of an electric field that is directed radially outward from the charge. To find the electric potential at a distance of r from the charge, we begin with the general expression for potential difference, Equation 20.3:

$$V_B - V_A = -\int_A^B \vec{\mathbf{E}} \cdot d\vec{\mathbf{s}}$$

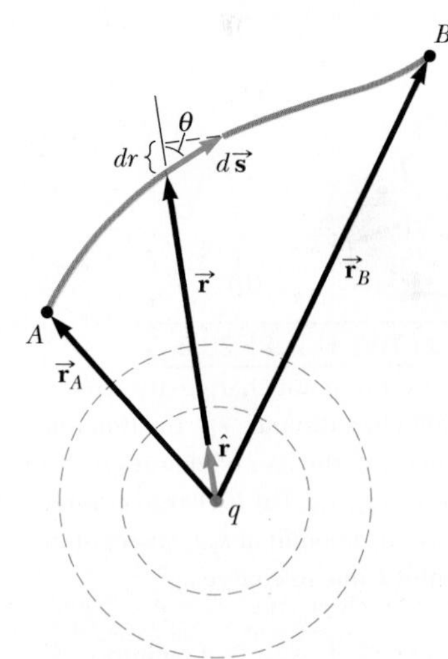

FIGURE 20.5 The potential difference between points A and B due to a point charge q depends *only* on the initial and final radial coordinates r_A and r_B. The two dashed circles represent cross-sections of spherical equipotential surfaces.

Because the electric field due to the point charge is given by $\vec{\mathbf{E}} = k_e q\hat{\mathbf{r}}/r^2$ (Eq. 19.5), where $\hat{\mathbf{r}}$ is a unit vector directed from the charge toward the field point, the quantity $\vec{\mathbf{E}} \cdot d\vec{\mathbf{s}}$ can be expressed as

$$\vec{\mathbf{E}} \cdot d\vec{\mathbf{s}} = k_e \frac{q}{r^2} \hat{\mathbf{r}} \cdot d\vec{\mathbf{s}}$$

The dot product $\hat{\mathbf{r}} \cdot d\vec{\mathbf{s}} = ds \cos\theta$, where θ is the angle between $\hat{\mathbf{r}}$ and $d\vec{\mathbf{s}}$ as in Figure 20.5. Furthermore, note that $ds \cos\theta$ is the projection of $d\vec{\mathbf{s}}$ onto $\vec{\mathbf{r}}$, so

PITFALL PREVENTION 20.3

SIMILAR EQUATION WARNING Be sure to avoid confusion between Equation 20.11 for the electric potential of a point charge and Equation 19.5 for the electric field of a point charge. The equations look very similar, but potential is proportional to $1/r$, whereas the field is proportional to $1/r^2$. The effect of a charge on the space surrounding it can be described in two ways. The charge sets up a vector electric field $\vec{\mathbf{E}}$, which is related to the force experienced by a test charge placed in the field. It also sets up a scalar potential V, which is related to the potential energy of the two-charge system when a test charge is placed in the field.

$ds \cos\theta = dr$. With these substitutions, we find that $\vec{\mathbf{E}} \cdot d\vec{\mathbf{s}} = (k_e q/r^2)\, dr$, so the expression for the potential difference becomes

$$V_B - V_A = -\int_{r_A}^{r_B} k_e \frac{q}{r^2}\, dr = -k_e q \int_{r_A}^{r_B} \frac{dr}{r^2} = \left.\frac{k_e q}{r}\right|_{r_A}^{r_B}$$

$$= k_e q \left[\frac{1}{r_B} - \frac{1}{r_A}\right] \qquad [20.10]$$

The line integral of $\vec{\mathbf{E}} \cdot d\vec{\mathbf{s}}$ is *independent* of the path between A and B, as it must be, because the electric field of a point charge is conservative.[2] Furthermore, Equation 20.10 expresses the important result that the potential difference between any two points A and B depends *only* on the *radial* coordinates r_A and r_B. As we learned in Section 20.1, it is customary to choose the reference of potential to be zero at $r_A = \infty$. With this choice, the electric potential due to a point charge at any distance r from the charge is

$$V = k_e \frac{q}{r} \qquad [20.11]$$

From this expression we see that V is constant on a spherical surface of radius r centered on the point charge. Hence, we conclude that **the equipotential surfaces for an isolated point charge consist of a family of spheres concentric with the charge** as shown in Figure 20.5. Note that the equipotential surfaces are perpendicular to the electric field lines, as is the case for a uniform electric field.

The electric potential at a point in space due to two or more point charges is obtained by applying the superposition principle. That is, the total potential at some point P due to multiple point charges is the sum of the potentials at P due to the individual charges. For a group of charges, we can write the total potential at P in the form

$$V = k_e \sum_i \frac{q_i}{r_i} \qquad [20.12]$$

where the potential is again taken to be zero at infinity and r_i is the distance from point P to the charge q_i. Note that the sum in Equation 20.12 is an *algebraic* sum of scalars rather than a *vector* sum (which is used to calculate the electric field of a group of charges, as in Eq. 19.6). Therefore, it is much easier to evaluate V for multiple charges than to evaluate $\vec{\mathbf{E}}$.

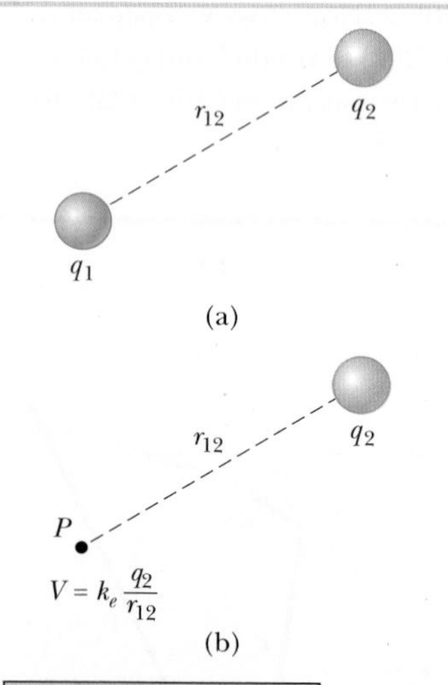

ACTIVE FIGURE **20.6**

(a) If two point charges are separated by a distance r_{12}, the potential energy of the pair of charges is given by $k_e q_1 q_2/r_{12}$. (b) If charge q_1 is removed, a potential $k_e q_2/r_{12}$ exists at point P due to charge q_2.

Physics Now™ Log into PhysicsNow at **www.pop4e.com** and go to Active Figure 20.6. You can move charge q_1 or point P and see the result on the electric potential energy of the system for part (a) and the electric potential due to charge q_2 for part (b).

We now consider the potential energy of a system of two charged particles. If V_2 is the electric potential at a point P due to charge q_2, the work an external agent must do to bring a second charge q_1 from infinity to P without acceleration is $q_1 V_2$. This work represents a transfer of energy into the system, and the energy appears in the system as potential energy U when the particles are separated by a distance r_{12} (Active Fig. 20.6a). We can therefore express the **electric potential energy of a pair of point charges** as

$$U = q_1 V_2 = k_e \frac{q_1 q_2}{r_{12}} \qquad [20.13]$$

Note that if the charges are of the same sign, U is positive, which is consistent because positive work must be done by an external agent on the system to bring the two charges near one another (because charges of the same sign repel). If the charges are of opposite sign, U is negative. Therefore, negative work is done by an external agent against the attractive force between the charges of opposite sign as they are brought near each other because a force must be applied opposite to the displacement to prevent q_1 from accelerating toward q_2.

[2]A conservative field is one that exerts a conservative force on an object placed within it. Both gravitational and electric fields are conservative.

In Active Figure 20.6b, we have removed the charge q_1. At the position that this charge previously occupied, point P, we can use Equations 20.2 and 20.13 to define a potential due to charge q_2 as $V = U/q_1 = k_e q_2/r_{12}$.

If the system consists of more than two charged particles, the total electric potential energy can be obtained by calculating U for every pair of charges and summing the terms algebraically. The total electric potential energy of a system of point charges is equal to the work required to bring the charges, one at a time, from an infinite separation to their final positions.

QUICK QUIZ 20.1 A spherical balloon contains a positively charged object at its center. **(i)** As the balloon is inflated to a greater volume while the charged object remains at the center, does the electric potential at the surface of the balloon **(a)** increase, **(b)** decrease, or **(c)** remain the same? **(ii)** Does the electric flux through the surface of the balloon **(a)** increase, **(b)** decrease, or **(c)** remain the same?

QUICK QUIZ 20.2 In Active Figure 20.6a, take q_1 to be a negative source charge and q_2 to be the test charge. **(i)** If q_2 is initially positive and is replaced with a charge of the same magnitude but negative, does the potential at the position of q_2 due to q_1 **(a)** increase, **(b)** decrease, or **(c)** remain the same? **(ii)** When q_2 is changed from positive to negative, does the potential energy of the two-charge system **(a)** increase, **(b)** decrease, or **(c)** remain the same?

INTERACTIVE EXAMPLE 20.3 The Electric Potential Due to Two Point Charges

A 2.00-μC point charge is located at the origin, and a second point charge of $-6.00\ \mu$C is located on the y axis at the position (0, 3.00) m as in Figure 20.7a.

A Find the total electric potential due to these charges at point P, whose coordinates are (4.00, 0) m.

Solution For two point charges, the sum in Equation 20.12 gives

$$V_P = k_e\left(\frac{q_1}{r_1} + \frac{q_2}{r_2}\right)$$

In this example, $q_1 = 2.00\ \mu$C, $r_1 = 4.00$ m, $q_2 = -6.00\ \mu$C, and $r_2 = 5.00$ m. Therefore, V_P reduces to

$$V_P = (8.99 \times 10^9\ \text{N}\cdot\text{m}^2/\text{C}^2) \times \left(\frac{2.00 \times 10^{-6}\ \text{C}}{4.00\ \text{m}} + \frac{-6.00 \times 10^{-6}\ \text{C}}{5.00\ \text{m}}\right)$$

$$= -6.29 \times 10^3\ \text{V}$$

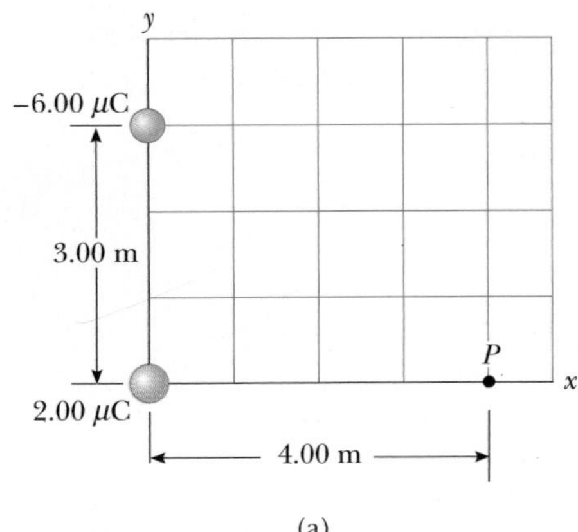

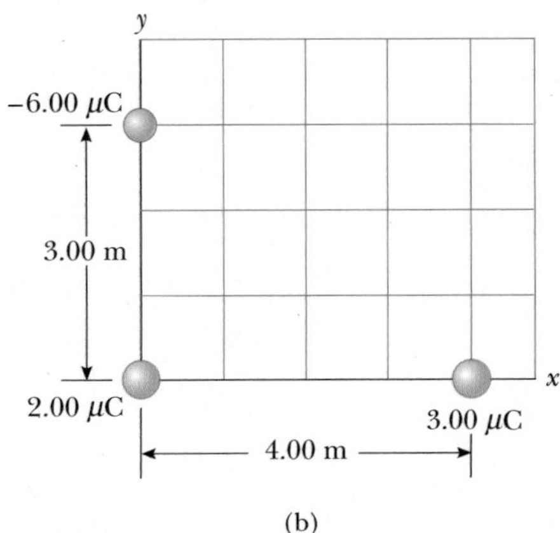

FIGURE 20.7 (Interactive Example 20.3) (a) The electric potential at point P due to the two point charges q_1 and q_2 is the algebraic sum of the potentials due to the individual charges. (b) How much work is done to bring a 3.00-μC charge from infinity to point P?

B How much work is required to bring a 3.00-μC point charge from infinity to the point P (Fig. 20.7b)?

Solution The work done is equal to the change in the system potential energy given by Equation 20.3:

$$W = \Delta U = q_3 \Delta V = q_3(V_P - 0)$$
$$= (3.00 \times 10^{-6}\text{ C})(-6.29 \times 10^3\text{ V})$$
$$= -18.9 \times 10^{-3}\text{ J}$$

The negative sign is because the 3.00-μC charge is attracted to the combination of q_1 and q_2, which has a net negative charge. The 3.00-μC charge would naturally move toward the other charges if released from infinity, so the external agent does not have to do anything to cause them to move together. To keep the charge from accelerating, however, the agent must apply a force *away* from point P. Therefore, the force exerted by the agent is opposite the displacement of the charge, leading to a negative value of the work. Positive work would have to be done by an external agent to remove the charge from P back to infinity.

PhysicsNow™ Log into PhysicsNow at **www.pop4e.com** and go to Interactive Example 20.3 to explore the value of the electric potential at point P and the electric potential energy of the system in Figure 20.7b.

20.4 OBTAINING ELECTRIC FIELD FROM ELECTRIC POTENTIAL

The electric field $\vec{\mathbf{E}}$ and the electric potential V are related by Equation 20.4, which shows how to find the potential if the electric field is known. We now show how to calculate the electric field if the electric potential is known in a certain region.

From Equation 20.3, we can express the potential difference dV between two points a distance ds apart as

$$dV = -\vec{\mathbf{E}} \cdot d\vec{\mathbf{s}} \qquad [20.14]$$

If the electric field has only *one* component—E_x, for example—then $\vec{\mathbf{E}} \cdot d\vec{\mathbf{s}} = E_x\,dx$. Therefore, Equation 20.14 becomes $dV = -E_x\,dx$, or

▮ **Relation between electric field and electric potential**

$$E_x = -\frac{dV}{dx}$$

That is, the electric field is equal to the negative of the derivative of the electric potential with respect to some coordinate. The potential change is zero for any displacement perpendicular to the electric field, which is consistent with the notion that equipotential surfaces are perpendicular to the field as in Figure 20.8a.

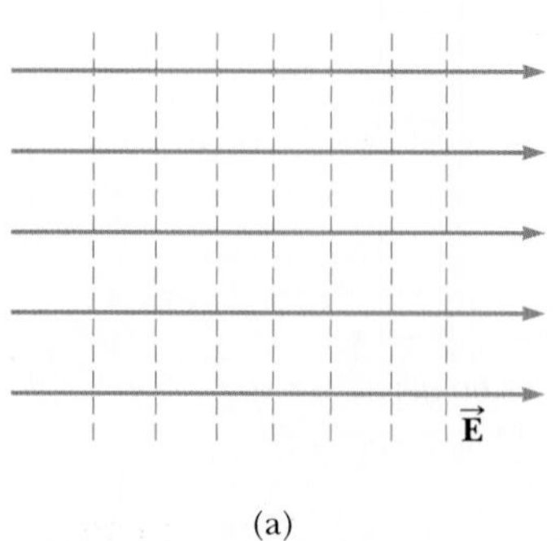

(a)

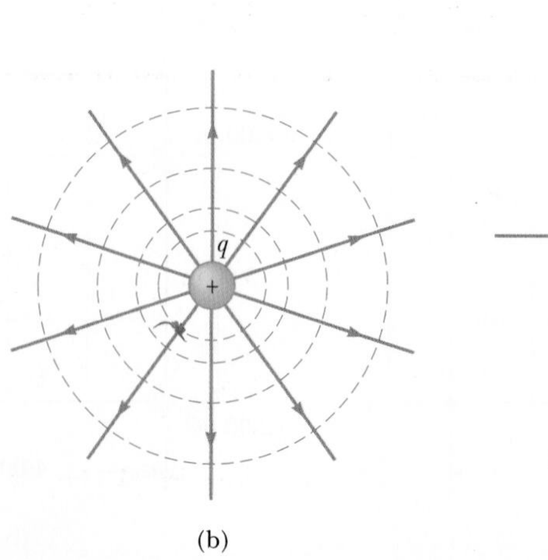

(b)

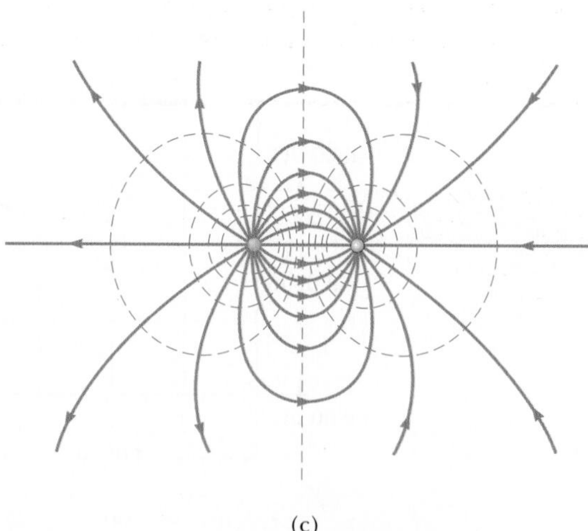

(c)

FIGURE 20.8 Equipotential surfaces (dashed blue lines) and electric field lines (brown lines) for (a) a uniform electric field produced by an infinite sheet of charge, (b) a point charge, and (c) an electric dipole. In all cases, the equipotential surfaces are *perpendicular* to the electric field lines at every point.

If the charge distribution has spherical symmetry such that the charge density depends only on the radial distance r, the electric field is radial. In this case, $\vec{\mathbf{E}} \cdot d\vec{\mathbf{s}} = E_r\,dr$, and we can express dV as $dV = -E_r\,dr$. Therefore,

$$E_r = -\frac{dV}{dr} \qquad [20.15]$$

For example, the potential of a point charge is $V = k_e q/r$. Because V is a function of r only, the potential function has spherical symmetry. Applying Equation 20.15, we find that the magnitude of the electric field due to the point charge is $E_r = k_e q/r^2$, a familiar result. Note that the potential changes only in the radial direction, not in a direction perpendicular to r. Therefore, V (like E_r) is a function only of r. Again, that is consistent with the idea that equipotential surfaces are perpendicular to field lines. In this case, the equipotential surfaces are a family of spheres concentric with the spherically symmetric charge distribution (Fig. 20.8b). The equipotential surfaces for the electric dipole are sketched in Figure 20.8c.

In general, the electric potential is a function of all three spatial coordinates. If V is given in terms of rectangular coordinates, the electric field components E_x, E_y, and E_z can be found from $V(x, y, z)$ as the partial derivatives

$$E_x = -\frac{\partial V}{\partial x} \qquad E_y = -\frac{\partial V}{\partial y} \qquad E_z = -\frac{\partial V}{\partial z} \qquad [20.16]$$

For example, if $V = 3x^2y + y^2 + yz$,

$$E_x = -\frac{\partial V}{\partial x} = -\frac{\partial}{\partial x}(3x^2y + y^2 + yz) = -\frac{\partial}{\partial x}(3x^2y) = -3y\frac{d}{dx}(x^2) = -6xy$$

$$E_y = -\frac{\partial V}{\partial y} = -\frac{\partial}{\partial y}(3x^2y + y^2 + yz) = -\left[3x^2\frac{d}{dy}(y) + \frac{d}{dy}(y^2) + z\frac{d}{dy}(y)\right]$$
$$= -3x^2 - 2y - z$$

$$E_z = -\frac{\partial V}{\partial z} = -\frac{\partial}{\partial z}(3x^2y + y^2 + yz) = -\frac{\partial}{\partial z}(yz) = -y\frac{d}{dz}(z) = -y$$

QUICK QUIZ 20.3 **(i)** In a certain region of space, the electric potential is zero everywhere along the x axis. From this information, we can conclude that the x component of the electric field in this region is **(a)** zero, **(b)** in the $+x$ direction, or **(c)** in the $-x$ direction. **(ii)** In a certain region of space, the electric field is zero. From this information, we can conclude that the electric potential in this region is **(a)** zero, **(b)** constant, **(c)** positive, or **(d)** negative.

EXAMPLE 20.4 The Electric Potential of a Dipole

An electric dipole consists of two charges of opposite sign but equal magnitude separated by a distance $2a$ as in Figure 20.9. The dipole is along the x axis and is centered at the origin.

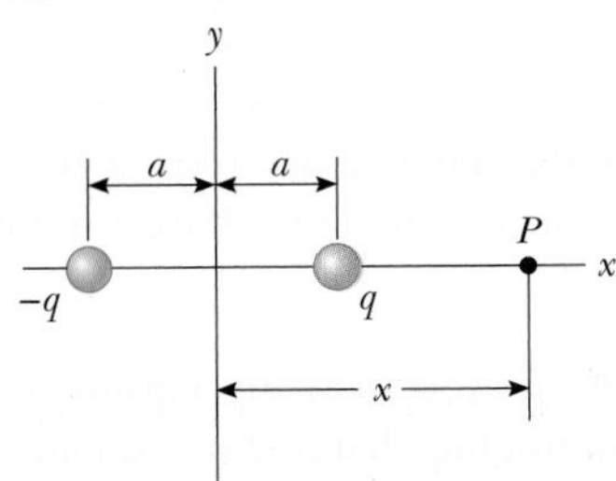

FIGURE 20.9 (Example 20.4) An electric dipole located on the x axis.

A Calculate the electric potential at any point P along the x axis.

Solution Using Equation 20.12, we have

$$V = k_e \sum_i \frac{q_i}{r_i} = k_e\left(\frac{q}{x-a} + \frac{-q}{x+a}\right) = \frac{2k_e qa}{x^2 - a^2}$$

B Calculate the electric field on this axis at points very far from the dipole.

Solution Using Equation 20.16 and the result from part A, we calculate the electric field at P:

$$E_x = -\frac{\partial V}{\partial x} = -\frac{d}{dx}\left(\frac{2k_e qa}{x^2 - a^2}\right) = -2k_e qa \frac{d}{dx}(x^2 - a^2)^{-1}$$
$$= (-2k_e qa)(-1)(x^2 - a^2)^{-2}(2x)$$
$$= \frac{4k_e qax}{(x^2 - a^2)^2}$$

If P is far from the dipole so that $x >> a$, then a^2 can be ignored in the term $x^2 - a^2$ and E_x becomes

$$E_x \approx \frac{4k_e qax}{x^4} = \frac{4k_e qa}{x^3} \quad (x >> a)$$

Comparing this result to that from Example 19.3, we see a factor of 2 difference between the results for the field far from the dipole. In the previous example, we were looking at the field along a line perpendicular to the line connecting the charges. As we see in Figure 19.11, the vertical components of the field cancel because the point at which we evaluate the field is equidistant from both charges. Therefore, only the very small horizontal components of the individual fields contribute to the total field. In this example, we are looking at the field along an extension of the line connecting the charges. For points along this line, the field vectors have components only along the line and the field vectors are in opposite directions. The point at which we evaluate the field, however, is necessarily closer to one charge than the other. As a result, the field is larger than that along the perpendicular direction by a factor of 2.

20.5 ELECTRIC POTENTIAL DUE TO CONTINUOUS CHARGE DISTRIBUTIONS

The **electric potential due to a continuous charge distribution** can be calculated in two ways. If the charge distribution is known, we can start with Equation 20.11 for the potential of a point charge. We then consider the potential due to a small charge element dq, modeling this element as a point charge (Fig. 20.10). The potential dV at some point P due to the charge element dq is

$$dV = k_e \frac{dq}{r} \quad [20.17]$$

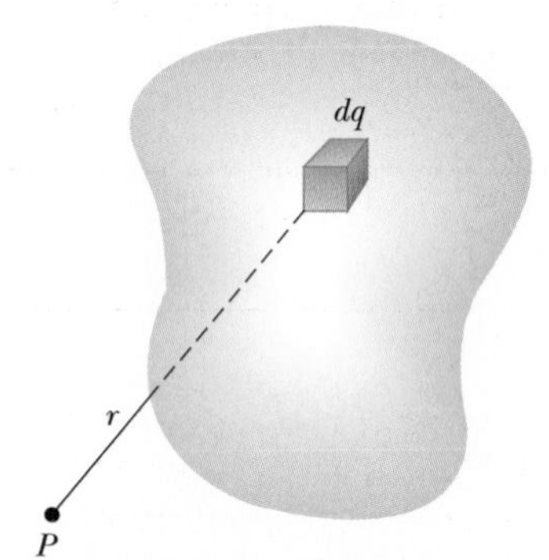

FIGURE 20.10 The electric potential at point P due to a continuous charge distribution can be calculated by dividing the charge distribution into elements of charge dq and summing the potential contributions over all elements.

where r is the distance from the charge element to P. To find the total potential at P, we integrate Equation 20.17 to include contributions from all elements of the charge distribution. Because each element is, in general, at a different distance from P and because k_e is a constant, we can express V as

$$V = k_e \int \frac{dq}{r} \quad [20.18]$$

In effect, we have replaced the sum in Equation 20.12 with an integral.

The second method for calculating the potential of a continuous charge distribution makes use of Equation 20.3. This procedure is useful when the electric field is already known from other considerations, such as Gauss's law. In this case, we substitute the electric field into Equation 20.3 to determine the potential difference between any two points. We then choose V to be zero at some convenient point. We shall illustrate both methods with examples.

PROBLEM-SOLVING STRATEGY Calculating Electric Potential

The following procedure is recommended for solving problems that involve the determination of an electric potential due to a charge distribution:

1. Conceptualize Think carefully about the individual charges or the charge distribution that you have in the problem and imagine what type of potential they would create so that you can establish the mental representation. Appeal to any symmetry in the arrangement of charges to help you visualize the potential.

2. Categorize Are you analyzing a group of individual charges or a continuous charge distribution? The answer to this question will tell you how to proceed in the *Analyze* step.

3. Analyze When working problems involving electric potential, remember that potential is a *scalar quantity,* so there are no vector components to consider. Therefore, when using the superposition principle to evaluate the electric potential at a point, simply take the algebraic sum of the potentials due to each charge. You must keep track of signs, however.

As with potential energy in mechanics, only *changes* in electric potential are significant; hence, the point where the potential is set at zero is arbitrary. When dealing with point charges or a finite-sized charge distribution, we usually define $V = 0$ to be at a point infinitely far from the charges. If the charge distribution itself extends to infinity, however, some other nearby point must be selected as the reference point.

(a) *If you are analyzing a group of individual charges:* Use the superposition principle. When several point charges are present, the resultant potential at a point in space is the *algebraic sum* of the individual potentials due to the individual charges (Eq. 20.12). Example 20.4 demonstrated this procedure.

(b) *If you are analyzing a continuous charge distribution:* Replace the sums for evaluating the total potential at some point P from individual charges by integrals (Eq. 20.18). The charge distribution is divided into infinitesimal elements of charge dq located at a distance of r from point P, and the sum is carried out by integrating over the entire charge distribution. An element is then treated as a point charge, so the potential at P due to the element is $dV = k_e\,(dq/r)$. The total potential at P is obtained by integrating dV over the entire charge distribution. For many problems, it is possible in performing the integration to express dq and r in terms of a single variable. To simplify the integration, it is important to give careful consideration to the geometry involved in the problem. Example 20.5 demonstrates such a procedure.

Another method that can be used to obtain the potential due to a finite continuous charge distribution is to start with the definition of the potential difference given by Equation 20.3. If $\vec{\mathbf{E}}$ is known or can be obtained easily (e.g., from Gauss's law), the line integral of $\vec{\mathbf{E}} \cdot d\vec{\mathbf{s}}$ can be evaluated. Example 20.6 uses this method.

4. Finalize Once you have determined your result, check to see if your potential is consistent with the mental representation and that it reflects any symmetry that you noted previously. Imagine varying parameters such as the distance of the observation point from the charges or the radius of any circular or spherical objects to see if the mathematical result changes in a reasonable way.

EXAMPLE 20.5 Potential Due to a Uniformly Charged Ring

Find the electric potential and electric field at a point P located on the axis of a uniformly charged ring of radius a and total charge Q. The plane of the ring is chosen perpendicular to the x axis (Fig. 20.11).

Solution Figure 20.11 helps us visualize the source of the potential and conceptualize what the potential might look like. We expect the potential to be symmetric around the x axis and to decrease for increasing values of x. We categorize this problem as one involving a continuous distribution of charge on the ring rather than a collection of individual charges. To analyze the problem, let us take P to be at a distance x from the center of the ring as in Figure 20.11. The charge element dq is at a distance equal to $r = \sqrt{x^2 + a^2}$ from point P. Hence, using Equation 20.18, we can express V as

$$V = k_e \int \frac{dq}{r} = k_e \int \frac{dq}{\sqrt{x^2 + a^2}}$$

In this case, each element dq is at the same distance from P. The term $\sqrt{x^2 + a^2}$ can therefore be removed from the integral, and V reduces to

$$V = \frac{k_e}{\sqrt{x^2 + a^2}} \int dq = \frac{k_e Q}{\sqrt{x^2 + a^2}}$$

Let us now address the electric field. The only variable in the expression for V is x. From the symmetry, we see that along the x axis $\vec{\mathbf{E}}$ can have only an x component. We can therefore use Equation 20.16 to find the magnitude of the electric field at P:

$$E_x = -\frac{\partial V}{\partial x} = -k_e Q \frac{d}{dx}(x^2 + a^2)^{-1/2}$$

$$= -k_e Q(-\tfrac{1}{2})(x^2 + a^2)^{-3/2}(2x)$$

$$= \frac{k_e Q x}{(x^2 + a^2)^{3/2}}$$

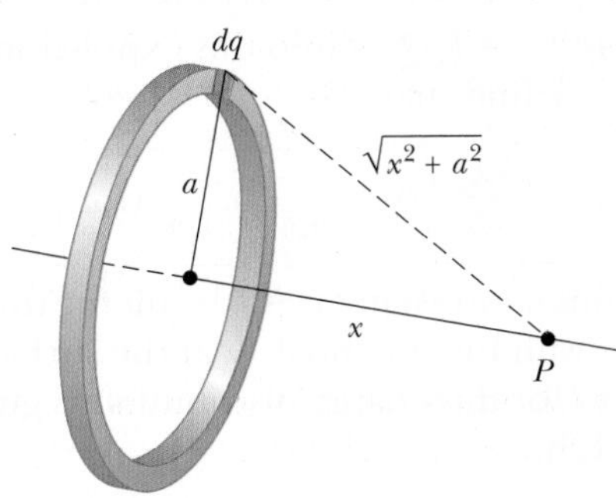

FIGURE 20.11 (Example 20.5) A uniformly charged ring of radius a, whose plane is perpendicular to the x axis. All elements dq of the ring are at the same distance from any point P on the x axis.

To finalize, note that V decreases as x increases, as we expected from our mental representation. If the point P is very far from the ring ($x >> a$), then a in the denominator of the expression for V can be ignored and $V \approx k_eQ/x$. This expression is just the one you would expect for a point charge. At large values of x, therefore, the charge distribution appears to be a point charge of magnitude Q as you should expect. Also notice that this result for the electric field agrees with that obtained by direct integration (see Example 19.5).

EXAMPLE 20.6 Potential of a Uniformly Charged Sphere

An insulating solid sphere of radius R has a total charge of Q, which is distributed uniformly throughout the volume of the sphere (Fig. 20.12a).

A Find the electric potential at a point outside the sphere, that is, for $r > R$. Take the potential to be zero at $r = \infty$.

Solution In Example 19.8, we found from Gauss's law that the magnitude of the electric field outside a spherically symmetric charge distribution is

$$E_r = k_e \frac{Q}{r^2} \qquad \text{(for } r > R\text{)}$$

where the field is directed radially outward when Q is positive. To obtain the potential at an exterior point, such as B in Figure 20.12a, we substitute this expression for E into Equation 20.4. Because $\vec{\mathbf{E}} \cdot d\vec{\mathbf{s}} = E_r\,dr$ in this case, we have

$$V_B = -\int_\infty^r E_r\,dr = -k_eQ\int_\infty^r \frac{dr}{r^2}$$

$$= k_e \frac{Q}{r} \qquad \text{(for } r > R\text{)}$$

Note that the result is identical to that for the electric potential due to a point charge. Because the potential must be continuous at $r = R$, we can use this expression to obtain the potential at the surface of the sphere. That is, the potential at a point such as C in Figure 20.12a is

$$V_C = k_e \frac{Q}{R} \qquad \text{(for } r = R\text{)}$$

B Find the potential at a point inside the charged sphere, that is, for $r < R$.

Solution In Example 19.8, we found that the electric field inside a uniformly charged sphere is

$$E_r = k_e \frac{Q}{R^3} r \qquad \text{(for } r < R\text{)}$$

We can use this result and Equation 20.3 to evaluate the potential difference $V_D - V_C$, where D is an interior point:

$$V_D - V_C = -\int_R^r E_r\,dr = -\frac{k_eQ}{R^3}\int_R^r r\,dr = \frac{k_eQ}{2R^3}(R^2 - r^2)$$

Substituting $V_C = k_eQ/R$ into this expression and solving for V_D, we find that

$$V_D = \frac{k_eQ}{2R}\left(3 - \frac{r^2}{R^2}\right) \qquad \text{(for } r < R\text{)}$$

At $r = R$, this expression gives a result for the potential that agrees with the potential V_C at the surface. A plot of V versus r for this charge distribution is given in Figure 20.12b.

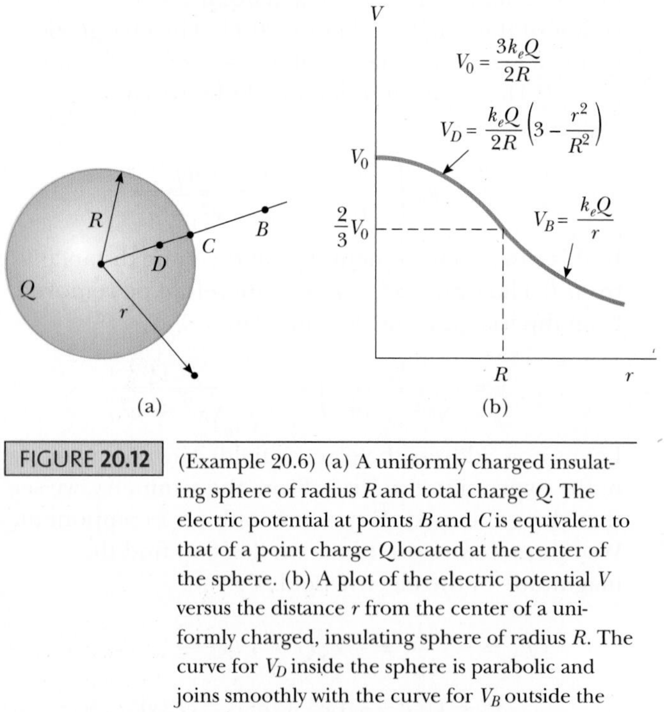

FIGURE 20.12 (Example 20.6) (a) A uniformly charged insulating sphere of radius R and total charge Q. The electric potential at points B and C is equivalent to that of a point charge Q located at the center of the sphere. (b) A plot of the electric potential V versus the distance r from the center of a uniformly charged, insulating sphere of radius R. The curve for V_D inside the sphere is parabolic and joins smoothly with the curve for V_B outside the sphere, which is a hyperbola. The potential has a maximum value V_0 at the center of the sphere.

20.6 ELECTRIC POTENTIAL OF A CHARGED CONDUCTOR

In Chapter 19, we found that when a solid conductor in electrostatic equilibrium carries a net charge, the charge resides on the outer surface of the conductor. Furthermore, we showed that the electric field just outside the surface of a conductor in equilibrium is perpendicular to the surface, whereas the field *inside* the conductor is zero.

We shall now show that **every point on the surface of a charged conductor in electrostatic equilibrium is at the same electric potential.** Consider two points A and B on the surface of a charged conductor as in Figure 20.13. Along a surface path connecting these points, $\vec{\mathbf{E}}$ is always perpendicular to the displacement $d\vec{\mathbf{s}}$; therefore, $\vec{\mathbf{E}} \cdot d\vec{\mathbf{s}} = 0$. Using this result and Equation 20.3, we conclude that the potential difference between A and B is necessarily zero. That is,

$$V_B - V_A = -\int_A^B \vec{\mathbf{E}} \cdot d\vec{\mathbf{s}} = 0$$

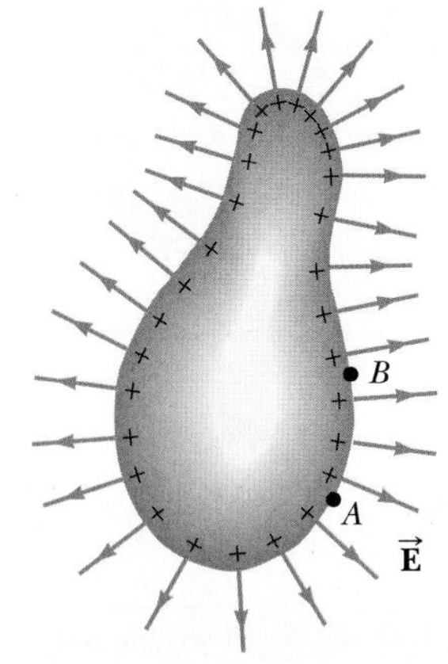

FIGURE 20.13 An arbitrarily shaped conductor with an excess positive charge. When the conductor is in electrostatic equilibrium, all the charge resides at the surface, $\vec{\mathbf{E}} = 0$ inside the conductor, and the electric field just outside the conductor is perpendicular to the surface. The potential is constant inside the conductor and is equal to the potential at the surface. The surface charge density is nonuniform.

This result applies to *any* two points on the surface. Therefore, V is constant everywhere on the surface of a charged conductor in equilibrium, so such a surface is an equipotential surface. Furthermore, because the electric field is zero inside the conductor, we conclude that the potential is constant everywhere inside the conductor and equal to its value at the surface. It follows that no work is required to move a test charge from the interior of a charged conductor to its surface.

For example, consider a solid metal sphere of radius R and total positive charge Q as in Figure 20.14a. The electric field outside the sphere has magnitude k_eQ/r^2 and points radially outward. Following Example 20.6, we see that the potential at the interior and surface of the sphere must be k_eQ/R relative to infinity. The potential outside the sphere is k_eQ/r. Figure 20.14b is a plot of the potential as a function of r, and Figure 20.14c shows the variations of the electric field with r.

When a net charge resides on a spherical conductor, the surface charge density is uniform as indicated in Figure 20.14a. If, however, the conductor is nonspherical as in Figure 20.13, the surface charge density is not uniform. To determine how the charge distributes on a nonspherical conductor, imagine a simplification model in which a nonspherical conductor is represented by the system shown in Figure 20.15. The system consists of two charged conducting spheres of radii r_1 and r_2, where $r_1 > r_2$, connected by a thin conducting wire. Imagine that the spheres are so far apart that the electric field of one does not influence the other (much farther apart than shown in Fig. 20.15). As a result, the electric field of each sphere can be modeled as that due to a spherically symmetric distribution of charge, which is the same as that due to a point charge.

Because the spheres are connected by a conducting wire, the entire system is a single conductor and all points must be at the same potential. In particular, the potentials at the surfaces of the two spheres must be equal. Using Equation 20.11 for the potential of a point charge, we set the potentials at the surfaces of the spheres equal:

$$k_e \frac{q_1}{r_1} = k_e \frac{q_2}{r_2} \quad \rightarrow \quad \frac{q_1}{q_2} = \frac{r_1}{r_2}$$

Therefore, the larger sphere has the larger amount of charge. Let us compare the surface charge densities on the two spheres, however:

$$\frac{\sigma_2}{\sigma_1} = \frac{\left(\dfrac{q_2}{4\pi r_2^{\,2}}\right)}{\left(\dfrac{q_1}{4\pi r_1^{\,2}}\right)} = \frac{q_2}{q_1}\frac{r_1^{\,2}}{r_2^{\,2}} = \frac{r_2}{r_1}\frac{r_1^{\,2}}{r_2^{\,2}} = \frac{r_1}{r_2}$$

Therefore, although the larger sphere has the larger total charge, the smaller sphere has the larger surface charge density, which leads to the fourth property

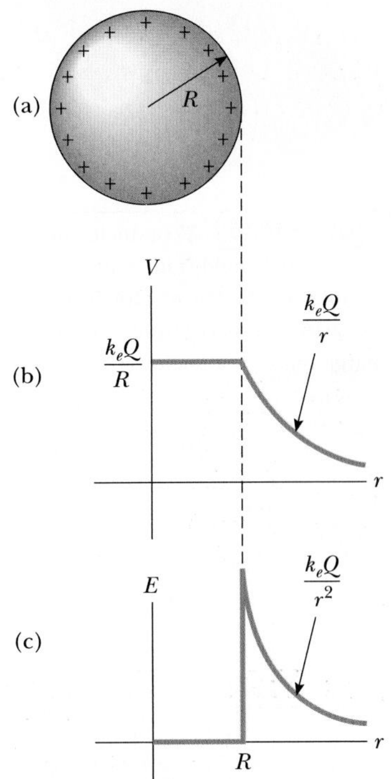

FIGURE 20.14 (a) The excess charge on a conducting sphere of radius R is uniformly distributed on its surface. (b) Electric potential versus distance r from the center of the charged conducting sphere. (c) Electric field versus distance r from the center of the charged conducting sphere.

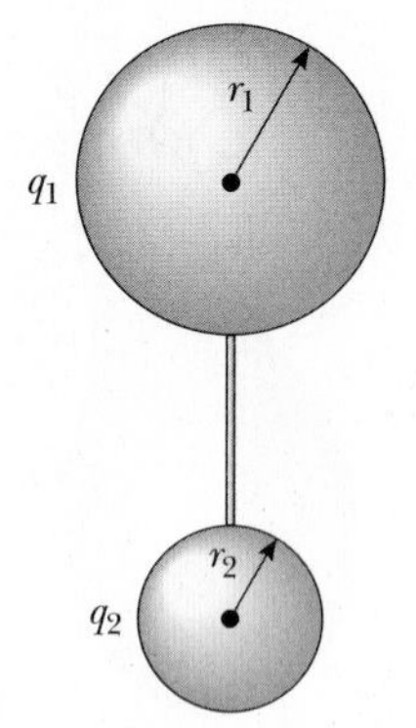

FIGURE 20.15 Two charged spherical conductors connected by a conducting wire. The spheres are at the same potential V.

listed in Section 19.11. Equation 19.25 tells us that the electric field near the surface of a conductor is proportional to the surface charge density. Therefore, the field near the smaller sphere is larger than the field close to the larger sphere.

We generalize this result by stating that **the electric field due to a charged conductor is large near convex surfaces of the conductor having small radii of curvature and is small near convex surfaces of the conductor having large radii of curvature.** A sharp point on a conductor is a region with an extremely small radius of curvature, so the field is very high near points on conductors.

■ Thinking Physics 20.1

Why is the end of a lightning rod pointed?

Reasoning The role of a lightning rod is to serve as a location at which the lightning strikes so that the charge delivered by the lightning will pass safely to the ground. If the lightning rod is pointed, the electric field due to charges moving between the rod and the ground is very strong near the point because the radius of curvature of the conductor is very small. This large electric field will greatly increase the likelihood that the return stroke will occur near the tip of the lightning rod rather than elsewhere. ■

A Cavity Within a Conductor in Equilibrium

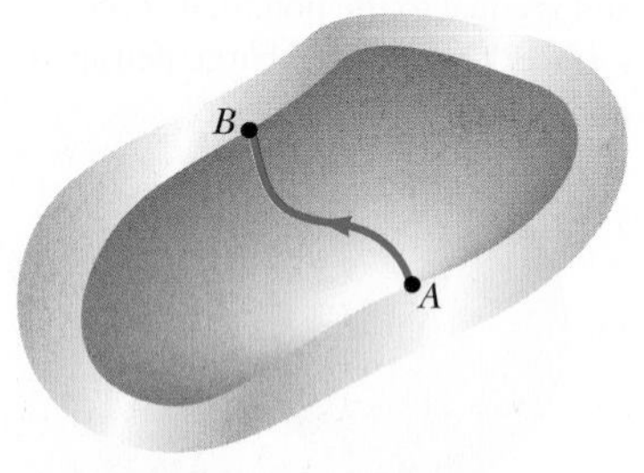

FIGURE 20.16 A conductor in electrostatic equilibrium containing an empty cavity. The electric field in the cavity is zero, regardless of the charge on the conductor.

Now consider a conductor of arbitrary shape containing a cavity as in Figure 20.16. Let us assume that no charges are inside the cavity. We shall show that **the electric field inside the cavity must be zero,** regardless of the charge distribution on the outside surface of the conductor. Furthermore, the field in the cavity is zero even if an electric field exists outside the conductor.

To prove this point, we remember that every point on the conductor is at the same potential and therefore any two points A and B on the surface of the cavity must be at the same potential. Now imagine that a field $\vec{\mathbf{E}}$ exists in the cavity and evaluate the potential difference $V_B - V_A$, defined by the expression

$$V_B - V_A = -\int_A^B \vec{\mathbf{E}} \cdot d\vec{\mathbf{s}}$$

where the path from A to B is within the cavity. Because $V_B - V_A = 0$, however, the integral must be zero regardless of the path chosen for the integration from A to B. The only way that the integral on the right side of the equation can be equal to zero for *all* possible paths within the cavity is for $\vec{\mathbf{E}}$ to be equal to zero at all points inside the cavity. Therefore, we conclude that a cavity surrounded by conducting walls is a field-free region as long as no charges are inside the cavity.

This result has some interesting applications. For example, it is possible to shield an electronic device or even an entire laboratory from external fields by surrounding it with conducting walls. Shielding is often necessary during highly sensitive electrical measurements. During a thunderstorm, the safest location is inside an automobile. Even if lightning strikes the car, the metal body guarantees that you will not receive a shock inside, where $\vec{\mathbf{E}} = 0$.

20.7 CAPACITANCE

As we continue with our discussion of electricity and, in later chapters, magnetism, we shall build *circuits* consisting of *circuit elements.* A circuit generally consists of a number of electrical components (circuit elements) connected together by conducting wires and forming one or more closed loops. These circuits can be considered as systems that exhibit a particular type of behavior. The first circuit element we shall consider is a **capacitor.**

In general, a capacitor consists of two conductors of any shape. Consider two conductors having a potential difference of ΔV between them. Let us assume that the conductors have charges of equal magnitude and opposite sign as in Figure 20.17. This situation can be accomplished by connecting two uncharged conductors to the terminals of a battery. Once that is done and the battery is disconnected, the charges remain on the conductors. We say that **the capacitor stores charge.**

The potential difference ΔV *across* the capacitor is the magnitude of the potential difference between the two conductors. This potential difference is proportional to the charge Q on the capacitor, which is defined as the magnitude of the charge on *either* of the two conductors. The **capacitance** C of a capacitor is defined as the ratio of the charge on the capacitor to the magnitude of the potential difference across the capacitor:

$$C \equiv \frac{Q}{\Delta V} \qquad [20.19]$$

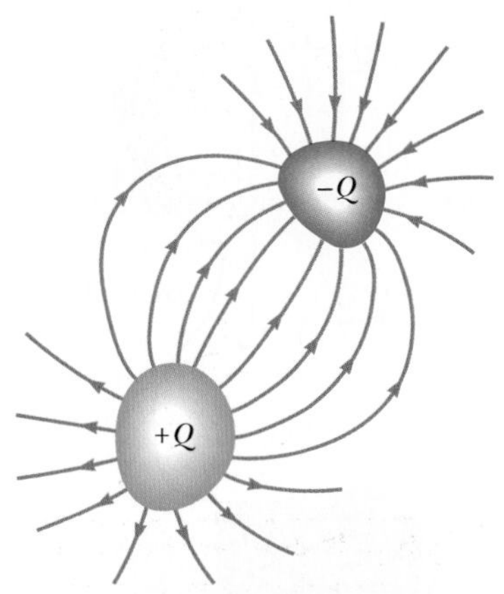

FIGURE 20.17 A capacitor consists of two conductors electrically isolated from each other and their surroundings. Once the capacitor is charged, the two conductors carry charges of equal magnitude but opposite sign.

By definition, **capacitance is always a positive quantity.** Because the potential difference is proportional to the charge, the ratio $Q/\Delta V$ is constant for a given capacitor. Equation 20.19 tells us that the capacitance of a system is a measure of the amount of charge that can be stored on the capacitor for a given potential difference.

From Equation 20.19, we see that capacitance has the SI units coulombs per volt, which is called a **farad** (F) in honor of Michael Faraday. The farad is a very large unit of capacitance. In practice, typical devices have capacitances ranging from microfarads to picofarads.

QUICK QUIZ 20.4 A capacitor stores charge Q at a potential difference ΔV. If the voltage applied by a battery to the capacitor is doubled to $2\,\Delta V$, **(a)** the capacitance falls to half its initial value and the charge remains the same, **(b)** the capacitance and the charge both fall to half their initial values, **(c)** the capacitance and the charge both double, or **(d)** the capacitance remains the same and the charge doubles.

The capacitance of a device depends on the geometric arrangement of the conductors. To illustrate this point, let us calculate the capacitance of an isolated spherical conductor of radius R and charge Q. (Based on the shape of the field lines from a single spherical conductor, we can model the second conductor as a concentric spherical shell of infinite radius.) Because the potential of the sphere is simply k_eQ/R (and $V = 0$ for the shell of infinite radius), the capacitance of the sphere is

$$C = \frac{Q}{\Delta V} = \frac{Q}{k_eQ/R} = \frac{R}{k_e} = 4\pi\epsilon_0 R \qquad [20.20]$$

(Remember from Section 19.4 that the Coulomb constant $k_e = 1/4\pi\epsilon_0$.) Equation 20.20 shows that the capacitance of an isolated charged sphere is proportional to the sphere's radius and is independent of both the charge and the potential difference.

The capacitance of a pair of oppositely charged conductors can be calculated in the following manner. A convenient charge of magnitude Q is assumed, and the potential difference is calculated using the techniques described in Section 20.5. One then uses $C = Q/\Delta V$ to evaluate the capacitance. As you might expect, the calculation is relatively straightforward if the geometry of the capacitor is simple.

Let us illustrate with two familiar geometries: parallel plates and concentric cylinders. In these examples, we shall assume that the charged conductors are separated by a vacuum. (The effect of a material between the conductors will be treated in Section 20.10.)

PITFALL PREVENTION 20.4

CAPACITANCE IS A CAPACITY To help you understand the concept of capacitance, think of similar notions that use a similar word. The *capacity* of a milk carton is the volume of milk it can store. The *heat capacity* of an object is the amount of energy an object can store per unit of temperature difference. The *capacitance* of a capacitor is the amount of charge the capacitor can store per unit of potential difference.

PITFALL PREVENTION 20.5

POTENTIAL DIFFERENCE IS ΔV, NOT V We use the symbol ΔV for the potential difference across a circuit element or a device because this notation is consistent with our definition of potential difference and with the meaning of the delta sign. It is a common but confusing practice to use the symbol V without the delta sign for a potential difference. Keep that in mind if you consult other texts.

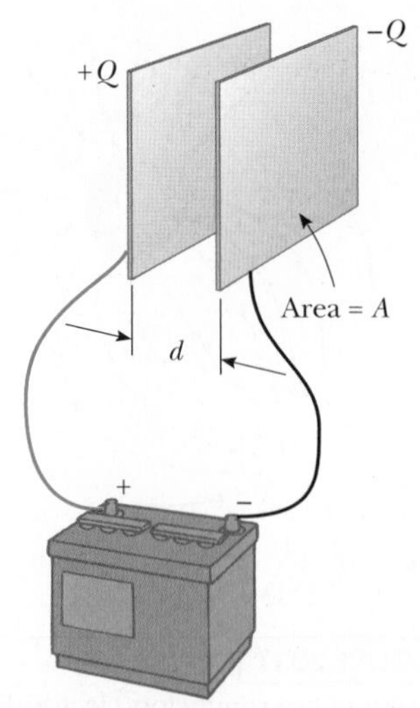

FIGURE 20.18 A parallel-plate capacitor consists of two parallel conducting plates, each of area A, separated by a distance d. When the capacitor is charged by connecting the plates to the terminals of a battery, the plates carry charges of equal magnitude but opposite sign.

The Parallel-Plate Capacitor

A parallel-plate capacitor consists of two parallel plates of equal area A separated by a distance d as in Figure 20.18. If the capacitor is charged, one plate has charge Q and the other, charge $-Q$. The magnitude of the charge per unit area on either plate is $\sigma = Q/A$. If the plates are very close together (compared with their length and width), we adopt a simplification model in which the electric field is uniform between the plates and zero elsewhere, as we discussed in Example 19.12. According to Example 19.12, the magnitude of the electric field between the plates is

$$E = \frac{\sigma}{\epsilon_0} = \frac{Q}{\epsilon_0 A}$$

Because the field is uniform, the potential difference across the capacitor can be found from Equation 20.6. Therefore,

$$\Delta V = Ed = \frac{Qd}{\epsilon_0 A}$$

Substituting this result into Equation 20.19, we find that the capacitance is

$$C = \frac{Q}{\Delta V} = \frac{Q}{Qd/\epsilon_0 A}$$

$$C = \frac{\epsilon_0 A}{d} \qquad [20.21]$$

That is, **the capacitance of a parallel-plate capacitor is proportional to the area of its plates and inversely proportional to the plate separation.**

As you can see from the definition of capacitance, $C = Q/\Delta V$, the amount of charge a given capacitor can store for a given potential difference across its plates increases as the capacitance increases. It therefore seems reasonable that a capacitor constructed from plates having large areas should be able to store a large charge.

A careful inspection of the electric field lines for a parallel-plate capacitor reveals that the field is uniform in the central region between the plates, but is nonuniform at the edges of the plates. Figure 20.19 shows a drawing and a photograph of the electric field pattern of a parallel-plate capacitor, showing the nonuniform field lines at the plates' edges. As long as the separation between the plates is small compared with the dimensions of the plates (unlike Fig. 20.19b), the edge effects can be ignored and we can use the simplification model in which the electric field is uniform everywhere between the plates.

PITFALL PREVENTION 20.6

Too many Cs Be sure not to confuse italic C for capacitance with regular C for the unit coulomb.

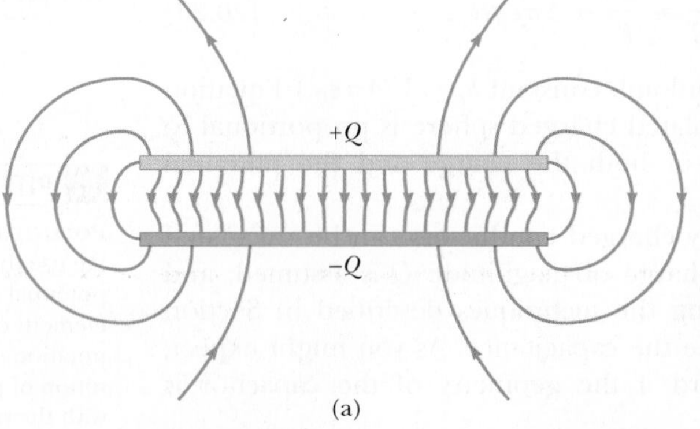

(a)

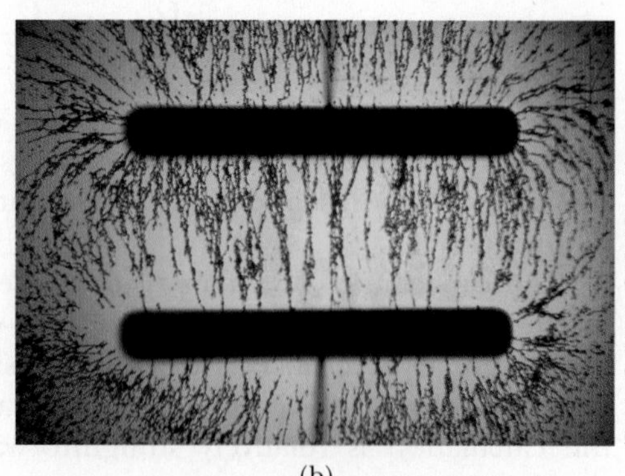
(b)

(Douglas C. Johnson/Cal Poly Pomona)

FIGURE 20.19 (a) The electric field between the plates of a parallel-plate capacitor is uniform near the center but nonuniform near the edges. (b) Electric field pattern of two oppositely charged conducting parallel plates. Small particles on an oil surface align with the electric field.

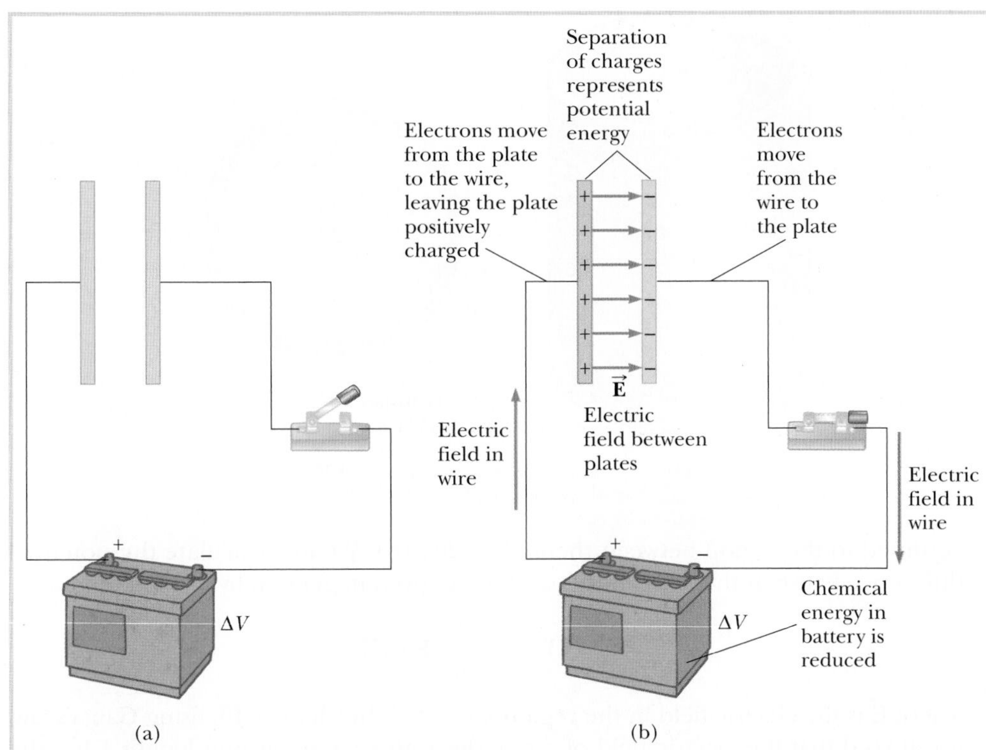
Separation of charges represents potential energy
Electrons move from the plate to the wire, leaving the plate positively charged
Electrons move from the wire to the plate
E
Electric field between plates
Electric field in wire
Electric field in wire
+
−
ΔV
+
−
ΔV
Chemical energy in battery is reduced
(a)
(b)

ACTIVE FIGURE 20.20
(a) A circuit consisting of a capacitor, a battery, and a switch. (b) When the switch is closed, the battery establishes an electric field in the wire that causes electrons to move from the left plate into the wire and into the right plate from the wire. As a result, a separation of charge exists on the plates, which represents an increase in electric potential energy of the system. This energy in the system of the circuit has been transformed from chemical energy in the battery.

PhysicsNow™ Log into PhysicsNow at **www.pop4e.com** and go to Active Figure 20.20 to adjust the battery voltage and see the result on the charge on the plates and the electric field between the plates.

Active Figure 20.20 shows a battery connected to a single parallel-plate capacitor with a switch in the circuit. Let us identify the circuit as a system. When the switch is closed, the battery establishes an electric field in the wires and charges flow between the wires and the capacitor. As that occurs, energy is transformed within the system. Before the switch is closed, energy is stored as chemical energy in the battery. This type of energy is associated with chemical bonds and is transformed during the chemical reaction that occurs within the battery when it is operating in an electric circuit. When the switch is closed, some of the chemical energy in the battery is converted to electric potential energy related to the separation of positive and negative charges on the plates. As a result, we can describe a capacitor as a device that stores energy as well as charge. We will explore this energy storage in more detail in Section 20.9.

EXAMPLE 20.7 **Parallel-Plate Capacitor**

A parallel-plate capacitor has an area $A = 2.00 \times 10^{-4}\ \text{m}^2$ and a plate separation $d = 1.00$ mm. Find its capacitance.

Solution From Equation 20.21, we find that

$$C = \frac{\epsilon_0 A}{d} = (8.85 \times 10^{-12}\ \text{C}^2/\text{N}\cdot\text{m}^2)\left(\frac{2.00 \times 10^{-4}\ \text{m}^2}{1.00 \times 10^{-3}\ \text{m}}\right)$$

$$= 1.77 \times 10^{-12}\ \text{F} = \boxed{1.77\ \text{pF}}$$

The Cylindrical Capacitor

A cylindrical capacitor consists of a cylindrical conductor of radius a and charge Q coaxial with a larger cylindrical shell of radius b and charge $-Q$ (Fig. 20.21a). Let us find the capacitance of this device if its length is ℓ. If we assume that ℓ is large compared with a and b, we can adopt a simplification model in which we ignore end effects. In this case, the field is perpendicular to the axis of the cylinders and is

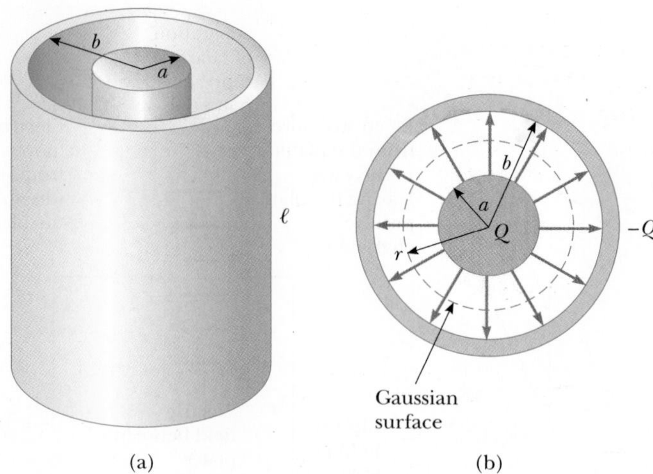

FIGURE 20.21 (a) A cylindrical capacitor consists of a solid cylindrical conductor of radius a and length ℓ surrounded by a coaxial cylindrical shell of radius b. (b) End view. The dashed line represents the end of the cylindrical gaussian surface of radius r.

confined to the region between them (Fig. 20.21b). We first calculate the potential difference between the two cylinders, which is given in general by

$$V_b - V_a = -\int_a^b \vec{\mathbf{E}} \cdot d\vec{\mathbf{s}}$$

where $\vec{\mathbf{E}}$ is the electric field in the region $a < r < b$. In Chapter 19, using Gauss's law, we showed that the electric field of a cylinder with charge per unit length λ has the magnitude $E = 2k_e\lambda/r$. The same result applies here because the outer cylinder does not contribute to the electric field inside it. Using this result and noting that the direction of $\vec{\mathbf{E}}$ is radially away from the inner cylinder in Figure 20.21b, we find that

$$V_b - V_a = -\int_a^b E_r\,dr = -2k_e\lambda \int_a^b \frac{dr}{r} = -2k_e\lambda \ln\left(\frac{b}{a}\right)$$

Substituting this result into Equation 20.19 and using that $\lambda = Q/\ell$, we find that

$$C = \frac{Q}{\Delta V} = \frac{Q}{\dfrac{2k_eQ}{\ell}\ln\left(\dfrac{b}{a}\right)} = \frac{\ell}{2k_e \ln\left(\dfrac{b}{a}\right)} \qquad [20.22]$$

where the magnitude of the potential difference between the cylinders is $\Delta V = |V_a - V_b| = 2k_e\lambda \ln(b/a)$, a positive quantity. Our result for C shows that the capacitance is proportional to the length of the cylinders. As you might expect, the capacitance also depends on the radii of the two cylindrical conductors. As an example, a coaxial cable consists of two concentric cylindrical conductors of radii a and b separated by an insulator. The cable carries currents in opposite directions in the inner and outer conductors. Such a geometry is especially useful for shielding an electrical signal from external influences. From Equation 20.22, we see that the capacitance per unit length of a coaxial cable is

$$\frac{C}{\ell} = \frac{1}{2k_e \ln\left(\dfrac{b}{a}\right)}$$

20.8 COMBINATIONS OF CAPACITORS

Two or more capacitors are often combined in electric circuits in different ways. The equivalent capacitance of certain combinations can be calculated using methods described in this section.

In studying electric circuits, we use a specialized simplified pictorial representation called a **circuit diagram.** Such a diagram uses **circuit symbols** to represent various circuit elements. The circuit symbols are connected by straight lines that represent the wires between the circuit elements. Figure 20.22 shows the circuit symbols for a capacitor, a battery, and an open switch. Notice that the circuit symbol for a capacitor consists of two parallel lines of equal length, representing the plates in a parallel-plate capacitor, and the lines in the battery symbol are of different lengths. The positive terminal of the battery is at the higher potential and is represented by the longer line in the battery symbol.

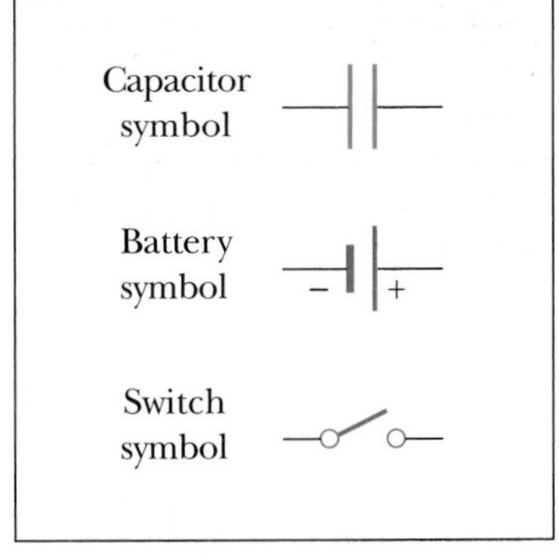

FIGURE 20.22 Circuit symbols for a capacitor, a battery, and an open switch. Note that capacitors are in blue, and batteries and switches are in red.

Parallel Combination

Two capacitors connected as shown in the pictorial representation in Active Figure 20.23a are known as a **parallel combination** of capacitors. Active Figure 20.23b shows the circuit diagram for this configuration. The left plates of both capacitors are connected by a conducting wire to the positive terminal of the battery, and both plates are therefore at the same potential as that of the battery terminal. Likewise, the right plates are connected to the negative terminal of the battery and are at the same potential as that terminal. The voltage applied across the combination is therefore the terminal voltage of the battery.[3] Furthermore, the voltage across *each capacitor* is the same as the terminal voltage of the battery.

When the capacitors are first connected in the circuit, electrons are transferred between the wires and the plates, causing the left plates to become positively charged and the right plates to become negatively charged. The flow of charge ceases when the voltage across the capacitors is equal to that across the battery terminals. At this point, the capacitors have reached their maximum charge. Let us call the maximum charges on the two capacitors Q_1 and Q_2. Then the *total charge* Q stored by the two capacitors is

$$Q = Q_1 + Q_2 \qquad [20.23]$$

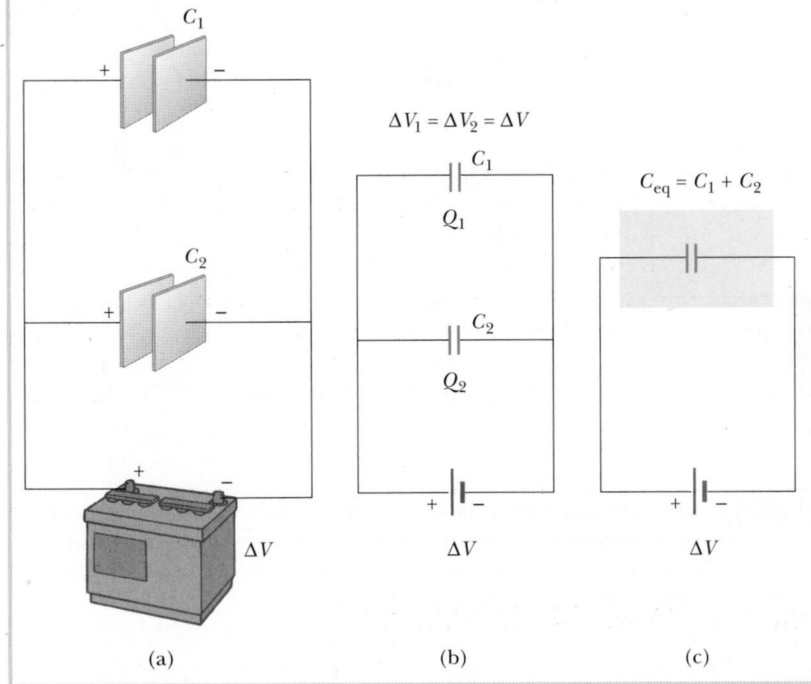

ACTIVE FIGURE 20.23 (a) A parallel combination of two capacitors connected to a battery. (b) The circuit diagram for the parallel combination. The potential difference is the same across each capacitor. (c) The equivalent capacitance is $C_{eq} = C_1 + C_2$.

PhysicsNow™ Log into PhysicsNow at www.pop4e.com and go to Active Figure 20.23 to adjust the battery voltage and the individual capacitances and see the resulting charges and voltages on the capacitors. You can combine up to four capacitors in parallel.

[3] In some situations, the parallel combination may be in a circuit with other circuit elements so that the potential difference across the combination is not that of a battery in the circuit, but must be determined by analyzing the entire circuit.

Suppose we wish to replace the two capacitors in Active Figure 20.23b with one equivalent capacitor having the capacitance C_{eq}. This equivalent capacitor (Active Fig. 20.23c) must have exactly the same result in the circuit as the original two. That is, it must store charge Q when connected to the battery. From Active Figure 20.23c, we see that the voltage across the equivalent capacitor is ΔV. Therefore, we have

$$Q = C_{eq}\,\Delta V$$

and, for the individual capacitors,

$$Q_1 = C_1\,\Delta V \qquad Q_2 = C_2\,\Delta V$$

Substitution of these relations into Equation 20.23 gives

$$C_{eq}\,\Delta V = C_1\,\Delta V + C_2\,\Delta V$$

or

$$C_{eq} = C_1 + C_2 \qquad \text{(parallel combination)} \qquad [20.24]$$

If we extend this treatment to three or more capacitors connected in parallel, the equivalent capacitance is

■ Equivalent capacitance of several capacitors in parallel

$$C_{eq} = C_1 + C_2 + C_3 + \cdots \qquad \text{(parallel combination)} \qquad [20.25]$$

Therefore, we see that **the equivalent capacitance of a parallel combination of capacitors is the algebraic sum of the individual capacitances and is larger than any of the individual capacitances.**

Series Combination

Now consider two capacitors connected in **series** as illustrated in Active Figure 20.24a. Active Figure 20.24b shows the circuit diagram. For this series combination of capacitors, **the magnitude of the charge is the same on all the plates.**

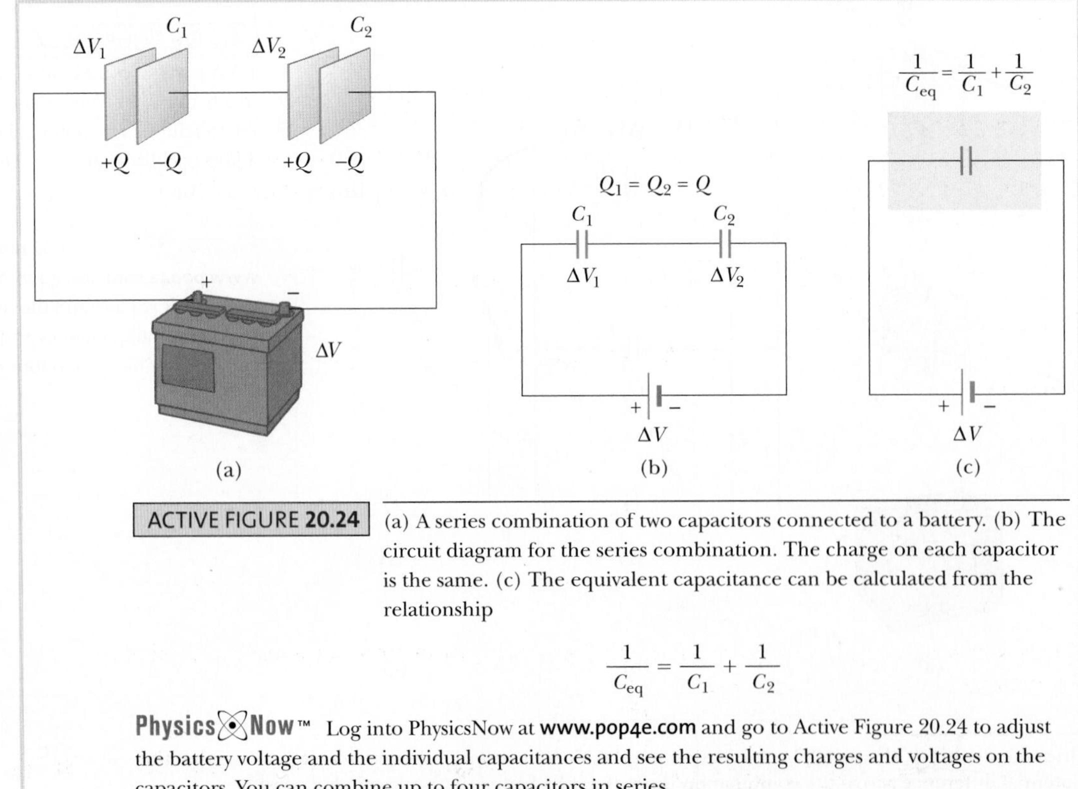

ACTIVE FIGURE 20.24 (a) A series combination of two capacitors connected to a battery. (b) The circuit diagram for the series combination. The charge on each capacitor is the same. (c) The equivalent capacitance can be calculated from the relationship

$$\frac{1}{C_{eq}} = \frac{1}{C_1} + \frac{1}{C_2}$$

PhysicsNow™ Log into PhysicsNow at **www.pop4e.com** and go to Active Figure 20.24 to adjust the battery voltage and the individual capacitances and see the resulting charges and voltages on the capacitors. You can combine up to four capacitors in series.

To see why that is true, let us consider the charge transfer process in some detail. We start with uncharged capacitors and follow what happens just after a battery is connected to the circuit. When the connection is made, the right plate of C_1 and the left plate of C_2 form an isolated conductor. Therefore, whatever negative charge enters one plate from the connecting wire must be equal to the positive charge of the other plate so as to maintain neutrality of the isolated conductor: that is the electric charge version of the isolated system model. As a result, both capacitors must have the same charge Q.

Suppose we wish to determine the capacitance of an equivalent capacitor that has the same effect in the circuit as the series combination. That is, as the equivalent capacitor is being charged, charge $-Q$ must enter its right plate from the wires and the charge on its left plate must be $+Q$. By applying the definition of capacitance to the circuit shown in Active Figure 20.24c, we have

$$\Delta V = \frac{Q}{C_{eq}} \qquad [20.26]$$

where ΔV is the potential difference between the terminals of the battery and C_{eq} is the equivalent capacitance.

Because the right plate of C_1 and the left plate of C_2 in Active Figure 20.24a form an isolated conductor, both plates are at the same potential V_i, where the i stands for the isolated conductor. The notation V_{left} represents the potential of the left plate of C_1, and V_{right} represents the potential of the right plate of C_2. Because these latter two plates are connected directly to the battery, the potential difference between them must be

$$\Delta V = V_{left} - V_{right}$$

If we add and subtract V_i to this equation, we have

$$\Delta V = (V_{left} - V_i) + (V_i - V_{right})$$

which we can write as

$$\Delta V = \Delta V_1 + \Delta V_2 \qquad [20.27]$$

where ΔV_1 and ΔV_2 are the potential differences across capacitors C_1 and C_2. In general, the potential difference across any number of capacitors in series is equal to the sum of the potential differences across the individual capacitors. Because $Q = C\,\Delta V$ can be applied to each capacitor, the potential difference across each is

$$\Delta V_1 = \frac{Q}{C_1} \qquad \Delta V_2 = \frac{Q}{C_2}$$

Substituting these expressions into Equation 20.27 and using Equation 20.26 to replace ΔV, we have

$$\frac{Q}{C_{eq}} = \frac{Q}{C_1} + \frac{Q}{C_2}$$

Canceling Q, we arrive at the relationship

$$\frac{1}{C_{eq}} = \frac{1}{C_1} + \frac{1}{C_2} \qquad \text{(series combination)} \qquad [20.28]$$

If this analysis is applied to three or more capacitors connected in series, the equivalent capacitance is found to be given by

$$\frac{1}{C_{eq}} = \frac{1}{C_1} + \frac{1}{C_2} + \frac{1}{C_3} + \cdots \qquad \text{(series combination)} \qquad [20.29]$$

■ Equivalent capacitance of several capacitors in series

which shows that **the inverse of the equivalent capacitance is the algebraic sum of the inverses of the individual capacitances and that the equivalent capacitance of a series combination is always less than any individual capacitance in the combination.**

QUICK QUIZ 20.5 Two capacitors are identical. They can be connected in series or in parallel. **(i)** If you want the *smallest* equivalent capacitance for the combination, **(a)** do you connect them in series, **(b)** do you connect them in parallel, or **(c)** do the combinations have the same capacitance? **(ii)** Each capacitor is charged to a voltage of 10 V. If you want the largest combined potential difference across the combination, **(a)** do you connect them in series, **(b)** do you connect them in parallel, or **(c)** do the combinations have the same potential difference?

INTERACTIVE EXAMPLE 20.8 Equivalent Capacitance

Find the equivalent capacitance between a and b for the combination of capacitors shown in Figure 20.25a. All capacitances are in microfarads.

Solution Using Equations 20.25 and 20.29, we reduce the combination step by step as indicated in the figure. The 1.0-μF and 3.0-μF capacitors are in parallel and combine according to $C_{eq} = C_1 + C_2$. Their equivalent capacitance is 4.0 μF. Likewise, the 2.0-μF and 6.0-μF capacitors are also in parallel and have an equivalent capacitance of 8.0 μF. The upper branch in Figure 20.25b now consists of two 4.0-μF capacitors in series, which combine according to

$$\frac{1}{C_{eq}} = \frac{1}{C_1} + \frac{1}{C_2} = \frac{1}{4.0\ \mu\text{F}} + \frac{1}{4.0\ \mu\text{F}} = \frac{1}{2.0\ \mu\text{F}}$$

$$C_{eq} = 2.0\ \mu\text{F}$$

Likewise, the lower branch in Figure 20.25b consists of two 8.0-μF capacitors in series, which give an equivalent of 4.0 μF. Finally, the 2.0-μF and 4.0-μF capacitors in Figure 20.25c are in parallel and have an equivalent capacitance of 6.0 μF. Hence, the equivalent capacitance of the circuit is 6.0 μF as shown in Figure 20.25d.

PhysicsNow™ Log into PhysicsNow at **www.pop4e.com** and go to Interactive Example 20.8 to practice reducing a combination of capacitors to a single equivalent capacitor.

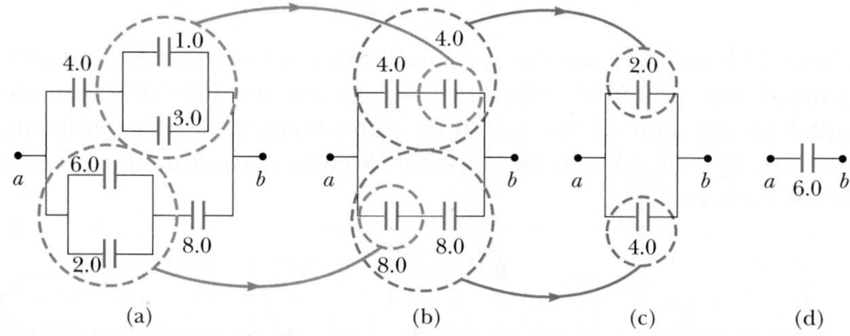

FIGURE 20.25 (Interactive Example 20.8) To find the equivalent combination of the capacitors in (a), the various combinations are reduced in steps as indicated in (b), (c), and (d), using the series and parallel rules described in the text. All capacitance values are in microfarads.

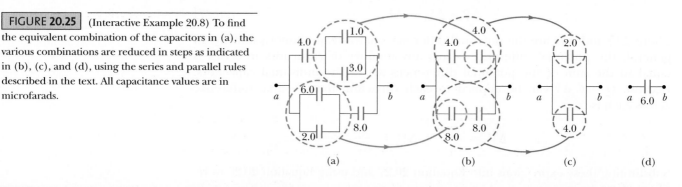

20.9 ENERGY STORED IN A CHARGED CAPACITOR

Almost everyone who works with electronic equipment has at some time verified that a capacitor can store energy. If the plates of a charged capacitor are connected by a conductor, such as a wire, charge transfers between the plates and the wire until the two plates are uncharged. The discharge can often be observed as a visible spark. If you accidentally touch the opposite plates of a charged capacitor, your fingers act as pathways by which the capacitor discharges, resulting in an electric shock. The degree of shock depends on the capacitance and the voltage applied to the capacitor. When high voltages are present, such as in the power supply of a television set, the shock can be fatal.

Consider a parallel-plate capacitor that is initially uncharged so that the initial potential difference across the plates is zero. Now imagine that the capacitor is connected to a battery and develops a charge of Q. The final potential difference across the capacitor is $\Delta V = Q/C$.

To calculate the energy stored in the capacitor, imagine charging the capacitor in a different way that achieves the same result. An external agent reaches in and grabs small bits of charge and transfers them from one plate to the other. Suppose q is the charge on the capacitor at some instant during this charging process. At the same instant, the potential difference across the capacitor is $\Delta V = q/C$. Now imagine that the external agent transfers an additional increment of charge dq from the plate of charge $-q$ to the plate of charge q (which is at the higher potential) by applying a force on the charge dq to move it through the electric field between the plates. The work required to transfer an increment of charge dq from one plate to the other is

$$dW = \Delta V\, dq = \frac{q}{C}\, dq$$

Therefore, the total work required to charge the capacitor from $q = 0$ to the final charge $q = Q$ is

$$W = \int_0^Q \frac{q}{C}\, dq = \frac{Q^2}{2C}$$

The capacitor can be modeled as a nonisolated system for this discussion. The work done by the external agent on the system in charging the capacitor appears as potential energy U stored in the capacitor. In reality, of course, this energy is not the result of mechanical work done by an external agent moving charge from one plate to the other, but is due to transformation of chemical energy in the battery. We have used a model of work done by an external agent that gives us a result that is also valid for the actual situation. Using $Q = C\Delta V$, the energy stored in a charged capacitor can be expressed in the following alternative forms:

$$U = \frac{Q^2}{2C} = \tfrac{1}{2}Q\,\Delta V = \tfrac{1}{2}C(\Delta V)^2 \qquad [20.30]$$

▪ Energy stored in a charged capacitor

This result applies to *any* capacitor, regardless of its geometry. In practice, the maximum energy (or charge) that can be stored is limited because electric discharge ultimately occurs between the plates of the capacitor at a sufficiently large value of ΔV. For this reason, capacitors are usually labeled with a maximum operating voltage.

For an object on an extended spring, the elastic potential energy can be modeled as being stored *in the spring*. Internal energy of a substance associated with its temperature is located *throughout the substance*. Where is the energy in a capacitor located? The energy stored in a capacitor can be modeled as being stored *in the electric field between the plates of the capacitor*. For a parallel-plate capacitor, the potential difference is related to the electric field through the relationship $\Delta V = Ed$. Furthermore, the capacitance is $C = \epsilon_0 A/d$. Substituting these expressions into Equation 20.30 gives

$$U = \tfrac{1}{2}\left(\frac{\epsilon_0 A}{d}\right)(Ed)^2 = \tfrac{1}{2}(\epsilon_0 Ad)E^2 \qquad [20.31]$$

Because the volume of a parallel-plate capacitor that is occupied by the electric field is Ad, the energy per unit volume $u = U/Ad$, called the **energy density,** is

$$u = \tfrac{1}{2}\epsilon_0 E^2 \qquad [20.32]$$

▪ Energy density in an electric field

Although Equation 20.32 was derived for a parallel-plate capacitor, the expression is generally valid. That is, **the energy density in any electric field is proportional to the square of the magnitude of the electric field at a given point.**

QUICK QUIZ 20.6 You have three capacitors and a battery. In which of the following combinations of the three capacitors will the maximum possible energy be stored when the combination is attached to the battery? **(a)** When in series the maximum amount is stored. **(b)** When parallel the maximum amount is stored. **(c)** Both combinations will store the same amount of energy.

■ Thinking Physics 20.2

You charge a capacitor and then remove it from the battery. The capacitor consists of large movable plates, with air between them. You pull the plates farther apart a small distance. What happens to the charge on the capacitor? To the potential difference? To the energy stored in the capacitor? To the capacitance? To the electric field between the plates? Is work done in pulling the plates apart?

Reasoning Because the capacitor is removed from the battery, charges on the plates have nowhere to go. Therefore, the charge on the capacitor remains the same as the plates are pulled apart. Because the electric field of large plates is independent of distance for uniform fields, the electric field remains constant. Because the electric field is a measure of the rate of change of potential with distance, the potential difference between the plates increases as the separation distance increases. Because the same charge is stored at a higher potential difference, the capacitance decreases. Because energy stored is proportional to both charge and potential difference, the energy stored in the capacitor increases. This energy must be transferred into the system from somewhere; the plates attract each other, so work is done by you on the system of two plates when you pull them apart. ■

INTERACTIVE EXAMPLE 20.9 Rewiring Two Charged Capacitors

Two capacitors with capacitances C_1 and C_2 (where $C_1 > C_2$) are charged to the same potential difference ΔV_i. The charged capacitors are removed from the battery, and their plates are connected as shown in Figure 20.26a. The switches S_1 and S_2 are then closed as in Figure 20.26b.

A Find the final potential difference ΔV_f between a and b after the switches are closed.

Solution Let us identify the left-hand plates of the capacitors as an isolated system because they are not connected to the right-hand plates by conductors. The charges on the left-hand plates before the switches are closed are

$$Q_{1i} = C_1\,\Delta V_i \qquad \text{and} \qquad Q_{2i} = -C_2\,\Delta V_i$$

The negative sign for Q_{2i} is necessary because the charge on the left plate of capacitor C_2 is negative. The total charge Q in the system is

$$(1)\quad Q = Q_{1i} + Q_{2i} = (C_1 - C_2)\Delta V_i$$

After the switches are closed, the electric charge version of the isolated system model tells us that the total charge on the left-hand plates remains the same:

$$(2)\quad Q = Q_{1f} + Q_{2f}$$

The charges will redistribute on the left-hand plates until the entire conductor in the system is at the same potential V_{left}. Similarly, charges will distribute on the two right-hand plates until the entire conductor in this system is at the same potential V_{right}. Therefore, the final potential difference $\Delta V_f = |V_{\text{left}} - V_{\text{right}}|$ across both capacitors is the same. To satisfy this requirement, the charges on the capacitors after the switches are closed are

$$(3)\quad Q_{1f} = C_1\,\Delta V_f$$

$$(4)\quad Q_{2f} = C_2\,\Delta V_f$$

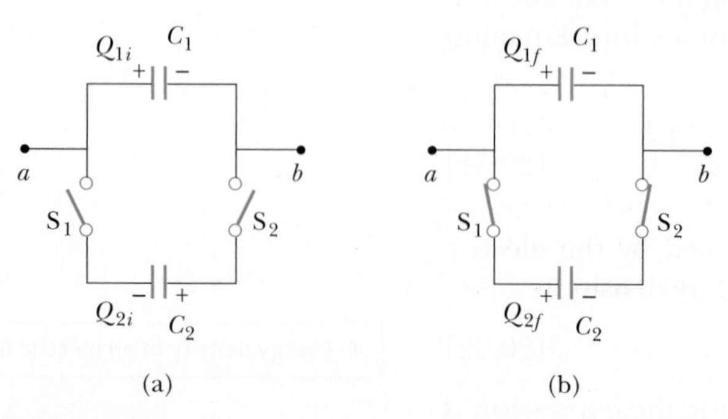

FIGURE 20.26 (Interactive Example 20.9) Two capacitors are connected with plates of opposite charge in contact.

Dividing these equations, we have

$$(5)\quad \frac{Q_{1f}}{Q_{2f}} = \frac{C_1\,\Delta V_f}{C_2\,\Delta V_f} = \frac{C_1}{C_2} \quad\rightarrow\quad Q_{1f} = \frac{C_1}{C_2}\,Q_{2f}$$

Combining (2) and (5) gives

$$Q = \frac{C_1}{C_2}\,Q_{2f} + Q_{2f} = Q_{2f}\left(1 + \frac{C_1}{C_2}\right)$$

$$(6)\quad Q_{2f} = Q\left(\frac{C_2}{C_1 + C_2}\right)$$

Using (5) and (6) to find Q_{1f}, we have

$$Q_{1f} = \frac{C_1}{C_2}\left[Q\left(\frac{C_2}{C_1 + C_2}\right)\right] = Q\left(\frac{C_1}{C_1 + C_2}\right)$$

Finally, we use (3) and (4) to find the voltage across each capacitor:

$$\Delta V_{1f} = \frac{Q_{1f}}{C_1} = \frac{Q\left(\dfrac{C_1}{C_1 + C_2}\right)}{C_1} = \boxed{\frac{Q}{C_1 + C_2}}$$

$$\Delta V_{2f} = \frac{Q_{2f}}{C_2} = \frac{Q\left(\dfrac{C_2}{C_1 + C_2}\right)}{C_2} = \boxed{\frac{Q}{C_1 + C_2}}$$

Notice that $\Delta V_{1f} = \Delta V_{2f} = \Delta V_f$, which is the expected result.

B Find the total energy stored in the capacitors before and after the switches are closed and the ratio of the final energy to the initial energy.

Solution Before the switches are closed, the total energy stored in the capacitors is

$$U_i = \tfrac{1}{2}C_1(\Delta V_i)^2 + \tfrac{1}{2}C_2(\Delta V_i)^2 = \tfrac{1}{2}(C_1 + C_2)(\Delta V_i)^2$$

After the switches are closed, the total energy stored in the capacitors is

$$U_f = \tfrac{1}{2}C_1(\Delta V_f)^2 + \tfrac{1}{2}C_2(\Delta V_f)^2 = \tfrac{1}{2}(C_1 + C_2)(\Delta V_f)^2$$

$$= \tfrac{1}{2}(C_1 + C_2)\left(\frac{Q}{C_1 + C_2}\right)^2 = \boxed{\frac{1}{2}\,\frac{Q^2}{C_1 + C_2}}$$

Using (1), this expression can be written as

$$U_f = \frac{1}{2}\,\frac{(C_1 - C_2)^2(\Delta V_i)^2}{C_1 + C_2}$$

Therefore, the ratio of the final energy stored to the initial energy stored is

$$\frac{U_f}{U_i} = \frac{\left(\dfrac{1}{2}\,\dfrac{(C_1 - C_2)^2(\Delta V_i)^2}{C_1 + C_2}\right)}{\tfrac{1}{2}(C_1 + C_2)(\Delta V_i)^2} = \boxed{\left(\frac{C_1 - C_2}{C_1 + C_2}\right)^2}$$

which shows that the final energy is less than the initial energy. Therefore, even though we correctly used an isolated system model for electric charge in this problem, we see that it is *not* an isolated system for energy. That begs the question as to how energy is transferred out of the system. The transfer mechanism is electromagnetic radiation, which may not be clear to you at this point but which will become clearer once we study the material in Chapter 24.

PhysicsNow™ Log into PhysicsNow at **www.pop4e.com** and go to Interactive Example 20.9 to explore this situation for various initial values of the voltage and the capacitance.

20.10 CAPACITORS WITH DIELECTRICS

A **dielectric** is an insulating material such as rubber, glass, or waxed paper. When a dielectric material is inserted between the plates of a capacitor, the capacitance increases. If the dielectric completely fills the space between the plates, the capacitance increases by the dimensionless factor κ, called the **dielectric constant** of the material.

The following experiment can be performed to illustrate the effect of a dielectric in a capacitor. Consider a parallel-plate capacitor of charge Q_0 and capacitance C_0 in the absence of a dielectric. The potential difference across the capacitor as measured by a voltmeter is $\Delta V_0 = Q_0/C_0$ (Fig. 20.27a). Notice that the capacitor circuit is *open;* that is, the plates of the capacitor are *not* connected to a battery and charge cannot flow through an ideal voltmeter. Hence, there is *no* path by which charge can flow and alter the charge on the capacitor. If a dielectric is now inserted between the plates as in Figure 20.27b, it is found that the voltmeter reading *decreases* by a factor of κ to the value ΔV, where

$$\Delta V = \frac{\Delta V_0}{\kappa}$$

Because $\Delta V < \Delta V_0$, we see that $\kappa > 1$.

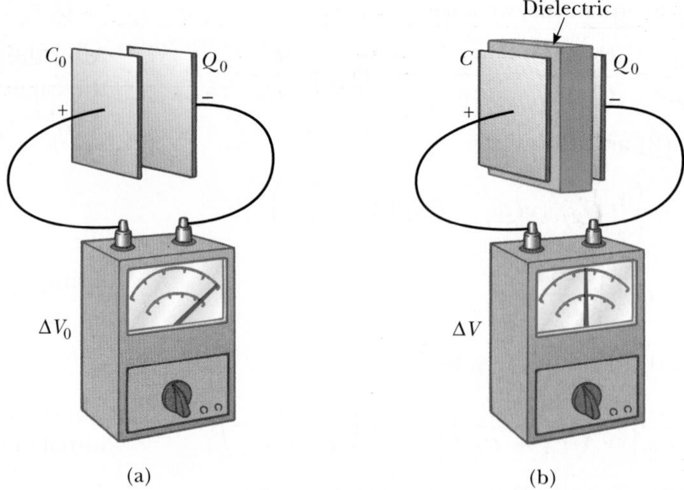

FIGURE 20.27 A charged capacitor (a) before and (b) after insertion of a dielectric between the plates. The charge on the plates remains unchanged, but the potential difference decreases from ΔV_0 to $\Delta V = \Delta V_0/\kappa$. Therefore, the capacitance *increases* from C_0 to κC_0.

Because the charge Q_0 on the capacitor *does not change,* we conclude that the capacitance must change to the value

$$C = \frac{Q_0}{\Delta V} = \frac{Q_0}{\Delta V_0/\kappa} = \kappa \frac{Q_0}{\Delta V_0}$$

$$C = \kappa C_0 \qquad [20.33]$$

where C_0 is the capacitance in the absence of the dielectric. That is, the capacitance *increases* by the factor κ when the dielectric completely fills the region between the plates.[4] For a parallel-plate capacitor, where $C_0 = \epsilon_0 A/d$, we can express the capacitance when the capacitor is filled with a dielectric as

$$C = \kappa \frac{\epsilon_0 A}{d} \qquad [20.34]$$

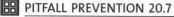

PITFALL PREVENTION 20.7

IS THE CAPACITOR CONNECTED TO A BATTERY? In problems in which you are modifying a capacitor (by insertion of a dielectric, for example), you must note whether modifications to the capacitor are being made while the capacitor is connected to a battery or after it is disconnected. If the capacitor remains connected to the battery, the voltage across the capacitor necessarily remains the same. If you disconnect the capacitor from the battery before making any modifications to the capacitor, the capacitor is an isolated system and its charge remains the same.

From this result, it would appear that the capacitance could be made very large by decreasing d, the distance between the plates. In practice, however, the lowest value of d is limited by the electric discharge that could occur through the dielectric medium separating the plates. For any given separation d, the maximum voltage that can be applied to a capacitor without causing a discharge depends on the **dielectric strength** (maximum electric field) of the dielectric, which for dry air is equal to 3×10^6 V/m. If the electric field in the medium exceeds the dielectric strength, the insulating properties break down and the medium begins to conduct. Most insulating materials have dielectric strengths and dielectric constants greater than those of air, as Table 20.1 indicates. Therefore, we see that a dielectric provides the following advantages:

- It increases the capacitance of a capacitor.
- It increases the maximum operating voltage of a capacitor.
- It may provide mechanical support between the conducting plates.

We can understand the effects of a dielectric by considering the polarization of molecules that we discussed in Section 19.3. Figure 20.28a shows polarized molecules of a dielectric in random orientations in the absence of an electric field. Figure 20.28b shows the polarization of the molecules when the dielectric is placed between the plates of the charged capacitor and the polarized molecules tend to line up parallel to the field lines. The plates set up an electric field $\vec{\mathbf{E}}_0$ in a direction to

[4]If another experiment is performed in which the dielectric is introduced while the potential difference is held constant by means of a battery, the charge increases to the value $Q = \kappa Q_0$. The additional charge is transferred from the connecting wires, and the capacitance still increases by the factor κ.

TABLE 20.1 Approximate Dielectric Constants and Dielectric Strengths of Various Materials at Room Temperature

Material	Dielectric Constant κ	Dielectric Strength[a] (10^6 V/m)
Air (dry)	1.000 59	3
Bakelite	4.9	24
Fused quartz	3.78	8
Mylar	3.2	7
Neoprene rubber	6.7	12
Nylon	3.4	14
Paper	3.7	16
Paraffin-impregnated paper	3.5	11
Polystyrene	2.56	24
Polyvinyl chloride	3.4	40
Porcelain	6	12
Pyrex glass	5.6	14
Silicone oil	2.5	15
Strontium titanate	233	8
Teflon	2.1	60
Vacuum	1.000 00	—
Water	80	—

[a]The dielectric strength equals the maximum electric field that can exist in a dielectric without electrical breakdown. Note that these values depend strongly on the presence of impurities and flaws in the materials.

the right in Figure 20.28b. In the body of the dielectric, a general homogeneity of charge exists, but look along the edges. There is a layer of negative charge along the left edge of the dielectric and a layer of positive charge along the right edge. These layers of charge can be modeled as additional charged parallel plates, as in Figure 20.28c. Because the polarity is opposite that of the real plates, these charges set up an induced electric field $\vec{\mathbf{E}}_{\text{ind}}$ directed to the left in the diagram that partially cancels the electric field due to the real plates. Therefore, for the charged capacitor removed from a battery, the electric field and hence the voltage between the plates is reduced by the introduction of the dielectric. The charge on the plates is stored at a lower potential difference, so the capacitance increases.

Types of Capacitors

Commercial capacitors are often made using metal foil interlaced with a dielectric such as thin sheets of paraffin-impregnated paper. These alternating layers of metal foil and dielectric are then rolled into the shape of a cylinder to form a small package (Fig. 20.29a). High-voltage capacitors commonly consist of interwoven metal plates immersed in silicone oil (Fig. 20.29b). Small capacitors are often constructed from ceramic materials. Variable capacitors (typically 10–500 pF) usually consist of

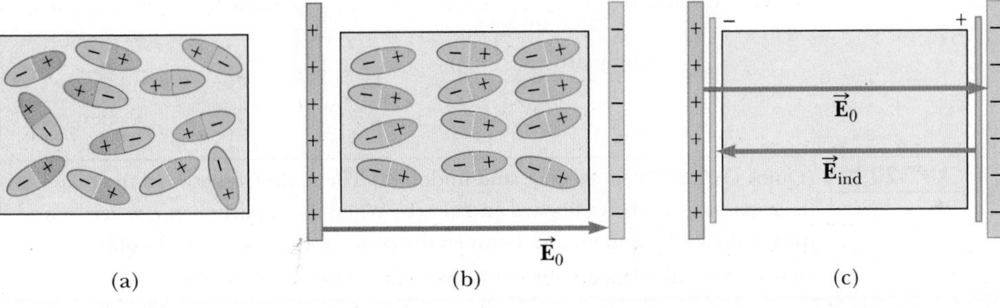

FIGURE 20.28 (a) Polar molecules are randomly oriented in the absence of an external electric field. (b) When an external electric field is applied, the molecules partially align with the field. (c) The charged edges of the dielectric can be modeled as an additional pair of parallel plates establishing an electric field $\vec{\mathbf{E}}_{\text{ind}}$ in the direction opposite to that of $\vec{\mathbf{E}}_0$.

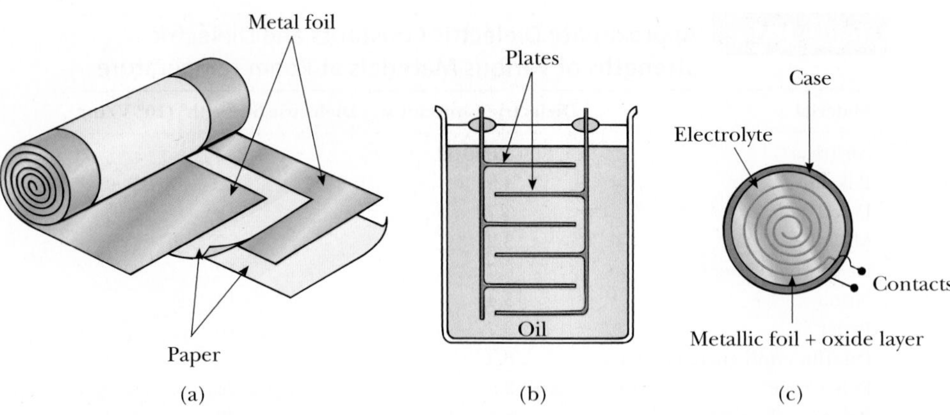

FIGURE 20.29 Three commercial capacitor designs. (a) A tubular capacitor whose plates are separated by paper and then rolled into a cylinder. (b) A high-voltage capacitor consists of many parallel plates separated by insulating oil. (c) An electrolytic capacitor.

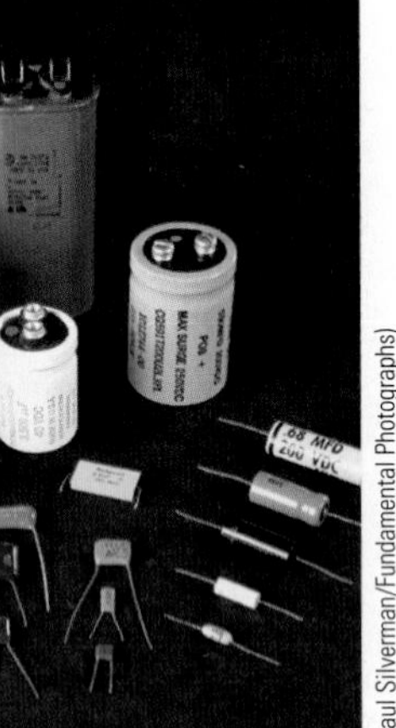

(Paul Silverman/Fundamental Photographs)

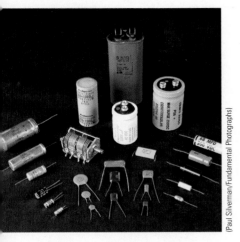

A collection of capacitors used in a variety of applications. ■

two interwoven sets of metal plates, one fixed and the other movable, with air as the dielectric.

An *electrolytic capacitor* is often used to store large amounts of charge at relatively low voltages. This device, shown in Figure 20.29c, consists of a metal foil in contact with an electrolyte, a solution that conducts electricity by virtue of the motion of ions contained in the solution. When a voltage is applied between the foil and the electrolyte, a thin layer of metal oxide (an insulator) is formed on the foil, and this layer serves as the dielectric. Very large capacitance values can be attained because the dielectric layer is very thin.

When electrolytic capacitors are used in circuits, they must be installed with the proper polarity. If the polarity of the applied voltage is opposite what is intended, the oxide layer will be removed and the capacitor will not be able to store charge.

QUICK QUIZ 20.7 If you have ever tried to hang a picture, you know it can be difficult to locate a wooden stud in which to anchor your nail or screw. A carpenter's electric stud finder will locate the studs in a wall. It consists of a parallel-plate capacitor, with the plates next to each other, as shown in Figure 20.30. Does the capacitance increase or decrease when the device is moved over a stud?

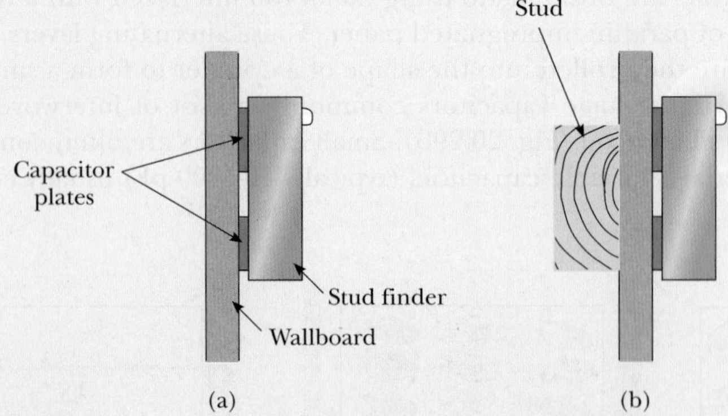

FIGURE 20.30 (Quick Quiz 20.7) An electric stud finder. (a) The materials between the plates of the capacitor are the wallboard and air. (b) When the capacitor moves across a stud in the wall, the materials between the plates are wallboard and wood. The change in the dielectric constant causes a signal light to illuminate.

EXAMPLE 20.10 A Paper-Filled Capacitor

A parallel-plate capacitor has plates of dimensions 2.0 cm × 3.0 cm separated by a 1.0-mm thickness of paper.

A Find the capacitance of this device.

Solution Because $\kappa = 3.7$ for paper (see Table 20.1), we have

$$C = \kappa \frac{\epsilon_0 A}{d}$$

$$= 3.7(8.85 \times 10^{-12} \text{ C}^2/\text{N} \cdot \text{m}^2) \left(\frac{6.0 \times 10^{-4} \text{ m}^2}{1.0 \times 10^{-3} \text{ m}} \right)$$

$$= 20 \times 10^{-12} \text{ F} = 20 \text{ pF}$$

B What is the maximum charge that can be placed on the capacitor?

Solution From Table 20.1, we see that the dielectric strength of paper is 16×10^6 V/m. Because the thickness of the paper is 1.0 mm, the maximum voltage that can be applied before breakdown is

$$\Delta V_{\text{max}} = E_{\text{max}} d = (16 \times 10^6 \text{ V/m})(1.0 \times 10^{-3} \text{ m})$$
$$= 16 \times 10^3 \text{ V}$$

Hence, the maximum charge is

$$Q_{\text{max}} = C\,\Delta V_{\text{max}} = (20 \times 10^{-12} \text{ F})(16 \times 10^3 \text{ V}) = 0.32\ \mu\text{C}$$

EXAMPLE 20.11 Energy Stored Before and After

A parallel-plate capacitor is charged with a battery to a charge Q_0 as in Figure 20.31a. The battery is then removed and a slab of material that has a dielectric constant κ is inserted between the plates as in Figure 20.31b. Find the energy stored in the capacitor before and after the dielectric is inserted.

Solution From Equation 20.30, the energy stored in the capacitor in the absence of the dielectric is

$$U_0 = \frac{Q_0^2}{2C_0}$$

After the battery is removed and the dielectric is inserted between the plates, the charge on the capacitor remains the same because the unconnected capacitor is an isolated system. Hence, the energy stored in the presence of the dielectric is

$$U = \frac{Q_0^2}{2C}$$

The capacitance in the presence of the dielectric, however, is given by $C = \kappa C_0$, so U becomes

$$U = \frac{Q_0^2}{2\kappa C_0} = \frac{U_0}{\kappa}$$

Because $\kappa > 1$, we see that the final energy is less than the initial energy by the factor $1/\kappa$.

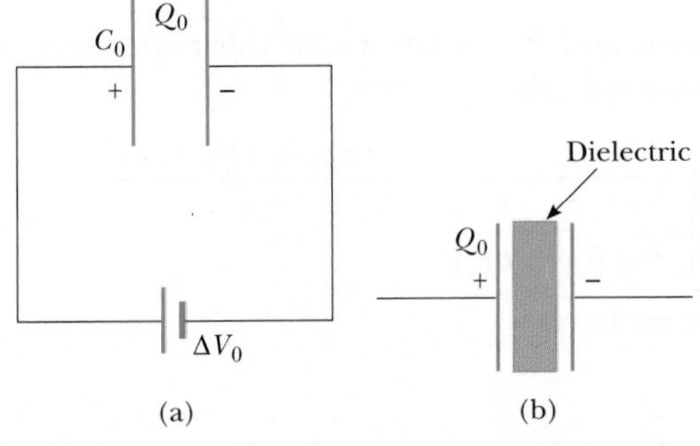

FIGURE 20.31 (Example 20.11) (a) A battery charges up a parallel-plate capacitor. (b) The battery is removed and a slab of dielectric material is inserted between the plates.

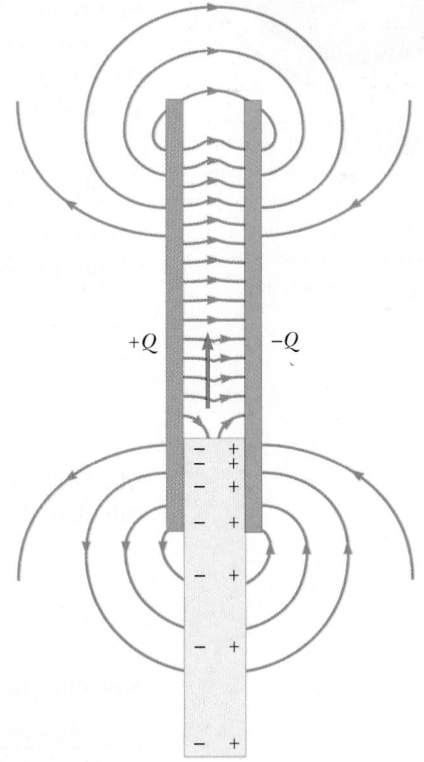

FIGURE 20.32 (Example 20.11) When a dielectric approaches an empty capacitor, the charge distributions induced on the edges cause an attraction between the dielectric and the capacitor.

This missing energy can be accounted for as follows. We identify the system as the capacitor and the dielectric. As the dielectric is brought near the capacitor so that electric field lines from the plates pass through the dielectric, the molecules of the dielectric become polarized. The edges of the dielectric take on a charge opposite to the plate nearest the edge as in Figure 20.32. Therefore, an attractive force exists between the dielectric and the plates. If the dielectric were released, it would be pulled into the plates and would pass through the plates with a kinetic energy, emerging on the other side and exhibiting oscillatory motion. If an external agent such as your hand holds the dielectric, allowing it to move into the plates at constant speed, the agent is doing negative work on the system. The displacement of the dielectric is into the plates, but the applied force is away from the plates. This work represents a decrease in energy for the system, so the energy change is $\Delta U = U - U_0$. Therefore, the unconnected capacitor and dielectric form an isolated system for electric charge, but when considering energy, it is a nonisolated system.

20.11 THE ATMOSPHERE AS A CAPACITOR

CONTEXT CONNECTION

Negative plate (Earth's surface)

Positive plate (charges in atmosphere)

FIGURE 20.33 The atmospheric capacitor. The Earth's surface serves as the negative plate, and the positive plate is modeled at a height in the atmosphere that represents positive charges spread through the atmosphere.

In the Context Connection of Chapter 19, we mentioned some processes occurring on the surface of the Earth and in the atmosphere that result in charge distributions. These processes result in a negative charge on the Earth's surface and positive charges distributed throughout the air.

This separation of charge can be modeled as a capacitor. The surface of the Earth is one plate and the positive charge in the air is the other plate. The positive charge in the atmosphere is not all located at one height but is spread throughout the atmosphere. Therefore, the position of the upper plate must be modeled, based on the charge distribution. Models of the atmosphere show that an appropriate effective height of the upper plate is about 5 km from the surface. The model atmospheric capacitor is shown in Figure 20.33.

Considering the charge distribution on the surface of the Earth to be spherically symmetric, we can use the result from Example 20.6 to claim that the potential at a point above the Earth's surface is

$$V = k_e \frac{Q}{r}$$

where Q is the charge on the surface. The potential difference between the plates of our atmospheric capacitor is

$$\begin{aligned}\Delta V &= \frac{Q}{4\pi\epsilon_0}\left(\frac{1}{r_{\text{surface}}} - \frac{1}{r_{\text{upper plate}}}\right) \\ &= \frac{Q}{4\pi\epsilon_0}\left(\frac{1}{R_E} - \frac{1}{R_E + h}\right) = \frac{Q}{4\pi\epsilon_0}\left(\frac{h}{R_E(R_E + h)}\right)\end{aligned}$$

where R_E is the radius of the Earth and $h = 5$ km. From this expression, we can calculate the capacitance of the atmospheric capacitor:

$$C = \frac{Q}{\Delta V} = \frac{Q}{\dfrac{Q}{4\pi\epsilon_0}\left[\dfrac{h}{R_E(R_E + h)}\right]} = \frac{4\pi\epsilon_0 R_E(R_E + h)}{h}$$

Substituting the numerical values, we have

$$\begin{aligned}C &= \frac{4\pi\epsilon_0 R_E(R_E + h)}{h} \\ &= \frac{4\pi(8.85 \times 10^{-12}\ \text{C}^2/\text{N}\cdot\text{m}^2)(6.4 \times 10^3\ \text{km})(6.4 \times 10^3\ \text{km} + 5\ \text{km})}{5\ \text{km}}\left(\frac{1000\ \text{m}}{1\ \text{km}}\right) \\ &\approx 0.9\ \text{F}\end{aligned}$$

This result is extremely large, compared with the *picofarads* and *microfarads* that are typical values for capacitors in electrical circuits, especially for a capacitor having plates that are 5 km apart! We shall use this model of the atmosphere as a capacitor in our Context Conclusion, in which we calculate the number of lightning strikes on the Earth in one day.

SUMMARY

PhysicsNow™ Take a practice test by logging into PhysicsNow at www.pop4e.com and clicking on the Pre-Test link for this chapter.

When a positive test charge q_0 is moved between points A and B in an electric field $\vec{\mathbf{E}}$, the **change in potential energy** of the charge–field system is

$$\Delta U = -q_0 \int_A^B \vec{\mathbf{E}} \cdot d\vec{\mathbf{s}} \qquad [20.1]$$

The **potential difference** ΔV between points A and B in an electric field $\vec{\mathbf{E}}$ is defined as the change in potential energy divided by the test charge q_0:

$$\Delta V = \frac{\Delta U}{q_0} = -\int_A^B \vec{\mathbf{E}} \cdot d\vec{\mathbf{s}} \qquad [20.3]$$

where **electric potential** V is a scalar and has the units joules per coulomb, defined as 1 **volt** (V).

The potential difference between two points A and B in a uniform electric field $\vec{\mathbf{E}}$ is

$$\Delta V = -\vec{\mathbf{E}} \cdot \Delta\vec{\mathbf{r}} \qquad [20.8]$$

where $\Delta\vec{\mathbf{r}}$ is the displacement vector between A and B.

Equipotential surfaces are surfaces on which the electric potential remains constant. Equipotential surfaces are *perpendicular* to electric field lines.

The electric potential due to a point charge q at a distance r from the charge is

$$V = k_e \frac{q}{r} \qquad [20.11]$$

The electric potential due to a group of point charges is obtained by summing the potentials due to the individual charges. Because V is a scalar, the sum is a simple algebraic operation.

The **electric potential energy of a pair of point charges** separated by a distance r_{12} is

$$U = k_e \frac{q_1 q_2}{r_{12}} \qquad [20.13]$$

which represents the work required to bring the charges from an infinite separation to the separation r_{12}. The potential energy of a distribution of point charges is obtained by summing terms like Equation 20.13 over *all pairs* of particles.

If the electric potential is known as a function of coordinates x, y, and z, the components of the electric field can be obtained by taking the negative derivative of the potential with respect to the coordinates. For example, the x component of an electric field in the x direction is

$$E_x = -\frac{\partial V}{\partial x} \qquad [20.16]$$

The **electric potential due to a continuous charge distribution** is

$$V = k_e \int \frac{dq}{r} \qquad [20.18]$$

Every point on the surface of a charged conductor in electrostatic equilibrium is at the same potential. Furthermore, the potential is constant everywhere inside the conductor and is equal to its value at the surface.

A capacitor is a device for storing charge. A charged capacitor consists of two equal and oppositely charged conductors with a potential difference ΔV between them. The **capacitance** C of any capacitor is defined as the ratio of the magnitude of the charge Q on either conductor to the magnitude of the potential difference ΔV:

$$C \equiv \frac{Q}{\Delta V} \qquad [20.19]$$

The SI units of capacitance are coulombs per volt, or the **farad** (F), and 1 F = 1 C/V.

If two or more capacitors are connected in parallel, the potential differences across them must be the same. The equivalent capacitance of a parallel combination of capacitors is

$$C_{\text{eq}} = C_1 + C_2 + C_3 + \cdots \qquad [20.25]$$

If two or more capacitors are connected in series, the charges on them are the same and the equivalent capacitance of the series combination is given by

$$\frac{1}{C_{\text{eq}}} = \frac{1}{C_1} + \frac{1}{C_2} + \frac{1}{C_3} + \cdots \qquad [20.29]$$

Energy is required to charge a capacitor because the charging process is equivalent to transferring charges from one conductor at a lower potential to another conductor at a higher potential. The electric potential energy U stored in the capacitor is

$$U = \frac{Q^2}{2C} = \tfrac{1}{2}Q\,\Delta V = \tfrac{1}{2}C(\Delta V)^2 \qquad [20.30]$$

When a dielectric material is inserted between the plates of a capacitor, the capacitance generally increases by the dimensionless factor κ, called the **dielectric constant.** That is,

$$C = \kappa C_0 \qquad [20.33]$$

where C_0 is the capacitance in the absence of the dielectric.

QUESTIONS

☐ = answer available in the *Student Solutions Manual and Study Guide*

1. Distinguish between electric potential and electric potential energy.
2. A negative charge moves in the direction of a uniform electric field. Does the potential energy of the charge–field system increase or decrease? Does the charge move to a position of higher or lower potential?
3. Give a physical explanation showing that the potential energy of a pair of charges with the same sign is positive, whereas the potential energy of a pair of charges with opposite signs is negative.
4. Explain why, under static conditions, all points in a conductor must be at the same electric potential.
5. Why is it important to avoid sharp edges or points on conductors used in high-voltage equipment?
6. How would you shield an electronic circuit or laboratory from stray electric fields? Why does that method work?
7. Study Figure 19.4 and the accompanying text discussion of charging by induction. When the grounding wire is touched to the rightmost point on the sphere in Figure 19.4c, electrons are drained away from the sphere to leave the sphere positively charged. Instead, suppose the grounding wire is touched to the leftmost point on the sphere. Will electrons still drain away, moving closer to the negatively charged rod as they do so? What kind of charge, if any, will remain on the sphere?
8. The plates of a capacitor are connected to a battery. What happens to the charge on the plates if the connecting wires are removed from the battery? What happens to the charge if the wires are removed from the battery and connected to each other?
9. One pair of capacitors is connected in parallel, whereas an identical pair is connected in series. Which pair would be more dangerous to handle after being connected to the same battery? Explain.
10. If you are given three different capacitors C_1, C_2, C_3, how many different combinations of capacitance can you produce?
11. If the potential difference across a capacitor is doubled, by what factor does the energy stored change?
12. Because the charges on the plates of a parallel-plate capacitor are opposite in sign, they attract each other. Hence, it would take positive work to increase the plate separation. What type of energy in the system changes due to the external work done in this process?
13. It is possible to obtain large potential differences by first charging a group of capacitors connected in parallel and then activating a switch arrangement that in effect disconnects the capacitors from the charging source and from each other and reconnects them in a series arrangement. The group of charged capacitors is then discharged in series. What is the maximum potential difference that can be obtained in this manner by using ten capacitors each of 500 μF and a charging source of 800 V?
14. Assume that you want to increase the maximum operating voltage of a parallel-plate capacitor. Describe how you can do so for a fixed plate separation.
15. If you were asked to design a capacitor in which small size and large capacitance were required, what factors would be important in your design?
16. Explain why a dielectric increases the maximum operating voltage of a capacitor although the physical size of the capacitor does not change.

PROBLEMS

1, 2, 3 = straightforward, intermediate, challenging
☐ = full solution available in the *Student Solutions Manual and Study Guide*
PhysicsNow™ = coached problem with hints available at www.pop4e.com
= computer useful in solving problem
= paired numerical and symbolic problems
= biomedical application

Section 20.1 ▪ Potential Difference and Electric Potential

1. (a) Calculate the speed of a proton that is accelerated from rest through a potential difference of 120 V. (b) Calculate the speed of an electron that is accelerated through the same potential difference.
2. How much work is done (by a battery, generator, or some other source of potential difference) in moving Avogadro's number of electrons from an initial point where the electric potential is 9.00 V to a point where the potential is -5.00 V? (The potential in each case is measured relative to a common reference point.)

Section 20.2 ▪ Potential Differences in a Uniform Electric Field

3. A uniform electric field of magnitude 250 V/m is directed in the positive x direction. A $+12.0$-μC charge moves from the origin to the point $(x, y) = (20.0$ cm, 50.0 cm$)$. (a) What is the change in the potential energy of the charge–field system? (b) Through what potential difference does the charge move?
4. The difference in potential between the accelerating plates in the electron gun of a TV picture tube is about

25 000 V. If the distance between these plates is 1.50 cm, what is the magnitude of the uniform electric field in this region?

5. PhysicsNow™ An electron moving parallel to the x axis has an initial speed of 3.70×10^6 m/s at the origin. Its speed is reduced to 1.40×10^5 m/s at the point $x = 2.00$ cm. Calculate the potential difference between the origin and that point. Which point is at the higher potential?

6. **Review problem.** A block having mass m and positive charge Q is connected to an insulating spring having constant k. The block lies on a frictionless, insulating horizontal track, and the system is immersed in a uniform electric field of magnitude E, directed as shown in Figure P20.6. If the block is released from rest when the spring is unstretched (at $x = 0$), (a) by what maximum amount does the spring expand? (b) What is the equilibrium position of the block? (c) Show that the block's motion is simple harmonic and determine its period. (d) Repeat part (a), assuming that the coefficient of kinetic friction between block and surface is μ_k.

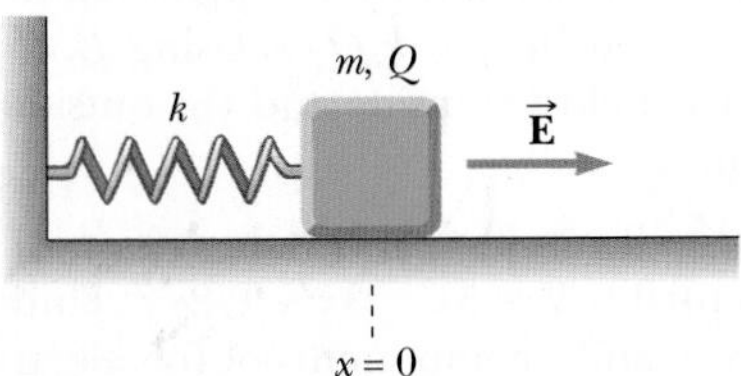

FIGURE P20.6

Section 20.3 ■ Electric Potential and Electric Potential Energy Due to Point Charges

Note: Unless stated otherwise, assume that the reference level of potential is $V = 0$ at $r = \infty$.

7. (a) Find the potential at a distance of 1.00 cm from a proton. (b) What is the potential difference between two points that are 1.00 cm and 2.00 cm from a proton? (c) Repeat parts (a) and (b) for an electron.

8. Given two 2.00-μC charges as shown in Figure P20.8 and a positive test charge $q = 1.28 \times 10^{-18}$ C at the origin, (a) what is the net force exerted by the two 2.00-μC charges on the test charge q? (b) What is the electric field at the origin due to the two 2.00-μC charges? (c) What is the electrical potential at the origin due to the two 2.00-μC charges?

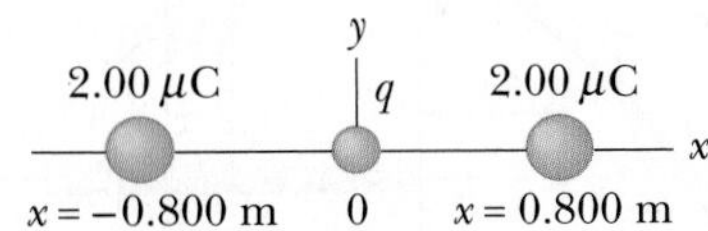

FIGURE P20.8

9. A charge $+q$ is at the origin. A charge $-2q$ is at $x = 2.00$ m on the x axis. (a) For what finite value(s) of x is the electric field zero? (b) For what finite value(s) of x is the electric potential zero?

10. *Compare this problem with Problem 19.55.* Four identical point charges ($q = +10.0$ μC) are located on the corners of a rectangle as shown in Figure P19.55. The dimensions of the rectangle are $L = 60.0$ cm and $W = 15.0$ cm. Calculate the change in electric potential energy of the system as the charge at the lower left corner in Figure P19.55 is brought to this position from infinitely far away. Assume that the other three charges remain fixed in position.

11. The three charges in Figure P20.11 are at the vertices of an isosceles triangle. Calculate the electric potential at the midpoint of the base, taking $q = 7.00$ μC.

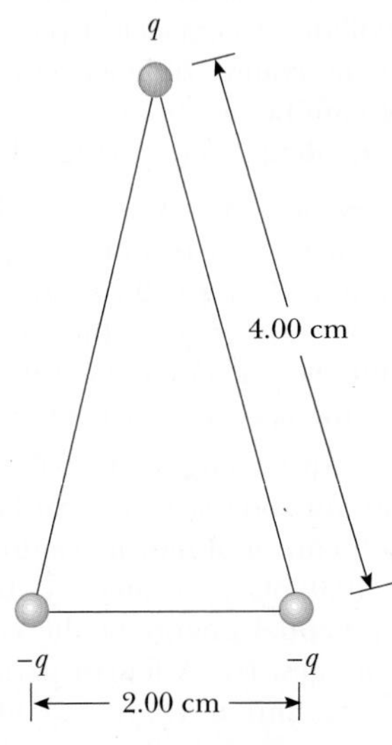

FIGURE P20.11

12. *Compare this problem with Problem 19.14.* Two point charges each of magnitude 2.00 μC are located on the x axis. One is at $x = 1.00$ m and the other is at $x = -1.00$ m. (a) Determine the electric potential on the y axis at $y = 0.500$ m. (b) Calculate the change in electric potential energy of the system as a third charge of -3.00 μC is brought from infinitely far away to a position on the y axis at $y = 0.500$ m.

13. PhysicsNow™ Show that the amount of work required to assemble four identical point charges of magnitude Q at the corners of a square of side s is $5.41k_eQ^2/s$.

14. Two charged particles create influences at the origin, described by the expressions

$$8.99 \times 10^9\ \text{N}\cdot\text{m}^2/\text{C}^2 \left[-\frac{7.00 \times 10^{-9}\ \text{C}}{(0.070\ 0\ \text{m})^2} \cos 70.0°\hat{\mathbf{i}} - \frac{7.00 \times 10^{-9}\ \text{C}}{(0.070\ 0\ \text{m})^2} \sin 70.0°\hat{\mathbf{j}} + \frac{8.00 \times 10^{-9}\ \text{C}}{(0.030\ 0\ \text{m})^2}\hat{\mathbf{j}} \right]$$

and

$$8.99 \times 10^9\ \text{N}\cdot\text{m}^2/\text{C}^2 \left[\frac{7.00 \times 10^{-9}\ \text{C}}{0.070\ 0\ \text{m}} - \frac{8.00 \times 10^{-9}\ \text{C}}{0.030\ 0\ \text{m}} \right]$$

(a) Identify the locations of the particles and the charges on them. (b) Find the force on a -16.0-nC charge placed at the origin and (c) the work required to move this third charge to the origin from a very distant point.

15. **Review problem.** Two insulating spheres have radii 0.300 cm and 0.500 cm, masses 0.100 kg and 0.700 kg, and uniformly distributed charges -2.00 μC and 3.00 μC. They are released from rest when their centers are separated by

1.00 m. (a) How fast will each be moving when they collide? (*Suggestion:* Consider conservation of energy and of linear momentum.) (b) If the spheres were conductors, would the speeds be greater or less than those calculated in part (a)? Explain.

16. **Review problem.** Two insulating spheres have radii r_1 and r_2, masses m_1 and m_2, and uniformly distributed charges $-q_1$ and q_2. They are released from rest when their centers are separated by a distance d. (a) How fast is each moving when they collide? (*Suggestion:* Consider conservation of energy and conservation of linear momentum.) (b) If the spheres were conductors, would their speeds be greater or less than those calculated in part (a)? Explain.

17. *Compare this problem with Problem 19.26.* Three equal positive charges q are at the corners of an equilateral triangle of side a as shown in Figure P19.26. (a) At what point, if any, in the plane of the charges is the electric potential zero? (b) What is the electric potential at the point P due to the two charges at the base of the triangle?

18. Two particles, with charges of 20.0 nC and -20.0 nC, are placed at the points with coordinates (0, 4.00 cm) and (0, -4.00 cm) as shown in Figure P20.18. A particle with charge 10.0 nC is located at the origin. (a) Find the electric potential energy of the configuration of the three fixed charges. (b) A fourth particle, with a mass of 2.00×10^{-13} kg and a charge of 40.0 nC, is released from rest at the point (3.00 cm, 0). Find its speed after it has moved freely to a very large distance away.

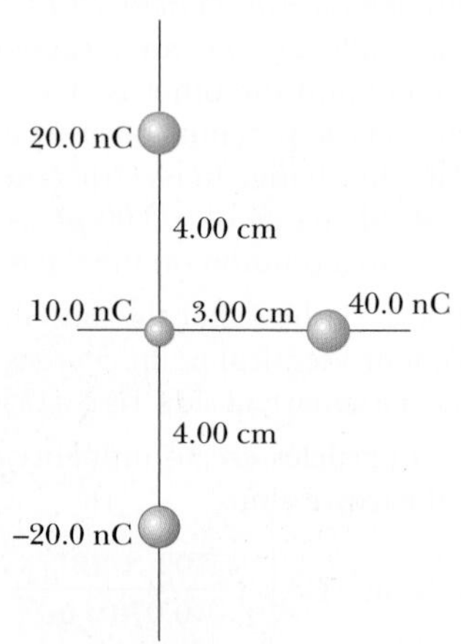

FIGURE **P20.18**

19. **Review problem.** A light, unstressed spring has length d. Two identical particles, each with charge q, are connected to the opposite ends of the spring. The particles are held stationary a distance d apart and are then released at the same time. The system then oscillates on a horizontal frictionless table. The spring has a bit of internal kinetic friction, so the oscillation is damped. The particles eventually stop vibrating when the distance between them is $3d$. Find the increase in internal energy that appears in the spring during the oscillations. Assume that the system of the spring and two charges is isolated.

20. In 1911, Ernest Rutherford and his assistants Hans Geiger and Ernest Marsden conducted an experiment in which they scattered alpha particles from thin sheets of gold. An alpha particle, having charge $+2e$ and mass 6.64×10^{-27} kg, is a product of certain radioactive decays. The results of the experiment led Rutherford to the idea that most of the mass of an atom is in a very small nucleus, with electrons in orbit around it, in his planetary model of the atom. Assume that an alpha particle, initially very far from a gold nucleus, is fired with a velocity of 2.00×10^7 m/s directly toward the nucleus (charge $+79e$). How close does the alpha particle get to the nucleus before turning around? Assume that the gold nucleus remains stationary.

Section 20.4 ▪ Obtaining Electric Field From Electric Potential

21. The potential in a region between $x = 0$ and $x = 6.00$ m is $V = a + bx$, where $a = 10.0$ V and $b = -7.00$ V/m. Determine (a) the potential at $x = 0$, 3.00 m, and 6.00 m; and (b) the magnitude and direction of the electric field at $x = 0$, 3.00 m, and 6.00 m.

22. The electric potential inside a charged spherical conductor of radius R is given by $V = k_eQ/R$ and outside the potential is given by $V = k_eQ/r$. Using $E_r = -dV/dr$, derive the electric field (a) inside and (b) outside this charge distribution.

23. **PhysicsNow™** Over a certain region of space, the electric potential is $V = 5x - 3x^2y + 2yz^2$. Find the expressions for the x, y, and z components of the electric field over this region. What is the magnitude of the field at the point P that has coordinates $(1, 0, -2)$ m?

Section 20.5 ▪ Electric Potential Due to Continuous Charge Distributions

24. Consider a ring of radius R with the total charge Q spread uniformly over its perimeter. What is the potential difference between the point at the center of the ring and a point on its axis a distance $2R$ from the center?

25. A rod of length L (Fig. P20.25) lies along the x axis with its left end at the origin. It has a nonuniform charge density $\lambda = \alpha x$, where α is a positive constant. (a) What are the units of α? (b) Calculate the electric potential at A.

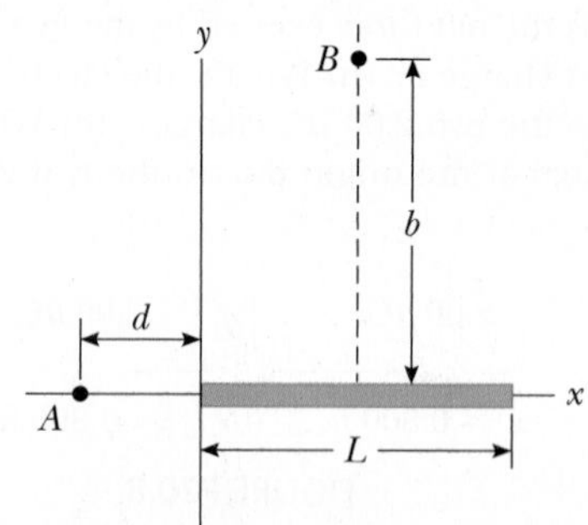

FIGURE **P20.25** Problems 20.25 and 20.26.

26. For the arrangement described in Problem 20.25, calculate the electric potential at point B that lies on the perpendicular bisector of the rod a distance b above the x axis.

27. *Compare this problem with Problem 19.21.* A uniformly charged insulating rod of length 14.0 cm is bent into the shape of a

semicircle as shown in Figure P19.21. The rod has a total charge of $-7.50\ \mu C$. Find the electric potential at O, the center of the semicircle.

Section 20.6 ▪ Electric Potential of a Charged Conductor

28. How many electrons should be removed from an initially uncharged spherical conductor of radius 0.300 m to produce a potential of 7.50 kV at the surface?

29. PhysicsNow™ A spherical conductor has a radius of 14.0 cm and charge of 26.0 μC. Calculate the electric field and the electric potential (a) $r = 10.0$ cm, (b) $r = 20.0$ cm, and (c) $r = 14.0$ cm from the center.

30. Electric charge can accumulate on an airplane in flight. You may have observed needle-shaped metal extensions on the wing tips and tail of an airplane. Their purpose is to allow charge to leak off before much of it accumulates. The electric field around the needle is much larger than around the body of the airplane and can become large enough to produce dielectric breakdown of the air, discharging the airplane. To model this process, assume that two charged spherical conductors are connected by a long conducting wire and that a charge of 1.20 μC is placed on the combination. One sphere, representing the body of the airplane, has a radius of 6.00 cm, and the other, representing the tip of the needle, has a radius of 2.00 cm. (a) What is the electric potential of each sphere? (b) What is the electric field at the surface of each sphere?

Section 20.7 ▪ Capacitance

31. (a) How much charge is on each plate of a 4.00-μF capacitor when it is connected to a 12.0-V battery? (b) If this same capacitor is connected to a 1.50-V battery, what charge is stored?

32. Two conductors having net charges of $+10.0\ \mu C$ and $-10.0\ \mu C$ have a potential difference of 10.0 V between them. (a) Determine the capacitance of the system. (b) What is the potential difference between the two conductors if the charges on each are increased to $+100\ \mu C$ and $-100\ \mu C$?

33. An isolated charged conducting sphere of radius 12.0 cm creates an electric field of 4.90×10^4 N/C at a distance 21.0 cm from its center. (a) What is its surface charge density? (b) What is its capacitance?

34. A variable air capacitor used in a radio tuning circuit is made of N semicircular plates each of radius R and positioned a distance d from its neighbors, to which it is electrically connected. As shown in the opening photograph on page 642 and modeled in Figure P20.34, a second identical set of plates is enmeshed with its plates halfway between those of the first set. The second set can rotate as a unit. Determine the capacitance as a function of the angle of rotation θ, where $\theta = 0$ corresponds to the maximum capacitance.

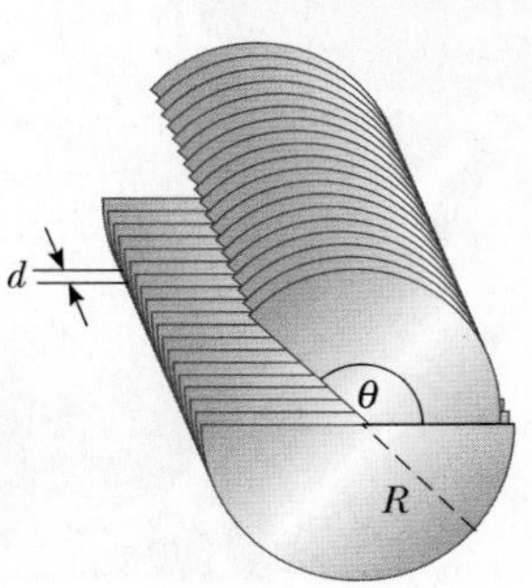

FIGURE **P20.34**

35. An air-filled capacitor consists of two parallel plates, each with an area of 7.60 cm^2, separated by a distance of 1.80 mm. A 20.0-V potential difference is applied to these plates. Calculate (a) the electric field between the plates, (b) the surface charge density, (c) the capacitance, and (d) the charge on each plate.

36. A 50.0-m length of coaxial cable has an inner conductor that has a diameter of 2.58 mm and carries a charge of 8.10 μC. The surrounding conductor has an inner diameter of 7.27 mm and a charge of $-8.10\ \mu C$. (a) What is the capacitance of this cable? (b) What is the potential difference between the two conductors? Assume that the region between the conductors is air.

37. A small object of mass m carries a charge q and is suspended by a thread between the vertical plates of a parallel-plate capacitor. The plate separation is d. If the thread makes an angle θ with the vertical, what is the potential difference between the plates?

38. A *spherical capacitor* consists of a spherical conducting shell of radius b and charge $-Q$ that is concentric with a smaller conducting sphere of radius a and charge $+Q$ (Fig. P20.38). (a) Show that its capacitance is

$$C = \frac{ab}{k_e(b-a)}$$

(b) Show that as b approaches infinity, the capacitance approaches the value $a/k_e = 4\pi\epsilon_0 a$.

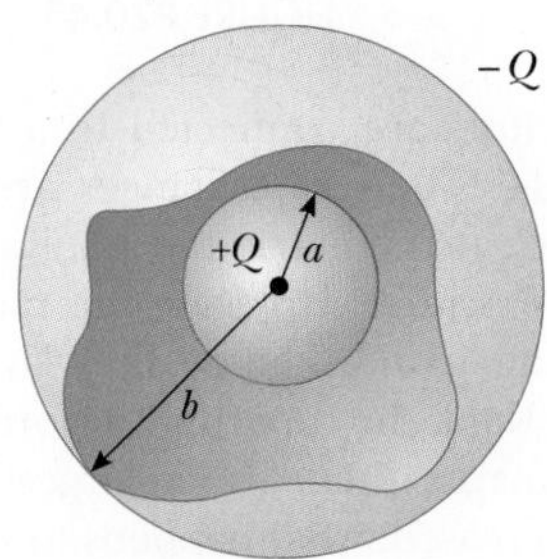

FIGURE **P20.38**

Section 20.8 ▪ Combinations of Capacitors

39. Two capacitors, $C_1 = 5.00\ \mu F$ and $C_2 = 12.0\ \mu F$, are connected in parallel, and the resulting combination is connected to a 9.00-V battery. (a) What is the equivalent capacitance of the combination? What are (b) the potential difference across each capacitor and (c) the charge stored on each capacitor?

40. The two capacitors of Problem 20.39 are now connected in series and to a 9.00-V battery. Find (a) the equivalent capacitance of the combination, (b) the potential difference across each capacitor, and (c) the charge on each capacitor.

41. Physics Now™ Four capacitors are connected as shown in Figure P20.41. (a) Find the equivalent capacitance between points a and b. (b) Calculate the charge on each capacitor, taking $\Delta V_{ab} = 15.0$ V.

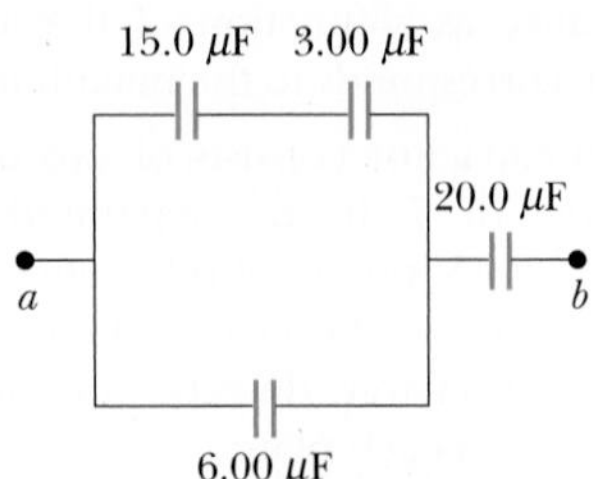

FIGURE **P20.41**

42. Two capacitors when connected in parallel give an equivalent capacitance of C_p and an equivalent capacitance of C_s when connected in series. What is the capacitance of each capacitor?

43. Consider the circuit shown in Figure P20.43, where $C_1 = 6.00$ μF, $C_2 = 3.00$ μF, and $\Delta V = 20.0$ V. Capacitor C_1 is first charged by the closing of switch S_1. Switch S_1 is then opened, and the charged capacitor is connected to the uncharged capacitor by the closing of S_2. Calculate the initial charge acquired by C_1 and the final charge on each capacitor.

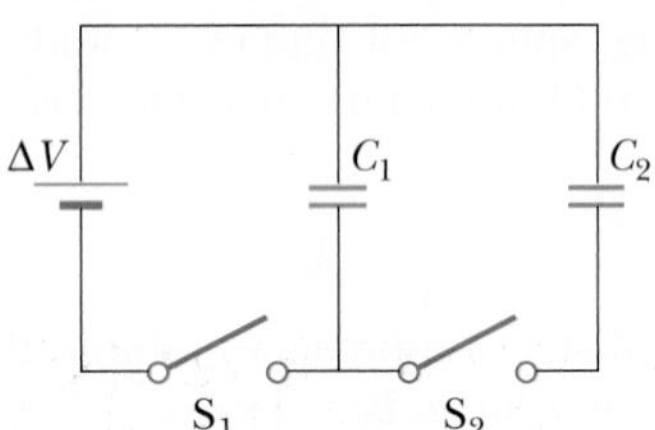

FIGURE **P20.43**

44. Three capacitors are connected to a battery as shown in Figure P20.44. Their capacitances are $C_1 = 3C$, $C_2 = C$, and $C_3 = 5C$. (a) What is the equivalent capacitance of this set of capacitors? (b) State the ranking of the capacitors according to the charge they store, from largest to smallest. (c) Rank the capacitors according to the potential differences across them, from largest to smallest. (d) If now C_3 is increased, what happens to the charge stored by each of the capacitors?

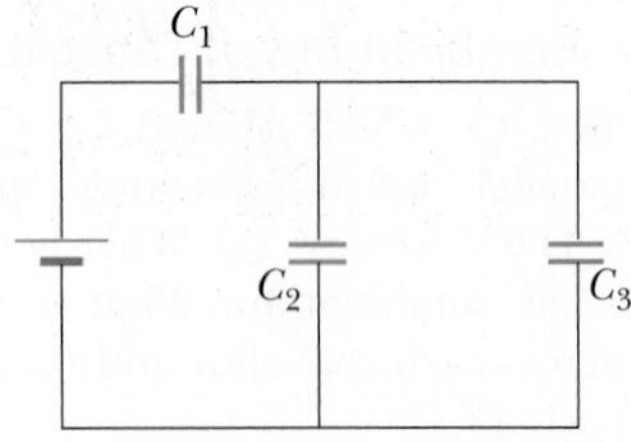

FIGURE **P20.44**

45. According to its design specification, the timer circuit delaying the closing of an elevator door is to have a capacitance of 32.0 μF between two points A and B. (a) When one circuit is being constructed, the inexpensive but durable capacitor installed between these two points is found to have capacitance 34.8 μF. To meet the specification, one additional capacitor can be placed between the two points. Should it be in series or in parallel with the 34.8-μF capacitor? What should be its capacitance? (b) The next circuit comes down the assembly line with capacitance 29.8 μF between A and B. What additional capacitor should be installed in series or in parallel in that circuit to meet the specification?

46. Find the equivalent capacitance between points a and b in the combination of capacitors shown in Figure P20.46.

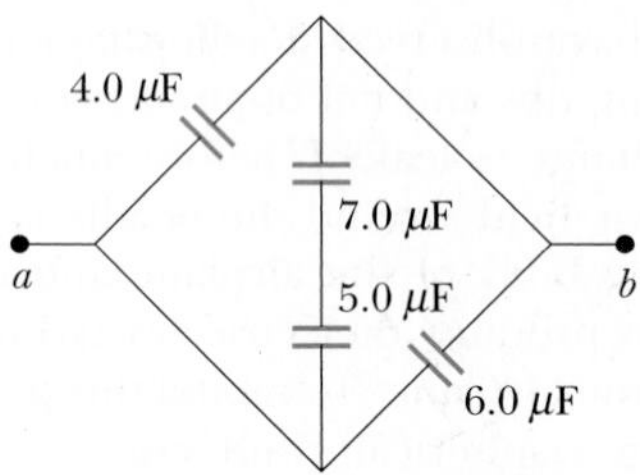

FIGURE **P20.46**

Section 20.9 ▪ Energy Stored in a Charged Capacitor

47. (a) A 3.00-μF capacitor is connected to a 12.0-V battery. How much energy is stored in the capacitor? (b) If the capacitor had been connected to a 6.00-V battery, how much energy would have been stored?

48. The immediate cause of many deaths is ventricular fibrillation, an uncoordinated quivering of the heart as opposed to proper beating. An electric shock to the chest can cause momentary paralysis of the heart muscle, after which the heart will sometimes start organized beating again. A *defibrillator* (Fig. P20.48) is a device that applies a strong electric shock to the chest over a time interval of a few milliseconds. The device contains a capacitor of several microfarads, charged to several thousand volts. Electrodes called paddles, about 8 cm across and coated with conducting paste, are held against the chest on both sides of the heart.

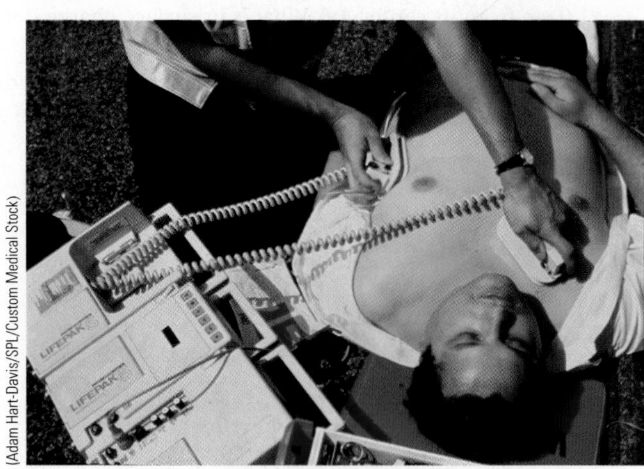

(Adam Hart-Davis/SPL/Custom Medical Stock)

FIGURE **P20.48** A defibrillator in use.

Their handles are insulated to prevent injury to the operator, who calls "Clear!" and pushes a button on one paddle to discharge the capacitor through the patient's chest. Assume that an energy of 300 J is to be delivered from a 30.0-μF capacitor. To what potential difference must it be charged?

49. Two capacitors, $C_1 = 25.0\ \mu$F and $C_2 = 5.00\ \mu$F, are connected in parallel and charged with a 100-V power supply. (a) Draw a circuit diagram and calculate the total energy stored in the two capacitors. (b) What potential difference would be required across the same two capacitors connected in series so that the combination stores the same energy as in part (a)? Draw a circuit diagram of this circuit.

50. As a person moves about in a dry environment, electric charge accumulates on the person's body. Once it is at high voltage, either positive or negative, the body can discharge via sometimes noticeable sparks and shocks. Consider a human body well separated from ground, with the typical capacitance 150 pF. (a) What charge on the body will produce a potential of 10.0 kV? (b) Sensitive electronic devices can be destroyed by electrostatic discharge from a person. A particular device can be destroyed by a discharge releasing an energy of 250 μJ. To what voltage on the body does this energy correspond?

51. A parallel-plate capacitor has a charge Q and plates of area A. What force acts on one plate to attract it toward the other plate? Because the electric field between the plates is $E = Q/A\epsilon_0$, you might think that the force is $F = QE = Q^2/A\epsilon_0$. This conclusion is wrong because the field E includes contributions from both plates, and the field created by the positive plate cannot exert any force on the positive plate. Show that the force exerted on each plate is actually $F = Q^2/2\epsilon_0 A$. (*Suggestion:* Let $C = \epsilon_0 A/x$ for an arbitrary plate separation x; then require that the work done in separating the two charged plates be $W = \int F\,dx$.) The force exerted by one charged plate on another is sometimes used in a machine shop to hold a workpiece stationary.

52. A uniform electric field $E = 3\ 000$ V/m exists within a certain region. What volume of space contains an energy equal to 1.00×10^{-7} J? Express your answer in cubic meters and in liters.

Section 20.10 ■ Capacitors with Dielectrics

53. Determine (a) the capacitance and (b) the maximum potential difference that can be applied to a Teflon-filled parallel-plate capacitor having a plate area of 1.75 cm^2 and plate separation of 0.040 0 mm.

54. (a) How much charge can be placed on a capacitor with air between the plates before it breaks down if the area of each of the plates is 5.00 cm^2? (b) Find the maximum charge assuming polystyrene is used between the plates instead of air.

55. A commercial capacitor is to be constructed as shown in Figure 20.29a. This particular capacitor is made from two strips of aluminum foil separated by a strip of paraffin-coated paper. Each strip of foil and paper is 7.00 cm wide. The foil is 0.004 00 mm thick, and the paper is 0.025 0 mm thick and has a dielectric constant of 3.70. What length should the strips have if a capacitance of 9.50×10^{-8} F is desired before the capacitor is rolled up? (Adding a second strip of paper and rolling the capacitor effectively doubles its capacitance by allowing charge storage on both sides of each strip of foil.)

56. The supermarket sells rolls of aluminum foil, of plastic wrap, and of waxed paper. Describe a capacitor made from supermarket materials. Compute order-of-magnitude estimates for its capacitance and its breakdown voltage.

57. A parallel-plate capacitor in air has a plate separation of 1.50 cm and a plate area of 25.0 cm^2. The plates are charged to a potential difference of 250 V and disconnected from the source. The capacitor is then immersed in distilled water. Determine (a) the charge on the plates before and after immersion, (b) the capacitance and potential difference after immersion, and (c) the change in energy of the capacitor. Assume that the liquid is an insulator.

Section 20.11 ■ Context Connection—The Atmosphere as a Capacitor

58. Lightning can be studied with a Van de Graaff generator, essentially consisting of a spherical dome on which charge is continuously deposited by a moving belt. Charge can be added until the electric field at the surface of the dome becomes equal to the dielectric strength of air (3×10^6 V/m). Any more charge leaks off in sparks as shown in Figure P20.58. Assume that the dome has a diameter of 30.0 cm and is surrounded by dry air. (a) What is the maximum potential of the dome? (b) What is the maximum charge on the dome?

(E. R. Degginer/H. Armstrong Roberts)

FIGURE **P20.58**

59. **Review problem.** A certain storm cloud has a potential of 1.00×10^8 V relative to a tree. If, during a lightning storm, 50.0 C of charge is transferred through this potential difference and 1.00% of the energy is absorbed by the tree, how much sap in the tree can be boiled away? Model the sap as water initially at 30.0°C. Water has a specific heat of

4 186 J/kg · °C, a boiling point of 100 °C, and a latent heat of vaporization of 2.26×10^6 J/kg.

Additional Problems

60. **Review problem.** From a large distance away, a particle of mass 2.00 g and charge 15.0 μC is fired at $21.0\hat{\mathbf{i}}$ m/s straight toward a second particle, originally stationary but free to move, with mass 5.00 g and charge 8.50 μC. (a) At the instant of closest approach, both particles will be moving at the same velocity. Explain why. (b) Find this velocity. (c) Find the distance of closest approach. (d) Find the velocities of both particles after they separate again.

61. The liquid-drop model of the atomic nucleus suggests that high-energy oscillations of certain nuclei can split the nucleus into two unequal fragments plus a few neutrons. The fission products acquire kinetic energy from their mutual Coulomb repulsion. Calculate the electric potential energy (in electron volts) of two spherical fragments from a uranium nucleus having the following charges and radii: $38e$ and 5.50×10^{-15} m; $54e$ and 6.20×10^{-15} m. Assume that the charge is distributed uniformly throughout the volume of each spherical fragment and that just before separating each fragment is at rest and their surfaces are in contact. The electrons surrounding the nucleus can be ignored.

62. The Bohr model of the hydrogen atom states that the single electron can exist only in certain allowed orbits around the proton. The radius of each Bohr orbit is $r = n^2(0.052\,9 \text{ nm})$ where $n = 1, 2, 3, \ldots$. Calculate the electric potential energy of a hydrogen atom when the electron is in the (a) first allowed orbit, with $n = 1$; (b) second allowed orbit, with $n = 2$; and (c) when the electron has escaped from the atom, with $r = \infty$. Express your answers in electron volts.

63. Calculate the work that must be done to charge a spherical shell of radius R to a total charge Q.

64. A Geiger–Mueller tube is a radiation detector that essentially consists of a closed, hollow metal cylinder (the cathode) of inner radius r_a and a coaxial cylindrical wire (the anode) of radius r_b (Fig. P20.64). The charge per unit length on the anode is λ, and the charge per unit length on the cathode is $-\lambda$. A gas fills the space between the electrodes. When a high-energy elementary particle passes through this space, it can ionize an atom of the gas. The strong electric field makes the resulting ion and electron accelerate in opposite directions. They strike other molecules of the gas to ionize them, producing an avalanche of electrical discharge. The pulse of electric current between the wire and the cylinder is counted by an external circuit. (a) Show that the magnitude of the potential difference between the wire and the cylinder is

$$\Delta V = 2k_e\lambda \ln\left(\frac{r_a}{r_b}\right)$$

(b) Show that the magnitude of the electric field in the space between cathode and anode is given by

$$E = \frac{\Delta V}{\ln(r_a/r_b)}\left(\frac{1}{r}\right)$$

where r is the distance from the axis of the anode to the point where the field is to be calculated.

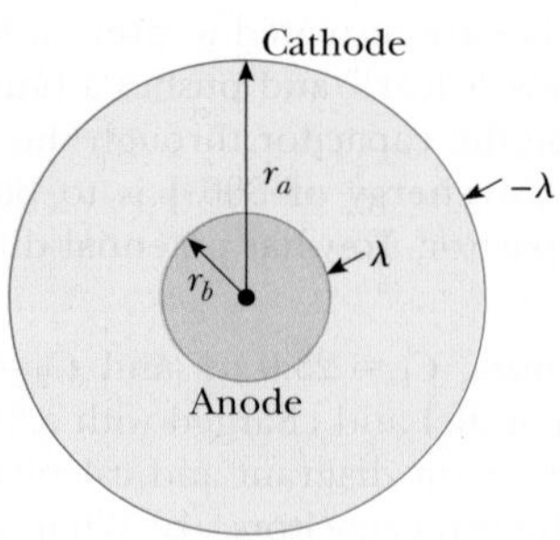

FIGURE **P20.64** Problems 20.64, 20.65, and 20.66.

65. Assume that the internal diameter of the Geiger–Mueller tube described in Problem 20.64 is 2.50 cm and that the wire along the axis has a diameter of 0.200 mm. The dielectric strength of the gas between the central wire and the cylinder is 1.20×10^6 V/m. Use the result of Problem 20.64 to calculate the maximum potential difference that can be applied between the wire and the cylinder before breakdown occurs in the gas.

66. The results of Problem 20.64 apply also to an electrostatic precipitator (Figs. P20.64 and P20.66). This pollution-control device consists of a vertical cylindrical duct with a wire along its axis at a high negative voltage. Corona discharge ionizes the air around the wire to produce free electrons and positive and negative molecular ions. The electrons and negative ions accelerate outward. As air passes through the cylinder, the dirt particles become electrically charged by collisions and ion capture. They are then swept out of the air by the horizontal electric field between the wire and the cylinder. In a particular case, an applied voltage $\Delta V = V_a - V_b = 50.0$ kV is to produce an electric field of magnitude 5.50 MV/m at the surface of the central wire. Assume that the outer cylindrical wall has uniform radius $r_a = 0.850$ m. (a) What should be the radius r_b of the central wire? You will need to solve a transcendental equation. (b) What is the magnitude of the electric field at the outer wall?

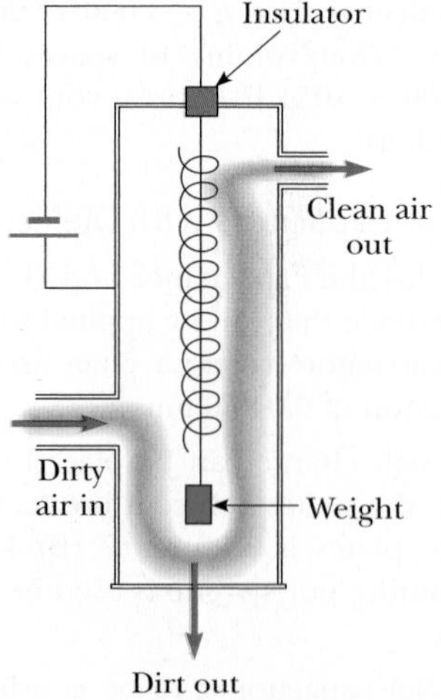

FIGURE **P20.66**

67. A model of a red blood cell portrays the cell as a capacitor with two spherical plates. It is a positively charged conducting liquid sphere of area A, separated by an insulating membrane of thickness t from the surrounding neg-

atively charged conducting fluid. Tiny electrodes introduced into the cell show a potential difference of 100 mV across the membrane. Take the membrane's thickness as 100 nm and its dielectric constant as 5.00. (a) Assume that a typical red blood cell has a mass of 1.00×10^{-12} kg and density 1 100 kg/m^3. Calculate its volume and its surface area. (b) Find the capacitance of the cell. (c) Calculate the charge on the surfaces of the membrane. How many electronic charges does this charge represent? (*Suggestion:* The chapter text models the Earth's atmosphere as a capacitor with two spherical plates.)

68. Four balls, each with mass m, are connected by four nonconducting strings to form a square with side a as shown in Figure P20.68. The assembly is placed on a horizontal, nonconducting, frictionless surface. Balls 1 and 2 each have charge q, and balls 3 and 4 are uncharged. Find the maximum speed of balls 3 and 4 after the string connecting balls 1 and 2 is cut.

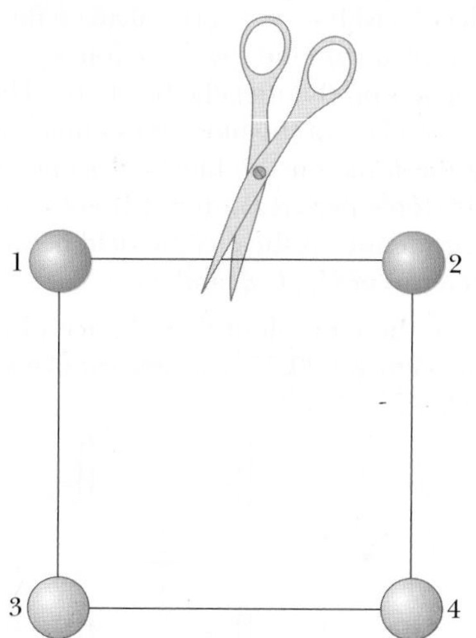

FIGURE **P20.68**

69. The x axis is the symmetry axis of a stationary, uniformly charged ring of radius R and charge Q (Fig. P20.69). A particle with charge Q and mass M is located at the center of the ring. When it is displaced slightly, the point charge accelerates along the x axis to infinity. Show that the ultimate speed of the point charge is

$$v = \left(\frac{2k_eQ^2}{MR}\right)^{1/2}$$

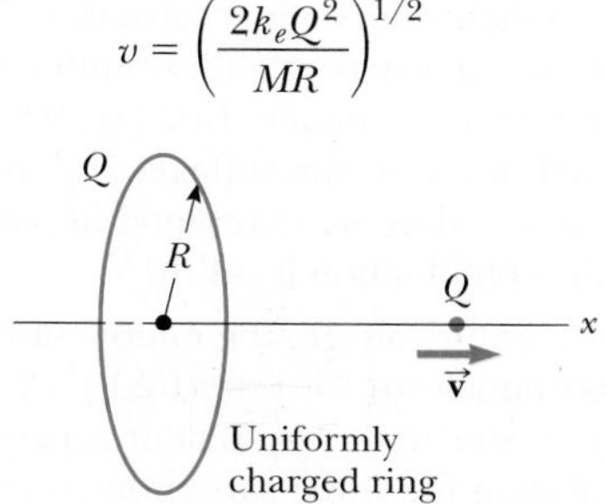

FIGURE **P20.69**

70. An electric dipole is located along the y axis as shown in Figure P20.70. The magnitude of its electric dipole moment is defined as $p = 2qa$. (a) At a point P, which is far from the dipole ($r \gg a$), show that the electric potential is

$$V = \frac{k_e p \cos \theta}{r^2}$$

(b) Calculate the radial component E_r and the perpendicular component E_θ of the associated electric field. Note that $E_\theta = -(1/r)(\partial V/\partial\theta)$. Do these results seem reasonable for $\theta = 90°$ and $0°$? for $r = 0$? (c) For the dipole arrangement shown, express V in terms of Cartesian coordinates using $r = (x^2 + y^2)^{1/2}$ and

$$\cos \theta = \frac{y}{(x^2 + y^2)^{1/2}}$$

Using these results and again taking $r \gg a$, calculate the field components E_x and E_y.

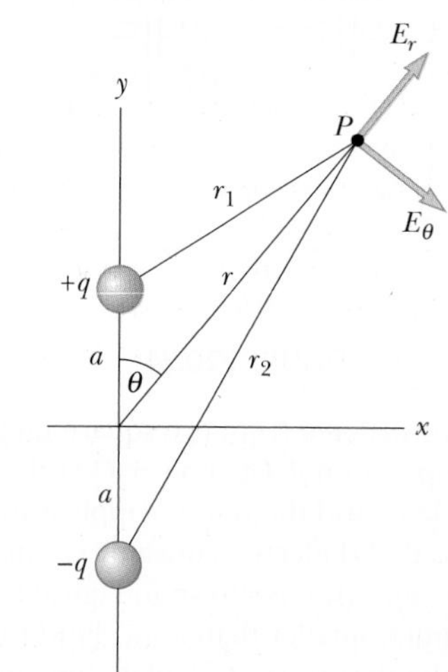

FIGURE **P20.70**

71. Two large, parallel metal plates are oriented horizontally and separated by a distance $3d$. A grounded conducting wire joins them, and initially each plate carries no charge. Now a third identical plate carrying charge Q is inserted between the two plates, parallel to them and located a distance d from the upper plate, as shown in Figure P20.71. (a) What induced charge appears on each of the two original plates? (b) What potential difference appears between the middle plate and each of the other plates? Let A represent the area of each plate.

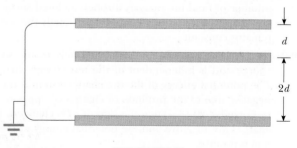

FIGURE **P20.71**

72. A 2.00-nF parallel-plate capacitor is charged to an initial potential difference $\Delta V_i = 100$ V and then isolated. The dielectric material between the plates is mica, with a dielectric constant of 5.00. (a) How much work is required to withdraw the mica sheet? (b) What is the potential difference of the capacitor after the mica is withdrawn?

73. A parallel-plate capacitor is constructed using a dielectric material whose dielectric constant is 3.00 and whose dielectric strength is 2.00×10^8 V/m. The desired capacitance is 0.250 μF, and the capacitor must withstand a maximum potential difference of 4 000 V. Find the minimum area of the capacitor plates.

74. A 10.0-μF capacitor is charged to 15.0 V. It is next connected in series with an uncharged 5.00-μF capacitor. The series combination is finally connected across a 50.0-V battery as diagrammed in Figure P20.74. Find the new potential differences across the 5.00-μF and 10.0-μF capacitors.

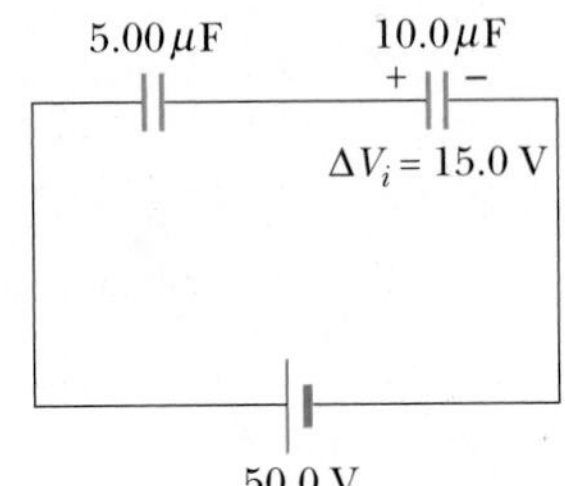

FIGURE P20.74

75. A capacitor is constructed from two square metallic plates of sides ℓ and separation d. Charges $+Q$ and $-Q$ are placed on the plates and the power supply is then removed. A material of dielectric constant κ is inserted a distance x into the capacitor as shown in Figure P20.75. Assume that d is much smaller than x. (a) Find the equivalent capacitance of the device. (b) Calculate the energy stored in the capacitor. (c) Find the direction and magnitude of the force exerted by the plates on the dielectric. (d) Obtain a numerical value for the force when $x = \ell/2$, assuming that $\ell = 5.00$ cm, $d = 2.00$ mm, the dielectric is glass ($\kappa = 4.50$), and the capacitor was charged to 2 000 V before the dieletric was inserted. (*Suggestion:* The system can be considered as two capacitors connected in parallel.)

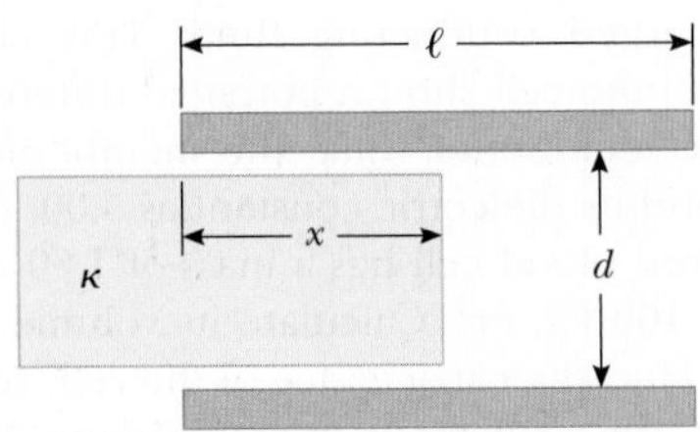

FIGURE P20.75 Problems 20.75 and 20.76.

76. Two square plates of sides ℓ are placed parallel to each other with separation d as suggested in Figure P20.75. You may assume that d is much less than ℓ. The plates carry uniformly distributed static charges $+Q_0$ and $-Q_0$. A block of metal has width ℓ, length ℓ, and thickness slightly less than d. It is inserted a distance x into the space between the plates. The charges on the plates are not disturbed as the block slides in. In a static situation, a metal prevents an electric field from penetrating inside it. The metal can be thought of as a perfect dielectric, with $\kappa \rightarrow \infty$. (a) Calculate the stored energy as a function of x. (b) Find the direction and magnitude of the force that acts on the metallic block. (c) The area of the advancing front face of the block is essentially equal to ℓd. Considering the force on the block as acting on this face, find the stress (force per area) on it. (d) For comparison, express the energy density in the electric field between the capacitor plates in terms of Q_0, ℓ, d, and ϵ_0.

77. Determine the equivalent capacitance of the combination shown in Figure P20.77. (*Suggestion:* Consider the symmetry involved.)

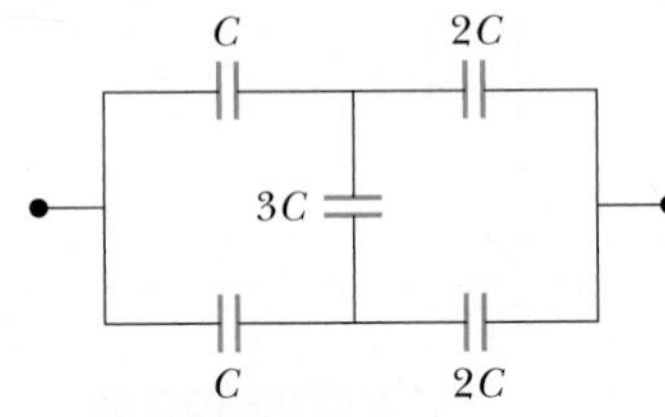

FIGURE P20.77

ANSWERS TO QUICK QUIZZES

20.1 (i), (b). The electric potential is inversely proportion to the radius (see Eq. 20.11). (ii), (c). Because the same number of field lines passes through a closed surface of any shape or size, the electric flux through the surface remains constant.

20.2 (i), (c). The potential is established only by the source charge and is independent of the test charge. (ii), (a). The potential energy of the two-charge system is initially negative due to the products of charges of opposite sign in Equation 20.13. When the sign of q_2 is changed, both charges are negative and the potential energy of the system is positive.

20.3 (i), (a). If the potential is constant (zero in this case), its derivative along this direction is zero. (ii), (b). If the electric field is zero, there is no change in the electric potential and it must be constant. This constant value *could be* zero, but it does not *have to be* zero.

20.4 (d). The capacitance is a property of the physical system and does not vary with applied voltage. According to Equation 20.19, if the voltage is doubled, the charge is doubled.

20.5 (i), (a). When connecting capacitors in series, the inverse of the capacitances add, resulting in a smaller overall equivalent capacitance. (ii), (a). When capacitors are connected in series, the voltages add, for a total of 20 V in this case. If they are combined in parallel, the voltage across the combination is still 10 V.

20.6 (b). For a given voltage, the energy stored in a capacitor is proportional to C: $U = C(\Delta V)^2/2$. Therefore, you want to maximize the equivalent capacitance. You do so by connecting the three capacitors in parallel so that the capacitances add.

20.7 Increase. The dielectric constant of wood (and of all other insulating materials, for that matter) is greater than 1; therefore, the capacitance increases (Eq. 20.33). This increase is sensed by the stud finder's special circuitry, which causes an indicator on the device to light up.

CHAPTER 21

Current and Direct Current Circuits

These power lines transfer energy from the electrical power company to homes and businesses. The energy is transferred at a very high voltage, possibly hundreds of thousands of volts in some cases. Even though that makes power lines very dangerous, the high voltage results in less loss of power due to resistance in the wires. We will study both resistance and power in this chapter.

CHAPTER OUTLINE

Thus far, our discussion of electrical phenomena has focused on charges at rest, or the study of *electrostatics.* We shall now consider situations involving electric charges in motion. The term *electric current,* or simply *current,* is used to describe the flow of charge through some region of space. Most practical applications of electricity involve electric currents. For example, in a flashlight, charges flow in the filament of the lightbulb after the switch is turned on. In most common situations, the flow of charge takes place in a conductor, such as a copper wire. It is also possible, however, for currents to exist outside a conductor. For instance, a beam of electrons in a TV picture tube constitutes a current in which charge flows through a vacuum.

In Chapter 20, we introduced the notion of a *circuit.* As we continue our investigations into circuits in this chapter, we introduce the *resistor* as a new circuit element.

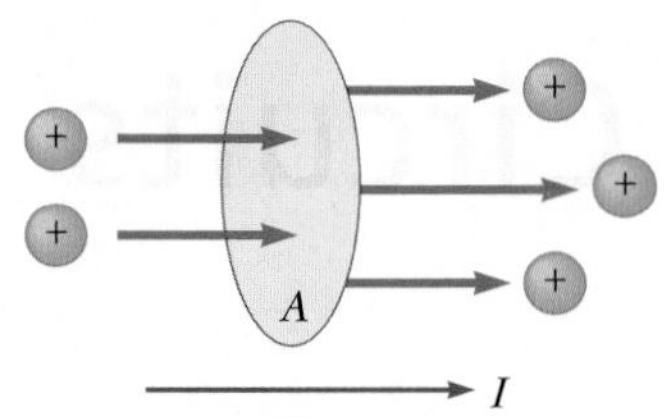

FIGURE 21.1 Charges in motion through an area A. The time rate at which charge flows through the area is defined as the current I. The direction of the current is the direction in which positive charges flow when free to do so.

21.1 ELECTRIC CURRENT

Whenever charge is flowing, an **electric current** is said to exist. To define current mathematically, suppose charged particles are moving perpendicular to a surface of area A as in Figure 21.1. (This area could be the cross-sectional area of a wire, for example.) The current is defined as **the rate at which electric charge flows through this surface.** If ΔQ is the amount of charge that passes through this area in a time interval Δt, the average current I_{avg} over the time interval is the ratio of the charge to the time interval:

$$I_{\text{avg}} = \frac{\Delta Q}{\Delta t} \qquad [21.1]$$

It is possible for the rate at which charge flows to vary in time. We define the **instantaneous current** I as the limit of the preceding expression as Δt goes to zero:

■ Electric current

$$I \equiv \lim_{\Delta t \to 0} \frac{\Delta Q}{\Delta t} = \frac{dQ}{dt} \qquad [21.2]$$

The SI unit of current is the **ampere** (A):

$$1 \text{ A} = 1 \text{ C/s} \qquad [21.3]$$

That is, 1 A of current is equivalent to 1 C of charge passing through a surface in 1 s.

The particles flowing through a surface as in Figure 21.1 can be charged positively or negatively, or we can have two or more types of particles moving, with charges of both signs in the flow. **Conventionally, we define the direction of the current as the direction of flow of positive charge,** regardless of the sign of the actual charged particles in motion.[1] In a common conductor such as copper, the current is physically due to the motion of the negatively charged electrons. Therefore, when we speak of current in such a conductor, **the direction of the current is opposite the direction of flow of electrons.** On the other hand, if one considers a beam of positively charged protons in a particle accelerator, the current is in the direction of motion of the protons. In some cases—gases and electrolytes, for example—the current is the result of the flow of both positive and negative charged particles. It is common to refer to a moving charged particle (whether it is positive or negative) as a mobile **charge carrier.** For example, the charge carriers in a metal are electrons.

PITFALL PREVENTION 21.1

Current flow is redundant The phrase *current flow* is commonly used, although it is strictly incorrect, because current *is* a flow (of charge). This terminology is similar to the phrase *heat transfer,* which is also redundant because heat *is* a transfer (of energy). We will avoid the phrase *current flow* and speak of *charge flow* or *flow of charge.*

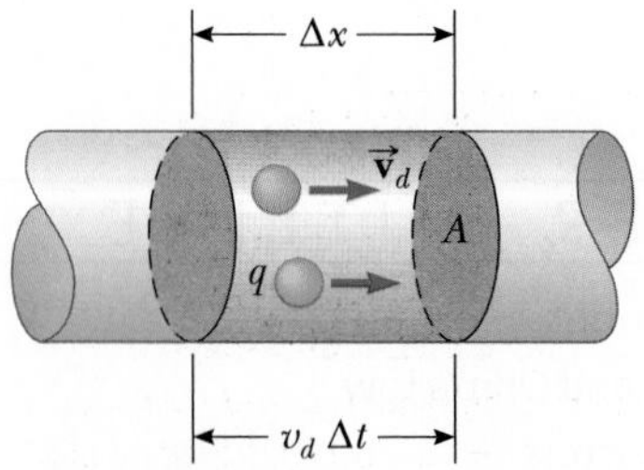

FIGURE 21.2 A section of a uniform conductor of cross-sectional area A. The mobile charge carriers move with an average speed v_d along the wire, and the displacement they experience in this direction in a time interval Δt is $\Delta x = v_d \Delta t$. If we choose Δt to be the time interval during which the charges are displaced, on the average, by the length of the cylinder, the number of carriers in the section of length Δx is $nAv_d \Delta t$, where n is the number of carriers per unit volume.

We now build a structural model that will allow us to relate the macroscopic current to the motion of the charged particles. Consider identical charged particles moving in a conductor of cross-sectional area A (Fig. 21.2). The volume of a section of the conductor of length Δx (the gray region shown in Fig. 21.2) is $A \Delta x$. If n represents the number of mobile charge carriers per unit volume (in other words, the charge carrier density), the number of carriers in the gray section is $nA \Delta x$. Therefore, the total charge ΔQ in this section is

$$\Delta Q = \text{number of carriers in section} \times \text{charge per carrier} = (nA\,\Delta x)q$$

where q is the charge on each carrier. If the carriers move with an average velocity component v_d in the x direction (along the wire), the displacement they experience in this direction in a time interval Δt is $\Delta x = v_d \Delta t$. The speed v_d of the charge carrier along the wire is an average speed called the **drift speed.** Let us choose Δt to be the time interval required for the charges in the cylinder to move through a displacement whose magnitude is equal to the length of the cylinder. This time

[1]Even though we discuss a direction for current, current is not a vector. As we shall see later in the chapter, currents add algebraically and not vectorially.

interval is also that required for all the charges in the cylinder to pass through the circular area at one end. With this choice, we can write ΔQ in the form

$$\Delta Q = (nAv_d\,\Delta t)\,q$$

If we divide both sides of this equation by Δt, we see that the average current in the conductor is

■ Current in terms of microscopic parameters

$$I_{\text{avg}} = \frac{\Delta Q}{\Delta t} = nqv_dA \qquad [21.4]$$

Equation 21.4 relates a macroscopically measured average current to the microscopic origin of the current: the density of charge carriers n, the charge per carrier q, and the drift speed v_d.

QUICK QUIZ 21.1 Consider positive and negative charges moving horizontally through the four regions shown in Figure 21.3. Rank the currents in these four regions, from lowest to highest.

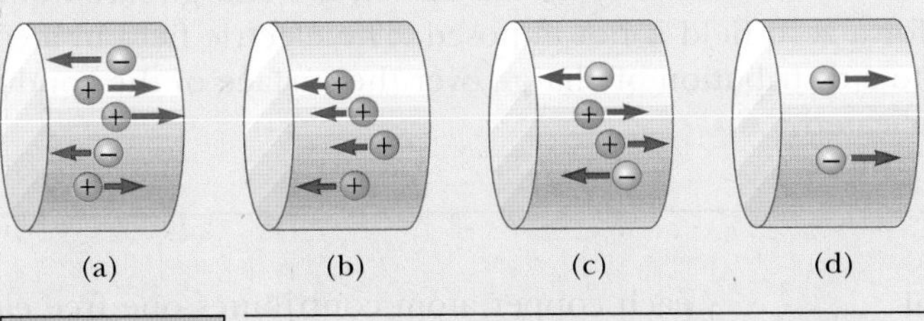

FIGURE 21.3 (Quick Quiz 21.1) Four groups of charges move through a region.

Let us investigate further the notion of drift speed. We have identified drift speed as an average speed along the wire, but the charge carriers are by no means moving in a straight line with speed v_d. Consider a conductor in which the charge carriers are free electrons. In the absence of a potential difference across the conductor, these electrons undergo random motion similar to that of gas molecules in the structural model of kinetic theory that we studied in Chapter 16. This random motion is related to the temperature of the conductor. The electrons undergo repeated collisions with the metal atoms, and the result is a complicated zigzag motion (Active Fig. 21.4). When a potential difference is applied across the conductor, an electric field is established in the conductor. The electric field exerts an electric force on the electrons (Eq. 19.4). This force accelerates the electrons and hence produces a current. The motion of the electrons due to the electric force is superimposed on their random motion to provide an average velocity whose magnitude is the drift speed.

When electrons make collisions with metal atoms during their motion, they transfer energy to the atoms. This energy transfer causes an increase in the vibrational energy of the atoms and a corresponding increase in the temperature of the conductor.[2] This process involves all three types of energy storage in the continuity equation for energy, Equation 6.20. If we consider the system to be the electrons, the metal atoms, and the electric field (which is established by an external source such as a battery), the energy at the instant when the potential difference is applied across the conductor is electric potential energy associated with the field and the electrons. This energy is transformed by work done by the field on the electrons to kinetic energy of electrons. When the electrons strike the metal atoms, some of the kinetic energy is transferred to the atoms, which adds to the internal energy of the system.

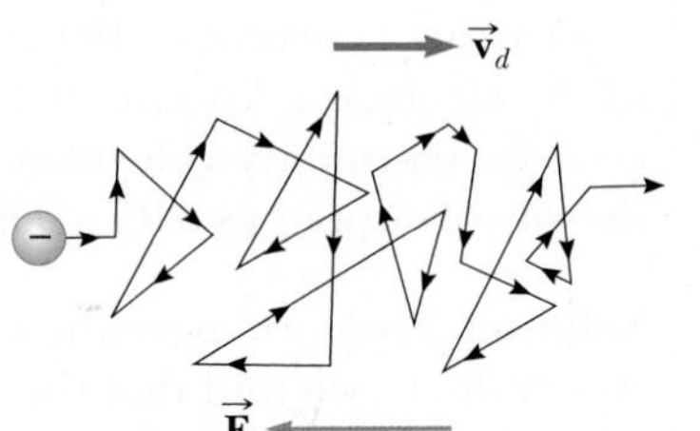

ACTIVE FIGURE 21.4 A schematic representation of the zigzag motion of a charge carrier in a conductor. The changes in direction are due to collisions with atoms in the conductor. Note that the net motion of electrons is opposite the direction of the electric field. Because of the acceleration of the charge carriers due to the electric force, the paths are actually parabolic. The drift speed, however, is much smaller than the average speed, so the parabolic shape is not visible on this scale.

Physics Now™ Log into PhysicsNow at **www.pop4e.com** and go to Active Figure 21.4 to adjust the electric field to see the resulting effect on the motion of an electron.

[2]This increase in temperature is sometimes called *Joule heating*, but that term is a misnomer because there is no heat involved. We will not use this wording.

The **current density** J in the conductor is defined as the current per unit area. From Equation 21.4, the current density is

$$J \equiv \frac{I}{A} = nqv_d \qquad [21.5]$$

where J has the SI units amperes per square meter.

Thinking Physics 21.1

In Chapter 19, we claimed that the electric field inside a conductor is zero. In the preceding discussion, however, we have used the notion of an electric field in a conducting wire that exerts electric forces on electrons, causing them to move with a drift velocity. Is this notion inconsistent with Chapter 19?

Reasoning The electric field is zero only in a conductor in *electrostatic equilibrium,* that is, a conductor in which the charges are at rest after having moved to equilibrium positions. In a current-carrying conductor, the charges are not at rest, so the requirement for a zero field is not imposed. The electric field in a conductor in a circuit is due to a distribution of charge over the surface of the conductor that can be quite complicated.[3] ■

EXAMPLE 21.1 Drift Speed in a Copper Wire

The 12-gauge copper wire in a typical residential building has a cross-sectional area of $3.31 \times 10^{-6}\ \text{m}^2$. If it carries a current of 10.0 A, what is the drift speed of the electrons? Assume that each copper atom contributes one free electron to the current. Take the density of copper as $8.95\ \text{g/cm}^3$.

Solution From the periodic table of the elements in Appendix C, we find that the molar mass of copper is 63.5 g/mol. Knowing the density of copper enables us to calculate the volume occupied by 1 mol of copper:

$$V = \frac{M}{\rho} = \frac{63.5\ \text{g/mol}}{8.95\ \text{g/cm}^3} = 7.09\ \text{cm}^3/\text{mol}$$

Recall that one mole of any substance contains Avogadro's number of atoms, 6.02×10^{23} atoms. Because each copper atom contributes one free electron to the current, the density of charge carriers is

$$n = \frac{6.02 \times 10^{23}\ \text{electrons}}{7.09\ \text{cm}^3}\left(\frac{1.00 \times 10^6\ \text{cm}^3}{1\ \text{m}^3}\right)$$
$$= 8.49 \times 10^{28}\ \text{electrons/m}^3$$

From Equation 21.4, we find that the drift speed is

$$v_d = \frac{I}{nqA}$$
$$= \frac{10.0\ \text{C/s}}{(8.49 \times 10^{28}\ \text{m}^{-3})(1.60 \times 10^{-19}\ \text{C})(3.31 \times 10^{-6}\ \text{m}^2)}$$
$$= 2.22 \times 10^{-4}\ \text{m/s}$$

PITFALL PREVENTION 21.2

ELECTRONS ARE AVAILABLE EVERYWHERE Let us emphasize the point being made here: *Electrons do not have to travel from the light switch to the light for the light to operate.* Electrons already in the filament of the lightbulb move in response to the electric field set up by the battery. Note also that the role of a battery is not to provide electrons to the circuit. It establishes the electric field that exerts a force on electrons already in the wires and elements of the circuit.

Example 21.1 shows that typical drift speeds in conductors are very small. In fact, the drift speed is much smaller than the average speed between collisions. For instance, electrons traveling with the drift speed calculated in Example 21.1 would take about 75 min to travel 1 m! In view of this low speed, you might wonder why a light turns on almost instantaneously when a switch is thrown. In a conductor, the electric field that drives the free electrons is established in the conductor almost instantaneously. Therefore, when you flip a light switch, the electric force that causes the electrons to start moving in the wire with a drift speed begins immediately. Electrons already in the filament of the lightbulb begin to move in response to this force, and the lightbulb begins to emit light.

[3]See Chapter 6 in R. Chabay and B. Sherwood, *Electric and Magnetic Interactions,* (New York: Wiley, 1995) for details on this charge distribution.

21.2 RESISTANCE AND OHM'S LAW

The drift speed of electrons in a current-carrying wire is related to the electric field in the wire. If the field is increased, the electric force on the electrons is stronger and the drift speed increases. We shall show in Section 21.4 that this relationship is linear and that the drift speed is directly proportional to the electric field. For a uniform field in a conductor of uniform cross-section, the potential difference across the conductor is proportional to the electric field as in Equation 20.6. Therefore, when a potential difference ΔV is applied across the ends of a metallic conductor as in Figure 21.5, the current in the conductor is found to be proportional to the applied voltage; that is, $I \propto \Delta V$. We can write this proportionality as $\Delta V = IR$, where R is called the **resistance** of the conductor. We define this resistance according to the equation we have just written, as the ratio of the voltage across the conductor to the current it carries:

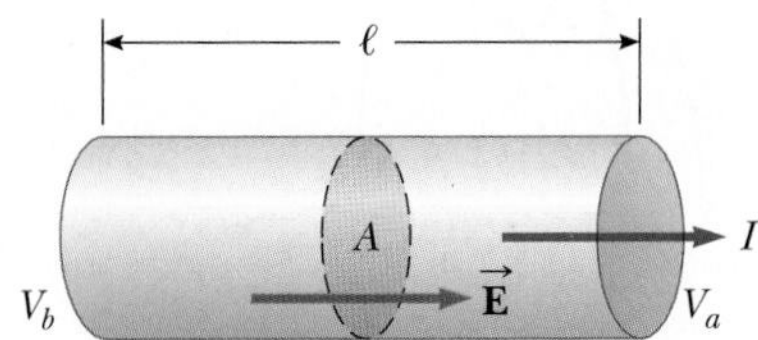

FIGURE 21.5 A uniform conductor of length ℓ and cross-sectional area A. A potential difference $V_b - V_a$ is maintained across the conductor so that an electric field $\vec{\mathbf{E}}$ exists in the conductor, and this field produces a current I that is proportional to the potential difference.

$$R \equiv \frac{\Delta V}{I} \qquad [21.6]$$

■ Definition of resistance

Resistance has the SI units volts per ampere, called **ohms** (Ω). Therefore, if a potential difference of 1 V across a conductor produces a current of 1 A, the resistance of the conductor is 1 Ω. As another example, if an electrical appliance connected to a 120-V source carries a current of 6.0 A, its resistance is 20 Ω.

Resistance is the quantity that determines the current that results due to a voltage in a simple circuit. For a fixed voltage, if the resistance increases, the current decreases. If the resistance decreases, the current increases.

It might be useful for you to build a mental model for current, voltage, and resistance by comparing these concepts to analogous concepts for the flow of water in a river. As water flows downhill in a river of constant width and depth, the rate of flow of water (analogous to current) depends on the angle that the river bottom makes with the horizontal (analogous to voltage) and on the width and depth as well as on the effects of rocks, the riverbank, and other obstructions (analogous to resistance). Likewise, electric current in a uniform conductor depends on the applied voltage and the resistance of the conductor is caused by collisions of the electrons with atoms in the conductor.

For many materials, including most metals, experiments show that the resistance is constant over a wide range of applied voltages. This behavior is known as **Ohm's law** after Georg Simon Ohm (1787–1854), who was the first to conduct a systematic study of electrical resistance.

Many individuals call Equation 21.6 Ohm's law, but this terminology is incorrect. This equation is simply the definition of resistance, and it provides an important relationship between voltage, current, and resistance. Ohm's law is *not* a fundamental law of nature, but a behavior that is valid only for certain materials and devices, and only over a limited range of conditions. Materials or devices that obey Ohm's law, and hence that have a constant resistance over a wide range of voltages, are said to be **ohmic** (Fig. 21.6a). Materials or devices that do not obey Ohm's law are

PITFALL PREVENTION 21.3

WE'VE SEEN SOMETHING LIKE EQUATION 21.6 BEFORE In Chapter 4, we introduced Newton's second law, $\Sigma F = ma$, for a net force on an object of mass m. It can be written as

$$m = \frac{\Sigma F}{a}$$

In Chapter 4, we defined mass as *resistance to a change in motion in response to an external force.* Mass as resistance to changes in motion is analogous to electrical resistance to charge flow, and Equation 21.6 is analogous to the form of Newton's second law above. Each equation states that the resistance (electrical or mechanical) is equal to (1) ΔV, the cause of current or (2) ΣF, the cause of changes in motion, divided by the result, (1) a charge flow, quantified by current I, or (2) a change in motion, quantified by acceleration a.

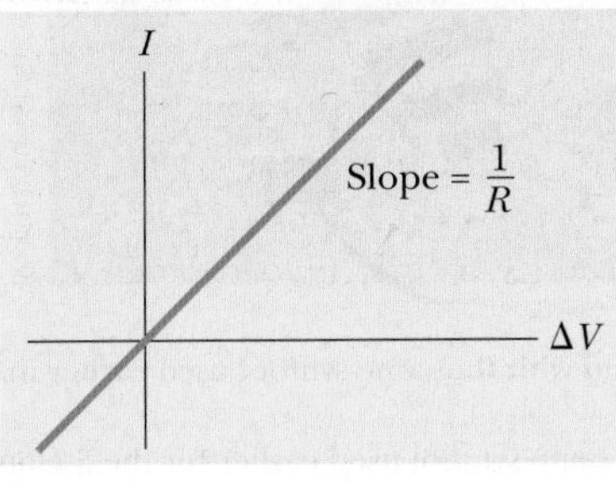

(a)

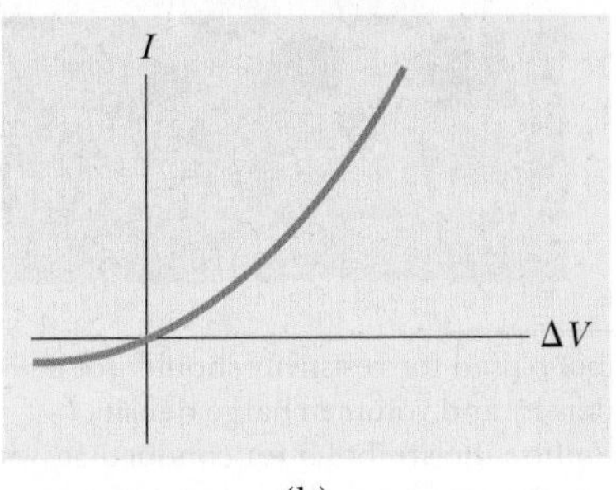

(b)

FIGURE 21.6 (a) The current–potential difference curve for an ohmic material. The curve is linear, and the slope is equal to the inverse of the resistance of the conductor. (b) A nonlinear current–potential difference curve for a semiconducting diode. This device does not obey Ohm's law.

nonohmic. One common semiconducting device that is nonohmic is the *diode,* a circuit element that acts like a one-way valve for current. Its resistance is small for currents in one direction (positive ΔV) and large for currents in the reverse direction (negative ΔV) as shown in Figure 21.6b. Most modern electronic devices have nonlinear current–voltage relationships; their operation depends on the particular ways they violate Ohm's law.

QUICK QUIZ 21.2 In Figure 21.6b, as the applied voltage increases, does the resistance of the diode **(a)** increase, **(b)** decrease, or **(c)** remain the same?

A **resistor** is a simple circuit element that provides a specified resistance in an electrical circuit. The symbol for a resistor in circuit diagrams is a zigzag red line (—/\/\/\—). We can express Equation 21.6 in the form

$$\Delta V = IR \tag{21.7}$$

This equation tells us that the voltage across a resistor is the product of the resistance and the current in the resistor.

The resistance of an ohmic conducting wire is found to be proportional to its length ℓ and inversely proportional to its cross-sectional area A. That is,

■ Resistance of a uniform material of resistivity ρ along a length ℓ

$$R = \rho \frac{\ell}{A} \tag{21.8}$$

where the constant of proportionality ρ is called the **resistivity** of the material,[4] which has the unit ohm meter ($\Omega \cdot \text{m}$). To understand this relationship between resistance and resistivity, note that every ohmic material has a characteristic resistivity, a parameter that depends on the properties of the material and on temperature. On the other hand, as you can see from Equation 21.8, the resistance of a particular conductor depends on its size and shape as well as on the resistivity of the material. Table 21.1 provides a list of resistivities for various materials measured at 20°C.

The inverse of the resistivity is defined[5] as the **conductivity** σ. Hence, the resistance of an ohmic conductor can be expressed in terms of its conductivity as

$$R = \frac{\ell}{\sigma A} \tag{21.9}$$

where $\sigma = 1/\rho$.

Equation 21.9 shows that the resistance of a conductor is proportional to its length and inversely proportional to its cross-sectional area, similar to the flow of

PITFALL PREVENTION 21.4

RESISTANCE AND RESISTIVITY Resistivity is a property of a *substance,* whereas resistance is a property of an *object.* We have seen similar pairs of variables before. For example, density is a property of a substance, whereas mass is a property of an object. Equation 21.8 relates resistance to resistivity, and we have seen a previous equation (Eq. 1.1) that relates mass to density.

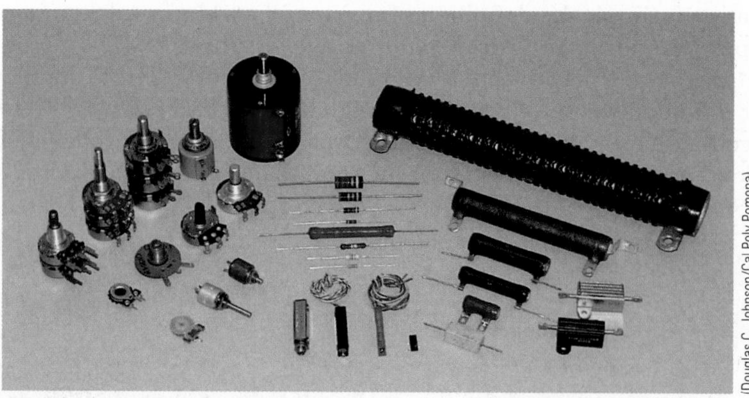
(Douglas C. Johnson/Cal Poly Pomona)

An assortment of resistors used in electric circuits. ■

[4]The symbol ρ used for resistivity should not be confused with the same symbol used earlier in the text for mass density and volume charge density.

[5]Do not confuse the symbol σ for conductivity with the same symbol used earlier for the Stefan–Boltzmann constant and surface charge density.

TABLE 21.1 Resistivities and Temperature Coefficients of Resistivity for Various Materials

Material	Resistivity[a] ($\Omega \cdot m$)	Temperature Coefficient α [$(°C)^{-1}$]
Silver	1.59×10^{-8}	3.8×10^{-3}
Copper	1.7×10^{-8}	3.9×10^{-3}
Gold	2.44×10^{-8}	3.4×10^{-3}
Aluminum	2.82×10^{-8}	3.9×10^{-3}
Tungsten	5.6×10^{-8}	4.5×10^{-3}
Iron	10×10^{-8}	5.0×10^{-3}
Platinum	11×10^{-8}	3.92×10^{-3}
Lead	22×10^{-8}	3.9×10^{-3}
Nichrome[b]	1.50×10^{-6}	0.4×10^{-3}
Carbon	3.5×10^{-5}	-0.5×10^{-3}
Germanium	0.46	-48×10^{-3}
Silicon	640	-75×10^{-3}
Glass	10^{10} to 10^{14}	
Hard rubber	$\sim 10^{13}$	
Sulfur	10^{15}	
Quartz (fused)	75×10^{16}	

[a]All values are at 20°C.
[b]Nichrome is a nickel–chromium alloy commonly used in heating elements.

liquid through a pipe. As the length of the pipe is increased and the pressure difference between the ends of the pipe is held constant, the pressure difference between any two points separated by a fixed distance decreases and there is less force pushing the element of fluid between these points through the pipe. As its cross-sectional area is increased, the pipe can transport more fluid in a given time interval, so its resistance drops.

As another analogy between electrical circuits and our previous studies, let us combine Equations 21.6 and 21.9:

$$R = \frac{\ell}{\sigma A} = \frac{\Delta V}{I} \quad \rightarrow \quad I = \sigma A \frac{\Delta V}{\ell} \quad \rightarrow \quad \frac{q}{\Delta t} = \sigma A \frac{\Delta V}{\ell}$$

where q is the amount of charge transferred in a time interval Δt. Let us compare this equation to Equation 17.35 for conduction of energy through a slab of material of area A, length ℓ, and thermal conductivity k, which we reproduce below:

$$\mathcal{P} = kA \frac{(T_h - T_c)}{L} \quad \rightarrow \quad \frac{Q}{\Delta t} = kA \frac{\Delta T}{L}$$

In this equation, Q is the amount of energy transferred by heat in a time interval Δt.

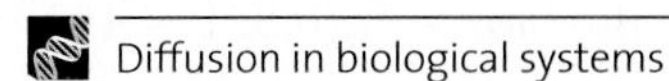
Diffusion in biological systems

Another analogy arises in an example that is important in biochemical applications. *Fick's law* describes the rate of transfer of a chemical solute through a solvent by the process of *diffusion.* This transfer occurs because of a difference in concentration of the solute (mass of solute per volume) between the two locations. Fick's law is as follows:

$$\frac{n}{\Delta t} = DA \frac{\Delta C}{L}$$

where $n/\Delta t$ is the rate of flow of the solute in moles per second, A is the area through which the solute moves, and L is the length over which the concentration difference is ΔC. The concentration is measured in moles per cubic meter. The parameter D is a diffusion constant (with units of meters squared per second) that describes the rate of diffusion of a solute through the solvent and is similar in nature to electrical or thermal conductivity. Fick's law has important applications in describing the transport of molecules across biological membranes.

(SuperStock)

FIGURE 21.7 The colored bands on a resistor represent a code for determining resistance. The first two colors give the first two digits in the resistance value. The third color represents the power of ten for the multiplier of the resistance value. The last color is the tolerance of the resistance value. As an example, the four colors on the circled resistors are red (= 2), black (= 0), orange (= 10^3), and gold (= 5%), and so the resistance value is $20 \times 10^3\ \Omega = 20\ \text{k}\Omega$ with a tolerance value of 5% = 1 kΩ. (The values for the colors come from Table 21.2.)

TABLE 21.2 Color Code for Resistors

Color	Number	Multiplier	Tolerance
Black	0	1	
Brown	1	10^1	
Red	2	10^2	
Orange	3	10^3	
Yellow	4	10^4	
Green	5	10^5	
Blue	6	10^6	
Violet	7	10^7	
Gray	8	10^8	
White	9	10^9	
Gold		10^{-1}	5%
Silver		10^{-2}	10%
Colorless			20%

All three of the preceding equations have exactly the same mathematical form. Each has a time rate of change on the left, and each has the product of a conductivity, an area, and a ratio of a difference in a variable to a length on the right. This type of equation is a *transport equation* used when we transport energy, charge, or moles of matter. The difference in the variable on the right side of each equation is what drives the transport. A temperature difference drives energy transfer by heat, a potential difference drives a transfer of charge, and a concentration difference drives a transfer of matter.

Most electric circuits use resistors to control the current level in the various parts of the circuit. Two common types of resistors are the *composition* resistor containing carbon and the *wire-wound* resistor, which consists of a coil of wire. Resistors are normally color-coded to give their values in ohms, as shown in Figure 21.7 and Table 21.2.

INTERACTIVE EXAMPLE 21.2 The Resistance of Nichrome Wire

A Calculate the resistance per unit length of a 22-gauge Nichrome wire, which has a radius of 0.321 mm.

Solution The cross-sectional area of this wire is

$$A = \pi r^2 = \pi(0.321 \times 10^{-3}\ \text{m})^2 = 3.24 \times 10^{-7}\ \text{m}^2$$

The resistivity of Nichrome is $1.50 \times 10^{-6}\ \Omega \cdot \text{m}$ (Table 21.1). We use Equation 21.8 to find the resistance per unit length:

$$\frac{R}{\ell} = \frac{\rho}{A} = \frac{1.50 \times 10^{-6}\ \Omega \cdot \text{m}}{3.24 \times 10^{-7}\ \text{m}^2} = 4.63\ \Omega/\text{m}$$

B If a potential difference of 10 V is maintained across a 1.0-m length of the Nichrome wire, what is the current in the wire?

Solution Because a 1.0-m length of this wire has a resistance of 4.63 Ω, we have

$$I = \frac{\Delta V}{R} = \frac{10\ \text{V}}{4.63\ \Omega} = 2.2\ \text{A}$$

Note from Table 21.1 that the resistivity of Nichrome wire is two orders of magnitude larger than that of copper. A copper wire of the same radius would have a resistance per unit length of only 0.052 Ω/m. A 1.0-m length of copper wire of the same radius would carry the same current (2.2 A) with an applied voltage of only 0.11 V.

Because of its high resistivity and its resistance to oxidation, Nichrome is often used for heating elements in toasters, irons, and electric heaters.

PhysicsNow™ Explore the resistance of different materials by logging into PhysicsNow at **www.pop4e.com** and going to Interactive Example 21.2.

Change in Resistivity with Temperature

Resistivity depends on a number of factors, one of which is temperature. For most metals, resistivity increases approximately linearly with increasing temperature over a limited temperature range according to the expression

■ Variation of resistivity with temperature

$$\rho = \rho_0[1 + \alpha(T - T_0)] \qquad [21.10]$$

where ρ is the resistivity at some temperature T (in degrees Celsius), ρ_0 is the resistivity at some reference temperature T_0 (usually 20°C), and α is called the **temperature coefficient of resistivity** (not to be confused with the average coefficient of linear expansion α in Chapter 16). From Equation 21.10, we see that α can be expressed as

■ Temperature coefficient of resistivity

$$\alpha = \frac{1}{\rho_0}\frac{\Delta\rho}{\Delta T} \qquad [21.11]$$

where $\Delta\rho = \rho - \rho_0$ is the change in resistivity in the temperature interval $\Delta T = T - T_0$.

The resistivities and temperature coefficients of certain materials are listed in Table 21.1. Note the enormous range in resistivities, from very low values for good conductors, such as copper and silver, to very high values for good insulators, such as glass and rubber. An ideal, or "perfect," conductor would have zero resistivity, and an ideal insulator would have infinite resistivity.

Because resistance is proportional to resistivity according to Equation 21.8, the temperature variation of the resistance can be written as

■ Variation of resistance with temperature

$$R = R_0[1 + \alpha(T - T_0)] \qquad [21.12]$$

Precise temperature measurements are often made using this property, as shown in Example 21.3.

QUICK QUIZ 21.3 When does a lightbulb carry more current: **(a)** just after it is turned on and the glow of the metal filament is increasing or **(b)** after it has been on for a few seconds and the glow is steady?

EXAMPLE 21.3 A Platinum Resistance Thermometer

A resistance thermometer, which measures temperature by measuring the change in resistance of a conductor, is made from platinum and has a resistance of 50.0 Ω at 20.0°C. When immersed in a vessel containing melting indium, its resistance increases to 76.8 Ω. Assuming that the resistance varies linearly with temperature over the temperature range in question, what is the melting point of indium?

Solution Solving Equation 21.12 for ΔT and obtaining α from Table 21.1, we have

$$\Delta T = \frac{R - R_0}{\alpha R_0} = \frac{76.8\ \Omega - 50.0\ \Omega}{[3.92 \times 10^{-3}\ (°\text{C})^{-1}](50.0\ \Omega)}$$
$$= 137°\text{C}$$

Because $T_0 = 20.0°\text{C}$, we find that $T = 157°\text{C}$.

21.3 SUPERCONDUCTORS

For several metals, resistivity is nearly proportional to temperature as shown in Figure 21.8. In reality, however, there is always a nonlinear region at very low temperatures, and the resistivity usually approaches some finite value near absolute zero (see the magnified inset in Fig. 21.8). This residual resistivity near absolute zero is due primarily to collisions of electrons with impurities and to imperfections in the metal. In contrast, the high temperature resistivity (the linear region) is dominated

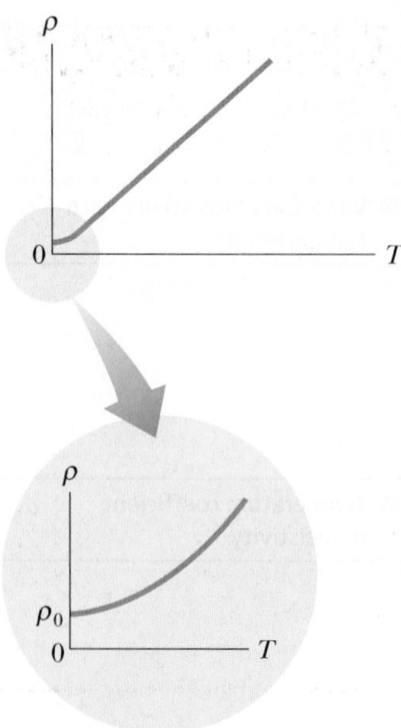

FIGURE 21.8 Resistivity versus temperature for a normal metal, such as copper. The curve is linear over a wide range of temperatures, and ρ increases with increasing temperature. As T approaches absolute zero (inset), the resistivity approaches a finite value ρ_0.

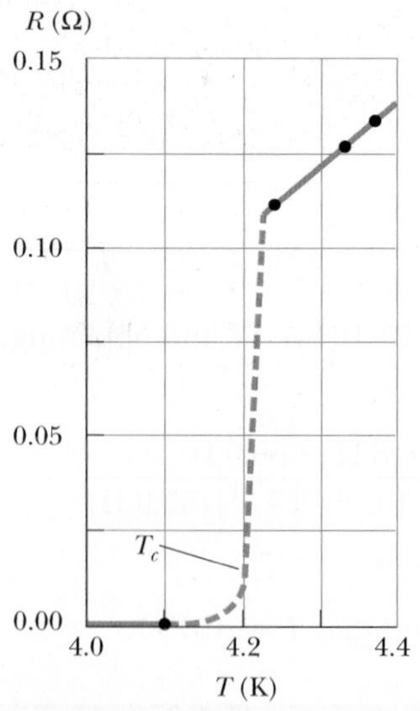

FIGURE 21.9 Resistance versus temperature for a sample of mercury. The graph follows that of a normal metal above the critical temperature T_c. The resistance drops to zero at T_c, which is 4.2 K for mercury.

by collisions of electrons with the vibrating metal atoms. We shall describe this process in more detail in Section 21.4.

There is a class of metals and compounds for which the resistivity goes to zero below a certain **critical temperature** T_c. These materials are known as **superconductors.** The resistance–temperature graph for a superconductor follows that of a normal metal at temperatures above T_c. When the temperature reaches T_c, the resistance of the sample drops suddenly to zero (Fig. 21.9). This phenomenon was discovered by Dutch physicist Heike Kamerlingh Onnes in 1911 as he worked with mercury, which is a superconductor below 4.2 K. Recent measurements have shown that the resistivities of superconductors below T_c are less than $4 \times 10^{-25}\ \Omega \cdot \text{m}$, which is about 10^{17} times smaller than the resistivity of copper and considered to be zero in practice.

Today, thousands of superconductors are known. Such common metals as aluminum, tin, lead, zinc, and indium are superconductors. Table 21.3 lists the critical temperatures of several superconductors. The value of T_c is sensitive to chemical composition, pressure, and crystalline structure. It is interesting to note that copper, silver, and gold, which are excellent conductors at room temperatures, do not exhibit superconductivity.

One truly remarkable feature of superconductors is that once a current is set up in them, it persists *without any applied voltage* (because $R = 0$). In fact, steady currents have been observed to persist in superconducting loops for several years with no apparent decay!

An important development in physics that created much excitement in the scientific community in the latter part of the twentieth century is the discovery of high-temperature copper-oxide-based superconductors. The excitement began with a 1986 publication by J. Georg Bednorz and K. Alex Müller, scientists at the IBM Zurich Research Laboratory in Switzerland, in which they reported evidence for superconductivity at a temperature near 30 K in an oxide of barium, lanthanum, and copper. Bednorz and Müller were awarded the Nobel Prize in Physics in 1987 for their remarkable discovery. Shortly thereafter, a new family of compounds was open for investigation, and research activity in the field of superconductivity proceeded vigorously. In early 1987, groups at the University of Alabama at Huntsville and the University of Houston announced the discovery of superconductivity at about 92 K in an oxide of yttrium, barium, and copper ($YBa_2Cu_3O_7$). Late in 1987, teams of scientists from Japan and the United States reported superconductivity at 105 K in an oxide of bismuth, strontium, calcium, and copper. More recently, scientists have reported superconductivity at temperatures as high as 134 K in a compound containing mercury. At this point, one cannot rule out the possibility of room-temperature superconductivity, and the search for novel superconducting materials continues. It is an important search both for scientific reasons and because practical applications become more probable and widespread as the critical temperature is raised.

An important and useful application is superconducting magnets in which the magnetic field magnitudes are about ten times greater than those of the best normal electromagnets. (We will study magnetism in Chapter 22.) Such superconducting magnets are being considered as a means of storing energy. The idea of using superconducting power lines for transmitting power efficiently is also receiving some consideration. Modern superconducting electronic devices consisting of two thin-film superconductors separated by a thin insulator have been constructed. They include magnetometers (magnetic-field measuring devices) and various microwave devices.

21.4 A STRUCTURAL MODEL FOR ELECTRICAL CONDUCTION

In Section 21.1, a structural model of electrical conduction was developed by relating the macroscopic current to the drift speed of microscopic charge carriers in a material. This section expands that model by introducing the microscopic origin of resistance. Once the model is completed, we shall compare its predictions to experimental measurements.

Consider a conductor as a regular array of atoms containing free electrons (sometimes called *conduction* electrons). Such electrons are free to move through the conductor (as we learned in our discussion of drift speed in Section 21.1) and are approximately equal in number to the number of atoms in the conductor. In the absence of an electric field, the free electrons move in random directions with average speeds on the order of 10^6 m/s. The situation is similar to the motion of gas molecules confined in a vessel that we studied in kinetic theory in Chapter 16. In fact, the conduction electrons in a metal are often called an *electron gas.*

Conduction electrons are not totally free because they are confined to the interior of the conductor and undergo frequent collisions with the array of atoms. The collisions are the predominant mechanism contributing to the resistivity of a metal at normal temperatures. Note that there is no current in a conductor in the absence of an electric field because the average velocity of the free electrons is zero. On the average, just as many electrons move in one direction as in the opposite direction, so there is no net flow of charge.

The situation is modified, however, when an electric field is applied to the metal. In addition to random thermal motion, the free electrons drift slowly in a direction opposite that of the electric field, with an average drift speed of v_d, which is much less (typically 10^{-4} m/s; see Example 21.1) than the average speed between collisions (typically 10^6 m/s).

In our structural model, we shall assume that the excess kinetic energy acquired by the electrons in the electric field is lost to the conductor in the collision process. The energy given up to the atoms in the collisions increases the total vibrational energy of the atoms, causing the conductor to warm up. The model also assumes that an electron's motion after a collision is independent of its motion before the collision.

Given this basis for our model, we now take the first step toward obtaining an expression for the drift speed. When a mobile, charged particle of mass m and charge q is subjected to an electric field $\vec{\mathbf{E}}$, it experiences a force $q\vec{\mathbf{E}}$ (Eq. 19.4). For electrons in a metal, $\vec{\mathbf{F}}_e = -e\vec{\mathbf{E}}$. The motion of the electron can be determined from Newton's second law, $\sum \vec{\mathbf{F}} = m_e\vec{\mathbf{a}}$. The acceleration of the electron is

$$\vec{\mathbf{a}} = \frac{\sum \vec{\mathbf{F}}}{m_e} = \frac{\vec{\mathbf{F}}_e}{m_e} = \frac{-e\vec{\mathbf{E}}}{m_e} \qquad [21.13]$$

The acceleration, which occurs for only a short time interval between collisions, changes the velocity of the electron. Because the force is constant, the acceleration is constant, and we can model the electron as a particle under constant acceleration. If $\vec{\mathbf{v}}_0$ is the velocity of the electron just after a collision, at which we define the time as $t = 0$, the velocity of the electron at time t is

$$\vec{\mathbf{v}} = \vec{\mathbf{v}}_0 + \vec{\mathbf{a}}t = \vec{\mathbf{v}}_0 - \frac{e\vec{\mathbf{E}}}{m_e}t \qquad [21.14]$$

The motion of the electron through the metal is characterized by a very large number of collisions per second. Consequently, we consider the average value of $\vec{\mathbf{v}}$ over a time interval long compared with the time interval between collisions, which gives us the drift velocity $\vec{\mathbf{v}}_d$. Because the velocity of the electron after a collision is assumed to be independent of its velocity before the collision, the initial velocities are randomly distributed in direction, so the average value of $\vec{\mathbf{v}}_0$ is zero. In the second term on the right of Equation 21.14, the charge, electric field, and mass are all constant. Therefore, the only factor affected by the averaging process is the time t. The average value of this term is $(-e\vec{\mathbf{E}}/m_e)\tau$, where τ is the *average time interval between collisions.* Therefore, Equation 21.14 becomes, after the averaging process,

■ Drift velocity

$$\vec{\mathbf{v}}_d = \frac{-e\vec{\mathbf{E}}}{m_e}\tau \qquad [21.15]$$

TABLE 21.3

Critical Temperatures for Various Superconductors

Material	T_c (K)
$HgBa_2Ca_2Cu_3O_8$	134
Tl–Ba–Ca–Cu–O	125
Bi–Sr–Ca–Cu–O	105
$YBa_2Cu_3O_7$	92
Nb_3Ge	23.2
Nb_3Sn	18.05
Nb	9.46
Pb	7.18
Hg	4.15
Sn	3.72
Al	1.19
Zn	0.88

(Courtesy of IBM Research Laboratory)

A small, permanent magnet levitated above a disk of the superconductor $YBa_2Cu_3O_7$, which is at 77 K. This levitation is one the phenomena related to the lack of resistance in the superconductor. ■

Substituting the magnitude of this drift velocity (the drift speed) into Equation 21.4, we have

$$I = nev_d A = ne\left(\frac{eE}{m_e}\tau\right)A = \frac{ne^2E}{m_e}\tau A \qquad [21.16]$$

According to Equation 21.6, the current is related to the macroscopic variables of potential difference and resistance:

$$I = \frac{\Delta V}{R}$$

Incorporating Equation 21.8, we can write this expression as

$$I = \frac{\Delta V}{\left(\rho\frac{\ell}{A}\right)} = \frac{\Delta V}{\rho\ell}A$$

In the conductor, the electric field is uniform, so we use Equation 20.6, $\Delta V = E\ell$, to substitute for the magnitude of the potential difference across the conductor:

$$I = \frac{E\ell}{\rho\ell}A = \frac{E}{\rho}A \qquad [21.17]$$

Setting the two expressions for the current, Equations 21.16 and 21.17, equal, we solve for the resistivity:

■ Resistivity in terms of microscopic parameters

$$I = \frac{ne^2E}{m_e}\tau A = \frac{E}{\rho}A \quad \rightarrow \quad \rho = \frac{m_e}{ne^2\tau} \qquad [21.18]$$

According to this structural model, resistivity does not depend on the electric field or, equivalently, on the potential difference, but depends only on fixed parameters associated with the material and the electron. This feature is characteristic of a conductor obeying Ohm's law. The model shows that the resistivity can be calculated from a knowledge of the density of the electrons, their charge and mass, and the average time interval τ between collisions. This time interval is related to the average distance between collisions ℓ_{avg} (the *mean free path*) and the average speed v_{avg} through the expression[6]

$$\tau = \frac{\ell_{avg}}{v_{avg}} \qquad [21.19]$$

EXAMPLE 21.4 Electron Collisions in Copper

A Using the data and results from Example 21.1 and the structural model of electron conduction, estimate the average time interval between collisions for electrons in copper at 20°C.

Solution From Equation 21.18 we see that

$$\tau = \frac{m_e}{ne^2\rho}$$

where $\rho = 1.7 \times 10^{-8}\ \Omega\cdot\text{m}$ for copper and the carrier density $n = 8.49 \times 10^{28}$ electrons/m^3 for the wire described in Example 21.1. Substitution of these values into the expression above gives

$$\tau = \frac{9.11 \times 10^{-31}\ \text{kg}}{(8.49 \times 10^{28}\ \text{m}^{-3})(1.6 \times 10^{-19}\ \text{C})^2(1.7 \times 10^{-8}\ \Omega\cdot\text{m})}$$
$$= 2.5 \times 10^{-14}\ \text{s}$$

Note that this result is a very short time interval and that the electrons make a very large number of collisions per second.

B Assuming that the average speed for free electrons in copper is 1.6×10^6 m/s and using the result from part A, calculate the mean free path for electrons in copper.

[6] Recall that the average speed of a group of particles depends on the temperature of the group (Chapter 16) and is not the same as the drift speed v_d.

Solution Using Equation 21.19, we find

$$\ell_{\text{avg}} = v_{\text{avg}}\tau = (1.6 \times 10^6 \text{ m/s})(2.5 \times 10^{-14} \text{ s})$$
$$= 4.0 \times 10^{-8} \text{ m}$$

which is equivalent to 40 nm (compared with atomic spacings of about 0.2 nm). Therefore, although the time interval between collisions is very short, the electrons travel about 200 atomic distances before colliding with an atom.

Although this structural model of conduction is consistent with Ohm's law, it does not correctly predict the values of resistivity or the behavior of the resistivity with temperature. For example, the results of classical calculations for v_{avg} using the ideal gas model for the electrons are about a factor of ten smaller than the actual values, which results in incorrect predictions of values of resistivity from Equation 21.18. Furthermore, according to Equations 21.18 and 21.19, the temperature variation of the resistivity is predicted to vary as v_{avg}, which according to an ideal-gas model (Chapter 16, Eq. 16.22) is proportional to $\sqrt{T}$. This behavior is in disagreement with the linear dependence of resistivity with temperature for pure metals (Fig. 21.8a). Because of these incorrect predictions, we must modify our structural model. We shall call the model that we have developed so far the *classical* model for electrical conduction. To account for the incorrect predictions of the classical model, we will develop it further into a *quantum mechanical* model, which we shall describe briefly.

We discussed two important simplification models in earlier chapters, the particle model and the wave model. Although we discussed these two simplification models separately, quantum physics tells us that this separation is not so clear-cut. As we shall discuss in detail in Chapter 28, particles have wave-like properties. The predictions of some models can only be matched to experimental results if the model includes the wave-like behavior of particles. The structural model for electrical conduction in metals is one of these cases.

Let us imagine that the electrons moving through the metal have wave-like properties. If the array of atoms in a conductor is regularly spaced (that is, periodic), the wave-like character of the electrons makes it possible for them to move freely through the conductor and a collision with an atom is unlikely. For an idealized conductor, no collisions would occur, the mean free path would be infinite, and the resistivity would be zero. Electrons are scattered only if the atomic arrangement is irregular (not periodic), as a result of structural defects or impurities, for example. At low temperatures, the resistivity of metals is dominated by scattering caused by collisions between the electrons and impurities. At high temperatures, the resistivity is dominated by scattering caused by collisions between the electrons and the atoms of the conductor, which are continuously displaced as a result of thermal agitation, destroying the perfect periodicity. The thermal motion of the atoms makes the structure irregular (compared with an atomic array at rest), thereby reducing the electron's mean free path.

Although it is beyond the scope of this text to show this modification in detail, the classical model modified with the wave-like character of the electrons results in predictions of resistivity values that are in agreement with measured values and predicts a linear temperature dependence. When discussing the hydrogen atom in Chapter 11, we had to introduce some quantum notions to understand experimental observations such as atomic spectra. Likewise, we had to introduce quantum notions in Chapter 17 to understand the temperature behavior of molar specific heats of gases. Here we have another case in which quantum physics is necessary for the model to agree with experiment. Although classical physics can explain a tremendous range of phenomena, we continue to see hints that quantum physics must be incorporated into our models. We shall study quantum physics in detail in Chapters 28 through 31.

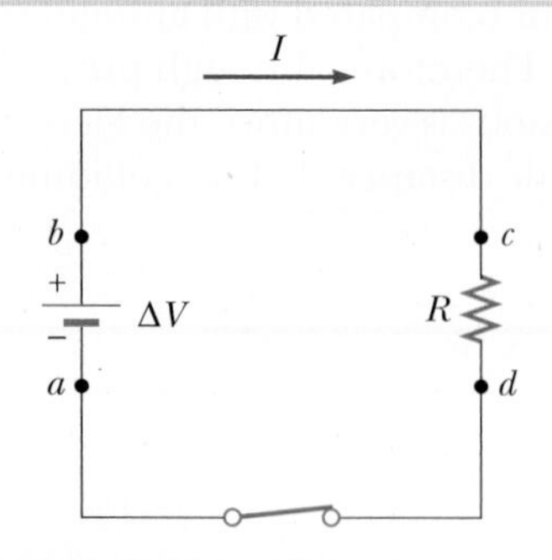

ACTIVE FIGURE 21.10

A circuit consisting of a resistor of resistance R and a battery having a potential difference ΔV across its terminals. Positive charge flows in the clockwise direction.

PhysicsNow™ Log into PhysicsNow at **www.pop4e.com** and go to Active Figure 21.10 to adjust the battery voltage and the resistance to see the resulting current in the circuit and power delivered to the resistor.

21.5 ELECTRIC ENERGY AND POWER

In Section 21.1, we discussed the energy transformations occurring in a circuit. If a battery is used to establish an electric current in a conductor, there is a continuous transformation of chemical energy in the battery to kinetic energy of the electrons to internal energy in the conductor, resulting in an increase in the temperature of the conductor.

In typical electric circuits, energy is transferred from a source, such as a battery, to some device, such as a lightbulb or a radio receiver by electrical transmission (T_{ET} in Eq. 6.20). Let us determine an expression that will allow us to calculate the rate of this energy transfer. First, consider the simple circuit in Active Figure 21.10, where we imagine that energy is being delivered to a resistor. Because the connecting wires also have resistance, some energy is delivered to the wires and some energy to the resistor. Unless noted otherwise, we will adopt a simplification model in which the resistance of the wires is so small compared with the resistance of the circuit element that we ignore the energy delivered to the wires.

Let us now analyze the energetics of the circuit in which a battery is connected to a resistor of resistance R as in Active Figure 21.10. Imagine following a positive quantity of charge Q around the circuit from point a through the battery and resistor and back to a. Point a is a reference point at which the potential is defined as zero. We identify the entire circuit as our system. As the charge moves from a to b through the battery whose potential difference is ΔV, the electrical potential energy of the system increases by the amount $Q\,\Delta V$, whereas the chemical energy in the battery decreases by the same amount. (Recall from Chapter 20 that $\Delta U = q\,\Delta V$.) As the charge moves from c to d through the resistor, however, the system loses this electrical potential energy during collisions with atoms in the resistor. In this process, the energy is transformed to internal energy corresponding to increased vibrational motion of the atoms in the resistor. Because we have neglected the resistance of the interconnecting wires, no energy transformation occurs for paths bc and da. When the charge returns to point a, the net result is that some of the chemical energy in the battery has been delivered to the resistor and resides in the resistor as internal energy associated with molecular vibration.

The resistor is normally in contact with air, so its increased temperature results in a transfer of energy by heat into the air. In addition, there will be thermal radiation from the resistor, representing another means of escape for the energy. After some time interval has passed, the resistor remains at a constant temperature because the input of energy from the battery is balanced by the output of energy by heat and radiation. Some electrical devices include *heat sinks*[7] connected to parts of the circuit to prevent these parts from reaching dangerously high temperatures. Heat sinks are pieces of metal with many fins. The high thermal conductivity of the metal provides a rapid transfer of energy by heat away from the hot component and the large number of fins provides a large surface area in contact with the air, so energy can transfer by radiation and into the air by heat at a high rate.

Let us consider now the rate at which the system loses electric potential energy as the charge Q passes through the resistor:

$$\frac{dU}{dt} = \frac{d}{dt}(Q\,\Delta V) = \frac{dQ}{dt}\Delta V = I\,\Delta V$$

where I is the current in the circuit. Of course, the system regains this potential energy when the charge passes through the battery, at the expense of chemical energy in the battery. The rate at which the system loses potential energy as the charge passes through the resistor is equal to the rate at which the system gains internal energy in the resistor. Therefore, the **power** $\mathcal{P}$, representing the rate at which energy

PITFALL PREVENTION 21.5

MISCONCEPTIONS ABOUT CURRENT Several common misconceptions are associated with current in a circuit like that in Active Figure 21.10. One is that current comes out of one terminal of the battery and is then "used up" as it passes through the resistor. According to this approach, there is current in only one part of the circuit. The correct understanding, however, is that the current is the same *everywhere* in the circuit. A related misconception has the current coming out of the resistor being smaller than that going in because some of the current is "used up." Another misconception has current coming out of both terminals of the battery, in opposite directions, and then "clashing" in the resistor, delivering the energy in this manner. We know that is not the case because the charges flow in the same rotational sense at *all* points in the circuit. Be sure your conceptual understanding of current is valid.

[7]This terminology is another misuse of the word *heat* that is ingrained in our common language.

is delivered to the resistor, is

$$\mathcal{P} = I\,\Delta V \qquad [21.20]$$

■ Power delivered to a device

We have developed this result by considering a battery delivering energy to a resistor. Equation 21.20, however, can be used to determine the power transferred from a voltage source to *any* device carrying a current I and having a potential difference ΔV between its terminals.

Using Equation 21.20 and that $\Delta V = IR$ for a resistor, we can express the power delivered to the resistor in the alternative forms

$$\mathcal{P} = I^2R = \frac{(\Delta V)^2}{R} \qquad [21.21]$$

The SI unit of power is the watt, introduced in Chapter 6. If you analyze the units in Equations 21.20 and 21.21, you will see that the result of the calculation provides a watt as the unit. The power delivered to a conductor of resistance R is often referred to as an *I^2R loss.*

As we learned in Section 6.8, the unit of energy your electric company uses to calculate energy transfer, the kilowatt-hour, is the amount of energy transferred in 1 h at the constant rate of 1 kW. Because 1 W = 1 J/s, we have

$$1 \text{ kWh} = (1.0 \times 10^3 \text{ W})(3\,600 \text{ s}) = 3.6 \times 10^6 \text{ J} \qquad [21.22]$$

QUICK QUIZ 21.4 For the two lightbulbs shown in Figure 21.11, rank the currents at points *a* through *f*, from largest to smallest.

■ Thinking Physics 21.2

Two lightbulbs A and B are connected across the same potential difference as in Figure 21.11. The electric input powers to the lightbulbs are shown. Which lightbulb has the higher resistance? Which carries the greater current?

Reasoning Because the voltage across each lightbulb is the same and the rate of energy delivered to a resistor is $\mathcal{P} = (\Delta V)^2/R$, the lightbulb with the lower resistance exhibits the higher rate of energy transfer. In this case, the resistance of A is larger than that for B. Furthermore, because $\mathcal{P} = I\,\Delta V$, we see that the current carried by B is larger than that of A. ■

■ Thinking Physics 21.3

When is a lightbulb more likely to fail, just after it is turned on or after it has been on for a while?

Reasoning When the switch is closed, the source voltage is immediately applied across the lightbulb. As the voltage is applied across the cold filament when the lightbulb is first turned on, the resistance of the filament is low. Therefore, the current is high and a relatively large amount of energy is delivered to the bulb per unit time interval. This causes the temperature of the filament to rise rapidly, resulting in thermal stress on the filament that makes it likely to fail at that moment. As the filament warms up in the absence of failure, its resistance rises and the current falls. As a result, the rate of energy delivered to the lightbulb falls. The thermal stress on the filament is reduced so that the failure is less likely to occur after the bulb has been on for a while. ■

PITFALL PREVENTION 21.6

CHARGES DO NOT MOVE ALL THE WAY AROUND A CIRCUIT The movement of a charge around the circuit is not what happens in a circuit, unless you wait for a very long time. Due to the very low magnitude of the drift velocity, it might take *hours* for a single electron to make one complete trip around the circuit. In terms of understanding the energy transfer in a circuit, however, it is useful to *imagine* a charge moving all the way around the circuit.

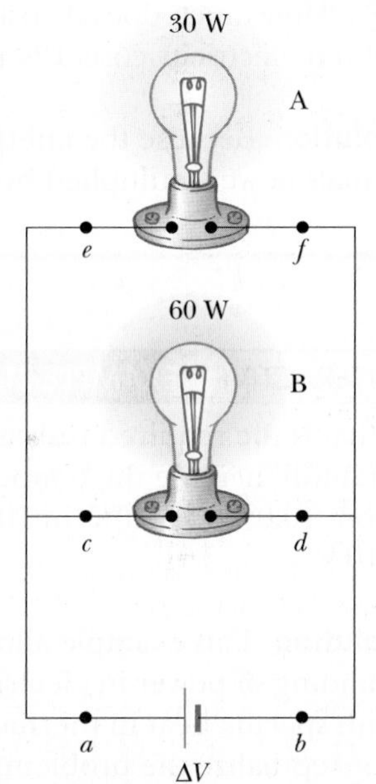

FIGURE 21.11 (Quick Quiz 21.4 and Thinking Physics 21.2) Two lightbulbs connected across the same potential difference.

PITFALL PREVENTION 21.7

ENERGY IS NOT "DISSIPATED" In some books, you may see Equation 21.20 described as the power "dissipated in" a resistor, suggesting that energy disappears. Instead, we say energy is "delivered to" a resistor. The notion of *dissipation* arises because a warm resistor will expel energy by radiation and heat, and energy delivered by the battery leaves the circuit. (It does not disappear!)

EXAMPLE 21.5 Electrical Rating of a Lightbulb

A lightbulb is rated at 120 V/75 W, which means that at its intended operating voltage of 120 V it has energy delivered to it at a rate of 75.0 W. The lightbulb is powered by a 120-V direct-current power supply.

A Find the current in the lightbulb and its resistance.

Solution Because the power rating of the lightbulb is 75.0 W and the operating voltage is 120 V, we can use $\mathcal{P} = I\,\Delta V$ to find the current:

$$I = \frac{\mathcal{P}}{\Delta V} = \frac{75.0 \text{ W}}{120 \text{ V}} = 0.625 \text{ A}$$

Using $\Delta V = IR$, the resistance is calculated to be

$$R = \frac{\Delta V}{I} = \frac{120 \text{ V}}{0.625 \text{ A}} = 192\ \Omega$$

B How much does it cost to operate the lightbulb for 24 h if electricity costs 12¢ per kilowatt-hour?

Solution Because the energy delivered to the lightbulb equals power multiplied by time interval, the amount of energy you must pay for, expressed in kWh, is

$$\text{Energy} = (0.075 \text{ kW})(24 \text{ h}) = 1.8 \text{ kWh}$$

If energy is purchased at 12¢ per kWh, the cost is

$$\text{Cost} = (1.8 \text{ kWh})(\$0.12/\text{kWh}) = \$0.22$$

That is, it costs 22¢ to operate the lightbulb for one day. This cost is a small amount, but when larger and more complex devices are used, the costs go up rapidly.

Demands on energy supplies have made it necessary to be aware of the energy requirements of electric devices, not only because they are becoming more expensive to operate but also because, with the dwindling of the coal and oil resources that ultimately supply us with electrical energy, increased awareness of conservation becomes necessary. Every electric appliance has a label that contains the information needed to calculate the power requirements of the appliance. The power consumption in watts is often stated directly, as on a lightbulb. In other cases, the amount of current in the device and the voltage at which it operates are given. This information and Equation 21.20 are sufficient to calculate the operating cost of any electric device.

INTERACTIVE EXAMPLE 21.6 Linking Electricity and Thermodynamics

What is the required resistance of an immersion heater that will increase the temperature of 1.50 kg of water from 10.0°C to 50.0°C in 10.0 min while operating at 110 V?

Solution This example allows us to link our new understanding of power in electricity with our experience with specific heat in thermodynamics in Chapter 17. To conceptualize the problem, we need to realize that an immersion heater is a resistor that is inserted into a container of water. As energy is delivered to the immersion heater, raising its temperature, energy leaves the surface of the resistor by heat, going into the water. When the immersion heater reaches a constant temperature, the rate of energy delivered to the resistance by electrical transmission is equal to the rate of energy delivered by heat to the water.

As a simplification model, we ignore the initial time interval during which the temperature of the resistor increases, and we also ignore any variation of resistance with temperature. Therefore, we imagine a constant rate of energy transfer for the entire 10.0 min. We categorize this problem as one in which energy is delivered to the resistor by electrical transmission and then to the water by heat. To analyze the problem, we set the rate of energy delivered to the resistor equal to the rate of energy entering the water:

$$\mathcal{P} = \frac{(\Delta V)^2}{R} = \frac{Q}{\Delta t}$$

where Q represents an amount of energy transfer by heat into the water and Equation 21.21 expresses the electrical power. The amount of energy transfer by heat necessary to raise the temperature of the water is given by Equation 17.3, $Q = mc\,\Delta T$. Therefore,

$$\frac{(\Delta V)^2}{R} = \frac{mc\,\Delta T}{\Delta t} \quad \rightarrow \quad R = \frac{(\Delta V)^2\,\Delta t}{mc\,\Delta T}$$

Substituting the values given in the statement of the problem gives

$$R = \frac{(110 \text{ V})^2(600 \text{ s})}{(1.50 \text{ kg})(4\,186 \text{ J/kg}\cdot{}^\circ\text{C})(50.0^\circ\text{C} - 10.0^\circ\text{C})}$$

$$= 28.9\ \Omega$$

To finalize this problem, let us compare the power and the cost of operation of the immersion heater to the lightbulb in Example 21.5. The power of the immersion heater is found from Equation 21.21:

$$\mathcal{P} = \frac{(\Delta V)^2}{R} = \frac{(110 \text{ V})^2}{28.9\ \Omega} = 419 \text{ W}$$

which is significantly larger than the power of the lightbulb in Example 21.5. The energy transferred to the heater during the operation time of 10.0 min is

$$\Delta E = \mathcal{P}\,\Delta t = (419\text{ W})(10.0\text{ min})\left(\frac{1\text{ h}}{60.0\text{ min}}\right)$$

$$= 69.8\text{ Wh} = 0.069\,8\text{ kWh}$$

If the energy is purchased at an estimated price of 12.0¢ per kilowatt-hour, the cost is

$$\text{Cost} = (0.069\,8\text{ kWh})(\$0.120/\text{kWh})$$

$$= \$0.008\,38 = 0.838¢$$

Even though the power rating is higher for the heater than for the lightbulb, it costs much less to operate the heater. Of course, the primary factor in this comparison is that the heater is operated for 10.0 min, whereas the lightbulb in Example 21.5 is operated for 24 h.

PhysicsNow™ Explore the heating of water by logging into PhysicsNow at **www.pop4e.com** and going to Interactive Example 21.6.

21.6 SOURCES OF emf

The entity that maintains the constant voltage in Figure 21.12 is called a **source of emf.**[8] Sources of emf are any devices (such as batteries and generators) that increase the potential energy of a circuit system by maintaining a potential difference between points in the circuit while charges move through the circuit. One can think of a source of emf as a "charge pump." The emf $\mathcal{E}$ of a source describes the work done per unit charge, and hence the SI unit of emf is the volt.

At this point, you may wonder why we need to define a second quantity, emf, with the volt as a unit when we have already defined the potential difference. To see the need for this new quantity, consider the circuit shown in Figure 21.12, consisting of a battery connected to a resistor. We shall assume that the connecting wires have no resistance. We might be tempted to claim that the potential difference across the battery terminals (the terminal voltage) equals the emf of the battery. A real battery, however, always has some **internal resistance** r. As a result, the terminal voltage is not equal to the emf, as we shall show.

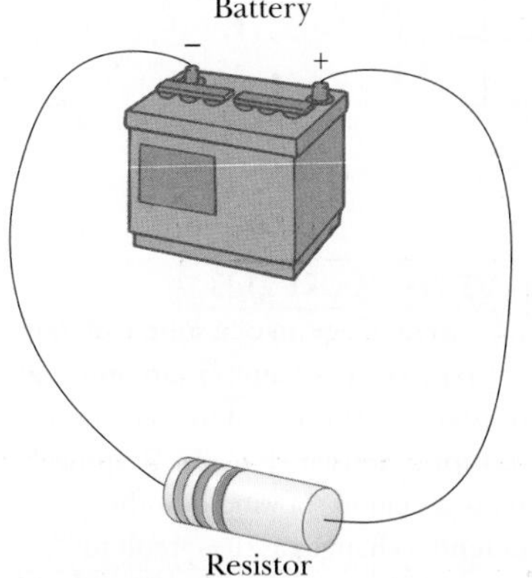

FIGURE 21.12 A circuit consisting of a resistor connected to the terminals of a battery.

The circuit shown in Figure 21.12 can be described by the circuit diagram in Active Figure 21.13a. The battery within the dashed rectangle is modeled as an ideal, zero-resistance source of emf $\mathcal{E}$ in series with the internal resistance r. Now imagine moving from a to b in Active Figure 21.13a. As you pass from the negative to the positive terminal within the source of emf the potential increases by $\mathcal{E}$. As you move through the resistance r, however, the potential decreases by an amount Ir, where I is the current in the circuit. Therefore, the terminal voltage $\Delta V = V_b - V_a$ of the battery is[9]

$$\Delta V = \mathcal{E} - Ir \qquad [21.23]$$

Note from this expression that $\mathcal{E}$ is equivalent to the **open-circuit voltage, that is, the terminal voltage when the current is zero.** Active Figure 21.13b is a graphical representation of the changes in potential as the circuit is traversed clockwise. By inspecting Active Figure 21.13a, we see that the terminal voltage ΔV must also equal the potential difference across the external resistance R, often called the **load resistance;** that is, $\Delta V = IR$. Combining this expression with Equation 21.23, we see that

$$\mathcal{E} = IR + Ir \qquad [21.24]$$

[8]The term *emf* was originally an abbreviation for *electromotive force,* but it is not a force, so the long form is discouraged. The name electromotive force was used early in the study of electricity before the understanding of batteries was as sophisticated as it is today.

[9]The terminal voltage in this case is less than the emf by the amount Ir. In some situations, the terminal voltage may *exceed* the emf by the amount Ir. Such a situation occurs when the direction of the current is *opposite* that of the emf, as when a battery is being charged by another source of emf.

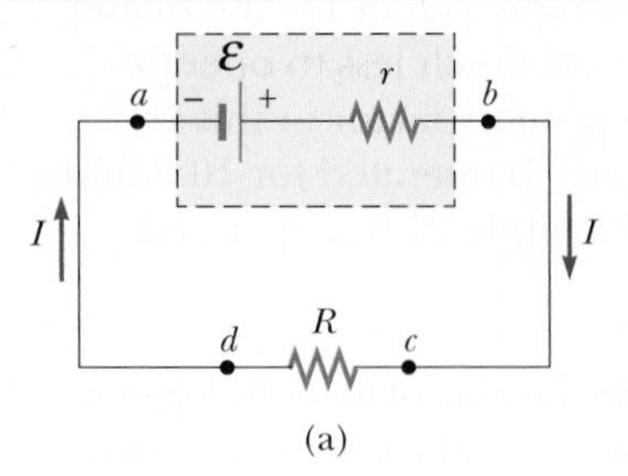

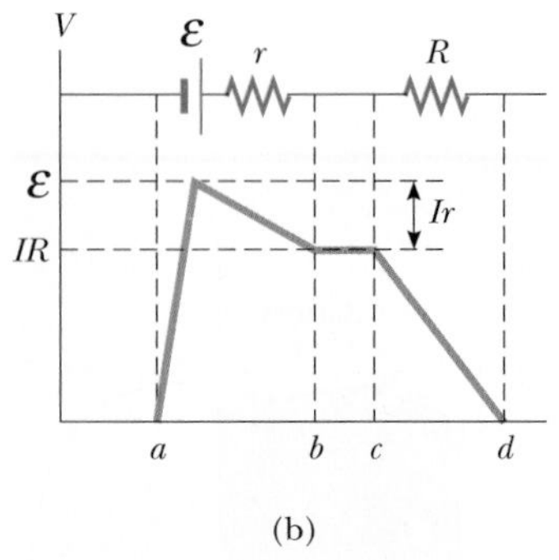

ACTIVE FIGURE 21.13
(a) Circuit diagram of a source of emf $\mathcal{E}$ (in this case, a battery) with internal resistance r, connected to an external resistor of resistance R. (b) Graphical representation showing how the potential changes as the circuit in (a) is traversed clockwise.

Physics Now™ Log into PhysicsNow at **www.pop4e.com** and go to Active Figure 21.13 to adjust the emf and resistances r and R to see the effect on the current and on the graph in (b).

PITFALL PREVENTION 21.8

WHAT IS CONSTANT IN A BATTERY? Notice that Equation 21.25 shows us that the current in the circuit depends on the resistance connected to the battery. It is a common misconception that a battery is a source of constant current. Equation 21.25 clearly shows that to be not true. It is also not true that a battery is a source of constant terminal voltage. Equation 21.23 shows that to be not true. **A battery is a source of constant emf.**

■ Equivalent resistance of resistors in series

Solving for the current gives

$$I = \frac{\mathcal{E}}{R + r} \quad [21.25]$$

which shows that the current in this simple circuit depends on both the resistance R external to the battery and the internal resistance r. If R is much greater than r, we can adopt a simplification model in which we neglect r in our analysis. In many circuits, we shall adopt this simplification model.

If we multiply Equation 21.24 by the current I, we have

$$I\mathcal{E} = I^2R + I^2r$$

This equation tells us that the total power output $I\mathcal{E}$ of the source of emf is equal to the rate I^2R at which energy is delivered to the load resistance plus the rate I^2r at which energy is delivered to the internal resistance. If $r \ll R$, much more of the energy from the battery is delivered to the load resistance than stays in the battery, although the amount of energy is relatively small because the load resistance is large, resulting in a small current. If $r \gg R$, a significant fraction of the energy from the source of emf stays in the battery package because it is delivered to the internal resistance. For example, if a wire is simply connected between the terminals of a flashlight battery, the battery becomes warm. This warming represents the transfer of energy from the source of emf to the internal resistance, where it appears as internal energy associated with temperature. Problem 21.57 explores the conditions under which the largest amount of energy is transferred from the battery to the load resistor.

21.7 RESISTORS IN SERIES AND IN PARALLEL

When two or more resistors are connected together end to end as in Active Figure 21.14a, they are said to be in **series.** (Compare this configuration to capacitors in series in Active Figure 20.24.) In a series connection, if an amount of charge Q exits resistor R_1, charge Q must also enter the second resistor R_2. Otherwise, charge will accumulate on the wire between the resistors. Therefore, the same amount of charge passes through both resistors in a given time interval and the currents are the same in both resistors.

Because the potential difference between a and b in the circuit diagram of Active Figure 21.14b equals IR_1 and the potential difference between b and c equals IR_2, the potential difference between a and c is

$$\Delta V = IR_1 + IR_2 = I(R_1 + R_2)$$

The potential difference across the battery is also applied to the equivalent resistance in Active Figure 21.14c:

$$\Delta V = IR_{\text{eq}}$$

where we have indicated that the equivalent resistance has the same effect on the circuit because it results in the same current in the battery as the combination of resistors. Combining these equations, we see that we can replace the two resistors in series with a single equivalent resistance whose value is the *sum* of the individual resistances:

$$\Delta V = IR_{\text{eq}} = I(R_1 + R_2) \quad \rightarrow \quad R_{\text{eq}} = R_1 + R_2 \quad [21.26]$$

The equivalent resistance of three or more resistors connected in series is simply

$$R_{\text{eq}} = R_1 + R_2 + R_3 + \cdots \quad [21.27]$$

Therefore, **the equivalent resistance of a series connection of resistors is the algebraic sum of the individual resistances and is always greater than any individual resistance.**

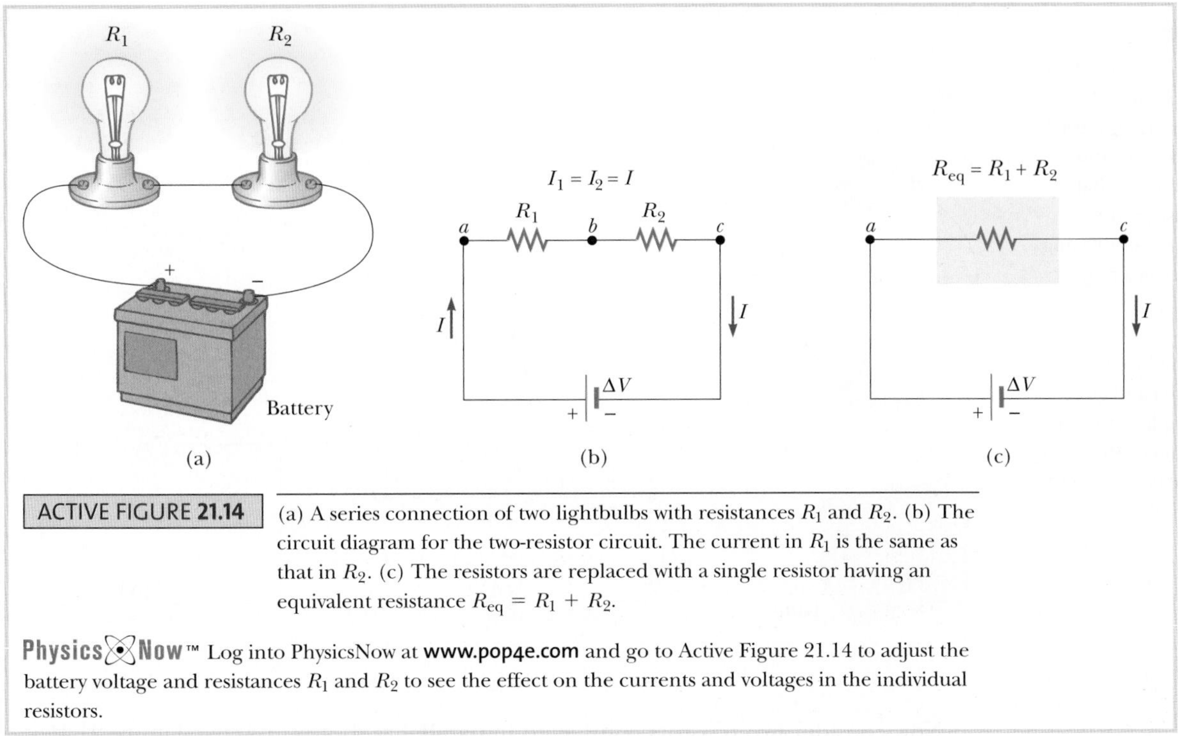

ACTIVE FIGURE 21.14 (a) A series connection of two lightbulbs with resistances R_1 and R_2. (b) The circuit diagram for the two-resistor circuit. The current in R_1 is the same as that in R_2. (c) The resistors are replaced with a single resistor having an equivalent resistance $R_{eq} = R_1 + R_2$.

PhysicsNow™ Log into PhysicsNow at **www.pop4e.com** and go to Active Figure 21.14 to adjust the battery voltage and resistances R_1 and R_2 to see the effect on the currents and voltages in the individual resistors.

Looking back at Equation 21.25, the denominator is the simple algebraic sum of the external and internal resistances, which is consistent with the internal and external resistances being in series in Active Figure 21.13a.

Note that if the filament of one lightbulb in Active Figure 21.14a were to fail,[10] the circuit would no longer be complete (an open-circuit condition would exist) and the second bulb would also go out.

PITFALL PREVENTION 21.9

Local and Global Changes A local change in one part of a circuit may result in a global change throughout the circuit. For example, if a single resistance is changed in a circuit containing several resistors and batteries, the currents in all resistors and batteries, the terminal voltages of all batteries, and the voltages across all resistors may change as a result.

QUICK QUIZ 21.5 If a piece of wire were used to connect points *b* and *c* in Active Figure 21.14b, does the brightness of lightbulb R_1 **(a)** increase, **(b)** decrease, or **(c)** remain the same?

QUICK QUIZ 21.6 With the switch in the circuit of Figure 21.15a closed, there is no current in R_2 because the current has an alternate zero-resistance path through the switch. There is current in R_1, and this current is measured with the ammeter (a device for measuring current) at the right side of the circuit. If the switch is opened (Fig. 21.15b), current exists in R_2. What happens to the reading on the ammeter when the switch is opened? **(a)** The reading goes up. **(b)** The reading goes down. **(c)** The reading does not change.

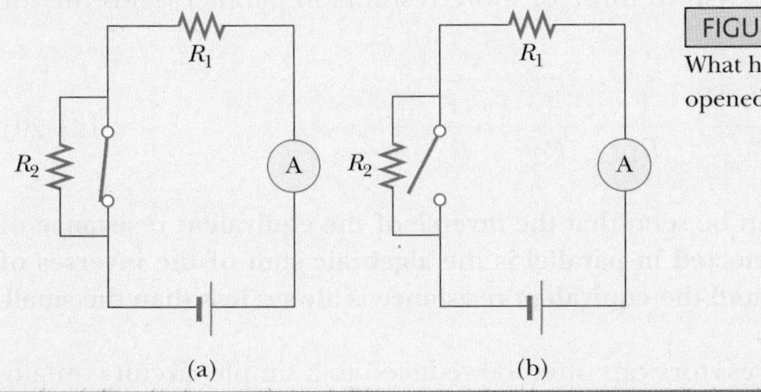

FIGURE 21.15 (Quick Quiz 21.6) What happens when the switch is opened?

[10]We will describe the end of the life of a lightbulb by saying that *the filament fails* rather than by saying that the lightbulb "burns out." The word *burn* suggests a combustion process, which is not what occurs in a lightbulb. When a filament fails, it breaks, so that the bulb can no longer carry a current.

ACTIVE FIGURE 21.16

(a) A parallel connection of two lightbulbs with resistances R_1 and R_2. (b) The circuit diagram for the two-resistor circuit. The potential difference across R_1 is the same as that across R_2. (c) The resistors are replaced with a single resistor having an equivalent resistance given by Equation 21.29.

Physics Now™ Log into PhysicsNow at **www.pop4e.com** and go to Active Figure 21.16 to adjust the battery voltage and resistances R_1 and R_2 to see the effect on the currents and voltages in the individual resistors.

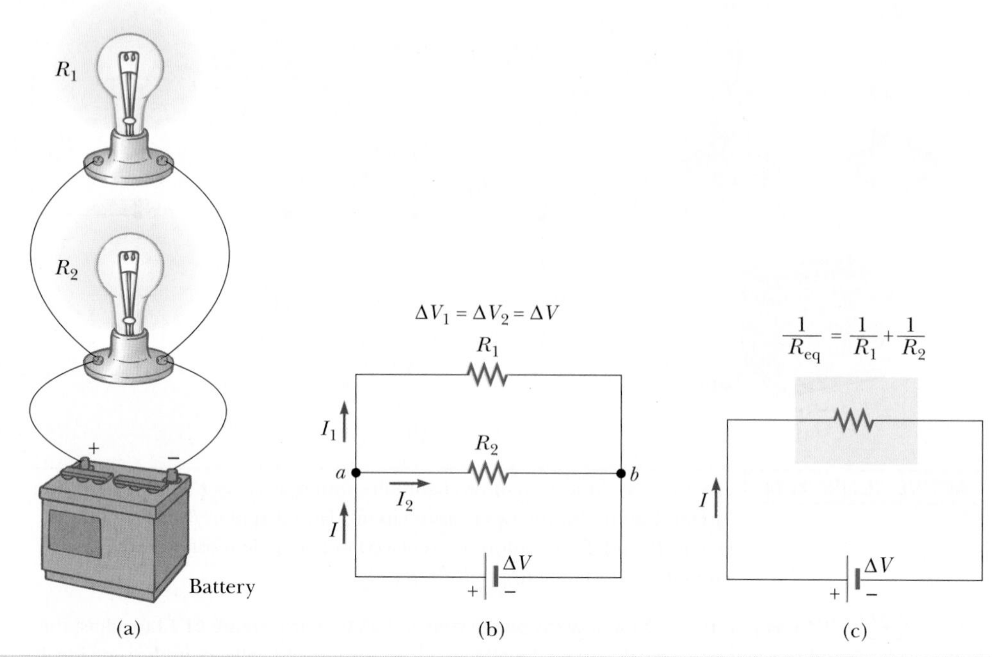

PITFALL PREVENTION 21.10

CURRENT DOES NOT TAKE THE PATH OF LEAST RESISTANCE You may have heard a phrase like "current takes the path of least resistance." This wording is a reference to a parallel combination of current paths such that there are two or more paths for the current to take. The phrase is incorrect, however. The current takes *all* paths. Those paths with lower resistance will have large currents, but even very high-resistance paths will carry *some* of the current.

Now consider two resistors connected in **parallel** as shown in Active Figure 21.16a. In this case, the potential differences across the resistors are equal because each resistor is connected directly across the battery terminals. The currents are generally not the same, however. When the charges reach point a (called a *junction*) in the circuit diagram in Active Figure 21.16b, the current splits into two parts, with I_1 in R_1 and I_2 in R_2. If R_1 is greater than R_2, then I_1 is less than I_2. Because charge must be conserved, the current I that enters point a must equal the total current leaving point a:

$$I = I_1 + I_2$$

Because the potential differences across the resistors are the same, $I = \Delta V/R$ gives

$$I = I_1 + I_2 = \frac{\Delta V}{R_1} + \frac{\Delta V}{R_2} = \Delta V\left(\frac{1}{R_1} + \frac{1}{R_2}\right) = \frac{\Delta V}{R_{eq}}$$

where R_{eq} is an equivalent single resistance that has the same effect on the circuit; that is, it causes the same current in the battery (Active Fig. 21.16c). From this result, we see that the equivalent resistance of two resistors in parallel is given by

$$\frac{1}{R_{eq}} = \frac{1}{R_1} + \frac{1}{R_2} \qquad [21.28]$$

An extension of this analysis to three or more resistors in parallel yields the following general expression:

■ Equivalent resistance of resistors in parallel

$$\frac{1}{R_{eq}} = \frac{1}{R_1} + \frac{1}{R_2} + \frac{1}{R_3} + \cdots \qquad [21.29]$$

From this expression, it can be seen that **the inverse of the equivalent resistance of two or more resistors connected in parallel is the algebraic sum of the inverses of the individual resistances, and the equivalent resistance is always less than the smallest resistance in the group.**

A circuit consisting of resistors can often be reduced to a simple circuit containing only one resistor. To do so, examine the initial circuit and replace any resistors in series or any in parallel with equivalent resistances using Equations 21.27 and 21.29. Draw a sketch of the new circuit after these changes have been made. Examine the new circuit and replace any new series or parallel combinations that now exist. Continue this process until a single equivalent resistance is found for the entire circuit. (That may not be possible; if not, see the techniques of Section 21.8.)

If the current in or the potential difference across a resistor in the initial circuit is to be found, start with the final circuit and gradually work your way back through the equivalent circuits. Find currents and voltages across resistors using $\Delta V = IR$ and your understanding of series and parallel combinations.

Household circuits are always wired so that the electrical devices are connected in parallel as in Active Figure 21.16a. In this manner, each device operates independently of the others so that if one is switched off, the others remain on. For example, if one of the lightbulbs in Active Figure 21.16a were removed from its socket, the other would continue to operate. Equally important, each device operates on the same voltage. If devices were connected in series, the voltage applied to the combination would divide among the devices, so the voltage applied to any one device would depend on how many devices were in the combination.

In many household circuits, circuit breakers are used in series with other circuit elements for safety purposes. A circuit breaker is designed to switch off and open the circuit at some maximum current (typically 15 A or 20 A) whose value depends on the nature of the circuit. If a circuit breaker were not used, excessive currents caused by turning on many devices could result in excessive temperatures in wires and, perhaps, cause a fire. In older home construction, fuses were used in place of circuit breakers. When the current in a circuit exceeds some value, the conductor in a fuse melts and opens the circuit. The disadvantage of fuses is that they are destroyed in the process of opening the circuit, whereas circuit breakers can be reset.

QUICK QUIZ 21.7 With the switch in the circuit of Figure 21.17a open, no current exists in R_2. Current does exist in R_1, and this current is measured with the ammeter at the right side of the circuit. If the switch is closed (Fig. 21.17b), current exists in R_2. What happens to the reading on the ammeter when the switch is closed? **(a)** The reading goes up. **(b)** The reading goes down. **(c)** The reading does not change.

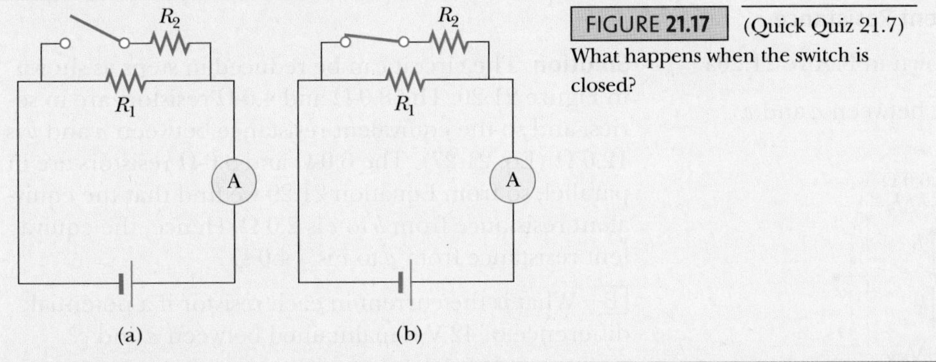

FIGURE 21.17 (Quick Quiz 21.7) What happens when the switch is closed?

QUICK QUIZ 21.8 **(i)** In Active Figure 21.14b, imagine that we add a third resistor in series with the first two. Does the current in the battery **(a)** increase, **(b)** decrease, or **(c)** remain the same? Does the terminal voltage of the battery **(d)** increase, **(e)** decrease, or **(f)** remain the same? **(ii)** In Active Figure 21.16b, imagine that we add a third resistor in parallel with the first two. Does the current in the battery **(a)** increase, **(b)** decrease, or **(c)** remain the same? Does the terminal voltage of the battery **(d)** increase, **(e)** decrease, or **(f)** remain the same?

Thinking Physics 21.4

Compare the brightnesses of the four identical lightbulbs in Figure 21.18. What happens if bulb A fails so that it cannot conduct? What if bulb C fails? What if bulb D fails ?

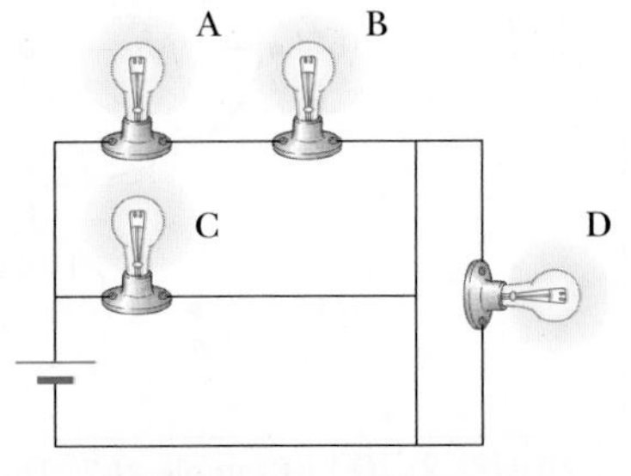

FIGURE 21.18 (Thinking Physics 21.4) What happens to the lightbulbs if one fails?

Reasoning Bulbs A and B are connected in series across the battery, whereas bulb C is connected by itself. Therefore, the terminal voltage of the battery is split between bulbs A and B. As a result, bulb C will be brighter than bulbs A and B, which should be equally as bright as each other. Bulb D has a wire connected across it. Therefore, there is no potential difference across bulb D and it does not glow at all. If bulb A fails, bulb B goes out but bulb C stays lit. If bulb C fails, there is no effect on the other bulbs. If bulb D fails, the event is undetectable because bulb D was not glowing initially. ■

Thinking Physics 21.5

Figure 21.19 illustrates how a three-way lightbulb is constructed to provide three levels of light intensity. The socket of the lamp is equipped with a three-way switch for selecting different light intensities. The bulb contains two filaments. Why are the filaments connected in parallel? Explain how the two filaments are used to provide three different light intensities.

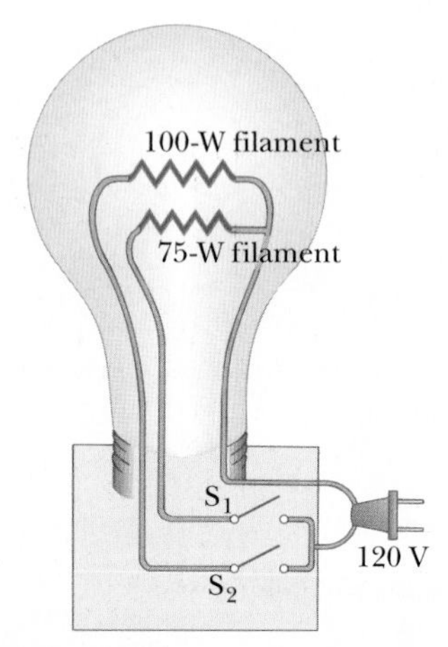

FIGURE 21.19 (Thinking Physics 21.5) A three-way lightbulb.

Reasoning If the filaments were connected in series and one of them were to fail, there would be no current in the bulb and the bulb would give no illumination, regardless of the switch position. When the filaments are connected in parallel, however, and one of them (say the 75-W filament) fails, the bulb still operates in some switch positions because there is current in the other (100-W) filament. The three light intensities are made possible by selecting one of three values of filament resistance, using a single value of 120 V for the applied voltage. The 75-W filament offers one value of resistance, the 100-W filament offers a second value, and the third resistance is obtained by combining the two filaments in parallel. When switch S_1 is closed and switch S_2 is opened, only the 75-W filament carries current. When switch S_1 is open and switch S_2 is closed, only the 100-W filament carries current. When both switches are closed, both filaments carry current, and a total illumination corresponding to 175 W is obtained. ■

EXAMPLE 21.7 Find the Equivalent Resistance

Four resistors are connected as shown in Figure 21.20a.

A Find the equivalent resistance between a and c.

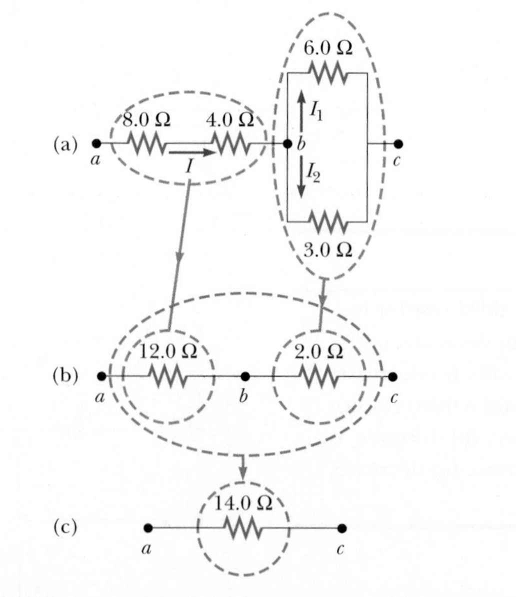

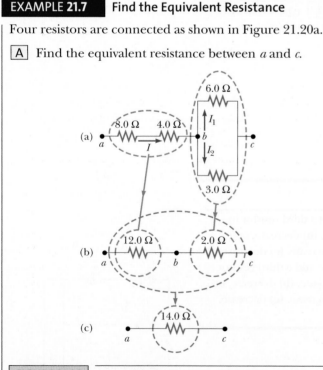

FIGURE 21.20 (Example 21.7) The four resistors shown in (a) can be reduced in steps to an equivalent 14.0-Ω resistor shown in (c).

Solution The circuit can be reduced in steps as shown in Figure 21.20. The 8.0-Ω and 4.0-Ω resistors are in series, and so the equivalent resistance between a and b is 12.0 Ω (Eq. 21.27). The 6.0-Ω and 3.0-Ω resistors are in parallel, so from Equation 21.29 we find that the equivalent resistance from b to c is 2.0 Ω. Hence, the equivalent resistance from a to c is 14.0 Ω.

B What is the current in each resistor if a potential difference of 42 V is maintained between a and c?

Solution Using $\Delta V = IR$ and the results from part A, we have

$$I = \frac{\Delta V_{ac}}{R_{\text{eq}}} = \frac{42\text{ V}}{14.0\ \Omega} = 3.0\text{ A}$$

The current I in the 8.0-Ω and 4.0-Ω resistors is the same because the resistors are in series. At the junction at b, the current splits. Part of it (I_1) is in the 6.0-Ω resistor, and part (I_2) is in the 3.0-Ω resistor. Because the potential differences ΔV_{bc} across these resistors are the same (they are in parallel), we see that $\Delta V_{bc} = IR = (6.0\ \Omega)I_1 = (3.0\ \Omega)I_2$, or $I_2 = 2.0I_1$. Using this result and that $I_1 + I_2 = 3.0$ A, we find that $I_1 =$ 1.0 A and $I_2 =$ 2.0 A. We could have guessed this result by noting that the current in the 3.0-Ω resistor has to be twice the current in the 6.0-Ω resistor in view of their relative resistances and that the same potential difference appears across each of them.

As a final check, note that $\Delta V_{bc} = (6.0\ \Omega)I_1 = (3.0\ \Omega)I_2 = 6.0$ V and $\Delta V_{ab} = (12.0\ \Omega)I = 36$ V; therefore, $\Delta V_{ac} = \Delta V_{ab} + \Delta V_{bc} = 42$ V, as it must.

INTERACTIVE EXAMPLE 21.8 Three Resistors in Parallel

Three resistors are connected in parallel as in Figure 21.21. A potential difference of 18.0 V is maintained between points *a* and *b*.

A Find the current in each resistor.

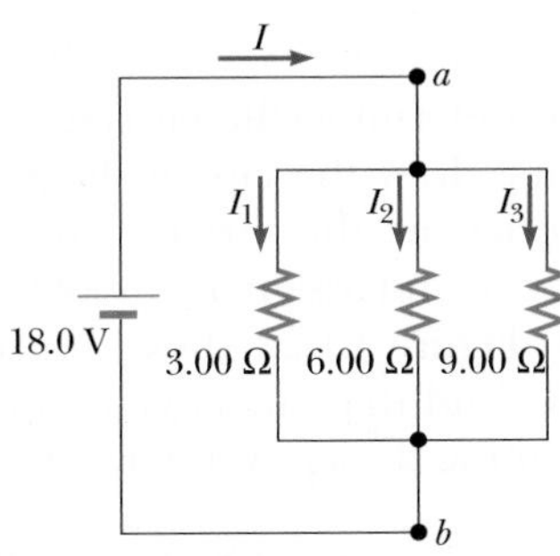

FIGURE 21.21 (Interactive Example 21.8) Three resistors connected in parallel. The voltage across each resistor is 18.0 V.

Solution The resistors are in parallel, and the potential difference across each is 18.0 V. Applying $\Delta V = IR$ to each resistor gives

$$I_1 = \frac{\Delta V}{R_1} = \frac{18.0\text{ V}}{3.00\ \Omega} = \boxed{6.00\text{ A}}$$

$$I_2 = \frac{\Delta V}{R_2} = \frac{18.0\text{ V}}{6.00\ \Omega} = \boxed{3.00\text{ A}}$$

$$I_3 = \frac{\Delta V}{R_3} = \frac{18.0\text{ V}}{9.00\ \Omega} = \boxed{2.00\text{ A}}$$

B Calculate the power delivered to each resistor and the total power delivered to the three resistors.

Solution Applying $\mathcal{P} = I^2R$ to each resistor gives

$$3.00\text{-}\Omega\text{:}\quad \mathcal{P}_1 = I_1^2R_1 = (6.00\text{ A})^2(3.00\ \Omega) = \boxed{108\text{ W}}$$

$$6.00\text{-}\Omega\text{:}\quad \mathcal{P}_2 = I_2^2R_2 = (3.00\text{ A})^2(6.00\ \Omega) = \boxed{54.0\text{ W}}$$

$$9.00\text{-}\Omega\text{:}\quad \mathcal{P}_3 = I_3^2R_3 = (2.00\text{ A})^2(9.00\ \Omega) = \boxed{36.0\text{ W}}$$

which shows that the smallest resistor receives the most power. You can also use $\mathcal{P} = (\Delta V)^2/R$ to find the power delivered to each resistor. Summing the three quantities gives a total power of 198 W.

C Calculate the equivalent resistance of the combination of three resistors.

Solution We can use Equation 21.29 to find R_{eq}:

$$\frac{1}{R_{eq}} = \frac{1}{3.00\ \Omega} + \frac{1}{6.00\ \Omega} + \frac{1}{9.00\ \Omega} = \frac{11}{18.0\ \Omega}$$

$$R_{eq} = \frac{18.0\ \Omega}{11} = \boxed{1.64\ \Omega}$$

We can check this answer using the battery voltage and the total current from part A:

$$R_{eq} = \frac{\Delta V}{I_{tot}} = \frac{18.0\text{ V}}{6.00\text{ A} + 3.00\text{ A} + 2.00\text{ A}} = 1.64\ \Omega$$

PhysicsNow™ By logging into PhysicsNow at **www.pop4e.com** and going to Interactive Example 21.8, you can explore different configurations of the battery and resistors.

21.8 KIRCHHOFF'S RULES

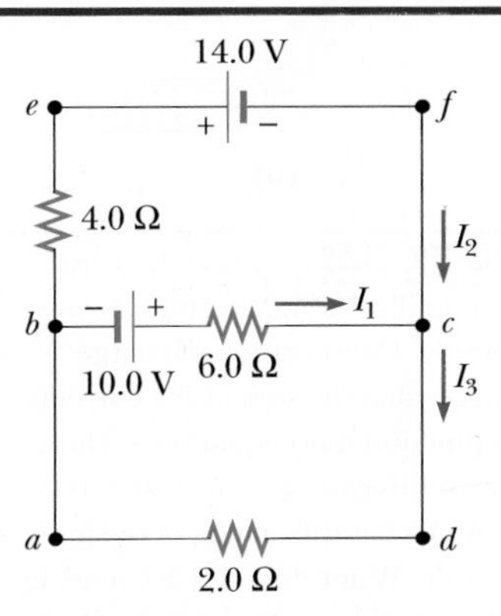

FIGURE 21.22 A circuit that cannot be simplified by using the rules for series and parallel resistors.

As indicated in the preceding section, some simple circuits can be analyzed using $\Delta V = IR$ and the rules for series and parallel combinations of resistors. Resistors, however, can be connected so that the circuits formed cannot be reduced to a single equivalent resistor. Consider the circuit in Figure 21.22, for example. If either battery were removed from this circuit, the resistors could be combined with the techniques of Section 21.7. With both batteries present, however, that cannot be done.

The procedure for analyzing such circuits is greatly simplified by the use of two simple rules called **Kirchhoff's rules**:

(AIP ESVA/W. F. Meggers Collection)

GUSTAV KIRCHHOFF (1824–1887)
Kirchhoff, a professor at Heidelberg, Germany, and Robert Bunsen invented the spectroscope and founded the science of spectroscopy, which led to atomic spectra such as those seen in Chapter 11. They discovered the elements cesium and rubidium and invented astronomical spectroscopy. Kirchhoff formulated another Kirchhoff's rule, namely, "a cool substance will absorb light of the same wavelengths that it emits when hot."

- At any junction, the sum of the currents must equal zero:

$$\sum_{\text{junction}} I = 0$$

This rule is often referred to as the **junction rule.** In Figure 21.22, there are junctions at b and c.

- The sum of the potential differences across each element around any closed circuit loop must be zero:

$$\sum_{\text{loop}} \Delta V = 0$$

This rule is usually called the **loop rule.** In Figure 21.22, we can identify three loops: *abcda, aefda,* and *befcb.*

Kirchhoff's rules are generally used to determine the current in each element in the circuit. In using these rules, we first draw the circuit diagram and assume a direction for the current in each device in the circuit. We draw an arrow representing that direction next to the device and assign a symbol to each independent current, such as I_1, I_2, and so on. Figure 21.22 shows the three different currents that exist in this circuit. Keep in mind that currents in devices connected in series are the same, so the currents in these devices will have the same assigned symbol.

The junction rule is a statement of **conservation of charge.** The amount of charge that enters a given point in a circuit in a time interval must also leave that point in the same time interval because charge cannot build up or disappear at a point. Currents with a direction into the junction are entered into the junction rule as $+I$, whereas currents with a direction out of a junction are entered as $-I$. If we apply the rule to the junction in Figure 21.23a, we have

$$I_1 - I_2 - I_3 = 0$$

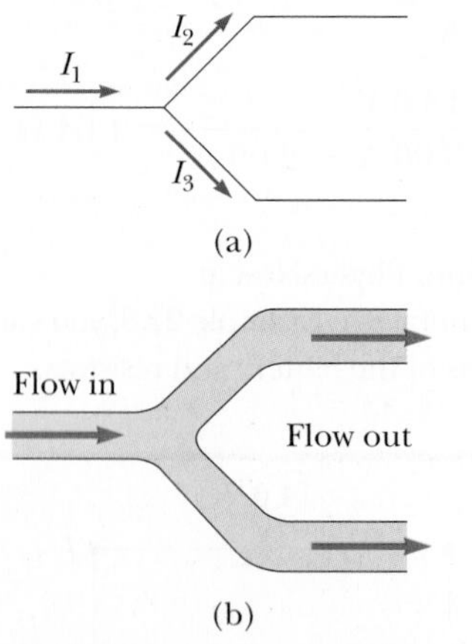

FIGURE 21.23 (a) A schematic diagram illustrating Kirchhoff's junction rule. Conservation of charge requires that the sum of the currents at a junction must equal zero. Therefore, in this case, $I_1 - I_2 - I_3 = 0$. (b) A mechanical analog of the junction rule. Water does not accumulate at the junction, so the amount of water flowing out of the branches on the right must equal the amount flowing into the single branch on the left.

Figure 21.23b represents a hydraulic analog to this situation in which water flows through a branched pipe with no leaks. The flow rate into the pipe equals the total flow rate out of the two branches.

The loop rule is equivalent to the law of **conservation of energy.** Suppose a charge moves around any closed loop in a circuit[11] (the charge starts and ends at the same point). In this case, the circuit must gain as much energy as it loses. In this isolated system model for the system of the circuit, no energy is transferred across the boundary of the system (ignoring energy transfer by radiation and heat into the air from warm circuit elements), but energy transformations do occur within the system. The energy of the circuit may decrease due to a potential drop $-IR$ as a charge moves through a resistor or as a result of having the charge move in the reverse direction through an emf. In the latter case, electric potential energy is converted to chemical energy as the battery is charged. The potential energy increases when the charge moves through a battery in the same direction as the emf.

Another approach to understanding the loop rule is to remember the definition of a conservative force from Chapter 7. One of the mathematical behaviors of a conservative force is that the work done by a such a force when a member of the system moves around a closed path is zero. A loop in a circuit is a closed path. If we imagine moving a charge around a loop, the total work done by the conservative electric force must be zero. The total work is the sum of positive and negative works as the charge passes through various circuit elements. Because work is related

[11]Remember that this situation is *not* what happens; a charge might take hours to traverse a loop. In terms of analyzing the circuit in terms of energy, however, we can build a mental model in which we *imagine* taking a charge all the way around the circuit.

to potential energy changes and because potential energy changes are related to potential differences (Eq. 20.3), that the sum of all the works is zero is equivalent to the sum of all the potential differences being zero, which is Kirchhoff's loop rule.

As an aid in applying the loop rule, the following sign conventions are used. We have already drawn arrows for currents on our diagram and have assigned symbols to the currents to apply the junction rule. To set up the sign conventions, we choose a direction around each loop that we imagine carrying a positive charge, clockwise or counterclockwise. Therefore, for any device, there will be two directions that we need to consider, one for our chosen current and one for our chosen travel through the device. The sign conventions for potential differences for resistors and batteries based on these two directions are summarized in Figure 21.24, where it is assumed that travel is from point a toward point b:

- If a resistor is traversed in the direction of the current, the potential difference across the resistor is $-IR$ (Fig. 21.24a).
- If a resistor is traversed in the direction *opposite* the current, the potential difference across the resistor is $+IR$ (Figure 21.24b).
- If a source of emf is traversed in the direction of the emf (from $-$ to $+$ on the terminals), the potential difference is $+\mathcal{E}$ (Fig. 21.24c).
- If a source of emf is traversed in the direction opposite the emf (from $+$ to $-$ on the terminals), the potential difference is $-\mathcal{E}$ (Fig. 21.24d).

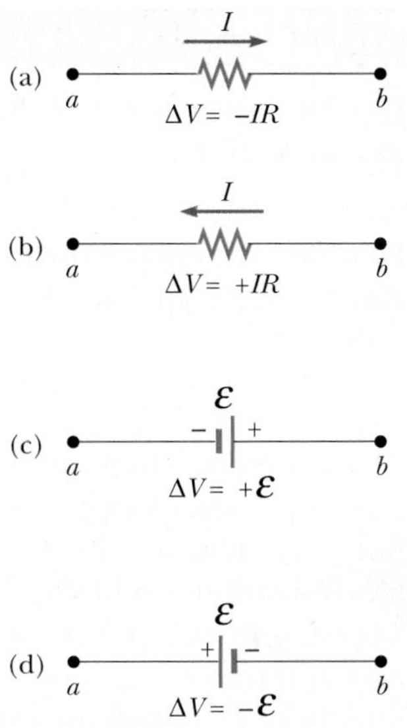

FIGURE 21.24 Rules for determining the potential differences across a resistor and a battery. (The battery is assumed to have no internal resistance.) Each circuit element is traversed from a to b.

There are limitations on the use of the junction rule and the loop rule. You may use the junction rule as often as needed, as long as each time you write an equation you include in it a current that has not been used in a previous junction rule equation. In general, the number of times the junction rule can be used is one fewer than the number of junction points in the circuit. The loop rule can be used as often as needed, as long as a new circuit element (a resistor or battery) or a new current appears in each new equation. In general, **the number of independent equations you need must equal the number of unknown currents to solve a particular circuit problem.**

PROBLEM-SOLVING STRATEGY Kirchhoff's Rules

The following procedure is recommended for solving problems that involve circuits that cannot be reduced by the rules for combining resistors in series or parallel.

1. Conceptualize Study the circuit diagram and make sure that you recognize all elements in the circuit. Identify the polarity of each battery and try to imagine the directions in which the current would exist through the batteries.

2. Categorize Determine whether the circuit can be reduced by means of combining series and parallel resistors. If so, use the techniques of Section 21.7. If not, apply Kirchhoff's rules according to step 3 below.

3. Analyze Assign labels to all the known quantities and assign symbols to all the unknown quantities. You must assign *directions* to the currents in each part of the circuit. Although the assignment of current directions is arbitrary, you must adhere *rigorously* to the directions you assign when you apply Kirchhoff's rules.

Apply the junction rule (Kirchhoff's first rule) to all junctions in the circuit except one. Now apply the loop rule (Kirchhoff's second rule) to as many loops in the circuit as are needed to obtain, in combination with the equations from the junction rule, as many equations as there are unknowns. To apply this rule, you must choose a direction in which to travel around the loop (either clockwise or counterclockwise) and correctly identify the change in potential as you cross each element. Watch out for signs!

Solve the equations simultaneously for the unknown quantities.

4. Finalize Check your numerical answers for consistency. Do not be alarmed if any of the resulting currents have a negative value; if so, you have guessed the direction of that current incorrectly, but *its magnitude will be correct.*

INTERACTIVE EXAMPLE 21.9 **Applying Kirchhoff's Rules**

A Find the currents I_1, I_2, and I_3 in the circuit shown in Figure 21.22.

Solution We choose the directions of the currents as in Figure 21.22. Applying Kirchhoff's first rule to junction *c* gives

$$(1) \quad I_1 + I_2 - I_3 = 0$$

There are three loops in the circuit: *abcda, befcb,* and *aefda* (the outer loop). We need only two loop equations to determine the unknown currents. The third loop equation would give no new information. Applying Kirchhoff's second rule to loops *abcda* and *befcb* and traversing these loops in the clockwise direction, we obtain the expressions

$$(2) \quad \text{Loop } abcda\text{:} \quad 10.0 \text{ V} - (6.0\ \Omega)I_1 - (2.0\ \Omega)I_3 = 0$$

$$(3) \quad \text{Loop } befcb\text{:} \quad -14.0 \text{ V} - 10.0 \text{ V} + (6.0\ \Omega)I_1 - (4.0\ \Omega)I_2 = 0$$

Note that in loop *befcb,* a positive sign is obtained when traversing the 6.0-Ω resistor because the direction of the path is opposite the direction of I_1. Loop *aefda* gives $-14.0 \text{ V} - (2.0\ \Omega)I_3 - (4.0\ \Omega)I_2 = 0$, which is just the sum of (2) and (3).

Expressions (1), (2), and (3) represent three independent equations with three unknowns. We can solve the problem as follows. Dropping the units for simplicity and substituting I_3 from (1) into (2) gives

$$10.0 - 6.0I_1 - 2.0(I_1 + I_2) = 0$$

$$10.0 = 8.0I_1 + 2.0I_2 \quad (4)$$

Dividing each term in (3) by 2 and rearranging the equation gives

$$-12.0 = -3.0I_1 + 2.0I_2 \quad (5)$$

Subtracting (5) from (4) eliminates I_2, giving

$$22.0 = 11.0I_1$$

$$I_1 = 2.0 \text{ A}$$

Using this value of I_1 in (5) gives a value for I_2:

$$2.0I_2 = 3.0I_1 - 12.0 = 3.0(2.0) - 12.0 = -6.0$$

$$I_2 = -3.0 \text{ A}$$

Finally, $I_3 = I_1 + I_2 = -1.0$ A. Hence, the currents have the values

$$I_1 = 2.0 \text{ A} \qquad I_2 = -3.0 \text{ A} \qquad I_3 = -1.0 \text{ A}$$

That I_2 and I_3 are negative indicates only that we chose the wrong directions for these currents. The numerical values, however, are correct.

B Find the potential difference between points *b* and *c*.

Solution In traversing the path from *b* to *c* along the central branch, we have

$$V_c - V_b = +10.0 \text{ V} - (6.0\ \Omega)I_1$$

$$= +10.0 \text{ V} - (6.0\ \Omega)(2.0 \text{ A}) = -2.0 \text{ V}$$

PhysicsNow™ By logging into PhysicsNow at www.pop4e.com and going to Interactive Example 21.9, you can practice applying Kirchhoff's rules.

21.9 *RC* CIRCUITS

So far, we have been concerned with circuits with constant currents, or *steady-state circuits.* We now consider circuits containing capacitors in which the currents may vary in time.

Charging a Capacitor

Consider the series circuit shown in Active Figure 21.25a. Let us assume that the capacitor is initially uncharged. No current exists when switch S is open (Active Fig. 21.25b). If the switch is thrown closed at $t = 0$, charges begin to flow, setting up a current in the circuit,[12] and the capacitor begins to charge (Active Fig. 21.25c). Note that during the charging, charges do not jump across the plates of the capacitor because the gap between the plates represents an open circuit. Instead, due to the electric field in the wires established by the battery, electrons move into the top plate from the wires and out of the bottom plate into the wires until the capacitor is

[12]By "a current in the circuit," we mean current in all parts of the circuit *except* for the region between the plates of the capacitor.

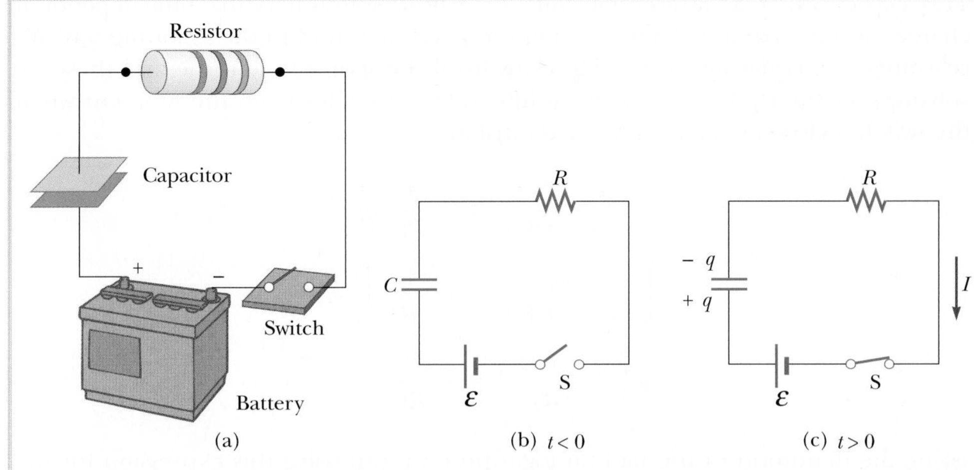

ACTIVE FIGURE 21.25
(a) A capacitor in series with a resistor, switch, and battery. (b) Circuit diagram representing this system at time $t < 0$, before the switch is closed. (c) Circuit diagram at time $t > 0$, after the switch has been closed.

PhysicsNow™ Log into PhysicsNow at **www.pop4e.com** and go to Active Figure 21.25 to adjust the values of R and C to see the effect on the charging of the capacitor.

fully charged. The value of the maximum charge depends on the emf of the battery. Once the maximum charge is reached, the current in the circuit is zero.

To put this discussion on a quantitative basis, let us apply Kirchhoff's second rule to the circuit after the switch is closed. In our sign conventions, we did not specify a convention for the potential difference across a capacitor. From our study of capacitors in Chapter 20, however, it should be clear that carrying a positive charge across a capacitor from − to + would represent an increase in potential energy for the circuit, a positive potential difference. Traversing the capacitor in the opposite direction would correspond to a decrease in potential energy, a negative potential difference.

Choosing clockwise as our direction around the circuit in Active Figure 21.25 and applying the sign convention for capacitors that we have just discussed, we have

$$\mathcal{E} - \frac{q}{C} - IR = 0 \qquad [21.30]$$

where $-q/C$ is the potential difference across the capacitor and $-IR$ is the potential difference across the resistor consistent with our direction of travel. Note that q and I are *instantaneous* values of the charge and current, respectively, as the capacitor is charged.

We can use Equation 21.30 to find the initial current in the circuit and the maximum charge on the capacitor. At $t = 0$, when the switch is closed, the charge on the capacitor is zero, and from Equation 21.30, we find that the initial current in the circuit I_0 is a maximum and equal to

$$I_0 = \frac{\mathcal{E}}{R} \qquad [21.31]$$

At this time, **the potential difference is entirely across the resistor.** Later, when the capacitor is charged to its maximum value Q, charges cease to flow, the current in the circuit is zero, and **the potential difference is entirely across the capacitor.** Substituting $I = 0$ into Equation 21.30 yields the following expression for Q:

$$Q = C\mathcal{E} \quad \text{(maximum charge)} \qquad [21.32]$$

To determine analytical expressions for the time dependence of the charge and current, we must solve Equation 21.30. To do so, let us substitute $I = dq/dt$ and rearrange the equation:

$$\frac{dq}{dt} = \frac{\mathcal{E}}{R} - \frac{q}{RC} = \frac{C\mathcal{E} - q}{RC}$$

This expression is a differential equation whose solution is the time-dependent charge on the capacitor. An expression for q may be found in the following way. We rearrange the equation by placing terms involving q on the left side and those involving t on the right side. Then we integrate both sides from the moment when the switch is closed to an arbitrary later instant:

$$\frac{dq}{(q - C\mathcal{E})} = -\frac{1}{RC}\,dt$$

$$\int_0^q \frac{dq}{(q - C\mathcal{E})} = -\frac{1}{RC}\int_0^t dt$$

$$\ln\left(\frac{q - C\mathcal{E}}{-C\mathcal{E}}\right) = -\frac{t}{RC}$$

Using the definition of the natural logarithm, we can solve this expression for the charge on the capacitor as a function of time:

$$\frac{q - C\mathcal{E}}{-C\mathcal{E}} = e^{-t/RC}$$

■ Charge versus time for a charging capacitor

$$q(t) = C\mathcal{E}[1 - e^{-t/RC}] = Q[1 - e^{-t/RC}] \qquad [21.33]$$

where e is the base of the natural logarithm (*not* the charge on the electron!) and $Q = C\mathcal{E}$ is the maximum charge on the capacitor.

An expression for the current as a function of time may be found by differentiating Equation 21.33 with respect to time. Using $I = dq/dt$, we obtain

$$I(t) = \frac{\mathcal{E}}{R}e^{-t/RC} \qquad [21.34]$$

where $\mathcal{E}/R$ is the initial current in the circuit.

Plots of charge and current versus time are shown in Figure 21.26. Note that the charge is zero at $t = 0$ and approaches the maximum value of $C\mathcal{E}$ as $t \to \infty$ (Fig. 21.26a). Furthermore, the current has its maximum value of $I_0 = \mathcal{E}/R$ at $t = 0$ and decays exponentially to zero as $t \to \infty$ (Fig. 21.26b). The quantity RC that appears in the exponential of Equations 21.33 and 21.34 is called the **time constant** τ of the circuit. It represents the time interval during which the current decreases to $1/e$ of its initial value; that is, at the end of the time interval τ, $I = e^{-1}I_0 = 0.368I_0$. After the time interval 2τ, $I = e^{-2}I_0 = 0.135I_0$, and so forth. Likewise, in a time interval τ the charge increases from zero to $C\mathcal{E}[1 - e^{-1}] = 0.632C\mathcal{E}$.

The energy decrease of the battery during the charging process is the product of the total charge and the emf, $Q\mathcal{E} = C\mathcal{E}^2$. After the capacitor is fully charged, the energy stored in it is $\frac{1}{2}Q\mathcal{E} = \frac{1}{2}C\mathcal{E}^2$, which is just half the energy decrease of the battery. It is left to an end-of-chapter problem to show that the remaining half of the energy supplied by the battery appears as internal energy in the resistor (Problem 21.58).

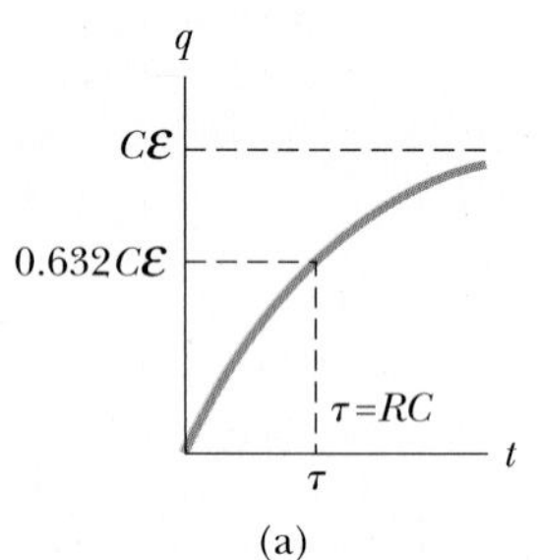

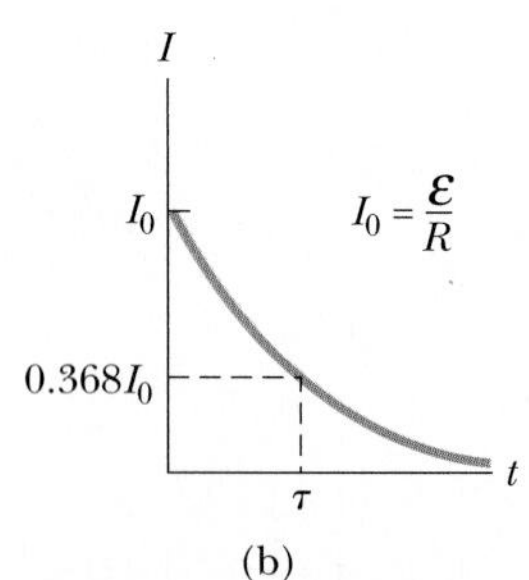

FIGURE 21.26 (a) Plot of capacitor charge versus time for the circuit shown in Active Figure 21.25. After a time interval equal to one time constant τ has passed, the charge is 63.2% of the maximum value $C\mathcal{E}$. The charge approaches its maximum value as t approaches infinity. (b) Plot of current versus time for the RC circuit shown in Active Figure 21.25. The current has its maximum value $I_0 = \mathcal{E}/R$ at $t = 0$ and decays to zero exponentially as t approaches infinity. After a time interval equal to one time constant τ has passed, the current is 36.8% of its initial value.

Discharging a Capacitor

Now consider the circuit in Active Figure 21.27, consisting of a capacitor with an initial charge Q, a resistor, and a switch. When the switch is open (Active Fig. 21.27a), a potential difference of Q/C exists across the capacitor and zero potential difference exists across the resistor because $I = 0$. If the switch is thrown closed at $t = 0$, the capacitor begins to discharge through the resistor. At some time during the discharge, the current in the circuit is I and the charge on the capacitor is q (Active Fig. 21.27b).

The circuit of Active Figure 21.27 is the same as the circuit of Active Figure 21.25 except for the absence of the battery. Therefore, we modify the Kirchhoff's rule expression in Equation 21.30 by dropping the emf from the equation:

$$-\frac{q}{C} - IR = 0 \qquad [21.35]$$

Because $I = dq/dt$, Equation 21.35 becomes

$$-R\frac{dq}{dt} = \frac{q}{C}$$

$$\frac{dq}{q} = -\frac{1}{RC}\,dt$$

In $I = dq/dt$, dq is negative, because the charge on the discharging capacitor is decreasing; therefore, I has a negative value. This is indicated in Active Figure 21.27 by the reversal of the current arrow compared to Active Figure 21.25. Furthermore, in Equation 21.37 below, the current will come out to have an explicit negative value. Integrating this expression from the moment the switch is closed, at which time $q = Q$, to an arbitrary later instant gives

$$\int_Q^q \frac{dq}{q} = -\frac{1}{RC}\int_0^t dt$$

$$\ln\left(\frac{q}{Q}\right) = -\frac{t}{RC}$$

$$q(t) = Qe^{-t/RC} \qquad [21.36]$$

Differentiating Equation 21.36 with respect to time gives the current as a function of time:

$$I(t) = \frac{dq}{dt} = -I_0e^{-t/RC} \qquad [21.37]$$

where the initial current is $I_0 = Q/RC$. Therefore, we see that both the charge on the capacitor and the current decay exponentially at a rate characterized by the time constant $\tau = RC$. The negative sign in Equation 21.37 indicates the direction of the current, which is opposite to the direction during the charging process.

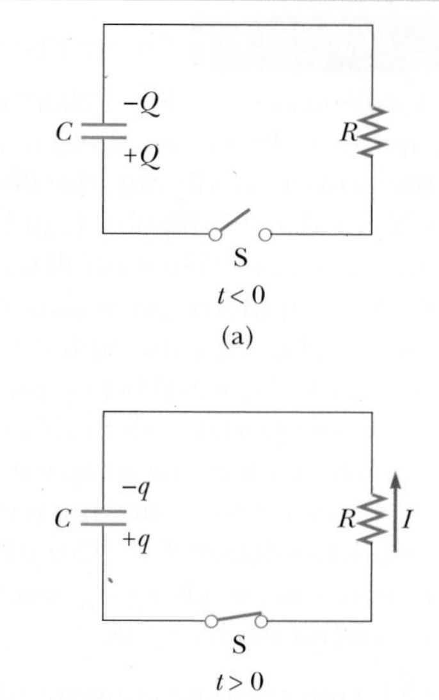

ACTIVE FIGURE 21.27
(a) A charged capacitor connected to a resistor and a switch, which is open for $t < 0$. (b) After the switch is closed at $t = 0$, a current that decreases in magnitude with time is set up in the direction shown and the charge on the capacitor decreases exponentially with time.

PhysicsNow™ Log into PhysicsNow at **www.pop4e.com** and go to Active Figure 21.27 to adjust the values of R and C to see the effect on the discharging of the capacitor.

QUICK QUIZ 21.9 Consider the circuit in Active Figure 21.25a and assume that the battery has no internal resistance. **(i)** Just after the switch is closed, the potential difference across which of the following is equal to the emf of the battery? **(a)** C **(b)** R **(c)** neither C nor R **(ii)** After a very long time, the potential difference across which of the following is equal to the emf of the battery? **(a)** C **(b)** R **(c)** neither C nor R

■ Thinking Physics 21.6

Many roadway construction sites have flashing yellow lights to warn motorists of possible dangers. What causes the lightbulbs to flash?

Reasoning A typical circuit for such a flasher is shown in Figure 21.28. The lamp L is a gas-filled lamp that acts as an open circuit until a large potential difference causes an electrical discharge in the gas, which gives off a bright light. During this discharge, charges flow through the gas between the electrodes of the lamp. After switch S is closed, the battery charges up the capacitor of capacitance C. At the beginning, the current is high and the charge on the capacitor is low, so most of the potential difference appears across the resistance R. As the capacitor charges, more potential difference appears across it, reflecting the lower current and therefore lower potential difference across the resistor. Eventually, the potential difference across the capacitor reaches a value at which the lamp will conduct, causing a flash. This discharges the capacitor through the lamp and the process of charging begins again. The period between flashes can be adjusted by changing the time constant of the RC circuit. ■

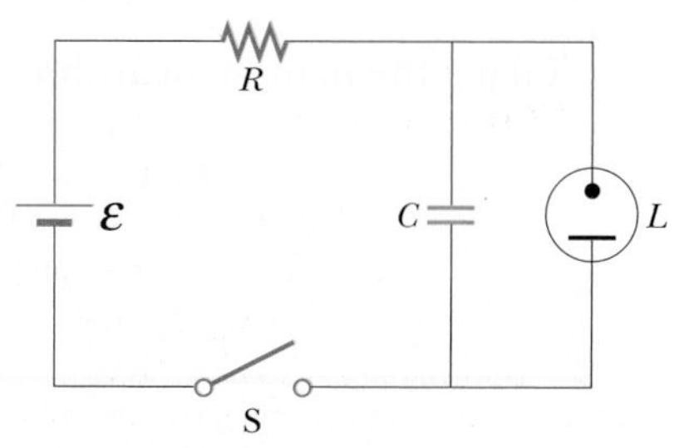

FIGURE 21.28 (Thinking Physics 21.6) The RC circuit in a roadway construction flasher. When the switch is closed, the charge on the capacitor increases until the voltage across the capacitor (and across the flash lamp) is high enough for the lamp to flash, discharging the capacitor.

EXAMPLE 21.10 Charging a Defibrillator

A *defibrillator* (see Fig. P20.48 on page 678) can store energy in the electric field of a large capacitor. Under the proper conditions, the defibrillator can be used to stop cardiac fibrillation (random contractions) in heart attack victims. When fibrillation occurs, the heart produces a rapid, irregular pattern of beats. A fast discharge of energy through the heart can return the organ to its normal beat pattern. Emergency medical teams use portable defibrillators that contain batteries capable of charging a capacitor to a high voltage.

Consider the following parameters for the *RC* circuit in a defibrillator: $C = 32.0\ \mu\text{F}$ and $R = 47.0\ \text{k}\Omega$. The circuitry in the charging system applies 5 000 V to the *RC* circuit to charge it.

A Find the time constant of the circuit, the maximum charge on the capacitor, the maximum current in the circuit during the charging process, and the charge and current as a function of time.

Solution The time constant of the circuit is $\tau = RC = (47.0 \times 10^3\ \Omega)(32.0 \times 10^{-6}\ \text{F}) = \boxed{1.50\ \text{s}}$.
The maximum charge on the capacitor is

$Q = C\mathcal{E} = (32.0 \times 10^{-6}\ \text{F})(5\ 000\ \text{V}) = \boxed{0.160\ \text{C}}$.
The maximum current in the circuit is
$I_0 = \mathcal{E}/R = (5\ 000\ \text{V})/(47.0 \times 10^3\ \Omega) = \boxed{0.106\ \text{A}}$.
Using these values and Equations 21.33 and 21.34, we find that

$$q(t) = \boxed{(0.160\ \text{C})\ [1 - e^{-t/1.50}]}$$

$$I(t) = \boxed{(0.106\ \text{A})\ e^{-t/1.50}}$$

B Find the energy stored in the capacitor when it is fully charged.

Solution Using Equation 20.30, we have

$$U = \tfrac{1}{2}C(\Delta V)^2 = \tfrac{1}{2}(32.0 \times 10^{-6}\ \text{C})(5\ 000\ \text{V})^2 = \boxed{400\ \text{J}}$$

Note that the time constant τ of 1.50 s means that several seconds are required until the capacitor is close to fully charged. Therefore, after an unsuccessful attempt to defibrillate a patient's heart by delivering the stored energy to the chest, the emergency personnel must wait several seconds for the capacitor to charge before trying again.

EXAMPLE 21.11 Discharging a Capacitor in an *RC* Circuit

Consider a capacitor *C* being discharged through a resistor *R* as in Active Figure 21.27.

A After how many time constants is the charge on the capacitor one fourth of its initial value?

Solution The charge on the capacitor varies with time according to Equation 21.36, $q(t) = Qe^{-t/RC}$. To find the time at which the charge q has dropped to one fourth of its initial value, we substitute $q(t) = Q/4$ into this expression and solve for t:

$$\tfrac{1}{4}Q = Qe^{-t/RC}$$

$$\tfrac{1}{4} = e^{-t/RC}$$

Taking the natural logarithm of both sides, we find that

$$-\ln 4 = -\frac{t}{RC}$$

$$t = RC \ln 4 = \boxed{1.39RC}$$

B The energy stored in the capacitor decreases with time as it discharges. After how many time constants is this stored energy one fourth of its initial value?

Solution Using Equations 20.30 and 21.36, we can express the energy stored in the capacitor at any time t as

$$U = \frac{q^2}{2C} = \frac{Q^2}{2C}e^{-2t/RC} = U_0 e^{-2t/RC}$$

where U_0 is the initial energy stored in the capacitor. Similar to part A, we now set $U = U_0/4$ and solve for t:

$$\tfrac{1}{4}U_0 = U_0 e^{-2t/RC}$$

$$\tfrac{1}{4} = e^{-2t/RC}$$

Again, taking the natural logarithm of both sides and solving for t gives

$$t = \tfrac{1}{2}RC \ln 4 = \boxed{0.693RC}$$

21.10 THE ATMOSPHERE AS A CONDUCTOR

When discussing capacitors with air between the plates in Chapter 20, we adopted the simplification model that air was a perfect insulator. Although that was a good model for typical potential differences encountered in capacitors, we know that it is possible for a current to exist in air. Lightning is a dramatic example of this possibility, but a more mundane example is the common spark that you might receive upon bringing your finger near a doorknob after rubbing your feet across a carpet.

Let us analyze the process that occurs in electrical discharge, which is the same for lightning and the doorknob spark except for the size of the current. Whenever a strong electric field exists in air, it is possible for the air to undergo electrical breakdown in which the effective resistivity of the air drops dramatically and the air becomes a conductor. At any given time, due to cosmic ray collisions and other events, air contains a number of ionized molecules (Fig. 21.29a). For a relatively weak electric field, such as the fair-weather electric field, these ions and freed electrons accelerate slowly due to the electric force. They collide with other molecules with no effect and eventually neutralize as a freed electron ultimately finds an ion and combines with it. In a strong electric field such as that associated with a thunderstorm, however, the freed electrons can accelerate to very high speeds (Fig. 21.29b) before making a collision with a molecule (Fig. 21.29c). If the field is strong enough, the electron may have enough energy to ionize the molecule in this collision (Fig. 21.29d). Now there are two electrons to be accelerated by the field, and each can strike another molecule at high speed (Fig. 21.29e). The result is a very rapid increase in the number of charge carriers available in the air and a corresponding decrease in resistance of the air. Therefore, there can be a large current in the air that tends to neutralize the charges that established the initial potential difference, such as the charges in the cloud and on the ground. When that happens, we have lightning.

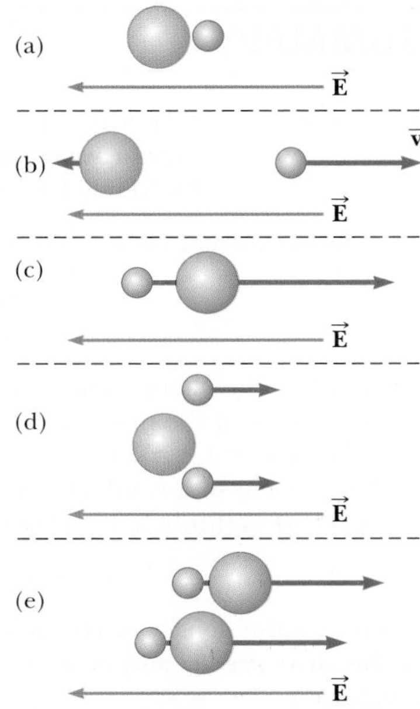

FIGURE 21.29 The anatomy of a spark. (a) A molecule is ionized as a result of a random event. (b) The ion accelerates slowly and the electron accelerates rapidly due to the force from the electric field. (c) The accelerated electron approaches another molecule at high speed. (d) The new molecule is ionized, and the original electron and the new electron accelerate rapidly. (e) These electrons approach other molecules, freeing two more electrons, and an avalanche of ionization proceeds.

Typical currents during lightning strikes can be very high. While the stepped leader is making its way toward the ground, the current is relatively modest, in the range of 200 to 300 A. This current is large compared with typical household currents but small compared with peak currents in lightning discharges. Once the connection is made between the stepped leader and the return stroke, the current rises rapidly to a typical value of 5×10^4 A. Considering that typical potential differences between cloud and ground in a thunderstorm can be measured in hundreds of thousands of volts, the power during a lightning stroke is measured in billions of watts. Much of the energy in the stroke is delivered to the air, resulting in a rapid temperature increase and the resultant flash of light and sound of thunder.

Even in the absence of a thundercloud, there is a flow of charge through the air. The ions in the air make the air a conductor, although not a very good one. Atmospheric measurements indicate a typical potential difference across our atmospheric capacitor (Section 20.11) of about 3×10^5 V. As we shall show in the Context 6 Conclusion, the total resistance of the air between the plates in the atmospheric capacitor is about 300 Ω. Therefore, the average fair-weather current in the air is

$$I = \frac{\Delta V}{R} = \frac{3 \times 10^5 \text{ V}}{300\ \Omega} \approx 1 \times 10^3 \text{ A}$$

A number of simplifying assumptions were made in these calculations, but this result is on the right order of magnitude for the global current. Although the result might seem surprisingly large, remember that this current is spread out over the entire surface area of the Earth. Therefore, the average fair-weather current density is

$$J = \frac{I}{A} = \frac{I}{4\pi r^2} = \frac{1 \times 10^3 \text{ A}}{4\pi(6.4 \times 10^6 \text{ m})^2} \approx 2 \times 10^{-12} \text{ A/m}^2$$

In comparison, the current density in a lightning strike is on the order of 10^5 A/m^2.

The fair-weather current and the lightning current are in opposite directions. The fair-weather current delivers positive charge to the ground, whereas lightning delivers negative charge. These two effects are in balance,[13] which is the principle that we shall use to estimate the average number of lightning strikes on the Earth in the Context Conclusion.

[13]There are a number of other effects, too, but we will adopt a simplification model in which these are the only two effects. For more information, see E. A. Bering, A. A. Few, and J. R. Benbrook, "The Global Electric Circuit," *Physics Today*, October 1998, pp. 24–30.

SUMMARY

PhysicsNow™ Take a practice test by logging into PhysicsNow at www.pop4e.com and clicking on the Pre-Test link for this chapter.

The **electric current** I in a conductor is defined as

$$I \equiv \frac{dQ}{dt} \qquad [21.2]$$

where dQ is the charge that passes through a cross-section of the conductor in the time interval dt. The SI unit of current is the ampere (A); 1 A = 1 C/s.

The current in a conductor is related to the motion of the charge carriers through the relationship

$$I_{\text{avg}} = nqv_d A \qquad [21.4]$$

where n is the density of charge carriers, q is their charge, v_d is the **drift speed,** and A is the cross-sectional area of the conductor.

The **resistance** R of a conductor is defined as the ratio of the potential difference across the conductor to the current:

$$R \equiv \frac{\Delta V}{I} \qquad [21.6]$$

The SI units of resistance are volts per ampere, defined as ohms (Ω); 1 Ω = 1 V/A.

If the resistance is independent of the applied voltage, the conductor obeys **Ohm's law,** and conductors that have a constant resistance over a wide range of voltages are said to be **ohmic.**

If a conductor has a uniform cross-sectional area A and a length ℓ, its resistance is

$$R = \rho \frac{\ell}{A} \qquad [21.8]$$

where ρ is called the **resistivity** of the material from which the conductor is made. The inverse of the resistivity is defined as the **conductivity** $\sigma = 1/\rho$.

The resistivity of a conductor varies with temperature in an approximately linear fashion; that is,

$$\rho = \rho_0[1 + \alpha(T - T_0)] \qquad [21.10]$$

where ρ_0 is the resistivity at some reference temperature T_0 and α is the **temperature coefficient of resistivity.**

In a classical model of electronic conduction in a metal, the electrons are treated as molecules of a gas. In the absence of an electric field, the average velocity of the electrons is zero. When an electric field is applied, the electrons move (on the average) with a drift velocity $\vec{\mathbf{v}}_d$, which is opposite the electric field:

$$\vec{\mathbf{v}}_d = \frac{-e\vec{\mathbf{E}}}{m_e}\tau \qquad [21.15]$$

where τ is the average time interval between collisions with the atoms of the metal. The resistivity of the material according to this model is

$$\rho = \frac{m_e}{ne^2\tau} \qquad [21.18]$$

where n is the number of free electrons per unit volume.

If a potential difference ΔV is maintained across a circuit element, the **power,** or the rate at which energy is delivered to the circuit element, is

$$\mathcal{P} = I\Delta V \qquad [21.20]$$

Because the potential difference across a resistor is $\Delta V = IR$, we can express the power delivered to a resistor in the form

$$\mathcal{P} = I^2R = \frac{(\Delta V)^2}{R} \qquad [21.21]$$

The **emf** of a battery is the voltage across its terminals when the current is zero. Because of the voltage drop across the **internal resistance** r of a battery, the **terminal voltage** of the battery is less than the emf when a current exists in the battery.

The **equivalent resistance** of a set of resistors connected in **series** is

$$R_{\text{eq}} = R_1 + R_2 + R_3 + \cdots \qquad [21.27]$$

The **equivalent resistance** of a set of resistors connected in **parallel** is given by

$$\frac{1}{R_{\text{eq}}} = \frac{1}{R_1} + \frac{1}{R_2} + \frac{1}{R_3} + \cdots \qquad [21.29]$$

Circuits involving more than one loop are analyzed using two simple rules called **Kirchhoff's rules:**

- At any junction, the sum of the currents must equal zero:

$$\sum_{\text{junction}} I = 0$$

- The sum of the potential differences across each element around any closed circuit loop must be zero:

$$\sum_{\text{loop}} \Delta V = 0$$

For the junction rule, current in a direction into a junction is $+I$, whereas current with a direction away from a junction is $-I$.

For the loop rule, when a resistor is traversed in the direction of the current, the change in potential ΔV across the resistor is $-IR$. If a resistor is traversed in the direction opposite the current, $\Delta V = +IR$.

If a source of emf is traversed in the direction of the emf (negative to positive), the change in potential is $+\mathcal{E}$. If it is traversed opposite the emf (positive to negative), the change in potential is $-\mathcal{E}$.

If a capacitor is charged with a battery of emf $\mathcal{E}$ through a resistance R, the charge on the capacitor and the current in the circuit vary in time according to the expressions

$$q(t) = Q[1 - e^{-t/RC}] \qquad [21.33]$$

$$I(t) = \frac{\mathcal{E}}{R}e^{-t/RC} \qquad [21.34]$$

where $Q = C\mathcal{E}$ is the maximum charge on the capacitor. The product RC is called the **time constant** of the circuit.

If a charged capacitor is discharged through a resistance R, the charge and current decrease exponentially in time according to the expressions

$$q(t) = Qe^{-t/RC} \quad [21.36]$$

$$I(t) = -I_0 e^{-t/RC} \quad [21.37]$$

where $I_0 = Q/RC$ is the initial current in the circuit and Q is the initial charge on the capacitor.

QUESTIONS

☐ = answer available in the *Student Solutions Manual and Study Guide*

1. In an analogy between electric current and automobile traffic flow, what would correspond to charge? What would correspond to current?
2. What factors affect the resistance of a conductor?
3. Two wires A and B of circular cross-section are made of the same metal and have equal lengths, but the resistance of wire A is three times greater than that of wire B. What is the ratio of their cross-sectional areas? How do their radii compare?
4. What would happen to the drift velocity of the electrons in a wire and to the current in the wire if the electrons could move freely without resistance through the wire?
5. Use the atomic theory of matter to explain why the resistance of a material should increase as its temperature increases.
6. Explain how a current can persist in a superconductor without any applied voltage.
7. If charges flow very slowly through a metal, why does it not require several hours for a light to come on when you throw a switch?
8. Two lightbulbs both operate from 120 V. One has a power of 25 W and the other 100 W. Which lightbulb has higher resistance? Which lightbulb carries more current?
9. Car batteries are often rated in ampere-hours. Does this rating designate the amount of current, power, energy, or charge that can be drawn from the battery?
10. When resistors are connected in series, which of the following would be the same for each resistor: potential difference, current, power?
11. When resistors are connected in parallel, which of the following would be the same for each resistor: potential difference, current, power?
12. A *short circuit* is a path of very low resistance in a circuit in parallel with some other part of the circuit. Discuss the effect of the short circuit on the portion of the circuit it parallels. Use a lamp with a frayed cord as an example.
13. Why is it possible for a bird to sit on a high-voltage wire without being electrocuted?
14. If electric power is transmitted over long distances, the resistance of the wires becomes significant. Why? Which method of transmission would result in less energy wasted: high current and low voltage or low current and high voltage? Explain your answer.
15. Referring to Figure Q21.15, describe what happens to the lightbulb after the switch is closed. Assume that the capacitor has a large capacitance and is initially uncharged, and assume that the lightbulb illuminates when connected directly across the battery terminals.

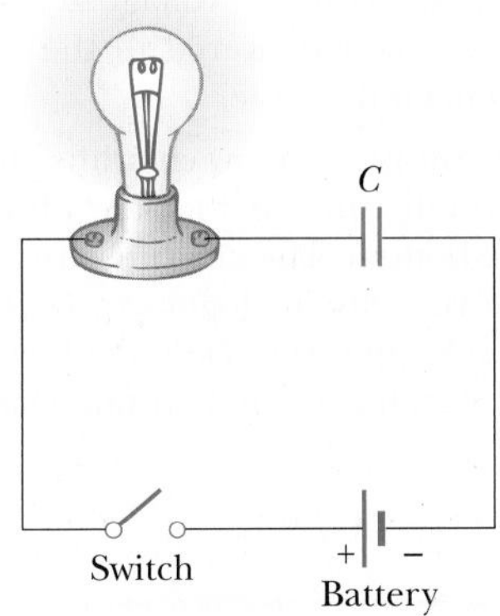

FIGURE Q21.15

16. Are the two headlights of a car wired in series or in parallel? How can you tell?
17. Embodied in Kirchhoff's rules are two conservation laws. What are they?
18. Figure Q21.18 shows a series combination of three lightbulbs, each rated at 120 V. From top to bottom, their power ratings are 60 W, 75 W, and 200 W. Why is the 60-W bulb the brightest and the 200-W bulb the dimmest? Which bulb has the greatest resistance? How would their intensities differ if they were connected in parallel?

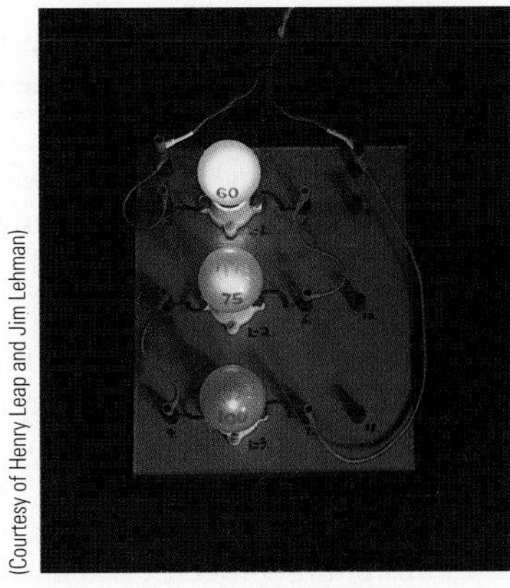

(Courtesy of Henry Leap and Jim Lehman)

FIGURE Q21.18

19. A student claims that the second lightbulb in series is less bright than the first because the first lightbulb uses up some of the current. How would you respond to this statement?
20. So that your grandmother can listen to *A Prairie Home Companion,* you take her bedside radio to the hospital where she is staying. You are required to have a maintenance

worker test it for electrical safety. Finding that it develops 120 V on one of its knobs, he does not let you take it up to your grandmother's room. She complains that she has had the radio for many years and that nobody has ever gotten a shock from it. You end up having to buy a new plastic radio. Is that fair? Will the old radio be safe back in her bedroom?

21. A series circuit consists of three identical lamps connected to a battery as shown in Figure Q21.21. When the switch S is closed, what happens (a) to the intensities of lamps A and B, (b) to the intensity of lamp C, (c) to the current in the circuit, and (d) to the voltage across the three lamps? (e) Does the power delivered to the circuit increase, decrease, or remain the same?

22. A ski resort consists of a few chairlifts and several interconnected downhill runs on the side of a mountain, with a lodge at the bottom. The chairlifts are analogous to batteries, and the runs are analogous to resistors. Describe how two runs can be in series. Describe how three runs can be in parallel. Sketch a junction of one chairlift and two runs. State Kirchhoff's junction rule for ski resorts. One of the skiers happens to be carrying a sky-diver's altimeter. She never takes the same set of chairlifts and runs twice, but keeps passing you at the fixed location where you are working. State Kirchhoff's loop rule for ski resorts.

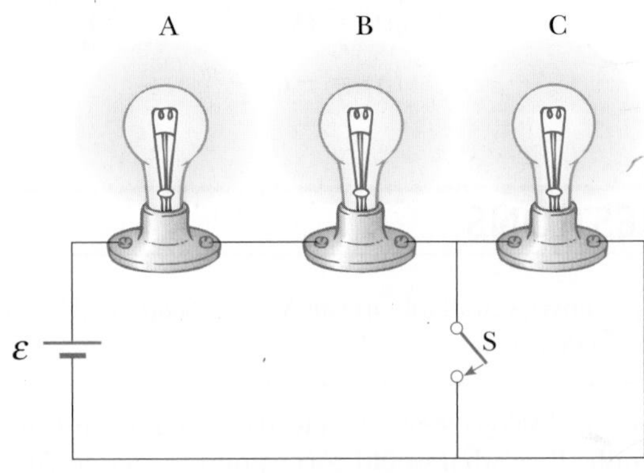

FIGURE **Q21.21**

PROBLEMS

1, 2, 3 = straightforward, intermediate, challenging
☐ = full solution available in the *Student Solutions Manual and Study Guide*
Physics Now™ = coached problem with hints available at www.pop4e.com
= computer useful in solving problem
= paired numerical and symbolic problems
= biomedical application

Section 21.1 ■ Electric Current

1. In a particular cathode-ray tube, the measured beam current is 30.0 μA. How many electrons strike the tube screen every 40.0 s?

2. A small sphere that carries a charge q is whirled in a circle at the end of an insulating string. The angular frequency of revolution is ω. What average current does this revolving charge represent?

3. **Physics Now™** Suppose the current in a conductor decreases exponentially with time according to the equation $I(t) = I_0e^{-t/\tau}$, where I_0 is the initial current (at $t = 0$) and τ is a constant having dimensions of time. Consider a fixed observation point within the conductor. (a) How much charge passes this point between $t = 0$ and $t = \tau$? (b) How much charge passes this point between $t = 0$ and $t = 10\tau$? (c) How much charge passes this point between $t = 0$ and $t = \infty$?

4. The quantity of charge q (in coulombs) that has passed through a surface of area 2.00 cm^2 varies with time according to the equation $q = 4t^3 + 5t + 6$, where t is in seconds. (a) What is the instantaneous current across the surface at $t = 1.00$ s? (b) What is the value of the current density?

5. An aluminum wire having a cross-sectional area of 4.00×10^{-6} m^2 carries a current of 5.00 A. Find the drift speed of the electrons in the wire. The density of aluminum is 2.70 g/cm^3. Assume that one conduction electron is supplied by each atom.

Section 21.2 ■ Resistance and Ohm's Law

6. A lightbulb has a resistance of 240 Ω when operating with a potential difference of 120 V across it. What is the current in the lightbulb?

7. **Physics Now™** A 0.900-V potential difference is maintained across a 1.50-m length of tungsten wire that has a cross-sectional area of 0.600 mm^2. What is the current in the wire?

8. Suppose you wish to fabricate a uniform wire out of 1.00 g of copper. If the wire is to have a resistance of $R = 0.500$ Ω and if all the copper is to be used, what will be (a) the length and (b) the diameter of this wire?

9. An aluminum wire with a diameter of 0.100 mm has a uniform electric field of 0.200 V/m imposed along its entire length. The temperature of the wire is 50.0°C. Assume one free electron per atom. (a) Use the information in Table 21.1 and determine the resistivity. (b) What is the current density in the wire? (c) What is the total current in the wire? (d) What is the drift speed of the conduction electrons? (e) What potential difference must exist between the ends of a 2.00-m length of the wire to produce the stated electric field?

10. While taking photographs in Death Valley on a day when the temperature is 58.0°C, Bill Hiker finds that a certain voltage applied to a copper wire produces a current of 1.000 A. Bill then travels to Antarctica and applies the same voltage to the same wire. What current does he register there if the temperature is $-88.0°C$? Assume that no change occurs in the wire's shape and size.

11. **Review problem.** An aluminum rod has a resistance of 1.234 Ω at 20.0°C. Calculate the resistance of the rod at 120°C by accounting for the changes in both the resistivity and the dimensions of the rod.

Section 21.4 ■ A Structural Model for Electrical Conduction

12. If the current carried by a conductor is doubled, what happens to (a) the charge carrier density, (b) the current density, (c) the electron drift velocity, and (d) the average time interval between collisions?

13. If the magnitude of the drift velocity of free electrons in a copper wire is 7.84×10^{-4} m/s, what is the electric field in the conductor?

Section 21.5 ■ Electric Energy and Power

14. A toaster is rated at 600 W when connected to a 120-V source. What current does the toaster carry, and what is its resistance?

15. In a hydroelectric installation, a turbine delivers 1 500 hp to a generator, which in turn transfers 80.0% of the mechanical energy out by electrical transmission. Under these conditions, what current does the generator deliver at a terminal potential difference of 2 000 V?

16. One rechargeable battery of mass 15.0 g delivers to a CD player an average current of 18.0 mA at 1.60 V for 2.40 h before the battery needs to be recharged. The recharger maintains a potential difference of 2.30 V across the battery and delivers a charging current of 13.5 mA for 4.20 h. (a) What is the efficiency of the battery as an energy storage device? (b) How much internal energy is produced in the battery during one charge–discharge cycle? (c) If the battery is surrounded by ideal thermal insulation and has an overall effective specific heat of $975\ \text{J/kg}\cdot°\text{C}$, by how much will its temperature increase during the cycle?

17. Suppose a voltage surge produces 140 V for a moment. By what percentage does the power output of a 120-V, 100-W lightbulb increase? Assume that its resistance does not change.

18. An 11.0-W energy-efficient fluorescent lamp is designed to produce the same illumination as a conventional 40.0-W incandescent lightbulb. How much money does the user of the energy-efficient lamp save during 100 h of use? Assume a cost of \$0.080 0/kWh for energy from the electric company.

19. A certain toaster has a heating element made of Nichrome wire. When the toaster is first connected to a 120-V source (and the wire is at a temperature of 20.0°C), the initial current is 1.80 A. The current begins to decrease as the heating element warms up, however. When the toaster reaches its final operating temperature, the current drops to 1.53 A. (a) Find the power delivered to the toaster when it is at its operating temperature. (b) What is the final temperature of the heating element?

20. We estimate that 270 million plug-in electric clocks are in the United States, approximately one clock for each person. The clocks convert energy at the average rate 2.50 W. To supply this energy, how many metric tons of coal are burned per hour in coal-fired electric generating plants that are, on average, 25.0% efficient? The heat of combustion for coal is 33.0 MJ/kg.

21. The cost of electricity varies widely through the United States; \$0.120/kWh is one typical value. At this unit price, calculate the cost of (a) leaving a 40.0-W porch light on for two weeks while you are on vacation, (b) making a piece of dark toast in 3.00 min with a 970-W toaster, and (c) drying a load of clothes in 40.0 min in a 5 200-W dryer.

22. An office worker uses an immersion heater to warm 250 g of water in a light, covered insulated cup from 20°C to 100°C in 4.00 min. In electrical terms, the heater is a Nichrome resistance wire connected to a 120-V power supply. Specify a diameter and a length that the wire can have. Can it be made from less than 0.5 cm^3 of Nichrome? You may assume that the wire is at 100°C throughout the time interval.

23. An electric car is designed to run off a bank of 12.0-V batteries with total energy storage of 2.00×10^7 J. (a) If the electric motor draws 8.00 kW, what is the current delivered to the motor? (b) If the electric motor draws 8.00 kW as the car moves at a steady speed of 20.0 m/s, how far will the car travel before it is "out of juice"?

24. Make an order-of-magnitude estimate of the cost of one person's routine use of a hair dryer for 1 yr. If you do not use a hair dryer yourself, observe or interview someone who does. State the quantities you estimate and their values.

Section 21.6 ■ Sources of emf

25. A battery has an emf of 15.0 V. The terminal voltage of the battery is 11.6 V when it is delivering 20.0 W of power to an external load resistor R. (a) What is the value of R? (b) What is the internal resistance of the battery?

26. Two 1.50-V batteries—with their positive terminals in the same direction—are inserted in series into the barrel of a flashlight. One battery has an internal resistance of 0.255 Ω and the other an internal resistance of 0.153 Ω. When the switch is closed, a current of 600 mA occurs in the lamp. (a) What is the lamp's resistance? (b) What fraction of the chemical energy transformed appears as internal energy in the batteries?

Section 21.7 ■ Resistors in Series and in Parallel

27. (a) Find the equivalent resistance between points a and b in Figure P21.27. (b) A potential difference of 34.0 V is applied between points a and b. Calculate the current in each resistor.

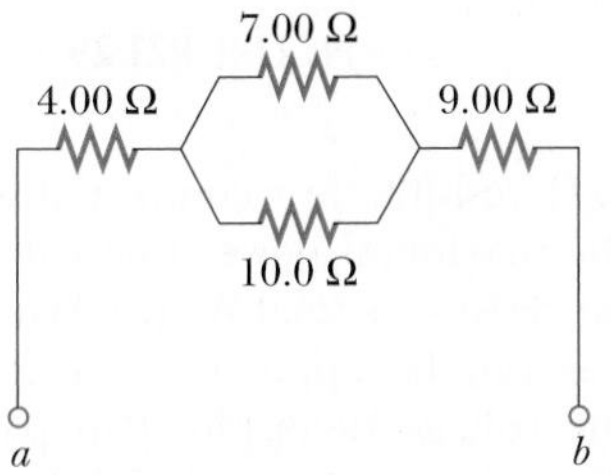

7.00 Ω
4.00 Ω
9.00 Ω
10.0 Ω
a
b

FIGURE **P21.27**

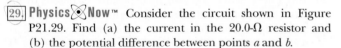

28. For the purpose of measuring the electric resistance of shoes through the body of the wearer to a metal ground plate, the American National Standards Institute (ANSI) specifies the circuit shown in Figure P21.28. The potential difference ΔV across the 1.00-MΩ resistor is measured with a high-resistance voltmeter. The resistance of the person's body is negligible by comparison. (a) Show that the resistance of the footwear is given by

$$R_{\text{shoes}} = 1.00\ \text{M}\Omega\left(\frac{50.0\ \text{V} - \Delta V}{\Delta V}\right)$$

(b) In a medical test, a current through the human body should not exceed 150 μA. Can the current delivered by the ANSI-specified circuit exceed 150 μA? To decide, consider a person standing barefoot on the ground plate.

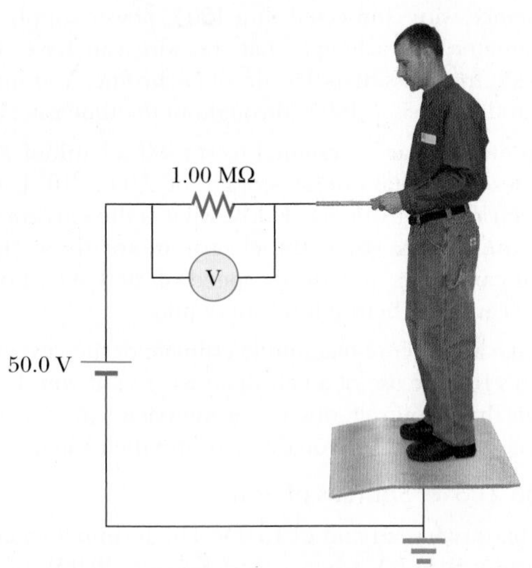

FIGURE P21.28

29. PhysicsNow™ Consider the circuit shown in Figure P21.29. Find (a) the current in the 20.0-Ω resistor and (b) the potential difference between points a and b.

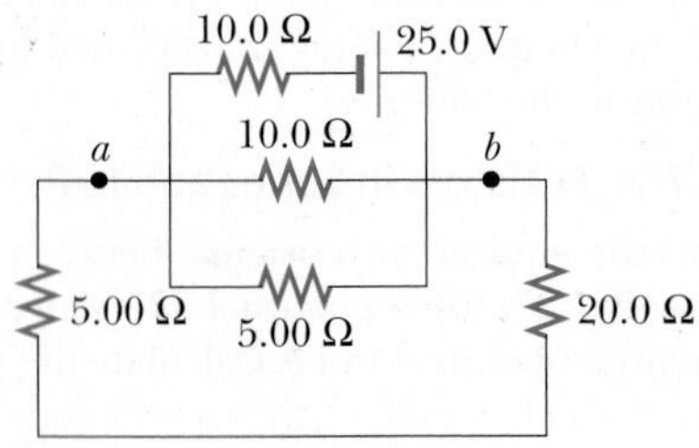

FIGURE P21.29

30. Three 100-Ω resistors are connected as shown in Figure P21.30. The maximum power that can safely be delivered to any one resistor is 25.0 W. (a) What is the maximum voltage that can be applied to the terminals a and b? (b) For the voltage determined in part (a), what is the power delivered to each resistor? What is the total power delivered?

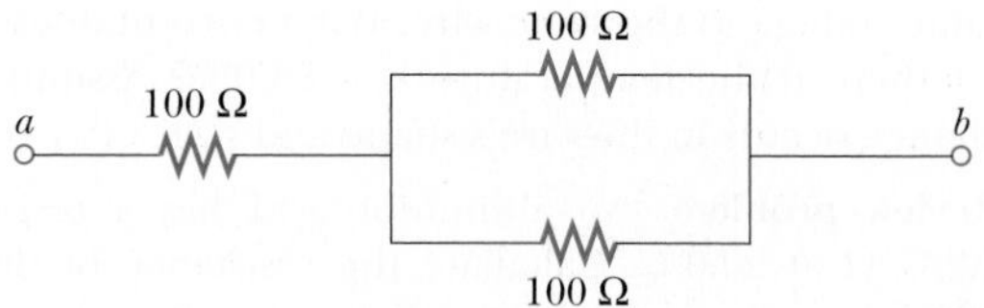

FIGURE P21.30

31. Calculate the power delivered to each resistor in the circuit shown in Figure P21.31.

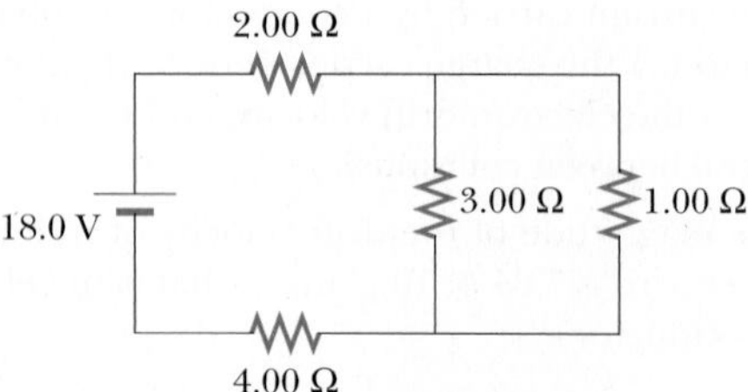

FIGURE P21.31

32. Four resistors are connected to a battery as shown in Figure P21.32. The current in the battery is I, the battery emf is $\mathcal{E}$, and the resistor values are $R_1 = R$, $R_2 = 2R$, $R_3 = 4R$, and $R_4 = 3R$. (a) Rank the resistors according to the potential difference across them, from largest to smallest. Note any cases of equal potential differences. (b) Determine the potential difference across each resistor in terms of $\mathcal{E}$. (c) Rank the resistors according to the current in them, from largest to smallest. Note any cases of equal currents. (d) Determine the current in each resistor in terms of I. (e) If R_3 is increased, what happens to the current in each of the resistors? (f) In the limit that $R_3 \to \infty$, what are the new values of the current in each resistor in terms of I, the original current in the battery?

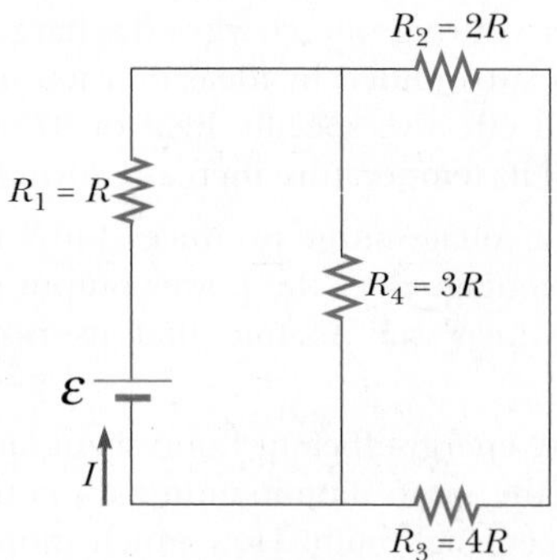

FIGURE P21.32

33. A young man has moved into his own apartment. His possessions include a canister vacuum cleaner marked 535 W at 120 V and a Volkswagen Beetle, which he wishes to clean. He must leave the car in a parking lot far from the building, so he needs an extension cord 15.0 m long to plug in the vacuum cleaner. You may assume that the vacuum cleaner has constant resistance. (a) If the resistance of each of the two conductors in an inexpensive cord is 0.900 Ω, what is the actual power delivered to the vacuum cleaner? (b) If instead the power is to be at least 525 W,

what must be the diameter of each of two identical copper conductors in the cord he buys? (c) Repeat part (b) if the power is to be at least 532 W. (*Suggestion:* A symbolic solution can simplify the calculations.)

Section 21.8 ▪ Kirchhoff's Rules

Note: The currents are not necessarily in the direction shown for some circuits.

34. The ammeter shown in Figure P21.34 reads 2.00 A. Find I_1, I_2, and $\mathcal{E}$.

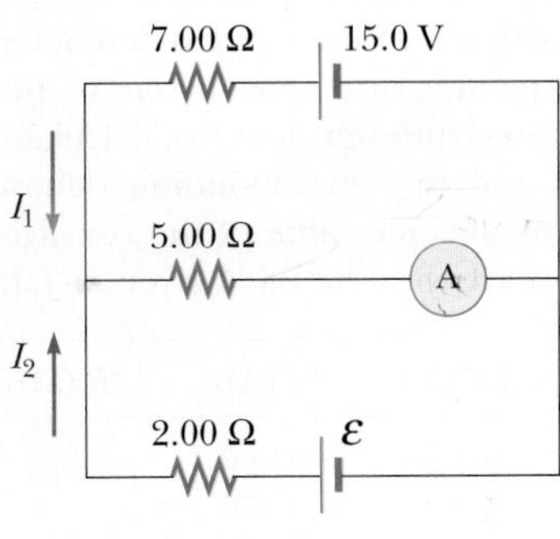

FIGURE **P21.34**

35. **Physics Now™** Determine the current in each branch of the circuit shown in Figure P21.35.

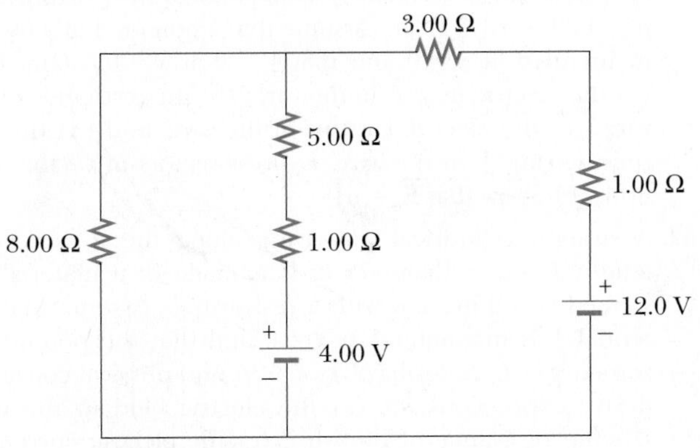

FIGURE **P21.35** Problems 21.35, 21.36, and 21.37.

36. In Figure P21.35, show how to add just enough ammeters to measure every different current. Show how to add just enough voltmeters to measure the potential difference across each resistor and across each battery.

37. The circuit considered in Problem 21.35 and shown in Figure P21.35 is connected for 2.00 min. (a) Find the energy delivered by each battery. (b) Find the energy delivered to each resistor. (c) Identify the net energy transformation that occurs in the operation of the circuit and the total amount of energy transformed.

38. The following equations describe an electric circuit:

$$-(220\ \Omega)I_1 + 5.80\text{ V} - (370\ \Omega)I_2 = 0$$

$$+(370\ \Omega)I_2 + (150\ \Omega)I_3 - 3.10\text{ V} = 0$$

$$I_1 + I_3 - I_2 = 0$$

(a) Draw a diagram of the circuit. (b) Calculate the unknowns and identify the physical meaning of each unknown.

39. Taking $R = 1.00\text{ k}\Omega$ and $\mathcal{E} = 250\text{ V}$ in Figure P21.39, determine the direction and magnitude of the current in the horizontal wire between a and e.

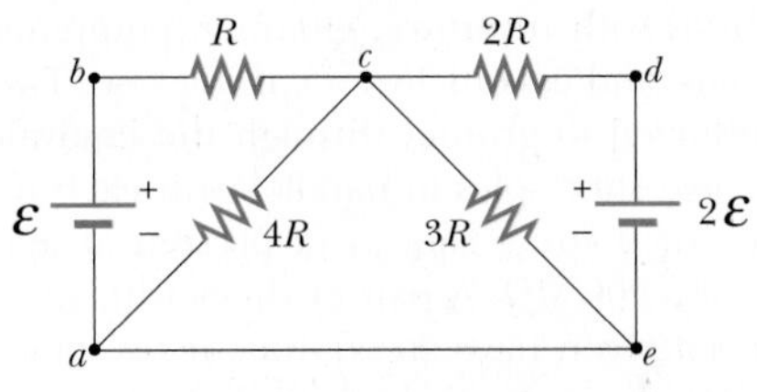

FIGURE **P21.39**

40. A dead battery is charged by connecting it to the live battery of another car with jumper cables (Fig. P21.40). Determine the current in the starter and in the dead battery.

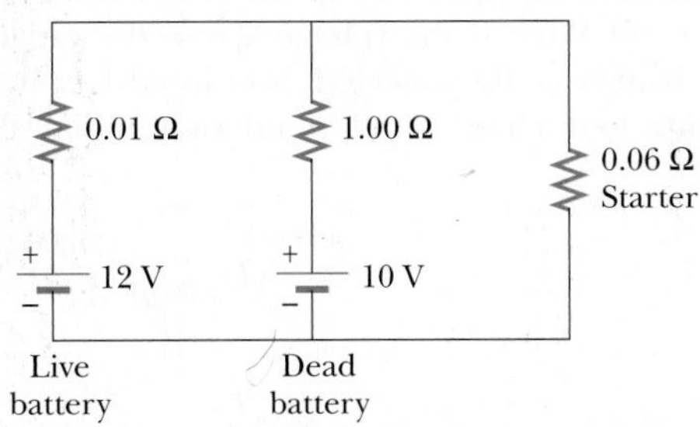

FIGURE **P21.40**

Section 12.9 ▪ *RC* Circuits

41. **Physics Now™** Consider a series *RC* circuit (see Fig. 21.25) for which $R = 1.00\text{ M}\Omega$, $C = 5.00\ \mu\text{F}$, and $\mathcal{E} = 30.0\text{ V}$. Find (a) the time constant of the circuit and (b) the maximum charge on the capacitor after the switch is closed. (c) Find the current in the resistor 10.0 s after the switch is closed.

42. A 2.00-nF capacitor with an initial charge of 5.10 μC is discharged through a 1.30-kΩ resistor. (a) Calculate the current in the resistor 9.00 μs after the resistor is connected across the terminals of the capacitor. (b) What charge remains on the capacitor after 8.00 μs? (c) What is the maximum current in the resistor?

43. In the circuit of Figure P21.43, the switch S has been open for a long time. It is then suddenly closed. Determine the time constant (a) before the switch is closed and (b) after the switch is closed. (c) Let the switch be closed at $t = 0$. Determine the current in the switch as a function of time.

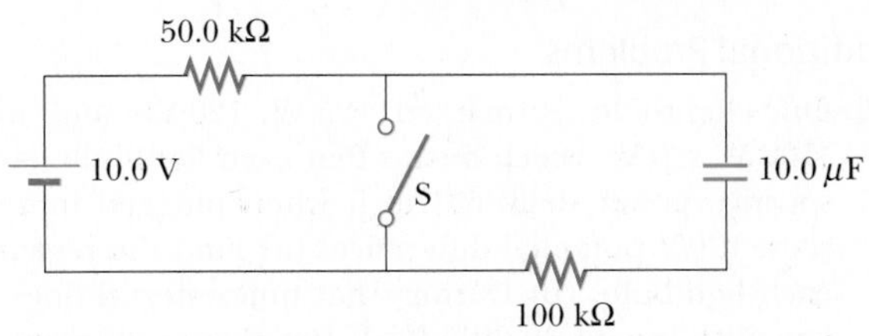

FIGURE **P21.43**

44. In places such as a hospital operating room and a factory for electronic circuit boards, electric sparks must be avoided. A person standing on a grounded floor and touching nothing else can typically have a body capacitance of 150 pF, in parallel with a foot capacitance of 80.0 pF produced by the dielectric soles of his or her shoes. The person acquires static electric charge from interactions with furniture, clothing, equipment, packaging materials, and essentially everything else. The static charge is conducted to ground through the equivalent resistance of the two shoe soles in parallel with each other. A pair of rubber-soled street shoes can present an equivalent resistance of 5 000 MΩ. A pair of shoes with special static-dissipative soles can have an equivalent resistance of 1.00 MΩ. Consider the person's body and shoes as forming an RC circuit with the ground. (a) How long does it take the rubber-soled shoes to reduce a 3 000-V static charge to 100 V? (b) How long does it take the static-dissipative shoes to do the same thing?

45. The circuit in Figure P21.45 has been connected for a long time. (a) What is the voltage across the capacitor? (b) If the battery is disconnected, how long does it take the capacitor to discharge to one tenth of its initial voltage?

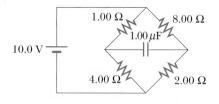

FIGURE **P21.45**

46. A 10.0-μF capacitor is charged by a 10.0-V battery through a resistance R. The capacitor reaches a potential difference of 4.00 V at the instant 3.00 s after charging begins. Find R.

Section 12.10 ▪ Context Connection—The Atmosphere as a Conductor

47. A current density of 6.00×10^{-13} A/m^2 exists in the atmosphere at a location where the electric field is 100 V/m. Calculate the electrical conductivity of the Earth's atmosphere in this region.

48. Assume that global lightning on the Earth constitutes a constant current of 1.00 kA between the ground and an atmospheric layer at potential 300 kV. (a) Find the power of terrestrial lightning. (b) For comparison, find the power of sunlight falling on the Earth. Sunlight has an intensity of 1 370 W/m^2 above the atmosphere. Sunlight falls perpendicularly on the circular projected area that the Earth presents to the Sun.

Additional Problems

49. One lightbulb is marked "25 W 120 V" and another "100 W 120 V," which means that each lightbulb has its respective power delivered to it when plugged into a constant 120-V potential difference. (a) Find the resistance of each lightbulb. (b) During what time interval does 1.00 C pass into the dim lightbulb? Is the charge different in any way upon its exit from the lightbulb versus its entry? (c) In what time interval does 1.00 J pass into the dim lightbulb? By what mechanisms does this energy enter and exit the lightbulb? (d) Find how much it costs to run the dim lightbulb continuously for 30.0 days, assuming that the electric company sells its product at \$0.070 0 per kWh. What product *does* the electric company sell? What is its price for one SI unit of this quantity?

50. An experiment is conducted to measure the electrical resistivity of Nichrome in the form of wires with different lengths and cross-sectional areas. For one set of measurements, a student uses 30-gauge wire, which has a cross-sectional area of 7.30×10^{-8} m^2. The student measures the potential difference across the wire and the current in the wire with a voltmeter and an ammeter, respectively. For each of the measurements given in the table taken on wires of three different lengths, calculate the resistance of the wires and the corresponding values of the resistivity. What is the average value of the resistivity? How does this value compare with the value given in Table 21.1?

L (m)	ΔV (V)	I (A)	R (Ω)	ρ (Ω·m)
0.540	5.22	0.500		
1.028	5.82	0.276		
1.543	5.94	0.187		

51. A straight cylindrical wire lying along the x axis has a length of 0.500 m and a diameter of 0.200 mm. It is made of a material described by Ohm's law with a resistivity of $\rho = 4.00 \times 10^{-8}$ Ω·m. Assume that a potential of 4.00 V is maintained at $x = 0$ and that $V = 0$ at $x = 0.500$ m. Find (a) the electric field $\vec{\mathbf{E}}$ in the wire, (b) the resistance of the wire, (c) the electric current in the wire, and (d) the current density $\vec{\mathbf{J}}$ in the wire. Express vectors in vector notation. (e) Show that $\vec{\mathbf{E}} = \rho\vec{\mathbf{J}}$.

52. A straight cylindrical wire lying along the x axis has a length L and a diameter d. It is made of a material described by Ohm's law with a resistivity ρ. Assume that potential V is maintained at $x = 0$ and that the potential is zero at $x = L$. In terms of L, d, V, ρ, and physical constants, derive expressions for (a) the electric field in the wire, (b) the resistance of the wire, (c) the electric current in the wire, and (d) the current density in the wire. Express vectors in vector notation. (e) Prove that $\vec{\mathbf{E}} = \rho\vec{\mathbf{J}}$.

53. An electric heater is rated at 1 500 W, a toaster at 750 W, and an electric grill at 1 000 W. The three appliances are connected to a common 120-V household circuit. (a) How much current does each draw? (b) Is a circuit with a 25.0-A circuit breaker sufficient in this situation? Explain your answer.

54. An oceanographer is studying how the ion concentration in sea water depends on depth. She does so by lowering into the water a pair of concentric metallic cylinders (Fig. P21.54) at the end of a cable and taking data to determine the resistance between these electrodes as a function of depth. The water between the two cylinders forms a cylindrical shell of inner radius r_a, outer radius r_b, and length L much larger than r_b. The scientist applies a potential difference ΔV between the inner and outer surfaces,

producing an outward radial current I. Let ρ represent the resistivity of the water. (a) Find the resistance of the water between the cylinders in terms of L, ρ, r_a, and r_b. (b) Express the resistivity of the water in terms of the measured quantities L, r_a, r_b, ΔV, and I.

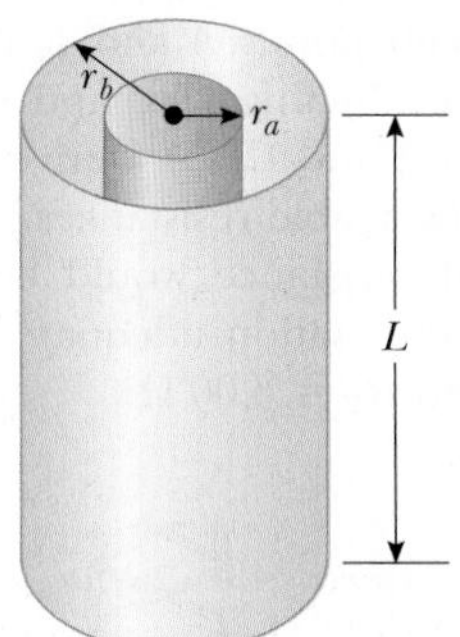

FIGURE **P21.54**

55. Four 1.50-V AA batteries in series are used to power a transistor radio. If the batteries can move a charge of 240 C, how long will they last if the radio has a resistance of 200 Ω?

56. A battery has an emf of 9.20 V and an internal resistance of 1.20 Ω. What resistance across the battery will extract from it (a) a power of 12.8 W and (b) a power of 21.2 W?

57. A battery has an emf $\mathcal{E}$ and internal resistance r. A variable load resistor R is connected across the terminals of the battery. (a) Determine the value of R such that the potential difference across the terminals is a maximum. (b) Determine the value of R so that the current in the circuit is a maximum. (c) Determine the value of R so that the power delivered to the load resistor is a maximum. Choosing the load resistance for maximum power transfer is a case of what is called *impedance matching* in general. Impedance matching is important in shifting gears on a bicycle, in connecting a loudspeaker to an audio amplifier, in connecting a battery charger to a bank of solar photoelectric cells, and in many other applications.

58. A battery is used to charge a capacitor through a resistor as shown in Figure 21.25. Show that half the energy supplied by the battery appears as internal energy in the resistor and that half is stored in the capacitor.

59. The values of the components in a simple series RC circuit containing a switch (Fig. 21.25) are $C = 1.00\ \mu$F, $R = 2.00 \times 10^6$ Ω, and $\mathcal{E} = 10.0$ V. At the instant 10.0 s after the switch is closed, calculate (a) the charge on the capacitor, (b) the current in the resistor, (c) the rate at which energy is being stored in the capacitor, and (d) the rate at which energy is being delivered by the battery.

60. The switch in Figure P21.60a closes when $\Delta V_c > 2\Delta V/3$ and opens when $\Delta V_c < \Delta V/3$. The voltmeter reads a voltage as plotted in Figure P21.60b. What is the period T of the waveform in terms of R_A, R_B, and C?

61. Switch S has been closed for a long time, and the electric circuit shown in Figure P21.61 carries a constant current. Take $C_1 = 3.00\ \mu$F, $C_2 = 6.00\ \mu$F, $R_1 = 4.00$ kΩ, and $R_2 = 7.00$ kΩ. The power delivered to R_2 is 2.40 W. (a) Find the charge on C_1. (b) Now the switch is opened. After many milliseconds, by how much has the charge on C_2 changed?

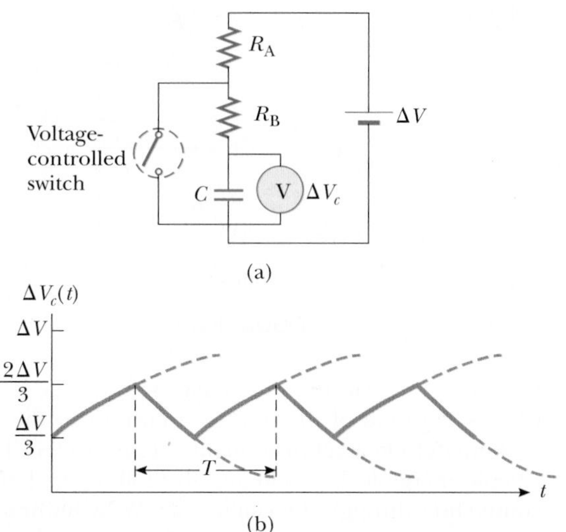

FIGURE **P21.60**

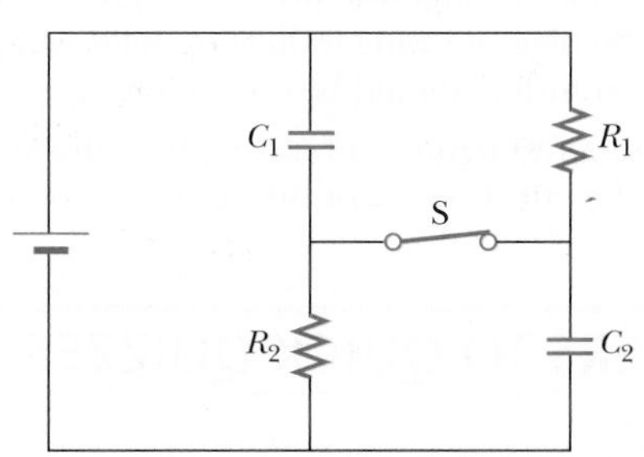

FIGURE **P21.61**

62. The circuit shown in Figure P21.62 is set up in the laboratory to measure an unknown capacitance C with the use of a voltmeter of resistance $R = 10.0$ MΩ and a battery whose emf is 6.19 V. The data given in the table are the measured voltages across the capacitor as a function of time, where $t = 0$ represents the instant at which the switch is opened. (a) Construct a graph of $\ln(\mathcal{E}/\Delta V)$ versus t and perform a linear least-squares fit to the data. (b) From the slope of your graph, obtain a value for the time constant of the circuit and a value for the capacitance.

ΔV (V)	t (s)	$\ln(\mathcal{E}/\Delta V)$
6.19	0	
5.55	4.87	
4.93	11.1	
4.34	19.4	
3.72	30.8	
3.09	46.6	
2.47	67.3	
1.83	102.2	

63. Four resistors are connected in parallel across a 9.20-V battery. They carry currents of 150 mA, 45.0 mA, 14.00 mA, and 4.00 mA. (a) If the resistor with the largest resistance is replaced with one having twice the resistance, what is the ratio of the new current in the battery to the original current? (b) If instead the resistor with the smallest

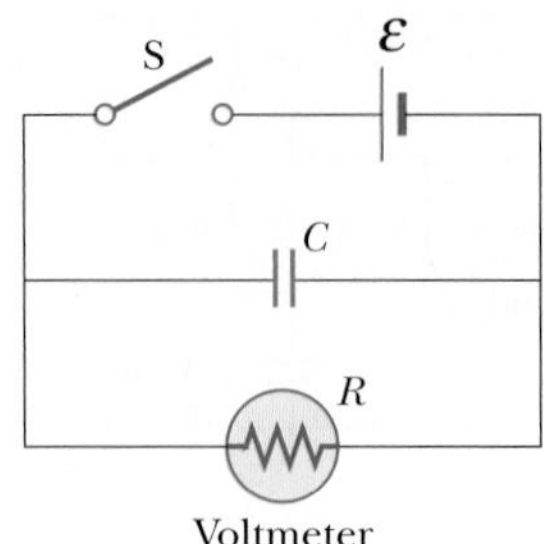

FIGURE **P21.62**

resistance is replaced with one having twice the resistance, what is the ratio of the new total current to the original current? (c) On a February night, energy leaves a house by several energy leaks, including the following: 1 500 W by conduction through the ceiling, 450 W by infiltration (air flow) around the windows, 140 W by conduction through the basement wall above the foundation sill, and 40.0 W by conduction through the plywood door to the attic. To produce the biggest saving in heating bills, which one of these energy transfers should be reduced first?

64. The student engineer of a campus radio station wishes to verify the effectiveness of the lightning rod on the antenna mast (Fig. P21.64). The unknown resistance R_x is between points C and E. Point E is a true ground but is inaccessible for direct measurement because this stratum is several meters below the Earth's surface. Two identical rods are driven into the ground at A and B, introducing an unknown resistance R_y. The procedure is as follows. Measure resistance R_1 between points A and B, then connect A and B with a heavy conducting wire and measure resistance R_2 between points A and C. (a) Derive an equation for R_x in terms of the observable resistances, R_1 and R_2. (b) A satisfactory ground resistance would be $R_x < 2.00\ \Omega$. Is the grounding of the station adequate if measurements give $R_1 = 13.0\ \Omega$ and $R_2 = 6.00\ \Omega$?

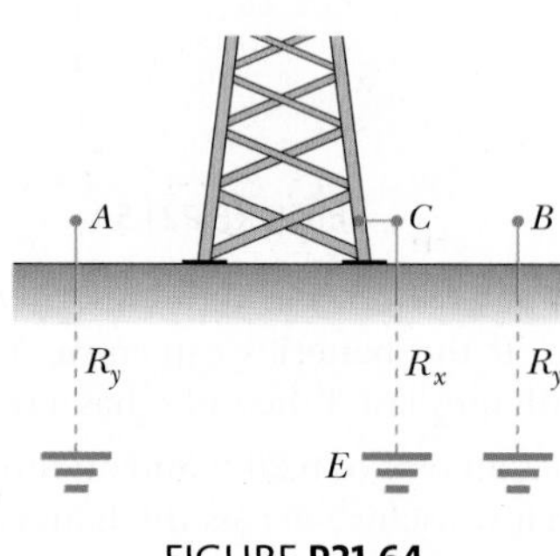

FIGURE **P21.64**

ANSWERS TO QUICK QUIZZES

21.1. (d), (b) = (c), (a). The current in part (d) is equivalent to two positive charges moving to the left. Parts (b) and (c) each represent four charges moving in the same direction because negative charges moving to the left are equivalent to positive charges moving to the right. The current in part (a) is equivalent to five positive charges moving to the right.

21.2. (b). According to Equation 21.6, resistance is the ratio of voltage across a device to current in the device. In Figure 21.6b, a line drawn from the origin to a point on the curve will have a slope equal to $I/\Delta V$, which is the inverse of resistance. As ΔV increases, the slope of this line also increases, so the resistance decreases.

21.3 (a). When the filament is at room temperature, its resistance is low and hence the current is relatively large. As the filament warms up, its resistance increases and the current decreases.

21.4 $I_a = I_b > I_c = I_d > I_e = I_f$. Charge constituting the current I_a leaves the positive terminal of the battery and then splits to flow through the two lightbulbs; therefore, $I_a = I_c + I_e$. Because the potential difference ΔV is the same across the two lightbulbs and because the power delivered to a device is $\mathcal{P} = I\,\Delta V$, the 60-W lightbulb with the higher power rating must carry the greater current. Because charge does not accumulate in the lightbulbs, we know that the same amount of charge flowing into a lightbulb from the left has to flow out on the right; consequently $I_c = I_d$ and $I_e = I_f$. The two currents leaving the lightbulbs recombine to form the current back into the battery, $I_f + I_d = I_b$.

21.5 (a). Connecting b to c "shorts out" lightbulb R_2 and changes the total resistance of the circuit from $R_1 + R_2$ to just R_1. Because the resistance of the circuit has decreased (and the potential difference supplied by the battery does not change), the current in the circuit increases.

21.6 (b). When the switch is opened, resistors R_1 and R_2 are in series, so the total circuit resistance is larger than when the switch was closed. As a result, the current drops.

21.7 (a). When the switch is closed, resistors R_1 and R_2 are in parallel, so the total circuit resistance is smaller than when the switch was open. As a result, the current increases.

21.8 (i), (b), (d). Adding another series resistor increases the total resistance of the circuit and therefore reduces the current in the circuit. The potential difference across the battery terminals increases because the reduced current results in a smaller voltage decrease across the internal resistance. (ii), (a), (e). If a third resistor were connected in parallel, the total resistance of the circuit would decrease and the current in the battery would increase. The potential difference across the terminals would decrease because the increased current results in a greater voltage drop across the internal resistance.

21.9 (i), (b). Just after the switch is closed, there is no charge on the capacitor, so there is no voltage across it. Charges begin to flow in the circuit to charge up the capacitor, so all the voltage $\Delta V = IR$ appears across the resistor. (ii), (a). After a long time, the capacitor is fully charged and the current drops to zero. Therefore, the battery voltage is now entirely across the capacitor.

Determining the Number of Lightning Strikes

Now that we have investigated the principles of electricity, let us respond to our central question for the *Lightning* Context:

How can we determine the number of lightning strikes on the Earth in a typical day?

We must combine several ideas from our knowledge of electricity to perform this calculation. In Chapter 20, the atmosphere was modeled as a capacitor. Such modeling was first done by Lord Kelvin, who modeled the ionosphere as the positive plate several tens of kilometers above the Earth's surface. More sophisticated models have shown the effective height of the positive plate to be the 5 km that we used in our earlier calculation.

The Atmospheric Capacitor Model

The plates of the atmospheric capacitor are separated by a layer of air containing a large number of free ions that can carry current. Air is a good insulator; measurements show that the resistivity of air is about $3 \times 10^{13}\ \Omega \cdot \text{m}$. Let us calculate the resistance of the air between our capacitor plates. The shape of the resistor is that of a spherical shell between the plates of the atmospheric capacitor. The length of 5 km, however, is very short compared with the radius of 6 400 km. Therefore, we can ignore the spherical shape and approximate the resistor as a 5-km slab of flat material whose area is the surface area of the Earth. Using Equation 21.8,

$$R = \rho \frac{\ell}{A} = (3 \times 10^{13}\ \Omega \cdot \text{m}) \frac{5 \times 10^3\ \text{m}}{4\pi(6.4 \times 10^6\ \text{m})^2} \approx 3 \times 10^2\ \Omega$$

The charge on the atmospheric capacitor can pass from the upper plate to the ground by electric current in the air between the plates. Thus, we can model the atmosphere as an RC circuit, using the capacitance found in Chapter 20, and the resistance connecting the plates calculated above (Fig. 1). The time constant for this RC circuit is

$$\tau = RC = (0.9\ \text{F})(3 \times 10^2\ \Omega) \approx 3 \times 10^2\ \text{s} = 5\ \text{min}$$

Thus, the charge on the atmospheric capacitor should fall to $e^{-1} = 37\%$ of its original value after only 5 min! After 30 min, less than 0.3% of the charge would remain! Why doesn't that happen? What keeps the atmospheric capacitor charged? The answer is *lightning*. The processes occurring in cloud charging result in lightning strikes that deliver negative charge to the ground to replace that neutralized by the flow of charge through the air. On the average, a net charge on the atmospheric capacitor results from a balance between these two processes.

Now, let's use this balance to numerically answer our central question. We first address the charge on the atmospheric capacitor. In Chapter 19, we mentioned a

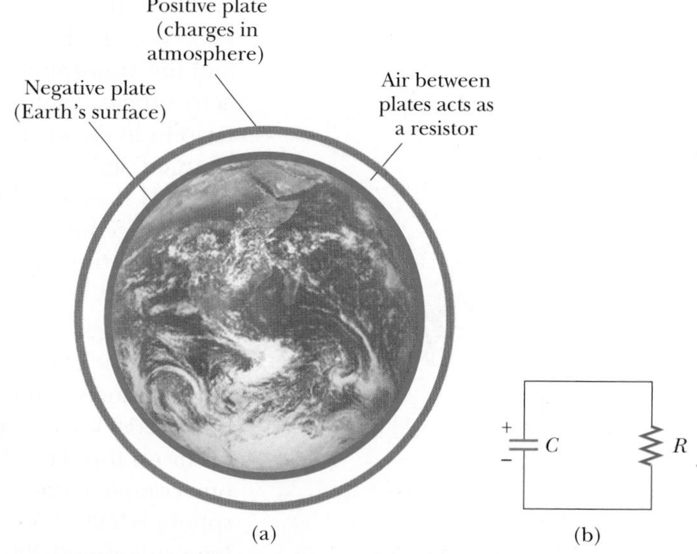

FIGURE 1 (a) The atmosphere can be modeled as a capacitor, with conductive air between the plates. (b) We can imagine an equivalent RC circuit for the atmosphere, with the natural discharge of the capacitor in balance with the charging of the capacitor by lightning.

charge of 5×10^5 C that is spread over the surface of the Earth, which is the charge on the atmospheric capacitor.

A typical lightning strike delivers about 25 C of negative charge to the ground in the process of charging the capacitor. Dividing the charge on the capacitor by the charge per lightning strike tells us the number of lightning strikes required to charge the capacitor:

$$\text{Number of lightning strikes} = \frac{\text{total charge}}{\text{charge per lightning strike}}$$

$$= \frac{5 \times 10^5 \text{ C}}{25 \text{ C per strike}} \approx 2 \times 10^4 \text{ lightning strikes}$$

According to our calculation for the RC circuit, the atmospheric capacitor almost completely discharges through the air in about 30 min. Thus, 2×10^4 lightning strikes must occur every 30 min, or 4×10^4/h, to keep the charging and discharging processes in balance. Multiplying by the number of hours in a day gives us

$$\text{Number of lightning strikes per day} = (4 \times 10^4 \text{ strikes/h})\left(\frac{24 \text{ h}}{1 \text{ d}}\right)$$

$$\approx 1 \times 10^6 \text{ strokes/day}$$

Despite the simplifications that we have adopted in our calculations, this number is on the right order of magnitude for the actual number of lightning strikes on the Earth in a typical day: 1 million!

Problems

1. Consider the atmospheric capacitor described in the text, with the ground as one plate and positive charges in the atmosphere as the other. On one particular day, the capacitance of the atmospheric capacitor is 0.800 F. The effective plate separation distance is 4.00 km, and the resistivity of the air between the plates is $2.00 \times 10^{13}\ \Omega \cdot \text{m}$. If no lightning events occur, the capacitor will discharge through the air. If a charge of 4.00×10^4 C is on the atmospheric capacitor at time $t = 0$, at what later time is the charge reduced (a) to 2.00×10^4 C, (b) to 5.00×10^3 C, and (c) to zero?
2. Consider this alternative line of reasoning to estimate the number of lightning strikes on the Earth in one day. Using the charge on the Earth of 5.00×10^5 C and the atmospheric capacitance of 0.9 F, we find that the potential difference across the capacitor is $\Delta V = Q/C = 5.00 \times 10^5 \text{ C}/0.9 \text{ F} \approx 6 \times 10^5$ V. The leakage current in the air is $I = \Delta V/R = 6 \times 10^5 \text{ V}/300\ \Omega \approx 2$ kA. To keep the capacitor charged, lightning should deliver the same net current in the opposite direction. (a) If each lightning strike delivers 25 C of charge to the ground, what is the average time interval between lightning strikes so that the average current due to lightning is 2 kA? (b) Using this average time interval between lightning strikes, calculate the number of lightning strikes in one day.
3. Consider again the atmospheric capacitor discussed in the text. (a) Assume that atmospheric conditions are such that, for one complete day, the lower 2.50 km of the air between the capacitor plates has resistivity $2.00 \times 10^{13}\ \Omega \cdot \text{m}$ and the upper 2.50 km has resistivity $0.500 \times 10^{13}\ \Omega \cdot \text{m}$. How many lightning strikes occur on this day? (b) Assume that atmospheric conditions are such that, for one complete day, resistivity of the air between the plates in the southern hemisphere is $2.00 \times 10^{13}\ \Omega \cdot \text{m}$ and the resistivity between the plates in the northern hemisphere is $0.200 \times 10^{13}\ \Omega \cdot \text{m}$. How many lightning strikes occur on this day?

CONTEXT 7

Magnetic Levitation Vehicles

All commercial long-distance ground transportation currently operating in the United States is subject to the force of friction between wheels and a roadway or a track. Recall that friction is a nonconservative force that transforms kinetic energy into internal energy. As discussed in Chapters 16 through 18 on thermodynamics, this internal energy is wasted.

Magnetic levitation (maglev) vehicles are suspended by magnetic forces and therefore do not make physical contact with a roadway or track. This suspension eliminates mechanical friction with the track, the primary cause of transformation of kinetic energy to internal energy. There is still a friction force from the surrounding air that will transform some of the kinetic energy.

Robert Goddard, of rocket fame, published a story in 1907 that describes many features of magnetic levitation. He also published a paper in *Scientific American* in 1909 describing a magnetically levitated vehicle operating in a tunnel between Boston and New York City. Emile Bachelet, a French engineer, published a paper in 1912 describing a magnetically levitated vehicle for delivering mail. He received a patent for his invention, but it required far too much power to be practical.

After these early ideas, no significant progress in magnetic levitation was made until the 1960s. At that time, advances in superconducting magnets spurred new interest in magnetic levitation because of the possible savings in energy costs over previous designs such as Bachelet's. In 1963, a physicist at Brookhaven National Laboratory proposed a system using superconducting magnets. Within a few years, projects were underway at Stanford University, MIT, Raytheon, Ford Motor Company, the University of Toronto, and McGill University. Projects were also initiated shortly thereafter in Japan, Germany, and England.

(© Transrapid International, Berlin, Germany)

FIGURE 1 Because a magnetic levitation (maglev) vehicle is not subject to mechanical friction with rails, it can reach very high speeds. This photo shows the German Transrapid maglev vehicle in operation. Of all proposed models of maglev, the Transrapid is furthest along in its development.

Despite this promising start by several U.S. companies and universities, federal funding for maglev research in the United States ended in 1975. Research in other countries continued, primarily in Germany and Japan. These studies and a variety of full-scale test vehicles have shown that maglev technology is very successful. Research in maglev has seen a modest revival in the United States following the National Maglev Initiative signed into law in 1991, but the United States remains far behind Germany and Japan.

The German maglev project is called the Transrapid. It has undergone extensive testing in Germany. In December 2003, it realized a major milestone in having the first commercial Transrapid line open for business in Shanghai,

(© Transrapid International, Berlin, Germany)

FIGURE 2 The German Transrapid in commercial operation in Shanghai, China. Recent tests have shown that this vehicle can travel at speeds of more than 500 km/h.

China. Additional proposals call for the Transrapid to be incorporated into transportation projects in Pittsburgh; Los Angeles; between Baltimore and Washington, D.C.; and between Anaheim, California, and Las Vegas, Nevada.

The Japanese maglev vehicle is dubbed the MLX01 (ML for maglev, X for "experimental"). This vehicle holds the world speed record for a maglev vehicle at 581 km/h, achieved in December 2003 with technicians on board. The MLX01 is currently in the final phase of testing before the Japanese Ministry of Transport decides whether to proceed with commercial development.

In addition to the energy savings in a maglev vehicle associated with the reduction of friction, there are other benefits. One is reduced environmental impact compared with a traditional railroad because of the absence of emissions. Furthermore, the reliability under various weather conditions such as snow and rain is enhanced because the motion is not dependent on a coefficient of friction. In this Context, we shall investigate the physics of magnetic fields and electromagnetism and apply these principles to understanding the processes of lifting, propelling, and braking a maglev vehicle. Two primary mechanisms—the *attractive* and *repulsive* models—are the basis of current research and development efforts. We shall study each of these models so we can respond to our central question:

(© Hashimoto Noburu/Corbis SYGMA)

FIGURE 3 The Japanese MLX01 test vehicle. Although this vehicle differs in technology from the German maglev vehicles, it also can travel at very high speeds and currently holds the world record for a maglev vehicle.

How can we lift, propel, and brake a vehicle with magnetic forces?

CHAPTER 22

Magnetic Forces and Magnetic Fields

Magnetic fingerprinting allows fingerprints to be seen on surfaces that otherwise would not allow prints to be lifted. The powder spread on the surface is coated with an organic material that adheres to the greasy residue in a fingerprint. A magnetic "brush" removes the excess powder and makes the fingerprint visible.

(James King-Holmes/Photo Researchers, Inc.)

CHAPTER OUTLINE

The list of technological applications of magnetism is very long. For instance, large electromagnets are used to pick up heavy loads in scrap yards. Magnets are used in such devices as meters, motors, and loudspeakers. Magnetic tapes are routinely used in sound and video recording equipment and for computer data storage. Intense magnetic fields generated by superconducting magnets are currently being used as a means of containing plasmas at temperatures on the order of 10^8 K used in controlled nuclear fusion research.

As we investigate magnetism in this chapter, we shall find that the subject cannot be divorced from electricity. For example, magnetic fields affect moving electric charges, and moving charges produce magnetic fields. This close association between electricity and magnetism will justify their union into *electromagnetism* that we explore in this chapter and the next.

22.1 HISTORICAL OVERVIEW

Many historians of science believe that the compass, which uses a magnetic needle, was used in China as early as the 13th century B.C., its invention being of Arab or Indian origin. The phenomenon of magnetism was known to the Greeks as early as about 800 B.C. They discovered that certain stones, made of a material now called *magnetite* (Fe_3O_4), attracted pieces of iron.

In 1269, Pierre de Maricourt (c. 1220–?) mapped out the directions taken by a magnetized needle when it was placed at various points on the surface of a spherical natural magnet. He found that the directions formed lines that encircled the sphere and passed through two points diametrically opposite each other, which he called the **poles** of the magnet. Subsequent experiments have shown that every magnet, regardless of its shape, has two poles, called **north** (N) and **south** (S), that exhibit forces on each other in a manner analogous to electric charges. That is, similar poles (N–N or S–S) repel each other and dissimilar poles (N–S) attract each other. The poles received their names because of the behavior of a magnet in the presence of the Earth's magnetic field. If a bar magnet is suspended from its midpoint by a piece of string so that it can swing freely in a horizontal plane, it rotates until its "north" pole points to the north geographic pole of the Earth (which is a south magnetic pole) and its "south" pole points to the Earth's south geographic pole. (The same idea is used to construct a simple compass.)

In 1600, William Gilbert (1544–1603) extended these experiments to a variety of materials. Using the fact that a compass needle orients in preferred directions, Gilbert suggested that magnets are attracted to land masses. In 1750, John Michell (1724–1793) used a torsion balance to show that magnetic poles exert attractive or repulsive forces on each other and that these forces vary as the inverse square of their separation. Although the force between two magnetic poles is similar to the force between two electric charges, an important difference exists. Electric charges can be isolated (witness the electron and proton), whereas magnetic poles cannot be isolated. That is, **magnetic poles are always found in pairs.** No matter how many times a permanent magnet is cut, each piece always has a north pole and a south pole. (Some theories speculate that magnetic monopoles—isolated north or south poles—may exist in nature, and attempts to detect them currently make up an active experimental field of investigation. None of these attempts has yet proven successful, however.)

The relationship between magnetism and electricity was discovered in 1819 when, while preparing for a lecture demonstration, Danish scientist Hans Christian Oersted found that an electric current in a wire deflected a nearby compass needle. Shortly thereafter, André-Marie Ampère (1775–1836) deduced quantitative laws of magnetic force between current-carrying conductors. He also suggested that electric current loops of molecular size are responsible for *all* magnetic phenomena.

In the 1820s, Faraday and, independently, Joseph Henry (1797–1878) identified further connections between electricity and magnetism. They showed that an electric current could be produced in a circuit either by moving a magnet near the circuit or by changing the current in a nearby circuit. Their observations demonstrated that a changing magnetic field produces an electric field. Years later, theoretical work by James Clerk Maxwell showed that the reverse is also true: a changing electric field gives rise to a magnetic field.

In this chapter, we shall investigate the effects of constant magnetic fields on charges and currents, and study the sources of magnetic fields. In the next chapter, we shall explore the effects of magnetic fields that vary in time.

(North Wind Picture Archives)

HANS CHRISTIAN OERSTED (1777–1851)

Oersted, a Danish physicist, is best known for observing that a compass needle deflects when placed near a wire carrying a current. This important discovery was the first evidence of the connection between electric and magnetic phenomena. Oersted was also the first to prepare pure aluminum.

22.2 THE MAGNETIC FIELD

In earlier chapters, we described the interaction between charged objects in terms of electric fields. Recall that an electric field surrounds any stationary electric charge. The region of space surrounding a *moving* charge includes a **magnetic field** in addition to the electric field. A magnetic field also surrounds any material with permanent magnetism. We find that the magnetic field is a vector field, as is the electric field.

To describe any type of vector field, we must define its magnitude and its direction. The direction of the magnetic field vector $\vec{\mathbf{B}}$ at any location is the direction in which the north pole of a compass needle points at that location. Active Figure 22.1 shows how the magnetic field of a bar magnet can be traced with the aid of a compass, defining a **magnetic field line,** similar in many ways to the electric field lines we studied in Chapter 19. Several magnetic field lines of a bar magnet traced out in this manner are shown in the two-dimensional pictorial representation in Active Figure 22.1. Magnetic field patterns can be displayed by small iron filings placed in the vicinity of a magnet, as in Figure 22.2.

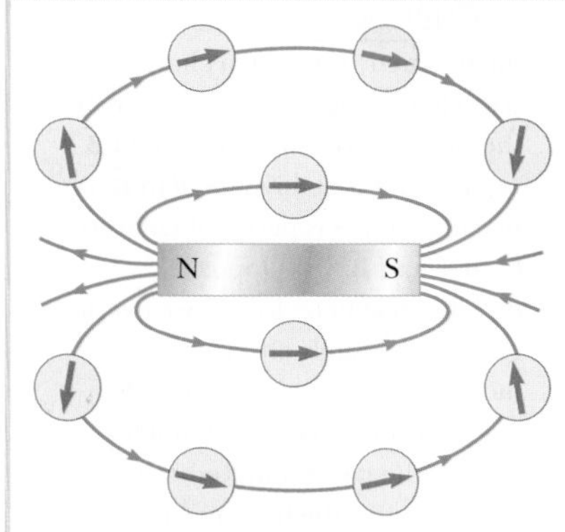

ACTIVE FIGURE 22.1
A small compass can be used to trace the magnetic field lines of a bar magnet.

Physics Now™ Log into PhysicsNow at **www.pop4e.com** and go to Active Figure 22.1 to move the compass around and trace the field lines for yourself.

We can quantify the magnetic field $\vec{\mathbf{B}}$ by using our model of a particle in a field. The existence of a magnetic field at some point in space can be determined by measuring the **magnetic force** $\vec{\mathbf{F}}_B$ exerted on an appropriate test particle placed at that point. This process is the same one we followed in defining the electric field in Chapter 19. Our test particle will be an electrically charged particle such as a proton. If we perform such an experiment, we find the following results:

- The magnetic force $\vec{\mathbf{F}}_B$ is proportional to the charge q of the particle as well as to the speed v of the particle.
- When a charged particle moves parallel to the magnetic field vector, the magnetic force $\vec{\mathbf{F}}_B$ on the charge is zero.
- When the velocity vector makes an angle θ with the magnetic field, the magnetic force acts in a direction perpendicular to both $\vec{\mathbf{v}}$ and $\vec{\mathbf{B}}$; that is, the magnetic force is perpendicular to the plane formed by $\vec{\mathbf{v}}$ and $\vec{\mathbf{B}}$ (Fig. 22.3a).
- The magnetic force on a negative charge is directed opposite to the force on a positive charge moving in the same direction (Fig. 22.3b).
- If the velocity vector makes an angle θ with the magnetic field, the magnitude of the magnetic force is proportional to sin θ.

These results show that the magnetic force on a particle is more complicated than the electric force. The magnetic force is distinctive because it depends on the

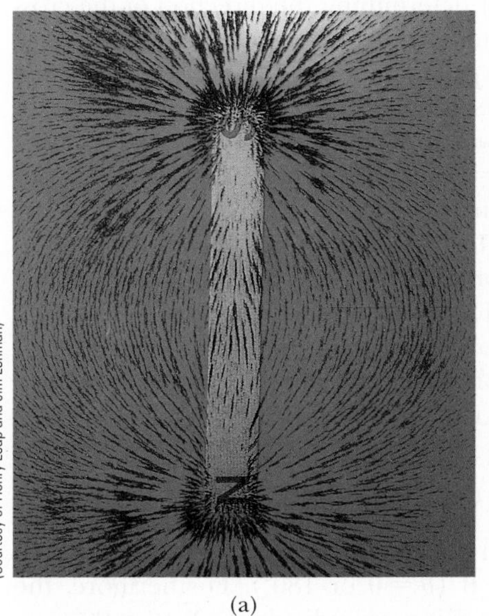

(a)

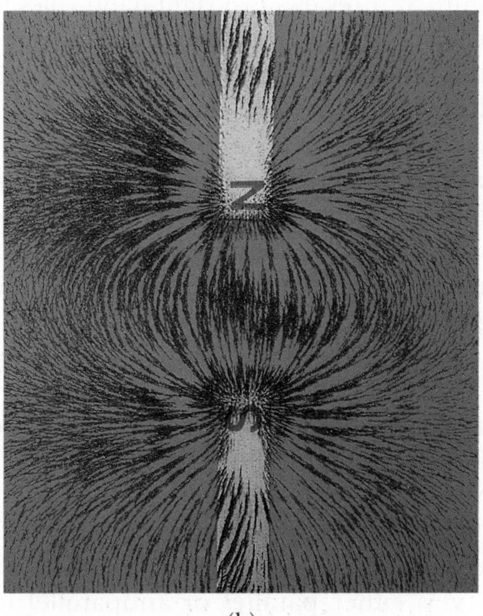

(b)

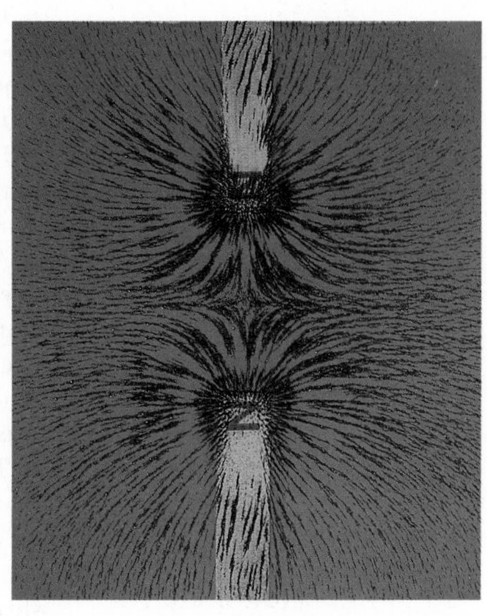

(c)

(Courtesy of Henry Leap and Jim Lehman)

FIGURE 22.2 (a) Magnetic field patterns surrounding a bar magnet as displayed with iron filings. (b) Magnetic field patterns between *dissimilar* poles of two bar magnets. (c) Magnetic field pattern between *similar* poles of two bar magnets.

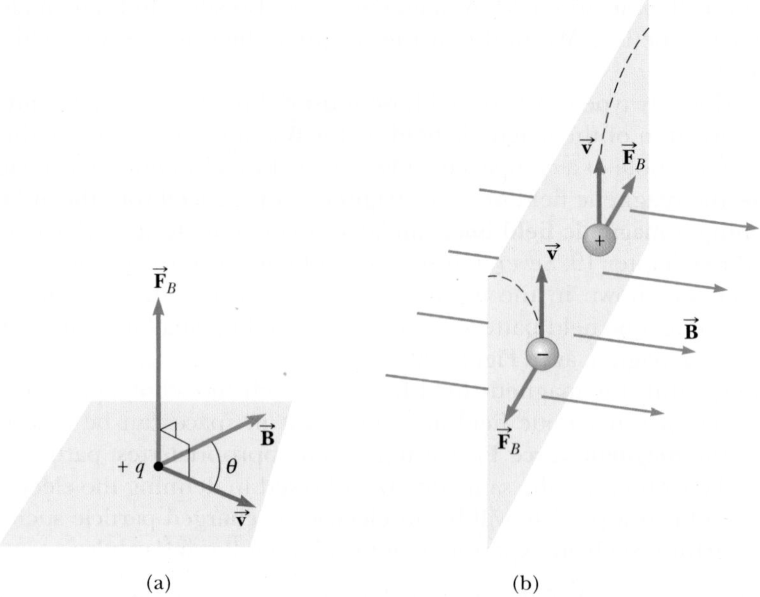

FIGURE 22.3 The direction of the magnetic force on a charged particle moving with a velocity $\vec{v}$ in the presence of a magnetic field $\vec{B}$. (a) When $\vec{v}$ is at an angle θ to $\vec{B}$, the magnetic force is perpendicular to both $\vec{v}$ and $\vec{B}$. (b) Oppositely directed magnetic forces are exerted on two oppositely charged particles moving with the same velocity in a magnetic field. The broken lines suggest the paths followed by the particles after the instant shown in the figure.

velocity of the particle and because its direction is perpendicular to both $\vec{v}$ and $\vec{B}$. Despite this complicated behavior, these observations can be summarized in a compact way by writing the magnetic force in the form

Magnetic force on a charged particle moving in a magnetic field

$$\vec{F}_B = q\vec{v} \times \vec{B} \qquad [22.1]$$

where the direction of the magnetic force is that of $\vec{v} \times \vec{B}$, which, by definition of the cross product, is perpendicular to both $\vec{v}$ and $\vec{B}$. Equation 22.1 is analogous to Equation 19.4, $\vec{F}_e = q\vec{E}$, but is clearly more complicated. We can regard Equation 22.1 as an operational definition of the magnetic field at a point in space. The SI unit of magnetic field is the **tesla** (T), where

$$1\ \text{T} = 1\ \text{N}\cdot\text{s/C}\cdot\text{m}$$

Figure 22.4 reviews two right-hand rules for determining the direction of the cross product $\vec{v} \times \vec{B}$ and determining the direction of $\vec{F}_B$. The rule in Figure 22.4a depends on our right-hand rule for the cross product in Figure 10.13. You point the four fingers of your right hand along the direction of $\vec{v}$ with the palm facing $\vec{B}$ and curl them toward $\vec{B}$. The extended thumb, which is at a right angle to the fingers, points in the direction of $\vec{v} \times \vec{B}$. Because $\vec{F}_B = q\vec{v} \times \vec{B}$, $\vec{F}_B$ is in the direction of your thumb if q is positive and opposite the direction of your thumb if q is negative.

A second rule is shown in Figure 22.4b. Here the thumb points in the direction of $\vec{v}$ and the extended fingers in the direction of $\vec{B}$. Now, the force $\vec{F}_B$ on a positive charge extends outward from your palm. The advantage of this rule is that the force on the charge is in the direction that you would push on something with your hand, outward from your palm. The force on a negative charge is in the opposite direction. Feel free to use either of these two right-hand rules.

The magnitude of the magnetic force is

$$F_B = |q|vB\sin\theta \qquad [22.2]$$

where θ is the angle between $\vec{v}$ and $\vec{B}$. From this expression, we see that F_B is zero when $\vec{v}$ is either parallel or antiparallel to $\vec{B}$ ($\theta = 0$ or $180°$). Furthermore, the force has its maximum value $F_B = |q|vB$ when $\vec{v}$ is perpendicular to $\vec{B}$ ($\theta = 90°$).

There are important differences between electric and magnetic forces on charged particles:

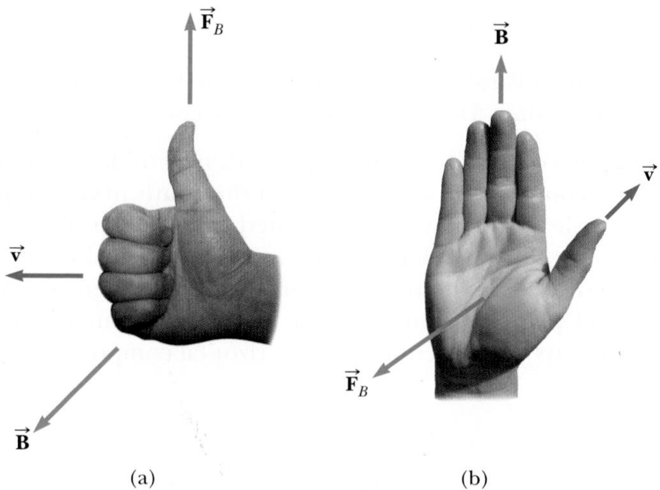

FIGURE 22.4 Two right-hand rules for determining the direction of the magnetic force $\vec{\mathbf{F}}_B = q\vec{\mathbf{v}} \times \vec{\mathbf{B}}$ acting on a particle with charge q moving with a velocity $\vec{\mathbf{v}}$ in a magnetic field $\vec{\mathbf{B}}$. (a) In this rule, the fingers point in the direction of $\vec{\mathbf{v}}$, with $\vec{\mathbf{B}}$ coming out of your palm, so that you can curl your fingers in the direction of $\vec{\mathbf{B}}$. The direction of $\vec{\mathbf{v}} \times \vec{\mathbf{B}}$, and the force on a positive charge, is the direction in which the thumb points. (b) In this rule, the vector $\vec{\mathbf{v}}$ is in the direction of your thumb and $\vec{\mathbf{B}}$ is in the direction of your fingers. The force $\vec{\mathbf{F}}_B$ on a positive charge is in the direction of your palm, as if you are pushing the particle with your hand.

- The electric force is always parallel or antiparallel to the direction of the electric field, whereas the magnetic force is perpendicular to the magnetic field.
- The electric force acts on a charged particle independent of the particle's velocity, whereas the magnetic force acts on a charged particle only when the particle is in motion and the force is proportional to the velocity.
- The electric force does work in displacing a charged particle, whereas the magnetic force associated with a constant magnetic field does *no* work when a charged particle is displaced.

This last statement is true because when a charge moves in a constant magnetic field, the magnetic force is always *perpendicular* to the displacement. That is, for a small displacement $d\vec{\mathbf{s}}$ of a particle, the work done by the magnetic force on the particle is $dW = \vec{\mathbf{F}}_B \cdot d\vec{\mathbf{s}} = (\vec{\mathbf{F}}_B \cdot \vec{\mathbf{v}})\,dt = 0$ because the magnetic force is a vector perpendicular to $\vec{\mathbf{v}}$. From this property and the work–kinetic energy theorem, we conclude that the kinetic energy of a charged particle *cannot* be altered by a constant magnetic field alone. In other words, when a charge moves with a velocity of $\vec{\mathbf{v}}$, an applied magnetic field can alter the direction of the velocity vector, but it cannot change the speed of the particle.

In Figures 22.3 and 22.4, we used green arrows to represent magnetic field vectors, which will be the convention in this book. In Active Figure 22.1, we represented the magnetic field of a bar magnet with green field lines. Studying magnetic fields presents a complication that we avoided in electric fields. In our study of electric fields, we drew all electric field vectors in the plane of the page or used perspective to represent them directed at an angle to the page. The cross product in Equation 22.1 requires us to think in three dimensions for problems in magnetism. Thus, in addition to drawing vectors pointing left or right and up or down, we will need a method of drawing vectors into or out of the page. These methods of representing the vectors are illustrated in Figure 22.5. A vector coming out of the page is represented by a dot, which we can think of as the tip of the arrowhead representing the vector coming through the paper toward us (Fig. 22.5a). A vector going into the page is represented by a cross, which we can think of as the tail feathers of an arrow going into the page (Fig. 22.5b). This depiction can be used for any type of vector we will encounter: magnetic field, velocity, force, and so on.

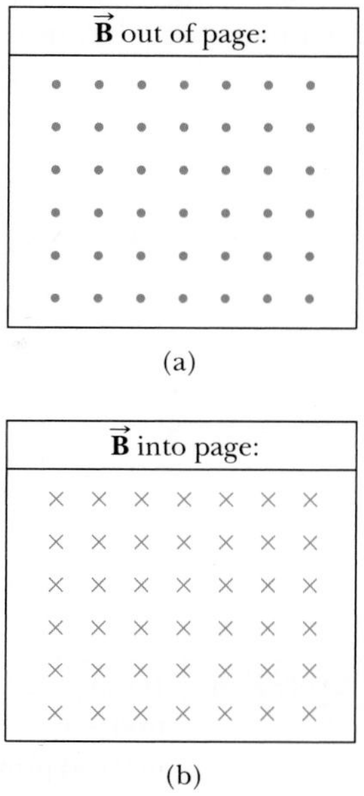

$\vec{\mathbf{B}}$ out of page:

(a)

$\vec{\mathbf{B}}$ into page:

(b)

FIGURE 22.5 (a) Magnetic field lines coming out of the paper are indicated by dots, representing the tips of arrows coming outward. (b) Magnetic field lines going into the paper are indicated by crosses, representing the feathers of arrows going inward.

QUICK QUIZ 22.1 An electron moves in the plane of this paper toward the top of the page. A magnetic field is also in the plane of the page and directed toward the right. What is the direction of the magnetic force on the electron? **(a)** toward the top of the page **(b)** toward the bottom of the page **(c)** toward the left edge of the page **(d)** toward the right edge of the page **(e)** upward out of the page **(f)** downward into the page

■ Thinking Physics 22.1

On a business trip to Australia, you take along your U.S.-made compass that you used in your Boy Scout days. Does this compass work correctly in Australia?

Reasoning Using the compass in Australia presents no problem. The north pole of the magnet in the compass will be attracted to the south magnetic pole near the north geographic pole, just as it was in the United States. The only difference in the magnetic field lines is that they have an upward component in Australia, whereas they have a downward component in the United States. When you hold the compass in a horizontal plane, it cannot detect the vertical component of the field, however; it only displays the direction of the horizontal component of the magnetic field. ■

EXAMPLE 22.1 An Electron Moving in a Magnetic Field

An electron in a television picture tube moves toward the front of the tube with a speed of 8.0×10^6 m/s along the *x* axis (Fig. 22.6). The neck of the tube is surrounded by a coil of wire that creates a magnetic field of magnitude 0.025 T, directed at an angle of 60° to the *x* axis and lying in the *xy* plane. Calculate the magnetic force on and acceleration of the electron.

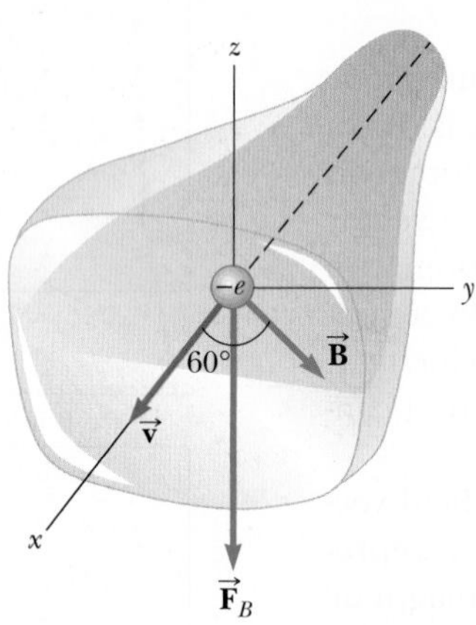

FIGURE 22.6 (Example 22.1) The magnetic force $\vec{\mathbf{F}}_B$ on the electron is in the negative *z* direction when $\vec{\mathbf{v}}$ and $\vec{\mathbf{B}}$ lie in the *xy* plane.

Solution Using Equation 22.2, we find the magnitude of the magnetic force:

$$F_B = |q|vB\sin\theta$$
$$= (1.60 \times 10^{-19}\ \text{C})(8.0 \times 10^6\ \text{m/s})(0.025\ \text{T})(\sin 60°)$$
$$= 2.8 \times 10^{-14}\ \text{N}$$

Because $\vec{\mathbf{v}} \times \vec{\mathbf{B}}$ is in the positive *z* direction (from the right-hand rule) and the charge is negative, $\vec{\mathbf{F}}_B$ is in the negative *z* direction.

Once we have determined the magnetic force, we have a Chapter 4 problem because the electron is a particle under a net force and the acceleration is determined from Newton's second law. The mass of the electron is $m_e = 9.1 \times 10^{-31}$ kg, and so its acceleration is

$$a = \frac{F_B}{m_e} = \frac{2.8 \times 10^{-14}\ \text{N}}{9.1 \times 10^{-31}\ \text{kg}} = 3.1 \times 10^{16}\ \text{m/s}^2$$

in the negative *z* direction.

22.3 MOTION OF A CHARGED PARTICLE IN A UNIFORM MAGNETIC FIELD

In Section 22.2, we found that the magnetic force acting on a charged particle moving in a magnetic field is perpendicular to the velocity of the particle and that, consequently, the work done on the particle by the magnetic force is zero. Consider now the special case of a positively charged particle moving in a uniform magnetic field when the initial velocity vector of the particle is perpendicular to the field. Let us assume that the direction of the magnetic field is into the page. Active Figure 22.7 shows that the particle moves in a circular path whose plane is perpendicular to the magnetic field.

The particle moves in this way because the magnetic force $\vec{\mathbf{F}}_B$ is perpendicular to $\vec{\mathbf{v}}$ and $\vec{\mathbf{B}}$ and has a constant magnitude qvB. As the force changes the direction of $\vec{\mathbf{v}}$, the direction of $\vec{\mathbf{F}}_B$ changes continuously as in Active Figure 22.7. Because $\vec{\mathbf{F}}_B$ always points toward the center of the circle, the particle can be modeled as being in uniform circular motion. As Active Figure 22.7 shows, the rotation is counterclockwise for a positive charge in a magnetic field directed into the page. If q were negative, the rotation would be clockwise. We can use Newton's second law to determine the radius of the circular path:

$$\sum F = F_B = ma$$

$$qvB = \frac{mv^2}{r}$$

$$r = \frac{mv}{qB} \qquad [22.3]$$

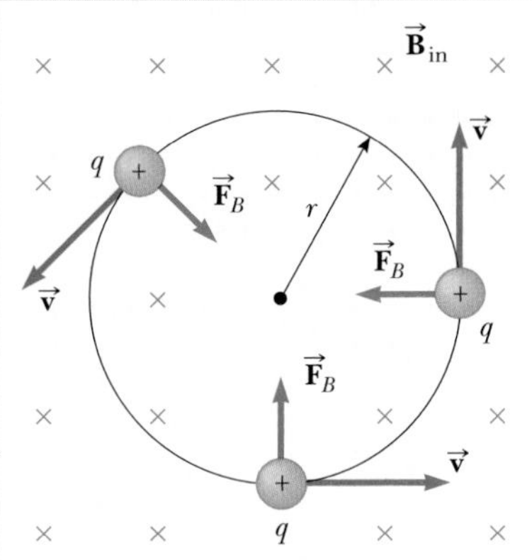

ACTIVE FIGURE 22.7
When the velocity of a charged particle is perpendicular to a uniform magnetic field, the particle moves in a circular path in a plane perpendicular to $\vec{\mathbf{B}}$. The magnetic force $\vec{\mathbf{F}}_B$ acting on the charge is always directed toward the center of the circle.

PhysicsNow™ Log into PhysicsNow at **www.pop4e.com** and go to Active Figure 22.7. You can adjust the mass, speed, and charge of the particle and the magnitude of the magnetic field to observe the resulting circular motion.

That is, the radius of the path is proportional to the linear momentum mv of the particle and inversely proportional to the magnitude of the charge on the particle and to the magnitude of the magnetic field. The angular speed of the particle is (from Eq. 10.10)

$$\omega = \frac{v}{r} = \frac{qB}{m} \qquad [22.4]$$

The period of the motion (the time interval required for the particle to complete one revolution) is equal to the circumference of the circular path divided by the speed of the particle:

$$T = \frac{2\pi r}{v} = \frac{2\pi}{\omega} = \frac{2\pi m}{qB} \qquad [22.5]$$

These results show that the angular speed of the particle and the period of the circular motion do not depend on the translational speed of the particle or the radius of the orbit for a given particle in a given uniform magnetic field. The angular speed ω is often referred to as the **cyclotron frequency** because charged particles circulate at this angular speed in one type of accelerator called a *cyclotron,* discussed in Section 22.4.

If a charged particle moves in a uniform magnetic field with its velocity at some arbitrary angle to $\vec{\mathbf{B}}$, its path is a helix. For example, if the field is in the x direction as in Active Figure 22.8, there is no component of force on the particle in the x direction. As a result, $a_x = 0$, and so the x component of velocity of the particle remains constant. The magnetic force $q\vec{\mathbf{v}} \times \vec{\mathbf{B}}$ causes the components v_y and v_z to change in time, however, and the resulting motion of the particle is a helix having its axis parallel to the magnetic field. The projection of the path onto the yz plane (viewed along the x axis) is a circle. (The projections of the path onto the xy and xz planes are sinusoids!) Equations 22.3 to 22.5 still apply provided that v is replaced by $v_\perp = \sqrt{v_y^2 + v_z^2}$.

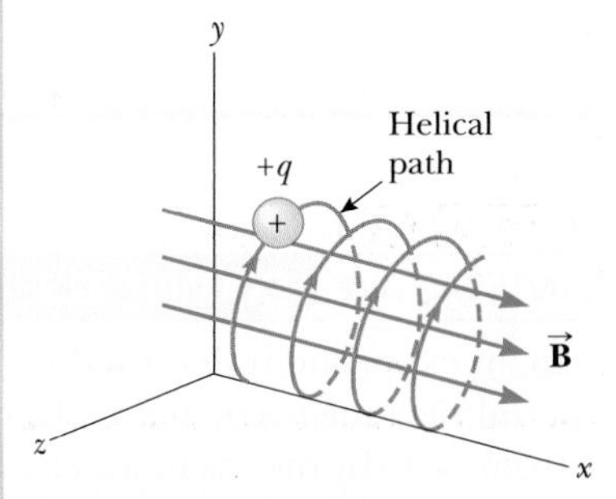

ACTIVE FIGURE 22.8
A charged particle having a velocity vector with a component parallel to a uniform magnetic field moves in a helical path.

PhysicsNow™ Log into PhysicsNow at **www.pop4e.com** and go to Active Figure 22.8. You can adjust the x component of the velocity of the particle and observe the resulting helical motion.

QUICK QUIZ 22.2 **(i)** A charged particle is moving perpendicular to a magnetic field in a circle with a radius r. The magnitude of the magnetic field is increased. Compared with the initial radius of the circular path, is the radius of the new path **(a)** smaller, **(b)** larger, or **(c)** equal in size? **(ii)** An identical particle enters the field, with $\vec{\mathbf{v}}$ perpendicular to $\vec{\mathbf{B}}$, but with a higher speed v than the first particle. Compared with the radius of the circle for the first particle in the same magnetic field, is the radius of the circle for the second particle **(a)** smaller, **(b)** larger, or **(c)** equal in size?

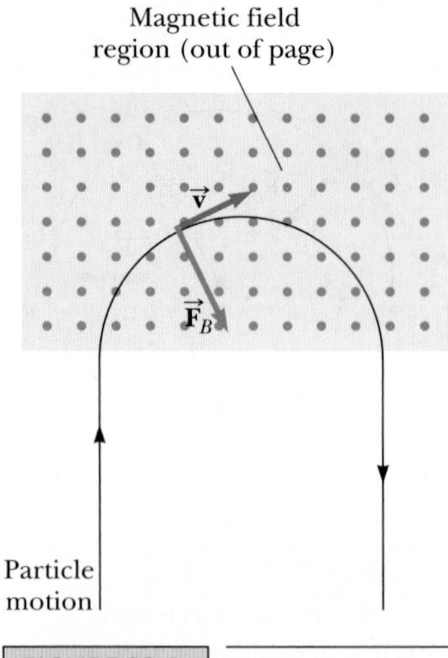

FIGURE 22.9 (Thinking Physics 22.2) A positively charged particle enters a region of magnetic field directed out of the page.

Thinking Physics 22.2

Suppose a uniform magnetic field exists in a finite region of space as in Figure 22.9. Can you inject a charged particle into this region and have it stay trapped in the region by the magnetic force?

Reasoning Consider separately the components of the particle velocity parallel and perpendicular to the field lines in the region. For the component parallel to the field lines, no force is exerted on the particle and it continues to move with the parallel component until it leaves the region of the magnetic field. Now consider the component perpendicular to the field lines. This component results in a magnetic force that is perpendicular to both the field lines and the velocity component. As discussed earlier, if the force acting on a charged particle is always perpendicular to its velocity, the particle moves in a circular path. Thus, the particle follows half of a circular arc and exits the field on the other side of the circle, as shown in Figure 22.9. Therefore, a particle injected into a uniform magnetic field cannot stay trapped in the field region. ■

EXAMPLE 22.2 A Proton Moving Perpendicular to a Uniform Magnetic Field

A proton is moving in a circular orbit of radius 14.0 cm in a uniform 0.350-T magnetic field directed perpendicular to the velocity of the proton.

A Find the translational speed of the proton.

Solution From Equation 22.3, we find that

$$v = \frac{qBr}{m_p} = \frac{(1.60 \times 10^{-19}\text{ C})(0.350\text{ T})(14.0 \times 10^{-2}\text{ m})}{1.67 \times 10^{-27}\text{ kg}}$$

$$= 4.69 \times 10^{6}\text{ m/s}$$

B Find the period of the circular motion of the proton.

Solution From Equation 22.5,

$$T = \frac{2\pi m_p}{qB} = \frac{2\pi(1.67 \times 10^{-27}\text{ kg})}{(1.60 \times 10^{-19}\text{ C})(0.350\text{ T})}$$

$$= 1.87 \times 10^{-7}\text{ s}$$

INTERACTIVE EXAMPLE 22.3 Bending an Electron Beam

In an experiment designed to measure the strength of a uniform magnetic field, electrons are accelerated from rest (by means of an electric field) through a potential difference of 350 V. After leaving the region of the electric field, the electrons enter a magnetic field and travel along a curved path because of the magnetic force exerted on them. The radius of the path is measured to be 7.50 cm. Figure 22.10 shows such a curved beam of electrons.

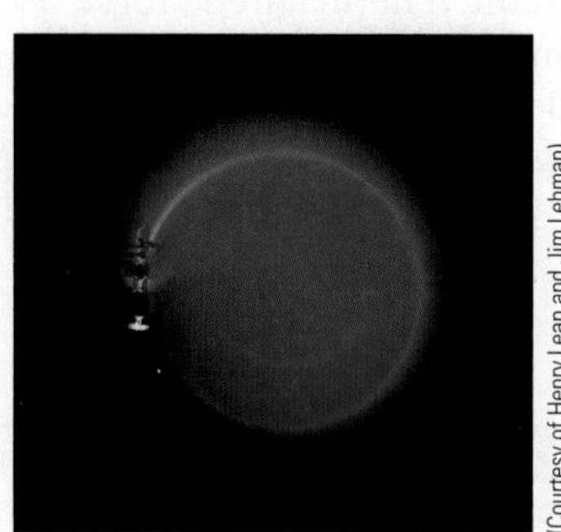

FIGURE 22.10 (Interactive Example 22.3) The bending of an electron beam in a magnetic field.

A Assuming that the magnetic field is perpendicular to the beam, what is the magnitude of the field?

Solution The drawing in Active Figure 22.7 and the photograph in Figure 22.10 help us conceptualize the circular motion of the electrons. We categorize this problem as one in which we will use our understanding of uniform circular motion along with our knowledge of the magnetic force. Looking at Equation 22.3, we see that we need the speed v of the electron if we are to find the magnetic field magnitude, and v is not given. Consequently, we must find the speed of the electron based on the potential difference through which it is accelerated. We can therefore also categorize this problem as one in which we must apply conservation of mechanical energy for an isolated system. We begin to analyze the problem by finding the electron speed. For the isolated electron–electric field system, the loss of potential energy as the electron moves through the 350-V potential difference appears as an increase in the kinetic energy of the electron. Because $K_i = 0$ and $K_f = \frac{1}{2}m_e v^2$, we have

$$\Delta K + \Delta U = 0 \quad \rightarrow \quad \tfrac{1}{2}m_e v^2 + (-e)\,\Delta V = 0$$

$$v = \sqrt{\frac{2e\,\Delta V}{m_e}} = \sqrt{\frac{2(1.60 \times 10^{-19}\text{ C})(350\text{ V})}{9.11 \times 10^{-31}\text{ kg}}}$$

$$= 1.11 \times 10^7\text{ m/s}$$

Now, using Equation 22.3, we find that

$$B = \frac{m_e v}{er} = \frac{(9.11 \times 10^{-31}\text{ kg})(1.11 \times 10^7\text{ m/s})}{(1.60 \times 10^{-19}\text{ C})(0.075\text{ m})}$$

$$= 8.4 \times 10^{-4}\text{ T}$$

B What is the angular speed of the electrons?

Solution Using Equation 22.4, we find that

$$\omega = \frac{v}{r} = \frac{1.11 \times 10^7\text{ m/s}}{0.075\text{ m}} = 1.5 \times 10^8\text{ rad/s}$$

To finalize this problem, note that the angular speed can be written as $\omega = (1.5 \times 10^8\text{ rad/s})(1\text{ rev}/2\pi\text{ rad}) = 2.4 \times 10^7\text{ rev/s}$. The electrons travel around the circle 24 million times per second! This very high speed is consistent with what we found in part A.

PhysicsNow™ By logging into PhysicsNow at www.pop4e.com and going to Interactive Example 22.3, you can investigate the relationship between the radius of the circular path of the electrons and the magnetic field.

22.4 APPLICATIONS INVOLVING CHARGED PARTICLES MOVING IN A MAGNETIC FIELD

A charge moving with velocity $\vec{\mathbf{v}}$ in the presence of an electric field $\vec{\mathbf{E}}$ and a magnetic field $\vec{\mathbf{B}}$ experiences both an electric force $q\vec{\mathbf{E}}$ and a magnetic force $q\vec{\mathbf{v}} \times \vec{\mathbf{B}}$. The total force, called the **Lorentz force,** acting on the charge is therefore the vector sum,

$$\vec{\mathbf{F}} = q\vec{\mathbf{E}} + q\vec{\mathbf{v}} \times \vec{\mathbf{B}} \qquad [22.6]$$

In this section, we look at three applications involving particles experiencing the Lorentz force.

Velocity Selector

In many experiments involving moving charged particles, it is important to have particles that all move with essentially the same velocity. That can be achieved by applying a combination of an electric field and a magnetic field oriented as shown in Active Figure 22.11a. A uniform electric field is directed vertically downward (in the plane of the page in Active Fig. 22.11a), and a uniform magnetic field is applied perpendicular to the electric field (into the page in Active Fig. 22.11a). Particles moving through this region will experience the Lorentz force, given by Equation 22.6. For a positively charged particle, the magnetic force $q\vec{\mathbf{v}} \times \vec{\mathbf{B}}$ is upward and the electric force $q\vec{\mathbf{E}}$ is downward. When the magnitudes of the two fields are chosen so that $qE = qvB$, the particle is in equilibrium (Active Fig. 22.11b) and moves in a straight horizontal line through the region of the fields. From $qE = qvB$ we find that

$$v = \frac{E}{B} \qquad [22.7]$$

Only those particles having this speed are undeflected as they move through the perpendicular electric and magnetic fields and pass through a small opening at the end of the device. The magnetic force exerted on particles moving faster than this

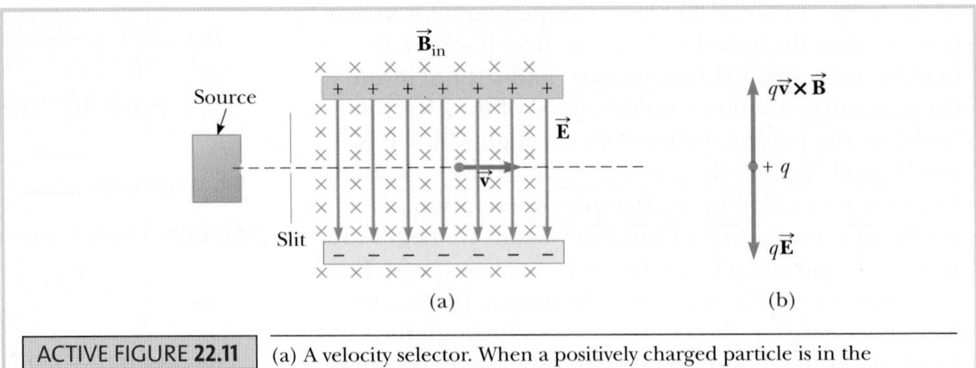

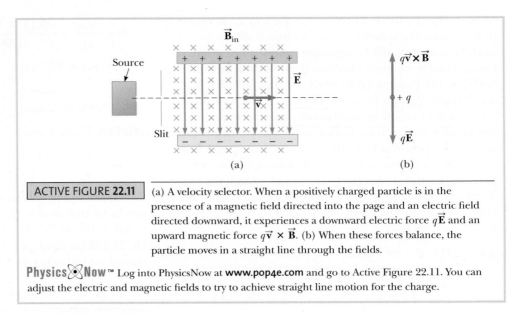

ACTIVE FIGURE 22.11 (a) A velocity selector. When a positively charged particle is in the presence of a magnetic field directed into the page and an electric field directed downward, it experiences a downward electric force $q\vec{E}$ and an upward magnetic force $q\vec{v} \times \vec{B}$. (b) When these forces balance, the particle moves in a straight line through the fields.

PhysicsNow™ Log into PhysicsNow at **www.pop4e.com** and go to Active Figure 22.11. You can adjust the electric and magnetic fields to try to achieve straight line motion for the charge.

speed is stronger than the electric force, and these particles are deflected upward. Those moving slower are deflected downward.

The Mass Spectrometer

A **mass spectrometer** separates ions according to their mass-to-charge ratio. In one version, known as the *Bainbridge mass spectrometer,* a beam of ions first passes through a velocity selector and then enters a second region with no electric field and a uniform magnetic field $\vec{B}_0$ that has the same direction as the magnetic field in the selector (Active Fig. 22.12). On entering the second magnetic field, the ions move in a semicircle of radius r before striking a detector array at P. If the ions are positively charged, the beam deflects upward as in Active Figure 22.12. If the ions are negatively charged, the beam deflects downward. From Equation 22.3, we can express the ratio m/q as

$$\frac{m}{q} = \frac{rB_0}{v}$$

Using Equation 22.7, we find that

$$\frac{m}{q} = \frac{rB_0 B}{E} \qquad [22.8]$$

ACTIVE FIGURE 22.12 A mass spectrometer. Positively charged particles are sent first through a velocity selector and then into a region where the magnetic field $\vec{B}_0$ causes the particles to move in a semicircular path and strike a detector array at P.

PhysicsNow™ Log into PhysicsNow at **www.pop4e.com** and go to Active Figure 22.12 to predict where particles will strike the detector array.

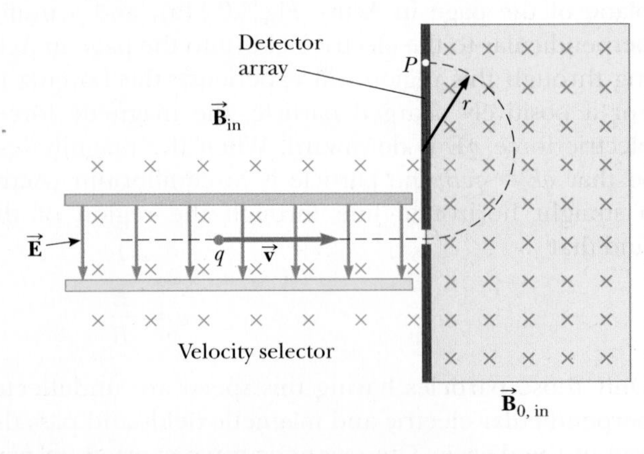

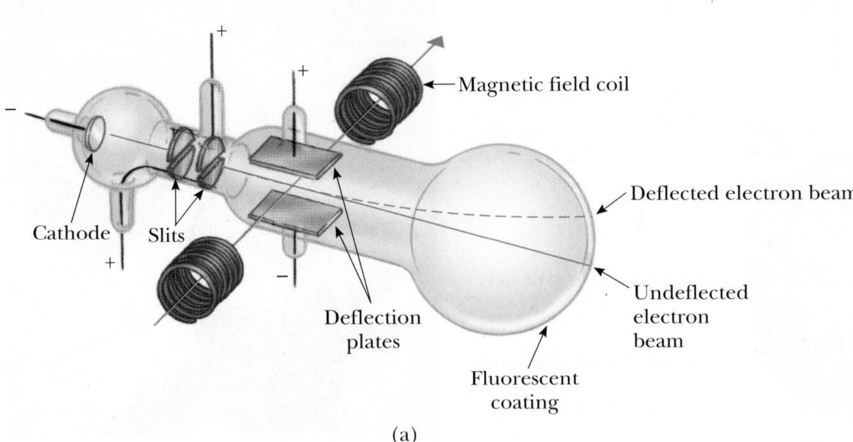

(a)

(b)

FIGURE 22.13 (a) Thomson's apparatus for measuring e/m_e. Electrons are accelerated from the cathode, pass through two slits, and are deflected by both an electric field and a magnetic field (directed perpendicular to the electric field). The electrons then strike a fluorescent screen. (b) J. J. Thomson *(left)* in the Cavendish Laboratory, University of Cambridge. It is interesting to note that the man on the right, Frank Baldwin Jewett, is a distant relative of John W. Jewett Jr., coauthor of this text.

Therefore, m/q can be determined by measuring the radius of curvature and knowing the field magnitudes B, B_0, and E. In practice, one usually measures the masses of various isotopes of a given ion, with the ions all carrying the same charge q. In this way, the mass ratios can be determined even if q is unknown.

A variation of this technique was used by J. J. Thomson (1856–1940) in 1897 to measure the ratio e/m_e for electrons. Figure 22.13a shows the basic apparatus he used. Electrons are accelerated from the cathode and pass through two slits. They then drift into a region of perpendicular electric and magnetic fields. The magnitudes of the two fields are first adjusted to produce an undeflected beam. When the magnetic field is turned off, the electric field produces a measurable beam deflection that is recorded on the fluorescent screen. From the size of the deflection and the measured values of E and B, the charge-to-mass ratio can be determined. The results of this crucial experiment represent the discovery of the electron as a fundamental particle of nature.

The Cyclotron

A **cyclotron** can accelerate charged particles to very high speeds. Both electric and magnetic forces play a key role in its operation. The energetic particles produced are used to bombard atomic nuclei and thereby produce nuclear reactions of interest to researchers. A number of hospitals use cyclotron facilities to produce radioactive substances for diagnosis and treatment as well as beams of high-energy particles for treating cancer. For example, retinoblastoma, a cancer of the eye, can be treated with a series of beams of protons from a cyclotron.

 Use of cyclotrons in medicine

A schematic drawing of a cyclotron is shown in Figure 22.14a. The charges move inside two hollow metal semicircular containers, D_1 and D_2, referred to as *dees* because they are shaped like the letter D. A high-frequency alternating potential difference is applied to the dees, and a uniform magnetic field is directed perpendicular to them. A positive ion released at P near the center of the magnet moves in a semicircular path in one dee (indicated by the dashed red line in the drawing) and arrives back at the gap in a time interval $T/2$, where T is the time interval needed to make one complete trip around the two dees, given by Equation 22.5. The frequency of the applied potential difference is chosen so that the polarity of the dees is reversed during the time interval in which the ion travels around one dee. If the applied potential difference is adjusted such that D_2 is at a lower electric potential

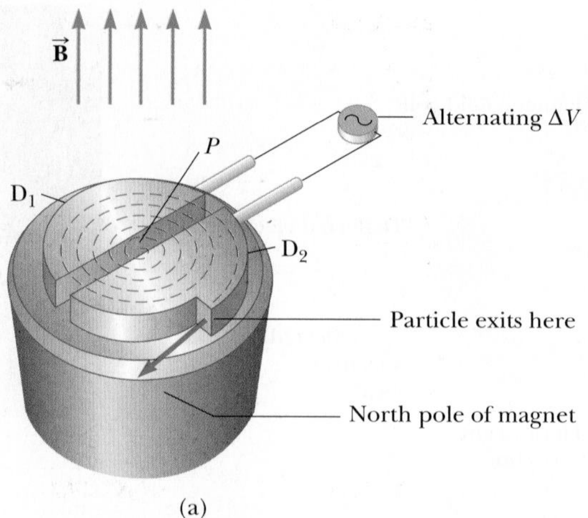

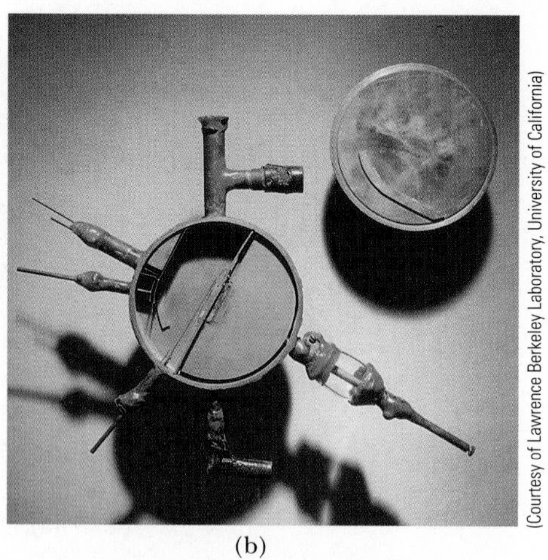

(Courtesy of Lawrence Berkeley Laboratory, University of California)

FIGURE 22.14 (a) A cyclotron consists of an ion source at P, two hollow sections called dees, D_1 and D_2, across which an alternating potential difference is applied, and a uniform magnetic field. (The south pole of the magnet is not shown.) The red dashed curved lines represent the path of the particles. (b) The first cyclotron, invented by E. O. Lawrence and M. S. Livingston in 1934.

than D_1 by an amount ΔV, the ion accelerates across the gap to D_2 and its kinetic energy increases by an amount $q\Delta V$. It then moves around D_2 in a semicircular path of larger radius (because its speed has increased). After a time interval $T/2$, it again arrives at the gap between the dees. By this time, the polarity across the dees has reversed again and the ion is given another "kick" across the gap. The motion continues so that for each half-circle trip, the ion gains additional kinetic energy equal to $q\,\Delta V$. When the radius of its path is nearly that of the dees, the energetic ion leaves the system through the exit slit. It is important to note that the operation of the cyclotron is based on T being independent of the speed of the ion and the radius of its circular path (Eq. 22.5).

We can obtain an expression for the kinetic energy of the ion when it exits from the cyclotron in terms of the radius R of the dees. From Equation 22.3 we know that $v = qBR/m$. Hence, the kinetic energy is

$$K = \tfrac{1}{2}mv^2 = \frac{q^2B^2R^2}{2m} \qquad [22.9]$$

When the energy of the ions in a cyclotron exceeds about 20 MeV, relativistic effects come into play. For this reason, the moving ions do not remain in phase with the applied potential difference. Some accelerators solve this problem by modifying the frequency of the applied potential difference so that it remains in phase with the moving ions.

PITFALL PREVENTION 22.1

THE CYCLOTRON IS NOT STATE-OF-THE-ART TECHNOLOGY The cyclotron is important historically because it was the first particle accelerator to achieve very high particle speeds. Cyclotrons are still used, for example, to accelerate particles to high speeds in medical applications. The cyclotron is also important as an application of the ideas in this discussion. Most accelerators built recently and currently in use in research, however, are not cyclotrons. They work on a different principle and are generally called *synchrotrons*.

22.5 MAGNETIC FORCE ON A CURRENT-CARRYING CONDUCTOR

Because a magnetic force is exerted on a single charged particle when it moves through an external magnetic field, it should not surprise you to find that a current-carrying wire also experiences a magnetic force when placed in an external magnetic field because the current represents a collection of many charged particles in motion. Hence, the resultant magnetic force on the wire is due to the sum of the individual magnetic forces on the charged particles. The force on the particles is transmitted to the "bulk" of the wire through collisions with the atoms making up the wire.

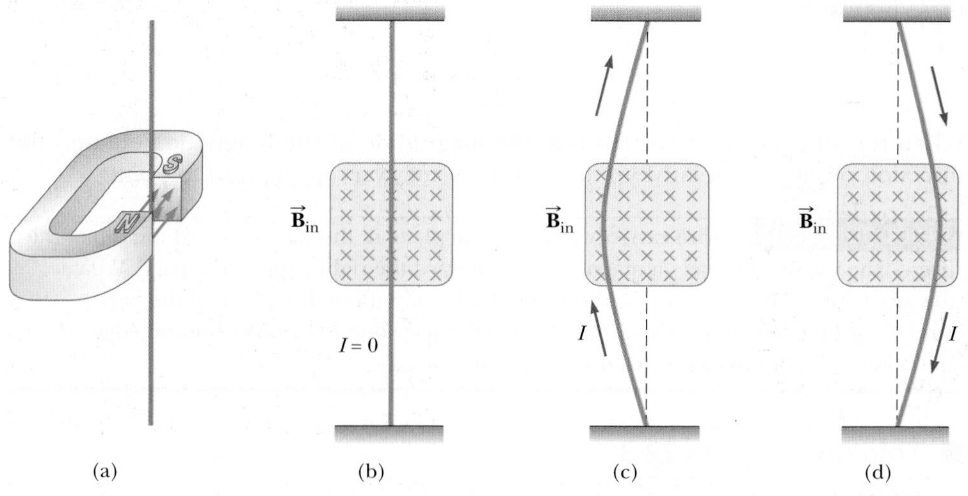

FIGURE 22.15 (a) A wire suspended vertically between the poles of a magnet. (b) The setup shown in (a) as seen looking at the south pole of the magnet so that the magnetic field (green crosses) is directed into the page. When no current is flowing in the wire, it remains vertical. (c) When the current is upward, the wire deflects to the left. (d) When the current is downward, the wire deflects to the right.

The magnetic force on a current-carrying conductor can be demonstrated by hanging a wire between the poles of a magnet as in Figure 22.15, where the magnetic field is directed into the page. The wire deflects to the left or right when a current is passed through it.

Let us quantify this discussion by considering a straight segment of wire of length ℓ and cross-sectional area A, carrying a current I in a uniform external magnetic field $\vec{\mathbf{B}}$ as in Figure 22.16. As a simplification model, we shall ignore the high-speed zigzag motion of the charges in the wire (which is valid because the net velocity associated with this motion is zero) and assume that the charges simply move with the drift velocity $\vec{\mathbf{v}}_d$. The magnetic force on a charge q moving with drift velocity $\vec{\mathbf{v}}_d$ is $q\vec{\mathbf{v}}_d \times \vec{\mathbf{B}}$. To find the total magnetic force on the wire segment, we multiply the magnetic force on one charge by the number of charges in the segment. Because the volume of the segment is $A\ell$, the number of charges in the segment is $nA\ell$, where n is the number of charges per unit volume. Hence, the total magnetic force on the wire of length ℓ is

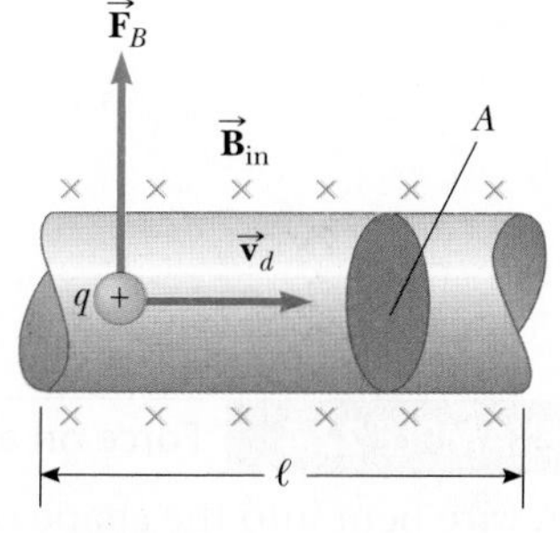

FIGURE 22.16 A section of a wire containing moving charges in a magnetic field $\vec{\mathbf{B}}$. The magnetic force on each charge is $q\vec{\mathbf{v}}_d \times \vec{\mathbf{B}}$, and the net force on a segment of length ℓ is $I\vec{\boldsymbol{\ell}} \times \vec{\mathbf{B}}$.

$$\vec{\mathbf{F}}_B = (q\vec{\mathbf{v}}_d \times \vec{\mathbf{B}})nA\ell$$

This equation can be written in a more convenient form by noting that, from Equation 21.4, the current in the wire is $I = nqv_dA$. Therefore, $\vec{\mathbf{F}}_B$ can be expressed as

$$\vec{\mathbf{F}}_B = I\vec{\boldsymbol{\ell}} \times \vec{\mathbf{B}} \qquad [22.10]$$

■ **Magnetic force on a current-carrying conductor**

where $\vec{\boldsymbol{\ell}}$ is a vector in the direction of the current I; the magnitude of $\vec{\boldsymbol{\ell}}$ equals the length of the segment. Note that this expression applies only to a straight segment of wire in a uniform external magnetic field.

Now consider an arbitrarily shaped wire of uniform cross-section in an external magnetic field as in Figure 22.17. It follows from Equation 22.10 that the magnetic force on a very small segment of the wire of length ds in the presence of an external field $\vec{\mathbf{B}}$ is

$$d\vec{\mathbf{F}}_B = I\,d\vec{\mathbf{s}} \times \vec{\mathbf{B}} \qquad [22.11]$$

where $d\vec{\mathbf{s}}$ is a vector representing the length segment, with its direction the same as that of the current, and $d\vec{\mathbf{F}}_B$ is directed out of the page for the directions assumed in Figure 22.17. We can consider Equation 22.11 as an alternative definition of $\vec{\mathbf{B}}$ to Equation 22.1. That is, the field $\vec{\mathbf{B}}$ can be defined in terms of a measurable force on a current element, where the force is a maximum when $\vec{\mathbf{B}}$ is perpendicular to the element and zero when $\vec{\mathbf{B}}$ is parallel to the element.

To obtain the total magnetic force $\vec{\mathbf{F}}_B$ on a length of the wire between arbitrary points a and b, we integrate Equation 22.11 over the length of the wire between

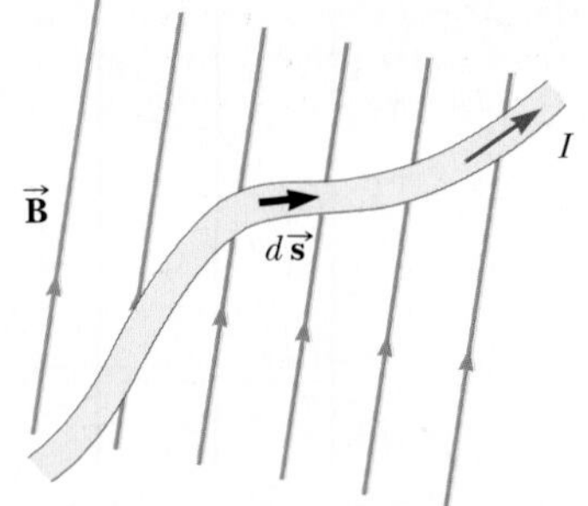

FIGURE 22.17 A wire segment of arbitrary shape carrying a current I in a magnetic field $\vec{\mathbf{B}}$ experiences a magnetic force. The force on any length element $d\vec{\mathbf{s}}$ is $I\,d\vec{\mathbf{s}} \times d\vec{\mathbf{B}}$ and is directed out of the page.

these points:

$$\vec{\mathbf{F}}_B = I\int_a^b d\vec{\mathbf{s}} \times \vec{\mathbf{B}} \qquad [22.12]$$

When this integration is carried out, the magnitude of the magnetic field and the direction of the field relative to the vector $d\vec{\mathbf{s}}$ may vary from point to point.

QUICK QUIZ 22.3 A wire carries current in the plane of this paper toward the top of the page. The wire experiences a magnetic force toward the right edge of the page. What is the direction of the magnetic field causing this force? **(a)** in the plane of the page and toward the left edge **(b)** in the plane of the page and toward the bottom edge **(c)** upward out of the page **(d)** downward into the page

Thinking Physics 22.3

In a lightning stroke, negative charge rapidly moves from a cloud to the ground. In what direction is a lightning stroke deflected by the Earth's magnetic field?

Reasoning The downward flow of negative charge in a lightning stroke is equivalent to an upward-moving current. Thus, the vector $d\vec{\mathbf{s}}$ is upward, and the magnetic field vector has a northward component. According to the cross product of the length element and magnetic field vectors (Eq. 22.11), the lightning stroke would be deflected to the *west.* ■

EXAMPLE 22.4 Force on a Semicircular Conductor

A wire bent into the shape of a semicircle of radius R forms a closed circuit and carries a current I. The circuit lies in the xy plane, and a uniform magnetic field is present along the positive y axis as in Figure 22.18. Find the magnetic force on the straight portion of the wire and on the curved portion.

Solution The force on the straight portion of the wire has a magnitude $F_1 = I\ell B = 2IRB$ because $\ell = 2R$ and the wire is perpendicular to $\vec{\mathbf{B}}$. The direction of $\vec{\mathbf{F}}_1$ is out of the paper because $\vec{\boldsymbol{\ell}} \times \vec{\mathbf{B}}$ is outward. (That is, $\vec{\boldsymbol{\ell}}$ is to the right, in the direction of the current, and so by the rule of cross products, $\vec{\boldsymbol{\ell}} \times \vec{\mathbf{B}}$ is outward.)

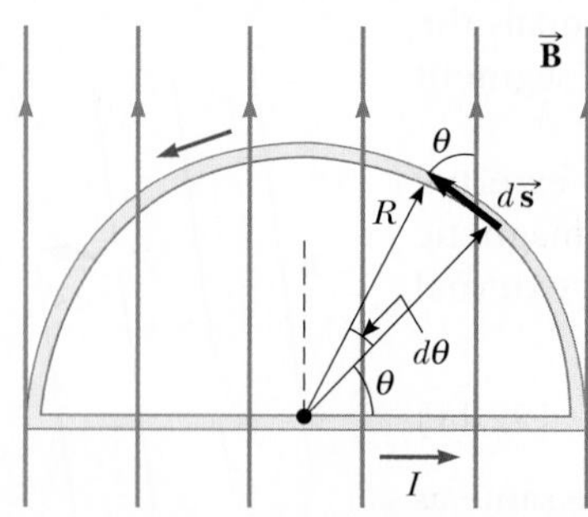

FIGURE 22.18 (Example 22.4) The net force on a closed current loop in a uniform magnetic field is zero. For the loop shown here, the force on the straight portion is $2IRB$ and out of the page, whereas the force on the curved portion is $2IRB$ and into the page.

To find the magnetic force on the curved part, we first write an expression for the magnetic force $d\vec{\mathbf{F}}_2$ on the element $d\vec{\mathbf{s}}$. If θ is the angle between $\vec{\mathbf{B}}$ and $d\vec{\mathbf{s}}$ in Figure 22.18, the magnitude of $d\vec{\mathbf{F}}_2$ is

$$dF_2 = I|d\vec{\mathbf{s}} \times \vec{\mathbf{B}}| = IB\sin\theta\, ds$$

To integrate this expression, we express ds in terms of θ. Because $s = R\theta$, $ds = R\,d\theta$, and the expression for dF_2 can be written as

$$dF_2 = IRB\sin\theta\, d\theta$$

To obtain the total magnetic force F_2 on the curved portion, we integrate this expression to account for contributions from all elements. Note that the direction of the magnetic force on every element is the same: into the paper (because $d\vec{\mathbf{s}} \times \vec{\mathbf{B}}$ is inward). Therefore, the resultant magnetic force $\vec{\mathbf{F}}_2$ on the curved wire must also be into the paper. Integrating dF_2 over the limits $\theta = 0$ to $\theta = \pi$ (i.e., the entire semicircle) gives

$$F_2 = IRB\int_0^{\pi}\sin\theta\, d\theta = IRB\Big[-\cos\theta\Big]_0^{\pi}$$

$$= -IRB(\cos\pi - \cos 0) = -IRB(-1-1) = 2IRB$$

Because $F_2 = 2IRB$ and the vector $\vec{\mathbf{F}}_2$ is directed into the paper and because the force on the straight wire has magnitude $F_1 = 2IRB$ and is out of the paper, we see that the net magnetic force on the closed loop is zero.

22.6 TORQUE ON A CURRENT LOOP IN A UNIFORM MAGNETIC FIELD

In the preceding section, we showed how a magnetic force is exerted on a current-carrying conductor when the conductor is placed in an external magnetic field. Starting at this point, we shall show that a *torque* is exerted on a current loop placed in a magnetic field. The results of this analysis are of great practical value in the design of motors and generators.

Consider a rectangular loop carrying a current I in the presence of a uniform external magnetic field *in the plane of the loop* as in Figure 22.19a. The magnetic forces on sides ① and ③, of length b, are zero because these wires are parallel to the field; hence, $d\vec{s} \times \vec{B} = 0$ for these sides. Nonzero magnetic forces act on sides ② and ④, however, because these sides are oriented perpendicular to the field. The magnitude of these forces is

$$F_2 = F_4 = IaB$$

We see that the net force on the loop is zero. The direction of $\vec{F}_2$, the magnetic force on side ②, is out of the paper, and that of $\vec{F}_4$, the magnetic force on side ④, is into the paper. If we view the loop from side ③ as in Figure 22.19b, we see the forces on ② and ④ directed as shown. If we assume that the loop is pivoted so that it can rotate about an axis perpendicular to the page and passing through point O, we see that these two magnetic forces produce a net torque about this axis that rotates the loop clockwise. The magnitude of the torque, which we will call τ_{max}, is

$$\tau_{max} = F_2 \frac{b}{2} + F_4 \frac{b}{2} = (IaB)\frac{b}{2} + (IaB)\frac{b}{2} = IabB$$

where the moment arm about this axis is $b/2$ for each force. Because the area of the loop is $A = ab$, the magnitude of the torque can be expressed as

$$\tau_{max} = IAB \qquad [22.13]$$

Remember that this torque occurs only when the field $\vec{B}$ is parallel to the plane of the loop. The sense of the rotation is clockwise when the loop is viewed as in Figure 22.19b. If the current were reversed, the magnetic forces would reverse their directions and the rotational tendency would be counterclockwise.

Now suppose the uniform magnetic field makes an angle θ with a line perpendicular to the plane of the loop as in Active Figure 22.20. For convenience, we shall assume that $\vec{B}$ is perpendicular to sides ② and ④. (The end view of these sides is shown in Active Fig. 22.20.) In this case, the magnetic forces on sides ① and ③

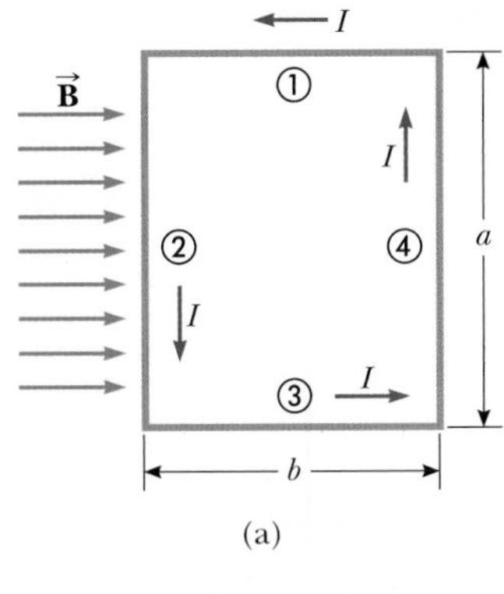

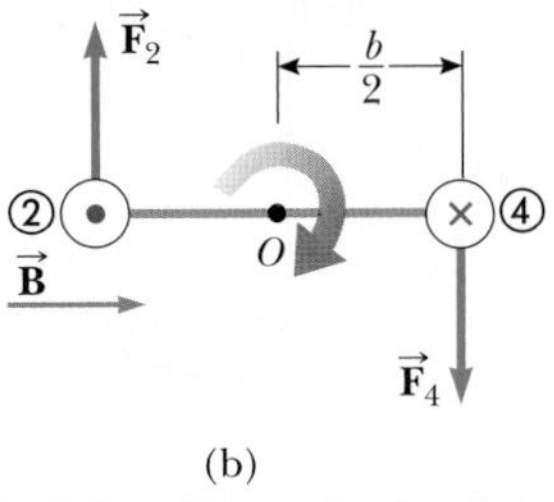

FIGURE 22.19 (a) Overhead view of a rectangular current loop in a uniform magnetic field. No magnetic forces are exerted on sides ① and ③ because these sides are parallel to $\vec{B}$. Forces are exerted on sides ② and ④, however. (b) Edge view of the loop sighting down ② and ④ shows that the forces $\vec{F}_2$ and $\vec{F}_4$ exerted on these sides create a torque that tends to rotate the loop clockwise. The purple dot in the left circle represents current in wire ② coming toward you; the purple × in the right circle represents current in wire ④ moving away from you.

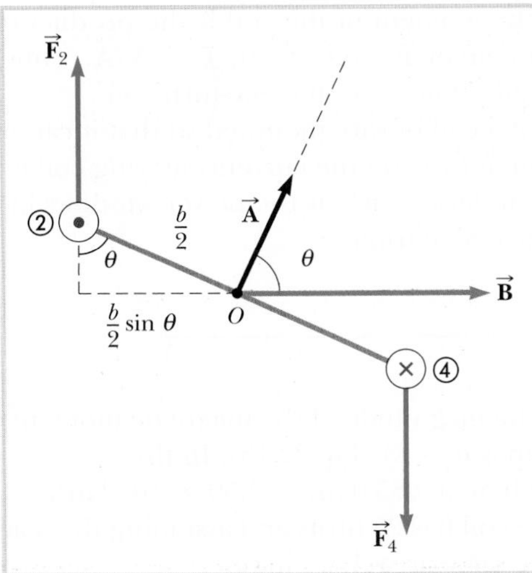

ACTIVE FIGURE 22.20
An end view of the loop in Figure 22.19b rotated through an angle with respect to the magnetic field. If $\vec{B}$ is at an angle θ with respect to vector $\vec{A}$, which is perpendicular to the plane of the loop, the torque is $IAB \sin \theta$.

Physics Now™ Log into PhysicsNow at **www.pop4e.com** and go to Active Figure 22.20. You can choose the current in the loop, the magnetic field, and the initial orientation of the loop and observe the subsequent motion.

cancel each other and produce no torque because they have the same line of action. The magnetic forces $\vec{\mathbf{F}}_2$ and $\vec{\mathbf{F}}_4$ acting on sides ② and ④, however, both produce a torque about an axis through the center of the loop. Referring to Active Figure 22.20, we note that the moment arm of $\vec{\mathbf{F}}_2$ about this axis is $(b/2)\sin\theta$. Likewise, the moment arm of $\vec{\mathbf{F}}_4$ is also $(b/2)\sin\theta$. Because $F_2 = F_4 = IaB$, the net torque τ has the magnitude

$$\begin{aligned}\tau &= F_2\frac{b}{2}\sin\theta + F_4\frac{b}{2}\sin\theta \\ &= (IaB)\left(\frac{b}{2}\sin\theta\right) + (IaB)\left(\frac{b}{2}\sin\theta\right) = IabB\sin\theta \\ &= IAB\sin\theta\end{aligned}$$

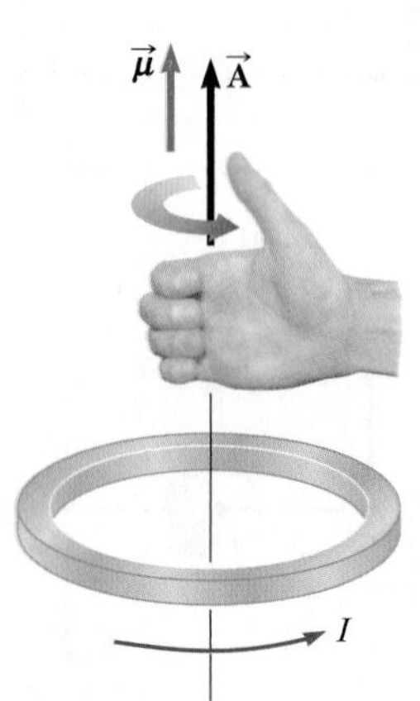

FIGURE 22.21 Right-hand rule for determining the direction of the vector $\vec{\mathbf{A}}$. The direction of the magnetic moment $\vec{\boldsymbol{\mu}}$ is the same as the direction of $\vec{\mathbf{A}}$.

where $A = ab$ is the area of the loop. This result shows that the torque has its maximum value IAB (Eq. 22.13) when the field is parallel to the plane of the loop ($\theta = 90°$) and is zero when the field is perpendicular to the plane of the loop ($\theta = 0$). As we see in Active Figure 22.20, the loop tends to rotate in the direction of decreasing values of θ (i.e., so that the normal to the plane of the loop rotates toward the direction of the magnetic field). A convenient vector expression for the torque is

$$\vec{\boldsymbol{\tau}} = I\vec{\mathbf{A}} \times \vec{\mathbf{B}} \quad [22.14]$$

where $\vec{\mathbf{A}}$, a vector perpendicular to the plane of the loop (Active Fig. 22.20), has a magnitude equal to the area of the loop. The sense of $\vec{\mathbf{A}}$ is determined by the right-hand rule illustrated in Figure 22.21. When the four fingers of the right hand are curled in the direction of the current in the loop, the thumb points in the direction of $\vec{\mathbf{A}}$. The product $I\vec{\mathbf{A}}$ is defined to be the **magnetic dipole moment** $\vec{\boldsymbol{\mu}}$ (often simply called the "magnetic moment") of the loop:

▪ Magnetic moment of a current loop

$$\vec{\boldsymbol{\mu}} = I\vec{\mathbf{A}} \quad [22.15]$$

The SI unit of magnetic dipole moment is the ampere-meter^2 ($\text{A}\cdot\text{m}^2$). Using this definition, the torque can be expressed as

▪ Torque on a current loop

$$\vec{\boldsymbol{\tau}} = \vec{\boldsymbol{\mu}} \times \vec{\mathbf{B}} \quad [22.16]$$

Although the torque was obtained for a particular orientation of $\vec{\mathbf{B}}$ with respect to the loop, Equation 22.16 is valid for any orientation. Furthermore, although the torque expression was derived for a rectangular loop, the result is valid for a loop of any shape. Once the torque is determined, the motion of the coil can be modeled as a rigid object under a net torque, which was studied in Chapter 10.

If a coil consists of N turns of wire, each carrying the same current and each having the same area, the total magnetic moment of the coil is the product of the number of turns and the magnetic moment for one turn, $\vec{\boldsymbol{\mu}} = NI\vec{\mathbf{A}}$. Thus, the torque on an N-turn coil is N times greater than that on a one-turn coil.

A common electric motor consists of a coil of wire mounted so that it can rotate in the field of a permanent magnet. The torque on the current-carrying coil is used to rotate a shaft that drives a mechanical device such as the power windows in your car, your household fan, or your electric hedge trimmer.

EXAMPLE 22.5 The Magnetic Moment and Torque on a Coil

A rectangular coil of dimensions 5.40 cm × 8.50 cm consists of 25 turns of wire. The coil carries a current of 15.0 mA.

A Calculate the magnitude of its magnetic moment.

Solution The magnitude of the magnetic moment of a current loop is $\mu = IA$ (Eq. 22.15). In this case, $A = (0.054\,0\text{ m})(0.085\,0\text{ m}) = 4.59 \times 10^{-3}\text{ m}^2$. Because the coil has 25 turns and assuming that each turn has the same area A, we have

$$\mu_{\text{coil}} = NIA = (25)(15.0 \times 10^{-3}\ \text{A})(4.59 \times 10^{-3}\ \text{m}^2)$$
$$= 1.72 \times 10^{-3}\ \text{A}\cdot\text{m}^2$$

B Suppose a uniform magnetic field of magnitude 0.350 T is applied parallel to the plane of the loop. What is the magnitude of the torque acting on the loop?

Solution The torque is given by Equation 22.16, $\vec{\boldsymbol{\tau}} = \vec{\boldsymbol{\mu}} \times \vec{\mathbf{B}}$. In this case, $\vec{\mathbf{B}}$ is perpendicular to $\vec{\boldsymbol{\mu}}_{\text{coil}}$, so

$$\tau = \mu_{\text{coil}} B = (1.72 \times 10^{-3}\ \text{A}\cdot\text{m}^2)(0.350\ \text{T})$$
$$= 6.02 \times 10^{-4}\ \text{N}\cdot\text{m}$$

22.7 THE BIOT–SAVART LAW

In the previous sections, we investigated the result of placing an object in an existing magnetic field. When a moving charge is placed in the field, it experiences a magnetic force. A current-carrying wire placed in the field also experiences a magnetic force; a current loop in the field experiences a torque.

Now we shift our thinking and investigate the *source* of the magnetic field. Oersted's 1819 discovery (Section 22.1) that an electric current in a wire deflects a nearby compass needle indicates that a current acts as a source of a magnetic field. From their investigations on the force between a current-carrying conductor and a magnet in the early 19th century, Jean-Baptiste Biot and Félix Savart arrived at an expression for the magnetic field at a point in space in terms of the current that produces the field. No point currents exist comparable to point charges (because we must have a complete circuit for a current to exist). Hence, we must investigate the magnetic field due to an infinitesimally small element of current that is part of a larger current distribution. Suppose the current distribution is a wire carrying a steady current I as in Figure 22.22. The **Biot–Savart law** says that the magnetic field $d\vec{\mathbf{B}}$ at point P created by an element of infinitesimal length ds of the wire has the following properties:

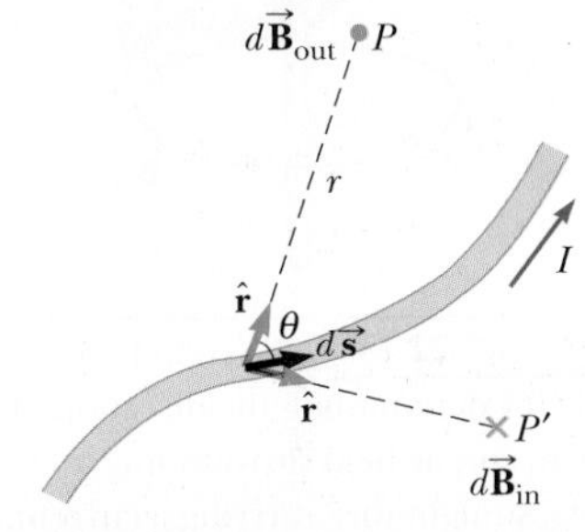

FIGURE 22.22 The magnetic field $d\vec{\mathbf{B}}$ at a point P due to a current I through a length element $d\vec{\mathbf{s}}$ is given by the Biot–Savart law. The field is out of the page at P and into the page at P'. (Both P and P' are in the plane of the page.)

- The vector $d\vec{\mathbf{B}}$ is perpendicular both to $d\vec{\mathbf{s}}$ (which is in the direction of the current) and to the unit vector $\hat{\mathbf{r}}$ directed from the element toward P.
- The magnitude of $d\vec{\mathbf{B}}$ is inversely proportional to r^2, where r is the distance from the element to P.
- The magnitude of $d\vec{\mathbf{B}}$ is proportional to the current I and to the length ds of the element.
- The magnitude of $d\vec{\mathbf{B}}$ is proportional to $\sin\theta$, where θ is the angle between $d\vec{\mathbf{s}}$ and $\hat{\mathbf{r}}$.

The **Biot–Savart law** can be summarized in the following compact form:

$$d\vec{\mathbf{B}} = k_m \frac{I\, d\vec{\mathbf{s}} \times \hat{\mathbf{r}}}{r^2} \qquad [22.17]$$

where k_m is a constant that in SI units is exactly $10^{-7}\ \text{T}\cdot\text{m/A}$. The constant k_m is usually written $\mu_0/4\pi$, where μ_0 is another constant, called the **permeability of free space:**

$$\frac{\mu_0}{4\pi} = k_m = 10^{-7}\ \text{T}\cdot\text{m/A} \qquad [22.18]$$

$$\mu_0 = 4\pi k_m = 4\pi \times 10^{-7}\ \text{T}\cdot\text{m/A} \qquad [22.19]$$

■ Permeability of free space

Hence, the Biot–Savart law, Equation 22.17, can also be written

$$d\vec{\mathbf{B}} = \frac{\mu_0}{4\pi} \frac{I\, d\vec{\mathbf{s}} \times \hat{\mathbf{r}}}{r^2} \qquad [22.20]$$

■ Biot–Savart law

PITFALL PREVENTION 22.2

THE BIOT–SAVART LAW When you are applying the Biot–Savart law, it is important to recognize that the magnetic field described in these calculations is the **field *due to* a given current-carrying conductor.** This magnetic field is not to be confused with any *external* field that may be applied to the conductor from some other source.

It is important to note that the Biot–Savart law gives the magnetic field at a point only for a small length element of the conductor. We identify the product $I\,d\vec{\mathbf{s}}$ as a **current element.** To find the total magnetic field $\vec{\mathbf{B}}$ at some point due to a conductor of finite size, we must sum contributions from all current elements making up the conductor. That is, we evaluate $\vec{\mathbf{B}}$ by integrating Equation 22.20 over the entire conductor.

There are two similarities between the Biot–Savart law of magnetism and Equation 19.7 for the electric field of a charge distribution, and there are two important differences. The current element $I\,d\mathbf{s}$ produces a magnetic field, and the charge element dq produces an electric field. Furthermore, the magnitude of the magnetic field varies as the inverse square of the distance from the current element, as does the electric field due to a charge element. The directions of the two fields are quite different, however. The electric field due to a charge element is radial; in the case of a positive point charge, $\vec{\mathbf{E}}$ is directed away from the charge. The magnetic field due to a current element is perpendicular to both the current element and the radius vector. Hence, if the conductor lies in the plane of the page, as in Figure 22.22, $d\vec{\mathbf{B}}$ points out of the page at the point P and into the page at P'. Another important difference is that an electric field can be a result either of a single charge or a distribution of charges, but a magnetic field can only be a result of a current distribution.

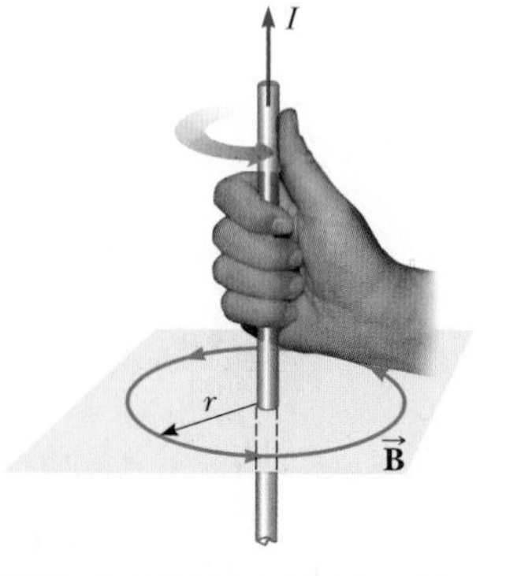

FIGURE 22.23 The right-hand rule for determining the direction of the magnetic field surrounding a long, straight wire carrying a current. Note that the magnetic field lines form circles around the wire. The magnitude of the magnetic field at a distance r from the wire is given by Equation 22.21.

Figure 22.23 shows a convenient right-hand rule for determining the direction of the magnetic field due to a current. Note that the field lines generally encircle the current. In the case of current in a long, straight wire, the field lines form circles that are concentric with the wire and are in a plane perpendicular to the wire. If the wire is grasped in the right hand with the thumb in the direction of the current, the fingers will curl in the direction of $\vec{\mathbf{B}}$.

Although the magnetic field due to an infinitely long, current-carrying wire can be calculated using the Biot–Savart law (Problem 22.52), in Section 22.9 we use a different method to show that the magnitude of this field at a distance r from the wire is

▪ Magnetic field due to a long, straight wire

$$B = \frac{\mu_0 I}{2\pi r} \qquad [22.21]$$

QUICK QUIZ 22.4 Consider the current in the length of wire shown in Figure 22.24. Rank the points A, B, and C, in terms of magnitude of the magnetic field due to the current in the length element $d\vec{\mathbf{s}}$ shown, from greatest to least.

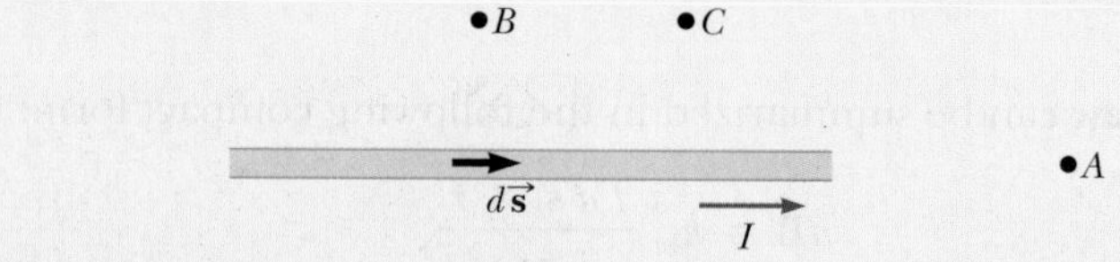

FIGURE 22.24 (Quick Quiz 22.4) Where is the magnetic field the greatest?

INTERACTIVE EXAMPLE 22.6 Magnetic Field on the Axis of a Circular Current Loop

Consider a circular loop of wire of radius R located in the yz plane and carrying a steady current I as in Figure 22.25. Calculate the magnetic field at an axial point P a distance x from the center of the loop.

Solution In this situation, note that any element $d\vec{\mathbf{s}}$ is perpendicular to $\hat{\mathbf{r}}$. Furthermore, all elements around the loop are at the same distance r from P, where $r^2 = x^2 + R^2$. Hence, the magnitude of $d\vec{\mathbf{B}}$ due to the element $d\vec{\mathbf{s}}$ is

$$dB = \frac{\mu_0 I}{4\pi}\frac{|d\vec{\mathbf{s}} \times \hat{\mathbf{r}}|}{r^2} = \frac{\mu_0 I}{4\pi}\frac{ds}{(x^2 + R^2)} \qquad [22.22]$$

The direction of the magnetic field $d\vec{\mathbf{B}}$ due to the element $d\vec{\mathbf{s}}$ is perpendicular to the plane formed by $\hat{\mathbf{r}}$ and $d\vec{\mathbf{s}}$ as in Figure 22.25. The vector $d\vec{\mathbf{B}}$ can be resolved into a component dB_x, along the x axis, and a component dB_y, which is perpendicular to the x axis. When the components dB_y are summed over the whole loop, the result

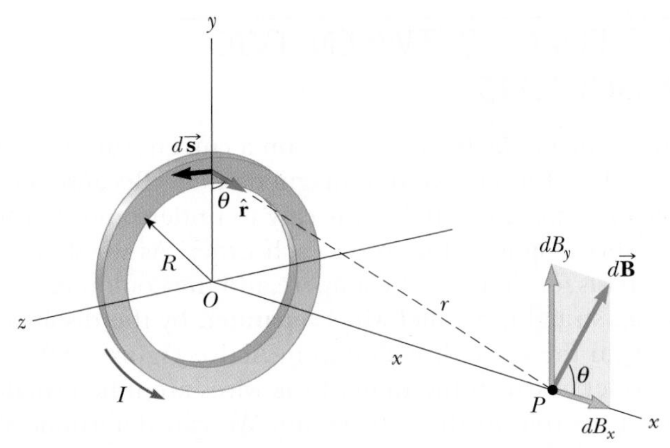

FIGURE 22.25 (Interactive Example 22.6) The geometry for calculating the magnetic field at a point P lying on the axis of a current loop. By symmetry, the total field $\vec{\mathbf{B}}$ is along this axis.

is zero. That is, by symmetry, any element on one side of the loop sets up a component dB_y that cancels the component set up by an element diametrically opposite it.

For these reasons, the resultant field at P must be along the x axis and can be found by integrating the components $dB_x = dB\cos\theta$, where this expression is obtained from resolving the vector $d\vec{\mathbf{B}}$ into its components as shown in Figure 22.25. That is, $\vec{\mathbf{B}} = B_x\hat{\mathbf{i}}$, where

$$B_x = \oint dB\cos\theta = \frac{\mu_0 I}{4\pi}\oint \frac{ds\cos\theta}{x^2 + R^2}$$

and the integral must be taken over the entire loop. Because θ, x, and R are constants for all elements of the loop and because $\cos\theta = R/(x^2 + R^2)^{1/2}$, we obtain

$$B_x = \frac{\mu_0 IR}{4\pi(x^2 + R^2)^{3/2}}\oint ds = \frac{\mu_0 IR^2}{2(x^2 + R^2)^{3/2}} \qquad [22.23]$$

where we have used that $\oint ds = 2\pi R$ (the circumference of the loop).

To find the magnetic field at the center of the loop, we set $x = 0$ in Equation 22.23. At this special point, we have

$$B = \frac{\mu_0 I}{2R} \qquad (\text{at } x = 0) \qquad [22.24]$$

It is also interesting to determine the behavior of the magnetic field far from the loop, that is, when x is large compared with R. In this case, we can ignore the term R^2 in the denominator of Equation 22.23 and we find that

$$B \approx \frac{\mu_0 IR^2}{2x^3} \qquad (\text{for } x >> R) \qquad [22.25]$$

Because the magnitude of the magnetic dipole moment μ of the loop is defined as the product of the current and the area (Eq. 22.15), $\mu = I(\pi R^2)$, we can express Equation 22.25 in the form

$$B = \frac{\mu_0}{2\pi}\frac{\mu}{x^3} \qquad [22.26]$$

This result is similar in form to the expression for the electric field due to an electric dipole, $E = k_e(2qa)/y^3 = k_e p/y^3$ (Example 19.3), where p is the electric dipole moment. The pattern of the magnetic field lines for a circular loop is shown in Figure 22.26a. For clarity, the lines are drawn only for one plane that contains the axis of the loop. Note that the field-line pattern is axially symmetric and looks like the pattern around a bar magnet, shown in Figure 22.26c.

PhysicsNow™ By logging into PhysicsNow at www.pop4e.com and going to Interactive Example 22.6, you can explore the field for different loop radii.

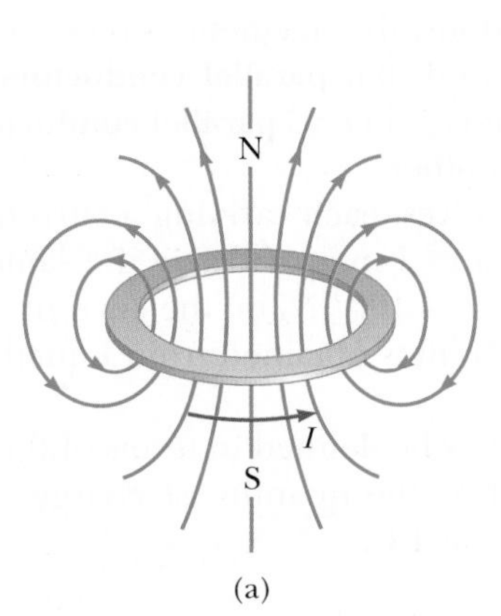

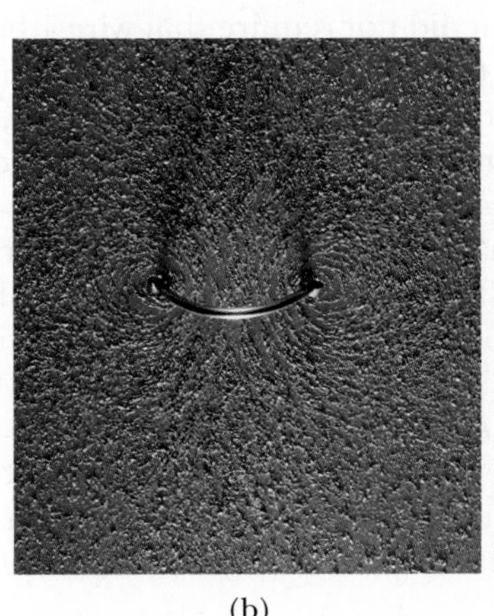

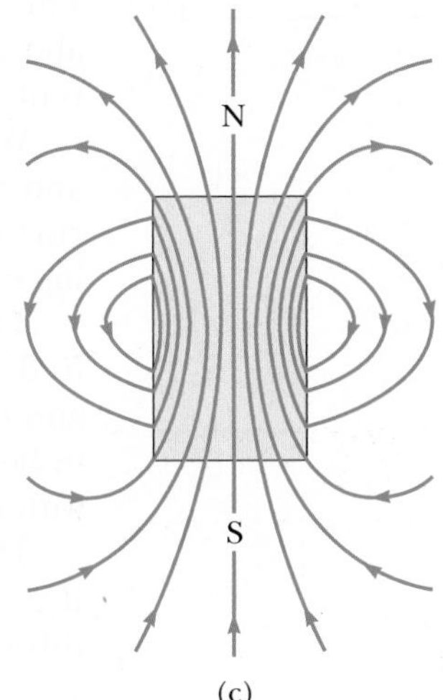

FIGURE 22.26 (Interactive Example 22.6) (a) Magnetic field lines surrounding a current loop. (b) Magnetic field lines surrounding a current loop displayed with iron filings. (c) Magnetic field lines surrounding a bar magnet. Note the similarity between this line pattern and that of a current loop.

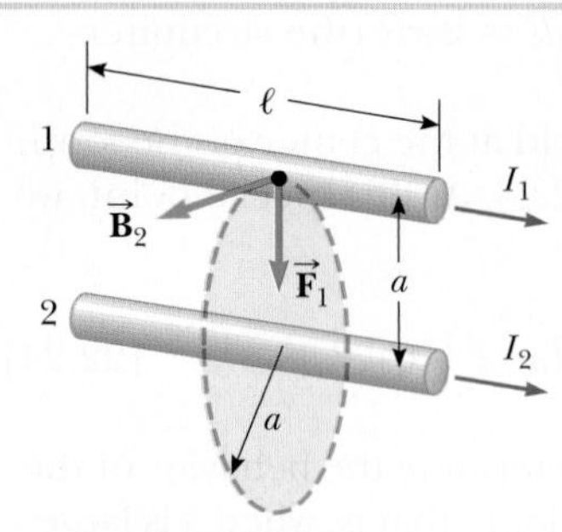

ACTIVE FIGURE 22.27

Two parallel wires that each carry a steady current exert a force on each other. The field $\vec{\mathbf{B}}_2$ due to the current in wire 2 exerts a force of magnitude $F_1 = I_1 \ell B_2$ on wire 1. The force is attractive if the currents are parallel (as shown) and repulsive if the currents are antiparallel.

Physics Now™ Log into PhysicsNow at www.pop4e.com and go to Active Figure 22.27. You can adjust the currents in the wires and the distance between them to see the effect on the force.

22.8 THE MAGNETIC FORCE BETWEEN TWO PARALLEL CONDUCTORS

In Section 22.5, we described the magnetic force that acts on a current-carrying conductor when the conductor is placed in an external magnetic field. Because a current in a conductor sets up its own magnetic field, it is easy to understand that two current-carrying conductors exert magnetic forces on each other. As we shall see, such forces can be used as the basis for defining the ampere and the coulomb.

Consider two infinitely long, straight, parallel wires separated by the distance a and carrying currents I_1 and I_2 in the same direction as in Active Figure 22.27. We shall adopt a simplification model in which the radii of the wires are much smaller than a so that the radius plays no role in the calculation. We can determine the force on one wire due to the magnetic field set up by the other wire. Wire 2, which carries current I_2, sets up a magnetic field $\vec{\mathbf{B}}_2$ at the position of wire 1. The direction of $\vec{\mathbf{B}}_2$ is perpendicular to the wire as shown in Active Figure 22.27. According to Equation 22.10, the magnetic force on a length ℓ of wire 1 is $\vec{\mathbf{F}}_1 = I_1 \vec{\ell} \times \vec{\mathbf{B}}_2$. Because $\vec{\ell}$ is perpendicular to $\vec{\mathbf{B}}_2$, the magnitude of $\vec{\mathbf{F}}_1$ is $F_1 = I_1 \ell B_2$. Because the field due to wire 2 is given by Equation 22.21, we see that

$$F_1 = I_1 \ell B_2 = I_1 \ell \left(\frac{\mu_0 I_2}{2\pi a} \right) = \frac{\ell \mu_0 I_1 I_2}{2\pi a}$$

We can rewrite this expression in terms of the force per unit length as

$$\frac{F_1}{\ell} = \frac{\mu_0 I_1 I_2}{2\pi a}$$

The direction of $\vec{\mathbf{F}}_1$ is downward, toward wire 2, because $\vec{\ell} \times \vec{\mathbf{B}}_2$ is downward. If one considers the field set up at wire 2 due to wire 1, the force $\vec{\mathbf{F}}_2$ on wire 2 is found to be equal in magnitude and opposite in direction to $\vec{\mathbf{F}}_1$. That is what one would expect because Newton's third law must be obeyed. Thus, we can drop the force subscript so that the magnetic force per unit length exerted by each long current-carrying wire on the other is

■ Magnetic force per unit length between parallel current-carrying wires

$$\frac{F}{\ell} = \frac{\mu_0 I_1 I_2}{2\pi a} \qquad [22.27]$$

This equation also applies if one of the wires is of finite length. In the discussion above, we used the equation for the magnetic field of an infinite wire carrying current I_2, but did not require that wire 1 be of infinite length.

When the currents are in opposite directions, the magnetic forces are reversed and the wires repel each other. Hence, we find that **parallel conductors carrying currents in the same direction attract each other,** whereas **parallel conductors carrying currents in opposite directions repel each other.**

The magnetic force between two parallel wires, each carrying a current, is used to define the **ampere:** If two long, parallel wires 1 m apart carry the same current and the force per unit length on each wire is 2×10^{-7} N/m, the current is defined to be 1 A. The numerical value of 2×10^{-7} N/m is obtained from Equation 22.27, with $I_1 = I_2 = 1$ A and $a = 1$ m.

The SI unit of charge, the **coulomb,** can now be defined in terms of the ampere: If a conductor carries a steady current of 1 A, the quantity of charge that flows through a cross-section of the conductor in 1 s is 1 C.

QUICK QUIZ 22.5 A loose spiral spring is hung from the ceiling and a large current is sent through it. Do the coils **(a)** move closer together, **(b)** move farther apart, or **(c)** not move at all?

22.9 AMPÈRE'S LAW

A simple experiment first carried out by Oersted in 1820 clearly demonstrates that a current-carrying conductor produces a magnetic field. In this experiment, several compass needles are placed in a horizontal plane near a long vertical wire as in Active Figure 22.28a. When the wire carries no current, all needles point in the same direction (that of the Earth's magnetic field), as one would expect. When the wire carries a strong, steady current, however, the needles all deflect in a direction tangent to the circle as in Active Figure 22.28b. These observations show that the direction of $\vec{\mathbf{B}}$ is consistent with the right-hand rule described in Section 22.7. When the current is reversed, the needles in Active Figure 22.28b also reverse.

Because the needles point in the direction of $\vec{\mathbf{B}}$, we conclude that the lines of $\vec{\mathbf{B}}$ form circles about the wire, as discussed in Section 22.7. By symmetry, the magnitude of $\vec{\mathbf{B}}$ is the same everywhere on a circular path that is centered on the wire and lies in a plane perpendicular to the wire. By varying the current and distance from the wire, one finds that $\vec{\mathbf{B}}$ is proportional to the current and inversely proportional to the distance from the wire.

In Chapter 19, we investigated Gauss's law, which is a relationship between an electric charge and the electric field it produces. Gauss's law can be used to determine the electric field in highly symmetric situations. We now consider an analogous relationship in magnetism between a current and the magnetic field it produces. This relationship can be used to determine the magnetic field created by a highly symmetric current distribution.

Let us evaluate the product $\vec{\mathbf{B}} \cdot d\vec{\mathbf{s}}$ for a small length element $d\vec{\mathbf{s}}$ on the circular path[1] centered on the wire in Active Figure 22.28b. Along this path, the vectors $d\vec{\mathbf{s}}$

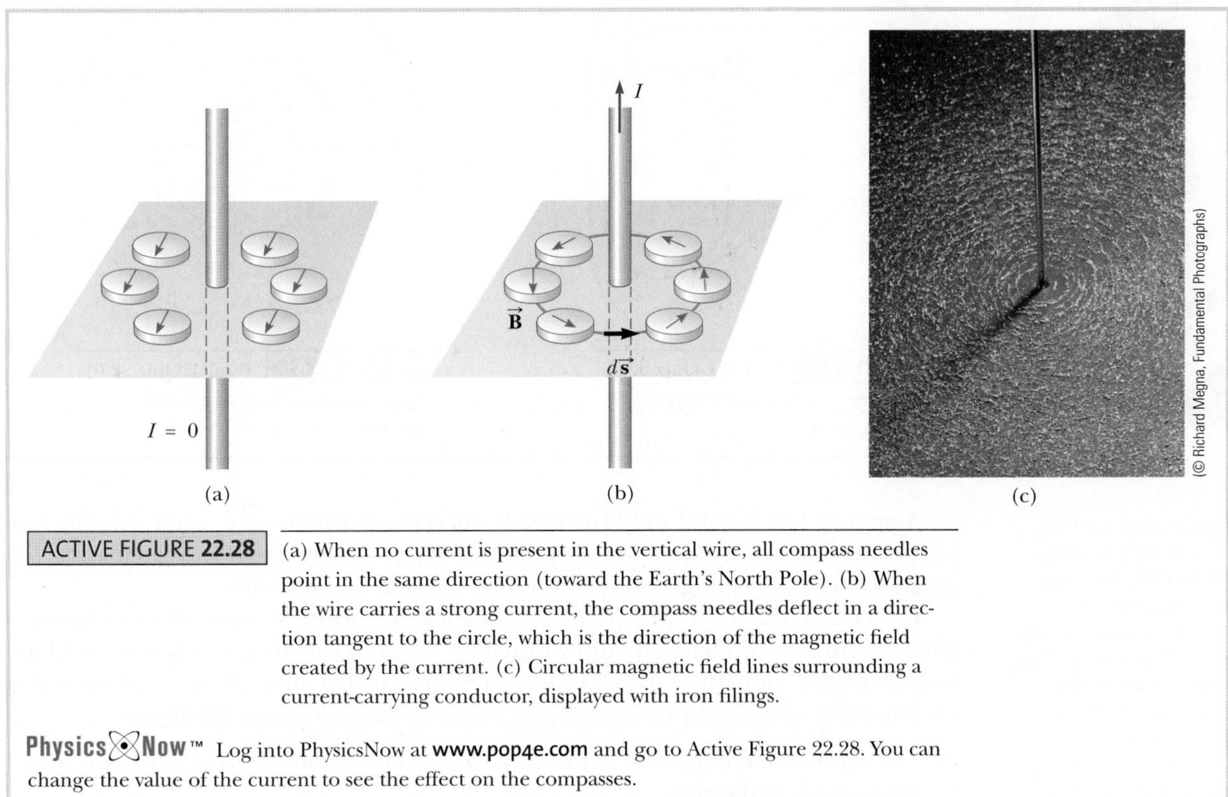

ACTIVE FIGURE 22.28 (a) When no current is present in the vertical wire, all compass needles point in the same direction (toward the Earth's North Pole). (b) When the wire carries a strong current, the compass needles deflect in a direction tangent to the circle, which is the direction of the magnetic field created by the current. (c) Circular magnetic field lines surrounding a current-carrying conductor, displayed with iron filings.

PhysicsNow™ Log into PhysicsNow at **www.pop4e.com** and go to Active Figure 22.28. You can change the value of the current to see the effect on the compasses.

[1] You may wonder why we would choose to do this evaluation. The origin of Ampère's law is in 19th-century science, in which a "magnetic charge" (the supposed analog to an isolated electric charge) was imagined to be moved around a circular field line. The work done on the charge was related to $\vec{\mathbf{B}} \cdot d\vec{\mathbf{s}}$, just like the work done moving an electric charge in an electric field is related to $\vec{\mathbf{E}} \cdot d\vec{\mathbf{s}}$. Thus, Ampère's law, a valid and useful principle, arose from an erroneous and abandoned work calculation!

and $\vec{\mathbf{B}}$ are parallel at each point, so $\vec{\mathbf{B}} \cdot d\vec{\mathbf{s}} = B\,ds$. Furthermore, by symmetry, $\vec{\mathbf{B}}$ is constant in magnitude on this circle and is given by Equation 22.21. Therefore, the sum of the products $B\,ds$ over the closed path, which is equivalent to the line integral of $\vec{\mathbf{B}} \cdot d\vec{\mathbf{s}}$, is

$$\oint \vec{\mathbf{B}} \cdot d\vec{\mathbf{s}} = B \oint ds = \frac{\mu_0 I}{2\pi r}(2\pi r) = \mu_0 I \qquad [22.28]$$

where $\oint ds = 2\pi r$ is the circumference of the circle.

This result, known as **Ampère's law,** was calculated for the special case of a circular path surrounding a wire. It can, however, also be applied in the general case in which a steady current passes through the area surrounded by an arbitrary closed path. That is, Ampère's law says that the line integral of $\vec{\mathbf{B}} \cdot d\vec{\mathbf{s}}$ around any closed path equals $\mu_0 I$, where I is the total steady current passing through any surface bounded by the closed path:

■ Ampère's law

$$\oint \vec{\mathbf{B}} \cdot d\vec{\mathbf{s}} = \mu_0 I \qquad [22.29]$$

QUICK QUIZ 22.6 (a) Rank the values of $\oint \vec{\mathbf{B}} \cdot d\vec{\mathbf{s}}$ for the closed paths in Figure 22.29, from smallest to largest. (b) Rank the values of $\oint \vec{\mathbf{B}} \cdot d\vec{\mathbf{s}}$ for the closed paths in Figure 22.30, from smallest to largest.

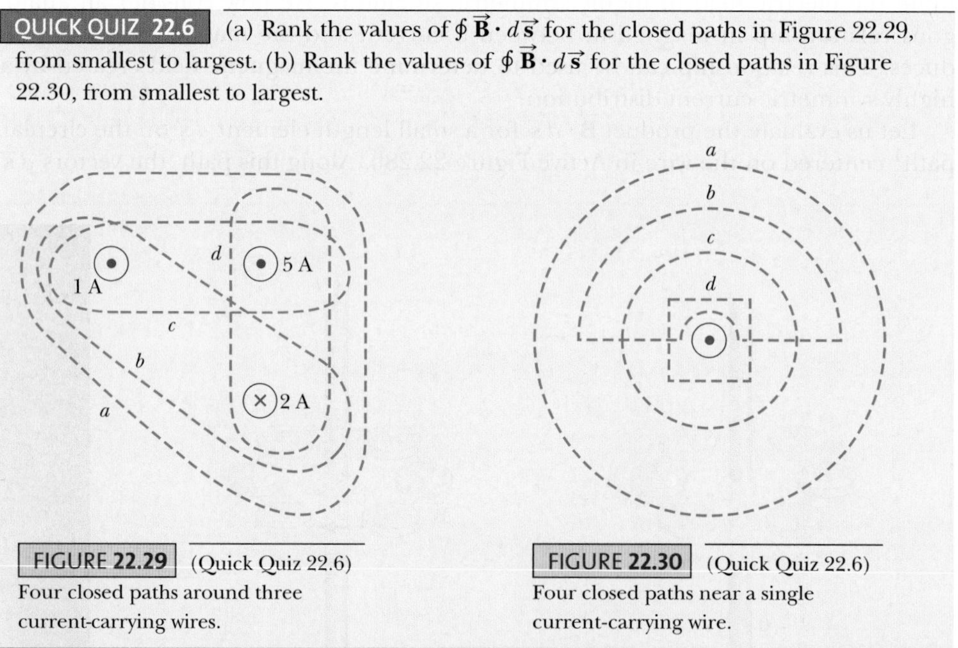

FIGURE 22.29 (Quick Quiz 22.6) Four closed paths around three current-carrying wires.

FIGURE 22.30 (Quick Quiz 22.6) Four closed paths near a single current-carrying wire.

(Leonard de Selva/CORBIS)

ANDRÉ-MARIE AMPÈRE
(1775–1836)

Ampère, a Frenchman, is credited with the discovery of electromagnetism, the relationship between electric currents and magnetic fields. Ampère's genius, particularly in mathematics, became evident by the time he was 12 years old; his personal life, however, was filled with tragedy. His father, a wealthy city official, was guillotined during the French Revolution, and his wife died young, in 1803. Ampère died at the age of 61 of pneumonia. His judgment of his life is clear from the epitaph he chose for his gravestone: *Tandem Felix* (Happy at Last).

Ampère's law is valid only for steady currents. Furthermore, even though Ampère's law is *true* for all current configurations, it is only *useful* for calculating the magnetic fields of configurations with high degrees of symmetry.

In Section 19.10, we provided some conditions to be sought when defining a gaussian surface. Similarly, to apply Equation 22.29 to calculate a magnetic field, we must determine a path of integration (sometimes called an *amperian loop*) such that each portion of the path satisfies one or more of the following conditions:

1. The value of the magnetic field can be argued by symmetry to be constant over the portion of the path.
2. The dot product in Equation 22.29 can be expressed as a simple algebraic product $B\,ds$ because $\vec{\mathbf{B}}$ and $d\vec{\mathbf{s}}$ are parallel.
3. The dot product in Equation 2.29 is zero because $\vec{\mathbf{B}}$ and $d\vec{\mathbf{s}}$ are perpendicular.
4. The magnetic field can be argued to be zero at all points on the portion of the path.

The following examples illustrate some symmetric current configurations for which Ampère's law is useful.

EXAMPLE 22.7 The Magnetic Field Created by a Long Current-Carrying Wire

A long, straight wire of radius R carries a steady current I that is uniformly distributed through the cross-section of the wire (Fig. 22.31). Calculate the magnetic field a distance r from the center of the wire in the regions $r \geq R$ and $r < R$.

Solution As mentioned in Section 22.7, we could use the Biot–Savart law to solve this problem, but Ampère's law provides a much simpler solution. For $r \geq R$, let us choose path 1 in Figure 22.31, a circle of radius r centered on the wire. From symmetry, we see that $\vec{\mathbf{B}}$ must be constant in magnitude—condition 1—and parallel to $d\vec{\mathbf{s}}$—condition 2—at every point on this circle. Because the total current passing through the plane of the circle is I, Ampère's law applied to the circular path gives

$$\oint \vec{\mathbf{B}} \cdot d\vec{\mathbf{s}} = B \oint ds = B(2\pi r) = \mu_0 I$$

$$B = \frac{\mu_0 I}{2\pi r} \qquad (\text{for } r \geq R)$$

which is the result (Eq. 22.21) referred to in Section 22.7.

Now consider the interior of the wire, where $r < R$. We choose the circular path 2 shown in Figure 22.31. Here the current I' passing through the plane of the circle is less than the total current I. Because the current is uniform over the cross-section of the wire, the fraction of the current enclosed by the circle of radius $r < R$ must equal the ratio of the area πr^2 enclosed by circular path 2 and the cross-sectional area πR^2 of the wire:

$$\frac{I'}{I} = \frac{\pi r^2}{\pi R^2}$$

$$I' = \frac{r^2}{R^2} I$$

Following the same procedure as for circular path 1, we apply Ampère's law to circular path 2:

$$\oint \vec{\mathbf{B}} \cdot d\vec{\mathbf{s}} = B(2\pi r) = \mu_0 I' = \mu_0 \left(\frac{r^2}{R^2} I \right)$$

$$B = \left(\frac{\mu_0 I}{2\pi R^2} \right) r \qquad (\text{for } r < R) \qquad [22.30]$$

The magnitude of the magnetic field versus r for this configuration is sketched in Figure 22.32. Note that inside the wire $B \rightarrow 0$ as $r \rightarrow 0$. This result is similar in form to that of the electric field inside a uniformly charged rod.

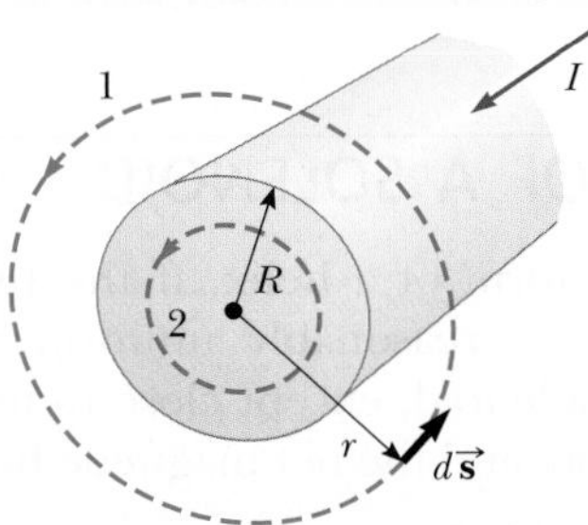

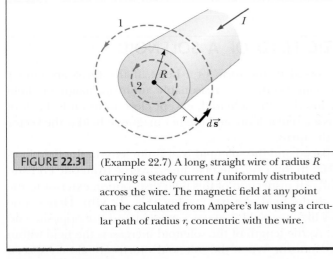

FIGURE 22.31 (Example 22.7) A long, straight wire of radius R carrying a steady current I uniformly distributed across the wire. The magnetic field at any point can be calculated from Ampère's law using a circular path of radius r, concentric with the wire.

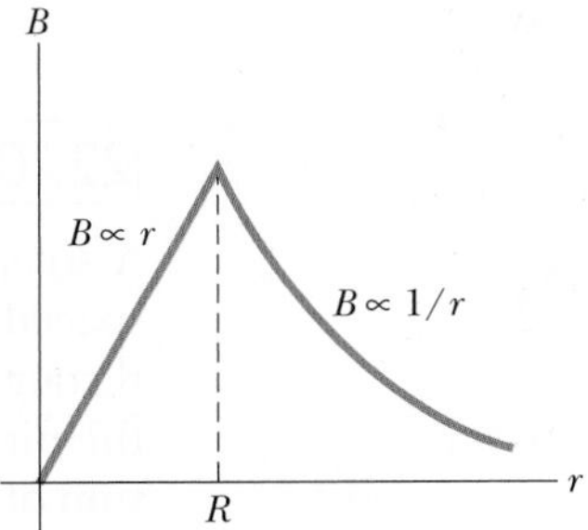

FIGURE 22.32 (Example 22.7) Magnitude of the magnetic field versus r for the wire described in Figure 22.31. The field is proportional to r inside the wire and varies as $1/r$ outside the wire.

EXAMPLE 22.8 The Magnetic Field Created by a Toroid

A device called a *toroid* (Fig. 22.33) is often used to create an almost uniform magnetic field in some enclosed area. The device consists of a conducting wire wrapped around a ring (a *torus*) made of a nonconducting material. For a toroid having N closely spaced turns of wire and air in the torus, calculate the magnetic field in the region occupied by the torus, a distance r from the center.

Solution To calculate the field inside the toroid, we evaluate the line integral of $\vec{\mathbf{B}} \cdot d\vec{\mathbf{s}}$ over the circular amperian loop of radius r in the plane of Figure 22.33. By symmetry, we see that conditions 1 and 2 apply: the magnetic field is constant in magnitude on this circle and tangent to it, so $\vec{\mathbf{B}} \cdot d\vec{\mathbf{s}} = B\,ds$.

Furthermore, note that the closed path surrounds a circular area through which N loops of wire pass, each of which carries a current I. The right side of Equation 22.29 is therefore $\mu_0 NI$ in this case. Ampère's law applied to the circular path gives

$$\oint \vec{\mathbf{B}} \cdot d\vec{\mathbf{s}} = B \oint ds = B(2\pi r) = \mu_0 NI$$

$$B = \frac{\mu_0 NI}{2\pi r} \qquad [22.31]$$

This result shows that B varies as $1/r$ and hence is nonuniform within the coil. If r is large compared with the cross-sectional radius a of the torus, however, the field is approximately uniform inside the coil.

For an ideal toroid in which the turns are closely spaced, the external magnetic field is close to zero. It is not exactly zero, however. In Figure 22.33, imagine the radius r of the amperian loop to be either smaller than b or larger than c. In either case, the loop encloses zero net current, so $\oint \vec{\mathbf{B}} \cdot d\vec{\mathbf{s}} = 0$. We might be tempted to claim that this expression proves that $\vec{\mathbf{B}} = 0$, but it does not. Consider the amperian loop on the right side of the toroid in Figure 22.33. The plane of this loop is perpendicular to the page and the toroid passes through the loop. As charges enter the toroid as indicated by the current directions in Figure 22.33, they work their way counterclockwise around the toroid. Thus, a current passes through the perpendicular amperian loop! This current is small, but it is not zero. As a result, the toroid acts as a current loop and produces a weak external field of the form shown in Figure 22.26a. The reason that $\oint \vec{\mathbf{B}} \cdot d\vec{\mathbf{s}} = 0$ for the amperian loops of radius $r < b$ and $r > c$ in the plane of the page is that the field lines are perpendicular to $d\vec{\mathbf{s}}$ (condition 3), *not* because $\vec{\mathbf{B}} = 0$ (condition 4).

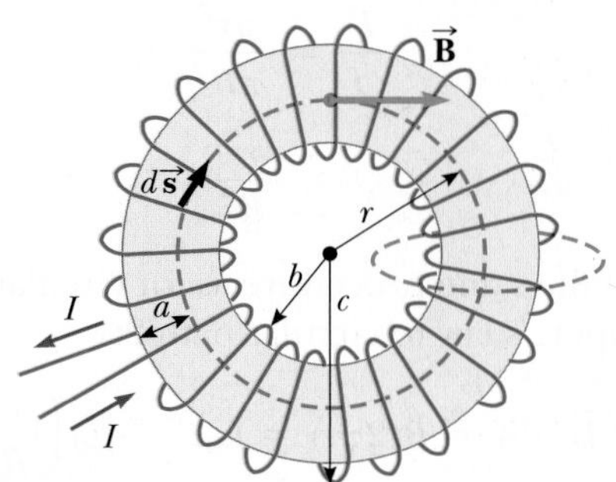

FIGURE 22.33 (Example 22.8) A toroid consisting of many turns of wire wrapped around a doughnut-shaped structure (called a torus). If the coils are closely spaced, the field in the interior of the toroid is tangent to the dashed circle and varies as $1/r$. The dimension a is the cross-sectional radius of the torus. The field outside the toroid is very small and can be described by using the amperian loop at the right side, perpendicular to the page.

22.10 THE MAGNETIC FIELD OF A SOLENOID

A solenoid is a long wire wound in the form of a helix. If the turns are closely spaced, this configuration can produce a reasonably uniform magnetic field throughout the volume enclosed by the solenoid, except close to its ends. Each of the turns can be modeled as a circular loop, and the net magnetic field is the vector sum of the fields due to all the turns.

If the turns are closely spaced and the solenoid is of finite length, the field lines are as shown in Figure 22.34a. In this case, the field lines diverge from one end and converge at the opposite end. An inspection of this field distribution exterior to the solenoid shows a similarity to the field of a bar magnet (Fig. 22.34b). Hence, one end of the solenoid behaves like the north pole of a magnet and the opposite end behaves like the south pole. As the length of the solenoid increases, the field within it becomes more and more uniform. When the solenoid's turns are closely spaced and its length is large compared with its radius, it approaches the case of an *ideal solenoid.* For an ideal solenoid, the field outside the solenoid is negligible and the field inside is uniform. We will use the ideal solenoid as a simplification model for a real solenoid.

If we consider the amperian loop perpendicular to the page in Figure 22.35, surrounding the ideal solenoid, we see that it it encloses a small current as the charges in the wire move coil by coil along the length of the solenoid. Thus, there is a

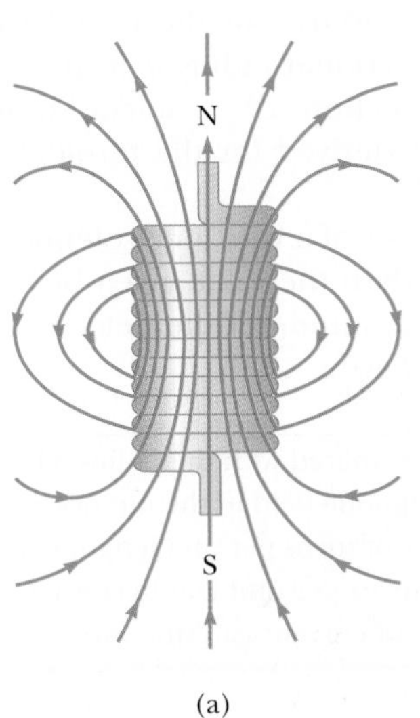

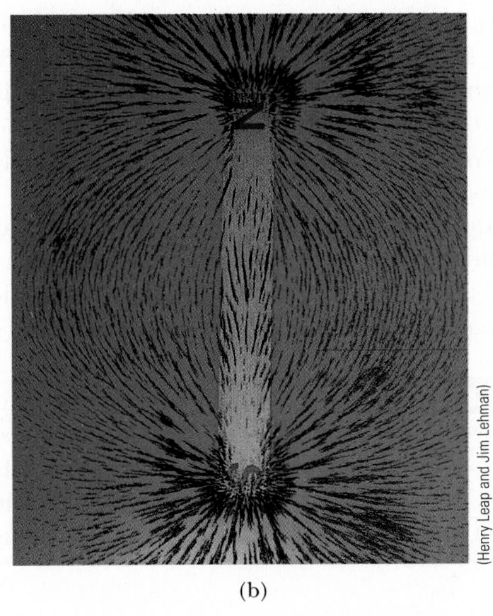

FIGURE 22.34 (a) Magnetic field lines for a tightly wound solenoid of finite length carrying a steady current. The field in the space enclosed by the solenoid is nearly uniform and strong. Note that the field lines resemble those of a bar magnet and that the solenoid effectively has north and south poles. (b) The magnetic field pattern of a bar magnet, displayed with iron filings.

nonzero magnetic field outside the solenoid. It is a weak field, with circular field lines, like those due to a line of current as in Figure 22.23. For an ideal solenoid, it is the only field external to the solenoid. We can eliminate this field in Figure 22.35 by adding a second layer of turns of wire outside the first layer. If the first layer of turns is wrapped so that the turns progress from the bottom of Figure 22.35 to the top and the second layer has turns progressing from the top to the bottom, the net current along the axis is zero.

We can use Ampère's law to obtain an expression for the magnetic field inside an ideal solenoid. A longitudinal cross-section of part of our ideal solenoid (Fig. 22.35) carries current I. Here, $\vec{\mathbf{B}}$ inside the ideal solenoid is uniform and parallel to the axis. Consider a rectangular path of length ℓ and width w as shown in Figure 22.35. We can apply Ampère's law to this path by evaluating the integral of $\vec{\mathbf{B}} \cdot d\vec{\mathbf{s}}$ over each of the four sides of the rectangle. The contribution along side 3 is zero because the magnetic field lines are perpendicular to the path in this region, which matches condition 3 in Section 22.9. The contributions from sides 2 and 4 are both zero because $\vec{\mathbf{B}}$ is perpendicular to $d\vec{\mathbf{s}}$ along these paths, both inside and outside the solenoid. Side 1, whose length is ℓ, gives a contribution to the integral because $\vec{\mathbf{B}}$ along this portion of the path is constant in magnitude and parallel to $d\vec{\mathbf{s}}$, which matches conditions 1 and 2. The integral over the closed rectangular path therefore has the value

$$\oint \vec{\mathbf{B}} \cdot d\vec{\mathbf{s}} = \int_{\text{side 1}} \vec{\mathbf{B}} \cdot d\vec{\mathbf{s}} = B \int_{\text{side 1}} ds = B\ell$$

The right side of Ampère's law involves the *total* current that passes through the surface bounded by the path of integration. In our case, the total current through the rectangular path equals the current through each turn of the solenoid multiplied by the number of turns enclosed by the path of integration. If N is the number of turns in the length ℓ, the total current through the rectangle equals NI. Ampère's law applied to this path therefore gives

$$\oint \vec{\mathbf{B}} \cdot d\vec{\mathbf{s}} = B\ell = \mu_0 NI$$

$$B = \mu_0 \frac{N}{\ell} I = \mu_0 nI \qquad [22.32]$$

where $n = N/\ell$ is the number of turns *per unit length* (not to be confused with N, the number of turns).

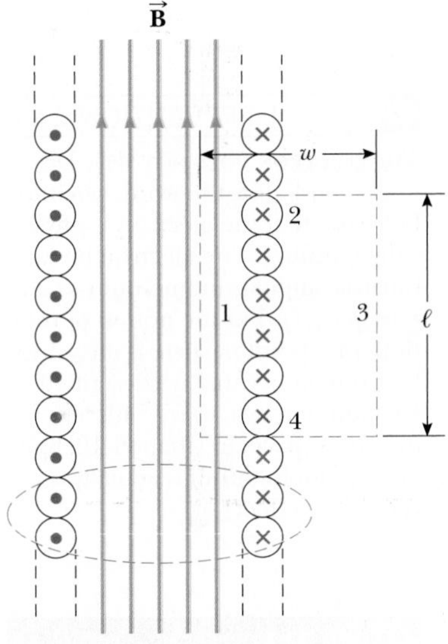

FIGURE 22.35 Cross-sectional view of an ideal solenoid, where the interior magnetic field is uniform and the exterior field is close to zero. Ampère's law applied to the circular path near the bottom whose plane is perpendicular to the page can be used to show that there is a weak field outside the solenoid. Ampère's law applied to the rectangular dashed path in the plane of the page can be used to calculate the magnitude of the interior field.

■ **Magnetic field inside a long solenoid**

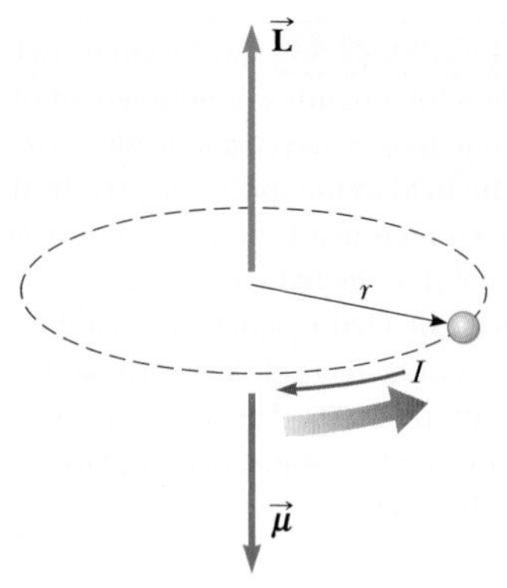

FIGURE 22.36 An electron moving in a circular orbit of radius r has an angular momentum $\vec{\mathbf{L}}$ in one direction and a magnetic moment $\vec{\mu}$ in the opposite direction. The motion of the electron in the direction of the gray arrow results in a current in the direction shown.

We also could obtain this result in a simpler manner by reconsidering the magnetic field of a toroidal coil (Example 22.8). If the radius r of the toroidal coil containing N turns is large compared with its cross-sectional radius a, a short section of the toroidal coil approximates a short section of a solenoid, with $n = N/2\pi r$. In this limit, we see that Equation 22.31 derived for the toroidal coil agrees with Equation 22.32.

Equation 22.32 is valid only for points near the center of a very long solenoid. As you might expect, the field near each end is smaller than the value given by Equation 22.32. At the very end of a long solenoid, the magnitude of the field is about one-half that of the field at the center (see Problem 22.46).

QUICK QUIZ 22.7 Consider a solenoid that is very long compared with the radius. Of the following choices, the most effective way to increase the magnetic field in the interior of the solenoid is to **(a)** double its length, keeping the number of turns per unit length constant, **(b)** reduce its radius by half, keeping the number of turns per unit length constant, or **(c)** overwrap the entire solenoid with an additional layer of current-carrying wire.

PITFALL PREVENTION 22.3

THE ELECTRON DOES NOT SPIN Do not be misled by the word *spin* into believing that the electron is physically spinning. The electron has an intrinsic angular momentum *as if it were spinning*, but the notion of rotation for a point particle is meaningless; remember that we described rotation of a *rigid object*, with an extent in space, in Chapter 10. Spin angular momentum is actually a relativistic effect.

22.11 MAGNETISM IN MATTER

The magnetic field produced by a current in a coil of wire gives a hint about what causes certain materials to exhibit strong magnetic properties. To understand why some materials are magnetic, it is instructive to begin this discussion with the Bohr structural model of the atom, in which electrons are assumed to move in circular orbits about the much more massive nucleus. Figure 22.36 shows the angular momentum associated with the electron. In the Bohr model, each electron, with its charge of magnitude 1.6×10^{-19} C, circles the atom once in about 10^{-16} s. If we divide the electronic charge by this time interval, we find that the orbiting electron is equivalent to a current of 1.6×10^{-3} A. Each orbiting electron is therefore viewed as a tiny current loop with a corresponding magnetic moment. Because the charge of the electron is negative, the magnetic moment is directed opposite to the angular momentum as shown in Figure 22.36.

In most substances, the magnetic moment of one electron in an atom is canceled by that of another electron in the atom, orbiting in the opposite direction. The net result is that **the magnetic effect produced by the orbital motion of the electrons is either zero or very small for most materials.**

In addition to its orbital angular momentum, an electron has an **intrinsic angular momentum,** called **spin,** which also contributes to its magnetic moment. The spin of an electron is an angular momentum separate from its orbital angular momentum, just as the spin of the Earth is separate from its orbital motion about the Sun. Even if the electron is at rest, it still has an angular momentum associated with spin. We shall investigate spin more deeply in Chapter 29.

In atoms or ions containing multiple electrons, many electrons are paired up with their spins in opposite directions, an arrangement that results in a cancellation of the spin magnetic moments. An atom with an odd number of electrons, however, must have at least one "unpaired" electron and a corresponding spin magnetic moment. The net magnetic moment of the atom leads to various types of magnetic behavior. The magnetic moments of several atoms and ions are listed in Table 22.1.

TABLE 22.1

Magnetic Moments of Some Atoms and Ions

Atom or Ion	Magnet Moment per Atom or Ion (10^{-24} J/T)
H	9.27
He	0
Ne	0
Fe	2.06
Co	16.0
Ni	5.62
Gd	65.8
Dy	92.7
Co^{2+}	44.5
Ni^{2+}	29.7
Fe^{2+}	50.1
Ce^{3+}	19.8
Yb^{3+}	37.1

Ferromagnetic Materials

Iron, cobalt, nickel, gadolinium, and dysprosium are strongly magnetic materials and are said to be **ferromagnetic.** Ferromagnetic substances, used to fabricate permanent magnets, contain atoms with spin magnetic moments that tend to align parallel to each other even in a weak external magnetic field. Once the moments are aligned, the substance remains magnetized after the external field is removed.

This permanent alignment is due to strong coupling between neighboring atoms, which can only be understood using quantum physics.

All ferromagnetic materials contain microscopic regions called **domains,** within which all magnetic moments are aligned. The domains range from about 10^{-12} to 10^{-8} m^3 in volume and contain 10^{17} to 10^{21} atoms. The boundaries between domains having different orientations are called **domain walls.** In an unmagnetized sample, the domains are randomly oriented so that the net magnetic moment is zero as in Figure 22.37a. When the sample is placed in an external magnetic field, domains with magnetic moment vectors initially oriented along the external field grow in size at the expense of other domains, which results in a magnetized sample, as in Figures 22.37b and 22.37c. When the external field is removed, the sample may retain most of its magnetism.

The extent to which a ferromagnetic substance retains its magnetism is described by its classification as being magnetically **hard** or **soft.** Soft magnetic materials, such as iron, are easily magnetized but also tend to lose their magnetism easily. When a soft magnetic material is magnetized and the external magnetic field is removed, thermal agitation produces domain motion and the material quickly returns to an unmagnetized state. In contrast, hard magnetic materials, such as cobalt and nickel, are difficult to magnetize but tend to retain their magnetism, and domain alignment persists in them after the external magnetic field is removed. Such hard magnetic materials are referred to as **permanent magnets.** Rare-earth permanent magnets, such as samarium–cobalt, are now regularly used in industry.

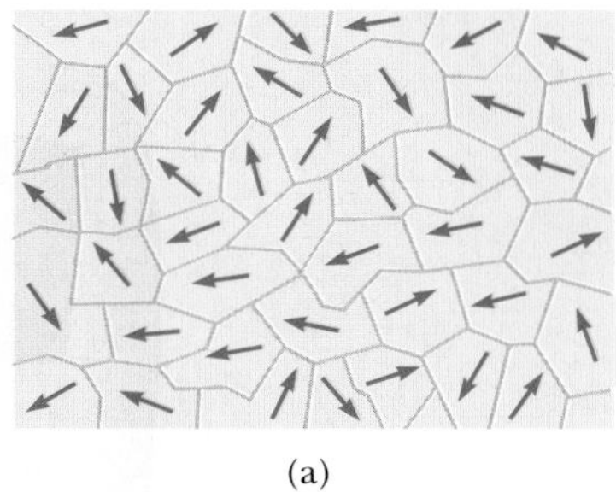

(a)

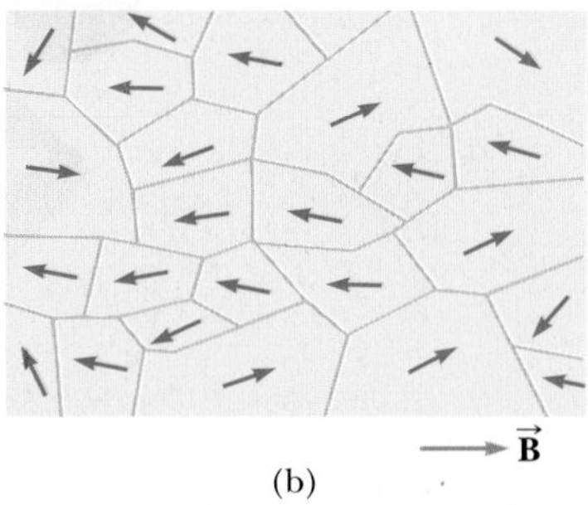

(b)

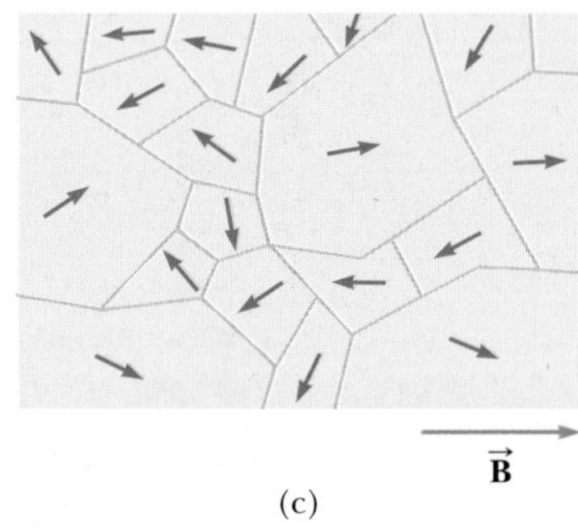

(c)

FIGURE 22.37 (a) Random orientation of atomic magnetic dipoles in the domains of an unmagnetized substance. (b) When an external field $\vec{B}$ is applied, the domains with components of magnetic moment in the same direction as $\vec{B}$ grow larger. (c) As the field is made even stronger, the domains with magnetic moment vectors not aligned with the external field become very small.

22.12 THE ATTRACTIVE MODEL FOR MAGNETIC LEVITATION

CONTEXT CONNECTION

A number of designs have been developed for magnetic levitation. In this section, we shall describe one design model called the *electromagnetic system* (EMS). This model is conceptually simple because it depends only on the attractive force between magnets and ferromagnetic materials. It has some technological complications, however. The EMS system is used in the German Transrapid design.

In an EMS system, the magnets supporting the vehicle are located below the track because the attractive force between these magnets and those in the track must lift the vehicle upward. A diagram of the German Transrapid system is shown in Figure 22.38.

The electromagnets attached to the vehicle are attracted to the steel rail, lifting the car. One disadvantage of this system is the *instability* of the vehicle caused by the variation of the magnetic force with distance. If the vehicle rises slightly, the magnet moves closer to the rail and the strength of the attractive force increases. As a result, the vehicle continues to move upward until the magnet makes contact with the rail. Conversely, if the vehicle drops slightly, the force decreases and the vehicle continues to drop. For these reasons, this system requires a proximity detector and electronic controls that adjust the magnetizing current to keep the vehicle at a constant position relative to the rail.

Figure 22.39 shows a typical method for controlling the separation between the magnets and the rails. The proximity detector is a device that uses magnetic induction (which we shall study in Chapter 23) to measure the magnet–rail separation. If the vehicle drops so that the levitation magnet moves farther from the rail, the detector causes the power supply to send more current to the magnet, pulling the vehicle back up. If the magnet rises, the decreased separation distance is detected and the power supply sends less current to the magnet so that the vehicle drops downward.

Another disadvantage of the EMS system is the relatively small separation between the levitating magnets and the track, about 10 mm. This small separation requires careful tolerance in track layout and curvature and steadfast maintenance of the track against problems with snow, ice, and temperature changes.

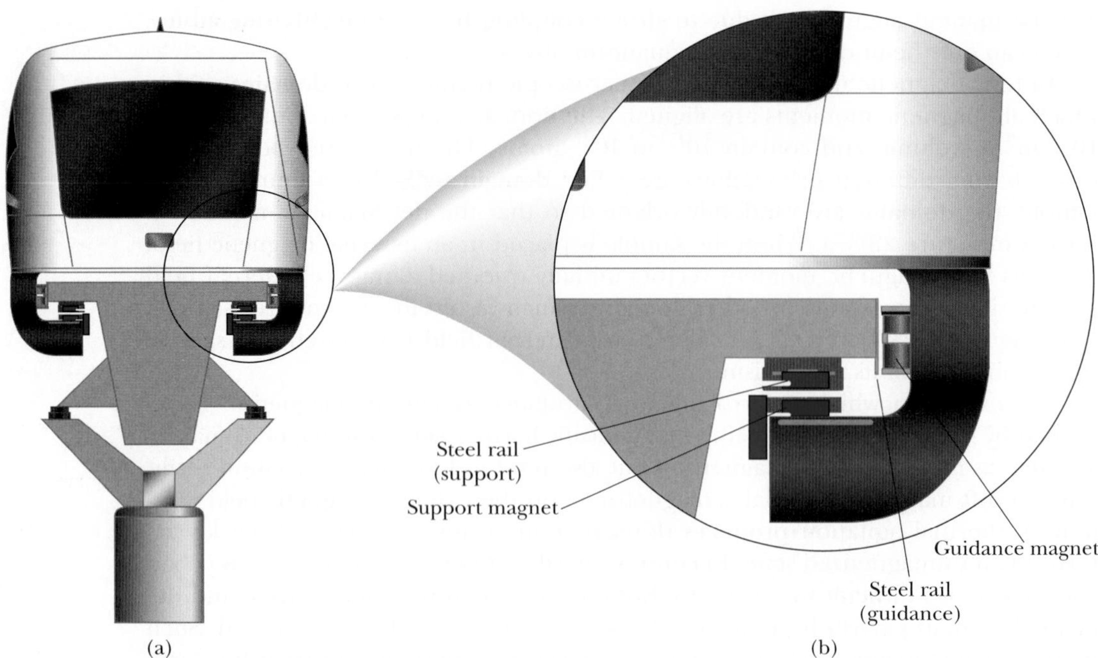

FIGURE 22.38 (a) A front view of a German Transrapid vehicle, showing it hovering above the track. (b) A close-up view of the support and guidance mechanisms. The attractive force between the support magnet and the steel rail lifts the vehicle upward. A second steel rail and the associated guidance magnet keep the vehicle laterally centered on the track. For more information, visit the Transrapid web site at www.transrapid.de/en/.

A major advantage of the EMS system is that the levitation is independent of speed so that wheels are not required; the vehicle is levitated even when stopped at a station. Wheels are still required, however, for an emergency "landing" system if a loss of power occurs.

The Transrapid system has undergone extensive testing in Germany and has achieved speeds of more than 450 km/h. As mentioned in the Context introduction,

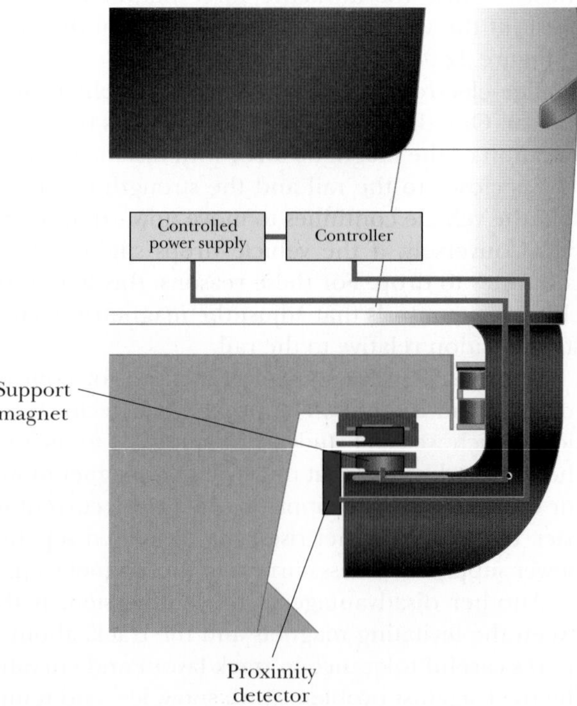

FIGURE 22.39 The control system for maintaining a fixed separation distance between the magnets and the track. The proximity detector signals the controller if the separation distance changes. The controlled power supply changes the current in the support magnet to counteract the change in the separation distance.

the Transrapid has entered commercial utilization in China, with a 30-km long track between Long Yang Station in Shanghai and Pudong International Airport. The station-to-station travel time is about 15 min, which is a significant reduction in time from that for a bus or a taxi. During the commissioning phase of this line, which spanned the year 2003, the vehicle achieved a speed of 501 km/h. During this phase, hundreds of thousands of visitors traveled on the line, including Chinese Premier Zhu Rongji and visiting German Chancellor Gerhard Schroeder. Scheduled commercial operations on this line began on December 29, 2003.

SUMMARY

The **magnetic force** that acts on a charge q moving with velocity $\vec{\mathbf{v}}$ in an external **magnetic field** $\vec{\mathbf{B}}$ is

$$\vec{\mathbf{F}}_B = q\vec{\mathbf{v}} \times \vec{\mathbf{B}} \qquad [22.1]$$

This force is in a direction perpendicular both to the velocity of the particle and to the magnetic field and given by the right-hand rules shown in Figure 22.4. The magnitude of the magnetic force is

$$F_B = |q|vB \sin \theta \qquad [22.2]$$

where θ is the angle between $\vec{\mathbf{v}}$ and $\vec{\mathbf{B}}$.

A particle with mass m and charge q moving with velocity $\vec{\mathbf{v}}$ perpendicular to a uniform magnetic field $\vec{\mathbf{B}}$ follows a circular path of radius

$$r = \frac{mv}{qB} \qquad [22.3]$$

If a straight conductor of length ℓ carries current I, the magnetic force on that conductor when placed in a uniform external magnetic field $\vec{\mathbf{B}}$ is

$$\vec{\mathbf{F}}_B = I\vec{\boldsymbol{\ell}} \times \vec{\mathbf{B}} \qquad [22.10]$$

where $\vec{\boldsymbol{\ell}}$ is in the direction of the current and $|\vec{\boldsymbol{\ell}}| = \ell$, the length of the conductor.

If an arbitrarily shaped wire carrying current I is placed in an external magnetic field, the magnetic force on a very small length element $d\vec{\mathbf{s}}$ is

$$d\vec{\mathbf{F}}_B = I\, d\vec{\mathbf{s}} \times \vec{\mathbf{B}} \qquad [22.11]$$

To determine the total magnetic force on the wire, one must integrate Equation 22.11 over the wire.

The **magnetic dipole moment** $\vec{\boldsymbol{\mu}}$ of a loop carrying current I is

$$\vec{\boldsymbol{\mu}} = I\vec{\mathbf{A}} \qquad [22.15]$$

where $\vec{\mathbf{A}}$ is perpendicular to the plane of the loop and $|\vec{\mathbf{A}}|$ is equal to the area of the loop. The SI unit of $\vec{\boldsymbol{\mu}}$ is the ampere-meter squared, or $\text{A} \cdot \text{m}^2$.

The torque $\vec{\boldsymbol{\tau}}$ on a current loop when the loop is placed in a uniform external magnetic field $\vec{\mathbf{B}}$ is

$$\vec{\boldsymbol{\tau}} = \vec{\boldsymbol{\mu}} \times \vec{\mathbf{B}} \qquad [22.16]$$

The **Biot–Savart law** says that the magnetic field $d\vec{\mathbf{B}}$ at a point P due to a wire element $d\vec{\mathbf{s}}$ carrying a steady current I is

$$d\vec{\mathbf{B}} = \frac{\mu_0}{4\pi} \frac{I\, d\vec{\mathbf{s}} \times \hat{\mathbf{r}}}{r^2} \qquad [22.20]$$

where $\mu_0 = 4\pi \times 10^{-7}\ \text{T} \cdot \text{m/A}$ is the **permeability of free space** and r is the distance from the element to the point P. To find the total field at P due to a current distribution, one must integrate this vector expression over the entire distribution.

The magnitude of the magnetic field at a distance r from a long, straight wire carrying current I is

$$B = \frac{\mu_0 I}{2\pi r} \qquad [22.21]$$

The field lines are circles concentric with the wire.

The magnetic force per unit length between two parallel wires (at least one of which is long) separated by a distance a and carrying currents I_1 and I_2 has the magnitude

$$\frac{F}{\ell} = \frac{\mu_0 I_1 I_2}{2\pi a} \qquad [22.27]$$

The force is attractive if the currents are in the same direction and repulsive if they are in opposite directions.

Ampère's law says that the line integral of $\vec{\mathbf{B}} \cdot d\vec{\mathbf{s}}$ around any closed path equals $\mu_0 I$, where I is the total steady current passing through any surface bounded by the closed path:

$$\oint \vec{\mathbf{B}} \cdot d\vec{\mathbf{s}} = \mu_0 I \qquad [22.29]$$

Using Ampère's law, one finds that the fields inside a toroidal coil and solenoid are

$$B = \frac{\mu_0 NI}{2\pi r} \quad \text{(toroid)} \qquad [22.31]$$

$$B = \mu_0 \frac{N}{\ell} I = \mu_0 nI \quad \text{(solenoid)} \qquad [22.32]$$

where N is the total number of turns and n is the number of turns per unit length.

QUESTIONS

☐ = answer available in the *Student Solutions Manual and Study Guide*

1. Two charged particles are projected into a magnetic field perpendicular to their velocities. If the charges are deflected in opposite directions, what can you say about them?
2. List several similarities and differences between electric and magnetic forces.
3. The electron beam in Figure Q22.3 is projected to the right. The beam deflects downward in the presence of a magnetic field produced by a pair of current-carrying coils. (a) What is the direction of the magnetic field? (b) What would happen to the beam if the current in the coils were reversed?

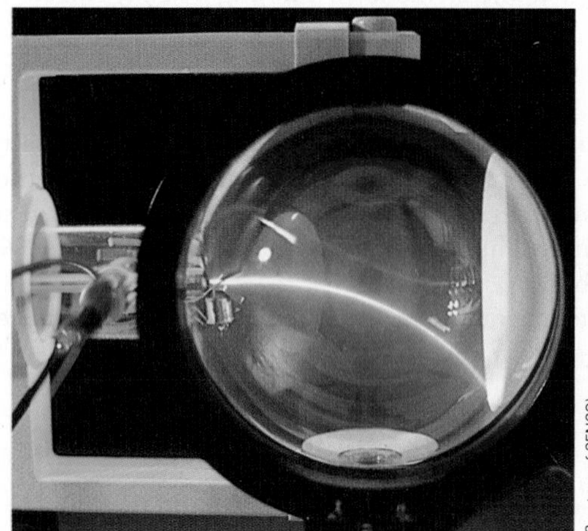

FIGURE Q22.3 Bending of a beam of electrons in a magnetic field.

4. A current-carrying conductor experiences no magnetic force when placed in a certain manner in a uniform magnetic field. Explain.
5. Is it possible to orient a current loop in a uniform magnetic field such that the loop does not tend to rotate? Explain.
6. Explain why it is not possible to determine the charge and the mass of a charged particle separately by measuring accelerations produced by electric and magnetic forces on the particle.
7. Charged particles from outer space, called cosmic rays, strike the Earth more frequently near the poles than near the equator. Why?
8. A *bubble chamber* is a device used for observing tracks of particles that pass through the chamber, which is immersed in a magnetic field. If some of the tracks are spirals and others are straight lines, what can you say about the particles?
9. Explain why two parallel wires carrying currents in opposite directions repel each other.
10. Parallel current-carrying wires exert magnetic forces on each other. What about perpendicular wires? Imagine two such wires oriented perpendicular to each other and almost touching. Does a magnetic force exist between the wires?
11. A hollow copper tube carries a current along its length. Why is $\vec{\mathbf{B}} = 0$ inside the tube? Is $\vec{\mathbf{B}}$ nonzero outside the tube?
12. Describe the change in the magnetic field in the space enclosed by a solenoid carrying a steady current I if (a) the length of the solenoid is doubled but the number of turns remains the same and (b) the number of turns is doubled but the length remains the same.
13. A magnet attracts a piece of iron. The iron can then attract another piece of iron. On the basis of domain alignment, explain what happens in each piece of iron.
14. The "north" pole of a bar magnet is attracted toward the geographic north pole of the Earth. Yet, similar poles repel. What is the way out of this dilemma?
15. Why does hitting a magnet with a hammer cause the magnetism to be reduced?
16. Should the surface of a computer disk be made from a hard or a soft ferromagnetic substance?
17. Figure Q22.17 shows two permanent magnets, each having a hole through its center. Note that the upper magnet is levitated above the lower one. (a) How does this situation occur? (b) What purpose does the pencil serve? (c) What can you say about the poles of the magnets from this observation? (d) If the upper magnet were inverted, what do you suppose would happen?

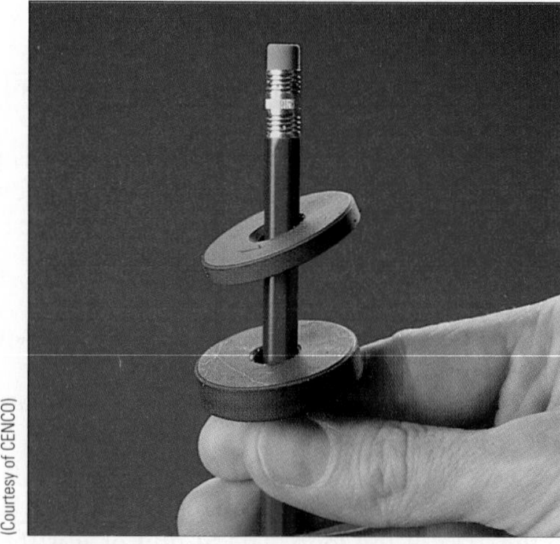

FIGURE Q22.17 Magnetic levitation using two ceramic magnets.

PROBLEMS

1, 2, 3 = straightforward, intermediate, challenging
□ = full solution available in the *Student Solutions Manual and Study Guide*
Physics Now™ = coached problem with hints available at www.pop4e.com
= computer useful in solving problem
= paired numerical and symbolic problems
= biomedical application

Section 22.2 ■ The Magnetic Field

1. Physics Now™ Determine the initial direction of the deflection of charged particles as they enter the magnetic fields as shown in Figure P22.1.

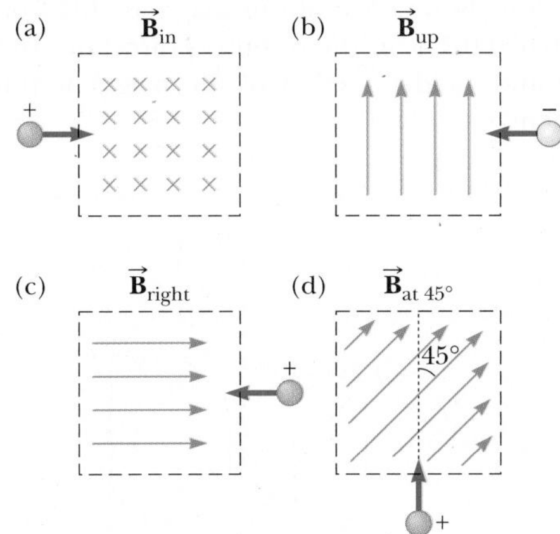

FIGURE **P22.1**

2. Consider an electron near the Earth's equator. In which direction does it tend to deflect if its velocity is directed (a) downward, (b) northward, (c) westward, or (d) southeastward?

3. A proton travels with a speed of 3.00×10^6 m/s at an angle of 37.0° with the direction of a magnetic field of 0.300 T in the $+y$ direction. What are (a) the magnitude of the magnetic force on the proton and (b) its acceleration?

4. An electron is accelerated through 2 400 V from rest and then enters a uniform 1.70-T magnetic field. What are (a) the maximum and (b) the minimum values of the magnetic force this charge can experience?

5. At the equator, near the surface of the Earth, the magnetic field is approximately 50.0 μT northward and the electric field is about 100 N/C downward in fair weather. Find the gravitational, electric, and magnetic forces on an electron in this environment, assuming that the electron has an instantaneous velocity of 6.00×10^6 m/s directed to the east.

6. A proton moves with a velocity of $\vec{\mathbf{v}} = (2\hat{\mathbf{i}} - 4\hat{\mathbf{j}} + \hat{\mathbf{k}})$ m/s in a region in which the magnetic field is $\vec{\mathbf{B}} = (\hat{\mathbf{i}} + 2\hat{\mathbf{j}} - 3\hat{\mathbf{k}})$ T. What is the magnitude of the magnetic force this charge experiences?

Section 22.3 ■ Motion of a Charged Particle in a Uniform Magnetic Field

7. **Review problem.** One electron collides elastically with a second electron initially at rest. After the collision, the radii of their trajectories are 1.00 cm and 2.40 cm. The trajectories are perpendicular to a uniform magnetic field of magnitude 0.044 0 T. Determine the energy (in keV) of the incident electron.

8. **Review problem.** An electron moves in a circular path perpendicular to a constant magnetic field of magnitude 1.00 mT. The angular momentum of the electron about the center of the circle is 4.00×10^{-25} J·s. Determine (a) the radius of the circular path and (b) the speed of the electron.

9. A cosmic-ray proton in interstellar space has an energy of 10.0 MeV and executes a circular orbit having a radius equal to that of Mercury's orbit around the Sun (5.80×10^{10} m). What is the magnetic field in that region of space?

Section 22.4 ■ Applications Involving Charged Particles Moving in a Magnetic Field

10. A velocity selector consists of electric and magnetic fields described by the expressions $\vec{\mathbf{E}} = E\hat{\mathbf{k}}$ and $\vec{\mathbf{B}} = B\hat{\mathbf{j}}$, with $B =$ 15.0 mT. Find the value of E such that a 750-eV electron moving along the positive x axis is undeflected.

11. Consider the mass spectrometer shown schematically in Active Figure 22.12. The magnitude of the electric field between the plates of the velocity selector is 2 500 V/m, and the magnetic field in both the velocity selector and the deflection chamber has a magnitude of 0.035 0 T. Calculate the radius of the path for a singly charged ion having a mass $m = 2.18 \times 10^{-26}$ kg.

12. A cyclotron designed to accelerate protons has an outer radius of 0.350 m. The protons are emitted nearly at rest from a source at the center and are accelerated through 600 V each time they cross the gap between the dees. The dees are between the poles of an electromagnet where the field is 0.800 T. (a) Find the cyclotron frequency. (b) Find the speed at which protons exit the cyclotron and (c) their maximum kinetic energy. (d) How many revolutions does a proton make in the cyclotron? (e) For what time interval does one proton accelerate?

13. The picture tube in a television uses magnetic deflection coils rather than electric deflection plates. Suppose an electron beam is accelerated through a 50.0-kV potential difference and then moves through a region of uniform magnetic field 1.00 cm wide. The screen is located 10.0 cm from the center of the coils and is 50.0 cm wide. When the field is turned off, the electron beam hits the center of the screen. What field magnitude is necessary to deflect the beam to the side of the screen? Ignore relativistic corrections.

14. The *Hall effect* finds important application in the electronics industry. It is used to find the sign and density of the carriers of electric current in semiconductor chips. The arrangement is shown in Figure P22.14. A semiconducting block of thickness t and width d carries a current I in the x direction. A uniform magnetic field B is applied in the y direction. If the charge carriers are positive, the magnetic force deflects them in the z direction. Positive charge accumulates on the top surface of the sample and negative charge on the bottom surface, creating a downward electric field. In equilibrium, the downward electric force on the charge carriers balances the upward magnetic force and the carriers move through the sample without deflection. The *Hall voltage* $\Delta V_H = V_c - V_a$ between the top and bottom surfaces is measured, and the density of the charge carriers can be calculated from it. (a) Demonstrate that if the charge carriers are negative the Hall voltage will be negative. Hence, the Hall effect reveals the sign of the charge carriers, so the sample can be classified as p-type (with positive majority charge carriers) or n-type (with negative). (b) Determine the number of charge carriers per unit volume n in terms of I, t, B, ΔV_H, and the magnitude q of the charge carrier.

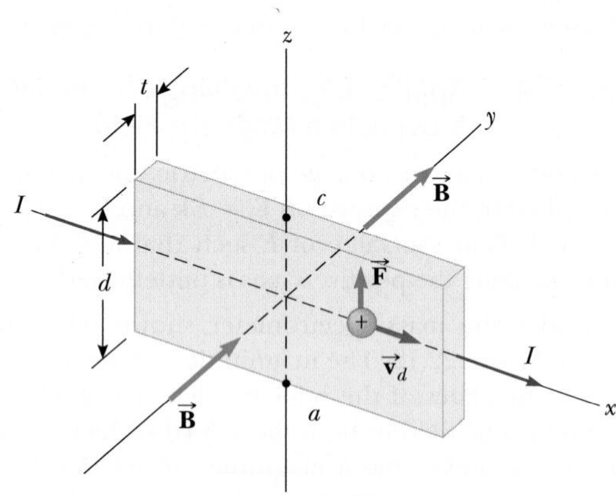

FIGURE **P22.14**

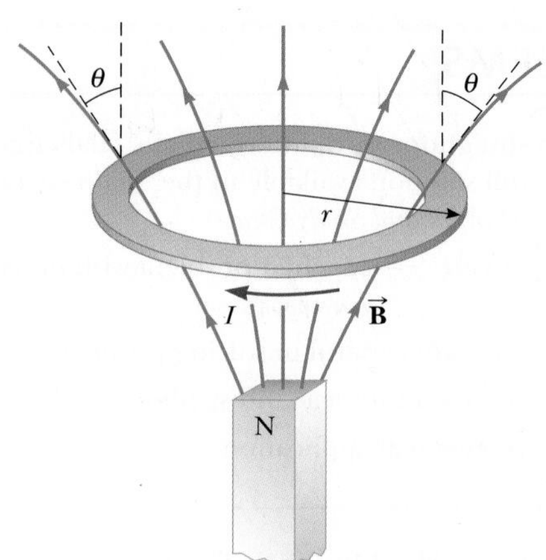

FIGURE **P22.17**

Section 22.5 ■ Magnetic Force on a Current-Carrying Conductor

15. A wire carries a steady current of 2.40 A. A straight section of the wire is 0.750 m long and lies along the x axis within a uniform magnetic field, $\vec{\mathbf{B}} = 1.60\hat{\mathbf{k}}$ T. If the current is in the $+x$ direction, what is the magnetic force on the section of wire?

16. A wire 2.80 m in length carries a current of 5.00 A in a region where a uniform magnetic field has a magnitude of 0.390 T. Calculate the magnitude of the magnetic force on the wire assuming that the angle between the magnetic field and the current is (a) 60.0°, (b) 90.0°, and (c) 120°.

17. **Physics Now™** *A nonuniform magnetic field exerts a net force on a magnetic dipole.* A strong magnet is placed under a horizontal conducting ring of radius r that carries current I as shown in Figure P22.17. If the magnetic field $\vec{\mathbf{B}}$ makes an angle θ with the vertical at the ring's location, what are the magnitude and direction of the resultant force on the ring?

18. In Figure P22.18, the cube is 40.0 cm on each edge. Four straight segments of wire—*ab*, *bc*, *cd*, and *da*—form a closed loop that carries a current $I = 5.00$ A, in the direction shown. A uniform magnetic field of magnitude $B = 0.020\,0$ T is in the positive y direction. Determine the magnitude and direction of the magnetic force on each segment.

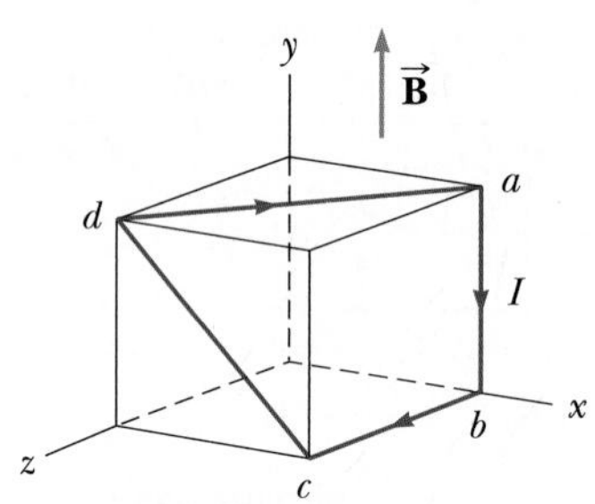

FIGURE **P22.18**

Section 22.6 ■ Torque on a Current Loop in a Uniform Magnetic Field

19. A current of 17.0 mA is maintained in a single circular loop of 2.00 m circumference. A magnetic field of 0.800 T is directed parallel to the plane of the loop. (a) Calculate the magnetic moment of the loop. (b) What is the magnitude of the torque exerted by the magnetic field on the loop?

20. A current loop with magnetic dipole moment $\vec{\boldsymbol{\mu}}$ is placed in a uniform magnetic field $\vec{\mathbf{B}}$, with its moment making angle θ with the field. With the arbitrary choice of $U = 0$ for $\theta = 90°$, prove that the potential energy of the dipole-field system is $U = -\vec{\boldsymbol{\mu}} \cdot \vec{\mathbf{B}}$.

21. **Physics Now™** A rectangular coil consists of $N = 100$ closely wrapped turns and has dimensions $a = 0.400$ m and $b = 0.300$ m. The coil is hinged along the y axis, and its plane makes an angle $\theta = 30.0°$ with the x axis (Fig. P22.21). What is the magnitude of the torque exerted on the coil by a uniform magnetic field $B = 0.800$ T directed along the x axis when the current is $I = 1.20$ A in the direction shown? What is the expected direction of rotation of the coil?

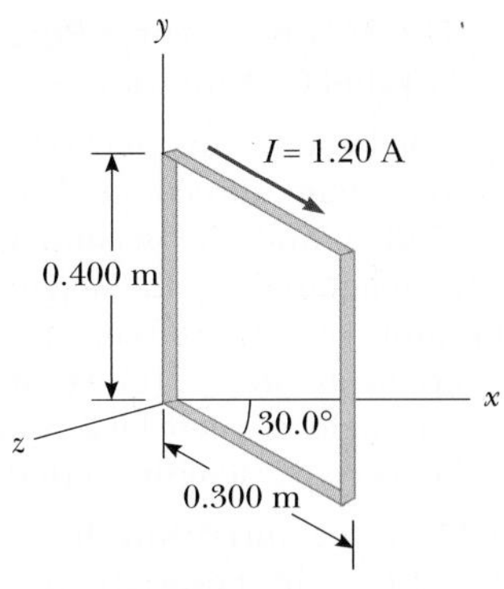

FIGURE **P22.21**

22. The rotor in a certain electric motor is a flat, rectangular coil with 80 turns of wire and dimensions 2.50 cm by 4.00 cm. The rotor rotates in a uniform magnetic field of 0.800 T. When the plane of the rotor is perpendicular to the direction of the magnetic field, it carries a current of 10.0 mA. In this orientation, the magnetic moment of the rotor is directed opposite the magnetic field. The rotor then turns through one-half revolution. This process is repeated to cause the rotor to turn steadily at 3 600 rev/min. (a) Find the maximum torque acting on the rotor. (b) Find the peak power output of the motor. (c) Determine the amount of work performed by the magnetic field on the rotor in every full revolution. (d) What is the average power of the motor?

Section 22.7 ■ The Biot–Savart Law

23. In Niels Bohr's 1913 model of the hydrogen atom, an electron circles the proton at a distance of 5.29×10^{-11} m with a speed of 2.19×10^6 m/s. Compute the magnitude of the magnetic field that this motion produces at the location of the proton.

24. A lightning bolt may carry a current of 1.00×10^4 A for a short time interval. What is the resulting magnetic field 100 m from the bolt? Assume that the bolt extends far above and below the point of observation.

25. **Physics Now™** Determine the magnetic field at a point *P* located a distance *x* from the corner of an infinitely long wire bent at a right angle as shown in Figure P22.25. The wire carries a steady current *I*.

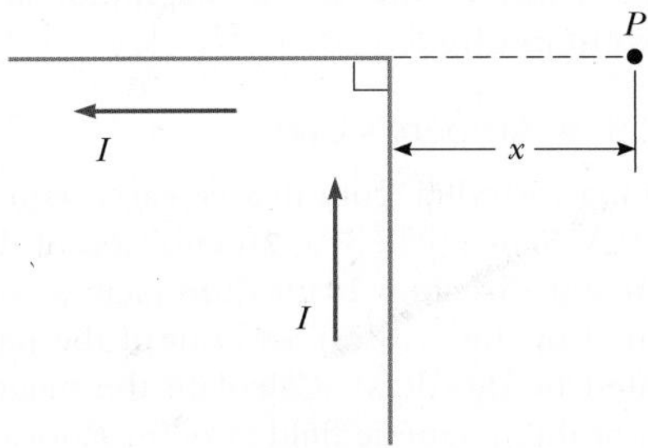

FIGURE **P22.25**

26. Calculate the magnitude of the magnetic field at a point 100 cm from a long, thin conductor carrying a current of 1.00 A.

27. A conductor consists of a circular loop of radius *R* and two straight, long sections as shown in Figure P22.27. The wire lies in the plane of the paper and carries a current *I*. Find an expression for the vector magnetic field at the center of the loop.

FIGURE **P22.27**

28. Consider a flat, circular current loop of radius *R* carrying current *I*. Choose the *x* axis to be along the axis of the loop, with the origin at the center of the loop. Plot a graph of the ratio of the magnitude of the magnetic field at coordinate *x* to that at the origin, for $x = 0$ to $x = 5R$. It may be useful to use a programmable calculator or a computer to solve this problem.

29. Two very long, straight, parallel wires carry currents that are directed perpendicular to the page as shown in Figure P22.29. Wire 1 carries a current I_1 into the page (in the $-z$ direction) and passes through the x axis at $x = +a$. Wire 2 passes through the x axis at $x = -2a$ and carries an unknown current I_2. The total magnetic field at the origin due to the current-carrying wires has the magnitude $2\mu_0 I_1/(2\pi a)$. The current I_2 can have either of two possible values. (a) Find the value of I_2 with the smaller magnitude, stating it in terms of I_1 and giving its direction. (b) Find the other possible value of I_2.

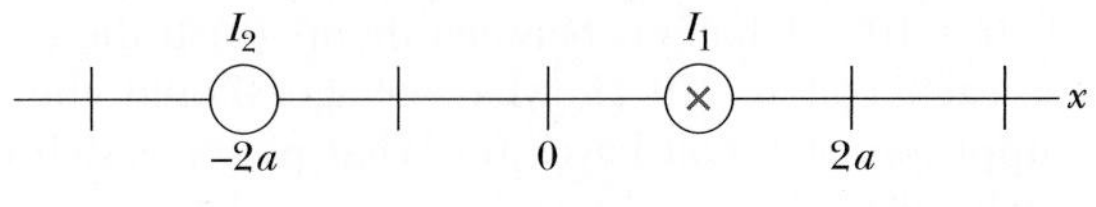

FIGURE **P22.29**

30. One very long wire carries current 30.0 A to the left along the *x* axis. A second very long wire carries current 50.0 A to the right along the line ($y = 0.280$ m, $z = 0$). (a) Where in the plane of the two wires is the total magnetic field equal to zero? (b) A particle with a charge of $-2.00\ \mu$C is moving with a velocity of $150\hat{\mathbf{i}}$ Mm/s along the line ($y = 0.100$ m, $z = 0$). Calculate the vector magnetic force acting on the particle. (c) A uniform electric field is applied to allow this particle to pass through this region undeflected. Calculate the required vector electric field.

31. A current path shaped as shown in Figure P22.31 produces a magnetic field at *P*, the center of the arc. If the arc subtends an angle of 30.0° and the radius of the arc is 0.600 m, what are the magnitude and direction of the field produced at *P* if the current is 3.00 A?

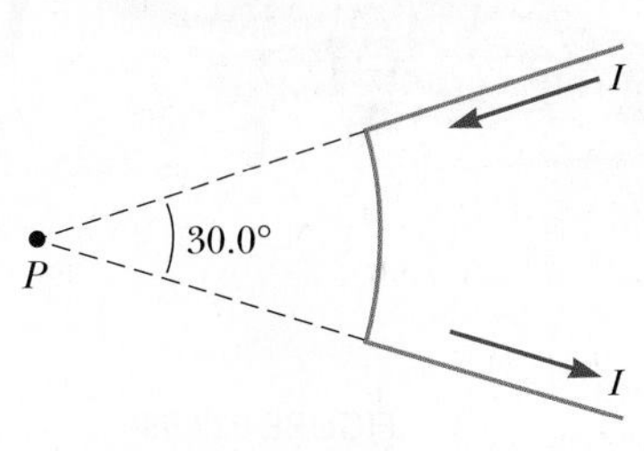

FIGURE **P22.31**

32. Three long, parallel conductors carry currents of $I = 2.00$ A. Figure P22.32 is an end view of the conductors, with each current coming out of the page. Taking $a = 1.00$ cm, determine the magnitude and direction of the magnetic field at points A, B, and C.

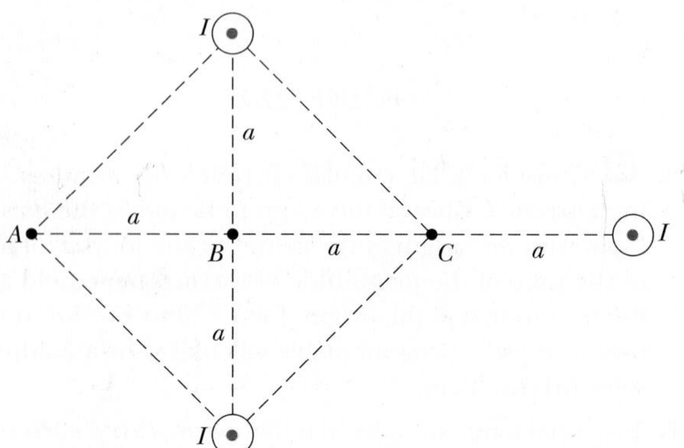

FIGURE **P22.32**

33. In studies of the possibility of migrating birds using the Earth's magnetic field for navigation, birds have been fitted with coils as "caps" and "collars" as shown in Figure P22.33. (a) If the identical coils have radii of 1.20 cm and are 2.20 cm apart, with 50 turns of wire apiece, what current should they both carry to produce a magnetic field of 4.50×10^{-5} T halfway between them? (b) If the resistance of each coil is 210 Ω, what voltage should the battery supplying each coil have? (c) What power is delivered to each coil?

FIGURE **P22.33**

Section 22.8 ■ The Magnetic Force Between Two Parallel Conductors

34. Two long, parallel conductors, separated by 10.0 cm, carry currents in the same direction. The first wire carries current $I_1 = 5.00$ A and the second carries $I_2 = 8.00$ A. (a) What is the magnitude of the magnetic field created by I_1 at the location of I_2? (b) What is the force per unit length exerted by I_1 on I_2? (c) What is the magnitude of the magnetic field created by I_2 at the location of I_1? (d) What is the force per length exerted by I_2 on I_1?

35. In Figure P22.35, the current in the long, straight wire is $I_1 = 5.00$ A and the wire lies in the plane of the rectangular loop, which carries the current $I_2 = 10.0$ A. The dimensions are $c = 0.100$ m, $a = 0.150$ m, and $\ell = 0.450$ m. Find the magnitude and direction of the net force exerted on the loop by the magnetic field created by the wire.

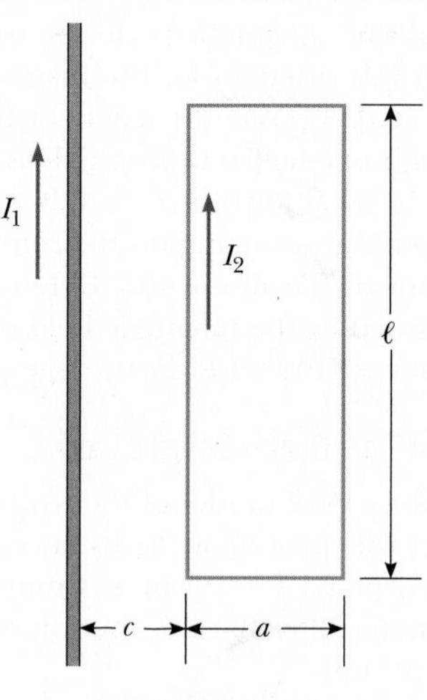

FIGURE **P22.35**

36. Three long wires (wire 1, wire 2, and wire 3) hang vertically. The distance between wire 1 and wire 2 is 20.0 cm. On the left, wire 1 carries an upward current of 1.50 A. To the right, wire 2 carries a downward current of 4.00 A. Wire 3 is located such that when it carries a certain current, each wire experiences no net force. Find (a) the position of wire 3 and (b) the magnitude and direction of the current in wire 3.

Section 22.9 ■ Ampère's Law

37. Four long, parallel conductors carry equal currents of $I = 5.00$ A. Figure P22.37 is an end view of the conductors. The current direction is into the page at points A and B (indicated by the crosses) and out of the page at C and D (indicated by the dots). Calculate the magnitude and direction of the magnetic field at point P, located at the center of the square of edge length 0.200 m.

38. A long, straight wire lies on a horizontal table and carries a current of 1.20 μA. In a vacuum, a proton moves parallel to the wire (opposite the current) with a constant speed of 2.30×10^4 m/s at a distance d above the wire. Determine the value of d. You may ignore the magnetic field due to the Earth.

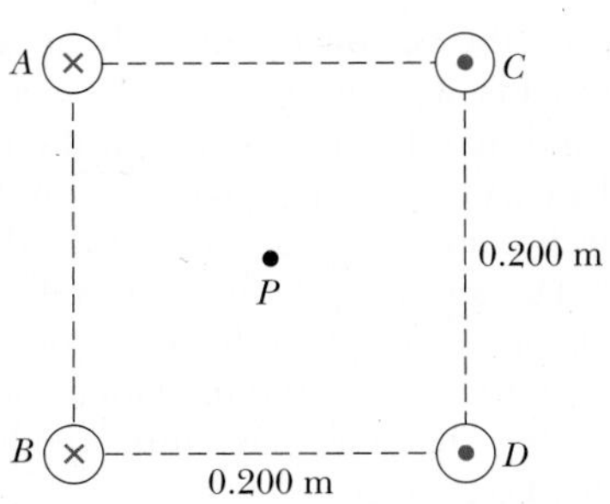

FIGURE P22.37

39. Physics⊗Now™ A packed bundle of 100 long, straight, insulated wires forms a cylinder of radius $R = 0.500$ cm. (a) If each wire carries 2.00 A, what are the magnitude and direction of the magnetic force per unit length acting on a wire located 0.200 cm from the center of the bundle? (b) Would a wire on the outer edge of the bundle experience a force greater or smaller than the value calculated in part (a)?

40. The magnetic field 40.0 cm away from a long, straight wire carrying current 2.00 A is 1.00 μT. (a) At what distance is it 0.100 μT? (b) At one instant, the two conductors in a long household extension cord carry equal 2.00-A currents in opposite directions. The two wires are 3.00 mm apart. Find the magnetic field 40.0 cm away from the middle of the straight cord, in the plane of the two wires. (c) At what distance is it one-tenth as large? (d) The center wire in a coaxial cable carries current 2.00 A in one direction and the sheath around it carries current 2.00 A in the opposite direction. What magnetic field does the cable create at points outside?

41. The magnetic coils of a tokamak fusion reactor are in the shape of a toroid having an inner radius of 0.700 m and an outer radius of 1.30 m. The toroid has 900 turns of large-diameter wire, each of which carries a current of 14.0 kA. Find the magnitude of the magnetic field inside the toroid along (a) the inner radius and (b) the outer radius.

42. Consider a column of electric current in a plasma (ionized gas). Filaments of current within the column are magnetically attracted to one another. They can crowd together to yield a very great current density and a very strong magnetic field in a small region. Sometimes the current can be cut off momentarily by this *pinch effect.* (In a metallic wire, a pinch effect is not important because the current-carrying electrons repel one another with electric forces.) The pinch effect can be demonstrated by making an empty aluminum can carry a large current parallel to its axis. Let R represent the radius of the can and I the upward current, uniformly distributed over its curved wall. Determine the magnetic field (a) just inside the wall and (b) just outside. (c) Determine the pressure on the wall.

43. Niobium metal becomes a superconductor when cooled below 9 K. Its superconductivity is destroyed when the surface magnetic field exceeds 0.100 T. Determine the maximum current a 2.00-mm-diameter niobium wire can carry and remain superconducting, in the absence of any external magnetic field.

Section 22.10 ▪ The Magnetic Field of a Solenoid

44. A single-turn square loop of wire, 2.00 cm on each edge, carries a clockwise current of 0.200 A. The loop is inside a solenoid, with the plane of the loop perpendicular to the magnetic field of the solenoid. The solenoid has 30 turns/cm and carries a clockwise current of 15.0 A. Find the force on each side of the loop and the torque acting on the loop.

45. What current is required in the windings of a long solenoid that has 1 000 turns uniformly distributed over a length of 0.400 m to produce at the center of the solenoid a magnetic field of magnitude 1.00×10^{-4} T?

46. Consider a solenoid of length ℓ and radius R, containing N closely spaced turns and carrying a steady current I. (a) In terms of these parameters, find the magnetic field at a point along the axis as a function of distance a from the end of the solenoid. (b) Show that as ℓ becomes very long, B approaches $\mu_0 NI/2\ell$ at each end of the solenoid.

47. A solenoid 10.0 cm in diameter and 75.0 cm long is made from copper wire of diameter 0.100 cm, with very thin insulation. The wire is wound onto a cardboard tube in a single layer, with adjacent turns touching each other. To produce a field of 8.00 mT at the center of the solenoid, what power must be delivered to the solenoid?

Section 22.11 ▪ Magnetism in Matter

48. In Bohr's 1913 model of the hydrogen atom, the electron is in a circular orbit of radius 5.29×10^{-11} m and its speed is 2.19×10^{6} m/s. (a) What is the magnitude of the magnetic moment due to the electron's motion? (b) If the electron moves in a horizontal circle, counterclockwise as seen from above, what is the direction of this magnetic moment vector?

49. The magnetic moment of the Earth is approximately 8.00×10^{22} A·m^2. (a) If it were caused by the complete magnetization of a huge iron deposit, how many unpaired electrons would participate? (b) At two unpaired electrons per iron atom, how many kilograms of iron would that correspond to? (Iron has a density of 7 900 kg/m^3 and approximately 8.50×10^{28} iron atoms/m^3.)

Section 22.12 ▪ Context Connection—The Attractive Model for Magnetic Levitation

50. The following represents a crude model for levitating a commercial transportation vehicle. Suppose the levitation is achieved by mounting small electrically charged spheres below the vehicle. The spheres pass through a magnetic field established by permanent magnets placed along the track. Let us assume that the permanent magnets produce a uniform magnetic field of 0.1 T at the location of the spheres and that an electronic control system maintains a charge of 1 μC on each sphere. The vehicle has a mass of 5×10^{4} kg and travels at a speed of 400 km/h. How many charged spheres are required to support the weight of the vehicle at this speed? Your answer should suggest that this design would not be practical as a means of magnetic levitation.

51. Data for the Transrapid maglev system show that the input electric power required to operate the vehicle is on the order of 10^{2} kW. (a) Assume that the Transrapid vehicle

moves at 400 km/h. Approximately how much energy, in joules, is used for each mile of travel for the vehicle? (b) Calculate the energy per mile used by an automobile that achieves 20 mi/gal. The energy available from gasoline is approximately 40 MJ/kg, a typical automobile engine efficiency is 20%, and the density of gasoline is 754 kg/m^3. (c) Considering 1 passenger in the automobile and 100 on the Transrapid vehicle, the energy per mile necessary for each passenger in the Transrapid is what fraction of that for an automobile?

Additional Problems

52. Consider a thin, straight wire segment carrying a constant current I and placed along the x axis as shown in Figure P22.52. (a) Use the Biot–Savart law to show that the total magnetic field at the point P, located a distance a from the wire, is

$$B = \frac{\mu_0 I}{4\pi a}(\cos\theta_1 - \cos\theta_2)$$

(b) Assuming that the wire is infinitely long, show that the result in part (a) gives a magnetic field that agrees with that obtained by using Ampère's law in Example 22.7.

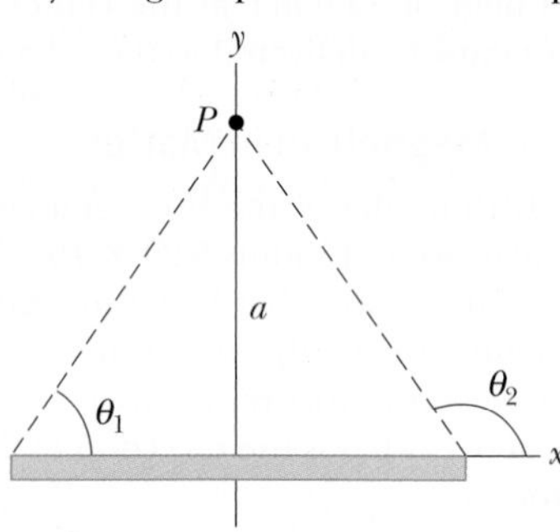

FIGURE P22.52

53. An infinite sheet of current lying in the yz plane carries a surface current of density $\vec{\mathbf{J}}_s$. The current is in the y direction, and J_s represents the current per unit length measured along the z axis. Figure P22.53 is an edge view of the sheet. Find the magnetic field near the sheet. (*Suggestion:* Use Ampère's law and evaluate the line integral for a rectangular path around the sheet, represented by the dashed line in Fig. P22.53.)

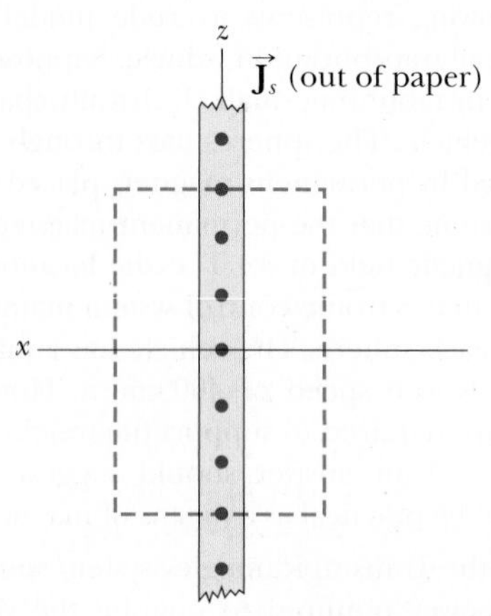

FIGURE P22.53

54. Assume that the region to the right of a certain vertical plane contains a vertical magnetic field of magnitude 1.00 mT and that the field is zero in the region to the left of the plane. An electron, originally traveling perpendicular to the boundary plane, passes into the region of the field. (a) Determine the time interval required for the electron to leave the "field-filled" region, noting that its path is a semicircle. (b) Find the kinetic energy of the electron assuming that the maximum depth of penetration into the field is 2.00 cm.

55. Heart–lung machines and artificial kidney machines employ blood pumps. A mechanical pump can mangle blood cells. Figure P22.55 represents an electromagnetic pump. The blood is confined to an electrically insulating tube, cylindrical in practice but represented as a rectangle of width w and height h. The simplicity of design makes the pump dependable. The blood is easily kept uncontaminated; the tube is simple to clean or cheap to replace. Two electrodes fit into the top and bottom of the tube. The potential difference between them establishes an electric current through the blood, with current density J over a section of length L. A perpendicular magnetic field exists in the same region. (a) Explain why this arrangement produces on the liquid a force that is directed along the length of the pipe. (b) Show that the section of liquid in the magnetic field experiences a pressure increase JLB. (c) After the blood leaves the pump, is it charged? Is it current-carrying? Is it magnetized? The same magnetic pump can be used for any fluid that conducts electricity, such as liquid sodium in a nuclear reactor.

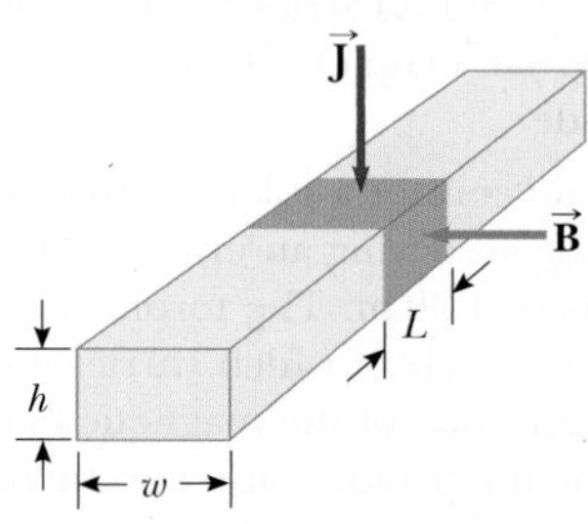

FIGURE P22.55

56. A 0.200-kg metal rod carrying a current of 10.0 A glides on two horizontal rails 0.500 m apart. What vertical magnetic field is required to keep the rod moving at a constant speed if the coefficient of kinetic friction between the rod and rails is 0.100?

57. A positive charge $q = 3.20 \times 10^{-19}$ C moves with a velocity $\vec{\mathbf{v}} = (2\hat{\mathbf{i}} + 3\hat{\mathbf{j}} - \hat{\mathbf{k}})$ m/s through a region where both a uniform magnetic field and a uniform electric field exist. (a) Calculate the total force on the moving charge (in unit-vector notation), taking $\vec{\mathbf{B}} = (2\hat{\mathbf{i}} + 4\hat{\mathbf{j}} + \hat{\mathbf{k}})$ T and $\vec{\mathbf{E}} = (4\hat{\mathbf{i}} - \hat{\mathbf{j}} - 2\hat{\mathbf{k}})$ V/m. (b) What angle does the force vector make with the positive x axis?

58. Protons having a kinetic energy of 5.00 MeV are moving in the positive x direction and enter a magnetic field $\vec{\mathbf{B}} = 0.050\,0\hat{\mathbf{k}}$ T directed out of the plane of the page and extending from $x = 0$ to $x = 1.00$ m as shown in Figure P22.58. (a) Calculate the y component of the protons'

momentum as they leave the magnetic field. (b) Find the angle α between the initial velocity vector of the proton beam and the velocity vector after the beam emerges from the field. Ignore relativistic effects and note that $1 \text{ eV} = 1.60 \times 10^{-19}$ J.

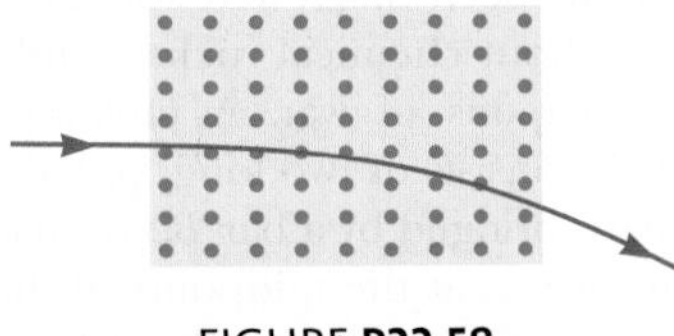

FIGURE **P22.58**

59. A handheld electric mixer contains an electric motor. Model the motor as a single flat, compact, circular coil carrying electric current in a region where a magnetic field is produced by an external permanent magnet. You need consider only one instant in the operation of the motor. (We will consider motors again in Chapter 23.) The coil moves because the magnetic field exerts torque on the coil as described in Section 22.6. Make order-of-magnitude estimates of the magnetic field, the torque on the coil, the current in it, its area, and the number of turns in the coil, so that they are related according to Equation 22.16. Note that the input power to the motor is electric, given by $\mathcal{P} = I\Delta V$, and the useful output power is mechanical, $\mathcal{P} = \tau\omega$.

60. A cyclotron is sometimes used for carbon dating as will be described in Chapter 30. Carbon-14 and carbon-12 ions are obtained from a sample of the material to be dated and are accelerated in the cyclotron. If the cyclotron has a magnetic field of magnitude 2.40 T, what is the difference in cyclotron frequencies for the two ions?

61. A uniform magnetic field of magnitude 0.150 T is directed along the positive x axis. A positron moving at 5.00×10^6 m/s enters the field along a direction that makes an angle of 85.0° with the x axis (Fig. P22.61). The motion of the particle is expected to be a helix as described in Section 22.3. Calculate (a) the pitch p and (b) the radius r of the trajectory.

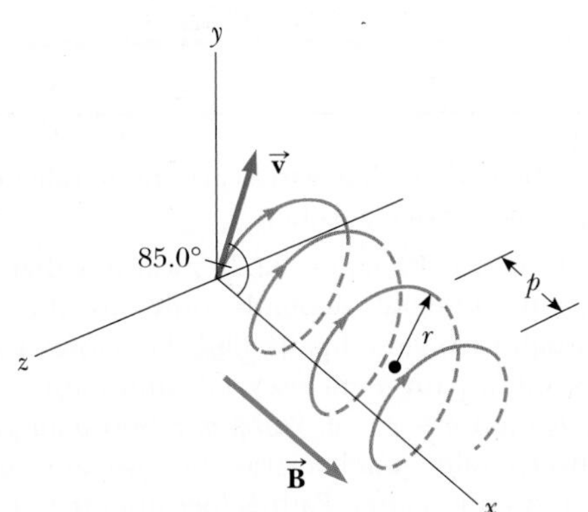

FIGURE **P22.61**

62. A heart surgeon monitors the flow rate of blood through an artery using an electromagnetic flowmeter (Fig. P22.62). Electrodes A and B make contact with the outer surface of the blood vessel, which has interior diameter 3.00 mm. (a) For a magnetic field magnitude of 0.040 0 T, an emf of 160 μV appears between the electrodes. Calculate the speed of the blood. (b) Verify that electrode A is positive as shown. Does the sign of the emf depend on whether the mobile ions in the blood are predominantly positively or negatively charged? Explain.

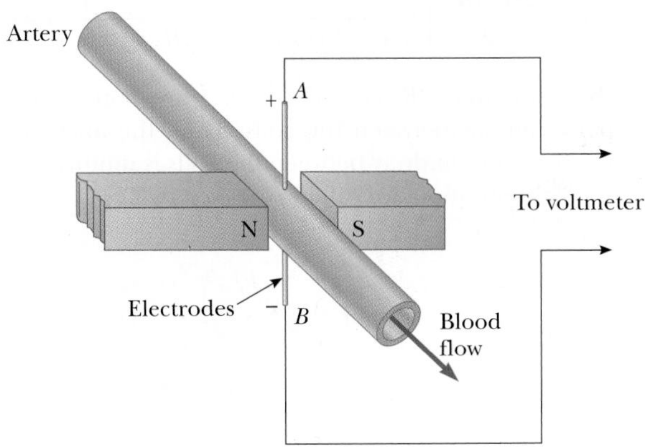

FIGURE **P22.62**

63. A very long, thin strip of metal of width w carries a current I along its length as shown in Figure P22.63. Find the magnetic field at the point P in the diagram. The point P is in the plane of the strip at distance b away from it.

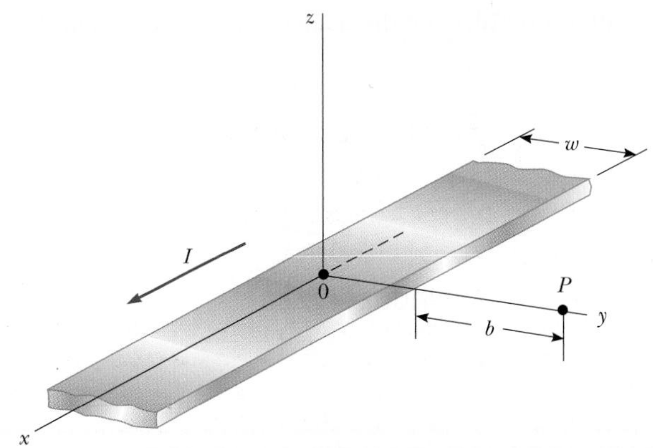

FIGURE **P22.63**

64. The magnitude of the Earth's magnetic field at either pole is approximately 7.00×10^{-5} T. Suppose the field fades away, before its next reversal. Scouts, sailors, and conservative politicians around the world join together in a program to replace the field. One plan is to use a current loop around the equator, without relying on magnetization of any materials inside the Earth. Determine the current that would generate such a field if this plan were carried out. (Take the radius of the Earth as $R_E = 6.37 \times 10^6$ m.)

65. A nonconducting ring of radius R is uniformly charged with a total positive charge q. The ring rotates at a constant angular speed ω about an axis through its center, perpendicular to the plane of the ring. What is the magnitude of the magnetic field on the axis of the ring a distance $R/2$ from its center?

66. Two circular coils of radius R, each with N turns, are perpendicular to a common axis. The coil centers are a distance R apart. Each coil carries a steady current I in the same direction as shown in Figure P22.66. (a) Show that the magnetic field on the axis at a distance x from the center of one coil is

$$B = \frac{N\mu_0 IR^2}{2}\left[\frac{1}{(R^2 + x^2)^{3/2}} + \frac{1}{(2R^2 + x^2 - 2Rx)^{3/2}}\right]$$

(b) Show that dB/dx and d^2B/dx^2 are both zero at the point midway between the coils. Thus, the magnetic field in the region midway between the coils is uniform. Coils in this configuration are called *Helmholtz coils.*

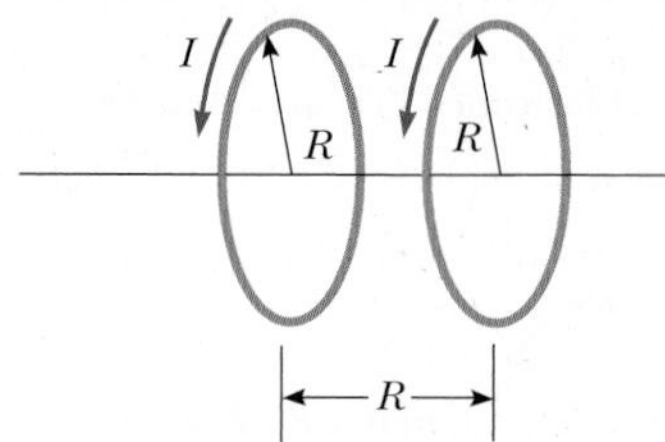

FIGURE **P22.66**

67. Two circular loops are parallel, coaxial, and almost in contact, 1.00 mm apart (Fig. P22.67). Each loop is 10.0 cm in radius. The top loop carries a clockwise current of 140 A. The bottom loop carries a counterclockwise current of 140 A. (a) Calculate the magnetic force exerted by the bottom loop on the top loop. (b) The upper loop has a mass of 0.021 0 kg. Calculate its acceleration, assuming that the only forces acting on it are the force in part (a) and the gravitational force. (*Suggestion:* Think about how one loop looks to a bug perched on the other loop.)

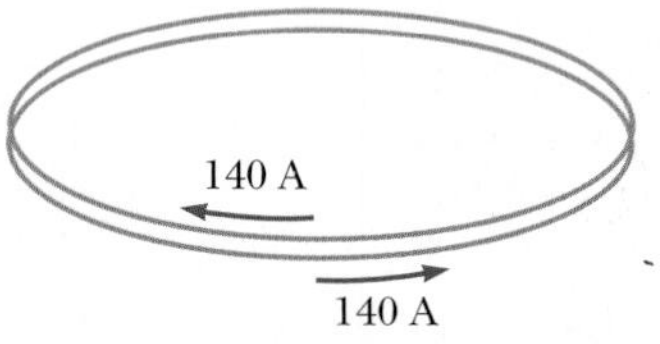

140 A

140 A

FIGURE **P22.67**

68. *Rail guns* have been suggested for launching projectiles into space without chemical rockets and for ground-to-air antimissile weapons of war. A tabletop model rail gun (Fig. P22.68) consists of two long, parallel, horizontal rails 3.50 cm apart, bridged by a bar BD of mass 3.00 g. The bar is originally at rest at the midpoint of the rails and is free to slide without friction. When the switch is closed, electric current is quickly established in the circuit $ABCDEA$. The rails and bar have low electric resistance, and the current is limited to a constant 24.0 A by the power supply. (a) Find the magnitude of the magnetic field 1.75 cm from a single very long, straight wire carrying current 24.0 A. (b) Find the magnitude and direction of the magnetic field at point C in the diagram, the midpoint of the bar, immediately after the switch is closed. (*Suggestion:* Consider what conclusions you can draw from the Biot–Savart law.) (c) At other points along the bar BD, the field is in the same direction as at point C but is larger in magnitude. Assume that the average effective magnetic field along BD is five times larger than the field at C. With this assumption, find the magnitude and direction of the force on the bar. (d) Find the acceleration of the bar when it is in motion. (e) Does the bar move with constant acceleration? (f) Find the velocity of the bar after it has traveled 130 cm to the end of the rails.

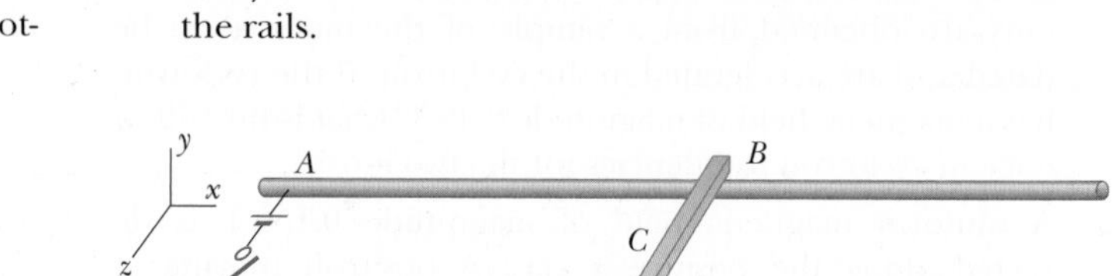

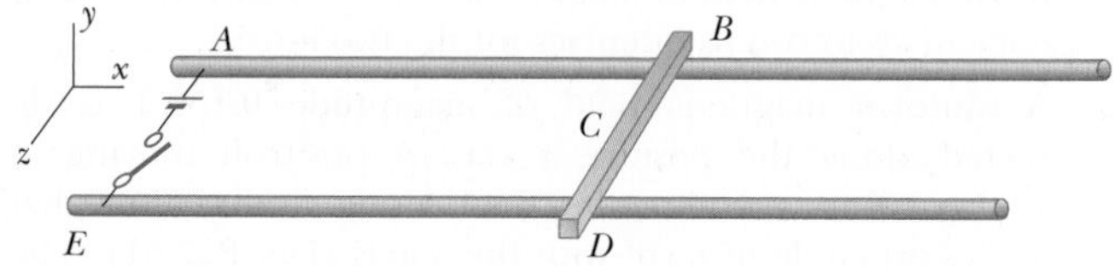

FIGURE **P22.68**

ANSWERS TO QUICK QUIZZES

22.1 (e). The right-hand rule gives the direction. Be sure to account for the negative charge on the electron.

22.2 (i), (a). The magnetic force on the particle increases in proportion to B. The result is a smaller radius, as we can see from Equation 22.3. (ii), (b). The magnetic force on the particle increases in proportion to v, but the centripetal acceleration increases according to the square of v. The result is a larger radius, as we can see from Equation 22.3.

22.3 (c). The right-hand rule is used to determine the direction of the magnetic field.

22.4 B, C, A. Point B is closest to the current element. Point C is farther away, and the field is further reduced by the $\sin\theta$ factor in the cross product $d\vec{s} \times \hat{r}$. The field at A is zero because $\theta = 0$.

22.5 (a). The coils act like wires carrying parallel currents and hence attract one another.

22.6 (a) b, d, a, c. Equation 22.29 indicates that the value of the line integral depends only on the net current through each closed path. Path b encloses 1 A, path d encloses 3 A, path a encloses 4 A, and path c encloses 6 A. (b) b, then $a = c = d$. Paths a, c, and d all give the same nonzero value $\mu_0 I$ because the size and shape of the paths do not matter. Path b does not enclose the current, and hence its line integral is zero.

22.7 (c). The magnetic field in a very long solenoid is independent of its length or radius. Overwrapping with an additional layer of wire increases the number of turns per unit length.

Faraday's Law and Inductance

In a commercial electric power plant, large generators transform energy that is transferred out of the plant by electrical transmission. These generators use *magnetic induction* to generate a potential difference when coils of wire in the generator are rotated in a magnetic field. The source of energy to rotate the coils might be falling water, burning fossil fuels, or a nuclear reaction.

(Michael Melford/Getty Images)

CHAPTER OUTLINE

Our studies in electromagnetism so far have been concerned with the electric fields due to stationary charges and the magnetic fields produced by moving charges. This chapter introduces a new type of electric field, one that is due to a changing magnetic field.

As we learned in Section 19.1, experiments conducted by Michael Faraday in England in the early 1800s and independently by Joseph Henry in the United States showed that an electric current can be induced in a circuit by a changing magnetic field. The results of those experiments led to a very basic and important law of electromagnetism known as *Faraday's law of induction.* Faraday's law explains how generators, as well as other practical devices, work.

23.1 FARADAY'S LAW OF INDUCTION

We begin discussing the concepts in this chapter by considering a simple experiment that builds on material presented in Chapter 22. Imagine that a straight metal wire resides in a uni-

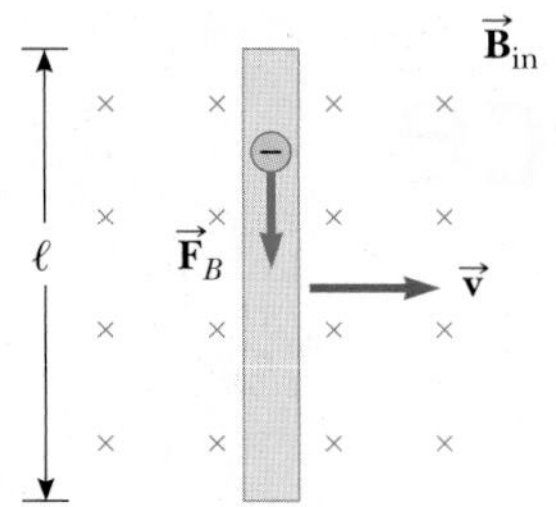

FIGURE 23.1 A straight electrical conductor of length ℓ moving with a velocity $\vec{\mathbf{v}}$ through a uniform magnetic field $\vec{\mathbf{B}}$ directed perpendicular to $\vec{\mathbf{v}}$. A current is induced in the conductor due to the magnetic force on charged particles in the conductor.

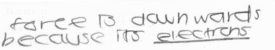

[illegible] s in Figure 23.1. Within the wire, [illegible] ow moved with a velocity $\vec{\mathbf{v}}$ toward [illegible] c force acts on the electrons in the [illegible] force on the electrons is downward in Figure 23.1 (remember that the electrons carry a negative charge). Because this direction is along the wire, the electrons move along the wire in response to this force. Thus, a *current* is produced in the wire as it moves through a magnetic field!

Let us consider another simple experiment that demonstrates that an electric current can be produced by a magnetic field. Consider a loop of wire connected to a sensitive ammeter, a device that measures current, as illustrated in Active Figure 23.2. If a magnet is moved toward the loop, the ammeter needle deflects in one direction as in Active Figure 23.2a. When the magnet is held stationary as in Active Figure 23.2b, the needle is not deflected. If the magnet is moved away from the loop as in Active Figure 23.2c, the ammeter needle deflects in the opposite direction from that caused by the motion of the magnet toward the ammeter. Finally, if the magnet is held stationary and the coil is moved either toward or away from it, the needle deflects. From these observations comes the conclusion that **an electric current is set up in the coil as long as relative motion occurs between the magnet and the coil.**

These results are quite remarkable when we consider that a current exists in a wire even though no batteries are connected to the wire. We call such a current an **induced current,** and it is produced by an **induced emf.**

ACTIVE FIGURE 23.2

(a) When a magnet is moved toward a loop of wire connected to a sensitive ammeter, the ammeter needle deflects as shown, indicating that a current is induced in the loop.
(b) When the magnet is held stationary, no current is induced in the loop, even when the magnet is inside the loop. (c) When the magnet is moved away from the loop, the ammeter needle deflects in the opposite direction, indicating that the induced current is opposite that shown in (a).

Physics Now™ By logging into PhysicsNow at **www.pop4e.com** and going to Active Figure 23.2, you can move the magnet and observe the current in the ammeter.

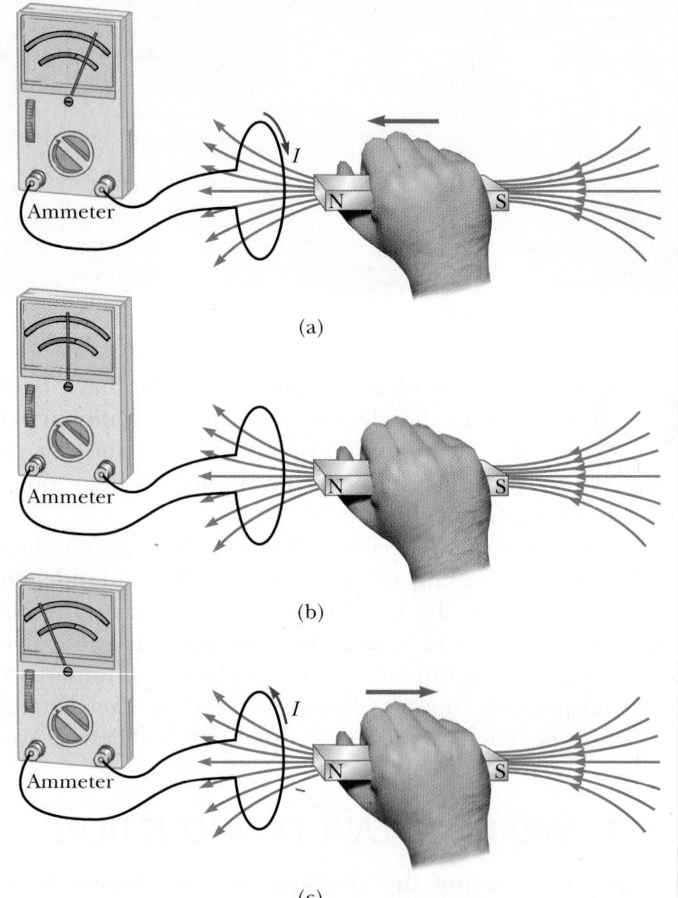

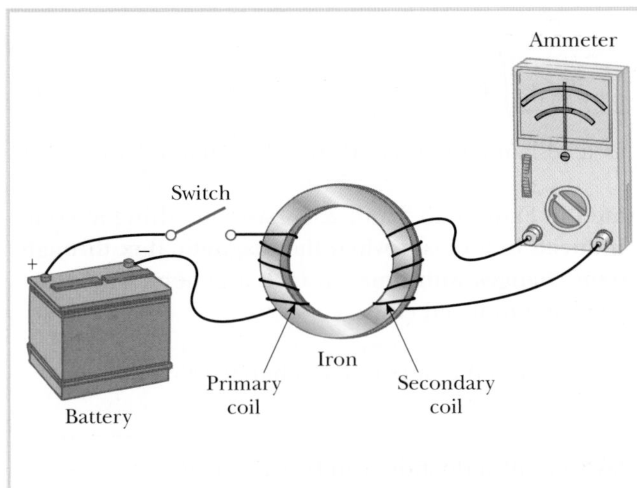

ACTIVE FIGURE 23.3 Faraday's experiment. When the switch in the primary circuit is closed, the ammeter needle in the secondary circuit deflects momentarily. The emf induced in the secondary circuit is caused by the changing magnetic field through the secondary coil.

Physics Now™ By logging into PhysicsNow at **www.pop4e.com** and going to Active Figure 23.3, you can open and close the switch and observe the current in the ammeter.

(By kind permission of the President and Council of the Royal Society)

MICHAEL FARADAY (1791–1867)

Faraday, a British physicist and chemist, is often regarded as the greatest experimental scientist of the 1800s. His many contributions to the study of electricity include the invention of the electric motor, the electric generator, and the transformer as well as the discovery of electromagnetic induction and the laws of electrolysis. Greatly influenced by his religious beliefs, he refused to work on the development of poison gas for the British military.

Another experiment, first conducted by Faraday, is illustrated in Active Figure 23.3. Part of the apparatus consists of a coil of insulated wire connected to a switch and a battery. We shall refer to this coil as the *primary coil* of wire and to the corresponding circuit as the primary circuit. The coil is wrapped around an iron ring to intensify the magnetic field produced by the current through the coil. A second coil of insulated wire at the right is also wrapped around the iron ring and is connected to a sensitive ammeter. We shall refer to this coil as the *secondary coil* and to the corresponding circuit as the secondary circuit. The secondary circuit has no battery, and the secondary coil is not electrically connected to the primary coil. The purpose of this apparatus is to detect any current that might be generated in the secondary circuit by a change in the magnetic field produced by the primary circuit.

Initially, you might guess that no current would ever be detected in the secondary circuit. Something quite surprising happens, however, when the switch in the primary circuit is opened or thrown closed. At the instant the switch is thrown closed, the ammeter needle deflects in one direction and then returns to zero. When the switch is opened, the ammeter needle deflects in the opposite direction and then again returns to zero. Finally, the ammeter reads zero when the primary circuit carries a steady current.

As a result of these observations, Faraday concluded that **an electric current can be produced by a time-varying magnetic field.** A current cannot be produced by a steady magnetic field. In the experiment shown in Active Figure 23.2, the changing magnetic field is a result of the relative motion between the magnet and the loop of wire. As long as the motion persists, the current is maintained. In the experiment shown in Active Figure 23.3, the current produced in the secondary circuit occurs for only an instant after the switch is closed while the magnetic field acting on the secondary coil builds from its zero value to its final value. In effect, the secondary circuit behaves as though a source of emf were connected to it for an instant. It is customary to say that an emf is induced in the secondary circuit by the changing magnetic field produced by the current in the primary circuit.

To quantify such observations, we define a quantity called **magnetic flux.** The flux associated with a magnetic field is defined in a similar manner to the electric flux (Section 19.8) and is proportional to the number of magnetic field lines passing through an area. Consider an element of area dA on an arbitrarily shaped open surface as in Figure 23.4. If the magnetic field at the location of this element is $\vec{\mathbf{B}}$, the magnetic flux through the element is $\vec{\mathbf{B}} \cdot d\vec{\mathbf{A}}$, where $d\vec{\mathbf{A}}$ is a vector perpendicular to the surface whose magnitude equals the area dA. Hence, the total magnetic

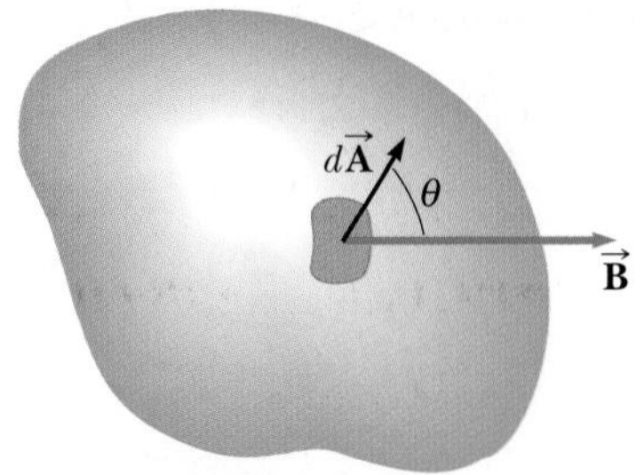

FIGURE 23.4 The magnetic flux through an area element $d\vec{\mathbf{A}}$ is given by $\vec{\mathbf{B}} \cdot d\vec{\mathbf{A}} = B\,dA\cos\theta$. Note that vector $d\vec{\mathbf{A}}$ is perpendicular to the surface.

flux Φ_B through the surface is

■ Magnetic flux

$$\Phi_B = \int \vec{\mathbf{B}} \cdot d\vec{\mathbf{A}} \qquad [23.1]$$

The SI unit of magnetic flux is a tesla-meter squared, which is named the *weber* (Wb); 1 Wb = 1 T · m^2.

The two experiments illustrated in Figures 23.2 and 23.3 have one thing in common. In both cases, **an emf is induced in a circuit when the magnetic flux through the surface bounded by the circuit changes with time.** In fact, a general statement summarizes such experiments involving induced emfs:

PITFALL PREVENTION 23.1

INDUCED EMF REQUIRES A CHANGE IN FLUX Remember that the *existence* of a magnetic flux through an area is not sufficient to create an induced emf. A *change* in the magnetic flux must occur for an emf to be induced.

> The emf induced in a circuit is equal to the time rate of change of magnetic flux through the circuit.

This statement, known as **Faraday's law of induction,** can be written as

■ Faraday's law

$$\mathcal{E} = -\frac{d\Phi_B}{dt} \qquad [23.2]$$

where Φ_B is the magnetic flux through the surface bounded by the circuit and is given by Equation 23.1. The negative sign in Equation 23.2 will be discussed in Section 23.3. If the circuit is a coil consisting of N identical and concentric loops and if the field lines pass through all loops, the induced emf is

$$\mathcal{E} = -N\frac{d\Phi_B}{dt} \qquad [23.3]$$

The emf is increased by the factor N because all the loops are in series, so the emfs in the individual loops add to give the total emf.

Suppose the magnetic field is uniform over the area A bounded by a loop lying in a plane as in Figure 23.5. In this case, the magnetic flux through the loop is

$$\Phi_B = \int \vec{\mathbf{B}} \cdot d\vec{\mathbf{A}} = \int B\,dA\cos\theta = B\cos\theta \int dA = BA\cos\theta$$

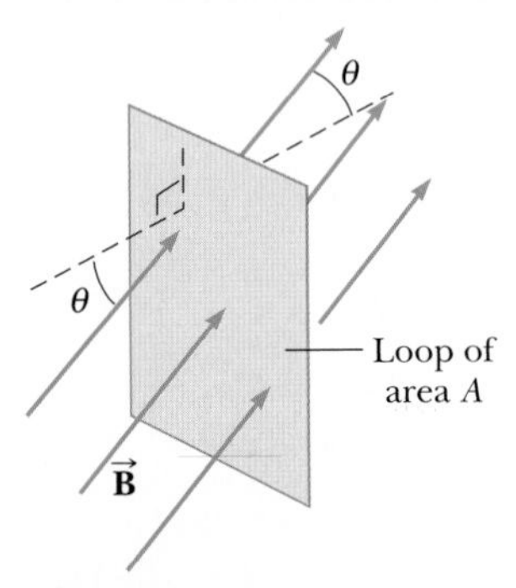

FIGURE 23.5 A conducting loop that encloses an area A in the presence of a uniform magnetic field $\vec{\mathbf{B}}$. The angle between $\vec{\mathbf{B}}$ and the normal to the loop is θ.

Hence, the induced emf is

$$\mathcal{E} = -\frac{d}{dt}(BA\cos\theta) \qquad [23.4]$$

This expression shows that an emf can be induced in a circuit by changing the magnetic flux in several ways: (1) the magnitude of $\vec{\mathbf{B}}$ can vary with time, (2) the area A of the circuit can change with time, (3) the angle θ between $\vec{\mathbf{B}}$ and the normal to the plane can change with time, and (4) any combination of these changes can occur.

An interesting application of Faraday's law is the production of sound in an electric guitar (Fig. 23.6). The coil in this case, called the *pickup coil,* is placed near the vibrating guitar string, which is made of a metal that can be magnetized. A permanent magnet inside the coil magnetizes the portion of the string nearest the coil. When the string vibrates at some frequency, its magnetized segment produces a changing magnetic flux through the coil. The changing flux induces an emf in the coil that is fed to an amplifier. The output of the amplifier is sent to the loudspeakers, which produce the sound waves we hear.

QUICK QUIZ 23.1 A circular loop of wire is held in a uniform magnetic field, with the plane of the loop perpendicular to the field lines. Which of the following will *not* cause a current to be induced in the loop? **(a)** crushing the loop **(b)** rotating the loop about an axis perpendicular to the field lines **(c)** keeping the orientation of the loop fixed and moving it along the field lines **(d)** pulling the loop out of the field

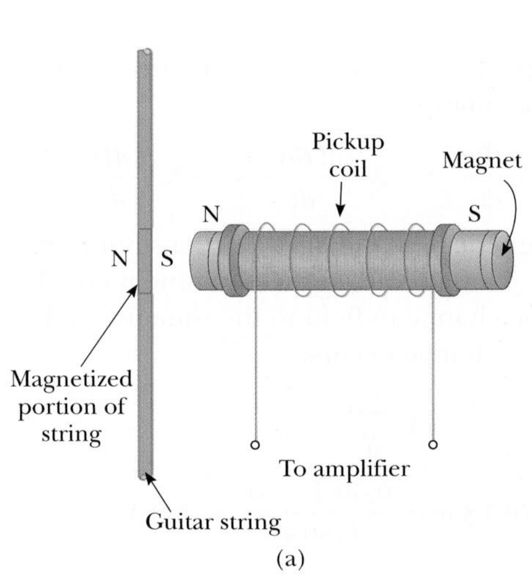

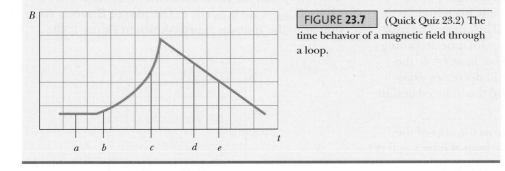

FIGURE 23.6 (a) In an electric guitar, a vibrating magnetized string induces an emf in a pickup coil. (b) The pickups (the circles beneath the metallic strings) of this electric guitar detect the vibrations of the strings and send this information through an amplifier and into speakers. (A switch on the guitar allows the musician to select which set of six pickups is used.)

QUICK QUIZ 23.2 Figure 23.7 shows a graphical representation of the field magnitude versus time for a magnetic field that passes through a fixed loop and that is oriented perpendicular to the plane of the loop. The magnitude of the magnetic field at any time is uniform over the area of the loop. Rank the magnitudes of the emf generated in the loop at the five instants indicated, from largest to smallest.

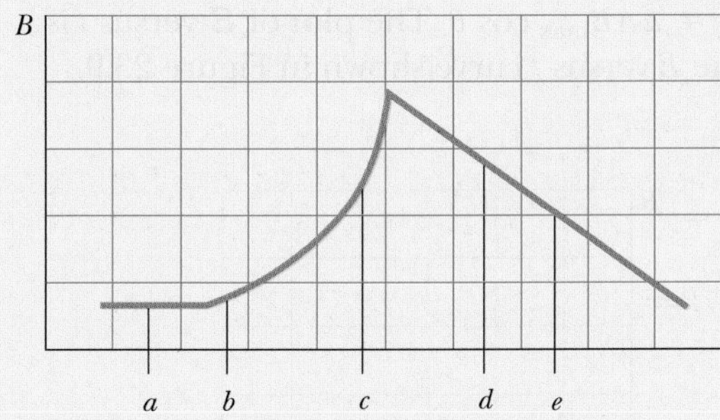

FIGURE 23.7 (Quick Quiz 23.2) The time behavior of a magnetic field through a loop.

■ Thinking Physics 23.1

The ground fault interrupter (GFI) is a safety device that protects users of electric power against electric shock when they touch appliances. Its essential parts are shown in Figure 23.8. How does the operation of a GFI make use of Faraday's law?

Reasoning Wire 1 leads from the wall outlet to the appliance being protected, and wire 2 leads from the appliance back to the wall outlet. An iron ring surrounds the two wires. A sensing coil wrapped around part of the iron ring activates a circuit breaker when changes in magnetic flux occur. Because the currents in the two wires are in opposite directions during normal operation of the appliance, the net magnetic field through the sensing coil due to the currents is zero. A change in magnetic flux through the sensing coil can happen, however, if one of the wires on the appliance loses its insulation and accidentally touches the metal case of the appliance, providing a direct path to ground. When such a short to ground occurs, a net magnetic flux occurs through the sensing coil that alternates in time because household current is alternating. This changing flux produces an induced voltage in the coil, which in turn triggers a circuit breaker, stopping the current before it reaches a level that might be harmful to the person using the appliance. ■

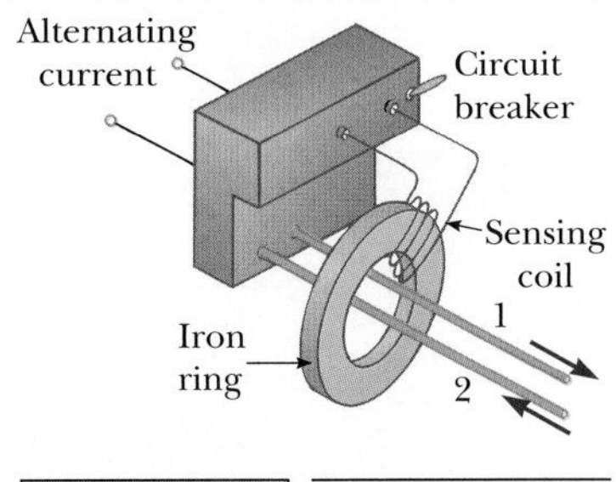

FIGURE 23.8 (Thinking Physics 23.1) Essential components of a ground fault interrupter.

EXAMPLE 23.1 **One Way to Induce an emf in a Coil**

A coil is wrapped with 200 turns of wire on the perimeter of a square frame with sides of 18 cm. Each turn has the same area, equal to that of the frame, and the total resistance of the coil is 2.0 Ω. A magnetic field is perpendicular to the plane of the coil and has the same magnitude at all points within the area of the coil at any time. If the field magnitude changes at a constant rate from 0 to 0.50 T in a time of 0.80 s, find the magnitude of the induced emf in the coil while the field is changing.

Solution Because the field is uniform across the area of the coil and perpendicular to the turns of wire, the magnetic flux at any time is simply the product of the magnitude of the field and the area of the turns, and Equation 23.3 becomes

$$\mathcal{E} = -N\frac{d\Phi_B}{dt} = -N\frac{d(BA)}{dt} = -NA\frac{dB}{dt}$$

Because the magnetic field changes at a constant rate, the derivative of the field with respect to time is equal to the ratio of the change in field to the time interval during which that change occurs:

$$|\mathcal{E}| = NA\frac{dB}{dt} = NA\frac{\Delta B}{\Delta t}$$

$$= (200)(0.18\text{ m})^2\frac{0.50\text{ T} - 0}{0.80\text{ s}} = 4.0\text{ V}$$

EXAMPLE 23.2 **An Exponentially Decaying B Field**

A plane loop of wire of area A is placed in a region where the magnetic field is at a fixed angle θ to the normal to the plane and has the same magnitude at all points within the area of the coil at any time. The magnitude of the magnetic field varies with time according to the expression $B = B_{\text{max}}e^{-at}$. That is, at $t = 0$, the field is B_{max}, and for $t > 0$, the field decreases exponentially with time (Fig. 23.9). Find the induced emf in the loop as a function of time.

Solution Because $\vec{\mathbf{B}}$ is uniform across the area of the coil, the magnetic flux through the loop at time $t > 0$ is

$$\Phi_B = BA\cos\theta = AB_{\text{max}}\cos\theta e^{-at}$$

Because the coefficient AB_{max} and the parameter a are constants, the induced emf from Equation 23.2 is

$$\mathcal{E} = -\frac{d\Phi_B}{dt} = -AB_{\text{max}}\cos\theta\frac{d}{dt}e^{-at}$$

$$= aAB_{\text{max}}\cos\theta e^{-at}$$

That is, the induced emf decays exponentially with time. Note that the maximum emf occurs at $t = 0$, where $\mathcal{E}_{\text{max}} = aAB_{\text{max}}\cos\theta$. The plot of $\mathcal{E}$ versus t is similar to the B versus t curve shown in Figure 23.9.

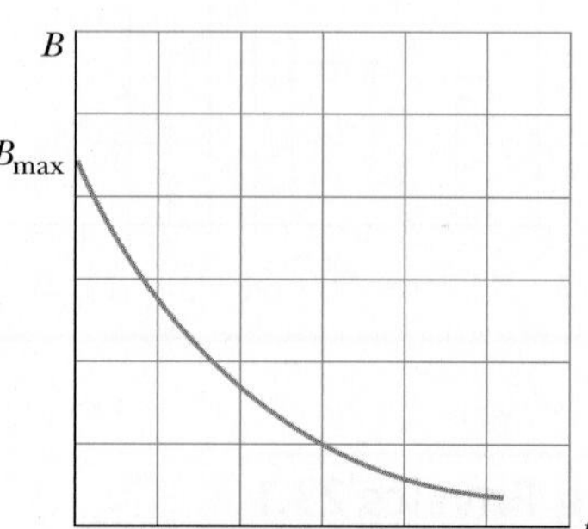

FIGURE 23.9 (Example 23.2) Exponential decrease in the magnitude of the magnetic field with time. The induced emf and induced current vary with time in the same way.

23.2 MOTIONAL emf

Examples 23.1 and 23.2 are cases in which an emf is produced in a circuit when the magnetic field changes with time. In this section, we describe **motional emf,** in which an emf is induced in a conductor moving through a magnetic field. This is the situation described in Figure 23.1 at the beginning of Section 23.1.

Consider a straight conductor of length ℓ moving with constant velocity through a uniform magnetic field directed into the page as in Figure 23.10. For simplicity, we shall assume that the conductor is moving with a velocity that is perpendicular to the field. The electrons in the conductor experience a force along the conductor with magnitude $|\vec{\mathbf{F}}_B| = |q\vec{\mathbf{v}} \times \vec{\mathbf{B}}| = qvB$. According to Newton's second law, the electrons accelerate in response to this force and move along the wire. Once the electrons move to the lower end of the wire, they accumulate there, leaving a net positive charge at the upper end. As a result of this charge separation, an electric field $\vec{\mathbf{E}}$ is produced within the conductor. The charge at the ends of the conductor builds up until the magnetic force qvB on an electron in the conductor is balanced by the electric force qE on the electron as shown in Figure 23.10. At this point, charge stops flowing. In this situation, the zero net force on an electron allows us to relate the electric field to the magnetic field:

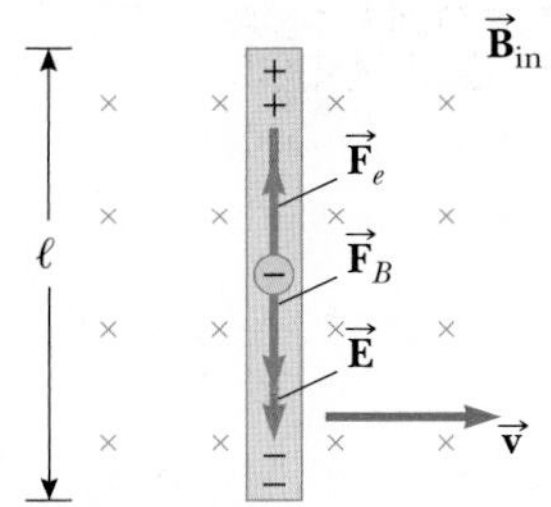

FIGURE 23.10 A straight electrical conductor of length ℓ moving with a velocity $\vec{\mathbf{v}}$ through a uniform magnetic field $\vec{\mathbf{B}}$ directed perpendicular to $\vec{\mathbf{v}}$.

$$\sum \vec{\mathbf{F}} = \vec{\mathbf{F}}_e - \vec{\mathbf{F}}_B = 0 \quad \rightarrow \quad qE = qvB \quad \rightarrow \quad E = vB$$

Because the electric field produced in the conductor is uniform, it is related to the potential difference across the ends of the conductor according to the relation $\Delta V = E\ell$ (Section 20.2). Thus,

$$\Delta V = E\ell = B\ell v$$

where the upper end is at a higher potential than the lower end. Therefore, **a potential difference is maintained as long as the conductor is moving through the magnetic field. If the motion is reversed, the polarity of ΔV is also reversed.**

An interesting situation occurs if we now consider what happens when the moving conductor is part of a closed circuit. Consider a circuit consisting of a conducting bar of length ℓ sliding along two fixed parallel conducting rails as in Active Figure 23.11a. For simplicity, we assume that the moving bar has zero electrical resistance and that the stationary part of the circuit has a resistance R. A uniform and constant magnetic field $\vec{\mathbf{B}}$ is applied perpendicular to the plane of the circuit.

As the bar is pulled to the right with a velocity $\vec{\mathbf{v}}$ under the influence of an applied force $\vec{\mathbf{F}}_{app}$, free charges in the bar experience a magnetic force along the length of the bar. Because the moving bar is part of a complete circuit, a continuous current is established in the circuit. In this case, the rate of change of magnetic flux through the loop and the accompanying induced emf across the moving bar are proportional to the change in loop area as the bar moves through the magnetic field.

Because the area of the circuit at any instant is ℓx, the magnetic flux through the circuit is

$$\Phi_B = B\ell x$$

where x is the width of the circuit, a parameter that changes with time. Using Faraday's law, we find that the induced emf is

$$\mathcal{E} = -\frac{d\Phi_B}{dt} = -\frac{d}{dt}(B\ell x) = -B\ell\frac{dx}{dt} = -B\ell v \qquad [23.5]$$

Because the resistance of the circuit is R, the magnitude of the induced current is

$$I = \frac{|\mathcal{E}|}{R} = \frac{B\ell v}{R} \qquad [23.6]$$

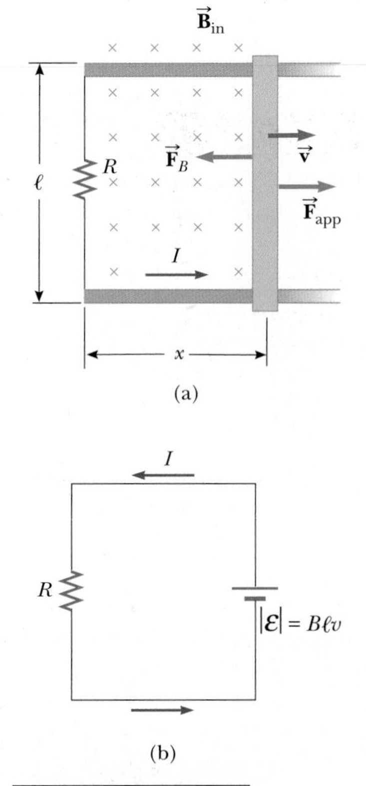

ACTIVE FIGURE 23.11
(a) A conducting bar sliding with a velocity $\vec{\mathbf{v}}$ along two conducting rails under the action of an applied force $\vec{\mathbf{F}}_{app}$. (b) The equivalent circuit diagram for the pictorial representation in (a).

PhysicsNow™ By logging into PhysicsNow at **www.pop4e.com** and going to Active Figure 23.11, you can adjust the applied force, the magnetic field, and the resistance to see the effects on the motion of the bar.

The equivalent circuit diagram for this example is shown in Active Figure 23.11b. The moving bar is behaving like a battery in that it is a source of emf as long as the bar continues to move.

Let us examine this situation using energy considerations in the nonisolated system model, with the system being the entire circuit. Because the circuit has no battery, you might wonder about the origin of the induced current and the energy delivered to the resistor. Note that the external force $\vec{\mathbf{F}}_{app}$ does work on the conductor, thereby moving charges through a magnetic field, which causes the charges to move along the conductor with some average drift velocity. Hence, a current is established. From the viewpoint of the continuity equation for energy (Eq. 6.20), the total work done on the system by the applied force while the bar moves with constant speed must equal the increase in internal energy in the resistor during this time interval. (This statement assumes that the energy stays in the resistor; in reality, energy leaves the resistor by heat and electromagnetic radiation.)

As the conductor of length ℓ moves through the uniform magnetic field $\vec{\mathbf{B}}$, it experiences a magnetic force $\vec{\mathbf{F}}_B$ of magnitude $I\ell B$ (Eq. 22.10), where I is the current induced due to its motion. The direction of this force is opposite the motion of the bar, or to the left in Active Figure 23.11a.

If the bar is to move with a *constant* velocity, the applied force $\vec{\mathbf{F}}_{\text{app}}$ must be equal in magnitude and opposite in direction to the magnetic force, or to the right in Active Figure 23.11a. (If the magnetic force acted in the direction of motion, it would cause the bar to accelerate once it was in motion, thereby increasing its speed. This state of affairs would represent a violation of the principle of energy conservation.) Using Equation 23.6 and that $F_{\text{app}} = F_B = I\ell B$, we find that the power delivered by the applied force is

$$\mathcal{P} = F_{\text{app}}v = (I\ell B)v = \frac{B^2\ell^2v^2}{R} = \left(\frac{B\ell v}{R}\right)^2 R = I^2R \qquad [23.7]$$

This power is equal to the rate at which energy is delivered to the resistor, as we expect.

QUICK QUIZ 23.3 You wish to move a rectangular loop of wire into a region of uniform magnetic field at a given speed so as to induce an emf in the loop. The plane of the loop must remain perpendicular to the magnetic field lines. In which orientation should you hold the loop while you move it into the region of magnetic field so as to generate the largest emf? **(a)** with the long dimension of the loop parallel to the velocity vector **(b)** with the short dimension of the loop parallel to the velocity vector **(c)** either way because the emf is the same regardless of orientation

QUICK QUIZ 23.4 In Active Figure 23.11, a given applied force of magnitude F_{app} results in a constant speed v and a power input $\mathcal{P}$. Imagine that the force is increased so that the constant speed of the bar is doubled to $2v$. Under these conditions, what are the new force and the new power input? **(a)** $2F$ and $2\mathcal{P}$ **(b)** $4F$ and $2\mathcal{P}$ **(c)** $2F$ and $4\mathcal{P}$ **(d)** $4F$ and $4\mathcal{P}$

INTERACTIVE EXAMPLE 23.3 Motional emf Induced in a Rotating Bar

A conducting bar of length ℓ rotates with a constant angular speed ω about a pivot at one end. A uniform magnetic field $\vec{\mathbf{B}}$ is directed perpendicular to the plane of rotation as in Figure 23.12. Find the emf induced between the ends of the bar.

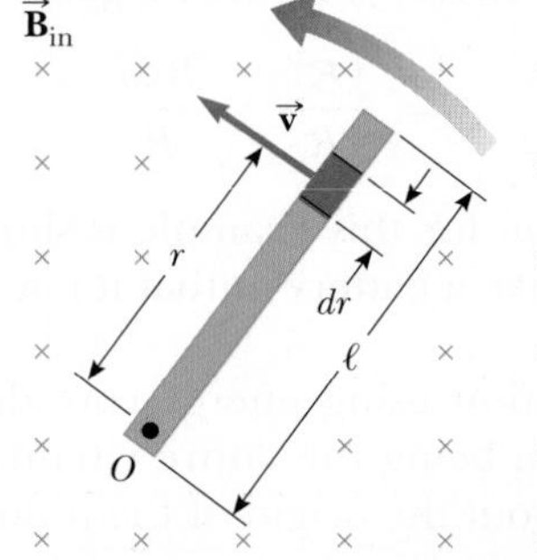

FIGURE 23.12 (Interactive Example 23.3) A conducting bar rotating about a pivot at one end in a uniform magnetic field that is perpendicular to the plane of rotation. An emf is induced between the ends of the bar.

Solution Consider a segment of the bar of length dr whose velocity is $\vec{\mathbf{v}}$. According to Equation 23.5, the magnitude of the emf induced in a conductor of length dr moving perpendicular to a field $\vec{\mathbf{B}}$ is

$$(1) \quad d\mathcal{E} = Bv\,dr$$

Each segment of the bar is moving perpendicular to $\vec{\mathbf{B}}$, so an emf is generated across each segment, the value of which is given by (1). Summing the emfs induced across all elements, which are in series, gives the magnitude of the total emf between the ends of the bar. That is,

$$\mathcal{E} = \int Bv\,dr$$

To integrate this expression, note that the linear speed of an element is related to the angular speed ω through the relationship $v = r\omega$ (Eq. 10.10). Because B and ω are constants, we therefore find that

$$\mathcal{E} = B\int_0^{\ell} v\,dr = B\omega \int_0^{\ell} r\,dr = \tfrac{1}{2}B\omega\ell^2$$

PhysicsNow™ By logging into PhysicsNow at www.pop4e.com and going to Interactive Example 23.3, you can explore the induced emf for different angular speeds and field magnitudes.

INTERACTIVE EXAMPLE 23.4 A Sliding Bar in a Magnetic Field

The conducting bar illustrated in Figure 23.13 moves on two frictionless, parallel rails in the presence of a uniform magnetic field directed into the page. The bar has mass m and its length is ℓ. The bar is given an initial velocity $\vec{\mathbf{v}}_i$ to the right and is released at $t = 0$.

A Using Newton's laws, find the velocity of the bar as a function of time.

Solution Conceptualize this situation as follows. As the bar slides to the right in Figure 23.13, a counterclockwise current is established in the circuit consisting of the bar, the rails, and the resistor. The upward current in the bar results in a magnetic force to the left on the bar as shown in the figure. As a result, the bar will slow down, so our mathematical solution should demonstrate that. The text of the question already categorizes this problem as one using Newton's laws. To analyze the problem, we determine from Equation 22.10 that the magnetic force is $F_B = -I\ell B$, where the negative sign indicates that the retarding force is to the left. Because this force is the *only* horizontal force acting on the bar, Newton's second law applied to motion in the horizontal direction gives

$$F_x = ma = m\frac{dv}{dt} = -I\ell B$$

From Equation 23.6, we know that $I = B\ell v/R$, and so we can write this expression as

$$m\frac{dv}{dt} = -\frac{B^2\ell^2}{R}v$$

$$\frac{dv}{v} = -\left(\frac{B^2\ell^2}{mR}\right)dt$$

Integrating this equation using the initial condition that $v = v_i$ at $t = 0$, we find that

$$\int_{v_i}^{v}\frac{dv}{v} = \frac{-B^2\ell^2}{mR}\int_0^t dt$$

$$\ln\left(\frac{v}{v_i}\right) = -\left(\frac{B^2\ell^2}{mR}\right)t = -\frac{t}{\tau}$$

where the constant $\tau = mR/B^2\ell^2$. From this result, we see that the velocity can be expressed in the exponential form

$$(1)\quad v = v_i e^{-t/\tau}$$

To finalize the problem, note that this expression for v indicates that the velocity of the bar decreases with time under the action of the magnetic retarding force, as we expect from our conceptualization of the problem.

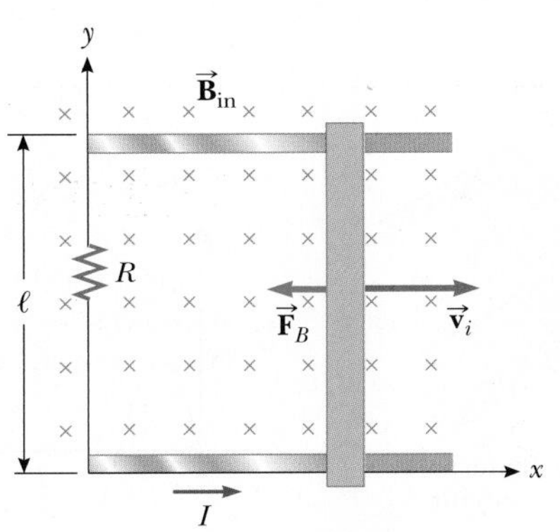

FIGURE 23.13 (Interactive Example 23.4) A conducting bar of length ℓ sliding on two fixed conducting rails is given an initial velocity $\vec{\mathbf{v}}_i$ in the positive x direction.

B Show that the same result is reached by using an energy approach.

Solution The wording of the text immediately categorizes this problem as one in energy conservation. Consider the sliding bar as one system possessing kinetic energy, which decreases because energy is transferring *out* of the system by electrical transmission through the rails. The resistor is another system possessing internal energy, which rises because energy is transferring *into* this system. Because energy is not leaving the combination of two systems, the rate of energy transfer out of the bar equals the rate of energy transfer into the resistor. Therefore,

$$\mathcal{P}_{\text{resistor}} = -\mathcal{P}_{\text{bar}}$$

where the negative sign is necessary because energy is leaving the bar and $\mathcal{P}_{\text{bar}}$ is a negative number. Substituting for the electrical power delivered to the resistor and the time rate of change of kinetic energy for the bar, we have

$$I^2R = -\frac{d}{dt}\left(\tfrac{1}{2}mv^2\right)$$

Using Equation 23.6 for the current and carrying out the derivative, we find that

$$\frac{B^2\ell^2v^2}{R} = -mv\frac{dv}{dt}$$

Rearranging terms gives

$$\frac{dv}{v} = -\left(\frac{B^2\ell^2}{mR}\right)dt$$

To finalize this part of the problem, note that this expression is the same one that we generated in part A, so the solution for v will be the same.

PhysicsNow™ By logging into PhysicsNow at www.pop4e.com and going to Interactive Example 23.4, you can study the motion of the bar after it is released.

The Alternating-Current Generator

The alternating-current (AC) generator is a device in which energy is transferred in by work and out by electrical transmission. A simplified pictorial representation of an AC generator is shown in Active Figure 23.14a. It consists of a coil of wire rotated in an external magnetic field by some external agent, which is the work input. In commercial power plants, the energy required to rotate the loop can be derived from a variety of sources. In a hydroelectric plant, for example, falling water directed against the blades of a turbine produces the rotary motion; in a coal-fired plant, the high temperature produced by burning the coal is used to convert water to steam and this steam is directed against turbine blades. As the loop rotates, the magnetic flux through it changes with time, inducing an emf and a current in a circuit connected to the coil.

Suppose the coil has N turns, all of the same area A, and suppose the coil rotates with a constant angular speed ω about an axis perpendicular to the magnetic field. If θ is the angle between the magnetic field and the direction perpendicular to the plane of the coil, the magnetic flux through the loop at any time t is given by

$$\Phi_B = BA\cos\theta = BA\cos\omega t$$

where we have used the relationship between angular position and a constant angular speed, $\theta = \omega t$. (See Eq. 10.7 and set the angular acceleration α equal to zero.)

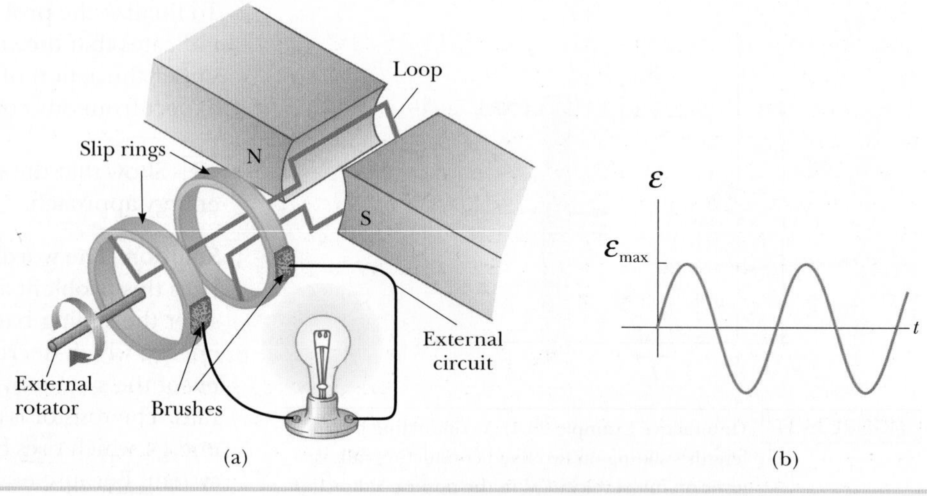

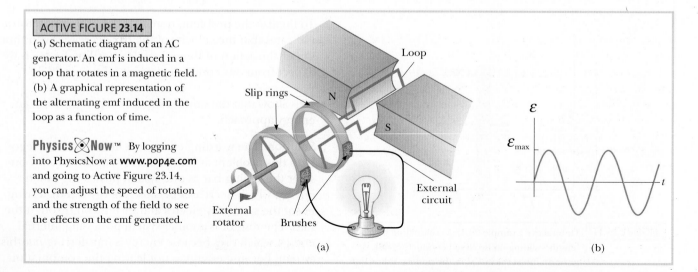

ACTIVE FIGURE 23.14
(a) Schematic diagram of an AC generator. An emf is induced in a loop that rotates in a magnetic field. (b) A graphical representation of the alternating emf induced in the loop as a function of time.

PhysicsNow™ By logging into PhysicsNow at www.pop4e.com and going to Active Figure 23.14, you can adjust the speed of rotation and the strength of the field to see the effects on the emf generated.

Hence, the induced emf in the coil is

$$\mathcal{E} = -N\frac{d\Phi_B}{dt} = -NAB\frac{d}{dt}(\cos\omega t) = NAB\omega\sin\omega t \qquad [23.8]$$

This result shows that the emf varies sinusoidally with time as shown in Active Figure 23.14b. From Equation 23.8, we see that the maximum emf has the value $\mathcal{E}_{max} = NAB\omega$, which occurs when $\omega t = 90°$ or $270°$. In other words, $\mathcal{E} = \mathcal{E}_{max}$ when the magnetic field is in the plane of the coil, and the time rate of change of flux is a maximum. In this position, the velocity vector for a wire in the loop is perpendicular to the magnetic field vector. Furthermore, the emf is *zero* when $\omega t = 0$ or $180°$—that is, when $\vec{\mathbf{B}}$ is perpendicular to the plane of the coil—and the time rate of change of flux is zero. In this orientation, the velocity vector for a wire in the loop is parallel to the magnetic field vector.

The sinusoidally varying emf in Equation 23.8 is the source of *alternating current* delivered to customers of electrical utility companies. It is called **AC voltage** as opposed to the DC voltage from a source such as a battery.

23.3 LENZ'S LAW

Let us now address the negative sign in Faraday's law. When a change occurs in the magnetic flux, the direction of the induced emf and induced current can be found from **Lenz's law:**

> The polarity of the induced emf in a loop is such that it produces a current whose magnetic field opposes the change in magnetic flux through the loop. That is, the induced current is in a direction such that the induced magnetic field attempts to maintain the original flux through the loop.

Notice that no equation is associated with Lenz's law. The law is in words only and provides a means for determining the direction of the current in a circuit when a magnetic change occurs.

PITFALL PREVENTION 23.2

INDUCED CURRENT OPPOSES THE CHANGE The induced current in a circuit opposes the *change* in the magnetic field, not the field itself. Therefore, in some cases the magnetic field due to the induced current is in the same direction as the changing external magnetic field. Such is the case if the external magnetic field is decreasing in magnitude, for example.

■ Thinking Physics 23.2

A transformer (Fig. 23.15) consists of a pair of coils wrapped around an iron form. When AC voltage is applied to one coil, the *primary,* the magnetic field lines cutting through the other coil, the *secondary,* induce an emf. (This arrangement is used in Faraday's experiment shown in Active Fig. 23.3.) By varying the number of turns of wire on each coil, the AC voltage in the secondary can be made larger or smaller than that in the primary. Clearly, this device cannot work with DC voltage. What's more, if DC voltage is applied, the primary coil sometimes overheats and burns. Why?

Reasoning When a current exists in the primary coil, the magnetic field lines from this current pass through the coil itself. Therefore, any change in the current causes a change in the magnetic field that in turn induces a current in the same coil. According to Lenz's law, this current is in the direction opposite the original current. The result is that when an AC voltage is applied, the opposing emf due to Lenz's law limits the current in the coil to a low value. If DC voltage is applied, no opposing emf occurs and the current can rise to a higher value. This increased current causes the temperature of the coil to rise, to the point at which the insulation on the wire sometimes burns. ■

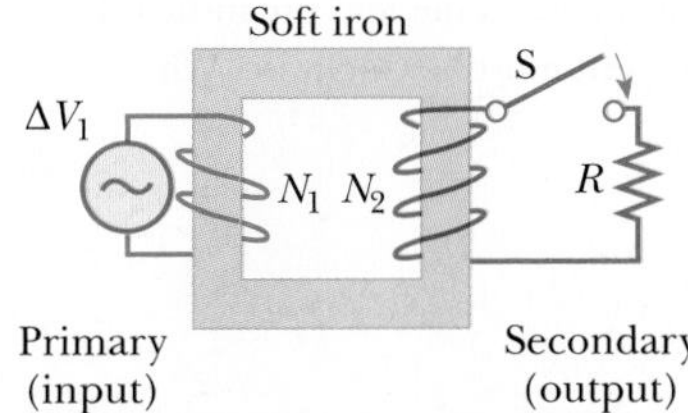

FIGURE 23.15 (Thinking Physics 23.2) An ideal transformer consists of two coils of wire wound on the same iron core. An alternating voltage ΔV_1 is applied to the primary coil, and the output voltage ΔV_2 appears across the resistance R.

To attain a better understanding of Lenz's law, let us return to the example of a bar moving to the right on two parallel rails in the presence of a uniform magnetic field directed into the page (Fig. 23.16a). As the bar moves to the right, the magnetic flux through the circuit increases with time because the area of the loop increases. Lenz's law says that the induced current must be in such a direction that the magnetic

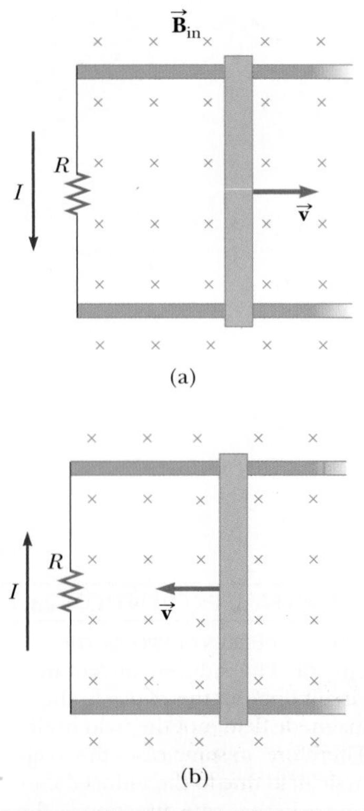

FIGURE 23.16 (a) As the conducting bar slides on the two fixed conducting rails, the flux due to the magnetic field directed inward through the area enclosed by the loop increases in time. By Lenz's law, the induced current must be counterclockwise so as to produce a counteracting magnetic field directed outward from the page. (b) When the bar moves to the left, the induced current must be clockwise. Why?

field *it* produces opposes the *change* in the magnetic flux of the external magnetic field. Because the flux is due to an external field *into* the page and is increasing, the induced current, if it is to oppose the change, must produce a magnetic field through the circuit *out* of the page. Hence, the induced current must be counterclockwise when the bar moves to the right to give a counteracting field out of the page in the region inside the loop. (Use the right-hand rule to verify this direction.) If the bar is moving to the left, as in Figure 23.16b, the magnetic flux through the loop decreases with time. Because the magnetic field is into the page, the induced current has to be clockwise to produce a magnetic field into the page inside the loop. In either case, the induced current attempts to maintain the original flux through the circuit.

Let us examine this situation from the viewpoint of energy considerations. Suppose the bar is given a slight push to the right. In the preceding analysis, we found that this motion leads to a counterclockwise current in the loop. What happens if we incorrectly assume that the current is clockwise? For a clockwise current I, the direction of the magnetic force $I\ell B$ on the sliding bar would be to the right. According to Newton's second law, this force would accelerate the rod and increase its speed, which in turn would cause the area of the loop to increase more rapidly. This increase would increase the induced current, which would increase the force, which would increase the current, and so on. In effect, the system would acquire energy with no additional energy input. This result is clearly inconsistent with all experience and with the continuity equation for energy. Thus, we are forced to conclude that the current must be counterclockwise.

Consider another situation, one in which a bar magnet is moved to the right toward a stationary loop of wire as in Figure 23.17a. As the magnet moves toward the loop, the magnetic flux through the loop increases with time. To counteract this increase in flux due to a magnetic field directed toward the right, the induced current produces a magnetic field to the left as in Figure 23.17b; hence, the induced current is in the direction shown. Therefore, the left face of the current loop is a north pole and the right face is a south pole.

If the magnet is moved to the left as in Figure 23.17c, the magnetic field through the loop, which is toward the right, decreases with time. Under these circumstances,

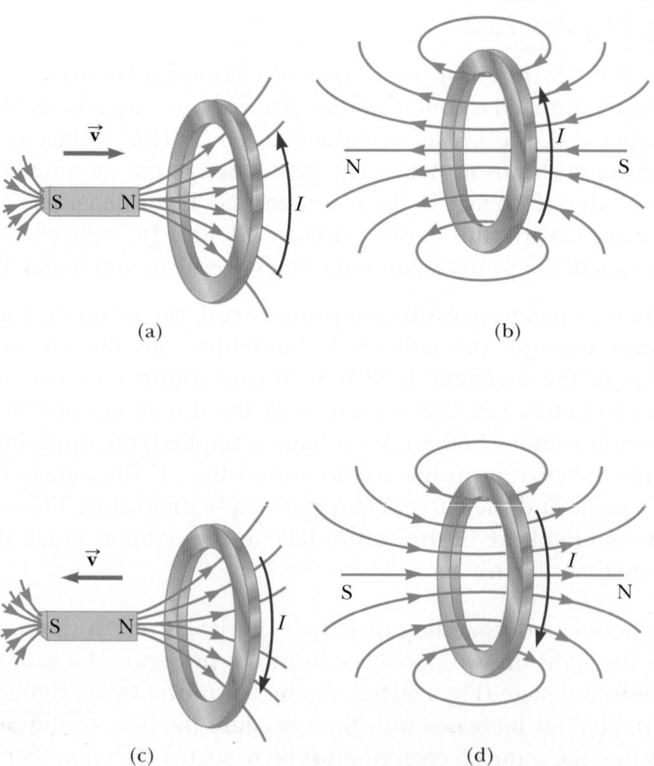

FIGURE 23.17 (a) When the magnet is moved toward the stationary conducting loop, a current is induced in the direction shown. (b) This induced current produces its own magnetic field that is directed to the left within the loop to counteract the increasing external flux. (c) When the magnet is moved away from the stationary conducting loop, a current is induced in the direction shown. (d) This induced current produces its own magnetic field that is directed to the right within the loop to counteract the decreasing external flux.

the induced current in the loop sets up a magnetic field through the loop from left to right in an effort to maintain a constant flux. Hence, the direction of the induced current in the loop is as shown in Figure 23.17d. In this case, the left face of the loop is a south pole and the right face is a north pole.

QUICK QUIZ 23.5 In equal-arm balances from the early 20th century (Fig. 23.18), it is sometimes observed that an aluminum sheet hangs from one of the arms and passes between the poles of a magnet, which causes the oscillations of the equal arm balance to decay rapidly. In the absence of such magnetic braking, the oscillation might continue for a very long time and the experimenter would have to wait to take a reading. Why do the oscillations decay? **(a)** The aluminum sheet is attracted to the magnet. **(b)** Currents in the aluminum sheet set up a magnetic field that opposes the oscillations. **(c)** Aluminum is ferromagnetic.

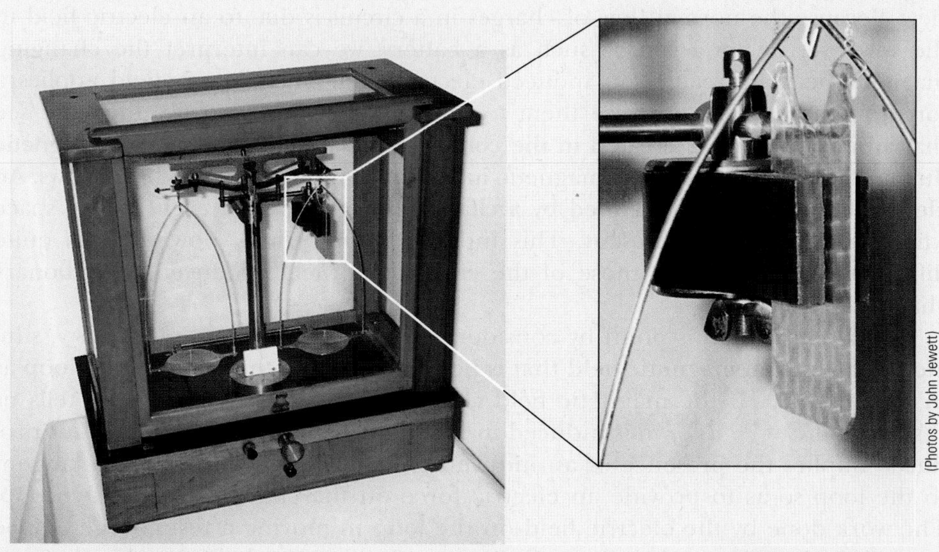

FIGURE 23.18 (Quick Quiz 23.5) In an old-fashioned equal-arm balance, an aluminum sheet hangs between the poles of a magnet.

EXAMPLE 23.5 Application of Lenz's Law

A coil of wire is placed near an electromagnet as in Figure 23.19a.

A Find the direction of the induced current in the coil at the instant the switch is closed.

Solution When the switch is closed, the situation changes from a condition in which no magnetic flux occurs through the coil to one in which flux does occur due to a magnetic field in the direction shown in Figure 23.19b. To counteract this increase in magnetic flux, the coil must set up a field from left to right in the figure, which requires a current directed as shown in Figure 23.19b.

B Find the direction of the induced current in the coil after the switch has been closed for several seconds.

Solution After the switch has been closed for several seconds, the magnetic flux through the loop does not change. Hence, the induced current is zero.

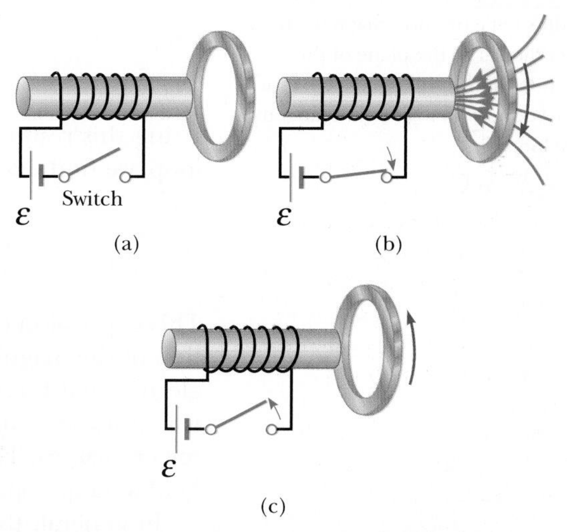

FIGURE 23.19 (Example 23.5) A current in the ring is induced when the switch is opened or closed.

C Find the direction of the induced current in the coil when the switch is opened.

Solution Opening the switch causes the magnetic field to change from a condition in which magnetic field lines pass through the coil from right to left to a condition of zero field. The induced current must then be as shown in Figure 23.19c so as to set up its own magnetic field from right to left.

23.4 INDUCED emfs AND ELECTRIC FIELDS

We have seen that a changing magnetic flux induces an emf and a current in a conducting loop. We can also interpret this phenomenon from another point of view. Because the normal flow of charges in a circuit is due to an electric field in the wires set up by a source such as a battery, we can interpret the changing magnetic field as creating an induced electric field. This electric field applies a force on the charges to cause them to move. With this approach, then, we see that **an electric field is created in the conductor as a result of changing magnetic flux.** In fact, the law of electromagnetic induction can be interpreted as follows: **An electric field is always generated by a changing magnetic flux,** even in free space where no charges are present. This induced electric field, however, has quite different properties from those of the electrostatic field produced by stationary charges.

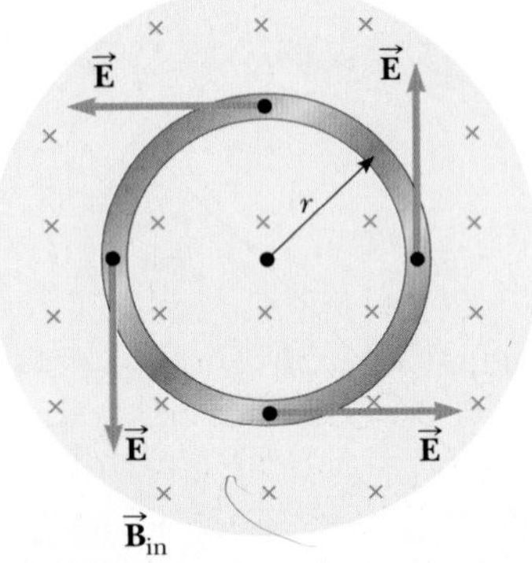

FIGURE 23.20 A conducting loop of radius r in a uniform magnetic field perpendicular to the plane of the loop. If $\vec{\mathbf{B}}$ changes in time, an electric field is induced in a direction tangent to the loop.

Let us illustrate this point by considering a conducting loop of radius r, situated in a uniform magnetic field that is perpendicular to the plane of the loop as in Figure 23.20. If the magnetic field changes with time, Faraday's law tells us that an emf $\mathcal{E} = -d\Phi_B/dt$ is induced in the loop. The induced current thus produced implies the presence of an induced electric field $\vec{\mathbf{E}}$ that must be tangent to the loop so as to provide an electric force on the charges around the loop. The work done by the electric field on the loop in moving a test charge q once around the loop is equal to $W = q\mathcal{E}$. Because the magnitude of the electric force on the charge is qE, the work done by the electric field can also be expressed from Equation 6.12 as $W = \int \vec{\mathbf{F}} \cdot d\vec{\mathbf{r}} = qE(2\pi r)$, where $2\pi r$ is the circumference of the loop. These two expressions for the work must be equal; therefore, we see that

$$q\mathcal{E} = qE(2\pi r)$$

$$E = \frac{\mathcal{E}}{2\pi r}$$

Using this result along with Faraday's law and that $\Phi_B = BA = B\pi r^2$ for a circular loop, we find that the induced electric field can be expressed as

$$E = -\frac{1}{2\pi r}\frac{d\Phi_B}{dt} = -\frac{1}{2\pi r}\frac{d}{dt}(B\pi r^2) = -\frac{r}{2}\frac{dB}{dt}$$

This expression can be used to calculate the induced electric field if the time variation of the magnetic field is specified. The negative sign indicates that the induced electric field $\vec{\mathbf{E}}$ results in a current that opposes the change in the magnetic field. It is important to understand that **this result is also valid in the absence of a conductor or charges.** That is, the same electric field is induced by the changing magnetic field in empty space.

In general, the magnitude of the emf for any closed path can be expressed as the line integral of $\vec{\mathbf{E}} \cdot d\vec{\mathbf{s}}$ over that path (Eq. 20.3). Hence, the general form of

Faraday's law of induction is

$$\mathcal{E} = \oint \vec{\mathbf{E}} \cdot d\vec{\mathbf{s}} = -\frac{d\Phi_B}{dt} \qquad [23.9]$$

■ Faraday's law in general form

It is important to recognize that **the induced electric field $\vec{\mathbf{E}}$ that appears in Equation 23.9 is a nonconservative field that is generated by a changing magnetic field.** We call it a nonconservative field because the work done in moving a charge around a closed path (the loop in Fig. 23.20) is not zero. This type of electric field is very different from an electrostatic field.

QUICK QUIZ 23.6 In a region of space, a magnetic field is uniform over space but increases at a constant rate. This changing magnetic field induces an electric field that **(a)** increases in time, **(b)** is conservative, **(c)** is in the direction of the magnetic field, or **(d)** has a constant magnitude.

■ Thinking Physics 23.3

In studying electric fields, we noted that electric field lines begin on positive charges and end on negative charges. Do *all* electric field lines begin and end on charges?

Reasoning The statement that electric field lines begin and end on charges is true only for *electrostatic* fields, that is, electric fields due to stationary charges. Electric field lines due to changing magnetic fields form closed loops, with no beginning and no end, and are independent of the presence of charges. ■

EXAMPLE 23.6 Electric Field Induced by a Changing Magnetic Field in a Solenoid

A long solenoid of radius R has n turns of wire per unit length and carries a time-varying current that varies sinusoidally as $I = I_{max} \cos \omega t$, where I_{max} is the maximum current and ω is the angular frequency of the AC source (Fig. 23.21).

A Determine the magnitude of the induced electric field outside the solenoid, a distance $r > R$ from its long central axis.

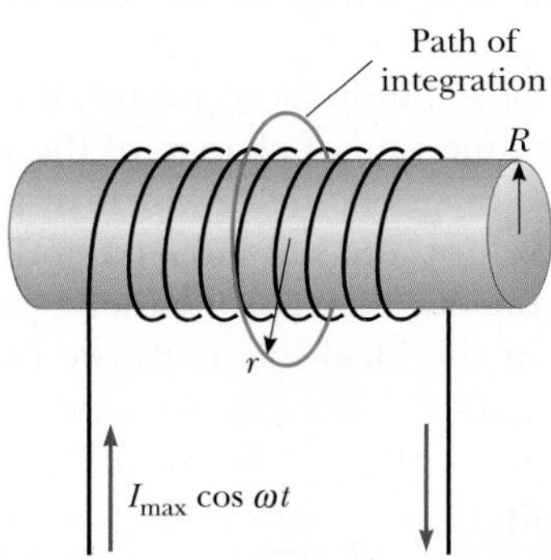

FIGURE 23.21 (Example 23.6) A long solenoid carrying a time-varying current given by $I = I_{max} \cos \omega t$. An electric field is induced both inside and outside the solenoid.

Solution First consider an external point and take the path for our line integral to be a circle of radius r centered on the solenoid as illustrated in Figure 23.21. By symmetry, we see that the magnitude of $\vec{\mathbf{E}}$ is constant on this path and that $\vec{\mathbf{E}}$ is tangent to it. The magnetic flux through the area enclosed by the path is $\int \vec{\mathbf{B}} \cdot d\vec{\mathbf{A}} = BA = B\pi R^2$; hence, Equation 23.9 gives

$$(1) \quad \oint \vec{\mathbf{E}} \cdot d\vec{\mathbf{s}} = -\frac{d}{dt}(B\pi R^2) = -\pi R^2 \frac{dB}{dt}$$

Based on the symmetry of the situation,

$$(2) \quad \oint \vec{\mathbf{E}} \cdot d\vec{\mathbf{s}} = \oint E\,ds = E\oint ds = E(2\pi r)$$

Setting these two expressions equal, we find that

$$(3) \quad E = -\frac{R^2}{2r}\frac{dB}{dt}$$

The magnetic field inside a long solenoid is given by Equation 22.32, $B = \mu_0 n I$. When we substitute $I = I_{max} \cos \omega t$ into this equation and then substitute the result into (3), we find that

$$E = -\frac{R^2}{2r}\frac{d}{dt}(\mu_0 n I_{\max} \cos \omega t)$$

$$= \frac{\mu_0 n I_{\max} \omega R^2}{2r} \sin \omega t \qquad (\text{for } r > R)$$

Hence, the electric field varies sinusoidally with time, and its amplitude falls off as $1/r$ outside the solenoid. According to the Ampère-Maxwell law, which we will study in Section 24.1, the changing electric field creates an additional contribution to the magnetic field. At high frequencies, an altogether new phenomenon can occur. The electric and magnetic fields, each supporting the other, can constitute an electromagnetic wave radiated by the solenoid, as we will study in Chapter 24.

B What is the magnitude of the induced electric field inside the solenoid, a distance r from its axis?

Solution For an interior point ($r < R$), the flux passing through the area bounded by a path of integration is given by $B\pi r^2$. Thus, the analogs to (1), (2), and (3) become

$$(4) \quad \oint \vec{\mathbf{E}} \cdot d\vec{\mathbf{s}} = -\frac{d}{dt}(B\pi r^2) = \pi r^2 \frac{dB}{dt}$$

$$(5) \quad \oint \vec{\mathbf{E}} \cdot d\vec{\mathbf{s}} = \oint E\, ds = E \oint ds = E(2\pi r)$$

$$(6) \quad E = -\frac{r}{2}\frac{dB}{dt}$$

Substituting the expression for the magnetic field into (6) gives

$$E = -\frac{r}{2}\frac{d}{dt}(\mu_0 n I_{\max} \cos \omega t)$$

$$= \frac{\mu_0 n I_{\max} \omega}{2} r \sin \omega t \qquad (\text{for } r < R)$$

This expression shows that the amplitude of the electric field induced inside the solenoid by the changing magnetic field varies linearly with r and varies sinusoidally with time.

23.5 SELF-INDUCTANCE

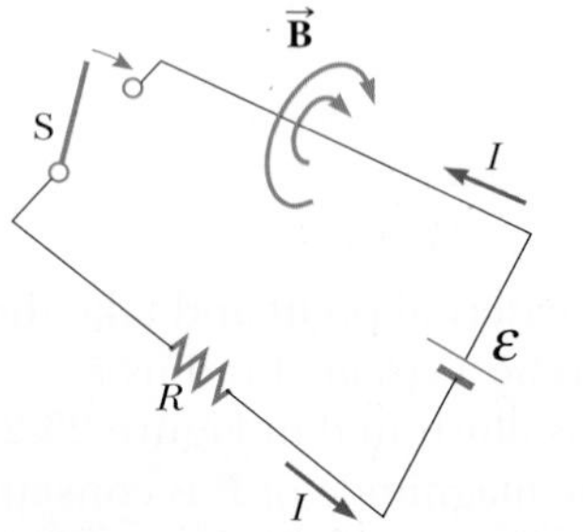

Consider an isolated circuit consisting of a switch, a resistor, and a source of emf as in Figure 23.22. The circuit diagram is represented in perspective so that we can see the orientations of some of the magnetic field lines due to the current in the circuit. When the switch is closed, the current doesn't immediately jump from zero to its maximum value $\mathcal{E}/R$; the law of electromagnetic induction (Faraday's law) describes the actual behavior. As the current increases with time, the magnetic flux through the loop of the circuit itself due to the current also increases with time. This increasing magnetic flux *from* the circuit induces an emf *in* the circuit (sometimes referred to as a *back emf*) that opposes the change in the net magnetic flux through the loop of the circuit. By Lenz's law, the induced electric field in the wires must therefore be opposite the direction of the current, and the opposing emf results in a *gradual* increase in the current. This effect is called *self-induction* because the changing magnetic flux through the circuit arises from the circuit itself. The emf set up in this case is called a **self-induced emf.**

FIGURE 23.22 After the switch is closed, the current produces a magnetic flux through the area enclosed by the loop of the circuit. As the current increases toward its final value, this magnetic flux changes with time and induces an emf in the loop.

To obtain a quantitative description of self-induction, we recall from Faraday's law that the induced emf is the negative time rate of change of the magnetic flux. The magnetic flux is proportional to the magnetic field, which in turn is proportional to the current in the circuit. Therefore, **the self-induced emf is always proportional to the time rate of change of the current.** For a closely spaced coil of N turns of fixed geometry (a toroidal coil or the ideal solenoid), we can express this proportionality as follows:

■ Self-induced emf

$$\mathcal{E}_L = -N\frac{d\Phi_B}{dt} = -L\frac{dI}{dt} \qquad [23.10]$$

where L is a proportionality constant, called the **inductance** of the coil, that depends on the geometric features of the coil and other physical characteristics. From this expression, we see that the inductance of a coil containing N

turns is

$$L = \frac{N\Phi_B}{I} \quad [23.11]$$

where it is assumed that the same magnetic flux passes through each turn. Later we shall use this equation to calculate the inductance of some special coil geometries.

From Equation 23.10, we can also write the inductance as the ratio

$$L = -\frac{\mathcal{E}_L}{dI/dt} \quad [23.12]$$

which is usually taken to be the defining equation for the inductance of any coil, regardless of its shape, size, or material characteristics. If we compare Equation 23.10 with Equation 21.6, $R = \Delta V/I$, we see that resistance is a measure of opposition to current, whereas inductance is a measure of opposition to the *change* in current.

The SI unit of inductance is the **henry (H),** which, from Equation 23.12, is seen to be equal to 1 volt-second per ampere:

$$1\text{ H} = 1\text{ V}\cdot\text{s/A}$$

As we shall see, **the inductance of a coil depends on its geometry.** Because inductance calculations can be quite difficult for complicated geometries, the examples we shall explore involve simple situations for which inductances are easily evaluated.

(North Wind Picture Archives)

JOSEPH HENRY (1797–1878)
Henry, an American physicist, became the first director of the Smithsonian Institution and first president of the Academy of Natural Science. He improved the design of the electromagnet and constructed one of the first motors. He also discovered the phenomenon of self-induction but failed to publish his findings. The unit of inductance, the henry, is named in his honor.

EXAMPLE 23.7 Inductance of a Solenoid

Consider a uniformly wound solenoid having N turns and length ℓ.

A Find the inductance of the solenoid. Assume that ℓ is long compared with the radius and that the core of the solenoid is air.

Solution Because ℓ is long compared with the radius, we can model the solenoid as an ideal solenoid. In this case, the interior magnetic field is uniform and given by Equation 22.32:

$$B = \mu_0 n I = \mu_0 \frac{N}{\ell} I$$

where $n = N/\ell$ is the number of turns per unit length. The magnetic flux through each turn is

$$\Phi_B = BA = \mu_0 \frac{NA}{\ell} I$$

where A is the cross-sectional area of the solenoid. Using this expression and Equation 23.11, we find that

$$L = \frac{N\Phi_B}{I} = \frac{\mu_0 N^2 A}{\ell}$$

which shows that L depends on the geometry of the solenoid and is proportional to the square of the number of turns. Because $N = n\ell$, we can also express the result in the form

$$L = \mu_0 \frac{(n\ell)^2}{\ell} A = \mu_0 n^2 A\ell = \mu_0 n^2 V$$

where $V = A\ell$ is the volume of the solenoid.

B Calculate the inductance of a solenoid containing 300 turns if the length of the solenoid is 25.0 cm and its cross-sectional area is $4.00\text{ cm}^2 = 4.00 \times 10^{-4}\text{ m}^2$.

Solution Using the expression for L from part A, we find that

$$L = \frac{\mu_0 N^2 A}{\ell}$$

$$= (4\pi \times 10^{-7}\text{ T}\cdot\text{m/A}) \frac{(300)^2(4.00 \times 10^{-4}\text{ m}^2)}{25.0 \times 10^{-2}\text{ m}}$$

$$= 1.81 \times 10^{-4}\text{ T}\cdot\text{m}^2/\text{A} = 0.181\text{ mH}$$

C Calculate the self-induced emf in the solenoid described in part B if the current through it is decreasing at the rate of 50.0 A/s.

Solution Using Equation 23.10 and given that $dI/dt = -50.0$ A/s, we have

$$\mathcal{E}_L = -L\frac{dI}{dt} = -(1.81 \times 10^{-4}\text{ H})(-50.0\text{ A/s})$$

$$= 9.05\text{ mV}$$

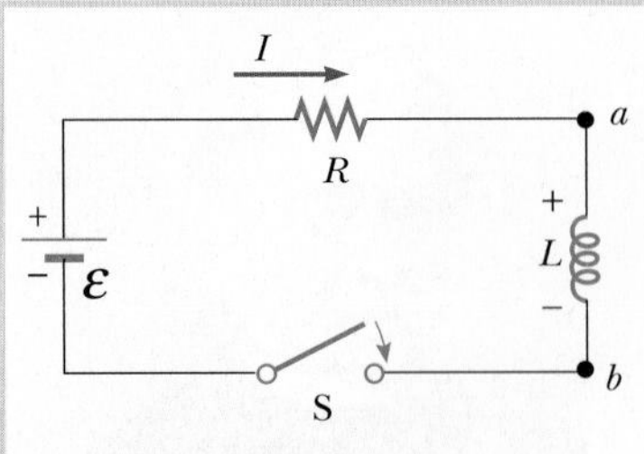

ACTIVE FIGURE 23.23
A series *RL* circuit. As the current increases toward its maximum value, an emf that opposes the increasing current is induced in the inductor.

PhysicsNow™ By logging into PhysicsNow at **www.pop4e.com** and going to Active Figure 23.23, you can adjust the values of *R* and *L* to see the effect on the current. A graphical display as in Active Figure 23.24 is available.

23.6 *RL* CIRCUITS

A circuit that contains a coil, such as a solenoid, has a self-inductance that prevents the current from increasing or decreasing instantaneously. A circuit element whose main purpose is to provide inductance in a circuit is called an **inductor.** The circuit symbol for an inductor is —ooo—. As a simplification model, we shall always assume that the self-inductance of the remainder of the circuit is negligible compared with that of any inductors in the circuit. In addition, any resistance in the inductor is assumed to be combined with other resistance in the circuit, so we model the inductor as having zero resistance.

Consider the circuit shown in Active Figure 23.23, consisting of a resistor, an inductor, a switch, and a battery. The internal resistance of the battery will be ignored as a further simplification model. Suppose the switch S is thrown closed at $t = 0$. The current begins to increase, and, due to the increasing current, the inductor produces an emf that opposes the increasing current. The back emf produced by the inductor is

$$\mathcal{E}_L = -L\frac{dI}{dt}$$

Because the current is increasing, dI/dt is positive; therefore $\mathcal{E}_L$ is negative, which corresponds to the potential drop occurring from a to b across the inductor. For this reason, point a is at a higher potential than point b as illustrated in Active Figure 23.23.

We can apply Kirchhoff's loop rule to this circuit. If we begin at the battery and travel clockwise, we have

$$\mathcal{E} - IR - L\frac{dI}{dt} = 0 \qquad [23.13]$$

where IR is the voltage across the resistor. The potential difference across the inductor is given a negative sign because its emf is in the opposite sense to that of the battery. We must now look for a solution to this differential equation, which is a mathematical representation of the behavior of the *RL* circuit. It is similar to Equation 21.30 for the *RC* circuit.

To obtain a mathematical solution of Equation 23.13, it is convenient to change variables by letting $x = (\mathcal{E}/R) - I$ so that $dx = -dI$. With these substitutions, Equation 23.13 can be written as

$$\frac{\mathcal{E}}{R} - I - \frac{L}{R}\frac{dI}{dt} = x + \frac{L}{R}\frac{dx}{dt} = 0$$

$$\frac{dx}{x} = -\frac{R}{L}\,dt$$

Integrating this last expression from an initial instant $t = 0$ to some later time t gives

$$\int_{x_i}^{x}\frac{dx}{x} = -\frac{R}{L}\int_0^t dt \quad\rightarrow\quad \ln\frac{x}{x_i} = -\frac{R}{L}t$$

Taking the antilog of this result gives

$$x = x_i e^{-Rt/L}$$

The value of x at $t = 0$ is expressed as $x_i = \mathcal{E}/R$ because $I = 0$ at $t = 0$. Hence, the preceding expression is equivalent to

$$\frac{\mathcal{E}}{R} - I = \frac{\mathcal{E}}{R}e^{-Rt/L}$$

$$I = \frac{\mathcal{E}}{R}(1 - e^{-Rt/L})$$

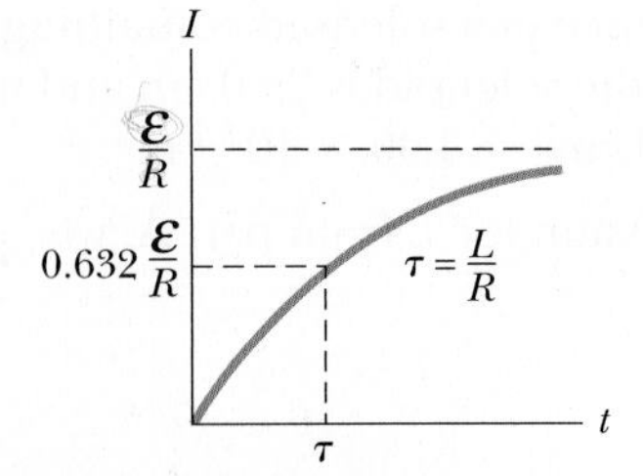

ACTIVE FIGURE 23.24
Plot of current versus time for the *RL* circuit shown in Active Figure 23.23. The switch is open for $t < 0$ and then closed at $t = 0$, and the current increases toward its maximum value $\mathcal{E}/R$. The time constant τ is the time interval required for I to reach 63.2% of its maximum value.

PhysicsNow™ By logging into PhysicsNow at **www.pop4e.com** and going to Active Figure 23.24, you can observe the graph develop after the switch in Active Figure 23.23 is closed.

This expression represents the solution of Equation 23.13, the current as a function of time. It can also be written as

$$I(t) = \frac{\mathcal{E}}{R}(1 - e^{-t/\tau}) \qquad [23.14]$$

where τ is the **time constant** of the *RL* circuit:

$$\tau = \frac{L}{R} \qquad [23.15]$$

■ Time constant of the *RL* circuit

It can be shown that the dimension of τ is time. Physically, τ is the time interval required for the current to reach $(1 - e^{-1}) = 0.632$ of its final value $\mathcal{E}/R$.

Active Figure 23.24 plots current versus time. Note that $I = 0$ at $t = 0$ and that the final steady-state value of the current, which occurs as $t \rightarrow \infty$, is $\mathcal{E}/R$. This result can be seen by setting dI/dt equal to zero in Equation 23.13 (in steady state, the change in the current is zero) and solving for the current. Thus, we see that the current rises very rapidly initially and then gradually approaches the maximum value $\mathcal{E}/R$ as $t \rightarrow \infty$. Notice that the final current does not involve L because the inductor has no effect on the circuit (ignoring any resistance associated with it) if the current is not changing.

Taking the first time derivative of Equation 23.14, we obtain

$$\frac{dI}{dt} = \frac{\mathcal{E}}{L}e^{-t/\tau} \qquad [23.16]$$

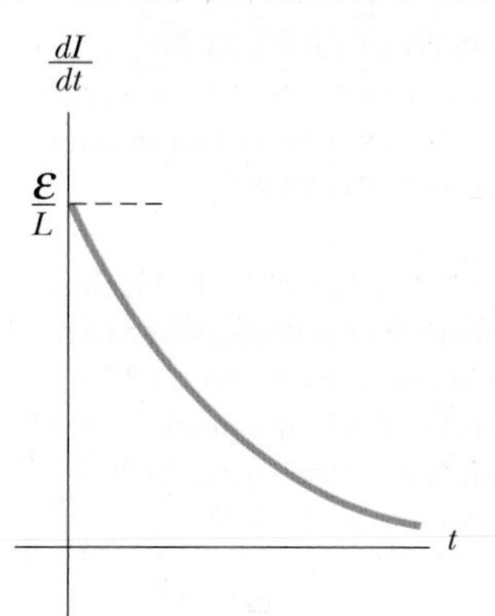

FIGURE 23.25 Plot of dI/dt versus time for the *RL* circuit shown in Active Figure 23.23. The time rate of change of current is a maximum at $t = 0$, which is the instant at which the switch is closed. The rate decreases exponentially with time as I increases toward its maximum value.

From this equation, we see that the rate of change of current dI/dt is a *maximum* (equal to $\mathcal{E}/L$) at $t = 0$ and falls off exponentially to zero as $t \rightarrow \infty$ (Fig. 23.25).

Now consider the *RL* circuit arranged as shown in Active Figure 23.26. The curved lines on the switch S represent a switch that is connected either to *a* or *b* at all times. (If the switch is connected to neither *a* nor *b*, the current in the circuit suddenly stops.) Suppose the switch has been set at position *a* long enough to allow the current to reach its equilibrium value $\mathcal{E}/R$. In this situation, the circuit is described by the outer loop in Active Figure 23.26. If the switch is thrown from *a* to *b*, the circuit is now described by just the right-hand loop in Active Figure 23.26. Thus, we have a circuit with no battery ($\mathcal{E} = 0$). Applying Kirchhoff's loop rule to the right-hand loop at the instant the switch is thrown from *a* to *b*, we obtain

$$IR + L\frac{dI}{dt} = 0 \qquad [23.17]$$

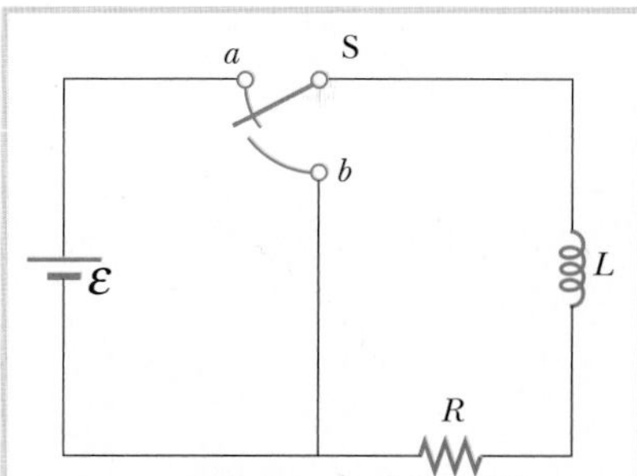

ACTIVE FIGURE 23.26 An *RL* circuit. When the switch S is in position *a*, the battery is in the circuit. When the switch is thrown to position *b*, the battery is no longer part of the circuit. The switch is designed so that it is never open, which would cause the current to stop.

PhysicsNow™ By logging into PhysicsNow at **www.pop4e.com** and going to Active Figure 23.26, you can adjust the values of *R* and *L* to see the effect on the current. A graphical display as in Active Figure 23.27 is available.

It is left to Problem 23.34 to show that the solution of this differential equation is

$$I(t) = \frac{\mathcal{E}}{R}e^{-t/\tau} \qquad [23.18]$$

where the current at $t = 0$ is $I_i = \mathcal{E}/R$ and $\tau = L/R$.

The graph of current versus time (Active Fig. 23.27) for the circuit of Active Figure 23.26 shows that the current is continuously decreasing with time, as one would expect. Furthermore, the slope dI/dt is always negative and has its maximum magnitude at $t = 0$. The negative slope signifies that $\mathcal{E}_L = -L(dI/dt)$ is now *positive.*

QUICK QUIZ 23.7 The circuit in Figure 23.28 includes a power source that provides a sinusoidal voltage. Thus, the magnetic field in the inductor is constantly changing. The inductor is a simple air-core solenoid. The switch in the circuit is closed and the lightbulb

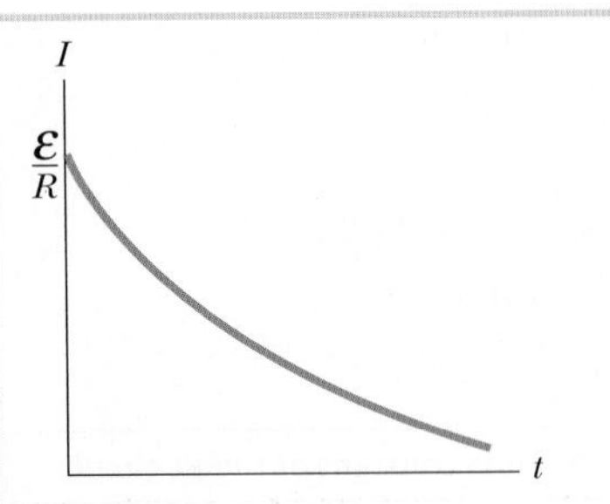

ACTIVE FIGURE 23.27
Current versus time for the right-hand loop of the circuit shown in Active Figure 23.26.

Physics Now™ By logging into PhysicsNow at **www.pop4e.com** and going to Active Figure 23.27, you can observe this graph develop after the switch in Active Figure 23.26 is thrown to position *b*.

glows steadily. An iron rod is inserted into the interior of the solenoid, which increases the magnitude of the magnetic field in the solenoid. As that happens, the brightness of the lightbulb **(a)** increases, **(b)** decreases, or **(c)** is unaffected.

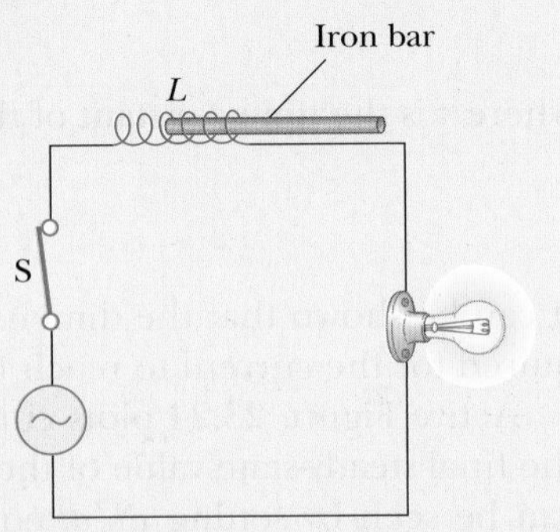

FIGURE 23.28 (Quick Quiz 23.7) A lightbulb is powered by an AC source with an inductor in the circuit. When the iron bar is inserted into the coil, what happens to the brightness of the lightbulb?

QUICK QUIZ 23.8 Two circuits like the one shown in Active Figure 23.26 are identical except for the value of L. In circuit A, the inductance of the inductor is L_A, and in circuit B, it is L_B. The switch has been in position *b* for both circuits for a long time. At $t = 0$, the switch is thrown to *a* in both circuits. At $t = 10$ s, the switch is thrown to *b* in both circuits. The resulting graphical representation of the current as a function of time is shown in Figure 23.29. Assuming that the time constant of each circuit is much less than 10 s, which of the following is true? (a) $L_A > L_B$. (b) $L_A < L_B$. (c) There is not enough information to determine the relative values.

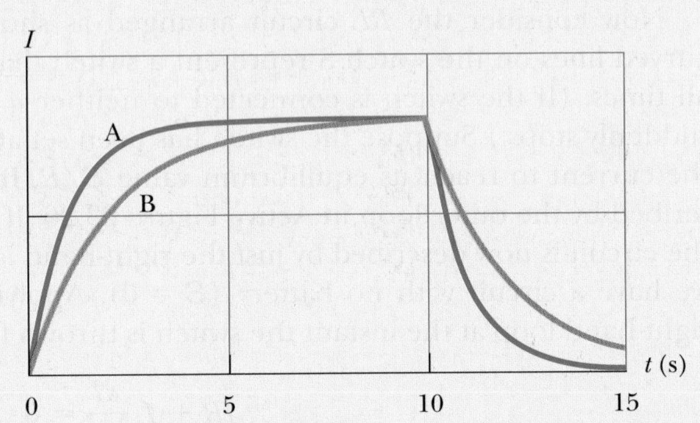

FIGURE 23.29 (Quick Quiz 23.8) Current–time graphs for two circuits with different inductances.

INTERACTIVE EXAMPLE 23.8 Time Constant of an *RL* Circuit

Consider the *RL* circuit in Figure 23.30a.

A Find the time constant of the circuit.

Solution The time constant is given by Equation 23.15:

$$\tau = \frac{L}{R} = \frac{30.0 \times 10^{-3}\ \text{H}}{6.00\ \Omega} = 5.00\ \text{ms}$$

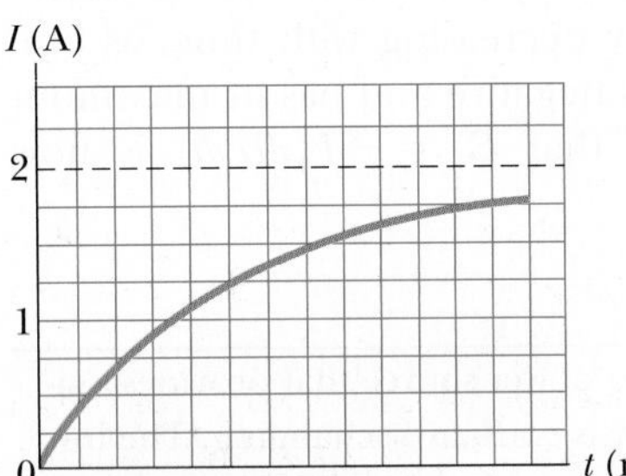

FIGURE 23.30 (Interactive Example 23.8) (a) The switch in this *RL* circuit is open for $t < 0$ and then closed at $t = 0$. (b) A graph of the current versus time for the circuit in (a).

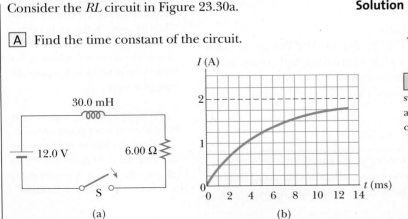

B The switch in Figure 23.30a is closed at $t = 0$. Calculate the current in the circuit at $t = 2.00$ ms.

Solution Using Equation 23.14 for the current as a function of time (with t and τ in milliseconds), we find that at $t = 2.00$ ms,

$$I = \frac{\mathcal{E}}{R}(1 - e^{-t/\tau}) = \frac{12.0\ \text{V}}{6.00\ \Omega}(1 - e^{-0.400}) = 0.659\ \text{A}$$

A plot of Equation 23.14 for this circuit is given in Figure 23.30b.

PhysicsNow™ By logging into PhysicsNow at www.pop4e.com and going to Interactive Example 23.8, you can explore the time behavior of the current in the circuit.

23.7 ENERGY STORED IN A MAGNETIC FIELD

In the preceding section, we found that the induced emf set up by an inductor prevents a battery from establishing an instantaneous current. Part of the energy supplied by the battery goes into internal energy in the resistor, and the remaining energy is stored in the inductor. If we multiply each term in Equation 23.13 by the current I and rearrange the expression, we have

$$I\mathcal{E} = I^2R + LI\frac{dI}{dt} \qquad [23.19]$$

This expression tells us that the rate $I\mathcal{E}$ at which energy is supplied by the battery equals the sum of the rate I^2R at which energy is delivered to the resistor and the rate $LI\,(dI/dt)$ at which energy is delivered to the inductor. Thus, Equation 23.19 is simply an expression of energy conservation for the isolated system of the circuit. (Actually, energy can leave the circuit by thermal conduction into the air and by electromagnetic radiation, so the system need not be completely isolated.) If we let U_B denote the energy stored in the inductor at any time, the rate dU_B/dt at which energy is delivered to the inductor can be written as

$$\frac{dU_B}{dt} = LI\frac{dI}{dt}$$

To find the total energy stored in the inductor at any instant, we can rewrite this expression as $dU_B = LI\,dI$ and integrate:

$$U_B = \int_0^{U_B} dU_B = \int_0^I LI\,dI$$

$$U_B = \tfrac{1}{2}LI^2 \qquad [23.20]$$

■ Energy stored in an inductor

where L is constant and so has been removed from the integral. Equation 23.20 represents the energy stored in the magnetic field of the inductor when the current is I.

Equation 23.20 is similar to the equation for the energy stored in the electric field of a capacitor, $U_E = \frac{1}{2}C(\Delta V)^2$ (Eq. 20.29). In either case, we see that energy from a battery is required to establish a field and that energy is stored in the field. In the case of the capacitor, we can conceptually relate the energy stored in the capacitor to the electric potential energy associated with the separated charge on the plates. We have not discussed a magnetic analogy to electric potential energy, so the storage of energy in an inductor is not as easy to conceptualize.

To argue that energy is stored in an inductor, consider the circuit in Figure 23.31a, which is the same circuit as in Active Figure 23.26, with the addition of a switch S_2 across the resistor R. With switch S_1 set to position a and S_2 closed as shown, a current is established in the inductor. Now, as in Figure 23.31b, switch S_1 is thrown to position b. The current persists in this (ideally) resistance-free and battery-free circuit (the right-hand loop in Fig. 23.31b), consisting of only the inductor and a conducting path between its ends. There is no current in the resistor (because the path around it through S_2 is resistance free), so no energy is being delivered to

PITFALL PREVENTION 23.3

COMPARE ENERGY IN A CAPACITOR, RESISTOR, AND INDUCTOR We have now seen three circuit elements to which we can transfer energy. Keep in mind the difference in energy transfer mechanisms. A capacitor stores a given amount of energy for a fixed charge on its plates. Further energy is delivered to the capacitor as a current in the wires connected to the capacitor delivers more charge to the plates. An inductor stores a given amount of energy if the current remains constant. Further energy is delivered to the inductor by increasing the current. A resistor acts differently because the energy is not stored as potential energy but rather is transformed to internal energy. Energy continues to be delivered to the resistor as long as it carries a current.

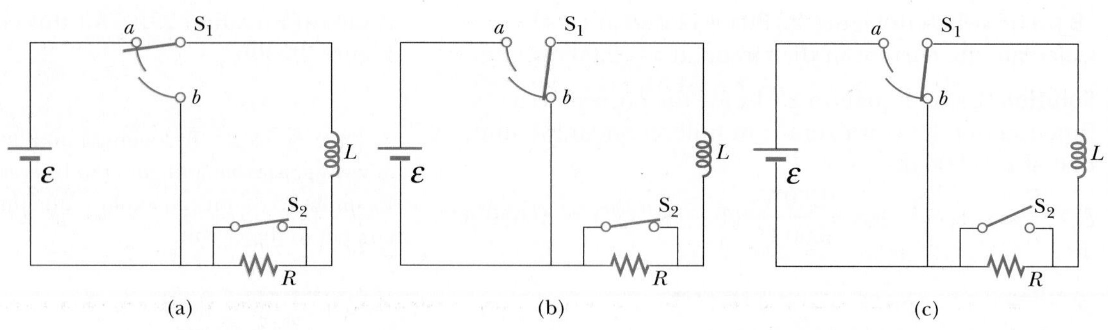

FIGURE 23.31 An RL circuit used for conceptualizing energy storage in an inductor. (a) With the switches as shown, the battery establishes a current through the inductor. (b) Switch S_1 is thrown to position b. Because the ends of the inductor are connected by a resistance-free path, the current continues to flow through the inductor. (c) Switch S_2 is opened, adding the resistor to the circuit, and energy is delivered to the resistor. This energy can only have been stored in the inductor because that is the only other element in the circuit.

it. The next step is to open switch S_2 as shown in Figure 23.31c, which puts the resistor into the circuit. There is now current in the resistor, and energy is delivered to the resistor. Where is the energy coming from? The only other element in the circuit previous to opening switch S_2 was the inductor. Energy must therefore have been stored in the inductor and is now being delivered to the resistor.

Now let us determine the energy per unit volume, or energy density, stored in a magnetic field. For simplicity, consider a solenoid whose inductance is $L = \mu_0 n^2 A\ell$ (see Example 23.7). The magnetic field of the solenoid is $B = \mu_0 nI$. Substituting the expression for L and $I = B/\mu_0 n$ into Equation 23.20 gives

$$U_B = \tfrac{1}{2}LI^2 = \tfrac{1}{2}\mu_0 n^2 A\ell\left(\frac{B}{\mu_0 n}\right)^2 = \frac{B^2}{2\mu_0}(A\ell) \qquad [23.21]$$

Because $A\ell$ is the volume of the solenoid, the energy stored per unit volume in a magnetic field—in other words, the *magnetic energy density*—is

▪ Magnetic energy density

$$u_B = \frac{U_B}{A\ell} = \frac{B^2}{2\mu_0} \qquad [23.22]$$

Although Equation 23.22 was derived for the special case of a solenoid, **it is valid for any region of space in which a magnetic field exists.** Note that it is similar to the equation for the energy per unit volume stored in an electric field, given by $\frac{1}{2}\epsilon_0 E^2$ (Eq. 20.32). In both cases, the energy density is proportional to the square of the magnitude of the field.

QUICK QUIZ 23.9 You are performing an experiment that requires the highest possible energy density in the interior of a very long solenoid. Which of the following increases the energy density? (More than one choice may be correct.) **(a)** increasing the number of turns per unit length on the solenoid **(b)** increasing the cross-sectional area of the solenoid **(c)** increasing only the length of the solenoid while keeping the number of turns per unit length fixed **(d)** increasing the current in the solenoid

EXAMPLE 23.9 What Happens to the Energy in the Inductor?

Consider once again the RL circuit shown in Active Figure 23.26 in which switch S is thrown to position b at $t = 0$. Recall that the current in the right-hand loop decays exponentially with time according to the expression $I = I_i e^{-t/\tau}$, where $I_i = \mathcal{E}/R$ is the initial current in the circuit and $\tau = L/R$ is the time constant. Let us show explicitly that all the energy stored in the magnetic field of the inductor is transferred to the resistor.

Solution The rate at which energy is transferred to the resistor is I^2R, where I is the instantaneous current. Using I from Equation 23.18,

$$\mathcal{P} = I^2R = (I_i e^{-Rt/L})^2 R = I_i^2 R e^{-2Rt/L}$$

To find the total energy transferred to the resistor, we integrate this expression over the limits $t = 0$ to $t = \infty$ (because it takes an infinite time for the current to reach zero):

$$(1)\quad E = \int_0^\infty \mathcal{P}\, dt = \int_0^\infty I_i^2 R e^{-2Rt/L}\, dt = I_i^2 R \int_0^\infty e^{-2Rt/L}\, dt$$

If we identify the variable $x = 2Rt/L$ so that $dx = (2R/L)\, dt$, we can rewrite (1) as,

$$E = I_i^2 R\left(\frac{L}{2R}\right)\int_0^\infty e^{-x}\, dx = \frac{I_i^2 L}{2}\left(-e^{-x}\right)\Big|_0^\infty$$

$$= \frac{I_i^2 L}{2}[0 - (-1)] = \tfrac{1}{2} I_i^2 L$$

Note that this expression is equal to the initial energy stored in the magnetic field of the inductor, given by Equation 23.20, as we set out to prove.

EXAMPLE 23.10 The Coaxial Cable

A long coaxial cable consists of two concentric cylindrical conductors of radii a and b and length ℓ as in Figure 23.32. The inner conductor is assumed to be a thin cylindrical shell. The conductors carry current I in opposite directions.

A Calculate the self-inductance L of this cable.

Solution To obtain L, we must know the magnetic flux through any cross-section between the two conductors. From Ampère's law (Eq. 22.29), we know that the magnetic field between the conductors is $B = \mu_0 I/2\pi r$. The magnetic field is zero outside the conductors ($r > b$) because the net current through a circular path surrounding both wires is zero. The magnetic field is zero inside the inner conductor because it is hollow and no current flows within a radius $r < a$.

The magnetic field is perpendicular to the light blue rectangle of length ℓ and width $(b - a)$, the cross-section of interest. Dividing this rectangle into strips of width dr, such as the dark blue strip in Figure 23.32, we see that the area of each strip is $\ell\, dr$ and the flux through each strip is $B\, dA = B\ell\, dr$. Hence, the total magnetic flux through the entire cross-section is

$$\Phi_B = \int B\, dA = \int_a^b \frac{\mu_0 I}{2\pi r}\ell\, dr = \frac{\mu_0 I\ell}{2\pi}\int_a^b \frac{dr}{r}$$

$$= \frac{\mu_0 I\ell}{2\pi}\ln\left(\frac{b}{a}\right)$$

Using this result, we find that the self-inductance of the cable is

$$L = \frac{\Phi_B}{I} = \frac{\mu_0 \ell}{2\pi}\ln\left(\frac{b}{a}\right)$$

B Calculate the total energy stored in the magnetic field of the cable.

Solution Using Equation 23.20 and the results to part A gives

$$U_B = \tfrac{1}{2} L I^2 = \frac{\mu_0 \ell I^2}{4\pi}\ln\left(\frac{b}{a}\right)$$

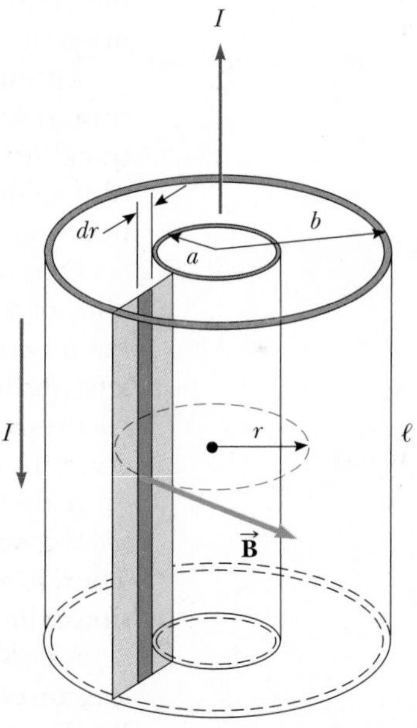

FIGURE 23.32 (Example 23.10) Section of a long coaxial cable. The inner and outer conductors carry equal currents in opposite directions.

23.8 THE REPULSIVE MODEL FOR MAGNETIC LEVITATION

CONTEXT CONNECTION

In Chapter 22, we considered a model for magnetic levitation that is based on the attractive force between a magnet and a rail made of magnetic material. The second major model for magnetically levitated vehicles is the EDS (*electrodynamic system*) model. The EDS model is used in Japan Railways's magnetic levitation system. This system uses superconducting magnets, unlike the conventional room-temperature magnets used in the German Transrapid. The result is improved energy efficiency. There is promise for even better efficiency in the future if higher-temperature superconductors are developed.

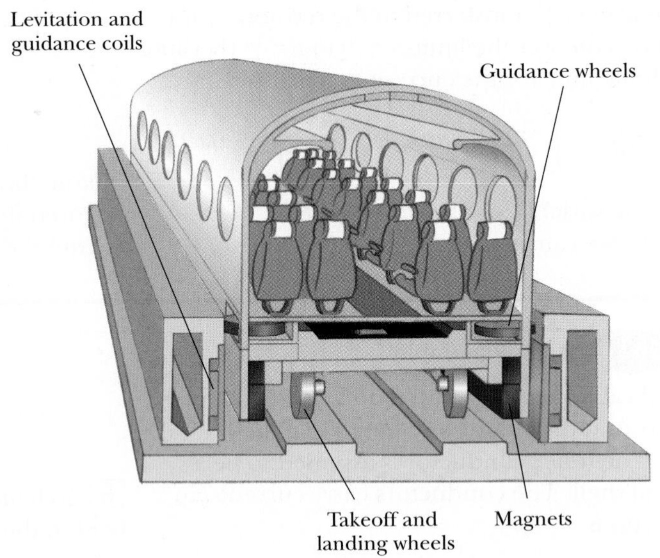

FIGURE 23.33 Schematic diagram of the levitation system for the Japanese maglev vehicle. The magnets induce currents in the coils at the side of the track so that, by Lenz's law, the repulsive force pushes the vehicle upward.

In this model, we appeal to Lenz's law. In the simplest form of the model, the vehicle carries a magnet. As the magnet passes over a metal plate that runs along the center of the track, currents are induced in the plate that tend to oppose the original change. The result is a *repulsive* force, which lifts the vehicle.

Although the idea of inducing a current in a metal plate is a valid concept, it represents a large expense in terms of the amount of metal required for a long track. Another technique is used in Japan's maglev vehicle. In this vehicle, the current is induced by magnets passing by coils located on the side of the railway channel. A schematic illustration of such a vehicle is given in Figure 23.33.

One of the disadvantages of the EMS model discussed in Chapter 22 is the instability of the attractive force, requiring feedback electronics. The EDS model, however, has a *natural* stabilizing feature. If the vehicle drops, the repulsion becomes stronger and pushes the vehicle back up. If the vehicle rises, the force decreases and the vehicle drops back down. Another advantage of the EDS system is the larger separation of about 10 cm between track and vehicle, as opposed to 10 mm in the EMS model.

A disadvantage of the EDS system is that *levitation only exists while the vehicle is moving* because it depends on Faraday's law; that is, a magnetic *change* must occur. Therefore, the vehicle must have landing wheels for stopping and starting at stations; these wheels are indicated in Figure 23.33.

Another disadvantage of the EDS system is that the induced currents result in a *drag* force as well as a lift force. The drag force requires more power for propulsion. It is larger than the lift force for small speeds, but the drag force maximizes at some speed and then begins to decrease. The lift force continues to increase as the speed increases. Therefore, it is advantageous to travel at high speeds, but the significant drag force at low speeds must be overcome every time the vehicle starts up.

The Japanese maglev project is jointly developed by the Central Japan Railway Co., the Railway Technical Research Institute, and the Japan Railway Construction, Transport, and Technology Agency. Exhaustive tests have been performed on five generations of maglev vehicles, beginning with the four-seater ML100, built in 1972 to celebrate Japan Railways's 100th anniversary. Current tests are being performed on the sixth-generation vehicle, the MLX01, a multicar train that can carry more than 100 passengers in its commercial form. A 43-km test line between Sakaigawa and Akiyama in Yamanashi Prefecture was opened in 1997. As mentioned in the Context introduction, the MLX01 holds the world speed record for magnetic levitation vehicles at 581 km/h. The Yamanashi Test Line is funded by the Japanese government, with the intention of final confirmation of maglev feasibility and commercial operation within the next few years. Once the Japanese system enters commercial operation, it will be interesting to watch the competition between it and the German Transrapid system! ■

SUMMARY

Physics Now™ Take a practice test by logging into PhysicsNow at **www.pop4e.com** and clicking on the Pre-Test link for this chapter.

The **magnetic flux** through a surface associated with a magnetic field $\vec{\mathbf{B}}$ is

$$\Phi_B = \int \vec{\mathbf{B}} \cdot d\vec{\mathbf{A}} \quad [23.1]$$

where the integral is over the surface.

Faraday's law of induction states that the emf induced in a circuit is directly proportional to the time rate of change of magnetic flux through the circuit:

$$\mathcal{E} = -N\frac{d\Phi_B}{dt} \quad [23.3]$$

where N is the number of turns and Φ_B is the magnetic flux through each turn.

When a conducting bar of length ℓ moves through a magnetic field $\vec{\mathbf{B}}$ with a velocity $\vec{\mathbf{v}}$ so that $\vec{\mathbf{v}}$ is perpendicular to $\vec{\mathbf{B}}$, the emf induced in the bar (called the **motional emf**) is

$$\mathcal{E} = -B\ell v \quad [23.5]$$

Lenz's law states that the induced current and induced emf in a conductor are in such a direction as to oppose the change that produced them.

A general form of Faraday's law of induction is

$$\mathcal{E} = \oint \vec{\mathbf{E}} \cdot d\vec{\mathbf{s}} = -\frac{d\Phi_B}{dt} \quad [23.9]$$

where $\vec{\mathbf{E}}$ is a nonconservative electric field produced by the changing magnetic flux.

When the current in a coil changes with time, an emf is induced in the coil according to Faraday's law. The **self-induced emf** is described by the expression

$$\mathcal{E}_L = -L\frac{dI}{dt} \quad [23.10]$$

where L is the *inductance* of the coil. Inductance is a measure of the opposition of a device to a change in current.

The **inductance** of a coil is

$$L = \frac{N\Phi_B}{I} \quad [23.11]$$

where Φ_B is the magnetic flux through the coil and N is the total number of turns. Inductance has the SI unit the **henry** (H), where $1\text{ H} = 1\text{ V}\cdot\text{s/A}$.

If a resistor and inductor are connected in series to a battery of emf $\mathcal{E}$ as shown in Active Figure 23.23 and a switch in the circuit is closed at $t = 0$, the current in the circuit varies with time according to the expression

$$I(t) = \frac{\mathcal{E}}{R}(1 - e^{-t/\tau}) \quad [23.14]$$

where $\tau = L/R$ is the **time constant** of the RL circuit.

If the battery is removed from an RL circuit as in Active Figure 23.26 with the switch thrown to position b, the current decays exponentially with time according to the expression

$$I(t) = \frac{\mathcal{E}}{R}e^{-t/\tau} \quad [23.18]$$

where $\mathcal{E}/R$ is the initial current in the circuit.

The energy stored in the magnetic field of an inductor carrying a current I is

$$U_B = \tfrac{1}{2}LI^2 \quad [23.20]$$

The energy per unit volume (or energy density) at a point where the magnetic field is B is

$$u_B = \frac{B^2}{2\mu_0} \quad [23.22]$$

QUESTIONS

☐ = answer available in the *Student Solutions Manual and Study Guide*

1. A loop of wire is placed in a uniform magnetic field. For what orientation of the loop is the magnetic flux a maximum? For what orientation is the flux zero?

2. A bar magnet is held above a loop of wire in a horizontal plane as shown in Figure Q23.2. The south end of the magnet is toward the loop of wire. The magnet is dropped toward the loop. Find the direction of the current in the resistor (a) while the magnet is falling toward the loop and (b) after the magnet has passed through the loop and is moving away from it.

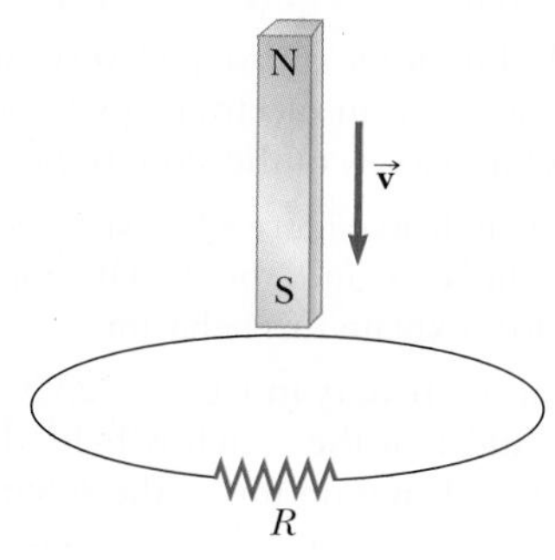

FIGURE Q23.2

3. As the bar in Figure Q23.3 moves to the right, an electric field is set up directed downward in the bar. Explain why the electric field would be upward if the bar were moving to the left.

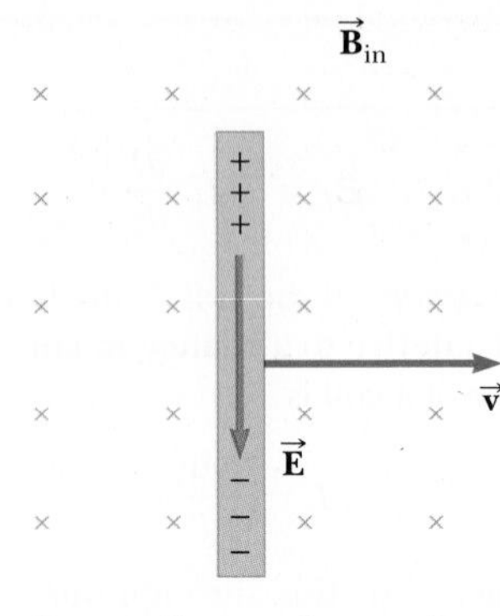

FIGURE Q23.3 Questions 23.3 and 23.4.

4. As the bar in Figure Q23.3 moves perpendicular to the field, is an external force required to keep it moving with constant speed?

5. The bar in Figure Q23.5 moves on rails to the right with a velocity $\vec{v}$, and the uniform, constant magnetic field is directed out of the page. Why is the induced current clockwise? If the bar were moving to the left, what would be the direction of the induced current?

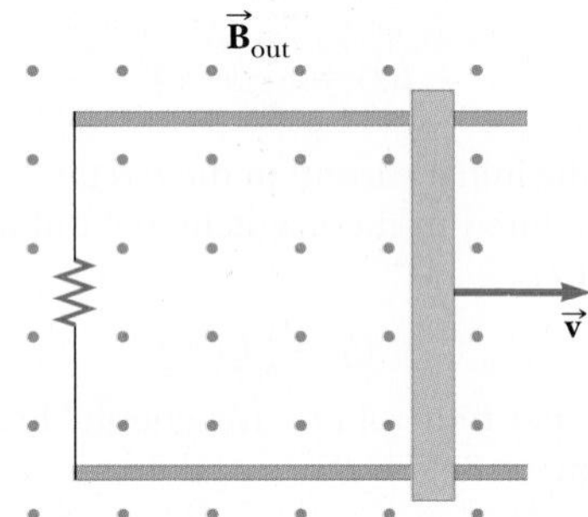

FIGURE Q23.5 Questions 23.5 and 23.6.

6. Explain why an applied force is necessary to keep the bar in Figure Q23.5 moving with a constant speed.

7. In a hydroelectric dam, how is the energy produced that is then transferred out by electrical transmission? That is, how is the energy of motion of the water converted to energy that is transmitted by AC electricity?

8. A piece of aluminum is dropped vertically downward between the poles of an electromagnet. Does the magnetic field affect the velocity of the aluminum?

9. When the switch in Figure Q23.9a is closed, a current is set up in the coil and the metal ring springs upward (Fig. Q23.9b). Explain this behavior.

10. Assume that the battery in Figure Q23.9a is replaced by an AC source and that the switch is held closed. If it is held down, the metal ring on top of the solenoid becomes hot. Why?

11. Find the direction of the current in the resistor in Figure Q23.11 (a) at the instant the switch is closed, (b) after the switch has been closed for several minutes, and (c) at the instant the switch is opened.

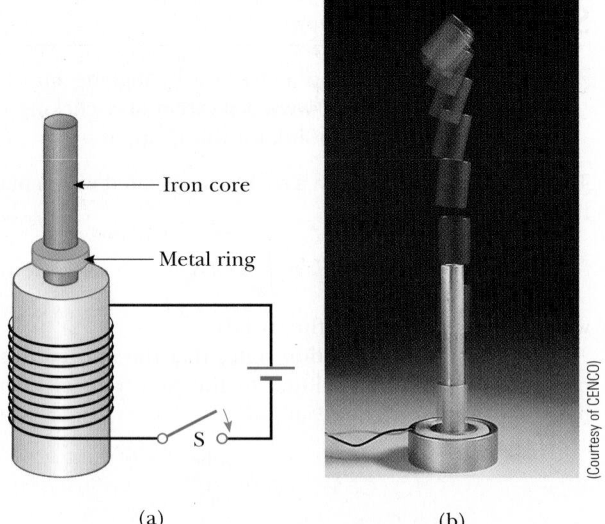

FIGURE Q23.9 Questions 23.9 and 23.10.

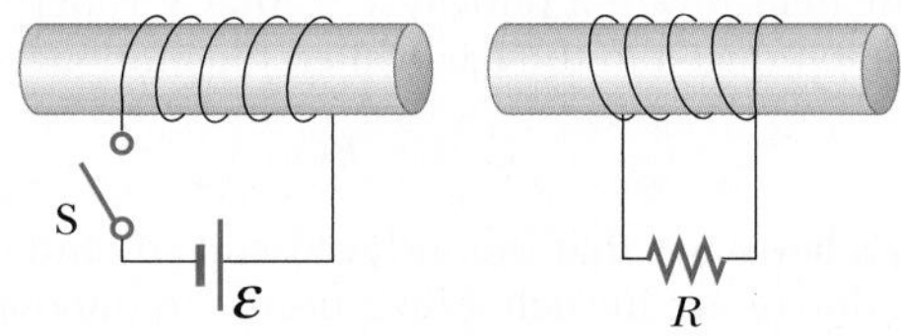

FIGURE Q23.11

12. An emf is induced between the wingtips of an airplane because of its motion in the Earth's magnetic field. Can this emf be used to power a light in the passenger compartment? Explain your answer.

13. Section 7.3 defined conservative and nonconservative forces. Section 20.1 stated that an electric charge creates an electric field that produces a conservative force. Argue now that induction creates an electric field that produces a nonconservative force.

14. What parameters affect the inductance of a coil? Does the inductance of a coil depend on the current in the coil?

15. Suppose the switch in Figure Q23.15 has been closed for a long time and is suddenly opened. Does the current instantaneously drop to zero? Why does a spark appear at the switch contacts at the moment the switch is opened?

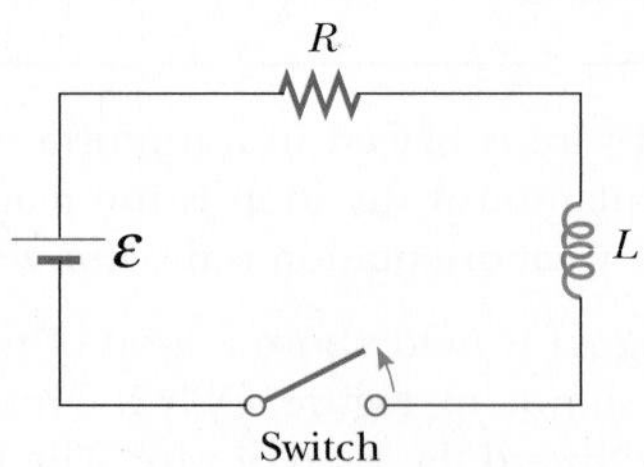

FIGURE Q23.15

16. Consider this thesis: "Joseph Henry, America's first professional physicist, caused the most recent basic change in the human view of the Universe when he discovered self-

induction during a school vacation at the Albany Academy about 1830. Before that time, one could think of the Universe as composed of just one thing: matter. The energy that temporarily maintains the current after a battery is removed from a coil, on the other hand, is not energy that belongs to any chunk of matter. It is energy in the massless magnetic field surrounding the coil. With Henry's discovery, Nature forced us to admit that the Universe consists of fields as well as matter." Argue for or against the statement. What in your view makes up the Universe?

17. If the current in an inductor is doubled, by what factor does the stored energy change?

18. Discuss the similarities between the energy stored in the electric field of a charged capacitor and the energy stored in the magnetic field of a current-carrying coil.

19. What is the inductance of two inductors connected in series? Does it matter if they are solenoids or toroids?

20. Can an object exert a force on itself? When a coil induces an emf in itself, does it exert a force on itself?

PROBLEMS

1, 2, 3 = straightforward, intermediate, challenging
☐ = full solution available in the *Student Solutions Manual and Study Guide*
PhysicsNow™ = coached problem with hints available at www.pop4e.com
= computer useful in solving problem
= paired numerical and symbolic problems
= biomedical application

Section 23.1 ■ Faraday's Law of Induction
Section 23.3 ■ Lenz's Law

1. A flat loop of wire consisting of a single turn of cross-sectional area 8.00 cm^2 is perpendicular to a magnetic field that increases uniformly in magnitude from 0.500 T to 2.50 T in 1.00 s. What is the resulting induced current if the loop has a resistance of 2.00 Ω?

2. A 25-turn circular coil of wire has diameter 1.00 m. It is placed with its axis along the direction of the Earth's magnetic field of 50.0 μT, and then in 0.200 s it is flipped 180°. An average emf of what magnitude is generated in the coil?

3. **PhysicsNow™** A strong electromagnet produces a uniform magnetic field of 1.60 T over a cross-sectional area of 0.200 m^2. We place a coil having 200 turns and a total resistance of 20.0 Ω around the electromagnet. We then smoothly reduce the current in the electromagnet until it reaches zero in 20.0 ms. What is the current induced in the coil?

4. An aluminum ring of radius r_1 and resistance R is placed around the top of a long air-core solenoid with n turns per meter and smaller radius r_2 as shown in Figure P23.4. Assume that the axial component of the field produced by the solenoid over the area of the end of the solenoid is one-half as strong as at the solenoid's center. Assume that the solenoid produces negligible field outside its cross-sectional area. The current in the solenoid is increasing at a rate of $\Delta I/\Delta t$. (a) What is the induced current in the ring? (b) At the center of the ring, what is the magnetic field produced by the induced current in the ring? (c) What is the direction of this field?

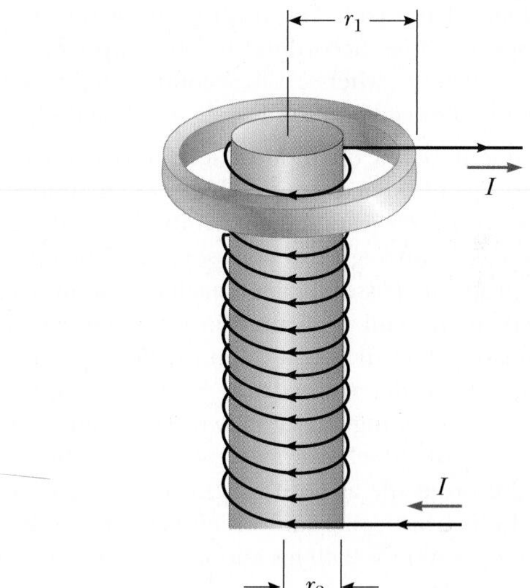

FIGURE P23.4

5. (a) A loop of wire in the shape of a rectangle of width w and length L and a long, straight wire carrying a current I lie on a tabletop as shown in Figure P23.5. (a) Determine the magnetic flux through the loop due to the current I. (b) Suppose the current is changing with time according to $I = a + bt$, where a and b are constants. Determine the emf that is induced in the loop if $b = 10.0$ A/s, $h = 1.00$ cm, $w = 10.0$ cm, and $L = 100$ cm. What is the direction of the induced current in the rectangle?

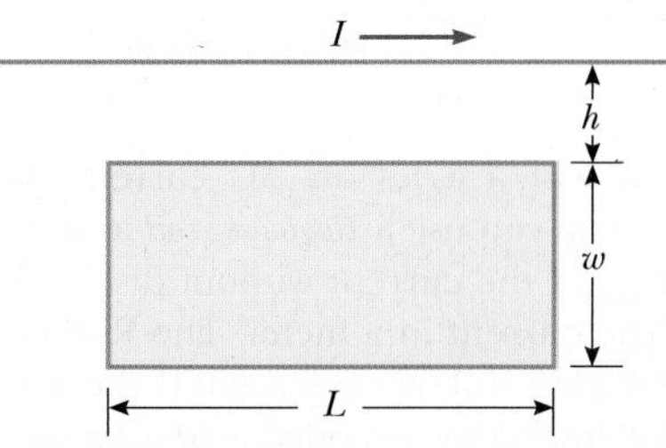

FIGURE P23.5 Problems 23.5 and 23.59.

6. A coil of 15 turns and radius 10.0 cm surrounds a long solenoid of radius 2.00 cm and 1.00×10^3 turns/m (Fig. P23.6). The current in the solenoid changes as $I = (5.00\text{ A}) \sin(120t)$. Find the induced emf in the 15-turn coil as a function of time.

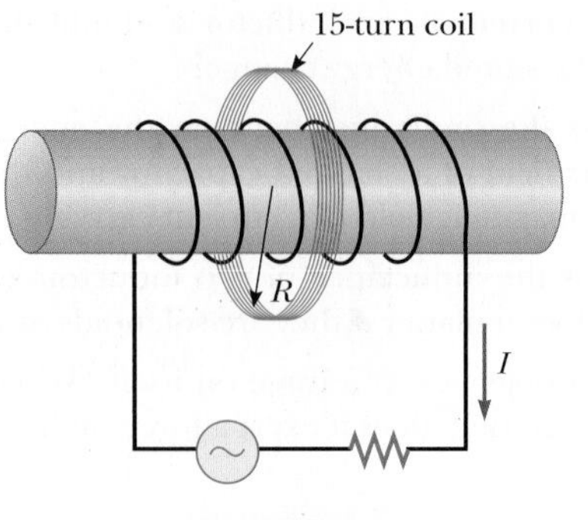

FIGURE P23.6

7. A 30-turn circular coil of radius 4.00 cm and resistance 1.00 Ω is placed in a magnetic field directed perpendicular to the plane of the coil. The magnitude of the magnetic field varies in time according to the expression $B = 0.010\,0t + 0.040\,0t^2$, where t is in seconds and B is in teslas. Calculate the induced emf in the coil at $t = 5.00$ s.

8. An instrument based on induced emf has been used to measure projectile speeds up to 6 km/s. A small magnet is imbedded in the projectile as shown in Figure P23.8. The projectile passes through two coils separated by a distance d. As the projectile passes through each coil, a pulse of emf is induced in the coil. The time interval between pulses can be measured accurately with an oscilloscope, and thus the speed can be determined. (a) Sketch a graph of ΔV versus t for the arrangement shown. Consider a current that flows counterclockwise as viewed from the starting point of the projectile as positive. On your graph, indicate which pulse is from coil 1 and which is from coil 2. (b) If the pulse separation is 2.40 ms and $d = 1.50$ m, what is the projectile speed?

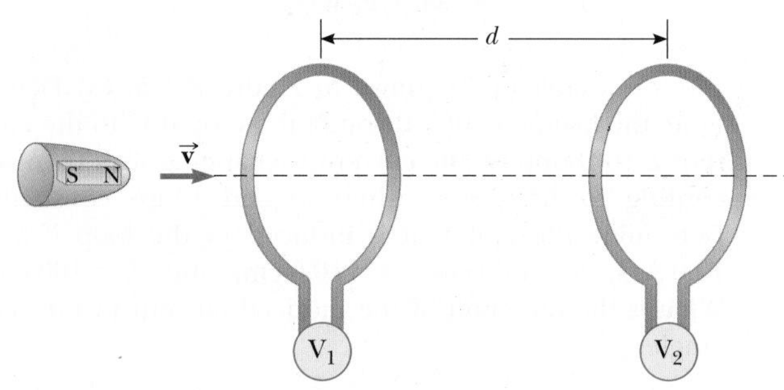

FIGURE P23.8

9. When a wire carries an AC current with a known frequency, you can use a *Rogowski coil* to determine the amplitude I_{max} of the current without disconnecting the wire to shunt the current in a meter. The Rogowski coil, shown in Figure P23.9, simply clips around the wire. It consists of a toroidal conductor wrapped around a circular return cord. The toroid has n turns per unit length and a cross-sectional area A. The current to be measured is given by $I(t) = I_{max} \sin \omega t$. (a) Show that the amplitude of the emf induced in the Rogowski coil is $\mathcal{E}_{max} = \mu_0 n A \omega I_{max}$. (b) Explain why the wire carrying the unknown current need not be at the center of the Rogowski coil and why the coil will not respond to nearby currents that it does not enclose.

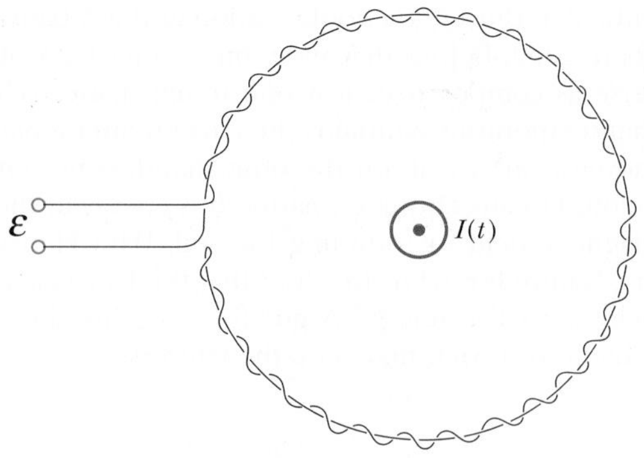

FIGURE P23.9

10. A piece of insulated wire is shaped into a figure eight as shown in Figure P23.10. The radius of the upper circle is 5.00 cm and that of the lower circle is 9.00 cm. The wire has a uniform resistance per unit length of 3.00 Ω/m. A uniform magnetic field is applied perpendicular to the plane of the two circles, in the direction shown. The magnetic field is increasing at a constant rate of 2.00 T/s. Find the magnitude and direction of the induced current in the wire.

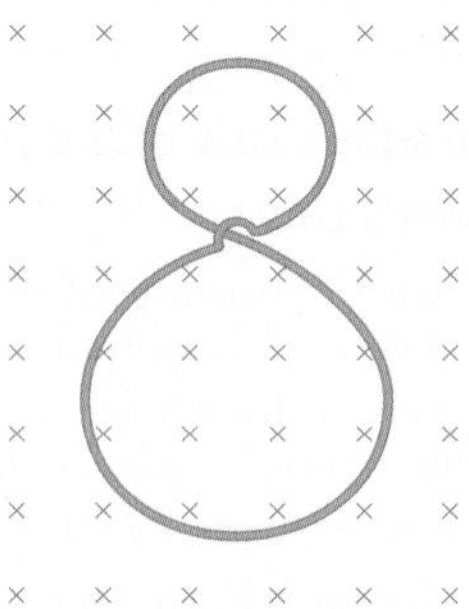

FIGURE P23.10

Section 23.2 ■ Motional emf
Section 23.3 ■ Lenz's Law

Note: Problem 22.62 can be assigned with this section.

11. An automobile has a vertical radio antenna 1.20 m long. The automobile travels at 65.0 km/h on a horizontal road where the Earth's magnetic field is 50.0 μT directed toward the north and downward at an angle of 65.0° below the horizontal. (a) Specify the direction that the automobile should move to generate the maximum motional emf in the antenna, with the top of the antenna positive relative to the bottom. (b) Calculate the magnitude of this induced emf.

12. Consider the arrangement shown in Figure P23.12. Assume that $R = 6.00$ Ω, $\ell = 1.20$ m, and a uniform 2.50-T magnetic field is directed into the page. At what speed should the bar be moved to produce a current of 0.500 A in the resistor?

13. Figure P23.12 shows a top view of a bar that can slide without friction. The resistor is 6.00 Ω, and a 2.50-T magnetic field is directed perpendicularly downward, into the paper.

Let $\ell = 1.20$ m. (a) Calculate the applied force required to move the bar to the right at a constant speed of 2.00 m/s. (b) At what rate is energy delivered to the resistor?

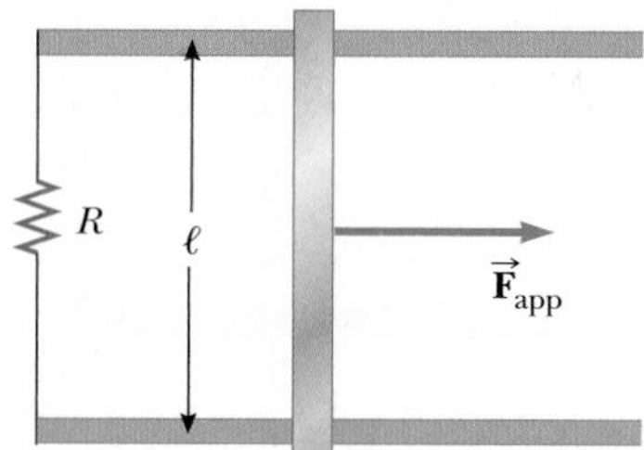

FIGURE **P23.12** Problems 23.12, 23.13, 23.14, and 23.15.

14. A conducting rod of length ℓ moves on two horizontal, frictionless rails as shown in Figure P23.12. If a constant force of 1.00 N moves the bar at 2.00 m/s through a magnetic field $\vec{\mathbf{B}}$ that is directed into the page, (a) what is the current through the 8.00-Ω resistor R? (b) What is the rate at which energy is delivered to the resistor? (c) What is the mechanical power delivered by the force $\vec{\mathbf{F}}_{app}$?

15. A metal rod of mass m slides without friction along two parallel horizontal rails, separated by a distance ℓ and connected by a resistor R, as shown in Figure P23.12. A uniform vertical magnetic field of magnitude B is applied perpendicular to the plane of the paper. The applied force shown in the figure acts only for a moment, to give the rod a speed v. In terms of m, ℓ, R, B, and v, find the distance the rod will then slide as it coasts to a stop.

16. Very large magnetic fields can be produced using a procedure called *flux compression.* A metallic cylindrical tube of radius R is placed coaxially in a long solenoid of somewhat larger radius. The space between the tube and the solenoid is filled with a highly explosive material. When the explosive is set off, it collapses the tube to a cylinder of radius $r < R$. If the collapse happens very rapidly, induced current in the tube maintains the magnetic flux nearly constant inside the tube. If the initial magnetic field in the solenoid is 2.50 T and $R/r = 12.0$, what maximum value of magnetic field can be achieved?

17. The *homopolar generator,* also called the *Faraday disk,* is a low-voltage, high-current electric generator. It consists of a rotating conducting disk with one stationary brush (a sliding electrical contact) at its axle and another at a point on its circumference as shown in Figure P23.17. A magnetic field is applied perpendicular to the plane of the disk. Assume that the field is 0.900 T, the angular speed is 3 200 rev/min, and the radius of the disk is 0.400 m. Find the emf generated between the brushes. When superconducting coils are used to produce a large magnetic field, a homopolar generator can have a power output of several megawatts. Such a generator is useful, for example, in purifying metals by electrolysis. If a voltage is applied to the output terminals of the generator, it runs in reverse as a *homopolar motor* capable of providing great torque, useful in ship propulsion.

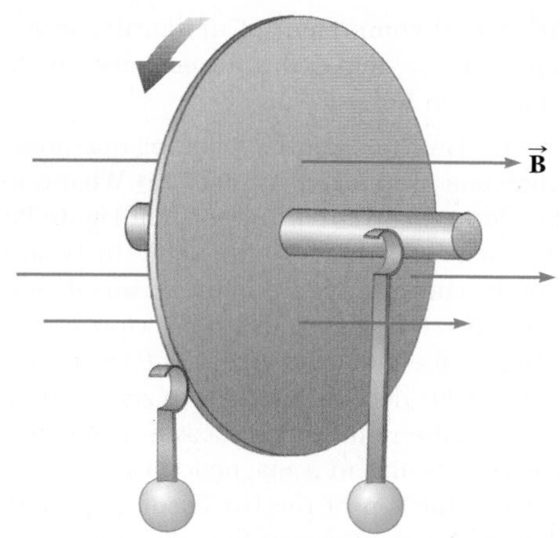

FIGURE **P23.17**

18. **Review problem.** A flexible metallic wire with linear density 3.00×10^{-3} kg/m is stretched between two fixed clamps 64.0 cm apart and held under tension 267 N. A magnet is placed near the wire as shown in Figure P23.18. Assume that the magnet produces a uniform field of 4.50 mT over a 2.00-cm length at the center of the wire and a negligible field elsewhere. The wire is set vibrating at its fundamental (lowest) frequency. The section of the wire in the magnetic field moves with a uniform amplitude of 1.50 cm. Find (a) the frequency and (b) the amplitude of the electromotive force induced between the ends of the wire.

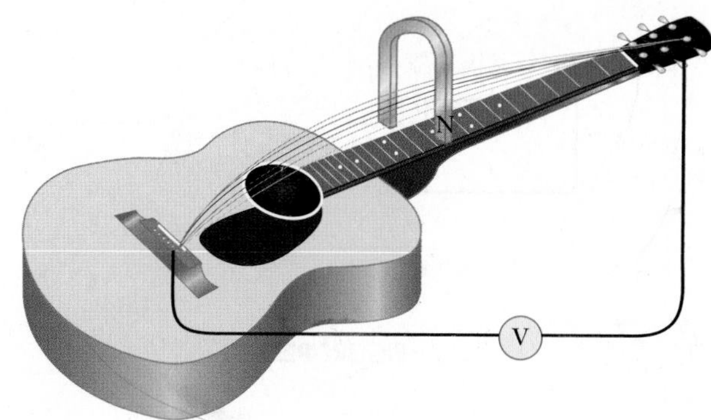

FIGURE **P23.18**

19. A helicopter (Fig. P23.19) has blades of length 3.00 m, extending out from a central hub and rotating at 2.00 rev/s.

(Ross Harrison/Getty Images)

FIGURE **P23.19**

If the vertical component of the Earth's magnetic field is 50.0 μT, what is the emf induced between the blade tip and the center hub?

20. Use Lenz's law to answer the following questions concerning the direction of induced currents. (a) What is the direction of the induced current in resistor R in Figure P23.20a when the bar magnet is moved to the left? (b) What is the direction of the current induced in the resistor R immediately after the switch S in Figure P23.20b is closed? (c) What is the direction of the induced current in R when the current I in Figure P23.20c decreases rapidly to zero? (d) A copper bar is moved to the right while its axis is maintained in a direction perpendicular to a magnetic field as shown in Figure P31.28d. If the top of the bar becomes positive relative to the bottom, what is the direction of the magnetic field?

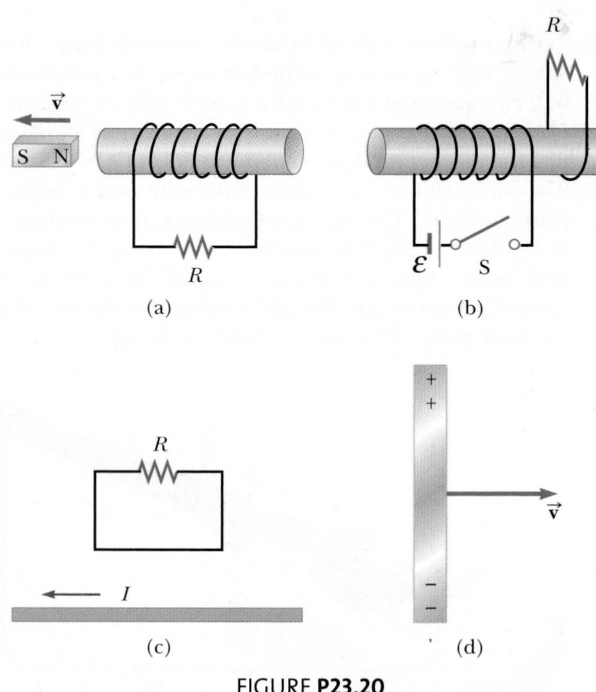

FIGURE **P23.20**

21. Physics Now™ A conducting rectangular loop of mass M, resistance R, and dimensions w by ℓ falls from rest into a magnetic field $\vec{B}$ as shown in Figure P23.21. During the time interval before the top edge of the loop reaches the field, the loop approaches a terminal speed v_T. (a) Show that

$$v_T = \frac{MgR}{B^2w^2}$$

(b) Why is v_T proportional to R? (c) Why is it inversely proportional to B^2?

22. A rectangular coil with resistance R has N turns, each of length ℓ and width w as shown in Figure P23.22. The coil moves into a uniform magnetic field $\vec{B}$ with constant velocity $\vec{v}$. What are the magnitude and direction of the total magnetic force on the coil (a) as it enters the magnetic field, (b) as it moves within the field, and (c) as it leaves the field?

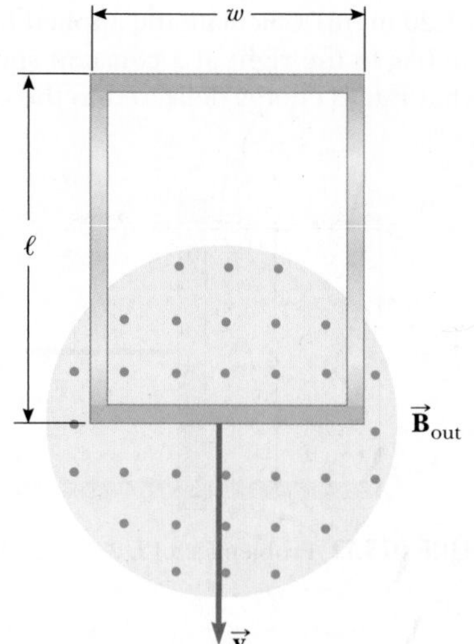

FIGURE **P23.21**

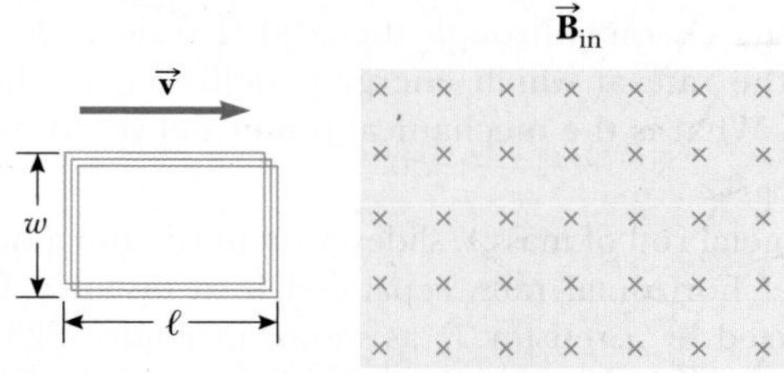

FIGURE **P23.22**

23. Physics Now™ A coil of area 0.100 m^2 is rotating at 60.0 rev/s with the axis of rotation perpendicular to a 0.200-T magnetic field. (a) If the coil has 1 000 turns, what is the maximum emf generated in it? (b) What is the orientation of the coil with respect to the magnetic field when the maximum induced voltage occurs?

24. A long solenoid, with its axis along the x axis, consists of 200 turns per meter of wire that carries a steady current of 15.0 A. A coil is formed by wrapping 30 turns of thin wire around a circular frame that has a radius of 8.00 cm. The coil is placed inside the solenoid and mounted on an axis that is a diameter of the coil and that coincides with the y axis. The coil is then rotated with an angular speed of 4.00π rad/s. (The plane of the coil is in the yz plane at $t = 0$.) Determine the emf generated in the coil as a function of time.

Section 23.4 Induced emfs and Electric Fields

25. A magnetic field directed into the page changes with time according to $B = (0.030\,0t^2 + 1.40)$ T, where t is in seconds. The field has a circular cross-section of radius $R = 2.50$ cm (Fig. P23.25). What are the magnitude and direction of the electric field at point P_1 when $t = 3.00$ s and $r_1 = 0.020\,0$ m?

26. For the situation shown in Figure P23.25, the magnetic field changes with time according to the expression $B = (2.00t^3 - 4.00t^2 + 0.800)$ T and $r_2 = 2R = 5.00$ cm. (a) Calculate the magnitude and direction of the force

exerted on an electron located at point P_2 when $t = 2.00$ s. (b) At what time is this force equal to zero?

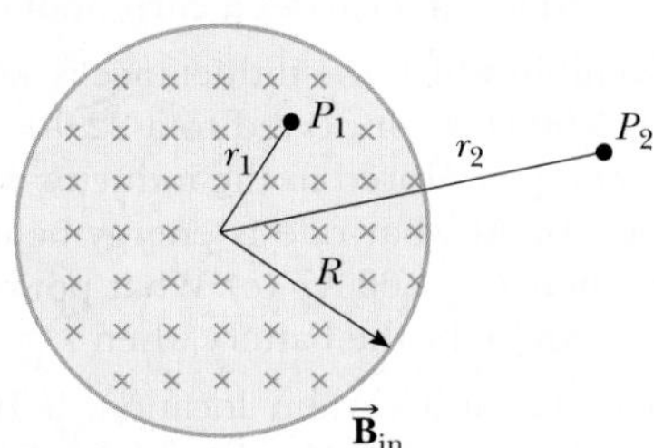

FIGURE **P23.25** Problems 23.25 and 23.26.

Section 23.5 ■ Self-Inductance

27. A coil has an inductance of 3.00 mH, and the current in it changes from 0.200 A to 1.50 A in a time interval of 0.200 s. Find the magnitude of the average induced emf in the coil during this time interval.

28. A coiled telephone cord forms a spiral with 70 turns, a diameter of 1.30 cm, and an unstretched length of 60.0 cm. Determine the self-inductance of one conductor in the unstretched cord.

29. **PhysicsNow™** A 10.0-mH inductor carries a current $I = I_{max} \sin \omega t$, with $I_{max} = 5.00$ A and $\omega/2\pi = 60.0$ Hz. What is the self-induced emf as a function of time?

30. An emf of 24.0 mV is induced in a 500-turn coil at an instant when the current is 4.00 A and is changing at the rate of 10.0 A/s. What is the magnetic flux through each turn of the coil?

31. The current in a 90.0-mH inductor changes with time as $I = 1.00t^2 - 6.00t$ (in SI units). Find the magnitude of the induced emf at (a) $t = 1.00$ s and (b) $t = 4.00$ s. (c) At what time is the emf zero?

32. A toroid has a major radius R and a minor radius r, and is tightly wound with N turns of wire as shown in Figure P23.32. If $R >> r$, the magnetic field in the region enclosed by the wire of the torus, of cross-sectional area $A = \pi r^2$, is essentially the same as the magnetic field of a solenoid that has been bent into a large circle of radius R. Modeling the field as the uniform field of a long solenoid, show that the self-inductance of such a toroid is approximately

$$L \approx \frac{\mu_0 N^2 A}{2\pi R}$$

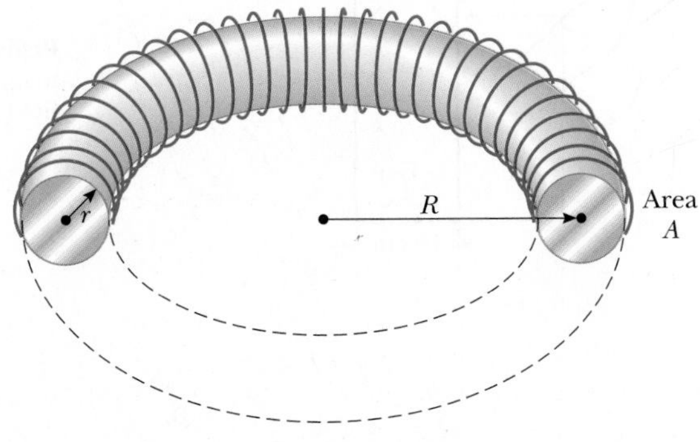

FIGURE **P23.32**

(An exact expression of the inductance of a toroid with a rectangular cross-section is derived in Problem 23.60.)

Section 23.6 ■ *RL* Circuits

33. A 12.0-V battery is connected into a series circuit containing a 10.0-Ω resistor and a 2.00-H inductor. In what time interval will the current reach (a) 50.0% and (b) 90.0% of its final value?

34. Show that $I = I_i e^{-t/\tau}$ is a solution of the differential equation

$$IR + L\frac{dI}{dt} = 0$$

where $\tau = L/R$ and I_i is the current at $t = 0$.

35. Consider the circuit in Figure P23.35, taking $\mathcal{E} = 6.00$ V, $L = 8.00$ mH, and $R = 4.00$ Ω. (a) What is the inductive time constant of the circuit? (b) Calculate the current in the circuit 250 μs after the switch is closed. (c) What is the value of the final steady-state current? (d) How long does it take the current to reach 80.0% of its maximum value?

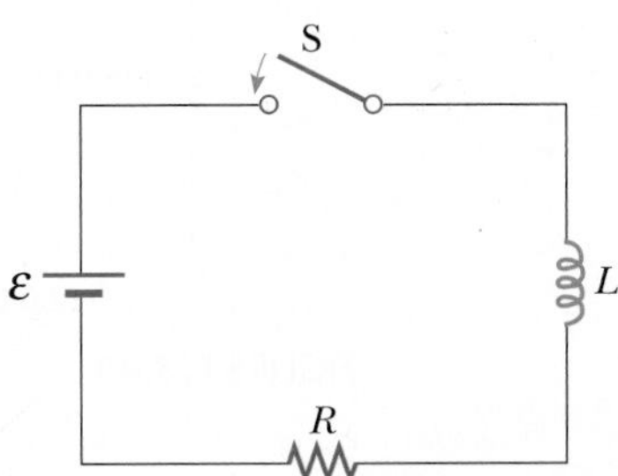

FIGURE **P23.35** Problems 23.35, 23.36, and 23.38.

36. For the *RL* circuit shown in Figure P23.35, let the inductance be 3.00 H, the resistance 8.00 Ω, and the battery emf 36.0 V. (a) Calculate the ratio of the potential difference across the resistor to the voltage across the inductor when the current is 2.00 A. (b) Calculate the voltage across the inductor when the current is 4.50 A.

37. A circuit consists of a coil, a switch, and a battery, all in series. The internal resistance of the battery is negligible compared with that of the coil. The switch is originally open. It is thrown closed, and after a time interval Δt, the current in the circuit reaches 80.0% of its final value. The switch remains closed for a time interval much longer than Δt. Then the battery is disconnected and the terminals of the coil are connected together to form a short circuit. (a) After an equal additional time interval Δt elapses, the current is what percentage of its maximum value? (b) At the moment $2\Delta t$ after the coil is short-circuited, the current in the coil is what percentage of its maximum value?

38. When the switch in Figure P23.35 is closed, the current takes 3.00 ms to reach 98.0% of its final value. If $R = 10.0$ Ω, what is the inductance?

39. The switch in Figure P23.39 is open for $t < 0$ and then closed at time $t = 0$. Find the current in the inductor and the current in the switch as functions of time thereafter.

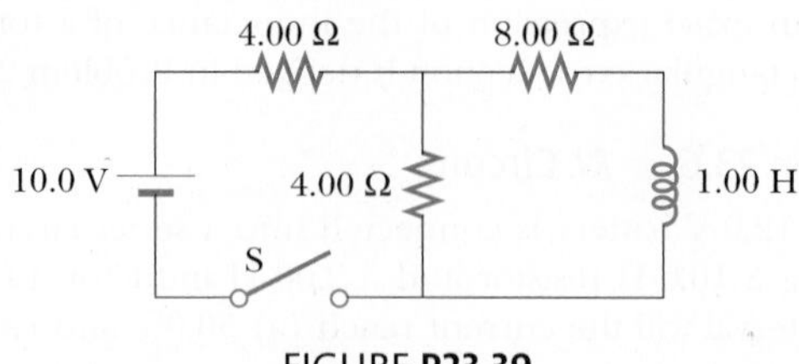

FIGURE P23.39

40. One application of an RL circuit is the generation of time-varying high voltage from a low-voltage source as shown in Figure P23.40. (a) What is the current in the circuit a long time after the switch has been in position a? (b) Now the switch is thrown quickly from a to b. Compute the initial voltage across each resistor and across the inductor. (c) How much time elapses before the voltage across the inductor drops to 12.0 V?

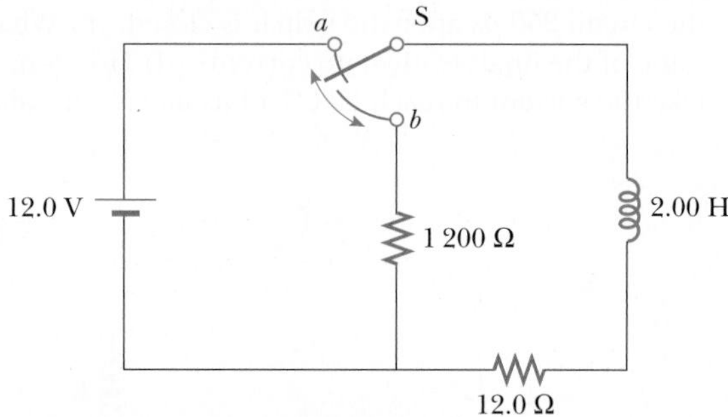

FIGURE P23.40

41. PhysicsNow™ A 140-mH inductor and a 4.90-Ω resistor are connected with a switch to a 6.00-V battery as shown in Figure P23.41. (a) If the switch is thrown to the left (connecting the battery), how much time elapses before the current reaches 220 mA? (b) What is the current in the inductor 10.0 s after the switch is closed? (c) Now the switch is quickly thrown from a to b. How much time elapses before the current falls to 160 mA?

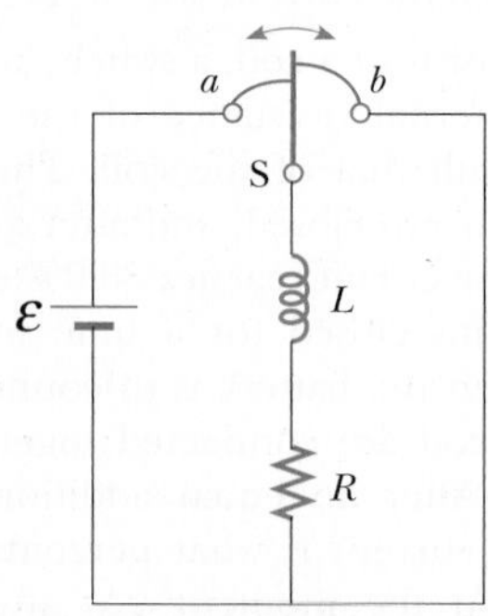

FIGURE P23.41

Section 23.7 ■ Energy Stored in a Magnetic Field

42. The magnetic field inside a superconducting solenoid is 4.50 T. The solenoid has an inner diameter of 6.20 cm and a length of 26.0 cm. Determine (a) the magnetic energy density in the field and (b) the energy stored in the magnetic field within the solenoid.

43. An air-core solenoid with 68 turns is 8.00 cm long and has a diameter of 1.20 cm. How much energy is stored in its magnetic field when it carries a current of 0.770 A?

44. An RL circuit in which the inductance is 4.00 H and the resistance is 5.00 Ω is connected to a 22.0-V battery at $t = 0$. (a) What energy is stored in the inductor when the current is 0.500 A? (b) At what rate is energy being stored in the inductor when $I = 1.00$ A? (c) What power is being delivered to the circuit by the battery when $I = 0.500$ A?

45. On a clear day at a certain location, a 100-V/m vertical electric field exists near the Earth's surface. At the same place, the Earth's magnetic field has a magnitude of 0.500×10^{-4} T. Compute the energy densities of the two fields.

Section 23.8 ■ Context Connection—The Repulsive Model for Magnetic Levitation

46. The following is a crude model for levitating a commercial transportation vehicle using Faraday's law. Assume that magnets are used to establish regions of horizontal magnetic field across the track as shown in Figure P23.46. Rectangular loops of wire are mounted on the vehicle so that the lower 20 cm of each loop passes into these regions of magnetic field. The upper portion of each loop contains a 25-Ω resistor. As the leading edge of a loop passes into the magnetic field, a current is induced in the loop as shown in the figure. The magnetic force on this current in the bottom of the loop, of length 10 cm, results in an upward force on the vehicle. (By electronic timing, a switch is opened in the loop before the loop's leading edge leaves the region of magnetic field, so a current is not induced in the opposite direction to apply a downward force on the vehicle.) The vehicle has a mass of 5×10^4 kg and travels at a speed of 400 km/h. If the vehicle has 100 loops carrying current at any moment, what is the approximate magnitude of the magnetic field required to levitate the vehicle? Assume that the magnetic force acts over the entire 10-cm length of the horizontal wire. Your answer should suggest that this design is impractical for magnetic levitation.

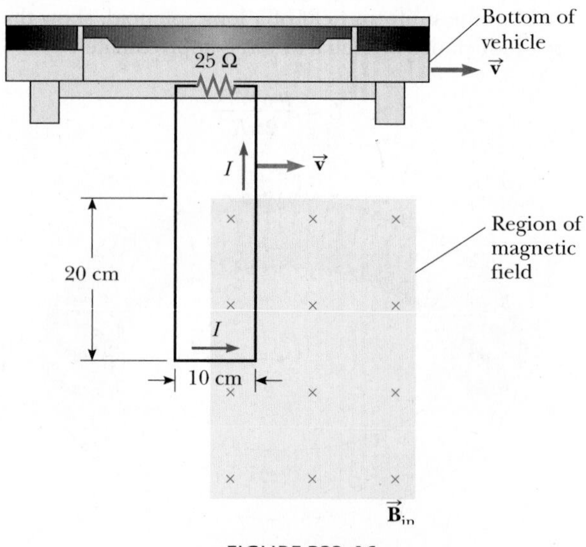

FIGURE P23.46

47. *The Meissner effect.* Compare this problem with Problem 20.76 on the force attracting a perfect dielectric into a strong electric field. A fundamental property of a type I superconducting material is *perfect diamagnetism,* or demonstration of the *Meissner effect,* illustrated in the photograph of the levitating magnet on page 693 and described as follows. The superconducting material has $\vec{\mathbf{B}} = 0$ everywhere inside it. If a sample of the material is placed into an externally produced magnetic field or if it is cooled to become superconducting while it is in a magnetic field, electric currents appear on the surface of the sample. The currents have precisely the strength and orientation required to make the total magnetic field zero throughout the interior of the sample. The following problem will help you understand the magnetic force that can then act on the superconducting sample.

    A vertical solenoid with a length of 120 cm and a diameter of 2.50 cm consists of 1 400 turns of copper wire carrying a counterclockwise current of 2.00 A as shown in Figure P23.47a. (a) Find the magnetic field in the vacuum inside the solenoid. (b) Find the energy density of the magnetic field. Note that the units J/m^3 of energy density are the same as the units N/m^2 of pressure. (c) Now a superconducting bar 2.20 cm in diameter is inserted partway into the solenoid. Its upper end is far outside the solenoid, where the magnetic field is negligible. The lower end of the bar is deep inside the solenoid. Identify the direction required for the current on the curved surface of the bar so that the total magnetic field is zero within the bar. The field created by the supercurrents is sketched in Figure P23.47b, and the total field is sketched in Figure P23.47c. (d) The field of the solenoid exerts a force on the current in the superconductor. Identify the direction of the force on the bar. (e) Calculate the magnitude of the force by multiplying the energy density of the solenoid field times the area of the bottom end of the superconducting bar.

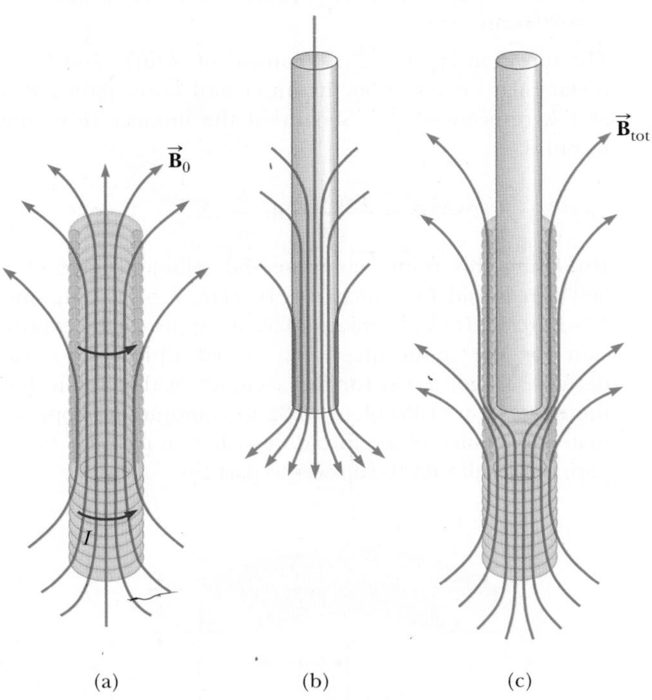

FIGURE **P23.47**

Additional Problems

48. Figure P23.48 is a graph of the induced emf versus time for a coil of N turns rotating with angular speed ω in a uniform magnetic field directed perpendicular to the axis of rotation of the coil. Copy this sketch (on a larger scale) and on the same set of axes show the graph of emf versus t (a) if the number of turns in the coil is doubled, (b) if instead the angular speed is doubled, and (c) if the angular speed is doubled while the number of turns in the coil is halved.

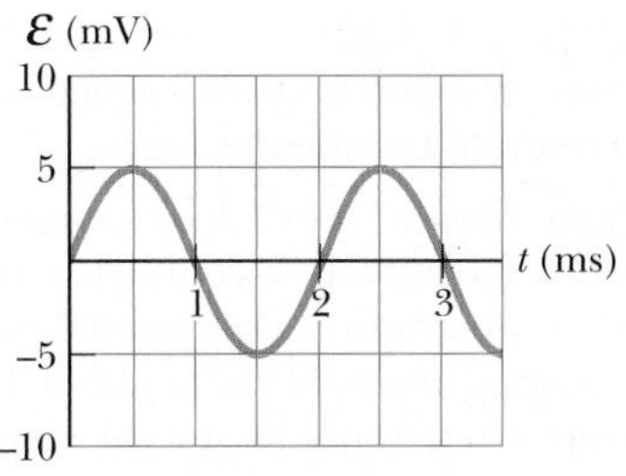

FIGURE **P23.48**

49. A steel guitar string vibrates (Figure 23.6). The component of magnetic field perpendicular to the area of a pickup coil nearby is given by

$$B = 50.0 \text{ mT} + (3.20 \text{ mT}) \sin(2\pi\, 523\, t/\text{s})$$

The circular pickup coil has 30 turns and radius 2.70 mm. Find the emf induced in the coil as a function of time.

50. Strong magnetic fields are used in such medical procedures as magnetic resonance imaging. A technician wearing a brass bracelet enclosing area 0.005 00 m^2 places her hand in a solenoid whose magnetic field is 5.00 T directed perpendicular to the plane of the bracelet. The electrical resistance around the circumference of the bracelet is 0.020 0 Ω. An unexpected power failure causes the field to drop to 1.50 T in a time of 20.0 ms. Find (a) the current induced in the bracelet and (b) the power delivered to the bracelet. (*Note:* As this problem implies, you should not wear any metal objects when working in regions of strong magnetic fields.)

51. Suppose you wrap wire onto the core from a roll of cellophane tape to make a coil. Describe how you can use a bar magnet to produce an induced voltage in the coil. What is the order of magnitude of the emf you generate? State the quantities you take as data and their values.

52. A bar of mass m, length d, and resistance R slides without friction in a horizontal plane, moving on parallel rails as shown in Figure P23.52. A battery that maintains a constant emf $\mathcal{E}$ is connected between the rails, and a constant

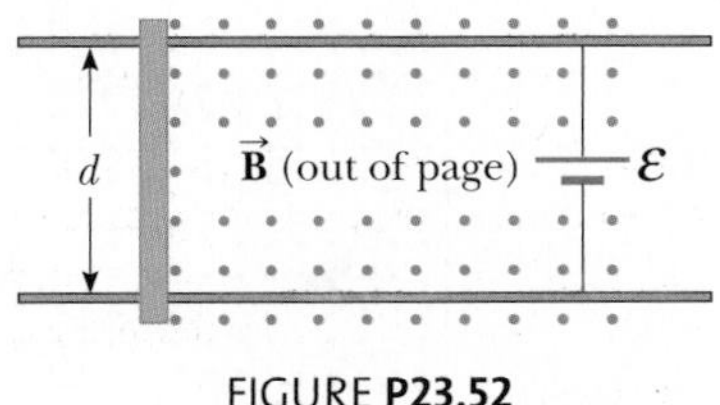

FIGURE **P23.52**

magnetic field $\vec{\mathbf{B}}$ is directed perpendicularly to the plane of the page. Assuming that the bar starts from rest, show that at time t it moves with a speed

$$v = \frac{\mathcal{E}}{Bd}(1 - e^{-B^2d^2t/mR})$$

53. **Review problem.** A particle with a mass of 2.00×10^{-16} kg and a charge of 30.0 nC starts from rest, is accelerated by a strong electric field, and is fired from a small source inside a region of uniform constant magnetic field 0.600 T. The velocity of the particle is perpendicular to the field. The circular orbit of the particle encloses a magnetic flux of 15.0 μWb. (a) Calculate the speed of the particle. (b) Calculate the potential difference through which the particle accelerated inside the source.

54. An *induction furnace* uses electromagnetic induction to produce eddy currents in a conductor, thereby raising the conductor's temperature. Commercial units operate at frequencies ranging from 60 Hz to about 1 MHz and deliver powers from a few watts to several megawatts. Induction heating can be used for warming a metal pan on a kitchen stove. It can also be used for welding in a vacuum enclosure so as to avoid oxidation and contamination of the metal. At high frequencies, induced currents occur only near the surface of the conductor—this phenomenon is the "skin effect." By creating an induced current for a short time interval at an appropriately high frequency, one can heat a sample down to a controlled depth. For example, the surface of a farm tiller can be tempered to make it hard and brittle for effective cutting while keeping the interior metal soft and ductile to resist breakage.

To explore induction heating, consider a flat conducting disk of radius R, thickness b, and resistivity ρ. A sinusoidal magnetic field $B_{max} \cos \omega t$ is applied perpendicular to the disk. Assume that the field is uniform in space and that the frequency is so low that the skin effect is not important. Assume that the eddy currents occur in circles concentric with the disk. (a) Calculate the average power delivered to the disk. By what factor does the power change (b) when the amplitude of the field doubles, (c) when the frequency doubles, and (d) when the radius of the disk doubles?

55. The magnetic flux through a metal ring varies with time t according to $\Phi_B = 3(at^3 - bt^2)\,\text{T}\cdot\text{m}^2$, with $a = 2.00\ \text{s}^{-3}$ and $b = 6.00\ \text{s}^{-2}$. The resistance of the ring is 3.00 Ω. Determine the maximum current induced in the ring during the interval from $t = 0$ to $t = 2.00$ s.

56. Figure P23.56 shows a stationary conductor whose shape is similar to the letter **e**. The radius of its circular portion is $a = 50.0$ cm. It is placed in a constant magnetic field of 0.500 T directed out of the page. A straight conducting rod, 50.0 cm long, is pivoted about point O and rotates with a constant angular speed of 2.00 rad/s. (a) Determine the induced emf in the loop POQ. Note that the area of the loop is $\theta a^2/2$. (b) If all the conducting material has a resistance per length of 5.00 Ω/m, what is the induced current in the loop POQ at the instant 0.250 s after point P passes point Q?

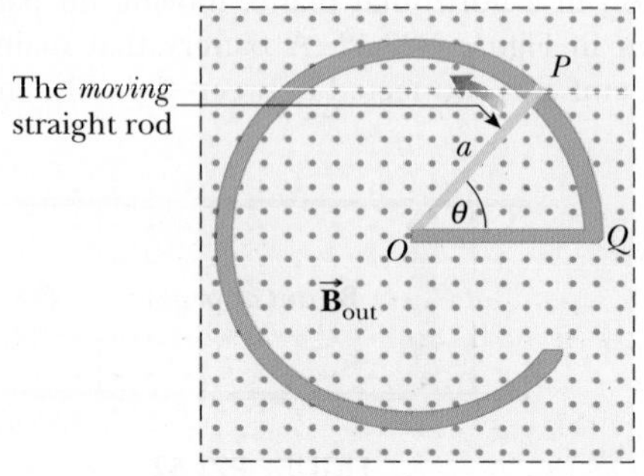

FIGURE **P23.56**

57. A *betatron* accelerates electrons to energies in the MeV range by means of electromagnetic induction. Electrons in a vacuum chamber are held in a circular orbit by a magnetic field perpendicular to the orbital plane. The magnetic field is gradually increased to induce an electric field around the orbit. (a) Show that the electric field is in the correct direction to make the electrons speed up. (b) Assume that the radius of the orbit remains constant. Show that the average magnetic field over the area enclosed by the orbit must be twice as large as the magnetic field at the circumference of the circle.

58. To monitor the breathing of a hospital patient, a thin belt is girded around the patient's chest. The belt is a 200-turn coil. When the patient inhales, the area encircled by the coil increases by 39.0 cm². The magnitude of the Earth's magnetic field is 50.0 μT and makes an angle of 28.0° with the plane of the coil. Assuming that a patient takes 1.80 s to inhale, find the average induced emf in the coil during this time.

59. A long, straight wire carries a current that is given by $I = I_{max} \sin(\omega t + \phi)$. The wire lies in the plane of a rectangular coil of N turns of wire as shown in Figure P23.5. The quantities I_{max}, ω, and ϕ are all constants. Determine the emf induced in the coil by the magnetic field created by the current in the straight wire. Assume that $I_{max} =$ 50.0 A, $\omega = 200\pi\ \text{s}^{-1}$, $N = 100$, $h = w = 5.00$ cm, and $L = 20.0$ cm.

60. The toroid in Figure P23.60 consists of N turns and has a rectangular cross-section. Its inner and outer radii are a and b, respectively. (a) Show that the inductance of the toroid is

$$L = \frac{\mu_0 N^2 h}{2\pi} \ln \frac{b}{a}$$

(b) Using this result, compute the self-inductance of a 500-turn toroid for which $a = 10.0$ cm, $b = 12.0$ cm, and $h = 1.00$ cm. (c) In Problem 23.32, an approximate expression for the inductance of a toroid with $R >> r$ was derived. To get a feel for the accuracy of that result, use the expression in Problem 23.32 to compute the approximate inductance of the toroid described in part (b). Compare the result with the answer to part (b).

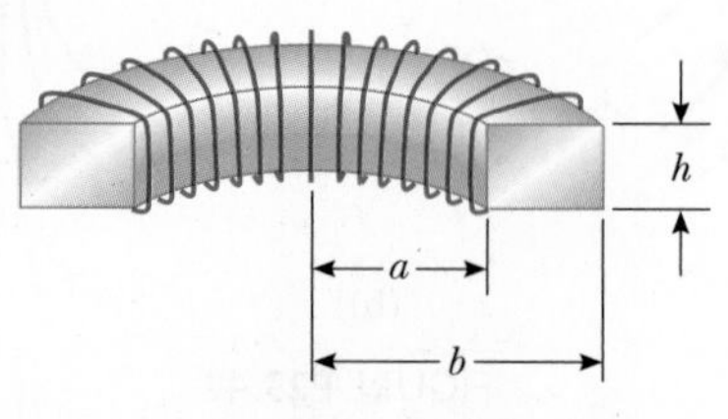

FIGURE **P23.60**

61. (a) A flat, circular coil does not really produce a uniform magnetic field in the area it encloses. Nonetheless, estimate the self-inductance of a flat, compact, circular coil, with radius R and N turns, by assuming that the field at its center is uniform over its area. (b) A circuit on a laboratory table consists of a 1.5-volt battery, a 270-Ω resistor, a switch, and three 30-cm-long patch cords connecting them. Suppose the circuit is arranged to be circular. Think of it as a flat coil with one turn. Compute the order of magnitude of its self-inductance and (c) of the time constant describing how fast the current increases when you close the switch.

62. To prevent damage from arcing in an electric motor, a discharge resistor is sometimes placed in parallel with the armature. If the motor is suddenly unplugged while running, this resistor limits the voltage that appears across the armature coils. Consider a 12.0-V DC motor with an armature that has a resistance of 7.50 Ω and an inductance of 450 mH. Assume that the magnitude of the self-induced emf in the armature coils is 10.0 V when the motor is running at normal speed. (The equivalent circuit for the armature is shown in Fig. P23.62.) Calculate the maximum resistance R that limits the voltage across the armature to 80.0 V when the motor is unplugged.

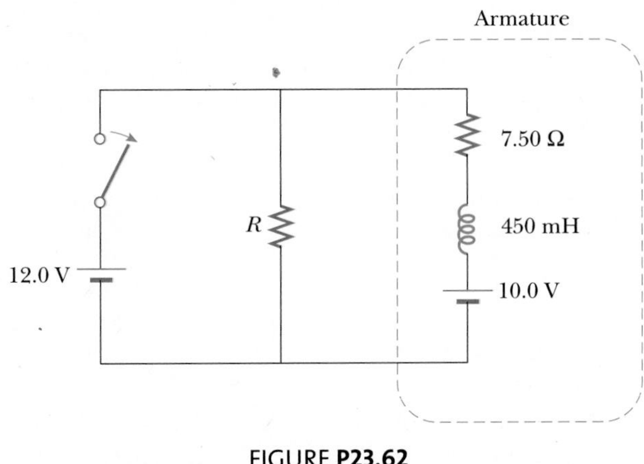

FIGURE **P23.62**

Review problems. Problems 23.63 through 23.65 apply ideas from this chapter and earlier chapters to some properties of superconductors, which were introduced in Section 21.3.

63. *The resistance of a superconductor.* In an experiment carried out by S. C. Collins between 1955 and 1958, a current was maintained in a superconducting lead ring for 2.50 yr with no observed loss. If the inductance of the ring was 3.14×10^{-8} H and the sensitivity of the experiment was 1 part in 10^9, what was the maximum resistance of the ring? (*Suggestion:* Treat this problem as a decaying current in an RL circuit and recall that $e^{-x} \approx 1 - x$ for small x.)

64. A novel method of storing energy has been proposed. A huge underground superconducting coil, 1.00 km in diameter, would be fabricated. It would carry a maximum current of 50.0 kA through each winding of a 150-turn Nb_3Sn solenoid. (a) If the inductance of this huge coil were 50.0 H, what would be the total energy stored? (b) What would be the compressive force per meter length acting between two adjacent windings 0.250 m apart?

65. *Superconducting power transmission.* The use of superconductors has been proposed for power transmission lines. A single coaxial cable (Fig. P23.65) could carry 1.00×10^3 MW (the output of a large power plant) at 200 kV, DC, over a distance of 1 000 km without loss. An inner wire of radius 2.00 cm, made from the superconductor Nb_3Sn, carries the current I in one direction. A surrounding superconducting cylinder, of radius 5.00 cm, would carry the return current I. In such a system, what is the magnetic field (a) at the surface of the inner conductor and (b) at the inner surface of the outer conductor? (c) How much energy would be stored in the space between the conductors in a 1 000-km superconducting line? (d) What is the pressure exerted on the outer conductor?

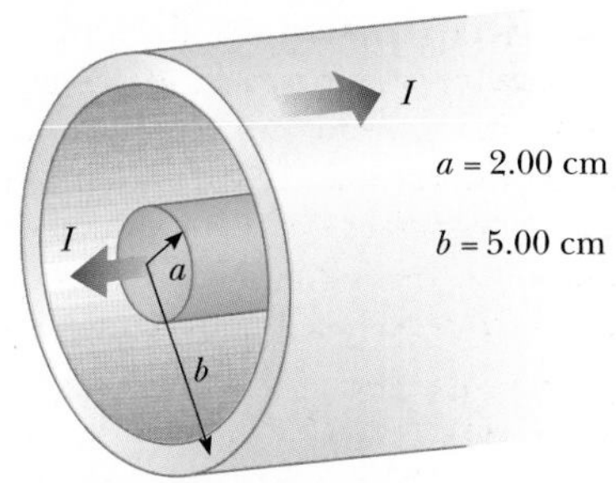

FIGURE **P23.65**

ANSWERS TO QUICK QUIZZES

23.1 (c). In all cases except this one, there is a change in the magnetic flux through the loop.

23.2 c, $d = e$, b, a. The magnitude of the emf is proportional to the rate of change of the magnetic flux. For the situation described, the rate of change of magnetic flux is proportional to the rate of change of the magnetic field. This rate of change is the slope of the graph in Figure 23.7. The magnitude of the slope is largest at c. Points d and e are on a straight line, so the slope is the same at each point. Point b represents a point of relatively small slope, and a is at a point of zero slope because the curve is horizontal at that point.

23.3 (b). According to Equation 23.5, because B and v are fixed, the emf depends only on the length of the wire moving in the magnetic field. Thus, you want the long dimension moving through the magnetic field lines so that it is perpendicular to the velocity vector. In this case, the short dimension is parallel to the velocity vector.

23.4 (c). The force on the wire is of magnitude $F_{app} = F_B = I\ell B$, with I given by Equation 23.6. Thus, the force is proportional to the speed and the force doubles. Because $\mathcal{P} = F_{app}v$, the doubling of the force *and* the speed results in the power being four times as large.

23.5 (b). When the aluminum sheet moves between the poles of the magnet, circular currents called *eddy currents* are established in the aluminum. According to Lenz's law, these currents are in a direction so as to oppose the original change, which is the movement of the aluminum sheet in the magnetic field. Thus, the effect of the eddy currents is create a force opposite to the velocity. This *magnetic braking* causes the oscillations of the equal arm balance to settle down, and a reading of the mass can take place. Magnetic damping has an advantage over frictional damping in that the magnetic damping force goes exactly to zero as the speed goes to zero. On the other hand, if mechanical friction were used to damp the oscillation of the balance beam, the speed might go to zero at a final position other than zero.

23.6 (d). The constant rate of change of B will result in a constant rate of change of the magnetic flux. According to Equation 23.9, if $d\Phi_B/dt$ is constant, $\vec{\mathbf{E}}$ is constant in magnitude.

23.7 (b). When the iron rod is inserted into the solenoid, the inductance of the coil increases. As a result, more potential difference appears across the coil than before. Consequently, less potential difference appears across the lightbulb, so the bulb is dimmer.

23.8 (b). Figure 23.29 shows that circuit B has the larger time constant because in this circuit it takes longer for the current to reach its maximum value and then longer for this current to drop back down to zero after the switch is thrown to *b*. Equation 23.15 indicates that, for equal resistances R_A and R_B, the condition $\tau_B > \tau_A$ means that $L_A < L_B$.

23.9 (a), (d). Because the energy density depends on the magnitude of the magnetic field, we must increase the magnetic field to increase the energy density. For a solenoid, $B = \mu_0 nI$, where n is the number of turns per unit length. In (a), we increase n to increase the magnetic field. In (b), the change in cross-sectional area has no effect on the magnetic field. In (c), increasing the length but keeping n fixed has no effect on the magnetic field. Increasing the current in (d) increases the magnetic field in the solenoid.

Lifting, Propelling, and Braking the Vehicle

Now that we have investigated the principles of electromagnetism, let us respond to our central question for the *Magnetic Levitation Vehicles* Context:

How can we lift, propel, and brake a vehicle with magnetic forces?

We have addressed two mechanisms for lifting the vehicle, one in each of the preceding chapters. Once the vehicle is suspended above the track, we must control its speed with propulsion and braking. We shall discuss these processes in this Context Conclusion.

Magnetic Propulsion

If you have a toy car made of iron, you can imagine how to propel the car with a magnet. You hold the magnet near the front of the car so that it exerts an attractive force on the car, causing it to accelerate from rest. As the car moves, you move the magnet in the same direction, not allowing the magnet and the car to touch. In essence, the car "chases" the moving magnet due to the attractive force.

You could increase the propelling force on the car by mounting a bar magnet on it, with the north pole near the front of the car and the south pole near the back. Now, you use two other bar magnets, one with its south pole in front of the car and another with its south pole near the back of the car. The magnet on the car is attracted to the front magnet and repelled by the rear magnet. In this manner, the car experiences a strong propelling force.

Such is the fundamental idea behind magnetic propulsion. Electromagnets are mounted on the magnetic levitation vehicle, and a series of propulsion coils are placed along the side of the track as shown in Figure 1. The polarity of the magnets on the vehicle remains constant. An electrical signal is passed through the coils such that each magnet on the vehicle "sees" a coil with opposite polarity ahead of it and a coil with the same polarity behind it. As a result, the vehicle experiences an attractive force from the coil ahead of it and a repulsive force from the coil behind it, both forces pushing the vehicle in the forward direction. Figure 1

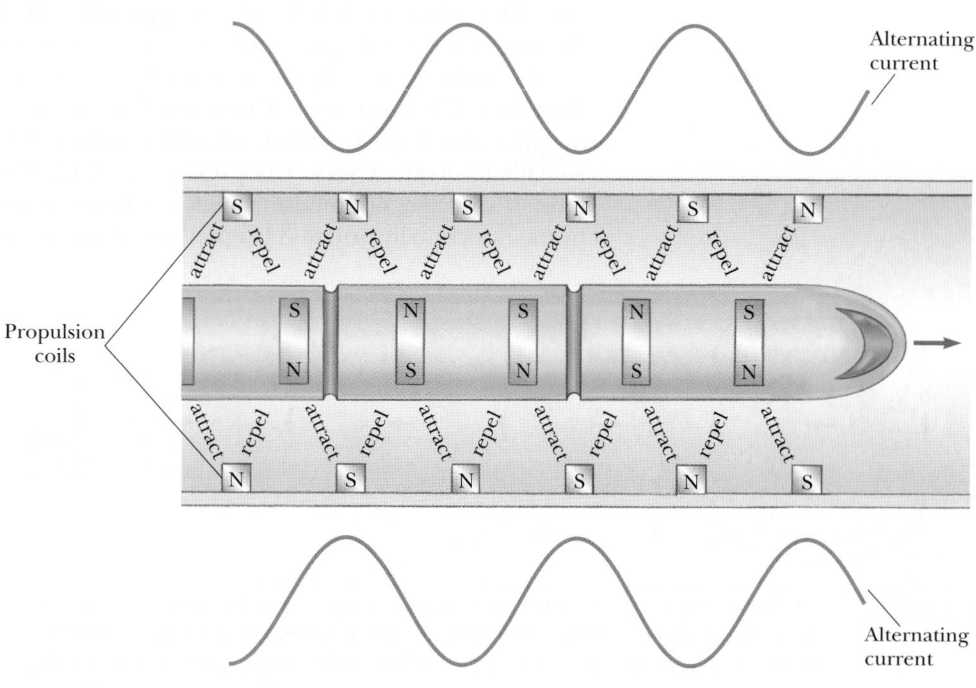

FIGURE 1 Magnetic propulsion. Magnets on the vehicle are attracted and repelled by electromagnetic poles created by a sinusoidal current passed through coils on the side of the track. As the wave of current moves along the track, the vehicle moves with it.

shows this situation at a single instant of time. As the sinusoidal wave in Figure 1 moves along the track, the vehicle "chases" it due to the magnetic forces.

Magnetic Braking

Electromagnetic transportation has the added benefit of a built-in braking mechanism. Lenz's law tells us that a magnetic change induces a current that acts to oppose the original change, which represents a natural braking mechanism. For example, if an aluminum plate is dropped between the poles of a very strong magnet, the plate drops slowly because currents established in the plate experience a magnetic force that opposes the fall.

In the case of magnetic levitation vehicles, imagine that the propulsion system is deactivated so that the vehicle begins to coast. The relative motion of magnets and coils in the train and track induces currents that slow the train down, according to Lenz's law. The propulsion system can be used in combination with this magnetic braking for complete control over the stopping process.

This principle is not new. Railroads operating in the Alps at the beginning of the 20th century included mechanisms for connecting the electric drive motor to a resistance when the train moved downhill. As the motion of the train causes the motor to turn, the motor acts as a generator. The generator produces a current in the resistance, resulting in a back emf. Because this back emf opposes the original change (the rotation of the motor), it controls the motion of the train down the hill. This same principle is now used in hybrid vehicles (see Context 1) so that the braking process feeds energy back to the battery.

A Third Levitation Model: the Inductrack

In Chapters 22 and 23, we discussed models for magnetic levitation that are under study in Germany and Japan, both of which require strong electromagnets. At Lawrence Livermore Laboratory in California, scientists are working on a magnetic levitation scheme that involves permanent magnets. The scheme is called the *Inductrack*. One attractive feature of this approach is that no energy is required to power the magnets, resulting in savings for energy costs to operate the system.

The Inductrack system uses a *Halbach array* of magnets. Figure 2 shows a one-dimensional Halbach array. Underneath the array, the magnetic field lines of adjacent magnets with poles oriented vertically combine with those with poles oriented horizontally to create a very strong magnetic field. Above the array, the magnetic field lines of vertically oriented magnets are in the opposite direction from those created by the horizontally oriented magnets, resulting in a weak field in this region.

The Halbach array, attached to the underside of the vehicle, passes over a series of coils of wire and induces currents in these coils. By Faraday's law, similar to the situation in the electrodynamic model of Section 23.8, the magnetic field resulting from the current in the coils exerts a repulsive force on the magnets, creating a levitation force on the vehicle. A levitation force of 40 metric tons per square meter can be achieved using high-field alloy magnets. A working model of the Induc-

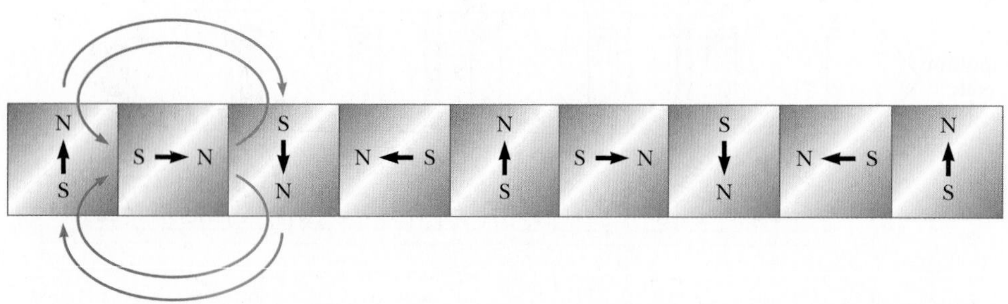

FIGURE 2 A Halbach array of magnets. A field line from the north pole of the leftmost magnet to the south pole of the third magnet is clockwise. The field line above the array from the north pole of the second magnet to the south pole of the same magnet is counterclockwise. The result is a relatively weak field above the array. Below the array, the field line from the north pole of the third magnet to the south pole of the leftmost magnet is clockwise. The field line below the array from the second magnet is also clockwise. The combination of these field lines results in a strong magnetic field below the array.

(a)

(b)

(Courtesy of Lawrence Livermore National Laboratory)

FIGURE 3 (a) General Atomics has built a full-scale Inductrack test vehicle, shown here on the test track in San Diego, California. (b) A drawing of the front end and magnetic components of a maglev vehicle using the Inductrack approach.

track system successfully propelled and levitated a 22-kg vehicle over a 20-m test track. Using federal funding, General Atomics in San Diego is developing the Inductrack technology as it moves toward commercial applications. Experiments with full-scale test vehicles are currently in progress on a 120-m test track (Fig. 3a), in preparation for development of a full-scale vehicle (Fig. 3b).

What does the future of magnetic levitation transportation hold? Will the Inductrack become the system of choice, or will it be the Japanese or German system? Or will all three coexist? At this point, it is impossible to predict, and we invite you to watch the newspapers for further developments!

Problems

1. Assume that the vehicle shown in Figure 1 is moving at 400 km/h. The distance between adjacent magnets on the vehicle is 10.0 m. What is the frequency of the alternating current in the coils on the side of the track required to propel the vehicle?
2. Figure 4 represents an electromagnetic brake that uses eddy currents. An electromagnet hangs from a railroad car near one rail. To stop the car, a large current is sent through the coils of the electromagnet. The moving electromagnet induces eddy currents in the rails, whose fields oppose the change in the field of the electromagnet. The magnetic fields of the eddy currents exert force on the current in the electromagnet, thereby slowing the car. The direction of the car's motion and the direction of the current in the electromagnet are shown correctly in the picture. Determine which of the eddy currents shown on the rails is correct. Explain your answer.

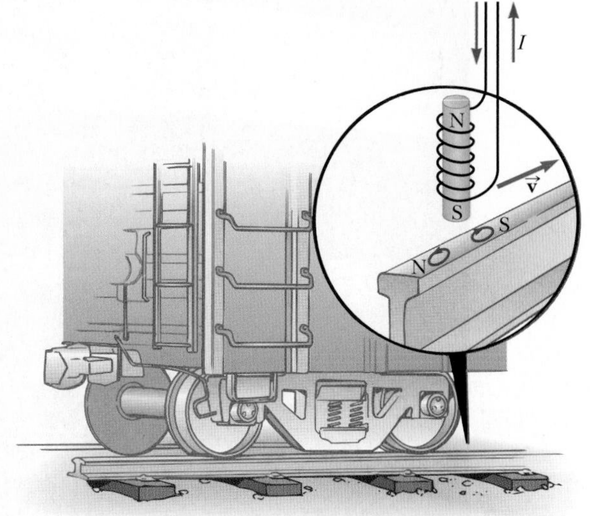

FIGURE 4 Magnets on a moving railroad car induce small current loops in the rail, called eddy currents.

CONTEXT 8

Lasers

The invention of the laser was popularly credited to Arthur L. Schawlow and Charles H. Townes for many years after their publication of a proposal for the laser in a 1958 issue of *Physical Review.* Schawlow and Townes received a patent for the device in 1959. In 1960, the first laser was built and operated by Theodore Maiman. This device used a ruby crystal to create the laser light, which was emitted in pulses from the end of a ruby cylinder. A flash lamp was used to excite the laser action.

In 1977, the first victory in a 30-year-long legal battle was completed in which Gordon Gould, who was a graduate student at Columbia University in the late 1950s, received a patent for inventing the laser in 1957 as well as coining the term. Believing erroneously that he had to have a working prototype before he could file for a patent, he did not file until later in 1959 than had Schawlow and Townes. Gould's legal battles ended in 1987. By this time, Gould's technology was being widely used in industry and medicine. His victory finally resulted in his control of patent rights to perhaps 90% of the lasers used and sold in the United States.

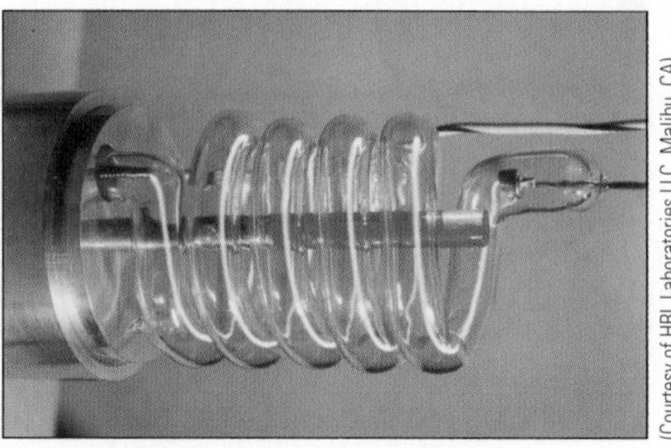
(Courtesy of HRL Laboratories LLC, Malibu, CA)

FIGURE 2 Photograph of an early ruby laser, showing the flash lamp (glass helix) surrounding the ruby rod (red cylinder).

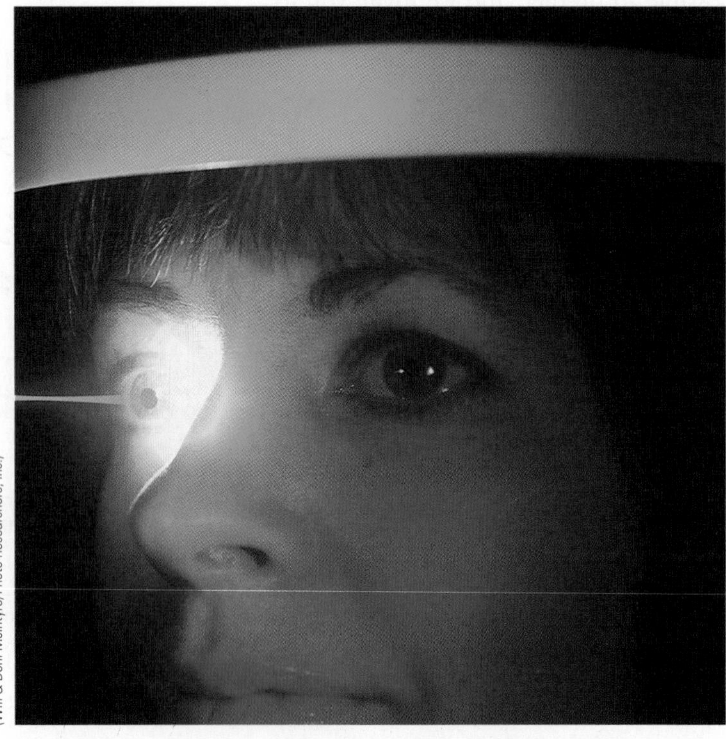
(Will & Deni McIntyre/Photo Researchers, Inc.)

FIGURE 1 A laser is used on a human eye to perform a surgical procedure. The word *laser* is an acronym meaning light amplification by stimulated emission of radiation.

Since the development of the first device, laser technology has experienced tremendous growth. Lasers that cover wavelengths in the infrared, visible, and ultraviolet regions are now available. Various types of lasers use solids, liquids, and gases as the active medium. Although the original laser emitted light over a very narrow range around a fixed wavelength, tunable lasers are now available, in which the wavelength can be varied.

The laser is an omnipresent technological tool in our daily life. Applications include surgical "welding" of detached retinas, precision surveying and length measurement, a potential source for inducing nuclear fusion reactions, precision cutting of metals and other materials, and telephone communication along optical fibers.

(© Lawrence Manning/CORBIS)

FIGURE 3 The original ruby laser emitted red light, as did many lasers developed soon afterward. Today, lasers are available in a variety of colors and various regions of the electromagnetic spectrum. In this photograph, a green laser is used to perform scientific research.

We also use lasers to read information from compact discs for use in audio entertainment and computer applications. Digital videodiscs use lasers to read video information. Lasers are used in retail stores to read price and inventory information from product labels. In the laboratory, lasers can be used to trap atoms and cool them to millikelvins above absolute zero and to move microscopic biological organisms around harmlessly.

These and other applications are possible because of the unique characteristics of laser light. In addition to its being almost monochromatic, laser light is also highly directional and can therefore be sharply focused to produce regions of extreme intensity.

In this Context, we shall investigate the physics of electromagnetic radiation and optics and apply the principles to an understanding of the behavior of laser light and its applications. A major focus of our study will be on the technology of optical fibers and how they are used in industry and medicine. We shall study the nature of light as we respond to our central question:

(Philippe Plailly/SPL/Photo Researchers, Inc.)

FIGURE 4 This robotic device, one of the many technological uses of lasers in our society, carries laser scissors, which can cut up to 50 layers of fabric at a time.

(© David Frazier/Stone/Getty Images)

FIGURE 5 A supermarket scanner uses light from a laser to identify products being purchased. The reflections from the bar code on the package are read and entered into the computer to determine the price of the item.

What is special about laser light, and how is it used in technological applications?

CHAPTER 24

Electromagnetic Waves

Electromagnetic waves cover a broad spectrum of wavelengths, with waves in various wavelength ranges having distinct properties. These photos of the Crab Nebula show different structure for observations made with waves of various wavelengths. The photos (clockwise starting from the upper left) were taken with x-rays, unpolarized visible light, radio waves, and visible light passing through a polarizing filter.

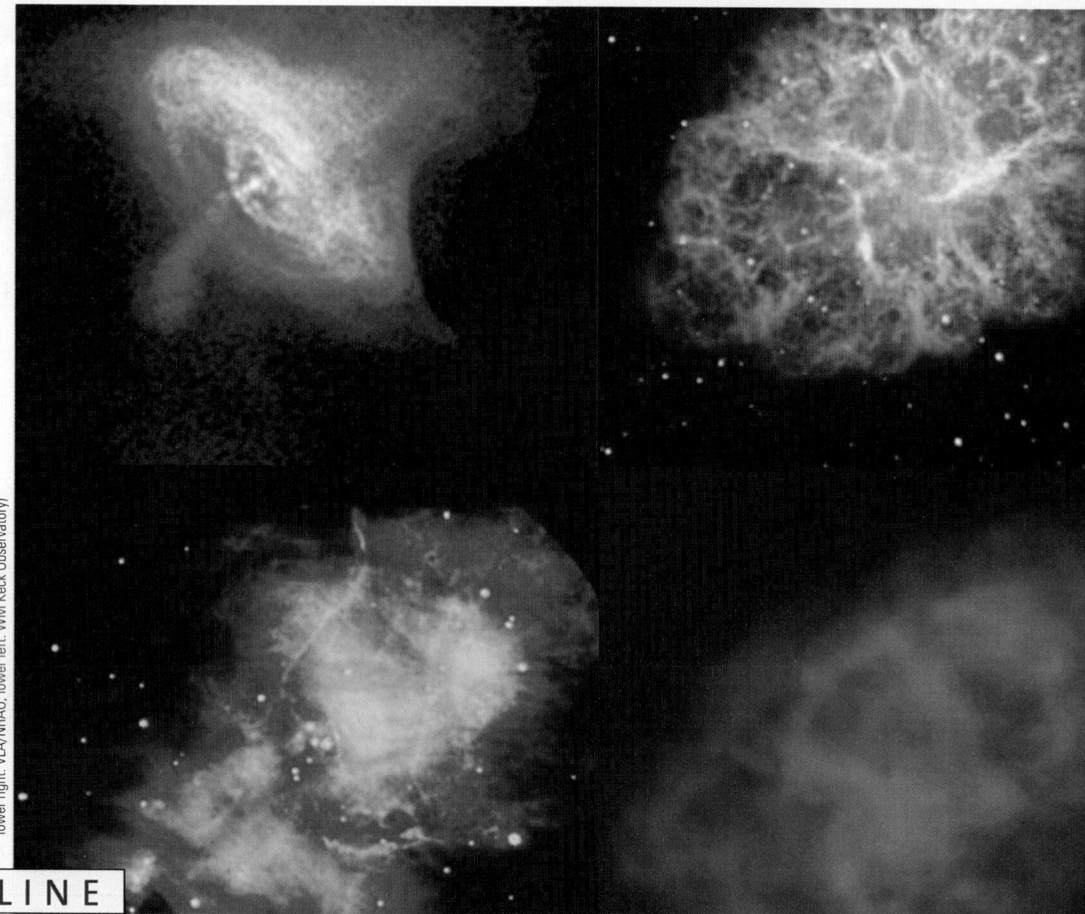

(Upper left: NASA/CXC/SAO; upper right: Palomar Observatory; lower right: VLA/NRAO; lower left: WM Keck Observatory)

CHAPTER OUTLINE

Although we are not always aware of their presence, electromagnetic waves permeate our environment. In the form of visible light, they enable us to view the world around us with our eyes. Infrared waves from the surface of the Earth warm our environment, radio-frequency waves carry our favorite radio entertainment, microwaves cook our food and are used in radar communication systems, and the list goes on and on. The waves described in Chapter 13 are mechanical waves, which require a medium through which to propagate. Electromagnetic waves, in contrast, can propagate through a vacuum. Despite this difference between mechanical and electromagnetic waves, much of the behavior in the wave models of Chapters 13 and 14 is similar for electromagnetic waves.

The purpose of this chapter is to explore the properties of electromagnetic waves. The fundamental laws of electricity and magnetism—Maxwell's equations—form the basis of all electromagnetic phenomena. One of these equations predicts that a time-varying electric field produces a magnetic field just as a time-varying magnetic field produces an electric field. From this generalization, Maxwell provided the final important link between electric and magnetic fields. The most dramatic prediction of his equations is the existence of electromagnetic waves that propagate through empty space with the speed of light. This discovery led to many practical applications, such as radio and television, and to the realization that light is one form of electromagnetic radiation.

24.1 DISPLACEMENT CURRENT AND THE GENERALIZED AMPÈRE'S LAW

We have seen that charges in motion, or currents, produce magnetic fields. When a current-carrying conductor has high symmetry, we can calculate the magnetic field using Ampère's law, given by Equation 22.29:

$$\oint \vec{\mathbf{B}} \cdot d\vec{\mathbf{s}} = \mu_0 I$$

where **the line integral is over any closed path through which the conduction current passes** and the conduction current is defined by $I = dq/dt$.

In this section, we shall use the term *conduction current* to refer to the type of current that we have already discussed, that is, current carried by charged particles in a wire. We use this term to differentiate this current from a different type of current we will introduce shortly. **Ampère's law in this form is valid only if the conduction current is continuous in space.** Maxwell recognized this limitation and modified Ampère's law to include all possible situations.

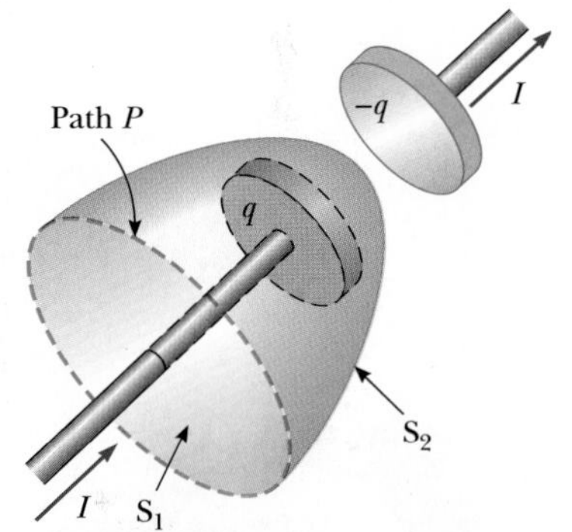

FIGURE 24.1 Two surfaces S_1 and S_2 near the plate of a capacitor are bounded by the same path P. The conduction current in the wire passes only through S_1, which leads to a contradiction in Ampère's law that is resolved only if one postulates a displacement current through S_2.

This limitation can be understood by considering a capacitor being charged as in Figure 24.1. When conduction current exists in the wires, the charge on the plates changes, but **no conduction current exists between the plates.** Consider the two surfaces S_1 (a circle, shown in blue) and S_2 (a paraboloid, in orange, passing between the plates) in Figure 24.1 bounded by the same path P. Ampère's law says that the line integral of $\vec{\mathbf{B}} \cdot d\vec{\mathbf{s}}$ around this path must equal $\mu_0 I$, where I is the conduction current through *any* surface bounded by the path P.

When the path P is considered as bounding S_1, the right-hand side of Equation 22.29 is $\mu_0 I$ because the conduction current passes through S_1 while the capacitor is charging. When the path bounds S_2, however, the right-hand side of Equation 22.29 is zero because no conduction current passes through S_2. Therefore, a contradictory situation arises because of the discontinuity of the current! Maxwell solved this problem by postulating an additional term on the right side of Equation 22.29, called the **displacement current** I_d, defined as

■ Displacement current

$$I_d \equiv \epsilon_0 \frac{d\Phi_E}{dt} \qquad [24.1]$$

Recall that Φ_E is the flux of the electric field, defined as $\Phi_E \equiv \oint \vec{\mathbf{E}} \cdot d\vec{\mathbf{A}}$ (Eq. 19.20). (The word *displacement* here does not have the same meaning as in Chapter 2; it is historically entrenched in the language of physics, however, so we continue to use it.)

Equation 24.1 is interpreted as follows. As the capacitor is being charged (or discharged), the changing electric field between the plates may be considered as equivalent to a current between the plates that acts as a continuation of the conduction current in the wire. When the expression for the displacement current given by Equation 24.1 is added to the conduction current on the right side of Ampère's law, the difficulty represented in Figure 24.1 is resolved. No matter what surface

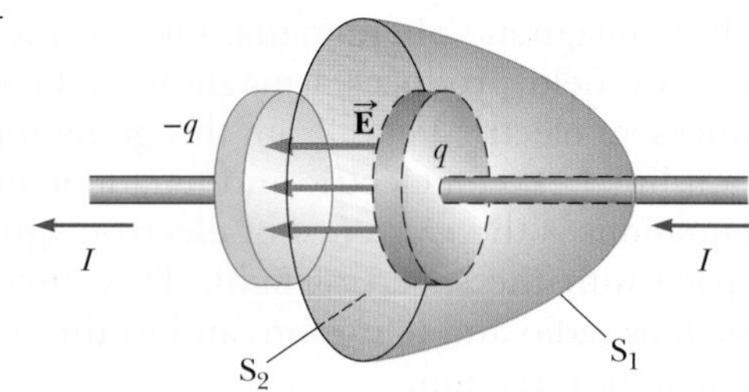

FIGURE 24.2 Because it exists only in the wires attached to the capacitor plates, the conduction current $I = dq/dt$ passes through the curved surface S_1 but not the flat surface S_2. Only the displacement current $I_d = \epsilon_0 d\Phi_E/dt$ passes through S_2. The two currents must be equal for continuity.

bounded by the path P is chosen, either conduction current or displacement current passes through it. With this new notion of displacement current, we can express the general form of Ampère's law (sometimes called the **Ampère–Maxwell law**) as[1]

■ Ampère–Maxwell law

$$\oint \vec{\mathbf{B}} \cdot d\vec{\mathbf{s}} = \mu_0(I + I_d) = \mu_0 I + \mu_0 \epsilon_0 \frac{d\Phi_E}{dt} \qquad [24.2]$$

The meaning of this expression can be understood by referring to Figure 24.2. The electric flux through S_2 (a circle, shown in gray, between the plates) is $\Phi_E = \oint \vec{\mathbf{E}} \cdot d\vec{\mathbf{A}} = EA$, where A is the area of the capacitor plates and E is the magnitude of the uniform electric field between the plates. If q is the charge on the plates at any instant, $E = q/\epsilon_0 A$ (Section 20.7). Therefore, the electric flux through S_2 is simply

$$\Phi_E = EA = \frac{q}{\epsilon_0}$$

Hence, the displacement current I_d through S_2 is

$$I_d = \epsilon_0 \frac{d\Phi_E}{dt} = \frac{dq}{dt} \qquad [24.3]$$

That is, the displacement current through S_2 is precisely equal to the conduction current I through S_1! The central point of this formalism is that **magnetic fields are produced both by conduction currents and by changing electric fields.** This result is a remarkable example of theoretical work by Maxwell and of his major contributions in advancing the understanding of electromagnetism.

(North Wind Picture Archives)

James Clerk Maxwell
(1831–1879)

Scottish theoretical physicist Maxwell developed the electromagnetic theory of light and the kinetic theory of gases, and he explained the nature of color vision and of Saturn's rings. His successful interpretation of electromagnetic fields produced the field equations that bear his name. Formidable mathematical ability combined with great insight enabled Maxwell to lead the way in the study of electromagnetism and kinetic theory. He died of cancer before he was 50.

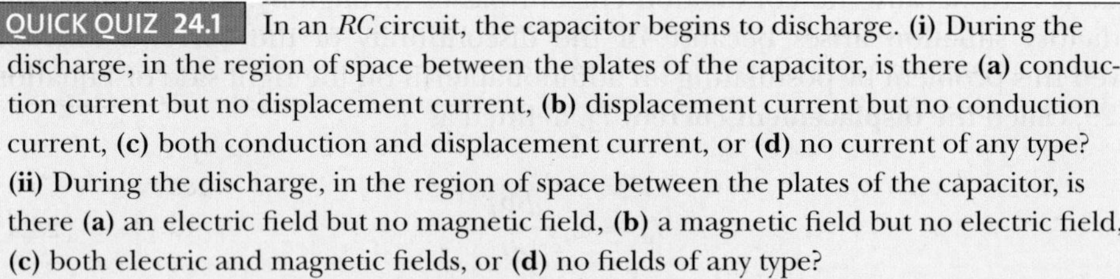

QUICK QUIZ 24.1 In an RC circuit, the capacitor begins to discharge. **(i)** During the discharge, in the region of space between the plates of the capacitor, is there **(a)** conduction current but no displacement current, **(b)** displacement current but no conduction current, **(c)** both conduction and displacement current, or **(d)** no current of any type? **(ii)** During the discharge, in the region of space between the plates of the capacitor, is there **(a)** an electric field but no magnetic field, **(b)** a magnetic field but no electric field, **(c)** both electric and magnetic fields, or **(d)** no fields of any type?

24.2 MAXWELL'S EQUATIONS

In this section, we gather together four equations from our studies in recent chapters that as a group can be regarded as the theoretical basis of all electric and magnetic fields. These relationships, known as Maxwell's equations after James Clerk Maxwell, are as fundamental to electromagnetic phenomena as Newton's laws are

[1]Strictly speaking, this expression is valid only in a vacuum. If a magnetic material is present, a magnetizing current must also be included on the right side of Equation 24.2 to make Ampère's law fully general.

to mechanical phenomena. In fact, the theory developed by Maxwell was more far reaching than even he imagined because it was shown by Albert Einstein in 1905 to be in agreement with the special theory of relativity. As we shall see, Maxwell's equations represent laws of electricity and magnetism that have already been discussed. The equations have additional important consequences, however, in that they predict the existence of electromagnetic waves (traveling patterns of electric and magnetic fields) that travel in vacuum with a speed of $c = 1/\sqrt{\epsilon_0\mu_0} = 3.00 \times 10^8$ m/s, the speed of light. Furthermore, Maxwell's equations show that electromagnetic waves are radiated by accelerating charges, as we discussed in Chapter 17 with regard to thermal radiation.

For simplicity, we present **Maxwell's equations** as applied to free space, that is, in the absence of any dielectric or magnetic material. The four equations are

■ Maxwell's equations

$$\oint \vec{\mathbf{E}} \cdot d\vec{\mathbf{A}} = \frac{q}{\epsilon_0} \qquad [24.4]$$

$$\oint \vec{\mathbf{B}} \cdot d\vec{\mathbf{A}} = 0 \qquad [24.5]$$

$$\oint \vec{\mathbf{E}} \cdot d\vec{\mathbf{s}} = -\frac{d\Phi_B}{dt} \qquad [24.6]$$

$$\oint \vec{\mathbf{B}} \cdot d\vec{\mathbf{s}} = \mu_0 I + \epsilon_0\mu_0 \frac{d\Phi_E}{dt} \qquad [24.7]$$

Equation 24.4 is **Gauss's law,** which states that **the total electric flux through any closed surface equals the net charge inside that surface divided by ϵ_0** (Section 19.9). This law describes how charges create electric fields.

Equation 24.5, which can be considered **Gauss's law for magnetism, says that the net magnetic flux through a closed surface is zero.** That is, the number of magnetic field lines entering a closed volume must equal the number leaving that volume. This law implies that magnetic field lines cannot begin or end at any point. If they did, it would mean that isolated magnetic monopoles existed at those points. That isolated magnetic monopoles have not been observed in nature can be taken as a basis of Equation 24.5.

Equation 24.6 is **Faraday's law of induction** (Eq. 23.9), which describes how a changing magnetic field creates an electric field. This law states that **the line integral of the electric field around any closed path (which equals the emf) equals the rate of change of magnetic flux through any surface area bounded by that path.**

Equation 24.7 is the generalized form of Ampère's law. It describes how both an electric current and a changing electric field create a magnetic field. That is, **the line integral of the magnetic field around any closed path is determined by the net current and the rate of change of electric flux through any surface bounded by that path.**

Once the electric and magnetic fields are known at some point in space, the force those fields exert on a particle of charge q can be calculated from the expression

■ Lorentz force

$$\vec{\mathbf{F}} = q\vec{\mathbf{E}} + q\vec{\mathbf{v}} \times \vec{\mathbf{B}} \qquad [24.8]$$

which is called the **Lorentz force** (Section 22.4). Maxwell's equations and this force law give a complete description of all classical electromagnetic interactions.

Note the interesting symmetry of Maxwell's equations. In a region of space free of charges, so that $q = 0$, Equations 24.4 and 24.5 are symmetric in that the surface integral of $\vec{\mathbf{E}}$ or $\vec{\mathbf{B}}$ over a closed surface equals zero. Furthermore, in a region free of conduction currents so that $I = 0$, Equations 24.6 and 24.7 are symmetric in that the line integrals of $\vec{\mathbf{E}}$ and $\vec{\mathbf{B}}$ around a closed path are related to the rate of change of magnetic flux and electric flux, respectively.

24.3 ELECTROMAGNETIC WAVES

In his unified theory of electromagnetism, Maxwell showed that time-dependent electric and magnetic fields satisfy a linear wave equation. (The linear wave equation for mechanical waves is Equation 13.19.) The most significant outcome of this theory is the prediction of the existence of **electromagnetic waves.**

Maxwell's equations predict that an electromagnetic wave consists of oscillating electric and magnetic fields. The changing fields induce each other, which maintains the propagation of the wave; a changing electric field induces a magnetic field, and a changing magnetic field induces an electric field. The $\vec{\mathbf{E}}$ and $\vec{\mathbf{B}}$ vectors are perpendicular to each other, and to the direction of propagation, as shown in Active Figure 24.3a at one instant of time and one point in space. The direction of propagation is the direction of the vector product $\vec{\mathbf{E}} \times \vec{\mathbf{B}}$, which we shall explore more fully in Section 24.5. Active Figure 24.3b shows how the electric and magnetic fields vary in phase sinusoidally along the x axis in the simplest type of electromagnetic wave. We will discuss this sinusoidal behavior shortly. As time progresses, imagine the construction in Active Figure 24.3b moving to the right along the x axis. That is what happens in an electromagnetic wave, with the movement taking place at the speed of light c.

To understand the prediction of electromagnetic waves, let us focus our attention on an electromagnetic wave traveling in the x direction. For this wave, the electric field $\vec{\mathbf{E}}$ is in the y direction and the magnetic field $\vec{\mathbf{B}}$ is in the z direction as in Active Figure 24.3. Waves in which the electric and magnetic fields are restricted to being parallel to certain directions are said to be **linearly polarized waves.** Furthermore, let us assume that at any point in space in Active Figure 24.3, the magnitudes E and B of the fields depend on x and t only, not on the y or z coordinates.

Let us also imagine that the source of the electromagnetic waves is such that a wave radiated from *any* position in the yz plane (not just from the origin as might be suggested by Active Fig. 24.3) propagates in the x direction and that all such waves are emitted in phase. If we define a **ray** as the line along which a wave travels, all rays for these waves are parallel. This whole collection of waves is often called a **plane wave.** A surface connecting points of equal phase on all waves, which we call a **wave front,** is a geometric plane. In comparison, a point source of radiation sends waves out in all directions. A surface connecting points of equal

PITFALL PREVENTION 24.1

WHAT IS "A" WAVE? A sticky point in these types of discussions is what we mean by a *single* wave. We could define one wave as that which is emitted by a single charged particle. In practice, however, the word *wave* represents both the emission from a *single point* ("wave radiated from any position in the *yz* plane") and the collection of waves from *all points* on the source ("plane wave"). You should be able to use this term in both ways and to understand its meaning from the context.

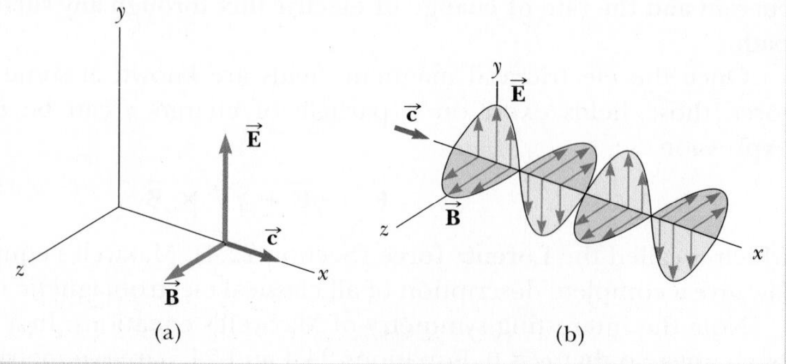

ACTIVE FIGURE 24.3 (a) The fields in an electromagnetic wave traveling at velocity $\vec{\mathbf{c}}$ in the positive x direction at one point on the x axis. These fields depend only on x and t. (b) Representation of a sinusoidal electromagnetic wave moving in the positive x direction with a speed c.

PhysicsNow™ Log into PhysicsNow at **www.pop4e.com** and go to Active Figure 24.3 to observe the wave in part (b). In addition, you can take a "snapshot" of the wave at an instant of time and investigate the electric and magnetic fields at that instant.

phase for this situation is a sphere, so we call the radiation from a point source a **spherical wave.**

To generate the prediction of electromagnetic waves, we start with Faraday's law, Equation 24.6:

$$\oint \vec{\mathbf{E}} \cdot d\vec{\mathbf{s}} = -\frac{d\Phi_B}{dt}$$

Let us assume that a plane electromagnetic wave is traveling in the x direction, with the electric field $\vec{\mathbf{E}}$ in the positive y direction and the magnetic field **B** in the positive z direction.

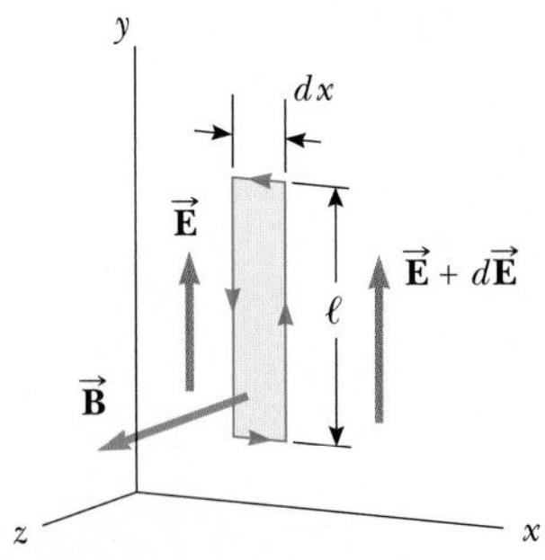

FIGURE 24.4 As a plane wave moving in the $+x$ direction passes through a rectangular path of width dx lying in the xy plane, the electric field in the y direction varies from $\vec{\mathbf{E}}$ to $\vec{\mathbf{E}} + d\vec{\mathbf{E}}$. This construction allows us to evaluate the line integral of $\vec{\mathbf{E}}$ over the perimeter of the rectangle.

Consider a rectangle of width dx and height ℓ lying in the xy plane as in Figure 24.4. To apply Equation 24.6, we first evaluate the line integral of $\vec{\mathbf{E}} \cdot d\vec{\mathbf{s}}$ around this rectangle. The contributions from the top and bottom of the rectangle are zero because $\vec{\mathbf{E}}$ is perpendicular to $d\mathbf{s}$ for these paths. We can express the electric field on the right side of the rectangle as[2]

$$E(x + dx, t) \approx E(x, t) + \left.\frac{dE}{dx}\right]_{t \text{ constant}} dx = E(x, t) + \frac{\partial E}{\partial x} dx$$

whereas the field on the left side of the rectangle is simply $E(x, t)$. The line integral over this rectangle is therefore approximately

$$\oint \vec{\mathbf{E}} \cdot d\vec{\mathbf{s}} = [E(x + dx, t)]\ell - [E(x, t)]\ell \approx \ell \left(\frac{\partial E}{\partial x}\right) dx \quad [24.9]$$

Because the magnetic field is in the z direction, the magnetic flux through the rectangle of area $\ell\ dx$ is approximately $\Phi_B = B\ell\ dx$. (This expression assumes that dx is very small compared with the wavelength of the wave so that B is uniform over the width dx.) Taking the time derivative of the magnetic flux at the location of the rectangle on the x axis gives

$$\frac{d\Phi_B}{dt} = \ell\ dx \left.\frac{dB}{dt}\right]_{x \text{ constant}} = \ell \frac{\partial B}{\partial t} dx \quad [24.10]$$

Substituting Equations 24.9 and 24.10 into Equation 24.6 gives

$$\ell \left(\frac{\partial E}{\partial x}\right) dx = -\ell \frac{\partial B}{\partial t} dx$$

$$\frac{\partial E}{\partial x} = -\frac{\partial B}{\partial t} \quad [24.11]$$

We can derive a second equation by starting with Maxwell's fourth equation in empty space (Eq. 24.7). In this case, we evaluate the line integral of $\vec{\mathbf{B}} \cdot d\vec{\mathbf{s}}$ around a rectangle lying in the xz plane and having width dx and length ℓ as in Figure 24.5. Using the sense of the integration shown and noting that the magnetic field changes from $B(x, t)$ to $B(x + dx, t)$ over the width dx, we find that

$$\oint \vec{\mathbf{B}} \cdot d\vec{\mathbf{s}} = [B(x, t)]\ell - [B(x + dx, t)]\ell = -\ell \left(\frac{\partial B}{\partial x}\right) dx \quad [24.12]$$

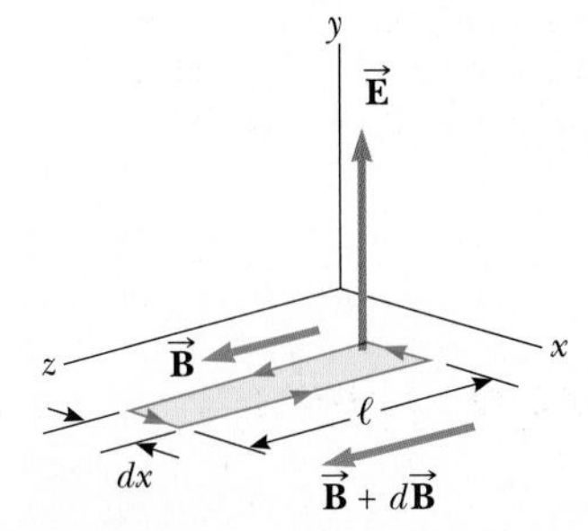

FIGURE 24.5 As a plane wave moving in the $+x$ direction passes through a rectangular path of width dx lying in the xz plane, the magnetic field in the z direction varies from $\vec{\mathbf{B}}$ to $\vec{\mathbf{B}} + d\vec{\mathbf{B}}$. This construction allows us to evaluate the line integral of $\vec{\mathbf{B}}$ over the perimeter of the rectangle.

The electric flux through the rectangle is $\Phi_E = E\ell\ dx$, which when differentiated with respect to time gives

$$\frac{d\Phi_E}{dt} = \ell \left(\frac{\partial E}{\partial t}\right) dx \quad [24.13]$$

[2]Because dE/dx in this equation is expressed as the change in E with x at a given instant t, dE/dx is equivalent to the partial derivative $\partial E/\partial x$. Likewise, we will shortly require dB/dt, which means the change in B with time at a particular position x, and so we can replace dB/dt by $\partial B/\partial t$.

Substituting Equations 24.12 and 24.13 into Equation 24.7 gives

$$-\ell\left(\frac{\partial B}{\partial x}\right)dx = \epsilon_0\mu_0\ell\left(\frac{\partial E}{\partial t}\right)dx$$

$$\frac{\partial B}{\partial x} = -\epsilon_0\mu_0\frac{\partial E}{\partial t} \qquad [24.14]$$

Taking the derivative of Equation 24.11 with respect to x and combining it with Equation 24.14 gives

$$\frac{\partial^2 E}{\partial x^2} = -\frac{\partial}{\partial x}\left(\frac{\partial B}{\partial t}\right) = -\frac{\partial}{\partial t}\left(\frac{\partial B}{\partial x}\right) = -\frac{\partial}{\partial t}\left(-\epsilon_0\mu_0\frac{\partial E}{\partial t}\right)$$

■ Electric field wave equation for electromagnetic waves in free space

$$\frac{\partial^2 E}{\partial x^2} = \epsilon_0\mu_0\frac{\partial^2 E}{\partial t^2} \qquad [24.15]$$

In the same manner, taking a derivative of Equation 24.14 with respect to x and combining it with Equation 24.11, we find that

■ Magnetic field wave equation for electromagnetic waves in free space

$$\frac{\partial^2 B}{\partial x^2} = \epsilon_0\mu_0\frac{\partial^2 B}{\partial t^2} \qquad [24.16]$$

Equations 24.15 and 24.16 both have the form of a linear wave equation (Eq. 13.19). As indicated in Chapter 13, such an equation is a mathematical representation of the traveling wave model. In this discussion, Equations 24.15 and 24.16 represent traveling electromagnetic waves. In the linear wave equation, the coefficient of the time derivative is the inverse of the speed of the waves. Therefore, these electromagnetic waves travel with a speed c of

■ The speed of electromagnetic waves

$$c = \frac{1}{\sqrt{\epsilon_0\mu_0}} \qquad [24.17]$$

Substituting $\epsilon_0 = 8.854\,19 \times 10^{-12}\ \mathrm{C^2/N \cdot m^2}$ and $\mu_0 = 4\pi \times 10^{-7}\ \mathrm{T \cdot m/A}$ in Equation 24.17, we find that $c = 2.997\,92 \times 10^8$ m/s. Because this speed is precisely the same as the speed of light in empty space,[3] one is led to believe (correctly) that **light is an electromagnetic wave.**

The simplest wave solutions to Equations 24.15 and 24.16 are those for which the field amplitudes E and B vary with x and t according to the expressions

$$E = E_{\max}\cos(kx - \omega t) \qquad [24.18]$$

$$B = B_{\max}\cos(kx - \omega t) \qquad [24.19]$$

In these expressions, $E_{\max}$ and $B_{\max}$ are the maximum values of the fields, the wave number $k = 2\pi/\lambda$, where λ is the wavelength, and the angular frequency $\omega = 2\pi f$, where f is the frequency. Active Figure 24.3b represents a view at one instant of a sinusoidal electromagnetic wave moving in the positive x direction.

Because electromagnetic waves are described by the traveling wave model, we can adopt another mathematical representation from the model, first seen in Equation 13.11 for mechanical waves. It is the relationship between wave speed, wavelength, and frequency for sinusoidal waves, $v = \lambda f$, which we can write for sinusoidal electromagnetic waves as

$$c = \lambda f \qquad [24.20]$$

The electric and magnetic fields of a plane electromagnetic wave are perpendicular to each other and to the direction of propagation. Therefore, **electromagnetic**

[3]Because of the redefinition of the meter in 1983, the speed of light is now a *defined* quantity with an *exact* value of $c = 2.997\,924\,58 \times 10^8$ m/s.

waves are transverse waves. The transverse mechanical waves studied in Chapter 13 exhibited physical displacements of the elements of the medium that were perpendicular to the direction of propagation of the wave. Electromagnetic waves do not require a medium for propagation, so there are no elements to be displaced. The transverse nature of an electromagnetic wave is represented by the direction of the field vectors with respect to the direction of propagation.

Taking partial derivatives of Equations 24.18 (with respect to x) and 24.19 (with respect to t), and substituting into Equation 24.11, we find that

$$\frac{E_{\text{max}}}{B_{\text{max}}} = c$$

Substituting from Equations 24.18 and 24.19 gives

$$\frac{E}{B} = c \qquad [24.21]$$

That is, **at every instant the ratio of the electric field to the magnetic field of an electromagnetic wave equals the speed of light.**

Finally, electromagnetic waves obey the **superposition principle** because the differential equations involving E and B are linear equations. For example, the resultant electric field magnitude of two waves coinciding in space with their $\vec{\mathbf{E}}$ vectors parallel can be found by simply adding the individual expressions for E given by Equation 24.18.

Doppler Effect for Light

In Section 13.8, we studied the Doppler effect for sound waves, in which the apparent frequency of the sound changes due to motion of the source or the observer. Light also exhibits a Doppler effect, which is demonstrated in astronomical observations by the wavelength shift of spectral lines from receding galaxies. This movement of spectral lines is toward the red end of the spectrum. It is therefore called the *red shift* and is evidence that other galaxies are moving away from us. (See Section 31.12 for more evidence of the expanding universe.)

The equation for the Doppler effect for light is not the same equation as that for sound for the following reason. For waves requiring a medium, the speeds of the source and observer can be separately measured with respect to a third entity, the medium. In the Doppler effect for sound, these two speeds are that of the source and that of the observer relative to the air. Because light does not require a medium, no third entity exists. Therefore, we cannot identify separate speeds for the source and observer. Only their relative speed can be identified. As a result, a different equation must be used, one that contains only this single speed. This equation can be generated from the laws of relativity and is found to be

■ Doppler effect for electromagnetic waves

$$f' = f\sqrt{\frac{c + v}{c - v}} \qquad [24.22]$$

where v is the relative speed between the source and the observer, c is the speed of light, f' is the frequency of light detected by the observer, and f is the frequency emitted by the source. For galaxies receding away from the Earth, v is entered into this equation as a negative number so that $f' < f$, which results in an apparent wavelength λ' such that $\lambda' > \lambda$. Therefore, the light should shift toward the red end of the spectrum, which is what is observed in the red shift.

INTERACTIVE EXAMPLE 24.1 An Electromagnetic Wave

A sinusoidal electromagnetic wave of frequency 40.0 MHz travels in free space in the x direction as in Figure 24.6.

A Determine the wavelength and period of the wave.

Solution Because $c = \lambda f$ and we know that $f = 40.0 \text{ MHz} = 4.00 \times 10^7 \text{ Hz}$, we have

$$\lambda = \frac{c}{f} = \frac{3.00 \times 10^8 \text{ m/s}}{4.00 \times 10^7 \text{ Hz}} = 7.50 \text{ m}$$

The period T of the wave equals the inverse of the frequency, so

$$T = \frac{1}{f} = \frac{1}{4.00 \times 10^7 \text{ Hz}} = 2.50 \times 10^{-8} \text{ s}$$

B At some point and at some instant, the electric field has its maximum value of 750 N/C and is along the y axis. Calculate the magnitude and direction of the magnetic field at this position and time.

Solution From Equation 24.21, we see that

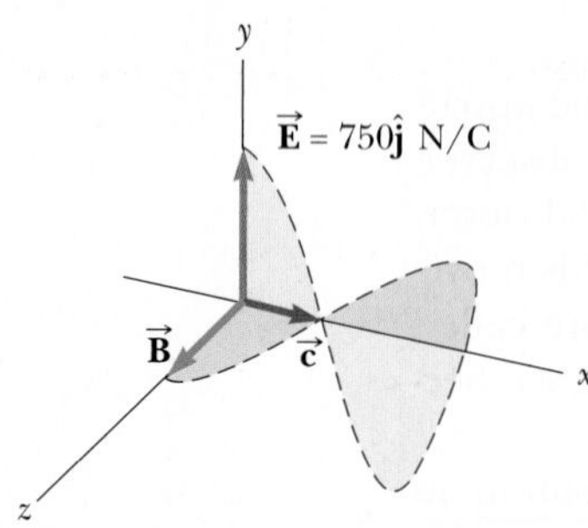

FIGURE 24.6 (Interactive Example 24.1) At some instant, a plane electromagnetic wave moving in the x direction has a maximum electric field of 750 N/C in the positive y direction. The corresponding magnetic field at that point has a magnitude E/c and is in the z direction.

$$B_{\text{max}} = \frac{E_{\text{max}}}{c} = \frac{750 \text{ N/C}}{3.00 \times 10^8 \text{ m/s}} = 2.50 \times 10^{-6} \text{ T}$$

Because $\vec{\mathbf{E}}$ and $\vec{\mathbf{B}}$ must be perpendicular to each other and $\vec{\mathbf{E}} \times \vec{\mathbf{B}}$ must be in the direction of wave propagation (x in this case), we conclude that $\vec{\mathbf{B}}$ is in the z direction.

C Write expressions for the space–time variation of the electric and magnetic field components for this wave.

Solution We can apply Equations 24.18 and 24.19 directly:

$$E = E_{\text{max}} \cos(kx - \omega t) = (750 \text{ N/C}) \cos(kx - \omega t)$$

$$B = B_{\text{max}} \cos(kx - \omega t) = (2.50 \times 10^{-6} \text{ T}) \cos(kx - \omega t)$$

where

$$\omega = 2\pi f = 2\pi(4.00 \times 10^7 \text{ Hz}) = 2.51 \times 10^8 \text{ rad/s}$$

$$k = \frac{2\pi}{\lambda} = \frac{2\pi}{7.50 \text{ m}} = 0.838 \text{ rad/m}$$

D An observer on the x axis, far to the right in Figure 24.6, moves to the left along the x axis at $0.500c$. What frequency does this observer measure for the electromagnetic wave?

Solution We use Equation 24.22 for the Doppler effect for light:

$$f' = f\sqrt{\frac{c + v}{c - v}} = 40.0 \text{ MHz}\sqrt{\frac{c + (+0.500c)}{c - (+0.500c)}}$$

$$= 69.3 \text{ MHz}$$

We have substituted v as a positive number because the observer is moving toward the source.

PhysicsNow™ Explore electromagnetic waves of different frequencies by logging into PhysicsNow at **www.pop4e.com** and going to Interactive Example 24.1.

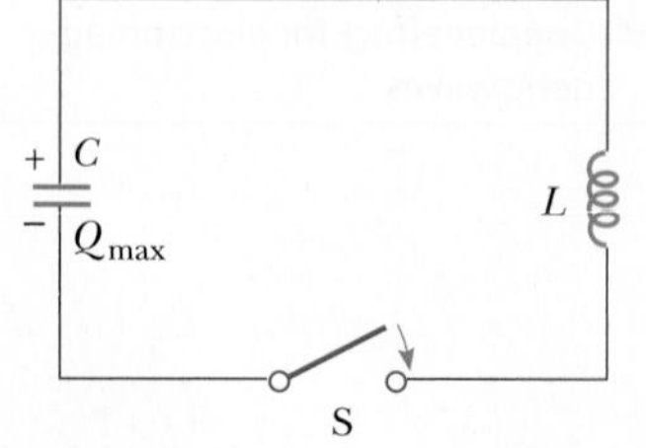

FIGURE 24.7 A simple LC circuit. The capacitor has an initial charge Q_{max}, and the switch is closed at $t = 0$.

24.4 HERTZ'S DISCOVERIES

In 1888, Heinrich Rudolf Hertz was the first to generate and detect electromagnetic waves in a laboratory setting. To appreciate the details of his experiment, let us first examine the properties of an LC circuit. In such a circuit, a charged capacitor is connected to an inductor as in Figure 24.7. When the switch is closed, both the current in the circuit and the charge on the capacitor oscillate in a manner closely related to our simple harmonic motion model in Chapter 12. If resistance is ignored, no energy is transformed to internal energy.

Let us investigate these oscillations in a way similar to our energy analysis of the simple harmonic motion model in Chapter 12. We assume that the capacitor has an initial charge of Q_{max} and that the switch is closed at $t = 0$. When the capacitor is fully charged, the total energy in the circuit is stored in the electric field of the capacitor and is equal to $Q_{max}^2/2C$. At this time, the current is zero and so no energy is stored in the inductor. As the capacitor begins to discharge, the energy stored in its electric field decreases. At the same time, the current increases and an amount of energy equal to $\frac{1}{2}LI^2$ is now stored in the magnetic field of the inductor. Therefore, energy is transferred from the electric field of the capacitor to the magnetic field of the inductor. When the capacitor is fully discharged, it stores no energy. At this time, the current reaches its maximum value and all the energy is stored in the inductor. The current continues in the same direction and begins to decrease in magnitude. While that occurs, the capacitor charges with polarity opposite to its previous polarity until the current stops and the capacitor is fully charged again. The process then repeats in the reverse direction. The energy continues to transfer between the inductor and the capacitor, corresponding to oscillations of both current and charge.

A representation of this energy transfer is shown in Active Figure 24.8. As mentioned, the behavior of the circuit is analogous to that of the oscillating block–spring system studied in Chapter 12. The potential energy $\frac{1}{2}kx^2$ stored in a stretched spring is analogous to the potential energy $Q_{max}^2/2C$ stored in the capacitor. The kinetic energy $\frac{1}{2}mv^2$ of the moving block is analogous to the magnetic energy $\frac{1}{2}LI^2$ stored in the inductor, which requires the presence of moving charges. In Active Figure 24.8a, all the energy is stored as electric potential energy in the capacitor at $t = 0$ (because $I = 0$), just like all the energy in a block–spring system is initially stored as potential energy in the spring if it is stretched and released at $t = 0$. In Active Figure 24.8b, all the energy is stored as magnetic energy $\frac{1}{2}LI_{max}^2$ in the inductor, where I_{max} is the maximum current. Active Figures 24.8c and 24.8d show subsequent quarter-cycle situations in which the energy is all electric or all magnetic. At intermediate points, part of the energy is electric and part is magnetic.

We now describe an alternative approach to the analogy between the LC circuit and the block–spring system of Chapter 12. Recall Equation 12.3, which is the differential equation describing the position of the block (modeled as a particle) in the simple harmonic motion model:

$$\frac{d^2x}{dt^2} = -\frac{k}{m}x$$

Applying Kirchhoff's loop rule to the circuit in Figure 24.7 gives

$$\frac{Q}{C} + L\frac{dI}{dt} = 0$$

Because $I = dQ/dt$, we can rewrite this equation as

$$\frac{Q}{C} = -L\frac{d}{dt}\left(\frac{dQ}{dt}\right) \rightarrow \frac{d^2Q}{dt^2} = -\frac{1}{LC}Q \qquad [24.23]$$

This equation has exactly the same mathematical form as Equation 12.3 for the block–spring system. Therefore, we conclude that the charge in the circuit will oscillate in a way analogous to a block on a spring.

In Chapter 12, we recognized the coefficient of x in Equation 12.3 as the square of the angular frequency (Eq. 12.4):

$$\omega^2 = \frac{k}{m}$$

Because of the identical mathematical form of the equation describing the LC circuit (Eq. 24.23), we can identify the coefficient of Q as the square of the angular

(The Bettman Archive/CORBIS)

HEINRICH RUDOLF HERTZ
(1857–1894)

German physicist Hertz made his greatest discovery—radio waves—in 1887. After finding that the speed of a radio wave is the same as that of light, he showed that radio waves, like light waves, could be reflected, refracted, and diffracted. Hertz died of blood poisoning at age 36. During his short life, he made many contributions to science. The hertz, equal to one complete vibration or cycle per second, is named after him.

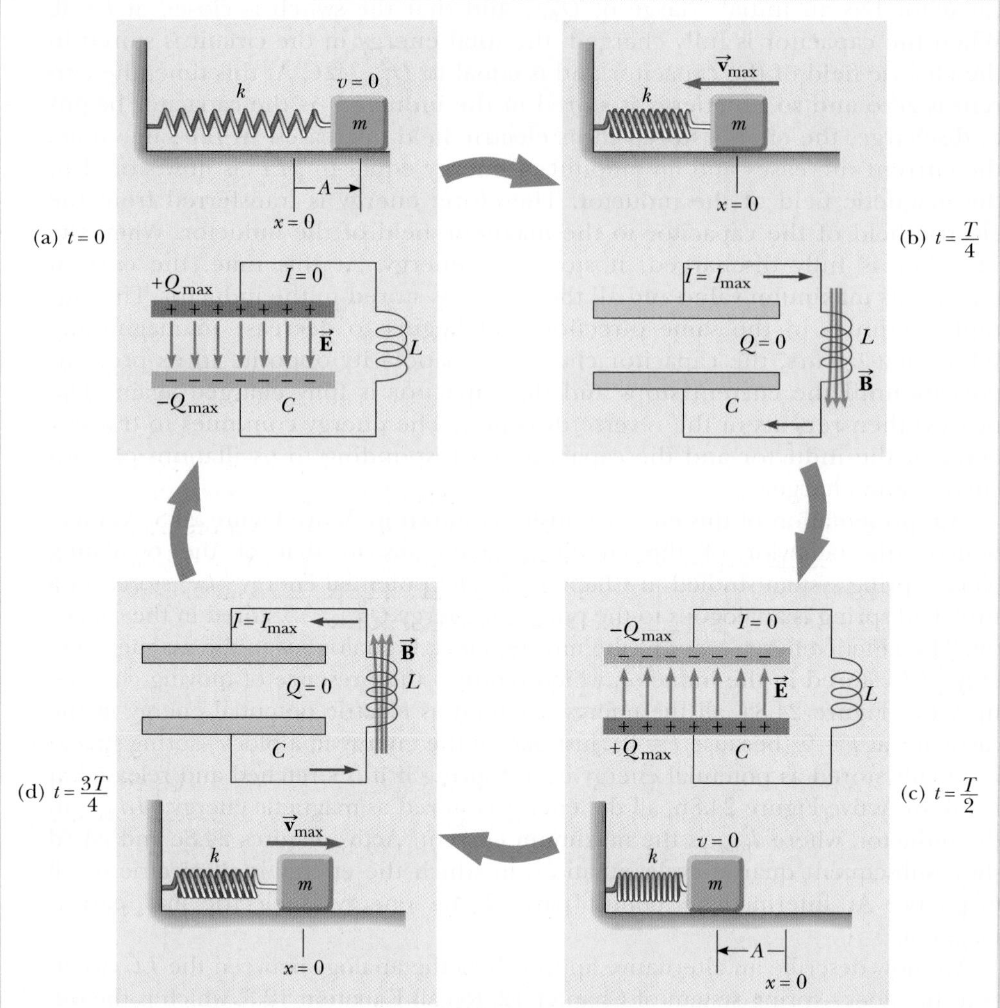

ACTIVE FIGURE 24.8

The conditions in a resistanceless LC circuit are shown at quarter-cycle intervals during its oscillation. Associated with each image of the circuit is the mechanical analog, the block–spring oscillating system. (a) At $t = 0$, the capacitor has a charge Q_{max} and there is an electric field between the plates. Because no current exists at this instant, there is no magnetic field in the inductor. In the mechanical system, the block of mass m is at its maximum displacement from equilibrium, with a speed of zero. (b) One quarter of a cycle later, the charge on the capacitor has reached zero and the current has its maximum value I_{max}, causing a magnetic field of maximum magnitude in the inductor. The block in the mechanical system is passing through $x = 0$ with maximum speed. (c) After another quarter cycle, the capacitor has charged up to its maximum value, with opposite polarity to that in (a). The mechanical system is similar to that in (a) except that the spring is at maximum compression rather than extension. (d) Circuit conditions are similar to those in (b) but with current in the opposite direction. The mechanical system is similar to that in (b) but with the direction of the velocity reversed. One quarter of a cycle later, the circuit and the mechanical system return to the conditions in (a), ready to begin a new cycle.

PhysicsNow™ Log into PhysicsNow at **www.pop4e.com** and go to Active Figure 24.8 to adjust the values of C and L and see the effect on the oscillating circuit. The block on the spring oscillates in a mechanical analog of the electrical oscillations. A graphical display of charge and current is available, as is an energy bar graph.

frequency:

$$\omega^2 = \frac{1}{LC}$$

Therefore, the frequency of oscillation of an LC circuit, called the *resonance frequency*, is

$$f_0 = \frac{1}{2\pi\sqrt{LC}} \qquad [24.24]$$

■ Resonance frequency of an *LC* circuit

The circuit Hertz used in his investigations of electromagnetic waves is shown schematically in Figure 24.9. A large coil of wire called an induction coil is connected to two metal spheres with a narrow gap between them to form a capacitor. Oscillations are initiated in the circuit by sending short voltage pulses via the coil to the spheres, initially charging one positive, the other negative. Based on the values of L and C in Hertz's circuit, the frequency of oscillation is $f \approx 100$ MHz. This circuit is called a transmitter because it produces electromagnetic waves.

Hertz placed a second circuit, the receiver, several meters from the transmitter circuit. This receiver circuit, which consisted of a single loop of wire connected to two spheres, had its own effective inductance, capacitance, and natural frequency of oscillation. Hertz found that energy was being sent from the transmitter to the receiver when the resonance frequency of the receiver was adjusted to match that of the transmitter.[4] The energy transfer was detected when the voltage across the spheres in the receiver circuit became high enough to cause sparks to appear in the air gap separating the spheres. Hertz's experiment is analogous to the mechanical phenomenon in which one tuning fork responds to acoustic vibrations from an identical vibrating fork. In the case of the tuning fork, the energy transfer from one fork to another is by means of sound, whereas the transfer mechanism is electromagnetic radiation for Hertz's apparatus.

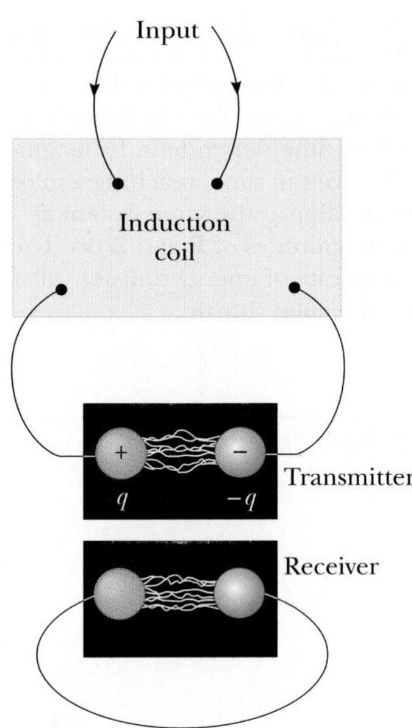

FIGURE 24.9 Schematic diagram of Hertz's apparatus for generating and detecting electromagnetic waves. The transmitter consists of two spherical electrodes connected to an induction coil, which provides short voltage surges to the spheres, setting up oscillations in the discharge. The receiver is a nearby loop containing a second spark gap.

Hertz assumed that the energy transferred from the transmitter to the receiver was carried in the form of waves, which are now known to have been electromagnetic waves. In a series of experiments, he also showed that the radiation generated by the transmitter exhibited the wave properties of interference, diffraction, reflection, refraction, and polarization. As we shall see shortly, all these properties are exhibited by light. Therefore, it became evident that the waves observed by Hertz had properties similar to those of light waves and differed only in frequency and wavelength.

Perhaps Hertz's most convincing experiment was his measurement of the speed of the waves from the transmitter. Waves of known frequency from the transmitter were reflected from a metal sheet so that a pattern of nodes and antinodes was set up, much like the standing wave pattern on a stretched string. As we saw in our discussion of standing waves (Chapter 14), the distance between nodes is $\lambda/2$, so Hertz was able to determine the wavelength λ. Using the relationship $v = f\lambda$, Hertz found that v was close to 3.00×10^8 m/s, the known speed of visible light. Therefore, Hertz's experiments provided the first evidence in support of Maxwell's theory.

■ Thinking Physics 24.1

In radio transmission, a radio wave serves as a carrier wave and the sound wave is superimposed on the carrier wave. In amplitude modulation (AM radio), the amplitude of the carrier wave varies according to the sound wave. (The word *modulate* means "to change.") In frequency modulation (FM radio), the frequency of the carrier wave varies according to the sound wave. The navy sometimes uses flashing

[4]Following Hertz's discoveries, Guglielmo Marconi succeeded in developing this phenomenon into a practical, long-range communication system, radio.

lights to send Morse code to neighboring ships, a process that is similar to radio broadcasting. Is this process AM or FM? What is the carrier frequency? What is the signal frequency? What is the broadcasting antenna? What is the receiving antenna?

Reasoning The flashing of the light according to Morse code is a drastic amplitude modulation because the amplitude is changing between a maximum value and zero. In this sense, it is similar to the on-and-off binary code used in computers and compact discs. The carrier frequency is that of the visible light, on the order of 10^{14} Hz. The signal frequency depends on the skill of the signal operator, but is on the order of a few hertz, as the light is flashed on and off. The broadcasting antenna for this modulated signal is the filament of the lightbulb in the signal source. The receiving antenna is the eye. ■

PITFALL PREVENTION 24.2

AN INSTANTANEOUS VALUE The Poynting vector given by Equation 24.25 is time-dependent. Its magnitude varies in time, reaching a maximum value at the same instant as the magnitudes of $\vec{\mathbf{E}}$ and $\vec{\mathbf{B}}$ do. The *average* rate of energy transfer will be calculated shortly.

24.5 ENERGY CARRIED BY ELECTROMAGNETIC WAVES

In Section 13.6, we found that mechanical waves carry energy. Electromagnetic waves also carry energy, and as they propagate through space they can transfer energy to objects placed in their path. This notion was introduced in Chapter 6 when we discussed the transfer mechanisms in the continuity equation for energy, and it was noted again in Chapter 17 in the discussion of thermal radiation. The rate of flow of energy in an electromagnetic wave is described by a vector $\vec{\mathbf{S}}$, called the **Poynting vector,** defined by the expression

■ Poynting vector

$$\vec{\mathbf{S}} = \frac{1}{\mu_0}\vec{\mathbf{E}} \times \vec{\mathbf{B}} \qquad [24.25]$$

The magnitude of the Poynting vector represents the rate at which energy flows through a unit surface area perpendicular to the flow and its direction is along the direction of wave propagation (Fig. 24.10). Therefore, the Poynting vector represents *power per unit area.* The SI units of the Poynting vector are $\text{J/s}\cdot\text{m}^2 = \text{W/m}^2$.

As an example, let us evaluate the magnitude of $\vec{\mathbf{S}}$ for a plane electromagnetic wave. We have $|\vec{\mathbf{E}} \times \vec{\mathbf{B}}| = EB$ because $\vec{\mathbf{E}}$ and $\vec{\mathbf{B}}$ are perpendicular to each other. In this case,

$$S = \frac{EB}{\mu_0} \qquad [24.26]$$

Because $B = E/c$, we can also express the magnitude as

$$S = \frac{E^2}{\mu_0 c} = \frac{cB^2}{\mu_0}$$

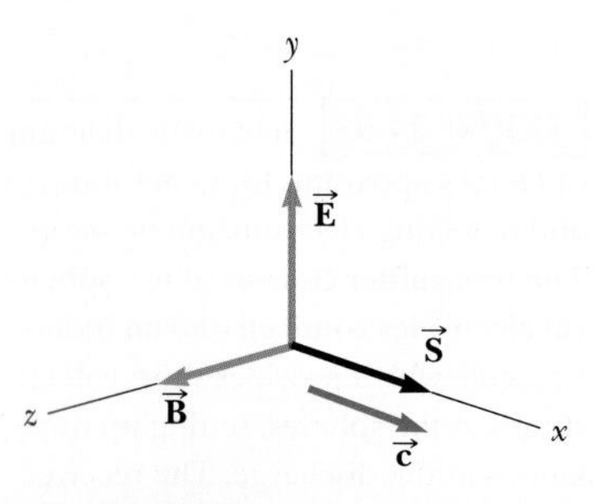

FIGURE 24.10 The Poynting vector $\vec{\mathbf{S}}$ for an electromagnetic wave is along the direction of wave propagation.

These equations for S apply at any instant of time.

What is of more interest for a sinusoidal electromagnetic wave (Eqs. 24.18 and 24.19) is the time average of S over one or more cycles, which is the **intensity** I. When this average is taken, we obtain an expression involving the time average of $\cos^2(kx - \omega t)$, which equals $\frac{1}{2}$. Therefore, the average value of S (or the intensity of the wave) is

■ Intensity of electromagnetic radiation

$$I = S_{\text{avg}} = \frac{E_{\text{max}}B_{\text{max}}}{2\mu_0} = \frac{E_{\text{max}}^2}{2\mu_0 c} = \frac{cB_{\text{max}}^2}{2\mu_0} \qquad [24.27]$$

Recall that the energy per unit volume u_E, which is the instantaneous energy density associated with an electric field (Section 20.9), is given by Equation 20.32:

$$u_E = \tfrac{1}{2}\epsilon_0 E^2 \qquad [24.28]$$

and that the instantaneous energy density u_B associated with a magnetic field (Section 23.7) is given by Equation 23.22:

$$u_B = \frac{B^2}{2\mu_0} \qquad [24.29]$$

Because E and B vary with time for an electromagnetic wave, the energy densities also vary with time. Using the relationships $B = E/c$ and $c = 1/\sqrt{\epsilon_0\mu_0}$, Equation 24.29 becomes

$$u_B = \frac{(E/c)^2}{2\mu_0} = \frac{\epsilon_0\mu_0}{2\mu_0}E^2 = \tfrac{1}{2}\epsilon_0 E^2$$

Comparing this result with the expression for u_E, we see that

$$u_B = u_E$$

That is, **for an electromagnetic wave, the instantaneous energy density associated with the magnetic field equals the instantaneous energy density associated with the electric field.** Therefore, in a given volume the energy is equally shared by the two fields.

The **total instantaneous energy density** u is equal to the sum of the energy densities associated with the electric and magnetic fields:

$$u = u_E + u_B = \epsilon_0 E^2 = \frac{B^2}{\mu_0}$$

■ Total instantaneous energy density of an electromagnetic wave

When this expression is averaged over one or more cycles of an electromagnetic wave, we again obtain a factor of $\frac{1}{2}$. Therefore, the total average energy per unit volume of an electromagnetic wave is

$$u_{avg} = \epsilon_0(E^2)_{avg} = \tfrac{1}{2}\epsilon_0 E_{max}^2 = \frac{B_{max}^2}{2\mu_0} \qquad [24.30]$$

■ Average energy density of an electromagnetic wave

Comparing this result with Equation 24.27 for the average value of S, we see that

$$I = S_{avg} = cu_{avg} \qquad [24.31]$$

In other words, **the intensity of an electromagnetic wave equals the average energy density multiplied by the speed of light.**

QUICK QUIZ 24.2 An electromagnetic wave propagates in the $-y$ direction. The electric field at a point in space is momentarily oriented in the $+x$ direction. Is the magnetic field at that point momentarily oriented in the **(a)** $-x$ direction, **(b)** $+y$ direction, **(c)** $+z$ direction, or **(d)** $-z$ direction?

QUICK QUIZ 24.3 Which of the following quantities does not vary in time for plane electromagnetic waves? **(a)** magnitude of the Poynting vector **(b)** energy density u_E **(c)** energy density u_B **(d)** intensity I

EXAMPLE 24.2 Fields Due to a Point Source

A point source of electromagnetic radiation has an average power output of 800 W. Calculate the maximum values of the electric and magnetic fields at a point 3.50 m from the source.

Solution For waves propagating uniformly from a point source, the energy of the wave at a distance r from the source is distributed over the surface area of an imaginary sphere of radius r. Therefore, the intensity of the radiation at a point on the sphere is

$$I = \frac{\mathcal{P}_{avg}}{4\pi r^2}$$

where $\mathcal{P}_{avg}$ is the average power output of the source and $4\pi r^2$ is the area of the sphere centered on the source. Because the intensity of an electromagnetic wave is also given by Equation 24.27, we have

$$I = \frac{\mathcal{P}_{avg}}{4\pi r^2} = \frac{E_{max}^2}{2\mu_0 c}$$

Solving for E_{max} gives us

$$E_{max} = \sqrt{\frac{\mu_0 c\mathcal{P}_{avg}}{2\pi r^2}}$$

$$= \sqrt{\frac{(4\pi \times 10^{-7}\ \text{T}\cdot\text{m/A})(3.00 \times 10^8\ \text{m/s})(800\ \text{W})}{2\pi(3.50\ \text{m})^2}}$$

$$= 62.6\ \text{V/m}$$

We calculate the maximum value of the magnetic field using this result and Equation 24.21:

$$B_{max} = \frac{E_{max}}{c} = \frac{62.6\ \text{V/m}}{3.00 \times 10^8\ \text{m/s}} = 2.09 \times 10^{-7}\ \text{T}$$

24.6 MOMENTUM AND RADIATION PRESSURE

Electromagnetic waves transport linear momentum as well as energy. Hence, it follows that pressure is exerted on a surface when an electromagnetic wave impinges on it. In what follows, let us assume that the electromagnetic wave strikes a surface at normal incidence and transports a total energy U to a surface in a time interval Δt. If the surface absorbs all the incident energy U in this time, Maxwell showed that the total momentum $\vec{\mathbf{p}}$ delivered to this surface has a magnitude

■ Momentum delivered to a perfectly absorbing surface

$$p = \frac{U}{c} \quad \text{(complete absorption)} \qquad [24.32]$$

The pressure exerted on the surface is defined as force per unit area F/A. Let us combine this definition with Newton's second law:

$$P = \frac{F}{A} = \frac{1}{A}\frac{dp}{dt}$$

If we now replace p, the momentum transported to the surface by radiation, from Equation 24.32, we have

$$P = \frac{1}{A}\frac{dp}{dt} = \frac{1}{A}\frac{d}{dt}\left(\frac{U}{c}\right) = \frac{1}{c}\frac{(dU/dt)}{A}$$

We recognize $(dU/dt)/A$ as the rate at which energy is arriving at the surface per unit area, which is the magnitude of the Poynting vector. Therefore, the radiation pressure P exerted on the perfectly absorbing surface is

■ Radiation pressure exerted on a perfect absorbing surface

$$P = \frac{S}{c} \quad \text{(complete absorption)} \qquad [24.33]$$

An absorbing surface for which all the incident energy is absorbed (none is reflected) is called a **black body.** A more detailed discussion of a black body will be presented in Chapter 28.

If the surface is a perfect reflector, the momentum delivered in a time interval Δt for normal incidence is twice that given by Equation 24.32, or $p = 2U/c$. That is, a momentum U/c is delivered first by the incident wave and then again by the reflected wave, a situation analogous to a ball colliding elastically with a wall.[5] Finally, the radiation pressure exerted on a perfect reflecting surface for normal incidence of the wave is twice that given by Equation 24.33, or $P = 2S/c$.

Although radiation pressures are very small (about $5 \times 10^{-6}\ \text{N/m}^2$ for direct sunlight), they have been measured using torsion balances such as the one shown in Figure 24.11. Light is allowed to strike either a mirror or a black disk, both of which are suspended from a fine fiber. Light striking the black disk is completely absorbed, so all its momentum is transferred to the disk. Light striking the mirror (normal incidence) is totally reflected and hence the momentum transfer is twice as great as that transferred to the disk. The radiation pressure is determined by measuring the angle through which the horizontal connecting rod rotates. The apparatus must be placed in a high vacuum to eliminate the effects of air currents.

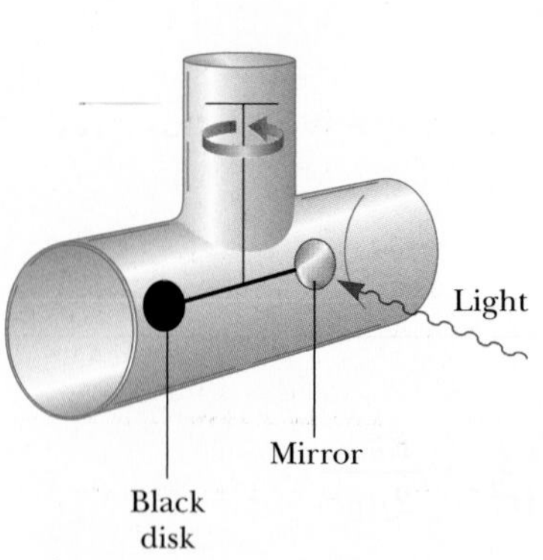

FIGURE 24.11 An apparatus for measuring the pressure exerted by light. In practice, the system is contained in a high vacuum.

QUICK QUIZ 24.4 In an apparatus such as that in Figure 24.11, suppose the black disk is replaced by one with half the radius. Which of the following are different after the disk is replaced? **(a)** radiation pressure on the disk **(b)** radiation force on the disk **(c)** radiation momentum delivered to the disk in a given time interval

[5] For *oblique* incidence, the momentum transferred is $2U \cos \theta / c$ and the pressure is given by $P = 2S \cos^2 \theta / c$, where θ is the angle between the normal to the surface and the direction of propagation.

■ Thinking Physics 24.2

A large amount of dust occurs in the interplanetary space in the Solar System. Although this dust can theoretically have a variety of sizes, from molecular size upward, very little of it is smaller than about 0.2 μm in our Solar System. Why? (*Hint:* The Solar System originally contained dust particles of all sizes.)

Reasoning Dust particles in the Solar System are subject to two forces: the gravitational force toward the Sun and the force from radiation pressure due to sunlight, which is away from the Sun. The gravitational force is proportional to the cube of the radius of a spherical dust particle because it is proportional to the particle's mass. The radiation force is proportional to the square of the radius because it depends on the circular cross-section of the particle. For large particles, the gravitational force is larger than the force from radiation pressure. For small particles, less than about 0.2 μm, the larger force from radiation pressure sweeps these particles out of the Solar System. ■

EXAMPLE 24.3 Solar Energy

The Sun delivers about 1 000 W/m^2 of energy to the Earth's surface.

A Calculate the total power incident on a roof of dimensions 8.00 m × 20.0 m.

Solution The Poynting vector has an average magnitude $I = S_{\text{avg}} = 1\,000\ \text{W/m}^2$, which represents the power per unit area. Assuming that the radiation is incident normal to the roof, we can find the power for the whole roof:

$$\mathcal{P} = IA = (1\,000\ \text{W/m}^2)(8.00 \times 20.0\ \text{m}^2)$$
$$= 1.60 \times 10^5\ \text{W}$$

If solar energy could all be converted to electric energy, it would provide more than enough power for the average home. Solar energy is not easily harnessed, however, and the prospects for large-scale conversion are not as bright as they may appear from this simple calculation. For example, the conversion efficiency from solar to electric energy is typically 10% for photovoltaic cells. Solar energy has other practical problems that must also be considered, such as overcast days, geographic location, and energy storage.

B Determine the radiation pressure and radiation force on the roof, assuming that the roof covering is a perfect absorber.

Solution Using Equation 24.33 with $I = 1\,000\ \text{W/m}^2$, we find that the average radiation pressure is

$$P = \frac{I}{c} = \frac{1\,000\ \text{W/m}^2}{3.00 \times 10^8\ \text{m/s}} = 3.33 \times 10^{-6}\ \text{N/m}^2$$

Because pressure equals force per unit area, this value of P corresponds to a radiation force of

$$F = PA = (3.33 \times 10^{-6}\ \text{N/m}^2)(160\ \text{m}^2)$$
$$= 5.33 \times 10^{-4}\ \text{N}$$

INTERACTIVE EXAMPLE 24.4 Pressure from a Laser Pointer

Many people giving presentations use a laser pointer to direct the attention of their audience. If a 3.0-mW pointer creates a spot that is 2.0 mm in diameter, determine the radiation pressure on a screen that reflects 70% of the light striking it. The power of 3.0 mW is a time-averaged power.

Solution In conceptualizing this problem, we certainly do not expect the pressure to be very large. We categorize this problem as one in which we calculate the radiation pressure by using something like Equation 24.33, but which is complicated by the 70% reflection. To analyze the problem, we begin by determining the Poynting vector of the beam. We divide the time-averaged power delivered via the electromagnetic wave by the cross-sectional area of the beam. Thus,

$$I = \frac{\mathcal{P}}{A} = \frac{\mathcal{P}}{\pi r^2} = \frac{3.0 \times 10^{-3}\ \text{W}}{\pi\left(\dfrac{2.0 \times 10^{-3}\ \text{m}}{2}\right)^2}$$
$$= 9.6 \times 10^2\ \text{W/m}^2$$

Now we can determine the radiation pressure from the laser beam. A completely reflected beam would apply an average pressure of $P_{\text{avg}} = 2S_{\text{avg}}/c$. We can model the actual reflection as follows. Imagine that the surface absorbs the beam, resulting in pressure

$P_{avg} = S_{avg}/c$. Then the surface emits the beam, resulting in additional pressure $P_{avg} = S_{avg}/c$. If the surface emits only a fraction f of the beam (so that f is the amount of the incident beam reflected), the pressure due to the emitted beam is $P_{avg} = fS_{avg}/c$. Therefore, the total pressure on the surface due to absorption and re-emission (reflection) is

$$P_{avg} = \frac{S_{avg}}{c} + f\frac{S_{avg}}{c} = (1+f)\frac{S_{avg}}{c}$$

For a beam that is 70% reflected, the pressure is

$$P = (1+0.70)\frac{9.6 \times 10^2 \text{ W/m}^2}{3.0 \times 10^8 \text{ m/s}} = 5.4 \times 10^{-6} \text{ N/m}^2$$

To finalize the problem, consider first the magnitude of the Poynting vector. It is about the same as the intensity of sunlight at the Earth's surface. (Therefore, it is not safe to shine the beam of a laser pointer into a person's eyes; that may be more dangerous than looking directly at the Sun.) To finalize further, note that the pressure has an extremely small value, as expected. (Recall from Section 15.1 that atmospheric pressure is approximately 10^5 N/m^2.)

PhysicsNow™ Log into PhysicsNow at **www.pop4e.com** and go to Interactive Example 24.4 to investigate the pressure on the screen for various laser and screen parameters.

Space Sailing

When imagining a trip to another planet, we normally think of traditional rocket engines that convert chemical energy in fuel carried on the spacecraft to kinetic energy of the spacecraft. An interesting alternative to this approach is that of **space sailing.** A space-sailing craft includes a very large sail that reflects light. The motion of the spacecraft depends on pressure from light, that is, the force exerted on the sail by the reflection of light from the Sun. Calculations performed (before U.S. government budget cutbacks shelved early space-sailing projects) showed that sailing craft could travel to and from the planets in times similar to those for traditional rockets, but for less cost.

Calculations show that the radiation force from the Sun on a practical sailcraft with large sails could be equal to or slightly larger than the gravitational force on the sailcraft. If these two forces are equal, the sailcraft can be modeled as a particle in equilibrium because the inward gravitational force of the Sun balances the outward force exerted by the light from the Sun. If the sailcraft has an initial velocity in some direction away from the Sun, it would move in a straight line under the action of these two forces, with no necessity for fuel. A traditional spacecraft with its rocket engines turned off, on the other hand, would slow down as a result of the gravitational force on it due to the Sun. Both the force on the sail and the gravitational force from the Sun fall off as the inverse square of the Sun–sailcraft separation. Therefore, in theory, the straight-line motion of the sailcraft would continue forever with no fuel requirement.

By using just the motion imparted to a sailcraft by the Sun, the craft could reach Alpha Centauri in about 10 000 years. This time interval can be reduced to 30 to 100 years using a *beamed power system.* In this concept, light from the Sun is gathered by a transformation device in orbit around the Earth and is converted to a laser beam or microwave beam aimed at the sailcraft. The force from this intense beam of radiation increases the acceleration of the craft, and the transit time is significantly reduced. Calculations indicate that the sailcraft could achieve design speeds of up to 20% of the speed of light using this technique.

24.7 THE SPECTRUM OF ELECTROMAGNETIC WAVES

Electromagnetic waves travel through vacuum with speed c, frequency f, and wavelength λ. The various types of electromagnetic waves, all produced by accelerating charges, are shown in Figure 24.12. Note the wide range of frequencies and wavelengths. Let us briefly describe the wave types shown in Figure 24.12.

Radio waves are the result of charges accelerating, for example, through conducting wires in a radio antenna. They are generated by such electronic devices as LC oscillators and are used in radio and television communication systems.

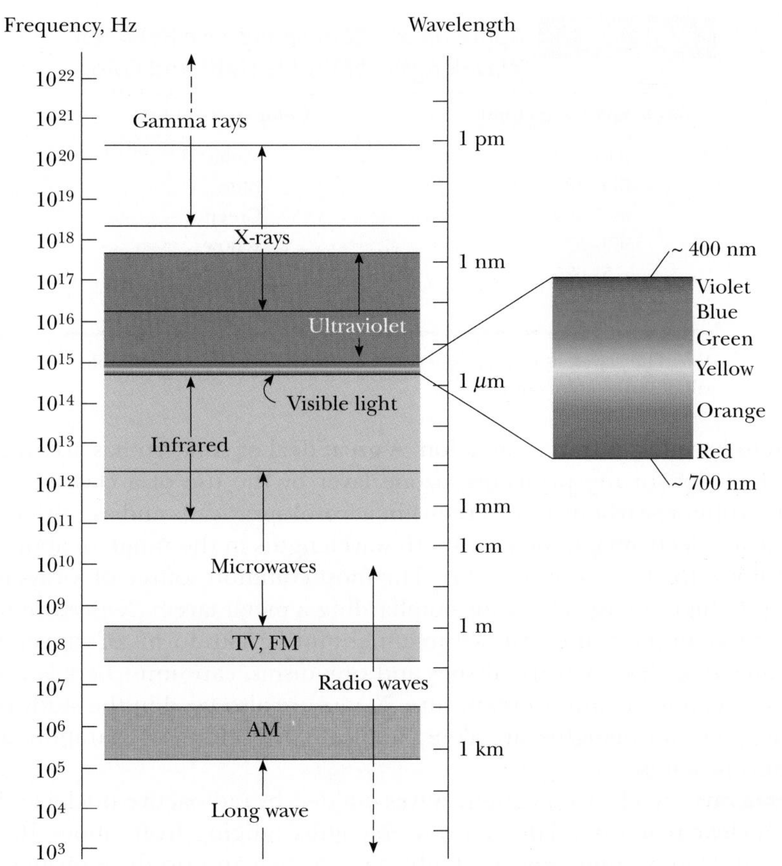

FIGURE 24.12 The electromagnetic spectrum. Note the overlap between adjacent wave types. The expanded view to the right shows details of the visible spectrum.

(Ron Chapple/Getty Images)

Wearing sunglasses that do not block ultraviolet (UV) light is worse for your eyes than wearing no sunglasses. The lenses of any sunglasses absorb some visible light, thus causing the wearer's pupils to dilate. If the glasses do not also block UV light, more damage may be done to the eye's lens because of the dilated pupils. If you wear no sunglasses at all, your pupils are contracted, you squint, and much less UV light enters your eyes. High-quality sunglasses block nearly all the eye-damaging UV light. ■

Microwaves (short-wavelength radio waves) have wavelengths ranging between about 1 mm and 30 cm and are also generated by electronic devices. Because of their short wavelengths, they are well suited for radar systems used in aircraft navigation and for studying the atomic and molecular properties of matter. Microwave ovens are a domestic application of these waves.

Infrared waves have wavelengths ranging from about 1 mm to the longest wavelength of visible light, 7×10^{-7} m. These waves, produced by objects at room temperature and by molecules, are readily absorbed by most materials. Infrared radiation has many practical and scientific applications, including physical therapy, infrared photography, and vibrational spectroscopy. Your remote control for your TV, VCR, or DVD player likely uses an infrared beam to communicate with the video device.

Visible light, the most familiar form of electromagnetic waves, is that part of the spectrum the human eye can detect. Light is produced by hot objects like lightbulb filaments and by the rearrangement of electrons in atoms and molecules. The wavelengths of visible light are classified by color, ranging from violet ($\lambda \approx 4 \times 10^{-7}$ m) to red ($\lambda \approx 7 \times 10^{-7}$ m). The eye's sensitivity is a function of wavelength and is a maximum at a wavelength of about 5.5×10^{-7} m (yellow–green). Table 24.1 provides approximate correspondences between the wavelength of visible light and the color assigned to it by humans. Light is the basis of the science of optics and optical instruments, to be discussed in Chapters 25 through 27.

Ultraviolet light covers wavelengths ranging from about 4×10^{-7} m (400 nm) down to 6×10^{-10} m (0.6 nm). The Sun is an important source of ultraviolet waves, which are the main cause of suntans and sunburns. Atoms in the stratosphere absorb most of the ultraviolet waves from the Sun (which is fortunate because ultraviolet waves in large quantities have harmful effects on humans). One important constituent of the stratosphere is ozone (O_3), which results from reactions of oxygen with ultraviolet radiation. This ozone shield converts lethal high-energy ultraviolet

PITFALL PREVENTION 24.3

Heat Rays Infrared rays are often called "heat rays." This terminology is a misnomer. Although infrared radiation is used to raise or maintain temperature, as in the case of keeping food warm with "heat lamps" at a fast-food restaurant, all wavelengths of electromagnetic radiation carry energy that can cause the temperature of a system to increase. As an example, consider using your microwave oven to bake a potato, whose temperature increases because of microwaves.

TABLE 24.1 Approximate Correspondence Between Wavelengths of Visible Light and Color

Wavelength Range (nm)	Color Description
400–430	Violet
430–485	Blue
485–560	Green
560–575	Yellow
575–625	Orange
625–700	Red

Note: The wavelength ranges here are approximate. Different people will describe colors differently.

radiation to harmless infrared radiation. A great deal of concern has arisen concerning the depletion of the protective ozone layer by the use of a class of chemicals called chlorofluorocarbons (e.g., Freon) in aerosol spray cans and as refrigerants.

X-rays are electromagnetic waves with wavelengths in the range of about 10^{-8} m (10 nm) down to 10^{-13} m (10^{-4} nm). The most common source of x-rays is the acceleration of high-energy electrons bombarding a metal target. X-rays are used as a diagnostic tool in medicine and as a treatment for certain forms of cancer. Because x-rays damage or destroy living tissues and organisms, care must be taken to avoid unnecessary exposure and overexposure. X-rays are also used in the study of crystal structure; x-ray wavelengths are comparable to the atomic separation distances ($\approx$ 0.1 nm) in solids.

Gamma rays are electromagnetic waves emitted by radioactive nuclei and during certain nuclear reactions. They have wavelengths ranging from about 10^{-10} m to less than 10^{-14} m. Gamma rays are highly penetrating and produce serious damage when absorbed by living tissues. Consequently, those working near such dangerous radiation must be protected with heavily absorbing materials, such as layers of lead.

QUICK QUIZ 24.5 In many kitchens, a microwave oven is used to cook food. The frequency of the microwaves is on the order of 10^{10} Hz. The wavelengths of these microwaves are on the order of **(a)** kilometers, **(b)** meters, **(c)** centimeters, or **(d)** micrometers.

QUICK QUIZ 24.6 A radio wave of frequency on the order of 10^5 Hz is used to carry a sound wave with a frequency on the order of 10^3 Hz. The wavelength of this radio wave is on the order of **(a)** kilometers, **(b)** meters, **(c)** centimeters, or **(d)** micrometers.

▪ Thinking Physics 24.3

The center of sensitivity of our eyes is close to the same frequency as the center of the wavelength distribution of light from the Sun. Is that an amazing coincidence?

 Center of eyesight sensitivity

Reasoning It is not a coincidence; rather, it is the result of biological evolution. Humans have evolved so as to be most visually sensitive to the wavelengths that are strongest from the Sun. It is an interesting conjecture to imagine aliens from another planet, with a Sun with a different temperature, arriving at Earth. Their eyes would have the center of sensitivity at different wavelengths than ours. How would their vision of the Earth compare with ours? ▪

24.8 POLARIZATION

As we learned in Section 24.3, the electric and magnetic vectors associated with an electromagnetic wave are perpendicular to each other and also to the direction of wave propagation as shown in Active Figure 24.3. The phenomenon of polarization described in this section is a property that specifies the directions of the electric and magnetic fields associated with an electromagnetic wave.

An ordinary beam of light consists of a large number of waves emitted by the atoms of the light source. Each atom produces a wave with its own orientation of the electric field $\vec{E}$, corresponding to the direction of vibration in the atom. The direction of polarization of the electromagnetic wave is defined to be the direction in which $\vec{E}$ is vibrating. Because all directions of vibration are possible in a group of atoms emitting a beam of light, however, the resultant beam is a superposition of waves produced by the individual atomic sources. The result is an **unpolarized** light wave, represented schematically in Figure 24.13a. The direction of wave propagation in this figure is perpendicular to the page. The figure suggests that *all* directions of the electric field vector lying in a plane perpendicular to the direction of propagation are equally probable.

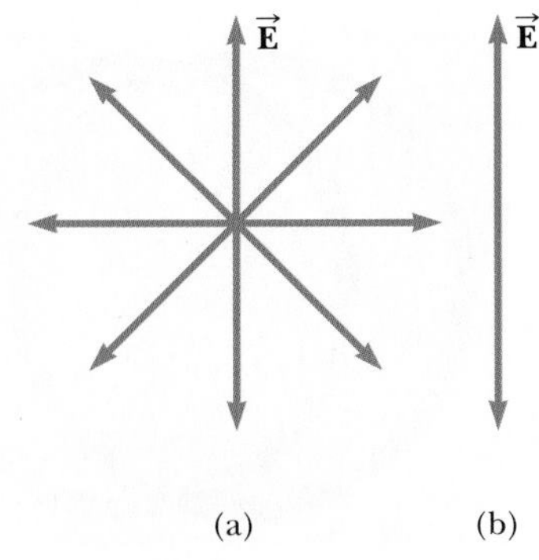

FIGURE 24.13 (a) An unpolarized light beam viewed along the direction of propagation (perpendicular to the page). The time-varying electric field vector can be in any direction in the plane of the page with equal probability. (b) A linearly polarized light beam with the time-varying electric field vector in the vertical direction.

A wave is said to be **linearly polarized** if the orientation of $\vec{E}$ is the same for all individual waves *at all times* at a particular point as suggested in Figure 24.13b. (Sometimes such a wave is described as **plane polarized.**) The wave described in Active Figure 24.3 is an example of a wave linearly polarized along the y axis. As the field propagates in the x direction, $\vec{E}$ is always along the y axis. The plane formed by $\vec{E}$ and the direction of propagation is called the **plane of polarization** of the wave. In Active Figure 24.3, the plane of polarization is the xy plane. It is possible to obtain a linearly polarized wave from an unpolarized wave by removing from the unpolarized wave all components of electric field vectors except those that lie in a single plane.

The most common technique for polarizing light is to send it through a material that passes only components of electric field vectors that are parallel to a characteristic direction of the material called the **polarizing direction.** In 1938, E. H. Land discovered such a material, which he called **Polaroid,** that polarizes light through selective absorption by oriented molecules. This material is fabricated in thin sheets of long-chain hydrocarbons, which are stretched during manufacture so that the molecules align. After a sheet is dipped into a solution containing iodine, the molecules become good electric conductors. The conduction, however, takes place primarily along the hydrocarbon chains because the valence electrons of the molecules can move easily only along the chains (valence electrons are "free" electrons that can readily move through the conductor). As a result, the molecules readily *absorb* light whose electric field vector is parallel to their length and *transmit* light whose electric field vector is perpendicular to their length. It is common to refer to the direction perpendicular to the molecular chains as the **transmission axis.** An ideal polarizer passes the components of electric vectors that are parallel to the transmission axis. Components perpendicular to the transmission axis are absorbed. If light passes through several polarizers, whatever is transmitted has the plane of polarization parallel to the polarizing direction of the last polarizer through which it passed.

Let us now obtain an expression for the intensity of light that passes through a polarizing material. In Active Figure 24.14, an unpolarized light beam is incident

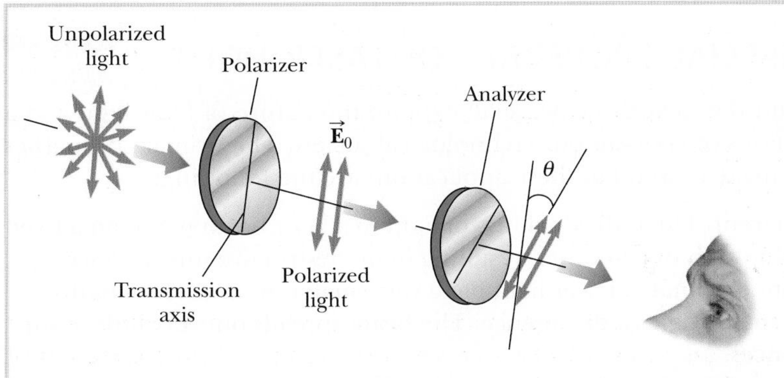

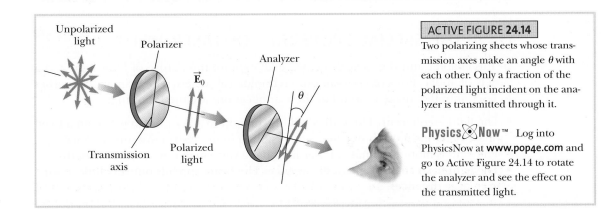

ACTIVE FIGURE 24.14 Two polarizing sheets whose transmission axes make an angle θ with each other. Only a fraction of the polarized light incident on the analyzer is transmitted through it.

PhysicsNow™ Log into PhysicsNow at **www.pop4e.com** and go to Active Figure 24.14 to rotate the analyzer and see the effect on the transmitted light.

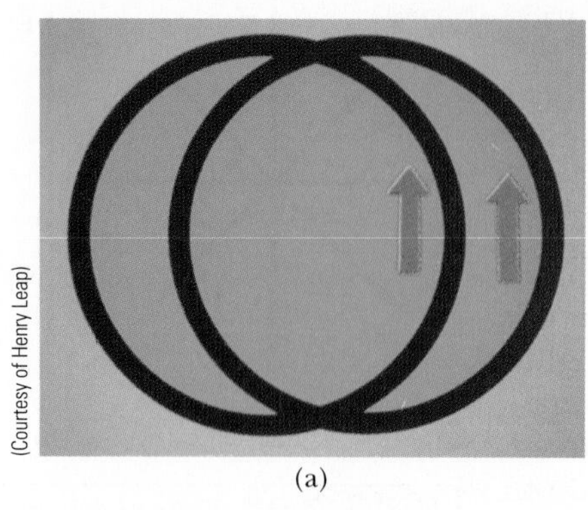
(a)

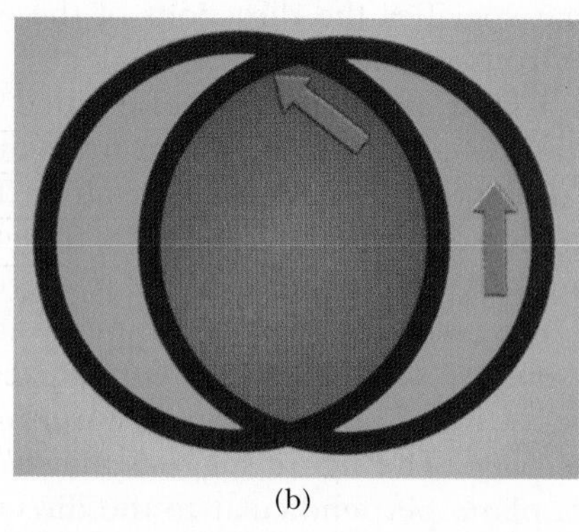
(b)

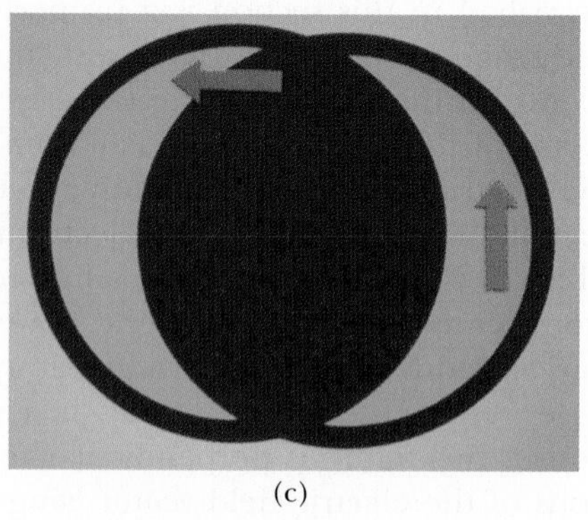
(c)

(Courtesy of Henry Leap)

FIGURE 24.15 The intensity of light transmitted through two polarizers depends on the relative orientation of their transmission axes. (a) The transmitted light has maximum intensity when the transmission axes are aligned with each other. (b) The transmitted light intensity diminishes when the transmission axes are at an angle of 45° with each other. (c) The transmitted light intensity is a minimum when the transmission axes are perpendicular to each other.

on the first polarizing sheet, called the **polarizer,** where the transmission axis is as indicated. The light that passes through this sheet is polarized vertically, and the transmitted electric field vector is $\vec{\mathbf{E}}_0$. A second polarizing sheet, called the **analyzer,** intercepts this beam with its transmission axis at an angle of θ to the axis of the polarizer. The component of $\vec{\mathbf{E}}_0$ that is perpendicular to the axis of the analyzer is completely absorbed, and the component parallel to that axis is $E_0 \cos \theta$. We know from Equation 24.27 that the transmitted intensity varies as the *square* of the transmitted amplitude, so we conclude that the intensity of the transmitted (polarized) light varies as

■ Malus's law

$$I = I_0 \cos^2 \theta \qquad [24.34]$$

where I_0 is the intensity of the polarized wave incident on the analyzer. This expression, known as **Malus's law,** applies to any two polarizing materials whose transmission axes are at an angle of θ to each other. From this expression, note that the transmitted intensity is a maximum when the transmission axes are parallel ($\theta = 0$ or 180°) and zero (complete absorption by the analyzer) when the transmission axes are perpendicular to each other. This variation in transmitted intensity through a pair of polarizing sheets is illustrated in Figure 24.15. Because the average value of $\cos^2 \theta$ is $\frac{1}{2}$, the intensity of initially unpolarized light is reduced by a factor of one half as the light passes through a single ideal polarizer.

QUICK QUIZ 24.7 A polarizer for microwaves can be made as a grid of parallel metal wires about a centimeter apart. Is the electric field vector for microwaves transmitted through this polarizer **(a)** parallel or **(b)** perpendicular to the metal wires?

24.9 THE SPECIAL PROPERTIES OF LASER LIGHT

In this chapter and the next three, we shall explore the nature of laser light and a variety of applications of lasers in our technological society. The primary properties of laser light that make it useful in these applications are the following:

- The light is coherent. The individual rays of light in a laser beam maintain a fixed phase relationship with one another, resulting in no destructive interference.
- The light is monochromatic. Laser light has a very small range of wavelengths.
- The light has a small angle of divergence. The beam spreads out very little, even over large distances.

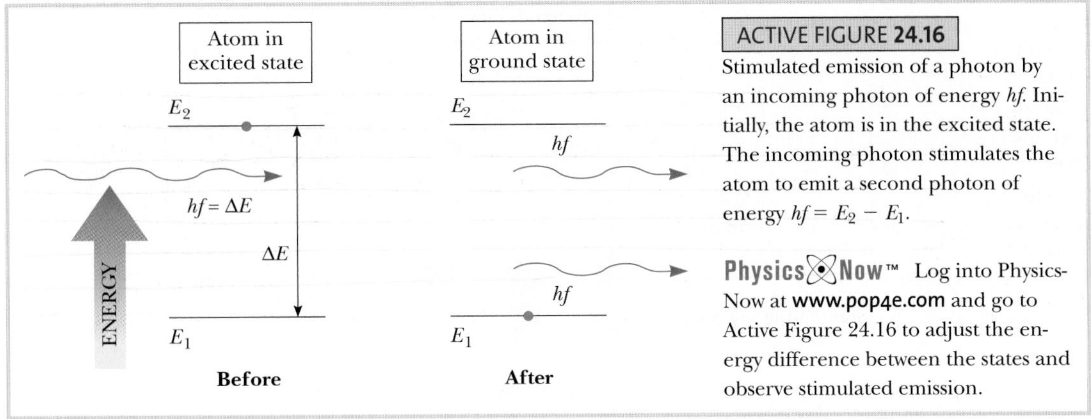

ACTIVE FIGURE 24.16 Stimulated emission of a photon by an incoming photon of energy hf. Initially, the atom is in the excited state. The incoming photon stimulates the atom to emit a second photon of energy $hf = E_2 - E_1$.

Physics Now™ Log into PhysicsNow at **www.pop4e.com** and go to Active Figure 24.16 to adjust the energy difference between the states and observe stimulated emission.

To understand the origin of these properties, let us combine our knowledge of atomic energy levels from Chapter 11 with some special requirements for the atoms that emit laser light.

As we found in Chapter 11, the energies of an atom are quantized. We used a semigraphical representation called an *energy level diagram* in that chapter to help us understand the quantized energies in an atom. The production of laser light depends heavily on the properties of these energy levels in the atoms, the source of the laser light.

The word *laser* is an acronym for **l**ight **a**mplification by **s**timulated **e**mission of **r**adiation. The full name indicates one of the requirements for laser light, that the process of **stimulated emission** must occur to achieve laser action.

Suppose an atom is in the excited state E_2 as in Active Figure 24.16 and a photon with energy $hf = E_2 - E_1$ is incident on it. The incoming photon can stimulate the excited atom to return to the ground state and thereby emit a second photon having the same energy hf and traveling in the same direction. Note that the incident photon is not absorbed, so after the stimulated emission, two identical photons exist: the incident photon and the emitted photon. The emitted photon is in phase with the incident photon. These photons can stimulate other atoms to emit photons in a chain of similar processes. The many photons produced in this fashion are the source of the intense, coherent light in a laser.

For the stimulated emission to result in laser light, we must have a buildup of photons in the system. The following three conditions must be satisfied to achieve this buildup:

- The system must be in a state of **population inversion.** More atoms must be in an excited state than in the ground state. Atoms in the ground state can absorb photons, raising them to the excited state. The population inversion assures that we have more emission of photons from excited atoms than absorption by atoms in the ground state.
- The excited state of the system must be a *metastable state,* which means that its lifetime must be long compared with the usually short lifetime of excited states, which is typically 10^{-8} s. In this case, stimulated emission is likely to occur before spontaneous emission. The energy of a metastable state is indicated with an asterisk, E^*.
- The emitted photons must be confined in the system long enough to enable them to stimulate further emission from other excited atoms, which is achieved by using reflecting mirrors at the ends of the system. One end is made totally reflecting, and the other is slightly transparent to allow the laser beam to escape (Fig. 24.17).

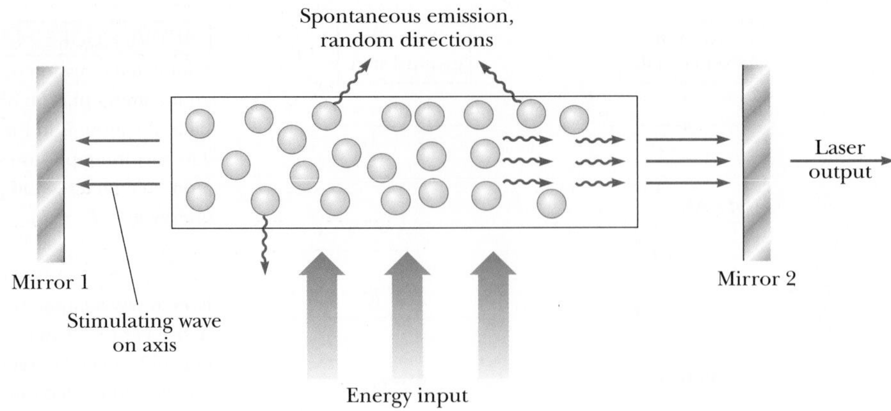

FIGURE 24.17 A schematic of a laser design. The tube contains atoms, which represent the active medium. An external source of energy (optical, electric, etc.) is needed to "pump" the atoms to excited energy states. The parallel end mirrors confine the photons to the tube. Mirror 2 is slightly transparent so that laser light leaves the tube through this mirror.

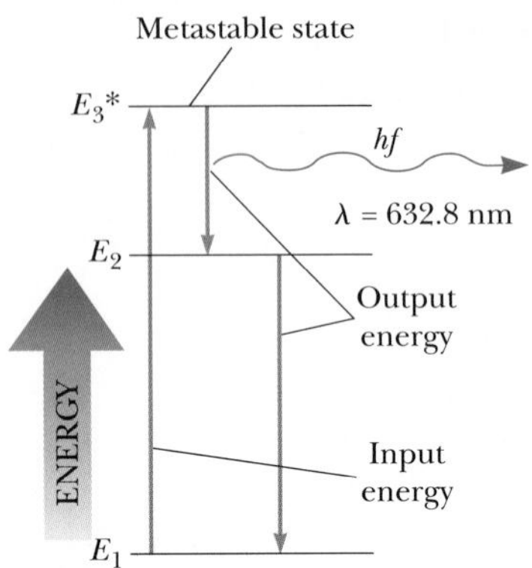

FIGURE 24.18 Energy level diagram for a neon atom in a helium–neon laser. The atom emits 632.8-nm photons through stimulated emission in the transition $E_3^* \rightarrow E_2$, which is the source of coherent light in the laser.

One device that exhibits stimulated emission of radiation is the helium–neon gas laser. Figure 24.18 is an energy level diagram for the neon atom in this system. The mixture of helium and neon is confined to a glass tube that is sealed at the ends by mirrors. A voltage applied across the tube causes electrons to sweep through the tube, colliding with the atoms of the gases and raising them into excited states. Neon atoms are excited to state E_3^* through this process and also as a result of collisions with excited helium atoms. Stimulated emission occurs as the neon atoms make a transition to state E_2 and neighboring excited atoms are stimulated. The result is the production of coherent light at a wavelength of 632.8 nm.

An exciting area of research and technological applications began in the 1990s with the development of *laser trapping* of atoms (Fig. 24.19). One scheme, called *optical molasses* and developed by Steven Chu of Stanford University and his colleagues, involves focusing six laser beams onto a small region in which atoms are to be trapped. Each pair of lasers is along one of the x, y, and z axes and emits light in opposite directions (Fig. 24.20). The frequency of the laser light is tuned to be just below the absorption frequency of the subject atom. Imagine that an atom has been placed into the trap region and moves along the positive x axis toward the laser that is emitting light toward it (the right-hand laser on the x axis in Fig. 24.20). Because the atom is moving, the light from the laser appears Doppler shifted upward in frequency in the reference frame of the atom. This shift creates a match between the Doppler-shifted laser frequency and the absorption frequency of the atom, and the atom absorbs photons.[6] The momentum carried by these photons results in the atom being pushed back to the center of the trap. By incorporating six lasers, the atoms are pushed back into the trap regardless of which way they move along any axis.

In 1986, Chu developed *optical tweezers* in which a single tightly focused laser beam can be used to trap and manipulate small particles. In combination with microscopes, optical tweezers have opened up many new possibilities for biologists. Optical tweezers have been used to manipulate live bacteria without damage, move chromosomes within a cell nucleus, and measure the elastic properties of a single DNA molecule. Chu shared the 1997 Nobel Prize in Physics with Claude Cohen-Tannoudji (Collège de France) and William Phillips (National Institute of Standards and Technology) for the development of the techniques of optical trapping.

An extension of laser trapping, *laser cooling*, is due to the reduction of the normal high speeds of the atoms when they are restricted to the region of the trap. As a result, the temperature of the collection of atoms can be reduced to a few

(Courtesy of Mark Helfer/National Institute of Standards and Technology)

FIGURE 24.19 A staff member of the National Institute of Standards and Technology views a sample of trapped sodium atoms (the small yellow dot in the center of the vacuum chamber) cooled to a temperature of less than 1 mK.

[6]The laser light traveling in the same direction as the atom (from the left-hand laser on the x axis in Fig. 24.20) is Doppler shifted further downward in frequency, so no absorption occurs. Therefore, the atom is not pushed out of the trap by the diametrically opposed laser.

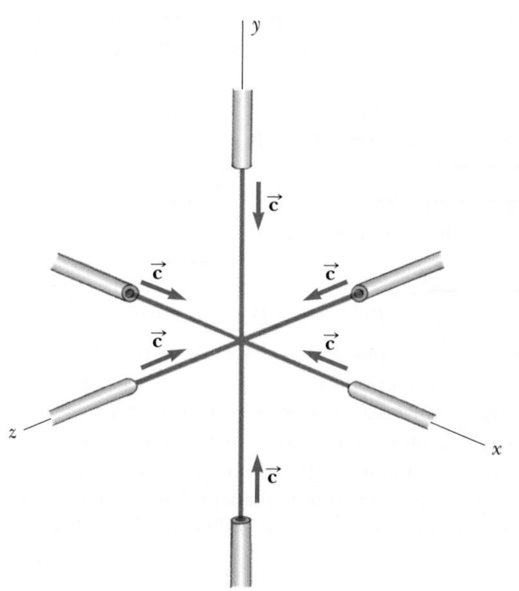

FIGURE 24.20 An optical trap for atoms is formed at the intersection point of six counterpropagating laser beams along mutually perpendicular axes. The frequency of the laser light is tuned to be just below that for absorption by the trapped atoms. If an atom moves away from the trap, it absorbs the Doppler-shifted laser light and the momentum of the light pushes the atom back into the trap.

nanokelvins. This laser cooling allows scientists to study the behavior of atoms at extremely low temperatures.

In the 1920s, Satyendra Nath Bose (1894–1974) was studying photons and investigating collections of identical photons, which can all be in the same quantum state. Einstein followed up on the work of Bose and predicted that a collection of atoms could all be in the same quantum state if the temperature were low enough. The proposed collection of atoms is called a *Bose–Einstein condensate.* In 1995, using laser cooling supplemented with evaporative cooling, the first Bose–Einstein condensate was created in the laboratory by Eric Cornell and Carl Wieman, who won the 2001 Nobel Prize in Physics for their work. Many laboratories are now creating Bose–Einstein condensates and studying their properties and possible applications. One interesting result was reported by a Harvard University group led by Lene Vestergaard Hau in 2001. She and her colleagues announced that they were able to bring a light pulse to a complete stop by using a Bose–Einstein condensate.[7]

We have explored general properties of laser light in this chapter. In the Context Connection of Chapter 25, we shall explore the technology of optical fibers, in which lasers are used in a variety of applications. ■

SUMMARY

Physics Now™ Take a practice test by logging into PhysicsNow at www.pop4e.com and clicking on the Pre-Test link for this chapter.

Displacement current I_d is defined as

$$I_d \equiv \epsilon_0 \frac{d\Phi_E}{dt} \qquad [24.1]$$

and represents an effective current through a region of space in which an electric field is changing in time.

When used with the Lorentz force law ($\vec{\mathbf{F}} = q\vec{\mathbf{E}} + q\vec{\mathbf{v}} \times \vec{\mathbf{B}}$), **Maxwell's equations** describe *all* electromagnetic phenomena:

$$\oint \vec{\mathbf{E}} \cdot d\vec{\mathbf{A}} = \frac{q}{\epsilon_0} \qquad [24.4]$$

$$\oint \vec{\mathbf{B}} \cdot d\vec{\mathbf{A}} = 0 \qquad [24.5]$$

$$\oint \vec{\mathbf{E}} \cdot d\vec{\mathbf{s}} = -\frac{d\Phi_B}{dt} \qquad [24.6]$$

$$\oint \vec{\mathbf{B}} \cdot d\vec{\mathbf{s}} = \mu_0 I + \epsilon_0 \mu_0 \frac{d\Phi_E}{dt} \qquad [24.7]$$

Electromagnetic waves, which are predicted by Maxwell's equations, have the following properties:

[7]C. Liu, Z. Dutton, C. H. Behroozi, and L. V. Hau, "Observation of coherent optical information storage in an atomic medium using halted light pulses," *Nature*, 409, 490–493, January 25, 2001.

- The electric and magnetic fields satisfy the following wave equations, which can be obtained from Maxwell's third and fourth equations:

$$\frac{\partial^2 E}{\partial x^2} = \epsilon_0 \mu_0 \frac{\partial^2 E}{\partial t^2} \qquad [24.15]$$

$$\frac{\partial^2 B}{\partial x^2} = \epsilon_0 \mu_0 \frac{\partial^2 B}{\partial t^2} \qquad [24.16]$$

- Electromagnetic waves travel through a vacuum with the speed of light $c = 3.00 \times 10^8$ m/s, where

$$c = \frac{1}{\sqrt{\epsilon_0 \mu_0}} \qquad [24.17]$$

- The electric and magnetic fields of an electromagnetic wave are perpendicular to each other and perpendicular to the direction of wave propagation; hence, electromagnetic waves are transverse waves. The electric and magnetic fields of a sinusoidal plane electromagnetic wave propagating in the positive x direction can be written

$$E = E_{\text{max}} \cos(kx - \omega t) \qquad [24.18]$$

$$B = B_{\text{max}} \cos(kx - \omega t) \qquad [24.19]$$

where ω is the angular frequency of the wave and k is the angular wave number. These equations represent special solutions to the wave equations for $\vec{\mathbf{E}}$ and $\vec{\mathbf{B}}$.
- The instantaneous magnitudes of $\vec{\mathbf{E}}$ and $\vec{\mathbf{B}}$ in an electromagnetic wave are related by the expression

$$\frac{E}{B} = c \qquad [24.21]$$

- Electromagnetic waves carry energy. The rate of flow of energy crossing a unit area is described by the **Poynting vector** $\vec{\mathbf{S}}$, where

$$\vec{\mathbf{S}} = \frac{1}{\mu_0} \vec{\mathbf{E}} \times \vec{\mathbf{B}} \qquad [24.25]$$

The average value of the Poynting vector for a plane electromagnetic wave has the magnitude

$$I = S_{\text{avg}} = \frac{E_{\text{max}} B_{\text{max}}}{2\mu_0} = \frac{E_{\text{max}}^2}{2\mu_0 c} = \frac{c B_{\text{max}}^2}{2\mu_0} \qquad [24.27]$$

The average power per unit area (intensity) of a sinusoidal plane electromagnetic wave equals the average value of the Poynting vector taken over one or more cycles.
- Electromagnetic waves carry momentum and hence can exert pressure on surfaces. If an electromagnetic wave whose intensity is I is completely absorbed by a surface on which it is normally incident, the radiation pressure on that surface is

$$P = \frac{S}{c} \quad \text{(complete absorption)} \qquad [24.33]$$

If the surface totally reflects a normally incident wave, the pressure is doubled.

The **electromagnetic spectrum** includes waves covering a broad range of frequencies and wavelengths.

When polarized light of intensity I_0 is incident on a polarizing film, the light transmitted through the film has an intensity equal to $I_0 \cos^2 \theta$, where θ is the angle between the transmission axis of the polarizing film and the electric field vector of the incident light.

QUESTIONS

☐ = answer available in the *Student Solutions Manual and Study Guide*

1. Radio station announcers often advertise "instant news." If they mean that you can hear the news the instant they speak it, is their claim true? About how long would it take for a message to travel across this country by radio waves, assuming that the waves could be detected at this range?

2. When light (or other electromagnetic radiation) travels across a given region, what is it that oscillates? What is it that is transported?

3. What is the fundamental source of electromagnetic radiation?

4. Does a wire connected to the terminals of a battery emit electromagnetic waves? Explain.

5. If you charge a comb by running it through your hair and then hold the comb next to a bar magnet, do the electric and magnetic fields produced constitute an electromagnetic wave?

6. List as many similarities and differences between sound waves and light waves as you can.

7. In the LC circuit shown in Figure 24.7, the charge on the capacitor is sometimes zero, but at such instants the current in the circuit is not zero. How is that possible?

8. Describe the physical significance of the Poynting vector.

9. Before the advent of cable television and satellite dishes, city dwellers often used "rabbit ears" atop their sets (Fig. Q24.9). Certain orientations of the receiving antenna on a television set give better reception than others. Furthermore, the best orientation varies from station to station. Explain.

10. Often when you touch the indoor antenna on a radio or television receiver, the reception instantly improves. Why?

11. What does a radio wave do to the charges in the receiving antenna to provide a signal for your car radio?

12. An empty plastic or glass dish being removed from a microwave oven is cool to the touch. How can that be possible? (Assume that your electric bill has been paid.)

FIGURE **Q24.9** Question 24.9 and Problem 24.57.

13. Suppose a creature from another planet had eyes that were sensitive to infrared radiation. Describe what the alien would see if it looked around the room you are now in. In particular, what would be bright and what would be dim?

14. Why should an infrared photograph of a person look different from a photograph taken with visible light?

15. A welder must wear protective glasses and clothing to prevent eye damage and sunburn. What does this practice imply about the nature of the light produced by the welding?

16. A home microwave oven uses electromagnetic waves with a wavelength of about 12.2 cm. Some 2.4-GHz cordless telephones suffer noisy interference when a microwave oven is used nearby. Locate the waves used by both devices on the electromagnetic spectrum. Do you expect them to interfere with each other?

17. Why is stimulated emission so important in the operation of a laser?

18. For a given incident energy of an electromagnetic wave, why is the radiation pressure on a perfectly reflecting surface twice as great as that on a perfect absorbing surface?

PROBLEMS

1, 2, 3 = straightforward, intermediate, challenging
☐ = full solution available in the *Student Solutions Manual and Study Guide*
Physics Now™ = coached problem with hints available at www.pop4e.com
= computer useful in solving problem
= paired numerical and symbolic problems
= biomedical application

Section 24.1 ■ Displacement Current and the Generalized Ampère's Law

1. Consider the situation shown in Figure P24.1. An electric field of 300 V/m is confined to a circular area 10.0 cm in diameter and directed outward perpendicular to the plane of the figure. If the field is increasing at a rate of 20.0 V/m · s, what are the direction and magnitude of the magnetic field at the point P, 15.0 cm from the center of the circle?

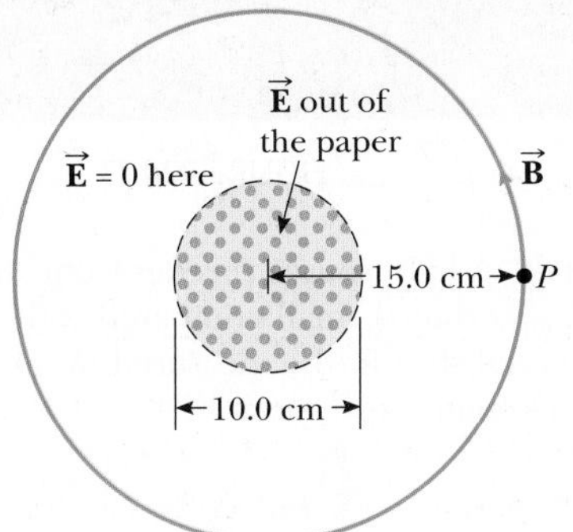

FIGURE **P24.1**

Section 24.2 ■ Maxwell's Equations

2. A very long, thin rod carries electric charge with the linear density 35.0 nC/m. It lies along the x axis and moves in the x direction at a speed of 15.0 Mm/s. (a) Find the electric field the rod creates at the point (0, 20.0 cm, 0). (b) Find the magnetic field it creates at the same point. (c) Find the force exerted on an electron at this point, moving with a velocity of $(240\hat{\mathbf{i}})$ Mm/s.

3. A proton moves through a uniform electric field $\vec{\mathbf{E}} = 50\hat{\mathbf{j}}$ V/m and a uniform magnetic field $\vec{\mathbf{B}} = (0.200\hat{\mathbf{i}} + 0.300\hat{\mathbf{j}} + 0.400\hat{\mathbf{k}})$ T. Determine the acceleration of the proton when it has a velocity $\vec{\mathbf{v}} = 200\hat{\mathbf{i}}$ m/s.

Section 24.3 ■ Electromagnetic Waves

Note: Assume that the medium is vacuum unless specified otherwise.

4. (a) The distance to the North Star, Polaris, is approximately 6.44×10^{18} m. If Polaris were to burn out today, in what year would we see it disappear? (b) How long does it take for sunlight to reach the Earth? (c) How long does it take for a microwave radar signal to travel from the Earth

to the Moon and back? (d) How long does it take for a radio wave to travel once around the Earth in a great circle, close to the planet's surface? (e) How long does it take for light to reach you from a lightning stroke 10.0 km away?

5. The speed of an electromagnetic wave traveling in a transparent nonmagnetic substance is $v = 1/\sqrt{\kappa\mu_0\epsilon_0}$, where κ is the dielectric constant of the substance. Determine the speed of light in water, which has a dielectric constant at optical frequencies of 1.78.

6. An electromagnetic wave in vacuum has an electric field amplitude of 220 V/m. Calculate the amplitude of the corresponding magnetic field.

7. **PhysicsNow™** Figure 24.3 shows a plane electromagnetic sinusoidal wave propagating in the x direction. Suppose the wavelength is 50.0 m and the electric field vibrates in the xy plane with an amplitude of 22.0 V/m. Calculate (a) the frequency of the wave and (b) the magnitude and direction of $\vec{\mathbf{B}}$ when the electric field has its maximum value in the negative y direction. (c) Write an expression for $\vec{\mathbf{B}}$ with the correct unit vector, with numerical values for B_{max}, k, and ω, and with its magnitude in the form

$$B = B_{max}\cos(kx - \omega t)$$

8. In SI units, the electric field in an electromagnetic wave is described by

$$E_y = 100\sin(1.00 \times 10^7 x - \omega t)$$

Find (a) the amplitude of the corresponding magnetic field oscillations, (b) the wavelength λ, and (c) the frequency f.

9. Verify by substitution that the following equations are solutions to Equations 24.15 and 24.16, respectively:

$$E = E_{max}\cos(kx - \omega t)$$
$$B = B_{max}\cos(kx - \omega t)$$

10. **Review problem.** A standing wave interference pattern is set up by radio waves between two metal sheets 2.00 m apart. That is the shortest distance between the plates that will produce a standing wave pattern. What is the fundamental frequency?

11. A microwave oven is powered by an electron tube called a magnetron, which generates electromagnetic waves of frequency 2.45 GHz. The microwaves enter the oven and are reflected by the walls. The standing wave pattern produced in the oven can cook food unevenly, with hot spots in the food at antinodes and cool spots at nodes, so a turntable is often used to rotate the food and distribute the energy. If a microwave oven intended for use with a turntable is instead used with a cooking dish in a fixed position, the antinodes can appear as burn marks on foods such as carrot strips or cheese. The separation distance between the burns is measured to be 6 cm ± 5%. From these data, calculate the speed of the microwaves.

12. **Review problem.** An alien civilization occupies a brown dwarf, nearly stationary relative to the Sun, several lightyears away. The extraterrestrials have come to love original broadcasts of *I Love Lucy*, on our television channel 2, at carrier frequency 57.0 MHz. Their line of sight to us is in the plane of the Earth's orbit. Find the difference between the highest and lowest frequencies they receive due to the Earth's orbital motion around the Sun.

13. Police radar detects the speed of a car (Fig. P24.13) as follows. Microwaves of a precisely known frequency are broadcast toward the car. The moving car reflects the microwaves with a Doppler shift. The reflected waves are received and combined with an attenuated version of the transmitted wave. Beats occur between the two microwave signals. The beat frequency is measured. (a) For an electromagnetic wave reflected back to its source from a mirror approaching at speed v, show that the reflected wave has frequency

$$f = f_{source}\frac{c + v}{c - v}$$

where f_{source} is the source frequency. (b) When v is much less than c, the beat frequency is much smaller than the transmitted frequency. In this case, use the approximation $f + f_{source} \approx 2f_{source}$ and show that the beat frequency can be written as $f_{beat} = 2v/\lambda$. (c) What beat frequency is measured for a car speed of 30.0 m/s if the microwaves have frequency 10.0 GHz? (d) If the beat frequency measurement is accurate to ± 5 Hz, how accurate is the speed measurement?

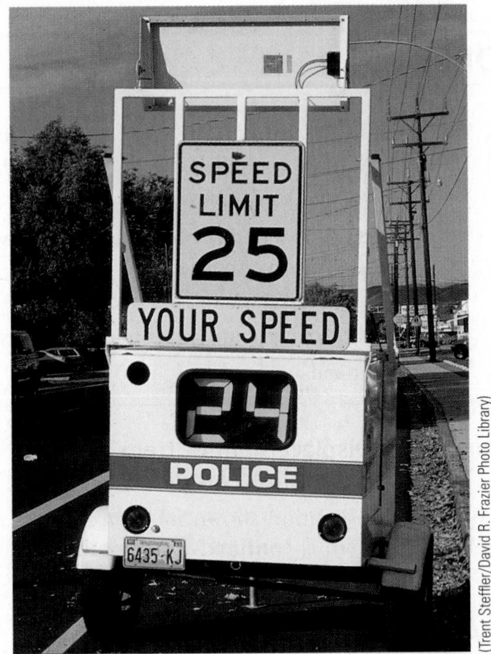

(Trent Steffler/David R. Frazier Photo Library)

FIGURE **P24.13**

14. *The red shift.* A light source recedes from an observer with a speed v_{source} that is small compared with c. (a) Show that the fractional shift in the measured wavelength is given by the approximate expression

$$\frac{\Delta\lambda}{\lambda} \approx \frac{v_{source}}{c}$$

This phenomenon is known as the red shift because the visible light is shifted toward the red. (b) Spectroscopic measurements of light at $\lambda = 397$ nm coming from a galaxy in Ursa Major reveal a red shift of 20.0 nm. What is the recessional speed of the galaxy?

15. A physicist drives through a stop light. When he is pulled over, he tells the police officer that the Doppler shift made the red light of wavelength 650 nm appear green to him, with a wavelength of 520 nm. The police officer writes out a traffic citation for speeding. How fast was the physicist traveling, according to his own testimony?

16. A Doppler weather radar station broadcasts a pulse of radio waves at frequency 2.85 GHz. From a relatively small batch of raindrops at bearing 38.6° east of north, the station receives a reflected pulse after 180 μs with a frequency shifted upward by 254 Hz. From a similar batch of raindrops at bearing 39.6° east of north, the station receives a reflected pulse after the same time delay, with a frequency shifted downward by 254 Hz. These pulses have the highest and lowest frequencies the station receives. (a) Calculate the radial velocity components of both batches of raindrops. (b) Assume that these raindrops are swirling in a uniformly rotating vortex. Find the angular speed of their rotation.

Section 24.4 ■ Hertz's Discoveries

17. A fixed inductance $L = 1.05\ \mu$H is used in series with a variable capacitor in the tuning section of a radiotelephone on a ship. What capacitance tunes the circuit to the signal from a transmitter broadcasting at 6.30 MHz?

18. Calculate the inductance of an LC circuit that oscillates at 120 Hz when the capacitance is 8.00 μF.

19. The switch in Figure P24.19 is connected to point a for a long time. After the switch is thrown to point b, what are (a) the frequency of oscillation of the LC circuit, (b) the maximum charge that appears on the capacitor, (c) the maximum current in the inductor, and (d) the total energy the circuit possesses at $t = 3.00$ s?

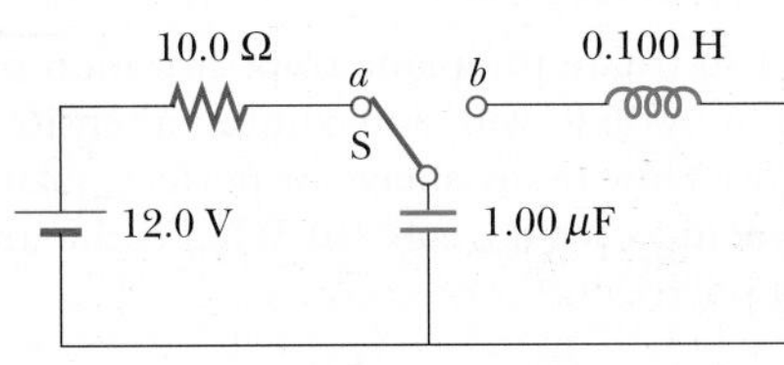

FIGURE **P24.19**

Section 24.5 ■ Energy Carried by Electromagnetic Waves

20. How much electromagnetic energy per cubic meter is contained in sunlight if the intensity of sunlight at the Earth's surface under a fairly clear sky is 1 000 W/m^2?

21. What is the average magnitude of the Poynting vector 5.00 miles from a radio transmitter broadcasting isotropically (equally in all directions) with an average power of 250 kW?

22. An AM radio station broadcasts isotropically (equally in all directions) with an average power of 4.00 kW. A dipole receiving antenna 65.0 cm long is at a location 4.00 miles from the transmitter. Compute the amplitude of the emf that is induced by this signal between the ends of the receiving antenna.

23. **PhysicsNow™** A community plans to build a facility to convert solar radiation to electrical power. The community requires 1.00 MW of power, and the system to be installed has an efficiency of 30.0% (that is, 30.0% of the solar energy incident on the surface is converted to useful energy that can power the community). What must be the effective area of a perfectly absorbing surface used in such an installation if sunlight has a constant intensity of 1 000 W/m^2?

24. One of the weapons considered for the "Star Wars" antimissile system is a laser that could destroy ballistic missiles. When a high-power laser is used in the Earth's atmosphere, the electric field can ionize the air, turning it into a conducting plasma that reflects the laser light. In dry air at 0°C and 1 atm, electric breakdown occurs for fields with amplitudes above about 3.00 MV/m. (a) What laser beam intensity will produce such a field? (b) At this maximum intensity, what power can be delivered in a cylindrical beam of diameter 5.00 mm?

25. **PhysicsNow™** The filament of an incandescent lamp has a 150-Ω resistance and carries a direct current of 1.00 A. The filament is 8.00 cm long and 0.900 mm in radius. (a) Calculate the Poynting vector at the surface of the filament, associated with the static electric field producing the current and the current's static magnetic field. (b) Find the magnitude of the static electric and magnetic fields at the surface of the filament.

26. In a region of free space, the electric field at an instant of time is $\vec{\mathbf{E}} = (80.0\hat{\mathbf{i}} + 32.0\hat{\mathbf{j}} - 64.0\hat{\mathbf{k}})$ N/C and the magnetic field is $\vec{\mathbf{B}} = (0.200\hat{\mathbf{i}} + 0.0800\hat{\mathbf{j}} + 0.290\hat{\mathbf{k}})\ \mu$T. (a) Show that the two fields are perpendicular to each other. (b) Determine the Poynting vector for these fields.

27. At what distance from a 100-W electromagnetic wave point source does $E_{max} = 15.0$ V/m?

28. Consider a bright star in our night sky. Assume that its power output is 4.00×10^{28} W, about 100 times that of the Sun, and that its distance is 20.0 ly. (a) Find the intensity of the starlight at the Earth. (b) Find the power of the starlight that the Earth intercepts.

Section 24.6 ■ Momentum and Radiation Pressure

29. A 15.0-mW helium-neon laser ($\lambda = 632.8$ nm) emits a beam of circular cross section with a diameter of 2.00 mm. (a) Find the maximum electric field in the beam. (b) What total energy is contained in a 1.00-m length of the beam? (c) Find the momentum carried by a 1.00-m length of the beam.

30. A possible means of space flight is to place a perfectly reflecting aluminized sheet into orbit around the Earth and then use the light from the Sun to push this "solar sail." Suppose a sail of area 6.00×10^5 m^2 and mass 6 000 kg is placed in orbit facing the Sun. (a) What force is exerted on the sail? (b) What is the sail's acceleration? (c) How long does it take the sail to reach the Moon, 3.84×10^8 m away? Ignore all gravitational effects, assume that the acceleration calculated in part (b) remains constant, and assume a solar intensity of 1 370 W/m^2.

Section 24.7 ■ The Spectrum of Electromagnetic Waves

31. *This just in!* An important news announcement is transmitted by radio waves to people sitting next to their radios 100 km

from the station and by sound waves to people sitting across the newsroom 3.00 m from the newscaster. Who receives the news first? Explain. Take the speed of sound in air to be 343 m/s.

32. Classify waves with frequencies of 2 Hz, 2 kHz, 2 MHz, 2 GHz, 2 THz, 2 PHz, 2 EHz, 2 ZHz, and 2 YHz on the electromagnetic spectrum. Classify waves with wavelengths of 2 km, 2 m, 2 mm, 2 μm, 2 nm, 2 pm, 2 fm, and 2 am.

33. The human eye is most sensitive to light having a wavelength of 5.50×10^{-7} m, which is in the green–yellow region of the visible electromagnetic spectrum. What is the frequency of this light?

34. Compute an order-of-magnitude estimate for the frequency of an electromagnetic wave with wavelength equal to (a) your height and (b) the thickness of this sheet of paper. How is each wave classified on the electromagnetic spectrum?

35. What are the wavelengths of electromagnetic waves in free space that have frequencies of (a) 5.00×10^{19} Hz and (b) 4.00×10^{9} Hz?

36. A diathermy machine, used in physiotherapy, generates electromagnetic radiation that gives the effect of "deep heat" when absorbed in tissue. One assigned frequency for diathermy is 27.33 MHz. What is the wavelength of this radiation?

37. **Review problem.** Accelerating charges radiate electromagnetic waves. Calculate the wavelength of radiation produced by a proton in a cyclotron with a radius of 0.500 m and magnetic field of 0.350 T.

38. Twelve VHF television channels (Channels 2 through 13) lie in the range of frequencies between 54.0 MHz and 216 MHz. Each channel is assigned a width of 6.0 MHz, with the two ranges 72.0–76.0 MHz and 88.0–174 MHz reserved for non-TV purposes. (Channel 2, for example, lies between 54.0 and 60.0 MHz.) Calculate the broadcast wavelength range for (a) Channel 4, (b) Channel 6, and (c) Channel 8.

39. Suppose you are located 180 m from a radio transmitter. (a) How many wavelengths are you from the transmitter if the station calls itself 1 150 AM? (The AM band frequencies are in kilohertz.) (b) What if this station is 98.1 FM? (The FM band frequencies are in megahertz.)

40. A radar pulse returns to the transmitter-receiver after a total travel time of 4.00×10^{-4} s. How far away is the object that reflected the wave?

Section 24.8 ■ Polarization

41. Plane-polarized light is incident on a single polarizing disk with the direction of $\vec{\mathbf{E}}_0$ parallel to the direction of the transmission axis. Through what angle should the disk be rotated so that the intensity in the transmitted beam is reduced by a factor of (a) 3.00, (b) 5.00, and (c) 10.0?

42. Unpolarized light passes through two Polaroid sheets. The axis of the first is vertical and that of the second is at 30.0° to the vertical. What fraction of the incident light is transmitted?

43. In Figure P24.43, suppose the transmission axes of the left and right polarizing disks are perpendicular to each other. In addition, let the center disk be rotated on the common axis with an angular speed ω. Show that if unpolarized light is incident on the left disk with an intensity $I_{\max}$, the intensity of the beam emerging from the right disk is

$$I = \frac{1}{16} I_{\max}(1 - \cos 4\omega t)$$

Hence, the intensity of the emerging beam is modulated at a rate that is four times the rate of rotation of the center disk. [*Suggestion:* Use the trigonometric identities $\cos^2 \theta = (1 + \cos 2\theta)/2$ and $\sin^2 \theta = (1 - \cos 2\theta)/2$, and recall that $\theta = \omega t$.]

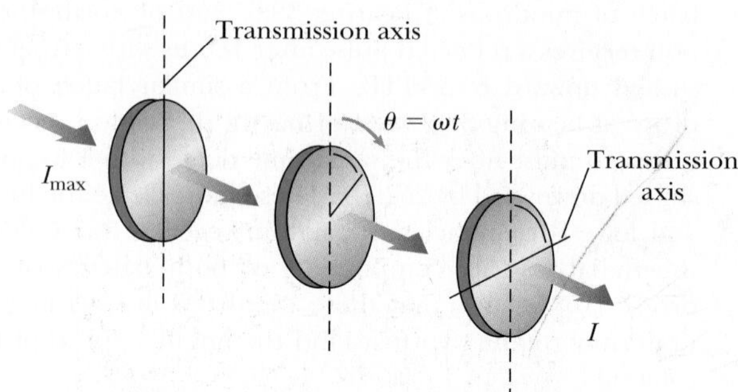

FIGURE **P24.43**

44. Two handheld radio transceivers with dipole antennas are separated by a large fixed distance. If the transmitting antenna is vertical, what fraction of the maximum received power will appear in the receiving antenna when it is inclined from the vertical by (a) 15.0°, (b) 45.0°, and (c) 90.0°?

45. Two polarizing sheets are placed together with their transmission axes crossed so that no light is transmitted. A third sheet is inserted between them with its transmission axis at an angle of 45.0° with respect to each of the other axes. Find the fraction of incident unpolarized light intensity transmitted by the three-sheet combination. (Assume each polarizing sheet is ideal.)

46. You want to rotate the plane of polarization of a polarized light beam by 45.0° with a maximum intensity reduction of 10.0%. (a) How many sheets of perfect polarizers do you need to achieve your goal? (b) What is the angle between adjacent polarizers?

Section 24.9 ■ Context Connection—The Special Properties of Laser Light

47. Figure P24.47 shows portions of the energy level diagrams of the helium and neon atoms in a helium–neon laser. An

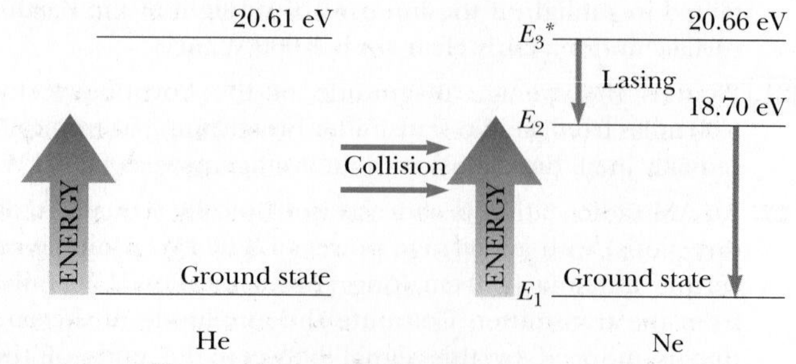

FIGURE **P24.47**

electrical discharge excites the He atom from its ground state to its excited state of 20.61 eV. The excited He atom collides with a Ne atom in its ground state and excites this atom to the state at 20.66 eV. Lasing action takes place for electron transitions from $E_3{}^*$ to E_2 in the Ne atoms. From the data in the figure, show that the wavelength of the red He–Ne laser light is approximately 633 nm.

48. High-power lasers in factories are used to cut through cloth and metal (Fig. P24.48). One such laser has a beam diameter of 1.00 mm and generates an electric field having an amplitude of 0.700 MV/m at the target. Find (a) the amplitude of the magnetic field produced, (b) the intensity of the laser, and (c) the power delivered by the laser.

(Philippe Plailly/SPL/Photo Researchers)

FIGURE **P24.48** A laser cutting device mounted on a robot arm is being used to cut through a metallic plate.

49. A neodymium–yttrium–aluminum garnet laser used in eye surgery emits a 3.00-mJ pulse in 1.00 ns, focused to a spot 30.0 μm in diameter on the retina. (a) Find (in SI units) the power per unit area at the retina. (This quantity is called the *irradiance* in the optics industry.) (b) What energy is delivered to an area of molecular size, taken as a circular area 0.600 nm in diameter?

50. The carbon dioxide laser is one of the most powerful developed. The energy difference between the two laser levels is 0.117 eV. Determine the frequency and wavelength of the radiation emitted by this laser. In what portion of the electromagnetic spectrum is this radiation?

51. PhysicsNow™ A ruby laser delivers a 10.0-ns pulse of 1.00 MW average power. If the photons have a wavelength of 694.3 nm, how many are contained in the pulse?

52. A pulsed ruby laser emits light at 694.3 nm. For a 14.0-ps pulse containing 3.00 J of energy, find (a) the physical length of the pulse as it travels through space and (b) the number of photons in it. (c) Assuming that the beam has a circular cross-section of 0.600 cm diameter, find the number of photons per cubic millimeter.

53. **Review problem.** Figure 24.17 represents the light bouncing between two mirrors in a laser cavity as two traveling waves. These traveling waves moving in opposite directions constitute a standing wave. If the reflecting surfaces are metallic films, the electric field has nodes at both ends. The electromagnetic standing wave is analogous to the standing string wave represented in Figure 14.9. (a) Assume that a helium–neon laser has precisely flat and parallel mirrors 35.124 103 cm apart. Assume that the active medium can efficiently amplify only light with wavelengths between 632.808 40 nm and 632.809 80 nm. Find the number of components that constitute the laser light, and the wavelength of each component, precise to eight digits. (b) Find the root-mean-square speed for a neon atom at 120°C. (c) Show that at this temperature the Doppler effect for light emission by moving neon atoms should realistically make the bandwidth of the light amplifier larger than the 0.001 40 nm assumed in part (a).

54. The number N of atoms in a particular state is called the population of that state. This number depends on the energy of that state and the temperature. In thermal equilibrium, the population of atoms in a state of energy E_n is given by a Boltzmann distribution expression

$$N = N_g e^{-(E_n - E_g)/k_B T}$$

where T is the absolute temperature and N_g is the population of the ground state, of energy E_g. For simplicity, we assume that each energy level has only one quantum state associated with it. (a) Before the power is switched on, the neon atoms in a laser are in thermal equilibrium at 27.0°C. Find the equilibrium ratio of the populations of the states $E_3{}^*$ and E_2 shown in Figure 24.18. Lasers operate by a clever artificial production of a "population inversion" between the upper and lower atomic energy states involved in the lasing transition. Thus, more atoms are in the upper excited state than in the lower one. Consider the helium–neon laser transition at 632.8 nm. Assume that 2% more atoms occur in the upper state than in the lower. (b) To demonstrate how unnatural such a situation is, find the temperature for which the Boltzmann distribution describes a 2.00% population inversion. (c) Why does such a situation not occur naturally?

Additional Problems

55. Assume that the intensity of solar radiation incident on the cloudtops of the Earth is 1 370 W/m^2. (a) Calculate the total power radiated by the Sun, taking the average Earth–Sun separation to be 1.496×10^{11} m. (b) Determine the maximum values of the electric and magnetic fields in the sunlight at the Earth's location.

56. The intensity of solar radiation at the top of the Earth's atmosphere is 1 370 W/m^2. Assuming that 60% of the incoming solar energy reaches the Earth's surface and assuming that you absorb 50% of the incident energy, make an order-of-magnitude estimate of the amount of solar energy you absorb in a 60-min sunbath.

57. PhysicsNow™ **Review problem.** In the absence of cable input or a satellite dish, a TV set can use a dipole-receiving antenna for VHF channels and a loop antenna for UHF channels (Fig. Q24.9). The UHF antenna produces an emf from the changing magnetic flux through the loop. The TV station broadcasts a signal with a frequency f, and the signal has an electric-field amplitude E_{max} and a magnetic-field amplitude B_{max} at the location of the receiving antenna. (a) Using Faraday's law, derive an expression for the amplitude of the emf that appears in a single-turn circular loop antenna with a radius r, which is small compared with the wavelength of the wave. (b) If the electric field in the signal points vertically, what orientation of the loop gives the best reception?

58. One goal of the Russian space program is to illuminate dark northern cities with sunlight reflected to the Earth from a 200-m diameter mirrored surface in orbit. Several smaller prototypes have already been constructed and put into orbit. (a) Assume that sunlight with intensity 1 370 W/m^2 falls on the mirror nearly perpendicularly and that the atmosphere of the Earth allows 74.6% of the energy of sunlight to pass though it in clear weather. What is the power received by a city when the space mirror is reflecting light to it? (b) The plan is for the reflected sunlight to cover a circle of diameter 8.00 km. What is the intensity of light (the average magnitude of the Poynting vector) received by the city? (c) This intensity is what percentage of the vertical component of sunlight at St. Petersburg in January, when the sun reaches an angle of 7.00° above the horizon at noon?

59. A dish antenna having a diameter of 20.0 m receives (at normal incidence) a radio signal from a distant source as shown in Figure P24.59. The radio signal is a continuous sinusoidal wave with amplitude $E_{max} = 0.200\ \mu$V/m. Assume that the antenna absorbs all the radiation that falls on the dish. (a) What is the amplitude of the magnetic field in this wave? (b) What is the intensity of the radiation received by this antenna? (c) What is the power received by the antenna? (d) What force do the radio waves exert on the antenna?

FIGURE **P24.59**

60. A handheld cellular telephone operates in the 860- to 900-MHz band and has a power output of 0.600 W from an antenna 10.0 cm long (Fig. P24.60). (a) Find the average magnitude of the Poynting vector 4.00 cm from the antenna, at the location of a typical person's head. Assume that the antenna emits energy with cylindrical wave fronts. (The actual radiation from antennas follows a more complicated pattern.) (b) The ANSI/IEEE C95.1-1991 maximum exposure standard is 0.57 mW/cm^2 for persons living near cellular telephone base stations, who would be continuously exposed to the radiation. Compare the answer to part (a) with this standard.

(Amos Morgan/Getty Images)

FIGURE **P24.60**

61. In 1965, Arno Penzias and Robert Wilson discovered the cosmic microwave radiation left over from the Big Bang expansion of the Universe. Suppose the energy density of this background radiation is 4.00×10^{-14} J/m^3. Determine the corresponding electric field amplitude.

62. A linearly polarized microwave of wavelength 1.50 cm is directed along the positive x axis. The electric field vector has a maximum value of 175 V/m and vibrates in the xy plane. (a) Assume that the magnetic field component of the wave can be written as $B = B_{max} \sin(kx - \omega t)$ and give values for B_{max}, k, and ω. Also determine in which plane the magnetic field vector vibrates. (b) Calculate the average magnitude of the Poynting vector for this wave. (c) What radiation pressure would this wave exert if it were directed at normal incidence onto a perfectly reflecting sheet? (d) What acceleration would be imparted to a 500-g sheet (perfectly reflecting and at normal incidence) with dimensions of 1.00 m × 0.750 m?

63. **Review problem.** A 1.00-m-diameter mirror focuses the Sun's rays onto an absorbing plate 2.00 cm in radius, which holds a can containing 1.00 L of water at 20.0°C. (a) If the solar intensity is 1.00 kW/m^2, what is the intensity on the absorbing plate? (b) What are the maximum magnitudes of the fields $\vec{\mathbf{E}}$ and $\vec{\mathbf{B}}$? (c) If 40.0% of the energy is absorbed, how long does it take to bring the water to its boiling point?

64. A microwave source produces pulses of 20.0-GHz radiation, with each pulse lasting 1.00 ns. A parabolic reflector with a face area of radius 6.00 cm is used to focus the microwaves into a parallel beam of radiation as shown in Figure P24.64. The average power during each pulse is 25.0 kW. (a) What is the wavelength of these microwaves? (b) What is the total energy contained in each pulse?

(c) Compute the average energy density inside each pulse. (d) Determine the amplitude of the electric and magnetic fields in these microwaves. (e) Assuming that this pulsed beam strikes an absorbing surface, compute the force exerted on the surface during the 1.00-ns duration of each pulse.

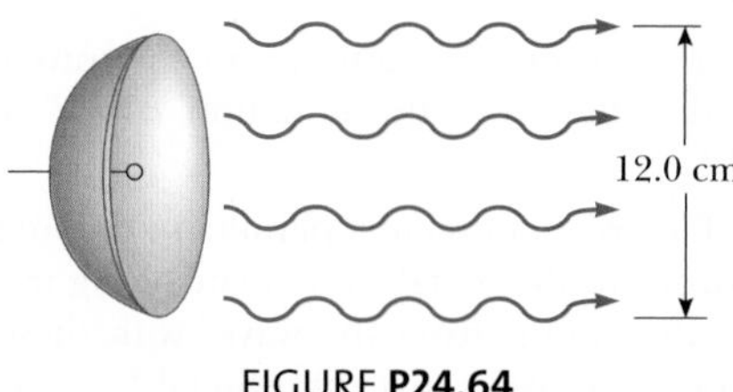

FIGURE **P24.64**

65. The electromagnetic power radiated by a nonrelativistic moving point charge q having an acceleration a is

$$\mathcal{P} = \frac{q^2 a^2}{6\pi\epsilon_0 c^3}$$

where ϵ_0 is the permittivity of free space and c is the speed of light in vacuum. (a) Show that the right side of this equation has units of watts. (b) An electron is placed in a constant electric field of magnitude 100 N/C. Determine the acceleration of the electron and the electromagnetic power radiated by this electron. (c) If a proton is placed in a cyclotron with a radius of 0.500 m and a magnetic field of magnitude 0.350 T, what electromagnetic power does this proton radiate?

Review problems. Section 17.10 discussed electromagnetic radiation as a mode of energy transfer. Problems 66 through 68 use ideas introduced both there and in this chapter.

66. Eliza is a black cat with four black kittens: Penelope, Rosalita, Sasha, and Timothy. Eliza's mass is 5.50 kg, and each kitten has mass 0.800 kg. One cool night, all five sleep snuggled together on a mat, with their bodies forming one hemisphere. (a) Assuming that the purring heap has uniform density 990 kg/m^3, find the radius of the hemisphere. (b) Find the area of its curved surface. (c) Assume that the surface temperature is uniformly 31.0°C and the emissivity is 0.970. Find the intensity of radiation emitted by the cats at their curved surface and (d) the radiated power from this surface. (e) You may think of the emitted electromagnetic wave as having a single predominant frequency (of 31.2 THz). Find the amplitude of the electric field just outside the surface of the cozy pile and (f) the amplitude of the magnetic field. (g) Are the sleeping cats charged? Are they current-carrying? Are they magnetic? Are they a radiation source? Do they glow in the dark? Give an explanation for your answers so that they do not seem contradictory. (h) The next night, the kittens all sleep alone, curling up into separate hemispheres like their mother. Find the total radiated power of the family. (For simplicity, we ignore throughout the cats' absorption of radiation from the environment.)

67. (a) An elderly couple has a solar water heater installed on the roof of their house (Fig. P24.67). The heater consists of a flat closed box with extraordinarily good thermal insulation. Its interior is painted black, and its front face is made of insulating glass. Assume that its emissivity for visible light is 0.900 and its emissivity for infrared light is 0.700. Assume that light from the noon Sun is incident perpendicular to the glass with an intensity of 1 000 W/m^2 and that no water enters or leaves the box. Find the steady-state temperature of the interior of the box. (b) The homeowners build an identical box with no water tubes. It lies flat on the ground in front of the house. They use it as a cold frame where they plant seeds in early spring. Assuming that the same noon Sun is at an elevation angle of 50.0°, find the steady-state temperature of the interior of this box when its ventilation slots are tightly closed.

FIGURE **P24.67**

68. The study of Creation suggests a Creator with an inordinate fondness for beetles and for small red stars. A small red star radiates electromagnetic waves with power 6.00×10^{23} W, which is only 0.159% of the luminosity of the Sun. Consider a spherical planet in a circular orbit around this star. Assume that the emissivity of the planet is equal for infrared and for visible light. Assume that the planet has a uniform surface temperature. Identify the projected area over which the planet absorbs starlight and the radiating area of the planet. If beetles thrive at a temperature of 310 K, what should be the radius of the planet's orbit?

69. An astronaut, stranded in space 10.0 m from her spacecraft and at rest relative to it, has a mass (including equipment) of 110 kg. Because she has a 100-W light source that forms a directed beam, she considers using the beam as a photon rocket to propel herself continuously toward the spacecraft. (a) Calculate the time interval required for her to reach the spacecraft by this method. (b) Assume, instead, that she throws the light source in the direction away from the spacecraft. The mass of the light source is 3.00 kg and, after being thrown, it moves at 12.0 m/s relative to the recoiling astronaut. After what time interval will the astronaut reach the spacecraft?

ANSWERS TO QUICK QUIZZES

24.1 (i), (b). There can be no conduction current because there is no conductor between the plates. There is a time-varying electric field because of the decreasing charge on the plates, and the time-varying electric flux represents a displacement current. (ii), (c). There is a time-varying electric field because of the decreasing charge on the plates. This time-varying electric field produces a magnetic field.

24.2 (c). The $\vec{\mathbf{B}}$ field must be in the $+z$ direction so that the Poynting vector is directed along the $-y$ direction.

24.3 (d). The first three choices are instantaneous values and vary in time. The intensity is an average value over a full cycle.

24.4 (b), (c). The radiation pressure (a) does not change because pressure is force per unit area. In (b), the smaller disk absorbs less radiation, resulting in a smaller force. For the same reason, the momentum in (c) is reduced.

24.5 (c). The order of magnitude of the wavelengths can be found either from the equation $c = \lambda f$ or from Figure 24.12.

24.6 (a). The order of magnitude of the wavelengths can be found either from the equation $c = \lambda f$ or from Figure 24.12.

24.7 (b). Electric field vectors parallel to the metal wires cause electrons in the metal to oscillate along the wires. Therefore, the energy from the waves with these electric field vectors is transferred to the metal by accelerating these electrons and is eventually transformed to internal energy through the resistance of the metal. Waves with electric field vectors perpendicular to the metal wires are not able to accelerate electrons and pass through.

CHAPTER 25

Reflection and Refraction of Light

(Patrick J. Endres/Visuals Unlimited)

This photograph of a rainbow shows a distinct secondary rainbow with the colors reversed. The appearance of the rainbow depends on three optical phenomena discussed in this chapter: reflection, refraction, and dispersion.

CHAPTER OUTLINE

The preceding chapter serves as a bridge between electromagnetism and the area of physics called *optics*. Now that we have established the wave nature of electromagnetic radiation, we shall study the behavior of visible light and apply what we learn to all electromagnetic radiation. Our emphasis in this chapter will be on the behavior of light as it encounters an interface between two media.

So far, we have focused on the wave nature of light and discussed it in terms of our wave simplification model. As we learn more about the behavior of light, however, we shall return to our particle simplification model, especially as we incorporate the notions of quantum physics, beginning in Chapter 28. As we discuss in Section 25.1, a long historical debate took place between proponents of wave and particle models for light.

25.1 THE NATURE OF LIGHT

We encounter light every day, as soon as we open our eyes in the morning. This everyday experience involves a phenomenon that is actually quite complicated. Since the beginning of this book, we have discussed both the particle model and the wave model as simplification models to help us gain understanding of physical phenomena. Both of these models have been applied to the behavior of light. Until the beginning of the 19th century, most scientists thought that light was a stream of particles emitted by a light source. According to this model, the light particles stimulated the sense of sight on entering the eye. The chief architect of this particle model of light was Isaac Newton. The model provided a simple explanation of some known experimental facts concerning the nature of light—namely, the laws of reflection and refraction—to be discussed in this chapter.

Most scientists accepted the particle model of light. During Newton's lifetime, however, another model was proposed—a model that views light as having wave-like properties. In 1678, a Dutch physicist and astronomer, Christiaan Huygens, showed that a wave model of light can also explain the laws of reflection and refraction. The wave model did not receive immediate acceptance for several reasons. All the waves known at the time (sound, water, and so on) traveled through a medium, but light from the Sun could travel to Earth through empty space. Even though experimental evidence for the wave nature of light was discovered by Francesco Grimaldi (1618–1663) around 1660, most scientists rejected the wave model for more than a century and adhered to Newton's particle model due, for the most part, to Newton's great reputation as a scientist.

The first clear and convincing demonstration of the wave nature of light was provided in 1801 by Englishman Thomas Young (1773–1829), who showed that under appropriate conditions, light exhibits interference behavior. That is, light waves emitted by a single source and traveling along two different paths can arrive at some point, combine, and cancel each other by destructive interference. Such behavior could not be explained at that time by a particle model, because scientists could not imagine how two or more particles could come together and cancel one another. Additional developments during the 19th century led to the general acceptance of the wave model of light.

A critical development concerning the understanding of light was the work of James Clerk Maxwell, who in 1865 mathematically predicted that light is a form of high-frequency electromagnetic wave. As discussed in Chapter 24, Hertz in 1887 provided experimental confirmation of Maxwell's theory by producing and detecting other electromagnetic waves. Furthermore, Hertz and other investigators showed that these waves exhibited reflection, refraction, and all the other characteristic properties of waves.

Although the electromagnetic wave model seemed to be well established and could explain most known properties of light, some experiments could not be explained by the assumption that light was a wave. The most striking of these was the *photoelectric effect,* discovered by Hertz, in which electrons are ejected from a metal when its surface is exposed to light. We shall explore this experiment in detail in Chapter 28.

In view of these developments, light must be regarded as having a dual nature. **In some cases, light acts like a wave, and in others, it acts like a particle.** The classical electromagnetic wave model provides an adequate explanation of light propagation and interference, whereas the photoelectric effect and other experiments involving the interaction of light with matter are best explained by assuming that light is a particle. Light is light, to be sure. The question "Is light a wave or a particle?" is inappropriate; in some experiments, we measure its wave properties; in other experiments, we measure its particle properties. This curious dual nature of light may be unsettling at this point, but it will be clarified when we introduce the notion of a *quantum particle.* The photon, a particle of light, is our first example of a quantum particle, which we shall explore more fully in Chapter 28. Until then, we focus our attention on the properties of light that can be satisfactorily explained with the wave model.

25.2 THE RAY MODEL IN GEOMETRIC OPTICS

In the beginning of our study of optics, we shall use a simplification model called the **ray model** or the **ray approximation.** A **ray** is a straight line drawn along the direction of propagation of a single wave, showing the path of the wave as it travels through space. The ray approximation involves geometric models based on these straight lines. Phenomena explained with the ray approximation do not depend explicitly on the wave nature of light, other than its propagation along a straight line.

A set of light waves can be represented by wave fronts (defined in Section 24.3) as illustrated in the pictorial representation in Figure 25.1 for a plane wave, which was introduced in Section 24.3. The definition of a wave front requires that the rays are perpendicular to the wave front at every location in space.

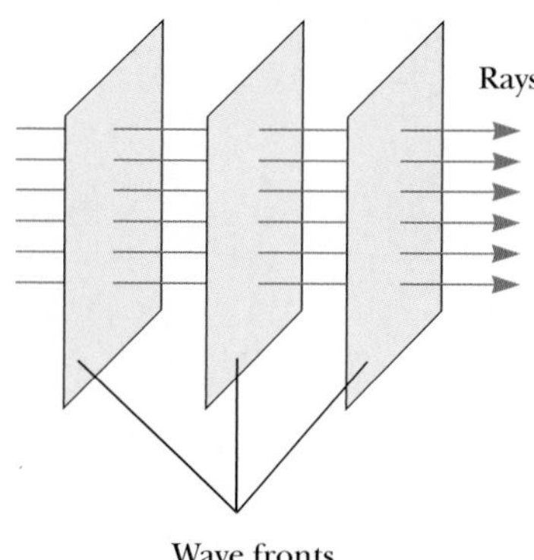

FIGURE 25.1 A plane wave propagating to the right. Note that the rays, which always point in the direction of wave motion, are straight lines perpendicular to the wave fronts.

If a plane wave meets a barrier containing an opening whose size is large relative to the wavelength as in Active Figure 25.2a, the individual waves emerging from the opening continue to move in a straight line (apart from some small edge effects); hence, the ray approximation continues to be valid. If the size of the opening is on the order of the wavelength as in Active Figure 25.2b, the waves (and, consequently, the rays we draw) spread out from the opening in all directions. We say that the incoming plane wave undergoes *diffraction* as it passes through the opening. If the opening is small relative to the wavelength, the diffraction is so strong that the opening can be approximated as a point source of waves (Active Fig. 25.2c). Thus, diffraction is more pronounced as the ratio d/λ approaches zero.

Suppose the opening is a circle of diameter d. The ray approximation assumes that $\lambda \ll d$ so that we do not concern ourselves with diffraction effects, which depend on the full wave nature of light. We shall delay studying diffraction until Chapter 27. The ray approximation is used in the current chapter and in Chapter 26. The material in these chapters is often called *geometric optics.* The ray

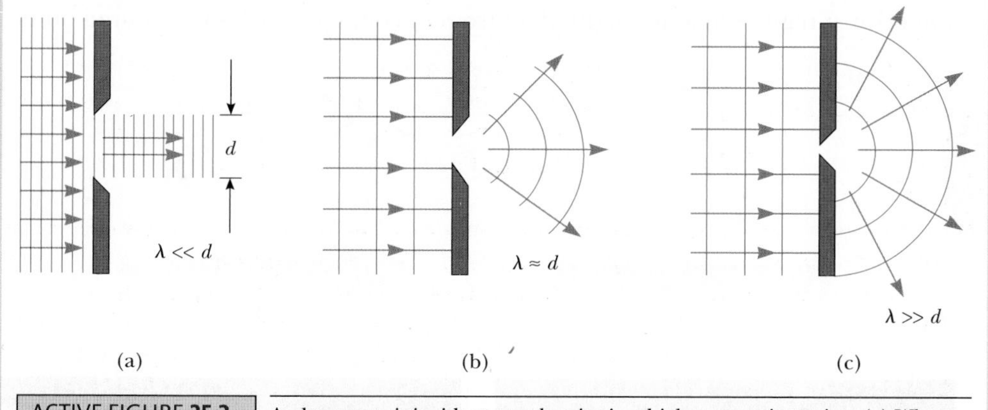

ACTIVE FIGURE 25.2 A plane wave is incident on a barrier in which an opening exists. (a) When the wavelength of the light is much smaller than the size of the opening, almost no observable diffraction takes place and the ray approximation remains valid. (b) When the wavelength of the light is comparable to the size of the opening, diffraction becomes significant. (c) When the wavelength of the light is much larger than the size of the opening, the opening behaves as a point source emitting spherical waves.

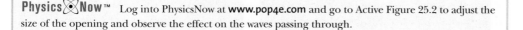

PhysicsNow™ Log into PhysicsNow at **www.pop4e.com** and go to Active Figure 25.2 to adjust the size of the opening and observe the effect on the waves passing through.

approximation is very good for the study of mirrors, lenses, prisms, and associated optical instruments, such as telescopes, cameras, and eyeglasses.

25.3 THE WAVE UNDER REFLECTION

In Chapter 13, we introduced a one-dimensional version of the model of a wave under reflection by considering waves on strings. When such a wave meets a discontinuity between strings representing different wave speeds, some of the energy is reflected and some of the energy is transmitted. In that discussion, the waves are constrained to move along the one-dimensional string. In this discussion of optics, we are not subject to that restriction. Light waves can move in three dimensions.

Figure 25.3 shows several rays of light incident on a surface. Unless the surface is perfectly absorbing, some portion of the light is reflected from the surface. (The transmitted portion will be discussed in Section 25.4.) If the surface is very smooth, the reflected rays are parallel as indicated in Figure 25.3a. Reflection of light from such a smooth surface is called **specular reflection.** If the reflecting surface is rough as in Figure 25.3b, it reflects the rays in various directions. Reflection from a rough surface is known as **diffuse reflection.** A surface behaves as a smooth surface as long as the surface variations are small compared with the wavelength of the incident light. For example, light passes through the small holes in a microwave oven door, allowing you to see the interior because the holes are large relative to the wavelengths of visible light. The large-wavelength microwaves, however, reflect from the door as if it were a solid piece of metal.

Figures 25.3c and 25.3d are photographs of specular reflection and diffuse reflection using laser light, made visible by dust in the air, which scatters the light toward the camera. The reflected laser beam is clearly visible in Figure 25.3c. In Figure 25.3d, the diffuse reflection has caused the incident beam to be reflected in many directions so that no clear outgoing beam is visible.

Specular reflection is necessary for the formation of clear images from reflecting surfaces, a topic we shall investigate in Chapter 26. Figure 25.4 shows an image resulting from specular reflection from a smooth water surface. If the water surface were rough, diffuse reflection would occur and the reflected image would not be visible.

Both types of reflection can occur from a road surface that you observe when you drive at night. On a dry night, light from oncoming vehicles is scattered off the

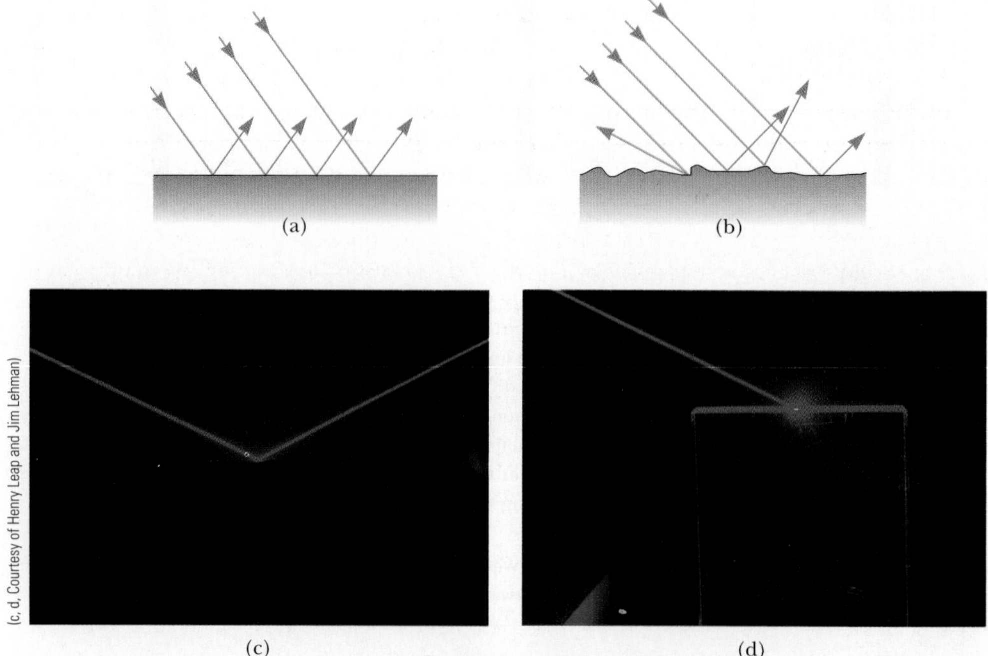

FIGURE 25.3 Schematic representation of (a) specular reflection, in which the reflected rays are all parallel, and (b) diffuse reflection, in which the reflected rays travel in scattered directions. (c) and (d) Photographs of specular and diffuse reflection using laser light.

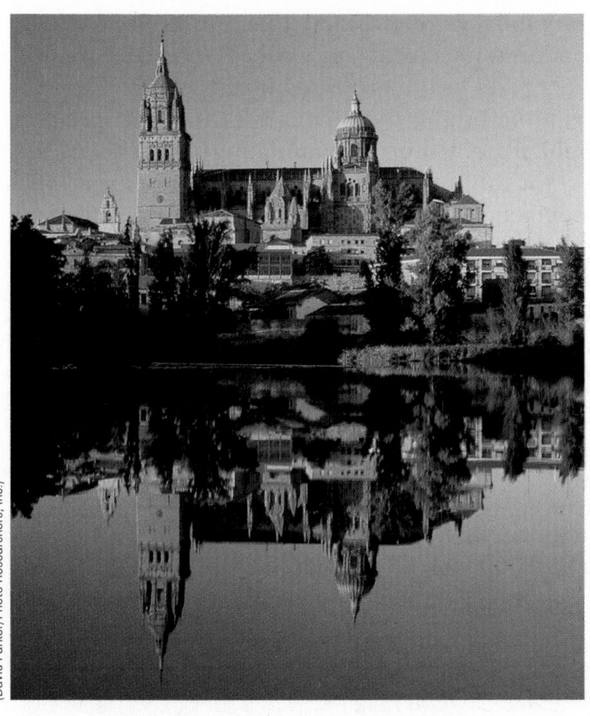
(David Parker/Photo Researchers, Inc.)

FIGURE 25.4 This photograph, taken in Salamanca, Spain, shows the reflection of the New Cathedral in the Tormes River. Because the water is so calm, the reflection is specular.

road in different directions (diffuse reflection) and the road is quite visible. On a rainy night, the small irregularities in the road surface are filled with water. Because the water surface is smooth, the light undergoes specular reflection and the glare from reflected light makes the road less visible.

Let us now develop the mathematical representation for the wave under reflection model. Consider a light ray that travels in air and is incident at an angle on a flat, smooth surface as in Active Figure 25.5. The incident and reflected rays make angles of θ_1 and θ_1', respectively, with a line drawn normal to the surface at the point where the incident ray strikes the surface. Experiments show that the incident ray, the normal to the surface, and the reflected ray all lie in the same plane and that **the angle of reflection equals the angle of incidence:**

$$\theta_1' = \theta_1 \qquad [25.1]$$

Equation 25.1 is called the **law of reflection.** By convention, the angles of incidence and reflection are measured from the normal to the surface rather than from the surface itself.

In diffuse reflection, the law of reflection is obeyed *with respect to the local normal.* Because of the roughness of the surface, the local normal varies significantly from one location to another. In this book, we shall concern ourselves only with specular reflection and shall use the term *reflection* to mean specular reflection.

As you might guess from Equation 25.1 and the figures we have seen so far, geometric models are used extensively in the study of optics. As we represent physical situations with geometric constructions, the mathematics of triangles and the principles of trigonometry will find many applications.

The path of a light ray is reversible. For example, the ray in Active Figure 25.5 travels from the upper left, reflects from the mirror, and then moves toward a point at the upper right. If the ray originated at the same point at the upper right, it would follow the same path in reverse to reach the same point at the upper left. This reversible property will be useful when we set up geometric constructions for finding the paths of light rays.

A practical application of the law of reflection is the digital projection of movies, television shows, and computer presentations. A digital projector makes use of an optical semiconductor chip called a *digital micromirror device.* This device contains an

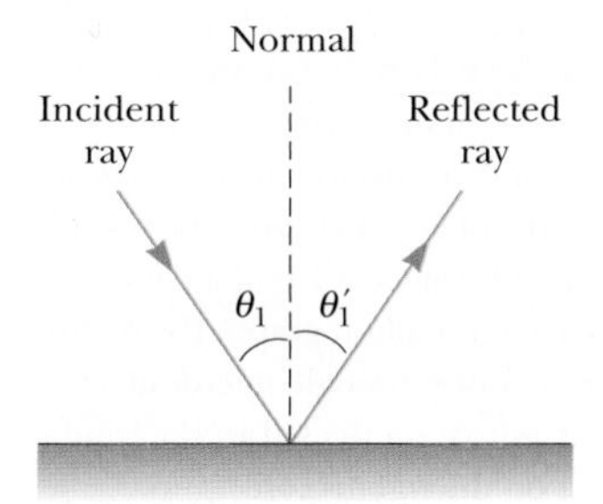

ACTIVE FIGURE 25.5
According to the law of reflection, $\theta_1 = \theta_1'$. The incident ray, the reflected ray, and the normal all lie in the same plane.

PhysicsNow™ Log into PhysicsNow at **www.pop4e.com** and go to Active Figure 25.5 to vary the incident angle and see the effect on the reflected ray.

PITFALL PREVENTION 25.1

Subscript notation We use the subscript 1 to refer to parameters for the light in the initial medium. When light travels from one medium to another, we use the subscript 2 for the parameters associated with the light in the new medium. In the current discussion, the light stays in the same medium, so we only have to use subscripts 1.

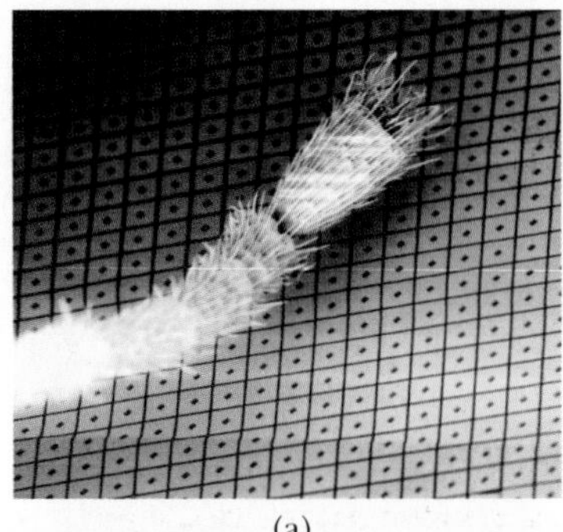

(a)

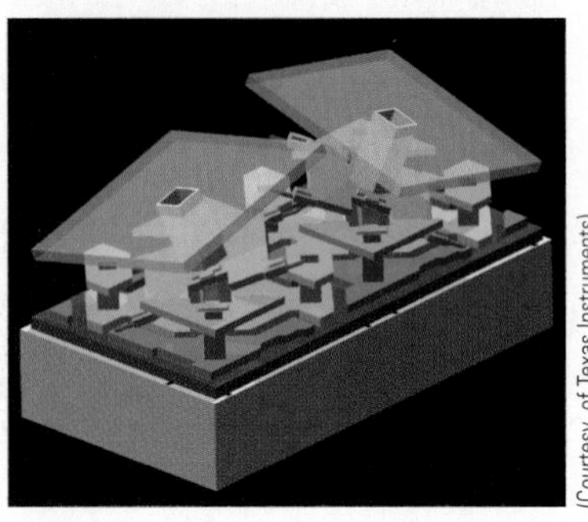

(b)

(Courtesy of Texas Instruments)

FIGURE 25.6 (a) An array of mirrors on the surface of a digital micromirror device. Each mirror has an area of about 16 μm^2. To provide a sense of scale, the leg of an ant appears in the photograph. (b) A close-up view of two single micromirrors. The mirror on the left is "on," and the one on the right is "off."

array of more than one million tiny mirrors (Fig. 25.6a) that can be individually tilted by means of signals to an address electrode underneath the edge of the mirror. Each mirror corresponds to a pixel in the projected image. When the pixel corresponding to a given mirror is to be bright, the mirror is in the "on" position and is oriented so as to reflect light from a source illuminating the array to the screen (Fig. 25.6b). When the pixel for this mirror is to be dark, the mirror is "off" and is tilted so that the light is reflected away from the screen. The brightness of the pixel is determined by the total time interval during which the mirror is in the "on" position during the display of one image.

Digital movie projectors use three micromirror devices, one for each of the primary colors red, blue, and green, so that movies can be displayed with up to 35 trillion colors. Because there is no physical storage mechanism for the movie, a digital movie does not degrade with time as does film. Furthermore, because the movie is entirely in the form of computer software, it can be delivered to theaters by means of satellites, optical discs, or optical fiber networks.

Several movies have been projected digitally to audiences and polls show that 85 percent of the viewers describe the image quality as "excellent." The first all-digital movie, from cinematography to postproduction to projection, was *Star Wars Episode II: Attack of the Clones* in 2002.

QUICK QUIZ 25.1 In the movies, you sometimes see an actor looking in a mirror and you can see his face in the mirror. During the filming of this scene, what does the actor see in the mirror? **(a)** his face **(b)** your face **(c)** the director's face **(d)** the movie camera **(e)** impossible to determine

■ Thinking Physics 25.1

When looking through a glass window to the outdoors at night, you sometimes see a *double* image of yourself. Why?

Reasoning Reflection occurs whenever light encounters an interface between two optical media. For the glass in the window, two such interfaces exist. The first is the inner surface of the glass and the second is the outer surface. Each interface results in an image. ■

INTERACTIVE EXAMPLE 25.1 The Double-Reflected Light Ray

Two mirrors make an angle of 120° with each other as in Figure 25.7. A ray is incident on mirror M_1 at an angle of 65° to the normal. Find the direction of the ray after it is reflected from mirror M_2.

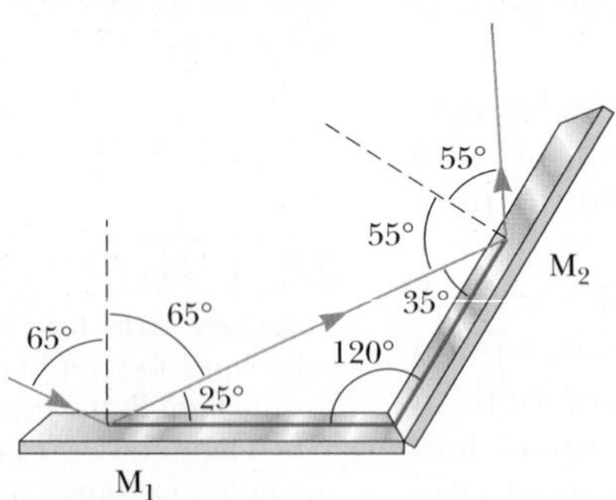

FIGURE 25.7 (Interactive Example 25.1) Mirrors M_1 and M_2 make an angle of 120° with each other.

Solution From the law of reflection, we know that the first reflected ray also makes an angle of 65° with the normal. Thus, this ray makes an angle of 90° − 65°, or 25°, with the horizontal. We identify the geometric model triangle as the triangle made by the first reflected ray and the two mirrors in Figure 25.7. The first reflected ray makes an angle of 35° with M_2 (because the sum of the interior angles of any triangle is 180°). Thus, this ray makes an angle of 55° with the normal to M_2. Hence, from the law of reflection, the second reflected ray makes an angle of 55° with the normal to M_2.

PhysicsNow™ Log into PhysicsNow at **www.pop4e.com** and go to Interactive Example 25.1 to investigate this reflection situation for various mirror angles.

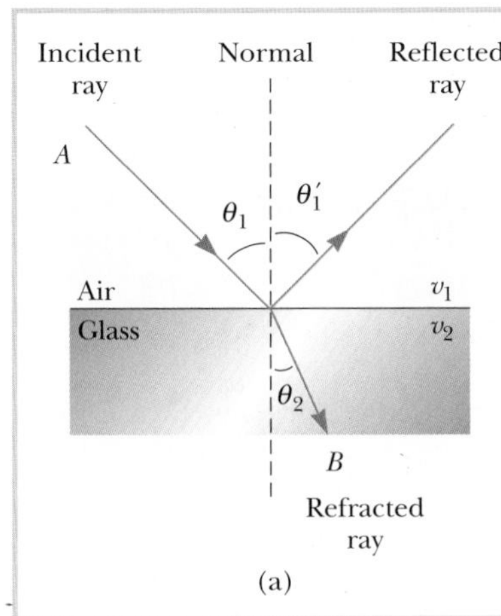

(a)

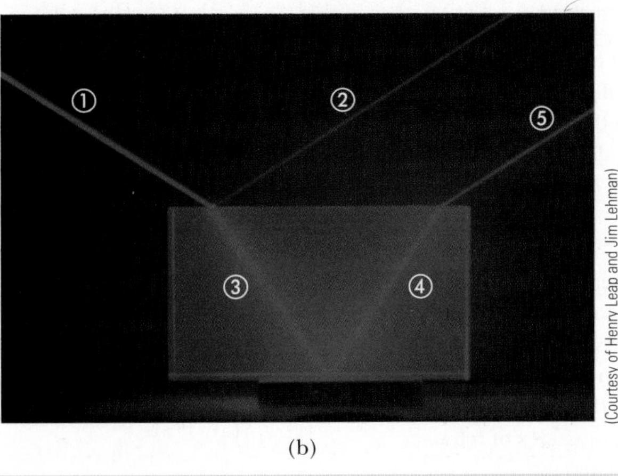

(b)

(Courtesy of Henry Leap and Jim Lehman)

ACTIVE FIGURE 25.8

(a) A light ray obliquely incident on an air–glass interface. The refracted ray is deviated toward the normal because $v_2 < v_1$. All rays and the normal lie in the same plane. (b) (Quick Quiz 25.2) Of light rays ② through ⑤, which are reflected and which are refracted?

PhysicsNow™ Log into PhysicsNow at **www.pop4e.com** and go to Active Figure 25.8 to vary the incident angle and see the effect on the reflected and refracted rays.

25.4 THE WAVE UNDER REFRACTION

Referring again to our discussion of string waves in Chapter 13, we discussed that some of the energy of a wave incident on a discontinuity in the string is transmitted through the discontinuity. As a light wave moves through three dimensions, understanding the transmitted light wave involves new principles that we now discuss.

When a ray of light traveling through a transparent medium is obliquely incident on a boundary leading into another transparent medium as in Active Figure 25.8a, part of the ray is reflected but part is transmitted into the second medium. The ray that enters the second medium experiences a change in direction at the boundary and is said to undergo **refraction. The incident ray, the reflected ray, and the refracted ray all lie in the same plane.** The **angle of refraction** θ_2 in Active Figure 25.8a depends on the properties of the two media and on the angle of incidence through the relationship

$$\frac{\sin \theta_2}{\sin \theta_1} = \frac{v_2}{v_1} = \text{constant} \qquad [25.2]$$

where v_1 is the speed of light in medium 1 and v_2 is the speed of light in medium 2. Equation 25.2 is a mathematical representation of the wave under refraction model, although we find a more commonly used form in Equation 25.7.

The path of a light ray through a refracting surface is reversible, as was the case for reflection. For example, the ray in Active Figure 25.8a travels from point *A* to point *B*. If the ray originated at *B*, it would follow the same path in reverse to reach point *A*. In the latter case, however, the reflected ray would be in the glass.

QUICK QUIZ 25.2 If beam ① is the incoming beam in Active Figure 25.8b, which of the other four red lines are reflected beams and which are refracted beams?

Equation 25.2 shows that when light moves from a material in which its speed is high to a material in which its speed is lower, the angle of refraction θ_2 is less than the angle of incidence. The refracted ray therefore deviates toward the normal as shown in Active Figure 25.9a. If the ray moves from a material in which it travels slowly to a material in which it travels more rapidly, θ_2 is greater than θ_1, so the ray deviates away from the normal as shown in Active Figure 25.9b.

The behavior of light as it passes from air into another substance and then re-emerges into air is often a source of confusion to students. Why is this behavior so different from other occurrences in our daily lives? When light travels in air, its speed is $c = 3.0 \times 10^8$ m/s; on entry into a block of glass, its speed is reduced to approximately 2.0×10^8 m/s. When the light re-emerges into air, its speed increases to its

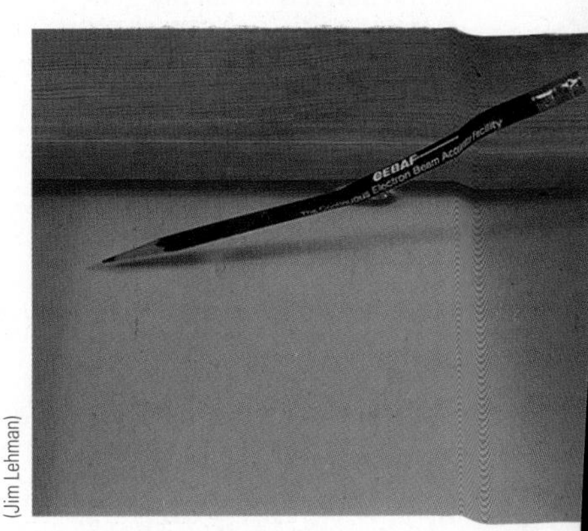

(Jim Lehman)

The pencil partially immersed in water appears bent because light from the lower part of the pencil is refracted as it travels across the boundary between water and air. ■

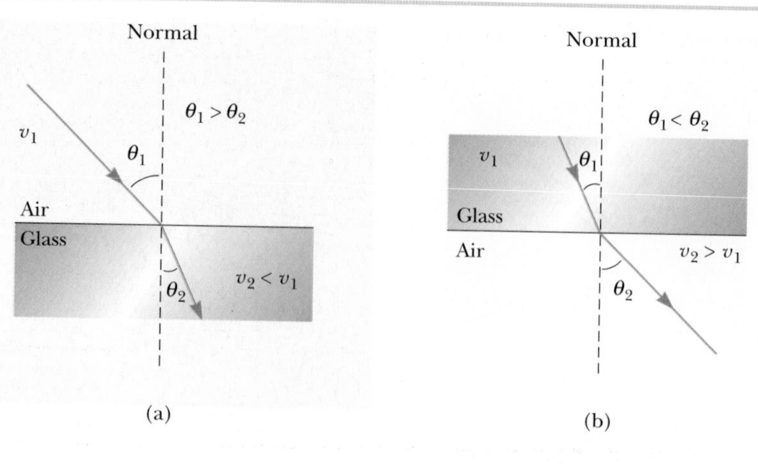

ACTIVE FIGURE 25.9 (a) When the light ray moves from air into glass, its path deviates toward the normal. (b) When the ray moves from glass into air, its path deviates away from the normal.

Physics Now™ Log into Physics-Now at **www.pop4e.com** and go to Active Figure 25.9 to see light passing through three layers of material. You can vary the incident angle and see the effect on the refracted rays for a variety of values of the index of refraction of the three materials.

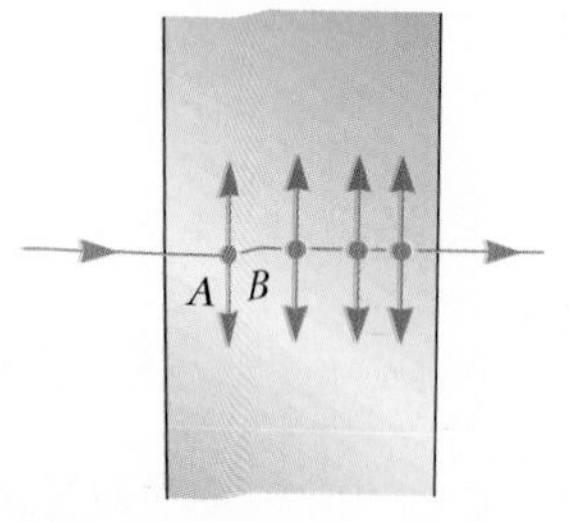

FIGURE 25.10 Light passing from one atom to another in a medium. The dots are atoms, and the vertical arrows represent their oscillations.

PITFALL PREVENTION 25.2

***n* IS NOT AN INTEGER HERE** We have seen *n* used in Chapter 11 to indicate the quantum number of a Bohr orbit and in Chapter 14 to indicate the standing wave mode on a string or in an air column. In those cases, *n* was an integer. The index of refraction *n* is *not* an integer.

original value 3.0×10^8 m/s. This process is very different from what happens, for example, when a bullet is fired through a block of wood. In that case, the speed of the bullet is reduced as it moves through the wood because some of its original energy is used to tear apart the fibers of the wood. When the bullet enters the air again, it emerges at a speed lower than that with which it entered the block of wood.

To see why light behaves as it does, consider Figure 25.10, which represents a beam of light entering a piece of glass from the left. Once inside the glass, the light may encounter an atom, represented by point *A* in the figure. Let us assume that light is absorbed by the atom, causing it to oscillate (a detail represented by the double-headed arrows in the drawing). The oscillating atom then radiates (emits) the beam of light toward an atom at point *B*, where the light is again absorbed. The details of these absorptions and emissions are best explained in terms of quantum physics, a subject we shall study in Chapter 28. For now, think of the process as one in which the light passes from one atom to another through the glass. (The situation is somewhat analogous to a relay race in which a baton is passed between runners on the same team.) Although light travels from one atom to another through the empty space between the atoms with a speed of $c = 3.0 \times 10^8$ m/s, the absorptions and emissions of light by the atoms require time to occur. Therefore, the *average* speed of light through the glass is lower than *c*. Once the light emerges into the air, the absorptions and emissions cease and the light's average speed returns to its original value.[1] Thus, whether the light is inside the material or outside, it always travels through vacuum with the same speed.

Light passing from one medium to another is refracted because the average speed of light is different in the two media. In fact, **light travels at its maximum speed in vacuum.** It is convenient to define the **index of refraction** *n* of a medium to be the ratio

■ Index of refraction

$$n \equiv \frac{\text{speed of light in vacuum}}{\text{average speed of light in the medium}} = \frac{c}{v} \qquad [25.3]$$

From this definition, we see that the index of refraction is a dimensionless number greater than or equal to unity because *v* in a medium is less than *c*. Furthermore, *n* is equal to unity for vacuum. The indices of refraction for various substances are listed in Table 25.1.

[1]As an analogy, consider a subway entering a city at a constant speed *v* and then stopping at several stations in the downtown region of the city. Although the subway may achieve the instantaneous speed *v* between stations, the *average* speed across the city is less than *v*. Once the subway leaves the city and makes no stops, it moves again at a constant speed *v*. The analogy, as is the case with many analogies, is not perfect because the subway requires time to accelerate to the speed *v* between stations, whereas light achieves speed *c* immediately as it travels between atoms.

TABLE 25.1 Indices of Refraction for Various Substances

Substance	Index of Refraction	Substance	Index of Refraction
Solids at 20° C		**Liquids at 20° C**	
Cubic zirconia	2.20	Benzene	1.501
Diamond (C)	2.419	Carbon disulfide	1.628
Fluorite (CaF_2)	1.434	Carbon tetrachloride	1.461
Fused quartz (SiO_2)	1.458	Corn syrup	2.21
Gallium phosphide	3.50	Ethyl alcohol	1.361
Glass, crown	1.52	Glycerin	1.473
Glass, flint	1.66	Water	1.333
Ice (H_2O)	1.309	**Gases at 0°C, 1 atm**	
Polystyrene	1.49	Air	1.000 293
Sodium chloride (NaCl)	1.544	Carbon dioxide	1.000 45

Note: All values are for light having a wavelength of 589 nm in vacuum.

As a wave travels from one medium to another, its frequency does not change. Let us first consider this notion for waves passing from a light string to a heavier string. If the frequencies of the incident and transmitted waves on the two strings at the junction point were different, the strings could not remain tied together because the joined ends of the two pieces of string would not move up and down in unison!

For a light wave passing from one medium to another, the frequency also remains constant. To see why, consider Figure 25.11. Wave fronts pass an observer at point A in medium 1 with a certain frequency and are incident on the boundary between medium 1 and medium 2. The frequency at which the wave fronts pass an observer at point B in medium 2 must equal the frequency at which they arrive at point A. If that were not the case, the wave fronts would either pile up at the boundary or be destroyed or created at the boundary. Because this situation does not occur, the frequency must be a constant as a light ray passes from one medium into another.

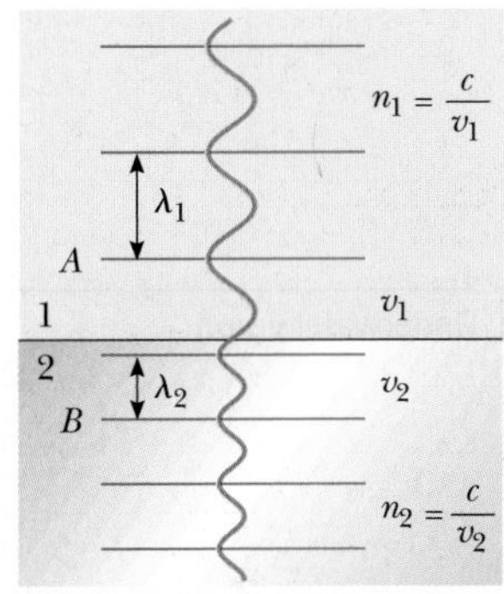

FIGURE 25.11 As a wave front moves from medium 1 to medium 2, its wavelength changes but its frequency remains constant.

Therefore, because the relation $v = f\lambda$ (Eq. 13.11) must be valid in both media and because $f_1 = f_2 = f$, we see that

$$v_1 = f\lambda_1 \quad \text{and} \quad v_2 = f\lambda_2$$

Because $v_1 \neq v_2$, it follows that $\lambda_1 \neq \lambda_2$. A relationship between index of refraction and wavelength can be obtained by dividing these two equations and making use of the definition of the index of refraction given by Equation 25.3:

$$\frac{\lambda_1}{\lambda_2} = \frac{v_1}{v_2} = \frac{c/n_1}{c/n_2} = \frac{n_2}{n_1} \qquad [25.4]$$

which gives

$$\lambda_1 n_1 = \lambda_2 n_2 \qquad [25.5]$$

It follows from Equation 25.5 that the index of refraction of any medium can be expressed as the ratio

$$n = \frac{\lambda_0}{\lambda_n} \qquad [25.6]$$

where λ_0 is the wavelength of light in vacuum and λ_n is the wavelength in the medium whose index of refraction is n.

We are now in a position to express Equation 25.2 in an alternative form. If we combine Equation 25.3 and Equation 25.2, we find that

$$n_1 \sin\theta_1 = n_2 \sin\theta_2 \qquad [25.7]$$

■ Law of refraction (Snell's law)

PITFALL PREVENTION 25.3

An Inverse Relationship The index of refraction is *inversely* proportional to the wave speed. As the wave speed v decreases, the index of refraction n increases. Thus, the higher the index of refraction of a material, the more it *slows down* light from its speed in vacuum. The more the light slows down, the more θ_2 differs from θ_1 in Equation 25.7.

This equation is the **law of refraction** and is the mathematical representation of the wave under refraction model. The experimental discovery of this relationship is usually credited to Willebrord Snell (1591–1626) and is therefore known as **Snell's law.**[2] Equation 25.7 is the conventional form of the law of refraction used in optics, expressed in terms of n values rather than speeds as in Equation 25.2.

QUICK QUIZ 25.3 Light passes from a material with index of refraction 1.3 into one with index of refraction 1.2. Compared with the incident ray, what happens to the refracted ray? **(a)** It bends toward the normal. **(b)** It is undeflected. **(c)** It bends away from the normal.

QUICK QUIZ 25.4 As light from the Sun enters the atmosphere, it refracts due to the small difference between the speeds of light in air and in vacuum. The *optical* length of the day is defined as the time interval between the instant when the top of the Sun is just visibly observed above the horizon to the instant at which the top of the Sun just disappears below the horizon. The *geometric* length of the day is defined as the time interval between the instant when a geometric straight line drawn from the observer to the top of the Sun just clears the horizon to the instant at which this line just dips below the horizon. Which is longer, **(a)** the optical length of a day or **(b)** the geometric length of a day?

■ Thinking Physics 25.2

Why do face masks make vision clearer under water? A face mask includes a flat piece of glass; the mask does not have lenses like those in eyeglasses.

Underwater vision

Reasoning The refraction necessary for focused viewing in the eye occurs at the air–cornea interface. The lens of the eye only performs some fine-tuning of this image, allowing for accommodation for objects at various distances. When the eye is opened underwater, the interface is water–cornea rather than air–cornea. Thus, the light from the scene is not focused on the retina and the scene is blurry. The face mask simply provides a layer of air in front of the eyes so that the air–cornea interface is re-established and the refraction is correct to focus the light on the retina. ■

EXAMPLE 25.2 Refraction in a Material

A beam of light of wavelength 550 nm traveling in air is incident on a slab of transparent material. The incident beam makes an angle of 40.0° with the normal, and the refracted beam makes an angle of 26.0° with the normal.

A Find the index of refraction of the material.

Solution We conceptualize the problem by looking again at Active Figure 25.9. Because the refracted angle is smaller than the incident angle, the situation is described by Active Figure 25.9a. The statement of the problem tells us to categorize this problem as one involving the model of a wave under refraction. To analyze the problem, we note that the index of refraction of air can be approximated as $n_1 = 1.00$. Snell's law of refraction (see Eq. 25.7) with these data gives

$$n_1 \sin\theta_1 = n_2 \sin\theta_2$$

$$n_2 = n_1 \frac{\sin\theta_1}{\sin\theta_2} = (1.00)\frac{\sin 40.0°}{\sin 26.0°}$$

$$= \frac{0.643}{0.438} = 1.47$$

B Find the speed of light in the material.

Solution The speed of light in the material can be easily obtained from Equation 25.3:

[2]The same law was deduced from the particle theory of light in 1637 by René Descartes (1596–1650) and hence is known as *Descartes's law* in France.

$$v = \frac{c}{n} = \frac{3.00 \times 10^8 \text{ m/s}}{1.47} = \boxed{2.04 \times 10^8 \text{ m/s}}$$

C What is the wavelength of the light in the material?

Solution We use Equation 25.6 to calculate the wavelength in the material, noting that we are given the wavelength in vacuum to be $\lambda_0 = 550$ nm:

$$\lambda_n = \frac{\lambda_0}{n} = \frac{550 \text{ nm}}{1.47} = \boxed{374 \text{ nm}}$$

To finalize the problem, note that the wavelength in the material in part C is shorter than that in vacuum. That is consistent with the concept of the wave slowing down in the material, as evidenced by the speed calculated in part B; the wave doesn't travel as far during one period of its oscillation.

INTERACTIVE EXAMPLE 25.3 Light Passing Through a Slab

A light beam passes from medium 1 to medium 2, with the latter being a thick slab of material whose index of refraction is n_2 (Fig. 25.12).

A Show that the emerging beam is parallel to the incident beam.

Solution First, let us apply Snell's law to the upper surface:

$$(1) \quad \sin\theta_2 = \frac{n_1}{n_2}\sin\theta_1$$

Applying Snell's law to the lower surface gives

$$(2) \quad \sin\theta_3 = \frac{n_2}{n_1}\sin\theta_2$$

Substituting (1) into (2) gives

$$\sin\theta_3 = \frac{n_2}{n_1}\left(\frac{n_1}{n_2}\sin\theta_1\right) = \sin\theta_1$$

Thus, $\theta_3 = \theta_1$, and so the layer does not alter the direction of the beam. It does, however, produce a lateral displacement d of the beam as shown in Figure 25.12.

B What if the thickness t of the slab is doubled? Does the lateral displacement d also double?

Solution Consider the magnification of the area of the light path within the slab in Figure 25.12b. The distance a is the hypotenuse of two right triangles. From the gold triangle, we see that

$$a = \frac{t}{\cos\theta_2}$$

and from the blue triangle, we see that

$$d = a\sin\gamma = a\sin(\theta_1 - \theta_2)$$

Combining these equations, we have

$$d = \frac{t}{\cos\theta_2}\sin(\theta_1 - \theta_2)$$

For a given incident angle θ_1, the refracted angle θ_2 is determined solely by the index of refraction, so the lateral displacement d is proportional to t. If the thickness doubles, so does the lateral displacement.

PhysicsNow™ By logging into PhysicsNow at **www.pop4e.com** and going to Interactive Example 25.3 you can explore refraction through slabs of various thicknesses.

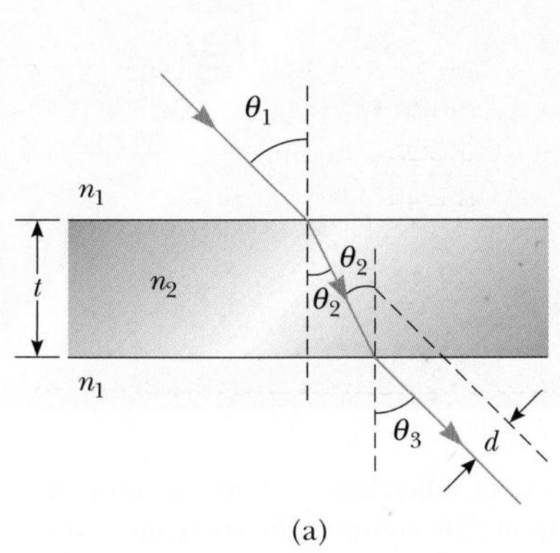

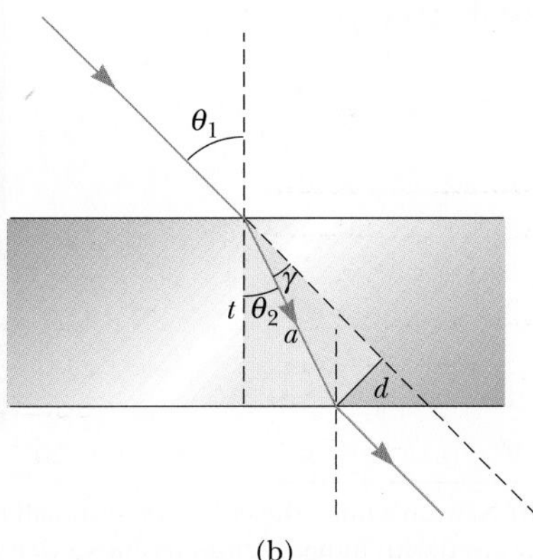

FIGURE 25.12 (Interactive Example 25.3) When light passes through a flat slab of material, the emerging beam is parallel to the incident beam and therefore $\theta_1 = \theta_3$. The dashed line parallel to the ray coming out the bottom of the slab represents the path the light would take if the slab were not there. (b) A magnification of the area of the light path inside the slab.

25.5 DISPERSION AND PRISMS

In the preceding section, we developed Snell's law, which incorporates the index of refraction of a material. In Table 25.1, we presented index of refraction values for a number of materials. If we make careful measurements, however, we find that the value of the index of refraction in anything but vacuum depends on the wavelength of light. The dependence of the index of refraction on wavelength, which results from the dependence of the wave speed on wavelength, is called **dispersion.** Figure 25.13 is a graphical representation of this variation in index of refraction with wavelength. Because n is a function of wavelength, Snell's law indicates that **the angle of refraction when light enters a material depends on the wavelength of the light.** As we see from Figure 25.13, the index of refraction for a material generally decreases with increasing wavelength in the visible range. Thus, violet light ($\lambda \approx 400$ nm) refracts more than red light ($\lambda \approx 650$ nm) when passing from air into a material.

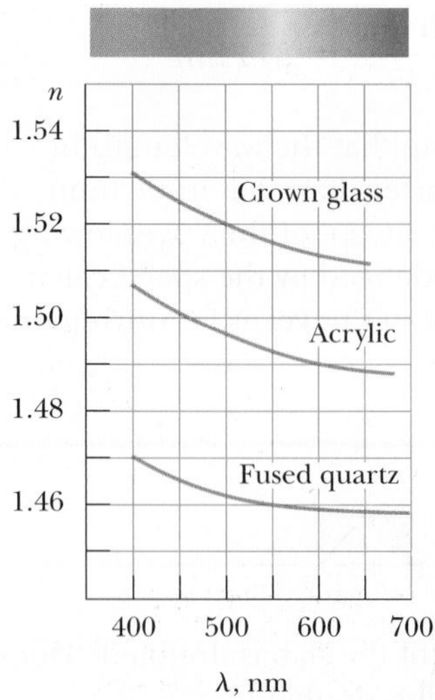

FIGURE 25.13 Variation of index of refraction with vacuum wavelength for three materials.

To understand the effects of dispersion on light, consider what happens when light strikes a prism as in Figure 25.14. The apex angle Φ of the prism is defined as shown in the figure. A ray of light of a single wavelength that is incident on the prism from the left emerges in a direction deviated from its original direction of travel by an angle of deviation δ that depends on the apex angle and the index of refraction of the prism material. Now suppose a beam of white light (a combination of all visible wavelengths) is incident on a prism. Because of dispersion, the different colors refract through different angles of deviation, and the rays that emerge from the second face of the prism spread out in a series of colors known as a **visible spectrum** as shown in Figure 25.15. These colors, in order of decreasing wavelength, are red, orange, yellow, green, blue, and violet.[3] Violet light deviates the most, red light deviates the least, and the remaining colors in the visible spectrum fall between these extremes.

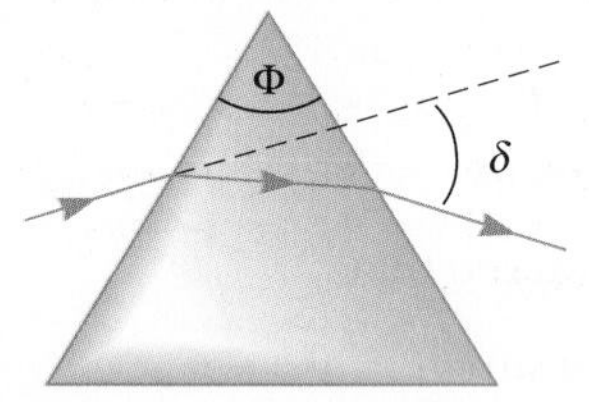

FIGURE 25.14 A prism refracts single-wavelength light and deviates the light through an angle δ. The apex angle Φ is the angle between the sides of the prism through which the light enters and leaves.

The dispersion of light into a spectrum is demonstrated most vividly in nature through the formation of a rainbow, often seen by an observer positioned between the Sun and a rain shower. To understand how a rainbow is formed, consider Active Figure 25.16. A ray of light passing overhead strikes a spherical drop of water in the

PITFALL PREVENTION 25.4

A RAINBOW OF MANY LIGHT RAYS Pictorial representations such as Active Figure 25.16 are subject to misinterpretation. The figure shows one ray of light entering the raindrop and undergoing reflection and refraction, exiting the raindrop in a range of 40° to 42° from the entering ray. This figure might be interpreted incorrectly as meaning that *all* light entering the raindrop exits in this small range of angles. In reality, light exits the raindrop over a much larger range of angles, from 0° to 42°. A careful analysis of the reflection and refraction from the spherical raindrop shows that the range of 40° to 42° is where the *highest intensity light* exits the raindrop.

FIGURE 25.15 White light enters a glass prism at the upper left. A reflected beam of light comes out of the prism just below the incoming beam. The beam moving toward the lower right shows distinct colors. Different colors are refracted at different angles because the index of refraction of the glass depends on wavelength. Violet light deviates the most; red light deviates the least.

(David Parker/Science Photo Library/Photo Researchers, Inc.)

[3]In Newton's time, the colors we now call teal and blue were called blue and indigo. Your "blue jeans" are dyed with indigo. A mnemonic device for remembering the colors of the spectrum is the acronym ROYGBIV, from the first letters of the colors: red, orange, yellow, green, blue, indigo, violet. Some individuals think of this acronym as the name of a person, Roy G. Biv!

atmosphere and is refracted and reflected as follows. It is first refracted at the front surface of the drop, with the violet light deviating the most and the red light the least. At the back surface of the drop, the light is reflected and returns to the front surface, where it again undergoes refraction as it moves from water into air.

Because light enters the front surface of the raindrop at all locations, there is a range of exit angles for the light leaving the raindrop after reflecting from the back surface. A careful analysis of the spherical shape of the water drop, however, shows that the exit angle of highest light intensity is 42° for the red light and 40° for the violet light. Thus, the light from the raindrop seen by the observer is brightest for these angles, and the observer sees a rainbow. Figure 25.17 shows the geometry for the observer. The colors of the rainbow are seen in a range of 40° to 42° from the antisolar direction, which is exactly 180° from the Sun. If red light is seen coming from a raindrop high in the sky, the violet light from this drop passes over the observer's head and is not seen. Thus, the portion of the rainbow in the vicinity of this drop is red. The violet portion of the rainbow seen by an observer is supplied by drops lower in the sky, which send violet light to the observer's eyes and red light below the eyes.

The opening photograph for this chapter shows a *double rainbow.* The secondary rainbow is fainter than the primary rainbow, and its colors are reversed. The secondary rainbow arises from light that makes two reflections from the interior surface before exiting the raindrop. In the laboratory, rainbows have been observed in which the light makes more than 30 reflections before exiting the water drop. Because each reflection involves some loss of light due to refraction out of the water drop, the intensity of these higher-order rainbows is very small.

Sunlight

40° 42°

R V

V

R

ACTIVE FIGURE 25.16 Path of sunlight through a spherical raindrop. Light following this path contributes to the visible rainbow.

Physics Now™ Log into PhysicsNow at **www.pop4e.com** and go to Active Figure 25.16 to vary the point at which the sunlight enters the raindrop and verify that the angles shown are the maximum angles.

QUICK QUIZ 25.5 In dispersive materials, the angle of refraction for a light ray depends on the wavelength of the light. True or false: The angle of reflection from the surface of the material depends on the wavelength.

25.6 HUYGENS'S PRINCIPLE

In this section, we introduce a geometric construction proposed by Huygens in 1678. Huygens assumed that light consists of waves rather than a stream of particles. He had no knowledge of the electromagnetic character of light. Nevertheless, his geometric model is adequate for understanding many practical aspects of the propagation of light.

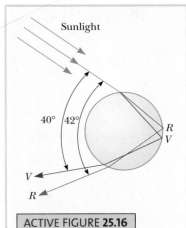

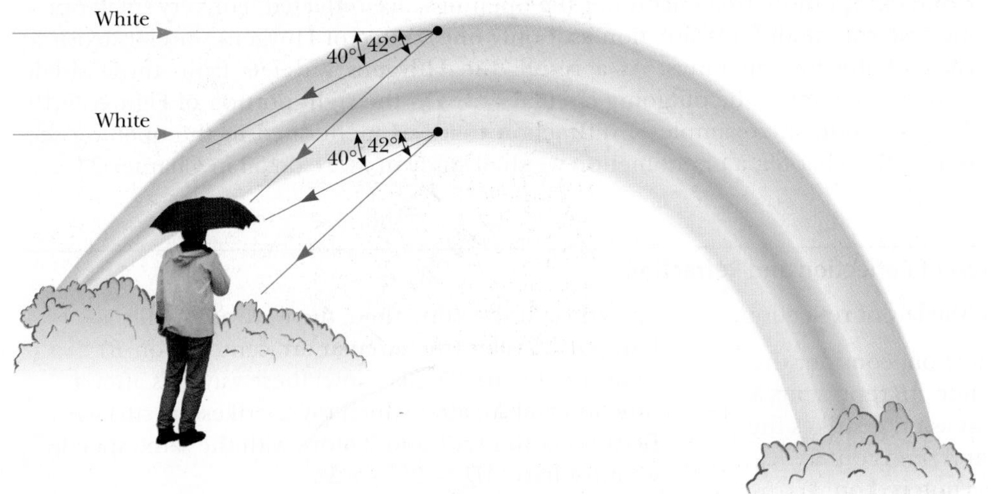

FIGURE 25.17 The formation of a rainbow.

(Courtesy of Rijksmuseum voor de Geschiedenis der Natuurwetenschappen. Courtesy AIP Niels Bohr Library)

CHRISTIAAN HUYGENS (1629–1695)

Huygens, a Dutch physicist and astronomer, is best known for his contributions to the fields of optics and dynamics. To Huygens, light was a type of vibratory motion, spreading out and producing the sensation of sight when impinging on the eye.

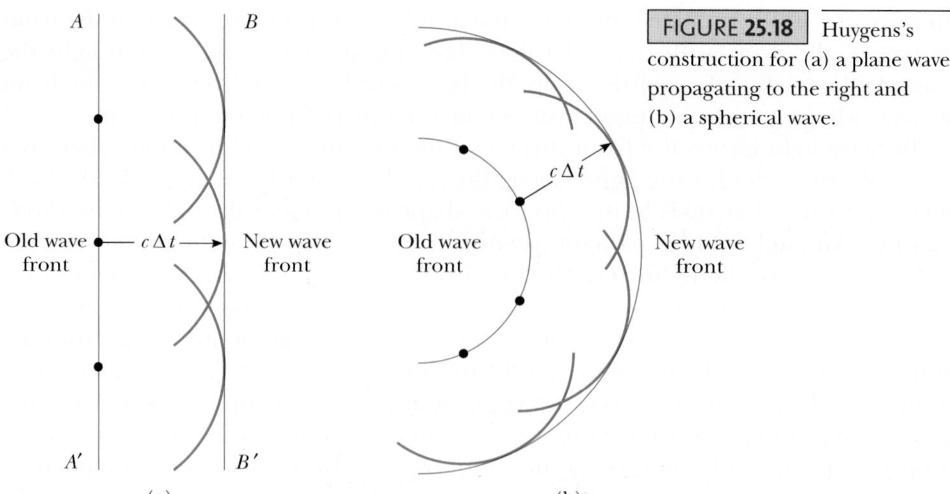

FIGURE 25.18 Huygens's construction for (a) a plane wave propagating to the right and (b) a spherical wave.

Huygens's principle is a geometric model that allows us to determine the position of a wave front from a knowledge of an earlier wave front. In Huygens's construction, all points on a given wave front are taken as point sources for the production of spherical secondary waves, called *wavelets,* that propagate outward with speeds characteristic of waves in that medium. After some time interval has elapsed, the new position of the wave front is the surface tangent to the wavelets.

Figure 25.18 illustrates two simple examples of a Huygens's principle construction. First, consider a plane wave moving through free space as in Figure 25.18a. At $t = 0$, the wave front is indicated by the plane labeled AA'. Each point on this wave front is a point source for a wavelet. Showing three of these points, we draw arcs of circles, each of radius $c\,\Delta t$, where c is the speed of light in free space and Δt is the time interval during which the wave propagates. The surface drawn tangent to the wavelets is the plane BB', which is parallel to AA'. This plane is the wave front at the end of the time interval Δt. In a similar manner, Figure 25.18b shows Huygens's construction for an outgoing spherical wave.

A convincing demonstration of the existence of Huygens wavelets is obtained with water waves in a shallow tank (called a ripple tank) as in Figure 25.19. Plane waves produced to the left of the slits emerge to the right of the slits as two-dimensional circular waves propagating outward. In the plane wave, each point on the wave front acts as a source of circular waves on the two-dimensional water surface. At a later time, the tangent of the circular wave fronts remains a straight line. As the wave front encounters the barrier, however, waves at all points on the wave front, except those that encounter the openings, are reflected. For very small openings, we can model this situation as if only one source of Huygens wavelets exists at each of the two openings. As a result, the Huygens wavelets from those single sources are seen as the outgoing circular waves in the right portion of Figure 25.19. This is a dramatic example of diffraction that was mentioned in the opening section of this chapter, a phenomenon we shall study in more detail in Chapter 27.

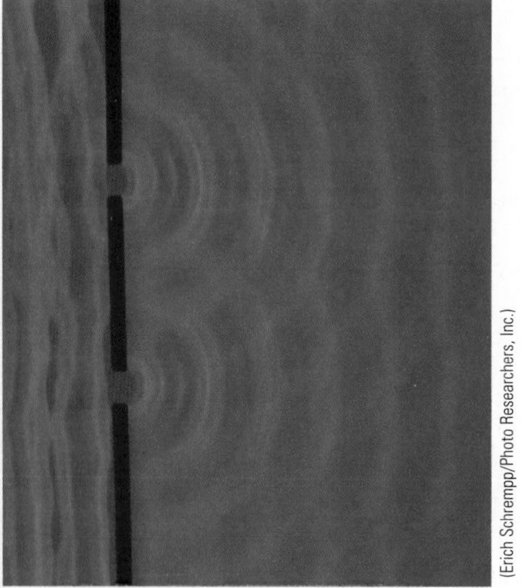

(Erich Schrempp/Photo Researchers, Inc.)

FIGURE 25.19 Water waves in a ripple tank demonstrate Huygens wavelets. A plane wave is incident on a barrier with two small openings. The openings act as sources of circular wavelets.

EXAMPLE 25.4 Deriving the Laws of Reflection and Refraction

Use Huygens's principle to derive the law of reflection.

Solution To derive the law of reflection, consider the rays shown in Figure 25.20a. The line AB represents a wave front of the incident light just as ray 1 strikes the surface. At this instant, the wave at A sends out a Huygens wavelet (the circular arc centered on A) toward D. At the same time, the wave at B emits a Huygens wavelet (the circular arc centered on B) toward C. Figure 25.20a shows these wavelets after a time interval Δt, after which ray 2 strikes the surface. Because both rays 1 and 2 move with the same speed, we must have $AD = BC = c\,\Delta t$.

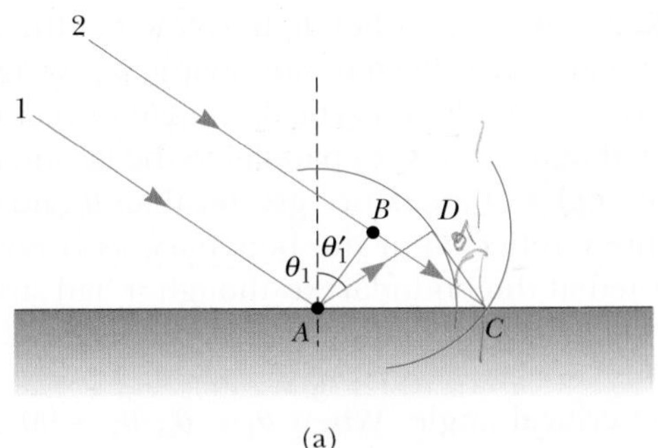

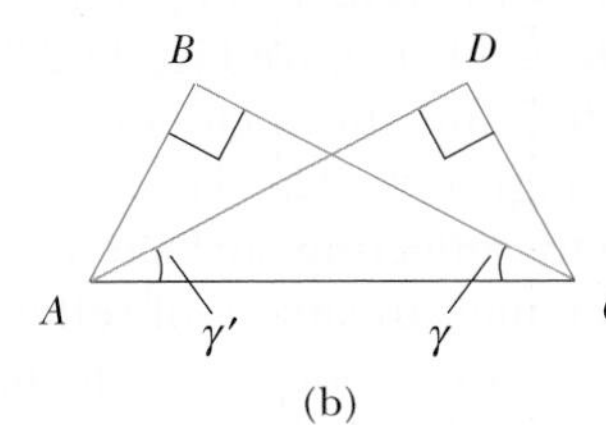

FIGURE 25.20 (Example 25.4) (a) Huygens's construction for proving the law of reflection. At the instant ray 1 strikes the surface, it sends out a Huygens wavelet from *A* and ray 2 sends out a Huygens wavelet from *B*. We choose a radius of the wavelet to be $c\,\Delta t$, where Δt is the time interval for ray 2 to travel from *B* to *C*. (b) Triangle *ADC* is congruent with triangle *ABC*.

The remainder of our analysis depends on geometry, as summarized in Figure 25.20b, in which we isolate the triangles *ABC* and *ADC*. Note that these two triangles are congruent because they have the same hypotenuse *AC* and because $AD = BC$. From Figure 25.20b, we have

$$\cos\gamma = \frac{BC}{AC} \quad \text{and} \quad \cos\gamma' = \frac{AD}{AC}$$

where, comparing Figures 25.20a and 25.20b, we see that $\gamma = 90° - \theta_1$ and $\gamma' = 90° - \theta_1'$. Because $AD = BC$,

$$\cos\gamma = \cos\gamma'$$

Therefore,

$$\gamma = \gamma'$$

$$90° - \theta_1 = 90° - \theta_1'$$

and

$$\theta_1 = \theta_1'$$

which is the law of reflection.

B Use Huygens's principle to derive the law of refraction.

Solution For the law of refraction, consider the geometric construction shown in Figure 25.21. We focus our attention on the instant ray 1 strikes the surface and the subsequent time interval until ray 2 strikes the surface. During this time interval, the wave at *A* sends out a Huygens wavelet (the arc centered on *A*) toward *D*. In the same time interval, the wave at *B* sends out a Huygens wavelet (the arc centered on *B*) toward *C*. Because these two wavelets travel through different media, the radii of the wavelets are different. The radius of the wavelet from *A* is $AD = v_2\,\Delta t$, where v_2 is the wave speed in the second medium. The radius of the wavelet from *B* is $BC = v_1\,\Delta t$, where v_1 is the wave speed in the original medium.

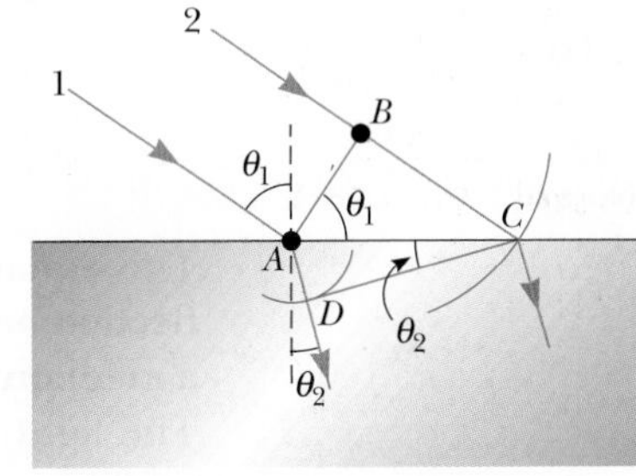

FIGURE 25.21 (Example 25.4) Huygens's construction for proving Snell's law of refraction. At the instant ray 1 strikes the surface, it sends out a Huygens wavelet from *A* and ray 2 sends out a Huygens wavelet from *B*. The two wavelets have different radii because they travel in different media.

From triangles *ABC* and *ADC*, we find that

$$\sin\theta_1 = \frac{BC}{AC} = \frac{v_1\,\Delta t}{AC} \quad \text{and} \quad \sin\theta_2 = \frac{AD}{AC} = \frac{v_2\,\Delta t}{AC}$$

If we divide the first equation by the second, we obtain

$$\frac{\sin\theta_1}{\sin\theta_2} = \frac{v_1}{v_2}$$

From Equation 25.3, however, we know that $v_1 = c/n_1$ and $v_2 = c/n_2$. Therefore,

$$\frac{\sin\theta_1}{\sin\theta_2} = \frac{c/n_1}{c/n_2} = \frac{n_2}{n_1}$$

$$n_1 \sin\theta_1 = n_2 \sin\theta_2$$

which is Snell's law of refraction.

25.7 TOTAL INTERNAL REFLECTION

An interesting effect called **total internal reflection** can occur when light travels from a medium with a high index of refraction to one with a lower index of refraction. Consider a light ray traveling in medium 1 and meeting the boundary between media 1 and 2, where $n_1 > n_2$ (Active Fig. 25.22a). Various possible directions of the ray are indicated by rays 1 through 5. The refracted rays are bent away

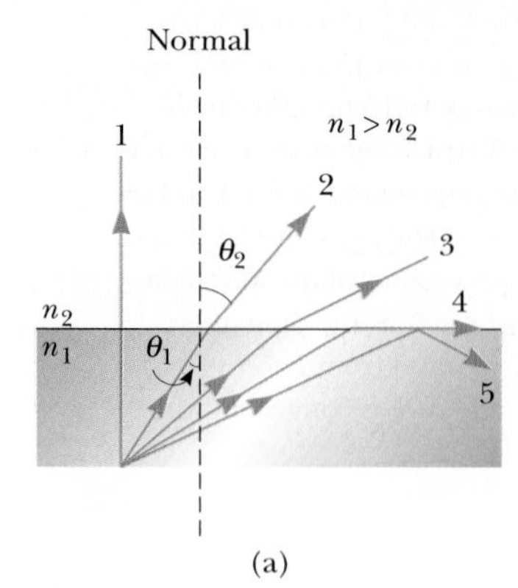

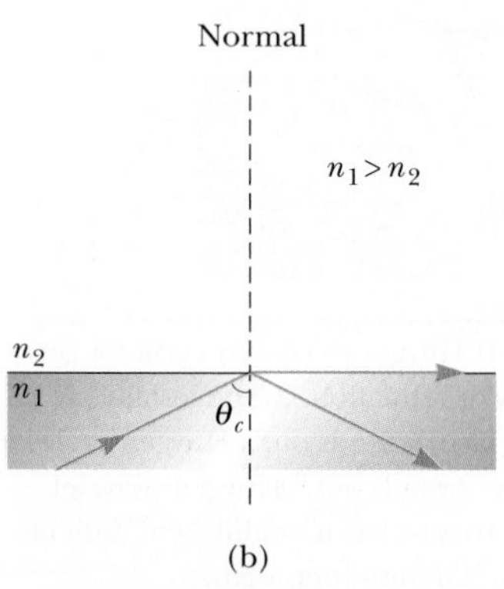

ACTIVE FIGURE 25.22
(a) Rays travel from a medium of index of refraction n_1 into a medium of index of refraction n_2, where $n_1 > n_2$. As the angle of incidence increases, the angle of refraction θ_2 increases until θ_2 is 90° (ray 4). For even larger angles of incidence, total internal reflection occurs (ray 5). (b) The angle of incidence producing an angle of refraction equal to 90° is the critical angle θ_c.

PhysicsNow™ Log into PhysicsNow at **www.pop4e.com** and go to Active Figure 25.22 to vary the incident angle and see the effect on the refracted ray and the distribution of incident energy between the reflected and refracted rays.

from the normal because $n_1 > n_2$. (Remember that when light refracts at the interface between the two media, it is also partially reflected. For simplicity, we ignore these reflected rays here, except for ray 5.) At some particular angle of incidence θ_c, called the **critical angle,** the refracted light ray moves parallel to the boundary so that $\theta_2 = 90°$ (Active Fig. 25.22b). For angles of incidence greater than θ_c, no ray is refracted and the incident ray is entirely reflected at the boundary, as is ray 5 in Active Figure 25.22a. This ray is reflected at the boundary as though it had struck a perfectly reflecting surface. It obeys the law of reflection; that is, the angle of incidence equals the angle of reflection.

We can use Snell's law to find the critical angle. When $\theta_1 = \theta_c$, $\theta_2 = 90°$, and Snell's law (Eq. 25.7) gives

$$n_1 \sin \theta_c = n_2 \sin 90° = n_2$$

$$\sin \theta_c = \frac{n_2}{n_1} \quad (\text{for } n_1 > n_2) \qquad [25.8]$$

This equation can be used only when n_1 is greater than n_2. That is, **total internal reflection occurs only when light travels from a medium of high index of refraction to a medium of lower index of refraction.** That is why the word *internal* is in the name. The light must initially be *inside* a material of higher index of refraction than the medium outside the material. If n_1 were less than n_2, Equation 25.8 would give $\sin \theta_c > 1$, which is meaningless because the sine of an angle can never be greater than unity.

The critical angle for total internal reflection is small when n_1 is considerably larger than n_2. Examples of this situation are diamond ($n = 2.42$ and $\theta_c = 24°$) and crown glass ($n = 1.52$ and $\theta_c = 41°$), where the angles given correspond to light refracting from the material into air. Total internal reflection combined with proper faceting causes diamonds and crystal glass to sparkle when observed in light.

QUICK QUIZ 25.6 **(i)** In Figure 25.23, five light rays enter a glass prism from the left. How many of these rays undergo total internal reflection at the slanted surface of the prism? **(a)** 1 **(b)** 2 **(c)** 3 **(d)** 4 **(e)** 5 **(ii)** Suppose the prism in Figure 25.23 can be rotated in the plane of the paper. For *all five* rays to experience total internal reflection from the slanted surface, should the prism be rotated **(a)** clockwise or **(b)** counterclockwise?

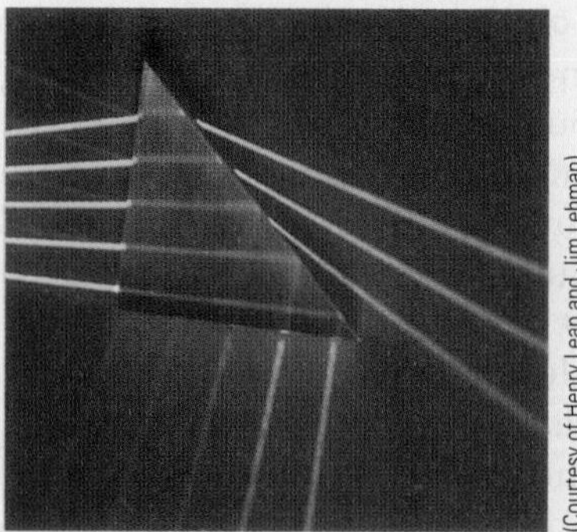

FIGURE 25.23 (Quick Quiz 25.6) Five nonparallel rays of light enter a glass prism from the left.

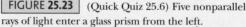

QUICK QUIZ 25.7 A beam of white light is incident on a crown glass–air interface as shown in Active Figure 25.22. The incoming beam is rotated clockwise, so the incident angle θ increases. Because of dispersion in the glass, some colors of light experience total internal reflection (ray 4 in Active Fig. 25.22a) before other colors, so the beam refracting out of the glass is no longer white. What is the last color to refract out of the upper surface? **(a)** violet **(b)** green **(c)** red **(d)** impossible to determine

EXAMPLE 25.5 A View from the Fish's Eye

A Find the critical angle for a water–air boundary if the index of refraction of water is 1.33.

Solution Applying Equation 25.8, we find the critical angle to be

$$\sin \theta_c = \frac{n_2}{n_1} = \frac{1}{1.33} = 0.752$$

$$\theta_c = 48.8°$$

B What if a fish in a still pond looks upward toward the water's surface at different angles relative to the surface as in Figure 25.24? What does it see?

Solution Examine Active Figure 25.22a. Because the path of a light ray is reversible, light traveling from medium 2 into medium 1 in Active Figure 25.22a follows the paths shown, but in the *opposite* direction. A fish looking upward toward the water surface as in Figure 25.24 can see out of the water if it looks toward the surface at an angle less than the critical angle. Thus, for example, when the fish's line of vision makes an angle of 40° with the normal to the surface, light from above the water reaches the fish's eye. At 48.8°, the critical angle for water, the light has to skim along the water's surface before being refracted to the fish's eye; at this angle, the fish can in principle see the whole shore of the pond. At angles greater than the critical angle, the light reaching the fish comes by means of internal reflection at the surface. Thus, at 60°, the fish sees a reflection of the bottom of the pond.

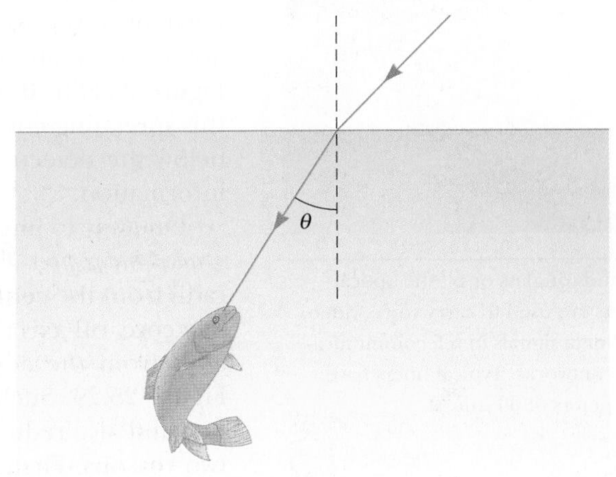

FIGURE 25.24 (Example 25.5) A fish looks upward toward the surface of the water.

25.8 OPTICAL FIBERS

CONTEXT CONNECTION

An interesting application of total internal reflection is the use of glass or transparent plastic rods to "pipe" light from one place to another. In the communication industry, digital pulses of laser light move along these light pipes, carrying information at an extremely high rate. In this Context Connection, we investigate the physics of this technological advance.

Figure 25.25 shows light traveling within a narrow transparent rod. The light is limited to traveling within the rod, even around gentle curves, as the result of successive total internal reflections. Such a light pipe can be flexible if thin fibers—called **optical fibers**—are used rather than thick rods. If a bundle of parallel optical fibers is used to construct an optical transmission line, images can be transferred from one point to another as we shall discuss in the Context Connection of Chapter 26. Typical diameters for optical fibers are measured in tens of micrometers.

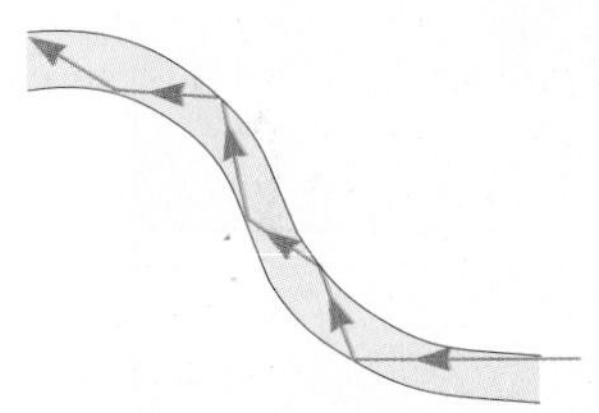

FIGURE 25.25 Light travels in a curved transparent rod by multiple internal reflections.

A typical optical fiber consists of a transparent core surrounded by a *cladding*, a material that has a lower index of refraction than the core. The combination may be surrounded by a plastic *jacket* to prevent mechanical damage. Figure 25.26 shows a cutaway view of this construction. Because the index of refraction of the cladding is less than that of the core, light traveling in the core experiences total internal reflection if it arrives at the interface between the core and the cladding at an angle of incidence that exceeds the critical angle. In this case, light "bounces" along the core of the optical fiber, losing very little of its intensity as it travels.

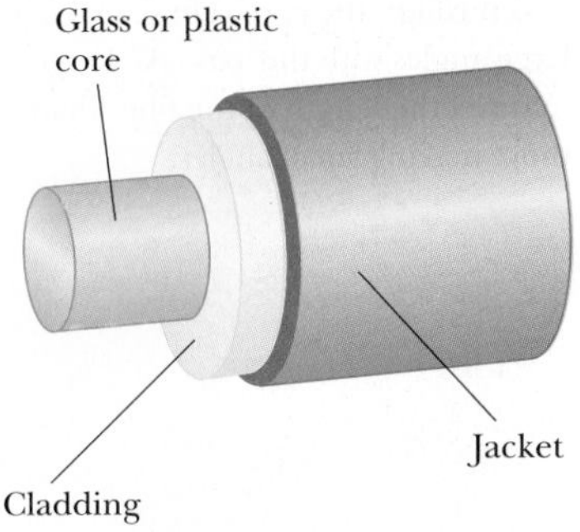

FIGURE 25.26 The construction of an optical fiber. Light travels in the core, which is surrounded by a cladding and a protective jacket.

Figure 25.27 shows a cross-sectional view from the side of a simple type of optical fiber known as a *multimode, stepped index fiber*. The term *stepped index* refers to the

Strands of glass or plastic optical fibers are used to carry voice, video, and data signals in telecommunication networks. Typical fibers have diameters of 60 μm. ■

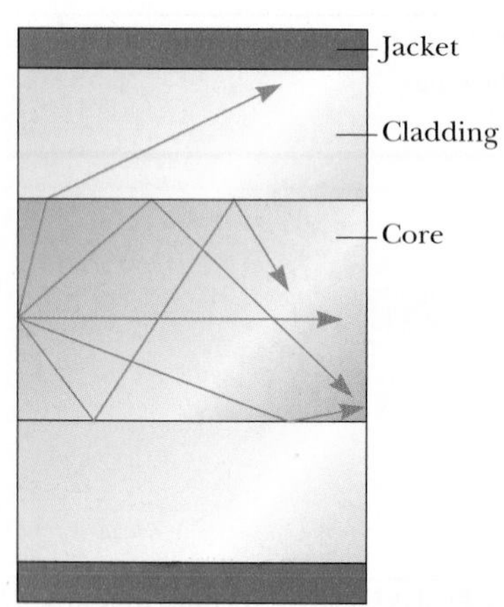

FIGURE 25.27 A multimode, stepped index optical fiber. Light rays entering over a wide range of angles pass through the core. Those making large angles with the axis take longer to travel the length of the fiber than those making small angles.

discontinuity in index of refraction between the core and the cladding, and *multimode* means that light entering the fiber at many angles is transmitted. This type of fiber is acceptable for transmitting signals over a short distance but not long distances because a digital pulse spreads with distance. Let us imagine that we input a perfectly rectangular pulse of laser light to the core of the optical fiber. Active Figure 25.28a shows the idealized time behavior of the laser light intensity for the input pulse. The laser light intensity rises instantaneously to its highest value, stays constant for the duration of the pulse, and then instantaneously drops to zero. The light from the pulse entering along the axis in Figure 25.27 travels the shortest distance and arrives at the other end first. The other light paths represent longer distance of travel because of the angled bounces. As a result, the light from the pulse arrives at the other end over a longer period and the pulse is spread out as in Active Figure 25.28b. If a series of pulses represents zeroes and ones for a binary signal, this spreading could cause the pulses to overlap or might reduce the peak intensity below the detection threshold; either situation would result in obliteration of the information.

One way to improve optical transmission in such a situation is to use a *multimode, graded index fiber.* This fiber has a core whose index of refraction is smaller at larger radii from the center as suggested by the shading in Figure 25.29. With a graded index core, off-axis rays of light experience continuous refraction and curve gradually away from the edges and back toward the center as shown by the light path in Figure 25.29. Such curving reduces the transit time through the fiber for off-axis rays and also reduces the spreading out of the pulse. The transit time is reduced for two reasons. First, the path length is reduced, and second, much of the time the wave travels in the lower index of refraction region, where the speed of light is higher than at the center.

The spreading effect in Active Figure 25.28 can be further reduced and almost eliminated by designing the fiber with two changes from the multimode, stepped index fiber in Figure 25.27. The core is made very small so that all paths within it are more nearly the same length, and the difference in index of refraction between core and cladding is made relatively small so that off-axis rays enter the cladding and are absorbed. These changes are suggested in Active Figure 25.30. This kind of fiber is called a *single-mode, stepped index fiber.* It can carry information at high bit rates because the pulses are minimally spread out.

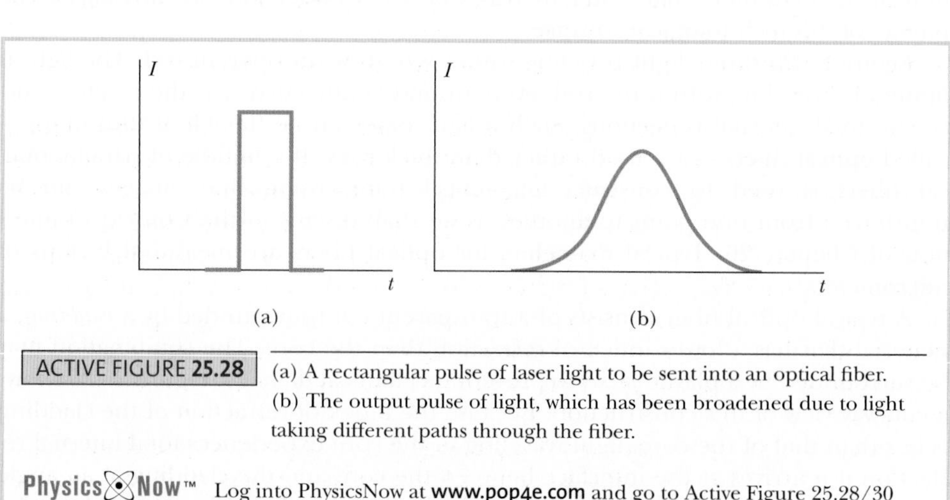

ACTIVE FIGURE 25.28 (a) A rectangular pulse of laser light to be sent into an optical fiber. (b) The output pulse of light, which has been broadened due to light taking different paths through the fiber.

PhysicsNow™ Log into PhysicsNow at **www.pop4e.com** and go to Active Figure 25.28/30 to see the variation in the pulse shape as changes to the optical fiber in Active Figure 25.30 are made.

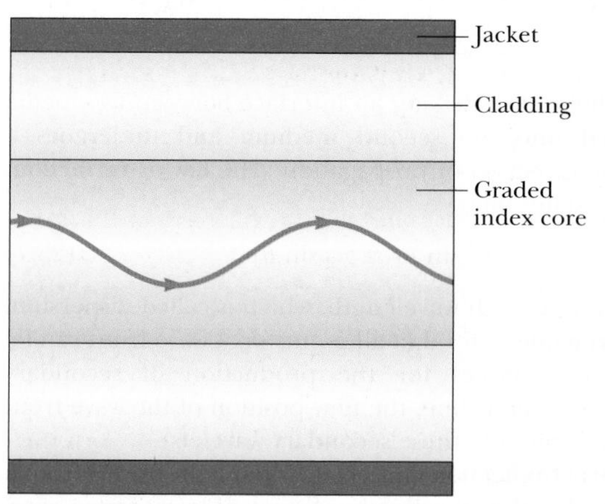

FIGURE 25.29 A multimode, graded index optical fiber. Because the index of refraction of the core varies radially, off-axis light rays follow curved paths through the core.

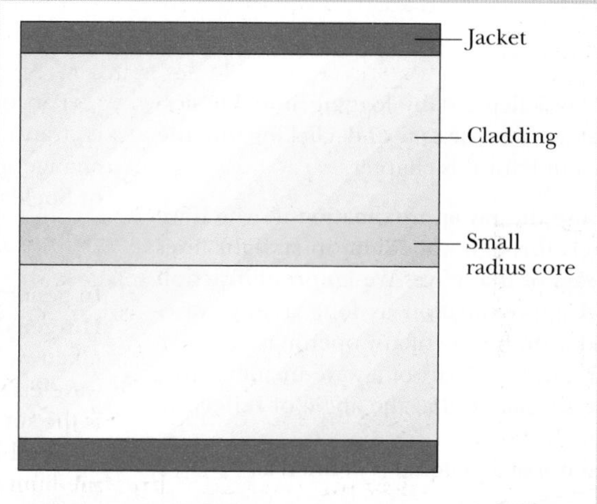

ACTIVE FIGURE 25.30 A single-mode, stepped index optical fiber. The small radius of the core and the small difference between the indices of refraction of the core and cladding reduce the broadening of light pulses.

PhysicsNow™ Log into PhysicsNow at **www.pop4e.com** and go to Active Figure 25.28/30 to make changes to the optical fiber and see the variation in the pulse shape in Active Figure 25.28.

In reality, the material of the core is not perfectly transparent. Some absorption and scattering occurs as the light travels down the fiber. Absorption transforms energy being transferred by electromagnetic radiation into increased internal energy in the fiber. Scattering causes light to strike the core–cladding interface at angles less than the critical angle for total internal reflection, resulting in some loss in the cladding or jacket. Even with these problems, optical fibers can transmit about 95% of the input energy over a kilometer. The problems are minimized by using as long a wavelength as possible for which the core material is transparent. The scattering and absorption centers then appear as small as possible to the waves and minimize the probability of interaction. Much of optical fiber communication occurs with light from infrared lasers, having wavelengths of about 1 300 nm.

The field of developing applications for optical fibers is called **fiber optics.** One common application is the use of optical fibers in telecommunications because the fibers can carry a much higher volume of telephone calls, or other forms of communication, than electric wires. Optical fibers are also used in "smart buildings." In this application, sensors are located at various points within a building and an optical fiber carries laser light to the sensor, which reflects it back to a control system. If any distortion occurs in the building due to earthquake or other causes, the intensity of the reflected light from the sensor changes and the control system locates the point of distortion by identifying the particular sensor involved.

A single optical fiber can carry a digital signal, as we already described. If it is desired for optical fibers to carry an image of a scene, it is necessary to use a bundle of optical fibers. A popular use of such bundles is in the use of *fiberscopes* in medicine. In the Context Connection of Chapter 26, we shall investigate these devices. ■

SUMMARY

Physics Now™ Take a practice test by logging into PhysicsNow at www.pop4e.com and clicking on the Pre-Test link for this chapter

In geometric optics, we use the **ray approximation** in which we assume that a wave travels through a medium in straight lines in the direction of the rays of that wave. We ignore diffraction effects, which is a good approximation as long as the wavelength is short compared with the size of any openings.

The **law of reflection** states that part of a wave incident on a surface reflects from the surface so that the angle of reflection θ_1' equals the angle of incidence θ_1.

The **index of refraction** n of a material is defined as

$$n \equiv \frac{c}{v} \quad [25.3]$$

where c is the speed of light in a vacuum and v is the speed of light in the material.

Part of a light wave striking an interface between two media is transmitted into the second medium and undergoes a change in the direction of propagation. The **law of refraction,** or **Snell's law,** states that

$$n_1 \sin \theta_1 = n_2 \sin \theta_2 \quad [25.7]$$

In general, n varies with wavelength, which is called **dispersion. Huygens's principle** states that all points on a wave front can be taken as point sources for the production of secondary wavelets. At some later time, the new position of the wave front is the surface tangent to these secondary wavelets.

Total internal reflection can occur when light travels from a medium of high index of refraction to one of lower index of refraction. The **critical angle** of incidence θ_c for which total internal reflection occurs at an interface is

$$\sin \theta_c = \frac{n_2}{n_1} \quad (\text{for } n_1 > n_2) \quad [25.8]$$

QUESTIONS

☐ = answer available in the *Student Solutions Manual and Study Guide*

1. Light of wavelength λ is incident on a slit of width d. Under what conditions is the ray approximation valid? Under what circumstances does the slit produce enough diffraction to make the ray approximation invalid?

2. The display windows of some department stores are slanted slightly inward at the bottom to decrease the glare from streetlights or the Sun, which would make it difficult for shoppers to see the display inside. Sketch a light ray reflecting from such a window to show how this design works.

3. The rectangular aquarium sketched in Figure Q25.3 contains only one goldfish. When the fish is near a corner of the tank and is viewed along a direction that make an equal angle with two adjacent faces, the observer sees two fish mirroring each other, as shown. Explain this observation.

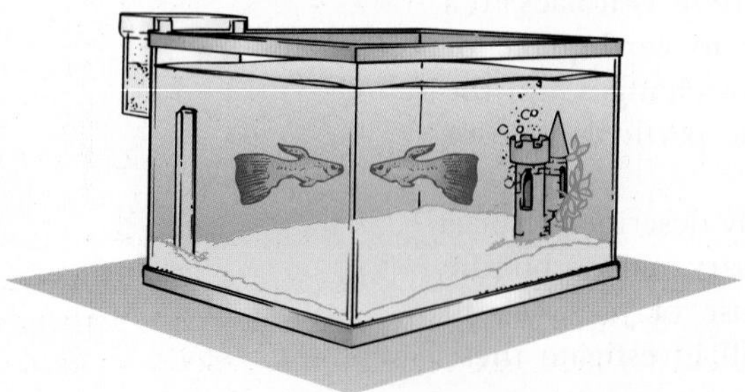

FIGURE Q25.3

4. Sound waves have much in common with light waves, including the properties of reflection and refraction. Give examples of these phenomena for sound waves.

5. ☐ As light travels from one medium to another, does the wavelength of the light change? Does the frequency change? Does the speed change? Explain.

6. A laser beam passing through a nonhomogeneous sugar solution follows a curved path. Explain.

7. ☐ Explain why a diamond sparkles more than a glass crystal of the same shape and size.

8. Why does a diamond show flashes of color when observed under white light?

9. ☐ Explain why a diamond loses most of its sparkle when it is submerged in carbon disulfide and why an imitation diamond of cubic zirconia loses all its sparkle in corn syrup.

10. Describe an experiment in which total internal reflection is used to determine the index of refraction of a medium.

11. ☐ When two colors of light (X and Y) are sent through a glass prism, X is bent more than Y. Which color travels more slowly in the prism?

12. Is it possible to have total internal reflection for light incident from air on water? Explain.

13. Total internal reflection is applied in the periscope of a submarine to let the user "see around corners." In this device, two prisms are arranged as shown in Figure Q25.13 so that an incident beam of light follows the path shown. Parallel tilted silvered mirrors could be used, but glass prisms with no silvered surfaces give higher light throughput. Propose a reason for the higher efficiency.

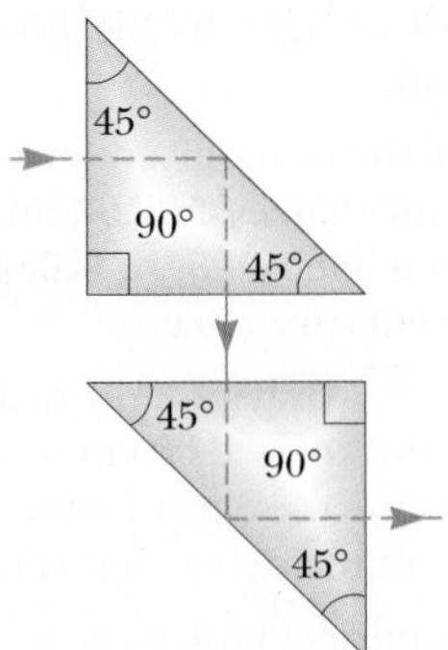

FIGURE **Q25.13**

14. At one restaurant, a worker uses colored chalk to write the daily specials on a blackboard illuminated with a spotlight. At another restaurant, a worker writes with colored grease pencils on a flat, smooth sheet of transparent acrylic plastic with index of refraction 1.55. The plastic panel hangs in front of a piece of black felt. Small, bright electric lights are installed all along the edges of the plastic sheet, inside an opaque channel. Figure Q25.14 shows a cutaway view. Explain why viewers at both restaurants see the letters shining against a black background. Explain why the sign at the second restaurant may use less energy from the electric company. What would be a good choice for the index of refraction of the material in the grease pencils?

FIGURE **Q25.14**

15. How is it possible that a complete circle of a rainbow can sometimes be seen from an airplane? With a stepladder, a lawn sprinkler, and a sunny day, how can you show the complete circle to children?

16. Under what conditions is a mirage formed? On a hot day, what are we seeing when we observe "water on the road"?

PROBLEMS

1, 2, 3 = straightforward, intermediate, challenging

☐ = full solution available in the *Student Solutions Manual and Study Guide*

PhysicsNow™ = coached problem with hints available at www.pop4e.com

= computer useful in solving problem

= paired numerical and symbolic problems

= biomedical application

Section 25.2 ▪ The Ray Model in Geometric Optics
Section 25.3 ▪ The Wave Under Reflection
Section 25.4 ▪ The Wave Under Refraction

Note: You may look up indices of refraction in Table 25.1.

1. The two mirrors illustrated in Figure P25.1 meet at a right angle. The beam of light in the vertical plane P strikes mirror 1 as shown. (a) Determine the distance the reflected light beam travels before striking mirror 2. (b) In what direction does the light beam travel after being reflected from mirror 2?

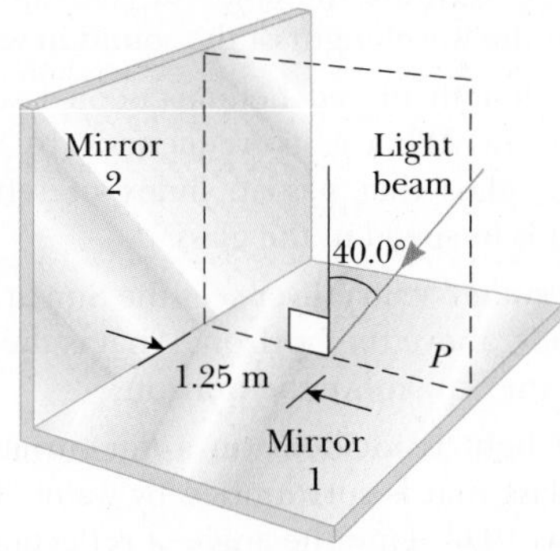

FIGURE **P25.1**

2. Two flat, rectangular mirrors, both perpendicular to a horizontal sheet of paper, are set edge to edge with their reflecting surfaces perpendicular to each other. (a) A light ray in the plane of the paper strikes one of the mirrors at an arbitrary angle of incidence θ_1. Prove that the final direction of the ray, after reflection from both mirrors, is opposite to its initial direction. In a clothing store, such a pair of mirrors shows you an image of yourself as others see you, with no apparent right–left reversal. (b) Now assume that the paper is replaced with a third flat mirror, touching edges with the other two and perpendicular to both. The set of three mirrors is called a *corner-cube reflector.* A ray of light is incident from any direction within the octant of space bounded by the reflecting surfaces. Argue

that the ray will reflect once from each mirror and that its final direction will be opposite to its original direction. The *Apollo 11* astronauts placed a panel of corner cube reflectors on the Moon. Analysis of timing data taken with the panel reveals that the radius of the Moon's orbit is increasing at the rate of 3.8 cm/yr as it loses kinetic energy because of tidal friction.

3. How many times will the incident beam shown in Figure P25.3 be reflected by each of the parallel mirrors?

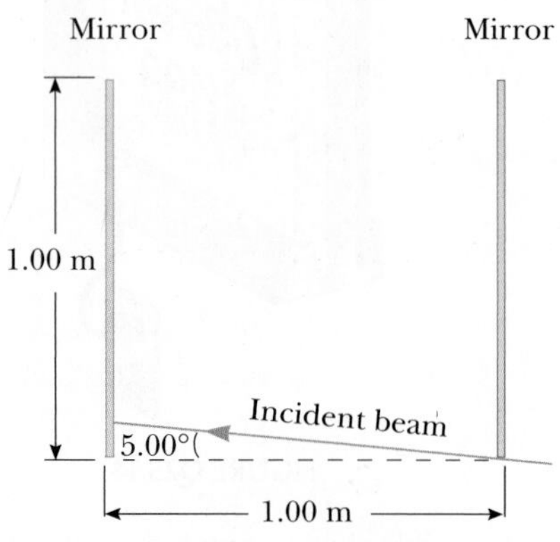

FIGURE **P25.3**

4. A narrow beam of sodium yellow light, with wavelength 589 nm in vacuum, is incident from air onto a smooth water surface at an angle of incidence of 35.0°. Determine the angle of refraction and the wavelength of the light in water.

5. *Compare this problem with Problem 25.4.* A plane sound wave in air at 20°C, with wavelength 589 mm, is incident on a smooth surface of water at 25°C, at an angle of incidence of 3.50°. Determine the angle of refraction for the sound wave and the wavelength of the sound in water.

6. The wavelength of red helium–neon laser light in air is 632.8 nm. (a) What is its frequency? (b) What is its wavelength in glass that has an index of refraction of 1.50? (c) What is its speed in the glass?

7. An underwater scuba diver sees the Sun at an apparent angle of 45.0° above the horizon. What is the actual elevation angle of the Sun above the horizon?

8. A ray of light is incident on a flat surface of a block of crown glass that is surrounded by water. The angle of refraction is 19.6°. Find the angle of reflection.

9. A laser beam with vacuum wavelength 632.8 nm is incident from air onto a block of glass as shown in Active Figure 25.8b. The line of sight of the photograph is perpendicular to the plane in which the light moves. Find the (a) speed, (b) frequency, and (c) wavelength of the light in the glass. The glass is not necessarily either of the types listed in Table 25.1. (*Suggestion:* Use a protractor.)

10. A laser beam is incident at an angle of 30.0° from the vertical onto a solution of corn syrup in water. The beam is refracted to 19.24° from the vertical. (a) What is the index of refraction of the corn syrup solution? Assume that the light is red, with vacuum wavelength 632.8 nm. Find its (b) wavelength, (c) frequency, and (d) speed in the solution.

11. Find the speed of light in (a) flint glass, (b) water, and (c) cubic zirconia.

12. A light ray initially in water enters a transparent substance at an angle of incidence of 37.0°, and the transmitted ray is refracted at an angle of 25.0°. Calculate the speed of light in the transparent substance.

13. PhysicsNow™ A ray of light strikes a flat block of glass ($n = 1.50$) of thickness 2.00 cm at an angle of 30.0° with the normal. Trace the light beam through the glass and find the angles of incidence and refraction at each surface.

14. An opaque cylindrical tank with an open top has a diameter of 3.00 m and is completely filled with water. When the afternoon Sun reaches an angle of 28.0° above the horizon, sunlight ceases to illuminate any part of the bottom of the tank. How deep is the tank?

15. Unpolarized light in vacuum is incident onto a sheet of glass with index of refraction n. The reflected and refracted rays are perpendicular to each other. Find the angle of incidence. This angle is called *Brewster's angle* or the *polarizing angle.* In this situation, the reflected light is linearly polarized, with its electric field restricted to be perpendicular to the plane containing the rays and the normal.

16. A narrow beam of ultrasonic waves reflects off the liver tumor in Figure P25.16. The speed of the wave is 10.0% less in the liver than in the surrounding medium. Determine the depth of the tumor.

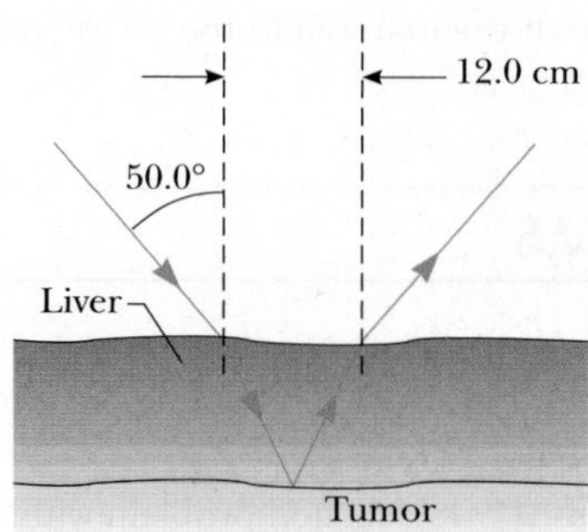

FIGURE **P25.16**

17. When the light illustrated in Figure P25.17 passes through the glass block, it is shifted laterally by the distance d. Taking $n = 1.50$, find the value of d.

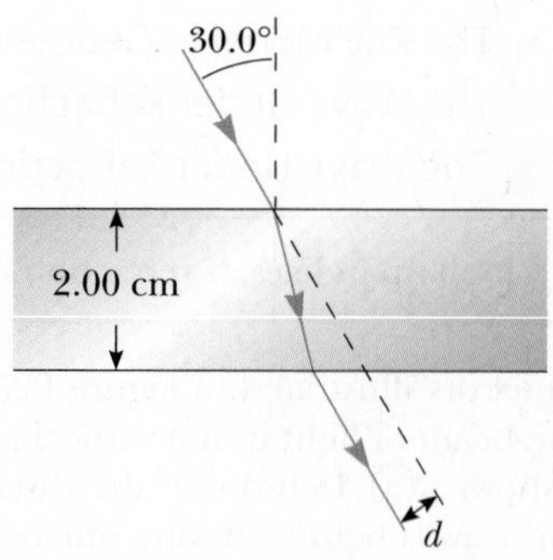

FIGURE **P25.17** Problems 25.17 and 25.18.

18. Find the time interval required for the light to pass through the glass block described in Problem 25.17.

19. The light beam shown in Figure P25.19 makes an angle of 20.0° with the normal line NN' in the linseed oil. Determine the angles θ and θ'. (*Note:* The index of refraction of linseed oil is 1.48.)

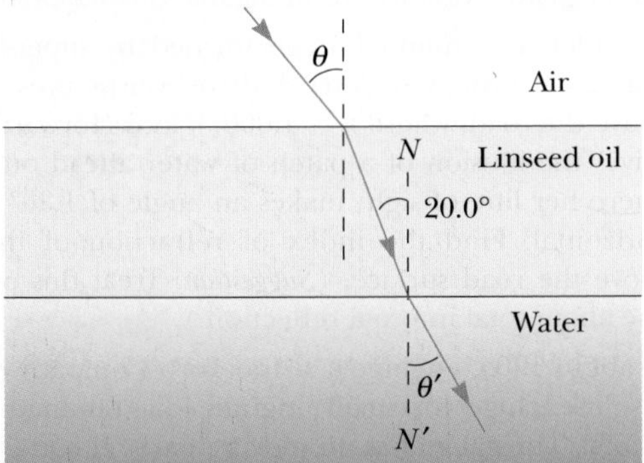

FIGURE **P25.19**

20. A digital video disc records information in a spiral track about 1 μm wide. The track consists of a series of pits in the information layer (see Fig. P25.20a) that scatter light from a laser beam sharply focused on them. The laser shines in through transparent plastic of thickness t = 1.20 mm and index of refraction 1.55. Assume that the width of the laser beam at the information layer must be a = 1.00 μm to read from just one track and not from its neighbors (Fig. P25.20b). Assume that the width of the beam as it enters the transparent plastic from below is w = 0.700 mm. A lens makes the beam converge into a cone with an apex angle $2\theta_1$ before it enters the disk. Find the incidence angle θ_1 of the light at the edge of the conical beam. Note that this design is relatively immune to small dust particles degrading the video quality. Particles on the plastic surface would have to be as large as 0.7 mm to obscure the beam.

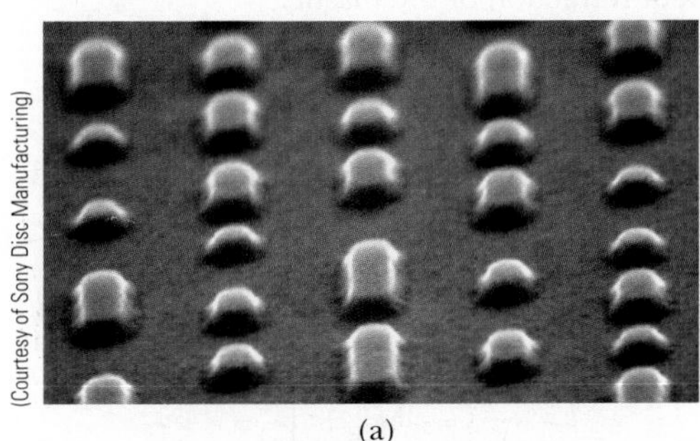

(a)

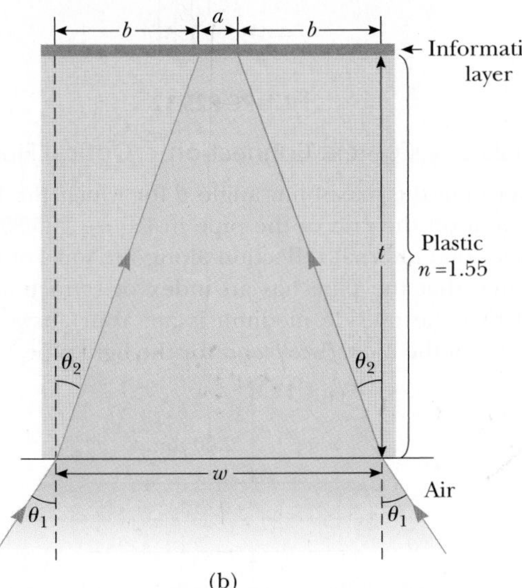

(b)

FIGURE **P25.20** (a) A micrograph of a DVD surface showing pits along each track. (b) Cross-section of a cone-shaped laser beam used to read a DVD.

21. When you look through a window, by what time interval is the light you see delayed by having to go through glass instead of air? Make an order-of-magnitude estimate on the basis of data you specify. By how many wavelengths is it delayed?

22. The reflecting surfaces of two intersecting flat mirrors are at an angle θ ($0° < \theta < 90°$) as shown in Figure P25.22. For a light ray that strikes the horizontal mirror, show that the emerging ray will intersect the incident ray at an angle $\beta = 180° - 2\theta$.

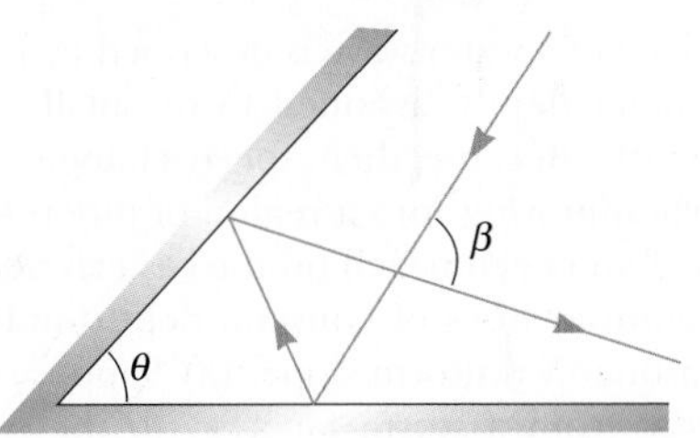

FIGURE **P25.22**

Section 25.5 ■ Dispersion and Prisms

23. A narrow white light beam is incident on a block of fused quartz at an angle of 30.0°. Find the angular width of the light beam inside the quartz.

24. A ray of light strikes the midpoint of one face of an equiangular glass prism (n = 1.50) at an angle of incidence of 30.0°. Trace the path of the light ray through the glass and find the angles of incidence and refraction at each surface.

25. A triangular glass prism with apex angle Φ = 60.0° has an index of refraction n = 1.50 (Fig. P25.25). What is the smallest angle of incidence θ_1 for which a light ray can emerge from the other side?

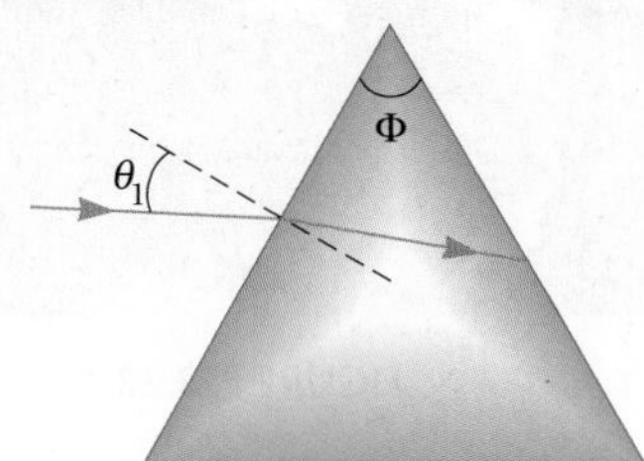

FIGURE **P25.25** Problems 25.25 and 25.26.

26. A triangular glass prism with apex angle Φ has index of refraction n. (See Fig. P25.25.) What is the smallest angle of incidence θ_1 for which a light ray can emerge from the other side?

27. **Physics Now™** The index of refraction for violet light in silica flint glass is 1.66 and that for red light is 1.62. What is the angular dispersion of visible light passing through a prism of apex angle 60.0° if the angle of incidence is 50.0°? (See Fig. P25.27.)

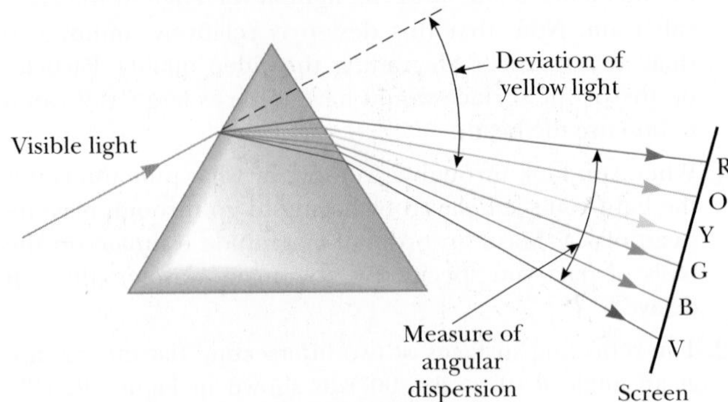

FIGURE **P25.27**

Section 25.6 ▪ Huygens's Principle

28. The speed of a water wave is described by $v = \sqrt{gd}$, where d is the water depth, assumed to be small compared to the wavelength. Because their speed changes, water waves refract when moving into a region of different depth. Sketch a map of an ocean beach on the eastern side of a landmass. Show contour lines of constant depth under water, assuming reasonably uniform slope. (a) Suppose waves approach the coast from a storm far away to the north–northeast. Demonstrate that the waves will move nearly perpendicular to the shoreline when they reach the beach. (b) Sketch a map of a coastline with alternating bays and headlands as suggested in Figure P25.28. Again make a reasonable guess about the shape of contour lines of constant depth. Suppose waves approach the coast, carrying energy with uniform density along originally straight wavefronts. Show that the energy reaching the coast is concentrated at the headlands and has lower intensity in the bays.

(Ray Atkeson/Image Archive)

FIGURE **P25.28**

Section 25.7 ▪ Total Internal Reflection

29. For 589-nm light, calculate the critical angle for the following materials surrounded by air: (a) diamond, (b) flint glass, and (c) ice.

30. A room contains air in which the speed of sound is 343 m/s. The walls of the room are made of concrete, in which the speed of sound is 1 850 m/s. (a) Find the critical angle for total internal reflection of sound at the concrete–air boundary. (b) In which medium must the sound be traveling to undergo total internal reflection? (c) "A bare concrete wall is a highly efficient mirror for sound." Give evidence for or against this statement.

31. Consider a common mirage formed by super-heated air just above a roadway. A truck driver whose eyes are 2.00 m above the road, where $n = 1.000\,3$, looks forward. She perceives the illusion of a patch of water ahead on the road, where her line of sight makes an angle of 1.20° below the horizontal. Find the index of refraction of the air just above the road surface. (*Suggestion:* Treat this problem as one about total internal reflection.)

32. In about 1965, engineers at the Toro Company invented a gasoline gauge for small engines, diagrammed in Figure P25.32. The gauge has no moving parts. It consists of a flat slab of transparent plastic fitting vertically into a slot in the cap on the gas tank. None of the plastic has a reflective coating. The plastic projects from the horizontal top down nearly to the bottom of the opaque tank. Its lower edge is cut with facets making angles of 45° with the horizontal. A lawn mower operator looks down from above and sees a boundary between bright and dark on the gauge. The location of the boundary, across the width of the plastic, indicates the quantity of gasoline in the tank. Explain how the gauge works. Explain the design requirements, if any, for the index of refraction of the plastic.

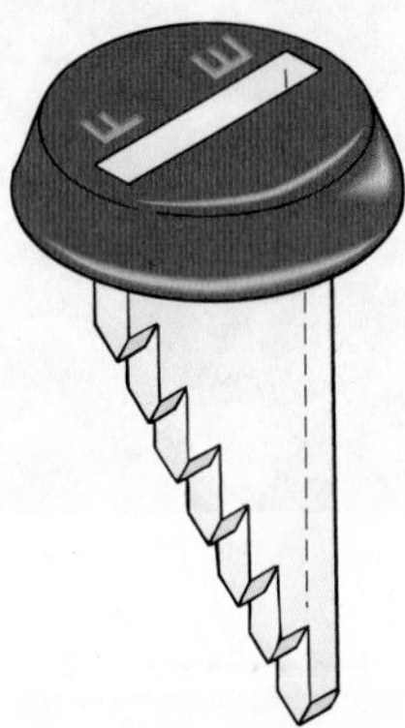

FIGURE **P25.32**

Section 25.8 ▪ Context Connection—Optical Fibers

33. Determine the maximum angle θ for which the light rays incident on the end of the pipe in Figure P25.33 are subject to total internal reflection along the walls of the pipe. Assume that the pipe has an index of refraction of 1.36 and that the outside medium is air. Your answer defines the size of the *cone of acceptance* for the light pipe.

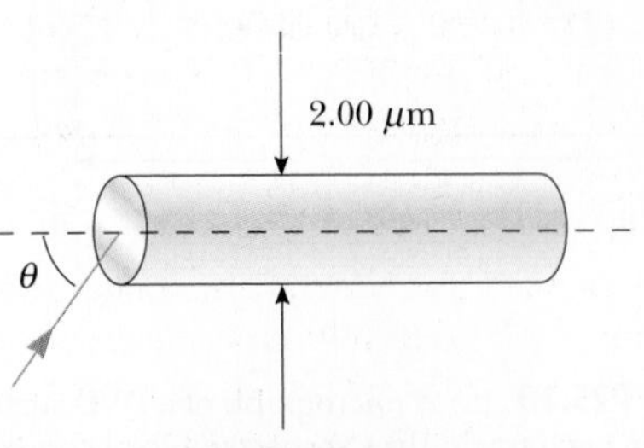

FIGURE **P25.33**

34. A glass fiber ($n = 1.50$) is submerged in water ($n = 1.33$). What is the critical angle for light to stay inside the optical fiber?

35. **PhysicsNow™** A laser beam strikes one end of a slab of material as shown in Figure P25.35. The index of refraction of the slab is 1.48. Determine the number of internal reflections of the beam before it emerges from the opposite end of the slab.

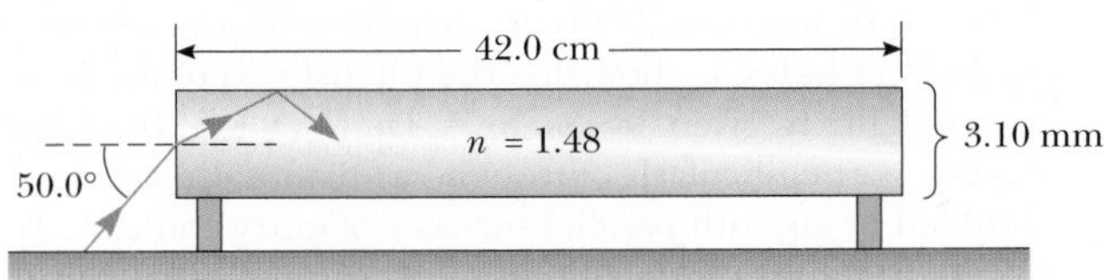

FIGURE **P25.35**

36. An optical fiber has index of refraction n and diameter d. It is surrounded by air. Light is sent into the fiber along its axis as shown in Figure P25.36. (a) Find the smallest outside radius R permitted for a bend in the fiber if no light is to escape. (b) Does the result for part (a) predict reasonable behavior as d approaches zero? As n increases? As n approaches 1? (c) Evaluate R assuming the fiber diameter is 100 μm and its index of refraction is 1.40.

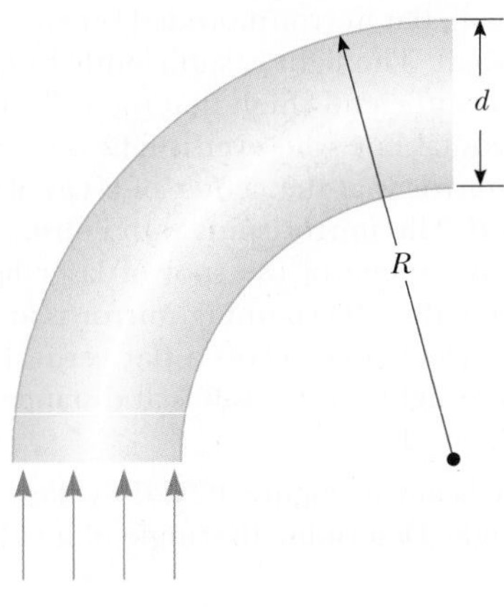

FIGURE **P25.36**

Additional Problems

37. Three sheets of plastic have unknown indices of refraction. Sheet 1 is placed on top of sheet 2, and a laser beam is directed onto the sheets from above so that it strikes the interface at an angle of 26.5° with the normal. The refracted beam in sheet 2 makes an angle of 31.7° with the normal. The experiment is repeated with sheet 3 on top of sheet 2, and, with the same angle of incidence, the refracted beam makes an angle of 36.7° with the normal. If the experiment is repeated again with sheet 1 on top of sheet 3, what is the expected angle of refraction in sheet 3? Assume the same angle of incidence.

38. Figure P25.38 shows a desk ornament globe containing a photograph. The flat photograph is in air, inside a vertical slot located behind a water-filled compartment having the shape of one half of a cylinder. Suppose you are looking at the center of the photograph and then rotate the globe about a vertical axis. You find that the center of the photograph disappears when you rotate the globe beyond a certain maximum angle (Fig. P25.38b). Account for this phenomenon and calculate the maximum angle. Briefly describe what you see when you turn the globe beyond this angle.

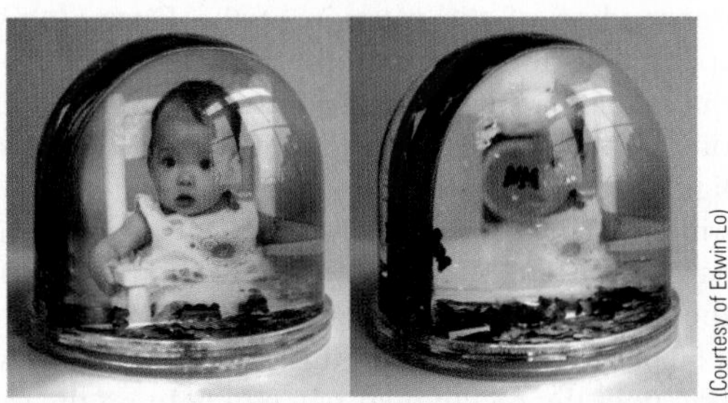

FIGURE **P25.38**

39. A light ray enters the atmosphere of a planet where it descends vertically to the surface a distance h below. The index of refraction where the light enters the atmosphere is 1.000, and it increases linearly to the surface where it has the value n. (a) How long does it take the ray to traverse this path? (b) Compare this time interval to that required in the absence of an atmosphere.

40. (a) Consider a horizontal interface between air above and glass of index 1.55 below. Draw a light ray incident from the air at angle of incidence 30.0°. Determine the angles of the reflected and refracted rays and show them on the diagram. (b) Now suppose the light ray is incident from the glass at angle of incidence 30.0°. Determine the angles of the reflected and refracted rays and show all three rays on a new diagram. (c) For rays incident from the air onto the air–glass surface, determine and tabulate the angles of reflection and refraction for all the angles of incidence at 10.0° intervals from 0° to 90.0°. (d) Do the same for light rays coming up to the interface through the glass.

41. **PhysicsNow™** A small light fixture is on the bottom of a swimming pool, 1.00 m below the surface. The light emerging from the water forms a circle on the still water surface. What is the diameter of this circle?

42. One technique for measuring the angle of a prism is shown in Figure P25.42. A parallel beam of light is directed on the angle so that parts of the beam reflect from opposite sides. Show that the angular separation of the two reflected beams is given by $B = 2A$.

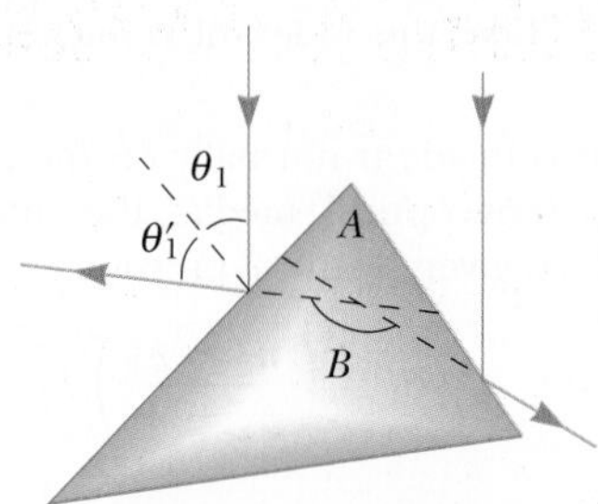

FIGURE **P25.42**

43. The walls of a prison cell are perpendicular to the four cardinal compass directions. On the first day of spring, light from the rising Sun enters a rectangular window in the eastern wall. The light traverses 2.37 m horizontally to shine perpendicularly on the wall opposite the window. A young prisoner observes the patch of light moving across this western wall and for the first time forms his own understanding of the rotation of the Earth. (a) With what speed does the illuminated rectangle move? (b) The prisoner holds a small, square mirror flat against the wall at one corner of the rectangle of light. The mirror reflects light back to a spot on the eastern wall close beside the window. How fast does the smaller square of light move across that wall? (c) Seen from a latitude of 40.0° north, the rising Sun moves through the sky along a line making a 50.0° angle with the southeastern horizon. In what direction does the rectangular patch of light on the western wall of the prisoner's cell move? (d) In what direction does the smaller square of light on the eastern wall move?

44. Figure P25.44 shows a top view of a square enclosure. The inner surfaces are plane mirrors. A ray of light enters a small hole in the center of one mirror. (a) At what angle θ must the ray enter so as to exit through the hole after being reflected once by each of the other three mirrors? (b) Are there other values of θ for which the ray can exit after multiple reflections? If so, make a sketch of one of the ray's paths.

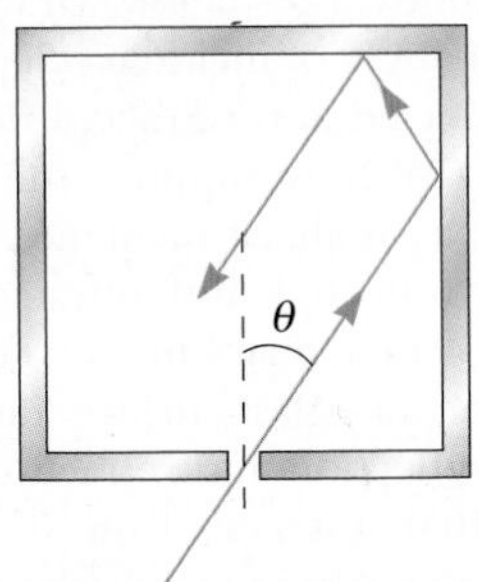

FIGURE **P25.44**

45. A hiker stands on an isolated mountain peak near sunset and observes a rainbow caused by water droplets in the air 8.00 km away. The valley is 2.00 km below the mountain peak and entirely flat. What fraction of the complete circular arc of the rainbow is visible to the hiker? (See Fig. 25.17.)

46. A 4.00-m-long pole stands vertically in a lake having a depth of 2.00 m. The Sun is 40.0° above the horizontal. Determine the length of the pole's shadow on the bottom of the lake. Take the index of refraction for water to be 1.33.

47. When light is incident normally on the interface between two transparent optical media, the intensity of the reflected light is given by the expression

$$S_1' = \left(\frac{n_2 - n_1}{n_2 + n_1}\right)^2 S_1$$

In this equation, S_1 represents the average magnitude of the Poynting vector in the incident light (the incident intensity), S_1' is the reflected intensity, and n_1 and n_2 are the refractive indices of the two media. (a) What fraction of the incident intensity is reflected for 589-nm light normally incident on an interface between air and crown glass? (b) Does it matter in part (a) whether the light is in the air or in the glass as it strikes the interface?

48. Refer to Problem 25.47 for its description of the reflected intensity of light normally incident on an interface between two transparent media. (a) For light normally incident on an interface between vacuum and a transparent medium of index n, show that the intensity S_2 of the transmitted light is given by $S_2/S_1 = 4n/(n + 1)^2$. (b) Light travels perpendicularly through a diamond slab, surrounded by air, with parallel surfaces of entry and exit. Apply the transmission fraction in part (a) to find the approximate overall transmission through the slab of diamond, as a percentage. Ignore light reflected back and forth within the slab.

49. This problem builds upon the results of Problems 25.47 and 25.48. Light travels perpendicularly through a diamond slab, surrounded by air, with parallel surfaces of entry and exit. The intensity of the transmitted light is what fraction of the incident intensity? Include the effects of light reflected back and forth inside the slab.

50. Builders use a leveling instrument with the beam from a fixed helium–neon laser reflecting in a horizontal plane from a small, flat mirror mounted on an accurately vertical rotating shaft. The light is sufficiently bright and the rotation rate is sufficiently high that the reflected light appears as a horizontal line wherever it falls on a wall. (a) Assume that the mirror is at the center of a circular grain elevator of radius R. The mirror spins with constant angular speed ω_m. Find the speed of the spot of laser light on the wall. (b) Assume that the spinning mirror is at a perpendicular distance d from point O on a flat vertical wall. When the spot of laser light on the wall is at distance x from point O, what is its speed?

51. The light beam in Figure P25.51 strikes surface 2 at the critical angle. Determine the angle of incidence θ_1.

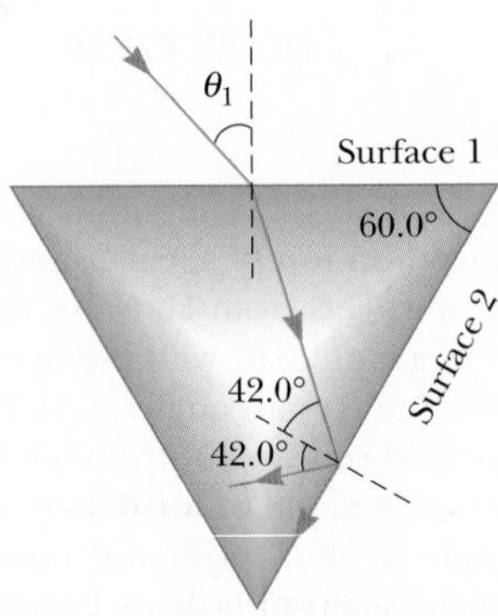

FIGURE **P25.51**

52. Refer to Quick Quiz 25.4. By how much does the duration of an optical day exceed that of a geometric day? Model the Earth's atmosphere as uniform, with index of refraction 1.000 293, a sharply defined upper surface, and depth 8 614 m. Assume that the observer is at the Earth's equator

so that the apparent path of the rising and setting Sun is perpendicular to the horizon.

53. **Physics Now™** A light ray of wavelength 589 nm is incident at an angle θ on the top surface of a block of polystyrene as shown in Figure P25.53. (a) Find the maximum value of θ for which the refracted ray undergoes total internal reflection at the left vertical face of the block. Repeat the calculation for cases in which the polystyrene block is immersed in (b) water and (c) carbon disulfide.

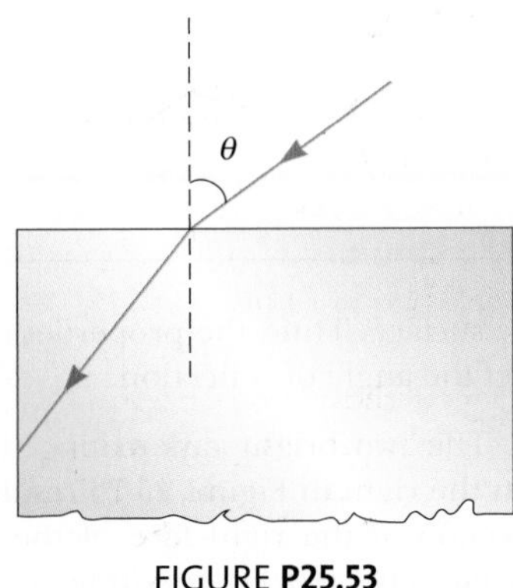

FIGURE **P25.53**

54. A ray of light passes from air into water. For its deviation angle $\delta = |\theta_1 - \theta_2|$ to be 10.0°, what must be its angle of incidence?

55. A shallow glass dish is 4.00 cm wide at the bottom as shown in Figure P25.55. When an observer's eye is placed as shown, the observer sees the edge of the bottom of the empty dish. When this dish is filled with water, the observer sees the center of the bottom of the dish. Find the height of the dish.

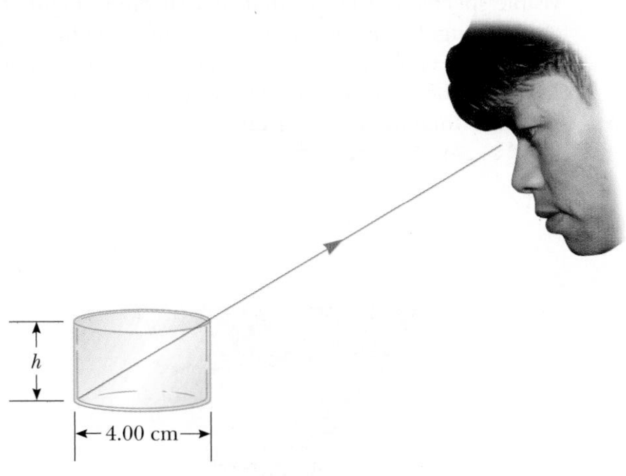

FIGURE **P25.55**

56. A material having an index of refraction n is surrounded by a vacuum and is in the shape of a quarter circle of radius R (Fig. P25.56). A light ray parallel to the base of the material is incident from the left at a distance L above the base and emerges from the material at the angle θ. Determine an expression for θ.

57. A transparent cylinder of radius $R = 2.00$ m has a mirrored surface on its right half as shown in Figure P25.57. A light ray traveling in air is incident on the left side of the cylinder. The incident light ray and exiting light ray are parallel, and $d = 2.00$ m. Determine the index of refraction of the material.

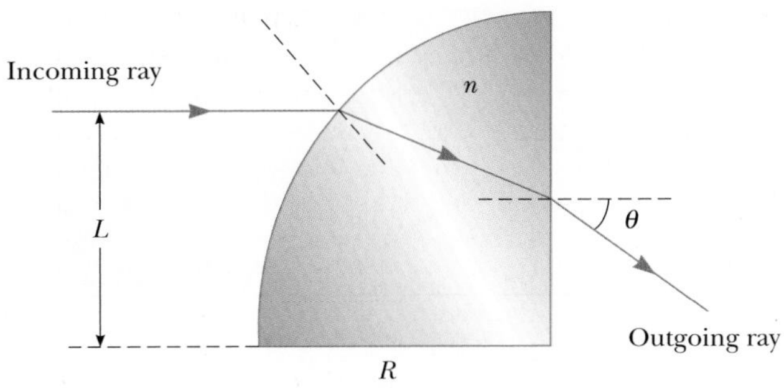

FIGURE **P25.56**

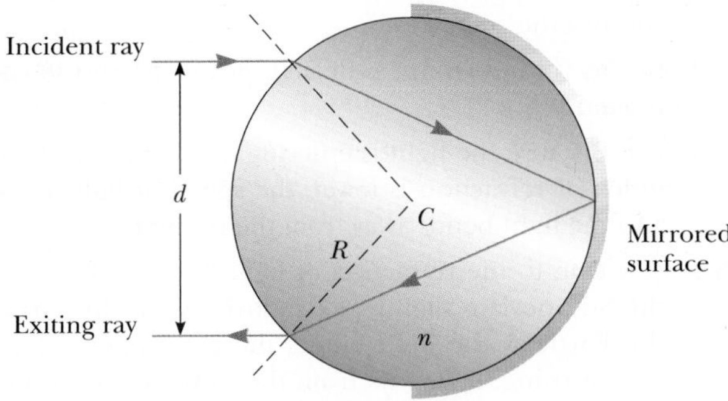

FIGURE **P25.57**

58. Students allow a narrow beam of laser light to strike a water surface. They arrange to measure the angle of refraction for selected angles of incidence and record the data shown in the accompanying table. Use the data to verify Snell's law of refraction by plotting the sine of the angle of incidence versus the sine of the angle of refraction. Use the resulting plot to deduce the index of refraction of water.

Angle of Incidence (degrees)	Angle of Refraction (degrees)
10.0	7.5
20.0	15.1
30.0	22.3
40.0	28.7
50.0	35.2
60.0	40.3
70.0	45.3
80.0	47.7

59. A light ray enters a rectangular block of plastic at an angle $\theta_1 = 45.0°$ and emerges at an angle $\theta_2 = 76.0°$ as shown in Figure P25.59. (a) Determine the index of refraction of the plastic. (b) If the light ray enters the plastic at a point $L = 50.0$ cm from the bottom edge, how long does it take the light ray to travel through the plastic?

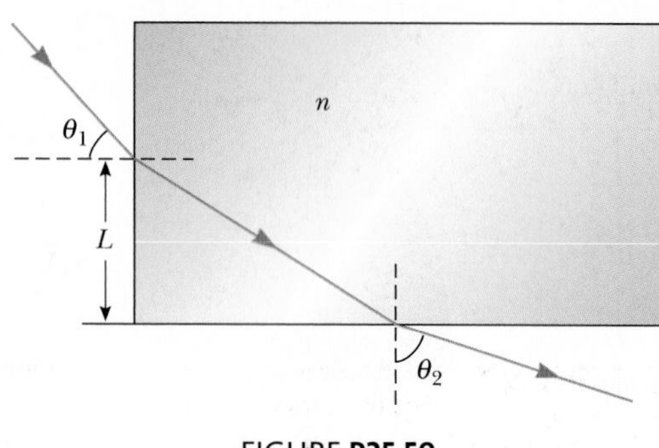

FIGURE **P25.59**

60. **Review problem.** A mirror is often "silvered" with aluminum. By adjusting the thickness of the metallic film, one can make a sheet of glass into a mirror that reflects anything between say 3% and 98% of the incident light, transmitting the rest. Prove that it is impossible to construct a "one-way mirror" that would reflect 90% of the electromagnetic waves incident from one side and reflect 10% of those incident from the other side. (*Suggestion:* Use Clausius's statement of the second law of thermodynamics.)

ANSWERS TO QUICK QUIZZES

25.1 (d). The light rays from the actor's face must reflect from the mirror and into the camera. If these light rays are reversed, light from the camera reflects from the mirror into the actor's eyes.

25.2 Beams ② and ④ are reflected; beams ③ and ⑤ are refracted.

25.3 (c). Because the light is entering a material in which the index of refraction is lower, the speed of light is higher and the light bends away from the normal.

25.4 (a). Due to the refraction of light by air, light rays from the Sun deviate slightly downward toward the surface of the Earth as the light enters the atmosphere. Thus, in the morning, light rays from the upper edge of the Sun arrive at your eyes before the geometric line from your eyes to the top of the Sun clears the horizon. In the evening, light rays from the top of the Sun continue to arrive at your eyes even after the geometric line from your eyes to the top of the Sun dips below the horizon.

25.5 False. There is no dependence of the angle of reflection on wavelength because the light does not enter deeply into the material during reflection; rather, it reflects from the surface. Thus, the properties of the material do not affect the angle of reflection.

25.6 (i), (b). The two bright rays exiting the bottom of the prism on the right in Figure 25.23 result from total internal reflection at the right face of the prism. Note that there is no refracted light exiting the slanted side for these rays. The light from the other three rays is divided into reflected and refracted parts. (ii), (b). Counterclockwise rotation of the prism will cause the rays to strike the slanted side of the prism at a larger angle. When all five rays strike at an angle larger than the critical angle, they will all undergo total internal reflection.

25.7 (c). When the outgoing beam approaches the direction parallel to the straight side, the incident angle is approaching the critical angle for total internal reflection. The index of refraction for light at the violet end of the visible spectrum is larger than that at the red end. Thus, as the outgoing beam approaches the straight side, the violet light experiences total internal reflection first, followed by the other colors. The red light is the last to experience total internal reflection.

CHAPTER 26

Image Formation by Mirrors and Lenses

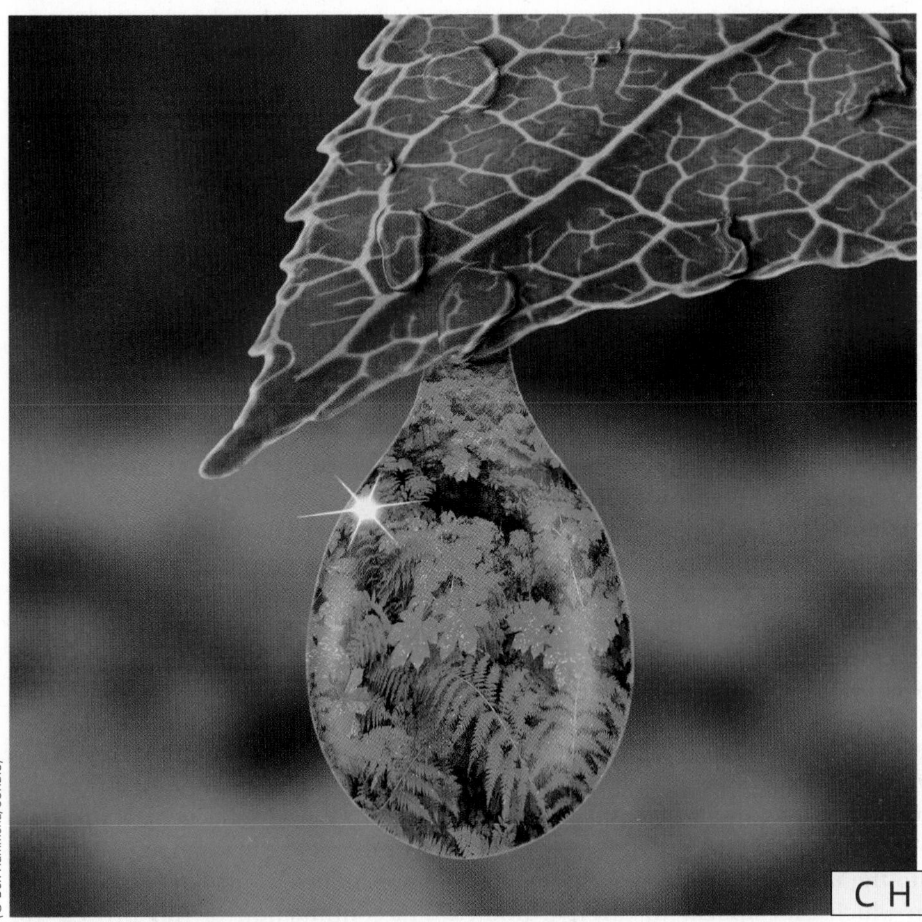

(© Don Hammond/CORBIS)

The light rays coming from the leaves in the background of this scene did not form a focused image on the film of the camera that took this photograph. Consequently, the background appears very blurry. Light rays passing though the raindrop, however, have been altered so as to form a focused image of the background leaves on the film. In this chapter, we investigate the formation of images as light rays reflect from mirrors and refract through lenses.

CHAPTER OUTLINE

This chapter is concerned with the images formed when light interacts with flat and curved surfaces. We find that images of an object can be formed by reflection or by refraction and that mirrors and lenses work because of these phenomena.

Images formed by reflection and refraction are used in a variety of everyday devices, such as the rearview mirror in your car, a shaving or makeup mirror, a camera, your eyeglasses, and a magnifying glass. In addition, more scientific devices, such as telescopes and microscopes, take advantage of the image formation principles discussed in this chapter.

We shall make extensive use of geometric models developed from the principles of reflection and refraction. Such constructions allow us to develop mathematical representations for the image locations of various types of mirrors and lenses.

26.1 IMAGES FORMED BY FLAT MIRRORS

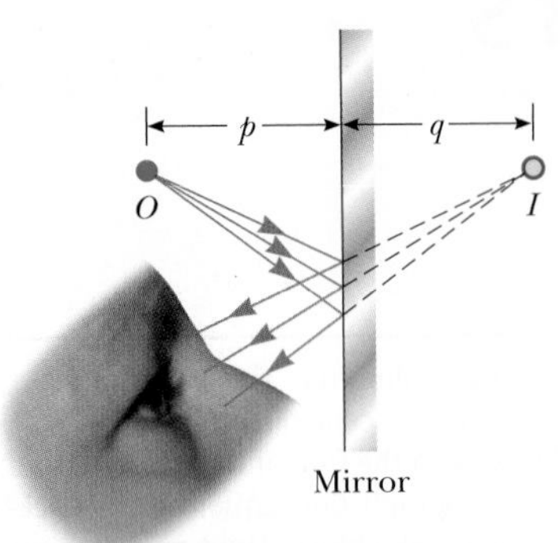

FIGURE 26.1 An image formed by reflection from a flat mirror. The image point I is located behind the mirror at a distance q, which is equal to the object distance p.

We begin by considering the simplest possible mirror, the flat mirror. Consider a point source of light[1] placed at O in Figure 26.1, a distance p in front of a flat mirror. The distance p is called the **object distance.** Light rays leave the source and are reflected from the mirror. Upon reflection, the rays continue to diverge (spread apart). The dashed lines in Figure 26.1 are extensions of the diverging rays back to a point of intersection at I. The diverging rays appear to the viewer to come from the point I behind the mirror. Point I is called the **image** of the object at O. Regardless of the system under study, we always locate images by extending diverging rays back to a point at which they intersect.[2] **Images are located either at a point from which rays of light *actually* diverge or at a point from which they *appear* to diverge.** Because the rays in Figure 26.1 appear to originate at I, which is a distance q behind the mirror, that is the location of the image. The distance q is called the **image distance.**

Images are classified as **real** or **virtual. A real image is formed when light rays pass through and diverge from the image point; a virtual image is formed when the light rays do not pass through the image point but only appear to diverge from that point.** The image formed by the mirror in Figure 26.1 is virtual. The image of an object seen in a flat mirror is *always* virtual. Real images can be displayed on a screen (as at a movie), but virtual images cannot be displayed on a screen. We shall see an example of a real image in Section 26.2.

Active Figure 26.2 is an example of a specialized pictorial representation, called a **ray diagram,** that is very useful in studies of mirrors and lenses. In a ray diagram, a small number of the myriad rays leaving a point source are drawn, and the location of the image is found by applying the laws of reflection (and refraction, in the case of refracting surfaces and lenses) to these rays. A carefully drawn ray diagram allows us to build a geometric model so that geometry and trigonometry can be used to solve a problem mathematically.

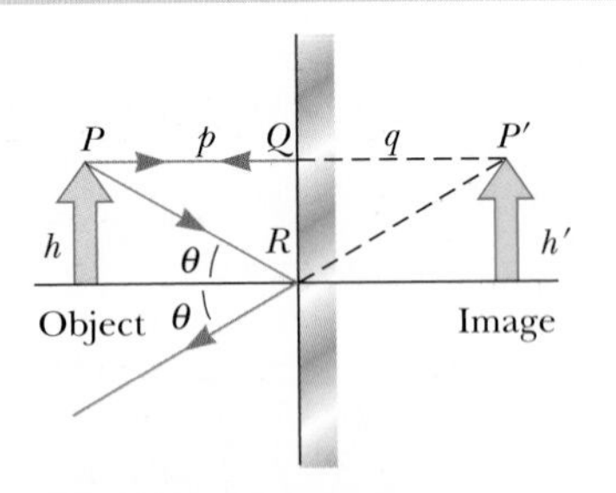

ACTIVE FIGURE 26.2 Geometric construction used to locate the image of an object placed in front of a flat mirror. Because the triangles PQR and $P'QR$ are congruent, $p = |q|$ and $h = h'$.

Physics Now™ Log into PhysicsNow at www.pop4e.com and go to Active Figure 26.2 to move the object and see the effect on the image.

We can use the simple geometry in Active Figure 26.2 to examine the properties of the images of extended objects formed by flat mirrors. Let us locate the image of the tip of the blue arrow. To find out where the image is formed, it is necessary to follow at least two rays of light as they reflect from the mirror. One of those rays starts at P, follows the horizontal path PQ to the mirror, and reflects back on itself. The second ray follows the oblique path PR and reflects at the same angle according to the law of reflection. We can extend the two reflected rays back to the point from which they appear to diverge, point P'. A continuation of this process for points other than P on the object would result in an image (drawn as a yellow arrow) to the right of the mirror. These rays and the extensions of the rays allow us to build a geometric model for the image formation based on triangles PQR and $P'QR$. Because these two triangles are identical, $PQ = P'Q$, or $p = |q|$. (We use the absolute value notation because, as we shall see shortly, a sign convention is associated with the values of p and q.) Hence, we conclude that **the image formed by an object placed in front of a flat mirror is as far behind the mirror as the object is in front of the mirror.**

Our geometric model also shows that the object height h equals the image height h'. We define the **lateral magnification** (or simply the **magnification**) M of an

PITFALL PREVENTION 26.1

MAGNIFICATION DOES NOT NECESSARILY IMPLY ENLARGEMENT For optical elements other than flat mirrors, the magnification defined in Equation 26.1 can result in a number with a magnitude larger *or* smaller than 1. Thus, despite the cultural usage of the word *magnification* to mean *enlargement,* the image could be smaller than the object. We shall see examples of such a situation in this chapter.

[1]We imagine the object to be a point source of light. It could actually *be* a point source, such as a very small lightbulb, but more often is a single point on some extended object that is illuminated from the exterior by a light source. Thus, the reflected light leaves the point on the object as if the point were a source of light.

[2]Your eyes and brain interpret diverging light rays as originating at the point from which the rays diverge. Your eye–brain system can detect the rays only *as they enter your eye* and has no access to information about what experiences the rays underwent before reaching your eyes. Thus, even though the light rays did not *actually originate* at point I, they enter the eye *as if they had,* and I is the point at which your brain locates the object.

image as follows:

$$M \equiv \frac{\text{image height}}{\text{object height}} = \frac{h'}{h} \qquad [26.1]$$

■ Magnification of an image

which is a general definition of the magnification for any type of image formed by a mirror or lens. Because $h' = h$ in this case, $M = 1$ for a flat mirror. We also note that the image is **upright** because the image arrow points in the same direction as the object arrow. An upright image is indicated mathematically by a positive value of the magnification. (Later we discuss situations in which *inverted* images, with negative magnifications, are formed.)

Finally, note that a flat mirror produces an image having an *apparent* left–right reversal. This reversal can be seen by standing in front of a mirror and raising your right hand. The image you see raises its left hand. Likewise, your hair appears to be parted on the opposite side, and a mole on your right cheek appears to be on your left cheek.

This reversal is not *actually* a left–right reversal. Imagine, for example, lying on your left side on the floor, with your body parallel to the mirror surface. Now, your head is on the left and your feet are on the right as you face the mirror. If you shake your feet, the image does not shake its head! If you raise your right hand, however, the image raises its left hand. Thus, it again appears like a left–right reversal, but in an up–down direction!

The apparent left–right reversal is actually a *front–back* reversal caused by the light rays going forward toward the mirror and then reflecting back from it. Figure 26.3 shows a person's right hand and its image in a flat mirror. Notice that no left–right reversal takes place; rather, the thumbs on both the real hand and the image are on the left side. It is the front–back reversal that makes the image of the right hand appear similar to the real left hand at the left side of the photograph.

An interesting experience with front–back reversal is to stand in front of a mirror while holding an overhead transparency in front of you so that you can read the writing on the transparency. You are also able to read the writing on the image of the transparency. You might have had a similar experience if you have a transparent decal with words on it on the rear window of your car. If the decal is placed so that it can be read from outside the car, you can also read it when looking into your rearview mirror from the front seat.

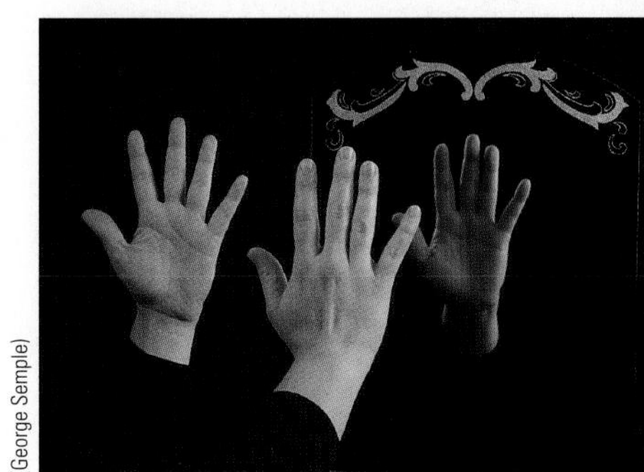

(George Semple)

FIGURE 26.3 The image in the mirror of a person's right hand is reversed front to back, which makes the image in the mirror appear to be a left hand. Notice that the thumb is on the left side of both real hands and on the left side of the image. That the thumb is not on the right side of the image indicates that the reversal is not left–right.

QUICK QUIZ 26.1 In the overhead view of Figure 26.4, the image of the stone seen by observer 1 is at C. At which of the five points A, B, C, D, or E does observer 2 see the image?

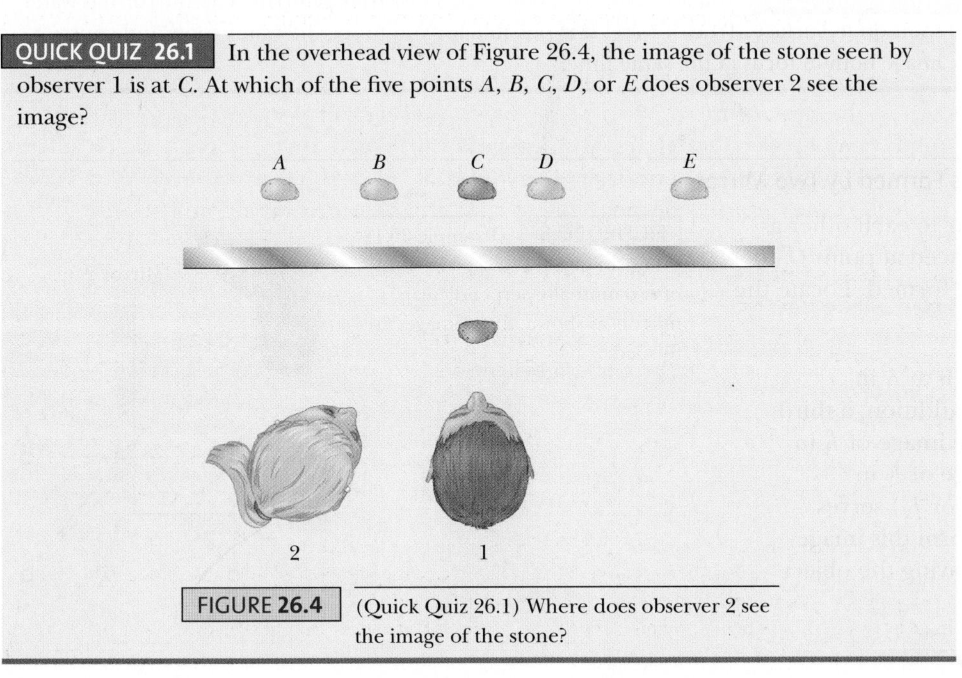

FIGURE 26.4 (Quick Quiz 26.1) Where does observer 2 see the image of the stone?

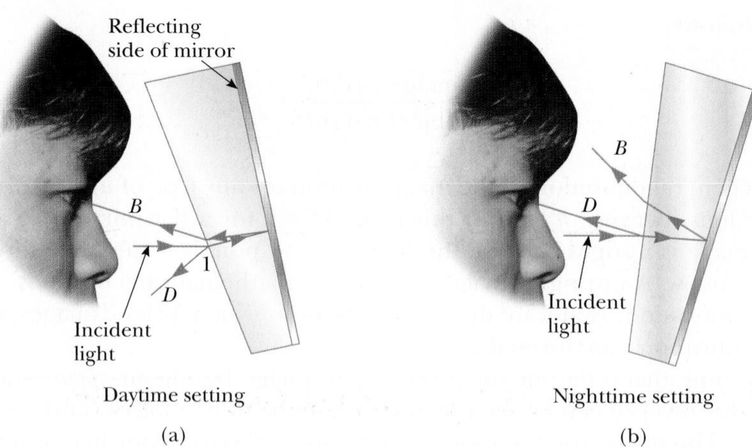

FIGURE 26.5 (Thinking Physics 26.1) (a) Daytime and (b) nighttime settings of a rearview mirror in an automobile.

Thinking Physics 26.1

Most rearview mirrors on cars have a day setting and a night setting. The night setting greatly diminishes the intensity of the image so that lights from trailing vehicles do not blind the driver. How does such a mirror work?

Reasoning Figure 26.5 represents a cross-sectional view of the mirror for the two settings. The mirror is a wedge of glass with a reflecting surface on the back side. When the mirror is in the day setting, as in Figure 26.5a, the light from an object behind the car strikes the mirror at point 1. Most of the light enters the wedge, is refracted, and reflects from the back surface to return to the front surface, where it is refracted again as it re-enters the air as ray B (for *bright*). In addition, a small portion of the light is reflected at the front surface as indicated by ray D (for *dim*). This dim reflected light is responsible for the image observed when the mirror is in the night setting, as in Figure 26.5b. In this case, the wedge is rotated so that the path followed by the bright light (ray B) does not lead to the eye. Instead, the dim light reflected from the front surface travels to the eye, and the brightness of trailing headlights does not become a hazard. ■

QUICK QUIZ 26.2 You are standing about 2 m away from a mirror. The mirror has water spots on its surface. True or false: It is possible for you to see the water spots and your image both in focus at the same time.

EXAMPLE 26.1 Multiple Images Formed by Two Mirrors

Two flat mirrors are perpendicular to each other as in Figure 26.6, and an object is placed at point O. In this situation, multiple images are formed. Locate the positions of these images.

Solution The image of the object is at I_1 in mirror 1 and at I_2 in mirror 2. In addition, a third image is formed at I_3, which is the image of I_1 in mirror 2 or, equivalently, the image of I_2 in mirror 1. That is, the image at I_1 (or I_2) serves as the object for I_3. Note that to form this image at I_3, the rays reflect twice after leaving the object at O.

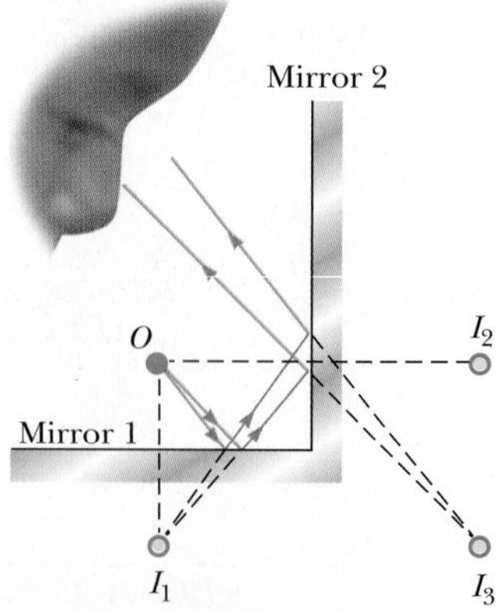

FIGURE 26.6 (Example 26.1) When an object is placed in front of two mutually perpendicular mirrors as shown, three images are formed.

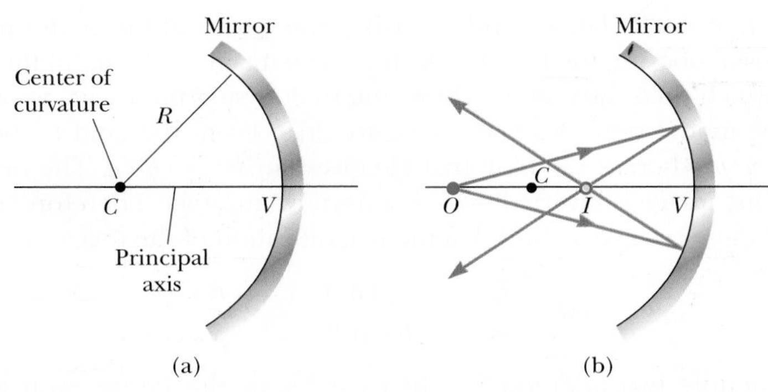

FIGURE 26.7 (a) A concave mirror of radius R. The center of curvature C is located on the principal axis. (b) A point source of light placed at O in front of a concave spherical mirror of radius R, where O is any point on the principal axis farther than R from the mirror surface, forms a real image at I. If the rays diverge from O at small angles, they all reflect through the same image point.

26.2 IMAGES FORMED BY SPHERICAL MIRRORS

In Section 26.1, we investigated images formed by a flat reflecting surface. In this section, we will explore images formed by curved mirrors, either from a concave surface of the mirror or a convex surface.

Concave Mirrors

A **spherical mirror,** as its name implies, has the shape of a segment of a sphere. Figure 26.7a shows the cross-section of a spherical mirror with its reflecting surface represented by the solid curved line. Such a mirror in which light is reflected from the inner, concave surface is called a **concave mirror.** The mirror's radius of curvature is R, and its center of curvature is at point C. Point V is the center of the spherical segment, and a line drawn from C to V is called the **principal axis** of the mirror.

Now consider a point source of light placed at point O in Figure 26.7b, on the principal axis and outside point C. Two diverging rays that originate at O are shown. After reflecting from the mirror, these rays converge and meet at I, the image point. They then continue to diverge from I as if a source of light existed there. Therefore, if your eyes detect the rays diverging from point I, you would claim that a light source is located at that point.

This example is the second one we have seen of rays diverging from an image point. Because the light rays pass through the image point in this case, unlike the situation in Active Figure 26.2, the image in Figure 26.7b is a real image.

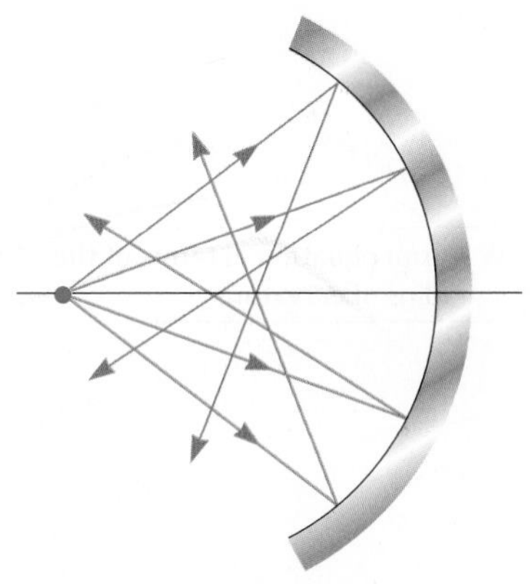

FIGURE 26.8 Rays diverging from an object at large angles from the principal axis reflect from a spherical concave mirror to intersect the principal axis at different points, resulting in a blurred image.

In what follows, we shall adopt a simplification model that assumes that all rays diverging from an object make small angles with the principal axis. Such rays, called **paraxial rays,** always reflect through the image point as in Figure 26.7b. Rays that make large angles with the principal axis as in Figure 26.8 converge at other points on the principal axis, producing a blurred image.

We can use a geometric model based on the ray diagram in Figure 26.9 to calculate the image distance q if we know the object distance p and radius of curvature R. By convention, these distances are measured from point V. Figure 26.9 shows two of the many light rays leaving the tip of the object. One ray passes through the center of curvature C of the mirror, hitting the mirror perpendicular to the mirror surface and

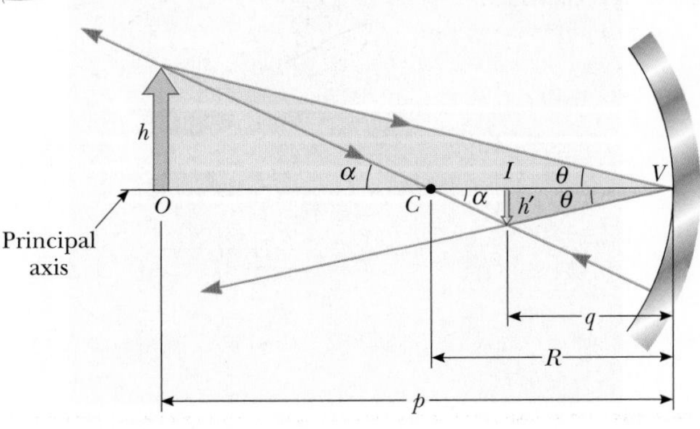

FIGURE 26.9 The image formed by a spherical concave mirror when the object O lies outside the center of curvature C. This geometric construction is used to derive Equation 26.4.

reflecting back on itself. The second ray strikes the mirror at the center point V and reflects as shown, obeying the law of reflection. The image of the tip of the arrow is at the point at which these two reflected rays intersect. Using these rays, we identify the gold and blue model right triangles in Figure 26.9. From the gold triangle, we see that $\tan\theta = h/p$, whereas the blue triangle gives $\tan\theta = -h'/q$. The negative sign signifies that the image is inverted, so h' is a negative number. Therefore, from Equation 26.1 and these results, we find that the magnification of the image is

$$M = \frac{h'}{h} = \frac{-q\tan\theta}{p\tan\theta} = -\frac{q}{p} \qquad [26.2]$$

We can identify two additional right triangles in the figure, with a common point at C and with angle α. These triangles tell us that

$$\tan\alpha = \frac{h}{p-R} \quad \text{and} \quad \tan\alpha = -\frac{h'}{R-q}$$

from which we find that

$$\frac{h'}{h} = -\frac{R-q}{p-R} \qquad [26.3]$$

If we compare Equations 26.2 and 26.3, we see that

$$\frac{R-q}{p-R} = \frac{q}{p}$$

Algebra reduces this expression to

■ Mirror equation in terms of the radius of curvature

$$\frac{1}{p} + \frac{1}{q} = \frac{2}{R} \qquad [26.4]$$

which is called the **mirror equation.** It is applicable only to the paraxial ray simplification model.

If the object is very far from the mirror—that is, if the object distance p is large compared with R, so that p can be said to approach infinity—$1/p \to 0$, and we see from Equation 26.4 that $q \approx R/2$. In other words, **when the object is very far from the mirror, the image point is halfway between the center of curvature and the center of the mirror** as in Figure 26.10a. The rays are essentially parallel in this figure because only those few rays traveling parallel to the axis from the distant object encounter the mirror. Rays not parallel to the axis miss the mirror. Figure 26.10b shows an experimental setup of this situation, demonstrating the crossing of the light rays at a single point. The point at which the parallel rays intersect after reflecting from the mirror is called the **focal point** of the mirror. The focal point is a

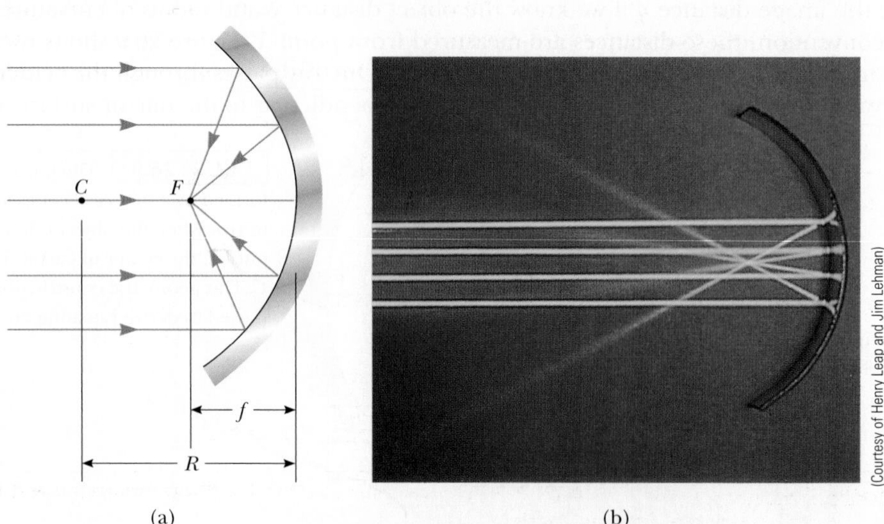

FIGURE 26.10 (a) Light rays from a distant object ($p \approx \infty$) reflect from a concave mirror through the focal point F. In this case, the image distance $q \approx R/2 = f$, where f is the focal length of the mirror. (b) Reflection of parallel rays from a concave mirror.

distance f from the mirror, called the **focal length.** The focal length is a parameter associated with the mirror and is given by

$$f = \frac{R}{2} \quad [26.5]$$

■ Focal length of a mirror

The mirror equation can therefore be expressed in terms of the focal length:

$$\frac{1}{p} + \frac{1}{q} = \frac{1}{f} \quad [26.6]$$

■ Mirror equation in terms of focal length

This equation is the commonly used mirror equation, in terms of the focal length of the mirror rather than its radius of curvature, as in Equation 26.4. We shall see how to use this equation in examples that follow shortly.

PITFALL PREVENTION 26.2

THE *FOCAL* POINT IS NOT THE *FOCUS* POINT The focal point *is usually not* the point at which the light rays focus to form an image. The focal point is determined solely by the curvature of the mirror; it does not depend on the location of the object at all. In general, an image forms at a point different from the focal point of a mirror (or a lens). The *only* exception is when the object is located infinitely far away from the mirror.

Convex Mirrors

Figure 26.11 shows the formation of an image by a **convex mirror,** a mirror that is silvered so that light is reflected from the outer, convex surface. Convex mirrors are sometimes called **diverging mirrors** because the rays from any point on an object diverge after reflection as though they were coming from some point behind the mirror. The image in Figure 26.11 is virtual rather than real because it lies behind the mirror at the point from which the reflected rays appear to diverge. In general, as shown in the figure, the image formed by a convex mirror is always upright, virtual, and smaller than the object.

We can set up a geometric model for a convex mirror using the ray diagram in Figure 26.11. The equations developed for concave mirrors can also be used with convex mirrors if we adhere to a particular sign convention. Let us refer to the region in which light rays move as the *front side* of the mirror and the other side, where virtual images are formed, as the *back side.* For example, in Figures 26.9 and 26.11, the side to the left of the mirror is the front side and that to the right of the mirror is the back side. Table 26.1 summarizes the sign conventions for all the necessary quantities. Notice in particular that we handle a convex mirror by assigning it a negative focal length. With this convention, the mirror equation for a convex mirror is the same as that for a concave mirror, Equation 26.6.

One entry in Table 26.1 that may appear strange is a "virtual object." A virtual object will only occur in some situations when combining two or more optical elements as we shall see in Section 26.4.

(Courtesy of Thomson Consumer Electronics).

A satellite-dish antenna is a concave reflector for television signals from a satellite in orbit around the Earth. The signals are carried by microwaves that, because the satellite is so far away, are parallel when they arrive at the dish. These waves reflect from the dish and are focused on the receiver at the focal point of the dish. ■

Constructing Ray Diagrams for Mirrors

We have been using the specialized pictorial representations called ray diagrams to help us locate images for flat and curved mirrors. Let us now formalize the procedure for drawing accurate ray diagrams. To construct such a diagram, we must know the position of the object and the locations of the focal point and center of

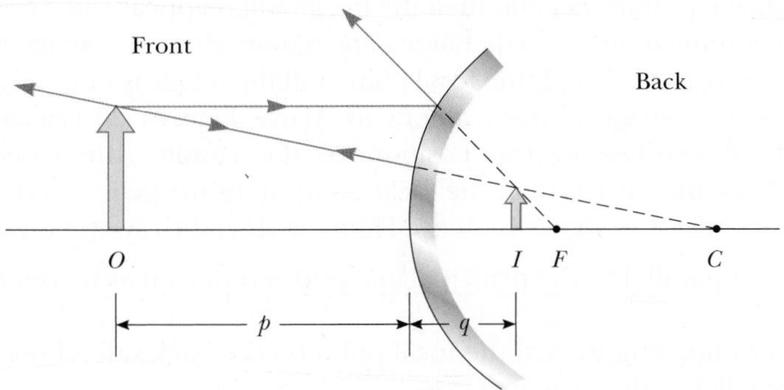

FIGURE 26.11 Formation of an image by a spherical convex mirror. The image formed by the real object is virtual and upright.

TABLE 26.1 Sign Conventions for Mirrors

Quantity	Positive when . . .	Negative when . . .
Object location (p)	object is in front of mirror (real object).	object is in back of mirror (virtual object).
Image location (q)	image is in front of mirror (real image).	image is in back of mirror (virtual image).
Image height (h')	image is upright.	image is inverted.
Focal length (f) and radius (R)	mirror is concave.	mirror is convex.
Magnification (M)	image is upright.	image is inverted.

PITFALL PREVENTION 26.3

WATCH YOUR SIGNS Success in working mirror problems (as well as problems involving refracting surfaces and thin lenses) is largely determined by proper sign choices when substituting into the equations. The best way to become adept at these problems is to work a multitude of them on your own. Watching your instructor or reading the example problems is no substitute for practice.

curvature of the mirror. We will construct three rays in the examples shown in Active Figure 26.12. Only two rays are necessary to locate the image, but we will include a third as a check. In each part of the figure, the right-hand portion shows a photograph of the situation described by the ray diagram in the left-hand portion. All three rays start from the same object point; in these examples, the top of the arrow is chosen as the starting point. For the concave mirrors in Active Figure 26.12a and 26.12b, the rays are drawn as follows:

- Ray 1 is drawn parallel to the principal axis and is reflected back through the focal point F. (This ray would be a light path followed by light from an object infinitely far from the mirror.)
- Ray 2 is drawn through the focal point (or as if coming from the focal point if $p < f$ as in Active Fig. 26.12b). It is reflected parallel to the principal axis. (This ray would be a light path followed by light from an object at the focal point and is the reverse of a ray approaching the mirror from an object infinitely far away.)
- Ray 3 is drawn through the center of curvature C and is reflected back on itself. (This ray follows the law of reflection for light incident along the normal to the surface.)

The image point obtained in this fashion must always agree with the value of q calculated from the mirror equation. With concave mirrors, note what happens as the object is moved closer to the mirror from infinity. The real, inverted image in Active Figure 26.12a moves to the left as the object approaches the mirror. When the object is at the center of curvature, the object and image are at the same distance from the mirror and are the same size. When the object is at the focal point, the image is infinitely far to the left. (Check these last three sentences with the mirror equation!)

When the object lies between the focal point and the mirror surface as in Active Figure 26.12b, the image is virtual, upright, and located on the back side of the mirror. The image is also larger than the object in this case. This situation illustrates the principle behind a shaving mirror or a makeup mirror. Your face is located closer to the concave mirror than the focal point, so you see an enlarged, upright image of your face, to assist you with shaving or applying makeup. If you have such a mirror, look into it and move your face farther from the mirror. Your head will pass through a point at which the image is indistinct and then the image will reappear with your face upside down as you continue to move farther away. The region where the image is indistinct is where your head passes through the focal point and the image is infinitely far away.

Notice that the image of the camera in Active Figures 26.12a and 26.12b is upside down. Regardless of the position of the candle, the camera remains farther away from the mirror than the focal point, so its image is inverted.

For a convex mirror as shown in Active Figure 26.12c, the rays are drawn as follows:

- Ray 1 is drawn parallel to the principal axis and is reflected as if coming from the focal point F.
- Ray 2 is drawn heading toward the focal point on the back side of the mirror. It is reflected parallel to the principal axis.
- Ray 3 is drawn heading toward the center of curvature C on the back side of the mirror and is reflected back on itself.

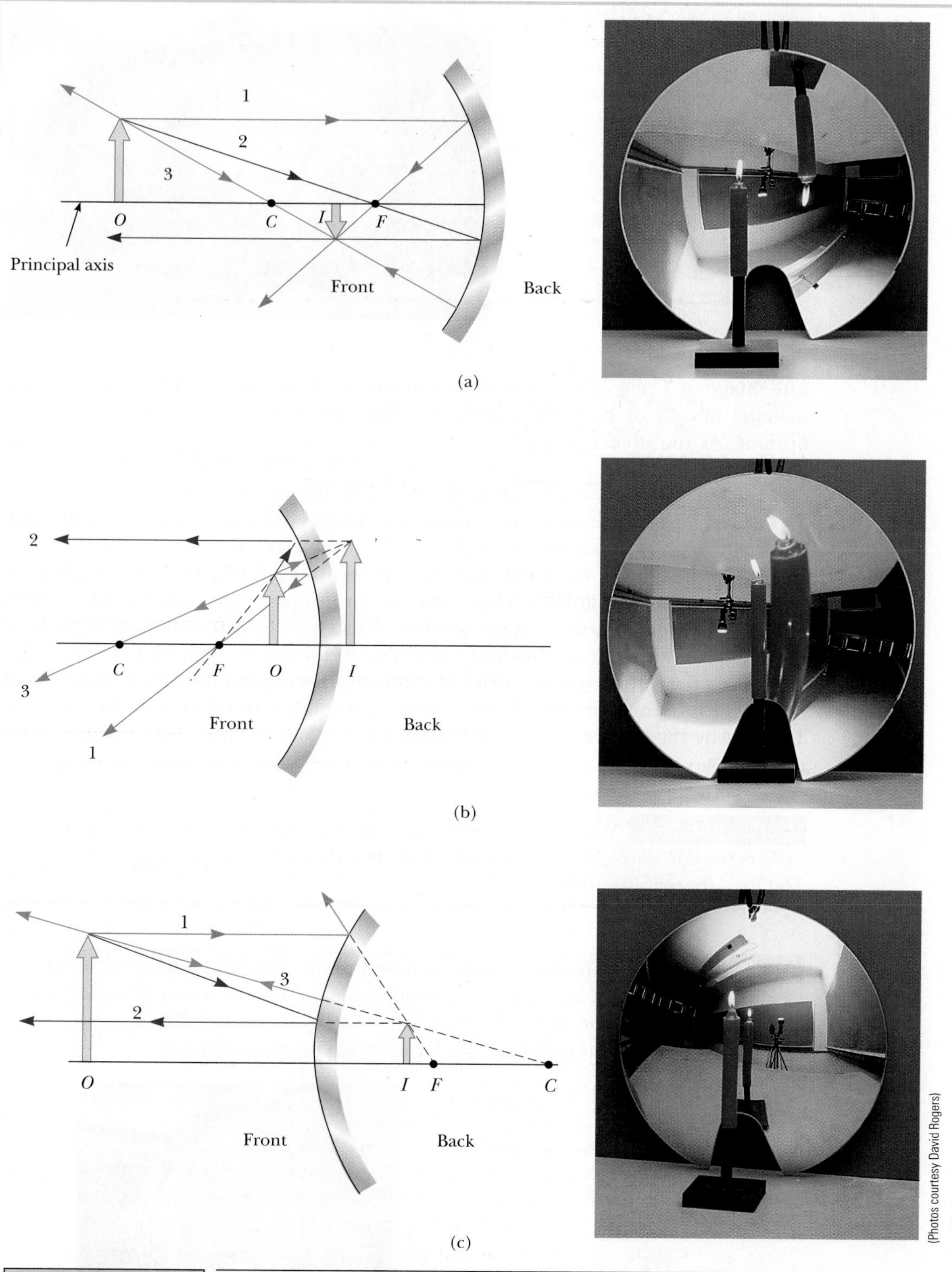

ACTIVE FIGURE 26.12 Ray diagrams for spherical mirrors, along with corresponding photographs of the images of a candle as the object. (a) When the object is located so that the focal point lies between the object and a concave mirror surface, the image is real and inverted. (b) When the object is located between the focal point and a concave mirror surface, the image is virtual, upright, and enlarged. (c) When the object is in front of a spherical convex mirror, the image is virtual, upright, and reduced in size.

PhysicsNow™ Log into PhysicsNow at **www.pop4e.com** and go to Active Figure 26.12 to move the objects and change the focal lengths of the mirrors to see the effect on the images.

FIGURE 26.13 An approaching truck is seen in a convex mirror on the right side of an automobile. Because the image is reduced in size, the truck appears to be farther away than it actually is. Notice also that the image of the truck is in focus but that the frame of the mirror is not, which demonstrates that the image is not at the same location as the mirror surface. (© Bo Zaunders/CORBIS)

The image of a real object in a convex mirror is always virtual and upright. Notice that the images of both the candle and the camera in Active Figure 26.12c are upright. As the object distance increases, the virtual image becomes smaller and approaches the focal point as p approaches infinity. You should construct other diagrams to verify how the image position varies with object position.

Convex mirrors are often used as security devices in large stores, where they are hung at a high position on the wall. The large field of view of the store is made smaller by the convex mirror so that store personnel can observe possible shoplifting activity in several aisles at once. Mirrors on the passenger side of automobiles are also often made with a convex surface. This type of mirror allows a wider field of view behind the automobile to be available to the driver (Fig. 26.13) than is the case with a flat mirror. These mirrors introduce a perceptual distortion, however, in that they cause cars behind the viewer to appear smaller and therefore farther away. That is why these mirrors carry the inscription, "Objects in this mirror are closer than they appear."

QUICK QUIZ 26.3 You wish to reflect sunlight from a mirror onto some paper under a pile of wood to start a fire. Which would be the best choice for the type of mirror? **(a)** flat **(b)** concave **(c)** convex

QUICK QUIZ 26.4 Consider the image in the mirror in Figure 26.14. Based on the appearance of this image, what conclusion would you make? **(a)** The mirror is concave and the image is real. **(b)** The mirror is concave and the image is virtual. **(c)** The mirror is convex and the image is real. **(d)** The mirror is convex and the image is virtual.

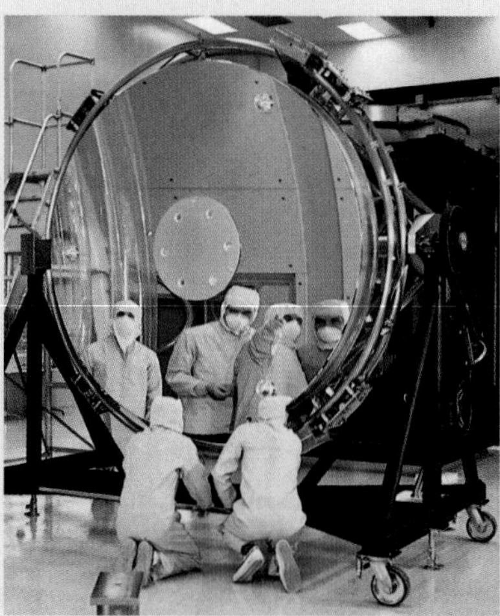

FIGURE 26.14 (Quick Quiz 26.4) What type of mirror is this one? (NASA)

INTERACTIVE EXAMPLE 26.2 The Image Formed by a Concave Mirror

A concave spherical mirror has a focal length of 10.0 cm.

A Find the location of the image for an object distance of 25.0 cm and describe the image.

Solution For an object distance of 25.0 cm, we find the image distance using the mirror equation:

$$\frac{1}{p} + \frac{1}{q} = \frac{1}{f}$$

$$\frac{1}{25.0 \text{ cm}} + \frac{1}{q} = \frac{1}{10.0 \text{ cm}}$$

$$q = \boxed{16.7 \text{ cm}}$$

The magnification is given by Equation 26.2:

$$M = -\frac{q}{p} = -\frac{16.7 \text{ cm}}{25.0 \text{ cm}} = -0.668$$

The magnitude of M less than unity tells us that the image is smaller than the object. The negative sign for M tells us that the image is inverted. Finally, because q is positive, the image is located on the front side of the mirror and is real. This situation is pictured in Active Figure 26.12a.

B Find the location of the image for an object distance of 10.0 cm and describe the image.

Solution When the object distance is 10.0 cm, the object is located at the focal point. Substituting the values $p = 10.0$ cm and $f = 10.0$ cm into the mirror equation, we find that

$$\frac{1}{10.0 \text{ cm}} + \frac{1}{q} = \frac{1}{10.0 \text{ cm}}$$

$$q = \boxed{\infty}$$

Thus, we see that light rays originating from an object located at the focal point of a mirror are reflected so that the image is formed at an infinite distance from the mirror; that is, the rays travel parallel to one another after reflection.

C Find the location of the image for an object distance of 5.00 cm and describe the image.

Solution When the object is at the position $p = 5.00$ cm, it lies between the focal point and the mirror surface. In this case, the mirror equation gives

$$\frac{1}{5.00 \text{ cm}} + \frac{1}{q} = \frac{1}{10.0 \text{ cm}}$$

$$q = \boxed{-10.0 \text{ cm}}$$

The negative value for q tells us the image is virtual and located on the back side of the mirror. The magnification is

$$M = -\frac{q}{p} = -\left(\frac{-10.0 \text{ cm}}{5.00 \text{ cm}}\right) = 2.00$$

From this value, we see that the image is larger than the object by a factor of 2.00. The positive sign for M indicates that the image is upright (see Active Fig. 26.12b).

Note the characteristics of the images formed by a concave spherical mirror. When the object is farther from the mirror than the focal point, the image is inverted and real; with the object at the focal point, the image is formed at infinity; with the object between the focal point and mirror surface, the image is upright and virtual.

PhysicsNow™ Investigate the image formed for various object positions and mirror focal lengths by logging into PhysicsNow at **www.pop4e.com** and going to Interactive Example 26.2.

INTERACTIVE EXAMPLE 26.3 The Image Formed by a Convex Mirror

An object 3.00 cm high is placed 20.0 cm from a convex mirror having a focal length of 8.00 cm.

A Find the position of the final image.

Solution Because the mirror is convex, its focal length is negative. To find the image position, we use the mirror equation:

$$\frac{1}{p} + \frac{1}{q} = \frac{1}{f} = \frac{1}{-8.00 \text{ cm}}$$

$$\frac{1}{q} = -\frac{1}{8.00 \text{ cm}} - \frac{1}{20.0 \text{ cm}}$$

$$q = \boxed{-5.71 \text{ cm}}$$

The negative value of q indicates that the image is virtual, or behind the mirror, as in Active Figure 26.12c.

B Find the height of the image.

Solution The magnification is

$$M = -\frac{q}{p} = -\left(\frac{-5.71 \text{ cm}}{20.0 \text{ cm}}\right) = 0.286$$

The image is upright because M is positive. Its height is

$$h' = Mh = (0.286)(3.00 \text{ cm}) = \boxed{0.858 \text{ cm}}$$

PhysicsNow™ Investigate the image formed for various object positions and mirror focal lengths by logging into PhysicsNow at **www.pop4e.com** and going to Interactive Example 26.3.

26.3 IMAGES FORMED BY REFRACTION

In this section, we describe how images are formed by the refraction of rays at the surface of a transparent material. We shall apply the law of refraction and use the simplification model in which we consider only paraxial rays.

Consider two transparent media with indices of refraction n_1 and n_2, where the boundary between the two media is a spherical surface with radius of curvature R (Fig. 26.15). We shall assume that the object at point O is in the medium with index of refraction n_1. As we shall see, all paraxial rays are refracted at the spherical surface and converge to a single point I, the image point.

Let us proceed by considering the geometric construction in Figure 26.16, which shows a single ray leaving point O and passing through point I. Snell's law applied to this refracted ray gives

$$n_1 \sin \theta_1 = n_2 \sin \theta_2$$

Because the angles θ_1 and θ_2 are small for paraxial rays, we can use the approximation $\sin \theta \approx \theta$ (angles in radians). Therefore, Snell's law becomes

$$n_1\theta_1 = n_2\theta_2$$

Now we make use of geometric model triangles and recall that an exterior angle of any triangle equals the sum of the two opposite interior angles. Applying this rule to the triangles *OPC* and *PIC* in Figure 26.16 gives

$$\theta_1 = \alpha + \beta$$

$$\beta = \theta_2 + \gamma$$

If we combine the last three equations and eliminate θ_1 and θ_2, we find that

$$n_1\alpha + n_2\gamma = (n_2 - n_1)\beta \qquad [26.7]$$

In the small angle approximation, $\tan \theta \approx \theta$, and so from Figure 26.16 we can write the approximate relations

$$\tan \alpha \approx \alpha \approx \frac{d}{p} \qquad \tan \beta \approx \beta \approx \frac{d}{R} \qquad \tan \gamma \approx \gamma \approx \frac{d}{q}$$

where d is the distance shown in Figure 26.16. We substitute these equations into Equation 26.7 and divide through by d to give

■ Images formed by a refracting surface

$$\frac{n_1}{p} + \frac{n_2}{q} = \frac{n_2 - n_1}{R} \qquad [26.8]$$

Because this expression does not involve any angles, all paraxial rays leaving an object at distance p from the refracting surface will be focused at the same distance q from the surface on the back side.

By setting up a geometric construction with an object and a refracting surface, we can show that the magnification of an image due to a refracting surface is

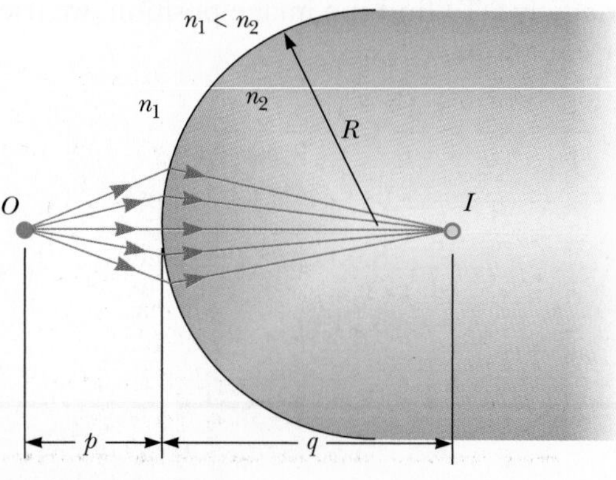

FIGURE 26.15 An image formed by refraction at a spherical surface. Rays making small angles with the principal axis diverge from a point object at O and are refracted through the image point I.

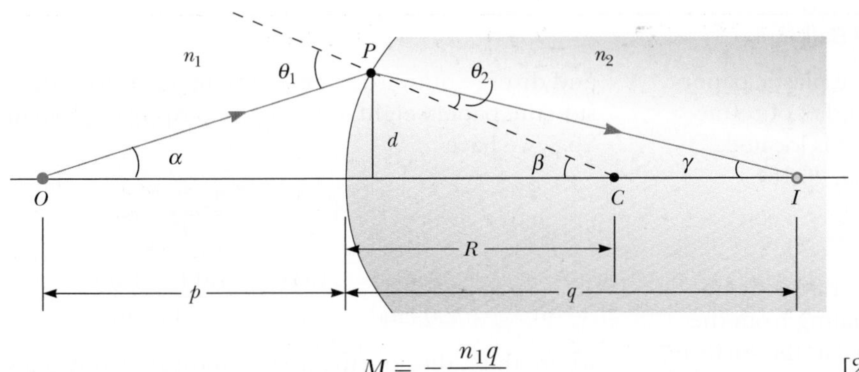

FIGURE 26.16 Geometry used to derive Equation 26.8, assuming that $n_1 < n_2$.

$$M = -\frac{n_1 q}{n_2 p} \qquad [26.9]$$

■ Magnification of an image formed by a refracting surface

As with mirrors, we must use a sign convention if we are to apply Equations 26.8 and 26.9 to a variety of circumstances. Note that real images are formed on the side of the surface that is *opposite* the side from which the light comes. That is in contrast to mirrors, for which real images are formed on the side where the light originates. Therefore, **the sign conventions for spherical refracting surfaces are similar to the conventions for mirrors, recognizing the change in sides of the surface for real and virtual images.** For example, in Figure 26.16, p, q, and R are all positive.

The sign conventions for spherical refracting surfaces are summarized in Table 26.2. The same conventions will be used for thin lenses discussed in the next section. As with mirrors, we assume that the front of the refracting surface is the side from which the light approaches the surface.

Flat Refracting Surfaces

If the refracting surface is flat, R approaches infinity and Equation 26.8 reduces to

$$\frac{n_1}{p} = -\frac{n_2}{q}$$

or

$$q = -\frac{n_2}{n_1} p \qquad [26.10]$$

From Equation 26.10, we see that the sign of q is opposite that of p. Thus, **the image formed by a flat refracting surface is on the same side of the surface as the object.** This situation is illustrated in Active Figure 26.17 for the case in which n_1 is greater than n_2, where a virtual image is formed between the object and the surface. Note that the refracted ray bends *away* from the normal in this case because $n_1 > n_2$.

The value of q given by Equation 26.10 is always smaller in magnitude than p when $n_1 > n_2$. This fact indicates that the image of an object located within a material with higher index of refraction than that of the material from which it is viewed is always closer to the flat refracting surface than the object. Thus, transparent bodies of water such as streams and swimming pools always appear shallower than they are because the image of the bottom of the body of water is closer to the surface than the bottom is in reality.

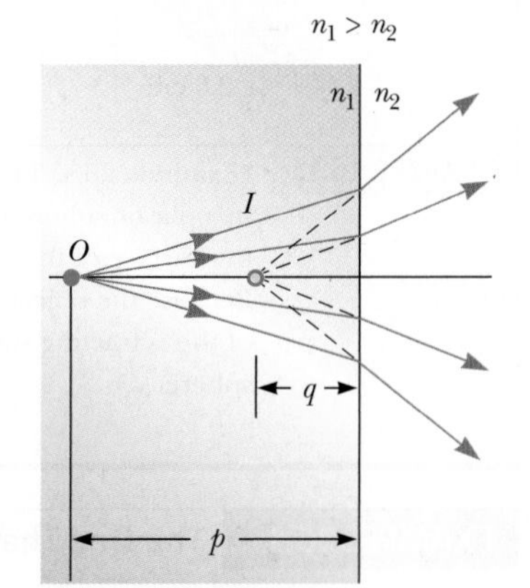

ACTIVE FIGURE 26.17 The image formed by a flat refracting surface is virtual; that is, it forms on the same side of the refracting surface as the object. All rays are assumed to be paraxial.

Physics Now™ Log into PhysicsNow at **www.pop4e.com** and go to Active Figure 26.17 to move the object and see the effect on the location of the image.

TABLE 26.2 Sign Conventions for Refracting Surfaces

Quantity	Positive when . . .	Negative when . . .
Object location (p)	object is in front of surface (real object).	object is in back of surface (virtual object).
Image location (q)	image is in back of surface (real image).	image is in front of surface (virtual image).
Image height (h')	image is upright.	image is inverted.
Radius (R)	center of curvature is in back of surface.	center of curvature is in front of surface.

EXAMPLE 26.4 Gaze into the Crystal Ball

A set of coins is embedded in a spherical plastic paperweight having a radius of 3.0 cm. The index of refraction of the plastic is $n_1 = 1.50$. One coin is located 2.0 cm from the edge of the sphere (Fig. 26.18).

A Find the position of the coin's image.

Solution Because $n_1 > n_2$, where $n_2 = 1.00$ is the index of refraction for air, the rays originating from the coin are refracted away from the normal at the surface and diverge outward. Hence, the image is formed inside the paperweight and is virtual. Applying Equation 26.8, we have

$$\frac{n_1}{p} + \frac{n_2}{q} = \frac{n_2 - n_1}{R}$$

$$\frac{1.50}{2.0\text{ cm}} + \frac{1.00}{q} = \frac{1.00 - 1.50}{-3.0\text{ cm}}$$

where the radius of curvature is indicated as negative because the center of curvature is in front of the concave surface (see Table 26.2). Solving for q gives

$$q = \boxed{-1.7\text{ cm}}$$

The negative sign indicates that the image is in the same medium as the object (the side of the incident light), in agreement with our ray diagram. Because the light rays do not pass through the image point, the image is virtual. The coin appears to be closer to the paperweight surface than it actually is.

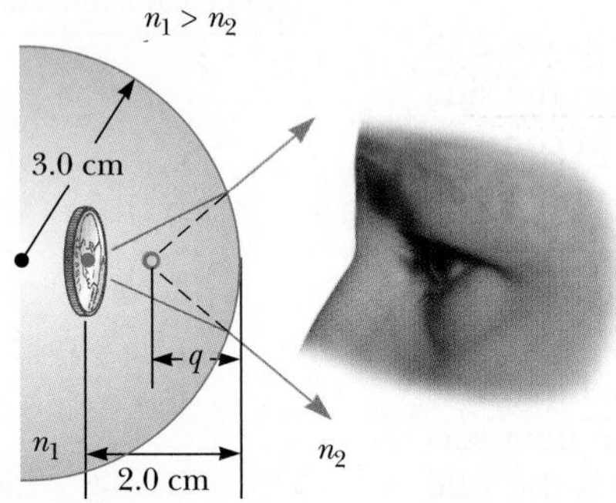

FIGURE 26.18 (Example 26.4) Light rays from a coin embedded in a plastic sphere form a virtual image between the surface of the object and the sphere surface. Because the object is inside the sphere, the front of the refracting surface is the *interior* of the sphere.

B What is the magnification of the image?

Solution Using Equation 26.9, we have

$$M = -\frac{n_1 q}{n_2 p} = -\frac{(1.50)(-1.7\text{ cm})}{(1.00)(2.0\text{ cm})} = \boxed{1.28}$$

Thus, the image is 28% larger than the actual object.

EXAMPLE 26.5 The One That Got Away

A small fish is swimming at a depth d below the surface of a pond (Fig. 26.19).

A What is the apparent depth of the fish as viewed from directly overhead?

Solution In this example, the refracting surface is flat, so R is infinite. Therefore, we can use Equation 26.10 to determine the location of the image. Using that $n_1 = 1.33$ for water and $p = d$ gives

$$q = -\frac{n_2}{n_1}p = -\frac{1.00}{1.33}d = \boxed{-0.752d}$$

Again, because q is negative, the image is virtual as indicated in Figure 26.19a. The apparent depth is approximately three-fourths the actual depth.

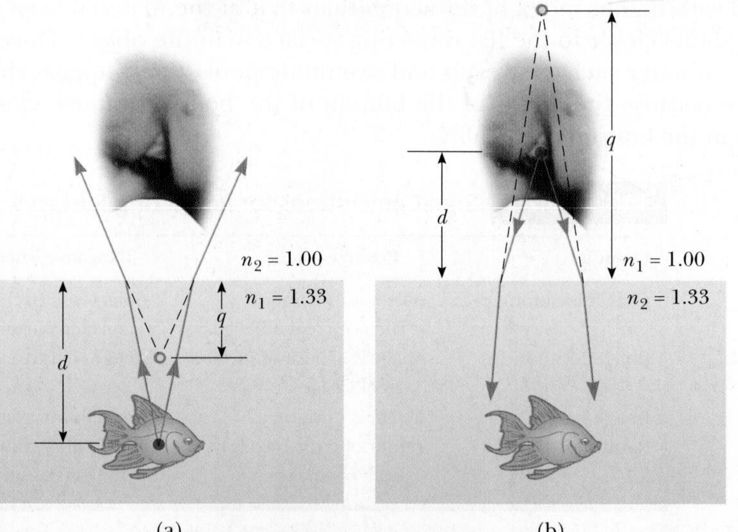

FIGURE 26.19 (Example 26.5) (a) The apparent depth q of the fish is less than the true depth d. All rays are assumed to be paraxial. (b) Your face appears to the fish to be higher above the surface than it is.

B If your face is a distance d above the water surface, at what apparent distance above the surface does the fish see your face?

Solution The light rays from your face are shown in Figure 26.19b. Because the rays refract toward the normal, your face will appear higher above the surface than it actually is. Using Equation 26.10,

$$q = -\frac{n_2}{n_1}p = -\frac{1.33}{1.00}d = -1.33d$$

The negative sign indicates that the image is in the medium from which the light originated, which is the air above the water.

26.4 THIN LENSES

A typical **thin lens** consists of a piece of glass or plastic, ground so that its two surfaces are either segments of spheres or planes. Lenses are commonly used in optical instruments such as cameras, telescopes, and microscopes to form images by refraction.

Figure 26.20 shows some representative shapes of lenses. These lenses have been placed in two groups. Those in Figure 26.20a are thicker at the center than at the rim, and those in Figure 26.20b are thinner at the center than at the rim. The lenses in the first group are examples of **converging lenses,** and those in the second group are called **diverging lenses.** The reason for these names will become apparent shortly.

As with mirrors, it is convenient to define a point called the **focal point** for a lens. For example, in Figure 26.21a, a group of rays parallel to the principal axis passes through the focal point after being converged by the lens. The distance from the focal point to the lens is again called the **focal length** f. **The focal length is the image distance that corresponds to an infinite object distance.**

To avoid the complications arising from the thickness of the lens, we adopt a simplification model called the **thin lens approximation, in which the thickness of the lens is assumed to be negligible.** As a result, it makes no difference whether we take the focal length to be the distance from the focal point to the surface of the lens or from the focal point to the center of the lens because the difference in

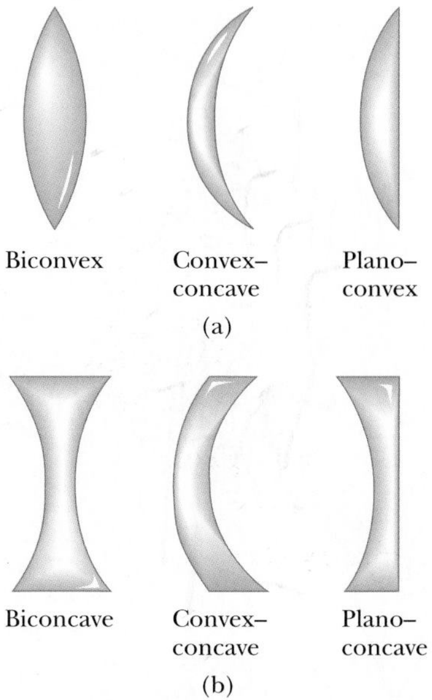

FIGURE 26.20 Cross sectional shapes of various lenses. (a) Converging lenses have a positive focal length and are thickest at the middle. (b) Diverging lenses have a negative focal length and are thickest at the edges.

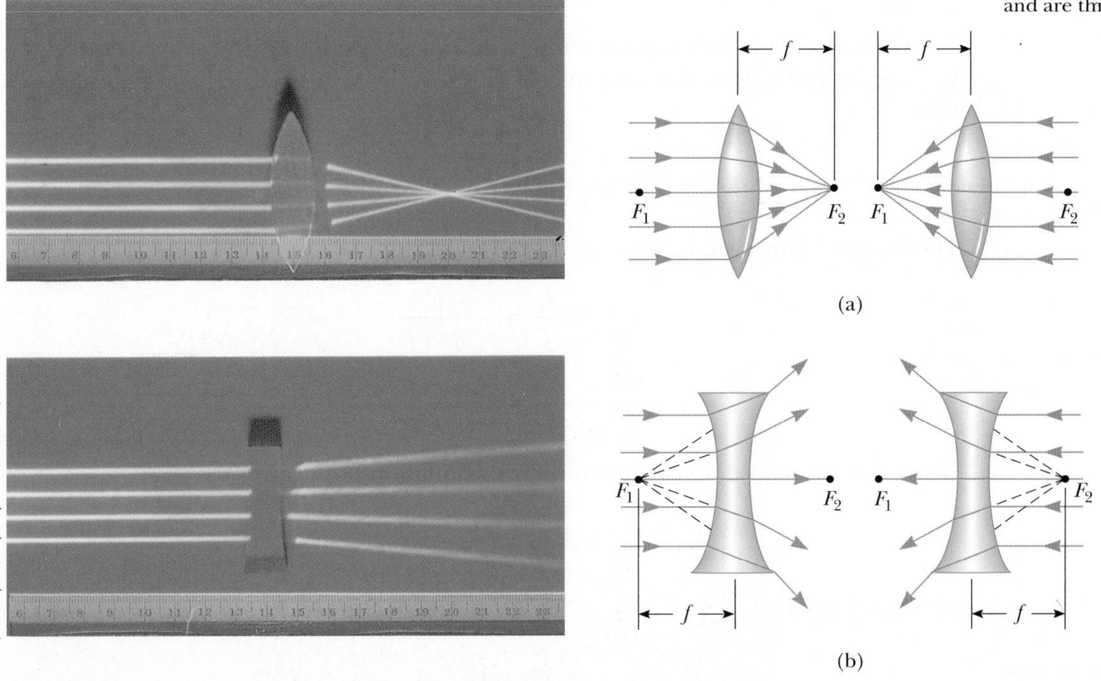

FIGURE 26.21 (*Left*) Effects of a converging lens (*top*) and a diverging lens (*bottom*) on parallel rays. (*Right*) Light rays passing through (a) a converging lens and (b) a diverging lens. The focal length is the same for light rays passing through a given lens in either direction. Both focal points F_1 and F_2 are the same distance from the lens.

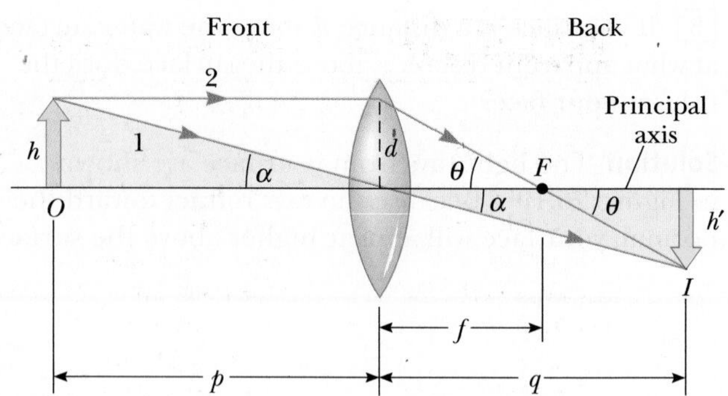

FIGURE 26.22 A geometric construction for developing the thin lens equation.

these two lengths is assumed to be negligible. (We will draw lenses in the diagrams with a thickness so that they can be seen.) A thin lens has one focal length and *two* focal points as illustrated in Figure 26.21, corresponding to parallel light rays traveling from the left or right.

Rays parallel to the axis diverge after passing through a lens of the shape shown in Figure 26.21b. In this case, the focal point is defined as the point from which the diverging rays appear to originate, as in Figure 26.21b. Figures 26.21a and 26.21b indicate why the names *converging* and *diverging* are applied to these lenses in Figure 26.20.

Consider now the ray diagram in Figure 26.22. Ray 1 passes through the center of the lens. Ray 2 is parallel to the principal axis of the lens (the horizontal axis passing through the center of the lens), and as a result it passes through the focal point F after refraction. The point at which these two rays intersect is the image point.

The tangent of the angle α can be found by using the blue and gold geometric model triangles in Figure 26.22:

$$\tan \alpha = \frac{h}{p} \quad \text{and} \quad \tan \alpha = -\frac{h'}{q}$$

from which

$$M = \frac{h'}{h} = -\frac{q}{p} \qquad [26.11]$$

Thus, the equation for magnification of an image by a lens is the same as the equation for magnification due to a mirror (Eq. 26.2). We also note from Figure 26.22 that

$$\tan \theta = \frac{d}{f} \quad \text{and} \quad \tan \theta = -\frac{h'}{q-f}$$

The height d, however, is the same as h. Therefore,

$$\frac{h}{f} = -\frac{h'}{q-f}$$

$$\frac{h'}{h} = -\frac{q-f}{f}$$

Using this expression in combination with Equation 26.11 gives us

$$\frac{q}{p} = \frac{q-f}{f}$$

which reduces to

■ **Thin lens equation**

$$\frac{1}{p} + \frac{1}{q} = \frac{1}{f} \qquad [26.12]$$

TABLE 26.3 Sign Conventions for Thin Lenses

Quantity	Positive when . . .	Negative when . . .
Object location (p)	object is in front of lens (real object).	object is in back of lens (virtual object).
Image location (q)	image is in back of lens (real image).	image is in front of lens (virtual image).
Image height (h')	image is upright.	image is inverted.
R_1 and R_2	center of curvature is in back of lens.	center of curvature is in front of lens.
Focal length (f)	a converging lens.	a diverging lens.

This equation, called the **thin lens equation** (which is identical to the mirror equation, Eq. 26.6), can be used with either converging or diverging lenses if we adhere to a set of sign conventions. Figure 26.23 is useful for obtaining the signs of p and q. (As with mirrors, we call the side from which the light approaches the *front* of the lens.) The complete sign conventions for lenses are provided in Table 26.3. Note that **a converging lens has a positive focal length** under this convention and **a diverging lens has a negative focal length.** Hence, the names *positive* and *negative* are often given to these lenses.

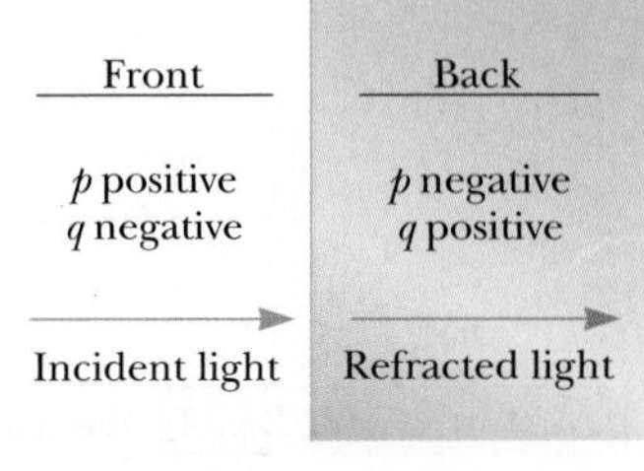

FIGURE 26.23 A diagram for obtaining the signs of p and q for a thin lens or a refracting surface.

The focal length for a lens in air is related to the curvatures of its surfaces and to the index of refraction n of the lens material by

■ Lens makers' equation

$$\frac{1}{f} = (n - 1)\left(\frac{1}{R_1} - \frac{1}{R_2}\right) \qquad [26.13]$$

where R_1 is the radius of curvature of the front surface and R_2 is the radius of curvature of the back surface. Equation 26.13 enables us to calculate the focal length from the known properties of the lens. It is called the **lens makers' equation.** Table 26.3 includes the sign conventions for determining the signs of the radii R_1 and R_2.

Ray Diagrams for Thin Lenses

Our specialized pictorial representations called ray diagrams are very convenient for locating the image of a thin lens or system of lenses. They should also help clarify the sign conventions we have already discussed. Active Figure 26.24 illustrates this method for three single-lens situations. To locate the image of a converging lens (Active Figs. 26.24a and 26.24b), the following three rays are drawn from the top of the object:

- Ray 1 is drawn parallel to the principal axis. After being refracted by the lens, this ray passes through the focal point on the back side of the lens.
- Ray 2 is drawn through the center of the lens and continues in a straight line.
- Ray 3 is drawn through the focal point on the front side of the lens (or as if coming from the focal point if $p < f$, as in Active Fig. 26.24b) and emerges from the lens parallel to the principal axis.

To locate the image of a diverging lens (Active Fig. 26.24c), the following three rays are drawn from the top of the object:

- Ray 1 is drawn parallel to the principal axis. After being refracted by the lens, this ray emerges directed away from the focal point on the front side of the lens.
- Ray 2 is drawn through the center of the lens and continues in a straight line.
- Ray 3 is drawn in the direction toward the focal point on the back side of the lens and emerges from the lens parallel to the principal axis.

In these ray diagrams, the point of intersection of *any two* of the rays can be used to locate the image. The third ray serves as a check of construction.

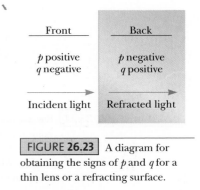

PITFALL PREVENTION 26.4

LENS HAVE TWO FOCAL POINTS BUT ONE FOCAL LENGTH A lens has a focal point on each side, front and back. There is, however, only one focal length for a thin lens. Each of the two focal points is located the same distance from the lens (Fig. 26.21), as can be seen mathematically by interchanging R_1 and R_2 in Equation 26.13 (and changing the signs of the radii because back and front have been interchanged). As a result, the lens forms an image of an object at the same point if it is turned around. In practice, that might not happen because real lenses are not infinitesimally thin.

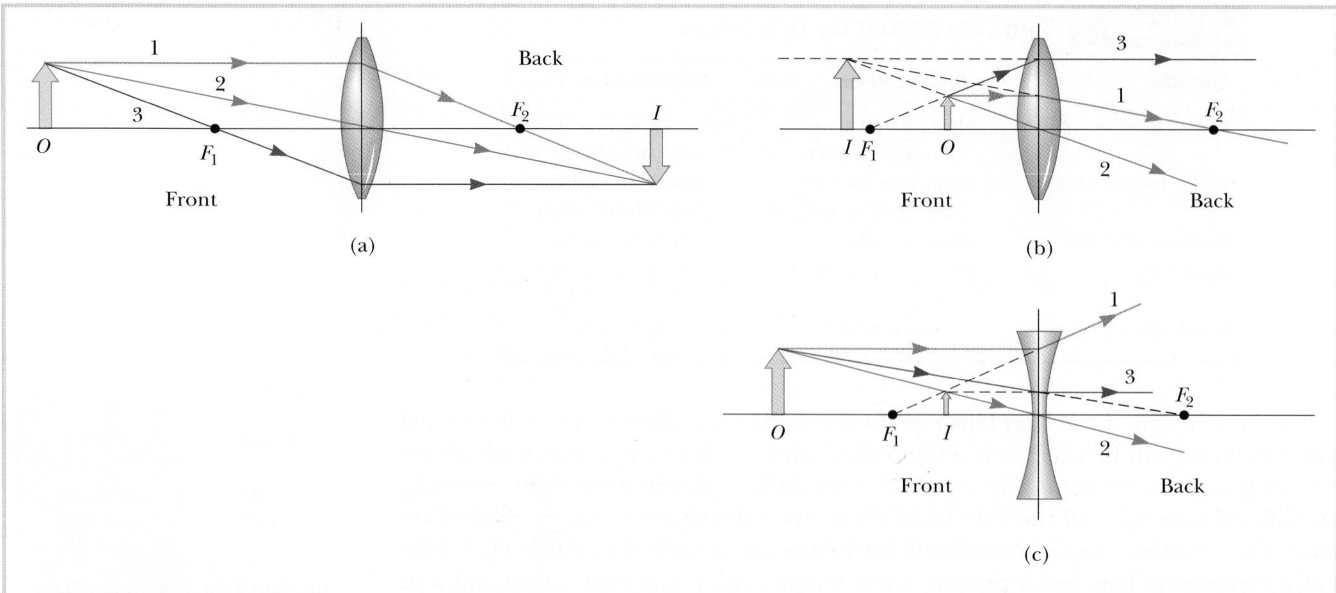

ACTIVE FIGURE 26.24 Ray diagrams for locating the image formed by a thin lens. (a) When the object is in front of and outside the focal point F_1 of a converging lens, the image is real, inverted, and on the back side of the lens. (b) When the object is between F_1 and a converging lens, the image is virtual, upright, larger than the object and on the front side of the lens. (c) When an object is anywhere in front of a diverging lens, the image is virtual, upright, smaller than the object and is on the front side of the lens.

Physics Now™ Log into PhysicsNow at **www.pop4e.com** and go to Active Figure 26.24 to move the objects and change the focal lengths of the lenses to see the effect on the images.

For the converging lens in Active Figure 26.24a where the object is *outside* the front focal point ($p > f$), the image is real and inverted and is located on the back side of the lens. This diagram would be representative of a movie projector, for which the film is the object, the lens is in the projector, and the image is projected on a large screen for the audience to watch. The film is placed in the projector with the scene upside down so that the inverted image is right side up for the audience.

When the object is *inside* the front focal point ($p < f$) as in Active Figure 26.24b, the image is virtual and upright. When used in this way, the lens is acting as a magnifying glass, providing an enlarged upright image for closer study of an object. The object might be a stamp, a fingerprint, or a printed page for someone with failing eyesight.

For the diverging lens of Active Figure 26.24c, the image is virtual and upright for all object locations. A diverging lens is used in a security peephole in a door to give a wide-angle view. Nearsighted individuals use diverging eyeglass lenses or contact lenses. Another use is for a panoramic lens for a camera (although a sophisticated camera "lens" is actually a combination of several lenses). A diverging lens in this application creates a small image of a wide field of view.

QUICK QUIZ 26.5 What is the focal length of a pane of window glass? **(a)** zero **(b)** infinity **(c)** the thickness of the glass **(d)** impossible to determine

QUICK QUIZ 26.6 If you cover the top half of the lens in Active Figure 26.24a with a piece of paper, which of the following happens to the appearance of the image of the object? **(a)** The bottom half disappears. **(b)** The top half disappears. **(c)** The entire image is visible but dimmer. **(d)** There is no change. **(e)** The entire image disappears.

Corrective lenses on diving masks

Thinking Physics 26.2

Diving masks often have lenses built into the glass for divers who do not have perfect vision. This kind of mask allows the individual to dive without the necessity for glasses because the lenses in the faceplate perform the necessary refraction to provide clear vision. Normal glasses have lenses that are curved on both the front and rear surfaces. The lenses in a diving mask faceplate often only have curved surfaces on the *inside* of the glass. Why is this design desirable?

Reasoning The main reason for curving only the inner surface of the lenses in the diving mask faceplate is so that the diver can see clearly when looking at objects straight ahead while underwater *and* in the air. Consider light rays approaching the mask along a normal to the plane of the faceplate. If curved surfaces were on both the front and the back of the diving lens on the faceplate, refraction would occur at each surface. The lens could be designed so that these two refractions would give clear vision while the diver is in air. When the diver is underwater, however, the refraction between the water and the glass at the first interface is now different because the index of refraction of water is different from that of air. Thus, the vision would not be clear underwater.

By making the outer surface of the lens flat, light is not refracted at normal incidence to the faceplate at the outer surface *in either air or water;* all the refraction occurs at the inner glass–air surface. Thus, the same refractive correction exists in water and in air, and the diver can see clearly in both environments. ■

EXAMPLE 26.6 The Lens Makers' Equation

The biconvex lens of Figure 26.25 has an index of refraction of 1.50. The radius of curvature of the front surface is 10 cm and that of the back surface is 15 cm. Find the focal length of the lens.

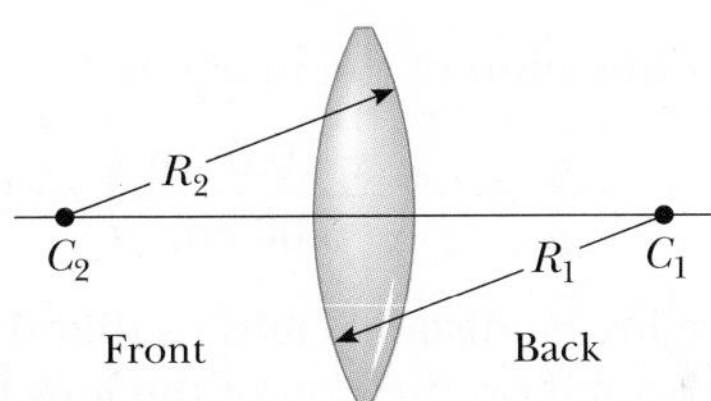

FIGURE 26.25 (Example 26.6) This lens has two curved surfaces with radii of curvature R_1 and R_2.

Solution From the sign conventions in Table 26.3 we find that $R_1 = +10$ cm and $R_2 = -15$ cm. Thus, using the lens makers' equation, we have

$$\frac{1}{f} = (n-1)\left(\frac{1}{R_1} - \frac{1}{R_2}\right)$$

$$= (1.50-1)\left(\frac{1}{10\text{ cm}} - \frac{1}{-15\text{ cm}}\right)$$

$$f = 12\text{ cm}$$

INTERACTIVE EXAMPLE 26.7 The Image Formed by a Converging Lens

A A converging lens of focal length 10.0 cm forms an image of an object placed 30.0 cm from the lens. Construct a ray diagram, find the image distance, and describe the image.

Solution First we construct a ray diagram as shown in Figure 26.26a. The diagram shows that we should expect a real, inverted, smaller image to be formed on the back side of the lens. The thin lens equation, Equation 26.12, can be used to find the image distance:

$$\frac{1}{p} + \frac{1}{q} = \frac{1}{f}$$

$$\frac{1}{30.0\text{ cm}} + \frac{1}{q} = \frac{1}{10.0\text{ cm}}$$

$$q = 15.0\text{ cm}$$

The positive sign for the image distance tells us that the image is indeed real and on the back side of the lens. The magnification of the image is

$$M = -\frac{q}{p} = -\frac{15.0\text{ cm}}{30.0\text{ cm}} = -0.500$$

Thus, the image is reduced in size by one half, and the negative sign for M tells us that the image is inverted.

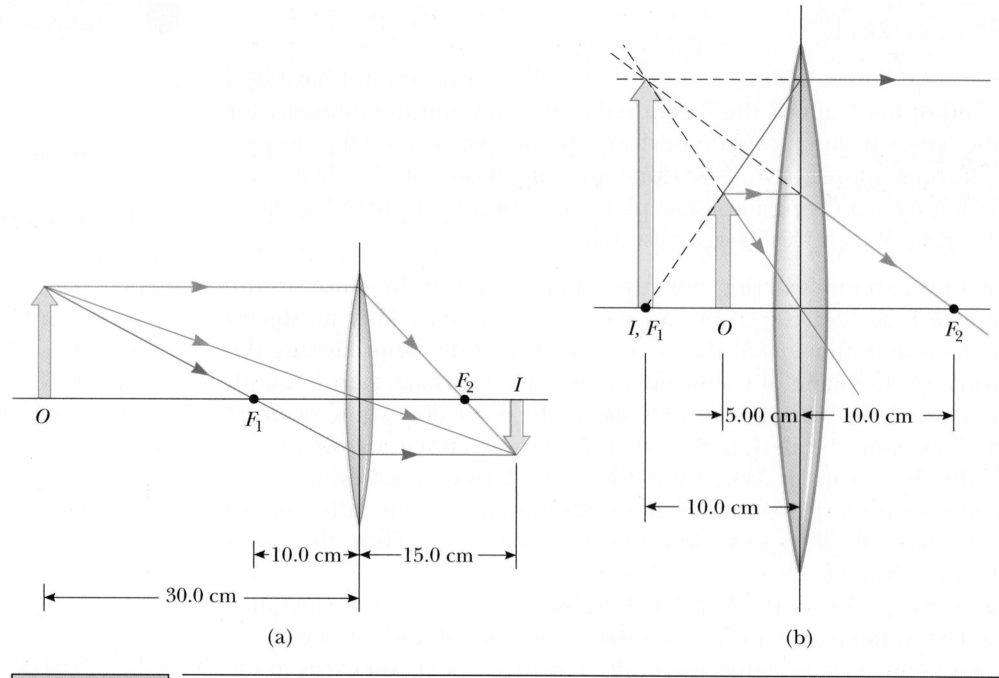

FIGURE 26.26 (Interactive Example 26.7) An image is formed by a converging lens. (a) The object is farther from the lens than the focal point. (b) The object is closer to the lens than the focal point.

B The object is now placed 10.0 cm from the lens. Construct a ray diagram, find the image distance, and describe the image.

Solution No calculation is necessary for this case because we know that when the object is placed at the focal point, the image is formed at infinity. That is verified by substituting $p = 10.0$ cm into the lens equation.

C Finally, the object is placed 5.00 cm from the lens. Construct a ray diagram, find the image distance, and describe the image.

Solution We now move inside the focal point. The ray diagram in Figure 26.26b shows that in this case the lens acts as a magnifying glass; that is, the image is magnified, upright, on the same side of the lens as the object, and virtual. Because the object distance is 5.00 cm, the thin lens equation gives us

$$\frac{1}{5.00 \text{ cm}} + \frac{1}{q} = \frac{1}{10.0 \text{ cm}}$$

$$q = -10.0 \text{ cm}$$

and the magnification of the image is

$$M = -\frac{q}{p} = -\left(\frac{-10.0 \text{ cm}}{5.00 \text{ cm}}\right) = 2.00$$

The negative image distance tells us that the image is virtual and formed on the side of the lens from which the light is incident, the front side. The image is enlarged, and the positive sign for M tells us that the image is upright.

PhysicsNow™ Investigate the image formed for various object positions and lens focal lengths by logging into PhysicsNow at **www.pop4e.com** and going to Interactive Example 26.7.

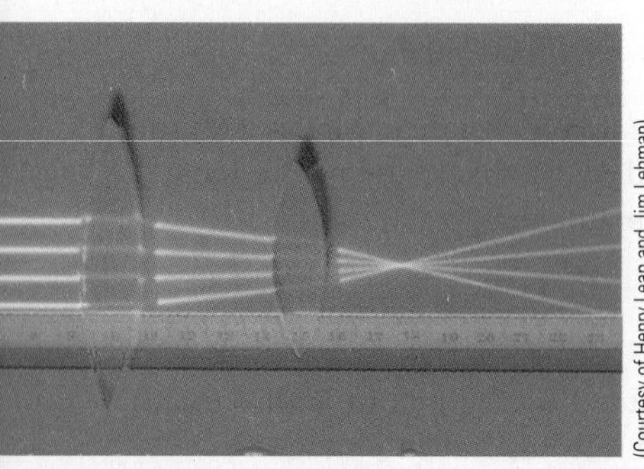
(Courtesy of Henry Leap and Jim Lehman)

Light from a distant object brought into focus by two converging lenses. ■

Combinations of Thin Lenses

If two thin lenses are used to form an image, the system can be treated in the following manner. The position of the image of the first lens is calculated as though the second lens were not present. The light then approaches the second lens *as if* it had originally come from the image formed by the first lens. Hence, the image of the first lens is treated as the object of the second lens. The image of the second lens is the final image of the system. If the image of the first lens lies on the back side of the second lens, the image is treated as a *virtual object* for the second lens (i.e., p is negative). The same procedure can be extended to a system of three or more lenses. The overall magnification of a system of thin lenses equals the *product* of the magnifications of the separate lenses.

INTERACTIVE EXAMPLE 26.8 Where Is the Final Image?

Two thin converging lenses of focal lengths 10.0 cm and 20.0 cm are separated by 20.0 cm as in Figure 26.27a. An object is placed 30.0 cm to the left of the first lens. Find the position and magnification of the final image.

Solution Conceptualize by imagining light rays passing through the first lens and forming a real image (because $p > f$) in the absence of the second lens. Figure 26.27b shows these light rays forming the inverted image I_1. Once the light rays converge to the image point, they do not stop. They continue through the image point and interact with the second lens. The rays leaving the image point behave in the same way as the rays leaving an object. Thus, the image of the first lens serves as the object of the second lens. We categorize this problem as one in which we apply the thin lens equation to the two lenses, but in stepwise fashion. To analyze the problem, we first draw a ray diagram (Fig. 26.27b) showing where the image from the first lens falls and how it acts as the object for the second lens. The location of the image formed by lens 1 is found from the thin lens equation:

$$\frac{1}{p_1} + \frac{1}{q_1} = \frac{1}{f}$$

$$\frac{1}{30.0 \text{ cm}} + \frac{1}{q_1} = \frac{1}{10.0 \text{ cm}}$$

$$q_1 = 15.0 \text{ cm}$$

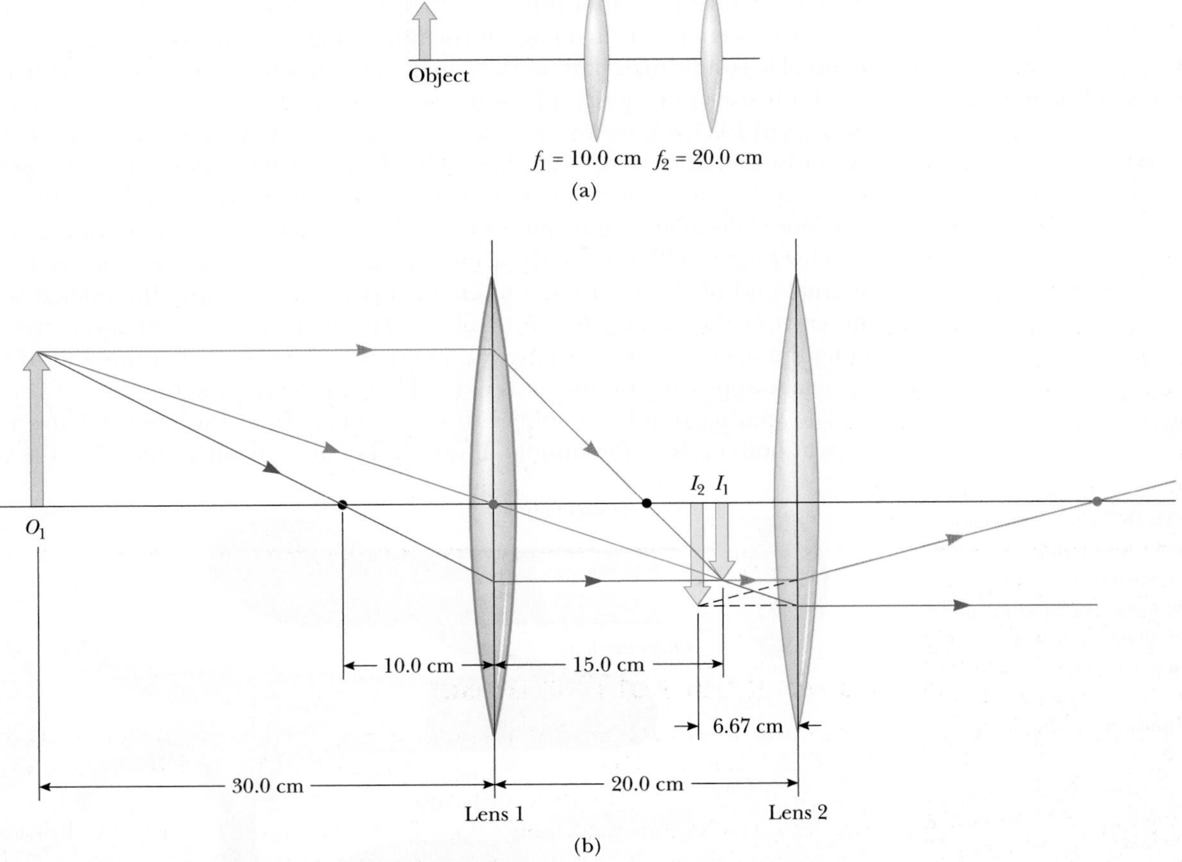

FIGURE 26.27 (Interactive Example 26.8) (a) A combination of two converging lenses. (b) The ray diagram showing the location of the final image due to the combination of lenses. The black dots are the focal points of lens 1, and the red dots are the focal points of lens 2.

where q_1 is measured from the first lens. The magnification of this image is

$$M_1 = -\frac{q_1}{p_1} = -\frac{15.0 \text{ cm}}{30.0 \text{ cm}} = -0.500$$

The image formed by this lens acts as the object for the second lens. The object distance for the second lens is 20.0 cm − 15.0 cm = 5.00 cm from the second lens. We again apply the thin lens equation to find the location of the final image:

$$\frac{1}{5.00 \text{ cm}} + \frac{1}{q_2} = \frac{1}{20.0 \text{ cm}}$$

$$q_2 = -6.67 \text{ cm}$$

Therefore, the final image lies 6.67 cm to the left of the second lens. The magnification of the second image is

$$M_2 = -\frac{q_2}{p_2} = -\frac{(-6.67 \text{ cm})}{5.00 \text{ cm}} = +1.33$$

The total magnification M of the image due to the two lenses is the product

$$M = M_1 M_2 = (-0.500)(1.33) = -0.667$$

To finalize the problem, note that the negative sign on the overall magnification indicates that the final image is inverted with respect to the initial object. That the absolute value of the magnification is less than 1 tells us that the final image is smaller than the object. That q_2 is negative tells us that the final image is on the front, or left, side of lens 2. All these conclusions are consistent with the ray diagram in Figure 26.27b.

PhysicsNow™ Investigate the image formed by a combination of lenses by logging into PhysicsNow at **www.pop4e.com** and going to Interactive Example 26.8.

26.5 MEDICAL FIBERSCOPES

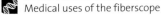

CONTEXT CONNECTION

Electromagnetic radiation in medicine

Electromagnetic radiation has played a role in the transfer of information in medicine for decades. Of particular interest is the ability to gain information about the relatively inaccessible regions inside the body without using invasive procedures such as surgery. An early advance in this area was the use of x-rays to create shadowy images of bones and other internal structures. In this section, we consider advances that have been made in image formation using optical fibers in medical instruments. These advances have in turn opened up new uses for lasers in medicine.

Medical uses of the fiberscope

The first use of optical fibers in medicine appeared with the invention of the *fiberscope* in 1957. Figure 26.28 indicates the construction of a fiberscope, which consists of two bundles of optical fibers. The *illuminating bundle* is an *incoherent* bundle, meaning that no effort is made to match the relative positions of the fibers at the two ends. This matching is not necessary because the sole purpose of this bundle is to deliver light to illuminate the scene. A lens (called the *objective lens*) is used at the internal end of the fiberscope to create a real image of the illuminated scene on the ends of the *viewing bundle* of fibers. The light from the image is transmitted along the fibers to the viewing end. An eyepiece lens is used at this end to magnify the image appearing on the ends of the fibers in the viewing bundle.

The viewing bundle is coherent, so the fibers have the same relative relationships at both ends of the bundle. If one end of an individual fiber is at the very top

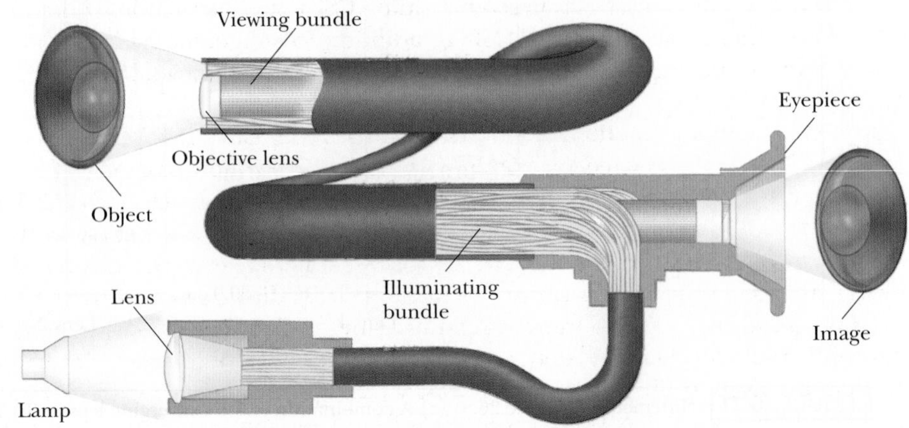

FIGURE 26.28 The construction of a fiberscope for viewing the interior of the body. The objective lens forms a real image of the scene on the end of a bundle of optical fibers. This image is carried to the other end of the bundle, where an eyepiece lens is used to magnify the image for the physician.

of the eyepiece end of the bundle, the other end of the fiber must be at the very top of the interior end of the bundle. This alignment is necessary because each fiber in the viewing bundle collects light from a particular part of the real image of the scene formed by the objective lens on the ends of the fibers. That part of the scene's image must appear in the correct place with all the parts at the other end for the image to make sense!

The diameter of such a fiberscope can be as small as 1 mm and still provide excellent optical imaging of the scene to be viewed. Therefore, the fiberscope can be inserted through very small surgical openings in the skin and threaded through narrow areas such as arteries. Fiber densities are currently about 10 000 fibers for a 1-mm-diameter scope. Resolution is as high as 70 μm.

As another example, a fiberscope can be passed through the esophagus and into the stomach to enable a physician to look for ulcers. The resulting image can be viewed directly by the physician through the eyepiece lens, but most often it is displayed on a television monitor, captured on film, or digitized for computer storage and display.

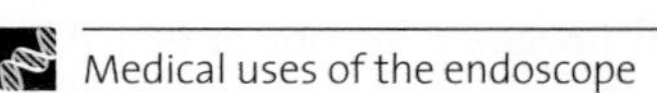
Medical uses of the endoscope

Endoscopes are fiberscopes with additional channels besides those for the illuminating and viewing fibers. These channels may be used for

withdrawing fluids
introducing fluids
vacuum suction
wire manipulators
scalpels for cutting tissue
needles for injections
lasers for surgical applications

Because these additional channels require more room, endoscopes range from 2 to 15 mm in diameter. Despite this larger size, however, endoscopes can be used to perform surgery within the body using incisions that are much smaller than those in traditional surgery.

Lasers are used with endoscopes in a variety of medical diagnostic and treatment procedures. As a diagnostic example, the dependence on wavelength of the amount of reflection from a surface allows a fiberscope to be used to make a direct measurement of the blood's oxygen content. Using two laser sources, red light and infrared light are both sent into the blood through optical fibers. Hemoglobin reflects a known fraction of infrared light, regardless of the oxygen carried. Thus, the measurement of the infrared reflection gives a total hemoglobin count. Red light is reflected much more by hemoglobin carrying oxygen than by hemoglobin that does not. Therefore, the amount of red laser light reflected allows a measurement of the ability of the patient's blood to carry oxygen.

Use of lasers in treating hydrocephalus

Lasers are used to treat medical conditions such as *hydrocephalus,* which occurs in about 0.1% of births. This condition involves an increase in intracranial pressure due to an overproduction of cerebrospinal fluid (CSF), an obstruction of the flow of CSF, or insufficient absorption of CSF. In addition to congenital hydrocephalus, the condition can be acquired later in life due to trauma to the head, brain tumors, or other factors.

The older treatment method for obstructive hydrocephalus involved placing a shunt (tube) between ventricular chambers in the brain to allow passage of CSF. A new alternative is *laser-assisted ventriculostomy,* in which a new pathway for CSF is made with an infrared laser beam and an endoscope having a spherical end as shown in Figure 26.29. As the laser beam strikes the spherical end, refraction at the spherical surface causes light waves to spread out in all directions as if the end of the endoscope were a point source of radiation. The result is a rapid decrease in intensity with distance from the sphere, avoiding damage to vital structures in the brain that are close to the area in which a new passageway is to be made. The surface of the spherical end is coated with an infrared radiation-absorbing material,

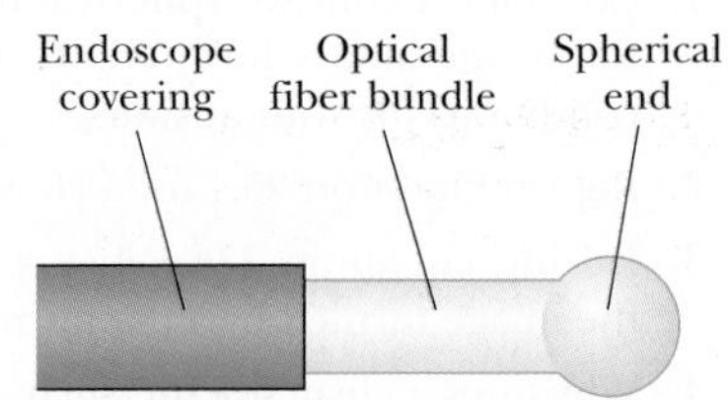

FIGURE 26.29 An endoscope probe used to open new passageways for cerebrospinal fluid in the treatment of hydrocephalus. Laser light raises the temperature of the sphere and radiates from the sphere to provide energy to tissues for cutting the new passageway.

and the absorbed laser energy raises the temperature of the sphere. As the sphere is placed in contact with the location of the desired passageway, the combination of the high temperature and laser radiation leaving the sphere burns a new passageway for the CSF. This treatment requires much less recovery time as well as significantly less postoperative care than that associated with the placement of shunts.

In Chapter 27, we shall investigate another application of lasers—the technology of *holography*—that has grown tremendously in recent years. In holography, three-dimensional images of objects are recorded on film. ■

SUMMARY

PhysicsNow™ Take a practice test by logging into PhysicsNow at **www.pop4e.com** and clicking on the Pre-Test link for this chapter.

An **image** of an object is a point from which light either diverges or seems to diverge after interacting with a mirror or lens. If light passes through the image point, the image is a **real image.** If light only appears to diverge from the image point, the image is a **virtual image.**

In the **paraxial ray** simplification model, the object distance p and image distance q for a spherical mirror of radius R are related by the **mirror equation**

$$\frac{1}{p} + \frac{1}{q} = \frac{2}{R} = \frac{1}{f} \qquad [26.4, 26.6]$$

where $f = R/2$ is the **focal length** of the mirror.

The **magnification** M of a mirror or lens is defined as the ratio of the image height h' to the object height h:

$$M = \frac{h'}{h} = -\frac{q}{p} \qquad [26.2, 26.11]$$

An image can be formed by refraction from a spherical surface of radius R. The object and image distances for refraction from such a surface are related by

$$\frac{n_1}{p} + \frac{n_2}{q} = \frac{n_2 - n_1}{R} \qquad [26.8]$$

where the light is incident from the medium of index of refraction n_1 and is refracted in the medium whose index of refraction is n_2.

For a thin lens, and in the paraxial ray approximation, the object and image distances are related by the **thin lens equation:**

$$\frac{1}{p} + \frac{1}{q} = \frac{1}{f} \qquad [26.12]$$

The **focal length** f of a thin lens in air is related to the curvature of its surfaces and to the index of refraction n of the lens material by

$$\frac{1}{f} = (n-1)\left(\frac{1}{R_1} - \frac{1}{R_2}\right) \qquad [26.13]$$

Converging lenses have positive focal lengths, and **diverging lenses** have negative focal lengths.

QUESTIONS

☐ = answer available in the *Student Solutions Manual and Study Guide*

1. Consider a concave spherical mirror with a real object. Is the image always inverted? Is the image always real? Give conditions for your answers.
2. Repeat Question 26.1 for a convex spherical mirror.
3. Do the equations $1/p + 1/q = 1/f$ or $M = -q/p$ apply to the image formed by a flat mirror? Explain your answer.
4. Why does a clear stream, such as a creek, always appear to be shallower than it actually is? By how much is its depth apparently reduced?
5. Consider the image formed by a thin converging lens. Under what conditions is the image (a) inverted, (b) upright, (c) real, (d) virtual, (e) larger than the object, and (f) smaller than the object?
6. Repeat Question 26.5 for a thin diverging lens.
7. Use the lens makers' equation to verify the sign of the focal length of each of the lenses in Figure 26.20.
8. If a solid cylinder of glass or clear plastic is placed above the words LEAD OXIDE and viewed from above as shown in Figure Q26.8, the LEAD appears inverted but the OXIDE does not. Explain.

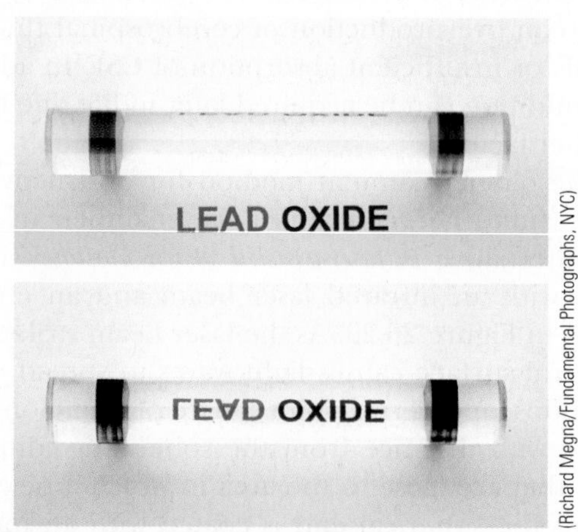

FIGURE Q26.8

9. Consider a spherical concave mirror with the object located to the left of the mirror beyond the focal point. Using ray diagrams, show that the image moves to the left as the object approaches the focal point.

10. Explain why a fish in a spherical goldfish bowl appears larger than it really is.

11. Why do some emergency vehicles have the symbol ƎƆИA⅃UBMA written on the front?

12. Lenses used in eyeglasses, whether converging or diverging, are always designed so that the middle of the lens curves away from the eye, like the center lenses of Figure 26.20a and 26.20b. Why?

13. In Active Figure 26.24a, assume that the blue object arrow is replaced by one that is much taller than the lens. How many rays from the object will strike the lens? How many principal rays can be drawn in a ray diagram?

14. In a Jules Verne novel, a piece of ice is shaped to form a magnifying lens to focus sunlight to start a fire. Is that possible?

15. Explain this statement: "The focal point of a lens is the location of the image of a point object at infinity." Discuss the notion of infinity in real terms as it applies to object distances. Based on this statement, can you think of a "quick and dirty" method for determining the focal length of a converging lens?

16. Discuss the proper position of a photographic slide relative to the lens in a slide projector. What type of lens must the slide projector have?

17. A solar furnace can be constructed by using a concave mirror to reflect and focus sunlight into a furnace enclosure. What factors in the design of the reflecting mirror would guarantee very high temperatures?

18. Figure Q26.18 shows a lithograph by M. C. Escher titled *Hand with Reflection Sphere (Self-Portrait in Spherical Mirror).* Escher had this to say about the work:

> The picture shows a spherical mirror, resting on a left hand. But as a print is the reverse of the original drawing on stone, it was my right hand that you see depicted. (Being left-handed, I needed my left hand to make the drawing.) Such a globe reflection collects almost one's whole surroundings in one disk-shaped image. The whole room, four walls, the floor, and the ceiling, everything, albeit distorted, is compressed into that one small circle. Your own head, or more exactly the point between your eyes, is the absolute center. No matter how you turn or twist yourself, you can't get out of that central point. You are immovably the focus, the unshakable core, of your world.

Comment on the accuracy of Escher's description.

FIGURE **Q26.18**

19. You can make a corner reflector by placing three flat mirrors in the corner of a room where the ceiling meets the walls. Show that no matter where you are in the room, you can see yourself reflected in the mirrors, upside down.

PROBLEMS

1, 2, 3 = straightforward, intermediate, challenging

☐ = full solution available in the *Student Solutions Manual and Study Guide*

PhysicsNow™ = coached problem with hints available at www.pop4e.com

= computer useful in solving problem

= paired numerical and symbolic problems

= biomedical application

Section 26.1 ■ Images Formed by Flat Mirrors

1. Does your bathroom mirror show you older or younger than you actually are? Compute an order-of-magnitude estimate for the age difference, based on data you specify.

2. In a church choir loft, two parallel walls are 5.30 m apart. The singers stand against the north wall. The organist faces the south wall, sitting 0.800 m away from it. To enable her to see the choir, a flat mirror 0.600 m wide is mounted on the south wall, straight in front of her. What width of the north wall can she see? (*Suggestion:* Draw a top-view diagram to justify your answer.)

3. Determine the minimum height of a vertical flat mirror in which a person 5′10″ in height can see his or her full image. (A ray diagram would be helpful.)

4. Two flat mirrors have their reflecting surfaces facing each other, with the edge of one mirror in contact with an edge of the other, so that the angle between the mirrors is α. When an object is placed between the mirrors, a number of images are formed. In general, if the angle α is such that $n\alpha = 360°$, where n is an integer, the number of images formed is $n - 1$. Graphically, find all the image positions for the case $n = 6$ when a point object is between the mirrors (but not on the angle bisector).

5. A person walks into a room with two flat mirrors on opposite walls, which produce multiple images. When the person is 5.00 ft from the mirror on the left wall and 10.0 ft from the mirror on the right wall, find the distance from the person to the first three images seen in the mirror on the left.

6. A periscope (Fig. P26.6) is useful for viewing objects that cannot be seen directly. It finds use in submarines and in watching golf matches or parades from behind a crowd of people. Suppose the object is a distance p_1 from the upper mirror and the two flat mirrors are separated by a distance h. (a) What is the distance of the final image from the lower mirror? (b) Is the final image real or virtual? (c) Is it upright or inverted? (d) What is its magnification? (e) Does it appear to be left–right reversed?

FIGURE **P26.6**

Section 26.2 ■ Images Formed by Spherical Mirrors

7. A concave spherical mirror has a radius of curvature of 20.0 cm. Find the location of the image for object distances of (a) 40.0 cm, (b) 20.0 cm, and (c) 10.0 cm. For each case, state whether the image is real or virtual and upright or inverted. Find the magnification in each case.

8. At an intersection of hospital hallways, a convex mirror is mounted high on a wall to help people avoid collisions. The mirror has a radius of curvature of 0.550 m. Locate and describe the image of a patient 10.0 m from the mirror. Determine the magnification.

9. **PhysicsNow™** A spherical convex mirror (Fig. P26.9) has a radius of curvature with a magnitude of 40.0 cm. Determine the position of the virtual image and the magnification for object distances of (a) 30.0 cm and (b) 60.0 cm. (c) Are the images upright or inverted?

10. A large church has a niche in one wall. On the floor plan it appears as a semicircular indentation of radius 2.50 m. A worshiper stands on the center line of the niche, 2.00 m out from its deepest point, and whispers a prayer. Where is the sound concentrated after reflection from the back wall of the niche?

11. A concave mirror has a radius of curvature of 60.0 cm. Calculate the image position and magnification of an object placed in front of the mirror at distances of (a) 90.0 cm and (b) 20.0 cm. (c) Draw ray diagrams to obtain the image characteristics in each case.

FIGURE **P26.9** Convex mirrors, often used for security in department stores, provide wide-angle viewing.

12. A dentist uses a mirror to examine a tooth. The tooth is 1.00 cm in front of the mirror, and the image is formed 10.0 cm behind the mirror. Determine (a) the mirror's radius of curvature and (b) the magnification of the image.

13. A certain Christmas tree ornament is a silver sphere having a diameter of 8.50 cm. Determine an object location for which the size of the reflected image is three-fourths the size of the object. Use a principal-ray diagram to arrive at a description of the image.

14. (a) A concave mirror forms an inverted image four times larger than the object. Find the focal length of the mirror, assuming that the distance between object and image is 0.600 m. (b) A convex mirror forms a virtual image half the size of the object. Assuming that the distance between image and object is 20.0 cm, determine the radius of curvature of the mirror.

15. To fit a contact lens to a patient's eye, a *keratometer* can be used to measure the curvature of the front surface of the eye, the cornea. This instrument places an illuminated object of known size at a known distance p from the cornea. The cornea reflects some light from the object, forming an image of the object. The magnification M of the image is measured by using a small viewing telescope that allows comparison of the image formed by the cornea with a second calibrated image projected into the field of view by a prism arrangement. Determine the radius of curvature of the cornea for the case $p = 30.0$ cm and $M = 0.013\,0$.

16. An object 10.0 cm tall is placed at the zero mark of a meter stick. A spherical mirror located at some point on the meter stick creates an image of the object that is upright, 4.00 cm tall, and located at the 42.0-cm mark of the meter stick. (a) Is the mirror convex or concave? (b) Where is the mirror? (c) What is the mirror's focal length?

17. A spherical mirror is to be used to form, on a screen located 5.00 m from the object, an image five times the size of the object. (a) Describe the type of mirror required. (b) Where should the mirror be positioned relative to the object?

18. A dedicated sports car enthusiast polishes the inside and outside surfaces of a hubcap that is a section of a sphere. When she looks into one side of the hubcap, she sees an image of her face 30.0 cm in back of the hubcap. She then flips the hubcap over and sees another image of her face 10.0 cm in back of the hubcap. (a) How far is her face from the hubcap? (b) What is the radius of curvature of the hubcap?

19. You unconsciously estimate the distance to an object from the angle it subtends in your field of view. This angle θ in radians is related to the linear height of the object h and to the distance d by $\theta = h/d$. Assume that you are driving a car and that another car, 1.50 m high, is 24.0 m behind you. (a) Suppose your car has a flat passenger-side rearview mirror, 1.55 m from your eyes. How far from your eyes is the image of the car following you? (b) What angle does the image subtend in your field of view? (c) Suppose instead your car has a convex rearview mirror with a radius of curvature of magnitude 2.00 m (Fig. 26.13 and Fig. P26.19). How far from your eyes is the image of the car behind you? (d) What angle does the image subtend at your eyes? (e) Based on its angular size, how far away does the following car appear to be?

THE FAR SIDE® BY GARY LARSON

OBJECTS IN MIRROR ARE
CLOSER THAN THEY APPEAR

FIGURE P26.19

20. **Review problem.** A ball is dropped at $t = 0$ from rest 3.00 m directly above the vertex of a concave mirror that has a radius of curvature of 1.00 m and lies in a horizontal plane. (a) Describe the motion of the ball's image in the mirror. (b) At what time do the ball and its image coincide?

Section 26.3 ■ Images Formed by Refraction

21. A cubical block of ice 50.0 cm on a side is placed on a level floor over a speck of dust. Find the location of the image of the speck as viewed from above. The index of refraction of ice is 1.309.

22. A flint glass plate ($n = 1.66$) rests on the bottom of an aquarium tank. The plate is 8.00 cm thick (vertical dimension) and is covered with a layer of water ($n = 1.33$) 12.0 cm deep. Calculate the apparent thickness of the plate as viewed from straight above the water.

23. A glass sphere ($n = 1.50$) with a radius of 15.0 cm has a tiny air bubble 5.00 cm above its center. The sphere is viewed looking down along the extended radius containing the bubble. What is the apparent depth of the bubble below the surface of the sphere?

24. A simple model of the human eye ignores its lens entirely. Most of what the eye does to light happens at the outer surface of the transparent cornea. Assume that this surface has a radius of curvature of 6.00 mm and that the eyeball contains just one fluid with a refractive index of 1.40. Prove that a very distant object will be imaged on the retina, 21.0 mm behind the cornea. Describe the image.

25. One end of a long glass rod ($n = 1.50$) is formed into a convex surface with a radius of curvature of 6.00 cm. An object is located in air along the axis of the rod. Find the image positions corresponding to object distances of (a) 20.0 cm, (b) 10.0 cm, and (c) 3.00 cm from the end of the rod.

26. A goldfish is swimming at 2.00 cm/s toward the front wall of a rectangular aquarium. What is the apparent speed of the fish measured by an observer looking in from outside the front wall of the tank? The index of refraction of water is 1.33.

Section 26.4 ■ Thin Lenses

27. PhysicsNow™ The left face of a biconvex lens has a radius of curvature of magnitude 12.0 cm, and the right face has a radius of curvature of magnitude 18.0 cm. The index of refraction of the glass is 1.44. (a) Calculate the focal length of the lens. (b) Calculate the focal length the lens has after is turned around to interchange the radii of curvature of the two faces.

28. A contact lens is made of plastic with an index of refraction of 1.50. The lens has an outer radius of curvature of +2.00 cm and an inner radius of curvature of +2.50 cm. What is the focal length of the lens?

29. A thin lens has a focal length of 25.0 cm. Locate and describe the image when the object is placed (a) 26.0 cm and (b) 24.0 cm in front of the lens.

30. A converging lens has a focal length of 20.0 cm. Locate the image for object distances of (a) 40.0 cm, (b) 20.0 cm, and (c) 10.0 cm. For each case, state whether the image is real or virtual and upright or inverted. Find the magnification in each case.

31. PhysicsNow™ The nickel's image in Figure P26.31 has twice the diameter of the nickel and is 2.84 cm from the lens. Determine the focal length of the lens.

32. An object located 32.0 cm in front of a lens forms an image on a screen 8.00 cm behind the lens. (a) Find the focal length of the lens. (b) Determine the magnification. (c) Is the lens converging or diverging?

FIGURE **P26.31**

33. Suppose an object has thickness dp so that it extends from object distance p to $p + dp$. Prove that the thickness dq of its image is given by $(-q^2/p^2)\,dp$. Then the longitudinal magnification is $dq/dp = -M^2$, where M is the lateral magnification.

34. The projection lens in a certain slide projector is a single thin lens. A slide 24.0 mm high is to be projected so that its image fills a screen 1.80 m high. The slide-to-screen distance is 3.00 m. (a) Determine the focal length of the projection lens. (b) How far from the slide should the lens of the projector be placed so as to form the image on the screen?

35. An object is located 20.0 cm to the left of a diverging lens having a focal length $f = -32.0$ cm. Determine (a) the location and (b) the magnification of the image. (c) Construct a ray diagram for this arrangement.

36. The use of a lens in a certain situation is described by the equation

$$\frac{1}{p} + \frac{1}{-3.50p} = \frac{1}{7.50\text{ cm}}$$

Determine (a) the object distance and (b) the image distance. (c) Use a ray diagram to obtain a description of the image. (d) Identify a practical device described by the given equation and write the statement of a problem for which the equation appears in the solution.

37. An antelope is at a distance of 20.0 m from a converging lens of focal length 30.0 cm. The lens forms an image of the animal. If the antelope runs away from the lens at a speed of 5.00 m/s, how fast does the image move? Does the image move toward or away from the lens?

38. Figure P26.38 shows a thin glass ($n = 1.50$) converging lens for which the radii of curvature are $R_1 = 15.0$ cm and $R_2 = -12.0$ cm. To the left of the lens is a cube having a face area of 100 cm^2. The base of the cube is on the axis of the lens, and the right face is 20.0 cm to the left of the lens. (a) Determine the focal length of the lens. (b) Draw the image of the square face formed by the lens. What type of geometric figure is it? (c) Determine the area of the image.

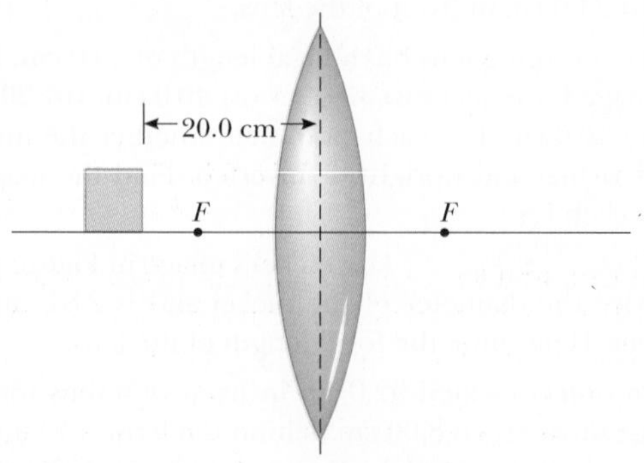

FIGURE **P26.38**

39. An object is at a distance d to the left of a flat screen. A converging lens with focal length $f < d/4$ is placed between object and screen. (a) Show that two lens positions exist that form an image on the screen and determine how far these positions are from the object. (b) How do the two images differ from each other?

40. Figure P26.40 diagrams a cross-section of a camera. It has a single lens of focal length 65.0 mm, which is to form an image on the film at the back of the camera. Suppose the position of the lens has been adjusted to focus the image of a distant object. How far and in what direction must the lens be moved to form a sharp image of an object that is 2.00 m away?

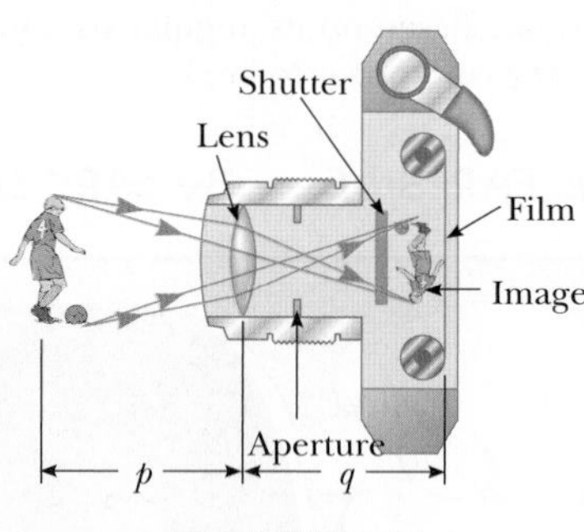

FIGURE **P26.40**

Section 26.5 ■ Context Connection—Medical Fiberscopes

41. You are designing an endoscope for use inside an air-filled body cavity. A lens at the end of the endoscope will form an image covering the end of a bundle of optical fibers. This image will then be carried by the optical fibers to an eyepiece lens at the outside end of the fiberscope. The radius of the bundle is 1.00 mm. The scene within the body that is to appear within the image fills a circle of radius 6.00 cm. The lens will be located 5.00 cm from the tissues you wish to observe. (a) How far should the lens be located from the end of an optical fiber bundle? (b) What is the focal length of the lens required?

42. Consider the endoscope probe used for treating hydrocephalus and shown in Figure 26.29. The spherical end, with refractive index 1.50, is attached to an optical fiber bundle of radius 1.00 mm, which is smaller than the radius of the sphere. The center of the spherical end is on the central axis of the bundle. Consider laser light that travels precisely parallel to the central axis of the bundle and then refracts out from the surface of the sphere into air. (a) In Figure 26.29, does light that refracts out of the sphere and travels toward the upper right come from the top half of the sphere or from the bottom half of the sphere? (b) If laser light that travels along the edge of the optical fiber bundle refracts out of the sphere tangent to the surface of the sphere, what is the radius of the sphere? (c) Find the angle of deviation of the ray considered in part (b), that is, the angle by which its direction changes as it leaves the sphere. (d) Show that the ray considered in part (b) has a greater angle of deviation than any other ray. Show that the light from all parts of the optical fiber bundle does not refract out of the sphere with spherical symmetry, but rather fills a cone around the forward direction. Find the angular diameter of the cone. (e) In reality, however, laser light can diverge from the sphere with approximate spherical symmetry. What considerations that we have not addressed will lead to this approximate spherical symmetry in practice?

Additional Problems

43. The distance between an object and its upright image is d. If the magnification is M, what is the focal length of the lens being used to form the image?

44. The lens and mirror in Figure P26.44 have focal lengths of $+80.0$ cm and -50.0 cm, respectively. An object is placed 1.00 m to the left of the lens as shown. Locate the final image, formed by light that has gone through the lens twice. State whether the image is upright or inverted, and determine the overall magnification.

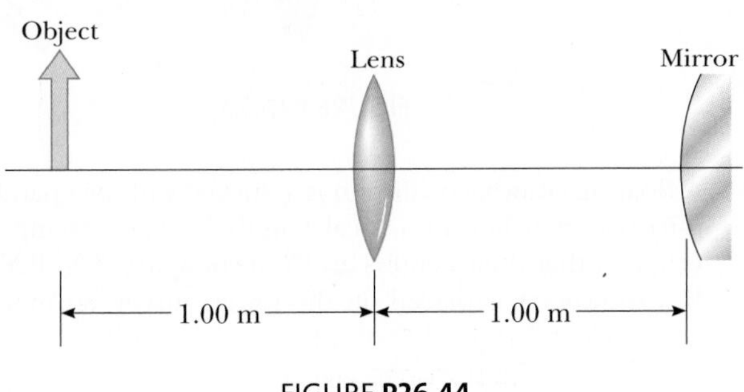

FIGURE **P26.44**

45. A real object is located at the zero end of a meter stick. A large concave mirror at the 100-cm end of the meter stick forms an image of the object at the 70.0-cm position. A small convex mirror placed at the 20.0-cm position forms a final image at the 10.0-cm point. What is the radius of curvature of the convex mirror?

46. Derive the lens makers' equation as follows. Consider an object in vacuum at $p_1 = \infty$ from a first refracting surface of radius of curvature R_1. Locate its image. Use this image as the object for the second refracting surface, which has nearly the same location as the first because the lens is thin. Locate the final image, proving it is at the image distance q_2 given by

$$\frac{1}{q_2} = (n-1)\left(\frac{1}{R_1} - \frac{1}{R_2}\right)$$

47. A *zoom lens* system is a combination of lenses that produces a variable magnification while maintaining fixed object and image positions. The magnification is varied by moving one or more lenses along the axis. Although multiple lenses are used in practice to obtain high-quality images, the effect of zooming in on an object can be demonstrated with a simple two-lens system. An object, two converging lenses, and a screen are mounted on an optical bench. The first lens, which is to the right of the object, has a focal length of 5.00 cm, and the second lens, which is to the right of the first lens, has a focal length of 10.0 cm. The screen is to the right of the second lens. Initially, an object is situated at a distance of 7.50 cm to the left of the first lens, and the image formed on the screen has a magnification of $+1.00$. (a) Find the distance between the object and the screen. (b) Both lenses are now moved along their common axis, while the object and the screen maintain fixed positions, until the image formed on the screen has a magnification of $+3.00$. Find the displacement of each lens from its initial position in part (a). Can the lenses be displaced in more than one way?

48. The object in Figure P26.48 is midway between the lens and the mirror. The mirror's radius of curvature is 20.0 cm, and the lens has a focal length of -16.7 cm. Considering only the light that leaves the object and travels first toward the mirror, locate the final image formed by this system. Is this image real or virtual? Is it upright or inverted? What is the overall magnification?

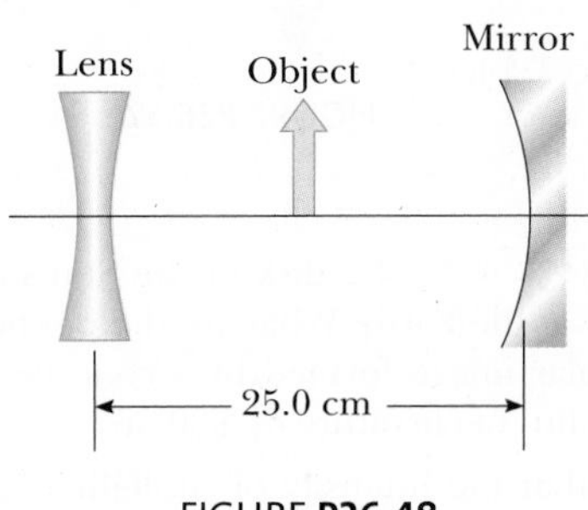

FIGURE **P26.48**

49. **PhysicsNow™** A parallel beam of light enters a glass hemisphere perpendicular to the flat face as shown in Figure P26.49. The magnitude of the radius is 6.00 cm, and the index of refraction is 1.560. Determine the point at which the beam is focused. (Assume paraxial rays.)

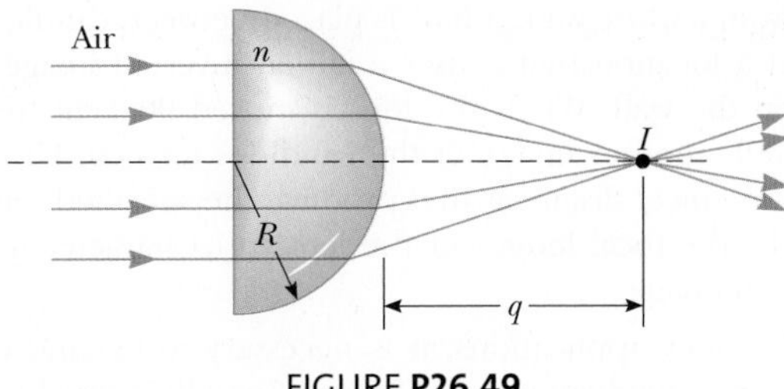

FIGURE **P26.49**

50. **Review problem.** A spherical lightbulb of diameter 3.20 cm radiates light equally in all directions, with power 4.50 W. (a) Find the light intensity at the surface of the lightbulb. (b) Find the light intensity 7.20 m away from the center of the lightbulb. (c) At this 7.20-m distance, a lens is set up with its axis pointing toward the lightbulb. The lens has a circular face with a diameter 15.0 cm and has a focal

length of 35.0 cm. Find the diameter of the image of the bulb. (d) Find the light intensity at the image.

51. An object is placed 12.0 cm to the left of a diverging lens of focal length − 6.00 cm. A converging lens of focal length 12.0 cm is placed a distance d to the right of the diverging lens. Find the distance d so that the final image is at infinity. Draw a ray diagram for this case.

52. An observer to the right of the mirror–lens combination shown in Figure P26.52 sees two real images that are the same size and in the same location. One image is upright and the other is inverted. Both images are 1.50 times larger than the object. The lens has a focal length of 10.0 cm. The lens and mirror are separated by 40.0 cm. Determine the focal length of the mirror. Do not assume that the figure is drawn to scale.

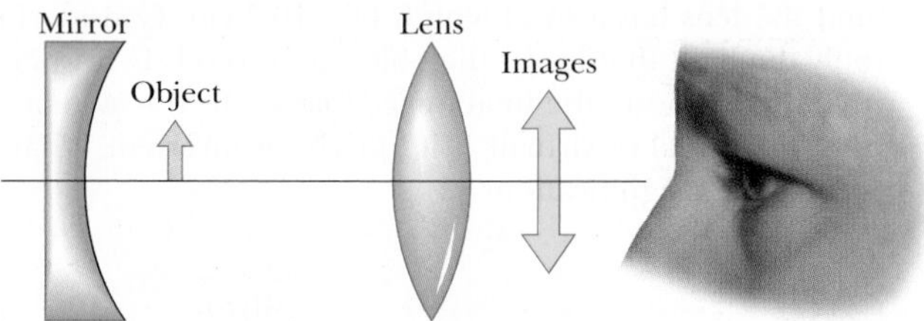

FIGURE **P26.52**

53. Physics⊗Now™ The disk of the Sun subtends an angle of 0.533° at the Earth. What are the position and diameter of the solar image formed by a concave spherical mirror with a radius of curvature of 3.00 m?

54. Assume that the intensity of sunlight is 1.00 kW/m² at a particular location. A highly reflecting concave mirror is to be pointed toward the Sun to produce a power of at least 350 W at the image. (a) Find the required radius R_a of the circular face area of the mirror. (b) Now suppose the light intensity is to be at least 120 kW/m² at the image. Find the required relationship between R_a and the radius of curvature R of the mirror. The disk of the Sun subtends an angle of 0.533° at the Earth.

55. In a darkened room, a burning candle is placed 1.50 m from a white wall. A lens is placed between candle and wall at a location that causes a larger, inverted image to form on the wall. When the lens is moved 90.0 cm toward the wall, another image of the candle is formed. Find (a) the two object distances that produce the specified images and (b) the focal length of the lens. (c) Characterize the second image.

56. In many applications, it is necessary to expand or to decrease the diameter of a beam of parallel rays of light. This change can be made by using a converging lens and a diverging lens in combination. Suppose you have a converging lens of focal length 21.0 cm and a diverging lens of focal length −12.0 cm. How can you arrange these lenses to increase the diameter of a beam of parallel rays? By what factor will the diameter increase?

57. The lens makers' equation applies to a lens immersed in a liquid if n in the equation is replaced by n_2/n_1. Here n_2 refers to the refractive index of the lens material and n_1 is that of the medium surrounding the lens. (a) A certain lens has focal length 79.0 cm in air and refractive index 1.55. Find its focal length in water. (b) A certain mirror has focal length 79.0 cm in air. Find its focal length in water.

58. Figure P26.58 shows a thin converging lens for which the radii of curvature are R_1 = 9.00 cm and R_2 = − 11.0 cm. The lens is in front of a concave spherical mirror with the radius of curvature $|R|$ = 8.00 cm. (a) Assume that its focal points F_1 and F_2 are 5.00 cm from the center of the lens. Determine its index of refraction. (b) The lens and mirror are 20.0 cm apart, and an object is placed 8.00 cm to the left of the lens. Determine the position of the final image and its magnification as seen by the eye in the figure. (c) Is the final image inverted or upright? Explain.

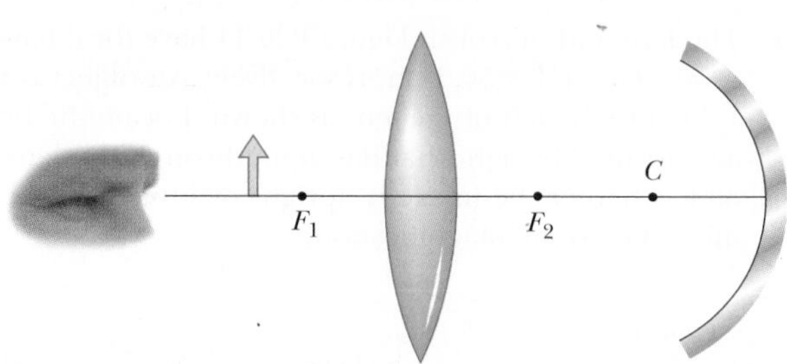

FIGURE **P26.58**

59. A floating strawberry illusion is achieved with two parabolic mirrors, each having a focal length 7.50 cm, facing each other so that their centers are 7.50 cm apart (Fig. P26.59). If a strawberry is placed on the lower mirror, an image of

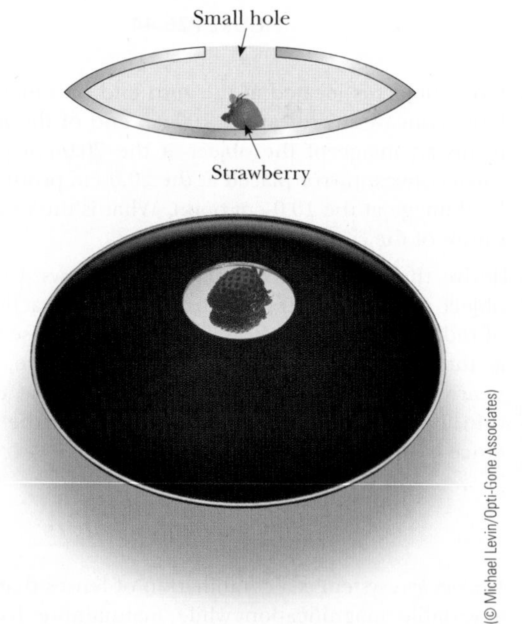

(© Michael Levin/Opti-Gone Associates)

FIGURE **P26.59**

the strawberry is formed at the small opening at the center of the top mirror. Show that the final image is formed at that location and describe its characteristics. (*Note:* A very startling effect is to shine a flashlight beam on this image. Even at a glancing angle, the incoming light beam is seemingly reflected from the image! Do you understand why?)

60. An object 2.00 cm high is placed 40.0 cm to the left of a converging lens having a focal length of 30.0 cm. A diverging lens with a focal length of − 20.0 cm is placed 110 cm to the right of the converging lens. (a) Determine the position and magnification of the final image. (b) Is the image upright or inverted? (c) Repeat parts (a) and (b) for the case where the second lens is a converging lens having a focal length of + 20.0 cm.

ANSWERS TO QUICK QUIZZES

26.1 At *C*. A ray traced from the stone to the mirror and then to observer 2 looks like this illustration:

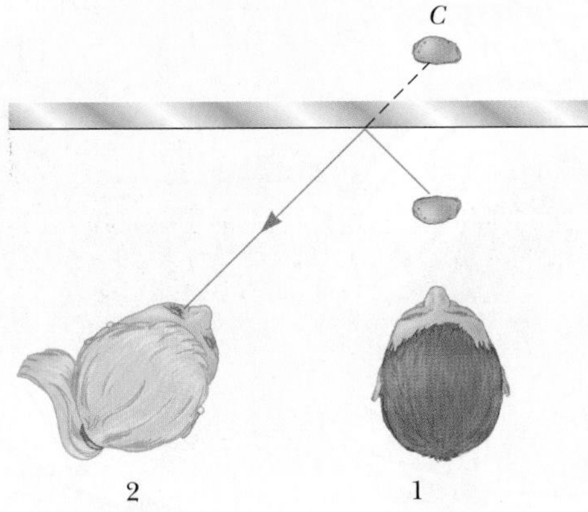

26.2 False. The water spots are 2 m away from you and your image is 4 m away. You cannot focus your eyes on both at the same time.

26.3 (b). A concave mirror will focus the light from a large area of the mirror onto a small area of the paper, resulting in a very high power input to the paper.

26.4 (b). A convex mirror always forms an image with a magnification less than one, so the mirror must be concave. In a concave mirror, only virtual images are upright. This particular photograph is of the Hubble Space Telescope primary mirror.

26.5 (b). Because the flat surfaces of the pane have infinite radii of curvature, Equation 26.13 indicates that the focal length is also infinite. Parallel rays striking the pane focus at infinity, which means that they remain parallel after passing through the glass.

26.6 (c). The entire image is visible but has half the intensity. Each point on the object is a source of rays that travel in all directions. Thus, light from all parts of the object goes through all parts of the lens and forms an image. If you block part of the lens, you are blocking some of the rays, but the remaining ones still come from all parts of the object.

CHAPTER 27

Wave Optics

Interference in soap bubbles. The colors are due to interference between light rays reflected from the front and back surfaces of the thin film of soap making up the bubble. The color depends on the thickness of the film, ranging from black where the film is thinnest to red where it is thickest.

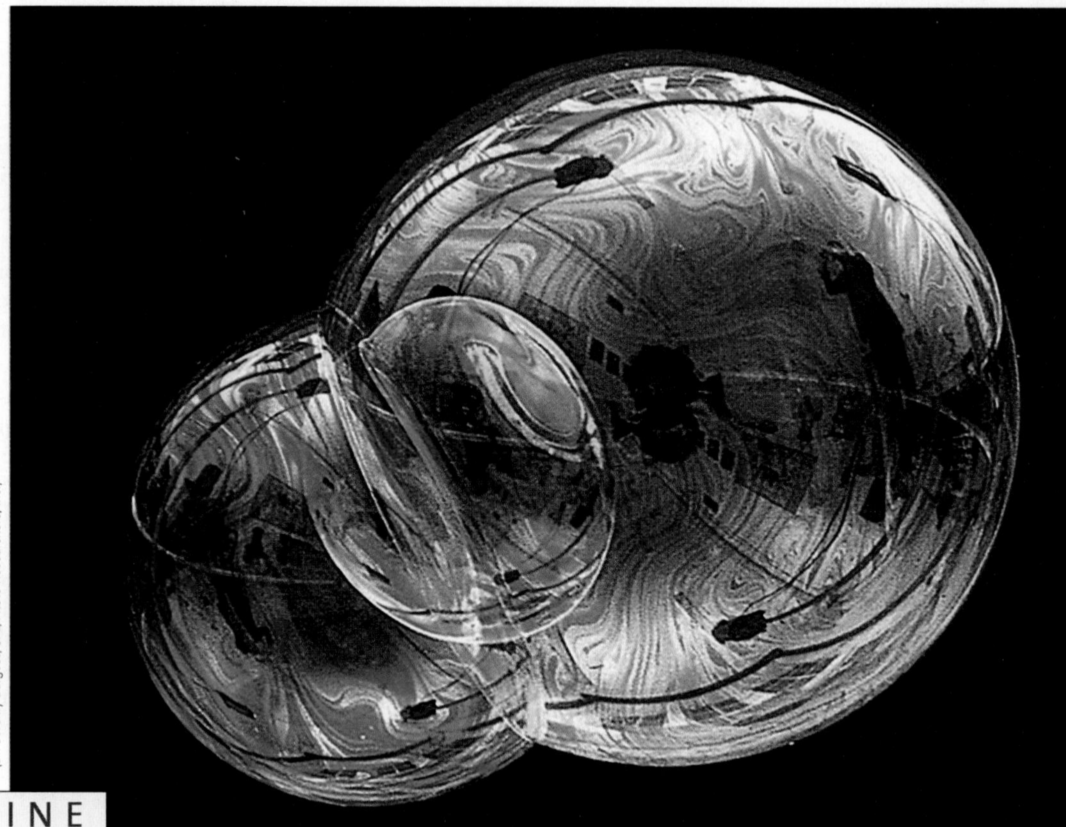

(Dr. Jeremy Burgess/SPL/Photo Researchers, Inc.)

CHAPTER OUTLINE

In Chapters 25 and 26, we used the ray approximation to examine what happens when light reflects from a surface or refracts into a new medium. We used the general term *geometric optics* for these discussions. This chapter is concerned with **wave optics,** a subject that addresses the optical phenomena of interference and diffraction. These phenomena cannot be adequately explained with the ray approximation. We must address the wave nature of light to be able to understand these phenomena.

We introduced the concept of wave interference in Chapter 14 for one-dimensional waves. This phenomenon depends on the principle of superposition, which tells us that when two or more traveling mechanical waves combine at a given point, the resultant displacement of the elements of the medium at that point is the sum of the displacements due to the individual waves.

We shall see the full richness of the waves in interference model in this chapter as we apply it to light. We used one-dimensional waves on strings to introduce interference in Figures 14.1 and 14.2. As we discuss the interference of light waves, two major changes from this previous discussion must be noted. First, we shall no longer focus on one-dimensional waves, so we must build geometric models to analyze the situation in two or three dimensions. Second, we shall study electromagnetic waves rather than mechanical waves. Therefore, the principle of superposition needs to be cast in terms of addition of field vectors rather than displacements of the elements of the medium.

27.1 CONDITIONS FOR INTERFERENCE

In our discussion of wave interference for mechanical waves in Chapter 14, we found that two waves can add together constructively or destructively. In constructive interference between waves, the amplitude of the resultant wave is greater than that of either individual wave, whereas in destructive interference, the resultant amplitude is less than that of either individual wave. Electromagnetic waves also undergo interference. Fundamentally, all interference associated with electromagnetic waves arises as a result of combining the electric and magnetic fields that constitute the individual waves.

In Figure 14.4, we described a device that allows interference to be observed for sound waves. Interference effects in visible electromagnetic waves are not easy to observe because of their short wavelengths (from about 4×10^{-7} to 7×10^{-7} m). Two sources producing two waves of identical wavelengths are needed to create interference. To produce a stable interference pattern, however, the individual waves must maintain a constant phase relationship with one another; they must be **coherent.** As an example, the sound waves emitted by two side-by-side loudspeakers driven by a single amplifier can produce interference because the two loudspeakers respond to the amplifier in the same way at the same time.

If two separate light sources are placed side by side, no interference effects are observed because the light waves from one source are emitted independently of the other source; hence, the emissions from the two sources do not maintain a constant phase relationship with each other over the time of observation. An ordinary light source undergoes random changes in time intervals less than a nanosecond. Therefore, the conditions for constructive interference, destructive interference, or some intermediate state are maintained only for such short time intervals. The result is that no interference effects are observed because the eye cannot follow such rapid changes. Such light sources are said to be **incoherent.**

27.2 YOUNG'S DOUBLE-SLIT EXPERIMENT

A common method for producing two coherent light sources is to use a monochromatic source to illuminate a barrier containing two small openings (usually in the shape of slits). The light emerging from the two slits is coherent because a single source produces the original light beam and the two slits serve only to separate the original beam into two parts (which, after all, is what was done to the sound signal from the side-by-side loudspeakers at the end of the preceding section). Any random change in the light emitted by the source occurs in both beams at the same time, and, as a result, interference effects can be observed when the light from the two slits arrives at a viewing screen.

If the light traveled only in its original direction after passing through the slits as shown in Figure 27.1a, the waves would not overlap and no interference pattern would be seen. Instead, as we have discussed in our treatment of Huygens's principle (Section 25.6), the waves spread out from the slits as shown in Figure 27.1b. In other words, the light deviates from a straight-line path and enters the region that would otherwise be shadowed. As noted in Section 25.2, this divergence of light from its initial line of travel is called **diffraction.**

Interference in light waves from two sources was first demonstrated by Thomas Young in 1801. A schematic diagram of the apparatus that Young used is shown in

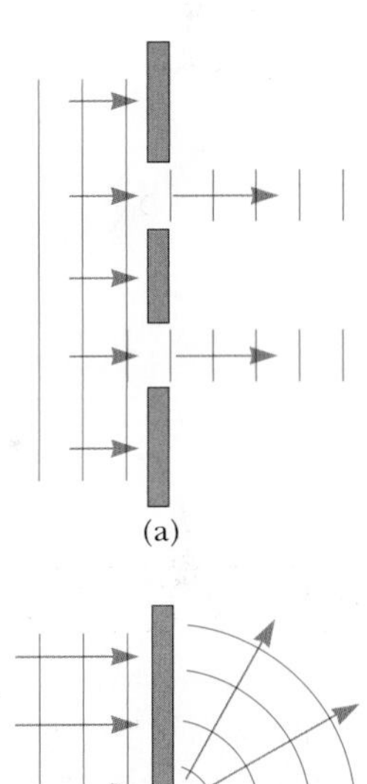
(a)

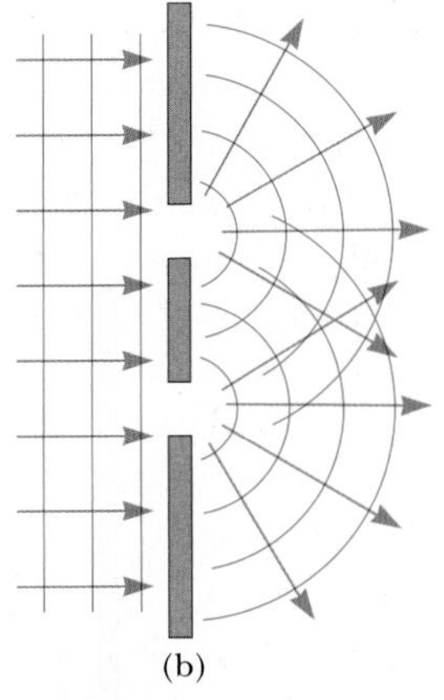
(b)

FIGURE 27.1 (a) If light waves did not spread out after passing through the slits, no interference would occur. (b) The light waves from the two slits overlap as they spread out, filling what we expect to be shadowed regions with light and producing interference fringes on a screen placed to the right of the slits.

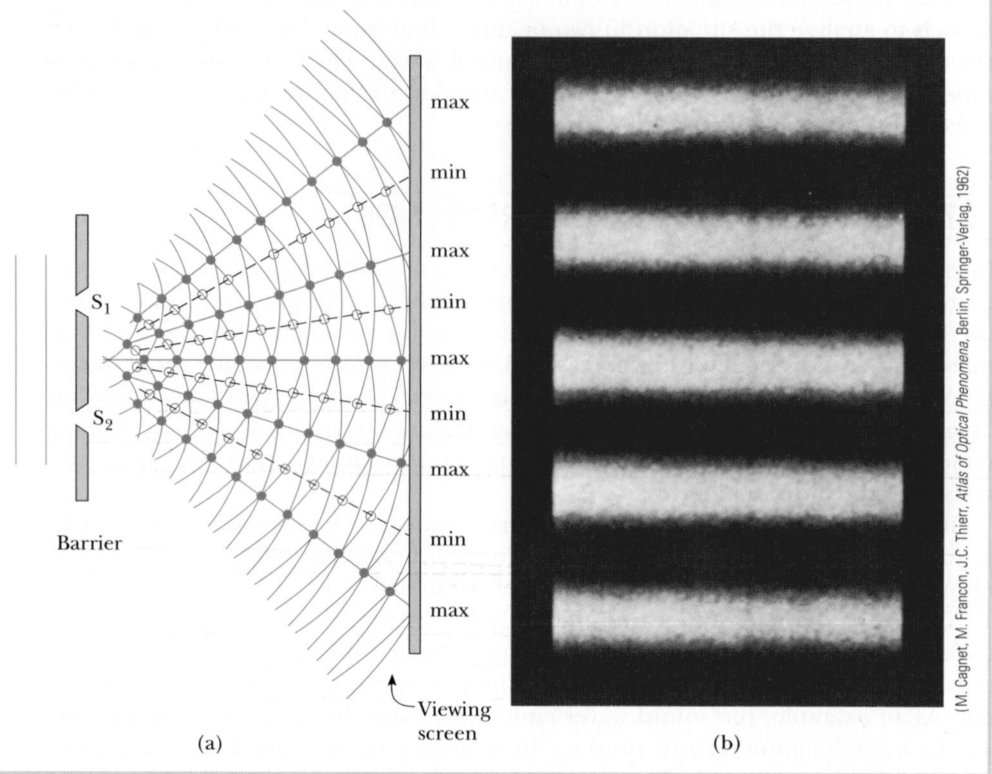

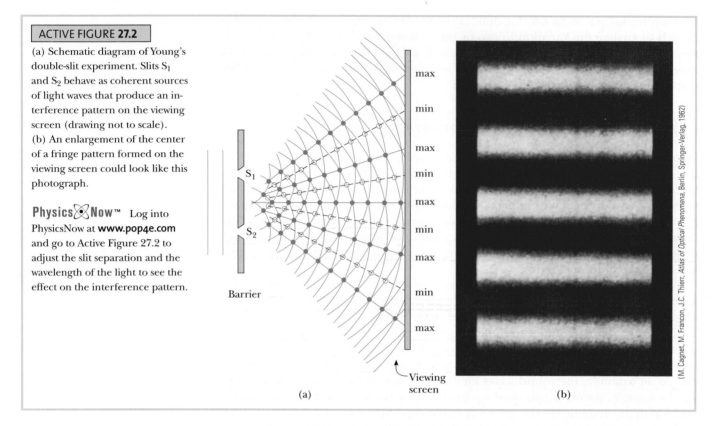

ACTIVE FIGURE 27.2
(a) Schematic diagram of Young's double-slit experiment. Slits S_1 and S_2 behave as coherent sources of light waves that produce an interference pattern on the viewing screen (drawing not to scale). (b) An enlargement of the center of a fringe pattern formed on the viewing screen could look like this photograph.

PhysicsNow™ Log into PhysicsNow at **www.pop4e.com** and go to Active Figure 27.2 to adjust the slit separation and the wavelength of the light to see the effect on the interference pattern.

Active Figure 27.2a. Plane light waves arrive at a barrier that contains two parallel slits S_1 and S_2. These two slits serve as a pair of coherent light sources because waves emerging from them originate from the same wave front and therefore maintain a constant phase relationship. The light from S_1 and S_2 produces on a viewing screen a visible pattern of bright and dark parallel bands called **fringes** (Active Fig. 27.2b). When the light from S_1 and that from S_2 both arrive at a point on the screen such that constructive interference occurs at that location, a bright fringe appears. When the light from the two slits combines destructively at any location on the screen, a dark fringe results.

Figure 27.3 is a schematic diagram that allows us to generate a mathematical representation by modeling the interference as if waves combine at the viewing screen.[1]

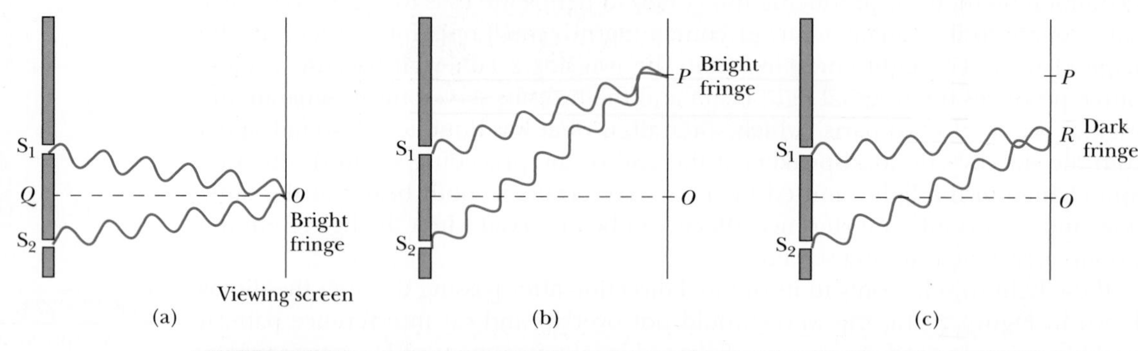

FIGURE 27.3 (a) Constructive interference occurs at point O when the waves combine. (b) Constructive interference also occurs at point P. (c) Destructive interference occurs at R because the wave from the upper slit falls half a wavelength behind the wave from the lower slit. (All figures not to scale.)

[1]The interference occurs everywhere between the slits and the screen, not only at the screen. See Thinking Physics 27.1. The model we have proposed will give us a valid result.

In Figure 27.3a, two waves leave the two slits in phase and strike the screen at the central point *O*. Because these waves travel equal distances, they arrive in phase at *O*. As a result, constructive interference occurs at this location and a bright fringe is observed. In Figure 27.3b, the two light waves again start in phase, but the lower wave has to travel one wavelength farther to reach point *P* on the screen. Because the lower wave falls behind the upper one by exactly one wavelength, they still arrive in phase at *P*. Hence, a second bright fringe appears at this location. Now consider point *R* located between *O* and *P* in Figure 27.3c. At this location, the lower wave has fallen half a wavelength behind the upper wave when they arrive at the screen. Hence, the trough from the lower wave overlaps the crest from the upper wave, giving rise to destructive interference at *R*. For this reason, one observes a dark fringe at this location.

Young's double-slit experiment is the prototype for many interference effects. Interference of waves occurs relatively commonly in technological applications, so this phenomenon represents an important analysis model to understand. In the next section, we develop the mathematical representation for interference of light.

27.3 LIGHT WAVES IN INTERFERENCE

We can obtain a quantitative description of Young's experiment with the help of a geometric model constructed from Figure 27.4a. The viewing screen is located a perpendicular distance *L* from the slits S_1 and S_2, which are separated by a distance *d*. Consider point *P* on the screen. Angle θ is measured from a line perpendicular to the screen from the midpoint between the slits and a line from the midpoint to point *P*. We identify r_1 and r_2 as the distances the waves travel from slit to screen. Let us assume that the source is monochromatic. Under these conditions, the waves emerging from S_1 and S_2 have the same wavelength and amplitude and are in phase. The light intensity on the screen at *P* is the result of the superposition of the light coming from both slits. Note from the geometric model triangle in gold in Figure 27.4a that a wave from the lower slit travels farther than a wave from the upper slit by an amount δ. This distance is called the **path difference.**

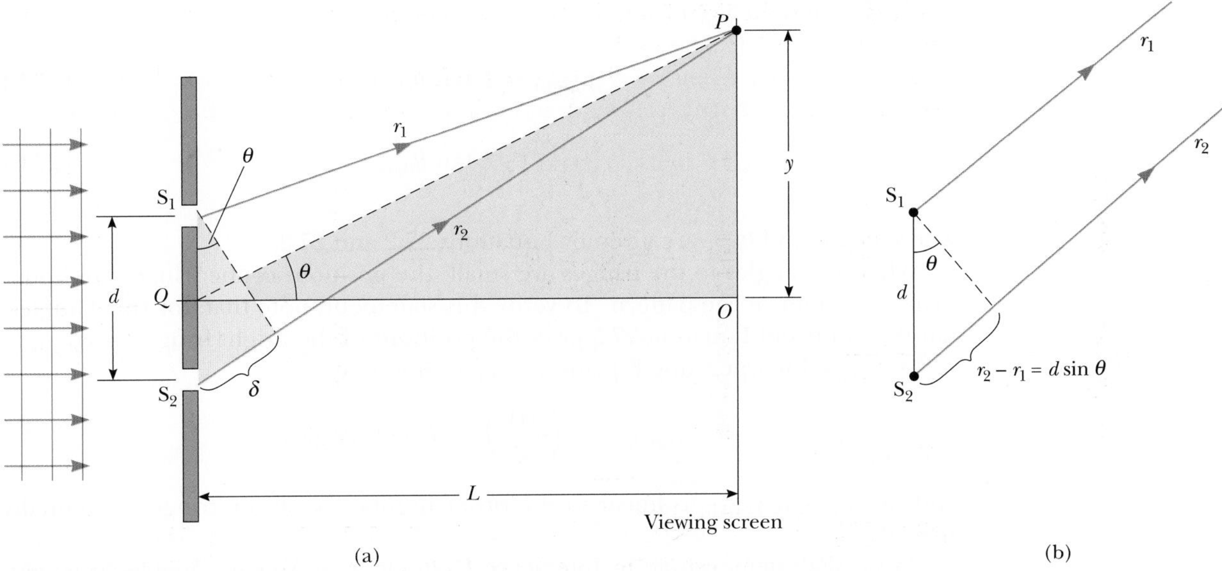

FIGURE 27.4 (a) Geometric construction for describing Young's double-slit experiment (not to scale). (b) When we assume that r_1 is parallel to r_2, the path difference between the two rays is $\delta = r_2 - r_1 = d \sin \theta$. For this approximation to be valid, it is essential that $L \gg d$.

If L is much greater than d, the two paths are very close to being parallel. We shall adopt a simplification model in which the two paths are exactly parallel. In this case, from Figure 27.4b, we see that

■ Path difference

$$\delta = r_2 - r_1 = d \sin \theta \qquad [27.1]$$

In Figure 27.4a, the condition $L \gg d$ is not satisfied because the figure is not to scale; in Figure 27.4b, the rays leave the slits as if the condition is satisfied. As noted earlier, the value of this path difference determines whether the two waves are in phase or out of phase when they arrive at P. If the path difference is either zero or some integral multiple of the wavelength, the two waves are in phase at P and **constructive interference** results. The condition for bright fringes at P is therefore

■ Conditions for constructive interference for two slits

$$\delta = d \sin \theta_{\text{bright}} = m\lambda \qquad (m = 0, \pm 1, \pm 2, \ldots) \qquad [27.2]$$

The number m is an integer called the **order number.** The central bright fringe at $\theta_{\text{bright}} = 0$ is associated with the order number $m = 0$ and is called the **zeroth-order maximum.** The first maximum on either side, for which $m = \pm 1$, is called the **first-order maximum,** and so forth.

Similarly, when the path difference is an odd multiple of $\lambda/2$, the two waves arriving at P are 180° out of phase and give rise to **destructive interference.** Therefore, the condition for dark fringes at P is

■ Conditions for destructive interference for two slits

$$\delta = d \sin \theta_{\text{dark}} = (m + \tfrac{1}{2})\lambda \qquad (m = 0, \pm 1, \pm 2, \ldots) \qquad [27.3]$$

These equations provide the *angular* positions of the fringes. It is also useful to obtain expressions for the *linear* positions measured along the screen from O to P. From the geometric model triangle OPQ in Figure 27.4a, we see that

$$\tan \theta = \frac{y}{L} \qquad [27.4]$$

Using this result, we can see that the linear positions of bright and dark fringes are given by

$$y_{\text{bright}} = L \tan \theta_{\text{bright}} \qquad [27.5]$$

$$y_{\text{dark}} = L \tan \theta_{\text{dark}} \qquad [27.6]$$

where θ_{bright} and θ_{dark} are given by Equations 27.2 and 27.3.

When the angles to the fringes are small, the positions of the fringes are linear near the center of the pattern. To verify this statement, note that, for small angles, $\tan \theta \approx \sin \theta$ and Equation 27.5 gives the positions of the bright fringes as $y_{\text{bright}} = L \sin \theta_{\text{bright}}$. Incorporating Equation 27.2, we find that

$$y_{\text{bright}} = L\left(\frac{m\lambda}{d}\right) \qquad \text{(small angles)}$$

and we see that y_{bright} is linear in the order number m, so the fringes are equally spaced.

As we shall demonstrate in Interactive Example 27.1, Young's double-slit experiment provides a method for measuring the wavelength of light. In fact, Young used this technique to make the first measurement of the wavelength of light. Young's experiment gave the wave model of light a great deal of credibility. Today we still use the phenomenon of interference to describe many observations of wave-like behavior.

QUICK QUIZ 27.1 Which of the following will cause the fringes in a two-slit interference pattern to move farther apart? **(a)** decreasing the wavelength of the light **(b)** decreasing the screen distance L **(c)** decreasing the slit spacing d **(d)** immersing the entire apparatus in water

■ Thinking Physics 27.1

Consider a double-slit experiment in which a laser beam is passed through a pair of very closely spaced slits and a clear interference pattern is displayed on a distant screen. Now suppose you place smoke particles between the double slit and the screen. With the presence of the smoke particles, will you see the effects of the interference in the space between the slits and the screen, or will you only see the effects on the screen?

Reasoning You see the effects in the area filled with smoke. Bright beams of light are directed toward the bright areas on the screen, and dark regions are directed toward the dark areas on the screen. The geometrical construction shown in Figure 27.4a is important for developing the mathematical description of interference. It is subject to misinterpretation, however, because it might suggest that the interference can only occur at the position of the screen. A better diagram for this situation is Active Figure 27.2a, which shows *paths* of destructive and constructive interference all the way from the slits to the screen. These paths are made visible by the smoke. ■

INTERACTIVE EXAMPLE 27.1 Measuring the Wavelength of Laser Light

A laser is used to illuminate a double slit. The distance between the two slits is 0.030 mm. A viewing screen is separated from the double slit by 1.2 m. The second-order bright fringe ($m = 2$) is 5.1 cm from the center line.

A Determine the wavelength of the laser light.

Solution Because the distance between the screen and the slits is much larger than the slit separation, Equation 27.2 is a valid mathematical representation of this situation. Incorporating Equation 27.5, with $m = 2$, $y_2 = 5.1 \times 10^{-2}$ m, $L = 1.2$ m, and $d = 3.0 \times 10^{-5}$ m, we have

$$\lambda = \frac{d \sin \theta_{\text{bright}}}{m} = \frac{d \sin\left(\tan^{-1} \dfrac{y_{\text{bright}}}{L}\right)}{m}$$

$$= \frac{(3.0 \times 10^{-5}\text{ m}) \sin\left(\tan^{-1} \dfrac{5.1 \times 10^{-2}\text{ m}}{1.2\text{ m}}\right)}{2}$$

$$= 6.4 \times 10^{-7}\text{ m} = \boxed{6.4 \times 10^{2}\text{ nm}}$$

B Calculate the distance between adjacent bright fringes near the center of the interference pattern.

Solution The position of the $m = 2$ fringe, 5.1 cm, is much smaller than the screen distance, 1.2 m. Therefore, the angular positions of the fringes near the center of the pattern are small. Consequently, these fringes are equally spaced, so the distance between fringes can be found by dividing the distance between the $m = 0$ and $m = 2$ fringes by 2:

$$\Delta y = \frac{y_2 - y_0}{2} = \frac{5.1\text{ cm} - 0}{2} = \boxed{2.6\text{ cm}}$$

PhysicsNow™ Investigate the double-slit interference pattern by logging into PhysicsNow at **www.pop4e.com** and going to Interactive Example 27.1.

(From M. Cagnet, M. Francon, and J. C. Thierr, Atlas of Optical Phenomena, Berlin, Springer-Verlag, 1962)

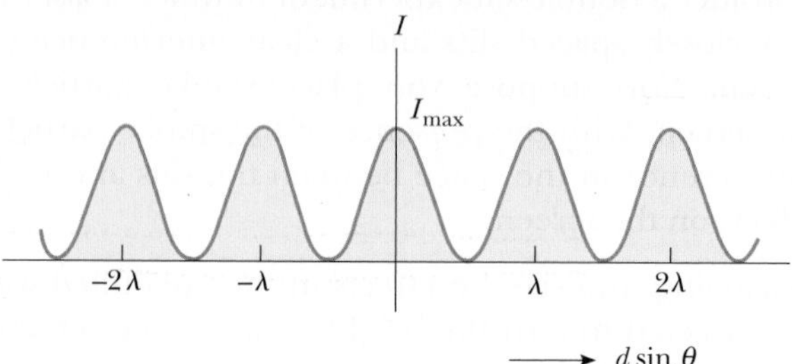

FIGURE 27.5 Light intensity versus $d \sin \theta$ for the double-slit interference pattern when the screen is far from the two slits ($L >> d$).

Intensity Distribution of the Double-Slit Interference Pattern

We shall now discuss briefly the distribution of light intensity I (the energy delivered by the light per unit area per unit time) associated with the double-slit interference pattern. Again, suppose the two slits represent coherent sources of sinusoidal waves. In this case, the two waves have the same angular frequency ω and a constant phase difference ϕ. Although the waves have equal phase at the slits, their phase difference ϕ at P depends on the path difference $\delta = r_2 - r_1 = d \sin \theta$. Because a path difference of λ corresponds to a phase difference of 2π rad, we can establish the equality of the ratios:

$$\frac{\delta}{\phi} = \frac{\lambda}{2\pi}$$

■ Phase difference

$$\phi = \frac{2\pi}{\lambda}\delta = \frac{2\pi}{\lambda} d \sin \theta \qquad [27.7]$$

This equation tells us how the phase difference ϕ depends on the angle θ.

Although we shall not prove it here, a careful analysis of the electric fields arriving at the screen from the two very narrow slits shows that the **time-averaged light intensity** at a given angle θ is

$$I_{avg} = I_{max} \cos^2\left(\frac{\pi d \sin \theta}{\lambda}\right) \qquad [27.8]$$

where I_{max} is the intensity at point O in Figure 27.4a, directly behind the midpoint between the slits. Intensity versus $d \sin \theta$ is plotted in Figure 27.5.

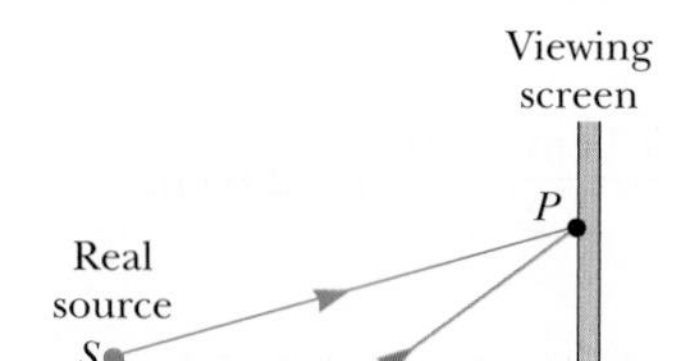

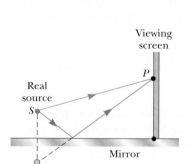

FIGURE 27.6 Lloyd's mirror. An interference pattern is produced on a screen at point P as a result of the combination of the direct ray (blue) and the reflected ray (brown). The reflected ray undergoes a phase change of 180°.

27.4 CHANGE OF PHASE DUE TO REFLECTION

Young's method of producing two coherent light sources involves illuminating a pair of slits with a single source. Another simple arrangement for producing an interference pattern with a single light source is known as *Lloyd's mirror.* A light source is placed at point S close to a mirror as illustrated in Figure 27.6. Waves can reach the point P either by the direct path SP or by the indirect path involving reflection

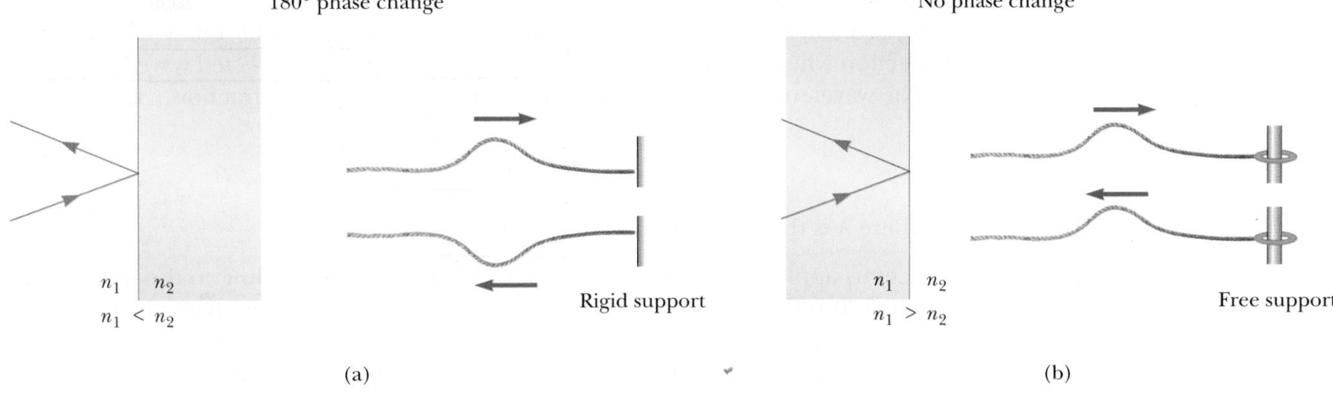

FIGURE 27.7 (a) For $n_1 < n_2$, a light ray traveling in medium 1 and reflected from the surface of medium 2 undergoes a 180° phase change. The same thing happens with a reflected pulse traveling along a string fixed at one end. (b) For $n_1 > n_2$, a light ray traveling in medium 1 undergoes no phase change when reflected from the surface of medium 2. The same is true of a reflected wave pulse on a string whose supported end is free to move.

from the mirror. The reflected ray strikes the screen as if it originated from a source at S' located below the mirror.

At points far from the source, one would expect an interference pattern due to waves from S and S', just as is observed for two real coherent sources at these points. An interference pattern is indeed observed. The positions of the dark and bright fringes, however, are *reversed* relative to the pattern of two real coherent sources (Young's experiment) because the coherent sources at S and S' differ in phase by 180°. This 180° phase change is produced on reflection. In general, an electromagnetic wave undergoes a phase change of 180° on reflection from a medium of higher index of refraction than the one in which it is traveling.

It is useful to draw an analogy between reflected light waves and the reflections of a transverse wave on a stretched string when the wave meets a boundary (Section 13.5) as in Figure 27.7. The reflected pulse on a string undergoes a phase change of 180° when it is reflected from a rigid end, and no phase change when it is reflected from a free end, as illustrated in Figures 13.12 and 13.13. If the boundary is between two strings, the transmitted wave exhibits no phase change. Similarly, an electromagnetic wave undergoes a 180° phase change when reflected from the boundary of a medium of higher index of refraction than the one in which it is traveling. There is no phase change for the reflected ray when the wave is incident on a boundary leading to a medium of lower index of refraction. In either case, the transmitted wave exhibits no phase change.

27.5 INTERFERENCE IN THIN FILMS

Interference effects can be observed in many situations in which one beam of light is split and then recombined. A common occurrence is the appearance of colored bands in a film of oil on water or in a soap bubble illuminated with white light. The colors in these situations result from the interference of waves reflected from the opposite surfaces of the film.

Consider a film of uniform thickness t and index of refraction n as in Figure 27.8. We adopt a simplification model in which the light ray is incident on the film from above and nearly normal to the surface of the film. Two rays are reflected from the film, one from the upper surface and one from the lower surface after the refracted ray has traveled through the film. Because the film is thin and has parallel sides, the reflected rays are parallel. Hence, rays reflected from the top surface can interfere with rays reflected from the bottom surface. To determine whether the reflected rays interfere constructively or destructively, we first note the following facts:

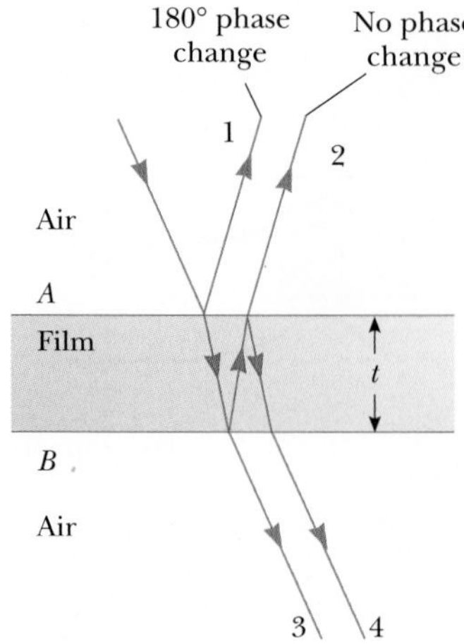

FIGURE 27.8 Interference in light reflected from a thin film is due to a combination of rays 1 and 2 reflected from the upper and lower surfaces of the film. Rays 3 and 4 lead to interference effects for light transmitted through the film.

- An electromagnetic wave traveling from a medium of index of refraction n_1 toward a medium of index of refraction n_2 undergoes a 180° phase change on reflection when $n_2 > n_1$. No phase change occurs in the reflected wave if $n_2 < n_1$.
- The wavelength λ_n of light in a medium whose index of refraction n is

$$\lambda_n = \frac{\lambda}{n} \qquad [27.9]$$

where λ is the wavelength of light in free space.

Let us apply these rules to the film of Figure 27.8. According to the first rule, ray 1, which is reflected from the upper surface (A), undergoes a phase change of 180° with respect to the incident wave. Ray 2, which is reflected from the lower surface (B), undergoes no phase change with respect to the incident wave. Therefore, ignoring the path difference for now, outgoing ray 1 is 180° out of phase with respect to ray 2, a phase difference that is equivalent to a path difference of $\lambda_n/2$. We must also consider, however, that ray 2 travels an extra distance approximately equal to $2t$ before the waves recombine. The *total* phase difference arises from a combination of the path difference and the 180° phase change on reflection. For example, if $2t = \lambda_n/2$, rays 1 and 2 will recombine in phase and constructive interference will result. In general, the condition for constructive interference is

■ Condition for constructive interference in thin films

$$2t = (m + \tfrac{1}{2})\lambda_n \qquad (m = 0, 1, 2, \ldots) \qquad [27.10]$$

This condition takes into account two factors: (a) the difference in optical path length for the two rays (the term $m\lambda_n$) and (b) the 180° phase change on reflection (the term $\lambda_n/2$). Because $\lambda_n = \lambda/n$, we can write Equation 27.10 in the form

$$2nt = (m + \tfrac{1}{2})\lambda \qquad (m = 0, 1, 2, \ldots) \qquad [27.11]$$

If the extra distance $2t$ traveled by ray 2 corresponds to a multiple of λ_n, the two waves will combine out of phase and destructive interference results. The general equation for destructive interference is

■ Condition for destructive interference in thin films

$$2nt = m\lambda \qquad (m = 0, 1, 2, \ldots) \qquad [27.12]$$

The preceding conditions for constructive and destructive interference are valid when the medium above the top surface of the film is the same as the medium below the bottom surface. The surrounding medium may have a refractive index less than or greater than that of the film. In either case, the rays reflected from the two surfaces will be out of phase by 180°. The conditions are also valid if different media are above and below the film and if both have n less than or larger than that of the film.

If the film is placed between two different media, one with $n < n_{\text{film}}$ and the other with $n > n_{\text{film}}$, the conditions for constructive and destructive interference are reversed. In this case, either a phase change of 180° takes place for both ray 1 reflecting from surface A and ray 2 reflecting from surface B, or no phase change occurs for either ray; hence, the net change in relative phase due to the reflections is zero.

Rays 3 and 4 in Figure 27.8 lead to interference effects in the light transmitted through the thin film. The analysis of these effects is similar to that of the reflected light.

QUICK QUIZ 27.2 In a laboratory accident, you spill two liquids onto water, neither of which mixes with the water. They both form thin films on the water surface. When the films become very thin as they spread, you observe that one film becomes bright and the other dark in reflected light. The film that appears dark **(a)** has an index of refraction higher than that of water, **(b)** has an index of refraction lower than that of water, **(c)** has an index of refraction equal to that of water, or **(d)** has an index of refraction lower than that of the bright film.

QUICK QUIZ 27.3 One microscope slide is placed on top of another with their left edges in contact and a human hair under the right edge of the upper slide. As a result, a wedge of air exists between the slides. An interference pattern results when monochromatic light is incident on the wedge. At the left edges of the slides, what kind of fringe is there? **(a)** a dark fringe **(b)** a bright fringe **(c)** impossible to determine

PROBLEM-SOLVING STRATEGY Thin-Film Interference

The following suggestions should be kept in mind while working thin-film interference problems:

1. Conceptualize Think about what is going on physically in the problem. Identify the light source and the location of the observer.

2. Categorize Confirm that you should use the techniques for thin film interference by identifying the thin film causing the interference.

3. Analyze The type of interference that occurs is determined by the phase relationship between the portion of the wave reflected at the upper surface of the film and the portion reflected at the lower surface. Phase differences between the two portions of the wave have two causes: (a) differences in the distances traveled by the two portions and (b) phase changes occurring on reflection. *Both* causes must be considered when determining which type of interference occurs. If the media above and below the film both have index of refraction larger than that of the film or if both indices are smaller, use Equation 27.11 for constructive interference and Equation 27.12 for destructive interference. If the film is located between two different media, one with $n < n_{\text{film}}$ and the other with $n > n_{\text{film}}$, reverse these two equations for constructive and destructive interference.

4. Finalize Inspect your final results to see if they make sense physically and are of an appropriate size.

EXAMPLE 27.2 Interference in a Soap Film

Calculate the minimum thickness of a soap film ($n = 1.33$) that results in constructive interference in reflected light if the film is illuminated with light whose wavelength in free space is 600 nm.

Solution The minimum film thickness for constructive interference in the reflected light corresponds to $m = 0$ in Equation 27.11, which gives $2nt = \lambda/2$, or

$$t = \frac{\lambda}{4n} = \frac{600 \text{ nm}}{4(1.33)} = 113 \text{ nm}$$

INTERACTIVE EXAMPLE 27.3 Nonreflecting Coatings for Solar Cells

Semiconductors such as silicon are used to fabricate solar cells, devices that absorb energy by electromagnetic radiation (e.g., sunlight), resulting in a potential difference so that the cell can transfer energy to a device by electrical transmission. Solar cells are often coated with a transparent thin film, such as silicon monoxide (SiO, $n = 1.45$), to minimize reflective losses from the surface. Suppose a silicon solar cell ($n = 3.5$) is coated with a thin film of silicon monoxide for this purpose (Fig. 27.9). Determine the minimum film thickness that produces the least reflection at a wavelength of 550 nm, which is the center of the visible spectrum.

Solution The reflected light is a minimum when rays 1 and 2 in Figure 27.9 meet the condition of destructive interference. Note that both rays undergo a 180° phase change on reflection in this case, one from the upper surface and one from the lower surface. Hence, the net change in phase is zero due to reflection, and the condition for a reflection minimum requires a path difference of $\lambda_n/2$; thus, $2t = \lambda/2n$. Therefore, the required thickness is

$$t = \frac{\lambda}{4n} = \frac{550 \text{ nm}}{4(1.45)} = 94.8 \text{ nm}$$

Typically, such antireflecting coatings reduce the reflective loss from 30% (with no coating) to 10% (with coating), thereby increasing the cell's efficiency because more light is available to provide energy to the cell. In reality, the coating is never perfectly

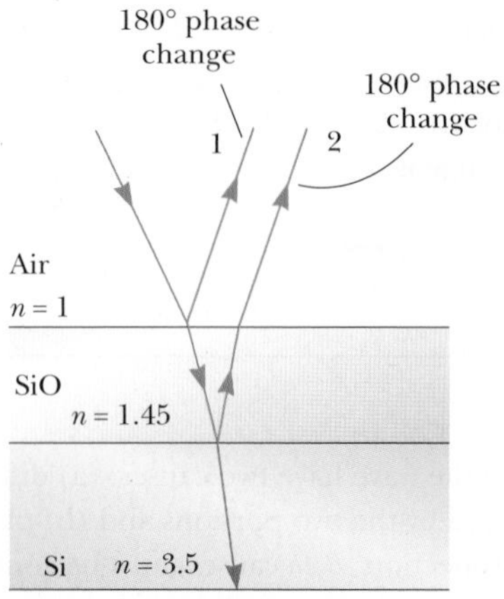

FIGURE 27.9 (Interactive Example 27.3) Reflective losses from a silicon solar cell are minimized by coating it with a thin film of silicon monoxide.

FIGURE 27.10 (Interactive Example 27.3) This camera lens has several coatings (of different thicknesses) that minimize reflection of light waves having wavelengths near the center of the visible spectrum. As a result, the little light that is reflected by the lens has a greater proportion of the far ends of the spectrum and appears purple.

nonreflecting for all light because the required thickness is wavelength dependent and the incident light covers a wide range of wavelengths.

Glass lenses used in cameras and other optical instruments are usually coated with a transparent thin film, such as magnesium fluoride (MgF_2), to reduce or eliminate unwanted reflection. The result is the enhancement of the transmission of light through the lenses. Figure 27.10 shows such a camera lens. Notice that the light reflecting from the lens is tinged purple. The coating on the lens is designed with a thickness such that light near the center of the visible spectrum experiences little reflection. Light near the ends of the spectrum is reflected from the coating. The combination of red and violet light from the ends of the spectrum provides the purple tinge.

PhysicsNow™ Investigate the interference for various film properties by logging into PhysicsNow at **www.pop4e.com** and going to Interactive Example 27.3.

EXAMPLE 27.4 Interference in a Wedge-Shaped Film

A thin, wedge-shaped film of refractive index n is illuminated with monochromatic light of wavelength λ as illustrated in Figure 27.11a. Describe the interference pattern observed for this case.

Solution The interference pattern is that of a thin film of variable thickness surrounded by air. Hence, the pattern is a series of alternating bright and dark parallel bands. A dark band corresponding to destructive interference appears at point O (where the path length difference is zero) because the ray reflected from the first surface undergoes a 180° phase change but the one reflected from the second surface does not. According to Equation 27.12, other dark bands appear when $2nt = m\lambda$, so that $t_1 = \lambda/2n$, $t_2 = \lambda/n$, $t_3 = 3\lambda/2n$, and so on. Similarly, bright bands are observed when the thickness satisfies the condition $2nt = (m + \frac{1}{2})\lambda$, corresponding to thicknesses of $\lambda/4n$, $3\lambda/4n$, $5\lambda/4n$, and so on. If white light is used, bands of different colors are observed at different points, corresponding to the different wavelengths of light. This situation is shown in the soap film in Figure 27.11b.

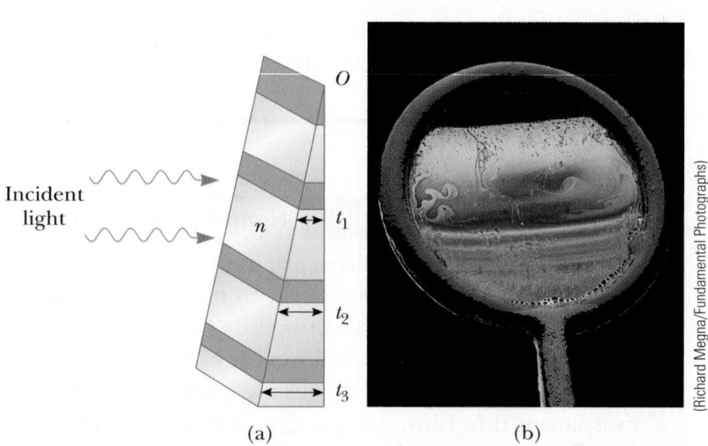

FIGURE 27.11 (Example 27.4) (a) Interference bands in reflected light can be observed by illuminating a wedge-shaped film with monochromatic light. The darker areas correspond to regions at which rays cancel due to destructive interference. (b) Interference in a vertical film of variable thickness. The top of the film appears darkest where the film is thinnest.

27.6 DIFFRACTION PATTERNS

In Sections 25.2 and 27.2, we discussed briefly the phenomenon of **diffraction,** and now we shall investigate this phenomenon more fully for light waves. In general, diffraction occurs when waves pass through small openings, around obstacles, or by sharp edges.

We might expect that the light passing through one such small opening would simply result in a broad region of light on a screen due to the spreading of the light as it passes through the opening. We find something more interesting, however. A **diffraction pattern** consisting of light and dark areas is observed, somewhat similar to the interference patterns discussed earlier. For example, when a narrow slit is placed between a distant light source (or a laser beam) and a screen, the light produces a diffraction pattern like that in Figure 27.12. The pattern consists of a broad, intense central band (called the **central maximum**), flanked by a series of narrower, less intense additional bands (called **side maxima**) and a series of dark bands (or **minima**).

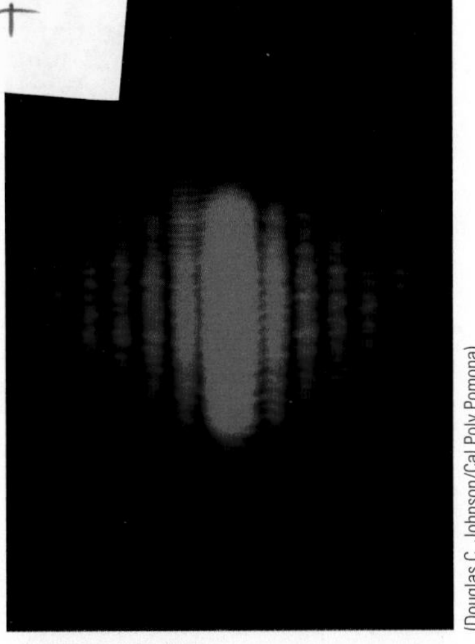

(Douglas C. Johnson/Cal Poly Pomona)

FIGURE 27.12 The diffraction pattern that appears on a screen when light passes through a narrow vertical slit. The pattern consists of a broad central band and a series of less intense and narrower side bands.

Figure 27.13 shows the shadow of a penny, which displays bright and dark rings of a diffraction pattern. The bright spot at the center (called the *Arago bright spot* after its discoverer, Dominique Arago) can be explained using the wave theory of light. Waves that diffract from all points on the edge of the penny travel the same distance to the midpoint on the screen. Thus, the midpoint is a region of constructive interference and a bright spot appears. In contrast, from the viewpoint of geometric optics, the center of the pattern would be completely screened by the penny, and so an approach that does not include the wave nature of light would not predict a central bright spot.

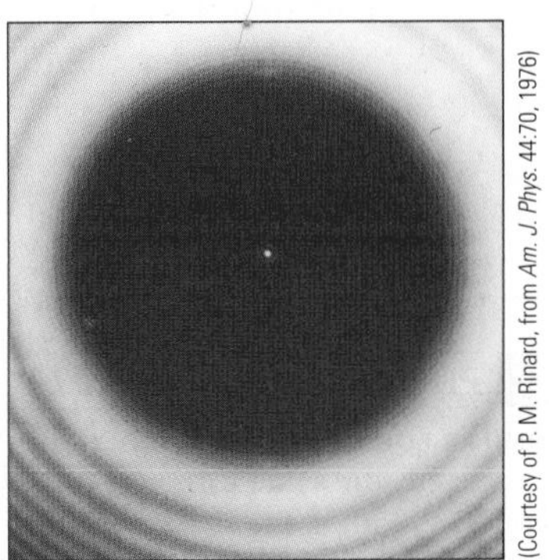

(Courtesy of P. M. Rinard, from *Am. J. Phys.* 44:70, 1976)

FIGURE 27.13 Diffraction pattern of a penny, taken with the penny midway between screen and source.

Let us consider a common situation, that of light passing through a narrow opening modeled as a slit and projected onto a screen. As a simplification model, we assume that the observing screen is far from the slit so that the rays reaching the screen are approximately parallel. This situation can also be achieved experimentally by using a converging lens to focus the parallel rays on a nearby screen. In this model, the pattern on the screen is called a **Fraunhofer diffraction pattern.**[2]

Active Figure 27.14a shows light entering a single slit from the left and diffracting as it propagates toward a screen. Active Figure 27.14b is a photograph of a

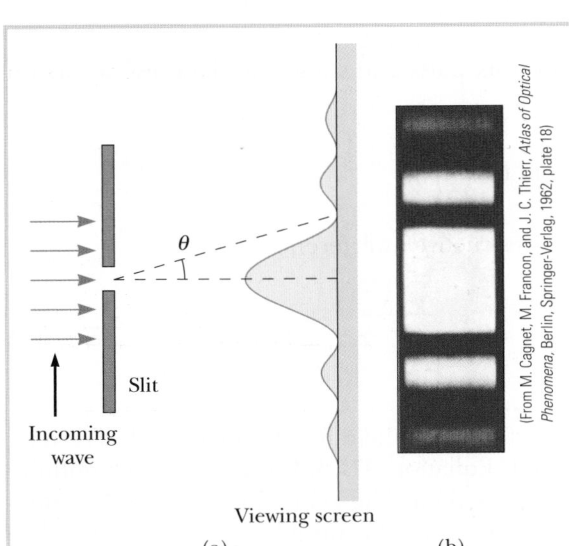

ACTIVE FIGURE 27.14

(a) Fraunhofer diffraction pattern of a single slit. The pattern consists of a central bright region flanked by much weaker maxima alternating with dark bands. (Drawing not to scale.)

(b) Photograph of a single-slit Fraunhofer diffraction pattern.

Physics Now™ Log into PhysicsNow at **www.pop4e.com** and go to Active Figure 27.14 to adjust the slit width and the wavelength of the light to see the effect on the diffraction pattern.

PITFALL PREVENTION 27.1

DIFFRACTION VERSUS DIFFRACTION PATTERN The word *diffraction* refers to the general behavior of waves spreading out as they pass through a slit. We used diffraction in explaining the existence of an interference pattern. A *diffraction pattern* is actually a misnomer, but it is deeply entrenched in the language of physics. We describe here the diffraction pattern seen on a screen when a single slit is illuminated. In reality, it is another interference pattern. The interference is between parts of the incident light illuminating different regions of the slit.

[2]If the screen were brought close to the slit (and no lens is used), the pattern is a *Fresnel* diffraction pattern. The Fresnel pattern is more difficult to analyze, so we shall restrict our discussion to Fraunhofer diffraction.

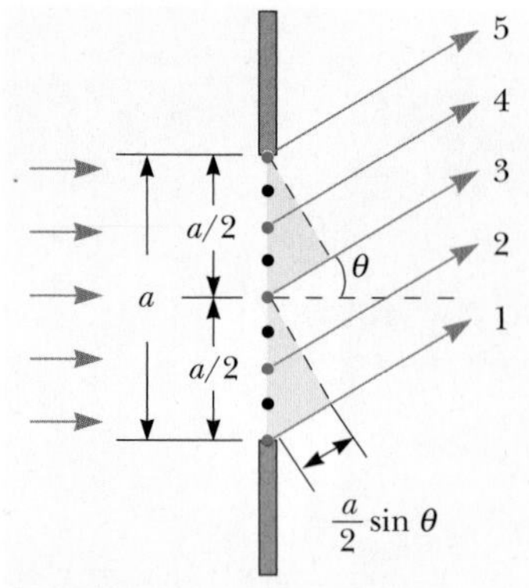

FIGURE 27.15 Diffraction of light by a narrow slit of width a. Each portion of the slit acts as a point source of waves. The path difference between rays 1 and 3, rays 2 and 4, or rays 3 and 5 is $(a/2)\sin\theta$. (Drawing not to scale.)

single-slit Fraunhofer diffraction pattern. A bright fringe is observed along the axis at $\theta = 0$, with alternating dark and bright fringes on each side of the central bright fringe.

Until now, we assumed that slits act as point sources of light. In this section, we shall determine how their finite widths are the basis for understanding the nature of the Fraunhofer diffraction pattern produced by a single slit. We can deduce some important features of this problem by examining waves coming from various portions of the slit as shown in the geometric model of Figure 27.15. According to Huygens's principle, **each portion of the slit acts as a source of waves. Hence, light from one portion of the slit can interfere with light from another portion,** and the resultant intensity on the screen depends on the direction θ.

To analyze the diffraction pattern, it is convenient to divide the slit into two halves as in Figure 27.15. All the waves that originate at the slit are in phase. Consider waves 1 and 3, which originate at the bottom and center of the slit, respectively. To reach the same point on the viewing screen, wave 1 travels farther than wave 3 by an amount equal to the path difference $(a/2)\sin\theta$, where a is the width of the slit. Similarly, the path difference between waves 3 and 5 is also $(a/2)\sin\theta$. If the path difference between two waves is exactly one half of a wavelength (corresponding to a phase difference of 180°), the two waves cancel each other and destructive interference results. That is true, in fact, for any two waves that originate at points separated by half the slit width because the phase difference between two such points is 180°. Therefore, waves from the upper half of the slit interfere *destructively* with waves from the lower half of the slit when

$$\frac{a}{2}\sin\theta = \frac{\lambda}{2}$$

or when

$$\sin\theta = \frac{\lambda}{a}$$

If we divide the slit into four parts rather than two and use similar reasoning, we find that the screen is also dark when

$$\sin\theta = \frac{2\lambda}{a}$$

Likewise, we can divide the slit into six parts and show that darkness occurs on the screen when

$$\sin\theta = \frac{3\lambda}{a}$$

Therefore, the general condition for destructive interference is

Condition for destructive interference in a diffraction pattern

$$\sin\theta_{\text{dark}} = m\frac{\lambda}{a} \qquad (m = \pm 1, \pm 2, \pm 3, \ldots) \qquad [27.13]$$

Equation 27.13 gives the values of θ for which the diffraction pattern has zero intensity, that is, a dark fringe is formed. Equation 27.13, however, tells us nothing about the variation in intensity along the screen. The general features of the intensity distribution are shown in Figure 27.16: a broad central bright fringe flanked by much weaker, alternating bright fringes. The various dark fringes (points of zero intensity) occur at the values of θ that satisfy Equation 27.13. The position of the points of constructive interference lie approximately halfway between the dark fringes. Note that the central bright fringe is twice as wide as the weaker maxima.

PITFALL PREVENTION 27.2

SIMILAR EQUATIONS Equation 27.13 has exactly the same form as Equation 27.2, with d, the slit separation, used in Equation 27.2 and a, the slit width, in Equation 27.13. Keep in mind, however, that Equation 27.2 describes the *bright* regions in a two-slit interference pattern, whereas Equation 27.13 describes the *dark* regions in a single-slit diffraction pattern. Furthermore, $m = 0$ does not represent a dark fringe in the diffraction pattern.

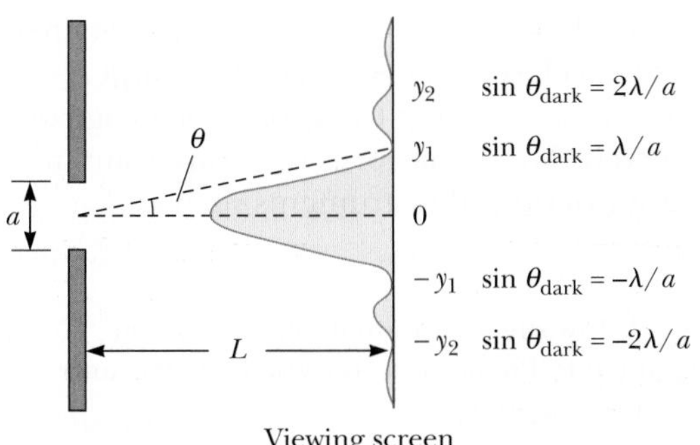

FIGURE 27.16 Light intensity distribution for the Fraunhofer diffraction pattern from a single slit of width a. The positions of two minima on each side of the central maximum are labeled. (Drawing not to scale.)

QUICK QUIZ 27.4 **(i)** Suppose the slit width in Figure 27.16 is made half as wide. The central bright fringe **(a)** becomes wider, **(b)** remains the same, or **(c)** becomes narrower. **(ii)** From the same choices, what happens to the central bright fringe when the wavelength of the light is made half as great?

■ Thinking Physics 27.2

If a classroom door is open slightly, you can hear sounds coming from the hallway. Yet you cannot see what is happening in the hallway. What accounts for the difference?

Reasoning The space between the slightly open door and the wall is acting as a single slit for waves. Sound waves have wavelengths larger than the slit width, so sound is effectively diffracted by the opening and spread throughout the room. The sound is then reflected from walls, floor, and ceiling, further distributing the sound throughout the room. Light wavelengths are much smaller than the slit width, so virtually no diffraction for the light occurs. You must have a direct line of sight to detect the light waves. ■

INTERACTIVE EXAMPLE 27.5 Where Are the Dark Fringes?

Light of wavelength 580 nm is incident on a slit of width 0.300 mm. The observing screen is 2.00 m from the slit.

A Find the positions of the first dark fringes and the width of the central bright fringe.

Solution The problem statement cues us to conceptualize a single-slit diffraction pattern similar to that in Figure 27.16. We categorize it as a straightforward application of our discussion of single-slit diffraction patterns. To analyze the problem, note that the two dark fringes that flank the central bright fringe correspond to $m = \pm 1$ in Equation 27.13. Hence, we find that

$$\sin\theta_{dark} = \pm\frac{\lambda}{a} = \pm\frac{5.80\times10^{-7}\text{ m}}{0.300\times10^{-3}\text{ m}}$$
$$= \pm 1.933\times10^{-3}$$

From the triangle in Figure 27.16, note that $\tan\theta_{dark} = y_1/L$. Because θ_{dark} is very small, we can use the approximation $\sin\theta_{dark} \approx \tan\theta_{dark}$; thus, $\sin\theta_{dark} \approx y_1/L$.

Therefore, the positions of the first minima measured from the central axis are given by

$$y_1 \approx L\sin\theta_{dark} = (2.00\text{ m})(\pm1.933\times10^{-3})$$
$$= \pm 3.87\times10^{-3}\text{ m}$$

The positive and negative signs correspond to the dark fringes on either side of the central bright fringe. Hence, the width of the central bright fringe is equal to $2|y_1| = 7.74\times10^{-3}\text{ m} = 7.74\text{ mm}$. To finalize this problem, note that this value is much greater than the width of the slit. We finalize further by exploring what happens if we change the slit width in part B.

B What if the slit width is increased by an order of magnitude to 3.00 mm? What happens to the diffraction pattern?

Solution Based on Equation 27.13, we expect that the angles at which the dark bands appear will decrease as a increases. Thus, the diffraction pattern narrows. For $a = 3.00$ mm, the sines of the angles θ_{dark} for the $m = \pm 1$ dark fringes are

$$\sin\theta_{\text{dark}} = \pm\frac{\lambda}{a} = \pm\frac{5.80 \times 10^{-7}\text{ m}}{3.00 \times 10^{-3}\text{ m}}$$
$$= \pm 1.933 \times 10^{-4}$$

The positions of the first minima measured from the central axis are given by

$$y_1 \approx L \sin\theta_{\text{dark}} = (2.00\text{ m})(\pm 1.933 \times 10^{-4})$$
$$= \pm 3.87 \times 10^{-4}\text{ m}$$

and the width of the central bright fringe is equal to $2|y_1| = 7.74 \times 10^{-4}\text{ m} = 0.774\text{ mm}$. Notice that this result is *smaller* than the width of the slit.

In general, for large values of a, the various maxima and minima are so closely spaced that only a large, central bright area resembling the geometric image of the slit is observed. This behavior is very important in the performance of optical instruments such as telescopes.

PhysicsNow™ Investigate the single-slit diffraction pattern by logging into PhysicsNow at **www.pop4e.com** and going to Interactive Example 27.5.

27.7 RESOLUTION OF SINGLE-SLIT AND CIRCULAR APERTURES

Imagine you are driving in the middle of a dark desert at night, along a road that is perfectly straight and flat for many kilometers. You see another vehicle coming toward you from a distance. When the vehicle is far away, you might be unable to determine whether it is an automobile with two headlights or a motorcycle with one. As it approaches you, at some point you will be able to distinguish the two headlights and determine that it is an automobile. Once you are able to see two separate headlights, you describe the light sources as being **resolved.**

The ability of optical systems to distinguish between closely spaced objects is limited because of the wave nature of light. To understand this limitation, consider Figure 27.17, which shows two light sources far from a narrow slit. The sources can be considered as two point sources S_1 and S_2 that are incoherent. For example, they could be two distant stars observed through the aperture of a telescope tube. If no diffraction occurred, one would observe two distinct bright spots (or images) on the screen at the right in the figure. Because of diffraction, however, each source is imaged as a bright central region flanked by weaker bright and dark bands. What is observed on the screen is the sum of two diffraction patterns: one from S_1 and the other from S_2.

If the two sources are far enough apart to ensure that their central maxima do not overlap as in Figure 27.17a, their images can be distinguished and are said to be resolved. If the sources are close together, however, as in Figure 27.17b, the two

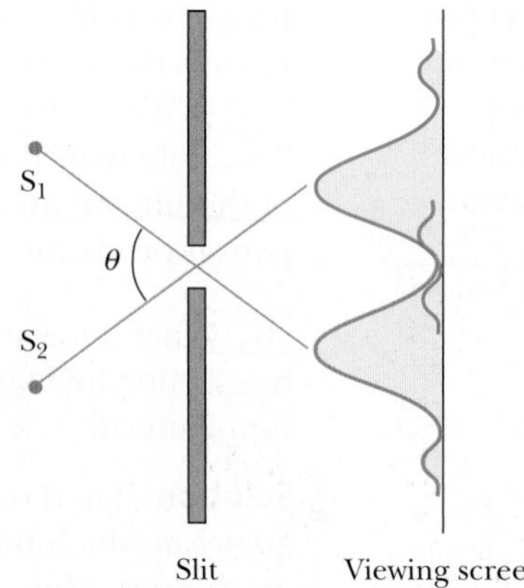

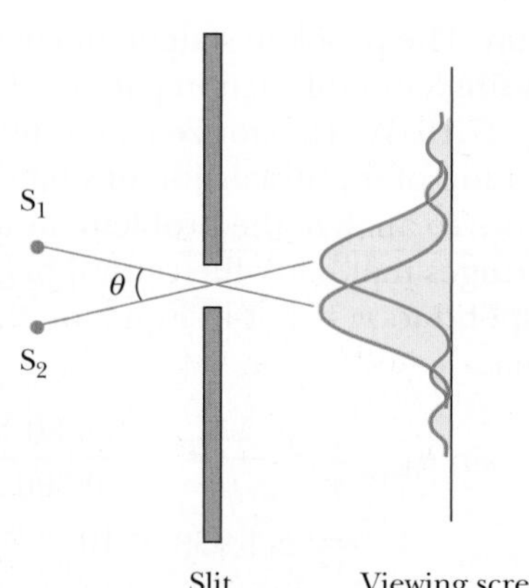

FIGURE 27.17 Two point sources far from a narrow slit each produce a diffraction pattern. (a) The angle subtended by the sources at the slit is large enough for the diffraction patterns to be distinguishable. (b) The angle subtended by the sources is so small that their diffraction patterns overlap and the images are not well resolved. (Note that the angles are greatly exaggerated. The drawings are not to scale.)

central maxima may overlap and the sources are not resolved. To decide when two sources are resolved, the following condition is often used:

When the central maximum of the diffraction pattern of one source falls on the first minimum of the diffraction pattern of another source, the sources are said to be just resolved. This limiting condition of resolution is known as **Rayleigh's criterion.**

■ Rayleigh's criterion

Figure 27.18 shows the diffraction patterns from circular apertures for three situations. When the objects are far apart, they are well resolved (Fig. 27.18a). They are just resolved when their angular separation satisfies Rayleigh's criterion (Fig. 27.18b). Finally, the sources are not resolved in Figure 27.18c.

From Rayleigh's criterion, we can determine the minimum angular separation $\theta_{\min}$ subtended by the sources at a slit such that the sources are just resolved. In Section 27.4, we found that the first minimum in a single-slit diffraction pattern occurs at the angle that satisfies the relationship

$$\sin \theta = \frac{\lambda}{a}$$

where a is the width of the slit. According to Rayleigh's criterion, this expression gives the smallest angular separation for which the two sources are resolved. Because $\lambda \ll a$ in most situations, $\sin \theta$ is small and we can use the approximation $\sin \theta \approx \theta$. Therefore, the limiting angle of resolution for a slit of width a is

$$\theta_{\min} = \frac{\lambda}{a} \qquad [27.14]$$

■ Limiting angle of resolution for a slit

where $\theta_{\min}$ is expressed in radians. Hence, the angle subtended by the two sources at the slit must be *greater* than λ/a if the sources are to be resolved.

Many optical systems use circular apertures rather than slits. The diffraction pattern of a circular aperture, as seen in Figure 27.18, consists of a central circular

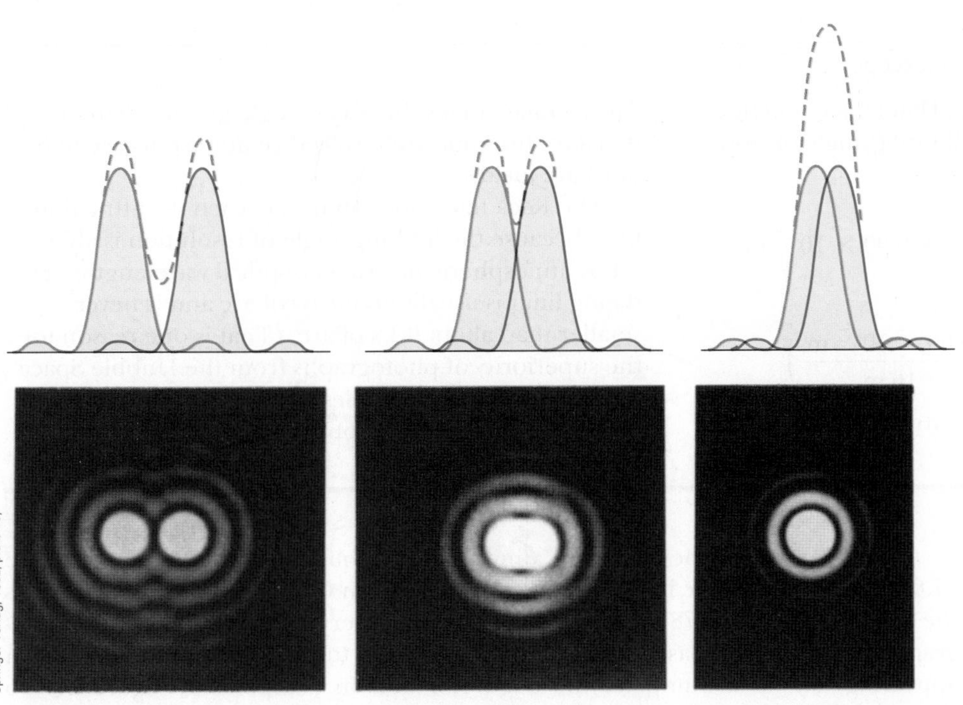

(From M. Cagnet, M. Francon, and J. C. Thierr, *Atlas of Optical Phenomena*, Berlin, Springer-Verlag, 1962, plate 16)

FIGURE 27.18 Individual diffraction patterns of two point sources (solid curves) and the resultant pattern (dashed curves) for various angular separations of the sources. In each case, the dashed curve is the sum of the two solid curves. (a) The sources are far apart, and the images are well resolved. (b) The sources are closer together such that the patterns satisfy Rayleigh's criterion, and the images are just resolved. (c) The sources are so close together that their images are not resolved.

bright disk surrounded by progressively fainter rings. Analysis shows that the limiting angle of resolution of the circular aperture is

■ Limiting angle of resolution for a circular aperture

$$\theta_{\min} = 1.22 \frac{\lambda}{D} \qquad [27.15]$$

where D is the diameter of the aperture. Note that Equation 27.15 is similar to Equation 27.14 except for the factor of 1.22, which arises from a mathematical analysis of diffraction from a circular aperture. This equation is related to the difficulty we had seeing the two headlights at the beginning of this section. When observing with the eye, D in Equation 27.15 is the diameter of the pupil. The diffraction pattern formed when light passes through the pupil causes the difficulty in resolving the headlights.

Another example of the effect of diffraction on resolution for circular apertures is the astronomical telescope. The end of the tube through which the light passes is circular, so the ability of the telescope to resolve light from closely spaced stars is limited by the diameter of this opening.

QUICK QUIZ 27.5 Suppose you are observing a binary star with a telescope and are having difficulty resolving the two stars. You decide to use a colored filter to maximize the resolution. (A filter of a given color transmits only that color of light.) What color filter should you choose? **(a)** blue **(b)** green **(c)** yellow **(d)** red

■ Thinking Physics 27.3

Cats' eyes have pupils that can be modeled as vertical slits. At night, are cats more successful in resolving headlights on a distant car or vertically separated lights on the mast of a distant boat?

Reasoning The effective slit width in the vertical direction of the cat's eye is larger than that in the horizontal direction. Thus, the eye has more resolving power for lights separated in the vertical direction and would be more effective at resolving the mast lights on the boat. ■

EXAMPLE 27.6 Resolution of a Telescope

The Keck telescope at Mauna Kea, Hawaii, has an effective diameter of 10 m. What is its limiting angle of resolution for 600-nm light?

Solution Because $D = 10$ m and $\lambda = 6.00 \times 10^{-7}$ m, Equation 27.15 gives

$$\theta_{\min} = 1.22 \frac{\lambda}{D} = 1.22 \left(\frac{6.00 \times 10^{-7}\ \text{m}}{10\ \text{m}} \right)$$

$$= 7.3 \times 10^{-8}\ \text{rad} \approx 0.015\ \text{s of arc}$$

Any two stars that subtend an angle greater than or equal to this value are resolved (if atmospheric conditions are ideal).

The Keck telescope can never reach its diffraction limit because the limiting angle of resolution is always set by atmospheric blurring at optical wavelengths. This seeing limit is usually about 1 s of arc and is never smaller than about 0.1 s of arc. (That is one reason for the superiority of photographs from the Hubble Space Telescope, which views celestial objects from an orbital position above the atmosphere.)

As an example of the effects of atmospheric turbulence discussed in Example 27.6, consider telescopic images of Pluto and its moon Charon. Figure 27.19a shows the image taken in 1978 that represents the discovery of Charon. In this photograph from an Earth-based telescope, atmospheric turbulence results in Charon appearing only as a bump on the edge of Pluto. In comparison, Figure 27.19b shows a photograph taken with the Hubble Space Telescope in 1994. Without the problems of atmospheric turbulence, Pluto and its moon are clearly resolved.

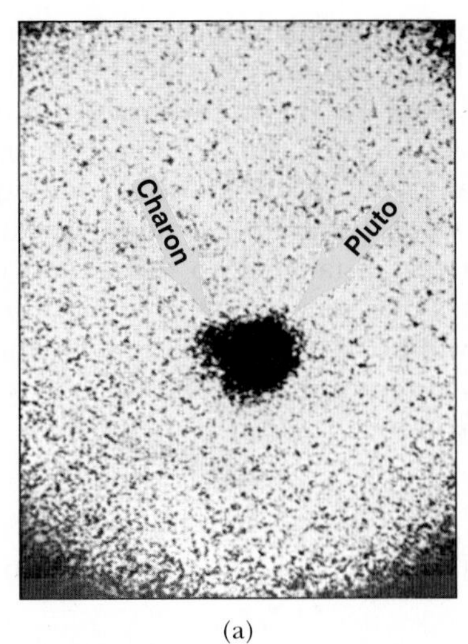

(a)

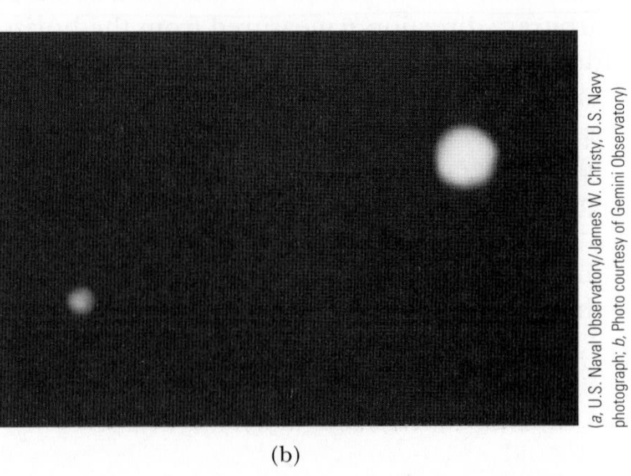

(b)

(a, U.S. Naval Observatory/James W. Christy, U.S. Navy photograph; b, Photo courtesy of Gemini Observatory)

FIGURE 27.19 (a) The photograph on which Charon, the moon of Pluto, was discovered in 1978. From an Earth-based telescope, atmospheric turbulence results in Charon appearing only as a subtle bump on the edge of Pluto. (b) A Hubble Space Telescope photo of Pluto and Charon, clearly resolving the two objects.

27.8 THE DIFFRACTION GRATING

The **diffraction grating,** a useful device for analyzing light sources, consists of a large number of equally spaced parallel slits. A grating can be made by cutting parallel, equally spaced grooves on a glass or metal plate with a precision ruling machine. In a *transmission grating,* the spaces between lines are transparent to the light and hence act as separate slits. In a *reflection grating,* the spaces between lines are highly reflective. Gratings with many lines very close to one another can have very small slit spacings. For example, a grating ruled with 5 000 lines/cm has a slit spacing of $d = (1/5\,000)$ cm $= 2 \times 10^{-4}$ cm.

Figure 27.20 shows a pictorial representation of a section of a flat diffraction grating. A plane wave is incident from the left, normal to the plane of the grating. The pattern observed on the screen at the right in Figure 27.20 is the result of the

PITFALL PREVENTION 27.3

A DIFFRACTION GRATING IS AN INTERFERENCE GRATING As with the term *diffraction pattern, diffraction grating* is a misnomer but is deeply entrenched. The diffraction grating depends on diffraction in the same way as the double slit, spreading the light so that light from different slits can interfere. It would be more correct to call it an *interference grating.* It is unlikely, however, that you will hear anything other than *diffraction grating* for this device.

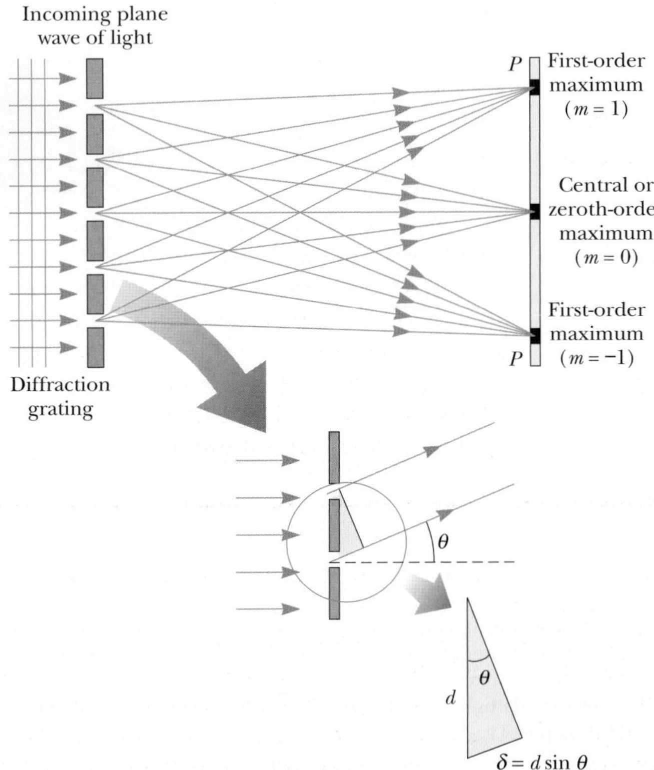

FIGURE 27.20 Side view of a diffraction grating. The slit separation is d and the path difference between adjacent slits is $d \sin \theta$.

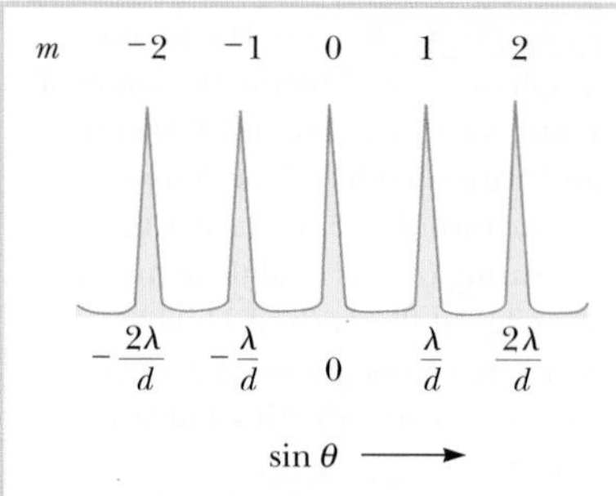

ACTIVE FIGURE 27.21
Intensity versus sin θ for a diffraction grating. The zeroth-, first-, and second-order maxima are shown.

Physics Now™ Log into PhysicsNow at **www.pop4e.com** and go to Active Figure 27.21 to choose the number of slits to be illuminated to see the effect on the interference pattern.

combined effects of interference and diffraction. Each slit produces diffraction, and the diffracted beams interfere with one another to produce the final pattern. Each slit acts as a source of waves, and all waves start at the slits in phase. For some arbitrary direction θ measured from the horizontal, however, the waves must travel different path lengths before reaching a particular point on the screen. From Figure 27.20, note that the path difference between waves from any two adjacent slits is equal to $d \sin \theta$. (We assume once again that the distance L to the screen is much larger than d.) If this path difference equals one wavelength or some integral multiple of a wavelength, waves from all slits will be in phase at the screen and a bright line will be observed. When the light is incident normally on the plane of the grating, the condition for *maxima* in the interference pattern at the angle θ is therefore[3]

$$d \sin \theta_{\text{bright}} = m\lambda \qquad (m = 0, 1, 2, 3, \ldots) \tag{27.16}$$

This expression can be used to calculate the wavelength from a knowledge of the grating spacing d and the angle of deviation θ. If the incident radiation contains several wavelengths, the mth-order maximum for each wavelength occurs at an angle determined from Equation 27.16. All wavelengths are mixed together at $\theta = 0$, corresponding to $m = 0$.

The intensity distribution for a diffraction grating is shown in Active Figure 27.21. If the source contains various wavelengths, a spectrum of lines at different positions for different order numbers will be observed. Note the sharpness of the principal maxima and the broad range of dark areas, which are in contrast to the broad, bright fringes characteristic of the two-slit interference pattern (see Fig. 27.5).

A simple arrangement for measuring the wavelength of light is shown in Active Figure 27.22. This arrangement is called a *diffraction grating spectrometer.* The light to be analyzed passes through a slit,[4] and a parallel beam of light exits from the collimator perpendicular to the grating. The diffracted light leaves the grating and

ACTIVE FIGURE 27.22
Diagram of a diffraction grating spectrometer. The collimated beam incident on the grating is spread into its various wavelength components with constructive interference for a particular wavelength occurring at the angles θ_{bright} that satisfy the equation $d \sin \theta_{\text{bright}} = m\lambda$, where $m = 0, 1, 2, \ldots$.

Physics Now™ Log into PhysicsNow at **www.pop4e.com** and go to Active Figure 27.22 to use the spectrometer to observe constructive interference for various wavelengths.

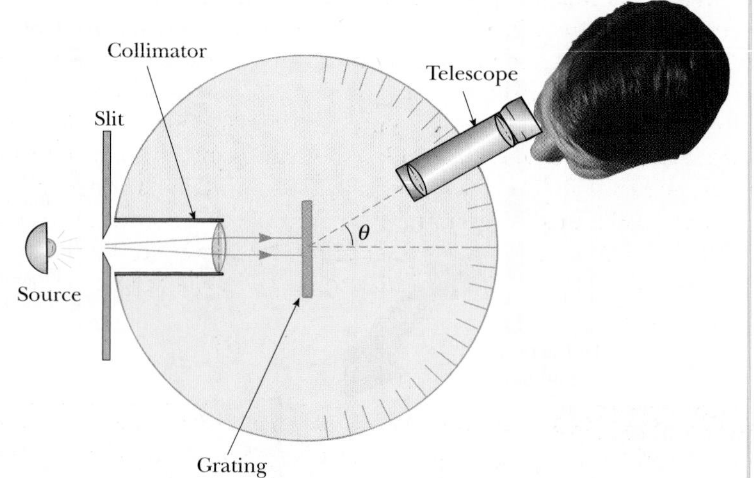

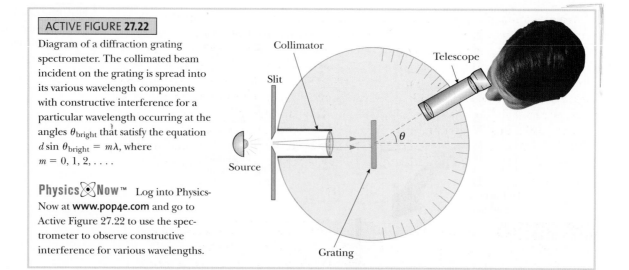

[3]Notice that this equation is identical to Equation 27.2. This equation can be used for a number of slits from two to any number N. The intensity distribution will change with the number of slits, but the locations of the maxima are the same.

[4]A long, narrow slit enables us to observe *line* spectra in the light coming from atomic and molecular systems, as discussed in Chapter 11.

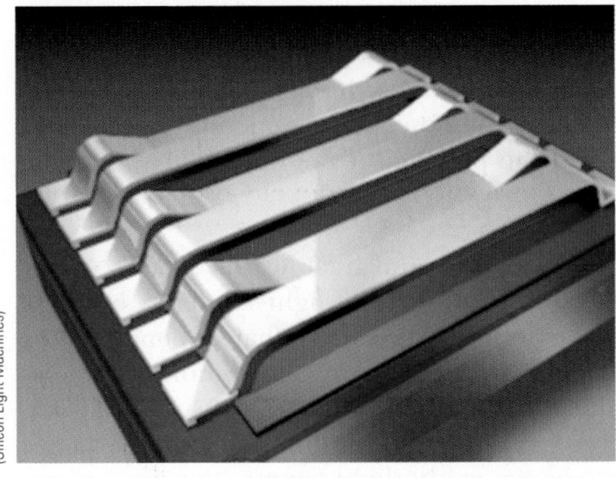
(Silicon Light Machines)

FIGURE 27.23 A small portion of a grating light valve. The alternating reflective ribbons at different levels act as a diffraction grating, offering very high speed control of the direction of light toward a digital display device.

exhibits constructive interference at angles that satisfy Equation 27.16. A telescope is used to view the image of the slit. The wavelength can be determined by measuring the precise angles at which the images of the slit appear for the various orders.

The spectrometer is a useful tool in *atomic spectroscopy*, in which the light from an atom is analyzed to find the wavelength components. These wavelength components can be used to identify the atom as discussed in Section 11.5. We will investigate atomic spectra further in Chapter 29.

Another application of diffraction gratings is in the recently developed *grating light valve* (GLV), which may compete in the near future in video projection with the digital micromirror devices (DMD) discussed in Section 25.3. The grating light valve consists of a silicon microchip fitted with an array of parallel silicon nitride ribbons coated with a thin layer of aluminum (Fig. 27.23). Each ribbon is about 20 μm long and about 5 μm wide and is separated from the silicon substrate by an air gap on the order of 100 nm. With no voltage applied, all ribbons are at the same level. In this situation, the array of ribbons acts as a flat surface, specularly reflecting incident light.

When a voltage is applied between a ribbon and the electrode on the silicon substrate, an electric force pulls the ribbon downward, closer to the substrate. Alternate ribbons can be pulled down, while those in between remain in the higher configuration. As a result, the array of ribbons acts as a diffraction grating, such that the constructive interference for a particular wavelength of light can be directed toward a screen or other optical display system. By using three such devices, one each for red, blue, and green light, full color display is possible.

The GLV tends to be simpler to fabricate and higher in resolution than comparable DMD devices. On the other hand, DMD devices have already made an entry into the market. It will be interesting to watch this technology competition in future years.

QUICK QUIZ 27.6 Ultraviolet light of wavelength 350 nm is incident on a diffraction grating with slit spacing d and forms an interference pattern on a screen a distance L away. The angular positions θ_{bright} of the interference maxima are large. The locations of the bright fringes are marked on the screen. Now red light of wavelength 700 nm is used with a diffraction grating to form another diffraction pattern on the screen. The bright fringes of this pattern will be located at the marks on the screen if **(a)** the screen is moved to a distance $2L$ from the grating, **(b)** the screen is moved to a distance $L/2$ from the grating, **(c)** the grating is replaced with one of slit spacing $2d$, **(d)** the grating is replaced with one of slit spacing $d/2$, or **(e)** nothing is changed

FIGURE 27.24 (Thinking Physics 27.4) A compact disc observed under white light. The colors observed in the reflected light and their intensities depend on the orientation of the disc relative to the eye and relative to the light source.

Thinking Physics 27.4

White light reflected from the surface of a compact disc has a multicolored appearance as shown in Figure 27.24. Furthermore, the observation depends on the orientation of the disc relative to the eye and the position of the light source. Explain how that works.

Reasoning The surface of a compact disc has a spiral track with a spacing of approximately 1 μm that acts as a reflection grating. The light scattered by these closely spaced tracks interferes constructively in directions that depend on the wavelength and on the direction of the incident light. Any one section of the disc serves as a diffraction grating for white light, sending beams of constructive interference for different colors in different directions. The different colors you see when viewing one section of the disc change as the light source, the disc, or you move to change the angle of incidence or the viewing angle. ■

INTERACTIVE EXAMPLE 27.7 The Orders of a Diffraction Grating

Monochromatic light from a helium–neon laser (λ = 632.8 nm) is incident normally on a diffraction grating containing 6 000 lines/cm. Find the angles at which the first-order, second-order, and third-order maxima can be observed.

Solution First, we calculate the slit separation, which is equal to the inverse of the number of lines per cm:

$$d = (1/6\,000)\text{ cm} = 1.667 \times 10^{-4}\text{ cm} = 1\,667\text{ nm}$$

For the first-order maximum (m = 1), we find that

$$\sin\theta_1 = \frac{\lambda}{d} = \frac{632.8\text{ nm}}{1\,667\text{ nm}} = 0.379\,6$$

$$\theta_1 = 22.31°$$

For m = 2, we find that

$$\sin\theta_2 = \frac{2\lambda}{d} = \frac{2(632.8\text{ nm})}{1\,667\text{ nm}} = 0.759\,2$$

$$\theta_2 = 49.39°$$

For m = 3, we find that $\sin\theta_3$ = 1.139. Because $\sin\theta$ cannot exceed unity, this result does not represent a realistic solution. Hence, only zeroth-, first-, and second-order maxima are observed for this situation.

PhysicsNow™ Investigate the interference pattern from a diffraction grating by logging into PhysicsNow at www.pop4e.com and going to Interactive Example 27.7.

27.9 DIFFRACTION OF X-RAYS BY CRYSTALS

In principle, the wavelength of any electromagnetic wave can be determined if a grating of the proper spacing (on the order of λ) is available. **X-rays,** discovered in 1895 by Wilhelm Roentgen (1845–1923), are electromagnetic waves with very short wavelengths (on the order of 10^{-10} m = 0.1 nm). In 1913, Max von Laue (1879–1960) suggested that the regular array of atoms in a crystal, whose spacing is known to be about 10^{-10} m, could act as a three-dimensional diffraction grating for x-rays. Subsequent experiments confirmed his prediction. The observed diffraction patterns are complicated because of the three-dimensional nature of the crystal. Nevertheless, x-ray diffraction is an invaluable technique for elucidating crystalline structures and for understanding the structure of matter.

Figure 27.25 is one experimental arrangement for observing x-ray diffraction from a crystal. A collimated beam of x-rays with a continuous range of wavelengths is incident on a crystal. The diffracted beams are very intense in certain directions, corresponding to constructive interference from waves reflected from layers of atoms in the crystal. The diffracted beams, which can be detected by a photographic film,

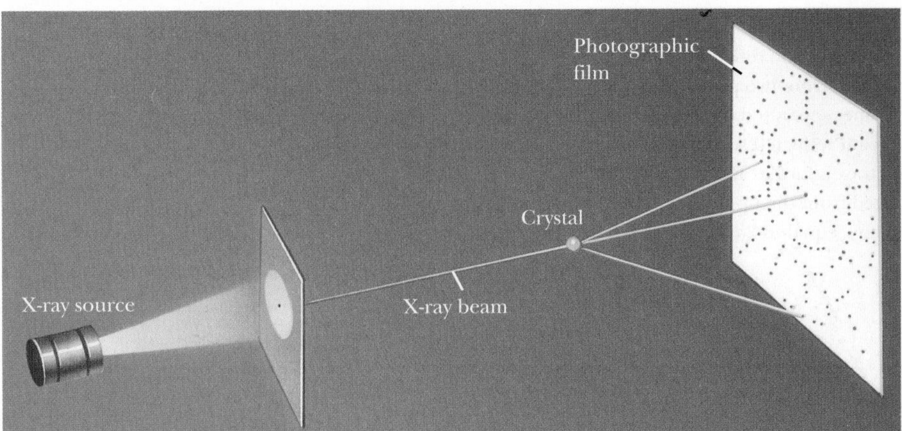

FIGURE 27.25 Schematic diagram of the technique used to observe the diffraction of x-rays by a crystal. The array of spots formed on the film is called a Laue pattern.

form an array of spots known as a *Laue pattern,* as in Figure 27.26a. One can deduce the crystalline structure by analyzing the positions and intensities of the various spots in the pattern. Figure 27.26b shows a Laue pattern from a crystalline enzyme, using a wide range of wavelengths so that a swirling pattern results.

Laue pattern of a crystalline enzyme

The arrangement of atoms in a crystal of NaCl is shown in Figure 27.27. The red spheres represent Na^+ ions, and the blue spheres represent Cl^- ions. Each unit cell (the geometric shape that repeats through the crystal) contains four Na^+ and four Cl^- ions. The unit cell is a cube whose edge length is a.

The ions in a crystal lie in various planes as shown in Figure 27.28. Suppose an incident x-ray beam makes an angle θ with one of the planes as in Figure 27.28. (Note that the angle θ is traditionally measured from the reflecting surface rather than from the normal, as in the case of the law of reflection in Chapter 25.) The beam can be reflected from both the upper plane and the lower one; the geometric construction in Figure 27.28, however, shows that the beam reflected from the lower surface travels farther than the beam reflected from the upper surface. The path difference between the two beams is $2d \sin \theta$, where d is the distance between the planes. The two beams reinforce each other (constructive interference) when this path difference equals some integral multiple of the wavelength λ. The same is

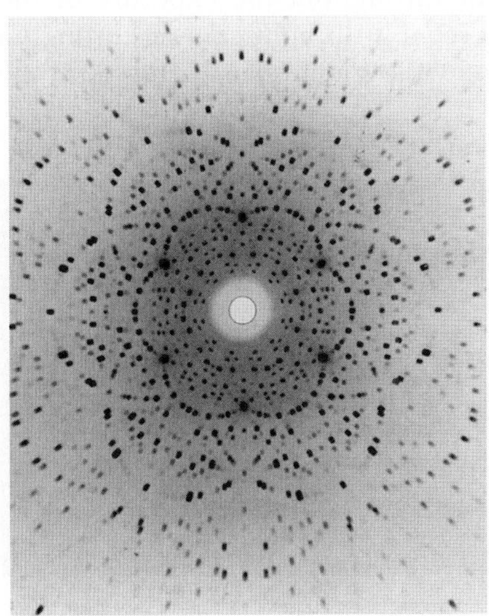

(a)

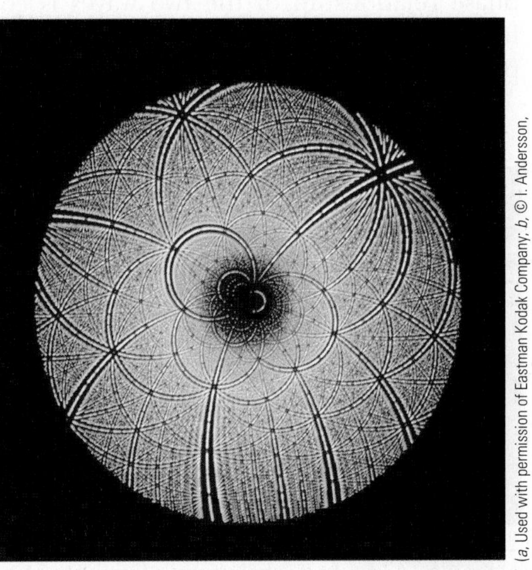

(b)

(a, Used with permission of Eastman Kodak Company; b, © I. Andersson, Oxford Molecular Biophysics Laboratory/Photo Researchers, Inc.)

FIGURE 27.26 (a) A Laue pattern of a single crystal of the mineral beryl (beryllium aluminum silicate). (b) A Laue pattern of the enzyme Rubisco, produced with a wide-band x-ray spectrum. This enzyme is present in plants and takes part in the process of photosynthesis. The Laue pattern is used to determine the crystal structure of Rubisco.

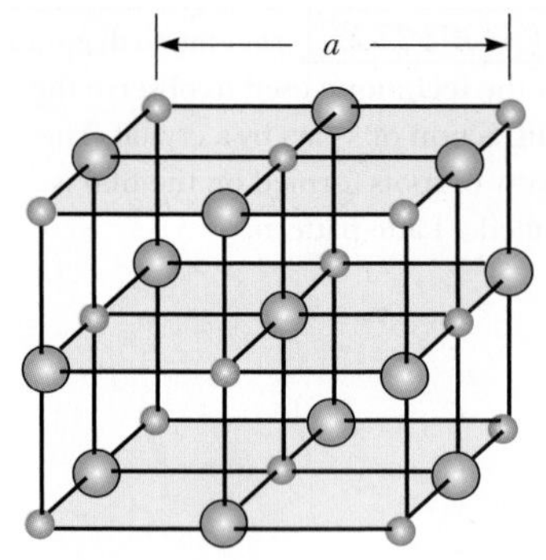

FIGURE 27.27 Crystalline structure of sodium chloride (NaCl). The blue spheres represent Cl^- ions, and the red spheres represent Na^+ ions.

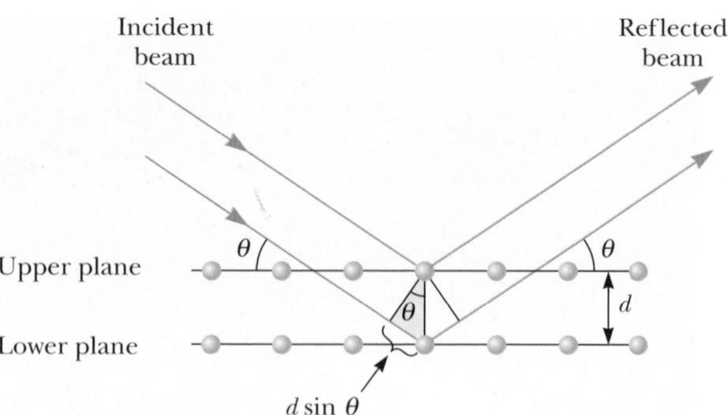

FIGURE 27.28 A two-dimensional description of the reflection of an x-ray beam from two parallel crystalline planes separated by a distance d. The beam reflected from the lower plane travels farther than the one reflected from the upper plane by a distance equal to $2d \sin \theta$.

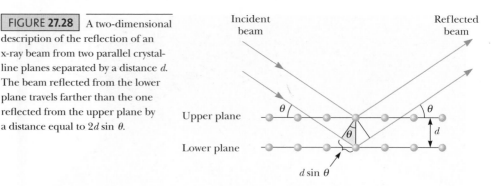

true of reflection from the entire family of parallel planes. Hence, the condition for constructive interference (maxima in the reflected wave) is

$$2d \sin \theta = m\lambda \qquad (m = 1, 2, 3, \ldots) \qquad [27.17]$$

This condition is known as **Bragg's law** after W. Lawrence Bragg (1890–1971), who first derived the relationship. If the wavelength and diffraction angle are measured, Equation 27.17 can be used to calculate the spacing between atomic planes.

27.10 HOLOGRAPHY

CONTEXT CONNECTION

One interesting application of the laser is **holography,** the production of three-dimensional images of objects. The physics of holography was developed by Dennis Gabor (1900–1979) in 1948, for which he was awarded the 1971 Nobel Prize in Physics. The requirement of coherent light for holography, however, delayed the realization of holographic images from Gabor's work until the development of lasers in the 1960s. Figure 27.29 shows a hologram and the three-dimensional character of its image.

Figure 27.30 shows how a hologram is made. Light from the laser is split into two parts by a half-silvered mirror at B. One part of the beam reflects off the object to be photographed and strikes an ordinary photographic film. The other half of the beam is diverged by lens L_2, reflects from mirrors M_1 and M_2, and finally strikes the film. The two beams overlap to form an extremely complicated interference pattern on the film. Such an interference pattern can be produced only if the phase relationship of the two waves is constant throughout the exposure of the film. This condition is met by illuminating the scene with light coming through a pinhole or with coherent laser radiation. The hologram records not only the

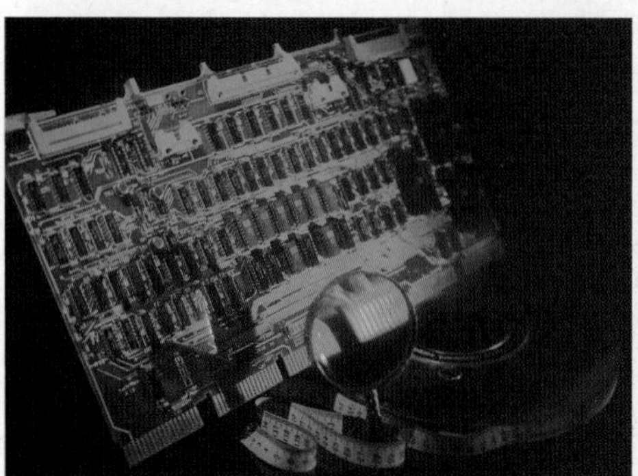

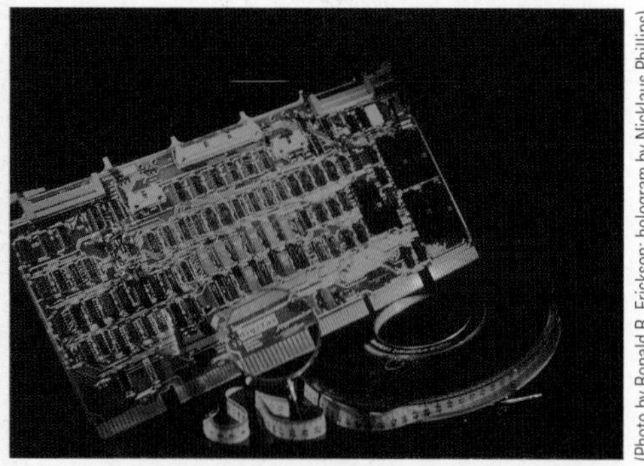

(Photo by Ronald R. Erickson; hologram by Nicklaus Phillips)

FIGURE 27.29 In this hologram, a circuit board is shown from two different views. Notice the difference in the appearance of the measuring tape and the view through the magnifying lens.

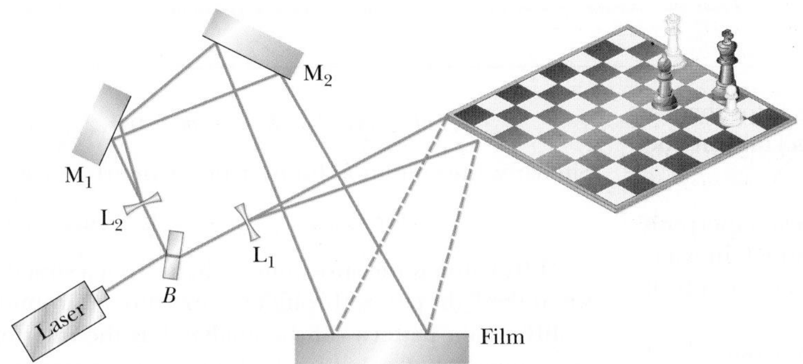

FIGURE 27.30 Experimental arrangement for producing a hologram.

intensity of the light scattered from the object (as in a conventional photograph), but also the phase difference between the reference beam and the beam scattered from the object. This phase difference results in an interference pattern that produces an image with full three-dimensional perspective.

In a normal photographic image, a lens is used to focus the image so that each point on the object corresponds to a single point on the film. Notice that no lens is used in Figure 27.30 to focus the light onto the film. Thus, light from each point on the object reaches *all* points on the film. As a result, each region of the photographic film on which the hologram is recorded contains information about all illuminated points on the object, which leads to a remarkable result: If a small section of the hologram is cut from the film, the complete image can be formed from this small piece!

A hologram is best viewed by allowing coherent light to pass through the developed film as one looks back along the direction from which the beam comes. The interference pattern on the film acts as a diffraction grating. Figure 27.31 shows two rays of light striking the film and passing through. For each ray, the $m = 0$ and $m = \pm 1$ rays in the diffraction pattern are shown emerging from the right side of the film. Notice that the $m = +1$ rays converge to form a real image of the scene, which is not the image that is normally viewed. By extending the light rays corresponding to $m = -1$ back behind the film, we see that there is a virtual image located there, with light coming from it in exactly the same way that light came from the actual object when the film was exposed. This image is the one we see by looking through the holographic film.

Holograms are finding a number of applications in displays and in precision measurements. You may have a hologram on your credit card. This special type of hologram is called a *rainbow hologram,* designed to be viewed in reflected white light.

Holograms represent a means of storing visual information using lasers. In the Context Conclusion, we will investigate means of using lasers to store digital information that can be converted into sound waves or video displays.

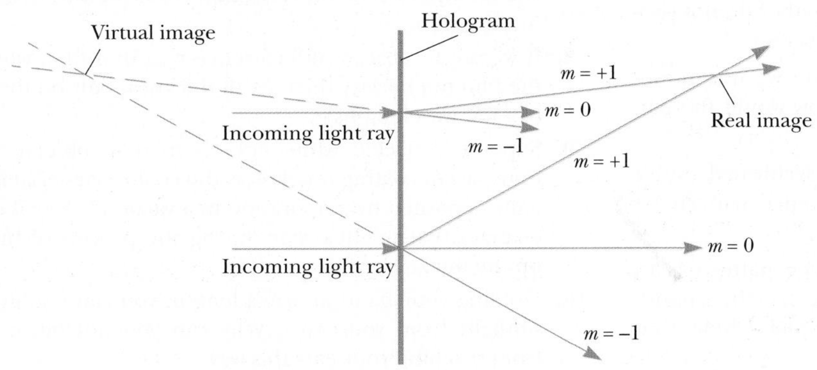

FIGURE 27.31 Two light rays strike a hologram at normal incidence. For each ray, outgoing rays corresponding to $m = 0$ and $m = \pm 1$ are shown. If the $m = -1$ rays are extended backward, a virtual image of the object photographed in the hologram exists on the front side of the hologram.

SUMMARY

PhysicsNow™ Take a practice test by logging into PhysicsNow at www.pop4e.com and clicking on the Pre-Test link for this chapter.

Interference of light waves is the result of the linear superposition of two or more waves at a given point. A sustained interference pattern is observed if (1) the sources have identical wavelengths and (2) the sources are coherent.

In Young's double-slit experiment, two slits separated by a distance d are illuminated by a monochromatic light source. An interference pattern consisting of bright and dark fringes is observed on a screen that is a distance of $L \gg d$ from the slits. The condition for **constructive interference** is

$$\delta = d \sin \theta_{\text{bright}} = m\lambda \qquad (m = 0, \pm 1, \pm 2, \ldots) \qquad [27.2]$$

The number m is called the **order number** of the fringe.

The condition for **destructive interference** is

$$\delta = d \sin \theta_{\text{dark}} = (m + \tfrac{1}{2})\lambda \qquad (m = 0, \pm 1, \pm 2, \ldots) \qquad [27.3]$$

The **time-averaged light intensity** of the double-slit interference pattern is

$$I_{\text{avg}} = I_{\max} \cos^2 \left(\frac{\pi d \sin \theta}{\lambda} \right) \qquad [27.8]$$

where $I_{\max}$ is the maximum intensity on the screen.

An electromagnetic wave traveling from a medium with an index of refraction n_1 toward a medium with index of refraction n_2 undergoes a 180° phase change on reflection when $n_2 > n_1$. No phase change occurs in the reflected wave if $n_2 < n_1$.

The condition for constructive interference in a film of thickness t and refractive index n with the same medium on both sides of the film is given by

$$2nt = (m + \tfrac{1}{2})\lambda \qquad (m = 0, 1, 2, \ldots) \qquad [27.11]$$

Similarly, the condition for destructive interference is

$$2nt = m\lambda \qquad (m = 0, 1, 2, \ldots) \qquad [27.12]$$

Diffraction is the spreading of light from a straight-line path when the light passes through an aperture or around obstacles. A **diffraction pattern** can be analyzed as the interference of a large number of coherent Huygens sources spread across the aperture.

The diffraction pattern produced by a single slit of width a on a distant screen consists of a central, bright maximum and alternating bright and dark regions of much lower intensities. The angles θ at which the diffraction pattern has *zero* intensity are given by

$$\sin \theta_{\text{dark}} = m\frac{\lambda}{a} \qquad (m = \pm 1, \pm 2, \pm 3, \ldots) \qquad [27.13]$$

Rayleigh's criterion, which is a limiting condition of resolution, says that two images formed by an aperture are just distinguishable if the central maximum of the diffraction pattern for one image falls on the first minimum of the other image. The limiting angle of resolution for a slit of width a is given by $\theta_{\min} = \lambda/a$, and the limiting angle of resolution for a circular aperture of diameter D is given by $\theta_{\min} = 1.22\lambda/D$.

A **diffraction grating** consists of a large number of equally spaced, identical slits. The condition for intensity maxima in the interference pattern of a diffraction grating for normal incidence is

$$d \sin \theta_{\text{bright}} = m\lambda \qquad (m = 0, 1, 2, 3, \ldots) \qquad [27.16]$$

where d is the spacing between adjacent slits and m is the order number of the diffraction maximum.

QUESTIONS

☐ = answer available in the *Student Solutions Manual and Study Guide*

1. What is the necessary condition on the path length difference between two waves that interfere (a) constructively and (b) destructively?

2. Explain why two flashlights held close together do not produce an interference pattern on a distant screen.

3. In Young's double-slit experiment, why do we use monochromatic light? If white light is used, how would the pattern change?

4. If Young's double-slit experiment were performed under water, how would the observed interference pattern be affected?

5. A simple way to observe an interference pattern is to look at a distant light source through a stretched handkerchief or an opened umbrella. Explain how that works.

6. A certain oil film on water appears brightest at the outer regions, where it is thinnest. From this information, what can you say about the index of refraction of oil relative to that of water?

7. As a soap bubble evaporates, it appears black just before it breaks. Explain this phenomenon in terms of the phase changes that occur on reflection from the two surfaces of the soap film.

8. If we are to observe interference in a thin film, why must the film not be very thick (with thickness only on the order of a few wavelengths)?

9. Suppose reflected white light is used to observe a thin transparent coating on glass as the coating material is gradually deposited by evaporation in a vacuum. Describe color changes that might occur during the process of building up the thickness of the coating.

10. Holding your hand at arm's length, you can readily block sunlight from your eyes. Why can you not block sound from reaching your ears this way?

11. Why can you hear around corners, but not see around corners?

12. When you receive a chest x-ray at a hospital, the rays pass through a series of parallel ribs in your chest. Do the ribs act as a diffraction grating for x-rays?

13. Describe the change in width of the central maximum of the single-slit diffraction pattern as the width of the slit is made narrower.

FIGURE Q27.14

14. John William Strutt, Lord Rayleigh (1842–1919), is known as the last person to understand all of physics and all of mathematics. He invented an improved foghorn. To warn ships of a coastline, a foghorn should radiate sound in a wide horizontal sheet over the ocean's surface. It should not waste energy by broadcasting sound upward. It should not emit sound downward because the water in front of the foghorn would reflect that sound upward. Rayleigh's foghorn trumpet is shown in Figure Q27.14. Is it installed in the correct orientation? Decide whether the long dimension of the rectangular opening should be horizontal or vertical, and argue for your decision.

15. A laser produces a beam a few millimeters wide, with uniform intensity across its width. A hair is stretched vertically across the front of the laser to cross the beam. How is the diffraction pattern it produces on a distant screen related to that of a vertical slit equal in width to the hair? How could you determine the width of the hair from measurements of its diffraction pattern?

16. A radio station serves listeners in a city to the northeast of its broadcast site. It broadcasts from three adjacent towers on a mountain ridge, along a line running east and west. Show that by introducing time delays among the signals the individual towers radiate, the station can maximize net intensity in the direction toward the city (and in the opposite direction) and minimize the signal transmitted in other directions. The towers together are said to form a *phased array*.

PROBLEMS

1, 2, 3 = straightforward, intermediate, challenging
☐ = full solution available in the *Student Solutions Manual and Study Guide*
Physics Now™ = coached problem with hints available at www.pop4e.com
= computer useful in solving problem
= paired numerical and symbolic problems
= biomedical application

Section 27.1 ■ Conditions for Interference
Section 27.2 ■ Young's Double-Slit Experiment
Section 27.3 ■ Light Waves in Interference

Note: Problems 14.8, 14.10, and 14.11 in Chapter 14 can be assigned with this section.

1. A Young's interference experiment is performed with monochromatic light. The separation between the slits is 0.500 mm, and the interference pattern on a screen 3.30 m away shows the first side maximum 3.40 mm from the center of the pattern. What is the wavelength?

2. In a location where the speed of sound is 354 m/s, a 2 000-Hz sound wave impinges on two slits 30.0 cm apart. (a) At what angle is the first maximum located? (b) If the sound wave is replaced by 3.00-cm microwaves, what slit separation gives the same angle for the first maximum? (c) If the slit separation is 1.00 μm, what frequency of light gives the same first maximum angle?

3. Two radio antennas separated by 300 m as shown in Figure P27.3 simultaneously broadcast identical signals at the same wavelength. A radio in a car traveling due north

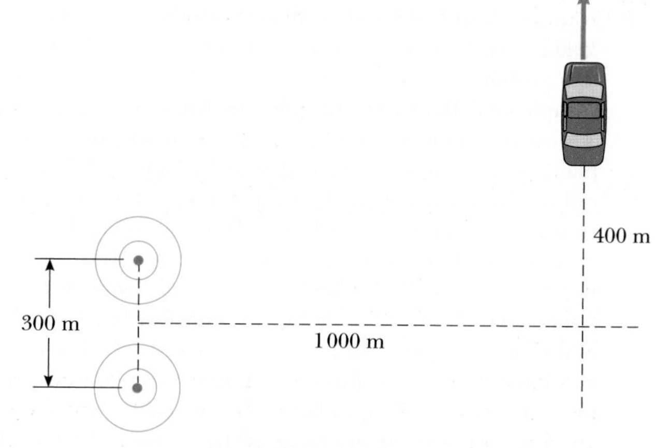

FIGURE P27.3

receives the signals. (a) If the car is at the position of the second maximum, what is the wavelength of the signals? (b) How much farther must the car travel to encounter the next minimum in reception?

4. The two speakers of a boom box are 35.0 cm apart. A single oscillator makes the speakers vibrate in phase at a frequency of 2.00 kHz. At what angles, measured from the perpendicular bisector of the line joining the speakers, would a distant observer hear maximum sound intensity? Minimum sound intensity? (Take the speed of sound as 340 m/s.)

5. **PhysicsNow**™ Young's double-slit experiment is performed with 589-nm light and a distance of 2.00 m between the slits and the screen. The tenth interference minimum is observed 7.26 mm from the central maximum. Determine the spacing of the slits.

6. A riverside warehouse has two open doors as shown in Figure P27.6. Its walls are lined with sound-absorbing material. A boat on the river sounds its horn. To person A the sound is loud and clear. To person B the sound is barely audible. The principal wavelength of the sound waves is 3.00 m. Assuming that person B is at the position of the first minimum, determine the distance between the doors, center to center.

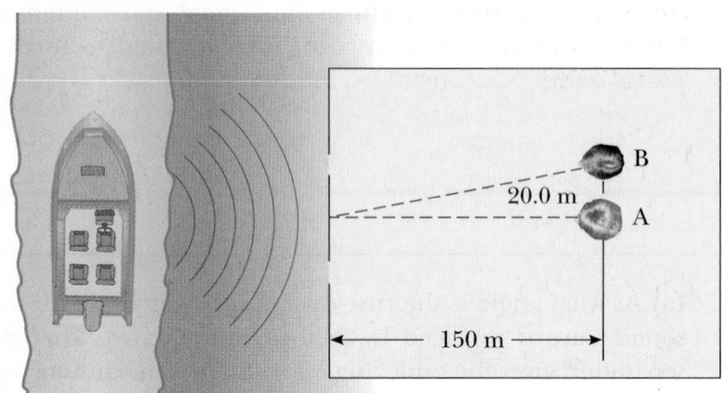

FIGURE **P27.6**

7. Two slits are separated by 0.320 mm. A beam of 500-nm light strikes the slits, producing an interference pattern. Determine the number of maxima observed in the angular range $-30.0° < \theta < 30.0°$.

8. Young's double-slit experiment underlies the *instrument landing system* used to guide aircraft to safe landings when the visibility is poor. Although real systems are more complicated than the example described here, they operate on the same principles. A pilot is trying to align her plane with a runway, as suggested in Figure P27.8a. Two radio antennas A_1 and A_2 are positioned adjacent to the runway, separated by 40.0 m. The antennas broadcast unmodulated coherent radio waves at 30.0 MHz. (a) Find the wavelength of the waves. The pilot "locks onto" the strong signal radiated along an interference maximum and steers the plane to keep the received signal strong. If she has found the central maximum, the plane will have just the right heading to land when it reaches the runway. (b) Suppose instead the plane is flying along the first side maximum (Fig. P27.8b). How far to the side of the runway centerline will the plane be when it is 2.00 km from the antennas? (c) It is possible to tell the pilot that she is on the wrong maximum by sending out two signals from each antenna and equipping the aircraft with a two-channel receiver. The ratio of the two frequencies must not be the ratio of small integers (such as 3/4). Explain how this two-frequency system would work and why it would not necessarily work if the frequencies were related by an integer ratio.

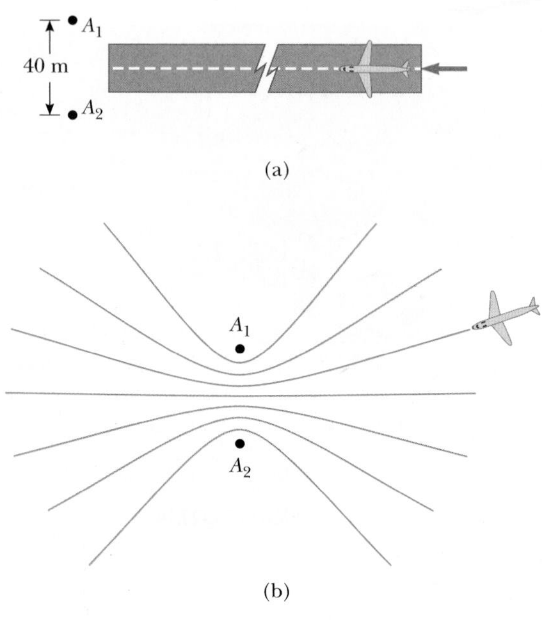

FIGURE **P27.8**

9. In Figure 27.4, let $L = 1.20$ m and $d = 0.120$ mm and assume that the slit system is illuminated with monochromatic 500-nm light. Calculate the phase difference between the two wave fronts arriving at P when (a) $\theta = 0.500°$ and (b) $y = 5.00$ mm. (c) What is the value of θ for which the phase difference is 0.333 rad? (d) What is the value of θ for which the path difference is $\lambda/4$?

10. Coherent light rays of wavelength λ strike a pair of slits separated by distance d at an angle θ_1 as shown in Figure P27.10. Assume that an interference maximum is formed at an angle θ_2 a great distance from the slits. Show that $d(\sin\theta_2 - \sin\theta_1) = m\lambda$, where m is an integer.

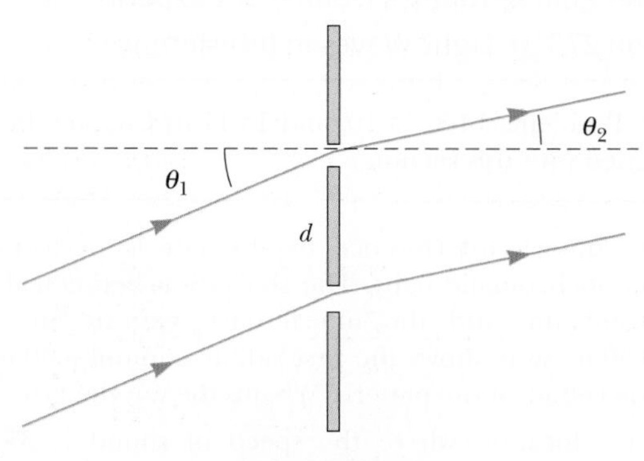

FIGURE **P27.10**

11. In Figure 27.4, let $L = 120$ cm and $d = 0.250$ cm. The slits are illuminated with coherent 600-nm light. Calculate the distance y above the central maximum for which the average intensity on the screen is 75.0% of the maximum.

12. The intensity on the screen at a certain point in a double-slit interference pattern is 64.0% of the maximum value. (a) What minimum phase difference (in radians) between sources produces this result? (b) Express this phase difference as a path difference for 486.1-nm light.

13. Two slits are separated by 0.180 mm. An interference pattern is formed on a screen 80.0 cm away by 656.3-nm light. Calculate the fraction of the maximum intensity 0.600 cm above the central maximum.

Section 27.4 ■ Change of Phase Due to Reflection

Section 27.5 ■ Interference in Thin Films

14. A soap bubble ($n = 1.33$) is floating in air. If the thickness of the bubble wall is 115 nm, what is the wavelength of the light that is most strongly reflected?

15. An oil film ($n = 1.45$) floating on water is illuminated by white light at normal incidence. The film is 280 nm thick. Find (a) the color of the light in the visible spectrum most strongly reflected and (b) the color of the light in the spectrum most strongly transmitted. Explain your reasoning.

16. A possible means for making an airplane invisible to radar is to coat the plane with an antireflective polymer. If radar waves have a wavelength of 3.00 cm and the index of refraction of the polymer is $n = 1.50$, how thick would you make the coating?

17. A material having an index of refraction of 1.30 is used as an antireflective coating on a piece of glass ($n = 1.50$). What should be the minimum thickness of this film to minimize reflection of 500-nm light?

18. Astronomers observe the chromosphere of the Sun with a filter that passes the red hydrogen spectral line of wavelength 656.3 nm, called the H_α line. The filter consists of a transparent dielectric of thickness d held between two partially aluminized glass plates. The filter is held at a constant temperature. (a) Find the minimum value of d that produces maximum transmission of perpendicular H_α light if the dielectric has index of refraction 1.378. (b) If the temperature of the filter increases above the normal value, what happens to the transmitted wavelength? (Its index of refraction does not change significantly.) (c) The dielectric will also pass what near-visible wavelength? One of the glass plates is colored red to absorb this light.

19. PhysicsNow™ An air wedge is formed between two glass plates separated at one edge by a very fine wire as shown in Figure P27.19. When the wedge is illuminated from above by 600-nm light and viewed from above, 30 dark fringes are observed. Calculate the radius of the wire.

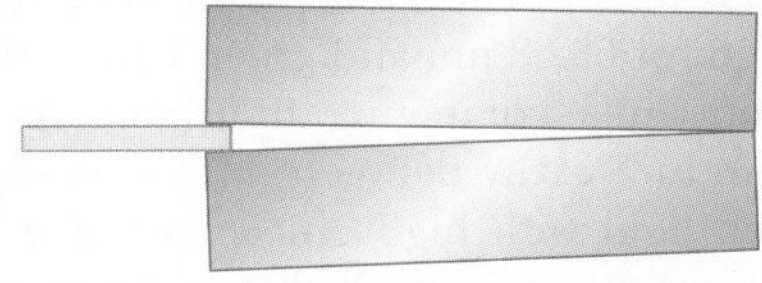

FIGURE **P27.19**

Section 27.6 ■ Diffraction Patterns

20. Helium–neon laser light ($\lambda = 632.8$ nm) is sent through a 0.300-mm-wide single slit. What is the width of the central maximum on a screen 1.00 m from the slit?

21. PhysicsNow™ A screen is placed 50.0 cm from a single slit, which is illuminated with 690-nm light. If the distance between the first and third minima in the diffraction pattern is 3.00 mm, what is the width of the slit?

22. A beam of monochromatic green light is diffracted by a slit of width 0.550 mm. The diffraction pattern forms on a wall 2.06 m beyond the slit. The distance between the positions of zero intensity on both sides of the central bright fringe is 4.10 mm. Calculate the wavelength of the light.

23. Coherent microwaves of wavelength 5.00 cm enter a long, narrow window in a building otherwise essentially opaque to the microwaves. If the window is 36.0 cm wide, what is the distance from the central maximum to the first-order minimum along a wall 6.50 m from the window?

24. Sound with a frequency 650 Hz from a distant source passes through a doorway 1.10 m wide in a sound-absorbing wall. Find the number and approximate directions of the diffraction-maximum beams radiated into the space beyond.

25. A beam of laser light of wavelength 632.8 nm has a circular cross-section 2.00 mm in diameter. A rectangular aperture is to be placed in the center of the beam so that when the light falls perpendicularly on a wall 4.50 m away, the central maximum fills a rectangle 110 mm wide and 6.00 mm high. The dimensions are measured between the minima bracketing the central maximum. Find the required width and height of the aperture.

Section 27.7 ■ Resolution of Single-Slit and Circular Apertures

26. The pupil of a cat's eye narrows to a vertical slit of width 0.500 mm in daylight. What is the angular resolution for horizontally separated mice? Assume that the average wavelength of the light is 500 nm.

27. PhysicsNow™ A helium–neon laser emits light that has a wavelength of 632.8 nm. The circular aperture through which the beam emerges has a diameter of 0.500 cm. Estimate the diameter of the beam 10.0 km from the laser.

28. Narrow, parallel, glowing gas-filled tubes in a variety of colors form block letters to spell out the name of a nightclub. Adjacent tubes are all 2.80 cm apart. The tubes forming one letter are filled with neon and radiate predominantly red light with a wavelength of 640 nm. For another letter, the tubes emit predominantly violet light at 440 nm. The pupil of a dark-adapted viewer's eye is 5.20 mm in diameter. If she is in a certain range of distances away, the viewer can resolve the separate tubes of one color but not the other. Which color is easier to resolve? The viewer's distance must be in what range for her to resolve the tubes of only one of these two colors?

29. The Impressionist painter Georges Seurat created paintings with an enormous number of dots of pure pigment, each of which was approximately 2.00 mm in diameter.

The idea was to have colors such as red and green next to each other to form a scintillating canvas (Fig. P27.29). Outside what distance would one be unable to discern individual dots on the canvas? (Assume that $\lambda = 500$ nm and that the pupil diameter is 4.00 mm.)

FIGURE P27.29 *Sunday Afternoon on the Isle of La Grande Jatte,* by Georges Seurat.

30. A spy satellite can consist essentially of a large-diameter concave mirror forming an image on a digital-camera detector and sending the picture to a ground receiver by radio waves. In effect, it is an astronomical telescope in orbit, looking down instead of up. Can a spy satellite read a license plate? Can it read the date on a dime? Argue for your answers by making an order-of-magnitude calculation, specifying the data you estimate.

31. A circular radar antenna on a Coast Guard ship has a diameter of 2.10 m and radiates at a frequency of 15.0 GHz. Two small boats are located 9.00 km away from the ship. How close together could the boats be and still be detected as two objects?

Section 27.8 ■ The Diffraction Grating

Note: In the following problems, assume that the light is incident normally on the gratings.

32. Light from an argon laser strikes a diffraction grating that has 5 310 grooves per centimeter. The central and first-order principal maxima are separated by 0.488 m on a wall 1.72 m from the grating. Determine the wavelength of the laser light.

33. **Physics Now™** The hydrogen spectrum has a red line at 656 nm and a blue line at 434 nm. What are the angular separations between two spectral lines obtained with a diffraction grating that has 4 500 grooves/cm?

34. A helium–neon laser ($\lambda = 632.8$ nm) is used to calibrate a diffraction grating. If the first-order maximum occurs at 20.5°, what is the spacing between adjacent grooves in the grating?

35. A grating with 250 grooves/mm is used with an incandescent light source. Assume that the visible spectrum ranges in wavelength from 400 to 700 nm. In how many orders can one see (a) the entire visible spectrum and (b) the short-wavelength region?

36. Show that whenever white light is passed through a diffraction grating of any spacing size, the violet end of the continuous visible spectrum in third order always overlaps with red light at the other end of the second-order spectrum.

37. A refrigerator shelf is an array of parallel wires with uniform spacing of 1.30 cm between centers. In air at 20°C, ultrasound with a frequency of 37.2 kHz from a distant source falls perpendicularly on the shelf. Find the number of diffracted beams leaving the other side of the shelf. Find the direction of each beam.

Section 27.9 ■ Diffraction of X-Rays by Crystals

38. Potassium iodide (KI) has the same crystalline structure as NaCl, with atomic planes separated by 0.353 nm. A monochromatic x-ray beam shows a first-order diffraction maximum when the grazing angle is 7.60°. Calculate the x-ray wavelength.

39. If the interplanar spacing of NaCl is 0.281 nm, what is the predicted angle at which 0.140-nm x-rays are diffracted in a first-order maximum?

40. In water of uniform depth, a wide pier is supported on pilings in several parallel rows 2.80 m apart. Ocean waves of uniform wavelength roll in, moving in a direction that makes an angle of 80.0° with the rows of posts. Find the three longest wavelengths of waves that will be strongly reflected by the pilings.

Section 27.10 ■ Context Connection—Holography

41. A wide beam of laser light with a wavelength of 632.8 nm is directed through several narrow parallel slits, separated by 1.20 mm, and falls on a sheet of photographic film 1.40 m away. The exposure time is chosen so that the film stays unexposed everywhere except at the central region of each bright fringe. (a) Find the distance between these interference maxima. The film is printed as a transparency; it is opaque everywhere except at the exposed lines. Next, the same beam of laser light is directed through the transparency and is allowed to fall on a screen 1.40 m beyond. (b) Argue that several narrow parallel bright regions, separated by 1.20 mm, will appear on the screen as real images of the original slits. If at last the screen is removed, light will diverge from the images of the original slits with the same reconstructed wave fronts as the original slits produced. (*Suggestion:* You may find it useful to draw a diagram similar to Fig. 27.20. A similar train of thought, at a soccer game, led Dennis Gabor to the invention of holography.)

42. A helium–neon laser can produce a green laser beam instead of red. Refer to Figure 24.18, which omits some energy levels between E_2 and E_1. After a population inversion is established, neon atoms make a variety of downward transitions in falling from the state labeled E_3* down eventually to level E_1. The atoms emit both red light with a wavelength of 632.8 nm and green light with a wavelength of 543 nm in a competing transition. Assume that the atoms are in a cavity between mirrors designed to reflect the green light with high efficiency but to allow the red light to leave the cavity immediately. Then stimulated emission can lead to the buildup of a collimated beam of green light between the mirrors, having a greater intensity than does the red light. A small fraction of the green light can be permitted to escape by transmission through one mirror to constitute the radiated laser beam. The mirrors forming the resonant cavity are not made of shiny metal, but of layered dielectrics, say silicon dioxide and titanium dioxide. (a) How thick a layer of silicon dioxide, between layers of titanium dioxide, would minimize reflection of the red light? (b) What should be the thickness of a similar but separate layer of silicon dioxide to maximize reflection of the green light?

Additional Problems

43. **Review problem.** This problem extends the result of Problem 14.11. Figure P27.43 shows two adjacent vibrating balls dipping into a tank of water. At distant points they produce an interference pattern as diagrammed in Figure 27.2. Let λ represent the wavelength of the ripples. Show that the two sources produce a standing wave along the line segment, of length d, between them. In terms of λ and d, find the number of nodes and the number of antinodes in the standing wave. Find the number of zones of constructive and of destructive interference in the interference pattern far away from the sources. Each line of destructive interference springs from a node in the standing wave, and each line of constructive interference springs from an antinode.

(Courtesy of CENCO Scientific Company)

FIGURE **P27.43**

44. Raise your hand and hold it flat. Think of the space between your index finger and your middle finger as one slit, and think of the space between middle finger and ring finger as a second slit. (a) Consider the interference resulting from sending coherent visible light perpendicularly through this pair of openings. Compute an order-of-magnitude estimate for the angle between adjacent zones of constructive interference. (b) To make the angles in the interference pattern easy to measure with a plastic protractor, you should use an electromagnetic wave with frequency of what order of magnitude? How is this wave classified on the electromagnetic spectrum?

45. **Review problem.** A flat piece of glass is held stationary and horizontal above the flat top end of a 10.0-cm-long vertical metal rod that has its lower end rigidly fixed. The thin film of air between the rod and glass is observed to be bright by reflected light when it is illuminated by light of wavelength 500 nm. As the temperature is slowly increased by 25.0°C, the film changes from bright to dark and back to bright 200 times. What is the coefficient of linear expansion of the metal?

46. Laser light with a wavelength of 632.8 nm is directed through one slit or two slits and allowed to fall on a screen 2.60 m beyond. Figure P27.46 shows the pattern on the screen, nearly actual size, with a centimeter rule below it. Did the light pass through one slit or two slits? If one, find its width. If two, find the distance between their centers.

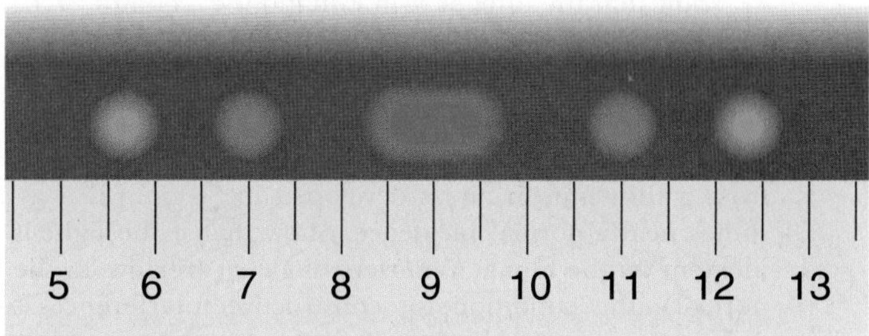

FIGURE **P27.46**

47. Interference effects are produced at point P on a screen as a result of direct rays from a 500-nm source and reflected rays from the mirror as shown in Figure P27.47. Assume that the source is 100 m to the left of the screen and 1.00 cm above the mirror. Find the distance y to the first dark band above the mirror.

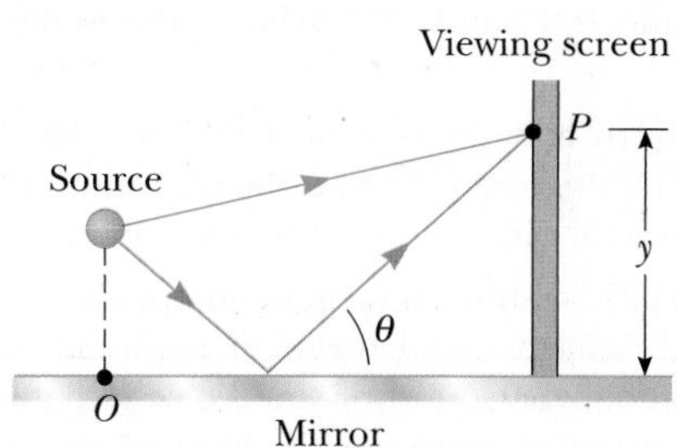

FIGURE **P27.47**

48. The waves from a radio station can reach a home receiver by two paths. One is a straight-line path from transmitter to home, a distance of 30.0 km. The second path is by reflection from the ionosphere (a layer of ionized air molecules high in the atmosphere). Assume that this reflection takes place at a point midway between receiver and transmitter and that the wavelength broadcast by the radio station is 350 m. Find the minimum height of the ionospheric layer that produces destructive interference between the direct and reflected beams. (Assume that no phase change occurs on reflection.)

49. Many cells are transparent and colorless. Structures of great interest in biology and medicine can be practically invisible to ordinary microscopy. An *interference microscope* reveals a difference in refractive index as a shift in interference fringes to indicate the size and shape of cell structures. The idea is exemplified in the following problem.

An air wedge is formed between two glass plates in contact along one edge and slightly separated at the opposite edge. When the plates are illuminated with monochromatic light from above, the reflected light has 85 dark fringes. Calculate the number of dark fringes that appear if water ($n = 1.33$) replaces the air between the plates.

50. (a) Both sides of a uniform film that has index of refraction n and thickness d are in contact with air. For normal incidence of light, an intensity minimum is observed in the reflected light at λ_2 and an intensity maximum is observed at λ_1, where $\lambda_1 > \lambda_2$. Assuming that no intensity minima are observed between λ_1 and λ_2, show that the integer m in Equations 27.11 and 27.12 is given by $m = \lambda_1/2(\lambda_1 - \lambda_2)$. (b) Determine the thickness of the film, assuming that $n = 1.40$, $\lambda_1 = 500$ nm, and $\lambda_2 = 370$ nm.

51. The condition for constructive interference by reflection from a thin film in air as developed in Section 27.5 assumes nearly normal incidence. Show that if the light is incident on the film at a nonzero angle ϕ_1 (relative to the normal), the condition for constructive interference is $2nt \cos \theta_2 = (m + \frac{1}{2})\lambda$, where θ_2 is the angle of refraction.

52. A soap film ($n = 1.33$) is contained within a rectangular wire frame. The frame is held vertically so that the film drains downward and forms a wedge with flat faces. The thickness of the film at the top is essentially zero. The film is viewed in reflected white light with near-normal incidence, and the first violet ($\lambda = 420$ nm) interference band is observed 3.00 cm from the top edge of the film. (a) Locate the first red ($\lambda = 680$ nm) interference band. (b) Determine the film thickness at the positions of the violet and red bands. (c) What is the wedge angle of the film?

53. Light from a helium–neon laser ($\lambda = 632.8$ nm) is incident on a single slit. What is the maximum width of the slit for which no diffraction minima are observed?

54. Figure P27.54 shows a megaphone in use. Construct a theoretical description of how a megaphone works. You may assume that the sound of your voice radiates just through the opening of your mouth. Most of the information in speech is carried not in a signal at the fundamental frequency, but in noises and in harmonics, with frequencies of a few thousand hertz. Does your theory allow any prediction that is simple to test?

(© Doug Pensigner/Getty Images)

FIGURE P27.54

55. **Review problem.** A beam of 541-nm light is incident on a diffraction grating that has 400 grooves/mm. (a) Determine the angle of the second-order ray. (b) If the entire apparatus is immersed in water, what is the new second-order angle of diffraction? (c) Show that the two diffracted rays of parts (a) and (b) are related through the law of refraction.

56. The *Very Large Array* (VLA) is a set of 27 radio telescope dishes in Caton and Socorro counties, New Mexico (Fig P27.56). The antennas can be moved apart on railroad tracks, and their combined signals give the resolving power of a synthetic aperture 36.0 km in diameter. (a) If the detectors are tuned to a frequency of 1.40 GHz, what is the angular resolution of the VLA? (b) Clouds of hydrogen radiate at this frequency. What must be the separation distance of two clouds at the center of the galaxy, 26 000 lightyears away, if they are to be resolved? (c) As the telescope looks up, a circling hawk looks down. Find the angular resolution of the hawk's eye. Assume that that the hawk is most sensitive to green light having wavelength 500 nm and that it has a pupil of diameter 12.0 mm. (d) A mouse is on the ground 30.0 m below. By what distance must the mouse's whiskers be separated if the hawk can resolve them?

(© Danny Lehman)

FIGURE P27.56 A rancher in New Mexico rides past one of the 27 radio telescopes that make up the Very Large Array (VLA).

57. Light of wavelength 500 nm is incident normally on a diffraction grating. If the third-order maximum of the diffraction pattern is observed at 32.0°, (a) what is the number of rulings per centimeter for the grating? (b) Determine the total number of primary maxima that can be observed in this situation.

58. Iridescent peacock feathers are shown in Figure P27.58a. The surface of one microscopic barbule is composed of transparent keratin that supports rods of dark brown melanin in a regular lattice, represented in Figure P27.58b. (Your fingernails are made of keratin, and melanin is the dark pigment giving color to human skin.) In a portion of the feather that can appear turquoise, assume that the melanin rods are uniformly separated by 0.25 μm, with air between them. (a) Explain how this

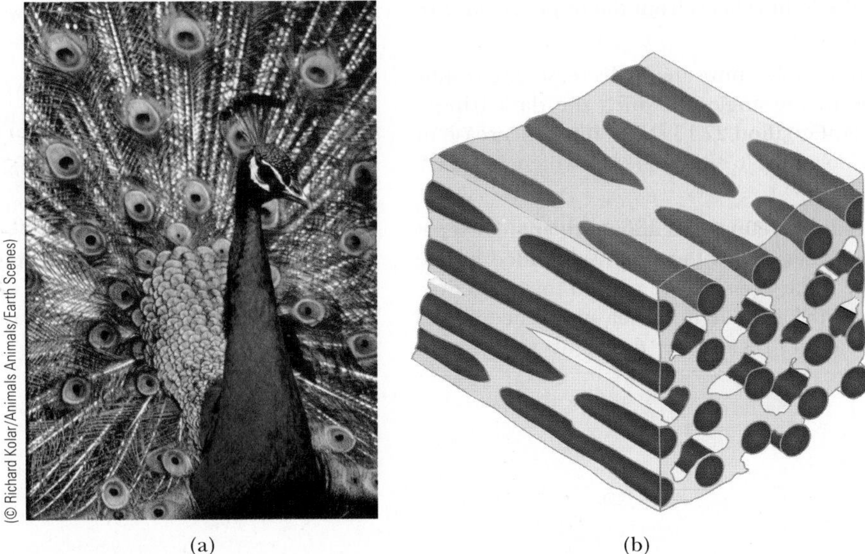

FIGURE P27.58 (a) Iridescence in peacock feathers. (b) Microscopic section of a feather showing dark melanin rods in a pale keratin matrix.

structure can appear blue–green when it contains no blue or green pigment. (b) Explain how it can also appear violet if light falls on it in a different direction. (c) Explain how it can present different colors to your two eyes at the same time, a characteristic of iridescence. (d) A compact disc can appear to be any color of the rainbow. Explain why this portion of the feather cannot appear yellow or red. (e) What could be different about the array of melanin rods in a portion of the feather that does appear to be red?

59. A beam of bright red light of wavelength 654 nm passes through a diffraction grating. Enclosing the space beyond the grating is a large screen forming one half of a cylinder centered on the grating, with its axis parallel to the slits in the grating. Fifteen bright spots appear on the screen. Find the maximum and minimum possible values for the slit separation in the diffraction grating.

60. A *pinhole camera* has a small circular aperture of diameter D. Light from distant objects passes through the aperture into an otherwise dark box, falling on a screen located a distance L away. If D is too large, the display on the screen will be fuzzy because a bright point in the field of view will send light onto a circle of diameter slightly larger than D. On the other hand, if D is too small, diffraction will blur the display on the screen. The screen shows a reasonably sharp image if the diameter of the central disk of the diffraction pattern, specified by Equation 27.15, is equal to D at the screen. (a) Show that for monochromatic light with plane wave fronts and $L \gg D$, the condition for a sharp view is fulfilled if $D^2 = 2.44\,\lambda L$. (b) Find the optimum pinhole diameter for 500-nm light projected onto a screen 15.0 cm away.

61. Two wavelengths λ and $\lambda + \Delta\lambda$ (with $\Delta\lambda \ll \lambda$) are incident on a diffraction grating. Show that the angular separation between the spectral lines in the mth-order spectrum is

$$\Delta\theta = \frac{\Delta\lambda}{\sqrt{(d/m)^2 - \lambda^2}}$$

where d is the slit spacing and m is the order number.

ANSWERS TO QUICK QUIZZES

27.1 (c). Equation 27.2 shows that decreasing λ will decrease the angle θ_{bright} and bring the fringes closer together. Equation 27.5 shows that decreasing L decreases y_{bright} and brings the fringes closer together. Immersing the apparatus in water decreases the wavelength so that the fringes move closer together.

27.2 (a). One of the materials has a higher index of refraction than water, the other lower. For the material with a higher index of refraction, there is a 180° phase shift for the light reflected from the upper surface but no such phase change from the lower surface because the index of refraction for water on the other side is lower than that of the film. Thus, the two reflections are out of phase and interfere destructively.

27.3 (a). At the left edge, the air wedge has zero thickness and the only contribution to the interference is the 180°

phase shift as the light reflects from the upper surface of the glass slide.

27.4 (i), (a). Equation 27.13 shows that a decrease in a results in an increase in the angles at which the dark fringes appear. (ii), (c). Equation 27.13 shows that a decrease in λ results in a decrease in the angles at which the dark fringes appear.

27.5 (a). We would like to reduce the minimum angular separation for two objects below the angle subtended by the two stars in the binary system. We can do that by reducing the wavelength of the light, which in essence makes the aperture larger, relative to the light wavelength, increasing the resolving power. Thus, we should choose a blue filter.

27.6 (c). With the doubled wavelength, the pattern will be wider. Choices (a) and (d) make the pattern even wider. From Equation 27.16, we see that choice (b) causes $\sin\theta_{\text{bright}}$ to be twice as large. Because we cannot use the small angle approximation, however, a doubling of $\sin\theta_{\text{bright}}$ is not the same as a doubling of θ_{bright}, which would translate to a doubling of the position of a maximum along the screen. If we only consider small-angle maxima, choice (b) would work, but it does not work in the large-angle case.

Using Lasers to Record and Read Digital Information

We have now investigated the principles of optics and can respond to our central question for the *Lasers* Context:

What is special about laser light and how is it used in technological applications?

In the Context Connections in Chapters 24 to 27, we discussed two primary technological applications of lasers: optical fibers and holography. In this Context Conclusion, we will choose one more from the vast number of possibilities, the storage and retrieval of information on compact discs (as well as CD-ROMs and digital video, or versatile, discs, DVDs).

The storage of information in a small volume of space is a goal toward which humans have worked for several decades. In the early days of computing, information was stored on punched cards. This method seems humorous in today's world, especially because the area taken up by laying the cards representing a page of text out on a table was larger than the original page of text.

The magnetic disc recording and storage technique introduced in the 1950s allowed a reduction in space over that taken up by the original data. The beginning of optical storage occurred in the 1970s with the introduction of videodiscs. These plastic discs included encoded pits representing the analog information associated with a video signal. A laser, focused by lenses to a spot about 1 micrometer (μm) in diameter, is used to read the data. When the laser light reflects off the flat area of the disc, the light is reflected back into the system and is detected. When the light encounters a pit, some of it is scattered. The light reflected from the bottom of the pit interferes destructively with that reflected from the surface, and very little of the incident light finds its way back to the detection system.

The next step in the optical recording story involves the digital revolution, exemplified by the introduction of the compact disc, or CD. The reading of the disc is similar to that of the videodisc, but the information is stored in a *digital* format. Musical CDs were rapidly accepted by the public with much more enthusiasm than videodiscs. Shortly after the introduction of CDs, plans were announced to market an optical disc for storage of information for computers, the CD-ROM.

Digital Recording

In digital recording, information is converted to binary code (ones and zeros), similar to the dots and dashes of Morse code. First, the waveform of the sound is *sampled,* typically at the rate of 44 100 times per second. Figure 1 illustrates this process. The sampling frequency is much higher than the upper range of hearing, about 20 000 Hz, so all audible frequencies of sound are sampled at this rate. During each sampling, the pressure of the wave is measured and converted to a voltage. Thus, there are 44 100 numbers associated with each second of the sound being sampled.

These measurements are then converted to *binary numbers,* which are numbers expressed to base 2 rather than base 10. Table 1 shows some sample binary numbers.

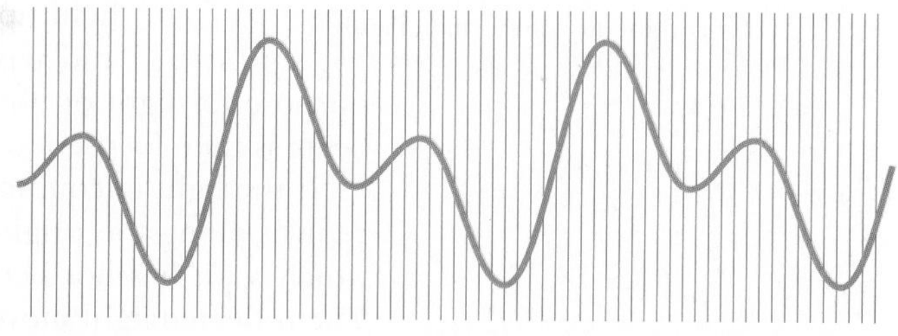

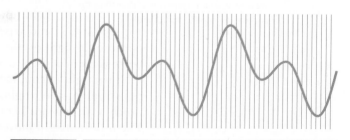

FIGURE 1 Sound is digitized by sampling the sound waveform at periodic intervals. During each interval, a number is recorded for the average voltage during the interval. The sampling rate shown here is much slower than the actual sampling rate of 44 100 per second.

TABLE 1 Sample Binary Numbers

Number in Base 10	Number in Binary	Sum
1	0000000000000001	1
2	0000000000000010	2 + 0
3	0000000000000011	2 + 1
10	0000000000001010	8 + 0 + 2 + 0
37	0000000000100101	32 + 0 + 0 + 4 + 0 + 1
275	0000000100010011	256 + 0 + 0 + 0 + 16 + 0 + 0 + 2 + 1

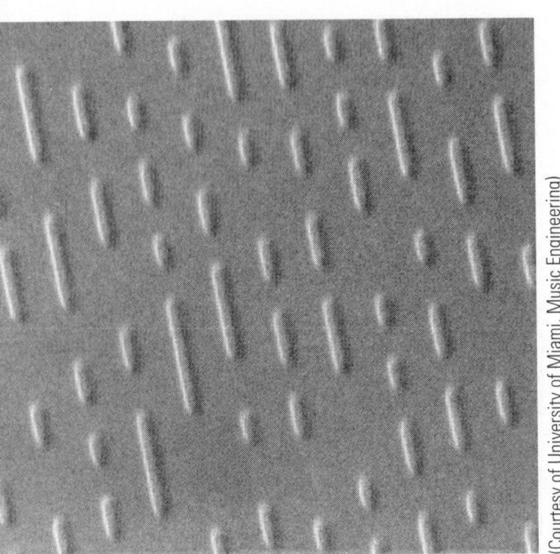
(Courtesy of University of Miami, Music Engineering)

FIGURE 2 The surface of a compact disc, showing the pits. Transitions between pits and lands correspond to ones. Regions without transitions correspond to zeros.

Generally, voltage measurements are recorded in 16-bit "words," where each bit is a one or a zero. Thus, the number of different voltage levels that can be assigned codes is $2^{16} = 65\ 536$. The number of bits in 1 second of sound is $16 \times 44\ 100 = 705\ 600$. These strings of ones and zeros, in 16-bit words, are recorded on the surface of a CD.

Figure 2 shows a magnification of the surface of a CD. There are two types of areas that are detected by the laser playback system: *lands* and *pits.* The lands are untouched regions of the disc surface that are highly reflective. The pits are areas that have been burned into the surface by a recording laser. The playback system, described below, converts the pits and lands into binary ones and zeros.

The binary numbers read from the CD are converted back to voltages, and the waveform is reconstructed as shown in Figure 3. Because the sampling rate is so high—44 100 voltage readings each second—the step-wise nature of the reconstructed waveform is not evident in the sound.

The advantage of digital recording is in the high fidelity of the sound. With analog recording, any small imperfection in the record surface or the recording equipment can cause a distortion of the waveform. If all peaks of a maximum in a waveform are clipped off so as to be only 90% as high, for example, there will be a major effect on the spectrum of the sound in an analog recording. With digital recording, however, it takes a major imperfection to turn a one into a zero. If an imperfection causes the magnitude of a one to be 90% of the original value, it still registers as a one and there is no distortion. Another advantage of digital recording is that the information is extracted optically, so there is no mechanical wear on the disc.

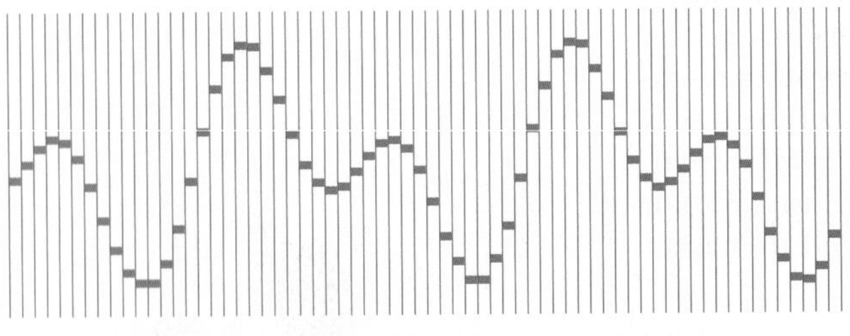

FIGURE 3 The reconstruction of the sound wave sampled in Figure 1. Notice that the reconstruction is step-wise, rather than the continuous waveform in Figure 1.

Digital Playback

Figure 4 shows the detection system of a CD player. The optical components are mounted on a track (not shown in the figure) that rolls radially so that the system can access all regions of the disc. The laser is located near the bottom of the figure, directing its light upward. The light is collimated by a lens into a parallel beam and passes through a beam splitter. The beam splitter serves no purpose for light on the way up, but it is important for the return light. The laser beam is then focused to a very small spot on the disc by the objective lens.

If the light encounters a pit in the disc, the light is scattered and very little light returns along the original path. If the light encounters a flat region of the disc at which a pit has not been recorded, the light reflects back along its original path. The reflected light moves downward in the diagram, arriving at the beam splitter so that it is partially reflected to the right. Lenses focus the beam, which is then detected by the photocell.

The playback system samples the reflected light 705 600 times per second. When the laser moves from a pit to a land or from a land to a pit, the reflected light changes during the sampling and the bit is recorded as a one. If there is no change during the sampling, the bit is recorded as a zero. The electronic circuitry in the CD player converts the series of zeros and ones back into an audible signal. This technology can also be used to store video information on a disc, leading to the rapid growth of DVD in the later years of the 20th century.

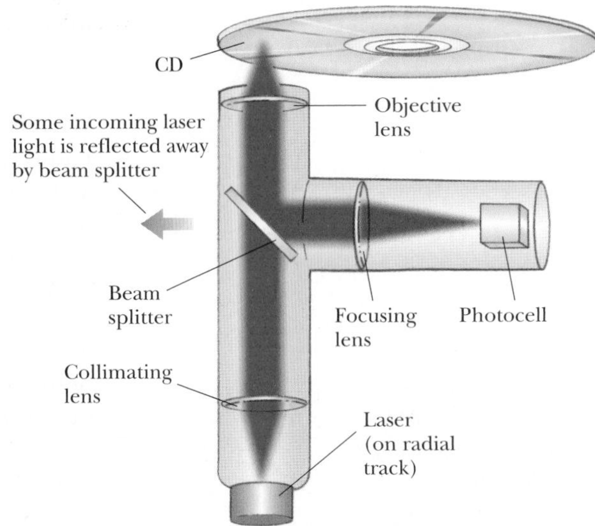

FIGURE 4 The detection system of a compact disc player. The laser (bottom) sends a beam of light upward. Laser light reflected back from the disc and then reflected to the right by the beam splitter enters a photocell. The digital information entering the photocell as pulses of light is converted to audio information.

Problems

1. Compact disc (CD) and digital video disc (DVD) players use interference to generate a strong signal from a tiny bump. The depth of a pit is chosen to be one quarter of the wavelength of the laser light used to read the disc. Then light reflected from the pit and light reflected from the adjoining land differ in path length traveled by one-half wavelength, to interfere destructively at the detector. As the disc rotates, the light intensity drops significantly whenever light is reflected from near a pit edge. The space between the leading and trailing edges of a pit determines the time interval between the fluctuations. The series of time intervals is decoded into a series of zeros and ones that carries the stored information. Assume that infrared light with a wavelength of 780 nm in vacuum is used in a CD player. The disc is coated with plastic having a refractive index of 1.50. What should be the depth of each pit? A DVD player uses light of a shorter wavelength, and the pit dimensions are correspondingly smaller, one factor that results in greater storage capacity on a DVD compared with a CD.
2. The laser in a CD player must precisely follow the spiral track, along which the distance between one loop of the spiral and the next is only about 1.25 μm. A feedback mechanism lets the player know if the laser drifts off the track so that the player can steer it back again. Figure 5 shows how a diffraction grating is used to provide information to keep the beam on track. The laser light passes through a diffraction grating just before it reaches the disc. The strong central maximum of the diffraction pattern is used to read the information in the track of pits. The two first-order side maxima are used for steering. The grating is designed so that the first-order maxima fall on the flat surfaces on both sides of the information track. Both side beams are reflected into their own detectors. As long as both beams are reflecting from smooth, nonpitted surfaces they are detected with constant high intensity. If the main beam wanders off the track, however, one of the side beams will begin to strike pits on the information track and the reflected light will diminish. This change is used with an electronic circuit to guide the beam back to the desired location. Assume that the laser light has a wavelength of 780 nm and that the diffraction grating is positioned 6.90 μm from the disc. Assume that the first-order beams are to fall on the disc 0.400 μm on either side of the information track. What should be the number of grooves per millimeter in the grating?
3. The speed with which the surface of a compact disc passes the laser is 1.3 m/s. What is the average length of the audio track on a CD associated with each bit of the audio information?
4. Consider the photograph of the compact disc surface in Figure 2. Audio data undergoes complicated processing to reduce a variety of errors in reading the data. Therefore, an audio "word" is not laid out linearly on the disc. Suppose data

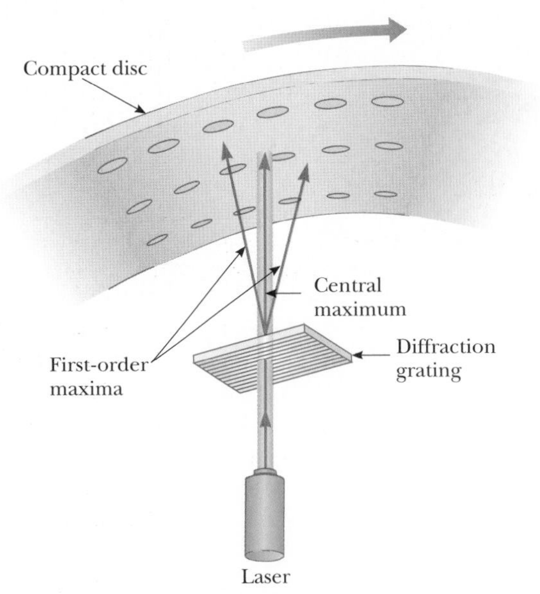

FIGURE 5 A tracking system in a CD player.

has been read from the disc, the error coding has been removed, and the resulting audio word is

1 0 1 1 1 0 1 1 1 0 1 1 1 0 1 1

What is the decimal number represented by this 16-bit word?

5. Lasers are also used in the recording process for a *magnetooptical disc.* To record a pit, its location on the ferromagnetic layer of the disc must be raised above a minimum temperature called the Curie temperature. Imagine that the surface moves past the laser at a speed on the order of 1 m/s and that the pit is modeled as a cylinder 1 μm deep with a radius of 1 μm. The ferromagnetic material has the following properties: its Curie temperature is 600 K, its specific heat is 300 J/kg·°C, and its density is 2×10^3 kg/m^3. What is the order of magnitude of the intensity of the laser beam necessary to raise the pit above the Curie temperature?

CONTEXT 9

The Cosmic Connection

In this final Context, we investigate the principles included in the area of physics commonly called *modern physics*. Modern physics encompasses the revolution in physics that commenced at the beginning of the 20th century. We began our discussion of modern physics in Chapter 9 in our study of relativity. Other aspects of modern physics—including atomic spectra and the Bohr model in Chapter 11, quantization of angular momentum and energy in Chapter 11, black holes in Chapter 11, black bodies in Chapter 24, and the discussion of the photon in Chapter 24—have appeared at various locations throughout the book.

In this book, we stress the importance of models in understanding physical phenomena. At the turn of the 20th century, classical physics was well established and provided many principles on which models for phenomena could be built. Many experimental observations, however, could not be brought into agreement with theory using classical models. Attempts to apply the laws of classical physics to atomic systems were consistently unsuccessful in making accurate predictions of the behavior of matter on the atomic scale. Various phenomena such as blackbody radiation, the photoelectric effect, and the emission of sharp spectral lines by atoms in a gas discharge could not be understood within the framework of classical physics. Between 1900 and 1930, however, new models collectively called *quantum physics* or *quantum mechanics* were highly successful in explaining the behavior of atoms, molecules, and nuclei. Like relativity, quantum physics requires a modification of our ideas concerning the physical world. Quantum mechanics does not, however, directly contradict or invalidate classical mechanics. As with relativity, the equations of quantum physics reduce to classical equations in the appropriate realm, that is, when the quantum equations are used to describe macroscopic systems.

An extensive study of quantum physics is certainly beyond the scope of this book and therefore this Context is simply an introduction to its underlying ideas. One of the true successes of quantum physics is the connection it makes between microscopic phenomena and the structure and evolution of the Universe. Ironically, recent developments in physics that probe smaller and smaller scales allow us to advance our understanding of the larger and larger systems that are familiar to us. This connection

(Comstock Royalty Free/Getty)

FIGURE 1 A person works on a personal digital assistant. The appearance of information on the display is due to the behavior of microscopic electrons in the circuitry of the microprocessor.

(a)

(b)

(1987 Anglo-Australian Observatory, photo by David Malin)

FIGURE 2 Supernova 1987A. (a) The region of the Tarantula Nebula (lower right) of the Large Magellanic Cloud before the supernova. (b) The supernova appears at the upper left on February 24, 1987. An understanding of this cosmic explosion is found in the interactions between the microscopic particles within the nucleus.

FIGURE 3 An image taken by the Hubble Space Telescope in January 2004 of galaxy AM 0644-741, called a "ring galaxy." Such a galaxy is formed from a collision with a second galaxy, called the intruder. The intruder punches through the center of the target galaxy, leaving a yellow nucleus in the case of AM 0644-741. The surrounding ring is expanding, similar to a ripple expanding outward from a disturbance in a pond. In the chaos in the ring, gas clouds collide and collapse gravitationally into new stars of large mass and high temperature, emitting light that is strong in the blue part of the visible spectrum. Several other galaxies are also visible in this photograph. Across the entire sky, it is estimated that the Hubble Space Telescope can detect 100 billion galaxies. It is also estimated that this is a very small fraction of all the galaxies in the visible part of the Universe. To develop a theory of the origin of this tremendously large system, we need to understand quarks, the most fundamental theorized particles.

between the small and the large is the theme of this Context.

Let us consider some examples of macroscopic systems and their connection to the behavior of microscopic particles. Consider your experiences with common electronic devices that are used today to view information on a liquid crystal display: handheld calculators, personal digital assistants (PDAs), cell phones, and video monitors. The events you observe—the appearance of numbers, to-do lists, or photographs on an LCD display—are macroscopic, but what controls these macroscopic events? They are controlled by a microprocessor within the electronic device. The operation of the microprocessor depends on the behavior of electrons within the solid-state material in an integrated circuit chip. The design and manufacture of the macroscopic electronic device are not possible without an understanding of the behavior of the electrons.

As a second example, a supernova explosion is clearly a macroscopic event; it is a star with a radius on the order of billions of meters undergoing a violent event. We have been able to advance our understanding of such events by studying the atomic nucleus, which is on the order of 10^{-15} m in size.

If we imagine an even larger system than a star—the entire Universe—we can advance our understanding of its origin by thinking about particles even smaller than the nucleus. Consider the constituents of protons and neutrons, called *quarks*. Models based on quarks provide further understanding of a theory of the origin of the Universe called the *Big Bang*. In this Context, we shall study both quarks and the Big Bang.

It seems that the larger the system we wish to investigate, the smaller are the particles whose behavior we must understand! We shall explore this relationship and study the principles of quantum physics as we respond to our central question:

How can we connect the physics of microscopic particles to the physics of the Universe?

CHAPTER 28

Quantum Physics

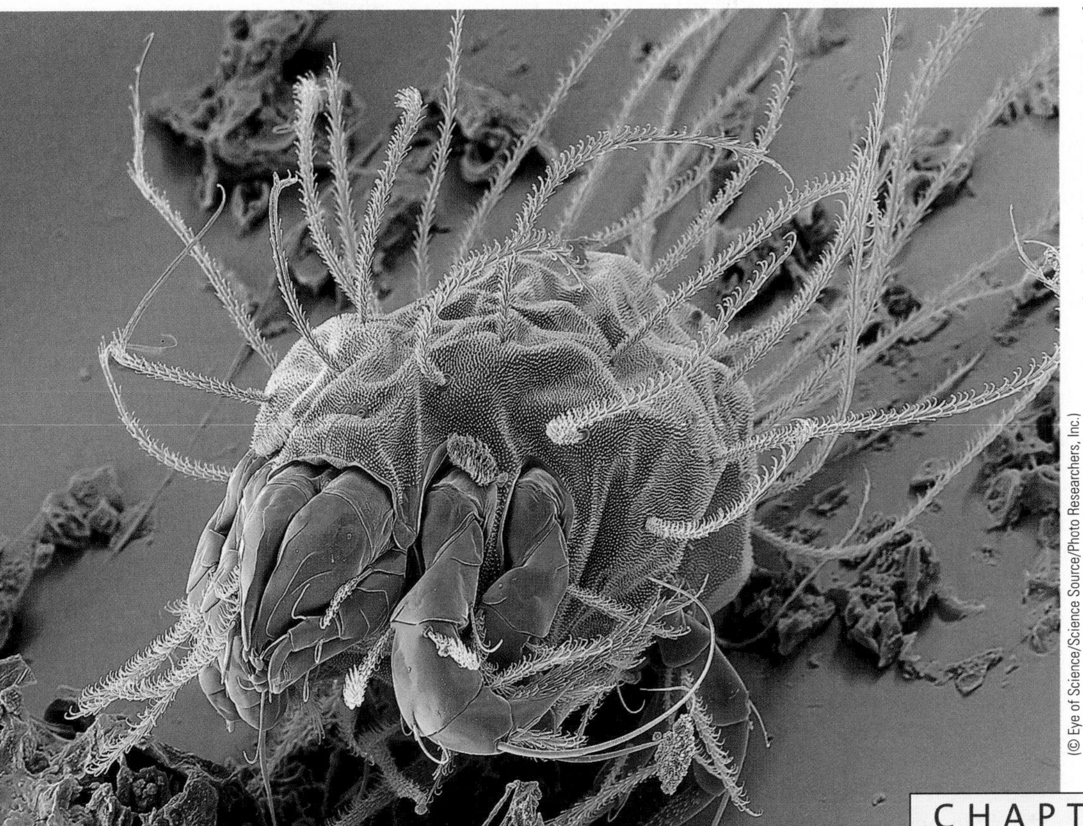

A color-enhanced electron microscope photograph shows significant detail of a storage mite, *Lepidoglyphus destructor.* The mite is so small, with a maximum length of 0.75 mm, that ordinary microscopes do not reveal minute anatomical details. The operation of the electron microscope is based on the wave nature of electrons, a central feature in quantum physics.

CHAPTER OUTLINE

In the earlier chapters of this book, we focused on the physics of particles. The particle model was a simplification model that allowed us to ignore the unnecessary details of an object when studying its behavior. We later combined particles into additional simplification models of systems and rigid objects. In Chapter 13, we introduced the wave as yet another simplification model and found that we could understand the motion of vibrating strings and the intricacies of sound by studying simple waves. In Chapters 24 to 27, we found that the wave model for light helped us understand many phenomena associated with optics.

It is hoped that you now have confidence in your abilities to analyze problems in the very different worlds of particles and waves. Your confidence may have been shaken somewhat by the discussion at the beginning of Chapter 25 in which we indicated that light has both wave-like and particle-like behaviors.

In this chapter, we return to this dual nature of light and study it in more detail. This study leads to two final analysis models: the quantum particle and the quantum particle under boundary conditions. A careful analysis of these two models shows that particles and waves are not as unrelated as you might expect.

PITFALL PREVENTION 28.1

Expect to be challenged If the discussions of quantum physics in this chapter seem strange and confusing to you, it's because your whole life experience has taken place in the macroscopic world, where quantum effects are not evident.

28.1 BLACKBODY RADIATION AND PLANCK'S THEORY

As we discussed in Chapter 17, an object at any temperature emits energy referred to as **thermal radiation.** The characteristics of this radiation depend on the temperature and properties of the surface of the object. If the surface is at room temperature, the wavelengths of the thermal radiation are primarily in the infrared region and hence are not observed by the eye. As the temperature of the surface increases, the object eventually begins to glow red. At sufficiently high temperatures, the object appears to be white, as in the glow of the hot tungsten filament of a lightbulb. A careful study of thermal radiation shows that it consists of a continuous distribution of wavelengths from all portions of the electromagnetic spectrum.

From a classical viewpoint, thermal radiation originates from accelerated charged particles near the surface of the object. The thermally agitated charges can have a distribution of accelerations, which accounts for the continuous spectrum of radiation emitted by the object. By the end of the 19th century, it had become apparent that this classical explanation of thermal radiation was inadequate. The basic problem was in understanding the observed distribution of wavelengths in the radiation emitted by an ideal object called a black body. As mentioned in Chapter 24, a **black body** is an ideal system that absorbs all radiation incident on it. A good approximation of a black body is a small hole leading to the inside of a hollow object as shown in Figure 28.1. The nature of the radiation emitted from the hole depends only on the temperature of the cavity walls.

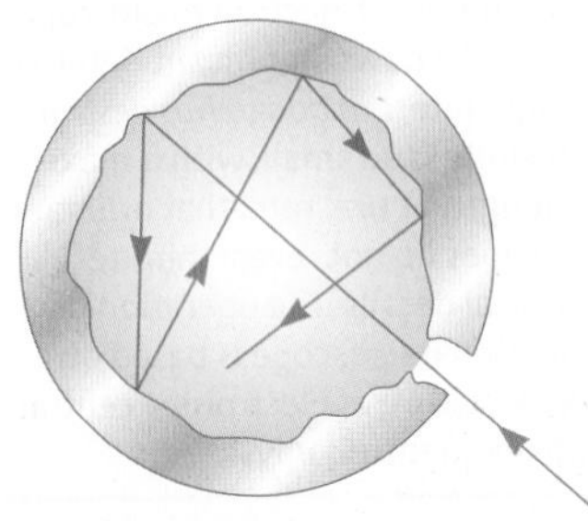

FIGURE 28.1 The opening to the cavity inside a hollow object is a good approximation of a black body. Light entering the small opening strikes the interior walls, where some is absorbed and some is reflected at a random angle. The cavity walls re-radiate at wavelengths corresponding to their temperature. Some of the energy from these standing waves can leave through the opening.

The wavelength distribution of radiation from cavities was studied extensively in the late 19th century. Experimental data for the distribution of energy in **blackbody radiation** at three temperatures are shown in Active Figure 28.2. The distribution of radiated energy varies with wavelength and temperature. Two regular features of the distribution were noted in these experiments.

1. **The total power of emitted radiation increases with temperature.** We discussed this feature briefly in Chapter 17, where we introduced **Stefan's law,** Equation 17.36, for the power emitted from a surface of area A and temperature T:

■ Stefan's law

$$\mathcal{P} = \sigma A e T^4$$

For a black body, the emissivity is $e = 1$ exactly.

2. **The peak of the wavelength distribution shifts to shorter wavelengths as the temperature increases.** This shift was found experimentally to obey the following relationship, called **Wien's displacement law:**

■ Wien's displacement law

$$\lambda_{\max} T = 2.898 \times 10^{-3} \text{ m} \cdot \text{K} \qquad [28.1]$$

where $\lambda_{\max}$ is the wavelength at which the curve peaks and T is the absolute temperature of the surface emitting the radiation.

(Corbis)

The glow emanating from the spaces between these hot charcoal briquettes is, to a close approximation, blackbody radiation. The color of the light depends on the temperature of the briquettes. ■

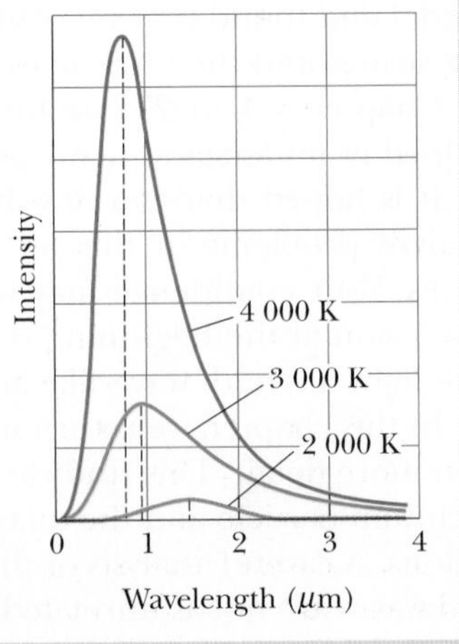

ACTIVE FIGURE 28.2 Intensity of blackbody radiation versus wavelength at three temperatures. Note that the amount of radiation emitted (the area under a curve) increases with increasing temperature. The visible range of wavelengths is between 0.4 μm and 0.7 μm. Therefore, the 4 000-K curve has a peak that is near the visible range and represents an object that would glow with a yellowish-white appearance. At about 6 000 K, the peak is in the center of the visible wavelengths and the object appears white.

Physics Now™ By logging into PhysicsNow at www.pop4e.com and going to Active Figure 28.2, you can adjust the temperature of the black body and study the radiation emitted from it.

A successful theoretical model for blackbody radiation must predict the shape of the curve in Active Figure 28.2, the temperature dependence expressed in Stefan's law, and the shift of the peak with temperature described by Wien's displacement law. Early attempts to use classical ideas to explain the shapes of the curves in Active Figure 28.2 failed. Figure 28.3 shows an experimental plot of the blackbody radiation spectrum (red curve) together with the curve predicted by classical theory (blue curve). At long wavelengths, classical theory is in good agreement with the experimental data. At short wavelengths, however, major disagreement exists between classical theory and experiment. This disagreement is often called the **ultraviolet catastrophe.** (This "catastrophe"—infinite energy—occurs as the wavelength approaches zero; the word "ultraviolet" was applied because ultraviolet wavelengths are short.)

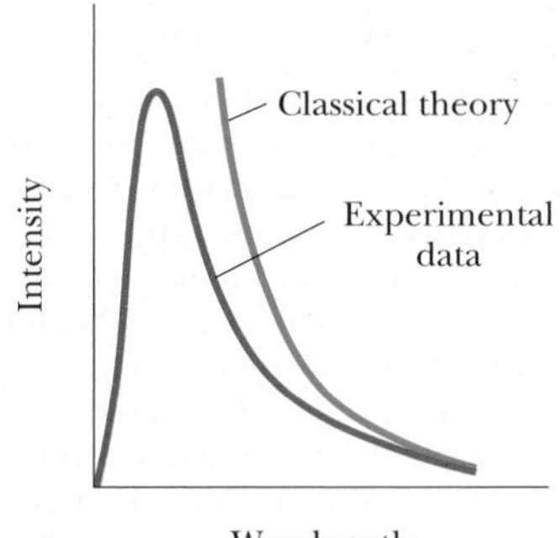

FIGURE 28.3 Comparison of the experimental results with the curve predicted by classical theory for the distribution of blackbody radiation.

In 1900, Max Planck developed a structural model for blackbody radiation that leads to a theoretical equation for the wavelength distribution that is in complete agreement with experimental results at all wavelengths. In his model, which represented the dawn of **quantum physics,** Planck imagined that oscillators exist at the surface of the black body, related to the charges within the molecules. He made two bold and controversial assumptions concerning the nature of these oscillators:

- The energy of the oscillator is quantized; that is, it can have only certain *discrete* amounts of energy E_n given by

$$E_n = nhf \qquad [28.2]$$

  where n is a positive integer called a **quantum number,**[1] f is the frequency of oscillation of the oscillator, and h is **Planck's constant,** first introduced in Chapter 11. Because the energy of each oscillator can have only discrete values given by Equation 28.2, we say that the energy is **quantized.** Each discrete energy value corresponds to a different **quantum state,** represented by the quantum number n. When the oscillator is in the $n = 1$ quantum state, its energy is hf; when it is in the $n = 2$ quantum state, its energy is $2hf$; and so on.
- The oscillators emit or absorb energy in discrete units. They emit or absorb these energy units by making a transition from one quantum state to another, similar to the transitions discussed in the Bohr model in Chapter 11. The entire energy difference between the initial and final states in the transition is emitted as a single quantum of radiation. If the transition is from one state to an adjacent state—say, from the $n = 3$ state to the $n = 2$ state—Equation 28.2 shows that the amount of energy radiated by the oscillator is

$$E = hf \qquad [28.3]$$

MAX PLANCK (1858–1947)

Planck introduced the concept of a "quantum of action" (Planck's constant, h) in an attempt to explain the spectral distribution of blackbody radiation, which laid the foundations for quantum theory. In 1918, he was awarded the Nobel Prize in Physics for this discovery of the quantized nature of energy.

An oscillator radiates or absorbs energy only when it changes quantum states. If it remains in one quantum state, no energy is absorbed or emitted. Figure 28.4 shows the quantized energy levels and allowed transitions proposed by Planck.

These assumptions may not sound bold to you because we have seen them in the Bohr model of the hydrogen atom in Chapter 11. It is important to keep in mind, however, that the Bohr model was not introduced until 1913, whereas Planck made his assumptions in 1900. The key point in Planck's theory is the radical assumption of quantized energy states. This development marked the birth of the quantum theory. Using these assumptions, Planck was able to generate a theoretical expression for the wavelength distribution that agreed remarkably well with the experimental curves in Active Figure 28.2. When Planck presented his theory, most scientists (including Planck!) did not consider the quantum concept realistic. It was believed to be a mathematical trick that happened to predict the correct results. Hence, Planck and others continued to search for what they believed to be a more rational explanation of blackbody radiation. Subsequent developments, however,

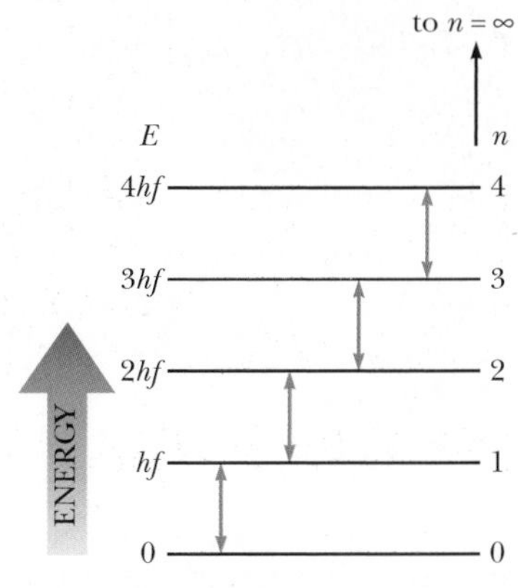

FIGURE 28.4 Allowed energy levels for an oscillator with a natural frequency f. Allowed transitions are indicated by the double-headed arrows.

[1]We first introduced the notion of a quantum number for microscopic systems in Section 11.5, in which we incorporated it into the Bohr model of the hydrogen atom. We put it in bold again here because it is an important notion for the remaining chapters in this book.

PITFALL PREVENTION 28.2

***n* IS AGAIN AN INTEGER** In the preceding chapters on optics, we used the symbol n for the index of refraction, which was not an integer. We are now using n again in the manner in which it was used in Chapter 11 to indicate the quantum number of a Bohr orbit and in Chapter 14 to indicate the standing wave mode on a string or in an air column. In quantum physics, n is often used as an integer quantum number to identify a particular quantum state of a system.

 The ear thermometer

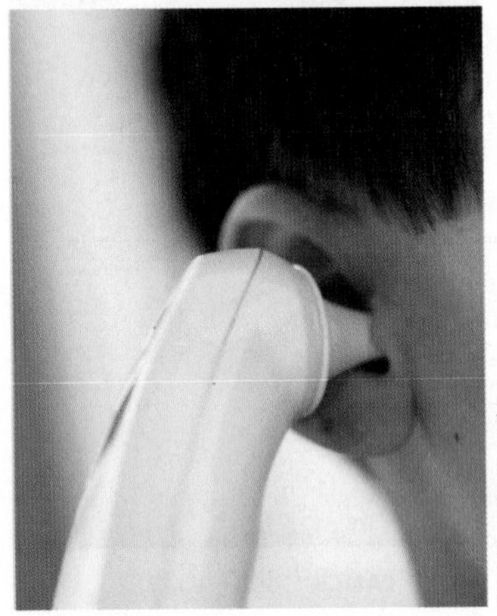
(Photodisc/Getty Images)

FIGURE 28.5 An ear thermometer measures a patient's temperature by detecting the intensity of infrared radiation leaving the eardrum.

showed that a theory based on the quantum concept (rather than on classical concepts) was required to explain a number of other phenomena at the atomic level.

We don't see quantum effects on an everyday basis because the energy change in a macroscopic system due to a transition between adjacent states is such a small fraction of the total energy of the system that we could never expect to detect the change. (See Example 28.2 for a numerical example.) Therefore, even though changes in the energy of a macroscopic system are indeed quantized and proceed by small quantum jumps, our senses perceive the decrease as continuous. **Quantum effects become important and measurable only on the submicroscopic level of atoms and molecules.** Furthermore, quantum results must blend smoothly with classical results when the quantum number becomes large. This statement is known as the **correspondence principle.**

You may have had your body temperature measured at the doctor's office by an *ear thermometer*, which can read your temperature in a matter of seconds (Fig. 28.5). This type of thermometer measures the amount of infrared radiation emitted by the eardrum in a fraction of a second. It then converts the amount of radiation into a temperature reading. This thermometer is very sensitive because temperature is raised to the fourth power in Stefan's law. Problem 28.1 allows you to explore the sensitivity of this device.

QUICK QUIZ 28.1 Figure 28.6 shows two stars in the constellation Orion. Betelgeuse appears to glow red, whereas Rigel looks blue in color. Which star has a higher surface temperature? **(a)** Betelgeuse **(b)** Rigel **(c)** both have the same surface temperature **(d)** impossible to determine

(John Chumack/Photo Researchers, Inc.)

FIGURE 28.6 (Quick Quiz 28.1) Which star is hotter?

■ Thinking Physics 28.1

You are observing a yellow candle flame, and your laboratory partner claims that the light from the flame is atomic in origin. You disagree, claiming that the candle flame is hot, so the radiation must be thermal in origin. Before this disagreement leads to fisticuffs, how could you determine who is correct?

Reasoning A simple determination could be made by observing the light from the candle flame through a diffraction grating spectrometer, which was discussed in Section 27.8. If the spectrum of the light is continuous, it is thermal in origin. If the spectrum shows discrete lines, it is atomic in origin. The results of the experiment show that the light is primarily thermal in origin and originates in the hot particles of soot in the candle flame. ■

EXAMPLE 28.1 Thermal Radiation from the Human Body

The temperature of your skin is approximately 35°C.

A What is the peak wavelength of the radiation it emits?

Solution From Wien's displacement law (Eq. 28.1), we have

$$\lambda_{max}T = 2.898 \times 10^{-3}\ \text{m}\cdot\text{K}$$

Solving for λ_{max} and noting that 35°C corresponds to an absolute temperature of 308 K, we have

$$\lambda_{max} = \frac{2.898 \times 10^{-3}\ \text{m}\cdot\text{K}}{308\ \text{K}} = 9.41\ \mu\text{m}$$

This radiation is in the infrared region of the spectrum.

B What total power is emitted by your skin, assuming that it emits like a black body?

Solution We need to make an estimate of the surface area of your skin. If we model your body as a rectangular box of height 2 m, width 0.3 m, and depth 0.2 m, the total surface area is

$$A = 2(2\ \text{m})(0.3\ \text{m}) + 2(2\ \text{m})(0.2\ \text{m}) + 2(0.2\ \text{m})(0.3\ \text{m}) \approx 2\ \text{m}^2$$

Therefore, from Stefan's law, we have

$$\mathcal{P} = \sigma AeT^4 \approx (5.7 \times 10^{-8}\ \text{W/m}^2\cdot\text{K}^4)(2\ \text{m}^2)(1)(308\ \text{K})^4 \approx 10^3\ \text{W}$$

C Based on your answer to part B, why don't you glow as brightly as several lightbulbs?

Solution The answer to part B indicates that your skin is radiating energy at approximately the rate as that which enters ten 100-W lightbulbs by electrical transmission. You are not visibly glowing, however, because most of this radiation is in the infrared range, as we found in part A, and our eyes are not sensitive to infrared radiation.

EXAMPLE 28.2 The Quantized Oscillator

A 2.0-kg block is attached to a massless spring of force constant $k = 25$ N/m. The spring is stretched 0.40 m from its equilibrium position and released.

A Find the total energy and frequency of oscillation according to classical calculations.

Solution Because of our study of oscillating blocks in Chapter 12, this problem is easy to conceptualize. The phrase "according to classical calculations" tells us that we should categorize this part of the problem as a classical analysis of the oscillator. To analyze the problem, we know that the total energy of a simple harmonic oscillator having an amplitude A is $\frac{1}{2}kA^2$ (Eq. 12.21). Therefore,

$$E = \tfrac{1}{2}kA^2 = \tfrac{1}{2}(25\ \text{N/m})(0.40\ \text{m})^2 = 2.0\ \text{J}$$

The frequency of oscillation is, from Equation 12.14,

$$f = \frac{1}{2\pi}\sqrt{\frac{k}{m}} = \frac{1}{2\pi}\sqrt{\frac{25\ \text{N/m}}{2.0\ \text{kg}}} = 0.56\ \text{Hz}$$

B Assuming that the energy is quantized, find the quantum number n for the system.

Solution This part of the problem is categorized as a quantum analysis of the oscillator. To analyze the problem, we note that the energy of the oscillator is quantized according to Equation 28.2. Therefore,

$$E_n = nhf = n(6.63 \times 10^{-34}\ \text{J}\cdot\text{s})(0.56\ \text{Hz}) = 2.0\ \text{J}$$

Solving for n,

$$n = \frac{2.0\ \text{J}}{(6.63 \times 10^{-34}\ \text{J}\cdot\text{s})(0.56\ \text{Hz})} = 5.4 \times 10^{33}$$

C How much energy is carried away when the oscillator makes a transition to the next lowest quantum state?

Solution The energy difference between adjacent quantum states is $\Delta E = hf$. Therefore, the energy carried away is

$$E = hf = (6.63 \times 10^{-34}\ \text{J}\cdot\text{s})(0.56\ \text{Hz}) = 3.7 \times 10^{-34}\ \text{J}$$

To finalize the problem, note that the energy carried away in part C due to a transition between adjacent states is a small fraction of the total energy of the oscillator (about one part in ten million billion billion billion, or $1{:}10^{34}$!). Consequently, we do not see the quantized nature of the oscillator for such a large quantum number as we found in part B, in agreement with the correspondence principle.

28.2 THE PHOTOELECTRIC EFFECT

Blackbody radiation was historically the first phenomenon to be explained with a quantum model. In the latter part of the 19th century, at the same time as data were being taken on thermal radiation, experiments showed that light incident on certain metallic surfaces causes electrons to be emitted from the surfaces. As mentioned in Section 25.1, this phenomenon, first discovered by Hertz, is known as the **photoelectric effect.** The emitted electrons are called **photoelectrons.**[2]

Active Figure 28.7 is a schematic diagram of a photoelectric effect apparatus. An evacuated glass or quartz tube contains a metal plate E connected to the negative terminal of a battery. Another metal plate C is maintained at a positive potential by the battery. When the tube is kept in the dark, the ammeter reads zero, indicating that there is no current in the circuit. When light of the appropriate wavelength shines on plate E, however, a current is detected by the ammeter, indicating a flow of charges across the gap between E and C. This current arises from electrons emitted from the negative plate E (the emitter) and collected at the positive plate C (the collector).

Active Figure 28.8, a graphical representation of the results of a photoelectric experiment, plots the photoelectric current versus the potential difference ΔV between E and C for two light intensities. For large positive values of ΔV, the current reaches a maximum value. In addition, the current increases as the incident light intensity increases, as you might expect. Finally, when ΔV is negative—that is, when the battery polarity is reversed to make E positive and C negative—the current drops because many of the photoelectrons emitted from E are repelled by the negative collecting plate C. Only those electrons ejected from the metal with a kinetic energy greater than $e|\Delta V|$ will reach C, where e is the magnitude of the charge on the electron. When the magnitude of ΔV is equal to ΔV_s, the **stopping potential,** no electrons reach C and the current is zero.

Let us consider the combination of the electric field between the plates and an electron ejected from plate E with the maximum kinetic energy to be an isolated system. Suppose this electron stops just as it reaches plate C. Applying the isolated

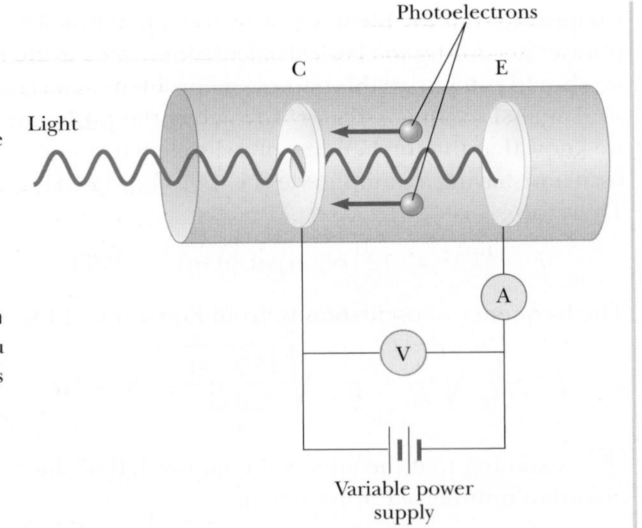

ACTIVE FIGURE 28.7
A circuit diagram for studying the photoelectric effect. When light strikes the plate E (the emitter), photoelectrons are ejected from the plate. Electrons moving from plate E to plate C (the collector) constitute a current in the circuit.

Physics Now™ By logging into PhysicsNow at **www.pop4e.com** and going to Active Figure 28.7, you can observe the motion of electrons for various frequencies and voltages.

[2]Photoelectrons are not different from other electrons. They are given this name solely because of their ejection from the metal by photons in the photoelectric effect.

system model, the total energy of the system must be conserved:

$$E_f = E_i \quad \rightarrow \quad K_f + U_f = K_i + U_i$$

where the initial configuration of the system refers to the instant that the electron leaves the metal with the maximum possible kinetic energy K_{max} and the final configuration is when the electron stops just before touching plate C. If we define the electric potential energy of the system in the initial configuration to be zero, we have

$$0 + (-e)(-\Delta V_s) = K_{max} + 0$$

$$K_{max} = e\,\Delta V_s \qquad [28.4]$$

This equation allows us to measure K_{max} experimentally by measuring the voltage at which the current drops to zero.

The following are several features of the photoelectric effect in which the predictions made by a classical approach are compared, using the wave model for light, with the experimental results. Notice the strong contrast between the predictions and the results.

1. Dependence of photoelectron kinetic energy on light intensity
   Classical prediction: Electrons should absorb energy continuously from the electromagnetic waves. A more intense light should transfer energy into the metal faster, and the electrons should be ejected with more kinetic energy.
   Experimental result: The maximum kinetic energy of the photoelectrons is *independent* of light intensity. This result is shown in Active Figure 28.8 by both curves falling to zero at the *same* negative voltage.
2. Time interval between incidence of light and ejection of photoelectrons
   Classical prediction: For very weak light, a measurable time interval should pass between the incidence of the light and the ejection of an electron. This time interval is required for the electron to absorb the incident radiation before it acquires enough energy to escape from the metal.
   Experimental result: Electrons are emitted from the surface almost *instantaneously* (less than 10^{-9} s after the surface is illuminated), even at very low light intensities.
3. Dependence of ejection of electrons on light frequency
   Classical prediction: Electrons should be ejected at any frequency of the incident light, as long as the intensity is high enough, because energy is being transferred to the metal regardless of the frequency.
   Experimental result: No electrons are emitted if the incident light frequency falls below some **cutoff frequency** f_c, which is characteristic of the material being illuminated. No electrons are ejected below this cutoff frequency *regardless* of how intense the light is.
4. Dependence of photoelectron kinetic energy on light frequency
   Classical prediction: No relationship should exist between the frequency of the light and the electron kinetic energy. The kinetic energy should be related to the intensity of the light.
   Experimental result: The maximum kinetic energy of the photoelectrons increases with increasing light frequency.

Notice that *all four* predictions of the classical model are incorrect. A successful explanation of the photoelectric effect was given by Einstein in 1905, the same year he published his special theory of relativity. As part of a general paper on electromagnetic radiation, for which he received the Nobel Prize in Physics in 1921, Einstein extended Planck's concept of quantization to electromagnetic waves. He assumed that light (or any other electromagnetic wave) of frequency f can be considered to be a stream of quanta, regardless of the source of the radiation. Today we call these quanta **photons.** Each photon has an energy E given by Equation 28.3, $E = hf$, and moves in a vacuum at the speed of light c, where is $c = 3.00 \times 10^8$ m/s.

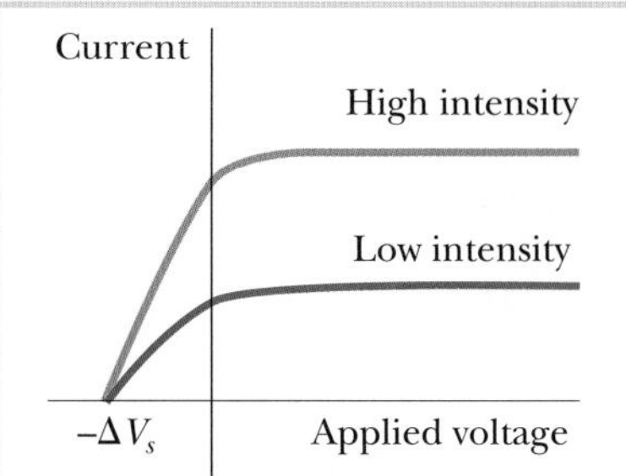

ACTIVE FIGURE 28.8

Photoelectric current versus applied potential difference for two light intensities. The current increases with intensity but reaches a saturation level for large values of ΔV. At voltages equal to or more negative than $-\Delta V_s$, where ΔV_s is the stopping potential, the current is zero.

PhysicsNow™ By logging into PhysicsNow at **www.pop4e.com** and going to Active Figure 28.8, you can sweep through the voltage range and observe the current curve for different intensities of radiation.

QUICK QUIZ 28.2 While standing outdoors on a dark night, you are subjected to the following four types of electromagnetic radiation: yellow light from a sodium street lamp, radio waves from a nearby AM radio station, radio waves from a nearby FM radio station, and microwaves from a nearby antenna of a communications system. Rank these types of waves in terms of increasing photon energy, lowest first.

In Einstein's model, a photon of the incident light gives *all* its energy hf to a *single* electron in the metal. Therefore, the absorption of energy by the electrons is not a continuous process as envisioned in the wave model, but rather a discontinuous process in which energy is delivered to the electrons in bundles. The energy transfer is accomplished via a one-photon/one-electron event.

Electrons ejected from the surface of the metal possess the maximum kinetic energy K_{max}. According to Einstein, the maximum kinetic energy for these liberated electrons is

▮ Photoelectric effect equation

$$K_{\text{max}} = hf - \phi \qquad [28.5]$$

where ϕ is called the **work function** of the metal. **The work function represents the minimum energy with which an electron is bound in the metal** and is on the order of a few electron volts. Table 28.1 lists selected values.

TABLE 28.1

Work Functions of Selected Metals

Metal	ϕ (eV)
Na	2.46
Al	4.08
Cu	4.70
Zn	4.31
Ag	4.73
Pt	6.35
Pb	4.14
Fe	4.50

Equation 28.5 is a statement of the continuity equation for energy, Equation 6.20, from Chapter 6:

$$\Delta E_{\text{system}} = \sum T$$

We imagine the system to consist of an electron that is to be ejected and the remainder of the metal, and then apply the nonisolated system model for energy. Energy is transferred into the system by electromagnetic radiation, the photon. The system has two types of energy: the potential energy of the metal–electron system and the kinetic energy of the electron. Therefore, we can write the continuity equation as

$$\Delta K + \Delta U = T_{\text{ER}} \qquad [28.6]$$

The energy transfer is that of the photon, $T_{\text{ER}} = hf$. During the process, the kinetic energy of the electron increases from zero to its final value, which we assume to be the maximum possible value K_{max}. The potential energy of the system increases because the electron is pulled away from the metal to which it is attracted. We define the potential energy of the system when the electron is outside the metal as zero. The potential energy of the system when the electron is in the metal is $U = -\phi$, where ϕ is the work function. Therefore, the increase in potential energy when the electron is removed from the metal is the work function ϕ. Substituting these energies into Equation 28.6, we have

$$K_{\text{max}} + \phi = hf$$

which is the same as Equation 28.5. If the electron makes collisions with other electrons or metal ions as it is being ejected, some of the incoming energy is transferred to the metal and the electron is ejected with less kinetic energy than K_{max}.

With the photon model of light, one can explain the observed features of the photoelectric effect that cannot be understood using classical concepts:

1. Dependence of photoelectron kinetic energy on light intensity

   That K_{max} is independent of the light intensity can be understood with the following argument. The maximum kinetic energy of any one electron, which equals $hf - \phi$, depends only on the light frequency and the work function, not on the light intensity. If the light intensity is doubled, the number of photons arriving per unit time interval is doubled, which doubles the rate at which photoelectrons are emitted. The maximum possible kinetic energy of any one emitted electron, however, is unchanged.

2. Time interval between incidence of light and ejection of photoelectrons
   That the electrons are emitted almost instantaneously is consistent with the photon model of light, in which the incident energy appears in small packets and the interaction between photons and electrons is one to one. Therefore, for very weak incident light, very few photons may arrive per unit time interval, but each one has sufficient energy to eject an electron immediately.
3. Dependence of ejection of electrons on light frequency
   That the effect is not observed below a certain cutoff frequency follows because the photon must have energy greater than the work function ϕ to eject an electron. If the energy of an incoming photon does not satisfy this requirement, an electron cannot be ejected from the surface, regardless of light intensity.
4. Dependence of photoelectron kinetic energy on light frequency
   That K_{max} increases with increasing frequency is easily understood with Equation 28.5.

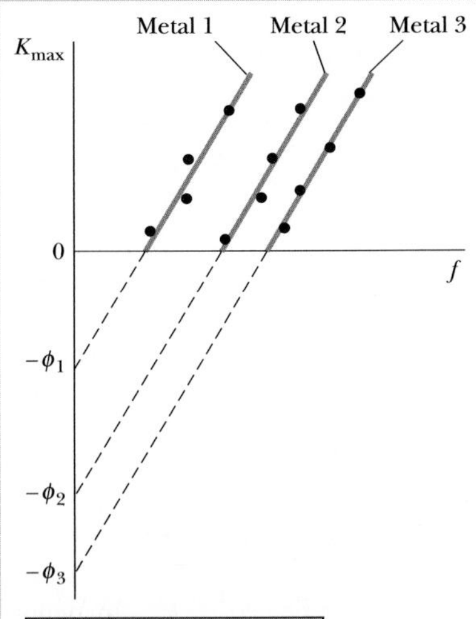

ACTIVE FIGURE **28.9**

A plot of results for K_{max} of photoelectrons versus frequency of incident light in a typical photoelectric effect experiment. Photons with frequency less than the cutoff frequency for a given metal do not have sufficient energy to eject an electron from the metal.

Physics Now™ By logging into PhysicsNow at **www.pop4e.com** and going to Active Figure 28.9, you can sweep through the frequency range and observe the curve for different target metals.

Einstein's theoretical result (Eq. 28.5) predicts a linear relationship between the maximum electron kinetic energy K_{max} and the light frequency f. Experimental observation of such a linear relationship would be a final confirmation of Einstein's theory. Indeed, such a linear relationship is observed as sketched in Active Figure 28.9. The slope of the curves for all metals is Planck's constant h. The absolute value of the intercept on the vertical axis is the work function ϕ, which varies from one metal to another. The intercept on the horizontal axis is the cutoff frequency, which is related to the work function through the relation $f_c = \phi/h$. This cutoff frequency corresponds to a **cutoff wavelength** of

$$\lambda_c = \frac{c}{f_c} = \frac{c}{\phi/h} = \frac{hc}{\phi} \qquad [28.7]$$

where c is the speed of light. Light with wavelength *greater* than λ_c incident on a material with a work function of ϕ does not result in the emission of photoelectrons.

The combination hc occurs often when relating the energy of a photon to its wavelength. A common shortcut to use in solving problems is to express this combination in useful units according to the numerical value

$$hc = 1\ 240 \text{ eV}\cdot\text{nm}$$

One of the first practical uses of the photoelectric effect was as a detector in a light meter of a camera. Light reflected from the object to be photographed strikes a photoelectric surface in the meter, causing it to emit photoelectrons that then pass through a sensitive ammeter. The magnitude of the current in the ammeter depends on the light intensity.

The phototube, another early application of the photoelectric effect, acts much like a switch in an electric circuit. It produces a current in the circuit when light of sufficiently high frequency falls on a metal plate in the phototube, but produces no current in the dark. Phototubes were used in burglar alarms and in the detection of the soundtrack on motion picture film. Modern semiconductor devices have now replaced older devices based on the photoelectric effect.

The photoelectric effect is used today in the operation of photomultiplier tubes. Figure 28.10 shows the structure of such a device. A photon striking the photocathode ejects an electron by means of the photoelectric effect. This electron is accelerated across the potential difference between the photocathode and the first *dynode,* shown as being at $+200$ V relative to the photocathode in Figure 28.10. This high-energy electron strikes the dynode and ejects several more electrons. This process is repeated through a series of dynodes at ever higher potentials until an electrical pulse is produced as millions of electrons strike the last dynode. Thus, the tube is

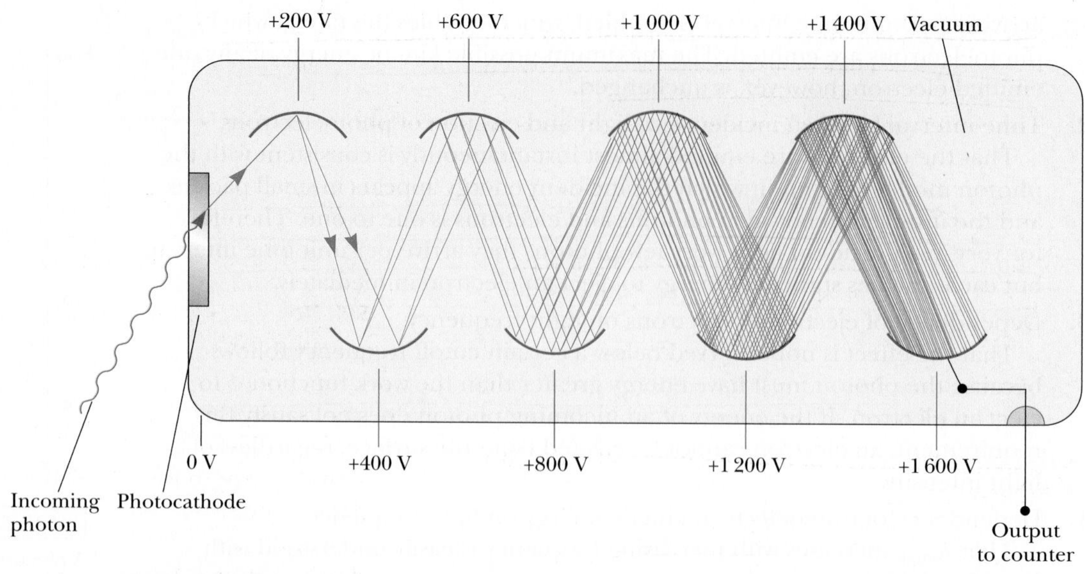

FIGURE 28.10 The multiplication of electrons in a photomultiplier tube.

called a *multiplier* because one photon at the input has resulted in millions of electrons at the output.

The photomultiplier tube is used in nuclear detectors to detect the presence of gamma rays emitted from radioactive nuclei, which we will study in Chapter 30. It is also used in astronomy in a technique called *photoelectric photometry.* In this technique, the light collected by a telescope from a single star is allowed to fall on a photomultiplier tube for a time interval. The tube measures the total light energy during the time interval, which can then be converted to a luminosity of the star.

The photomultiplier tube is being replaced in many astronomical observations with a *charge-coupled device* (CCD), which is the same device that is used in a digital camera. In this device, an array of pixels are formed on the silicon surface of an integrated circuit. When the surface is exposed to light from an astronomical scene through a telescope or a terrestrial scene through a digital camera, electrons generated by the photoelectric effect are caught in "traps" beneath the surface. The number of electrons is related to the intensity of the light striking the surface. A signal processor measures the number of electrons associated with each pixel and converts this information into a digital code that a computer can use to reconstruct and display the scene.

The *electron bombardment CCD camera* allows higher sensitivity than a conventional CCD. In this device, electrons ejected from a photocathode by the photoelectric effect are accelerated through a high voltage before striking a CCD array. The higher energy of the electrons results in a very sensitive detector of low-intensity radiation.

The explanation of the photoelectric effect with a quantum model, combined with Planck's quantum model for blackbody radiation, laid a strong foundation for further investigation into quantum physics. In the next section, we present a third experimental result that provides further strong evidence of the quantum nature of light.

QUICK QUIZ 28.3 Consider one of the curves in Active Figure 28.8. Suppose the intensity of the incident light is held fixed but its frequency is increased. The stopping potential in Active Figure 28.8 **(a)** remains fixed, **(b)** moves to the right, or **(c)** moves to the left.

QUICK QUIZ 28.4 Suppose classical physicists had come up with the idea of predicting the appearance of a plot of K_{max} versus f as in Active Figure 28.9. What would their expected plot look like, based on the wave model for light?

INTERACTIVE EXAMPLE 28.3 **The Photoelectric Effect for Sodium**

A sodium surface is illuminated with light of wavelength 300 nm. The work function for sodium metal is 2.46 eV.

A Find the maximum kinetic energy of the ejected photoelectrons.

Solution The energy of photons in the illuminating light beam is

$$E = hf = \frac{hc}{\lambda} = \frac{1\,240 \text{ eV}\cdot\text{nm}}{300 \text{ nm}} = 4.13 \text{ eV}$$

Using Equation 28.5 gives

$$K_{max} = hf - \phi = 4.13 \text{ eV} - 2.46 \text{ eV} = 1.67 \text{ eV}$$

B Find the cutoff wavelength for sodium.

Solution The cutoff wavelength can be calculated from Equation 28.7:

$$\lambda_c = \frac{hc}{\phi} = \frac{1\,240 \text{ eV}\cdot\text{nm}}{2.46 \text{ eV}} = 504 \text{ nm}$$

This wavelength is in the yellow–green region of the visible spectrum.

PhysicsNow™ Investigate the photoelectric effect for different materials and different wavelengths of light by logging into PhysicsNow at **www.pop4e.com** and going to Interactive Example 28.3.

28.3 THE COMPTON EFFECT

In 1919, Einstein proposed that a photon of energy E carries a momentum equal to $E/c = hf/c$. In 1923, Arthur Holly Compton carried Einstein's idea of photon momentum further with the **Compton effect.**

Prior to 1922, Compton and his coworkers had accumulated evidence that showed that the classical wave theory of light failed to explain the scattering of x-rays from electrons. According to classical theory, incident electromagnetic waves of frequency f_0 should have two effects: (1) the electrons should accelerate in the direction of propagation of the x-ray by radiation pressure (see Section 24.6), and (2) the oscillating electric field should set the electrons into oscillation at the apparent frequency of the radiation as detected by the moving electron. The apparent frequency detected by the electron differs from f_0 due to the Doppler effect (see Section 24.3) because the electron absorbs as a moving particle. The electron then re-radiates as a moving particle, exhibiting another Doppler shift in the frequency of emitted radiation.

Because different electrons move at different speeds, depending on the amount of energy absorbed from the electromagnetic waves, the scattered wave frequency at a given angle should show a distribution of Doppler-shifted values. Contrary to this prediction, Compton's experiment showed that, at a given angle, only *one* frequency of radiation was observed that was different from that of the incident radiation. Compton and his coworkers realized that the scattering of x-ray photons from electrons could be explained by treating photons as point-like particles with energy hf and momentum hf/c and by assuming that the energy and momentum of the isolated system of the photon and the electron are conserved in a collision. By doing so, Compton was adopting a particle model for something that was well known as a wave, as had Einstein in his explanation of the photoelectric effect. Figure 28.11 shows the quantum picture of the exchange of momentum and energy between an individual x-ray photon and an electron. In the classical model, the electron is

(Courtesy of AIP Niels Bohr Library)

ARTHUR HOLLY COMPTON (1892–1962)

Compton measured and explained the effect named for him at the University of Chicago in 1923 and shared the 1927 Nobel Prize in Physics. He went on to demonstrate that cosmic rays are charged particles and to direct research on producing plutonium for nuclear weapons.

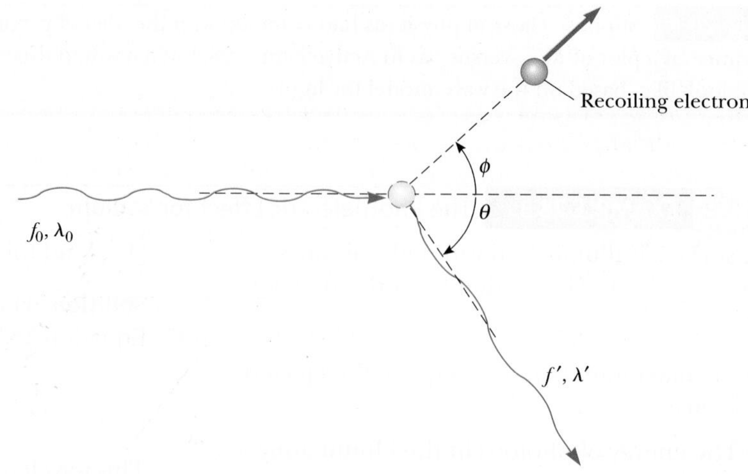

FIGURE 28.11 The quantum model for x-ray scattering from an electron. The collision of the photon with the electron displays the particle-like nature of the photon.

pushed along the direction of propagation of the incident x-ray by radiation pressure. In the quantum model in Figure 28.11, the electron is scattered through an angle ϕ with respect to this direction as if it were a billiard-ball type collision.

Figure 28.12 is a schematic diagram of the apparatus used by Compton. In his original experiment, Compton measured how scattered x-ray intensity depends on wavelength at various scattering angles. The incident beam consisted of monochromatic x-rays of wavelength $\lambda_0 = 0.071$ nm. The experimental plots of intensity versus wavelength obtained by Compton for four scattering angles are shown in Figure 28.13. They show two peaks, one at λ_0 and the other at a longer wavelength λ'. The peak at λ_0 is caused by x-rays scattered from electrons that are tightly bound to the target atoms, and the shifted peak at λ' is caused by x-rays scattered from free electrons in the target. In his analysis, Compton predicted that the shifted peak should depend on scattering angle θ as

■ **Compton shift equation**

$$\lambda' - \lambda_0 = \frac{h}{m_e c}(1 - \cos\theta) \qquad [28.8]$$

In this expression, known as the **Compton shift equation,** m_e is the mass of the electron; $h/m_e c$ is called the **Compton wavelength** λ_C for the electron and has the value

■ **Compton wavelength**

$$\lambda_C = \frac{h}{m_e c} = 0.002\,43 \text{ nm} \qquad [28.9]$$

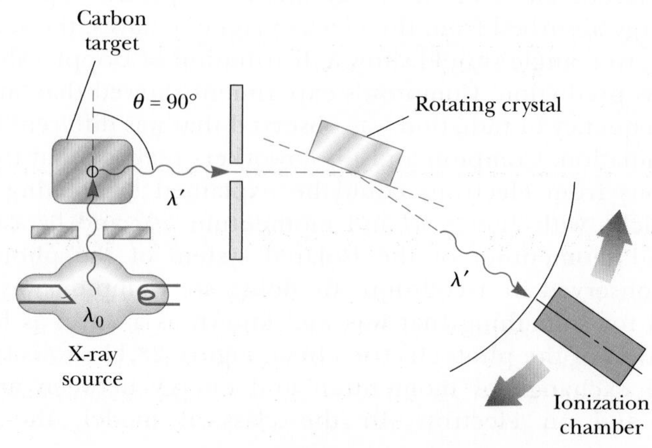

FIGURE 28.12 Schematic diagram of Compton's apparatus. Photons are scattered through 90° from a carbon target. The wavelength is measured with a rotating crystal spectrometer using Bragg's law (Section 27.9).

Compton's measurements were in excellent agreement with the predictions of Equation 28.8. They were the first experimental results to convince most physicists of the fundamental validity of the quantum theory!

The Compton effect should be kept in mind by x-ray technicians working in hospitals and radiology laboratories. X-rays directed into the patient's body are Compton scattered by electrons in the body in all directions. Equation 28.8 shows that the scattered wavelength is still well within the x-ray region so that these scattered x-rays can damage human tissue. In general, technicians operate the x-ray machine from behind an absorbing wall to avoid exposure to the scattered x-rays. Furthermore, when dental x-rays are taken, a lead apron is placed over the patient to reduce the absorption of scattered x-rays by other parts of the patient's body.

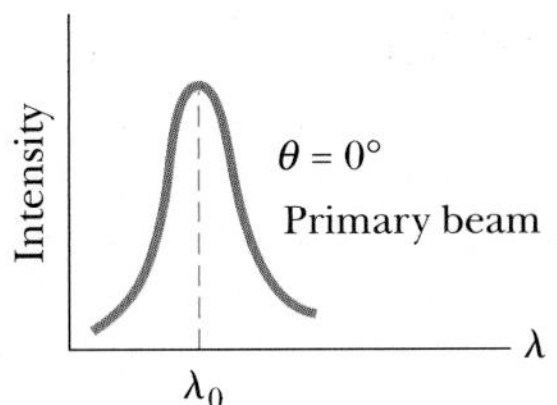

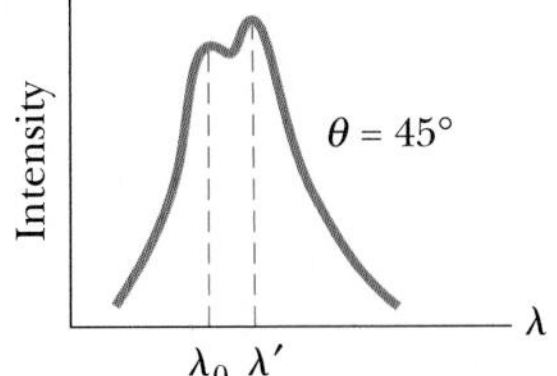

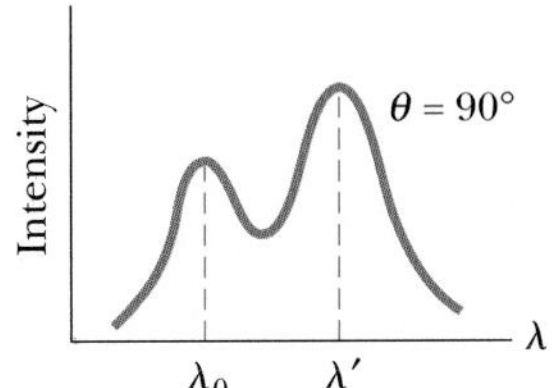

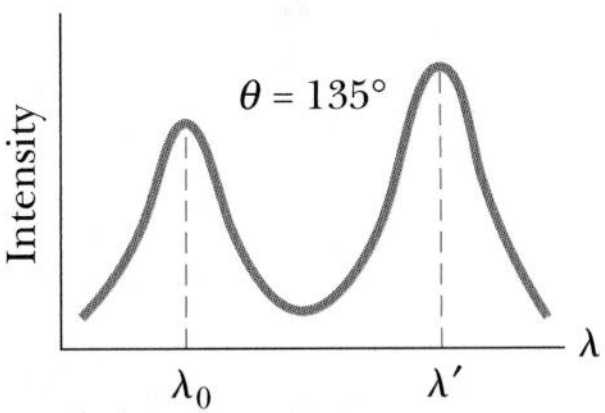

FIGURE 28.13 Scattered x-ray intensity versus wavelength for Compton scattering at $\theta = 0°$, 45°, 90°, and 135°.

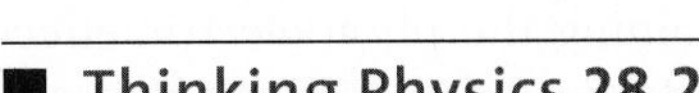

■ Thinking Physics 28.2

The Compton effect involves a change in wavelength as photons are scattered through different angles. Suppose we illuminate a piece of material with a beam of light and then view the material from different angles relative to the beam of light. Will we see a *color change* corresponding to the change in wavelength of the scattered light?

Reasoning Visible light scattered by the material undergoes a change in wavelength, but the change is far too small to detect as a color change. The largest possible change in wavelength, at 180° scattering, is twice the Compton wavelength, about 0.005 nm, which represents a change of less than 0.001% of the wavelength of red light. The Compton effect is only detectable for wavelengths that are very short to begin with, so the Compton wavelength is an appreciable fraction of the incident wavelength. As a result, the usual radiation for observing the Compton effect is in the x-ray range of the electromagnetic spectrum. ■

INTERACTIVE EXAMPLE 28.4 Compton Scattering at 45°

X-rays of wavelength $\lambda_0 = 0.200\ 000$ nm are scattered from a block of material. The scattered x-rays are observed at an angle of 45.0° to the incident beam. Calculate the wavelength of the x-rays scattered at this angle.

Solution The shift in wavelength of the scattered x-rays is given by Equation 28.8:

$$\Delta\lambda = \frac{h}{m_e c}(1 - \cos\theta)$$

$$= \frac{6.626 \times 10^{-34}\ \text{J}\cdot\text{s}}{(9.11 \times 10^{-31}\ \text{kg})(3.00 \times 10^{8}\ \text{m/s})}(1 - \cos 45°)$$

$$= 7.10 \times 10^{-13}\ \text{m} = 0.000\ 710\ \text{nm}$$

Hence, the wavelength of the scattered x-rays at this angle is

$$\lambda' = \Delta\lambda + \lambda_0 = \boxed{0.200\ 710\ \text{nm}}$$

Physics Now™ Study Compton scattering for different angles by logging into PhysicsNow at **www.pop4e.com** and going to Interactive Example 28.4.

28.4 PHOTONS AND ELECTROMAGNETIC WAVES

The agreement between experimental measurements and theoretical predictions based on quantum models for phenomena such as the photoelectric effect and the Compton effect offers clear evidence that when light and matter interact, the light behaves as if it were composed of particles with energy *hf* and momentum *hf/c*. An obvious question at this point is, "How can light be considered a photon when it exhibits wave-like properties?" We describe light in terms of photons having energy and momentum, which are parameters of the particle model. Remember, however, that light and other electromagnetic waves exhibit interference and diffraction effects, which are consistent only with the wave model.

Which model is correct? Is light a wave or a particle? The answer depends on the phenomenon being observed. Some experiments can be explained better, or solely, with the photon model, whereas others are best described, or can only be described, with a wave model. The end result is that **we must accept both models and admit that the true nature of light is not describable in terms of any single classical picture.** Hence, **light has a dual nature in that it exhibits both wave and particle characteristics.** You should recognize, however, that the same beam of light that can eject photoelectrons from a metal can also be diffracted by a grating. In other words, the **particle model and the wave model of light complement each other.**

The success of the particle model of light in explaining the photoelectric effect and the Compton effect raises many other questions. Because a photon is a particle, what is the meaning of its "frequency" and "wavelength," and which determines its energy and momentum? Is light in some sense simultaneously a wave and a particle? Although photons have no rest energy, can some simple expression describe the effective mass of a "moving" photon? If a "moving" photon has mass, do photons experience gravitational attraction? What is the spatial extent of a photon, and how does an electron absorb or scatter one photon? Some of these questions can be answered, but others demand a view of atomic processes that is too pictorial and literal. Furthermore, many of these questions stem from classical analogies such as colliding billiard balls and water waves breaking on a shore. Quantum mechanics gives light a more fluid and flexible nature by treating the particle model and wave model of light as both necessary and complementary. Neither model can be used exclusively to describe all properties of light. A complete understanding of the observed behavior of light can be attained only if the two models are combined in a complementary manner. Before discussing this combination in more detail, we now turn our attention from electromagnetic waves to the behavior of entities that we have called particles.

28.5 THE WAVE PROPERTIES OF PARTICLES

We feel quite comfortable in adopting a particle model for matter because we have studied such concepts as conservation of energy and momentum for particles as well as extended objects. It might therefore be even more difficult to accept that *matter* also has a dual nature!

In 1923, in his doctoral dissertation, Louis Victor de Broglie postulated that **because photons have wave and particle characteristics, perhaps all forms of matter have wave as well as particle properties.** This postulate was a highly revolutionary idea with no experimental confirmation at that time. According to de Broglie, an electron in motion exhibits both wave and particle characteristics. De Broglie explained the source of this assertion in his 1929 Nobel Prize acceptance speech:

> On the one hand the quantum theory of light cannot be considered satisfactory since it defines the energy of a light corpuscle by the equation $E = hf$ containing the frequency f. Now a purely corpuscular theory contains nothing that enables us to define a frequency; for this reason alone, therefore, we are compelled, in the case of light, to introduce the idea of a corpuscle and that of periodicity simultaneously. On the other hand, determination of the stable motion of electrons in the atom introduces integers, and up to this point the only phenomena involving integers in physics were those of interference and of normal modes of vibration. This fact suggested to me the idea that electrons too could not be considered simply as corpuscles, but that periodicity must be assigned to them also.

(AIP Niels Bohr Library)

LOUIS DE BROGLIE (1892–1987)

A French physicist, de Broglie was awarded the Nobel Prize in Physics in 1929 for his prediction of the wave nature of electrons.

In Chapter 9, we found that the relationship between energy and momentum for a photon is $p = E/c$. We also know from Equation 28.3 that the energy of a photon is $E = hf = hc/\lambda$. Therefore, the momentum of a photon can be expressed as

$$p = \frac{E}{c} = \frac{hf}{c} = \frac{hc}{c\lambda} = \frac{h}{\lambda}$$

From this equation, we see that the photon wavelength can be specified by its momentum: $\lambda = h/p$. De Broglie suggested that material particles of momentum p should also have wave properties and a corresponding wavelength. Because the magnitude of the momentum of a nonrelativistic particle of mass m and speed v is $p = mv$, the **de Broglie wavelength** of that particle is[3]

$$\lambda = \frac{h}{p} = \frac{h}{mv} \qquad [28.10]$$

■ De Broglie wavelength of a particle

Furthermore, in analogy with photons, de Broglie postulated that particles obey the Einstein relation $E = hf$, so the frequency of a particle is

$$f = \frac{E}{h} \qquad [28.11]$$

■ Frequency of a particle

The dual nature of matter is apparent in these last two equations because each contains both particle concepts (p and E) and wave concepts (λ and f). That these relationships are established experimentally for photons makes the de Broglie hypothesis that much easier to accept.

The Davisson–Germer Experiment

De Broglie's proposal that any kind of particle exhibits both wave and particle properties was first regarded as pure speculation. If particles such as electrons had wave-like properties, under the correct conditions they should exhibit diffraction effects. In 1927, three years after de Broglie published his work, C. J. Davisson and L. H. Germer of the United States succeeded in measuring the wavelength of electrons. Their important discovery provided the first experimental confirmation of the wave nature of particles proposed by de Broglie.

Interestingly, the intent of the initial Davisson–Germer experiment was not to confirm the de Broglie hypothesis. In fact, the discovery was made by accident (as is often the case). The experiment involved the scattering of low-energy electrons ($\approx$ 54 eV) projected toward a nickel target in a vacuum. During one experiment, the nickel surface was badly oxidized because of an accidental break in the vacuum system. After the nickel target was heated in a flowing stream of hydrogen to remove the oxide coating, electrons scattered by it exhibited intensity maxima and minima at specific angles. The experimenters finally realized that the nickel had formed large crystal regions on heating and that the regularly spaced planes of atoms in the crystalline regions served as a diffraction grating (Section 27.8) for electrons.

Shortly thereafter, Davisson and Germer performed more extensive diffraction measurements on electrons scattered from single-crystal targets. Their results showed conclusively the wave nature of electrons and confirmed the de Broglie relation $p = h/\lambda$. A year later in 1928, G. P. Thomson of Scotland observed electron diffraction patterns by passing electrons through very thin gold foils. Diffraction patterns have since been observed for helium atoms, hydrogen atoms, and neutrons. Hence, the wave nature of particles has been established in a variety of ways.

PITFALL PREVENTION 28.3

What's Waving? If particles have wave properties, what's waving? You are familiar with waves on strings, which are very concrete. Sound waves are more abstract, but you are likely comfortable with them. Electromagnetic waves are even more abstract, but at least they can be described in terms of physical variables, electric and magnetic fields. Waves associated with particles are very abstract and cannot be associated with a physical variable. Later in this chapter, we will describe the wave associated with a particle in terms of probability.

[3]The de Broglie wavelength for a particle moving at any speed v, including relativistic speeds, is $\lambda = h/\gamma mv$, where $\gamma = (1 - v^2/c^2)^{-1/2}$.

QUICK QUIZ 28.5 An electron and a proton both moving at nonrelativistic speeds have the same de Broglie wavelength. Which of the following are also the same for the two particles? **(a)** speed **(b)** kinetic energy **(c)** momentum **(d)** frequency

QUICK QUIZ 28.6 We have discussed two wavelengths associated with the electron, the Compton wavelength and the de Broglie wavelength. Which is an actual *physical* wavelength associated with the electron? **(a)** the Compton wavelength **(b)** the de Broglie wavelength **(c)** both wavelengths **(d)** neither wavelength

EXAMPLE 28.5 The Wavelength of an Electron

Calculate the de Broglie wavelength for an electron ($m_e = 9.11 \times 10^{-31}$ kg) moving with a speed of 1.00×10^7 m/s.

Solution Equation 28.10 gives

$$\lambda = \frac{h}{m_e v}$$

$$= \frac{6.626 \times 10^{-34}\ \text{J}\cdot\text{s}}{(9.11 \times 10^{-31}\ \text{kg})(1.00 \times 10^7\ \text{m/s})}$$

$$= 7.27 \times 10^{-11}\ \text{m}$$

This wavelength corresponds to that of typical x-rays in the electromagnetic spectrum. Furthermore, note that the calculated wavelength is on the order of the spacing of atoms in a crystalline substance such as sodium chloride.

EXAMPLE 28.6 The Wavelength of a Rock

A rock of mass 50.0 g is thrown with a speed of 40.0 m/s. What is its de Broglie wavelength?

Solution From Equation 28.10, we have

$$\lambda = \frac{h}{mv} = \frac{6.626 \times 10^{-34}\ \text{J}\cdot\text{s}}{(50.0 \times 10^{-3}\ \text{kg})(40.0\ \text{m/s})}$$

$$= 3.31 \times 10^{-34}\ \text{m}$$

This wavelength is much smaller than any aperture through which the rock could possibly pass. Thus, we could not observe diffraction effects, and as a result the wave properties of large-scale objects cannot be observed.

EXAMPLE 28.7 An Accelerated Charge

A particle of charge q and mass m is accelerated from rest through a potential difference ΔV. Assuming that the particle moves with a nonrelativistic speed, find its de Broglie wavelength.

Solution We apply the isolated system model for the particle and the electric field associated with the potential difference. The total energy of the system must be conserved:

$$K_f + U_f = K_i + U_i$$

where the initial configuration of the system refers to the instant the particle begins to move from rest and the final configuration is when the particle reaches its final speed after accelerating through the potential difference ΔV. If we define the electric potential energy of the system in the initial configuration to be zero, we have

$$\tfrac{1}{2}mv^2 + q(-\Delta V) = 0 + 0$$

where the negative sign indicates that a positive charge accelerates in the direction of decreasing potential. Because $p = mv$, we can express this equation in the form

$$\frac{p^2}{2m} = q\,\Delta V \qquad \text{or} \qquad p = \sqrt{2mq\,\Delta V}$$

Substituting this expression for p in the de Broglie relation $\lambda = h/p$ gives

$$\lambda = \frac{h}{\sqrt{2mq\,\Delta V}}$$

The Electron Microscope

The electron microscope

A practical device that relies on the wave characteristics of electrons is the **electron microscope.** A *transmission* electron microscope, used for viewing flat, thin samples, is shown in Figure 28.14. In many respects, it is similar to an optical microscope, but the electron microscope has a much greater resolving power because it can accelerate electrons to very high kinetic energies, giving them very short wavelengths. No microscope can resolve details that are significantly smaller than the wavelength of the waves used to illuminate the object. Typically, the wavelengths of electrons are about 100 times shorter than those of the visible light used in optical microscopes. As a result, an electron microscope with ideal lenses would be able to distinguish details about 100 times smaller than those distinguished by an optical microscope. (Electromagnetic radiation of the same wavelength as the electrons in an electron microscope is in the x-ray region of the spectrum.)

The electron beam in an electron microscope is controlled by electrostatic or magnetic deflection, which acts on the electrons to focus the beam and form an image. Rather than examining the image through an eyepiece as in an optical microscope, the viewer looks at an image formed on a monitor or other type of display screen. The photograph at the beginning of this chapter shows the amazing detail available with an electron microscope.

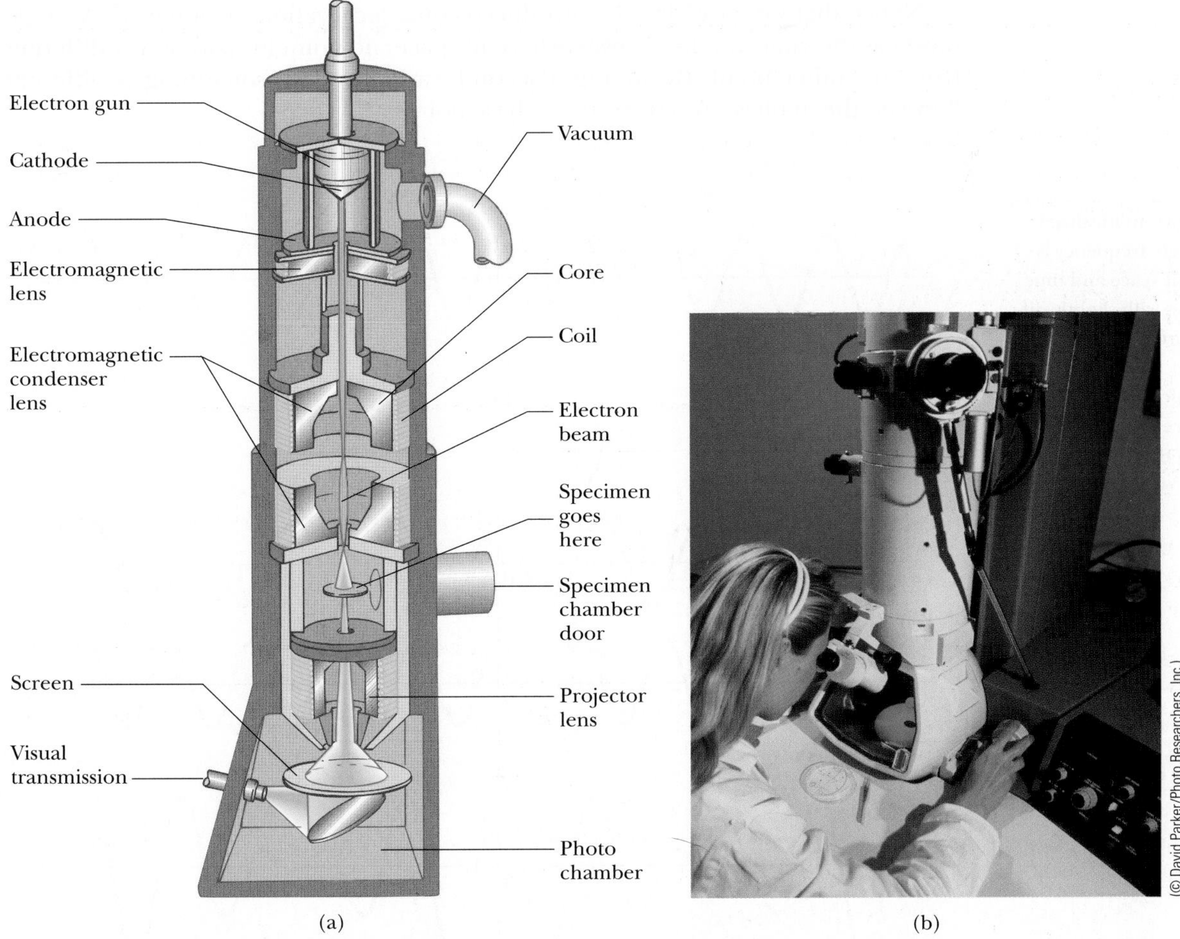

FIGURE 28.14 (a) Diagram of a transmission electron microscope for viewing a thinly sectioned sample. The "lenses" that control the electron beam are magnetic deflection coils. (b) An electron microscope in use.

28.6 THE QUANTUM PARTICLE

The discussions presented in previous sections may be quite disturbing because we considered the particle and wave models to be distinct in earlier chapters. The notion that both light and material particles have both particle and wave properties does not fit with this distinction. We have experimental evidence, however, that this dual nature is just what we must accept. This acceptance leads to a new simplification model, the **quantum particle model.** In this model, entities have both particle and wave characteristics, and we must choose one appropriate behavior—particle or wave—to understand a particular phenomenon.

In this section, we shall investigate this model, which might bring you more comfort with this idea. As we shall demonstrate, we can construct from waves an entity that exhibits properties of a particle.

Let us first review the characteristics of ideal particles and waves. An ideal particle has zero size. As mentioned in Section 13.2, an ideal wave has a single frequency and is infinitely long. Therefore, an essential identifying feature of a particle that differentiates it from a wave is that it is *localized* in space. Let us show that we can build a localized entity from infinitely long waves. Imagine drawing one wave along the x axis, with one of its crests located at $x = 0$, as in Figure 28.15a. Now, draw a second wave, of the same amplitude but a different frequency, with one of its crests also at $x = 0$. The result of the superposition of these two waves is a *beat* because the waves are alternately in phase and out of phase. (Beats were discussed in Section 14.6.) Figure 28.15b shows the results of superposing these two waves.

Notice that we have already introduced some localization by doing so. A single wave has the same amplitude everywhere in space; no point in space is any different from any other point. By adding a second wave, however, something is different between the in-phase and the out-of-phase points.

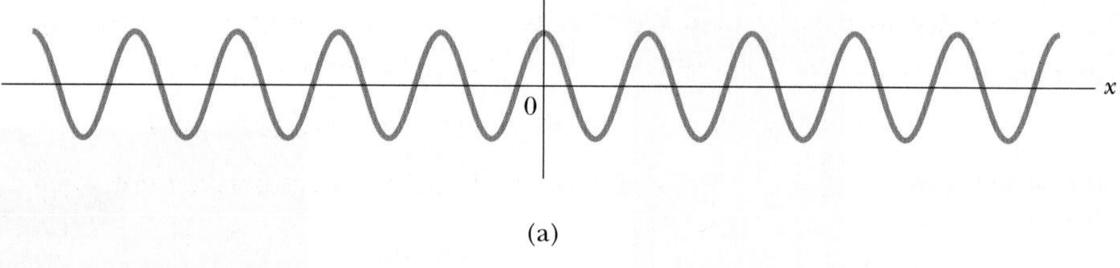

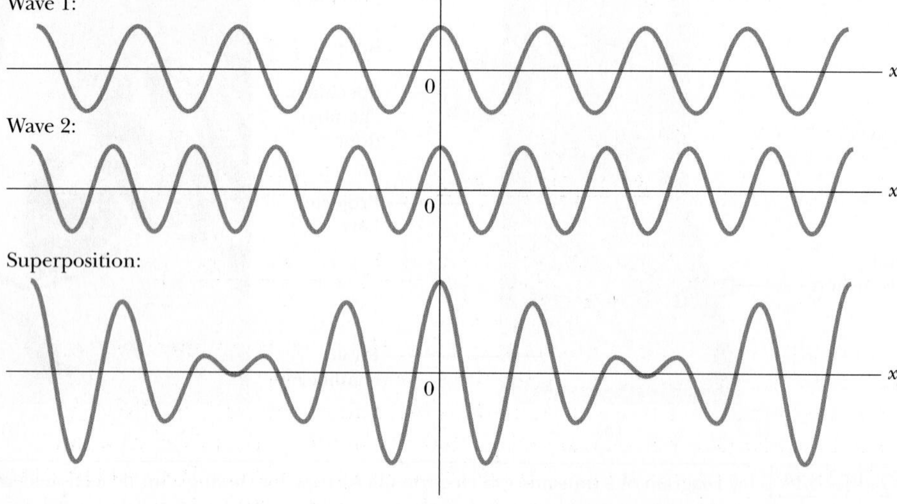

FIGURE 28.15 (a) An idealized wave of an exact single frequency is the same throughout space and time. (b) If two ideal waves with slightly different frequencies are combined, beats result (Section 14.6). The regions of space at which there is constructive interference are different from those at which there is destructive interference.

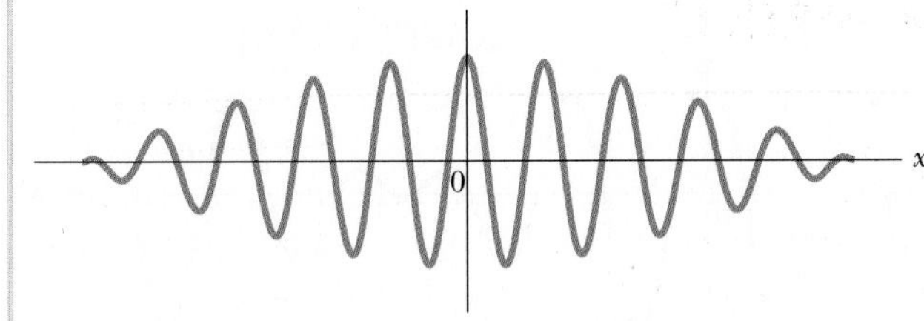

ACTIVE FIGURE 28.16 If a large number of waves are combined, the result is a wave packet, which represents a particle.

Physics Now™ Choose the number of waves to add together and observe the resulting wave packet by logging into PhysicsNow at **www.pop4e.com** and going to Active Figure 28.16.

Now imagine that more and more waves are added to our original two, each new wave having a new frequency. Each new wave is added so that one of its crests is at $x = 0$. The result at $x = 0$ is that all the waves add constructively. When we consider a large number of waves, the probability of a positive value of a wave function at any point x is equal to the probability of a negative value and destructive interference occurs *everywhere* except near $x = 0$, where we superposed all the crests. The result is shown in Active Figure 28.16. The small region of constructive interference is called a **wave packet.** This wave packet is a localized region of space that is different from all other regions, because the result of the superposition of the waves everywhere else is zero. We can identify the wave packet as a particle because it has the localized nature of what we have come to recognize as a particle!

The localized nature of this entity is the *only* characteristic of a particle that was generated with this process. We have not addressed how the wave packet might achieve such particle characteristics as mass, electric charge, spin, and so on. Therefore, you may not be completely convinced that we have built a particle. As further evidence that the wave packet can represent the particle, let us show that the wave packet has another characteristic of a particle.

Let us return to our combination of only two waves so as to make the mathematical representation simple. Consider two waves with equal amplitudes but different frequencies f_1 and f_2. We can represent the waves mathematically as

$$y_1 = A\cos(k_1x - \omega_1t) \quad \text{and} \quad y_2 = A\cos(k_2x - \omega_2t)$$

where, as in Chapter 13, $\omega = 2\pi f$ and $k = 2\pi/\lambda$. Using the superposition principle, we add the waves:

$$y = y_1 + y_2 = A\cos(k_1x - \omega_1t) + A\cos(k_2x - \omega_2t)$$

It is convenient to write this expression in a form that uses the trigonometric identity

$$\cos a + \cos b = 2\cos\left(\frac{a-b}{2}\right)\cos\left(\frac{a+b}{2}\right)$$

Letting $a = k_1x - \omega_1t$ and $b = k_2x - \omega_2t$, we find that

$$y = 2A\cos\left[\frac{(k_1x - \omega_1t) - (k_2x - \omega_2t)}{2}\right]\cos\left[\frac{(k_1x - \omega_1t) + (k_2x - \omega_2t)}{2}\right]$$

$$= \left[2A\cos\left(\frac{\Delta k}{2}x - \frac{\Delta\omega}{2}t\right)\right]\cos\left(\frac{k_1+k_2}{2}x - \frac{\omega_1+\omega_2}{2}t\right) \qquad [28.12]$$

The second cosine factor represents a wave with a wave number and frequency equal to the averages of the values for the individual waves.

The factor in brackets represents the envelope of the wave as shown in Active Figure 28.17. Notice that this factor also has the mathematical form of a wave. **This envelope of the combination can travel through space with a different speed than the individual waves.** As an extreme example of this possibility, imagine combining two identical waves moving in opposite directions. The two waves move with the

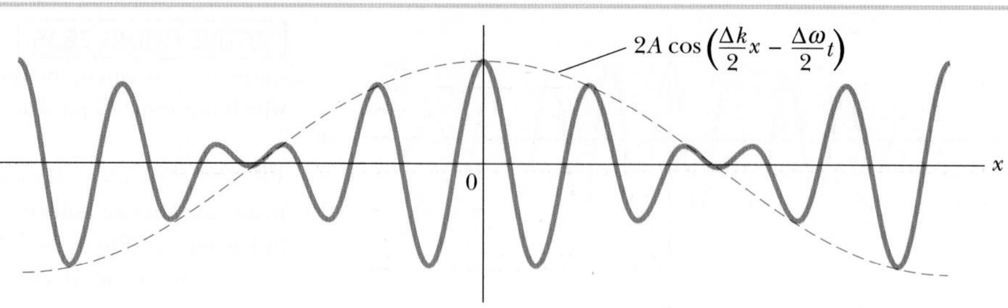

ACTIVE FIGURE 28.17 The beat pattern of Figure 28.15b, with an envelope function (blue curve) superimposed.

Physics Now™ Observe the movement of the waves and the envelope by logging into PhysicsNow at **www.pop4e.com** and going to Active Figure 28.17.

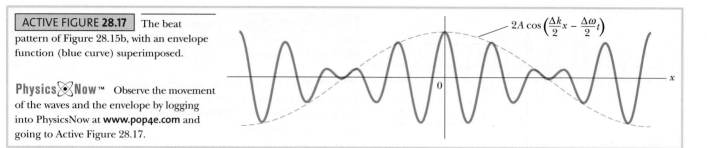

same speed, but the envelope has a speed of *zero* because we have built a standing wave, which we studied in Section 14.3.

For an individual wave, the speed is given by Equation 13.10:

■ Phase speed for a wave

$$v_{\text{phase}} = \frac{\omega}{k}$$

It is called the **phase speed** because it is the rate of advance of a crest on a single wave, which is a point of fixed phase. This equation can be interpreted as the following: the phase speed of a wave is the ratio of the coefficient of the time variable t to the coefficient of the space variable x in the equation for the wave, $y = A\cos(kx - \omega t)$.

The factor in brackets in Equation 28.12 is of the form of a wave, so it moves with a speed given by this same ratio:

$$v_g = \frac{\text{coefficient of time variable } t}{\text{coefficient of space variable } x} = \frac{(\Delta\omega/2)}{(\Delta k/2)} = \frac{\Delta\omega}{\Delta k}$$

The subscript g on the speed indicates that it is commonly called the **group speed,** or the speed of the wave packet (the *group* of waves) that we have built. We have generated this expression for a simple addition of two waves. For a superposition of a very large number of waves to form a wave packet, this ratio becomes a derivative:

■ Group speed for a wave packet

$$v_g = \frac{d\omega}{dk} \qquad [28.13]$$

Let us multiply the numerator and the denominator by $\hbar$, where $\hbar = h/2\pi$:

$$v_g = \frac{\hbar\, d\omega}{\hbar\, dk} = \frac{d(\hbar\omega)}{d(\hbar k)} \qquad [28.14]$$

We look at the terms in the parentheses in the numerator and denominator in this equation separately. For the numerator

$$\hbar\omega = \frac{h}{2\pi}(2\pi f) = hf = E$$

For the denominator,

$$\hbar k = \frac{h}{2\pi}\left(\frac{2\pi}{\lambda}\right) = \frac{h}{\lambda} = p$$

Therefore, Equation 28.14 can be written as

$$v_g = \frac{d(\hbar\omega)}{d(\hbar k)} = \frac{dE}{dp} \qquad [28.15]$$

Because we are exploring the possibility that the envelope of the combined waves represents the particle, consider a free particle moving with a speed u that is small compared with that of light. The energy of the particle is its kinetic energy:

$$E = \frac{1}{2}mu^2 = \frac{p^2}{2m}$$

Differentiating this equation with respect to p, where $p = mu$, gives

$$v_g = \frac{dE}{dp} = \frac{d}{dp}\left(\frac{p^2}{2m}\right) = \frac{1}{2m}(2p) = u \qquad [28.16]$$

Therefore, the group speed of the wave packet is identical to the speed of the particle that it is modeled to represent! Thus, we have further confidence that the wave packet is a reasonable way to build a particle.

QUICK QUIZ 28.7 As an analogy to wave packets, consider an "automobile packet" that occurs near the scene of an accident on a freeway. The phase speed is analogous to the speed of individual automobiles as they move through the backup caused by the accident. The group speed can be identified as the speed of the leading edge of the packet of cars. For the automobile packet, is the group speed **(a)** the same as the phase speed, **(b)** less than the phase speed, or **(c)** greater than the phase speed?

28.7 THE DOUBLE-SLIT EXPERIMENT REVISITED

One way to crystallize our ideas about the electron's wave–particle duality is to consider a hypothetical experiment in which electrons are fired at a double slit. Consider a parallel beam of monoenergetic electrons that is incident on a double slit as in Figure 28.18. We shall assume that the slit widths are small compared with the electron wavelength, so we need not worry about diffraction maxima and minima as discussed for light in Section 27.6. An electron detector is positioned far from the slits at a distance much greater than the separation distance d of the slits. If the detector collects electrons for a long enough time interval, one finds a typical wave interference pattern for the counts per minute, or probability of arrival of

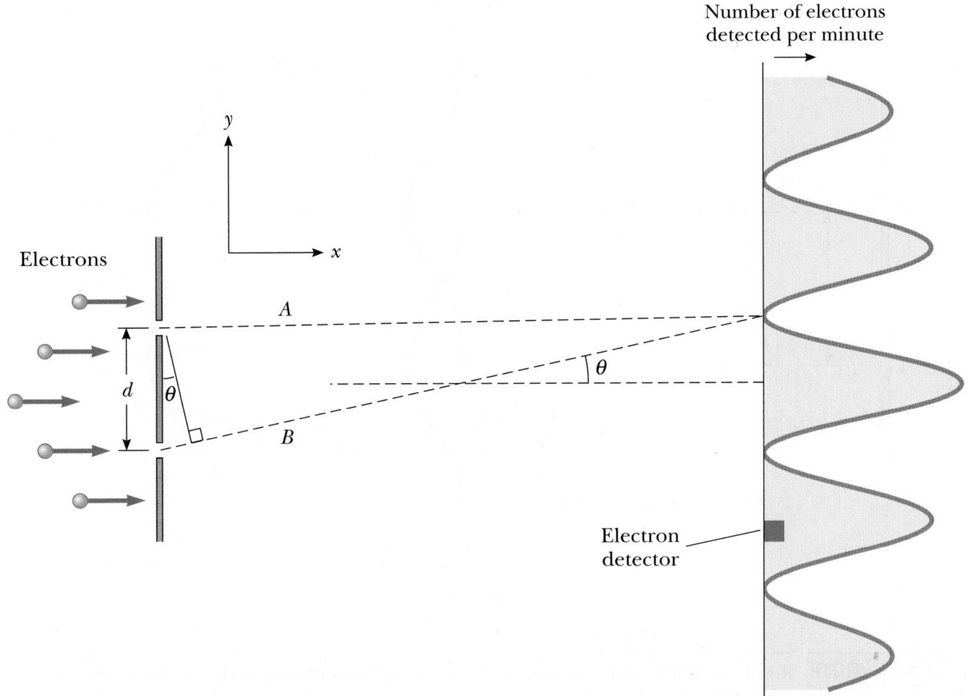

FIGURE 28.18 Electron interference. The slit separation d is much greater than the individual slit widths and much less than the distance between the slit and the detector. The electron detector is movable along the y direction in the drawing and so can detect electrons diffracted at different values of θ. The detector acts like the "viewing screen" of Young's double-slit experiment with light discussed in Chapter 27.

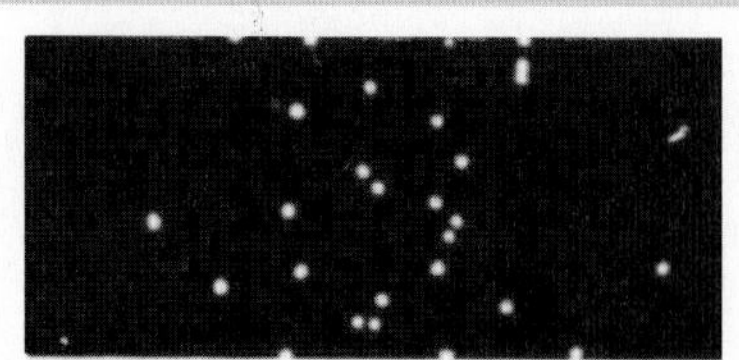

(a) After 28 electrons

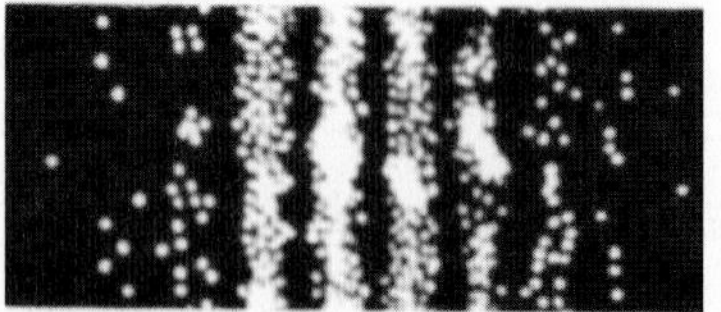

(b) After 1 000 electrons

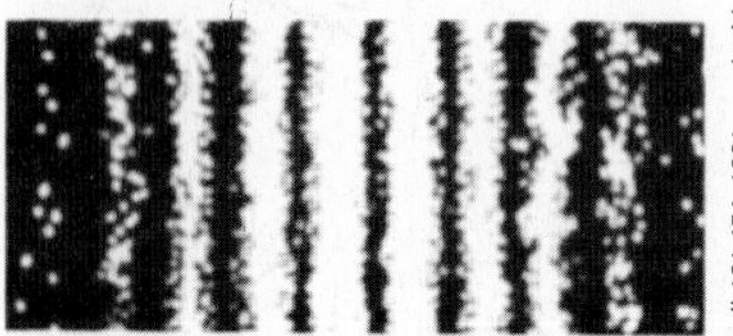

(c) After 10 000 electrons

(d) Two-slit electron pattern

(a–d, From C. Jönsson, *Zeitschrift fur Physik* 161:454, 1961; used with permission Springer Verlag)

ACTIVE FIGURE 28.19
(a, b, c) Computer-simulated interference patterns for a beam of electrons incident on a double slit. (d) Photograph of a double-slit interference pattern produced by electrons.

Physics Now™ By logging into PhysicsNow at **www.pop4e.com** and going to Active Figure 28.19, you can watch the interference pattern develop over time and see how it is destroyed by the action of keeping track of which slit an electron goes through.

electrons. Such an interference pattern would not be expected if the electrons behaved as classical particles. It is clear that electrons are interfering, which is a distinct wave-like behavior.

If the experiment is carried out at lower electron beam intensities, the interference pattern is still observed if the time interval of the measurement is sufficiently long as illustrated by the computer-simulated patterns in Active Figure 28.19. Note that the interference pattern becomes clearer as the number of electrons reaching the screen increases.

If one imagines a single electron in the beam producing in-phase "wavelets" as it reaches one of the slits, the waves in interference model (Section 27.3) can be used to find the angular separation θ between the central probability maximum and its neighboring minimum. The minimum occurs when the path length difference between A and B is half a wavelength, or when

$$d \sin \theta = \frac{\lambda}{2}$$

Because an electron's wavelength is given by $\lambda = h/p_x$, we see that for small θ,

$$\sin \theta \approx \theta = \frac{h}{2p_x d}$$

Thus, the dual nature of the electron is clearly shown in this experiment. **The electrons are detected as particles at a localized spot at some instant of time, but the probability of arrival at that spot is determined by finding the intensity of two interfering waves.**

Let us now look at this experiment from another point of view. If one slit is covered during the experiment, a symmetric curve peaked around the center of the open slit is observed; it is the central maximum of the single-slit diffraction pattern. Plots of the counts per minute (probability of arrival of electrons) with the lower or upper slit closed are shown as blue curves in the central part of Figure 28.20.

If another experiment is now performed with slit 2 of Figure 28.20 blocked half of the time and then slit 1 blocked during the remaining time, the accumulated pattern of counts per minute shown by the blue curve on the right side of the

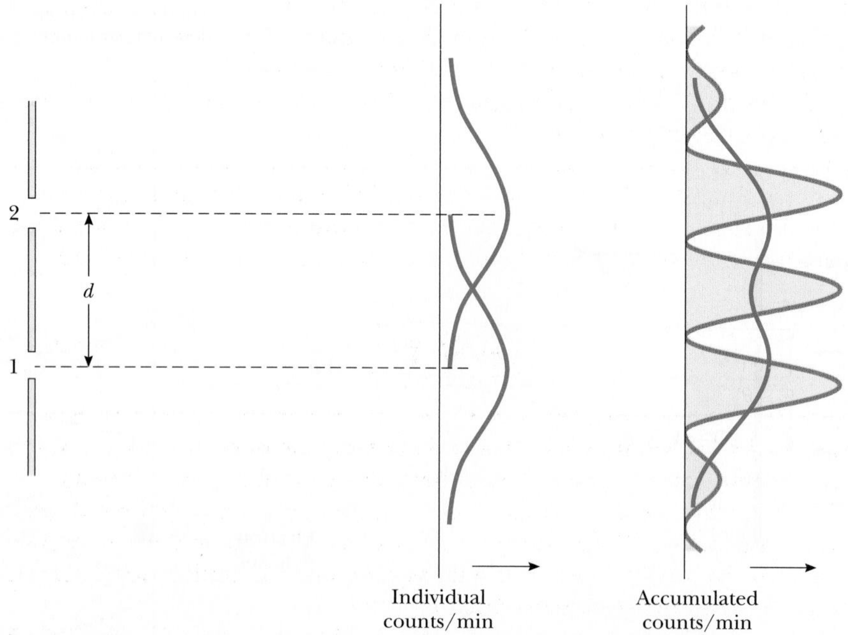

FIGURE 28.20 Results of the two-slit electron diffraction experiment with each slit closed half the time (blue). The result with both slits open is shown in brown.

figure is completely different from the case with both slits open (brown curve). A maximum probability of arrival of an electron at $\theta = 0$ no longer exists. In fact, **the interference pattern has been lost, and the accumulated result is simply the sum of the individual results.** When only one slit is open at a time, we know that the electron has the same localizability and indivisibility at the slits as we measure at the detector because the electron clearly goes through slit 1 or slit 2. Therefore, the total must be analyzed as the sum of those electrons that come through slit 1 and those that come through slit 2.

When both slits are open, it is tempting to assume that the electron goes through either slit 1 or slit 2 and that the counts per minute are again given by the combination of the single-slit distributions. We know, however, that the experimental results indicated by the brown interference pattern in Figure 28.20 contradict this assumption. Hence, our assumption that the electron is localized and goes through only one slit when both slits are open must be wrong (a painful conclusion!).

To interpret these results, we are forced to conclude that **an electron interacts with both slits simultaneously.** If we attempt to determine experimentally which slit the electron goes through, the act of measuring destroys the interference pattern. It is impossible to determine which slit the electron goes through. In effect, **we can say only that the electron passes through both slits!** The same arguments apply to photons.

If we restrict ourselves to a pure particle model, it is an uncomfortable notion that the electron can be present at both slits at once. From the quantum particle model, however, the particle can be considered to be built from waves that exist throughout space. Therefore, the wave components of the electron are present at both slits at the same time, and this model leads to a more comfortable interpretation of this experiment.

(Courtesy of the University of Hamburg)

WERNER HEISENBERG (1901–1976)

A German theoretical physicist, Heisenberg made many significant contributions to physics, including his famous uncertainty principle, for which he received the Nobel Prize in Physics in 1932; the development of an abstract model of quantum mechanics called matrix mechanics; the prediction of two forms of molecular hydrogen; and theoretical models of the nucleus.

28.8 THE UNCERTAINTY PRINCIPLE

Whenever one measures the position or velocity of a particle at any instant, experimental uncertainties are built into the measurements. According to classical mechanics, there is no fundamental barrier to an ultimate refinement of the apparatus or experimental procedures. In other words, it is possible, in principle, to make such measurements with arbitrarily small uncertainty. Quantum theory predicts, however, that **it is fundamentally impossible to make simultaneous measurements of a particle's position and momentum with infinite accuracy.**

In 1927, Werner Heisenberg introduced this notion, which is now known as the **Heisenberg uncertainty principle:**

If a measurement of the position of a particle is made with uncertainty Δx and a simultaneous measurement of its momentum is made with uncertainty Δp_x, the product of the two uncertainties can never be smaller than $\hbar/2$:

$$\Delta x \, \Delta p_x \geq \frac{\hbar}{2} \qquad [28.17]$$

■ **Uncertainty principle for momentum and position**

That is, **it is physically impossible to simultaneously measure the exact position and exact momentum of a particle.** Heisenberg was careful to point out that the inescapable uncertainties Δx and Δp_x do not arise from imperfections in practical measuring instruments. Furthermore, they do not arise due to any perturbation of the system that we might cause in the measuring process. Rather, **the uncertainties arise from the quantum structure of matter.**

To understand the uncertainty principle, consider a particle for which we know the wavelength *exactly*. According to the de Broglie relation $\lambda = h/p$, we would know the momentum to infinite accuracy, so $\Delta p_x = 0$.

PITFALL PREVENTION 28.4

THE UNCERTAINTY PRINCIPLE Some students incorrectly interpret the uncertainty principle as meaning that a measurement interferes with the system. For example, if an electron is observed in a hypothetical experiment using an optical microscope, the photon used to see the electron collides with it and makes it move, giving it an uncertainty in momentum. That is *not* the idea of the uncertainty principle. The uncertainty principle is independent of the measurement process and is grounded in the wave nature of matter.

In reality, as we have mentioned, a single-wavelength wave would exist throughout space. Any region along this wave is the same as any other region (see Fig. 28.15a). If we were to ask, "Where is the particle that this wave represents?" no special location in space along the wave could be identified with the particle because all points along the wave are the same. Therefore, we have *infinite* uncertainty in the position of the particle and we know nothing about where it is. Perfect knowledge of the momentum has cost us all information about the position.

In comparison, now consider a particle with some uncertainty in momentum so that a range of values of momentum are possible. According to the de Broglie relation, the result is a range of wavelengths. Therefore, the particle is not represented by a single wavelength, but a combination of wavelengths within this range. This combination forms a wave packet as we discussed in Section 28.6 and illustrated in Active Figure 28.16. Now, if we are asked to determine the location of the particle, we can only say that it is somewhere in the region defined by the wave packet because a distinct difference exists between this region and the rest of space. Therefore, by losing some information about the momentum of the particle, we have gained information about its position.

If we were to lose all information about the momentum, we would be adding together waves of all possible wavelengths. The result would be a wave packet of zero length. Therefore, if we know nothing about the momentum, we know exactly where the particle is.

The mathematical form of the uncertainty principle argues that the product of the uncertainties in position and momentum will always be larger than some minimum value. This value can be calculated from the types of arguments discussed earlier, which result in the value of $\hbar/2$ in Equation 28.17.

Another form of the uncertainty principle can be generated by reconsidering Active Figure 28.16. Imagine that the horizontal axis is time rather than spatial position x. We can then make the same arguments that we made about knowledge of wavelength and position in the time domain. The corresponding variables would be frequency and time. Because frequency is related to the energy of the particle by $E = hf$, the uncertainty principle in this form is

▮ Uncertainty principle for energy and time

$$\Delta E \, \Delta t \geq \frac{\hbar}{2} \qquad [28.18]$$

This form of the uncertainty principle suggests that energy conservation can appear to be violated by an amount ΔE as long as it is only for a short time interval Δt consistent with Equation 28.18. We shall use this notion to estimate the rest energies of particles in Chapter 31.

EXAMPLE 28.8 Locating an Electron

The speed of an electron is measured to be 5.00×10^3 m/s $\pm$ 0.003%. Within what limits could one determine the position of this electron along the direction of its velocity vector?

Solution The momentum of the electron is

$$p = m_e v = (9.11 \times 10^{-31}\ \text{kg})(5.00 \times 10^3\ \text{m/s})$$
$$= 4.56 \times 10^{-27}\ \text{kg}\cdot\text{m/s}$$

Because the uncertainty is 0.003% of this value, we have

$$\Delta p = 0.000\,03p = 1.37 \times 10^{-31}\ \text{kg}\cdot\text{m/s}$$

The minimum uncertainty in position can now be calculated by using this value of Δp and Equation 28.17:

$$\Delta x \geq \frac{\hbar}{2\,\Delta p} = \frac{1.05 \times 10^{-34}\ \text{J}\cdot\text{s}}{2(1.37 \times 10^{-31}\ \text{kg}\cdot\text{m/s})}$$
$$= 0.384 \times 10^{-3}\ \text{m} = 0.384\ \text{mm}$$

28.9 AN INTERPRETATION OF QUANTUM MECHANICS

We have been introduced to some new and strange ideas so far in this chapter. In an effort to understand the concepts of quantum physics better, let us investigate another bridge between particles and waves. We first think about electromagnetic radiation from the particle point of view. The probability per unit volume of finding a photon in a given region of space at an instant of time is proportional to the number of photons per unit volume at that time:

$$\frac{\text{probability}}{V} \propto \frac{N}{V}$$

The number of photons per unit volume is proportional to the intensity of the radiation:

$$\frac{N}{V} \propto I$$

Now, we form the bridge to the wave model by recalling that the intensity of electromagnetic radiation is proportional to the square of the electric field amplitude for the electromagnetic wave (Section 24.5):

$$I \propto E^2$$

Equating the beginning and the end of this string of proportionalities, we have

$$\frac{\text{probability}}{V} \propto E^2 \qquad [28.19]$$

Therefore, for electromagnetic radiation, the probability per unit volume of finding a particle associated with this radiation (the photon) is proportional to the square of the amplitude of the wave associated with the particle.

Recognizing the wave–particle duality of both electromagnetic radiation and matter, we should suspect a parallel proportionality for a material particle. That is, the probability per unit volume of finding the particle is proportional to the square of the amplitude of a wave representing the particle. In Section 28.5 we learned that there is a de Broglie wave associated with every particle. The amplitude of the de Broglie wave associated with a particle is not a measurable quantity (because the wave function representing a particle is generally a complex function, as we discuss below). In contrast, the electric field is a measurable quantity for an electromagnetic wave. The matter analog to Equation 28.19 relates the square of the wave's amplitude to the probability per unit volume of finding the particle. As a result, we call the amplitude of the wave associated with the particle the **probability amplitude,** or the **wave function,** and give it the symbol Ψ. In general, the complete wave function Ψ for a system depends on the positions of all the particles in the system and on time; therefore, it can be written $\Psi(\vec{\mathbf{r}}_1, \vec{\mathbf{r}}_2, \vec{\mathbf{r}}_3, \ldots, \vec{\mathbf{r}}_j, \ldots, t)$, where $\vec{\mathbf{r}}_j$ is the position vector of the jth particle in the system. For many systems of interest, including all those in this text, the wave function Ψ is mathematically separable in space and time and can be written as a product of a space function ψ for one particle of the system and a complex time function:[4]

■ Space- and time-dependent wave function Ψ

$$\Psi(\vec{\mathbf{r}}_1, \vec{\mathbf{r}}_2, \vec{\mathbf{r}}_3, \ldots, \vec{\mathbf{r}}_j, \ldots, t) = \psi(\vec{\mathbf{r}}_j)e^{-i\omega t} \qquad [28.20]$$

where ω $(= 2\pi f)$ is the angular frequency of the wave function and $i = \sqrt{-1}$.

[4]The standard form of a complex number is $a + ib$. The notation $e^{i\theta}$ is equivalent to the standard form as follows:

$$e^{i\theta} = \cos\theta + i\sin\theta$$

Therefore, the notation $e^{-i\omega t}$ in Equation 28.20 is equivalent to $\cos(-\omega t) + i\sin(-\omega t) = \cos\omega t - i\sin\omega t$.

PITFALL PREVENTION 28.5

THE WAVE FUNCTION BELONGS TO A SYSTEM The common language in quantum mechanics is to associate a wave function with a particle. The wave function, however, is determined by the particle and its interaction with its environment, so it more rightfully belongs to a system. In many cases, the particle is the only part of the system that experiences a change, which is why the common language has developed. You will see examples in the future in which it is more proper to think of the system wave function rather than the particle wave function.

For any system in which the potential energy is time-independent and depends only on the positions of particles within the system, the important information about the system is contained within the space part of the wave function. The time part is simply the factor $e^{-i\omega t}$. Therefore, the understanding of ψ is the critical aspect of a given problem.

The wave function ψ is often complex-valued. The quantity $|\psi|^2 = \psi^*\psi$, where ψ^* is the complex conjugate[5] of ψ, is always real and positive and is proportional to the probability per unit volume of finding a particle at a given point at some instant. The wave function contains within it all the information that can be known about the particle.

This probability interpretation of the wave function was first suggested by Max Born (1882–1970) in 1928. In 1926, Erwin Schrödinger (1887–1961) proposed a wave equation that describes the manner in which the wave function changes in space and time. The *Schrödinger wave equation,* which we shall examine in Section 28.12, represents a key element in the theory of quantum mechanics.

In Section 28.5, we found that the de Broglie equation relates the momentum of a particle to its wavelength through the relation $p = h/\lambda$. If an ideal free particle has a precisely known momentum p_x, its wave function is a sinusoidal wave of wavelength $\lambda = h/p_x$ and the particle has equal probability of being at any point along the x axis. The wave function for such a free particle moving along the x axis can be written as

■ Wave function for a free particle

$$\psi(x) = Ae^{ikx} \qquad [28.21]$$

where $k = 2\pi/\lambda$ is the angular wave number and A is a constant amplitude.[6]

Although we cannot measure ψ, we can measure the quantity $|\psi|^2$, the absolute square of ψ, which can be interpreted as follows. If ψ represents a single particle, $|\psi|^2$—called the **probability density**—is the relative probability per unit volume that the particle will be found at any given point in the volume. This interpretation can also be stated in the following manner. If dV is a small volume element surrounding some point, the probability of finding the particle in that volume element is $|\psi|^2\,dV$. In this section, we deal only with one-dimensional systems, where the particle must be located along the x axis, and we therefore replace dV with dx. In this case, the probability $P(x)\,dx$ that the particle will be found in the infinitesimal interval dx around the point x is

$$P(x)\,dx = |\psi|^2\,dx \qquad [28.22]$$

Because the particle must be somewhere along the x axis, the sum of the probabilities over all values of x must be 1:

■ Normalization condition on ψ

$$\int_{-\infty}^{\infty} |\psi|^2\,dx = 1 \qquad [28.23]$$

Any wave function satisfying Equation 28.23 is said to be **normalized.** Normalization is simply a statement that the particle exists at some point at all times.

Although it is not possible to specify the position of a particle with complete certainty, it is possible through $|\psi|^2$ to specify the probability of observing it in a small

[5] For a complex number $z = a + ib$, the complex conjugate is found by changing i to $-i$: $z^* = a - ib$. The product of a complex number and its complex conjugate is always real and positive: $z^*z = (a - ib)(a + ib) = a^2 - (ib)^2 = a^2 - (i)^2b^2 = a^2 + b^2$.

[6] For the free particle, the full wave function, based on Equation 28.20, is

$$\Psi(x, t) = Ae^{ikx}e^{-i\omega t} = Ae^{i(kx-\omega t)} = A[\cos(kx - \omega t) + i\sin(kx - \omega t)]$$

The real part of this wave function has the same form as the waves that we added together to form wave packets in Section 28.6.

region surrounding a given point. **The probability of finding the particle in the arbitrarily sized interval $a \leq x \leq b$ is**

$$P_{ab} = \int_a^b |\psi|^2 \, dx \qquad [28.24]$$

The probability P_{ab} is the area under the curve of $|\psi|^2$ versus x between the points $x = a$ and $x = b$ as in Figure 28.21.

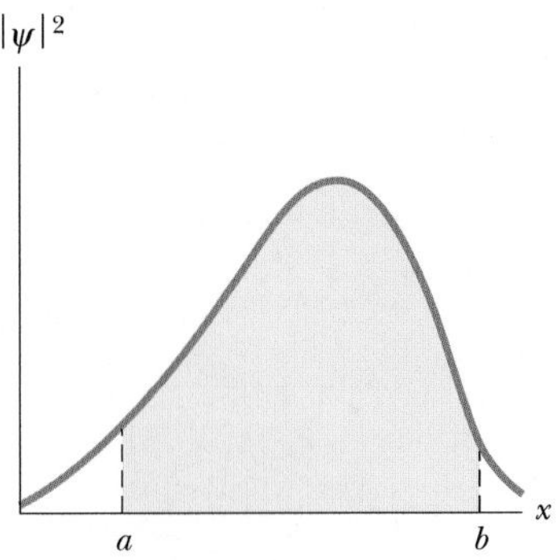

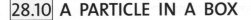

FIGURE 28.21 The probability of a particle being in the interval $a \leq x \leq b$ is the area under the probability density curve from a to b.

Experimentally, the probability is finite of finding a particle in an interval near some point at some instant. The value of that probability must lie between the limits 0 and 1. For example, if the probability is 0.3, there is a 30% chance of finding the particle in the interval.

The wave function ψ satisfies a wave equation, just as the electric field associated with an electromagnetic wave satisfies a wave equation that follows from Maxwell's equations. The wave equation satisfied by ψ is the Schrödinger equation (Section 28.12), and ψ can be computed from it. Although ψ is not a measurable quantity, all the measurable quantities of a particle, such as its energy and momentum, can be derived from a knowledge of ψ. For example, once the wave function for a particle is known, it is possible to calculate the average position at which you would find the particle after many measurements. This average position is called the **expectation value** of x and is defined by the equation

Expectation value for position x

$$\langle x \rangle \equiv \int_{-\infty}^{\infty} \psi^* x \psi \, dx \qquad [28.25]$$

where brackets $\langle \ \rangle$ are used to denote expectation values. Furthermore, one can find the expectation value of any function $f(x)$ associated with the particle by using the following equation:

$$\langle f(x) \rangle \equiv \int_{-\infty}^{\infty} \psi^* f(x) \psi \, dx \qquad [28.26]$$

28.10 A PARTICLE IN A BOX

In this section, we shall apply some of the ideas we have developed to a sample problem. Let us choose a simple problem: a particle confined to a one-dimensional region of space, called the *particle in a box* (even though the "box" is one-dimensional!). From a classical viewpoint, if a particle is confined to bouncing back and forth along the x axis between two impenetrable walls as in the pictorial representation in Figure 28.22a, its motion is easy to describe. If the speed of the particle is v, the magnitude of its momentum mv remains constant, as does its kinetic energy. Classical physics places no restrictions on the values of a particle's momentum and energy. The quantum mechanics approach to this problem is quite different and requires that we find the appropriate wave function consistent with the conditions of the situation.[7]

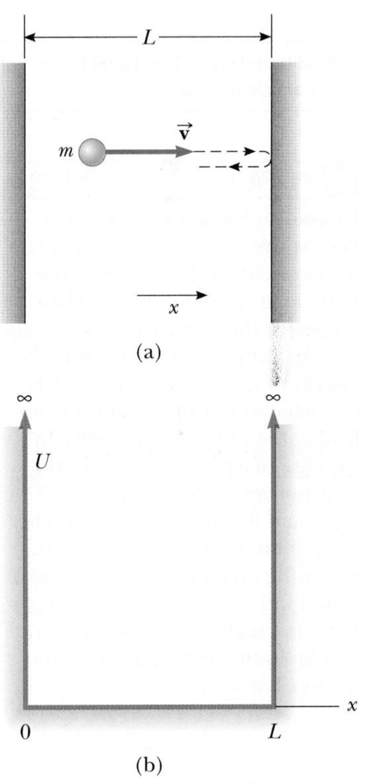

FIGURE 28.22 (a) A particle of mass m and velocity $\vec{v}$, confined to bouncing between two impenetrable walls separated by a distance L. (b) The potential energy function for the system.

Because the walls are impenetrable, the probability of finding the particle outside the box is zero, so the wave function $\psi(x)$ must be zero for $x < 0$ and for $x > L$, where L is the distance between the two walls. A mathematical condition for any wave function is that it must be continuous in space.[8] Therefore, if ψ is zero outside the walls, it must also be zero *at* the walls; that is, $\psi(0) = 0$ and $\psi(L) = 0$. Only those wave functions that satisfy this condition are allowed.

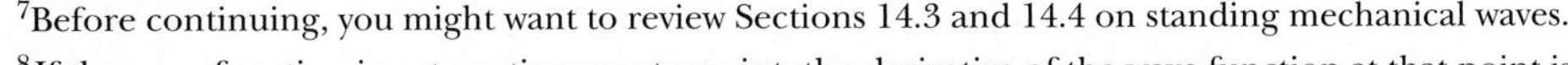

[7]Before continuing, you might want to review Sections 14.3 and 14.4 on standing mechanical waves.

[8]If the wave function is not continuous at a point, the derivative of the wave function at that point is infinite. This issue leads to problems in the Schrödinger equation, for which the wave function is a solution and which is discussed in Section 28.12.

Figure 28.22b shows a graphical representation of the particle in a box problem, which graphs the potential energy of the particle–environment system as a function of the position of the particle. When the particle is inside the box, the potential energy of the system does not depend on the particle's location and we can choose its value to be zero. Outside the box, we have to ensure that the wave function is zero. We can do so by defining the potential energy of the system as infinitely large if the particle were outside the box. Because kinetic energy is necessarily positive, the only way a particle could be outside the box is if the system has an infinite amount of energy.

The wave function for a particle in the box can be expressed as a real sinusoidal function:[9]

$$\psi(x) = A \sin\left(\frac{2\pi x}{\lambda}\right) \qquad [28.27]$$

This wave function must satisfy the boundary conditions at the walls. The boundary condition at $x = 0$ is satisfied already because the sine function is zero when $x = 0$. For the boundary condition at $x = L$, we have

$$\psi(L) = 0 = A \sin\left(\frac{2\pi L}{\lambda}\right)$$

which can only be true if

$$\frac{2\pi L}{\lambda} = n\pi \quad \rightarrow \quad \lambda = \frac{2L}{n} \qquad [28.28]$$

where $n = 1, 2, 3, \ldots$. Therefore, only certain wavelengths for the particle are allowed! Each of the allowed wavelengths corresponds to a quantum state for the system, and n is the quantum number. Expressing the wave function in terms of the quantum number n, we have

■ Allowed wave functions for a particle in a box

$$\psi(x) = A \sin\left(\frac{2\pi x}{\lambda}\right) = A \sin\left(\frac{2\pi x}{2L/n}\right) = A \sin\left(\frac{n\pi x}{L}\right) \qquad [28.29]$$

Active Figures 28.23a and 28.23b are graphical representations of ψ versus x and $|\psi|^2$ versus x for $n = 1$, 2, and 3 for the particle in a box. Note that although ψ can be positive or negative, $|\psi|^2$ is always positive. Because $|\psi|^2$ represents a probability density, a negative value for $|\psi|^2$ is meaningless.

Further inspection of Active Figure 28.23b shows that $|\psi|^2$ is zero at the boundaries, satisfying our boundary condition. In addition, $|\psi|^2$ is zero at other points, depending on the value of n. For $n = 2$, $|\psi|^2 = 0$ at $x = L/2$; for $n = 3$, $|\psi|^2 = 0$ at $x = L/3$ and $x = 2L/3$. The number of zero points increases by one each time the quantum number increases by one.

Because the wavelengths of the particle are restricted by the condition $\lambda = 2L/n$, the magnitude of the momentum of the particle is also restricted to specific values that we can find from the expression for the de Broglie wavelength, Equation 28.10:

$$p = \frac{h}{\lambda} = \frac{h}{2L/n} = \frac{nh}{2L}$$

From this expression, we find that the allowed values of the energy, which is simply the kinetic energy of the particle, are

■ Allowed energies for a particle in a box

$$E_n = \tfrac{1}{2}mv^2 = \frac{p^2}{2m} = \frac{(nh/2L)^2}{2m}$$

$$= \left(\frac{h^2}{8mL^2}\right)n^2 \qquad n = 1, 2, 3, \ldots \qquad [28.30]$$

PITFALL PREVENTION 28.6

Reminder: Energy belongs to a system We describe Equation 28.30 as representing the energy of the particle; it is commonly used language for the particle in a box problem. In reality, we are analyzing the energy of the *system* of the particle and whatever environment is establishing the impenetrable walls. In the case of a particle in a box, the only nonzero type of energy is kinetic and it belongs to the particle. In general, energies that we calculate using quantum physics are associated with a system of interacting particles, such as the electron and proton in the hydrogen atom studied in Chapter 11.

[9]We show that this function is the correct one explicitly in Section 28.12.

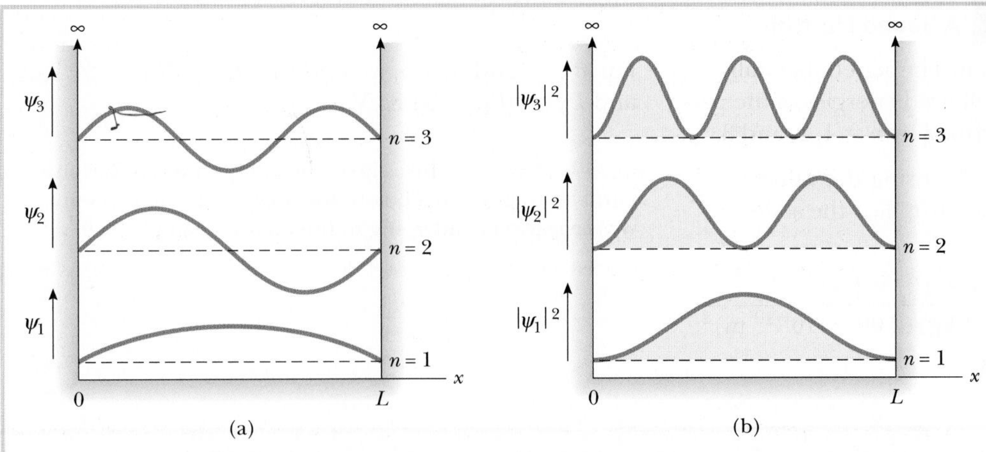

ACTIVE FIGURE **28.23** The first three allowed states for a particle confined to a one-dimensional box. The states are shown superimposed on the potential energy function of Figure 28.22b. (a) The wave functions ψ for $n = 1, 2$, and 3. (b) The probability densities $|\psi|^2$ for $n = 1, 2$, and 3. The wave functions and probability densities are plotted vertically from separate axes that are offset vertically for clarity. The positions of these axes on the potential energy function suggest the relative energies of the states, but the positions are not shown to scale.

PhysicsNow™ By logging into PhysicsNow at **www.pop4e.com** and going to Active Figure 28.23, you can measure the probability of a particle being between two points for the three quantum states in the figure.

As we see from this expression, **the energy of the particle is quantized,** similar to our quantization of energy in the hydrogen atom in Chapter 11. The lowest allowed energy corresponds to $n = 1$, for which $E_1 = h^2/8mL^2$. Because $E_n = n^2E_1$, the excited states corresponding to $n = 2, 3, 4, \ldots$ have energies given by $4E_1$, $9E_1$, $16E_1$, ...

Active Figure 28.24 is an energy level diagram[10] describing the energy values of the allowed states. Note that the state $n = 0$, for which E would be equal to zero, is not allowed. Thus, according to quantum mechanics, the particle can never be at rest. The least energy it can have, corresponding to $n = 1$, is called the **zero-point energy.** This result is clearly contradictory to the classical viewpoint, in which $E = 0$ is an acceptable state, as are all positive values of E.

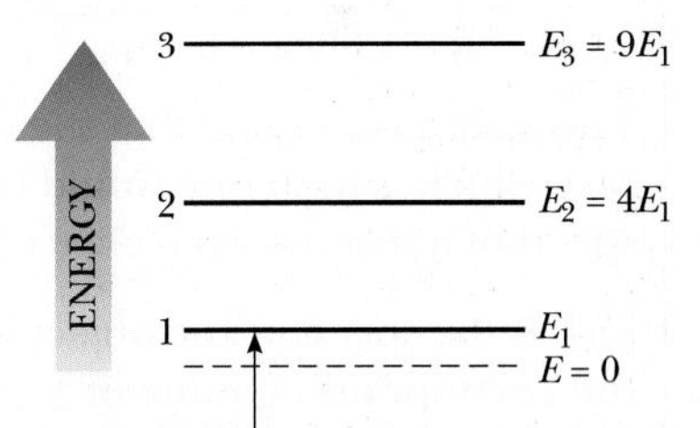

ACTIVE FIGURE **28.24**

Energy level diagram for a particle confined to a one-dimensional box of length L. The lowest allowed energy is $E_1 = h^2/8mL^2$.

PhysicsNow™ Adjust the length of the box and the mass of the particle to see the effect on the energy levels by logging into PhysicsNow at **www.pop4e.com** and going to Active Figure 28.24.

QUICK QUIZ **28.8** Redraw Active Figure 28.23b, the probability of finding a particle at a particular location in a box, on the basis of expectations from classical physics rather than quantum physics.

QUICK QUIZ **28.9** A particle is in a box of length L. Suddenly, the length of the box is increased to $2L$. What happens to the energy levels shown in Active Figure 28.24? **(a)** Nothing happens; they are unaffected. **(b)** They move farther apart. **(c)** They move closer together.

[10] We introduced the energy level diagram as a specialized semigraphical representation in Chapter 11.

INTERACTIVE EXAMPLE 28.9 A Bound Electron

An electron is confined between two impenetrable walls 0.200 nm apart. Determine the allowed energies of the particle for the quantum states described by $n = 1, 2,$ and 3.

Solution We apply Equation 28.30, using the value $m_e = 9.11 \times 10^{-31}$ kg for the electron. For the state described by $n = 1$, we have

$$E_1 = \frac{h^2}{8m_eL^2} = \frac{(6.63 \times 10^{-34}\,\text{J}\cdot\text{s})^2}{8(9.11 \times 10^{-31}\,\text{kg})(2.00 \times 10^{-10}\,\text{m})^2}$$
$$= 1.51 \times 10^{-18}\,\text{J} = 9.42\ \text{eV}$$

For $n = 2$ and $n = 3$, we find that $E_2 = 4E_1 = 37.7$ eV and $E_3 = 9E_1 = 84.8$ eV.

PhysicsNow™ Investigate the energy levels of various particles trapped in a box by logging into PhysicsNow at www.pop4e.com and going to Interactive Example 28.9.

EXAMPLE 28.10 Energy Quantization for a Macroscopic Object

A 1.00-mg object is confined between two rigid walls separated by 1.00 cm.

A Calculate its minimum speed.

Solution The minimum speed corresponds to the state for which $n = 1$. Using Equation 28.30 with $n = 1$ gives the zero-point energy:

$$E_1 = \frac{h^2}{8mL^2} = \frac{(6.63 \times 10^{-34}\,\text{J}\cdot\text{s})^2}{8(1.00 \times 10^{-6}\,\text{kg})(1.00 \times 10^{-2}\,\text{m})^2}$$
$$= 5.49 \times 10^{-58}\,\text{J}$$

Because $E = K = \frac{1}{2}mv^2$, we can find v as follows:

$$\tfrac{1}{2}mv^2 = 5.49 \times 10^{-58}\,\text{J}$$

$$v = \left[\frac{2(5.49 \times 10^{-58}\,\text{J})}{1.00 \times 10^{-6}\,\text{kg}}\right]^{1/2} = 3.31 \times 10^{-26}\ \text{m/s}$$

This speed is so small that the object appears to be at rest, which is what one would expect for the zero-point speed of a macroscopic object.

B If the speed of the object is 3.00×10^{-2} m/s, find the corresponding value of n.

Solution The kinetic energy of the object is

$$K = \tfrac{1}{2}mv^2 = \tfrac{1}{2}(1.00 \times 10^{-6}\,\text{kg})(3.00 \times 10^{-2}\,\text{m/s})^2$$
$$= 4.50 \times 10^{-10}\,\text{J}$$

Because $E_n = n^2E_1$ and $E_1 = 5.49 \times 10^{-58}$ J, we find that

$$K = E_n = n^2E_1 = 4.50 \times 10^{-10}\,\text{J}$$

$$n = \left(\frac{4.50 \times 10^{-10}\,\text{J}}{E_1}\right)^{1/2} = \left(\frac{4.50 \times 10^{-10}\,\text{J}}{5.49 \times 10^{-58}\,\text{J}}\right)^{1/2}$$
$$= 9.05 \times 10^{23}$$

This value of n is so large that we would never be able to distinguish the quantized nature of the energy levels. That is, the difference in energy between the two states $n_1 = 9.05 \times 10^{23}$ and $n_2 = (9.05 \times 10^{23}) + 1$ is too small to be detected experimentally. Like Example 28.2, this example illustrates the working of the correspondence principle; that is, as m or L become large, the quantum description must agree with the classical result.

28.11 THE QUANTUM PARTICLE UNDER BOUNDARY CONDITIONS

The particle in a box discussed in Section 28.10 is an example of how all quantum problems can be addressed. To begin, consider Equation 28.28 and compare it with Equation 14.6. The allowed wavelengths for the particle in a box are identical to the allowed wavelengths for mechanical waves on a string fixed at both ends. In both the string wave and the particle wave, we apply **boundary conditions** to determine the allowed states of the system. For the string fixed at both ends, the boundary condition is that the displacement of the string at the boundaries is zero. For the particle in a box, the probability amplitude at the boundaries is zero. In both cases, the result is quantized wavelengths. In the case of the vibrating string, wavelength is related to the frequency, so we have a set of harmonics, or quantized frequencies, given by Equation 14.8. In the case of the particle in a box, we also have quantized frequencies. We can go further in this case, however, because the frequency is related to the energy through $E = hf$, and we generate a set of quantized *energies*.

The quantization of energy for a quantum particle is therefore no more surprising than the quantization of frequencies for the vibrating guitar string. The essential feature of the analysis model of the quantum particle under boundary conditions is the recognition that **an interaction of a particle with its environment represents one or more boundary conditions and, if the interaction restricts the particle to a finite region of space, results in quantization of the energy of the system.** Because particles have wave-like characteristics, the allowed quantum states of a system are those in which the boundary conditions on the wave function representing the system are satisfied.

The only quantization of energy we have seen before this chapter is that of the hydrogen atom in Chapter 11. In that case, the electric force between the proton and the electron creates a constraint that requires the electron and the proton to stay near each other (assuming that we have not supplied enough energy to ionize the atom). This constraint results in boundary conditions that limit the energies of the atom to those corresponding to specific allowed wave functions.

In general, boundary conditions are related to the coordinates describing the problem. For the particle in a box, we required a zero value of the wave function at two values of x. In the case of the hydrogen atom, the problem is best presented in *spherical coordinates.* These coordinates are an extension of the polar coordinates introduced in Chapter 1 and consist of a radial coordinate r and two angular coordinates. Boundary conditions on the wave function related to r are that the radial part of the wave function must approach zero as $r \rightarrow \infty$ (so that the wave function can be normalized) and remain finite as $r \rightarrow 0$. A boundary condition on the wave function related to an angular coordinate is that adding 2π to the angle must return the wave function to the same value because an addition of 2π results in the same angular position. The generation of the wave function and application of the boundary conditions for the hydrogen atom are beyond the scope of this book. We shall, however, examine the behavior of some of the wave functions in Section 29.3.

PITFALL PREVENTION 28.7

QUANTUM STATES ARE NOT NECESSARILY STANDING WAVES Although there are many similarities between the quantization of states for quantum systems and the quantization of frequencies for waves on a string (Chapter 14), quantum states are not necessarily standing waves. There may be no stationary "nodes" and no sinusoidal shape associated with a quantum wave function under boundary conditions. Systems more complicated than the particle in a box will have more complicated wave functions, and some boundary conditions will not lead to zeros of the wave function at fixed points.

28.12 THE SCHRÖDINGER EQUATION

In Section 24.3, we discussed a wave equation for electromagnetic radiation. The waves associated with particles also satisfy a wave equation. We might guess that the wave equation for material particles is different from that associated with photons because material particles have a nonzero rest energy. The appropriate wave equation was developed by Schrödinger in 1926. In analyzing the behavior of a quantum system, the approach is to determine a solution to this equation and then apply the appropriate boundary conditions to the solution. The solution yields the allowed wave functions and energy levels of the system under consideration. Proper manipulation of the wave function then enables one to calculate all measurable features of the system.

(AIP Emilio Segrè Visual Archives)

ERWIN SCHRÖDINGER (1887–1961)

An Austrian theoretical physicist, Schrödinger is best known as the creator of quantum mechanics. He also produced important papers in the fields of statistical mechanics, color vision, and general relativity. Schrödinger did much to hasten the universal acceptance of quantum theory by demonstrating the mathematical equivalence between his wave mechanics and the more abstract matrix mechanics developed by Heisenberg.

The Schrödinger equation as it applies to a particle of mass m confined to moving along the x axis and interacting with its environment through a potential energy function $U(x)$ is

$$-\frac{\hbar^2}{2m}\frac{d^2\psi}{dx^2} + U\psi = E\psi \qquad [28.31]$$

■ Time-independent Schrödinger equation

where E is the total energy of the system (particle and environment). Because this equation is independent of time, it is commonly referred to as the **time-independent**

Schrödinger equation. (We shall not discuss the time-dependent Schrödinger equation, whose solution is Ψ, Eq. 28.20, in this text.)

This equation is consistent with the energy version of the isolated system model. The system is the particle and its environment. Problem 28.44 shows, both for a free particle and a particle in a box, that the first term in the Schrödinger equation reduces to the kinetic energy of the particle multiplied by the wave function. Therefore, Equation 28.31 tells us that the total energy is the sum of the kinetic energy and the potential energy and that the total energy is a constant: $K + U = E =$ constant.

In principle, if the potential energy $U(x)$ for the system is known, one can solve Equation 28.31 and obtain the wave functions and energies for the allowed states of the system. Because U may vary with position, it may be necessary to solve the equation separately for various regions. In the process, the wave functions for the different regions must join smoothly at the boundaries and we require that $\psi(x)$ be *continuous*. Furthermore, so that $\psi(x)$ obeys the normalization condition, we require that $\psi(x)$ approach zero as x approaches $\pm\infty$. Finally, $\psi(x)$ must be *single-valued* and $d\psi/dx$ must also be continuous[11] for finite values of $U(x)$.

The task of solving the Schrödinger equation may be very difficult, depending on the form of the potential energy function. As it turns out, the Schrödinger equation has been extremely successful in explaining the behavior of atomic and nuclear systems, whereas classical physics has failed to do so. Furthermore, when quantum mechanics is applied to macroscopic objects, the results agree with classical physics, as required by the correspondence principle.

The Particle in a Box via the Schrödinger Equation

To see how the Schrödinger equation is applied to a problem, let us return to our particle in a one-dimensional box of width L (see Fig. 28.22) and analyze it with the Schrödinger equation. In association with Figure 28.22b, we discussed the potential energy diagram that describes the problem. A potential energy diagram such as this one is a useful representation for understanding and solving problems with the Schrödinger equation.

Because of the shape of the curve in Figure 28.22b, the particle in a box is sometimes said to be in a **square well,**[12] where a **well** is an upward-facing region of the curve in a potential energy diagram. (A downward-facing region is called a *barrier,* which we shall investigate in Section 28.13.)

In the region $0 < x < L$, where $U = 0$, we can express the Schrödinger equation in the form

$$\frac{d^2\psi}{dx^2} = -\frac{2mE}{\hbar^2}\psi = -k^2\psi \qquad [28.32]$$

where

$$k = \frac{\sqrt{2mE}}{\hbar} \qquad [28.33]$$

The solution to Equation 28.32 is a function whose second derivative is the negative of the same function multiplied by a constant k^2. We recognize both the sine and cosine functions as satisfying this requirement. Therefore, the most general solution

---

[11]If $d\psi/dx$ were not continuous, we would not be able to evaluate $d^2\psi/dx^2$ in Equation 28.31 at the point of discontinuity.

[12]It is called a square well even if it has a rectangular shape in a potential energy diagram.

to the equation is a linear combination of both solutions:

$$\psi(x) = A \sin kx + B \cos kx$$

where A and B are constants determined by the boundary conditions.

Our first boundary condition is that $\psi(0) = 0$:

$$\psi(0) = A \sin 0 + B \cos 0 = 0 + B = 0$$

Therefore, our solution reduces to

$$\psi(x) = A \sin kx$$

The second boundary condition, $\psi(L) = 0$, when applied to the reduced solution, gives

$$\psi(L) = A \sin kL = 0$$

which is satisfied only if kL is an integral multiple of π, that is, if $kL = n\pi$, where n is an integer. Because $k = \sqrt{2mE}/\hbar$, we have

$$kL = \frac{\sqrt{2mE}}{\hbar} L = n\pi$$

For each integer choice for n, this equation determines a quantized energy E_n. Solving for the allowed energies E_n gives

$$E_n = \left(\frac{h^2}{8mL^2}\right) n^2 \qquad [28.34]$$

which are identical to the allowed energies in Equation 28.30.

Substituting the values of k in the wave function, the allowed wave functions $\psi_n(x)$ are given by

$$\psi_n(x) = A \sin\left(\frac{n\pi x}{L}\right) \qquad [28.35]$$

This wave function agrees with Equation 28.29.

Normalizing this relationship shows that $A = \sqrt{(2/L)}$. (See Problem 28.46.) Therefore, the normalized wave function is

$$\psi(x) = \sqrt{\frac{2}{L}} \sin\left(\frac{n\pi x}{L}\right) \qquad [28.36]$$

The notion of trapping particles in potential wells is used in the burgeoning field of **nanotechnology,** which refers to the design and application of devices having dimensions ranging from 1 to 100 nm. The fabrication of these devices often involves manipulating single atoms or small groups of atoms to form structures such as the quantum corral in Figure 28.25.

One area of nanotechnology of interest to researchers is the **quantum dot.** The quantum dot, a small region that is grown in a silicon crystal, acts as a potential well. This region can trap electrons into states with quantized energies. The wave functions for a particle in a quantum dot look similar to those in Active Figure 28.23a if L is on the order of nanometers. The storage of binary information using quantum dots is an active field of research. A simple binary scheme would involve associating a one with a quantum dot containing an electron and a zero with an empty dot. Other schemes involve cells of multiple dots such that arrangements of electrons among the dots correspond to ones and zeros. Several research laboratories are studying the properties and potential applications of quantum dots. Information should be forthcoming from these laboratories at a steady rate in the next few years.

FIGURE 28.25 This photograph is a demonstration of a quantum corral consisting of a ring of 48 iron atoms located on a copper surface. The diameter of the ring is 143 nm, and the photograph was obtained using a low-temperature scanning tunneling microscope (STM) as mentioned in Section 28.13. Corrals and other structures are able to confine surface electron waves. The study of such structures will play an important role in determining the future of small electronic devices.

EXAMPLE 28.11 The Expectation Values for the Particle in a Box

A particle of mass m is confined to a one-dimensional box between $x = 0$ and $x = L$. Find the expectation value of the position x of the particle for a state with quantum number n.

Solution Using Equation 28.25 and the wave function in Equation 28.36, we can set up the expectation value for x:

$$\langle x \rangle = \int_{-\infty}^{\infty} \psi^* x \psi \, dx = \int_0^L x \left[\sqrt{\frac{2}{L}} \sin\left(\frac{n\pi x}{L}\right)\right]^2 dx$$

$$= \frac{2}{L} \int_0^L x \sin^2\left(\frac{n\pi x}{L}\right) dx$$

where we have reduced the limits on the integral to 0 to L because the value of the wave function is zero elsewhere. Evaluating the integral by consulting an integral table or by mathematical integration[13] gives

$$\langle x \rangle = \frac{2}{L}\left[\frac{x^2}{4} - \frac{x \sin\left(2\frac{n\pi x}{L}\right)}{4\frac{n\pi}{L}} - \frac{\cos\left(2\frac{n\pi x}{L}\right)}{8\left(\frac{n\pi}{L}\right)^2}\right]_0^L$$

$$= \frac{2}{L}\left(\frac{L^2}{4}\right) = \frac{L}{2}$$

Notice that the expectation value is right at the center of the box, which we would expect from the symmetry of the square of the wave function about the center (see Active Fig. 28.23b). Because the squares of all wave functions are symmetric about the midpoint, the expectation value does not depend on n.

28.13 TUNNELING THROUGH A POTENTIAL ENERGY BARRIER

Consider the potential energy function shown in Figure 28.26, in which the potential energy of the system is zero everywhere except for a region of width L where the potential energy has a constant value of U. This type of potential energy function is called a **square barrier,** and U is called the **barrier height.** A very interesting and peculiar phenomenon occurs when a moving particle encounters such a barrier of finite height and width. Consider a particle of energy $E < U$ that is incident on the barrier from the left (see Fig. 28.26). Classically, the particle is reflected by the barrier. If the particle were to exist in region II, its kinetic energy would be negative, which is not allowed classically. Therefore, region II, and in turn region III, are both classically *forbidden* to the particle incident from the left. According to quantum mechanics, however, **all regions are accessible to the particle, regardless of its energy.** (Although all regions are accessible, the probability of the particle being in a region that is classically forbidden is very low.) According to the uncertainty principle, the particle can be within the barrier as long as the time interval during which it is in the barrier is short and consistent with Equation 28.18. If the barrier is

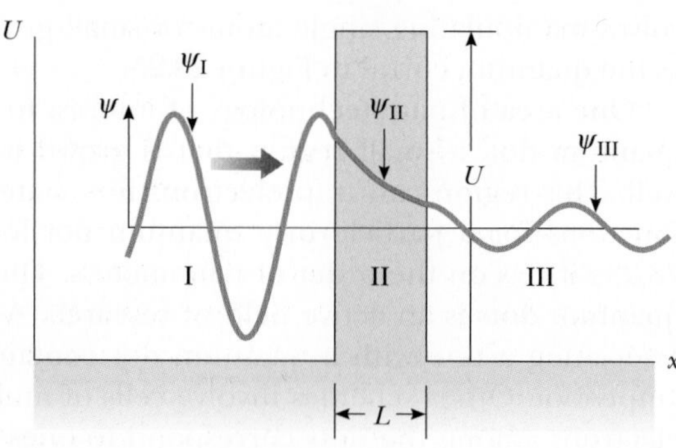

FIGURE 28.26 Wave function ψ for a particle incident from the left on a barrier of height U and width L. The wave function is sinusoidal in regions I and III but exponentially decaying in region II. The wave function is plotted vertically from an axis positioned at the energy of the particle.

[13] To integrate this function, first replace $\sin^2(\pi x/L)$ with $\frac{1}{2}(1 - \cos 2\pi x/L)$ (Table B.3 in Appendix B). That step will allow $\langle x \rangle$ to be expressed as two integrals. The second integral can then be evaluated by partial integration (Section B.7 in Appendix B).

relatively narrow, this short time interval can allow the particle to move across the barrier. Therefore, it is possible for us to understand the passing of the particle through the barrier with the help of the uncertainty principle.

Let us approach this situation using a mathematical representation. The Schrödinger equation has valid solutions in all three regions I, II, and III. The solutions in regions I and III are sinusoidal as in Equation 28.21. In region II, the solution is exponential. Applying the boundary conditions that the wave functions in the three regions must join smoothly at the boundaries, we find that a full solution can be found such as that represented by the curve in Figure 28.26. Therefore, Schrödinger's equation and the boundary conditions are satisfied, which tells us mathematically that such a process can theoretically occur according to the quantum theory.

Because the probability of locating the particle is proportional to $|\psi|^2$, we conclude that the chance of finding the particle beyond the barrier in region III is nonzero. This result is in complete disagreement with classical physics. The movement of the particle to the far side of the barrier is called **tunneling** or **barrier penetration.**

The probability of tunneling can be described with a **transmission coefficient** T and a **reflection coefficient** R. **The transmission coefficient represents the probability that the particle penetrates to the other side of the barrier, and the reflection coefficient is the probability that the particle is reflected by the barrier.** Because the incident particle is either reflected or transmitted, we require that $T + R = 1$. An approximate expression for the transmission coefficient, obtained when $T \ll 1$ (a very wide barrier or a very high barrier, that is, $U \gg E$), is

$$T \approx e^{-2CL} \qquad [28.37]$$

where

$$C = \frac{\sqrt{2m(U - E)}}{\hbar} \qquad [28.38]$$

According to quantum physics, Equation 28.37 tells us that T can be nonzero, which is in contrast to the classical point of view that requires that $T = 0$. That we experimentally observe the phenomenon of tunneling provides further confidence in the principles of quantum physics.

Figure 28.26 shows the wave function of a particle tunneling through a barrier in one dimension. A similar wave function having spherical symmetry describes the barrier penetration of a particle leaving a radioactive nucleus, which we will study in Chapter 30. The wave function exists both inside and outside the nucleus, and its amplitude is constant in time. In this way, the wave function correctly describes the small but constant probability that the nucleus will decay. The moment of decay cannot be predicted. In general, quantum mechanics implies that the future is indeterminate. (This feature is in contrast to classical mechanics, from which the trajectory of an object can be calculated to arbitrarily high precision from precise knowledge of its initial position and velocity and of the forces exerted on it.) We must conclude that the fundamental laws of nature are probabilistic.

A radiation detector can be used to show that a nucleus decays by radiating a particle at a particular moment and in a particular direction. To point out the contrast between this experimental result and the wave function describing it, Schrödinger imagined a box containing a cat, a radioactive sample, a radiation counter, and a vial of poison. When a nucleus in the sample decays, the counter triggers the administration of lethal poison to the cat. Quantum mechanics correctly predicts the probability of finding the cat dead when the box is opened. Before the box is opened, does the animal have a wave function describing it as a fractionally dead cat, with some chance of being alive?

This question is currently under investigation, never with actual cats, but sometimes with interference experiments building upon the experiment described in

PITFALL PREVENTION 28.8

"HEIGHT" ON AN ENERGY DIAGRAM The word *height* (as in *barrier height*) refers to an energy in discussions of barriers in potential energy diagrams. For example, we might say the height of the barrier is 10 eV. On the other hand, the barrier *width* refers to our traditional usage of such a word. It is an actual physical length measurement between the two locations of the vertical sides of the barrier.

Section 28.7. Does the act of measurement change the system from a probabilistic to a definite state? When a particle emitted by a radioactive nucleus is detected at one particular location, does the wave function describing the particle drop instantaneously to zero everywhere else in the Universe? (Einstein called such a state change a "spooky action at a distance.") Is there a fundamental difference between a quantum system and a macroscopic system? The answers to these questions are basically unknown.

QUICK QUIZ 28.10 Which of the following changes would increase the probability of transmission of a particle through a potential barrier? (You may choose more than one answer.) **(a)** decreasing the width of the barrier **(b)** increasing the width of the barrier **(c)** decreasing the height of the barrier **(d)** increasing the height of the barrier **(e)** decreasing the kinetic energy of the incident particle **(f)** increasing the kinetic energy of the incident particle

INTERACTIVE EXAMPLE 28.12 Transmission Coefficient for an Electron

A 30-eV electron is incident on a square barrier of height 40 eV.

A What is the probability that the electron will tunnel through if the barrier width is 1.0 nm?

Solution Let us assume that the probability of transmission is low so that we can use the approximation in Equation 28.37. For the given barrier height and electron energy, the quantity $U - E$ has the value

$$U - E = (40 \text{ eV} - 30 \text{ eV}) = 10 \text{ eV} = 1.6 \times 10^{-18} \text{ J}$$

Using Equation 28.38, the quantity $2CL$ is

$$2CL = 2\frac{\sqrt{2(9.11 \times 10^{-31} \text{ kg})(1.6 \times 10^{-18} \text{ J})}}{1.054 \times 10^{-34} \text{ J}\cdot\text{s}}(1.0 \times 10^{-9} \text{ m})$$
$$= 32.4$$

Therefore, the probability of tunneling through the barrier is

$$T \approx e^{-2CL} = e^{-32.4} = 8.5 \times 10^{-15}$$

That is, the electron has only about 1 chance in 10^{14} to tunnel through the 1.0-nm-wide barrier.

B What is the probability that the electron will tunnel through if the barrier width is 0.10 nm?

Solution For $L = 0.10$ nm, we find $2CL = 3.24$, and

$$T \approx e^{-2CL} = e^{-3.24} = 0.039$$

This result shows that the electron has a relatively high probability, about 4%, compared with 10^{-12}% in part A, of penetrating the 0.10-nm barrier. Notice an important behavior that leads to effective practical applications for tunneling: that reducing the width of the barrier by only one order of magnitude increases the probability of tunneling by about 12 orders of magnitude!

PhysicsNow™ Investigate the tunneling of particles through barriers by logging into PhysicsNow at www.pop4e.com and going to Interactive Example 28.12.

Applications of Tunneling

As we have seen, tunneling is a quantum phenomenon, a result of the wave nature of matter. Many applications may be understood only on the basis of tunneling.

- **Alpha decay.** One form of radioactive decay is the emission of alpha particles (the nuclei of helium atoms) by unstable, heavy nuclei (Chapter 30). For an alpha particle to escape from the nucleus, it must penetrate a barrier whose height is several times larger than the energy of the nucleus–alpha particle system. The barrier is due to a combination of the attractive nuclear force (discussed in Chapter 30) and the Coulomb repulsion (discussed in detail in Chapter 19) between the alpha particle and the rest of the nucleus. Occasionally, an alpha particle tunnels through the barrier, which explains the basic mechanism for this type of decay and the large variations in the mean lifetimes of various radioactive nuclei.
- **Nuclear fusion.** The basic reaction that powers the Sun and, indirectly, almost everything else in the solar system is fusion, which we will study in Chapter 30. In

one step of the process that occurs at the core of the Sun, protons must approach each other to within such a small distance that they fuse to form a deuterium nucleus. According to classical physics, these protons cannot overcome and penetrate the barrier caused by their mutual electrical repulsion. Quantum-mechanically, however, the protons are able to tunnel through the barrier and fuse together.

- **Scanning tunneling microscope.** The scanning tunneling microscope, or STM, is a remarkable device that uses tunneling to create images of surfaces with resolution comparable to the size of a single atom. A small probe with a very fine tip is made to scan very close to the surface of a specimen. A tunneling current is maintained between the probe and specimen; the current (which is related to the probability of tunneling) is very sensitive to the barrier height (which is related to the separation between the tip and specimen) as seen in Interactive Example 28.12. Maintaining a constant tunneling current produces a feedback signal that is used to raise and lower the probe as the surface is scanned. Because the vertical motion of the probe follows the contour of the specimen's surface, an image of the surface is obtained. The image of the quantum corral shown in Figure 28.25 is made with a scanning tunneling microscope.

28.14 THE COSMIC TEMPERATURE

CONTEXT CONNECTION

Now that we have introduced the concepts of quantum physics for microscopic particles and systems, let us see how we can connect these concepts to processes occurring on a cosmic scale. For our first such connection, consider the Universe as a system. It is widely believed that the Universe began with a cataclysmic explosion called the **Big Bang,** first mentioned in Chapter 5. Because of this explosion, all the material in the Universe is moving apart. This expansion causes a Doppler shift in radiation left over from the Big Bang such that the wavelength of the radiation lengthens. In the 1940s, Ralph Alpher, George Gamow, and Robert Hermann developed a structural model of the Universe in which they predicted that the thermal radiation from the Big Bang should still be present and that it should now have a wavelength distribution consistent with a black body with a temperature of a few kelvins.

In 1965, Arno Penzias and Robert Wilson of Bell Telephone Laboratories were measuring radiation from the Milky Way galaxy using a special 20-ft antenna as a radio telescope. They noticed a consistent background "noise" of radiation in the signals from the antenna. Despite their great efforts to test alternative hypotheses for the origin of the noise in terms of interference from the Sun, an unknown source in the Milky Way, structural problems in the antenna, and even the presence of pigeon droppings in the antenna, none of the hypotheses was sufficient to explain the noise.

What Penzias and Wilson were detecting was the thermal radiation from the Big Bang. That it was detected by their system regardless of the direction of the antenna was consistent with the radiation being spread throughout the Universe, as the Big Bang model predicts. A measurement of the intensity of this radiation suggested that the temperature associated with the radiation was about 3 K, consistent with Alpher, Gamow, and Hermann's prediction from the 1940s. Although the measured intensity was consistent with their prediction, the measurement was taken at only a single wavelength. Full agreement with the model of the Universe as a black body would come only if measurements at many wavelengths demonstrated a distribution in wavelengths consistent with Active Figure 28.2.

In the years following Penzias and Wilson's discovery, other researchers made measurements at different wavelengths. In 1989, the COBE (*CO*smic *B*ackground *E*xplorer) satellite was launched by NASA and added critical measurements at wavelengths below 0.1 cm. The results of these measurements are shown in

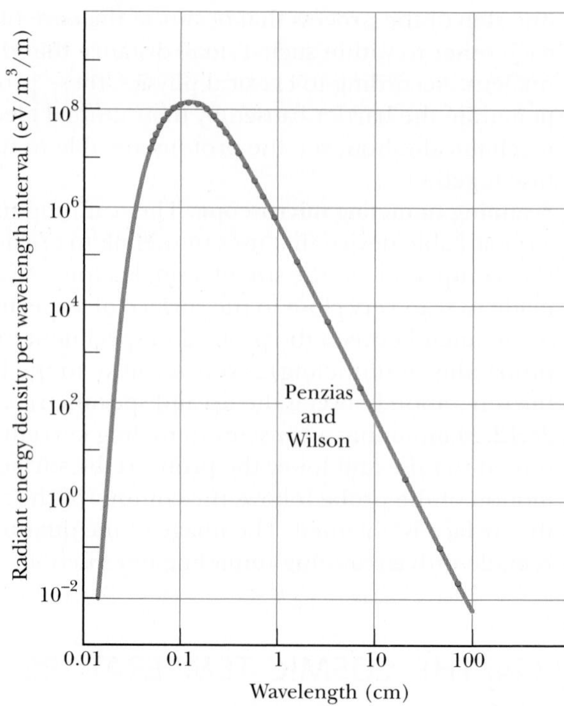

FIGURE 28.27 Theoretical blackbody wavelength distribution (brown curve) and measured data points (blue) for radiation from the Big Bang. Most of the data were collected from the Cosmic Background Explorer (COBE) satellite. The datum of Penzias and Wilson is indicated.

Figure 28.27. The series of measurements taken since 1965 are consistent with thermal radiation associated with a temperature of 2.7 K. The whole story of the cosmic temperature is a remarkable example of science at work: building a model, making a prediction, taking measurements, and testing the measurements against the predictions.

The first chapter of our *Cosmic Connection* Context describes the first example of this connection. By studying the thermal radiation from microscopic vibrating objects, we learn something about the origin of our Universe. In Chapter 29, we shall see more examples of this fascinating connection.

SUMMARY

PhysicsNow™ Take a practice test by logging into PhysicsNow at www.pop4e.com and clicking on the Pre-Test link for this chapter.

The characteristics of **blackbody radiation** cannot be explained by classical concepts. Planck introduced the first model of **quantum physics** when he argued that the atomic oscillators responsible for this radiation exist only in discrete **quantum states.**

In the **photoelectric effect,** electrons are ejected from a metallic surface when light is incident on that surface. Einstein provided a successful explanation of this effect by extending Planck's quantum theory to the electromagnetic field. In this model, light is viewed as a stream of particles called **photons,** each with energy $E = hf$, where f is the frequency and h is **Planck's constant.** The maximum kinetic energy of the ejected photoelectron is given by

$$K_{\max} = hf - \phi \quad [28.5]$$

where ϕ is the **work function** of the metal.

X-rays striking a target are scattered at various angles by electrons in the target. A shift in wavelength is observed for the scattered x-rays, and the phenomenon is known as the **Compton effect.** Classical physics does not correctly explain the experimental results of this effect. If the x-ray is treated as a photon, conservation of energy and momentum applied to the isolated system of the photon and the electron yields for the Compton shift the expression

$$\lambda' - \lambda_0 = \frac{h}{m_e c}(1 - \cos\theta) \quad [28.8]$$

where m_e is the mass of the electron, c is the speed of light, and θ is the scattering angle.

Every object of mass m and momentum p has wave-like properties, with a **de Broglie wavelength** given by the relation

$$\lambda = \frac{h}{p} = \frac{h}{mv} \quad [28.10]$$

The wave–particle duality is the basis of the **quantum particle model.** It can be interpreted by imagining a particle to be made up of a combination of a large number of waves. These waves interfere constructively in a small region of space called a **wave packet.**

The **uncertainty principle** states that if a measurement of position is made with uncertainty Δx and a *simultaneous* measurement of momentum is made with uncertainty Δp_x, the product of the two uncertainties can never be less than $\hbar/2$:

$$\Delta x \, \Delta p_x \geq \frac{\hbar}{2} \quad [28.17]$$

The uncertainty principle is a natural outgrowth of the wave packet model.

Particles are represented by a **wave function** $\psi(x, y, z)$. The **probability density** that a particle will be found at a point is $|\psi|^2$. If the particle is confined to moving along the x axis, the probability that it will be located in an interval dx is given by $|\psi|^2 \, dx$. Furthermore, the wave function must be **normalized:**

$$\int_{-\infty}^{\infty} |\psi|^2 \, dx = 1 \quad [28.23]$$

The measured position x of the particle, averaged over many trials, is called the **expectation value** of x and is defined by

$$\langle x \rangle \equiv \int_{-\infty}^{\infty} \psi^* x \psi \, dx \quad [28.25]$$

If a particle of mass m is confined to moving in a one-dimensional box of width L whose walls are perfectly rigid, the allowed wave functions for the particle are

$$\psi(x) = A \sin\left(\frac{n\pi x}{L}\right) \quad [28.29]$$

where n is an integer quantum number starting at 1. The particle has a well-defined wavelength λ whose values are such that the width L of the box is equal to an integral number of half wavelengths, that is, $L = n\lambda/2$. The energies of a particle in a box are quantized and are given by

$$E_n = \left(\frac{h^2}{8mL^2}\right) n^2 \qquad n = 1, 2, 3, \ldots \quad [28.30]$$

Quantum systems generally involve a particle under some constraints imposed by the environment, which is the basis of the quantum particle under **boundary conditions** model. In this model, the wave function for a system is found and application of boundary conditions on the system allows a determination of the allowed energies and unknown constants in the wave function.

The wave function must satisfy the Schrödinger equation. The **time-independent Schrödinger equation** for a particle confined to moving along the x axis is

$$-\frac{\hbar^2}{2m}\frac{d^2\psi}{dx^2} + U\psi = E\psi \quad [28.31]$$

where E is the total energy of the system and U is the potential energy of the system.

When a particle of energy E meets a barrier of height U, where $E < U$, the particle has a finite probability of penetrating the barrier. This process, called **tunneling,** is the basic mechanism that explains the operation of the scanning tunneling microscope and the phenomenon of alpha decay in some radioactive nuclei.

QUESTIONS

☐ = answer available in the *Student Solutions Manual and Study Guide*

1. What assumptions did Planck make in dealing with the problem of blackbody radiation? Discuss the consequences of these assumptions.
2. Which is more likely to cause sunburn because individual molecules in skin cells absorb more energy: (a) infrared light, (b) visible light, or (c) ultraviolet light?
3. If the photoelectric effect is observed for one metal, can you conclude that the effect will also be observed for another metal under the same conditions? Explain.
4. How does the Compton effect differ from the photoelectric effect?
5. Why does the existence of a cutoff frequency in the photoelectric effect favor a particle theory for light over a wave theory?
6. Suppose a photograph were made of a person's face using only a few photons. Would the result be simply a very faint image of the face? Explain your answer.
7. An x-ray photon is scattered by an electron. What happens to the frequency of the scattered photon relative to that of the incident photon?
8. Is light a wave or a particle? Support your answer by citing specific experimental evidence.
9. Is an electron a wave or a particle? Support your answer by citing some experimental results.
10. Why was the demonstration of electron diffraction by Davisson and Germer an important experiment?
11. If matter has a wave nature, why is this wave-like characteristic not observable in our daily experiences?
12. An electron and a proton are accelerated from rest through the same potential difference. Which particle has the longer wavelength?
13. In describing the passage of electrons through a slit and arriving at a screen, physicist Richard Feynman said that "electrons arrive in lumps, like particles, but the probability of arrival of these lumps is determined as the intensity of the waves would be. It is in this sense that the electron behaves sometimes like a particle and sometimes like a wave." Elaborate on this point in your own words. For a further discussion of this point, see R. Feynman, *The Character of Physical Law* (Cambridge, MA: MIT Press, 1980), Chapter 6.
14. *Blacker than black, brighter than white.* (a) Take a large, closed, empty cardboard box. Cut a slot a few millimeters wide in one side. Use black pens, markers, and black material to make some stripes next to the slot as shown in Figure Q28.14a. Inspect them with care and choose which is blackest (the figure does not show enough contrast to

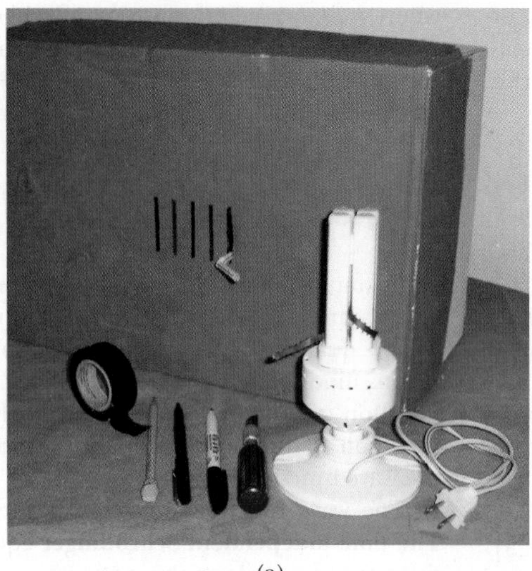

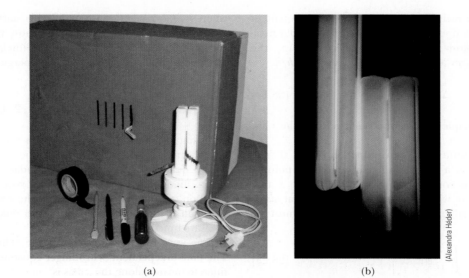

(a)

(b)

(Alexandra Héder)

FIGURE Q28.14

reveal which it is). Explain why it is blackest. (b) Locate an intricately shaped compact fluorescent light fixture. Look at it through dark glasses and describe where it appears brightest. Explain why it is brightest there. Figure Q28.14b shows two such light fixtures held near each other. [*Suggestion:* Gustav Kirchhoff, professor at Heidelberg and master of the obvious, gave the same answer to part (a) as you likely will. His answer to part (b) would begin as follows. When electromagnetic radiation falls on its surface, an object reflects some fraction r of the energy and absorbs the rest. Whether the fraction reflected is 0.8 or 0.001, the fraction absorbed is $a = 1 - r$. Suppose the object and its surroundings are at the same temperature. The energy the object absorbs joins its fund of internal energy, but the second law of thermodynamics implies that the absorbed energy cannot raise the object's temperature. It does not produce a temperature increase because the object's energy budget has one more term: energy radiated. . . . You still have to make the observations and answer parts (a) and (b), but you can incorporate some of Kirchhoff's ideas into your answer if you wish.]

15. For a particle in a box, the probability density at certain points is zero as seen in Active Figure 28.23b. Does that imply that the particle cannot move across these points? Explain.

PROBLEMS

1, 2, 3 = straightforward, intermediate, challenging

☐ = full solution available in the *Student Solutions Manual and Study Guide*

Physics Now™ = coached problem with hints available at www.pop4e.com

= computer useful in solving problem

= paired numerical and symbolic problems

= biomedical application

Section 28.1 ■ Blackbody Radiation and Planck's Theory

1. With young children and the elderly, use of a traditional fever thermometer has risks of bacterial contamination and tissue perforation. The radiation thermometer shown in Figure 28.5 works fast and avoids most risks. The instrument measures the power of infrared radiation from the ear canal. This cavity is accurately described as a black body and is close to the hypothalamus, the body's temperature control center. Take normal body temperature as 37.0°C. If the body temperature of a feverish patient is 38.3°C, what is the percentage increase in radiated power from his ear canal?

2. The radius of our Sun is 6.96×10^{8} m, and its total power output is 3.85×10^{26} W. (a) Assuming that the Sun's surface emits as a black body, calculate its surface temperature. (b) Using the result of part (a), find λ_{max} for the Sun.

3. Figure P28.3 shows the spectrum of light emitted by a firefly. Determine the temperature of a black body that

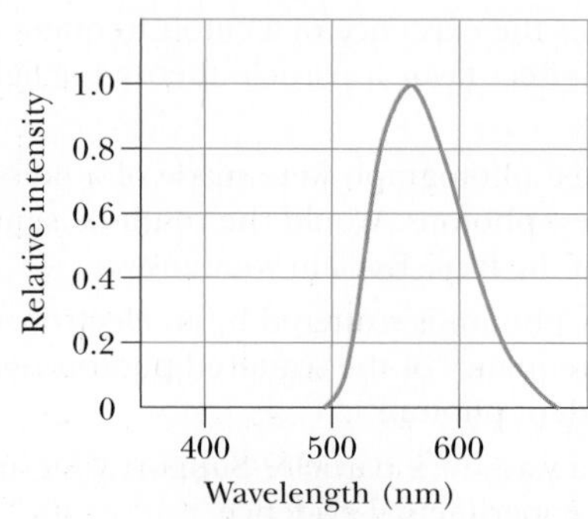

FIGURE P28.3

would emit radiation peaked at the same wavelength. Based on your result, would you say that firefly radiation is blackbody radiation?

4. Calculate the energy, in electron volts, of a photon whose frequency is (a) 620 THz, (b) 3.10 GHz, and (c) 46.0 MHz. (d) Determine the corresponding wavelengths for these photons and state the classification of each on the electromagnetic spectrum.

5. An FM radio transmitter has a power output of 150 kW and operates at a frequency of 99.7 MHz. How many photons per second does the transmitter emit?

6. The average threshold of dark-adapted (scotopic) vision is 4.00×10^{-11} W/m^2 at a central wavelength of 500 nm. If light having this intensity and wavelength enters the eye and the pupil is open to its maximum diameter of 8.50 mm, how many photons per second enter the eye?

7. A simple pendulum has a length of 1.00 m and a mass of 1.00 kg. The amplitude of oscillations of the pendulum is 3.00 cm. Estimate the quantum number for the pendulum.

8. **Review problem.** This problem is about how strongly matter is coupled to radiation, the subject with which quantum mechanics began. For a very simple model, consider a solid iron sphere 2.00 cm in radius. Assume that its temperature is always uniform throughout its volume. (a) Find the mass of the sphere. (b) Assume that it is at 20°C and has emissivity 0.860. Find the power with which it is radiating electromagnetic waves. (c) If it were alone in the Universe, at what rate would its temperature be changing? (d) Assume that Wien's law describes the sphere. Find the wavelength λ_{max} of electromagnetic radiation it emits most strongly. Although it emits a spectrum of waves having all different wavelengths, model its whole power output as carried by photons of wavelength λ_{max}. Find (e) the energy of one photon and (f) the number of photons it emits each second. The answer to part (f) gives an indication of how fast the object is emitting and also absorbing photons when it is in thermal equilibrium with its surroundings at 20°C.

Section 28.2 ▪ The Photoelectric Effect

9. Molybdenum has a work function of 4.20 eV. (a) Find the cutoff wavelength and cutoff frequency for the photoelectric effect. (b) What is the stopping potential if the incident light has a wavelength of 180 nm?

10. Electrons are ejected from a metallic surface with speeds ranging up to 4.60×10^5 m/s when light with a wavelength of 625 nm is used. (a) What is the work function of the surface? (b) What is the cutoff frequency for this surface?

11. Two light sources are used in a photoelectric experiment to determine the work function for a particular metal surface. When green light from a mercury lamp (λ = 546.1 nm) is used, a stopping potential of 0.376 V reduces the photocurrent to zero. (a) Based on this measurement, what is the work function for this metal? (b) What stopping potential would be observed when using the yellow light from a helium discharge tube (λ = 587.5 nm)?

12. From the scattering of sunlight, J. J. Thomson calculated the classical radius of the electron as having a value of 2.82×10^{-15} m. Sunlight with an intensity of 500 W/m^2 falls on a disk with this radius. Calculate the time interval required to accumulate 1.00 eV of energy. Assume that light is a classical wave and that the light striking the disk is completely absorbed. How does your result compare with the observation that photoelectrons are emitted promptly (within 10^{-9} s)?

13. **Review problem.** An isolated copper sphere of radius 5.00 cm, initially uncharged, is illuminated by ultraviolet light of wavelength 200 nm. What charge will the photoelectric effect induce on the sphere? The work function for copper is 4.70 eV.

Section 28.3 ▪ The Compton Effect

14. Calculate the energy and momentum of a photon of wavelength 700 nm.

15. X-rays having an energy of 300 keV undergo Compton scattering from a target. The scattered rays are detected at 37.0° relative to the incident rays. Find (a) the Compton shift at this angle, (b) the energy of the scattered x-ray, and (c) the energy of the recoiling electron.

16. A 0.110-nm photon collides with a stationary electron. After the collision, the electron moves forward and the photon recoils backward. Find the momentum and the kinetic energy of the electron.

17. PhysicsNow™ A 0.001 60-nm photon scatters from a free electron. For what (photon) scattering angle does the recoiling electron have kinetic energy equal to the energy of the scattered photon?

18. After a 0.800-nm x-ray photon scatters from a free electron, the electron recoils at 1.40×10^6 m/s. (a) What was the Compton shift in the photon's wavelength? (b) Through what angle was the photon scattered?

Section 28.4 ▪ Photons and Electromagnetic Waves

19. An electromagnetic wave is called *ionizing radiation* if its photon energy is larger than about 10.0 eV so that a single photon has enough energy to break apart an atom. With reference to Figure 24.12, identify what regions of the electromagnetic spectrum fit this definition of ionizing radiation and what do not.

Section 28.5 ▪ The Wave Properties of Particles

20. Calculate the de Broglie wavelength for a proton moving with a speed of 1.00×10^6 m/s.

21. (a) An electron has kinetic energy 3.00 eV. Find its wavelength. (b) A photon has energy 3.00 eV. Find its wavelength.

22. In the Davisson–Germer experiment, 54.0-eV electrons were diffracted from a nickel lattice. If the first maximum in the diffraction pattern was observed at ϕ = 50.0° (Fig. P28.22), what was the lattice spacing a between the vertical

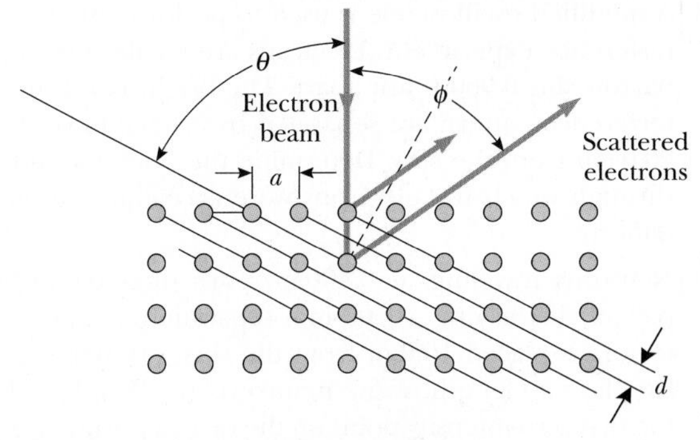

FIGURE **P28.22**

rows of atoms in the figure? (It is not the same as the spacing between the horizontal rows of atoms.)

23. Physics Now™ The nucleus of an atom is on the order of 10^{-14} m in diameter. For an electron to be confined to a nucleus, its de Broglie wavelength would have to be on this order of magnitude or smaller. (a) What would be the kinetic energy of an electron confined to this region? (b) Make also an order-of-magnitude estimate of the electric potential energy of a system of an electron inside an atomic nucleus. Would you expect to find an electron in a nucleus? Explain.

24. After learning about de Broglie's hypothesis that particles of momentum p have wave characteristics with wavelength $\lambda = h/p$, an 80.0-kg student has grown concerned about being diffracted when passing through a 75.0-cm-wide doorway. Assume that significant diffraction occurs when the width of the diffraction aperture is less than 10.0 times the wavelength of the wave being diffracted. (a) Determine the maximum speed at which the student can pass through the doorway so as to be significantly diffracted. (b) With that speed, how long will it take the student to pass through the doorway if it is in a wall 15.0 cm thick? Compare your result to the currently accepted age of the Universe, which is 4×10^{17} s. (c) Should this student worry about being diffracted?

25. The resolving power of a microscope depends on the wavelength used. If one wished to "see" an atom, a resolution of approximately 1.00×10^{-11} m would be required. (a) If electrons are used (in an electron microscope), what minimum kinetic energy is required for the electrons? (b) If photons are used, what minimum photon energy is needed to obtain the required resolution?

Section 28.6 ■ The Quantum Particle

26. Consider a freely moving quantum particle with mass m and speed u. Its energy is $E = K = \frac{1}{2}mu^2$. Determine the phase speed of the quantum wave representing the particle and show that it is different from the speed at which the particle transports mass and energy.

27. For a free relativistic quantum particle moving with speed v, the total energy is $E = hf = \hbar\omega = \sqrt{p^2c^2 + m^2c^4}$ and the momentum is $p = h/\lambda = \hbar k = \gamma mv$. For the quantum wave representing the particle, the group speed is $v_g = d\omega/dk$. Prove that the group speed of the wave is the same as the speed of the particle.

Section 28.7 ■ The Double-Slit Experiment Revisited

28. A modified oscilloscope is used to perform an electron interference experiment. Electrons are incident on a pair of narrow slits 0.060 0 μm apart. The bright bands in the interference pattern are separated by 0.400 mm on a screen 20.0 cm from the slits. Determine the potential difference through which the electrons were accelerated to give this pattern.

29. Neutrons traveling at 0.400 m/s are directed through a pair of slits having a 1.00-mm separation. An array of detectors is placed 10.0 m from the slits. (a) What is the de Broglie wavelength of the neutrons? (b) How far off axis is the first zero-intensity point on the detector array? (c) When a neutron reaches a detector, can we say which slit the neutron passed through? Explain.

30. In a certain vacuum tube, electrons evaporate from a hot cathode at a slow, steady rate and accelerate from rest through a potential difference of 45.0 V. Then they travel 28.0 cm as they pass through an array of slits and fall on a screen to produce an interference pattern. If the beam current is below a certain value, only one electron at a time will be in flight in the tube. What is this value? In this situation, the interference pattern still appears, showing that each individual electron can interfere with itself.

Section 28.8 ■ The Uncertainty Principle

31. An electron ($m_e = 9.11 \times 10^{-31}$ kg) and a bullet ($m =$ 0.020 0 kg) each have a velocity with a magnitude of 500 m/s, accurate to within 0.010 0%. Within what limits could we determine the position of the objects along the direction of the velocity?

32. Suppose Fuzzy, a quantum-mechanical duck, lives in a world in which $h = 2\pi$ J·s. Fuzzy has a mass of 2.00 kg and is initially known to be within a pond 1.00 m wide. (a) What is the minimum uncertainty in the component of the duck's velocity parallel to the width of the pond? (b) Assuming that this uncertainty in speed prevails for 5.00 s, determine the uncertainty in the duck's position after this time interval.

33. An air rifle is used to shoot 1.00-g particles at 100 m/s through a hole of diameter 2.00 mm. How far from the rifle must an observer be to see the beam spread by 1.00 cm because of the uncertainty principle? Compare this answer with the diameter of the visible Universe (2×10^{26} m).

34. A π^0 meson is an unstable particle produced in high-energy particle collisions. Its rest energy is about 135 MeV, and it exists for an average lifetime of only 8.70×10^{-17} s before decaying into two gamma rays. Using the uncertainty principle, estimate the fractional uncertainty $\Delta m/m$ in its mass determination.

35. A woman on a ladder drops small pellets toward a point target on the floor. (a) Show that, according to the uncertainty principle, the average miss distance must be at least

$$\Delta x_f = \left(\frac{2\hbar}{m}\right)^{1/2}\left(\frac{2H}{g}\right)^{1/4}$$

where H is the initial height of each pellet above the floor and m is the mass of each pellet. Assume that the spread in impact points is given by $\Delta x_f = \Delta x_i + (\Delta v_x)t$. (b) If $H = 2.00$ m and $m = 0.500$ g, what is Δx_f?

Section 28.9 ■ An Interpretation of Quantum Mechanics

36. The wave function for a particle is

$$\psi(x) = \sqrt{\frac{a}{\pi(x^2 + a^2)}}$$

for $a > 0$ and $-\infty < x < +\infty$. Determine the probability that the particle is located somewhere between $x = -a$ and $x = +a$.

37. A free electron has a wave function

$$\psi(x) = Ae^{i(5.00\times10^{10}x)}$$

where x is in meters. Find (a) its de Broglie wavelength, (b) its momentum, and (c) its kinetic energy in electron volts.

Section 28.10 ■ A Particle in a Box

38. An electron that has an energy of approximately 6 eV moves between rigid walls 1.00 nm apart. Find (a) the quantum number n for the energy state that the electron occupies and (b) the precise energy of the electron.

39. **Physics Now™** An electron is contained in a one-dimensional box of length 0.100 nm. (a) Draw an energy level diagram for the electron for levels up to $n = 4$. (b) Find the wavelengths of all photons that can be emitted by the electron in making downward transitions that could eventually carry it from the $n = 4$ state to the $n = 1$ state.

40. The nuclear potential energy that binds protons and neutrons in a nucleus is often approximated by a square well. Imagine a proton confined in an infinitely high square well of length 10.0 fm, a typical nuclear diameter. Calculate the wavelength and energy associated with the photon emitted when the proton moves from the $n = 2$ state to the ground state. In what region of the electromagnetic spectrum does this wavelength belong?

41. A photon with wavelength λ is absorbed by an electron confined to a box. As a result, the electron moves from state $n = 1$ to $n = 4$. (a) Find the length of the box. (b) What is the wavelength of the photon emitted in the transition of that electron from the state $n = 4$ to the state $n = 2$?

Section 28.11 ■ The Quantum Particle Under Boundary Conditions

Section 28.12 ■ The Schrödinger Equation

42. The wave function of a particle is given by

$$\psi(x) = A\cos(kx) + B\sin(kx)$$

where A, B, and k are constants. Show that ψ is a solution of the Schrödinger equation (Eq. 28.31), assuming the particle is free ($U = 0$), and find the corresponding energy E of the particle.

43. Show that the wave function $\psi = Ae^{i(kx-\omega t)}$ is a solution to the Schrödinger equation (Eq. 28.31) where $k = 2\pi/\lambda$ and $U = 0$.

44. Prove that the first term in the Schrödinger equation, $-(\hbar^2/2m)(d^2\psi/dx^2)$, reduces to the kinetic energy of the particle multiplied by the wave function (a) for a freely moving particle, with the wave function given by Equation 28.21, and (b) for a particle in a box, with the wave function given by Equation 28.36.

45. A particle in an infinitely deep square well has a wave function given by

$$\psi_2(x) = \sqrt{\frac{2}{L}}\sin\left(\frac{2\pi x}{L}\right)$$

for $0 \le x \le L$ and zero otherwise. (a) Determine the expectation value of x. (b) Determine the probability of finding the particle near $L/2$ by calculating the probability that the particle lies in the range $0.490L \le x \le 0.510L$. (c) Determine the probability of finding the particle near $L/4$ by calculating the probability that the particle lies in the range $0.240L \le x \le 0.260L$. (d) Argue that the result of part (a) does not contradict the results of parts (b) and (c).

46. The wave function for a particle confined to moving in a one-dimensional box is

$$\psi(x) = A\sin\left(\frac{n\pi x}{L}\right)$$

Use the normalization condition on ψ to show that

$$A = \sqrt{\frac{2}{L}}$$

(*Suggestion:* Because the box length is L, the wave function is zero for $x < 0$ and for $x > L$, so the normalization condition, Equation 28.23, reduces to $\int_0^L |\psi|^2\,dx = 1$.)

47. The wave function of an electron is

$$\psi(x) = \sqrt{\frac{2}{L}}\sin\left(\frac{2\pi x}{L}\right)$$

Calculate the probability of finding the electron between $x = 0$ and $x = L/4$.

48. A particle of mass m moves in a potential well of length $2L$. The potential energy is infinite for $x < -L$ and for $x > +L$. Inside the region $-L < x < L$, the potential energy is given by

$$U(x) = \frac{-\hbar^2 x^2}{mL^2(L^2 - x^2)}$$

In addition, the particle is in a stationary state that is described by the wave function $\psi(x) = A(1 - x^2/L^2)$ for $-L < x < +L$ and by $\psi(x) = 0$ elsewhere. (a) Determine the energy of the particle in terms of $\hbar$, m, and L. (*Suggestion:* Use the Schrödinger equation, Eq. 28.31.) (b) Show that $A = (15/16L)^{1/2}$. (c) Determine the probability that the particle is located between $x = -L/3$ and $x = +L/3$.

Section 28.13 ■ Tunneling Through a Potential Energy Barrier

49. An electron with kinetic energy $E = 5.00$ eV is incident on a barrier with thickness $L = 0.200$ nm and height $U = 10.0$ eV (Fig. P28.49). What is the probability that the electron (a) will tunnel through the barrier and (b) will be reflected?

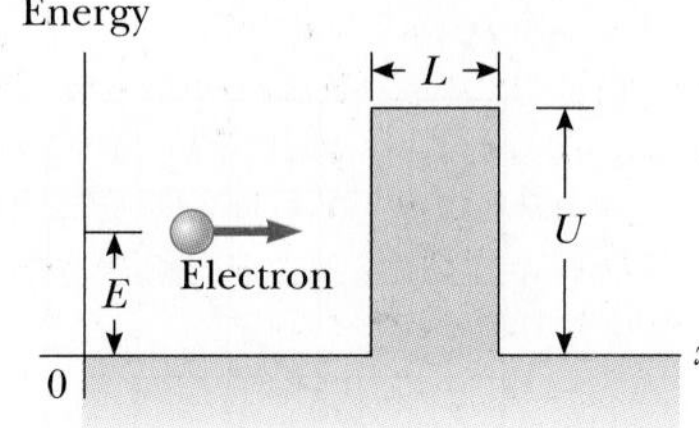

FIGURE P28.49 Problems 28.49 and 28.50.

50. An electron having total energy $E = 4.50$ eV approaches a rectangular energy barrier with $U = 5.00$ eV and $L = 950$ pm as shown in Figure P28.49. Classically, the electron cannot pass through the barrier because $E < U$. Quantum-mechanically, however, the probability of tunneling is not zero. Calculate this probability, which is the transmission coefficient.

51. An electron has a kinetic energy of 12.0 eV. The electron is incident upon a rectangular barrier of height 20.0 eV and thickness 1.00 nm. By what factor would the electron's probability of tunneling through the barrier increase if the electron absorbs all the energy of a photon of green light (with wavelength 546 nm) just as it reaches the barrier?

Section 28.14 ▪ Context Connection—The Cosmic Temperature

Problems 24.14 and 24.59 in Chapter 24 can be assigned with this section.

52. **Review problem.** A star moving away from the Earth at $0.280c$ emits radiation that we measure to be most intense at the wavelength 500 nm. Determine the surface temperature of this star.

53. The cosmic background radiation is blackbody radiation from a source at a temperature of 2.73 K. (a) Determine the wavelength at which this radiation has its maximum intensity. (b) In what part of the electromagnetic spectrum is the peak of the distribution?

54. Find the intensity of the cosmic background radiation, emitted by the fireball of the Big Bang at a temperature of 2.73 K.

Additional Problems

55. **Review problem.** Design an incandescent lamp filament. Specify the length and radius a tungsten wire can have to radiate electromagnetic waves with power 75.0 W when its ends are connected across a 120-V power supply. Assume that its constant operating temperature is 2 900 K and that its emissivity is 0.450. Assume that it takes in energy only by electrical transmission and loses energy only by electromagnetic radiation. From Table 21.1, you may take the resistivity of tungsten at 2 900 K as $5.6 \times 10^{-8}\,\Omega\cdot\text{m} \times [1 + (4.5 \times 10^{-3}/°\text{C})(2\,607\,°\text{C})] = 7.13 \times 10^{-7}\,\Omega\cdot\text{m}$.

56. Figure P28.56 shows the stopping potential versus the incident photon frequency for the photoelectric effect for sodium. Use the graph to find (a) the work function, (b) the ratio h/e, and (c) the cutoff wavelength. The data are taken from R. A. Millikan, *Physical Review* 7:362 (1916).

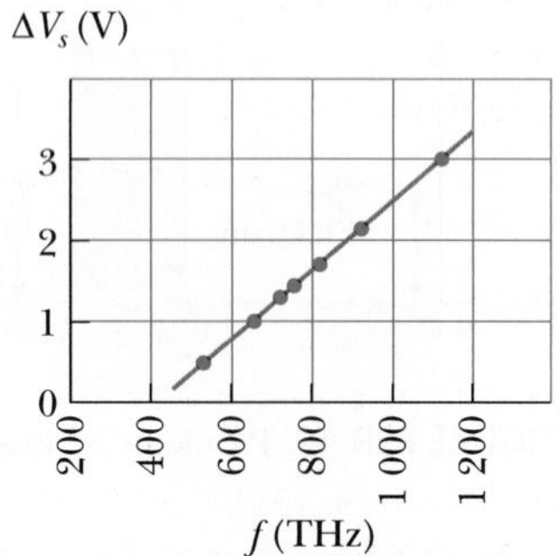

FIGURE P28.56

57. PhysicsNow™ The following table shows data obtained in a photoelectric experiment. (a) Using these data, make a graph similar to Active Figure 28.9 that plots as a straight line. From the graph, determine (b) an experimental value for Planck's constant (in joule-seconds) and (c) the work function (in electron volts) for the surface. (Two significant figures for each answer are sufficient.)

Wavelength (nm)	Maximum Kinetic Energy of Photoelectrons (eV)
588	0.67
505	0.98
445	1.35
399	1.63

58. **Review problem.** Photons of wavelength λ are incident on a metal. The most energetic electrons ejected from the metal are bent into a circular arc of radius R by a magnetic field having a magnitude B. What is the work function of the metal?

59. Johnny Jumper's favorite trick is to step out of his 16th-story window and fall 50.0 m into a pool. A news reporter takes a picture of 75.0-kg Johnny just before he makes a splash, using an exposure time of 5.00 ms. Find (a) Johnny's de Broglie wavelength at this moment, (b) the uncertainty of his kinetic energy measurement during such a period of time, and (c) the percent error caused by such an uncertainty.

60. A particle of mass 2.00×10^{-28} kg is confined to a one-dimensional box of length 1.00×10^{-10} m. For $n = 1$, what are (a) the particle's wavelength, (b) its momentum, and (c) its ground-state energy?

61. PhysicsNow™ An electron is represented by the time-independent wave function

$$\psi(x) = \begin{cases} Ae^{-\alpha x} & \text{for } x > 0 \\ Ae^{+\alpha x} & \text{for } x < 0 \end{cases}$$

(a) Sketch the wave function as a function of x. (b) Sketch the probability density representing the likelihood that the electron is found between x and $x + dx$. (c) Only an infinite value of potential energy could produce the discontinuity in the derivative of the wave function at $x = 0$. Aside from this feature, argue that $\psi(x)$ can be a physically reasonable wave function. (d) Normalize the wave function. (e) Determine the probability of finding the electron somewhere in the range

$$x_1 = -\frac{1}{2\alpha} \quad \text{to} \quad x_2 = \frac{1}{2\alpha}$$

62. Particles incident from the left are confronted with a step in potential energy shown in Figure P28.62. Located at $x = 0$, the step has a height U. The particles have energy $E > U$. Classically, we would expect all the particles to continue on, although with reduced speed. According to quantum mechanics, a fraction of the particles are reflected at the barrier. (a) Prove that the reflection coefficient R for this case is

$$R = \frac{(k_1 - k_2)^2}{(k_1 + k_2)^2}$$

where $k_1 = 2\pi/\lambda_1$ and $k_2 = 2\pi/\lambda_2$ are the wave numbers for the incident and transmitted particles. Proceed as follows. Show that the wave function $\psi_1 = Ae^{ik_1x} + Be^{-ik_1x}$ satisfies the Schrödinger equation in region 1, for $x < 0$. Here Ae^{ik_1x} represents the incident beam and Be^{-ik_1x} represents the reflected particles. Show that $\psi_2 = Ce^{ik_2x}$ satisfies the Schrödinger equation in region 2, for $x > 0$. Impose the boundary conditions $\psi_1 = \psi_2$ and $d\psi_1/dx = d\psi_2/dx$ at $x = 0$ to find the relationship between B and A. Then evaluate $R = B^2/A^2$. (b) A particle that has kinetic energy $E = 7.00$ eV is incident from a region where the potential energy is zero onto one in which $U = 5.00$ eV. Find its probability of being reflected and its probability of being transmitted.

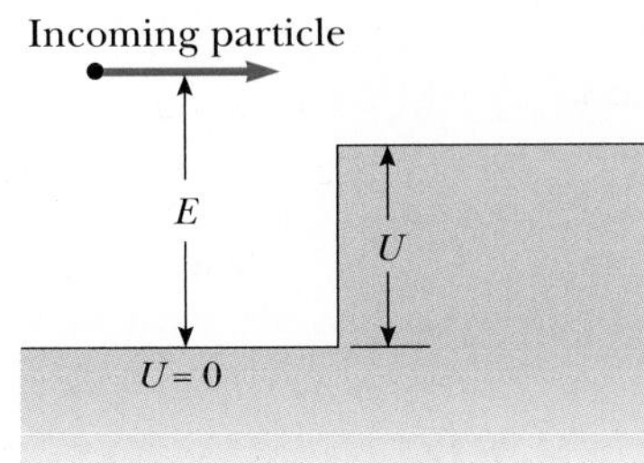

FIGURE **P28.62**

63. For a particle described by a wave function $\psi(x)$, the expectation value of a physical quantity $f(x)$ associated with the particle is defined by

$$\langle f(x)\rangle \equiv \int_{-\infty}^{\infty} \psi^* f(x)\psi\, dx$$

For a particle in a one-dimensional box extending from $x = 0$ to $x = L$, show that

$$\langle x^2\rangle = \frac{L^2}{3} - \frac{L^2}{2n^2\pi^2}$$

64. A particle of mass m is placed in a one-dimensional box of length L. Assume that the box is so small that the particle's motion is *relativistic*, so $K = p^2/2m$ is not valid. (a) Derive an expression for the kinetic energy levels of the particle. (b) Assume that the particle is an electron in a box of length $L = 1.00 \times 10^{-12}$ m. Find its lowest possible kinetic energy. By what percent is the nonrelativistic equation in error? (*Suggestion:* See Eq. 9.18.)

65. Imagine that a particle has a wave function

$$\psi(x) = \begin{cases} \sqrt{\dfrac{2}{a}}\, e^{-x/a} & \text{for } x > 0 \\ 0 & \text{for } x < 0 \end{cases}$$

(a) Find and sketch the probability density. (b) Find the probability that the particle will be at any point where $x < 0$. (c) Show that ψ is normalized, and then find the probability that the particle will be found between $x = 0$ and $x = a$.

ANSWERS TO QUICK QUIZZES

28.1 (b). A very hot star has a peak in the blackbody intensity distribution curve at wavelengths shorter than the visible. As a result, more blue light is emitted than red light.

28.2 AM radio, FM radio, microwaves, sodium light. The order of photon energy is the same as the order of frequency. See Figure 24.12 for a pictorial representation of electromagnetic radiation in order of frequency.

28.3 (c). When the frequency is increased, the photons each carry more energy, so a stopping potential larger in magnitude is required for the current to fall to zero.

28.4 Classical physics predicts that light of sufficient intensity causes emission of photoelectrons, independent of frequency and without a cutoff frequency. Also, the greater the intensity, the larger the maximum kinetic energy of the electrons, with some time delay in emission at low intensities. Therefore, the classical expectation (which did not match experiment) yields a graph that looks like the one at the bottom of the left column.

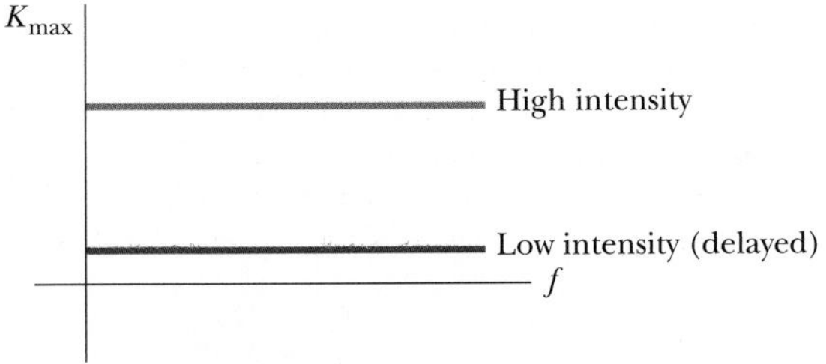

28.5 (c). According to Equation 28.10, two particles with the same de Broglie wavelength have the same momentum $p = mv$. If the electron and proton have the same momentum, they cannot have the same speed (a) because of the difference in their masses. For the same reason, remembering that $K = p^2/2m$, they cannot have the same kinetic energy (b). Because the particles have different kinetic energies, Equation 28.11 tells us that the particles do not have the same frequency (d).

28.6 (b). The Compton wavelength (Eq. 28.9) is a combination of constants and has no relation to the motion of the electron. The de Broglie wavelength (Eq. 28.10) is associated with the motion of the electron through its momentum.

28.7 (b). The group speed is zero because the leading edge of the packet remains fixed at the location of the accident.

28.8 Classically, we expect the particle to bounce back and forth between the two walls at constant speed. Therefore, we are as likely to find it on the left side of the box as in the middle, on the right side, or anywhere else inside the box. Our graph of probability density versus x would therefore be a horizontal line, with a total area under the line of unity, as shown on the next page.

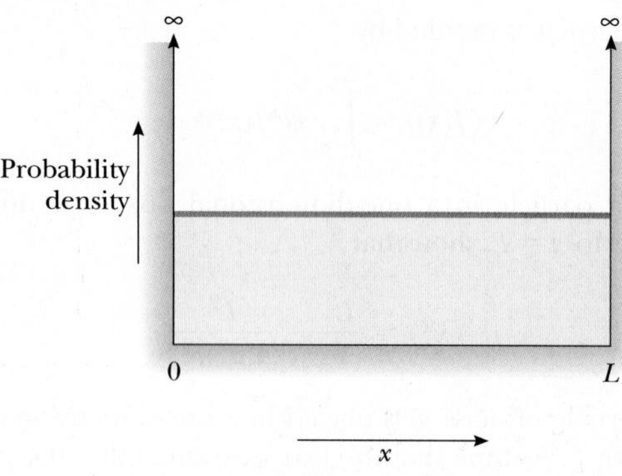

28.9 (c). According to Equation 28.30, if the length L is increased, all quantized energies become smaller. Therefore, the energy levels move closer together. As L becomes macroscopic, the energy levels are so close together that we do not observe the quantized behavior.

28.10 (a), (c), (f). Decreasing the barrier height and increasing the particle energy both reduce the value of C in Equation 28.38, increasing the transmission coefficient in Equation 28.37. Decreasing the width L of the barrier increases the transmission coefficient in Equation 28.37.

CHAPTER 29

Atomic Physics

This fireworks display shows several different colors. The colors are determined by the types of atoms in the material burning in the explosion. Bright white light often comes from oxidizing magnesium or aluminum. Red light often comes from strontium and yellow from sodium. Blue light is more difficult to achieve, but can be obtained by burning a mixture of copper powder, copper chloride, and hexachloroethane. The emission of light from atoms is an important clue that allows us to learn about the structure of the atom.

CHAPTER OUTLINE

In Chapter 28, we introduced some of the basic concepts and techniques used in quantum physics along with their applications to various simple systems. This chapter describes the application of quantum physics to more sophisticated structural models of atoms than we have seen previously.

We studied the hydrogen atom in Chapter 11 using Bohr's semiclassical approach. In this chapter, we shall analyze the hydrogen atom with a full quantum model. Although the hydrogen atom is the simplest atomic system, it is an especially important system to understand, for several reasons:

- Much of what we learn about the hydrogen atom, with its single electron, can be extended to such single-electron ions as He^+ and Li^{2+}.
- The hydrogen atom is an ideal system for performing precise tests of theory against experiment and for improving our overall understanding of atomic structure.

- The quantum numbers used to characterize the allowed states of hydrogen can be used to qualitatively describe the allowed states of more complex atoms. This characterization enables us to understand the periodic table of the elements, which is one of the greatest triumphs of quantum physics.
- The basic ideas about atomic structure must be well understood before we attempt to deal with the complexities of molecular structures and the electronic structures of solids.

29.1 EARLY STRUCTURAL MODELS OF THE ATOM

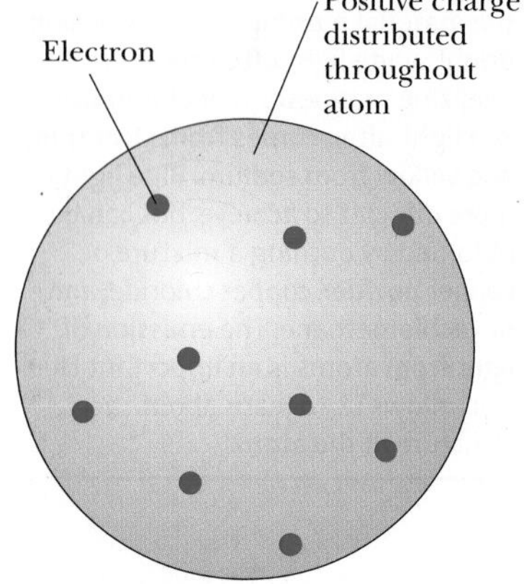

FIGURE 29.1 Thomson's model of the atom: negatively charged electrons in a volume of continuous positive charge.

The structural model of the atom in Newton's day described the atom as a tiny, hard, indestructible sphere, a particle model that ignored any internal structure of the atom. Although this model was a good basis for the kinetic theory of gases (Chapter 16), new structural models had to be devised when later experiments revealed the electrical nature of atoms. J. J. Thomson suggested a structural model that describes the atom as a continuous volume of positive charge with electrons embedded throughout it (Fig. 29.1).

In 1911, Ernest Rutherford and his students Hans Geiger and Ernst Marsden performed a critical experiment that showed that Thomson's model could not be correct. In this experiment, a beam of positively charged alpha particles was projected into a thin metal foil as in Figure 29.2a. Most of the particles passed through the foil as if it were empty space, which is consistent with the Thomson model. Some of the results of the experiment, however, were astounding. Many alpha particles were deflected from their original direction of travel through large angles. Some particles were even deflected backward, reversing their direction of travel. When Geiger informed Rutherford of these results, Rutherford wrote, "It was quite the most incredible event that has ever happened to me in my life. It was almost as incredible as if you fired a 15-inch [artillery] shell at a piece of tissue paper and it came back and hit you."

Such large deflections were not expected on the basis of Thomson's model. According to this model, a positively charged alpha particle would never come close enough to a sufficiently large concentration of positive charge to cause any large-angle deflections. Rutherford explained his astounding results with a new

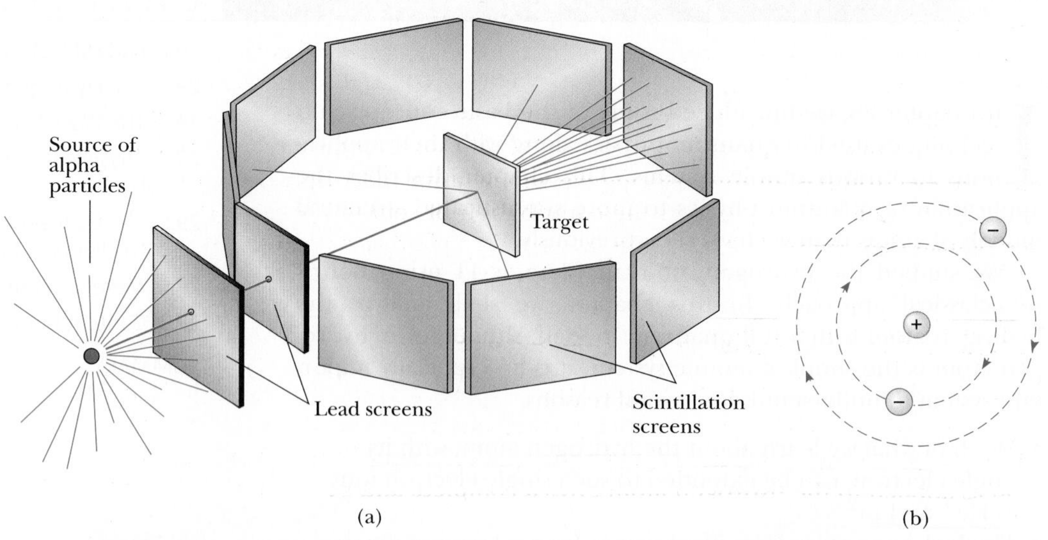

FIGURE 29.2 (a) Rutherford's technique for observing the scattering of alpha particles from a thin foil target. The source is a naturally occurring radioactive substance, such as radium. (b) Rutherford's planetary model of the atom.

structural model: he assumed that the positive charge was concentrated in a region that was small relative to the size of the atom. He called this concentration of positive charge the **nucleus** of the atom. Any electrons belonging to the atom were assumed to be outside the nucleus. To explain why these electrons were not pulled into the nucleus by the attractive electric force, Rutherford imagined that the electrons move in orbits about the nucleus in the same manner as the planets orbit the Sun, as in Figure 29.2b.

There are two basic difficulties with Rutherford's planetary structural model. As we saw in Chapter 11, an atom emits discrete characteristic frequencies of electromagnetic radiation and no others; the Rutherford model is unable to explain this phenomenon. A second difficulty is that Rutherford's electrons experience a centripetal acceleration. According to Maxwell's equations in electromagnetism, charges orbiting with frequency f experience centripetal acceleration and therefore should radiate electromagnetic waves of frequency f. Unfortunately, this classical model leads to disaster when applied to the atom. As the electron radiates energy from the electron–proton system, the radius of the orbit of the electron steadily decreases and its frequency of revolution increases. Energy is continuously transferred out of the system by electromagnetic radiation. As a result, the energy of the system decreases, resulting in the decay of the orbit of the electron. This decrease in total energy leads to an increase in the kinetic energy of the electron,[1] an ever-increasing frequency of emitted radiation, and a rapid collapse of the atom as the electron plunges into the nucleus (Fig. 29.3).

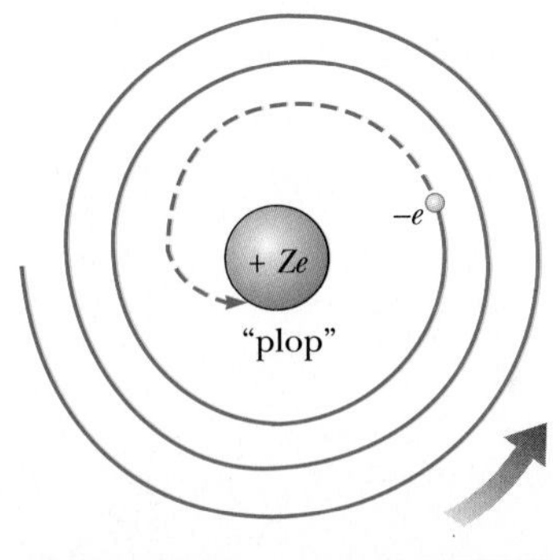

FIGURE 29.3 The classical model of the nuclear atom. Because the accelerating electron radiates energy, the orbit decays until the electron falls into the nucleus.

The stage was set for Bohr! To circumvent the erroneous predictions of the Rutherford model—electrons falling into the nucleus and a continuous emission spectrum from elements—Bohr postulated that classical radiation theory does not hold for atomic-sized systems. He overcame the problem of an atom that continuously loses energy by applying Planck's ideas of quantized energy levels to orbiting atomic electrons. Therefore, as described in Section 11.5, Bohr postulated that electrons in atoms are generally confined to stable, nonradiating orbits called stationary states. Furthermore, he applied Einstein's concept of the photon to arrive at an expression for the frequency of radiation emitted when the atom makes a transition from one stationary state to another.

One of the first indications that the Bohr theory needed modification arose when improved spectroscopic techniques were used to examine the spectral lines of hydrogen. It was found that many of the lines in the Balmer and other series were not single lines at all. Instead, each was a group of closely spaced lines. An additional difficulty arose when it was observed that, in some situations, some single spectral lines were split into three closely spaced lines when the atoms were placed in a strong magnetic field. The Bohr model cannot explain this phenomenon.

Efforts to explain these difficulties with the Bohr model led to improvements in the structural model of the atom. One of the changes introduced was the concept that the electron has an intrinsic angular momentum called *spin,* which we introduced in Chapter 22 in terms of the contribution of spin to the magnetic properties of materials. We shall discuss spin in more detail in this chapter.

29.2 THE HYDROGEN ATOM REVISITED

A quantum treatment of the hydrogen atom requires a solution to the Schrödinger equation (Eq. 28.31), with U being the electric potential energy of the electron–proton system. The full mathematical solution of the Schrödinger equation as applied to the hydrogen atom gives a complete and beautiful description of

[1]As an orbital system that interacts via an inverse square force law loses energy, the kinetic energy of the orbiting object increases but the potential energy of the system decreases by a larger amount; thus, the change in the total energy of the system is negative.

the atom's properties. The mathematical procedures that make up the solution are beyond the scope of this text, however, and so the details shall be omitted. The solutions for some states of hydrogen will be discussed, together with the quantum numbers used to characterize allowed stationary states. We also discuss the physical significance of the quantum numbers.

Let us outline the steps we take in developing a quantum structural model for the hydrogen atom. We apply the quantum particle under boundary conditions model by solving the Schrödinger equation and then applying boundary conditions to the solution to determine the allowed wave functions and energies of the atom. For the particle in a one-dimensional box in Section 28.10, we found that the imposition of boundary conditions generated a single quantum number. For the three-dimensional system of the hydrogen atom, the application of boundary conditions in each dimension introduces a quantum number, so the model will generate three quantum numbers. We also find the need for a fourth quantum number, representing the spin, that cannot be extracted from the Schrödinger equation.

To set up the Schrödinger equation, we must first specify the potential energy function for the system. For the hydrogen atom, this function is

$$U(r) = -k_e \frac{e^2}{r} \qquad [29.1]$$

where k_e is the Coulomb constant and r is the radial distance between the proton (situated at $r = 0$) and the electron.

The formal procedure for solving the problem of the hydrogen atom is to substitute $U(r)$ into the Schrödinger equation and find appropriate solutions to the equation. We did that for the particle in a box in Section 28.12. The current problem is more complicated, however, because it is three dimensional and because U is not constant. In addition, U depends on the radial coordinate r rather than a Cartesian coordinate x, y, or z. As a result, we must use spherical coordinates. We shall not attempt to carry out these solutions because they are quite complicated. Rather, we shall simply describe their properties and some of their implications with regard to atomic structure.

When the boundary conditions are applied to the solutions of the Schrödinger equation, we find that the energies of the allowed states for the hydrogen atom are

▪ Allowed energies for the hydrogen atom

$$E_n = -\left(\frac{k_e e^2}{2a_0}\right)\frac{1}{n^2} = -\frac{13.606 \text{ eV}}{n^2} \qquad n = 1, 2, 3, \ldots \qquad [29.2]$$

where a_0 is the Bohr radius. This result is in precise agreement with the Bohr model and with observed spectral lines, which is a triumph for both the Bohr approach and the quantum approach! Note that the allowed energies in our model depend only on the quantum number n, called the **principal quantum number.**

PITFALL PREVENTION 29.1

ENERGY DEPENDS ON *n* ONLY FOR HYDROGEN The statement after Equation 29.2 that the energy depends only on the quantum number n is a simplification model. The energy levels for all atoms depend primarily on n, but also depend to a lesser degree on other quantum numbers, especially for heavier atoms.

The imposition of boundary conditions also leads to two new quantum numbers that do not appear in the Bohr model. The quantum number ℓ is called the **orbital quantum number,** and m_ℓ is called the **orbital magnetic quantum number.** Although n is related to the energy of the atom, the quantum numbers ℓ and m_ℓ are related to the angular momentum of the atom as described in Section 29.4. From the solution to the Schrödinger equation, we find the following allowed values for these three quantum numbers:

- n is an integer that can range from 1 to ∞.

For a particular value of n,

- ℓ is an integer that can range from 0 to $n - 1$.

For a particular value of ℓ,

- m_ℓ is an integer that can range from $-\ell$ to ℓ.

TABLE 29.1 Three Quantum Numbers for the Hydrogen Atom

Quantum Number	Name	Allowed Values	Number of Allowed States
n	Principal quantum number	$1, 2, 3, \ldots$	Any number
ℓ	Orbital quantum number	$0, 1, 2, \ldots, n-1$	n
m_ℓ	Orbital magnetic quantum number	$-\ell, -\ell+1, \ldots, 0, \ldots, \ell-1, \ell$	$2\ell+1$

Table 29.1 summarizes the rules for determining the allowed values of ℓ and m_ℓ for a given value of n.

For historical reasons, all states with the same principal quantum number are said to form a **shell.** Shells are identified by the letters K, L, M, . . . , which designate the states for which $n = 1, 2, 3, \ldots$. Likewise, all states with given values of n and ℓ are said to form a **subshell.** Based on early practices in spectroscopy, the letters[2] $s, p, d, f, g, h, \ldots$ are used to designate the subshells for which $\ell = 0, 1, 2, 3, 4, 5, \ldots$. For example, the subshell designated by $3p$ has the quantum numbers $n = 3$ and $\ell = 1$; the $2s$ subshell has the quantum numbers $n = 2$ and $\ell = 0$. These notations are summarized in Table 29.2.

States with quantum numbers that violate the rules given in Table 29.1 cannot exist because they do not satisfy the boundary conditions on the wave function of the system. For instance, a $2d$ state, which would have $n = 2$ and $\ell = 2$, cannot exist; the highest allowed value of ℓ is $n - 1$, or 1 in this case. Therefore, for $n = 2$, $2s$ and $2p$ are allowed states but $2d$, $2f$, . . . are not. For $n = 3$, the allowed subshells are $3s$, $3p$, and $3d$.

TABLE 29.2 Atomic Shell and Subshell Notations

n	Shell Symbol	ℓ	Subshell Symbol
1	K	0	s
2	L	1	p
3	M	2	d
4	N	3	f
5	O	4	g
6	P	5	h
.		.	
.		.	
.		.	

QUICK QUIZ 29.1 How many possible subshells are there for the $n = 4$ level of hydrogen? **(a)** 5 **(b)** 4 **(c)** 3 **(d)** 2 **(e)** 1

QUICK QUIZ 29.2 When the principal quantum number is $n = 5$, how many different values of **(a)** ℓ and **(b)** m_ℓ are possible?

EXAMPLE 29.1 The $n = 2$ Level of Hydrogen

For a hydrogen atom, determine the number of allowed states corresponding to the principal quantum number $n = 2$ and calculate the energies of these states.

Solution When $n = 2$, ℓ can be 0 or 1. For $\ell = 0$, m_ℓ can only be 0; for $\ell = 1$, m_ℓ can be -1, 0, or 1. Hence, we have one allowed state designated as the $2s$ state associated with the quantum numbers $n = 2$, $\ell = 0$, and $m_\ell = 0$, and three states designated as $2p$ states for which the quantum numbers are $n = 2$, $\ell = 1$, $m_\ell = -1$; $n = 2$, $\ell = 1$, $m_\ell = 0$; and $n = 2$, $\ell = 1$, $m_\ell = 1$, for a total of four states.

Because all these states have the same principal quantum number, they also have the same energy, which can be calculated with Equation 29.2, with $n = 2$:

$$E_2 = -\frac{13.606 \text{ eV}}{2^2} = -3.401 \text{ eV}$$

29.3 THE WAVE FUNCTIONS FOR HYDROGEN

The potential energy of the hydrogen atom depends only on the radial distance r between nucleus and electron. We therefore expect that some of the allowed states for this atom can be represented by wave functions that depend only on r, which

[2]These seemingly strange letter designations come from descriptions of spectral lines in the early history of spectroscopy: s—sharp; p—principal; d—diffuse; f—fine. After s, p, d, and f, the subsequent letters follow alphabetically from f.

indeed is the case. (Other wave functions depend on r and on the angular coordinates.) The simplest wave function for the hydrogen atom describes the $1s$ state and is designated $\psi_{1s}(r)$:

■ **Wave function for hydrogen in its ground state**

$$\psi_{1s}(r) = \frac{1}{\sqrt{\pi a_0^3}} e^{-r/a_0} \quad [29.3]$$

where a_0 is the Bohr radius and the wave function as given is normalized. This wave function satisfies the boundary conditions mentioned in Section 28.11; that is, ψ_{1s} approaches zero as $r \to \infty$ and remains finite as $r \to 0$. Because ψ_{1s} depends only on r, it is spherically symmetric. In fact, **all s states have spherical symmetry.**

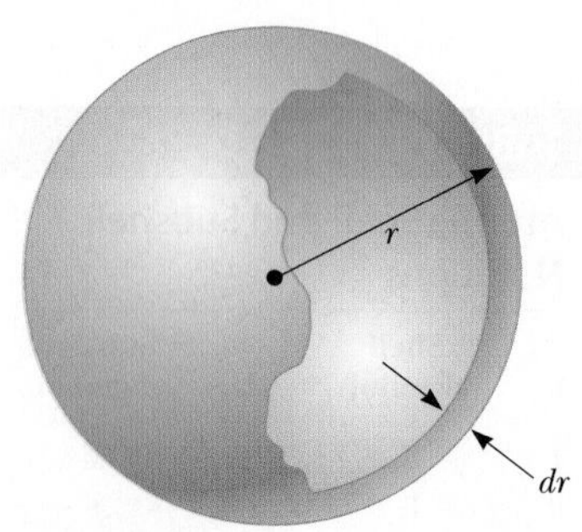

FIGURE 29.4 A spherical shell of radius r and thickness dr has a volume equal to $4\pi r^2\,dr$.

Recall that the probability of finding the electron in any region is equal to an integral of the probability density $|\psi|^2$ over the region, if ψ is normalized. The probability density for the $1s$ state is

$$|\psi_{1s}|^2 = \left(\frac{1}{\pi a_0^3}\right) e^{-2r/a_0} \quad [29.4]$$

The probability of finding the electron in a volume element dV is $|\psi|^2\,dV$. It is convenient to define the **radial probability density function** $P(r)$ as the probability per unit radial distance of finding the electron in a spherical shell of radius r and thickness dr. The volume of such a shell equals its surface area $4\pi r^2$ multiplied by the shell thickness dr (Fig. 29.4), so that

$$P(r)\,dr = |\psi|^2\,dV = |\psi|^2 4\pi r^2\,dr \quad [29.5]$$

$$P(r) = 4\pi r^2 |\psi|^2 \quad [29.6]$$

Substituting Equation 29.4 into Equation 29.6 gives the radial probability density function for the hydrogen atom in its ground state:

■ **Radial probability density for the 1s state of hydrogen**

$$P_{1s}(r) = \left(\frac{4r^2}{a_0^3}\right) e^{-2r/a_0} \quad [29.7]$$

A graphical representation of the function $P_{1s}(r)$ versus r is presented in Figure 29.5a. The peak of the curve corresponds to the most probable value of r for this particular state. The spherical symmetry of the distribution function is shown in Figure 29.5b.

In Example 29.2, we show that the most probable value of r for the ground state of hydrogen equals the Bohr radius a_0. It turns out that the average value of r for the ground state of hydrogen is $\frac{3}{2}a_0$, which is 50% larger than the most probable value of r. (See Problem 29.45.) The reason that the average value is larger than the most probable value lies in the asymmetry in the radial distribution function shown in Figure 29.5a. According to quantum mechanics, the atom has no sharply defined

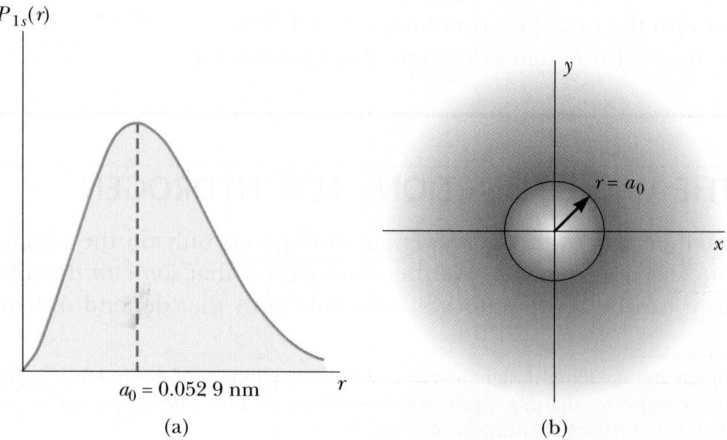

FIGURE 29.5 (a) The probability density of finding the electron as a function of distance from the nucleus for the hydrogen atom in the $1s$ (ground) state. Note that the probability has its maximum value when r equals the Bohr radius a_0. (See Example 29.2.) (b) The cross-section in the xy plane of the spherical electronic charge distribution for the hydrogen atom in its $1s$ state.

boundary. The probability distribution in Figure 29.5a suggests that the charge of the electron is extended throughout a diffuse region of space, commonly referred to as an **electron cloud.** This electron cloud model is quite different from the Bohr model, which places the electron at a fixed distance from the nucleus. Figure 29.5b shows the probability density of the electron in a hydrogen atom in the 1*s* state as a function of position in the *xy* plane. The darkest portion of the distribution appears at $r = a_0$, corresponding to the most probable value of r for the electron.

For an atom in a quantum state that is a solution to the Schrödinger equation, the electron cloud structure remains the same, on the average, over time. Therefore, *the atom does not radiate when it is in one particular quantum state.* This fact removes the problem that plagued the Rutherford model, in which the atom continuously radiates until the electron spirals into the nucleus. Because no change occurs in the charge structure in the electron cloud, the atom does not radiate. Radiation occurs only when a transition is made, so the structure of the electron cloud changes in time.

The next simplest wave function for the hydrogen atom is the one corresponding to the 2*s* state ($n = 2$, $\ell = 0$). The normalized wave function for this state is

$$\psi_{2s}(r) = \frac{1}{4\sqrt{2\pi}}\left(\frac{1}{a_0}\right)^{3/2}\left[2 - \frac{r}{a_0}\right]e^{-r/2a_0} \qquad [29.8]$$

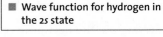

Wave function for hydrogen in the 2*s* state

Like the ψ_{1s} function, ψ_{2s} depends only on r and is spherically symmetric. The energy corresponding to this state is $E_2 = -(13.6 \text{ eV}/4) = -3.4 \text{ eV}$. This energy level represents the first excited state of hydrogen.

A plot of the radial probability density function for this state in comparison to the 1*s* state is shown in Active Figure 29.6. The plot for the 2*s* state has two peaks. In this case, the most probable value corresponds to that value of r that corresponds to the highest value of P_{2s}, which is at $r \approx 5a_0$. An electron in the 2*s* state would be much farther from the nucleus (on the average) than an electron in the 1*s* state.

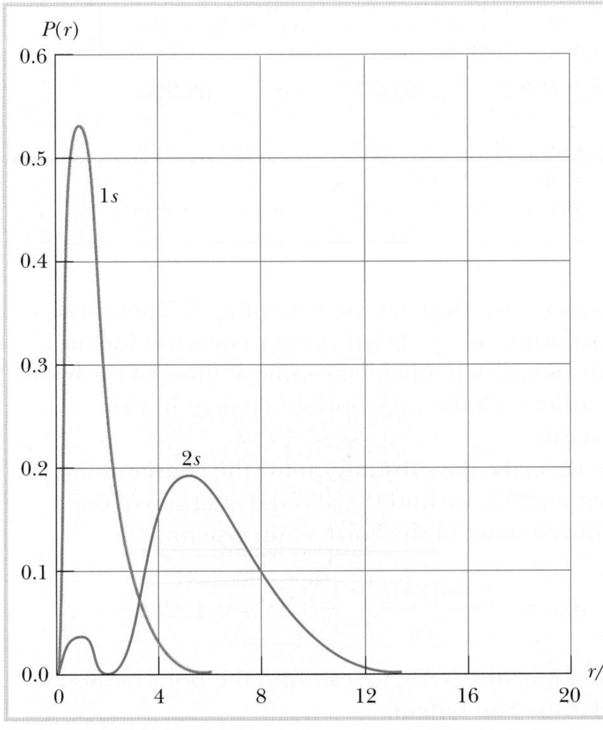

ACTIVE FIGURE 29.6

The radial probability density function versus r/a_0 for the 1*s* and 2*s* states of the hydrogen atom.

Physics Now™ By logging into PhysicsNow at **www.pop4e.com** and going to Active Figure 29.6, you can choose values of r/a_0 and find the probability that the electron is located between two values.

EXAMPLE 29.2 The Ground State of Hydrogen

Calculate the most probable value of r for a hydrogen atom in its ground state.

Solution We conceptualize a hydrogen atom as having a single electron and proton. Because the statement of the problem asks for the "most probable value of r," we categorize this problem as one in which we use the quantum approach. (In the Bohr atom, the electron moves in an orbit with an *exact* value of r.) Therefore, our conceptualization should include the electron cloud image of the electron rather than the well-defined orbits of the Bohr model. To analyze the problem, we note that the most probable value of r corresponds to the peak of the plot of $P_{1s}(r)$ versus r. Because the slope of the curve at this point is zero, we can evaluate the most probable value of r by setting $dP_{1s}/dr = 0$ and solving for r. Using Equation 29.7, we find that

$$\frac{dP_{1s}(r)}{dr} = \frac{d}{dr}\left[\left(\frac{4r^2}{a_0{}^3}\right)e^{-2r/a_0}\right] = 0$$

Carrying out the derivative operation and simplifying the expression, we have

$$e^{-2r/a_0}\frac{d}{dr}(r^2) + r^2\frac{d}{dr}(e^{-2r/a_0}) = 0$$

$$2re^{-2r/a_0} + r^2\left(-\frac{2}{a_0}\right)e^{-2r/a_0} = 0$$

$$(1)\quad 2r\left[1 - \left(\frac{r}{a_0}\right)\right]e^{-2r/a_0} = 0$$

This expression is satisfied if

$$1 - \left(\frac{r}{a_0}\right) = 0 \quad \rightarrow \quad r = a_0$$

To finalize the problem, notice that although the quantum model differs from the Bohr model in that the electron has a finite probability of being at *any* distance from the nucleus, the most probable distance is the same as the orbital radius in the Bohr model! Note also that (1) is satisfied at $r = 0$ and as $r \rightarrow \infty$. These are points of *minimum* probability, which is equal to zero, as seen in Figure 29.5a.

EXAMPLE 29.3 Probabilities for the Electron in Hydrogen

Calculate the probability that the electron in the ground state of hydrogen will be found outside the Bohr radius.

Solution The probability is found by integrating the radial probability density $P_{1s}(r)$ for this state from the Bohr radius a_0 to ∞. Using Equation 29.7, we have

$$P = \int_{a_0}^{\infty} P_{1s}(r)\, dr = \frac{4}{a_0{}^3}\int_{a_0}^{\infty} r^2 e^{-2r/a_0}\, dr$$

We can put the integral in dimensionless form by changing variables from r to $z = 2r/a_0$. Noting that $z = 2$ when $r = a_0$ and that $dr = (a_0/2)\, dz$, we find that

$$P = \tfrac{1}{2}\int_2^{\infty} z^2 e^{-z}\, dz = -\tfrac{1}{2}(z^2 + 2z + 2)e^{-z}\Big|_2^{\infty}$$

$$P = 5e^{-2} = 0.677 \quad \text{or} \quad 67.7\%$$

EXAMPLE 29.4 The Quantized Solar System

Consider the Schrödinger equation for the Earth and the Sun as a system of two particles interacting via the gravitational force. What is the quantum number of the system with the Earth in its present orbit?

Solution The potential energy function for the system is

$$U(r) = -G\frac{M_E M_S}{r}$$

where M_E is the mass of the Earth and M_S is the mass of the Sun. Comparing this expression with Equation 29.1 for the hydrogen atom, $U(r) = -k_e e^2/r$, we see that it has the same mathematical form and that the constant $GM_E M_S$ plays the role of $k_e e^2$. Therefore, the solutions to the Schrödinger equation for the Earth–Sun system will be the same as those of the hydrogen atom with the appropriate change in the constants.

If we make the substitution for the constants in Equation 29.2, we find the allowed energies of the quantized states of the Earth–Sun system:

$$E_n = -\left(\frac{GM_E M_S}{2a_0}\right)\frac{1}{n^2} \qquad n = 1, 2, 3, \ldots$$

From Equation 11.23, we can find the Bohr radius for the Earth–Sun system:

$$a_0 = \frac{\hbar^2}{M_E(GM_EM_S)} = \frac{\hbar^2}{GM_E^2M_S}$$

$$= \frac{(1.055 \times 10^{-34}\ \text{J}\cdot\text{s})^2}{(6.67 \times 10^{-11}\ \text{N}\cdot\text{m}^2/\text{kg}^2)(5.98 \times 10^{24}\ \text{kg})^2(1.99 \times 10^{30}\ \text{kg})}$$

$$= 2.22 \times 10^{-104}\ \text{m}$$

Therefore, evaluating the allowed energies of the system, we have

$$E_n = -\left(\frac{GM_EM_S}{2a_0}\right)\frac{1}{n^2}$$

$$= -\frac{(6.67 \times 10^{-11}\ \text{N}\cdot\text{m}^2/\text{kg}^2)(5.98 \times 10^{24}\ \text{kg})(1.99 \times 10^{30}\ \text{kg})}{2(2.22 \times 10^{-104}\ \text{m})\,n^2}$$

$$= -\frac{1.79 \times 10^{148}\ \text{J}}{n^2} \qquad n = 1, 2, 3, \ldots$$

We now evaluate the energy of the Earth–Sun system from Equation 11.10, assuming a circular orbit:

$$E_n = -\frac{GM_EM_S}{2r}$$

$$= -\frac{(6.67 \times 10^{-11}\ \text{N}\cdot\text{m}^2/\text{kg}^2)(5.98 \times 10^{24}\ \text{kg})(1.99 \times 10^{30}\ \text{kg})}{2(1.50 \times 10^{11}\ \text{m})}$$

$$= -2.65 \times 10^{33}\ \text{J}$$

Finally, we find the quantum number associated with this state:

$$E_n = -\frac{1.79 \times 10^{148}\ \text{J}}{n^2}$$

$$n = \sqrt{\frac{-1.79 \times 10^{148}\ \text{J}}{E_n}} = \sqrt{\frac{-1.79 \times 10^{148}\ \text{J}}{-2.65 \times 10^{33}\ \text{J}}}$$

$$= 2.60 \times 10^{57}$$

This result is a tremendously large quantum number. Therefore, according to the correspondence principle, classical mechanics describes the Earth's motion as well as quantum mechanics does. The energies of quantum states for adjacent values of n are so close together that we do not see the quantized nature of the energy. For example, if the Earth were to move into the next higher quantum state, calculations show that it would be farther from the Sun by a distance on the order of 10^{-80} m. Even on a nuclear scale of 10^{-15} m, that value is undetectable.

29.4 PHYSICAL INTERPRETATION OF THE QUANTUM NUMBERS

As discussed in Section 29.2, the energy of a particular state in our model depends on the principal quantum number. Now let us see what the other three quantum numbers contribute to the physical nature of our quantum structural model of the atom.

The Orbital Quantum Number ℓ

If a particle moves in a circle of radius r, the magnitude of its angular momentum relative to the center of the circle is $L = mvr$. The direction of $\vec{\mathbf{L}}$ is perpendicular to the plane of the circle, and the sense of $\vec{\mathbf{L}}$ is given by a right-hand rule.[3] According to classical physics, L can have any value. The Bohr model of hydrogen, however, postulates that the angular momentum is restricted to integer multiples of $\hbar$; that is, $mvr = n\hbar$. This model must be modified because it predicts (incorrectly) that the ground state of hydrogen ($n = 1$) has one unit of angular momentum. Our quantum model shows that the lowest value of the orbital quantum number, which is related to the orbital momentum, is $\ell = 0$, which corresponds to zero angular momentum.

According to the quantum model, an atom in a state whose principal quantum number is n can take on the following *discrete* values for the magnitude of the **orbital angular momentum** vector:[4]

[3]See Sections 10.8 and 10.9 for a review of this material on angular momentum.

[4]Equation 29.9 on the next page is a direct result of the mathematical solution of the Schrödinger equation and the application of angular boundary conditions. This development, however, is beyond the scope of this text and will not be presented.

■ Allowed values of L

$$|\vec{\mathbf{L}}| = L = \sqrt{\ell(\ell + 1)}\,\hbar \qquad \ell = 0, 1, 2, \ldots, n - 1 \qquad [29.9]$$

That L can be zero in this model points out the difficulties inherent in any attempt to describe results based on quantum mechanics in terms of a purely particle-like model. We cannot think in terms of electrons traveling in well-defined orbits of circular shape or any other shape, for that matter. It is more consistent with the probability notions of quantum physics to imagine the electron smeared out in space in an electron cloud, with the "density" of the cloud highest where the probability is highest. In the quantum mechanical interpretation, the electron cloud for the $L = 0$ state is spherically symmetric and has no fundamental axis of rotation.

EXAMPLE 29.5 Calculating L for a p State

Calculate the magnitude of the orbital angular momentum for a p state of hydrogen.

Solution Because we know that $\hbar = 1.055 \times 10^{-34}\,\mathrm{J \cdot s}$, we can use Equation 29.9 to calculate L. With $\ell = 1$ for a p state, we have

$$L = \sqrt{\ell(\ell + 1)}\hbar = \sqrt{2}\hbar = 1.49 \times 10^{-34}\,\mathrm{J \cdot s}$$

This value is extremely small relative to that of the orbital angular momentum of a macroscopic system, such as the Earth orbiting the Sun, which is about $2.7 \times 10^{40}\,\mathrm{J \cdot s}$. The quantum number that describes L for macroscopic systems, such as the Earth and the Sun, is so large that the separation between adjacent states cannot be measured. We do not see quantized angular momentum for macroscopic systems. Once again, the correspondence principle is upheld.

The Magnetic Orbital Quantum Number m_ℓ

We have seen in the preceding discussion that the magnitude of the orbital angular momentum is quantized. Because angular momentum is a vector, its direction must also be specified. An orbiting electron can be considered an effective current loop with a corresponding magnetic moment. Such a moment placed in a magnetic field $\vec{\mathbf{B}}$ will interact with the field.

Suppose a weak magnetic field is applied to an atom and we define the direction of the field as the z axis. According to quantum mechanics, we find a startling result in that the *direction* of the angular momentum vector relative to the z axis is quantized! Once a z axis is specified, the angular momentum vector can only point in certain directions with respect to this axis. That the direction of $\vec{\mathbf{L}}$ is quantized is often referred to as **space quantization** because we are quantizing a *direction* rather than a *magnitude*.

The quantization of the direction of $\vec{\mathbf{L}}$ is described by giving the allowed z components of the vector. The magnetic orbital quantum number m_ℓ specifies the allowed values of L_z according to the expression

■ Allowed values of L_z

$$L_z = m_\ell \hbar \qquad [29.10]$$

Let us look at the possible orientations of $\vec{\mathbf{L}}$ for a given value of ℓ. Recall that m_ℓ can have values ranging from $-\ell$ to ℓ. If $\ell = 0$, then $L = 0$ and there is no vector for which to consider a direction. If $\ell = 1$, then the possible values of m_ℓ are -1, 0, and 1, so L_z may be $-\hbar$, 0, or $\hbar$. If $\ell = 2$, m_ℓ can be -2, -1, 0, 1, or 2, corresponding to L_z values of $-2\hbar$, $-\hbar$, 0, $\hbar$, or $2\hbar$, and so on.

A useful specialized pictorial representation for understanding space quantization is commonly called a **vector model.** A vector model for $\ell = 2$ is shown in Figure 29.7a. Note that $\vec{\mathbf{L}}$ **can never be aligned parallel or antiparallel to the z axis** because L_z must be smaller than the magnitude of the angular momentum $\vec{\mathbf{L}}$. The vector $\vec{\mathbf{L}}$ can be *perpendicular* to the z axis, which is the case if $m_\ell = 0$. From a

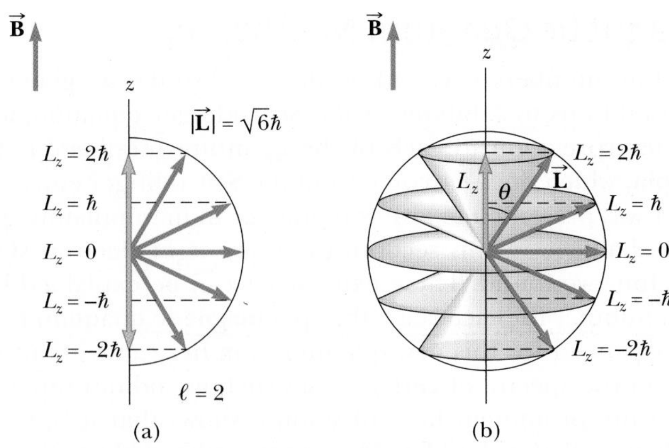

FIGURE 29.7 A vector model for $\ell = 2$. (a) The allowed projections of the orbital angular momentum $\vec{\mathbf{L}}$ relative to a magnetic field that defines the z direction. (b) The orbital angular momentum vector $\vec{\mathbf{L}}$ lies on the surface of a cone.

three-dimensional viewpoint, $\vec{\mathbf{L}}$ can lie on the surfaces of cones that make angles θ with the z axis as shown in Figure 29.7b. From the figure, we see that θ is also quantized and that its values are specified through a relation based on a geometric model triangle with the $\vec{\mathbf{L}}$ vector as the hypotenuse and the z component as one leg of the triangle:

$$\cos\theta = \frac{L_z}{|\vec{\mathbf{L}}|} = \frac{m_\ell}{\sqrt{\ell(\ell+1)}} \qquad [29.11]$$

Note that m_ℓ is never greater than ℓ, so m_ℓ is always smaller than $\sqrt{\ell(\ell+1)}$ and therefore θ can never be zero, consistent with our restriction on $\vec{\mathbf{L}}$ not being parallel to the z axis.

Because of the uncertainty principle, $\vec{\mathbf{L}}$ does not point in a specific direction but rather lies somewhere on a cone as mentioned above. If $\vec{\mathbf{L}}$ had a definite direction, all three components L_x, L_y, and L_z would be exactly specified. For the moment, let us assume this case to be true and let us suppose the electron moves in the xy plane, so the uncertainty $\Delta z = 0$. Because the electron moves in the xy plane, $p_z = 0$. Thus, p_z is *precisely* known, so $\Delta p_z = 0$. The product of these two uncertainties is $\Delta z\,\Delta p_z = 0$, but that is in violation of the uncertainty principle, which requires that $\Delta z\,\Delta p_z \geq \hbar/2$. In reality, only the magnitude of $\vec{\mathbf{L}}$ and one component (which is traditionally chosen as L_z) can have definite values at the same time. In other words, quantum mechanics allows us to specify L and L_z but not L_x and L_y. Because the direction of $\vec{\mathbf{L}}$ is constantly changing, the average values of L_x and L_y are zero and L_z maintains a fixed value $m_\ell\hbar$.

QUICK QUIZ 29.3 Sketch a vector model (shown in Fig. 29.7 for $\ell = 2$) for $\ell = 1$.

INTERACTIVE EXAMPLE 29.6 Space Quantization for Hydrogen

For the hydrogen atom in the $\ell = 3$ state, calculate the magnitude of $\vec{\mathbf{L}}$ and the allowed values of L_z and θ.

Solution We use Equation 29.9 with $\ell = 3$:

$$L = |\vec{\mathbf{L}}| = \sqrt{\ell(\ell+1)}\,\hbar = \sqrt{3(3+1)}\,\hbar = 2\sqrt{3}\,\hbar$$

The allowed values of L_z are $L_z = m_\ell\hbar$ with $m_\ell = -3, -2, -1, 0, 1, 2,$ and 3:

$$L_z = -3\hbar, -2\hbar, -\hbar, 0, \hbar, 2\hbar, 3\hbar$$

Finally, we use Equation 29.11 to calculate the allowed values of θ. Because $L = 2\sqrt{3}\hbar$, we have

$$\cos\theta = \frac{m_\ell}{2\sqrt{3}}$$

Substitution of the allowed values of m_ℓ gives

$$\cos\theta = \pm 0.866, \pm 0.577, \pm 0.289, 0$$

$$\theta = 30.0°, 54.8°, 73.2°, 90.0°, 107°, 125°, 150°$$

PhysicsNow™ Log into PhysicsNow at **www.pop4e.com** and go to Interactive Example 29.6 to practice evaluating the angular momentum for various quantum states of the hydrogen atom.

The Spin Magnetic Quantum Number m_s

The three quantum numbers n, ℓ, and m_ℓ discussed so far are generated by applying boundary conditions to solutions of the Schrödinger equation, and we can assign a physical interpretation to each of the quantum numbers. Let us now consider **electron spin,** which does *not* come from the Schrödinger equation.

Example 29.1 was presented to give you practice in manipulating quantum numbers, but, as we shall see in this section, there are *eight* electron states for $n = 2$ rather than the four we found. These extra states can be explained by requiring a fourth quantum number for each state, the **spin magnetic quantum number** m_s.

Evidence of the need for this new quantum number came about because of an unusual feature in the spectra of certain gases such as sodium vapor. Close examination of one of the prominent lines of sodium shows that it is, in fact, two very closely spaced lines called a doublet. The wavelengths of these lines occur in the yellow region at 589.0 nm and 589.6 nm. In 1925, when this doublet was first noticed, atomic models could not explain it. To resolve this dilemma, Samuel Goudsmit and George Uhlenbeck, following a suggestion by the Austrian physicist Wolfgang Pauli, proposed a new quantum number, called the spin quantum number. The origin of this fourth quantum number was shown by Arnold Sommerfeld and Paul Dirac to lie in the relativistic properties of the electron, which requires four quantum numbers to describe it in four-dimensional space-time.

PITFALL PREVENTION 29.2

THE ELECTRON IS NOT SPINNING Keep in mind Pitfall Prevention 22.3: that the electron is *not* physically spinning. Although the analogy between the spin of the Earth and the electron spin is conceptually useful, it should not be taken literally. The spin of the Earth is a physical rotation. Electron spin is a purely quantum effect that gives the electron an angular momentum as if it were physically spinning.

To describe the spin quantum number, it is convenient (but incorrect!) to think of the electron as spinning on its axis as it orbits the nucleus in a planetary model, just as the Earth spins on its axis as it orbits the Sun. The direction in which the spin angular momentum vector can point is quantized; it can have only two directions as shown in Figure 29.8. If the direction of spin is as shown in Figure 29.8a, the electron is said to have "spin up." If the direction of spin is as shown in Figure 29.8b, the electron is said to have "spin down." In the presence of a magnetic field, the energy of the system (the electron and the magnetic field) is slightly different for the two spin directions, and this energy difference accounts for the sodium doublet. The quantum numbers associated with electron spin are $m_s = \frac{1}{2}$ for the spin-up state and $m_s = -\frac{1}{2}$ for the spin-down state. As we shall see in Example 29.7, this added quantum number doubles the number of allowed states specified by the quantum numbers n, ℓ, and m_ℓ.

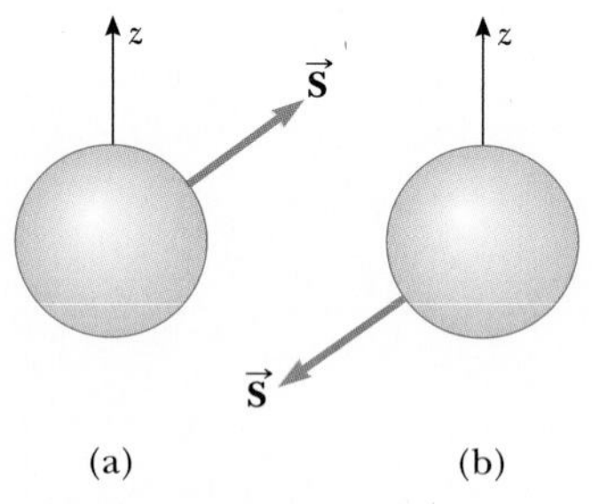

FIGURE 29.8 The spin of an electron can be either (a) up or (b) down relative to a specified z axis. The spin can never be aligned with the axis.

In 1921, Otto Stern and Walther Gerlach performed an experiment (Fig. 29.9) that detected the effects of the force on a magnetic moment in a nonuniform magnetic field. The experiment demonstrated that the angular momentum of an atom is quantized. In their experiment, a beam of neutral silver atoms was sent through a

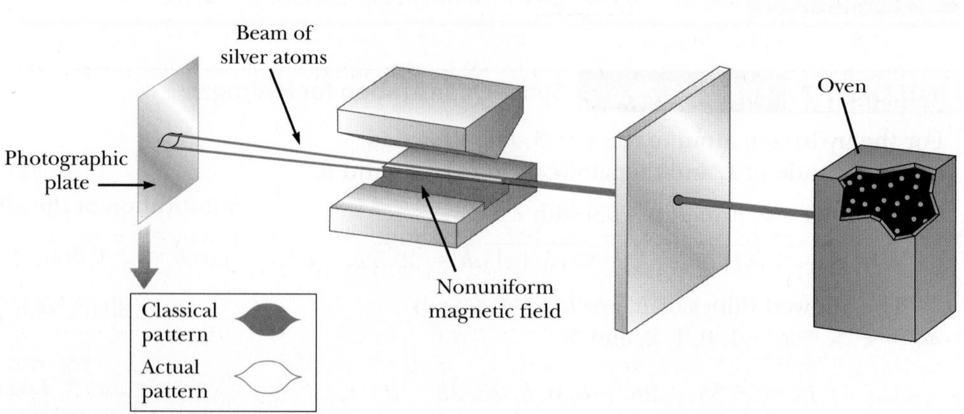

FIGURE 29.9 The apparatus used by Stern and Gerlach to verify space quantization. A beam of neutral silver atoms is split into two components by a nonuniform magnetic field as shown by the actual pattern in the box.

nonuniform magnetic field. In such a situation, the atoms experience a force (in the vertical direction in Fig. 29.9) due to their magnetic moments in this field. Classically, we would expect the beam to be spread out into a continuous distribution on the photographic plate in Figure 29.9 because all possible directions of the atomic magnetic moments are allowed. Stern and Gerlach found, however, that the beam split into two *discrete* components. The experiment was repeated using other atoms, and in each case the beam split into two or more discrete components.

These results are clearly inconsistent with the prediction of a classical model. According to a quantum model, however, the direction of the total angular momentum of the atom, and hence the direction of its magnetic moment, is quantized. Therefore, the deflected beam has an integral number of discrete components, and the number of components determines the number of possible values of μ_z. Because the Stern–Gerlach experiment showed discrete beams, space quantization was at least qualitatively verified.

(Courtesy of AIP Niels Bohr Library, Margarethe Bohr Collection)

Wolfgang Pauli and Niels Bohr watch a spinning top.

For the moment, let us assume that the angular momentum of the atom is due to the orbital angular momentum.[5] Because μ_z is proportional to m_ℓ, the number of possible values of μ_z is $2\ell + 1$. Furthermore, because ℓ is an integer, the number of values of μ_z is always odd. This prediction was not consistent with the observations of Stern and Gerlach, who observed two components, an even number, in the deflected beam of silver atoms. Therefore, although the Stern–Gerlach experiment demonstrated space quantization, the number of components was not consistent with the quantum model developed at that time.

In 1927, T. E. Phipps and J. B. Taylor repeated the Stern–Gerlach experiment using a beam of hydrogen atoms. This experiment is important because it deals with an atom with a single electron in its ground state, for which the quantum model makes reliable predictions. At room temperature, almost all hydrogen atoms are in the ground state. Recall that $\ell = 0$ for hydrogen in its ground state, and so $m_\ell = 0$. Hence, from the orbital angular momentum approach, one would not expect the beam to be deflected by the field at all because μ_z would be zero. The beam in the Phipps–Taylor experiment, however, was again split into two components. On the basis of this result, one can conclude only one thing: that there is some contribution to the angular momentum of the atom and its magnetic moment other than the orbital angular momentum.

As we learned earlier, Goudsmit and Uhlenbeck had proposed that the electron has an intrinsic angular momentum, spin, apart from its orbital angular momentum. In other words, the total angular momentum of the electron in a particular electronic state contains both an orbital contribution $\vec{\mathbf{L}}$ and a spin contribution $\vec{\mathbf{S}}$. A quantum number s exists for spin that is analogous to ℓ for orbital angular momentum. The value of s for an electron, however, is *always* $s = \frac{1}{2}$, unlike ℓ, which varies for different states of the atom.

Like $\vec{\mathbf{L}}$, the **spin angular momentum** vector $\vec{\mathbf{S}}$ must obey the rules of the quantum model. In analogy with Equation 29.9, the **magnitude of the spin angular momentum $\vec{\mathbf{S}}$** for the electron is

$$S = \sqrt{s(s+1)}\hbar = \frac{\sqrt{3}}{2}\hbar \qquad [29.12]$$

■ Spin angular momentum of an electron

This result is the only allowed value for the magnitude of the spin angular momentum vector for an electron, so we usually do not include s in a list of quantum numbers describing states of the atom. Like orbital angular momentum, spin angular momentum is quantized in space as described in Figure 29.10. It can have two orientations, specified by the spin magnetic quantum number m_s, where m_s has two possible values, $\pm\frac{1}{2}$. In analogy with Equation 29.10, the z component of spin angular

[5]The Stern–Gerlach experiment was performed in 1921, before spin was hypothesized, so orbital angular momentum was the only type of angular momentum in the quantum model at the time.

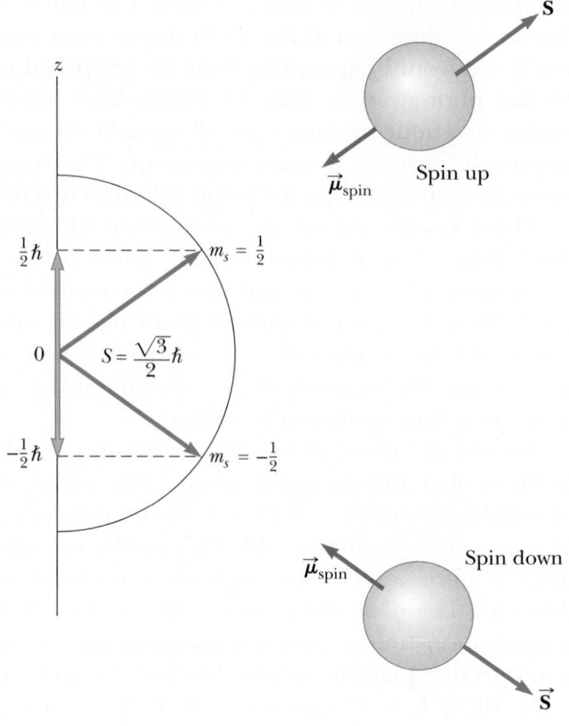

FIGURE 29.10 Spin angular momentum $\vec{\mathbf{S}}$ exhibits space quantization. This figure shows the two allowed orientations of the spin angular momentum vector $\vec{\mathbf{S}}$ and the spin magnetic moment vector $\vec{\boldsymbol{\mu}}_{spin}$ for a spin-$\frac{1}{2}$ particle such as the electron.

momentum is

$$S_z = m_s\hbar = \pm \tfrac{1}{2}\hbar \qquad [29.13]$$

The two values $\pm\hbar/2$ for S_z correspond to the two possible orientations for $\vec{\mathbf{S}}$ shown in Figure 29.10. The quantum number m_s is listed as the fourth quantum number describing a particular state of the atom.

The spin magnetic moment $\vec{\boldsymbol{\mu}}_s$ of the electron is related to its spin angular momentum $\vec{\mathbf{S}}$ by the expression

$$\vec{\boldsymbol{\mu}}_s = -\frac{e}{m_e}\vec{\mathbf{S}} \qquad [29.14]$$

Because $S_z = \pm\frac{1}{2}\hbar$, the z component of the spin magnetic moment can have the values

$$\mu_{sz} = \pm\frac{e\hbar}{2m_e} \qquad [29.15]$$

The quantity $e\hbar/2m_e$ is called the **Bohr magneton** μ_B and has the numerical value 9.274×10^{-24} J/T.

Today physicists explain the outcome of the Stern–Gerlach experiment as follows. The observed moments for both silver and hydrogen are due to spin angular momentum alone and not to orbital angular momentum. (The hydrogen atom in its ground state has $\ell = 0$; for silver, used in the Stern–Gerlach experiment, the net orbital angular momentum for all the electrons is $|\vec{\mathbf{L}}| = 0$.) A single-electron atom such as hydrogen has its electron spin quantized in the magnetic field in such a way that its z component of spin angular momentum is either $\frac{1}{2}\hbar$ or $-\frac{1}{2}\hbar$, corresponding to $m_s = \pm\frac{1}{2}$. Electrons with spin $+\frac{1}{2}$ are deflected in one direction by the nonuniform magnetic field, and those with spin $-\frac{1}{2}$ are deflected in the opposite direction.

The Stern–Gerlach experiment provided two important results. First, it verified the concept of space quantization. Second, it showed that spin angular momentum exists even though this property was not recognized until long after the experiments were performed.

Thinking Physics 29.1

Does the Stern–Gerlach experiment differentiate between orbital angular momentum and spin angular momentum?

Reasoning A magnetic force on the magnetic moment arises from both orbital angular momentum and spin angular momentum. In this sense, the experiment does not differentiate between the two. The number of components on the screen does tell us something, however, because orbital angular momenta are described by an integral quantum number ℓ, whereas spin angular momentum depends on a half-integral quantum number s. If an odd number of components occur on the screen, three possibilities arise: the atom has (1) orbital angular momentum only, (2) an even number of electrons with spin angular momentum, or (3) a combination of orbital angular momentum and an even number of electrons with spin angular momentum. If an even number of components occurs on the screen, at least one unpaired spin angular momentum exists, possibly in combination with orbital angular momentum. The only numbers of components for which we can specify the type of angular momentum are one component (no orbital, no spin) and two components (spin of one electron). Once we see more than two components multiple possibilities arise because of various combinations of $\vec{\mathbf{L}}$ and $\vec{\mathbf{S}}$. ■

EXAMPLE 29.7 Putting a Spin on Hydrogen

For a hydrogen atom, determine the quantum numbers associated with the possible states that correspond to the principal quantum number $n = 2$.

Solution With the results from Example 29.1 and the addition of the spin quantum number, we have the possibilities given in the table to the right. Therefore, there are eight possible states.

n	ℓ	m_ℓ	m_s	Subshell	Shell	Number of States in Subshell
2	0	0	$\frac{1}{2}$	$2s$	L	2
2	0	0	$-\frac{1}{2}$			
2	1	1	$\frac{1}{2}$	$2p$	L	6
2	1	1	$-\frac{1}{2}$			
2	1	0	$\frac{1}{2}$			
2	1	0	$-\frac{1}{2}$			
2	1	-1	$\frac{1}{2}$			
2	1	-1	$-\frac{1}{2}$			

29.5 THE EXCLUSION PRINCIPLE AND THE PERIODIC TABLE

The quantum model generated from the Schrödinger equation is based on the hydrogen atom, which is a system consisting of one electron and one proton. As soon as we consider the next atom, helium, we introduce complicating factors. The two electrons in helium both interact with the nucleus, so we can define a potential energy function for those interactions. They also interact with each other, however. The line of action of the electron–nucleus interaction is along a line between the electron and the nucleus. The line of action of the electron–electron interaction is along the line between the two electrons, which is different from that of the electron–nucleus interaction. Thus, the Schrödinger equation is extremely difficult to solve. As we consider atoms with more and more electrons, the possibility of an algebraic solution of the Schrödinger equation becomes hopeless.

We find, however, that despite our inability to solve the Schrödinger equation, **we can use the same four quantum numbers developed for hydrogen for the electrons in heavier atoms.** We are not able to calculate the quantized energy levels easily, but we can gain information about the levels from theoretical models and experimental measurements.

PITFALL PREVENTION 29.3

QUANTUM NUMBERS DESCRIBE A SYSTEM The common usage is to assign the quantum numbers to an electron. Remember, however, that these quantum numbers arise from the Schrödinger equation, which involves a potential energy function for the *system* consisting of the electron and the nucleus. Therefore, it is more *proper* to assign the quantum numbers to the atom, but it is more *popular* to assign them to an electron. We will follow this latter usage because it is so common, but keep the notion of the system in the back of your mind.

(AIP Emilio Segrè Visual Archives, Goudsmit Collection)

WOLFGANG PAULI (1900–1958)

An extremely talented Austrian theoretical physicist, Pauli made important contributions in many areas of modern physics. Pauli gained public recognition at the age of 21 with a masterful review article on relativity, which is still considered one of the finest and most comprehensive introductions to the subject. Other major contributions were the discovery of the exclusion principle, the explanation of the connection between particle spin and statistics, and theories of relativistic quantum electrodynamics, the neutrino hypothesis, and the hypothesis of nuclear spin.

Because a quantum state in any atom is specified by four quantum numbers, n, ℓ, m_ℓ, and m_s, an obvious and important question is, "How many electrons in an atom can have a particular set of quantum numbers?" Pauli provided an answer in 1925 in a powerful statement known as the **exclusion principle:**

> No two electrons in an atom can ever be in the same quantum state; that is, no two electrons in the same atom can have the same set of quantum numbers.

It is interesting that if this principle were not valid, every atom would radiate energy by means of photons and end up with all electrons in the lowest energy state. The chemical behavior of the elements would be grossly modified because this behavior depends on the electronic structure of atoms. Nature as we know it would not exist! In reality, we can view the electronic structure of complex atoms as a succession of filled levels increasing in energy, where the outermost electrons are primarily responsible for the chemical properties of the element.

Imagine building an atom by forming the nucleus and then filling in the available quantum states with electrons until the atom is neutral. We shall use the common language here that "electrons go into available states." Keep in mind, however, that the states are those of the *system* of the atom. As a general rule, the order of filling of an atom's subshells with electrons is as follows. Once one subshell is filled, the next electron goes into the vacant subshell that is lowest in energy.

Before we discuss the electronic configurations of some elements, it is convenient to define an **orbital** as the state of an electron characterized by the quantum numbers n, ℓ, and m_ℓ. From the exclusion principle, we see that **at most two electrons can be in any orbital.** One of these electrons has $m_s = +\frac{1}{2}$ and the other has $m_s = -\frac{1}{2}$. Because each orbital is limited to two electrons, the numbers of electrons that can occupy the shells are also limited.

Table 29.3 shows the allowed quantum states for an atom up to $n = 3$. Each square in the bottom row of the table represents one orbital, with the ↑ arrows representing $m_s = +\frac{1}{2}$ and the ↓ arrows representing $m_s = -\frac{1}{2}$. The $n = 1$ shell can accommodate only two electrons because only one orbital is allowed with $m_\ell = 0$. The $n = 2$ shell has two subshells, with $\ell = 0$ and $\ell = 1$. The $\ell = 0$ subshell is limited to only two electrons because $m_\ell = 0$. The $\ell = 1$ subshell has three allowed orbitals, corresponding to $m_\ell = 1$, 0, and -1. Because each orbital can accommodate two electrons, the $\ell = 1$ subshell can hold six electrons (and the $n = 2$ shell can hold eight). The $n = 3$ shell has three subshells and nine orbitals and can accommodate up to 18 electrons. In general, each shell can accommodate up to $2n^2$ electrons.

The results of the exclusion principle can be illustrated by an examination of the electronic arrangement in a few of the lighter atoms. For example, **hydrogen** has only one electron, which, in its ground state, can be described by either of two sets of quantum numbers: 1, 0, 0, $+\frac{1}{2}$ or 1, 0, 0, $-\frac{1}{2}$. The electronic configuration of this atom is often designated as $1s^1$. The notation $1s$ refers to a state for which $n = 1$ and $\ell = 0$, and the superscript indicates that one electron is present in the s subshell.

Neutral **helium** has two electrons. In the ground state, the quantum numbers for these two electrons are 1, 0, 0, $+\frac{1}{2}$ and 1, 0, 0, $-\frac{1}{2}$. No other combinations of

PITFALL PREVENTION 29.4

THE EXCLUSION PRINCIPLE IS MORE GENERAL The exclusion principle stated here is a limited form of the more general exclusion principle, which states that no two *fermions,* which are *all* particles with half-integral spin $\frac{1}{2}, \frac{3}{2}, \frac{5}{2}, \ldots$ can be in the same quantum state. The present form is satisfactory for our discussions of atomic physics, and we will discuss the general form further in Chapter 31.

TABLE 29.3 Allowed Quantum States for an Atom Up to $n = 3$

n	1	2				3								
ℓ	0	0	1			0	1			2				
m_ℓ	0	0	1	0	−1	0	1	0	−1	2	1	0	−1	−2
m_s	↑↓	↑↓	↑↓	↑↓	↑↓	↑↓	↑↓	↑↓	↑↓	↑↓	↑↓	↑↓	↑↓	↑↓

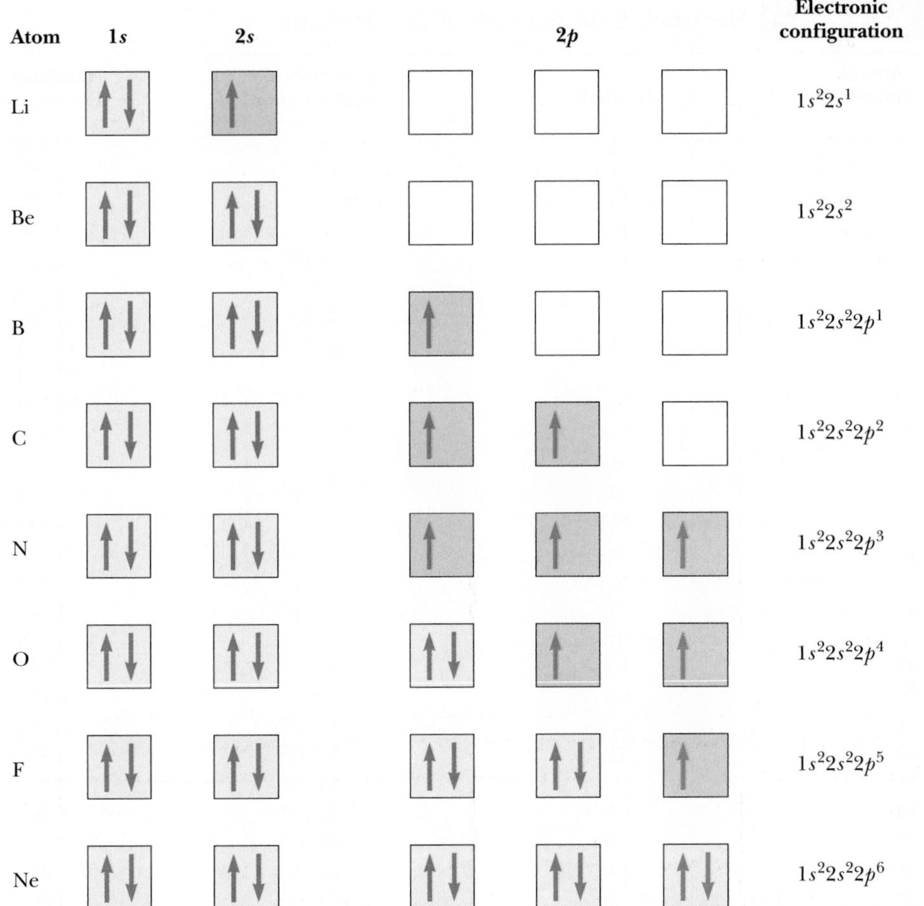

Atom
1s
2s
2p
Electronic configuration
Li
Be
B
C
N
O
F
Ne
$1s^22s^1$
$1s^22s^2$
$1s^22s^22p^1$
$1s^22s^22p^2$
$1s^22s^22p^3$
$1s^22s^22p^4$
$1s^22s^22p^5$
$1s^22s^22p^6$

FIGURE 29.11 The filling of electronic states must obey both the exclusion principle and Hund's rules.

quantum numbers are possible for this level, and we say that the K shell is filled. The electronic configuration of helium is designated as $1s^2$.

The electronic configurations of some successive elements are given in Figure 29.11. Neutral **lithium** has three electrons. In the ground state, two of them are in the $1s$ subshell and the third is in the $2s$ subshell because this subshell is lower in energy than the $2p$ subshell. (In addition to the simple dependence of E on n in Eq. 29.2, there is an additional dependence on ℓ, which will be addressed in Section 29.6.) Hence, the electronic configuration for lithium is $1s^22s^1$.

Note that the electronic configuration of **beryllium,** with its four electrons, is $1s^22s^2$, and **boron** has a configuration of $1s^22s^22p^1$. The $2p$ electron in boron may be described by one of six sets of quantum numbers, corresponding to six states of equal energy.

Carbon has six electrons, and a question arises concerning how to assign the two $2p$ electrons. Do they go into the same orbital with paired spins (↑↓), or do they occupy different orbitals with unpaired spins (↑↑ or ↓↓)? Experimental data show that the lowest energy configuration is the latter, where the spins are unpaired. Hence, the two $2p$ electrons in carbon and the three $2p$ electrons in nitrogen have unpaired spins in the ground state (see Fig. 29.11). The general rules that govern such situations throughout the periodic table are called **Hund's rules.** The rule appropriate for elements like carbon is that **when an atom has orbitals of equal energy, the order in which they are filled by electrons is such that a maximum number of electrons will have unpaired spins.** Some exceptions to this rule occur in elements having subshells close to being filled or half-filled.

A complete list of electronic configurations is provided in the tabular representation in Table 29.4. An early attempt to find some order among the elements was

TABLE 29.4 Electronic Configuration of the Elements

Atomic Number Z	Symbol	Ground-State Configuration	Ionization Energy (eV)
1	H	$1s^1$	13.595
2	He	$1s^2$	24.581
3	Li	[He] $2s^1$	5.39
4	Be	$2s^2$	9.320
5	B	$2s^22p^1$	8.296
6	C	$2s^22p^2$	11.256
7	N	$2s^22p^3$	14.545
8	O	$2s^22p^4$	13.614
9	F	$2s^22p^5$	17.418
10	Ne	$2s^22p^6$	21.559
11	Na	[Ne] $3s^1$	5.138
12	Mg	$3s^2$	7.644
13	Al	$3s^23p^1$	5.984
14	Si	$3s^23p^2$	8.149
15	P	$3s^23p^3$	10.484
16	S	$3s^23p^4$	10.357
17	Cl	$3s^23p^5$	13.01
18	Ar	$3s^23p^6$	15.755
19	K	[Ar] $4s^1$	4.339
20	Ca	$4s^2$	6.111
21	Sc	$3d^14s^2$	6.54
22	Ti	$3d^24s^2$	6.83
23	V	$3d^34s^2$	6.74
24	Cr	$3d^54s^1$	6.76
25	Mn	$3d^54s^2$	7.432
26	Fe	$3d^64s^2$	7.87
27	Co	$3d^74s^2$	7.86
28	Ni	$3d^84s^2$	7.633
29	Cu	$3d^{10}4s^1$	7.724
30	Zn	$3d^{10}4s^2$	9.391
31	Ga	$3d^{10}4s^24p^1$	6.00
32	Ge	$3d^{10}4s^24p^2$	7.88
33	As	$3d^{10}4s^24p^3$	9.81
34	Se	$3d^{10}4s^24p^4$	9.75
35	Br	$3d^{10}4s^24p^5$	11.84
36	Kr	$3d^{10}4s^24p^6$	13.996
37	Rb	[Kr] $5s^1$	4.176
38	Sr	$5s^2$	5.692
39	Y	$4d^15s^2$	6.377
40	Zr	$4d^25s^2$	
41	Nb	$4d^45s^1$	6.881
42	Mo	$4d^55s^1$	7.10
43	Tc	$4d^65s^1$	7.228
44	Ru	$4d^75s^1$	7.365
45	Rh	$4d^85s^1$	7.461
46	Pd	$4d^{10}$	8.33
47	Ag	$4d^{10}5s^1$	7.574
48	Cd	$4d^{10}5s^2$	8.991
49	In	$4d^{10}5s^25p^1$	
50	Sn	$4d^{10}5s^25p^2$	7.342
51	Sb	$4d^{10}5s^25p^3$	8.639
52	Te	$4d^{10}5s^25p^4$	9.01
53	I	$4d^{10}5s^25p^5$	10.454
54	Xe	$4d^{10}5s^25p^6$	12.127
55	Cs	[Xe] $6s^1$	3.893
56	Ba	$6s^2$	5.210
57	La	$5d^16s^2$	5.61

TABLE 29.4 Electronic Configuration of the Elements *(Continued)*

Atomic Number Z	Symbol	Ground-State Configuration	Ionization Energy (eV)
58	Ce	$4f^15d^16s^2$	6.54
59	Pr	$4f^36s^2$	5.48
60	Nd	$4f^46s^2$	5.51
61	Pm	$4f^56s^2$	
62	Fm	$4f^66s^2$	5.6
63	Eu	$4f^76s^2$	5.67
64	Gd	$4f^75d^16s^2$	6.16
65	Tb	$4f^96s^2$	6.74
66	Dy	$4f^{10}6s^2$	
67	Ho	$4f^{11}6s^2$	
68	Er	$4f^{12}6s^2$	
69	Tm	$4f^{13}6s^2$	
70	Yb	$4f^{14}6s^2$	6.22
71	Lu	$4f^{14}5d^16s^2$	6.15
72	Hf	$4f^{14}5d^26s^2$	7.0
73	Ta	$4f^{14}5d^36s^2$	7.88
74	W	$4f^{14}5d^46s^2$	7.98
75	Re	$4f^{14}5d^56s^2$	7.87
76	Os	$4f^{14}5d^66s^2$	8.7
77	Ir	$4f^{14}5d^76s^2$	9.2
78	Pt	$4f^{14}5d^96s^1$	8.88
79	Au	[Xe, $4f^{14}5d^{10}$] $6s^1$	9.22
80	Hg	$6s^2$	10.434
81	Tl	$6s^26p^1$	6.106
82	Pb	$6s^26p^2$	7.415
83	Bi	$6s^26p^3$	7.287
84	Po	$6s^26p^4$	8.43
85	At	$6s^26p^5$	
86	Rn	$6s^26p^6$	10.745
87	Fr	[Rn] $7s^1$	
88	Ra	$7s^2$	5.277
89	Ac	$6d^17s^2$	6.9
90	Th	$6d^27s^2$	
91	Pa	$5f^26d^17s^2$	
92	U	$5f^36d^17s^2$	4.0
93	Np	$5f^46d^17s^2$	
94	Pu	$5f^67s^2$	
95	Am	$5f^77s^2$	
96	Cm	$5f^76d^17s^2$	
97	Bk	$5f^97s^2$	
98	Cf	$5f^{10}7s^2$	
99	Es	$5f^{11}7s^2$	
100	Fm	$5f^{12}7s^2$	
101	Md	$5f^{13}7s^2$	
102	No	$5f^{14}7s^2$	
103	Lr	$5f^{14}7s^27p^1$	
104	Rf	$5f^{14}6d^27s^2$	
105	Db	$5f^{14}6d^37s^2$	
106	Sg	$5f^{14}6d^47s^2$	
107	Bh	$5f^{14}6d^57s^2$	
108	Hs	$5f^{14}6d^67s^2$	
109	Mt	$5f^{14}6d^77s^2$	
110	Ds	$5f^{14}6d^97s^1$	

Note: The bracket notation is used as a shorthand method to avoid repetition in indicating inner-shell electrons. Therefore, [He] represents $1s^2$, [Ne] represents $1s^22s^22p^6$, [Ar] represents $1s^22s^22p^63s^23p^6$, and so on. Configurations for elements above $Z = 102$ are tentative.

made by a Russian chemist, Dmitri Mendeleev, in 1871. He developed a tabular representation of the elements, which has become one of the most important, as well as well-recognized, tools of science. He arranged the atoms in a table (similar to that in Appendix C) according to their atomic masses and chemical similarities. Thus was born the first **periodic table of the elements.** The first table Mendeleev proposed contained many blank spaces, and he boldly stated that the gaps were there only because the elements had not yet been discovered. By noting the columns in which these missing elements should be located, he was able to make rough predictions about their chemical properties. Within 20 years of Mendeleev's announcement, the missing elements were indeed discovered. The predictions made possible by this table represent an excellent example of the power of presenting information in an alternative representation.

The elements in the periodic table are arranged so that all those in a vertical column have similar chemical properties. For example, consider the elements in the last column: He (helium), Ne (neon), Ar (argon), Kr (krypton), Xe (xenon), and Rn (radon). The outstanding characteristic of all these elements is that they do not normally take part in chemical reactions; that is, they do not readily join with other atoms to form molecules. They are therefore called *inert gases.*

We can partially understand this behavior by looking at the electronic configurations in Table 29.4. The element helium is one in which the electronic configuration is $1s^2$; in other words, one shell is filled. Additionally, it is found that the energy associated with this filled shell is considerably lower than the energy of the next available level, the $2s$ level. Next, look at the electronic configuration for neon, $1s^22s^22p^6$. Again, the outermost shell is filled, and a gap in energy occurs between the $2p$ level and the $3s$ level. Argon has the configuration $1s^22s^22p^63s^23p^6$. Here, the $3p$ subshell is filled, and a gap in energy arises between the $3p$ subshell and the $3d$ subshell. We could continue this procedure through all the inert gases; the pattern remains the same. An inert gas is formed when either a shell or a subshell is filled and a gap in energy occurs before the next possible level is encountered.

If we consider the column to the left of the inert gases in the periodic table, we find a group of elements called the *halogens*: fluorine, chlorine, bromine, iodine, and astatine. At room temperature, fluorine and chlorine are gases, bromine is a liquid, and iodine and astatine are solids. In each of these atoms, the outer subshell is one electron short of being filled. As a result, the halogens are chemically very active, readily accepting an electron from another atom to form a closed shell. The halogens tend to from strong ionic bonds with atoms at the other side of the periodic table. In a halogen lightbulb, bromine or iodine atoms combine with tungsten atoms evaporated from the filament and return them to the filament, resulting in a longer-lasting bulb. In addition, the filament can be operated at a higher temperature than in ordinary lightbulbs, giving a brighter and whiter light.

At the left side of the periodic table, the Group I elements consist of hydrogen and the *alkali metals,* lithium, sodium, potassium, rubidium, cesium, and francium. Each of these atoms contains one electron in a subshell outside of a closed subshell. Therefore, these elements easily form positive ions because the lone electron is bound with a relatively low energy and is easily removed. Thus, the alkali metal atoms are chemically active and form very strong bonds with halogen atoms. For example, table salt, NaCl, is a combination of an alkali metal and a halogen. Because the outer electron is weakly bound, pure alkali metals tend to be good electrical conductors, although, because of their high chemical activity, pure alkali metals are not generally found in nature.

Table 29.4 also lists the ionization energies for certain elements. It is interesting to plot ionization energy versus the atomic number Z as in Figure 29.12. Note the pattern of differences in atomic numbers between the peaks in the graph: 8, 8, 18, 18, 32. This pattern follows from the Pauli exclusion principle and helps explain why the elements repeat their chemical properties in groups. For example, the

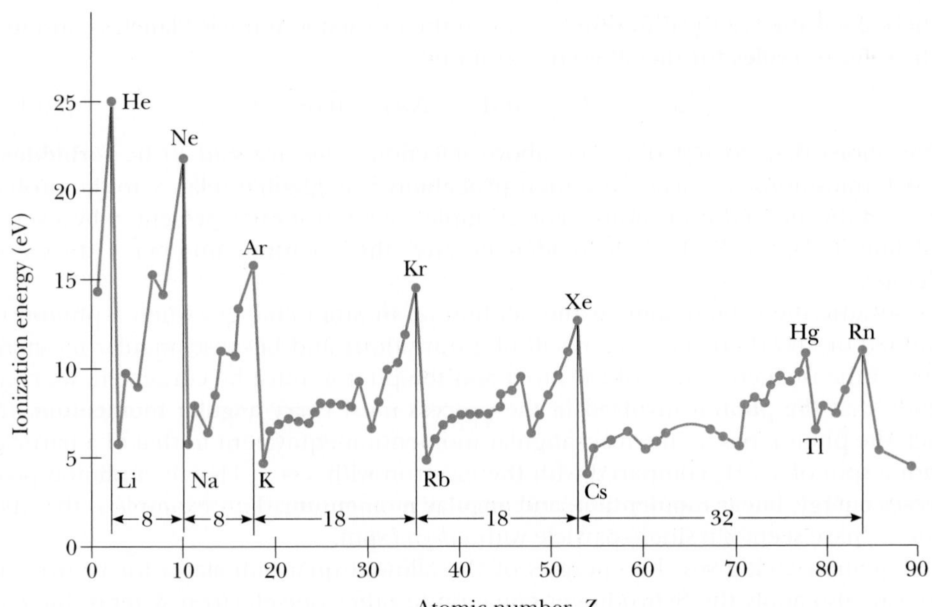

FIGURE 29.12 Ionization energy of the elements versus atomic number.

peaks at $Z = 2$, 10, 18, and 36 correspond to the elements He, Ne, Ar, and Kr, which have filled shells. These elements have similar chemical behavior.

QUICK QUIZ 29.4 Rank the energy necessary to remove the outermost electron from the following three elements, smallest to largest: lithium, potassium, cesium.

29.6 MORE ON ATOMIC SPECTRA: VISIBLE AND X-RAY

In Chapter 11, we briefly discussed the origin of the spectral lines for hydrogen and hydrogen-like ions. Recall that an atom in an excited state will emit electromagnetic radiation if it makes a transition to a lower energy state.

The energy level diagram for hydrogen is shown in Figure 29.13. This semigraphical representation is different from Figure 11.20 in that the individual states corresponding to different values of ℓ within a given value of n are spread out horizontally. Figure 29.13 shows only those states up to $\ell = 2$; the shells from $n = 4$ upward would have more sets of states to the right, which are not shown.

The diagonal lines in Figure 29.13 represent allowed transitions between stationary states. Whenever an atom makes a transition from a higher energy state to a lower one, a photon of light is emitted. The frequency of this photon is $f = \Delta E/h$,

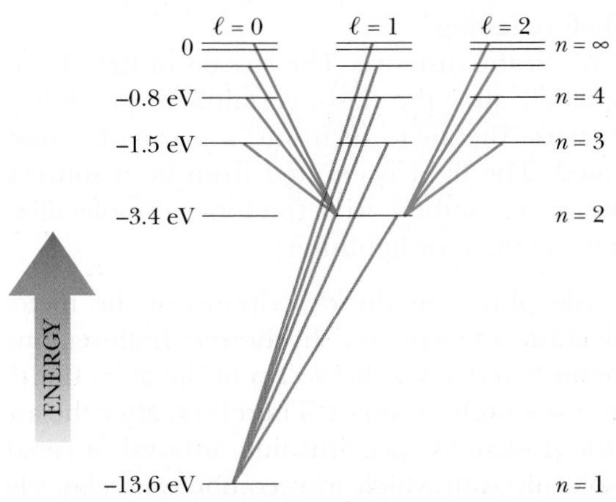

FIGURE 29.13 Some allowed electronic transitions for hydrogen, represented by the colored lines. These transitions must obey the selection rule $\Delta\ell = \pm 1$.

where ΔE is the energy difference between the two states and h is Planck's constant. The **selection rules** for the allowed transitions are

■ Selection rules for allowed atomic transitions

$$\Delta \ell = \pm 1 \quad \text{and} \quad \Delta m_\ell = 0 \text{ or } \pm 1 \qquad [29.16]$$

Transitions that do not obey the above selection rules are said to be **forbidden.** (Such transitions can occur, but their probability is negligible relative to the probability of the allowed transitions.) For example, any transition represented by a vertical line in Figure 29.13 is forbidden because the quantum number ℓ does not change.

Because the orbital angular momentum of an atom changes when a photon is emitted or absorbed (i.e., as a result of a transition) and because angular momentum of the isolated system of the atom and the photon must be conserved, we conclude that **the photon involved in the process must carry angular momentum.** In fact, the photon has an intrinsic angular momentum equivalent to that of a particle with a spin of $s = 1$, compared with the electron with $s = \frac{1}{2}$. Hence, **a photon possesses energy, linear momentum, and angular momentum.** This example is the first one we have seen of a single particle with *integral* spin.

Equation 29.2 gives the energies of the allowed quantum states for hydrogen. We can also apply the Schrödinger equation to other one-electron systems, such as the He^+ and Li^{++} ions. The primary difference between these ions and the hydrogen atom is the different number of protons Z in the nucleus. The result is a generalization of Equation 29.2 for these other one-electron systems:

$$E_n = -\frac{(13.606 \text{ eV})\, Z^2}{n^2} \qquad [29.17]$$

For outer electrons in multielectron atoms, the nuclear charge Ze is largely canceled or shielded by the negative charge of the inner-core electrons. Hence, the outer electrons interact with a net charge that is reduced below the actual charge of the nucleus. (According to Gauss's law, the electric field at the position of an outer electron depends on the net charge of the nucleus and the electrons closer to the nucleus.) The expression for the allowed energies for multielectron atoms has the same form as Equation 29.17, with Z replaced by an effective atomic number Z_{eff}. That is,

$$E_n \approx -\frac{(13.6 \text{ eV})\, Z_{\text{eff}}^2}{n^2} \qquad [29.18]$$

where Z_{eff} depends on n and ℓ.

■ Thinking Physics 29.2

A physics student is watching a meteor shower in the early morning hours. She notices that the streaks of light from the meteoroids entering the very high regions of the atmosphere last for up to 2 or 3 s before fading.

She also notices a lightning storm off in the distance. The streaks of light from the lightning fade away almost immediately after the flash, certainly in much less than 1 s. Both lightning and meteors cause the air to turn into a plasma because of the very high temperatures generated. The light is emitted from both sources when the stripped electrons in the plasma recombine with the ionized molecules. Why would this light last longer for meteors than for lightning?

Reasoning The answer lies in the subtle phrase in the description of the meteoroids "entering the very high regions of the atmosphere." In the very high regions of the atmosphere, the pressure of the air is very low. The *density* of the air is therefore very low, so molecules of the air are relatively far apart. Therefore, after the air is ionized by the passing meteoroid, the probability per unit time interval of freed electrons encountering an ionized molecule with which to recombine is relatively low. As a result, the recombination process for all freed electrons occurs over a relatively long time interval, measured in seconds.

On the other hand, lightning occurs in the lower regions of the atmosphere (the troposphere) where the pressure and density are relatively high. After the ionization by the lightning flash, the freed electrons and ionized molecules are much closer together than in the upper atmosphere. The probability per unit time interval of a recombination is much higher, and the time interval for the recombination of all the electrons and ions to occur is much shorter. ■

X-Ray Spectra

X-rays are emitted from a metal target that is being bombarded by high-energy electrons. If we consider the target as the system, the continuity equation for energy (see Eq. 6.20) for this process can be written as

$$\Delta E_{\text{system}} = \sum T \quad \rightarrow \quad \Delta E_{\text{int}} = T_{\text{MT}} + T_{\text{ER}}$$

The matter-transfer term on the right-hand side represents the process by which energy enters the target; it travels with the electron. The second term on the right has a negative value and represents the transfer of energy out of the system by x-rays. On the left, there is an increase in internal energy of the target, which recognizes that only a fraction of the incoming energy leaves as x-rays. A large fraction of the incoming energy results in an increase in temperature of the target.

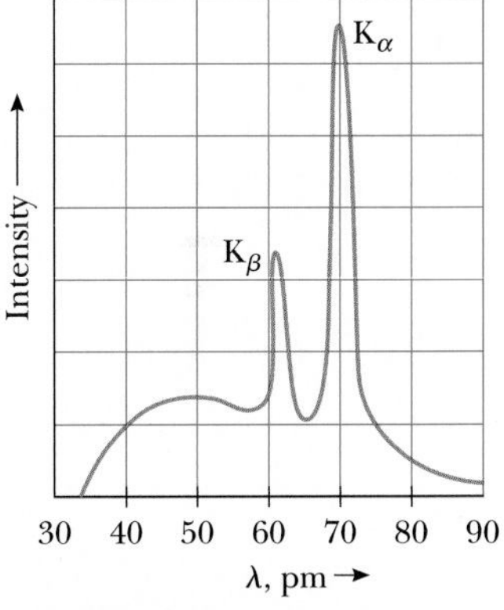

FIGURE 29.14 The x-ray spectrum of a metal target consists of a broad continuous spectrum *(bremsstrahlung)* plus a number of sharp lines, which are due to *characteristic x-rays.* The data shown were obtained when 37-keV electrons bombarded a molybdenum target.

The **x-ray spectrum** typically consists of a broad continuous band and a series of sharp lines that depend on the type of material used for the target as shown in Figure 29.14. In Chapter 24, we mentioned that an accelerated electric charge emits electromagnetic radiation. The x-rays we see in Figure 29.14 are the result of the slowing down of high-energy electrons as they strike the target. It may take several interactions with the atoms of the target before the electron loses all its kinetic energy. The amount of kinetic energy lost in any given interaction can vary from zero up to the entire kinetic energy of the electron. Therefore, the wavelength of radiation from these interactions lies in a continuous range from some minimum value up to infinity. It is this general slowing down of the electrons that provides the continuous curve in Figure 29.14, which shows the cutoff of x-rays below a minimum wavelength value that depends on the kinetic energy of the incoming electrons. X-radiation with its origin in the slowing down of electrons is called **bremsstrahlung,** German for "braking radiation."

The discrete lines in Figure 29.14, called **characteristic x-rays** and discovered in 1908, have a different origin. Their origin remained unexplained until the details of atomic structure were understood. The first step in the production of characteristic x-rays occurs when a bombarding electron collides with a target atom. The incoming electron must have sufficient energy to remove an inner-shell electron from the atom. The vacancy created in the shell is filled when an electron in a higher shell drops down into the shell containing the vacancy. The time interval required for this to happen is very short, less than 10^{-9} s. As usual, this transition is accompanied by the emission of a photon whose energy equals the difference in energy between the two shells. Typically, the energy of such transitions is greater than 1 000 eV, and the emitted x-ray photons have wavelengths in the range of 0.01 to 1 nm.

Let us assume that the incoming electron has dislodged an atomic electron from the innermost shell, the K shell. If the vacancy is filled by an electron dropping from the next higher shell, the L shell, the photon emitted in the process has an energy corresponding to the K_α line on the curve of Figure 29.14. If the vacancy is filled by an electron dropping from the M shell, the line produced is called the K_β line. In this notation, the letter K represents the final shell into which the electron drops and the subscript provides a Greek letter corresponding to the number of the shell above the final shell in which the electron originates. Therefore, K_α indicates that the final shell is the K shell, whereas the initial shell is the first shell above K (because α is the first letter in the Greek alphabet), which is the L shell.

Other characteristic x-ray lines are formed when electrons drop from upper shells to vacancies in shells other than the K shell. For example, L lines are produced when vacancies in the L shell are filled by electrons dropping from higher shells. An L_α line is produced as an electron drops from the M shell to the L shell, and an L_β line is produced by a transition from the N shell to the L shell.

Although multielectron atoms cannot be analyzed exactly using either the Bohr model or the Schrödinger equation, we can apply our knowledge of Gauss's law from Chapter 19 to make some surprisingly accurate estimates of expected x-ray energies and wavelengths. Consider an atom of atomic number Z in which one of the two electrons in the K shell has been ejected. Imagine that we draw a gaussian sphere just inside the most probable radius of the L electrons. The electric field at the position of the L electrons is a combination of that due to the nucleus, the single K electron, the other L electrons, and the outer electrons. The wave functions of the outer electrons are such that they have a very high probability of being farther from the nucleus than the L electrons are. Therefore, they are much more likely to be outside the gaussian surface than inside and, on the average, do not contribute significantly to the electric field at the position of the L electrons. The effective charge inside the gaussian surface is the positive nuclear charge and one negative charge due to the single K electron. If we ignore the interactions between L electrons, a single L electron behaves as if it experiences an electric field due to a charge enclosed by the gaussian surface of $(Z-1)e$. The nuclear charge is in effect shielded by the electron in the K shell such that Z_{eff} in Equation 29.18 is $Z-1$. For higher-level shells, the nuclear charge is shielded by electrons in all the inner shells.

We can now use Equation 29.18 to estimate the energy associated with an electron in the L shell:

$$E_{\text{L}} \approx -(Z-1)^2 \frac{13.6\ \text{eV}}{2^2}$$

The final state of the atom after it makes the transition is such that there are two electrons in the K shell. We can use a similar argument by drawing a gaussian surface just inside the most probable radius for one K electron. The energy associated with one of these electrons is approximately that of a one-electron atom with the nuclear charge reduced by the negative charge of the other electron. Therefore,

$$E_{\text{K}} \approx -(Z-1)^2(13.6\ \text{eV}) \qquad [29.19]$$

As we show in Example 29.8, the energy of the atom with an electron in an M shell can be estimated in a similar fashion. Taking the energy difference between the initial and final levels, the energy and wavelength of the emitted photon can then be calculated.

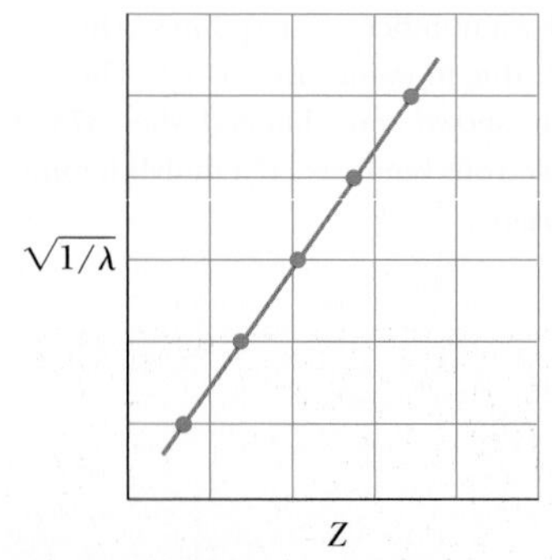

FIGURE 29.15 A Moseley plot of $\sqrt{1/\lambda}$ versus Z, where λ is the wavelength of the K_α x-ray line of the element with atomic number Z.

In 1914, Henry G. J. Moseley plotted $\sqrt{1/\lambda}$ versus the Z values for a number of elements, where λ is the wavelength of the K_α line of each element. He found that the curve is a straight line as in Figure 29.15. This finding is consistent with rough calculations of the energy levels given by Equation 29.19. From this plot, Moseley was able to determine the Z values of some missing elements, which provided a periodic chart in excellent agreement with the known chemical properties of the elements.

QUICK QUIZ 29.5 What are the initial and final shells for an M_β line in an x-ray spectrum?

QUICK QUIZ 29.6 In an x-ray tube, as you increase the energy of the electrons striking the metal target, **(i)** the wavelengths of the characteristic x-rays **(a)** increase, **(b)** decrease, or **(c)** do not change and **(ii)**, the minimum wavelength of the bremsstrahlung **(a)** increases, **(b)** decreases, or **(c)** does not change.

EXAMPLE 29.8 Estimating the Energy of an X-Ray

Estimate the energy of the characteristic x-ray emitted from a tungsten target when an electron drops from an M shell ($n = 3$ state) to a vacancy in the K shell ($n = 1$ state).

Solution The atomic number for tungsten is $Z = 74$. Using Equation 29.19, we see that the energy associated with the electron in the K shell is approximately

$$E_K \approx -(74 - 1)^2(13.6\text{ eV}) = -7.2 \times 10^4\text{ eV}$$

An electron in the M shell is subject to an effective nuclear charge that depends on the number of electrons in the $n = 1$ and $n = 2$ states, which shield the nucleus. Because eight electrons are in the $n = 2$ state and one electron is in the $n = 1$ state, nine electrons shield the nucleus, and so $Z_{\text{eff}} = Z - 9$. Hence, the energy of the M shell, following Equation 29.18, is approximately

$$E_M \approx -\frac{(13.6\text{ eV})(74 - 9)^2}{(3)^2} \approx -6.4 \times 10^3\text{ eV}$$

The emitted x-ray therefore has an energy equal to $E_M - E_K \approx -6.4 \times 10^3\text{ eV} - (-7.2 \times 10^4\text{ eV}) \approx 6.6 \times 10^4\text{ eV} = 66\text{ keV}$. Consultation of x-ray tables shows that the M–K transition energies in tungsten vary from 66.9 to 67.7 keV, where the range of energies is due to slightly different energy values for states of different ℓ. Therefore, our estimate differs from the midpoint of this experimentally measured range by about 2%.

29.7 ATOMS IN SPACE

CONTEXT CONNECTION

We have spent quite a bit of time on the hydrogen atom in this chapter. Let us now consider hydrogen atoms located in space. Because hydrogen is the most abundant element in the Universe, its role in astronomy and cosmology is very important.

Let us begin by considering pictures of some nebulae you might have seen in an astronomy text, such as Figure 29.16. Time-exposure photographs of these objects show a variety of colors. What causes the colors in these clouds of gas and grains of dust? Let us imagine a cloud of hydrogen atoms in space near a very hot star. The high-energy photons from the star can interact with the hydrogen atoms, either raising them to a high-energy state or ionizing them. As the atoms fall back to the lower states, many atoms emit the Balmer series of wavelengths. Therefore, these atoms provide red, green, blue, and violet colors to the nebula, corresponding to the colors seen in the hydrogen spectrum in Chapter 11.

In practice, nebulae are classified into three groups depending on the transitions occurring in the hydrogen atoms. **Emission nebulae** (Fig. 29.16a) are near a hot star, so hydrogen atoms are excited by light from the star as described above.

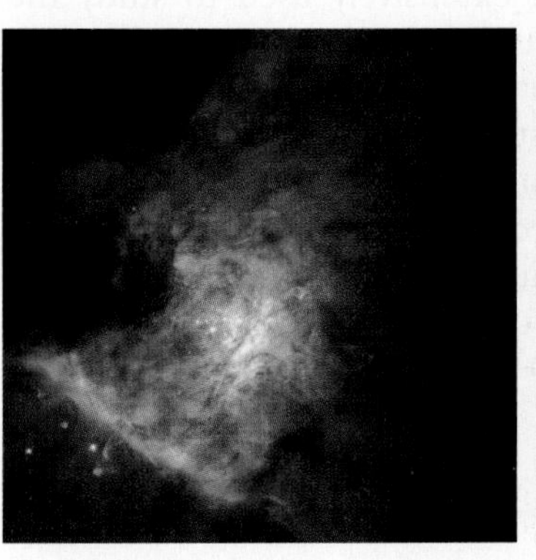

(a)

(b)

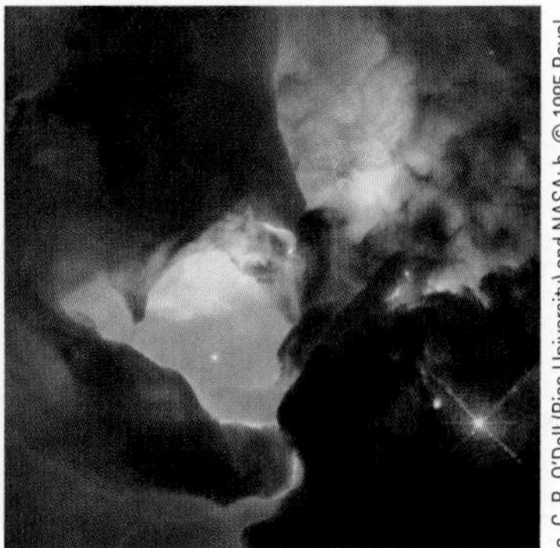

(c)

(a, C. R. O'Dell (Rice University) and NASA; b, © 1985 Royal Observatory Edinburgh/Anglo-Australian Observatory by David Malin; c, A. Caulet (ST-ECF, ESA) and NASA)

FIGURE 29.16 Types of astronomical nebulae. (a) The central part of the Orion Nebula represents an emission nebula, from which colored light is emitted from atoms. (b) The Pleiades. The clouds of light surrounding the stars represent a reflection nebula, from which starlight is reflected by dust particles. (c) The Lagoon Nebula shows the effects of a dark nebula, in which clouds of dust block starlight and appear as a dark silhouette against the light from stars farther away.

Therefore, the light from an emission nebula is dominated by discrete emission spectral lines and contains colors. **Reflection nebulae** (Fig. 29.16b) are near a cool star. In these cases, most of the light from the nebula is the starlight reflected from larger grains of material in the nebula rather than emitted by excited atoms. Therefore, the spectrum of the light from the nebula is the same as that from the star: an absorption spectrum with dark lines corresponding to atoms and ions in the outer regions of the star. The light from these nebulae tends to appear white. Finally, **dark nebulae** (Fig. 29.16c) are not close to a star. Therefore, little radiation is available to excite atoms or reflect from grains of dust. As a result, the material in these nebulae screens out light from stars beyond them, and they appear as black patches against the brightness of the more distant stars.

In addition to hydrogen, some other atoms and ions in space are raised to higher energy states by radiation from stars and proceed to emit various colors. Some of the more prominent colors are violet (373 nm) from the O^+ ion and green (496 nm and 501 nm) from the O^{++} ion. Helium and nitrogen also provide strong colors.

In our discussion of the quantum numbers for the hydrogen atom, we claimed that two states are possible in the $1s$ shell, corresponding to up or down spin, and that these two states are equivalent in energy. If we modify our structural model to include the spin of the proton, however, we find that the two atomic states corresponding to the electron spin are not the same in energy. The state in which the electron and proton spins are parallel is slightly higher in energy than the state in which they are antiparallel. The energy difference is only 5.9×10^{-6} eV. Because these two states differ in energy, it is possible for the atom to make a transition between the states. If the transition is from the parallel state to the antiparallel state, a photon is emitted, with energy equal to the difference in energy between the states. The wavelength of this photon is

$$\lambda = \frac{c}{f} = \frac{hc}{hf} = \frac{hc}{E} = \frac{1\,240\text{ eV}\cdot\text{nm}}{5.9 \times 10^{-6}\text{ eV}}\left(\frac{10^{-9}\text{ m}}{1\text{ nm}}\right)$$
$$= 0.21\text{ m} = 21\text{ cm}$$

This radiation is called, for obvious reasons, **21-cm radiation.** It is radiation with a wavelength that is identifiable with the hydrogen atom. Therefore, by looking for this radiation in space, we can detect hydrogen atoms. Furthermore, if the wavelength of the observed radiation is not equal to 21 cm, we can infer that it has been Doppler shifted due to relative motion between the Earth and the source. This Doppler shift can then be used to measure the relative speed of the source toward or away from the Earth. This technique has been extensively used to study the hydrogen distribution in the Milky Way galaxy and to detect the presence of spiral arms in our galaxy, similar to the spiral arms in other galaxies.

Our study of atomic physics allows us to understand an important connection between the microscopic world of quantum physics and the macroscopic Universe. Atoms throughout the Universe act as transmitters of information to us about the local conditions. In Chapter 30, which deals with nuclear physics, we shall see how our understanding of microscopic processes helps us understand the local conditions at the center of a star.

SUMMARY

PhysicsNow™ Take a practice test by logging into PhysicsNow at www.pop4e.com and clicking on the Pre-Test link for this chapter.

The methods of quantum mechanics can be applied to the hydrogen atom using the appropriate potential energy function $U(r) = -k_e e^2/r$ in the Schrödinger equation. The solution to this equation yields the wave functions for the allowed states and the allowed energies, given by

$$E_n = -\left(\frac{k_e e^2}{2a_0}\right)\frac{1}{n^2} = -\frac{13.606\text{ eV}}{n^2} \qquad n = 1, 2, 3, \ldots \qquad [29.2]$$

which is precisely the result obtained in the Bohr theory. The allowed energy depends only on the **principal quantum number** n. The allowed wave functions depend on three quantum numbers, n, ℓ, and m_ℓ, where ℓ is the **orbital quantum number** and m_ℓ is the **orbital magnetic quantum number.** The restrictions on the quantum numbers are as follows:

$$n = 1, 2, 3, \ldots$$
$$\ell = 0, 1, 2, \ldots, (n-1)$$
$$m_\ell = -\ell, -\ell+1, \ldots, \ell-1, \ell$$

All states with the same principal quantum number n form a **shell,** identified by the letters K, L, M, . . . (corresponding to $n = 1, 2, 3, \ldots$). All states with the same values of both n and ℓ form a **subshell,** designated by the letters *s, p, d, f,* . . . (corresponding to $\ell = 0, 1, 2, 3, \ldots$).

An atom in a state characterized by a specific n can have the following values of **orbital angular momentum** L:

$$|\vec{\mathbf{L}}| = L = \sqrt{\ell(\ell+1)}\hbar \qquad \ell = 0, 1, 2, \ldots, n-1 \qquad [29.9]$$

The allowed values of the projection of the angular momentum vector $\vec{\mathbf{L}}$ along the z axis are given by

$$L_z = m_\ell \hbar \qquad [29.10]$$

where m_ℓ is restricted to integer values lying between $-\ell$ and ℓ. Only discrete values of L_z are allowed, and they are determined by the restrictions on m_ℓ. This quantization of L_z is referred to as **space quantization.**

To describe a quantum state of the hydrogen atom completely, it is necessary to include a fourth quantum number m_s, called the **spin magnetic quantum number.** This quantum number can have only two values, $\pm\frac{1}{2}$. In effect, this additional quantum number doubles the number of allowed states specified by the quantum numbers n, ℓ, and m_ℓ.

The electron has an intrinsic angular momentum called **spin angular momentum.** That is, the total angular momentum of an atom can have two contributions: one arising from the spin of the electron ($\vec{\mathbf{S}}$) and one arising from the orbital motion of the electron ($\vec{\mathbf{L}}$).

Electronic spin can be described by a quantum number $s = \frac{1}{2}$. The **magnitude of the spin angular momentum** is

$$S = \frac{\sqrt{3}}{2}\hbar \qquad [29.12]$$

and the z component of $\vec{\mathbf{S}}$ is

$$S_z = m_s\hbar = \pm\tfrac{1}{2}\hbar \qquad [29.13]$$

The magnetic moment $\vec{\boldsymbol{\mu}}_s$ associated with the spin angular momentum of an electron is

$$\vec{\boldsymbol{\mu}}_s = -\frac{e}{m_e}\vec{\mathbf{S}} \qquad [29.14]$$

The z component of $\vec{\boldsymbol{\mu}}_s$ can have the values

$$\mu_{sz} = \pm\frac{e\hbar}{2m_e} \qquad [29.15]$$

The quantity $e\hbar/2m_e$ is called the **Bohr magneton** μ_B and has the numerical value 9.274×10^{-24} J/T.

The **exclusion principle** states that no two electrons in an atom can have the same set of quantum numbers n, ℓ, m_ℓ, and m_s. Using this principle, one can determine the electronic configuration of the elements. This procedure serves as a basis for understanding atomic structure and the chemical properties of the elements.

The allowed electronic transitions between any two states in an atom are governed by the **selection rules**

$$\Delta\ell = \pm 1 \qquad \text{and} \qquad \Delta m_\ell = 0 \text{ or } \pm 1 \qquad [29.16]$$

The **x-ray spectrum** of a metal target consists of a set of sharp characteristic lines superimposed on a broad, continuous spectrum. **Bremsstrahlung** is x-radiation with its origin in the slowing down of high-energy electrons as they encounter the target. **Characteristic x-rays** are emitted by atoms when an electron undergoes a transition from an outer shell into an electron vacancy in one of the inner shells.

QUESTIONS

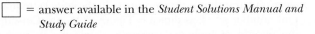

☐ = answer available in the *Student Solutions Manual and Study Guide*

1. According to Bohr's model of the hydrogen atom, what is the uncertainty in the radial coordinate of the electron? What is the uncertainty in the radial component of the velocity of the electron? In what way does the model violate the uncertainty principle?
2. Why are three quantum numbers needed to describe the state of a one-electron atom (ignoring spin)?
3. Compare the Bohr theory and the Schrödinger treatment of the hydrogen atom. Comment on the total energy and orbital angular momentum.
4. Discuss why the term *electron cloud* is used to describe the electronic arrangement in the quantum mechanical model of an atom.
5. Why is the direction of the orbital angular momentum of an electron opposite to that of its magnetic moment?
6. Why is a *nonuniform* magnetic field used in the Stern–Gerlach experiment?
7. Could the Stern–Gerlach experiment be performed with ions rather than neutral atoms? Explain.
8. Describe some experiments that support the conclusion that the spin magnetic quantum number for electrons can only have the values $\pm\frac{1}{2}$.
9. Discuss some of the consequences of the exclusion principle.
10. How is it possible that electrons, whose positions are described by a probability distribution around a nucleus, can exist in atoms with states of *definite* energy (e.g., 1*s*, 2*p*, 3*d*, . . .)?

11. Why do lithium, potassium, and sodium exhibit similar chemical properties?

12. An energy of about 21 eV is required to excite an electron in a helium atom from the $1s$ state to the $2s$ state. The same transition for the He^+ ion requires approximately twice as much energy. Explain.

13. The absorption or emission spectrum of a gas consists of lines that broaden as the density of gas molecules increases. Why do you suppose that occurs?

14. It is easy to understand how two electrons (one spin up, one spin down) can fill the $1s$ shell for a helium atom. How is it possible that eight more electrons can fit into the $2s$, $2p$ level to complete the $1s^22s^22p^6$ shell for a neon atom?

15. In 1914, Henry G. J. Moseley was able to define the atomic number of an element from its characteristic x-ray spectrum. How was that possible? (*Suggestion:* See Figs. 29.14 and 29.15.)

16. (a) "As soon as I define a particular direction as the z axis, precisely one half of the electrons in this part of the Universe have their magnetic moment vectors oriented at 54.735 61° to that axis, and all the rest have their magnetic moments at 125.264 39°." Argue for or against this statement. (b) "The Universe is not simply stranger than we suppose; it is stranger than we *can* suppose." Argue for or against this statement.

17. A message reads, "*All your base are belong to us!*" Argue for or against the view that a scientific discovery is like a communication from an utterly alien source, needing interpretation and susceptible to misunderstanding. Argue for or against the view that the human mind is not necessarily well adapted to understand the Universe. Argue for or against the view that education in science is the best preparation for life in a rapidly changing world.

PROBLEMS

1, 2, 3 = straightforward, intermediate, challenging
☐ = full solution available in the *Student Solutions Manual and Study Guide*
Physics Now™ = coached problem with hints available at www.pop4e.com
= computer useful in solving problem
= paired numerical and symbolic problems
= biomedical application

Section 29.1 ▪ Early Structural Models of the Atom

1. Physics Now™ According to classical physics, a charge e moving with an acceleration a radiates at a rate

$$\frac{dE}{dt} = -\frac{1}{6\pi\epsilon_0}\frac{e^2a^2}{c^3}$$

(a) Show that an electron in a classical hydrogen atom (see Fig. 29.3) spirals into the nucleus at a rate

$$\frac{dr}{dt} = -\frac{e^4}{12\pi^2\epsilon_0{}^2r^2m_e{}^2c^3}$$

(b) Find the time interval over which the electron will reach $r = 0$, starting from $r_0 = 2.00 \times 10^{-10}$ m.

2. **Review problem.** In the Rutherford scattering experiment, 4.00-MeV alpha particles (^{4}He nuclei containing 2 protons and 2 neutrons) scatter off gold nuclei (containing 79 protons and 118 neutrons). Assume that a particular alpha particle makes a direct head-on collision with the gold nucleus and scatters backward at 180°. Determine (a) the distance of closest approach of the alpha particle to the gold nucleus and (b) the maximum force exerted on the alpha particle. Assume that the gold nucleus remains fixed throughout the entire process.

3. (a) Calculate the angular momentum of the Moon due to its orbital motion about the Earth. In your calculation, use 3.84×10^8 m as the average Earth–Moon distance and 2.36×10^6 s as the period of the Moon in its orbit. (b) Assume that the Moon's angular momentum is described by Bohr's assumption $mvr = n\hbar$. Determine the corresponding quantum number. (c) By what fraction would the Earth–Moon distance have to be increased to raise the quantum number by 1?

4. (a) An isolated atom of a certain element emits light of wavelength 520 nm when the atom falls from its fifth excited state into its second excited state. The atom emits a photon of wavelength 410 nm when it drops from its sixth excited state into its second excited state. Find the wavelength of the light radiated when the atom makes a transition from its sixth to its fifth excited state. (b) Solve the same problem again in symbolic terms. Letting λ_{BA} represent the wavelength emitted in the transition B to A and λ_{CA} the shorter wavelength emitted in the transition C to A, find λ_{CB}. This problem exemplifies the *Ritz combination principle,* an empirical rule formulated in 1908.

Section 29.2 ▪ The Hydrogen Atom Revisited

5. The Balmer series for the hydrogen atom corresponds to electronic transitions that terminate in the state with quantum number $n = 2$ as shown in Figure P29.5. (a) Consider the photon of longest wavelength; determine its energy and wavelength. (b) Consider the spectral line of shortest wavelength; find its photon energy and wavelength.

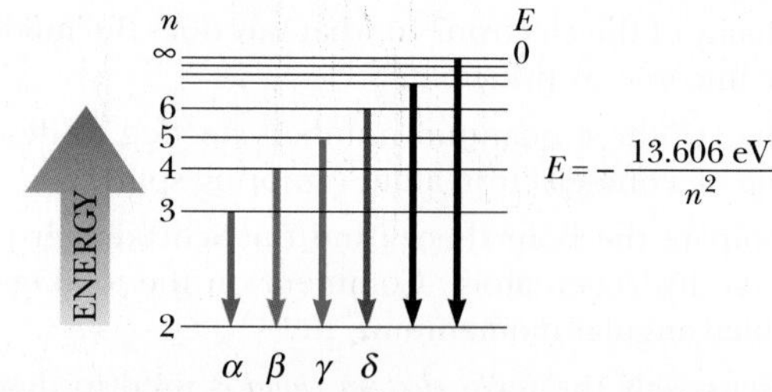

FIGURE P29.5 An energy level diagram for hydrogen showing the Balmer series (not drawn to scale).

6. A photon with energy 2.28 eV is barely capable of causing a photoelectric effect when it strikes a sodium plate. Suppose the photon is instead absorbed by hydrogen. Find (a) the minimum n for a hydrogen atom that can be ionized by such a photon and (b) the speed of the released electron far from the nucleus.

7. A general expression for the energy levels of one-electron atoms and ions is

$$E_n = -\frac{\mu k_e^2 q_1^2 q_2^2}{2\hbar^2 n^2}$$

where k_e is the Coulomb constant, q_1 and q_2 are the charges of the electron and the nucleus, and μ is the reduced mass of the atom, given by $\mu = m_1 m_2/(m_1 + m_2)$ where m_1 is the mass of the electron and m_2 is the mass of the nucleus. In Problem 29.5 we found that the wavelength for the $n = 3$ to $n = 2$ transition of the hydrogen atom is 656.3 nm (visible red light). What are the wavelengths for this same transition in (a) positronium, which consists of an electron and a positron, and (b) singly ionized helium? (*Note:* A positron is a positively charged electron.)

8. Ordinary hydrogen gas is a mixture of two kinds of atoms (isotopes) containing either one- or two-particle nuclei. These isotopes are hydrogen-1 with a proton nucleus and hydrogen-2, called deuterium, with a deuteron nucleus. A deuteron is one proton and one neutron bound together. Hydrogen-1 and deuterium have identical chemical properties, but can be separated via an ultracentrifuge or by other methods. Their emission spectra show lines of the same colors at very slightly different wavelengths. (a) Use the equation given in Problem 29.7 to show that the difference in wavelength, between the hydrogen-1 and deuterium spectral lines associated with a particular electron transition, is given by

$$\lambda_H - \lambda_D = \left(1 - \frac{\mu_H}{\mu_D}\right)\lambda_H$$

(b) Evaluate the wavelength difference for the Balmer alpha line of hydrogen, with wavelength 656.3 nm, emitted by an atom making a transition from an $n = 3$ state to an $n = 2$ state. Harold Urey observed this wavelength difference in 1931, confirming his discovery of deuterium.

9. An electron of momentum p is at a distance r from a stationary proton. The electron has kinetic energy $K = p^2/2m_e$. The atom has potential energy $U = -k_e e^2/r$ and total energy $E = K + U$. If the electron is bound to the proton to form a hydrogen atom, its average position is at the proton, but the uncertainty in its position is approximately equal to the radius r of its orbit. The electron's average vector momentum is zero, but its average squared momentum is approximately equal to the squared uncertainty in its momentum, as given by the uncertainty principle. Treating the atom as a one-dimensional system, (a) estimate the uncertainty in the electron's momentum in terms of r. (b) Estimate the electron's kinetic, potential, and total energies in terms of r. (c) The actual value of r is the one that *minimizes the total energy*, resulting in a stable atom. Find that value of r and the resulting total energy. Compare your answer with the predictions of the Bohr theory.

Section 29.3 ■ The Wave Functions for Hydrogen

10. Plot the wave function $\psi_{1s}(r)$ (see Eq. 29.3) and the radial probability density function $P_{1s}(r)$ (see Eq. 29.7) for hydrogen. Let r range from 0 to $1.5a_0$, where a_0 is the Bohr radius.

11. The ground-state wave function for a hydrogen atom is

$$\psi_{1s}(r) = \frac{1}{\sqrt{\pi a_0^3}} e^{-r/a_0}$$

where r is the radial coordinate of the electron and a_0 is the Bohr radius. (a) Show that the wave function as given is normalized. (b) Find the probability of locating the electron between $r_1 = a_0/2$ and $r_2 = 3a_0/2$.

12. The wave function for the $2p$ state of hydrogen is

$$\psi_{2p} = \frac{1}{\sqrt{3}(2a_0)^{3/2}} \frac{r}{a_0} e^{-r/2a_0}$$

What is the most likely distance from the nucleus to find an electron in the $2p$ state?

13. **Physics Now™** For a spherically symmetric state of a hydrogen atom, the Schrödinger equation in spherical coordinates is

$$-\frac{\hbar^2}{2m}\left(\frac{d^2\psi}{dr^2} + \frac{2}{r}\frac{d\psi}{dr}\right) - \frac{k_e e^2}{r}\psi = E\psi$$

Show that the $1s$ wave function for hydrogen,

$$\psi_{1s}(r) = \frac{1}{\sqrt{\pi a_0^3}} e^{-r/a_0}$$

satisfies the Schrödinger equation.

14. In an experiment, electrons are fired at a sample of neutral hydrogen atoms and observations are made of how the incident particles scatter. A large set of trials can be thought of as containing 1 000 observations of the electron in the ground state of a hydrogen atom being momentarily at a distance $a_0/2$ from the nucleus. How many times is the atomic electron observed at a distance $2a_0$ from the nucleus in this set of trials?

Section 29.4 ■ Physical Interpretation of the Quantum Numbers

15. List the possible sets of quantum numbers for the hydrogen atom associated with (a) the $3d$ subshell and (b) the $3p$ subshell.

16. Calculate the orbital angular momentum for a hydrogen atom in (a) the $4d$ state and (b) the $6f$ state.

17. If a hydrogen atom has orbital angular momentum 4.714×10^{-34} J·s, what is the orbital quantum number for the state of the atom?

18. A hydrogen atom is in its fifth excited state, with principal quantum number 6. The atom emits a photon with a wavelength of 1 090 nm. Determine the maximum possible orbital angular momentum of the atom after emission.

19. **Physics Now™** How many sets of quantum numbers are possible for a hydrogen atom for which (a) $n = 1$, (b) $n = 2$, (c) $n = 3$, (d) $n = 4$, and (e) $n = 5$? Check your results to show that they agree with the general rule

that the number of sets of quantum numbers for a shell is equal to $2n^2$.

20. Find all possible values of L, L_z, and θ for a hydrogen atom in a $3d$ state.

21. (a) Find the mass density of a proton, modeling it as a solid sphere of radius 1.00×10^{-15} m. (b) Consider a classical model of an electron as a solid sphere with the same density as the proton. Find its radius. (c) Imagine that this electron possesses spin angular momentum $I\omega = \hbar/2$ because of classical rotation about the z axis. Determine the speed of a point on the equator of the electron and (d) compare this speed to the speed of light.

22. An electron is in the N shell. Determine the maximum value the z component of its angular momentum could have.

23. The ρ^- meson has a charge of $-e$, a spin quantum number of 1, and a mass 1 507 times that of the electron. The possible values for its spin magnetic quantum number are -1, 0, and 1. Imagine that the electrons in atoms were replaced by ρ^- mesons. List the possible sets of quantum numbers for ρ^- mesons in the $3d$ subshell.

Section 29.5 ▪ The Exclusion Principle and the Periodic Table

24. (a) Write out the electronic configuration for the ground state of oxygen ($Z = 8$). (b) Write out a set of possible values for the quantum numbers n, ℓ, m_ℓ, and m_s for each electron in oxygen.

25. As we go down the periodic table, which subshell is filled first, the $3d$ or the $4s$ subshell? Which electronic configuration has a lower energy: $[\text{Ar}]3d^4 4s^2$ or $[\text{Ar}]3d^5 4s^1$? Which has the greater number of unpaired spins? Identify this element and discuss Hund's rule in this case. (*Note:* The notation [Ar] represents the filled configuration for argon.)

26. Devise a table similar to that shown in Figure 29.11 for atoms containing 11 through 19 electrons. Use Hund's rule and educated guesswork.

27. A certain element has its outermost electron in a $3p$ subshell. It has valence $+3$ because it has 3 more electrons than a certain inert gas. What element is it?

28. Two electrons in the same atom both have $n = 3$ and $\ell = 1$. (a) List the quantum numbers for the possible states of the atom. (b) How many states would be possible if the exclusion principle were inoperative?

29. **PhysicsNow™** (a) Scanning through Table 29.4 in order of increasing atomic number, note that the electrons usually fill the subshells in such a way that those subshells with the lowest values of $n + \ell$ are filled first. If two subshells have the same value of $n + \ell$, the one with the lower value of n is generally filled first. Using these two rules, write the order in which the subshells are filled through $n + \ell = 7$. (b) Predict the chemical valence for the elements that have atomic numbers 15, 47, and 86 and compare your predictions with the actual valences (which may be found in a chemistry text).

30. For a neutral atom of element 110, what would be the probable ground-state electronic configuration?

31. **Review problem.** For an electron with magnetic moment $\vec{\mu}_s$ in a magnetic field $\vec{\mathbf{B}}$, the result of Problem 22.20 in Chapter 22 shows the following. The electron-field system can be in a higher energy state with the z component of the magnetic moment of the electron opposite to the field or in a lower energy state with the z component of the magnetic moment in the direction of the field. The difference in energy between the two states is $2\mu_B B$.

Under high resolution, many spectral lines are observed to be doublets. The most famous of these are the two yellow lines in the spectrum of sodium (the D lines), with wavelengths of 588.995 nm and 589.592 nm. Their existence was explained in 1925 by Goudsmit and Uhlenbeck, who postulated that an electron has intrinsic spin angular momentum. When the sodium atom is excited with its outermost electron in a $3p$ subshell, the orbital motion of the outermost electron creates a magnetic field. The atom's energy is somewhat different depending on whether the electron is itself spin-up or spin-down in this field. Then the photon energy the atom radiates as it falls back into its ground state depends on the energy of the excited state. Calculate the magnitude of the internal magnetic field mediating this so-called spin-orbit coupling.

Section 29.6 ▪ More on Atomic Spectra: Visible and X-ray

32. (a) Determine the possible values of the quantum numbers ℓ and m_ℓ for the He^+ ion in the state corresponding to $n = 3$. (b) What is the energy of this state?

33. If you wish to produce 10.0-nm x-rays in the laboratory, what is the minimum voltage you must use in accelerating the electrons?

34. In x-ray production, electrons are accelerated through a high voltage ΔV and then decelerated by striking a target. Show that the shortest wavelength of an x-ray that can be produced is

$$\lambda_{\min} = \frac{1\,240 \text{ nm} \cdot \text{V}}{\Delta V}$$

35. Use the method illustrated in Example 29.8 to calculate the wavelength of the x-ray emitted from a molybdenum target ($Z = 42$) when an electron moves from the L shell ($n = 2$) to the K shell ($n = 1$).

36. The K series of the discrete x-ray spectrum of tungsten contains wavelengths of 0.018 5 nm, 0.020 9 nm, and 0.021 5 nm. The K-shell ionization energy is 69.5 keV. Determine the ionization energies of the L, M, and N shells. Draw a diagram of the transitions.

37. The wavelength of characteristic x-rays in the K_β line from a particular source is 0.152 nm. Determine the material in the target.

Section 29.7 ▪ Context Connection—Atoms in Space

38. In interstellar space, atomic hydrogen produces the sharp spectral line called the 21-cm radiation, which astronomers find most helpful in detecting clouds of hydrogen between stars. This radiation is useful because it is the only signal cold hydrogen emits and because interstellar dust that obscures visible light is transparent to these radio waves. The radiation is not generated by an electron transition between energy states characterized by different values of n.

Instead, in the ground state ($n = 1$), the electron and proton spins may be parallel or antiparallel, with a resultant slight difference in these energy states. (a) Which condition has the higher energy? (b) More precisely, the line has wavelength 21.11 cm. What is the energy difference between the states? (c) The average lifetime in the excited state is about 10^7 yr. Calculate the associated uncertainty in energy of the excited energy level.

39. **Review problem.** Refer to Section 24.3. Prove that the Doppler shift in wavelength of electromagnetic waves is described by

$$\lambda' = \lambda\sqrt{\frac{1 + v/c}{1 - v/c}}$$

where λ' is the wavelength measured by an observer moving at speed v away from a source radiating waves of wavelength λ.

40. Astronomers observe a series of spectral lines in the light from a distant galaxy. On the hypothesis that the lines form the Lyman series for a (new?!) one-electron atom, they start to construct the energy level diagram shown in Figure P29.40, which gives the wavelengths of the first four lines and the short-wavelength limit of this series. Based on this information, calculate (a) the energies of the ground state and first four excited states for this one-electron atom and (b) the wavelengths of the first three lines and the short-wavelength limit in the Balmer series for this atom. (c) Show that the wavelengths of the first four lines and the short wavelength limit of the Lyman series for the hydrogen atom are all 60.0% of the wavelengths for the Lyman series in the one-electron atom described in part (b). (d) Based on this observation, explain why this atom could be hydrogen.

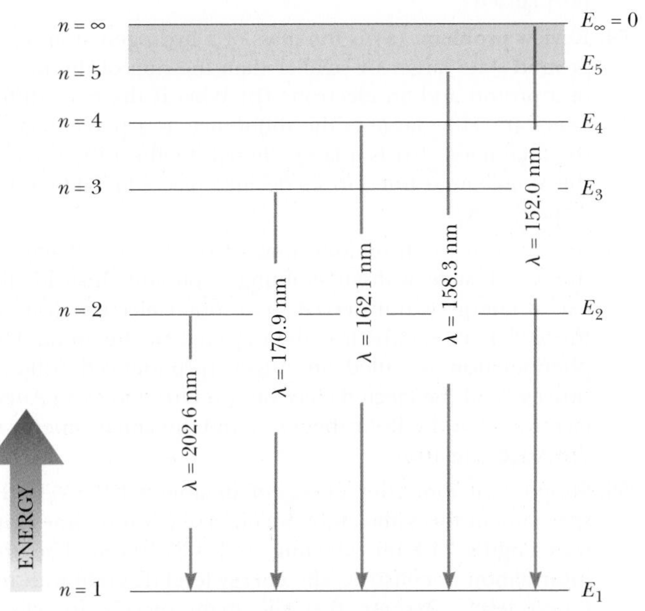

FIGURE **P29.40**

41. **Physics Now™** A distant quasar is moving away from the Earth at such high speed that the blue 434-nm H_γ line of hydrogen is observed at 510 nm, in the green portion of the spectrum (Fig. P29.41). (a) How fast is the quasar receding? You may use the result of Problem 29.39. (b) Edwin Hubble discovered that all objects outside the local group of galaxies are moving away from us, with speeds proportional to their distances. Hubble's law is expressed as $v = HR$, where Hubble's constant has the approximate value $H = 17 \times 10^{-3}$ m/s · ly. Determine the distance from the Earth to this quasar.

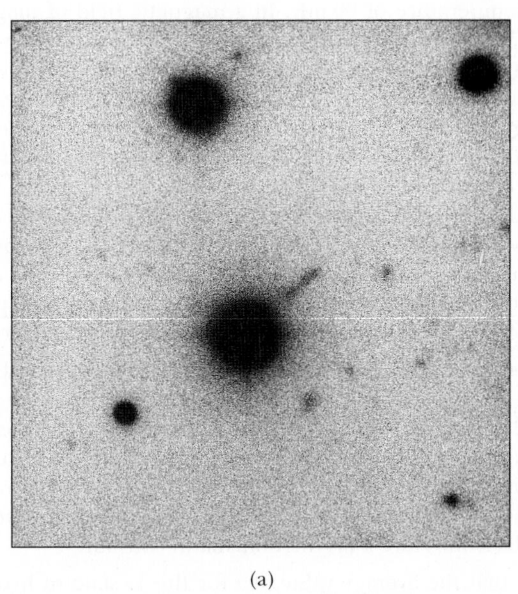

(a)

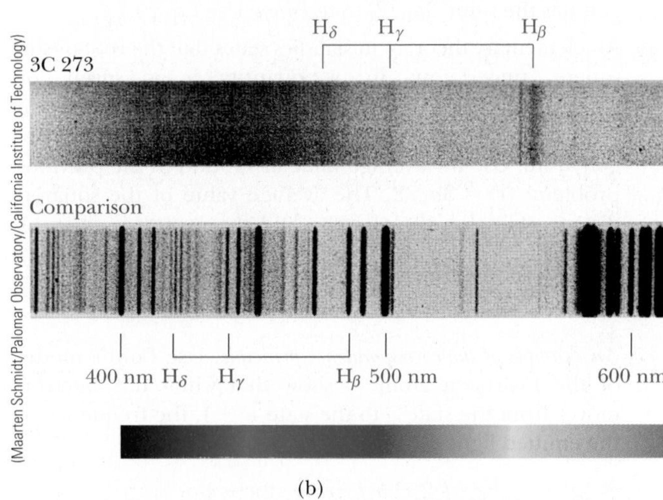

(b)

FIGURE **P29.41** (a) Image of the quasar 3C273. (b) Spectrum of the quasar above a comparison spectrum emitted by stationary hydrogen and helium atoms. Both parts of the figure are printed as black-and-white photographic negatives to reveal detail.

Additional Problems

42. (a) If a hydrogen atom makes a transition from the $n = 4$ state to the $n = 2$ state, determine the wavelength of the photon created in the process. (b) Assuming that the atom was initially at rest, determine the recoil speed of the hydrogen atom when it emits this photon.

43. LENINGRAD, 1930—Four years after the publication of the Schrödinger equation, Lev Davidovich Landau, age 23, solved the equation for a charged particle moving in a uniform magnetic field. A single electron moving perpendicular to a field $\vec{\mathbf{B}}$ can be considered as a model atom without a nucleus, or as the irreducible quantum limit of the cyclotron. Landau proved that its energy is quantized in uniform steps of $e\hbar B/m_e$.
CAMBRIDGE, MA, 1999—Gerald Gabrielse trapped a single electron in an evacuated centimeter-size metal can cooled to a temperature of 80 mK. In a magnetic field of magnitude 5.26 T, the electron circulated for hours in its lowest energy level, generating a measurable signal as it moved. (a) Evaluate the size of a quantum jump in the electron's energy. (b) For comparison, evaluate k_BT as a measure of the energy available to the electron in blackbody radiation from the walls of its container. (c) Microwave radiation was introduced to excite the electron. Calculate the frequency and wavelength of the photon that the electron absorbs as it jumps to its second energy level. Measurement of the resonant absorption frequency verified the theory and permitted precise determination of properties of the electron.

44. Review problem. (a) How much energy is required to cause a hydrogen atom to move from the $n = 1$ state to the $n = 2$ state? (b) Suppose the atom gains this energy through collisions with other hydrogen atoms at a high temperature. At what temperature would the average atomic kinetic energy $3k_BT/2$ be great enough to excite the electron? Here k_B is the Boltzmann constant.

45. Show that the average value of r for the $1s$ state of hydrogen has the value $3a_0/2$. (*Suggestion:* Use Eq. 29.7.)

46. An elementary theorem in statistics states that the root-mean-square uncertainty in a quantity r is given by $\Delta r = \sqrt{\langle r^2\rangle - \langle r\rangle^2}$. Evaluate the uncertainty in the radial position of the electron in the ground state of the hydrogen atom. Use the average value of r found in the previous problem: $\langle r\rangle = 3a_0/2$. The average value of the squared distance between the electron and the proton is given by

$$\langle r^2\rangle = \int_{\text{all space}} |\psi|^2 r^2\, dV = \int_0^\infty P(r)r^2\, dr$$

47. *An example of the correspondence principle.* Use Bohr's model of the hydrogen atom to show that when the electron moves from the state n to the state $n-1$, the frequency of the emitted light is

$$f = \left(\frac{2\pi^2 m_e k_e^2 e^4}{h^3 n^2}\right)\frac{2n-1}{(n-1)^2}$$

Show that as $n \to \infty$, this expression varies as $1/n^3$ and reduces to the classical frequency one expects the atom to emit. (*Suggestion:* To calculate the classical frequency, note that the frequency of revolution is $v/2\pi r$, where r is given by Eq. 11.22.)

48. Example 29.2 calculates the most probable value for the radial coordinate r of the electron in the ground state of a hydrogen atom. Problem 29.45 shows that the average value is $\langle r\rangle = 3a_0/2$. For comparison with these modal and mean values, find the median value of r. Proceed as follows. (a) Derive an expression for the probability, as a function of r, that the electron in the ground state of hydrogen will be found outside a sphere of radius r centered on the nucleus. (b) Make a graph of the probability as a function of r/a_0. Choose values of r/a_0 ranging from 0 to 4.00 in steps of 0.250. (c) Find the value of r for which the probability of finding the electron outside a sphere of radius r is equal to the probability of finding the electron inside this sphere. You must solve a transcendental equation numerically, and your graph is a good starting point.

49. Suppose a hydrogen atom is in the $2s$ state, with its wave function given by Equation 29.8. Taking $r = a_0$, calculate values for (a) $\psi_{2s}(a_0)$, (b) $|\psi_{2s}(a_0)|^2$, and (c) $P_{2s}(a_0)$.

50. The states of matter are solid, liquid, gas, and plasma. Plasma can be described as a gas of charged particles or a gas of ionized atoms. Most of the matter in the Solar System is plasma (throughout the interior of the Sun). In fact, most of the matter in the Universe is plasma; so is a candle flame. Use the information in Figure 29.12 to make an order-of-magnitude estimate for the temperature to which a typical chemical element must be raised to turn into plasma by ionizing most of the atoms in a sample. Explain your reasoning.

51. Assume that three identical uncharged particles of mass m and spin $\frac{1}{2}$ are contained in a one-dimensional box of length L. What is the ground-state energy of this system?

52. The force on a magnetic moment μ_z in a nonuniform magnetic field B_z is given by $F_z = \mu_z(dB_z/dz)$. If a beam of silver atoms travels a horizontal distance of 1.00 m through such a field and each atom has a speed of 100 m/s, how strong must be the field gradient dB_z/dz to deflect the beam 1.00 mm?

53. (a) Show that the most probable radial position for an electron in the $2s$ state of hydrogen is $r = 5.236a_0$. (b) Show that the wave function given by Equation 29.8 is normalized.

54. Review problem. (a) Is the mass of a hydrogen atom in its ground state larger or smaller than the sum of the masses of a proton and an electron? (b) What is the mass difference? (c) How large is the difference as a percentage of the total mass? (d) Is it large enough to affect the value of the atomic mass listed to six decimal places in Table A.3 in Appendix A?

55. An electron in chromium moves from the $n = 2$ state to the $n = 1$ state without emitting a photon. Instead, the excess energy is transferred to an outer electron (one in the $n = 4$ state), which is then ejected by the atom. This phenomenon is called an Auger (pronounced "ohjay") process, and the ejected electron is referred to as an Auger electron. Use the Bohr theory to find the kinetic energy of the Auger electron.

56. Suppose the ionization energy of an atom is 4.10 eV. In the spectrum of this same atom, we observe emission lines with wavelengths 310 nm, 400 nm, and 1 377.8 nm. Use this information to construct the energy level diagram with the fewest levels. Assume that the higher levels are closer together.

57. For hydrogen in the $1s$ state, what is the probability of finding the electron farther than $2.50a_0$ from the nucleus?

58. All atoms have the same size, to an order of magnitude. (a) To show that, estimate the diameters for aluminum (with molar mass 27.0 g/mol and density 2.70 g/cm^3) and uranium (molar mass 238 g/mol and density 18.9 g/cm^3). (b) What do the results imply about the wave functions for inner-shell electrons as we progress to higher and higher atomic mass atoms? (*Suggestion:* The molar volume is approximately D^3N_A, where D is the atomic diameter and N_A is Avogadro's number.)

59. In the technique known as electron spin resonance (ESR), a sample containing unpaired electrons is placed in a magnetic field. Consider the simplest situation, in which only one electron is present and therefore only two energy states are possible, corresponding to $m_s = \pm\frac{1}{2}$. In ESR, the absorption of a photon causes the electron's spin magnetic moment to flip from the lower energy state to the higher energy state. According to the result of Problem 22.20 in Chapter 22, the change in energy is $2\mu_B B$. (The lower energy state corresponds to the case where the z component of the magnetic moment $\vec{\mu}_{spin}$ is aligned with the magnetic field, and the higher energy state is the case where the z component of $\vec{\mu}_{spin}$ is aligned opposite to the field.) What is the photon frequency required to excite an ESR transition in a 0.350-T magnetic field?

60. Show that the wave function for a hydrogen atom in the $2s$ state

$$\psi_{2s}(r) = \frac{1}{4\sqrt{2\pi}}\left(\frac{1}{a_0}\right)^{3/2}\left(2 - \frac{r}{a_0}\right)e^{-r/2a_0}$$

satisfies the spherically symmetric Schrödinger equation given in Problem 29.13.

61. **Review problem.** Steven Chu, Claude Cohen-Tannoudji, and William Phillips received the 1997 Nobel Prize in Physics for "the development of methods to cool and trap atoms with laser light." One part of their work was with a beam of atoms (mass $\sim 10^{-25}$ kg) that move at a speed on the order of 1 km/s, similar to the speed of molecules in air at room temperature. An intense laser light beam tuned to a visible atomic transition (assume 500 nm) is directed straight into the atomic beam. That is, the atomic beam and the light beam are traveling in opposite directions. An atom in the ground state immediately absorbs a photon. Total system momentum is conserved in the absorption process. After a lifetime on the order of 10^{-8} s, the excited atom radiates by spontaneous emission. It has an equal probability of emitting a photon in any direction. Therefore, the average "recoil" of the atom is zero over many absorption and emission cycles. (a) Estimate the average deceleration of the atomic beam. (b) What is the order of magnitude of the distance over which the atoms in the beam will be brought to a halt?

62. Find the average (expectation) value of $1/r$ in the $1s$ state of hydrogen. Note that the general expression is given by

$$\langle 1/r \rangle = \int_{\text{all space}} |\psi|^2(1/r)\, dV = \int_0^\infty P(r)\,(1/r)\, dr$$

Is the result equal to the inverse of the average value of r?

ANSWERS TO QUICK QUIZZES

29.1 (b). The number of subshells is the same as the number of allowed values of ℓ. The allowed values of ℓ for $n = 4$ are $\ell = 0$, 1, 2, and 3, so there are four subshells.

29.2 (a) Five values (0, 1, 2, 3, 4) of ℓ and (b) nine different values ($-4, -3, -2, -1, 0, 1, 2, 3, 4$) of m_ℓ as follows:

ℓ	m_ℓ
0	0
1	$-1, 0, 1$
2	$-2, -1, 0, 1, 2$
3	$-3, -2, -1, 0, 1, 2, 3$
4	$-4, -3, -2, -1, 0, 1, 2, 3, 4$

29.3 The vector model for $\ell = 1$ is shown at the top of the right column.

29.4 Cesium, potassium, lithium. The higher the value of Z, the closer to zero is the energy associated with the outermost electron and the smaller is the ionization energy.

29.5 Final: M. Initial: O (because the subscript β indicates that the initial shell is the second shell higher than M).

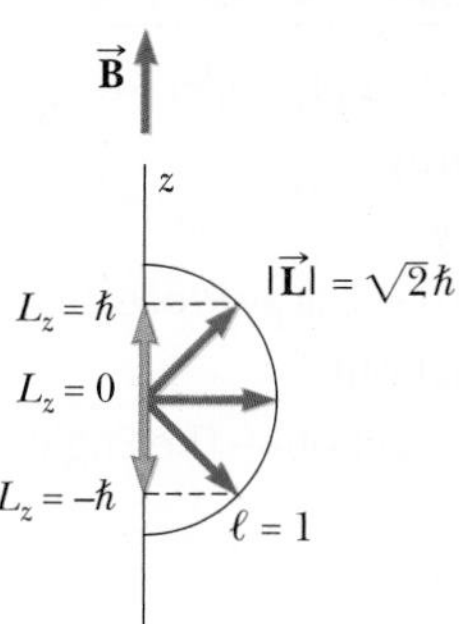

29.6 (i), (c). The wavelengths of the characteristic x-rays are determined by the separation between energy levels in the atoms of the target, which is unrelated to the energy with which electrons are fired at the target. The only dependence is that the incoming electrons must have enough energy to eject an atomic electron from an inner shell. (ii), (b). The minimum wavelength of the bremsstrahlung is associated with the highest-energy photon. This photon comes from an electron striking the target and giving up all its energy to electromagnetic radiation in one collision. Therefore, higher-energy incoming electrons will result in higher-energy photons with shorter wavelengths.

CHAPTER 30

Nuclear Physics

The Ice Man, discovered in 1991 when an Italian glacier melted enough to expose his remains. His possessions, particularly his tools, have shed light on the way people lived in the Bronze Age. A dating technique using radioactive carbon-14 was used to determine how long ago this person lived.

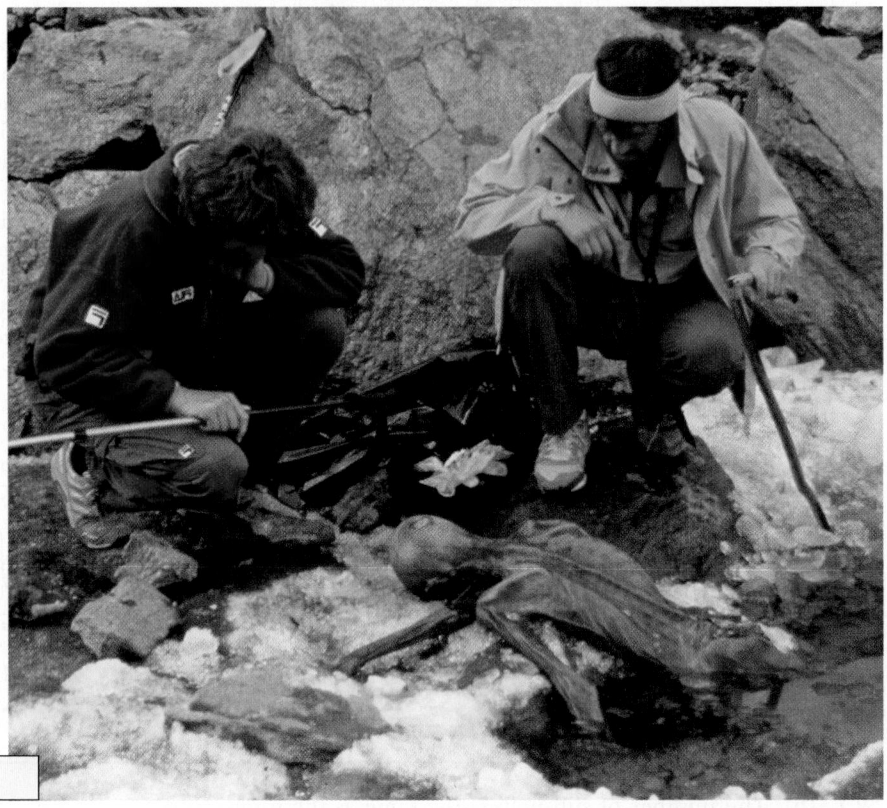
(Paul Hanny/Gamma Liaison)

CHAPTER OUTLINE

In 1896, the year that marked the birth of nuclear physics, Antoine-Henri Becquerel (1852–1908) introduced the world of science to radioactivity in uranium compounds by accidentally discovering that uranyl potassium sulfate crystals emit an invisible radiation that can darken a photographic plate when the plate is covered to exclude light. After a series of experiments, he concluded that the radiation emitted by the crystals was of a new type, one that requires no external stimulation and is so penetrating that it can darken protected photographic plates and ionize gases.

A great deal of research followed as scientists attempted to understand the radiation emitted by radioactive nuclei. Pioneering work by Rutherford showed that the radiation was of three types, which he called alpha, beta, and gamma rays. Later experiments showed that alpha rays are helium nuclei, beta rays are electrons or related particles called positrons, and gamma rays are high-energy photons.

As we saw in Section 29.1, the 1911 experiments of Rutherford established that the nucleus of an atom has a very small volume and that most of the atomic mass is contained in the nucleus.

Furthermore, such studies demonstrated a new type of force, the nuclear force, first introduced in Section 5.5, that is predominant at distances on the order of 10^{-15} m and essentially zero at distances larger than that.

In this chapter, we discuss the structure of the atomic nucleus. We shall describe the basic properties of nuclei, nuclear forces, nuclear binding energy, the phenomenon of radioactivity, and nuclear reactions.

30.1 SOME PROPERTIES OF NUCLEI

In the commonly accepted structural model of the nucleus, all nuclei are composed of two types of particles: protons and neutrons. The only exception is the ordinary hydrogen nucleus, which is a single proton with no neutrons. In describing the atomic nucleus, we identify the following integer quantities:

- The **atomic number** Z (introduced in Chapter 29) equals the number of protons in the nucleus (the atomic number is sometimes called the charge number).
- The **neutron number** N equals the number of neutrons in the nucleus.
- The **mass number** A equals the number of nucleons (neutrons plus protons) in the nucleus. That is, $A = N + Z$.

In representing nuclei, it is convenient to have a symbolic representation that shows how many protons and neutrons are present. The symbol used is ${}^{A}_{Z}\text{X}$, where X represents the chemical symbol for the element. For example, ${}^{56}_{26}\text{Fe}$ (iron) has a mass number of 56 and an atomic number of 26; therefore, it contains 26 protons and 30 neutrons. When no confusion is likely to arise, we omit the subscript Z because the chemical symbol can always be used to determine Z. Therefore, ${}^{56}_{26}\text{Fe}$ is the same as ${}^{56}\text{Fe}$ and can also be expressed as "iron-56."

The nuclei of all atoms of a particular element contain the same number of protons but often contain different numbers of neutrons. Nuclei that are related in this way are called **isotopes. The isotopes of an element have the same Z value but different N and A values.** The natural abundances of isotopes can differ substantially. For example, ${}^{11}_{6}\text{C}$, ${}^{12}_{6}\text{C}$, ${}^{13}_{6}\text{C}$, and ${}^{14}_{6}\text{C}$ are four isotopes of carbon. The natural abundance of the ${}^{12}_{6}\text{C}$ isotope is about 98.9%, whereas that of the ${}^{13}_{6}\text{C}$ isotope is only about 1.1%. (${}^{11}_{6}\text{C}$ and ${}^{14}_{6}\text{C}$ exist in trace amounts.) Even the simplest element, hydrogen, has isotopes: ${}^{1}_{1}\text{H}$, the ordinary hydrogen nucleus; ${}^{2}_{1}\text{H}$, deuterium; and ${}^{3}_{1}\text{H}$, tritium. Some isotopes do not occur naturally but can be produced in the laboratory through nuclear reactions.

QUICK QUIZ 30.1 (i) Consider the following three nuclei: ${}^{12}\text{C}$, ${}^{13}\text{N}$, ${}^{14}\text{O}$. What is the same for these three nuclei? **(a)** number of protons **(b)** number of neutrons **(c)** number of nucleons. **(ii)** Consider the following three nuclei: ${}^{12}\text{N}$, ${}^{13}\text{N}$, ${}^{14}\text{N}$. From the same list of choices, what is the same for these three nuclei? **(iii)** Consider the following three nuclei: ${}^{14}\text{C}$, ${}^{14}\text{N}$, ${}^{14}\text{O}$. From the same list of choices, what is the same for these three nuclei?

Charge and Mass

The proton carries a single positive charge $+e$ and the electron carries a single negative charge $-e$, where $e = 1.60 \times 10^{-19}$ C. The neutron is electrically neutral, as its name implies. Because the neutron has no charge, it was difficult to detect with early experimental apparatus and techniques. Today we can detect neutrons relatively easily with modern detection devices.

A convenient unit for measuring mass on a nuclear scale is the **atomic mass unit** u. This unit is defined in such a way that the atomic mass of the isotope ${}^{12}_{6}\text{C}$ is exactly 12 u, where $1\text{ u} = 1.660\,539 \times 10^{-27}$ kg. The proton and neutron each

TABLE 30.1 Masses of Selected Particles in Various Units

Particle	Mass kg	Mass u	Mass MeV/c^2
Proton	$1.672\,62 \times 10^{-27}$	1.007 276	938.28
Neutron	$1.674\,93 \times 10^{-27}$	1.008 665	939.57
Electron	$9.109\,39 \times 10^{-31}$	$5.48\,579 \times 10^{-4}$	0.510 999
${}^1_1$H atom	$1.673\,53 \times 10^{-27}$	1.007 825	938.783
${}^4_2$He atom	$6.646\,48 \times 10^{-27}$	4.002 603	3 728.40
${}^{12}_{6}$C atom	$1.992\,65 \times 10^{-27}$	12.000 000	11 177.9

have a mass of approximately 1 u, and the electron has a mass that is only a small fraction of an atomic mass unit:

Mass of proton = 1.007 276 u
Mass of neutron = 1.008 665 u
Mass of electron = 0.000 548 6 u

Because the rest energy of a particle is given by $E_R = mc^2$ (Section 9.7), it is often convenient to express the atomic mass unit in terms of its rest energy equivalent. For one atomic mass unit, we have

$$E_R = mc^2 = (1.660\,539 \times 10^{-27}\ \text{kg})(2.997\,92 \times 10^8\ \text{m/s})^2 = 931.494\ \text{MeV}/c^2$$

where we have used the conversion $1\ \text{eV} = 1.602\,176 \times 10^{-19}\ \text{J}$. Using this equivalence, nuclear physicists often express mass in terms of the unit MeV/c^2. The masses of several simple particles are given in Table 30.1. The masses and some other properties of selected isotopes are provided in Table A.3 in Appendix A.

The Size of Nuclei

The size and structure of nuclei were first investigated in the scattering experiments of Rutherford, discussed in Section 29.1. Using the principle of conservation of energy, Rutherford found an expression for how close an alpha particle moving directly toward the nucleus can approach the nucleus before being turned around by Coulomb repulsion.

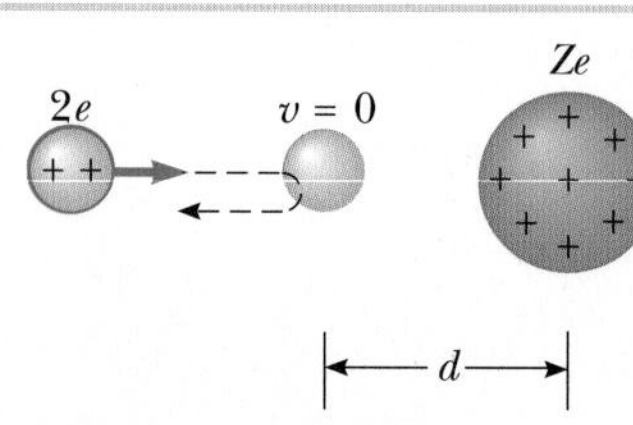

ACTIVE FIGURE 30.1
An alpha particle on a head-on collision course with a nucleus of charge *Ze*. Because of the Coulomb repulsion between charges of the same sign, the alpha particle approaches to a distance *d* from the target nucleus, called the distance of closest approach.

PhysicsNow™ By logging into PhysicsNow at **www.pop4e.com** and going to Active Figure 30.1, you can adjust the atomic number of the target nucleus and the kinetic energy of the alpha particle. Then observe the approach of the alpha particle toward the nucleus.

Let us consider the system of the incoming alpha particle ($Z = 2$) and the nucleus (arbitrary Z), and apply the energy version of the isolated system model. Because the nucleus is assumed to be much more massive than the alpha particle, we identify the kinetic energy of the system as the kinetic energy of the alpha particle alone. When the alpha particle and the nucleus are far apart, we can approximate the potential energy of the system as zero. If the collision is head-on, the alpha particle stops momentarily at some point (Active Fig. 30.1) and the energy of the system is entirely potential. Therefore, the initial kinetic energy of the incoming alpha particle is converted completely to electric potential energy of the system when the particle stops:

$$\tfrac{1}{2}mv^2 = k_e \frac{q_1 q_2}{r} = k_e \frac{(2e)(Ze)}{d}$$

where d is the distance of closest approach, Z is the atomic number of the target nucleus, and we have used the nonrelativistic expression for kinetic energy because speeds of alpha particles from radioactive decay are small relative to c. Solving for d, we find that

$$d = \frac{4k_e Z e^2}{mv^2}$$

From this expression, Rutherford found that alpha particles approached to within 3.2×10^{-14} m of a nucleus when the foil was made of gold. Based on this calculation and his analysis of results for collisions that were not head-on, Rutherford argued that the radius of the gold nucleus must be less than this value. For silver atoms, the distance of closest approach was found to be 2×10^{-14} m. From these results, Rutherford reached his conclusion that the positive charge in an atom is concentrated in a small sphere called the nucleus, whose radius is no greater than about 10^{-14} m. Note that this radius is on the order of 10^{-4} of the Bohr radius, corresponding to a nuclear volume which is on the order of 10^{-12} of the volume of a hydrogen atom. The nucleus is an incredibly small part of the atom! Because such small lengths are common in nuclear physics, a convenient unit of length is the *femtometer* (fm), sometimes called the **fermi,** defined as

$$1 \text{ fm} \equiv 10^{-15} \text{ m}$$

FIGURE 30.2 A nucleus can be modeled as a cluster of tightly packed spheres, each of which is a nucleon.

Since the time of Rutherford's scattering experiments, a multitude of other experiments have shown that most nuclei can be geometrically modeled as being approximately spherical with an average radius of

Radius of a nucleus

$$r = r_0 A^{1/3} \quad [30.1]$$

where A is the mass number and r_0 is a constant equal to 1.2×10^{-15} m. Because the volume of a sphere is proportional to the cube of the radius, it follows from Equation 30.1 that the volume of a nucleus (assumed to be spherical) is directly proportional to A, the total number of nucleons, which suggests that **all nuclei have nearly the same density.** Nucleons combine to form a nucleus as though they were tightly packed spheres (Fig. 30.2).

EXAMPLE 30.1 **Nuclear Volume and Density**

A Find an approximate expression for the mass of a nucleus of mass number A.

Solution The mass of the proton is approximately equal to that of the neutron. Therefore, if the mass of one of these particles is m, the mass of the nucleus is approximately Am.

B Find an expression for the volume of this nucleus in terms of the mass number.

Solution Assuming that the nucleus is spherical and using Equation 30.1, we find that the volume is

$$V = \tfrac{4}{3}\pi r^3 = \tfrac{4}{3}\pi r_0{}^3 A$$

C Find a numerical value for the density of this nucleus.

Solution The nuclear density is

$$\rho_n = \frac{m_{\text{nucleus}}}{V} = \frac{Am}{\frac{4}{3}\pi r_0{}^3 A} = \frac{3m}{4\pi r_0{}^3}$$

$$= \frac{3(1.67 \times 10^{-27}\text{ kg})}{4\pi(1.2 \times 10^{-15}\text{ m})^3} = 2.3 \times 10^{17}\text{ kg/m}^3$$

Recalling that the density of water is 10^3 kg/m^3, note that the nuclear density is about 2.3×10^{14} times greater than that of water!

Nuclear Stability

Because the nucleus consists of a closely packed collection of protons and neutrons, you might be surprised that it can exist at all. The very large repulsive electrostatic forces between protons in close proximity should cause the nucleus to fly apart. Nuclei are stable, however, because of the presence of another force, the **nuclear force** (see Section 5.5). This short-range force (it is nonzero only for particle separations less than about 2 fm) is an attractive force that acts between all nuclear particles. The nuclear force also acts between pairs of neutrons and between neutrons and protons.

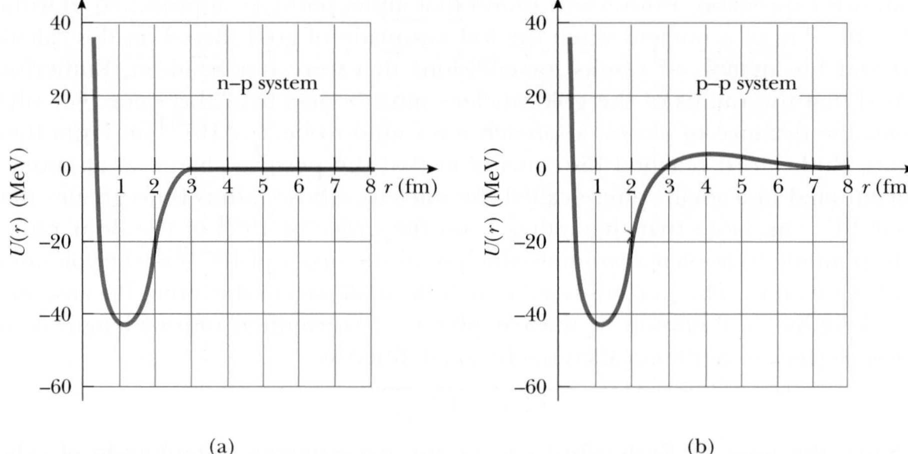

FIGURE 30.3 (a) Potential energy versus separation distance for the neutron–proton system. (b) Potential energy versus separation distance for the proton–proton system. The difference in the two curves is due to the Coulomb repulsion in the case of the proton–proton interaction. To display the difference in the curves on this scale, the height of the peak for the proton–proton curve has been exaggerated by a factor of 10.

The nuclear force dominates the Coulomb repulsive force within the nucleus (at short ranges). If that were not the case, stable nuclei would not exist. Moreover, the nuclear force is independent of charge. In other words, the forces associated with the proton–proton, proton–neutron, and neutron–neutron interactions are the same, apart from the additional repulsive Coulomb force for the proton–proton interaction.

Evidence for the limited range of nuclear forces comes from scattering experiments and from studies of nuclear binding energies, which we shall discuss shortly. The short range of the nuclear force is shown in the neutron–proton (n–p) potential energy plot of Figure 30.3a obtained by scattering neutrons from a target containing hydrogen. The depth of the n–p potential energy well is 40 to 50 MeV, and a strong repulsive component prevents the nucleons from approaching much closer than 0.4 fm.

The nuclear force does not affect electrons, enabling energetic electrons to serve as point-like probes of the charge density of nuclei. The charge independence of the nuclear force also means that the main difference between the n–p and p–p interactions is that the p–p potential energy consists of a *superposition* of nuclear and Coulomb interactions as shown in Figure 30.3b. At distances less than 2 fm, the p–p and n–p potential energies are nearly identical, but for distances greater than this, the p–p potential has a positive energy barrier with a maximum at 4 fm.

About 260 stable nuclei exist; hundreds of other nuclei have been observed but are unstable. A useful graphical representation in nuclear physics is a plot of N versus Z for stable nuclei as shown in Figure 30.4. Note that light nuclei are stable if they contain equal numbers of protons and neutrons—that is, if $N = Z$—but heavy nuclei are stable if $N > Z$. This behavior can be partially understood by recognizing that as the number of protons increases, the strength of the Coulomb force increases, which tends to break the nucleus apart. As a result, more neutrons are needed to keep the nucleus stable because neutrons experience only the attractive nuclear force. Eventually, when $Z = 83$, the repulsive forces between protons cannot be compensated by the addition of more neutrons. Elements that contain more than 83 protons do not have stable nuclei.

Interestingly, most stable nuclei have even values of A. In fact, certain values of Z and N correspond to nuclei with unusually high stability. These values of Z and N, called **magic numbers,** are

$$Z \text{ or } N = 2, 8, 20, 28, 50, 82, 126 \qquad [30.2]$$

For example, the helium nucleus (two protons and two neutrons), which has $Z = 2$ and $N = 2$, is very stable. This stability is reminiscent of the chemical stability of inert gases and suggests quantized nuclear energy levels, which we indeed find to

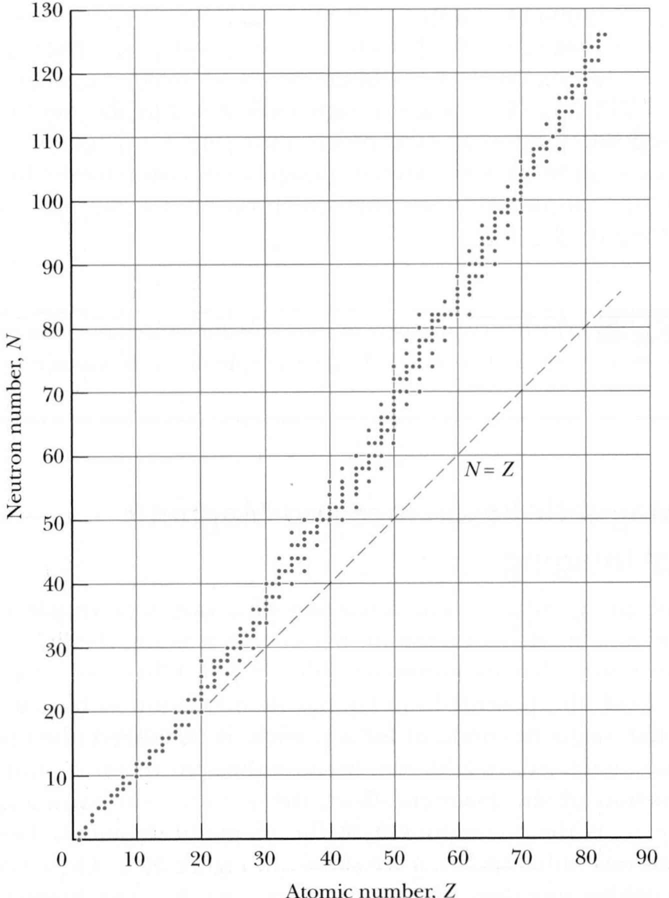

FIGURE 30.4 Neutron number N versus atomic number Z for the stable nuclei (blue dots). These nuclei lie in a narrow band called the line of stability. The dashed line corresponds to the condition $N = Z$.

be the case. Some structural models of the nucleus predict a shell structure similar to that for the atom.

Nuclear Spin and Magnetic Moment

In Chapter 29, we discussed that an electron has an intrinsic angular momentum called spin. Protons and neutrons, like electrons, also have an intrinsic angular momentum. Furthermore, a nucleus has a net intrinsic angular momentum that arises from the individual spins of the protons and neutrons. This angular momentum must obey the same quantum rules as orbital angular momentum and spin (Section 29.4). Therefore, the magnitude of the **nuclear angular momentum** is due to the combination of all nucleons and is equal to $\sqrt{I(I+1)}\,\hbar$, where I is called the **nuclear spin quantum number** and may be an integer or a half-integer. The maximum component of the nuclear angular momentum projected along any direction is $I\hbar$. Figure 30.5 illustrates the possible orientations of the nuclear spin and its projections along the z axis for the case where $I = \frac{3}{2}$.

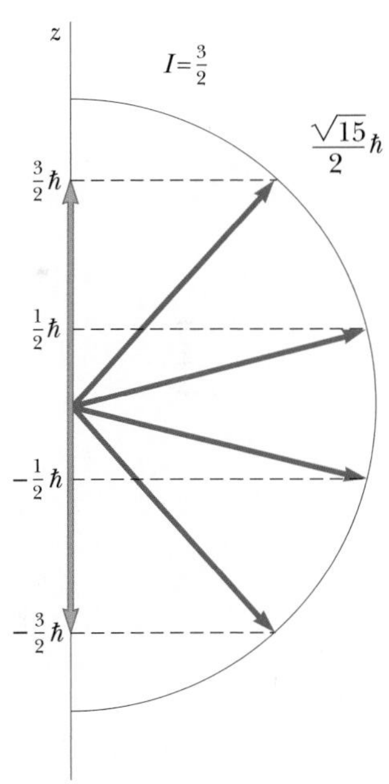

FIGURE 30.5 A vector model showing possible orientations of the nuclear spin angular momentum vector and its projections along the z axis for the case $I = \frac{3}{2}$.

The nuclear angular momentum has a nuclear magnetic moment associated with it. The magnetic moment of a nucleus is measured in terms of the **nuclear magneton** μ_n, a unit of magnetic moment defined as

■ Nuclear magneton

$$\mu_n \equiv \frac{e\hbar}{2m_p} = 5.05 \times 10^{-27}\,\mathrm{J/T} \qquad [30.3]$$

This definition is analogous to Equation 29.15 for the z component of the spin magnetic moment for an electron, which is the Bohr magneton μ_B. Note that μ_n is smaller than μ_B by a factor of about 2 000 because of the large difference in masses of the proton and electron.

The magnetic moment of a free proton is $2.792\,8\mu_n$. Unfortunately, no general theory of nuclear magnetism explains this value. Another surprising point is that a neutron, despite having no electric charge, also has a magnetic moment, which has a value of $-1.913\,5\mu_n$. The negative sign indicates that the neutron's magnetic moment is opposite its spin angular momentum. Such a magnetic moment for a neutral particle suggests that we need to design a structural model for the neutron that explains such an observation. This structural model, the *quark model*, will be discussed in Chapter 31.

QUICK QUIZ 30.2 Which do you expect to show very little variation among different isotopes of an element? **(a)** atomic mass **(b)** nuclear spin magnetic moment **(c)** chemical behavior

Nuclear Magnetic Resonance and Magnetic Resonance Imaging

The potential energy of a system consisting of a magnetic dipole moment in a magnetic field is $-\vec{\boldsymbol{\mu}} \cdot \vec{\mathbf{B}}$. When the direction of $\vec{\boldsymbol{\mu}}$ is along the field, the potential energy of the system has its minimum value $-\mu B$. When the direction of $\vec{\boldsymbol{\mu}}$ is opposite the field, the potential energy has its maximum value μB. Because the direction of the magnetic moment for a particle is quantized, the energies of the system are also quantized. In addition, because the spin vector cannot align exactly with the direction of the magnetic field, the extreme values of the energy are $\pm \mu_z B$, where μ_z is the z component of the magnetic moment. The two energy states for a nucleus with a spin of $\frac{1}{2}$ are shown in Figure 30.6. These states are often called **spin states** because they differ in energy as a result of the direction of the spin.

It is possible to observe transitions between these two spin states in a sample using a technique known as **nuclear magnetic resonance** (NMR). A constant magnetic field changes the energy associated with the spin states, splitting them apart in energy (Fig. 30.6). In addition, the sample is irradiated with electromagnetic waves in the radio range of the electromagnetic spectrum. When the frequency of the radio waves is adjusted such that the photon energy matches the separation energy between spin states, the photon is absorbed by a nucleus in the ground state, raising the nucleus–magnetic field system to the higher-energy spin state. The result is a net absorption of energy by the system, which is detected by the experimental control and measurement system. A diagram of the apparatus used to detect an NMR signal is illustrated in Figure 30.7. The absorbed energy is supplied by the oscillator producing the radio waves. Nuclear magnetic resonance and a related technique called electron spin resonance are extremely important methods for studying nuclear and atomic systems and how these systems interact with their surroundings.

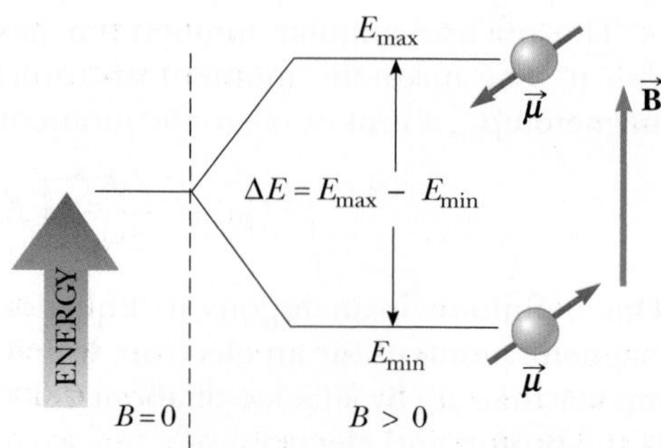

FIGURE 30.6 A nucleus with spin $\frac{1}{2}$ can occupy one of two energy states when placed in an external magnetic field. The lower energy state E_{min} corresponds to the case where the spin is aligned with the field as much as possible according to quantum mechanics, and the higher energy state E_{max} corresponds to the case where the spin is opposite the field as much as possible.

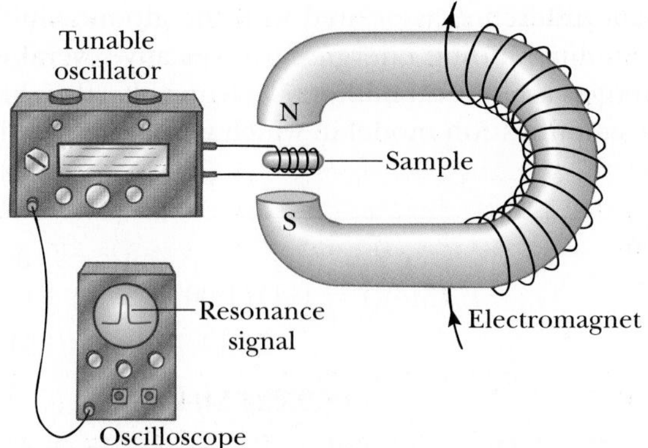

FIGURE 30.7 Experimental arrangement for nuclear magnetic resonance. The radio-frequency magnetic field created by the coil surrounding the sample and provided by the variable-frequency oscillator is perpendicular to the constant magnetic field created by the electromagnet. When the nuclei in the sample meet the resonance condition, the nuclei absorb energy from the radio-frequency field of the coil, and this absorption changes the characteristics of the circuit in which the coil is included. Most modern NMR spectrometers use superconducting magnets at fixed field strengths and operate at frequencies of approximately 200 MHz.

A widely used medical diagnostic technique called **MRI,** for **magnetic resonance imaging,** is based on nuclear magnetic resonance. In MRI, the patient is placed inside a large solenoid that supplies a spatially varying magnetic field. Because of the variation in the magnetic field across the patient's body, protons in hydrogen atoms in water molecules in different parts of the body have different splittings in energy between spin states, and the resonance signal can be used to provide information on the positions of the protons. A computer is used to analyze the position information to provide data for constructing a final image. An MRI scan showing incredible detail in internal body structure is shown in Figure 30.8. The main advantage of MRI over other imaging techniques in medical diagnostics is that it does not cause damage to cellular structures as x-rays do. Photons associated with the radio-frequency signals used in MRI have energies of only about 10^{-7} eV. Because molecular bond strengths are much larger (on the order of 1 eV), the radio-frequency radiation cannot cause cellular damage. In comparison, x-rays or γ-rays have energies ranging from 10^4 to 10^6 eV and can cause considerable cellular damage. Therefore, despite some individuals' fears of the word *nuclear* associated with magnetic resonance imaging, the radio-frequency radiation involved is overwhelmingly safer than x-rays!

 Magnetic resonance imaging

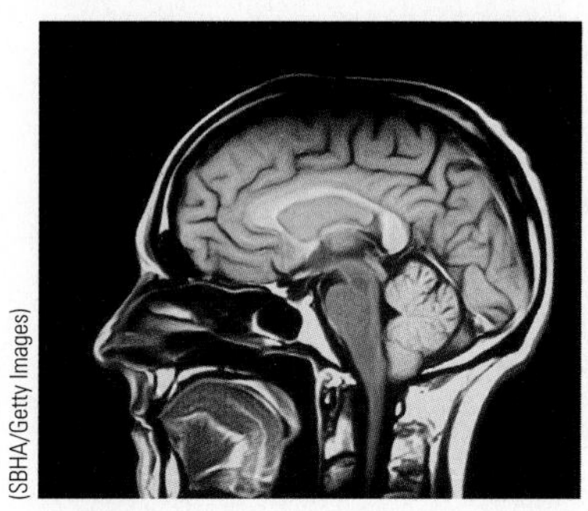

(SBHA/Getty Images)

FIGURE 30.8 A color-enhanced MRI scan of a human brain.

30.2 BINDING ENERGY

It is found that the mass of a nucleus is always less than the sum of the masses of its nucleons. Because mass is a manifestation of energy, **the total rest energy of the bound system (the nucleus) is less than the combined rest energy of the separated nucleons.** This difference in energy is called the **binding energy** E_b of the nucleus and represents the energy that must be added to a nucleus to break it apart into its components:

$$E_b(\text{MeV}) = [ZM(\text{H}) + Nm_n - M(^A_Z\text{X})] \times 931.494 \text{ MeV/u} \qquad [30.4]$$

■ Binding energy of a nucleus

where $M(\text{H})$ is the atomic mass of the neutral hydrogen atom, $M(^A_Z\text{X})$ represents the atomic mass of an atom of the isotope ^A_ZX, m_n is the mass of the neutron, and the masses are all in atomic mass units. Note that the mass of the Z electrons included in $M(\text{H})$ cancels with the mass of the Z electrons included in the term

PITFALL PREVENTION 30.1

Binding Energy When separate nucleons are combined to form a nucleus, the rest energy of the system is reduced. Therefore, the change in energy is negative. The absolute value of this change is called the binding energy. This difference in sign may be a source of confusion. For example, an *increase* in binding energy corresponds to a *decrease* in the rest energy of the system.

$M(^A_Z\text{X})$ within a small difference associated with the atomic binding energy of the electrons. Because atomic binding energies are typically several electron volts and nuclear binding energies are several million electron volts, this difference is negligible, and we adopt a simplification model in which we ignore this difference.

EXAMPLE 30.2 The Binding Energy of the Deuteron

Calculate the binding energy of the deuteron (the nucleus of a deuterium atom), which consists of a proton and a neutron, given that the atomic mass of deuterium is 2.014 102 u.

Solution From Table 30.1, we see that the mass of the hydrogen atom, representing the proton, is $M(\text{H}) = 1.007\,825$ u and that the neutron mass $m_n = 1.008\,665$ u. Therefore,

$$E_b(\text{MeV}) = [(1)(1.007\,825\text{ u}) + (1)(1.008\,665\text{ u}) - 2.014\,102\text{ u}] \times 931.494\text{ MeV/u}$$

$$= 2.224\text{ MeV}$$

This result tells us that separating a deuteron into its constituent proton and neutron requires adding 2.224 MeV of energy to the deuteron. One way of supplying the deuteron with this energy is by bombarding it with energetic particles.

A plot of binding energy per nucleon E_b/A as a function of mass number for various stable nuclei is shown in Figure 30.9. Note that the curve has a maximum in the vicinity of $A = 60$, corresponding to isotopes of iron, cobalt, and nickel. That is, nuclei having mass numbers either greater or less than 60 are not as strongly bound as those near the middle of the periodic table. The higher values of binding energy per nucleon near $A = 60$ imply that energy is released when a heavy nucleus splits, or *fissions,* into two lighter nuclei. Energy is released in fission because the nucleons in each product nucleus are more tightly bound to one another than are the nucleons in the original nucleus. The important process of fission and a second important process of *fusion,* in which energy is released as light nuclei combine, are considered in detail in Section 30.6.

The binding energy per nucleon in Figure 30.9 is approximately constant at 8 MeV for $A > 20$. In this case, the nuclear forces between a particular nucleon and all the other nucleons in the nucleus are said to be *saturated;* that is, a particular

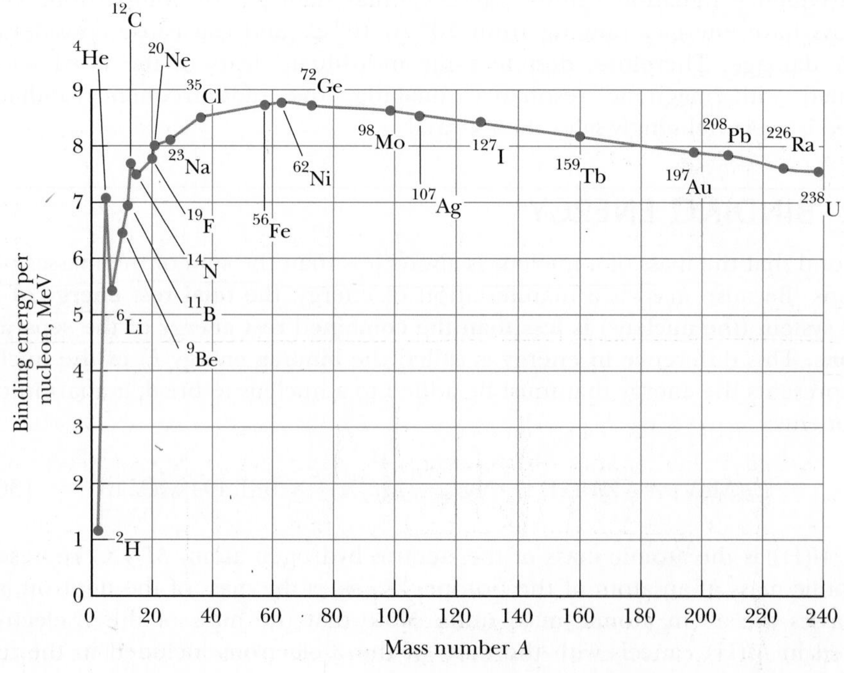

FIGURE 30.9 Binding energy per nucleon versus mass number for nuclei that lie along the line of stability in Figure 30.4. Some representative nuclei appear as blue dots with labels. (Nuclei to the right of ^{208}Pb are unstable. The curve represents the binding energy for the most stable isotopes.)

nucleon interacts with only a limited number of other nucleons because of the short-range character of the nuclear force. These other nucleons can be viewed as being the nearest neighbors in the closely packed structure shown in Figure 30.2.

Figure 30.9 provides insight into fundamental questions about the origin of the chemical elements. In the early life of the Universe, there were only hydrogen and helium. Clouds of cosmic gas coalesced under gravitational forces to form stars. As a star ages, it produces heavier elements from the lighter elements contained within it, beginning by fusing hydrogen atoms to form helium. This process continues as the star becomes older, generating atoms having larger and larger atomic numbers. The nuclide $^{62}_{28}$Ni has the largest binding energy per nucleon of 8.794 5 MeV/nucleon. It takes additional energy to create elements in a star with mass numbers larger than 62 because of their lower binding energies per nucleon. This energy comes from the supernova explosion that occurs at the end of some large stars' lives. Therefore, all the heavy atoms in your body were produced from the explosions of ancient stars. You are literally made of stardust!

■ Thinking Physics 30.1

Figure 30.9 shows a graph of the average amount of energy necessary to remove a nucleon from the nucleus. Figure 29.12 shows the energy necessary to remove an electron from an atom. Why does Figure 30.9 show an *approximately constant* amount of energy necessary to remove a nucleon (above about $A = 20$), but Figure 29.12 shows *widely varying* amounts of energy necessary to remove an electron from the atom?

Reasoning In the case of Figure 30.9, the approximately constant value of the nuclear binding energy is a result of the short-range nature of the nuclear force. A given nucleon interacts only with its few nearest neighbors, rather than with all the nucleons in the nucleus. Therefore, no matter how many nucleons are present in the nucleus, removing one nucleon involves separating it only from its nearest neighbors. The energy to do so is therefore approximately independent of how many nucleons are present.

On the other hand, the electric force holding the electrons to the nucleus in an atom is a long-range force. An electron in the atom interacts with all the protons in the nucleus. When the nuclear charge increases, a stronger attraction occurs between the nucleus and the electrons. As a result, as the nuclear charge increases, more energy is necessary to remove an electron, as demonstrated by the upward tendency of the ionization energy in Figure 29.12 for each period. ■

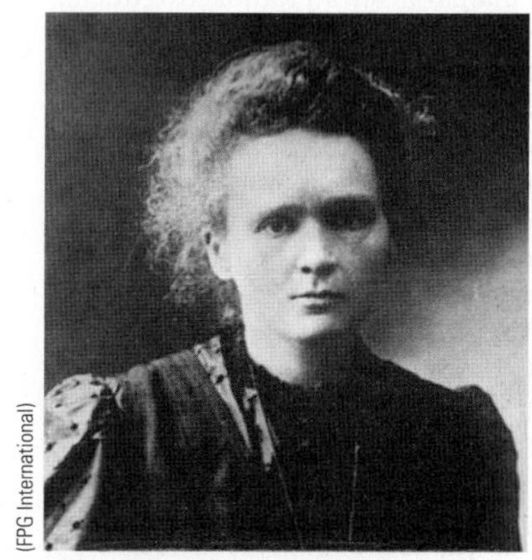

(FPG International)

Marie Curie (1867–1934)
A Polish scientist, Marie Curie shared the Nobel Prize in Physics in 1903 with her husband, Pierre, and with Henri Becquerel for their work on spontaneous radioactivity and the radiation emitted by radioactive substances. She wrote "I persist in believing that the ideas that then guided us are the only ones which can lead to the true social progress. We cannot hope to build a better world without improving the individual. Toward this end, each of us must work toward his own highest development, accepting at the same time his share of responsibility in the general life of humanity."

30.3 RADIOACTIVITY

At the beginning of this chapter, we discussed the discovery of radioactivity by Becquerel, which indicated that nuclei emit particles and radiation. This spontaneous emission was soon to be called **radioactivity.**

The most significant investigations of this phenomenon were conducted by Marie Curie and Pierre Curie (1859–1906). After several years of careful and laborious chemical separation processes on tons of pitchblende, a radioactive ore, the Curies reported the discovery of two previously unknown elements, both of which were radioactive, named polonium and radium. Subsequent experiments, including Rutherford's famous work on alpha particle scattering, suggested that radioactivity was the result of the decay, or disintegration, of unstable nuclei.

Three types of radiation can be emitted by a radioactive substance: alpha (α) rays, where the emitted particles are ^{4}He nuclei; beta (β) rays, in which the emitted particles are either electrons or positrons; and gamma (γ) rays, in which the emitted rays are high-energy photons. A **positron** is a particle similar to the electron in all respects except that it has a charge of $+e$ (the positron is said to be the **antiparticle**

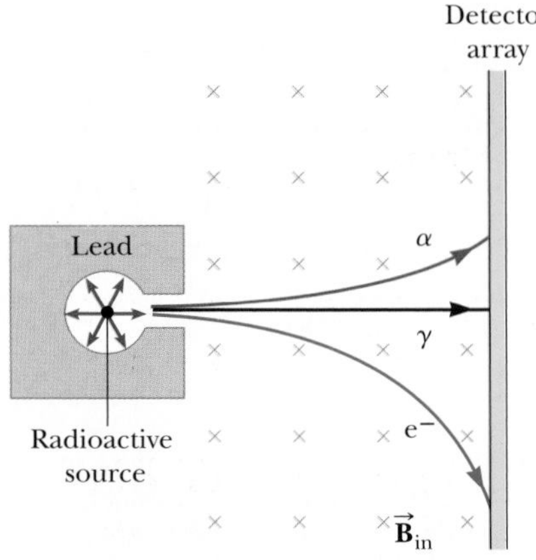

FIGURE 30.10 The radiation from radioactive sources can be separated into three components by using a magnetic field to deflect the charged particles. The detector array at the right records the events. The gamma ray is not deflected by the magnetic field.

of the electron; we shall discuss antiparticles further in Chapter 31). The symbol e^- is used to designate an electron and e^+ designates a positron.

It is possible to distinguish these three forms of radiation using the scheme illustrated in Figure 30.10. The radiation from a variety of radioactive samples is directed into a region with a magnetic field. The radiation is separated into three components by the magnetic field, two bending in opposite directions and the third experiencing no change in direction. From this simple observation, one can conclude that the radiation of the undeflected beam carries no charge (the gamma ray), the component deflected upward corresponds to positively charged particles (alpha particles), and the component deflected downward corresponds to negatively charged particles (e^-). If the beam includes positrons (e^+), these particles are deflected upward with a different radius of curvature from that of the alpha particles.

The three types of radiation have quite different penetrating powers. Alpha particles barely penetrate a sheet of paper, beta particles can penetrate a few millimeters of aluminum, and gamma rays can penetrate several centimeters of lead.

The rate at which a decay process occurs in a radioactive sample is proportional to the number of radioactive nuclei present in the sample (i.e., those nuclei that have not yet decayed). This dependence is similar to the behavior of population growth in that the rate at which babies are born is proportional to the number of people currently alive. If N is the number of radioactive nuclei present at some instant, the rate of change of N is

$$\frac{dN}{dt} = -\lambda N \qquad [30.5]$$

where λ is called either the **decay constant** or the **disintegration constant** and has a different value for different nuclei. The negative sign indicates that dN/dt is a negative number; that is, N decreases in time.

If we write Equation 30.5 in the form

$$\frac{dN}{N} = -\lambda\, dt$$

we can integrate from an arbitrary initial instant $t = 0$ to a later time t:

$$\int_{N_0}^{N} \frac{dN}{N} = -\lambda \int_0^t dt$$

$$\ln\left(\frac{N}{N_0}\right) = -\lambda t$$

■ Number of undecayed nuclei as a function of time

$$N = N_0 e^{-\lambda t} \qquad [30.6]$$

The constant N_0 represents the number of undecayed radioactive nuclei at $t = 0$. We have seen exponential behaviors before, for example, with the discharging of a capacitor in Section 21.9. Based on these experiences, we can identify the inverse of the decay constant $1/\lambda$ as the time interval required for the number of undecayed nuclei to fall to $1/e$ of its original value. Therefore, $1/\lambda$ is the **time constant** for this decay, similar to the time constants we investigated for the decay of the current in an RC circuit in Section 21.9 and an RL circuit in Section 23.6.

The **decay rate** R is obtained by differentiating Equation 30.6 with respect to time:

$$R = \left|\frac{dN}{dt}\right| = N_0 \lambda e^{-\lambda t} = R_0 e^{-\lambda t} \qquad [30.7]$$

where $R = N\lambda$ and $R_0 = N_0\lambda$ is the decay rate at $t = 0$. The decay rate of a sample is often referred to as its **activity.** Note that both N and R decrease exponentially

PITFALL PREVENTION 30.2

Rays or Particles? Early in the history of nuclear physics, the term *radiation* was used to describe the emanations from radioactive nuclei. We now know that two out of the three types, namely alpha radiation and beta radiation, involve the emission of particles. Even though these particles are not examples of electromagnetic radiation, the use of the term *radiation* for all three types is deeply entrenched in our language. We will use this term because of its wide usage in the physics community.

PITFALL PREVENTION 30.3

Notation Warning In Section 30.1, we introduced the symbol N as an integer representing the number of neutrons in a nucleus. In this discussion, the symbol N represents the number of undecayed nuclei in a radioactive sample remaining after some time interval. As you read further, be sure to consider the context to determine the appropriate meaning for the symbol N.

with time. The plot of N versus t in Active Figure 30.11 illustrates the exponential decay law.

A common unit of activity for a radioactive sample is the **curie** (Ci), defined as

■ The curie

$$1 \text{ Ci} \equiv 3.7 \times 10^{10} \text{ decays/s}$$

This unit was selected as the original unit of activity because it is the approximate activity of 1 g of radium. The SI unit of activity is called the **becquerel** (Bq):

■ The becquerel

$$1 \text{ Bq} \equiv 1 \text{ decay/s}$$

Therefore, 1 Ci = 3.7×10^{10} Bq. The most commonly used units of activity are millicuries (mCi) and microcuries (μCi).

A useful parameter for characterizing radioactive decay is the **half-life** $T_{1/2}$. **The half-life of a radioactive substance is the time interval required for half of a given number of radioactive nuclei to decay.** Setting $N = N_0/2$ and $t = T_{1/2}$ in Equation 30.6 gives

$$\frac{N_0}{2} = N_0 e^{-\lambda T_{1/2}}$$

Writing this equation in the form $e^{\lambda T_{1/2}} = 2$ and taking the natural logarithm of both sides, we have

■ Relationship between half-life and decay constant

$$T_{1/2} = \frac{\ln 2}{\lambda} = \frac{0.693}{\lambda} \qquad [30.8]$$

which is a convenient expression relating the half-life to the decay constant. Note that after a time interval of one half-life, $N_0/2$ radioactive nuclei remain (by definition); after two half-lives, half of these have decayed and $N_0/4$ radioactive nuclei remain; after three half-lives, $N_0/8$ remain; and so on. In general, after n half-lives, the number of radioactive nuclei remaining is $N_0/2^n$.

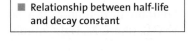

PITFALL PREVENTION 30.4

HALF-LIFE It is *not* true that all the original nuclei have decayed after two half-lives! In one half-life, *half of those nuclei that are left* will decay.

QUICK QUIZ 30.3 On your birthday, you measure the activity of a sample of ^{210}Bi, which has a half-life of 5.01 days. The activity you measure is 1.000 μCi. What is the activity of this sample on your next birthday? **(a)** 1.000 μCi **(b)** 0 **(c)** $\sim$0.2 μCi **(d)** $\sim$0.01 μCi **(e)** $\sim 10^{-22}$ μCi

QUICK QUIZ 30.4 Suppose you have a pure radioactive material with a half-life of $T_{1/2}$. You begin with N_0 undecayed nuclei of the material at $t = 0$. At $t = \frac{1}{2}T_{1/2}$, how many of the nuclei *have decayed*? **(a)** $\frac{1}{4}N_0$ **(b)** $\frac{1}{2}N_0$ **(c)** $\frac{3}{4}N_0$ **(d)** $0.707N_0$ **(e)** $0.293N_0$

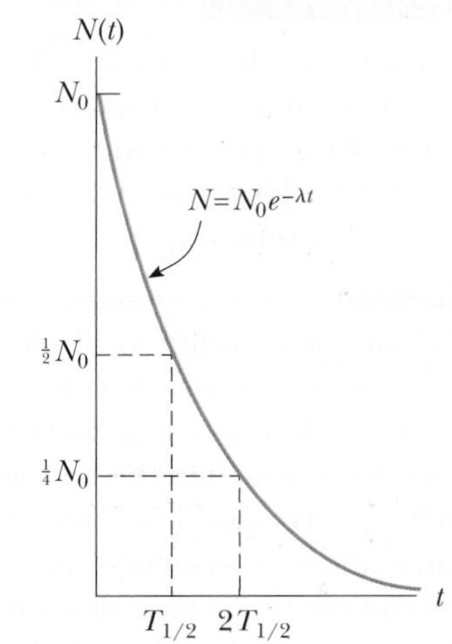

ACTIVE FIGURE **30.11**
Plot of the exponential decay law for radioactive nuclei. The vertical axis represents the number of undecayed radioactive nuclei present at any time t, and the horizontal axis is time. The time $T_{1/2}$ is the half-life of the sample.

Physics Now™ By logging into PhysicsNow at www.pop4e.com and going to Active Figure 30.11, you can observe the decay curves for nuclei with varying half-lives.

■ Thinking Physics 30.2

The isotope $^{14}_{6}$C is radioactive and has a half-life of 5 730 years. If you start with a sample of 1 000 carbon-14 nuclei, how many remain (have not decayed) after 17 190 yr?

Reasoning In 5 730 yr, half the sample will have decayed, leaving 500 radioactive $^{14}_{6}$C nuclei. In another 5 730 yr (for a total elapsed time of 11 460 yr), the number remaining is 250 nuclei. After another 5 730 yr (total of 17 190 yr), 125 remain.

These numbers represent ideal circumstances. Radioactive decay is an averaging process over a very large number of atoms, and the actual outcome depends on statistics. Our original sample in this example contained only 1 000 nuclei, certainly not a very large number when we are dealing with atoms, for which we measure the numbers in macroscopic samples in terms of Avogadro's number. Therefore, if we were actually to count the number remaining after one half-life for this small sample, it probably would not be exactly 500. ■

INTERACTIVE EXAMPLE 30.3 The Activity of Radium

The half-life of the radioactive nucleus radium-226, ${}^{226}_{88}\text{Ra}$, is 1.6×10^3 yr.

A What is the decay constant λ of this nucleus?

Solution We can calculate λ using Equation 30.8 and that

$$T_{1/2} = 1.6 \times 10^3 \text{ yr}\left(\frac{3.16 \times 10^7 \text{ s}}{1 \text{ yr}}\right)$$
$$= 5.0 \times 10^{10} \text{ s}$$

Therefore,

$$\lambda = \frac{0.693}{T_{1/2}} = \frac{0.693}{5.0 \times 10^{10} \text{ s}} = 1.4 \times 10^{-11} \text{ s}^{-1}$$

Note that this result is also the probability that any single ${}^{226}_{88}\text{Ra}$ nucleus will decay in a time interval of 1 second.

B If a sample contains 3.0×10^{16} ${}^{226}_{88}\text{Ra}$ nuclei at $t = 0$, determine its activity in curies at this time.

Solution By definition (Eq. 30.7), R_0, the activity at $t = 0$, is λN_0, where N_0 is the number of radioactive nuclei present at $t = 0$. With $N_0 = 3.0 \times 10^{16}$, we have

$$R_0 = \lambda N_0 = (1.4 \times 10^{-11} \text{ s}^{-1})(3.0 \times 10^{16})$$
$$= (4.2 \times 10^5 \text{ Bq})\left(\frac{1 \text{ Ci}}{3.7 \times 10^{10} \text{ Bq}}\right)$$
$$= 11 \mu\text{Ci}$$

C What is the activity after the sample is 2.0×10^3 yr old?

Solution We use Equation 30.7 and that $2.0 \times 10^3 \text{ yr} = 6.3 \times 10^{10}$ s:

$$R = R_0 e^{-\lambda t}$$
$$= (11 \mu\text{Ci})e^{-(1.4 \times 10^{-11} \text{ s}^{-1})(6.3 \times 10^{10} \text{ s})}$$
$$= 4.7 \mu\text{Ci}$$

Physics Now™ By logging into PhysicsNow at **www.pop4e.com** and going to Interactive Example 30.3, you can practice evaluating the parameters for radioactive decay of various isotopes of radium.

EXAMPLE 30.4 A Radioactive Isotope of Iodine

A sample of the isotope ${}^{131}\text{I}$, which has a half-life of 8.04 days, has an activity of 5.0 mCi at the time of shipment. Upon receipt in a medical laboratory, the activity is 4.2 mCi. How much time has elapsed between the two measurements?

Solution To conceptualize this problem, consider that the sample is continuously decaying as it is in transit. The decrease in the activity is 16% during the time interval between shipment and receipt, so we expect the elapsed time to be less than the half-life of 8.04 d. The stated activity corresponds to many decays per second, so N is large and we can categorize this problem as one in which we can use our statistical analysis of radioactivity. To analyze the problem, we use Equation 30.7 in the form

$$\frac{R}{R_0} = e^{-\lambda t}$$

where the sample is shipped at $t = 0$, at which time the activity is R_0. Taking the natural logarithm of each side, we have

$$\ln\left(\frac{R}{R_0}\right) = -\lambda t$$

$$t = -\frac{1}{\lambda}\ln\left(\frac{R}{R_0}\right) \quad (1)$$

To find λ, we use Equation 30.8:

$$\lambda = \frac{0.693}{T_{1/2}} = \frac{0.693}{8.04 \text{ d}} = 8.62 \times 10^{-2} \text{ d}^{-1} \quad (2)$$

Substituting (2) into (1) gives

$$t = -\left(\frac{1}{8.62 \times 10^{-2} \text{ d}^{-1}}\right)\ln\left(\frac{4.2 \text{ mCi}}{5.0 \text{ mCi}}\right) = 2.0 \text{ d}$$

To finalize this problem, note that this value is indeed less than the half-life, as we expected. This problem demonstrates the difficulty in shipping radioactive samples with short half-lives. If the shipment were to be delayed by several days, only a small fraction of the sample would remain upon receipt. This difficulty can be addressed by shipping a combination of isotopes in which the desired isotope is the product of a decay occurring within the sample. It is possible for the desired isotope to be in *equilibrium,* in which case it is created at the same rate as it decays. Therefore, the amount of the desired isotope remains constant during the shipping process. Upon receipt, the desired isotope can be separated from the rest of the sample, and its decay from the initial activity begins upon receipt rather than upon shipment.

30.4 THE RADIOACTIVE DECAY PROCESSES

When one nucleus changes into another without external influence, the process is called **spontaneous decay.** As we stated in Section 30.3, a radioactive nucleus spontaneously decays by one of three processes: alpha decay, beta decay, or gamma decay. Active Figure 30.12 shows a close-up view of a portion of Figure 30.4 from $Z = 65$ to $Z = 80$. The blue circles are the stable nuclei seen in Figure 30.4. In addition, unstable nuclei above and below the line of stability for each value of Z are shown. Above the line of stability, the red circles show unstable nuclei that are neutron-rich and undergo a beta decay process in which an electron is emitted. Below the blue circles are green circles corresponding to proton-rich unstable nuclei that primarily undergo a beta decay process in which a positron is emitted or a competing process called electron capture. Beta decay and electron capture are described in more detail below. Further below the line of stability (with a few exceptions) are yellow circles that represent very proton-rich nuclei for which the primary decay mechanism is alpha decay, which we will discuss first.

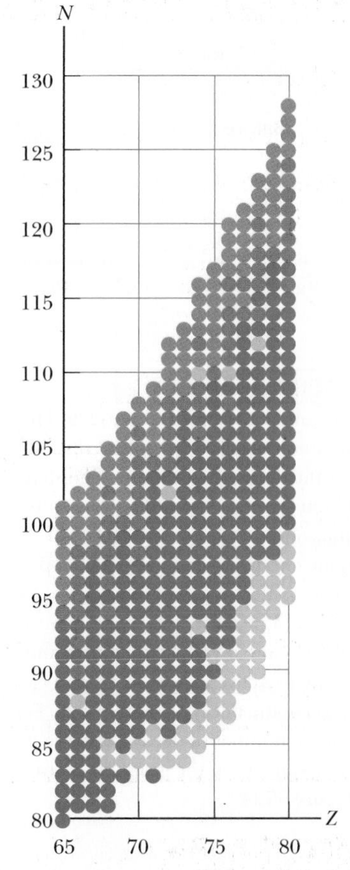

ACTIVE FIGURE **30.12**
A close-up view of the line of stability in Figure 30.4 from $Z = 65$ to $Z = 80$. The blue dots represent stable nuclei as in Figure 30.4. The other colored dots represent unstable isotopes above and below the line of stability, with the color of the dot indicating the primary means of decay.

Physics Now™ Study the decay modes and decay energies by logging into PhysicsNow at **www.pop4e.com** and going to Active Figure 30.12. Click on any of the colored dots to view information about the decay.

Alpha Decay

If a nucleus emits an alpha particle ($^4_2$He) in a spontaneous decay, it loses two protons and two neutrons. Therefore, N decreases by 2, Z decreases by 2, and A decreases by 4. The **alpha decay** can be written with a symbolic representation as

$$^A_Z\text{X} \rightarrow {}^{A-4}_{Z-2}\text{Y} + {}^4_2\text{He} \qquad [30.9]$$

where X is called the **parent nucleus** and Y the **daughter nucleus.** As general rules, (1) the sum of the mass numbers must be the same on both sides of the symbolic representation and (2) the sum of the atomic numbers must be the same on both sides. As examples, ^{238}U and ^{226}Ra are both alpha emitters and decay according to the schemes

$$^{238}_{92}\text{U} \rightarrow {}^{234}_{90}\text{Th} + {}^4_2\text{He} \qquad [30.10]$$

$$^{226}_{88}\text{Ra} \rightarrow {}^{222}_{86}\text{Rn} + {}^4_2\text{He} \qquad [30.11]$$

The half-life for ^{238}U decay is 4.47×10^9 years, and the half-life for ^{226}Ra decay is 1.60×10^3 years. In both cases, note that the mass number A of the daughter nucleus is 4 less than that of the parent nucleus. Likewise, the atomic number Z is reduced by 2.

The decay of ^{226}Ra is shown in Active Figure 30.13. In addition to the rules for the mass number and the atomic number, the total energy of the system must be conserved in the decay. If we call M_X the mass of the parent nucleus, M_Y the mass of the daughter nucleus, and M_α the mass of the alpha particle, we can define the **disintegration energy** Q:

$$Q \equiv (M_X - M_Y - M_\alpha)c^2 \qquad [30.12]$$

Note that the value of Q will be in joules if the masses are in kilograms and $c = 3.00 \times 10^8$ m/s. When the nuclear masses are expressed in the more convenient atomic mass unit u, however, the value of Q can be calculated in MeV units using the expression

$$Q = (M_X - M_Y - M_\alpha) \times 931.494 \text{ MeV/u} \qquad [30.13]$$

The disintegration energy Q represents the decrease in binding energy of the system and appears in the form of kinetic energy of the daughter nucleus and the alpha particle. In this nuclear example of the energy version of the isolated system model, no energy is entering or leaving the system. The energy in the system simply transforms from rest energy to kinetic energy, and Equation 30.13 gives the amount of energy transformed in the process. This quantity is sometimes referred to as the ***Q* value** of the nuclear reaction.

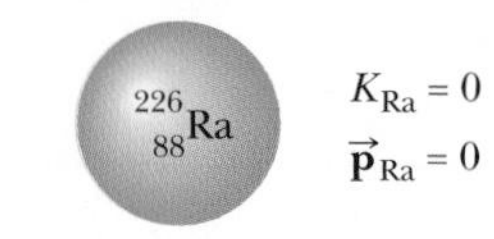

Before decay

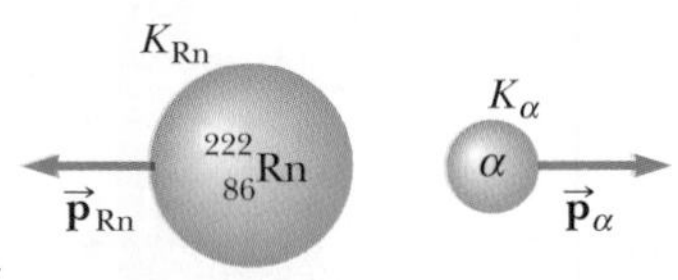

After decay

ACTIVE FIGURE 30.13

The alpha decay of radium-226. The radium nucleus is initially at rest. After the decay, the radon nucleus has kinetic energy K_{Rn} and momentum $\vec{p}_{Rn}$, and the alpha particle has kinetic energy K_α and momentum $\vec{p}_\alpha$.

PhysicsNow™ By logging into PhysicsNow at **www.pop4e.com** and going to Active Figure 30.13, you can observe the decay of radium-226. For a large number of decays, observe the development of the graph in Active Figure 30.14.

In addition to energy conservation, we can also apply the momentum version of the isolated system model to the decay. Because momentum of the isolated system must be conserved, the lighter alpha particle moves with a much higher speed than the daughter nucleus after the decay occurs. As a result, most of the available kinetic energy is associated with the alpha particle. Generally, light particles carry off most of the energy in nuclear decays.

Equation 30.13 suggests that the alpha particles are emitted with a discrete energy. Such an energy is calculated in Example 30.5. In practice, we find that alpha particles are emitted with a *set* of discrete energies (Active Fig. 30.14), with the *maximum* value calculated as in Example 30.5. This set of energies occurs because the energy of the nucleus is quantized, similar to the quantized energies in an atom. In Equation 30.13, we assume that the daughter nucleus is left in the ground state. If the daughter nucleus is left in an excited state, however, less energy is available for the decay and the alpha particle is emitted with less than the maximum kinetic energy. That the alpha particles have a discrete set of energies is direct evidence for the quantization of energy in the nucleus. This quantization is consistent with the model of a quantum particle under boundary conditions because the nucleons are quantum particles and they are subject to the constraints imposed by their mutual forces.

Finally, it is interesting to note that if one assumes that ^{238}U (or other alpha emitters) decays by emitting protons and neutrons, the mass of the decay products exceeds that of the parent nucleus, corresponding to negative Q values. Because that cannot occur for an isolated system, such spontaneous decays do not occur.

QUICK QUIZ 30.5 Which of the following is the correct daughter nucleus associated with the alpha decay of $^{157}_{72}$Hf? **(a)** $^{153}_{72}$Hf **(b)** $^{153}_{70}$Yb **(c)** $^{157}_{70}$Yb

EXAMPLE 30.5 The Energy Liberated When Radium Decays

The ^{226}Ra nucleus undergoes alpha decay according to Equation 30.11. Calculate the Q value for this process.

Solution Using Equation 30.13 and the mass values in Table A.3 in Appendix A, we see that

$$Q = [M(^{226}\text{Ra}) - M(^{222}\text{Rn}) - M(^{4}\text{He})] \times 931.494 \text{ MeV/u}$$

$$= (226.025\,403 \text{ u} - 222.017\,570 \text{ u} - 4.002\,603 \text{ u}) \times 931.494 \text{ MeV/u}$$

$$= (0.005\,230 \text{ u}) \times (931.494 \text{ MeV/u}) = \boxed{4.87 \text{ MeV}}$$

It is left to Problem 30.49 to show that the kinetic energy of the alpha particle is about 4.8 MeV, whereas that of the recoiling daughter nucleus is only about 0.1 MeV.

PITFALL PREVENTION 30.5

ANOTHER Q We have seen the symbol Q before, but in this section we introduced a brand new meaning for this symbol: the disintegration energy. It is neither heat nor charge, for which we have used Q before.

We now turn to a structural model for the mechanism of alpha decay that allows some understanding of the decay process. Imagine that the alpha particle forms within the parent nucleus so that the parent nucleus is modeled as a system consisting of the alpha particle and the remaining daughter nucleus. Figure 30.15 is a graphical representation of the potential energy of this system as a function of the separation distance r between the alpha particle and the daughter nucleus. The distance R is the range of the nuclear force. The curve represents the combined effects of (1) the repulsive Coulomb force, which describes the curve for $r > R$, and (2) the attractive nuclear force, which causes the energy curve to be negative for $r < R$. As we saw in Example 30.5, a typical disintegration energy is a few MeV, which is the approximate kinetic energy of the emitted alpha particle, represented by the lower dotted line in Figure 30.15. According to classical physics, the alpha particle is trapped in the potential well. How, then, does it ever escape from the nucleus?

The answer to this question was provided by Gamow and, independently, Ronald Gurney and Edward Condon in 1928, using quantum mechanics. The view of quantum mechanics is that there is always some probability that the particle can *tunnel* through the barrier as we discussed in Section 28.13. Our model of the potential energy curve, combined with the possibility of tunneling, predicts that the probability of tunneling should increase as the particle energy increases because of the narrowing of the barrier for higher energies. This increased probability should be reflected as an increased activity and consequently a shorter half-life. Experimental data show just this relationship: nuclei with higher alpha particle energies have shorter half-lives. If the potential energy curve in Figure 30.15 is modeled as a series of square barriers whose heights vary with particle separation according to the curve, we can generate a theoretical relationship between particle energy and half-life that is in excellent agreement with the experimental results. This particular application of modeling and quantum physics is a very effective demonstration of the power of these approaches.

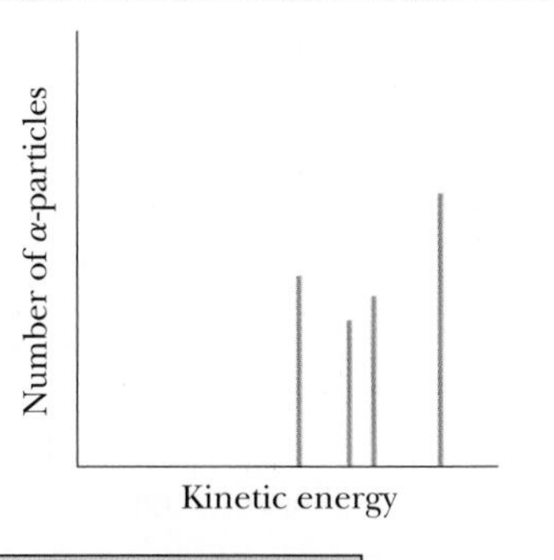

ACTIVE FIGURE 30.14 Distribution of alpha particle energies in a typical alpha decay. The energies of the alpha particles are discrete.

Physics Now™ By logging into PhysicsNow at **www.pop4e.com** and going to Active Figure 30.14, you can observe the development of this graph for the decay in Active Figure 30.13.

Beta Decay

When a radioactive nucleus undergoes **beta decay,** the daughter nucleus has the same number of nucleons as the parent nucleus, but the atomic number is changed by 1:

$$ {}^{A}_{Z}\mathrm{X} \rightarrow {}^{\;\;A}_{Z+1}\mathrm{Y} + \mathrm{e}^{-} \quad \text{(incomplete expression)} \qquad [30.14]$$

$$ {}^{A}_{Z}\mathrm{X} \rightarrow {}^{\;\;A}_{Z-1}\mathrm{Y} + \mathrm{e}^{+} \quad \text{(incomplete expression)} \qquad [30.15]$$

Again, note that nucleon number and total charge are both conserved in these decays. As we shall see later, however, these processes are not described completely by these expressions. We shall explain this incomplete description shortly.

The electron or positron involved in these decays is created within the nucleus as an initial step in the decay process. For example, during beta-minus decay, a neutron in the nucleus is transformed into a proton and an electron:

$$\mathrm{n} \rightarrow \mathrm{p} + \mathrm{e}^{-} \quad \text{(incomplete expression)}$$

For beta-plus decay, we have a proton transformed into a neutron and a positron:

$$\mathrm{p} \rightarrow \mathrm{n} + \mathrm{e}^{+} \quad \text{(incomplete expression)}$$

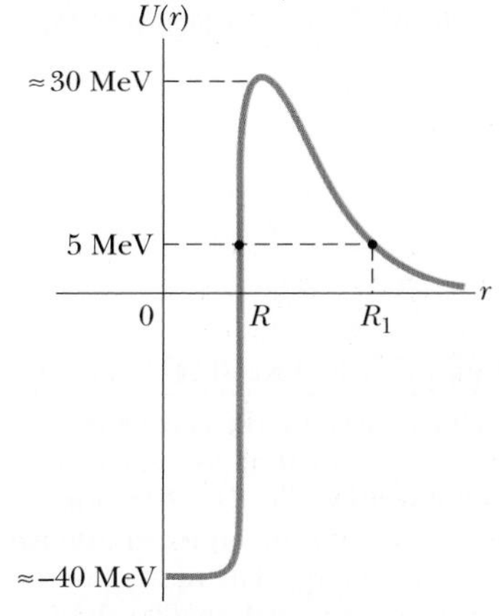

FIGURE 30.15 Potential energy versus separation distance for a system consisting of an alpha particle and a daughter nucleus. Classically, the energy associated with the alpha particle is not sufficiently large to overcome the energy barrier and so the particle should not be able to escape the nucleus. In reality, the alpha particle does escape by tunneling through the barrier.

Outside the nucleus, this latter process will not occur because the neutron and electron have more total mass than the proton. This process can occur within the nucleus, however, because we consider the rest energy changes of the entire nuclear system, not just the individual particles. In beta-plus decay, the process $\mathrm{p} \rightarrow \mathrm{n} + \mathrm{e}^{+}$ does indeed result in a decrease in the mass of the nucleus, so the process does occur spontaneously.

As with alpha decay, the energy of the isolated system of the nucleus and the emitted particle must be conserved in beta decay. Experimentally, one finds that the beta particles are emitted over a continuous range of energies (Active Fig. 30.16), unlike alpha particles, which are emitted with discrete energies (Active Fig. 30.14). The kinetic energy increase of the system must be balanced by the decrease in rest energy of the system; either of these changes is the Q value. Because all decaying nuclei have the same initial mass, however, **the Q value must be the same for each decay.** Then why do the emitted electrons have a range of kinetic energies? The energy version of the isolated system model seems to make an incorrect prediction! Further experimentation shows that, according to the decay processes given by Equations 30.14 and 30.15, the angular momentum (spin) and linear momentum versions of the isolated system model fail, too, and neither angular momentum nor linear momentum of the system is conserved!

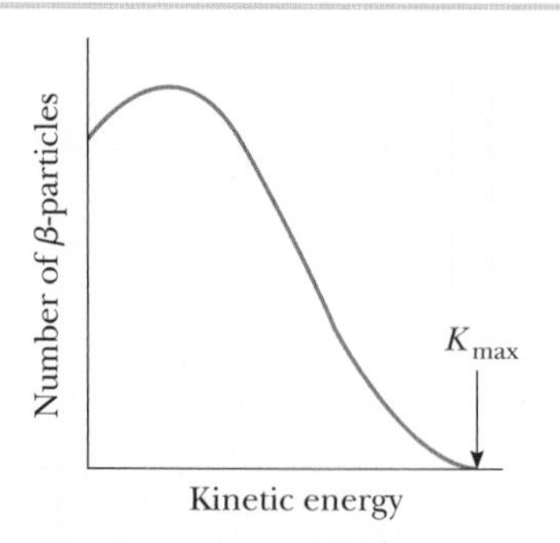

ACTIVE FIGURE 30.16
Distribution of beta particle energies in a typical beta decay. All energies are observed up to a maximum value. Compare this continuous distribution of energies to the discrete distribution of alpha particle energies in Active Figure 30.14.

PhysicsNow™ By logging into PhysicsNow at **www.pop4e.com** and going to Active Figure 30.16, you can observe the development of this graph for the decay in Active Figure 30.17a.

Clearly, the structural model for beta decay must differ from that for alpha decay. After a great deal of experimental and theoretical study, Pauli proposed in 1930 that a third particle must be involved in the decay to account for the "missing" energy and momentum. Enrico Fermi later named this particle the **neutrino** (little neutral one) because it has to be electrically neutral and have little or no rest energy. Although it eluded detection for many years, the neutrino (symbolized by ν) was finally detected experimentally in 1956 by Frederick Reines and Clyde Cowan. It has the following properties:

- It has zero electric charge.
- Its mass is much smaller than that of the electron. Recent experiments show that the mass of the neutrino is not 0 but is less than 2.8 eV/c^2.
- It has a spin of $\frac{1}{2}$, which allows the law of conservation of angular momentum to be satisfied in beta decay.
- It interacts very weakly with matter and is therefore very difficult to detect.

We can now write the beta decay processes (Eqs. 30.14 and 30.15) in their correct form:

$$^{A}_{Z}\text{X} \rightarrow {}^{A}_{Z+1}\text{Y} + \text{e}^- + \overline{\nu} \quad \text{(complete expression)} \qquad [30.16]$$

$$^{A}_{Z}\text{X} \rightarrow {}^{A}_{Z-1}\text{Y} + \text{e}^+ + \nu \quad \text{(complete expression)} \qquad [30.17]$$

where $\overline{\nu}$ represents the **antineutrino,** the antiparticle to the neutrino. We shall discuss antiparticles further in Chapter 31. For now, it suffices to say that **a neutrino is emitted in positron decay, and an antineutrino is emitted in electron decay.** The spin of the neutrino allows angular momentum to be conserved in the decay processes. Despite its small mass, the neutrino does carry momentum, which allows linear momentum to be conserved.

The decays of the neutron and proton within the nucleus are more properly written as

$$\text{n} \rightarrow \text{p} + \text{e}^- + \overline{\nu} \quad \text{(complete expression)}$$

$$\text{p} \rightarrow \text{n} + \text{e}^+ + \nu \quad \text{(complete expression)}$$

PITFALL PREVENTION 30.6

Mass Number of the Electron An alternative notation for an electron is the symbol $_{-1}^{0}$e. This notation does not imply that the electron has zero rest energy. The mass of the electron is so much smaller than that of the lightest nucleon, however, that we approximate it as zero in the context of nuclear decays and reactions.

As examples of beta decay, we can write the decay schemes for carbon-14 and nitrogen-12:

$$^{14}_{6}\text{C} \rightarrow {}^{14}_{7}\text{N} + \text{e}^- + \overline{\nu} \quad \text{(complete expression)} \qquad [30.18]$$

$$^{12}_{7}\text{N} \rightarrow {}^{12}_{6}\text{C} + \text{e}^+ + \nu \quad \text{(complete expression)} \qquad [30.19]$$

Active Figure 30.17 shows a pictorial representation of the decays described by Equations 30.18 and 30.19.

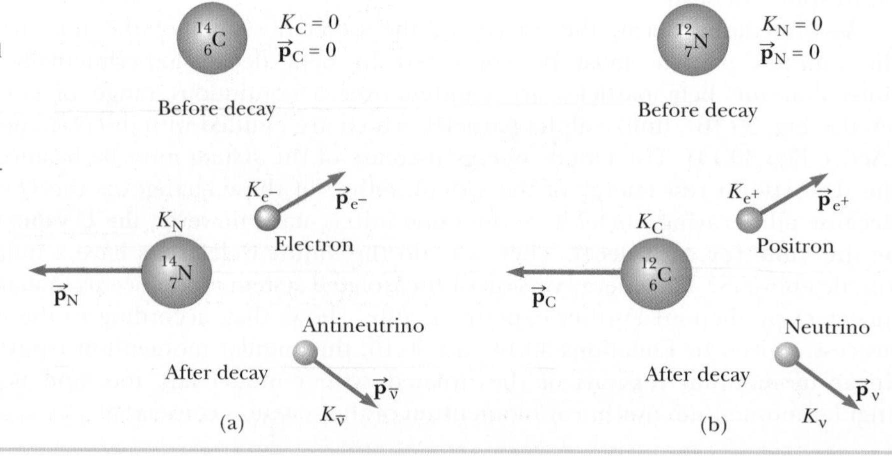

ACTIVE FIGURE 30.17
(a) The beta decay of carbon-14. The final products of the decay are the nitrogen-14 nucleus, an electron and an antineutrino. (b) The beta decay of nitrogen-12. The final products of the decay are the carbon-12 nucleus, a positron and a neutrino.

PhysicsNow™ By logging into PhysicsNow at **www.pop4e.com** and going to Active Figure 30.17, you can observe the decay of carbon-14. For a large number of decays, observe the development of the graph in Active Figure 30.16.

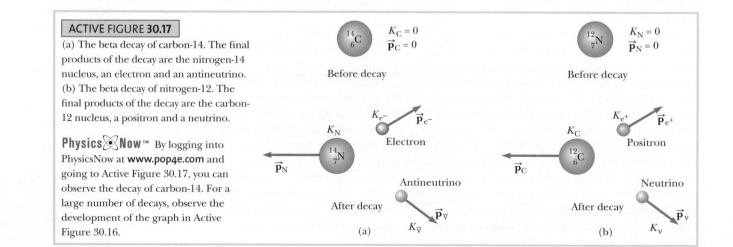

In beta-plus decay, the final system consists of the daughter nucleus, the ejected positron and neutrino, and an electron shed from the atom to neutralize the daughter atom. In some cases, this process represents an overall increase in rest energy, so it does not occur. There is an alternative process that allows some proton-rich nuclei to decay and become more stable. This process, called **electron capture,** occurs when a parent nucleus captures one of its own orbital electrons and emits a neutrino. The final product after decay is a nucleus whose charge is $Z - 1$:

$$^{A}_{Z}\text{X} + \text{e}^{-} \rightarrow \, _{Z-1}^{A}\text{Y} + \nu \qquad [30.20]$$

■ **Electron capture process**

In most cases, an inner K-shell electron is captured, a process referred to as **K capture.** In this process, the only outgoing particles are the neutrino and x-ray photons, originating in higher-shell electrons falling into the vacancy left by the captured K electron.

QUICK QUIZ 30.6 Which of the following is the correct daughter nucleus associated with the beta decay of $^{184}_{72}\text{Hf}$? **(a)** $^{183}_{72}\text{Hf}$ **(b)** $^{183}_{73}\text{Ta}$ **(c)** $^{184}_{73}\text{Ta}$

Carbon Dating

 Carbon dating

The beta decay of ^{14}C given by Equation 30.18 is commonly used to date organic samples. Cosmic rays (high-energy particles from outer space) in the upper atmosphere cause nuclear reactions that create ^{14}C. The ratio of ^{14}C to ^{12}C in the carbon dioxide molecules of our atmosphere has a constant value of about 1.3×10^{-12}. All living organisms have the same ratio of ^{14}C to ^{12}C because they continuously exchange carbon dioxide with their surroundings. When an organism dies, however, it no longer absorbs ^{14}C from the atmosphere, and so the ratio of ^{14}C to ^{12}C decreases as the result of the beta decay of ^{14}C, which has a half-life of 5 730 yr. It is therefore possible to determine the age of a biological sample by measuring its activity per unit mass due to the decay of ^{14}C. Using carbon dating, samples of wood, charcoal, bone, and shell have been identified as having lived 1 000 to 25 000 yr ago.

A particularly interesting example is the dating of the Dead Sea Scrolls, a group of manuscripts discovered by a shepherd in 1947 (Fig. 30.18). Translation showed them to be religious documents, including most of the books of the Old Testament. Because of their historical and religious significance, scholars wanted to know their age. Carbon dating applied to the material in which they were wrapped established their age at approximately 1 950 yr.

(© CORBIS)

ENRICO FERMI (1901–1954)

An Italian physicist who immigrated to the United States to escape the Fascists, Fermi was awarded the Nobel Prize in Physics in 1938 for producing the transuranic elements by neutron irradiation and for his discovery of nuclear reactions brought about by slow neutrons. He made many other outstanding contributions to physics including his theory of beta decay, the free electron theory of metals, and the development of the world's first fission reactor in 1942. Fermi was truly a gifted theoretical and experimental physicist. He was also well known for his ability to present physics in a clear and exciting manner. He wrote, "Whatever Nature has in store for mankind, unpleasant as it may be, men must accept, for ignorance is never better than knowledge."

(Corbis SYGMA)

(a)

(M. Milner/Corbis SYGMA)

(b)

FIGURE 30.18 (a) A fragment of the Dead Sea Scrolls, which were discovered in the caves in the photograph (b). The packing material of the scrolls was analyzed by carbon dating to determine their age.

■ Thinking Physics 30.3

In 1991, a German tourist discovered the well-preserved remains of the Ice Man trapped in a glacier in the Italian Alps, shown in the opening photograph for this chapter. Radioactive dating of a sample of the Ice Man revealed an age of 5 300 yr. Why did scientists date the sample using the isotope ^{14}C rather than ^{11}C, a beta emitter with a half-life of 20.4 min?

Reasoning ^{14}C has a long half-life of 5 730 yr, so the fraction of ^{14}C nuclei remaining after one half-life is high enough to measure accurate changes in the sample's activity. The ^{11}C isotope, which has a very short half-life, is not useful because its activity decreases to a vanishingly small value over 5 300 yr, making it impossible to detect.

An isotope used to date a sample must be present in a known amount in the sample when it is formed. As a general rule, the isotope chosen to date a sample should also have a half-life with the same order of magnitude as the age of the sample. If the half-life is much less than the age of the sample, there won't be enough activity left to measure because almost all the original radioactive nuclei will have decayed. If the half-life is much greater than the age of the sample, the reduction in activity that has taken place since the sample died will be too small to measure. ■

INTERACTIVE EXAMPLE 30.6 Radioactive Dating

A piece of charcoal of mass 25.0 g is found in the ruins of an ancient city. The sample shows a ^{14}C activity of 250 decays/min. How long has the tree from which this charcoal came been dead?

Solution We begin by calculating the decay constant for ^{14}C, which has a half-life of 5 730 yr:

$$\lambda = \frac{0.693}{T_{1/2}} = \frac{0.693}{(5\,730\text{ yr})(3.16 \times 10^{7}\text{ s/yr})}$$
$$= 3.83 \times 10^{-12}\text{ s}^{-1}$$

The number of ^{14}C nuclei can be calculated in two steps. First, the number of ^{12}C nuclei in 25 g of carbon is

$$N(^{12}\text{C}) = \frac{6.02 \times 10^{23}\text{ nuclei/mol}}{12.0\text{ g/mol}}(25.0\text{ g})$$
$$= 1.25 \times 10^{24}\text{ nuclei}$$

Assuming that the initial ratio of ^{14}C to ^{12}C was 1.3×10^{-12}, we see that the number of ^{14}C nuclei in 25.0 g *before* decay is

$$N_0(^{14}\text{C}) = (1.3 \times 10^{-12})(1.25 \times 10^{24})$$
$$= 1.6 \times 10^{12}\text{ nuclei}$$

Hence, the initial activity of the sample is

$$R_0 = \lambda N_0 = (3.83 \times 10^{-12}\text{ s}^{-1})(1.6 \times 10^{12}\text{ nuclei})$$
$$= 6.13\text{ decays/s} = 368\text{ decays/min}$$

We can now calculate the age of the charcoal using Equation 30.7, which relates the activity R at any time t to the initial activity R_0:

$$R = R_0 e^{-\lambda t} \quad \text{or} \quad e^{-\lambda t} = \frac{R}{R_0}$$

Because it is given that $R = 250$ decays/min and because we found that $R_0 = 368$ decays/min, we can calculate t by taking the natural logarithm of both sides of the last equation:

$$-\lambda t = \ln\left(\frac{R}{R_0}\right) = \ln\left(\frac{250}{368}\right) = -0.39$$

$$t = \frac{0.39}{\lambda} = \frac{0.39}{3.83 \times 10^{-12}\text{ s}^{-1}}$$
$$= 1.0 \times 10^{11}\text{ s} = 3.2 \times 10^{3}\text{ yr}$$

PhysicsNow™ Practice using carbon dating on samples by logging into PhysicsNow at **www.pop4e.com** and going to Interactive Example 30.6.

Gamma Decay

Very often, a nucleus that undergoes radioactive decay is left in an excited quantum state. The nucleus can then undergo a second decay, **a gamma decay,** to a lower state, perhaps to the ground state, by emitting a photon:

■ Gamma decay

$$^{A}_{Z}\text{X}^{*} \rightarrow {}^{A}_{Z}\text{X} + \gamma \qquad [30.21]$$

TABLE 30.2 Various Decay Pathways

Alpha decay	^A_ZX	$\rightarrow$	$^{A-4}_{Z-2}Y + ^4_2He$
Beta decay (e^-)	^A_ZX	$\rightarrow$	$_{Z+1}^{\;\;A}Y + e^- + \bar{\nu}$
Beta decay (e^+)	^A_ZX	$\rightarrow$	$_{Z-1}^{\;\;A}Y + e^+ + \nu$
Electron capture	$^A_ZX + e^-$	$\rightarrow$	$_{Z-1}^{\;\;A}Y + \nu$
Gamma decay	$^A_ZX^*$	$\rightarrow$	$^A_ZX + \gamma$

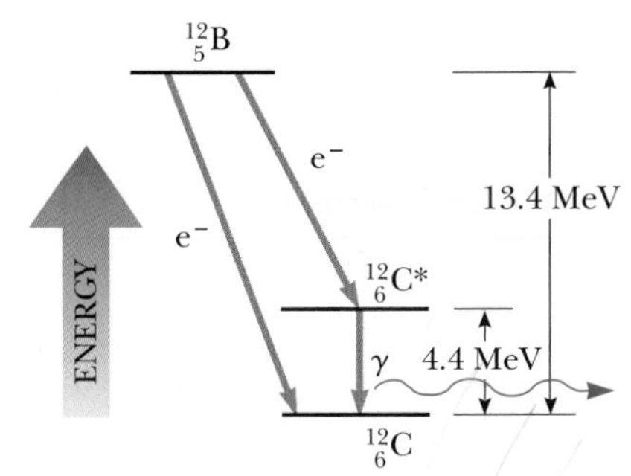

FIGURE 30.19 An energy level diagram showing the initial nuclear state of a ^{12}B nucleus and two possible lower-energy states of the ^{12}C nucleus. The beta decay of the ^{12}B nucleus can result in either of two situations, with the ^{12}C nucleus in the ground state or in the excited state, in which case the nucleus is denoted as $^{12}C^*$. In the latter case, the beta decay to $^{12}C^*$ is followed by a gamma decay to ^{12}C as the excited nucleus makes a transition to the ground state.

where X* indicates a nucleus in an excited state. The typical half-life of an excited nuclear state is 10^{-10} s. Photons emitted in such a de-excitation process are called **gamma rays.** Such photons have very high energy (on the order of 1 MeV or higher) relative to the energy of visible light (on the order of a few electron volts). Recall from Chapter 29 that the energy of photons emitted (or absorbed) by an atom equals the difference in energy between the two atomic quantum states involved in the transition. Similarly, a gamma ray photon has an energy hf that equals the energy difference ΔE between two nuclear quantum states. When a nucleus decays by emitting a gamma ray, it ends up in a lower state, but its atomic mass A and atomic number Z do not change.

A nucleus may reach an excited state as the result of a violent collision with another particle. It is more common, however, for a nucleus to be in an excited state after it has undergone an alpha or beta decay. The following sequence of events represents a typical situation in which gamma decay occurs:

$$^{12}_5B \rightarrow ^{12}_6C^* + e^- + \bar{\nu} \qquad [30.22]$$

$$^{12}_6C^* \rightarrow ^{12}_6C + \gamma \qquad [30.23]$$

Figure 30.19 shows the decay scheme for ^{12}B, which undergoes beta decay with a half-life of 20.4 ms to either of two levels of ^{12}C. It can either (1) decay directly to the ground state of ^{12}C by emitting a 13.4-MeV electron or (2) undergo beta-minus decay to an excited state of $^{12}C^*$, followed by gamma decay to the ground state. The latter process results in the emission of a 9.0-MeV electron and a 4.4-MeV photon. Table 30.2 summarizes the pathways by which radioactive nuclei undergo decay.

30.5 NUCLEAR REACTIONS

In Section 30.4, we discussed the processes by which nuclei can *spontaneously* change to another nucleus by undergoing a radioactive decay process. It is also possible to change the structures and properties of nuclei by bombarding them with energetic particles. Such changes are called **nuclear reactions.** In 1919, Rutherford was the first to observe nuclear reactions, using naturally occurring radioactive sources for the bombarding particles. Since then, thousands of nuclear reactions have been observed following the development of charged-particle accelerators in the 1930s. With today's advanced technology in particle accelerators and particle detectors, it is possible to achieve particle energies of more than 1 000 GeV = 1 TeV. These high-energy particles are used to create new particles whose properties are helping solve the mysteries of the nucleus.

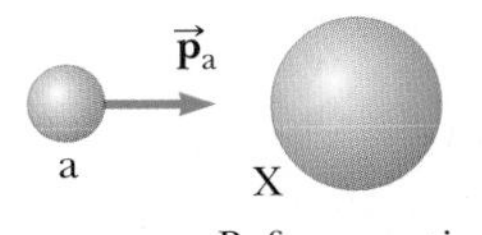

Before reaction

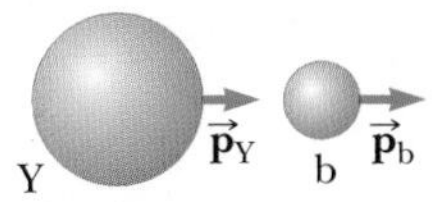

After reaction

FIGURE 30.20 A nuclear reaction. Before the reaction, an incoming particle a moves toward a target nucleus X. After the reaction, the target nucleus has changed to nucleus Y and an outgoing particle b moves away from the reaction site.

Consider a reaction (Fig. 30.20) in which a target nucleus X is bombarded by an incoming particle a, resulting in a different nucleus Y and an outgoing particle b:

■ Nuclear reaction

$$a + X \rightarrow Y + b \qquad [30.24]$$

Sometimes this reaction is written in the equivalent symbolic representation

$$X(a, b)Y$$

In the preceding section, the Q value, or disintegration energy, associated with radioactive decay was defined as the change in the rest energy, which is the amount of the rest energy transformed to kinetic energy during the decay process. In a similar way, we define the **reaction energy** Q associated with a nuclear reaction as the **total change in rest energy that results from the reaction:**

■ Reaction energy Q

$$Q = (M_a + M_X - M_Y - M_b)c^2 \qquad [30.25]$$

A reaction for which Q is positive is called **exothermic.** After the reaction, the transformed rest energy appears as an increase in kinetic energy of Y and b over that of a and X.

A reaction for which Q is negative is called **endothermic** and represents an increase in rest energy. An endothermic reaction will not occur unless the bombarding particle has a kinetic energy greater than $|Q|$. The minimum kinetic energy of the incoming particle necessary for such a reaction to occur is called the **threshold energy.** The threshold energy is larger than $|Q|$ because we must also conserve linear momentum in the isolated system of the initial and final particles. If an incoming particle has just energy $|Q|$, enough energy is present to increase the rest energy of the system, but none is left over for kinetic energy of the final particles, that is, nothing is moving after the reaction. Therefore, the incoming particle has momentum before the reaction but there is no momentum of the system afterward, which is a violation of the law of conservation of momentum.

If particles a and b in a nuclear reaction are identical so that X and Y are also necessarily identical, the reaction is called a **scattering event.** If the kinetic energy of the system (a and X) before the event is the same as that of the system (b and Y) after the event, it is classified as *elastic scattering.* If the kinetic energies of the system before and after the event are not the same, the reaction is described as *inelastic scattering.* In this case, the difference in energy is accounted for by the target nucleus being raised to an excited state by the event. The final system now consists of b and an excited nucleus Y*, and eventually it will become b, Y, and γ, where γ is the gamma-ray photon that is emitted when the system returns to the ground state. This elastic and inelastic terminology is identical to that used in describing collisions between macroscopic objects (Section 8.3).

In addition to energy and momentum, the total charge and total number of nucleons must be conserved in the system of particles for a nuclear reaction. For example, consider the reaction $^{19}\text{F}(\text{p}, \alpha)^{16}\text{O}$, which has a Q value of 8.124 MeV. We can show this reaction more completely as

$$^{1}_{1}\text{H} + ^{19}_{9}\text{F} \rightarrow ^{16}_{8}\text{O} + ^{4}_{2}\text{He}$$

We see that the total number of nucleons before the reaction $(1 + 19 = 20)$ is equal to the total number after the reaction $(16 + 4 = 20)$. Furthermore, the total charge $(Z = 10)$ is the same before and after the reaction.

30.6 THE ENGINE OF THE STARS

CONTEXT CONNECTION

One of the important features of nuclear reactions is that much more energy is released (i.e., converted from rest energy) than in normal chemical reactions such as in the burning of fossil fuels. Let us look back at our binding energy curve (see Fig. 30.9) and consider two important nuclear reactions that relate to that curve. If a heavy nucleus at the right of the graph splits into two lighter nuclei, the total binding energy within the system increases, representing energy released from the nuclei. This type of reaction was observed and reported in 1939 by Otto Hahn and Fritz Strassman. This reaction, known as **fission,** was of great scientific and political interest at the time of World War II because of the development of the first nuclear weapon.

In the fission reaction, a fissionable nucleus (the target nucleus X), which is often ^{235}U, absorbs a slowly moving neutron (the incoming particle a) and the nucleus splits into two smaller nuclei (two nuclei Y_1 and Y_2), releasing energy and

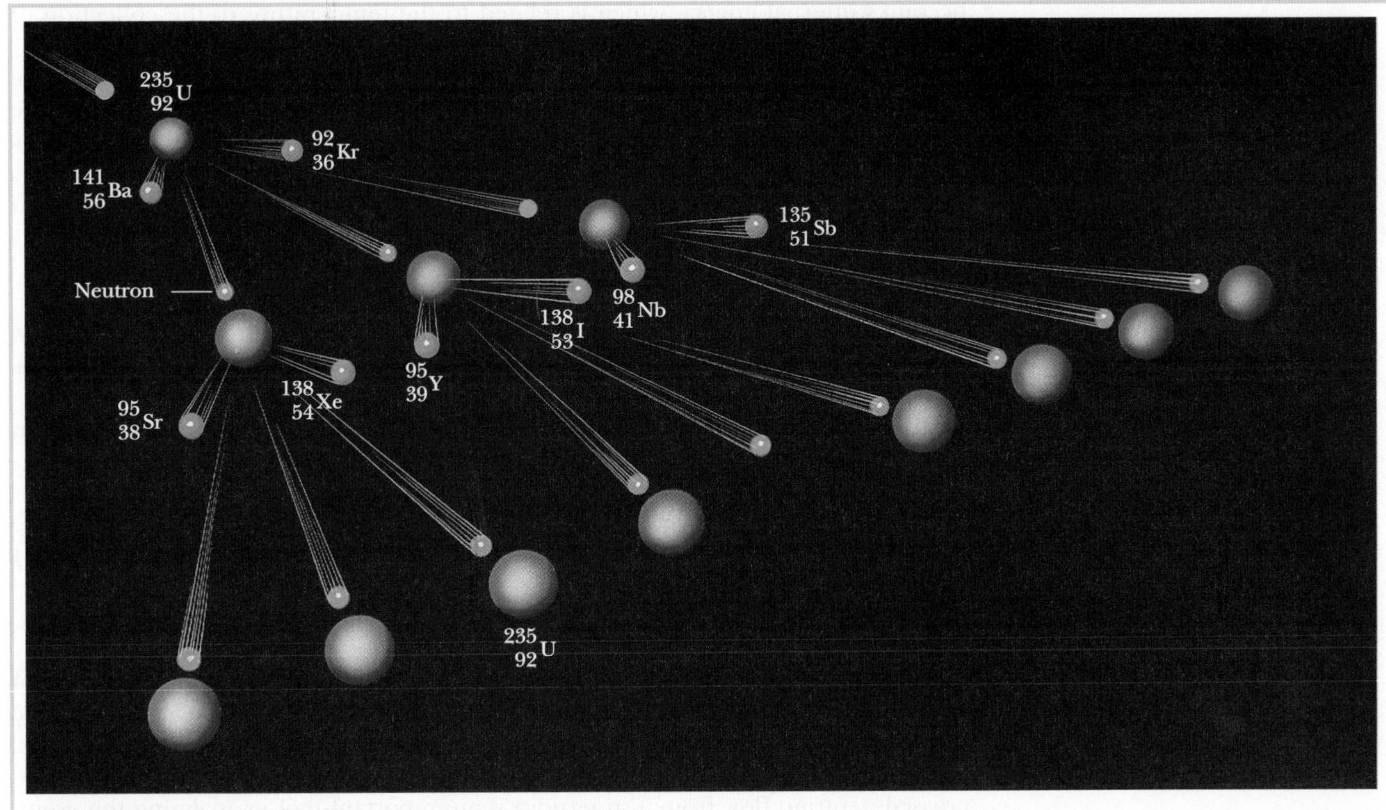

ACTIVE FIGURE 30.21 A nuclear chain reaction initiated by the capture of a neutron by a ^{235}U nucleus.

Physics Now™ By logging into PhysicsNow at **www.pop4e.com** and going to Active Figure 30.21, you can observe the chain reaction.

more neutrons (several particles b). These neutrons can then go on to be absorbed within other nuclei, causing other fissions. With no means of control, the result is a chain reaction explosion as suggested by Active Figure 30.21. With proper control, the fission process is used in nuclear power generating stations.

Examining the other end of the binding energy curve, we see that we could also increase the binding energy of the system and release energy by combining two light nuclei. This process of **fusion** is made difficult because the nuclei must overcome a very strong Coulomb repulsion before they become close enough together to fuse. One way to assist the nuclei in overcoming this repulsion is to cause them to move with very high kinetic energy by raising the system of nuclei to a very high temperature. If the density of nuclei is high also, the probability of nuclei colliding is high and fusion can occur. The technological problem of creating very high temperatures and densities is a major challenge in the area of Earth-based controlled fusion research.

At some natural locations (e.g., the cores of stars), the necessary high temperatures and densities exist. Consider a collection of gas and dust somewhere in the Universe to be an isolated system. What happens as this system collapses under its own gravitational attraction? Energy of the system is conserved, and the gravitational potential energy associated with the separated particles decreases while the kinetic energy of the particles increases, just like a falling ball with cosmic particles "falling" into a gravitational center. As the falling particles collide with the particles that have already fallen into the central region of collapse, their kinetic energy is distributed to the other particles by collisions and randomized; it becomes internal energy, which is related to the temperature of the collection of particles.

If the temperature and density of the system's core rise to the point where fusion can occur, the system becomes a star. The primary constituent of the Universe is hydrogen, so the fusion reaction at the center of a star combines hydrogen nuclei—protons—into helium nuclei. A common reaction process for stars with relatively cool cores ($T < 15 \times 10^6$ K) is the **proton–proton cycle.** In the first step of the process, two protons combine to form deuterium:

$$^1_1\text{H} + ^1_1\text{H} \rightarrow ^2_1\text{H} + \text{e}^+ + \nu$$

Notice the implicit ^2_2H nucleus that is formed but that does not appear in the reaction equation. This nucleus is highly unstable and decays very rapidly by beta-plus decay to the deuterium nucleus, a positron and a neutrino.

In the next step, the deuterium nucleus undergoes fusion with another proton to form a helium-3 nucleus:

$$^1_1\text{H} + ^2_1\text{H} \rightarrow ^3_2\text{He} + \gamma$$

Finally, two helium-3 nuclei formed in such reactions can fuse to form helium-4 and two protons:

$$^3_2\text{He} + ^3_2\text{He} \rightarrow ^4_2\text{He} + ^1_1\text{H} + ^1_1\text{H}$$

The net result of this cycle has been the joining of four protons to form a helium-4 nucleus, with the release of energy that eventually leaves the star as electromagnetic radiation from its surface. In addition, notice that the reaction releases neutrinos, which serve as a signal for beta decay occurring within the star. The observation of increased neutrino flow from a supernova is an important tool in analyzing the event.

For stars with hotter cores ($T > 15 \times 10^6$ K), another process, called the **carbon cycle,** dominates. At such high temperatures, hydrogen nuclei can fuse into nuclei heavier than helium such as carbon. In the first of six steps in the cycle, a carbon nucleus fuses with a proton to form nitrogen:

$$^1_1\text{H} + ^{12}_6\text{C} \rightarrow ^{13}_7\text{N}$$

The nitrogen nucleus is proton-rich and undergoes beta-plus decay:

$$^{13}_7\text{N} \rightarrow ^{13}_6\text{C} + \text{e}^+ + \nu$$

The resulting carbon-13 nucleus fuses with another proton, with the emission of a gamma ray:

$$^1_1\text{H} + ^{13}_6\text{C} \rightarrow ^{14}_7\text{N} + \gamma$$

The nitrogen-14 fuses with another proton, with more gamma emission:

$$^1_1\text{H} + ^{14}_7\text{N} \rightarrow ^{15}_8\text{O} + \gamma$$

The oxygen nucleus undergoes beta-plus decay:

$$^{15}_8\text{O} \rightarrow ^{15}_7\text{N} + \text{e}^+ + \nu$$

Finally, the nitrogen-15 fuses with another proton:

$$^1_1\text{H} + ^{15}_7\text{N} \rightarrow ^{12}_6\text{C} + ^4_2\text{He}$$

Notice that the net effect of this process is to combine four protons into a helium nucleus, just like the proton–proton cycle. The carbon-12 with which we began the process is returned at the end, so it acts only as a catalyst to the process and is not consumed.

Depending on its mass, a star transforms energy in its core at a rate between 10^{23} and 10^{33} W. The energy transformed from the rest energy of the nuclei in the core is transferred outward through the surrounding layers by matter transfer in two forms. First, neutrinos carry energy directly through these layers to space because these particles interact only weakly with matter. Second, energy carried by photons from the core is absorbed by the gases in layers outside the core and slowly works its way to the surface by convection. This energy is eventually radiated from the surface of the star by electromagnetic radiation, mostly in the infrared, visible, and ultraviolet regions of the electromagnetic spectrum. The weight of the layers outside the core keeps the core from exploding. The whole system of a star is stable as long as the supply of hydrogen in the core lasts.

In the previous chapters, we presented examples of the applications of quantum physics and atomic physics to processes in space. In this chapter, we have seen that nuclear processes also have an important role in the cosmos. The formation of stars is a critical process in the development of the Universe. The energy provided by stars is crucial to life on planets such as the Earth. In our next, and final, chapter, we shall discuss the processes that occur on an even smaller scale, the scale of *elementary particles.* We shall find again that looking at a smaller scale allows us to advance our understanding of the largest scale system, the Universe.

SUMMARY

PhysicsNow™ Take a practice test by logging into PhysicsNow at www.pop4e.com and clicking on the Pre-Test link for this chapter.

A nuclear species can be represented by ${}^A_Z\text{X}$, where A is the **mass number,** the total number of nucleons, and Z is the **atomic number,** the total number of protons. The total number of neutrons in a nucleus is the **neutron number** N, where $A = N + Z$. Elements with the same Z but different A and N values are called **isotopes.**

Assuming that a nucleus is spherical, its radius is

$$r = r_0 A^{1/3} \qquad [30.1]$$

where $r_0 = 1.2$ fm.

Nuclei are stable because of the **nuclear force** between nucleons. This short-range force dominates the Coulomb repulsive force at distances of less than about 2 fm and is independent of charge.

Light nuclei are most stable when the number of protons equals the number of neutrons. Heavy nuclei are most stable when the number of neutrons exceeds the number of protons. In addition, many stable nuclei have Z and N values that are both even. Nuclei with unusually high stability have Z or N values of 2, 8, 20, 28, 50, 82, and 126, called **magic numbers.**

Nuclei have an intrinsic angular momentum (spin) of magnitude $\sqrt{I(I+1)}\,\hbar$, where I is the **nuclear spin quantum number.** The magnetic moment of a nucleus is measured in terms of the **nuclear magneton** μ_n, where

$$\mu_n \equiv \frac{e\hbar}{2m_p} = 5.05 \times 10^{-27}\ \text{J/T} \qquad [30.3]$$

The difference in mass between the separate nucleons and the nucleus containing these nucleons, when multiplied by c^2, gives the **binding energy** E_b of the nucleus. We can calculate the binding energy of any nucleus ${}^A_Z\text{X}$ using the expression

$$E_b(\text{MeV}) = [ZM(\text{H}) + Nm_n - M({}^A_Z\text{X})] \times 931.494\ \text{MeV/u} \qquad [30.4]$$

Radioactive processes include alpha decay, beta decay, and gamma decay. An alpha particle is a ${}^4\text{He}$ nucleus, a beta particle is either an electron (e^-) or a positron (e^+), and a gamma particle is a high-energy photon.

If a radioactive material contains N_0 radioactive nuclei at $t = 0$, the number N of nuclei remaining at time t is

$$N = N_0 e^{-\lambda t} \qquad [30.6]$$

where λ is the **decay constant,** or **disintegration constant.** The **decay rate,** or **activity,** of a radioactive substance is given by

$$R = \left|\frac{dN}{dt}\right| = N_0 \lambda e^{-\lambda t} = R_0 e^{-\lambda t} \qquad [30.7]$$

where $R_0 = N_0\lambda$ is the activity at $t = 0$. The **half-life** $T_{1/2}$ is defined as the time interval required for half of a given number of radioactive nuclei to decay, where

$$T_{1/2} = \frac{\ln 2}{\lambda} = \frac{0.693}{\lambda} \qquad [30.8]$$

Alpha decay can occur because according to quantum mechanics some nuclei have barriers that can be penetrated by the alpha particles (the tunneling process). This process is energetically more favorable for those nuclei having large excesses of neutrons. A nucleus can undergo **beta decay** in two ways. It can emit either an electron (e^-) and an antineutrino ($\bar{\nu}$) or a positron (e^+) and a neutrino (ν). In the **electron capture** process, the nucleus of an atom absorbs one of its own electrons (usually from the K shell) and emits a neutrino. In **gamma decay,** a nucleus in an excited state decays to its ground state and emits a gamma ray.

Nuclear reactions can occur when a target nucleus X is bombarded by a particle a, resulting in a nucleus Y and an outgoing particle b:

$$\text{a} + \text{X} \rightarrow \text{Y} + \text{b} \quad \text{or} \quad \text{X(a, b)Y} \qquad [30.24]$$

The rest energy transformed to kinetic energy in such a reaction, called the **reaction energy** Q, is

$$Q = (M_\text{a} + M_\text{X} - M_\text{Y} - M_\text{b})c^2 \qquad [30.25]$$

A reaction for which Q is positive is called **exothermic.** A reaction for which Q is negative is called **endothermic.** The minimum kinetic energy of the incoming particle necessary for such a reaction to occur is called the **threshold energy.**

QUESTIONS

☐ = answer available in the *Student Solutions Manual and Study Guide*

1. In Rutherford's experiment, assume that an alpha particle is headed directly toward the nucleus of an atom. Why doesn't the alpha particle make physical contact with the nucleus?
2. Why are very heavy nuclei unstable?
3. Why do nearly all the naturally occurring isotopes lie above the $N = Z$ line in Figure 30.4?
4. Explain why nuclei that are well off the line of stability in Figure 30.4 tend to be unstable.
5. From Table A.3 in Appendix A, identify the four stable nuclei that have magic numbers in both Z and N.
6. If a nucleus has a half-life of 1 year, does that mean that its whole life is 2 years? Will it be completely decayed after 2 years? Explain.
7. Two samples of the same radioactive nuclide are prepared. Sample A has twice the initial activity of sample B. How does the half-life of A compare with the half-life of B? After each has passed through five half-lives, what is the ratio of their activities?
8. "If no more people were to be born, the law of population growth would strongly resemble the radioactive decay law." Discuss this statement.
9. If a nucleus such as ^{226}Ra initially at rest undergoes alpha decay, which has more kinetic energy after the decay, the alpha particle or the daughter nucleus?
10. Can a nucleus emit alpha particles that have different energies? Explain.
11. Suppose it could be shown that the cosmic ray intensity at the Earth's surface was much greater 10 000 years ago. How would this difference affect what we accept as valid carbon-dated values of the age of ancient samples of once-living matter?
12. Explain why many heavy nuclei undergo alpha decay but do not spontaneously emit neutrons or protons.
13. Do all natural events have causes? Is the Universe intelligible? Give reasons for your answers. (*Note:* You may wish to consider again Question 5.17 in Chapter 5 on whether the future is determinate.)
14. Discuss the similarities and differences between fusion and fission.
15. *And swift, and swift past comprehension*
Turn round Earth's beauty and her might.
The heavens blaze in alternation
With deep and chill and rainy night.
In mighty currents foams the ocean
Up from the rocks' abyssal base,
With rock and sea torn into motion
In ever-swift celestial race.
Corrosive, choking smoke is spraying.
Above infernos, lava flies.
A perilous bridge, the land is swaying
Between them and the gaping skies.
And tempests bluster in a contest
From sea to land, from land to sea.
In rage they forge a chain around us
Of primal meaning, energy.
There flames a lightning disaster
Before the thunder, in its way.
But all Your servants honor, Master,
The gentle order of Your day.

Johann Wolfgang von Goethe wrote the song of the archangels in *Faust* half a century before the law of conservation of energy was recognized. Students often find it useful to think of a list of several "forms of energy," from kinetic to nuclear. Argue for or against the view that these lines of poetry make an obvious or oblique reference to every form of energy and energy transfer.

PROBLEMS

1, 2, 3 = straightforward, intermediate, challenging
☐ = full solution available in the *Student Solutions Manual and Study Guide*
Physics Now™ = coached problem with hints available at www.pop4e.com
= computer useful in solving problem
= paired numerical and symbolic problems
= biomedical application

Note: Atomic masses are listed in Table A.3 in Appendix A.

Section 30.1 ▪ Some Properties of Nuclei

1. What is the order of magnitude of the number of protons in your body? Of the number of neutrons? Of the number of electrons?
2. **Review problem.** Singly ionized carbon is accelerated through 1 000 V and passed into a mass spectrometer to determine the isotopes present (see Chapter 22). The magnitude of the magnetic field in the spectrometer is 0.200 T. (a) Determine the orbit radii for the ^{12}C and the ^{13}C isotopes as they pass through the field. (b) Show that the ratio of radii may be written in the form

$$\frac{r_1}{r_2} = \sqrt{\frac{m_1}{m_2}}$$

and verify that your radii in part (a) agree with this equation.
3. In a Rutherford scattering experiment, alpha particles having kinetic energy of 7.70 MeV are fired toward a gold nucleus. (a) Use energy conservation to determine the distance of closest approach between the alpha particle and gold nucleus. Assume that the nucleus remains at rest.

(b) Calculate the de Broglie wavelength for the 7.70-MeV alpha particle and compare it with the distance obtained in part (a). (c) Based on this comparison, why is it proper to treat the alpha particle as a particle and not as a wave in the Rutherford scattering experiment?

4. Find the radius of (a) a nucleus of $^{4}_{2}$He and (b) a nucleus of $^{238}_{92}$U.

5. A star ending its life with a mass of two times the mass of the Sun is expected to collapse, combining its protons and electrons to form a neutron star. Such a star could be thought of as a gigantic atomic nucleus. If a star of mass $2 \times 1.99 \times 10^{30}$ kg collapsed into neutrons ($m_n = 1.67 \times 10^{-27}$ kg), what would its radius be? (Assume that $r = r_0 A^{1/3}$.)

6. **Review problem.** What would be the gravitational force exerted by each of two golf balls on the other, if they were made of nuclear matter? Assume that each has a 4.30-cm diameter and that the balls are 1.00 m apart.

7. The radio frequency at which a nucleus displays resonance absorption between spin states is called the Larmor precessional frequency and is given by

$$f = \frac{\Delta E}{h} = \frac{2\mu B}{h}$$

Calculate the Larmor frequency for (a) free neutrons in a magnetic field of 1.00 T, (b) free protons in a magnetic field of 1.00 T, and (c) free protons in the Earth's magnetic field at a location where the magnitude of the field is 50.0 μT.

Section 30.2 ■ Binding Energy

8. Calculate the binding energy per nucleon for (a) ^{2}H, (b) ^{4}He, (c) ^{56}Fe, and (d) ^{238}U.

9. **Physics Now™** A pair of nuclei for which $Z_1 = N_2$ and $Z_2 = N_1$ are called *mirror isobars* (the atomic and neutron numbers are interchanged). Binding energy measurements on these nuclei can be used to obtain evidence of the charge independence of nuclear forces (i.e., proton–proton, proton–neutron, and neutron–neutron nuclear forces are equal). Calculate the difference in binding energy for the two mirror isobars $^{15}_{8}$O and $^{15}_{7}$N. The electric repulsion among eight protons rather than seven accounts for the difference.

10. Nuclei having the same mass numbers are called *isobars.* The isotope $^{139}_{57}$La is stable. The radioactive isobar $^{139}_{59}$Pr is located below the line of stable nuclei in Figure 30.4 and decays by e^+ emission. Another radioactive isobar of $^{139}_{57}$La, $^{139}_{55}$Cs, decays by e^- emission and is located above the line of stable nuclei in Figure 30.4. (a) Which of these three isobars has the highest neutron-to-proton ratio? (b) Which has the greatest binding energy per nucleon? (c) Which do you expect to be heavier, $^{139}_{59}$Pr or $^{139}_{55}$Cs?

11. Using the graph in Figure 30.9, estimate how much energy is released when a nucleus of mass number 200 fissions into two nuclei each of mass number 100.

Section 30.3 ■ Radioactivity

12. The half-life of ^{131}I is 8.04 days. On a certain day, the activity of an iodine-131 sample is 6.40 mCi. What is its activity 40.2 days later?

13. **Physics Now™** A freshly prepared sample of a certain radioactive isotope has an activity of 10.0 mCi. After 4.00 h, its activity is 8.00 mCi. (a) Find the decay constant and half-life. (b) How many atoms of the isotope were contained in the freshly prepared sample? (c) What is the sample's activity 30.0 h after it is prepared?

14. A sample of radioactive material contains 1.00×10^{15} atoms and has an activity of 6.00×10^{11} Bq. What is its half-life?

15. What time interval elapses while 90.0% of the radioactivity of a sample of $^{72}_{33}$As disappears as measured by its activity? The half-life of $^{72}_{33}$As is 26 h.

16. A radioactive nucleus has half-life $T_{1/2}$. A sample containing these nuclei has initial activity R_0. Calculate the number of nuclei that decay during the interval between the times t_1 and t_2.

17. In an experiment on the transport of nutrients in the root structure of a plant, two radioactive nuclides X and Y are used. Initially 2.50 times more nuclei of type X are present than of type Y. Just three days later there are 4.20 times more nuclei of type X than of type Y. Isotope Y has a half-life of 1.60 d. What is the half-life of isotope X?

18. (a) The daughter nucleus formed in radioactive decay is often radioactive. Let N_{10} represent the number of parent nuclei at time $t = 0$, $N_1(t)$ the number of parent nuclei at time t, and λ_1 the decay constant of the parent. Suppose the number of daughter nuclei at time $t = 0$ is zero, let $N_2(t)$ be the number of daughter nuclei at time t, and let λ_2 be the decay constant of the daughter. Show that $N_2(t)$ satisfies the differential equation

$$\frac{dN_2}{dt} = \lambda_1 N_1 - \lambda_2 N_2$$

(b) Verify by substitution that this differential equation has the solution

$$N_2(t) = \frac{N_{10}\lambda_1}{\lambda_1 - \lambda_2}(e^{-\lambda_2 t} - e^{-\lambda_1 t})$$

This equation is the law of successive radioactive decays. (c) ^{218}Po decays into ^{214}Pb with a half-life of 3.10 min, and ^{214}Pb decays into ^{214}Bi with a half-life of 26.8 min. On the same axes, plot graphs of $N_1(t)$ for ^{218}Po and $N_2(t)$ for ^{214}Pb. Let N_{10} = 1 000 nuclei, and choose values of t from 0 to 36 min in 2-min intervals. The curve for ^{214}Pb at first rises to a maximum and then starts to decay. At what instant t_m is the number of ^{214}Pb nuclei a maximum? (d) By applying the condition for a maximum $dN_2/dt = 0$, derive a symbolic equation for t_m in terms of λ_1 and λ_2. Does the value obtained in part (c) agree with this equation?

Section 30.4 ■ The Radioactive Decay Processes

19. Find the energy released in the alpha decay

$$^{238}_{92}\text{U} \rightarrow {}^{234}_{90}\text{Th} + {}^{4}_{2}\text{He}$$

You will find Table A.3 useful.

20. Identify the missing nuclide or particle (X):

(a) X $\rightarrow$ $^{65}_{28}$Ni + γ
(b) $^{215}_{84}$Po $\rightarrow$ X + α

(c) $X \rightarrow {}^{55}_{26}\text{Fe} + e^+ + \nu$
(d) ${}^{109}_{48}\text{Cd} + X \rightarrow {}^{109}_{47}\text{Ag} + \nu$
(e) ${}^{14}_{7}\text{N} + {}^{4}_{2}\text{He} \rightarrow X + {}^{17}_{8}\text{O}$

21. A living specimen in equilibrium with the atmosphere contains one atom of ^{14}C (half-life = 5 730 yr) for every 7.7×10^{11} stable carbon atoms. An archeological sample of wood (cellulose, $C_{12}H_{22}O_{11}$) contains 21.0 mg of carbon. When the sample is placed inside a shielded beta counter with 88.0% counting efficiency, 837 counts are accumulated in one week. Assuming that the cosmic-ray flux and the Earth's atmosphere have not changed appreciably since the sample was formed, find the age of the sample.

22. A ^{3}H nucleus beta decays into ^{3}He by creating an electron and an antineutrino according to the reaction

$${}^{3}_{1}\text{H} \rightarrow {}^{3}_{2}\text{He} + e^- + \bar{\nu}$$

The symbols in this reaction refer to nuclei. Write the reaction referring to neutral atoms by adding one electron to both sides. Then use Table A.3 to determine the total energy released in this reaction.

23. The nucleus ${}^{15}_{8}\text{O}$ decays by electron capture. The nuclear reaction is written

$${}^{15}_{8}\text{O} + e^- \rightarrow {}^{15}_{7}\text{N} + \nu$$

(a) Write the process going on for a single particle within the nucleus. (b) Write the decay process referring to neutral atoms. (c) Determine the energy of the neutrino. Disregard the daughter's recoil.

24. Enter the correct isotope symbol in each open square in Figure P30.24, which shows the sequences of decays in the natural radioactive series starting with the long-lived isotope uranium-235 and ending with the stable nucleus lead-207.

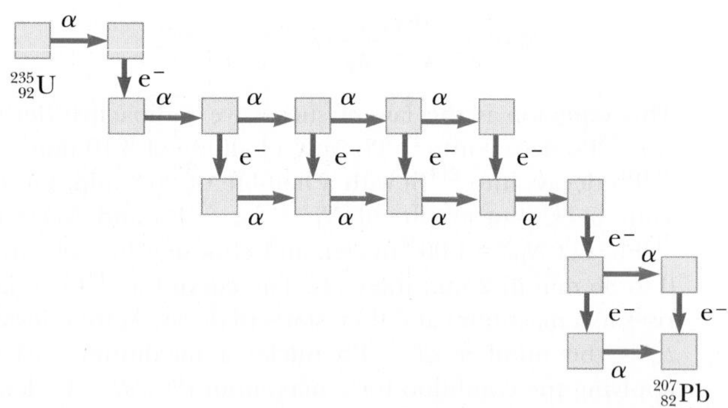

FIGURE **P30.24**

25. *Indoor air pollution.* Uranium is naturally present in rock and soil. At one step in its series of radioactive decays, ^{238}U produces the chemically inert gas radon-222, with a half-life of 3.82 days. The radon seeps out of the ground to mix into the atmosphere, typically making open air radioactive with activity 0.3 pCi/L. In homes, ^{222}Rn can be a serious pollutant, accumulating to reach much higher activities in enclosed spaces. If the radon activity exceeds 4 pCi/L, the Environmental Protection Agency suggests taking action to reduce it, such as by reducing infiltration of air from the ground. (a) Convert the activity 4 pCi/L to units of becquerel per cubic meter. (b) How many ^{222}Rn atoms are in one cubic meter of air displaying this activity? (c) What fraction of the mass of the air does the radon constitute?

Section 30.5 ■ Nuclear Reactions

Note: Problem 20.61 in Chapter 20 can be assigned with this section.

26. Identify the unknown nuclei and particles X and X′ in the following nuclear reactions:
(a) $X + {}^{4}_{2}\text{He} \rightarrow {}^{24}_{12}\text{Mg} + {}^{1}_{0}\text{n}$
(b) ${}^{235}_{92}\text{U} + {}^{1}_{0}\text{n} \rightarrow {}^{90}_{38}\text{Sr} + X + 2{}^{1}_{0}\text{n}$
(c) $2{}^{1}_{1}\text{H} \rightarrow {}^{2}_{1}\text{H} + X + X'$

27. PhysicsNow™ Natural gold has only one isotope, ${}^{197}_{79}\text{Au}$. If natural gold is irradiated by a flux of slow neutrons, electrons are emitted. (a) Write the reaction equation. (b) Calculate the maximum energy of the emitted electrons.

28. A beam of 6.61 MeV protons is incident on a target of ${}^{27}_{13}\text{Al}$. Those protons that collide with a target nucleus produce the reaction

$$p + {}^{27}_{13}\text{Al} \rightarrow {}^{27}_{14}\text{Si} + n$$

(${}^{27}_{14}\text{Si}$ has mass 26.986 705 u.) Ignoring any recoil of the product nucleus, determine the kinetic energy of the emerging neutrons.

29. **Review problem.** Suppose enriched uranium containing 3.40% of the fissionable isotope ${}^{235}_{92}\text{U}$ is used as fuel for a ship. The water exerts an average friction force of magnitude 1.00×10^5 N on the ship. How far can the ship travel per kilogram of fuel? Assume that the energy released per fission event is 208 MeV and that the ship's engine has an efficiency of 20.0%.

30. (a) The following fission reaction is typical of those occurring in a nuclear electric generating station:

$${}^{1}_{0}\text{n} + {}^{235}_{92}\text{U} \rightarrow {}^{141}_{56}\text{Ba} + {}^{92}_{36}\text{Kr} + 3({}^{1}_{0}\text{n})$$

Find the energy released. The required masses are

$$M({}^{1}_{0}\text{n}) = 1.008\,665 \text{ u}$$
$$M({}^{235}_{92}\text{U}) = 235.043\,923 \text{ u}$$
$$M({}^{141}_{56}\text{Ba}) = 140.914\,4 \text{ u}$$
$$M({}^{92}_{36}\text{Kr}) = 91.926\,2 \text{ u}$$

(b) What fraction of the initial mass of the system is transformed?

31. PhysicsNow™ It has been estimated that on the order of 10^9 tons of natural uranium is available at concentrations exceeding 100 parts per million, of which 0.7% is the fissionable isotope ^{235}U. Assume that all the world's energy use (7×10^{12} J/s) were supplied by ^{235}U fission in conventional nuclear reactors, releasing 208 MeV for each reaction. How long would the supply last? The estimate of uranium supply is taken from K. S. Deffeyes and I. D. MacGregor, "World Uranium Resources," *Scientific American* 242(1):66, 1980.

32. Of all the hydrogen in the oceans, 0.030 0% of the mass is deuterium. The oceans have a volume of 317 million mi^3. (a) If nuclear fusion were controlled and all the deuterium in the oceans were fused to $^4_2$He, how many joules of energy would be released? (b) World power consumption is about 7.00×10^{12} W. If consumption were 100 times greater, how many years would the energy calculated in part (a) last?

Section 30.6 ■ Context Connection—The Engine of the Stars

33. The Sun radiates energy at the rate of 3.85×10^{26} W. Suppose the net reaction

$$4(^1_1\text{H}) + 2(^{\ 0}_{-1}\text{e}) \rightarrow {}^4_2\text{He} + 2\nu + \gamma$$

accounts for all the energy released. Calculate the number of protons fused per second.

34. In addition to the proton–proton cycle, the carbon cycle, first proposed by Hans Bethe in 1939, is another cycle by which energy is released in stars as hydrogen is converted to helium. The carbon cycle requires higher temperatures than the proton–proton cycle. The series of reactions is

$$\begin{aligned}
^{12}\text{C} + {}^1\text{H} &\rightarrow {}^{13}\text{N} + \gamma \\
^{13}\text{N} &\rightarrow {}^{13}\text{C} + \text{e}^+ + \nu \\
\text{e}^+ + \text{e}^- &\rightarrow 2\gamma \\
^{13}\text{C} + {}^1\text{H} &\rightarrow {}^{14}\text{N} + \gamma \\
^{14}\text{N} + {}^1\text{H} &\rightarrow {}^{15}\text{O} + \gamma \\
^{15}\text{O} &\rightarrow {}^{15}\text{N} + \text{e}^+ + \nu \\
\text{e}^+ + \text{e}^- &\rightarrow 2\gamma \\
^{15}\text{N} + {}^1\text{H} &\rightarrow {}^{12}\text{C} + {}^4\text{He}
\end{aligned}$$

(a) Assuming that the proton–proton cycle requires a temperature of 1.5×10^7 K, estimate by proportion the temperature required for the carbon cycle. (b) Calculate the Q value for each step in the carbon cycle and the overall energy released. (c) Do you think that the energy carried off by the neutrinos is deposited in the star? Explain.

35. Consider the two nuclear reactions

$$\begin{aligned}
\text{A} + \text{B} &\rightarrow \text{C} + \text{E} \qquad \text{(I)} \\
\text{C} + \text{D} &\rightarrow \text{F} + \text{G} \qquad \text{(II)}
\end{aligned}$$

(a) Show that the net disintegration energy for these two reactions ($Q_{\text{net}} = Q_{\text{I}} + Q_{\text{II}}$) is identical to the disintegration energy for the net reaction

$$\text{A} + \text{B} + \text{D} \rightarrow \text{E} + \text{F} + \text{G}$$

(b) One chain of reactions in the proton–proton cycle in the Sun's core is

$$\begin{aligned}
^1_1\text{H} + {}^1_1\text{H} &\rightarrow {}^2_1\text{H} + {}^0_1\text{e} + \nu \\
^0_1\text{e} + {}^{\ 0}_{-1}\text{e} &\rightarrow 2\gamma \\
^1_1\text{H} + {}^2_1\text{H} &\rightarrow {}^3_2\text{He} + \gamma \\
^1_1\text{H} + {}^3_2\text{He} &\rightarrow {}^4_2\text{He} + {}^0_1\text{e} + \nu \\
^0_1\text{e} + {}^{\ 0}_{-1}\text{e} &\rightarrow 2\gamma
\end{aligned}$$

Based on part (a), what is Q_{net} for this sequence?

36. After determining that the Sun has existed for hundreds of millions of years but before the discovery of nuclear physics, scientists could not explain why the Sun has continued to burn for such a long time. For example, if it were a coal fire, it would have burned up in about 3 000 yr. Assume that the Sun, whose mass is 1.99×10^{30} kg, originally consisted entirely of hydrogen and that its total power output is 3.85×10^{26} W. (a) Assuming the energy-generating mechanism of the Sun is the fusion of hydrogen into helium via the net reaction

$$4(^1_1\text{H}) + 2(\text{e}^-) \rightarrow {}^4_2\text{He} + 2\nu + \gamma$$

calculate the energy (in joules) given off by this reaction. (b) Determine how many hydrogen atoms constitute the Sun. Take the mass of one hydrogen atom to be 1.67×10^{-27} kg. (c) If the total power output remains constant, after what time interval will all the hydrogen be converted into helium, making the Sun die? The actual projected lifetime of the Sun is about 10 billion years, because only the hydrogen in a relatively small core is available as a fuel. Only in the core are temperatures and densities high enough for the fusion reaction to be self-sustaining.

37. Carbon detonations are powerful nuclear reactions that temporarily tear apart the cores inside massive stars late in their lives. These blasts are produced by carbon fusion, which requires a temperature of about 6×10^8 K to overcome the strong Coulomb repulsion between carbon nuclei. (a) Estimate the repulsive energy barrier to fusion, using the temperature required for carbon fusion. (In other words, what is the average kinetic energy of a carbon nucleus at 6×10^8 K?) (b) Calculate the energy (in MeV) released in each of these "carbon-burning" reactions:

$$\begin{aligned}
^{12}\text{C} + {}^{12}\text{C} &\rightarrow {}^{20}\text{Ne} + {}^4\text{He} \\
^{12}\text{C} + {}^{12}\text{C} &\rightarrow {}^{24}\text{Mg} + \gamma
\end{aligned}$$

(c) Calculate the energy (in kWh) given off when 2.00 kg of carbon completely fuses according to the first reaction.

38. A theory of nuclear astrophysics proposes that all the elements heavier than iron are formed in supernova explosions ending the lives of massive stars. Assume that at the time of the explosion the amounts of ^{235}U and ^{238}U were equal. How long ago did the star(s) explode that released the elements that formed our Earth? The present $^{235}\text{U}/^{238}\text{U}$ ratio is 0.007 25. The half-lives of ^{235}U and ^{238}U are 0.704×10^9 yr and 4.47×10^9 yr.

Additional Problems

39. (a) One method of producing neutrons for experimental use is bombardment of light nuclei with alpha particles. In the method used by James Chadwick in 1932, alpha particles emitted by polonium are incident on beryllium nuclei:

$$^4_2\text{He} + {}^9_4\text{Be} \rightarrow {}^{12}_{\ 6}\text{C} + {}^1_0\text{n}$$

What is the Q value? (b) Neutrons are also often produced by small particle accelerators. In one design, deuterons accelerated in a Van de Graaff generator bombard other deuterium nuclei:

$$^2_1\text{H} + {}^2_1\text{H} \rightarrow {}^3_2\text{He} + {}^1_0\text{n}$$

Is this reaction exothermic or endothermic? Calculate its Q value.

40. As part of his discovery of the neutron in 1932, Chadwick determined the mass of the newly identified particle by firing a beam of fast neutrons, all having the same speed, at two different targets and measuring the maximum recoil speeds of the target nuclei. The maximum speeds arise when an elastic head-on collision occurs between a neutron and a stationary target nucleus. (a) Represent the masses and final speeds of the two target nuclei as m_1, v_1, m_2, and v_2 and assume that Newtonian mechanics applies. Show that the neutron mass can be calculated from the equation

$$m_n = \frac{m_1 v_1 - m_2 v_2}{v_2 - v_1}$$

(b) Chadwick directed a beam of neutrons (produced from a nuclear reaction) on paraffin, which contains hydrogen. The maximum speed of the protons ejected was found to be 3.3×10^7 m/s. Because the velocity of the neutrons could not be determined directly, a second experiment was performed using neutrons from the same source and nitrogen nuclei as the target. The maximum recoil speed of the nitrogen nuclei was found to be 4.7×10^6 m/s. The masses of a proton and a nitrogen nucleus were taken as 1 u and 14 u, respectively. What was Chadwick's value for the neutron mass?

41. When the nuclear reaction represented by Equation 30.24 is endothermic, the reaction energy Q is negative. For the reaction to proceed, the incoming particle must have a minimum energy called the threshold energy, E_{th}. Some fraction of the energy of the incident particle is transferred to the compound nucleus to conserve momentum. Therefore, E_{th} must be greater than Q. (a) Show that

$$E_{\text{th}} = -Q\left(1 + \frac{M_{\text{a}}}{M_{\text{X}}}\right)$$

(b) Calculate the threshold energy of the incident alpha particle in the reaction

$${}^4_2\text{He} + {}^{14}_7\text{N} \rightarrow {}^{17}_8\text{O} + {}^1_1\text{H}$$

42. (a) Find the radius of the ${}^{12}_6\text{C}$ nucleus. (b) Find the force of repulsion between a proton at the surface of a ${}^{12}_6\text{C}$ nucleus and the remaining five protons. (c) How much work (in MeV) has to be done to overcome this electric repulsion to put the last proton into the nucleus? (d) Repeat parts (a), (b), and (c) for ${}^{238}_{92}\text{U}$.

43. (a) Why is the beta decay $\text{p} \rightarrow \text{n} + \text{e}^+ + \nu$ forbidden for a free proton? (b) Why is the same reaction possible if the proton is bound in a nucleus? For example, the following reaction occurs:

$${}^{13}_7\text{N} \rightarrow {}^{13}_6\text{C} + \text{e}^+ + \nu$$

(c) How much energy is released in the reaction given in part (b)? [*Suggestion:* Add seven electrons to both sides of the reaction to write it for neutral atoms. You may use the masses $m(\text{e}^+) = 0.000\,549$ u, $M({}^{13}\text{C}) = 13.003\,355$ u, and $M({}^{13}\text{N}) = 13.005\,739$ u.]

44. The activity of a radioactive sample was measured over 12 h, with the net count rates shown in the table.

Time (h)	Counting Rate (counts/min)
1.00	3 100
2.00	2 450
4.00	1 480
6.00	910
8.00	545
10.0	330
12.0	200

(a) Plot the logarithm of counting rate as a function of time. (b) Determine the decay constant and half-life of the radioactive nuclei in the sample. (c) What counting rate would you expect for the sample at $t = 0$? (d) Assuming the efficiency of the counting instrument to be 10.0%, calculate the number of radioactive atoms in the sample at $t = 0$.

45. When, after a reaction or disturbance of any kind, a nucleus is left in an excited state, it can return to its normal (ground) state by emission of a gamma-ray photon (or several photons). This process is illustrated by Equation 30.21. The emitting nucleus must recoil to conserve both energy and momentum. (a) Show that the recoil energy of the nucleus is

$$E_r = \frac{(\Delta E)^2}{2Mc^2}$$

where ΔE is the difference in energy between the excited and ground states of a nucleus of mass M. (b) Calculate the recoil energy of the ${}^{57}\text{Fe}$ nucleus when it decays by gamma emission from the 14.4-keV excited state. For this calculation, take the mass to be 57 u. [*Suggestions:* When writing the equation for conservation of energy, use $(Mv)^2/2M$ for the kinetic energy of the recoiling nucleus. Also, assume that $hf \ll Mc^2$ and use the binomial expansion.]

46. After the sudden release of radioactivity from the Chernobyl nuclear reactor accident in 1986, the radioactivity of milk in Poland rose to 2 000 Bq/L due to iodine-131 present in the grass eaten by dairy cattle. Radioactive iodine, with half-life 8.04 days, is particularly hazardous because the thyroid gland concentrates iodine. The Chernobyl accident caused a measurable increase in thyroid cancers among children in Belarus. (a) For comparison, find the activity of milk due to potassium. Assume that one liter of milk contains 2.00 g of potassium, of which 0.011 7% is the isotope ${}^{40}\text{K}$ with half-life 1.28×10^9 yr. (b) After what time interval would the activity due to iodine fall below that due to potassium?

47. Europeans named a certain direction in the sky as between the horns of Taurus the Bull. On the day they named as A.D. July 4, 1054, a brilliant light appeared there. Europeans left no surviving record of the supernova, which could be seen in daylight for some days. As it faded it remained visible for years, dimming for a time with the 77.1-day half-life of the radioactive cobalt-56 that had been created in the explosion. (a) The remains of the star now form the Crab Nebula. (See Fig. 10.23 and the opening photographs of Chapter 24.) In it, the cobalt-56 has now decreased to what fraction of its original activity? (b) Suppose an American, of the people called the Anasazi, made a charcoal drawing of the supernova. The carbon-14 in the charcoal has now decayed to what fraction of its original activity?

48. In a piece of rock from the Moon, the ^{87}Rb content is assayed to be 1.82×10^{10} atoms per gram of material and the ^{87}Sr content is found to be 1.07×10^{9} atoms per gram. (a) Calculate the age of the rock. (b) Could the material in the rock actually be much older? What assumption is implicit in using the radioactive dating method? (The relevant decay is $^{87}\text{Rb} \rightarrow {}^{87}\text{Sr} + e^{-} + \overline{\nu}$. The half-life of the decay is 4.75×10^{10} yr.)

49. **Physics Now™** The decay of an unstable nucleus by alpha emission is represented by Equation 30.9. The disintegration energy Q given by Equation 30.12 must be shared by the alpha particle and the daughter nucleus to conserve both energy and momentum in the decay process. (a) Show that Q and K_α, the kinetic energy of the alpha particle, are related by the expression

$$Q = K_\alpha\left(1 + \frac{M_\alpha}{M}\right)$$

where M is the mass of the daughter nucleus. (b) Use the result of part (a) to find the energy of the alpha particle emitted in the decay of ^{226}Ra. (See Example 30.5 for the calculation of Q.)

50. *Student determination of the half-life of ^{137}Ba.* The radioactive barium isotope ^{137}Ba has a relatively short half-life and can be easily extracted from a solution containing its parent cesium (^{137}Cs). This barium isotope is commonly used in an undergraduate laboratory exercise for demonstrating the radioactive decay law. Undergraduate students using modest experimental equipment took the data presented in Figure P30.50. Determine the half-life for the decay of ^{137}Ba using their data.

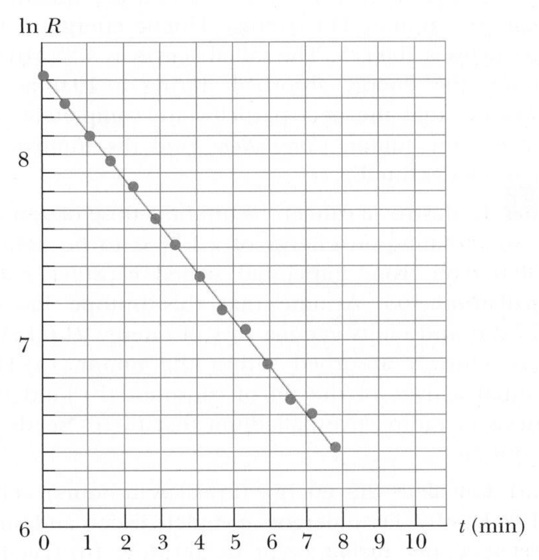

FIGURE **P30.50**

51. A small building has become accidentally contaminated with radioactivity. The longest-lived material in the building is strontium-90. ($^{90}_{38}$Sr has an atomic mass 89.907 7 u, and its half-life is 29.1 yr. It is particularly dangerous because it substitutes for calcium in bones.) Assume that the building initially contained 5.00 kg of this substance uniformly distributed throughout the building and that the safe level is defined as less than 10.0 decays/min (to be small in comparison to background radiation). How long will the building be unsafe?

52. *Lead shielding.* When gamma rays are incident on matter, the intensity of the gamma rays passing through the material varies with depth x as $I(x) = I_0 e^{-\mu x}$, where μ is the absorption coefficient and I_0 is the intensity of the radiation at the surface of the material. For 0.400-MeV gamma rays in lead, the absorption coefficient is 1.59 cm^{-1}. (a) Determine the "half-thickness" for lead, that is, the thickness of lead that would absorb half the incident gamma rays. (b) What thickness will reduce the radiation by a factor of 10^4?

53. *A thickness gauge.* When gamma rays are incident on matter, the intensity of the gamma rays passing through the material varies with depth x as $I(x) = I_0 e^{-\mu x}$, where μ is the absorption coefficient and I_0 is the intensity of the radiation at the surface of the material. For low-energy gamma rays in steel, take the absorption coefficient to be 0.720 mm^{-1}. (a) Determine the "half-thickness" for steel, that is, the thickness of steel that would absorb half the incident gamma rays. (b) In a steel mill, the thickness of sheet steel passing into a roller is measured by monitoring the intensity of gamma radiation reaching a detector below the rapidly moving metal from a small source just above the metal. If the thickness of the sheet changes from 0.800 mm to 0.700 mm, by what percentage will the gamma-ray intensity change?

54. During the manufacture of a steel engine component, radioactive iron (^{59}Fe) is included in the total mass of 0.200 kg. The component is placed in a test engine when the activity due to this isotope is 20.0 μCi. After a 1 000-h test period, some of the lubricating oil is removed from the engine and found to contain enough ^{59}Fe to produce 800 disintegrations/min/L of oil. The total volume of oil in the engine is 6.50 L. Calculate the total mass worn from the engine component per hour of operation. (The half-life of ^{59}Fe is 45.1 days.)

55. *Neutron activation analysis* is a method for chemical analysis at the level of isotopes. When a sample is irradiated by neutrons, radioactive atoms are produced continuously and then decay according to their characteristic half-lives. (a) Assume that one species of radioactive nuclei is produced at a constant rate R and that its decay is described by the conventional radioactive decay law. Defining $t = 0$ as the time irradiation begins, show that the number of radioactive atoms accumulated at time t is

$$N = \frac{R}{\lambda}(1 - e^{-\lambda t})$$

(b) What is the maximum number of radioactive atoms that can be produced?

56. On August 6, 1945, the United States dropped on Hiroshima a nuclear bomb that released 5×10^{13} J of energy, equivalent to that from 12 000 tons of TNT. The fission of one $^{235}_{92}$U nucleus releases an average of 208 MeV. Estimate (a) the number of nuclei fissioned and (b) the mass of this $^{235}_{92}$U.

57. **Review problem.** A nuclear power plant operates by using the energy released in nuclear fission to convert 20°C water into 400°C steam. How much water could theoretically be converted to steam by the complete fissioning of 1.00 gram of ^{235}U at 200 MeV/fission?

58. **Review problem.** The first nuclear bomb was a fissioning mass of plutonium-239, exploded in the Trinity test, before dawn on July 16, 1945, at Alamogordo, New Mexico. Enrico Fermi was 14 km away, lying on the ground facing away from the bomb. After the whole sky had flashed with unbelievable brightness, Fermi stood up and began dropping bits of paper to the ground. They first fell at his feet in the calm and silent air. As the shock wave passed, about 40 s after the explosion, the paper then in flight jumped about 5 cm away from ground zero. (a) Assume that the shock wave in air propagated equally in all directions without absorption. Find the change in volume of a sphere of radius 14 km as it expands by 5 cm. (b) Find the work $P\Delta V$ done by the air in this sphere on the next layer of air farther from the center. (c) Assume that the shock wave carried on the order of one tenth of the energy of the explosion. Make an order-of-magnitude estimate of the bomb yield. (d) One ton of exploding trinitrotoluene (TNT) releases energy 4.2 GJ. What was the order of magnitude of the energy of the Trinity test in equivalent tons of TNT? The dawn revealed the mushroom cloud. Fermi's immediate knowledge of the bomb yield agreed with that determined days later by analysis of elaborate measurements.

59. About 1 of every 3 300 water molecules contains one deuterium atom. (a) If all the deuterium nuclei in 1 L of water are fused in pairs according to the D–D reaction $^2\text{H} + {}^2\text{H} \rightarrow {}^3\text{He} + \text{n} + 3.27$ MeV, how much energy in joules is liberated? (b) Burning gasoline produces about 3.40×10^7 J/L. Compare the energy obtainable from the fusion of the deuterium in 1 L of water with the energy liberated from the burning of 1 L of gasoline.

60. The alpha-emitter polonium-210 ($^{210}_{84}$Po) is used in a nuclear energy source on a spacecraft (Fig. P30.60). Determine the initial power output of the source. Assume that it contains 0.155 kg of ^{210}Po and that the efficiency for conversion of radioactive decay energy to energy transferred by electrical transmission is 1.00%.

(Courtesy of NASA Ames)

FIGURE P30.60 The *Pioneer 10* spacecraft leaves the Solar System. It carries radioactive power supplies at the ends of two booms. Solar panels would not work far from the Sun.

61. Natural uranium must be processed to produce uranium enriched in ^{235}U for bombs and power plants. The processing yields a large quantity of nearly pure ^{238}U as a by-product, called "depleted uranium." Because of its high mass density, it is used in armor-piercing artillery shells. (a) Find the edge dimension of a 70.0-kg cube of ^{238}U. The density of uranium is 18.7×10^3 kg/m^3. (b) The isotope ^{238}U has a long half-life of 4.47×10^9 yr. As soon as one nucleus decays, it begins a relatively rapid series of 14 steps that together constitute the net reaction

$$^{238}_{92}\text{U} \rightarrow 8(^4_2\text{He}) + 6(^{\;0}_{-1}\text{e}) + {}^{206}_{82}\text{Pb} + 6\,\bar{\nu} + Q_{\text{net}}$$

Find the net decay energy. (Refer to Table A.3.) (c) Argue that a radioactive sample with decay rate R and decay energy Q has power output $\mathcal{P} = QR$. (d) Consider an artillery shell with a jacket of 70.0 kg of ^{238}U. Find its power output due to the radioactivity of the uranium and its daughters. Assume that the shell is old enough that the daughters have reached steady-state amounts. Express the power in joules per year. (e) A 17-year-old soldier of mass 70.0 kg works in an arsenal where many such artillery shells are stored. Assume that his radiation exposure is limited to absorbing 45.5 mJ per year per kilogram of body mass. Find the net rate at which he can absorb energy of radiation, in joules per year.

62. A sealed capsule containing the radiopharmaceutical phosphorus-32 ($^{32}_{15}$P), an e$^-$ emitter, is implanted into a patient's tumor. The average kinetic energy of the beta particles is 700 keV. The initial activity is 5.22 MBq. Determine the energy absorbed during a 10.0-day period. Assume that the beta particles are completely absorbed within the tumor. (*Suggestion:* Find the number of beta particles emitted.)

63. To destroy a cancerous tumor, a dose of gamma radiation totaling an energy of 2.12 J is to be delivered in 30.0 days from implanted sealed capsules containing palladium-103. Assume that this isotope has half-life 17.0 d and emits gamma rays of energy 21.0 keV, which are entirely absorbed within the tumor. (a) Find the initial activity of the set of capsules. (b) Find the total mass of radioactive palladium that these "seeds" should contain.

64. (a) Calculate the energy (in kilowatt-hours) released if 1.00 kg of ^{239}Pu undergoes complete fission and the energy released per fission event is 200 MeV. (b) Calculate the energy (in electron volts) released in the deuterium–tritium

fusion reaction

$$^2_1\text{H} + ^3_1\text{H} \rightarrow ^4_2\text{He} + ^1_0\text{n}$$

(c) Calculate the energy (in kilowatt-hours) released if 1.00 kg of deuterium undergoes fusion according to this reaction. (d) Calculate the energy (in kilowatt-hours) released by the combustion of 1.00 kg of coal if each $C + O_2 \rightarrow CO_2$ reaction yields 4.20 eV. (e) List advantages and disadvantages of each of these methods of energy generation.

ANSWERS TO QUICK QUIZZES

30.1 (i),(b). The value of $N = A - Z$ is the same for all three nuclei. (ii), (a). The value of Z is the same for all three nuclei because they are all nuclei of nitrogen. (iii), (c). The value of A is the same for all three nuclei, as seen by the unchanging preceding superscript.

30.2 (c). Isotopes of a given element correspond to nuclei with different numbers of neutrons. The result is different masses of the atom, and different magnetic moments because the neutron, despite being uncharged, has a magnetic moment. The chemical behavior, however, is governed by the electrons. All isotopes of a given element have the same number of electrons and therefore the same chemical behavior.

30.3 (e). A year of 365 days is equivalent to 365 d/5.01 d $\approx$ 73 half-lives. Therefore, the activity will be reduced after one year to approximately $(1/2)^{73}(1.000\ \mu\text{Ci}) \sim 10^{-22}\ \mu\text{Ci}$.

30.4 (e). The time we are interested in is half of a half-life. Therefore, the number of *remaining* nuclei is $(\frac{1}{2})^{1/2}N_0 = (1/\sqrt{2})N_0 = 0.707N_0$. The number of nuclei that *have decayed* is $N_0 - 0.707N_0 = 0.293N_0$.

30.5 (b). In alpha decay, the atomic number decreases by two and the atomic mass number decreases by four.

30.6 (c). In e^- decay, the atomic number increases by one and the atomic mass number stays fixed. None of the choices is consistent with e^+ decay, so we assume that the decay must be by e^-.

CHAPTER 31

Particle Physics

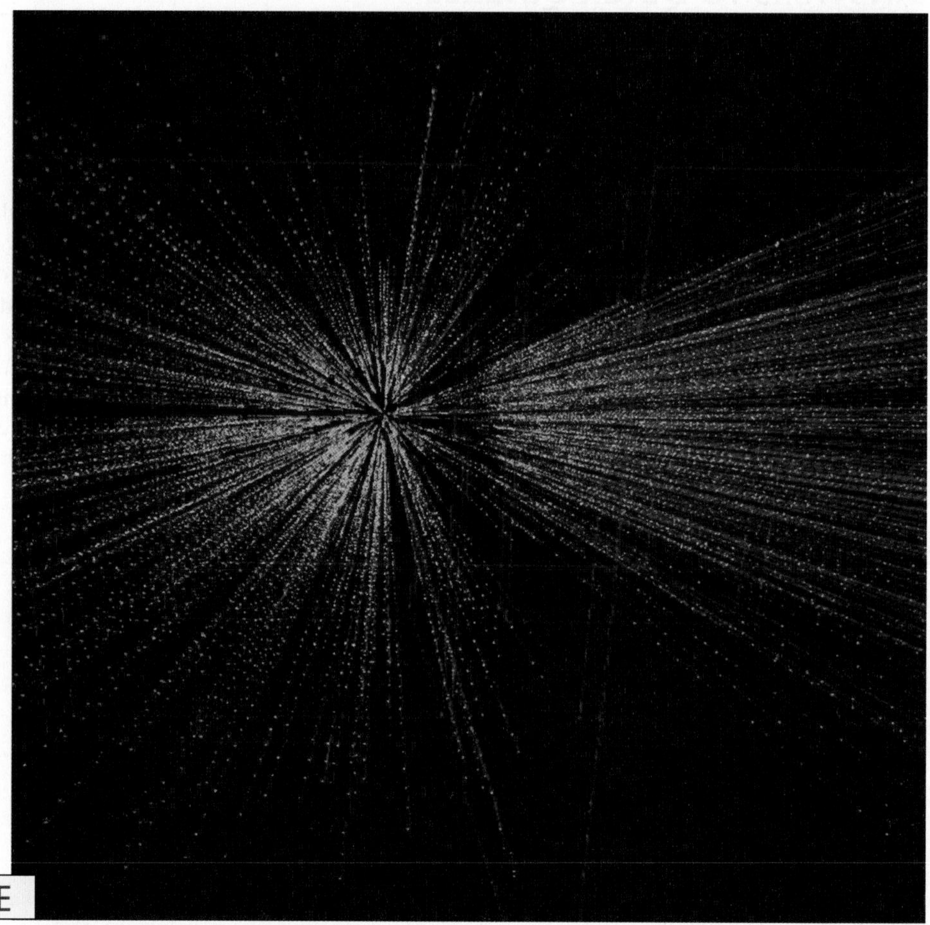

In this image from the NA49 experiment at CERN, hundreds of subatomic particles are created in the collision of high-energy nuclei with a lead target. The aim of the experiment is to create a quark-gluon plasma, in which the force that normally locks quarks within protons and neutrons is broken.

(Courtesy of CERN)

CHAPTER OUTLINE

In the early chapters of this book, we discussed the particle model, which treats an object as a particle of zero size with no structure. Some behaviors of objects, such as thermal expansion, can be understood by modeling the object as a collection of particles: atoms. In these models, any internal structure of the atom is ignored. We could not ignore the internal structure of the atom to understand such phenomena as atomic spectra, however. Modeling the hydrogen atom as a system of an electron in orbit about a particle-like nucleus helped in this regard (Section 11.5). In Chapter 30, however, we could not model the nucleus as a particle and ignore its structure to understand behavior such as nuclear stability and radioactive decay. We had to model the nucleus as a collection of smaller particles, nucleons. What about these nuclear constituents, the protons and neutrons? Can we apply the particle model to these entities? As we shall see, even protons and neutrons have structure, which leads to a puzzling question. As we continue to investigate the structure of smaller and smaller "particles," will we ever reach a level at which the building blocks are truly and completely described by the particle model?

In this concluding chapter, we explore this question by examining the properties and classifications of the various known subatomic particles and the fundamental interactions that govern their behavior. We also discuss the current model of elementary particles, in which all matter is believed to be constructed from only two families of particles: quarks and leptons.

The word *atom* is from the Greek *atomos,* which means "indivisible." At one time, atoms were thought to be the indivisible constituents of matter; that is, they were regarded as elementary particles. After 1932, physicists viewed all matter as consisting of only three constituent particles: electrons, protons, and neutrons. (The neutron was observed and identified in 1932.) With the exception of the free neutron (as opposed to a neutron within a nucleus), these particles are very stable. Beginning in 1945, many new particles were discovered in experiments involving high-energy collisions between known particles. These new particles are characteristically very unstable and have very short half-lives, ranging between 10^{-6} s and 10^{-23} s. So far, more than 300 of these unstable, temporary particles have been catalogued.

Since the 1930s, many powerful particle accelerators have been constructed throughout the world, making it possible to observe collisions of highly energetic particles under controlled laboratory conditions so as to reveal the subatomic world in finer detail. Until the 1960s, physicists were bewildered by the large number and variety of subatomic particles being discovered. They wondered if the particles were like animals in a zoo, having no systematic relationship connecting them, or whether a pattern was emerging that would provide a better understanding of the elaborate structure in the subnuclear world. Since that time, physicists have advanced our knowledge of the structure of matter tremendously by developing a structural model in which most of the particles in the ever-growing particle zoo are made of smaller particles called quarks. Therefore, protons and neutrons, for example, are not truly elementary but are systems of tightly bound quarks.

31.1 THE FUNDAMENTAL FORCES IN NATURE

As we learned in Chapter 5, all natural phenomena can be described by four fundamental forces between particles. In order of decreasing strength, they are the **strong** force, the **electromagnetic** force, the **weak** force, and the **gravitational** force. In current models, the electromagnetic and weak forces are considered to be two manifestations of a single interaction, the **electroweak force,** as discussed in Section 31.11.

The **nuclear force,** as we mentioned in Chapter 30, holds nucleons together. It is very short range and is negligible for separations greater than about 2 fm (about the size of the nucleus). The electromagnetic force, which binds atoms and molecules together to form ordinary matter, has about 10^{-2} times the strength of the nuclear force. It is a long-range force that decreases in strength as the inverse square of the separation between interacting particles. The weak force is a short-range force that accounts for radioactive decay processes such as beta decay, and its strength is only about 10^{-5} times that of the nuclear force. Finally, the gravitational force is a long-range force that has a strength of only about 10^{-41} times that of the nuclear force. Although this familiar interaction is the force that holds the planets, stars, and galaxies together, its effect on elementary particles is negligible.

■ Field particles

In modern physics, interactions between particles are often described in terms of a structural model that involves the exchange of **field particles,** or **quanta.** In the case of the familiar electromagnetic interaction, for instance, the field particles are photons. In the language of modern physics, we say that the electromagnetic force is *mediated* by photons and that photons are the quanta of the electromagnetic field. Likewise, the nuclear force is mediated by field particles called **gluons,** the weak

TABLE 31.1 Fundamental Forces

Force	Relative Strength	Range of Force	Mediating Field Particle	Mass of Field Particle (GeV/c^2)
Nuclear	1	Short (~1 fm)	Gluon	0
Electromagnetic	10^{-2}	∞	Photon	0
Weak	10^{-5}	Short ($\sim 10^{-3}$ fm)	$W^{\pm}$, Z^0 bosons	80.4, 80.4, 91.2
Gravitational	10^{-41}	∞	Graviton	0

force is mediated by particles called the W and Z **bosons** (in general, all particles with integral spin are called *bosons*), and the gravitational force is mediated by quanta of the gravitational field called **gravitons.** These forces, their ranges, and their relative strengths are summarized in Table 31.1.

31.2 POSITRONS AND OTHER ANTIPARTICLES

In the 1920s, English theoretical physicist Paul Adrien Maurice Dirac developed a version of quantum mechanics that incorporated special relativity. Dirac's theory explained the origin of electron spin and its magnetic moment. It also presented a major difficulty, however. Dirac's relativistic wave equation required solutions corresponding to negative energy states even for free electrons. If negative energy states existed, however, one would expect an electron in a state of positive energy to make a rapid transition to one of these states, emitting a photon in the process. Dirac avoided this difficulty by postulating a structural model in which all negative energy states are filled. The electrons occupying these negative energy states are collectively called the *Dirac sea.* Electrons in the Dirac sea are not directly observable because the Pauli exclusion principle does not allow them to react to external forces; there are no states available to which an electron can make a transition in response to an external force. Therefore, an electron in such a state acts as an isolated system unless an interaction with the environment is strong enough to excite the electron to a positive energy state. Such an excitation causes one of the negative energy states to be vacant, as in Figure 31.1, leaving a hole in the sea of filled states. (Notice that positive energy states exist only for $E > m_ec^2$, representing the rest energy of the electron. Similarly, negative energy states exist only for $E < -m_ec^2$.) *The hole can react to external forces and is observable.* The hole reacts in a way similar to that of the electron, except that it has a positive charge. It is the **antiparticle** to the electron.

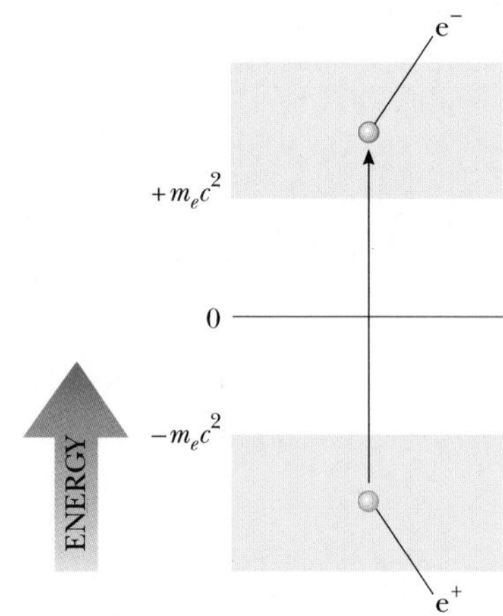

FIGURE 31.1 Dirac's model for the existence of antielectrons (positrons). The states lower in energy than $-m_ec^2$ are filled with electrons (the Dirac sea). One of these electrons can make a transition out of its state only if it is provided with energy equal to or larger than $2m_ec^2$. That leaves a vacancy in the Dirac sea, which can behave as a particle identical to the electron except for its positive charge.

The profound implication of this model is that *every particle has a corresponding antiparticle.* The antiparticle has the same mass as the particle, but the opposite charge. For example, the electron's antiparticle, called a **positron,** has a mass of 0.511 MeV/c^2 and a positive charge of 1.60×10^{-19} C.

Carl Anderson (1905–1991) observed and identified the positron in 1932, and in 1936 he was awarded the Nobel Prize in Physics for that achievement. Anderson discovered the positron while examining tracks in a cloud chamber created by electron-like particles of positive charge. (A cloud chamber contains a gas that has been supercooled to just below its usual condensation point. An energetic radioactive particle passing through ionizes the gas and leaves a visible track. These early experiments used cosmic rays—mostly energetic protons passing through interstellar space—to initiate high-energy reactions in the upper atmosphere, which resulted in the production of positrons at ground level.) To discriminate between positive and negative charges, Anderson placed the cloud chamber in a magnetic field, causing moving charged particles to follow curved paths as discussed in Section 22.3. He

PITFALL PREVENTION 31.1

ANTIPARTICLES An antiparticle is not identified solely on the basis of opposite charge; even neutral particles have antiparticles, which are defined in terms of other properties, such as spin magnetic moment.

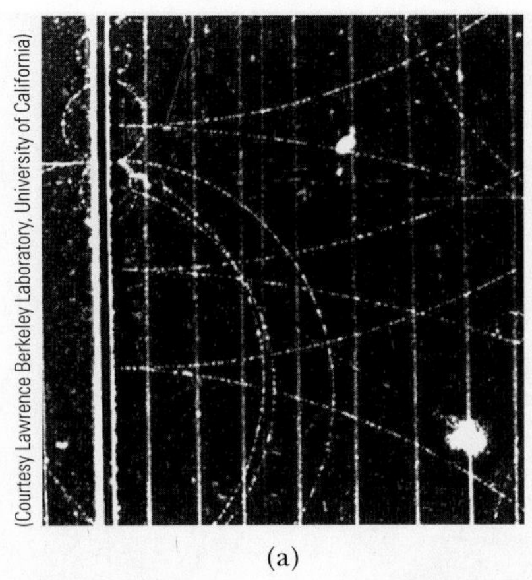

(a)

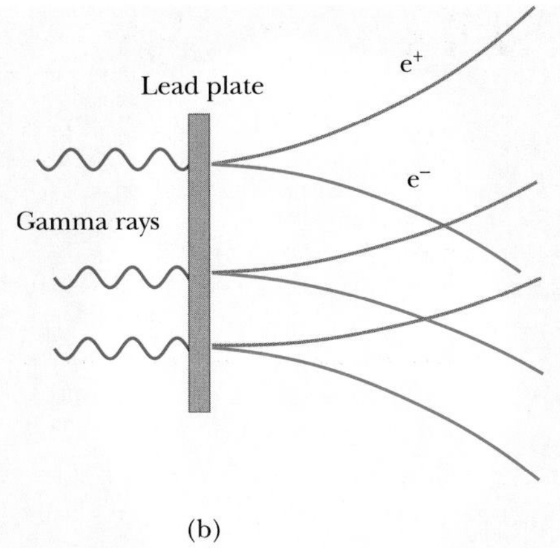

(b)

FIGURE 31.2 (a) Bubble-chamber tracks of electron–positron pairs produced by 300-MeV gamma rays striking a lead plate. (b) Sketch of the pertinent pair-production events. Note that the positrons deflect upward and the electrons deflect downward in an applied magnetic field.

noted that some of the electron-like tracks deflected in a direction corresponding to a positively charged particle.

Since Anderson's discovery, the positron has been observed in a number of experiments. A common process for producing positrons is **pair production.** In this process, a gamma-ray photon with sufficiently high energy interacts with a nucleus and an electron–positron pair is created. In the Dirac sea model, an electron in a negative energy state is excited to a positive energy state, resulting in a new observable electron and a hole, which is the positron. Because the total rest energy of the electron–positron pair is $2m_ec^2 = 1.022$ MeV, the photon must have at least this much energy to create an electron–positron pair. Therefore, energy in the form of a gamma-ray photon is converted to rest energy in accordance with Einstein's relationship $E_R = mc^2$. We can use the isolated system model to describe this process. The energy of the system of the photon and the nucleus is conserved and transformed to rest energy of the electron and positron, kinetic energy of these particles, and some small amount of kinetic energy associated with the nucleus. Figure 31.2a shows tracks of electron–positron pairs created by 300-MeV gamma rays striking a lead plate.

PAUL ADRIEN MAURICE DIRAC (1902–1984)

British physicist Dirac was instrumental in the understanding of antimatter and the unification of quantum mechanics and relativity. He made many contributions to the development of quantum physics and cosmology. In 1933, he won the Nobel Prize in Physics.

QUICK QUIZ 31.1 Given the identification of the particles in Figure 31.2b, what is the direction of the external magnetic field in Figure 31.2a? **(a)** into the page **(b)** out of the page **(c)** impossible to determine

The reverse process can also occur. Under the proper conditions, an electron and positron can annihilate each other to produce two gamma-ray photons (see Thinking Physics 31.1) that have a combined energy of at least 1.022 MeV:

$$e^- + e^+ \rightarrow 2\gamma$$

 Positron-emission tomography (PET)

Electron–positron annihilation is used in the medical diagnostic technique called *positron-emission tomography* (PET). The patient is injected with a glucose solution containing a radioactive substance that decays by positron emission, and the material is carried by the blood throughout the body. A positron emitted during a decay event in one of the radioactive nuclei in the glucose solution annihilates with an electron in the immediately surrounding tissue, resulting in two gamma-ray photons emitted in opposite directions. A gamma detector surrounding the patient pinpoints the source of the photons and, with the assistance of a computer, displays an image of the sites at which the glucose accumulates. (Glucose is metabolized rapidly in cancerous tumors and accumulates in these sites, providing a strong signal for a PET detector system.) The images from a PET scan can indicate a wide

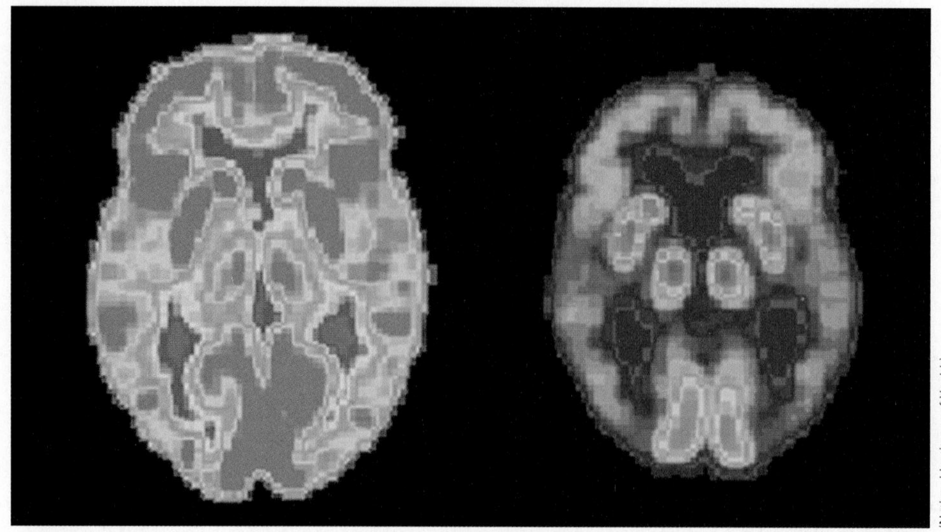

FIGURE 31.3 PET scans of the brain of a healthy older person (*left*) and that of a patient suffering from Alzheimer's disease (*right*). Lighter regions contain higher concentrations of radioactive glucose, indicating higher metabolism rates and therefore increased brain activity.

variety of disorders in the brain, including Alzheimer's disease (Fig. 31.3). In addition, because glucose metabolizes more rapidly in active areas of the brain, a PET scan can indicate which areas of the brain are involved when the patient is engaging in such activities as language use, music, or vision.

Prior to 1955, on the basis of the Dirac theory, it was expected that every particle has a corresponding antiparticle, but antiparticles such as the antiproton and antineutron had not been detected experimentally. Because the relativistic Dirac theory had some failures (it predicted the wrong-size magnetic moment for the photon) as well as many successes, it was important to determine whether the antiproton really existed. In 1955, a team led by Emilio Segrè (1905–1989) and Owen Chamberlain (b. 1920) used the Bevatron particle accelerator at the University of California–Berkeley to produce antiprotons and antineutrons. They therefore established with certainty the existence of antiparticles. For this work, Segrè and Chamberlain received the Nobel Prize in Physics in 1959. It is now established that **every particle has a corresponding antiparticle with equal mass and spin, and with charge, magnetic moment, and strangeness of equal magnitude but opposite sign.** (The property of strangeness is explained in Section 31.6.) The only exception to these rules for particles and antiparticles are the neutral photon, pion, and eta, each of which is its own antiparticle.

An intriguing aspect of the existence of antiparticles is that if we replace every proton, neutron, and electron in an atom with its antiparticle, we can create a stable antiatom; combinations of antiatoms should form antimolecules and eventually antiworlds. As far as we know, everything would behave in the same way in an antiworld as in our world. In principle, it is possible that some distant antimatter galaxies exist, separated from normal-matter galaxies by millions of lightyears. Unfortunately, because the photon is its own antiparticle, the light emitted from an antimatter galaxy is no different from that from a normal-matter galaxy, so astronomical observations cannot determine if the galaxy is composed of matter or antimatter. Although no evidence of antimatter galaxies exists at present, it is awe-inspiring to imagine the cosmic spectacle that would result if matter and antimatter galaxies were to collide: a gigantic eruption of jets of annihilation radiation, transforming the entire galactic mass into energetic particles fleeing the collision point.

■ Thinking Physics 31.1

When an electron and a positron meet at low speed in free space, why are two 0.511-MeV gamma rays produced rather than one gamma ray with an energy of 1.022 MeV?

Reasoning Gamma rays are photons, and photons carry momentum. We apply the momentum version of the isolated system model to the system, which consists initially of the electron and positron. If the system, assumed to be at rest, transformed to only one photon, momentum would not be conserved because the initial momentum of the electron–positron system is zero, whereas the final system consists of a single photon of energy 1.022 MeV and nonzero momentum. On the other hand, the two gamma-ray photons travel in *opposite* directions, so the total momentum of the final system—two photons—is zero, and momentum is conserved. ■

31.3 MESONS AND THE BEGINNING OF PARTICLE PHYSICS

In the mid-1930s, physicists had a fairly simple view of the structure of matter. The building blocks were the proton, the electron, and the neutron. Three other particles were known or had been postulated at the time: the photon, the neutrino, and the positron. These six particles were considered the fundamental constituents of matter. With this marvelously simple picture of the world, however, no one was able to answer an important question. Because many protons in proximity in any nucleus should strongly repel one another due to their positive charges, what is the nature of the force that holds the nucleus together? Scientists recognized that this mysterious force, which we now call the nuclear force, must be much stronger than anything encountered in nature up to that time.

In 1935, Japanese physicist Hideki Yukawa proposed the first theory to successfully explain the nature of the nuclear force, an effort that later earned him the Nobel Prize in Physics. To understand Yukawa's theory, it is useful to first recall that in the modern structural model of electromagnetic interactions, charged particles interact by exchanging photons. Yukawa used this idea to explain the nuclear force by proposing a new particle whose exchange between nucleons in the nucleus produces the nuclear force. Furthermore, he established that the range of the force is inversely proportional to the mass of this particle and predicted that the mass would be about 200 times the mass of the electron. Because the new particle would have a mass between that of the electron and that of the proton, it was called a meson (from the Greek *meso,* meaning "middle").

In an effort to substantiate Yukawa's predictions, physicists began an experimental search for the meson by studying cosmic rays entering the Earth's atmosphere. In 1937, Anderson and his collaborators discovered a particle of mass 106 MeV/c^2, about 207 times the mass of the electron. Subsequent experiments showed that the particle interacted very weakly with matter, however, and hence could not be the carrier of the nuclear force. The puzzling situation inspired several theoreticians to propose that two mesons existed with slightly different masses. This idea was confirmed by the discovery in 1947 of the **pi** (π) **meson,** or simply **pion,** by Cecil Frank Powell (1903–1969) and Giuseppe P. S. Occhialini (1907–1993). The particle discovered by Anderson in 1937, the one thought to be Yukawa's meson, is not really a meson. (We shall discuss the requirements for a particle to be a meson in Section 31.4). Instead, it takes part in the weak and electromagnetic interactions only and is now called the **muon** (μ). We first discussed the muon in Section 9.4, with regard to time dilation.

The pion, Yukawa's carrier of the nuclear force, comes in three varieties corresponding to three charge states: π^+, π^-, and π^0. The π^+ and π^- particles have masses of 139.6 MeV/c^2, and the π^0 particle has a mass of 135.0 MeV/c^2. Pions and muons are very unstable particles. For example, the π^-, which has a mean lifetime of 2.6×10^{-8} s, first decays to a muon and an antineutrino. The muon, which has a mean lifetime of 2.2 μs, then decays into an electron, a neutrino, and

PITFALL PREVENTION 31.2

THE NUCLEAR FORCE AND THE STRONG FORCE The nuclear force that we discussed in Chapter 30 and continue to discuss here was originally called the strong force. Once the quark theory (Section 31.9) was established, however, the phrase *strong force* was identified as the force between quarks. Currently, the strong force is associated both with the force between quarks and the force between particles made up of quarks. If those particles happen to be neutrons and protons, the strong force is often called the nuclear force because these particles make up the nucleus.

(UPI/Corbis-Bettman)

HIDEKI YUKAWA (1907–1981)

Japanese physicist Yukawa was awarded the Nobel Prize in Physics in 1949 for predicting the existence of mesons. This photograph of him at work was taken in 1950 in his office at Columbia University. Yukawa came to Columbia in 1949 after spending the early part of his career in Japan.

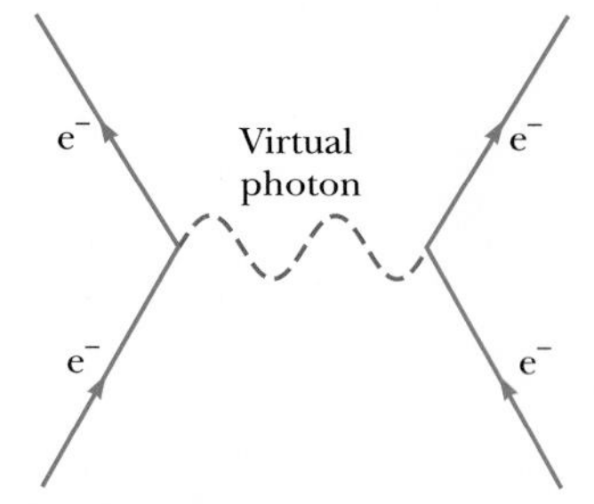

FIGURE 31.4 Feynman diagram representing a photon mediating the electromagnetic force between two electrons.

(© Shelly Gazin/CORBIS)

RICHARD FEYNMAN (1918–1988)

Inspired by Dirac, Feynman developed quantum electrodynamics, the theory of the interaction of light and matter on a relativistic and quantum basis. Feynman won the Nobel Prize in Physics in 1965. The prize was shared by Feynman, Julian Schwinger, and Sin Itiro Tomonaga. Early in his career, Feynman was a leading member of the team developing the first nuclear weapon in the Manhattan Project. Toward the end of his career, he worked on the commission investigating the 1986 *Challenger* tragedy and demonstrated the effects of cold temperatures on the rubber O-rings used in the space shuttle.

an antineutrino:

$$\pi^- \rightarrow \mu^- + \bar{\nu} \qquad [31.1]$$
$$\mu^- \rightarrow e^- + \nu + \bar{\nu}$$

Note that for chargeless particles (as well as some charged particles such as the proton), a bar over the symbol indicates an antiparticle.

The interaction between two particles can be represented in a simple qualitative graphical representation called a **Feynman diagram,** developed by American physicist Richard P. Feynman. Figure 31.4 is such a diagram for the electromagnetic interaction between two electrons approaching each other. A Feynman diagram is a qualitative graph of time in the vertical direction versus space in the horizontal direction. It is qualitative in the sense that the actual values of time and space are not important, but the overall appearance of the graph provides a representation of the process. The time evolution of the process can be approximated by starting at the bottom of the diagram and moving your eyes upward.

In the simple case of the electron–electron interaction in Figure 31.4, a photon is the field particle that mediates the electromagnetic force between the electrons. Notice that the entire interaction is represented in such a diagram as if it occurs at a single point in time. Therefore, the paths of the electrons appear to undergo a discontinuous change in direction at the moment of interaction. This representation is correct on a microscopic level over a time interval that includes the exchange of one photon. It is different from the paths produced over the much longer interval during which we watch the interaction from a macroscopic point of view. In this case, the paths would be curved (as in Fig. 31.2) due to the continuous exchange of large numbers of field particles, illustrating another aspect of the qualitative nature of Feynman diagrams.

In the electron–electron interaction, the photon, which transfers energy and momentum from one electron to the other, is called a *virtual photon* because it vanishes during the interaction without having been detected. In Chapter 28, we discussed that a photon has energy $E = hf$, where f is its frequency. Consequently, for a system of two electrons initially at rest, the system has energy $2m_ec^2$ before a virtual photon is released and energy $2m_ec^2 + hf$ after the virtual photon is released (plus any kinetic energy of the electron resulting from the emission of the photon). Is that a violation of the law of conservation of energy for an isolated system? No; this process does *not* violate the law of conservation of energy because the virtual photon has a very short lifetime Δt that makes the uncertainty in the energy $\Delta E \approx \hbar/2\,\Delta t$ of the system consisting of two electrons and the photon greater than the photon energy. Therefore, within the constraints of the uncertainty principle, the energy of the system is conserved.

Now consider a pion exchange between a proton and a neutron according to Yukawa's model (Fig. 31.5a). The energy ΔE_R needed to create a pion of mass m_π is given by Einstein's equation $\Delta E_R = m_\pi c^2$. As with the photon in Figure 31.4, the very existence of the pion would appear to violate the law of conservation of energy

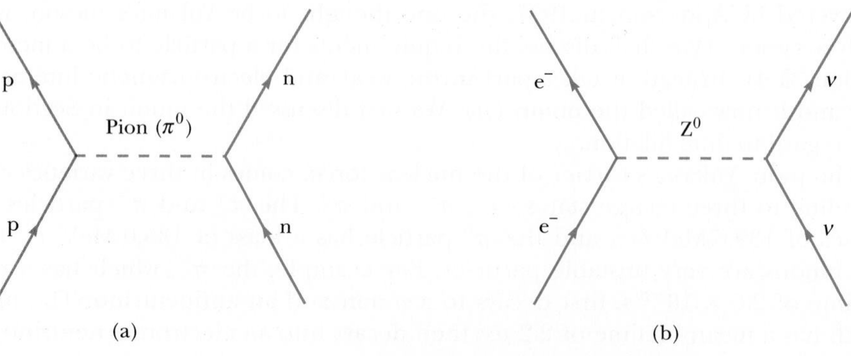

FIGURE 31.5 (a) Feynman diagram representing a proton and a neutron interacting via the nuclear force with a neutral pion mediating the force. (This model is *not* the most fundamental model for nucleon interaction.) (b) Feynman diagram for an electron and a neutrino interacting via the weak force with a Z^0 boson mediating the force.

if the particle existed for a time greater than $\Delta t \approx \hbar / 2\,\Delta E_R$ (from the uncertainty principle), where Δt is the time interval required for the pion to transfer from one nucleon to the other. Therefore,

$$\Delta t \approx \frac{\hbar}{2\,\Delta E_R} = \frac{\hbar}{2 m_\pi c^2} \qquad [31.2]$$

Because the pion cannot travel faster than the speed of light, the maximum distance d it can travel in a time interval Δt is $c\,\Delta t$. Therefore, using Equation 31.2 and $d = c\,\Delta t$, we find this maximum distance to be

$$d = c\,\Delta t \approx \frac{\hbar}{2 m_\pi c} \qquad [31.3]$$

From Chapter 30, we know that the range of the nuclear force is on the order of 10^{-15} fm. Using this value for d in Equation 31.3, we estimate the rest energy of the pion to be

$$m_\pi c^2 \approx \frac{\hbar c}{2d} = \frac{(1.055 \times 10^{-34}\,\mathrm{J \cdot s})(3.00 \times 10^{8}\,\mathrm{m/s})}{2(1 \times 10^{-15}\,\mathrm{m})}$$
$$= 1.6 \times 10^{-11}\,\mathrm{J} \approx 100\,\mathrm{MeV}$$

which corresponds to a mass of 100 MeV/c^2 (approximately 250 times the mass of the electron). This value is in reasonable agreement with the observed pion mass.

The concept we have just described is quite revolutionary. In effect, it says that a system of two nucleons can change into two nucleons plus a pion as long as it returns to its original state in a very short time interval. (Remember that this model is the older, historical one, which assumes that the pion is the field particle for the nuclear force.) Physicists often say that a nucleon undergoes *fluctuations* as it emits and absorbs pions. As we have seen, these fluctuations are a consequence of a combination of quantum mechanics (through the uncertainty principle) and special relativity (through Einstein's mass–energy relationship $E_R = mc^2$).

This section has dealt with the particles that mediate the nuclear force, pions, and the mediators of the electromagnetic force, photons. **Current ideas indicate that the nuclear force is more fundamentally described as an average or residual effect of the force between quarks,** as will be explained in Section 31.10. The graviton, which is the mediator of the gravitational force, has yet to be observed. The $W^\pm$ and Z^0 particles that mediate the weak force were discovered in 1983 by Italian physicist Carlo Rubbia (b. 1934) and his associates using a proton–antiproton collider. Rubbia and Simon van der Meer (b. 1925), both at CERN (European Organization for Nuclear Research), shared the 1984 Nobel Prize in Physics for the detection and identification of the $W^\pm$ and Z^0 particles and the development of the proton–antiproton collider. In this accelerator, protons and antiprotons undergo head-on collisions with each other. In some of the collisions, $W^\pm$ and Z^0 particles are produced, which in turn are identified by their decay products. Figure 31.5b shows a Feynman diagram for a weak interaction mediated by a Z^0 boson.

31.4 CLASSIFICATION OF PARTICLES

All particles other than field particles can be classified into two broad categories, *hadrons* and *leptons*. The criterion for separating these particles into categories is whether or not they interact via a force called the **strong force.** This force (discussed in Section 31.10) increases with separation distance, similar to the force exerted by a stretched spring. The nuclear force between nucleons in a nucleus is a particular manifestation of the strong force, but, as mentioned in Pitfall Prevention 31.2, we will use the term *strong force* in general to refer to any interaction between particles made up of more elementary units called quarks. (Today it is believed that hadrons are not elementary particles, but rather are composed of more elementary

TABLE 31.2 Some Particles and Their Properties

Category	Particle Name	Symbol	Anti-particle	Mass (MeV/c^2)	B	L_e	L_μ	L_τ	S	Lifetime(s)	Principal Decay Modes[a]
Leptons	Electron	e^-	e^+	0.511	0	+1	0	0	0	Stable	
	Electron–neutrino	ν_e	$\overline{\nu}_e$	$< 7\ \text{eV}/c^2$	0	+1	0	0	0	Stable	
	Muon	μ^-	μ^+	105.7	0	0	+1	0	0	2.20×10^{-6}	$e^-\overline{\nu}_e\nu_\mu$
	Muon–neutrino	ν_μ	$\overline{\nu}_\mu$	< 0.3	0	0	+1	0	0	Stable	
	Tau	τ^-	τ^+	1 784	0	0	0	+1	0	$<4 \times 10^{-13}$	$\mu^-\overline{\nu}_\mu\nu_\tau$, $e^-\overline{\nu}_e\nu_\tau$
	Tau–neutrino	ν_τ	$\overline{\nu}_\tau$	< 30	0	0	0	+1	0	Stable	
Hadrons											
Mesons	Pion	π^+	π^-	139.6	0	0	0	0	0	2.60×10^{-8}	$\mu^+\nu_\mu$
		π^0	Self	135.0	0	0	0	0	0	0.83×10^{-16}	2γ
	Kaon	K^+	K^-	493.7	0	0	0	0	+1	1.24×10^{-8}	$\mu^+\nu_\mu$, $\pi^+\pi^0$
		$K_s{}^0$	$\overline{K}_s{}^0$	497.7	0	0	0	0	+1	0.89×10^{-10}	$\pi^+\pi^-$, $2\pi^0$
		$K_L{}^0$	$\overline{K}_L{}^0$	497.7	0	0	0	0	+1	5.2×10^{-8}	$\pi^\pm e^\mp\overline{\nu}_e$, $3\pi^0$ $\pi^\pm\mu^\mp\overline{\nu}_\mu$
	Eta	η	Self	548.8	0	0	0	0	0	$<10^{-18}$	2γ, $3\pi^0$
		η'	Self	958	0	0	0	0	0	2.2×10^{-21}	$\eta\pi^+\pi^-$
Baryons	Proton	p	$\overline{p}$	938.3	+1	0	0	0	0	Stable	
	Neutron	n	$\overline{n}$	939.6	+1	0	0	0	0	614	$pe^-\overline{\nu}_e$
	Lambda	Λ^0	$\overline{\Lambda}^0$	1 115.6	+1	0	0	0	−1	2.6×10^{-10}	$p\pi^-$, $n\pi^0$
	Sigma	Σ^+	$\overline{\Sigma}^-$	1 189.4	+1	0	0	0	−1	0.80×10^{-10}	$p\pi^0$, $n\pi^+$
		Σ^0	$\overline{\Sigma}^0$	1 192.5	+1	0	0	0	−1	6×10^{-20}	$\Lambda^0\gamma$
		Σ^-	$\overline{\Sigma}^+$	1 197.3	+1	0	0	0	−1	1.5×10^{-10}	$n\pi^-$
	Delta	Δ^{++}	$\overline{\Delta}^{--}$	1 230	+1	0	0	0	0	6×10^{-24}	$p\pi^+$
		Δ^+	$\overline{\Delta}^-$	1 231	+1	0	0	0	0	6×10^{-24}	$p\pi^0$, $n\pi^+$
		Δ^0	$\overline{\Delta}^0$	1 232	+1	0	0	0	0	6×10^{-24}	$n\pi^0$, $p\pi^-$
		Δ^-	$\overline{\Delta}^+$	1 234	+1	0	0	0	0	6×10^{-24}	$n\pi^-$
	Xi	Ξ^0	$\overline{\Xi}^0$	1 315	+1	0	0	0	−2	2.9×10^{-10}	$\Lambda^0\pi^0$
		Ξ^-	Ξ^+	1 321	+1	0	0	0	−2	1.64×10^{-10}	$\Lambda^0\pi^-$
	Omega	Ω^-	Ω^+	1 672	+1	0	0	0	−3	0.82×10^{-10}	$\Xi^-\pi^0$, $\Xi^0\pi^-$, Λ^0K^-

[a]Notations in this column such as $p\pi^-$, $n\pi^0$ mean two possible decay modes. In this case, the two possible decays are $\Lambda^0 \rightarrow p + \pi^-$ and $\Lambda^0 \rightarrow n + \pi^0$.

units called quarks. We shall discuss quarks in Section 31.9.) Table 31.2 provides a summary of the properties of some of these particles.

Hadrons

Particles that interact through the strong force are called **hadrons.** The two classes of hadrons—*mesons* and *baryons*—are distinguished by their masses and spins.

Mesons all have zero or integer spin (0 or 1).[1] As indicated in Section 31.3, the origin of the name comes from the expectation that Yukawa's proposed meson mass would lie between the mass of the electron and the mass of the proton. Several meson masses do lie in this range, although there are heavier mesons that have masses larger than that of the proton.

All mesons are known to decay into final products including electrons, positrons, neutrinos, and photons. The pions are the lightest of the known mesons; they have masses of about 140 MeV/c^2 and a spin of 0. Another is the K meson, with a mass of approximately 500 MeV/c^2 and a spin of 0.

[1]Thus, the particle discovered by Anderson in 1937, the muon, is not a meson; the muon has spin $\frac{1}{2}$. It belongs in the *lepton* classification described shortly.

Baryons, the second class of hadrons, have masses equal to or greater than the proton mass (*baryon* means "heavy" in Greek), and their spins are always an odd half-integer value ($\frac{1}{2}$ or $\frac{3}{2}$). Protons and neutrons are baryons, as are many other particles. With the exception of the proton, all baryons decay in such a way that the end products include a proton. For example, the baryon called the Ξ hyperon decays to the Λ^0 baryon in about 10^{-10} s. The Λ^0 baryon then decays to a proton and a π^- in approximately 3×10^{-10} s.

Today it is believed that hadrons are not elementary particles, but rather are composed of more elementary units called quarks. We shall discuss quarks in Section 31.9.

Leptons

Leptons (from the Greek *leptos,* meaning "small" or "light") are a group of particles that participate in the electromagnetic (if charged) and weak interactions. All leptons have spins of $\frac{1}{2}$. Unlike hadrons, which have size and structure, **leptons appear to be truly elementary particles with no structure.**

Quite unlike hadrons, the number of known leptons is small. Currently, scientists believe that only six leptons exist: the electron, the muon, and the tau, e^-, μ^-, τ^-, and a neutrino associated with each, ν_e, ν_μ, ν_τ. The tau lepton, discovered in 1975, has a mass equal to about twice that of the proton. Direct experimental evidence for the neutrino associated with the tau was announced by the Fermi National Accelerator Laboratory (Fermilab) in July 2000. Each of these six leptons has an antiparticle.

Current studies indicate that neutrinos may have a small but nonzero mass. If they do have mass, they cannot travel at the speed of light. Also, so many neutrinos exist that their combined mass may be sufficient to cause all the matter in the Universe to eventually collapse to infinite density and then explode and create a completely new Universe! We shall discuss this concept in more detail in Section 31.12.

31.5 CONSERVATION LAWS

We have seen the importance of conservation laws for isolated systems many times in earlier chapters and have solved problems using conservation of energy, linear momentum, angular momentum, and electric charge. Conservation laws are important in understanding why certain decays and reactions occur but others do not. In general, our familiar conservation laws provide us with a set of rules that all processes must follow.

Certain new conservation laws have been identified through experimentation and are important in the study of elementary particles. The members of the isolated system change identity during a decay or reaction. The initial particles before the decay or reaction are different from the final particles afterward.

Baryon Number

Experimental results tell us that whenever a baryon is created in a nuclear reaction or decay, an antibaryon is also created. This scheme can be quantified by assigning a baryon number $B = +1$ for all baryons, $B = -1$ for all antibaryons, and $B = 0$ for all other particles. Therefore, the **law of conservation of baryon number** states that **whenever a reaction or decay occurs, the sum of the baryon numbers of the system before the process must equal the sum of the baryon numbers after the process.** An equivalent statement is that the net number of baryons remains constant in any process.

■ Conservation of baryon number

If baryon number is absolutely conserved, the proton must be absolutely stable. For example, a decay of the proton to a positron and a neutral pion would satisfy conservation of energy, momentum, and electric charge. Such a decay has never been observed, however. At present, we can say only that the proton has a half-life of at least 10^{33} years (the estimated age of the Universe is only 10^{10} years). Therefore, it is extremely unlikely that one would see a given proton undergo a decay process. If we collect a huge number of protons, however, perhaps we might see *some* proton in the collection undergo a decay, as addressed in Interactive Example 31.2.

QUICK QUIZ 31.2 Consider the following decay: $n \rightarrow \pi^+ + \pi^- + \mu^+ + \mu^-$. What conservation laws are violated by this decay? **(a)** energy **(b)** electric charge **(c)** baryon number **(d)** angular momentum **(e)** no conservation laws

QUICK QUIZ 31.3 Consider the following decay: $n \rightarrow p + \pi^-$. What conservation laws are violated by this decay? **(a)** energy **(b)** electric charge **(c)** baryon number **(d)** angular momentum **(e)** no conservation laws

EXAMPLE 31.1 Checking Baryon Numbers

A Use the law of conservation of baryon number to determine whether the reaction $p + n \rightarrow p + p + n + \bar{p}$ can occur.

Solution The left side of the equation gives a total baryon number of $1 + 1 = 2$. The right side gives a total baryon number of $1 + 1 + 1 + (-1) = 2$. Therefore, baryon number is conserved and the reaction can occur (provided the incoming proton has sufficient kinetic energy so that energy conservation is satisfied).

B Use the law of conservation of baryon numbers to determine whether the reaction $p + n \rightarrow p + p + \bar{p}$ can occur.

Solution The left side of the equation gives a total baryon number of $1 + 1 = 2$; the right side, however, gives $1 + 1 + (-1) = 1$. Because baryon number is not conserved, the reaction cannot occur.

INTERACTIVE EXAMPLE 31.2 Detecting Proton Decay

Measurements taken at the Super Kamiokande neutrino detection facility in Japan (Fig. 31.6) indicate that the half-life of protons is at least 10^{33} years.

A Estimate how long we would have to watch, on average, to see a proton in a glass of water decay.

Solution To conceptualize the problem, imagine the number of protons in a glass of water. Although this number is huge, we know that the probability of a single proton undergoing decay is small, so we would expect to wait a long time before observing a decay. Because a half-life is provided in the problem, we categorize this problem as one in which we can apply our statistical analysis techniques from Section 30.3. To analyze the problem, let us estimate that a glass contains about 250 g of water. The number of molecules of water is

$$\frac{(250 \text{ g})(6.02 \times 10^{23} \text{ molecules/mol})}{18 \text{ g/mol}} = 8.4 \times 10^{24} \text{ molecules}$$

Each water molecule contains one proton in each of its two hydrogen atoms plus eight protons in its oxygen atom, for a total of ten. Therefore, 8.4×10^{25} protons are in the glass of water. The decay constant is given by Equation 30.8:

$$\lambda = \frac{0.693}{T_{1/2}} = \frac{0.693}{10^{33} \text{ yr}} = 6.9 \times 10^{-34} \text{ yr}^{-1}$$

This result is the probability that any one proton will decay in a year. The probability that *any* proton in our glass of water will decay in the one-year interval is (Eqs. 30.5 and 30.7)

$$R = (8.4 \times 10^{25})(6.9 \times 10^{-34} \text{ yr}^{-1}) = 5.8 \times 10^{-8} \text{ yr}^{-1}$$

To finalize this part of the problem, note that we have to watch our glass of water for $1/R \approx$ 17 million years! This answer is indeed a long time, as we suspected.

B The Super Kamiokande neutrino facility contains 50 000 metric tons of water. Estimate the average time interval between detected proton

(Courtesy of ICRR [Institute for Cosmic Ray Research], University of Tokyo)

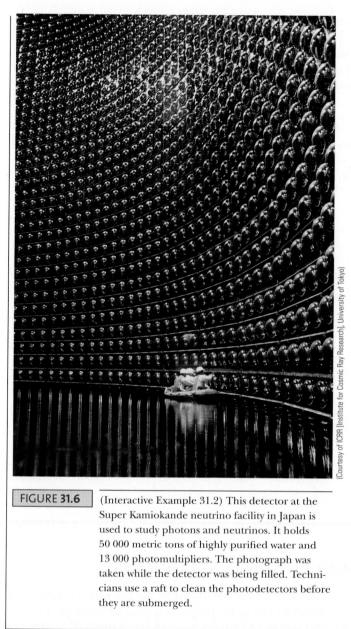

FIGURE 31.6 (Interactive Example 31.2) This detector at the Super Kamiokande neutrino facility in Japan is used to study photons and neutrinos. It holds 50 000 metric tons of highly purified water and 13 000 photomultipliers. The photograph was taken while the detector was being filled. Technicians use a raft to clean the photodetectors before they are submerged.

decays in this much water if the half-life of a proton is 10^{33} yr.

Solution We find the ratio of the number of molecules in 50 000 metric tons of water to that in the glass of water in part A, which will be same as the ratio of masses:

$$\frac{N_{\text{Kamiokande}}}{N_{\text{glass}}} = \frac{m_{\text{Kamiokande}}}{m_{\text{glass}}} = \frac{50\ 000 \text{ metric ton}}{250 \text{ g}}\left(\frac{1\ 000 \text{ kg}}{1 \text{ metric ton}}\right)\left(\frac{1\ 000 \text{ g}}{1 \text{ kg}}\right) = 2.0 \times 10^8$$

$$N_{\text{Kamiokande}} = (2.0 \times 10^8)N_{\text{glass}} = (2.0 \times 10^8)(8.4 \times 10^{24} \text{ molecules}) = 1.7 \times 10^{33} \text{ molecules}$$

Each of these molecules contains ten protons. The probability that one of these protons will decay in one year is

$$R = (10)(1.7 \times 10^{33})(6.9 \times 10^{-34} \text{ yr}^{-1}) \approx 12 \text{ yr}^{-1}$$

To finalize this part of the problem, note that the average time interval between decays is about one twelfth of a year, or approximately one month. This result is much shorter than the time interval in part A due to the tremendous amount of water in the detector facility.

PhysicsNow™ Practice the statistics of proton decay by logging into PhysicsNow at www.pop4e.com and going to Interactive Example 31.2.

Lepton Number

From observations of commonly occurring decays of the electron, muon, and tau, we arrive at three conservation laws involving lepton numbers, one for each variety of lepton. The **law of conservation of electron-lepton number** states that **the sum of the electron-lepton numbers of the system before a reaction or decay must equal the sum of the electron-lepton numbers after the reaction or decay.**

■ Conservation of electron-lepton number

The electron and the electron neutrino are assigned a positive electron-lepton number $L_e = +1$, the antileptons e^+ and $\bar{\nu}_e$ are assigned a negative electron-lepton number $L_e = -1$; all others have $L_e = 0$. For example, consider the decay of the neutron

$$\text{n} \rightarrow \text{p} + \text{e}^- + \bar{\nu}_e$$

Before the decay, the electron-lepton number is $L_e = 0$; after the decay, it is $0 + 1 + (-1) = 0$. Therefore, the electron-lepton number is conserved. It is important to recognize that the baryon number must also be conserved; which can easily be checked by noting that before the decay $B = +1$ and after the decay B is $+1 + 0 + 0 = +1$.

Similarly, when a decay involves muons, the muon-lepton number L_μ is conserved. The μ^- and the ν_μ are assigned positive numbers, $L_\mu = +1$, the antimuons μ^+ and $\bar{\nu}_\mu$ are assigned negative numbers, $L_\mu = -1$; all others have $L_\mu = 0$. Finally, the tau-lepton number L_τ is conserved, and similar assignments can be made for the tau lepton and its neutrino.

QUICK QUIZ 31.4 Consider the following decay: $\pi^0 \rightarrow \mu^- + e^+ + \nu_\mu$. What conservation laws are violated by this decay? **(a)** energy **(b)** angular momentum **(c)** electric charge **(d)** baryon number **(e)** electron-lepton number **(f)** muon-lepton number **(g)** tau-lepton number **(h)** no conservation laws

QUICK QUIZ 31.5 Suppose a claim is made that the decay of the neutron is given by $n \rightarrow p + e^-$. What conservation laws are violated by this decay? **(a)** energy **(b)** angular momentum **(c)** electric charge **(d)** baryon number **(e)** electron-lepton number **(f)** muon-lepton number **(g)** tau-lepton number **(h)** no conservation laws

EXAMPLE 31.3 Checking Lepton Numbers

A Use the law of conservation of electron-lepton number to determine if the decay scheme $\mu^- \rightarrow e^- + \bar{\nu}_e + \nu_\mu$ can occur.

Solution Because this decay involves a muon and an electron, L_μ and L_e must both be conserved. Before the decay, $L_\mu = +1$ and $L_e = 0$. After the decay, $L_\mu = 0 + 0 + 1 = +1$ and $L_e = +1 + (-1) + 0 = 0$. Therefore, both numbers are conserved, and on this basis the decay is possible.

B Use the law of conservation of electron-lepton number to determine if the decay scheme $\pi^+ \rightarrow \mu^+ + \nu_\mu + \nu_e$ can occur.

Solution Before the decay, $L_\mu = 0$ and $L_e = 0$. After the decay, $L_\mu = -1 + 1 + 0 = 0$, but $L_e = 0 + 0 + 1 = 1$. Therefore, the decay is not possible because electron-lepton number is not conserved.

31.6 STRANGE PARTICLES AND STRANGENESS

Many particles discovered in the 1950s were produced by the nuclear interaction of pions with protons and neutrons in the atmosphere. A group of these particles—the kaon (K), lambda (Λ), and sigma (Σ) particles—exhibited unusual properties in production and decay and hence were called *strange particles.*

One unusual property is that these particles are always produced in pairs. For example, when a pion collides with a proton, two neutral strange particles are produced with high probability:

$$\pi^- + p \rightarrow \Lambda^0 + K^0$$

On the other hand, the reaction $\pi^- + p \rightarrow n^0 + K^0$ in which only one of the final particles is strange never occurs, even though no conservation laws known in the 1950s are violated and the energy of the pion is sufficient to initiate the reaction.

The second peculiar feature of strange particles is that, although they are produced by the strong force at a high rate, they do not decay at a very high rate into particles that interact via the strong force. Instead, they decay very slowly, which is characteristic of the weak interaction as shown in Table 31.1. Their half-lives are in the range 10^{-10} s to 10^{-8} s; most other particles that interact via the strong force have very short lifetimes, on the order of 10^{-20} s or less.

Such observations indicate the necessity to make modifications in our model. To explain these unusual properties of strange particles, a new quantum number S, called **strangeness,** was introduced into our model of elementary particles, together with a new conservation law. The strangeness numbers for some particles are given in Table 31.2. The production of strange particles in pairs is handled by assigning $S = +1$ to one of the particles and $S = -1$ to the other. All nonstrange particles are assigned strangeness $S = 0$. The **law of conservation of strangeness** states that **whenever a reaction or decay occurs via the strong force, the sum of the strangeness numbers of the system before the process must equal the sum of the strangeness numbers after the process.**

■ Conservation of strangeness

The low decay rate of strange particles can be explained by assuming that the nuclear and electromagnetic interactions obey the law of conservation of strangeness, but the weak interaction does not. Because the decay reaction involves the loss of one strange particle, it violates strangeness conservation and hence proceeds slowly via the weak interaction.

EXAMPLE 31.4 **Is Strangeness Conserved?**

A Determine whether the following reaction occurs on the basis of conservation of strangeness.

$$\pi^0 + \text{n} \rightarrow \text{K}^+ + \Sigma^-$$

Solution From Table 31.2, we see that the initial system has strangeness $S = 0 + 0 = 0$. Because the strangeness of the K^+ is $S = +1$ and the strangeness of the Σ^- is $S = -1$, the strangeness of the final sysstem is $+1 - 1 = 0$. Therefore, strangeness is conserved and the reaction is allowed.

B Show that the following reaction does not conserve strangeness.

$$\pi^- + \text{p} \rightarrow \pi^- + \Sigma^+$$

Solution The initial system has strangeness $S = 0 + 0 = 0$, and the final system has strangeness $S = 0 + (-1) = -1$. Therefore, strangeness is not conserved.

31.7 MEASURING PARTICLE LIFETIMES

The bewildering array of entries in Table 31.2 leaves one yearning for firm ground. In fact, it is natural to wonder about an entry, for example, that shows a particle (Σ^0) that exists for 10^{-20} s and has a mass of 1 192.5 MeV/c^2. How is it possible to detect a particle that exists for only 10^{-20} s?

Most particles are unstable and are created in nature only rarely, in cosmic ray showers. In the laboratory, however, large numbers of these particles are created in controlled collisions between high-energy particles and a suitable target. The incident particles must have very high energy, and it takes a considerable time interval for electromagnetic fields to accelerate particles to high energies. Therefore, stable charged particles such as electrons or protons generally make up the incident beam. Similarly, targets must be simple and stable, and the simplest target, hydrogen, serves nicely as both target (the proton) and detector.

Figure 31.7 shows a typical event in which hydrogen in a bubble chamber served as both target source and detector. (A bubble chamber is a device in which the tracks of charged particles are made visible in liquid hydrogen that is maintained near its boiling point.) Many parallel tracks of negative pions are visible entering the photograph from the bottom. As the labels in the inset drawing show, one of the pions has hit a stationary proton in the hydrogen, producing two strange particles, Λ^0 and K^0, according to the reaction

$$\pi^- + \text{p} \rightarrow \Lambda^0 + \text{K}^0$$

Neither neutral strange particle leaves a track, but their subsequent decays into charged particles can be clearly seen as indicated in Figure 31.7. A magnetic field directed into the plane of the photograph causes the track of each charged particle to curve, and from the measured curvature one can determine the particle's charge

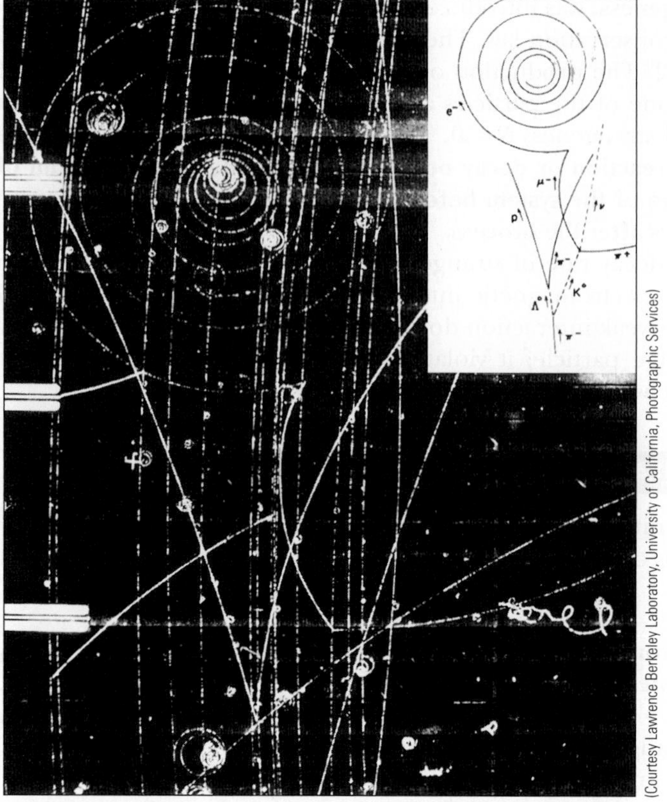

(Courtesy Lawrence Berkeley Laboratory, University of California, Photographic Services)

FIGURE 31.7 This bubble-chamber photograph shows many events, and the inset is a drawing of identified tracks. The strange particles Λ^0 and K^0 are formed at the bottom as the π^- interacts with a proton according to $\pi^- + p \rightarrow \Lambda^0 + K^0$. (Note that the neutral particles leave no tracks, as indicated by the dashed lines.) The Λ^0 and K^0 then decay according to $\Lambda^0 \rightarrow \pi^- + p$ and $K^0 \rightarrow \pi^0 + \mu^- + \bar{\nu}_\mu$.

and linear momentum. If the mass and momentum of the incident particle are known, we can then usually calculate the product particle mass, kinetic energy, and speed from conservation of momentum and energy. Finally, by combining a product particle's speed with a measurable decay track length, we can calculate the product particle's lifetime. Figure 31.7 shows that sometimes one can use this lifetime technique even for a neutral particle, which leaves no track. As long as the beginning and end points of the missing track are known as well as the particle speed, one can infer the missing track length and find the lifetime of the neutral particle.

Resonance Particles

With clever experimental technique and much effort, decay track lengths as short as 10^{-6} m can be measured. Thus, lifetimes as short as 10^{-16} s can be measured for high-energy particles traveling at about the speed of light. We arrive at this result by assuming that a decaying particle travels 1 μm at a speed of $0.99c$ in the reference frame of the laboratory, yielding a lifetime of $\Delta t_{\text{lab}} = 1 \times 10^{-6}\ \text{m}/0.99c \approx 3.4 \times 10^{-15}$ s. This result is not our final one, however, because we must account for the relativistic effects of time dilation. Because the proper lifetime Δt_p as measured in the decaying particle's reference frame is shorter than the laboratory frame value Δt_{lab} by a factor of $\sqrt{1 - (v^2/c^2)}$ (see Eq. 9.6), we can calculate the proper lifetime:

$$\Delta t_p = \Delta t_{\text{lab}} \sqrt{1 - \frac{v^2}{c^2}} = (3.4 \times 10^{-15}\ \text{s}) \sqrt{1 - \frac{(0.99c)^2}{c^2}} = 4.8 \times 10^{-16}\ \text{s}$$

Unfortunately, even with Einstein's help, the best answer we can obtain with the track length method is several orders of magnitude away from lifetimes of 10^{-20} s. How then can we detect the presence of particles that exist for time intervals like 10^{-20} s? For such short-lived particles, known as **resonance particles,** all we can do is infer their masses, their lifetimes, and, indeed, their very existence from data on their decay products.

31.8 FINDING PATTERNS IN THE PARTICLES

A tool scientists use to help understand nature is the detection of patterns in data. One of the best examples of the use of this tool is the development of the periodic table, which provides fundamental understanding of the chemical behavior of the elements. The periodic table explains how more than a hundred elements can be formed from three particles: the electron, proton, and neutron. The number of observed particles and resonances observed by particle physicists is even larger than the number of elements. Is it possible that a small number of entities could exist from which all these particles could be built? Motivated by the success of the periodic table, let us explore the historical search for patterns among the particles.

Many classification schemes have been proposed for grouping particles into families. Consider, for instance, the baryons listed in Table 31.2 that have spins of $\frac{1}{2}$: p, n, Λ^0, Σ^+, Σ^0, Σ^-, Ξ^0, and Ξ^-. If we plot strangeness versus charge for these baryons using a sloping coordinate system, as in Figure 31.8a, we observe a fascinating pattern. Six of the baryons form a hexagon, and the remaining two are at the hexagon's center.[2]

As a second example, consider the following nine spin-zero mesons listed in Table 31.2: π^+, π^0, π^-, K^+, K^0, K^-, η, η', and the antiparticle $\overline{K}^0$. Figure 31.8b is a plot of strangeness versus charge for this family. Again, a hexagonal pattern emerges. In this case, each particle on the perimeter of the hexagon lies opposite its antiparticle, and the remaining three (which form their own antiparticles) are at its center. These and related symmetric patterns were developed independently in 1961 by Murray Gell-Mann and Yuval Ne'eman (b. 1925). Gell-Mann called the patterns the **eightfold way,** after the eightfold path to nirvana in Buddhism.

MURRAY GELL-MANN (b. 1929)
American physicist Gell-Mann was awarded the Nobel Prize in Physics in 1969 for his theoretical studies dealing with subatomic particles.

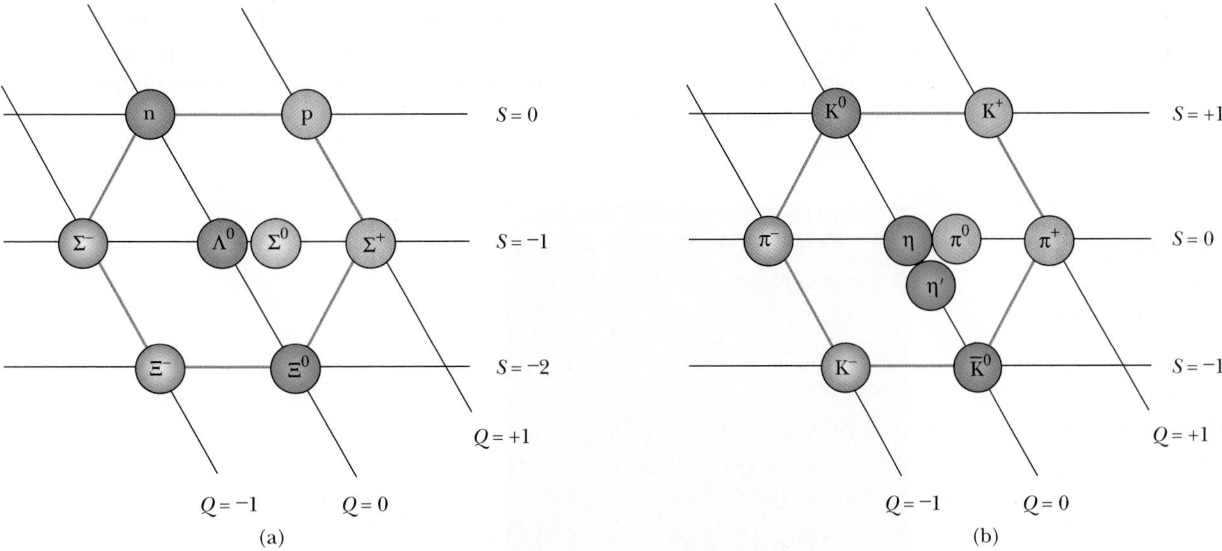

FIGURE 31.8 (a) The hexagonal eightfold-way pattern for the eight spin-$\frac{1}{2}$ baryons. This strangeness-versus-charge plot uses a sloping axis for charge number Q and a horizontal axis for strangeness S. (b) The eightfold-way pattern for the nine spin-zero mesons.

[2]The reason for the sloping coordinate system is so that a *regular* hexagon is formed, one with equal sides. If a normal orthogonal coordinate system is used, the pattern still appears, but the hexagonal shape does not have equal sides. Try it!

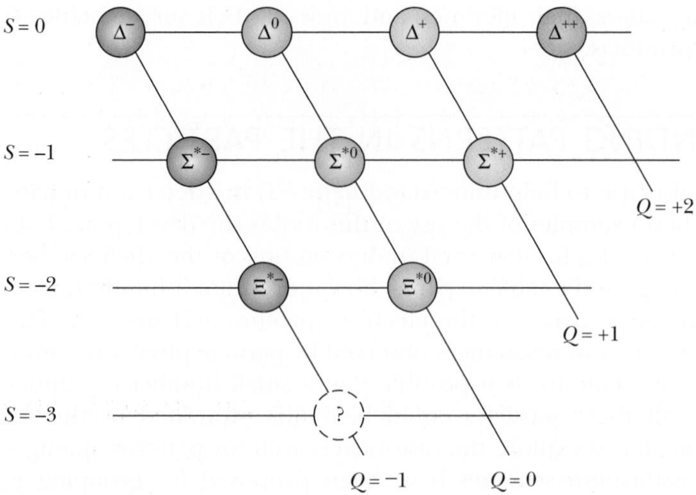

FIGURE 31.9 The pattern for the higher-mass, spin-$\frac{3}{2}$ baryons known at the time the pattern was proposed. The three Σ^* and two Ξ^* particles are excited states of the corresponding spin-$\frac{1}{2}$ particles in Figure 31.8. These excited states have higher mass and spin $\frac{3}{2}$. The absence of a particle in the bottom position was evidence of a new particle yet to be discovered, the Ω^-.

Groups of baryons and mesons can be displayed in many other symmetric patterns within the framework of the eightfold way. For example, the family of spin-$\frac{3}{2}$ baryons known in 1961 contains nine particles arranged in a pattern like that of the pins in a bowling alley as in Figure 31.9. [The particles Σ^{*+}, Σ^{*0}, Σ^{*-}, Ξ^{*0}, and Ξ^{*-} are excited states of the particles Σ^+, Σ^0, Σ^-, Ξ^0, and Ξ^-. In these higher-energy states, the spins of the three quarks (see Section 31.9) making up the particle are aligned so that the total spin of the particle is $\frac{3}{2}$.] When this pattern was proposed, an empty spot occurred in it (at the bottom position), corresponding to a particle that had never been observed. Gell-Mann predicted that the missing particle, which he called the omega minus (Ω^-), should have spin $\frac{3}{2}$, charge -1, strangeness -3, and rest energy of approximately 1 680 MeV. Shortly thereafter, in 1964, scientists at the Brookhaven National Laboratory found the missing particle through careful analyses of bubble-chamber photographs (Fig. 31.10) and confirmed all its predicted properties.

The prediction of the missing particle from the eightfold way has much in common with the prediction of missing elements in the periodic table. Whenever a vacancy occurs in an organized pattern of information, experimentalists have a guide for their investigations.

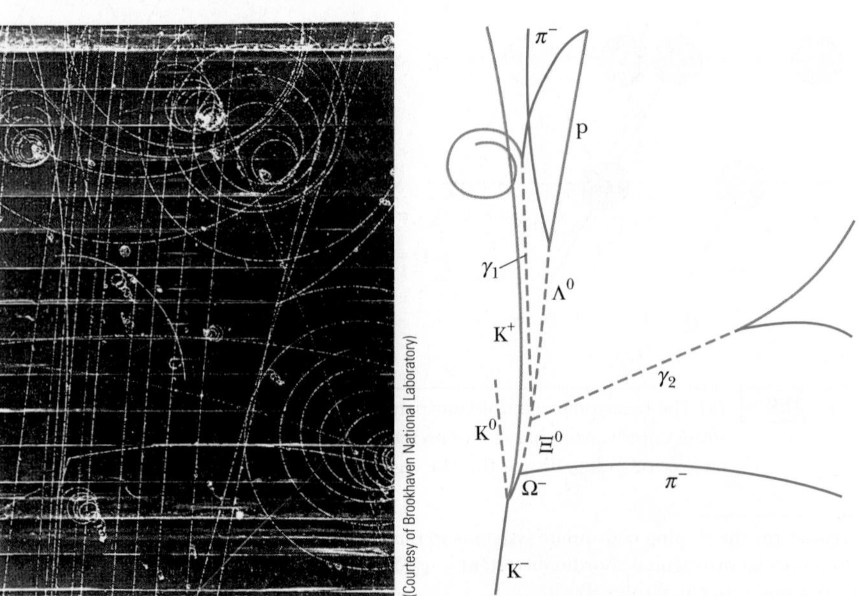

FIGURE 31.10 Discovery of the Ω^- particle. The photograph on the left shows the original bubble-chamber tracks. The drawing on the right isolates the tracks of the important events. The K^- particle at the bottom collides with a proton to produce the first detected Ω^- particle plus a K^0 and a K^+.

31.9 QUARKS

As we have noted, leptons appear to be truly elementary particles because they occur in a small number of types, have no measurable size or internal structure, and do not seem to break down to smaller units. Hadrons, on the other hand, are complex particles having size and structure. The existence of the eightfold-way patterns suggests that hadrons have a more elemental substructure. Furthermore, we know that hundreds of types of hadrons exist and that many of them decay into other hadrons. These facts strongly suggest that hadrons cannot be truly elementary. In this section, we show that the complexity of hadrons can be explained by a simple substructure.

The Original Quark Model: A Structural Model for Hadrons

In 1963, Gell-Mann and George Zweig (b. 1937) independently proposed that hadrons have a more elemental substructure. According to their structural model, all hadrons are composite systems of two or three fundamental constituents called **quarks** (pronounced to rhyme with *forks*). (Gell-Mann borrowed the word *quark* from the passage "Three quarks for Muster Mark" in James Joyce's *Finnegan's Wake.*) The model proposes that three types of quarks exist, designated by the symbols u, d, and s. They are given the arbitrary names **up, down,** and **strange.** The various types of quarks are called **flavors.** Baryons consist of three quarks, and mesons consist of a quark and an antiquark. Active Figure 31.11 is a pictorial representation of the quark composition of several hadrons.

An unusual property of quarks is that they carry a fractional electronic charge. The u, d, and s quarks have charges of $+\frac{2}{3}e$, $-\frac{1}{3}e$, and $-\frac{1}{3}e$, respectively, where e is the elementary charge 1.6×10^{-19} C. These and other properties of quarks and antiquarks are given in Table 31.3. Notice that quarks have spin $\frac{1}{2}$, which means that all quarks are *fermions,* defined as any particle having half-integral spin. As Table 31.3 shows, associated with each quark is an antiquark of opposite charge, baryon number, and strangeness.

The composition of all hadrons known when Gell-Mann and Zweig presented their models can be completely specified by three simple rules:

- A meson consists of one quark and one antiquark, giving it a baryon number of 0, as required.
- A baryon consists of three quarks.
- An antibaryon consists of three antiquarks.

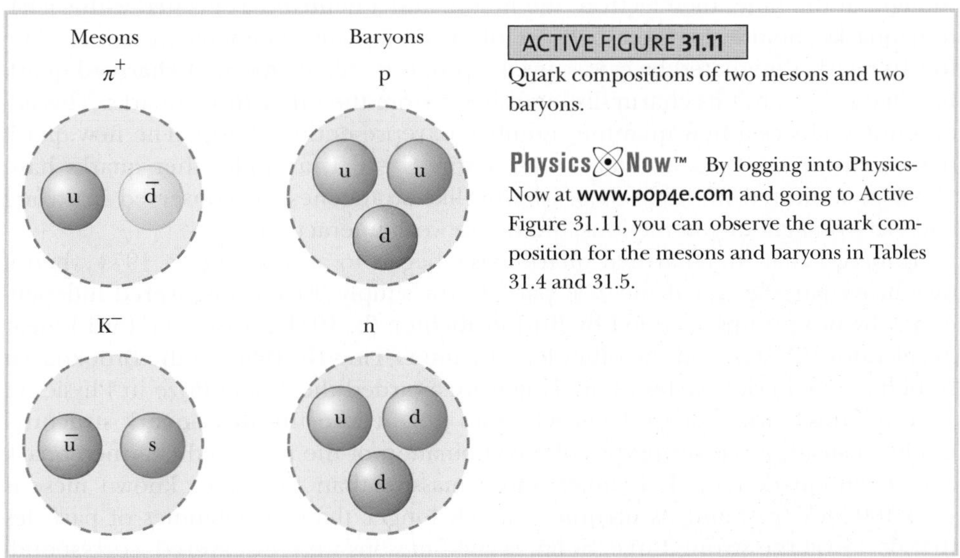

ACTIVE FIGURE 31.11
Quark compositions of two mesons and two baryons.

Physics Now™ By logging into PhysicsNow at **www.pop4e.com** and going to Active Figure 31.11, you can observe the quark composition for the mesons and baryons in Tables 31.4 and 31.5.

TABLE 31.3 Properties of Quarks and Antiquarks

Quarks

Name	Symbol	Spin	Charge	Baryon Number	Strangeness	Charm	Bottomness	Topness
Up	u	$\frac{1}{2}$	$+\frac{2}{3}e$	$\frac{1}{3}$	0	0	0	0
Down	d	$\frac{1}{2}$	$-\frac{1}{3}e$	$\frac{1}{3}$	0	0	0	0
Strange	s	$\frac{1}{2}$	$-\frac{1}{3}e$	$\frac{1}{3}$	-1	0	0	0
Charmed	c	$\frac{1}{2}$	$+\frac{2}{3}e$	$\frac{1}{3}$	0	$+1$	0	0
Bottom	b	$\frac{1}{2}$	$-\frac{1}{3}e$	$\frac{1}{3}$	0	0	$+1$	0
Top	t	$\frac{1}{2}$	$+\frac{2}{3}e$	$\frac{1}{3}$	0	0	0	$+1$

Antiquarks

Name	Symbol	Spin	Charge	Baryon Number	Strangeness	Charm	Bottomness	Topness
Anti-up	$\overline{\text{u}}$	$\frac{1}{2}$	$-\frac{2}{3}e$	$-\frac{1}{3}$	0	0	0	0
Anti-down	$\overline{\text{d}}$	$\frac{1}{2}$	$+\frac{1}{3}e$	$-\frac{1}{3}$	0	0	0	0
Anti-strange	$\overline{\text{s}}$	$\frac{1}{2}$	$+\frac{1}{3}e$	$-\frac{1}{3}$	$+1$	0	0	0
Anti-charmed	$\overline{\text{c}}$	$\frac{1}{2}$	$-\frac{2}{3}e$	$-\frac{1}{3}$	0	-1	0	0
Anti-bottom	$\overline{\text{b}}$	$\frac{1}{2}$	$+\frac{1}{3}e$	$-\frac{1}{3}$	0	0	-1	0
Anti-top	$\overline{\text{t}}$	$\frac{1}{2}$	$-\frac{2}{3}e$	$-\frac{1}{3}$	0	0	0	-1

The theory put forth by Gell-Mann and Zweig is referred to as the *original quark model.*

QUICK QUIZ 31.6 Using a coordinate system like that in Figure 31.8, draw an eightfold-way diagram for the three quarks in the original quark model.

Charm and Other Developments

Although the original quark model was highly successful in classifying particles into families, some discrepancies were evident between predictions of the model and certain experimental decay rates. It became clear that the structural model needed to be modified to remove these discrepancies. Consequently, several physicists proposed a fourth quark in 1967. They argued that if four leptons exist (as was thought at the time: the electron, the muon, and a neutrino associated with each), four quarks should also exist because of an underlying symmetry in nature. The fourth quark, designated by c, was given a property called **charm.** A **charmed** quark has charge $+\frac{2}{3}e$, but its charm distinguishes it from the other three quarks. This addition introduces a new quantum number *C*, representing charm. The new quark has charm $C = +1$, its antiquark has charm $C = -1$, and all other quarks have $C = 0$ as indicated in Table 31.3. Charm, like strangeness, is conserved in strong and electromagnetic interactions, but not in weak interactions.

Evidence that the charmed quark exists began to accumulate in 1974 when a new heavy particle called the J/Ψ particle (or simply Ψ) was discovered independently by two groups, one led by Burton Richter (b. 1931) at the Stanford Linear Accelerator (SLAC), and the other led by Samuel Ting (b. 1936) at the Brookhaven National Laboratory. Richter and Ting were awarded the Nobel Prize in Physics in 1976 for this work. The J/Ψ particle does not fit into the three-quark structural model; instead, it has properties of a combination of the proposed charmed quark and its antiquark ($c\overline{c}$). It is much more massive than the other known mesons ($\sim 3\ 100\ \text{MeV}/c^2$), and its lifetime is much longer than the lifetimes of particles that decay via the strong force. Soon, related mesons were discovered, corresponding to such quark combinations as $\overline{c}d$ and $c\overline{d}$, which all have large masses and long

TABLE 31.4 Quark Composition of Mesons

		Antiquarks									
		$\bar{b}$		$\bar{c}$		$\bar{s}$		$\bar{d}$		$\bar{u}$	
Quarks	b	Υ	($\bar{b}$b)	B_c^-	($\bar{c}$b)	$\bar{B}_s^0$	($\bar{s}$b)	$\bar{B}^0$	($\bar{d}$b)	B^-	($\bar{u}$b)
	c	B_c^+	($\bar{b}$c)	J/Ψ	($\bar{c}$c)	D_s^+	($\bar{s}$c)	D^+	($\bar{d}$c)	D^0	($\bar{u}$c)
	s	B_s^0	($\bar{b}$s)	D_s^-	($\bar{c}$s)	η, η'	($\bar{s}$s)	$\bar{K}^0$	($\bar{d}$s)	K^-	($\bar{u}$s)
	d	B^0	($\bar{b}$d)	D^-	($\bar{c}$d)	K^0	($\bar{s}$d)	π^0, η, η'	($\bar{d}$d)	π^-	($\bar{u}$d)
	u	B^+	($\bar{b}$u)	$\bar{D}^0$	($\bar{c}$u)	K^+	($\bar{s}$u)	π^+	($\bar{d}$u)	π^0, η, η'	($\bar{u}$u)

Note: The top quark does not form mesons because it decays too quickly.

lifetimes. The existence of these new mesons provided firm evidence for the fourth quark flavor.

In 1975, researchers at Stanford University reported strong evidence for the tau (τ) lepton with a mass of 1 784 MeV/c^2. It is the fifth type of lepton to be discovered, which led physicists to propose that more flavors of quarks may exist, based on symmetry arguments similar to those leading to the proposal of the charmed quark. These proposals led to more elaborate quark models and the prediction of two new quarks: **top** (t) and **bottom** (b). To distinguish these quarks from the original four, quantum numbers called *topness* and *bottomness* (with allowed values +1, 0, −1) are assigned to all quarks and antiquarks (Table 31.3). In 1977, researchers at the Fermi National Laboratory, under the direction of Leon Lederman (b. 1922), reported the discovery of a very massive new meson Υ whose composition is considered to be b$\bar{b}$, providing evidence for the bottom quark. In March 1995, researchers at Fermilab announced the discovery of the top quark (supposedly the last of the quarks to be found), with a mass of 173 GeV/c^2.

Table 31.4 lists the quark compositions of mesons formed from the up, down, strange, charmed, and bottom quarks. Table 31.5 shows the quark combinations for the baryons listed in Table 31.2. Note that only two flavors of quarks, u and d, are contained in all hadrons encountered in ordinary matter (protons and neutrons).

You are probably wondering if such discoveries will ever end. How many "building blocks" of matter really exist? At present, physicists believe that the fundamental particles in nature are six quarks and six leptons (together with their antiparticles)

TABLE 31.5 Quark Composition of Several Baryons

Particle	Quark Composition
p	uud
n	udd
Λ^0	uds
Σ^+	uus
Σ^0	uds
Σ^-	dds
Δ^{++}	uuu
Δ^+	uud
Δ^0	udd
Δ^-	ddd
Ξ^0	uss
Ξ^-	dss
Ω^-	sss

Note: Some baryons have the same quark composition, such as the p and the Δ^+ and the n and the Δ^0. In these cases, the Δ particles are considered to be excited states of the proton and neutron.

TABLE 31.6 The Elementary Particles and Their Rest Energies and Charges

Particle	Rest Energy	Charge
Quarks		
u	360 MeV	$+\frac{2}{3}e$
d	360 MeV	$-\frac{1}{3}e$
s	540 MeV	$-\frac{1}{3}e$
c	1 500 MeV	$+\frac{2}{3}e$
b	5 GeV	$-\frac{1}{3}e$
t	173 GeV	$+\frac{2}{3}e$
Leptons		
e^-	511 keV	$-e$
μ^-	105.7 MeV	$-e$
τ^-	1 784 MeV	$-e$
ν_e	<7 eV	0
ν_μ	<0.3 MeV	0
ν_τ	<30 MeV	0

listed in Table 31.6 and the field particles listed in Table 31.1. Table 31.6 lists the rest energies and charges of the quarks and leptons.

Despite extensive experimental effort, no isolated quark has ever been observed. Physicists now believe that quarks are permanently confined inside hadrons because of the strong force, which prevents them from escaping. Current efforts are under way to form a **quark-gluon plasma,** a state of matter in which the quarks are freed from neutrons and protons. In 2000, scientists at CERN announced evidence for a quark-gluon plasma formed by colliding lead nuclei. Experiments continue at CERN as well as at the Relativistic Heavy Ion Collider (RHIC) at Brookhaven to verify the production of a quark-gluon plasma.

■ Thinking Physics 31.2

We have seen a law of conservation of *lepton number* and a law of conservation of *baryon number*. Why isn't there a law of conservation of *meson number*?

Reasoning We can argue from the point of view of creating particle–antiparticle pairs from available energy. (Review pair production in Section 31.2.) If energy is converted to rest energy of a lepton–antilepton pair, no net change occurs in lepton number because the lepton has a lepton number of $+1$ and the antilepton -1. Energy can also be transformed into rest energy of a baryon–antibaryon pair. The baryon has baryon number $+1$, the antibaryon -1, and no net change in baryon number occurs.

Now, however, suppose energy is transformed into rest energy of a quark–antiquark pair. By definition in quark theory, a quark–antiquark pair *is a meson.* Therefore, we have created a meson from energy because no meson existed before, now one does. Therefore, meson number is not conserved. With more energy, we can create more mesons, with no restriction from a conservation law other than that of energy. ■

31.10 COLORED QUARKS

Shortly after the concept of quarks was proposed, scientists recognized that certain particles had quark compositions that violated the Pauli exclusion principle. As noted in Pitfall Prevention 29.4 in Chapter 29, all fermions obey the exclusion principle. Because all quarks are fermions with spin $\frac{1}{2}$, they are expected to follow the exclusion principle. One example of a particle that appears to violate the exclusion principle is the Ω^- (sss) baryon that contains three s quarks having parallel spins, giving it a total spin of $\frac{3}{2}$. Other examples of baryons that have identical quarks with parallel spins are the Δ^{++} (uuu) and the Δ^- (ddd). To resolve this problem, in 1965 Moo-Young Han (b. 1934) and Yoichiro Nambu (b. 1921) suggested a modification of the structural model of quarks in which quarks possess a new property called **color** or **color charge.** This property is similar in many respects to electric charge except that it occurs in three varieties called **red, green,** and **blue.** The antiquarks have the colors **antired, antigreen,** and **antiblue.** To satisfy the exclusion principle, all three quarks in a baryon must have different colors. Just as a combination of actual colors of light can produce the neutral color white, a combination of three quarks with different colors is also described as white, or colorless. A meson consists of a quark of one color and an antiquark of the corresponding anticolor. The result is that baryons and mesons are always colorless (or white).

Although the concept of color in the quark model was originally conceived to satisfy the exclusion principle, it also provided a better theory for explaining certain experimental results. For example, the modified theory correctly predicts the lifetime of the π^0 meson. The theory of how quarks interact with one another is called **quantum chromodynamics,** or QCD, to parallel quantum electrodynamics (the theory of interaction between electric charges). In QCD, the quark is said to carry a **color charge,** in analogy to electric charge. The strong force between quarks is often called the **color force.**

The color force between quarks is analogous to the electric force between charges; like colors repel and opposite colors attract. Therefore, two green quarks repel each other, but a green quark is attracted to an antigreen quark. The attraction between quarks of opposite color to form a meson ($q\bar{q}$) is indicated in Figure 31.12a. Differently colored quarks also attract one another, but with less strength than opposite colors of quark and antiquark. For example, a cluster of red, blue, and green quarks all attract one another to form a baryon as indicated in Figure 31.12b. Therefore, every baryon contains three quarks of three different colors.

As stated earlier, the strong force between quarks is carried by massless particles that travel at the speed of light called **gluons.** According to QCD, there are eight gluons, all carrying two color charges, a color and an anticolor such as a "blue–antired" gluon. When a quark emits or absorbs a gluon, its color changes. For example, a blue quark that emits a blue–antired gluon becomes a red quark, and a red quark that absorbs this gluon becomes a blue quark.

Figure 31.13a shows the interaction between a neutron and a proton by means of Yukawa's pion, in this case a π^-. In Figure 31.13a, the charged pion carries charge from one nucleon to the other, so the nucleons change identities and the

PITFALL PREVENTION 31.3

Color is not really color The description of color for a quark has nothing to do with visual sensation from light. It is simply a convenient name for a property analogous to electric charge, except that we need to combine three types of this property to achieve neutrality.

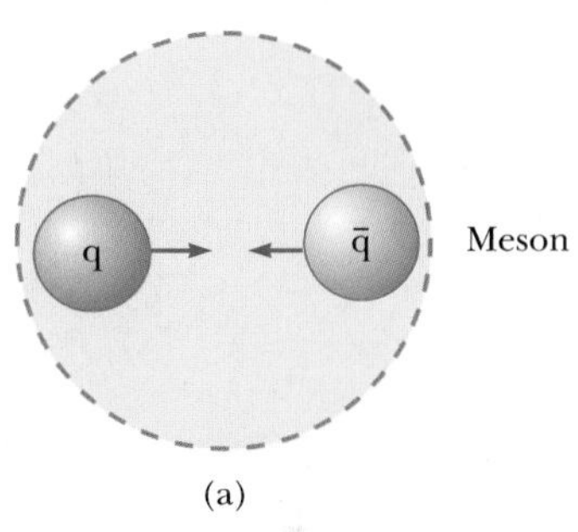

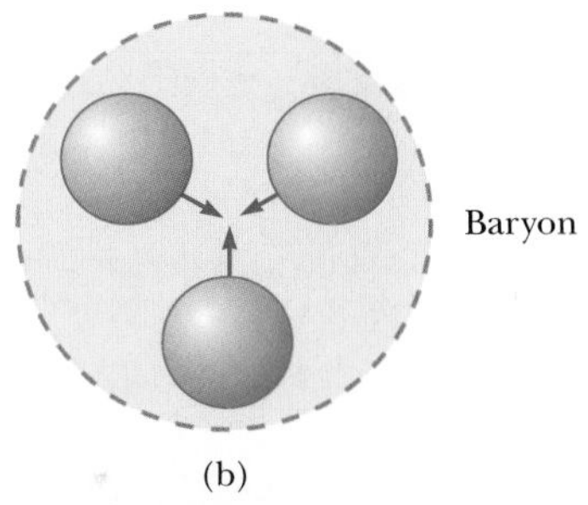

FIGURE 31.12 (a) A green quark is attracted to an antigreen quark, forming a meson whose quark structure is ($q\bar{q}$). (b) Three quarks of different colors attract one another to form a baryon.

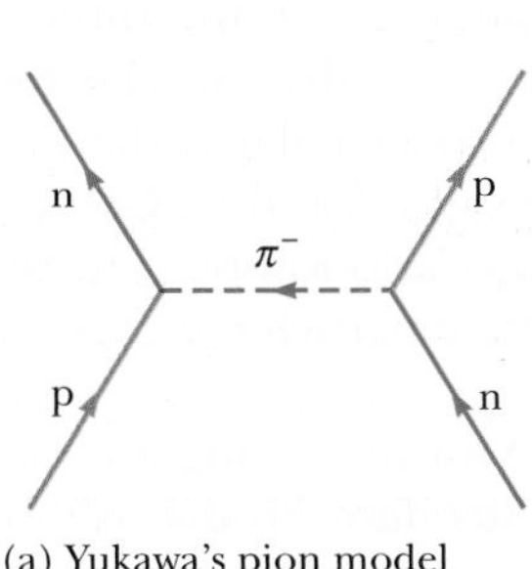

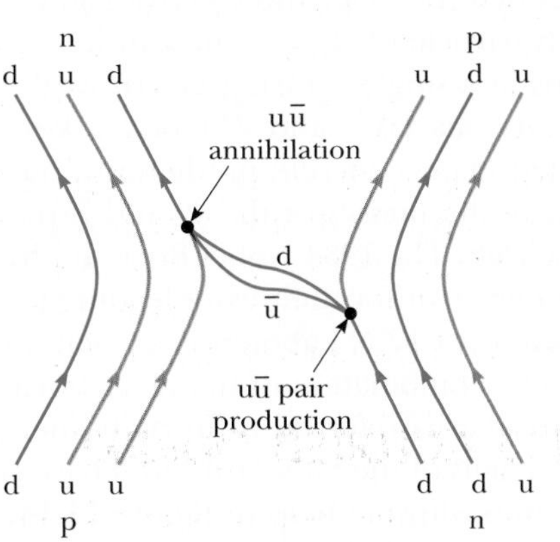

FIGURE 31.13 (a) A nuclear interaction between a proton and a neutron explained in terms of Yukawa's pion-exchange model. Because the pion carries charge, the proton and neutron switch identities. (b) The same interaction, explained in terms of quarks and gluons. Note that the exchanged $\bar{u}d$ quark pair makes up a π^- meson.

proton becomes a neutron and the neutron becomes a proton. (This process differs from Fig. 31.5, in which the field particle is a π^0, resulting in no transfer of charge from one nucleon to the other.)

Let us look at the same interaction from the viewpoint of the quark model shown in Figure 31.13b. In this Feynman diagram, the proton and neutron are represented by their quark constituents. Each quark in the neutron and proton is continuously emitting and absorbing gluons. The energy of a gluon can result in the creation of quark–antiquark pairs. This is similar to the creation of electron–positron pairs in pair production, which we investigated in Section 31.2. When the neutron and proton approach to within 1 to 2 fm of each other, these gluons and quarks can be exchanged between the two nucleons, and such exchanges produce the strong force. Figure 31.13b depicts one possibility for the process shown in Figure 31.13a. A down quark in the neutron on the right emits a gluon. The energy of the gluon is then transformed to create a $u\bar{u}$ pair. The u quark stays within the nucleon (which has now changed to a proton), and the recoiling d quark and the $\bar{u}$ antiquark are transmitted to the proton on the left side of the diagram. Here the $\bar{u}$ annihilates a u quark within the proton and the d is captured. Therefore, the net effect is to change a u quark to a d quark, and the proton has changed to a neutron.

As the d quark and $\bar{u}$ antiquark in Figure 31.13 transfer between the nucleons, the d and $\bar{u}$ exchange gluons with each other and can be considered to be bound to each other by means of the strong force. If we look back at Table 31.4, we see that this combination is a π^-, which is Yukawa's field particle! Therefore, the quark model of interactions between nucleons is consistent with the pion-exchange model.

31.11 THE STANDARD MODEL

Scientists now believe that there are three classifications of truly elementary particles: leptons, quarks, and field particles. These three particles are further classified as either fermions or bosons. Quarks and leptons have spin $\frac{1}{2}$ and hence are fermions, whereas the field particles have integral spin of 1 or higher and are bosons.

Recall from Section 31.1 that the weak force is believed to be mediated by the W^+, W^-, and Z^0 bosons. These particles are said to have *weak charge* just as quarks have color charge. Therefore, each elementary particle can have mass, electric charge, color charge, and weak charge. Of course, one or more of these could be zero.

In 1979, Sheldon Glashow (b. 1932), Abdus Salam (1926–1996), and Steven Weinberg (b. 1933) won the Nobel Prize in Physics for developing a theory that unified the electromagnetic and weak interactions. This **electroweak theory** postulates that the weak and electromagnetic interactions have the same strength at very high particle energies. The two interactions are viewed as two different manifestations of a single unifying electroweak interaction. The photon and the three massive bosons ($W^\pm$ and Z^0) play a key role in the electroweak theory. The theory makes many concrete predictions, but perhaps the most spectacular is the prediction of the masses of the W and Z particles at about 82 GeV/c^2 and 93 GeV/c^2, respectively. The 1984 Nobel Prize in Physics was awarded to Carlo Rubbia and Simon van der Meer for their work leading to the discovery of these particles at these energies at the CERN Laboratory in Geneva, Switzerland.

The combination of the electroweak theory and QCD for the strong interaction form what is referred to in high-energy physics as the **Standard Model.** Although the details of the Standard Model are complex, its essential ingredients can be summarized with the help of Figure 31.14. (The Standard Model does not include the gravitational force at present; we include gravity in Fig. 31.14, however, because physicists hope to eventually incorporate this force into a unified theory.) This

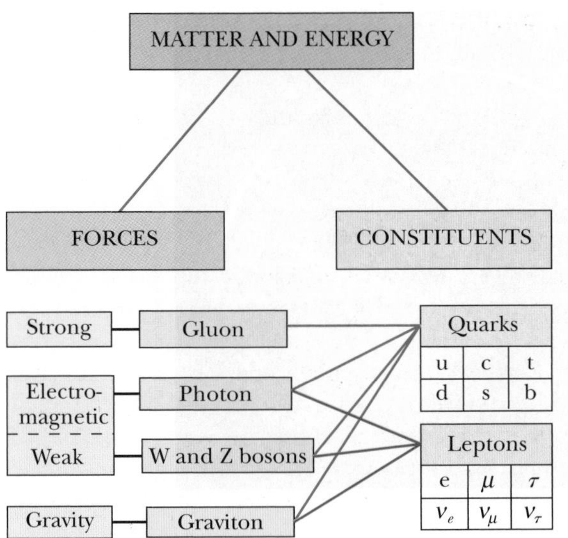

FIGURE 31.14 The Standard Model of particle physics.

diagram shows that quarks participate in all the fundamental forces and that leptons participate in all except the strong force.

The Standard Model does not answer all questions. A major question that is still unanswered is why, of the two mediators of the electroweak interaction, the photon has no mass but the W and Z bosons do. Because of this mass difference, the electromagnetic and weak forces are quite distinct at low energies but become similar at very high energies, when the rest energy is negligible relative to the total energy. The behavior as one goes from high to low energies is called *symmetry breaking* because the forces are similar, or symmetric, at high energies but are very different at low energies. The nonzero rest energies of the W and Z bosons raise the question of the origin of particle masses. To resolve this problem, a hypothetical particle called the **Higgs boson,** which provides a mechanism for breaking the electroweak symmetry, has been proposed. The Standard Model, modified to include the Higgs mechanism, provides a logically consistent explanation of the massive nature of the W and Z bosons. Unfortunately, the Higgs boson has not yet been found, but physicists know that its rest energy should be less than 1 TeV. To determine whether the Higgs boson exists, two quarks of at least 1 TeV of energy must collide. Calculations show, however, that this process requires injecting 40 TeV of energy within the volume of a proton.

Scientists are convinced that because the energy available in conventional accelerators using fixed targets is too limited, it is necessary to build colliding-beam accelerators called **colliders.** The concept of colliders is straightforward. Particles with equal masses and kinetic energies, traveling in opposite directions in an accelerator ring, collide head-on to produce the required reaction and the formation of new particles. Because the total momentum of the isolated system of interacting particles is zero, all their kinetic energy is available for the reaction. The Large Electron–Positron (LEP) Collider at CERN (Fig. 31.15), near Geneva, Switzerland, and the Stanford Linear Collider in California collide both electrons and positrons. The Super Proton Synchrotron at CERN accelerates protons and antiprotons to energies of 270 GeV. The world's highest energy proton accelerator, the Tevatron located at Fermilab in Illinois, produces protons at almost 1 000 GeV (1 TeV). CERN expects a 2007 completion date for the Large Hadron Collider (LHC), a proton–proton collider that will provide a center of mass energy of 14 TeV and allow an exploration of Higgs boson physics. The accelerator will be constructed in the same 27-km circumference tunnel now housing the LEP collider, and many countries will participate in the project.

(Courtesy of CERN)

FIGURE 31.15 A view from inside the Large Electron–Positron (LEP) Collider tunnel, which is 27 km in circumference.

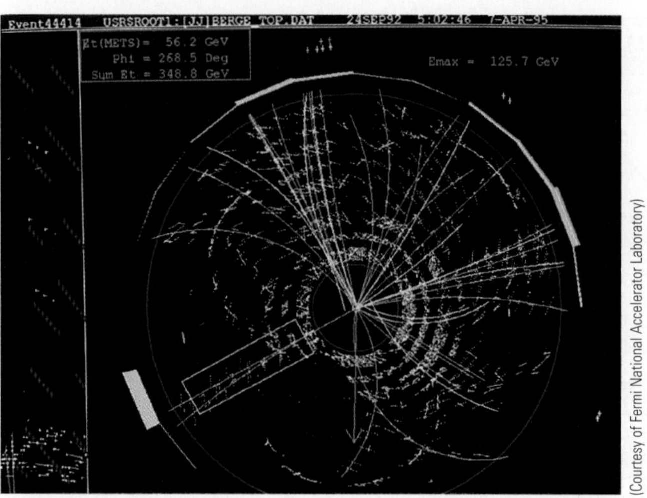

FIGURE 31.16 Computers at Fermilab create a pictorial representation such as this one of the paths of particles after a collision.

(Courtesy of Fermi National Accelerator Laboratory)

In addition to increasing energies in modern accelerators, detection techniques have become increasingly sophisticated. Figure 31.16 shows the computer-generated pictorial representation of the tracks of particles after a collision from a modern particle detector.

■ Thinking Physics 31.3

Consider a car making a head-on collision with an identical car moving in the opposite direction at the same speed. Compare that collision with one of the cars making a collision with the second car at rest. In which collision is the transformation of kinetic energy to other forms larger? How does this example relate to particle accelerators?

Reasoning In the head-on collision with both cars moving, conservation of momentum for the system of two cars requires that the cars come to rest during the collision. Therefore, *all* the original kinetic energy is transformed to other forms. In the collision between a moving car and a stationary car, the cars are still moving with reduced speed after the collision, in the direction of the initially moving car. Therefore, *only part* of the kinetic energy is transformed to other forms.

This example suggests the importance of colliding beams in a particle accelerator as opposed to firing a beam into a stationary target. When particles moving in opposite directions collide, all the kinetic energy is available for transformation into other forms, which in this case is the creation of new particles. When a beam is fired into a stationary target, only part of the energy is available for transformation, so higher mass particles cannot be created. ■

31.12 INVESTIGATING THE SMALLEST SYSTEM TO UNDERSTAND THE LARGEST

In this section, we shall describe further one of the most fascinating theories in all science—the Big Bang theory of the creation of the Universe, introduced in the Context Connection of Chapter 28—and the experimental evidence that supports it. This theory of cosmology states that the Universe had a beginning and, further, that the beginning was so cataclysmic that it is impossible to look back beyond it. According to this theory, the Universe erupted from a singularity with infinite density about 15 to 20 billion years ago. The first few fractions of a second after the Big Bang saw such extremes of energy that all four fundamental forces of physics were believed to be unified and all matter was contained in a quark-gluon plasma.

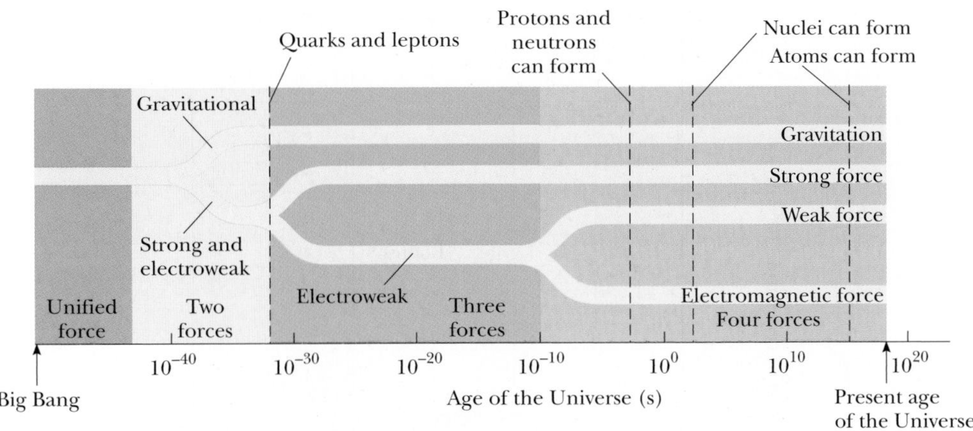

FIGURE 31.17 A brief history of the Universe from the Big Bang to the present. The four forces became distinguishable during the first nanosecond. Following that, all the quarks combined to form particles that interact via the strong force. The leptons remained separate, however, and exist as individually observable particles to this day.

The evolution of the four fundamental forces from the Big Bang to the present is shown in Figure 31.17. During the first 10^{-43} s (the ultrahot epoch, $T \sim 10^{32}$ K), it is presumed that the strong, electroweak, and gravitational forces were joined to form a completely unified force. In the first 10^{-35} s following the Big Bang (the hot epoch, $T \sim 10^{29}$ K), gravity broke free of this unification while the strong and electroweak forces remained unified. During this period, particle energies were so great ($> 10^{16}$ GeV) that very massive particles as well as quarks, leptons, and their antiparticles existed. Then, after 10^{-35} s, the Universe rapidly expanded and cooled (the warm epoch, $T \sim 10^{29} - 10^{15}$ K), and the strong and electroweak forces parted company. As the Universe continued to cool, the electroweak force split into the weak force and the electromagnetic force about 10^{-10} s after the Big Bang.

After a few minutes, protons condensed out of the plasma. For half an hour the Universe underwent thermonuclear detonation, exploding like a hydrogen bomb and producing most of the helium nuclei that now exist. The Universe continued to expand and its temperature dropped. Until about 700 000 years after the Big Bang, the Universe was dominated by radiation. Energetic radiation prevented matter from forming single hydrogen atoms because collisions would instantly ionize any atoms that happened to form. Photons experienced continuous Compton scattering from the vast numbers of free electrons, resulting in a Universe that was opaque to radiation. By the time the Universe was about 700 000 years old, it had expanded and cooled to about 3 000 K, and protons could bind to electrons to form neutral hydrogen atoms. Because of the quantized energies of the atoms, far more wavelengths of radiation were not absorbed by atoms than were, and the Universe suddenly became transparent to photons. Radiation no longer dominated the Universe, and clumps of neutral matter steadily grew, first atoms, followed by molecules, gas clouds, stars, and finally galaxies.

Evidence for the Expanding Universe

In Chapter 28, we discussed the observation of blackbody radiation by Penzias and Wilson that represents the leftover glow from the Big Bang. We discuss here additional relevant astronomical observations. Vesto Melvin Slipher (1875–1969), an American astronomer, reported that most nebulae are receding from the Earth at speeds up to several million miles per hour. Slipher was one of the first to use the methods of Doppler shifts in spectral lines to measure galactic speeds.

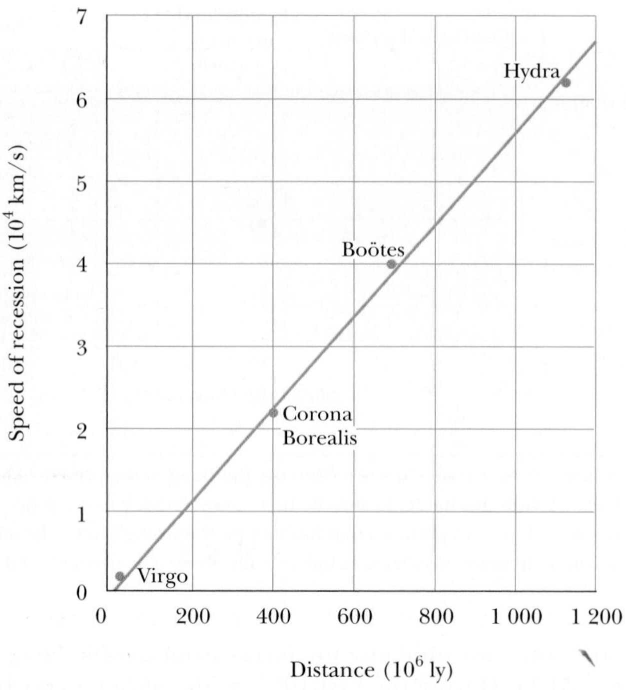

FIGURE 31.18 Hubble's law. The speed of recession is directly proportional to distance. Data points for four galaxies are shown here.

In the late 1920s, Edwin P. Hubble (1889–1953) made the bold assertion that the whole Universe is expanding. From 1928 to 1936, he and Milton Humason (1891–1972) toiled at the Mount Wilson Observatory in California to prove this assertion until they reached the limits of that 100-in. telescope. The results of this work and its continuation on a 200-in. telescope in the 1940s showed that the speeds of galaxies increase in direct proportion to their distance R from us (Fig. 31.18). This linear relationship, known as **Hubble's law,** may be written as

Hubble's law

$$v = HR \qquad [31.7]$$

where H, called the **Hubble parameter,** has the approximate value

$$H \approx 17 \times 10^{-3}\ \text{m/(s}\cdot\text{ly)}$$

EXAMPLE 31.5 Recession of a Quasar

A quasar is an object that appears similar to a star and that is very distant from the Earth. Its speed can be measured from Doppler shift measurements in the light it emits.

A A certain quasar recedes from the Earth at a speed of $0.55c$. How far away is it?

Solution We can find the distance from Hubble's law:

$$R = \frac{v}{H} = \frac{(0.55)(3.00 \times 10^8\ \text{m/s})}{17 \times 10^{-3}\ \text{m/(s}\cdot\text{ly)}} = 9.7 \times 10^9\ \text{ly}$$

B Suppose we assume that the quasar has moved at this speed ever since the Big Bang. With this assumption, estimate the age of the Universe.

Solution We approximate the distance from the Earth to the quasar as the distance that the quasar has moved from the singularity since the Big Bang. We can then find the time interval from a calculation as performed in Chapter 2: $\Delta t = \Delta x/v = R/v = 1/H \approx 18$ billion years, which is in approximate agreement with other calculations.

Will the Universe Expand Forever?

In the 1950s and 1960s, Allan R. Sandage (b. 1926) used the 200-in. telescope at the Mount Palomar Observatory in California to measure the speeds of galaxies at distances of up to 6 billion lightyears from the Earth. These measurements showed

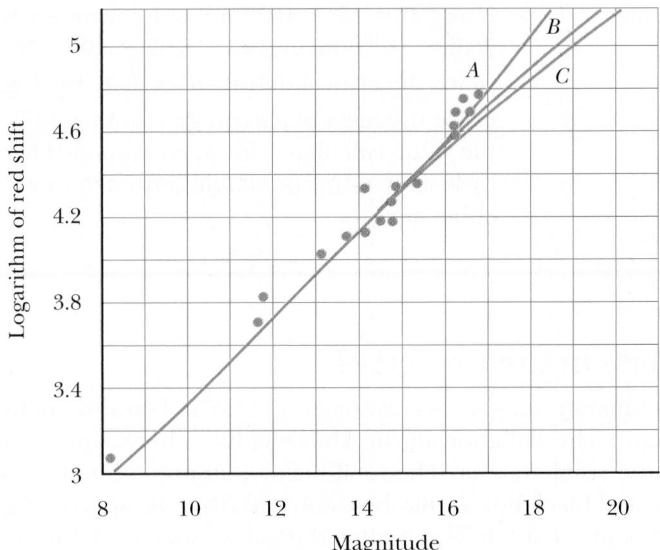

FIGURE 31.19 Red shift, or speed of recession, versus magnitude (which is related to brightness) of 18 faint galaxy clusters. Significant scatter of the data occurs, so the extrapolation of the curve to the upper right is uncertain. Curve *A* is the trend suggested by the six faintest clusters. Curve *C* corresponds to a Universe having a constant rate of expansion. If more data are taken and the complete set of data indicates a curve that falls between *B* and *C*, the expansion will slow but never stop. If the data fall to the left of *B*, expansion will eventually stop and the Universe will begin to contract.

that these very distant galaxies were moving about 10 000 km/s faster than Hubble's law predicted. According to this result, the Universe must have been expanding more rapidly 1 billion years ago, and consequently the expansion is slowing (Fig. 31.19). Today, astronomers and physicists are trying to determine the rate of slowing.

If the average mass density of atoms in the Universe is less than some critical density (about 3 atoms/m^3), the galaxies will slow in their outward rush but still escape to infinity. If the average density exceeds the critical value, the expansion will eventually stop and contraction will begin, possibly leading to a new superdense state and another expansion. In this scenario, we have an **oscillating Universe.**

EXAMPLE 31.6 The Critical Density of the Universe

Estimate the critical mass density ρ_c of the Universe, using energy considerations.

Solution Figure 31.20 shows a large section of the Universe with radius R, containing galaxies with a total mass M. Let us apply the isolated system model to an escaping galaxy and the section of the Universe; a galaxy of mass m and speed v at R will just escape to infinity with zero speed if the sum of its kinetic energy and the gravitational potential energy of the system is zero. The Universe may be infinite in extent, but a theorem such as the gravitational form of Gauss's law implies that only the mass inside the sphere contributes to the gravitational potential energy of the system of the sphere and the galaxy. Therefore,

$$E_{\text{total}} = 0 = K + U = \tfrac{1}{2}mv^2 - \frac{GmM}{R}$$

$$\tfrac{1}{2}mv^2 = \frac{Gm\frac{4}{3}\pi R^3\rho_c}{R}$$

$$(1)\quad v^2 = \frac{8\pi G}{3}R^2\rho_c$$

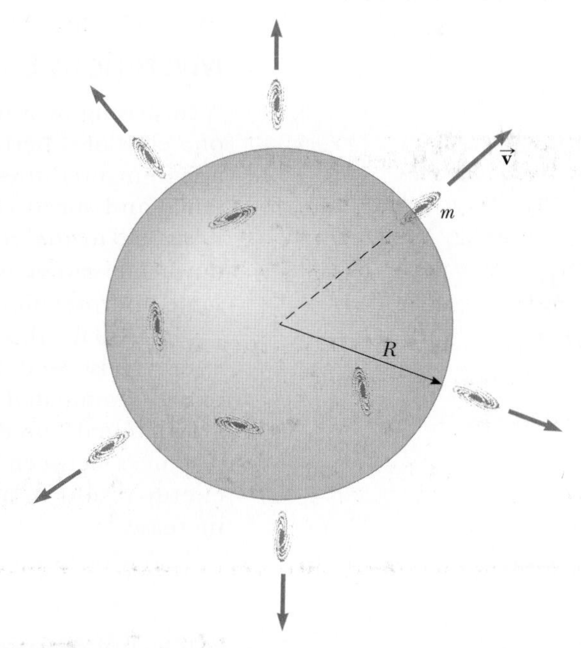

FIGURE 31.20 (Example 31.6) The galaxy labeled with mass m is escaping from a large cluster of galaxies contained within a spherical volume of radius R. Only the mass within the sphere slows the escaping galaxy.

Because the galaxy of mass m obeys the Hubble law, $v = HR$, (1) becomes

$$H^2 = \frac{8\pi G}{3}\rho_c$$

$$(2)\quad \rho_c = \frac{3H^2}{8\pi G}$$

Using $H = 17 \times 10^{-3}$ m/(s·ly), where 1 ly = 9.46×10^{12} km, and $G = 6.67 \times 10^{-11}$ N·m^2/kg^2 yields the critical density $\rho_c = 6 \times 10^{-30}$ g/cm^3. Because the mass of a hydrogen atom is 1.67×10^{-24} g, the value calculated for ρ_c corresponds to 3×10^{-6} hydrogen atoms per cubic centimeter or 3 atoms per cubic meter.

Missing Mass in the Universe?

The luminous matter in galaxies averages out to a Universe density of about 5×10^{-33} g/cm^3. The radiation in the Universe has a mass equivalent of approximately 2% of the visible matter. The total mass of all nonluminous matter (such as interstellar gas and black holes) may be estimated from the speeds of galaxies orbiting one another in a cluster. The higher the galaxy speeds, the more mass in the cluster. Measurements on the Coma cluster of galaxies indicate that the amount of nonluminous matter is 20 to 30 times the amount of luminous matter present in stars and luminous gas clouds. Yet even this large invisible component of dark matter, if extrapolated to the Universe as a whole, leaves the observed mass density a factor of 10 less than ρ_c. The deficit, called *missing mass,* has been the subject of intense theoretical and experimental work. Exotic particles such as axions, photinos, and superstring particles have been suggested as candidates for the missing mass. More mundane proposals argue that the missing mass is present in certain galaxies as neutrinos. In fact, neutrinos are so abundant that a tiny neutrino rest energy on the order of only 20 eV would furnish the missing mass and "close" the Universe. Therefore, current experiments designed to measure the rest energy of the neutrino will affect predictions for the future of the Universe, showing a clear connection between one of the smallest pieces of the Universe and the Universe as a whole!

Mysterious Energy in the Universe?

A surprising twist in the story of the Universe arose in 1998 with the observation of a class of supernovae that have a fixed absolute brightness. By combining the apparent brightness and the redshift of light from these explosions, their distance and speed of recession of the Earth can be determined. These observations led to the conclusion that the expansion of the Universe is not slowing down but rather is accelerating! Observations by other groups also led to the same interpretation.

To explain this acceleration, physicists have proposed *dark energy,* which is energy possessed by the vacuum of space. In the early life of the Universe, gravity dominated over the dark energy. As the Universe expanded and the gravitational force between galaxies became smaller because of the great distances between them, the dark energy became more important. The dark energy results in an effective repulsive force that causes the expansion rate to increase.[3]

[3]For a discussion of dark energy, see S. Perlmutter, "Supernovae, Dark Energy, and the Accelerating Universe," *Physics Today,* 56(4): 53–60, April 2003.

Although we have some degree of certainty about the beginning of the Universe, we are uncertain about how the story will end. Will the Universe keep on expanding forever, or will it someday collapse and then expand again, perhaps in an endless series of oscillations? Results and answers to these questions remain inconclusive, and the exciting controversy continues.

SUMMARY

Physics Now™ Take a practice test by logging into Physics-Now at www.pop4e.com and clicking on the Pre-Test link for this chapter.

There are four fundamental forces in nature: **strong, electromagnetic, weak,** and **gravitational.** The strong force is the force between quarks. A residual effect of the strong force is the **nuclear force** between nucleons that keeps the nucleus together. The weak force is responsible for beta decay. The electromagnetic and weak forces are now considered to be manifestations of a single force called the **electroweak force.** Every fundamental interaction is mediated by the exchange of **field particles.** The electromagnetic interaction is mediated by the photon; the weak interaction is mediated by the $W^{\pm}$ and Z^0 **bosons;** the gravitational interaction is mediated by **gravitons;** the strong interaction is mediated by **gluons.**

An **antiparticle** and a particle have the same mass, but opposite charge, and other properties may have opposite values such as lepton number and baryon number. It is possible to produce particle–antiparticle pairs in nuclear reactions if the available energy is greater than $2mc^2$, where m is the mass of the particle (or antiparticle).

Particles other than field particles are classified as hadrons or leptons. **Hadrons** interact through the strong force. They have size and structure and are not elementary particles. Hadrons are of two types, baryons and mesons. **Mesons** have baryon number zero and have either zero or integral spin. **Baryons,** which generally are the most massive particles, have nonzero baryon number and a spin of $\frac{1}{2}$ or $\frac{3}{2}$. The neutron and proton are examples of baryons.

Leptons have no structure or size and are considered truly elementary. They interact through the weak and electromagnetic forces. The six leptons are the electron e^-, the muon μ^-, the tau τ^-; and their neutrinos ν_e, ν_μ, and ν_τ.

In all reactions and decays, quantities such as energy, linear momentum, angular momentum, electric charge, baryon number, and lepton number are strictly conserved. Certain particles have properties called **strangeness** and **charm.** These unusual properties are conserved only in those reactions and decays that occur via the strong force.

Theories in elementary particle physics have postulated that all hadrons are composed of smaller units known as **quarks.** Quarks have fractional electric charge and come in six "flavors": **up** (u), **down** (d), **strange** (s), **charmed** (c), **top** (t), and **bottom** (b). Each baryon contains three quarks, and each meson contains one quark and one antiquark.

According to the theory of **quantum chromodynamics,** quarks have a property called **color charge,** and the strong force between quarks is referred to as the **color force.**

QUESTIONS

☐ = answer available in the *Student Solutions Manual and Study Guide*

1. Name the four fundamental forces and the field particle that mediates each.

2. Describe the quark model of hadrons, including the properties of quarks.

3. What are the differences between hadrons and leptons?

4. Describe the properties of baryons and mesons and the important differences between them.

5. Particles known as resonances have very short lifetimes, on the order of 10^{-23} s. From this information, would you guess that they are hadrons or leptons? Explain.

6. Kaons all decay into final states that contain no protons or neutrons. What is the baryon number of kaons?

7. Two protons in a nucleus interact via the nuclear interaction. Are they also subject to the weak interaction?

8. The Ξ^0 particle decays by the weak interaction according to the decay mode $\Xi^0 \rightarrow \Lambda^0 + \pi^0$. Would you expect this decay to be fast or slow? Explain.

9. Identify the particle decays in Table 31.2 that occur by the weak interaction. Justify your answers.

10. Identify the particle decays in Table 31.2 that occur by the electromagnetic interaction. Justify your answers.

11. Discuss the following conservation laws: energy, linear momentum, angular momentum, electric charge, baryon number, lepton number, and strangeness. Are all these laws based on fundamental properties of nature? Explain.

12. An antibaryon interacts with a meson. Can a baryon be produced in such an interaction? Explain.

13. Describe the essential features of the Standard Model of particle physics.

14. How many quarks are in each of the following: (a) a baryon, (b) an antibaryon, (c) a meson, (d) an antimeson? How do you explain that baryons have half-integral spins, whereas mesons have spins of 0 or 1? (*Note:* Quarks have spin $\frac{1}{2}$.)

15. In the theory of quantum chromodynamics, quarks come in three colors. How would you justify the statement that "all baryons and mesons are colorless"?

16. Which baryon did Murray Gell-Mann predict in 1961? What is the quark composition of this particle?

17. What is the quark composition of the Ξ^- particle? (See Table 31.5.)

18. The W and Z bosons were first produced at CERN in 1983 by causing a beam of protons and a beam of antiprotons to meet at high energy. Why was this discovery important?

19. How did Edwin Hubble determine in 1928 that the Universe is expanding?

20. Neutral atoms did not exist until hundreds of thousands of years after the Big Bang. Why?

21. What does the infinite range of the electromagnetic and gravitational interactions tell you about the masses of the photon and the graviton?

22. If high-energy electrons, with deBroglie wavelengths smaller than the size of the nucleus, are scattered from nuclei, the behavior of the electrons is consistent with scattering from very dense structures much smaller in size than the nucleus—quarks. Is this experiment similar to another classic experiment that detected small structures in atoms? Explain.

23. Observations of galaxies outside our Local Group show that they are all moving away from us. Is it therefore correct to propose that we are at the center of the Universe?

PROBLEMS

1, 2, 3 = straightforward, intermediate, challenging
☐ = full solution available in the *Student Solutions Manual and Study Guide*

Physics Now™ = coached problem with hints available at www.pop4e.com

= computer useful in solving problem

= paired numerical and symbolic problems

= biomedical application

Section 31.1 ■ The Fundamental Forces in Nature

Section 31.2 ■ Positrons and Other Antiparticles

1. A photon produces a proton–antiproton pair according to the reaction $\gamma \rightarrow p + \bar{p}$. What is the minimum possible frequency of the photon? What is its wavelength?

2. At some time in your past or future, you may find yourself in a hospital to have a PET scan. The acronym stands for *positron-emission tomography*. In the procedure, a radioactive element that undergoes e^+ decay is introduced into your body. The equipment detects the gamma rays that result from pair annihilation when the emitted positron encounters an electron in your body's tissue. Suppose you receive an injection of glucose containing on the order of 10^{10} atoms of ^{14}O. Assume that the oxygen is uniformly distributed through 2 L of blood after 5 min. What will be the order of magnitude of the activity of the oxygen atoms in 1 cm^3 of the blood?

3. Model a penny as 3.10 g of copper. Consider an anti-penny minted from 3.10 g of copper anti-atoms, each with 29 positrons in orbit around a nucleus comprising 29 antiprotons and 34 or 36 antineutrons. (a) Find the energy released if the two coins collide. (b) Find the value of this energy at the unit price of

\$0.14/kWh, a representative retail rate for energy from the electric company.

4. Two photons are produced when a proton and antiproton annihilate each other. In the reference frame in which the center of mass of the proton–antiproton system is stationary, what are the minimum frequency and corresponding wavelength of each photon?

5. A photon with an energy $E_\gamma = 2.09$ GeV creates a proton–antiproton pair in which the proton has a kinetic energy of 95.0 MeV. What is the kinetic energy of the antiproton? ($m_pc^2 = 938.3$ MeV.)

Section 31.3 ■ Mesons and the Beginning of Particle Physics

6. Occasionally, high-energy muons collide with electrons and produce two neutrinos according to the reaction $\mu^+ + e^- \rightarrow 2\nu$. What kind of neutrinos are they?

7. PhysicsNow™ One of the mediators of the weak interaction is the Z^0 boson, with mass 91.2 GeV/c^2. Use this information to find the order of magnitude of the range of the weak interaction.

8. Calculate the range of the force that might be produced by the virtual exchange of a proton.

9. PhysicsNow™ A neutral pion at rest decays into two photons according to

$$\pi^0 \rightarrow \gamma + \gamma$$

Find the energy, momentum, and frequency of each photon.

10. When a high-energy proton or pion traveling near the speed of light collides with a nucleus, it travels an average distance of 3×10^{-15} m before interacting. From this information, find the order of magnitude of the time interval required for the strong interaction to occur.

11. A free neutron beta decays by creating a proton, an electron, and an antineutrino according to the reaction $n \rightarrow p + e^- + \bar{\nu}$. Imagine that a free neutron were to decay by creating a proton and electron according to the reaction

$$n \rightarrow p + e^-$$

and assume that the neutron is initially at rest in the laboratory. (a) Determine the energy released in this reaction. (b) Determine the speeds of the proton and electron after the reaction. (Energy and momentum are conserved in the reaction.) (c) Is either of these particles moving at a relativistic speed? Explain.

Section 31.4 ■ Classification of Particles

12. Identify the unknown particle on the left side of the reaction

$$? + p \rightarrow n + \mu^+$$

13. Name one possible decay mode (see Table 31.2) for Ω^+, $\overline{K}_S{}^0$, $\overline{\Lambda}{}^0$, and $\overline{n}$.

Section 31.5 ■ Conservation Laws

14. Each of the following reactions is forbidden. Determine a conservation law that is violated for each reaction.
(a) $p + \bar{p} \rightarrow \mu^+ + e^-$ (b) $\pi^- + p \rightarrow p + \pi^+$
(c) $p + p \rightarrow p + \pi^+$ (d) $p + p \rightarrow p + p + n$
(e) $\gamma + p \rightarrow n + \pi^0$

15. (a) Show that baryon number and charge are conserved in the following reactions of a pion with a proton.

$$(1) \quad \pi^+ + p \rightarrow K^+ + \Sigma^+$$
$$(2) \quad \pi^+ + p \rightarrow \pi^+ + \Sigma^+$$

(b) The first reaction is observed, but the second never occurs. Explain.

16. The first of the following two reactions can occur, but the second cannot. Explain.

$$K_S{}^0 \rightarrow \pi^+ + \pi^- \quad \text{(can occur)}$$
$$\Lambda^0 \rightarrow \pi^+ + \pi^- \quad \text{(cannot occur)}$$

17. PhysicsNow™ The following reactions or decays involve one or more neutrinos. In each case, supply the missing neutrino (ν_e, ν_μ, or ν_τ) or antineutrino.
(a) $\pi^- \rightarrow \mu^- + ?$ (b) $K^+ \rightarrow \mu^+ + ?$
(c) $? + p \rightarrow n + e^+$ (d) $? + n \rightarrow p + e^-$
(e) $? + n \rightarrow p + \mu^-$ (f) $\mu^- \rightarrow e^- + ? + ?$

18. A $K_S{}^0$ particle at rest decays into a π^+ and a π^-. What will be the speed of each of the pions? The mass of the $K_S{}^0$ is 497.7 MeV/c^2, and the mass of each π is 139.6 MeV/c^2.

19. PhysicsNow™ Determine which of the following reactions can occur. For those that cannot occur, determine the conservation law (or laws) violated.
(a) $p \rightarrow \pi^+ + \pi^0$ (b) $p + p \rightarrow p + p + \pi^0$
(c) $p + p \rightarrow p + \pi^+$ (d) $\pi^+ \rightarrow \mu^+ + \nu_\mu$
(e) $n \rightarrow p + e^- + \bar{\nu}_e$ (f) $\pi^+ \rightarrow \mu^+ + n$

20. (a) Show that the proton-decay reaction

$$p \rightarrow e^+ + \gamma$$

cannot occur because it violates conservation of baryon number. (b) Imagine that this reaction does occur and that the proton is initially at rest. Determine the energy and momentum of the positron and photon after the reaction. (*Suggestion:* Recall that energy and momentum must be conserved in the reaction.) (c) Determine the speed of the positron after the reaction.

21. Determine the type of neutrino or antineutrino involved in each of the following processes.
(a) $\pi^+ \rightarrow \pi^0 + e^+ + ?$ (b) $? + p \rightarrow \mu^- + p + \pi^+$
(c) $\Lambda^0 \rightarrow p + \mu^- + ?$ (d) $\tau^+ \rightarrow \mu^+ + ? + ?$

Section 31.6 ■ Strange Particles and Strangeness

22. The neutral meson ρ^0 decays by the strong interaction into two pions: $\rho^0 \rightarrow \pi^+ + \pi^-$, half-life 10^{-23} s. The neutral kaon also decays into two pions: $K_S{}^0 \rightarrow \pi^+ + \pi^-$, half-life 10^{-10} s. How do you explain the difference in half-lives?

23. Determine whether or not strangeness is conserved in the following decays and reactions.
(a) $\Lambda^0 \rightarrow p + \pi^-$ (b) $\pi^- + p \rightarrow \Lambda^0 + K^0$
(c) $\bar{p} + p \rightarrow \bar{\Lambda}^0 + \Lambda^0$ (d) $\pi^- + p \rightarrow \pi^- + \Sigma^+$
(e) $\Xi^- \rightarrow \Lambda^0 + \pi^-$ (f) $\Xi^0 \rightarrow p + \pi^-$

24. For each of the following forbidden decays, determine which conservation law is violated.
(a) $\mu^- \rightarrow e^- + \gamma$ (b) $n \rightarrow p + e^- + \nu_e$
(c) $\Lambda^0 \rightarrow p + \pi^0$ (d) $p \rightarrow e^+ + \pi^0$
(e) $\Xi^0 \rightarrow n + \pi^0$

25. Which of the following processes are allowed by the strong interaction, the electromagnetic interaction, the weak interaction, or no interaction at all?
(a) $\pi^- + p \rightarrow 2\eta$ (b) $K^- + n \rightarrow \Lambda^0 + \pi^-$
(c) $K^- \rightarrow \pi^- + \pi^0$ (d) $\Omega^- \rightarrow \Xi^- + \pi^0$
(e) $\eta \rightarrow 2\gamma$

26. Identify the conserved quantities in the following processes.
(a) $\Xi^- \rightarrow \Lambda^0 + \mu^- + \nu_\mu$ (b) $K_S{}^0 \rightarrow 2\pi^0$
(c) $K^- + p \rightarrow \Sigma^0 + n$ (d) $\Sigma^0 \rightarrow \Lambda^0 + \gamma$
(e) $e^+ + e^- \rightarrow \mu^+ + \mu^-$ (f) $\bar{p} + n \rightarrow \bar{\Lambda}^0 + \Sigma^-$

27. Fill in the missing particle. Assume that (a) occurs via the strong interaction and that (b) and (c) involve the weak interaction.
(a) $K^+ + p \rightarrow ? + p$ (b) $\Omega^- \rightarrow ? + \pi^-$
(c) $K^+ \rightarrow ? + \mu^+ + \nu_\mu$

Section 31.7 ■ Measuring Particle Lifetimes

28. The particle decay $\Sigma^+ \rightarrow \pi^+ + n$ is observed in a bubble chamber. Figure P31.28 represents the curved tracks of the particles Σ^+ and π^+, and the invisible track of the neutron, in the presence of a uniform magnetic field of 1.15 T directed out of the page. The measured radii of curvature are 1.99 m for the Σ^+ particle and 0.580 m for the π^+ particle. (a) Find the momenta of the Σ^+ and the π^+ particles in units of MeV/c. (b) The angle between the momenta of the Σ^+ and the π^+ particles at the moment of decay is 64.5°. Find the momentum of the neutron. (c) Calculate the total energy of the π^+ particle, and of the neutron, from their known masses (m_π = 139.6 MeV/c^2, m_n = 939.6 MeV/c^2) and the relativistic energy-momentum relation. What is the total energy of the Σ^+ particle? (d) Calculate the mass and speed of the Σ^+ particle.

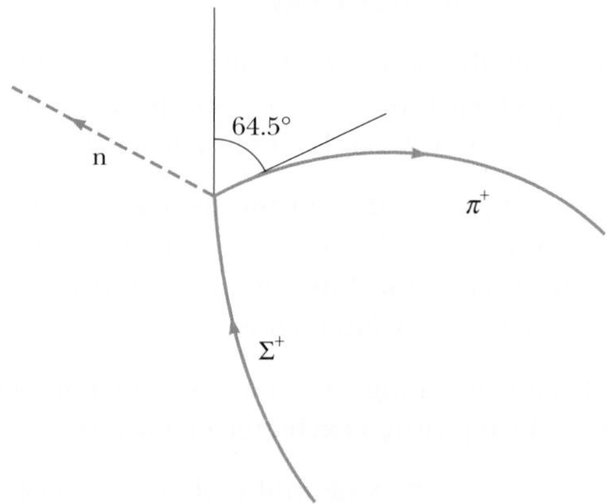

FIGURE **P31.28**

29. If a $K_S{}^0$ meson at rest decays in 0.900×10^{-10} s, how far will a $K_S{}^0$ meson travel if it is moving at 0.960c?

30. A particle of mass m_1 is fired at a stationary particle of mass m_2, and a reaction takes place in which new particles are created out of the incident kinetic energy. Taken together, the product particles have total mass m_3. The minimum kinetic energy that the bombarding particle must have to induce the reaction is called the threshold energy. At this energy, the kinetic energy of the products is a minimum, so the fraction of the incident kinetic energy that is available to create new particles is a maximum. This situation occurs when all the product particles have the same velocity; then the particles have no kinetic energy of motion relative to one another. (a) By using conservation of relativistic energy and momentum, and the relativistic energy-momentum relation, show that the threshold energy is given by

$$K_{\min} = \frac{[m_3{}^2 - (m_1 + m_2)^2]c^2}{2m_2}$$

Calculate the threshold energy for each of the following reactions:

(b) $p + p \rightarrow p + p + p + \bar{p}$
(One of the initial protons is at rest. Antiprotons are produced.)

(c) $\pi^- + p \rightarrow K^0 + \Lambda^0$
(The proton is at rest. Strange particles are produced.)

(d) $p + p \rightarrow p + p + \pi^0$
(One of the initial protons is at rest. Pions are produced.)

(e) $p + \bar{p} \rightarrow Z^0$
(One of the initial particles is at rest. Z^0 particles (mass 91.2 GeV/c^2) are produced.)

Section 31.8 ▪ Finding Patterns in the Particles

Section 31.9 ▪ Quarks

Section 31.10 ▪ Colored Quarks

Section 31.11 ▪ The Standard Model

Note: Problem 9.59 in Chapter 9 can be assigned with Section 31.11.

31. (a) Find the number of electrons and the number of each species of quark in 1 L of water. (b) Make an order-of-magnitude estimate of the number of each kind of fundamental matter particle in your body. State your assumptions and the quantities you take as data. Note that the t and b quarks were sometimes called "truth" and "beauty."

32. The quark composition of the proton is uud and that of the neutron is udd. Show that in each case the charge, baryon number, and strangeness of the particle equal, respectively, the sums of these numbers for the quark constituents.

33. Imagine that binding energies could be ignored. Find the masses of the u and d quarks from the masses of the proton and neutron.

34. The quark compositions of the K^0 and Λ^0 particles are $\bar{s}d$ and uds, respectively. Show that the charge, baryon number, and strangeness of these particles equal, respectively, the sums of these numbers for the quark constituents.

35. Analyze each reaction in terms of constituent quarks.
(a) $\pi^- + p \rightarrow K^0 + \Lambda^0$
(b) $\pi^+ + p \rightarrow K^+ + \Sigma^+$
(c) $K^- + p \rightarrow K^+ + K^0 + \Omega^-$
(d) $p + p \rightarrow K^0 + p + \pi^+ + ?$
In the last reaction, identify the mystery particle.

36. The text states that the reaction $\pi^- + p \rightarrow K^0 + \Lambda^0$ occurs with high probability, whereas the reaction $\pi^- + p \rightarrow K^0 + n$ never occurs. Analyze these reactions at the quark level. Show that the first reaction conserves the total number of each type of quark and that the second reaction does not.

37. A Σ^0 particle traveling through matter strikes a proton; then a Σ^+ and a gamma ray emerge, as well as a third particle. Use the quark model of each to determine the identity of the third particle.

38. Identify the particles corresponding to the quark combinations (a) suu, (b) $\bar{u}d$, (c) $\bar{s}d$, and (d) ssd.

39. What is the electrical charge of the baryons with the quark compositions (a) $\bar{u}\bar{u}\bar{d}$ and (b) $\bar{u}\bar{d}\bar{d}$? What are these baryons called?

Section 31.12 ▪ Context Connection—Investigating the Smallest System to Understand the Largest

Note: Problem 24.14 in Chapter 24, Problems 29.39 and 29.41 in Chapter 29, and Problems 28.53 and 28.54 in Chapter 28 can be assigned with this section.

40. Imagine that all distances expand at a rate described by the Hubble constant of 17.0×10^{-3} m/s · ly. (a) At what rate would the 1.85-m height of a basketball player be increasing? (b) At what rate would the distance between the Earth and the Moon be increasing? In fact, gravitation and other forces prevent the Hubble's-law expansion from taking place except in systems larger than clusters of galaxies.

41. Using Hubble's law, find the wavelength of the 590-nm sodium line emitted from galaxies (a) 2.00×10^6 ly away from the Earth, (b) 2.00×10^8 ly away, and (c) 2.00×10^9 ly away. You may use the result of Problem 29.39 in Chapter 29.

42. The various spectral lines observed in the light from a distant quasar have longer wavelengths λ_n' than the wavelengths λ_n measured in light from a stationary source. Here n is an index taking different values for different spectral lines. The fractional change in wavelength toward the red is the same for all spectral lines. That is, the redshift parameter Z defined by

$$Z = \frac{\lambda_n' - \lambda_n}{\lambda_n}$$

is common to all spectral lines for one object. In terms of Z, determine (a) the speed of recession of the quasar and (b) the distance from the Earth to this quasar. Use the result of Problem 29.39 in Chapter 29 and Hubble's law.

43. Assume that dark matter exists throughout space with a uniform density of 6.00×10^{-28} kg/m^3. (a) Find the amount of such dark matter inside a sphere centered on the Sun, having the Earth's orbit as its equator. (b) Would the gravitational field of this dark matter have a measurable effect on the Earth's revolution?

44. It is mostly your roommate's fault. Nosy astronomers have discovered enough junk and clutter in your dorm room to constitute the missing mass required to close the Universe. After observing your floor, closet, bed, and computer files, they extrapolate to slobs in other galaxies and calculate the average density of the observable Universe as $1.20\rho_c$. How many times larger will the Universe become before it begins to collapse? That is, by what factor will the distance between remote galaxies increase in the future?

45. The early Universe was dense with gamma-ray photons of energy $\sim k_B T$ and at such a high temperature that protons and antiprotons were created by the process $\gamma \rightarrow \text{p} + \bar{\text{p}}$ as rapidly as they annihilated each other. As the Universe cooled in adiabatic expansion, its temperature fell below a certain value and proton pair production became rare. At that time slightly more protons than antiprotons existed, and essentially all the protons in the Universe today date from that time. (a) Estimate the order of magnitude of the temperature of the Universe when protons condensed out. (b) Estimate the order of magnitude of the temperature of the Universe when electrons condensed out.

46. If the average density of the Universe is small compared with the critical density, the expansion of the Universe described by Hubble's law proceeds with speeds that are nearly constant over time. (a) Prove that in this case the age of the Universe is given by the inverse of Hubble's constant. (b) Calculate $1/H$ and express it in years.

47. Assume that the average density of the Universe is equal to the critical density. (a) Prove that the age of the Universe is given by $2/3H$. (b) Calculate $2/3H$ and express it in years.

48. Hubble's law can be stated in vector form as $\vec{\mathbf{v}} = H\vec{\mathbf{R}}$. Outside the local group of galaxies, all objects are moving away from us with velocities proportional to their displacements from us. In this form, it sounds as if our location in the Universe is specially privileged. Prove that Hubble's law would be equally true for an observer elsewhere in the Universe. Proceed as follows. Assume that we are at the origin of coordinates, that one galaxy cluster is at location $\vec{\mathbf{R}}_1$ and has velocity $\vec{\mathbf{v}}_1 = H\vec{\mathbf{R}}_1$ relative to us, and that another galaxy cluster has position vector $\vec{\mathbf{R}}_2$ and velocity $\vec{\mathbf{v}}_2 = H\vec{\mathbf{R}}_2$. Suppose the speeds are nonrelativistic. Consider the frame of reference of an observer in the first of these galaxy clusters. Show that our velocity relative to her, together with the position vector of our galaxy cluster relative to hers, satisfies Hubble's law. Show that the position and velocity of cluster 2 relative to cluster 1 satisfy Hubble's law.

Additional Problems

49. Review problem. Supernova Shelton 1987A, located about 170 000 ly from the Earth, is estimated to have emitted a burst of neutrinos carrying energy $\sim 10^{46}$ J

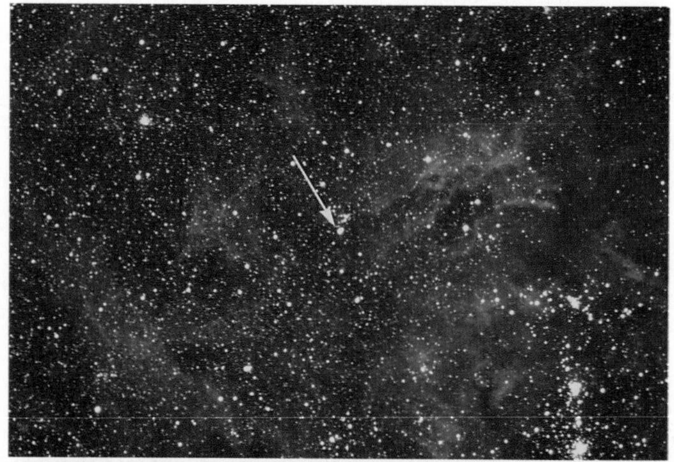

(Anglo-Australian Telescope Board)

FIGURE P31.49 (Problems 31.49 and 31.50) The giant star catalogued as Sanduleak $-69°$ 202 in the "before" picture (*top*) became Supernova Shelton 1987A in the "after" picture (*bottom*).

(Fig. P31.49). Suppose the average neutrino energy was 6 MeV and your mother's body presented cross-sectional area 5 000 cm^2. To an order of magnitude, how many of these neutrinos passed through your mother?

50. The most recent naked-eye supernova was Supernova Shelton 1987A (Fig. P31.49). It was 170 000 ly away in the next galaxy to ours, the Large Magellanic Cloud. About 3 h before its optical brightening was noticed, two continuously running neutrino detection experiments simultaneously registered the first neutrinos from an identified source other than the Sun. The Irvine-Michigan-Brookhaven experiment in a salt mine in Ohio registered eight neutrinos over a 6-s period, and the Kamiokande II experiment in a zinc mine in Japan counted eleven neutrinos in 13 s. (Because the supernova is far south in the sky, these neutrinos entered the detectors from below. They passed through the Earth before they were by chance absorbed by nuclei in the detectors.) The neutrino energies were between about 8 MeV and 40 MeV. If neutrinos have no mass, neutrinos of all energies should travel together at the speed of light, and the data are consistent with this possibility. The arrival times could show scatter simply because neutrinos were created at different moments as the core of the star collapsed into a neutron star. If neutrinos have nonzero mass, lower-energy neutrinos should move comparatively slowly. The data are consistent with a 10 MeV neutrino requiring at most about 10 s more than a photon would require to travel from the supernova to us. Find the upper limit that this observation sets on the mass of a neutrino. (Other evidence sets an even tighter limit.)

51. **PhysicsNow™** The energy flux carried by neutrinos from the Sun is estimated to be on the order of 0.4 W/m^2 at the Earth's surface. Estimate the fractional mass loss of the Sun over 10^9 yr due to the emission of neutrinos. (The mass of the Sun is 2×10^{30} kg. The Earth–Sun distance is 1.5×10^{11} m.)

52. Name at least one conservation law that prevents each of the following reactions: (a) $\pi^- + \text{p} \rightarrow \Sigma^+ + \pi^0$, (b) $\mu^- \rightarrow \pi^- + \nu_e$, (c) $\text{p} \rightarrow \pi^+ + \pi^+ + \pi^-$.

53. Assume that the half-life of free neutrons is 614 s. What fraction of a group of free thermal neutrons with kinetic energy 0.040 0 eV will decay before traveling a distance of 10.0 km?

54. Two protons approach each other head-on, each with 70.4 MeV of kinetic energy, and engage in a reaction in which a proton and positive pion emerge at rest. What third particle, obviously uncharged and therefore difficult to detect, must have been created?

55. Determine the kinetic energies of the proton and pion resulting from the decay of a Λ^0 at rest:

$$\Lambda^0 \rightarrow \text{p} + \pi^-$$

56. A rocket engine for space travel using photon drive and matter–antimatter annihilation has been suggested. Suppose the fuel for a short-duration burn consists of N protons and N antiprotons, each with mass m. (a) Assume that all the fuel is annihilated to produce photons. When the photons are ejected from the rocket, what momentum can be imparted to it? (b) If half of the protons and antiprotons annihilate each other and the energy released is used to eject the remaining particles, what momentum could be given to the rocket? Which scheme results in the greatest change in speed for the rocket?

57. An unstable particle, initially at rest, decays into a proton (rest energy 938.3 MeV) and a negative pion (rest energy 139.6 MeV). A uniform magnetic field of 0.250 T exists perpendicular to the velocities of the created particles. The radius of curvature of each track is found to be 1.33 m. What is the mass of the original unstable particle?

58. A gamma-ray photon strikes a stationary electron. Determine the minimum gamma-ray energy to make this reaction occur:

$$\gamma + \text{e}^- \rightarrow \text{e}^- + \text{e}^- + \text{e}^+$$

59. Two protons approach each other with velocities of equal magnitude in opposite directions. What is the minimum kinetic energy of each of the protons if they are to produce a π^+ meson at rest in the following reaction?

$$\text{p} + \text{p} \rightarrow \text{p} + \text{n} + \pi^+$$

60. A Σ^0 particle at rest decays according to

$$\Sigma^0 \rightarrow \Lambda^0 + \gamma$$

Find the gamma-ray energy.

61. **Review problem.** Use the Boltzmann distribution function e^{-E/k_BT} to calculate the temperature at which 1.00% of a population of photons will have energy greater than 1.00 eV. The energy required to excite an atom is on the order of 1 eV. Therefore, as the temperature of the Universe fell below the value you calculate, neutral atoms could form from plasma and the Universe became transparent. The cosmic background radiation represents our vastly red-shifted view of the opaque fireball of the Big Bang as it was at this time and temperature. The fireball surrounds us; we are embers.

62. A π-meson at rest decays according to $\pi^- \rightarrow \mu^- + \bar{\nu}_\mu$. What is the energy carried off by the neutrino? (Assume

that the neutrino has no mass and moves off with the speed of light. Take $m_\pi c^2 = 139.6$ MeV and $m_\mu c^2 = 105.7$ MeV.)

63. Identify the mediators for the two interactions described in the Feynman diagrams shown in Figure P31.63.

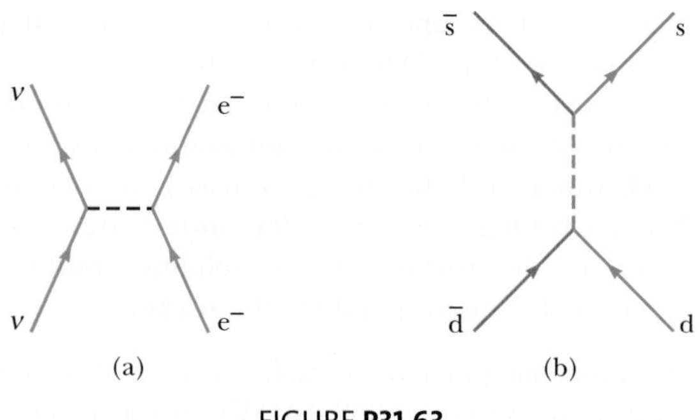

FIGURE **P31.63**

64. What processes are described by the Feynman diagrams in Figure P31.64? What is the exchanged particle in each process?

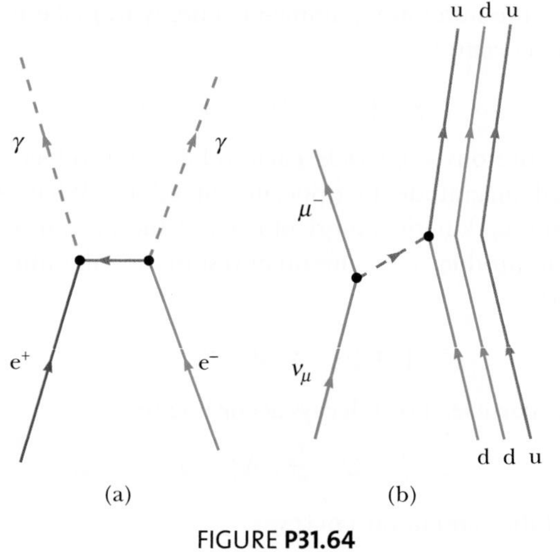

FIGURE **P31.64**

65. The cosmic rays of highest energy are mostly protons, accelerated by unknown sources. Their spectrum shows a cutoff at an energy on the order of 10^{20} eV. Above that energy, a proton will interact with a photon of cosmic microwave background radiation to produce mesons, for example, according to

$$p + \gamma \rightarrow p + \pi^0$$

Demonstrate this fact by taking the following steps. (a) Find the minimum photon energy required to produce this reaction in the reference frame where the total momentum of the photon–proton system is zero. The reaction was observed experimentally in the 1950s with photons of a few hundred MeV. (b) Use Wien's displacement law to find the wavelength of a photon at the peak of the blackbody spectrum of the primordial microwave background radiation, with a temperature of 2.73 K. (c) Find the energy of this photon. (d) Consider the reaction in part (a) in a moving reference frame so that the photon is the same as that in part (c). Calculate the energy of the proton in this frame, which represents the Earth reference frame.

ANSWERS TO QUICK QUIZZES

31.1 (a). The right-hand rule for the positive particle tells you that this direction is the one that leads to a force directed toward the center of curvature of the path.

31.2 (c), (d). There is a baryon, the neutron, on the left of the reaction, but no baryon on the right. Therefore, baryon number is not conserved. The neutron has spin $\frac{1}{2}$. On the right side of the reaction, the pions each have integral spin, and the combination of two muons must also have integral spin. Therefore, the total spin of the particles on the right-hand side is integral and angular momentum is not conserved.

31.3 (a). The sum of the proton and pion masses is larger than the mass of the neutron, so energy conservation is violated.

31.4 (b), (e), (f). The pion on the left has integral spin, whereas the three spin-$\frac{1}{2}$ leptons on the right must result in a total spin that is half-integral. Therefore, angular momentum (b) is not conserved. There is an electron on the right but no lepton on the left, so electron lepton number (e) is not conserved. There are no muons on the left, but a muon and its neutrino on the right (both with $L_\mu = +1$). Therefore, muon lepton number (f) is not conserved.

31.5 (b), (e). There is one spin-$\frac{1}{2}$ particle on the left and two on the right, so angular momentum is not conserved. There are no leptons on the left and an electron on the right, so electron lepton number is not conserved.

31.6 The diagram would look like this one:

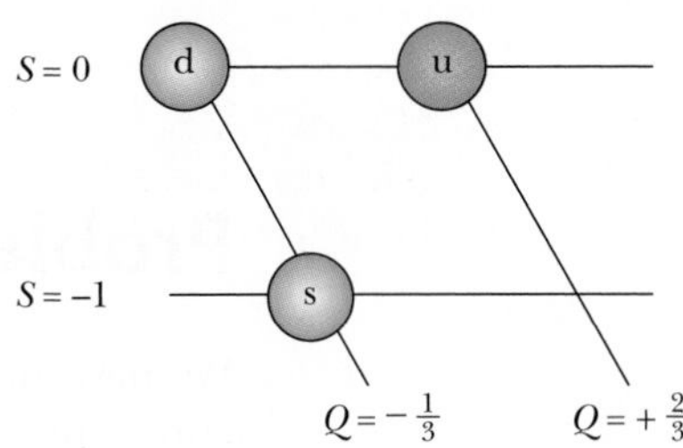

Problems and Perspectives

We have now investigated the principles of quantum physics and have seen many connections to our central question for the *Cosmic Connection* Context:

How can we connect the physics of microscopic particles to the physics of the Universe?

While particle physicists have been exploring the realm of the very small, cosmologists have been exploring cosmic history back to the first second of the Big Bang. Observation of events that occur when two particles collide in an accelerator is essential in reconstructing the early moments in cosmic history. The key to understanding the early Universe is first to understand the world of elementary particles. Cosmologists and physicists now find that they have many common goals and are joining hands to attempt to understand the physical world at its most fundamental level.

Problems

We have made great progress in understanding the Universe and its underlying structure, but a multitude of questions remain unanswered. Why does so little antimatter exist in the Universe? Do neutrinos have a small rest energy, and if so, how do they contribute to the "dark matter" of the Universe? Is there "dark energy" in the Universe? Is it possible to unify the strong and electroweak forces in a logical and consistent manner? Can gravity be unified with the other forces? Why do quarks and leptons form three similar but distinct families? Are muons the same as electrons (apart from their difference in mass), or do they have other subtle differences that have not been detected? Why are some particles charged and others neutral? Why do quarks carry a fractional charge? What determines the masses of the fundamental constituents? Can isolated quarks exist? Do leptons and quarks have a substructure?

String Theory: A New Perspective

Let us briefly discuss one current effort at answering some of these questions by proposing a new perspective on particles. As you read this book, you may recall starting off with the particle model and doing quite a bit of physics with it. In the *Earthquakes* Context, we introduced the wave model, and more physics was used to investigate the properties of waves. We used a wave model for light in the *Lasers* Context. Early in this Context, however, we saw the need to return to the particle model for light. Furthermore, we found that material particles had wave-like characteristics. The quantum particle model of Chapter 28 allowed us to build particles out of waves, suggesting that a wave is the fundamental entity. In Chapter 31, however, we discussed the elementary particles as the fundamental entities. It seems as if we cannot make up our mind! In some sense, that is true because the wave–particle duality is still an area of active research. In this final Context Conclusion, we shall discuss a current research effort to build particles out of waves and vibrations.

String theory is an effort to unify the four fundamental forces by modeling all particles as various quantized vibrational modes of a single entity, an incredibly small string. The typical length of such a string is on the order of 10^{-35} m, called the **Planck length.** We have seen quantized modes before with the frequencies of vibrating guitar strings in Chapter 14 and the quantized energy levels of atoms in

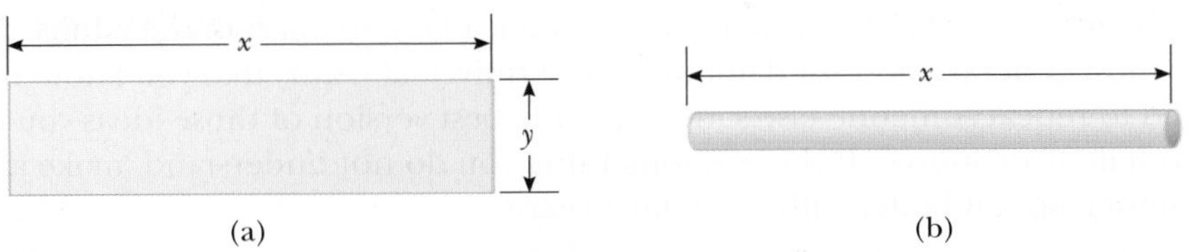

FIGURE 1 (a) A piece of paper is cut into a rectangular shape. As a rectangle, the shape has two dimensions. (b) The paper is rolled up into a soda straw. From far away, it appears to be one-dimensional. The curled-up second dimension is not visible when viewed from a distance that is large compared with the diameter of the straw.

Chapter 29. In string theory, each quantized mode of vibration of the string corresponds to a different elementary particle in the Standard Model.

One complicating factor in string theory is that it requires space–time to have ten dimensions. Despite the theoretical and conceptual difficulties in dealing with ten dimensions, string theory holds promise in incorporating gravity with the other forces. Four of the ten dimensions are visible to us—three space dimensions and one time dimension—and the other six are *compactified*. In other words, the six dimensions are curled up so tightly that they are not visible in the macroscopic world.

As an analogy, consider a soda straw. We can build a soda straw by cutting a rectangular piece of paper (Fig. 1a), which clearly has two dimensions, and rolling it up into a small tube (Fig. 1b). From far away, the soda straw looks like a one-dimensional straight line. The second dimension has been curled up and is not visible. String theory claims that six space–time dimensions are curled up in an analogous way, with the curling on the size of the Planck length and impossible to see from our viewpoint.

Another complicating factor with string theory is that it is difficult for string theorists to guide experimentalists in how and what to look for in an experiment. The Planck length is so incredibly small that direct experimentation on strings is impossible. Until the theory has been further developed, string theorists are restricted to applying the theory to known results and testing for consistency.

One of the predictions of string theory is called **supersymmetry** (SUSY), which suggests that every elementary particle has a superpartner that has not yet been observed. It is believed that supersymmetry is a broken symmetry (like the broken electroweak symmetry at low energies) and that the masses of the superpartners are above our current capabilities of detection by accelerators. Some theorists claim that the mass of superpartners is the missing mass discussed in the Context Conclusion of Chapter 31. Keeping with the whimsical trend in naming particles and their properties that we saw in Chapter 31, superpartners are given names such as the *squark* (the superpartner to a quark), the *selectron* (electron), and the *gluinos* (gluon).

Other theorists are working on **M-theory,** which is an 11-dimensional theory based on membranes rather than strings. In a way reminiscent of the correspondence principle, M-theory is claimed to reduce to string theory if one compactifies from 11 dimensions to 10.

The questions that we listed at the beginning of this Context Conclusion go on and on. Because of the rapid advances and new discoveries in the field of particle physics, by the time you read this book some of these questions may be resolved and other new questions may emerge.

Question

1. **Review question.** A girl and her grandmother grind corn while the woman tells the girl stories about what is most important. A boy keeps crows away from ripening corn while his grandfather sits in the shade and explains to him the Universe and his place in it. What the children do not understand this year they

will better understand next year. Now you must take the part of the adults. State the most general, most fundamental, most universal truths that you know. If you need to repeat someone else's ideas, get the best version of those ideas you can and state your source. If there is something you do not understand, make a plan to understand it better within the next year.

Problem

1. Classical general relativity views the structure of space–time as deterministic and well defined down to arbitrarily small distances. On the other hand, quantum general relativity forbids distances smaller than the Planck length given by $L = (\hbar G/c^3)^{1/2}$. (a) Calculate the value of the Planck length. The quantum limitation suggests that after the Big Bang, when all the presently observable section of the Universe was contained within a point-like singularity, nothing could be observed until that singularity grew larger than the Planck length. Because the size of the singularity grew at the speed of light, we can infer that no observations were possible during the time interval required for light to travel the Planck length. (b) Calculate this time interval, known as the Planck time T, and compare it with the ultrahot epoch mentioned in the text. (c) Does this answer suggest that we may never know what happened between the time $t = 0$ and the time $t = T$?

The Meaning of Success

To earn the respect of intelligent people and to win the affection of children;
To appreciate the beauty in nature and all that surrounds us;
To seek out and nurture the best in others;
To give the gift of yourself to others without the slightest thought of return, for it is in giving that we receive;
To have accomplished a task, whether it be saving a lost soul, healing a sick child, writing a book, or risking your life for a friend;
To have celebrated and laughed with great joy and enthusiasm and sung with exultation;
To have hope even in times of despair, for as long as you have hope, you have life;
To love and be loved;
To be understood and to understand;
To know that even one life has breathed easier because you have lived;
This is the meaning of success.

Ralph Waldo Emerson and modified by Ray Serway

Tables

TABLE A.1 Conversion Factors

Length

	m	cm	km	in.	ft	mi
1 meter	1	10^2	10^{-3}	39.37	3.281	6.214×10^{-4}
1 centimeter	10^{-2}	1	10^{-5}	0.393 7	3.281×10^{-2}	6.214×10^{-6}
1 kilometer	10^3	10^5	1	3.937×10^4	3.281×10^3	0.621 4
1 inch	2.540×10^{-2}	2.540	2.540×10^{-5}	1	8.333×10^{-2}	1.578×10^{-5}
1 foot	0.304 8	30.48	3.048×10^{-4}	12	1	1.894×10^{-4}
1 mile	1 609	1.609×10^5	1.609	6.336×10^4	5 280	1

Mass

	kg	g	slug	u
1 kilogram	1	10^3	6.852×10^{-2}	6.024×10^{26}
1 gram	10^{-3}	1	6.852×10^{-5}	6.024×10^{23}
1 slug	14.59	1.459×10^4	1	8.789×10^{27}
1 atomic mass unit	1.660×10^{-27}	1.660×10^{-24}	1.137×10^{-28}	1

Note: 1 metric ton = 1 000 kg.

Time

	s	min	h	day	yr
1 second	1	1.667×10^{-2}	2.778×10^{-4}	1.157×10^{-5}	3.169×10^{-8}
1 minute	60	1	1.667×10^{-2}	6.994×10^{-4}	1.901×10^{-6}
1 hour	3 600	60	1	4.167×10^{-2}	1.141×10^{-4}
1 day	8.640×10^4	1 440	24	1	2.738×10^{-5}
1 year	3.156×10^7	5.259×10^5	8.766×10^3	365.2	1

Speed

	m/s	cm/s	ft/s	mi/h
1 meter per second	1	10^2	3.281	2.237
1 centimeter per second	10^{-2}	1	3.281×10^{-2}	2.237×10^{-2}
1 foot per second	0.304 8	30.48	1	0.681 8
1 mile per hour	0.447 0	44.70	1.467	1

Note: 1 mi/min = 60 mi/h = 88 ft/s.

Force

	N	lb
1 newton	1	0.224 8
1 pound	4.448	1

(Continued)

TABLE A.1 Conversion Factors *(Continued)*

Work, Energy, Heat

	J	ft · lb	eV
1 joule	1	0.737 6	6.242×10^{18}
1 foot-pound	1.356	1	8.464×10^{18}
1 electron volt	1.602×10^{-19}	1.182×10^{-19}	1
1 calorie	4.186	3.087	2.613×10^{19}
1 British thermal unit	1.055×10^{3}	7.779×10^{2}	6.585×10^{21}
1 kilowatt-hour	3.600×10^{6}	2.655×10^{6}	2.247×10^{25}

	cal	Btu	kWh
1 joule	0.238 9	9.481×10^{-4}	2.778×10^{-7}
1 foot-pound	0.323 9	1.285×10^{-3}	3.766×10^{-7}
1 electron volt	3.827×10^{-20}	1.519×10^{-22}	4.450×10^{-26}
1 calorie	1	3.968×10^{-3}	1.163×10^{-6}
1 British thermal unit	2.520×10^{2}	1	2.930×10^{-4}
1 kilowatt-hour	8.601×10^{5}	3.413×10^{2}	1

Pressure

	Pa	atm
1 pascal	1	9.869×10^{-6}
1 atmosphere	1.013×10^{5}	1
1 centimeter mercury[a]	1.333×10^{3}	1.316×10^{-2}
1 pound per square inch	6.895×10^{3}	6.805×10^{-2}
1 pound per square foot	47.88	4.725×10^{-4}

	cm Hg	lb/in.^{2}	lb/ft^2
1 pascal	7.501×10^{-4}	1.450×10^{-4}	2.089×10^{-2}
1 atmosphere	76	14.70	2.116×10^{3}
1 centimeter mercury[a]	1	0.194 3	27.85
1 pound per square inch	5.171	1	144
1 pound per square foot	3.591×10^{-2}	6.944×10^{-3}	1

[a]At 0°C and at a location where the free-fall acceleration has its "standard" value, 9.806 65 m/s^2.

TABLE A.2 Symbols, Dimensions, and Units of Physical Quantities

Quantity	Common Symbol	Unit[a]	Dimensions[b]	Unit in Terms of Base SI Units
Acceleration	$\vec{\mathbf{a}}$	m/s^2	L/T^2	m/s^2
Amount of substance	n	MOLE		mol
Angle	θ, ϕ	radian (rad)	1	
Angular acceleration	$\vec{\boldsymbol{\alpha}}$	rad/s^2	T^{-2}	s^{-2}
Angular frequency	ω	rad/s	T^{-1}	s^{-1}
Angular momentum	$\vec{\mathbf{L}}$	kg · m^2/s	ML^2/T	kg · m^2/s
Angular velocity	$\vec{\boldsymbol{\omega}}$	rad/s	T^{-1}	s^{-1}
Area	A	m^2	L^2	m^2
Atomic number	Z			

(Continued)

TABLE A.2 Symbols, Dimensions, and Units of Physical Quantities *(Continued)*

Quantity	Common Symbol	Unit[a]	Dimensions[b]	Unit in Terms of Base SI Units
Capacitance	C	farad (F)	Q^2T^2/ML^2	$A^2 \cdot s^4/kg \cdot m^2$
Charge	q, Q, e	coulomb (C)	Q	$A \cdot s$
Charge density				
Line	λ	C/m	Q/L	$A \cdot s/m$
Surface	σ	C/m^2	Q/L^2	$A \cdot s/m^2$
Volume	ρ	C/m^3	Q/L^3	$A \cdot s/m^3$
Conductivity	σ	$1/\Omega \cdot m$	Q^2T/ML^3	$A^2 \cdot s^3/kg \cdot m^3$
Current	I	AMPERE	Q/T	A
Current density	$\vec{\mathbf{J}}$	A/m^2	Q/TL^2	A/m^2
Density	ρ	kg/m^3	M/L^3	kg/m^3
Dielectric constant	κ			
Electric dipole moment	$\vec{\mathbf{p}}$	$C \cdot m$	QL	$A \cdot s \cdot m$
Electric field	$\vec{\mathbf{E}}$	V/m	ML/QT^2	$kg \cdot m/A \cdot s^3$
Electric flux	Φ_E	$V \cdot m$	ML^3/QT^2	$kg \cdot m^3/A \cdot s^3$
Electromotive force	$\mathcal{E}$	volt (V)	ML^2/QT^2	$kg \cdot m^2/A \cdot s^3$
Energy	E, U, K	joule (J)	ML^2/T^2	$kg \cdot m^2/s^2$
Entropy	S	J/K	$ML^2/T^2 \cdot K$	$kg \cdot m^2/s^2 \cdot K$
Force	$\vec{\mathbf{F}}$	newton (N)	ML/T^2	$kg \cdot m/s^2$
Frequency	f	hertz (Hz)	T^{-1}	s^{-1}
Heat	Q	joule (J)	ML^2/T^2	$kg \cdot m^2/s^2$
Inductance	L	henry (H)	ML^2/Q^2	$kg \cdot m^2/A^2 \cdot s^2$
Length	ℓ, L	METER	L	m
Displacement	$\Delta x, \Delta\vec{\mathbf{r}}$			
Distance	d, h			
Position	$x, y, z, \vec{\mathbf{r}}$			
Magnetic dipole moment	$\vec{\boldsymbol{\mu}}$	$N \cdot m/T$	QL^2/T	$A \cdot m^2$
Magnetic field	$\vec{\mathbf{B}}$	tesla (T) (= Wb/m^2)	M/QT	$kg/A \cdot s^2$
Magnetic flux	Φ_B	weber (Wb)	ML^2/QT	$kg \cdot m^2/A \cdot s^2$
Mass	m, M	KILOGRAM	M	kg
Molar specific heat	C	$J/mol \cdot K$		$kg \cdot m^2/s^2 \cdot mol \cdot K$
Moment of inertia	I	$kg \cdot m^2$	ML^2	$kg \cdot m^2$
Momentum	$\vec{\mathbf{p}}$	$kg \cdot m/s$	ML/T	$kg \cdot m/s$
Period	T	s	T	s
Permeability of free space	μ_0	N/A^2 (= H/m)	ML/Q^2	$kg \cdot m/A^2 \cdot s^2$
Permittivity of free space	ϵ_0	$C^2/N \cdot m^2$ (= F/m)	Q^2T^2/ML^3	$A^2 \cdot s^4/kg \cdot m^3$
Potential	V	volt (V) (= J/C)	ML^2/QT^2	$kg \cdot m^2/A \cdot s^3$
Power	$\mathcal{P}$	watt (W) (= J/s)	ML^2/T^3	$kg \cdot m^2/s^3$
Pressure	P	pascal (Pa) (= N/m^2)	M/LT^2	$kg/m \cdot s^2$
Resistance	R	ohm (Ω) (= V/A)	ML^2/Q^2T	$kg \cdot m^2/A^2 \cdot s^3$
Specific heat	c	$J/kg \cdot K$	$L^2/T^2 \cdot K$	$m^2/s^2 \cdot K$
Speed	v	m/s	L/T	m/s
Temperature	T	KELVIN	K	K
Time	t	SECOND	T	s
Torque	$\vec{\boldsymbol{\tau}}$	$N \cdot m$	ML^2/T^2	$kg \cdot m^2/s^2$
Velocity	$\vec{\mathbf{v}}$	m/s	L/T	m/s
Volume	V	m^3	L^3	m^3
Wavelength	λ	m	L	m
Work	W	joule (J) (= $N \cdot m$)	ML^2/T^2	$kg \cdot m^2/s^2$

[a]The base SI units are given in uppercase letters.

[b]The symbols M, L, T, and Q denote mass, length, time, and charge, respectively.

TABLE A.3 Table of Atomic Masses

Atomic Number Z	Element	Symbol	Chemical Atomic Mass (u)	Mass Number (*indicates radioactive) A	Atomic Mass (u)	Percent Abundance	Half-Life (if radioactive) $T_{1/2}$
0	(Neutron)	n		1*	1.008 665		10.4 min
1	Hydrogen	H	1.007 94	1	1.007 825	99.988 5	
	Deuterium	D		2	2.014 102	0.011 5	
	Tritium	T		3*	3.016 049		12.33 yr
2	Helium	He	4.002 602	3	3.016 029	0.000 137	
				4	4.002 603	99.999 863	
				6*	6.018 888		0.81 s
3	Lithium	Li	6.941	6	6.015 122	7.5	
				7	7.016 004	92.5	
				8*	8.022 487		0.84 s
4	Beryllium	Be	9.012 182	7*	7.016 929		53.3 days
				9	9.012 182	100	
				10*	10.013 534		1.5×10^6 yr
5	Boron	B	10.811	10	10.012 937	19.9	
				11	11.009 306	80.1	
				12*	12.014 352		0.020 2 s
6	Carbon	C	12.010 7	10*	10.016 853		19.3 s
				11*	11.011 434		20.4 min
				12	12.000 000	98.93	
				13	13.003 355	1.07	
				14*	14.003 242		5 730 yr
				15*	15.010 599		2.45 s
7	Nitrogen	N	14.006 7	12*	12.018 613		0.011 0 s
				13*	13.005 739		9.96 min
				14	14.003 074	99.632	
				15	15.000 109	0.368	
				16*	16.006 101		7.13 s
				17*	17.008 450		4.17 s
8	Oxygen	O	15.999 4	14*	14.008 595		70.6 s
				15*	15.003 065		122 s
				16	15.994 915	99.757	
				17	16.999 132	0.038	
				18	17.999 160	0.205	
				19*	19.003 579		26.9 s
9	Fluorine	F	18.998 403 2	17*	17.002 095		64.5 s
				18*	18.000 938		109.8 min
				19	18.998 403	100	
				20*	19.999 981		11.0 s
				21*	20.999 949		4.2 s
10	Neon	Ne	20.179 7	18*	18.005 697		1.67 s
				19*	19.001 880		17.2 s
				20	19.992 440	90.48	
				21	20.993 847	0.27	
				22	21.991 385	9.25	
				23*	22.994 467		37.2 s
11	Sodium	Na	22.989 77	21*	20.997 655		22.5 s
				22*	21.994 437		2.61 yr
				23	22.989 770	100	
				24*	23.990 963		14.96 h
12	Magnesium	Mg	24.305 0	23*	22.994 125		11.3 s
				24	23.985 042	78.99	
				25	24.985 837	10.00	

(Continued)

TABLE A.3 Table of Atomic Masses *(Continued)*

Atomic Number Z	Element	Symbol	Chemical Atomic Mass (u)	Mass Number (*indicates radioactive) A	Atomic Mass (u)	Percent Abundance	Half-Life (if radioactive) $T_{1/2}$
(12)	Magnesium			26	25.982 593	11.01	
				27*	26.984 341		9.46 min
13	Aluminum	Al	26.981 538	26*	25.986 892		7.4×10^5 yr
				27	26.981 539	100	
				28*	27.981 910		2.24 min
14	Silicon	Si	28.085 5	28	27.976 926	92.229 7	
				29	28.976 495	4.683 2	
				30	29.973 770	3.087 2	
				31*	30.975 363		2.62 h
				32*	31.974 148		172 yr
15	Phosphorus	P	30.973 761	30*	29.978 314		2.50 min
				31	30.973 762	100	
				32*	31.973 907		14.26 days
				33*	32.971 725		25.3 days
16	Sulfur	S	32.066	32	31.972 071	94.93	
				33	32.971 458	0.76	
				34	33.967 869	4.29	
				35*	34.969 032		87.5 days
				36	35.967 081	0.02	
17	Chlorine	Cl	35.452 7	35	34.968 853	75.78	
				36*	35.968 307		3.0×10^5 yr
				37	36.965 903	24.22	
18	Argon	Ar	39.948	36	35.967 546	0.336 5	
				37*	36.966 776		35.04 days
				38	37.962 732	0.063 2	
				39*	38.964 313		269 yr
				40	39.962 383	99.600 3	
				42*	41.963 046		33 yr
19	Potassium	K	39.098 3	39	38.963 707	93.258 1	
				40*	39.963 999	0.011 7	1.28×10^9 yr
				41	40.961 826	6.730 2	
20	Calcium	Ca	40.078	40	39.962 591	96.941	
				41*	40.962 278		1.0×10^5 yr
				42	41.958 618	0.647	
				43	42.958 767	0.135	
				44	43.955 481	2.086	
				46	45.953 693	0.004	
				48	47.952 534	0.187	
21	Scandium	Sc	44.955 910	41*	40.969 251		0.596 s
				45	44.955 910	100	
22	Titanium	Ti	47.867	44*	43.959 690		49 yr
				46	45.952 630	8.25	
				47	46.951 764	7.44	
				48	47.947 947	73.72	
				49	48.947 871	5.41	
				50	49.944 792	5.18	
23	Vanadium	V	50.941 5	48*	47.952 254		15.97 days
				50*	49.947 163	0.250	1.5×10^{17} yr
				51	50.943 964	99.750	
24	Chromium	Cr	51.996 1	48*	47.954 036		21.6 h
				50	49.946 050	4.345	

(Continued)

TABLE A.3 Table of Atomic Masses *(Continued)*

Atomic Number Z	Element	Symbol	Chemical Atomic Mass (u)	Mass Number (*indicates radioactive) A	Atomic Mass (u)	Percent Abundance	Half-Life (if radioactive) $T_{1/2}$
(24)	Chromium			52	51.940 512	83.789	
				53	52.940 654	9.501	
				54	53.938 885	2.365	
25	Manganese	Mn	54.938 049	54*	53.940 363		312.1 days
				55	54.938 050	100	
26	Iron	Fe	55.845	54	53.939 615	5.845	
				55*	54.938 298		2.7 yr
				56	55.934 942	91.754	
				57	56.935 399	2.119	
				58	57.933 280	0.282	
				60*	59.934 077		1.5×10^6 yr
27	Cobalt	Co	58.933 200	59	58.933 200	100	
				60*	59.933 822		5.27 yr
28	Nickel	Ni	58.693 4	58	57.935 348	68.076 9	
				59*	58.934 351		7.5×10^4 yr
				60	59.930 790	26.223 1	
				61	60.931 060	1.139 9	
				62	61.928 349	3.634 5	
				63*	62.929 673		100 yr
				64	63.927 970	0.925 6	
29	Copper	Cu	63.546	63	62.929 601	69.17	
				65	64.927 794	30.83	
30	Zinc	Zn	65.39	64	63.929 147	48.63	
				66	65.926 037	27.90	
				67	66.927 131	4.10	
				68	67.924 848	18.75	
				70	69.925 325	0.62	
31	Gallium	Ga	69.723	69	68.925 581	60.108	
				71	70.924 705	39.892	
32	Germanium	Ge	72.61	70	69.924 250	20.84	
				72	71.922 076	27.54	
				73	72.923 459	7.73	
				74	73.921 178	36.28	
				76	75.921 403	7.61	
33	Arsenic	As	74.921 60	75	74.921 596	100	
34	Selenium	Se	78.96	74	73.922 477	0.89	
				76	75.919 214	9.37	
				77	76.919 915	7.63	
				78	77.917 310	23.77	
				79*	78.918 500		$\leq 6.5 \times 10^4$ yr
				80	79.916 522	49.61	
				82*	81.916 700	8.73	1.4×10^{20} yr
35	Bromine	Br	79.904	79	78.918 338	50.69	
				81	80.916 291	49.31	
36	Krypton	Kr	83.80	78	77.920 386	0.35	
				80	79.916 378	2.28	
				81*	80.916 592		2.1×10^5 yr
				82	81.913 485	11.58	
				83	82.914 136	11.49	
				84	83.911 507	57.00	
				85*	84.912 527		10.76 yr
				86	85.910 610	17.30	

(Continued)

TABLE A.3 Table of Atomic Masses *(Continued)*

Atomic Number Z	Element	Symbol	Chemical Atomic Mass (u)	Mass Number (*indicates radioactive) A	Atomic Mass (u)	Percent Abundance	Half-Life (if radioactive) $T_{1/2}$
37	Rubidium	Rb	85.467 8	85	84.911 789	72.17	
				87*	86.909 184	27.83	4.75×10^{10} yr
38	Strontium	Sr	87.62	84	83.913 425	0.56	
				86	85.909 262	9.86	
				87	86.908 880	7.00	
				88	87.905 614	82.58	
				90*	89.907 738		29.1 yr
39	Yttrium	Y	88.905 85	89	88.905 848	100	
40	Zirconium	Zr	91.224	90	89.904 704	51.45	
				91	90.905 645	11.22	
				92	91.905 040	17.15	
				93*	92.906 476		1.5×10^{6} yr
				94	93.906 316	17.38	
				96	95.908 276	2.80	
41	Niobium	Nb	92.906 38	91*	90.906 990		6.8×10^{2} yr
				92*	91.907 193		3.5×10^{7} yr
				93	92.906 378	100	
				94*	93.907 284		2×10^{4} yr
42	Molybdenum	Mo	95.94	92	91.906 810	14.84	
				93*	92.906 812		3.5×10^{3} yr
				94	93.905 088	9.25	
				95	94.905 842	15.92	
				96	95.904 679	16.68	
				97	96.906 021	9.55	
				98	97.905 408	24.13	
				100	99.907 477	9.63	
43	Technetium	Tc		97*	96.906 365		2.6×10^{6} yr
				98*	97.907 216		4.2×10^{6} yr
				99*	98.906 255		2.1×10^{5} yr
44	Ruthenium	Ru	101.07	96	95.907 598	5.54	
				98	97.905 287	1.87	
				99	98.905 939	12.76	
				100	99.904 220	12.60	
				101	100.905 582	17.06	
				102	101.904 350	31.55	
				104	103.905 430	18.62	
45	Rhodium	Rh	102.905 50	103	102.905 504	100	
46	Palladium	Pd	106.42	102	101.905 608	1.02	
				104	103.904 035	11.14	
				105	104.905 084	22.33	
				106	105.903 483	27.33	
				107*	106.905 128		6.5×10^{6} yr
				108	107.903 894	26.46	
				110	109.905 152	11.72	
47	Silver	Ag	107.868 2	107	106.905 093	51.839	
				109	108.904 756	48.161	
48	Cadmium	Cd	112.411	106	105.906 458	1.25	
				108	107.904 183	0.89	
				109*	108.904 986		462 days
				110	109.903 006	12.49	
				111	110.904 182	12.80	

(Continued)

TABLE A.3 Table of Atomic Masses *(Continued)*

Atomic Number Z	Element	Symbol	Chemical Atomic Mass (u)	Mass Number (*indicates radioactive) A	Atomic Mass (u)	Percent Abundance	Half-Life (if radioactive) $T_{1/2}$
(48)	Cadmium			112	111.902 757	24.13	
				113*	112.904 401	12.22	9.3×10^{15} yr
				114	113.903 358	28.73	
				116	115.904 755	7.49	
49	Indium	In	114.818	113	112.904 061	4.29	
				115*	114.903 878	95.71	4.4×10^{14} yr
50	Tin	Sn	118.710	112	111.904 821	0.97	
				114	113.902 782	0.66	
				115	114.903 346	0.34	
				116	115.901 744	14.54	
				117	116.902 954	7.68	
				118	117.901 606	24.22	
				119	118.903 309	8.59	
				120	119.902 197	32.58	
				121*	120.904 237		55 yr
				122	121.903 440	4.63	
				124	123.905 275	5.79	
51	Antimony	Sb	121.760	121	120.903 818	57.21	
				123	122.904 216	42.79	
				125*	124.905 248		2.7 yr
52	Tellurium	Te	127.60	120	119.904 020	0.09	
				122	121.903 047	2.55	
				123*	122.904 273	0.89	1.3×10^{13} yr
				124	123.902 820	4.74	
				125	124.904 425	7.07	
				126	125.903 306	18.84	
				128*	127.904 461	31.74	$>8 \times 10^{24}$ yr
				130*	129.906 223	34.08	$\leq 1.25 \times 10^{21}$ yr
53	Iodine	I	126.904 47	127	126.904 468	100	
				129*	128.904 988		1.6×10^{7} yr
54	Xenon	Xe	131.29	124	123.905 896	0.09	
				126	125.904 269	0.09	
				128	127.903 530	1.92	
				129	128.904 780	26.44	
				130	129.903 508	4.08	
				131	130.905 082	21.18	
				132	131.904 145	26.89	
				134	133.905 394	10.44	
				136*	135.907 220	8.87	$\geq 2.36 \times 10^{21}$ yr
55	Cesium	Cs	132.905 45	133	132.905 447	100	
				134*	133.906 713		2.1 yr
				135*	134.905 972		2×10^{6} yr
				137*	136.907 074		30 yr
56	Barium	Ba	137.327	130	129.906 310	0.106	
				132	131.905 056	0.101	
				133*	132.906 002		10.5 yr
				134	133.904 503	2.417	
				135	134.905 683	6.592	
				136	135.904 570	7.854	
				137	136.905 821	11.232	
				138	137.905 241	71.698	
57	Lanthanum	La	138.905 5	137*	136.906 466		6×10^{4} yr
				138*	137.907 107	0.090	1.05×10^{11} yr

(Continued)

TABLE A.3 Table of Atomic Masses *(Continued)*

Atomic Number Z	Element	Symbol	Chemical Atomic Mass (u)	Mass Number (*indicates radioactive) A	Atomic Mass (u)	Percent Abundance	Half-Life (if radioactive) $T_{1/2}$
(57)	Lanthanum			139	138.906 349	99.910	
58	Cerium	Ce	140.116	136	135.907 144	0.185	
				138	137.905 986	0.251	
				140	139.905 434	88.450	
				142*	141.909 240	11.114	$>5 \times 10^{16}$ yr
59	Praseodymium	Pr	140.907 65	141	140.907 648	100	
60	Neodymium	Nd	144.24	142	141.907 719	27.2	
				143	142.909 810	12.2	
				144*	143.910 083	23.8	2.3×10^{15} yr
				145	144.912 569	8.3	
				146	145.913 112	17.2	
				148	147.916 888	5.7	
				150*	149.920 887	5.6	$>1 \times 10^{18}$ yr
61	Promethium	Pm		143*	142.910 928		265 days
				145*	144.912 744		17.7 yr
				146*	145.914 692		5.5 yr
				147*	146.915 134		2.623 yr
62	Samarium	Sm	150.36	144	143.911 995	3.07	
				146*	145.913 037		1.0×10^{8} yr
				147*	146.914 893	14.99	1.06×10^{11} yr
				148*	147.914 818	11.24	7×10^{15} yr
				149*	148.917 180	13.82	$>2 \times 10^{15}$ yr
				150	149.917 272	7.38	
				151*	150.919 928		90 yr
				152	151.919 728	26.75	
				154	153.922 205	22.75	
63	Europium	Eu	151.964	151	150.919 846	47.81	
				152*	151.921 740		13.5 yr
				153	152.921 226	52.19	
				154*	153.922 975		8.59 yr
				155*	154.922 889		4.7 yr
64	Gadolinium	Gd	157.25	148*	147.918 110		75 yr
				150*	149.918 656		1.8×10^{6} yr
				152*	151.919 788	0.20	1.1×10^{14} yr
				154	153.920 862	2.18	
				155	154.922 619	14.80	
				156	155.922 120	20.47	
				157	156.923 957	15.65	
				158	157.924 100	24.84	
				160	159.927 051	21.86	
65	Terbium	Tb	158.925 34	159	158.925 343	100	
66	Dysprosium	Dy	162.50	156	155.924 278	0.06	
				158	157.924 405	0.10	
				160	159.925 194	2.34	
				161	160.926 930	18.91	
				162	161.926 795	25.51	
				163	162.928 728	24.90	
				164	163.929 171	28.18	
67	Holmium	Ho	164.930 32	165	164.930 320	100	
				166*	165.932 281		1.2×10^{3} yr
68	Erbium	Er	167.6	162	161.928 775	0.14	
				164	163.929 197	1.61	

(Continued)

TABLE A.3 Table of Atomic Masses *(Continued)*

Atomic Number Z	Element	Symbol	Chemical Atomic Mass (u)	Mass Number (*indicates radioactive) A	Atomic Mass (u)	Percent Abundance	Half-Life (if radioactive) $T_{1/2}$
(68)	Erbium			166	165.930 290	33.61	
				167	166.932 045	22.93	
				168	167.932 368	26.78	
				170	169.935 460	14.93	
69	Thulium	Tm	168.934 21	169	168.934 211	100	
				171*	170.936 426		1.92 yr
70	Ytterbium	Yb	173.04	168	167.933 894	0.13	
				170	169.934 759	3.04	
				171	170.936 322	14.28	
				172	171.936 378	21.83	
				173	172.938 207	16.13	
				174	173.938 858	31.83	
				176	175.942 568	12.76	
71	Lutecium	Lu	174.967	173*	172.938 927		1.37 yr
				175	174.940 768	97.41	
				176*	175.942 682	2.59	3.78×10^{10} yr
72	Hafnium	Hf	178.49	174*	173.940 040	0.16	2.0×10^{15} yr
				176	175.941 402	5.26	
				177	176.943 220	18.60	
				178	177.943 698	27.28	
				179	178.945 815	13.62	
				180	179.946 549	35.08	
73	Tantalum	Ta	180.947 9	180	179.947 466	0.012	
				181	180.947 996	99.988	
74	Tungsten (Wolfram)	W	183.84	180	179.946 706	0.12	
				182	181.948 206	26.50	
				183	182.950 224	14.31	
				184	183.950 933	30.64	
				186	185.954 362	28.43	
75	Rhenium	Re	186.207	185	184.952 956	37.40	
				187*	186.955 751	62.60	4.4×10^{10} yr
76	Osmium	Os	190.23	184	183.952 491	0.02	
				186*	185.953 838	1.59	2.0×10^{15} yr
				187	186.955 748	1.96	
				188	187.955 836	13.24	
				189	188.958 145	16.15	
				190	189.958 445	26.26	
				192	191.961 479	40.78	
				194*	193.965 179		6.0 yr
77	Iridium	Ir	192.217	191	190.960 591	37.3	
				193	192.962 924	62.7	
78	Platinum	Pt	195.078	190*	189.959 930	0.014	6.5×10^{11} yr
				192	191.961 035	0.782	
				194	193.962 664	32.967	
				195	194.964 774	33.832	
				196	195.964 935	25.242	
				198	197.967 876	7.163	
79	Gold	Au	196.966 55	197	196.966 552	100	
80	Mercury	Hg	200.59	196	195.965 815	0.15	
				198	197.966 752	9.97	
				199	198.968 262	16.87	
				200	199.968 309	23.10	
				201	200.970 285	13.18	

(Continued)

TABLE A.3 Table of Atomic Masses *(Continued)*

Atomic Number Z	Element	Symbol	Chemical Atomic Mass (u)	Mass Number (*indicates radioactive) A	Atomic Mass (u)	Percent Abundance	Half-Life (if radioactive) $T_{1/2}$
(80)	Mercury			202	201.970 626	29.86	
				204	203.973 476	6.87	
81	Thallium	Tl	204.383 3	203	202.972 329	29.524	
				204*	203.973 849		3.78 yr
				205	204.974 412	70.476	
		(Ra E″)		206*	205.976 095		4.2 min
		(Ac C″)		207*	206.977 408		4.77 min
		(Th C″)		208*	207.982 005		3.053 min
		(Ra C″)		210*	209.990 066		1.30 min
82	Lead	Pb	207.2	202*	201.972 144		5×10^4 yr
				204*	203.973 029	1.4	$\geq 1.4 \times 10^{17}$ yr
				205*	204.974 467		1.5×10^7 yr
				206	205.974 449	24.1	
				207	206.975 881	22.1	
				208	207.976 636	52.4	
		(Ra D)		210*	209.984 173		22.3 yr
		(Ac B)		211*	210.988 732		36.1 min
		(Th B)		212*	211.991 888		10.64 h
		(Ra B)		214*	213.999 798		26.8 min
83	Bismuth	Bi	208.980 38	207*	206.978 455		32.2 yr
				208*	207.979 727		3.7×10^5 yr
				209	208.980 383	100	
		(Ra E)		210*	209.984 105		5.01 days
		(Th C)		211*	210.987 258		2.14 min
				212*	211.991 272		60.6 min
		(Ra C)		214*	213.998 699		19.9 min
				215*	215.001 832		7.4 min
84	Polonium	Po		209*	208.982 416		102 yr
		(Ra F)		210*	209.982 857		138.38 days
		(Ac C′)		211*	210.986 637		0.52 s
		(Th C′)		212*	211.988 852		0.30 μs
		(Ra C′)		214*	213.995 186		164 μs
		(Ac A)		215*	214.999 415		0.001 8 s
		(Th A)		216*	216.001 905		0.145 s
		(Ra A)		218*	218.008 966		3.10 min
85	Astatine	At		215*	214.998 641		≈ 100 μs
				218*	218.008 682		1.6 s
				219*	219.011 297		0.9 min
86	Radon	Rn					
		(An)		219*	219.009 475		3.96 s
		(Tn)		220*	220.011 384		55.6 s
		(Rn)		222*	222.017 570		3.823 days
87	Francium	Fr					
		(Ac K)		223*	223.019 731		22 min
88	Radium	Ra					
		(Ac X)		223*	223.018 497		11.43 days
		(Th X)		224*	224.020 202		3.66 days
		(Ra)		226*	226.025 403		1 600 yr
		(Ms Th_1)		228*	228.031 064		5.75 yr
89	Actinium	Ac		227*	227.027 747		21.77 yr
		(Ms Th_2)		228*	228.031 015		6.15 h
90	Thorium	Th	232.038 1				

(Continued)

TABLE A.3 **Table of Atomic Masses** *(Continued)*

Atomic Number Z	Element	Symbol	Chemical Atomic Mass (u)	Mass Number (*indicates radioactive) A	Atomic Mass (u)	Percent Abundance	Half-Life (if radioactive) $T_{1/2}$
(90)	Thorium	(Rd Ac)		227*	227.027 699		18.72 days
		(Rd Th)		228*	228.028 731		1.913 yr
				229*	229.031 755		7 300 yr
		(Io)		230*	230.033 127		75.000 yr
		(UY)		231*	231.036 297		25.52 h
		(Th)		232*	232.038 050	100	1.40×10^{10} yr
		(UX_1)		234*	234.043 596		24.1 days
91	Protactinium	Pa	231.035 88	231*	231.035 879		32.760 yr
		(Uz)		234*	234.043 302		6.7 h
92	Uranium	U	238.028 9	232*	232.037 146		69 yr
				233*	233.039 628		1.59×10^5 yr
				234*	234.040 946	0.005 5	2.45×10^5 yr
		(Ac U)		235*	235.043 923	0.720 0	7.04×10^8 yr
				236*	236.045 562		2.34×10^7 yr
		(UI)		238*	238.050 783	99.274 5	4.47×10^9 yr
93	Neptunium	Np		235*	235.044 056		396 days
				236*	236.046 560		1.15×10^5 yr
				237*	237.048 167		2.14×10^6 yr
94	Plutonium	Pu		236*	236.046 048		2.87 yr
				238*	238.049 553		87.7 yr
				239*	239.052 156		2.412×10^4 yr
				240*	240.053 808		6 560 yr
				241*	241.056 845		14.4 yr
				242*	242.058 737		3.73×10^6 yr
				244*	244.064 198		8.1×10^7 yr

Sources: Chemical atomic masses are from T. B. Coplen, "Atomic Weights of the Elements 1999," a technical report to the International Union of Pure and Applied Chemistry, and published in *Pure and Applied Chemistry* 73(4), 667–683, 2001. Atomic masses of the isotopes are from G. Audi and A. H. Wapstra, "The 1995 Update to the Atomic Mass Evaluation," *Nuclear Physics* A595, vol. 4, 409–480, December 25, 1995. Percent abundance values are from K. J. R. Rosman and P. D. P. Taylor, "Isotopic Compositions of the Elements 1999," a technical report to the International Union of Pure and Applied Chemistry, and published in *Pure and Applied Chemistry* 70(1), 217–236, 1998.

Mathematics Review

This appendix in mathematics is intended as a brief review of operations and methods. Early in this course, you should be totally familiar with basic algebraic techniques, analytic geometry, and trigonometry. The sections on differential and integral calculus are more detailed and are intended for those students who have difficulty applying calculus concepts to physical situations.

B.1 SCIENTIFIC NOTATION

Many quantities that scientists deal with often have very large or very small values. The speed of light, for example, is about 300 000 000 m/s, and the ink required to make the dot over an *i* in this textbook has a mass of about 0.000 000 001 kg. Obviously, it is very cumbersome to read, write, and keep track of such numbers. We avoid this problem by using a method dealing with powers of the number ten:

$$10^0 = 1$$

$$10^1 = 10$$

$$10^2 = 10 \times 10 = 100$$

$$10^3 = 10 \times 10 \times 10 = 1\,000$$

$$10^4 = 10 \times 10 \times 10 \times 10 = 10\,000$$

$$10^5 = 10 \times 10 \times 10 \times 10 \times 10 = 100\,000$$

and so on. The number of zeros corresponds to the power to which ten is raised, called the **exponent** of ten. For example, the speed of light, 300 000 000 m/s, can be expressed as 3×10^8 m/s.

In this method, some representative numbers smaller than unity are the following:

$$10^{-1} = \frac{1}{10} = 0.1$$

$$10^{-2} = \frac{1}{10 \times 10} = 0.01$$

$$10^{-3} = \frac{1}{10 \times 10 \times 10} = 0.001$$

$$10^{-4} = \frac{1}{10 \times 10 \times 10 \times 10} = 0.000\,1$$

$$10^{-5} = \frac{1}{10 \times 10 \times 10 \times 10 \times 10} = 0.000\,01$$

In these cases, the number of places the decimal point is to the left of the digit 1 equals the value of the (negative) exponent. Numbers expressed as some power of ten multiplied by another number between one and ten are said to be in **scientific notation.** For example, the scientific notation for 5 943 000 000 is 5.943×10^9 and that for 0.000 083 2 is 8.32×10^{-5}.

When numbers expressed in scientific notation are being multiplied, the following general rule is very useful:

$$10^n \times 10^m = 10^{n+m} \qquad \text{[B.1]}$$

where n and m can be *any* numbers (not necessarily integers). For example, $10^2 \times 10^5 = 10^7$. The rule also applies if one of the exponents is negative: $10^3 \times 10^{-8} = 10^{-5}$.

When dividing numbers expressed in scientific notation, note that

$$\frac{10^n}{10^m} = 10^n \times 10^{-m} = 10^{n-m} \qquad \text{[B.2]}$$

Exercises

With help from the preceding rules, verify the answers to the following equations.

1. $86\,400 = 8.64 \times 10^4$
2. $9\,816\,762.5 = 9.816\,762\,5 \times 10^6$
3. $0.000\,000\,039\,8 = 3.98 \times 10^{-8}$
4. $(4.0 \times 10^8)(9.0 \times 10^9) = 3.6 \times 10^{18}$
5. $(3.0 \times 10^7)(6.0 \times 10^{-12}) = 1.8 \times 10^{-4}$
6. $\dfrac{75 \times 10^{-11}}{5.0 \times 10^{-3}} = 1.5 \times 10^{-7}$
7. $\dfrac{(3 \times 10^6)(8 \times 10^{-2})}{(2 \times 10^{17})(6 \times 10^5)} = 2 \times 10^{-18}$

B.2 ALGEBRA

Some Basic Rules

When algebraic operations are performed, the laws of arithmetic apply. Symbols such as x, y, and z are usually used to represent unspecified quantities, called the **unknowns.**

First, consider the equation

$$8x = 32$$

If we wish to solve for x, we can divide (or multiply) each side of the equation by the same factor without destroying the equality. In this case, if we divide both sides by 8, we have

$$\frac{8x}{8} = \frac{32}{8}$$

$$x = 4$$

Next consider the equation

$$x + 2 = 8$$

In this type of expression, we can add or subtract the same quantity from each side. If we subtract 2 from each side, we have

$$x + 2 - 2 = 8 - 2$$

$$x = 6$$

In general, if $x + a = b$, then $x = b - a$.

Now consider the equation

$$\frac{x}{5} = 9$$

If we multiply each side by 5, we are left with x on the left by itself and 45 on the right:

$$\left(\frac{x}{5}\right)(5) = 9 \times 5$$

$$x = 45$$

In all cases, *whatever operation is performed on the left side of the equality must also be performed on the right side.*

The following rules for multiplying, dividing, adding, and subtracting fractions should be recalled, where a, b, c, and d are four numbers:

	Rule	Example
Multiplying	$\left(\frac{a}{b}\right)\left(\frac{c}{d}\right) = \frac{ac}{bd}$	$\left(\frac{2}{3}\right)\left(\frac{4}{5}\right) = \frac{8}{15}$
Dividing	$\frac{(a/b)}{(c/d)} = \frac{ad}{bc}$	$\frac{2/3}{4/5} = \frac{(2)(5)}{(4)(3)} = \frac{10}{12}$
Adding	$\frac{a}{b} \pm \frac{c}{d} = \frac{ad \pm bc}{bd}$	$\frac{2}{3} - \frac{4}{5} = \frac{(2)(5) - (4)(3)}{(3)(5)} = -\frac{2}{15}$

Exercises

In the following exercises, solve for x:

		Answers
1.	$a = \frac{1}{1 + x}$	$x = \frac{1 - a}{a}$
2.	$3x - 5 = 13$	$x = 6$
3.	$ax - 5 = bx + 2$	$x = \frac{7}{a - b}$
4.	$\frac{5}{2x + 6} = \frac{3}{4x + 8}$	$x = -\frac{11}{7}$

Powers

When powers of a given quantity x are multiplied, the following rule applies:

$$x^n x^m = x^{n+m} \quad \text{[B.3]}$$

For example, $x^2 x^4 = x^{2+4} = x^6$.

When dividing the powers of a given quantity, the rule is

$$\frac{x^n}{x^m} = x^{n-m} \quad \text{[B.4]}$$

For example, $x^8/x^2 = x^{8-2} = x^6$.

A power that is a fraction, such as $\frac{1}{3}$, corresponds to a root as follows:

$$x^{1/n} = \sqrt[n]{x} \quad \text{[B.5]}$$

For example, $4^{1/3} = \sqrt[3]{4} = 1.5874$. (A scientific calculator is useful for such calculations.)

Finally, any quantity x^n raised to the mth power is

$$(x^n)^m = x^{nm} \qquad \text{[B.6]}$$

Table B.1 summarizes the rules of exponents.

TABLE B.1 Rules of Exponents

$x^0 = 1$
$x^1 = x$
$x^n x^m = x^{n+m}$
$x^n / x^m = x^{n-m}$
$x^{1/n} = \sqrt[n]{x}$
$(x^n)^m = x^{nm}$

Exercises

Verify the following equations.

1. $3^2 \times 3^3 = 243$
2. $x^5 x^{-8} = x^{-3}$
3. $x^{10}/x^{-5} = x^{15}$
4. $5^{1/3} = 1.709\,975$ (Use your calculator.)
5. $60^{1/4} = 2.783\,158$ (Use your calculator.)
6. $(x^4)^3 = x^{12}$

Factoring

Some useful formulas for factoring an equation are the following:

$$ax + ay + az = a(x + y + x) \qquad \text{common factor}$$

$$a^2 + 2ab + b^2 = (a + b)^2 \qquad \text{perfect square}$$

$$a^2 - b^2 = (a + b)(a - b) \qquad \text{differences of squares}$$

Quadratic Equations

The general form of a quadratic equation is

$$ax^2 + bx + c = 0 \qquad \text{[B.7]}$$

where x is the unknown quantity and a, b, and c are numerical factors referred to as **coefficients** of the equation. This equation has two roots, given by

$$x = \frac{-b \pm \sqrt{b^2 - 4ac}}{2a} \qquad \text{[B.8]}$$

If $b^2 \geq 4ac$, the roots are real.

EXAMPLE B.1

The equation $x^2 + 5x + 4 = 0$ has the following roots corresponding to the two signs of the square-root term:

$$x = \frac{-5 \pm \sqrt{5^2 - (4)(1)(4)}}{2(1)} = \frac{-5 \pm \sqrt{9}}{2} = \frac{-5 \pm 3}{2}$$

$$x_+ = \frac{-5 + 3}{2} = -1 \qquad x_- = \frac{-5 - 3}{2} = -4$$

where x_+ refers to the root corresponding to the positive sign and x_- refers to the root corresponding to the negative sign.

Exercises

Solve the following quadratic equations.

		Answers	
1.	$x^2 + 2x - 3 = 0$	$x_+ = 1$	$x_- = -3$
2.	$2x^2 - 5x + 2 = 0$	$x_+ = 2$	$x_- = \frac{1}{2}$
3.	$2x^2 - 4x - 9 = 0$	$x_+ = 1 + \sqrt{22}/2$	$x_- = 1 - \sqrt{22}/2$

Linear Equations

A linear equation has the general form

$$y = mx + b \quad \text{[B.9]}$$

where m and b are constants. This equation is referred to as being linear because the graph of y versus x is a straight line as shown in Figure B.1. The constant b, called the **y-intercept,** represents the value of y at which the straight line intersects the y axis. The constant m is equal to the **slope** of the straight line. If any two points on the straight line are specified by the coordinates (x_1, y_1) and (x_2, y_2) as in Figure B.1, the slope of the straight line can be expressed as

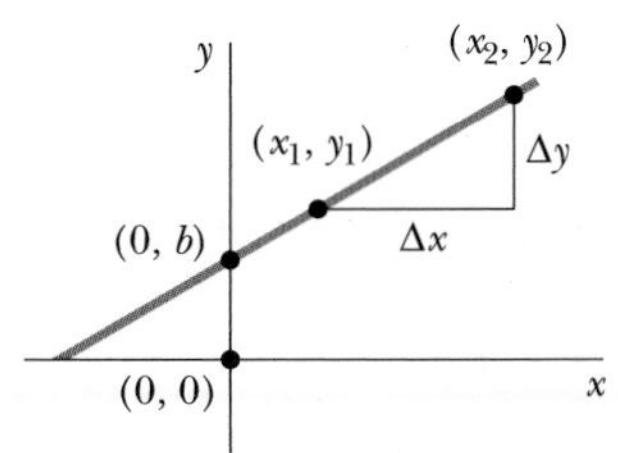

FIGURE B.1 A straight line graphed on an x-y coordinate system. The slope of the line is the ratio of Δy to Δx.

$$\text{Slope} = \frac{y_2 - y_1}{x_2 - x_1} = \frac{\Delta y}{\Delta x} \quad \text{[B.10]}$$

Note that m and b can have either positive or negative values. If $m > 0$, the straight line has a *positive* slope as in Figure B.1. If $m < 0$, the straight line has a *negative* slope. In Figure B.1, both m and b are positive. Three other possible situations are shown in Figure B.2.

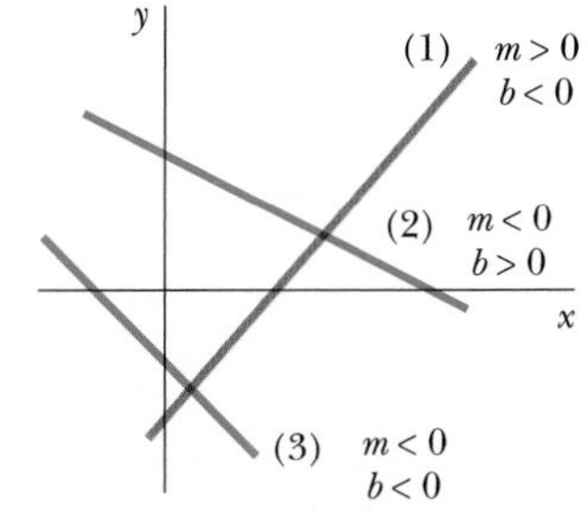

FIGURE B.2 The brown line has a positive slope and a negative y-intercept. The blue line has a negative slope and a positive y-intercept. The green line has a negative slope and a negative y-intercept.

Exercises

1. Draw graphs of the following straight lines: (a) $y = 5x + 3$, (b) $y = -2x + 4$, (c) $y = -3x - 6$.
2. Find the slopes of the straight lines described in Exercise 1.

Answers (a) 5 (b) -2 (c) -3

3. Find the slopes of the straight lines that pass through the following sets of points: (a) $(0, -4)$ and $(4, 2)$, (b) $(0, 0)$ and $(2, -5)$, (c) $(-5, 2)$ and $(4, -2)$.

Answers (a) $\frac{3}{2}$ (b) $-\frac{5}{2}$ (c) $-\frac{4}{9}$

Solving Simultaneous Linear Equations

Consider the equation $3x + 5y = 15$, which has two unknowns, x and y. Such an equation does not have a unique solution. For example, note that $(x = 0, y = 3)$, $(x = 5, y = 0)$, and $\left(x = 2, y = \frac{9}{5}\right)$ are all solutions to this equation.

If a problem has two unknowns, a unique solution is possible only if we have *two* equations. In general, if a problem has n unknowns, its solution requires n equations. To solve two simultaneous equations involving two unknowns, x and y, we solve one of the equations for x in terms of y and substitute this expression into the other equation.

EXAMPLE B.2

Solve the two simultaneous equations

$$(1) \quad 5x + y = -8$$

$$(2) \quad 2x - 2y = 4$$

Solution From (2), $x = y + 2$. Substitution of this equation into (1) gives

$$5(y + 2) + y = -8$$

$$6y = -18$$

$$y = -3$$

$$x = y + 2 = -1$$

Alternative Solution Multiply each term in (1) by the factor 2 and add the result to (2):

$$10x + 2y = -16$$

$$\underline{2x - 2y = 4}$$

$$12x \qquad = -12$$

$$x = -1$$

$$y = x - 2 = -3$$

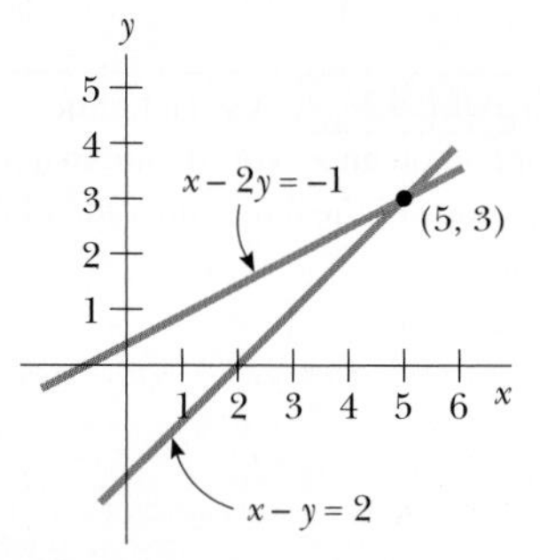

FIGURE B.3 A graphical solution for two linear equations.

Two linear equations containing two unknowns can also be solved by a graphical method. If the straight lines corresponding to the two equations are plotted in a conventional coordinate system, the intersection of the two lines represents the solution. For example, consider the two equations

$$x - y = 2$$

$$x - 2y = -1$$

These equations are plotted in Figure B.3. The intersection of the two lines has the coordinates $x = 5$ and $y = 3$, which represents the solution to the equations. You should check this solution by the analytical technique discussed earlier.

Exercises

Solve the following pairs of simultaneous equations involving two unknowns.

		Answers
1.	$x + y = 8$ $x - y = 2$	$x = 5, y = 3$
2.	$98 - T = 10a$ $T - 49 = 5a$	$T = 65, a = 3.27$
3.	$6x + 2y = 6$ $8x - 4y = 28$	$x = 2, y = -3$

Logarithms

Suppose a quantity x is expressed as a power of some quantity a:

$$x = a^y \qquad [B.11]$$

The number a is called the **base** number. The **logarithm** of x with respect to the base a is equal to the exponent to which the base must be raised to satisfy the expression $x = a^y$:

$$y = \log_a x \qquad [B.12]$$

Conversely, the **antilogarithm** of y is the number x:

$$x = \text{antilog}_a\, y \qquad [B.13]$$

In practice, the two bases most often used are base 10, called the *common* logarithm base, and base $e = 2.718\,282$, called Euler's constant or the *natural* logarithm base. When common logarithms are used,

$$y = \log_{10} x \qquad (\text{or } x = 10^y) \qquad \text{[B.14]}$$

When natural logarithms are used,

$$y = \ln x \qquad (\text{or } x = e^y) \qquad \text{[B.15]}$$

For example, $\log_{10} 52 = 1.716$, so $\text{antilog}_{10}\, 1.716 = 10^{1.716} = 52$. Likewise, $\ln 52 = 3.951$, so $\text{antiln}\, 3.951 = e^{3.951} = 52$.

In general, you can convert between base 10 and base e with the equality

$$\ln x = (2.302\,585) \log_{10} x \qquad \text{[B.16]}$$

Finally, some useful properties of logarithms are the following:

$$\left.\begin{aligned} \log(ab) &= \log a + \log b \\ \log(a/b) &= \log a - \log b \\ \log(a^n) &= n \log a \end{aligned}\right\} \text{ any base}$$

$$\ln e = 1$$

$$\ln e^a = a$$

$$\ln\left(\frac{1}{a}\right) = -\ln a$$

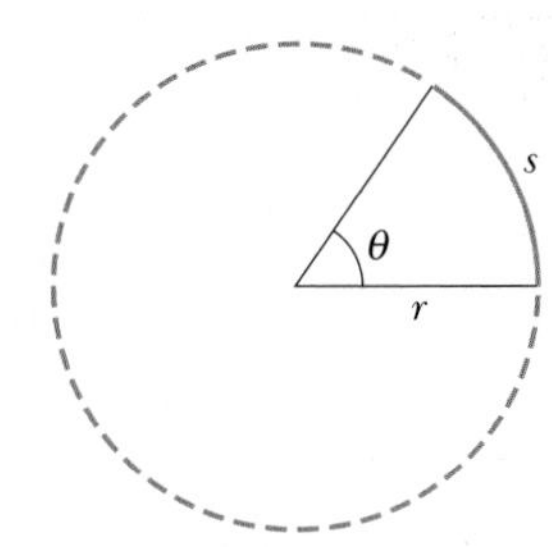

FIGURE B.4 The angle θ in radians is the ratio of the arc length s to the radius r of the circle.

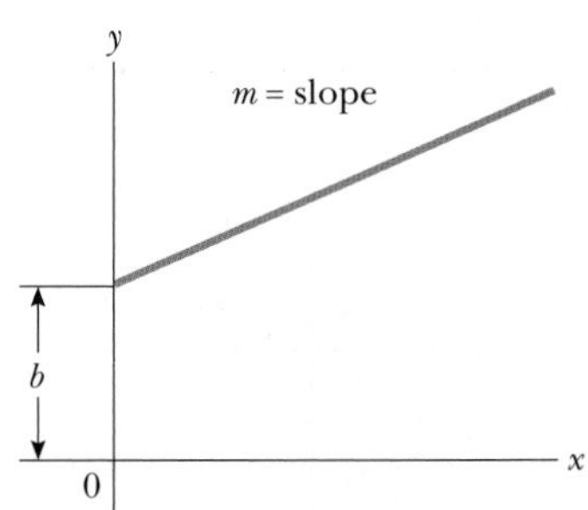

FIGURE B.5 A straight line with a slope of m and a y-intercept of b.

B.3 GEOMETRY

The **distance** d between two points having coordinates (x_1, y_1) and (x_2, y_2) is

$$d = \sqrt{(x_2 - x_1)^2 + (y_2 - y_1)^2} \qquad \text{[B.17]}$$

Radian measure: The arc length s of a circular arc (Fig. B.4) is proportional to the radius r for a fixed value of θ (in radians):

$$s = r\theta$$
$$\theta = \frac{s}{r} \qquad \text{[B.18]}$$

Table B.2 gives the **areas** and **volumes** for several geometric shapes used throughout this text.

The equation of a **straight line** (Fig. B.5) is

$$y = mx + b \qquad \text{[B.19]}$$

where b is the y-intercept and m is the slope of the line.

The equation of a **circle** of radius R centered at the origin is

$$x^2 + y^2 = R^2 \qquad \text{[B.20]}$$

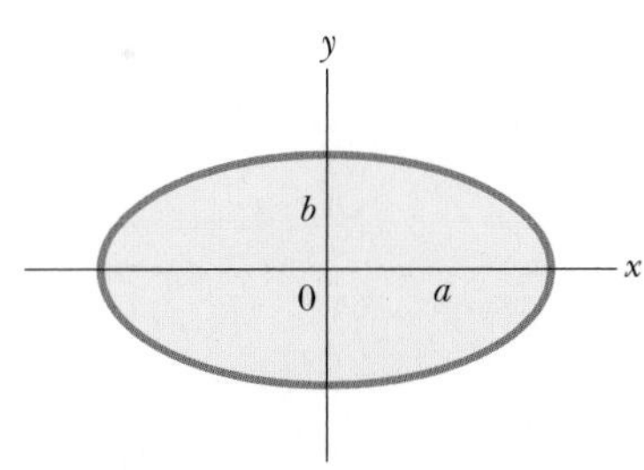

FIGURE B.6 An ellipse with semimajor axis a and semiminor axis b.

The equation of an **ellipse** having the origin at its center (Fig. B.6) is

$$\frac{x^2}{a^2} + \frac{y^2}{b^2} = 1 \qquad \text{[B.21]}$$

where a is the length of the semimajor axis (the longer one) and b is the length of the semiminor axis (the shorter one).

The equation of a **parabola** the vertex of which is at $y = b$ (Fig. B.7) is

$$y = ax^2 + b \qquad \text{[B.22]}$$

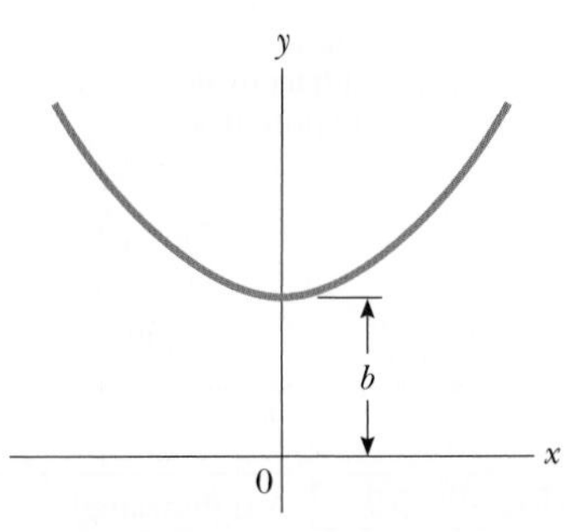

FIGURE B.7 A parabola.

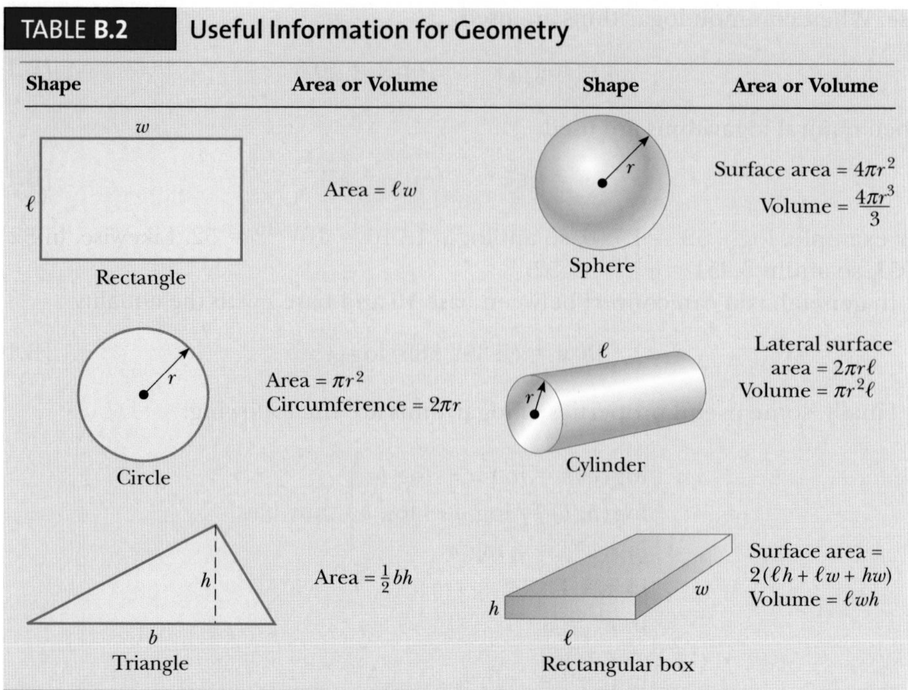

TABLE B.2 Useful Information for Geometry

Shape	Area or Volume	Shape	Area or Volume
Rectangle (ℓ, w)	Area = ℓw	Sphere (r)	Surface area = $4\pi r^2$ Volume = $\frac{4\pi r^3}{3}$
Circle (r)	Area = πr^2 Circumference = $2\pi r$	Cylinder (r, ℓ)	Lateral surface area = $2\pi r\ell$ Volume = $\pi r^2\ell$
Triangle (b, h)	Area = $\frac{1}{2}bh$	Rectangular box (ℓ, w, h)	Surface area = $2(\ell h + \ell w + hw)$ Volume = ℓwh

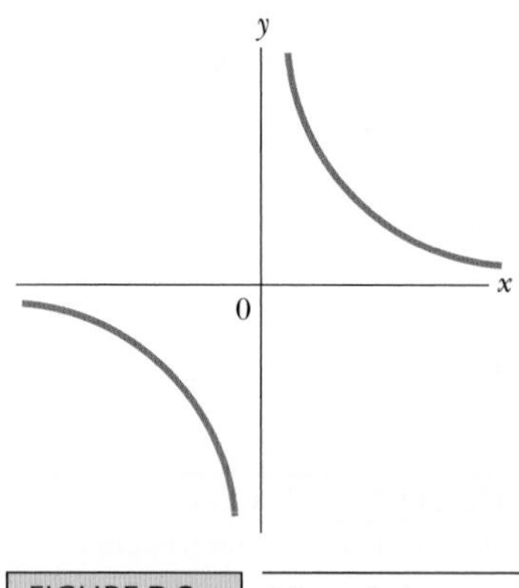

FIGURE **B.8** A hyperbola.

The equation of a **rectangular hyperbola** (Fig. B.8) is

$$xy = \text{constant} \qquad \text{[B.23]}$$

B.4 TRIGONOMETRY

That portion of mathematics based on the special properties of the right triangle is called trigonometry. By definition, a right triangle is a triangle containing a 90° angle. Consider the right triangle shown in Figure B.9, where side a is opposite the angle θ, side b is adjacent to the angle θ, and side c is the hypotenuse of the triangle. The three basic trigonometric functions defined by such a triangle are the sine (sin), cosine (cos), and tangent (tan) functions. In terms of the angle θ, these functions are defined by

$$\sin\theta = \frac{\text{side opposite } \theta}{\text{hypotenuse}} = \frac{a}{c} \qquad \text{[B.24]}$$

$$\cos\theta = \frac{\text{side adjacent to } \theta}{\text{hypotenuse}} = \frac{b}{c} \qquad \text{[B.25]}$$

$$\tan\theta = \frac{\text{side opposite } \theta}{\text{side adjacent to } \theta} = \frac{a}{b} \qquad \text{[B.26]}$$

a = opposite side
b = adjacent side
c = hypotenuse

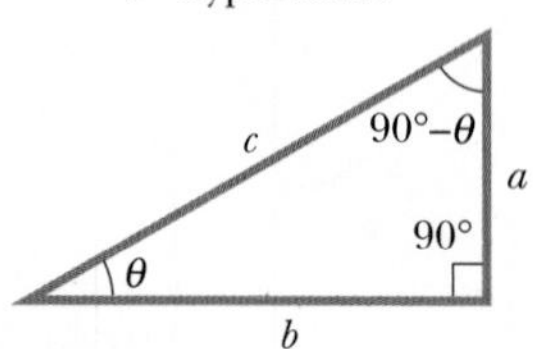

FIGURE **B.9** A right triangle, used to define the basic functions of trigonometry.

The Pythagorean theorem provides the following relationship among the sides of a right triangle:

$$c^2 = a^2 + b^2 \qquad \text{[B.27]}$$

From the preceding definitions and the Pythagorean theorem, it follows that

$$\sin^2\theta + \cos^2\theta = 1$$

$$\tan\theta = \frac{\sin\theta}{\cos\theta}$$

TABLE B.3 **Some Trigonometric Identities**

$\sin^2\theta + \cos^2\theta = 1$	$\csc^2\theta = 1 + \cot^2\theta$
$\sec^2\theta = 1 + \tan^2\theta$	$\sin^2\dfrac{\theta}{2} = \frac{1}{2}(1 - \cos\theta)$
$\sin 2\theta = 2\sin\theta\cos\theta$	$\cos^2\dfrac{\theta}{2} = \frac{1}{2}(1 + \cos\theta)$
$\cos 2\theta = \cos^2\theta - \sin^2\theta$	$1 - \cos\theta = 2\sin^2\dfrac{\theta}{2}$
$\tan 2\theta = \dfrac{2\tan\theta}{1 - \tan^2\theta}$	$\tan\dfrac{\theta}{2} = \sqrt{\dfrac{1 - \cos\theta}{1 + \cos\theta}}$

$\sin(A \pm B) = \sin A\cos B \pm \cos A\sin B$
$\cos(A \pm B) = \cos A\cos B \mp \sin A\sin B$
$\sin A \pm \sin B = 2\sin\left[\frac{1}{2}(A \pm B)\right]\cos\left[\frac{1}{2}(A \mp B)\right]$
$\cos A + \cos B = 2\cos\left[\frac{1}{2}(A + B)\right]\cos\left[\frac{1}{2}(A - B)\right]$
$\cos A - \cos B = 2\sin\left[\frac{1}{2}(A + B)\right]\sin\left[\frac{1}{2}(B - A)\right]$

The cosecant, secant, and cotangent functions are defined by

$$\csc\theta = \frac{1}{\sin\theta} \qquad \sec\theta = \frac{1}{\cos\theta} \qquad \cot\theta = \frac{1}{\tan\theta}$$

The following relationships are derived directly from the right triangle shown in Figure B.9:

$$\sin\theta = \cos(90° - \theta)$$
$$\cos\theta = \sin(90° - \theta)$$
$$\cot\theta = \tan(90° - \theta)$$

Some properties of trigonometric functions are

$$\sin(-\theta) = -\sin\theta$$
$$\cos(-\theta) = \cos\theta$$
$$\tan(-\theta) = -\tan\theta$$

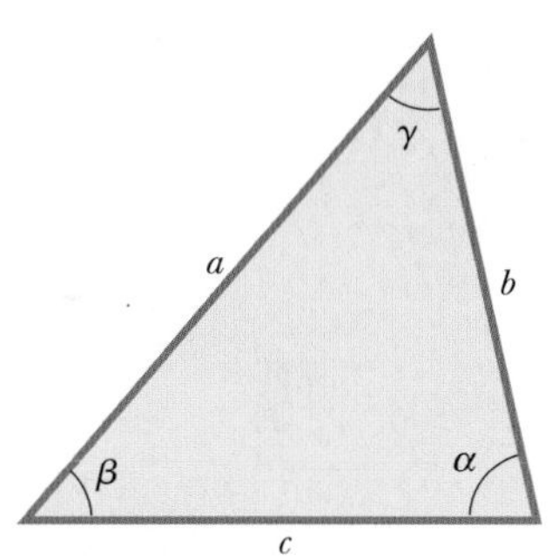

FIGURE B.10 An arbitrary, non-right triangle.

The following relationships apply to *any* triangle, as shown in Figure B.10:

$$\alpha + \beta + \gamma = 180°$$

Law of cosines
$$\begin{cases} a^2 = b^2 + c^2 - 2bc\cos\alpha \\ b^2 = a^2 + c^2 - 2ac\cos\beta \\ c^2 = a^2 + b^2 - 2ab\cos\gamma \end{cases}$$

Law of sines
$$\frac{a}{\sin\alpha} = \frac{b}{\sin\beta} = \frac{c}{\sin\gamma}$$

Table B.3 lists a number of useful trigonometric identities.

EXAMPLE B.3

Consider the right triangle in Figure B.11 in which $a = 2.00$, $b = 5.00$, and c is unknown. From the Pythagorean theorem we have

$$c^2 = a^2 + b^2 = 2.00^2 + 5.00^2 = 4.00 + 25.0 = 29.0$$

$$c = \sqrt{29.0} = 5.39$$

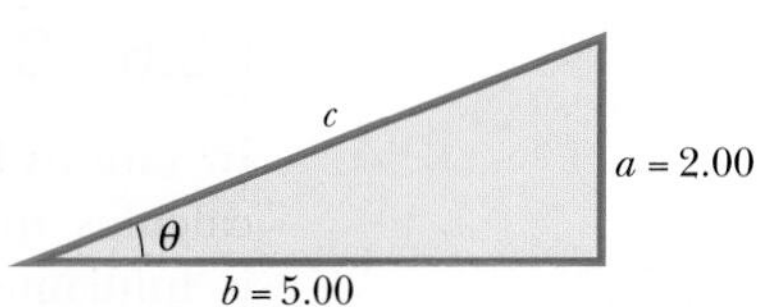

FIGURE B.11 (Example B.3)

To find the angle θ, note that

$$\tan\theta = \frac{a}{b} = \frac{2.00}{5.00} = 0.400$$

Using a calculator, we find that

$$\theta = \tan^{-1}(0.400) = 21.8°$$

where $\tan^{-1}(0.400)$ is the notation for "angle whose tangent is 0.400," sometimes written as arctan(0.400).

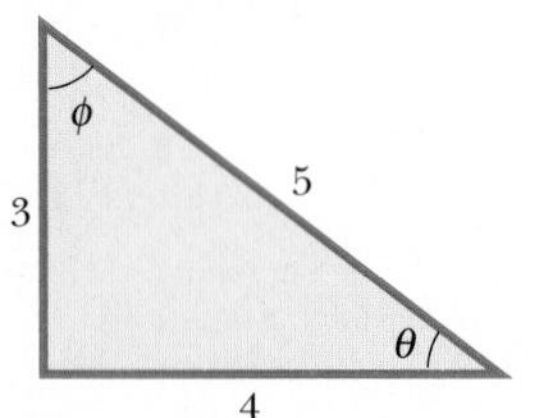

FIGURE B.12 (Exercise 1)

Exercises

1. In Figure B.12, identify (a) the side opposite θ and (b) the side adjacent to ϕ and then find (c) $\cos\theta$, (d) $\sin\phi$, and (e) $\tan\phi$.

Answers (a) 3 (b) 3 (c) $\frac{4}{5}$ (d) $\frac{4}{5}$ (e) $\frac{4}{3}$

2. In a certain right triangle, the two sides that are perpendicular to each other are 5.00 m and 7.00 m long. What is the length of the third side?

Answer 8.60 m

3. A right triangle has a hypotenuse of length 3.0 m, and one of its angles is 30°. (a) What is the length of the side opposite the 30° angle? (b) What is the length of the side adjacent to the 30° angle?

Answers (a) 1.5 m (b) 2.6 m

B.5 SERIES EXPANSIONS

$$(a+b)^n = a^n + \frac{n}{1!}a^{n-1}b + \frac{n(n-1)}{2!}a^{n-2}b^2 + \cdots$$

$$(1+x)^n = 1 + nx + \frac{n(n-1)}{2!}x^2 + \cdots$$

$$e^x = 1 + x + \frac{x^2}{2!} + \frac{x^3}{3!} + \cdots$$

$$\ln(1 \pm x) = \pm x - \tfrac{1}{2}x^2 \pm \tfrac{1}{3}x^3 - \cdots$$

$$\left.\begin{aligned} \sin x &= x - \frac{x^3}{3!} + \frac{x^5}{5!} - \cdots \\ \cos x &= 1 - \frac{x^2}{2!} + \frac{x^4}{4!} - \cdots \\ \tan x &= x + \frac{x^3}{3} + \frac{2x^5}{15} + \cdots \qquad |x| < \frac{\pi}{2} \end{aligned}\right\} x \text{ in radians}$$

For $x \ll 1$, the following approximations can be used:[1]

$$(1+x)^n \approx 1 + nx \qquad \sin x \approx x$$

$$e^x \approx 1 + x \qquad \cos x \approx 1$$

$$\ln(1 \pm x) \approx \pm x \qquad \tan x \approx x$$

B.6 DIFFERENTIAL CALCULUS

In various branches of science, it is sometimes necessary to use the basic tools of calculus, invented by Newton, to describe physical phenomena. The use of calculus is fundamental in the treatment of various problems in Newtonian mechanics,

[1] The approximations for the functions $\sin x$, $\cos x$, and $\tan x$ are for $x \le 0.1$ rad.

electricity, and magnetism. In this section, we simply state some basic properties and "rules of thumb" that should be a useful review to the student.

First, a **function** must be specified that relates one variable to another (e.g., a coordinate as a function of time). Suppose one of the variables is called y (the dependent variable) and the other x (the independent variable). We might have a function relationship such as

$$y(x) = ax^3 + bx^2 + cx + d$$

If a, b, c, and d are specified constants, y can be calculated for any value of x. We usually deal with continuous functions, that is, those for which y varies "smoothly" with x.

The **derivative** of y with respect to x is defined as the limit, as Δx approaches zero, of the slopes of chords drawn between two points on the y versus x curve. Mathematically, we write this definition as

$$\frac{dy}{dx} = \lim_{\Delta x \to 0} \frac{\Delta y}{\Delta x} = \lim_{\Delta x \to 0} \frac{y(x + \Delta x) - y(x)}{\Delta x} \quad \text{[B.28]}$$

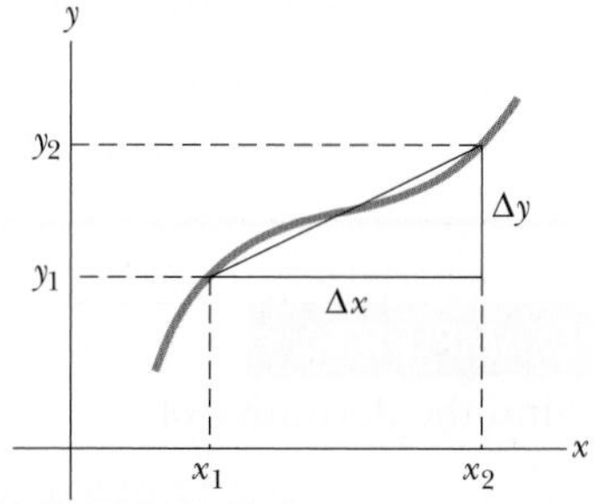

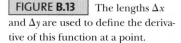

FIGURE B.13 The lengths Δx and Δy are used to define the derivative of this function at a point.

where Δy and Δx are defined as $\Delta x = x_2 - x_1$ and $\Delta y = y_2 - y_1$ (Fig. B.13). It is important to note that dy/dx *does not* mean dy divided by dx, but rather is simply a notation of the limiting process of the derivative as defined by Equation B.28.

A useful expression to remember when $y(x) = ax^n$, where a is a *constant* and n is *any* positive or negative number (integer or fraction), is

$$\frac{dy}{dx} = nax^{n-1} \quad \text{[B.29]}$$

If $y(x)$ is a polynomial or algebraic function of x, we apply Equation B.29 to *each* term in the polynomial and take $d[\text{constant}]/dx = 0$. In Examples 4 through 7, we evaluate the derivatives of several functions.

Special Properties of the Derivative

A. Derivative of the product of two functions If a function $f(x)$ is given by the product of two functions—say, $g(x)$ and $h(x)$—the derivative of $f(x)$ is defined as

$$\frac{d}{dx} f(x) = \frac{d}{dx}[g(x)h(x)] = g\frac{dh}{dx} + h\frac{dg}{dx} \quad \text{[B.30]}$$

B. Derivative of the sum of two functions If a function $f(x)$ is equal to the sum of two functions, the derivative of the sum is equal to the sum of the derivatives:

$$\frac{d}{dx} f(x) = \frac{d}{dx}[g(x) + h(x)] = \frac{dg}{dx} + \frac{dh}{dx} \quad \text{[B.31]}$$

C. Chain rule of differential calculus If $y = f(x)$ and $x = g(z)$, then dy/dz can be written as the product of two derivatives:

$$\frac{dy}{dz} = \frac{dy}{dx}\frac{dx}{dz} \quad \text{[B.32]}$$

D. The second derivative The second derivative of y with respect to x is defined as the derivative of the function dy/dx (the derivative of the derivative). It is usually written as

$$\frac{d^2y}{dx^2} = \frac{d}{dx}\left(\frac{dy}{dx}\right) \quad \text{[B.33]}$$

EXAMPLE B.4

Suppose $y(x)$ (that is, y as a function of x) is given by

$$y(x) = ax^3 + bx + c$$

where a and b are constants. Then it follows that

$$y(x + \Delta x) = a(x + \Delta x)^3 + b(x + \Delta x) + c$$
$$= a(x^3 + 3x^2\Delta x + 3x\Delta x^2 + \Delta x^3) + b(x + \Delta x) + c$$

so

$$\Delta y = y(x + \Delta x) - y(x) = a(3x^2\Delta x + 3x\Delta x^2 + \Delta x^3) + b\Delta x$$

Substituting this equation into Equation B.28 gives

$$\frac{dy}{dx} = \lim_{\Delta x \to 0} \frac{\Delta y}{\Delta x} = \lim_{\Delta x \to 0} a[3x^2 + 3x\Delta x + \Delta x^2] + b$$

$$\frac{dy}{dx} = 3ax^2 + b$$

EXAMPLE B.5

Find the derivative of

$$y(x) = 8x^5 + 4x^3 + 2x + 7$$

Solution By applying Equation B.29 to each term independently and remembering that d/dx (constant) $= 0$, we have

$$\frac{dy}{dx} = 8(5)x^4 + 4(3)x^2 + 2(1)x^0 + 0$$

$$\frac{dy}{dx} = 40x^4 + 12x^2 + 2$$

EXAMPLE B.6

Find the derivative of $y(x) = x^3/(x + 1)^2$ with respect to x.

Solution We can rewrite this function as $y(x) = x^3(x + 1)^{-2}$ and apply Equation B.30:

$$\frac{dy}{dx} = (x + 1)^{-2}\frac{d}{dx}(x^3) + x^3\frac{d}{dx}(x + 1)^{-2}$$
$$= (x + 1)^{-2}\,3x^2 + x^3(-2)(x + 1)^{-3}$$

$$\frac{dy}{dx} = \frac{3x^2}{(x + 1)^2} - \frac{2x^3}{(x + 1)^3}$$

EXAMPLE B.7

A useful formula that follows from Equation B.30 is the derivative of the quotient of two functions. Show that

$$\frac{d}{dx}\left[\frac{g(x)}{h(x)}\right] = \frac{h\dfrac{dg}{dx} - g\dfrac{dh}{dx}}{h^2}$$

Solution We can write the quotient as gh^{-1} and then apply Equations B.29 and B.30:

$$\frac{d}{dx}\left(\frac{g}{h}\right) = \frac{d}{dx}(gh^{-1}) = g\frac{d}{dx}(h^{-1}) + h^{-1}\frac{d}{dx}(g)$$
$$= -gh^{-2}\frac{dh}{dx} + h^{-1}\frac{dg}{dx}$$
$$= \frac{h\dfrac{dg}{dx} - g\dfrac{dh}{dx}}{h^2}$$

Some of the more commonly used derivatives of functions are listed in Table B.4.

B.7 INTEGRAL CALCULUS

We think of integration as the inverse of differentiation. As an example, consider the expression

$$f(x) = \frac{dy}{dx} = 3ax^2 + b \qquad \text{[B.34]}$$

which was the result of differentiating the function

$$y(x) = ax^3 + bx + c$$

in Example 4. We can write Equation B.34 as $dy = f(x)\,dx = (3ax^2 + b)\,dx$ and obtain $y(x)$ by "summing" over all values of x. Mathematically, we write this inverse operation

$$y(x) = \int f(x)\,dx$$

For the function $f(x)$ given by Equation B.34, we have

$$y(x) = \int (3ax^2 + b)\ dx = ax^3 + bx + c$$

where c is a constant of the integration. This type of integral is called an *indefinite integral* because its value depends on the choice of c.

A general **indefinite integral** $I(x)$ is defined as

$$I(x) = \int f(x)\,dx \qquad \text{[B.35]}$$

where $f(x)$ is called the *integrand* and $f(x) = dI(x)/dx$.

For a *general continuous* function $f(x)$, the integral can be described as the area under the curve bounded by $f(x)$ and the x axis, between two specified values of x, say, x_1 and x_2, as in Figure B.14.

The area of the blue element in Figure B.14 is approximately $f(x_i)\ \Delta x_i$. If we sum all these area elements between x_1 and x_2 and take the limit of this sum as $\Delta x_i \rightarrow 0$, we obtain the *true* area under the curve bounded by $f(x)$ and the x axis, between the limits x_1 and x_2:

$$\text{Area} = \lim_{\Delta x_i \rightarrow 0} \sum_i f(x_i)\Delta x_i = \int_{x_1}^{x_2} f(x)\ dx \qquad \text{[B.36]}$$

Integrals of the type defined by Equation B.36 are called **definite integrals.**

One common integral that arises in practical situations has the form

$$\int x^n\ dx = \frac{x^{n+1}}{n+1} + c \qquad (n \neq -1) \qquad \text{[B.37]}$$

This result is obvious because differentiation of the right-hand side with respect to x gives $f(x) = x^n$ directly. If the limits of the integration are known, this integral becomes a *definite integral* and is written

$$\int_{x_1}^{x_2} x^n\ dx = \frac{x^{n+1}}{n+1}\bigg|_{x_1}^{x_2} = \frac{x_2^{\,n+1} - x_1^{\,n+1}}{n+1} \qquad (n \neq -1) \qquad \text{[B.38]}$$

TABLE B.4 Derivatives for Several Functions

$\dfrac{d}{dx}(a) = 0$
$\dfrac{d}{dx}(ax^n) = nax^{n-1}$
$\dfrac{d}{dx}(e^{ax}) = ae^{ax}$
$\dfrac{d}{dx}(\sin ax) = a\cos ax$
$\dfrac{d}{dx}(\cos ax) = -a\sin ax$
$\dfrac{d}{dx}(\tan ax) = a\sec^2 ax$
$\dfrac{d}{dx}(\cot ax) = -a\csc^2 ax$
$\dfrac{d}{dx}(\sec x) = \tan x\sec x$
$\dfrac{d}{dx}(\csc x) = -\cot x\csc x$
$\dfrac{d}{dx}(\ln ax) = \dfrac{1}{x}$

Note: The letters a and n are constants.

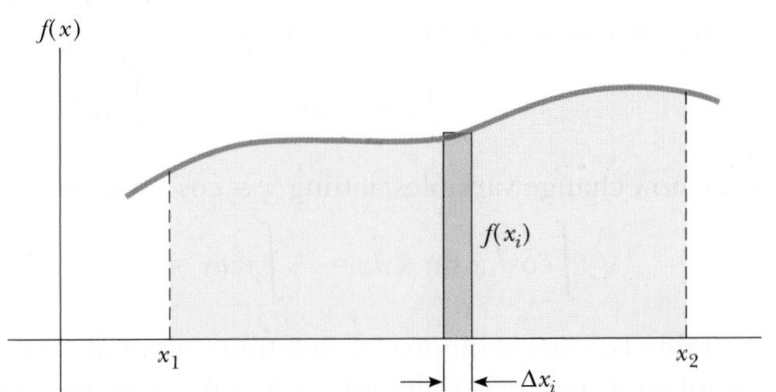

FIGURE B.14 The definite integral of a function is the area under the curve of the function between the limits x_1 and x_2.

EXAMPLES

1. $\displaystyle\int_0^a x^2\,dx = \frac{x^3}{3}\Big]_0^a = \frac{a^3}{3}$

2. $\displaystyle\int_0^b x^{3/2}\,dx = \frac{x^{5/2}}{5/2}\Big]_0^b = \tfrac{2}{5}b^{5/2}$

3. $\displaystyle\int_3^5 x\,dx = \frac{x^2}{2}\Big]_3^5 = \frac{5^2-3^2}{2} = 8$

Partial Integration

Sometimes it is useful to apply the method of *partial integration* (also called "integrating by parts") to evaluate certain integrals. The method uses the property that

$$\int u\,dv = uv - \int v\,du \qquad \text{[B.39]}$$

where u and v are *carefully* chosen so as to reduce a complex integral to a simpler one. In many cases, several reductions have to be made. Consider the function

$$I(x) = \int x^2 e^x\,dx$$

which can be evaluated by integrating by parts twice. First, if we choose $u = x^2$, $v = e^x$, we obtain

$$\int x^2 e^x\,dx = \int x^2\,d(e^x) = x^2 e^x - 2\int e^x x\,dx + c_1$$

Now, in the second term, choose $u = x$, $v = e^x$, which gives

$$\int x^2 e^x\,dx = x^2 e^x - 2xe^x + 2\int e^x\,dx + c_1$$

or

$$\int x^2 e^x\,dx = x^2 e^x - 2xe^x + 2e^x + c_2$$

The Perfect Differential

Another useful method to remember is the use of the *perfect differential,* in which we look for a change of variable such that the differential of the function is the differential of the independent variable appearing in the integrand. For example, consider the integral

$$I(x) = \int \cos^2 x \sin x\,dx$$

This integral becomes easy to evaluate if we rewrite the differential as $d(\cos x) = -\sin x\,dx$. The integral then becomes

$$\int \cos^2 x \sin x\,dx = -\int \cos^2 x\,d(\cos x)$$

If we now change variables, letting $y = \cos x$, we obtain

$$\int \cos^2 x \sin x\,dx = -\int y^2 dy = -\frac{y^3}{3} + c = -\frac{\cos^3 x}{3} + c$$

Table B.5 lists some useful indefinite integrals. Table B.6 gives Gauss's probability integral and other definite integrals. A more complete list can be found in

TABLE B.5 **Some Indefinite Integrals (An arbitrary constant should be added to each of these integrals.)**

$\int x^n\,dx = \frac{x^{n+1}}{n+1}$ (provided $n \neq -1$)	$\int \ln ax\,dx = (x \ln ax) - x$
$\int \frac{dx}{x} = \int x^{-1}\,dx = \ln x$	$\int xe^{ax}\,dx = \frac{e^{ax}}{a^2}(ax - 1)$
$\int \frac{dx}{a + bx} = \frac{1}{b}\ln(a + bx)$	$\int \frac{dx}{a + be^{cx}} = \frac{x}{a} - \frac{1}{ac}\ln(a + be^{cx})$
$\int \frac{x\,dx}{a + bx} = \frac{x}{b} - \frac{a}{b^2}\ln(a + bx)$	$\int \sin ax\,dx = -\frac{1}{a}\cos ax$
$\int \frac{dx}{x(x + a)} = -\frac{1}{a}\ln\frac{x + a}{x}$	$\int \cos ax\,dx = \frac{1}{a}\sin ax$
$\int \frac{dx}{(a + bx)^2} = -\frac{1}{b(a + bx)}$	$\int \tan ax\,dx = -\frac{1}{a}\ln(\cos ax) = \frac{1}{a}\ln(\sec ax)$
$\int \frac{dx}{a^2 + x^2} = \frac{1}{a}\tan^{-1}\frac{x}{a}$	$\int \cot ax\,dx = \frac{1}{a}\ln(\sin ax)$
$\int \frac{dx}{a^2 - x^2} = \frac{1}{2a}\ln\frac{a + x}{a - x}$ $\quad (a^2 - x^2 > 0)$	$\int \sec ax\,dx = \frac{1}{a}\ln(\sec ax + \tan ax) = \frac{1}{a}\ln\left[\tan\left(\frac{ax}{2} + \frac{\pi}{4}\right)\right]$
$\int \frac{dx}{x^2 - a^2} = \frac{1}{2a}\ln\frac{x - a}{x + a}$ $\quad (x^2 - a^2 > 0)$	$\int \csc ax\,dx = \frac{1}{a}\ln(\csc ax - \cot ax) = \frac{1}{a}\ln\left(\tan\frac{ax}{2}\right)$
$\int \frac{x\,dx}{a^2 \pm x^2} = \pm\frac{1}{2}\ln(a^2 \pm x^2)$	$\int \sin^2 ax\,dx = \frac{x}{2} - \frac{\sin 2ax}{4a}$
$\int \frac{dx}{\sqrt{a^2 - x^2}} = \sin^{-1}\frac{x}{a} = -\cos^{-1}\frac{x}{a}$ $\quad (a^2 - x^2 > 0)$	$\int \cos^2 ax\,dx = \frac{x}{2} + \frac{\sin 2ax}{4a}$
$\int \frac{dx}{\sqrt{x^2 \pm a^2}} = \ln\left(x + \sqrt{x^2 \pm a^2}\right)$	$\int \frac{dx}{\sin^2 ax} = -\frac{1}{a}\cot ax$
$\int \frac{x\,dx}{\sqrt{a^2 - x^2}} = -\sqrt{a^2 - x^2}$	$\int \frac{dx}{\cos^2 ax} = \frac{1}{a}\tan ax$
$\int \frac{x\,dx}{\sqrt{x^2 \pm a^2}} = \sqrt{x^2 \pm a^2}$	$\int \tan^2 ax\,dx = \frac{1}{a}(\tan ax) - x$
$\int \sqrt{a^2 - x^2}\,dx = \frac{1}{2}\left(x\sqrt{a^2 - x^2} + a^2\sin^{-1}\frac{x}{a}\right)$	$\int \cot^2 ax\,dx = -\frac{1}{a}(\cot ax) - x$
$\int x\sqrt{a^2 - x^2}\,dx = -\frac{1}{3}(a^2 - x^2)^{3/2}$	$\int \sin^{-1} ax\,dx = x(\sin^{-1} ax) + \frac{\sqrt{1 - a^2x^2}}{a}$
$\int \sqrt{x^2 \pm a^2}\,dx = \frac{1}{2}\left[x\sqrt{x^2 \pm a^2} \pm a^2\ln\left(x + \sqrt{x^2 \pm a^2}\right)\right]$	$\int \cos^{-1} ax\,dx = x(\cos^{-1} ax) - \frac{\sqrt{1 - a^2x^2}}{a}$
$\int x\left(\sqrt{x^2 \pm a^2}\right)dx = \frac{1}{3}(x^2 \pm a^2)^{3/2}$	$\int \frac{dx}{(x^2 + a^2)^{3/2}} = \frac{x}{a^2\sqrt{x^2 + a^2}}$
$\int e^{ax}\,dx = \frac{1}{a}e^{ax}$	$\int \frac{x\,dx}{(x^2 + a^2)^{3/2}} = -\frac{1}{\sqrt{x^2 + a^2}}$

various handbooks, such as *The Handbook of Chemistry and Physics* (Boca Raton, FL: CRC Press, published annually).

B.8 PROPAGATION OF UNCERTAINTY

In laboratory experiments, a common activity is to take measurements that act as raw data. These measurements are of several types—length, time interval, temperature, voltage, and so on—and are taken by a variety of instruments. Regardless of the measurement and the quality of the instrumentation, **there is always uncertainty associated with a physical measurement.** This uncertainty is a combination of that associated with the instrument and that related to the system being measured.

TABLE B.6 Gauss's Probability Integral and Other Definite Integrals

$$\int_0^\infty x^n e^{-ax}\,dx = \frac{n!}{a^{n+1}}$$

$$I_0 = \int_0^\infty e^{-ax^2}\,dx = \tfrac{1}{2}\sqrt{\frac{\pi}{a}} \qquad \text{(Gauss's probability integral)}$$

$$I_1 = \int_0^\infty x e^{-ax^2}\,dx = \frac{1}{2a}$$

$$I_2 = \int_0^\infty x^2 e^{-ax^2}\,dx = -\frac{dI_0}{da} = \tfrac{1}{4}\sqrt{\frac{\pi}{a^3}}$$

$$I_3 = \int_0^\infty x^3 e^{-ax^2}\,dx = -\frac{dI_1}{da} = \frac{1}{2a^2}$$

$$I_4 = \int_0^\infty x^4 e^{-ax^2}\,dx = \frac{d^2 I_0}{da^2} = \tfrac{3}{8}\sqrt{\frac{\pi}{a^5}}$$

$$I_5 = \int_0^\infty x^5 e^{-ax^2}\,dx = \frac{d^2 I_1}{da^2} = \frac{1}{a^3}$$

$$\vdots$$

$$I_{2n} = (-1)^n \frac{d^n}{da^n} I_0$$

$$I_{2n+1} = (-1)^n \frac{d^n}{da^n} I_1$$

An example of the former is the inability to determine exactly the position of a length measurement between the lines on a meter stick. An example of uncertainty related to the system being measured is the variation of temperature within a sample of water so that a single temperature for the sample is difficult to determine.

Uncertainties can be expressed in two ways. **Absolute uncertainty** refers to an uncertainty expressed in the same units as the measurement. Thus, the length of a computer disk label might be expressed as (5.5 ± 0.1) cm. The uncertainty of ± 0.1 cm by itself is not descriptive enough for some purposes, however. This uncertainty is large if the measurement is 1.0 cm, but it is small if the measurement is 100 m. To give a more descriptive account of the uncertainty, **fractional uncertainty** or **percent uncertainty** is used. In this type of description, the uncertainty is divided by the actual measurement. Therefore, the length of the computer disk label could be expressed as

$$\ell = 5.5\,\text{cm} \pm \frac{0.1\,\text{cm}}{5.5\,\text{cm}} = 5.5\,\text{cm} \pm 0.018 \qquad \text{(fractional uncertainty)}$$

or as

$$\ell = 5.5\,\text{cm} \pm 1.8\% \qquad \text{(percent uncertainty)}$$

When combining measurements in a calculation, the percent uncertainty in the final result is generally larger than the uncertainty in the individual measurements. This **propagation of uncertainty** is one of the challenges of experimental physics.

Some simple rules can provide a reasonable estimate of the uncertainty in a calculated result.

Multiplication and division: When measurements with uncertainties are multiplied or divided, add the *percent uncertainties* to obtain the percent uncertainty in the result.

Example: The Area of a Rectangular Plate

$$A = \ell w = (5.5\text{ cm} \pm 1.8\%) \times (6.4\text{ cm} \pm 1.6\%) = 35\text{ cm}^2 \pm 3.4\%$$
$$= (35 \pm 1)\text{ cm}^2$$

Addition and subtraction: When measurements with uncertainties are added or subtracted, add the *absolute uncertainties* to obtain the absolute uncertainty in the result.

Example: A Change in Temperature

$$\Delta T = T_2 - T_1 = (99.2 \pm 1.5)^\circ\text{C} - (27.6 \pm 1.5)^\circ\text{C} = (71.6 \pm 3.0)^\circ\text{C}$$
$$= 71.6^\circ\text{C} \pm 4.2\%$$

Powers: If a measurement is taken to a power, the percent uncertainty is multiplied by that power to obtain the percent uncertainty in the result.

Example: The Volume of a Sphere

$$V = \tfrac{4}{3}\pi r^3 = \tfrac{4}{3}\pi(6.20\text{ cm} \pm 2.0\%)^3 = 998\text{ cm}^3 \pm 6.0\%$$
$$= (998 \pm 60)\text{ cm}^3$$

For complicated calculations, many uncertainties are added together. This can cause the uncertainty in the final result to be undesirably large. Experiments should be designed such that calculations are as simple as possible.

Notice that uncertainties in a calculation always add. As a result, an experiment involving a subtraction should be avoided if possible, especially if the measurements being subtracted are close together. The result of such a calculation is a small difference in the measurements and uncertainties that add together. It is possible that the uncertainty in the result could be larger than the result itself!

APPENDIX C

Periodic Table of the Elements

Symbol: Ca — Atomic number: 20
Atomic mass†: 40.078
Electron configuration: $4s^2$

Group I	Group II	Transition elements						
H 1 1.007 9 $1s$								
Li 3 6.941 $2s^1$	**Be** 4 9.0122 $2s^2$							
Na 11 22.990 $3s^1$	**Mg** 12 24.305 $3s^2$							
K 19 39.098 $4s^1$	**Ca** 20 40.078 $4s^2$	**Sc** 21 44.956 $3d^14s^2$	**Ti** 22 47.867 $3d^24s^2$	**V** 23 50.942 $3d^34s^2$	**Cr** 24 51.996 $3d^54s^1$	**Mn** 25 54.938 $3d^54s^2$	**Fe** 26 55.845 $3d^64s^2$	**Co** 27 58.933 $3d^74s^2$
Rb 37 85.468 $5s^1$	**Sr** 38 87.62 $5s^2$	**Y** 39 88.906 $4d^15s^2$	**Zr** 40 91.224 $4d^25s^2$	**Nb** 41 92.906 $4d^45s^1$	**Mo** 42 95.94 $4d^55s^1$	**Tc** 43 (98) $4d^55s^2$	**Ru** 44 101.07 $4d^75s^1$	**Rh** 45 102.91 $4d^85s^1$
Cs 55 132.91 $6s^1$	**Ba** 56 137.33 $6s^2$	57–71*	**Hf** 72 178.49 $5d^26s^2$	**Ta** 73 180.95 $5d^36s^2$	**W** 74 183.84 $5d^46s^2$	**Re** 75 186.21 $5d^56s^2$	**Os** 76 190.23 $5d^66s^2$	**Ir** 77 192.2 $5d^76s^2$
Fr 87 (223) $7s^1$	**Ra** 88 (226) $7s^2$	89–103**	**Rf** 104 (261) $6d^27s^2$	**Db** 105 (262) $6d^37s^2$	**Sg** 106 (266)	**Bh** 107 (264)	**Hs** 108 (269)	**Mt** 109 (268)

*Lanthanide series	**La** 57 138.91 $5d^16s^2$	**Ce** 58 140.12 $5d^14f^16s^2$	**Pr** 59 140.91 $4f^36s^2$	**Nd** 60 144.24 $4f^46s^2$	**Pm** 61 (145) $4f^56s^2$	**Sm** 62 150.36 $4f^66s^2$
**Actinide series	**Ac** 89 (227) $6d^17s^2$	**Th** 90 232.04 $6d^27s^2$	**Pa** 91 231.04 $5f^26d^17s^2$	**U** 92 238.03 $5f^36d^17s^2$	**Np** 93 (237) $5f^46d^17s^2$	**Pu** 94 (244) $5f^66d^07s^2$

Note: Atomic mass values given are averaged over isotopes in the percentages in which they exist in nature.
†For an unstable element, mass number of the most stable known isotope is given in parentheses.
††Elements 111, 112, and 114 have not yet been named
†††For a description of the atomic data, visit **physics.nist.gov/atomic**.

			Group III	Group IV	Group V	Group VI	Group VII	Group 0
							H 1 1.007 9 $1s^1$	**He** 2 4.002 6 $1s^2$
			B 5 10.811 $2p^1$	**C** 6 12.011 $2p^2$	**N** 7 14.007 $2p^3$	**O** 8 15.999 $2p^4$	**F** 9 18.998 $2p^5$	**Ne** 10 20.180 $2p^6$
			Al 13 26.982 $3p^1$	**Si** 14 28.086 $3p^2$	**P** 15 30.974 $3p^3$	**S** 16 32.066 $3p^4$	**Cl** 17 35.453 $3p^5$	**Ar** 18 39.948 $3p^6$
Ni 28 58.693 $3d^84s^2$	**Cu** 29 63.546 $3d^{10}4s^1$	**Zn** 30 65.39 $3d^{10}4s^2$	**Ga** 31 69.723 $4p^1$	**Ge** 32 72.61 $4p^2$	**As** 33 74.922 $4p^3$	**Se** 34 78.96 $4p^4$	**Br** 35 79.904 $4p^5$	**Kr** 36 83.80 $4p^6$
Pd 46 106.42 $4d^{10}$	**Ag** 47 107.87 $4d^{10}5s^1$	**Cd** 48 112.41 $4d^{10}5s^2$	**In** 49 114.82 $5p^1$	**Sn** 50 118.71 $5p^2$	**Sb** 51 121.76 $5p^3$	**Te** 52 127.60 $5p^4$	**I** 53 126.90 $5p^5$	**Xe** 54 131.29 $5p^6$
Pt 78 195.08 $5d^96s^1$	**Au** 79 196.97 $5d^{10}6s^1$	**Hg** 80 200.59 $5d^{10}6s^2$	**Tl** 81 204.38 $6p^1$	**Pb** 82 207.2 $6p^2$	**Bi** 83 208.98 $6p^3$	**Po** 84 (209) $6p^4$	**At** 85 (210) $6p^5$	**Rn** 86 (222) $6p^6$
Ds 110 (271)	111†† (272)	112†† (285)		114†† (289)				

Eu 63 151.96 $4f^76s^2$	**Gd** 64 157.25 $5d^14f^76s^2$	**Tb** 65 158.93 $5d^14f^86s^2$	**Dy** 66 162.50 $4f^{10}6s^2$	**Ho** 67 164.93 $4f^{11}6s^2$	**Er** 68 167.26 $4f^{12}6s^2$	**Tm** 69 168.93 $4f^{13}6s^2$	**Yb** 70 173.04 $4f^{14}6s^2$	**Lu** 71 174.97 $5d^14f^{14}6s^2$
Am 95 (243) $5f^76d^07s^2$	**Cm** 96 (247) $5f^76d^17s^2$	**Bk** 97 (247) $5f^86d^17s^2$	**Cf** 98 (251) $5f^{10}6d^07s^2$	**Es** 99 (252) $5f^{11}6d^07s^2$	**Fm** 100 (257) $5f^{12}6d^07s^2$	**Md** 101 (258) $5f^{13}6d^07s^2$	**No** 102 (259) $6d^07s^2$	**Lr** 103 (262) $6d^17s^2$

APPENDIX D

SI Units

TABLE D.1 SI Base Units

	SI Base Unit	
Base Quantity	**Name**	**Symbol**
Length	meter	m
Mass	kilogram	kg
Time	second	s
Electric current	ampere	A
Temperature	kelvin	K
Amount of substance	mole	mol
Luminous intensity	candela	cd

TABLE D.2 Some Derived SI Units

Quantity	**Name**	**Symbol**	**Expression in Terms of Base Units**	**Expression in Terms of Other SI Units**
Plane angle	radian	rad	m/m	
Frequency	hertz	Hz	s^{-1}	
Force	newton	N	$kg \cdot m/s^2$	J/m
Pressure	pascal	Pa	$kg/m \cdot s^2$	N/m^2
Energy; work	joule	J	$kg \cdot m^2/s^2$	$N \cdot m$
Power	watt	W	$kg \cdot m^2/s^3$	J/s
Electric charge	coulomb	C	$A \cdot s$	
Electric potential	volt	V	$kg \cdot m^2/A \cdot s^3$	W/A
Capacitance	farad	F	$A^2 \cdot s^4/kg \cdot m^2$	C/V
Electric resistance	ohm	w	$kg \cdot m^2/A^2 \cdot s^3$	V/A
Magnetic flux	weber	Wb	$kg \cdot m^2/A \cdot s^2$	$V \cdot s$
Magnetic field	tesla	T	$kg/A \cdot s^2$	
Inductance	henry	H	$kg \cdot m^2/A^2 \cdot s^2$	$T \cdot m^2/A$

Nobel Prizes

All Nobel Prizes in Physics are listed (and marked with a P), as well as relevant Nobel Prizes in Chemistry (C). The key dates for some of the scientific work are supplied; they often antedate the prize considerably.

1901 (P) *Wilhelm Roentgen* for discovering x-rays (1895).

1902 (P) *Hendrik A. Lorentz* for predicting the Zeeman effect and *Pieter Zeeman* for discovering the Zeeman effect, the splitting of spectral lines in magnetic fields.

1903 (P) *Antoine-Henri Becquerel* for discovering radioactivity (1896) and *Pierre Curie* and *Marie Curie* for studying radioactivity.

1904 (P) *Lord Rayleigh* for studying the density of gases and discovering argon.

(C) *William Ramsay* for discovering the inert gas elements helium, neon, xenon, and krypton, and placing them in the periodic table.

1905 (P) *Philipp Lenard* for studying cathode rays, electrons (1898–1899).

1906 (P) *J. J. Thomson* for studying electrical discharge through gases and discovering the electron (1897).

1907 (P) *Albert A. Michelson* for inventing optical instruments and measuring the speed of light (1880s).

1908 (P) *Gabriel Lippmann* for making the first color photographic plate, using interference methods (1891).

(C) *Ernest Rutherford* for discovering that atoms can be broken apart by alpha rays and for studying radioactivity.

1909 (P) *Guglielmo Marconi* and *Carl Ferdinand Braun* for developing wireless telegraphy.

1910 (P) *Johannes D. van der Waals* for studying the equation of state for gases and liquids (1881).

1911 (P) *Wilhelm Wien* for discovering Wien's law giving the peak of a blackbody spectrum (1893).

(C) *Marie Curie* for discovering radium and polonium (1898) and isolating radium.

1912 (P) *Nils Dalén* for inventing automatic gas regulators for lighthouses.

1913 (P) *Heike Kamerlingh Onnes* for the discovery of superconductivity and liquefying helium (1908).

1914 (P) *Max T. F. von Laue* for studying x-rays from their diffraction by crystals, showing that x-rays are electromagnetic waves (1912).

(C) *Theodore W. Richards* for determining the atomic weights of 60 elements, indicating the existence of isotopes.

1915 (P) *William Henry Bragg* and *William Lawrence Bragg*, his son, for studying the diffraction of x-rays in crystals.

1917 (P) *Charles Barkla* for studying atoms by x-ray scattering (1906).

1918 (P) *Max Planck* for discovering energy quanta (1900).

1919 (P) *Johannes Stark* for discovering the Stark effect, the splitting of spectral lines in electric fields (1913).

1920 (P) *Charles-Édouard Guillaume* for discovering invar, a nickel–steel alloy with low coefficient of expansion.

(C) *Walther Nernst* for studying heat changes in chemical reactions and formulating the third law of thermodynamics (1918).

1921 (P) *Albert Einstein* for explaining the photoelectric effect and for his services to theoretical physics (1905).

(C) *Frederick Soddy* for studying the chemistry of radioactive substances and discovering isotopes (1912).

1922 (P) *Niels Bohr* for his model of the atom and its radiation (1913).

(C) *Francis W. Aston* for using the mass spectrograph to study atomic weights, thus discovering 212 of the 287 naturally occurring isotopes.

1923 (P) *Robert A. Millikan* for measuring the charge on an electron (1911) and for studying the photoelectric effect experimentally (1914).

1924 (P) *Karl M. G. Siegbahn* for his work in x-ray spectroscopy.

1925 (P) *James Franck* and *Gustav Hertz* for discovering the Franck–Hertz effect in electron–atom collisions.

1926 (P) *Jean-Baptiste Perrin* for studying Brownian motion to validate the discontinuous structure of matter and measure the size of atoms.

1927 (P) *Arthur Holly Compton* for discovering the Compton effect on x-rays, their change in wavelength when they collide with matter (1922), and *Charles T. R. Wilson* for inventing the cloud chamber, used to study charged particles (1906).

1928 (P) *Owen W. Richardson* for studying the thermionic effect and electrons emitted by hot metals (1911).

1929 (P) *Louis Victor de Broglie* for discovering the wave nature of electrons (1923).

1930 (P) *Chandrasekhara Venkata Raman* for studying Raman scattering, the scattering of light by atoms and molecules with a change in wavelength (1928).

1932 (P) *Werner Heisenberg* for creating quantum mechanics (1925).

1933 (P) *Erwin Schrödinger* and *Paul A. M. Dirac* for developing wave mechanics (1925) and relativistic quantum mechanics (1927).

(C) *Harold Urey* for discovering heavy hydrogen, deuterium (1931).

1935 (P) *James Chadwick* for discovering the neutron (1932).

(C) *Irène Joliot-Curie* and *Frédéric Joliot-Curie* for synthesizing new radioactive elements.

1936 (P) *Carl D. Anderson* for discovering the positron in particular and antimatter in general (1932) and *Victor F. Hess* for discovering cosmic rays.

(C) *Peter J. W. Debye* for studying dipole moments and diffraction of x-rays and electrons in gases.

1937 (P) *Clinton Davisson* and *George Thomson* for discovering the diffraction of electrons by crystals, confirming de Broglie's hypothesis (1927).

1938 (P) *Enrico Fermi* for producing the transuranic radioactive elements by neutron irradiation (1934–1937).

1939 (P) *Ernest O. Lawrence* for inventing the cyclotron.

1943 (P) *Otto Stern* for developing molecular-beam studies (1923) and using them to discover the magnetic moment of the proton (1933).

1944 (P) *Isidor I. Rabi* for discovering nuclear magnetic resonance in atomic and molecular beams.

(C) *Otto Hahn* for discovering nuclear fission (1938).

1945 (P) *Wolfgang Pauli* for discovering the exclusion principle (1924).

1946 (P) *Percy W. Bridgman* for studying physics at high pressures.

1947 (P) *Edward V. Appleton* for studying the ionosphere.

1948 (P) *Patrick M. S. Blackett* for studying nuclear physics with cloud-chamber photographs of cosmic-ray interactions.

1949 (P) *Hideki Yukawa* for predicting the existence of mesons (1935).

1950 (P) *Cecil F. Powell* for developing the method of studying cosmic rays with photographic emulsions and discovering new mesons.

1951 (P) *John D. Cockcroft* and *Ernest T. S. Walton* for transmuting nuclei in an accelerator (1932).

(C) *Edwin M. McMillan* for producing neptunium (1940) and *Glenn T. Seaborg* for producing plutonium (1941) and further transuranic elements.

1952 (P) *Felix Bloch* and *Edward Mills Purcell* for discovering nuclear magnetic resonance in liquids and gases (1946).

1953 (P) *Frits Zernike* for inventing the phase-contrast microscope, which uses interference to provide high contrast.

1954 (P) *Max Born* for interpreting the wave function as a probability (1926) and other quantum-mechanical discoveries and *Walther Bothe* for developing the coincidence method to study subatomic particles (1930–1931), producing, in particular, the particle interpreted by Chadwick as the neutron.

1955 (P) *Willis E. Lamb Jr.*, for discovering the Lamb shift in the hydrogen spectrum (1947) and *Polykarp Kusch* for determining the magnetic moment of the electron (1947).

1956 (P) *John Bardeen, Walter H. Brattain,* and *William Shockley* for inventing the transistor (1956).

1957 (P) *T.-D. Lee* and *C.-N. Yang* for predicting that parity is not conserved in beta decay (1956).

1958 (P) *Pavel A. Čerenkov* for discovering Čerenkov radiation (1935) and *Ilya M. Frank* and *Igor Tamm* for interpreting it (1937).

1959 (P) *Emilio G. Segrè* and *Owen Chamberlain* for discovering the antiproton (1955).

1960 (P) *Donald A. Glaser* for inventing the bubble chamber to study elementary particles (1952).

(C) *Willard Libby* for developing radiocarbon dating (1947).

1961 (P) *Robert Hofstadter* for discovering internal structure in protons and neutrons and *Rudolf L. Mössbauer* for discovering the Mössbauer effect of recoilless gamma-ray emission (1957).

1962 (P) *Lev Davidovich Landau* for studying liquid helium and other condensed matter theoretically.

1963 (P) *Eugene P. Wigner* for applying symmetry principles to elementary-particle theory and *Maria Goeppert Mayer* and *J. Hans D. Jensen* for studying the shell model of nuclei (1947).

1964 (P) *Charles H. Townes, Nikolai G. Basov,* and *Alexandr M. Prokhorov* for developing masers (1951–1952) and lasers.

1965 (P) *Sin-itiro Tomonaga, Julian S. Schwinger,* and *Richard P. Feynman* for developing quantum electrodynamics (1948).

1966 (P) *Alfred Kastler* for his optical methods of studying atomic energy levels.

1967 (P) *Hans Albrecht Bethe* for discovering the routes of energy production in stars (1939).

1968 (P) *Luis W. Alvarez* for discovering resonance states of elementary particles.

1969 (P) *Murray Gell-Mann* for classifying elementary particles (1963).

1970 (P) *Hannes Alfvén* for developing magnetohydrodynamic theory and *Louis Eugène Félix Néel* for discovering antiferromagnetism and ferrimagnetism (1930s).

1971 (P) *Dennis Gabor* for developing holography (1947).

(C) *Gerhard Herzberg* for studying the structure of molecules spectroscopically.

1972 (P) *John Bardeen, Leon N. Cooper,* and *John Robert Schrieffer* for explaining superconductivity (1957).

1973 (P) *Leo Esaki* for discovering tunneling in semiconductors, *Ivar Giaever* for discovering tunneling in superconductors, and *Brian D. Josephson* for predicting the Josephson effect, which involves tunneling of paired electrons (1958–1962).

1974 (P) *Anthony Hewish* for discovering pulsars and *Martin Ryle* for developing radio interferometry.

1975 (P) *Aage N. Bohr, Ben R. Mottelson,* and *James Rainwater* for discovering why some nuclei take asymmetric shapes.

1976 (P) *Burton Richter* and *Samuel C. C. Ting* for discovering the J/psi particle, the first charmed particle (1974).

1977 (P) *John H. Van Vleck, Nevill F. Mott,* and *Philip W. Anderson* for studying solids quantum-mechanically.

(C) *Ilya Prigogine* for extending thermodynamics to show how life could arise in the face of the second law.

1978 (P) *Arno A. Penzias* and *Robert W. Wilson* for discovering the cosmic background radiation (1965) and *Pyotr Kapitsa* for his studies of liquid helium.

1979 (P) *Sheldon L. Glashow, Abdus Salam,* and *Steven Weinberg* for developing the theory that unified the weak and electromagnetic forces (1958–1971).

1980 (P) *Val Fitch* and *James W. Cronin* for discovering CP (charge-parity) violation (1964), which possibly explains the cosmological dominance of matter over antimatter.

1981 (P) *Nicolaas Bloembergen* and *Arthur L. Schawlow* for developing laser spectroscopy and *Kai M. Siegbahn* for developing high-resolution electron spectroscopy (1958).

1982 (P) *Kenneth G. Wilson* for developing a method of constructing theories of phase transitions to analyze critical phenomena.

1983 (P) *William A. Fowler* for theoretical studies of astrophysical nucleosynthesis and *Subramanyan Chandrasekhar* for studying physical processes of importance to stellar structure and evolution, including the prediction of white dwarf stars (1930).

1984 (P) *Carlo Rubbia* for discovering the W and Z particles verifying the electroweak unification, and *Simon van der Meer* for developing the method of stochastic cooling of the CERN beam that allowed the discovery (1982–1983).

1985 (P) *Klaus von Klitzing* for the quantized Hall effect, relating to conductivity in the presence of a magnetic field (1980).

1986 (P) *Ernst Ruska* for inventing the electron microscope (1931) and *Gerd Binnig* and *Heinrich Rohrer* for inventing the scanning-tunneling electron microscope (1981).

1987 (P) *J. Georg Bednorz* and *Karl Alex Müller* for the discovery of high-temperature superconductivity (1986).

1988 (P) *Leon M. Lederman, Melvin Schwartz,* and *Jack Steinberger* for a collaborative experiment that led to the development of a new tool for studying the weak nuclear force, which affects the radioactive decay of atoms.

1989 (P) *Norman Ramsay* for various techniques in atomic physics and *Hans Dehmelt* and *Wolfgang Paul* for the development of techniques for trapping single-charge particles.

1990 (P) *Jerome Friedman, Henry Kendall,* and *Richard Taylor* for experiments important to the development of the quark model.

1991 (P) *Pierre-Gilles de Gennes* for discovering that methods developed for studying order phenomena in simple systems can be generalized to more complex forms of matter, in particular to liquid crystals and polymers.

1992 (P) *George Charpak* for developing detectors that trace the paths of evanescent subatomic particles produced in particle accelerators.

1993 (P) *Russell Hulse* and *Joseph Taylor* for discovering evidence of gravitational waves.

1994 (P) *Bertram N. Brockhouse* and *Clifford G. Shull* for pioneering work in neutron scattering.

1995 (P) *Martin L. Perl* and *Frederick Reines* for discovering the tau particle and the neutrino, respectively.

1996 (P) *David M. Lee, Douglas C. Osheroff,* and *Robert C. Richardson* for developing a superfluid using helium-3.

1997 (P) *Steven Chu, Claude Cohen-Tannoudji,* and *William D. Phillips* for developing methods to cool and trap atoms with laser light.

1998 (P) *Robert B. Laughlin, Horst L. Störmer,* and *Daniel C. Tsui* for discovering a new form of quantum fluid with fractionally charged excitations.

1999 (P) *Gerardus 'T Hooft* and *Martinus J. G. Veltman* for studies in the quantum structure of electroweak interactions in physics.

2000 (P) *Zhores I. Alferov* and *Herbert Kroemer* for developing semiconductor heterostructures used in high-speed electronics and optoelectronics and *Jack St. Clair Kilby* for participating in the invention of the integrated circuit.

2001 (P) *Eric A. Cornell, Wolfgang Ketterle,* and *Carl E. Wieman* for the achievement of Bose–Einstein condensation in dilute gases of alkali atoms.

2002 (P) *Raymond Davis Jr.* and *Masatoshi Koshiba* for the detection of cosmic neutrinos and *Riccardo Giacconi* for contributions to astrophysics that led to the discovery of cosmic x-ray sources.

2003 *Alexei A. Abrikosov, Vitaly L. Ginzburg,* and *Anthony J. Leggett* for pioneering contributions to the theory of superconductors and superfluids.

2004 *David J. Gross, H. David Politzer,* and *Frank Wilczeck* for the discovery of asymptotic freedom in the theory of the strong interaction.

Answers to Odd-Numbered Problems

Chapter 1

1. 5.52×10^3 kg/m^3, between the density of aluminum and that of iron and greater than the densities of typical surface rocks

3. $4\pi\rho(r_2{}^3 - r_1{}^3)/3$

5. No

7. (b) only

9. (a) 0.071 4 gal/s (b) 2.70×10^{-4} m^3/s (c) 1.03 h

11. 667 lb/s

13. 151 μm

15. 2.86 cm

17. (a) 2.07 mm (b) 8.62×10^{13} times as large

19. $\sim 10^6$ balls

21. $\sim 10^2$ tuners

23. (a) 3 (b) 4 (c) 3 (d) 2

25. 31 556 926.0 s

27. 5.2 m^3, 3%

29. 108° and 288°

31. 3.46 or $-$ 3.46

33. (a) 2.24 m (b) 2.24 m at 26.6°

35. (a) r, $180° - \theta$ (b) $2r$, $180° + \theta$ (c) $3r$, $-\theta$

37. (a) 10.0 m (b) 15.7 m (c) 0

39. Approximately 420 ft at $-$ 3°

41. 47.2 units at 122°

43. 196 cm at 345°

45. (a) $2.00\hat{\mathbf{i}} - 6.00\hat{\mathbf{j}}$ (b) $4.00\hat{\mathbf{i}} + 2.00\hat{\mathbf{j}}$ (c) 6.32 (d) 4.47 (e) 288°; 26.6°

47. 240 m at 237°

49. (a) 10.4 cm (b) 35.5°

51. (a) $8.00\hat{\mathbf{i}} + 12.0\hat{\mathbf{j}} - 4.00\hat{\mathbf{k}}$ (b) $2.00\hat{\mathbf{i}} + 3.00\hat{\mathbf{j}} - 1.00\hat{\mathbf{k}}$ (c) $-24.0\hat{\mathbf{i}} - 36.0\hat{\mathbf{j}} + 12.0\hat{\mathbf{k}}$

53. (a) $49.5\hat{\mathbf{i}} + 27.1\hat{\mathbf{j}}$ (b) 56.4 units at 28.7°

55. 70.0 m

57. 0.141 nm

59. 4.50 m^2

61. 0.449%

63. (a) 0.529 cm/s (b) 11.5 cm/s

65. $\sim 10^{11}$ stars

67. (a) 185 N at 77.8° from the $+x$ axis (b) $(-39.3\hat{\mathbf{i}} - 181\hat{\mathbf{j}})$ N

69. (a) (10.0 m, 16.0 m)

71. (a) $\vec{\mathbf{R}}_1 = a\hat{\mathbf{i}} + b\hat{\mathbf{j}}$; $R_1 = \sqrt{a^2 + b^2}$ (b) $\vec{\mathbf{R}}_2 = a\hat{\mathbf{i}} + b\hat{\mathbf{j}} + c\hat{\mathbf{k}}$; $R_2 = \sqrt{a^2 + b^2 + c^2}$

Chapter 2

1. (a) 2.30 m/s (b) 16.1 m/s (c) 11.5 m/s

3. (a) 5 m/s (b) 1.2 m/s (c) $-$ 2.5 m/s (d) $-$3.3 m/s (e) 0

5. (a) $-$ 2.4 m/s (b) $-$ 3.8 m/s (c) 4.0 s

7. (b) $v_{t\,=\,5.0\,\text{s}} = 23$ m/s, $v_{t\,=\,4.0\,\text{s}} = 18$ m/s, $v_{t\,=\,3.0\,\text{s}} = 14$ m/s, $v_{t\,=\,2.0\,\text{s}} = 9.0$ m/s (c) 4.6 m/s^2 (d) 0

9. 5.00 m

11. (a) 20.0 m/s, 5.00 m/s (b) 262 m

13. (a) 2.00 m (b) $-$ 3.00 m/s (c) $-$ 2.00 m/s^2

15. (a) 1.3 m/s^2 (b) 2.0 m/s^2 at 3 s (c) at $t = 6$ s and for $t > 10$ s (d) $-$ 1.5 m/s^2 at 8 s

17. (a) 6.61 m/s (b) $-$ 0.448 m/s^2

19. $-$ 16.0 cm/s^2

21. (a) 20.0 s (b) no

23. 3.10 m/s

25. (a) 35.0 s (b) 15.7 m/s

27. yes; 212 m, 11.4 s

29. (a) 29.4 m/s (b) 44.1 m

31. (a) 10.0 m/s up (b) 4.68 m/s down

33. (a) 7.82 m (b) 0.782 s

35. (b) 7.4 m/s^2 and 2.1 m/s^2 (c) 48 m and 170 m (d) 2.74 s

37. (a) 70.0 mi/h·s = 31.3 m/s^2 = 3.19g (b) 321 ft = 97.8 m

39. (a) $-$ 202 m/s^2 (b) 198 m

41. 2.74×10^5 m/s^2, which is 2.79×10^4 g

43. (a) 3.00 m/s (b) 6.00 s (c) $-$ 0.300 m/s^2 (d) 2.05 m/s

45. 1.60 m/s^2

47. (a) 41.0 s (b) 1.73 km (c) $-$ 184 m/s

49. (a) 5.43 m/s^2 and 3.83 m/s^2 (b) 10.9 m/s and 11.5 m/s (c) Maggie by 2.62 m

51. (a) 3.00 s (b) – 15.3 m/s (c) 31.4 m/s down and 34.8 m/s down

53. (c) v_{boy}^2/h, 0 (d) v_{boy}, 0

55. (a) 26.4 m (b) 6.82%

57. $0.577v$

Chapter 3

1. (a) 4.87 km at 209° from east (b) 23.3 m/s (c) 13.5 m/s at 209°

3. (a) $(0.800\hat{\mathbf{i}} - 0.300\hat{\mathbf{j}})\text{m/s}^2$ (b) 339° (c) $(360\hat{\mathbf{i}} - 72.7\hat{\mathbf{j}})\text{m}$, – 15.2°

5. (a) $\vec{\mathbf{r}} = (5.00t\,\hat{\mathbf{i}} + 1.50t^2\,\hat{\mathbf{j}})\text{m}$, $\vec{\mathbf{v}} = (5.00\hat{\mathbf{i}} + 3.00t\,\hat{\mathbf{j}})\text{m/s}$ (b) (10.0 m, 6.00 m), 7.81 m/s

7. (a) $3.34\hat{\mathbf{i}}$ m/s (b) – 50.9°

9. 12.0 m/s

11. 22.4° or 89.4°

13. 67.8°

15. (a) The ball clears by 0.889 m while (b) descending

17. (a) 18.1 m/s (b) 1.13 m (c) 2.79 m

19. 9.91 m/s

21. $\tan^{-1}[(2gh)^{1/2}/v]$

23. 377 m/s^2

25. 10.5 m/s, 219 m/s^2 inward

27. 7.58×10^3 m/s, 5.80×10^3 s

29. 1.48 m/s^2 inward and 29.9° backward

31. (a) 13.0 m/s^2 (b) 5.70 m/s (c) 7.50 m/s^2

33. 2.02×10^3 s; 21.0% longer

35. 153 km/h at 11.3° north of west

37. 15.3 m

39. $0.975g$

41. (a) 101 m/s (b) 32 700 ft (c) 20.6 s (d) 180 m/s

43. 54.4 m/s^2

45. (a) 41.7 m/s (b) 3.81 s (c) $(34.1\hat{\mathbf{i}} - 13.4\hat{\mathbf{j}})\text{m/s}$; 36.7 m/s

47. 10.7 m/s

49. (a) 6.80 km (b) 3.00 km vertically above the impact point (c) 66.2°

51. (a) 20.0 m/s, 5.00 s (b) $(16.0\hat{\mathbf{i}} - 27.1\hat{\mathbf{j}})\text{m/s}$ (c) 6.53 s (d) $24.5\hat{\mathbf{i}}$ m

53. (a) 22.9 m/s (b) 360 m from the base of the cliff (c) $\vec{\mathbf{v}} = (114\hat{\mathbf{i}} - 44.3\hat{\mathbf{j}})\text{m/s}$

55. (a) 1.52 km (b) 36.1 s (c) 4.05 km

57. (a) 43.2 m (b) $(9.66\hat{\mathbf{i}} - 25.6\hat{\mathbf{j}})\text{m/s}$

59. 4.00 km/h

61. Safe distances are less than 270 m and greater than 3.48×10^3 m from the western shore.

Chapter 4

1. (a) 1/3 (b) 0.750 m/s^2

3. $(6.00\hat{\mathbf{i}} + 15.0\hat{\mathbf{j}})$ N; 16.2 N

5. (a) $(2.50\hat{\mathbf{i}} + 5.00\hat{\mathbf{j}})$ N (b) 5.59 N

7. (a) 5.00 m/s^2 at 36.9° (b) 6.08 m/s^2 at 25.3°

9. (a) 534 N down (b) 54.5 kg

11. 2.55 N for an 88.7-kg person

13. (a) 3.64×10^{-18} N (b) 8.93×10^{-30} N is 408 billion times smaller

15. (a) $\sim 10^{-22}$ m/s^2 (b) $\sim 10^{-23}$ m

17. (a) 15.0 lb up (b) 5.00 lb up (c) 0

21. (a) From a free-body diagram of the forces on the bit of string touching the weight hanger we have $\Sigma F_y = 0$: $-F_g + T\sin\theta = 0$, so $T = F_g/\sin\theta$. The force the child feels gets smaller, changing from T to $T\cos\theta$ when the counterweight hangs from the string. On the other hand, the kite does not notice what you are doing and the tension in the main part of the string stays constant. You do not need a level because you learned in physics lab to sight to a horizontal line in a building. Share with the parents your estimate of the experimental uncertainty, which you made by thinking critically about the measurement, repeating trials, practicing in advance, and looking for variations and improvements in technique, including using other observers. You will then be glad to have the parents themselves repeat your measurements. (b) 1.79 N

23. (a) $a = g\tan\theta$ (b) 4.16 m/s^2

25. 100 N and 204 N

27. 8.66 N east

29. 3.73 m

31. A is in compression 3.83 kN and B is in tension 3.37 kN

33. 950 N

35. (a) $F_x > 19.6$ N (b) $F_x \leq -78.4$ N

(c)

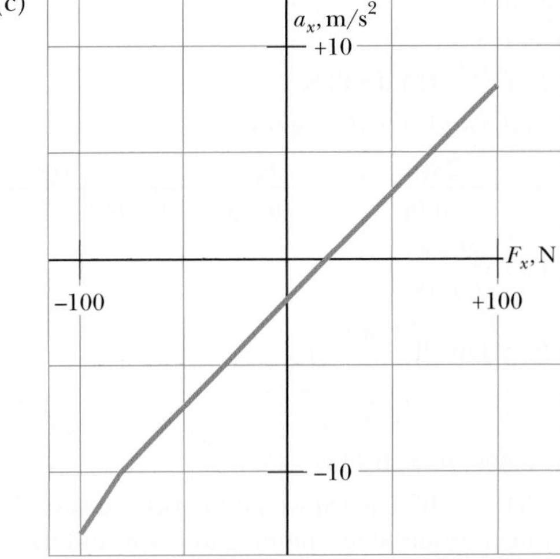

37. (a) 706 N (b) 814 N (c) 706 N (d) 648 N

39. (a) Removing mass (b) 13.7 mi/h·s

41. (a)

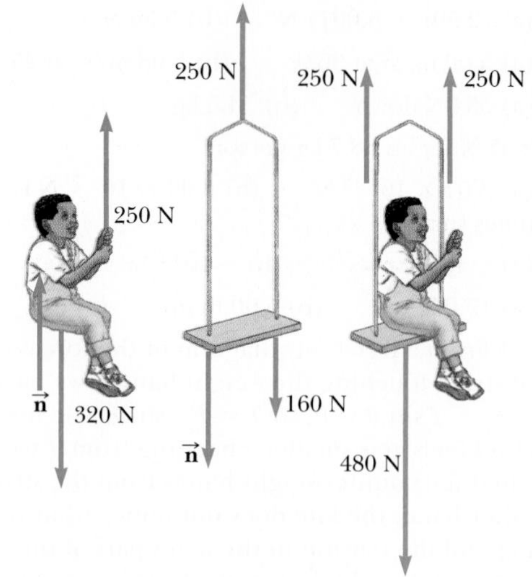

(b) 0.408 m/s^2 (c) 83.3 N

43. (a) 2.00 m/s^2 forward (b) 4.00 N forward on 2 kg, 6.00 N forward on 3 kg, 8.00 N forward on 4 kg (c) 14.0 N between 2 kg and 3 kg, 8.00 N between 4 kg and 3 kg (d) The 3-kg block models the heavy block of wood. The contact force on your back is represented by Q, which is much less than F. The difference between F and Q is the net force causing acceleration of the 5-kg pair of objects. The acceleration is real and nonzero but lasts for so short a time interval that it is never associated with a large velocity. The frame of the building and your legs exert forces, small compared with the hammer blow, to bring the partition, block, and you to rest again over a time interval large compared with the duration of the hammer blow.

45. (a) $Mg/2$, $Mg/2$, $Mg/2$, $3Mg/2$, Mg (b) $Mg/2$

47. $(M + m_1 + m_2)(m_2 g/m_1)$

49. (c) 3.56 N

51. 1.16 cm

53. (a) 30.7° (b) 0.843 N

55. $mg \sin\theta \cos\theta \hat{\mathbf{i}} + (M + m\cos^2\theta) g\hat{\mathbf{j}}$

57. (a) $T_1 = \dfrac{2mg}{\sin\theta_1}$, $T_2 = \dfrac{mg}{\sin\theta_2} = \dfrac{mg}{\sin[\tan^{-1}(\frac{1}{2}\tan\theta_1)]}$,

$$T_3 = \frac{2\,mg}{\tan\theta_1}$$

(b) $\theta_2 = \tan^{-1}\left(\dfrac{\tan\theta_1}{2}\right)$

Chapter 5

1. $\mu_s = 0.306$; $\mu_k = 0.245$

3. (a) 3.34 (b) The car would flip over backwards; or the wheels would skid, spinning in place, and the time would increase.

5. (a) 1.11 s (b) 0.875 s

7. $\mu_s = 0.727$, $\mu_k = 0.577$

9. (a) 1.78 m/s^2 (b) 0.368 (c) 9.37 N (d) 2.67 m/s

11. (a)

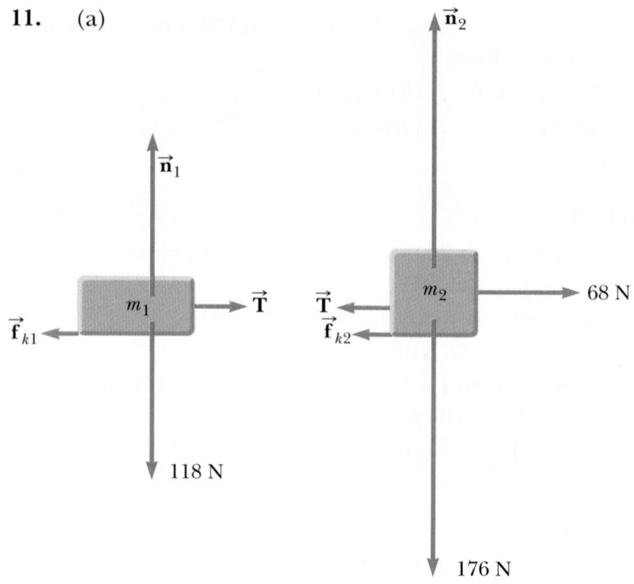

(b) 27.2 N, 1.29 m/s^2

13. any value between 31.7 N and 48.6 N

15. any speed up to 8.08 m/s

17. $v \leq 14.3$ m/s

19. (a) 68.6 N toward the center of the circle and 784 N up (b) 0.857 m/s^2

21. No. The jungle-lord needs a vine of tensile strength 1.38 kN.

23. 3.13 m/s

25. (a) 32.7 s^{-1} (b) 9.80 m/s^2 down (c) 4.90 m/s^2 down

27. (a) 1.47 N·s/m (b) 2.04×10^{-3} s (c) 2.94×10^{-2} N

29. (a) $0.034\,7 \text{ s}^{-1}$ (b) 2.50 m/s (c) $a = -cv$

31. 2.97 nN

33. 0.613 m/s^2 toward the Earth

35. -0.212 m/s^2

37. (a) $M = 3m\sin\theta$ (b) $T_1 = 2mg\sin\theta$, $T_2 = 3mg\sin\theta$

(c) $a = \dfrac{g\sin\theta}{1 + 2\sin\theta}$

(d) $T_1 = 4mg\sin\theta\left(\dfrac{1+\sin\theta}{1+2\sin\theta}\right)$

$$T_2 = 6mg\sin\theta\left(\frac{1+\sin\theta}{1+2\sin\theta}\right)$$

(e) $M_{\text{max}} = 3m(\sin\theta + \mu_s\cos\theta)$

(f) $M_{\text{min}} = 3m(\sin\theta - \mu_s\cos\theta)$

(g) $T_{2,\text{max}} - T_{2,\text{min}} = (M_{\text{max}} - M_{\text{min}})g = 6\mu_s mg\cos\theta$

39. (b)

θ	0	15°	30°	45°	60°
P (N)	40.0	46.4	60.1	94.3	260

41. (a) 0.087 1 (b) 27.4 N

43. (a) 2.13 s (b) 1.67 m

45. (a)

$$v_{\text{min}} = \sqrt{\frac{Rg(\tan\theta - \mu_s)}{1 + \mu_s\tan\theta}}$$

$$v_{\text{max}} = \sqrt{\frac{Rg(\tan\theta + \mu_s)}{1 - \mu_s\tan\theta}}$$

(b) $\mu_s = \tan\theta$
(c) $8.57\ \text{m/s} \leq v \leq 16.6\ \text{m/s}$

47. 0.835 rev/s

49. (b) 732 N down at the equator and 735 N down at the poles

51. (a) $1.58\ \text{m/s}^2$ (b) 455 N (c) 329 N (d) 397 N upward and 9.15° inward

53. 2.14 rev/min

55. (b) 2.54 s; 23.6 rev/min

57. (a) 0.013 2 m/s (b) 1.03 m/s (c) 6.87 m/s

59. 12.8 N

Chapter 6

1. (a) 31.9 J (b) 0 (c) 0 (d) 31.9 J

3. − 4.70 kJ

5. 5.33 W

7. (a) 16.0 J (b) 36.9°

9. (a) 11.3° (b) 156° (c) 82.3°

11. (a) 7.50 J (b) 15.0 J (c) 7.50 J (d) 30.0 J

13. (a) 0.938 cm (b) 1.25 J

15. (a) 575 N/m (b) 46.0 J

17. 12.0 J

19. (b) mgR

21. (a) 1.20 J (b) 5.00 m/s (c) 6.30 J

23. (a) 60.0 J (b) 60.0 J

25. 878 kN up

27. 0.116 m

29. (a) 650 J (b) 588 J (c) 0 (d) 0 (e) 62.0 J (f) 1.76 m/s

31. (a) − 168 J (b) 184 J (c) 500 J (d) 148 J (e) 5.65 m/s

33. 2.04 m

35. 875 W

37. $46.2

39. (a) 423 mi/gal (b) 776 mi/gal

41. 830 N

43. 2.92 m/s

45. (a) $(2 + 24t^2 + 72t^4)$ J (b) $12t\ \text{m/s}^2$; $48t$ N (c) $(48t + 288t^3)$ W (d) 1 250 J

47. (a) $\dfrac{mgnhh_s}{v + nh_s}$ (b) $\dfrac{mgvh}{v + nh_s}$

49. 7.37 N/m

51. (b) 240 W

53. (a) 4.12 m (b) 3.35 m

55. 1.68 m/s

57. -1.37×10^{-21} J

59. 0.799 J

61. (a) 2.17 kW (b) 58.6 kW

Chapter 7

1. (a) 259 kJ, 0, − 259 kJ (b) 0, − 259 kJ, − 259 kJ

3. 22.0 kW

5. (a) $v = (3gR)^{1/2}$ (b) 0.098 0 N down

7. 1.84 m

9. (a) 4.43 m/s (b) 5.00 m

11. (b) 60.0°

13. (a) 1.24 kW (b) 20.9%

15. (a) 125 J (b) 50.0 J (c) 66.7 J (d) Nonconservative; the work done depends on the path.

17. 10.2 m

19. (a) 22.0 J, 40.0 J (b) Yes; the total mechanical energy changes.

21. 26.5 m/s

23. 3.74 m/s

25. (a) − 160 J (b) 73.5 J (c) 28.8 N (d) 0.679

27. (a) 1.40 m/s (b) 4.60 cm after release (c) 1.79 m/s

29. (a) 0.381 m (b) 0.143 m (c) 0.371 m

31. (a) 40.0 J (b) −40.0 J (c) 62.5 J

33. (A/r^2) away from the other particle

35. (a) -4.77×10^9 J (b) 569 N (c) 569 N up

37. 2.52×10^7 m

39. (a) + at Ⓑ, − at Ⓓ, 0 at Ⓐ, Ⓒ, and Ⓔ (b) Ⓒ stable; Ⓐ and Ⓔ unstable

(c)

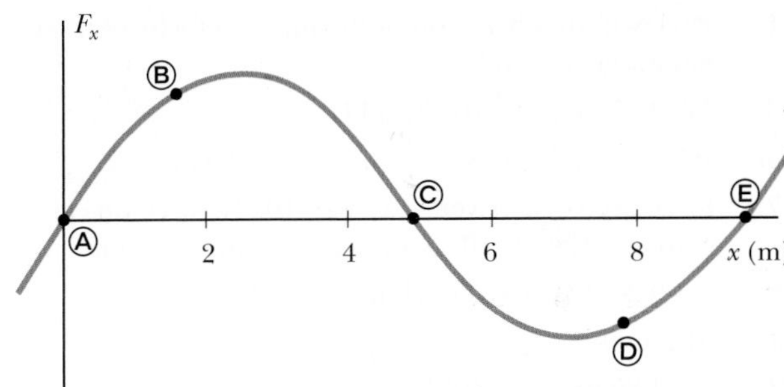

41. (b)

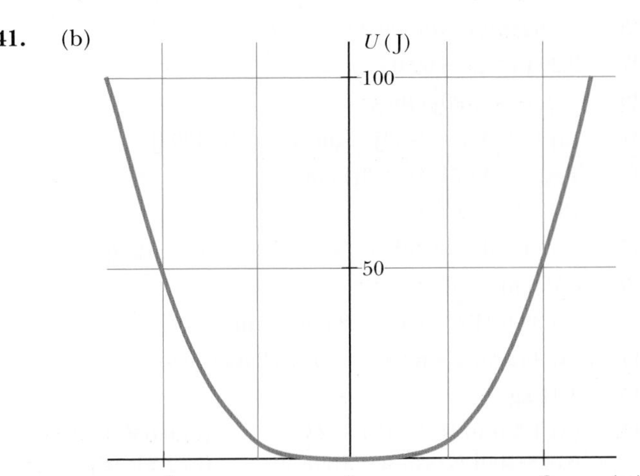

Equilibrium at $x = 0$. (c) 0.823 m/s

43. 0.27 MJ/kg for a battery. 17 MJ/kg for hay is 63 times larger. 44 MJ/kg for gasoline is 2.6 times larger still. 142 MJ/kg for hydrogen is 3.2 times larger than that.

45. ~10^3 W peak or ~10^2 W sustainable

47. $(8gh/15)^{1/2}$

49. (a) 0.225 J (b) $\Delta E_{mech} = -0.363$ J (c) No; the normal force changes in a complicated way.

51. 0.328

53. 1.24 m/s

55. (a) 0.400 m (b) 4.10 m/s (c) The block stays on the track.

57. (a) 6.15 m/s (b) 9.87 m/s

59. (a) 11.1 m/s (b) 19.6 m/s^2 upward (c) 2.23×10^3 N upward (d) 1.01×10^3 J (e) 5.14 m/s (f) 1.35 m (g) 1.39 s

63. (a) 14.1 m/s (b) − 7.90 kJ (c) 800 N (d) 771 N (e) 1.57 kN up

Context 1 Conclusion

1. (a) 315 kJ (b) 220 kJ (c) 187 kJ (d) 127 kJ (e) 14.0 m/s (f) 40.5% (g) 187 kJ

Chapter 8

1. (a) $(9.00\hat{\mathbf{i}} - 12.0\hat{\mathbf{j}})$ kg·m/s (b) 15.0 kg·m/s at 307°

3. 40.5 g

5. (a) 6.00 m/s toward the left (b) 8.40 J

7. (a) 13.5 N·s (b) 9.00 kN (c) 18.0 kN

9. 260 N normal to the wall

11. 15.0 N in the direction of the initial velocity of the exiting water stream

13. (a) 2.50 m/s (b) 37.5 kJ

15. (a) $v_{gx} = 1.15$ m/s (b) $v_{px} = -0.346$ m/s

17. force on truck driver = 1.78×10^3 N; force on car driver = 8.89×10^3 N in the opposite direction

19. (a) 0.284 (b) 115 fJ and 45.4 fJ

21. 91.2 m/s

23. (a) 4.85 m/s (b) 8.41 m

25. orange: $v_i \cos\theta$; yellow: $v_i \sin\theta$

27. 2.50 m/s at − 60.0°

29. $(3.00\hat{\mathbf{i}} - 1.20\hat{\mathbf{j}})$ m/s

31. (a) $(-9.33\hat{\mathbf{i}} - 8.33\hat{\mathbf{j}})$ Mm/s (b) 439 fJ

33. $\vec{\mathbf{r}}_{CM} = (11.7\hat{\mathbf{i}} + 13.3\hat{\mathbf{j}})$ cm

35. (b) 3.57×10^8 J

37. (a) $(1.40\hat{\mathbf{i}} + 2.40\hat{\mathbf{j}})$ m/s (b) $(7.00\hat{\mathbf{i}} + 12.0\hat{\mathbf{j}})$ kg·m/s

39. 0.700 m

41. (a) 39.0 MN (b) 3.20 m/s^2 up

43. (a) 442 metric tons (b) 19.2 metric tons

45. 4.41 kg

47. (a) $1.33\hat{\mathbf{i}}$ m/s (b) $-235\hat{\mathbf{i}}$ N (c) 0.680 s (d) $-160\hat{\mathbf{i}}$ N·s and $+160\hat{\mathbf{i}}$ N·s (e) 1.81 m (f) 0.454 m (g) − 427 J (h) + 107 J (i) Equal friction forces act through different distances on person and cart to do different amounts of work on them. The total work on both together, − 320 J, becomes + 320 J of extra internal energy in this perfectly inelastic collision.

49. (a) 2.07 m/s^2 (b) 3.88 m/s

51. (a) − 0.667 m/s (b) 0.952 m

53. (a) $-0.256\hat{\mathbf{i}}$ m/s and $0.128\hat{\mathbf{i}}$ m/s (b) $-0.064\ 2\hat{\mathbf{i}}$ m/s and 0 (c) 0 and 0

55. $2v_i$ and 0

57. (a) $m/M = 0.403$ (b) no changes; no difference

59. (a) 3.75 kg·m/s^2 to the right (b) 3.75 N to the right (c) 3.75 N (d) 2.81 J (e) 1.41 J (f) Friction between sand and belt causes half of the input work to appear as extra internal energy.

Chapter 9

5. $0.866c$

7. (a) 25.0 yr (b) 15.0 yr (c) 12.0 ly

9. 1.54 ns

11. $0.800c$

13. (a) 20.0 m (b) 19.0 m (c) $0.312c$

15. (a) 21.0 yr (b) 14.7 ly (c) 10.5 ly (d) 35.7 yr

17. (a) 17.4 m (b) 3.30°

19. (a) 2.50×10^8 m/s (b) 4.97 m (c) -1.33×10^{-8} s

21. $0.960c$

23. (a) 2.73×10^{-24} kg·m/s (b) 1.58×10^{-22} kg·m/s (c) 5.64×10^{-22} kg·m/s

25. 4.50×10^{-14}

27. $0.285c$

29. (a) 0.582 MeV (b) 2.45 MeV

31. (a) 3.07 MeV (b) $0.986c$

33. (a) 938 MeV (b) 3.00 GeV (c) 2.07 GeV

35. (a) $0.979c$ (b) $0.065\ 2c$ (c) $0.914c = 274$ Mm/s (d) $0.999\ 999\ 97c$; $0.948c$; $0.052\ 3c = 15.7$ Mm/s

39. 4.08 MeV and 29.6 MeV

41. 4.28×10^9 kg/s

43. 1.02 MeV

45. (a) 3.87 km/s (b) -8.36×10^{-11} (c) 5.29×10^{-10} (d) $+4.46 \times 10^{-10}$

47. (a) $v/c = 1 - 1.12 \times 10^{-10}$ (b) 6.00×10^{27} J (c) \$$2.17 \times 10^{20}$

49. (a) a few hundred seconds (b) ~ 10^8 km

51. 0.712%

53. (a) $0.946c$ (b) 0.160 ly (c) 0.114 yr (d) 7.50×10^{22} J

55. yes, with 18.8 m to spare

57. (b) For u small compared to c, the relativistic expression agrees with the classical expression. As u approaches c, the acceleration approaches zero, so the object can never reach or surpass the speed of light.
(c) Perform $\int(1 - u^2/c^2)^{-3/2}du = (qE/m)\int dt$ to obtain $u = qEct(m^2c^2 + q^2E^2t^2)^{-1/2}$ and then $\int dx = \int qEct(m^2c^2 + q^2E^2t^2)^{-1/2}dt$ to obtain $x = (c/qE)[(m^2c^2 + q^2E^2t^2)^{1/2} - mc]$.

63. (a) The refugees conclude that Tau Ceti exploded 16.0 yr before the Sun.
(b) A stationary observer at the midpoint concludes that they exploded simultaneously.

Chapter 10

1. (a) 5.00 rad, 10.0 rad/s, 4.00 rad/s^2 (b) 53.0 rad, 22.0 rad/s, 4.00 rad/s^2
3. (a) 5.24 s (b) 27.4 rad
5. 50.0 rev
7. (a) 7.27×10^{-5} rad/s (b) 2.57×10^4 s = 428 min
9. (a) 126 rad/s (b) 3.77 m/s (c) 1.26 km/s^2 (d) 20.1 m
11. (a) 0.605 m/s (b) 17.3 rad/s (c) 5.82 m/s (d) The crank length is unnecessary.
13. 0.572
17. (a) $\sqrt{\dfrac{2(m_1 - m_2)gh}{m_1 + m_2 + I/R^2}}$ (b) $\sqrt{\dfrac{2(m_1 - m_2)gh}{m_1R^2 + m_2R^2 + I}}$
19. 24.5 m/s
21. -3.55 N·m
23. $\vec{\tau} = (2.00\hat{\mathbf{k}})$ N·m
27. $[(m_1 + m_b)d + m_1\ell/2]/m_2$
29. (a) 1.04 kN at 60.0° (b) $(370\hat{\mathbf{i}} + 900\hat{\mathbf{j}})$ N
31. (a) $T = F_g(L + d)/\sin\theta(2L + d)$ (b) $R_x = F_g(L + d)\cot\theta/(2L + d)$; $R_y = F_gL/(2L + d)$
33. (a) 21.6 kg·m^2 (b) 3.60 N·m (c) 52.4 rev
35. 21.5 N
37. (a) 118 N and 156 N (b) 1.17 kg·m^2
39. (a) 11.4 N, 7.57 m/s^2, 9.53 m/s down (b) 9.53 m/s
41. $(60.0\hat{\mathbf{k}})$ kg·m^2/s
43. (a) 0.433 kg·m^2/s (b) 1.73 kg·m^2/s
45. (a) $\omega_f = \omega_i I_1/(I_1 + I_2)$ (b) $I_1/(I_1 + I_2)$
47. (a) 0.360 rad/s counterclockwise (b) 99.9 J
49. (a) 7.20×10^{-3} kg·m^2/s (b) 9.47 rad/s
51. 5.99×10^{-2} J
53. (a) 500 J (b) 250 J (c) 750 J
55. (a) 2.38 m/s. Its weight is insufficient to provide the centripetal acceleration. (b) 4.31 m/s (c) The ball does not reach the top of the loop.
57. 131 s
59. (a) $(3g/L)^{1/2}$ (b) $3g/2L$ (c) $-\frac{3}{2}g\hat{\mathbf{i}} - \frac{3}{4}g\hat{\mathbf{j}}$ (d) $-\frac{3}{2}Mg\hat{\mathbf{i}} + \frac{1}{4}Mg\hat{\mathbf{j}}$
61. (a) $\sqrt{\dfrac{2mgd\sin\theta + kd^2}{I + mR^2}}$ (b) 1.74 rad/s
67. (a) 61.2 J (b) 50.8 J
69. (a) Mvd (b) Mv^2 (c) Mvd (d) $2v$ (e) $4Mv^2$ (f) $3Mv^2$
71. $T = 2.71$ kN, $R_x = 2.65$ kN
73. (a) 20.1 cm to the left of the front edge; $\mu_k = 0.571$ (b) 0.501 m
75. (a) 133 N (b) $n_A = 429$ N and $n_B = 257$ N (c) $R_x = 133$ N and $R_y = -257$ N
77. $\frac{3}{8}F_g$

Chapter 11

1. 2.67×10^{-7} m/s^2
3. 7.41×10^{-10} N
5. (a) 4.39×10^{20} N toward the Sun (b) 1.99×10^{20} N away from the Sun (c) 3.55×10^{22} N toward the Sun
7. $\rho_M/\rho_E = 2/3$
9. (a) 7.61 cm/s^2 (b) 363 s (c) 3.08 km (d) 28.9 m/s at 72.9° below the horizontal
11. $\vec{\mathbf{g}} = 2MGr(r^2 + a^2)^{-3/2}$ toward the center of mass
13. (a) 4.23×10^7 m (b) 0.285 s
15. 1.90×10^{27} kg
17. 1.26×10^{32} kg
19. After 3.93 yr, Mercury would be farther from the Sun than Pluto.
21. (a) 1.84×10^9 kg/m^3 (b) 3.27×10^6 m/s^2 (c) -2.08×10^{13} J
23. 1.78 km
25. 1.66×10^4 m/s
29. 1.58×10^{10} J
31. (b) 1.00×10^7 m (c) 1.00×10^4 m/s
33. (a) 0.980 (b) 127 yr (c) -2.13×10^{17} J
35. (a) 5 (b) no; no
37. (a) ii (b) i (c) ii and iii
39. (a) 0.212 nm (b) 9.95×10^{-25} kg·m/s (c) 2.11×10^{-34} kg·m^2/s (d) 3.40 eV (e) -6.80 eV (f) -3.40 eV
41. 4.42×10^4 m/s
43. (a) 29.3% (b) no change
45. 2.26×10^{-7}
47. (c) 1.85×10^{-5} m/s^2
49. $v = 492$ m/s
51. (a) 7.79 km/s (b) 7.85 km/s (c) -3.04 GJ (d) -3.08 GJ (e) loss = 46.9 MJ (f) A component of the Earth's gravity pulls forward on the satellite on its downward-banking trajectory.
53. (a) $m_2(2G/d)^{1/2}(m_1 + m_2)^{-1/2}$ and $m_1(2G/d)^{1/2}(m_1 + m_2)^{-1/2}$; relative speed $(2G/d)^{1/2}(m_1 + m_2)^{1/2}$ (b) 1.07×10^{32} J and 2.67×10^{31} J
55. (a) 200 Myr (b) $\sim 10^{41}$ kg; $\sim 10^{11}$ stars
57. $(GM_E/4R_E)^{1/2}$
61. $r_n = (0.106\text{ nm})n^2$, $E_n = -6.80\text{ eV}/n^2$, for $n = 1, 2, 3, \ldots$

Context 2 Conclusion

1. (a) 146 d (b) Venus 53.9° behind the Earth
3. (a) 2.95 km/s (b) 2.65 km/s (c) 10.7 km/s (d) 4.80 km/s

Chapter 12

1. (a) The motion repeats precisely. (b) 1.81 s (c) No, the force is not in the form of Hooke's law.
3. (a) 1.50 Hz, 0.667 s (b) 4.00 m (c) π rad (d) 2.83 m
5. (b) 18.8 cm/s, 0.333 s (c) 178 cm/s^2, 0.500 s (d) 12.0 cm
9. 40.9 N/m

11. (a) 40.0 cm/s, 160 cm/s^2 (b) 32.0 cm/s, $-$96.0 cm/s^2 (c) 0.232 s

13. 2.23 m/s

15. (a) 0.542 kg (b) 1.81 s (c) 1.20 m/s^2

17. (a) 28.0 mJ (b) 1.02 m/s (c) 12.2 mJ (d) 15.8 mJ

19. (a) E increases by a factor of 4 (b) v_{max} is doubled. (c) a_{max} is doubled. (d) The period is unchanged.

21. 2.60 cm and $-$2.60 cm

23. Assume simple harmonic motion: (a) 0.820 m/s (b) 2.57 rad/s^2 (c) 0.641 N More precisely: (a) 0.817 m/s (b) 2.54 rad/s^2 (c) 0.634 N

27. 0.944 kg · m^2

31. 1.00×10^{-3} s^{-1}

33. (a) 1.00 s (b) 5.09 cm

37. 1.74 Hz

39. If the cyclist goes over them at one certain speed, the washboard bumps can excite a resonance vibration of the bike, so large in amplitude as to make the rider lose control. $\sim 10^1$ m

41. 6.62 cm

43. 9.19×10^{13} Hz

45. (b) 1.04 m/s (c) four times larger, 3.40 m

47. $f = (2\pi L)^{-1}(gL + kh^2/M)^{1/2}$

49. (b) 1.23 Hz

51. (a) 3.00 s (b) 14.3 J (c) 25.5°

57. (a) 5.20 s (b) 2.60 s

Chapter 13

1. $y = 6\,[(x - 4.5t)^2 + 3]^{-1}$

3. 0.319 m

5. (a) $(3.33\hat{\mathbf{i}})$ m/s (b) $-$5.48 cm (c) 0.667 m, 5.00 Hz (d) 11.0 m/s

7. (a) 31.4 rad/s (b) 1.57 rad/m (c) $y = (0.120 \text{ m}) \sin(1.57x - 31.4t)$ (d) 3.77 m/s (e) 118 m/s^2

9. (a) $y = (8.00 \text{ cm}) \sin(7.85x + 6\pi t)$ (b) $y = (8.00 \text{ cm}) \sin(7.85x + 6\pi t - 0.785)$

11. (a) 0.021 5 m (b) 1.95 rad (c) 5.41 m/s (d) $y(x, t) = (0.021\,5 \text{ m}) \sin(8.38x + 80.0\pi t + 1.95)$

13. 80.0 N

15. 13.5 N

17. 0.329 s

19. 1.07 kW

21. 55.1 Hz

23. 5.56 km

25. (a) 23.2 cm (b) 1.38 cm

27. (a) 4.16 m (b) 0.455 μs (c) 0.158 mm

29. 5.81 m

31. $\Delta P = (0.200 \text{ N/m}^2) \sin(62.8x/\text{m} - 2.16 \times 10^4 t/\text{s})$

33. (a) 3.04 kHz (b) 2.08 kHz (c) 2.62 kHz; 2.40 kHz

35. 26.4 m/s

37. 19.3 m

39. (a) 0.364 m (b) 0.398 m (c) 941 Hz (d) 938 Hz

41. 184 km

43. $(Lm/Mg \sin\theta)^{1/2}$

45. 0.084 3 rad

49. (a) $\dfrac{\mu\omega^3}{2k} A_0^2 e^{-2bx}$ (b) $\dfrac{\mu\omega^3}{2k} A_0^2$ (c) e^{-2bx}

51. (a) $\mu_0 + (\mu_L - \mu_0)x/L$

55. 6.01 km

57. (a) 55.8 m/s (b) 2 500 Hz

59. The gap between bat and insect is closing at 1.69 m/s.

Chapter 14

1. (a) $-$1.65 cm (b) $-$6.02 cm (c) 1.15 cm

3. (a) $+x$, $-x$ (b) 0.750 s (c) 1.00 m

5. (a) 9.24 m (b) 600 Hz

7. 91.3°

9. (a) 156° (b) 0.058 4 cm

11. (a) To reach the receiver, waves from the more distant source must travel an extra distance $\Delta r = \lambda/2$ and interfere destructively with waves from the closer source. (b) It should move along the hyperbola represented by $9.00x^2 - 16.0y^2 = 144$.

13. at 0.089 1 m, 0.303 m, 0.518 m, 0.732 m, 0.947 m, and 1.16 m from one speaker

15. (a) 4.24 cm (b) 6.00 cm (c) 6.00 cm (d) 0.500 cm, 1.50 cm, 2.50 cm

17. 0.786 Hz, 1.57 Hz, 2.36 Hz, 3.14 Hz

19. 15.7 Hz

21. (a) reduced by $\frac{1}{2}$ (b) reduced by $1/\sqrt{2}$ (c) increased by $\sqrt{2}$

23. (a) 163 N (b) 660 Hz

25. $\dfrac{Mg}{4Lf^2 \tan\theta}$

27. (a) 0.357 m (b) 0.715 m

29. 57.9 Hz

31. n(206 Hz) for $n = 1$ to 9 and n(84.5 Hz) for $n = 2$ to 23

33. 50.0 Hz, 1.70 m

35. n(0.252 m) with $n = 1, 2, 3, \ldots$

37. (a) 350 m/s (b) 1.14 m

39. 5.64 beats/s

41. (a) 1.99 beats/s (b) 3.38 m/s

43. The second harmonic of E is close to the third harmonic of A, and the fourth harmonic of C$^\#$ is close to the fifth harmonic of A.

45. The condition for resonance is satisfied because the 12 h 24 min period of free oscillation agrees precisely with the period of the lunar excitation.

47. (a) 34.8 m/s (b) 0.977 m

49. 3.85 m/s away from the station and 3.77 m/s toward the station

51. 21.5 m

53. (a) 59.9 Hz (b) 20.0 cm

55. (a) $\frac{1}{2}$ (b) $[n/(n+1)]^2T$ (c) $\frac{9}{16}$

57. $y_1 + y_2 = 11.2 \sin(2.00x - 10.0t + 63.4°)$

59. (a) 78.9 N (b) 211 Hz

Context 3 Conclusion

1. 3.5 cm

2. The speed decreases by a factor of 25.

3. Station 1: 15:46:32; Station 2: 15:46:22; Station 3: 15:46:08, all with uncertainties of ±1 s

Chapter 15

1. 0.111 kg

3. 6.24 MPa

5. 1.62 m

7. 7.74×10^{-3} m^2

9. 271 kN horizontally backward

11. 2.31 lb

13. 10.5 m; no because some alcohol and water evaporate

15. 98.6 kPa

17. (a) 7.54 kg (b) 39.8 N (c) 41.9 N up (d) zero (e) The tension decreases and the normal force increases.

19. 0.258 N

21. (a) $1.017\,9 \times 10^3$ N down, $1.029\,7 \times 10^3$ N up (b) 86.2 N (c) 11.8 N

23. (a) 7.00 cm (b) 2.80 kg

25. 1 430 m^3

27. 1 250 kg/m^3 and 500 kg/m^3

29. 1.01 kJ

31. 12.8 kg/s

33. $2\sqrt{h(h_0 - h)}$

35. (a) 27.9 N (b) 3.32×10^4 kg (c) 7.26×10^4 Pa

37. 0.247 cm

39. (a) 1 atm + 15.0 MPa (b) 2.95 m/s (c) 4.34 kPa

41. 347 m/s

43. (a) 4.43 m/s (b) The siphon can be no higher than 10.3 m.

45. 12.6 m/s

47. 1.61×10^4 m^2

51. 0.604 m

55. The top scale reads $(1 - \rho_0/\rho_{Fe})m_{Fe}g$. The bottom scale reads $[m_b + m_0 + \rho_0 m_{Fe}/\rho_{Fe}]g$.

57. (a) 2.79 μm/s (b) 7.95 h (c) 8.88×10^3 m/s^2 (d) 31.6 s

59. 4.43 m/s

61. (a) 1.25 cm (b) 13.8 m/s

63. (a) 3.307 g (b) 3.271 g (c) 3.48×10^{-4} N

Context 4 Conclusion

1. 9.8×10^9 N·m

2. (a) 1.30 MPa (b) yes, but only with specialized equipment and techniques

3. (a) 0.42 m/s (b) greater

4. (a) 16 knots at 56° west of south (b) 47%

Chapter 16

1. (a) −273°C (b) 1.27 atm (c) 1.74 atm

3. (a) −320°F (b) 77.3 K

5. 1.54 km. The pipeline can be supported on rollers. Ω-shaped loops can be built between straight sections. They bend as the steel changes length.

7. 0.001 58 cm

9. (a) 0.176 mm (b) 8.78 μm (c) 0.093 0 cm^3

11. (a) 0.109 cm^2 (b) increase

13. (a) 99.4 cm^3 (b) 0.943 cm

15. 8.72×10^{11} atoms/s

17. (a) 400 kPa (b) 449 kPa

19. 1.50×10^{29} molecules

21. 472 K

23. (a) 7.13 m (b) The open end of the tube should be at the bottom after the bird surfaces so that the water can drain out. There is no other requirement. Air does not tend to bubble out of a narrow tube.

25. 4.39 kg

27. 3.55 L

29. $m_1 - m_2 = \dfrac{P_0 VM}{R}\left(\dfrac{1}{T_1} - \dfrac{1}{T_2}\right)$

31. 17.6 kPa

33. (a) 3.54×10^{23} atoms (b) 6.07×10^{-21} J (c) 1.35 km/s

35. (a) 8.76×10^{-21} J for both (b) 1.62 km/s for helium and 514 m/s for argon

39. (a) 2.37×10^4 K (b) 1.06×10^3 K

41. (a) −9.73°C/km (b) As rising air drops in temperature, water vapor in it condenses into liquid. It releases energy in this process to reduce the net temperature drop. (c) −4.60°C/km (d) 4.34 km (e) Dust aloft absorbs sunlight to raise the temperature there. *Mariner* occurred in dustier conditions.

43. 0.523 kg

45. (a) Expansion makes density drop. (b) 5×10^{-5}/°C

47. (a) $h = nRT/(mg + P_0A)$ (b) 0.661 m

49. We assume that $\alpha\,\Delta T$ is much less than 1.

51. (a) 0.340% (b) 0.480%

53. 2.74 m

55. (a) It increases. As the disk cools, its radius and hence its moment of inertia decrease. Conservation of angular momentum then requires that its angular speed increase. (b) 25.7 rad/s

57. (b) 1.33 kg/m^3

59. 1.12 atm

61. (d) 0.275 mm (e) The plate creeps down the roof each day by an amount given by the same expression.

63. 1.09×10^{-3}, 2.69×10^{-2}, 0.529, 1.00, 0.199, 1.01×10^{-41}, $1.25 \times 10^{-1\,082}$

Chapter 17

1. 0.281°C

3. 87.0°C

5. 29.6°C

7. (a) 16.1°C (b) 16.1°C

9. 23.6°C

11. 1.22×10^5 J

13. 0.294 g

15. (a) 0°C (b) 114 g

17. liquid lead at 805°C

19. − 1.18 MJ

21. − 466 J

23. $Q = -720$ J

25.

	Q	W	ΔE_{int}
BC	−	0	−
CA	−	+	−
AB	+	−	+

27. (a) 7.50 kJ (b) 900 K

29. − 3.10 kJ; 37.6 kJ

31. (a) 0.041 0 m^3 (b) + 5.48 kJ (c) − 5.48 kJ

33. (a) 3.46 kJ (b) 2.45 kJ (c) − 1.01 kJ

35. (a) 209 J (b) zero (c) 317 K

37. between 10^{-2}°C and 10^{-3}°C

39. 13.5PV

41. (a) 1.39 atm (b) 366 K, 253 K (c) 0, − 4.66 kJ, − 4.66 kJ

43. (a)

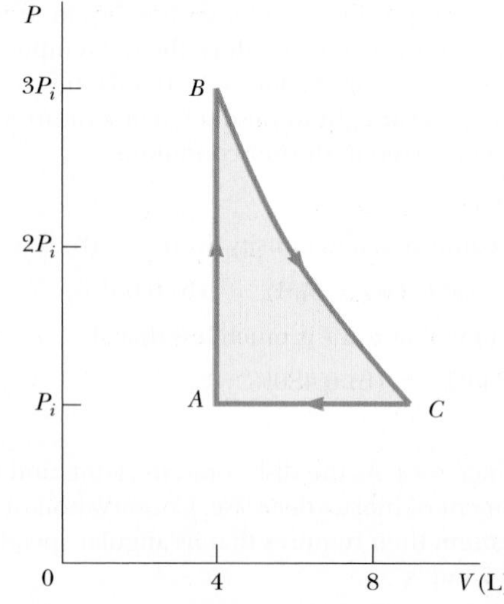

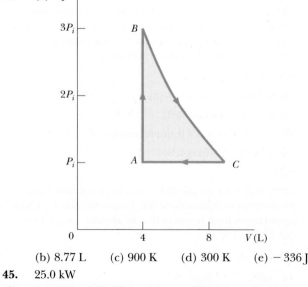

(b) 8.77 L (c) 900 K (d) 300 K (e) − 336 J

45. 25.0 kW

47. (a) 9.95 cal/K, 13.9 cal/K (b) 13.9 cal/K, 17.9 cal/K

49. 51.2°C

51. 3.85×10^{26} J/s

53. 74.8 kJ

55. 279 K = 6°C

57. (a) 0.964 kg or more (b) The test samples and the inner surface of the insulation can be preheated to 37.0°C as the box is assembled. Then nothing changes in temperature during the test period, and the masses of the test samples and insulation make no difference.

59. (a) 13.0°C (b) − 0.532 °C/s

61. $c = \mathcal{P}/\rho R \Delta T$

63. (a) 9.31×10^{10} J (b) -8.47×10^{12} J (c) 8.38×10^{12} J

65. 5.31 h

67. (a) 15.0 mg; block: $Q = 0$, $W = -5.00$ J, $\Delta E_{int} = 0$, $\Delta K = -5.00$ J; ice: $Q = 0$, $W = +5.00$ J; $\Delta E_{int} = 5.00$ J, $\Delta K = 0$ (b) 15.0 mg; block: $Q = 0$, $W = 0$, $\Delta E_{int} = 5.00$ J, $\Delta K = -5.00$ J; metal: $Q = 0$, $W = 0$, $\Delta E_{int} = 0$, $\Delta K = 0$ (c) 0.004 04°C; moving block: $Q = 0$, $W = -2.50$ J, $\Delta E_{int} = 2.50$ J, $\Delta K = -5.00$ J; stationary block: $Q = 0$, $W = +2.50$ J, $\Delta E_{int} = 2.50$ J, $\Delta K = 0$

69. 38.6 m^3/d

71. (a) 100 kPa, 66.5 L, 400 K; 5.82 kJ; 7.48 kJ; − 1.66 kJ (b) 133 kPa, 49.9 L, 400 K; 5.82 kJ; 5.82 kJ; 0 (c) 120 kPa, 41.6 L, 300 K; 0; − 909 J; + 909 J (d) 120 kPa, 43.3 L, 312 K; 722 J; 0; + 722 J

73. (a) 300 K (b) 1.00 atm

75. (a) 0.203 mol (b) $T_B = T_C = 900$ K, $V_C = 15.0$ L

(c, d)	P, atm	V, L	T, K	E_{int}, kJ
A	1.00	5.00	300	0.760
B	3.00	5.00	900	2.28
C	1.00	15.0	900	2.28
A	1.00	5.00	300	0.760

(e) Lock the piston in place and put the cylinder into an oven at 900 K. Keep the gas in the oven while gradually letting the gas expand to lift a load on the piston as far as it can. Move the cylinder from the oven back to the 300-K room and let the gas cool and contract.

(f, g)	Q, kJ	W, kJ	ΔE_{int}, kJ
AB	1.52	0	1.52
BC	1.67	− 1.67	0
CA	− 2.53	+ 1.01	− 1.52
ABCA	0.656	− 0.656	0

Chapter 18

1. (a) 6.94% (b) 335 J

3. (a) 10.7 kJ (b) 0.533 s

5. (a) 67.2% (b) 58.8 kW

7. (a) 741 J (b) 459 J

9. 0.330

11. (b) $1 - T_c/T_h$ (c) $(T_c + T_h)/2$ (d) $(T_hT_c)^{1/2}$

13. (a) 24.0 J (b) 144 J

15. (a) 2.93 (b) coefficient of performance for a refrigerator (c) $300 is twice as large as $150

19. 1.17 J

21. 72.2 J

23. 195 J/K

25. (a) isobaric (b) 402 kJ (c) 1.20 kJ/K

27. (a) 1 (b) 6

29. (a)

Result	Number of ways to draw
All R	1
2R, 1G	3
1R, 2G	3
All G	1

(b)

Result	Number of ways to draw
All R	1
4R, 1G	5
3R, 2G	10
2R, 3G	10
1R, 4G	5
All G	1

31. 1.02 kJ/K

33. $\sim 10^0$ W/K from metabolism; much more if you are using high-power electric appliances or an automobile, or if your taxes are paying for a war.

35. 0.507 J/K

37. (a) 5.2×10^{17} J (b) 1.8×10^3 s

39. (a) 5.00 kW (b) 763 W

41. 32.9 kJ

43. (a) $2nRT_i \ln 2$ (b) 0.273

45. 5.97×10^4 kg/s

49. (a) 4.11 kJ (b) 14.2 kJ (c) 10.1 kJ (d) 28.9%

51. (a) $10.5nRT_i$ (b) $8.50nRT_i$ (c) 0.190 (d) 0.833

53. (a) $nC_P \ln 3$ (b) Both ask for the change in entropy between the same two states of the same system. Entropy is a state variable. The change in entropy does not depend on path, but only on original and final states.

55. (a) $V_A = 1.97$ L, $V_B = 11.9$ L, $V_C = 32.8$ L, $V_D = 5.44$ L, $P_B = 4.14$ atm, $P_D = 6.03$ atm (b) 2.99 kJ (c) 0.333

57. 1.18 J/K

Context 5 Conclusion

1. 298 K

2. 60 km

3. (c) 336 K (d) The troposphere and stratosphere are too thick to be accurately modeled as having uniform temperatures. (e) 227 K (f) 107 (g) The multilayer model should be better for Venus than for the Earth. There are many layers, so the temperature of each can be reasonably uniform.

Chapter 19

1. (a) + 160 zC, 1.01 u (b) + 160 zC, 23.0 u (c) − 160 zC, 35.5 u (d) + 320 zC, 40.1 u (e) − 480 zC, 14.0 u (f) + 640 zC, 14.0 u (g) + 1.12 aC, 14.0 u (h) − 160 zC, 18.0 u

3. The force is $\sim 10^{26}$ N.

5. 0.872 N at 330°

7. (a) 2.16×10^{-5} N toward the other (b) 8.99×10^{-7} N away from the other

9. (a) 82.2 nN (b) 2.19 Mm/s

11. 1.82 m to the left of the negative charge

13. (a) $(-0.599\hat{\mathbf{i}} - 2.70\hat{\mathbf{j}})$ kN/C (b) $(-3.00\hat{\mathbf{i}} - 13.5\hat{\mathbf{j}})$ μN

15. (a) $5.91\, k_e q/a^2$ at 58.8° (b) $5.91\, k_e q^2/a^2$ at 58.8°

17. 1.59×10^6 N/C toward the rod

19. (a) $6.64\hat{\mathbf{i}}$ MN/C (b) $24.1\hat{\mathbf{i}}$ MN/C (c) $6.40\hat{\mathbf{i}}$ MN/C (d) $0.664\hat{\mathbf{i}}$ MN/C, taking the axis of the ring as the x axis

21. $-21.6\hat{\mathbf{i}}$ MN/C

23. (a) 2.00×10^{-10} C (b) 1.41×10^{-10} C (c) 5.89×10^{-11} C

25.

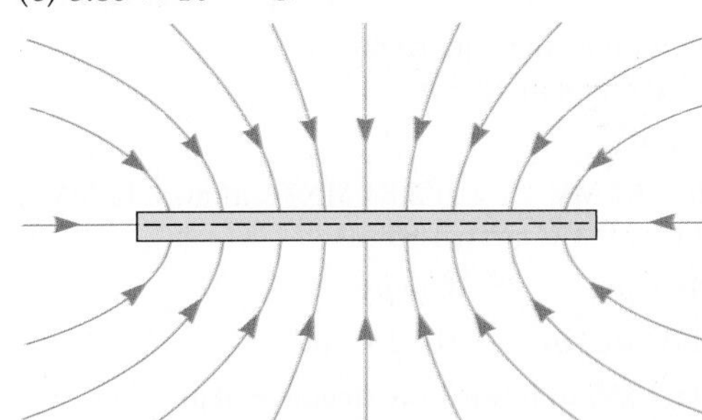

27. (a) 61.3 Gm/s^2 (b) 19.5 μs (c) 11.7 m (d) 1.20 fJ

29. (a) 111 ns (b) 5.68 mm (c) $(450\hat{\mathbf{i}} + 102\hat{\mathbf{j}})$ km/s

31. 4.14 MN/C

33. (a) $+Q/2\epsilon_0$ (b) $-Q/2\epsilon_0$

35. (a) 0 (b) 365 kN/C radially outward (c) 1.46 MN/C outward (d) 649 kN/C radially outward

37. (a) 913 nC (b) 0

39. $\vec{\mathbf{E}} = \rho r/2\epsilon_0$ away from the axis

41. 3.50 kN

43. $\vec{\mathbf{E}} = Q/2\epsilon_0 A$ vertically upward in each case if $Q > 0$

45. (a) 0 (b) 79.9 MN/C radially outward (c) 0 (d) 7.34 MN/C radially outward

47. (a) $-\lambda$, $+3\lambda$ (b) $3\lambda/2\pi\epsilon_0 r$ radially outward

49. (a) 80.0 nC/m^2 on each face (b) $9.04\hat{\mathbf{k}}$ kN/C (c) $-9.04\hat{\mathbf{k}}$ kN/C

51. 1.77×10^{-12} C/m^3, positive

53. possible only with a charge of + 51.3 μC at $x = -16.0$ cm

55. 40.9 N at 263°

57. 26.7 μC

59. (a) $\theta_1 = \theta_2$

61. (b) in the $+z$ direction

63. (a) σ/ϵ_0 away from both plates (b) 0 (c) σ/ϵ_0 away from both plates

65. (a) $\rho r/3\epsilon_0$; $Q/4\pi\epsilon_0 r^2$; 0; $Q/4\pi\epsilon_0 r^2$, all radially outward (b) $-Q/4\pi b^2$ and $+Q/4\pi c^2$

Chapter 20

1. (a) 152 km/s (b) 6.49 Mm/s

3. (a) $-600\ \mu$J (b) -50.0 V

5. 38.9 V; the origin

7. (a) 1.44×10^{-7} V (b) -7.19×10^{-8} V (c) -1.44×10^{-7} V, $+7.19 \times 10^{-8}$ V

9. (a) -4.83 m (b) 0.667 m and -2.00 m

11. -11.0 MV

15. (a) 10.8 m/s and 1.55 m/s (b) greater

17. (a) no point at a finite distance from the charges (b) $2k_eq/a$

19. $5k_eq^2/9d$

21. (a) 10.0 V, -11.0 V, -32.0 V (b) 7.00 N/C in the $+x$ direction

23. $\vec{\mathbf{E}} = (-5 + 6xy)\hat{\mathbf{i}} + (3x^2 - 2z^2)\hat{\mathbf{j}} - 4yz\hat{\mathbf{k}}$; 7.07 N/C

25. (a) coulombs per square meter (b) $k_e\alpha[L - d\ln(1 + L/d)]$

27. -1.51 MV

29. (a) 0, 1.67 MV (b) 5.84 MN/C away, 1.17 MV (c) 11.9 MN/C away, 1.67 MV

31. (a) 48.0 μC (b) 6.00 μC

33. (a) 1.33 μC/m^2 (b) 13.3 pF

35. (a) 11.1 kV/m toward the negative plate (b) 98.3 nC/m^2 (c) 3.74 pF (d) 74.7 pC

37. $mgd\tan\theta/q$

39. (a) 17.0 μF (b) 9.00 V (c) 45.0 μC and 108 μC

41. (a) 5.96 μF (b) 89.5 μC on 20 μF, 63.2 μC on 6 μF, 26.3 μC on 15 μF and on 3 μF

43. 120 μC; 80.0 μC and 40.0 μC

45. (a) 398 μF in series (b) 2.20 μF in parallel

47. (a) 216 μJ (b) 54.0 μJ

49. (a) circuit diagram:

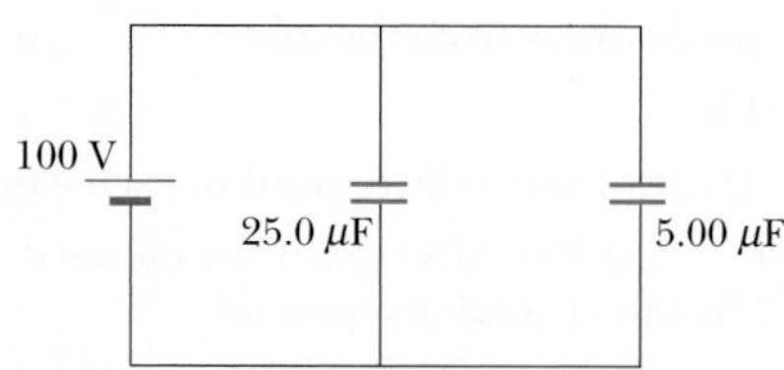

stored energy = 0.150 J

(b) potential difference = 268 V

circuit diagram:

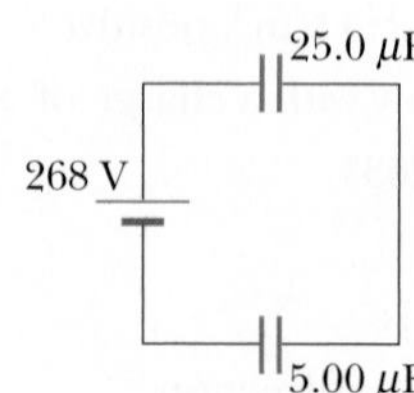

53. (a) 81.3 pF (b) 2.40 kV

55. 1.04 m

57. (a) 369 pC (b) 118 pF, 3.12 V (c) -45.5 nJ

59. 9.79 kg

61. 253 MeV

63 $k_eQ^2/2R$

65. 579 V

67. (a) volume 9.09×10^{-16} m^3, area 4.54×10^{-10} m^2 (b) 2.01×10^{-13} F (c) 2.01×10^{-14} C; 1.26×10^5 electronic charges

71. (a) $-2Q/3$ on upper plate, $-Q/3$ on lower plate (b) $2Qd/3\epsilon_0 A$

73. 0.188 m^2

75. (a) $\dfrac{\epsilon_0}{d}(\ell^2 + \ell x(\kappa - 1))$

(b) $\dfrac{Q^2 d}{2\epsilon_0(\ell^2 + \ell x(\kappa - 1))}$

(c) $\dfrac{Q^2 d\ell(\kappa - 1)}{2\epsilon_0(\ell^2 + \ell x(\kappa - 1))^2}$ to the right

(d) 205 μN to the right

77. $\frac{4}{3}C$

Chapter 21

1. 7.50×10^{15} electrons

3. (a) 0.632 $I_0\tau$ (b) 0.999 95 $I_0\tau$ (c) $I_0\tau$

5. 0.130 mm/s

7. 6.43 A

9. (a) 31.5 n$\Omega\cdot$m (b) 6.35 MA/m^2 (c) 49.9 mA (d) 659 μm/s (e) 0.400 V

11. 1.71 Ω

13. 0.181 V/m

15. 448 A

17. 36.1%

19. (a) 184 W (b) 461°C

21. (a) \$1.61 (b) \$0.005 82 (c) \$0.416

23. (a) 667 A (b) 50.0 km

25. (a) 6.73 Ω (b) 1.97 Ω

27. (a) 17.1 Ω (b) 1.99 A for 4 Ω and 9 Ω, 1.17 A for 7 Ω, 0.818 A for 10 Ω

29. (a) 227 mA (b) 5.68 V

31. 14.2 W to 2 Ω, 28.4 W to 4 Ω, 1.33 W to 3 Ω, 4.00 W to 1 Ω

33. (a) 470 W (b) 1.60 mm or more (c) 2.93 mm or more

35. 846 mA down in the 8-Ω resistor; 462 mA down in the middle branch; 1.31 A up in the right-hand branch

37. (a) -222 J and 1.88 kJ (b) 687 J, 128 J, 25.6 J, 616 J, 205 J (c) 1.66 kJ of chemical energy is transformed into internal energy

39. 50.0 mA from a to e

41. (a) 5.00 s (b) 150 μC (c) 4.06 μA

43. (a) 1.50 s (b) 1.00 s (c) $(200 + 100e^{-t/1.00\text{ s}})\ \mu$A

45. (a) 6.00 V (b) 8.29 μs

47. $6.00 \times 10^{-15}/\Omega \cdot \text{m}$

49. (a) 576 Ω, 144 Ω (b) 4.80 s. The charge is the same. It is at a location of lower potential energy. (c) 0.040 0 s. Energy entering by electrical transmission exits by heat and electromagnetic radiation. (d) \$1.26, energy at 1.94×10^{-8} \$/J

51. (a) $(8.00\hat{\mathbf{i}})$ V/m (b) 0.637 Ω (c) 6.28 A (d) $(200\hat{\mathbf{i}})$ MA/m^2

53. (a) 12.5 A, 6.25 A, 8.33 A (b) No; together they would require 27.1 A.

55. 2.22 h

57. (a) $R \rightarrow \infty$ (b) $R \rightarrow 0$ (c) $R = r$

59. (a) 9.93 μC (b) 33.7 nA (c) 334 nW (d) 337 nW

61. (a) 222 μC (b) increases by 444 μC

63. (a) 0.991 (b) 0.648 (c) Insulation should be added to the ceiling.

Context 6 Conclusion

1. (a) 87.0 s (b) 261 s (c) $t \rightarrow \infty$

2. (a) 0.01 s (b) 7×10^6

3. (a) 3×10^6 (b) 9×10^6

Chapter 22

1. (a) up (b) toward you, out of the plane of the paper (c) no deflection (d) into the plane of the paper

3. (a) 8.67×10^{-14} N (b) 5.19×10^{13} m/s^2

5. 8.93×10^{-30} N down, 1.60×10^{-17} N up, 4.80×10^{-17} N down

7. 115 keV

9. 7.88 pT

11. 0.278 m

13. 70.1 mT

15. $(-2.88\hat{\mathbf{j}})$ N

17. $2\pi r I B \sin\theta$ up

19. (a) 5.41 mA · m^2 (b) 4.33 mN · m

21. 9.98 N · m clockwise as seen looking down from above

23. 12.5 T

25. $\dfrac{\mu_0 I}{4\pi x}$ into the paper

27. $\left(1 + \dfrac{1}{\pi}\right)\dfrac{\mu_0 I}{2R}$ directed into the page

29. (a) $2I_1$ out of the page (b) $6I_1$ into the page

31. 261 nT into the page

33. (a) 21.5 mA (b) 4.51 V (c) 96.7 mW

35. $(-27.0\hat{\mathbf{i}})$ μN

37. 20.0 μT toward the bottom of the page

39. (a) 6.34 mN/m inward (b) greater

41. (a) 3.60 T (b) 1.94 T

43. 500 A

45. 31.8 mA

47. 207 W

49. (a) 8.63×10^{45} electrons (b) 4.01×10^{20} kg

51. (a) 1.4 MJ/mi (b) 5.7 MJ/mi (c) 1/400

53. $\vec{\mathbf{B}} = \dfrac{\mu_0 J_s}{2}\,\hat{\mathbf{k}}$ for $x > 0$ and $\vec{\mathbf{B}} = -\dfrac{\mu_0 J_s}{2}\,\hat{\mathbf{k}}$ for $x < 0$

55. (a) The electric current experiences a magnetic force. (c) no, no, no

57. (a) $(3.52\hat{\mathbf{i}} - 1.60\hat{\mathbf{j}})$ aN (b) 24.4°

59. $B \sim 10^{-1}$ T, $\tau \sim 10^{-1}$ N · m, $I \sim 10^{0}$ A, $A \sim 10^{-3}$ m^2, $N \sim 10^3$

61. (a) 1.04×10^{-4} m (b) 1.89×10^{-4} m

63. $\dfrac{\mu_0 I}{2\pi w} \ln\left(1 + \dfrac{w}{b}\right)\hat{\mathbf{k}}$

65. $\dfrac{\mu_0 q\omega}{2.5\pi R\sqrt{5}}$

67. (a) 2.46 N up (b) 107 m/s^2 up

Chapter 23

1. 0.800 mA

3. 160 A

5. (a) $(\mu_0 I L/2\pi)\ln(1 + w/h)$ (b) -4.80 μV; current is counterclockwise

7. 61.8 mV

9. (b) The emf induced in the coil is proportional to the line integral of the magnetic field around the circular axis of the toroid. By Ampère's law, this line integral depends only on the current the circle encloses.

11. (a) eastward (b) 458 μV

13. (a) 3.00 N to the right (b) 6.00 W

15. $mvR/B^2\ell^2$

17. 24.1 V with the outer contact positive

19. 2.83 mV

21. (b) Larger R makes current smaller, so the loop must travel faster to maintain equality of magnetic force and weight. (c) The magnetic force is proportional to the product of field and current, whereas the current is itself proportional to field. If B becomes two times smaller, the speed must become four times larger to compensate.

23. (a) 7.54 kV (b) The plane of the coil is parallel to $\vec{\mathbf{B}}$.

25. 1.80 mN/C upward and to the left, perpendicular to r_1

27. 19.5 mV

29. $-(18.8 \text{ V})\cos(377t)$

31. (a) 360 mV (b) 180 mV (c) 3.00 s

33. (a) 0.139 s (b) 0.461 s

35. (a) 2.00 ms (b) 0.176 A (c) 1.50 A (d) 3.22 ms

37. (a) 20.0% (b) 4.00%

39. $(500 \text{ mA})(1 - e^{-10t/\text{s}})$, $1.50 \text{ A} - (0.250 \text{ A})e^{-10t/\text{s}}$

41. (a) 5.66 ms (b) 1.22 A (c) 58.1 ms

43. 2.44 μJ

45. 44.2 nJ/m^3 for the $\vec{\mathbf{E}}$ field and 995 μJ/m^3 for the $\vec{\mathbf{B}}$ field

47. (a) 2.93 mT up (b) 3.42 Pa (c) clockwise (d) up (e) 1.30 mN

49. $-7.22 \text{ mV} \cos(2\pi 523\, t/\text{s})$

51. $\sim 10^{-4}$ V, by reversing a 20-turn coil of diameter 3 cm in 0.1 s in a field of 10^{-3} T

53. (a) 254 km/s (b) 215 V

55. 6.00 A

59. $(-87.1 \text{ mV})\cos(200\pi t + \phi)$

61. (a) $L \approx (\pi/2)N^2\mu_0 R$ (b) ~100 nH (c) ~1 ns

63. $3.97 \times 10^{-25}\ \Omega$

65. (a) 50.0 mT (b) 20.0 mT (c) 2.29 MJ (d) 318 Pa

Context 7 Conclusion

1. 5.56 Hz

2. Both are correct.

Chapter 24

1. 1.85 aT up

3. $(-2.87\hat{\mathbf{j}} + 5.75\hat{\mathbf{k}})$ Gm/s^2

5. 2.25×10^8 m/s

7. (a) 6.00 MHz (b) $(-73.3\hat{\mathbf{k}})$ nT (c) $\vec{\mathbf{B}} = [(-73.3\hat{\mathbf{k}})\text{ nT}]\cos(0.126x - 3.77 \times 10^7 t)$

11. 2.9×10^8 m/s $\pm$ 5%

13. (c) 2.00 kHz (d) $\pm$ 0.075 0 m/s $\approx$ 0.2 mi/h

15. $0.220c = 6.59 \times 10^7$ m/s

17. 608 pF

19. (a) 503 Hz (b) 12.0 μC (c) 37.9 mA (d) 72.0 μJ

21. 307 μW/m^2

23. 3.33×10^3 m^2

25. (a) 332 kW/m^2 radially inward (b) 1.88 kV/m and 222 μT

27. 5.16 m

29. (a) 1.90 kN/C (b) 50.0 pJ (c) 1.67×10^{-19} kg·m/s

31. The radio audience hears it 8.41 ms sooner.

33. 545 THz

35. (a) 6.00 pm (b) 7.50 cm

37. 56.2 m

39. (a) 0.690 wavelengths (b) 58.9 wavelengths

41. (a) 54.7° (b) 63.4° (c) 71.6°

45. $\frac{1}{8}$

49. (a) 4.24 PW/m^2 (b) 1.20 pJ = 7.50 MeV

51. 3.49×10^{16} photons

53. (a) three: 632.808 57 nm, 632.809 14 nm, and 632.809 71 nm (b) 697 m/s (c) For an atom moving away from the observer at the rms speed, the wavelength is increased by 0.001 47 nm. For an approaching atom, the wavelength is decreased by this amount. Many atoms are moving at speeds higher than the rms speed.

55. (a) 3.85×10^{26} W (b) 1.02 kV/m and 3.39 μT

57. (a) $2\pi^2 r^2 f B_{\max}\cos\theta$, where θ is the angle between the magnetic field and the normal to the loop (b) The loop should be in the vertical plane containing the line of sight to the transmitter.

59. (a) 6.67×10^{-16} T (b) 5.31×10^{-17} W/m^2 (c) 1.67×10^{-14} W (d) 5.56×10^{-23} N

61. 95.1 mV/m

63. (a) 625 kW/m^2 (b) 21.7 kN/C, 72.4 μT (c) 17.8 min

65. (b) 17.6 Tm/s^2, 1.75×10^{-27} W (c) 1.80×10^{-24} W

67. (a) 388 K (b) 363 K

69. (a) 22.6 h (b) 30.6 s

Chapter 25

1. (a) 1.94 m (b) 50.0° above the horizontal

3. six times from the mirror on the left and five times from the mirror on the right

5. 15.4°; 2.56 m

7. 19.5° above the horizon

9. (a) 2.0×10^8 m/s (b) 474 THz (c) 4.2×10^{-7} m

11. (a) 181 Mm/s (b) 225 Mm/s (c) 136 Mm/s

13. 30.0° and 19.5° at entry; 19.5° and 30.0° at exit

15. $\tan^{-1} n$

17. 3.88 mm

19. 30.4° and 22.3°

21. $\sim 10^{-11}$ s; between 10^3 and 10^4 wavelengths

23. 0.171°

25. 27.9°

27. 4.61°

29. (a) 24.4° (b) 37.0° (c) 49.8°

31. 1.000 08

33. 67.2°

35. 82 reflections

37. 23.1°

39. (a) $\frac{h}{c}\left(\frac{n + 1.00}{2}\right)$ (b) $\left(\frac{n + 1.00}{2}\right)$ times longer

41. 2.27 m

43. (a) 0.172 mm/s (b) 0.345 mm/s (c) northward at 50.0° below the horizontal (d) northward at 50.0° below the horizontal

45. 62.2%

47. (a) 0.042 6 = 4.26% (b) no difference

49. 70.6%

51. 27.5°

53. (a) It always happens. (b) 30.3° (c) It cannot happen.

55. 2.36 cm

57. 1.93

59. (a) 1.20 (b) 3.40 ns

Chapter 26

1. $\sim 10^{-9}$ s younger

3. 35.0 in.

5. 10.0 ft, 30.0 ft, 40.0 ft

7. (a) 13.3 cm, real and inverted, -0.333 (b) 20.0 cm, real and inverted, -1.00 (c) No image is formed.

9. (a) -12.0 cm; 0.400 (b) -15.0 cm; 0.250 (c) upright

11. (a) $q = 45.0$ cm; $M = -0.500$ (b) $q = -60.0$ cm; $M = 3.00$ (c) Image (a) is real, inverted, and diminished. Image (b) is virtual, upright, and enlarged.

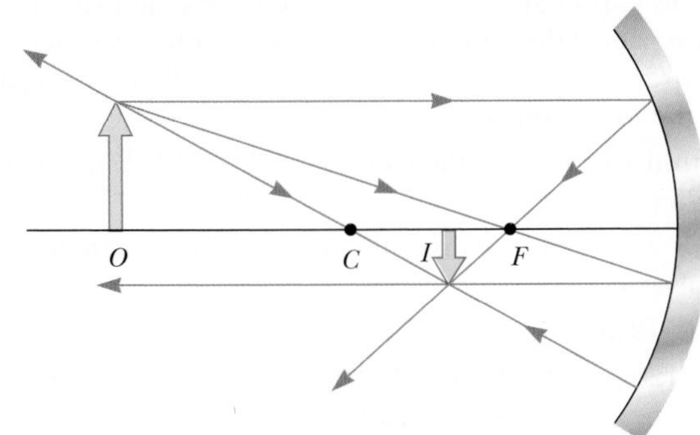

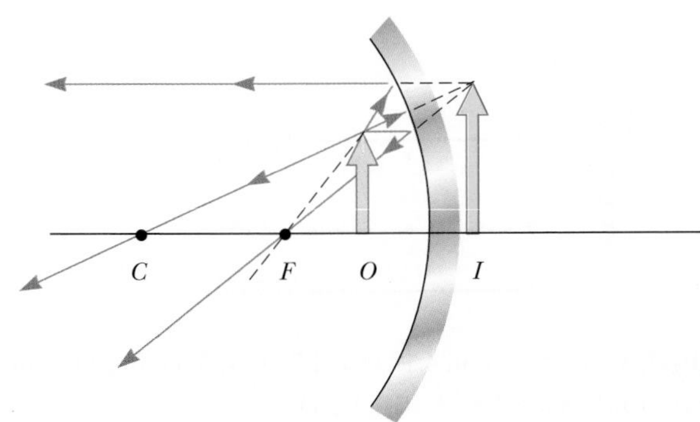

13. At 0.708 cm in front of the reflecting surface. Image is virtual, upright, and diminished.

15. 7.90 mm

17. (a) a concave mirror with radius of curvature 2.08 m (b) 1.25 m from the object

19. (a) 25.6 m (b) 0.058 7 rad (c) 2.51 m (d) 0.023 9 rad (e) 62.8 m from your eyes

21. 38.2 cm below the top surface of the ice

23. 8.57 cm

25. (a) 45.0 cm (b) -90.0 cm (c) -6.00 cm

27. (a) 16.4 cm (b) 16.4 cm

29. (a) 650 cm from the lens on the opposite side from the object; real, inverted, enlarged (b) 600 cm from the lens on the same side as the object; virtual, upright, enlarged

31. 2.84 cm

35. (a) -12.3 cm, to the left of the lens (b) 0.615 (c)

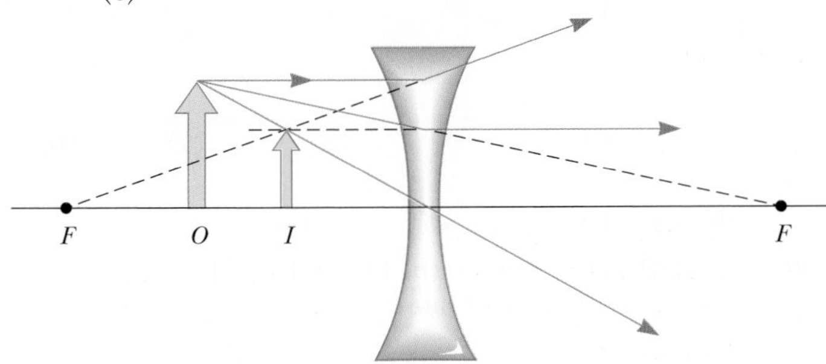

37. 1.16 mm/s toward the lens

39. (a) $p = \frac{d}{2} \pm \sqrt{\frac{d^2}{4} - fd}$ (b) Both images are real and inverted. One is enlarged, the other diminished.

41. (a) 0.833 mm (b) 0.820 mm

43. $f = \frac{-Md}{(1-M)^2}$ if $M < 1$, $f = \frac{Md}{(M-1)^2}$ if $M > 1$

45. -25.0 cm

47. (a) 67.5 cm (b) The lenses can be displaced in two ways. The first lens can be displaced 1.28 cm farther away from the object and the second lens 17.7 cm toward the object. Alternatively, the first lens can be displaced 0.927 cm toward the object and the second lens 4.44 cm toward the object.

49. 0.107 m to the right of the vertex of the hemispherical face

51. 8.00 cm

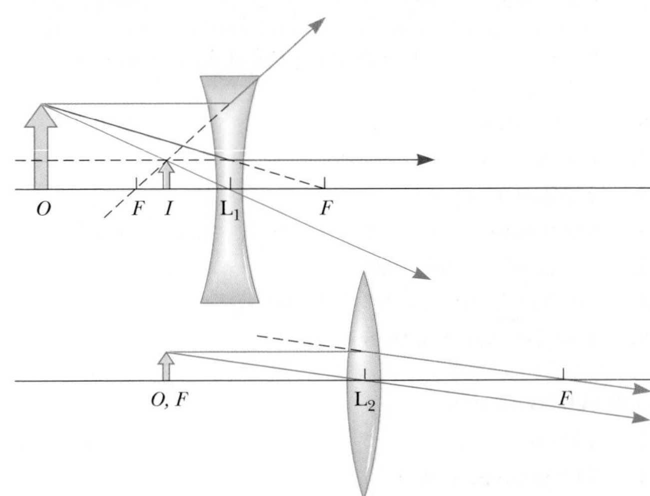

53. 1.50 m in front of the mirror; 1.40 cm (inverted)

55. (a) 30.0 cm and 120 cm (b) 24.0 cm (c) real, inverted, diminished with $M = -0.250$

57. (a) 263 cm (b) 79.0 cm

59. The image is real, inverted, and actual size.

Chapter 27

1. 515 nm

3. (a) 55.7 m (b) 124 m

5. 1.54 mm

7. 641

9. (a) 13.2 rad (b) 6.28 rad (c) 0.012 7° (d) 0.059 7°

11. 48.0 μm

13. 0.968

15. (a) green (b) violet

17. 96.2 m

19. 4.35 μm

21. 0.230 mm

23. 91.2 cm

25. 51.8 μm wide and 949 μm high

27. 3.09 m

29. 13.1 m

31. 105 m

33. 5.91° in first order, 13.2° in second order, 26.5° in third order

35. (a) 5 orders (b) 10 orders in the short-wavelength region

37. three, at 0° and at 45.2° to the right and left

39. 14.4°

41. (a) 0.738 mm (b) Individual waves from all the transparent zones will add crest-on-crest to interfere constructively at the slit images. The grating equation $d \sin \theta = m\lambda$ is satisfied at the slit images. Elsewhere on the screen destructive interference will prevent light from reaching the screen.

43. number of antinodes = number of constructive interference zones = 1 plus 2 times the greatest positive integer $\leq d/\lambda$; number of nodes = number of destructive interference zones = 2 times the greatest positive integer $< (d/\lambda + \frac{1}{2})$

45. $20.0 \times 10^{-6}\ °C^{-1}$

47. 2.50 mm

49. 113 dark fringes

53. 632.8 nm

55. (a) 25.6° (b) 19.0°

57. (a) $3.53 \times 10^3\ cm^{-1}$ (b) 11

59. $4.58\ \mu m < d < 5.23\ \mu m$

Context 8 Conclusion

1. 130 nm

2. 74.2 grooves/mm

3. 1.8 μm/bit

4. 48 059

5. $\sim 10^8\ W/m^2$

Chapter 28

1. 1.69%

3. About 5 200 K. A firefly cannot be at this temperature, so its light cannot be blackbody radiation.

5. 2.27×10^{30} photons/s

7. 1.32×10^{31}

9. (a) 296 nm, 1.01 PHz (b) 2.71 V

11. (a) 1.90 eV (b) 0.216 V

13. 8.41 pC

15. (a) 488 fm (b) 268 keV (c) 31.5 keV

17. 70.0°

19. By this definition, ionizing radiation is the ultraviolet light, x-rays and γ rays with wavelength shorter than 124 nm; that is, with frequency higher than 2.41×10^{15} Hz.

21. (a) 0.709 nm (b) 414 nm

23. (a) ~100 MeV or more (b) No. With kinetic energy much larger than the magnitude of its negative electric potential energy, the electron would immediately escape.

25. (a) 14.9 keV (b) 124 keV

29. (a) 993 nm (b) 4.96 mm (c) If its detection forms part of an interference pattern, the neutron must have passed through both slits. If we test to see which slit a particular neutron passes through, it will not form part of the interference pattern.

31. within 1.16 mm for the electron, 5.28×10^{-32} m for the bullet

33. 3.79×10^{28} m, 190 times the diameter of the visible Universe

35. (b) 519 am

37. (a) 126 pm (b) $5.27 \times 10^{-24}\ kg \cdot m/s$ (c) 95.5 eV

39. (a)

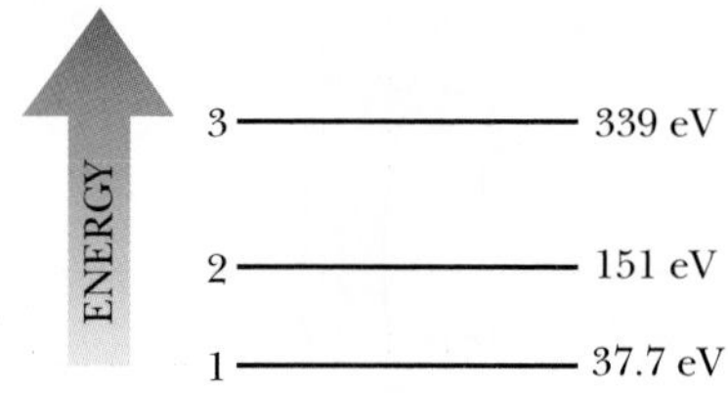

(b) 2.20 nm, 2.75 nm, 4.12 nm, 4.71 nm, 6.60 nm, 11.0 nm

41. (a) $(15h\lambda/8m_e c)^{1/2}$ (b) 1.25λ

45. (a) $L/2$ (b) 5.26×10^{-5} (c) 3.99×10^{-2} (d) The probability density has peaks around $L/4$ and $3L/4$, and a zero at $L/2$. Because the probability density is symmetric about $L/2$, the average experimental value has to be $L/2$.

47. 0.250

49. (a) 0.010 3 (b) 0.990

51. 85.9

53. (a) 1.06 mm (b) microwave

55. length 0.333 m, radius 19.8 μm

57. (a)

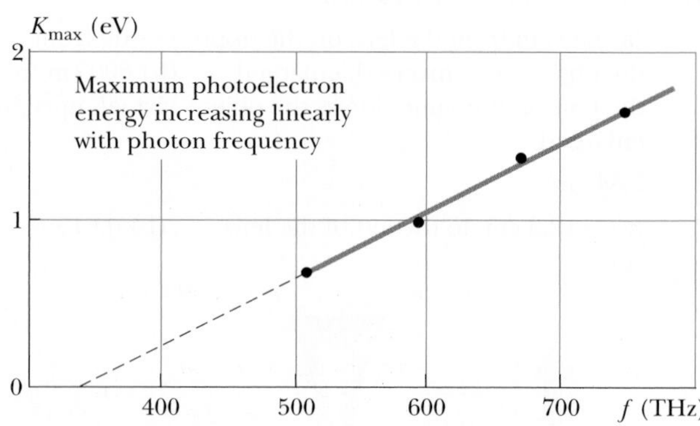

(b) $6.4 \times 10^{-34}\ J \cdot s \pm 8\%$ (c) 1.4

59. (a) 2.82×10^{-37} m (b) 1.06×10^{-32} J (c) 2.87×10^{-35} % or more

61. (a)

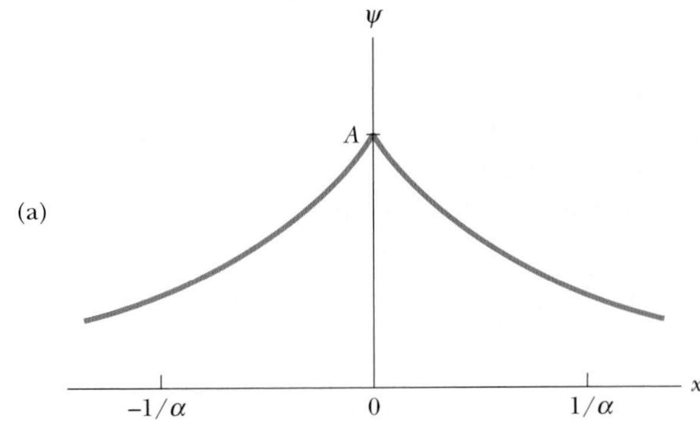

(b)

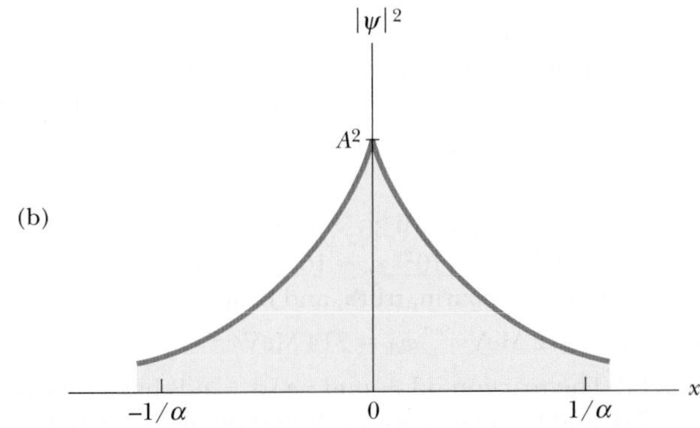

(c) The wave function is continuous. It shows localization by approaching zero as $x \to \pm\infty$. It is everywhere finite and can be normalized. (d) $A = \sqrt{\alpha}$ (e) 0.632

65. (a)

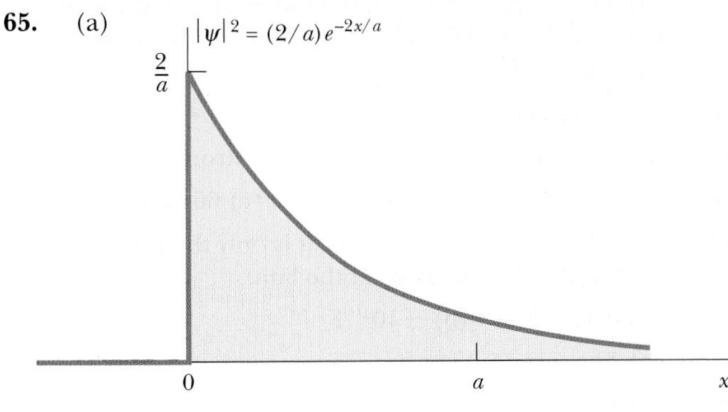

(b) 0 (c) 0.865

Chapter 29

1. (b) 0.846 ns

3. (a) 2.89×10^{34} kg·m²/s (b) 2.74×10^{68} (a) 7.30×10^{-69}

5. (a) 1.89 eV, 656 nm (b) 3.40 eV, 365 nm

7. (a) 1.31 μm (b) 164 nm

9. (a) $\Delta p \geq \hbar/2r$ (b) Choosing $p \approx \hbar/r$, we find that $E = K + U = \hbar^2/2m_er^2 - k_ee^2/r$. (c) $r = \hbar^2/m_ek_ee^2 = a_0$ and $E = -13.6$ eV, in agreement with the Bohr theory

11. (b) 0.497

13. It does, with $E = -k_ee^2/2a_0$.

15. (a)

n	ℓ	m_ℓ	m_s
3	2	2	$\frac{1}{2}$
3	2	2	$-\frac{1}{2}$
3	2	1	$\frac{1}{2}$
3	2	1	$-\frac{1}{2}$
3	2	0	$\frac{1}{2}$
3	2	0	$-\frac{1}{2}$
3	2	−1	$\frac{1}{2}$
3	2	−1	$-\frac{1}{2}$
3	2	−2	$\frac{1}{2}$
3	2	−2	$-\frac{1}{2}$

(b)

n	ℓ	m_ℓ	m_s
3	1	1	$\frac{1}{2}$
3	1	1	$-\frac{1}{2}$
3	1	0	$\frac{1}{2}$
3	1	0	$-\frac{1}{2}$
3	1	−1	$\frac{1}{2}$
3	1	−1	$-\frac{1}{2}$

17. $\ell = 4$

19. (a) 2 (b) 8 (c) 18 (d) 32 (e) 50

21. (a) 3.99×10^{17} kg/m³ (b) 81.7 am (c) 1.77 Tm/s (d) $5.91 \times 10^3\ c$

23. $n = 3$; $\ell = 2$; $m_\ell = -2, -1, 0, 1,$ or 2; $s = 1$; $m_s = -1$, 0, or 1, for a total of 15 states

25. The 4*s* subshell is filled first. We would expect [Ar]$3d^44s^2$ to have lower energy, but [Ar]$3d^54s^1$ has more unpaired spins and lower energy according to Hund's rule. It is the ground-state configuration of chromium.

27. aluminum

29. (a) 1*s*, 2*s*, 2*p*, 3*s*, 3*p*, 4*s*, 3*d*, 4*p*, 5*s*, 4*d*, 5*p*, 6*s*, 4*f*, 5*d*, 6*p*, 7*s* (b) Element 15 should have valence +5 or −3, and it does. Element 47 should have valence −1, but it has valence +1. Element 86 should be inert, and it is.

31. 18.4 T

33. 124 V

35. 0.072 5 nm

37. iron

41. (a) $0.160c$ (b) 2.82×10^9 ly

43. (a) 609 μeV (b) 6.9 μeV (c) 147 GHz, 2.04 mm

47. The classical frequency is $4\pi^2m_ek_e^2e^4/h^3n^3$.

49. (a) 1.57×10^{14} m$^{-3/2}$ (b) 2.47×10^{28} m^{-3} (c) 8.69×10^8 m^{-1}

51. $3h^2/4mL^2$

55. 5.39 keV

57. 0.125

59. 9.79 GHz

61. (a) $\sim -10^6$ m/s² (b) ~ 1 m

Chapter 30

1. $\sim 10^{28}$; $\sim 10^{28}$; $\sim 10^{28}$
3. (a) 29.5 fm (b) 5.18 fm (c) The wavelength is much less than the distance of closest approach.
5. 16.0 km
7. (a) 29.2 MHz (b) 42.6 MHz (c) 2.13 kHz
9. greater for $^{15}_{7}N$ by 3.54 MeV
11. 200 MeV
13. (a) 1.55×10^{-5}/s, 12.4 h (b) 2.39×10^{13} atoms (c) 1.88 mCi
15. 86.4 h
17. 2.66 d
19. 4.27 MeV
21. 9.96×10^{3} yr
23. (a) $e^- + p \rightarrow n + \nu$ (b) $^{15}_{8}O$ atom $\rightarrow$ $^{15}_{7}N$ atom $+ \nu$ (c) 2.75 MeV
25. (a) 148 Bq/m^3 (b) 7.05×10^{7} atoms/m^3 (c) 2.17×10^{-17}
27. (a) $^{197}_{79}Au + ^{1}_{0}n \rightarrow ^{198}_{80}Hg + ^{0}_{-1}e^- + \bar{\nu}$ (b) 7.89 MeV
29. 5.80 Mm
31. about 3 000 yr
33. 3.60×10^{38} protons/s
35. (b) 26.7 MeV
37. (a) 8×10^{4} eV (b) 4.62 MeV and 13.9 MeV (c) 1.03×10^{7} kWh
39. (a) 5.70 MeV (b) exothermic; 3.27 MeV
41. (b) 1.53 MeV
43. (a) conservation of energy (b) electric potential energy of the nucleus (c) 1.20 MeV
45. (b) 1.94 meV
47. (a) $\sim 10^{-1356}$ (b) 0.891
49. (b) 4.78 MeV
51. 1.66×10^{3} yr
53. (a) 0.963 mm (b) It increases by 7.47%
55. (b) R/λ
57. 2.56×10^{4} kg
59. (a) 2.65 GJ (b) The fusion energy is 78.0 times larger.
61. (a) 15.5 cm (b) 51.7 MeV (c) The number of decays per second is the decay rate R, and the energy released in each decay is Q. Then the energy released per unit time interval is $\mathcal{P} = QR$. (d) 227 kJ/yr (e) 3.18 J/yr
63. (a) 422 MBq (b) 153 ng

Chapter 31

1. 453 ZHz; 662 am
3. (a) 558 TJ (b) \$$2.17 \times 10^{7}$
5. 118 MeV
7. $\sim 10^{-18}$ m
9. 67.5 MeV, 67.5 MeV/c, 16.3 ZHz
11. (a) 0.782 MeV (b) $v_e = 0.919c$, $v_p = 380$ km/s (c) The electron is relativistic; the proton is not.
13. $\Omega^+ \rightarrow \overline{\Lambda}^0 + K^+$, $\overline{K}_S{}^0 \rightarrow \pi^+ + \pi^-$, $\overline{\Lambda}^0 \rightarrow \bar{p} + \pi^+$, $\bar{n} \rightarrow \bar{p} + e^+ + \nu_e$
15. (b) The second violates strangeness conservation.
17. (a) $\bar{\nu}_\mu$ (b) ν_μ (c) $\bar{\nu}_e$ (d) ν_e (e) ν_μ (f) $\bar{\nu}_e + \nu_\mu$
19. (a), (c), and (f) violate baryon number conservation. (b), (d), and (e) can occur. (f) violates muon-lepton number conservation.
21. (a) ν_e (b) ν_μ (c) $\bar{\nu}_\mu$ (d) $\nu_\mu + \bar{\nu}_\tau$
23. (b) and (c) conserve strangeness. (a), (d), (e), and (f) violate strangeness conservation.
25. (a) not allowed; violates conservation of baryon number (b) strong interaction (c) weak interaction (d) weak interaction (e) electromagnetic interaction
27. (a) K^+ (b) Ξ^0 (c) π^0
29. 9.26 cm
31. (a) 3.34×10^{26} e$^-$, 9.36×10^{26} u, 8.70×10^{26} d (b) $\sim 10^{28}$ e$^-$, $\sim 10^{29}$ u, $\sim 10^{29}$ d. You have zero strangeness, charm, truth, and beauty.
33. $m_u = 312$ MeV/c^2, $m_d = 314$ MeV/c^2
35. (a) The reaction $\bar{u}d + uud \rightarrow \bar{s}d + uds$ has a total of 1 u, 2 d, and 0 s quarks originally and finally. (b) The reaction $\bar{d}u + uud \rightarrow \bar{s}u + uus$ has a net of 3 u, 0 d, and 0 s before and after. (c) $\bar{u}s + uud \rightarrow \bar{s}u + \bar{s}d + sss$ shows conservation at 1 u, 1 d, and 1 s quark. (d) The process $uud + uud \rightarrow \bar{s}d + uud + \bar{d}u + uds$ nets 4 u, 2 d, and 0 s initially and finally; the mystery particle is a Λ^0 or a Σ^0.
37. a neutron, udd
39. (a) $-e$, antiproton (b) 0, antineutron
41. (a) 590.07 nm (b) 597 nm (c) 661 nm
43. (a) 8.41×10^{6} kg (b) No. It is only the fraction 4.23×10^{-24} of the mass of the Sun.
45. (a) $\sim 10^{13}$ K (b) $\sim 10^{10}$ K
47. (b) 11.8 Gyr
49. $\sim 10^{14}$
51. one part in 50 000 000
53. 0.407%
55. 5.35 MeV and 32.3 MeV
57. 1 116 MeV/c^2
59. 70.4 MeV
61. 2.52×10^{3} K
63. (a) Z^0 boson (b) gluon or photon
65. (a) 127 MeV (b) 1.06 mm (c) 1.17 meV (d) 58.1 EeV

Context 9 Conclusion

1. (a) 1.61×10^{-35} m (b) 5.38×10^{-44} s (c) yes

Credits

Photographs

This page constitutes an extension of the copyright page. We have made every effort to trace the ownership of all copyrighted material and to secure permission from copyright holders. In the event of any question arising as to the use of any material, we will be pleased to make the necessary corrections in future printings. Thanks are due to the following authors, publishers, and agents for permission to use the material indicated.

Invitation. **1:** © Stockbyte **3:** © David Parker Photo Researchers, Inc.

Chapter 1. **4:** Mark Wagner/Stone/Getty Images **6:** Courtesy of National Institute of Standards and Technology, U.S. Department of Commerce **10:** Phil Boorman/Getty Images **14:** Mack Henley/Visuals Unlimited

Context 1. **34:** Courtesy of The Exhibition Alliance, Hamilton, N.Y. **35:** top left, © Bettmann/CORBIS; bottom right, © Martin Bond/Photo Researchers, Inc. **36:** © Mehau Kulyk/Photo Researchers, Inc.

Chapter 2. **37:** Jean Y. Ruszniewski/Getty Images **56:** top left, North Wind Picture Archive; bottom, © 1993 James Sugar/Black Star **57:** George Semple **60:** George Lepp/Stone/Getty **66:** top left, Courtesy of the U.S. Air Force; top right, Photri, Inc., **67:** Courtesy Amtrak NEC Media Relations

Chapter 3. **69:** © Arndt/Premium Stock/PictureQuest **74:** © The Telegraph Colour Library/Getty Images **79:** Tony Duffy/Getty Images **89:** Frederick McKinney/FPG/Getty **90:** top left, Jed Jacobsohn/Getty Images; middle left, Bill Lee/Dembinsky Photo Associates; bottom left, Sam Sargent/Liaison International **91:** Courtesy of NASA **92:** Courtesy of NASA

Chapter 4. **96:** Steve Raymer/CORBIS **99:** Giraudon/Art Source **104:** NASA **105:** top right, John Gillmoure, The Stock Market **116:** Roger Viollet, Mill Valley, CA, University Science Books, 1982 **118:** © Tony Arruza/CORBIS

Chapter 5. **125:** Paul Hardy/CORBIS **134:** top, Robin Smith/Getty Images; bottom left, © Tom Carroll/Index Stock Imagery/Picture Quest **143:** Jump Run Productions/Image Bank **146:** a) Courtesy of GM; b) Courtesy of GM-Hummer **148:** Mike Powell/Getty Images **150:** Frank Cezus/FPG International

Chapter 6. **156:** Billy Hustace/Getty Images **170:** a–c, e, f) George Semple; d) Digital Vision/Getty Images **172:** Sinclair Stammers/Science Photo Library/Photo Researchers, Inc. **185:** Ron Chapple/FPG/Getty

Chapter 7. **188:** © Harold E. Edgerton/Courtesy of Palm Press, Inc. **213:** Gamma **219:** Engraving from Scientific American, July 1888

Context Conclusion 1. **220:** Courtesy of Honda Motor Co., Inc **222:** top left, Photo by Brent Romans/www.Edmunds.com; bottom left, © Adam Hart-Davis/Photo Researchers, Inc.

Context 2. **223:** Japanese Aerospace Exploration Agency (JAXA) **224:** top, courtesy of NASA/JPL; bottom, Courtesy of NASA/JPL/Cornell **225:** Pierre Mion/National Geographic Image Collection

Chapter 8. **226:** © Harold and Esther Edgerton Foundation 2002, courtesy of Palm Press, Inc. **232:** Courtesy of Saab **233:** Tim Wright/CORBIS **246:** Richard Megna, Fundamental Photographs **251:** © Bill Stormont/The Stock Market

Chapter 9. **259:** Emily Serway **262:** AIP Emilio Segré Visual Archives, Michelson Collection **264:** AIP Niels Bohr Library **288:** Courtesy of Garmin Ltd.

Chapter 10. **291:** Courtesy of Tourism Malaysia **317:** top left, © Stuart Franklin/Getty Images; bottom, David Malin, Anglo-Australian Observatory **320:** Courtesy of Henry Leap and Jim Lehman **327:** John Lawrence/Stone/Getty Images

Chapter 11. **337:** © 1987 Royal Observatory/Anglo-Australian Observatory, by David F. Mahlin from U.K. Schmidt plates **342:** Art Resource **346:** U.S. Space Command, NORAD **350:** bottom, H. Ford et al. & NASA **352:** K. W. Whitten, R. E. Davis, M. L. Peck, and G. G. Stanley, General Chemistry, 7th ed., Belmont, CA, Brooks/Cole, 2004 **353:** Photo courtesy of AIP Niels Bohr Library, Margarethe Bohr Collection **361:** Courtesy of NASA/JPL **365:** NASA

Context Conclusion 3. **371:** top right, Guillermo Aldana Espinosa/National Geographic Image Collection; bottom right, Joseph Sohm/ChromoSohm Inc./CORBIS **372:** T. Campion/SYGMA

Chapter 12. **373:** © Simon Kwong/Reuters/CORBIS **390:** Special Collections Division, Univ. of Wash. Libraries, Photo by Farquharson **394:** Telegraph Colour Library/FPG International

Chapter 13. **400:** Don Bonsey/Stone Getty **416:** © Howard Sochurek/Woodfin Camp & Associates, Inc. **419:** top right, Courtesy of U.S. Navy. Photo by Ensign John Gay; bottom right, Courtesy of the Educational Development Center, Newton, MA **427:** Joe McDonald/Visuals Unlimited **429:** Gregg Adams/Stone

Chapter 14. **432:** Kathy Ferguson Johnson/PhotoEdit/PictureQuest **433:** bottom right, Education Development Center, Newton, MA **434:** top right, Education Development Center, Newton, MA **438:** © 1991 Richard Megna/Fundamental Photographs **450:** Photographs courtesy of (a) and (b) Getty Images; (c) Photodisc Green/Getty Images **457:** top left, Courtesy Professor Thomas D. Rossing, Northern Illinois University; top right, Steve Bronstein/Image Bank/Getty Images; bottom left, Murray Greenberg

Context Conclusion 3. **459:** Paul X. Scott/SYGMA **460:** Photo by M. Hamada; used by permission from the Multidisciplinary Center for Earthquake Engineering Research, Univ. of Buffalo

Context 4. **462:** bottom left, Illustrations by Ken Marschall © 1992 from *Titanic: An Illustrated History,* a Hyperion/Madison Press Book); top right, Illustrations by Ken Marschall © 1992 from *Titanic: An Illustrated History,* a Hyperion/Madison Press Book **463:** top left, Courtesy of RMS Titanic, Inc.; top right, Courtesy of RMS Titanic, Inc.

Chapter 15. **464:** Norman Tobley/Taxi/Getty Images **466:** Royalty-free/CORBIS **468:** top right, David Frazier **470:** © Hulton Deutsch Collection/CORBIS **472:** Richard Megna/Fundamental Photographs **475:** Andy Sacks/Tony Stone Images/Getty Images **476:** Werner Wolff/Black Star **478:** © Bettmann/CORBIS **482:** Courtesy of The Father Browne S.J. Collection **483:** © TempSport/CORBIS **484:** Pamela Zilly/The Image Bank/Getty **488:** George Semple **489:** Stan Osolinski/Dembinsky Photo Associates **490:** bottom right, The Granger Collection **492:** Courtesy of Jeanne Maier

Context Conclusion 4. **493:** Illustrations by Ken Marschall © 1992 from *Titanic: An Illustrated History,* a Hyperion/Madison Press Book **494:** Illustrations by Ken Marschall © 1992 from *Titanic: An Illustrated History,* a Hyperion/Madison Press Book **495:** top, Illustrations by Ken Marschall © 1992 from *Titanic: An Illustrated History,* a Hyperion/Madison Press Book; bottom, E. Paul Oberlander/Woods Hole Oceanographic Institution

Context 5. **497:** © Tony Roberts/CORBIS **498:** Courtesy of NASA

Chapter 16. **499:** Lowell Georgia/CORBIS **502:** Charles D. Winters **505:** a) Frank Siteman, Stock/Boston; b) George Semple **506:** Wide World Photos **510:** © Peter Turnley/Corbis **513:** Courtesy of Niels Bohr Library, Lande Collection **523:** Courtesy of Central Scientific Company **528:** Charles D. Winters

Chapter 17. **531:** Jon Eisberg/Taxi/Getty **532:** top, By kind permission of the President and Council of the Royal Society; bottom, Charles D. Winters **554:** Courtesy of Dr. Albert A. Bartlett, University of Colorado, Boulder **557:** National Optical Astronomy Observatories/Photo Researchers, Inc. **568:** Nathan Bilow/Leo de Wys, Inc. **570:** AP/World Wide Photos

Chapter 18. **572:** Roger Harris/Photo Researchers, Inc. **573:** J. L. Charmet/SPL/Photo Researchers, Inc. **575:** J. L. Charmet/SPL/Photo Researchers, Inc. **590:** Courtesy of PASCO Scientific Company **594:** CORBIS/Stock Market

Context 6. **601:** Paul and LindaMarie Ambrose/Taxi/Getty Images **602:** left, M. Zhilin/M. Newman/Photo Researchers, Inc.; right, © Johnny Autery

Chapter 19. **603:** Courtesy of Resonance Research Corporation **604:** Charles D. Winters **607:** © 1968 Fundamental Photographs **608:** Photo courtesy of AIP Niels Bohr Library/E. Scott Barr Collection **617:** top, Douglas C. Johnson/Cal Poly Pomona; bottom, Douglas C. Johnson/Cal Poly Pomona **618:** Douglas C. Johnson/Cal Poly Pomona

Chapter 20. **642:** George Semple **658:** Douglas C. Johnson/Cal Poly Pomona **670:** Paul Silverman/Fundamental Photographs **678:** Adam-Hart Davis/SPL/Custom Medical Stock **679:** E. R. Degginer/H. Armstrong Roberts

Chapter 21. **683:** © Lester Lefkowitz/Image Bank/Getty Images **688:** Douglas C. Johnson/Cal Poly Pomona **690:** SuperStock **693:** Courtesy of IBM Research Laboratory **706:** AIP ESVA/W. F. Meggers Collection **715:** Courtesy of Henry Leap and Jim Lehman

Context 7 Opener. **725:** © Transrapid International, Berlin, Germany **726:** left, © Transrapid International, Berlin, Germany; right, © Hashimoto Noburu/Corbis SYGMA

Chapter 22. **727:** James King-Holmes/Photo Researchers, Inc. **728:** North Wind Picture Archives **729:** Courtesy of Henry Leap and Jim Lehman **734:** Courtesy of Henry Leap and Jim Lehman **737:** Bell Telephone Labs/Courtesy of Emilio Segré Visual Archives **738:** Courtesy of Lawrence Berkeley Laboratory, University of California **745:** © Richard Megna, Fundamental Photographs **747:** © Richard Megna, Fundamental Photographs **748:** Leonard de Selva/CORBIS **751:** Henry Leap and Jim Lehman **756:** left, Courtesy of CENCO; bottom right, Courtesy of CENCO

Chapter 23. **765:** Michael Melford/Getty Images **767:** By kind permission of the President and Council of the Royal Society **769:** Charles D. Winters **777:** Photos by John Jewett **781:** North Wind Picture Archives **790:** Courtesy of CENCO **793:** Ross Harrison/Getty Images

Context 7 Conclusion. **803:** a–b, Courtesy of Lawrence Livermore Laboratory

Context 8 Opener. **804:** bottom left, Will & Deni McIntyre/Photo Researchers, Inc.; top right, Courtesy of HRL Laboratories LLC, Malibu, CA **805:** top left, © Lawrence Manning/CORBIS; middle, © David Frazier/Stone/Getty Images; top right, Philippe Plailly/SPL/Photo Researchers, Inc.

Chapter 24. **806:** upper left, NASA/CXC/SAO; upper right, Palomar Observatory; lower right, VLA/NRAO; lower left, WM Keck Observatory **808:** North Wind Picture Archives **815:** The Bettman Archive/CORBIS **823:** Ron Chapple/Getty Images **828:** Courtesy of Mark Helfer/National Institute of Standards and Technology **831:** George Semple **832:** Trent Steffler/David R. Frazier Photo Library **835:** Philippe Plaily/SPL/Photo Researchers **836:** Amos Morgan/Getty Images **837:** © Bill Banaszewski/Visuals Unlimited

Chapter 25. **839:** Patrick J. Endres/Visuals Unlimited **842:** c,d) Courtesy of Henry Leap and Jim Lehman **843:** David Parker/Photo Researchers, Inc. **844:** a,b) Courtesy of Texas Instruments **845:** top, Courtesy of Henry Leap and Jim Lehman; bottom, Jim Lehman **850:** David Parker/Science Photo Library/Photo Researchers, Inc. **852:** top left, Courtesy of Rijksmuseum voor de Geschiedenis der Natuurwetenschappen. Courtesy of Niels Bohr Library; middle left, Erich Schrempp/Photo Researchers, Inc. **854:** Courtesy of Henry Leap and Jim Lehman **856:** Dennis O'Clair/Stone/Getty Images **861:** Courtesy of Sony Disc Manufacturing **862:** Ray Atkeson/Image Archive **863:** Courtesy of Edwin Lo

Chapter 26. **867:** © Don Hammond/CORBIS **869:** George Semple **872:** Henry Leap and Jim Lehman **873:** Courtesy of Thompson Consumer Electronics **875:** Photos courtesy of David Rogers **876:** top, © Bo Zaunders/CORBIS; bottom, NASA **881:** Courtesy of Henry Leap and Jim Lehman **886:** Courtesy of Henry Leap and Jim Lehman **890:** Richard Megna/Fundamental Photographs, NYC **891:** M. C. Escher/Cordon Art–Baarn–Holland. All rights reserved. **892:** © Paul Silverman 1990, Fundamental Photographs **896:** © Michael Levin/Optigone Associates

Chapter 27. **898:** Dr. Jeremy Burgess/SPL/Photo Researchers, Inc. **900:** M. Cagnet, M. Francon, J.C. Thierr, *Atlas of Optical Phenomena,* Berlin, Springer-Verlag, 1962 **904:** From M. Cagnet, M. Francon, and J.C. Thierr, *Atlas of Optical Phenomena,* Berlin, Springer-Verlag, 1962 **908:** top, Kristen Brochmann/Fundamental Photographs; bottom, Richard Megna/Fundamental Photographs **909:** top right, Douglas C. Johnson/Cal Poly Pomona; middle right, Courtesy of P. M. Rinard, from *Am. J. Phys.* 44: 70, 1976; bottom, From M. Cagnet, M. Francon, and J.C. Thierr, *Atlas of Optical Phenomena,* Berlin, Springer-Verlag, 1962, plate 18 **913:** From M. Cagnet, M. Francon, J.C. Thierr, *Atlas of Optical Phenomena,* Berlin, Springer-Verlag, 1962 **915:** a, U.S. Naval Observatory/James W. Christy, U.S. Navy photograph; b, Photo courtesy of Gemini Observatory **917:** Silicon Light Machines **918:** © Kristen Brochmann/Fundamental Photographs **919:** a, Used with permission of Eastman Kodak Company; b, © I. Anderson, Oxford Molecular Biophysics Laboratory/Photo Researchers, Inc. **920:** Ronald Erickson/Media Interface **926:** SuperStock **927:** Courtesy of CENCO Scientific Company **928:** bottom left, © Doug Pensigner/Getty Images; right, © Danny Lehman **929:** © Richard Kolar/Animals Animals/Earth Scenes

Context 8 Conclusion. **932:** Courtesy of University of Miami, Music Engineering

Context 9 Opener. **935:** top right, Comstock Royalty Free/Getty; bottom, 1987 Anglo-Australian Observatory, photo by David Malin **936:** NASA, ESA, and The Hubble Heritage Team

Chapter 28. **937:** © Eye of Science/Science Source/Photo Researchers, Inc. **938:** Corbis **939:** © Bettmann/CORBIS **940:** left, Photodisc/Getty Images; right, John Chumack/Photo Researchers, Inc. **947:** Courtesy of AIP Niels Bohr Library **950:** AIP Niels Bohr Library **953:** © David Parker/Photo Researchers, Inc. **958:** a–d, From C. Jonsson, *Zeitschrift fur Physik* 161:454, 1961; used with permission Springer-Verlag **959:** Courtesy of the University of Hamburg **967:** AIP Emilio Segré Visual Archives **969:** IBM Research, Almaden Research Center. Unauthorized use prohibited **976:** a–b, Alexandra Héder

Chapter 29. **983:** © Jeff Hunter/Image Bank/Getty Images **995:** Courtesy of AIP Niels Bohr Library, Margarethe Bohr Collection **998:** AIP Emilio Segré Visual Archives, Goudsmit Collection **1007:** a, C. R. O'Dell (Rice University) and NASA; b, © 1985 Royal Observatory Edinburgh/Anglo-Australian Observatory by David Malin; c, A. Caulet (ST-ECF, ESA) and NASA **1013:** Maarten Schmidt/Palomar Observatory/California Institute of Technology

Chapter 30. **1016:** Paul Hanny/Gamma Liaison **1023:** SBHA/Getty Images **1025:** FPG International **1033:** middle right, © CORBIS; bottom left, Corbis SYGMA; bottom right, M. Milner/Corbis SYGMA **1046:** Courtesy of NASA Ames

Chapter 31. **1048:** Courtesy of CERN **1051:** top left, Courtesy Lawrence Berkeley Laboratory, University of California; middle right, Courtesy of AIP Emilio Segré Visual Archives **1052:** National Institutes of Health **1053:** UPI/Corbis-Bettmann **1054:** © Shelly Gazin/CORBIS **1059:** Courtesy of ICRR (Institute for Cosmic Ray Research), University of Tokyo **1062:** Courtesy Lawrence Berkeley Laboratory, University of California, Photographic Services **1063:** AIP Meggers Gallery of Nobel Laureates **1064:** Courtesy of Brookhaven National Laboratory **1071:** Courtesy of CERN **1072:** Courtesy of Fermi National Accelerator Laboratory **1081:** Anglo-Australian Telescope Board

Tables and Illustrations

This page constitutes an extension of the copyright page. We have made every effort to trace the ownership of all copyrighted material and to secure permission from copyright holders. In the event of any question arising as to the use of any material, we will be pleased to make the necessary corrections in future printings. Thanks are due to the following authors, publishers, and agents for permission to use the material indicated.

Chapter 1. **33:** By permission of John L. Hart FLP, and Creators Syndicate, Inc.

Chapter 11. **363:** By permission of John L. Hart FLP, and Creators Syndicate, Inc.

Chapter 14. **452:** Adapted from C. A. Culver, *Musical Acoustics,* 4th Edition, New York, McGraw-Hill, 1956, p. 128. **453:** Adapted from C. A. Culver, Musical Acoustic, 4th Edition, New York, McGraw-Hill, 1956.

Chapter 26. **893:** The Far Side ® by Gary Larson © 1985 FarWorks, Inc. All rights reserved. Used with permission.

Chapter 29. **989:** From E. U. Condon and G. H. Shortley, *The Theory of Atomic Spectra,* Cambridge, Cambridge University Press, 1953, used with permission.

Index

Page numbers in *italics* indicate figures; page numbers followed by "n" indicate footnotes; page numbers followed by "t" indicate tables.

D

T

Standard Abbreviations and Symbols for Units

Symbol	Unit	Symbol	Unit
A	ampere	K	kelvin
u	atomic mass unit	kg	kilogram
atm	atmosphere	kmol	kilomole
Btu	British thermal unit	L	liter
C	coulomb	lb	pound
°C	degree Celsius	ly	lightyear
cal	calorie	m	meter
d	day	min	minute
eV	electron volt	mol	mole
°F	degree Fahrenheit	N	newton
F	farad	Pa	pascal
ft	foot	rad	radian
G	gauss	rev	revolution
g	gram	s	second
H	henry	T	tesla
h	hour	V	volt
hp	horsepower	W	watt
Hz	hertz	Wb	weber
in.	inch	yr	year
J	joule	Ω	ohm

Mathematical Symbols Used in the Text and Their Meaning

Symbol	Meaning
$=$	is equal to
$\equiv$	is defined as
$\neq$	is not equal to
$\propto$	is proportional to
$\sim$	is on the order of
$>$	is greater than
$<$	is less than
$\gg (\ll)$	is much greater (less) than
$\approx$	is approximately equal to
Δx	the change in x
$\sum_{i=1}^{N} x_i$	the sum of all quantities x_i from $i = 1$ to $i = N$
$\|x\|$	the magnitude of x (always a nonnegative quantity)
$\Delta x \rightarrow 0$	Δx approaches zero
$\frac{dx}{dt}$	the derivative of x with respect to t
$\frac{\partial x}{\partial t}$	the partial derivative of x with respect to t
$\int$	integral